Understanding Normal and Clinical Nutrition

UNDERSTANDING NORMAL AND CLINICAL NUTRITION

SIXTH EDITION

Eleanor Noss Whitney

Corinne Balog Cataldo

Sharon Rady Rolfes

WADSWORTH
THOMSON LEARNING™

Australia • Canada • Mexico • Singapore
Spain • United Kingdom • United States

WADSWORTH
THOMSON LEARNING™

Publisher: Peter Marshall
Development Editor: Elizabeth Howe
Assistant Editor: John Boyd
Editorial Assistant: Andrea Kesterke
Marketing Manager: Becky Tollerson
Marketing Assistant: Adam Hofmann
Advertising Project Manager: Brian Chaffee
Project Manager: Sandra Craig
Print/Media Buyer: Barbara Britton
Permissions Editor: Joohee Lee

Production: The Book Company
Text and Cover Designer: Norman Baugher
Photo Researcher: Myrna Engler
Copy Editor: Patricia Lewis
Illustrators: Impact Publications; J/B Woolsey Associates
Compositor: Parkwood Composition
Cover Image: © Kevin R. Morris / Corbis
Cover Printer: Phoenix Color
Printer: Quebecor/World, Versailles

Printed in the United States of America
2 3 4 5 6 7 05 04 03 02

Library of Congress Cataloging-in-Publication Data
Whitney, Eleanor Noss.
Understanding normal and clinical nutrition / Eleanor Noss Whitney, Corinne Balog Cataldo, Sharon Rady Rolfes.--6th ed.
p. ; cm.
Includes bibliographical references and index.
ISBN 0-534-58995-2
1. Nurition. 2. Diet therapy. 3. Dietetics. I. Title: Normal and clinical nutrition. II. Cataldo, Corinne Balog. III. Rolfes, Sharon Rady. IV. Title.
[DNLM: 1. Nutrition. 2. Diet Therapy. QU 145 W618ua2002]
QP141.W458 2002
613.2--dc21 2001026056

Wadsworth/Thomson Learning
10 Davis Drive
Belmont, CA 94002-3098
USA

For more information about our products, contact us:
Thomson Learning Academic Resource Center
1-800-423-0563
http://www.wadsworth.com

International Headquarters
Thomson Learning
International Division
290 Harbor Drive, 2nd Floor
Stamford, CT 06902-7477
USA

UK/Europe/Middle East/South Africa
Thomson Learning
Berkshire House
168-173 High Holborn
London WC1V 7AA
United Kingdom

Asia
Thomson Learning
60 Albert Street, #15-01
Albert Complex
Singapore 189969

Canada
Nelson Thomson Learning
1120 Birchmount Road
Toronto, Ontario M1K 5G4
Canada

To

All my children, stepchildren, and grandchildren, and to my loved ones who enrich my life today beyond measure: Council, Vicki, Arthur, Brad, John, David, Jarod, Mechelle, Brittany, Jacob, and all.

ELLIE

To

My husband and soulmate Mike Fogel, who lifts my spirits and brightens my days.

CORKIE

To

My parents, Tom and Gladys Rady, whose love and guidance throughout the years enabled me to fulfill my dreams.

SHARON

ABOUT THE AUTHORS

Eleanor Noss Whitney, Ph.D., received her B.A. in biology from Radcliffe College in 1960 and her Ph.D. in biology from Washington University, St. Louis, in 1970. Formerly on the faculty at Florida State University, and a dietitian registered with the American Dietetic Association, she now devotes full time to research, writing, and consulting. Her earlier publications include articles in *Science, Genetics,* and other journals. Her textbooks include *Understanding Nutrition, Nutrition Concepts and Controversies, Life Span Nutrition: Conception through Life, Nutrition and Diet Therapy,* and *Nutrition for Health and Health Care* for college students and *Making Life Choices* for high school students. Her most intense interests currently include energy conservation, solar energy uses, alternatively fueled vehicles, and ecosystem restoration.

Corinne Balog Cataldo, M.M.Sc., R.D., C.N.S.D., received her B.S. in community health nutrition from Georgia State University in 1976 and her M.M.Sc. in clinical dietetics from Emory University in 1979. She has worked in private practice in Atlanta, as a clinical dietitian and metabolic support nutritionist at Georgia Baptist Medical Center in Atlanta, as a faculty member and dietetic internship coordinator at Emory University, and as a nutritionist with the Infant Formula Council. She has made numerous presentations, and in addition to this book, she has written a manual on tube feedings and the books *Nutrition and Diet Therapy, Nutrition for Health and Health Care,* and *Understanding Clinical Nutrition.* She is a certified nutrition support dietitian.

Sharon Rady Rolfes, M.S., R.D., received her B.S. in psychology and criminology in 1974 and her M.S. in nutrition and food science in 1982 from Florida State University. She is a founding member of Nutrition and Health Associates, an information resource center that maintains an ongoing bibliographic database that tracks research in over 1000 nutrition-related topics. Her other publications include the textbooks *Understanding Nutrition, Understanding Clinical Nutrition, Life Span Nutrition: Conception through Life,* and *Nutrition for Health and Health Care* and a multimedia CD-ROM called *Nutrition Interactive.* In addition to writing, she also lectures at universities and at professional conferences and serves as a consultant for various educational projects. She maintains her registration as a dietitian and membership in the American Dietetic Association.

BRIEF CONTENTS

CONTENTS

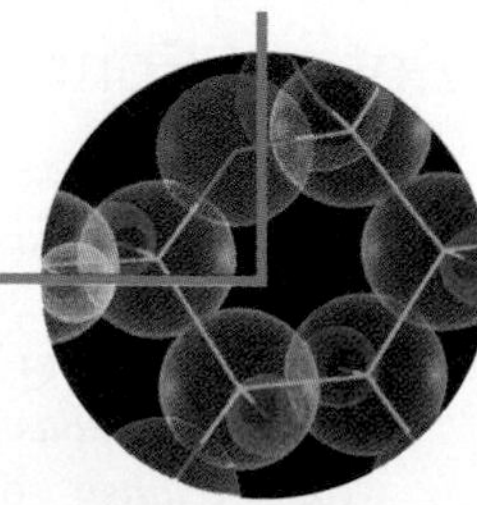

BOXES

HOW TO

CASE STUDIES

PREFACE

THE CONTENTS of the sixth edition of *Understanding Normal and Clinical Nutrition* reflect the dramatic changes that have transpired in nutrition science and health care delivery since the first edition was written many current research topics—such as phytochemicals, leptin, complementary therapies, and the metabolic syndrome—weren't even mentioned in the first edition. This edition discusses each of these topics, and more. As with each previous edition, every chapter has been substantially revised to reflect the many changes that have occurred in the field of nutrition. Even though the information is new, this edition maintains the same goals: to reveal the facts about nutrition, and to show readers how to apply this information to their daily lives and clinical practice.

• The Chapters • *Understanding Normal and Clinical Nutrition* presents the core information of an introductory nutrition course. Chapter 1 wastes no time in exploring why we eat the foods we do and continues with a brief overview of the nutrients, the science of nutrition, recommended nutrient intakes, assessment, and important relationships between diet and health. Chapter 2 describes the diet-planning principles and food guides used to create diets that support good health and includes instructions on how to read a food label. In Chapter 3, readers follow the journey of digestion and absorption as the body transforms foods into nutrients. Chapters 4 through 6 describe carbohydrates, fats, and proteins—their chemistry, roles in the body, and places in the diet. Then Chapter 7 shows how the body derives energy from these three nutrients. Chapters 8 and 9 continue the story with a look at the benefits and dangers of weight loss and weight gain. Chapters 10 through 13 complete the introductory lessons by describing the vitamins, the minerals, and water—their roles in the body, deficiency and toxicity symptoms, and sources. Chapters 14, 15, and 16 present the special nutrient needs of people through the life cycle—pregnancy and lactation; infancy, childhood, and adolescence; and adulthood and the later years.

The remaining chapters of the book focus on the nutrition care of clients with health problems. Chapter 17 describes the ways illnesses and their treatments alter nutrient needs and shows how health care professionals assess the affects of illness on nutrition status. Chapter 18 provides an in-depth look at potential interactions between nutrients and prescription medications, over-the-counter medications, herbs, and other dietary supplements. Chapter 19 discusses how health care professionals make plans for clients' nutrition care. Chapters 20 through 29 explore specific diseases. Chapters 20 and 21 examine health problems that affect the upper and lower GI tract. Chapters 22 and 23 explain special ways of feeding people who cannot eat conventional foods. Chapter 24 looks at the ways severe stresses affect metabolism and nutrient needs. Chapter 25 provides an overview of diabetes mellitus. Chapters 26 through 28 discuss disorders of the heart, blood vessels, lungs, kidneys, and liver. The last chapter describes the multiple effects of cancers and HIV infections on nutrition status.

• The Highlights • Every chapter is followed by a highlight. Each highlight provides readers with a review of a topic that relates to the companion chapter. New highlights in this edition feature functional foods, a quick and easy way to plan and prepare healthy meals, the immune system, multiple organ failure, the metabolic syndrome, dialysis, and gallstones.

• Special Features • The chapters in this edition have been designed with special features to enhance learning. For example, definitions are provided whenever new terms are introduced. These definitions often include pronunciations and derivations to facilitate understanding. A glossary at the end of the text includes all defined terms.

> **definition** (DEF-eh-NISH-en): the meaning of a word.
> - **de** = from
> - **finis** = boundary

New to this edition are health promotion icons in the first part of the book that call the reader's attention to the many connections between specific nutrients and health. Another new feature found in each of the first 16 chapters is a fascinating fact offered just for fun. These informative tidbits may not be on the exam, but they are sure to amuse readers and stimulate discussion. Also new to this edition and found in later chapters are "Think nutrition" reminders. These boxes prompt readers to be alert for nutrition problems when clients have specific illnesses or symptoms.

◆ Did you know that an adult has 10,000 tastebuds?◆

Many chapters include "How to" sections that guide readers through problem-solving tasks. For example, the "How to" in Chapter 1 shows readers how to calculate energy intake from the grams of carbohydrate, fat, and protein in a food; another "How to" in Chapter 26 describes how to help clients implement heart-healthy diets.

Several chapters in the first part of the book close with a "Making It Click" section. Later chapters include case studies and "Clinical Applications" sections. The problems posed in these sections enable readers to apply the chapter material to hypothetical situations. Readers who successfully master these exercises will be well prepared for "real-life" nutrition-related problems.

> **IN SUMMARY** Each major section within a chapter concludes with a summary paragraph that reviews the key concepts. Similarly, summary tables cue readers to important reviews.

Also featured in the early chapters of this edition are the Healthy People 2010 nutrition-related priorities, which are presented whenever their subjects are discussed. Healthy People 2010 is a report developed by the U.S. Department of Health and Human Services that establishes national objectives in health promotion and disease prevention for the year 2010.

> **HEALTHY PEOPLE 2010**
> These nutrition-related priorities are presented throughtout the text whenever their subjects are discussed.

How are you doing?

New to this edition are the "How are you doing?" questions at the end of each of the earlier chapters that reflect the *Dietary Guidelines*. These questions prompt readers to ponder their own eating habits and activity patterns. Later chapters include "Nutrition assessment checklists" that help readers evaluate the impact of various disorders on nutrition status by highlighting the medical, drug, nutrient intake, anthropometric, laboratory, and physical findings that are particularly relevant to a specific group of clients.

Nutrition Assessment Checklist

New to the later chapters of this edition are "Diet-Drug Interactions" boxes. These boxes describe the nutrition-related concerns of the medications used to treat the disorders described in the chapter.

Diet-Drug Interactions

Each chapter and many highlights also conclude with Nutrition on the Net—a list of websites and interactive Internet activities for further study of topics covered in the accompanying text. These listings do not imply an endorsement of the

Nutrition on the Net

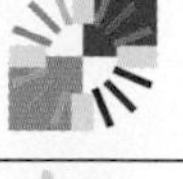

organizations or their programs. We have tried to provide reputable sources, but cannot be responsible for the content of these sites. (Read Highlight 1 to learn how to find reliable information on the Internet.)

Study Questions

Each chapter ends with study questions in essay and multiple-choice format. Study questions offer readers the opportunity to review the major concepts presented in the chapters in preparation for exams. The page numbers after each essay question refer readers to discussions that answer the question; multiple-choice answers appear on the last page of the chapter.

• The Appendixes • The appendixes are valuable references for a number of purposes. Appendix A summarizes background information on the hormonal and nervous systems, complementing Appendixes B and C on basic chemistry, the chemical structure of nutrients, and major metabolic pathways. Appendix D describes measures of protein quality. Appendix E provides supplemental information about nutrition assessment, and Appendix F lists nutrition resources, including websites. Appendix G presents the U.S. Exchange System. Appendix H is a 2000-item food composition table compiled from the latest nutrient database assembled by ESHA Research, Inc., of Salem, Oregon. Appendix I presents recommendations from the World Health Organization (WHO) and information for Canadians—the Choice System and guidelines to healthy eating and physical activity. Appendix J provides information about enteral formulas.

• The Inside Covers • The inside covers put commonly used information at your fingertips. The front covers present the current nutrient recommendations (introduced in Chapter 1); the inside back cover (left) features the Daily Values used on food labels (presented in Chapter 2) and a glossary of nutrient measures; and the inside back cover (right) shows the suggested weight ranges for various heights (discussed in Chapter 8). Aids to Calculations can be found on the last two pages of the book.

• Closing Comments • We have tried to keep the number of references manageable. Many statements that have appeared in previous editions with references now appear without them, but every statement is backed by research, and the authors will supply references upon request. We have not provided a separate list of suggested readings, but have tried to include references that will provide readers with additional details or a good overview of the subject. Nutrition is a fascinating subject, and we hope our enthusiasm for it comes through on every page.

ELEANOR NOSS WHITNEY
CORINNE BALOG CATALDO
SHARON RADY ROLFES
MAY 2001

ACKNOWLEDGMENTS

To produce a book requires the coordinated effort of a team of people—and, no doubt, each team member has another team of support people as well. We salute, with a big round of applause, everyone who has worked so diligently to ensure the quality of this book.

We thank our partner and friend, Linda DeBruyne, for her valuable contributions to the fitness and diet and health chapters, and Margaret Hedley for her attention to the Canadian information throughout the text and in Appendixes F and I. A million thank yous to Lynn Earnest for her careful attention to manuscript preparation and to Katie McMurry for her work on a multitude of other daily tasks. To Joohee Lee, a special thanks for her assistance in obtaining permissions. We also thank the many people who have prepared the ancillaries that accompany this text: Melaney Jones for writing and enhancing the Test Bank and Instructor's

Manual; Lori Turner, Mary Rhiner, and Margaret Hedley for preparing the Instructor's Manual; Charlene Hamilton for developing the Internet activities and electronic lecture presentations; and Lori Turner and Jana Kicklighter for authoring the Webtutor Student Study Guide and Linda DeBruyne for creating the clinical Internet activities. A big thank you to Elizabeth Hands, Bob Geltz, and their staff at ESHA for their meticulous effort in creating the food composition appendix, verifying the data in figures and tables, and developing the computerized diet analysis program that accompanies this book. Our special thanks to Peter Marshall for his continued support and insightful ideas; to Sandra Craig for her guidance of this revision from conception to conclusion; to Dusty Friedman for her diligent attention to the innumerable details involved in production; to Becky Tollerson, Joy Westberg, and Brian Chaffee for their talented marketing efforts; to Elizabeth Howe for her efficient coordination of reviews and other tasks; to John Boyd for his competent coordination of ancillaries; and to Andrea Kesterke for her willingness to fill in the gaps. We also thank Norman Baugher for designing the pages to enhance our work; John Woolsey and his team of artists for creating accurate and attractive artwork to complement our writing; Myrna Engler for selecting photographs and coordinating photography sessions to deliver nutrition messages beautifully; Pat Lewis for copyediting thousands of pages of manuscript; and Erin Taylor for composing a thorough and useful index. To the many others involved in production, marketing, and sales, we tip our hats in appreciation.

We are especially grateful to our associates, friends, and families for their continued encouragement and support. We also thank our many reviewers for their comments and contributions.

REVIEWERS OF UNDERSTANDING NORMAL AND CLINICAL NUTRITION

Sara Long Anderson
Southern Illinois University at Carbondale

Dean E. Bell
College of Eastern Utah

James W. Bailey
University of Tennessee

Wayne Billon
University of Southern Mississippi

M. Brian-Keber
University of San Francisco

Marjorie Busby
University of North Carolina, Chapel Hill

Eileen Monahan Chopnick
Widener University School of Nursing

Beth Clark
University of Maine, Augusta

Wendy Mueller-Cunningham
California State University, Sacramento

Colleen Duggan
Johnson Community College

Carmen Edwards
Midland College

Jamie Erskine
University of Northern Colorado

Amelia Finan
Anne Arundel Community College

Betty Forbes
West Virginia University

Denise T. Garner
York College of Pennsylvania

Judith Harr
University of San Francisco

Ann Hunter
Wichita State University

Irene Langille
Southern Alberta Institute of Technology

Robert D. Lee
Central Michigan University

Myrtle McCulloch
Georgetown University

Dawna Torres Mughal
Gannon University

Jean Nelson
University of Missouri, St. Louis

Emilia Papakonstantinou
University of Georgia

Amy R. Pemberton
Lamar University

Kathy Redding
Westark College

Ruth Reilly
University of New Hampshire

Tonia Reinhard
Wayne State University

Susan Swadener
California Polytechnic University, San Louis Obispo

Marilyn Tapia
Montana State University at Billings

Elizabeth Thompson
University of North Carolina, Chapel Hill

Connie B. Till
Orangeburg-Calhoun Technical College

Anne VanBeber
Texas Christian University

Understanding Normal and Clinical Nutrition

CHAPTER 1
An Overview of Nutrition

www.comstock.com

Welcome to the world of **nutrition.** Nutrition has played a significant role in your life, even from before your birth, although you may not always have been aware of it. And it will continue to affect you in major ways, depending on the **foods** you select.

Health promotion

Every day, several times a day, you make food choices that influence your body's health for better or worse. Each day's choices may benefit or harm your health only a little, but when these choices are repeated over years and decades, the rewards or consequences become major. That being the case, close attention to good eating habits now can bring health benefits later. Conversely, carelessness about food choices from youth on can be a major contributor to many of today's most prevalent chronic diseases◆ of later life, including heart disease and cancer. Of course, some people will become ill or die young no matter what choices they make, and others will live long lives despite making poor choices. For the large majority, however, the food choices they make each and every day will benefit or impair their health in proportion to the wisdom of those choices.

◆ In general, a **chronic** disease is one of long duration that progresses slowly or with little change. By comparison, an **acute** disease develops quickly, produces sharp symptoms, and runs a short course.

- chronos = time
- acute = sharp

Although most people realize that their food habits affect their health, they often choose foods for other reasons. After all, foods bring to the table a variety of pleasures, traditions, and associations as well as nourishment. The challenge, then, is to combine favorite foods and fun times with a nutritionally balanced **diet.**

FOOD CHOICES

People decide what to eat, when to eat, and even whether to eat in highly personal ways, often based on behavioral or social motives rather than on awareness of nutrition's importance to health. Fortunately, many different food choices can be healthy ones, but nutrition awareness helps to make them so.

© SuperStock

An enjoyable way to learn about other cultures is to taste their ethnic foods.

• Personal Preference • As you might expect, the number one reason people choose foods is taste—they like certain flavors.[1] Two widely shared preferences are for the sweetness of sugar and the tang of salt. Liking high-fat foods appears to be another universally common preference.[2] Other preferences might be for the hot peppers common in Mexican cooking or the curry spices of Indian cuisine. Recent research suggests that genetics may influence people's food preferences.[3]

• Habit • People sometimes select foods out of habit. They eat cereal every morning, for example, simply because they have always eaten cereal for breakfast. Eating a familiar food and not having to make any decisions can be comforting.

• Ethnic Heritage or Tradition • Among the strongest influences on food choices are ethnic heritage and tradition. People eat the foods they grew up eating. Every country, and in fact every region of a country, has its own typical foods and ways of combining them into meals. The "American diet" includes many ethnic foods from various countries, all adding variety to the diet. This is most evident when we eat out: 60 percent of our restaurants (excluding fast-food places) have an ethnic emphasis, most commonly Chinese, Italian, or Mexican.[4]

nutrition: the science of foods and the nutrients and other substances they contain, and of their actions within the body (including ingestion, digestion, absorption, transport, metabolism, and excretion). A broader definition includes the social, economic, cultural, and psychological implications of food and eating.

foods: products derived from plants or animals that can be taken into the body to yield energy and nutrients for the maintenance of life and the growth and repair of tissues.

diet: the foods and beverages a person eats and drinks.

• Social Interactions • Most people enjoy companionship while eating. Meals are social events, and the sharing of food is part of hospitality. Social customs almost compel people to accept food or drink offered by a host or shared by a group. When your friends are ordering pizza or going out for ice cream, how can you refuse?

• Availability, Convenience, and Economy • People eat foods that are accessible, quick and easy to prepare, and within their financial means. Consumers today value convenience especially highly and are willing to spend a little over half of their food budget on meals that require little, if any, further preparation.[5] They frequently eat out, bring home ready-to-eat meals, or have food delivered. Even when they venture into the kitchen, they want to prepare a meal in 15 to 20 minutes, using only four to six ingredients.[6] Such convenience limits food choices to the selections offered on menus and products designed for quick preparation.

• Positive and Negative Associations • People tend to like foods with happy associations—such as hot dogs at ball games or cake and ice cream at birthday parties. By the same token, people can attach intense and unalterable dislikes to foods that they ate when they felt sick or that were forced on them when they weren't hungry.[7] Parents may teach their children to like and dislike certain foods by using those foods as rewards or punishments.

© The Stock Market/Chuck Savage 2001

For many people, a special family dinner brings pleasant memories of the holidays.

• Emotional Comfort • Some people eat in response to emotional stimuli—for example, to relieve boredom or depression or to calm anxiety. A depressed person may choose to eat rather than to call a friend. A person who has returned home from an exciting evening out may unwind with a late-night snack. Eating in response to emotions can easily lead to overeating and obesity, but may be appropriate at times. For example, sharing food at times of bereavement serves both the giver's need to provide comfort and the receiver's need to be cared for and to interact with others, as well as to take nourishment.

• Values • Food choices may reflect people's religious beliefs, political views, or environmental concerns. For example, many Christians forgo meat during Lent, the period prior to Easter, and Jewish law includes an extensive set of dietary rules. A political activist may boycott fruit picked by migrant workers who have been exploited. People may buy vegetables from local farmers to save the fuel and environmental costs of foods shipped in from far away. They may also select foods packaged in containers that can be reused or recycled. Some consumers accept or reject foods that have been irradiated or genetically modified, depending on their understanding of these processes.

• Body Weight and Image • Sometimes people select certain foods and supplements that they believe will improve their physical appearance and avoid those they believe might be detrimental. Such decisions can be beneficial when based on sound nutrition and fitness knowledge, but undermine good health when based on faddism or carried to extremes, as Highlight 9 points out.

• **Nutrition** • Finally, of course, many consumers make food choices that will benefit health. Food manufacturers and restaurant chefs have responded to scientific findings linking health with nutrition by offering an abundant selection of health-promoting foods and beverages. In some cases, the health-promoting foods are as natural and familiar as oatmeal or tomatoes. In other cases, the foods have been processed or prepared in a way that provides health benefits, perhaps by lowering the fat contents (see Highlight 5). In still other cases, manufacturers have developed **functional foods**—products that contain physiologically active ingredients that provide health benefits (see Highlight 13).[8] Examples of functional foods include a margarine made with a certain plant sterol that lowers blood cholesterol and orange juice fortified with calcium to help build strong bones.

Consumers welcome these new foods into their diets, provided that the foods are reasonably priced, clearly labeled, easy to find in the grocery store, and convenient to prepare. These foods must also taste good—as good as the traditional choices.[9] Of course, a person need not eat any of these "special" foods to enjoy a healthy diet; many "regular" foods provide numerous health benefits as well (in other words, they too would be considered functional foods).

IN SUMMARY

A person selects foods for a variety of reasons. Whatever those reasons may be, food choices influence health. Individual food selections neither make nor break a diet's healthfulness, but the balance of foods selected over time can make an important difference to health. For this reason, people are wise to think "nutrition" when making their food choices.

© Bob Daemmrich/Stock Boston

Foods bring pleasure—and nutrients.

THE NUTRIENTS

Biologically speaking, people eat to receive nourishment. Do you ever think of yourself as a biological being made of carefully arranged atoms, molecules, cells, tissues, and organs? Are you aware of the activity going on within your body even as you sit still? The atoms, molecules, and cells of your body continually move and change, even though the structures of your tissues and organs and your external appearance remain relatively constant. Your skin, which has covered you since your birth, is replaced entirely by new cells every seven years. The fat beneath your skin is not the same fat that was there a year ago. Your oldest red blood cell is only 120 days old, and the entire lining of your digestive tract is renewed every 3 days. To maintain your "self," you must continually replenish, from foods, the **energy** and the **nutrients** you deplete in maintaining your body.

Nutrients in Foods and in the Body

Amazingly, the body can derive all the energy, structural materials, and regulating agents that it needs from the foods we eat. This section introduces the nutrients that foods bring to your body and shows how they participate in the dynamic processes that keep people alive and well.

• **Composition of Foods** • Chemical analysis of a food such as a tomato shows that it is composed primarily of water (95 percent). Most of the solid materials are carbohydrates, lipids,◆ and proteins. If you could remove these materials, you would find a tiny residue of vitamins, minerals, and other compounds. Water, carbohydrates, lipids, proteins, vitamins, and some of the minerals found in foods are nutrients—substances the body uses for the growth, maintenance, and repair of its tissues.

◆ As Chapter 5 explains, most lipids are fats.

functional foods: foods that contain physiologically active compounds that provide health benefits beyond their nutrient contributions.

energy: the capacity to do work. The energy in food is chemical energy. The body can convert this chemical energy to mechanical, electrical, or heat energy.

nutrients: chemical substances obtained from food and used in the body to provide energy, structural materials, and regulating agents to support growth, maintenance, and repair of the body's tissues. Nutrients may also reduce the risks of some diseases.

• **Composition of the Body** • A complete chemical analysis of your body would show that it is made of materials similar to those found in foods. A healthy 150-pound body contains about 90 pounds of water and about 30 pounds of fat. The other 30 pounds are mostly compounds containing protein and carbohydrate and the major minerals of the bones. Vitamins, other minerals, and incidental extras constitute a fraction of a pound.

TABLE 1-1 **Elements in the Six Classes of Nutrients**

Notice that organic nutrients contain carbon.

	Carbon	Hydrogen	Oxygen	Nitrogen	Minerals
Inorganic nutrients					
Minerals					✓
Water		✓	✓		
Organic nutrients					
Carbohydrates	✓	✓	✓		
Lipids (fats)	✓	✓	✓		
Proteins[a]	✓	✓	✓	✓	
Vitamins[b]	✓	✓	✓		

[a]Some proteins also contain the mineral sulfur.
[b]Some vitamins contain nitrogen; some contain minerals.

• **Chemical Composition of Nutrients** • The simplest of the nutrients are the minerals. Each mineral is a chemical element; its atoms are all alike. As a result, its identity never changes. Iron, for example, remains iron when a food is cooked, when a person eats the food, when iron becomes part of a red blood cell, when the cell is broken down, and when the iron is lost from the body by excretion. The next simplest nutrient is water, a compound made of two elements—hydrogen and oxygen. Minerals and water are **inorganic** nutrients—they contain no carbon.

The other four classes of nutrients (carbohydrates, lipids, proteins, and vitamins) are more complex. In addition to hydrogen and oxygen, they all contain carbon, an element found in all living things. They are therefore called **organic** compounds (meaning, literally, "alive"). Protein and some vitamins also contain nitrogen and may contain other elements as well (see Table 1-1).

• **Essential Nutrients** • The body can make some nutrients, but it cannot make all of them and it makes some in insufficient quantities to meet its needs. It must obtain these nutrients from foods. The nutrients that foods must supply are **essential nutrients.** When used to refer to nutrients, the word *essential* means more than just "necessary"; it means "needed from outside the body"—normally, from foods.

◆ Carbohydrate, fat, and protein are sometimes called **macronutrients** because they are relatively large molecules. In contrast, vitamins and minerals are **micronutrients.**

• **Nonnutrients** • This book focuses mostly on the nutrients, but foods contain other compounds as well—alcohols, **phytochemicals,** pigments, additives, and others. Some are beneficial, some are neutral, and a few are harmful. Later sections of the book touch on these **nonnutrients** and their significance.

The Energy-Yielding Nutrients

In the body, three of the organic nutrients can be used to provide energy: carbohydrate, fat, and protein.◆ In contrast to these **energy-yielding nutrients,** vitamins, minerals, and water do not yield energy in the human body.

• **Energy Measured in kCalories** • The energy released from carbohydrates, fats, and proteins can be measured in **calories**—tiny units of energy so small that a single apple provides tens of thousands of them. To ease calculations, energy is expressed in 1000-calorie metric units known as kilocalories (shortened

inorganic: not containing carbon or pertaining to living things.
• **in** = not

organic: a substance or molecule containing carbon-carbon bonds or carbon-hydrogen bonds.* Some farmers call their produce "organic" if it was grown without manufactured fertilizers and pesticides, but by the definition given here, all foods are organic.

essential nutrients: nutrients a person must obtain from food because the body cannot make them for itself in sufficient quantity to meet physiological needs; also called **indispensable nutrients.** About 40 nutrients are known to be essential for human beings.

phytochemicals (FIE-toe-KEM-ih-cals): nonnutrient compounds found in plant-derived foods that have biological activity in the body.
• **phyto** = plant

nonnutrients: compounds in foods that do not fit within the six classes of nutrients.

energy-yielding nutrients: the nutrients that break down to yield energy the body can use:
• Carbohydrate.
• Fat.
• Protein.

calories: units by which energy is measured. Food energy is measured in **kilocalories** (1000 calories equal 1 kilocalorie), abbreviated **kcalories** or **kcal.** A capitalized version is also sometimes used: **Calories.** One kcalorie is the amount of heat necessary to raise the temperature of 1 kilogram (kg) of water 1°C.

*This definition excludes coal, diamonds, and a few carbon-containing compounds that contain only a single carbon and no hydrogen, such as carbon dioxide (CO_2), calcium carbonate ($CaCO_3$), magnesium carbonate ($MgCO_3$), and sodium cyanide (NaCN).

HOW TO Think Metric

Like other scientists, nutrition scientists use metric units of measure. They measure food energy in kilocalories, people's height in centimeters, people's weight in kilograms, and the weights of foods and nutrients in grams, milligrams, or micrograms. For ease in using these measures, it helps to remember that the prefixes on the grams imply 1000. For example, a *kilo*gram is 1000 grams, a *milli*gram is 1/1000 of a gram, and a *micro*gram is 1/1000 of a milligram.

Most food labels and many recipe books provide "dual measures," listing both household measures, such as cups, quarts, and teaspoons, and metric measures, such as milliliters, liters, and grams. This practice gives people a chance to gradually learn to "think metric."

A person might begin to "think metric" by simply observing the measure—by noticing the amount of soda in a 2-liter bottle, for example. Through such experiences, a person can become familiar with a measure without having to do any conversions.

To facilitate communication, many members of the international scientific community have adopted a common system of measurement—the International System of Units (SI). In addition to using metric measures, the SI establishes common units of measurement. For example, the SI unit for measuring food energy is the joule (not the kcalorie). A joule is the amount of energy expended when 1 kilogram is moved 1 meter by a force of 1 newton. The joule is thus a measure of *work* energy, whereas the kcalorie is a measure of *heat* energy. While many scientists and journals report their findings in kilojoules (kJ), many others, particularly those in the United States, use kcalories. To convert energy measures from kcalories to kilojoules, multiply by 4.2. For example, a 50-kcalorie cookie provides 210 kilojoules:

$$50 \text{ kcal} \times 4.2 = 210 \text{ kJ}.$$

Exact conversion factors for these and other units of measure are in the Aids to Calculation section on the last two pages of the book.

Volume: Liters (L)

1 L = 1000 milliliters (mL).
0.95 L = 1 quart.
1 mL = 0.03 fluid ounces.
250 mL = 1 cup.

© Felicia Martinez/Photo Edit (both)

A liter of liquid is approximately one U.S. quart. (Four liters are only about 5 percent more than a gallon.)

One cup is about 250 milliliters; a half-cup of liquid is about 125 milliliters.

Weight: Grams (g)

1 g = 1000 milligrams (mg).
1 g = 0.04 ounces (oz).
1 oz = 28.35 g or ≈ 30 g.
100 g ≈ 3½ oz.
1 kilogram (kg) = 1000 g.
1 kg = 2.2 pounds (lb).
454 g = 1 lb.

A kilogram is slightly more than 2 lbs; conversely, a pound weighs about ½ kg.

© Thomas Harm, Tom Peterson/Quest Photographic Inc.

A half-cup of vegetables weighs about 100 grams.

© Tony Freeman/Photo Edit

A 5-pound bag of potatoes weighs about 2 kilograms, and a 176-pound person weighs 80 kilograms.

to kcalories, but commonly called "calories"). When you read in popular books or magazines that an apple provides "100 calories," understand that it means 100 kcalories. This book uses the term *kcalorie* and its abbreviation *kcal* throughout, as do other scientific books and journals. The "How to" on this page provides a few tips on how to "think metric."

A kcalorie is not a constituent of foods; it is a measure of the potential energy in foods. Thus to speak of "kcalories" in a cookie is technically incorrect, just as to speak of the inches in a person is incorrect. It is correct to speak of the *energy* a food provides (and of the *height* of a person).

• Energy from Foods • The amount of energy a food provides depends on how much carbohydrate, fat, and protein it contains. When completely broken down in the body, a gram of carbohydrate yields about 4 kcalories of energy; a gram of protein also yields 4 kcalories; and a gram of fat yields 9 kcalories (see Table 1-2). The "How to" below explains how to calculate the energy available from foods.

One other substance contributes energy: alcohol. Alcohol is not a nutrient because it interferes with the growth, maintenance, and repair of the body, but it does yield energy (7 kcalories per gram) when metabolized in the body.

Most foods contain all three energy-yielding nutrients, as well as water, vitamins, minerals, and other substances. Thus it is inaccurate to describe a food as its predominant nutrient—for example, to speak of meat as protein or of bread as carbohydrate. Meat and bread are *foods* rich in these nutrients. Meat contains water, fat, vitamins, and minerals as well as protein. Bread contains water, a trace of fat, a little protein, and some vitamins and minerals in addition to its carbohydrate. Only a few foods are exceptions to this rule, the common ones being sugar (pure carbohydrate) and oil (essentially pure fat).

TABLE 1-2 kCalorie Values of Energy Nutrients

Energy Nutrients	kCalories[a] (per gram)
Carbohydrate	4 kcal/g
Fat	9 kcal/g
Protein	4 kcal/g

Note: Alcohol contributes 7 kcalories per gram that can be used for energy, but it is not considered a nutrient because it interferes with the body's growth, maintenance, and repair.
[a]For those using kilojoules: 1 g carbohydrate = 17 kJ; 1 g protein = 17 kJ; 1 g fat = 37 kJ; and 1 g alcohol = 29 kJ.

• Energy in the Body • The body uses the energy-yielding nutrients to fuel all its activities. When the body uses carbohydrate, fat, or protein for energy, the bonds between the nutrient's atoms break. As the bonds break, they release energy.◆ Some of this energy is released as heat, but some is used to send electrical impulses through the brain and nerves, to synthesize body compounds, and to move muscles. Thus the energy from food supports every activity from quiet thought to vigorous sports.

◆ The processes by which nutrients are broken down to yield energy or rearranged into body structures are known as **metabolism** (defined and described further in Chapter 7).

If the body does not use these nutrients to fuel its current activities, it rearranges them into storage compounds (such as body fat), to be used between meals and overnight when fresh energy supplies run low. If more energy is consumed than expended, the result is an increase in energy stores and weight gain. Similarly, if less energy is consumed than expended, the result is a decrease in energy stores and weight loss.

When consumed in excess of energy need, alcohol, too, can be converted to body fat and stored. When alcohol contributes a substantial portion of the energy in a person's diet, the harm it does extends far beyond the problems of excess body fat. (Highlight 7 describes the effects of alcohol on health and nutrition.)

• Other Roles of Energy-Yielding Nutrients • In addition to providing energy, carbohydrates, fats, and proteins provide the raw materials for building the body's tissues and regulating its many activities. In fact, protein's role as a fuel source is relatively minor compared with both the other two nutrients and its other roles. Proteins are found in structures such as the muscles and skin and help to regulate activities such as digestion and energy metabolism.

HOW TO Calculate the Energy Available from Foods

To calculate the energy available from a food, multiply the number of grams of carbohydrate, protein, and fat by 4, 4, and 9, respectively. Then add the results together. For example, 1 slice of bread with 1 tablespoon of peanut butter on it contains 16 grams carbohydrate, 7 grams protein, and 9 grams fat:

$$16 \text{ g carbohydrate} \times 4 \text{ kcal/g} = 64 \text{ kcal.}$$
$$7 \text{ g protein} \times 4 \text{ kcal/g} = 28 \text{ kcal.}$$
$$9 \text{ g fat} \times 9 \text{ kcal/g} = 81 \text{ kcal.}$$
$$\text{Total} = 173 \text{ kcal.}$$

From this information, you can calculate the percentage of kcalories each of the energy nutrients contributes to the total. To determine the percentage of kcalories from fat, for example, divide the 81 fat kcalories by the total 173 kcalories:

$$81 \text{ fat kcal} \div 173 \text{ total kcal} = 0.468 \text{ (rounded to 0.47).}$$

Then multiply by 100 to get the percentage:

$$0.47 \times 100 = 47\%.$$

Health recommendations that urge people to limit fat intake to 30 percent of kcalories refer to the day's total energy intake, not to individual foods. Still, if the proportion of fat in each food choice throughout a day exceeds 30 percent of kcalories, then the day's total surely will, too. Knowing that this snack provides 47 percent of its kcalories from fat alerts a person to the need to make lower-fat selections at other times that day.

The Vitamins

The **vitamins** are also organic, but they do not provide energy. Instead, they facilitate the release of energy from carbohydrate, fat, and protein.

There are 13 different vitamins, each with its own special roles to play.* One vitamin enables the eyes to see in dim light, another helps protect the lungs from air

vitamins: organic, essential nutrients required in small amounts by the body for health.

*The water-soluble vitamins are vitamin C and the eight B vitamins: thiamin, riboflavin, niacin, vitamins B_6 and B_{12}, folate, biotin, and pantothenic acid. The fat-soluble vitamins are vitamins A, D, E, and K. The water-soluble vitamins are the subject of Chapter 10 and fat-soluble vitamins, of Chapter 11.

pollution, and still another helps make the sex hormones—among other things. When you cut yourself, one vitamin helps stop the bleeding and another helps repair the skin. Vitamins busily help replace old red blood cells and the lining of the digestive tract. Almost every action in the body requires the assistance of vitamins.

Vitamins can function only if they are intact, but because they are complex organic molecules, they are vulnerable to destruction by heat, light, and chemical agents. This is why the body handles them carefully, and why nutrition-wise cooks do, too. The strategies of cooking vegetables at moderate temperatures, using small amounts of water, and for short times all help to preserve the vitamins.

The Minerals

In the body, some **minerals** are put together in orderly arrays in such structures as bones and teeth. Minerals are also found in the fluids of the body and influence their properties. Whatever their roles, minerals do not yield energy.

Some 16 minerals are known to be essential in human nutrition.* Others are still being studied to determine whether they play significant roles in the human body. Still other minerals are *not* essential nutrients, but are important nevertheless because they are environmental contaminants that displace the nutrient minerals from their workplaces in the body, disrupting body functions. The problems caused by contaminant minerals are described in Chapter 13.

Because minerals are inorganic, they are indestructible and need not be handled with the special care that vitamins require. Minerals can, however, be bound by substances that interfere with the body's ability to absorb them. They can also be lost during food-refining processes or during cooking when they dissolve in water that is discarded.

Water

Water itself is an essential nutrient and naturally carries many minerals, which give it flavor.

Water, indispensable and abundant, provides the environment in which nearly all the body's activities are conducted. It participates in many metabolic reactions and supplies the medium for transporting vital materials to cells and waste products away from them. Water is discussed fully in Chapter 12, but it is mentioned in every chapter. If you watch for it, you cannot help but be impressed by water's participation in all life processes.

IN SUMMARY

Foods provide nutrients—substances that support the growth, maintenance, and repair of the body's tissues. The six classes of nutrients include:

- *Carbohydrates.*
- *Lipids.*
- *Proteins.*
- *Vitamins.*
- *Minerals.*
- *Water.*

Foods rich in the energy-yielding nutrients (carbohydrates, fats, and proteins) provide the major materials for building the body's tissues and yield energy for the body's use or storage. Energy is measured in kcalories. Vitamins, minerals, and water facilitate a variety of activities in the body. Without exaggeration, nutrients provide the physical and metabolic basis for nearly all that we are and all that we do.

minerals: inorganic elements. Some minerals are essential nutrients required in small amounts.

*The major minerals are calcium, phosphorus, potassium, sodium, chloride, magnesium, and sulfur. The trace minerals are iron, iodine, zinc, chromium, selenium, fluoride, molybdenum, copper, and manganese. Chapters 12 and 13 are devoted to the major and trace minerals, respectively.

THE SCIENCE OF NUTRITION

The science of nutrition is the study of the nutrients and other substances in foods and the body's handling of them. As sciences go, nutrition is a young one, but as you can see from the size of this book, much has happened in nutrition's short life. This section introduces the research methods scientists have used in uncovering the wonders of nutrition.

Craig M. Moore

Knowledge about the nutrients and their effects on health comes from scientific study.

Nutrition Research

Research always begins with a question. For example, "What foods or nutrients might protect against the common cold?" In search of an answer, scientists make educated guesses (hypotheses) and then systematically conduct research studies to test each hypothesis. Some examples of the various types of research studies are presented in Figure 1-1. Each type of study has advantages and disadvantages. Findings must be interpreted with an awareness of the study's limitations. (See Highlight 1 for a discussion of how to evaluate research findings.)

In attempting to discover whether a nutrient relieves symptoms or cures a disease, all research tries to answer the same kinds of questions. Research on vitamin C and the common cold illustrates particularly well what those questions

FIGURE 1-1 Research Designs

Epidemiological studies. Scientists observe how much and what kinds of foods a group of people eat and how healthy those people are. Their findings identify factors that might influence the incidence of a disease in various populations.

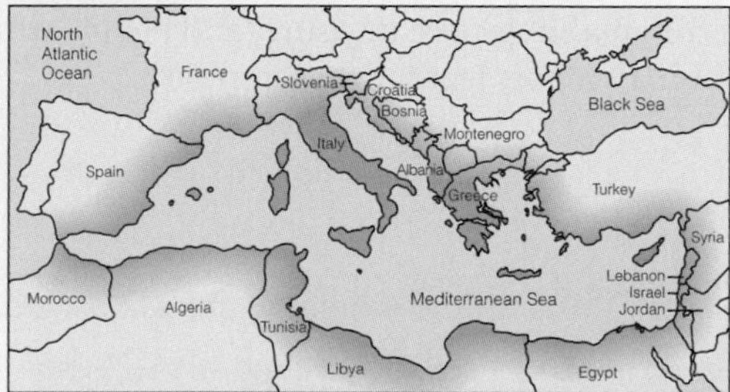

Example. The people of the Mediterranean region drink lots of wine, eat plenty of fat from olive oil, and have a lower incidence of heart disease than northern Europeans and North Americans.

Case-control studies. Researchers compare people who do and do not have a given condition such as a disease, closely matching them in age, gender, and other key variables so that differences in other factors will stand out. These differences may account for the condition in the group that has it.

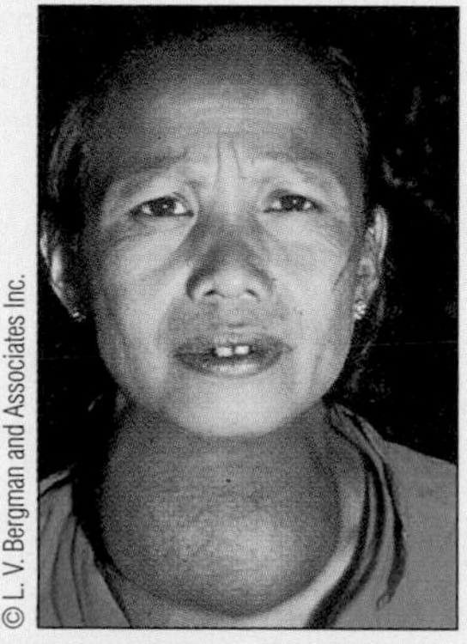
© L. V. Bergman and Associates Inc.

Example. People with goiter lack iodine in their diets.

Animal studies. Researchers feed animals special diets that provide or omit specific nutrients and then observe any changes in health. Such studies test possible disease causes and treatments in a laboratory where all conditions can be controlled.

© R. Benali-Liaison Gamma

Example. Mice fed a high-fat diet eat less food than mice given a lower-fat diet so that they receive the same number of kcalories—but the mice eating the fat-rich diet become severely obese.

Human intervention (or clinical) trials. Scientists ask people to adopt a new behavior (for example, eat a citrus fruit, take a vitamin C supplement, or exercise daily). These trials help determine the effectiveness of such interventions on the development or prevention of disease.

©2001 Photo Disc Inc.

Example. Heart disease risk factors improve when men receive fresh-squeezed orange juice daily for two months compared with those on a diet low in vitamin C—even when both groups follow a diet high in saturated fat.

◆ If the 3-million-year history of the human race were compressed into 24 hours, then scientific discoveries began about 12 seconds ago, and nutrition as an organized science emerged only during the last 3 to 6 seconds.◆

are. Relevant terms appear in boldface type in the text and are defined in the accompanying glossary.

• **Controls** • In studies examining the effectiveness of vitamin C, researchers typically divide the **subjects** into two groups. One group (the **experimental group**) receives a vitamin C supplement, and the other (the **control group**) does not. Researchers observe both groups to determine whether the vitamin C group has fewer or shorter colds than the control group. A number of pitfalls are inherent in an experiment of this kind and must be avoided.

First, each person must have an equal chance of being assigned to either the experimental group or the control group. This is accomplished by **randomization;** that is, the members are chosen from the same population by flipping a coin or some other method involving chance.

Importantly, the two groups of people must be similar and must have the same track record with respect to colds to rule out the possibility that an observed difference might have occurred anyway. If group A would have caught twice as many colds as group B anyway and group B happens to receive the treatment, then the findings prove nothing.

In experiments involving a nutrient, the diets of both groups must also be similar, especially with respect to that nutrient. If those in group B were receiving less vitamin C from their diet, this might cancel the effects of the supplement.

• **Sample Size** • To ensure that chance variation between the two groups does not influence the results, the groups must be large. If one member of a group of five people catches a bad cold by chance, he will pull the whole group's average toward bad colds; but if one member of a group of 500 catches a bad cold, she will not unduly affect the group average. Statistical methods are used to determine whether differences between groups of various sizes support a hypothesis.

• **Placebos** • If people take vitamin C for colds and *believe* it will cure them, their chances of recovery are improved. Taking anything believed to be beneficial hastens recovery in about half of all cases. This phenomenon, the effect of faith on healing, is known as the **placebo effect.** In experiments designed to determine vitamin C's effect on colds, this mind-body effect must be rigorously controlled. Severity of symptoms is often a subjective measure, and people who believe they are receiving treatment may report less severe symptoms.

GLOSSARY OF RESEARCH TERMS

blind experiment: an experiment in which the subjects do not know whether they are members of the experimental group or the control group.

control group: a group of individuals similar in all possible respects to the experimental group except for the treatment. Ideally, the control group receives a placebo while the experimental group receives a real treatment.

correlation (CORE-ee-LAY-shun): the simultaneous increase, decrease, or change in two variables. If A increases as B increases, or if A decreases as B decreases, the correlation is **positive.** (This does not mean that A causes B or vice versa.) If A increases as B decreases, or if A decreases as B increases, the correlation is **negative.** (This does not mean that A prevents B or vice versa.) Some third factor may account for both A and B.

double-blind experiment: an experiment in which neither the subjects nor the researchers know which subjects are members of the experimental group and which are serving as control subjects, until after the experiment is over.

experimental group: a group of individuals similar in all possible respects to the control group except for the treatment. The experimental group receives the real treatment.

peer review: a process in which a panel of scientists rigorously evaluates a research study to assure that the scientific method was followed.

placebo (pla-SEE-bo): an inert, harmless medication given to provide comfort and hope; a sham treatment used in controlled research studies.

placebo effect: the healing effect that faith in medicine, even medicine without pharmaceutical effects, often has.

randomization (RAN-dom-ih-ZAY-shun): a process of choosing the members of the experimental and control groups without bias.

replication (REP-lee-KAY-shun): repeating an experiment and getting the same results. The skeptical scientist, on hearing of a new, exciting finding, will ask, "Has it been replicated yet?" If it hasn't, the scientist will withhold judgment regarding the finding's validity.

subjects: the people or animals participating in a research project.

validity (va-LID-ih-tee): having the quality of being founded on fact or evidence.

variables: factors that change. A variable may depend on another variable (for example, a child's height depends on his age), or it may be independent (for example, a child's height does not depend on the color of her eyes). Sometimes both variables correlate with a third variable (a child's height and eye color both depend on genetics).

One way experimenters control for the placebo effect is to give pills to all participants; some receive pills containing vitamin C, and others receive a **placebo,** pills of similar appearance and taste containing an inactive ingredient. This way, the effects of faith will work equally in both groups. It is not necessary to convince all subjects that they are receiving vitamin C, but the extent of belief or unbelief must be the same in both groups. A study conducted under these conditions is called a **blind experiment**—that is, the subjects do not know (are blind to) whether they are members of the experimental group (receiving treatment) or the control group (receiving the placebo).

• Double Blind • When both the subjects and the researchers do not know which subjects are in which group, the study is called a **double-blind experiment.** Being fallible human beings and having an emotional investment in a successful outcome, researchers might record and interpret results with a bias in the expected direction. To prevent such distortions, the pills are coded by a third party, who does not reveal to the experimenters which subjects were in which group until all results have been recorded.

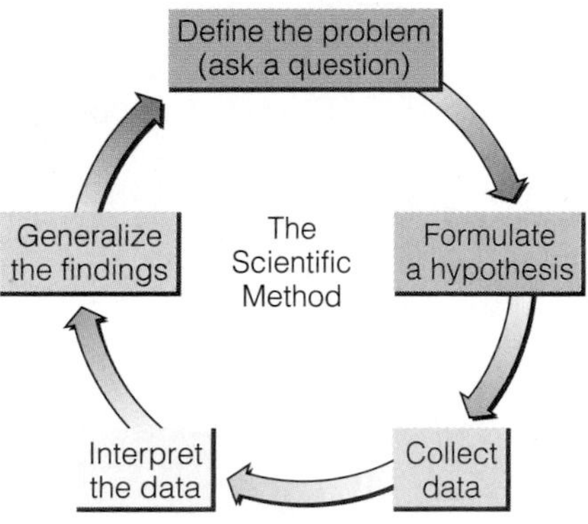

In conducting research, scientists follow these steps, which define the scientific method.

• Correlations and Causes • Researchers often examine the relationships between two or more **variables**—for example, daily vitamin C intake and the number of colds. Findings sometimes suggest no **correlation** between the two variables (regardless of the amount of vitamin C consumed, the number of colds remains the same). Other times, studies find either a **positive correlation** (the more vitamin C, the more colds) or a **negative correlation** (the more vitamin C, the fewer colds). Correlational evidence proves only that two variables are associated, not that one is the cause of the other. People often jump to conclusions when they learn of correlations, but the conclusions are often wrong. To prove that A causes B, scientists have to find evidence of the *mechanism*—that is, to catch A in the act of causing B, so to speak. Furthermore, other scientists must confirm or disprove the findings through **replication** before the results are accepted into the body of nutrition knowledge. Before the findings are published, they are subjected to **peer review**—a process whereby a panel of scientists evaluates the study to confirm that it followed standard scientific methods.

Research versus Rumors

In discussing these subtleties of experimental design, our intent is to show you what a far cry scientific **validity** is from the experience of your neighbor Mary (sample size, one; no control group), who says she takes vitamin C when she feels a cold coming on and "it works every time." She knows what she is taking, she has faith in its effectiveness, and she tends not to notice when it doesn't work. Before concluding that an experiment has shown that a nutrient cures a disease or alleviates a symptom, ask these questions:

- Was there similarity between the control group and the experimental group?
- Was the sample size large enough to rule out chance variation?
- Was a placebo effectively administered (blind)?
- Was the experiment double blind?

These characteristics of well-designed research have enabled scientists to study the actions of nutrients in the body. Such research has laid the foundation for quantifying how much of each nutrient the body needs.

IN SUMMARY Scientists learn about nutrition by conducting experiments that follow the protocol of scientific research. Researchers take care to establish similar control and experimental groups, large sample sizes, placebos, and blind treatments. Their findings must be reviewed and replicated by other scientists before being accepted as valid.

DIETARY REFERENCE INTAKES

Defining the amounts of energy, nutrients, and other dietary components that best support health is a huge task. For more than 60 years, nutrition experts have produced a set of energy and nutrient standards. These standards are revised periodically as new evidence becomes available. Current revisions maintain the original goal of protecting against nutrient deficiencies, but given the abundance of research now linking diet and health, that goal has been broadened to include supporting optimal activities within the body and preventing chronic diseases as well. Previous reports also focused narrowly on nutrients known to be essential. With recent research revealing the health benefits of other dietary components such as fiber and phytochemicals, their recommended intakes need to be addressed as well.

© Tony Freeman/PhotoEdit

To enhance your health, keep nutrition in mind when selecting foods.

To that end, a major revision of nutrient recommendations was implemented. These recommendations are called **Dietary Reference Intakes (DRI)** and reflect the collaborative efforts of researchers in both the United States and Canada. The first in a series of reports was published in 1997 and presents both the framework for developing the DRI and the revised recommendations for the five nutrients that play key roles in bone health—calcium, phosphorus, magnesium, vitamin D, and fluoride.[10] A second report came out in 1998 and features revised recommendations for the eight B vitamins and a related compound choline.[11] The third report, released in 2000, focuses on beta-carotene and the antioxidant nutrients—vitamin C, vitamin E, and selenium.[12] The most recent report, published in 2001, presents recommendations for vitamins A and K and for a dozen minerals.[13]◆ Recommendations for all these nutrients appear on the inside front cover.

◆ Look for upcoming DRI reports on energy and energy-yielding nutrients, fluids and their associated minerals, and other food components.

◆ The DRI reports are produced by the Food and Nutrition Board of the Institute of Medicine in cooperation with scientists from Canada.

Establishing Nutrient Recommendations

The DRI Committee◆ consists of highly qualified scientists who base their estimates of nutrient needs on careful examination and interpretation of scientific evidence. The next paragraphs discuss specific aspects of how the committee goes about establishing the values that make up the DRI:

- Estimated Average Requirements.
- Recommended Dietary Allowances.
- Adequate Intakes.
- Tolerable Upper Intake Levels.

• Estimated Average Requirements • The committee reviews hundreds of research studies to determine the **requirement** for a nutrient—how much is needed in the diet. The committee selects a different criterion for each nutrient based on its roles both in performing activities in the body and in reducing disease risks. From this information, the committee determines an **Estimated Average Requirement** for the nutrient—an amount that appears sufficient to maintain a specific body function in half of the population.

An examination of all the available data reveals that each person's body is unique and has its own set of requirements. Men differ from women, and needs change as a person grows from infancy through old age. For this reason, the committee clusters its recommendations for people into groups by age and gender. Even so, the exact requirements of people the same age and gender are likely to be different. For example, person A might need 40 units of the nutrient each day; person B might need 35; person C, 57. A look at enough individuals might reveal that their requirements fall into a symmetrical distribution, with most near the midpoint (shown in Figure 1-2 as 45 units) and only a few at the extremes.

Dietary Reference Intakes (DRI): a set of values for the dietary nutrient intakes of healthy people in the United States and Canada. These values are used for planning and assessing diets and include:

- Estimated Average Requirements.
- Recommended Dietary Allowances.
- Adequate Intakes.
- Tolerable Upper Intake Levels.

requirement: the lowest continuing intake of a nutrient that will maintain a specified criterion of adequacy.

Estimated Average Requirement: the amount of a nutrient that will maintain a specific biochemical or physiological function in half the people of a given age and gender group.

• Recommended Dietary Allowances (RDA) • Then the committee must decide what intake to recommend for everybody—the **Recommended Dietary Allowance (RDA).** Assuming the distribution shown in Figure 1-2, the Estimated Average Requirement (shown in the figure as 45 units) for each nutrient is probably closest to everyone's need. (Actually, the data for most nutrients other than protein have a much less symmetrical distribution.) But if people consumed exactly the average requirement of a given nutrient each day, half of the population would develop deficiencies of that nutrient; in Figure 1-2, person C would be among them. Recommendations should be set high enough above the Estimated Average Requirement to meet the needs of most healthy people.

FIGURE 1-2 **Estimated Average Requirements and Recommended Dietary Allowances Compared**

Number of people

A B C

20 30 40 50 60 70

Daily requirement for nutrient X (units/day)

Each square represents a person. Some people require only a small amount of the nutrient, and some require a lot, but most fall somewhere near the middle. The text discusses three of these people: A, B, and C.

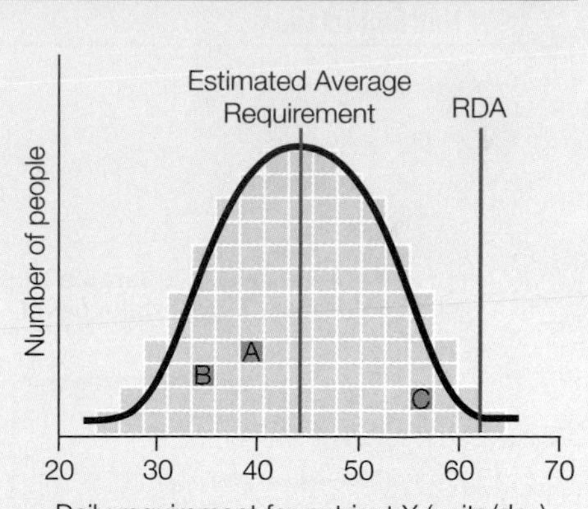

The RDA for a nutrient is set well above the Estimated Average Requirement. It covers about 98% of the population.

In this example, a reasonable RDA might be 63 units a day (see Figure 1-2). Such a point can be calculated mathematically so that it covers about 98 percent of a population. Almost everybody—including person C whose needs were higher than the average—would be covered if they met this dietary goal. Relatively few people's requirements would exceed this recommendation, and even then, they wouldn't exceed by much.

In contrast to the recommendations for nutrients, the value set for energy is not generous. It is set at the *mean* of the population's estimated requirements (see Figure 1-3), representing the average energy needs of a group of individuals, not the exact needs of any one individual. Enough energy is needed to sustain a healthy and active life, but too much energy leads to obesity. In contrast, in the case of vitamins and minerals, small amounts above the daily requirement do no harm, whereas amounts below the requirement lead to health problems. When people's vitamin and mineral intakes are consistently **deficient** (less than the requirement), their nutrient stores decline, and over time this decline leads to poor health and deficiency symptoms. Therefore, to ensure that the vitamin and mineral RDA meet the needs of as many people as possible, these RDA are set near the top end of the range of the population's estimated requirements.

FIGURE 1-3 **Recommended Intakes of Nutrients and Energy Compared**

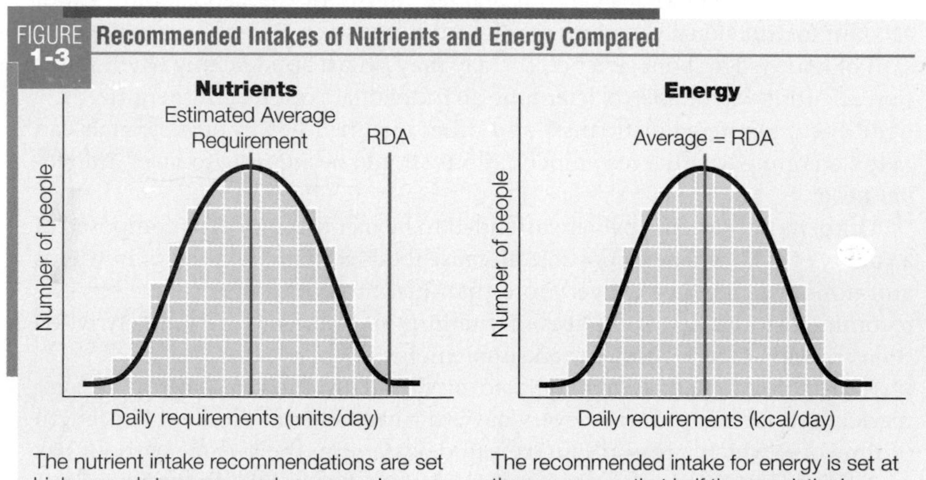

The nutrient intake recommendations are set high enough to cover nearly everyone's requirements (the boxes represent people).

The recommended intake for energy is set at the average so that half the population's requirements fall below and half above it.

Recommended Dietary Allowance (RDA): the average daily amount of a nutrient considered adequate to meet the known nutrient needs of practically all healthy people; a goal for dietary intake by individuals.

deficient: the amount of a nutrient below which almost all healthy people can be expected, over time, to experience deficiency symptoms.

FIGURE 1-4 Naive versus Accurate View of Nutrient Intakes

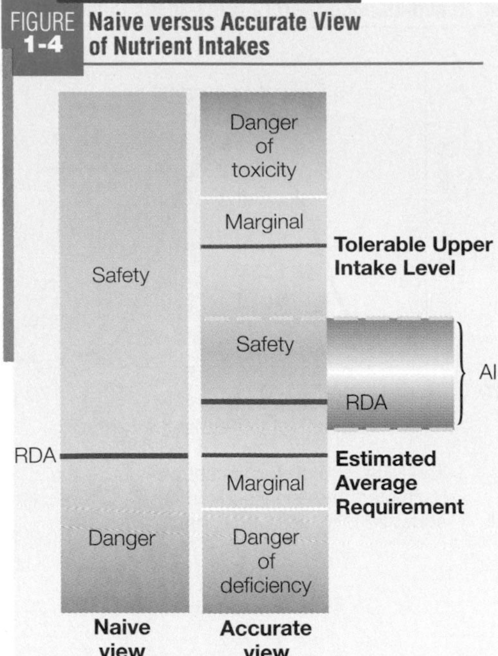

The RDA for a given nutrient represents a point that lies within a range of appropriate and reasonable intakes between toxicity and deficiency. The AI also falls within this range, but its determination is not as exact as an RDA's; the AI may cover more or fewer people than the RDA. Both of these recommendations are high enough to provide reserves in times of short-term dietary inadequacies, but not so high as to approach toxicity. Nutrient intakes above or below this range may be equally harmful.

◆ A **registered dietitian** is a college-educated food and nutrition specialist who is qualified to evaluate people's nutritional health and needs. See Highlight 1 for more on what constitutes a nutrition expert.

Adequate Intake (AI): the average amount of a nutrient that appears sufficient to maintain a specified criterion; a value used as a guide for nutrient intake when an RDA cannot be determined.

Tolerable Upper Intake Level: the maximum amount of a nutrient that appears safe for most healthy people and beyond which there is an increased risk of adverse health effects.

• Adequate Intakes (AI) • For some nutrients, there is insufficient scientific evidence to determine an Estimated Average Requirement (which is needed to set an RDA). In these cases, the committee establishes an **Adequate Intake (AI)** instead of an RDA. An AI reflects the average amount of a nutrient that a group of healthy people consumes. Like the RDA, the AI may be used as nutrient goals for individuals.

Although both RDA and AI serve as nutrient intake goals for individuals, their differences are noteworthy. An RDA for a given nutrient is based on enough scientific evidence to expect that the needs of almost all healthy people will be met. An AI, on the other hand, must rely more heavily on scientific judgments because sufficient evidence is lacking. The percentage of people covered by an AI is unknown; an AI is expected to exceed average requirements, but it may cover more or fewer people than an RDA would (if an RDA could be determined). For these reasons, AI values are more tentative than RDA. The table on the inside front cover identifies which nutrients have an RDA and which have an AI. Later chapters present the RDA and AI values for the vitamins and minerals.

• Tolerable Upper Intake Levels • The recommended intakes for nutrients are generous, and although they do not necessarily cover every individual for every nutrient, they probably should not be exceeded by much. People's tolerances for high doses of nutrients vary, and somewhere above the recommended intake is a **Tolerable Upper Intake Level** beyond which a nutrient is likely to become toxic.[14] It is naive to think of recommendations as minimum amounts. A more accurate view is to see a person's nutrient needs as falling within a range, with marginal and danger zones both below and above it (see Figure 1-4).

Upper levels are particularly useful in guarding against the overconsumption of nutrients, which is most likely to occur when people use supplements or fortified foods regularly. Later chapters discuss the dangers associated with excessively high intakes of vitamins and minerals, and the inside front cover presents a table that includes the upper-level values for selected nutrients.

Using Nutrient Recommendations

Although the intent of nutrient recommendations may seem simple enough, they are the subject of much misunderstanding and controversy. Perhaps the following facts will help put them in perspective. First, estimates of adequate energy and nutrient intakes apply to *healthy* people. They do not apply to malnourished people or to those with medical problems who may require supplemented or restricted intakes.

Second, these *recommendations* include a generous margin of safety. They are not minimum requirements, nor are they necessarily optimal intakes for all individuals. Recommendations can only target "most" of the people and cannot account for individual variations in nutrient needs—yet. Given the recent explosion of knowledge about genetics, the day may be fast approaching when nutrition scientists will be able to determine an individual's optimal nutrient needs.[15] Until then, registered dietitians◆ and other qualified health professionals can help determine whether recommendations should be adjusted to meet individual needs.

Third, most nutrient goals are intended to be met through diets composed of a variety of *foods* whenever possible. Because foods contain mixtures of nutrients and nonnutrients, they deliver more than just those nutrients covered by the recommendations. Excess intakes of vitamins and minerals are unlikely when their sources are foods rather than supplements.

Fourth, recommendations apply to *average* daily intakes. Meeting recommendations for every nutrient every day is difficult and unnecessary. The length of time over which a person's intake can deviate from the average without risk of deficiency or overdose varies for each nutrient, depending on the body's use and storage of the nutrient. For most nutrients (such as thiamin and vitamin C), deprivation would lead to rapid development of deficiency symptoms (within days or weeks); for others (such as vitamin A and vitamin B_{12}), deficiencies would develop much slower (over months or years).

Fifth, each of the four DRI categories serves a unique purpose. For example, the Estimated Average Requirements are most appropriately used to develop and evaluate nutrition programs for *groups* such as schoolchildren or military personnel. The RDA (or AI if an RDA is not available) can be used to set goals for *individuals*. Tolerable Upper Intake Levels help to keep nutrient intakes below the amounts that increase the risk of toxicity. With these understandings, professionals can use the DRI for a variety of purposes.

Comparing Nutrient Recommendations

At least 40 different nations and international organizations have published nutrient standards similar to those used in the United States and Canada. Slight differences may be apparent, reflecting differences both in the interpretation of the data from which the standards were derived and in the food habits and physical activities of the populations they serve.

Many countries use the recommendations developed by two international groups: the Food and Agriculture Organization (of the United Nations) and the World Health Organization.◆ The FAO/WHO recommendations are considered sufficient to maintain health in nearly all healthy people worldwide.

◆ Nutrient recommendations from FAO/WHO are provided in Appendix I.

IN SUMMARY

The Dietary Reference Intakes (DRI) are a set of four nutrient intake values that can be used to plan and evaluate diets for healthy people. The Estimated Average Requirement defines the amount of a nutrient that supports a specific function in the body for half of the population. The Recommended Dietary Allowance (RDA) is based on the Estimated Average Requirement and establishes a goal for dietary intake that will meet the needs of almost all healthy people. An Adequate Intake (AI) serves a similar purpose when an RDA cannot be determined. The Tolerable Upper Intake Level establishes the highest amount that appears safe for regular consumption.

NUTRITION ASSESSMENT

What happens when a person doesn't get enough of a nutrient or energy or gets too much? If the deficiency or excess is significant over time, the person exhibits signs of **malnutrition.** With a deficiency of energy, the person may display the symptoms of **undernutrition** by becoming extremely thin, losing muscle tissue, and becoming prone to infection and disease. With a deficiency of a nutrient, the person may experience skin rashes, depression, hair loss, bleeding gums, muscle spasms, night blindness, or other symptoms. With an excess of energy, the person may become obese and vulnerable to diseases associated with **overnutrition** such as heart disease and diabetes. With a sudden nutrient overdose, the person may experience hot flashes, yellowing skin, a rapid heart rate, low blood pressure, or other symptoms. Similarly, regular intakes in excess of needs may also have adverse effects.

Malnutrition symptoms are easy to miss. They resemble the symptoms of other diseases: diarrhea, skin rashes, pain, and the like. But a person who has learned how to use assessment techniques to detect malnutrition can tell when these conditions are caused by poor nutrition and can take steps to correct it. This discussion presents the basics of nutrition assessment; many more details are offered in Chapter 17 and in Appendix E.

malnutrition: any condition caused by excess or deficient food energy or nutrient intake or by an imbalance of nutrients.
- **mal** = bad

undernutrition: deficient energy or nutrients.

overnutrition: excess energy or nutrients.

Nutrition Assessment of Individuals

To prepare a **nutrition assessment,** the assessor, usually a registered dietitian or a physician trained in clinical nutrition, uses:

- Historical information.
- Anthropometric data.
- Physical examinations.
- Laboratory tests.

Each of these methods involves collecting data in various ways and interpreting each finding in relation to the others to create a total picture.

• Historical Information • One step in evaluating nutrition status is to obtain information about a person's history with respect to health status, socioeconomic status, drug use, and diet. The health history reflects a person's medical record and may reveal a disease that interferes with the person's ability to eat or the body's use of nutrients. The person's family history of major diseases is also noteworthy, especially for conditions such as heart disease that have a genetic tendency to run in families. Economic circumstances may show a financial inability to buy foods or inadequate kitchen facilities in which to prepare them. Social factors such as marital status, ethnic background, and educational level also influence food choices and nutrition status. A drug history may highlight possible diet-medication interactions that lead to nutrient deficiencies (as described in Chapter 18). A diet history can indicate whether the diet may be under- or oversupplying nutrients or energy.

◆ Chapter 17 describes the tools used to obtain food intake data: the 24-hour recall, usual intake record, food frequency questionnaire, and food record.

To take a diet history,◆ the assessor collects and analyzes data about the foods a person eats. The data may be collected by recording the foods the person has eaten over a period of 24 hours, three days, or a week or more or by asking what foods the person typically eats and how much of each. The days in the record have to be fairly typical of the person's diet, and portion sizes must be recorded accurately. To determine the amounts of nutrients consumed, the assessor usually enters the foods and their portion sizes into a computer using a diet analysis program. Alternatively, this step can be done manually by looking up each food in a table of food composition such as Appendix H in this book. Then the assessor compares the calculated nutrient intakes with recommended intakes.

An estimate of energy and nutrient intakes from a diet history, combined with other sources of information, can help confirm or rule out the *possibility* of suspected nutrition problems. A sufficient intake of a nutrient does not guarantee adequate nutrition status for an individual, and an insufficient intake does not always indicate a deficiency, but such findings warn of possible problems.

• Anthropometric Data • A second technique that may help to reveal nutrition problems is the taking of **anthropometric** measures such as height and weight. The assessor compares measurements taken on an individual with standards specific for gender and age or with previous measures on the same individual.

Measurements taken periodically and compared with previous measurements reveal patterns and indicate trends in a person's overall nutrition status, but they provide little information about specific nutrients. Instead, measurements out of line with expectations may reveal such problems as growth failure in children, wasting or swelling of body tissues in adults, and obesity—conditions that may reflect energy or nutrient deficiencies or excesses.

• Physical Examinations • A third nutrition assessment technique is a physical examination that looks for clues to poor nutrition status. Every part of the body that can be inspected can offer such clues: the hair, eyes, skin, posture, tongue, fingernails, and others. The examination requires skill, for many physical signs can reflect more than one nutrient deficiency or toxicity or even nonnutrition conditions. Like the other assessment techniques, a physical examination does not by itself point to firm conclusions. Instead, it reveals pos-

nutrition assessment: a comprehensive evaluation of a person's nutrition status, completed by a registered dietitian, using health, socioeconomic, drug, and diet histories; anthropometric measurements; physical examinations; and laboratory tests.

anthropometric (AN-throw-poe-MET-rick): relating to measurement of the physical characteristics of the body, such as height and weight.
- **anthropos** = human
- **metric** = measuring

sible nutrient imbalances for other assessment techniques to confirm, or it confirms data collected from other assessment measures.

• Laboratory Tests • A fourth way to detect a developing deficiency, imbalance, or toxicity is to take samples of blood or urine, analyze them in the laboratory, and compare the results with normal values for a similar population. A goal of nutrition assessment is to uncover early signs of malnutrition before symptoms appear, and laboratory tests are most useful for this purpose. In addition, they can confirm suspicions raised by other assessment methods.

• Iron, for Example • The mineral iron can be used to illustrate the stages in the development of a nutrient deficiency and the assessment techniques useful in detecting them. The **overt,** or outward, signs of an iron deficiency appear at the end of a long sequence of events. Figure 1-5 describes what happens in the body as a nutrient deficiency progresses and shows which assessment methods can reveal those changes.

First, too little iron gets into the body—either because iron is lacking in the person's diet (a **primary deficiency**) or because the person's body doesn't absorb or use iron normally (a **secondary deficiency**). A diet history provides clues to primary deficiencies; a health history provides clues to secondary deficiencies.

Then the body begins to use up its stores of iron. At this stage, the deficiency might be described as **subclinical.** It exists as a **covert** condition and might be detected by laboratory tests, but outward signs have not yet appeared.

Finally, iron stores are exhausted. Now, the body cannot make enough iron-containing red blood cells to replace those that are aging and dying. The iron in red blood cells normally carries oxygen to all the body's tissues. When iron is

A peek inside the mouth provides clues to a person's nutrition status.

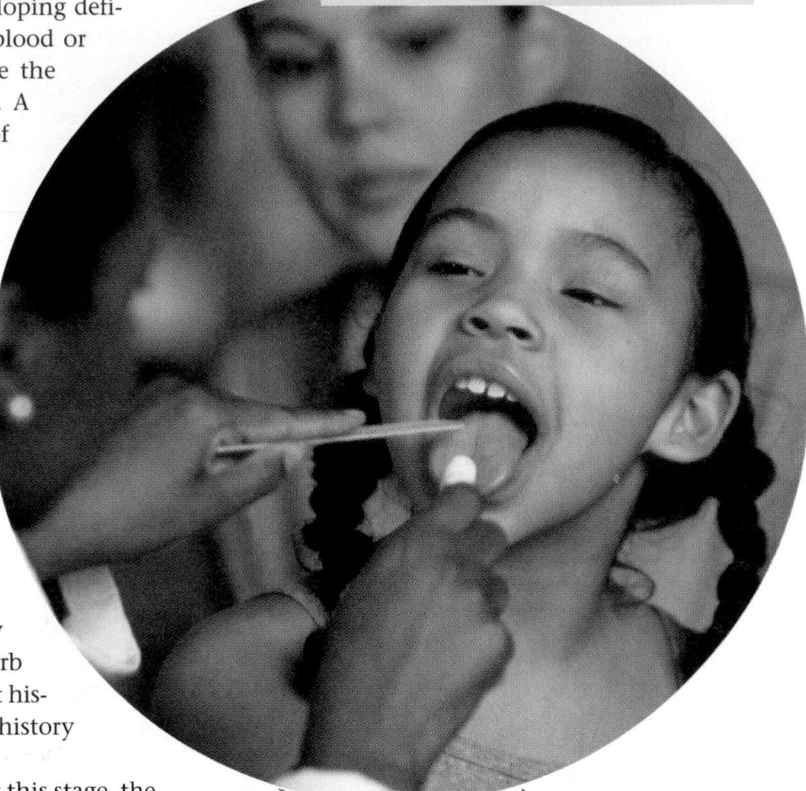

Tom & Dee Ann McCarthy/The Stock Market

FIGURE 1-5 Stages in the Development of a Nutrient Deficiency

Internal changes precede outward signs of deficiencies. As a corollary, signs of sickness need not appear before a person takes corrective measures. Tests can either reveal the presence of problems in the early stages or confirm that nutrient stores are adequate.

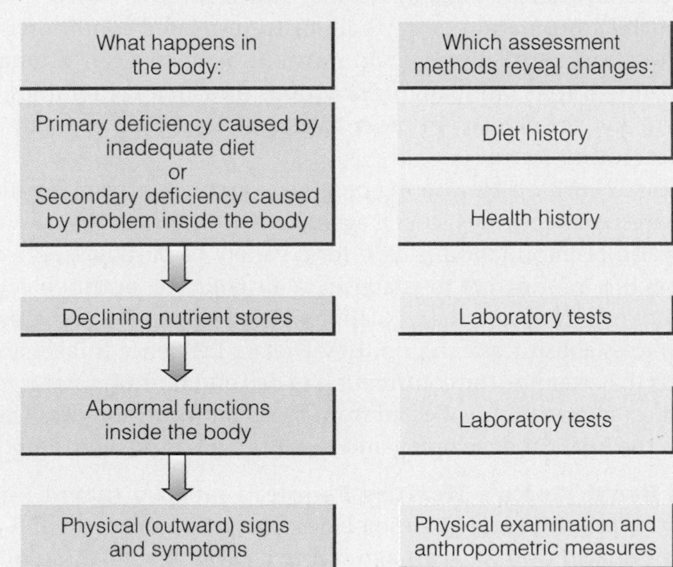

overt (oh-VERT): out in the open and easy to observe.
- **ouvrir** = to open

primary deficiency: a nutrient deficiency caused by inadequate dietary intake of a nutrient.

secondary deficiency: a nutrient deficiency caused by something other than an inadequate intake such as a disease condition that reduces absorption, accelerates use, hastens excretion, or destroys the nutrient.

subclinical deficiency: a deficiency in the early stages, before the outward signs have appeared.

covert (KOH-vert): hidden, as if under covers.
- **couvrir** = to cover

TABLE 1-3 Healthy People 2010 Nutrition and Overweight Objectives

- Increase the proportion of adults who are at a *healthy weight.*
- Reduce the proportion of adults who are *obese.*
- Reduce the proportion of children and adolescents who are *overweight* or *obese.*
- Reduce *growth retardation* among low-income children under age 5 years.
- Increase the proportion of persons aged 2 years and older who consume at least two daily servings of *fruit.*
- Increase the proportion of persons aged 2 years and older who consume at least three daily servings of *vegetables,* with at least one-third being dark green or orange vegetables.
- Increase the proportion of persons aged 2 years and older who consume at least six daily servings of *grain products,* with at least three being whole grains.
- Increase the proportion of persons aged 2 years and older who consume less than 10 percent of kcalories from *saturated fat.*
- Increase the proportion of persons aged 2 years and older who consume no more than 30 percent of kcalories from *total fat.*
- Increase the proportion of persons aged 2 years and older who consume 2400 mg or less of *sodium.*
- Increase the proportion of persons aged 2 years and older who meet dietary recommendations for *calcium.*
- Reduce *iron deficiency* among young children, females of childbearing age, and pregnant females.
- Reduce *anemia* among low-income pregnant females in their third trimester.
- Increase the proportion of children and adolescents aged 6 to 19 years whose intake of *meals and snacks at school* contributes to good overall dietary quality.
- Increase the proportion of worksites that offer *nutrition or weight management classes or counseling.*
- Increase the proportion of physician office visits made by patients with a diagnosis of cardiovascular disease, diabetes, or hyperlipidemia that include *counseling or education related to diet and nutrition.*
- Increase *food security* among U.S. households and in so doing reduce hunger.

Note: "Nutrition and Overweight" is one of 26 focus areas, each with numerous objectives. Several of the other focus areas have nutrition-related objectives, and these are presented in later chapters.
SOURCE: Healthy People 2010, www.healthypeople.gov, site updated on November 11, 2000, and accessed on November 29, 2000.

lacking, fewer red blood cells are made, the new ones are pale and small, and every part of the body feels the effects of an oxygen shortage. Now the overt symptoms of deficiency appear—weakness, fatigue, pallor, and headaches, reflecting the iron-deficient state of the blood. Physical examination would reveal these symptoms.

Nutrition Assessment of Populations

To assess a population's nutrition status, researchers conduct surveys using techniques similar to those used on individuals. The data collected are then used by various agencies for numerous purposes, including the development of national health goals.

• **National Nutrition Surveys** • One kind of survey—a **food consumption survey**—determines the kinds and amounts of foods people eat. Then researchers calculate the energy and nutrients in the foods and compare the amounts consumed with a standard. An example of this type of survey is the Continuing Survey of Food Intakes by Individuals (CSFII), popularly known as *What We Eat in America,* which gathered information from about 16,000 people using food intake records and questionnaires about their knowledge and attitudes about diet and health.

Another kind of survey—a **nutrition status survey**—examines the people themselves, using nutrition assessment methods. The National Health and Nutrition Examination Survey (NHANES) is an example of a nutrition status survey. The current NHANES is an ongoing project that gathers information from about 50,000 people using diet histories, anthropometric measurements, physical examinations, and laboratory tests.[16] The data provide valuable information on several nutrition-related conditions, such as growth retardation, heart disease, and nutrient deficiencies. National nutrition surveys often oversample high-risk groups (low-income families, infants and children, and the elderly) in order to glean an accurate estimate of their health and nutrition status.

The National Nutrition Monitoring program coordinates the many nutrition-related activities of the various federal agencies that conduct these surveys. The resulting wealth of information is used for a variety of purposes. For example, Congress uses this information to establish public policy on nutrition education, food assistance programs, and the regulation of the food supply. Scientists use the information to establish research priorities. Dietary Reference Intakes and other major reports that examine the contribution of diet and nutrition status to health also depend on information collected from these nutrition surveys. These data also provide the basis for developing and monitoring national health goals.

• **National Health Goals** • **Healthy People,** a program that identifies the nation's health priorities and guides policies that promote health and prevent disease, was initiated over 20 years ago. Table 1-3 lists the nutrition objectives for 2010.

food consumption survey: a survey that measures the amounts and kinds of foods people consume (using diet histories), estimates the nutrient intakes, and compares them with a standard.

nutrition status survey: a survey that evaluates people's nutrition status using diet histories, anthropometric measures, physical examinations, and laboratory tests.

Healthy People: a national public health initiative under the jurisdiction of the U.S. Department of Health and Human Services (DHHS) that identifies the most significant preventable threats to health and focuses efforts toward eliminating them.

At the start of each decade, the program sets goals for improving the nation's health during the following 10 years. The goals of Healthy People 2010 focus on "improving the quality of life and eliminating disparity in health among racial and ethnic groups."[17]

At the close of the twentieth century, the nation's progress toward meeting its Healthy People 2000 goals was mixed.[18] For almost 60 percent of the objectives, the population either met the target or was moving in the right direction. Successes included reductions in the incidence of food- and water-borne infections, oral and breast cancer, and infant mortality, for example. On the downside, the population was moving in the opposite direction of several key objectives, most notably for reducing overweight and increasing physical activity.

TABLE 1-4 Ten Leading Causes of Death in the United States

	Percentage of Total Deaths
1. **Heart disease.**	31.4
2. **Cancers.**	23.3
3. **Strokes.**	6.9
4. Chronic obstructive lung disease.	4.7
5. Accidents.	4.1
6. Pneumonia and influenza.	3.7
7. **Diabetes mellitus.**	2.7
8. Suicide.	1.3
9. Kidney disease.	1.1
10. Chronic liver disease.	1.1

Note: The diseases in bold type have relationships with diet.

IN SUMMARY

People become malnourished when they get too little or too much energy or nutrients. Deficiencies, excesses, and imbalances of nutrients lead to malnutrition diseases. To detect malnutrition in individuals, health care professionals use four nutrition assessment methods. Reviewing dietary data and health information may suggest a nutrition problem in its earliest stages. Laboratory tests may detect it before it becomes overt, whereas anthropometrics and physical examinations pick up on the problem only after it is causing symptoms. Similar assessment methods are used in surveys to measure people's food consumption and to evaluate the nutrition status of populations.

DIET AND HEALTH

Diet has always played a vital role in supporting health. Early nutrition research focused on identifying the nutrients in foods that would prevent such common diseases as rickets and scurvy, the vitamin D– and vitamin C–deficiency diseases. More recently, with nutrient deficiencies no longer a major health threat, nutrition research has focused on **chronic diseases** associated with energy and nutrient excesses. Today, overconsumption of foods—especially foods high in fats—is a major health concern for people in the United States.

Chronic Diseases

Table 1-4 lists the ten leading causes of death in the United States. These "causes" are stated as if a single condition such as heart disease caused death, but most chronic diseases arise from multiple factors over many years. A person who died of heart failure may have been overweight, had high blood pressure, been a cigarette smoker, and spent years eating a high-fat diet and getting too little exercise.

Of course, not all people who die of heart disease fit this description, nor do all people with these characteristics die of heart disease. People who are overweight might die from the complications of diabetes instead, or those who smoke might die of cancer. They might even die from something totally unrelated to any of these factors, such as an automobile accident. Still, statistical studies have shown that certain conditions and behaviors are linked to certain diseases.

Risk Factors for Chronic Diseases

Factors that increase or reduce the *risk* of developing chronic diseases are identified by analyzing statistical data. A strong association between a **risk factor** and a disease means that when the factor is present, the *likelihood* of developing the disease increases. It does not mean that all people with the risk factor will

chronic diseases: long-duration degenerative diseases characterized by deterioration of the body organs. Examples include heart disease, cancer, and diabetes.

risk factor: a condition or behavior associated with an elevated frequency of a disease but not proved to be causal. Risk factors for disease include overweight, cigarette smoking, alcohol abuse, high blood pressure, high blood cholesterol, high-fat diet, and physical inactivity.

develop the disease. Similarly, a lack of risk factors does not guarantee freedom from a given disease. On the average, though, the more risk factors in a person's life, the greater that person's chances of developing the disease. Conversely, the fewer risk factors in a person's life, the better the chances for good health.

• **Risk Factors Persist** • Risk factors tend to persist over time. Without intervention, a young adult with high blood pressure will most likely continue to have high blood pressure as an older adult, for example. Thus, to minimize the damage, early intervention is most effective.

• **Risk Factors Cluster** • Risk factors also tend to cluster. For example, a person who is overweight is likely to be physically inactive, to have high blood pressure, and to have high blood cholesterol—all risk factors associated with heart disease. Intervention that focuses on one risk factor often benefits the others as well. For example, physical activity can help reduce weight. Then both physical activity and weight loss will help to lower blood pressure and blood cholesterol.

• **Risk Factors in Perspective** • The most prominent factor contributing to death in the United States is tobacco use,◆ followed by diet and activity patterns, and alcohol use. Risk factors such as smoking, poor dietary habits, physical inactivity, and alcohol consumption are personal behaviors that can be changed. Decisions to not smoke, to eat a well-balanced diet, to engage in regular physical activity, and to drink alcohol in moderation (if at all) improve the likelihood that a person will enjoy good health. Other risk factors, such as genetics, gender, and age, also play important roles in the development of chronic diseases, but they cannot be changed. Health recommendations acknowledge the influence of such factors on the development of disease, but must focus on those that are changeable. For the two out of three Americans who do not smoke or drink alcohol excessively, the one choice that can influence long-term health prospects more than any other is diet.[19]

◆ Cigarette smoking is responsible for almost one of every five deaths each year.

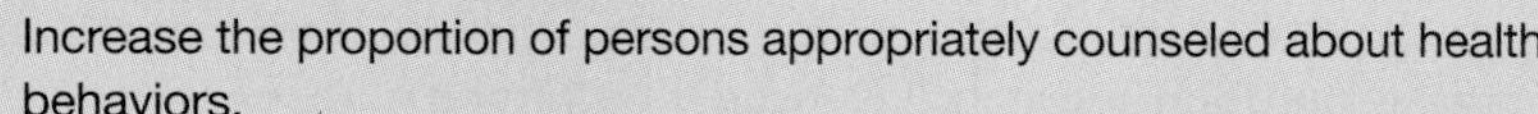

HEALTHY PEOPLE 2010

Increase the proportion of persons appropriately counseled about health behaviors.

IN SUMMARY

Within the range set by genetics, a person's choice of diet influences long-term health. Diet has no influence on some diseases, but is linked closely to others. Personal life choices, such as physical activity and tobacco or alcohol use, also affect health for the better or worse.

The next several chapters will provide many more details about nutrients and how they support health. Whenever appropriate, the discussion will show how diet influences each of today's major diseases. Dietary recommendations will appear again and again, as each nutrient's relationships with health are explored. Most people who follow the recommendations will benefit and can enjoy good health into their later years.

How are you doing?

- How often do you think of health and nutrition when making food choices?
- Which disease risk factors (listed on p. 19 in the definition) do you have?
- What lifestyle changes could you make to improve your chances of enjoying good health?

Nutrition on the Net

WEBSITES

Access these websites for further study of topics covered in this chapter.

- Find updates and quick links to these and other nutrition-related sites at our website: **www.wadsworth.com/nutrition**
- Search for "nutrition" at the U.S. Government health information site: **www.healthfinder.gov**
- Review the Dietary Reference Intakes: **www.nap.edu/readingroom** or **www2.nas.edu/fnb**
- Review nutrient recommendations from the Food and Agriculture Organization and the World Health Organization: **www.fao.org** and **www.who.org**
- View Healthy People 2010: **web.health.gov/healthypeople**
- Review the Canadian *National Plan of Action for Nutrition:* **www.hc-sc.gc.ca/datahpsb/npu**
- Learn about NHANES: **www.cdc.gov/nchs/nhanes.htm**
- Get information from the Food Surveys Research Group: **www.barc.usda.gov/bhnrc/foodsurvey**
- Visit the nutrition center of the Mayo Clinic: **www.mayohealth.org**

INTERNET ACTIVITIES

The explosive growth of the Internet has dramatically increased the amount of health-related information available to consumers. Some of this information is helpful, while some is worthless. Do you know how to tell the difference?

- Go to: **www.quackwatch.com**
- Scroll down to "General Observations."
- Click on "Quackery: 25 Ways to Spot It."
- Read the information on this page and identify the techniques that you have read or heard about personally.

How do you sort out the reliable information from the junk? Become an informed consumer.

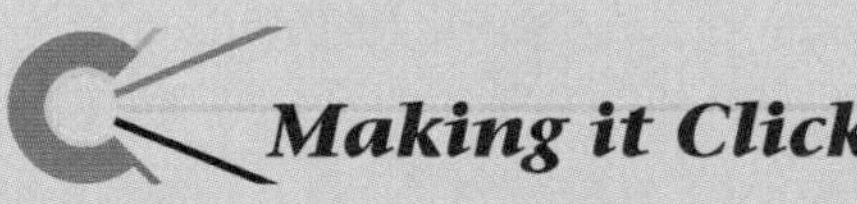

Making it Click

Several chapters end with problems to give you practice in doing simple nutrition-related calculations. They use hypothetical situations in order to teach a lesson that can provide answers (see p. 23). Once you have mastered these examples, you will be prepared to examine your own food choices. Be sure to show your calculations for each problem.

1. Calculate the energy provided by a food from its energy-nutrient contents. A cup of fried rice contains 5 grams protein, 30 grams carbohydrate, and 11 grams fat.
   a. How many kcalories does the rice provide from these energy nutrients?

      __________ = ___ kcal protein.
      __________ = ___ kcal carbohydrate.
      __________ = ___ kcal fat.
      Total = ___ kcal.

   b. What percentage of the energy in the fried rice comes from each of the energy-yielding nutrients?

      __________ = ___ % kcal from protein.
      __________ = ___ % kcal from carbohydrate.
      __________ = ___ % kcal from fat.
      Total = ___ %

      Note: The total should add up to 100%; 99% or 101% due to rounding is also acceptable.

   c. Calculate how many of the 146 kcalories provided by a 12-ounce can of beer come from alcohol, if the beer contains 1 gram protein and 13 grams carbohydrate. (Hint: The remaining kcalories derive from alcohol.)

      1 g protein = ___ kcal protein.
      13 g carbohydrate = ___ kcal carbohydrate.
      = ___ kcal alcohol.

      How many grams of alcohol does this represent? ___ g alcohol.

2. Even a little nutrition knowledge can help you identify some bogus claims. Consider an advertisement for a new "super supplement" that claims the product provides 15 grams protein and 10 kcalories per dose. Is this possible? ___ Why or why not? __________ = ___ kcal.

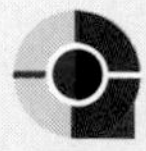

Study Questions

These questions will help you review this chapter. You will find the answers in the discussions on the pages provided.

1. Give several reasons (and examples) why people make the food choices that they do. (pp. 2–4)
2. What is a nutrient? Name the six classes of nutrients found in foods. What is an essential nutrient? (pp. 4–5)
3. Which nutrients are inorganic, and which are organic? Discuss the significance of that distinction. (pp. 5, 8)
4. Which nutrients yield energy, and how much energy do they yield per gram? How is energy measured? (pp. 5–7)
5. Describe how alcohol resembles nutrients. Why is alcohol not considered a nutrient? (p. 7)
6. What is the science of nutrition? Describe the types of research studies and methods used in acquiring nutrition information. (pp. 9–11)
7. Explain how variables might be correlational but not causal. (p. 11)
8. What are the DRI? Who develops the DRI? To whom do they apply? How are they used? In your description, identify the four categories of DRI and indicate how they are related. (pp. 12–15)
9. What judgment factors are involved in setting the energy and nutrient recommendations? (pp. 13–14)
10. What happens when people get either too little or too much energy or nutrients? Define malnutrition, undernutrition, and overnutrition. Describe the four methods used to detect energy and nutrient deficiencies and excesses. (pp. 15–17)
11. What methods are used in nutrition surveys? What kinds of information can these surveys provide? (pp. 18–19)
12. Describe risk factors and their relationships to disease. (pp. 19–20)

These questions will help you prepare for an exam. Answers can be found on p. 23.

1. When people eat the foods typical of their families or geographic region, their choices are influenced by:
   a. habit.
   b. nutrition.
   c. personal preference.
   d. ethnic heritage or tradition.
2. Both the human body and many foods are composed mostly of:
   a. fat.
   b. water.
   c. minerals.
   d. proteins.
3. The inorganic nutrients are:
   a. proteins and fats.
   b. vitamins and minerals.
   c. minerals and water.
   d. vitamins and proteins.
4. The energy-yielding nutrients are:
   a. fats, minerals, and water.
   b. minerals, proteins, and vitamins.
   c. carbohydrates, fats, and vitamins.
   d. carbohydrates, fats, and proteins.
5. Studies of populations that reveal correlations between dietary habits and disease incidence are:
   a. clinical trials.
   b. laboratory studies.
   c. case-control studies.
   d. epidemiological studies.
6. An experiment in which neither the researchers nor the subjects know who is receiving the treatment is known as:
   a. double blind.
   b. double control.
   c. blind variable.
   d. placebo control.
7. An RDA represents the:
   a. highest amount of a nutrient that appears safe for most healthy people.
   b. lowest amount of a nutrient that will maintain a specified criterion of adequacy.
   c. average amount of a nutrient considered adequate to meet the known nutrient needs of practically all healthy people.
   d. average amount of a nutrient that will maintain a specific biochemical or physiological function in half the people.
8. Historical information, physical examinations, laboratory tests, and anthropometric measures are:
   a. techniques used in diet planning.
   b. steps used in the scientific method.
   c. approaches used in disease prevention.
   d. methods used in a nutrition assessment.
9. A deficiency caused by an inadequate dietary intake is a(n):
   a. overt deficiency.
   b. covert deficiency.
   c. primary deficiency.
   d. secondary deficiency.
10. Behaviors such as smoking, dietary habits, physical activity, and alcohol consumption that influence the development of disease are known as:
    a. risk factors.
    b. chronic causes.
    c. preventive agents.
    d. disease descriptors.

References

1 K. Glanz and coauthors, Why Americans eat what they do: Taste, nutrition, cost, convenience, and weight control concerns as influences on food consumption, *Journal of the American Dietetic Association* 98 (1998): 1118–1126; Food Marketing Institute, *Trends in the United States: Consumer Attitudes and the Supermarket* (Chicago: Food Marketing Institute, 1996).

2 A. Drewnowski, Why do we like fat? *Journal of the American Dietetic Association* 97 (1997): S58–S62.

3 L. L. Birch, Development of food preferences, *Annual Review of Nutrition* 19 (1999): 41–62; M. B. M. van den Bree, L. J. Eaves, and J. T. Dwyer, Genetic and environmental influences on eating patterns of twins aged ≥50 y, *American Journal of Clinical Nutrition* 70 (1999): 456–465.

4 J. Cousminer and G. Hartman, Understanding America's regional taste preferences, *Food Technology* 50 (1996): 73–77.

5 F. Katz, "How nutritious?" meets "How convenient?" *Food Technology* 53 (1999): 44–50.

6 R. S. Gradgenett, Watching the trends helps focus nutrition education strategies, *Journal of the American Dietetic Association* 98 (1998): 1000.

7 B. Bower, Forbidden flavors: Scientists consider how disgusting tastes can linger surreptitiously in memory, *Science News* 151 (1997): 198.

8 International Life Sciences Institute, First International Conference on East-West Perspectives on Functional Foods, *Nutrition Reviews* 54 (1996): entire issue.

9 A. Drewnowski, Taste preferences and food intake, *Annual Review of Nutrition* 17 (1997): 237–253.

10 Committee on Dietary Reference Intakes, *Dietary Reference Intakes for Calcium, Phosphorus, Magnesium, Vitamin D, and Fluoride* (Washington, D.C.: National Academy Press, 1997).

11 Committee on Dietary Reference Intakes, *Dietary Reference Intakes for Thiamin, Riboflavin, Niacin, Vitamin B_6, Folate, Vitamin B_{12}, Pantothenic Acid, Biotin, and Choline* (Washington, D.C.: National Academy Press, 1998).

12 Committee on Dietary Reference Intakes, *Dietary Reference Intakes for Vitamin C, Vitamin E, Selenium, and Carotenoids* (Washington, D.C.: National Academy Press, 2000).

13 Committee on Dietary Reference Intakes, *Dietary Reference Intakes for Vitamin A, Vitamin K, Arsenic, Boron, Chromium, Copper, Iodine, Iron, Manganese, Molybdenum, Nickel, Silicon, Vanadium, and Zinc* (Washington, D.C.: National Academy Press, 2001).

14 W. Mertz, Risk assessment of essential trace elements: New approaches to setting Recommended Dietary Allowances and safety limits, *Nutrition Reviews* 53 (1995): 179–185.

15 C. D. Berndanier, Nutrient-gene interactions, *Nutrition Today* 35 (2000): 8–17.

16 S. S. Smith, NCHS launches latest National Health and Nutrition Examination Survey, *Public Health Reports* 114 (1999): 190–192.

17 D. Satcher, U.S. Surgeon General, as cited in C. Marwick, Healthy People 2010 Initiative launched, *Journal of American Medical Association* 283 (2000): 989–990.

18 D. S. Satcher, Healthy People at 2000, *Public Health Reports* 114 (1999): 563–564.

19 *The Surgeon General's Report on Nutrition and Health: Summary and Recommendations,* DHHS (PHS) publication no. 88-50211 (Washington, D.C.: Government Printing Office, 1988).

ANSWERS

Making it Click

1. a. 5 g protein × 4 kcal/g = 20 kcal protein.
   30 g carbohydrate × 4 kcal/g = 120 kcal carbohydrate.
   11 g fat × 9 kcal/g = 99 kcal fat.
   Total = 239 kcal.

   b. 20 kcal ÷ 239 kcal × 100 = 8.4% kcal from protein.
   120 kcal ÷ 239 kcal × 100 = 50.2% kcal from carbohydrate.
   99 kcal ÷ 239 kcal × 100 = 41.4% kcal from fat.
   Total = 100%.

   c. 1 g protein = 4 kcal protein.
   13 g carbohydrate = 52 kcal carbohydrate.
   146 total kcal − 56 kcal protein + carbohydrate = 90 kcal alcohol.
   90 kcal alcohol ÷ 7 g/kcal = 12.9 g alcohol.

2. No. 15 g protein × 4 kcal/g = 60 kcal.

Multiple Choice

1. d	2. b	3. c	4. d	5. d
6. a	7. c	8. d	9. c	10. a

NUTRITION INFORMATION—EXPERTS AND QUACKS, ON THE NET AND IN THE NEWS

PEOPLE LEARN about nutrition daily as they watch television, read newspapers, turn the pages of magazines, talk with friends, and search the Internet. They want to know how best to take care of themselves. In some cases, they are seeking miracles: tricks to help them lose weight, foods to forestall aging, and supplements to build muscles. People's heightened interest in nutrition and health translates into billions of dollars spent on services and products sold by both legitimate and fraudulent businesses. Although consumers who obtain legitimate products can improve their health, those enticed by **fraud** may lose their health, their savings, or both (boldface terms are defined in the glossary on p. 26). Ironically, nutrition **quackery** prevents people from attaining the health they seek by giving them false hope and delaying the implementation of effective strategies. Furthermore, the conflicting information that results from a mixture of science and quackery confuses consumers.[1]

Science and quackery may be easy to tell apart at the extremes, but much nutrition information lies between the extremes. How can people distinguish valid nutrition information from misinformation? One excellent approach is to notice *who* is purveying the information. The "who" behind the information is not always evident, though, especially in the world of electronic media. Consumers need to keep in mind that *people* develop CD-ROMs and create websites on the Internet, just as people write books and report the news.

This highlight begins by examining the unique potential and problems of relying on the Internet and the media for nutrition information. It continues with a discussion of how to identify reliable nutrition information that applies to all resources, including the Internet and the news.

The quality of nutrition information depends on the provider's knowledge and credentials.

NUTRITION ON THE NET

Got a question? The **Internet** has an answer. The Internet offers endless opportunities to obtain high-quality information, but it also delivers an abundance of incomplete, misleading, or inaccurate information.[2] Simply put: anyone can publish anything.

For experienced users with nutrition knowledge, results are just a mouse click away; not only is access easy, but the information is often more current than that obtainable from other sources. For others, though, answers lie tangled in a web of information overload and questionable reliability.

With hundreds of millions of **websites** on the **World Wide Web,** searching for nutrition information can be an overwhelming experience—much like walking into an enormous bookstore with millions of books, magazines, newspapers, and videos. And like a bookstore, the Internet offers no guarantees of the accuracy of the information found there—and much of it is pure fiction.

When using the Internet, keep in mind that the quality of information available covers a broad range. Just because you find it on the Net doesn't make it true. Websites must be evaluated for their accuracy, just like every other source. The "How to" on p. 25 provides tips for determining whether a website is reliable.

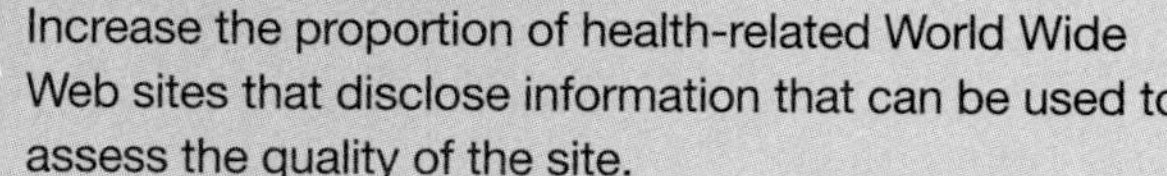

HEALTHY PEOPLE 2010

Increase the proportion of health-related World Wide Web sites that disclose information that can be used to assess the quality of the site.

To help users find reliable nutrition information on the Internet, Tufts University maintains an online rating and review guide called the Nutrition Navigator (navigator.tufts.edu). The ratings reflect the opinions of a panel of nutrition experts who have scored selected websites on the basis of their accuracy, depth, and ease of use. In addition to a rating, the Nutrition Navigator provides a review of the website's content and links to recommended sites. The Nutrition Navigator is an excellent site from which to launch your ventures into nutrition **cyberspace.**

NUTRITION IN THE NEWS

Consumers get most of their nutrition information from television news and magazine reports, which have heightened awareness of how diet influences the development of diseases. Coverage of nutrition in the news has benefited consumers as they learn to make lifestyle changes that will improve their health. Sometimes, however, magazine articles or television programs reporting on nutrition trends mislead consumers and create confusion. The high-protein diet craze, for example, was featured numerous times in national news magazines and on all major networks. No doubt, it was a hot topic and people wanted to know all about it (see Highlight 8 for coverage of the high-protein weight-loss diets). Unfortunately, many of these reports told a lopsided story based on a few testimonials. They did not present the results of research studies or a balance of expert opinions. Telling the whole story might not have been as entertaining, but it would have been more informative.

Tight deadlines and limited understanding sometimes make it difficult to provide a thorough report. Hungry for the latest news, the media often report scientific findings prematurely—without benefit of careful interpretation, replication, and review. Usually, the reports present findings from a single, recently released study, making the news current and controversial.[3] Consequently, the public receives diet and health news quickly, but not always in perspective. Pressure to write catchy headlines and sensational stories twists inconclusive findings into "meaningful discoveries."

As a result, "surprising new findings" seem to contradict one another, and consumers feel frustrated and betrayed. Occasionally, the reports are downright erroneous, but sometimes the apparent contradictions are simply the normal result of science at work. A single study contributes to the big picture, but when viewed alone, the image is distorted. To be meaningful, its conclusions must be presented cautiously within the context of other research findings.

People who do not understand how science operates may become distrustful as they try to learn nutrition from current news reports: "How am I supposed to know what to eat when the scientists themselves don't know?" General background knowledge about the science of nutrition is the best foundation a person can have for judging the validity of new nutrition information. (Congratulations on your decision to take this course.)

Determine Whether a Website Is Reliable

To determine whether a website offers reliable nutrition information, ask the following questions:

- **Who?** Who is responsible for the site? Look for the author's name and credentials. Is the person qualified to speak on nutrition?
- **When?** When was the site last updated? Because nutrition is an ever-changing science, sites need to be dated and updated frequently.
- **Where?** Where is the information coming from? The three letters following the dot in a Web address identify the site's affiliation. Addresses ending in "gov" (government), "edu" (educational institute), and "org" (organization) generally provide reliable information; "com" (commercial) sites represent businesses and, depending on their qualifications and integrity, may or may not offer dependable information.
- **Why?** Why is the site giving you this information? Is the site providing a public service or selling a product? Many commercial sites provide accurate information, but some do not. When money is the prime motivation, be aware that the information may be biased.

If you have answered all of the above questions with satisfaction, then ask this final question:

- **What?** What is the message, and is it in line with other reliable sources? Information that contradicts common knowledge should be questioned. Many reliable sites provide links to other sites to facilitate your quest for knowledge, but this provision alone does not guarantee a reputable intention. Be aware that any site can link to any other site without permission.

IDENTIFYING NUTRITION EXPERTS

Regardless of whether the medium is electronic, print, or video, consumers need to ask whether the person behind the information is qualified to speak on nutrition. If the creator of a website on the Internet recommends eating three pineapples a day to lose weight, a trainer at the gym praises a high-protein diet, or a health-store clerk suggests an herbal supplement, should you believe these people? Can you distinguish between accurate news reports and sensational programs on television? Have you noticed that many televised nutrition messages are presented by celebrities, fitness experts, psychologists, food editors, and chefs—that is, almost anyone except a **dietitian**?[4] When you are confused or need sound dietary advice, whom should you ask?

Physicians

Many people turn to their physicians for dietary advice, expecting them to know all about health-related matters. But are physicians the best sources of accurate and current information on nutrition? Only about one-fourth of all medical schools in the United States require students to take even one nutrition course; less than half provide an elective nutrition course.[5] Students attending these classes receive an average of 20 hours of nutrition instruction—an amount they, themselves, consider inadequate. By comparison, most students reading this text are taking a nutrition class that provides an average of 45 hours of instruction.

The **American Dietetic Association (ADA)** asserts that standardized nutrition education should be included in the curricula for all health care professionals: physician's assistants, dental hygienists, physical and occupational

GLOSSARY

accredited: approved; in the case of medical centers or universities, certified by an agency recognized by the U.S. Department of Education.

American Dietetic Association (ADA): the professional organization of dietitians in the United States. The Canadian equivalent is Dietitians of Canada, which operates similarly.

correspondence schools: schools that offer courses and degrees by mail. Some correspondence schools are accredited; others are *diploma mills.*

cyberspace: a term coined by William Gibson referring to the nonphysical place where all Internet activity occurs.

dietetic technician: a person who has completed a minimum of an associate's degree from an accredited university or college and an approved dietetic technician program that includes a supervised practice experience. See also *dietetic technician, registered (DTR).*

dietetic technician, registered (DTR): a dietetic technician who has passed a national examination and maintains registration through continuing professional education.

dietitian: a person trained in nutrition, food science, and diet planning. See also *registered dietitian.*

DTR: see *dietetic technician, registered.*

fraud or **quackery:** the promotion, for financial gain, of devices, treatments, services, plans, or products (including diets and supplements) that alter or claim to alter a human condition without proof of safety or effectiveness. (The word *quackery* comes from the term *quacksalver,* meaning a person who quacks loudly about a miracle product—a lotion or a salve.)

Internet (the Net): a worldwide network of millions of computers linked together to share information.

license to practice: permission under state or federal law, granted on meeting specified criteria, to use a certain title (such as dietitian) and offer certain services. **Licensed dietitians** may use the initials **LD** after their names.

misinformation: false or misleading information.

nutritionist: a person who specializes in the study of nutrition. Some nutritionists are registered dietitians, whereas others are self-described experts whose training is questionable. In states with responsible legislation, the term applies only to people who have MS or PhD degrees from properly accredited institutions.

public health dietitians: dietitians who specialize in providing nutrition services through organized community efforts.

RD: see *registered dietitian.*

registered dietitian (RD): a person who has completed a minimum of a bachelor's degree from an accredited university or college, has completed approved course work and a supervised practice program, has passed a national examination, and maintains registration through continuing professional education.

registration: listing; with respect to health professionals, listing with a professional organization that requires specific course work, experience, and passing of an examination.

websites: Internet resources composed of text and graphic files, each with a unique URL (Uniform Resource Locator) that names the site (for example, www.usda.gov).

World Wide Web (WWW, the Web): a graphical subset of the Internet.

therapists, social workers, and all others who provide services directly to clients.[6] When these professionals understand the relevance of nutrition in the treatment and prevention of disease and have command of reliable nutrition information, then all the people they serve will also be better informed.[7]

Most physicians appreciate the connections between health and nutrition. Those who have specialized in clinical nutrition are especially well qualified to speak on the subject. Few physicians, however, have the time or experience to develop diet plans and provide detailed diet instructions for clients. Often physicians wisely refer their clients to a qualified nutrition expert—a **registered dietitian (RD).**

HEALTHY PEOPLE 2010

Increase the proportion of physician office visits that provide or order nutrition counseling and educational services.

Registered Dietitians

A registered dietitian has the educational background necessary to deliver reliable nutrition advice and care. To become an RD, a person must earn an undergraduate degree requiring some 60 or so semester hours in nutrition, food science, and other related subjects; complete a year's clinical internship or the equivalent; pass a national examination administered by the ADA; and maintain up-to-date knowledge and **registration** by participating in required continuing education activities such as attending seminars, taking courses, or writing professional papers.

Dietitians perform a multitude of duties in many settings in most communities.[8] They work in the food industry, pharmaceutical companies, home health agencies, long-term care institutions, private practice, public health departments, research centers, education settings, fitness centers, and hospitals. Depending on their work settings, dietitians can assume a number of different job responsibilities and positions. In hospitals, administrative dietitians manage the foodservice system; clinical dietitians provide client care (see Table H1-1); and nutrition support team dietitians coordinate nutrition care with other health care professionals. In the food industry, dietitians conduct research, develop products, and market services.

Public health dietitians who work in government-funded agencies play a key role in delivering nutrition services to people in the community. Among their many roles, public health dietitians help plan, coordinate, and evaluate food assistance programs; act as consultants to other agencies; manage finances; and much more.

Other Dietary Employees

In some facilities, a **dietetic technician, registered (DTR)** assists registered dietitians in both administrative and clinical responsibilities.[9] A DTR has been educated and trained to work under the guidance of a registered dietitian.

TABLE H1-1 Responsibilities of a Clinical Dietitian

- Assesses clients' nutrition status.
- Determines clients' nutrient requirements.
- Monitors clients' nutrient intakes.
- Develops, implements, and evaluates clients' nutrition care plans.
- Counsels clients to cope with unique diet plans.
- Teaches clients and their families about nutrition needs and diet plans.
- Provides training for other dietitians, nurses, interns, and dietetics students.
- Serves as liaison between clients and the foodservice department.
- Communicates with physicians, nurses, pharmacists, and other health care professionals about clients' progress, needs, and treatments.
- Participates in professional activities to enhance knowledge and skill.

In addition to the **dietetic technician,** other dietary employees may include clerks, aides, cooks, porters, and other assistants. These dietary employees do not have extensive formal training in nutrition, and their ability to provide accurate information may be limited.

IDENTIFYING FAKE CREDENTIALS

In contrast to registered dietitians, thousands of people possess fake nutrition degrees and claim to be nutrition consultants or doctors of "nutrimedicine."[10] These and other such titles may sound meaningful, but most of these people lack the established credentials and training of an ADA-sanctioned dietitian. If you look closely, you can see signs of their fake expertise.

Consider educational background, for example. The minimal standards of education for a dietitian specify a bachelor of science (BS) degree in food science and human nutrition or related fields from an **accredited** college or university. Such a degree generally requires four to five years of study. In contrast, a fake nutrition expert may display a degree from a six-month correspondence course. Such a degree simply falls short. In some cases, schools posing as legitimate **correspondence schools** offer even less—they sell certificates to anyone who pays the fees. To obtain these "degrees," a candidate need not read any books or pass any examinations.

To guard educational quality, an accrediting agency recognized by the U.S. Department of Education (DOE) certifies that certain schools meet criteria established to ensure that an institution provides complete and accurate schooling. Unfortunately, fake nutrition degrees are available from schools "accredited" by more than 30 phony accrediting agencies.

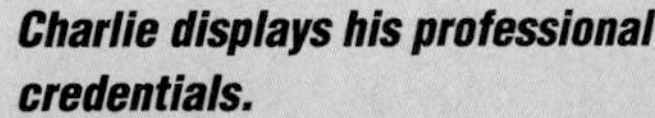

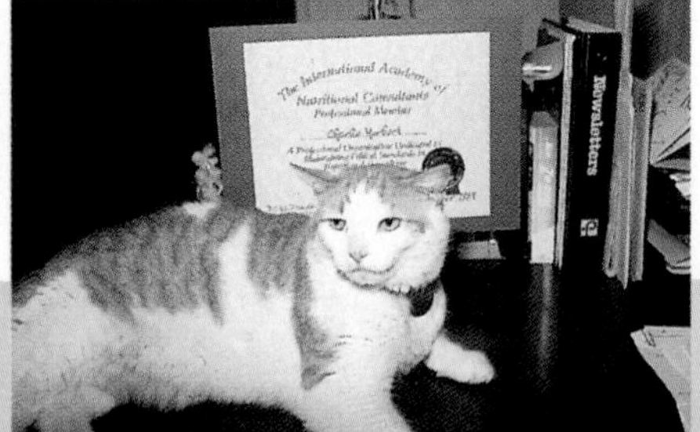

Charlie displays his professional credentials.

To dramatize the ease with which anyone can obtain a fake nutrition degree, one writer enrolled in a correspondence course for a fee of $82. She made every attempt to fail, intentionally answering all examination questions incorrectly. Even so, she received a "nutritionist" certificate at the end of the course. The "school" explained that it was sure she must have just misread the test.

In a similar stunt, Ms. Sassafras Herbert was named a "professional member" of a professional association. For her efforts, Sassafras has received a wallet card and is listed in a *Who's Who* publication that is distributed at health fairs and trade shows nationwide. Sassafras is a poodle; her master, Victor Herbert, MD, paid $50 to prove that she could be awarded these honors merely by sending in her name. Mr. Charlie Herbert, who is also a professional member of such an organization, is a cat.

Some states allow anyone to use the title dietitian or **nutritionist,** but others allow only RDs or people with certain graduate degrees to call themselves dietitians. Many states provide a further guarantee: the **license to practice.**[11] By requiring dietitians to be licensed, states identify people who have met minimal standards of education and experience.

By knowing what qualifies someone to speak on nutrition, consumers can determine whether that person's advice might be harmful or helpful. Don't be afraid to ask for credentials. Does the personal trainer at the gym have a degree in nutrition from an accredited university? Is the creator of a nutrition website an RD or otherwise qualified to write on nutrition? Have you seen the health-store clerk's license to practice as a dietitian? If not, seek a better-qualified source. After all, your health depends on it.

IDENTIFYING VALID INFORMATION

Where do nutrition experts get their information? As Chapter 1 explained, nutrition knowledge derives from scientific research.

Researchers conduct experiments and then record and analyze their results, exercising caution in their interpretation of the findings. For example, in an epidemiological study, scientists may use a specific segment of the population—say, men 18 to 30 years old. When the scientists draw conclusions, they are careful not to generalize the findings to all people. Similarly, scientists performing research studies using animals are cautious in applying their findings to human beings. Conclusions from any one research study are always tentative and take into account findings from studies conducted by other scientists as well. As evidence accumulates, scientists gain confidence about making recommendations

TABLE H1-2 Parts of a Research Article

- *Abstract.* The abstract provides a brief overview of the article.
- *Introduction.* The introduction clearly states the purpose of the current study by proposing a hypothesis.
- *Review of literature.* A comprehensive review of the literature reveals all that science has uncovered on the subject to date.
- *Methodology.* The methodology section defines key terms and describes the instruments and procedures used in conducting the study.
- *Results.* The results report the findings and may include tables and figures that summarize the information.
- *Conclusions.* The conclusions drawn are those supported by the data and reflect the original purpose as stated in the introduction. Usually, they answer a few questions and raise several more.
- *References.* The references reflect the investigator's knowledge of the subject and should include an extensive list of relevant studies (including key studies several years old as well as current ones).

that affect people's health and lives. Still, their statements are worded cautiously, as in "A diet high in fruits and vegetables *may* protect against *some* cancers."

Quite often, as they approach an answer to one research question, scientists raise several more questions, so future research projects are never lacking. Further scientific investigation then seeks to answer questions such as "What substance or substances within fruits and vegetables provide protection?" If those substances turn out to be the vitamins A and C found so abundantly in fresh produce, then, "How much vitamin A and C is needed to offer protection?" "How do these vitamins protect against cancer?" "Is it their action as antioxidant nutrients?" "If not, might it be another action or even another substance that accounts for the protection fruits and vegetables provide against cancer?" (Highlight 11 explores the answers to these questions and reviews recent research on antioxidant nutrients, phytochemicals, and disease.)

The findings from a research study are submitted to a board of reviewers composed of other scientists who rigorously evaluate the study to assure that the scientific method was followed—a process known as peer review. The reviewers critique the study's hypothesis, methodology, statistical significance, and conclusions (Table H1-2 describes the parts of a research article). If the reviewers consider the conclusions to be well supported by the evidence, they endorse the work for publication in a scientific journal where others can read it. This raises an important point regarding information found on the Internet: much gets published without the rigorous scrutiny of peer review.[12] Consequently, readers must assume greater responsibility for examining the data and conclusions presented—often without the benefit of journal citations. Until you feel confident in critically evaluating nutrition information, you would be wise to restrict your research to one of the many online peer-reviewed journals (see the "How to" on this page for selected website addresses).

Regardless of whether an article is presented on the Internet, on television, or in print, readers must evaluate the study and assess the findings in light of knowledge gleaned from other studies. Figure H1-1 provides examples of reliable nutrition information.

Even when a new finding is published or released to the media, it is still only preliminary and not very meaningful by itself. Other scientists will need to confirm or disprove

Find Credible Sources of Nutrition Information

Government agencies, volunteer associations, consumer groups, and professional organizations provide consumers with reliable health and nutrition information. Credible sources of nutrition information include:

- Nutrition and food science departments at a university or community college.
- Local agencies such as the health department or County Cooperative Extension Service.
- Government health agencies such as:
  - Department of Agriculture (USDA) — www.usda.gov
  - Department of Health and Human Services (DHHS) — www.os.dhhs.gov
  - Food and Drug Administration (FDA) — www.fda.gov
  - Health Canada — www.hc-sc.gc.ca/nutrition
- Volunteer health agencies such as:
  - American Cancer Society — www.cancer.org
  - American Diabetes Association — www.diabetes.org
  - American Heart Association — www.americanheart.org
- Reputable consumer groups such as:
  - American Council on Science and Health — www.acsh.org
  - Federal Consumer Information Center — www.pueblo.gsa.gov
  - International Food Information Council — ificinfo.health.org
- Professional health organizations such as:
  - American Dietetic Assocation — www.eatright.org
  - American Medical Association — www.ama-assn.org
  - Dietitians of Canada — www.dietitians.ca
- Journals such as:
  - *American Journal of Clinical Nutrition* — www.faseb.org/ajcn
  - *New England Journal of Medicine* — www.nejm.org
  - *Nutrition Reviews* — www.ilsi.org/pubs.html

Appendix F provides websites and addresses for these and other organizations.

FIGURE H1-1 Sources of Reliable Nutriton Information

REVIEWS
Articles that examine all the major work on a subject are published in review journals like* Nutrition Reviews. *These articles provide references to all of the original work reviewed.

The American Journal of CLINICAL NUTRITION
Official Journal of The American Society for Clinical Nutrition, Inc
FULL TEXT ONLINE AT WWW.AJCN.ORG
JANUARY 2001 • VOLUME 73 • NUMBER 1

EDITORIALS
ORIGINAL RESEARCH COMMUNICATIONS
Obesity and eating disorders
Lipids and cardiovascular risk
Nutritional status, dietary intake, and body composition
Carbohydrate metabolism and diabetes
Vitamins, minerals, and phytochemicals
Pregnancy and lactation
Digestion
Bone metabolism
REPORT OF A MEETING
Letters to the Editor
Book Reviews

JOURNALS
***Articles that present all the details of the methods, results, and conclusions of a particular study are published in journals like* the American Journal of Clinical Nutrition.**

INDEXES
An index of abstracts directs you to many research articles on a given subject. This index from* Biological Abstracts *lists recently published works on vitamin C.

ONLINE INDEXES
Several online indexes are available, but one of the best for nutrition research is the National Library of Medicine's MEDLINE, which provides an index of more than 9 million articles from almost 4000 journals. For free access, visit www.nlm.nih.gov/medlineplus or www.ncbi.nlm.nih.gov/PubMed

WEBSITES
Websites on the Internet developed by credible sources, such as those listed on p. 28, can provide valuable nutrition information and direct users to other resources. A quick link to many of these nutrition resources is available when you visit www.wadsworth.com/nutrition

the findings through replication. To be accepted into the body of nutrition knowledge, a finding must stand up to rigorous, repeated testing in experiments performed by several different researchers. What we "know" in nutrition results from years of replicating study findings. Communicating the latest finding in its proper context without distorting or oversimplifying the message is a challenge for scientists and journalists alike.[13]

With each report from scientists, the field of nutrition changes a little—each finding contributes another piece to the whole body of knowledge. People who know how science works understand that single findings, like single frames in a movie, are just small parts of a larger story. Over years, the picture of what is "true" in nutrition gradually changes, and modifications in recommendations then follow. Instead of eating 4 servings of fruits and vegetables as recommended by the old Four Food Group Plan, people are now encouraged to eat 2 to 4 servings of fruits and 3 to 5 servings of vegetables as suggested by the current Daily Food Guide (presented in Chapter 2).

Because science is a step-by-step, information-gathering and testing process, old research still has value. A hypothesis first advanced in 1960 that stands up to decades of validation has real strength. When it comes to scientific information, "new" does not necessarily mean "improved." In fact, any science report based on all new references is suspect, for truly strong research is based on a body of work conducted over many years. This is why, even in books published just this year, you will see references to old reports. Some studies have become classics: they were exciting when they first appeared, and they have stood up to the test of time.

IDENTIFYING MISINFORMATION

Nutrition is a hot topic, and scattered among the valid research findings are thousands of misleading and unfounded claims. How can a person identify nutrition **misinformation** and

FIGURE H1-2 Red Flags of Nutrition Quackery

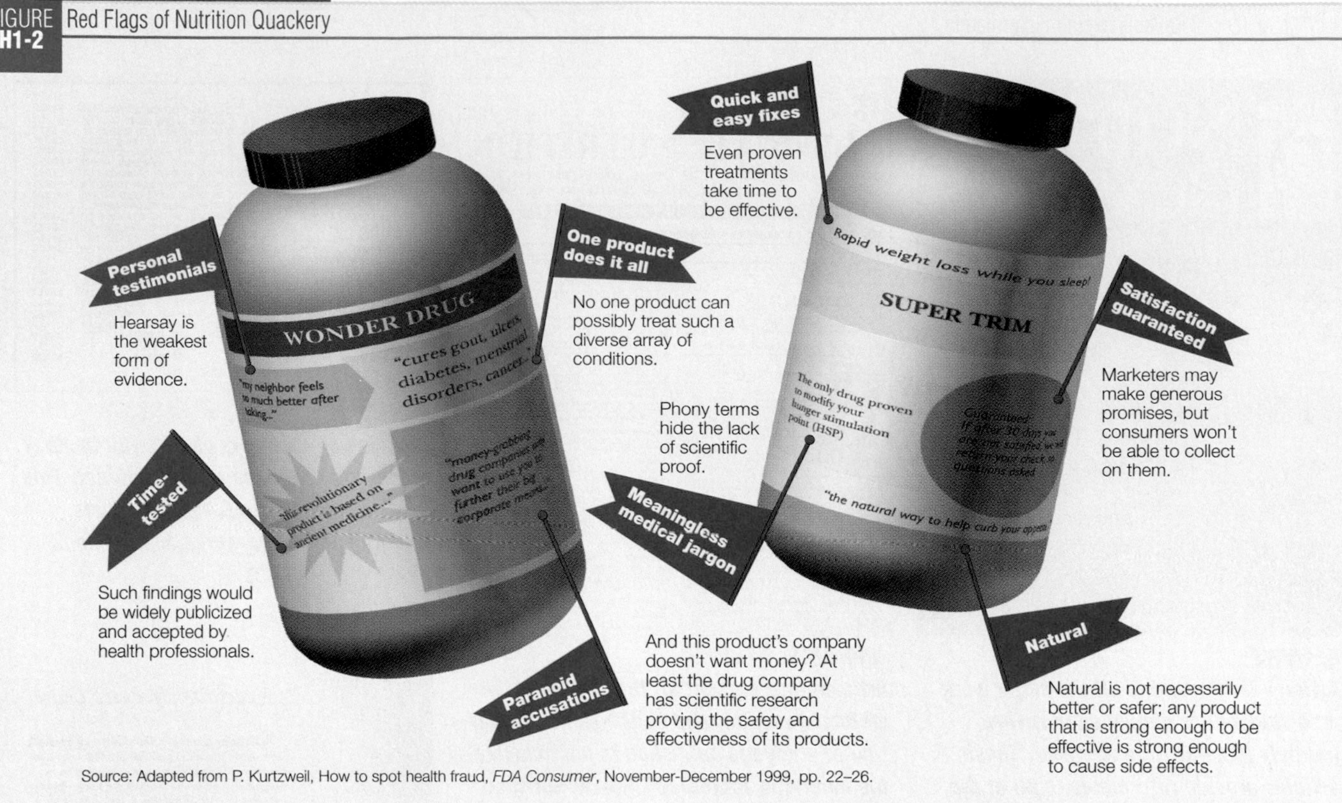

Source: Adapted from P. Kurtzweil, How to spot health fraud, *FDA Consumer*, November-December 1999, pp. 22–26.

quackery? Once upon a time, quacks rode into town in wooden wagons hawking snake oil for 50 cents a bottle to "cure what ails you," but those days are gone. Today's purveyors approach consumers in less obvious ways. They deliver their messages via the Internet, on glossy pages of magazines, in televised infomercials, and at social gatherings. The claims may look slick and sound logical, but they lack the research support found in nutrition science. Figure H1-2 presents red flags that alert consumers to nutrition misinformation.

Sales of unproven and dangerous products have always been a concern, but the Internet now provides merchants with an easy and inexpensive way to reach millions of customers around the world. Because of the difficulty in regulating the Internet, fraudulent and illegal sales of medical products have hit a bonanza.[14] As is the case with the air, no one owns the Internet, and similarly, no one has control over the pollution. Countries have different laws regarding sales of drugs, dietary supplements, and other health products, but applying these laws to the Internet marketplace is almost impossible. Even if illegal activities could be defined and identified, finding the person responsible for a particular website is not always possible. Websites can open and close in a blink of a cursor. Now, more than ever, consumers must heed the caution "Buyer beware."

In summary, when you hear nutrition news, consider its source. Ask yourself these two questions: Is the person purveying the information qualified to speak on nutrition? Is the information based on valid scientific research? If not, find a better source, for your health is your most precious asset.

Nutrition on the Net

•WEBSITES•

Access these websites for further study of topics covered in this highlight.

- Find updates and quick links to these and other nutrition-related sites at our website: **www.wadsworth.com/nutrition**
- Visit the National Council for Reliable Health Information: **www.ncrhi.org**
- Check the ratings and reviews of websites by Tufts University Nutrition Navigator: **navigator.tufts.edu**
- Find a registered dietitian in your area from the American Dietetic Association: **www.eatright.org**

- Find a nutrition professional in Canada from the Dietitians of Canada: **www.dietitians.ca**
- Find out whether a correspondence school is accredited from the Distance Education and Training Council's Accrediting Commission: **www.detc.org**
- Find out whether a school is properly accredited for a dietetics degree from the American Dietetic Association: **www.eatright.org/caade**
- Obtain a listing of accredited institutions, professionally accredited programs, and candidates for accreditation from the American Council on Education: **www.acenet.edu**
- Learn more about quackery from Stephen Barrett's Quackwatch: **www.quackwatch.com**
- Search "quackery" at the U.S. government health information site: **www.healthfinder.gov**

References

1 Position of The American Dietetic Association: Food and nutrition misinformation, *Journal of the American Dietetic Association* 95 (1995): 705–707.

2 W. M. Silberg, G. D. Lundberg, and R. A. Musacchio, Assessing, controlling, and assuring the quality of medical information on the Internet, *Journal of the American Medical Association* 277 (1997): 1244–1245.

3 N. S. Wellman and coauthors, Do we facilitate the scientific process and the development of dietary guidance when findings from single studies are publicized? An American Society for Nutritional Sciences Controversy Session Report, *American Journal of Clinical Nutrition* 70 (1999): 802–805.

4 T. MacLaren, Messages for the masses: Food and nutrition issues on television, *Journal of the American Dietetic Association* 97 (1997): 733–734.

5 J. A. Schulman, Nutrition education in medical schools: Trends and implications for health educators, *Med Ed Online,* www.med-ed-Online.org/f0000015.htm (accessed October 24, 2000).

6 Position of The American Dietetic Association: Nutrition education of health professionals, *Journal of the American Dietetic Association* 98 (1998): 343–346.

7 Intersociety Professional Nutrition Education Consortium, Bringing nutrition specialists into the mainstream: Rationale for the Intersociety Professional Nutrition Education Consortium, *American Journal of Clinical Nutrition* 68 (1998): 894–898.

8 C. Gopalan, Dietetics and nutrition: Impact of scientific advances and development, *Journal of the American Dietetic Association* 97 (1997): 737–741.

9 J. Arena and P. Walters, Do you know what a dietetic technician can do? A focus on clinical technicians and their expanded roles and responsibilities, *Journal of the American Dietetic Association* 97 (1997): S139–S141.

10 J. Raso, Nutrition-related "credentialing" organizations: The good, the bad, and the abysmal, *Priorities* 7 (1995): 31–34.

11 Update on state licensure laws and ADA regulatory remarks, *Journal of the American Dietetic Association* 97 (1997): 1251.

12 D. E. Kipp, J. D. Radel, and J. A. Hogue, The Internet and the nutritional scientist, *American Journal of Clinical Nutrition* 64 (1996): 659–662.

13 Advisory group convened by the Harvard School of Public Health and the International Food Information Council Foundation, Improving public understanding: Guidelines for communicating emerging science on nutrition, food safety, and health, *Journal of the National Cancer Institute* 90 (1998): 194–199.

14 A. A. Skolnick, WHO considers regulating ads, sale of medical products on Internet, *Journal of the American Medical Association* 278 (1997): 1723–1725.

CHAPTER 2 Planning a Healthy Diet

Chapter 1 explained that the body's many activities are supported by the array of nutrients delivered by the foods people eat. Food choices made over years influence the body's health, and consistently poor choices increase the risks of developing chronic diseases. This chapter shows how a person can select from the tens of thousands of foods available to create a diet that supports health. Fortunately, most foods provide several nutrients, so one trick for wise diet planning is to select a combination of foods that deliver a full array of nutrients. This chapter begins by introducing the diet-planning principles and dietary guidelines that assist people in selecting foods that will deliver nutrients without excess energy.

Health promotion

◆ In 1900, people in the United States chose from 500 or so different foods; a century later, they can choose from more than 50,000.◆

PRINCIPLES AND GUIDELINES

How well you nourish yourself does not depend on the selection of any one food. Instead it depends on the selection of many different foods at numerous meals over days, months, and years. Diet-planning principles and dietary guidelines are key concepts to keep in mind whenever you are selecting foods—whether shopping at the grocery store, choosing from a restaurant menu, or preparing a home-cooked meal.

Diet-Planning Principles

Diet planners have developed several ways to select foods. Whatever plan or combination of plans they use, though, they keep in mind the six basic diet-planning principles◆ listed in the margin.

• Adequacy • **Adequacy** means that the diet provides sufficient energy and enough of all the nutrients to meet the needs of healthy people. Take the essential nutrient iron, for example. Each day the body loses some iron, so people have to replace it by eating foods that contain iron. A person whose diet fails to provide enough iron-rich foods may develop the symptoms of iron-deficiency anemia: the person may feel weak, tired, and listless; have frequent headaches; and find that even the smallest amount of muscular work brings disabling fatigue. To prevent these deficiency symptoms, a person must include foods that supply adequate iron. The same is true for all the other essential nutrients introduced in Chapter 1.

• Balance • The art of balancing the diet involves using enough—but not too much—of each type of food. The essential minerals calcium and iron, taken together, illustrate the importance of dietary **balance.** Meats, fish, and poultry are rich in iron but poor in calcium. Conversely, milk and milk products are rich in calcium but poor in iron. In fact, milk (except breast milk) and milk products are so low in iron that overuse of these foods can lead to iron-deficiency anemia by displacing iron-rich foods from the diet. Yet milk is the single most nutritious food for infants and can be an important source of calcium for people of all ages. Use some meat or meat alternates for iron; use some milk and milk products for calcium; and save some space for other foods, too, since a diet consisting of milk and meat alone would not be adequate.◆ For the other nutrients, people need grains, vegetables, and fruits.

• kCalorie (Energy) Control • Clearly, designing an adequate, balanced diet without overeating requires careful planning. The discussion of weight control in Chapter 9 examines this issue in more detail, but the key to **kcalorie control** is to select foods of high **nutrient density.**

• Nutrient Density • To eat well without overeating, select foods that deliver the most nutrients for the least food energy. Consider foods containing calcium, for example. You can get about 300 milligrams of calcium from either 1½ ounces of cheddar cheese or 1 cup of nonfat milk, but the cheese

Diet-planning principles: ◆

- **A**dequacy.
- **B**alance.
- k**C**alorie (energy) control.
- Nutrient **D**ensity.
- **M**oderation.
- **V**ariety.

Balance in the diet helps to ensure adequacy. ◆

adequacy (dietary): providing all the essential nutrients, fiber, and energy in amounts sufficient to maintain health.

balance (dietary): providing foods of a number of types in proportion to each other, such that foods rich in some nutrients do not crowd out of the diet foods that are rich in other nutrients.

kcalorie (energy) control: management of food energy intake.

nutrient density: a measure of the nutrients a food provides relative to the energy it provides. The more nutrients and the fewer kcalories, the higher the nutrient density.

HOW TO Compare Foods Based on Nutrient Density

One way to evaluate foods is simply to notice their nutrient contribution *per serving:* 1 cup of milk provides 301 milligrams of calcium, and ½ cup of fresh, cooked turnip greens provides 99 milligrams. Thus a serving of milk offers three times as much calcium as a serving of turnip greens. To get 300 milligrams of calcium, a person could choose either 1 serving of milk or 3 servings (1½ cups) of turnip greens.

Another valuable way to evaluate foods is to consider their nutrient density—their nutrient contribution *per kcalorie*. Nonfat milk delivers 86 kcalories with its 301 milligrams of calcium. To calculate the nutrient density, divide milligrams by kcalories:

$$\frac{301 \text{ mg calcium}}{86 \text{ kcal}} = 3.5 \text{ mg per kcal.}$$

Do the same for the fresh turnip greens, which provide 15 kcalories with the 99 milligrams of calcium:

$$\frac{99 \text{ mg calcium}}{15 \text{ kcal}} = 6.6 \text{ mg per kcal.}$$

The more milligrams per kcalorie, the greater the nutrient density. Turnip greens are more calcium dense than milk. They provide more calcium *per kcalorie* than milk, but milk offers more calcium *per serving*. Both approaches offer valuable information, espeically when combined with a realistic appraisal. What matters most is which are you more likely to consume—1½ cups of turnip greens or 1 cup of milk? You can get 300 milligrams of calcium from either, but the greens will save you about 40 kcalories (the savings would be even greater if you use whole milk).

Keep in mind, too, that calcium is only one of the many nutrients that foods provide. Similar calculations for protein, for example, would show that nonfat milk provides more protein both *per kcalorie* and *per serving* than turnip greens—that is, milk is more protein dense. Combining variety with nutrient density helps to ensure the adequacy of all nutrients.

contributes about twice as much food energy (kcalories) as the milk. The nonfat milk, then, is twice as calcium dense as the cheddar cheese; it offers the same amount of calcium for half the kcalories. Both foods are excellent choices for adequacy's sake alone, but to achieve adequacy while controlling kcalories,◆ the nonfat milk is the better choice. The many bar graphs that appear in Chapters 10–13 highlight the most nutrient-dense choices, and the accompanying "How to" describes how to compare foods based on nutrient density.

Just like a person who has to pay for rent, food, clothes, and tuition on a tight budget, a person whose energy allowance is limited has to obtain iron, calcium, and all the other essential nutrients on a tight energy budget. To succeed, the person has to get many nutrients for each kcalorie "dollar." In the cola and grapes example (in the photo on p. 35), both provide about the same number of kcalories, but the grapes deliver many more nutrients. A person who makes nutrient-dense choices such as fruit over cola can meet daily nutrient needs on a lower energy budget.

Foods that are notably low in nutrient density—such as potato chips, candies, and colas—are sometimes called **empty-kcalorie foods.** The kcalories these foods provide are "empty" in that they deliver only energy (from sugar, fat, or both) with little, or no, protein, vitamins, or minerals.

◆ Nutrient density promotes adequacy and kcalorie control.

◆ Moderation contributes to adequacy, balance, and kcalorie control.

• **Moderation** • Foods rich in fat and sugar provide enjoyment and energy but relatively few nutrients. In addition, they promote weight gain when eaten in excess. A person practicing **moderation**◆ would eat such foods only on occasion and would regularly select foods low in fat and sugar, a practice that automatically improves nutrient density. Returning to the example of cheddar cheese and nonfat milk, the nonfat milk not only offers the same amount of calcium for less energy, but it contains far less fat than the cheese.

• **Variety** • A diet may have all of the virtues just described and still lack **variety,** if a person eats the same foods day after day. People should select foods from each of the food groups daily and vary their choices within each food group from day to day for several reasons. First, different foods within the same group contain different arrays of nutrients. Among the fruits, for example, strawberries are especially rich in vitamin C while cantaloupes are rich in vitamin A. Second, no food is guaranteed entirely free of substances that, in excess, could be harmful. The strawberries might contain trace amounts of one contaminant, the cantaloupes another. By alternating fruit choices, a person will ingest very little of either contaminant. Third, as the adage goes, variety is the spice of life. Even if a person eats beans frequently, the person can enjoy pinto beans in Mexican burritos today, garbanzo beans in Greek salad tomorrow, and baked beans with barbecued chicken on the weekend. Eating nutritious meals need never be boring.

empty-kcalorie foods: a popular term used to denote foods that contribute energy but lack protein, vitamins, and minerals.

moderation (dietary): providing enough but not too much of a substance.

variety (dietary): eating a wide selection of foods within and among the major food groups (the opposite of monotony).

FIGURE 2-1 *Dietary Guidelines for Americans:* The ABC's of Good Health

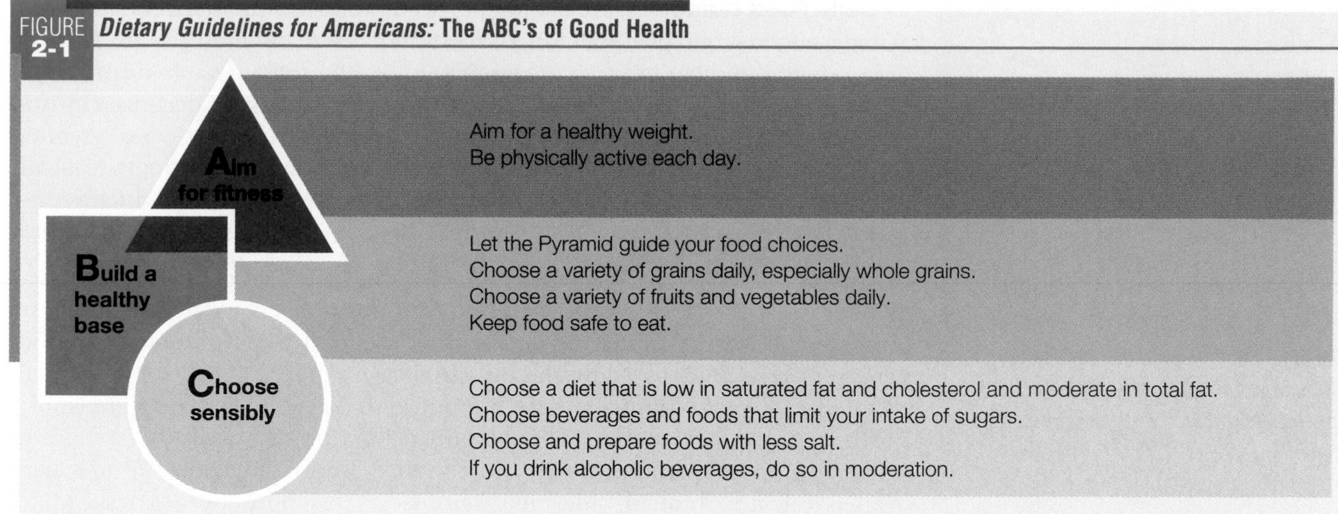

Note: These guidelines are intended for adults and healthy children ages 2 and older.
SOURCE: U.S. Department of Agriculture and U.S. Department of Health and Human Services, *Nutrition and Your Health: Dietary Guidelines for Americans,* Home and Garden Bulletin no. 232 (Washington, D.C.: 2000).

Dietary Guidelines for Americans

Figure 2-1 presents the 2000 *Dietary Guidelines for Americans.*[1] In general, the *Dietary Guidelines* answer the question, What should an individual eat to stay healthy? The first two guidelines encourage people to aim for fitness by combining sensible eating with regular physical activity to achieve and maintain a healthy weight. The next four guidelines urge people to build a healthy base by using the Food Guide Pyramid◆ in meal planning; choosing a variety of grains, vegetables, and fruits daily; and keeping foods safe. The last four guidelines encourage people to choose sensibly in their use of fats,◆ sugars, salt, and alcoholic beverages for those who partake. Together, these ten guidelines point the way toward better health. Table 2-1 (on p. 36) presents *Canada's Guidelines for Healthy Eating.*

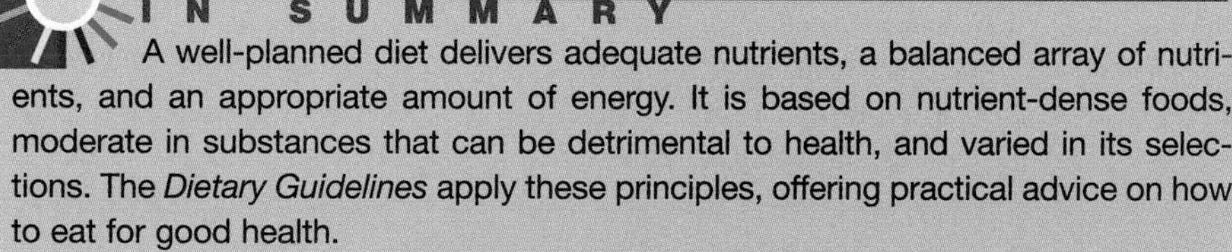

IN SUMMARY

A well-planned diet delivers adequate nutrients, a balanced array of nutrients, and an appropriate amount of energy. It is based on nutrient-dense foods, moderate in substances that can be detrimental to health, and varied in its selections. The *Dietary Guidelines* apply these principles, offering practical advice on how to eat for good health.

© Polara Studios Inc.

This cola and bunch of grapes illustrate nutrient density. Each provides about 150 kcalories, but the grapes offer a trace of protein, some vitamins, minerals, and fiber along with the energy; the cola beverage offers only "empty" kcalories from sugar without any other nutrients. Grapes, or any fruit for that matter, are more nutrient dense than cola beverages.

DIET-PLANNING GUIDES

To plan a diet that achieves all of the dietary ideals just outlined, a person needs tools as well as knowledge. Two of the tools most widely used for diet planning are food group plans and exchange lists.

◆ The Food Guide Pyramid is introduced on p. 37.

◆ Choose a diet that provides:
- ≤30% kcalories from fat.
- 15–20% kcalories from protein.
- 55–60% kcalories from carbohydrate.

Food Group Plans

Food group plans build a diet from clusters of foods that are similar in origin and nutrient content. Thus each group represents a set of nutrients that differs from the nutrients supplied by the other groups. Selecting foods from each of the groups eases the task of creating a balanced diet.

food group plans: diet-planning tools that sort foods of similar origin and nutrient content into groups and then specify that people should eat certain numbers of servings from each group.

© Thomas Harm/Tom Peterson/Quest Photographic Inc.

Flavored "beverages" or "drinks" may taste delicious, but they lack the nutrient richness of real milk products.

• Daily Food Guide • Figure 2-2 (on pp. 38–39) presents the USDA's Daily Food Guide, a food group plan that assigns foods to five◆ major food groups. The figure lists the number of servings recommended, the most notable nutrients of each group, the serving sizes, and the foods within each group categorized by nutrient density. It also includes an illustration of the USDA's Food Guide Pyramid, a pictorial representation of the Daily Food Guide. These guides apply to older children and adults only; young children have their own pyramid, which is presented in Chapter 15. Appendix I presents *Canada's Food Guide to Healthy Eating.*

• Notable Nutrients • The beauty of the Daily Food Guide lies in its simplicity and flexibility. For example, a person can substitute cheese for milk because both supply the key nutrients for the milk group. A person following a food group plan receives not only the nutrients each group is noted for, but small amounts of other nutrients and nonnutrients as well. For example, milk, cheese, and yogurt are notable for their calcium, protein, and riboflavin, but they also provide other nutrients. In contrast, a drink concocted from sugar, water, calcium, protein, and riboflavin lacks this nutrient richness. Milk, cheese, and yogurt are foundation foods; synthetic drinks are not.

• Miscellaneous Foods • Some foods—such as the synthetic drink just mentioned—do not fit into any of the food groups. Foods that are high in fat, sugar, or alcohol provide energy, but too few nutrients to hold a significant place in the diet. Such foods should be used sparingly and only after basic nutrient needs have been met by the foundation foods. Examples of "miscellaneous" foods include salad dressings, jams, and alcoholic beverages.

◆ Five food groups:
- Breads, cereals, and other grain products.
- Vegetables.
- Fruits.
- Meat, poultry, fish, and alternates.
- Milk, cheese, and yogurt.

• Mixtures of Foods • Some foods—such as casseroles, soups, and sandwiches—fall into two or more food groups. With a little practice, users can begin to see the number of servings represented by each food group. From the Daily Food Guide point of view, a chicken enchilada looks like one serving from each of four different food groups if it is made with a corn tortilla from the bread group; ½ cup chopped onion, pepper, and tomatoes from the vegetable group; 3 ounces of chicken from the meat group; and 1½ ounces of shredded cheese from the milk group.

• Nutrient Density • The Daily Food Guide provides a strong foundation for a healthy diet, but it fails to account for the fat and energy differences between foods within a single group—for example, between nonfat milk and ice cream, baked fish and hot dogs, green beans and french fries, apples and avocados, or bread and biscuits. Yet according to the Daily Food Guide, any of these substitutions would be acceptable. People who have low energy allowances are advised to select the most nutrient-dense foods within each group, whereas people with high energy needs may select some of the less nutrient-dense, higher-kcalorie foods. Notice that Figure 2-2 provides a key indicating which foods *within each group* are high, moderate, or low in nutrient density.

TABLE 2-1 Canada's Guidelines for Healthy Eating

- Enjoy a variety of foods.
- Emphasize cereals, breads, other grain products, vegetables, and fruits.
- Choose lower-fat dairy products, leaner meats, and foods prepared with little or no fat.
- Achieve and maintain a healthy body weight by enjoying regular physical activity and healthy eating.
- Limit salt, alcohol, and caffeine.

SOURCE: These guidelines derive from *Action Towards Healthy Eating: The Report of the Communications/Implementation Committee and Nutrition Recommendations . . . A Call for Action: Summary Report of the Scientific Review Committee and the Communications/Implementation Committee,* which are available from Branch Publications Unit, Health Services and Promotion Branch, Department of Health and Welfare, 5th Floor, Jeanne Mance Building, Ottawa, Ontario K1A 1B4.

• Recommended Servings • As mentioned earlier, all food groups offer valuable nutrients, and people should make selections from each group daily. The recommended numbers of daily servings are:

- 6 to 11 servings of breads and cereals.
- 3 to 5 servings of vegetables.
- 2 to 4 servings of fruits.
- 2 to 3 servings of meats and meat alternates.
- 2 servings of milk and milk products. (Teenagers and young adults, women who are pregnant or breastfeeding, and women past menopause are advised to have 3 servings, and teenagers who are pregnant or breastfeeding should have 4.)

The lower number of servings from each group provides about the right amount of food energy (about 1600 kcalories) for sedentary women and older adults. The middle of the range (about 2200 kcalories) is appropriate for most children, teenage girls, active women, and sedentary men. The upper end (about 2800 kcalories) meets the needs of teenage boys, active men, and very active women. Table 2-2 shows recommended numbers of servings for each of these three levels. Physical activity increases a person's energy allowance and permits the person to eat more foods, or higher-kcalorie foods, to supply the needed nutrients without gaining unwanted weight.

TABLE 2-2 **Recommended Servings for Different Energy Intakes**

Food Group	Energy[a] (kcal)		
	1600	*2200*	*2800*
Bread, especially whole grain	6	9	11
Vegetable	3	4	5
Fruit	2	3	4
Milk, preferably nonfat or low fat[b]	2–3	2–3	2–3
Meat, preferably lean or low fat	2 (5 oz)	2 (6 oz)	3 (7 oz)

SOURCE: Adapted from U.S. Department of Agriculture, Center for Nutrition Policy and Promotion, *The Food Guide Pyramid,* Home and Garden Bulletin no. 252, 1996.
[a]Choose low-fat and lean foods from the five major food groups and use foods from the fats, oils, and sweet group sparingly.
[b]Women who are pregnant or breastfeeding, teenagers and young adults, and women past menopause need 3 servings; teenagers who are pregnant or breastfeeding need 4 servings. In fact, given the 1997 DRI, which raised calcium recommendations, all individuals may need an additional serving from the milk group.

• **Serving Sizes** • What counts as a serving? The answer differs for each food group and for various foods within a group. For example, a serving of milk is 1 cup whereas a serving of fruit juice is ¾ cup. Furthermore, serving sizes may not represent the amounts people actually put on their plates. Standard serving sizes for the Food Guide Pyramid are generally smaller than people typically eat. For example, the Pyramid describes half of a 2-ounce bagel as one serving, yet most bagels today weigh in at 4 ounces or more—meaning that a person eating one of these bagels for breakfast is actually getting four or more bread servings, not one.

Figure 2-2 lists the serving sizes for standard foods within each group. When counting servings, be sure to consider the quantity consumed. For example, ½ cup of cooked rice is considered one serving. So, 1 cup of rice counts as 2 of the recommended 6 to 11 daily servings from the bread group. Similarly, ¼ cup counts as ½ serving.

• **Food Guide Pyramid** • The Food Guide Pyramid is a graphic depiction of the Daily Food Guide (see Figure 2-2 again). The illustration was designed to depict variety, moderation, and also proportions: the relative size of each section represents the number of daily servings recommended. The broad base at the bottom conveys the message that grains should be abundant and form the foundation of a healthy diet. Fruits and vegetables appear at the next level, showing that they have a slightly less prominent, but still important, place in the diet. Meats and milks appear in a smaller band near the top. A few servings of each can contribute valuable nutrients, such as protein, vitamins, and minerals, without too much fat and cholesterol. Fats, oils, and sweets occupy the tiny apex, indicating that they should be used sparingly.

Alcoholic beverages do not appear in the Pyramid, but they too should be limited. Items such as spices, coffee, tea, and diet soft drinks provide few, if any, nutrients, but can add flavor and pleasure to meals when used judiciously.

Tiny dots and triangles are sprinkled over the food groups to represent naturally occurring and added fats and added sugars, respectively. These symbols are meant to remind users that specific foods within the various groups are high in fats, sugars, or both, and so should be eaten in moderation.

The Daily Food Guide plan and Food Guide Pyramid emphasize grains, vegetables, and fruits—all plant foods. Some 75 percent of a day's servings should come from these three groups. This strategy helps all people obtain complex carbohydrates, fiber, vitamins, and minerals with little fat. It also eases diet planning for vegetarians.

• **Vegetarian Food Guide** • Vegetarian diets rely mainly on plant foods: grains, vegetables, **legumes,**◆ fruits, seeds, and nuts. Some vegetarian diets include eggs, milk products, or both. People who do not eat meats or milk products can still use the Daily Food Guide to create an adequate diet.[2] The food groups are similar, and the number of servings remains the same. Vegetarians

◆ **Legumes include a variety of beans and peas:**

- Black beans.
- Black-eyed peas.
- Garbanzo beans.
- Great northern beans.
- Kidney beans.
- Lentils.
- Navy beans.
- Peanuts.
- Pinto beans.
- Soybeans.
- Split peas.

legumes (lay-GYOOMS, LEG-yooms): plants of the bean and pea family, rich in high-quality protein compared with other plant-derived foods.

FIGURE 2-2 Daily Food Guide

Breads, Cereals, and Other Grain Products: 6 to 11 servings per day.

These foods are notable for their contributions of complex carbohydrates, riboflavin, thiamin, niacin, folate, iron, protein, magnesium, and fiber.
Serving = 1 slice bread; ½ c cooked cereal, rice, or pasta; 1 oz ready-to-eat cereal; ½ bun, bagel, or English muffin; 1 small roll, biscuit, or muffin; 3 to 4 small or 2 large crackers.

- ◆ Whole grains (wheat, oats, barley, millet, rye, bulgur, couscous, polenta), enriched breads, rolls, tortillas, cereals, bagels, rice, pastas (macaroni, spaghetti), air-popped corn.
- ◆ Pancakes, muffins, cornbread, crackers, cookies, biscuits, presweetened cereals, granola, taco shells, waffles, french toast.
- ◆ Croissants, fried rice, doughnuts, pastries, cakes, pies.

Vegetables: 3 to 5 servings per day (use dark green, leafy vegetables and legumes several times a week).

These foods are notable for their contributions of vitamin A, vitamin C, folate, potassium, magnesium, and fiber, and for their lack of fat and cholesterol.
Serving = ½ c cooked or raw vegetables; 1 c leafy raw vegetables; ½ c cooked legumes; ¾ c vegetable juice.

- ◆ Bamboo shoots, bok choy, bean sprouts, broccoli, brussels sprouts, cabbage, carrots, cauliflower, corn, cucumbers, eggplant, green beans, green peas, leafy greens (spinach, mustard, and collard greens), legumes, lettuce, mushrooms, okra, onions, peppers, potatoes, pumpkin, scallions, seaweed, snow peas, soybeans, tomatoes, water chestnuts, winter squash.
- ◆ Candied sweet potatoes.
- ◆ French fries, tempura vegetables, scalloped potatoes, potato salad.

Fruits: 2 to 4 servings per day.

These foods are notable for their contributions of vitamin A, vitamin C, potassium, and fiber, and for their lack of sodium, fat, and cholesterol.
Serving = typical portion (such as 1 medium apple, banana, or orange, ½ grapefruit, 1 melon wedge); ¾ c juice; ½ c berries; ½ c diced, cooked, or canned fruit; ¼ c dried fruit.

- ◆ Apricots, blueberries, cantaloupe, grapefruit, guava, oranges, kiwi, papaya, peaches, strawberries, plums, apples, bananas, pears, watermelon; unsweetened juices.
- ◆ Canned or frozen fruit (in syrup); sweetened juices.
- ◆ Dried fruit, coconut, avocados, olives.

Meat, Poultry, Fish, and Alternates: 2 to 3 servings per day.

Meat, poultry, and fish are notable for their contributions of protein, phosphorus, vitamin B_6, vitamin B_{12}, zinc, iron, niacin, and thiamin; legumes are notable for their protein, fiber, thiamin, folate, vitamin E, potassium, magnesium, iron, and zinc, and for their lack of fat and cholesterol.
Servings = 2 to 3 oz lean, cooked meat, poultry, or fish (total 5 to 7 oz per day); count 1 egg, ½ c cooked legumes, 4 oz tofu, or 2 tbs nuts, seeds, or peanut butter as 1 oz meat (or about ⅓ serving).

- ◆ Poultry (light meat, no skin), fish, shellfish, legumes, egg whites.
- ◆ Lean meat (fat-trimmed beef, lamb, pork); poultry (dark meat, no skin); ham; refried beans; whole eggs, tofu, tempeh.
- ◆ Hot dogs, luncheon meats, ground beef, peanut butter, nuts, sausage, bacon, fried fish or poultry, duck.

Key:
- ◆ Foods generally highest in nutrient density (good first choice).
- ◆ Foods moderate in nutrient density (reasonable second choice).
- ◆ Foods lowest in nutrient density (limit selections).

Milk, Cheese, and Yogurt: 2 servings per day; 3 servings per day for teenagers and young adults, pregnant/lactating women, women past menopause; 4 servings per day for pregnant/lactating teenagers.

These foods are notable for their contributions of calcium, riboflavin, protein, vitamin B_{12}, and, when fortified, vitamin D and vitamin A.

Serving = 1 c milk or yogurt; 2 oz process cheese food; 1½ oz cheese.

- ◆ Nonfat and 1% low-fat milk (and nonfat products such as buttermilk, cottage cheese, cheese, yogurt); fortified soy milk.
- ◆ 2% reduced-fat milk (and low-fat products such as yogurt, cheese, cottage cheese); chocolate milk; sherbet; ice milk.
- ◆ Whole milk (and whole-milk products such as cheese, yogurt); custard; milk shakes; ice cream.

Note: These serving recommendations were established before the 1997 DRI, which raised the recommended intake for calcium; meeting the calcium recommendation may require an additional serving from the milk, cheese, and yogurt group.

Fats, Sweets, and Alcoholic Beverages: Use sparingly.

These foods contribute sugar, fat, alcohol, and food energy (kcalories). They should be used sparingly because they provide food energy while contributing few nutrients. Miscellaneous foods not high in kcalories, such as spices, herbs, coffee, tea, and diet soft drinks, can be used freely.

- ◆ Foods high in fat include margarine, salad dressing, oils, lard, mayonnaise, sour cream, cream cheese, butter, gravy, sauces, potato chips, chocolate bars.
- ◆ Foods high in sugar include cakes, pies, cookies, doughnuts, sweet rolls, candy, soft drinks, fruit drinks, jelly, syrup, gelatin, desserts, sugar, and honey.
- ◆ Alcoholic beverages include wine, beer, and liquor.

Food Guide Pyramid

A Guide to Daily Food Choices
The breadth of the base shows that grains (breads, cereals, rice, and pasta) deserve most emphasis in the diet. The tip is smallest: use fats, oils, and sweets sparingly.

FIGURE 2-3 Ethnic Cuisines and Food Choices

	Bread	Vegetables	Fruit	Meat	Milk	Comment
Asian	Rice, noodles, millet	Amaranth, baby corn, bamboo shoots, chayote, bok choy, mung bean sprouts, sugar peas, straw mushrooms, water chestnuts	Carambola, guava, kumquat, lychee, persimmon	Soybeans, squid, tofu, duck eggs, pork, poultry	Soy milk	Asian people eating traditional foods consume more fiber and less fat than North Americans—dietary habits that support good health. The abundance of vegetables and soy in the diet also helps to protect against disease.
Mediterranean	Pita pocket bread, pastas, rice, couscous, polenta, bulgur	Eggplant, tomatoes, peppers, cucumbers, grape leaves	Olives, grapes, figs	Fish and other seafood, gyros, lamb, chicken, beef	Feta, mozzarella, and goat cheeses; yogurt	Among the dietary factors that may protect people in the Mediterranean region against heart disease are their high intakes of olive oil, red wines, and fruits and vegetables rich in antioxidants (see Highlight 11).
Mexican	Tortillas (corn or flour), taco shells, rice	Chayote, corn, jicama, tomato salsa, cactus, cassava, tomatoes	Guava, mango, papaya, avocado, plantain	Refried beans, fish, chicken, chorizo, beef, eggs	Cheese, custard	To lower the fat in Mexican meals, use sour cream lightly and skip the fried tortillas. Refry beans in a little olive oil instead of lard—or eat them whole or mashed, without added fat.

© Becky Luigart-Stayner/Corbis
© PhotoDisc Inc
© PhotoDisc Inc.

select *meat alternates* from the meat group—foods such as legumes, seeds, nuts, tofu, and, for those who eat them, eggs. Legumes and at least one cup of dark leafy greens help to supply the iron that meats usually provide. Vegetarians who do not drink cow's milk can use soy "milk"—a product made from soybeans that provides similar nutrients if it has been **fortified** with calcium, vitamin D, and vitamin B_{12}. Highlight 6 presents the Daily Food Guide for Vegetarians, defines vegetarian terms, and provides more information on vegetarian diet planning.

• **Ethnic Food Choices** • People can use the Food Guide Pyramid and still enjoy a diverse array of cuisines by sorting ethnic foods into their appropriate food groups. For example, a person eating Mexican foods would find tortillas in the bread group, jicama in the vegetable group, and guava in the fruit group. Figure 2-3 features foods unique to selected cuisines and shows how people can enjoy ethnic meals and still limit their energy and fat intakes.

• **Perceptions and Actual Intakes** • The Daily Food Guide and Food Guide Pyramid were developed to help people choose a balanced and healthful diet. Are these plans successful? Yes, they can help people select nutrient-rich diets.

fortified: the addition to a food of nutrients that were either not originally present or present in insignificant amounts. Fortification can be used to correct or prevent a widespread nutrient deficiency or to balance the total nutrient profile of a food.

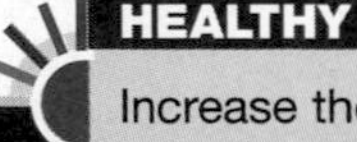

HEALTHY PEOPLE 2010

Increase the proportion of persons aged 2 years and older who consume at least two daily servings of fruit; at least three daily servings of vegetables, with at least one-third being dark green or orange vegetables; and at least six daily servings of grain products, with at least three being whole grains.

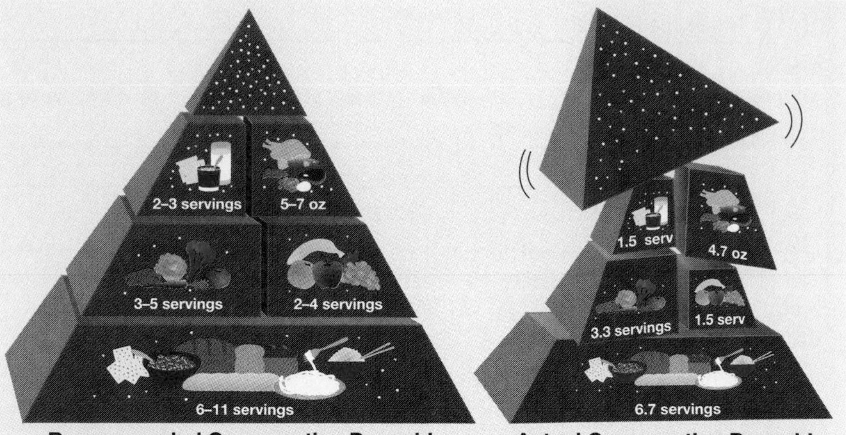

In fact, one survey reports that only adults who select the recommended number of servings from each of the five food groups meet recommendations for energy, fat, fiber, vitamins, and minerals.[3] Unfortunately, only 1 percent of the more than 8000 people surveyed made such selections. More commonly, they neglected the fruit, milk, and bread groups and, by doing so, raised their percentage of kcalories from fat and lowered their fiber, calcium, and zinc intakes.

Many adults *think* they are selecting foods that reflect the recommendations of the Food Guide Pyramid. In reality they are barely meeting, or falling short, of recommendations for all food groups, and eating too many fats, sweets, and oils.[4] In a sense, our pyramids are top-heavy and "tumbling." They need more support from each of the five food groups to build a balanced diet.

Recommended Consumption Pyramid **Actual Consumption Pyramid**

Compared with recommendations, actual consumption resembles a precariously built "tumbling" pyramid.

Exchange Lists

Food group plans are particularly well suited to help a person achieve dietary adequacy, balance, and variety. **Exchange lists**◆ provide additional help in achieving kcalorie control and moderation. Originally developed for people with diabetes, exchange systems have proved useful for general diet planning as well.

Appendix G gives complete details of the major exchange system used in the United States, and Appendix I provides details of the choice system used in Canada. ◆

Unlike the Daily Food Guide, which sorts foods primarily by their protein, vitamin, and mineral contents, the exchange system sorts foods into three main groups by their proportions of carbohydrate, fat, and protein. These three groups—the carbohydrate group, the fat group, and the meat and meat substitutes group (protein)—organize foods into several exchange lists (see Table 2-3 on p. 42). The carbohydrate group covers these exchange lists:

- Starch (cereals, grains, pasta, breads, crackers, snacks, starchy vegetables, and dried beans, peas, and lentils).
- Fruit.
- Milk (nonfat, reduced-fat, and whole).
- Other carbohydrates (desserts and snacks with added sugars and fats).
- Vegetables.

The fat group covers this exchange list:

- Fats.

The meat and meat substitutes group (protein) covers these exchange lists:

- Meat and meat substitutes (very lean, lean, medium-fat, and high-fat).

• Portion Sizes • The exchange system helps people control their energy intakes by paying close attention to portion sizes.◆ The portion sizes have been adjusted so that a portion of any food on a given list provides roughly the same amount of carbohydrate, fat, and protein and, therefore, total kcalories. Any food on a list can then be exchanged, or traded, for any other food on that same list without significantly affecting the diet's balance or total kcalories. For example, a person may select either 17 grapes or ½ grapefruit as one fruit portion, and either choice would provide roughly 60 kcalories. A whole grapefruit, however, would count as two portions.

Quick and easy estimates of portion sizes: ◆

- 1 small fruit is about the size of a tennis ball.
- 1 c cooked vegetables is about the size of your fist.
- ½ c ice cream is about the size of a racquetball.
- 3 oz meat is about the size of a deck of cards.
- 1 oz of cheese is about the size of four stacked dice.

The amount of food per serving in the exchange system is not always the same as in the Daily Food Guide, especially when it comes to meats. The exchange

exchange lists: diet-planning tools that organize foods by their proportions of carbohydrate, fat, and protein. Foods on any single list can be used interchangeably.

TABLE 2-3 The Exchange Lists

Group/Lists	Typical Item/Portion Size	Carbohydrate (g)	Protein (g)	Fat (g)	Energy[a] (kcal)
Carbohydrate Group					
Starch[b]	1 slice bread	15	3	1 or less	80
Fruit	1 small apple	15	—	—	60
Milk					
Nonfat	1 c nonfat milk	12	8	0–3	90
Reduced-fat	1 c reduced-fat milk	12	8	5	120
Whole	1 c whole milk	12	8	8	150
Other carbohydrates[c]	2 small cookies	15	varies	varies	varies
Vegetable	½ c cooked carrots	5	2	—	25
Meat and Meat Substitute Group[d]					
Meat					
Very lean	1 oz chicken (white meat, no skin)	—	7	0–1	35
Lean	1 oz lean beef	—	7	3	55
Medium-fat	1 oz ground beef	—	7	5	75
High-fat	1 oz pork sausage	—	7	8	100
Fat Group					
Fat	1 tsp butter	—	—	5	45

Note: The complete details of the U.S. exchange system are provided in Appendix G. Those of the Canadian system are shown in Appendix I.

[a]The energy value for each exchange list represents an approximate average for the group and does not reflect the precise number of grams of carbohydrate, protein, and fat. For example, a slice of bread contains 15 grams of carbohydrate (that's 60 kcalories), 3 grams protein (that's another 12 kcalories), and a little fat—rounded to 80 kcalories for ease in calculating. A half-cup of vegetables (not including starchy vegetables) contains 5 grams carbohydrate (20 kcalories) and 2 grams protein (8 more), which has been rounded down to 25 kcalories.

[b]The starch list includes cereals, grains, breads, crackers, snacks, starchy vegetables (such as corn, peas, and potatoes), and legumes (dried beans, peas, and lentils).

[c]The other carbohydrates list includes foods that contain added sugars and fats such as cakes, cookies, doughnuts, ice cream, potato chips, pudding, syrup, and frozen yogurt.

[d]The meat and meat substitutes list includes legumes, cheeses, and peanut butter.

system lists meats and most cheeses in single ounces; that is, one *exchange* of meat is 1 ounce, whereas one *serving* in the Daily Food Guide is 2 to 3 ounces. Calculating meat by the ounce encourages a person to keep close track of the exact amounts eaten. This in turn helps control energy and fat intakes. Be aware, too, that most people do not serve foods in carefully measured portions, nor do the amounts reflect the exchange system or Daily Food Guide serving sizes. Many restaurants, for example, offer 8- to 16-ounce steaks that are equivalent to four or five (2- to 3-ounce) *servings* of meat. Similarly, a bakery may sell muffins or bagels that are two to three times the size of a typical bread serving.

To apply the system successfully, users must become familiar with portion sizes. A convenient way to remember the portion sizes and energy values is to keep in mind a typical item from each list. The photos on pages 44–45 show the foods on each of the exchange lists and their accurate portion sizes.

© Polara Studios Inc.

It may look like one, but the large muffin counts as two servings.

• The Foods on the Lists • Foods do not always appear on the exchange list where you might first expect to find them. They are grouped according to their energy-nutrient contents rather than by their source (such as milks), their outward appearance, or their vitamin and mineral contents. For example, cheeses are grouped with meats because, like meats, cheeses contribute energy from protein and fat but provide negligible carbohydrate. (In the food group plan presented earlier, cheeses are classed with milk because they are milk products with similar calcium contents.)

For similar reasons, starchy vegetables such as corn, green peas, and potatoes are listed on the starch list in the exchange system, rather than with the vegetables. Likewise, olives are not classed as a "fruit" as a botanist would claim; they are classified as a "fat" because their fat content makes them more similar to butter than to berries. Bacon and nuts are also on the fat list to remind users of their high fat content. These groupings highlight the characteristics of foods that are significant to energy intake.

Users of the exchange lists learn to view mixtures of foods, such as casseroles and soups, as combinations of foods from different exchange lists. They also learn to interpret food labels with the exchange system in mind (see Figure 2-4).

• Controlling Energy and Fat • By assigning items like bacon and avocados to the fat list, the exchange system alerts consumers to foods that are unexpectedly high in fat. Even the starch list specifies which grain products contain added fat (such as biscuits, muffins, and waffles). In addition, the exchange system encourages users to think of nonfat milk as milk and of whole milk as milk with added fat, and to think of very lean meats as meats and of lean, medium-fat, and high-fat meats as meats with added fat. To that end, foods on the milk and meat lists are separated into categories based on their fat contents. The milk group is classed as nonfat, reduced-fat, and whole; the meat group as very lean, lean, medium-fat, and high-fat.

Control of food energy and fat intake can be highly successful with the exchange system. Exchange plans do not, however, guarantee adequate intakes of vitamins and minerals. Food group plans work better from that standpoint because the food groupings are based on similarities in vitamin-mineral content. In the exchange system, for example, meats are grouped with cheeses, yet the meats are iron-rich and calcium-poor, whereas the cheeses are iron-poor and calcium-rich. To take advantage of the strengths of both food group plans and exchange patterns, and to compensate for their weaknesses, diet planners often combine these two diet-planning tools.

FIGURE 2-4 Seeing Exchanges on a Food Label

HOME TASTE
Lasagna Dinner
WITH MEAT SAUCE

Nutrition Facts
Serving size 10 1/2 oz (298 g)
Servings per Package 1

Amount per serving		
Calories 361	Calories from Fat 117	
		% Daily Value
Total Fat 13 g		**20%**
Saturated Fat 8 g		**40%**
Cholesterol 87 mg		**29%**
Sodium 860 mg		**36%**
Total Carbohydrate 37 g		**12%**
Dietary fiber 0 g		
Sugars 8 g		
Protein 26 g		

Can you "see" these exchanges in the label above?

Exchange	Carbohydrate	Protein	Fat
2 starches	30 g	6 g	—
1 vegetable	5 g	2 g	—
3 medium-fat meats	—	21 g	15 g
Exchange totals	**35**	**29**	**15**
Label totals	**37**	**26**	**13**

Knowing that foods on the starch list provide 15 grams of carbohydrate and those on the vegetable list provide 5, you can count a lasagna dinner that provides 37 grams of carbohydrate as "2 starches and 1 vegetable"; knowing that foods on the meat list provide 7 grams of protein, you might count it as "3 meats"; the grams of fat suggest that the meat (and cheese) is probably medium-fat.

Combining Food Group Plans and Exchange Lists

A person may find that using a food group plan together with the exchange lists eases the task of choosing foods that provide all the nutrients. The food group plan ensures that all classes of nutritious foods are included, thus promoting adequacy, balance, and variety. The exchange system classifies the food selections by their energy-yielding nutrients, thus controlling energy and fat intakes.

Table 2-4 (on p. 46) shows how to use the Daily Food Guide plan together with the exchange lists to plan a diet. The Daily Food Guide ensures that a certain number of servings is chosen from each of the five food groups (see the first column of the table). The second column translates the number of servings (using the midpoint) into exchanges. With the addition of a small amount of fat, this sample diet plan provides about 1750 kcalories. Most people can meet their needs for all the nutrients within this reasonable energy allowance. The next step in diet planning is to assign the exchanges to meals and snacks. The final plan might look like the one in Table 2-5 (on p. 46).

Next, a person could begin to fill in the plan with real foods to create a menu (use Figure 2-5 and Appendix G or I). For example, the breakfast plan calls for 2 starch exchanges, 1 fruit exchange, and 1 nonfat milk exchange. A person might select a bowl of shredded wheat with banana slices and milk:

1 cup shredded wheat = 2 starch exchanges.
1 small banana = 1 fruit exchange.
1 cup nonfat milk = 1 milk exchange.

Or a bagel and a bowl of cantaloupe pieces topped with yogurt:

1 bagel = 2 starch exchanges.
⅓ cantaloupe melon = 1 fruit exchange.
¾ cup nonfat plain yogurt = 1 milk exchange.

To ensure an adequate and balanced diet, eat a variety of foods daily, choosing different foods from each group.

FIGURE 2-5 **The Exchange System: Example Foods, Portion Sizes, and Energy-Nutrient Contributions**

Starch
1 starch exchange is like:
1 slice bread.
¾ c ready-to-eat cereal.
½ c cooked pasta, rice noodles, or bulgur.
⅓ c cooked rice.
½ c cooked beans.[a]
½ c corn, peas, or yams.
1 small (3 oz) potato.
½ bagel, English muffin, or bun.
1 tortilla, waffle, roll, taco, or matzoh.
(1 starch = 15 g carbohydrate, 3 g protein, 0–1 g fat, and 80 kcal.)

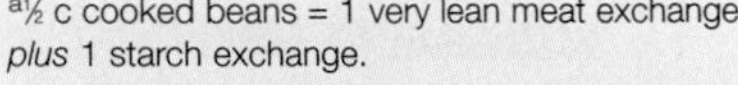

[a]½ c cooked beans = 1 very lean meat exchange *plus* 1 starch exchange.

Vegetables
1 vegetable exchange is like:
½ c cooked carrots, greens, green beans, brussels sprouts, beets, broccoli, cauliflower, or spinach.
1 c raw carrots, radishes, or salad greens.
1 lg tomato.
(1 vegetable = 5 g carbohydrate, 2 g protein, and 25 kcal.)

Fruits
1 fruit exchange is like:
1 small banana, nectarine, apple, or orange.
½ large grapefruit, pear, or papaya.
½ c orange, apple, or grapefruit juice.
17 small grapes.
⅓ cantaloupe (or 1 c cubes).
2 tbs raisins.
1½ dried figs.
3 dates.
1½ carambola (star fruit).
(1 fruit = 15 g carbohydrate and 60 kcal.)

Meat and substitutes (very lean)
1 very lean meat exchange is like:
1 oz chicken (white meat, no skin).
1 oz cod, flounder, or trout.
1 oz tuna (canned in water).
1 oz clams, crab, lobster, scallops, shrimp, or imitation seafood.
1 oz fat-free cheese.
½ c cooked beans, peas, or lentils.
¼ c nonfat or low-fat cottage cheese.
2 egg whites (or ¼ c egg substitute).
(1 very lean meat = 7 g protein, 0–1 g fat, and 35 kcal).

Meats and substitutes (lean)
1 lean meat exchange is like:
1 oz beef or pork tenderloin.
1 oz chicken (dark meat, no skin).
1 oz herring or salmon.
1 oz tuna (canned in oil, drained).
1 oz low-fat cheese or luncheon meats.
(1 lean meat = 7 g protein, 3 g fat, and 55 kcal.)

Meats and substitutes (medium-fat)
1 medium-fat meat exchange is like:
1 oz ground beef.
1 oz pork chop.
1 egg.
¼ c ricotta.
4 oz tofu.
(1 medium-fat meat = 7 g protein, 5 g fat, and 75 kcal.)

A person who wanted butter on the bagel could move a fat exchange or two from dinner to breakfast. If willing to use two fat exchanges at breakfast, the person could have pancakes with strawberries and milk:

4 small pancakes = 2 starch plus 2 fat exchanges.
1¼ cup strawberries = 1 fruit exchange.
1 cup nonfat milk = 1 milk exchange.

Other carbohydrates
1 other carbohydrates exchange is like:
2 small cookies.
1 small brownie or cake.
5 vanilla wafers.
1 granola bar.
½ c ice cream.
(1 other carbohydrate = 15 g carbohydrate and may be exchanged for 1 starch, 1 fruit, or 1 milk. Because many items on this list contain added sugar and fat, their fat and kcalorie values vary, and their portion sizes are small.)

Meats and substitutes (high-fat)
1 high-fat meat exchange is like:
1 oz pork sausage.
1 oz luncheon meat (such as bologna).
1 oz regular cheese (such as cheddar or swiss).
1 small hot dog (turkey or chicken).[b]
2 tbs peanut butter.[c]
(1 high-fat meat = 7 g protein, 8 g fat, and 100 kcal.)

[b]A beef or pork hot dog counts as 1 high-fat meat exchange *plus* 1 fat exchange.
[c]Peanut butter counts as 1 high-fat meat exchange *plus* 1 fat exchange.

Milks (nonfat and low-fat)
1 nonfat milk exchange is like:
1 c nonfat or 1% milk.
¾ c nonfat yogurt, plain.
1 c nonfat or low-fat buttermilk.
½ c evaporated nonfat milk.
⅓ c dry nonfat milk.
(1 nonfat milk = 12 g carbohydrate, 8 g protein, 0–3 g fat, and 90 kcal.)

Milks (reduced-fat)
1 reduced-fat milk exchange is like:
1 c 2% milk.
¾ c low-fat yogurt, plain.
(1 reduced-fat milk = 12 g carbohydrate, 8 g protein, 5 g fat, and 120 kcal.)

Milks (whole)
1 whole milk exchange is like:
1 c whole milk.
½ c evaporated whole milk.
(1 whole milk = 12 g carbohydrate, 8 g protein, 8 g fat, and 150 kcal.)

THE FAT GROUP

Fats
1 fat exchange is like:
1 tsp butter.
1 tsp margarine or mayonnaise (1 tbs reduced fat).
1 tsp any oil.
1 tbs salad dressing (2 tbs reduced fat).
8 large black olives.
10 large peanuts.
⅛ medium avocado.
1 slice bacon.
2 tbs shredded coconut.
1 tbs cream cheese (2 tbs reduced fat).
(1 fat = 5 g fat and 45 kcal.)

Note: Health recommendations urge people to limit their intakes of saturated fats; butter, bacon, coconut, and cream cheese contain saturated fats.

Then the person could move on to complete the menu for lunch, dinner, and snacks. (Table 2-5 includes a sample menu.) As you can see, we all make countless food-related decisions daily—whether we have a plan or not. Following a plan, like the Daily Food Guide, that incorporates health recommendations and diet-planning principles helps a person to make wise decisions. Highlight 2 offers a shortcut plan for those who haven't taken the time and effort to develop

TABLE 2-4 Diet Planning with the Exchange System Using the Daily Food Guide Pattern

Patterns from Daily Food Guide Plan	Selections Made Using the Exchange System	Energy Cost (kcal)
Grains (breads and cereals)—6 to 11 servings	Starch list—select 9 exchanges	720
Vegetables—3 to 5 servings	Vegetable list—select 4 exchanges	100
Fruits—2 to 4 servings	Fruit list—select 3 exchanges	180
Meat—2 to 3 servings[a]	Meat list—select 6 lean exchanges	330
Milk—2 servings	Milk list—select 2 nonfat exchanges	180
	Fat list—select 5 exchanges	225
Total		1735

[a]In the food group plan, 1 serving is 2 to 3 ounces; in the exchange system, 1 exchange is 1 ounce. The Daily Food Guide suggests that amounts should total 5 to 7 ounces of meat daily.

a nutritionally adequate, well-balanced diet for themselves. It's not perfect, but it's a start toward thinking "nutrition" when deciding what to eat.

From Guidelines to Groceries

Dietary recommendations emphasize foods low in fat such as grains, fruits, vegetables, lean meats, fish, poultry, and low-fat milk products. Only you can design such a diet for yourself, but how do you begin? Start with the foods you enjoy eating. Then try to make improvements, little by little. When shopping, think of the food groups, and choose nutrient-dense foods within each group.

• Breads, Cereals, and Other Grain Products • When shopping for grain products, you will find them described as *refined, enriched,* or *whole grain.* These terms refer to the milling process and the making of grain products, and they have different nutrition implications (see Figure 2-6). **Refined** foods may have lost many nutrients during processing; **enriched** products may have had some nutrients added back; and **whole-grain** products may be rich in all nutrients found in the original grain.

When it became a common practice to refine the wheat flour used for bread by milling it and throwing away the bran and the germ, consumers suffered a tragic loss of many nutrients. As a consequence, in the early 1940s Congress passed legislation requiring that all grain products that cross state lines be enriched with iron, thiamin, riboflavin, and niacin. In 1996 this legislation was amended to include folate, a vitamin considered essential in the prevention of

TABLE 2-5 A Sample Diet Plan and Menu

This diet plan is one of many possibilities. It follows the number of servings suggested by the Daily Food Guide and meets dietary recommendations to provide 55 to 60 percent of its kcalories from carbohydrate, 15 to 20 percent from protein, and less than 30 percent from fat.

Exchange	Breakfast	Lunch	Snack	Dinner	Snack
9 starch	2	2	1	3	1
4 vegetable				4	
3 fruit	1	1	1		
6 lean meat		2		4	
2 nonfat milk	1				1
5 fat		1		4	

A Sample Menu

Breakfast: Cereal with banana and milk

Lunch: Turkey sandwich and a small bunch of grapes

Snack: Popcorn and apple juice

Dinner: Spagetti with meat sauce; salad with sunflower seeds adn dressing; green beans; corn on the cob

Snack: Graham crackers and milk

FIGURE 2-6 **A Wheat Plant**

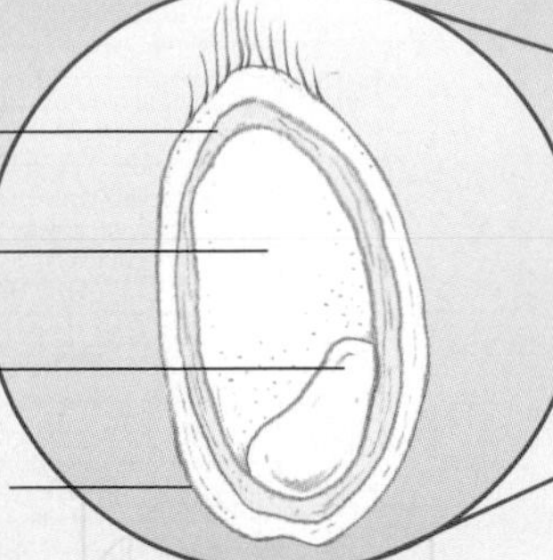

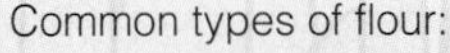

Common types of flour:

- *White flour*—an endosperm flour that has been refined and bleached for maximum softness and whiteness.
- *Unbleached flour*—a tan-colored endosperm flour with texture and nutritive qualities that approximate those of regular white flour.
- *Wheat flour*—any flour made from wheat, including white flour; wheat flour has been refined whereas *whole-wheat flour* has not.
- *Whole-wheat flour*—flour made from whole-wheat kernels; a whole-grain flour.

Refined white grain products contain only the endosperm. Even with nutrients added back, they are not as nutritious as whole-grain products, as the next figure shows.

some birth defects. Most grain products that have been refined, such as rice, wheat pastas like macaroni and spaghetti, and cereals (both cooked and ready-to-eat types), have subsequently been enriched,◆ and their labels say so.

Enrichment doesn't make a slice of bread rich in these added nutrients, but people who eat several slices a day obtain significantly more of these nutrients than they would from unenriched white bread. To a great extent, the enrichment of white flour helps to prevent deficiencies of these nutrients, but it fails to compensate for losses of many other nutrients and fiber. As Figure 2-7 shows, whole-grain items still outshine the enriched ones. Only *whole-grain* flour contains all of the nutritive portions of the grain. Whole-grain products, such as brown rice and oatmeal, not only provide more nutrients and fiber, but do not contain the added salt and sugar of flavored, processed rice or sweetened cereals.

Speaking of cereals, ready-to-eat breakfast cereals are the most highly fortified foods on the market. Like an enriched food, a *fortified* food◆ has had nutrients added during processing, but in a fortified food, the added nutrients may not have been present in the original product. Some breakfast cereals made from refined flour and fortified with high doses of vitamins and minerals are actually more like supplements disguised as cereals than they are like whole grains. They may be nutritious—with respect to the nutrients added—but they still may fail to convey the full spectrum of nutrients that a whole-

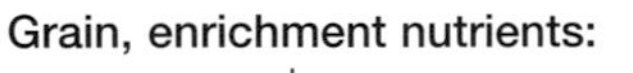

Grain, enrichment nutrients: ◆
- Iron.
- Thiamin.
- Riboflavin.
- Niacin.
- Folate.

The terms *fortified* and *enriched* may be used interchangeably.[5] ◆

refined: the process by which the coarse parts of a food are removed. When wheat is refined into flour, the bran, germ, and husk are removed, leaving only the endosperm.

enriched: the addition to a food of nutrients that were lost during processing so that the food will meet a specified standard.

whole grain: a grain milled in its entirety (all but the husk), not refined.

With its many grains (including wheat, rye, oats, corn, and rice) and types of foods (such as pastas, breads, and cereals), this group does more than its share for variety.

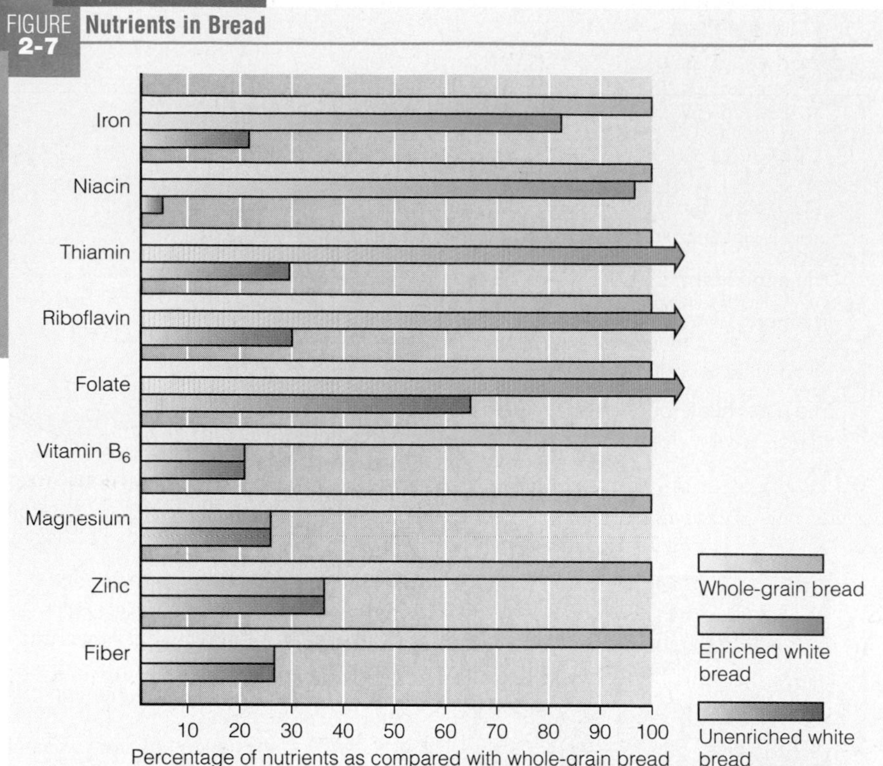

Whole-grain bread is more nutritious than other breads, even enriched bread. For iron, thiamin, riboflavin, niacin, and folate, enriched bread provides about the same quantities as whole-grain bread and significantly more than unenriched bread. For fiber and the other nutrients (both those shown here and those not shown), enriched bread provides less than whole-grain bread.

grain food or a mixture of such foods might provide. Furthermore, minerals (especially iron) are not as well absorbed from enriched foods as from naturally occurring sources.

• **Vegetables** • Choose fresh vegetables, especially dark green leafy and yellow-orange vegetables like spinach, broccoli, and sweet potatoes. Cooked or raw, vegetables are good sources of vitamins, minerals, and fiber. Frozen and canned vegetables without added salt are acceptable alternatives to fresh. To control fat, energy, and sodium intakes, limit butter, salad dressings, and salt on vegetables.

• **Legumes** • Choose often from the variety of legumes available. They are an economical, low-fat, nutrient- and fiber-rich food choice.

• **Fruit** • Choose fresh fruits often, especially citrus fruits and yellow-orange fruits like cantaloupes and apricots. Frozen, dried, and canned fruits without added sugar are acceptable alternatives to fresh. Fruits supply valuable vitamins, minerals, and fibers. They add flavors, colors, and textures to meals, and their natural sweetness makes them enjoyable as snacks or desserts.

Fruit juices are healthy beverages, but contain little dietary fiber compared with whole fruits. Whole fruits satisfy the appetite better than juices, thereby helping people to limit food energy intakes. Juices, on the other hand, are a good choice for people who need extra food energy. Be aware that sweetened fruit "drinks" or "ades" contain mostly water, sugar, and a little juice for flavor. Some may have been fortified with vitamin C, but lack any other significant nutritional value.

© 1998 Photo Disc Inc.

Combining legumes with foods from other food groups creates delicious meals.

© 1998 Photo Disc Inc.

Add rice to red beans for a hearty meal.

© Felicia Martinez/PhotoEdit

Enjoy a Greek salad topped with garbanzo beans for a little ethnic diversity.

© Michael Newman/PhotoEdit

A bit of meat and lots of spices turn kidney beans into chili con carne.

• **Meat, Fish, and Poultry** • Meat, fish, and poultry provide essential minerals, such as iron and zinc, and abundant B vitamins as well as protein. To buy and prepare these foods without excess energy, fat, and sodium takes a little knowledge and planning. When shopping in the meat department, choose fish, poultry, and lean cuts of beef and pork named "round" or "loin" (as in top round or pork tenderloin). As a guide, "prime" and "choice" cuts generally have more fat than "select" cuts. Restaurants usually serve prime cuts. Ground beef, even "lean" ground beef, derives most of its food energy from fat as Table 2-6 shows. Have the butcher trim and grind a lean round steak instead. Alternatively, **textured vegetable protein** can be used instead of ground beef in a casserole, spaghetti sauce, or chili, saving fat kcalories.

Weigh meat after it is cooked and the bones and fat are removed. In general, 4 ounces of raw meat is equal to about 3 ounces of cooked meat. Some examples of 3-ounce portions of meat include 1 medium pork chop, ½ chicken breast, or 1 steak or hamburger about the size of a deck of cards. To keep fat intake down, bake, roast, broil, grill, or braise meats (but do not fry them in fat); remove the skin from poultry after cooking; trim visible fat before cooking; and drain fat after cooking. Chapter 5 offers many additional strategies for lowering fat intake.

TABLE 2-6 Percent kCalories from Fat in Selected Meats

• Ground beef	
Regular	66%
Lean	57%
Extra lean	54%
• Ground turkey	51%
• Ground round	
(lean and trimmed)	27%

• **Milk** • Shoppers will find a variety of fortified foods in the dairy case. Examples are milk, to which vitamins A and D have been added, and soy milk,◆ to which calcium, vitamin D, and vitamin B_{12} have been added. In addition, shoppers may find **imitation foods** (such as cheese products) and **food substitutes** (such as egg substitutes). As food technology advances, many such foods offer nonfat and low-fat alternatives. For example, egg substitutes help people who want to reduce their fat and cholesterol intakes. Highlight 5 gives other examples.

◆ **Be aware that not all soy milks have been fortified. Read labels carefully.**

When shopping, choose nonfat◆ or low-fat milk, yogurt, and cheeses. Such selections help consumers lower their fat intake to 30 percent of their daily energy intake.[6] Milk products are important sources of calcium, but can provide too much sodium and fat if not selected with care.

◆
- **Nonfat** milk may also be called **fat-free, skim, zero-fat,** or **no-fat.**
- **Low-fat** milk refers to 1% milk.
- **Reduced-fat** milk refers to 2% milk; may also be called **less-fat.**

IN SUMMARY

Food group plans select from different families of similar foods to provide adequacy, balance, and variety in the diet. Exchange lists define portion sizes so that foods within a given group supply similar amounts of energy nutrients, thus helping to attain kcalorie control and moderation. Together, they make it easier to plan a diet that includes abundant grains, vegetables, legumes, and fruits and moderate amounts of meats and milk products. In making any food choice, remember to view the food in the context of your total diet. It is the combination of many different foods that provides the abundance of nutrients so essential to a healthy diet.

textured vegetable protein: processed soybean protein used in vegetarian products such as soy burgers.

imitation foods: foods that substitute for and resemble another food, but are nutritionally inferior to it with respect to vitamin, mineral, or protein content. If the substitute is not inferior to the food it resembles and if its name provides an accurate description of the product, it need not be labeled "imitation."

food substitutes: foods that are designed to replace other foods.

FOOD LABELS

Many consumers read food labels to help them select foods with less fat, saturated fat, cholesterol, and sodium and more complex carbohydrates and dietary fiber. Food labels appear on virtually all processed foods, and posters or brochures provide similar nutrition information for fresh meats, fruits, and vegetables (see Figure 2-8). A few foods need not carry nutrition labels: those contributing few nutrients, such as plain coffee, tea, and spices; those produced by small businesses; and those prepared and sold in the same establishment. Producers of some of these items, however, voluntarily use labels. Even markets selling nonpackaged items voluntarily present nutrient information, either in brochures or on signs posted at the point of purchase. Restaurants need not supply complete nutrition information for menu items unless claims such as "low fat" or "heart healthy" have been made.[7] When ordering such items, keep in mind that restaurants tend to serve extra-large portions—two to three times standard serving sizes. A "low-fat" ice cream, for example, may have only 3 grams of fat per ½ cup, but you may be served 2 cups for a total of 12 grams of fat and all their accompanying kcalories.

The Ingredient List

All foods must list all ingredients on the label in descending order of predominance by weight. Knowing that the first ingredient predominates by weight, consumers can glean much information. Compare these products, for example:

- An orange powder that contains "sugar, citric acid, orange flavor . . ." versus a juice that contains "water, tomato concentrate, concentrated juices of carrots, celery . . . ."
- A cereal that contains "puffed milled corn, sugar, corn syrup, molasses, salt . . ." versus one that contains "100 percent rolled oats."
- A canned fruit that contains "sugar, apples, water" versus one that contains simply "apples, water."

In each comparison, consumers can tell that the second product is the more nutrient dense.

Consumers read food labels to learn about the nutrient contents of a food or to compare similar foods.

© Bob Daemmrich Photography

Serving Sizes

Because labels present nutrient information per serving, they must identify the size of a serving. The Food and Drug Administration (FDA) has established specific serving sizes that reflect amounts that people customarily consume and requires that all labels for a given product use the same serving size. For example, the serving size for all ice creams is ½ cup and for all beverages, 8 fluid ounces. This facilitates comparison shopping. Consumers can see at a glance which brand has more or fewer kcalories or grams of fat. Standard serving sizes are expressed in both common household measures, such as cups, and metric measures, such as milliliters, to accommodate users of both types of measures (see Table 2-7 on p. 51).

When examining the nutrition facts on a food label, consumers need to consider how the serving size compares with the actual quantity eaten. If it is not the same, they will need to adjust the quantities accordingly. For example, if the serving size is four cookies and you only eat two, then you need to cut the

FIGURE 2-8 Example of a Food Label

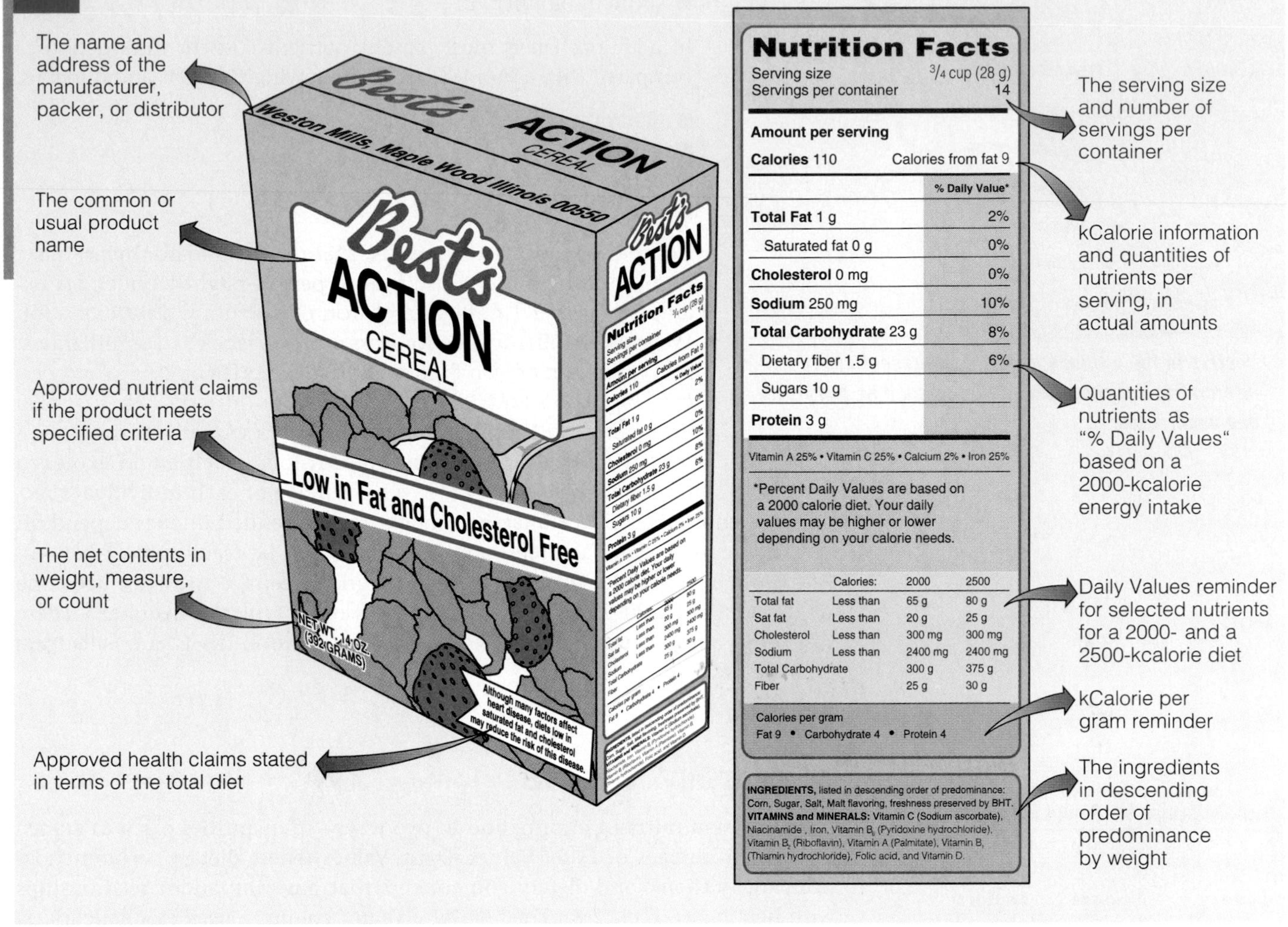

The name and address of the manufacturer, packer, or distributor

The common or usual product name

Approved nutrient claims if the product meets specified criteria

The net contents in weight, measure, or count

Approved health claims stated in terms of the total diet

Best's ACTION CEREAL

Weston Mills, Maple Wood Illinois 00550

Low in Fat and Cholesterol Free

NET WT. 14 OZ. (392 GRAMS)

Although many factors affect heart disease, diets low in saturated fat and cholesterol may reduce the risk of this disease.

Nutrition Facts

Serving size 3/4 cup (28 g)
Servings per container 14

Amount per serving	
Calories 110	Calories from fat 9
	% Daily Value*
Total Fat 1 g	2%
Saturated fat 0 g	0%
Cholesterol 0 mg	0%
Sodium 250 mg	10%
Total Carbohydrate 23 g	8%
Dietary fiber 1.5 g	6%
Sugars 10 g	
Protein 3 g	

Vitamin A 25% • Vitamin C 25% • Calcium 2% • Iron 25%

*Percent Daily Values are based on a 2000 calorie diet. Your daily values may be higher or lower depending on your calorie needs.

	Calories:	2000	2500
Total fat	Less than	65 g	80 g
Sat fat	Less than	20 g	25 g
Cholesterol	Less than	300 mg	300 mg
Sodium	Less than	2400 mg	2400 mg
Total Carbohydrate		300 g	375 g
Fiber		25 g	30 g

Calories per gram
Fat 9 • Carbohydrate 4 • Protein 4

INGREDIENTS, listed in descending order of predominance: Corn, Sugar, Salt, Malt flavoring, freshness preserved by BHT.
VITAMINS and MINERALS: Vitamin C (Sodium ascorbate), Niacinamide, Iron, Vitamin B_6 (Pyridoxine hydrochloride), Vitamin B_2 (Riboflavin), Vitamin A (Palmitate), Vitamin B_1 (Thiamin hydrochloride), Folic acid, and Vitamin D.

The serving size and number of servings per container

kCalorie information and quantities of nutrients per serving, in actual amounts

Quantities of nutrients as "% Daily Values" based on a 2000-kcalorie energy intake

Daily Values reminder for selected nutrients for a 2000- and a 2500-kcalorie diet

kCalorie per gram reminder

The ingredients in descending order of predominance by weight

nutrient and kcalorie values in half; similarly, if you eat eight cookies, then you need to double the values.

Be aware that serving sizes on food labels are not always the same as those of the Pyramid.[8] For example, a serving of rice on a food label is 1 cup, whereas in the Pyramid it is ½ cup. Unfortunately, this discrepancy, coupled with each person's own perception of standard serving sizes, sometimes creates confusion for consumers trying to follow recommendations.

Nutrition Facts

In addition to the serving size and the servings per container, the "Nutrition Facts" panel on a label shows the quantities per serving of the following:

- Total food energy (kcalories).
- Food energy from fat (kcalories).
- Total fat (grams).
- Saturated fat (grams).
- Cholesterol (milligrams).
- Sodium (milligrams).
- Total carbohydrate, including starch, sugar, and fiber (grams).
- Dietary fiber (grams).

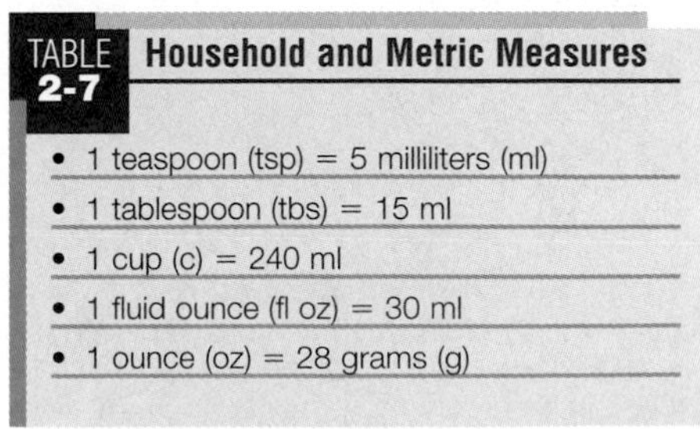
TABLE 2-7 Household and Metric Measures

- 1 teaspoon (tsp) = 5 milliliters (ml)
- 1 tablespoon (tbs) = 15 ml
- 1 cup (c) = 240 ml
- 1 fluid ounce (fl oz) = 30 ml
- 1 ounce (oz) = 28 grams (g)

Courtesy of the FDA

Posters in the produce department present nutrition information for nonpackaged items such as raw fruits and vegetables.

- Sugars (grams), including both those naturally present in and those added to the food.
- Protein (grams).

In addition, labels must present nutrient content information as compared with a standard for the following vitamins and minerals:

- Vitamin A.
- Vitamin C.
- Iron.
- Calcium.

Comparing nutrient amounts against a standard helps make them meaningful to consumers. A person might wonder, for example, whether 1 milligram of iron or calcium is a little or a lot. Well, the standard value for iron on food labels is 18 milligrams, so 1 milligram of iron is enough to take notice of: it is over 5 percent. But the standard value for calcium on food labels is 1000 milligrams, so 1 milligram of calcium is essentially nothing.

It would be nice for consumers if food labels could express a food's nutrient contents as a percentage of each individual's recommended intakes. Unfortunately, though, recommended intakes depend on age and gender. Manufacturers can't know who will be reading the label—an 8-year-old boy, a 70-year-old woman, or a pregnant teenage girl. They need one set of standard values to represent the needs of a "typical consumer." These standard values, developed by the FDA for use on food labels, are called the **Daily Values.**

The Daily Values

Labels present nutrient information in two ways—in quantities (such as grams) and as percentages of Daily Values. Daily Values reflect dietary recommendations for nutrients and dietary components that have important relationships with health (see Table 2-8). The "% Daily Value" column provides a ballpark estimate of how individual foods contribute to the total diet. It compares key nutrients in a serving of food with the daily goals of a person consuming 2000 kcalories. A 2000-kcalorie diet is considered about right for moderately active women, teenage girls, and sedentary men. Older adults, children, and sedentary women may need fewer kcalories. Most labels list, at the bottom, Daily Values for both a 2000-kcalorie and a 2500-kcalorie diet, but the "% Daily Value" column on all labels applies only to a 2000-kcalorie diet. A 2500-kcalorie diet is considered about right for many men, teenage boys, and active women. People who are exceptionally active may need still higher kcalorie intakes. Labels may also provide a reminder of the kcalories in a gram of carbohydrate, fat, and protein below the Daily Value information.

A person who consumes 2000 kcalories a day can simply add up all the "% Daily Values" for a particular nutrient to see if the day's diet fits with recommendations. If the "% Daily Values" total 100 percent, then recommendations are met. People who require more or less than 2000 kcalories daily must do some calculations to see how foods compare with their personal nutrition goals. They can use the calculation column in Table 2-8 or the suggestions presented in the accompanying "How to" feature.

Daily Values help consumers see easily whether a food contributes "a little" or "a lot" of a nutrient. For example, the "% Daily Value" column on a label of macaroni and cheese may say 20 percent for fat. This tells the consumer that each serving of this food contains about 20 percent of the day's allotted 65 grams of fat. A person consuming 2000 kcalories a day could simply keep track of the percentages of Daily Values from foods eaten in a day and try not to ex-

TABLE 2-8 Daily Values for Food Labels

Food Component	Amount	Calculation Factors
Fat	65 g	30% of kcalories
Saturated fat	20 g	10% of kcalories
Cholesterol	300 mg	—
Carbohydrate (total)	300 g	60% of kcalories
Fiber	25 g	11.5 g per 1000 kcalories
Protein	50 g	10% of kcalories
Sodium	2400 mg	—
Potassium	3500 mg	—
Vitamin C	60 mg	—
Vitamin A	1500 μg	—
Calcium	1000 mg	—
Iron	18 mg	—

Note: Daily values were established for adults and children over 4 years old. The values for energy-yielding nutrients are based on 2000 kcalories a day.

Daily Values (DV): reference values developed by the FDA specifically for use on food labels.

ceed 100 percent. Be aware that for some nutrients (such as fat and sodium) you will want to select foods with a low "% Daily Value" and for others (such as calcium and fiber) you will want a high "% Daily Value." To determine whether a particular food is a wise choice, a consumer needs to consider the other foods eaten during the day.

Daily Values also make it easy to compare foods. For example, a consumer might discover that frozen macaroni and cheese has a Daily Value for fat of 20 percent, whereas macaroni and cheese prepared from a boxed mix has a Daily Value of 15 percent. By comparing labels, consumers who are concerned about their fat intakes will be able to make informed decisions.

HOW TO Calculate Personal Daily Values

The Daily Values on food labels are designed for a 2000-kcalorie intake, but you can calculate a personal set of Daily Values based on your energy allowance. Consider a person with a 1500-kcalorie intake, for example. To calculate a daily goal for fat, multiply energy intake by 30 percent:

1500 kcal × 0.30 kcal from fat
= 450 kcal from fat.

The "kcalories from fat" are listed on food labels, so a person could then add all the "kcalories from fat" values for a day, using 450 as a goal. A person who preferred to count grams of fat could divide this 450 kcalories from fat by 9 kcalories per gram to determine the goal in grams:

450 kcal from fat ÷ 9 kcal/g
= 50 g fat.

Alternatively, a person could calculate that 1500 kcalories is 75 percent of the 2000–kcalorie intake used for Daily Values:

1500 kcal ÷ 2000 kcal = 0.75.
0.75 × 100 = 75%.

Then, instead of trying to achieve 100 percent of the Daily Value, a person consuming 1500 kcalories would aim for 75 percent. Similarly, a person consuming 2800 kcalories would aim for 140 percent:

2800 kcal ÷ 2000 kcal = 1.40 or 140%.

Table 2-8 includes a calculation column that can help you estimate your personal daily value for several nutrients.

Nutrient Claims

The FDA defines the **nutrient claims** a label may use to describe the contents of a product (see Table 2-9 on p. 54). Definitions include the conditions under which each term can be used. For example, in addition to having less than 2 milligrams of cholesterol, a "cholesterol-free" product may not contain more than 2 grams of saturated fat per serving.

Some descriptions *imply* that a food contains, or does not contain, a nutrient. Implied claims are prohibited unless they meet specified criteria. For example, a claim that a product "contains no oil" *implies* that the food contains no fat. If the product is truly fat-free, then it may make the no-oil claim, but if it contains another source of fat, such as butter, it may not.

Health Claims

Health claims◆ describe an association between a nutrient or food substance and a specific health problem. The FDA has approved several health claims based on specified criteria, including:

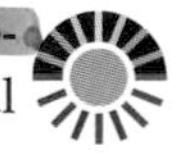

- The nutrient or food substance must be related to a disease or health condition for which most people or specific groups of people, such as the elderly, are at risk.
- The relationship between diet and health has been clearly established by scientific evidence.

Health claims on products must emphasize the importance of the total diet and not exaggerate the role of a particular food or diet in disease prevention. No one food possesses magical healing powers, and manufacturers must take care not to distort the roles of their products in promoting health.

Claims must be honest and balanced. For example, health claims can say that foods high in calcium "may" or "might" reduce the risk of osteoporosis. Claims must also explain that diseases develop in response to many factors. They may even mention beneficial factors, such as exercise. For example, a health claim may state that "Development of cancer depends on many factors. A diet low in total fat may reduce the risk of some cancers." The health claim is true, it acknowledges that diet is among many factors influencing disease development, and it is phrased in terms of total diet, not in terms of the particular product. Table 2-10 presents the criteria for health claims on labels that have been authorized.

◆ Health claims on supplement labels are presented in Highlight 10.

nutrient claims: statements that characterize the quantity of a nutrient in a food.

health claims: statements that characterize the relationship between a nutrient or other substance in a food and a disease or health-related condition.

TABLE 2-9 Terms on Food Labels

General Terms

Free: "nutritionally trivial" and unlikely to have a physiological consequence; synonyms include "without," "no," and "zero." A food that does not contain a nutrient naturally may make such a claim, but only as it applies to all similar foods (for example, "applesauce, a fat-free food").

Healthy: a food that is low in fat, saturated fat, cholesterol, and sodium and that contains at least 10% of the Daily Values for vitamin A, vitamin C, iron, calcium, protein, or fiber.

High: 20% or more of the Daily Value for a given nutrient per serving; synonyms include "rich in" or "excellent source."

Less: at least 25% less of a given nutrient or kcalories than the comparison food (see individual nutrients below); synonyms include "fewer" and "reduced."

Light or **lite:** any use of the term other than as defined below must specify what it is referring to (for example, "light in color" or "light in texture").

Low: an amount that would allow frequent consumption of a food without exceeding the Daily Value for the nutrient. A food that is naturally low in a nutrient may make such a claim, but only as it applies to all similar foods (for example, "fresh cauliflower, a low-sodium food"); synonyms include "little," "few," and "low source of."

More: at least 10% more of the Daily Value for a given nutrient than the comparison food; synonyms include "added" and "extra."

Good source of: product provides between 10 and 19% of the Daily Value for a given nutrient per serving.

Energy

kCalorie-free: fewer than 5 kcal per serving.

Light: one-third fewer kcalories than the comparison food.

Low kcalorie: 40 kcal or less per serving.

Reduced kcalorie: at least 25% fewer kcalories per serving than the comparison food.

Fat and Cholesterol[a]

Percent fat-free: may be used only if the product meets the definition of *low fat* or *fat-free* and must reflect the amount of fat in 100 g (for example, a food that contains 2.5 g of fat per 50 g can claim to be "95 percent fat free").

Fat-free: less than 0.5 g of fat per serving (and no added fat or oil); synonyms include "zero-fat," "no-fat," and "nonfat."

Low fat: 3 g or less fat per serving.

Less fat: 25% or less fat than the comparison food.

Saturated fat-free: less than 0.5 g of saturated fat and 0.5 g of *trans*-fatty acids per serving.

Low saturated fat: 1 g or less saturated fat per serving.

Less saturated fat: 25% or less saturated fat than the comparison food.

Cholesterol-free: less than 2 mg cholesterol per serving and 2 g or less saturated fat per serving.

Low cholesterol: 20 mg or less cholesterol per serving and 2 g or less saturated fat per serving.

Less cholesterol: 25% or less cholesterol than the comparison food (reflecting a reduction of at least 20 mg per serving), and 2 g or less saturated fat per serving.

Extra lean: less than 5 g of fat, 2 g of saturated fat, and 95 mg of cholesterol per serving and per 100 g of meat, poultry, and seafood.

Lean: less than 10 g of fat, 4.5 g of saturated fat, and 95 mg of cholesterol per serving and per 100 g of meat, poultry, and seafood.

Light: 50% or less of the fat than in the comparison food (for example, 50% less fat than our regular cookies).

Carbohydrates: Fiber and Sugar

High fiber: 5 g or more fiber per serving. A high-fiber claim made on a food that contains more than 3 g fat per serving and per 100 g of food must also declare total fat.

Sugar-free: less than 0.5 g of sugar per serving.

Sodium

Sodium-free and **salt-free:** less than 5 mg of sodium per serving.

Low sodium: 140 mg or less per serving.

Light: a low-kcalorie, low-fat food with a 50% reduction in sodium.

Light in sodium: no more than 50% of the sodium of the comparison food.

Very low sodium: 35 mg or less per serving.

[a]Foods containing more than 13 g total fat per serving or per 50 g of food must indicate those contents immediately after a cholesterol claim. As you can see, all cholesterol claims are prohibited when the food contains more than 2 g saturated fat per serving.

Consumer Education

Labels are valuable only if people know how to use them, and so the FDA has designed several programs to educate consumers. Consumers who understand how to read labels will be best able to apply the information to achieve and maintain healthful dietary practices.

Figure 2-9 (on p. 56) shows how the messages from the *Dietary Guidelines,* the Food Guide Pyramid, and food labels coordinate with each other. To help consumers understand and coordinate these messages, an alliance of health organizations, the food industry, and government agencies has developed an ed-

TABLE 2-10 **Food Label Health Claims and Their Criteria**

Health Claim	Criteria
Calcium and reduced risk of osteoporosis	• High in calcium (≥20% DV) • No more phosphorus than calcium
Sodium and reduced risk of hypertension	• Low in sodium (≤140 mg/serving)
Dietary saturated fat and cholesterol and reduced risk of coronary heart disease	• Low in saturated fat (≤1 g/serving) • Low in cholesterol (≤20 mg/serving) • Low in fat (≤3 g/serving)
Dietary fat and reduced risk of cancer	• Low in fat (≤3 g/serving)
Fiber-containing grain products, fruits, and vegetables and reduced risk of cancer	• Low in fat (≤3 g/serving) • Good source of dietary fiber (≥10% DV)
Fruits, vegetables, and grain products that contain fiber, particularly soluble fiber, and reduced risk of coronary heart disease	• Low in saturated fat (≤1 g/serving) • Low in cholesterol (≤20 mg/serving) • Low in fat (≤3 g/serving) • Soluble fiber (≥0.6 g/serving)
Fruits and vegetables and reduced risk of cancer	• Low in fat (≤3 g/serving) • Good source of vitamin A, vitamin C, or dietary fiber (≥10% DV)
Folate and reduced risk of neural tube defects	• Good source of folate (≥10% DV) • Limited in vitamin A and vitamin D (≤100% DV)
Sugar alcohols and reduced risk of tooth decay	• Sugar-free • Cannot lower dental plaque pH below 5.7 by bacterial fermentation
Soluble fiber from whole oats and from psyllium seed husk and reduced risk of heart disease	• Soluble fiber from oats (≥0.75 g/serving) or from psyllium seed husk (≥1.7 g/serving) • Low in saturated fat (≤1 g/serving) • Low in cholesterol (≤20 mg/serving) • Low in fat (≤3 g/serving)
Soy protein and reduced risk of heart disease	• Soy protein (6.25 g/serving) • Low in saturated fat (≤1 g/serving) • Low in cholesterol (≤20 mg/serving) • Low in fat (≤3 g/serving)
Whole grains and reduced risk of heart disease and certain cancers	• Whole-grain (≥51%) • Low in saturated fat (≤1 g/serving) • Low in cholesterol (≤20 mg/serving) • Low in fat (≤3 g/serving)
Plant sterol and plant stanol esters and heart disease	• Plant sterol esters (0.65 g/serving) or plant stanol esters (1.7 g/serving) • Low in saturated fat (≤1 g/serving) • Low in cholesterol (≤20 mg/serving) • Low in fat (≤3 g/serving)
Potassium and reduced risk of hypertension and stroke	• Good source of potassium (≥10% DV) • Low in sodium (≤140 mg/serving) • Low in saturated fat (≤1 g/serving) • Low in cholesterol (≤20 mg/serving) • Low in fat (≤3 g/serving)

Note: With the exception of sugar alcohols and dental caries, all other health claims must also meet two additional criteria. First, a food making a health claim must be a naturally good source (containing at least 10 percent of the Daily Value) of at least one of the following nutrients: vitamin A, vitamin C, iron, calcium, protein, or fiber. Second, foods are disqualified from making health claims if a standard serving contains more than 20 percent of the Daily Value for total fat, saturated fat, cholesterol, or sodium.

ucational program called "It's All About You." The program is designed to deliver simple messages that will motivate consumers to think positively about making reasonable changes in their eating and physical activity habits (see Table 2-11 on p. 57).

IN SUMMARY Food labels provide consumers with information they need to select foods that will help them meet their nutrition and health goals. Given labels with relevant information presented in a standardized, easy-to-read format, consumers are well prepared to plan and create healthful diets.

FIGURE 2-9 From Guidelines to Groceries

Dietary Guidelines	Food Guide Pyramid	Food Labels
Aim for a healthy weight.	Build a healthy base by eating vegetables, fruits, and grains. Choose foods low in fats and added sugars. Select sensible portion sizes.	Look for foods that describe their kcalorie contents as *free, low, reduced, light,* or *less.*
Be physically active each day.	Increase your physical activity.	
Let the Pyramid guide your food choices.	Choose at least the minimum number of servings from each food group.	Look for foods that describe their vitamin, mineral, or fiber contents as a *good source* or *high.*
Choose a variety of grains daily, especially whole grains.	Choose at least 6 servings of grain products daily.	Look for foods that describe their fiber contents as *high.* Aim for 100% of the Daily Value for fiber from a variety of sources. A heart disease health claim identifies grains with soluble fibers from oats and psyllium seeds. A cancer health claim identifies grains that are good sources of fiber and low in fat.
Choose a variety of fruits and vegetables daily.	Choose at least 2 servings of fruits and 3 servings of vegetables daily.	Look for foods that describe their fiber contents as *high.* Aim for 100% of the Daily Value for fiber from a variety of sources. A cancer health claim identifies fruits and vegetables that are low in fat and good sources of fiber, vitamin A, or vitamin C.
Keep food safe to eat.		Follow the *safe handling instructions* on packages of meat and other safety instructions, such as *keep refrigerated,* on packages of perishable foods.
Choose a diet that is low in saturated fat and cholesterol and moderate in total fat.	Limit foods in the tip of the Pyramid. Choose foods within each group that are low in fat, saturated fat, and cholesterol.	Look for foods that describe their fat, saturated fat, and cholesterol contents as *free, less, low, light, reduced, lean,* or *extra lean.* Keep your intake under 100% of the Daily Value for fat, saturated fat, and cholesterol. A heart disease or a cancer health claim identifies foods low in fat, saturated fat, and cholesterol.
Choose beverages and foods that limit your intake of sugars.	Limit foods in the tip of the Pyramid. Choose foods within each group that are low in added sugars.	Look for foods that describe their sugar contents as *free* or *reduced.* A tooth decay health claim identifies foods that contain sugar alcohols, but no sugar. A food may be high in sugar if its ingredients list begins with, or contains several of, the following: *sugar, sucrose, fructose, maltose, lactose, honey, syrup, corn syrup, high-fructose corn syrup, molasses,* or *fruit juice concentrate.*
Choose and prepare foods with less salt.	Choose foods within each group that are low in salt and sodium.	Look for foods that describe their salt and sodium contents as *free, low* or *reduced.* Keep your intake under 100% of the Daily Value for sodium. A high blood pressure health claim identifies foods low in sodium.
If you drink alcoholic beverages, do so in moderation.	Like other foods in the tip, use sparingly (no more than one drink a day for women and two drinks a day for men).	*Light* beverages contain fewer kcalories and less alcohol than regular versions.

TABLE 2-11 Messages from the "It's All About You" Campaign

Make healthy choices that fit your lifestyle so you can do the things you want to do.

- **Be realistic**
Make small changes over time in what you eat and the level of activity you do. After all, small steps work better than giant leaps. *Tip: Sprinkle shredded cheese on salads or pasta to boost your calcium intake.*
- **Be adventurous**
Expand your tastes to enjoy a variety of foods. *Tip: Try a new food or recipe once a month.*
- **Be flexible**
Go ahead and balance what you eat and the physical activity you do over several days. No need to worry about just one meal or one day. *Tip: If you eat ice cream, increase your physical activity for several days.*
- **Be sensible**
Enjoy all foods, just don't overdo it. *Tip: If your favorite food is high in fat, eat a smaller portion.*
- **Be active**
Walk the dog, don't just watch the dog walk. *Tip: Climb the stairs instead of taking the elevator or escalator.*

How are you doing?

- Do you eat at least the minimum number of servings from each of the five food groups daily?
- Do you try to vary your choices within each food group from day to day?
- What dietary changes could you make to improve your chances of enjoying good health?

Nutrition on the Net

WEBSITES

Access these websites for further study of topics covered in this chapter.

- Find updates and quick links to these and other nutrition-related sites at our website: **www.wadsworth.com/nutrition**
- Search for "diet" and "food labels" at the U.S. Government health information site: **www.healthfinder.gov**
- Learn more about the *Dietary Guidelines for Americans:* **health.gov/dietaryguidelines**
- View Canadian information on nutrition guidelines and food labels at: **www.hc-sc.gc.ca**
- Visit the Food Guide Pyramid section (including its ethnic/cultural pyramids) of the U.S. Department of Agriculture: **www.nal.usda.gov/fnic**
- See food pyramids for various ethnic groups at Oldways Preservation and Exchange Trust: **www.oldwayspt.org**
- Search for "exchange lists" at the American Diabetes Association: **www.diabetes.org**
- Learn more about food labeling from the Food and Drug Administration: **www.cfsan.fda.gov** or **vm.cfsan.fda.gov/label.html**
- Search for "food labels" at the International Food Information Center: **ificinfo.health.org**

INTERNET ACTIVITIES

Much exciting nutrition and health-related information is available on the Internet. As you begin your study of nutrition, what topics are you most interested in learning more about?

- Go to: **www.navigator.tufts.edu**
- Click on the "Hot Topics" button.
- Check out some of the topics on this page.
- Visit some of the web pages on the topic of your choice.

The Tufts University Nutrition Navigator provides timely nutrition information on a variety of topics. What new information did you find on your topic of interest?

Making it Click

These problems will give you practice in doing simple nutrition-related calculations. They use hypothetical situations in order to teach a lesson that can provide answers (see p. 60 for answers). Be sure to show your calculations for each problem.

1. *Read a food label.* Look at the cereal label in Figure 2-8 and answer the following questions:
   a. What is the size of a serving of cereal?
   b. How many kcalories are in a serving?
   c. How much fat is in a serving?
   d. How many kcalories does this represent?
   e. What percentage of the kcalories in this product comes from fat?
   f. What does this tell you?
   g. What is the % Daily Value for fat?
   h. What does this tell you?
   i. Does this cereal meet the criteria for a low-fat product (refer to Table 2-9).
   j. How much fiber is in a serving?
   k. Read the Daily Value chart on the lower section of the label. What is the Daily Value for fiber?
   l. What percentage of the Daily Value for fiber does a serving of the cereal contribute? Show the calculation the label-makers used to come up with the % Daily Value for fiber.
   m. What is the predominant ingredient in the cereal?
   n. Have any nutrients been added to this cereal (is it fortified)?
2. *Calculate a personal Daily Value.* The Daily Values on food labels are for people with a 2000-kcalorie intake.
   a. Suppose a person has a 1600-kcalorie energy allowance. Use the calculation factors listed in Table 2-8 to calculate a set of personal "Daily Values" based on 1600 kcalories. Show your calculations.
   b. Revise the % Daily Value chart of the cereal label in Figure 2-8 based on your "Daily Values" for a 1600-kcalorie diet.

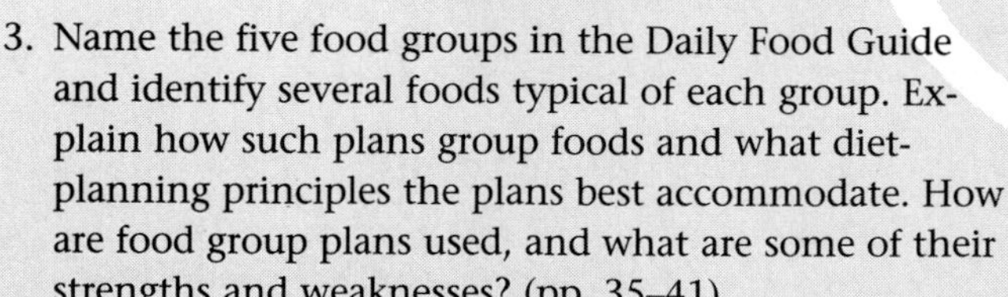

Study Questions

These questions will help you review the chapter. You will find the answers in the discussion on the pages provided.

1. Name the diet-planning principles and briefly describe how each principle helps in diet planning. (pp. 33–34)
2. What recommendations appear in the *Dietary Guidelines for Americans*? (p. 35)
3. Name the five food groups in the Daily Food Guide and identify several foods typical of each group. Explain how such plans group foods and what diet-planning principles the plans best accommodate. How are food group plans used, and what are some of their strengths and weaknesses? (pp. 35–41)
4. Name the exchange lists and identify a food typical of each list. Explain how the exchange system groups foods and what diet-planning principles the system best accommodates. How are exchange systems used,

and what are some of their strengths and weaknesses? (pp. 41–46)
5. Review the *Dietary Guidelines*. What types of grocery selections would you make to achieve those recommendations? (pp. 35, 46–49)
6. What information can you expect to find on a food label? How can this information help you choose between two similar products? (pp. 50–52)
7. What are the Daily Values? How can they help you meet health recommendations? (pp. 52–53)
8. What health claims have been approved by the FDA for use on labels? What criteria must all health claims meet? (pp. 53–55)

These questions will help you prepare for an exam. Answers can be found on p. 60.

1. The diet-planning principle that provides all the essential nutrients in sufficient amounts to support health is:
   a. balance.
   b. variety.
   c. adequacy.
   d. moderation.
2. A person who chooses a chicken leg that provides 0.5 milligrams of iron and 95 kcalories instead of two tablespoons of peanut butter that also provide 0.5 milligrams of iron but 188 kcalories is using the principle of nutrient:
   a. control.
   b. density.
   c. adequacy.
   d. moderation.
3. Which of the following is consistent with the *Dietary Guidelines for Americans*?
   a. Choose a diet restricted in fat and cholesterol.
   b. Balance the food you eat with physical activity.
   c. Choose a diet with plenty of milk products and meats.
   d. Eat an abundance of foods to ensure nutrient adequacy.
4. According to the Food Guide Pyramid, which food group provides the foundation of a healthy diet?
   a. vegetables
   b. milk, yogurt, and cheese
   c. breads, cereals, rice, and pasta
   d. meat, poultry, fish, dry beans, eggs, and nuts
5. Foods within a given food group of the Pyramid are similar in their contents of:
   a. energy.
   b. proteins and fibers.
   c. vitamins and minerals.
   d. carbohydrates and fats.
6. In the exchange system, each portion of food on any given list provides about the same amount of:
   a. energy.
   b. satiety.
   c. vitamins.
   d. minerals.
7. In the exchange system, corn and potatoes are on the:
   a. fruit list.
   b. starch list.
   c. vegetable list.
   d. meat alternate list.
8. Which of the following is *not* on the fat exchange list?
   a. bacon
   b. avocados
   c. black olives
   d. peanut butter
9. Enriched grain products are fortified with:
   a. fiber, folate, iron, niacin, and zinc.
   b. thiamin, iron, calcium, zinc, and sodium.
   c. iron, thiamin, riboflavin, niacin, and folate.
   d. folate, magnesium, vitamin B_6, zinc, and fiber.
10. Daily Values on food labels are based on a:
    a. 1500-kcalorie diet.
    b. 2000-kcalorie diet.
    c. 2500-kcalorie diet.
    d. 3000-kcalorie diet.

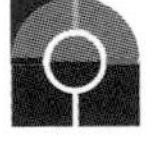

References

1 U.S. Department of Agriculture and U.S. Department of Health and Human Services, *Nutrition and Your Health: Dietary Guidelines for Americans,* Home and Garden Bulletin no. 232 (Washington, D.C.: 2000).

2 Position of The American Dietetic Association: Vegetarian diets, *Journal of the American Dietetic Association* 97 (1997): 1317–1319.

3 S. M. Krebs-Smith and coauthors, Characterizing food intake patterns of American adults, *American Journal of Clinical Nutrition* 65 (1997): 1264S–1268S.

4 A. K. Kant, Consumption of energy-dense, nutrient-poor foods by adult Americans: Nutrition and health implications. The third National Health and Nutrition Examination Survey, 1988–1994, *American Journal of Clinical Nutrition* 72 (2000): 929–936; K. S. Tippett, C. W. Enns, and A. J. Moshfegh, Food consumption surveys in the US

Department of Agriculture, *Nutrition Today* 34 (1999): 33–46.

5 As cited in 21 Code of Federal Regulations—Food and Drugs, Section 104.20, 45 *Federal Register* 6323, January 25, 1980, as amended in 58 *Federal Register* 2228, January 6, 1993.

6 H. H. C. Lee, S. A. Gerrior, and J. A. Smith, Energy, macronutrient, and food intakes in relation to energy compensation in consumers who drink different types of milk, *American Journal of Clinical Nutrition* 67 (1998): 616–623.

7 Foods in menu claims must meet FDA rule, *FDA Consumer,* October 1996, p. 5.

8 M. B. Hogbin and M. A. Hess, Public confusion over food portions and servings, *Journal of the American Dietetic Association* 99 (1999): 1209–1211.

ANSWERS

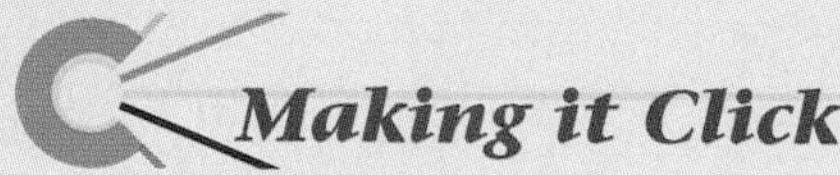

Making it Click

1. a. ¾ cup (28 g).
   b. 110 kcalories.
   c. 1 g fat.
   d. 9 kcalories.
   e. 9 kcal ÷ 110 kcal = 0.08.
   0.08 × 100 = 8%.
   f. This cereal derives 8 percent of its kcalories from fat.
   g. 2%.
   h. A serving of this cereal provides 2 percent of the 65 grams of fat recommended for a 2000-kcalorie diet.
   i. Yes.
   j. 1.5 g fiber.
   k. 25 g.
   l. 1.5 g ÷ 25 g = 0.06.
   0.06 × 100 = 6%.
   m. Corn.
   n. Yes.

2. a. Daily Values for 1600-kcalorie diet:

Fat: 1600 kcal × 0.30 = 480 kcal from fat.

480 kcal ÷ 9 kcal/g = 53 g fat.

Saturated fat: 1600 kcal × 0.10 = 160 kcal from saturated fat.

160 kcal ÷ 9 kcal/g = 18 g saturated fat.

Cholesterol: 300 mg.

Carbohydrate: 1600 kcal × 0.60 = 960 kcal from carbohydrate.

960 kcal ÷ 4 kcal/g = 240 g carbohydrate.

Fiber: 1600 kcal ÷ 1000 kcal = 1.6.

1.6 × 11.5 g = 18.4 g fiber.

Protein: 1600 kcal × 0.10 = 160 kcal from protein.

160 kcal ÷ 4 kcal/g = 40 g protein.

Sodium: 2400 mg.

Potassium: 3500 mg.

b.

Total fat	2%	(1 g ÷ 53 g)
Saturated fat	0%	(0 g ÷ 18 g)
Cholesterol	0%	(no calculation needed)
Sodium	10%	(no calculation needed)
Total carbohydrate	10%	(23 g ÷ 240 g)
Dietary fiber	8%	(1.5 g ÷ 18.4 g)

Multiple Choice

1. c	2. b	3. b	4. c	5. c
6. a	7. b	8. d	9. c	10. b

COLLEGE BOUND—AND HUNGRY

EVERY DAY, several times a day, we stop what we are doing to eat. That would not be a problem if the campus cafeteria were delivering gourmet meals at no cost, but that is not happening. More likely, you have a couple of dollars in your pocket and less than an hour to feed yourself before heading out the door again. You turn to your food supply, but can't quite figure out how to make a meal out of microwave popcorn and packets of catsup. What's a person to do?

Because we eat so frequently, it's easy for us to choose a meal without giving its nutritious merit any thought. Even when we want to make a healthy choice, we may not know which foods to select. Television, newspaper, and magazine reports that describe the health benefits of a particular nutrient tend to focus on "one disease–one nutrient" relationships. Although it is true that fiber helps defend against colon cancer and vitamin E helps protect against heart disease, focusing narrowly on these links oversimplifies the story. In reality, each nutrient may have connections with several diseases because its role in the body is not specific to a disease, but to a body function. Fiber—because it binds substances in the digestive tract—helps prevent cancer *and* control diabetes. Vitamin E—because it acts as an antioxidant—helps prevent both heart disease *and* cancer.

If the media reports convince you to get your fiber and vitamin E each day, you'll eat more fruits and vegetables for their fiber and cook with vegetable oils for the vitamin E. But what about the 40-something other nutrients we need each day? Upon learning that calcium helps to protect against osteoporosis, you might grab a quart of milk. Then to get your vitamin C, you might throw in a couple of oranges. Selecting one food for each nutrient though would quickly fill your shopping cart with dozens of foods and thousands of kcalories! Fortunately, most foods provide several nutrients, so one trick for wise diet planning is to select a combination of foods that deliver a full array of nutrients. Between classes and term papers, who has time to think about planning or preparing a meal? In less than the 20 minutes it takes to travel to a fast-food restaurant and order a cheeseburger, you can prepare a nutritious—and delicious—meal. The secret is learning to think "Food Pyramid." This highlight presents a shortcut to meal planning for people who want a reasonably healthy meal in a hurry.

As Chapter 2 explained, the Food Guide Pyramid clusters foods into five major food groups (as described on pp. 36–39). Ideally, you would create a meal using foods from most, if not all, of the five food groups. Making daily selections from each of the five food groups helps to ensure a well-balanced diet. For quick and easy menu planning purposes, however, it may be useful to think of the Pyramid's *three* main levels (see Figure H2-1). (The items on the very top of the Food Pyramid—such as mayonnaise, chips, and colas—are not included in this discussion because they should be used sparingly for flavor and fun, not as the main ingredients of a meal.) The first level covers the entire base of the Food Pyramid and is dedicated to the breads, cereals, and other grain products that provide the energy and foundation of a healthy diet. Fruits and vegetables share the second level, making important contributions of fiber, vitamins, and minerals. Meats and milks appear on the third level in slightly smaller sections, providing valuable protein, vitamins, and minerals.

When you're hungry and time is short, try combining at least one food from each of the three main levels of the Food Pyramid for a fast and healthy meal. Selecting foods from three or more different food groups helps to add variety and nutrient balance. It really is as easy as:

1. Selecting a serving or two of bread, cereal, rice, pasta, or another grain product from the bottom of the Pyramid.
2. Picking a fruit or vegetable (or two) from the middle.
3. And choosing a serving of milk (or soy milk), cheese, yogurt, meat, fish, poultry, eggs, legumes, tofu, seeds, or nuts from the next level up.

For example, you could make a healthful Greek tuna

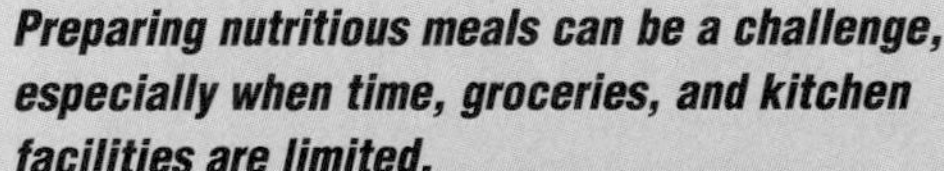

Preparing nutritious meals can be a challenge, especially when time, groceries, and kitchen facilities are limited.

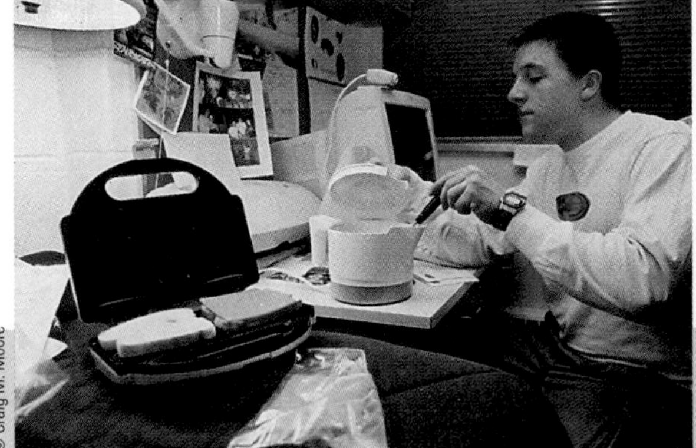

© Craig M. Moore

FIGURE H2-1 **A Simplified Pyramid**

For quick and easy meal planning, select foods from each of the Pyramid's three main levels.

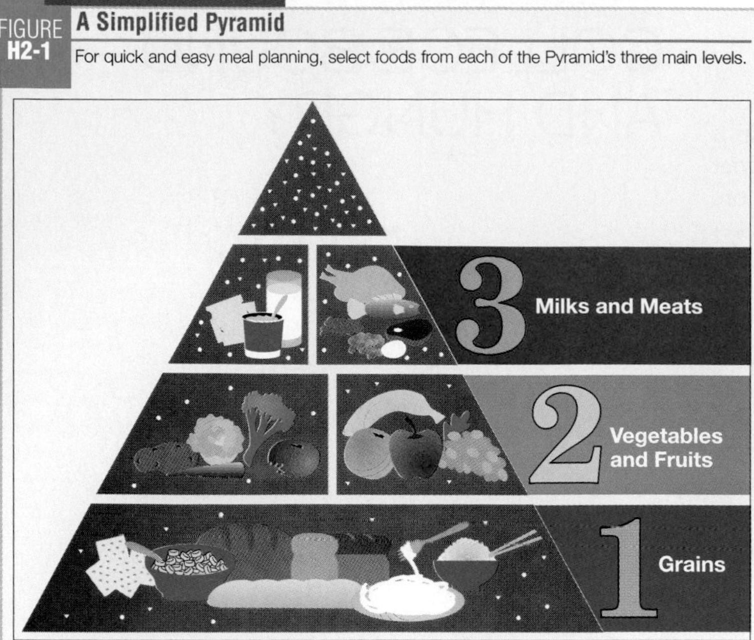

TABLE H2-1 Basic Cooking Tools

- Mixing bowl and spoon
- Cutting board
- Paring knife
- Can opener
- Sponge or dish rag and dish soap
- Dish towel or paper towels
- Measuring cups and spoons
- 2 qt microwave-safe casserole dish with lid
- 2 qt pot with lid
- 10" frying pan and spatula

salad sandwich by mixing together drained tuna (from level 3) with chopped tomatoes and cucumbers (from level 2), and plain yogurt (from level 3) and spooning it all into pita pocket bread (from level 1). Alternatively, combining pasta (level 1), broccoli (level 2), and chicken (level 3) creates a tasty meal. Figure H2-2 illustrates how recipe ingredients from each of the Pyramid's three main levels can be combined to create healthful meals. Preparing such meals is easy if your kitchen is equipped with a few basic cooking tools (see Table H2-1).

One more tip: try to include foods from all five good groups daily. Consistently selecting milk products from level 3, for example, would leave the diet short in iron, zinc, and other nutrients that meat and meat alternates deliver. Similarly, regularly neglecting the milk group in favor of the meat and meat alternate group would fail to provide enough calcium. Think about which food groups were missing from your last meal and include them in the next one. Better yet, think about how you might improve each meal nutritionally with one simple addition—a vitamin A–rich peach with the Greek tuna salad sandwich or a couple ounces of calcium-rich shredded cheese on top of the pasta, for example.

Each food group offers easy-to-prepare or ready-to-eat options. Keep them in mind when time is short. Grab a piece of bread. Crunch on carrot sticks. Peel a banana. Drink a glass of milk. Open a can of sardines. There are dozens of choices and variety is key to nutrient adequacy. (Chapter 15 offers several healthful snack ideas, using combinations from the food groups.)

It's amazing how many soups, salads, sandwiches, and other speedy suppers you can create from as few as three main ingredients. The combinations are endless, so be creative. Or look for recipes in cookbooks or magazines. Especially useful are recipes that include ingredients from each of the three main levels of the Pyramid; if they don't, improvise and revise them. The accompanying "How to" feature provides tips for changing recipes and shopping for ingredients. Bon appétit!

HOW TO *Make a Shopping List*

As you decide which meals you want to prepare, check the groceries you have on hand and make a list of items you will need. Your list may be fairly short if you keep staples such as rice and pasta on hand. To keep from running out of staple foods that you use regularly, post a shopping list on the refrigerator or bulletin board. When you open the next-to-last can of tuna, get down to the last inch of peanut butter, or have only one cup of rice left, add it to the list.

Keeping an ongoing grocery list is a great start, but you will still need to add a few items that are unique to the recipes you have selected. Even then, the list need not exceed your budget if you improvise a little. If you don't have pita pocket bread, wrap sandwich ingredients in flour tortillas or serve them open-faced on English muffins if that's what you have on hand. Similarly, if you don't like cucumbers or don't have a tomato, use "whatever you got"—grated carrots, diced onions, or black olives all add texture and flavor to a meal.

Finally, before heading to the grocery store, think of your day's meals. What do you like to eat for breakfast? If it's cereal and milk, put them on the list. If it's a banana and yogurt, write that down. Then consider your favorite options for lunch, supper, and snacks. Making a list will help ensure that you get home with the groceries you need to create nutritious meals in minutes.

FIGURE H2-2 **Examples of Meals Created with the Simplified Pyramid**

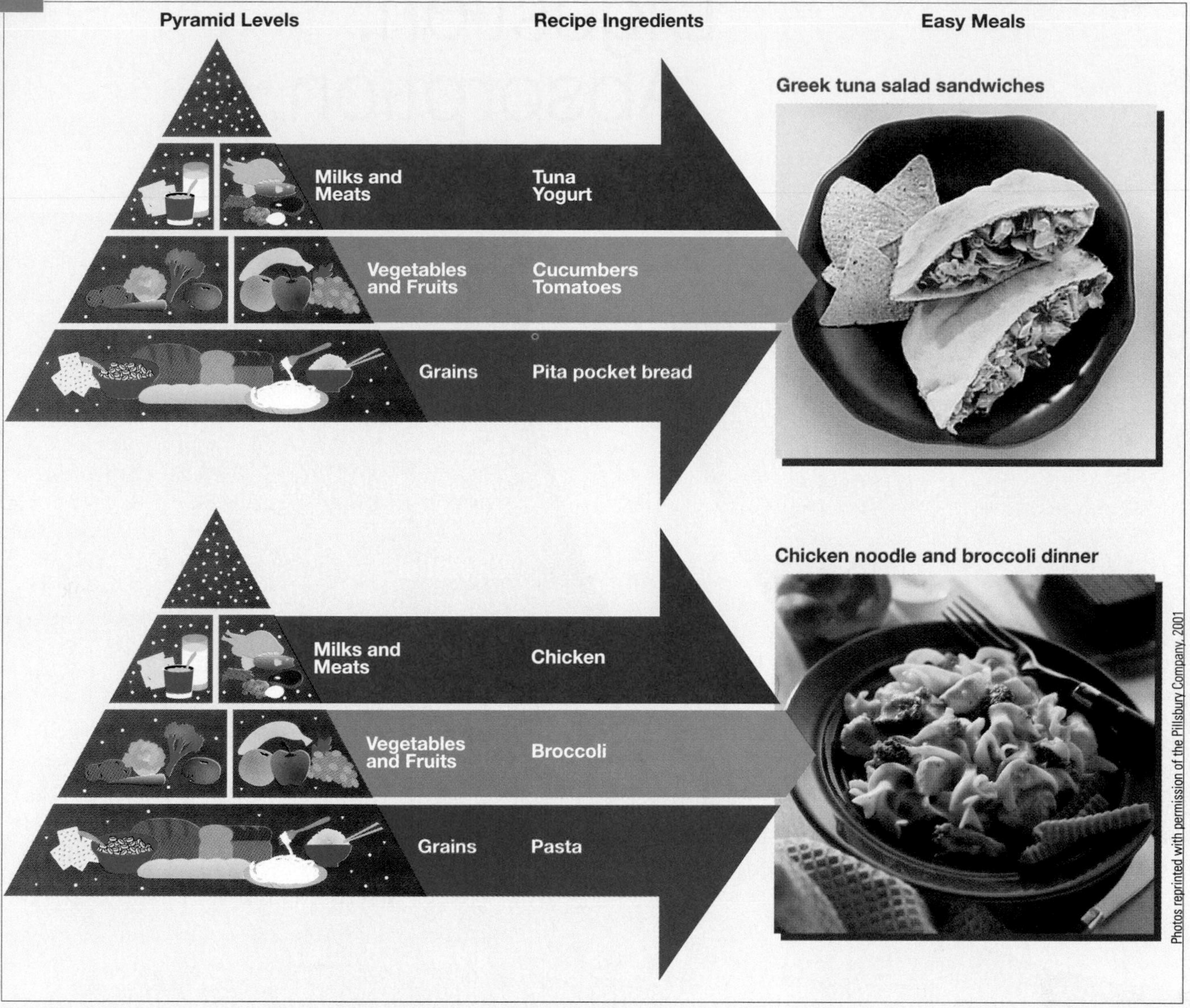

Pyramid Levels	Recipe Ingredients	Easy Meals
Milks and Meats	Tuna Yogurt	Greek tuna salad sandwiches
Vegetables and Fruits	Cucumbers Tomatoes	
Grains	Pita pocket bread	
Milks and Meats	Chicken	Chicken noodle and broccoli dinner
Vegetables and Fruits	Broccoli	
Grains	Pasta	

Nutrition on the Net

•WEBSITES•

Access these websites for further study of topics covered in this highlight.

- Find updates and quick links to these and other nutrition-related sites at our website: **www.wadsworth.com/nutrition**
- Check out the meal plans and recipe suggestions from Meals for You: **www.mymenus.com**
- Visit the Healthy Lifestyle section of the American Dietetic Association: **www.eatright.org**
- Find everything you need to plan a healthy diet: **www.cyberdiet.com**

CHAPTER 3 Digestion, Absorption, and Transport

www.comstock.com

Have you ever wondered what happens to the food you eat after you swallow it? Or how your body extracts nutrients from food? Have you ever marveled how it all just seems to happen? This chapter takes you on the journey that transforms the foods you eat into the nutrients featured in the later chapters. Then it follows the nutrients as they travel through the intestinal cells and into the body to do their work. This introduction presents a general overview of the processes common to all nutrients; later chapters discuss the specifics of digesting and absorbing individual nutrients.

1998 Photo Disc Inc.

The process of digestion transforms all kinds of foods into nutrients.

DIGESTION

Digestion is the body's ingenious way of breaking down foods into nutrients in preparation for **absorption.** In the process, it overcomes many obstacles for you without any conscious effort on your part. Consider these obstacles:

1. Human beings breathe, eat, and drink through their mouths. Air taken in through the mouth must go to the lungs; food and liquid must go to the stomach. The throat must be arranged so that swallowing and breathing don't interfere with each other.
2. Below the lungs lies the diaphragm, a dome of muscle that separates the upper half of the major body cavity from the lower half. Food must pass through this wall to reach the stomach.
3. The materials within the tract should be kept moving forward, slowly but steadily, at a pace that permits all reactions to reach completion.
4. To move through the system, food must be lubricated with water. Too much water would form a liquid that would flow too rapidly; too little water would form a paste too dry and compact to move at all. The amount of water must be regulated to keep the intestinal contents at the right consistency to move smoothly along.
5. When the digestive enzymes are breaking food down, they need it in finely divided form, suspended in enough water so that every particle is accessible. Once digestion is complete and the needed nutrients have been absorbed out of the tract into the body, the system must excrete the residue that remains, but excreting all the water along with the solid residue would be both wasteful and messy. Some water should be withdrawn, leaving a paste just solid enough to be smooth and easy to pass.
6. Once waste matter has reached the end of the tract, it must be excreted, but it would be inconvenient and embarrassing if this function occurred continuously. Provision must be made for periodic, voluntary evacuation.
7. The enzymes of the digestive tract are designed to digest carbohydrate, fat, and protein. The walls of the tract, composed of living cells, are also made of carbohydrate, fat, and protein. These cells need protection against the action of the powerful digestive juices that they secrete.

The following sections show how the body elegantly and efficiently handles these obstacles.

Anatomy of the Digestive Tract

The **gastrointestinal (GI) tract** is a flexible muscular tube from the mouth, through the esophagus, stomach, small intestine, large intestine, and rectum to the anus. Figure 3-1 traces the path followed by food from one end to the other. In a sense, the human body surrounds the GI tract. The inner space within the GI tract, called the **lumen,** is continuous from one end to the other (see the glossary on p. 67 for GI anatomy terms). Only when a nutrient or other substance penetrates the GI tract's wall does it enter the body proper; many materials pass through the GI tract without being digested or absorbed.

digestion: the process by which food is broken down into absorbable units.
- **digestion** = take apart

absorption: the passage of nutrients from the GI tract into either the blood or the lymph.
- **absorb** = suck in

gastrointestinal (GI) tract: the digestive tract. The principal organs are the stomach and intestines.
- **gastro** = stomach
- **intestinalis** = intestine

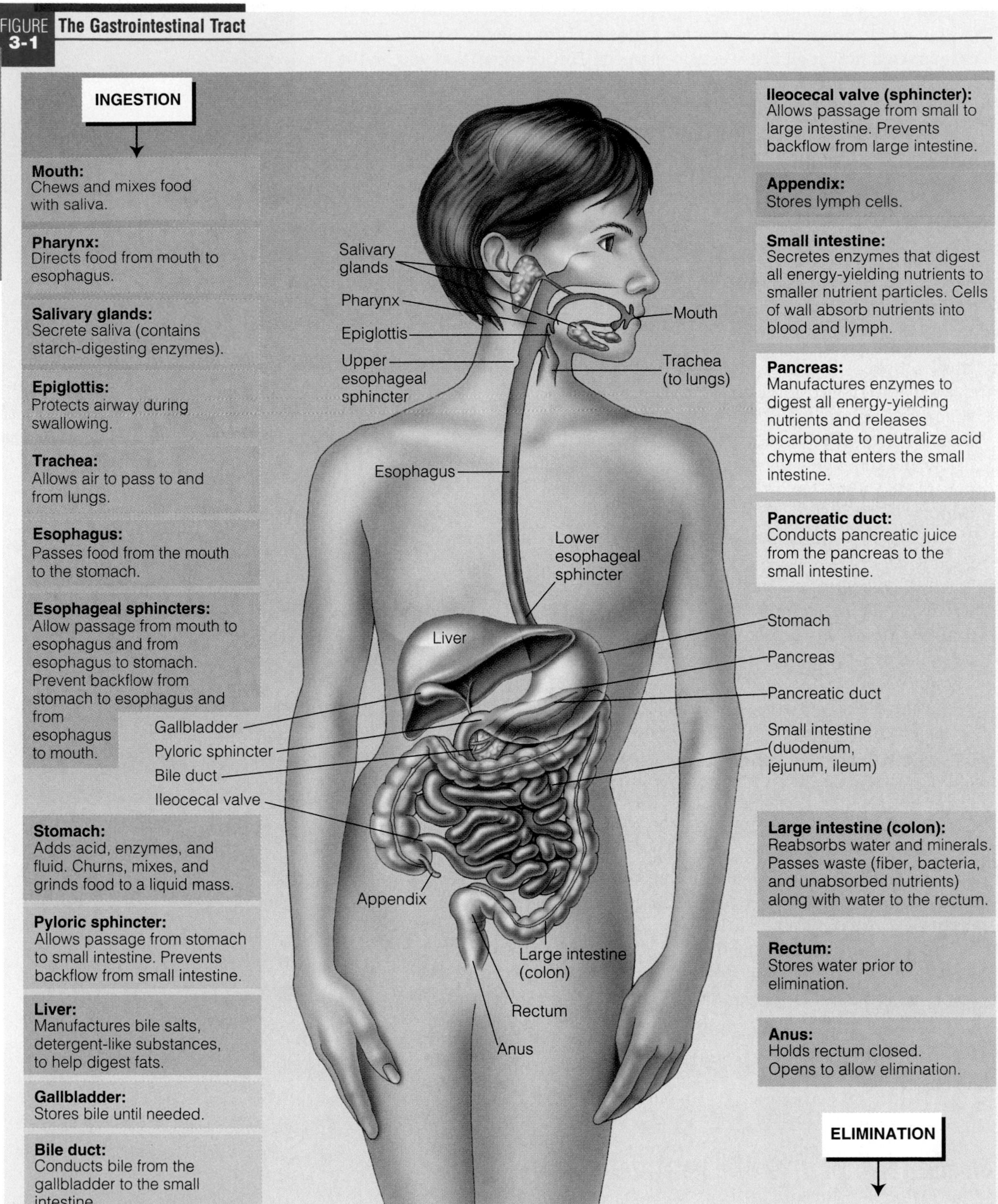

FIGURE 3-1 The Gastrointestinal Tract

◆ The process of chewing is called **mastication** (mass-tih-KAY-shun).

• **Mouth** • The process of digestion begins in the **mouth.** As you chew,◆ your teeth crush large pieces of food into smaller ones (see Figure 3-2 on p. 68), and fluids blend with these pieces to ease swallowing. Fluids also help dissolve the food so that you can taste it; only particles in solution can react with taste buds. The tongue allows you not only to taste food, but also to move food

GLOSSARY OF GI ANATOMY TERMS

These terms are listed in order from start to end of the digestive tract.

mouth: the oral cavity containing the tongue and teeth.

pharynx (FAIR-inks): the passageway leading from the nose and mouth to the larynx and esophagus, respectively.

epiglottis (epp-ee-GLOTT-iss): cartilage in the throat that guards the entrance to the trachea and prevents fluid or food from entering it when a person swallows.
- **epi** = upon (over)
- **glottis** = back of tongue

esophagus (ee-SOFF-ah-gus): the food pipe; the conduit from the mouth to the stomach.

esophageal sphincters (ee-SOF-ah-GEE-al SFINK-ters): sphincter muscles at the upper and lower ends of the esophagus. The *lower esophageal sphincter* is also called the *cardiac sphincter.*

sphincter (SFINK-ter): a circular muscle surrounding, and able to close, a body opening. Sphincters are found at specific points along the GI tract and regulate the flow of food particles.
- **sphincter** = band (binder)

stomach: a muscular, elastic, saclike portion of the digestive tract that grinds and churns swallowed food, mixing it with acid and enzymes to form chyme.

pyloric (pie-LORE-ic) **sphincter:** the circular muscle that separates the stomach from the small intestine and regulates the flow of partially digested food into the small intestine; also called *pylorus* or *pyloric valve.*
- **pylorus** = gatekeeper

gallbladder: the organ that stores and concentrates bile. When it receives the signal that fat is present in the duodenum, the gallbladder contracts and squirts bile through the bile duct into the duodenum.

pancreas: a gland that secretes digestive enzymes and juices into the duodenum.

small intestine: a 10-foot length of small-diameter intestine that is the major site of digestion of food and absorption of nutrients. Its segments are the duodenum, jejunum, and ileum.

lumen (LOO-men): the space within a vessel, such as the intestine.

duodenum (doo-oh-DEEN-um, doo-ODD-num): the top portion of the small intestine (about "12 fingers' breadth" long in ancient terminology).
- **duodecim** = twelve

jejunum (je-JOON-um): the first two-fifths of the small intestine beyond the duodenum.

ileum (ILL-ee-um): the last segment of the small intestine.

ileocecal (ill-ee-oh-SEEK-ul) **valve:** the sphincter separating the small and large intestines.

large intestine or **colon** (COAL-un): the lower portion of intestine that completes the digestive process. Its segments are the ascending colon, the transverse colon, the descending colon, and the sigmoid colon.
- **sigmoid** = shaped like the letter S (sigma in Greek)

appendix: a narrow blind sac extending from the beginning of the colon that stores lymph cells.

rectum: the muscular terminal part of the intestine, extending from the sigmoid colon to the anus.

anus (AY-nus): the terminal outlet of the GI tract.

around the mouth, facilitating chewing and swallowing. When you swallow a mouthful of food, it passes through the **pharynx,** a short tube that is shared by both the **digestive system** and the respiration system. To bypass the entrance to your lungs, the **epiglottis** closes off your air passages so that you don't choke when you swallow, thus resolving obstacle 1. (Choking is discussed on p. 85.) After a mouthful of food has been swallowed, it is called a **bolus.**

• Esophagus to the Stomach • The **esophagus** has a **sphincter** muscle at each end. During a swallow, the upper **esophageal sphincter** opens. The bolus then slides down the esophagus, which passes through a hole in the diaphragm (obstacle 2) to the **stomach.** The lower esophageal sphincter at the entrance to the stomach closes behind the bolus so that it proceeds forward and doesn't slip back into the esophagus (obstacle 3). The stomach retains the bolus for a while in its upper portion. Little by little, the stomach transfers the food to its lower portion, adds juices to it, and grinds it to a semiliquid mass called **chyme.** Then, bit by bit, the stomach releases the chyme through the **pyloric sphincter,** which opens into the **small intestine** and then closes behind the chyme.

• Small Intestine • At the top of the small intestine, the chyme bypasses the opening from the common bile duct, which is dripping fluids (obstacle 4) into the small intestine from two organs outside the GI tract—the **gallbladder** and the **pancreas.** The chyme travels on down the small intestine through its three segments—the **duodenum,** the **jejunum,** and the **ileum**—almost 10 feet of tubing coiled within the abdomen.[1]

• Large Intestine (Colon) • Having traveled the length of the small intestine, the chyme arrives at another sphincter (obstacle 3 again): the **ileocecal valve,** at the beginning of the **large intestine (colon)** in the lower right-hand side of the abdomen. As the chyme enters the colon, it passes another opening. Had it slipped into this opening, it would have ended up in the **appendix,** a blind

◆ Young adults have up to 10,000 taste buds.◆

digestive system: all the organs and glands associated with the ingestion and digestion of food.

bolus (BOH-lus): a portion; with respect to food, the amount swallowed at one time.
- **bolos** = lump

chyme (KIME): the semiliquid mass of partly digested food expelled by the stomach into the duodenum.
- **chymos** = juice

FIGURE 3-2 The Teeth

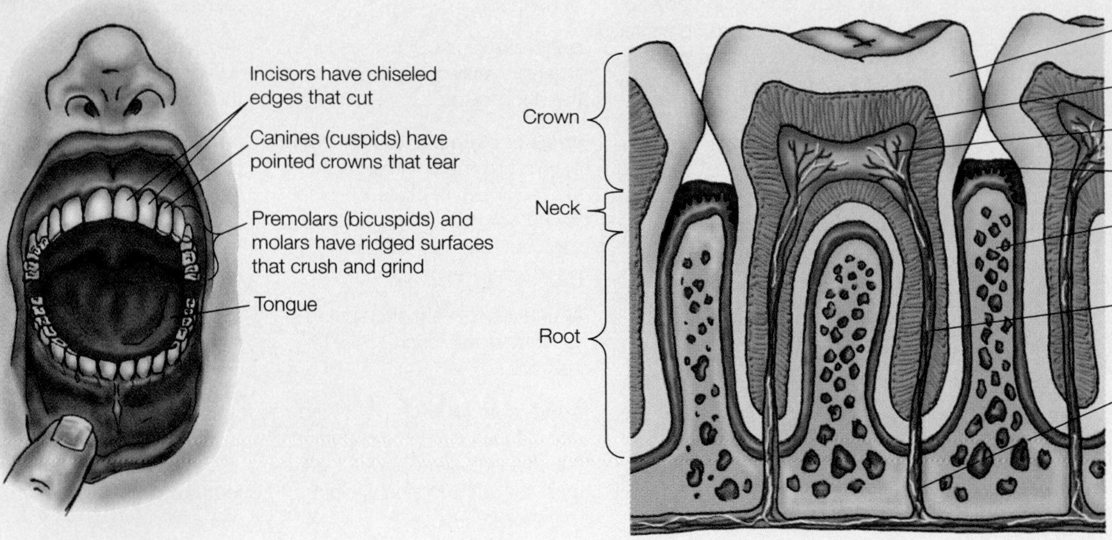

sac about the size of your little finger. The chyme bypasses this opening, however, and travels along the large intestine up the right-hand side of the abdomen, across the front to the left-hand side, down to the lower left-hand side, and finally below the other folds of the intestines to the back side of the body, above the **rectum.**

During the chyme's passage to the rectum, the colon withdraws water from it, leaving semisolid waste (obstacle 5). The strong muscles of the rectum and anal canal hold back this waste until it is time to defecate. Then the rectal muscles relax (obstacle 6), and the two sphincters of the **anus** open to allow passage of the waste.

The Muscular Action of Digestion

The first step in the reduction of food to a liquid takes place in the mouth, where chewing, the addition of saliva, and the action of the tongue reduce the food to a coarse mash. Then you swallow, and thereafter, you are generally unaware of all the activity that follows. As is the case with so much else that happens in the body, the muscles of the digestive tract meet internal needs without your having to exert any conscious effort. They keep things moving◆ at just the right pace, slow enough to get the job done and fast enough to make progress.

◆ The ability of the GI tract muscles to move is called their **motility** (moh-TIL-ah-tee).

• **Peristalsis** • The entire GI tract is ringed with circular muscles that can squeeze it tightly. Surrounding these rings of muscle are longitudinal muscles. When the rings tighten and the long muscles relax, the tube is constricted. When the rings relax and the long muscles tighten, the tube bulges. This action—called **peristalsis**—occurs continuously and pushes the intestinal contents along (obstacle 3 again). (If you have ever watched a lump of food pass along the body of a snake, you have a good picture of how these muscles work.)

The waves of contraction ripple along the GI tract at varying rates and intensities depending on the part of the GI tract and on whether food is present. For example, waves occur three times per minute in the stomach, but speed up to ten times per minute when chyme reaches the small intestine. When you

peristalsis (peri-STALL-sis): wavelike muscular contractions of the GI tract that push its contents along.
- **peri** = around
- **stellein** = wrap

have just eaten a meal, the waves are slow and continuous; when the GI tract is empty, the intestine is quiet except for periodic bursts of powerful rhythmic waves. Peristalsis, along with the sphincter muscles that surround the tract at key places, keeps things moving along.

• Stomach Action • The stomach has the thickest walls and strongest muscles of all the GI tract organs. In addition to the circular and longitudinal muscles, it has a third layer of diagonal muscles that also alternately contract and relax (see Figure 3-3). These three sets of muscles work to force the chyme downward, but the pyloric sphincter usually remains tightly closed, preventing the chyme from passing into the duodenum of the small intestine. As a result, the chyme is churned and forced down, hits the pyloric sphincter, and remains in the stomach. Meanwhile, the stomach wall releases juices. When the chyme is completely liquefied, the pyloric sphincter opens briefly, about three times a minute, to allow small portions of chyme through. At this point, the chyme no longer resembles food in the least.

FIGURE 3-3 **Stomach Muscles**

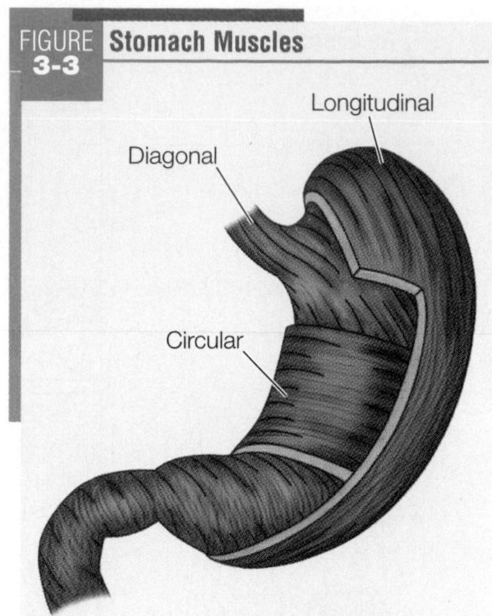

The stomach has three layers of muscles.

• Segmentation • The circular muscles of the intestines rhythmically contract and squeeze their contents. These contractions, called **segmentation,** mix the chyme and promote close contact with the digestive juices and the absorbing cells of the intestinal walls before letting the contents move slowly along.

• Sphincter Contractions • Sphincter muscles periodically open and close, allowing the contents of the GI tract to move along at a controlled pace (obstacle 3 again). At the top of the esophagus, the upper esophageal sphincter opens in response to swallowing. At the bottom of the esophagus, the lower esophageal sphincter (sometimes called the cardiac sphincter because of its proximity to the heart) prevents **reflux** of the stomach contents.[2] At the bottom of the stomach, the pyloric sphincter, which stays closed most of the time, holds the chyme in the stomach long enough so that it can be thoroughly mixed with gastric juice and liquefied. The pyloric sphincter also prevents the intestinal contents from backing up into the stomach. At the end of the small intestine, the ileocecal valve performs a similar function, emptying the contents of the small intestine into the large intestine. Finally, the tightness of the rectal muscle is a kind of safety device; together with the two sphincters of the anus, it prevents elimination until you choose to perform it voluntarily (obstacle 6). Figure 3-4 (on p. 70) illustrates how sphincter muscles contract and relax to close and open passageways.

The Secretions of Digestion

Have you ever wondered how people can eat differently yet have essentially the same body composition? It all comes down to the process of digestion, of rendering food—whatever kind of food it is to start with—into nutrients.

To break down food into small nutrients that the body can absorb, five different organs produce secretions: the salivary glands, the stomach, the pancreas, the liver (via the gallbladder), and the small intestine. These secretions enter the GI tract at various points along the way, bringing an abundance of water (obstacle 4) and a variety of enzymes.

Enzymes◆ are formally introduced in Chapter 6, but for now a simple definition will suffice. An enzyme is a protein that facilitates a chemical reaction—making a molecule, breaking a molecule, changing the arrangement of a molecule, or exchanging parts of molecules. As a **catalyst,** the enzyme itself remains unchanged. The enzymes involved in digestion facilitate a chemical reaction known as **hydrolysis—**the addition of water *(hydro)* to break *(lysis)* a molecule into smaller pieces. The glossary on p. 71 (top) identifies some of the common **digestive enzymes** and related terms; later chapters introduce specific enzymes. When learning about enzymes, it helps to know

◆ All enzymes and some hormones are proteins, but enzymes are not hormones. Enzymes facilitate the making and breaking of bonds in chemical reactions; hormones act as chemical messengers, sometimes regulating enzyme action.

segmentation (SEG-men-TAY-shun): a periodic squeezing or partitioning of the intestine at intervals along its length by its circular muscles.

reflux: a backward flow.
- **re** = back
- **flux** = flow

catalyst (CAT-uh-list): a compound that facilitates chemical reactions without itself being changed in the process.

FIGURE 3-4 An Example of a Sphincter Muscle

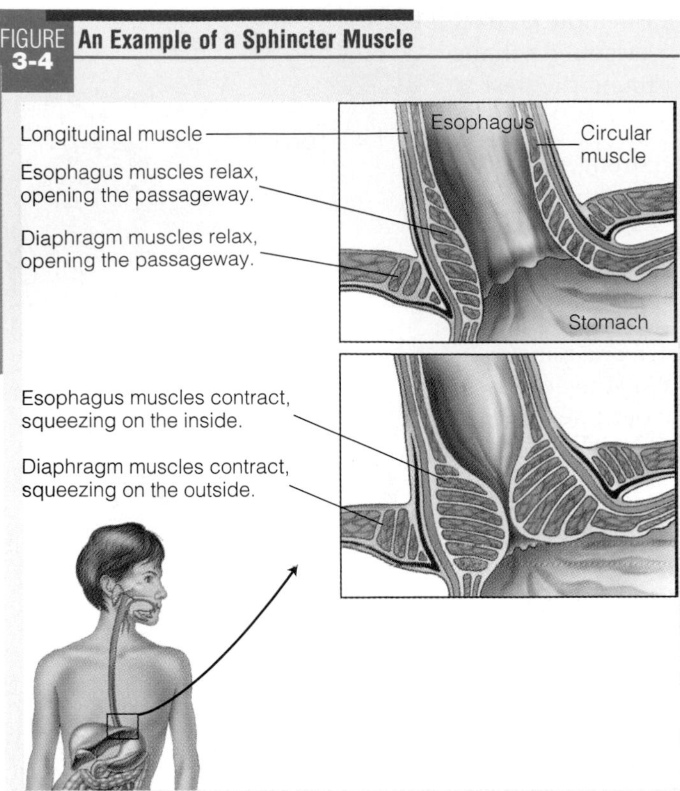

When the circular muscles of a sphincter contract, the passage closes; when they relax, the passage opens.

that the word ending *-ase* denotes an enzyme. Enzymes are often identified by the place they come from and the compounds they work on; *gastric lipase,* for example, is a stomach enzyme that acts on lipids, whereas *pancreatic lipase* comes from the pancreas (and also works on lipids).

• **Saliva** • The **salivary glands,** shown in Figure 3-5 on p. 71, squirt just enough **saliva** to moisten each mouthful of food so that it can pass easily down the esophagus (obstacle 4). (Digestive glands and their secretions appear in boldface type and are defined in the glossary at the bottom of p. 71.) The saliva contains water, salts, and enzymes that initiate the digestion of carbohydrates. Saliva also protects the teeth and the linings of the mouth, esophagus, and stomach from attack by substances that might harm them.

• **Gastric Juice** • Cells in the stomach secrete **gastric juice,** a mixture of water, enzymes, and **hydrochloric acid.** The acid is so strong that it causes the sensation of heartburn if it happens to reflux into the esophagus. Highlight 3, following this chapter, discusses heartburn, ulcers, and other common digestive problems.

The strong acidity of the stomach prevents bacterial growth and kills most bacteria that enter the body with food. It would destroy the cells of the stomach as well, but for their natural defenses. To protect themselves from gastric juice, the **goblet cells** of the stomach wall secrete **mucus,** a thick, slippery, white substance that coats the cells, protecting them from the acid and enzymes that might otherwise harm them (obstacle 7).

Figure 3-6 on p. 72 shows how the strength of acids is measured—in **pH** units. Note that the acidity of gastric juice registers below "2" on the pH scale—stronger than vinegar. The stomach enzymes work most efficiently in the stomach's strong acid, but the salivary enzymes, which are swallowed with food, do not work in acid this strong. Consequently, the salivary digestion of carbohydrate gradually ceases as the stomach acid penetrates each newly swallowed bolus of food. In fact, salivary enzymes become just other proteins to be digested.

• **Pancreatic Juice and Intestinal Enzymes** • By the time food leaves the stomach, digestion of all three energy nutrients (carbohydrates, fats, and proteins) has begun, and the action gains momentum in the small intestine. There the pancreas contributes digestive juices by way of ducts leading into the duodenum. The **pancreatic juice** contains enzymes that act on all three energy nutrients, and the cells of the intestinal wall also possess digestive enzymes on their surfaces.

In addition to enzymes, the pancreatic juice contains sodium **bicarbonate,** which is basic or alkaline—the opposite of the stomach's acid (review Figure 3-6). The pancreatic juice thus neutralizes the acid chyme arriving in the small intestine from the stomach. From this point on, the chyme remains at a neutral or slightly alkaline pH. The enzymes of both the intestine and the pancreas work best in this environment.

• **Bile** • **Bile** also flows into the duodenum. The **liver** continuously produces bile, which is then concentrated and stored in the gallbladder. The gallbladder squirts the bile into the duodenum when fat arrives there. Bile is not an

goblet cells: cells of the GI tract (and lungs) that secrete mucus.

pH: the unit of measure expressing a substance's acidity or alkalinity (Chapter 12 provides a more detailed definition).

GLOSSARY OF DIGESTIVE ENZYMES

digestive enzymes: proteins found in digestive juices that act on food substances, causing them to break down into simpler compounds.

-ase (ACE): a word ending denoting an enzyme. The word beginning often identifies the compounds the enzyme works on. Examples include:

- **Carbohydrase** (KAR-boe-HIGH-drase), an enzyme that hydrolyzes carbohydrates.
- **Lipase** (LYE-pase), an enzyme that hydrolyzes lipids (fats).
- **Protease** (PRO-tee-ase), an enzyme that hydrolyzes proteins.

hydrolysis (high-DROL-ih-sis): a chemical reaction in which a major reactant is split into two products, with the addition of a hydrogen atom (H) to one and a hydroxyl group (OH) to the other (from water, H_2O). (The noun is **hydrolysis;** the verb is **hydrolyze.**)

- **hydro** = water
- **lysis** = breaking

enzyme, but an **emulsifier** that brings fats into suspension in water so that enzymes can break them down into their component parts. Thanks to all these secretions, the three energy-yielding nutrients are digested in the small intestine (the summary on p. 72 provides a table of digestive secretions and their actions).

• Protective Factors • Both the small and the large intestine, being neutral in pH, permit the growth of bacteria (known as the intestinal flora). In fact, a healthy intestinal tract supports a thriving bacterial population that normally does the body no harm and may actually do some good. Bacteria in the GI tract produce a couple of vitamins, including a significant amount of vitamin K, although the amount is insufficient to meet the body's total need for that vitamin.

Provided that the normal intestinal flora are thriving, infectious bacteria have a hard time getting established and launching an attack on the system. Diet is one of several factors that influence the bacterial population and its environment.[3] In addition, secretions from the GI tract—saliva, mucus, gastric acid, and digestive enzymes—not only help with digestion, but also defend against foreign invaders. The GI tract also maintains several different kinds of defending cells that confer specific immunity against intestinal diseases such as inflammatory bowel disease.[4]

FIGURE 3-5 The Salivary Glands

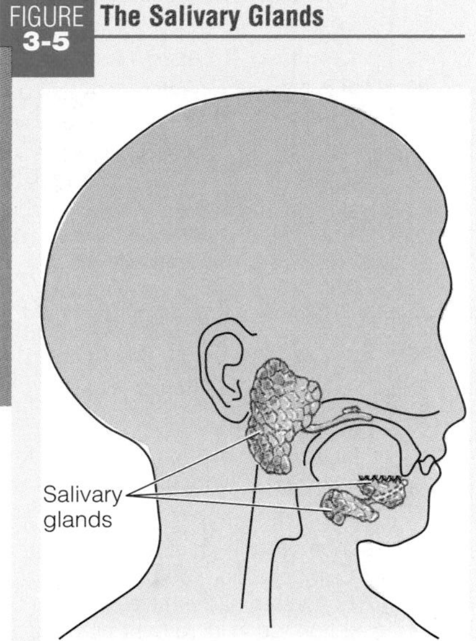

The salivary glands secrete saliva into the mouth and begin the digestive process. Given the short time food is in the mouth, salivary enzymes contribute little to digestion.

GLOSSARY OF DIGESTIVE GLANDS AND THEIR SECRETIONS

These terms are listed in order from start to end of the digestive tract.

gland: a cell or group of cells that secretes materials for special uses in the body. Glands may be **exocrine** (EKS-oh-crin) **glands,** secreting their materials "out" (into the digestive tract or onto the surface of the skin), or **endocrine** (EN-doe-crin) **glands,** secreting their materials "in" (into the blood).

- **exo** = outside
- **endo** = inside
- **krine** = to separate

salivary glands: exocrine glands that secrete saliva into the mouth.

saliva: the secretion of the salivary glands. Its principal enzyme begins carbohydrate digestion.

gastric glands: exocrine glands in the stomach wall that secrete gastric juice into the stomach.

- **gastro** = stomach

gastric juice: the digestive secretion of the gastric glands of the stomach.

hydrochloric acid: an acid composed of hydrogen and chloride atoms (HCl). The gastric glands normally produce this acid.

mucus (MYOO-kus): a slippery substance secreted by goblet cells of the GI lining (and other body linings) that protects the cells from exposure to digestive juices (and other destructive agents). The lining of the GI tract with its coat of mucus is a **mucous membrane.** (The noun is **mucus;** the adjective is **mucous.**)

liver: the organ that manufactures bile. The liver's many other functions are described in Chapter 7.

bile: an emulsifier that prepares fats and oils for digestion; an exocrine secretion made by the liver, stored in the gallbladder, and released into the small intestine when needed.

emulsifier (ee-MUL-sih-fire): a substance with both water-soluble and fat-soluble portions that promotes the mixing of oils and fats in a watery solution.

pancreatic (pank-ree-AT-ic) **juice:** the exocrine secretion of the pancreas, containing enzymes for the digestion of carbohydrate, fat, and protein as well as bicarbonate, a neutralizing agent. The juice flows from the pancreas into the small intestine through the pancreatic duct. (The pancreas also has an endocrine function, the secretion of insulin and other hormones.)

bicarbonate: an alkaline secretion of the pancreas, part of the pancreatic juice. (Bicarbonate also occurs widely in all cell fluids.)

FIGURE 3-6 The pH Scale

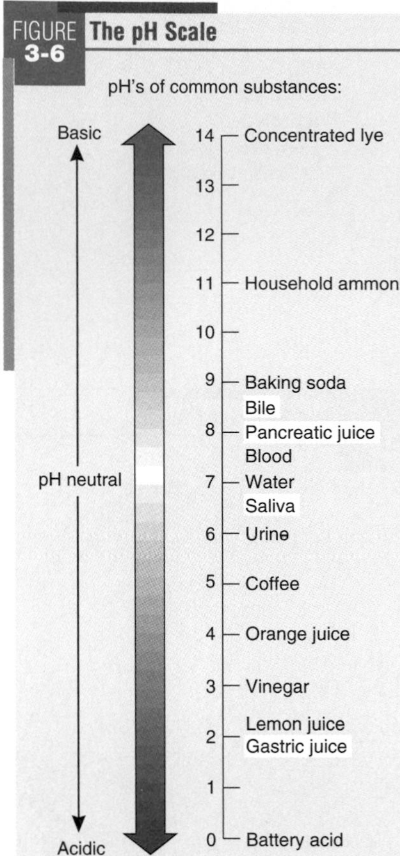

A substance's acidity or alkalinity is measured in pH units. The pH is the negative logarithm of the hydrogen ion concentration. Each increment presents a tenfold increase in concentration of hydrogen particles. For example, a pH of 2 is 1000 times stronger than a pH of 5.

The Final Stage

The story of how digestion prepares food for absorption is now nearly complete. The three energy-yielding nutrients—carbohydrate, fat, and protein—have been disassembled and are ready to be absorbed. Most of the other nutrients—vitamins, minerals, and water—need no such disassembly; some vitamins and minerals are altered slightly during digestion, but most are absorbed as they are. Undigested residues, such as some fibers, are not absorbed, but continue through the digestive tract, providing a semisolid mass that helps exercise the muscles and keeps them strong enough to perform peristalsis efficiently. Fiber also retains water, accounting for the **stools'** pasty consistency, and carries some bile acids, some minerals, and some additives and contaminants with it out of the body.

By the time the contents of the GI tract reach the end of the small intestine, little remains but water, a few dissolved salts and body secretions, and undigested materials such as fiber. These enter the large intestine (colon).

In the colon, intestinal bacteria degrade some of the fiber to simpler compounds, while the colon itself retrieves all materials that the body can recycle—water and dissolved salts (see Figure 3-7). The waste that is finally excreted has little or nothing of value left in it. The body has extracted all that it can use from the food. Figure 3-8 on p. 74 summarizes digestion by following a sandwich through the GI tract and into the body.

IN SUMMARY

As Figure 3-1 shows, food enters the mouth and travels down the esophagus and through the upper and lower esophageal sphincters to the stomach, then through the pyloric sphincter to the small intestine, on through the ileocecal valve to the large intestine, past the appendix to the rectum, ending at the anus. The wavelike contractions of peristalsis and the periodic squeezing of segmentation keep things moving at a reasonable pace. Along the way, secretions from the salivary glands, stomach, pancreas, liver (via the gallbladder), and small intestine deliver fluids and digestive enzymes.

Digestive Secretions

Organ or Gland	Target Organ	Secretion	Action
Salivary glands	Mouth	Saliva	Fluid eases swallowing; salivary enzyme breaks down **carbohydrate.**
Gastric glands	Stomach	Gastric juice	Fluid mixes with bolus; hydrochloric acid uncoils **proteins;** enzymes break down proteins; mucus protects stomach cells.
Pancreas	Small intestine	Pancreatic juice	Bicarbonate neutralizes acidic gastric juices; pancreatic enzymes break down **carbohydrates, fats,** and **proteins.**
Liver	Gallbladder	Bile	Bile stored until needed.
Gallbladder	Small intestine	Bile	Bile emulsifies **fat** so that enzymes can attack.
Intestinal glands	Small intestine	Intestinal juice	Intestinal enzymes break down **carbohydrate** and **protein** fragments; mucus protects the intestinal wall.

ABSORPTION

The problem of absorption: Given a meal that delivers millions of molecules of nutrients, provide a means by which all can enter the body simultaneously. Within three or four hours after you have eaten a dinner of beans and rice (or spinach lasagna, or steak and potatoes) with vegetable, salad, beverage, and

stools: waste matter discharged from the colon; also called **feces** (FEE-seez).

FIGURE 3-7 The Colon

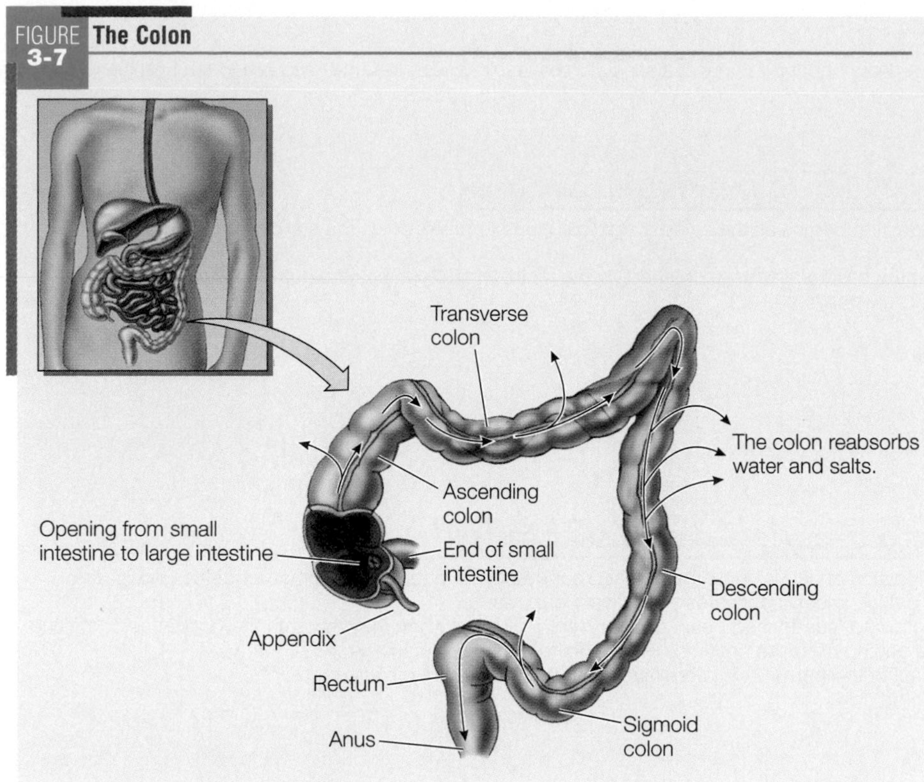

dessert, your body must find a way to absorb—one by one—some two hundred thousand million, million, million molecules derived from carbohydrate digestion; a comparable number of molecules derived from protein and fat digestion; and many vitamin and mineral molecules as well.

Most absorption takes place in the small intestine, one of the most elegantly designed organ systems in the body. In 10 feet of small intestine, it provides a surface area equivalent to a tennis court, which engulfs and absorbs the nutrient molecules. To remove the molecules rapidly and provide room for more to be absorbed, a rush of circulating blood continuously washes the underside of this surface, carrying the absorbed nutrients away to the liver and other parts of the body.

Anatomy of the Absorptive System

The inner surface of the small intestine looks smooth and slippery, but viewed through a microscope, it turns out to be wrinkled into hundreds of folds. Each fold in turn is contoured into thousands of fingerlike projections, as numerous as the hairs on velvet fabric. These small intestinal projections are the **villi.** A single villus, magnified still more, turns out to be composed of hundreds of cells, each covered with its own microscopic hairs, the **microvilli** (see Figure 3-9). In the crevices between the villi lie the **crypts**—tubular glands that secrete the intestinal juices into the small intestine.

The villi are in constant motion. Each villus is lined by a thin sheet of muscle, so it can wave, squirm, and wriggle like the tentacles of a sea anemone. Any nutrient molecule small enough to be absorbed is trapped among the microvilli that coat the cells and then drawn into the cells. Some partially digested nutrients are caught in the microvilli, digested further by enzymes there, and then absorbed into the cells. Figure 3-10 on p. 76 describes how nutrients are absorbed by simple diffusion, facilitated diffusion, or active transport. Later chapters provide details on specific nutrients. Before following nutrients through the body, we must look more closely at the digestive cells themselves.

villi (VILL-ee, VILL-eye): fingerlike projections from the folds of the small intestine; singular **villus.**

microvilli (MY-cro-VILL-ee, MY-cro-VILL-eye): tiny, hairlike projections on each cell of every villus that can trap nutrient particles and transport them into the cells; singular **microvillus.**

crypts (KRIPTS): tubular glands that lie between the intestinal villi and secrete intestinal juices into the small intestine.

FIGURE 3-8 **The Digestive Fate of a Sandwich**

To review the digestive processes, follow a peanut butter and banana sandwich on whole-wheat, sesame seed bread through the GI tract.

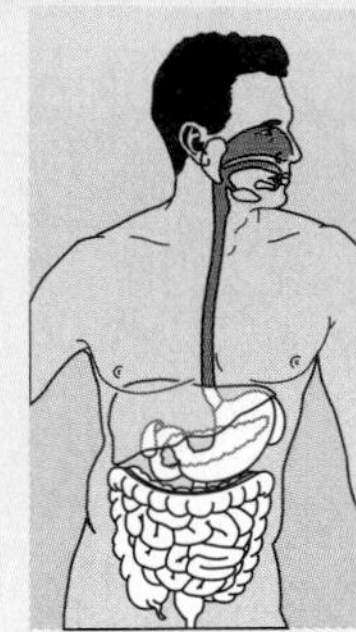

MOUTH: CHEWING AND SWALLOWING, WITH LITTLE DIGESTION

- **Carbohydrate** digestion begins as the salivary enzyme starts to break down the starch from bread and peanut butter.
- **Fiber** covering on the sesame seeds is crushed by the teeth, which exposes the nutrients inside the seeds to the upcoming digestive enzymes.

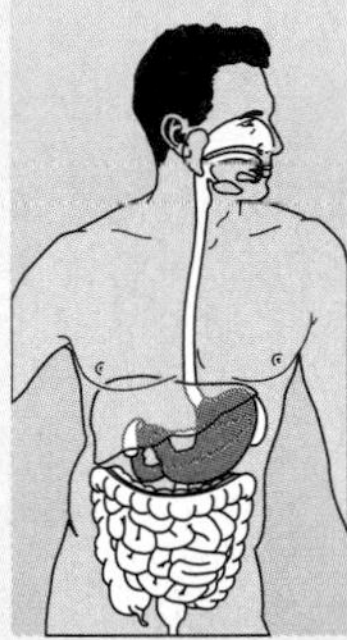

STOMACH: COLLECTING AND CHURNING, WITH SOME DIGESTION

- **Carbohydrate** digestion continues until the mashed sandwich has been mixed with the gastric juices; the stomach acid of the gastric juices inactivates the salivary enzyme.
- **Proteins** from the bread, seeds, and peanut butter begin to uncoil when they mix with the gastric acid, making them available to the gastric protease enzymes that begin to digest proteins.
- **Fat** from the peanut butter forms a separate layer on top of the watery mixture.

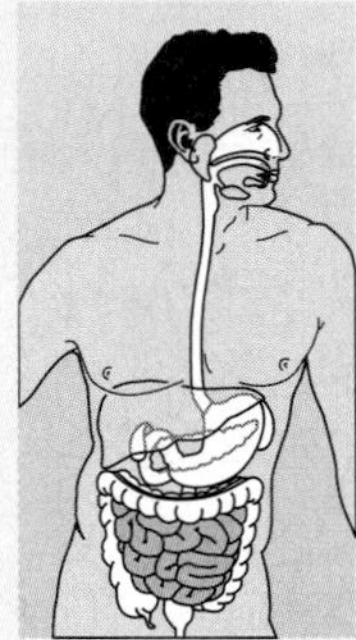

SMALL INTESTINE: DIGESTING AND ABSORBING

- **Sugars** from the banana require so little digestion that they begin to traverse the intestinal cells immediately on contact.
- **Starch** digestion picks up when the pancreas sends pancreatic enzymes to the small intestine via the pancreatic duct. Enzymes on the surfaces of the small intestinal cells complete the process of breaking down starch into small fragments that can be absorbed through the intestinal cell walls and into the blood.
- **Fat** from the peanut butter and seeds is emulsified with the watery digestive fluids by bile. Now the pancreatic and intestinal lipases can begin to break down the fat to smaller fragments that can be absorbed through the cells of the small intestinal wall and into the lymph.
- **Protein** digestion depends on the pancreatic and intestinal proteases. Small fragments of protein are liberated and absorbed through the cells of the small intestinal wall and into the blood.
- **Vitamins and minerals** are absorbed.

Note: Sugars and starches are members of the carbohydrate family.

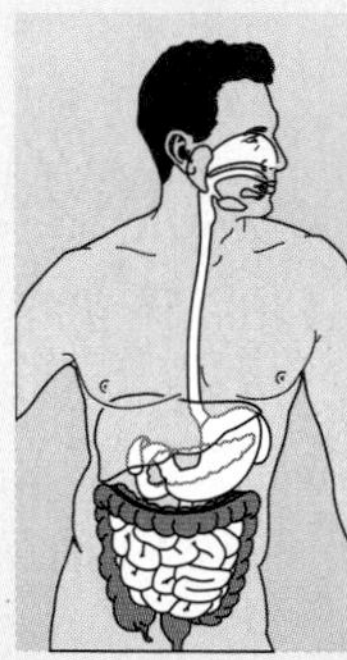

LARGE INTESTINE: REABSORBING AND ELIMINATING

- **Fluids and some minerals** are absorbed.
- **Some fibers** from the seeds, whole-wheat bread, peanut butter, and banana are partly digested by the bacteria living there, and some of these products are absorbed.
- **Most fibers** pass through the large intestine and are excreted as feces; some fat, cholesterol, and minerals bind to fiber and are also excreted.

A Closer Look at the Intestinal Cells

The cells of the villi are among the most amazing in the body, for they recognize and select the nutrients the body needs and regulate their absorption. A close look at these cells is worthwhile, because it will help to explode a common misconception about nutrition: that you have to do anything to ensure that your digestive tract does its job. Nothing could be further from the truth.

FIGURE 3-9 **The Small Intestinal Villi**

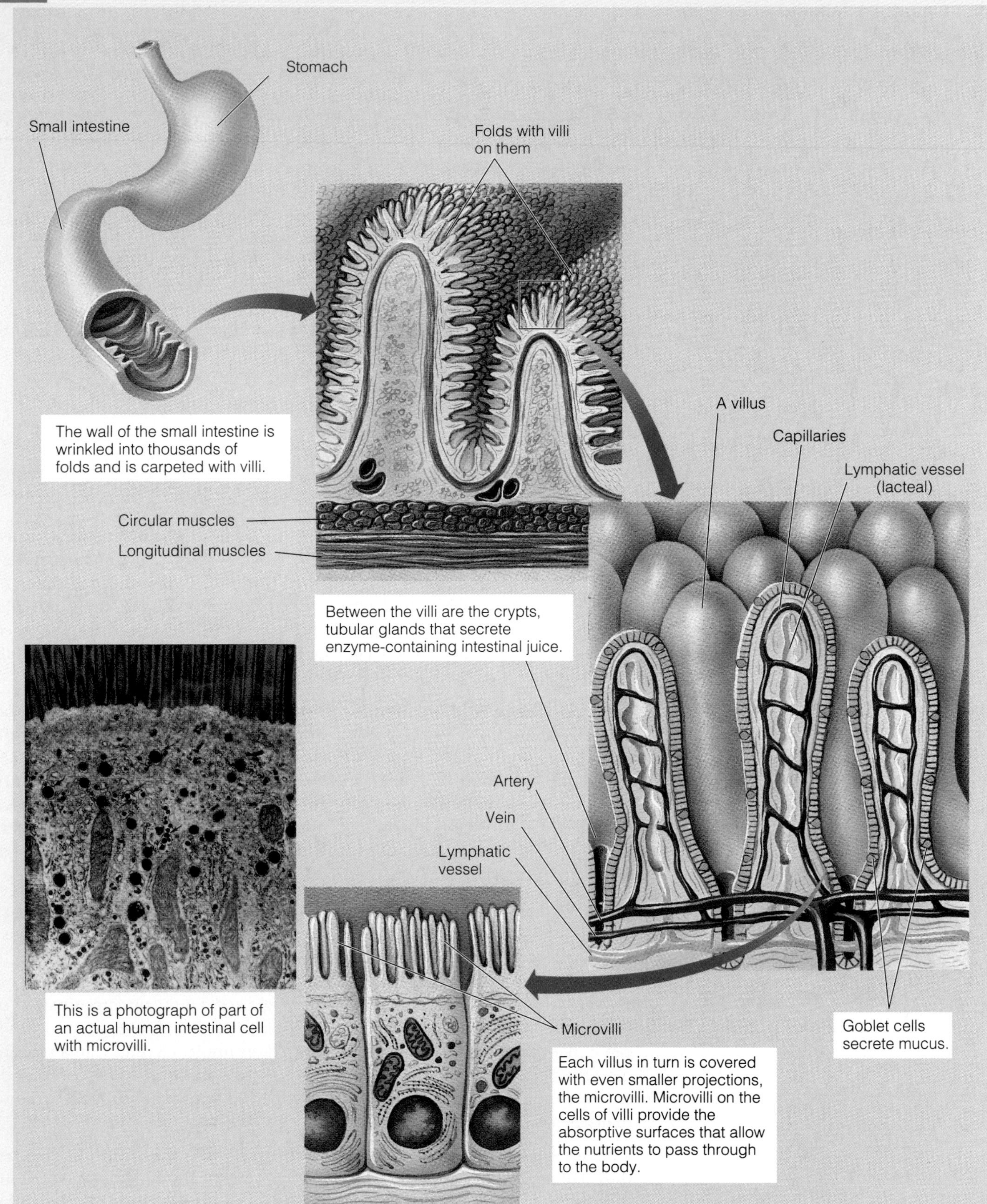

FIGURE 3-10 **Absorption of Nutrients**

Absorption of nutrients into intestinal cells typically occurs by simple diffusion, facilitated diffusion, or active transport.

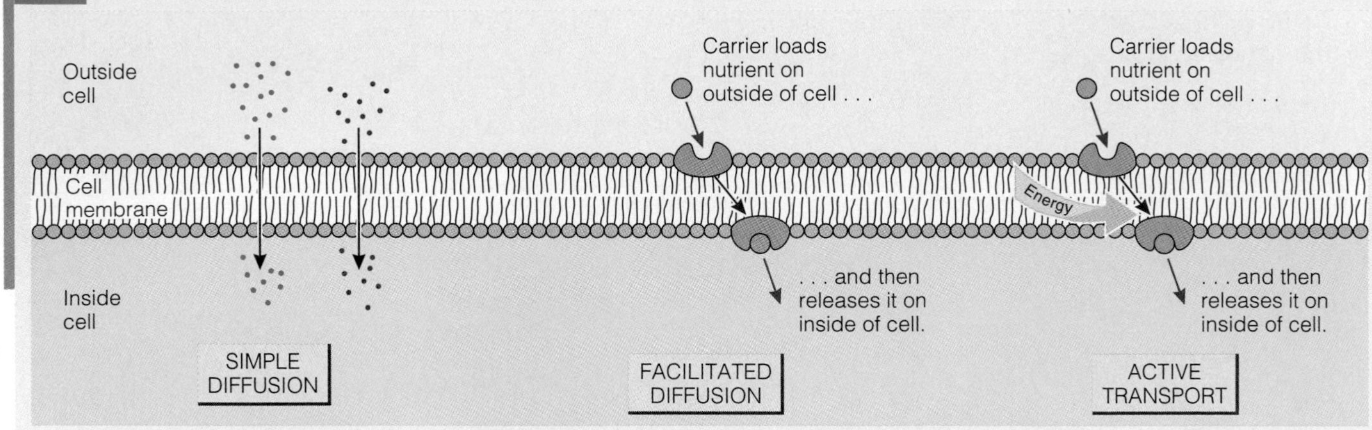

Some nutrients (such as water and small lipids) are absorbed by simple diffusion. They cross into intestinal cells freely.

Some nutrients (such as the water soluble vitamins) are absorbed by facilitated diffusion. They need a specific carrier to transport them from one side of the cell membrane to the other. (Alternatively, facilitated diffusion may occur when the carrier changes the cell membrane in such a way that the nutrients can pass through.)

Some nutrients (such as glucose and amino acids) must be absorbed actively. These nutrients move against a concentration gradient, which requires energy.

• **The Cells' Capabilities** • As already described, each cell of a villus is coated with thousands of microvilli, which project from the cell's membrane (review Figure 3-9). In these microvilli and in the membrane lie hundreds of different kinds of enzymes and "pumps," which recognize and act on different nutrients. Descriptions of specific enzymes and "pumps" for each nutrient are presented in the following chapters where appropriate, but the point here is that the cells are equipped to handle all kinds and combinations of foods and nutrients.

• **Specialization in the GI Tract** • A further refinement of the system is that the cells of successive portions of the intestinal tract are specialized to absorb different nutrients. The nutrients that are ready for absorption early are absorbed near the top of the tract; those that take longer to be digested are absorbed farther down. Registered dietitians and medical professionals who treat digestive disorders learn the specialized absorptive functions of different parts of the GI tract so that if one part becomes dysfunctional, the diet can be adjusted accordingly.

• **The Myth of "Food Combining"** • The idea that people should not eat certain food combinations (for example, fruit and meat) at the same meal, because the digestive system cannot handle more than one task at a time, is a myth. The art of "food combining" (which actually emphasizes "food separating") is based on this idea, and it represents faulty logic and a gross underestimation of the body's capabilities. In fact, the contrary is often true; foods eaten together can enhance each other's use by the body. For example, vitamin C in a pineapple or other citrus fruit can enhance the absorption of iron from a meal of chicken and rice or other iron-containing foods. Many other instances of mutually beneficial interactions are presented in later chapters.

• **Preparing Nutrients for Transport** • When a nutrient molecule has crossed the cell of a villus, it enters either the bloodstream or the lymphatic system. Both transport systems supply vessels to each villus, as shown in Figure 3-9. The water-soluble nutrients and the smaller products of fat digestion are released directly into the bloodstream. The larger fats and the fat-soluble vitamins are insoluble in water, however, and blood is mostly water. The intestinal cells assemble many of the products of fat digestion into larger molecules. These larger molecules cluster together with special proteins, forming chylomicrons.◆ These chylomicrons cannot pass into the capillaries and are released into the lymphatic system instead; the chylomicrons move through the lymph and later enter the bloodstream at a point near the heart.

◆ Chylomicrons (kye-lo-MY-cronz) are described in Chapter 5.

IN SUMMARY The many folds and villi of the small intestine dramatically increase its surface area, facilitating nutrient absorption. Nutrients pass through the cells of the villi and enter either the blood (if they are water soluble or small fat fragments) or the lymph (if they are fat soluble).

THE CIRCULATORY SYSTEMS

Once a nutrient has entered the bloodstream, it may be transported to any of the cells in the body, from the tips of the toes to the roots of the hair. The circulatory systems deliver nutrients wherever they are needed.

The Vascular System

The vascular, or blood circulatory, system is a closed system of vessels through which blood flows continuously in a figure eight, with the heart serving as a pump at the crossover point (see Figure 3-11 on p. 78). As the blood circulates through this system, it picks up and delivers materials as needed.

All the body tissues derive oxygen and nutrients from the blood and deposit carbon dioxide and other wastes into it. The lungs exchange carbon dioxide (which leaves the blood to be exhaled) and oxygen (which enters the blood to be delivered to all cells). The digestive system supplies the nutrients to be picked up. In the kidneys, wastes other than carbon dioxide are filtered out of the blood to be excreted in the urine.

Blood leaving the right side of the heart circulates through the lungs and then back to the left side of the heart. The left side of the heart then pumps the blood out through **arteries** to all systems of the body. The blood circulates in the **capillaries,** where it exchanges material with the cells, and then collects into **veins,** which return it again to the right side of the heart. In short, blood travels this simple route:

- Heart to arteries to capillaries to veins to heart.

The routing of the blood past the digestive system has a special feature. The blood is carried to the digestive system (as to all organs) by way of an artery, which (as in all organs) branches into capillaries to reach every cell. Blood leaving the digestive system, however, goes by way of a vein,◆ not back to the heart, but to another organ—the liver. This vein *again* branches into *capillaries* so that every cell of the liver also has access to the blood carried by the vein. Blood leaving the liver then *again* collects into a vein,◆ which returns to the heart.

The route is:

- Heart to arteries to capillaries (in intestines) to vein to capillaries (in liver) to vein to heart.

An anatomist studying this system knows there must be a reason for this special arrangement. The liver's placement ensures that it will be first to receive the materials absorbed from the GI tract. In fact, the liver has many jobs to do in preparing the absorbed nutrients for use by the body. It is the body's major metabolic organ.

◆ The vein that collects blood from the GI tract and conducts it to capillaries in the liver is the **portal vein.**
• portal = gateway

◆ The vein that collects blood from the liver capillaries and returns it to the heart is the **hepatic vein.**
• hepatic = liver

arteries: vessels that carry blood away from the heart.

capillaries (CAP-ill-aries): small vessels that branch from an artery. Capillaries connect arteries to veins. Exchange of oxygen, nutrients, and waste materials takes place across capillary walls.

veins (VANES): vessels that carry blood back to the heart.

FIGURE 3-11 **The Vascular System**

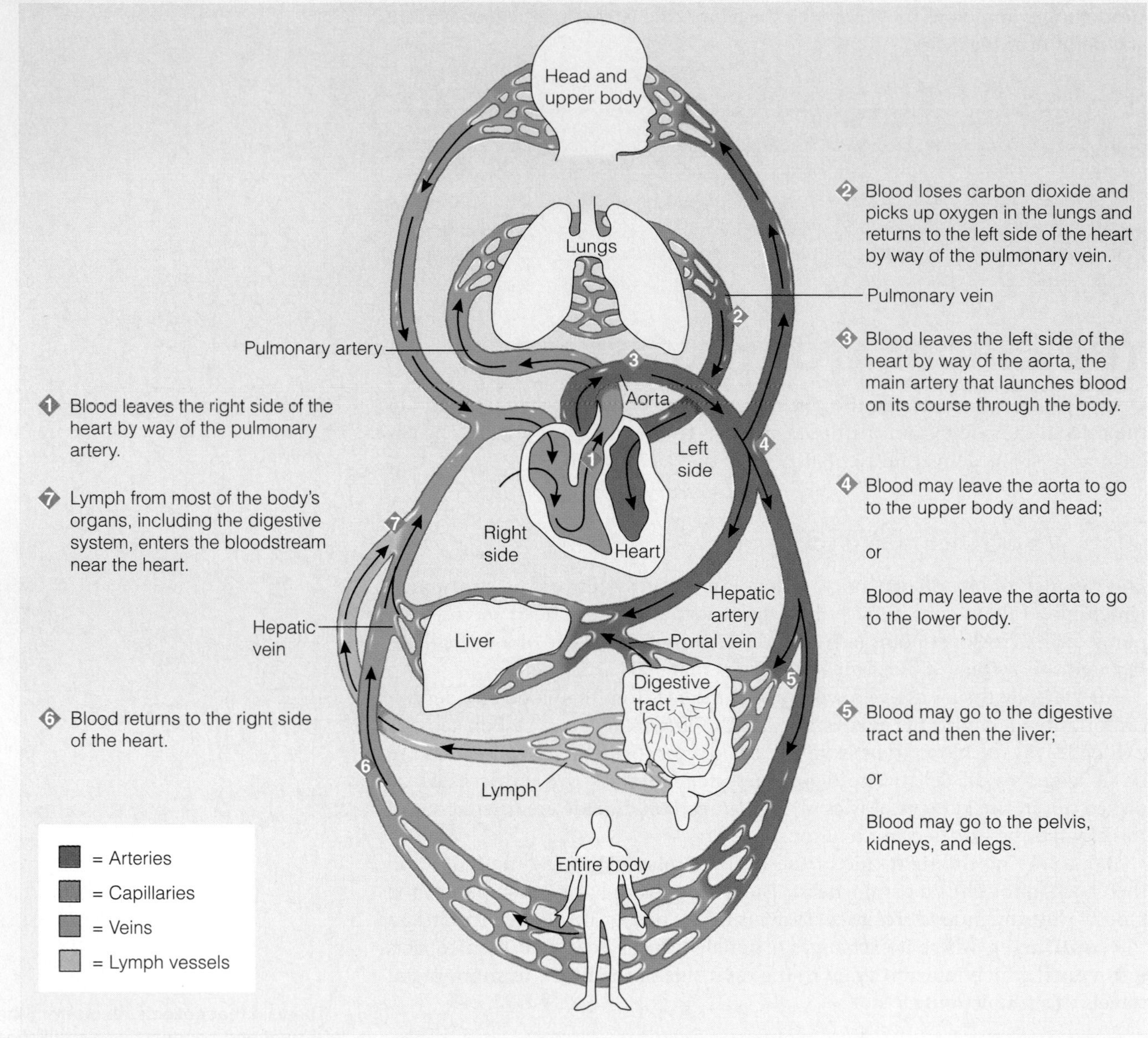
Head and upper body
Lungs
Pulmonary vein
Pulmonary artery
Aorta
Left side
Right side
Heart
Hepatic artery
Hepatic vein
Liver
Portal vein
Digestive tract
Lymph
Entire body
1 Blood leaves the right side of the heart by way of the pulmonary artery.
2 Blood loses carbon dioxide and picks up oxygen in the lungs and returns to the left side of the heart by way of the pulmonary vein.
3 Blood leaves the left side of the heart by way of the aorta, the main artery that launches blood on its course through the body.
4 Blood may leave the aorta to go to the upper body and head;
or
Blood may leave the aorta to go to the lower body.
5 Blood may go to the digestive tract and then the liver;
or
Blood may go to the pelvis, kidneys, and legs.
6 Blood returns to the right side of the heart.
7 Lymph from most of the body's organs, including the digestive system, enters the bloodstream near the heart.
= Arteries
= Capillaries
= Veins
= Lymph vessels

You might guess that, in addition, the liver serves as a gatekeeper to defend against substances that might harm the heart or brain. This is why, when people ingest poisons that succeed in passing the first barrier (the intestinal cells), the liver quite often suffers the damage—from the hepatitis virus, from drugs such as barbiturates or alcohol, from poisons, and from contaminants such as mercury. Perhaps, in fact, you have been undervaluing your liver, not knowing what heroic tasks it quietly performs for you. Figure 3-12 shows the liver's key position in nutrient transport.

lymphatic (lim-FAT-ic) **system:** a loosely organized system of vessels and ducts that convey fluids toward the heart. The GI part of the lymphatic system carries the products of digestion into the bloodstream.

lymph (LIMF): a clear yellowish fluid that is almost identical to blood except that it contains no red blood cells or platelets. Lymph from the GI tract transports fat and fat-soluble vitamins to the bloodstream via lymphatic vessels.

The Lymphatic System

The **lymphatic system** provides a one-way route for fluid from the tissue spaces to enter the blood. Unlike the vascular system, the lymphatic system has no pump; instead, **lymph** circulates between the cells of the body and collects

FIGURE 3-12 **The Liver**

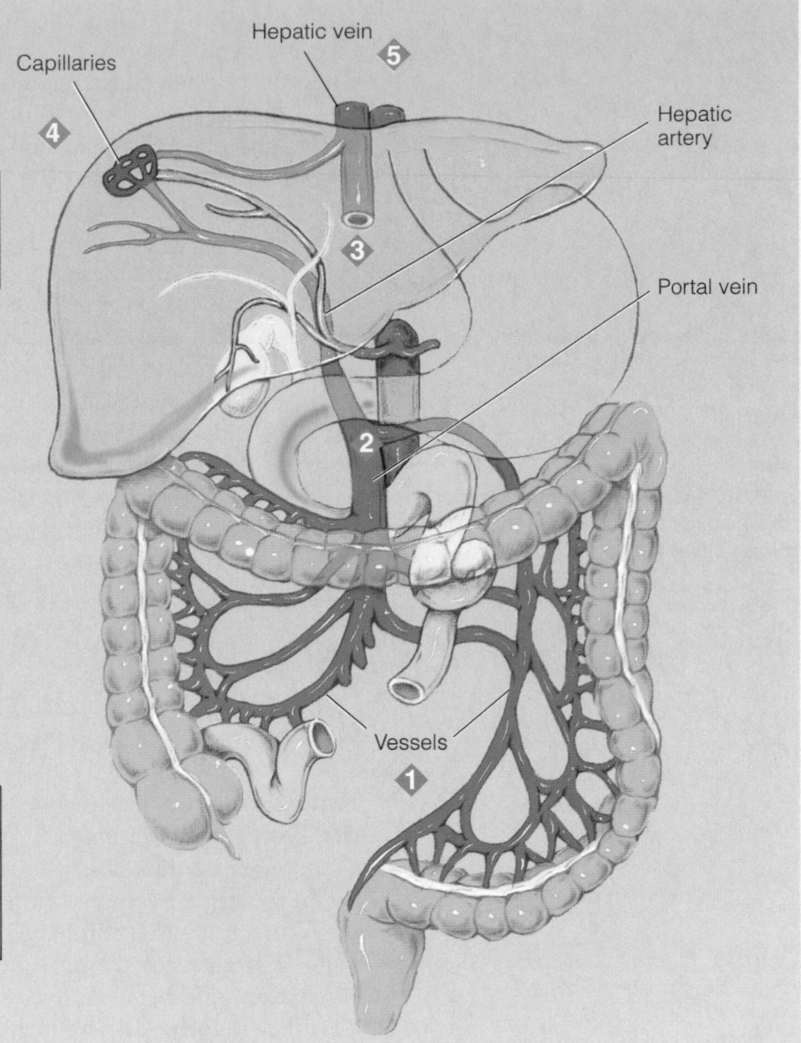

1. Vessels gather up nutrients and reabsorbed water and salts from all over the digestive tract.

Not shown here:
Parallel to these vessels (veins) are other vessels (arteries) that carry oxygen-rich blood from the heart to the intestines.

2. The vessels merge into the portal vein, which conducts all absorbed materials to the liver.

3. The hepatic artery brings a supply of freshly oxygenated blood (not loaded with nutrients) from the lungs to supply oxygen to the liver's own cells.

4. Capillaries branch all over the liver, making nutrients and oxygen available to all its cells and giving the cells access to blood from the digestive system.

5. The hepatic vein gathers up blood in the liver and returns it to the heart.

In contrast, nutrients absorbed into lymph do not go to the liver first. They go to the heart, which pumps them to all the body's cells. The cells remove the nutrients they need, and the liver then has to deal only with the remnants.

into tiny vessels. The fluid moves from one portion of the body to another as muscles contract and create pressure here and there. Ultimately, much of the lymph collects in a large duct behind the heart. This duct terminates in a vein◆ that conducts the lymph toward the heart. Thus materials from the GI tract that enter lymphatic vessels◆ (large fats and fat-soluble vitamins) ultimately enter the bloodstream, circulating through arteries, capillaries, and veins like the other nutrients, with a notable exception—they bypass the liver at first.

Once inside the vascular system, the nutrients can travel freely to any destination and can be taken into cells and used as needed. What becomes of them is described in later chapters.

◆ The duct that conveys lymph toward the heart is the **thoracic** (thor-ASS-ic) **duct.** The **subclavian** (sub-KLAY-vee-an) **vein** connects this duct with the right upper chamber of the heart, providing a passageway by which lymph can be returned to the vascular system.

◆ The lymphatic vessels of the intestine that take up nutrients and pass them to the lymph circulation are called **lacteals** (LACK-tee-als).

IN SUMMARY

Nutrients leaving the digestive system via the blood are routed directly to the liver before being transported to the body's cells. Those leaving via the lymphatic system eventually enter the vascular system, but bypass the liver at first.

REGULATION OF DIGESTION AND ABSORPTION

There is nothing random about digestion and absorption; they are coordinated in every detail. The ability of the digestive tract to handle its ever-changing contents routinely illustrates an important physiological principle that governs the way all living things function—the principle of **homeostasis.** Simply stated, conditions have to stay about the same for an organism to survive; if they deviate too far from the norm, the organism must "do something" to bring them back to normal. The body's regulation of digestion is one example of homeostatic regulation. The body also regulates its temperature, its blood pressure, and all other aspects of its blood chemistry in similar ways.

The following paragraphs describe the regulation of digestion and absorption in healthy adults, but many factors◆ can influence normal GI function. For example, peristalsis and sphincter action are poorly coordinated in newborns, so infants tend to "spit up" during the first several months of life. Older adults often experience constipation, in part because the intestinal wall loses strength and elasticity with age, which slows GI motility. Diseases can also interfere with digestion and absorption and often lead to malnutrition. Lack of nourishment, in general, and lack of certain dietary constituents such as fiber, in particular, alter the structure and function of GI cells. Quite simply, GI tract health depends on food.

Health promotion

◆ **Factors influencing GI function:**
- Physical immaturity.
- Aging.
- Illness.
- Nutrition.

◆ **Appendix A presents a brief summary of the body's hormonal system and nervous system.**

Gastrointestinal Hormones and Nerve Pathways

Two intricate and sensitive systems◆ coordinate all the digestive and absorptive processes: the hormonal (or endocrine) system and the nervous system. Even before the first bite of food is taken, the mere thought, sight, or smell of food can trigger a response from these systems.[5] Then, as food travels through the GI tract, it either stimulates or inhibits digestive secretions by way of messages that are carried from one section of the GI tract to another by both **hormones** and nerve pathways.

Notice that the kinds of regulation that will be described are all examples of *feedback* mechanisms. A certain condition demands a response. The response changes that condition, and the change then cuts off the response. Thus the system is self-corrective. Examples follow:

- *The stomach normally maintains a pH between 1.5 and 1.7. How does it stay that way?* Food entering the stomach stimulates cells in the stomach wall to release the hormone **gastrin.** Gastrin, in turn, stimulates the stomach glands to secrete the components of hydrochloric acid. When pH 1.5 is reached, the acid itself turns off the gastrin-producing cells. They stop releasing gastrin, and the glands stop producing hydrochloric acid. Thus the system adjusts itself.

  Nerve receptors in the stomach wall also respond to the presence of food and stimulate both the gastric glands to secrete juices and the muscles to contract. As the stomach empties, the receptors are no longer stimulated, the flow of juices slows, and the stomach quiets down.
- *The pyloric sphincter opens to let out a little chyme, then closes again. How does it know when to open and close?* When the pyloric sphincter relaxes, acidic chyme slips through. The cells of the pyloric muscle on the intestinal side sense the acid, causing the pyloric sphincter to close tightly. Only after the chyme has been neutralized by pancreatic bicarbonate and the juices surrounding the pyloric sphincter have become

homeostasis (HOME-ee-oh-STAY-sis): the maintenance of constant internal conditions (such as blood chemistry, temperature, and blood pressure) by the body's control systems. A homeostatic system is constantly reacting to external forces so as to maintain limits set by the body's needs.
- **homeo** = the same
- **stasis** = staying

hormones: chemical messengers. Hormones are secreted by a variety of glands in response to altered conditions in the body. Each hormone travels to one or more specific target tissues or organs, where it elicits a specific response to maintain homeostasis. In general, any gastrointestinal hormone may be called an **enterogastrone** (EN-ter-oh-GAS-trone), but the term refers specifically to any hormone that slows motility and inhibits gastric secretions.

gastrin: a hormone secreted by cells in the stomach wall. Target organ: the glands of the stomach. Response: secretion of gastric acid.

alkaline can the muscle relax again. This process ensures that the chyme will be released slowly enough to be neutralized as it flows through the small intestine. This is important, because the small intestine has less of a mucous coating than the stomach does and so is not as well protected from acid.

- *As the chyme enters the intestine, the pancreas adds bicarbonate to it so that the intestinal contents always remain at a slightly alkaline pH. How does the pancreas know how much to add?* The presence of chyme stimulates the cells of the duodenum wall to release the hormone **secretin** into the blood. When secretin reaches the pancreas, it stimulates the pancreas to release its bicarbonate-rich juices. Thus, whenever the duodenum signals that acidic chyme is present, the pancreas responds by sending bicarbonate to neutralize it. When the need has been met, the cells of the duodenal wall are no longer stimulated to release secretin, the hormone no longer flows through the blood, the pancreas no longer receives the message, and it stops sending pancreatic juice. Nerves also regulate pancreatic secretions.
- *Pancreatic secretions contain a mixture of enzymes to digest carbohydrate, fat, and protein. How does the pancreas know how much of each type of enzyme to provide?* This is one of the most interesting questions physiologists have asked. The question awaits final answer, but clearly the pancreas does know, somehow, what its owner has been eating, and it secretes enzyme mixtures tailored to deal with the food mixtures that have been arriving lately (over the last several days). Enzyme activity changes proportionately in response to the amounts of carbohydrate, fat, and protein in the diet.[6] If a person has been eating mostly carbohydrates, the pancreas makes and secretes mostly carbohydrases; if the person's diet has been high in fat, the pancreas produces more lipases; and so forth. Presumably, hormones from the GI tract, secreted in response to meals, keep the pancreas informed as to its digestive tasks. The day or two lag between the time a person's diet changes and the time digestion of the new diet becomes efficient explains why dietary changes can "upset digestion" and should be made gradually.
- *When fat is present in the intestine, the gallbladder contracts to squirt bile into the intestine to emulsify the fat. How does the gallbladder get the message that fat is present?* Fat in the intestine stimulates cells of the intestinal wall to release the hormone **cholecystokinin (CCK).** This hormone, traveling by way of the blood to the gallbladder, stimulates it to contract, releasing bile into the small intestine. Once the fat in the intestine is emulsified and enzymes have begun to work on it, the fat no longer provokes release of the hormone, and the message to contract is canceled. (By the way, fat emulsification can continue even after a diseased gallbladder has been surgically removed because the liver can deliver bile directly to the small intestine.)
- *Fat takes longer to digest than carbohydrate does. When fat is present, intestinal motility slows to allow time for its digestion. How does the intestine know when to slow down?* Cholecystokinin and **gastric-inhibitory peptide** slow GI tract motility. By slowing the digestive process, fat helps to maintain a pace that will allow all reactions to reach completion. Gastric-inhibitory peptide also inhibits gastric acid secretion. Hormonal and nervous mechanisms like these account for much of the body's ability to adapt to changing conditions.

Once a person has started to learn the answers to questions like these, it may be hard to stop. Some people devote their whole lives to the study of physiology. For now, however, these few examples will be enough to illustrate how all the processes throughout the digestive system are precisely and automatically regulated without any conscious effort.

To become part of your body, food must first be digested and absorbed.

secretin (see-CREET-in): a hormone produced by cells in the duodenum wall. Target organ: the pancreas. Response: secretion of bicarbonate-rich pancreatic juice.

cholecystokinin (coal-ee-sis-toe-KINE-in), or **CCK:** a hormone produced by cells of the intestinal wall. Target organ: the gallbladder. Response: release of bile and slowing of GI motility.

gastric-inhibitory peptide: a hormone produced by the intestine. Target organ: the stomach. Response: slowing of the secretion of gastric juices and of GI motility.

IN SUMMARY Digestion and absorption depend on the coordinated efforts of the hormonal system and the nervous system. Together, they regulate the processes of transforming foods into nutrients.

The System at Its Best

This chapter has described the anatomy of the digestive tract on several levels: the sequence of digestive organs, the cells and structures of the villi, and the selective machinery of the cell membranes. The intricate architecture of the digestive system makes it sensitive and responsive to conditions in its environment. Knowing the optimal conditions will help you to promote the best functioning of the system.

One indispensable condition is good health of the digestive tract itself. This health is affected by such lifestyle factors as sleep, physical activity, and state of mind. Adequate sleep allows for repair and maintenance of tissue and removal of wastes that might impair efficient functioning. Activity promotes healthy muscle tone. Mental state influences the activity of regulatory nerves and hormones; for healthy digestion, you should be relaxed and tranquil at mealtimes.

Another factor is the kind of meals you eat. Among the characteristics of meals that promote optimal absorption of nutrients are those mentioned in Chapter 2: balance, moderation, variety, and adequacy. Balance and moderation require having neither too much nor too little of anything. For example, too much fat is harmful, but some fat is needed to slow down intestinal motility, providing time for absorption of some of the nutrients that are slow to be absorbed.

Variety is important for many reasons, but partly because some food constituents interfere with nutrient absorption. For example, some compounds common in high-fiber foods such as whole-grain cereals, certain leafy green vegetables, and legumes bind with minerals. To some extent, then, the minerals in those foods may become unavailable. Not that these high-fiber foods are undesirable, but people who use cereals, leafy greens, and legumes to the exclusion of other foods may be obtaining fewer minerals from their diets than they would if their choices were more varied. They might want to exercise moderation in their use of these high-fiber foods.

As for adequacy—in a sense, this entire book is about dietary adequacy. But here, at the end of this chapter, is a good place to underline the interdependence of the nutrients. It could almost be said that every nutrient depends on every other. All the nutrients work together, and all are present in the cells of a healthy digestive tract. To maintain health and promote the functions of the GI tract, you should make balance, moderation, variety, and adequacy features of every day's menus.

How are you doing?

- Do you frequently overeat to the point of discomfort?
- Do you experience GI distress regularly?
- What changes can you make in your eating habits to promote GI health?

Nutrition on the Net

WEBSITES

Access these websites for further study of topics covered in this chapter.

- Find updates and quick links to these and other nutrition-related sites at our website: **www.wadsworth.com/nutrition**
- Visit the Center for Digestive Health and Nutrition: **www.gihealth.com**
- Visit the Digest This! section of the American College of Gastroenterology: **www.acg.gi.org**

INTERNET ACTIVITIES

If you are like most people, you take your digestive system for granted. As long as it is working well, you seldom think of it. Have you ever wondered what the inside of your GI tract looks like? Explore the following website to see what happens when things go wrong inside the GI tract.

- Go to: **www.gihealth.com**
- Click on "Patient Pamphlets."
- Click on "Look Inside Your Body."
- Explore the still pictures and video clips of the GI tract taken during endoscopy examinations.

Following general guidelines for a healthy lifestyle, including sensible eating and regular exercise, will help keep GI tract activities normal. What can you do to keep your GI tract in top condition?

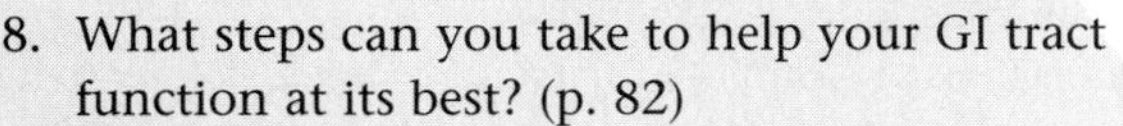

Study Questions

These questions will help you review the chapter. You will find the answers in the discussion on the pages provided.

1. Describe the obstacles associated with digesting food and the solutions offered by the human body. (pp. 65–70)
2. Describe the path food follows as it travels through the digestive system. Summarize the muscular actions that take place along the way. (pp. 67–69)
3. Name five organs that secrete digestive juices. How do the juices and enzymes facilitate digestion? (pp. 69–72)
4. Describe the problems associated with absorbing nutrients and the solutions offered by the small intestine. (pp. 73–77)
5. How is blood routed through the digestive system? Which nutrients enter the bloodstream directly? Which are first absorbed into the lymph? (pp. 77–79)
6. Describe how the body coordinates and regulates the processes of digestion and absorption. (p. 80)
7. How does the composition of the diet influence the functioning of the GI tract? (pp. 80–81)
8. What steps can you take to help your GI tract function at its best? (p. 82)

These questions will help you prepare for an exam. Answers can be found on p. 84.

1. The semiliquid, partially digested food that travels through the intestinal tract is called:
   a. bile.
   b. lymph.
   c. chyme.
   d. secretin.
2. The muscular contractions that move food through the GI tract are called:
   a. hydrolysis.
   b. sphincters.
   c. peristalsis.
   d. bowel movements.
3. The main function of bile is to:
   a. emulsify fats.
   b. catalyze hydrolysis.
   c. slow protein digestion.
   d. neutralize stomach acidity.
4. All blood leaving the GI tract travels first to the:
   a. heart.
   b. liver.

c. kidneys.
d. pancreas.

5. Which nutrients leave the GI tract by way of the lymphatic system?
   a. water and minerals
   b. proteins and minerals
   c. all vitamins and minerals
   d. fats and fat-soluble vitamins

6. Digestion and absorption are coordinated by the:
   a. pancreas and kidneys.
   b. liver and gallbladder.
   c. hormonal system and the nervous system.
   d. vascular system and the lymphatic system.

References

1 The length of the small intestine in living adults is almost 2½ times shorter than at death, when muscles are relaxed and elongated. *Review of Medical Physiology,* ed. W. F. Ganong (Norwalk, Conn.: Appleton & Lange, 1993), pp. 438–465; E. A. Shaffer, Digestive system, physiology, and biochemistry, in *Encyclopedia of Human Biology* (San Diego: Academic Press, 1991), p. 76.

2 R. K. Mittal and D. H. Balaban, The esophagogastric junction, *New England Journal of Medicine* 336 (1997): 924–932.

3 M. B. Roberfroid and coauthors, Colonic microflora: Nutrition and health, *Nutrition Reviews* 53 (1995): 127–130.

4 C. Galperin and E. Gershwin, Immunopathogenesis of gastrointestinal and hepatobiliary diseases, *Journal of the American Medical Association* 278 (1997): 1946–1955.

5 R. D. Mattes, Physiological responses to sensory stimulation by food: Nutritional implications, *Journal of the American Dietetic Association* 97 (1997): 406–410, 413.

6 M. Armand and coauthors, Dietary fat modulates gastric lipase activity in healthy humans, *American Journal of Clinical Nutrition* 62 (1995): 74–80; P. M. Brannon, Adaptation of the exocrine pancreas to diet, *Annual Review of Nutrition* 10 (1990): 85–105.

ANSWERS

Multiple Choice

1. c 2. c 3. a
4. b 5. d 6. c

COMMON DIGESTIVE PROBLEMS

THE FACTS of anatomy and physiology presented in Chapter 3 permit easy understanding of some common problems that occasionally arise in the digestive tract. Food may slip into the air passages instead of the esophagus, causing choking. Bowel movements may be loose and watery, as in diarrhea, or painful and hard, as in constipation. Some people complain about belching, while others are bothered by intestinal gas. Sometimes people develop medical problems such as an ulcer. This highlight describes some of the symptoms and strategies for preventing these common digestive problems (the glossary on p. 86 defines these terms).

CHOKING

A person chokes when a piece of food slips into the **trachea** and cuts off breathing (see Figure H3-1). Food can lodge so securely that it cuts off all air. No sound can be made because the **larynx** is in the trachea and makes sounds only when air is pushed across it. For this reason, it is imperative that everyone learn to recognize the international signal for choking (shown in Figure H3-2).

The choking scenario might read like this. A person is dining in a restaurant with friends. A chunk of food, usually meat, becomes lodged in his trachea so firmly that he cannot make a sound. Often he chooses to suffer alone rather than "make a scene in public." If he tries to communicate distress to his friends, he must depend on pantomime. The friends are bewildered by his antics and become terribly worried when he "faints" after a few minutes without air. They call for an ambulance, but by the time it arrives, he is dead from suffocation.

To help a person who is choking, first ask this critical question: "Can you make any sound at all?" If so, relax. You have time to decide what you can do to help. Whatever you do, don't hit him on the back—the particle may become lodged more firmly in his air passage. If the person cannot make a sound, shout for help and perform the **Heimlich maneuver** (described in Figure H3-2). You would do well to take a life-saving course and practice these techniques, for you will have no time for hesitation once you are called on to perform this death-defying act.

Almost any food can cause choking, although some are cited more often than others: tough meats, hot dogs, nuts, grapes, carrots, marshmallows, hard candies, popcorn, and peanut butter. These foods are particularly difficult for young children to safely chew and swallow. Each year, more than 300 children in the United States choke to death. Always remain alert to the dangers of choking whenever young children are eating. To prevent choking, cut food into small pieces, chew thoroughly before swallowing, don't talk or laugh with food in your mouth, and don't eat when breathing hard.

Taking the time to enjoy a meal can enhance physical and emotional health.

VOMITING

Another common digestive mishap is **vomiting.** Vomiting can be a symptom of many different diseases or may arise in situations that upset the body's equilibrium, such as air or sea travel. For whatever reason, the waves of peristalsis reverse direction, and the contents of the stomach are propelled up through the esophagus to the mouth and expelled.

If vomiting continues long enough or is severe enough, the reverse peristalsis will extend beyond the stomach and carry the contents of the duodenum, with its green bile, into the stomach and then up the esophagus. Although certainly unpleasant and wearying for the nauseated person, vomiting such as this is no cause for alarm. Vomiting is one of the body's adaptive mechanisms to rid itself of something irritating. The best advice is to rest and drink small amounts of fluids as tolerated until the nausea subsides.

A physician's care may be needed, however, when large quantities of fluid are lost from the GI tract, causing dehydration. With massive fluid loss from the GI tract, all of the body's other fluids redistribute themselves so that, eventually, fluid is taken from every cell of the body. Leaving the cells with the fluid are salts that are absolutely

GLOSSARY

acid controllers: medications used to prevent or relieve indigestion by suppressing production of acid in the stomach; also called **H2 blockers.** Common brands include Pepcid AC, Tagamet HB, Zantac 75, and Axid AR.

antacids: medications used to relieve indigestion by neutralizing acid in the stomach. Common brands include Alka-Seltzer, Maalox, Rolaids, and Tums.

belching: the expulsion of gas from the stomach through the mouth.

colitis (ko-LYE-tis): inflammation of the colon.

colonic irrigation: the popular, but potentially harmful practice of "washing" the large intestine with a powerful enema machine.

constipation: the condition of having infrequent or difficult bowel movements.

defecate (DEF-uh-cate): to move the bowels and eliminate waste.
- **defaecare** = to remove dregs

diarrhea: the frequent passage of watery bowel movements.

diverticula (dye-ver-TIC-you-la): sacs or pouches that develop in the weakened areas of the intestinal wall (like bulges in an inner tube where the tire wall is weak).
- **divertir** = to turn aside

diverticulosis (DYE-ver-tic-you-LOH-sis): the condition of having diverticula. About one in every six people in Western countries develops diverticulosis in middle or later life.
- **osis** = condition

diverticulitis (DYE-ver-tic-you-LYE-tis): infected or inflamed diverticula.
- **itis** = infection or inflammation

enemas: solutions inserted into the rectum and colon to stimulate a bowel movement and empty the lower large intestine.

gastroesophageal reflux: the backflow of stomach acid into the esophagus, causing damage to the cells of the esophagus and the sensation of heartburn.

heartburn: a burning sensation in the chest area caused by backflow of stomach acid into the esophagus.

Heimlich (HIME-lick) **maneuver (abdominal thrust maneuver):** a technique for dislodging an object from the trachea of a choking person (see Figure H3-2); named for the physician who developed it.

hemorrhoids (HEM-oh-royds): painful swelling of the veins surrounding the rectum.

hiccups (HICK-ups): repeated cough-like sounds and jerks that are produced when an involuntary spasm of the diaphragm muscle sucks air down the windpipe; also spelled *hiccoughs.*

indigestion: incomplete or uncomfortable digestion, usually accompanied by pain, nausea, vomiting, heartburn, intestinal gas, or belching.
- **in** = not

irritable bowel syndrome: an intestinal disorder of unknown cause. Symptoms include abdominal discomfort and cramping, diarrhea, constipation, or alternating diarrhea and constipation.

larynx: the voice box (see Figure H3-1).

peptic ulcer: an erosion in the mucous membrane of either the stomach (a gastric ulcer) or the duodenum (a duodenal ulcer).

trachea (TRAKE-ee-uh): the windpipe; the passageway from the mouth and nose to the lungs.

ulcer: an erosion in the topmost, and sometimes underlying, layers of cells in an area. See also *peptic ulcer.*

vomiting: expulsion of the contents of the stomach up through the esophagus to the mouth.

essential to the life of the cells, and they must be replaced, which is difficult while the vomiting continues. Intravenous feedings of saline and glucose are frequently necessary while the physician is diagnosing the cause of the vomiting and instituting corrective therapy.

In an infant, vomiting is likely to become serious early in its course, and a physician should be contacted soon after onset. Infants have more fluid between their body cells than adults do, so more fluid can move readily into the digestive tract and be lost from the body. Consequently, the body water of infants becomes depleted and their body salt balance upset faster than in adults.

Self-induced vomiting, such as occurs in bulimia nervosa, also has serious consequences. In addition to fluid and salt imbalances, repeated vomiting can cause irritation and infection of the pharynx, esophagus, and salivary glands; erosion of the teeth; and dental caries. The esophagus may rupture or tear, as may the stomach. Sometimes the eyes become red from pressure during vomiting. Bulimic behavior reflects underlying problems that require intervention. (Bulimia nervosa is discussed in Highlight 9.)

Projectile vomiting is also serious. The contents of the stomach are expelled

FIGURE H3-1 Normal Swallowing and Choking

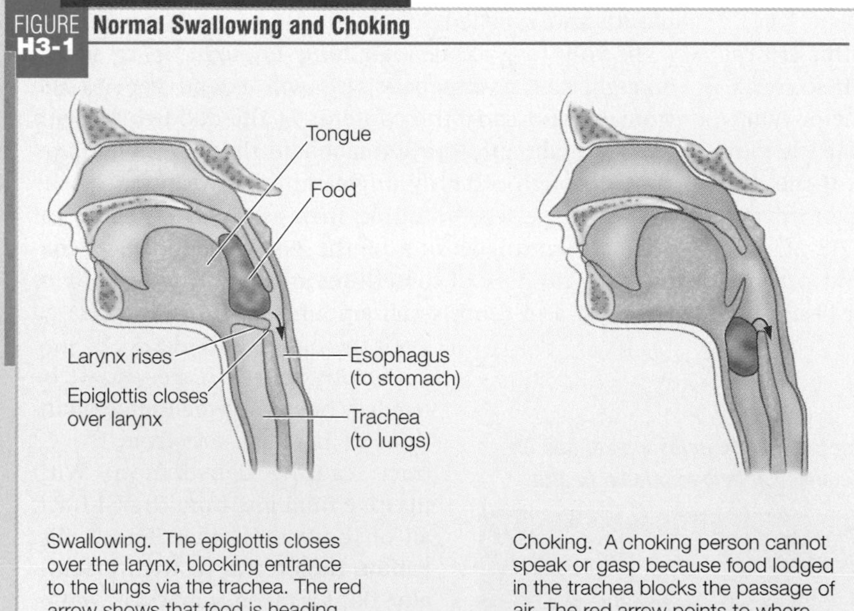

Swallowing. The epiglottis closes over the larynx, blocking entrance to the lungs via the trachea. The red arrow shows that food is heading down the esophagus normally.

Choking. A choking person cannot speak or gasp because food lodged in the trachea blocks the passage of air. The red arrow points to where the food should have gone to prevent choking.

with such force that they leave the mouth in a wide arc like a bullet leaving a gun. This type of vomiting requires immediate medical attention.

DIARRHEA

Diarrhea is characterized by frequent, loose, watery stools. Such stools indicate that the intestinal contents have moved too quickly through the intestines for fluid absorption to take place, or that water has been drawn from the cells lining the intestinal tract and added to the food residue. Like vomiting, diarrhea can lead to considerable fluid and salt losses, but the composition of the fluids is different. Stomach fluids lost in vomiting are highly acidic, whereas intestinal fluids lost in diarrhea are nearly neutral. When fluid losses require medical attention, correct replacement is crucial.

Diarrhea is a symptom of a variety of medical conditions and treatments. It may occur abruptly in a healthy person as a result of infections (such as food poisoning) or as a side effect of medications. When used in large quantities, food ingredients such as the sugar alternative sorbitol and the fat alternative olestra may also cause diarrhea in some people. If a food is responsible, then that food must be omitted from the diet, at least temporarily. If medication is responsible, a different medicine, when possible, or a different form (injectable versus oral, for example) may alleviate the problem.

Diarrhea may also occur as a result of disorders of the GI tract, such as irritable bowel syndrome or colitis. **Irritable bowel syndrome** is one of the most common GI disorders and is characterized by a disturbance in the motility of the GI tract. Dietary treatment hinges on identifying and avoiding individual foods that cause intolerance. For most people, a low-fat diet provided in small meals, with a gradual increase in fiber, is helpful. People with **colitis,** an inflammation of the large intestine, may also suffer from severe diarrhea. They often benefit from complete bowel rest and medication. If treatment fails, surgery to remove the colon and rectum may be necessary.

As you can see, treatment for diarrhea depends on its cause and its severity.[1] Mild diarrhea may remit without treatment; simply rest and drink fluids to replace losses. If diarrhea persists, though, especially in an infant, call a physician. Severe diarrhea can lead to dehydration and electrolyte imbalances. (Chapter 12 provides more information on dehydration and its therapy.)

CONSTIPATION

Like diarrhea, **constipation** describes a symptom, not a disease. Each person's GI tract has its own cycle of waste elimination, which depends on its owner's health, the type of food eaten, when it was eaten, and when the

FIGURE H3-2 **First Aid for Choking**

- The strategy most likely to succeed is abdominal thrusts, sometimes called the Heimlich maneuver.

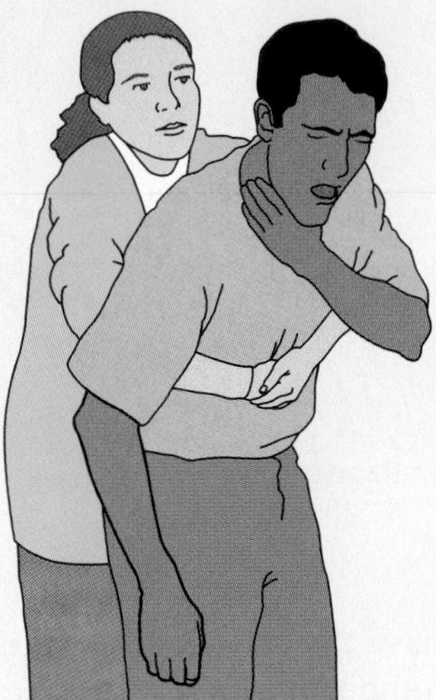

This universal signal for choking alerts others to the need for assistance. Stand behind the person, and wrap your arms around him. Place the thumb side of one fist snugly against his body, slightly above the navel and below the rib cage. Grasp your fist with your other hand and give him a sudden strong hug inward and upward. Repeat thrusts as necessary.

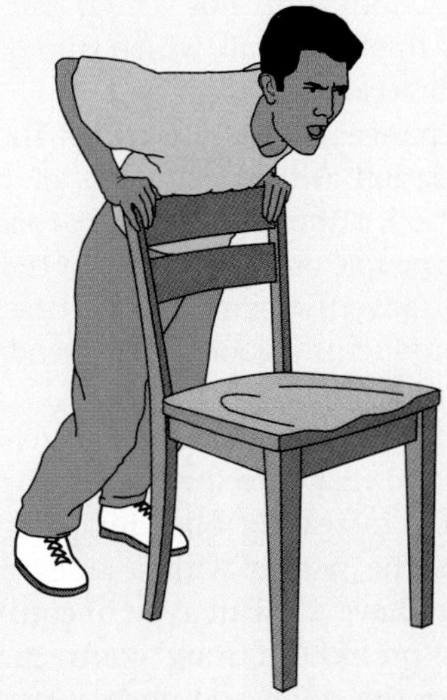

To self-administer first aid, place the thumb side of one fist slightly above the navel and below the rib cage, grasp the fist with your other hand, and then press inward and upward with a quick motion. If this is unsuccessful, quickly press your upper abdomen over any firm surface such as the back of a chair, a countertop, or a railing.

- If all else fails, open the mouth by grasping both the tongue and lower jaw and lifting. Then, and only if you can see the object, use your finger to sweep it out and begin rescue breathing.

person takes time to **defecate.** For some people, bowel movements occur daily; for others, several days may pass between movements. Only when people pass stools that are difficult or painful to expel or they experience a reduced frequency of bowel movements from their typical pattern are they constipated. Abdominal discomfort, headaches, backaches, and the passing of gas sometimes accompany constipation.

Often a person's lifestyle may cause constipation. Being too busy to respond to the defecation signal is a common complaint. If a person receives the signal to defecate and ignores it, the signal may not return for several hours. In the meantime, water continues to be withdrawn from the fecal matter, so when the person does defecate, the bowel movement is dry and hard. In such a case, a person's daily regimen may need to be revised to allow time to have a bowel movement when the body sends its signal. One possibility is to go to bed earlier in order to rise earlier, allowing ample time for a leisurely breakfast and a movement.

Another cause of constipation is lack of physical activity. Physical activity improves muscle tone, not just of the outer body, but also of the digestive tract.

Although constipation usually reflects lifestyle habits, in some cases it may be a side effect of medication or may reflect a medical problem such as tumors that are obstructing the passage of waste. If discomfort is associated with passing fecal matter, seek medical advice to rule out disease. Once this has been done, dietary or other measures for correction can be considered.

One dietary measure that may be appropriate is to increase dietary fiber. Fibers found in cereal products help to prevent constipation by increasing fecal mass. In the GI tract, fiber attracts water, creating soft, bulky stools that stimulate bowel contractions to push the contents along. These contractions strengthen the intestinal muscles. The improved muscle tone, together with the water content of the stools, eases elimination, reducing the pressure in the rectal veins and helping to prevent **hemorrhoids.** Chapter 4 provides more information on fiber's role in maintaining a healthy colon and reducing the risks of colon cancer and diverticulosis. **Diverticulosis** is a condition in which the intestinal walls develop bulges in weakened areas, most commonly in the colon (see Figure H3-3). These bulging pockets, known as **diverticula,** can worsen constipation, entrap feces, and become painfully infected and inflamed **(diverticulitis).** Treatment may require hospitalization, antibiotics, or surgery.

Drinking plenty of water in conjunction with eating high-fiber foods also helps with constipation. The increased bulk physically stimulates the upper GI tract, promoting peristalsis throughout.

Eating prunes can also be helpful. Prunes are high in fiber and also contain a laxative substance.* If a morning defecation is desired, a person can drink prune juice at bedtime; if the evening is preferred, the person can drink prune juice with breakfast.

FIGURE H3-3 **Diverticula**

Diverticula may develop anywhere along the GI tract, but are most common in the colon.

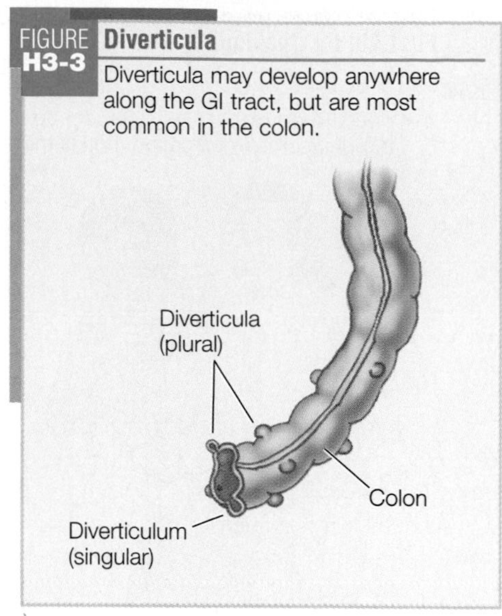

Honey can also have a laxative effect due to its incomplete absorption.[2] Although this characteristic may cause problems for people with irritable bowel syndrome, eating honey may be an easy and effective treatment for those who are constipated. Honey should never be fed to infants, however, because of the risk of botulism (as explained in Chapter 15).

Adding fat to the diet can relieve some constipation by stimulating the hormone cholecystokinin, which summons bile into the duodenum. Bile's high salt content draws water from the intestinal wall, which stimulates peristalsis and softens the fecal matter.

These suggested changes in lifestyle or diet should correct chronic constipation without the use of laxatives, **enemas,** or mineral oil, although television commercials often try to persuade people otherwise. One of the fallacies often perpetrated by advertisements is that one person's successful use of a product is a good recommendation for others to use that product.

As a matter of fact, even diet changes that relieve constipation for one person may increase the constipation of another. For instance, increasing fiber intake stimulates peristalsis and helps the person with a sluggish colon. Some people, though, have a spastic type of constipation, in which peristalsis promotes strong contractions that close off a segment of the colon and prevent passage; for these people, increasing fiber intake would be exactly the wrong thing to do.

A person who seems to need products such as laxatives should seek a physician's opinion. Advice from friends or alternative medicine practitioners may cause more harm than good. One potentially harmful but currently popular practice that is being promoted by some alternative medicine practitioners is **colonic irrigation**—the internal washing of the large intestine with a powerful enema machine. Such an extreme cleansing is not only unnecessary, but the force of the machine can rupture the intes-

*This substance is dihydroxyphenyl isatin.

tine. Less extreme practices can cause problems, too. Frequent use of laxatives and enemas can lead to dependency; upset the body's fluid, salt, and mineral balances; and, in the case of mineral oil, interfere with the absorption of fat-soluble vitamins. (Mineral oil dissolves the vitamins, but is not itself absorbed; instead, it leaves the body, carrying the vitamins with it.)

BELCHING AND GAS

Many people complain of problems that they attribute to excessive gas. For some, **belching** is the complaint. Others blame intestinal gas for abdominal discomforts and embarrassment. Most people believe that the problems occur after they eat certain foods. This may be the case with intestinal gas, but belching results from swallowing air. The best advice for belching seems to be to eat slowly, chew thoroughly, and relax while eating.

Everyone swallows a little bit of air with each mouthful of food, but people who eat too fast may swallow too much air and then have to belch. Ill-fitting dentures, carbonated beverages, and chewing gum can also contribute to the swallowing of air with resultant belching. Occasionally, belching can be a sign of a more serious disorder, such as gallbladder disease or a peptic ulcer.

People who eat or drink too fast may also trigger **hiccups,** the repeated spasms that produce a cough-like sound and jerky movement. Normally, hiccups soon subside and are of no medical significance, but they can be bothersome. The most effective cure is to hold the breath for as long as possible, which helps to relieve the spasms of the diaphragm.

While expelling gas can be a humiliating experience, it is quite normal. (People experiencing painful bloating from malabsorption diseases, however, require medical treatment.) Healthy people expel several hundred milliliters of gas several times a day. Almost all (99 percent) of the gases expelled—nitrogen, oxygen, hydrogen, methane, and carbon dioxide—are odorless. The remaining "volatile" gases are the infamous ones.

Foods that produce gas usually must be determined individually. The most common offenders are foods rich in the carbohydrates—sugars, starches, and fibers. When partially digested carbohydrates reach the large intestine, bacteria digest them, giving off gas as a byproduct. People can test foods suspected of forming gas by omitting them individually for a trial period and seeing if there is any improvement.

Beans, broccoli, cabbage, and onions produce gas in many people. People troubled by gas need to determine which foods bother them and then eat those foods in moderation.

HEARTBURN AND "ACID INDIGESTION"

Almost everyone has experienced heartburn at one time or another, usually soon after eating a meal. **Heartburn** is the painful sensation a person feels behind the breastbone that occurs when the lower esophageal sphincter fails to prevent the stomach contents from refluxing into the esophagus. This may happen if a person eats or drinks too much (or both). Tight clothing and even changes of position (lying down, bending over) can cause it, too, as can some medications and smoking. A defect of the sphincter muscle itself is a possible, but less common cause.

If the heartburn is not caused by an anatomical defect, treatment is fairly simple. To avoid such misery in the future, the person needs to learn to eat less at a sitting, chew food more thoroughly, and eat it more slowly.

As far as "acid indigestion" is concerned, recall from Chapter 3 that the strong acidity of the stomach is a desirable condition—television commercials for **antacids** and **acid controllers** notwithstanding. People who overeat or eat too quickly are likely to suffer from **indigestion.** The muscular reaction of the stomach to unchewed lumps or to being overfilled may be so violent that it causes regurgitation (reverse peristalsis). When this happens, overeaters may taste the stomach acid and feel pain. Responding to advertisements, they may reach for antacids or acid controllers. Both of these drugs were originally designed to treat GI illnesses such as ulcers. As is true of most over-the-counter medicines, antacids and acid controllers should be used only infrequently for occasional heartburn; they may mask or cause problems if used regularly, as the next section explains. Instead of self-medicating, people who suffer from frequent and regular bouts of heartburn and indigestion need to see a physician, who can prescribe specific medication to control **gastroesophageal reflux.** Without treatment, the repeated splashes of acid can severely damage the cells of the esophagus, creating a condition known as Barrett's esophagus. At that stage, the risk of cancer in the throat or esophagus increases dramatically. To repeat, if symptoms persist, see a doctor—don't self-medicate.

ULCERS

Ulcers of the stomach (gastric ulcers) or duodenum (duodenal ulcers) are another common digestive problem. (The term **peptic ulcer** includes both types.) An **ulcer** is an erosion of the top layer of cells from an area, such as the wall of the stomach or duodenum.

This erosion leaves the underlying layers of cells unprotected and exposed to gastric juices. The erosion may proceed until the gastric juices reach the capillaries that feed the area, leading to bleeding, and reach the nerves, causing pain. If GI bleeding is excessive, iron deficiency may develop.[3] If the erosion penetrates all the way through the GI lining, a life-threatening infection can develop.

Many people naively believe that an ulcer is caused by stress or spicy foods, but this is not the case—at least not at first.[4] The stomach lining in a healthy person is well protected by its mucous coat. What, then, causes ulcers to form?

Three major causes of ulcers have been identified: bacterial infection with *Helicobacter pylori,* the use of certain anti-inflammatory drugs such as ibuprofen and naproxen, and disorders that cause excessive gastric acid secretion. The cause of the ulcer dictates the type of medication used in treatment.[5] For example, people with ulcers caused by infection receive antibiotics, whereas those with ulcers caused by medicines discontinue their use. In addition, all treatment plans aim to relieve pain, heal the ulcer, and prevent recurrence.

Diet therapy once played a major role in ulcer treatment, but it no longer does. Current practice is simply to treat for infection, eliminate any food that routinely causes indigestion or pain, and avoid coffee and caffeine- and alcohol-containing beverages. Both regular and decaffeinated coffee stimulate acid secretion and so aggravate *existing* ulcers.

Ulcers and their treatments highlight the importance of not self-medicating when symptoms persist. People with *H. pylori* infection often take over-the-counter acid controllers to relieve the pain of their ulcers when they need physician-prescribed antibiotics instead. Suppressing gastric acidity not only fails to heal the ulcer, but actually worsens inflammation during an *H. pylori* infection.[6] Furthermore, *H. pylori* infection has been linked with stomach cancer as well, making prompt diagnosis and appropriate treatment most important.[7]

TABLE H3-1 Strategies to Prevent or Alleviate Common GI Problems

GI Problem	Strategies
Choking	• Take small bites of food. • Chew thoroughly before swallowing. • Don't talk or laugh with food in your mouth. • Don't eat when breathing hard.
Diarrhea	• Rest. • Drink fluids to replace losses. • Call for medical help if diarrhea persists.
Constipation	• Eat a high-fiber diet. • Drink plenty of fluids. • Exercise regularly. • Respond promptly to the urge to defecate.
Belching	• Eat slowly. • Chew thoroughly. • Relax while eating.
Intestinal gas	• Eat bothersome foods in moderation.
Heartburn	• Eat small meals. • Drink liquids between meals. • Sit up while eating. • Wait 1 hour after eating before lying down. • Wait 2 hours after eating before exercising. • Refrain from wearing tight-fitting clothing. • Avoid foods, beverages, and medications that aggravate your heartburn. • Refrain from smoking cigarettes. • Lose weight if overweight.
Ulcer	• Take medicine as prescribed by your physician. • Avoid coffee and caffeine- and alcohol-containing beverages. • Avoid foods that aggravate your ulcer. • Minimize aspirin use. • Refrain from smoking cigarettes.

Note: Chapters 20 and 21 present clinical interventions for disorders of the upper and lower GI tract.

Table H3-1 summarizes strategies to prevent or alleviate common GI problems. Many of these problems reflect hurried lifestyles. For this reason, many of their remedies require that people slow down and take the time to eat leisurely; chew food thoroughly to prevent choking, heartburn, and acid indigestion; rest until vomiting and diarrhea subside; and heed the urge to defecate. In addition, learn how to handle life's day-to-day problems and challenges without overreacting and becoming upset; learn how to relax, to get enough sleep, and to enjoy life. Remember, "what's eating you" may cause more GI distress than what you eat.

Nutrition on the Net

•WEBSITES•

Access these websites for further study of topics covered in this highlight.

- Find updates and quick links to these and other nutrition-related sites at our website: **www.wadsworth.com/nutrition**
- Search for "choking," "vomiting," "diarrhea," "constipation," "heartburn," "indigestion," and "ulcers" at the U.S. Government health information site: **www.healthfinder.gov**
- Visit the Center for Digestive Health and Nutrition: **www.gihealth.com**
- Visit the Digestive Diseases section of the National Institute of Diabetes, Digestive, and

Kidney Diseases: **www.niddk.nih.gov/health/health.htm**

- Visit the Digest This! section of the American College of Gastroenterology: **www.acg.gi.org**
- Learn more about *H. pylori* from the Helicobacter Foundation: **www.helico.com**

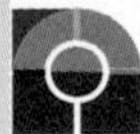

References

1 M. Donowitz, F. T. Kokke, and R. Saidi, Evaluation of patients with chronic diarrhea, *New England Journal of Medicine* 332 (1995): 725–729.

2 S. D. Ladas, D. N. Haritos, and S. A. Raptis, Honey may have a laxative effect on normal subjects because of incomplete fructose absorption, *American Journal of Clinical Nutrition* 62 (1995): 1212–1215.

3 R. Yip and coauthors, Pervasive occult gastrointestinal bleeding in an Alaska native population with prevalent iron deficiency: Role of *Helicobacter pylori* gastritis, *Journal of the American Medical Association* 277 (1997): 1135–1139.

4 Knowledge about causes of peptic ulcer disease—United States, March-April 1997, *Morbidity and Mortality Weekly Report* 46 (1997): 985–987.

5 M. S. Khuroo and coauthors, A comparison of omeprazole and placebo for bleeding peptic ulcer, *New England Journal of Medicine* 336 (1997): 1054–1058; A. H. Soll for the Practice Parameters Committee of the American College of Gastroenterology, Medical treatment of peptic ulcer disease: Practice guidelines, *Journal of the American Medical Association* 275 (1996): 622–629.

6 E. J. Kuipers and coauthors, Atrophic gastritis and *Helicobacter pylori* infection in patients with reflux esophagitis treated with omeprazole or fundoplication, *New England Journal of Medicine* 334 (1996): 1018–1022.

7 L. E. Hansson and coauthors, The risk of stomach cancer in patients with gastric or duodenal ulcer disease, *New England Journal of Medicine* 335 (1996): 242–249.

The Carbohydrates: Sugars, Starches, and Fibers

CHAPTER 4

A student, quietly studying a textbook, is seldom aware that within his brain cells, billions of glucose molecules are splitting each second to provide the energy that permits him to learn. Yet glucose provides nearly all of the energy the human brain uses daily. Similarly, a marathon runner, bursting across the finish line in an explosion of sweat and triumph, seldom gives thanks to the glycogen fuel her muscles have devoured to help her finish the race. Yet, together, glucose and its storage form glycogen provide about half of all the energy muscles and other body tissues use. The other half of the body's energy comes mostly from fat.

People don't eat glucose and glycogen directly; they eat foods rich in **carbohydrates.** Then their bodies convert the carbohydrates mostly into glucose for immediate energy and into glycogen for reserve energy.

Many people mistakenly think of carbohydrates as "fattening" and avoid them when trying to lose weight. Actually, such a strategy may be counterproductive. People can better control body weight by selecting high-carbohydrate, high-fiber foods and limiting fat-rich foods. All unrefined plant foods—whole grains, vegetables, legumes, and fruits—provide ample carbohydrate and fiber with little or no fat. Milk also contains carbohydrates. (So do shellfish and organ meats such as liver, but only a little.)

FIGURE 4-1 Atoms and Their Bonds

The four main types of atoms found in nutrients are hydrogen (H), oxygen (O), nitrogen (N), and carbon (C). Appendix B presents basic chemistry terms and relationships.

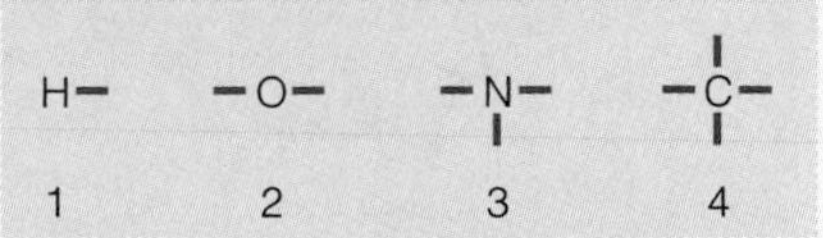

H–	–O–	–N– (with a third bond)	–C– (with four bonds)
1	2	3	4

Each atom has a characteristic number of bonds it can form with other atoms.

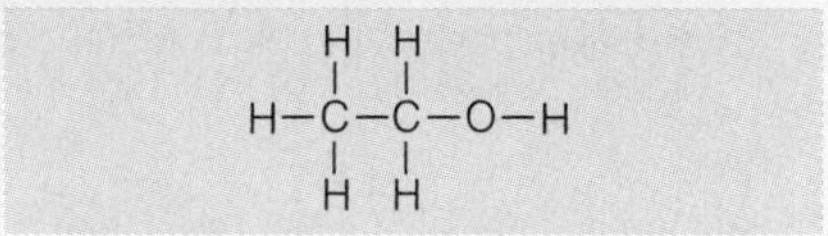

```
  H H
  | |
H-C-C-O-H
  | |
  H H
```

Ethyl alcohol, a simple molecule showing bonding.

THE CHEMIST'S VIEW OF CARBOHYDRATES

The dietary carbohydrate family includes the **simple carbohydrates** (the sugars) and the **complex carbohydrates** (the starches and fibers). The simple carbohydrates are those that chemists describe as:

- Monosaccharides—single sugars.
- Disaccharides—sugars composed of pairs of monosaccharides.

The complex carbohydrates are:

- Polysaccharides—large molecules composed of chains of monosaccharides.

To understand the structure of carbohydrates, look at the units of which they are made. The sugars most important in nutrition are the 6-carbon monosaccharides known as hexoses.◆ Each contains 6 carbon atoms, 12 hydrogens, and 6 oxygens (written in shorthand as $C_6H_{12}O_6$).

Each atom can form a certain number of chemical bonds with other atoms:

- Carbon atoms can form four bonds.
- Nitrogen atoms, three.
- Oxygen atoms, two.
- Hydrogen atoms, only one.

◆ **Most of the monosaccharides important in nutrition are hexoses, simple sugars with six atoms of carbon and the formula $C_6H_{12}O_6$.**
• hex = six

Chemists represent the bonds as lines between the chemical symbols (such as C, N, O, and H) that stand for the atoms (see Figure 4-1).

Atoms form molecules in ways that satisfy the bonding requirements of each atom. Figure 4-1 includes the structure of ethyl alcohol, the active ingredient of alcoholic beverages, as an example. The two carbons each have four bonds represented by lines; the oxygen has two; and each hydrogen has one bond connecting it to other atoms. An accurate drawing of a chemical structure must obey these rules because the laws of nature demand it.

IN SUMMARY The carbohydrates are made of carbon (C), oxygen (O), and hydrogen (H). Each of these atoms can form a specified number of chemical bonds: carbon forms four, oxygen forms two, and hydrogen forms one.

carbohydrates: compounds composed of carbon, oxygen, and hydrogen arranged as monosaccharides or multiples of monosaccharides. Most, but not all, carbohydrates have a ratio of one carbon molecule to one water molecule: $(CH_2O)_n$.
• **carbo** = carbon (C)
• **hydrate** = with water (H_2O)

simple carbohydrates (sugars): monosaccharides and disaccharides.

complex carbohydrates (starches and **fibers):** polysaccharides composed of straight or branched chains of monosaccharides.

FIGURE 4-2 Chemical Structure of Glucose

On paper, the structure of glucose has to be drawn flat, but in nature the five carbons and oxygen are roughly in a plane. The atoms attached to the ring carbons extend above and below the plane.

THE SIMPLE CARBOHYDRATES

The following list of the simple carbohydrates most important in nutrition symbolizes them as hexagons and pentagons of different colors. Three are monosaccharides:

- Glucose.
- Fructose.
- Galactose. 

Three are disaccharides:

- Maltose (glucose + glucose).
- Sucrose (glucose + fructose).
- Lactose (glucose + galactose).

Monosaccharides

The three **monosaccharides** important in nutrition all have the same numbers and kinds of atoms, but in different arrangements. These chemical differences account for the differing sweetness of the monosaccharides. A pinch of purified glucose on the tongue gives only a mild sweet flavor, and galactose hardly tastes sweet at all, but fructose is as intensely sweet as honey and, in fact, is the sugar primarily responsible for honey's sweetness.

• **Glucose** • Chemically, **glucose** is a larger and more complicated molecule than ethyl alcohol, but it obeys the same rules of chemistry: each carbon atom has four bonds; each oxygen, two bonds; and each hydrogen, one bond. Figure 4-2 illustrates the chemical structure of a glucose molecule.

The diagram of a glucose molecule shows all the relationships between the atoms and proves simple on examination, but chemists have adopted even simpler ways to depict chemical structures. Figure 4-3 shows that a chemical structure can combine or omit several symbols without losing the information it conveys.

Commonly known as blood sugar, glucose serves as an essential energy source for all the body's activities. Its significance to nutrition is tremendous. Glucose is one of the two sugars in every disaccharide and is the unit from which the polysaccharides are made almost exclusively. One of these polysaccharides, starch, is the chief food source of energy for the world's people; another, glycogen, is a major storage form of energy in the body. Glucose reappears frequently throughout this chapter and all those that follow.

• **Fructose** • **Fructose** is the sweetest of the sugars. Curiously, fructose has exactly the same chemical *formula* as glucose—$C_6H_{12}O_6$—but its *structure* differs (see Figure 4-4).◆ The arrangement of the atoms in fructose stimulates the taste buds on the tongue to produce the sweet sensation. Fructose occurs naturally in fruits and honey; other sources include products such as soft drinks, ready-to-cereals, and desserts that have been sweetened with high-fructose corn syrup (HFCS).

◆ **Fructose is shown as a pentagon, but it does have 6 carbon atoms. The ring contains 4 carbons and an oxygen; 2 carbons stick out from the ring (see Figure 4-4).**

• **Galactose** • Seldom occurring free in nature,◆ **galactose** binds with glucose to form the sugar in milk. Galactose has the same numbers and kinds of atoms as glucose and fructose in yet another arrangement. Figure 4-5 on p. 96 shows galactose beside a molecule of glucose for comparison.

◆ **Galactose occurs only as a part of lactose.**

Disaccharides

The **disaccharides** are pairs of the three monosaccharides just discussed. Glucose occurs in all three; the second member of the pair is either fructose, galac-

monosaccharides (mon-oh-SACK-uh-rides): carbohydrates of the general formula $C_nH_{2n}O_n$ that consist of a single ring. See Appendix C for the chemical structures of the monosaccharides.
- **mono** = one
- **saccharide** = sugar

glucose (GLOO-kose): a monosaccharide; sometimes known as blood sugar or **dextrose.**
- **ose** = carbohydrate
- = glucose

fructose (FRUK-tose or FROOK-tose): a monosaccharide. Sometimes known as fruit sugar or **levulose,** fructose is found abundantly in fruits, honey, and saps.
- **fruct** = fruit
- = fructose

galactose (ga-LAK-tose): a monosaccharide; part of the disaccharide lactose.
- = galactose

disaccharides (dye-SACK-uh-rides): pairs of monosaccharides linked together. See Appendix C for the chemical structures of the disaccharides.
- **di** = two

FIGURE 4-3 **Simplified Diagrams of Glucose**

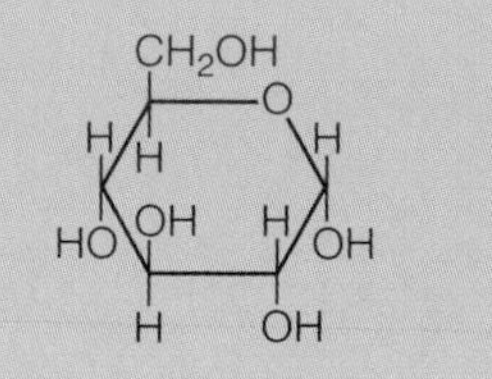

The lines representing some of the bonds and the carbons at the corners are not shown; the formula CH_2OH stands for the structure in Figure 4-2.

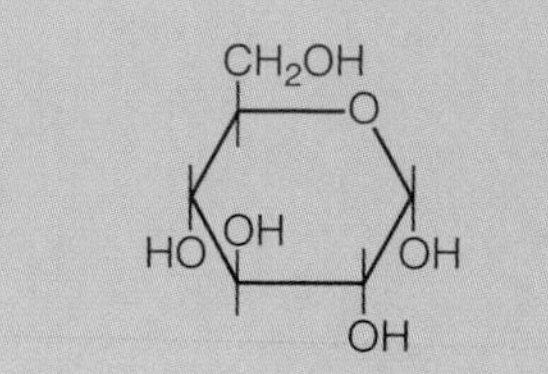

Now the single hydrogens are not shown, but lines still extend upward or downward from the ring to show where they belong.

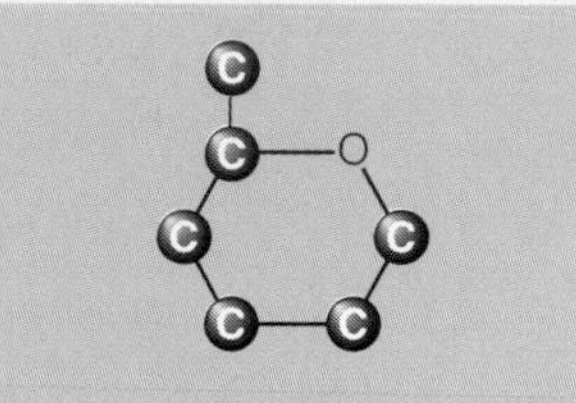

Another way to look at glucose is to notice that its six carbon atoms are all connected.

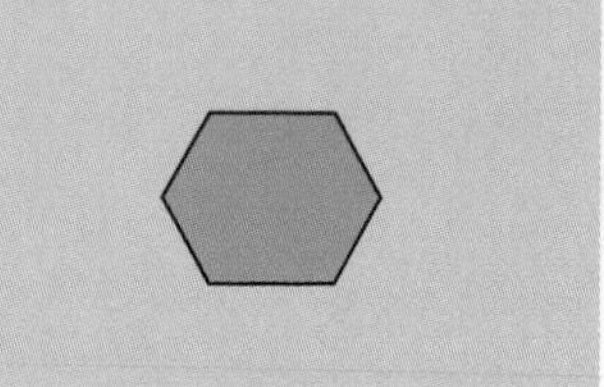

In this and other illustrations throughout this book, glucose is represented as a blue hexagon.

tose, or another glucose. These carbohydrates and all the other energy nutrients are put together and taken apart by similar chemical reactions: condensation and hydrolysis.

• **Condensation** • To make a disaccharide, a chemical reaction known as **condensation** links two monosaccharides together (see Figure 4-6 on p. 96). A hydroxyl (OH) group from one monosaccharide and a hydrogen atom (H) from the other combine to create a molecule of water (H_2O). The two originally separate monosaccharides link together with a single oxygen (O).

• **Hydrolysis** • To break a disaccharide in two, a chemical reaction known as hydrolysis◆ occurs (see Figure 4-7 on p. 97). A molecule of water splits to provide the H and OH needed to complete the resulting monosaccharides. Hydrolysis reactions commonly occur during digestion.

• **Maltose** • The disaccharide **maltose** consists of two glucose units. Maltose is produced whenever starch breaks down—as happens in plants when seeds germinate and in human beings during carbohydrate digestion. It also occurs during the fermentation process that yields alcohol. Maltose is only a minor constituent of a few foods.

• **Sucrose** • Fructose and glucose together form **sucrose.** Because the fructose is in a position accessible to the taste receptors, sucrose tastes sweet, accounting for some of the natural sweetness of fruits, vegetables, and grains. To make table sugar, sucrose is refined from the juices of sugarcane and sugar

Fruits package their simple sugars with fibers, vitamins, and minerals, making them a sweet and healthy snack.

◆ Reminder: A *hydrolysis* reaction splits a molecule into two, with H added to one and OH to the other (from water).

FIGURE 4-4 **Two Monosaccharides: Glucose and Fructose**

Glucose

Fructose

Can you see the similarities? If you learned the rules in Figure 4-3, you will be able to "see" 6 carbons (numbered), 12 hydrogens (those shown plus one at the end of each single line), and 6 oxygens in both these compounds.

condensation: a chemical reaction in which two reactants combine to yield a larger product.

maltose (MAWL-tose): a disaccharide composed of two glucose units; sometimes known as malt sugar.

- = maltose

sucrose (SUE-krose): a disaccharide composed of glucose and fructose; commonly known as table sugar, beet sugar, or cane sugar. Sucrose also occurs in many fruits and some vegetables and grains.

- **sucro** = sugar
-  = sucrose

FIGURE 4-5 Two Monosaccharides: Glucose and Galactose

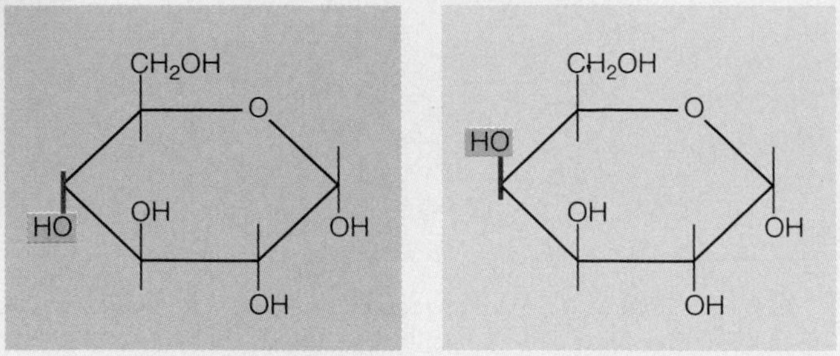

Glucose

Galactose

Notice the similarities and the difference (highlighted in red).

beets, then granulated. Depending on the extent to which it is refined, the product becomes the familiar brown, white, and powdered sugars available at grocery stores.

• **Lactose** • The combination of galactose and glucose makes the disaccharide **lactose,** the principal carbohydrate of milk. Known as milk sugar, lactose contributes about 5 percent of milk's weight. Depending on the milk's fat content, lactose contributes 30 to 50 percent of milk's energy.

IN SUMMARY

Six simple carbohydrates, or sugars, are important in nutrition. The three monosaccharides (glucose, fructose, and galactose) all have the same chemical formula ($C_6H_{12}O_6$), but their structures differ. The three disaccharides (sucrose, lactose, and maltose) are pairs of monosaccharides, each containing a glucose paired with one of the three monosaccharides. The sugars derive primarily from plants, except for lactose and its component galactose, which come from milk and milk products. Two monosaccharides can be linked together by a condensation reaction to form a disaccharide and water. A disaccharide, in turn, can be broken into its two monosaccharides by a hydrolysis reaction using water.

FIGURE 4-6 Condensation of Two Monosaccharides to Form a Disaccharide

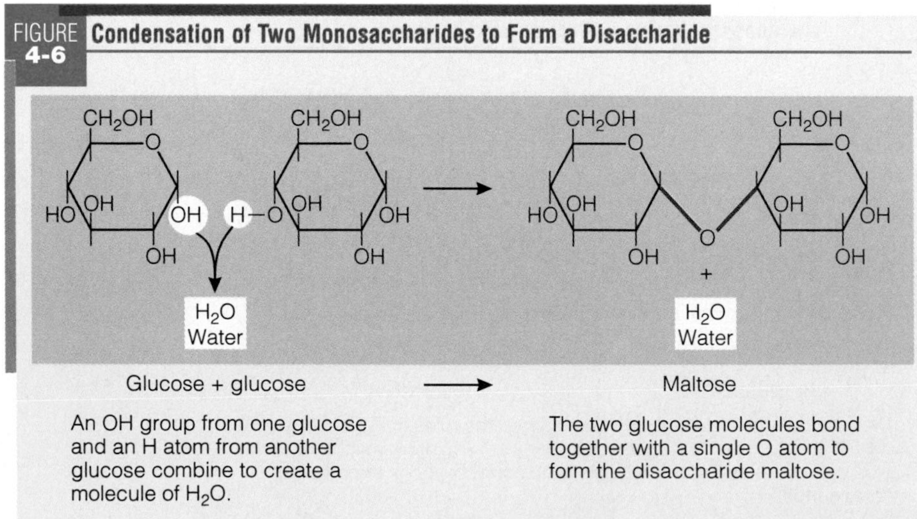

Glucose + glucose → Maltose

An OH group from one glucose and an H atom from another glucose combine to create a molecule of H_2O.

The two glucose molecules bond together with a single O atom to form the disaccharide maltose.

lactose (LAK-tose): a disaccharide composed of glucose and galactose; commonly known as milk sugar.

- **lact** = milk
-  = lactose

FIGURE 4-7 **Hydrolysis of a Disaccharide**

Hydrolysis occurs during digestion.

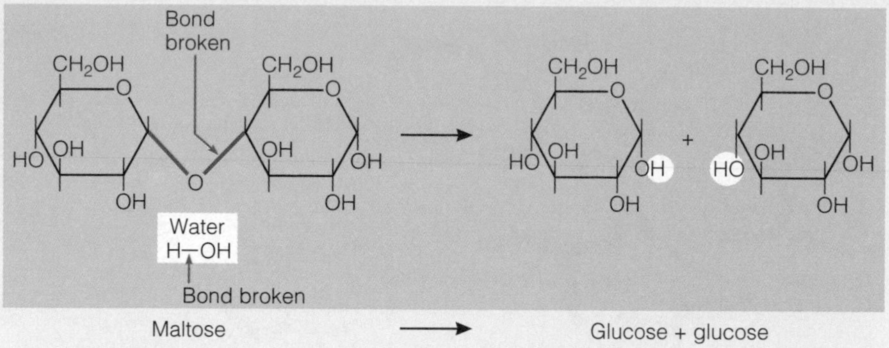

The disaccharide maltose splits into two glucose molecules with H added to one and OH to the other (from water).

THE COMPLEX CARBOHYDRATES

The simple carbohydrates are the sugars just mentioned—glucose, fructose, and galactose—either singly or paired with glucose. In contrast, the complex carbohydrates contain many glucose units and, in some cases, a few other monosaccharides strung together as **polysaccharides.** Three polysaccharides are important in nutrition: glycogen, starches, and fibers.

Glycogen is a storage form of energy in the animal body; starches play that role in plants; and fibers provide structure in stems, trunks, roots, leaves, and skins of plants. Both glycogen and starch are built of glucose units, but they are linked together differently. Fibers are composed of a variety of monosaccharides and other carbohydrate derivatives.

Glycogen

Glycogen is found only to a limited extent in meats and not at all in plants.* For this reason, glycogen is not a significant food source of carbohydrate, but it does perform an important role in the body. The human body stores much of its glucose as glycogen—many glucose molecules linked together in highly branched chains (see the left side of Figure 4-8, p. 98). This arrangement permits rapid hydrolysis. When the hormonal message "Release energy" arrives at the storage sites in a liver or muscle cell, enzymes respond by attacking all the many branches of each glycogen simultaneously, making a surge of glucose available.†

Starches

Just as the human body stores glucose as glycogen, plant cells store glucose as **starches**—long, branched or unbranched chains of hundreds or thousands of glucose molecules linked together (see the middle and right side of Figure 4-8 on p. 98). These giant molecules are packed side by side in grains such as wheat or rice, in tubers such as potatoes, and in legumes such as peas and beans. A cubic inch of food may contain as many as a million starch molecules. When you eat the plant, your body hydrolyzes the starch to glucose and uses the glucose for its own energy purposes.

*Glycogen in animal muscles rapidly hydrolyzes after slaughter.
†Normally, only the liver can return glucose *directly* from glycogen to the blood; muscle cells use glycogen internally to produce glucose. Muscle cells can restore the blood glucose level *indirectly*, however, as Chapter 7 explains.

polysaccharides: compounds composed of many monosaccharides linked together. An intermediate string of three to ten monosaccharides is an **oligosaccharide.**
- **poly** = many
- **oligo** = few

glycogen (GLY-co-gen): an animal polysaccharide composed of glucose; manufactured and stored in the liver and muscles as a storage form of glucose. Glycogen is not a significant food source of carbohydrate and is not counted as one of the complex carbohydrates in foods.
- **glyco** = glucose
- **gen** = gives rise to

starches: plant polysaccharides composed of glucose.

FIGURE 4-8 **Glycogen and Starch Molecules Compared (Small Segments)**

Notice the more highly branched the structure, the greater the number of ends from which glucose can be released. (These units would have to be magnified millions of times to appear at the size shown in this figure. For details of the chemical structures, see Appendix C.)

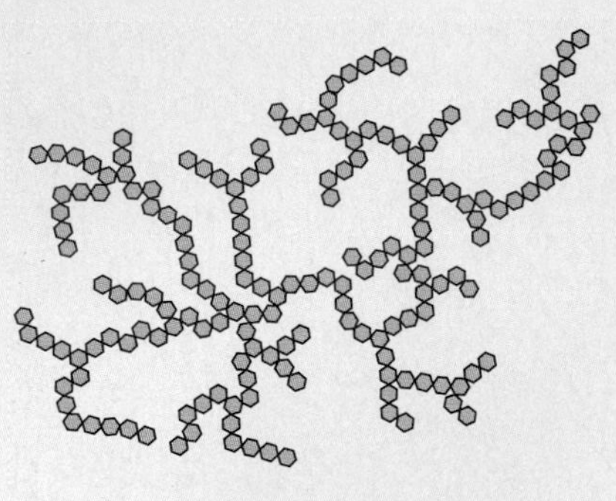

Glycogen

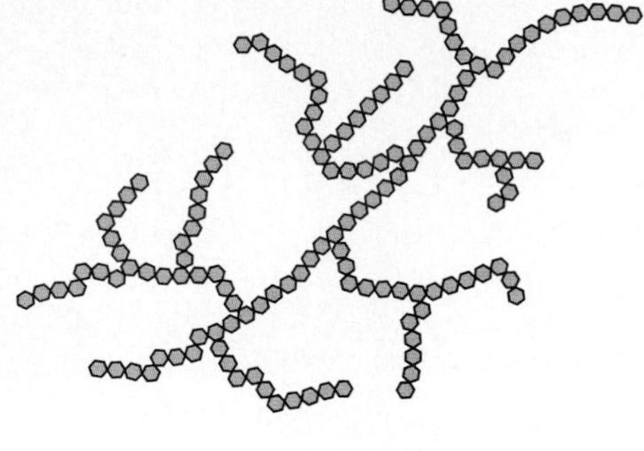

Starch (amylopectin)

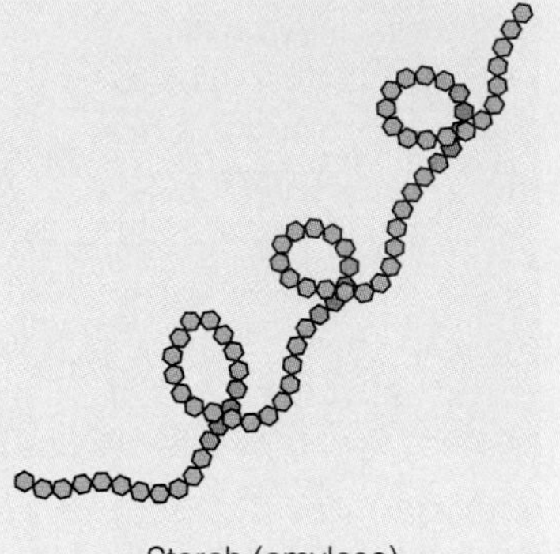

Starch (amylose)

A glycogen molecule contains hundreds of glucose units in long, highly branched chains.

A starch molecule contains hundreds of glucose molecules in either occasionally branched chains (amylopectin) or unbranched chains (amylose).

© Polara Studios Inc.

Major sources of starch include grains, legumes, and tubers (such as potatoes, yams, and cassava).

All starchy foods derive from plants. Grains are the richest food source of starch, providing much of the food energy for people all over the world—rice in Asia; wheat in Canada, the United States, and Europe; corn in much of Central and South America; and millet, rye, barley, and oats elsewhere.[1] Legumes and tubers, such as potatoes, yams, and cassava, are also important sources of starch.

Fibers

Fibers are the structural parts of plants and thus are found in all plant-derived foods—vegetables, fruits, grains, and legumes. Most fibers are polysaccharides. Starch, however, is not a fiber; in fact, fibers are often described as *nonstarch polysaccharides*. Nonstarch polysaccharides include cellulose, hemicelluloses, pectins, gums, and mucilages. Fibers also include some *nonpolysaccharides* such as lignins, cutins, and tannins.

Fibers differ from starches in that the bonds between their monosaccharides cannot be broken down by human digestive enzymes. Consequently, fibers do not contribute monosaccharides to the body. The bacteria of the GI tract can break some fibers down, however, and this is important to digestion and to health.

Each of the fibers has a different structure. Most contain monosaccharides, but differ in the types they contain and in the bonds that link the monosaccharides to each other. These differences produce diverse health effects.

• Cellulose • Cellulose is the primary constituent of plant cell walls and therefore occurs in all vegetables, fruits, and legumes. Like starch, cellulose is composed of glucose molecules connected in long chains. Unlike starch, however, the chains do not branch, and the bonds linking the glucose molecules together resist digestion by human enzymes (see Figure 4-9).

• Hemicelluloses • The hemicelluloses are the main constituent of cereal fibers. They are composed of various monosaccharide backbones with branching side

fibers: in plant foods, the *nonstarch polysaccharides* that are not digested by human digestive enzymes, although some are digested by GI tract bacteria. Fibers include cellulose, hemicelluloses, pectins, gums, and mucilages and the nonpolysaccharides lignins, cutins, and tannins.

chains of monosaccharides.* The many backbones and side chains make the hemicelluloses a diverse group; some are soluble, while others are insoluble.

• Pectins • All pectins consist of a backbone derived from carbohydrate with side chains of various monosaccharides. Commonly found in vegetables and fruits (especially citrus fruits and apples), pectins may be isolated and used by the food industry to thicken jelly, keep salad dressings from separating, and otherwise control texture and consistency. Pectins can perform these functions because they readily form gels in water.

• Gums and Mucilages • When cut, a plant secretes gums from the site of the injury. Like the other fibers, gums are composed of various monosaccharides and their derivatives. Gums such as *gum arabic* are used as additives by the food industry. Mucilages are similar to gums in structure; they include *guar* and *carrageenan,* which are added to foods as stabilizers.

• Lignin • This *nonpolysaccharide* fiber has a three-dimensional structure that gives it strength.† Because of its toughness, few of the foods that people eat contain much lignin. It occurs in the woody parts of vegetables such as carrots and the small seeds of fruits such as strawberries.

• Other Classifications of Fibers • Scientists classify fibers in several ways. The previous paragraphs classified them according to their chemical properties. Fibers can also be classified as **soluble fibers** or **insoluble fibers,** depending on their solubility in water. The effects of fibers on the body do not divide neatly along the lines of solubility, but some generalizations of significance to health can be made (see Table 4-1).

Some researchers classify fibers according to other physical properties that affect GI function and nutrient absorption. Physical properties of fibers include:

- *Water-holding capacity*—the capacity to capture water like a sponge, swelling and increasing the bulk of the intestines' contents.
- *Viscosity*—the capacity to form viscous, gel-like solutions.
- *Cation-exchange capacity*—the ability to bind minerals.
- *Bile-binding capacity*—the ability to bind to bile acids.

*In hemicelluloses, the most common backbone monosaccharides are xylose, mannose, and galactose; the common side chains are arabinose, glucuronic acid, and galactose (see Appendix C for structures).
†Lignins are polymers of several dozen molecules of phenol (an organic alcohol), with strong internal bonds that make them impervious to digestive enzymes.

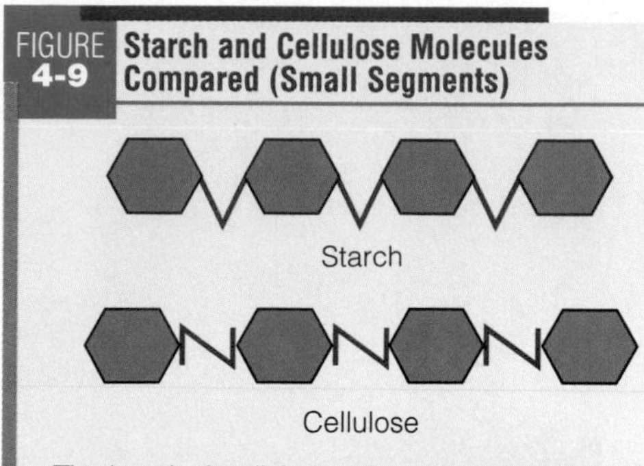

FIGURE 4-9 Starch and Cellulose Molecules Compared (Small Segments)

The bonds that link the glucose molecules together in cellulose are different from the bonds in starch (and glycogen). Human enzymes cannot digest cellulose. See Appendix C for chemical structures and descriptions of linkages.

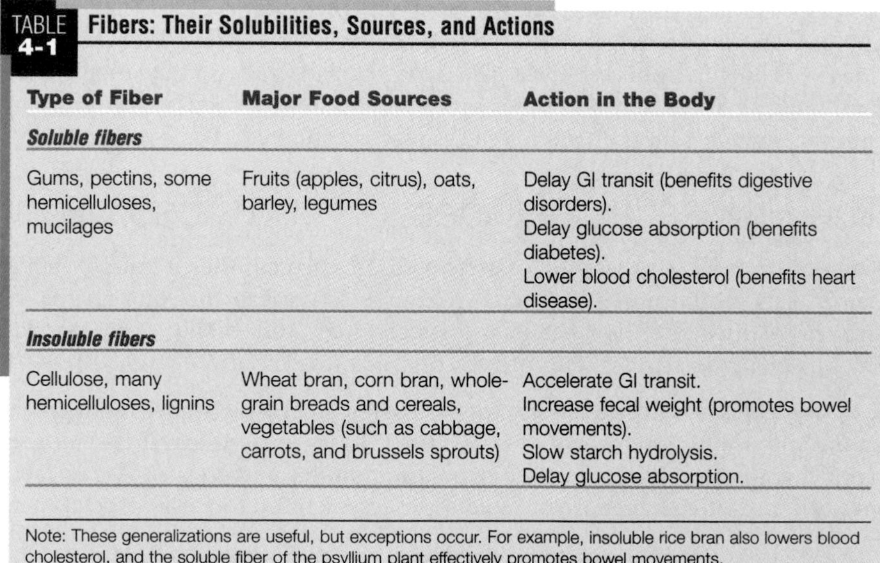

TABLE 4-1 Fibers: Their Solubilities, Sources, and Actions

Type of Fiber	Major Food Sources	Action in the Body
Soluble fibers		
Gums, pectins, some hemicelluloses, mucilages	Fruits (apples, citrus), oats, barley, legumes	Delay GI transit (benefits digestive disorders). Delay glucose absorption (benefits diabetes). Lower blood cholesterol (benefits heart disease).
Insoluble fibers		
Cellulose, many hemicelluloses, lignins	Wheat bran, corn bran, whole-grain breads and cereals, vegetables (such as cabbage, carrots, and brussels sprouts)	Accelerate GI transit. Increase fecal weight (promotes bowel movements). Slow starch hydrolysis. Delay glucose absorption.

Note: These generalizations are useful, but exceptions occur. For example, insoluble rice bran also lowers blood cholesterol, and the soluble fiber of the psyllium plant effectively promotes bowel movements.

soluble fibers: indigestible food components that dissolve in water to form a gel. An example is pectin from fruit, which is used to thicken jellies.

insoluble fibers: indigestible food components that do not dissolve in water. Examples include the tough, fibrous structures found in the strings of celery and the skins of corn kernels.

© Wartenberg/Picture Press/Corbis

When a person eats carbohydrate-rich foods, the body receives a valuable commodity—glucose.

- *Fermentability*—the extent to which bacteria in the GI tract can break down fibers to fragments that the body can use.*

Clearly, the fibers are a diverse group of compounds.

A compound not classed as a fiber but often found with it in foods is **phytic acid.** Because of this close association, researchers have been unable to determine whether it is the fiber, the phytic acid, or both, that binds with minerals, preventing their absorption. This binding presents a risk of mineral deficiencies, but the risk is minimal when fiber intake is reasonable and mineral intake adequate. The nutrition consequences of such mineral losses are described further in Chapters 12 and 13.

IN SUMMARY The complex carbohydrates are the polysaccharides (chains of monosaccharides): glycogen, starches, and fibers. Both glycogen and starch are storage forms of glucose—glycogen in the body, and starch in plants—and both yield energy for human use. The fibers also contain glucose (and other monosaccharides), but their bonds cannot be broken by human digestive enzymes, so they yield little, if any, energy. The accompanying table summarizes the carbohydrate family of compounds.

The Carbohydrate Family

Simple Carbohydrates (sugars)	Complex Carbohydrates
• Monosaccharides	• Polysaccharides
Glucose	Glycogen[a]
Fructose	Starches
Galactose	Fibers (nonstarch polysaccharides)
• Disaccharides	Soluble
Maltose	Insoluble
Sucrose	
Lactose	

[a]Glycogen is a complex carbohydrate (a polysaccharide), but not a *dietary* source of carbohydrate.

◆ The short chains of glucose units that result from the breakdown of starch are known as **dextrins.** The word sometimes appears on food labels because dextrins can be used as thickening agents in foods.

DIGESTION AND ABSORPTION OF CARBOHYDRATES

The ultimate goal of digestion and absorption of sugars and starches is to dismantle them into small molecules—chiefly glucose—that the body can absorb and use. The large starch molecules require extensive breakdown; the disaccharides need only to be broken once. The initial splitting begins in the mouth; the final splitting and absorption occur in the small intestine; and conversion to a common energy currency (glucose) takes place in the liver. The details follow.

The Processes of Digestion and Absorption

Figure 4-10 traces the digestion of carbohydrates through the GI tract. When a person eats foods containing starch, enzymes hydrolyze the long chains to shorter chains,◆ the short chains to disaccharides, and, finally, the disaccharides to monosaccharides. This process begins in the mouth.

• In the Mouth • In the mouth, vigorous chewing of high-fiber foods slows eating and stimulates the flow of saliva. The salivary enzyme **amylase** starts to work, hydrolyzing starch to shorter polysaccharides and to maltose. In fact, you can taste the change if you hold a piece of starchy food like a cracker in

phytic (FYE-tick) **acid:** a nonnutrient component of plant seeds; also called **phytate** (FYE-tate). Phytic acid occurs in the husks of grains, legumes, and seeds and is capable of binding minerals such as zinc, iron, calcium, magnesium, and copper in insoluble complexes in the intestine, which the body excretes unused.

amylase (AM-ih-lace): an enzyme that hydrolyzes amylose (a form of starch). Amylase is a carbohydrase, an enzyme that breaks down carbohydrates.

*Dietary fibers are fermented by bacteria in the colon to short-chain fatty acids, which are absorbed and metabolized by cells in the GI tract mucosa and liver (Chapter 5 describes fatty acids).

FIGURE 4-10 **Carbohydrate Digestion in the GI Tract**

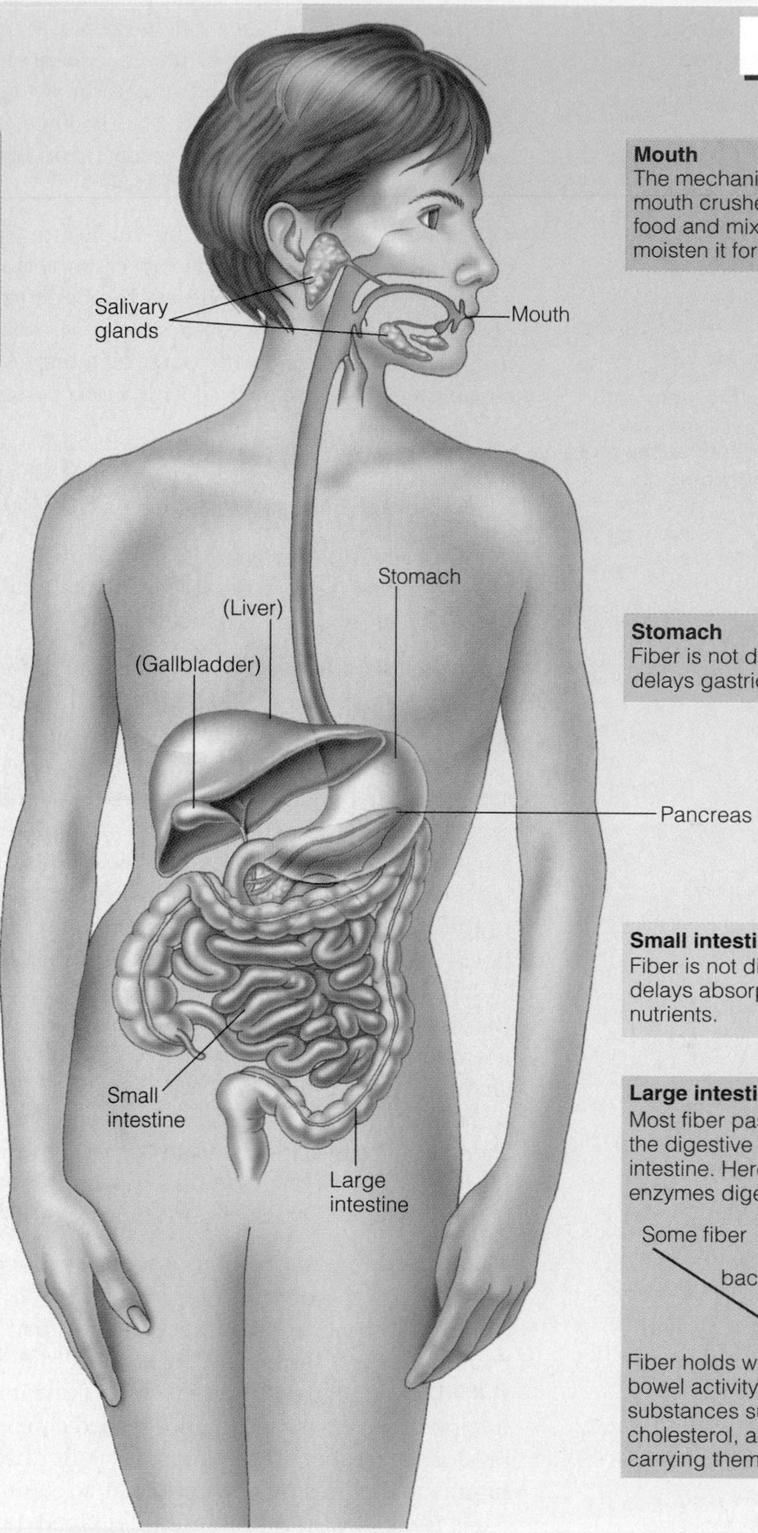

STARCH

Mouth and salivary glands
The salivary glands secrete saliva into the mouth to moisten the food. The salivary enzyme amylase begins digestion:

Starch —amylase→ small polysaccharides, maltose

Stomach
Stomach acid inactivates salivary enzymes, halting starch digestion.

Small intestine and pancreas
The pancreas produces an amylase that is released through the pancreatic duct into the small intestine:

Starch —pancreatic amylase→ small polysaccharides, disaccharides

Then disaccharidase enzymes on the surface of the small intestinal cells hydrolyze the disaccharides into monosaccharides:

Maltose —maltase→ glucose + glucose

Sucrose —sucrase→ fructose + glucose

Lactose —lactase→ galactose + glucose

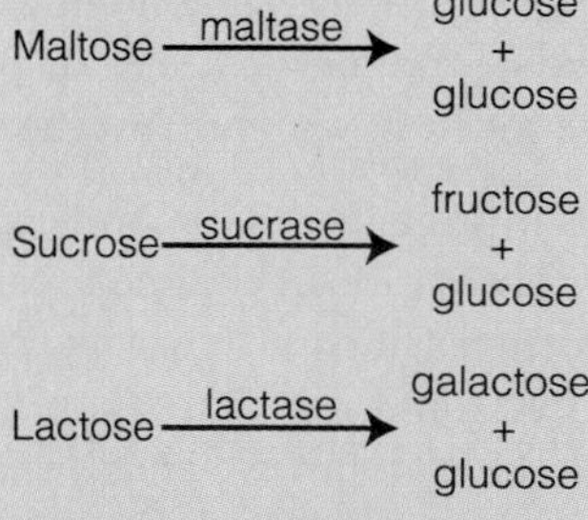

Intestinal cells absorb these monosaccharides.

FIBER

Mouth
The mechanical action of the mouth crushes and tears fiber in food and mixes it with saliva to moisten it for swallowing.

Stomach
Fiber is not digested, and it delays gastric emptying.

Small intestine
Fiber is not digested, and it delays absorption of other nutrients.

Large intestine
Most fiber passes intact through the digestive tract to the large intestine. Here, bacterial enzymes digest fiber:

Some fiber —bacterial enzymes→ fatty acids, gas

Fiber holds water; regulates bowel activity; and binds substances such as bile, cholesterol, and some minerals, carrying them out of the body.

your mouth for a few minutes without swallowing it—the cracker begins tasting sweeter as the enzyme acts on it. Because food is in the mouth for only a short time, very little digestion takes place there.

◆ Reminder: A *bolus* is a portion of food swallowed at one time.

• In the Stomach • The swallowed bolus◆ mixes with the stomach's acid and protein-digesting enzymes, which inactivate salivary amylase. Thus amylase cannot complete its job of starch digestion. To a small extent, the stomach's acid continues breaking down starch, but its juices contain no enzymes to digest carbohydrate. Fibers linger in the stomach and delay gastric emptying, thereby providing a feeling of fullness and **satiety.**

• In the Small Intestine • The small intestine performs most of the work of carbohydrate digestion. A major carbohydrate-digesting enzyme, pancreatic amylase, enters the intestine via the pancreatic duct and continues breaking down the polysaccharides to shorter glucose chains and disaccharides. The final step takes place on the outer membranes of the intestinal cells. There specific enzymes◆ dismantle specific disaccharides:

◆ Reminder: In general, the word ending *-ase* identifies an enzyme, and the word beginning identifies the molecule that the enzyme works on.

- **Maltase** breaks maltose into 2 glucose molecules.
- **Sucrase** breaks sucrose into 1 glucose and 1 fructose molecule.
- **Lactase** breaks lactose into 1 glucose and 1 galactose molecule.

At this point, all disaccharides contribute at least one glucose molecule to the body. Fructose and galactose can eventually become glucose after being processed in the liver, as explained later.

• In the Large Intestine • Within one to four hours after a meal, all the sugars and most of the starches have been digested. Only a small fraction of the starches and the indigestible fibers remain in the digestive tract.

The small fraction of starches that escapes digestion and absorption in the small intestine is known as **resistant starch.** Starch may resist digestion for several reasons, reflecting both the individual's efficiency in digesting starches and the food's physical properties.[2] Resistant starch is common in whole legumes, raw potatoes, and unripe bananas.[3] Because resistant starches remain in the large intestine, they promote bowel movements as fibers do, but unlike fibers, they do not lower blood cholesterol.[4]

Like resistant starches, fibers in the large intestine attract water, which softens the stools for passage without straining. Also, bacteria in the GI tract **ferment** both fibers and resistant starches. This process generates water, gas, and short-chain fatty acids (described in Chapter 5).* These short-chain fatty acids are absorbed in the colon and yield energy when metabolized. Metabolism of short-chain fatty acids occurs in both the intestine and the liver. Food fibers and resistant starches, therefore, do contribute some energy (about 2 kcalories per gram), depending on the extent to which they are broken down and absorbed.[5]

• Absorption into the Bloodstream • Glucose is unique in that it can be absorbed to some extent through the lining of the mouth, but for the most part, nutrient absorption takes place in the small intestine. Glucose and galactose traverse the cells lining the small intestine by active transport; fructose is absorbed by facilitated diffusion, which slows its entry and produces a smaller rise in blood glucose. Likewise, unbranched chains of starch are digested slowly and produce a smaller rise in blood glucose than branched chains, which have many more places for enzymes to attack and release glucose rapidly.

As the blood from the intestines circulates through the liver, cells take up fructose and galactose and convert them to other compounds, most often to glucose, as shown in Figure 4-11. Thus all disaccharides not only provide at least one glucose molecule directly, but they can also provide another one indirectly—through the conversion of fructose and galactose to glucose.

satiety (sah-TIE-eh-tee): the feeling of fullness and satisfaction that food brings (Chapter 8 provides a more detailed description).
- **sate** = to fill

maltase: an enzyme that hydrolyzes maltose.

sucrase: an enzyme that hydrolyzes sucrose.

lactase: an enzyme that hydrolyzes lactose.

resistant starch: starch that escapes digestion and absorption in the small intestine of healthy people.

ferment: to digest in the absence of oxygen.

*The short-chain fatty acids produced by GI bacteria are primarily acetic acid, propionic acid, and butyric acid.

FIGURE 4-11 Absorption of Monosaccharides

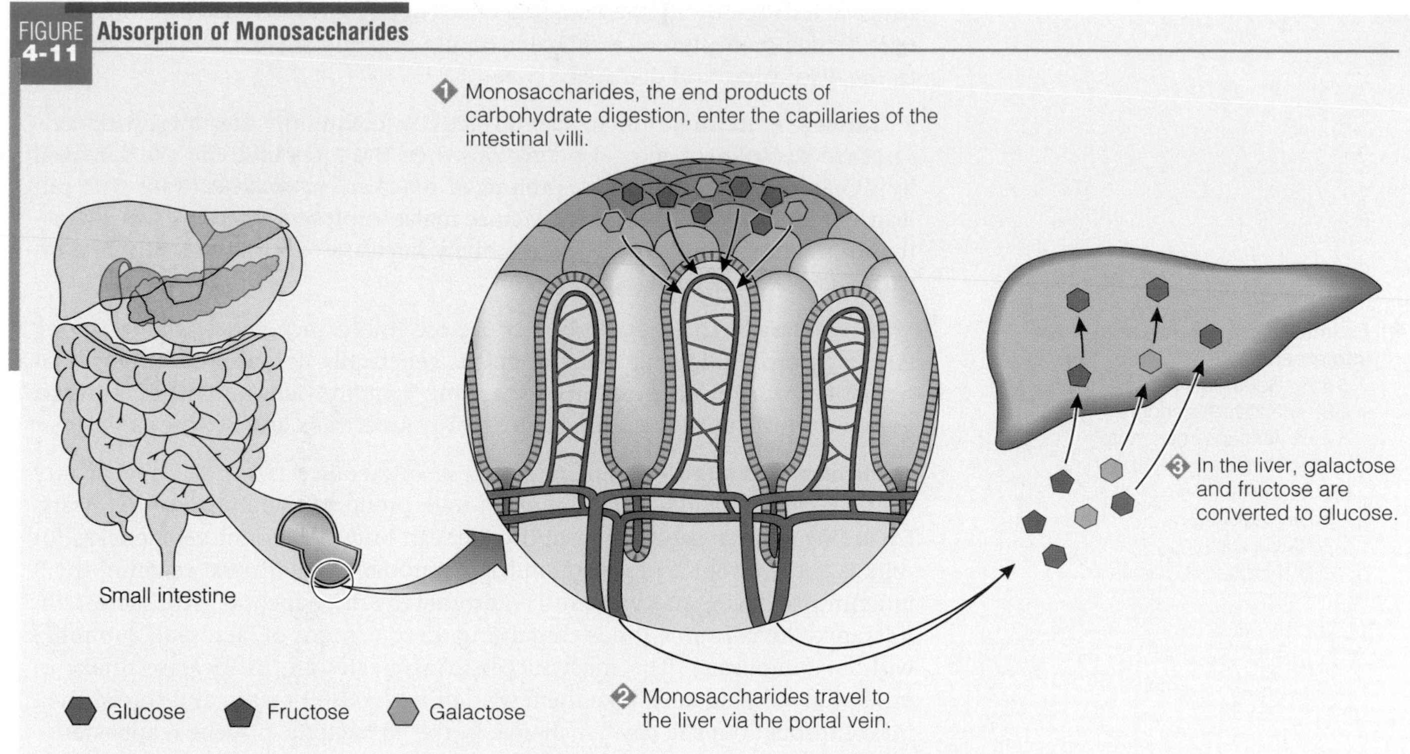

This description of the way the body receives carbohydrate should help explode a myth perpetrated by advertisers of high-sugar foods and beverages. They describe sugar as "quick energy" and imply that when you need a pick-me-up, you should reach for a candy bar and a cola beverage. Concentrated sugars do offer energy, but that's all. Clearly, a better choice for a pick-me-up is a carbohydrate-containing food that delivers vitamins, minerals, and fiber along with its energy. Next time you need an energy boost, why not have a delicious peanut butter and banana sandwich, a tall, cool glass of milk, and a fresh, juicy orange?

IN SUMMARY

In the digestion and absorption of carbohydrates, the body breaks down starches into disaccharides and disaccharides into monosaccharides; it then converts monosaccharides mostly to glucose to provide energy for the cells' work. The fibers help to regulate the passage of food through the GI system and slow the absorption of glucose, but contribute little, if any, energy.

Lactose Intolerance

Normally, the enzyme lactase ensures that the disaccharide lactose found in milk is both digested and absorbed efficiently. Lactase activity is highest immediately after birth, as befits an infant whose first and only food for a while will be breast milk or infant formula. In the great majority of the world's populations, lactase activity declines dramatically during childhood and adolescence to about 5 to 10 percent of the activity at birth.[6] Only a relatively small percentage (about 30 percent) of the people in the world retain enough lactase to digest and absorb lactose efficiently throughout adult life.

• **Symptoms** • When more lactose is consumed than the available lactase can handle, lactose molecules remain in the intestine undigested, attracting water and causing bloating, abdominal discomfort, and diarrhea—the symptoms of

lactose intolerance. The undigested lactose becomes food for intestinal bacteria, which multiply and produce irritating acid and gas, further contributing to the discomfort and diarrhea.

• **Causes** • As mentioned, lactase activity commonly declines with age. **Lactase deficiency** may also develop when the intestinal villi are damaged by disease, certain medicines, prolonged diarrhea, or malnutrition; this can lead to temporary or permanent lactose malabsorption, depending on the extent of the intestinal damage. In extremely rare cases, an infant is simply born with a lactase deficiency.

◆ **Estimated prevalence of lactose intolerance:**
- >80% Southeast Asians.
- 80% Native Americans.
- 75% African Americans.
- 70% Mediterranean peoples.
- 60% Inuits.
- 50% Hispanics.
- 20% Caucasians.
- <10% Northern Europeans.

• **Prevalence** • The prevalence◆ of lactose intolerance varies widely among ethnic groups, indicating that the trait is genetically determined.[7] The prevalence of lactose intolerance is lowest among Scandinavians and other northern Europeans and highest among native North Americans and Southeast Asians.

• **Dietary Changes** • Managing lactose intolerance requires some dietary changes, although total elimination of milk products usually is not necessary. Excluding all milk products from the diet can lead to nutrient deficiencies, for milk is a major source of several nutrients, notably the mineral calcium, the B vitamin riboflavin, and vitamin D. Fortunately, many people with lactose intolerance can consume foods containing up to 6 grams of lactose (½ cup milk) without symptoms.[8] The most successful strategies are to increase intake of milk products gradually, take them with other foods in meals, and spread their intake throughout the day.[9] A change in the GI bacteria, not the reappearance of the missing enzyme, accounts for the ability to adapt to milk products.[10]

◆ **Lactose in selected foods:**

Food	Lactose
Whole-wheat bread, 1 slice	0.5 g
Dinner roll, 1	0.5 g
Cheese, 1 oz	
Cheddar or American	0.5 g
Parmesan or cream	0.8 g
Doughnut (cake type), 1	1.2 g
Chocolate candy, 1 oz	2.3 g
Sherbet, 1 c	4.0 g
Cottage cheese (low-fat), 1 c	7.5 g
Ice cream, 1 c	9.0 g
Milk, 1 c	12.0 g
Yogurt (low-fat), 1 c	15.0 g

Note: Yogurt is often enriched with nonfat milk solids, which increase its lactose content to a level higher than milk's.

In many cases, lactose-intolerant people can tolerate fermented milk products such as yogurt and **acidophilus milk.**[11] The bacteria in these products digest lactose for their own use, thus reducing the lactose content. Even when the lactose content is equivalent to milk, yogurt produces fewer symptoms. Hard cheeses and cottage cheese are often well tolerated because most of the lactose is removed with the whey during manufacturing. Lactose continues to diminish as the cheese ages.

Many lactose-intolerant people use commercially prepared milk products that have been treated with an enzyme that breaks down the lactose. Alternatively, they take enzyme tablets with meals or add enzyme drops to their milk. The enzyme hydrolyzes much of the lactose in milk to glucose and galactose, which lactose-intolerant people can absorb without ill effects.

Because people's tolerance to lactose varies widely, lactose-restricted diets must be highly individualized. A completely lactose-free diet can be difficult because lactose appears not only in milk and milk products but also as an ingredient in many nondairy foods◆ such as breads, cereals, breakfast drinks, salad dressings, and cake mixes. People on strict lactose-free diets need to read labels and avoid foods that include milk, milk solids, whey (milk liquid), and casein (milk protein, which may contain traces of lactose). They also need to check all medications with the pharmacist because 20 percent of prescription drugs and 5 percent of over-the-counter medications contain lactose as a filler.

People who consume few or no milk products must take care to meet riboflavin, vitamin D, and calcium needs. Later chapters on the vitamins and minerals offer help with finding good nonmilk sources of these nutrients.

lactose intolerance: a condition that results from inability to digest the milk sugar lactose; characterized by bloating, gas, abdominal discomfort, and diarrhea. Lactose intolerance differs from milk allergy, which is caused by an immune reaction to the protein in milk.

lactase deficiency: a lack of the enzyme required to digest the disaccharide lactose into its component monosaccharides (glucose and galactose).

acidophilus (ASS-ih-DOF-ih-lus) **milk:** a cultured milk created by adding *Lactobacillus acidophilus*, a bacterium that breaks down lactose to glucose and galactose, producing a sweet, lactose-free product.

IN SUMMARY

Lactose intolerance is a common condition that occurs when there is insufficient lactase to digest the disaccharide lactose found in milk and milk products. Symptoms include GI distress. Because treatment requires limiting milk intake, other sources of riboflavin, vitamin D, and calcium must be included in the diet.

GLUCOSE IN THE BODY

The primary role of the available carbohydrates◆ in human nutrition is to supply the body's cells with glucose to deliver the indispensable commodity, energy. Starch contributes most to the body's glucose supply, but as explained earlier, any of the monosaccharides can also provide glucose.

Glucose plays the central role in carbohydrate metabolism. The next two sections provide an overview first of the pathways glucose can follow in the body and then of the ways the body regulates those pathways.

© Polara Studios Inc.

The carbohydrates of grains, vegetables, fruits, and legumes supply most of the energy in a healthful diet.

◆ Starches and sugars are called **available carbohydrates** because human digestive enzymes break them down for the body's use. In contrast, fibers are called **unavailable carbohydrates** because human digestive enzymes cannot break their bonds.

A Preview of Carbohydrate Metabolism

This brief discussion provides just enough information about carbohydrate metabolism to illustrate that the body needs and uses glucose as a chief energy nutrient. Chapter 7 provides a full description of energy metabolism.

• Storing Glucose as Glycogen • The liver stores one-third of the body's total glycogen and releases glucose as needed. During times of plenty, blood glucose rises, and liver cells link the excess glucose molecules into long, branching chains of glycogen. When blood glucose falls, the liver cells dismantle the glycogen into single molecules of glucose and release them into the bloodstream. Thus glucose becomes available to supply energy to the central nervous system and other organs regardless of whether the person has eaten recently. Muscle cells can also store glucose as glycogen (the other two-thirds), but they hoard most of their own supply, using it just for themselves during exercise.

Glycogen holds water and therefore is rather bulky. The body can store only enough glycogen to provide energy for relatively short periods of time—during exercise, a few hours' worth at most. For its long-term energy reserves, for use over days or weeks of food deprivation, the body uses its abundant, water-free fuel, fat, as Chapter 5 describes.

• Using Glucose for Energy • Glucose fuels the work of most of the body's cells. Inside a cell, enzymes break glucose in half. These halves can be put back together to make glucose, or they can be further broken down into smaller fragments (never again to be reassembled to form glucose). The small fragments can yield energy when broken down completely to carbon dioxide and water, or they can be reassembled, but only into units of body fat.

As mentioned, glycogen stores last only for hours, not for days. To keep providing glucose to meet the body's energy needs, a person has to eat dietary carbohydrate frequently. Yet people who do not always attend faithfully to their bodies' carbohydrate needs still survive. How do they manage without glucose from dietary carbohydrate? Do they simply draw energy from the other two energy-yielding nutrients, fat and protein? They do draw energy, but not simply.

• Making Glucose from Protein • Only glucose can provide energy for brain cells, other nerve cells, and developing red blood cells. Body protein can be converted to glucose to some extent, but protein has jobs of its own that no other nutrient can do. Body fat cannot be converted to glucose to any significant extent. Thus, when a person does not replenish depleted glycogen stores by eating carbohydrate, body proteins are dismantled to make glucose to fuel these special cells.

The conversion of protein to glucose is called **gluconeogenesis**—literally, the making of new glucose. Only adequate dietary carbohydrate can prevent this use of protein for energy, and this role of carbohydrate is known as its **protein-sparing action.**

• Making Ketone Bodies from Fat Fragments • An inadequate supply of carbohydrate combined with an accelerated breakdown of fat can shift the body's

gluconeogenesis (gloo-co-nee-oh-GEN-ih-sis): the making of glucose from a noncarbohydrate source (described in more detail in Chapter 7).

- **gluco** = glucose
- **neo** = new
- **genesis** = making

protein-sparing action: the action of carbohydrate (and fat) in providing energy that allows protein to be used for other purposes.

energy metabolism in a precarious direction. With less carbohydrate available for energy, more fat may be broken down, but not all the way to energy. Instead, the fat fragments combine with each other, forming **ketone bodies.** Muscles and other tissues can use ketone bodies for energy, but when their production exceeds their use, they accumulate in the blood, causing **ketosis,**◆ a condition that disturbs the body's normal **acid-base balance,** as Chapter 7 describes.

◆ Highlight 8 explores ketosis and low-carbohydrate diets further.

To spare body protein and prevent ketosis, the body needs 50 to 100 grams of carbohydrate a day. Dietary recommendations urge people to select abundantly from carbohydrate-rich foods to provide for this minimum need and considerably more.

• **Converting Glucose to Fat** • Given more carbohydrate than it needs to meet its energy needs and fill its glycogen stores to capacity, the body must find a way to store any extra glucose. The liver breaks it into smaller molecules and puts them together into the more permanent energy-storage compound—fat. Then the fat travels to the fatty tissues of the body for storage. Unlike the liver cells, which can store only about half a day's worth of glycogen, fat cells can store unlimited quantities of fat.

Even though excess carbohydrate can be converted to fat and stored, this is a relatively minor pathway. Storing carbohydrate as body fat is energetically expensive. Quite simply, the body uses more energy to convert dietary carbohydrate to body fat than it does to convert dietary fat to body fat. Consequently, body fat comes mainly from dietary fat. A balanced diet high in *complex* carbohydrates actually helps control body weight. Most foods rich in starch and fibers are so bulky and naturally so low in fat that when large quantities are eaten, they tend to crowd fat out of the diet. Since carbohydrate is less energy dense than fat (with only 4 kcalories to the gram compared with fat's 9), eating a diet high in complex carbohydrates usually reduces energy intake and supports weight control.

The Constancy of Blood Glucose

Every body cell depends on glucose for its fuel to some extent, and ordinarily, the cells of the brain and the rest of the nervous system depend *primarily* on glucose for their energy. The activities of these cells never cease, and they do not have the ability to store glucose. Day and night they continually draw on the supply of glucose in the fluid surrounding them. To maintain the supply, a steady stream of blood moves past these cells bringing more glucose from either the intestines (food) or the liver (via glycogen breakdown or glucose synthesis).

◆ Normal blood glucose: 80 to 120 mg/dL.

◆ Reminder: *Homeostasis* is the maintenance of constant internal conditions by the body's control systems.

• **Maintaining Glucose Homeostasis** • To function optimally, the body must maintain blood glucose within limits that permit the cells to nourish themselves. If blood glucose falls below normal,◆ the person may become dizzy and weak; if it rises above normal, the person may become fatigued. Left untreated, fluctuations to the extremes—either high or low—can be fatal.

• **The Regulating Hormones** • Blood glucose homeostasis◆ is regulated primarily by two hormones: insulin, which moves glucose from the blood into the cells, and glucagon, which brings glucose out of storage when necessary. Figure 4-12 depicts these hormonal regulators at work.

After a meal, as blood glucose rises, special cells of the pancreas respond by secreting **insulin** into the blood.* As the circulating insulin contacts the receptors on the body's other cells, the receptors respond by ushering glucose from the blood into the cells. Most of the cells take only the glucose they can use for energy right away, but the liver and muscle cells can assemble the small glucose units into long, branching chains of glycogen for storage. The liver cells can also convert glucose to fat for export to other cells. Thus high blood

*The *beta* (BAY-tuh) *cells,* one of several types of cells in the pancreas, secrete insulin in response to elevated blood glucose concentration.

ketone (KEE-tone) **bodies:** the product of the incomplete breakdown of fat when glucose is not available in the cells.

ketosis (kee-TOE-sis): an undesirably high concentration of ketone bodies in the blood and urine.

acid-base balance: the equilibrium in the body between acid and base concentrations; see Chapter 12.

insulin (IN-suh-lin): a hormone secreted by special cells in the pancreas in response to (among other things) increased blood glucose concentration. The primary role of insulin is to control the transport of glucose from the bloodstream into the muscle and fat cells.

FIGURE 4-12 Maintaining Blood Glucose Homeostasis

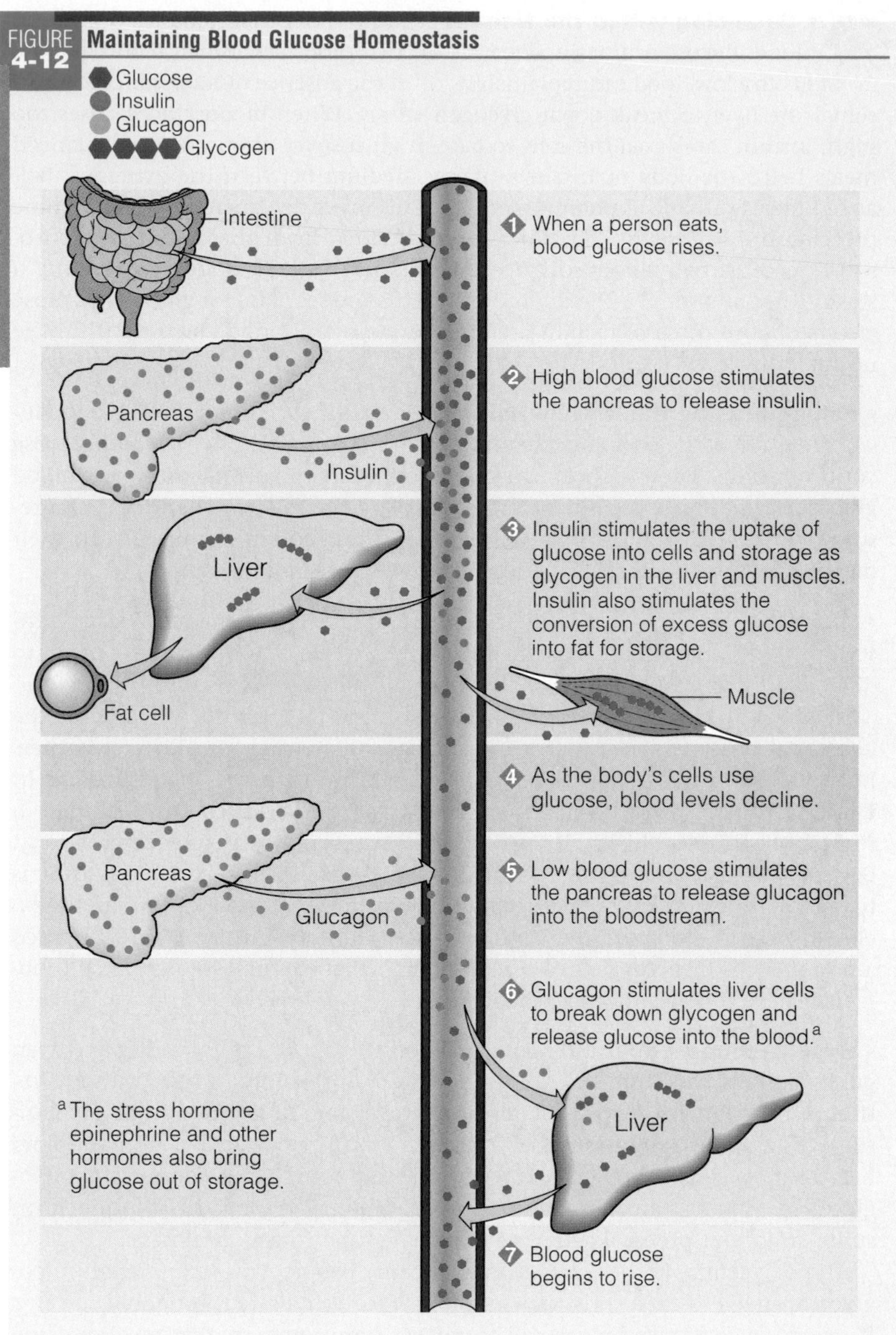

glucose returns to normal as excess glucose is stored as glycogen (which can be converted back to glucose) and fat (which cannot be).

When blood glucose falls (as occurs between meals), other special cells of the pancreas respond by secreting **glucagon** into the blood.* Glucagon raises blood glucose by signaling the liver to dismantle its glycogen stores and release glucose into the blood for use by all the other body cells.

Another hormone that calls glucose from the liver cells is the "fight-or-flight" hormone, **epinephrine.** When a person experiences stress, epinephrine acts quickly, ensuring that all the body cells have energy fuel in emergencies. Like glucagon, epinephrine works to release glucose from liver glycogen to the blood.

glucagon (GLOO-ka-gon): a hormone that is secreted by special cells in the pancreas in response to low blood glucose concentration and elicits release of glucose from storage.

epinephrine (EP-ih-NEFF-rin): a hormone of the adrenal gland that modulates the stress response; formerly called **adrenaline.**

*The *alpha cells* of the pancreas secrete glucagon in response to low blood glucose.

Health promotion

• Balancing within the Normal Range • The maintenance of normal blood glucose ordinarily depends on two processes. When blood glucose falls too low, food can replenish it, or in the absence of food, glucagon can signal the liver to break down glycogen stores. When blood glucose rises too high, insulin can signal the cells to take in glucose for energy. Eating balanced meals helps the body maintain a happy medium between the extremes. Balanced meals provide abundant complex carbohydrates, including fibers, some protein, and a little fat. The fibers and fat slow down the digestion and absorption of carbohydrate, so glucose enters the blood gradually, providing a steady, ongoing supply. Dietary protein elicits the secretion of glucagon, whose effects oppose those of insulin, helping to maintain blood glucose within the normal range.

• Falling outside the Normal Range • This influence of foods on blood glucose has given rise to the oversimplification that foods *govern* blood glucose concentrations. Foods do not; the body does. In some people, however, blood glucose regulation fails. When this happens, either of two conditions can result: diabetes or hypoglycemia. People with these conditions often plan their diets to help maintain their blood glucose within a normal range.

• Diabetes • In **diabetes,** blood glucose remains high after a meal because insulin is either inadequate or ineffective. Thus, while *blood* glucose is central to diabetes, *dietary* carbohydrates do not cause diabetes.

There are two main types of diabetes.[12] In **type 1 diabetes,** which is the less common type of diabetes, the pancreas fails to make insulin; researchers hold genetics, toxins, a virus, and a disordered immune system responsible. In **type 2 diabetes,** which is the more common type of diabetes, the cells fail to respond to insulin; this condition tends to occur as a consequence of obesity. Because obesity can precipitate type 2 diabetes, the best preventive measure is to maintain a healthy body weight. Concentrated sweets are not strictly excluded from the diabetic diet as they once were, but can be eaten in limited amounts with meals as part of a healthy diet. Diabetes and its associated problems receive full attention in Chapter 25.

• Hypoglycemia • In healthy people, blood glucose rises after eating and then gradually falls back into the normal range. The transition occurs without notice. In people with **hypoglycemia,** however, blood glucose drops dramatically, producing symptoms that mimic an anxiety attack: weakness, rapid heartbeat, sweating, anxiety, hunger, and trembling. Most commonly, hypoglycemia occurs as a consequence of poorly managed diabetes—too much insulin, strenuous physical activity, inadequate food intake, or illness.

Hypoglycemia in healthy people is rare. People who experience hypoglycemia need to eat a consistent amount of carbohydrate throughout the day. If several average-sized meals fail to relieve symptoms, smaller meals◆ eaten more frequently may help.

◆ Popular articles sometimes describe eating many small meals and snacks throughout the day as **grazing.**

• The Glycemic Effect • The term **glycemic effect** refers to the way blood glucose responds to foods: how quickly glucose is absorbed after a person eats, how high blood glucose rises, and how quickly it returns to normal. Slow absorption, a modest rise in blood glucose, and a smooth return to normal are desirable (a low glycemic effect); fast absorption, a surge in blood glucose, and an overreaction that plunges glucose below normal are undesirable (a high glycemic effect). Different foods have different effects on blood glucose depending on a number of factors working together, and the effect is not always what one might expect. Ice cream, for example, is a high-sugar food, but it produces less of a response than baked potatoes, a high-starch food. Table 4-2 compares selected foods based on their **glycemic index.**

Most relevant to real life, a food's glycemic effect differs depending on whether it is eaten alone or as part of a meal. In addition, eating small meals

diabetes (DYE-ah-BEE-teez): a disorder of carbohydrate metabolism resulting from inadequate or ineffective insulin.

type 1 diabetes: the less common type of diabetes in which the person produces no insulin at all; also known as **insulin-dependent diabetes mellitus (IDDM)** or **juvenile-onset diabetes** (because it frequently develops in childhood), although some cases arise in adulthood.

type 2 diabetes: the more common type of diabetes in which the fat cells resist insulin; also called **noninsulin-dependent diabetes mellitus (NIDDM)** or **adult-onset diabetes.** Type 2 usually progresses more slowly than type 1.

hypoglycemia (HIGH-po-gligh-SEE-me-ah): an abnormally low blood glucose concentration.

glycemic (gligh-SEEM-ic) **effect:** a measure of the extent to which a food, as compared with pure glucose, raises the blood glucose concentration and elicits an insulin response.

glycemic index: a method used to classify foods according to their potential for raising blood glucose.

TABLE 4-2 Glycemic Index of Selected Foods

Low (best)	Moderate (better)	High (good)
	Grains	
Pumpernickel bread	Sourdough or rye bread	White bread
Pasta		Short-grain rice
Bran cereals	Shredded wheat	Cornflakes
		Waffles
	Vegetables and Legumes	
Soybeans		Potatoes
Lentils		Carrots
Baked beans		
	Fruits	
Peaches	Banana	Watermelon
Apples (and apple juice)	Pineapple	
Oranges	Orange juice	
	Milk Products	
Milk	Ice cream	
Yogurt		
	Sweets	
Chocolate		Jelly beans
		Soft drinks
		Honey

Note: In an effort not to categorize foods as "good" or "bad," dietitians typically refer to high, moderate, and low glycemic index foods as "good," "better," and "best" choices, respectively.

frequently spreads glucose absorption across the day and thus offers similar metabolic advantages to eating foods with a low glycemic effect.

The rate of glucose absorption is particularly important to people with diabetes, who may benefit from limiting foods that produce too great a rise, or too sudden a fall, in blood glucose. Indeed, some studies have shown that taking the glycemic effect into account in meal planning is a practical way to improve glucose control. Others disagree. The usefulness of the glycemic index is surrounded by controversy as researchers debate whether selecting foods based on the glycemic index offers any health benefits.

Those in favor of using the glycemic index in meal planning claim that lowering the glycemic index of the diet reduces insulin secretion and improves glucose and lipid metabolism.[13] Consequently, a low glycemic diet may help prevent heart disease and diabetes.[14] It may also help prevent obesity. Fibers and other slowly digested carbohydrates prolong the presence of foods in the digestive tract, thus providing greater satiety and diminishing the insulin response, which can help with weight control.[15] In contrast, the rapid absorption of glucose from a high glycemic diet seems to promote overeating in overweight people.[16]

Those opposing the use of the glycemic index argue that it is too complicated to teach: relatively few foods have had their glycemic index determined, and this information is neither intuitively apparent nor provided on food labels. In addition, current guidelines already suggest many of the same choices: whole-grain breads, legumes, fruits, and milk products.[17] Perhaps most importantly, experts are concerned that instead of selecting carbohydrates based on *quality,* consumers will reduce their carbohydrate *quantity*—that is, they will adopt a low-carbohydrate diet instead of a low glycemic one.[18] (The problems associated with a low-carbohydrate diet are addressed in Highlight 8.)

Clearly, the issues surrounding the glycemic index are complex and need further elucidation. Until we get more answers, perhaps a reasonable approach is to include at least one low glycemic index food per meal.[19]

IN SUMMARY Dietary carbohydrates provide glucose that can be used by the cells for energy, stored by the liver and muscles as glycogen, or converted into fat if intakes exceed needs. All of the body's cells depend on glucose; those of the central nervous system are especially dependent on it. Without glucose, the body is forced to break down its protein tissues to make glucose and to alter energy metabolism to make ketone bodies from fats. Blood glucose regulation depends primarily on two pancreatic hormones: insulin to remove glucose from the blood into the cells when levels are high and glucagon to free glucose from glycogen stores and release it into the blood when levels are low. The glycemic effect describes how blood glucose responds to foods.

HEALTH EFFECTS AND RECOMMENDED INTAKES OF SUGARS

Ever since people first discovered honey and dates, they have enjoyed the sweetness of sugars. In the United States, the natural sugars of milk, fruits, vegetables, and grains account for about half of the sugar intake; the other half consists of sugars that have been refined and added to foods for a variety of purposes.◆ Added sugars assume various names: sucrose, invert sugar, corn sugar, corn syrups and solids, high-fructose corn syrup, and honey (see Table 4-3).

◆ As an additive, sugar:
- Enhances flavor.
- Supplies texture and color to baked goods.
- Provides fuel for fermentation, causing bread to rise or producing alcohol.
- Acts as a bulking agent in ice cream and baked goods.
- Acts as a preservative in jams.
- Balances the acidity of tomato- and vinegar-based products.

TABLE 4-3 Examples of Added Sugars

brown sugar: refined white sugar crystals to which manufacturers have added molasses syrup with natural flavor and color; 91 to 96 percent pure sucrose.

confectioners' sugar: finely powdered sucrose; 99.9 percent pure.

corn sweeteners: corn syrup and sugars derived from corn.

corn syrup: a syrup made from cornstarch that has been treated with acid, high temperatures, and enzymes that produce glucose, maltose, and dextrins. See also *high-fructose corn syrup (HFCS).*

dextrose: an older name for glucose.

granulated sugar: crystalline sucrose; 99.9 percent pure.

high-fructose corn syrup (HFCS): a syrup made from corn syrup that has been treated with an enzyme that changes glucose to the sweeter fructose; made especially for use in processed foods and beverages, where it is the predominant sweetener. HFCS is mostly fructose; glucose makes up the balance.

honey: sugar (mostly sucrose) formed from nectar gathered by bees. An enzyme splits the sucrose into glucose and fructose. Composition and flavor vary, but honey always contains a mixture of sucrose, fructose, and glucose.

invert sugar: a mixture of glucose and fructose formed by the hydrolysis of sucrose in a chemical process; sold only in liquid form and sweeter than sucrose. Invert sugar is used as a food additive to help preserve freshness and prevent shrinkage.

levulose: an older name for fructose.

maple sugar: a sugar (mostly sucrose) purified from the concentrated sap of the sugar maple tree.

molasses: the thick brown syrup produced during sugar refining. Molasses retains residual sugar and other by-products and a few minerals; blackstrap molasses contains significant amounts of calcium and iron—the iron comes from the *machinery* used to process the sugar.

raw sugar: the first crop of crystals harvested during sugar processing. Raw sugar cannot be sold in the United States because it contains too much filth (dirt, insect fragments, and the like). Sugar sold as "raw sugar" domestically has actually gone through over half of the refining steps.

turbinado (ter-bih-NOD-oh) **sugar:** sugar produced using the same refining process as white sugar, but without the bleaching and anti-caking treatment. Traces of molasses give turbinado its sandy color.

white sugar: pure sucrose or "table sugar," produced by dissolving, concentrating, and recrystallizing raw sugar.

The use of sweeteners in food manufacturing has risen steadily over the past several decades. Estimates of sugar consumption typically include all sweeteners used in the marketing system, including sugar lost or wasted, such as in the brine of sweet pickles or in jams or bakery goods that spoil before they can be eaten. They also include sugar used in pet foods and in fermentation. Estimates of *intake* indicate that on the average, each person consumes about 65 pounds of added sugar per year. This amount represents about 16 percent of daily energy intake, which exceeds current recommendations that concentrated sugars contribute no more than about 10 percent of energy intake.[20] When added sugars occupy this much of a diet, intakes from the five food groups fall below recommendations.[21]

Health Effects of Sugars

In moderate amounts, sugars add pleasure to meals without harming health. In excess, however, they can be detrimental in two ways. One, sugars can contribute to nutrient deficiencies by supplying energy (kcalories) without providing nutrients. Two, sugars contribute to tooth decay.

◆ Most teenagers consume soft drinks daily; 25 percent of them drink at least 2 cans a day, which is equivalent to more than ¾ cup of sugar.◆

• Nutrient Deficiencies • Empty-kcalorie foods that contain lots of added sugar such as cakes, candies, and sodas deliver glucose and energy with few, if any, other nutrients. By comparison, foods such as whole grains, vegetables, legumes, and fruits that contain some natural sugars and lots of starches and fibers deliver their glucose and energy along with protein, vitamins, and minerals.

A person spending 200 kcalories of a day's energy allowance on a 16-ounce soda gets little of value for those kcaloric "dollars." In contrast, a person using 200 kcalories on three slices of whole-wheat bread gets 9 grams of protein, 6 grams of fiber, plus several of the B vitamins with those kcalories. For the person who wants something sweet, perhaps a reasonable compromise would be to have two slices of bread with a teaspoon of jam on each. The amount of sugar a person can afford depends on how many kcalories are available beyond those needed to deliver indispensable vitamins and minerals.

With careful food selections, a person can obtain all the needed nutrients within an allowance of about 1500 kcalories. Some people have more generous energy allowances with which to "purchase" nutrients. For example, an active teenage boy may need as many as 4000 kcalories a day. If he eats mostly nutritious foods, then the "empty kcalories" of cola beverages may be an acceptable addition to his diet. On the other hand, an inactive older woman who is limited to fewer than 1500 kcalories a day cannot afford any but the most nutrient-dense foods.

Some people believe that because honey is a natural food, it is nutritious—or, at least, more nutritious than sugar. A look at their chemical structures reveals the truth. Honey, like table sugar, contains glucose and fructose. The primary difference is that in table sugar the two monosaccharides are bonded together as a disaccharide, whereas in honey some of them are free. Whether a person eats monosaccharides individually, as in honey, or linked together, as in table sugar, they end up the same way in the body: as glucose and fructose.

Honey does contain a few vitamins and minerals, but not many, as Table 4-4 on p. 112 shows. Honey is denser than crystalline sugar, too, so it provides more energy per spoon.

This is not to say that all sugar sources are alike, for some are more nutritious than others. Consider a fruit, say, an orange. The fruit may give you the same amounts of fructose and glucose and the same number of kcalories as a dose of sugar or honey, but the packaging is more valuable nutritionally. The fruit's sugars arrive in the body diluted in a large volume of water, packaged in fiber, and mixed with valuable minerals and vitamins.

As these comparisons illustrate, the significant difference between sugar sources is not between "natural" honey and "purified" sugar but between

TABLE 4-4 Sample Nutrients in Sugar and Other Foods

The indicated portion of any of these foods provides approximately 100 kcalories. Notice that for a similar number of kcalories and grams of carbohydrate, milk, legumes, fruits, grains, and vegetables offer more of the other nutrients than do the sugars.

	Size of 100 kcal Portion	Carbohydrate (g)	Protein (g)	Calcium (mg)	Iron (mg)	Vitamin A (µg)	Vitamin C (mg)
Foods							
Milk, 1% low-fat	1 c	12	8	300	0.1	144	2
Kidney beans	½ c	20	7	30	1.6	0	2
Apricots	6	24	2	30	1.1	554	22
Bread, whole wheat	1½ slices	20	4	30	1.9	0	0
Broccoli, cooked	2 c	20	12	188	2.2	696	148
Sugars							
Sugar, white	2 tbs	24	0	trace	trace	0	0
Molasses, blackstrap	2½ tbs	28	0	343	12.6	0	0.1
Cola beverage	1 c	26	0	6	trace	0	0
Honey	1½ tbs	26	trace	2	0.2	0	trace

concentrated sweets and the dilute, naturally occurring sugars that sweeten foods. You can suspect an exaggerated nutrition claim when someone asserts that one product is more nutritious than another because it contains honey.

Sugar can contribute to nutrient deficiencies only by displacing nutrients. For nutrition's sake, the appropriate attitude to take is not that sugar is "bad" and must be avoided, but that nutritious foods must come first. If the nutritious foods end up crowding sugar out of the diet, that is fine—but not the other way around. As always, the goals to seek are balance, variety, and moderation.

1 tsp honey = 22 kcal.
1 tsp sugar = 16 kcal.

You receive the same sugars from an orange as from honey, but the packaging makes a big nutrition difference.

• Dental Caries • Both sugars and starches begin breaking down to sugars in the mouth and so can contribute to tooth decay. Bacteria in the mouth ferment the sugars and in the process produce an acid that dissolves tooth enamel (see Figure 4-13). People can eat sugar without this happening, though, for much depends on how long foods stay in the mouth. Sticky foods stay on the teeth longer and keep yielding acid longer than foods that are readily cleared from the mouth. For that reason, sugar consumed quickly in a soft drink, for example, is less likely to cause **dental caries** than sugar in a pastry. By the same token, the sugar in sticky foods such as dried fruits is more detrimental than its quantity alone would suggest.

Another concern is how often people eat sugar. Bacteria produce acid for 20 to 30 minutes after each exposure. If a person eats three pieces of candy at one time, the teeth will be exposed to approximately 30 minutes of acid destruction. But, if the person eats three pieces at half-hour intervals, the time of exposure increases to 90 minutes. Likewise, slowly sipping a sugary soft drink may be more harmful than drinking quickly and clearing the mouth of sugar. Nonsugary foods can help remove sugar from tooth surfaces; hence, it is better to eat sugar with meals than between meals.

dental caries: decay of teeth.
- **caries** = rottenness

The development of caries depends on several factors: the bacteria that reside in **dental plaque,** the saliva that cleanses the mouth, the minerals that form the teeth, and the foods that remain after swallowing. For most people, good oral hygiene will prevent◆ dental caries.[22] In fact, regular brushing (twice a day, with a fluoride toothpaste) and flossing may be more effective in preventing dental caries than restricting sugary foods.[23]

FIGURE 4-13 Dental Caries

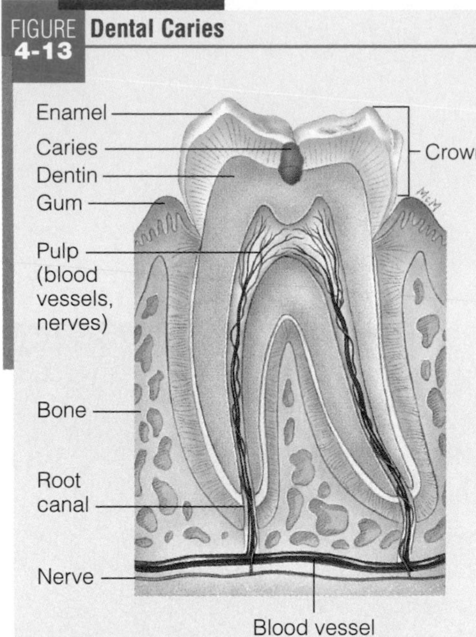

Caries begins when acid dissolves the enamel that covers the tooth. If not repaired, the decay may penetrate the dentin and spread into the pulp of the tooth, causing inflammation and an abscess.

Accusations against Sugars

Sugars have been blamed for a variety of other problems. The following paragraphs evaluate some of these accusations.

• Accusation: Sugar Causes Obesity • Foods high in added sugars are usually high in fat, too, so consumption of these foods increases total energy and fat intakes. Thus sugar contributes to obesity as a companion to fat, not as the sole cause of obesity—and obesity can occur without a high-sugar consumption.[24] The notion that eating sweet foods stimulates appetite and promotes overeating has not been supported by research.[25]

• Accusation: Sugar Causes Heart Disease • Researchers agree that unusually high doses of refined sugar can alter blood lipids◆ to favor heart disease.[26] This effect is most dramatic in "carbohydrate-sensitive" individuals—people who respond to sucrose with abnormally high insulin secretions, which promote the making of excess fat.[27] For most people, though, moderate sugar intakes do *not* elevate blood lipids.[28] To keep these findings in perspective, consider that heart disease correlates most closely with factors that have nothing to do with nutrition, such as smoking and genetics. Among dietary risk factors, several—such as total fats, saturated fats, cholesterol, and obesity—have much stronger associations with heart disease than do sugar intakes.

• Accusation: Sugar Causes Misbehavior in Children and Criminal Behavior in Adults • Sugar has been blamed for the misbehaviors of hyperactive children, delinquent adolescents, and lawbreaking adults. Such speculations have been based on personal stories and have not been confirmed by scientific research.[29] No scientific evidence supports a relationship between sugar and hyperactivity or other misbehaviors.[30] Chapter 15 provides accurate information on diet and children's behavior.

Recommended Intakes of Sugars

The *Dietary Guidelines* urge people to choose beverages and foods that limit sugar intake. Specifically, sugars should account for only 10 percent or less of the day's total energy intake. A person consuming 2000 kcalories a day, then, should receive no more then 200 kcalories◆ (that is, 50 grams or less) from concentrated sugars.

Food labels list the total grams of sugar a food provides. This total reflects both added sugars and those occurring naturally in foods. A food is likely to be high in sugars if its ingredient list starts with any of the sugars named in Table 4-3 on p. 110 or if it includes several of them. To help estimate sugar and energy intakes accurately, the list in the margin◆ shows what concentrated sweets are equivalent to 1 teaspoon of white sugar. These sugars all provide about 5 grams of carbohydrate and about 20 kcalories per teaspoon. Some are lower (16 kcalories for table sugar), while others are higher (22 kcalories for honey), but a 20-kcalorie average is an acceptable approximation. For a person who uses catsup liberally, it may help to remember that 1 tablespoon of catsup supplies about 1 teaspoon of sugar.

To prevent dental caries: ◆
- Eat sugary foods with meals.
- Limit between-meal snacks containing sugars and starches.
- Brush and floss teeth regularly.
- If brushing and flossing are not possible, at least rinse with water.

Lipids include fats and cholesterol, as Chapter 5 explains. ◆

For perspective, each of these concentrated sugars provides 200 kcal: ◆
- 16 oz soda.
- 2 oz jelly beans.
- ¾ c gelatin dessert.

1 tsp white sugar = ◆
- 1 tsp brown sugar.
- 1 tsp candy.
- 1 tsp corn sweetener or corn syrup.
- 1 tsp honey.
- 1 tsp jam or jelly.
- 1 tsp maple sugar or maple syrup.
- 1 tsp molasses.
- 1½ oz carbonated soda.
- 1 tbs catsup.

dental plaque: a gummy mass of bacteria that grows on teeth and can lead to dental caries and gum disease.

IN SUMMARY
Sugars pose no major health threat except for an increased risk of dental caries. Excessive intakes may displace needed nutrients and fiber; when accompanied by fat, sugars may contribute to obesity. A person deciding to limit daily sugar intake should recognize that not all sugars need to be restricted, just concentrated sweets, which are relatively empty of other nutrients and high in kcalories. Sugars that occur naturally in fruits, vegetables, and milk are acceptable.

HEALTH EFFECTS AND RECOMMENDED INTAKES OF STARCH AND FIBERS

Carbohydrates and fats are the two major sources of energy in the diet. When one is high, the other is usually low—and vice versa. The average fat intake in the United States is high compared with health recommendations. To lower fat intake and improve the balance between these two energy nutrients, people need to replace fatty foods with whole grains, vegetables, legumes, and fruits—foods noted for their starch, fibers, and naturally occurring sugars.

© Polara Studios Inc.

Foods rich in starch and fiber offer many health benefits.

Health Effects of Starch and Fibers

In addition to starch, fibers, and natural sugars, whole grains, vegetables, legumes, and fruits supply valuable vitamins and minerals and little or no fat. The following paragraphs describe some of the health benefits of diets that include a variety of these foods daily.

• **Weight Control** • Foods rich in complex carbohydrates tend to be low in fat and added sugars and can therefore promote weight loss by delivering less energy per bite, providing satiety, and delaying hunger.[32] As mentioned earlier, a high-fiber diet may also protect against obesity by virtue of its low glycemic effect.

Many weight-loss products on the market today contain bulk-inducing fibers such as methylcellulose, but buying pure fiber compounds like this is neither necessary nor advisable. To use fiber in a weight-loss plan, select fresh fruits, vegetables, legumes, and whole-grain foods. High-fiber foods not only add bulk to the diet, but are economical and nutritious. (A note of caution, though: on baked goods, read the label—many items, including bran muffins, may be high in fat and added sugars.)

• **Heart Disease** • High-carbohydrate diets, rich in whole grains, may protect against heart disease and stroke, although sorting out the exact reasons why can be difficult.[31] Such diets are low in animal fat and cholesterol and high in soluble fibers, vegetable proteins, and phytochemicals—all factors◆ associated with a lower risk of heart disease.

◆ The role of animal fat and cholesterol in heart disease is discussed in Chapter 5. The role of vegetable proteins in heart disease is presented in Chapter 6. The benefits of phytochemicals in disease prevention are featured in Highlight 11.

Foods rich in soluble fibers (such as oat bran, barley, and legumes) lower blood cholesterol by binding with bile acids, thus increasing their excretion. Consequently, the liver must use its cholesterol to make new bile acids. In addition, the bacterial by-products of fiber digestion in the colon also inhibit cholesterol synthesis in the liver. The net result is lower blood cholesterol.[33]

Several researchers have speculated that fiber may also exert its effect by displacing fats in the diet. Even when dietary fat is low, however, high intakes of soluble fibers exert a separate and significant cholesterol-lowering effect. In other words, a high-fiber diet helps to prevent heart disease independent of fat intake.

• **Cancer** • A high-carbohydrate diet, especially one that includes plenty of green and yellow vegetables and citrus fruits, protects against some types of cancer. Again, it is unclear whether the protection derives from the fiber, the vitamins, or phytochemicals.◆

Populations consuming high-fiber diets generally have lower rates of colon cancer than similar populations consuming low-fiber diets.[34] Many, but not all, research studies also suggest that increasing dietary fiber protects against colon cancer.[35] Fiber may help prevent colon cancer by diluting, binding, and rapidly removing potentially cancer-causing agents from the colon. Alternatively, the protective effect may be due to the fermentation of resistant starch and fiber in the colon, which lowers the pH.[36] A lower pH in the colon is associated with decreased colon cancer risks. Still another explanation is that the phytochemical lignin in dietary fiber defends cells against cancer-causing damage.[37]

The role antioxidant vitamins and phytochemicals play in cancer prevention is discussed in Highlight 11. ◆

• **Diabetes** • Populations eating high-carbohydrate diets often have low rates of diabetes, most likely because such diets are low in fat. High-carbohydrate, low-fat diets help control weight, and this is the most effective way to prevent the most common type of diabetes (type 2). Furthermore, when soluble fibers trap nutrients and delay their transit through the GI tract, glucose absorption is slowed, and this helps to prevent the glucose surge and rebound that seem to be associated with diabetes onset. High-fiber foods play a key role in reducing the risk of diabetes.[38]

• **GI Health** • Dietary fibers enhance the health of the large intestine. The healthier the intestinal walls, the better they can block absorption of unwanted constituents. Insoluble fibers such as cellulose (as in cereal brans, fruits, and vegetables) enlarge the stools, easing passage, and speed up transit time. In this way, the undigested fibers, together with the microbial growth they stimulate, help to alleviate or prevent constipation.

Taken with ample fluids, fibers help to prevent several GI disorders. Large, soft stools ease elimination for the rectal muscles and reduce the pressure in the lower bowel, making it less likely that rectal veins will swell (hemorrhoids). Fiber prevents compaction of the intestinal contents, which could obstruct the appendix and permit bacteria to invade and infect it (appendicitis). In addition, fiber stimulates the GI tract muscles so that they retain their strength and resist bulging out into pouches known as diverticula (illustrated in Figure H3-3 on p. 88).

• **Harmful Effects of Excessive Fiber Intake** • Despite fiber's benefits to health, a diet high in fiber also has a few drawbacks. A person who has a small capacity and eats mostly high-fiber foods may not be able to take in enough food energy or nutrients. The malnourished, the elderly, and children adhering to all-plant (vegan) diets are especially vulnerable to this problem.

Launching suddenly into a high-fiber diet can cause temporary bouts of abdominal discomfort, gas, and diarrhea and, more seriously, can obstruct the GI tract. To prevent such complications, a person adopting a high-fiber diet is advised to:

- Increase fiber intake gradually over several weeks to give the GI tract time to adapt.
- Drink lots of fluids to soften the fiber as it moves through the GI tract.
- Select fiber-rich foods from a variety of sources—fruits, vegetables, legumes, and whole-grain breads and cereals.

Insoluble fibers can limit the absorption of some nutrients by speeding the transit of foods through the GI tract and by binding to minerals. When mineral intake is adequate, however, a reasonable intake of high-fiber foods does not seem to compromise mineral balance.

Clearly, fiber is like all the nutrients in that "more" is only "better" up to a point. Again, the key words are balance, moderation, and variety.

IN SUMMARY

An adequate fiber intake:

- *Fosters weight control.*
- *Lowers blood cholesterol.*
- *Helps prevent colon cancer.*
- *Helps prevent and control diabetes.*
- *Helps prevent and alleviate hemorrhoids.*
- *Helps prevent appendicitis.*
- *Helps prevent diverticulosis.*

An excessive intake of fiber:

- *Displaces energy- and nutrient-dense foods.*
- *Causes intestinal discomfort and distention.*
- *Interferes with mineral absorption.*

Recommended Intakes of Starch and Fibers

Dietary recommendations suggest that carbohydrates provide more than half (55 to 60 percent)◆ of the energy requirement. A person consuming 2000 kcalories a day should therefore have 1100 to 1200 kcalories of carbohydrate, or about 275 to 300 grams. The Food and Drug Administration (FDA) used this 60 percent of kcalories guideline in establishing the Daily Value on food labels (300 grams of carbohydrate per day). For most people, this means increasing carbohydrate intake. To this end, the *Dietary Guidelines* encourage people to choose a variety of whole grains, vegetables, fruits, and legumes daily.

◆ **Quick and easy estimate: To attain 55 to 60% of energy from carbohydrate, look for 14 to 15 g of carbohydrate for each 100 kcal of food.**

Recommendations for fiber◆ suggest the same foods just mentioned: whole grains, vegetables, fruits, and legumes, which also provide minerals and vitamins. The FDA set a Daily Value on food labels for fiber at 25 grams or 11.5 grams per 1000-kcalorie energy intake. The American Dietetic Association suggests 20 to 35 grams of dietary fiber daily, which is about two times higher than the average intake in the United States.[39] An effective way to add fiber while cutting fat is to substitute plant sources of proteins (legumes) for animal sources (meats). Table 4-5 presents a list of fiber sources.

◆ **To increase your fiber intake:**
- Eat whole-grain cereals that contain ≥5 g fiber per serving for breakfast.
- Eat raw vegetables.
- Eat fruits (such as pears) and vegetables (such as potatoes) with their skins.
- Add legumes to soups, salads, and casseroles.
- Eat fresh and dried fruit for snacks.

As mentioned earlier, too much fiber is no better than too little. The World Health Organization recommends an upper limit of 40 grams of dietary fiber a day.

• **Choose Wisely** • In selecting high-fiber foods, keep in mind the principle of variety. The fibers in some foods lower cholesterol; those in other foods help promote GI tract health.

A diet following the Food Guide Pyramid, which includes 3 to 5 vegetable servings, 2 to 4 fruit servings, and 6 to 11 bread servings daily, can easily supply the recommended amount of carbohydrates and fiber. The "How to" feature on p. 118 describes an easy way to estimate the carbohydrate content of a meal.

• **Read Food Labels** • Food labels list the amount, in grams, of *total* carbohydrate—including starch, fibers, and sugars—per serving (see Figure 4-14). Fiber grams are also listed separately, as are the grams of sugars. (With this information, you could calculate starch grams by subtracting the grams of fibers and sugars from the total carbohydrate.) Sugars reflect both added sugars and those that occur naturally in foods. Total carbohydrate and dietary fiber are also expressed as "% Daily Values" for a person consuming 2000 kcalories; there is no Daily Value for sugars.

TABLE 4-5 Fiber in Selected Foods

Bread, Cereal, Rice, and Pasta Group

Whole-grain products provide about 1 to 2 grams (or more) of fiber per serving:

- 1 slice whole-wheat, pumpernickel, rye bread.
- 1 oz ready-to-eat cereal (100% bran cereals contain 10 grams or more).
- ½ c cooked barley, bulgur, grits, oatmeal.

Vegetable Group

Most vegetables contain about 2 to 3 grams of fiber per serving:

- 1 c raw bean sprouts.
- ½ c cooked broccoli, brussels sprouts, cabbage, carrots, cauliflower, collards, corn, eggplant, green beans, green peas, kale, mushrooms, okra, parsnips, potatoes, pumpkin, spinach, sweet potatoes, swiss chard, winter squash.
- ½ c chopped raw carrots, peppers.

Fruit Group

Fresh, frozen, and dried fruits have about 2 grams of fiber per serving:

- 1 medium apple, banana, kiwi, nectarine, orange, pear.
- ½ c applesauce, blackberries, blueberries, raspberries, strawberries.
- Fruit juices contain very little fiber.

Legumes

Many legumes provide about 8 grams of fiber per serving:

- ½ c cooked baked beans, black beans, black-eyed peas, kidney beans, navy beans, pinto beans.

Some legumes provide about 5 grams of fiber per serving:

- ½ c cooked garbanzo beans, great northern beans, lentils, lima beans, split peas.

Note: Appendix H provides fiber grams for over 2000 foods.

FIGURE 4-14 Carbohydrates on Food Labels

Food labels provide the quantities of total carbohydrate, dietary fiber, and sugars. Total carbohydrate and dietary fiber are also stated as "% Daily Values."

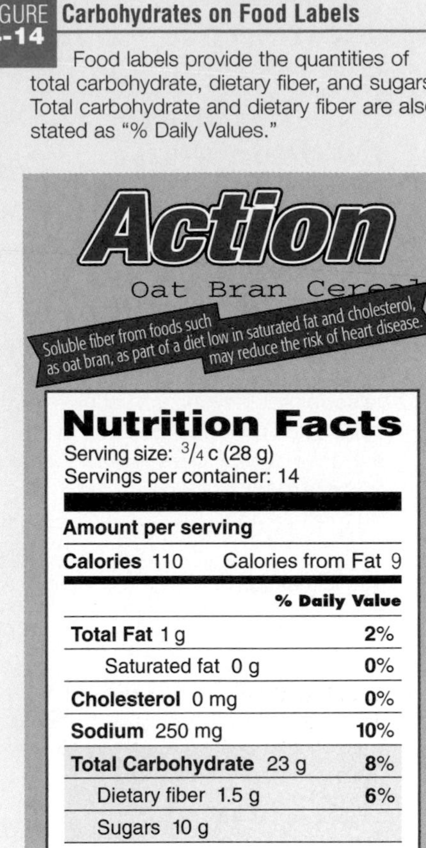

The FDA authorizes three health claims concerning fiber on food labels: one is for "fruits, vegetables, and grain products that contain fiber, particularly soluble fiber, and risk of coronary heart disease"; another is for "fiber-containing grain products, fruits, and vegetables and cancer"; and the most recent is for "soluble fiber from whole oats and heart disease." Other more general health claims address "fruits and vegetables and cancer" and "whole grains and heart disease and certain cancers." Chapter 2 describes the criteria foods must meet to bear these health claims.

IN SUMMARY

Clearly, a diet rich in complex carbohydrates—starches and fibers—supports efforts to control body weight and prevent heart disease, cancer, diabetes, and GI disorders. For these reasons, recommendations urge people to eat plenty of whole grains, vegetables, legumes, and fruits—enough to provide 55 to 60 percent of the daily energy intake from carbohydrate.

In today's world, there is one other reason why plant foods rich in complex carbohydrates and natural sugars are a better choice than animal foods or foods high in concentrated sweets: in general, less energy and resources are required to grow and process plant foods than to produce sugar or foods derived from animals. People who make environmentally-friendly food choices help to preserve the earth's ability to produce enough food to feed its many people.

HOW TO Use the Exchange System to Estimate Carbohydrate

The exchange system described in Chapter 2 provides a convenient way to estimate carbohydrate intake because the foods on each list have a similar carbohydrate content. To use the system, you need to know the carbohydrate value for each exchange list (see the table below) and the foods on that list with their portion sizes (review Figure 2-5 on pp. 44–45).

One Exchange	Carbohydrate (g)
Starch	15
Fruits	15
Other carbohydrates	15
Milks	12
Vegetables	5
Meats	—
Fats	—
Sugars (1 tsp)[a]	5

[a] The exchange system has no sugar list, but sugars do contribute to carbohydrates and energy intake.

Familiarity with portion sizes makes estimations easier. For example, it helps to recognize that this bowl of cereal contains 1 cup shredded wheat with 1 cup of milk and ½ banana. Then you can translate these portions into exchanges: 2 starches, 1 milk, and ½ fruit. Finally, you can calculate 15 grams of carbohydrate for each starch, 12 grams for the milk, and 8 grams for the ½ fruit.

Using the exchange system to estimate, this breakfast provides about 50 grams of carbohydrate. A computer diet analysis program came to a slightly higher conclusion (59 grams), as would a diet analysis using the values in Appendix H. Small variations between values arrived at differently may seem disconcerting, but remember that all are only estimates. Estimates save time; often only a ballpark figure is needed anyway.

Breakfast		Exchange	Carbohydrate (g) Estimate	Carbohydrate (g) Actual
1 c shredded wheat	=	2 starches	30	34
1 c milk	=	1 milk	12	12
½ banana	=	½ fruit	8	13
			50	59

Carbohydrate-containing foods appear in several exchange lists: starches, fruits, milks, "others" such as desserts and snacks, and vegetables. Sugars also contribute carbohydrate.

both, © Polara Studios Inc.

Most estimates of the nutrient contents of foods are rough but serviceable approximations. A "90-kcalorie potato" actually means a "90-kcalorie plus or minus about 20 percent potato," which makes it not significantly different from a 100-kcalorie potato. For most purposes, a variation of about 20 percent is considered reasonable, which is the difference between the values in this example.

Fiber appears in only the starches, vegetables, and fruits lists. To estimate fiber, remember that most items on these lists provide at least 2 grams of fiber per serving; some provide 3 or more. Knowing this, a reasonable fiber estimate for this breakfast (with 2 starches and ½ fruit) would be 5 to 7 grams—and a diet analysis report of 5 grams would agree.

Just a few calculations of this kind will give you a feel for the carbohydrate content of a diet. Once you are aware of the major carbohydrate-contributing foods you eat, you can return to thinking simply in terms of foods, developing a sense of how much of each is enough.

How are you doing?

- Do you eat at least 6 servings of grain products daily?
- Do you eat at least 5 servings of fruits and vegetables daily?
- Do you choose beverages and foods that limit your intake of sugars?

Nutrition on the Net

WEBSITES

Access these websites for further study of topics covered in this chapter.

- Find updates and quick links to these and other nutrition-related sites at our website: **www.wadsworth.com/nutrition**
- Search for "lactose intolerance" at the U.S. Government health information site: **www.healthfinder.gov**
- Search for "sugars" and "fiber" at the International Food Information Council site: **www.ificinfo.health.org**
- Learn more about dental caries from the American Dental Association and the National Institute of Dental Research: **www.ada.org** and **www.nidr.nih.gov**
- Learn more about diabetes from the American Diabetes Association, the Canadian Diabetes Association, and the National Diabetes Information Clearinghouse: **www.diabetes.org**, **www.diabetes.ca**, and **www.niddk.nih.gov**

INTERNET ACTIVITIES

The cause of diabetes is a mystery, although both genetics and environmental factors such as obesity and lack of exercise appear to play roles. Do you know what your risk is for developing this disease? Take the Diabetes Risk Test to find out.

- Go to: **www.diabetes.org**
- Click on "Take the Risk Test" located on the left side bar.
- Answer the questions and then click "Calculate."

These questions address the most common risk factors for the development of diabetes and your score suggests how strong your personal risk factors are. What did you learn about your risk of developing diabetes?

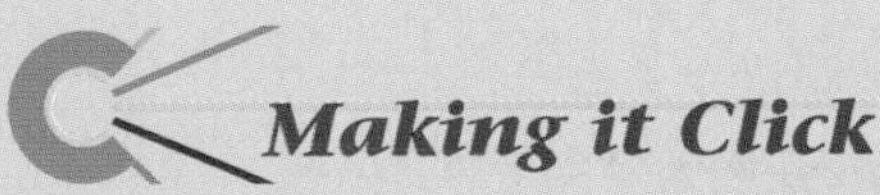

Making it Click

These problems use hypothetical situations to give you practice in doing simple nutrition-related calculations (see p. 122 for answers). Be sure to show your calculations for each problem.

Health recommendations suggest that 55 to 60 percent of the daily energy intake come from carbohydrates. Stating recommendations in terms of percentage of energy intake is meaningful only if energy intake is known. The following exercises illustrate this concept.

1. Calculate the carbohydrate intake (in grams) for a student who has a high carbohydrate intake (70 percent of energy intake) and a moderate energy intake (2000 kcalories a day).

   How does this carbohydrate intake compare to the Daily Value of 300 grams? To the "over half" recommendation?
2. Now consider a professor who eats half as much carbohydrate as the student (in grams) and has the same energy intake. What percentage does carbohydrate contribute to the daily intake?

   How does carbohydrate intake compare to the Daily Value of 300 grams? To the "over half" recommendation?
3. Now consider an athlete who eats twice as much carbohydrate (in grams) as the student and has a much higher energy intake (6000 kcalories a day). What percentage does carbohydrate contribute to this person's daily intake?

   How does carbohydrate intake compare to the Daily Value of 300 grams? To the "over half" recommendation?
4. One more example. In an attempt to lose weight, a person adopts a diet that provides 150 grams of carbohydrate per day and limits energy intake to 1000 kcalories. What percentage does carbohydrate contribute to this person's daily intake?

   How does this carbohydrate intake compare to the Daily Value of 300 grams? To the "over half" recommendation?

These exercises should convince you of the importance of examining actual intake as well the percentage of energy intake.

Study Questions

These questions will help you review the chapter. You will find the answers in the discussions on the pages provided.

1. Which carbohydrates are described as simple, and which are complex? (p. 93)
2. Describe the structure of a monosaccharide and name the three monosaccharides important in nutrition. Name the three disaccharides commonly found in foods and their component monosaccharides. In what foods are these sugars found? (pp. 94–96)
3. What happens in a condensation reaction? In a hydrolysis reaction? (p. 95)
4. Describe the structure of polysaccharides and name the ones important in nutrition. How are starch and glycogen similar, and how do they differ? How do the fibers differ from the other polysaccharides? (pp. 97–100)
5. Describe carbohydrate digestion and absorption. What role does fiber play in the process? (pp. 100–103)
6. What are the possible fates of glucose in the body? What is the protein-sparing action of carbohydrate? (pp. 105–106)
7. How does the body maintain its blood glucose concentration? What happens when the blood glucose concentration rises too high or falls too low? (pp. 106–109)
8. What are the health effects of sugars? What are the dietary recommendations regarding concentrated sugar intakes? (pp. 110–114)
9. What are the health effects of starches and fibers? What are the dietary recommendations regarding these complex carbohydrates? (pp. 114–116)
10. What foods provide starches and fibers? (p. 117)

These questions will help you prepare for an exam. Answers can be found on p. 122.

1. Carbohydrates are found in virtually all foods except:
   a. milks.
   b. meats.
   c. breads.
   d. fruits.
2. Disaccharides include:
   a. starch, glycogen, and fiber.
   b. amylose, pectin, and dextrose.
   c. sucrose, maltose, and lactose.
   d. glucose, galactose, and fructose.
3. The making of a disaccharide from two monosaccharides is an example of:
   a. digestion.
   b. hydrolysis.
   c. condensation.
   d. gluconeogenesis.
4. The storage form of glucose in the body is:
   a. insulin.
   b. maltose.
   c. glucagon.
   d. glycogen.
5. The significant difference between starch and cellulose is that:
   a. starch is a polysaccharide, but cellulose is not.
   b. animals can store glucose as starch, but not as cellulose.
   c. hormones can make glucose from cellulose, but not from starch.
   d. digestive enzymes can break the bonds in starch, but not in cellulose.
6. The ultimate goal of carbohydrate digestion and absorption is to yield:
   a. fibers.
   b. glucose.
   c. enzymes.
   d. amylase.
7. The enzyme that breaks a disaccharide into glucose and galactose is:
   a. amylase.
   b. maltase.
   c. sucrase.
   d. lactase.
8. With insufficient glucose in metabolism, fat fragments combine to form:
   a. dextrins.
   b. mucilages.
   c. phytic acids.
   d. ketone bodies.
9. What does the pancreas secrete when blood glucose rises? When blood glucose falls?
   a. insulin; glucagon
   b. glucagon; insulin
   c. insulin; glycogen
   d. glycogen; epinephrine
10. What percentage of the daily energy intake should come from carbohydrates?
    a. 15 to 20
    b. 25 to 30
    c. 45 to 50
    d. 55 to 60

References

1 A. M. Stephen and coauthors, Intake of carbohydrate and its components—International comparisons, trends over time, and effects of changing to low-fat diets, *American Journal of Clinical Nutrition* 62 (1995): 851S–867S.

2 K. M. Behall and J. C. Howe, Breath-hydrogen production and amylose content of the diet, *American Journal of Clinical Nutrition* 65 (1997): 1783–1789; K. R. Silvester, H. N. Englyst, and J. H. Cummings, Ileal recovery of starch from whole diets containing resistant starch measured in vitro and fermentation of ileal effluent, *American Journal of Clinical Nutrition* 62 (1995): 403–411.

3 P. A. Baghurst, K. I. Baghurst, and S. J. Record, Dietary fibre, non-starch polysaccharides and resistant starch—A review, *Supplement to Food Australia* 48 (1996): S1–S36.

4 M. L. A. Heijnen and coauthors, Neither raw nor retrograded resistant starch lowers fasting serum cholesterol concentrations in healthy normolipidemic subjects, *American Journal of Clinical Nutrition* 64 (1996): 312–318.

5 K. M. Behall and J. C. Howe, Contribution of fiber and resistant starch to metabolize energy, *American Journal of Clinical Nutrition* 62 (1995): 1158S–1160S.

6 M. Lee and S. D. Krasinski, Human adult-onset lactase decline: An update, *Nutrition Reviews* 56 (1998): 1–8.

7 F. L. Suarez and D. A. Savaiano, Diet, genetics, and lactose intolerance, *Food Technology* 51 (1997): 74–76.

8 T. H. Vesa, R. A. Korpela, and T. Sahi, Tolerance to small amounts of lactose in lactose maldigesters, *American Journal of Clinical Nutrition* 64 (1996): 197–201; S. R. Hertzler, B. L. Huynh, and D. A. Savaiano, How much lactose is low lactose? *Journal of the American Dietetic Association* 96 (1996): 243–246.

9 L. D. McBean and G. D. Miller, Allaying fears and fallacies about lactose intolerance, *Journal of the American Dietetic Association* 98 (1998): 671–676; F. L. Suarez and coauthors, Tolerance to the daily ingestion of two cups of milk by individuals claiming lactose intolerance, *American Journal of Clinical Nutrition* 65 (1997): 1502–1506; F. L. Suarez, D. A. Savaiano, and M. D. Levitt, A comparison of symptoms after the consumption of milk or lactose-hydrolyzed milk by people with self-reported severe lactose intolerance, *New England Journal of Medicine* 333 (1995): 1–4.

10 S. R. Hertzler and D. A. Savaiano, Colonic adaptation to daily lactose feeding in lactose maldigesters reduces lactose intolerance, *American Journal of Clinical Nutrition* 64 (1996): 232–236.

11 S. W. Rizkalla and coauthors, Chronic consumption of fresh but not heated yogurt improves breath-hydrogen status and short-chain fatty acid profiles: A controlled study in healthy men with or without lactose maldigestion, *American Journal of Clinical Nutrition* 72 (2000): 1474–1479.

12 The Expert Committee on the Diagnosis and Classification of Diabetes Mellitus, Report of the Expert Committee on the Diagnosis and Classification of Diabetes Mellitus, *Diabetes Care* 20 (1997): 1183–1197.

13 T. M. S. Wolever, Dietary recommendations for diabetes: High carbohydrate or high monounsaturated fat? *Nutrition Today* 34 (1999): 73–77.

14 K. L. Morris and M. B. Zemel, Glycemic index, cardiovascular disease, and obesity, *Nutrition Reviews* 57 (1999): 273–276.

15 D. S. Ludwig and coauthors, Dietary fiber, weight gain, and cardiovascular disease risk factors in young adults, *Journal of the American Medical Association* 282 (1999): 1539–1546.

16 D. S. Ludwig and coauthors, High glycemic index foods, overeating, and obesity, *Pediatrics* 103 (1999): e26 (www.pediatrics.org).

17 C. Beebe, Diets with a low glycemic index: Not ready for practice yet! *Nutrition Today* 34 (1999): 82–86.

18 E. Saltzman, The low glycemic index diet: Not yet ready for prime time, *Nutrition Reviews* 57 (1999): 297.

19 J. Brand-Miller and K. Foster-Powell, Diets with a low glycemic index: From theory to practice, *Nutrition Today* 34 (1999): 64–72; H. Katanas, Diets with a low glycemic index are ready for practice, *Nutrition Today* 34 (1999): 87–88.

20 J. F. Guthrie and J. F. Morton, Food sources of added sweeteners in the diets of Americans, *Journal of the American Dietetic Association* 100 (2000): 43–48, 51.

21 S. A. Bowman, Diets of individuals based on energy intakes from added sugars, *Family Economics and Nutrition Review* 12 (1999): 31–38.

22 K. G. König and J. M. Navia, Nutritional role of sugars in oral health, *American Journal of Clinical Nutrition* 62 (1995): 275S–283S.

23 S. Gibson and S. Williams, Dental caries in pre-school children: Associations with social class, toothbrushing habit and consumption of sugars and sugar-containing foods. Further analysis of data from the National Diet and Nutrition Survey of children aged 1.5–4.5 years, *Caries Research* 33 (1999): 101–113.

24 J. O. Hill and A. M. Prentice, Sugar and body weight regulation, *American Journal of Clinical Nutrition* 62 (1995): 264S–274S.

25 G. H. Anderson, Sugars, sweetness, and food intake, *American Journal of Clinical Nutrition* 62 (1995): 195S–202S.

26 L. C. Hudgins and coauthors, Human fatty acid synthesis is reduced after the substitution of dietary starch for sugar, *American Journal of Clinical Nutrition* 67 (1998): 631–639.

27 K. N. Frayn and S. M. Kingman, Dietary sugars and lipid metabolism in humans, *American Journal of Clinical Nutrition* 62 (1995): 250S–263S.

28 E. J. Parks and M. K. Hellerstein, Carbohydrate-induced hypertriacylglycerolemia: Historical perspective and review of biological mechanisms, *American Journal of Clinical Nutrition* 71 (2000): 412–433.

29 J. W. White and M. Wolraich, Effect of sugar on behavior and mental performance, *American Journal of Clinical Nutrition* 62 (1995): 242S–249S.

30 M. L. Wolraich, D. B. Wilson, and J. W. White, The effect of sugar on behavior

or cognition in children: A meta-analysis, *Journal of the American Medical Association* 274 (1995): 1617–1621.

31 S. Liu and coauthors, Whole-grain consumption and risk of ischemic stroke in women—A prospective study, *Journal of the American Medical Association* 284 (2000): 1534–1540; S. Liu and coauthors, Whole-grain consumption and risk of coronary heart disease: Results from the Nurses' Health Study, *American Journal of Clinical Nutrition* 70 (1999): 412–419; A. Wolk and coauthors, Long-term intake of dietary fiber and decreased risk of coronary heart disease among women, *Journal of the American Medical Association* 281 (1999): 1998–2004; J. L. Slavin and coauthors, Plausible mechanisms for the protectiveness of whole grains, *American Journal of Clinical Nutrition* 70 (1999): 459S–463S; M. B. Katan, Forever fiber, *Nutrition Reviews* 54 (1996): 253–257.

32 A. Sparti and coauthors, Effect of diets high or low in unavailable and slowly digestible carbohydrates on the pattern of 24-h substrate oxidation and feelings of hunger in humans, *American Journal of Clinical Nutrition* 72 (2000): 1461–1468.

33 L. Brown and coauthors, Cholesterol-lowering effects of dietary fiber: A meta-analysis, *American Journal of Clinical Nutrition* 69 (1999): 30–42.

34 B. S. Reddy, Role of dietary fiber in colon cancer: An overview, *American Journal of Medicine* 106 (1999): S16–S19; M. J. Hill, Cereals, cereal fibre and colorectal cancer risk: A review of the epidemiological literature, *European Journal of Cancer Prevention* 6 (1997): 219–225; L. LeMarchand and coauthors, Dietary fiber and colorectal cancer risk, *Epidemiology* 8 (1997): 658–665.

35 A. Schatzkin and coauthors, Lack of effect of a low-fat, high-fiber diet on the recurrence of colorectal adenomas, *New England Journal of Medicine* 342 (2000): 1149–1155; D. S. Alberts and coauthors, Lack of effect of a high-fiber cereal supplement on the recurrence of colorectal adenomas, *New England Journal of Medicine* 342 (2000): 1156–1162; F. Macrae, Wheat bran fiber and development of adenomatous polyps: Evidence from randomized, controlled clinical trials, *American Journal of Medicine* 106 (1999): S38–S42; D. Kritchevsky, Protective role of wheat bran fiber: Preclinical data, *American Journal of Medicine* 106 (1999): S28–S31; C. S. Fuchs and coauthors, Dietary fiber and risk of colorectal cancer and adenoma in women, *New England Journal of Medicine* 340 (1999): 169–176; E. Negri and coauthors, Fiber intake and risk of colorectal cancer, *Cancer Epidemiology, Biomarkers, and Prevention* 7 (1998): 667–671; J. Faivre and A. Giacosa, Primary prevention of colorectal cancer through fibre supplementation, *European Journal of Cancer Prevention* 7 (1998): S29–S32.

36 M. Noakes and coauthors, Effect of high-amylose starch and oat bran on metabolic variables and bowel function in subjects with hypertriglyceridemia, *American Journal of Clinical Nutrition* 64 (1996): 944–951.

37 F. J. Lu, L. H. Chu, and R. J. Gau, Free radical–scavenging properties of lignin, *Nutrition and Cancer* 30 (1998): 31–38; J. G. Erhardt and coauthors, A diet rich in fat and poor in dietary fiber increases the in vitro formation of reactive oxygen species in human feces, *Journal of Nutrition* 127 (1997): 706–709.

38 J. Salmerón and coauthors, Dietary fiber, glycemic load, and risk of non-insulin-dependent diabetes mellitus in women, *Journal of the American Medical Association* 277 (1997): 472–477.

39 Position of The American Dietetic Association: Health implications of dietary fiber, *Journal of the American Dietetic Association* 97 (1997): 1157–1159.

ANSWERS

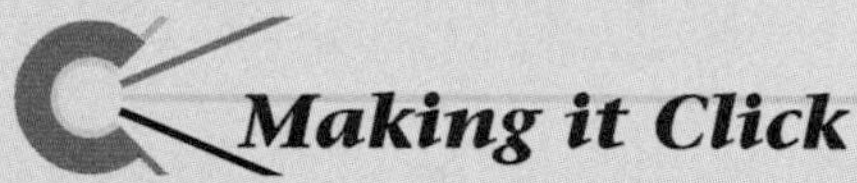

Making it Click

1. 0.7 × 2000 total kcal/day = 1400 kcal from carbohydrate/day.
   1400 kcal from carbohydrate ÷ 4 kcal/g = 350 g carbohydrate.

This carbohydrate intake is higher than the Daily Value and higher than the "over half" recommendation.

2. 350 g carbohydrate ÷ 2 = 175 g carbohydrate/day.
   175 g carbohydrate × 4 kcal/g = 700 kcal from carbohydrate.
   700 kcal from carbohydrate ÷ 2000 total kcal/day = 0.35.
   0.35 × 100 = 35% kcal from carbohydrate.

This carbohydrate intake is lower than the Daily Value and lower than the "over half" recommendation.

3. 350 g carbohydrate × 2 = 700 g carbohydrate/day.
   700 g carbohydrate × 4 kcal/g = 2800 kcal from carbohydrate.
   2800 kcal from carbohydrate ÷ 6000 total kcal/day = 0.466 (rounded to 0.47)
   0.47 × 100 = 47% kcal from carbohydrate.

This carbohydrate intake is higher than the Daily Value and lower than the "over half" recommendation.

4. 150 g carbohydrate × 4 kcal/g = 600 kcal from carbohydrate.
   600 kcal from carbohydrate ÷ 1000 total kcal/day = 0.60.
   0.60 × 100 = 60% kcal from carbohydrate.

This carbohydrate intake is lower than the Daily Value and higher than the "over half" recommendation.

Multiple Choice

1. b	2. c	3. c	4. d	5. d
6. b	7. d	8. d	9. a	10. d

ALTERNATIVES TO SUGAR

PEOPLE WHO want to limit their use of sugar may encounter two sets of alternative sweeteners. One set, the **artificial sweeteners,** provide virtually no energy and are sometimes referred to as nonnutritive sweeteners. The other set, the **sugar replacers,** yield energy and are sometimes referred to as **nutritive sweeteners.**

ARTIFICIAL SWEETENERS

Artificial sweeteners permit people to keep their sugar and energy intakes down, yet still enjoy the delicious sweet tastes of their favorite foods and beverages. The Food and Drug Administration (FDA) has approved the use of four artificial sweeteners—saccharin, aspartame, acesulfame potassium (acesulfame-K), and sucralose. Two others have petitioned the FDA and are awaiting approval—alitame and cyclamate. Table H4-1 and the glossary on p. 124 provide general details about each of these sweeteners.

Saccharin, acesulfame-K, and sucralose are not metabolized in the body; in contrast, the body digests aspartame as a protein. In fact, aspartame is *technically* classified as a nutritive sweetener because it yields energy (4 kcalories per gram, as does protein). But because so little is used, its energy contribution is negligible.

Some consumers have challenged the safety of using artificial sweeteners. Considering that all compounds are toxic at some dose, it is little surprise that large doses of artificial sweeteners (or their components or metabolic byproducts) have toxic effects. The question to ask is whether their ingestion is safe for human beings in quantities people normally use (and potentially abuse). The answer is yes, except in the special case described for aspartame later.

The FDA reached this conclusion after decades of extensive research. Such research is lacking for the herb **stevia,** a shrub whose leaves have long been used by the people of South America to sweeten their beverages. In the United States, stevia is sold in health-food stores as a dietary supplement. The FDA has reviewed the limited research on the use of stevia as an alternative to artificial sweeteners and found concerns regarding its effect on reproduction, cancer development, and energy metabolism. Used sparingly, stevia may do little harm, but the FDA could not approve its extensive and widespread use in the U.S. market. Canada, the European Union, and the United Nations have reached similar conclusions. That stevia can be sold as a dietary supplement, but not used as a food additive in the United States, highlights key differences in FDA regulations. Food additives must prove their safety and effectiveness before receiving FDA approval, whereas dietary supplements are not required to submit to any testing or receive any approval. (See Highlight 10 and Chapter 18 for information on herbs and other dietary supplements.)

The Safety of Saccharin

Saccharin, used for over 100 years in the United States, is currently used by some 50 million people—primarily in soft drinks, secondarily as a tabletop sweetener. Saccharin is rapidly excreted in the urine and does not accumulate in the body.

Questions about saccharin's safety surfaced in 1977, when experiments suggested that large doses of saccharin (equivalent to hundreds of cans of diet soda daily for a lifetime) increased the risk of bladder cancer in rats. The FDA proposed banning saccharin as a result. Public outcry in favor of saccharin was so loud, however, that Congress imposed a moratorium on the ban—a moratorium that was repeatedly extended until 1991, when the FDA withdrew its proposal to ban saccharin. Products containing saccharin must still carry a warning label: "Use of this product may be hazardous to your health. This product contains saccharin, which has been determined to cause cancer in laboratory animals."

Does saccharin cause cancer? The largest population study to date, involving 9000 men and women, showed overall that saccharin use did not increase the risk of cancer. Among certain small groups of the population, however, such as those who both smoked heavily and used saccharin, the risk of bladder cancer was slightly

People wanting to limit their sugar intake can find a variety of foods and beverages made with artificial sweetners.

TABLE H4-1 Artificial Sweeteners

Artificial Sweeteners	Relative Sweetness[a]	Energy (kcal/g)	Acceptable Daily Intake	Approved Uses
Approved Sweeteners				
Saccharin	450	0	5 mg/kg body weight	Tabletop sweeteners, wide range of foods, beverages, cosmetics, and pharmaceutical products
Aspartame	200	4[b]	50 mg/kg body weight[c] Warning to people with PKU: Contains phenylalanine	General purpose sweetener in all foods and beverages
Acesulfame-K	200	0	15 mg/kg body weight[d]	Tabletop sweeteners, puddings, gelatins, chewing gum, candies, baked goods, desserts, alcoholic beverages
Sucralose	600	0	5 mg/kg body weight	Carbonated beverages, dairy products, baked goods, coffee and tea, fruit spreads, syrups, tabletop sweeteners, chewing gum, frozen desserts, salad dressing
Sweeteners with Approval Pending				***Proposed Uses***
Alitame	2000	4[e]	—	Beverages, baked goods, tabletop sweeteners, frozen desserts
Cyclamate	30	0	—	Tabletop sweeteners, baked goods

[a]Relative sweetness is determined by comparing the approximate sweetness of a sugar substitute with the sweetness of pure sucrose, which has been defined as 1.0. Chemical structure, temperature, acidity, and other flavors of the foods in which the substance occurs all influence relative sweetness.
[b]Aspartame provides 4 kcalories per gram, as does protein, but because so little is used, its energy contribution is negligible. In powdered form it is sometimes mixed with lactose, however, so a 1-gram packet may provide 4 kcalories.
[c]Recommendations from the World Health Organization and in Europe and Canada limit aspartame intake to 40 milligrams per kilogram of body weight.
[d]Recommendations from the World Health Organization limit acesulfame-K intake to 9 milligrams per kilogram of body weight.
[e]Alitame provides 4 kcalories per gram, as does protein, but because so little is used, its energy contribution is negligible.

greater. Other studies involving more than 5000 people with bladder cancer showed no association between bladder cancer and saccharin use.[1] In 2000, saccharin was removed from the list of suspected cancer-causing substances.

Common sense dictates that consuming large amounts of any substance is probably not wise, but at current, moderate intake levels, saccharin appears to be safe for most people. It has been approved for use in more than 100 countries.

The Safety of Aspartame

Aspartame is one of the most studied of all food additives; extensive animal and human studies document its safety. Long-term consumption of aspartame is not associated with any adverse health effects.

The nutrients in aspartame may present a problem for certain people, however, and for this reason, aspartame also carries a warning on its label. Aspartame is a simple chemical compound made of components common to

GLOSSARY

Acceptable Daily Intake (ADI): the estimated amount of a sweetener that individuals can safely consume each day over the course of a lifetime without adverse effect. It includes a 100-fold safety factor.

acesulfame (AY-sul-fame) **potassium:** an artificial sweetener composed of an organic salt that has been approved in both the United States and Canada; also known as acesulfame-K because K is the chemical symbol for potassium.

alitame (AL-ih-tame): an artificial sweetener composed of two amino acids (alanine and aspartic acid); FDA approval pending.

artificial sweeteners: sugar substitutes that provide negligible, if any, energy; sometimes called **nonnutritive sweeteners.**

aspartame (ah-SPAR-tame or ASS-par-tame): an artificial sweetener composed of two amino acids (phenylalanine and aspartic acid); approved in both the United States and Canada.

cyclamate (SIGH-kla-mate): an artificial sweetener that is being considered for approval in the United States and is available in Canada as a tabletop sweetener, but not as an additive.

nutritive sweeteners: sweeteners that yield energy, including both sugars and sugar replacers.

saccharin (SAK-ah-ren): an artificial sweetener that has been approved in the United States. In Canada, approval for use in foods and beverages is pending; currently available only in pharmacies and only as a tabletop sweetener, not as an additive.

stevia (STEE-vee-ah): a South American shrub whose leaves are used as a sweetener; sold in the United States as a dietary supplement that provides sweetness without kcalories.

sucralose (SUE-kra-lose): an artificial sweetener approved for use in the United States.

sugar replacers: sugarlike compounds that can be derived from fruits or commercially produced from dextrose; also called **sugar alcohols** or **polyols.** Sugar alcohols are absorbed more slowly than other sugars and metabolized differently in the human body; they are not readily utilized by ordinary mouth bacteria. Examples are **maltitol, mannitol, sorbitol, xylitol, isomalt,** and **lactitol.**

FIGURE H4-1 **Structure of Aspartame**

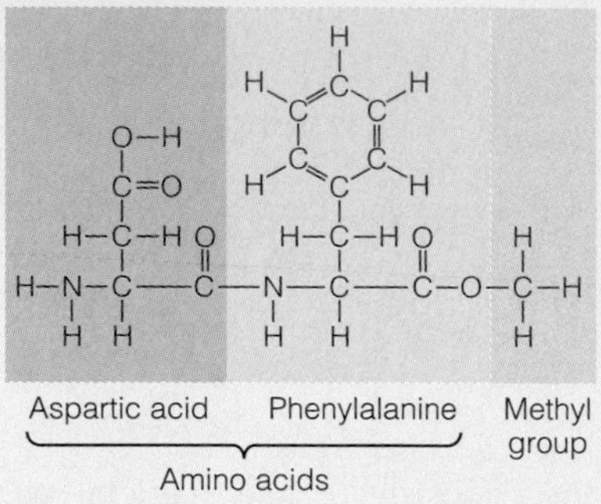

Aspartic acid | Phenylalanine | Methyl group

Amino acids

FIGURE H4-2 **Metabolism of Aspartame**

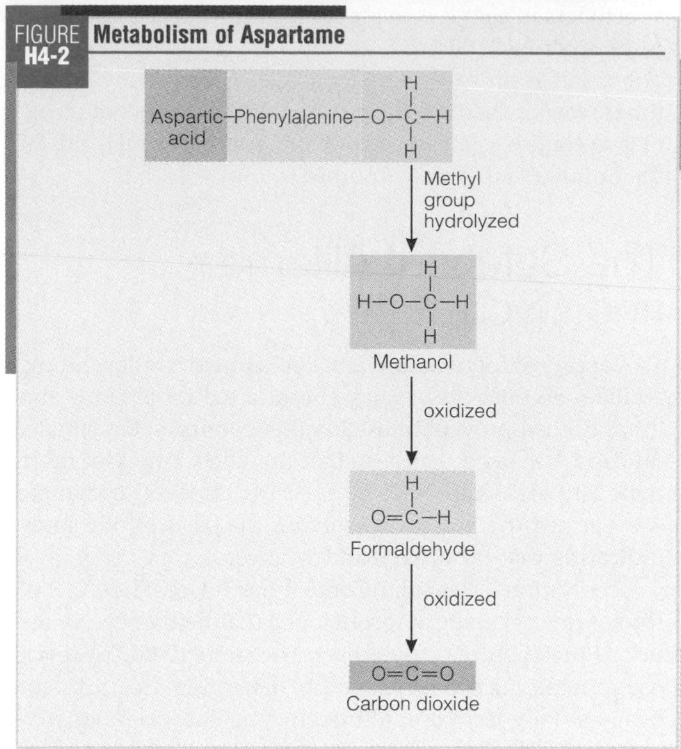

many foods: two amino acids (phenylalanine and aspartic acid) and a methyl group (CH3). Figure H4-1 shows its chemical structure. The flavors of the components give no clue to the combined effect; one of them tastes bitter, and the other is tasteless, but the combination creates a product that is 200 times sweeter than sucrose.

In the digestive tract, enzymes split aspartame into its three component parts. The body absorbs the two amino acids and uses them just as if they had come from food protein, which is made entirely of amino acids including these two.

Because this sweetener contributes phenylalanine, products containing aspartame must bear a warning label for people with the inherited disease phenylketonuria (PKU). As Highlight 22 explains, people with PKU are unable to dispose of any excess phenylalanine. The accumulation of phenylalanine and its by-products is toxic to the developing nervous system, causing irreversible brain damage. For this reason, all newborns in the United States are screened for PKU. The treatment for PKU is a special diet that must strike a balance, providing enough phenylalanine to support normal growth and health but not enough to cause harm. The question then is does aspartame raise blood phenylalanine high enough to be toxic to people with PKU? Apparently not. The little extra phenylalanine from aspartame poses only a small risk, even in heavy users.

Still, there is a compelling reason why children with PKU need to get all their phenylalanine from foods, and not from an artificial sweetener. The PKU diet excludes such protein- and nutrient-rich foods as milk, meat, fish, poultry, cheese, eggs, nuts, legumes, and many bread products. Only with difficulty can these children obtain the many essential nutrients—such as calcium, iron, and the B vitamins—found along with phenylalanine in these foods. To suggest that children with PKU squander any of their limited phenylalanine allowance on the purified phenylalanine of aspartame, which contributes none of the associated vitamins or minerals essential for good health and normal growth, would open the way for poor nutrition.

Setting aside the special case of PKU, is there any reason to be concerned about the products aspartame yields in the body? During metabolism, the methyl group momentarily becomes methyl alcohol (methanol)—a potentially toxic compound (see Figure H4-2). Then enzymes convert methanol to formaldehyde, another toxic compound. Finally, formaldehyde is broken down to carbon dioxide. Before aspartame could be approved, the quantities of these products generated during metabolism had to be determined; they were found to fall below the threshold at which they would cause harm. In fact, ounce for ounce, tomato juice yields six times as much methanol as a diet soda.

In conclusion, except for people with PKU, aspartame is safe. Some individuals may exhibit vague, but not dangerous, symptoms due to unusual sensitivity to aspartame, but it is generally safe. Like saccharin, aspartame has been approved for use in more than 100 countries.

The Safety of Acesulfame-K

The FDA approved **acesulfame-K** in 1988 after reviewing more than 90 safety studies conducted over 15 years. Some consumer groups believe that acesulfame-K causes tumors in rats and should not have been approved. The FDA counters that the tumors were not caused by the sweetener, but were typical of those commonly found in rat studies. Acesulfame-K has been approved for use in more than 60 countries.

The Safety of Sucralose

Sucralose is the "new kid on the block," having received FDA approval in 1998. The FDA approval came after a review of over 110 safety studies conducted on both animals and human beings. Sucralose is unique among the artificial sweeteners in that it is made from sugar that has had three of its hydroxyl (OH) groups replaced by chlorine atoms. The result is an exceptionally stable molecule that is 600 times sweeter than sugar. Because the body does not recognize sucralose as a carbohydrate, it passes through the GI tract undigested and unabsorbed.

The Safety of Alitame and Cyclamate

FDA approval for **alitame** and **cyclamate** is still pending. To date, no safety issues have been raised for alitame, and it has been approved for use in other countries. Cyclamate, on the other hand, has been battling safety issues for more than 50 years. Approved by the FDA in 1949, cyclamate was banned in 1969 principally on the basis of one study indicating that it caused bladder cancer in rats.

The National Research Council has reviewed dozens of studies on cyclamate and concluded that neither cyclamate nor its metabolites cause cancer. The council did, however, recommend further research to determine the risks for heavy or long-term use. Although cyclamate does not *initiate* cancer, it may *promote* cancer development once started. The FDA currently has no policy on substances that enhance the cancer-causing activities of other substances, but it is unlikely to approve cyclamate soon, if at all. Agencies in more than 50 other countries, including Canada, have approved cyclamate.

Acceptable Daily Intake

The amount of artificial sweetener considered safe for daily use is called the **Acceptable Daily Intake (ADI).** The ADI represents the level of consumption that, if maintained every day throughout a person's life, would still be considered safe by a wide margin.

For example, the ADI for aspartame is 50 milligrams per kilogram of body weight. That is, the FDA approved aspartame based on the assumption that no one would consume more than 50 milligrams per kilogram of body weight in a day. This maximum daily intake is indeed a lot: for a 150-pound adult, it adds up to 97 packets of Equal or 20 cans of soft drinks sweetened only with aspartame. The company that produces aspartame estimates that if all the sugar and saccharin in the U.S. diet were replaced with aspartame, 1 percent of the population would be consuming the FDA maximum. Most people who use aspartame consume less than 5 milligrams per kilogram of body weight per day. A young child who drinks four glasses of aspartame-sweetened beverages on a hot day and has five servings of other products with aspartame that day (such as pudding, chewing gum, cereal, gelatin, and frozen desserts) takes in the FDA maximum level. Although this presents no proven hazard, it seems wise to offer children other foods so as not to exceed the limit. Table H4-2 lists the average amounts of aspartame in some common foods.

TABLE H4-2 Average Aspartame Contents of Selected Foods

Food	Aspartame (mg)
12 oz diet soft drink	170
8 oz powdered drink	100
8 oz sugar-free fruit yogurt	124
4 oz gelatin dessert	80
1 packet sweetener	35

For persons choosing to use artificial sweeteners, the American Dietetic Association wisely advises that they be used in moderation and only as part of a well-balanced nutritious diet.[2] The dietary principles of both moderation and variety help to reduce the possible risks associated with any food.

Artificial Sweeteners and Weight Control

Many people eat and drink products sweetened with artificial sweeteners to help them control weight. Ironically, a few studies have reported that intense sweeteners, such as aspartame, may stimulate appetite, which could lead to weight gain.[3] Contradicting these reports, most studies find no change in feelings of hunger and no change in food intakes or body weight.[4] Adding to the confusion, some studies report lower energy intakes and greater weight losses when people eat or drink artificially sweetened products.[5]

When studying the effects of artificial sweeteners on food intake and body weight, researchers ask different questions and take different approaches. It matters, for example, whether the people used in a study are of a healthy weight and whether they are following a weight-loss diet. Motivations for using sweeteners differ, too, and this influences a person's actions. For example, a person might drink an artificially sweetened beverage now so as to be able to eat a high-kcalorie food later. This person's energy intake might stay the same or increase. On the other hand, a person trying to control food energy intake might drink an artificially sweetened beverage now and then choose a low-kcalorie food later. This plan would help reduce the person's energy intake.

In designing experiments on artificial sweeteners, researchers have to distinguish between the effects of sweetness and the effects of a particular substance. If a person is hungry shortly after eating an artificially sweetened snack, is that because the sweet taste (of all sweeteners, including sugars) stimulates appetite? Or is it because the artificial sweetener itself stimulates appetite? Research must also distinguish between the effects of food energy and the effects of the substance. If a person is hungry shortly after eating an artificially sweetened snack, is that because less food energy was available to satisfy hunger? Or is it because the artificial sweetener itself triggers hunger? Furthermore, if appetite is stimulated and a person feels hungry, does that actually lead to increased food intake?

Whether a person compensates for the energy reduction of artificial sweeteners either partially or fully depends on several factors. Using artificial sweeteners will not automatically lower energy intake; to control energy intake successfully, a person needs to make informed diet and activity decisions throughout the day (as Chapter 9 explains).

SUGAR REPLACERS

Some "sugar-free" or reduced-kcalorie products contain sugar replacers.* The term *sugar replacers* describes the sugar alcohols—mannitol, sorbitol, xylitol, maltitol, isomalt, and lactitol—that provide bulk and sweetness in cookies, hard candies, sugarless gums, jams, and jellies. These products claim to be "sugar-free" on their labels, but in this case, "sugar-free" does not mean free of kcalories. Sugar replacers do provide kcalories, but fewer than their carbohydrate cousins, the sugars. Table H4-3 includes their energy values, but a simple estimate can help consumers: divide grams by 2.[6] Sugar alcohols occur naturally in fruits and vegetables; they are also used by manufacturers as a low-energy bulk ingredient in many products.[7]

Sugar alcohols evoke a low glycemic response. The body absorbs sugar alcohols slowly; consequently, they are slower to enter the bloodstream than other sugars. Side effects such as gas, abdominal discomfort, and diarrhea, however, make them less attractive than the artificial sweeteners. For this reason, regulations require food labels to state that "Excess consumption may have a laxative effect" if reasonable consumption could result in the daily ingestion of 50 grams of a sugar alcohol.

*To minimize confusion, the American Diabetes Association prefers the term *sugar replacers* instead of "sugar alcohols" (which connotes alcohol), "bulk sweeteners" (which connotes fiber), or "sugar substitutes" (which connotes aspartame and saccharin).

TABLE H4-3 Sugar Replacers

Sugar Alcohols	Relative Sweetness[a]	Energy (kcal/g)	Approved Uses
Isomalt	0.5	2.0	Candies, chewing gum, ice cream, jams and jellies, frostings, beverages, baked goods
Lactitol	0.4	2.0	Candies, chewing gum, frozen dairy desserts, jams and jellies, frostings, baked goods
Maltitol	0.9	2.1	Particularly good for candy coating
Mannitol	0.7	1.6	Bulking agent, chewing gum
Sorbitol	0.5	2.6	Special dietary foods, candies, gums
Xylitol	1.0	2.4	Chewing gum, candies, pharmaceutical and oral health products

[a]Relative sweetness is determined by comparing the approximate sweetness of a sugar replacer with the sweetness of pure sucrose, which has been defined as 1.0. Chemical structure, temperature, acidity, and other flavors of the foods in which the substance occurs all influence relative sweetness.

The real benefit of using sugar replacers is that they do not contribute to dental caries. Bacteria in the mouth cannot metabolize sugar alcohols as rapidly as sugar. They are therefore valuable in chewing gums, breath mints, and other products that people keep in their mouths for a while.

The sugar replacers, like the artificial sweeteners, can occupy a place in the diet, and provided they are used in moderation, they will do no harm. In fact, they can help, both by providing an alternative to sugar for people with diabetes and by inhibiting caries-causing bacteria. People may find it appropriate to use all three sweeteners at times: artificial sweeteners, sugar replacers, and sugar itself.

Products containing sugar replacers may claim to "not promote tooth decay" if they meet FDA criteria for dental plaque activity.

Nutrition on the Net

•WEBSITES•

Access these websites for further study of topics covered in this highlight.

- Find updates and quick links to these and other nutrition-related sites at our website: **www.wadsworth.com/nutrition**
- Search for "artificial sweeteners" at the U.S. Government health information site: **www.healthfinder.gov**
- Search for "sweeteners" at the International Food Information Council site: **ificinfo.health.org**

References

1 Position of The American Dietetic Association: Use of nutritive and nonnutritive sweeteners, *Journal of the American Dietetic Association* 98 (1998): 580–587.

2 Position of The American Dietetic Association, 1998.

3 J. E. Blundell and P. J. Rogers, Sweet carbohydrate substitutes (intense sweeteners) and the control of appetite: Scientific issues, in *Appetites and Body Weight Regulation: Sugar, Fat, and Macronutrient Substitutes*, ed. J. D. Fernstrom and G. D. Miller (Boca Raton, Fla.: CRC Press, 1994), pp. 113–124.

4 S. J. Gatenby and coauthors, Extended use of foods modified in fat and sugar content: Nutrition implications in a free-living female population, *American Journal of Clinical Nutrition* 65 (1997): 1867–1873; A. Drewnowski, Intense sweeteners and the control of appetite, *Nutrition Reviews* 53 (1995): 1–7.

5 G. L. Blackburn and coauthors, The effect of aspartame as part of a multidisciplinary weight-control program on short- and long-term control of body weight, *American Journal of Clinical Nutrition* 65 (1997): 409–418; M. G. Tordoff and A. M. Alleva, Effect of drinking soda sweetened with aspartame or high-fructose corn syrup on food intake and body weight, *American Journal of Clinical Nutrition* 51 (1990): 963–969.

6 K. McNutt, What clients need to know about sugar replacers, *Journal of the American Dietetic Association* 100 (2000): 466-469.

7 F. R. J. Bonet, Undigestible sugars in food products, *American Journal of Clinical Nutrition* 59 (1994): 763S–769S.

The Lipids: Triglycerides, Phospholipids, and Sterols

CHAPTER 5

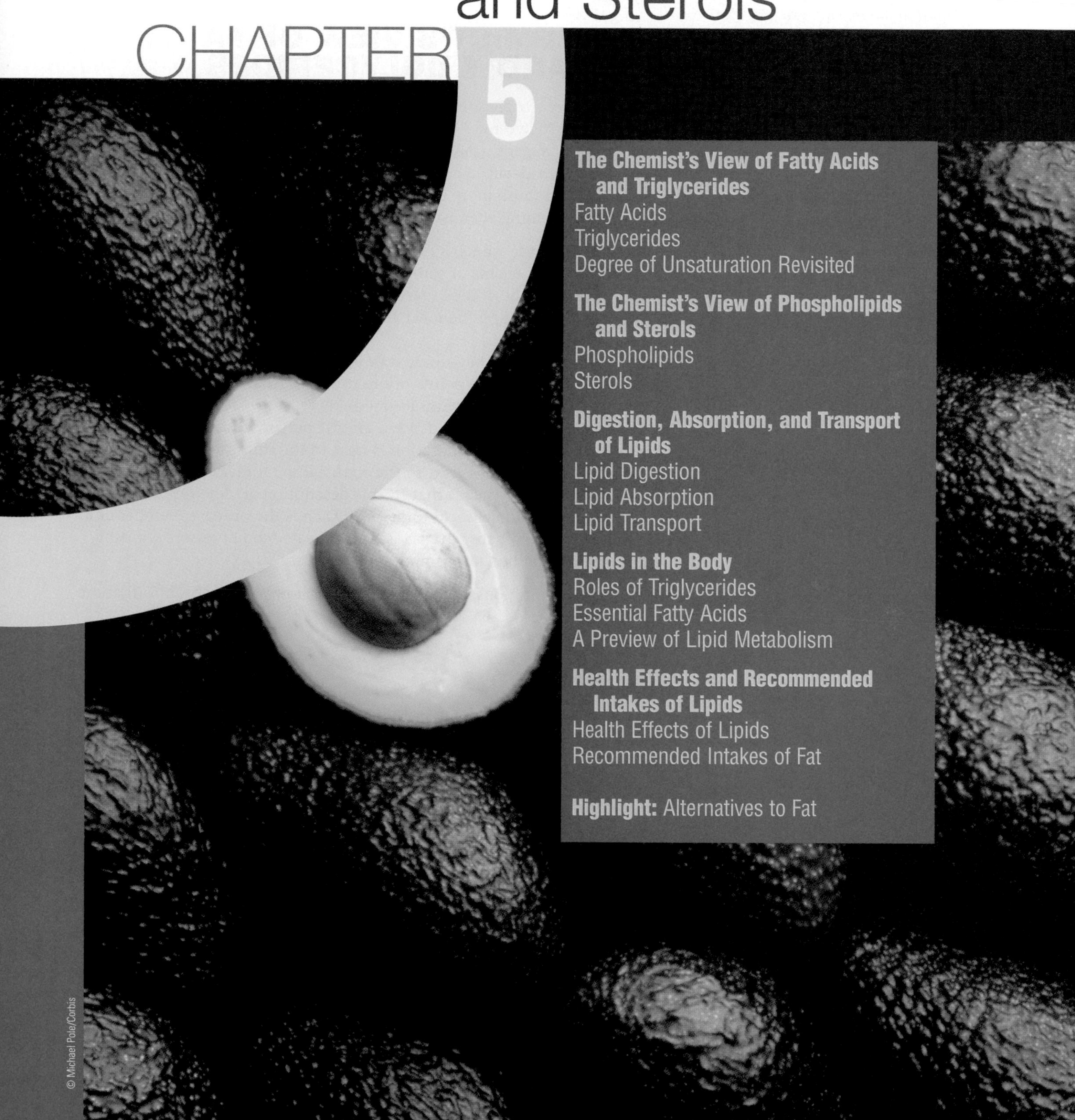

© Michael Pole/Corbis

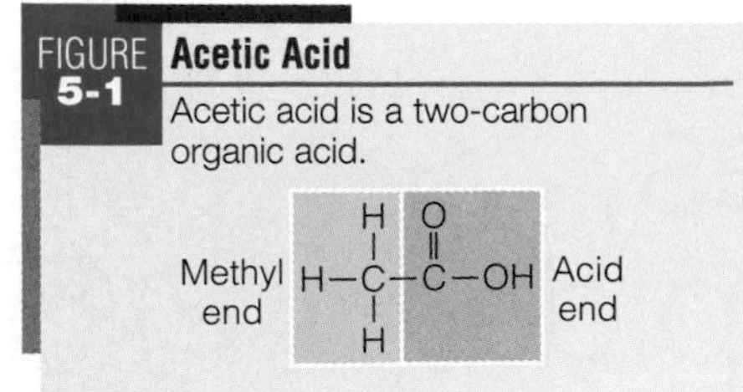

FIGURE 5-1 Acetic Acid

Acetic acid is a two-carbon organic acid.

◆ Of the lipids in foods, 95% are fats and oils (triglycerides), and 5% are other lipids (phospholipids and sterols). Of the lipids stored in the body, 99% are triglycerides.

Most people are surprised to learn that fat has some virtues. Only when people consume either too much or too little fat does ill health follow. It is true, though, that in our society of abundance, people are likely to encounter too much fat.

Fat is actually a subset of the class of nutrients known as **lipids,** but the term *fat* is often used to refer to all the lipids. The lipid family includes triglycerides (**fats** and **oils**), phospholipids, and sterols, all important to nutrition. The triglycerides◆ predominate, both in foods and in the body.

THE CHEMIST'S VIEW OF FATTY ACIDS AND TRIGLYCERIDES

Like carbohydrates, triglycerides are composed of carbon (C), hydrogen (H), and oxygen (O). Triglycerides have many more carbons and hydrogens in proportion to their oxygens, however, and so can supply more energy per gram (Chapter 7 provides details).

For people who think more easily in words than in chemical symbols, this *preview* of the upcoming chemistry may be helpful:

1. Every triglyceride contains one molecule of glycerol and three fatty acids (basically, chains of carbon atoms).
2. Fatty acids may be 4 to 24 (even numbers of) carbons long, the 18-carbon ones being the most common in foods and especially noteworthy in nutrition.
3. Fatty acids may also be saturated or unsaturated. Unsaturated fatty acids may have one or more points of unsaturation (that is, they may be monounsaturated or polyunsaturated).
4. Of special importance in nutrition are the polyunsaturated fatty acids whose *first* point of unsaturation is next to the third carbon (known as omega-3 fatty acids) or next to the sixth carbon (omega-6).
5. The 18-carbon fatty acids that fit this description are linolenic acid (omega-3) and linoleic acid (omega-6). Each is the primary member of a family of longer-chain fatty acids that regulate blood pressure, clotting, and other body functions important to health.

The paragraphs, definitions, and diagrams that follow present this information again in much more detail.

Fatty Acids

A **fatty acid** is an organic acid—a chain of carbon atoms with hydrogens attached—that has an acid group (COOH) at one end and a methyl group (CH_3) at the other end. The organic acid shown in Figure 5-1 is acetic acid, the compound that gives vinegar its sour taste. Acetic acid is the simplest such acid, with a "chain" only two carbon atoms long.

• The Length of the Carbon Chain • Most naturally occurring fatty acids contain even numbers of carbons in their chains—up to 24 carbons in length. This discussion begins with the 18-carbon fatty acids, which are abundant in our food supply. Stearic acid is the simplest of the 18-carbon fatty acids; the bonds between its carbons are all alike:

```
  H H H H H H H H H H H H H H H H H O
  | | | | | | | | | | | | | | | | | ‖
H-C-C-C-C-C-C-C-C-C-C-C-C-C-C-C-C-C-C-O-H
  | | | | | | | | | | | | | | | | |
  H H H H H H H H H H H H H H H H H
```

As you can see, stearic acid is 18 carbons long, and each atom meets the rules of chemical bonding described in Figure 4-1 on p. 93. The following structure

lipids: a family of compounds that includes triglycerides (fats and oils), phospholipids, and sterols.

fats: lipids in foods or the body; composed mostly of triglycerides.

oils: liquid fats (at room temperature).

fatty acid: an organic compound composed of a carbon chain with hydrogens attached and an acid group (COOH) at one end. The COOH group of an organic acid can be represented this way:

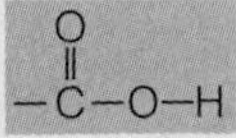

also depicts stearic acid, but in a simpler way, with each "corner" on the zigzag line representing a carbon atom with two attached hydrogens:

```
    H                                                   O
    |                                                   ‖
 H—C /\/\/\/\/\/\/\/\/\/\/\/\/\/\/\/\                   C—O—H
    |
    H
```

Stearic acid (simplified structure).

As mentioned, the carbon chains of fatty acids vary in length. The long-chain (12 to 24 carbons) fatty acids of meats and fish are most common in the diet.◆ Smaller amounts of medium-chain (6 to 10 carbons) and short-chain (fewer than 6 carbons) fatty acids also occur, primarily in dairy products.

Tables C-1 and C-2 in Appendix C provide the names, chain lengths, and sources of fatty acids commonly found in foods. ◆

• The Degree of Unsaturation • Stearic acid is a **saturated fatty acid** (terms that describe the saturation of fatty acids are defined in the glossary below). A saturated fatty acid is fully loaded with hydrogen atoms and contains only single bonds between its carbon atoms. If two hydrogens were missing from the middle of the carbon chain, the remaining structure might be:

```
   H H H H H H H H     H H H H H H H O
   | | | | | | | |     | | | | | | | ‖
 H—C—C—C—C—C—C—C—C—C—C—C—C—C—C—C—C—C—O—H
   | | | | | | | | | | | | | | | | |
   H H H H H H H H H H H H H H H H H
```

An impossible chemical structure.

Such a compound cannot exist, however, because two of the carbons have only three bonds each, and nature requires that every carbon have four bonds. The two carbons therefore form a double bond:

```
   H H H H H H H H     H H H H H H H O
   | | | | | | | |     | | | | | | | ‖
 H—C—C—C—C—C—C—C—C—C=C—C—C—C—C—C—C—C—O—H
   | | | | | | | | | | | | | | | | |
   H H H H H H H H H H H H H H H H H
```

Oleic acid, an 18-carbon monounsaturated fatty acid.

The same structure drawn more simply looks like this:*

```
    H                                        O
    |                                        ‖
 H—C /\/\/\/\/\/\/\/=\/\/\/\/\/\/\/\         C—O—H
    |
    H
```

Oleic acid (simplified structure).

The double bond is a **point of unsaturation.** Hence, a fatty acid like this—with two hydrogens missing and a double bond—is an ***un*saturated fatty acid.** This one is the 18-carbon ***mono*unsaturated fatty acid** oleic acid, which is abundant in olive oil.

*Remember that each "corner" on the zigzag line represents a carbon atom with two attached hydrogens. In addition, the actual shape bends at the double bonds and rotates around single bonds. These molecules, although drawn straight on paper, are constantly twisting and bending. At any given moment, they may be coiled, horseshoe shaped, or straight.

GLOSSARY OF SATURATION TERMS

These terms are listed in order from the most saturated to the most unsaturated.

saturated fatty acid: a fatty acid carrying the maximum possible number of hydrogen atoms—for example, stearic acid. A **saturated fat** is composed of triglycerides in which most of the fatty acids are saturated.

point of unsaturation: the double bond of a fatty acid, where hydrogen atoms can easily be added to the structure.

unsaturated fatty acid: a fatty acid that lacks hydrogen atoms and has at least one double bond between carbons (includes monounsaturated and polyunsaturated fatty acids). An **unsaturated fat** is composed of triglycerides in which most of the fatty acids are unsaturated.

monounsaturated fatty acid: a fatty acid that lacks two hydrogen atoms and has one double bond between carbons—for example, oleic acid. A **monounsaturated fat** is composed of triglycerides in which most of the fatty acids are monounsaturated.

- **mono** = one

polyunsaturated fatty acid (PUFA): a fatty acid that lacks four or more hydrogen atoms and has two or more double bonds between carbons—for example, linoleic acid (two double bonds) and linolenic acid (three double bonds). A **polyunsaturated fat** is composed of triglycerides in which most of the fatty acids are polyunsaturated.

- **poly** = many

A ***polyunsaturated* fatty acid** has two or more carbon-to-carbon double bonds. **Linoleic acid,** the 18-carbon fatty acid common in vegetable oils, lacks four hydrogens and has two double bonds:

Linoleic acid, an 18-carbon polyunsaturated fatty acid.

```
    H  H  H  H  H        H        H  H  H  H  H  H  H  O
    |  |  |  |  |        |        |  |  |  |  |  |  |  ‖
H—C—C—C—C—C—C=C—C—C=C—C—C—C—C—C—C—C—C—O—H
    |  |  |  |  |  |  |  |  |  |  |  |  |  |  |  |  |
    H  H  H  H  H  H  H  H  H  H  H  H  H  H  H  H  H
```

Drawn more simply, linoleic acid looks like this:

Linoleic acid (simplified structure).

```
    H
    |                                                      O
H—C                                                      ‖
    |                                                      C—O—H
    H
```

A fourth 18-carbon fatty acid is **linolenic acid,** which has three double bonds. Table 5-1 presents the 18-carbon fatty acids.

• The Location of Double Bonds • Fatty acids differ not only in the length of their chains and their degree of saturation, but also in the locations of their double bonds. Chemists identify polyunsaturated fatty acids by the position of the double bond nearest the methyl (CH_3) end of the carbon chain, which is described by an **omega** number. A polyunsaturated fatty acid with its first double bond three carbons away from the methyl end is an **omega-3 fatty acid,** for example (see Figure 5-2). Similarly, an **omega-6 fatty acid** is a polyunsaturated fatty acid with its first double bond six carbons away from the methyl end.

Triglycerides

Few fatty acids occur free in foods or in the body. Most often, they are incorporated into **triglycerides**—lipids composed of three fatty acids attached to a **glycerol.** (Figure 5-3 presents a glycerol molecule.) To make a triglyceride, a series of condensation reactions combine a hydrogen atom (H) from the glycerol and a hydroxyl (OH) group from a fatty acid, forming a molecule of water (H_2O) and leaving a bond between the other two molecules (see Figure 5-4). Most triglycerides contain a mixture of more than one type of fatty acid (see Figure 5-5 on p. 134).

Degree of Unsaturation Revisited

The chemistry of a fatty acid—whether it is short or long, saturated or unsaturated, with its first double bond here or there—influences the characteristics of foods and the health of the body. A later section of this chapter explains how these features affect health; this section describes how the degree of unsaturation influences the fats and oils in foods.

TABLE 5-1 18-Carbon Fatty Acids

Name	Notation[a]	Number of Double Bonds	Saturation	Common Food Sources
Stearic acid	18:0	0	Saturated	Most animal fats
Oleic acid	18:1	1	Monounsaturated	Olive, canola oils
Linoleic acid	18:2	2	Polyunsaturated	Sunflower, safflower, corn oils
Linolenic acid	18:3	3	Polyunsaturated	Soybean oils

[a]Chemists use a shorthand notation to describe fatty acids. The first number indicates the number of carbon atoms; the second, the number of double bonds.

linoleic (lin-oh-LAY-ick) **acid:** an essential fatty acid with 18 carbons and two double bonds (18:2).

linolenic (lin-oh-LEN-ick) **acid:** an essential fatty acid with 18 carbons and three double bonds (18:3).

omega: the last letter of the Greek alphabet (ω), used by chemists to refer to the position of the first double bond from the methyl end in a fatty acid.

omega-3 fatty acid: a polyunsaturated fatty acid in which the first double bond is three carbons away from the methyl (CH_3) end of the carbon chain.

omega-6 fatty acid: a polyunsaturated fatty acid in which the first double bond is six carbons from the methyl (CH_3) end of the carbon chain.

triglycerides (try-GLISS-er-rides): the chief form of fat in the diet and the major storage form of fat in the body; composed of a molecule of glycerol with three fatty acids attached; also called **triacylglycerols** (try-ay-seel-GLISS-er-ols).*

- **tri** = three
- **glyceride** = a compound of glycerol
- **acyl** = a carbon chain

glycerol (GLISS-er-ol): an alcohol composed of a three-carbon chain, which can serve as the backbone for a triglyceride.

- **ol** = alcohol

*Research scientists commonly use the term *triacyglycerols;* this book continues to use the more familiar term *triglycerides,* as do many other health and nutrition books and journals.

FIGURE 5-2 **Omega-3 and Omega-6 Fatty Acids Compared**

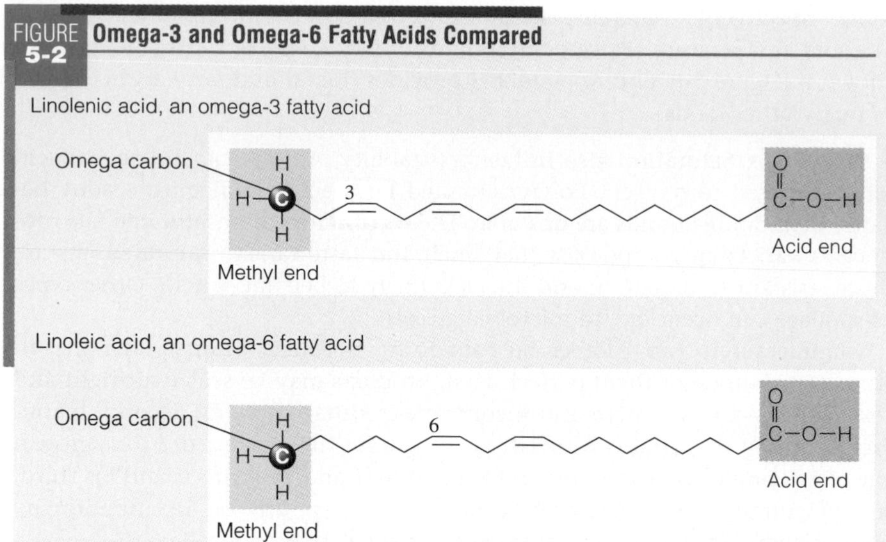

The omega number indicates the position of the first double bond in a fatty acid, counting from the methyl (CH_3) end. Thus an omega-3 fatty acid's first double bond occurs three carbons from the methyl end, and an omega-6 fatty acid's first double bond occurs six carbons from the methyl end. The members of an omega family may have different lengths and different numbers of double bonds, but the first double bond occurs at the same point in all of them.

FIGURE 5-3 **Glycerol**

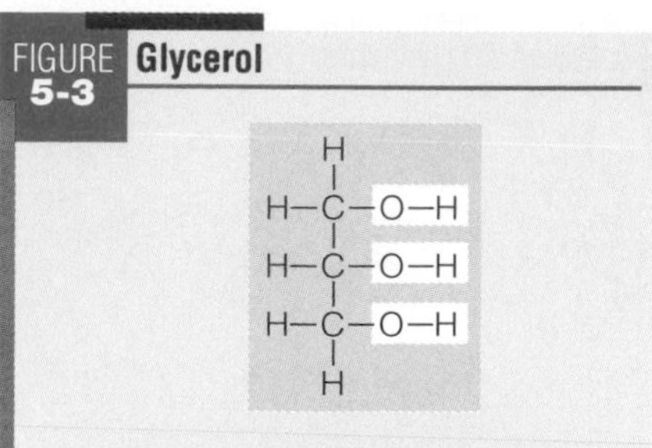

When glycerol is free, an OH group is attached to each carbon. When glycerol is part of a triglyceride, each carbon is attached to a fatty acid by a carbon-oxygen bond.

• **Firmness** • The degree of unsaturation influences the firmness of fats at room temperature. Generally speaking, the polyunsaturated vegetable oils are liquid at room temperature, and the more saturated animal fats are solid. Butter is harder than margarine because butter is more saturated than margarine; this is why people limiting their intakes of saturated fats use margarine. Not all vegetable oils are polyunsaturated, however. Cocoa butter and palm◆ and coconut oils are saturated even though they are of vegetable origin; they are firmer than most vegetable oils because of their saturation, but softer than most animal fats because of their short carbon chains (only 10 and 12 carbons

© Polara Studios Inc.

At room temperature, saturated fats (such as those found in butter) are solid, whereas unsaturated fats (such as those found in oil) are usually liquid.

◆ The food industry often refers to palm and coconut oils as the "tropical oils."

FIGURE 5-4 **Condensation of Glycerol and Fatty Acids to Form a Triglyceride**

To make a triglyceride, three fatty acids attach to glycerol in condensation reactions:

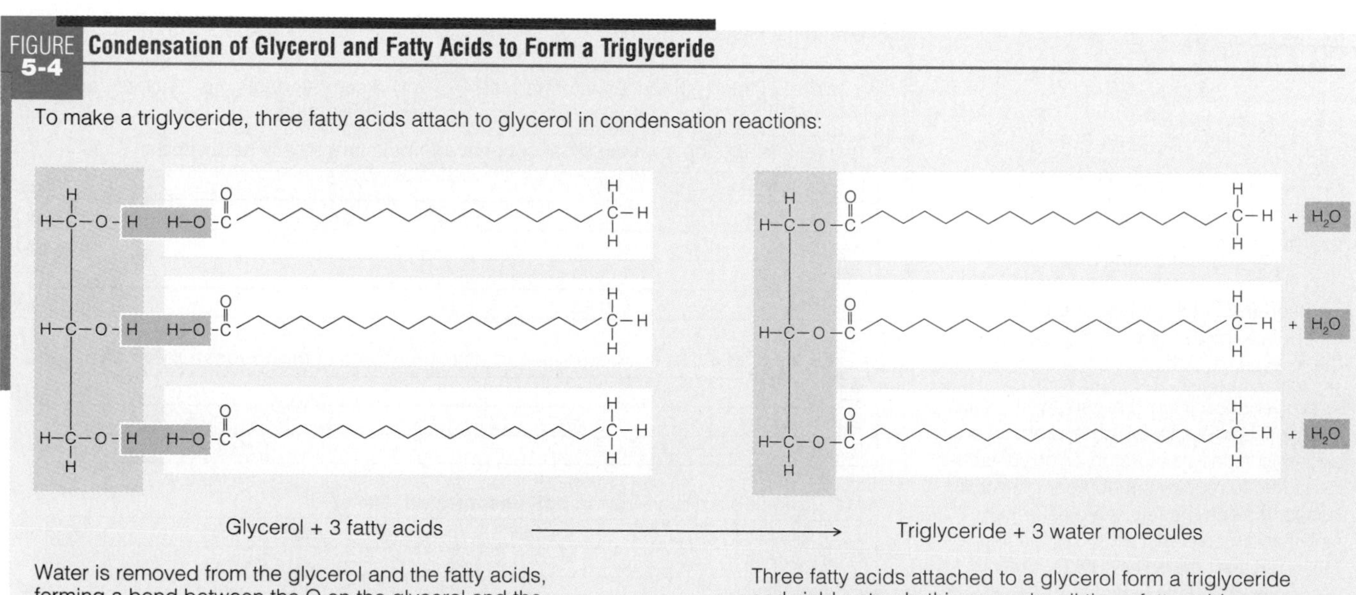

Water is removed from the glycerol and the fatty acids, forming a bond between the O on the glycerol and the C at the acid end of each fatty acid.

Three fatty acids attached to a glycerol form a triglyceride and yield water. In this example, all three fatty acids are stearic acid, but most often triglycerides contain mixtures of fatty acids (as shown in Figure 5-5).

FIGURE 5-5 A Mixed Triglyceride

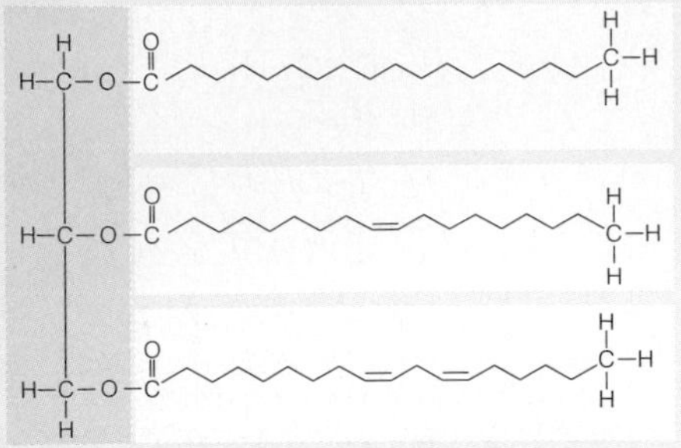

This mixed triglyceride includes a saturated fatty acid, a monounsaturated fatty acid, and a polyunsaturated fatty acid. Sometimes the chemical structure of a triglyceride is drawn with the second fatty acid to the left of the glycerol.

long, respectively). Generally, the shorter the carbon chain, the softer the fat is at room temperature. Fatty acid compositions of selected fats and oils are shown in Figure 5-6, and Appendix H provides the fat and fatty acid contents of many other foods.

• **Stability** • Saturation also influences stability. All fats can become rancid when exposed to oxygen. Polyunsaturated fatty acids spoil most readily because their double bonds are unstable. The **oxidation** of unsaturated fats produces a variety of compounds that smell and taste rancid; saturated fats are more resistant to oxidation and thus less likely to become rancid. Other types of spoilage can occur due to microbial growth.

Manufacturers can protect fat-containing products against rancidity in three ways—none of them perfect. First, products may be sealed air-tight and refrigerated—an expensive and inconvenient storage system. Second, manufacturers may add **antioxidants** to compete for the oxygen and thus protect the oil (examples are the additives BHA and BHT and vitamins C and E). Third, manufacturers may saturate some or all of the points of unsaturation by adding hydrogen molecules—a process known as hydrogenation.

• **Hydrogenation** • **Hydrogenation** offers two advantages: it protects against oxidation (which prolongs shelf life) and alters the texture of foods. When partially hydrogenated, vegetable oils become spreadable margarine. Hydrogenated fats make pie crusts flaky and puddings creamy. A disadvantage is that hydrogenation makes polyunsaturated fats more saturated (see Figure 5-7). Consequently, any health advantages of using polyunsaturated fats instead of saturated fats are lost in hydrogenation.

• ***Trans*-Fatty Acids** • Another disadvantage of hydrogenation is that some of the molecules that remain unsaturated after processing change shape from *cis* to *trans*. In nature, most unsaturated fatty acids are *cis*-fatty acids—meaning that the hydrogens next to the double bonds are on the same side of the carbon chain. Only a few (notably those found in milk and butter) are ***trans*-fatty acids**—meaning that the hydrogens next to the double bonds are on

oxidation (OKS-ee-day-shun): the process of a substance combining with oxygen.

antioxidants: compounds that protect others from oxidation by being oxidized themselves.

hydrogenation (high-dro-gen-AY-shun): a chemical process by which hydrogens are added to monounsaturated or polyunsaturated fats to reduce the number of double bonds, making the fats more saturated (solid) and more resistant to oxidation (protecting against rancidity). Hydrogenation produces *trans*-fatty acids.

***trans*-fatty acids:** fatty acids with an unusual configuration around the double bond.

FIGURE 5-6 Comparison of Dietary Fats

Most fats are a mixture of saturated, monounsaturated, and polyunsaturated fatty acids.

Saturated fats | Monounsaturated fats | Polyunsaturated fats, ω3 Linolenic acid, ω6 Linoleic acid

• Animal fats and the tropical oils of coconut and palm are mostly **saturated.**

Coconut oil (ω6)
Butter (ω6, ω3)
Beef tallow (ω6, ω3)
Palm oil (ω6)
Lard (ω6, ω3)

• Some vegetable oils, such as olive and canola, are rich in **monounsaturated** fatty acids.

Olive oil (ω6, ω3)
Canola oil (ω6, ω3)
Peanut oil (ω6)

• Many vegetable oils are rich in **polyunsaturated** fatty acids.

Safflower oil (ω6, ω3)
Sunflower oil (ω6)
Corn oil (ω6, ω3)
Soybean oil (ω6, ω3)
Cottonseed oil (ω6)

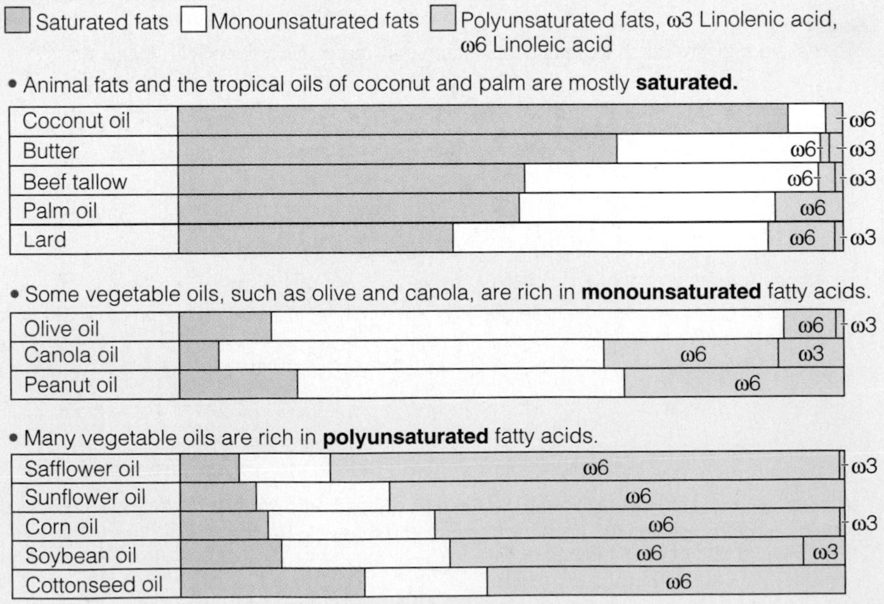

FIGURE 5-7 Hydrogenation

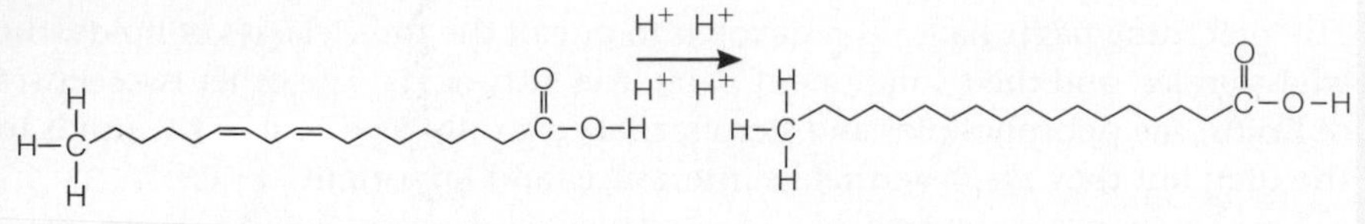

Polyunsaturated fatty acid — Hydrogenated (saturated) fatty acid

Double bonds carry a slightly negative charge and readily accept positively charged hydrogen atoms, creating a saturated fatty acid.

opposite sides of the carbon chain (see Figure 5-8). These arrangements result in different configurations for the fatty acids, and this difference affects function: in the body, *trans*-fatty acids behave more like saturated fats than like unsaturated fats. The relationship between *trans*-fatty acids and heart disease has been the subject of much recent research, as a later section describes.

IN SUMMARY The lipids important in nutrition are of three classes: triglycerides (commonly called fats), phospholipids, and sterols. The predominant lipids both in foods and in the body are triglycerides: glycerol backbones with three fatty acids attached. Fatty acids vary in the length of their carbon chains, their degrees of unsaturation, and the location of their double bond(s). Those that are fully loaded with hydrogens are saturated; those that are missing hydrogens and therefore have double bonds are unsaturated (monounsaturated or polyunsaturated). The vast majority of triglycerides contain more than one type of fatty acid. Fatty acid saturation affects fats' physical characteristics and storage properties. Hydrogenation, which makes polyunsaturated fats more saturated, gives rise to *trans*-fatty acids, altered fatty acids that may have health effects similar to those of saturated fatty acids.

FIGURE 5-8 *cis*- and *trans*-Fatty Acids Compared

Manufacturers rarely use total hydrogenation; most often a fat is partially hydrogenated, yielding a *trans*-monounsaturated fatty acid. This example shows the *cis* configuration for the 18-carbon monounsaturated fatty acid (oleic acid) and its corresponding *trans* configuration (elaidic acid).

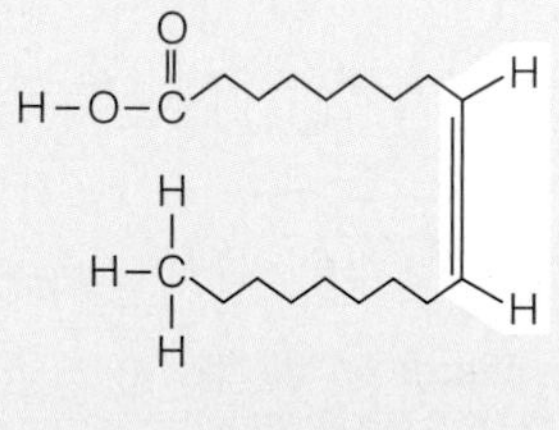

cis-fatty acid

A *cis*-fatty acid has its hydrogens on the same side of the double bond; *cis* molecules fold back into a U-like formation. Most naturally ocurring unsaturated fatty acids in foods are *cis*.

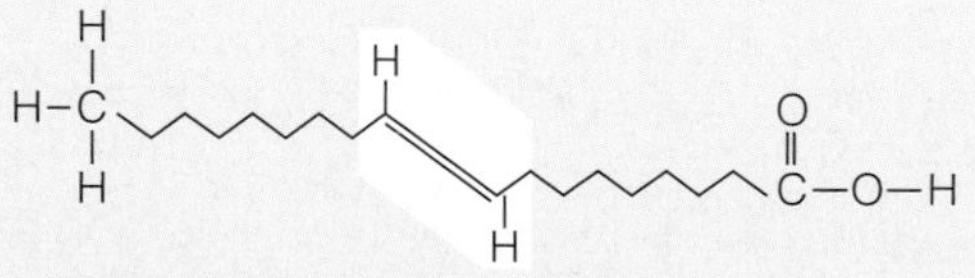

trans-fatty acid

A *trans*-fatty acid has its hydrogens on the opposite sides of the double bond; *trans* molecules are more linear. The *trans* form typically occurs in partially hydrogenated foods when hydrogen atoms shift around some double bonds and change the configuration from *cis* to *trans*.

THE CHEMIST'S VIEW OF PHOSPHOLIPIDS AND STEROLS

The preceding pages have been devoted to one of the three classes of lipids, the triglycerides, and their component parts, the fatty acids. The other two classes of lipids, the phospholipids and sterols, make up only 5 percent of the lipids in the diet, but they are nevertheless interesting and important.

Phospholipids

The best-known **phospholipid** is **lecithin.** A diagram of a lecithin molecule is shown in Figure 5-9. Notice that each lecithin has a backbone of glycerol with two of its three attachment sites occupied by fatty acids like those in triglycerides. The third site is occupied by a phosphate group and a molecule of **choline.** The fatty acids make phospholipids soluble in fat; the phosphate group allows them to dissolve in water. Such versatility enables the food industry to use phospholipids as emulsifiers to mix fats with water in such products as mayonnaise and candy bars.

• Phospholipids in Foods • In addition to the phospholipids used by the food industry as emulsifiers, phospholipids are also found in foods naturally. The richest food sources of lecithin are eggs, liver, soybeans, wheat germ, and peanuts.

• Roles of Phospholipids • The lecithins and other phospholipids are important constituents of cell membranes (see the margin drawing).◆ Because phospholipids can dissolve in both water and fat, they can help lipids move back and forth across the cell membranes into the watery fluids on both sides. They thus allow fat-soluble substances, including vitamins and hormones, to pass easily in and out of cells. The phospholipids also act as emulsifiers in the body, helping to keep fats suspended in the blood and body fluids.

Lecithin periodically receives attention in the popular press. Its fans claim that it is a major constituent of cell membranes (true), that all cells depend on the integrity of their membranes (true), and that consumers must therefore take lecithin supplements (false). The liver makes from scratch all the lecithin a person needs. As for lecithin taken as a supplement, the digestive enzyme lecithinase◆ in the intestine hydrolyzes most of it before it passes into the body fluids, so little

FIGURE 5-9 A Lecithin

From 2 fatty acids

The plus charge on the N is balanced by a negative ion—usually chloride.

From choline

From glycerol

From phosphate

This is one of the lecithins. Other lecithins have different fatty acids at the upper two positions. Notice that a molecule of lecithin is similar to a triglyceride but contains only two fatty acids. The third position is occupied by a phosphate group and a molecule of choline.

◆ **Phospholipids of a Cell Membrane**

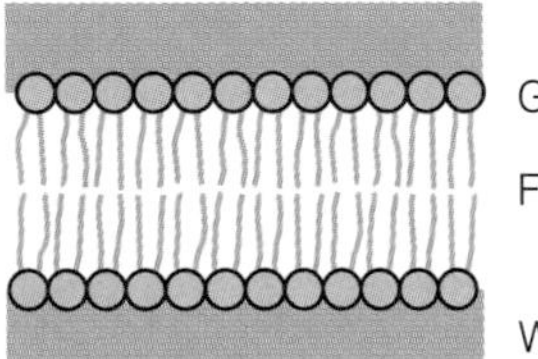

A cell membrane is made of phospholipids assembled into an orderly formation called a bilayer. The fatty acid "tails" orient themselves away from the watery fluid inside and outside of the cell. The glycerol and phosphate "heads" are attracted to the watery fluid.

◆ **Reminder: The word ending *-ase* denotes an enzyme. Hence, lecithinase is an enzyme that works on lecithin.**

phospholipid (FOS-foe-LIP-id): a compound similar to a triglyceride but having a phosphate group (a phosphorus-containing salt) and choline (or another nitrogen-containing compound) in place of one of the fatty acids.

lecithin (LESS-uh-thin): one of the phospholipids; a compound of glycerol to which are attached two fatty acids, a phosphate group, and a choline molecule. Both nature and the food industry use lecithin as an emulsifier to combine two ingredients that do not ordinarily mix, such as water and oil.

choline (KOH-leen): a nitrogen-containing compound found in foods as part of lecithin and other phospholipids.

lecithin reaches the body tissues intact. In other words, the lecithins are *not essential nutrients;* they are just another lipid. Like other lipids, they contribute 9 kcalories per gram—an unexpected "bonus" many people taking lecithin supplements fail to realize. Furthermore, large doses of lecithin may cause GI distress, sweating, salivation, and loss of appetite. Perhaps these symptoms are beneficial because they may warn people to stop self-dosing with lecithin.

IN SUMMARY Phospholipids, including lecithin, have a unique chemical structure that allows them to be soluble in both water and fat. In the body, phospholipids are part of cell membranes; the food industry uses phospholipids as emulsifiers.

Sterols

In addition to triglycerides and phospholipids, the lipids include the **sterols,** compounds with a multiple-ring structure.* The most famous sterol is **cholesterol;** Figure 5-10 shows its chemical structure.

• **Sterols in Foods** • Foods derived from both plants and animals contain sterols, but only those from animals contain cholesterol: meats, eggs, fish, poultry, and dairy products. Some people, confused about the distinction between dietary and blood cholesterol, have asked which foods contain the "good" cholesterol. "Good" cholesterol is not a type of cholesterol found in foods, but refers to the way the body transports cholesterol in the blood, as explained later (p. 143).

• **Roles of Sterols** • Many vitally important body compounds are sterols. Among them are bile acids, the sex hormones (such as testosterone), the adrenal hormones (such as cortisol), and vitamin D, as well as cholesterol itself. Cholesterol in the body can serve as the starting material for the synthesis of these compounds or as a structural component of cell membranes; more than 90 percent of all the body's cholesterol resides in the cells. Despite popular impressions to the contrary, cholesterol is not a villain lurking in some evil foods—it is a compound the body makes◆ and uses. Right now, as you read, your liver

◆ Cholesterol that is made in the body is **endogenous** (en-DOGDE-eh-nus), whereas cholesterol from outside the body (from foods) is **exogenous** (eks-ODGE-eh-nus).
- **endo** = within
- **gen** = arising
- **exo** = outside (the body)

*The four-ring core structure identifies a steroid; sterols are alcohol derivatives with a steroid ring structure.

FIGURE 5-10 **Cholesterol**

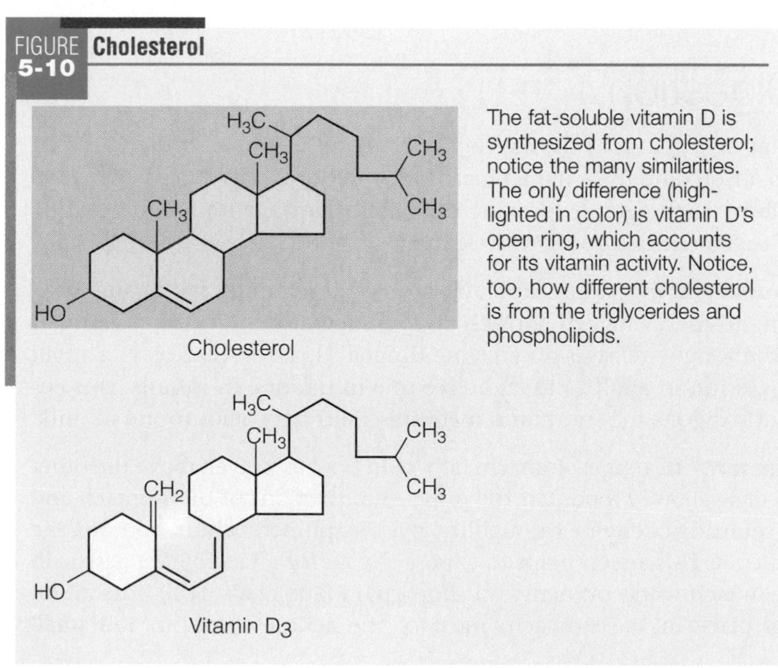

The fat-soluble vitamin D is synthesized from cholesterol; notice the many similarities. The only difference (highlighted in color) is vitamin D's open ring, which accounts for its vitamin activity. Notice, too, how different cholesterol is from the triglycerides and phospholipids.

sterols (STARE-ols or STEER-ols): compounds composed of C, H, and O atoms arranged in rings, like those of cholesterol, with any of a variety of side chains attached.

cholesterol (koh-LESS-ter-ol): one of the sterols containing four carbon rings and a carbon side chain.

◆ The liver makes cholesterol at the rate of perhaps 50,000,000,000,000,000 molecules per second.◆

is manufacturing cholesterol from fragments of carbohydrate, protein, and fat. In fact, the liver makes about 800 to 1500 milligrams of cholesterol per day, thus contributing much more to the body's total than does the diet.

Cholesterol's harmful effects in the body occur when it forms deposits in the artery walls. These deposits lead to **atherosclerosis,** a disease that causes heart attacks and strokes (Chapter 26 provides many more details).

IN SUMMARY

Sterols, including cholesterol, have a multiple-ring structure that differs from the other lipids. In the body, sterols include bile, vitamin D, and the sex hormones. Only animal-derived foods contain cholesterol. To summarize, the members of the lipid family include:

- **Triglycerides** (fats and oils), which are made of:
  - Glycerol (1 per triglyceride) and
  - Fatty acids (3 per triglyceride). Depending on the number of double bonds, fatty acids may be:
    - *Saturated,*
    - *Monounsaturated,* or
    - *Polyunsaturated.* Depending on the location of double bonds, polyunsaturated fatty acids may belong to families called:
      - *Omega-3* or
      - *Omega-6.*
- **Phospholipids** (such as lecithin) and
- **Sterols** (such as cholesterol).

DIGESTION, ABSORPTION, AND TRANSPORT OF LIPIDS

◆ Reminder: An enzyme that hydrolyzes lipids is called a *lipase; lingual* refers to the tongue.

Each day, the GI tract receives, on average, 50 to 100 grams of triglycerides, 4 to 8 grams of phospholipids, and 300 to 450 milligrams of cholesterol. The body faces a challenge in digesting and absorbing these lipids: getting at them. Fats are **hydrophobic**—that is, they tend to separate from the watery fluids of the GI tract—whereas the enzymes for digesting fats are **hydrophilic.** The challenge then is how the body mixes the fats into the watery fluids and then digests them.

atherosclerosis (ath-er-oh-scler-OH-sis): a type of artery disease characterized by accumulations of lipid-containing material on the inner walls of the arteries (see Chapter 26).
- **athero** = porridge or soft
- **scleros** = hard
- **osis** = condition

hydrophobic (high-dro-FOE-bick): a term referring to water-fearing, or non-water-soluble, substances; also known as **lipophilic** (fat loving).
- **hydro** = water
- **phobia** = fear
- **lipo** = lipid
- **phile** = friend

hydrophilic (high-dro-FIL-ick): a term referring to water-loving, or water-soluble, substances.

monoglycerides: molecules of glycerol with one fatty acid attached. A molecule of glycerol with two fatty acids attached is a **diglyceride.**

Lipid Digestion

The goal of fat digestion is to dismantle triglycerides into small molecules that the body can absorb and use—namely, **monoglycerides,** fatty acids, and glycerol. The following paragraphs provide the details, and Figure 5-11 traces the digestion of triglycerides through the GI tract.

• In the Mouth • Fat digestion starts off slowly in the mouth, with some hard fats beginning to melt when they reach body temperature. A salivary gland at the base of the tongue releases an enzyme (lingual lipase)◆ that plays a small role in fat digestion in adults and an active role in infants. In infants, this enzyme efficiently digests the short- and medium-chain fatty acids found in milk.

• In the Stomach • In a quiet stomach, fat would float as a layer above the other components of swallowed food. But the muscle contractions of the stomach and the periodic squirting of chyme through the pyloric sphincter churn and mix the stomach contents. This action helps to expose the fat for attack by the gastric lipase enzyme, which works primarily on short-chain fatty acids. Still, little fat digestion takes place in the stomach; most of the action occurs in the small intestine.

FIGURE 5-11 Triglyceride Digestion in the GI Tract

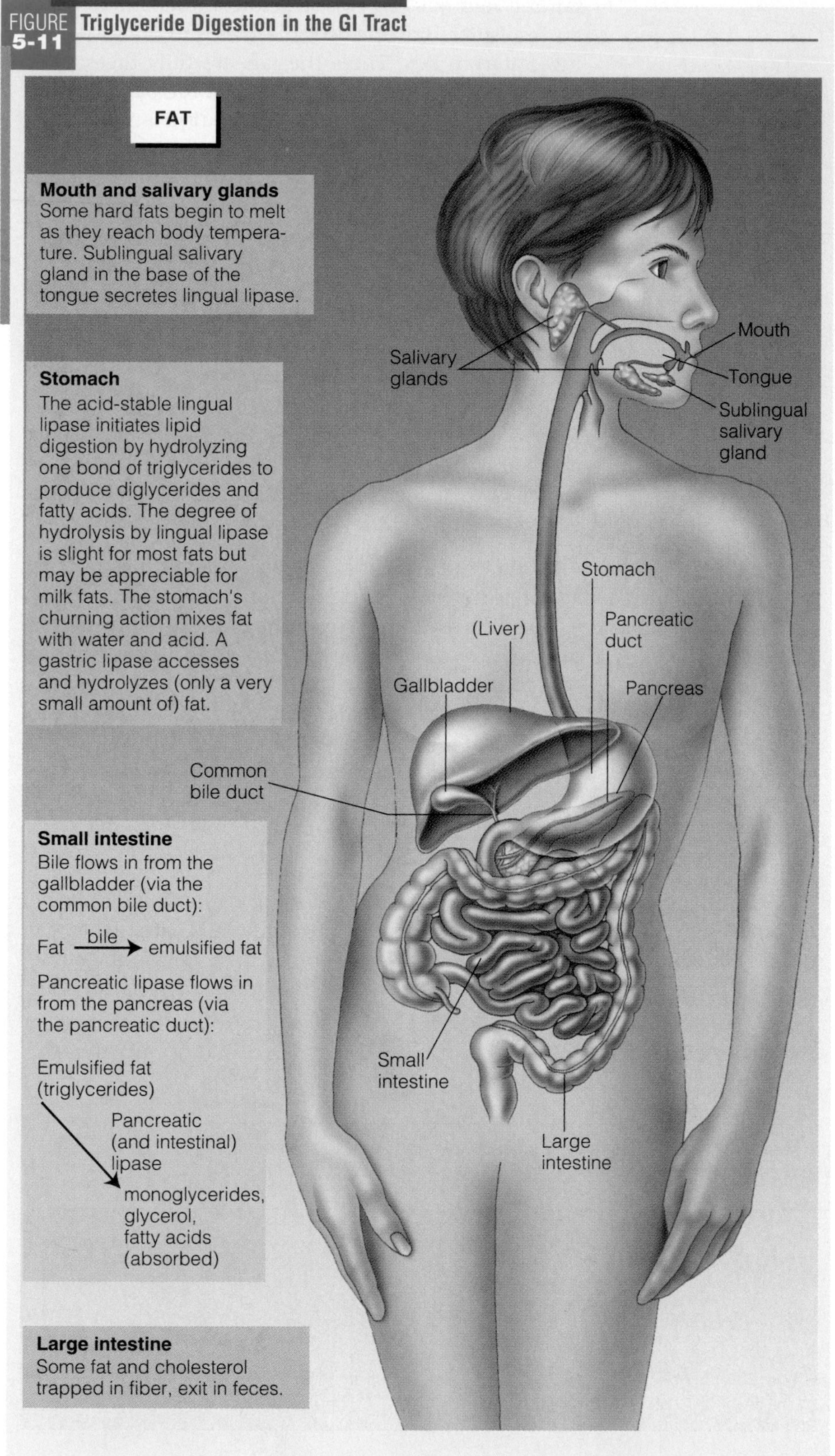

• **In the Small Intestine** • Fat in the small intestine triggers the release of the hormone cholecystokinin (CCK), which signals the gallbladder to release its stores of bile. (The liver manufactures bile acids from cholesterol, and the gallbladder stores the bile until it is needed.)

At one end of each bile acid are side chains of amino acids (units of protein) that are attracted to water, and at the other end is a sterol that is attracted to

FIGURE 5-12 **A Bile Acid**

Bile acid made from cholesterol

CH_3

HO CH—CH_2—CH_2—C—NH—CH_2—COOH

O

Bound to an amino acid from protein

HO H OH

This is one of several bile acids the liver makes from cholesterol. It is then bound to an amino acid to improve its ability to form micelles, spherical complexes of emulsified fat. Most bile acids occur as bile salts, usually in association with sodium, but sometimes with potassium or calcium. In addition to bile acids and bile salts, bile contains cholesterol, phospholipids (especially lecithin), antibodies, water, electrolytes, and bilirubin (a pigment resulting from the breakdown of heme).

fat (see Figure 5-12). This structure allows bile to act as an emulsifier, drawing fat molecules into the surrounding watery fluids. There the fats are fully digested as they encounter lipase enzymes from the pancreas and small intestine. The process of emulsification is diagrammed in Figure 5-13.

Most of the hydrolysis of triglycerides occurs in the small intestine. The major fat-digesting enzymes are pancreatic lipases; some intestinal lipases are also active. These enzymes remove one, then the other, of each triglyceride's outer fatty acids, leaving a monoglyceride. Occasionally, enzymes remove all three fatty acids, leaving a free molecule of glycerol. Hydrolysis of a triglyceride is shown in Figure 5-14.

Phospholipids are digested similarly—that is, their fatty acids are removed by hydrolysis. The two fatty acids and the remaining phospholipid fragment are then absorbed. Sterols can be absorbed as is; if any fatty acids are attached, they are first hydrolyzed off.

• **Bile's Routes** • After bile has entered the small intestine and emulsified fat, it has two possible destinations, illustrated in Figure 5-15 (on p. 142). Most of the bile is reabsorbed from the intestine and recycled. The other possibility is that some of the bile can be trapped by dietary fibers in the large intestine and carried out of the body with the feces. Because cholesterol is needed to make bile, the excretion of bile effectively reduces blood cholesterol. The fibers most effective at lowering blood cholesterol this way are the soluble pectins and gums commonly found in fruits, oats, and legumes.[1]

Lipid Absorption

Figure 5-16 (on p. 142) illustrates the absorption of lipids. Small molecules of digested triglycerides (glycerol and short- and medium-chain fatty acids) can diffuse easily into the intestinal cells; they are absorbed directly into the bloodstream. Larger molecules (the monoglycerides and long-chain fatty acids)

FIGURE 5-13 **Emulsification of Fat by Bile**

Like bile, detergents are emulsifiers and work the same way, which is why they are effective in removing grease spots from clothes. Molecule by molecule, the grease is dissolved out of the spot and suspended in the water, where it can be rinsed away.

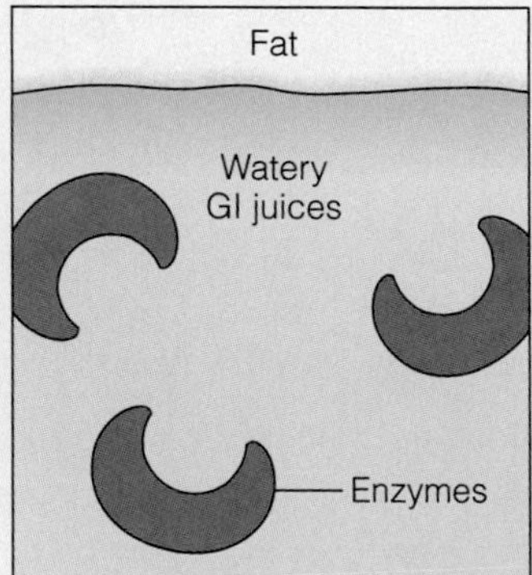

In the stomach, the fat and watery GI juices tend to separate. The enzymes are in the water and can't get at the fat.

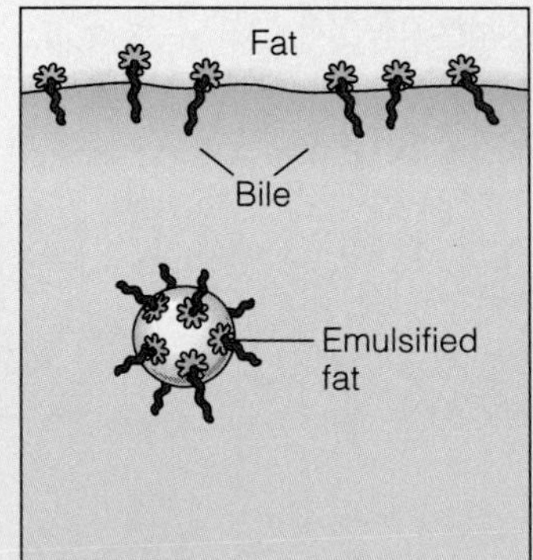

When fat enters the small intestine, the gallbladder secretes bile. Bile has an affinity for both fat and water, so it can bring the fat into the water.

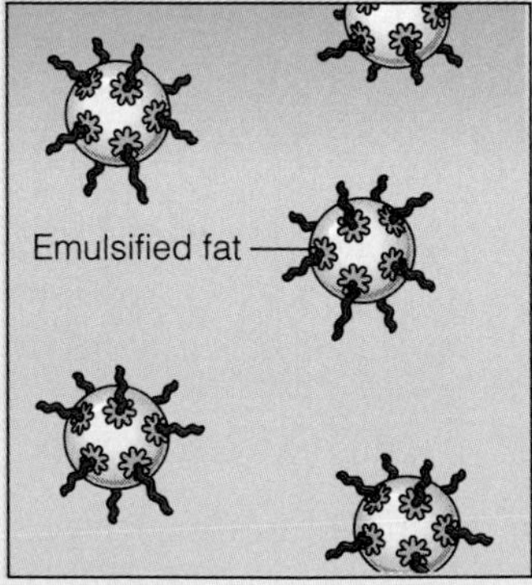

Bile's emulsifying action converts large fat globules into small droplets that repel each other.

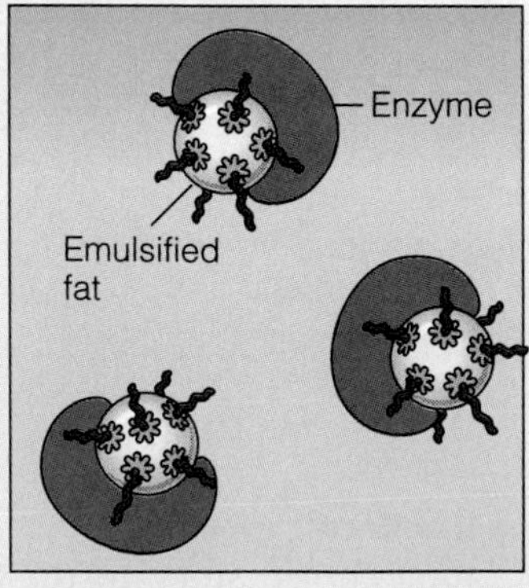

After emulsification, the enzymes have easy access to the fat droplets.

FIGURE 5-14 **Digestion (Hydrolysis) of a Triglyceride**

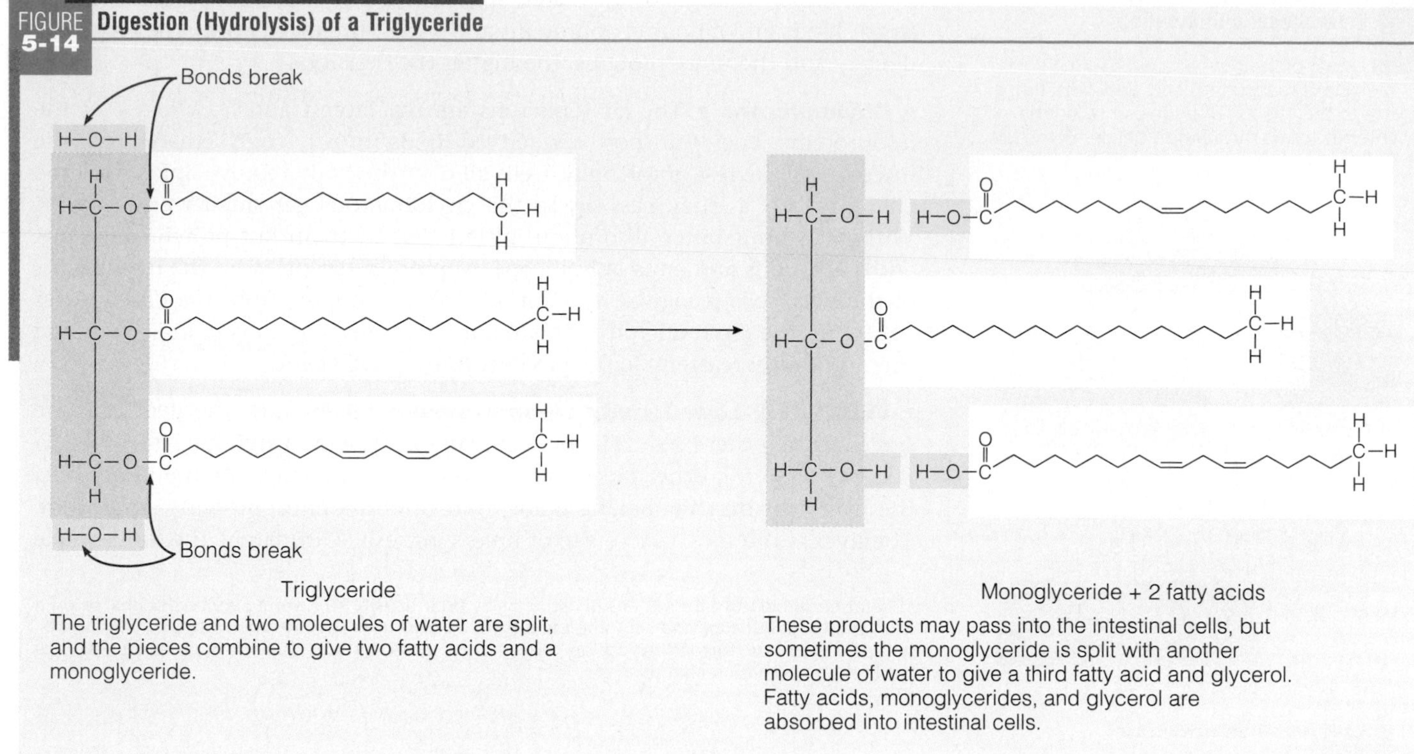

The triglyceride and two molecules of water are split, and the pieces combine to give two fatty acids and a monoglyceride.

These products may pass into the intestinal cells, but sometimes the monoglyceride is split with another molecule of water to give a third fatty acid and glycerol. Fatty acids, monoglycerides, and glycerol are absorbed into intestinal cells.

merge into spherical complexes, known as **micelles.** Micelles are emulsified fat droplets formed by molecules of bile surrounding monoglycerides and fatty acids. This configuration permits solubility in the watery digestive fluids and transportation to the intestinal cells. Upon arrival, the lipid contents of the micelles diffuse into the intestinal cells. Once inside, the monoglycerides and long-chain fatty acids are reassembled into new triglycerides.

Within the intestinal cells, the newly made triglycerides and the other large lipids (cholesterol and phospholipids) are packed into transport vehicles known as **chylomicrons.** The intestinal cells then release the chylomicrons into the lymphatic system. The chylomicrons glide through the lymph until they reach a point of entry into the bloodstream at the thoracic duct near the heart. (Recall that nutrients from the GI tract that enter the lymph system bypass the liver at first.) The blood can then carry these lipids to the rest of the body.

IN SUMMARY

The body makes special arrangements to digest and absorb lipids. It provides the emulsifier bile to make them accessible to the fat-digesting lipases that dismantle triglycerides, mostly to monoglycerides and fatty acids, for absorption by the intestinal cells. The intestinal cells assemble freshly absorbed lipids into chylomicrons, lipid packages with protein escorts, for transport so that cells all over the body may select needed lipids from them.

micelles (MY-cells): tiny spherical complexes of emulsified fat that arise during digestion. Each carries dozens of molecules of bile and fatty acids and/or monoglycerides.

chylomicrons (kye-lo-MY-cronz): the class of lipoproteins that transport lipids from the intestinal cells to the rest of the body.

lipoproteins (LIP-oh-PRO-teenz): clusters of lipids associated with proteins that serve as transport vehicles for lipids in the lymph and blood.

Lipid Transport

The chylomicrons are only one of several clusters of lipids and proteins that are used as transport vehicles for fats. As a group, these vehicles are known as **lipoproteins,** and they solve the body's problem of transporting fatty materials through the watery bloodstream. The body makes four main types of

FIGURE 5-15 Enterohepatic Circulation

Most of the bile released into the small intestine is reabsorbed and sent back to the liver to be reused. This cycle is called the **enterohepatic circulation** of bile. Some bile is excreted.

- **enteron** = intestine
- **hepat** = liver

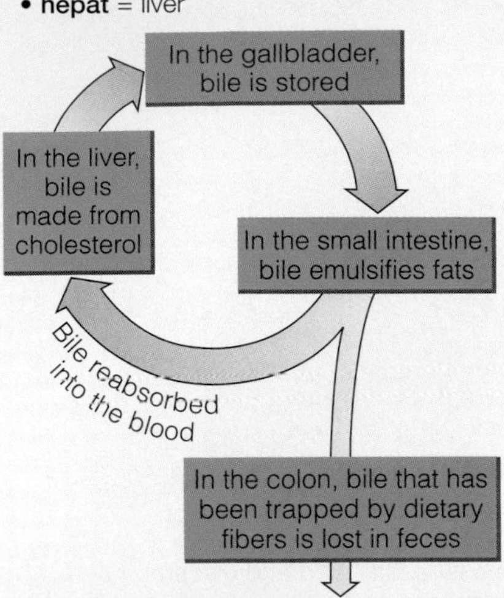

lipoproteins, distinguished by their size and density.* Each type contains different kinds and amounts of proteins and lipids: the more lipids, the lower the density and the more proteins, the higher the density.

• **Chylomicrons** • The chylomicrons are the largest and least dense of the lipoproteins. They transport *diet*-derived lipids (mostly triglycerides) from the intestine to the rest of the body. Cells all over the body remove lipids from the chylomicrons as they pass by, so the chylomicrons get smaller and smaller. Within 14 hours after absorption, little is left of them but protein remnants and a few odds and ends of lipid. Special protein receptors on the membranes of the liver cells recognize and remove these remnants from the blood. After collecting the chylomicron remnants, the liver cells first dismantle them and then promptly reassemble the pieces into new triglycerides.

• **VLDL (Very-Low-Density Lipoproteins)** • Meanwhile, the liver cells are synthesizing other lipids. The liver cells use fatty acids arriving in the blood to make cholesterol, other fatty acids, and other compounds. At the same time, the liver cells may be making lipids from carbohydrates, proteins, or alcohol. The liver is the most active site of lipid synthesis. Ultimately, the lipids made

*Chemists can identify the various lipoproteins by their density by layering a blood sample below a thick fluid in a test tube and spinning the tube in a centrifuge. The most buoyant particles (highest in lipids) rise to the top and have a low density; the densest particles (highest in proteins) remain at the bottom and have a high density.

FIGURE 5-16 Absorption of Lipids

The end products of fat digestion are mostly monoglycerides, some fatty acids, and very little glycerol. Their absorption differs depending on their size. (In reality, molecules of fatty acid are too small to see without a powerful microscope, while villi are visible to the naked eye.)

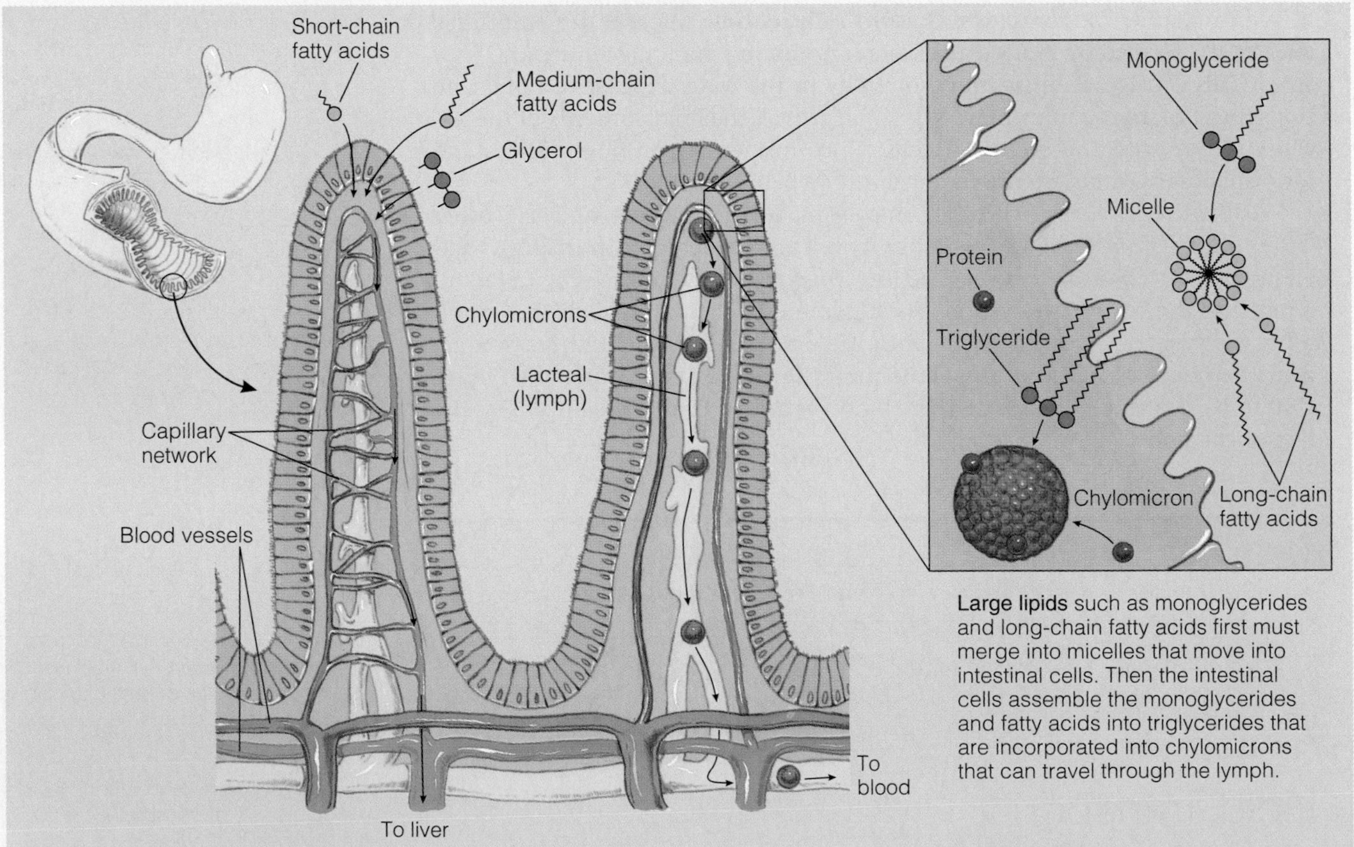

Large lipids such as monoglycerides and long-chain fatty acids first must merge into micelles that move into intestinal cells. Then the intestinal cells assemble the monoglycerides and fatty acids into triglycerides that are incorporated into chylomicrons that can travel through the lymph.

Glycerol and small lipids such as short- and medium-chain fatty acids can move directly into the bloodstream.

in the liver and those collected from chylomicron remnants are packaged with proteins as very-low-density lipoproteins **(VLDL)**◆ and shipped to other parts of the body.

As the VLDL travel through the body, cells remove triglycerides, causing the VLDL to shrink. As a VLDL loses triglycerides, the proportion of lipids shifts and the lipoprotein becomes more dense. The remaining cholesterol-rich lipoprotein eventually becomes a low-density lipoprotein **(LDL)**.* This transformation explains why LDL contain few triglycerides but are loaded with cholesterol.

◆ **Chylomicrons and VLDL transport triglycerides.**

• LDL (Low-Density Lipoproteins) • The LDL circulate throughout the body, making their contents available to the cells of all tissues—muscles, including the heart muscle; fat stores; the mammary glands; and others. The cells take triglycerides, cholesterol, and phospholipids to build new membranes, make hormones or other compounds, or store for later use. Special LDL receptors on the liver cells play a crucial role in the control of blood cholesterol concentrations by removing LDL from circulation.

• HDL (High-Density Lipoproteins) • Fat cells may release glycerol, fatty acids, cholesterol, and phospholipids to the blood. The liver makes high-density lipoprotein **(HDL)** packages to carry cholesterol◆ and other lipids from the cells back to the liver for recycling or disposal.

◆ **LDL and HDL transport cholesterol.**

• Health Implications • The distinction between LDL and HDL has implications for the health of the heart and blood vessels. The blood cholesterol linked to heart disease is LDL cholesterol. HDL also carry cholesterol, but elevated HDL represent cholesterol returning◆ from the rest of the body to the liver for breakdown and excretion.[2] High LDL cholesterol is associated with a high risk of heart attack, whereas high HDL cholesterol seems to have a protective effect. This is why some people refer to LDL as "bad," and HDL as "good," cholesterol.◆ Keep in mind that there is only *one* kind of cholesterol and that the differences between LDL and HDL reflect the *proportions* of lipids and proteins within them—not the type of cholesterol. The margin◆ lists factors that influence LDL and HDL, and Chapter 26 provides many more details.

Health promotion

◆ **The reverse transport of cholesterol is called the *scavenger pathway.***

◆ **To help you remember, think of elevated HDL as Healthy and elevated LDL as Less Healthy.**

◆ **Factors that improve the LDL-to-HDL ratio:**

- Weight control.
- Monounsaturated or polyunsaturated, instead of saturated, fat in the diet.
- Soluble fibers (see Chapter 4).
- Antioxidants (see Chapter 26).
- *Moderate* alcohol consumption.
- Physical activity.

IN SUMMARY

The liver packages lipids with proteins into lipoproteins for transport around the body. All four types of lipoproteins carry all classes of lipids (triglycerides, phospholipids, and cholesterol), but the chylomicrons are the largest and the highest in triglycerides; VLDL are smaller and are about half triglycerides; LDL are smaller still and are high in cholesterol; and HDL are the smallest and are rich in protein. Figure 5-17 on p. 144 shows the relative compositions and sizes of the lipoproteins.

LIPIDS IN THE BODY

The blood carries lipids to various sites around the body. Once they arrive at their destinations, the lipids can get to work providing energy, insulating against temperature extremes, protecting against shock, and building cell structures. This section provides an overview first of the roles of triglycerides and fatty acids and then of the metabolic pathways they can follow within the body's cells.

*Before becoming LDL, the VLDL are first transformed into intermediate-density lipoproteins (IDL), sometimes called VLDL remnants. Some IDL may be picked up by the liver and rapidly broken down; those IDL that remain in circulation pick up cholesterol and become LDL. Researchers debate whether IDL are simply transitional particles or a separate class of lipoproteins; normally, IDL do not accumulate in the blood.

VLDL (very-low-density lipoprotein): the type of lipoprotein made primarily by liver cells to transport lipids to various tissues in the body; composed primarily of triglycerides.

LDL (low-density lipoprotein): the type of lipoprotein derived from very-low-density lipoproteins (VLDL) as cells remove triglycerides from them; composed primarily of cholesterol.

HDL (high-density lipoprotein): the type of lipoprotein that transports cholesterol back to the liver from the cells; composed primarily of protein.

FIGURE 5-17 **Size and Compositions of the Lipoproteins**

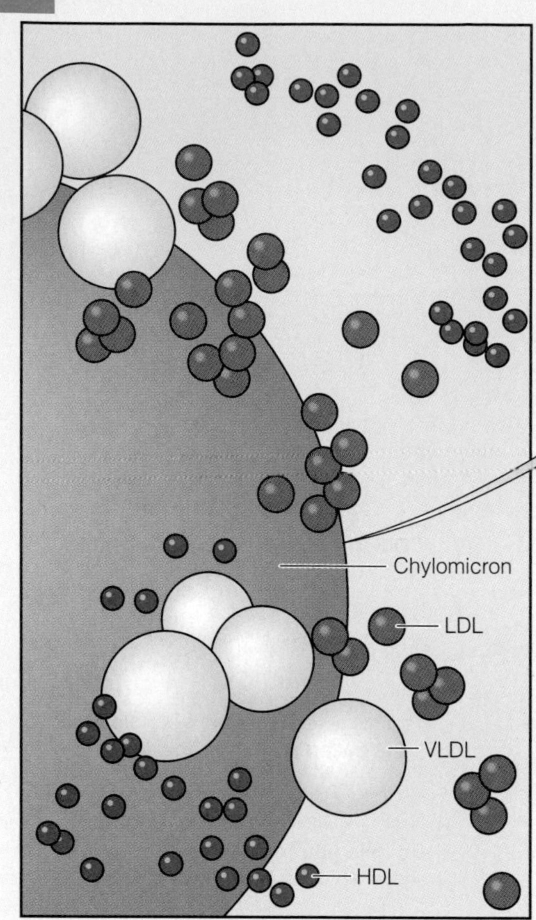

This solar system of lipoproteins shows their relative sizes. Notice how large the fat-filled chylomicron is compared with the others and how the others get progressively smaller as their proportion of fat declines and protein increases.

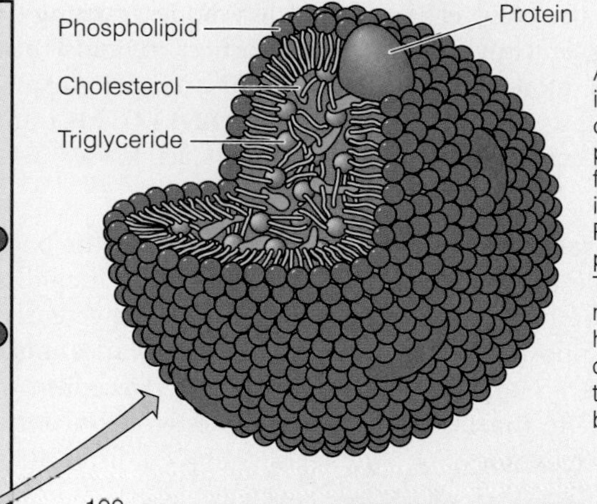

A typical lipoprotein contains an interior of triglycerides and cholesterol surrounded by phospholipids. The phospholipids' fatty acid "tails" point toward the interior, where the lipids are. Proteins near the outer ends of the phospholipids cover the structure. This arrangement of hydrophobic molecules on the inside and hydrophilic molecules on the outside allows lipids to travel through the watery fluids of the blood.

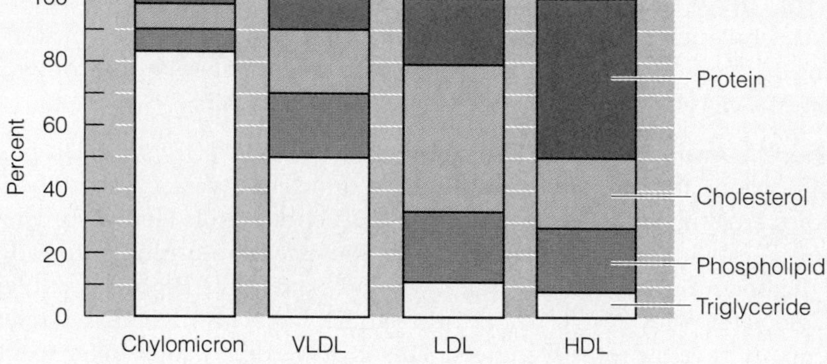

Chylomicrons contain so little protein and so much triglyceride that they are the lowest in density.

Very-low-density lipoproteins (VLDL) are half triglycerides, accounting for their low density.

Low-density lipoproteins (LDL) are half cholesterol, accounting for their implication in heart disease.

High-density lipoproteins (HDL) are half protein, accounting for their high density.

Thanks to the body's fat pads, a horseback ride causes no serious damage to internal organs.

Roles of Triglycerides

First and foremost, the triglycerides—either from food or the body's fat stores—provide the body with energy. When a person dances all night, her dinner's triglycerides provide the fuel to keep her moving; when a person loses his appetite, his stored triglycerides fuel much of his body's work until he can eat again.

Fat also insulates the body. Fat is a poor conductor of heat, so the layer of fat beneath the skin helps keep the body warm. Fat pads also serve as shock absorbers, supporting and cushioning the vital organs.

Fat also helps the body use its two other energy nutrients—carbohydrate and protein—efficiently. Fat fragments combine with glucose fragments during energy metabolism, and fat helps spare protein, providing energy so that protein can be used for other important tasks.

Essential Fatty Acids

The human body needs fatty acids, and it can make all but two of them—linoleic acid (the 18-carbon omega-6 fatty acid) and linolenic acid (the 18-carbon omega-

3 fatty acid). These two fatty acids must be supplied by the diet and are therefore called **essential fatty acids.** A simple definition of an essential nutrient has already been given: a nutrient that the body cannot make, or cannot make in sufficient quantities to meet its physiological needs. The cells do not possess the enzymes to make any of the omega-6 or omega-3 fatty acids from scratch; nor can they convert an omega-6 fatty acid to an omega-3 fatty acid or vice versa. They *can* start with the 18-carbon member of an omega family and make the longer fatty acids of that family by forming double bonds (desaturation) and lengthening the chain two carbons at a time (elongation), as shown in the margin drawing.◆ This is a slow process because the two families compete for the same enzymes. Therefore, the most effective way to maintain body supplies of all the omega-6 and omega-3 fatty acids is to obtain them directly from foods—most notably, from vegetable oils, seeds, nuts, and fish (see Table 5-2).

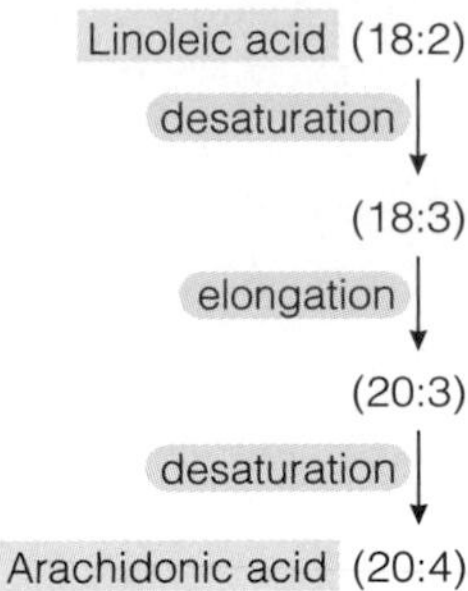

◆ Note: The first number indicates the number of carbons and the second, the number of double bonds. Similar reactions occur when the body makes EPA and DHA from linolenic acid.

• Linoleic Acid • Linoleic acid is the primary member of the omega-6 family. Given linoleic acid, the body can make other members of the omega-6 family—such as the 20-carbon polyunsaturated fatty acid, **arachidonic acid.** Should a linoleic acid deficiency develop,◆ arachidonic acid, and all other fatty acids that derive from linoleic acid, would also become essential and have to be obtained from the diet. Normally, vegetable oils and meats supply enough omega-6 fatty acids to meet the body's needs.

◆ **A nonessential nutrient (such as arachidonic acid) that must be supplied by the diet in special circumstances (as in a linoleic deficiency) is considered *conditionally* essential.**

• Linolenic Acid • Linolenic acid is the primary member of the omega-3 family.* Like linoleic acid, this 18-carbon acid cannot be made in the body and must be supplied by foods. Given dietary linolenic acid, the body can make the 20- and 22-carbon members of the omega-3 series, **EPA (eicosapentaenoic acid)** and **DHA (docosahexaenoic acid).** These omega-3 fatty acids are essential for normal growth and development, and they may play an important role in the prevention and treatment of heart disease, hypertension, arthritis, and cancer.

• Eicosanoids • The body uses the essential fatty acids to maintain the structural parts of cell membranes and to make substances known as **eicosanoids.** Eicosanoids are a diverse group of compounds that are sometimes described as "hormonelike," but they differ from hormones in important ways. For one, hormones are secreted in one location and travel to affect cells all over the

*This omega-3 linolenic acid is known as alpha-linolenic acid and is the fatty acid referred to in this discussion. Another fatty acid, also with 18 carbons and three double bonds, belongs to the omega-6 family and is known as gamma-linolenic acid.

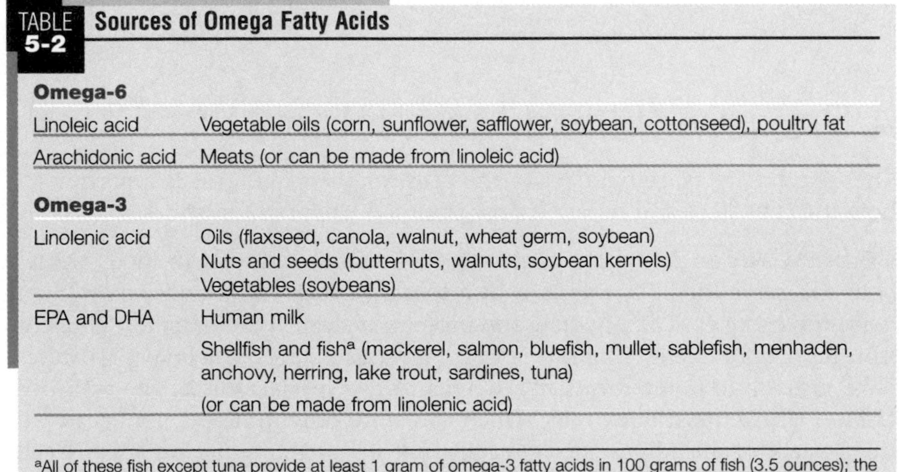

TABLE 5-2 Sources of Omega Fatty Acids

Omega-6	
Linoleic acid	Vegetable oils (corn, sunflower, safflower, soybean, cottonseed), poultry fat
Arachidonic acid	Meats (or can be made from linoleic acid)
Omega-3	
Linolenic acid	Oils (flaxseed, canola, walnut, wheat germ, soybean) Nuts and seeds (butternuts, walnuts, soybean kernels) Vegetables (soybeans)
EPA and DHA	Human milk Shellfish and fish[a] (mackerel, salmon, bluefish, mullet, sablefish, menhaden, anchovy, herring, lake trout, sardines, tuna) (or can be made from linolenic acid)

[a]All of these fish except tuna provide at least 1 gram of omega-3 fatty acids in 100 grams of fish (3.5 ounces); the fish oil content of each species varies with the season and site of harvest. Tuna provides fewer omega-3 fatty acids, but because it is commonly consumed, its contribution can be significant.

essential fatty acids: fatty acids needed by the body, but not made by it in amounts sufficient to meet physiological needs.

arachidonic (a-RACK-ih-DON-ic) **acid:** an omega-6 polyunsaturated fatty acid with 20 carbons and four double bonds (20:4); synthesized from linoleic acid.

eicosapentaenoic (EYE-cossa-PENTA-ee-NO-ick) **acid (EPA):** an omega-3 polyunsaturated fatty acid with 20 carbons and five double bonds (20:5); synthesized from linolenic acid.

docosahexaenoic (DOE-cossa-HEXA-ee-NO-ick) **acid (DHA):** an omega-3 polyunsaturated fatty acid with 22 carbons and six double bonds (22:6); synthesized from linolenic acid.

eicosanoids (eye-COSS-uh-noyds): derivatives of 20-carbon fatty acids; biologically active compounds that regulate blood pressure, blood clotting, and other body functions. They include *prostaglandins* (PROS-tah-GLAND-ins), *thromboxanes* (throm-BOX-ains), and *leukotrienes* (LOO-ko-TRY-eens).

body, whereas eicosanoids appear to affect only the cells in which they are made. For another, hormones elicit the same response from all their target cells, whereas eicosanoids often have different effects on different cells.

The actions of various eicosanoids sometimes oppose each other. One causes muscles to relax and blood vessels to dilate, while another causes muscles to contract and blood vessels to constrict, for example.

Certain eicosanoids participate in the immune response to injury and infection, producing fever, inflammation, and pain. Aspirin relieves these symptoms by slowing the synthesis of these eicosanoids.

Because eicosanoids help regulate blood pressure, blood clot formation, blood lipids, and the immune response, they play an important role in maintaining health. To make eicosanoids in sufficient quantities, the body needs the long-chain polyunsaturated omega fatty acids (review Table 5-2 on p. 145 for common food sources).

• **Fatty Acid Deficiencies** • Essential fatty acids should make up at least 3 percent of the day's energy intake, and most diets meet this minimum requirement more than adequately. Historically, deficiencies have developed only in infants and young children fed nonfat milk and low-fat diets or in hospital clients fed formulas that provided no polyunsaturated fatty acids for long periods of time. More recently, researchers have identified essential fatty acid deficiencies in people with chronic intestinal diseases.[3] Classic deficiency symptoms include growth retardation, reproductive failure, skin lesions, kidney and liver disorders, and subtle neurological and visual problems.

Interestingly, a deficiency of omega-3 fatty acids (EPA and DHA) may be associated with depression.[4] It is unclear, however, which comes first—whether inadequate intake alters brain activity or depression alters fatty acid metabolism. Answers remain tangled in a multitude of confounding factors.

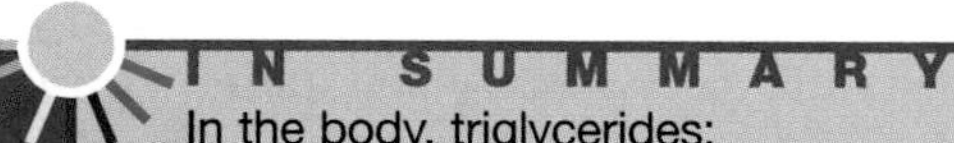

IN SUMMARY

In the body, triglycerides:

- *Provide an energy reserve when stored in the body's fat tissue.*
- *Insulate against temperature extremes.*
- *Protect organs against shock.*
- *Help the body use carbohydrate and protein efficiently.*

Linoleic acid (18 carbons, omega-6) and linolenic acid (18 carbons, omega-3) are essential nutrients. In addition to serving as structural parts of cell membranes, they make eicosanoids—powerful compounds that help regulate blood pressure, blood clot formation, and the immune response to injury and infection. Deficiencies are unlikely.

A Preview of Lipid Metabolism

The blood delivers triglycerides to the cells for their use. This is a preview of how the cells store and release energy from fat; Chapter 7 provides details.

• **Storing Fat as Fat** • The triglycerides, familiar as the fat in foods and as body fat, serve the body primarily as a source of fuel. Fat provides more than twice the energy of carbohydrate and protein, making it an extremely efficient storage form of energy. Unlike the liver's glycogen stores, the body's fat stores have virtually unlimited capacity, thanks to the special cells of the **adipose tissue.** Unlike most body cells, which can store only limited amounts of fat, the fat cells of the adipose tissue readily take up and store fat. An adipose cell is depicted in Figure 5-18.

Adipose cells have an enzyme on their surfaces—**lipoprotein lipase (LPL)**—that captures circulating triglycerides from lipoproteins passing by

adipose (ADD-ih-poce) **tissue:** the body's fat tissue; consists of masses of fat-storing cells.

lipoprotein lipase (LPL): an enzyme mounted on the surface of fat cells (and other cells) that hydrolyzes triglycerides passing by in the bloodstream and directs their parts into the cells, where they can be metabolized or reassembled for storage.

after meals. This enzyme hydrolyzes the triglycerides to fatty acids and monoglycerides and passes these products into the cells. Inside the cells, other enzymes reassemble the pieces into triglycerides again for storage. Triglycerides pack tightly together within adipose cells, storing a lot of energy in a relatively small space. Adipose cells always store fat after meals when a heavy traffic of chylomicrons and VLDL loaded with triglycerides passes by; they release it later whenever the blood needs replenishing.

FIGURE 5-18 **An Adipose Cell**

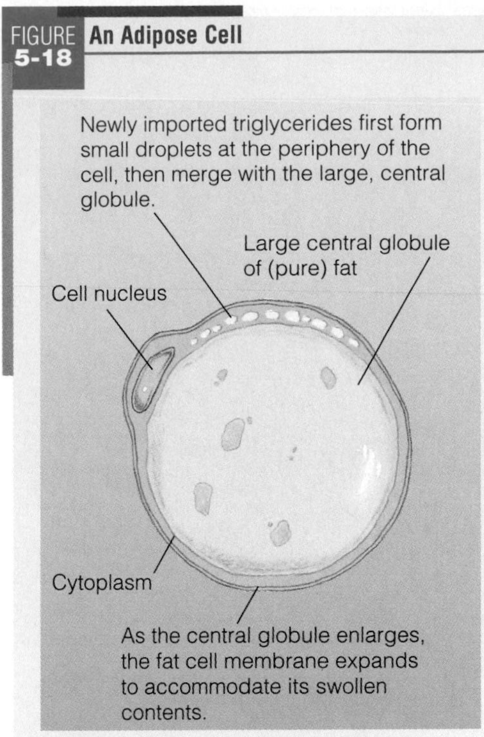

• Making Fat from Carbohydrate or Protein • Earlier, Figure 5-4 (on p. 133) showed how the body can make a triglyceride from glycerol and fatty acids. Fatty acids, in turn, can be made from two-carbon fragments derived from carbohydrate or protein. (Remember that most fatty acid carbon chains come in even numbers.) Thus glucose can be converted to body fat: enzymes break six-carbon glucose into two-carbon fragments and then combine them to make long-chain fatty acids. Enzymes can also convert some of the components of protein (certain amino acids) to fatty acids. The food source from which the body most easily makes fat for storage, though, is fat itself.

• Making Fat from Fat • To convert food fats to body fat, the body simply absorbs the parts and puts them (and others) together again in storage. It requires very little energy to do this. By comparison, to convert dietary carbohydrate to body fat, the body must break starches first into disaccharides and then into monosaccharides, absorb the monosaccharides, then dismantle them, and reassemble many of their fragments into fatty acid chains. Each conversion requires energy. Thus it costs less (energetically) to store dietary fat as body fat than to convert and store dietary carbohydrate as body fat.

• Using Fat for Energy • Fat supplies 60 percent of the body's ongoing energy needs during rest. During prolonged light to moderately intense exercise or extended periods of food deprivation, fat stores may make a slightly greater contribution to energy needs.

When cells demand energy, an enzyme **(hormone-sensitive lipase)** inside the adipose cells responds by dismantling stored triglycerides and releasing the glycerol and fatty acids directly into the blood. Energy-hungry cells anywhere in the body can then capture these compounds and take them through a series of chemical reactions to yield energy, carbon dioxide, and water.

A person who fasts (drinking only water) will rapidly metabolize body fat. A pound of body fat provides 3500 kcalories,◆ so you might think a fasting person who expends 2000 kcalories a day could lose more than half a pound of body fat each day.* Actually, the person has to obtain some energy from lean tissue because the brain, nervous system, and red blood cells need glucose. Also, fat needs carbohydrate or protein to break down completely. Even on a total fast, a person cannot lose more than half a pound of pure fat per day. Still, in conditions of enforced starvation—say, during a siege or a famine—a fatter person can survive longer than a thinner person thanks to this energy reserve.

1 lb body fat = 3500 kcal. ◆

Although fat provides energy during a fast, it can provide very little glucose to give energy to the brain and nerves. Only the small glycerol molecule can be converted to glucose; fatty acids cannot be. After prolonged glucose deprivation, brain and nerve cells develop the ability to derive about two-thirds of their energy from the ketone bodies that the body makes from fat fragments. Ketone bodies cannot sustain life by themselves, however. As Chapter 7 explains, fasting for too long will cause death, even if the person still has ample body fat.

hormone-sensitive lipase: an enzyme inside adipose cells that responds to the body's need for fuel by hydrolyzing triglycerides so that their parts (glycerol and fatty acids) escape into the general circulation and thus become available to other cells as fuel. The signals to which this enzyme responds include epinephrine and glucagon, which oppose insulin (see Chapter 4).

*The reader who knows that 1 pound = 454 grams and that 1 gram of fat = 9 kcalories may wonder why a pound of body fat does not equal 9 × 454 kcalories. The reason is that body fat contains some cell water and other materials; it is not quite pure fat.

© Bob Daemmrich

Fat supplies most of the energy during a long-distance run.

> **IN SUMMARY** The body can easily store unlimited amounts of fat if given excesses, and this body fat is used for energy when needed. The liver can also convert excess carbohydrate and protein into fat. Fat breakdown requires simultaneous carbohydrate breakdown for maximum efficiency; without carbohydrate, fats break down to ketone bodies.

HEALTH EFFECTS AND RECOMMENDED INTAKES OF LIPIDS

Of all the nutrients, fat is most often linked with chronic diseases. A high-fat diet raises the risks of heart disease, some types of cancer, hypertension, diabetes, and obesity.[5] Fortunately, the same recommendation can help with all of these health problems: eat less fat.

Health Effects of Lipids

Hearing a physician say, "Your blood lipid profile looks fine," is reassuring. The **blood lipid profile** reveals the concentrations of various lipids in the blood, notably triglycerides and cholesterol, and their lipoprotein carriers (VLDL, LDL, and HDL). This information alerts people to their disease risks and their need to change eating habits.

• Heart Disease • Most people realize that elevated blood cholesterol is a major risk factor◆ for **cardiovascular disease.** Cholesterol accumulates in the arteries, restricting blood flow and raising blood pressure. The consequences are deadly; in fact, heart disease is the nation's number one killer of adults. Blood cholesterol is often used to predict the likelihood of a person's suffering a heart attack or stroke; the higher the cholesterol, the earlier and more likely the tragedy.

◆ Other risk factors for heart disease include smoking, high blood pressure, diabetes, family history, gender, race, and obesity. Chapter 26 provides many more details about these risk factors, the development of heart disease, and dietary recommendations.

Commercials advertise products that are low in cholesterol, and magazine articles tell readers how to cut the cholesterol in their favorite recipes. What most people don't realize, though, is that *food* cholesterol does not raise *blood* cholesterol as dramatically as *saturated fat* does.

• Risks from Saturated Fats • Recall that LDL cholesterol raises the risk of heart disease. LDL concentrations respond to both the total amount and the type of fat in the diet.[6] Most often implicated in raising LDL cholesterol are the saturated fats, although not all saturated fats have the same cholesterol-raising effect. Most notable among the saturated fatty acids that raise blood cholesterol are lauric, myristic, and palmitic acids (12, 14, and 16 carbons, respectively).[7] In contrast, stearic acid (18 carbons) does not seem to raise blood cholesterol, although it does appear to increase the risk of heart disease similarly.[8] Furthermore, these saturated fatty acids typically appear in the same foods together, making such distinctions impractical in diet planning. Common sources of stearic acid and other saturated fats include red meat, high-fat milk products, and chocolate.[9]

blood lipid profile: results of blood tests that reveal a person's total cholesterol, triglycerides, and various lipoproteins. Desirable blood lipid profile:
- Total cholesterol: <200 mg/dL.
- LDL cholesterol: <130 mg/dL.
- HDL cholesterol: >35 mg/dL.
- Triglycerides: <200 mg/dL.

cardiovascular disease (CVD): a general term for all diseases of the heart and blood vessels. Atherosclerosis is the main cause of CVD. When the arteries that carry blood to the heart muscle become blocked, the heart suffers damage known as **coronary heart disease (CHD).**
- **cardio** = heart
- **vascular** = blood vessels

• Risks from *trans*-Fats • In the body, *trans*-fatty acids—even the monounsaturated ones—alter blood cholesterol the same way some saturated fats do: they raise LDL and lower HDL cholesterol.[10] Epidemiological studies also suggest an association between dietary *trans*-fatty acids and heart disease.[11] Evidence linking *trans*-fatty acids and cancer is generally less conclusive.[12]

Reports on *trans*-fatty acids have raised consumer doubts about whether margarine is, after all, a better choice than butter for heart health. The American Heart Association has stated that because butter is rich in both saturated fat and cholesterol while margarine is made from vegetable fat with no dietary cholesterol, margarine is still preferable to butter. Others disagree, claiming the occasional use of butter is preferable to the use of margarine and other products containing *trans*-fatty acids.[13] All things considered, health risks from saturated fatty acids appear to outweigh those from *trans*-fatty acids.[14] Replacing both saturated fats and *trans*-fats with monounsaturated and polyunsaturated fats may be the most effective strategy in preventing heart disease.[15]

© Tony Freeman/PhotoEdit

Whether you decide to use butter or margarine, remember to use them sparingly.

• Benefits from Monounsaturated Fats • The lower rates of heart disease among people in the Mediterranean region of the world are often attributed to their liberal use of olive oil, a rich source of monounsaturated fatty acids. Using monounsaturated fatty acids (as in olive oil and canola oil) instead of saturated fatty acids (as in butter and shortening) lowers blood levels of LDL cholesterol—even when energy and fat intakes remain the same.[16] Some research indicates that a diet rich in monounsaturated fatty acids may improve blood lipids even more favorably than a low-fat diet.[17]

• Benefits from Omega-3 Polyunsaturated Fats • Research on the different types of fat has spotlighted the beneficial effects of the omega-3 polyunsaturated fatty acids in lowering blood cholesterol and preventing heart disease.[18]◆ Inuit peoples of Alaska and Greenland enjoy relative freedom from heart disease despite high-energy, high-fat, and high-cholesterol diets. Why? Their foods derive primarily from fish and other marine animals that are rich in omega-3 fatty acids, particularly EPA and DHA (review Table 5-2 on p. 145 for a list of fish rich in EPA and DHA).

◆ Chapter 26 presents a few more details on the action of omega-3 fatty acids in preventing heart disease.

People who eat some fish each week can lower their blood cholesterol and their risk of heart attack and stroke.[19] A diet low in both fat and saturated fat, combined with regular fish consumption, improves a person's lipid profile. In addition to improving blood lipids, fish oils prevent blood clots and may also lower blood pressure, especially in people with hypertension or atherosclerosis.[20]

• Cancer • The evidence linking dietary fats with cancer◆ is less conclusive than for heart disease, but it does suggest an association between total fat and some types of cancers. Dietary fat seems not to *initiate* cancer development but to *promote* cancer once it has arisen.[21] Some studies report a relationship between specific cancers and saturated fats or dietary fat from animal sources (which is mostly saturated).[22] Thus health advice to reduce cancer risks parallels that given to reduce heart disease risks: reduce total fat, especially saturated fat, intake.

◆ Other risk factors for cancer include smoking, alcohol, and environmental contaminants. Chapter 29 provides more details about risk factors and the development of cancer.

The relationship between dietary fat and the risk of cancer differs for various types of cancers. In the case of breast cancer, evidence has been inconclusive.[23] Some studies indicate little or no association between dietary fat and cancer; others find that total *energy* intake and obesity are better predictors than percentage of kcalories from fat.[24] In the case of prostate cancer, there appears to be a strong association with fat.[25]

The relationship between dietary fat and the risk of cancer differs for various types of fats as well.[26] The association between cancer and fat appears to be due primarily to the saturated fat from meats; fat from milk or fish has not been implicated in cancer risk. In fact, eating fish seems to protect against some cancers.[27]

• Obesity • Fat contributes twice as many kcalories◆ per gram as either carbohydrate or protein. Consequently, people who eat high-fat diets regularly may exceed their energy needs and gain weight.

◆ Reminder: Fat is a more concentrated energy source than the other energy nutrients: 1 g carbohydrate or protein = 4 kcal, but 1 g fat = 9 kcal.

Furthermore, people who eat high-fat diets tend to store body fat efficiently and have more body fat than their energy intakes would predict. Later chapters revisit the issue of the fattening power of fat and conclude that low-fat, high-carbohydrate foods are most appropriate for satisfying hunger and controlling appetite.[28]

• Don't Overdo Fat Restriction • Although it is very difficult to do, some people actually manage to eat too little fat—to their detriment. Among them are people with eating disorders, described in Highlight 9. Most adults should

© Polara Studios Inc.

Enjoy low-fat foods for good heart health.

consume at least 15 percent of their energy intake from fat.[29] A person consuming 2000 kcalories a day should therefore have *at least* 300 kcalories of fat, or about 33 grams. As a practical guideline, it is wise to include the equivalent of at least a teaspoon of fat in every meal—a little peanut butter on toast or mayonnaise on tuna, for example. Dietary recommendations that limit fat were developed for healthy people over age two; Chapter 15 discusses the fat needs of infants and young children.

IN SUMMARY Health authorities single out high fat intakes as a major flaw in the North American diet: excess fat contributes to heart disease, obesity, and other health problems. High blood LDL cholesterol, specifically, poses a risk of heart disease, and high intakes of saturated fat contribute most to high LDL. High intakes of *trans*-fat also appear to raise LDL, although perhaps to a lesser extent. Cholesterol in foods presents much less of a risk. Omega-3 fatty acids appear to be protective. High-fat diets may also accelerate (but do not initiate) cancer development.

Recommended Intakes of Fat

In foods, triglycerides accompany protein in foods derived from animals, such as meat, fish, poultry, and eggs, and carbohydrate in foods derived from plants, such as avocados and coconuts. Triglycerides carry with them the four fat-soluble vitamins—A, D, E, and K—together with many of the compounds that give foods their flavor, texture, and palatability. Fat is responsible for the delicious aromas associated with sizzling bacon and hamburgers on the grill, onions being sautéed, or vegetables in a stir-fry. Of course, these wonderful characteristics lure people into eating too much from time to time.[30]

Some fat in the diet is essential for good health, but too much fat, especially saturated fat, increases the risks for chronic diseases. Dietary fat recommendations include the following:

- Reduce total fat intake to 30 percent or less of energy intake.
- Reduce saturated fat intake to less than 10 percent of energy intake.
- Reduce cholesterol intake to less than 300 milligrams daily.

A person consuming 2000 kcalories a day◆ should therefore have 600 kcalories or less from fat (roughly 65 grams). Of those fat kcalories, no more than 200 should come from saturated fats (roughly 20 grams).

◆ Daily Values for a 2000 kcal diet:
- Fat: 65 g.
- Saturated fat: 20 g.

HEALTHY PEOPLE 2010

Increase the proportion of persons aged 2 years and older who consume less than 10 percent of kcalories from saturated fat and no more than 30 percent of kcalories from total fat.

To meet dietary fat recommendations, many people have reduced their fat intakes. Fat intake peaked at more than 40 percent of daily kcalories in the 1950s and has fallen steadily ever since.[31] According to recent surveys, adults in the United States receive about 35 percent of their total energy from fat, with saturated fat contributing about 12 percent of the total.[32] Cholesterol intakes in the United States average 250 milligrams a day for women and 350 for men.

• **Reduce Total Fat Intake** • As the photos in Figure 5-19 show, fat accounts for a lot of the energy in foods, and removing the fat from foods cuts energy intake dramatically. To reduce dietary fat, eliminate fat as a seasoning and in cooking; remove the fat from high-fat foods; replace high-fat foods with low-

FIGURE 5-19 **Cutting Fat Cuts kCalories**

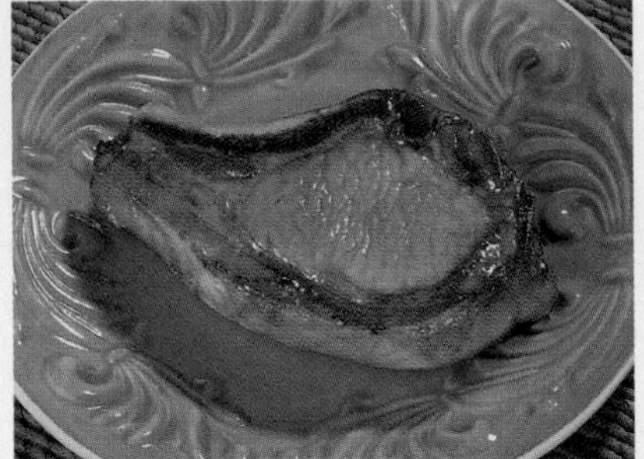

Pork chop with a half-inch of fat (275 kcal and 19 g fat).

Potato with 1 tbs butter and 1 tbs sour cream (350 kcal and 14 g fat).

Whole milk, 1 c (150 kcal and 8 g fat).

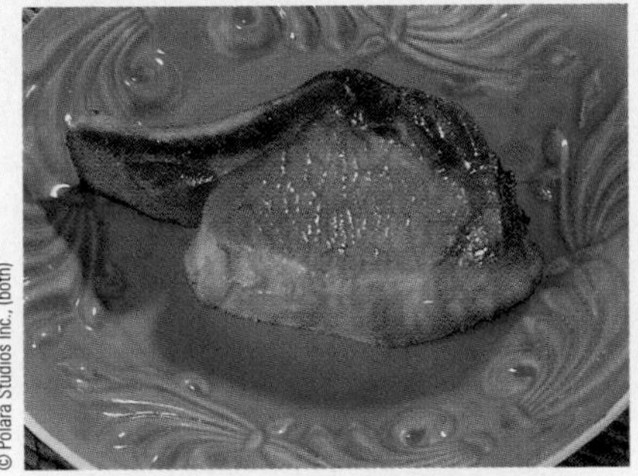

© Polara Studios Inc., (both)

Pork chop with fat trimmed off (165 kcal and 8 g fat).

© Polara Studios Inc., (both)

Plain potato (220 kcal and <1 g fat).

© Thomsas Ham and Tom Peterson/Quest Photographic Inc. (both)

Nonfat milk, 1 c (90 kcal and 1 g fat).

fat alternatives; and emphasize grains, fruits, and vegetables. The "How to" feature (on p. 152) provides additional tips for reducing fat in the diet, food group by food group.

• Reduce Saturated and *trans*-Fat Intake • Fats from animal sources are the main sources of saturated fats in most people's diets. Some vegetable fats (coconut and palm) and hydrogenated fats provide smaller amounts of saturated fats. Selecting poultry or fish and nonfat milk products helps to lower saturated fat intake and heart disease risk.[33] Using monounsaturated margarine and cooking oil is another simple change that can make a big difference.[34]

Limiting the intake of *trans*-fatty acids can also improve blood cholesterol. A food that lists partially hydrogenated oils among its first three ingredients usually contains substantial amounts of *trans*-fatty acids,◆ as well as some saturated fat. *Trans*-fatty acids make up approximately 7 percent of the fat intake in the U.S. diet.[35]

Major sources of *trans*-fatty acids: ◆
- Margarine (hard stick, soft tub).
- Imitation cheese.
- Cakes, cookies, doughnuts, pastry, crackers.
- Meats and dairy products.
- Snack chips.
- Peanut butter.
- Deep-fried foods.

• Reduce Cholesterol Intake • Recall that cholesterol is found only in animal products. Consequently, eating less fat from meat, eggs, and milk products will also help lower dietary cholesterol intake (as well as total and saturated fat intakes). Figure 5-20 (on p. 153) shows the cholesterol contents of selected foods. Many more foods, with their cholesterol contents, appear in Appendix H.

An egg contains just over 200 milligrams of cholesterol, all of it in the yolk. A person on a strict low-cholesterol diet must curtail the use of egg yolks, and food manufacturers have produced several nonfat, no-cholesterol egg substitutes. For most people trying to lower blood cholesterol, however, limiting saturated fat is more effective than limiting cholesterol intake. Eggs are a valuable part of the diet because they are inexpensive, useful in cooking, and a source

HOW TO Lower Fat Intake by Food Group

Meat, Fish, and Poultry

- Fat adds up quickly, even with lean meat; limit intake to about 6 ounces (cooked weight) daily.
- Choose fish, poultry, or lean cuts of pork or beef; look for unmarbled cuts named *round* or *loin* (eye of round, top round, bottom round, round tip, tenderloin, sirloin, center loin, and top loin).
- Trim the fat from pork and beef; remove the skin from poultry.
- Grill, roast, broil, bake, stir-fry, stew, or braise meats; don't fry. When possible, place meat on a rack so that fat can drain.
- Use lean ground turkey or lean ground beef in recipes; brown ground meats without added fat, then drain off fat.
- Refrigerate meat pan drippings and broth; when the fat solidifies, remove it and use the defatted broth in recipes.
- Select tuna, sardines, and other canned meats packed in water; rinse oil-packed items with hot water to remove much of the fat.
- Fill kabob skewers with lots of vegetables and slivers of meat; create main dishes and casseroles by combining a little meat, fish, or poultry with a lot of pasta, rice, or vegetables.
- Make meatless spaghetti sauces and casseroles; use legumes often.
- Eat a meatless meal or two daily.

Milk and Cheeses

- Switch from whole milk to reduced-fat, from reduced-fat to low-fat, and from low-fat to fat-free (nonfat).
- Use nonfat and low-fat cheeses (such as part-skim ricotta and low-fat mozzarella) instead of regular cheeses.
- Use nonfat or low-fat yogurt or sour cream instead of regular sour cream.
- Use evaporated nonfat milk instead of cream.
- Enjoy nonfat frozen yogurt, sherbet, or ice milk instead of ice cream.

Fruits and Vegetables

- Enjoy the natural flavor of steamed vegetables for dinner and fruits for dessert.
- Use butter-flavored granules on vegetables instead of butter or margarine.
- Use nonfat yogurt or nonfat salad dressing instead of sour cream, cheese, mayonnaise, or other sauces on vegetables and in casseroles.
- Select nonfat or low-fat salad dressings, or use herbs, lemon juice, and spices instead of regular salad dressing.
- Add a little water to thick, bottled salad dressing to dilute the amount of fat each serving provides.
- Eat at least two vegetables (in addition to a salad) with dinner.
- Snack on raw vegetables or fruits instead of high-fat items like potato chips.
- Buy frozen vegetables without sauce.

Breads and Cereals

- Use fruit butters, honey, or jellies on bread instead of butter or margarine.
- Select breads, cereals, and crackers that are low in fat (for example, bagels instead of croissants).
- Prepare pasta with a tomato sauce instead of a cheese or cream sauce.

Other Foods and Cooking Tips

- Use a nonstick pan or coat the pan lightly with vegetable oil.
- Use egg substitutes in recipes instead of whole eggs or use 2 egg whites in place of each whole egg.
- Use half the margarine, butter, or oil called for in a recipe. (The minimum amount of fat for muffins, quick breads, and biscuits is 1 to 2 tablespoons per cup of flour; for cakes and cookies, 2 tablespoons per cup.)
- Use less butter or margarine. Select the whipped types of butter, margarine, or cream cheese for use at the table; they contain half the kcalories of the regular types.
- Use butter replacers instead of butter.
- For sandwiches and salads, use spicy mustard, lemon juice, flavored vinegar, salsa, or the nonfat versions instead of regular mayonnaise, salad dressing, or sour cream.
- Use wine; lemon, orange, or tomato juice; herbs; spices; fruits; or broth instead of butter, margarine, or oil when cooking.
- Stir-fry in a small amount of oil; add moisture and flavor with broth, tomato juice, or wine.
- Use variety to enhance enjoyment of the meal: vary colors, textures, and temperatures—hot cooked versus cool raw foods—and use garnishes to complement food.

of high-quality protein. The American Heart Association approves an intake of up to four eggs a week and some research suggests that eating up to one egg a day is not detrimental for healthy people.[36]

• Balance Omega-3 and Omega-6 Intakes • The Committee on Dietary Reference Intakes is currently working on establishing a recommended intake for omega-3 and omega-6 fatty acids.[37] The 1990 Canadian RNI include specific amounts for both omega-3 and omega-6 fatty acids.* Many researchers believe the body's requirements depend on an optimal ratio of between 1 to 5 and 1 to 10 (omega-3 to omega-6 fatty acids); current intake is estimated at 1 to 9.8.[38]

*For omega-3 fatty acids, the RNI is 0.5 percent of total energy or 0.55 grams per 1000 kcalories; for omega-6 fatty acids, the RNI is 3 percent of total energy or 3.3 grams per 1000 kcalories. These recommendations represent a 1-to-6 ratio (omega-3 to omega-6).

FIGURE 5-20 Cholesterol in Selected Foods

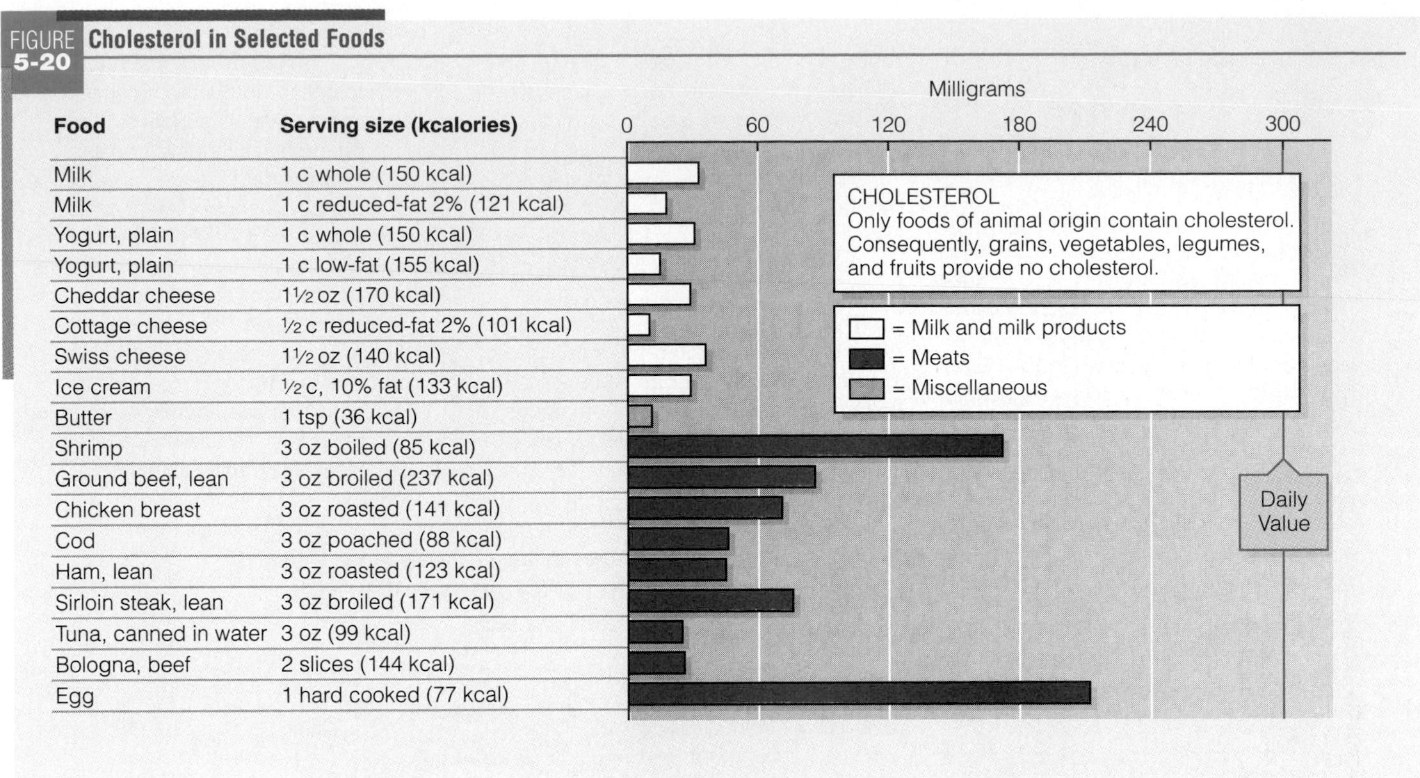

Food	Serving size (kcalories)
Milk	1 c whole (150 kcal)
Milk	1 c reduced-fat 2% (121 kcal)
Yogurt, plain	1 c whole (150 kcal)
Yogurt, plain	1 c low-fat (155 kcal)
Cheddar cheese	1½ oz (170 kcal)
Cottage cheese	½ c reduced-fat 2% (101 kcal)
Swiss cheese	1½ oz (140 kcal)
Ice cream	½ c, 10% fat (133 kcal)
Butter	1 tsp (36 kcal)
Shrimp	3 oz boiled (85 kcal)
Ground beef, lean	3 oz broiled (237 kcal)
Chicken breast	3 oz roasted (141 kcal)
Cod	3 oz poached (88 kcal)
Ham, lean	3 oz roasted (123 kcal)
Sirloin steak, lean	3 oz broiled (171 kcal)
Tuna, canned in water	3 oz (99 kcal)
Bologna, beef	2 slices (144 kcal)
Egg	1 hard cooked (77 kcal)

To obtain the right balance between omega-3 and omega-6 fatty acids, most people need to eat more fish and less meat. The American Heart Association recommends two 3-ounce servings of fatty fish a week.[39] Eating fish instead of meat supports heart health, especially when combined with physical activity. Even one fish meal a week may be enough to make a difference.[40] The fish may not even need to be rich in omega-3 fatty acids; one study found that farm-raised catfish (which is relatively low in omega-3 fatty acids) improves lipid profiles similarly to wild Alaskan salmon.[41] Fish provides many minerals (except iron) and vitamins and is leaner than most other animal-protein sources. When used in a weight-loss program, eating fish improves blood lipids even more effectively than either measure alone.[42]

Fish oil should come from fish, not from supplements. Fish oil supplements are not recommended for a number of reasons.* Perhaps most importantly, high intakes of fish oil increase bleeding time, interfere with wound healing, worsen diabetes, and impair immune function. Fish oil supplements are made from fish skins and livers, which may contain environmental contaminants. Fish oils also naturally contain large amounts of the two most potentially toxic vitamins, A and D. Lastly, supplements are expensive; money is better spent on foods that can provide a full array of nutrients. In addition to fish, other functional foods◆ are being developed to help consumers improve their omega-3 fatty acid intake. For example, hens fed flaxseed produce eggs rich in omega-3 fatty acids.[43] Including even one enriched egg in the diet daily can significantly increase a person's intake of omega-3 fatty acids and improve the omega-3 to omega-6 ratio.[44]

Reminder: Functional foods contain physiologically active compounds that provide health benefits beyond basic nutrition (see Highlight 13 for a full discussion). ◆

• Select Lean Meats and Nonfat Milks • Many foods that contain fat, saturated fat, and cholesterol—such as meats, milk, cheese, and eggs—also provide high-quality protein and valuable vitamins and minerals. They can be included in a healthy diet if a person selects lean and nonfat products and prepares them

*In Canada, fish oil supplements require a physician's prescription.

HOW TO Use the Exchange System to Estimate Fat

The exchange system is especially informative about the fats in foods. To use the exchange system, you need to know the fat value for each list (see the table below) and the foods on that list with their portion sizes (review Figure 2-5 on pp. 44–45). Two of the lists—vegetables and fruits—contain no fat. A third list—starch—provides only a little. (Starch foods prepared with fat are counted as one starch exchange *plus* a fat exchange.) So you only need to learn the fat values for three lists: milks, meats, and fats.

The milk list offers three fat values for nonfat, reduced-fat, and whole milk. Think of nonfat milk as milk and of reduced-fat and whole milk as milk with added fat.

The meat list offers four fat values for very lean, lean, medium-fat, and high-fat products. People are often surprised to learn how much fat comes from meats and cheeses. An ounce of lean meat or low-fat cheese supplies about half of its energy from fat (28 protein kcalories and 27 fat kcalories). An ounce of high-fat meat or most cheeses supplies 72 percent of its energy from fat (28 protein kcalories and 72 fat kcalories). As for the meat alternate, peanut butter, 2 tablespoons supply 76 percent of their energy from fat (32 protein kcalories and 144 fat kcalories). Note that one meat exchange is a single ounce; to use the exchange system, learn to recognize the number of ounces in a serving.

The fat list includes butter, margarine, and oil, of course, but it also includes bacon, olives, avocados, and many

One Exchange	Fat (g)
Milks	
Nonfat (and low-fat)	0–3
Reduced-fat	5
Whole	8
Meats	
Very lean	0–1
Lean	3
Medium-fat	5
High-fat	8
Fats	5
Starch	1 or less
Vegetables	—
Fruits	—
Other carbohydrates	varies

Fat-containing foods appear primarily in three of the exchanges lists: milks, meats, and fats.

0–3 g in 1 c nonfat milk

8 g in 1 oz high-fat meat

3 g in 1 oz lean meat

5 g in 1 tsp butter or margarine

5 g in 1 c reduced-fat (2%) milk

8 g in 1 c whole milk

0–1 g in 1 oz very lean meat

5 g in 1 oz medium-fat meat

Even well-balanced, healthy meals provide some fat. In this chicken stir-fry, only two teaspoons of oil were used in preparation, but 30 percent of the kcalories come from fat. The chicken and sesame seeds also contribute some fat.

using the suggestions outlined on p. 152. Figure 2-5 on pp. 44-45 shows examples of very lean, lean, medium-fat, and high-fat meats and of nonfat, low-fat, reduced-fat, and whole-milk products.

• Eat Plenty of Vegetables, Fruits, and Grains • Choosing vegetables, fruits, whole grains, and legumes also helps lower fat intake. Vegetables and fruits naturally contain no fat, and most grains contain only trace amounts. Some grain *products* such as fried taco shells, croissants, and granola cereal are high in fat, though, so consumers need to read food labels. Similarly, many people prepare grains and vegetables with added fat. Because fruits are often eaten without added fat, a diet that includes several servings of fruit can more readily meet the recommendation for 30 percent or less of kcalories from fat.[45]

Because a low-fat diet is usually rich in vegetables, fruits, whole grains, and legumes, it offers abundant vitamin C, folate, vitamin A, and dietary fiber—all important in supporting health. Consequently, such a diet protects against disease by both reducing fat and increasing nutrients. It also provides valuable phytochemicals that help defend against heart disease.

• Use Fats and Oils Sparingly • Practice moderation when using fats and oils such as butter, margarine, mayonnaise, and salad dressings. These foods offer much fat and little nourishment.

HOW TO (continued)

kinds of nuts. These foods are grouped together because a portion of any of them contains as much fat as a pat of butter and, like butter, offers negligible protein and carbohydrate. In Appendix G, the fat list is sorted into saturated, monounsaturated, and polyunsaturated groups, which helps people make heart-wise selections when choosing fat.

To estimate the fat in this meal, you first need to recognize that this spaghetti dinner is really 1 cup of pasta, with 1 cup of tomato sauce and 3 ounces of lean ground beef. Then you need to translate these portions into exchanges: 2 starches, 1 vegetable, and 3 lean meats, respectively. Ignore the vegetables in the salad, but count the ½ cup of garbanzo beans as 1 starch + 1 very lean meat, and the sunflower seeds and ranch dressing as 1 fat each.

Dinner	Exchange	Fat (g) Estimate	Fat (g) Actual
Salad:			
1 c raw spinach leaves, shredded carrots, and sliced mushrooms	= free	0	0
½ c garbanzo beans	= 1 starch and 1 very lean meat	2	2
1 tbs sunflower seeds	= 1 fat	5	4
1 tbs ranch salad dressing	= 1 fat	5	6
Entrée:			
Spaghetti with meat sauce			
1 c pasta (cooked)	= 2 starches	2	12
1 c tomato sauce	= 1 vegetable	—	
3 oz ground round	= 3 lean meats	9	
½ c green beans	= 1 vegetable	—	0
1 medium corn on the cob	= 1 starch	1	0
2 tsp butter	= 2 fats	10	8
Dessert:			
1⁄12 angel food cake	= 1 other carbohydrate	—	0
Beverage:			
1 c 1% low-fat milk	= 1 nonfat milk	3	3
		37	35

© Thomas Harm and Tom Peterson/Quest Photographic Inc.

Using the exchange system to estimate, this dinner provides about 37 grams of fat. A computer diet analysis program came to a similar conclusion (35 grams), as would a diet analysis using the values in Appendix H. To keep from underestimating fat intake, count "1 or less" and "0–1" as "1 gram of fat" and count "0–3" as 3 grams of fat.

• Look for Invisible Fat • *Visible* fat, such as butter, the oil in salad dressing, and the fat trimmed from meat, is easy to see. *Invisible* fat is less apparent and can be present in foods in surprising amounts. Invisible fat "marbles" a steak or is hidden in foods like nuts, cheese, avocados, and olives. Any *fried* food contains abundant fat: potato chips, french fries, fried wontons, and fried fish. Many *baked* goods, too, are high in fat: pie crusts, pastries, crackers, biscuits, cornbread, doughnuts, sweet rolls, cookies, and cakes. Most chocolate bars contain more fat energy than sugar energy. Even cream-of-mushroom soup prepared with water derives 66 percent of its energy from fat. Abundant fat lurks on salad bars, too, not only in the dressings, but also in the potato salad, the macaroni salad, the coleslaw, and the marinated beans that are mixed with oil-based dressings. Keep invisible fats in mind when making food selections.

© Polara Studios Inc.

Salad dressing can add more than 20 grams of fat to an otherwise low-fat salad.

• Choose Wisely • The *Dietary Guidelines* urge people to choose a diet low in saturated fat and cholesterol and moderate in total fat. A diet following the Food Guide Pyramid can support this goal if selections are made carefully. The "How to" feature above describes an easy way to estimate the fat content of a meal.

FIGURE 5-21 Fats on Food Labels

Food labels list the kcalories from fat and the quantities and Daily Values for fat, saturated fat, and cholesterol.

Information on polyunsaturated and monounsaturated fats is optional, but if it is provided, you can add the three types of fat together (0.5 + 0 + 1 = 1.5) and subtract from the total (4) to calculate *trans*-fat. In this example, there are 2.5 grams *trans*-fat (4 – 1.5 = 2.5).

If the ingredients list includes hydrogenated oils, you know the food contains *trans*-fat—you just don't know how much.

Consumers can find an abundant array of low-fat food choices. In many cases, they are familiar foods presented with less fat. Animals fed special diets produce lower-fat meats, milks, and eggs, which can improve a person's blood cholesterol with little effort.[46] Cuts of meat are often closely trimmed of fat. In the dairy case, nonfat and reduced-fat milk, yogurt, cheeses, and sour cream offer healthy alternatives to their higher-fat counterparts. Many processed foods such as salad dressings, crackers, chips, and cookies are available with little or no fat. Beyond lowering the fat content, manufacturers have developed margarines fortified with phytochemicals that lower blood cholesterol.*[47] (Highlight 13 explores these and other functional foods designed to support health.) Such choices make low-fat, heart-healthy eating easy. A simple switch to nonfat salad dressing can lower a person's average fat intake. Every little fat-saving step helps a person get closer to the 30 percent goal.

Many fat-free foods have been developed using proteins or carbohydrate derivatives. Highlight 5 examines some of these alternatives to fat.

• Read Food Labels • Labels list total fat, saturated fat, and cholesterol contents of foods in addition to fat kcalories per serving (see Figure 5-21). Because each package provides information for a single serving and serving sizes are standardized, consumers can easily compare similar products. Currently, labels do not provide information on *trans*-fatty acids, although the Food and Drug Administration (FDA) is considering it. Among the questions being contemplated is whether to list the amount of *trans*-fatty acids separately or together with saturated fatty acids. In the meantime, Figure 5-21 includes tips on finding the *trans*-fats on food labels.

Total fat, saturated fat, and cholesterol are also expressed as "% Daily Values" for a person consuming 2000 kcalories. People who are consuming more or less than 2000 kcalories daily can calculate their personal Daily Value for fat as described in the "How to" on p. 157.

Be aware that the "% Daily Value" for fat is not the same as "% kcalories from fat." Consider, for example, a piece of lemon meringue pie that provides 140 kcalories and 12 grams of fat. Because the Daily Value for fat is 65 grams for a 2000-kcalorie intake, 12 grams represent about 18 percent, or almost one-fifth, of the day's fat allowance (the pie's "% Daily Value" is 18).◆ Uninformed consumers may mistakenly believe that this food meets the guideline to limit fat to "30 percent kcalories," but it doesn't—for two reasons. First, the pie's 12 grams of fat contribute 108 of the 140 kcalories, for a total of 77 percent kcalories from fat. Second, the "30 percent kcalories from fat" guideline applies to a day's total intake, not to an individual food. (Of course, if every selection throughout the day exceeds 30 percent kcalories from fat, you can be certain that the day's total intake will, too.)

◆ **The photo caption on p. 157 provides calculations.**

Because recommendations apply to average daily intakes and not to individual food items, food labels do not provide "% kcalories from fat." Still, you can get an idea of whether a particular food is high or low in fat. To quickly compare the fat content of a food with recommendations, use the quick and easy estimate◆ that 3 grams of fat (27 kcalories of fat) represent about 30 percent of the kcalories in 100 kcalories of food. Alternatively, you can multiply the grams of fat in a serving by 30 and then compare that number to the kcalories.[48] If it is less, then the food has less than 30 percent kcalories from fat.

◆ **Quick and easy estimates:**
- A food is low in fat if it has: $\leq$3 g fat in 100 kcal food.
- A food is low in fat if: g fat $\times$ 30 $<$ kcal.

The FDA authorizes two health claims on labels concerning fat: one for "dietary saturated fat and cholesterol and risk of coronary heart disease" and one for "dietary fat and cancer." To make these claims, foods must meet specified criteria, as described in Chapter 2.

*Marketed under the brand names Benecol and Take Control.

Calculate a Personal Daily Value for Fat

The % Daily Value for fat on food labels is based on 65 grams. To know that you've met recommendations, you can either count grams until you reach 65, or add the "% Daily Values" until you reach 100 percent—if your energy intake is 2000 kcalories a day. If your energy intake is more or less, you have a couple of options.

You can calculate your personal daily fat allowance in grams. Multiply your total energy intake by 30 percent, then divide by 9. Suppose your energy intake is 1800 kcalories per day and your goal is 30 percent kcalories from fat:

1800 total kcal × 0.30 from fat = 540 fat kcal.
540 fat kcal ÷ 9 kcal/g = 60 g fat.

Another way to calculate your personal fat allowance is to cross out the last digit of your energy intake and divide by 3.[a] For example, 1800 kcalories becomes 180; then you divide by 3:

180 ÷ 3 = 60 g fat/day.

(In familiar measures, 60 grams of fat is about the same as ⅔ stick of butter or ¼ cup of oil.)

[a] K. McNutt, Fat traps, tips, and tricks, *Nutrition Today,* May/June 1992, pp. 47–49.

The accompanying table shows the numbers of grams of fat allowed per day for various energy intakes. With one of these numbers in mind, you can quickly evaluate the number of fat grams in foods you are considering eating.

Recommended Grams of Fat for Different Energy Intakes

Energy (kcal/day)	30% kCalories from Fat	Fat (g/day)
1200	360	40
1500	450	50
1800	540	60
1900 (RDA for women 51 years and over)	570	63
2000 (Daily Value for food labels)	600	65
2200 (RDA for women 19 to 50 years old)	660	73
2300 (RDA for men 51 years and over)	690	77
2600	780	87
2900 (RDA for men 19 to 50 years old)	870	97
3000	900	100

IN SUMMARY

In foods, triglycerides:

- *Deliver fat-soluble vitamins, energy, and essential fatty acids.*
- *Contribute to the sensory appeal of foods and stimulate appetite.*
- *Make foods tender.*

While some fat in the diet is necessary, health authorities recommend limiting total fat to 30 percent or less of energy intake; saturated fat to one-third of total fat, or 10 percent of energy intake; and cholesterol to less than 300 milligrams a day. They also recommend consuming relatively more monounsaturated and polyunsaturated fats, particularly omega-3 fatty acids from foods such as fish, not from supplements. Many purchasing and cooking strategies can help bring these goals within reach, and food labels make it easier to select foods consistent with these guidelines.

If people were to make only one change in their diets, they would be wise to limit their intakes of saturated fat. A second change might be to moderate their total fat. Chances are good that if saturated fat and total fat meet recommendations, then cholesterol intake will, too. Sometimes lowering fat intake can be difficult, though, because fats make foods taste delicious. To maintain good health, must a person give up all high-fat foods forever—never again to eat marbled steak, hollandaise sauce, or gooey chocolate cake? Not at all. These foods bring pleasure to a meal and can be enjoyed as part of a healthy diet when eaten in small quantities on occasion, but it is true that they are not everyday foods. The key word for fat is not deprivation, but moderation: appreciate the energy and enjoyment that fat provides, but take care not to exceed your needs.

% Daily Value for fat:
- ***12 g ÷ 65 g = 0.18 × 100 = 18%.***

% kCalories from fat:
- ***12 g × 9 kcal/g = 108 kcal.***

108 kcal ÷ 140 kcal = 77%.

How are you doing?

- Do you use polyunsaturated and monounsaturated vegetable oils instead of animal fats, hard margarines, and partially hydrogenated shortenings?
- Do you choose fat-free or low-fat milk products and lean meats, fish, and poultry regularly?
- Do you eat plenty of grain products, vegetables, legumes, and fruits daily?

Nutrition on the Net

WEBSITES

Access these websites for further study of topics covered in this chapter.

- Find updates and quick links to these and other nutrition-related sites at our website: **www.wadsworth.com/nutrition**
- Search for "cholesterol" and "dietary fat" at the U.S. Government health information site: **www.healthfinder.gov**
- Review the American Dietetic Association's *ABC's of Fats, Oils, and Cholesterol:* **www.eatright.org/nfs2.html**
- Search for "fat" at the International Food Information Council site: **www.ificinfo.health.org**

INTERNET ACTIVITIES

Heart disease is the number one killer in the United States, implicated in more deaths annually than any other disease. Take the American Heart Association risk assessment quiz to learn how your personal health characteristics contribute to your risk of developing heart disease.

- Go to: **www.americanheart.org**
- Click on "Risk Assessment" located under "Warning Signs."
- Check the boxes in the Risk Assessment quiz that apply to you and then click on "Get Results."

The results of this quiz give you some indication of your personal risks for heart disease and stroke. Some factors (such as genetics) are beyond your control, but others (described as "lifestyle risk factors" such as diet and physical activity) can be modified. What lifestyle changes might you make to reduce your personal risks for heart disease?

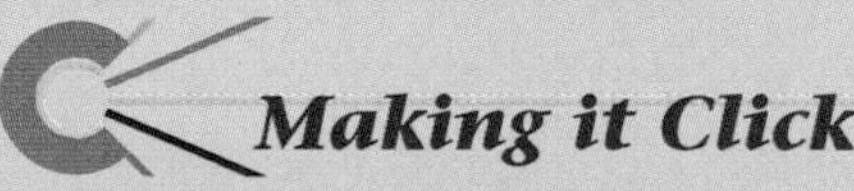

Making it Click

These problems will give you practice in doing simple nutrition-related calculations (see p. 161 for answers). Show your calculations for each problem.

1. Be aware of the fats in milks. Following are four categories of milk.

	Wt (g)	Fat (g)	Prot (g)	Carb (g)
Milk A (1 c)	244	8	8	12
Milk B (1 c)	244	5	8	12
Milk C (1 c)	244	3	8	12
Milk D (1 c)	244	0	8	12

   a. Based on *weight,* what percentage of each milk is fat (round off to a whole number)?
   b. How much energy from fat will a person receive from drinking 1 cup of each milk?
   c. How much total energy will the person receive from 1 cup of each milk?
   d. What percentage of the energy in each milk comes from fat?
   e. In the grocery store, how is each milk labeled?

2. Judge foods' fat contents by their labels.
   a. A food label says that one serving of the food contains 6.5 grams fat. What would the % Daily Value for fat be? What does the Daily Value you just calculated mean?

b. How many kcalories from fat does a serving contain? (Round off to the nearest whole number.)
c. If a *serving* of the food contains 200 kcalories, what percentage of the energy is from fat?

This example should show you how easy it is to evaluate foods' fat contents by reading labels and to see the difference between the % Daily Value and the percentage of kcalories from fat.

3. Now consider a piece of carrot cake. Remember that the Daily Value suggests 65 grams of fat as acceptable within a 2000-kcalorie diet. A serving of carrot cake provides 30 grams of fat. What percentage of the Daily Value is that? What does this mean?

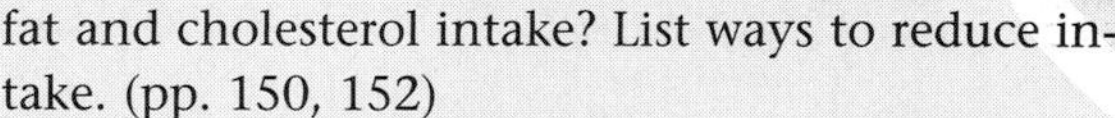

Study Questions

These questions will help you review the chapter. You will find the answers in the discussions on the pages provided.

1. Name three classes of lipids found in the body and in foods. What are some of their functions in the body? What features do fats bring to foods? (pp. 130, 144, 150)
2. What features distinguish fatty acids from each other? (pp. 130–132)
3. What does the term *omega* mean with respect to fatty acids? Describe the roles of the omega fatty acids in disease prevention. (pp. 132, 145–146, 149)
4. What are the differences between saturated, unsaturated, monounsaturated, and polyunsaturated fatty acids? Describe the structure of a triglyceride. (pp. 131–133)
5. What does hydrogenation do to fats? What are *trans*-fatty acids, and how do they influence heart disease? (pp. 134–135, 148–149)
6. How do phospholipids differ from triglycerides in structure? How does cholesterol differ? How do these differences in structure affect function? (pp. 136–138)
7. What roles do phospholipids perform in the body? What roles does cholesterol play in the body? (pp. 136–137)
8. Trace the steps in fat digestion, absorption, and transport. Describe the routes cholesterol takes in the body. (pp. 138–140)
9. What do lipoproteins do? What are the differences among the chylomicrons, VLDL, LDL, and HDL? (pp. 141–143)
10. Which of the fatty acids are essential? Name their chief dietary sources. (pp. 144–145)
11. How does excessive fat intake influence health? What factors influence LDL, HDL, and total blood cholesterol? (pp. 148–149)
12. What are the dietary recommendations regarding fat and cholesterol intake? List ways to reduce intake. (pp. 150, 152)
13. Which food lists of the exchange system supply fat in abundance? In moderation? Not at all? (pp. 153–155)
14. What is the Daily Value for fat (for a 2000-kcalorie diet)? What does this number represent? (pp. 156–157)

These questions will help you prepare for an exam. Answers can be found on p. 161.

1. Saturated fatty acids:
   a. are always 18 carbons long.
   b. have at least one double bond.
   c. are fully loaded with hydrogens.
   d. are always liquid at room temperature.
2. A triglyceride consists of:
   a. three glycerols attached to a lipid.
   d. three fatty acids attached to a glucose.
   c. three fatty acids attached to a glycerol.
   d. three phospholipids attached to a cholesterol.
3. The difference between *cis*- and *trans*-fatty acids is:
   a. the number of double bonds.
   b. the length of their carbon chains.
   c. the location of the first double bond.
   d. the configuration around the double bond.
4. Which of the following is *not* true? Lecithin is:
   a. an emulsifier.
   b. a phospholipid.
   c. an essential nutrient.
   d. a constituent of cell membranes.
5. Chylomicrons are produced in the:
   a. liver.
   b. pancreas.
   c. gallbladder.
   d. small intestine.
6. Transport vehicles for lipids are called:
   a. micelles.
   b. lipoproteins.

c. blood vessels.
d. monoglycerides.

7. The lipoprotein most associated with a high risk of heart disease is:
a. CHD.
b. HDL.
c. LDL.
d. LPL.

8. Which of the following is *not* true? Fats:
a. contain glucose.
b. provide energy.
c. protect against organ shock.
d. carry vitamins A, D, E, and K.

9. The essential fatty acids include:
a. stearic acid and oleic acid.
b. oleic acid and linoleic acid.
c. palmitic acid and linolenic acid.
d. linoleic acid and linolenic acid.

10. A person consuming 2200 kcalories a day who wants to meet the *Diet and Health* recommendations should limit daily fat intake to:
a. 25 grams or less.
b. 73 grams or less.
c. 98 grams or less.
d. 123 grams or less.

References

1 L. Brown and coauthors, Cholesterol-lowering effects of dietary fiber: A meta-analysis, *American Journal of Clinical Nutrition* 69 (1999): 30–42.

2 C. Bruce, R. A. Chouinard, Jr., and A. R. Tall, Plasma lipid transfer proteins, high-density lipoproteins, and reverse cholesterol transport, *Annual Review of Nutrition* 18 (1998): 297–330.

3 E. N. Siguel and R. H. Lerman, Prevalence of essential fatty acid deficiency in patients with chronic gastrointestinal disorders, *Metabolism* 45 (1996): 12–23.

4 K. A. Bruinsma and D. L. Taren, Dieting, essential fatty acid intake, and depression, *Nutrition Reviews* 58 (2000): 98–108.

5 L. H. Kuller, Dietary fat and chronic diseases: Epidemiologic overview, *Journal of the American Dietetic Association* 97 (1997): S9–S15.

6 R. H. Knopp and coauthors, Long-term cholesterol-lowering effects of 4 fat-restricted diets in hypercholesterolemic and combined hyperlipidemic men: The Dietary Alternatives Study, *Journal of the American Medical Association* 278 (1997): 1509–1515.

7 A. M. Salter and D. A. White, Effect of dietary fat on cholesterol metabolism: Regulation of plasma LDL concentrations, *Nutrition Research Reviews* 9 (1996): 241–257.

8 F. B. Hu and coauthors, Dietary saturated fats and their food sources in relation to the risk of coronary heart disease in women, *American Journal of Clinical Nutrition* 70 (1999): 1001–1008; S. M. Grundy, Influence of stearic acid on cholesterol metabolism relative to other long-chain fatty acids, *American Journal of Clinical Nutrition* 60 (1994): 986S–990S.

9 W. E. Connor, Harbingers of coronary heart disease: Dietary saturated fatty acids and cholesterol—Is chocolate benign because of its stearic acid content? *American Journal of Clinical Nutrition* 70 (1999): 951–952.

10 M. B. Katan, *Trans* fatty acids and plasma lipoproteins, *Nutrition Reviews* 58 (2000): 188–191; P. O. Kwiterovich, The effect of dietary fat, antioxidants, and pro-oxidants on blood lipids, lipoproteins, and atherosclerosis, *Journal of the American Dietetic Association* 97 (1997): S31–S41.

11 G. J. Nelson, Dietary fat, *trans* fatty acids, and risk of coronary heart diseases, *Nutrition Reviews* 56 (1998): 250–252; S. Shapiro, Do *trans* fatty acids increase the risk of coronary artery disease? A critique of the epidemiologic evidence, *American Journal of Clinical Nutrition* 66 (1997): 1011S–1017S; D. B. Allison, *Trans* fatty acids and coronary heart disease risk: Epidemiology, *American Journal of Clinical Nutrition* 62 (1995): 670S–678S.

12 C. Ip, Review of the effects of *trans* fatty acids, oleic acid, n-3 polyunsaturated fatty acids, and conjugated linoleic acid on mammary carcinogenesis in animals, *American Journal of Clinical Nutrition* 66 (1997): 1523S–1529S.

13 A. Ascherio and W. C. Willett, Health effects of *trans* fatty acids, *American Journal of Clinical Nutrition* 66 (1997): 1006S–1010S.

14 ASCN/AIN Task Force on *Trans* Fatty Acids, Position paper on *trans* fatty acids, *American Journal of Clinical Nutrition* 63 (1996): 663–670.

15 A. H. Lichtenstein and coauthors, Effects of different forms of dietary hydrogenated fats on serum lipoprotein cholesterol levels, *New England Journal of Medicine* 340 (1999): 1933–1940; F. B. Hu and coauthors, Dietary fat intake and the risk of coronary heart disease in women, *New England Journal of Medicine* 337 (1997): 1491–1499; M. B. Katan, High-oil compared with low-fat, high-carbohydrate diets in the prevention of ischemic heart disease, *American Journal of Clinical Nutrition* 66 (1997): 974S–979S.

16 C. Thomsen and coauthors, Differential effects of saturated and monounsaturated fatty acids on postprandial lipemia and incretin responses in healthy subjects, *American Journal of Clinical Nutrition* 69 (1999): 1135–1143; H. M. Roche and coauthors, Effect of long-term olive oil dietary intervention on postprandial triacylglycerol and factor VII metabolism, *American Journal of Clinical Nutrition* 68 (1998): 552–560; Kwiterovich, 1997.

17 P. M. Kris-Etherton and coauthors, High-monounsaturated fatty acid diets lower both plasma cholesterol and triacylglycerol concentrations, *American Journal of Clinical Nutrition* 70 (1999): 1009–1115.

18 C. von Schacky, n-3 Fatty acids and the prevention of coronary atherosclerosis, *American Journal of Clinical Nutrition* 71 (2000): 224S–227S; W. S. Harris, n-3 Fatty acids and serum lipoproteins: Human studies, *American Journal of Clinical Nutrition* 65 (1997): 1645S–1654S.

19 H. Iso and coauthors, Intake of fish and omega-3 fatty acids and risk of stroke in women, *Journal of the American Medical Association* 285 (2001): 304–312; M. L. Daviglus and coauthors, Fish consumption and the 30-year risk of fatal myocardial infarction, *New England Journal of Medicine* 336 (1997): 1046–1053; N. J. Stone, Fish consumption, fish oil, lipids, and coronary heart disease, *Circulation* 94 (1996): 2337–2340; R. F. Gillum, M. E. Mussolino, and J. H. Madans, The relationship between fish consumption and stroke incidence: The NHANES I epidemiologic follow-up study, *Archives of Internal Medicine* 156 (1996): 537–542.

20 P. J. Nestel, Fish oil and cardiovascular disease: Lipids and arterial function, *American Journal of Clinical Nutrition* 71 (2000): 228S–231S; S. L. Connor and W. E. Connor, Are fish oils beneficial in the prevention and treatment of coronary artery disease? *American Journal of Clinical Nutrition* 66 (1997): 1020S–1031S.

21 D. P. Rose, Dietary fatty acids and cancer, *American Journal of Clinical Nutrition* 66 (1997): 998S–1003S; J. H. Weisburger, Dietary fat and risk of chronic disease: Mechanistic insights from experimental studies, *Journal of the American Dietetic Association* 97 (1997): S16–S23.

22 B. C.-H. Chiu and coauthors, Diet and risk of non-Hodgkin lymphoma in older women, *Journal of the American Medical Association* 275 (1996): 1315–1321.

23 E. B. Feldman, Breast cancer risk and intake of fat, *Nutrition Reviews* 57 (1999): 353–356.

24 M. D. Holmes and coauthors, Association of dietary intake of fat and fatty acids with risk of breast cancer, *Journal of the American Medical Association* 281 (1999): 914–920; D. J. Hunter and coauthors, Cohort studies of fat intake and the risk of breast cancer—A pooled analysis, *New England Journal of Medicine* 334 (1996): 356–361; E. Barrett-Connor and N. J. Friedlander, Dietary fat, calories, and the risk of breast cancer in postmenopausal women: A prospective population-based study, *Journal of the American College of Nutrition* 12 (1993): 390–399.

25 J. A. Thomas, Diet, micronutrients, and the prostate gland, *Nutrition Reviews* 57 (1999): 95–103.

26 H. Senzaki and coauthors, Dietary effects of fatty acids on growth and metastasis of KPL-1 human breast cancer cells in vivo and in vitro, *Anticancer Research* 18 (1998): 1621–1627; N. R. Simonsen and coauthors, Tissue stores of individual monounsaturated fatty acids and breast cancer: The EURAMIC Study, *American Journal of Clinical Nutrition* 68 (1998): 134–141; D. Y. Kim, K. H. Chung, and J. H. Lee, Stimulatory effects of high-fat diets on colon cell proliferation depend on the type of dietary fat and site of the colon, *Nutrition and Cancer* 30 (1998): 118–123; L. Kohlmeier and coauthors, Adipose tissue, *trans* fatty acids and breast cancer in the European

ANSWERS

Making it Click

1. a. Milk A: 8 g fat ÷ 244 g total = 0.03; 0.03 × 100 = 3%.
   Milk B: 5 g fat ÷ 244 g total = 0.02; 0.02 × 100 = 2%.
   Milk C: 3 g fat ÷ 244 g total = 0.01; 0.01 × 100 = 1%.
   Milk D: 0 g fat ÷ 244 g total = 0.00; 0.00 × 100 = 0%.
   b. Milk A: 8 g fat × 9 kcal/g = 72 kcal from fat.
   Milk B: 5 g fat × 9 kcal/g = 45 kcal from fat.
   Milk C: 3 g fat × 9 kcal/g = 27 kcal from fat.
   Milk D: 0 g fat × 9 kcal/g = 0 kcal from fat.
   c. Milk A: (8 g fat × 9 kcal/g) + (8 g prot × 4 kcal/g) + (12 g carb × 4 kcal/g) = 152 kcal.
   Milk B: (5 g fat × 9 kcal/g) + (8 g prot × 4 kcal/g) + (12 g carb × 4 kcal/g) = 125 kcal.
   Milk C: (3 g fat × 9 kcal/g) + (8 g prot × 4 kcal/g) + (12 g carb × 4 kcal/g) = 107 kcal.
   Milk D: (0 g fat × 9 kcal/g) + (8 g prot × 4 kcal/g) + (12 g carb × 4 kcal/g) = 80 kcal.
   d. Milk A: 72 kcal from fat ÷ 152 total kcal = 0.47; 0.47 × 100 = 47%.
   Milk B: 45 kcal from fat ÷ 125 total kcal = 0.36; 0.36 × 100 = 36%.
   Milk C: 27 kcal from fat ÷ 107 total kcal = 0.25; 0.25 × 100 = 25%.
   Milk D: 0 kcal from fat ÷ 80 total kcal = 0.00; 0.00 × 100 = 0%.
   e. Milk A: whole.
   Milk B: reduced-fat, 2%, or less-fat.
   Milk C: low-fat or 1%.
   Milk D: nonfat, fat-free, skim, zero-fat, or no-fat.
2. a. 6.5 g ÷ 65 g = 0.1; 0.1 × 100 = 10%. A Daily Value of 10% means that one serving of this food contributes about ⅒ of the day's fat allotment.
   b. 6.5 g × 9 kcal/g = 58.5, rounded to 59 kcal from fat.
   c. (59 kcal from fat ÷ 200 kcal) × 100 = 30% kcalories from fat.
3. (30 g fat ÷ 65 g fat) × 100 = 46% of the Daily Value for fat; this means that almost half of the day's fat allotment would be used in this one dessert.

Multiple Choice

1. c	2. c	3. d	4. c	5. d
6. b	7. c	8. a	9. d	10. b

Community Multicenter Study on antioxidants, myocardial infarction, and breast cancer, *Cancer Epidemiology, Biomarkers & Prevention* 6 (1997): 705–710.

27 E. Fernandez and coauthors, Fish consumption and cancer risk, *American Journal of Clinical Nutrition* 70 (1999): 85–90.

28 B. J. Rolls, Carbohydrates, fats, and satiety, *American Journal of Clinical Nutrition* 61 (1995): 960S–967S.

29 WHO/FAO Joint Consultation, Fats and oils in human nutrition, *Nutrition Reviews* 53 (1995): 202–205.

30 A. Drewnowski, Taste preferences and food intake, *Annual Review of Nutrition* 17 (1997): 237–253.

31 S. M. Garn, From the Miocene to olestra: A historical perspective on fat consumption, *Journal of the American Dietetic Association* 97 (1997): S54–S57.

32 N. D. Ernst and coauthors, Consistency between US dietary fat intake and serum total cholesterol concentrations: The National Health and Nutrition Examination Surveys, *American Journal of Clinical Nutrition* 66 (1997): 965S–972S; National Center for Health Statistics, www.cdc.gov/nchs, site visited on November 6, 2000.

33 Hu and coauthors, 1999.

34 B. Matheson and coauthors, Effect on serum lipids of monounsaturated oil and margarine in the diet of an Antarctic expedition, *American Journal of Clinical Nutrition* 63 (1996): 933–938.

35 D. B. Allison and coauthors, Estimated intakes of *trans* fatty and other fatty acids in the US population, *Journal of the American Dietetic Association* 99 (1999): 166–174.

36 F. B. Hu and coauthors, A prospective study of egg consumption and risk of cardiovascular disease in men and women, *Journal of the American Medical Association* 281 (1999): 1387–1394.

37 A. P. Simopoulos, A. Leaf, and N. Salem, Workshop on the essentiality of and recommended dietary intakes for omega-6 and omega-3 fatty acids, *Nutrition Today* 35 (2000): 166–167.

38 P. M. Kris-Etherton and coauthors, Polyunsaturated fatty acids in the food chain in the United States, *American Journal of Clinical Nutrition* 71 (2000): 179S–188S.

39 AHA Dietary Guidelines, Published online on October 5, 2000 http://circ.ahajournals.org/cgi/content/full/4304635102).

40 C. M. Albert and coauthors, Fish consumption and risk of sudden cardiac death, *Journal of the American Medical Association* 279 (1998): 23–28; A. Ascherio and coauthors, Dietary intake of marine n-3 fatty acids, fish intake, and the risk of coronary disease among men, *New England Journal of Medicine* 332 (1995): 977–982.

41 D. K. Tidwell and coauthors, Comparison of the effects of adding fish high or low in n-3 fatty acids to a diet conforming to the Dietary Guidelines for Americans, *Journal of the American Dietetic Association* 93 (1993): 1124–1128.

42 T. A. Mori and coauthors, Dietary fish as a major component of a weight-loss diet: Effect on serum lipids, glucose, and insulin metabolism in overweight hypertensive subjects, *American Journal of Clinical Nutrition* 70 (1999): 817–825.

43 L. K. Ferrier and coauthors, α-Linoleic acid—and docosahexaenoic acid—enriched eggs frm hens fed flaxseed: Influence on blood lipids and platelet phospholipid fatty acids in humans, *American Journal of Clinical Nutrition* 62 (1995): 81–86.

44 D. J. Farrell, Enrichment of hen eggs with n-3 long-chain fatty acids and evaluation of enriched eggs in humans, *American Journal of Clinical Nutrition* 68 (1998): 538–544.

45 S. M. Krebs-Smith and coauthors, Characterizing food intake patterns of American adults, *American Journal of Clinical Nutrition* 65 (1997): 1264S–1268S.

46 M. Noakes and coauthors, Modifying the fatty acid profile of dairy products through feedlot technology lowers plasma cholesterol of humans consuming the products, *American Journal of Clinical Nutrition* 63 (1996): 42–46.

47 P. J. H. Jones and coauthors, Cholesterol-lowering efficacy of a sitostanol-containing phytosterol mixture with a prudent diet in hyperlipidemic men, *American Journal of Clinical Nutrition* 69 (1999): 1144–1150; M. A. Hallikainen and M. I. J. Uusitupa, Effects of 2 low-fat stanol ester-containing margarines on serum cholesterol concentrations as part of a low-fat diet in hypercholesterolemic subjects, *American Journal of Clinical Nutrition* 69 (1999): 403–410.

48 D. Green-Burgeson, Calculating fat the easy way, *Journal of the American Dietetic Association* 94 (1994): 256.

ALTERNATIVES TO FAT

As PEOPLE learn more about the health consequences of high-fat diets, they want to lower their fat intake, but they'd rather not give up their favorite foods. As the adage goes, they want to have their cake and eat it, too—both figuratively and literally.

Food chemists have been working for decades on ways to reduce the fat in foods. Juggling the needs of the human body, the taste perceptions of consumers, and the requirements of food preparation is a complex task. For the body, products must contribute little food energy, be nontoxic and completely excreted, and not rob the body of valuable fat-soluble nutrients. To satisfy consumers, products must be attractive, feel pleasant in the mouth, and have an acceptable flavor. Food manufacturers need a compound that remains stable while meeting a product's requirements for temperature, moisture, and texture. That's a tall order, but food chemists have mastered the task. Today shoppers can select from thousands of fat-free, low-fat, and reduced-fat products. Many bakery goods, cheeses, frozen desserts, and other products offer less than half a gram of fat in a serving.

Some techniques for reducing food fat are quite simple. For example, fat can be removed by skimming milk or trimming meats. Manufacturers can dilute fat by adding water or whipping in air. They use nonfat milk in creamy desserts and lean meats in frozen entrées. Sometimes they simply prepare the products differently. For example, fat-free potato chips can be baked instead of fried.

Other techniques to replace some or all of the fat in foods use ingredients that derive from either carbohydrate, protein, or fat. Because these ingredients are common dietary substances, companies can ask the Food and Drug Administration (FDA) to approve them as generally recognized as safe (GRAS) substances. The body may digest and absorb some of these substances, so they may contribute some energy, although significantly less than fat's 9 kcalories per gram.[1] The glossary and Table H5-1 (on p. 164) describe the various **fat replacers** in more detail.

CARBOHYDRATE-BASED FAT REPLACERS

Manufacturers have long used carbohydrate-based compounds to thicken and stabilize foods. Today carbohydrate-based ingredients are used to replace the fat in a variety of products. For example, maltodextrin, a carbohydrate derived from corn, melts when sprinkled on hot, moist foods such as baked potatoes, providing a flavor similar to that of butter or margarine.

Many carbohydrate-based products mimic the texture and feel of fat by forming gels when added to water. They are heat stable enough to be used in baking, but not in frying.

Some (such as dextrins and modified food starches) can be digested and so provide energy—up to 4 kcalories per gram. This represents less than half of the kcalories provided by fat, but kcalories nevertheless. Others (such as cellulose) are insoluble fibers that pass through the body undigested, thus yielding 0 kcalories. Better still is the contribution these fibers make to a person's daily intake. When used to replace 100 grams of fat, these products may save 900 kcalories *and* provide 10 grams of fiber.

PROTEIN-BASED FAT REPLACERS

Protein-based fat replacers are made from either egg white, soy, or milk proteins processed into mistlike particles that feel creamy on the tongue and taste like fat. As a result, they create the *perception* of fat without all the kcalories. In the body, they are digested and absorbed, contributing 1 to 4 kcalories per gram. Because 1 gram of a protein-based ingredient can replace 3 grams of fat, this represents a substantial kcalorie savings. Such substitutions can reduce the energy values of some foods dramatically. In some cases, though, such as fat-free frozen desserts, so much sugar is added that the kcalorie count of the fat-free product may be as high as in the regular ice cream it is intended to replace (see Table H5-2, p. 165).

Low-fat and nonfat foods offer as much flavor and enjoyment as their high-fat counterparts—for less fat and fewer calories.

GLOSSARY

fat replacers: ingredients that replace some or all of the functions of fat and may or may not provide energy. In this text, the term *fat replacer* is used interchangeably with **fat substitute,** which technically applies only to an ingredient that replaces all of the functions of fat and provides no energy.

olestra: a synthetic fat made from sucrose and fatty acids that provides 0 kcalories per gram; also known as **sucrose polyester.**

Replacing both the fat and the sugar in a product is difficult to do because both contribute to flavor, texture, and stability. Still, the ice cream's fat and cholesterol content have been reduced appreciably.

Some people, such as those who are allergic or sensitive to egg or milk proteins, may have to avoid these products. FDA regulations require a product's label to identify the source of its protein in the ingredients list.

FAT-BASED FAT REPLACERS

Fat-based fat replacers share many of the properties of regular fats, but provide fewer kcalories. Two of these products—caprenin and salatrim—are triglycerides that have been created with a specific combination of fatty acids attached to the glycerol molecule.[2] By using short-chain (8-carbon) fatty acids that provide fewer kcalories and a long-chain (22-carbon) fatty acid that is poorly absorbed, manufacturers have created a fat that provides 5 kcalories per gram instead of 9.

SYNTHETIC FAT REPLACERS

In January 1996, the FDA approved a fat replacer known as **olestra** for use as an additive in snack foods.[3] Olestra

TABLE H5-1 A Sampling of Fat Replacers

Fat Replacers	Energy (kcal/g)	Properties	Uses in Foods
Carbohydrate-Based Fat Replacers			
• *Fruit* purees and pastes of apples, bananas, cherries, plums, or prunes.	0	Replace bulk; add moisture and tenderness.	Baked goods, candy, dairy products.
• *Gels* derived from cellulose or starch.	0–4[a]	Add moisture; stabilize emulsions; mimic texture and feel of fat.	Dairy products, salad dressings, frozen desserts, sauces.
• *Gums* extracted from beans, sea vegetables, or other sources.	0–4	Add moisture; thicken; provide creamy texture; stabilize emulsions.	Salad dressings, processed meats, desserts.
• *Dextrins and maltodextrins* made from corn.	1–4	Form gels that mimic texture and feel of fat; thicken; stabilize emulsions.	Salad dressings, puddings, spreads, dairy products, frozen desserts, chips, baked goods, meat products, frostings, soups, butter-flavored "sprinkles" for melting on hot foods.
• *Grain based* fibers[b] made from oat flour; the seed hulls of oats, soybeans, peas, and rice; or the bran of corn or wheat.	0–4	Add fiber; provide creamy texture; add moisture; replace bulk of fat; can be used in baking but not in frying.	Dips, dressings, baked goods, cheese, ground beef, chocolate.
Fat-Based Fat Replacers			
• *Salatrim*[c] derived from **s**hort- **a**nd **l**ong-chain fatty **a**cid **tri**glyceride **m**olecules.	5	Same properties as fats; can be used in baking but not frying.	Chocolate coatings, dairy products, spreads, cookies, crackers, frozen desserts.
• *Caprenin* derived from a triglyceride with poorly absorbed fatty acids.	5	Same properties as fats; can be used in baking but not frying.	Soft candies, cocoa butter, confectionery coatings.
Protein-Based Fat Replacers			
• *Microparticulated protein*[d] processed from the proteins of soy, milk, or egg white into mistlike particles that roll over the tongue, making the protein feel and taste like fat.	1–4	Creamy texture; heat stable in some cooking and baking but not frying.	Dairy products, mayonnaise, salad dressings, spreads, baked goods, frozen desserts.
Synthetic Fat Replacers			
• *Olestra,*[e] an artificial fat made from sucrose and fatty acids; formerly called *sucrose polyester.*	0	Same properties as fats; heat stable in frying, cooking, and baking.	Potato, corn, tortilla chips; crackers; cheese puffs.

[a]Energy made available by intestinal bacteria.
[b]Trade names: Oatrim, Z-Trim, Nu-Trim, Trim-Choice
[c]Trade name: Benefat.
[d]Trade names: Simplesse and K-Blazer.
[e]Trade name: Olean.

TABLE H5-2 Fat Content of Regular Foods and Foods Prepared with Fat Replacers

Food	Fat (g)	Cholesterol (mg)	Energy (kcal)
Ice cream			
Super premium (½ c)	19	97	274
Regular (½ c)	7	30	135
Ice milk (½ c)	3	9	92
Frozen dessert			
Made with Simplesse (½ c)	<1	14	120
Made with Oatrim (½ c)	1	4	135
Butter (1 tsp)	4	11	36
Margarine (1 tsp)	4	0	34
Maltodextrin sprinkles (½ tsp)	<1	0	3
French fries			
Fried in vegetable oil	12.3	0	227
Fried in 75% olestra blend	3.1	0	144
Chicken			
Fried in vegetable oil	14.6	0	252
Fried in 75% olestra blend	8.6	0	198
Onion rings			
Fried in vegetable oil	19.5	0	315
Fried in 75% olestra blend	4.9	0	184

Note: Equivalent serving sizes were compared; in the case of butter and margarine, ½ teaspoon of sprinkles was compared with 1 teaspoon butter or margarine as per label directions.

is the first artificial fat to reach the market that can withstand the heat needed to fry foods such as potato chips or cheese puffs. Manufactured under the trade name Olean (pronounced oh-LEEN), olestra has been approved for use in snack foods such as potato chips, crackers, and tortilla chips.

Olestra's chemical structure is similar to that of a regular fat (a triglyceride) but with important differences. A triglyceride is composed of a glycerol molecule with three fatty acids attached, whereas olestra is made of a sucrose molecule with six to eight fatty acids attached. Enzymes in the digestive tract cannot break the bonds of olestra, so unlike [illegible] fatty acids, [illegible] the system u[illegible]s characteristic is responsib[illegible] the troubles of this nonfat [illegible]

Because o[illegible]tty acids common to shor[illegible]hares many of their physical [illegible]e, color, taste, heat stability, and shelf life. Consequently, olestra looks, feels, and tastes like dietary fat and can be used in frying, cooking, and baking. Best of all, olestra performs like a fat without losing flavor, adding kcalories, or raising blood lipids. Potato chips made with olestra deliver half the kcalories of regular chips and none of the fat. Does this sound too good to be true? It may be.

The FDA's evaluation of olestra's safety answered two questions. First, is olestra toxic? Second, does it affect either nutrient absorption or the health of the digestive tract?

Regarding possible toxicity, there is little controversy. Research on both animals and human beings supports the safety of olestra as a partial replacement for dietary fats and oils. Researchers conducting studies on animals have reported no evidence of either cancer or birth defects caused by olestra.

Olestra's effects on nutrient absorption and digestive tract health, however, raise concerns.[4] One of the positive attributes of fats is that they carry the fat-soluble vitamins A, D, E, and K with them into the body. When olestra passes through the digestive tract unabsorbed, it binds with some of these vitamins and carries them out of the body, robbing the person of these valuable nutrients. To compensate for these losses, the FDA has required the manufacturer to fortify olestra with vitamins A, D, E, and K. Saturating olestra with these vitamins blocks its ability to bind with the vitamins from other foods.

Olestra also sweeps other fat-soluble substances through the digestive system. Among those lost are the carotenoids, the colorful pigments found in fruits and vegetables.[5] Carotenoids act as antioxidants and may be important in protecting against a variety of diseases, including heart

Potato chips made with olestra deliver half of the kcalories of regular chips and none of the fat but may cause GI distress for some people.

disease, some cancers, and macular degeneration (an eye disorder that causes blurry vision and blindness). Because the relationships between carotenoids and disease prevention have not yet been proved, the FDA has required the manufacturer to conduct long-term studies on whether olestra use and carotenoid losses will impair health. Should this prove to be a problem, fortifying products is not a likely solution. If you consider that there are hundreds of carotenoids commonly found in foods and then think of the hundreds of *non*carotenoid substances that may also be beneficial to health, then you will quickly realize that fortification is not feasible. The FDA will review olestra's status periodically as new research findings become available.

Consumers may not perceive any immediate ill effects of a diet depleted of its fat-soluble vitamins and carotenoids, but they will surely notice the digestive distress that sometimes accompanies olestra consumption: cramps, gas, bloating, and diarrhea. For some people, eating as little as 2 ounces of olestra-containing potato chips produces "fecal urgency" (the immediate need for a bathroom) and "anal leakage" (noted by stained underwear); for others, the incidence and severity of GI symptoms are no different than when eating regular potato chips.[6] The FDA considers these symptoms "unpleasant" but not "medically significant." Many consumers seem to agree and have accepted such "annoyances" in trade for a bag of fat-free chips.

In approving olestra, the FDA required foods made with it to carry a label warning that "olestra may cause abdominal cramping and loose stools" and that it "inhibits the absorption of some vitamins and other nutrients." Those who read food labels may hesitate to purchase products with these warnings; others may not even notice the warnings.

Will people become slimmer by munching on chips and cookies made with artificial fats? Or will they simply feel free to eat more chips and cookies? Interestingly, one study found that women did eat more lunch after having eaten a low-fat yogurt than after a high-fat yogurt when the yogurt was labeled as such.[7] If they did not know which yogurt they had eaten, the women eating the low-fat yogurt consumed less lunch than those eating the high-fat yogurt. Simply knowing the fat content of the yogurt influenced the energy intake of their next meal. Other studies report that substituting olestra chips for regular chips reduces fat and energy intakes.[8]

Consumers need to keep in mind that low-fat and fat-free foods still deliver kcalories. Decades ago, consumers hailed the arrival of artificial sweeteners as a weight-loss wonder, but in reality, kcalories saved by using artificial sweeteners were readily replaced by kcalories from other foods. The alternatives to fat discussed in this highlight can help to lower fat intake and support weight loss only when they actually *replace* fat and energy in the diet.[9]

Nutrition on the Net

•WEBSITES•

Access these websites for further study of topics covered in this highlight.

- Find updates and quick links to these and other nutrition-related sites at our website: **www.wadsworth.com/nutrition**
- Search for "fat replacers" at the International Food Information Council site: **ificinfo.health.org**
- Search for "fat substitutes" and "olestra" in the foods section of the Food and Drug Administration's site: **www.fda.gov**

References

1 A. M. Miraglio, Nutrient substitutes and their energy values in fat substitutes and replacers, *American Journal of Clinical Nutrition* 62 (1995): 1175S–1179S.

2 S. J. Bell and coauthors, The new dietary fats in health and disease, *Journal of the American Dietetic Association* 97 (1997): 280–286.

3 Olestra: Approved with special labeling, *FDA Consumer,* April 1996, p. 11.

4 H. Blackburn, Olestra and the FDA, *New England Journal of Medicine* 334 (1996): 984–986.

5 J. A. Westrate and K. H. van het Hof, Sucrose polyester and plasma carotenoid concentrations in healthy subjects, *American Journal of Clinical Nutrition* 62 (1995): 591–597.

6 L. J. Cheskin and coauthors, Gastrointestinal symptoms following consumption of olestra or regular triglyceride potato chips: A controlled comparison, *Journal of the American Medical Association* 279 (1998): 150–152.

7 D. J. Shide and B. J. Rolls, Information about the fat content of preloads influences energy intake in healthy women, *Journal of the American Dietetic Association* 95 (1995): 993–998.

8 J. O. Hill and coauthors, Effects of 14 d of covert substitution of olestra for conventional fat on spontaneous food intake, *American Journal of Clinical Nutrition* 67 (1998): 1178–1185; D. L. Miller and coauthors, Effect of fat-free potato chips with and without nutrition labels on fat and energy intakes, *American Journal of Clinical Nutrition* 68 (1998): 282–290.

9 Position of The American Dietetic Association: Fat replacers, *Journal of the American Dietetic Association* 98 (1998): 463–468.

CHAPTER 6
Protein: Amino Acids

© Richard Fukuhara/Corbis

FIGURE 6-1 Amino Acid Structure

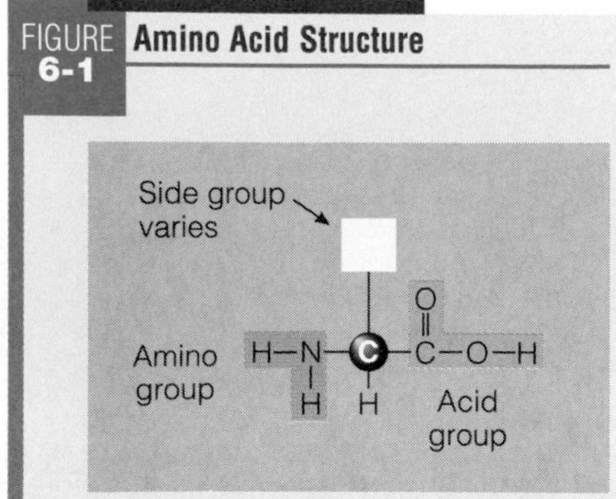

All amino acids have a carbon (known as the alpha-carbon), with an amino group (NH_2), an acid group (COOH), a hydrogen (H), and a side group attached. The side group is a unique chemical structure that differentiates one amino acid from another.

People commonly associate protein with strength and meat with protein. Consequently, they eat steak to build their muscles, but their thinking is only partly correct. Protein is a vital structural and working substance in all cells, not just muscle cells. Meat is a good source of protein, but so are milk, eggs, legumes, and many grains and vegetables. People who overvalue protein may overemphasize meat in their diets, sometimes at the expense of other, equally important nutrients and foods. Protein is important, but it is only one of the nutrients needed to maintain the body's health.

THE CHEMIST'S VIEW OF PROTEINS

Chemically, **proteins** contain the same atoms as carbohydrates and lipids—carbon (C), hydrogen (H), and oxygen (O)—but proteins also contain nitrogen (N) atoms. These nitrogen atoms give the name *amino* (nitrogen containing) to the amino acids—the links in the chains of proteins.

Amino Acids

All **amino acids** have the same basic structure—a central carbon (C) atom with a hydrogen (H), an amino group (NH_2), and an acid group (COOH) attached to it. Carbon atoms need to form four bonds,◆ though, so a fourth attachment is necessary. It is this fourth site that distinguishes each amino acid from the others. Attached to the carbon atom at the fourth bond is a distinct atom, or group of atoms, known as the *side group* or *side chain* (see Figure 6-1).

◆ **Reminder:**
- H forms 1 bond.
- O forms 2 bonds.
- N forms 3 bonds.
- C forms 4 bonds.

• Unique Side Groups • The side groups on amino acids vary from one amino acid to the next, making proteins more complex than either carbohydrates or lipids. A polysaccharide (starch, for example) may be several thousand units long, but every unit is a glucose molecule just like all the others. A protein, on the other hand, is made up of about 20 different amino acids, each with a different side group. Table 6-1 lists the amino acids most common in proteins.*

*Besides the 20 common amino acids, which can all be components of proteins, others do not occur in proteins, but can be found individually (for example, taurine and ornithine). Some amino acids occur in related forms (for example, proline can acquire an OH group to become hydroxyproline).

TABLE 6-1 Amino Acids

Proteins are made up of about 20 common amino acids. The first column lists the essential amino acids for human beings (those the body cannot make—that must be provided in the diet).

Essential Amino Acids		Nonessential Amino Acids	
Histidine	(HISS-tuh-deen)	Alanine	(AL-ah-neen)
Isoleucine	(eye-so-LOO-seen)	Arginine	(ARJ-ih-neen)
Leucine	(LOO-seen)	Asparagine	(ah-SPAR-ah-geen)
Lysine	(LYE-seen)	Aspartic acid	(ah-SPAR-tic acid)
Methionine	(meh-THIGH-oh-neen)	Cysteine	(SIS-teh-een)
Phenylalanine	(fen-il-AL-ah-neen)	Glutamic acid	(GLU-tam-ic acid)
Threonine	(THREE-oh-neen)	Glutamine	(GLU-tah-meen)
Tryptophan	(TRIP-toe-fan, TRIP-toe-fane)	Glycine	(GLY-seen)
		Proline	(PRO-leen)
Valine	(VAY-leen)	Serine	(SEER-een)
		Tyrosine	(TIE-roe-seen)

Note: In special cases, some nonessential amino acids may become conditionally essential (see the text).

proteins: compounds composed of carbon, hydrogen, oxygen, and nitrogen atoms, arranged into amino acids linked in a chain. Some amino acids also contain sulfur atoms.

amino (a-MEEN-oh) **acids:** building blocks of proteins. Each contains an amino group, an acid group, a hydrogen atom, and a distinctive side group, all attached to a central carbon atom.
- **amino** = containing nitrogen

The simplest amino acid, glycine, has a hydrogen atom as its side group. A slightly more complex amino acid, alanine, has an extra carbon with three hydrogen atoms. Other amino acids have more complex side groups (see Figure 6-2 for examples). Thus, although all amino acids share a common structure, they differ in size, shape, electrical charge, and other characteristics because of differences in these side groups.

• **Nonessential Amino Acids** • More than half of the amino acids are **nonessential,** meaning that the body can synthesize them for itself. Proteins in foods usually deliver these amino acids, but it is not essential that they do so. Given nitrogen to form the amino group and fragments from carbohydrate and fat to form the rest of the structure, the body can make any nonessential amino acid.

• **Essential Amino Acids** • There are nine amino acids that the human body either cannot make at all or cannot make in sufficient quantity to meet its needs. These nine amino acids must be supplied by the diet; they are **essential.** The first column in Table 6-1 presents the essential amino acids.

• **Conditionally Essential Amino Acids** • Sometimes a nonessential amino acid becomes essential under special circumstances. For example, the body normally uses the essential amino acid phenylalanine to make tyrosine (a nonessential amino acid). But if the diet fails to supply enough phenylalanine, or if the body cannot make the conversion for some reason (as happens in the inherited disease phenylketonuria), then tyrosine becomes **conditionally essential.**

Proteins

Cells link amino acids end-to-end in a virtually infinite variety of sequences to form thousands of different proteins. Each amino acid is connected to the next by a **peptide bond.**

• **Amino Acid Chains** • Condensation reactions connect amino acids, just as they combine monosaccharides to form disaccharides, and fatty acids with glycerol to form triglycerides. Two amino acids bonded together form a **dipeptide** (see Figure 6-3 on p. 170). By another such reaction, a third amino acid can be added to the chain to form a **tripeptide.** As additional amino acids join the chain, a **polypeptide** is formed. Most proteins are a few dozen to several hundred amino acids long. Figure 6-4 (on p. 170) provides an example—insulin.

• **Amino Acid Sequences** • If a person could walk along a carbohydrate molecule like starch, the first stepping stone would be a glucose. The next stepping

FIGURE 6-2 Examples of Amino Acids

Note that all amino acids have a common chemical structure but that each has a different side group. Appendix C presents the chemical structures of the 20 amino acids most common in proteins.

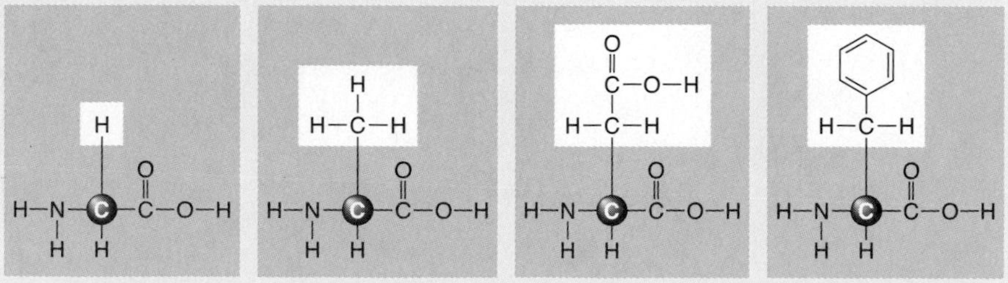

Glycine — Alanine — Aspartic acid — Phenylalanine

nonessential amino acids: amino acids that the body can synthesize (see Table 6-1).

essential amino acids: amino acids that the body cannot synthesize in amounts sufficient to meet physiological needs (see Table 6-1). Some researchers refer to essential amino acids as **indispensable** and to nonessential amino acids as **dispensable.**

conditionally essential amino acid: an amino acid that is normally nonessential, but must be supplied by the diet in special circumstances when the need for it exceeds the body's ability to produce it.

peptide bond: a bond that connects the acid end of one amino acid with the amino end of another, forming a link in a protein chain.

dipeptide (dye-PEP-tide): two amino acids bonded together.
- **di** = two
- **peptide** = amino acid

tripeptide: three amino acids bonded together.
- **tri** = three

polypeptide: many (ten or more) amino acids bonded together. An intermediate string of four to nine amino acids is an **oligopeptide** (OL-ee-go-PEP-tide).
- **poly** = many
- **oligo** = few

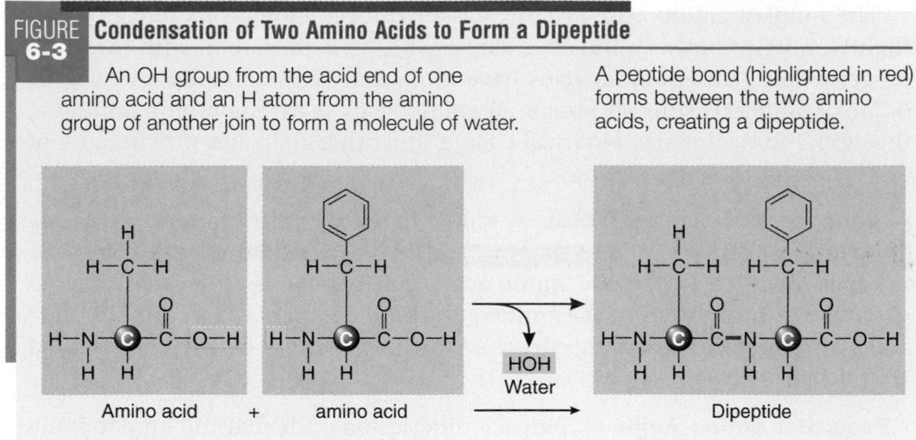

FIGURE 6-3 Condensation of Two Amino Acids to Form a Dipeptide

An OH group from the acid end of one amino acid and an H atom from the amino group of another join to form a molecule of water.

A peptide bond (highlighted in red) forms between the two amino acids, creating a dipeptide.

stone would also be a glucose, and it would be followed by a glucose, and yet another glucose. But if a person were to walk along a polypeptide chain, each stepping stone would be one of 20-odd different amino acids. The first stepping stone might be the amino acid methionine. The second might be an alanine. The third might be a glycine, and the fourth a tryptophan, and so on. Walking along another polypeptide path, a person might step on a phenylalanine, then a valine, and a glutamine. In other words, amino acid sequences within proteins vary.

The amino acids can act somewhat like the letters in an alphabet. If you had only the letter *G*, all you could write would be a string of Gs: G–G–G–G–G–G–G. But with 20 different letters available, you could create poems, songs, or novels. The 20 amino acids can be linked together in an even greater variety of sequences than are possible for letters in a word or words in a sentence. Thus the variety of possible sequences for polypeptide chains is tremendous.

• **Protein Shapes** • Polypeptide chains twist into a variety of complex, tangled shapes, depending on their amino acid sequences. The unique side group of each amino acid gives it characteristics that attract it to, or repel it from, the surrounding fluids and other amino acids. Some amino acid side groups carry electrical charges that are attracted to water molecules (they are hydrophilic). Other side groups are neutral and are repelled by water (they are hydrophobic). As amino acids are strung together to make a polypeptide, the chain folds so that its charged hydrophilic side groups are on the outer surface near water; the

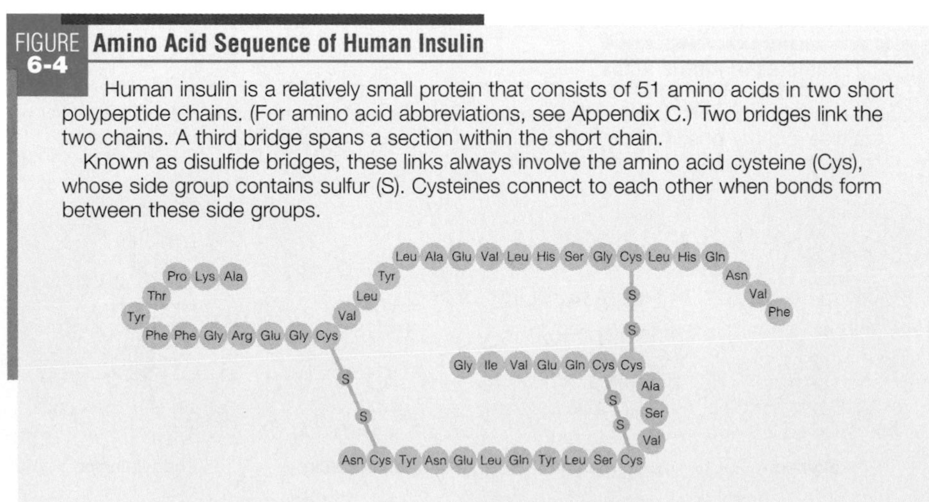

FIGURE 6-4 Amino Acid Sequence of Human Insulin

Human insulin is a relatively small protein that consists of 51 amino acids in two short polypeptide chains. (For amino acid abbreviations, see Appendix C.) Two bridges link the two chains. A third bridge spans a section within the short chain.

Known as disulfide bridges, these links always involve the amino acid cysteine (Cys), whose side group contains sulfur (S). Cysteines connect to each other when bonds form between these side groups.

neutral hydrophobic groups tuck themselves inside, away from water. The intricate, coiled shape the polypeptide finally assumes gives it maximum stability in the body's watery fluids.

• Protein Functions • The extraordinary and unique shapes of proteins enable them to perform their various tasks in the body. Some form hollow balls that can carry and store materials within them, and some, such as those of tendons, are more than ten times as long as they are wide, forming strong, rodlike structures. Some polypeptides are functioning proteins as they are; others need to associate with other polypeptides to form larger working complexes. Some proteins require minerals to activate them. One molecule of **hemoglobin**—the large, globular protein molecule that, by the billions, packs the red blood cells and carries oxygen—is made of four associated polypeptide chains, each holding the mineral iron (see Figure 6-5).

FIGURE 6-5 **The Structure of Hemoglobin**

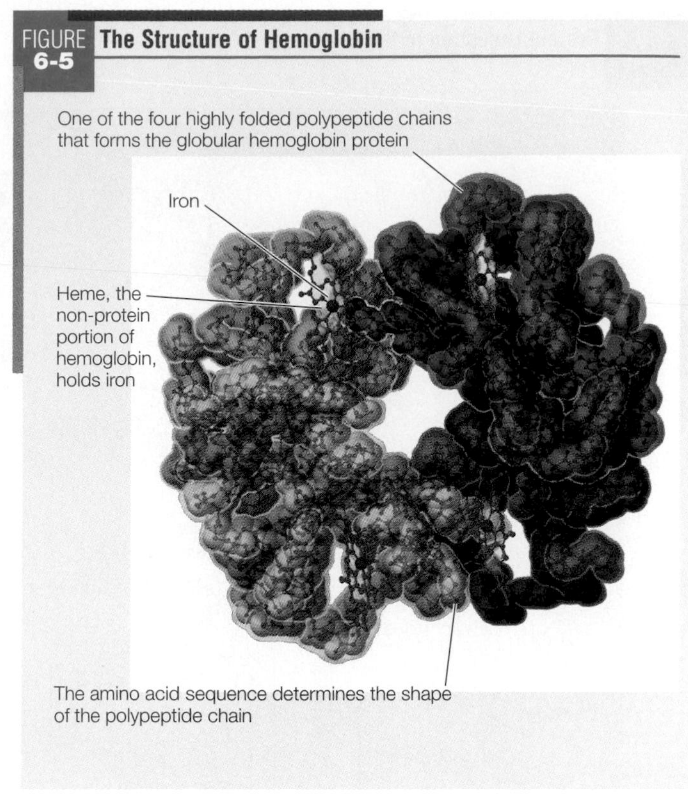

• Protein Denaturation • When proteins are subjected to heat, acid, or other conditions that disturb their stability, they undergo **denaturation**—that is, they uncoil and lose their shapes and, consequently, their ability to function. Past a certain point, denaturation is irreversible. Familiar examples of denaturation include the hardening of an egg when it is cooked, the curdling of milk when acid is added, and the stiffening of egg whites when they are whipped.

IN SUMMARY

Chemically speaking, proteins are more complex than carbohydrates or lipids, being made of some 20 different amino acids, 9 of which the body cannot make (they are essential). Each amino acid contains an amino group, an acid group, a hydrogen atom, and a distinctive side group. Cells link amino acids together in a series of condensation reactions to create proteins. The distinctive sequence of amino acids in each protein determines its unique shape and function.

DIGESTION AND ABSORPTION OF PROTEIN

Proteins in foods do not become body proteins directly. Instead, they supply the amino acids from which the body makes its own proteins. When a person eats foods containing protein, enzymes break the long polypeptide strands into shorter strands, the short strands into tripeptides and dipeptides, and, finally, the tripeptides and dipeptides into amino acids.

The Process of Digestion

Figure 6-6 (on p. 172) illustrates the digestion of protein through the GI tract. Proteins are crushed and moistened in the mouth, but the real action begins in the stomach.

• In the Stomach • The major event in the stomach is the partial breakdown (hydrolysis) of proteins. Hydrochloric acid uncoils (denatures) each protein's tangled strands so that digestive enzymes can attack the peptide bonds. The hydrochloric acid also converts the inactive◆ form of the enzyme pepsinogen to

◆ The inactive form of an enzyme is called a **proenzyme.**
• **pro** = before

hemoglobin (HE-moh-GLOW-bin): the globular protein of the red blood cells that carries oxygen from the lungs to the cells throughout the body.
- **hemo** = blood
- **globin** = globular protein

denaturation (dee-NAY-chur-AY-shun): the change in a protein's shape and consequent loss of its function brought about by heat, agitation, acid, base, alcohol, heavy metals, or other agents.

FIGURE 6-6 Protein Digestion in the GI Tract

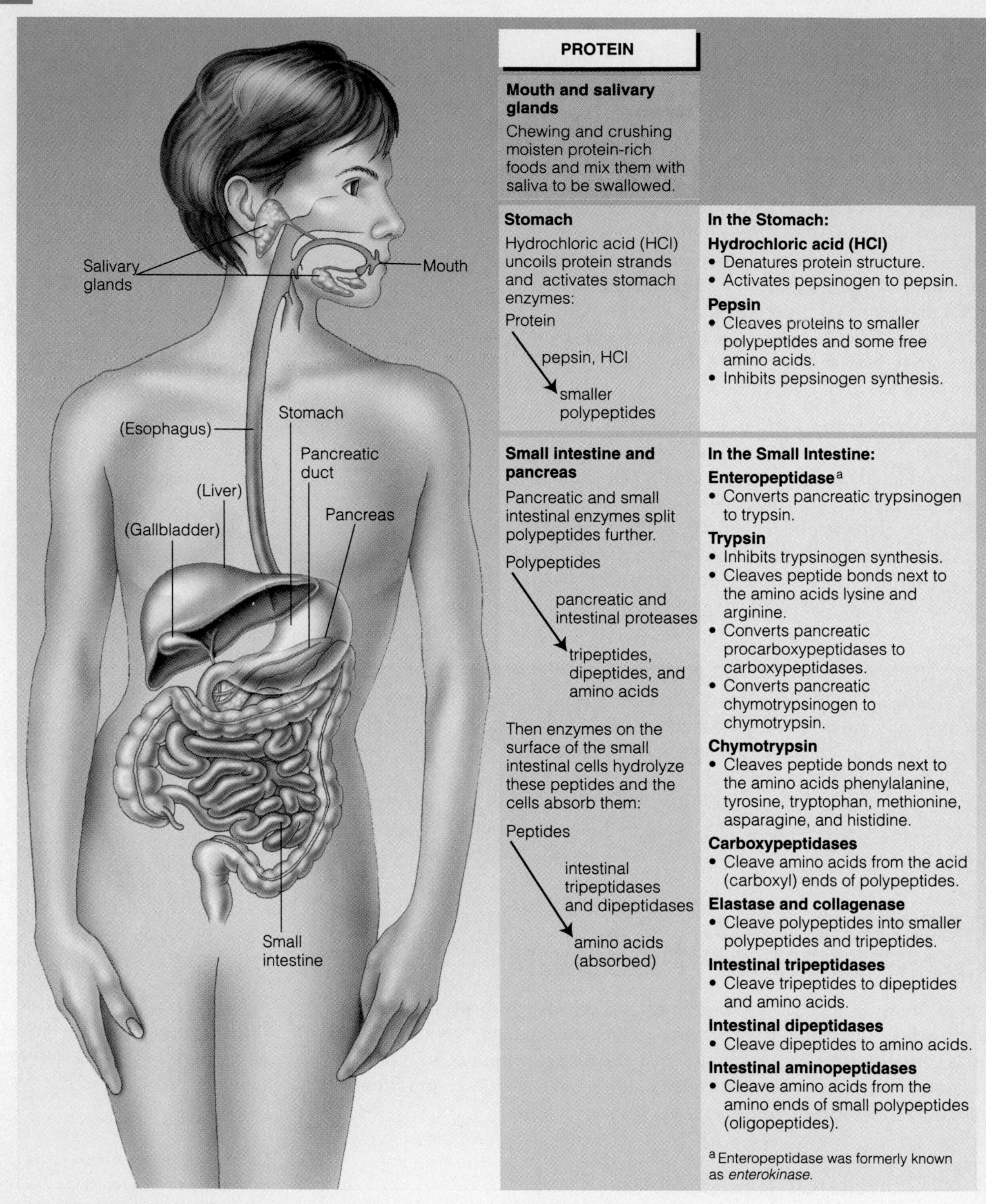

pepsin: a gastric enzyme that hydrolyzes protein. Pepsin is secreted in an inactive form, **pepsinogen,** which is activated by hydrochloric acid in the stomach.

proteases (PRO-tee-aces): enzymes that hydrolyze protein.

its active form, **pepsin.** Pepsin cleaves proteins—large polypeptides—into smaller polypeptides and some amino acids.

• In the Small Intestine • When polypeptides enter the small intestine, several pancreatic and intestinal **proteases** hydrolyze them further into short peptide chains (oligopeptides), tripeptides, dipeptides, and amino acids. Then

peptidase enzymes on the membrane surfaces of the intestinal cells split most of the dipeptides and tripeptides into single amino acids. Only a few peptides escape digestion and enter the blood intact. Figure 6-6 names the digestive enzymes for protein and describes their actions.

The Process of Absorption

A number of specific carriers transport amino acids (and some dipeptides and tripeptides) into the intestinal cells. Once inside the intestinal cells, amino acids may be used for energy or to synthesize needed compounds. Those not used by the intestinal cells are transported across the cell membrane into the surrounding fluid where they enter the capillaries on their way to the liver.

Some nutrition faddists fail to realize that most proteins are broken down to amino acids before absorption. They urge consumers to "Eat enzyme A. It will help you digest your food." Or "Don't eat food B. It contains enzyme C, which will digest cells in your body." In reality, though, enzymes in foods are digested, just as all proteins are. Even the digestive enzymes—which function optimally at their specific pH—are denatured and digested when the pH of their environment changes. (For example, the enzyme pepsin, which works best in the low pH of the stomach becomes inactive and digested when it enters the higher pH of the small intestine.)

Another misconception is that eating predigested proteins (amino acid supplements) saves the body from having to digest proteins and keeps the digestive system from "overworking." Such a belief grossly underestimates the body's abilities. As a matter of fact, the digestive system handles whole proteins *better* than predigested ones because it dismantles and absorbs the amino acids at rates that are optimal for the body's use. (The last section of this chapter discusses amino acid supplements further.)

IN SUMMARY

Digestion is facilitated mostly by the stomach's acid and enzymes, which first denature dietary proteins, then cleave them into smaller polypeptides and some amino acids. Pancreatic and intestinal enzymes split these polypeptides further, to oligo-, tri-, and dipeptides, and then split most of these to single amino acids. Then carriers in the membranes of intestinal cells transport the amino acids into the cells, where they are released into the bloodstream.

PROTEINS IN THE BODY

The human body contains an estimated 10,000 to 50,000 different kinds of proteins. Of these, about 1000 have been studied,◆ although with the recent surge in knowledge gained from sequencing the human genome,◆ this number is sure to grow rapidly. Only about 10 are described in this chapter—but these should be enough to illustrate the versatility, uniqueness, and importance of proteins. As you will see, each protein has a specific function and that function is determined during protein synthesis.

◆ The study of the body's proteins is called **proteomics.**

◆ Information gleaned from sequencing and mapping the human genome is expected to influence health and health care in the not too distant future.

Protein Synthesis

Each human being is unique because of minute differences in the body's proteins. These differences are determined by the amino acid sequences of proteins, which, in turn, are determined by genes. The following paragraphs describe in words the ways cells synthesize proteins; Figure 6-7 provides a pictorial description.

peptidase: a digestive enzyme that hydrolyzes peptide bonds. *Tripeptidases* cleave tripeptides; *dipeptidases* cleave dipeptides. *Endopeptidases* cleave peptide bonds within the chain to create smaller fragments, whereas *exopeptidases* cleave bonds at the ends to release free amino acids.

- **tri** = three
- **di** = two
- **endo** = within
- **exo** = outside

FIGURE 6-7 **Protein Synthesis**

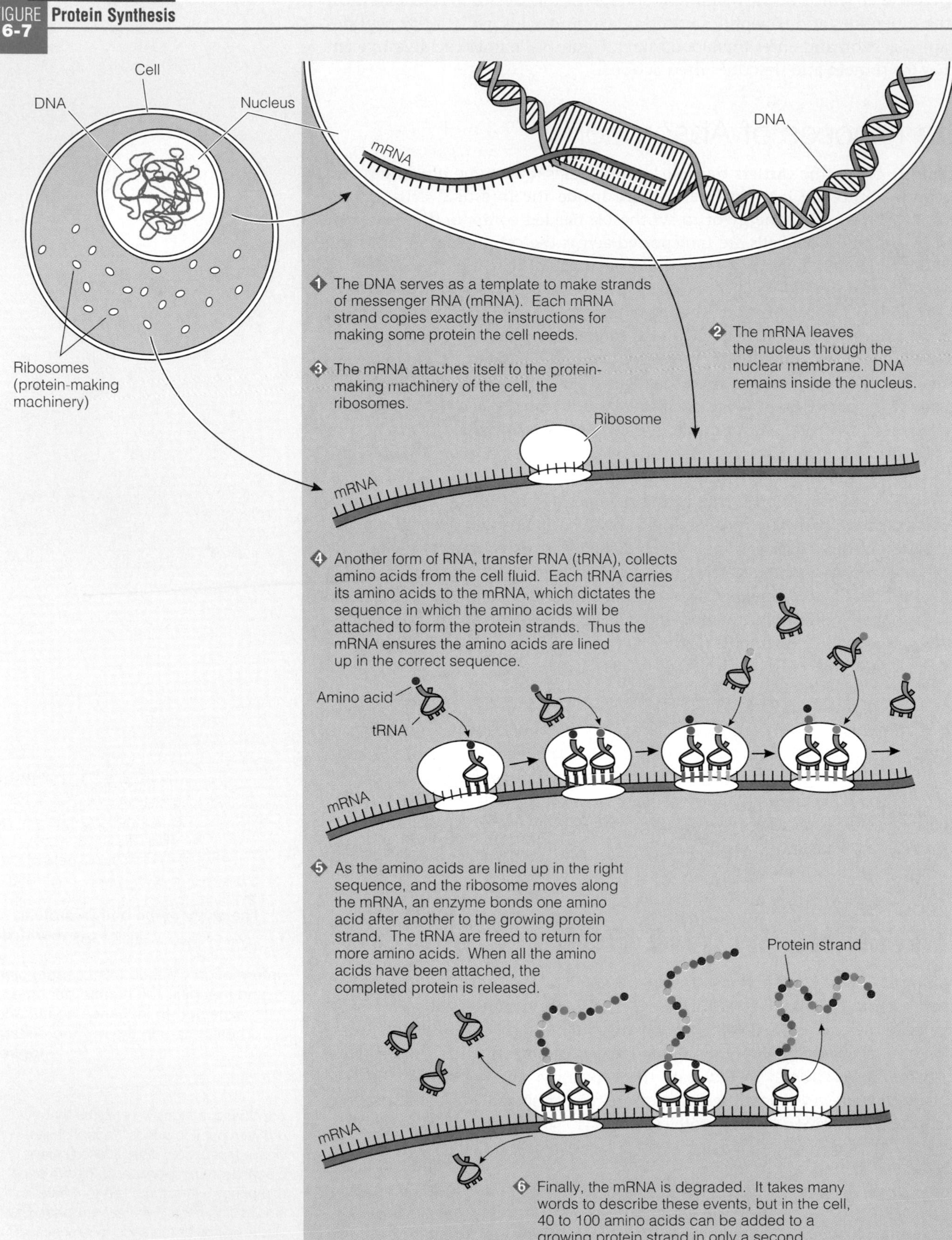

The instructions for making every protein in a person's body are transmitted by way of the genetic information received at conception. This body of knowledge, which is filed in the DNA within the nucleus of every cell, never leaves the nucleus.

• Delivering the Instructions • To inform a cell of the sequence of amino acids for a needed protein, a stretch of DNA serves as a template for making a strand of RNA that carries a code. Known as messenger RNA, this molecule escapes through the nuclear membrane. Messenger RNA seeks out and attaches itself to one of the ribosomes (a protein-making machine, which is itself composed of RNA and protein). Thus situated, messenger RNA presents its list, specifying the sequence in which the amino acids are to line up to make a strand of protein.

• Lining Up the Amino Acids • Other forms of RNA, called transfer RNA, collect amino acids from the cell fluid and bring them to the messenger. Each of the 20 amino acids has a specific transfer RNA. Thousands of transfer RNA, each carrying its amino acid, cluster around the ribosomes, awaiting their turn to unload. When the messenger's list calls for a specific amino acid, the transfer RNA carrying that amino acid moves into position. Then the next loaded transfer RNA moves into place and then the next and the next. In this way, the amino acids line up in the sequence that is called for, and enzymes bind them together. Finally, the completed protein strand is released, the messenger is degraded, and the transfer RNA are freed to return for other loads of amino acids.

• Sequencing Errors • The sequence of amino acids in each protein determines its shape, which supports a specific function. If a genetic error alters the amino acid sequence of a protein, or if a mistake is made in copying the sequence, an altered protein will result, sometimes with dramatic consequences. The protein hemoglobin offers one example of such a genetic variation. In a person with **sickle-cell anemia,◆** two of hemoglobin's four polypeptide chains (described earlier on p. 171) have the normal sequence of amino acids, but the other two chains do not—they have the amino acid valine in a position that is normally occupied by glutamic acid (see Figure 6-8). This single alteration in the amino acid sequence changes the character and shape of the protein so much that hemoglobin loses its ability to carry oxygen effectively.[1] The red blood cells filled with this abnormal hemoglobin stiffen into elongated sickle, or crescent, shapes instead of maintaining their normal pliable disc shape—hence the name, sickle-cell anemia. Sickle-cell anemia causes many medical problems and can be fatal.[2] Caring for children with sickle-cell anemia includes diligent attention to their water needs; dehydration can trigger a crisis.[3]

• Nutrients and Gene Expression • When a cell makes a protein as described earlier, scientists say that the gene for that protein has been "expressed." Cells can regulate gene expression to make the type of protein, in the amounts and at the rate, they need. Nearly all of the body's cells possess the genes for making all human proteins, but each type of cell makes only the proteins it needs. For example, only cells of the pancreas express the gene for insulin; in other cells, that gene is idle. Similarly, the cells of the pancreas do not make the protein hemoglobin, which is needed only by the red blood cells.

Recent research has unveiled some of the fascinating ways nutrients regulate gene expression and protein synthesis. These discoveries have begun to explain some of the relationships among nutrients, genes, and disease development. The benefits of polyunsaturated fatty acids in defending against heart disease, for example, are partially explained by their role in influencing gene expression for lipid enzymes. Later chapters provide additional examples of how nutrients influence gene expression.

FIGURE 6-8 Normal Red Blood Cells Compared with Sickle Cells

Normally, red blood cells are disc-shaped; in the inherited disorder sickle-cell anemia, red blood cells are sickle- or crescent-shaped. This alteration in shape occurs because valine replaces glutamic acid in the amino acid sequence of one of hemoglobin's polypeptide chain. As a result of this one alteration, the hemoglobin has a diminished capacity to carry oxygen.

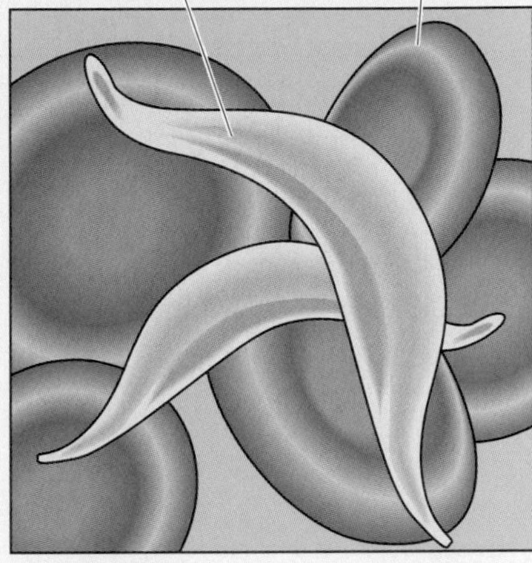

Amino acid sequence of normal hemoglobin:

Val–His–Leu–Thr–Pro–Glu–Glu

Amino acid sequence of sickle-cell hemoglobin:

Val–His–Leu–Thr–Pro–Val–Glu

◆ Anemia is not a disease, but a symptom of various diseases. In the case of sickle-cell anemia, a defect in the hemoglobin molecule changes the shape of the red blood cells. Later chapters describe how vitamin and mineral deficiencies change the size and color of the red blood cells. In all cases, the abnormal blood cells are unable to meet the body's oxygen demands.

sickle-cell anemia: a hereditary form of anemia characterized by abnormal sickle- or crescent-shaped red blood cells. Sickled cells interfere with oxygen transport and blood flow. Symptoms are precipitated by dehydration and insufficient oxygen (as may occur at high altitudes) and include hemolytic anemia (red blood cells burst), fever, and severe pain in the joints and abdomen.

IN SUMMARY Cells synthesize proteins according to the genetic information provided by the DNA in the nucleus of each cell. This information dictates the order in which amino acids must be linked together to form a given protein. Sequencing errors occasionally occur, sometimes with significant consequences.

Roles of Proteins

Whenever the body is growing, repairing, or replacing tissue, proteins are involved. Sometimes their role is to facilitate or to regulate; other times it is to become part of a structure. Versatility is a key feature of proteins.

◆ Mussels anchor themselves underwater with strong, sticky proteins that researchers hope to replicate for use as nontoxic, moisture-resistant glues in medicine and dentistry.◆

• **As Building Materials** • From the moment of conception, proteins form the building blocks of most body structures. For example, to build a bone or a tooth, cells first lay down a **matrix** of the protein **collagen** and then fill it with crystals of calcium, phosphorus, magnesium, fluoride, and other minerals.

Collagen also provides the material of ligaments and tendons and the strengthening glue between the cells of the artery walls that enables the arteries to withstand the pressure of the blood surging through them with each heartbeat. Also made of collagen are scars that knit the separated parts of torn tissues together.

Proteins are also needed for replacement. The life span of a skin cell is only about 30 days. As old skin cells shed, new cells made largely of protein grow from underneath to compensate. Cells in the deeper skin layers synthesize new proteins to go into hair and fingernails. Cells of the GI tract are replaced every three days. Both inside and outside, then, the body continuously deposits protein into new cells that replace those that have been lost.

© International eStock Photography/PhotoQuest

Growing children end each day with more bone, blood, muscle, and skin cells than they had at the beginning of the day.

• **As Enzymes** • Digestive enzymes have appeared in every chapter since Chapter 3, but digestion is only one of the many processes enzymes facilitate. **Enzymes** not only break down substances, they also build substances (such as bone) and transform one substance into another (amino acids into glucose, for example). Figure 6-9 diagrams a synthesis reaction.

An analogy may help to clarify the role of enzymes. Enzymes are comparable to the clergy and judges who make and dissolve marriages. When a minister marries two people, they become a couple, with a new bond between them. They are joined together—but the minister remains unchanged. The minister represents enzymes that synthesize large compounds from smaller ones. One minister can perform thousands of marriage ceremonies, just as one enzyme can perform billions of synthetic reactions.

Similarly, a judge who lets married couples separate may decree many divorces before retiring or dying. The judge represents enzymes that hydrolyze larger compounds to smaller ones; for example, the digestive enzymes. The point is that, like the minister and the judge, enzymes themselves are not altered by the reactions they facilitate. They are catalysts, permitting reactions to occur more quickly and efficiently than if substances depended on chance encounters alone.

• **As Hormones** • Cells can switch their protein machinery on or off in response to the body's needs. Often hormones do the switching, with marvelous

matrix (MAY-tricks): the basic substance that gives form to a developing structure; in the body, the formative cells from which teeth and bones grow.

collagen (KOL-ah-jen): the protein from which connective tissues such as scars, tendons, ligaments, and the foundations of bones and teeth are made.

enzymes: proteins that facilitate chemical reactions without being changed in the process; protein catalysts.

FIGURE 6-9 Enzyme Action

Each enzyme facilitates a specific chemical reaction. In this diagram, an enzyme enables two compounds to make a more complex structure, but the enzyme itself remains unchanged.

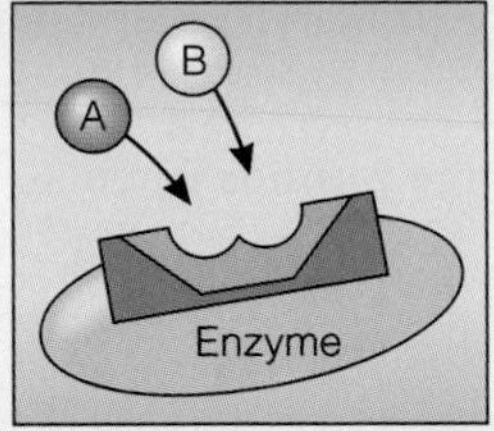

Two separate compounds, A and B, are attracted to the enzyme's active site, making a reaction likely.

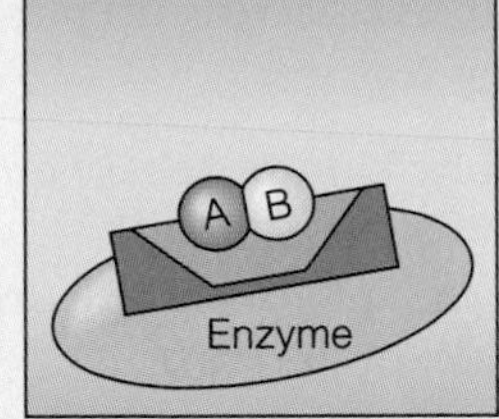

The enzyme forms a complex with A and B.

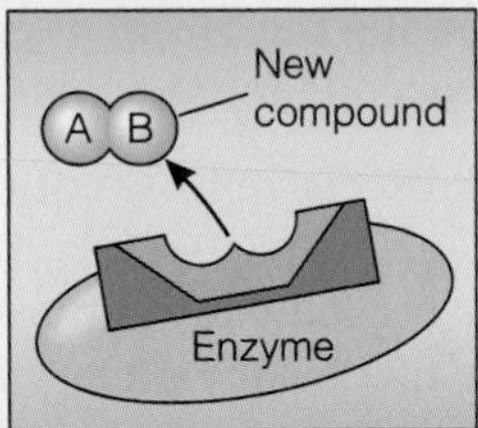

The enzyme is unchanged, but A and B have formed a new compound, AB.

precision. The body's many hormones are messenger molecules, and *some* hormones are proteins. Various glands in the body release hormones in response to changes in the internal environment. The blood carries the hormones from these glands to their target tissues, where they elicit the appropriate responses to restore normal conditions.

The hormone insulin provides a familiar example. When blood glucose rises, the pancreas releases its insulin. Insulin stimulates the transport proteins of the muscles and adipose tissue to pump glucose into the cells faster than it can leak out. (After acting on the message, the cells destroy the insulin.) Then, as blood glucose falls, the pancreas reduces its insulin output. Many other proteins act as hormones, regulating a variety of actions in the body (see Table 6-2 for examples).

• **As Regulators of Fluid Balance** • Proteins help to maintain the body's **fluid balance.** The body's fluids are contained within the cells (intracellular) or outside the cells (extracellular).◆ Extracellular fluids, in turn, can be found either in the spaces between the cells (interstitial) or within the blood vessels (intravascular). The fluid within the intravascular spaces of blood vessels is called plasma (essentially blood without its red blood cells). Fluids can flow freely between these compartments, but being large, proteins cannot. Proteins are trapped primarily within the cells and to a lesser extent in the plasma. Wherever proteins are, they attract water.

The exchange of materials between the blood and the cells takes place across the capillary walls, which allow the passage of fluids and a variety of materials—

◆ **Figure 12-1 in Chapter 12 illustrates a cell and its associated fluids.**

TABLE 6-2 Examples of Hormones and Their Actions

Hormones	Actions
Growth hormone	Promotes growth.
Insulin and glucagon	Regulate blood glucose (see Chapter 4).
Thyroxin	Regulates the body's metabolic rate (see Chapter 8).
Calcitonin and parathormone	Regulate blood calcium (see Chapter 12).
Antidiuretic hormone	Regulates fluid and electrolyte balance (see Chapter 12).

Note: *Hormones* are chemical messengers that are secreted by endocrine glands in response to altered conditions in the body. Each travels to one or more specific target tissues or organs, where it elicits a specific response. For descriptions of many hormones important in nutrition, see Appendix A.

fluid balance: maintenance of the proper types and amounts of fluid in each compartment of the body fluids (see also Chapter 12).

but usually not plasma proteins. Still some plasma proteins leak out of the capillaries into the interstitial fluid between the cells. These proteins cannot be reabsorbed back into the plasma; they normally reenter circulation via the lymph system. If plasma proteins enter the interstitial spaces faster than they can be cleared, fluid accumulates (because proteins attract water) and causes swelling. Swelling due to an excess of interstitial fluid is known as **edema.**

Among the causes of edema is a decrease in plasma proteins. Plasma proteins may decline for the following reasons:

- Excessive losses caused by kidney disease or large wounds (such as extensive burns).
- Inadequate synthesis caused by liver disease.
- Inadequate dietary intake of protein.

Whatever the cause of edema, the result is the same: a diminished capacity to deliver nutrients and oxygen to the cells and to remove wastes from them. As a consequence, cells fail to function adequately.

• As Acid-Base Regulators • Proteins also help to maintain the balance between **acids** and **bases** within the body fluids. Normal body processes continually produce acids and bases, which the blood carries to the kidneys and lungs for excretion. The challenge is to do this without upsetting the blood's acid-base balance.

In an acid solution, hydrogen ions abound; the more hydrogen ions, the more concentrated the acid. Proteins, which have negative charges on their surfaces, attract hydrogen ions, which have positive charges. By accepting and releasing hydrogen ions,◆ proteins maintain the acid-base balance of the blood and body fluids.

◆ Compounds that help keep a solution's acidity or alkalinity constant are called **buffers.**

The blood's acid-base balance is tightly controlled. The extremes of **acidosis** and **alkalosis** lead to coma and death, largely because they denature working proteins. Disturbing a protein's shape renders it useless. To give just one example, denatured hemoglobin loses its capacity to carry oxygen.

• As Transporters • Some proteins move about in the body fluids, carrying nutrients and other molecules. The protein hemoglobin carries oxygen from the lungs to the cells. The lipoproteins transport lipids around the body. Special transport proteins carry vitamins and minerals.

The transport of the mineral iron provides an especially good illustration of these proteins' specificity and precision. When iron enters an intestinal cell after a meal has been digested and absorbed, it is captured by a protein.◆ This protein will not release the iron unless the body needs it. Before leaving the intestinal cell to enter the bloodstream, iron is attached to a carrier protein. The carrier protein, in turn, can pass iron on to a storage protein in the bone marrow or other tissues, which will hold the iron until it is needed. When it is needed, iron is incorporated into proteins in the red blood cells and muscles that assist in oxygen transport and use.

◆ The protein in the cells of the intestinal wall is **ferritin;** the carrier protein, **transferrin;** the storage protein, **ferritin** again; the red blood cell protein, **hemoglobin;** and the muscle cell protein, **myoglobin.** Figure 13-1 illustrates how these protein carriers transport iron.

Some transport proteins reside in cell membranes and act as "pumps," picking up compounds on one side of the membrane and depositing them on the other as needed. Each transport protein is specific for a certain compound or group of related compounds. Figure 6-10 illustrates how a membrane-bound transport protein helps to maintain the sodium and potassium concentrations in the fluids inside and outside cells. The balance of these two minerals is critical to neural transmissions and muscle contractions; imbalances can cause irregular heartbeats, muscular weakness, kidney failure, and even death.

• As Antibodies • Proteins also defend the body against disease. A virus—whether it is one that causes flu, smallpox, measles, or the common cold—enters the cells and multiplies there. One virus may produce 100 replicas of itself within an hour or so. Each replica can then burst out and invade 100 different

edema (eh-DEEM-uh): the swelling of body tissue caused by excessive amounts of fluid in the interstitial spaces; seen in protein deficiency (among other conditions).

acids: compounds that release hydrogen ions in a solution.

bases: compounds that accept hydrogen ions in a solution.

acidosis (assi-DOE-sis): above-normal acidity in the blood and body fluids.

alkalosis (alka-LOE-sis): above-normal alkalinity (base) in the blood and body fluids.

FIGURE 6-10 A Membrane-bound Transport Protein

This transport protein resides within a cell membrane. It picks up substances on one side of the membrane and carries them to the other side without leaving the membrane. The substances being transported here are sodium and potassium. Maintaining a high concentration of potassium and a low concentration of sodium within the cells requires energy.

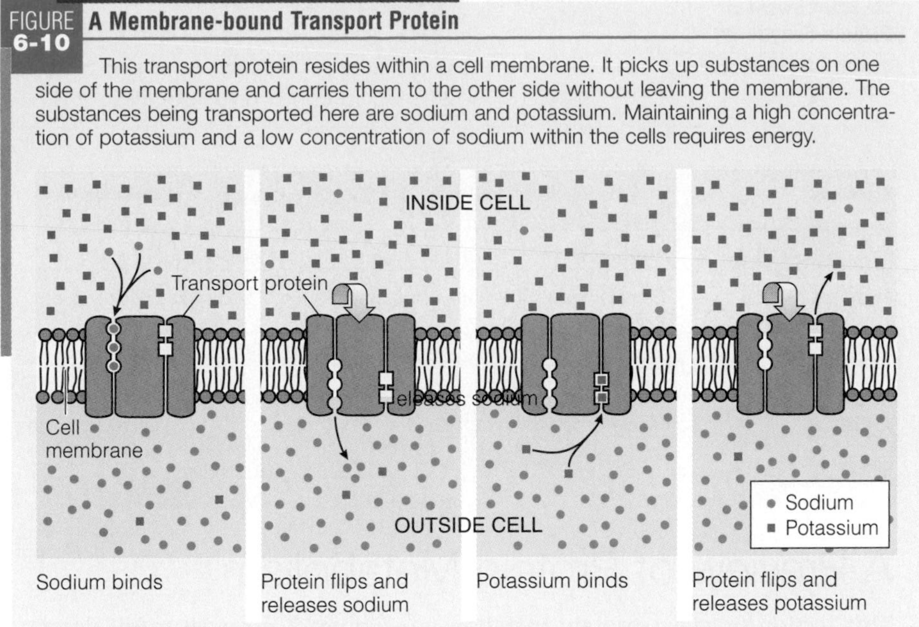

cells, soon yielding 10,000 virus particles, which invade 10,000 cells. Left free to do their worst, they will soon overwhelm the body with disease.

Fortunately, when the body detects those invading **antigens,**◆ it manufactures **antibodies,** giant protein molecules designed specifically to combat them. The antibodies work so swiftly and efficiently that in a normal, healthy individual, most diseases never have a chance to get started. Without sufficient protein, though, the body cannot maintain its army of antibodies to resist infectious diseases.

Each antibody is designed to destroy just one antigen. Once the body has manufactured antibodies against a particular antigen (such as the measles virus), it remembers how to make them. Consequently, the next time the body encounters that same antigen, it will produce antibodies even more quickly. In other words, the body develops a molecular memory, known as **immunity.**

• As a Source of Energy and Glucose • Even though proteins are needed to do the work that only they can perform, they will be sacrificed to provide energy◆ and glucose◆ if need be. Without energy, cells die; without glucose, the brain and nervous system falter. Chapter 7 provides many more details on energy metabolism.

• Other Roles • As mentioned earlier, proteins form integral parts of most body structures such as skin, muscles, and bones. They also participate in some of the body's most amazing activities such as blood clotting and vision. When a tissue is injured, a rapid chain of events leads to the production of fibrin, a stringy, insoluble mass of protein fibers that forms a clot from liquid blood. Later, more slowly, the protein collagen forms a scar to replace the clot and permanently heal the cut. The light-sensitive pigments in the cells of the retina are molecules of the protein opsin. Opsin responds to light by changing its shape, thus initiating the nerve impulses that convey the sense of sight to the brain.

IN SUMMARY

The protein functions discussed here are summarized in the table on p. 180. They are only a few of the many roles proteins play, but they convey some sense of the immense variety of proteins and their importance in the body.

(continued)

Chapter 15 describes food allergies—the immune system's response to food antigens. ◆

Reminder: Protein provides 4 kcal/g. Return to p. 7 for a refresher on how to calculate the protein kcalories from foods. ◆

Reminder: The making of glucose from noncarbohydrate sources such as amino acids is *gluconeogenesis*. ◆

antigens: substances that elicit the formation of antibodies or an inflammation reaction from the immune system. A bacterium, a virus, a toxin, and a protein in food that causes allergy are all examples of antigens.

antibodies: large proteins produced by the immune system in response to the invasion of the body by foreign molecules (usually proteins called antigens). Antibodies combine with and inactivate the foreign invaders, thus protecting the body.

immunity: the body's ability to recognize and eliminate foreign invaders; see Highlight 18.

Protein Functions

Growth and maintenance	Proteins form integral parts of most body structures such as skin, tendons, membranes, muscles, organs, and bones. As such, they support the growth and repair of body tissues.
Enzymes	Proteins facilitate chemical reactions.
Hormones	Proteins regulate body processes. (Some, but not all, hormones are made of protein.)
Fluid balance	Proteins help to maintain the volume and composition of body fluids.
Acid-base balance	Proteins help maintain the acid-base balance of body fluids by acting as buffers.
Transportation	Proteins transport substances, such as lipids, vitamins, minerals, and oxygen, around the body.
Antibodies	Proteins inactivate foreign invaders, thus protecting the body against diseases.
Energy	Proteins provide some fuel for the body's energy needs.

A Preview of Protein Metabolism

This section previews protein metabolism; Chapter 7 provides a full description. Cells have several metabolic options, depending on their protein and energy needs.

• **Protein Turnover and the Amino Acid Pool** • Within each cell, proteins are constantly being made and broken down, a process known as **protein turnover.** When proteins break down, they free amino acids to join the general circulation.◆ These amino acids mix with amino acids from dietary protein to form an **"amino acid pool"** within the cells and circulating blood. The rate of protein degradation and the amount of protein intake may vary, but the pattern of amino acids within the pool remains fairly constant. Regardless of their source, any of these amino acids can be used to make body proteins or other nitrogen-containing compounds, or they can be stripped of their nitrogen and used for energy (either immediately or stored as fat for later use).

◆ Amino acids (or proteins) that derive from within the body are **endogenous** (en-DODGE-eh-nus). In contrast, those that derive from foods are **exogenous** (eks-ODGE-eh-nus).
- endo = within
- gen = arising
- exo = outside (the body)

◆ Nitrogen equilibrium (zero nitrogen balance): N in = N out.

◆ Positive nitrogen: N in > N out.

◆ Negative nitrogen: N in < N out.

• **Nitrogen Balance** • Protein turnover and **nitrogen balance** go hand in hand. In healthy adults, protein synthesis balances with degradation, and protein intake from food balances with nitrogen excretion in the urine, feces, and sweat. When nitrogen intake equals nitrogen output, the person is in nitrogen equilibrium, or zero nitrogen balance.◆

If the body synthesizes more than it degrades and adds protein, nitrogen status becomes positive.◆ Nitrogen status is positive in growing infants and children, pregnant women, and people recovering from protein deficiency or illness; their nitrogen intake exceeds their nitrogen output. They are retaining protein in new tissues as they add blood, bone, skin, and muscle cells to their bodies.

If the body degrades more than it synthesizes and loses protein, nitrogen status becomes negative.◆ Nitrogen status is negative in people who are starving or suffering other severe stresses such as burns, injuries, infections, and fever; their nitrogen output exceeds their nitrogen intake. During these times, the body loses nitrogen as it breaks down muscle and other body proteins for energy. (Chapter 24 details the body's response to severe stresses.)

protein turnover: the degradation and synthesis of protein.

amino acid pool: the supply of amino acids derived from either food proteins or body proteins that collect in the cells and circulating blood and stand ready to be incorporated in proteins and other compounds or used for energy.

nitrogen balance: the amount of nitrogen consumed (N in) as compared with the amount of nitrogen excreted (N out) in a given period of time.*

*The genetic materials DNA and RNA contain nitrogen, but the quantity is insignificant compared with the amount in protein. The average amino acid weighs about 6.25 times as much as the nitrogen it contains, so scientists can estimate the amount of protein in a sample of food, body tissue, or other material by multiplying the weight of the nitrogen in it by 6.25.

• **Using Amino Acids to Make Proteins or Nonessential Amino Acids** • As mentioned, cells can assemble amino acids into the proteins they need to do their work. If a particular nonessential amino acid is not readily available, cells can make it from another amino acid. If an essential amino acid is missing, the body may break down some of its own proteins to obtain it.

• **Using Amino Acids to Make Other Compounds** • Cells can also use amino acids to make other compounds. For example, the amino acid tyrosine is used to make the **neurotransmitters** norepinephrine and epinephrine, which relay nervous system messages throughout the body. Tyrosine can also be made into the pigment melanin, which is responsible for brown hair, eye, and skin color, or into the hormone thyroxin, which helps to regulate the metabolic rate. For another example, the amino acid tryptophan serves as a precursor for the vitamin niacin and for serotonin, a neurotransmitter important in sleep regulation, appetite control, and sensory perception.

• Using Amino Acids for Energy • As mentioned earlier, when glucose or fatty acids are limited, cells are forced to use amino acids for energy and glucose. The body does not make a specialized storage form of protein as it does for carbohydrate and fat. Glucose is stored as glycogen in the liver and fat as triglycerides in adipose tissue, but protein in the body is available only as the working and structural components of the tissues. When the need arises, the body dismantles its tissue proteins and uses them for energy. Thus, over time, energy deprivation (starvation) always incurs wasting of lean body tissue as well as fat loss. An adequate supply of carbohydrates and fats spares◆ amino acids from being used for energy and allows them to perform their unique roles.

◆ **Reminder: The action of carbohydrate and fat in providing enough energy to allow amino acids to be used to build body proteins is known as the *protein-sparing action* of carbohydrate and fat.**

• Deaminating Amino Acids • When amino acids are broken down (as occurs when they are used for energy), they are first deaminated—stripped of their nitrogen-containing amino groups. **Deamination** produces ammonia, which the cells release into the bloodstream. The liver picks up the ammonia, converts it into urea◆ (a less toxic compound), and returns the urea to the blood. The kidneys filter urea out of the blood; thus the amino nitrogen ends up in the urine. The remaining carbon fragments may enter a number of metabolic pathways—for example, they may be used to make fat.

◆ **Urea metabolism is described in Chapter 7.**

• Using Amino Acids to Make Fat • If a person eats more protein than the body needs, the amino acids are deaminated, the nitrogen is excreted, and the remaining carbon fragments are converted to fat and stored for later use.* In this way, valuable, expensive, protein-rich foods can contribute to obesity.

IN SUMMARY

Proteins are constantly being synthesized and broken down as needed. The body's assimilation of amino acids into proteins and its release of amino acids via protein degradation and excretion can be tracked by measuring nitrogen balance, which should be positive during growth and steady in adulthood. An energy deficit or an inadequate protein intake may force the body to use amino acids as fuel, creating a negative nitrogen balance. Protein eaten in excess of need is degraded and stored as body fat.

PROTEIN IN FOODS

In the United States and Canada, where nutritious foods are abundant, people eat protein in such large quantities that they receive all the amino acids they need. In countries where food is scarce and the people eat only marginal amounts of protein-rich foods, however, the *quality* of the protein becomes crucial. It determines, in large part, how well the children grow and how well the adults maintain their health.

neurotransmitters: chemicals that are released at the end of a nerve cell when a nerve impulse arrives there. They diffuse across the gap to the next cell and alter the membrane of that second cell to either inhibit or excite it.

deamination (dee-AM-eh-NAY-shun): removal of the amino (NH_2) group from a compound such as an amino acid.

*Chemists sometimes classify amino acids according to the destinations of their carbon fragments after deamination. If the fragment leads to the production of glucose, the amino acid is called "glucogenic"; if it leads to the formation of ketone bodies, fats, and sterols, the amino acid is called "ketogenic." There is no sharp distinction between glucogenic and ketogenic amino acids, however. A few are both; most are considered glucogenic; only one (leucine) is clearly ketogenic.

Protein Quality

The protein quality of the diet is of great concern when making nutrition recommendations, especially in countries where malnutrition is widespread. Low-quality proteins fail to provide enough of all the essential amino acids needed to support the body's work.

• Limiting Amino Acids • To make proteins, a cell must have all the needed amino acids available simultaneously. The liver can produce any nonessential amino acid that may be in short supply so that the cells can continue linking amino acids into protein strands. If an essential amino acid is missing, though, a cell must dismantle its own proteins to obtain it. Therefore, to prevent protein breakdown, dietary protein must supply at least the nine essential amino acids plus enough nitrogen-containing amino groups and energy for the synthesis of the others. If the diet supplies too little of any essential amino acid, protein synthesis will be limited. The body makes whole proteins only; if one amino acid is missing, the others cannot form a "partial" protein. An essential amino acid supplied in less than the amount needed to support protein synthesis is called a **limiting amino acid.**

Black beans and rice, a favorite Hispanic combination, together provide a balanced array of amino acids.

© Polara Studios Inc.

• Complete Protein • A **complete protein** contains all the essential amino acids in relatively the same amounts as human beings require; it may or may not contain all the nonessential amino acids. Proteins that are low in an essential amino acid cannot, by themselves, support protein synthesis. Generally, proteins derived from animals (meat, fish, poultry, cheese, eggs, and milk) are complete, although gelatin is an exception (it lacks tryptophan and cannot support growth and health as a diet's sole protein). Proteins from plants (vegetables, grains, and legumes) have more diverse amino acid patterns, and some tend to be limiting in one or more essential amino acids. Some plant proteins (for example, corn protein) are notoriously incomplete. Others (for example, soy protein) are complete.

• Complementary Proteins • In general, plant proteins are of lower quality than animal proteins, and plants also offer less protein (per weight or measure of food). For this reason, many vegetarians improve the quality of proteins in their diets by combining plant-protein foods that have different but complementary amino acid patterns. This strategy is called **mutual supplementation,** and it yields **complementary proteins** that together contain all the essential amino acids in quantities sufficient to support health. The protein quality of the combination is greater than for either food alone (see Figure 6-11).

Many people have long believed that mutual supplementation at every meal is critical to protein nutrition. For most healthy vegetarians, though, it is not necessary to balance amino acids at each meal when protein intake is varied and energy intake is sufficient.[4] Vegetarians can receive all the amino acids they need over the course of a day, if they eat a variety of grains, legumes, seeds, nuts, and vegetables. Protein deficiency will develop, however, when fruits and certain vegetables make up the core of the diet, severely limiting both the *quantity* and *quality* of protein. Highlight 6 shows how to plan a nutritious vegetarian diet.

• Digestibility • Ideally, a complete protein is digestible enough to provide the amino acids needed for protein synthesis. Such a protein is a **high-quality protein. Protein digestibility** depends on such factors as a food protein's source and the other foods eaten with it. The di-

limiting amino acid: the essential amino acid found in the shortest supply relative to the amounts needed for protein synthesis in the body. Four amino acids are most likely to be limiting:
- Lysine.
- Methionine.
- Threonine.
- Tryptophan.

complete protein: a dietary protein containing all the essential amino acids in relatively the same amounts that human beings require. It may also contain nonessential amino acids.

mutual supplementation: the strategy of combining two protein foods in a meal so that each food provides the essential amino acid(s) lacking in the other. Mutual supplementation is the dietary strategy that brings complementary proteins together in a meal.

complementary proteins: two or more proteins whose amino acid assortments complement each other in such a way that the essential amino acids missing from one are supplied by the other.

high-quality protein: an easily digestible, complete protein.

protein digestibility: a measure of the amount of amino acids absorbed from a given protein intake.

gestibility of most animal proteins is high (90 to 99 percent); plant proteins are less digestible (70 to 90 percent), in part because they are contained within plant cell walls that resist digestion.

FIGURE 6-11 An Example of Mutual Supplementation

In general, legumes provide plenty of isoleucine (Ile) and lysine (Lys), but fall short in methionine (Met) and tryptophan (Trp). Grains have the opposite strengths and weaknesses, making them a perfect match for legumes.

	Ile	Lys	Met	Trp
Legumes				
Grains				
Together				

• Reference Protein • One of the most complete and digestible proteins is egg protein. Egg protein can be used as a standard for measuring protein quality; it is assigned a value of 100, and the quality of other food proteins is determined based on how they compared with egg. Such a standard is called a **reference protein.** In the early 1990s, the Food and Agriculture Organization (FAO) of the United Nations and the World Health Organization (WHO) established another standard for the reference protein: the essential amino acid requirements of preschool-age children. The rationale behind using the requirements of this age group is that if a protein will effectively support a young child's growth and development, then it will meet or exceed the requirements of older children and adults.

IN SUMMARY

A diet inadequate in any of the essential amino acids limits protein synthesis. The best guarantee of amino acid adequacy is to eat foods containing complete proteins or mixtures of foods containing incomplete but complementary proteins so that each can supply the amino acids missing in the other. Vegetarians can meet their protein needs by eating a variety of whole grains, legumes, seeds, nuts, and vegetables.

© Polara Studios Inc.

Vegetarians obtain their protein from whole grains, legumes, nuts, vegetables, and, in some cases, eggs and milk products.

Measures of Protein Quality

Researchers have developed several methods for evaluating the quality of food proteins and identifying high-quality proteins. The following paragraphs briefly describe these measures; Appendix D provides more detail.

• Amino Acid Scoring • Amino acid scoring evaluates a food protein's quality by determining its amino acid composition and comparing it with that of a reference protein. Scientists can easily identify the limiting amino acid—it is the one that falls shortest compared with the reference. If the food protein's limiting amino acid is 70 percent of the amount found in the reference protein, it receives a score of 70. Such calculations fail to estimate digestibility, however.

• Biological Value • The **biological value (BV)** of a protein measures its efficiency in supporting the body's needs. Scientists feed a given protein to experimental animals as the sole protein in their diet and measure the animals' retention of food nitrogen absorbed. The more nitrogen retained, the higher the protein quality. (Recall that when an essential amino acid is missing, protein synthesis stops, and the remaining amino acids are deaminated and the nitrogen excreted.) Egg protein has a BV of 100, indicating that 100 percent of the nitrogen absorbed is retained. Appendix D includes a table that lists the BV of selected foods.

• Net Protein Utilization • Like BV, **net protein utilization (NPU)** measures how efficiently a protein is used by the body. The difference is that NPU measures retention of food nitrogen *consumed* rather than food nitrogen absorbed (as in BV).

• Protein Efficiency Ratio • The **protein efficiency ratio (PER)** measures the weight gain of a growing animal and compares it to the animal's protein intake. Until recently, the PER was generally accepted in the United States

reference protein: a standard against which to measure the quality of other proteins.

amino acid scoring: a method of evaluating protein quality by comparing a protein's amino acid pattern with that of a reference protein; sometimes called **chemical scoring.**

biological value (BV): the amount of protein nitrogen that is retained from a given amount of protein nitrogen absorbed; a measure of protein quality.

net protein utilization (NPU): the amount of protein nitrogen that is retained from a given amount of protein nitrogen eaten; a measure of protein quality.

protein efficiency ratio (PER): a measure of protein quality assessed by determining how well a given protein supports weight gain in growing rats; used to establish the protein quality for infant formulas and baby foods.

and Canada as the official method for assessing protein quality, and it is still used to evaluate proteins for infants.

• PDCAAS • The **protein digestibility–corrected amino acid score,** or **PDCAAS,** compares the amino acid contents of a protein with human amino acid requirements and corrects for digestibility. The protein's amino acid score is determined as described earlier, and then it is compared against the amino acid requirements of preschool-age children. This comparison reveals the most limiting amino acid. The amino acid score is multiplied by the food's protein digestibility percentage to determine the PDCAAS. Appendix D provides an example of how to calculate the PDCAAS and a table that lists the PDCAAS values of selected foods.

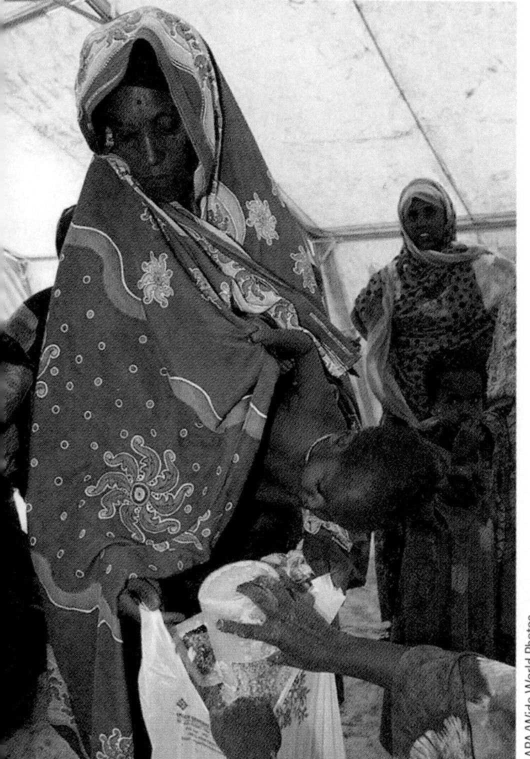

© APA/Wide World Photos

Donated food saves some people from starvation, but it is usually insufficient to meet nutrient needs or even to provide a full belly for every person who is hungry.

Protein Regulations for Food Labels

The Food and Drug Administration's (FDA) labeling regulations use the PDCAAS method to assess protein quality in foods intended for people over age one. For infant formulas and baby foods, the PER method using casein (the chief protein in milk) as a standard is used to measure protein quality.

All food labels must state the *quantity* of protein in grams. The "% Daily Value" for protein is not mandatory on all labels, but is required whenever a food makes a protein claim or is intended for consumption by children under four years old.* Whenever the Daily Value percentage is declared, researchers must determine the *quality* of the protein by using the PDCAAS method. Thus, when a % Daily Value is stated for protein, it reflects both quantity and quality.

IN SUMMARY

The quality of protein is measured by its amino acid content, its digestibility, and its ability to support growth. Such measures are of great importance in dealing with malnutrition worldwide, but in the United States and Canada, where protein deficiency is not common, protein quality scores of individual foods deserve little emphasis.

HEALTH EFFECTS AND RECOMMENDED INTAKES OF PROTEIN

As you know by now, protein is indispensable to life. It should come as no surprise that protein deficiency can have devastating effects on people's health. But like the other nutrients, protein in excess can also be harmful. This section examines the health effects and recommended intakes of protein.

Protein-Energy Malnutrition

When people are deprived of protein, energy, or both, the result is **protein-energy malnutrition (PEM).** Although PEM touches many adult lives, it most often strikes early in childhood. It is one of the most prevalent and devastating forms of malnutrition in the world, afflicting over 500 million children. Most of the 33,000 children who die each day are malnourished.[5]

Inadequate food intake leads to poor growth in children and to weight loss and wasting in adults. Children who are thin for their height may be suffering

protein digestibility–corrected amino acid score (PDCAAS): a measure of protein quality assessed by comparing the amino acid score of a food protein with the amino acid requirements of preschool-age children and then correcting for the true digestibility of the protein; recommended by the FAO/WHO and used to establish protein quality of foods for Daily Value percentages on food labels.

protein-energy malnutrition (PEM), also called **protein-kcalorie malnutrition (PCM):** a deficiency of protein, energy, or both, including kwashiorkor, marasmus, and instances in which they overlap.

*For labeling purposes, the Daily Values for protein are as follows: for infants, 14 grams; for children under age four, 16 grams; for older children and adults, 50 grams; for pregnant women, 60 grams; and for lactating women, 65 grams.

TABLE 6-3 **Features of Marasmus and Kwashiorkor in Children**

Separating PEM into two classifications oversimplifies the condition, but at the extremes, marasmus and kwashiorkor exhibit marked differences. Marasmus-kwashiorkor mix presents symptoms common to both marasmus and kwashiorkor. In all cases, children are likely to develop diarrhea, infections, and multiple nutrient deficiencies.

Marasmus	Kwashiorkor
Infancy (less than 2 yr)	Older infants and young children (1 to 3 yr)
Severe deprivation, or impaired absorption, of protein, energy, vitamins, and minerals	Inadequate protein intake or, more commonly, infections
Develops slowly; chronic PEM	Rapid onset; acute PEM
Severe weight loss	Some weight loss
Severe muscle wasting, with no body fat	Some muscle wasting, with retention of some body fat
Growth: <60% weight-for-age	Growth: 60 to 80% weight-for-age
No detectable edema	Edema
No fatty liver	Enlarged fatty liver
Anxiety, apathy	Apathy, misery, irritability, sadness
Good appetite possible	Loss of appetite
Hair is sparse, thin, and dry; easily pulled out	Hair is dry and brittle; easily pulled out; changes color; becomes straight
Skin is dry, thin, and easily wrinkles	Skin develops lesions

from **acute PEM** (recent severe food deprivation), whereas children who are short for their age have experienced **chronic PEM** (long-term food deprivation). Poor growth due to PEM is easy to overlook because a small child may look quite normal, but it is the most common sign of malnutrition.

PEM is most prevalent in Africa, Central America, South America, the Middle East, and East and Southeast Asia. In the United States, homeless people and those living in substandard housing in inner cities and rural areas have been diagnosed with PEM. In addition to those living in poverty, elderly people who live alone and adults who are addicted to drugs and alcohol are frequently victims of PEM. Adult PEM is also seen in people hospitalized with infections such as AIDS or tuberculosis; these infections deplete body proteins, demand extra energy, induce nutrient losses, and alter metabolic pathways. Furthermore, poor nutrient intake during hospitalization worsens malnutrition and impairs recovery.[6] PEM is also common in those suffering from the eating disorder anorexia nervosa (discussed in Highlight 9). Prevention emphasizes frequent, nutrient-dense, energy-dense meals and, equally important, resolution of the underlying causes of PEM—poverty, infections, and illness.

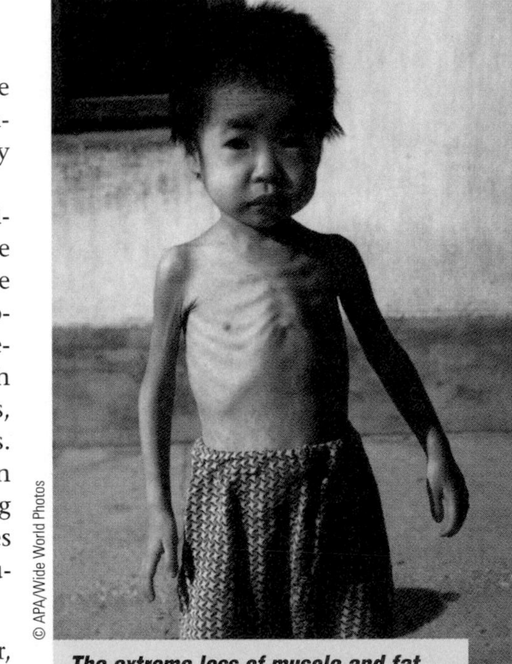

The extreme loss of muscle and fat characteristic of marasmus is apparent in this child's "matchstick" arms.

• Classifying PEM • PEM occurs in two forms: marasmus and kwashiorkor, which differ in their clinical features (see Table 6-3). The following paragraphs present three clinical syndromes—marasmus, kwashiorkor, and the combination of the two.

• Marasmus • Appropriately named from the Greek word meaning "dying away," **marasmus** reflects a severe deprivation of food over a long time (chronic PEM). It is therefore caused by an inadequate energy *and* protein intake (and by inadequate essential fatty acids, vitamins, and minerals as well). Marasmus occurs most commonly in children from 6 to 18 months of age in all the overpopulated urban slums of the world. Children in impoverished nations simply do not have enough to eat and subsist on diluted cereal drinks that supply scant energy and protein of low quality; such food can barely sustain life, much less support growth. Consequently, marasmic children look like little old people—just skin and bones.

Without adequate nutrition, muscles, including the heart, waste and weaken. Because the brain normally grows to almost its full adult size within the first two years of life, marasmus impairs brain development and learning ability. Reduced

acute PEM: protein-energy malnutrition caused by recent severe food restriction; characterized in children by thinness for height (wasting).

chronic PEM: protein-energy malnutrition caused by long-term food deprivation; characterized in children by short height for age (stunting).

marasmus (ma-RAZ-mus): a form of PEM that results from a severe deprivation, or impaired absorption, of energy, protein, vitamins, and minerals.

synthesis of key hormones slows metabolism and lowers body temperature. There is little or no fat under the skin to insulate against cold. Hospital workers find that children with marasmus need to be wrapped up and kept warm. They also need love because they have often been deprived of parental attention as well as food.

The starving child faces this threat to life by engaging in as little activity as possible—not even crying for food. The body musters all its forces to meet the crisis, so it cuts down on any expenditure of protein not needed for the functioning of the heart, lungs, and brain. Growth ceases; the child is no larger at age four than at age two. Enzymes are in short supply and the GI tract lining deteriorates. Consequently, the child can't digest and absorb what little food is eaten.

• Kwashiorkor • **Kwashiorkor** typically reflects a sudden and recent deprivation of food (acute PEM). Kwashiorkor was originally a Ghanaian word meaning "the evil spirit that infects the first child when the second child is born." When a mother who has been nursing her first child bears a second child, she weans the first child and puts the second one on the breast. The first child, suddenly switched from nutrient-dense, protein-rich breast milk to a starchy, protein-poor cereal, soon begins to sicken and die. Kwashiorkor typically sets in between 18 months and two years.

Kwashiorkor usually develops rapidly as a result of protein deficiency or, more commonly, is precipitated by an illness such as measles or other infection. Other factors may also contribute to the symptoms that accompany kwashiorkor.

The loss of weight and body fat is usually not as severe in kwashiorkor as in marasmus, but there may be some muscle wasting. Proteins and hormones that previously maintained fluid balance diminish, and fluid leaks into the interstitial spaces. The child's limbs and face become swollen with edema, a distinguishing feature of kwashiorkor. The lack of the protein carriers that transport fat out of the liver causes the belly to bulge with a fatty liver. The fatty liver lacks enzymes to clear metabolic toxins from the body, so their harmful effects are prolonged. Inflammation in response to these toxins and to infections further contributes to the edema that accompanies kwashiorkor.[7] Without sufficient tyrosine to make melanin, the child's hair loses its color; inadequate protein synthesis leaves the skin patchy and scaly, often with sores that fail to heal. The lack of proteins to carry or store iron leaves iron free. Unbound iron is common in children with kwashiorkor and may contribute to their illnesses and deaths by promoting bacterial growth and free-radical damage.[8] (Free-radical damage and severe stress are discussed fully in Highlight 11 and Chapter 24.)

The edema and enlarged liver characteristic of kwashiorkor are apparent in this child's swollen belly. Malnourished children commonly have an enlarged abdomen from parasites as well.

• Marasmus-Kwashiorkor Mix • The combination of marasmus and kwashiorkor is characterized by the edema of kwashiorkor with the wasting of marasmus. Most often, the child is suffering the effects of both malnutrition and infections. Some researchers believe that kwashiorkor and marasmus are two stages of the same disease. They point out that kwashiorkor and marasmus often exist side by side in the same community where children consume the same diet. They note that a child who has marasmus can later develop kwashiorkor. Some research indicates that marasmus represents the body's adaptation to starvation and that kwashiorkor develops when adaptation fails.

• Infections • In PEM, antibodies to fight off invading bacteria are degraded to provide amino acids for other uses, leaving the malnourished child vulnerable to infections. Blood proteins, including hemoglobin, are no longer synthesized, so the child becomes anemic and weak. **Dysentery,** an infection of the digestive tract, causes diarrhea, further depleting the body of nutrients. In the marasmic child, once infection sets in, kwashiorkor often follows.

The combination of infections, fever, fluid imbalances, and anemia often leads to heart failure and occasionally sudden death. Infections combined with malnutrition are responsible for two-thirds of the deaths of young children in

kwashiorkor (kwash-ee-OR-core, kwash-ee-or-CORE): a form of PEM that results either from inadequate protein intake or, more commonly, from infections.

dysentery (DISS-en-terry): an infection of the digestive tract that causes diarrhea.

developing countries. Measles, which might make a healthy child sick for a week or two, kills a child with PEM within two or three days.

• **Rehabilitation** • If caught in time, the life of a starving child may be saved with nutrition intervention. Diarrhea will have incurred dramatic fluid and mineral losses that will require careful correction to help raise the blood pressure and strengthen the heartbeat. After the first 24 to 48 hours, protein and food energy may be given in *small* quantities, with intakes *gradually* increased as tolerated. Severely malnourished adults, especially those with edema, recover better with an initial diet that is low in protein (9 percent kcalories from protein) than with a higher-protein diet (16 percent).[9]

Experts assure us that we possess the knowledge, technology, and resources to end hunger. Programs that tailor interventions to the local people and involve them in the process of identifying problems and devising solutions have the most success.[10] To win the war on hunger, those who have the food, technology, and resources must make fighting hunger a priority (see Highlight 16 for more on hunger).

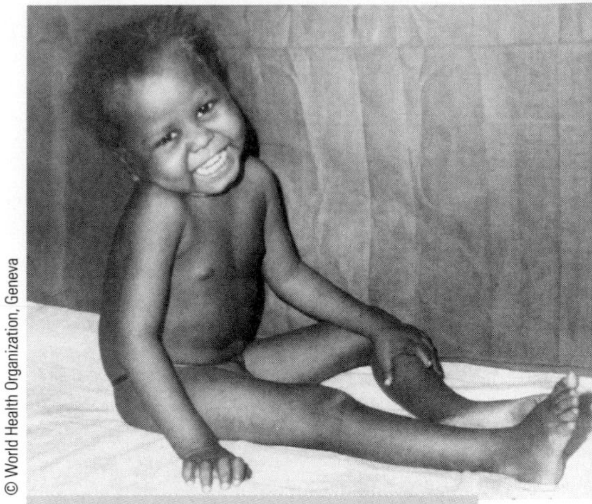

Given appropriate nutrition care, this child has successfully recovered from kwashiorkor.

Health Effects of Protein

Health promotion

While many of the world's people struggle to obtain enough food energy and protein, in developed countries both are so abundant that problems of excess are seen. Overconsumption of protein offers no benefits and may pose health risks.

The relationships between excess protein and chronic diseases are not clearly evident. Population studies have difficulty determining whether diseases correlate with animal proteins or with their accompanying saturated fats, for example. Studies that rely on data from vegetarians must sort out the many lifestyle factors, other than a "no-meat diet," that might explain relationships between protein and health.

• **Heart Disease** • As mentioned, foods rich in animal protein tend to be rich in saturated fats. Consequently, it is not surprising to find a correlation between animal-protein intake and heart disease, although no independent effect has been demonstrated. On the other hand, substituting soy protein for animal protein lowers blood cholesterol, especially in those with high blood cholesterol.[11]

Research suggests that elevated levels of the amino acid homocysteine may be an independent risk factor for heart disease.[12] Researchers do not yet understand the many factors that can raise homocysteine in the blood or whether elevated levels are a cause or an effect of heart disease.[13] Until they can determine the exact role homocysteine plays in heart disease, they are following several leads in pursuit of the answers.[14] Coffee's role in heart disease has been controversial, but research suggests it is among the most influential factors in raising homocysteine, which may explain some of the adverse health effects of heavy consumption.[15] Homocysteine is also elevated with suboptimal intakes of B vitamins and can usually be lowered with supplements of vitamin B_{12}, vitamin B_6, and folate.[16] Such research suggests that a high intake of these vitamins may reduce the risk of heart disease.[17]

In contrast to homocysteine, the amino acid arginine may be a protective factor for heart disease, slowing the progression of atherosclerosis.[18] The exact amount of arginine needed to defend against heart disease has not yet been determined, but it appears to be much greater than a healthy diet or a reasonable quantity of supplements alone can provide. If research confirms the benefits of arginine, look for manufacturers to begin producing functional foods enriched with this amino acid. In the meantime, it would be unwise for consumers to use supplements of arginine, or any other amino acid for that matter (as pp. 191–192 explain).

• **Cancer** • As in heart disease, the effects of protein and fats on cancers cannot be easily separated. Population studies suggest a correlation between high intakes of animal proteins and some types of cancer (notably, cancer of the colon, breast, kidneys, pancreas, and prostate).

• **Adult Bone Loss (Osteoporosis)** • Chapter 12 presents calcium metabolism, and Highlight 12 elaborates on the main factors that influence osteoporosis. This section briefly describes the controversy over whether a high protein intake accelerates bone loss. Researchers know that calcium excretion rises as protein intake increases. Whether excess protein depletes the bones of their chief mineral may depend upon the ratio of calcium intake to protein intake. An ideal ratio has not been established, but a young woman whose intake meets recommendations for both nutrients has a calcium-to-protein ratio of more than 20 to 1 (milligrams to grams), which probably provides adequate protection for the bones.[19] For most women in the United States, however, average calcium intakes are lower and protein intakes are higher, yielding a 9-to-1 ratio, which may produce calcium losses significant enough to compromise bone health. In establishing calcium recommendations, the Committee on Dietary Reference Intakes considered protein's effect on calcium retention, but did not find sufficient evidence to warrant an adjustment.[20]

Some research suggests that calcium losses occur when protein intake is high because such diets are typically low in fruits and vegetables and produce a lot of acid.[21] The skeleton responds to this acidity by giving up its calcium. Adding fruits and vegetables to a diet high in protein creates less acidity and decreases the loss of calcium in the urine—another good reason to "Eat 5 a Day."

• **Weight Control** • Protein-rich foods are often fat-rich foods that contribute to obesity with its accompanying health risks. As Highlight 8 explains, weight-loss gimmicks that encourage a high-protein diet are rarely effective over the long term; overweight people have better success with diets that provide adequate protein, minimal fat, and ample energy from carbohydrates. The higher a person's intake of protein-rich foods such as meat and milk, the more likely that fruits, vegetables, and grains will be crowded out, making the diet inadequate in other nutrients.

• **Kidney Disease** • Excretion of the end products of protein metabolism depends, in part, on an adequate fluid intake and healthy kidneys. A high protein intake increases the work of the kidneys.[22] In fact, one of the most effective ways to slow the progression of kidney disease is to restrict dietary protein.[23]

IN SUMMARY

Protein deficiencies arise from both energy-poor and protein-poor diets and lead to the devastating diseases of marasmus and kwashiorkor. Together, these diseases are known as PEM (protein-energy malnutrition), a major form of malnutrition causing death in children worldwide. Excesses of protein offer no advantage; in fact, overconsumption of protein-rich foods may incur health problems as well.

Recommended Intakes of Protein

As mentioned earlier, the body continuously breaks down and loses its proteins and cannot store amino acids. To replace protein, the body needs dietary protein for two reasons: first, food protein is the only source of the *essential* amino acids; and second, it is the only practical source of *nitrogen* with which to build the nonessential amino acids and other nitrogen-containing compounds the body needs.

Given recommendations that people's fat intakes should contribute 30 percent or less of total food energy, and carbohydrate, 55 percent or more, that leaves about 15 percent for protein. In a 2000-kcalorie diet, that represents 300 kcalories from protein, or 75 grams. Current intakes in the United States and Canada, though higher than this, do not seem to be high enough to cause harm. Health experts advise people to maintain moderate protein intakes—between the RDA and twice the RDA.

• **Protein RDA** • The protein RDA for adults is 0.8 grams per kilogram of healthy body weight per day. For infants and children, the RDA is higher. When compared to total energy intake, however, the protein RDA for infants and children is similar to that for adults as Table 6-4 shows. The RDA generously covers the needs for replacing worn-out tissue, so it increases for larger people; it also covers the needs for building new tissue during growth, so it increases for infants, children, and pregnant women. Protein recommendations for athletes◆ are somewhat higher than those for the general population.[24] The accompanying "How to" shows how to calculate your RDA for protein.

In setting the RDA, the committee assumes that people are healthy and do not have unusual metabolic needs for protein; that the protein eaten will be of mixed quality (from both high- and low-quality sources); and that the body will use the protein about as efficiently as it uses reference proteins. In addition, the committee assumes that the protein is consumed along with sufficient carbohydrate and fat to provide adequate energy and that other nutrients in the diet are adequate.

TABLE 6-4 Protein RDA as a Percentage of Energy RDA

When expressed as a percentage of energy intake, the protein requirement represents about 10 percent of the energy RDA.

Age (yr)	Protein RDA (g)	Protein RDA (g/kg)	Protein RDA (in kCalories) as a Percentage of Energy RDA (%)
0 to ½	13	2.2	8.0
½ to 1	14	1.6	6.5
1 to 3	16	1.2	4.9
4 to 6	24	1.1	5.3
7 to 10	28	1.0	5.6
Males			
11 to 14	45	1.0	7.2
15 to 18	59	0.9	7.9
19 to 24	58	0.8	8.0
25 to 50	63	0.8	8.7
51 +	63	0.8	11.0
Females			
11 to 14	46	1.0	8.4
15 to 18	44	0.8	8.0
19 to 24	46	0.8	8.4
25 to 50	50	0.8	9.1
51 +	50	0.8	10.5

• **Adequate Energy** • Note the qualification "adequate energy" in the preceding statement, and consider what happens if energy intake falls short of needs. An intake of 50 grams of protein provides 200 kcalories, which represents 10 percent of the total energy from protein, if the person receives 2000 kcalories a day. But if the person cuts energy intake drastically—to, say, 800 kcalories a day—then an intake of 200 kcalories from protein is suddenly 25 percent of the total; yet it's still the same amount of protein (number of grams). The protein intake is reasonable, but the energy intake is not; the low energy intake will force the body to use the protein to meet energy needs rather than to replace lost body protein. Similarly, if the person's energy intake is high—say, 4000 kcalories—the 50-gram protein intake will represent only 5 percent of the total; yet it *still* is a reasonable protein intake. Again, the energy intake is unreasonable for most people, but in this case, it will permit the protein to be used to meet the body's needs.

◆ Recommended protein intake for:
- Power athletes: 1.6–1.7 g/kg/day.
- Endurance athletes: 1.2–1.6 g/kg/day.

Be careful when judging protein (or carbohydrate or fat) intake as a percentage of energy. Always ascertain the number of grams as well, and compare it with the RDA or another standard stated in grams. A recommendation stated as a percentage of energy intake is useful only if the energy intake is within reason.

• **Protein in Abundance** • Most people in the United States and Canada receive much more protein than they need. Even athletes in training typically don't need to increase their protein intakes because the additional foods they eat to meet their high energy needs deliver protein as well. This is not surprising considering the abundance of food eaten and the central role meats hold in the North American diet. A single ounce of meat delivers about 7 grams of protein, so 8 ounces of meat alone supplies

Calculate Recommended Protein Intakes

To figure your protein RDA:
- Look up the healthy weight for a person of your height (inside back cover). If your present weight falls within that range, use it for the following calculations. If your present weight falls outside the range, use the midpoint of the healthy weight range as your reference weight.
- Convert pounds to kilograms, if necessary (pounds divided by 2.2 equals kilograms).
- Multiply kilograms by 0.8 to get your RDA in grams per day. (Males 15 to 18 years old, multiply by 0.9.) Example:

Weight = 150 lb.

150 lb ÷ 2.2 lb/kg = 68 kg (rounded off).

68 kg × 0.8 g/kg= 54 g protein (rounded off).

HOW TO Use the Exchange System to Estimate Protein

Exchange	Protein (g)
Milks	8
Meats	7
Starch	3
Vegetables	2
Fruits	—
Fats	—

The exchange system provides an easy way to estimate dietary protein. The foods on the milk and meat lists supply protein in abundance: a cup of milk provides 8 grams of protein; an ounce of meat, 7 grams. The starch and vegetable lists contribute small amounts of protein, but they can add up to significant quantities; fruits and fats provide no protein.

To estimate the protein in this meal consisting of a burrito, 1 cup of milk, and an apple, you first need to recognize that the burrito contains about ½ cup pinto beans and ½ ounce shredded cheese wrapped in a tortilla. Then you need to translate these portions into exchanges: 1½ meats, 1 starch, 1 milk, and 1 fruit, respectively.

Using the exchange system to estimate, this lunch provides about 22 grams of protein. A computer diet analysis program calculated the same. The exchange system sometimes over- or underestimates the protein contents of individual foods, but for most, its estimates of daily intakes are close. In any case, for nutrients eaten in such large quantities as protein, a difference of a few grams in a day's total is insignificant.

Lunch	Exchange	Protein (g) Estimate	Protein (g) Actual
½ c pinto beans	= 1 meat	7	
½ oz cheese	= ½ meat	4	14
1 tortilla	= 1 starch	3	
1 c milk	= 1 milk	8	8
1 apple	= 1 fruit	—	—
		22	22

© Polara Studios Inc. (both)

Milks and meats provide lots of protein; starch and vegetables contain a little; fruits and fats have none.

more than the RDA for an average-sized person. Besides meat, well-fed people eat many other nutritious foods, many of which also provide protein.

To illustrate how easy it is to overconsume protein, consider the *minimum* recommended servings for the Food Guide Pyramid. Six servings from the bread, cereal, rice, and pasta group provide about 18 grams of protein; 3 servings of vegetables deliver about 6 grams; 2 servings of milk offer 16 grams; and 2 servings (about 5 ounces) of meat contain about 35 grams. This totals 75 grams of protein—higher than recommendations for most people and yet still lower than the average intake of people in the United States. (The accompanying "How to" describes how to estimate protein in foods.)

Just think how much more protein people receive when they eat additional servings. No wonder most people in the United States and Canada get more protein than they need. If they have an adequate *food* intake, they have a more-than-adequate protein intake. The key diet-planning principle to emphasize for protein is moderation. Even though most people receive plenty of protein, some feel compelled to take supplements as well, as the next section describes.

IN SUMMARY Optimally, the diet will be adequate in energy from carbohydrate and fat and will deliver 0.8 grams of protein per kilogram of healthy body weight each day. U.S. and Canadian diets are typically more than adequate in this respect.

Protein and Amino Acid Supplements

Websites, health-food stores, and popular magazine articles advertise a wide variety of protein supplements, and people take these supplements for many different reasons, all of them unfounded. Athletes take them to build muscle. Dieters take them to spare their bodies' protein while losing weight. Women take them to strengthen their fingernails. People take individual amino acids, too—to cure herpes, to make themselves sleep better, to lose weight, and to relieve pain and depression.* Like many other magic solutions to health problems, protein and amino acid◆ supplements don't work these miracles, and they can be harmful.

◆ **Use of amino acids as dietary supplements is *inappropriate*, especially for:**
- All women of childbearing age.
- Pregnant or lactating women.
- Infants, children, and adolescents.
- Elderly people.
- People with inborn errors of metabolism that affect their bodies' handling of amino acids.
- Smokers.
- People on low-protein diets.
- People with chronic or acute mental or physical illnesses who take amino acids without medical supervision.

Muscle work builds muscle; protein supplements do not, and athletes do not need them. Instead, athletes need a well-balanced diet that provides sufficient dietary protein and adequate food energy. Food energy spares body protein; carbohydrate and fat serve this purpose equally well, and carbohydrate is safer. Fingernails are not affected by protein supplements, provided the diet is adequate. Furthermore, protein supplements are expensive, less completely digested than protein-rich foods, and, when used as replacements for such foods, often downright dangerous.

Single amino acids do not occur naturally in foods and offer no benefit to the body; in fact, they can be harmful. The body was not designed to handle the high concentrations and unusual combinations of amino acids found in supplements. An excess of one amino acid can create such a demand for a carrier that it prevents the absorption of another amino acid, leading to a deficiency. Those amino acids winning the competition enter in excess, creating the possibility of a toxicity. Toxicity of single amino acids in animal studies raises concerns about their use in human beings. Anyone considering taking amino acid supplements should check with a registered dietitian or physician first.

In two cases, recommendations for single amino acid supplements have led to widespread public use—lysine to prevent or relieve the infections that cause herpes cold sores on the mouth or genital organs, and tryptophan to relieve pain, depression, and insomnia. In both cases, enthusiastic popular reports preceded careful scientific experiments and health recommendations. A review of the research indicates that lysine may suppress herpes infections in some individuals and appears safe (up to 3 grams per day) when taken in divided doses with meals.[25]

Tryptophan is also effective with respect to pain and sleep, but its use for these purposes is still experimental. More than 1500 people who elected to take tryptophan supplements developed a rare blood disorder known as eosinophilia-myalgia syndrome (EMS). EMS is characterized by severe muscle pain, extremely high fever, and, in over three dozen cases, death. Treatment for EMS usually involves physical therapy and low doses of corticosteroids to relieve symptoms temporarily. Early evidence suggested that a major tryptophan processing plant may have introduced contaminants that caused the disease, but later research indicated that multiple factors were involved; the exact

*Canada allows single amino acid supplements to be sold as drugs or used as food additives.

causes of EMS remain unknown. The FDA issued a recall of all products containing tryptophan.

IN SUMMARY Normal, healthy people never need protein or amino acid supplements. It is safest to obtain lysine, tryptophan, and all other amino acids from protein-rich foods, eaten with carbohydrate to facilitate their use in the body. With all that we know about science, it is hard to improve on nature.

How are you doing?

- Do you receive enough, but not too much, protein daily?
- How often do you select plant-based protein foods?
- Do you think you need to take protein or amino acid supplements?

Nutrition on the Net

WEBSITES

Access these websites for further study of topics covered in this chapter.

- Find updates and quick links to these and other nutrition-related sites at our website: **www.wadsworth.com/nutrition**
- Learn more about sickle-cell anemia from the National Heart, Lung, and Blood Institute or the Sickle Cell Disease Association of America: **www.nhlbi.nih.gov** or **www.sicklecelldisease.org**
- Learn more about protein-energy malnutrition and world hunger from the World Health Organization Nutrition Programme: **www.who.ch/nut/prot**
- Highlight 16 offers many more websites on malnutrition and world hunger.
- Search for "amino acid supplements" at the National Council for Reliable Health Information site: **www.ncrhi.org**

INTERNET ACTIVITIES

People concerned about healthy eating sometimes consider following a vegetarian diet. Play the Vegetarian Resource Group's "Vegetarian Game" and test your knowledge of vegetarianism.

- Go to: **www.vrg.org/game**
- Click on "Play!"
- Follow the instructions for playing the game.
- Note your total score and the questions (if any) you answered incorrectly.

There are many reasons for eating less meat and reducing animal protein in the diet. How might a vegetarian diet affect the quality of your diet?

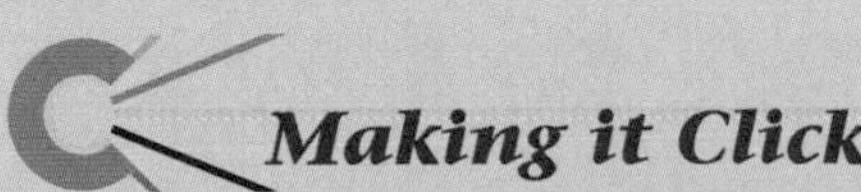

Making it Click

These problems will give you practice in doing simple nutrition-related calculations using hypothetical situations (see p. 194 for answers). Once you have mastered these examples, you will be prepared to examine your own protein needs. Be sure to show your calculations for each problem.

1. Compute recommended protein intakes for people of different sizes. Refer to the "How to" on p. 189 and compute the protein recommendation for the follow-

ing people. The intake for a woman 5 feet 8 inches tall is computed for you as an example.

A woman 5 feet 8 inches tall is 68 inches tall. From the table on the inside back cover, the midpoint in the green area for this woman is 144 pounds.

$$144 \text{ lb} \div 2.2 \text{ lb/kg} = 65 \text{ kg}.$$

$$0.8 \text{ g/kg} \times 65 \text{ kg} = 52 \text{ g protein per day}.$$

a. A woman 5 feet 1 inch tall.
a. A man (18 years) 6 feet 4 inches tall.

2. The chapter warns that recommendations based on percentage of energy intake are not always appropriate. Consider a man 26 years old who is 5 feet 10 inches tall, weighs 163 pounds, is moderately active, and eats 3500 kcalories/day with 10 percent of the kcalories from protein.
   a. What is this man's protein intake? Show your calculations.
   b. Is his protein intake appropriate? Too high? Too low? Justify your answer.

This exercise should help you develop a perspective on protein recommendations.

Study Questions

These questions will help you review the chapter. You will find the answers in the discussions on the pages provided.

1. How does the chemical structure of proteins differ from the structures of carbohydrates and fats? (pp. 168–170)
2. Describe the structure of amino acids, and explain how their sequence in proteins affects the proteins' shapes. What are essential amino acids? (pp. 168, 169–170)
3. Describe protein digestion and absorption. (pp. 171–173)
4. Describe protein synthesis. (pp. 173–175)
5. Describe some of the roles proteins play in the human body. (pp. 176–179)
6. What are enzymes? What roles do they play in chemical reactions? Describe the differences between enzymes and hormones. (pp. 176–177)
7. How does the body use amino acids? What is deamination? Define nitrogen balance. What conditions are associated with zero, positive, and negative balance? (pp. 180–181)
8. What factors affect the quality of dietary protein? What is a complete protein? (pp. 182–183)
9. How can vegetarians meet their protein needs without eating meat? (p. 182)
10. What are the health consequences of ingesting inadequate protein and energy? Describe marasmus and kwashiorkor. How can the two conditions be distinguished, and in what ways do they overlap? (pp. 184–186)
11. How might protein excess, or the type of protein eaten, influence health? (pp. 187–188)
12. What factors are considered in establishing recommended protein intakes? (p. 189)
13. Which food lists of the exchange system supply protein in abundance? In moderation? Not at all? (pp. 189–191)
14. What are the benefits and risks of taking protein and amino acid supplements? (pp. 191–192)

These questions will help you prepare for an exam. Answers can be found on p. 194.

1. Which part of its chemical structure differentiates one amino acid from another?
   a. its side group
   b. its acid group
   c. its amino group
   d. its double bonds
2. Isoleucine, leucine, and lysine are:
   a. proteases.
   b. polypeptides.
   c. essential amino acids.
   d. complementary proteins.
3. In the stomach, hydrochloric acid:
   a. denatures proteins and activates pepsin.
   b. hydrolyzes proteins and denatures pepsin.
   c. emulsifies proteins and releases peptidase.
   d. condenses proteins and facilitates digestion.
4. Proteins that facilitate chemical reactions are:
   a. buffers.
   b. enzymes.
   c. hormones.
   d. antigens.
5. If an essential amino acid that is needed to make a protein is unavailable, the cells must:
   a. deaminate another amino acid.
   b. substitute a similar amino acid.
   c. break down proteins to obtain it.
   d. synthesize the amino acid from glucose and nitrogen.
6. Eating two foods together so that each provides an amino acid that the other lacks is known as:
   a. dual deamination.
   b. random limitation.
   c. mutual supplementation.
   d. double complementation.

7. The protein efficiency ratio and PDCAAS are two methods used to:
   a. determine protein quality.
   b. assess protein-energy malnutrition.
   c. estimate the weight of nitrogen in a food.
   d. calculate the percentage kcalories from protein.
8. Marasmus develops from:
   a. too much fat clogging the liver.
   b. megadoses of amino acid supplements.
   c. inadequate protein and energy intake.
   d. excessive fluid intake causing edema.
9. The protein RDA for a healthy adult who weighs 180 pounds is:
   a. 50 milligrams/day.
   b. 65 grams/day.
   c. 180 grams/day.
   d. 2000 milligrams/day.
10. Which of these foods has the least protein per serving?
   a. rice
   b. broccoli
   c. pinto beans
   d. orange juice

References

1 H. F. Bunn, Pathogenesis and treatment of sickle cell disease, *New England Journal of Medicine* 337 (1997): 762–769.

2 S. T. Miller and coauthors, Prediction of adverse outcomes in children with sickle cell disease, *New England Journal of Medicine* 342 (2000): 83–89.

3 Committee on Genetics, Health supervision for children with sickle cell diseases and their families, *Pediatrics* 98 (1996): 467–472; E. M. Chiocca, Sickle cell crisis, *American Journal of Nursing* 96 (1996): 49.

4 Position of The American Dietetic Association: Vegetarian diets, *Journal of the American Dietetic Association* 97 (1997): 1317–1321.

5 D. G. Schroeder and R. Martorell, Enhancing child survival by preventing malnutrition, *American Journal of Clinical Nutrition* 65 (1997): 1080–1081.

6 D. H. Sullivan, S. Sun, and R. C. Walls, Protein-energy undernutrition among elderly hospitalized patients: A prospective study, *Journal of the American Medical Association* 281 (1999): 2013–2019.

7 R. W. Sauerwein and coauthors, Inflammatory mediators in children with protein-energy malnutrition, *American Journal of Clinical Nutrition* 65 (1997): 1534–1539.

8 W. S. Dempster and coauthors, Misplaced iron in kwashiorkor, *European Journal of Clinical Nutrition* 49 (1995): 208–210.

9 S. Collins, M. Myatt, and B. Golden, Dietary treatment of severe malnutrition in adults, *American Journal of Clinical Nutrition* 68 (1998): 193–199.

10 B. A. Underwood and S. Smitasiri, Micronutrient malnutrition: Policies and programs for control and their implications, *Annual Review of Nutrition* 19 (1999): 303–324; C. G. Victora and coauthors, Potential interventions for the prevention of childhood pneumonia in developing countries: Improving nutrition, *American Journal of Clinical Nutrition* 70 (1999): 309–320.

11 S. M. Potter, Soy protein and cardiovascular disease: The impact of bioactive components in soy, *Nutrition Reviews* 56 (1998): 231–235; J. W. Anderson, B. M. Johnstone, and M. E. Cook-Newell, Meta-analysis of the effects of soy protein intake on serum lipids, *New England Journal of Medicine*

ANSWERS

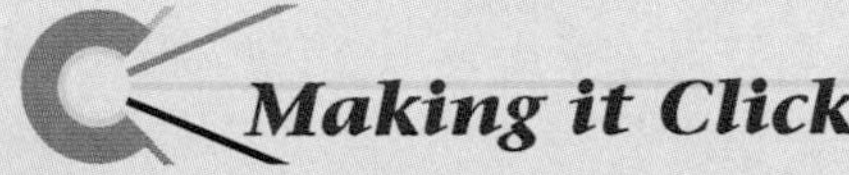

Making it Click

1. a. The midpoint weight for a woman 5 feet 1 inch tall is 116 lb.
   116 lb ÷ 2.2 lb/kg = 53 kg.
   0.8 g/kg × 53 kg = 42 g protein per day.

   b. The midpoint weight for a man 6 feet 4 inches is 180 lb.
   180 lb ÷ 2.2 lb/kg = 82 kg.
   He is 18 years old, so use 0.9 g/kg. 0.9 g/kg × 82 kg = 74 g protein per day.

2. a. 10% of 3500 kcal = 350 kcal from protein.
   350 kcal ÷ 4 kcal/g = 87.5 g protein per day.

   b. Using the RDA guideline of 0.8 g/kg, an appropriate protein intake for this man would be 59 g protein/day (163 lb ÷ 2.2 lb/kg = 74 kg; 0.8 g/kg × 74 kg = 59 g/day). His intake of 87.5 g protein per day falls between the RDA (59 g) and twice the RDA (2 × 59 g = 118 g), and so meets diet and health recommendations.

Multiple Choice

1. a	2. c	3. a	4. b	5. c
6. c	7. a	8. c	9. b	10. d

333 (1995): 276–282; D. Kritchevsky, Dietary protein, cholesterol, and atherosclerosis: A review of the early history, *Journal of Nutrition* 125 (1995): 589S–593S.

12 P. M. Ridker, Homocysteine and risk of cardiovascular disease among postmenopausal women, *Journal of the American Medical Association* 281 (1999): 1817–1821; J. Selhub and coauthors, Association between plasma homocysteine concentrations and extracranial carotid-artery stenosis, *New England Journal of Medicine* 332 (1995): 286–291; M. J. Stampfer and M. R. Malinow, Can lowering homocysteine levels reduce cardiovascular risk? *New England Journal of Medicine* 332 (1995): 328–329; J. B. Ubbink, Homocysteine—An atherogenic and a thrombogenic factor? *Nutrition Reviews* 53 (1995): 323–332.

13 L. Brattström and D. E. L. Wilcken, Homocysteine and cardiovascular disease: Cause or effect? *American Journal of Clinical Nutrition* 72 (2000): 315–323.

14 J. Selhub, Homocysteine metabolism, *Annual Review of Nutrition* 19 (1999): 217–246; J. S. Stamler and A. Slivka, Biological chemistry of thiols in the vasculature and in vascular-related disease, *Nutrition Reviews* 54 (1996): 1–30.

15 M. J. Grubben and coauthors, Unfiltered coffee increases plasma homocysteine concentrations in healthy volunteers: A randomized trial, *American Journal of Clinical Nutrition* 71 (2000): 480–484; O. Nygård and coauthors, Coffee consumption and plasma total homocysteine: The Hordeland Homocysteine Study, *American Journal of Clinical Nutrition* 65 (1997): 136–143.

16 P. F. Jacques and coauthors, The effect of folic acid fortification on plasma folate and total homocysteine concentrations, *New England Journal of Medicine* 340 (1999): 1449–1454; A. Chait and coauthors, Increased dietary micronutrients decrease serum homocysteine concentrations in patients at high risk of cardiovascular disease, *American Journal of Clinical Nutrition* 70 (1999): 881–887; I. A. Brouwer and coauthors, Low-dose folic acid supplementation decreases plasma homocysteine concentrations: A randomized trial, *American Journal of Clinical Nutrition* 69 (1999): 99–104; A. Brönstrup and coauthors, Effects of folic acid and combinations of folic acid and vitamin B-12 on plasma homocysteine concentrations in healthy, young women, *American Journal of Clinical Nutrition* 68 (1998): 1104–1110.

17 R. Meleady and I. Graham, Plasma homocysteine as a cardiovascular risk factor: Causal, consequential, or of no consequence? *Nutrition Reviews* 57 (1999): 299–305; E. B. Rimm and coauthors, Folate and vitamin B_6 from diet and supplements in relation to risk of coronary heart disease among women, *Journal of the American Medical Association* 279 (1998): 359–364.

18 A. J. Maxwell and J. P. Cooke, Cardiovascular effects of L-arginine, *Current Opinion in Nephrology and Hypertension* 7 (1998): 63–70; R. H. Boger and coauthors, Restoring vascular nitric oxide formation by L-arginine improves the symptoms of intermittent claudication in patients with peripheral arterial occlusive disease, *Journal of the American College of Cardiology* 32 (1998): 1336–1344.

19 R. P. Heaney, Excess dietary protein may not adversely affect bone, *Journal of Nutrition* 128 (1998): 1054–1057.

20 Committee on Dietary Reference Intakes, *Dietary Reference Intakes for Calcium, Phosphorus, Magnesium, Vitamin D, and Fluoride* (Washington, D.C.: National Academy Press, 1997), pp. 75–76.

21 U. S. Barzel and L. K. Massey, Excess dietary protein can adversely affect bone, *Journal of Nutrition* 128 (1998): 1051–1053.

22 A. R. Skov and coauthors, Changes in renal function during weight loss induced by high vs low-protein low-fat diets in overweight subjects, *International Journal of Obesity and Related Metabolic Disorders* 23 (1999): 1170–1177; E. Brändle, H. G. Sieberth, and R. E. Hautmann, Effect of chronic dietary protein intake on the renal function in healthy subjects, *European Journal of Clinical Nutrition* 50 (1996): 734–740.

23 M. T. Pedrini and coauthors, The effect of dietary protein restriction on the progression of diabetic and nondiabetic renal diseases: A meta-analysis, *Annals of Internal Medicine* 124 (1996): 627–632.

24 Position of The American Dietetic Association, Dietitians of Canada, and the American College of Sports Medicine: Nutrition and athletic performance, *Journal of the American Dietetic Association* 100 (2000): 1543–1556.

25 N. W. Flodin, The metabolic roles, pharmacology, and toxicology of lysine, *Journal of the American College of Nutrition* 16 (1997): 7–21.

VEGETARIAN DIETS

THE WAITER presents this evening's specials: a fresh spinach salad topped with mandarin oranges, raisins, and sunflower seeds, served with a bowl of pasta smothered in a mushroom and tomato sauce and topped with grated parmesan cheese. Then this one: a salad made of chopped parsley, scallions, celery, and tomatoes mixed with bulgur wheat and dressed with olive oil and lemon juice, served with a spinach and feta cheese pie. Do these meals sound good to you? Or is something missing . . . a pork chop or ribeye, perhaps?

Would vegetarian fare be acceptable to you some of the time? Most of the time? Ever? Perhaps it is helpful to recognize that dietary choices fall along a continuum—from one end, where people eat no meat or foods of animal origin, to the other end, where they eat generous quantities daily. Meat's place in the diet has been the subject of much research and controversy, as this highlight will reveal. One of the missions of this highlight, in fact, is to identify the *range* of meat intakes most compatible with health.

People who choose to exclude meat and other animal-derived foods from their diets today do so for many of the same reasons the Greek philosopher Pythagoras cited in the sixth century B.C.: physical health, ecological responsibility, and philosophical concerns. They might also cite world hunger issues, economic reasons, ethical concerns, or religious beliefs as motivating factors. Whatever their reasons—and even if they don't have a particular reason—people who exclude meat will be better prepared to plan well-balanced meals if they understand the nutrition and health implications of vegetarian diets.

Vegetarians generally are categorized, not by their motivations, but by the foods they choose to exclude (see the glossary on p. 197). Some exclude red meat only; some also exclude chicken or fish; others also exclude eggs; and still others exclude milk and milk products as well. As you will see, though, the foods a person *excludes* are not nearly as important as the foods a person *includes* in the diet. Vegetarian diets that include a variety of whole grains, vegetables, legumes, and fruits offer abundant complex carbohydrates and fibers, an assortment of vitamins and minerals, and little fat—characteristics that reflect current dietary recommendations aimed at promoting health and reducing obesity. This highlight examines the health benefits and potential problems of vegetarian diets and shows how to plan a well-balanced vegetarian diet.

HEALTH BENEFITS OF VEGETARIAN DIETS

Research on the health impacts of vegetarianism would be relatively easy if vegetarians differed from other people only in not eating meat. Many vegetarians, however, have adopted lifestyles that differentiate them from others: they typically maintain a healthy weight, use no tobacco or illicit drugs, use little (if any) alcohol, and are physically active.[1] Researchers must account for these lifestyle differences before they can determine which aspects of health correlate just with diet. Even then, *correlations* merely reveal what health factors *go with* the vegetarian diet, not what health effects may be *caused by* the diet. Without more evidence, conclusions remain tentative. Still, with all these qualifications, research findings suggest that well-planned vegetarian diets offer sound nutrition and health benefits to adults.[2]

Weight Control

In general, vegetarians maintain a healthier body weight than nonvegetarians. Lower body weights correlate with their high intakes of fiber and low intakes of animal fat.[3] Since obesity impairs health in a number of ways, this gives vegetarians a health advantage.

Well-planned vegetarian meals can provide adequate amounts of all the nutrients a person needs for good health.

Blood Pressure

Appropriate body weight helps to maintain a healthy blood pressure, as does a diet low in total fat and saturated fat and high in fiber, fruits, and vegetables.[4] Lifestyle factors also seem to influence blood pressure: smoking and alcohol intake raise blood pressure, and physical activity lowers it.

GLOSSARY

lactovegetarians: people who include milk and milk products, but exclude meat, poultry, fish, seafood, and eggs from their diets.
- **lacto** = milk

lacto-ovo-vegetarians: people who include milk, milk products, and eggs, but exclude meat, poultry, fish, and seafood from their diets.
- **ovo** = egg

macrobiotic diets: extremely restrictive diets limited to a few grains and vegetables; based on metaphysical beliefs and not on nutrition.

meat replacements: products formulated to look and taste like meat, fish, or poultry; usually made of textured vegetable protein.

omnivores: people who have no formal restriction on the eating of any foods.
- **omni** = all
- **vores** = to eat

tempeh (TEM-pay): a fermented soybean food, rich in protein and fiber.

textured vegetable protein: processed soybean protein used in vegetarian products such as soy burgers; see also *meat replacements.*

tofu (TOE-foo): a curd made from soybeans, rich in protein and often fortified with calcium; used in many Asian and vegetarian dishes in place of meat.

vegans (VEE-guns, VAY-guns, or VEJ-ans): people who exclude all animal-derived foods (including meat, poultry, fish, eggs, and dairy products) from their diets; also called **pure vegetarians, strict vegetarians,** or **total vegetarians.**

vegetarians: a general term used to describe people who exclude meat, poultry, fish, or other animal-derived foods from their diets.

Coronary Artery Disease

Fewer vegetarians than meat eaters suffer from diseases of the heart and arteries. The dietary factor most directly related to coronary artery disease is saturated animal fat, and in general, vegetarian diets are lower in total fat, saturated fat, and cholesterol than typical meat-based diets.[5] The fats common in plant-based diets—the monounsaturated fats of seeds and nuts and the polyunsaturated fats of vegetable oils—are not associated with heart disease.[6] Furthermore, vegetarian diets are generally higher in dietary fiber, another factor that helps control blood lipids and protect against heart disease.

When vegetarians are fed meat, which contains saturated fat, their blood lipid profiles change for the worse; when meat eaters are fed a low-fat vegetarian diet, their blood lipid profiles and blood pressure improve.[7] Similarly, vegetarians who eat fish and milk products have blood lipids between the low blood lipids of **vegans** and the higher blood lipids of nonvegetarians.[8]

Cancer

Seventh-Day Adventists, a religious group whose foodways center on a lacto-ovo-vegetarian diet, have a significantly lower mortality rate from cancer than the rest of the population, even after all the cancers attributed to smoking and alcohol are discounted.[9] Their low cancer rates may be due to their vegetarian diets; evidence is overwhelming that high intakes of fruits and vegetables reduce the risks of cancer.

Some scientific findings indicate that vegetarian diets are associated not only with lower cancer mortality in general, but with lower incidence of cancer at specific sites as well, most notably, colon cancer. People with colon cancer seem to eat more meat, more saturated fat, and fewer vegetables than others without cancer. High-protein, high-fat, low-fiber diets create an environment in the colon that promotes the development of cancer in some people. A high-meat diet has been associated with other cancers as well.[10]

In general, then, adults who eat vegetarian diets can reduce their risks of mortality and several chronic diseases, including obesity, high blood pressure, heart disease, and cancer.[11] But there is nothing mysterious or magical about the vegetarian diet; it simply includes ample fruits, vegetables, whole grains, and legumes—foods that are higher in fiber, richer in antioxidant vitamins, and lower in fats than meat-based diets.[12]

Some people find it easier to meet today's dietary recommendations for health by adopting a plant-based diet. They have decreased their use of animal products and increased their consumption of plant foods without any intention of "becoming a vegetarian." Such a plan offers many of the same health advantages of a vegetarian diet if it limits meat intake to the recommended 5 to 7 ounces daily and includes lean cuts, as well as abundant whole grains, fruits, and vegetables.

Conversely, both plant-based and meat-based diets can be detrimental to health when overloaded with fat. Vegetarians who dine on cheddar cheese, butter sauces, sour cream, and deep-fried vegetables invite the same health hazards as **omnivores** who overeat high-fat meats. And both diets, if not properly balanced, can lack nutrients. Poorly planned vegetarian diets typically lack iron, zinc, calcium, vitamin B_{12}, and vitamin D; without planning, the meat eater's diet may lack vitamin A, vitamin C, folate, and fiber, among others. Quite simply, the negative health aspects of any diet, including vegetarian diets, reflect poor diet planning. Careful attention to energy intake and specific problem nutrients can ensure adequacy.

VEGETARIAN DIET PLANNING

The vegetarian has the same meal-planning task as any other person—using a variety of foods that will deliver all the needed nutrients within an energy allowance that maintains a healthy body weight (as discussed in Chapter 2). An added challenge is to do so with fewer foods.

Vegetarians who include milk products and eggs can meet recommendations for most nutrients about as easily as nonvegetarians.[13] Such diets provide enough energy, protein, and other nutrients to support the health of adults and the growth of children and adolescents.

Vegetarians are often advised to follow the Daily Food Guide presented in Chapter 2 with a few modifications (see Table H6-1). Those who include milk products and eggs can follow the regular plan, using legumes and products made from them, such as peanut butter, **tempeh,** and **tofu,** in place of meat. Those who do not use milk can use soy milk and tofu fortified with calcium, vitamin D, and vitamin B_{12}. Dark green vegetables and legumes help meet iron and zinc needs.

A new food guide is being developed specifically for vegetarian diets.[14] At first glance, it looks like the familiar Food Guide Pyramid, but it actually has a large trapezoid bottom with a small triangle top. The bottom contains the five major plant-based food groups—whole grains, legumes, vegetables, fruits, and nuts and seeds. The top contains four optional groups—vegetable oils, dairy products, eggs, and sweets—that can be included at the user's discretion. This design is flexible enough that a variety of people can use it: people who have adopted diverse vegetarian diets, those who want to make the transition to a vegetarian diet, and those who simply want to include more plant-based meals in their diets. This vegetarian food guide also includes other lifestyle factors that contribute to good health: moderate exposure to sunlight, physical activity, and water intake. Figure H6-1 presents both the proposed vegetarian food guide pyramid and the USDA Food Guide Pyramid for comparison.

Most vegetarians easily obtain large quantities of the nutrients that are abundant in plant foods: thiamin, riboflavin, folate, and vitamins B_6, C, A, and E. These vegetarian guides help to ensure adequate intakes of the main nutrients vegetarian diets might otherwise lack: iron, zinc, calcium, vitamin B_{12}, and vitamin D.

TABLE H6-1 Daily Food Guide for Vegetarian Meal Planning

Food Group	Suggested Daily Servings	Serving Sizes
Breads, cereals, rice, pasta, and other grain products	6 to 11	1 slice bread ½ bun, bagel, or English muffin ½ c cooked cereal, rice, or pasta 1 oz ready-to-eat cereal
Vegetables	3 to 5[a]	½ c cooked or chopped raw vegetables 1 c raw leafy vegetables
Fruits	2 to 4	1 medium-sized piece fresh fruit ¾ c fruit juice ½ c canned, cooked, or chopped raw fruit ¼ c dried fruit
Legumes, nuts, seeds, eggs, and other meat substitutes	2 to 3	½ c cooked legumes ¼ c tofu or tempeh 1 c soy milk 2 tbs peanut butter, nuts, or seeds (these tend to be high in fat, so use sparingly) 1 egg or 2 egg whites
Milk, yogurt, cheese, and other milk products	2 to 3[b]	1 c milk 1 c yogurt 1½ oz cheese

[a]Include 1 cup of dark green vegetables daily to help meet iron requirements.
[b]People who do not use milk or milk products: use soy milk fortified with calcium, vitamin D, and vitamin B_{12}. Suggestions for other nonmilk calcium-rich food sources are provided in Chapter 12.
SOURCE: Adapted from Position of The American Dietetic Association: Vegetarian diets, *Journal of the American Dietetic Association* 97 (1997): 1320.

Protein

Protein is not the problem it was once thought to be for vegetarian diets. **Lacto-ovo-vegetarians** who use animal-derived foods such as milk and eggs receive high-quality proteins and are unlikely to develop protein deficiencies. Even those who adopt only plant-based diets are unlikely to develop protein deficiencies provided that energy intakes are adequate and the protein sources varied.[15] The proteins of whole grains, legumes, seeds, nuts, and vegetables can provide adequate amounts of all the amino acids. An advantage of many vegetarian protein foods is that they are generally lower in saturated fat than meats and are often higher in fiber and richer in some vitamins and minerals.

To ease meal preparation, vegetarians sometimes use **meat replacements** made of **textured vegetable protein** (soy protein). These foods are formulated to look and taste like meat, fish, or poultry. Many of these products are designed to match the known nutrient contents of animal-protein foods, but sometimes they fall short. A wise vegetarian does not rely on these products too heavily, but learns to use a variety of whole foods instead. Vegetarians may also use soybeans in the form of tofu (bean curds), to bolster protein intake.

Iron

Getting enough iron can be a problem even for meat eaters, and those who eat no meat must pay special attention to their iron intake. The iron in plant foods such as legumes, dark green leafy vegetables, iron-fortified cereals, and whole-grain breads and cereals is poorly absorbed. Fortunately, the body seems to adapt to a vegetarian diet by absorbing iron more efficiently.[16] Furthermore, iron absorption is enhanced by vitamin C, and vegetarians typically eat many vitamin C–rich fruits and vegetables. Consequently, vegetarians suffer no more iron deficiency than other people do.[17]

Zinc

Zinc is similar to iron in that meat is its richest food source and zinc from plant sources is not well absorbed. In addition, soy, which is commonly used as a meat alter-

FIGURE H6-1 Food Pyramids Compared

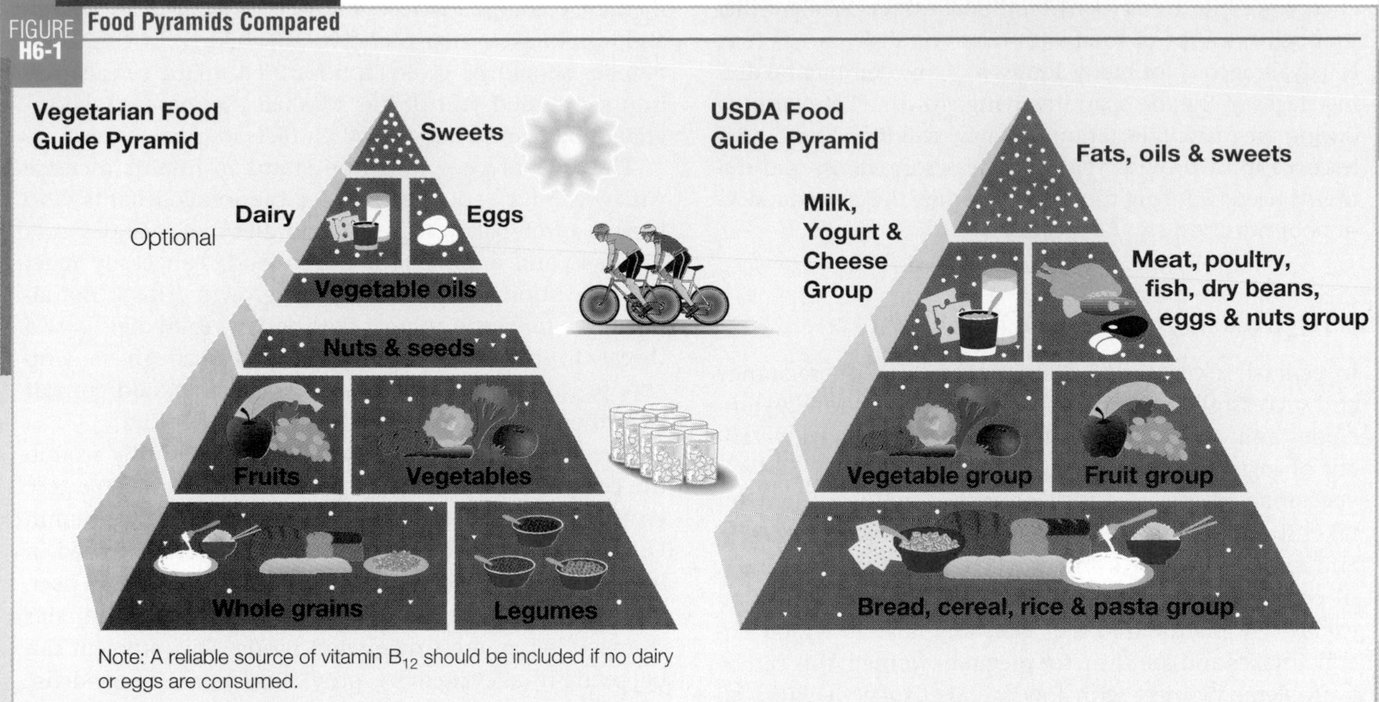

Note: A reliable source of vitamin B_{12} should be included if no dairy or eggs are consumed.

SOURCE: E. H. Haddad, J. Sabaté, and C. G. Whitten, Vegetarian food guide pyramid: A conceptual framework, *American Journal of Clinical Nutrition* 70 (1999): 615S–619S; U.S. Department of Agriculture, 1992.

nate in vegetarian meals, interferes with zinc absorption. Nevertheless, most vegetarian adults are not zinc deficient.[18] Perhaps the best advice to vegetarians regarding zinc is to eat a variety of nutrient-dense foods; include whole grains, nuts, and legumes such as black-eyed peas, pinto beans, and kidney beans; and maintain an adequate energy intake. For those who include seafood, oysters, crabmeat, and shrimp are rich in zinc.

Calcium

The calcium intakes of **lactovegetarians** are similar to those of the general population, but people who use no milk risk deficiency. Careful planners select calcium-rich foods, such as calcium-fortified juices or soy milk, in ample quantities regularly. This is especially important for children and adolescents. Soy formulas for infants are fortified with calcium and can be used in cooking, even for adults. Other good calcium sources include calcium-set tofu, some legumes, some green vegetables such as broccoli and turnip greens, some nuts such as almonds, and certain seeds such as sesame seeds.*[19] The choices should be varied because calcium absorption from some plant foods may be limited (as Chapter 12 explains).

Vitamin B_{12}

The requirement for vitamin B_{12} is small, but this vitamin is found only in animal-derived foods. Fermented soy products such as tempeh may contain some vitamin B_{12} from the bacteria that did the fermenting, but unfortunately, much of the vitamin B_{12} found in these products may be an inactive form. Seaweeds such as nori and chlorella supply some vitamin B_{12}, but not much, and excessive intakes can lead to iodine toxicity.[20] To defend against vitamin B_{12} deficiency, vegans must rely on vitamin B_{12}–fortified sources (such as soy milk or breakfast cereals) or supplements.

*Calcium salts are often added during processing to coagulate the tofu.

Vitamin D

For people who do not use vitamin D–fortified milk and do not receive enough exposure to sunlight to synthesize adequate vitamin D, supplements may be warranted. This is particularly important for children and older adults. In northern climates during winter months, young children on vegan diets can readily develop rickets, the vitamin D–deficiency disease.

VEGETARIAN DIETS THROUGH THE LIFE SPAN

Vegetarians who plan their diets carefully easily obtain all the nutrients they need to support good health. Achieving adequate energy and nutrient intakes may be difficult, however, for the vegan who excludes all animal products, and particularly for growing children and pregnant and lactating women. Foods of plant origin generally offer much less energy per bite than foods of animal origin. While a diet that delivers a lot of food with relatively little energy may be advantageous for many adults, it can be detrimental during stages of the life span involving growth. Diet planning during pregnancy, lactation, infancy, childhood, and adolescence must provide for the increases in energy and nutrients needed during those times—when the consequences of poor nutrition can be great.

Pregnancy and Lactation

In general, a vegetarian diet favors a healthy pregnancy and successful lactation if it provides adequate energy; includes milk and milk products; and contains a wide variety of legumes, cereals, fruits, and vegetables.[21] Many vegetarian women are well nourished, with nutrient intakes from diet alone exceeding the RDA for all vitamins and minerals except iron, which is low for most women. In contrast, vegan women who restrict themselves to an exclusively plant-based diet generally have low food energy intakes and are thin; for pregnant women, this can be a problem. Women with low prepregnancy weights and small weight gains during pregnancy jeopardize a healthy pregnancy.

Vegan diets that exclude all foods of animal origin may require supplementation with vitamin B_{12}, calcium, and vitamin D, or the addition of foods fortified with these nutrients. Infants of vegan parents may suffer spinal cord damage and develop severe psychomotor retardation due to a lack of vitamin B_{12} in the mother's diet during pregnancy.[22] Breastfed infants of vegan mothers have been reported to develop vitamin B_{12} deficiency and severe movement disorders.[23] Giving the infants vitamin B_{12} supplements corrects the blood and neurological symptoms of deficiency, as well as the structural abnormalities, but cognitive and language development delays may persist.[24] A vegan mother needs a regular source of vitamin B_{12}–fortified foods or a supplement that provides 2.6 micrograms daily.

A pregnant woman who cannot meet her calcium needs through diet alone may need 600 milligrams of supplemental calcium daily, taken with meals. Pregnant women who do not receive sufficient dietary vitamin D or enough exposure to sunlight may need a supplement that provides 10 micrograms daily.

Infancy

The newborn infant is a lactovegetarian. As long as the infant has access to sunlight as a source of vitamin D and to sufficient quantities of breast milk from a mother who eats an adequate diet, the infant will thrive during the early months.

Infants beyond about four months of age present a greater challenge in terms of meeting nutrient needs by way of vegetarian and, especially, vegan diets. Continued breastfeeding or formula feeding is recommended, but supplementary feedings are necessary to ensure adequate energy and iron intakes. Infants and young children in vegetarian families should be given iron-fortified infant cereals well into the second year of life. Mashed legumes and whole-grain foods can be added to their diets in place of meat.

The risks of poor nutrition status in infants increase with weaning and reliance on table foods. Infants who receive a well-balanced vegetarian diet that includes milk products and a variety of other foods can easily meet their nutritional requirements for growth. This is not always true for vegan infants. Restrictive vegan diets pose a threat to infants' health. Parents or caregivers who choose to feed their infants vegan diets should consult with their pediatrician and a registered dietitian.

The growth of vegan infants slows significantly around the time of transition from breast milk to solid foods.[25] Protein-energy malnutrition and deficiencies of vitamin D, vitamin B_{12}, iron, and calcium have been reported in infants fed vegan diets.[26] Vegan diets that are high in fiber, other complex carbohydrates, and water fill an infant's stomach before meeting energy needs. This problem can be partially alleviated by providing more energy-dense foods: nut butters, legumes, dried fruit spreads, and mashed avocado. Using soy formulas (or milk) fortified with calcium, vitamin B_{12}, and vitamin D and including vitamin C–containing foods at meals to enhance iron absorption will help prevent other nutrient deficiencies in vegan diets.

Childhood and Adolescence

Well-planned vegetarian diets, especially those that include eggs, milk, and milk products, can easily provide adequate nutrient intakes for growing children. The growth of vegetarian children is similar to that of their peers.[27]

Vegan diets, on the other hand, can fail to provide sufficient energy to support the growth of a child within a quantity of food small enough for the child to eat. A child's small stomach can hold only so much food, and a vegan child may feel full before eating enough to meet nutrient and energy needs. A vegan child's diet should emphasize cereals, legumes, and nuts to meet protein and energy needs in a small volume. Meat, which contains abundant protein, iron, and food energy in less bulk, supports the growth of children more efficiently. Compared with meat-eating children, vegan children tend to be shorter and lighter in weight; their low energy intakes can impair growth.[28]

When vegan children get their protein only from plant foods, they may need protein intakes higher than the RDA for normal growth and health. The standard protein recommendations may be inadequate to support the growth of vegan children, but specific recommendations have not been established.

Other nutritional concerns for vegans include vitamin B_{12}, calcium, and vitamin D. Children who were raised on vegan diets and then switched to more liberal diets have difficulty achieving an adequate vitamin B_{12} status even with a moderate consumption of animal products.[29] Adolescents following a vegan diet low in calcium and vitamin D have a reduced bone density, which may have implications for bone health later in life.[30]

CLOSING THOUGHTS

As you can see, vegetarianism is not a religion like Buddhism or Hinduism, but merely an eating plan that selects plant foods to deliver needed nutrients. The quality of the diet depends not on whether it includes meat, but on whether the other food choices are nutritionally sound. Variety is key to nutritional adequacy in a vegetarian diet. Restrictive plans, such as **macrobiotic diets,** that limit selections to a few grains and vegetables cannot possibly deliver a full array of nutrients.

Having learned some of the relationships between diet and health, many people may discover that their strategies for planning meals need to change. In the past, they decided what cut of beef, ham, pork, lamb, poultry, or fish to prepare and then filled in the menu with an accompanying "starch" (potato, rice, or noodles), salad or other vegetable, and bread. Now they fill their dinner plates with legumes, whole grains, vegetables, and fruits. Then they may add small quantities of milk products, eggs, lean meat, fish, or poultry.

Two meat servings of the size depicted here represent the maximum daily meat intake suggested by the Daily Food Guide as health promoting.

© Courtesy National Cattlemen's Beef Association

For the most part, it seems that nonmeat and low-meat diets can both support good health. The USDA Food Guide Pyramid suggests 5 to 7 ounces of meat, poultry, or fish a day. This amount of meat alone provides most of a person's daily recommended protein intake—and other foods together can provide a similar amount. Some researchers argue that this much meat eaten daily is not compatible with good health; if any meat is eaten, they suggest that it be eaten infrequently and in small portions.[31] With the evidence pointing to the health advantages of a plant-based diet, perhaps between 0 and 6 ounces of meat daily would best serve the needs of most people.

Keep in mind, too, that diet is only one factor influencing health. Whatever a diet consists of, its context is also important: no smoking; alcohol consumption in moderation, if at all; regular physical activity; adequate rest; and medical attention when needed all contribute to a healthy life. Establishing these healthy habits early in life seems to be the most important step one can take to reduce the risks of later diseases (as Highlight 15 explains).[32]

Nutrition on the Net

•WEBSITES•

Access these websites for further study of topics covered in this highlight.

- Find updates and quick links to these and other nutrition-related sites at our website: **www.wadsworth.com/nutrition**
- Search for "vegetarian" in the Foods section of the Food and Drug Administration's site: **www.fda.gov**
- Visit the Vegetarian Resource Group: **www.vrg.org**
- Review another vegetarian diet pyramid developed by Oldways Preservation & Exchange Trust: **www.oldwayspt.org**

References

1 P. Walter, Effects of vegetarian diets on aging and longevity, *Nutrition Reviews* 55 (1997): S61–S68.

2 Position of The American Dietetic Association: Vegetarian diets, *Journal of the American Dietetic Association* 97 (1997): 1317–1321.

3 P. N. Appleby and coauthors, Low body mass index in non-meat eaters: The possible roles of animal fat, dietary fibre and alcohol, *International*

Journal of Obesity and Related Metabolic Disorders 22 (1998): 454–460; T. Key and G. Davey, Prevalence of obesity is low in people who do not eat meat, *British Medical Journal* 313 (1996): 816–817.

4 L. J. Beilin and V. Burke, Vegetarian diet components, protein and blood pressure: Which nutrients are important? *Clinical and Experimental Pharmacology and Physiology* 22 (1995): 195–198.

5 J. I. Mann and coauthors, Dietary determinants of ischaemic heart disease in health conscious individuals, *Heart* 78 (1997): 450–455.

6 A. M. Coulston, The role of dietary fats in plant-based diets, *American Journal of Clinical Nutrition* 70 (1999): 512S–515S.

7 J. McDougall and coauthors, Rapid reduction of serum cholesterol and blood pressure by a twelve-day, very low fat, strictly vegetarian diet, *Journal of the American College of Nutrition* 14 (1995): 491–496.

8 P. N. Appleby and coauthors, The Oxford Vegetarian Study: An overview, *American Journal of Clinical Nutrition* 70 (1999): 525S–531S.

9 P. K. Mills and coauthors, Cancer incidence among California Seventh-Day Adventists, 1976–1982, *American Journal of Clinical Nutrition* 59 (1994): 1136S–1142S.

10 Z. Djuric and coauthors, Oxidative DNA damage levels in blood from women at high risk for breast cancer are associated with dietary intakes of meats, vegetables, and fruits, *Journal of the American Dietetic Association* 98 (1998): 524–528; B. C.-H. Chiu and coauthors, Diet and risk of non-Hodgkin lymphoma in older women, *Journal of the American Medical Association* 275 (1996): 1315–1321.

11 G. E. Fraser, Associations between diet and cancer, ischemic heart disease, and all-cause mortality in non-Hispanic white California Seventh-Day Adventists, *American Journal of Clinical Nutrition* 70 (1999): 532S–538S; T. J. A. Key and coauthors, Dietary habits and mortality in 11,000 vegetarians and health conscious people: Results of a 17 year follow up, *British Medical Journal* 313 (1996): 775–779.

12 M. Thorogood, The epidemiology of vegetarianism and health, *Nutrition Research Reviews* 8 (1995): 179–192; A.-L. Rauma and coauthors, Antioxidant status in long-term adherents to a strict uncooked vegan diet, *American Journal of Clinical Nutrition* 62 (1995): 1221–1227.

13 K. C. Janelle and S. I. Barr, Nutrient intakes and eating behavior score of vegetarian and nonvegetarian women, *Journal of the American Dietetic Association* 95 (1995): 180–189.

14 E. H. Haddad, J. Sabaté, and C. G. Whitten, Vegetarian food guide pyramid: A conceptual framework, *American Journal of Clinical Nutrition* 70 (1999): 615S–619S.

15 V. R. Young and P. L. Pellett, Plant proteins in relation to human protein and amino acid nutrition, *American Journal of Clinical Nutrition* 59 (1994): 1203S–1212S; Position of The American Dietetic Association, 1997.

16 J. R. Hunt and Z. K. Roughead, Nonheme-iron absorption, fecal ferritin excretion, and blood indexes of iron status in women consuming controlled lactoovovegetarian diets for 8 wk, *American Journal of Clinical Nutrition* 69 (1999): 944–952.

17 M. J. Ball and M. A. Bartlett, Dietary intake and iron status of Australian vegetarian women, *American Journal of Clinical Nutrition* 70 (1999): 353–358; W. J. Craig, Iron status of vegetarians, *American Journal of Clinical Nutrition* 59 (1994): 1233S–1237S.

18 R. J. Gibson, Content and bioavailability of trace elements in vegetarian diets, *American Journal of Clinical Nutrition* 59 (1994): 1223S–1232S.

19 C. M. Weaver and K. L. Plawecki, Dietary calcium: Adequacy of a vegetarian diet, *American Journal of Clinical Nutrition* 59 (1994): 1238S–1241S.

20 A.-L. Rauma and coauthors, Vitamin B-12 status of long-term adherents of a strict uncooked vegan diet ("Living Food Diet") is compromised, *Journal of Nutrition* 125 (1995): 2511–2515.

21 Position of The American Dietetic Association, 1997.

22 K. Lovblad and coauthors, Retardation of myelination due to dietary vitamin B_{12} deficiency: Cranial MRI findings, *Pediatric Radiology* 27 (1997): 155–158; U. von Schenck, C. Bender-Gotze, and B. Koletzko, Persistence of neurological damage induced by dietary vitamin B-12 deficiency in infancy, *Archives of Diseases in Childhood* 77 (1997): 137–139.

23 P. J. Grattan-Smith and coauthors, The neurological syndrome of infantile cobalamin deficiency: Developmental regression and involuntary movements, *Movement Disorders* 12 (1997): 39–46.

24 Von Schenck, 1997.

25 P. C. Dagnelie and W. A. van Staveren, Macrobiotic nutrition and child health: Results of a population-based, mixed-longitudinal cohort study in the Netherlands, *American Journal of Clinical Nutrition* 59 (1994): 1187S–1196S.

26 Dagnelie and van Staveren, 1994.

27 M. Hebbelinck, P. Clarys, and A. DeMalsche, Growth, development, and physical fitness of Flemish vegetarian children, adolescents, young adults, *American Journal of Clinical Nutrition* 70 (1999): 579S–585S; I. Nathan, A. F. Hackett, and S. Kirby, A longitudinal study of the growth of matched pairs of vegetarian and omnivorous children, aged 7–11 years, in the northwest of England, *European Journal of Clinical Nutrition* 51 (1997): 20–25.

28 T. A. B. Sanders and S. Reddy, Vegetarian diets and children, *American Journal of Clinical Nutrition* 59 (1994): 1176S–1181S.

29 M. van Dusseldorp and coauthors, Risk of persistent cobalamin deficiency in adolescents fed a macrobiotic diet in early life, *American Journal of Clinical Nutrition* 69 (1999): 664–671.

30 T. J. Parsons and coauthors, Reduced bone mass in Dutch adolescents fed a macrobiotic diet in early life, *Journal of Bone Mineral Research* 12 (1997): 1486–1494.

31 L. H. Kushi, E. B. Lenart, and W. C. Willett, Health implications of Mediterranean diets in light of contemporary knowledge. 2. Meat, wine, fats, and oils, *American Journal of Clinical Nutrition* 61 (1995): 1416S–1427S.

32 V. Fønnebø, The healthy Seventh-Day Adventist lifestyle: What is the Norwegian experience? *American Journal of Clinical Nutrition* 59 (1994): 1124S–1129S.

Metabolism: Transformations and Interactions

CHAPTER 7

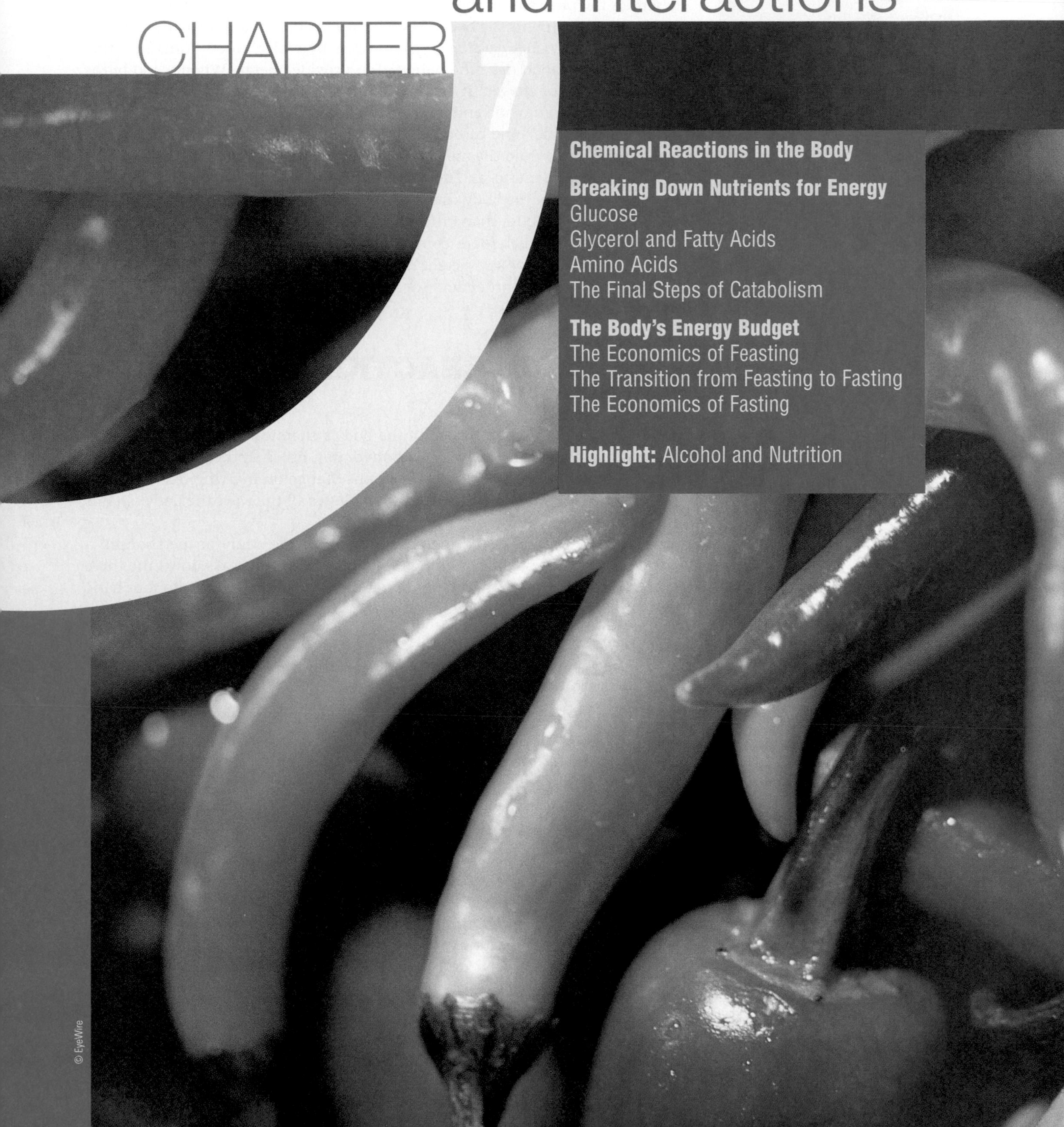

© EyeWire

© Photo Disc Inc.

We receive energy from the sun by way of the foods we eat.

Almost all living things depend on the sun's energy. Plants rely directly on the sun to provide the light energy that drives the reactions of **photosynthesis**—the process by which plants make carbohydrate from carbon dioxide and water. That energy from the sun is captured in the energy of chemical bonds that hold the atoms of sugars and starches together. We humans, and all other animals, use the sun indirectly, for we cannot photosynthesize. We depend on plants, or on animals that eat plants, for the food that provides us with energy.

This chapter answers the question, How do we obtain energy from foods? It describes the processes in the human body that *release* energy from the chemical bonds in nutrients the body uses for **fuel.** As the bonds break, they release energy in a controlled version of the process by which wood burns in a fire. Both wood and food have the potential to provide energy. When wood burns in the presence of oxygen, it generates heat and light (energy), steam (water), and some carbon dioxide and ash (waste). Similarly, during the body's metabolism, energy, water, and carbon dioxide are released.

Energy derived from the metabolism of fuel nutrients enables people to ride bicycles, compose music, and do everything else they do. By studying metabolism, we learn how the body uses foods to meet its needs and why some foods meet those needs better than others. Readers who are interested in weight control will discover which foods contribute most to body fat and which to select when trying to gain or lose weight safely. Physically active readers will discover which foods best support endurance activities and which to select when trying to build lean body mass.

CHEMICAL REACTIONS IN THE BODY

◆ Appendix B provides an overview of basic chemistry concepts.

Earlier chapters introduced some of the body's chemical reactions:◆ the making and breaking of the bonds in carbohydrates, lipids, and proteins. The sum of these and all the other chemical reactions that go on in living cells is known as **metabolism;** and *energy* metabolism includes all the ways the body obtains and spends energy from food.

Chapters 4, 5, and 6 laid the groundwork for the study of metabolism; a brief review may be helpful. During digestion, the body breaks down the three energy-yielding nutrients—carbohydrates, lipids, and proteins—into four basic units that can be absorbed into the blood:

- From carbohydrates—glucose.*
- From lipids (triglycerides)—glycerol and fatty acids.
- From proteins—amino acids.

Amino acids are primarily building blocks for proteins, but they can follow energy pathways if needed or if eaten in excess.

Look for these four basic units to appear again and again in the metabolic reactions described in this chapter. Alcohol also enters many of the metabolic pathways; Highlight 7 focuses on how alcohol disrupts metabolism and how the body handles it.

•Building Reactions—Anabolism • The cells can use the basic units of energy-yielding nutrients to build body compounds. Glucose molecules may be joined together to make glycogen chains. Glycerol and fatty acids may be assembled into triglycerides. Amino acids may be linked together to make proteins. Each of these reactions starts with small, simple compounds and uses them as building blocks to form larger, more complex structures. Such reac-

*This chapter features glucose because of its central role in carbohydrate metabolism, but the monosaccharides fructose and galactose also enter the metabolic pathways.

photosynthesis: the process by which green plants make carbohydrates from carbon dioxide and water using the green pigment chlorophyll to trap the sun's energy.
- **photo** = light
- **synthesis** = put together (making)

fuel: compounds that cells can use for energy. The major fuels include glucose, fatty acids, and amino acids; other fuels include ketone bodies, lactic acid, glycerol, and alcohol.

metabolism: the sum total of all the chemical reactions that go on in living cells. **Energy metabolism** includes all the reactions by which the body obtains and spends the energy from food.
- **metaballein** = change

FIGURE 7-1 **Anabolic and Catabolic Reactions Compared**

Note: You need not memorize a color code to understand the figures in this chapter, but you may find it helpful to know that blue is used for carbohydrates, yellow for fats, and red for proteins.

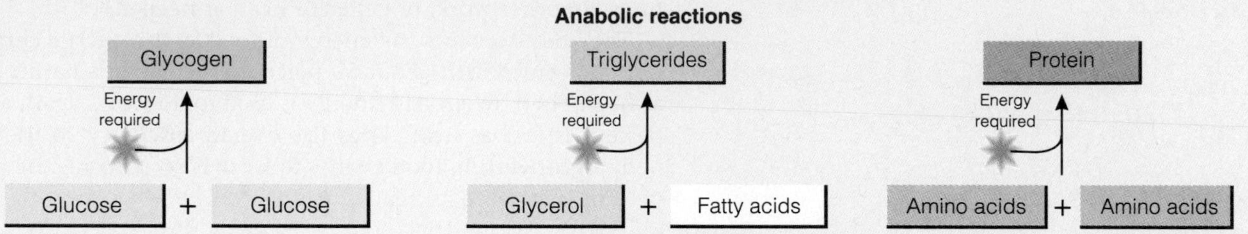

Anabolic reactions include the making of glycogen, triglycerides, and protein; these reactions require differing amounts of energy.

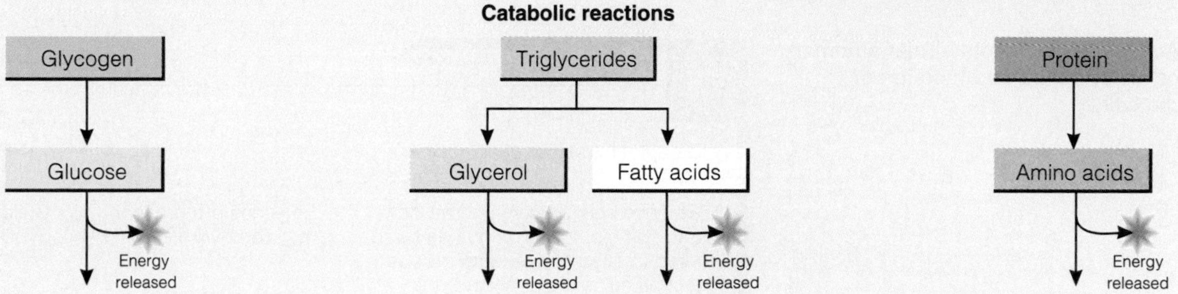

Catabolic reactions include the breakdown of glycogen, triglycerides, and protein; the further catabolism of glucose, glycerol, fatty acids, and amino acids releases differing amounts of energy.

tions involve doing work and so require energy. The building up of body compounds is known as **anabolism;** this book represents anabolic reactions, wherever possible, with "up" arrows in chemical diagrams (such as those shown in Figure 7-1).

• Breakdown Reactions—Catabolism • The breaking down of body compounds is known as **catabolism;** catabolic reactions usually release energy and are represented, wherever possible, by "down" arrows in chemical diagrams (as in Figure 7-1). Catabolic reactions include the breakdown of glycogen to glucose, of triglycerides to fatty acids and glycerol, and of proteins to amino acids. When the body needs energy, it breaks down any or all of these four basic units into even smaller units, as described later.

• The Transfer of Energy in Reactions • When a chemical bond breaks, energy can be released as heat, captured in another chemical bond, or both. Often, as one compound is broken apart, some of the energy is released as heat, and some is used to put together another compound. Such reactions, in which the breakdown of one compound provides energy for the building of another, are known as **coupled reactions.**

The energy released during catabolism is often captured by go-between molecules that can easily transfer that energy to other compounds. These molecules are sometimes called the body's "common energy currency," or "high-energy compounds." One such compound is **ATP (adenosine triphosphate).** The breakdown of energy-nutrient molecules is coupled to the making of many ATP molecules, which capture much of the released energy in their bonds.

ATP, as its name indicates, contains three phosphate groups (see Figure 7-2 on p. 206).◆ The energy in the bonds between each phosphate group is greater than the energy in most other chemical bonds. When energy is needed, hydrolysis reactions readily break these high-energy bonds, splitting off one or

ATP = A-P~P~P. ◆
(Each ~ denotes a "high-energy" bond.)

anabolism (an-ABB-o-lism): reactions in which small molecules are put together to build larger ones. Anabolic reactions require energy.
- **ana** = up

catabolism (ca-TAB-o-lism): reactions in which large molecules are broken down to smaller ones. Catabolic reactions usually release energy.
- **kata** = down

coupled reactions: pairs of chemical reactions in which energy released from the breakdown of one compound is used to create a bond in the formation of another compound.

ATP or **adenosine** (ah-DEN-oh-seen) **triphosphate** (try-FOS-fate): a common high-energy compound composed of a purine (adenine), a sugar (ribose), and three phosphate groups.

FIGURE 7-2 ATP (Adenosine Triphosphate)

ATP is one of the body's quick-energy molecules. Notice that the bonds connecting the three phosphate groups have been drawn as wavy lines, indicating a high-energy bond. When these bonds are broken, a large amount of energy is released.

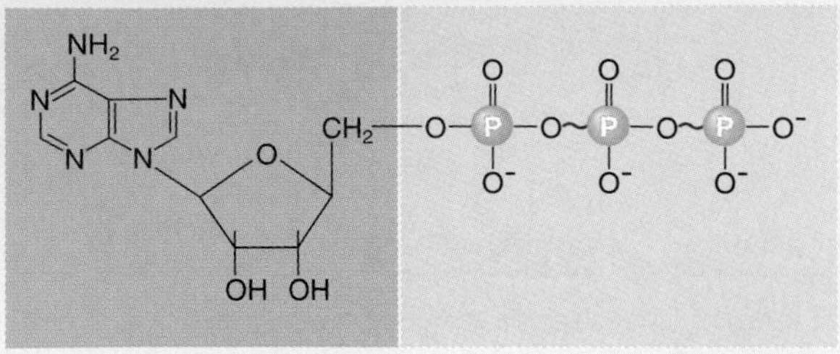

Adenosine + 3 phosphate groups

two phosphate groups and releasing their energy. These reactions, in turn, are coupled to other reactions that use that energy. Thus the body uses ATP to transfer the energy produced during catabolic reactions to power its anabolic reactions. Figure 7-3 illustrates how the body uses ATP to carry its energy currency, build body structures, do other work, or generate heat, as needed.

The body converts the energy of food to the energy currency of ATP molecules with about 50 percent efficiency, radiating the rest as heat. Then, when ATP energy is used to do work, again about 50 percent is lost as heat. Thus the overall efficiency of the human body in converting food energy to work is 25 percent; the other 75 percent is released as heat.

• **The Site of Reactions—Cells** • Metabolic work is going on all the time within all the body's trillions of cells.◆ Figure 7-4 depicts a typical cell and shows where the major reactions of energy pro-

◆ **Appendix A presents a brief summary of the structure and function of the cell.**

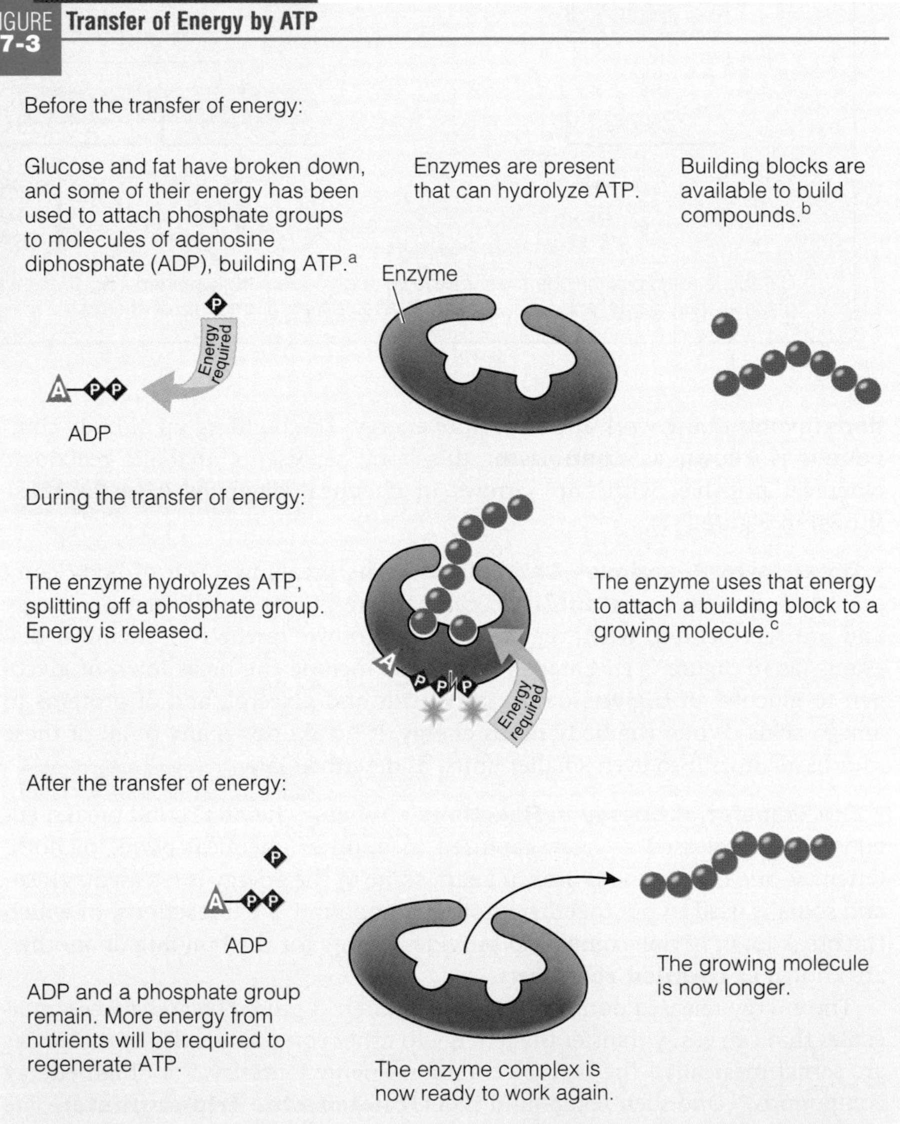

FIGURE 7-3 Transfer of Energy by ATP

[a]ADP (adenosine diphosphate) is lower in energy than ATP; AMP (adenosine monophosphate) is even lower.
[b]Compounds that ATP energy might be used to build include glycogen, fat, proteins, and hormones, among others.
[c]In all such reactions, half or more of the total original energy is lost as heat, accounting for the temperature-raising effect of metabolism. ATP can also break apart without doing work and release all of its energy as heat if needed. The breakdown of ATP to release heat only, with no energy captured for use in another reaction, is an *uncoupled* reaction.

FIGURE 7-4 A Typical Cell (Simplified Diagram)

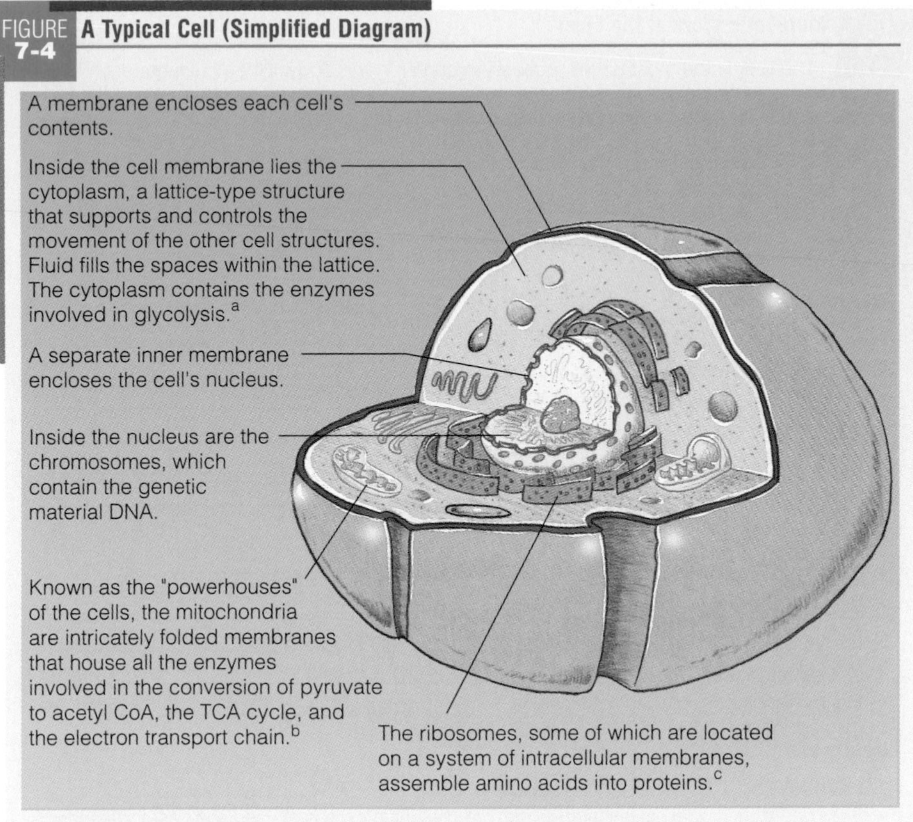

[a]Glycolysis is described on pp. 209–210.
[b]The conversion of pyruvate to acetyl CoA, TCA cycle, and electron transport chain are described on pp. 210, 219–220.
[c]Figure 6-7 on p. 174 describes protein synthesis.

duction take place. The type and extent of metabolic activity vary depending on the type of cell, but of all the body's cells, the liver cells are the most versatile and metabolically active. Table 7-1 (on p. 208) offers insights into the liver's work.

• The Helpers in Reactions—Enzymes and Coenzymes • Metabolic reactions almost always require enzymes to facilitate their action. In some cases, the enzymes need assistants to help them. Enzyme helpers are called **coenzymes.**

Coenzymes are small organic molecules that associate closely with most enzymes, but are not proteins themselves. The relationships between coenzymes and enzymes differ in detail, but one thing is true of all: without its coenzyme, an enzyme cannot function. Some of the B vitamins serve as coenzymes to the enzymes that release energy from glucose, glycerol, fatty acids, and amino acids. These B vitamin coenzymes stand alongside the metabolic pathways, so to speak, and help to keep the disassembly lines moving. Chapter 10 provides more details on the coenzyme actions of the B vitamins.

IN SUMMARY

During digestion the energy-yielding nutrients—carbohydrates, lipids, and proteins—are broken down to glucose, glycerol, fatty acids, and amino acids. Aided by enzymes and coenzymes, the cells use these products of digestion to build more complex compounds (anabolism) or break them down further to release energy (catabolism). The energy released during catabolism is often captured by high-energy compounds such as ATP.

coenzymes: small organic molecules that work with enzymes to facilitate the enzymes' activity. Many coenzymes have B vitamins as part of their structures (Figure 10-2 in Chapter 10 illustrates coenzyme action).
- **co** = with

TABLE 7-1 Metabolic Work of the Liver

The liver is the most active processing center in the body. When nutrients enter the body, the liver receives them first; then it metabolizes, packages, stores, or ships them out for use by other organs. When alcohol, drugs, or poisons enter the body, they are also sent directly to the liver; here they are detoxified and their by-products shipped out for excretion. An enthusiastic anatomy and physiology professor once remarked that given the many vital activities of the liver, we should express our feelings for others by saying, "I love you with all my liver," instead of with all my heart. Granted, this declaration lacks romance, but it makes a valid point. Here are just some of the many jobs performed by the liver.

Carbohydrates:

- Converts fructose and galactose to glucose.
- Makes and stores glycogen.
- Breaks down glycogen and releases glucose.
- Breaks down glucose for energy when needed.
- Makes glucose from some amino acids and glycerol when needed.
- Converts excess glucose to fatty acids.

Lipids:

- Builds and breaks down triglycerides, phospholipids, and cholesterol as needed.
- Breaks down fatty acids for energy when needed.
- Packages extra lipids in lipoproteins for transport to other body organs.
- Manufactures bile to send to the gallbladder for use in fat digestion.
- Makes ketone bodies when necessary.

Proteins:

- Manufactures nonessential amino acids that are in short supply.
- Removes from circulation amino acids that are present in excess of need and deaminates them or converts them to other amino acids.
- Removes ammonia from the blood and converts it to urea to be sent to the kidneys for excretion.
- Makes other nitrogen-containing compounds the body needs (such as bases used in DNA and RNA).
- Makes plasma proteins such as clotting factors.

Other:

- Detoxifies alcohol, other drugs, and poisons; prepares waste products for excretion.
- Helps dismantle old red blood cells and captures the iron for recycling.
- Stores most vitamins and many minerals.

To renew your appreciation for this remarkable organ, you might want to review Figure 3-12 on p. 79.

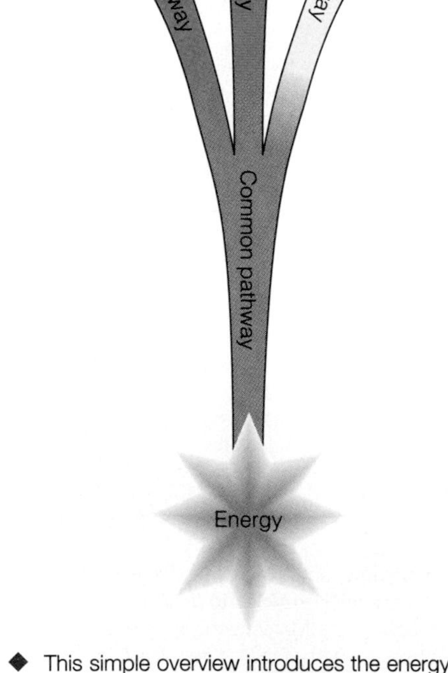

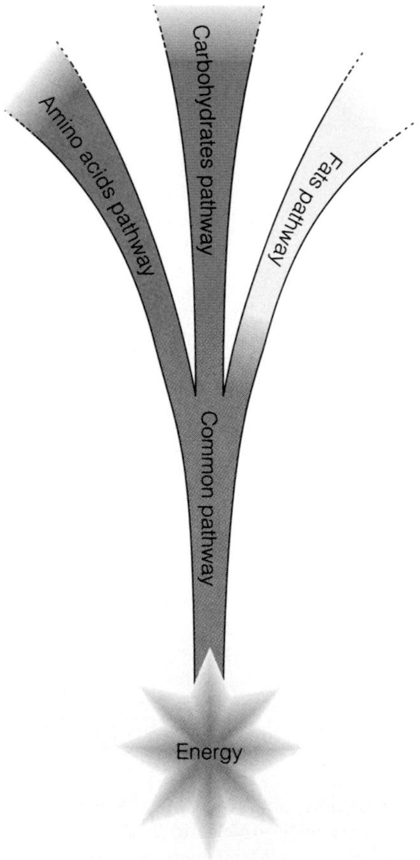

◆ This simple overview introduces the energy metabolism that is presented in the upcoming text and detailed in Figure 7-20 on p. 222.

BREAKING DOWN NUTRIENTS FOR ENERGY

With these introductory remarks in mind, it is time to enter a cell and follow the various paths that glucose, glycerol, fatty acids, and amino acids take to yield energy. As you will see, each starts down a different path, but they all reach a common destination.◆ At a certain point, they lose their individuality and most of their options—during catabolism all roads lead to energy.

Glucose, glycerol, fatty acids, and amino acids are the basic units derived from food, but a molecule of each of these compounds is made of still smaller units, the atoms—carbons, nitrogens, oxygens, and hydrogens. During catabolism, the body separates these atoms from one another. To follow this action, recall how many carbons are in the "backbones" of these compounds:

- Glucose has 6 carbons:

- Glycerol has 3 carbons:

- A fatty acid usually has an even number of carbons, commonly 18 carbons:

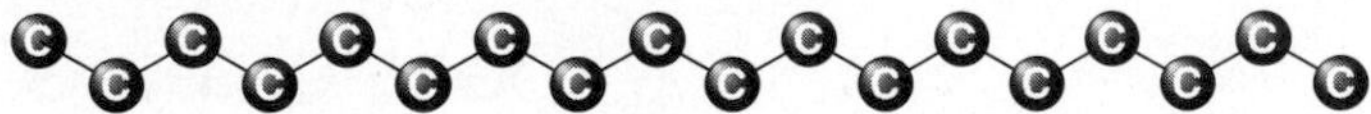

- An amino acid has 2, 3, or more carbons with a nitrogen attached:*

Full chemical structures and reactions appear both in the earlier chapters and in Appendix C; this chapter diagrams the reactions using just the compounds' carbon and nitrogen backbones.

What happens to these compounds inside cells can best be understood by starting with glucose. Two new names appear—**pyruvate** (a 3-carbon structure) and **acetyl CoA** (a 2-carbon structure with a coenzyme, **CoA,** attached)—and the rest of the story falls into place around them.† Two major points to notice in the following discussion:

- All compounds that can be converted to pyruvate can be used to make glucose.
- Compounds that are converted directly to acetyl CoA cannot make glucose.

Glucose

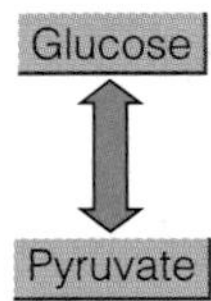

The first pathway glucose takes on its way to yield energy is called **glycolysis** (glucose splitting).‡ Figure 7-5 (on p. 210) shows a simplified drawing of glycolysis, which actually involves several steps and several enzymes (see Appendix C for details). Along the way, the 6-carbon glucose is split in half, forming two 3-carbon compounds. These 3-carbon compounds continue along the pathway until they are converted to pyruvate. Thus the net yield of one glucose molecule is two pyruvate molecules. If they continue breaking down, both pyruvate molecules will release much of their energy to form ATP molecules and some of their energy as heat.

• Glucose-to-Pyruvate, and Back Again • A cell can make glucose again from pyruvate in a process similar to the reversal of glycolysis. Making glucose requires energy, however, and a few different enzymes. Still, glucose is retrievable from pyruvate, so the arrows between glucose and pyruvate could point up as well as down.◆

• Glucose-to-Pyruvate, an Anaerobic Pathway • To start the process of splitting glucose to pyruvate, the cell must use a little energy, but then it produces more energy than it had to invest initially.§ No oxygen has been required thus far—that is, glycolysis is an **anaerobic** pathway. More energy can be released by taking pyruvate through additional metabolic reactions, but oxygen is needed for these reactions (they are **aerobic**). With sufficient oxygen, pyruvate molecules enter the mitochondria of the cell (review Figure 7-4) where they will be converted to acetyl CoA.

---

*The figures in this chapter usually show amino acids as compounds of 2, 3, or 5 carbons arranged in a straight line, but in reality amino acids may contain other numbers of carbons and assume other structural shapes (see Appendix C).

†The term *pyruvate* means a salt of *pyruvic acid.* (Throughout this book, the ending *-ate* is used interchangeably with *-ic acid;* for our purposes they mean the same thing.)

‡Glycolysis takes place in the cytoplasm of the cell (see Figure 7-4).

§The cell uses 2 ATP to begin the breakdown of glucose to pyruvate, but then gains 4 ATP for a net gain of 2 ATP.

pyruvate (PIE-roo-vate): pyruvic acid, a 3-carbon compound that plays a key role in energy metabolism.

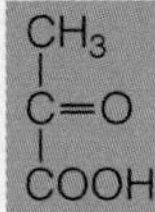

acetyl CoA (ASS-eh-teel, or ah-SEET-il, coh-AY): a 2-carbon compound (**acetate,** or **acetic acid,** shown in Figure 5-1 on p. 130) to which a molecule of CoA is attached.

CoA (coh-AY): coenzyme A; the coenzyme derived from the B vitamin pantothenic acid and central to energy metabolism.

glycolysis (gligh-COLL-ih-sis): the metabolic breakdown of glucose to pyruvate. Glycolysis does not require oxygen (anaerobic).

- **glyco** = glucose
- **lysis** = breakdown

anaerobic (AN-air-ROE-bic): not requiring oxygen.

- **an** = not

aerobic (air-ROE-bic): requiring oxygen.

FIGURE 7-5 **Glycolysis: The Glucose-to-Pyruvate Pathway (Anaerobic)**

Glycolysis occurs in anaerobic conditions (that is, it does not require oxygen). (If there is a shortage of oxygen, pyruvate is converted to lactic acid, as a later section of the text describes.)

Glycolysis begins with the splitting of the 6-carbon compound glucose into two interchangeable 3-carbon compounds. These compounds are converted to pyruvate in a series of reactions.

Any of the monosaccharides can enter the glycolysis pathway at various points.

Glycolysis ends with the production of pyruvate.

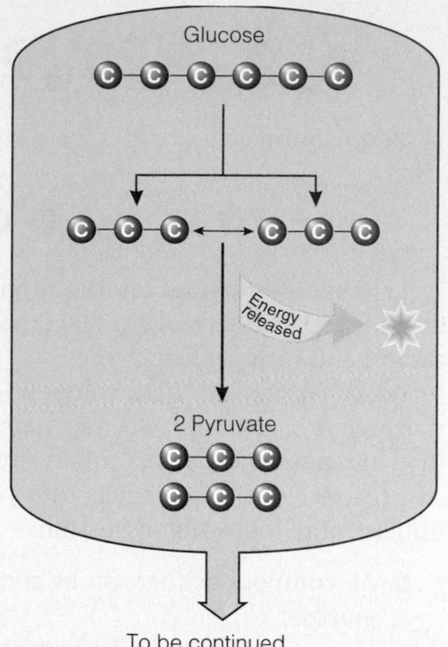

Note: These arrows point down indicating the breakdown of glucose to pyruvate during energy metabolism. Alternatively, the arrows could point up indicating the making of glucose from pyruvate, but that is not the focus of this discussion.

• **Pyruvate-to-Acetyl CoA** • If the cell needs energy and oxygen is available, it removes a carbon group (COOH) from the 3-carbon pyruvate to produce a 2-carbon compound that bonds with a molecule of CoA, becoming acetyl CoA. The carbon group from pyruvate becomes carbon dioxide, which is released into the blood, circulated to the lungs, and breathed out. Figure 7-6 diagrams the pyruvate-to-acetyl CoA reaction.

• **Glucose Retrieval via the Cori Cycle** • Alternatively, when less oxygen is available, pyruvate is converted to **lactic acid.** This anaerobic reaction occurs to a limited extent even at rest, but increases dramatically during high-intensity exercise—that is, whenever exertion exceeds the capacity of the heart and lungs to deliver oxygen to and clear carbon dioxide from the muscles. With limited oxygen available and limited carbon dioxide clearance, lactic acid accumulates in muscles, causing burning pain and fatigue. (To relieve this pain, relax the muscles frequently so that the circulating blood can carry the lactic acid away to the liver.) The liver can convert lactic acid to glucose in a recycling process called the **Cori cycle.** (Muscle cells cannot recycle lactic acid to glucose because they lack a necessary enzyme.)

• **Muscles' Needs for Oxygen** • Whenever carbohydrates, fats, or proteins are broken down to provide energy, oxygen is always ultimately involved in the process. The role of oxygen in metabolism is worth noticing, for it helps our understanding of physiology and metabolic reactions. The first pathway in glucose metabolism (glycolysis) yields some energy without oxygen (it is anaerobic), but anaerobic metabolism cannot be sustained for long. Conversely, the later pathways require oxygen (they are aerobic) and can be sustained for a long time. Aerobic metabolism yields by far the *most energy* and so is crucial for endurance activities.

lactic acid: lactate, a 3-carbon compound produced from pyruvate during anaerobic metabolism.

Cori cycle: the path from muscle glycogen to glucose to pyruvate to lactic acid (which travels to the liver) to glucose (which can travel back to the muscle) to glycogen; named after the scientist who elucidated this pathway.

• **Pyruvate-to-Acetyl CoA, an Irreversible Step** • The step from pyruvate to acetyl CoA is metabolically irreversible: a cell cannot retrieve the shed carbons from carbon dioxide to remake pyruvate, and then glucose. It is a one-way step and is therefore shown with only a "down" arrow in Figure 7-7 below. Notice that acetyl CoA can be used as a building block for fatty acids, but it cannot be used to make glucose or amino acids.

• **Acetyl CoA-to-Carbon Dioxide: The TCA Cycle** • Once made, acetyl CoA may take different metabolic paths, depending on the cell's needs. If the cell needs energy, acetyl CoA may proceed through a series of reactions known as the **TCA cycle** and **electron transport chain.** These reactions convert the 2-carbon acetyl CoA to two carbon dioxide molecules and free its coenzyme (CoA) to be reused (see Figure 7-8 on p. 212). In the process, much more energy is released than during glycolysis (more details are given later).

• **Acetyl CoA-to-Fat** • If energy is not needed, acetyl CoA will not enter the TCA cycle, but will be used to make fatty acids instead. This explains how carbohydrate, eaten in excess of the body's needs, can add to fat stores. As you will see, fat or protein eaten in excess of immediate energy needs can also take the same pathway to body fat.

FIGURE 7-6 Pyruvate-to-Acetyl CoA (Aerobic)

Each pyruvate loses a carbon as carbon dioxide and picks up a molecule of CoA, becoming acetyl CoA. The arrow goes only one way (down), because the step is not reversible. Result from 1 glucose: 2 carbon dioxide and 2 acetyl CoA.

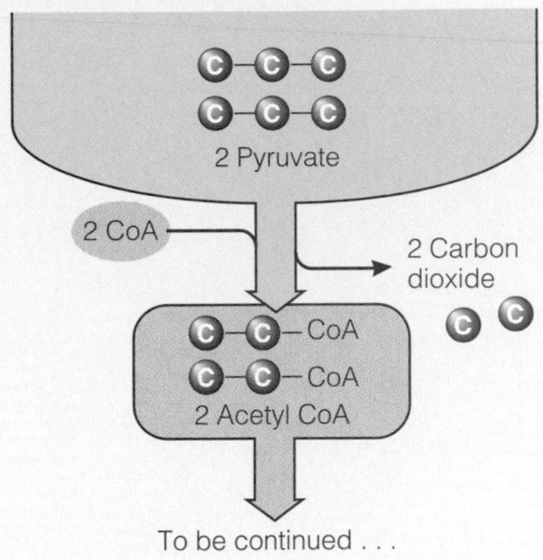

IN SUMMARY

Figure 7-9 on p. 213 combines the pathways presented in Figures 7-5, 7-6, and 7-8 and shows the whole sequence of steps in glucose breakdown. In summary, the main steps in the catabolism of glucose are:

Glucose to
pyruvate to
acetyl CoA to
carbon dioxide.

The first step—glucose-to-pyruvate—is anaerobic, but the later steps are aerobic and require oxygen. Keep in mind that glucose can be retrieved only from pyruvate or compounds earlier in the pathway. Once the commitment to acetyl CoA is made, glucose is not retrievable; acetyl CoA can go on to carbon dioxide, fat, or other compounds but not back to glucose.

Glycerol and Fatty Acids

Once glucose breakdown is understood, fat and protein breakdown are easily learned, for all three share a common metabolic pathway. Recall that triglycerides can break down to glycerol and fatty acids.

FIGURE 7-7 The Paths of Pyruvate and Acetyl CoA

Pyruvate may follow several reversible paths, but the path from pyruvate to acetyl CoA is irreversible.

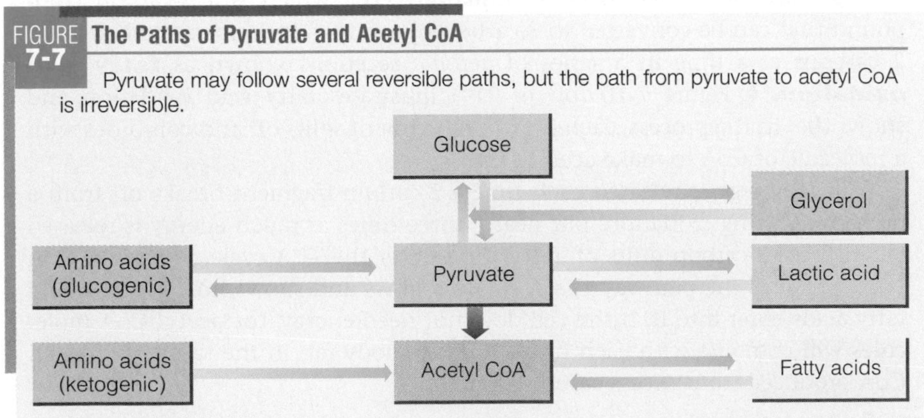

Note: Amino acids that can be used to make glucose are called *glucogenic;* amino acids that are converted to acetyl CoA are called *ketogenic.*

TCA cycle: tricarboxylic acid cycle; a series of metabolic reactions that break down molecules of acetyl CoA to carbon dioxide and hydrogen atoms (more details are provided later in the text); also called the **Krebs cycle** after the biochemist who elucidated its reactions.

electron transport chain (ETC): the final pathway in energy metabolism where the electrons from hydrogen are passed to oxygen and the energy released is trapped in the bonds of ATP.

FIGURE 7-8 The Breakdown of Acetyl CoA

The complete oxidation of acetyl CoA is accomplished through the reactions of the TCA (tricarboxylic acid) cycle and the electron transport chain. In the TCA cycle, the acetyl CoA carbons are converted to carbon dioxide. Each CoA returns to pick up another acetate (coming from glucose, glycerol, fatty acids, and amino acids).

The net result is that acetyl CoA splits, the carbons combine with oxygen, and the energy originally in the acetyl CoA is stored in ATP and similar compounds, thus becoming available for the body's use. Chapter 10 describes how the B vitamin coenzymes participate in these metabolic pathways. For more details, see the text and Appendix C.

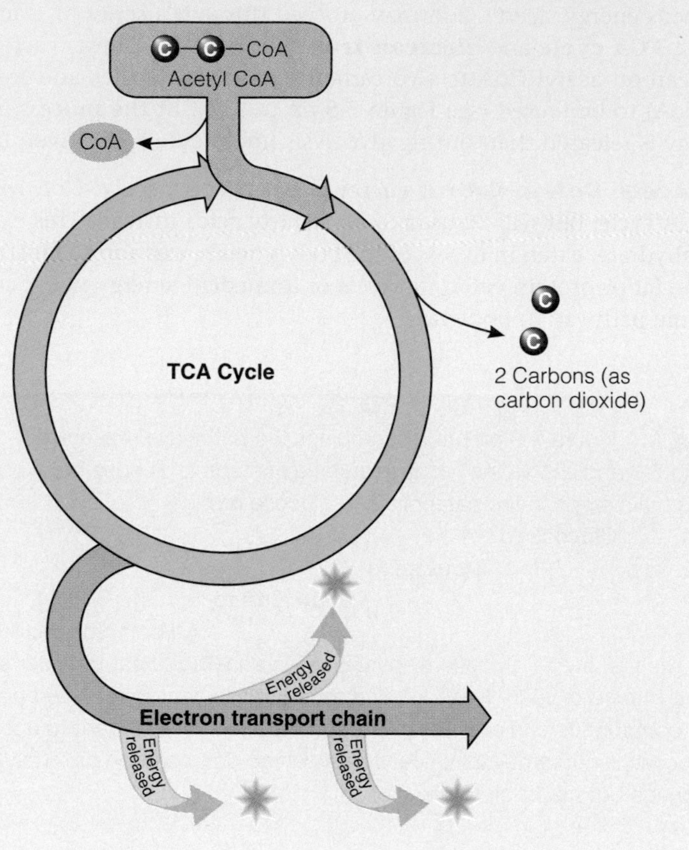

◆ 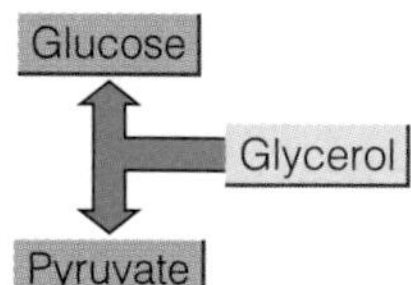

◆ Note: The *oxidation* of energy nutrients refers to the metabolic reactions that lead to the production of energy.

fatty acid oxidation: the metabolic breakdown of fatty acids to acetyl CoA; also called **beta oxidation.**

•Glycerol-to-Pyruvate • Glycerol (a 3-carbon compound like pyruvate, but with a different arrangement of H and OH on the C) is easily converted to another 3-carbon compound. This compound may go either "up" the pathway to form glucose or "down" to form pyruvate◆ and acetyl CoA and, finally, carbon dioxide.

•Fatty Acids-to-Acetyl CoA • Unlike glycerol, which is a 3-carbon compound that can be converted to 3-carbon pyruvate, fatty acids are taken apart 2 carbons at a time in a series of aerobic reactions known as **fatty acid oxidation.***◆ Figure 7-10 (on p. 214) illustrates fatty acid oxidation and shows that in the process, each 2-carbon fragment splits off and combines with a molecule of CoA to make acetyl CoA.

A little energy is released each time a 2-carbon fragment breaks off from a fatty acid during oxidation, but nearly three times as much energy is released when these 2-carbon units of acetyl CoA enter the TCA cycle. Figure 7-11 on p. 215 repeats the pathway that glucose follows and shows how glycerol and fatty acids enter into it. If the cell does not need energy, the acetyl CoA molecules will combine with each other to make body fat, in the same way acetyl CoA produced from excess carbohydrate does.

•Glucose Not Retrievable from Fatty Acids • Cells can make glucose from pyruvate and other 3-carbon compounds, such as glycerol, but they cannot

*Oxidation of fatty acids occurs in the mitochondria of the cells (see Figure 7-4).

FIGURE 7-9 **Glucose-to-Energy Pathway**

Through these processes, energy from glucose is made available to do the cells' work. Ultimately, glucose is completely disassembled to single-carbon fragments, and the fragments are combined with oxygen to form carbon dioxide. Much of the energy released is trapped and stored in ATP. Details of the TCA cycle and the electron transport chain are given later and in Appendix C.

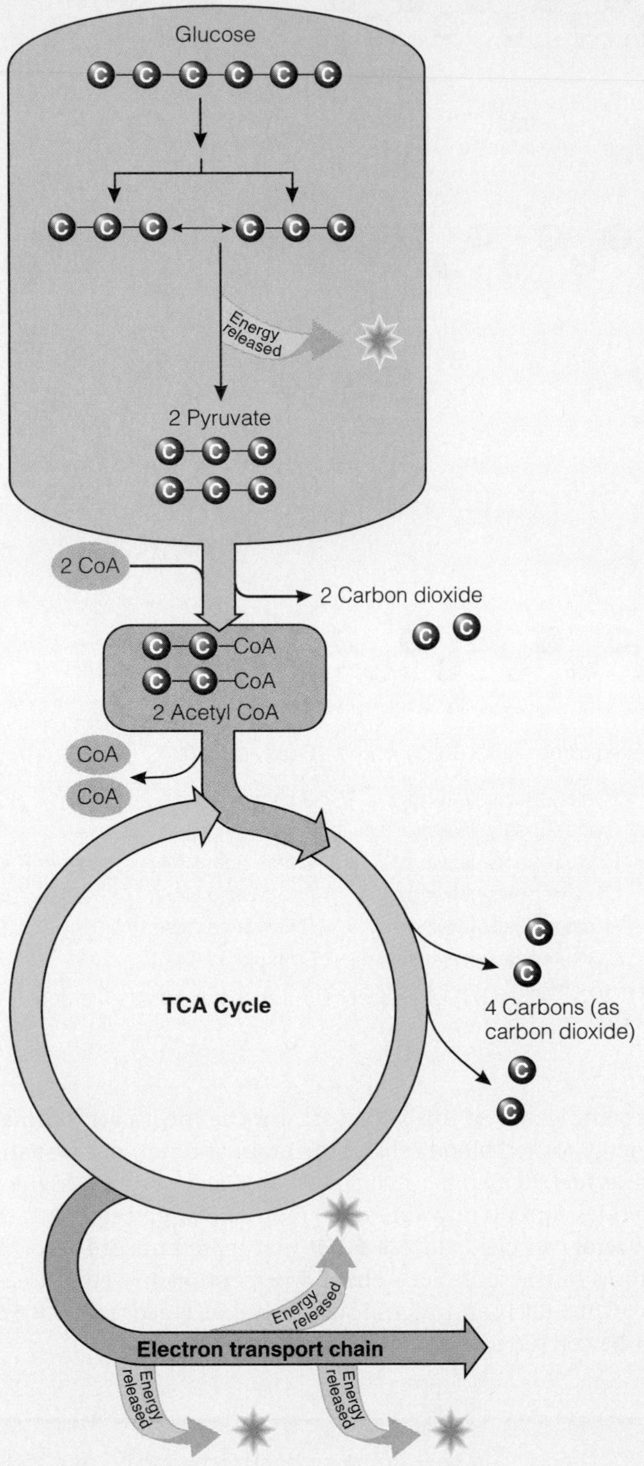

Glucose
Energy released
2 Pyruvate
2 CoA
2 Carbon dioxide
2 Acetyl CoA
CoA
CoA
TCA Cycle
4 Carbons (as carbon dioxide)
Energy released
Electron transport chain
Energy released
Energy released

make glucose from the 2-carbon fragments of fatty acids. In chemical diagrams, the arrow between pyruvate and acetyl CoA always points only one way—down—and fatty acid fragments enter the metabolic path below this arrow (review Figure 7-7 on p. 211). Thus fatty acids cannot be used to make glucose.

FIGURE 7-10 Fatty Acid Oxidation

During oxidation, fatty acids are taken apart to 2-carbon fragments that combine with CoA to make acetyl CoA. Fatty acid oxidation is a series of aerobic reactions (that is, it requires oxygen).

16-C fatty acid

The fatty acid is first activated by coenzyme A.

CoA

Energy released

Energy released

A little energy is released each time a carbon-carbon bond is cleaved.

CoA

Another CoA joins the chain, and the bond at the second carbon (the beta-carbon) weakens. Acetyl CoA splits off, leaving a fatty acid that is two carbons shorter.

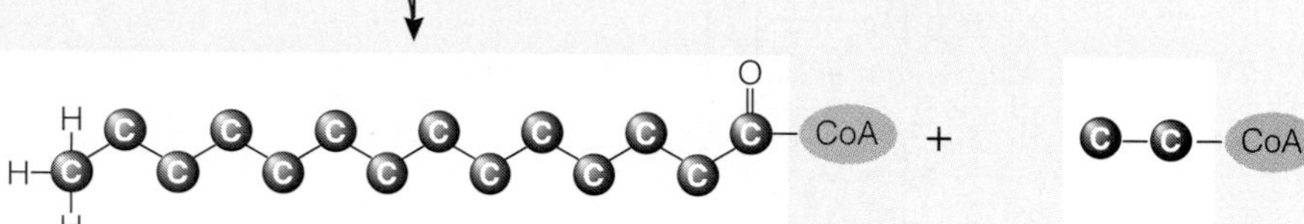

Net result from a 16-C fatty acid:	14-C fatty acid CoA	+	1 acetyl CoA
Cycle repeats, leaving:	12-C fatty acid CoA	+	2 acetyl CoA
Cycle repeats, leaving:	10-C fatty acid CoA	+	3 acetyl CoA
Cycle repeats, leaving:	8-C fatty acid CoA	+	4 acetyl CoA
Cycle repeats, leaving:	6-C fatty acid CoA	+	5 acetyl CoA
Cycle repeats, leaving:	4-C fatty acid CoA	+	6 acetyl CoA
Cycle repeats, leaving:	2-C fatty acid CoA*	+	7 acetyl CoA

The shorter fatty acid enters the pathway and the cycle repeats. The molecules of acetyl CoA enter the TCA cycle, yielding abundant energy.

*Notice that 2-C fatty acid CoA = acetyl CoA, so that the final yield from a 16-C fatty acid is 8 acetyl CoA.

The significance of this is that fat, for the most part, normally cannot provide energy for red blood cells or the brain and nervous system, which require glucose as fuel. Remember that almost all dietary fats are triglycerides, and that triglycerides contain only one small molecule of glycerol with three fatty acids. The glycerol can yield glucose,◆ but that represents only 3 of the 50 or so carbon atoms in the molecule—about 5 percent of its weight (see Figure 7-12 on p. 216). Thus fat is an insignificant source of glucose; about 95 percent of fat cannot be converted to glucose.

◆ **Reminder: The making of glucose from the glycerol of triglycerides (or from amino acids) is *gluconeogenesis.* About 5% of fat (the glycerol portion of a triglyceride) and most amino acids can be converted to glucose (review Figure 7-7 on p. 211).**

IN SUMMARY

The body can convert the small glycerol portion of a triglyceride to either pyruvate (and then glucose) or acetyl CoA. The fatty acids of a triglyceride, on the other hand, cannot make glucose, but they can provide acetyl CoA. Acetyl CoA from either source may then enter the TCA cycle to produce energy or combine with other molecules of acetyl CoA to make body fat.

FIGURE 7-11 **Fats-to-Energy Pathway**

Glycerol enters the metabolic path about midway between glucose and pyruvate and can be converted to either; fatty acids are broken down into 2-carbon fragments that combine with CoA to form acetyl CoA. Net from an 18-carbon fatty acid: 9 acetyl CoA molecules, which are converted to 18 carbon dioxide molecules. Notice that you have seen parts of this figure before—in Figure 7-9.

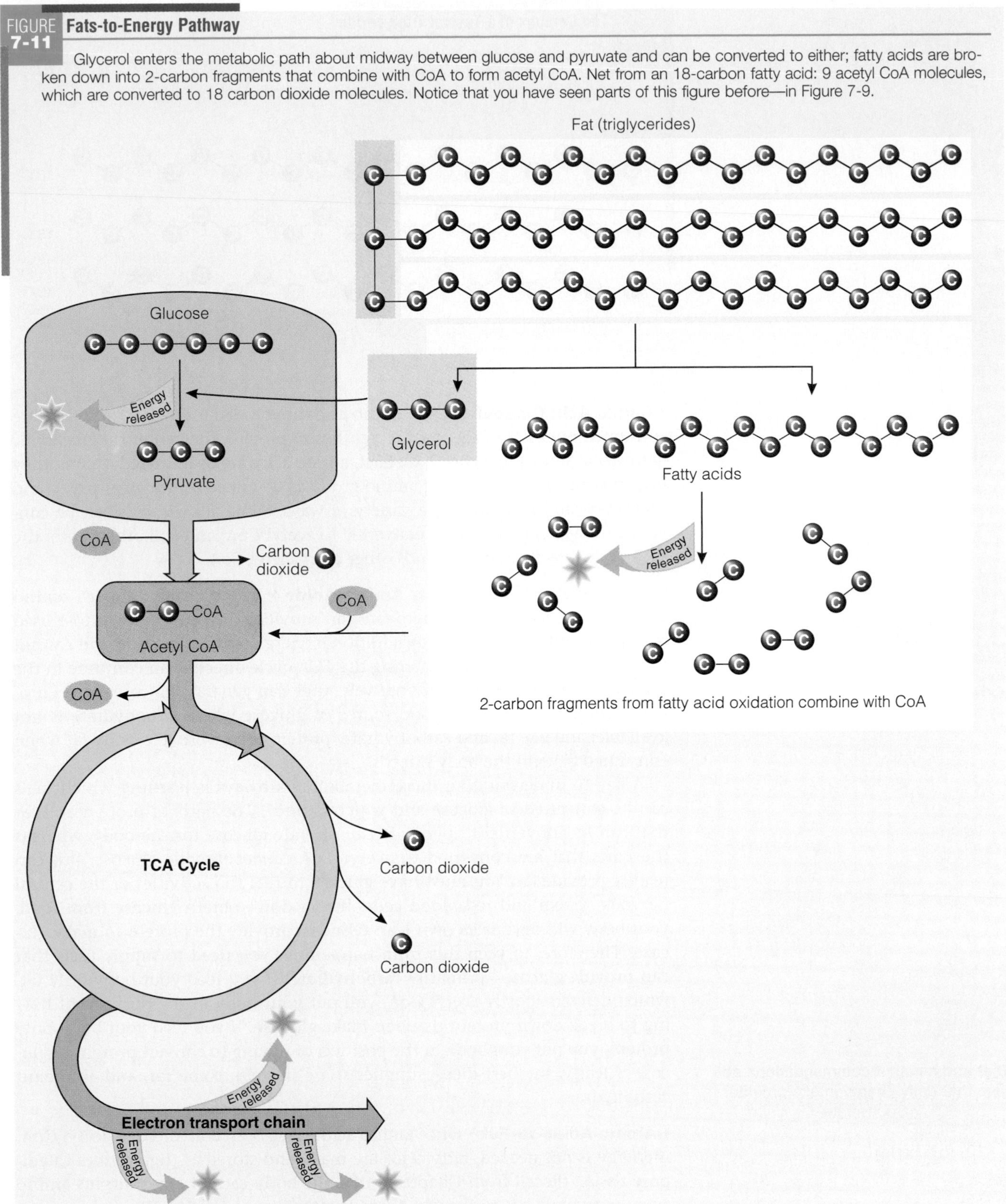

Amino Acids

The preceding two sections have shown how the breakdown of carbohydrate and fat provides energy for the body's use. One energy-yielding nutrient remains: protein or, rather, the amino acids of protein.

FIGURE 7-12 **The Carbons of a Typical Triglyceride**

A typical triglyceride contains only one small molecule of glycerol (3 C), but has three fatty acids (each about 18 C on the average, or about 54 C). Only the glycerol portion of a triglyceride can yield glucose, making fat an insignificant source.

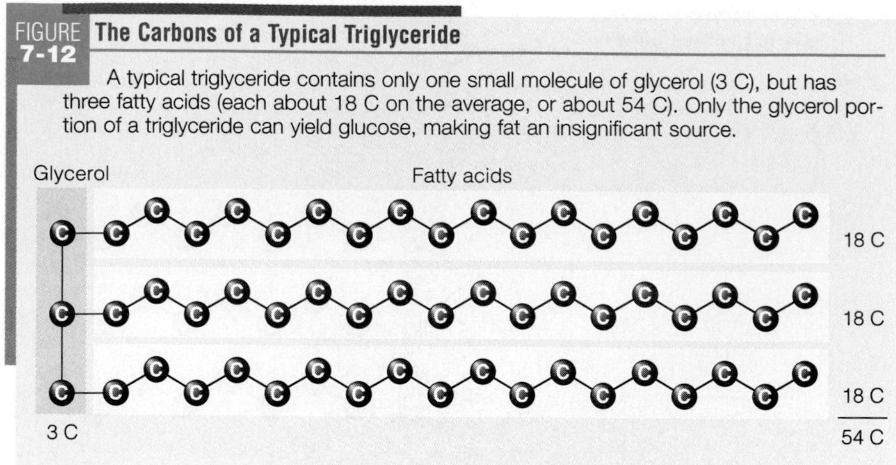

• Amino Acid Catabolism • If amino acids are needed for energy, or if they are consumed in excess of the need to synthesize protein, they enter the metabolic pathway as shown in Figure 7-13. First, amino acids are deaminated (that is, they lose their nitrogen-containing amino group as described on the next page), and then they are catabolized in a variety of ways. Some amino acids can be converted to pyruvate; others are converted to acetyl CoA; and still others enter the TCA cycle directly as compounds other than acetyl CoA.

• Glucose Retrievable from Amino Acids • As you might expect, amino acids that are used to make pyruvate can provide glucose, whereas those used to make acetyl CoA can provide additional energy or make body fat but cannot make glucose. Amino acids entering the TCA cycle directly can continue in the cycle and generate energy; alternatively, they can generate glucose.[1] Thus protein, unlike fat, is a fairly good source of glucose when carbohydrate is not available; and like fat and carbohydrate, protein is converted to body fat when consumed beyond the body's needs.

A key to understanding these metabolic pathways is learning which fuels can be converted to glucose and which cannot. The parts of protein and fat that can be converted to pyruvate *can* provide glucose for the body, whereas the parts that are converted to acetyl CoA *cannot* provide glucose, but can readily provide fat. You must have glucose to fuel the activities of the central nervous system and red blood cells. If you don't obtain glucose from food, your body will devour its own lean tissue to provide the protein to make glucose. Therefore, to keep this from happening, you need to supply fuels that can provide glucose—primarily carbohydrate. If you feed your body only fat, which delivers mostly acetyl CoA, you put your body in the position of having to break down protein tissue to make glucose. If you feed your body only protein, you put your body in the position of having to convert protein to glucose. Clearly, the best diet◆ supplies some protein, some fat, and abundant carbohydrate.

◆ **Diet and health recommendations advise that daily energy intake provide:**
- 55–60% carbohydrate.
- ≤30% fat.
- 10–15% protein.

• Amino Acids-to-Fat • Once amino acids have been converted to acetyl CoA, if energy is not needed, fatty acids are made and stored as triglycerides in adipose tissue. (Recall from Chapter 6 that the body cannot store surplus amino acids as such; it has to convert them to other compounds.) Thus protein can also add to fat stores if eaten in excess.

People who eat huge portions of meat and other protein-rich foods may wonder why they have weight problems. Not only does the fat in those foods lead to fat storage, but the protein can, too, when energy intake exceeds energy needs. Many fad weight-loss diets encourage high protein intakes based on the false assumption that protein builds only muscle, not fat (see Highlight 8 for more details).

FIGURE 7-13 **Amino Acids-to-Energy Pathway**

Notice that you have seen parts of this figure before—in Figures 7-9 and 7-11. Note: The arrows from pyruvate and the TCA cycle to amino acids are possible only for *nonessential* amino acids; remember, the body cannot make essential amino acids.

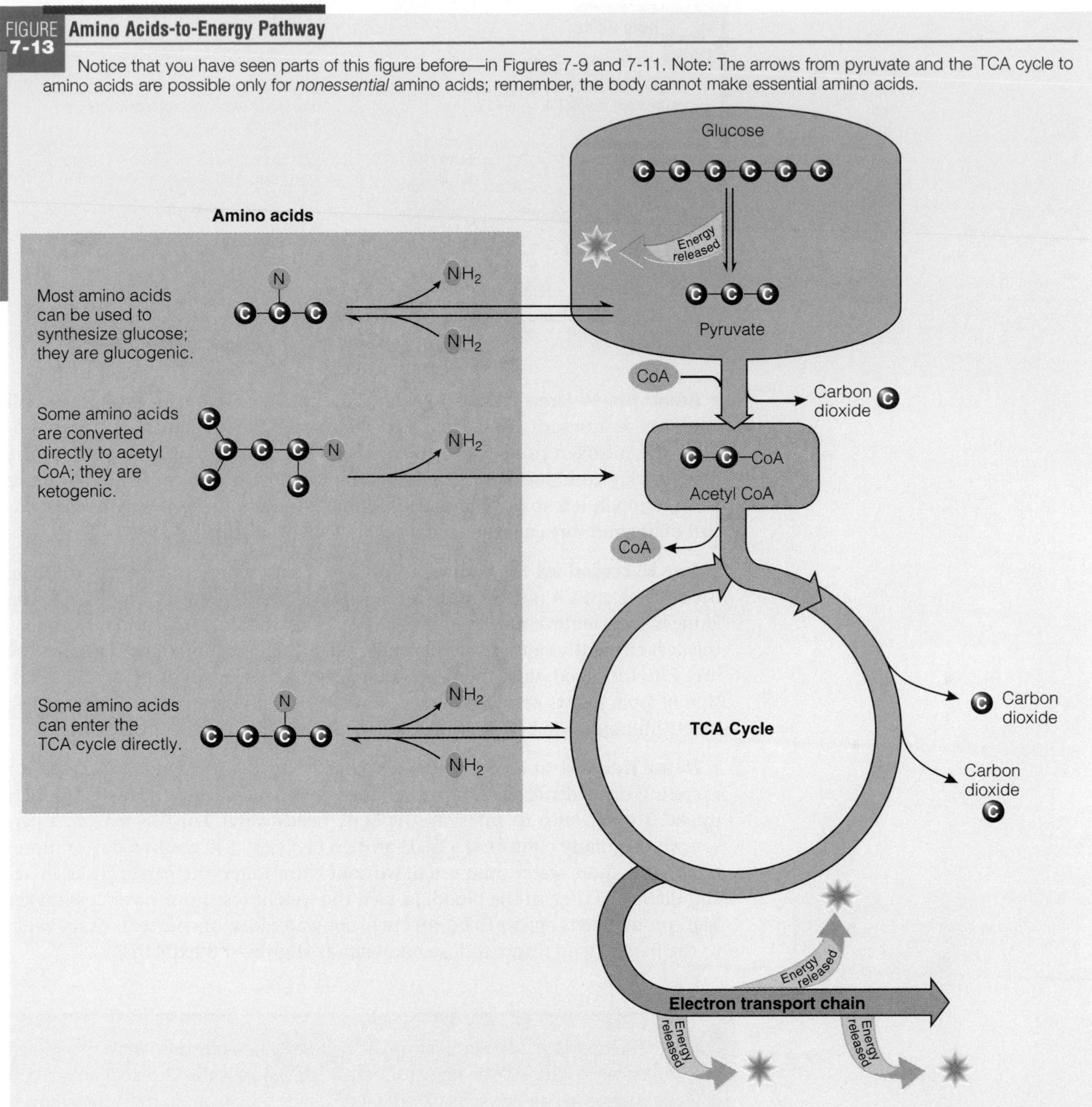

• **Deamination** • When amino acids are metabolized for energy or used to make fat, they must be deaminated first. Two products result from deamination. One is, of course, the carbon structure without its amino group—often a **keto acid** (see Figure 7-14 on p. 218). The other product is **ammonia,** a toxic compound chemically identical to the strong-smelling ammonia in bottled cleaning solutions. Ammonia is a base, and if the body produces larger quantities than it can handle, the blood's critical acid-base balance becomes upset.

• **Transamination** • As the discussion of protein in Chapter 6 pointed out, only some amino acids are essential; others can be made in the body, given a source of nitrogen. By transferring an amino group from one amino acid to its corresponding keto acid, cells can make a new amino acid and a new keto acid, as shown in Figure 7-15. Through many such **transamination** reactions, involving many different keto acids, the liver cells can synthesize the nonessential amino acids.

keto (KEY-toe) **acid:** an organic acid that contains a carbonyl group (C=O).

ammonia: a compound with the chemical formula NH_3; produced during the deamination of amino acids.

transamination (TRANS-am-ih-NAY-shun): the transfer of an amino group from one amino acid to a keto acid, producing a new nonessential amino acid and a new keto acid.

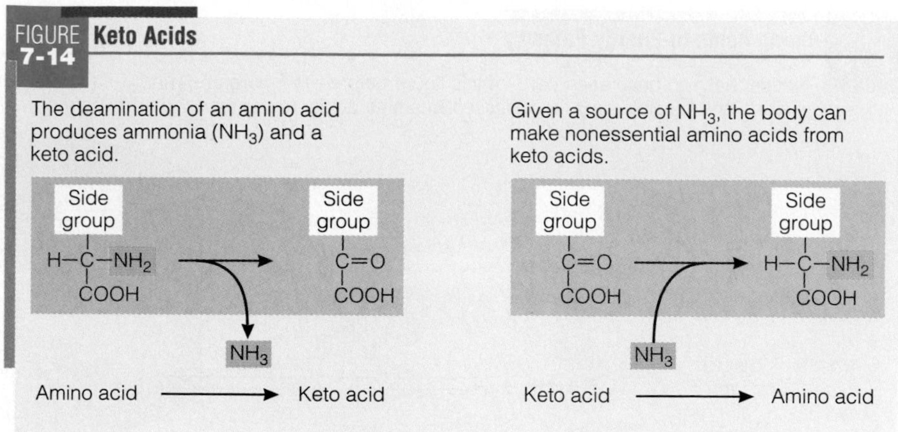

FIGURE 7-14 Keto Acids

The deamination of an amino acid produces ammonia (NH_3) and a keto acid.

Given a source of NH_3, the body can make nonessential amino acids from keto acids.

Side group
H–C–NH2
COOH
Side group
C=O
COOH
NH3
Amino acid → Keto acid
Side group
C=O
COOH
Side group
H–C–NH2
COOH
NH3
Keto acid → Amino acid

• **Ammonia-to-Urea in the Liver** • The liver continuously produces small amounts of ammonia in deamination reactions. Some of this ammonia provides the nitrogen needed for the synthesis of nonessential amino acids. The liver quickly combines any remaining ammonia with carbon dioxide to make **urea,** a much less toxic compound. Figure 7-16 provides a greatly oversimplified diagram of urea synthesis; details are shown in Appendix C.

• **Urea Excreted via the Kidneys** • Liver cells release urea into the blood, where it circulates until it passes through the kidneys (see Figure 7-17 on p. 220). The kidneys then remove urea from the blood for excretion in the urine. Normally, the liver efficiently captures all the ammonia, makes urea from it, and releases the urea into the blood; then the kidneys clear all the urea from the blood. This division of labor allows easy diagnosis of diseases of both organs. If the liver is sick, blood ammonia will be high; if the kidneys are sick, blood urea will be high.

• **Water Needed to Excrete Urea** • Urea is the body's principal vehicle for excreting unused nitrogen, and the amount produced increases with protein intake. To keep urea in solution, the body needs water. For this reason, a person who regularly consumes a high-protein diet (say, 100 grams a day or more) must drink more water than usual; without extra water, the person risks an accumulation of urea in the blood. In fact, the weight loss from water loss makes high-protein diets *appear* to be effective, but water loss, of course, is of no value to the person who wants to lose body fat (as Highlight 8 explains).

IN SUMMARY

The body can use some amino acids to produce glucose, while others can be used either to generate energy or to make fat. Before an amino acid enters one of these metabolic pathways, its nitrogen-containing amino group must be removed through deamination. Some of the nitrogen may be used to make nonessential amino acids and other nitrogen-containing compounds; the rest is cleared from the body via urea synthesis in the liver and excretion in the kidneys.

FIGURE 7-15 Transamination to Make a Nonessential Amino Acid

The body can transfer amino groups (NH_2) from an amino acid to a keto acid, forming a new nonessential amino acid and a new keto acid.

Transamination reactions require the vitamin B_6 coenzyme.

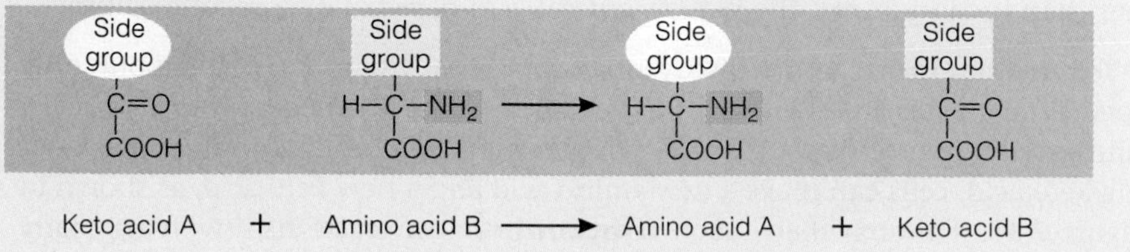

Side group
C=O
COOH
Side group
H–C–NH2
COOH
Side group
H–C–NH2
COOH
Side group
C=O
COOH
Keto acid A + Amino acid B → Amino acid A + Keto acid B

IN SUMMARY To review the ways the body can use the energy-yielding nutrients, see the summary table below. To obtain energy, the body uses glucose and fatty acids as its primary fuels, although it can use amino acids to provide energy if need be. To make glucose, the body can use all carbohydrates and most amino acids, but it can convert only 5 percent of fat (the glycerol portion) to glucose. To make body proteins, the body needs amino acids. It can use glucose to make some nonessential amino acids when nitrogen is available; it cannot use fats to make body proteins. Finally, when energy is consumed beyond the body's needs, the body can convert all three energy-yielding nutrients to fat for storage.

The Body's Use of Energy-Yielding Nutrients

Nutrient	Yields Energy	Yields Glucose	Yields Amino Acids and Body Proteins	Yields Fat Stores
Carbohydrates (glucose)	Yes	Yes	Yes—when nitrogen is available, can yield nonessential amino acids	Yes
Lipids (triglycerides)	Yes	No—glycerol provides minimal amount	No	Yes
Proteins (amino acids)	Yes—if needed	Yes—when carbohydrate is unavailable	Yes	Yes

FIGURE 7-16 Urea Synthesis

When amino nitrogen is stripped from amino acids, ammonia is produced. The liver detoxifies ammonia before releasing it into the bloodstream by combining it with another waste product, carbon dioxide, to produce urea. See Appendix C for details.

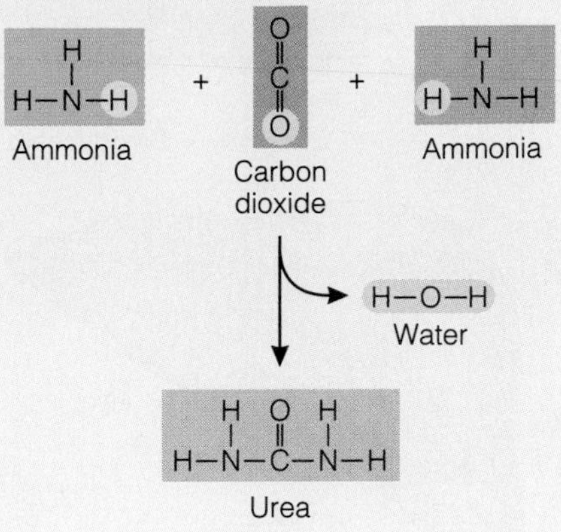

The Final Steps of Catabolism

Thus far the discussion has followed each of the energy-yielding nutrients down three different pathways. All lead to the point where acetyl CoA enters the TCA cycle.*

• **The TCA Cycle** • The TCA cycle serves as a busy traffic center through which these 2-carbon acetyl CoA molecules pass on their way to carbon dioxide, releasing their energy to other compounds as they go. The TCA cycle is called a cycle, but that doesn't mean it regenerates acetyl CoA. Acetyl CoA goes one way only—to carbon dioxide, releasing energy as it goes. The TCA cycle is a circular path, though, in the sense that a 4-carbon compound does cycle around and around.†

The first 4-carbon compound to enter the cycle is known as **oxaloacetate,** and it is made primarily from pyruvate, although it can be made from certain amino acids. Importantly, it cannot be made from fat. Thus a diet that provides ample carbohydrate ensures an adequate supply of oxaloacetate (because glucose produces pyruvate during glycolysis).

Oxaloacetate's role in replenishing the TCA cycle is critical. When oxaloacetate is insufficient, the TCA cycle slows down, and the cells face an energy crisis. That oxaloacetate must be available for acetyl CoA to enter the TCA cycle underscores the importance of carbohydrates in the diet. (Highlight 8 presents more information on the consequences of low-carbohydrate diets.)

*The TCA cycle reactions take place in the mitochondria of the cell (see Figure 7-4).

†Actually, the 4-carbon compound (oxaloacetate) does not cycle around as the same structure throughout; instead it starts a series of reactions. On picking up acetyl CoA, it becomes a 6-carbon compound. On dropping off carbon dioxide, it becomes a 5- and then a 4-carbon compound. Each reaction changes the structure slightly until finally the original 4-carbon compound forms again and picks up another acetyl CoA, starting the series of reactions over again.

The carbons that enter the cycle in acetyl CoA may not be the ones that are given off as carbon dioxide. In one of the steps of the cycle, a 6-carbon compound of the cycle becomes symmetrical, both ends being identical. Thereafter it loses carbons to carbon dioxide at one end or the other. Thus only half of the carbons from acetyl CoA are given off as carbon dioxide in any one turn of the cycle; the other half become part of the compound that returns to pick up another acetyl CoA. It is true to say, though, that for each acetyl CoA that enters the TCA cycle, 2 carbons are given off as carbon dioxide. It is also true that with each turn of the cycle the energy equivalent of one acetyl CoA is released.

urea (you-REE-uh): the principal nitrogen-excretion product of metabolism. Two ammonia fragments are combined with carbon dioxide to form urea.

oxaloacetate (OKS-ah-low-AS-eh-tate): a carbohydrate intermediate of the TCA cycle.

FIGURE 7-17 Urea Excretion

The liver and kidneys both play a role in disposing of excess nitrogen. Can you see why the person with liver disease has high blood ammonia, while the person with kidney disease has high blood urea? (Figure 12-2 provides details of how the kidneys work.)

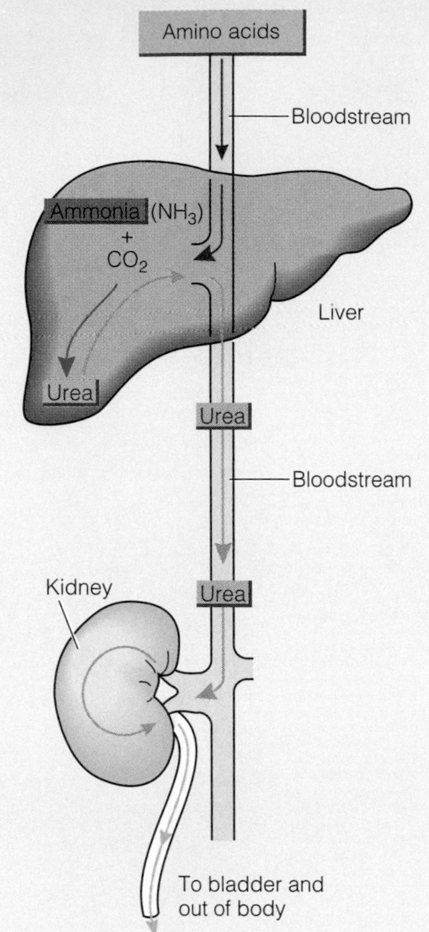

When oxaloacetate is available, it picks up acetyl CoA (a 2-carbon compound), drops off one carbon (as carbon dioxide), then another carbon (as carbon dioxide), and returns to pick up another acetyl CoA. As for the acetyl CoA, its carbons go only one way—to carbon dioxide (see Appendix C for additional details).

As acetyl CoA molecules break down to carbon dioxide, hydrogen atoms with their electrons are removed from the compounds in the cycle. Coenzymes made from the B vitamins niacin and riboflavin receive the hydrogens and their electrons and transfer them to the electron transport chain.

• The Electron Transport Chain • The electron transport chain (ETC) consists of a series of proteins that serve as electron "carriers." These carriers are mounted in sequence on a membrane inside the energy-generating structures within the cell known as mitochondria (review Figure 7-4). As each carrier receives electrons, it releases a little energy and passes the electrons on to the next carrier. While some of the energy is released as heat, much of it is captured in the bonds of ATP molecules. These electron-transferring molecules continue passing electrons and giving up energy until, at the end of the chain, any usable energy has been captured in the body's ATP molecules. In the last step, the low-energy electrons with their hydrogen atoms (H) combine with oxygen (O) from the lungs, forming water (H_2O), from which the body cannot extract any more energy. Everyone knows that oxygen is essential to life—now you understand why. The TCA cycle and the ETC represent the body's most efficient means of capturing the energy from nutrients and transferring it into the bonds of ATP. Figure 7-18 provides a simple diagram of the ETC; see Appendix C for details.

• The kCalories per Gram Secret Revealed • Of the three energy-yielding nutrients, fat provides the most energy per gram. The reason may be apparent from Figure 7-19, which compares a fatty acid molecule with a glucose molecule. Notice that nearly all the bonds in a fatty acid molecule are between carbons and hydrogens. Oxygen can be added to all of them (forming carbon dioxide with the carbons, and water with the hydrogens). As this happens, the energy in the bonds is released. In glucose, on the other hand, an oxygen is already bonded to each carbon; thus there is less potential for oxidation, and less energy is released when the remaining bonds are broken.

FIGURE 7-18 Electron Transport Chain

An important concept to remember is that an electron is not a fixed amount of energy. The electrons that bond the hydrogens to the B vitamin coenzymes have a relatively large amount of energy. In the series of reactions that follow, they lose this energy in small amounts, until at the end the electrons with their hydrogens (H) are attached to oxygen (O) to make water (H_2O). In some of the steps, the energy the electrons lose is captured into ATP in coupled reactions. Appendix C provides a more detailed explanation.

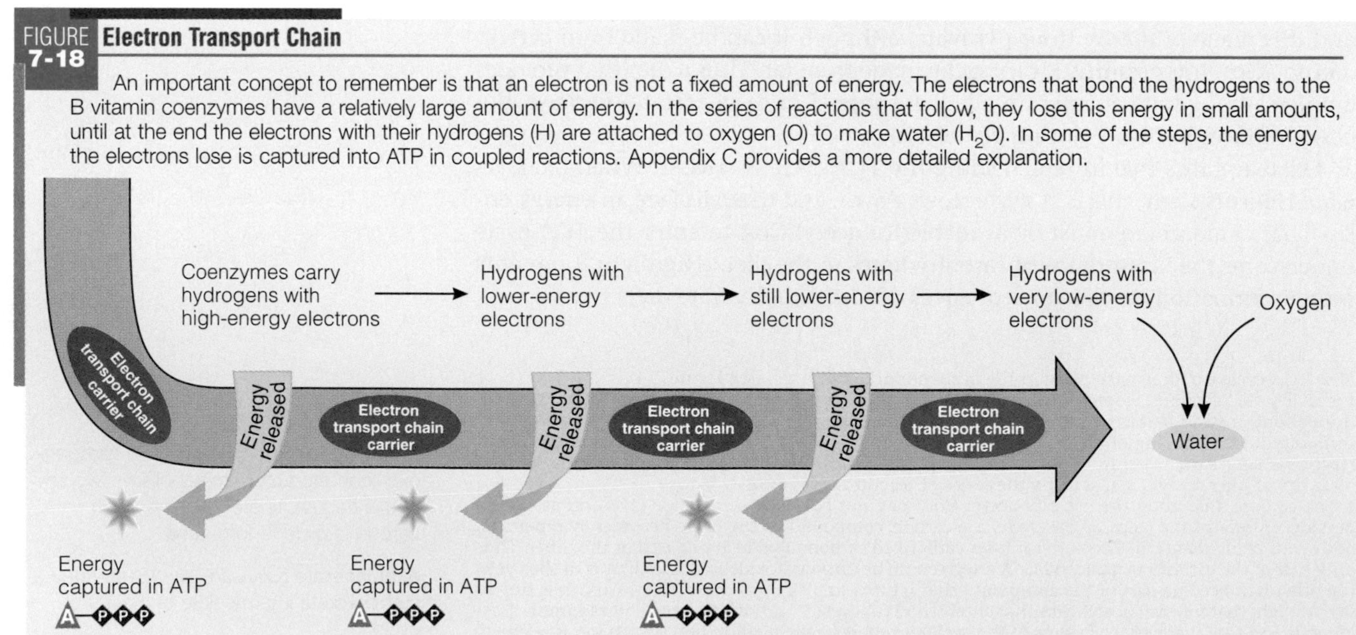

FIGURE 7-19 **Chemical Structures of a Fatty Acid and Glucose Compared**

To ease comparison, the structure shown here for glucose is not the ring structure shown in Chapter 4, but an alternative way of drawing its chemical structure.

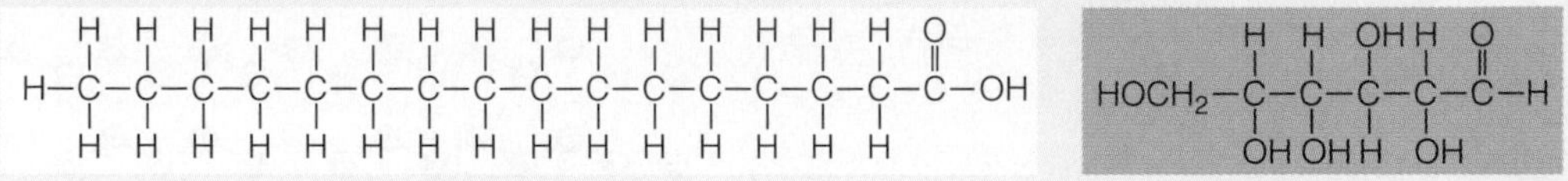

```
  H H H H H H H H H H H H H H H O
  | | | | | | | | | | | | | | | ||
H-C-C-C-C-C-C-C-C-C-C-C-C-C-C-C-C-OH
  | | | | | | | | | | | | | | |
  H H H H H H H H H H H H H H H
```

Fatty acid

```
        H  H  OH H  O
        |  |  |  |  ||
HOCH2-C--C--C--C--C-H
        |  |  |  |
        OH OH H  OH
```

Glucose

Because fat contains many carbon-hydrogen bonds that can be readily oxidized, it generates abundant ATP during oxidation. This explains why fat yields more kcalories per gram than carbohydrate or protein. (Remember that each ATP holds energy and that kcalories measure energy; thus the more ATP generated, the more kcalories have been collected.) For example, one glucose molecule will yield 38 ATP when completely oxidized. In comparison, one 16-carbon fatty acid molecule will yield 129 ATP when completely oxidized. Gram for gram, fat can provide much more energy than either of the other two energy-yielding nutrients, making it the body's preferred form of energy storage.

IN SUMMARY

After a balanced meal, the body handles the nutrients as follows. The digestion of carbohydrate yields glucose; some is stored as glycogen, and some is broken down to pyruvate and acetyl CoA to provide energy. The acetyl CoA can then enter the TCA cycle and ETC to provide more energy. The digestion of fat yields glycerol and fatty acids; some are reassembled and stored as fat, and others are broken down to acetyl CoA, enter the TCA cycle and ETC, and provide energy. The digestion of protein yields amino acids, some of which are used to build body protein. If there is a surplus, however, or if not enough carbohydrate and fat are available to meet energy needs, some amino acids are broken down through the same pathways as glucose to provide energy. Other amino acids enter directly into the TCA cycle, and these, too, can be broken down to yield energy. In summary, although carbohydrate, fat, and protein enter the TCA cycle by different routes, the energy-generating pathways that follow are common to all energy-yielding nutrients (see Figure 7-20).

THE BODY'S ENERGY BUDGET

Every day, a healthy diet delivers thousands of kcalories from foods, and the body uses most of them to do its work. As a result, body weight changes little, if at all. This remarkable achievement, which many people manage without even thinking about it, could be called the economy of maintenance. The body's energy budget is balanced. Some people, however, eat too much and get fat; others eat too little and get thin. The metabolic details have already been described; the next sections will review them from the perspective of the body fat gained or lost. The possible reasons why people gain or lose weight are explored in Chapter 8.

◆ If you live for 65 years or longer, you will have consumed more than 70,000 meals and disposed of 50 tons of foods.◆

The Economics of Feasting

When a person eats too much, metabolism favors fat formation. Fat cells enlarge and multiply regardless of whether the excess derives from protein, carbohydrate, or fat. The pathway from dietary fat to body fat, however, is the most direct (requiring only a few metabolic steps) and the most efficient

FIGURE 7-20 The Central Pathways of Energy Metabolism

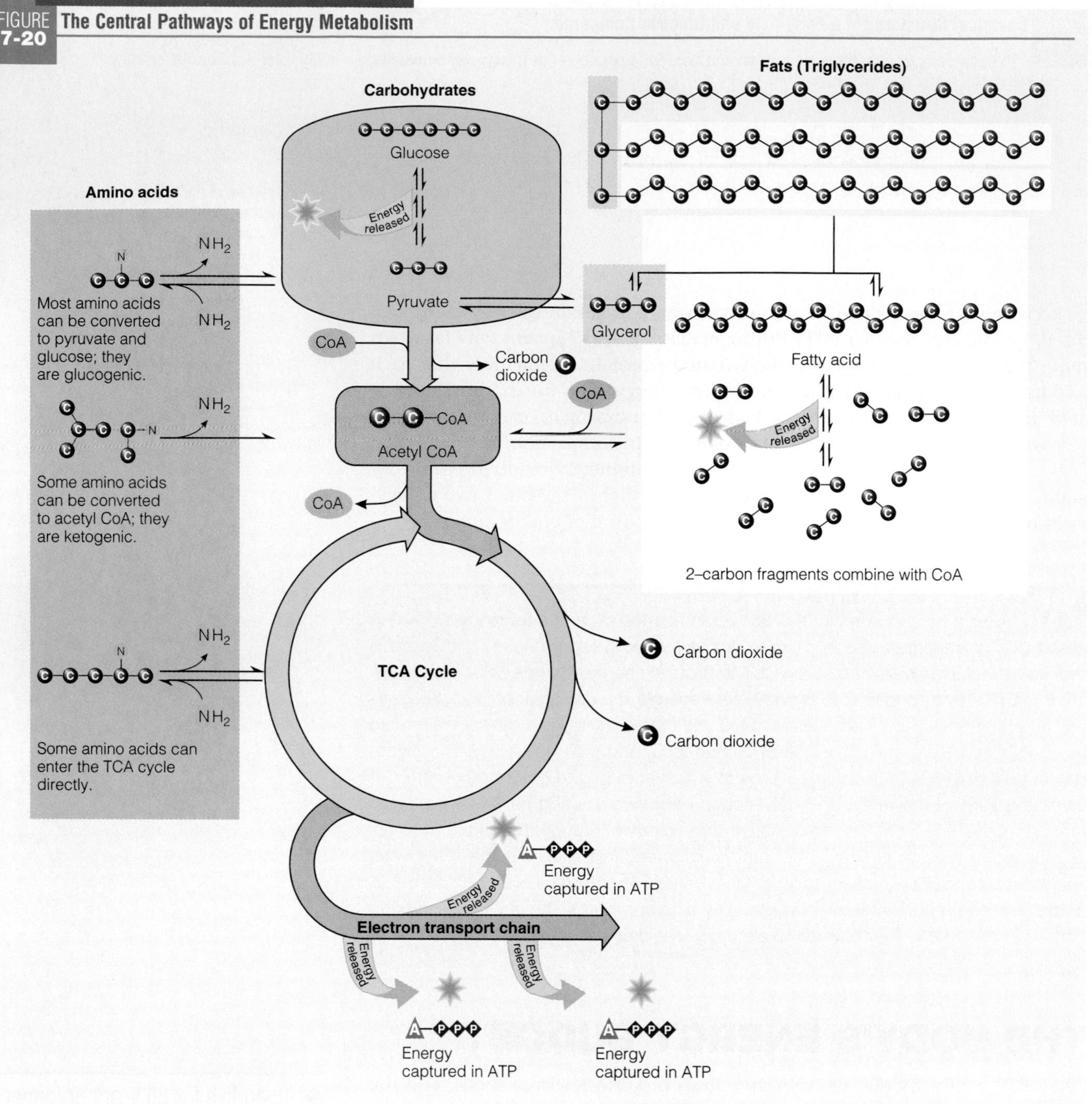

(costing only a few kcalories). To convert a dietary triglyceride to a triglyceride in adipose tissue, the body removes two of the fatty acids from the glycerol backbone, absorbs the parts, and puts them (and others) together again. By comparison, to convert a molecule of sucrose, the body has to split glucose from fructose, absorb them, dismantle them to pyruvate and acetyl CoA, assemble many acetyl CoA molecules into fatty acid chains, and finally attach fatty acids to a glycerol backbone to make a triglyceride for storage in adipose tissue. Quite simply, the body uses much less energy to convert dietary fat to body fat than it does to convert dietary carbohydrate to body fat. On average, storing excess energy from dietary fat in body fat uses only 5 percent of the ingested energy intake, but storing excess energy from dietary carbohydrate in body fat requires an expenditure of 25 percent of the ingested energy intake.

The pathways from excess protein and excess carbohydrate to body fat are not only indirect and inefficient, but also less preferred (having other priorities). Before entering fat storage, protein must first tend to its many roles in the body's lean tissues, and carbohydrate must fill the glycogen stores. Simply put, making fat is a low priority for these two nutrients. Still, if eaten in abundance, any of the energy-yielding nutrients can make fat.

This chapter has described each of the energy-yielding nutrients individually, but cells use a mixture of these fuels. How much of which nutrient is in the fuel mix depends, in part, on its availability from the diet. (The proportion of each fuel also depends on activity.) Dietary protein and dietary carbohydrate influence the mixture of fuel used during energy metabolism. Usually, protein's contribution to the fuel mix is relatively minor and fairly constant; protein oxidation increases only slightly, if at all, when protein is eaten in excess. Dietary carbohydrate, however, significantly enhances carbohydrate oxidation when eaten in excess. In contrast, fat oxidation does *not* respond to dietary fat intake, especially when dietary changes occur abruptly. The more protein or carbohydrate in the fuel mix, the less fat contributes to the fuel mix. Instead of being oxidized, fat accumulates in storage. Details follow.

People can enjoy bountiful meals such as this without storing body fat, provided that they spend as much energy as they take in.

• Surplus Protein • Recall from Chapter 6 that extra protein is not stored as protein in the body. Contrary to popular opinion, a person cannot grow muscle simply by overeating protein. Lean tissue such as muscle develops in response to a stimulus such as hormones or physical activity. When a person overeats protein, the body uses the surplus first by replacing normal daily losses and then by increasing protein oxidation slightly.[2] The body achieves protein balance this way, but any increase in protein oxidation displaces fat in the fuel mix. Any additional protein is then deaminated and the remaining carbons used to make fatty acids. Thus a person can grow fat by eating too much protein.

• Surplus Carbohydrate • Compared with protein, the proportions of carbohydrate and fat in the fuel mix change more dramatically when a person overeats. The body handles abundant carbohydrate by first storing it as glycogen, but glycogen storage areas are limited and fill quickly. Because maintaining glucose balance is critical, the body uses glucose frugally when the diet provides only small amounts and freely when stores are abundant. In other words, glucose oxidation rapidly adjusts to the dietary intake of carbohydrate.[3]

Excess glucose can be converted to fat directly, but this is a minor pathway.[4] As mentioned earlier, converting glucose to fat is energetically expensive and does not occur until after glycogen stores have been filled.[5] Even then, little or no new fat is made from carbohydrate.[6]

Nevertheless, excess dietary carbohydrate can lead to weight gain when extra carbohydrate displaces fat in the fuel mix.[7] When this occurs, carbohydrate spares both dietary fat and body fat from oxidation—an effect that may be more pronounced in overweight people than in lean people.[8] The net result: excess carbohydrate contributes to obesity or at least to the maintenance of an overweight body.

• Surplus Fat • Unlike excess protein and carbohydrate, which both enhance their own oxidation, eating too much fat does not promote fat oxidation.[9] Instead, excess dietary fat moves efficiently into the body's fat stores; almost all of the excess is stored.

IN SUMMARY

If energy intake exceeds the body's energy needs, the result will be weight gain—regardless of whether the excess intake is from protein, carbohydrate, or fat. The difference is that the body is much more efficient at storing energy when the excess derives from dietary fat.

The Transition from Feasting to Fasting

After a meal, glucose, glycerol, and fatty acids from foods are either used or stored. Later, as the body shifts from a fed state to a fasting one, it begins drawing on these stores. Glycogen and fat are released from storage to provide more glucose, glycerol, and fatty acids for energy.

Energy is needed all the time. Even when a person is asleep and totally relaxed, the cells of many organs are hard at work. In fact, this work—the cells' work that maintains all life processes without any conscious effort—represents about two-thirds to three-fourths of the total energy a person spends in a day. The small remainder is the work that a person's muscles perform voluntarily during waking hours.

The body's top priority is to meet the cells' needs for energy, and it normally does this by periodic refueling—that is, by eating several times a day. When food is not available, the body turns to its own tissues for other fuel sources. If people choose not to eat, we say they are fasting; if they have no choice, we say they are starving. The body makes no such distinction. In either case, the body is forced to switch to a wasting metabolism, drawing on its reserves of carbohydrate and fat and, within a day or so, on its vital protein tissues as well. Figure 7-21 shows the metabolic pathways operating in the body as it shifts from feasting (part A) to fasting (parts B and C).

The Economics of Fasting

During fasting, carbohydrate, fat, and protein are all eventually used for energy—fuel must be delivered to every cell. As the fast begins, glucose from

FIGURE 7-21 **Feasting and Fasting**

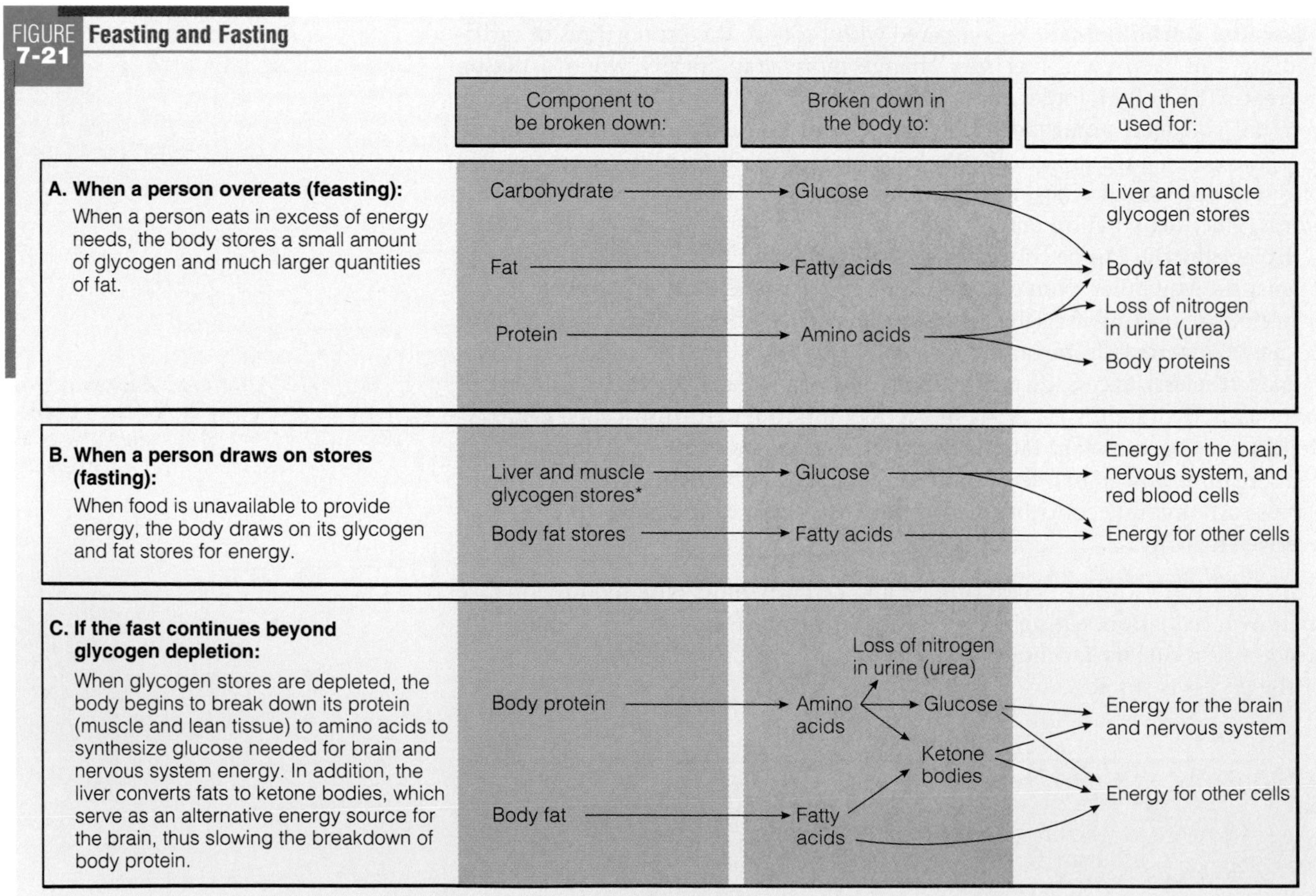

*The muscles' stored glycogen provides glucose only for the muscle in which the glycogen is stored.

the liver's stored glycogen and fatty acids from the adipose tissue's stored fat are both flowing into cells, then breaking down to yield acetyl CoA, and delivering energy to power the cells' work. Several hours later, however, most of the glucose is used up—liver glycogen is exhausted and blood glucose begins to fall. Low blood glucose serves as a signal that promotes further fat breakdown.

• Glucose Needed for the Brain • At this point, most of the cells are depending on fatty acids to continue providing their fuel. But red blood cells and the cells of the nervous system need glucose. Glucose is their major energy fuel, and even when other energy fuels are available, glucose must be present to permit the energy-metabolizing machinery of the nervous system to work. Normally, the brain and nerve cells consume about two-thirds of the total *glucose* used each day—about 400 to 600 kcalories' worth. About one-fifth to one-fourth of the *energy* the adult body uses when it is at rest is spent by the brain; in children, it can be up to one-half.

• Protein Called on to Meet Glucose Needs • The red blood cells' and brain's special requirements for glucose pose a problem for the fasting body. The body can use its stores of fat, which may be quite generous, to furnish most of its cells with energy, but the red blood cells are completely dependent on glucose, and the brain and nerves prefer energy in the form of glucose. For this reason, body protein tissues such as muscle and liver always break down to some extent during fasting. Amino acids that yield pyruvate can be used to make glucose; and to obtain them, body proteins must be broken down. The amino acids that can't be used to make glucose are used as an energy source for other body cells.

• Fat's Small Glucose Contribution from Glycerol • The breakdown of body protein is an expensive way to obtain glucose. In the first few days of a fast, body protein provides about 90 percent of the needed glucose; glycerol, about 10 percent. If body protein losses were to continue at this rate, death would ensue within three weeks, regardless of the quantity of fat a person had stored. Fortunately, fat breakdown also increases with fasting—in fact, fat breakdown almost doubles, providing energy for other body cells and glycerol for glucose production.[10]

• The Shift to Ketosis • As the fast continues, the body finds a way to use its fat to fuel the brain. It adapts by combining acetyl CoA fragments derived from fatty acids to produce an alternate energy source, ketone bodies (see Figure 7-22 on p. 226). Normally produced and used only in small quantities, ketone bodies◆ can provide fuel for some brain cells. Ketone body production rises until, after about ten days of fasting, it is meeting much of the nervous system's energy needs. Still, many areas of the brain rely exclusively on glucose, and to produce it, the body continues to sacrifice protein—albeit at a slower rate than in the early days of fasting.

◆ Reminder: *Ketone bodies* are compounds produced during the incomplete breakdown of fat when glucose is not available.

When ketone bodies contain an acid group (COOH), they are called keto acids. Small amounts of keto acids are a normal part of the blood chemistry; but when their concentration rises, the pH of the blood drops. This is ketosis, and it is a sign that the body's chemistry is going awry. Elevated blood ketones (ketonemia) sometimes spill into the urine (ketonuria). A fruity odor on the breath (known as acetone breath) develops, reflecting the presence of the ketone acetone.

• Suppression of Appetite • The starvation that produces ketosis also causes loss of appetite. Researchers have theorized that having a reduced appetite is an advantage to a person without access to food, because the search for food would be a waste of energy. When the person finds food and eats again, the body shifts out of ketosis, the hunger center gets the message that food is again available, and the appetite returns. This chain of events has served as justification for

FIGURE 7-22 **Ketone Body Formation**

1. The first step in the formation of ketone bodies is the condensation of two molecules of acetyl CoA and the removal of the CoA to form a compound that is converted to the first ketone body.

2. This ketone body may lose a molecule of carbon dioxide to become another ketone.

3. Or, the acetoacetate may add two hydrogens, becoming another ketone body (beta-hydroxybutyrate). See Appendix C for more details.

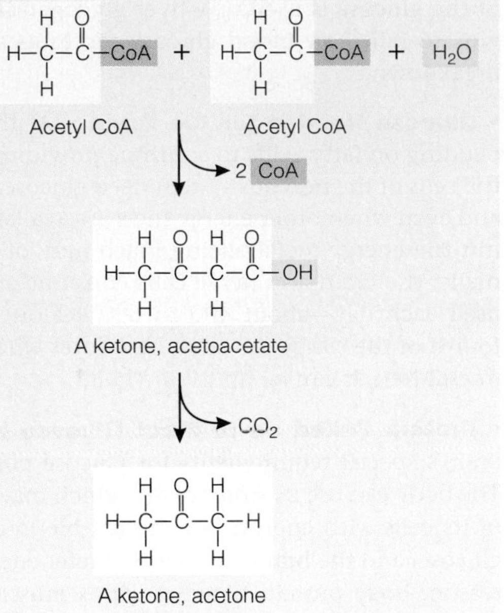

weight-loss routines that induce ketosis, such as fasting and low-carbohydrate diets. However, any kind of food restriction, with or without ketosis, leads a person to adapt by losing appetite. A well-balanced low-kcalorie diet can induce the same effect. Therefore ketosis-producing diets offer no special advantage in terms of appetite suppression, and because ketosis can disrupt the body's acid-base balance, other weight-loss regimens are preferred. Highlight 8 includes a discussion of the risks of ketosis-producing diets in its review of popular weight-loss diets.

• **Slowing of Metabolism** • When the body shifts to the use of ketone bodies, it simultaneously reduces its energy output and conserves both its fat and its lean tissue. As the lean (protein-containing) organ tissues shrink in mass, they perform less metabolic work, reducing energy expenditures. As the muscles waste, they can do less work and so demand less energy, reducing expenditures further. The hormones of fasting slow metabolism even further in the effort to conserve lean body tissues for as long as possible. Because of the slowed metabolism, the loss of fat falls to a bare minimum—less, in fact, than the fat that would be lost on a low-kcalorie diet. Thus, although *weight* loss during fasting may be quite dramatic, *fat* loss may be less than when at least some food is eaten.

• **Symptoms of Starvation** • The adaptations just described—slowing of energy output and reduction in fat loss—occur in the starving child, the hungry homeless adult, the fasting religious person, the adolescent with anorexia nervosa, and the malnourished hospital client. Such adaptations help to prolong their lives and explain the physical symptoms of energy deprivation: wasting, slowed metabolism, lowered body temperature, and reduced resistance to disease.

The body's adaptations to fasting are sufficient to maintain life for a long time. Mental alertness need not be diminished, and even some physical energy may remain unimpaired for a surprisingly long time. Still, fasting presents hazards. The same alterations in metabolism occur on a low-carbohydrate diet, as Highlight 8 explains.

IN SUMMARY

When fasting, the body makes a number of adaptations: increasing the breakdown of fat to provide energy for most of the cells, using glycerol and amino acids to make glucose for the red blood cells and central nervous system, producing ketones to fuel the brain, suppressing the appetite, and slowing metabolism. All of these measures conserve energy and minimize losses. In fact, metabolism slows to such an extent that the loss of fat eventually slows to less than would be achieved with a low-kcalorie diet.

This chapter has probed the intricate details of metabolism at the level of the cells, exploring the transformations of nutrients to energy and to storage compounds. Several chapters and highlights to come build on this information. The highlight that follows this chapter shows how alcohol disrupts normal metabolism. Chapter 8 describes how a person's intake and expenditure of energy are reflected in body weight and body composition. Chapter 9 examines the consequences of unbalanced energy budgets—overweight and underweight. Chapter 10 shows the vital roles the B vitamins play as coenzymes assisting all the metabolic pathways described here.

Indeed, the sun's energy sparks every move that we make. And our beautifully designed bodies make use of it in astonishing ways.

How are you doing?

- Are you impressed with how metabolically active the body's cells are?
- Do you eat more protein, carbohydrate, or fat than your body needs?
- Do you follow a low-carbohydrate diet that forces your body into ketosis?

Nutrition on the Net

INTERNET ACTIVITIES

Have you ever been confused about issues related to alcohol consumption? Have you ever thought that maybe you occasionally drink too much? Have you ever been in a car when the driver was drunk? Explore these issues and more in the following virtual tour.

- Go to: **www.health.org**
- Click on "Alcohol and Drug Facts" on the left side bar.
- Scroll down to "Information related to" and click on "College Students."
- Scroll to the bottom of the page and click on "What's driving you?"
- Click on the car to start the tour.

Issues related to young adults and alcohol consumption are complicated. Sometimes it's hard to know what to do about drinking situations. A virtual tour like this gives you a chance to experience decision making ahead of time in a neutral environment. Did you gain new insights about alcohol consumption?

Study Questions

These questions will help you review the chapter. You will find the answers in the discussions on the pages provided.

1. Define metabolism, anabolism, and catabolism; give an example of each. (pp. 204–205)
2. Name one of the body's quick-energy molecules, and describe how is it used. (pp. 205–206)
3. What are coenzymes, and what service do they provide in metabolism? (p. 207)
4. Name the four basic units, derived from foods, that are used by the body in metabolic transformations. How many carbons are in the "backbones" of each? (pp. 208–209)
5. Define aerobic and anaerobic metabolism. How does insufficient oxygen influence metabolism? (pp. 209–210)
6. How does the body dispose of excess nitrogen? (p. 218)
7. Summarize the main steps in the metabolism of glucose, glycerol, fatty acids, and amino acids. (pp. 221–222)
8. Describe how a surplus of the three energy nutrients contributes to body fat stores. (p. 223)
9. What adaptations does the body make during a fast? What are ketone bodies? Define ketosis. (pp. 224–226)
10. Distinguish between a loss of *fat* and a loss of *weight,* and describe how each might happen. (pp. 226–227)

These questions will help you prepare for an exam. Answers can be found on p. 229.

1. Hydrolysis is an example of a(n):
   a. coupled reaction.
   b. anabolic reaction.
   c. catabolic reaction.
   d. synthesis reaction.
2. During metabolism, released energy is captured and transferred by:
   a. enzymes.
   b. pyruvate.
   c. acetyl CoA.
   d. adenosine triphosphate.
3. Glycolysis:
   a. requires oxygen.
   b. generates abundant energy.
   c. converts glucose to pyruvate.
   d. produces ammonia as a by-product.
4. The pathway from pyruvate to acetyl CoA:
   a. produces lactic acid.
   b. is known as gluconeogenesis.
   c. is metabolically irreversible.
   d. requires more energy than it produces.
5. For complete oxidation, acetyl CoA enters:
   a. glycolysis and the TCA cycle.
   b. glycolysis and the Cori cycle.
   c. the TCA cycle and electron transport chain.
   d. the Cori cycle and electron transport chain.
6. Deamination of an amino acid produces:
   a. vitamin B_6 and energy.
   b. pyruvate and acetyl CoA.
   c. ammonia and a keto acid.
   d. carbon dioxide and water.
7. Before entering the TCA cycle, each of the energy-yielding nutrients is broken down to:
   a. ammonia.
   b. pyruvate.
   c. electrons.
   d. acetyl CoA.
8. The body stores energy for future use in:
   a. proteins.
   b. acetyl CoA.
   c. triglycerides.
   d. ketone bodies.
9. During a fast, when glycogen stores have been depleted, the body begins to synthesize glucose from:
   a. acetyl CoA.
   b. amino acids.
   c. fatty acids.
   d. ketone bodies.
10. During a fast, the body produces ketone bodies by:
    a. hydrolyzing glycogen.
    b. condensing acetyl CoA.
    c. transaminating keto acids.
    d. converting ammonia to urea.

References

1 J. L. Groff and S. S. Gropper, *Advanced Nutrition and Human Metabolism* (Belmont, Calif.: Wadsworth/Thomson Learning, 2000), p. 188.

2 S. A. Jebb and coauthors, Changes in macronutrient balance during over- and underfeeding assessed by 12-d continuous whole-body calorimetry, *American Journal of Clinical Nutrition* 64 (1996): 259–266.

3 T. J. Horton and coauthors, Fat and carbohydrate overfeeding in humans: Different effects on energy storage, *American Journal of Clinical Nutrition* 62 (1995): 19–29; P. S. Shetty and coauthors, Alterations in fuel selection and voluntary food intake in response to isoenergetic manipulation of glycogen stores in humans, *American Journal of Clinical Nutrition* 60 (1994): 534–543.

4 M. K. Hellerstein, Regulation of hepatic de novo lipogenesis in humans, *Annual Review of Nutrition* 16 (1996): 523–557.

5 J. P. Flatt, McCollum Award Lecture, 1995: Diet, lifestyle, and weight maintenance, *American Journal of Clinical Nutrition* 62 (1995): 820–836.

6 C. Prosperpi and coauthors, Ad libitum intake of a high-carbohydrate or high-fat diet in young men: Effects on nutrient balances, *American Journal of Clinical Nutrition* 66 (1997): 539–545; Jebb and coauthors, 1996; Horton and coauthors, 1995; B. Swinburn and E. Ravussin, Energy balance or fat balance? *American Journal of Clinical Nutrition* 57 (1993): 766S–771S.

7 Hellerstein, 1996.

8 I. Marques–Lopes and coauthors, Postprandial de novo lipogenesis and metabolic changes induced by a high-carbohydrate, low-fat meal in lean and overweight men, *American Journal of Clinical Nutrition* 73 (2001): 253–261.

9 E. Ravussin and A. Tataranni, Dietary fat and human obesity, *Journal of the American Dietetic Association* 97 (1997): S42–S46; J. P. Flatt, Use and storage of carbohydrate and fat, *American Journal of Clinical Nutrition* 61 (1995): 952S–959S; Horton and coauthors, 1995.

10 J. E. Ati, C. Beji, and J. Danguir, Increased fat oxidation during Ramadan fasting in healthy women: An adaptive mechanism for body weight maintenance, *American Journal of Clinical Nutrition* 62 (1995): 302–307; M. G. Carlson, W. L. Snead, and P. J. Campbell, Fuel and energy metabolism in fasting humans, *American Journal of Clinical Nutrition* 60 (1994): 29–36.

ANSWERS

Multiple Choice

1. c   2. d   3. c   4. c   5. c
6. c   7. d   8. c   9. b   10. b

ALCOHOL AND NUTRITION

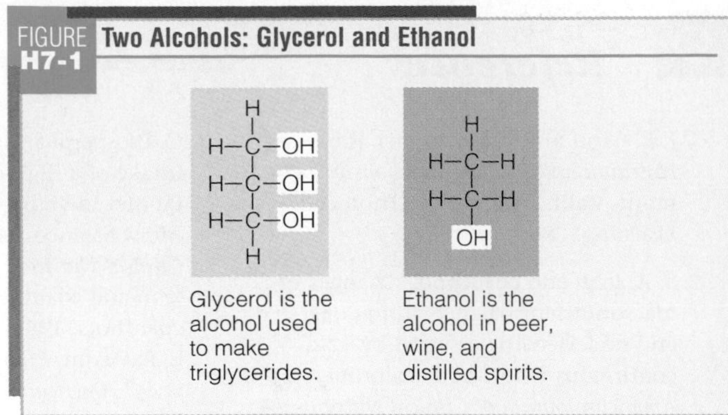

FIGURE H7-1 **Two Alcohols: Glycerol and Ethanol**

Glycerol is the alcohol used to make triglycerides.

Ethanol is the alcohol in beer, wine, and distilled spirits.

FROM BACKYARD barbecues to formal weddings, many social gatherings are occasions for offering beverages that contain alcohol, and people must choose whether to drink them. Most people who drink manage their relationships with alcohol relatively safely. Unfortunately, some 14 million people in the United States abuse alcohol to the point that their personal relationships, work, and health become impaired. With the understanding of metabolism gained from Chapter 7, you are in a position to understand how the body handles alcohol, how alcohol interferes with metabolism, and how alcohol impairs health and nutrition. (The potential benefits of *moderate* alcohol consumption are presented in Chapter 26.)

ALCOHOL IN BEVERAGES

To the chemist, **alcohol** refers to a class of organic compounds containing hydroxyl (OH) groups (the accompanying glossary defines alcohol and related terms). The glycerol to which fatty acids are attached in triglycerides is an example of an alcohol to a chemist. To most people, though, *alcohol* refers to the intoxicating ingredient in **beer, wine,** and **distilled spirits (hard liquor).** The chemist's name for this particular alcohol is *ethyl alcohol,* or **ethanol.** Glycerol has 3 carbons with 3 hydroxyl groups attached; ethanol has only 2 carbons and 1 hydroxyl group (see Figure H7-1). The remainder of this highlight talks about the particular alcohol, ethanol, but refers to it simply as *alcohol.*

Alcohols affect living things profoundly, partly because they act as lipid solvents. Their ability to dissolve lipids out of cell membranes allows alcohols to penetrate rapidly into cells, destroying cell structures and thereby killing the cells. For this reason, most alcohols are toxic in relatively small amounts; by the same token, because they kill microbial cells, they are useful as disinfectants.

Ethanol is less toxic than the other alcohols. Sufficiently diluted and taken in small enough doses, its action in the brain produces an effect that people seek—not with zero risk, but with a low enough risk (if the doses are low enough) to be tolerable. Used in this way, alcohol is a **drug**—that is, a substance that modifies body functions. Like all drugs, alcohol offers both benefits and hazards. It must be used with caution, if used at all.

Beer, wine, and liquor deliver different amounts of alcohol. The amount of alcohol in distilled liquor is stated as **proof:** 100 proof liquor is 50 percent alcohol, 80 proof is 40 percent alcohol, and so forth. Wine (at 8 to 14 percent) and beer (at 4 to 6 percent) have less alcohol than distilled liquor, although some fortified wines and beers have more alcohol than the regular varieties.

Taken in **moderation,** alcohol can be compatible with good health. The term *moderation* is important in describing alcohol use. How many drinks constitute moderate use, and how much is "a drink"? First, a **drink** is any alcoholic beverage that delivers ½ ounce of *pure ethanol:*

- 5 ounces of wine.
- 10 ounces of wine cooler.
- 12 ounces of beer.
- 1½ ounces of distilled liquor (80 proof whiskey, scotch, rum, or vodka).

Second, because people have different tolerances to alcohol, it is impossible to name an exact daily amount of alcohol that is appropriate for everyone. Authorities have attempted to set limits that are acceptable for most healthy people. An accepted definition of moderation is not more than two drinks a day for the average-sized man and not more than one drink a day for the average-sized woman. (Pregnant women are advised to

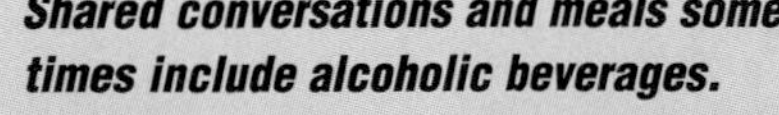
Shared conversations and meals sometimes include alcoholic beverages.

© 1998 Photo Disc Inc.

GLOSSARY

acetaldehyde (ass-et-AL-duh-hide): an intermediate in alcohol metabolism.

alcohol: a class of organic compounds containing hydroxyl (OH) groups.

alcohol abuse: a pattern of drinking that includes failure to fulfill work, school, or home responsibilities; drinking in situations that are physically dangerous (as in driving while intoxicated); recurring alcohol-related legal problems (as in aggravated assault charges); or continued drinking despite ongoing social problems that are caused by or worsened by alcohol.

alcohol dehydrogenase (dee-high-DROJ-eh-nayz): an enzyme active in the stomach and the liver that converts ethanol to acetaldehyde.

alcoholism: a pattern of drinking that includes a strong craving for alcohol, a loss of control and an inability to stop drinking once begun, withdrawal symptoms (nausea, sweating, shakiness, and anxiety) after heavy drinking, and the need for increasing amounts of alcohol in order to feel "high."

antidiuretic hormone (ADH): a hormone produced by the pituitary gland in response to dehydration (or a high sodium concentration in the blood). It stimulates the kidneys to reabsorb more water and therefore to excrete less. This ADH should not be confused with the enzyme alcohol dehydrogenase, which is sometimes also abbreviated ADH.

beer: an alcoholic beverage brewed by fermenting malt and hops.

cirrhosis (seer-OH-sis): advanced liver disease in which liver cells turn orange, die, and harden, permanently losing their function; often associated with alcoholism.

- **cirrhos** = an orange

distilled liquor: an alcoholic beverage made by fermenting and distilling grains; sometimes called *distilled spirits* or *hard liquor.*

drink: a dose of any alcoholic beverage that delivers ½ oz of pure ethanol:

- 5 oz of wine.
- 10 oz of wine cooler.
- 12 oz of beer.
- 1½ oz of hard liquor (80 proof whiskey, scotch, rum, or vodka).

drug: a substance that can modify one or more of the body's functions.

ethanol: a particular type of alcohol found in beer, wine, and distilled spirits; also called *ethyl alcohol* (see Figure H7-1). Ethanol is the most widely used—and abused—drug in our society. It is also the only legal, nonprescription drug that produces euphoria.

fatty liver: an early stage of liver deterioration seen in several diseases, including kwashiorkor and alcoholic liver disease. Fatty liver is characterized by an accumulation of fat in the liver cells.

fibrosis (fye-BROH-sis): an intermediate stage of liver deterioration seen in several diseases, including viral hepatitis and alcoholic liver disease. In fibrosis, the liver cells lose their function and assume the characteristics of connective tissue cells (fibers).

MEOS or **microsomal** (my-krow-SO-mal) **ethanol-oxidizing system:** a system of enzymes in the liver that oxidize not only alcohol, but also several classes of drugs.

moderation: in relation to alcohol consumption, not more than two drinks a day for the average-sized man and not more than one drink a day for the average-sized woman.

NAD (nicotinamide adenine dinucleotide): the main coenzyme form of the vitamin niacin. Its reduced form is NADH.

narcotic (nar-KOT-ic): drug that dulls the senses, induces sleep, and becomes addictive with prolonged use.

proof: a way of stating the percentage of alcohol in distilled liquor. Liquor that is 100 proof is 50% alcohol; 90 proof is 45%, and so forth.

wine: an alcoholic beverage made by fermenting grape juice.

abstain from alcohol, as Highlight 14 explains.) Notice that this advice is stated as a maximum, not as an average; seven drinks one night a week would not be considered moderate, even though one a day would be. Doubtless some people could consume slightly more; others could not handle nearly so much without risk. The amount a person can drink safely is highly individual, depending on genetics, health, gender, body composition, age, and family history.

HEALTHY PEOPLE 2010

Reduce average annual alcohol consumption.

ALCOHOL IN THE BODY

From the moment an alcoholic beverage enters the body, it is treated as if it has special privileges. Unlike foods, which require time for digestion, alcohol needs no digestion and is quickly absorbed. About 20 percent is absorbed directly across the walls of an empty stomach and can reach the brain within a minute. Consequently, a person can immediately feel euphoric when drinking, especially on an empty stomach.

When the stomach is full of food, alcohol has less chance of touching the walls and diffusing through, so its influence on the brain is slightly delayed. This information leads to a practical tip: eat snacks when drinking alcoholic beverages. Carbohydrate snacks slow alcohol absorption and high-fat snacks slow peristalsis, keeping the alcohol in the stomach longer. Salty snacks make a person thirsty; to quench thirst, drink water instead of more alcohol.

The stomach begins to break down alcohol with its **alcohol dehydrogenase** enzyme.[1] This action can reduce the amount of alcohol entering the blood by about 20 percent. Women produce less of this stomach enzyme than men; consequently, more alcohol reaches the intestine for absorption into the bloodstream. As a result, women absorb about one-third more alcohol than men of the same size who drink the same amount of alcohol. Consequently, they are more likely to become more intoxicated on less alcohol than men. These differences between men and women help explain why women have a lower alcohol tolerance and a lower recommendation for moderate intake.

Alcohol is rapidly absorbed in the small intestine. From this point on, alcohol receives VIP (Very Important Person) treatment: it gets absorbed and metabolized before most nutrients. Alcohol's priority status helps to ensure a speedy disposal and reflects two facts: alcohol cannot be stored in the body, and it is potentially toxic.

ALCOHOL ARRIVES IN THE LIVER

The capillaries of the digestive tract merge into veins that carry the alcohol-laden blood to the liver. These veins branch and rebranch into capillaries that touch every liver cell. Liver cells are the only other cells in the body that can make enough of the alcohol dehydrogenase enzyme to oxidize alcohol at an appreciable rate. The routing of blood through the liver cells gives them the chance to dispose of some alcohol before it moves on.

Alcohol affects every organ of the body, but the most dramatic evidence of its disruptive behavior appears in the liver. If liver cells could talk, they would describe alcohol as demanding, egocentric, and disruptive of the liver's efficient way of running its business. For example, liver cells normally prefer fatty acids as their fuel, and they like to package excess fatty acids into triglycerides and ship them out to other tissues. When alcohol is present, however, the liver cells are forced to metabolize alcohol and let the fatty acids accumulate, sometimes in huge stockpiles. Alcohol metabolism also permanently changes liver cell structure, impairing the liver's ability to metabolize fats. This explains why heavy drinkers develop fatty livers.

The liver can process about ½ ounce *ethanol* per hour (the amount in a typical drink), depending on the person's body size, previous drinking experience, food intake, and general health. This maximum rate of alcohol breakdown is set by the amount of alcohol dehydrogenase available. If more alcohol arrives at the liver than the enzymes can handle, the extra alcohol travels to all parts of the body, circulating again and again until liver enzymes are finally available to process it. Another practical tip derives from this information: drink slowly enough to allow the liver to keep up—no more than one drink per hour.

The amount of alcohol dehydrogenase enzyme present in the liver varies with individuals, depending on the genes they have inherited and on how recently they have eaten. Fasting for as little as a day forces the body to degrade its proteins, including the alcohol-processing enzymes, and this can slow the rate of alcohol metabolism by half. Drinking on an empty stomach thus causes the drinker to feel the effects more promptly for two reasons: rapid absorption and slowed breakdown. By maintaining higher blood alcohol concentrations for longer times, alcohol can anesthetize the brain more completely.

The alcohol dehydrogenase enzyme breaks down alcohol by removing hydrogens in two steps. (Figure H7-2 provides a simplified diagram of alcohol metabolism; Appendix C provides the chemical details.) In the first step, alcohol dehydrogenase oxidizes alcohol to **acetaldehyde.** High concentrations of acetaldehyde in the brain and other tissues are responsible for many of the punishing effects of **alcohol abuse.**

In the second step, a related enzyme, acetaldehyde dehydrogenase, converts acetaldehyde to acetate, which is then converted to acetyl CoA—the "crossroads" compound introduced in Chapter 7 that can enter the TCA cycle to generate energy. These reactions produce hydrogen ions (H^+). The B vitamin niacin (in its role as the coenzyme **NAD**) helpfully picks up these hydrogen ions (becoming NADH). Thus, whenever the body breaks down alcohol, NAD diminishes and NADH accumulates. (Chapter 10 presents information on NAD and the other coenzyme roles of the B vitamins.)

ALCOHOL DISRUPTS THE LIVER

During alcohol metabolism, the multitude of other metabolic processes for which NAD is required, including glycolysis, the TCA cycle, and the electron transport chain, falter. Its presence is sorely missed in these energy pathways because it is the chief carrier of the hydrogens that travel with their electrons along the electron transport chain. Without adequate NAD, the energy pathway cannot function. Traffic either backs up, or an alternate route

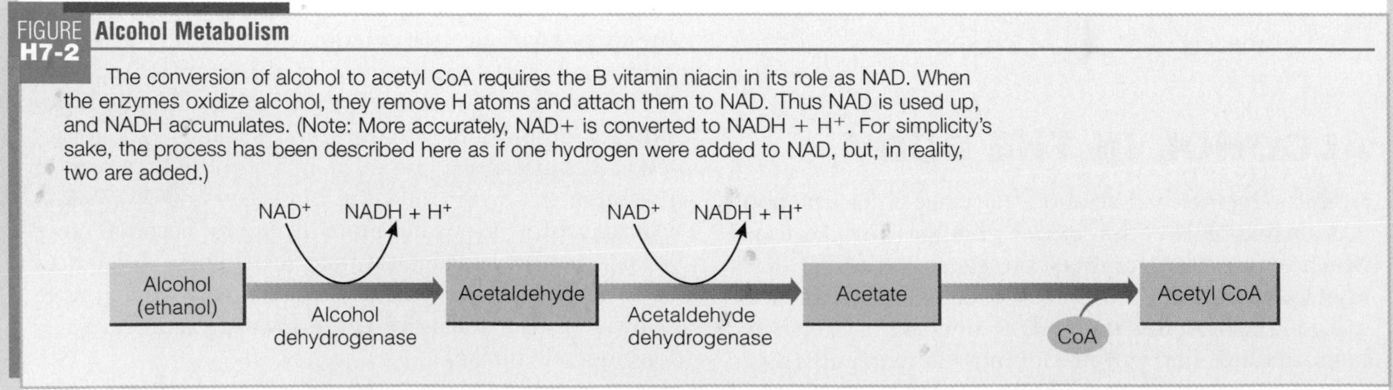

FIGURE H7-2 **Alcohol Metabolism**

The conversion of alcohol to acetyl CoA requires the B vitamin niacin in its role as NAD. When the enzymes oxidize alcohol, they remove H atoms and attach them to NAD. Thus NAD is used up, and NADH accumulates. (Note: More accurately, NAD+ is converted to NADH + H^+. For simplicity's sake, the process has been described here as if one hydrogen were added to NAD, but, in reality, two are added.)

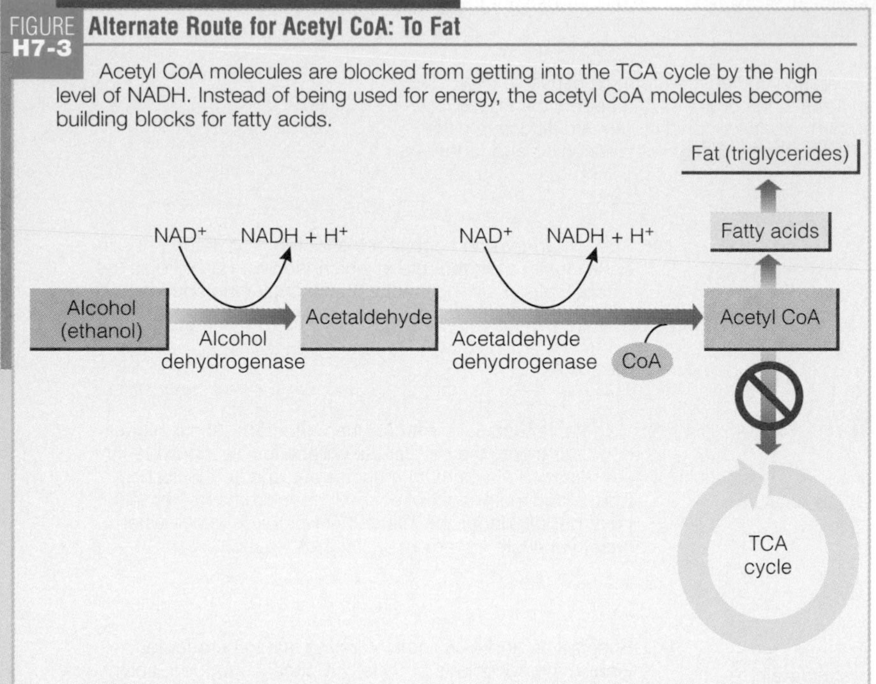

FIGURE H7-3 **Alternate Route for Acetyl CoA: To Fat**

Acetyl CoA molecules are blocked from getting into the TCA cycle by the high level of NADH. Instead of being used for energy, the acetyl CoA molecules become building blocks for fatty acids.

is taken. Such changes in the normal flow from glucose to energy have striking physical consequences.

For one, the accumulation of hydrogen ions during alcohol metabolism shifts the body's acid-base balance toward acid. For another, the accumulation of NADH slows the TCA cycle, so pyruvate and acetyl CoA build up. Excess acetyl CoA then takes the route to fatty acid synthesis (as Figure H7-3 illustrates), and fat clogs the liver.

As you might expect, a liver clogged with fat cannot function properly. Liver cells become less efficient at performing a number of tasks. Much of this inefficiency impairs a person's nutritional health in ways that cannot be corrected by diet alone. For example, the liver has difficulty activating vitamin D, as well as producing and releasing bile. To overcome such problems, a person needs to stop drinking alcohol.

The synthesis of fatty acids accelerates with exposure to alcohol. Fat accumulation can be seen in the liver after a single night of heavy drinking. **Fatty liver,** the first stage of liver deterioration seen in heavy drinkers, interferes with the distribution of nutrients and oxygen to the liver cells. If the condition lasts long enough, the liver cells will die and form fibrous scar tissue—the second stage of liver deterioration, called **fibrosis.** Some liver cells can regenerate with good nutrition and abstinence from alcohol, but in the most advanced stage, **cirrhosis,** damage is the least reversible (see Chapter 28).

The fatty liver has difficulty generating glucose from protein. The lack of glucose together with the overabundance of acetyl CoA sets the stage for ketosis. The body uses the acetyl CoA to make ketone bodies, which push the acid-base balance further toward acid.

Excess NADH also promotes the making of lactic acid from pyruvate. The conversion of pyruvate to lactic acid uses the hydrogens from NADH and restores some NAD, but a lactic acid buildup has serious consequences of its own—it adds still further to the body's acid burden and interferes with the excretion of another acid, uric acid, causing inflammation of the joints.

Alcohol alters both amino acid and protein metabolism. Synthesis of proteins important in the immune system slows down, weakening the body's defenses against infection. Protein deficiency can develop, both from a diminished synthesis of protein and from a poor diet. Normally, the cells would at least use the amino acids from the protein foods a person eats, but the drinker's liver deaminates the amino acids and uses the carbon fragments primarily to make fat or ketones. Eating well does not protect the drinker from protein depletion; a person has to stop drinking alcohol.

The liver's VIP treatment of alcohol affects its handling of drugs as well as nutrients. In addition to the dehydrogenase enzyme already described, the liver possesses an enzyme system that metabolizes *both* alcohol and several other types of drugs. Called the **MEOS (microsomal ethanol-oxidizing system),** this system handles about one-fifth of the total alcohol a person consumes. At high blood alcohol concentrations, however, or if repeatedly exposed to alcohol, the MEOS grows larger.

As a person's blood alcohol rises, alcohol competes with—and wins out over—other drugs whose metabolism relies on the MEOS. If a person drinks and uses another drug at the same time, the drug will be metabolized more slowly because the MEOS is busy disposing of alcohol. Consequently the drug cannot be handled until later; the dose may build up so that its effects are greatly amplified—sometimes to the point of being fatal.

In contrast, once a heavy drinker stops drinking and alcohol is no longer competing with other drugs, the enlarged MEOS metabolizes drugs much faster than before. As a result, determining the correct dosages of medications can be challenging. A skilled anesthesiologist always asks clients about their drinking patterns before sedating them.

This discussion has emphasized the major way that the blood is cleared of alcohol—metabolism by the liver—but there is another way. About 10 percent of the alcohol leaves the body through the breath and in the urine. This is the basis for the breath and urine tests for drunkenness. The amounts of alcohol in the breath and in the urine are in proportion to the amount still in the bloodstream and brain. In nearly all states, legal drunkenness is set at 0.10 percent or less, reflecting the relationship between alcohol use and industrial and traffic accidents.

FIGURE H7-4 Alcohol's Effects on the Brain

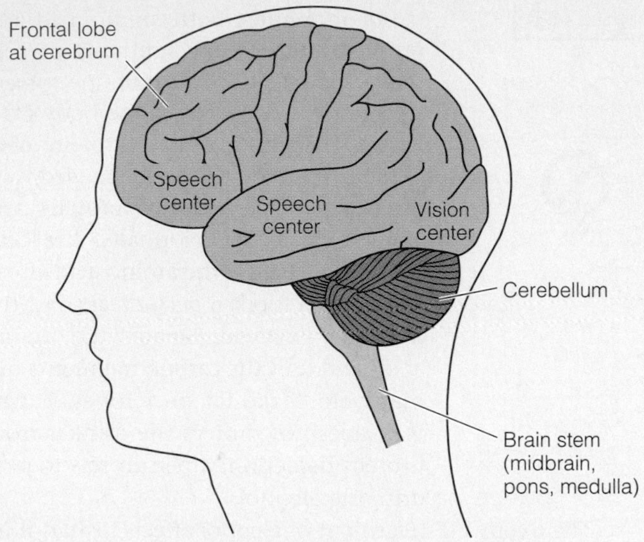

1. Judgment and reasoning centers are most sensitive to alcohol. When alcohol flows to the brain, it first sedates the frontal lobe, the reasoning part. As the alcohol molecules diffuse into the cells of these lobes, they interfere with reasoning and judgment.

2. Speech and vision centers are affected next. If the drinker drinks faster than the rate at which the liver can oxidize the alcohol, blood alcohol concentrations rise: the speech and vision centers of the brain become sedated.

3. Voluntary muscular control is then affected. At still higher concentrations, the cells in the cerebellum responsible for coordination of voluntary muscles are affected including those used in speech, eye, and limb movements. At this point people under the influence stagger or weave when they try to walk, or they may slur their speech.

4. Respiration and heart action are the last to be affected. Finally, the conscious brain is completely subdued, and the person passes out. Now the person can drink no more; this is fortunate because higher doses have an anesthetic effect that could reach the deepest brain centers, which control breathing and heartbeat, and the person could die.

ALCOHOL ARRIVES IN THE BRAIN

Alcohol is a **narcotic.** People used it for centuries as an anesthetic because it can deaden pain. But alcohol was a poor anesthetic because one could never be sure how much a person would need and how much would be a fatal dose. Consequently, new, more predictable anesthetics have replaced alcohol. However, alcohol continues to be used today as a kind of social anesthetic to help people relax or to relieve anxiety. People think that alcohol is a stimulant because it seems to relieve inhibitions. Actually, though, it accomplishes this by sedating *inhibitory* nerves, which are more numerous than excitatory nerves. Ultimately, alcohol acts as a depressant and affects all the nerve cells. Figure H7-4 describes alcohol's effects on the brain.

It is lucky that the brain centers respond to a rising blood alcohol concentration in the order described in Figure H7-4 because a person usually passes out before managing to drink a lethal dose. It is possible, though, to drink so fast that the effects of alcohol continue to accelerate after the person has passed out. Occasionally, a person dies from drinking enough to stop the heart before passing out. Table H7-1 shows the blood alcohol levels that correspond to progressively greater intoxication, and Table H7-2 shows the brain responses that occur at these blood levels.

Like liver cells, brain cells die with excessive exposure to alcohol. Liver cells may be replaced, but not all brain cells can regenerate. Thus some heavy drinkers suffer permanent brain damage.

People who drink alcoholic beverages may notice that they urinate more, but they may be unaware of the vicious cycle that results. Alcohol depresses production of **antidiuretic hormone (ADH),** a hormone produced by the pituitary gland that retains water. Loss of body water leads to thirst, and thirst leads to more drinking. The only fluid that will relieve dehydration is water, but the thirsty drinker may drink alcohol instead. This only worsens the problem. Such information provides another practical tip: drink water when thirsty and before each alcoholic drink. Drink an extra glass or two before going to bed.

Water loss is accompanied by the loss of important minerals. As Chapters 12 and 13 will explain, these min-

TABLE H7-1 Alcohol Doses and Blood Levels

Number of Drinks[a]	Percentage of Blood Alcohol by Body Weight				
	100 lb	*120 lb*	*150 lb*	*180 lb*	*200 lb*
2	0.08	0.06	0.05	0.04	0.04
4	0.15	0.13	0.10	0.08	0.08
6	0.23	0.19	0.15	0.13	0.11
8	0.30	0.25	0.20	0.17	0.15
12	0.45	0.36	0.30	0.25	0.23
14	0.52	0.42	0.35	0.34	0.27

Note: In some states driving under the influence is proved when an adult's blood contains 0.08 percent alcohol, and in others, 0.10. Many states have adopted a "zero-tolerance" policy for drivers under age 21, using 0.02 percent as the limit.
[a]Taken within an hour or so; each drink equivalent to ½ ounce pure ethanol.

erals are vital to the body's fluid balance and to many chemical reactions in the cells, including muscle action. Detoxification treatment includes restoration of mineral balance as quickly as possible.

ALCOHOL AND MALNUTRITION

For many moderate drinkers, alcohol does not suppress food intake and may actually stimulate appetite.[2] Moderate drinkers usually consume alcohol as *added* energy—on top of their normal food intake. In addition, alcohol in moderate doses is efficiently metabolized. Consequently, alcohol can contribute to body fat.[3] Metabolically, alcohol behaves like fat in promoting obesity; each ounce of alcohol represents about a half-ounce of fat. Alcohol's contribution to body fat is most evident in the central obesity that commonly accompanies alcohol consumption, popularly—and appropriately—known as the "beer belly."[4] Alcohol in heavy doses, though, is not efficiently metabolized, generating more heat than fat. Heavy drinkers usually consume alcohol as *substituted* energy—instead of their normal food intake. They tend to eat poorly and suffer malnutrition.

Alcohol is rich in energy (7 kcalories per gram), but as with pure sugar or fat, the kcalories are empty of nutrients. The more alcohol people drink, the less likely that they will eat enough food to obtain adequate nutrients. The more kcalories spent on alcohol, the fewer kcalories available to spend on nutritious foods. Table H7-3 shows the kcalorie amounts of typical alcoholic beverages.

Chronic alcohol abuse not only displaces nutrients from the diet, but also interferes with the body's metabolism of nutrients. Most dramatic is alcohol's effect on the B vitamin folate. The liver loses its ability to retain folate, and the kidneys increase their excretion of it. Alcohol abuse creates a folate deficiency that devastates digestive system function. The intestine normally releases and retrieves folate continuously, but it becomes damaged by folate deficiency and alcohol toxicity, so it fails to retrieve its own folate and misses any that may trickle in from food as well. Alcohol also interferes with the action of folate in converting homocysteine to the nonessential amino acid methionine. The result is an excess of homocysteine, which has been linked to heart disease, and an inadequate supply of methionine, which slows the production of new cells, especially the rapidly dividing cells of the intestine and the blood.[5] The combination of poor folate status and alcohol consumption has also been implicated in promoting colorectal cancer.[6]

The inadequate food intake and impaired nutrient absorption that accompany chronic alcohol abuse frequently lead to a deficiency of another B vitamin—thiamin. In fact, the cluster of thiamin-deficiency symptoms commonly seen in chronic **alcoholism** has its own name—the Wernicke-Korsakoff syndrome. This syndrome is characterized by paralysis of the eye muscles, poor muscle

TABLE H7-2 Alcohol Blood Levels and Brain Responses

Blood Alcohol Concentration	Effect on Brain
0.05	Impaired judgment, relaxed inhibitions, altered mood, increased heart rate
0.10	Impaired coordination, delayed reaction time, exaggerated emotions, impaired peripheral vision, impaired ability to operate a vehicle
0.15	Slurred speech, blurred vision, staggered walk, seriously impaired coordination and judgment
0.20	Double vision, inability to walk
0.30	Uninhibited behavior, stupor, confusion, inability to comprehend
0.40 to 0.60	Unconsciousness, shock, coma, death (cardiac or respiratory failure)

Note: Blood alcohol concentration depends on a number of factors, including alcohol in the beverage, the rate of consumption, the person's gender, and body weight. For example, a 100-pound female can become legally drunk (0.10 concentration) by drinking three beers in an hour, whereas a 220-pound male consuming that amount at the same rate would have a 0.05 blood alcohol concentration.

TABLE H7-3 kCalories in Alcoholic Beverages and Mixers

Beverage	Amount (oz)	Energy (kcal)
Beer		
Regular	12	150
Light	12	78–131
Nonalcoholic	12	32–82
Distilled liquor (gin, rum, vodka, whiskey)		
80 proof	1½	100
86 proof	1½	105
90 proof	1½	110
Liqueurs		
Coffee liqueur, 53 proof	1½	175
Coffee and cream liqueur, 34 proof	1½	155
Crème de menthe, 72 proof	1½	185
Mixers		
Club soda	12	0
Cola	12	150
Cranberry juice cocktail	8	145
Diet drinks	12	2
Ginger ale or tonic	12	125
Grapefruit juice	8	95
Orange juice	8	110
Tomato or vegetable juice	8	45
Wine		
Dessert	3½	110–135
Nonalcoholic	8	14
Red or rosé	3½	75
White	3½	70
Wine cooler	12	170

coordination, impaired memory, and damaged nerves; it is treated with thiamin supplements.

Acetaldehyde, an intermediate in alcohol metabolism, interferes with nutrient use, too. For example, acetaldehyde dislodges vitamin B_6 from its protective binding protein so that it is destroyed, causing a vitamin B_6 deficiency and, thereby, lowered production of red blood cells.

Malnutrition occurs not only because of lack of intake and altered metabolism, but because of direct toxic effects as well. Alcohol causes stomach cells to oversecrete both gastric acid and histamine, an immune system agent that produces inflammation. Beer in particular stimulates gastric acid secretion, irritating the stomach and esophagus linings and making them vulnerable to ulcer formation.

Nutrient deficiencies are virtually inevitable in alcohol abuse, not only because alcohol displaces food but also because alcohol directly interferes with the body's use of nutrients, making them ineffective even if they are present. Intestinal cells fail to absorb B vitamins, notably, thiamin, folate, and vitamin B_{12}. Liver cells lose efficiency in activating vitamin D. Cells in the retina of the eye, which normally process the alcohol form of vitamin A (retinol) to its aldehyde form needed in vision (retinal), find themselves processing ethanol to acetaldehyde instead.

Regardless of dietary intake, excessive drinking over a lifetime creates deficits of all the nutrients mentioned in this discussion and more. No diet can compensate for the damage caused by heavy alcohol consumption.

ALCOHOL'S SHORT-TERM EFFECTS

Heavy or binge drinking (defined as at least four to five drinks in a row) is widespread on college campuses and poses serious health and social consequences to drinkers and nondrinkers alike.*[7] In fact, binge drinking can kill: the respiratory center of the brain becomes anesthetized, and breathing stops. Acute alcohol intoxication can cause coronary artery spasms, leading to heart attacks.[8]

Binge drinking is most common among college students who live in a fraternity or sorority house, attend parties frequently, engage in other risky behaviors, and have a history of binge drinking in high school.[9] Compared with nondrinkers or moderate drinkers, people who frequently binge drink (at least three times within two weeks) are more likely to engage in unprotected sex, have multiple sex partners, damage property, and assault others.[10]

Binge drinkers skew the statistics on college campus alcohol use. The median number of drinks consumed by college students is 1.5 per week, but for binge drinkers, it is 14.5. Nationally, only 20 percent of all students are frequent binge drinkers; yet they account for two-thirds of all the alcohol students report consuming and most of the alcohol-related problems.[11]

Binge drinking is not limited to college campuses, of course, but that environment seems most accepting of such behavior despite its problems. Social acceptance may make it difficult for binge drinkers to recognize themselves as problem drinkers. For this reason, interventions must focus both on individuals and on the whole population. The damage alcohol causes only becomes worse if the pattern is not broken. Alcohol abuse sets in much more quickly in young people than in adults.[12] Those who start drinking at an early age more often suffer from alcoholism than people who start later on. Table H7-4 lists the key signs of alcoholism.

*This definition of binge drinking, without specification of time elapsed, is consistent with standard practice in alcohol research.

HEALTHY PEOPLE 2010

Reduce the proportion of persons engaging in binge drinking of alcoholic beverages.

ALCOHOL'S LONG-TERM EFFECTS

The most devastating long-term effect of alcohol is the damage done to a child whose mother abused alcohol

TABLE H7-4 Signs of Alcoholism

- Tolerance—the person needs higher and higher intakes of alcohol to achieve intoxication.
- Withdrawal—the person who stops drinking experiences anxiety, agitation, increased blood pressure, or seizures, or seeks alcohol to relieve these symptoms.
- Impaired control—the person intends to have 1 or 2 drinks, but has 9 or 10 instead, or the person tries to control or quit drinking, but fails.
- Disinterest—the person neglects important social, family, job, or school activities because of drinking.
- Time—the person spends a great deal of time obtaining and drinking alcohol or recovering from excessive drinking.
- Impaired ability—the person's intoxication or withdrawal symptoms interfere with work, school, or home.
- Problems—the person continues drinking despite physical hazards or medical, legal, psychological, family, employment, or school problems.

The presence of three or more of these conditions is required to make a diagnosis.

SOURCE: Adapted from *Diagnostic and Statistical Manual of Mental Disorders*, 4th ed. (Washington, D.C.: American Psychiatric Association, 1994).

TABLE H7-5 **Health Effects of Heavy Alcohol Consumption**

Health Problem	Effects of Alcohol
Arthritis	Increases the risk of inflamed joints.
Cancer	Increases the risk of cancer of the liver, pancreas, rectum, and breast; increases the risk of cancer of the mouth, pharynx, larynx, and esophagus, where alcohol interacts synergistically with tobacco.
Fetal alcohol syndrome	Causes physical and behavioral abnormalities in the fetus (see Highlight 14).
Heart disease	In heavy drinkers, raises blood pressure, blood lipids, and the risk of stroke and heart disease; when compared with those who abstain, heart disease risk is generally lower in light-to-moderate drinkers (see Chapter 26).
Hyperglycemia	Raises blood glucose.
Hypoglycemia	Lowers blood glucose, especially in people with diabetes.
Infertility	Increases the risks of menstrual disorders and spontaneous abortions (in women); suppresses luteinizing hormone (in women) and testosterone (in men).
Kidney disease	Enlarges the kidneys, alters hormone functions, and increases the risk of kidney failure.
Liver disease	Causes fatty liver, alcoholic hepatitis, and cirrhosis (see Chapter 28).
Malnutrition	Increases the risk of protein-energy malnutrition; low intakes of protein, calcium, iron, vitamin A, vitamin C, thiamin, vitamin B_6, and riboflavin; and impaired absorption of calcium, phosphorus, vitamin D, and zinc.
Nervous disorders	Causes neuropathy and dementia; impairs balance and memory.
Obesity	Increases energy intake, but is not a primary cause of obesity.
Psychological disturbances	Causes depression, anxiety, and insomnia.

Note: This list is by no means all-inclusive. Alcohol has direct toxic effects on all body systems.

during pregnancy. The devastating effects of alcohol on the unborn, and the message that pregnant women should not drink alcohol, are presented in Highlight 14.

For nonpregnant adults, a drink or two sets in motion many destructive processes in the body, but the next day's abstinence reverses them. As long as the doses are moderate, the time between them is ample, and nutrition is adequate, recovery is probably complete.

If the doses of alcohol are heavy and the time between them short, complete recovery cannot take place. Repeated onslaughts of alcohol gradually take a toll on all parts of the body (see Table H7-5). Compared with nondrinkers, heavy drinkers have significantly greater risks of dying from all causes.

PERSONAL STRATEGIES

One obvious option available to people attending social gatherings is to enjoy the conversation, eat the food, and drink nonalcoholic beverages. Several nonalcoholic beverages are available that mimic the look and taste of their alcoholic counterparts. For those who enjoy champagne or beer, sparkling ciders and beers without alcohol are available. Instead of drinking a cocktail, a person can sip tomato juice with a slice of lime and a stalk of celery or just a plain cola beverage. Any of these drinks can ease conversation.

The person who chooses to drink alcohol should sip each drink slowly with food. The alcohol should arrive at the liver cells slowly enough that the enzymes can handle the load. It is best to space drinks, too, allowing about an hour or so to metabolize each drink.

If you want to help sober up a friend who has had too much to drink, don't bother walking arm in arm around the block. Walking muscles have to work harder, but muscle cells can't metabolize alcohol; only liver cells can. Remember that each person has a limited amount of the alcohol dehydrogenase enzyme that clears the blood at a steady rate. Time alone will do the job.

Nor will it help to give your friend a cup of coffee. Caffeine is a stimulant, but it won't speed up alcohol metabolism. The police say ruefully, "If you give a drunk a cup of coffee, you'll just have a wide-awake drunk on your hands." Table H7-6 (on p. 238) presents other alcohol myths.

People who have passed out from drinking need 24 hours to sober up completely. Let them sleep, but watch over them. Encourage them to lie on their sides, instead of their backs. That way, if they vomit, they won't choke.

Don't drive too soon after drinking. The lack of glucose for the brain's function and the length of time needed to clear the blood of alcohol make alcohol's adverse effects linger long after its blood concentration has fallen. Driving coordination is still impaired the morning *after* a night of drinking, even if the drinking was moderate. Responsible aircraft pilots know that they must allow 24 hours for their bodies to clear alcohol completely, and they refuse to fly any sooner. The Federal Aviation Administration and major airlines enforce this rule.

Society also pays a high price when a drinker gets behind the wheel of an automobile—and that occurs an estimated 123 million times a year.[13] Traffic accidents are the number one cause of death among young people (ages 5 to 32), and 38 percent of all traffic fatalities involve alcohol.[14] On the average, a person is killed in an alcohol-related traffic accident every 30 minutes. This rate is actually lower than in past decades thanks to the educational efforts of MADD (Mothers Against Drunk

TABLE H7-6 Myths and Truths concerning Alcohol

Myth:	Hard liquors such as rum, vodka, and tequila are more harmful than wine and beer.
Truth:	The damage caused by alcohol depends largely on the *amount* consumed. Compared with hard liquor, beer and wine have relatively low percentages of alcohol, but they are often consumed in larger quantities.
Myth:	Consuming alcohol with raw seafood diminishes the likelihood of getting hepatitis.
Truth:	People have eaten contaminated oysters while drinking alcoholic beverages and not gotten as sick as those who were not drinking. But do not be misled: hepatitis is too serious an illness for anyone to depend on alcohol for protection.
Myth:	Alcohol stimulates the appetite.
Truth:	For some people, alcohol may stimulate appetite, but it seems to have the opposite effect in heavy drinkers. Heavy drinkers tend to eat poorly and suffer malnutrition.
Myth:	Drinking alcohol reduces the risk of heart disease.
Truth:	Moderate alcohol consumption is associated with a lower risk for heart disease in some people (see Chapter 26 for more details). Higher intakes, however, raise the risks for high blood pressure, stroke, heart disease, some cancers, accidents, violence, suicide, birth defects, and deaths in general. Furthermore, excessive alcohol consumption damages the liver, pancreas, brain, and heart. No authority recommends that nondrinkers begin drinking alcoholic beverages to obtain health benefits.
Myth:	Wine increases the body's absorption of minerals.
Truth:	Wine may increase the body's absorption of potassium, calcium, phosphorus, magnesium, and zinc, but the alcohol in wine also promotes the body's excretion of these minerals, so no benefit is gained.
Myth:	Alcohol is legal and, therefore, not a drug.
Truth:	Alcohol is legal for adults 21 years old and older, but it is also a drug—a substance that alters one or more of the body's functions.
Myth:	A shot of alcohol warms you up.
Truth:	Alcohol diverts blood flow to the skin making you *feel* warmer, but it actually cools the body.
Myth:	Wine and beer are mild; they do not lead to alcoholism.
Truth:	Alcoholism is not related to the kind of beverage, but rather to the quantity and frequency of consumption.
Myth:	Mixing different types of drinks gives you a hangover.
Truth:	Too much alcohol in any form produces a hangover.
Myth:	Alcohol is a stimulant.
Truth:	People think alcohol is a stimulant because it seems to relieve inhibitions, but it does so by depressing the activity of the brain. Alcohol is medically defined as a depressant drug.
Myth:	Beer is a great source of carbohydrate, vitamins, minerals, and fluids.
Truth:	Beer does provide some carbohydrate, but most of its kcalories come from alcohol. The few vitamins and minerals in beer cannot compete with rich food sources. And the diuretic effect of alcohol causes the body to lose more fluid in urine than is provided by the beer.

Driving), SADD (Students Against Drunk Driving or Students Against Destructive Decisions), the implementation of designated driver programs, a higher minimum drinking age in many states, and the severe legal consequences of driving while under the influence (DUI). In addition to traffic fatalities, alcohol use has been implicated in most of the other deaths among young people, including drownings, falls, suicides, and homicides.[15]

HEALTHY PEOPLE 2010

Reduce deaths and injuries caused by alcohol- and drug-related motor vehicle crashes.

Look again at the drawing of the brain in Figure H7-4 and note that when someone drinks, judgment fails first. Judgment might tell a person to limit alcohol consumption to two drinks at a party, but if the first drink takes judgment away, many more drinks may follow. The failure to stop drinking as planned, on repeated occasions, is a danger sign warning that the person should not drink at all. The accompanying Nutrition on the Net and Appendix F provide addresses for organizations that offer information about alcohol and alcohol abuse.

Ethanol interferes with a multitude of chemical and hormonal reactions in the body—many more than have been enumerated here. With heavy alcohol consumption, the potential for harm is great. The best way to escape the harmful effects of alcohol is, of course, to refuse alcohol altogether. If you do drink alcoholic beverages, do so with care, and in moderation.

Nutrition on the Net

•WEBSITES•

Access these websites for further study of topics covered in this highlight.

- Find updates and quick links to these and other nutrition-related sites at our website: **www.wadsworth.com/nutrition**
- Search for "alcohol" at the U.S. Government health site: **www.healthfinder.gov**

- Gather information on alcohol and drug abuse from the National Clearinghouse for Alcohol and Drug Information (NCADI): **www.health.org**
- Learn more about alcoholism and drug dependence from the National Council on Alcoholism and Drug Dependence (NCADD): **www.ncadd.org**
- Find help for a family alcohol problem from Alateen and Al-Anon Family support groups: **www.al-anon.alateen.org**
- Find help for an alcohol or drug problem from Alcoholics Anonymous (AA) or Narcotics Anonymous: **www.aa.org** or **www.wsoinc.com**
- Get tips for hosting a safe party from Mothers Against Drunk Driving (MADD): **www.madd.org/programs/safe_party.shtml**

References

1 H. K. Seitz and C. M. Oneta, Gastrointestinal alcohol dehydrogenase, *Nutrition Reviews* 56 (1998): 52–60.

2 M. S. Westerterp-Plantenga and C. R. Verwegen, The appetizing effect of an aperitif in overweight and normal-weight humans, *American Journal of Clinical Nutrition* 69 (1999): 205–212.

3 P. M. Suter, E. Häsler, and W. Vetter, Effects of alcohol on energy metabolism and body weight regulation: Is alcohol a risk factor for obesity? *Nutrition Reviews* 55 (1997): 157–171.

4 Y. Sakurai and coauthors, Relation of total and beverage-specific alcohol intake to body mass index and waist-to-hip ratio: A study of self-defense officials in Japan, *European Journal of Epidemiology* 13 (1997): 893–898; B. B. Duncan and coauthors, Association of the waist-to-hip ratio is different with wine than with beer or hard liquor consumption, *American Journal of Epidemiology* 142 (1995): 1034–1038.

5 M. L. Cravo and coauthors, Hyperhomocysteinemia in chronic alcoholism: Correlation with folate, vitamin B-12, and vitamin B-6 status, *American Journal of Clinical Nutrition* 63 (1996): 220–224.

6 Folate, alcohol, methionine, and colon cancer risk: Is there a unifying theme? *Nutrition Reviews* 52 (1994): 18–20.

7 H. Wechsler and coauthors, College binge drinking in the 1990s: A continuing problem—Results of the Harvard School of Public Health 1999 College Alcohol Study, *Journal of American College Health* 48 (2000): 199–210.

8 M. J. Williams, N. J. Restieaux, and C. J. Low, Myocardial infarction in young people with normal coronary arteries, *Heart* 79 (1998): 191–194.

9 P. W. Meilman, J. S. Leichliter, and C. A. Presley, Greeks and athletes: Who drinks more? *Journal of American College Health* 47 (1999): 187–190; B. E. Borsari and K. B. Carey, Understanding fraternity drinking: Five recurring themes in the literature, 1980–1998, *Journal of American College Health* 48 (1999): 30–37; H. Wechsler and coauthors, Binge drinking, tobacco, and illicit drug use and involvement in college athletes: A survey of students at 140 American colleges, *Journal of American College Health* 45 (1995): 195–200.

10 Wechsler and coauthors, 2000; K. A. Douglas and coauthors, Results from the 1995 National College Health Risk Behavior Survey, *Journal of American College Health* 46 (1997): 55–66; D. M. Ferguson and M. T. Lynskey, Alcohol misuse and adolescent sexual behaviors and risk taking, *Pediatrics* 98 (1996): 91–96.

11 H. Wechsler and coauthors, College alcohol use: A full or empty glass? *Journal of American College Health* 47 (1999): 247–252.

12 Committee on Substance Abuse, Alcohol use and abuse: A pediatric concern, *Pediatrics* 95 (1995): 439–442.

13 S. Liu and coauthors, Prevalence of alcohol-impaired driving: Results from a national self-reported survey of health behaviors, *Journal of the American Medical Association* 277 (1997): 122–125.

14 Alcohol involvement in fatal motor-vehicle crashes—United States, 1997–1998, *Morbidity and Mortality Weekly Report* 38 (1999): 1086–1087.

15 Committee on Substance Abuse, 1995.

CHAPTER 8
Energy Balance and Body Composition

The body's remarkable machinery can cope with many extremes of diet. As you have seen, it can convert both carbohydrate (glucose) and protein (amino acids) to fat. To some extent, it can convert amino acids to glucose. To a very limited extent, it can even convert fat (the glycerol portion) to glucose. But a grossly unbalanced diet imposes hardships on the body. If energy intake is too low or if too little carbohydrate or protein is supplied, the body must degrade its own lean tissue to meet its glucose and protein needs. If energy intake is too high or if fat is abundant, the body stores fat.

Overfatness◆ and underweight both result from unbalanced energy budgets. The simple picture is as follows. Overfat people have consumed more food energy than they have spent and have banked the surplus as body fat. To reduce body fat, they need to spend more energy than they take in from food. In contrast, underweight people have consumed too little food energy to support their bodies' activities and so have depleted their bodies' fat stores and possibly some of their lean tissues as well. To gain weight, these people need to take in more food energy than they expend. As you will see, though, the details of the body's weight regulation are quite complex. This chapter describes energy balance and body composition and examines the health problems associated with having too much or too little body fat; the next chapter presents strategies toward resolving these problems.

The term *overfat* refers to an excess of body fat, which is not necessarily the same as *overweight*, as a later section of the chapter explains. ◆

◆ A healthy person may take in close to a million kcalories a year and expend more than 99 percent of them, thus maintaining a relatively stable weight for years on end.◆

ENERGY BALANCE

People spend energy continuously and eat periodically to refuel. Ideally, their energy intakes cover their energy expenditures without too much excess. Excess energy is stored as fat, and stored fat is used for energy between meals. The amount of body fat a person deposits in, or withdraws from, "storage" on any given day depends on the energy balance for that day—the amount consumed (energy in) versus the amount expended (energy out). When a person is maintaining weight, energy in equals energy out.

Most people maintain a steady energy balance over time. On any given day, they may eat a little more or a little less than usual, and their weight may go up or down a pound or two, but for the most part, they stay in balance. When the balance shifts, their weight changes. For each 3500 kcalories ◆ eaten in excess, a pound of body fat is stored; similarly, a pound of fat is lost for each 3500 kcalories expended beyond those consumed.

A reasonable rate of weight loss for overweight people is ½ to 2 pounds a week,◆ or 10 percent of body weight over six months. For a person weighing 250 pounds, a 10 percent loss is 25 pounds, or about 1 pound a week for six months. Such gradual weight losses are more likely to be maintained than rapid losses. If food energy is restricted too severely, dieters lose lean tissue and may not receive enough nutrients. In addition, restrictive eating may set in motion the unhealthy cycle of restrictive dieting and binge eating.

Besides, quick changes in weight are not just changes in fat. Weight gained or lost rapidly includes some fat, large amounts of fluid, and some lean tissues such as muscles and bone minerals. (Because water constitutes about 60 percent of an adult's body weight, retention or loss of water influences body weight.) Even over the long term, the composition of weight gained or lost is normally about 75 percent fat and 25 percent lean. During starvation, losses of fat and lean are about equal. Invariably, though, *fat* gains and losses are gradual.

1 lb body fat = 3500 kcal. Body fat, or adipose tissue, is composed of a mixture of mostly fat, some protein, and water. A pound of body fat (454 g) is approximately 87% fat, or (454 × 0.87) 395 g, and 395 g × 9 kcal/g = 3555 kcal. ◆

Safe rate for weight loss: ◆
- ½ to 2 lb/week.
- 10% body weight/6 mo.

IN SUMMARY

When the energy consumed equals the energy expended, the person is in energy balance and body weight is stable. If more energy is taken in than is expended, the person gains weight. If more energy is spent than is taken in, the person loses weight. The next two sections examine the two sides of the energy-balance equation: energy in and energy out.

When energy in balances with energy out, a person's body weight is stable.

FIGURE 8-1 **Bomb Calorimeter**

When food is burned, the chemical bonds between the carbons and hydrogens are broken, and energy is released in the form of heat. The amount of heat generated provides a direct measure of the amount of energy stored in the food's chemical bonds.

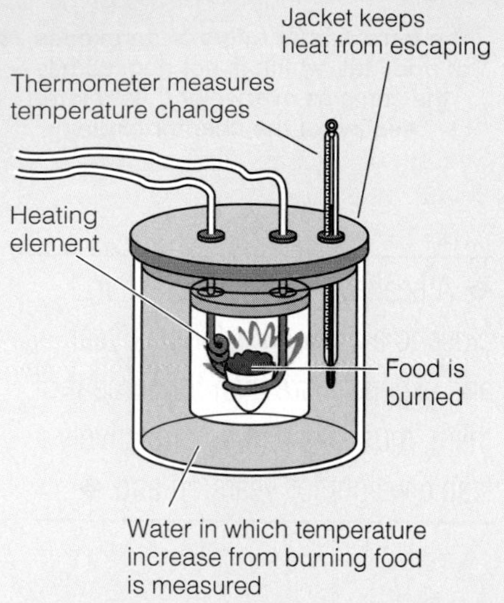

◆ **Reminder: A *kcalorie* is a unit of *heat* energy. As Chapter 1 mentioned, many scientists measure food energy in kilojoules instead. Conversion factors for these and other measures are in the Aids to Calculation section on the last two pages of the book.**

◆ **Food energy values can be determined by:**
- **Direct calorimetry,** which measures the amount of heat released.
- **Indirect calorimetry,** which measures the amount of oxygen consumed.

◆ **The number of kcalories that the body derives from a food, as contrasted with the number of kcalories determined by calorimetry, is the physiological fuel value.**

◆ **Reminder:**
- 1 g carbohydrate = 4 kcal.
- 1 g fat = 9 kcal.
- 1 g protein = 4 kcal.
- 1 g alcohol = 7 kcal.

bomb calorimeter (KAL-oh-RIM-eh-ter): an instrument that measures the heat energy released when foods are burned, thus providing an estimate of the potential energy of foods.
- **calor** = heat
- **metron** = measure

hunger: the physiological drive for food that initiates food-seeking behavior.

ENERGY IN: THE kCALORIES FOODS PROVIDE

Foods and beverages are the "energy in" part of the energy-balance equation. How much energy a person receives depends on the composition of the foods and beverages and on the amount the person eats and drinks.

Food Composition

To find out how many kcalories a food provides, a scientist can burn the food in a **bomb calorimeter** (see Figure 8-1). When the food burns, the chemical bonds between the carbon and hydrogen atoms break, releasing energy in the form of heat. The amount of heat given off provides a *direct* measure of the food's energy value (remember◆ that kcalories are units of heat energy). In addition to releasing heat, these reactions generate carbon dioxide and water—just as the body's cells do when they metabolize the energy-yielding nutrients. When the food burns and the chemical bonds break, the carbons (C) and hydrogens (H) combine with oxygen (O) to form carbon dioxide (CO_2) and water (H_2O). The amount of oxygen consumed gives an *indirect* measure◆ of the amount of energy released.

A bomb calorimeter measures the available energy in foods, but overstates the amount of energy that the human body◆ derives from foods. The body is less efficient than a calorimeter and cannot metabolize all of a food's energy-yielding nutrients all the way to carbon dioxide and water. Researchers can correct for this discrepancy mathematically to create useful tables of the energy values of foods (such as Appendix H). These values provide reasonable estimates, but do not reflect the *precise* amount of energy a person will derive from the foods consumed.

The energy values of foods can also be computed from the amounts of carbohydrate, fat, and protein (and alcohol, if present) in the foods.* For example, a food containing 12 grams of carbohydrate, 5 grams of fat, and 8 grams of protein would provide 48 carbohydrate kcalories, 45 fat kcalories, and 32 protein kcalories,◆ for a total of 125 kcalories. (To review how to calculate the energy available from foods, turn to p. 7.)

Food Intake

To achieve energy balance, the body must meet its needs without taking in too much or too little energy. Somehow the body decides how much and how often to eat—when to start eating and when to stop. As you will see, many signals initiate or delay eating.

• **Hunger** • People eat for a variety of reasons, most obviously (although not necessarily most commonly) because they are hungry. Most people recognize **hunger** as an irritating feeling that prompts thoughts of food and motivates them to start eating. In the body, hunger is the physiological response to a need for food triggered by chemical messengers originating and acting in the brain, primarily in the hypothalamus.[1] Hunger can be influenced by nutrients in the bloodstream, the size and composition of the preceding meal, customary eating patterns, climate (heat reduces food intake; cold increases it), exercise, hormones, and physical and mental diseases.

The stomach is ideally designed to handle periodic batches of food, and people typically eat meals at roughly four-hour intervals. Four hours after a meal,

*Some of the food energy values in the table of food composition in Appendix H were derived by bomb calorimetry, and many were calculated from their energy-yielding nutrient contents.

most, if not all, of the food has left the stomach and been absorbed by the intestine. Most people do not feel like eating again until the stomach is either empty or almost so. Even then, a person may not feel hungry for quite a while.

The body's hunger response adapts to accommodate changes in food intake. People who restrict their food intakes may feel pangs of hunger for the first few days, but these sensations diminish with time. After the body has adapted to less food, eating a large, energy-rich meal makes the person feel uncomfortable, in part because the stomach's capacity has diminished. People can adapt to eating excessive amounts of food as well, until a normal-size meal no longer feels satisfying. This observation may partly explain why obesity is on the rise: many people's stomachs have adapted to accommodate giant burgers, super-sized fries, huge candy bars, and 32-ounce soft drinks.

Receptors in the GI tract also adapt and change their responses depending on nutrient intake. After two weeks on a high-fat diet, for example, digestion and absorption of fat are accelerated; the GI tract has adapted to handle the increase in fat intake efficiently. Such findings have interesting implications for the development of obesity.

• **Appetite** • Hunger is only one of the signals determining whether a person will eat. **Appetite** also initiates eating. A person may experience appetite without hunger, for example, when presented with a hot piece of homemade apple pie after having eaten a large dinner. In contrast, a person may feel hungry but have no appetite for food when faced with unfamiliar foods, a stressful situation, or illness; in such circumstances, eating becomes a chore.

• **Satiation** • During the course of a meal, as food enters the GI tract and hunger diminishes, **satiation** develops. Receptors in the stomach stretch, and the person begins to feel too uncomfortable to continue eating. Nutrients in the small intestine trigger the release of GI hormones (such as cholecystokinin) and the stimulation of nerves. Together, gastric distention, nutrients in the small intestine, and GI hormones send messages about the amount of food eaten and the kinds of nutrients received to the hypothalamus. The response: satiation occurs and people stop eating.

• **Satiety** • After a meal, the feeling of **satiety** continues to suppress hunger and allows a person to not eat again for a while. Whereas *satiation* informs us of when to "stop eating," *satiety* maintains the signal to "not start eating again." Figure 8-2 on p. 244 summarizes the relationships among hunger, satiation, and satiety. Of course, people can override these signals, especially when presented with stressful situations or favorite foods.

• **Overriding Hunger and Satiety Signals** • Not surprisingly, eating can be triggered by signals other than hunger, even when the body does not need food. Some people experience food cravings when they are bored or anxious. In fact, they may eat in response to any kind of stress,◆ negative or positive. ("What do I do when I'm grieving? Eat. What do I do when I'm celebrating? Eat!") Some people respond to external cues such as the time of day ("It's time to eat") or the availability, sight, and taste of food ("I'd love a piece of chocolate even though I'm stuffed"). Being presented with a variety of foods also stimulates eating; people eating one food until satisfied may override this feeling and begin eating enthusiastically again when offered a fresh selection of different foods. Such behavior can easily lead to weight gain.

Eating can also be suppressed by signals other than satiety, even when a person is hungry. People with the eating disorder◆ anorexia nervosa, for example, use tremendous discipline to ignore the pangs of hunger. Some people simply cannot eat during times of stress, negative or positive. ("I'm too sad to eat. I'm too excited to eat!") Why some people overeat in response to stress and others cannot eat at all remains a bit of a mystery. Factors that appear to be involved include how the person perceives the stress and whether usual eating behaviors are restrained.

◆ Eating in response to arousal is called **stress eating.**

◆ Highlight 9 features eating disorders.

appetite: the integrated response to the sight, smell, thought, or taste of food that initiates or delays eating.

satiation (say-she-AY-shun): the feeling of satisfaction and fullness that occurs during a meal and halts eating. Satiation determines how much food is consumed during a meal.

satiety (sah-TIE-eh-tee): the feeling of satisfaction that occurs after a meal and inhibits eating until the next meal. Satiety determines how much time passes between meals.

FIGURE 8-2 Hunger, Satiation, and Satiety

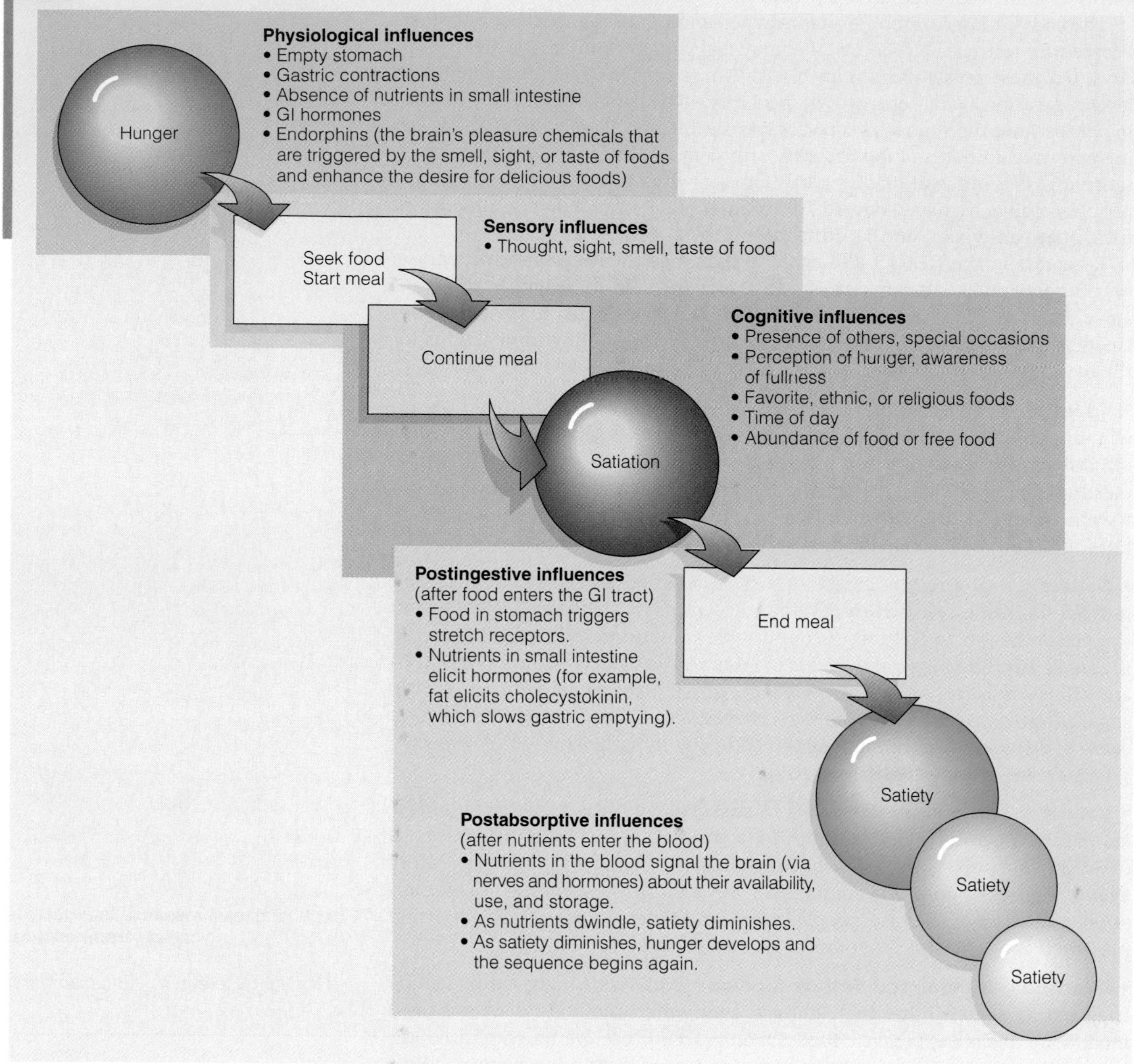

• Nutrients, Satiation, and Satiety • The extent to which foods produce satiation and sustain satiety depends in part on the nutrient composition of a meal. Of the three energy-yielding nutrients, protein is the most **satiating.** Foods rich in complex carbohydrates and fibers also effectively extend the duration of satiety by filling the stomach and delaying the absorption of nutrients. In contrast, fat has a weak satiating effect; consequently, eating high-fat foods leads to passive overconsumption.[2] High-fat foods offer flavor, which entices people to eat more, and energy density, which delivers more kcalories per bite, but produce little satiation during a meal and little satiety after a meal.

Eating high-fat foods while trying to limit energy intake can leave a person feeling hungry. People trying to manage their weight may find it easier to limit energy intake by selecting foods based on satiety. High-fat foods such as croissants, doughnuts, peanuts, and potato chips please the palate and stimulate the

satiating: having the power to suppress hunger and inhibit eating.

FIGURE 8-3 How Fat Influences Serving Sizes

For the same size serving, peanuts deliver more than 15 times the kcalories and 20 times the fat of popcorn.

Popcorn offers twice the satiety of peanuts. For the same number of kcalories, a person can have a few high-fat peanuts or almost 2 cups of high-fiber popcorn. (This comparison used oil-based popcorn; using air-popped popcorn would double the amount of popcorn in this example.)

appetite, but score low on satiety; by comparison, simple, whole foods such as potatoes, apples, oranges, whole-grain pastas, fish, and steak are highly satiating—and they provide a rich array of nutrients. Instead of feeling deprived eating small servings of high-fat foods, a person can feel satisfied eating large servings of high-protein and high-fiber foods. Serving size correlates directly with a food's satiety index, and Figure 8-3 illustrates how fat influences serving size.

• **Message Central—The Hypothalamus** • As you can see, eating is a complex behavior controlled by a variety of psychological, social, metabolic, and physiological factors. The **hypothalamus** appears to be the control center, integrating messages about energy intake, expenditure, and storage from other parts of the brain and from the mouth, GI tract, and liver. Some of these messages influence satiation, controlling the size of a meal; others influence satiety, determining the frequency of meals.

Dozens of chemicals in the brain participate in appetite control and energy balance.[3] By understanding the action of these brain chemicals, researchers may one day be able to control appetite. Their greatest challenge now is in sorting out the many actions of these brain chemicals. For example, one of these chemicals, **neuropeptide Y,** causes carbohydrate cravings, initiates eating, decreases energy expenditure, and increases fat storage—all factors favoring a positive energy balance and weight gain.

IN SUMMARY

A mixture of signals governs people's eating behaviors. Hunger, satiation, and satiety each result from several stimuli generated by the nervous and hormonal systems. Superimposed on these are complex factors involving emotions, habits, and other aspects of human behavior.

hypothalamus (high-po-THAL-ah-mus): a brain center that controls activities such as maintenance of water balance, regulation of body temperature, and control of appetite.

neuropeptide Y: a chemical produced in the brain that stimulates appetite, diminishes energy expenditure, and increases fat storage.

ENERGY OUT: THE kCALORIES THE BODY SPENDS

Chapter 7 explained that heat is released whenever the body breaks down carbohydrate, fat, or protein for energy and again when that energy is used to do

FIGURE 8-4 **Components of Energy Expenditure**

The amount of energy spent in a day differs for each individual, but in general, basal metabolism is the largest component of energy expenditure (60 to 65 percent), and the thermic effect of food is the smallest (only 10 percent). The amount spent in voluntary physical activities has the greatest variability, depending on a person's activity patterns.

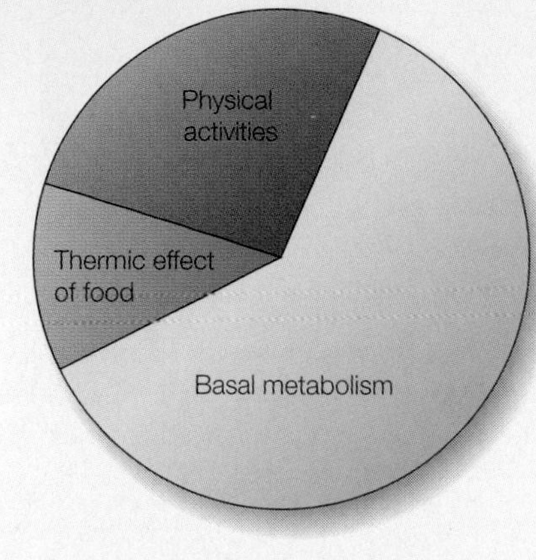

work. The work itself, as it is done, generates heat as well. The body's generation of heat is known as **thermogenesis,** and it can be measured to determine the amount of energy expended.◆ The total energy a body spends reflects three main categories of thermogenesis:

- Basal thermogenesis (basal metabolism).
- Exercise-induced thermogenesis (physical activity).
- Diet-induced thermogenesis (thermic effect of food).

A fourth category is sometimes involved:

- Adaptive thermogenesis (energy of adaptation).

◆ **Energy expenditure can be determined by:**
- **Direct calorimetry,** which measures the amount of heat released.
- **Indirect calorimetry,** which measures the amount of oxygen consumed and carbon dioxide expelled.

Components of Energy Expenditure

People spend energy when they are physically active, of course, but they also spend energy when they are resting quietly. In fact, quiet metabolic activities account for the lion's share of most people's energy expenditures, as Figure 8-4 shows.

• Basal Metabolism • About two-thirds of the energy the average person spends in a day supports the body's **basal metabolism.** Metabolic activities maintain the body temperature, keep the lungs inhaling and exhaling air, the bone marrow making new red blood cells, the heart beating 100,000 times a day, and the kidneys filtering wastes—in short, they support all the basic processes of life.

The **basal metabolic rate (BMR)** is the rate at which the body spends energy for these maintenance activities. The rate may vary dramatically from person to person and may vary for the same individual with a change in circumstance or physical condition. The rate is slowest when a person is sleeping undisturbed, but it is usually measured in a room with a comfortable temperature when the person is lying still after a restful sleep and is not digesting any food. This measure of energy output is called the **resting metabolic rate (RMR).**

In general, the more a person weighs, the more *total* energy is expended on basal metabolism, but the amount of energy *per pound* of body weight may be lower. For example, an adult's BMR might be 1500 kcalories per day and an infant's only 500, but compared to body weight, the infant's BMR is more than twice as fast. Similarly, a normal-weight adult may have a metabolic rate one and a half times that of an obese adult when compared to body weight.

Table 8-1 summarizes the factors that raise and lower the BMR. For the most part, the BMR is highest in people who are growing (children and pregnant women) and in those with considerable **lean body mass** (physically fit people and males). One way to increase the BMR then is to participate in endurance and strength-building activities regularly to maximize lean body mass. The BMR is also high in people who are tall and so have a large surface area for their weight, in people with fever or under stress, and in people with highly active thyroid glands.

The BMR declines during adulthood as lean body mass diminishes.[4] This change in body composition occurs, in part, because some hormones that influence metabolism become more, or less, active as a person ages. **Voluntary activities** tend to decline as well, bringing the average reduction in energy expenditure to about 5 percent per decade. The decline in the BMR that occurs when a person reduces voluntary activity reflects the loss of

thermogenesis: the generation of heat; used in physiology and nutrition studies as an index of how much energy the body is spending.

basal metabolism: the energy needed to maintain life when a body is at complete digestive, physical, and emotional rest.

basal metabolic rate (BMR): the rate of energy use for metabolism under specified conditions: after a 12-hour fast and restful sleep, without any physical activity or emotional excitement, and in a comfortable setting. It is usually expressed as kcalories per kilogram body weight per hour. (Table 8-3 on p. 249 provides equations for estimating BMR.)

resting metabolic rate (RMR): similar to the BMR, a measure of a person at rest in a comfortable setting, but with less stringent criteria for the number of hours fasting. Consequently, the RMR is slightly higher than the BMR.

lean body mass: the weight of the body minus the fat content.

voluntary activities: conscious and deliberate muscular work—walking, lifting, climbing, and other physical activities. In contrast, involuntary activities occur independently, without conscious will or knowledge—heart beating, lungs breathing, and other activities critical to maintaining life.

TABLE 8-1 Factors That Affect the BMR

Factor	Effect on BMR
Age	Lean body mass diminishes with age, slowing the BMR.[a]
Height	In tall, thin people, the BMR is higher.[b]
Growth	In children and pregnant women, the BMR is higher.
Body composition	The more lean tissue, the higher the BMR (which is why males usually have a higher BMR than females). The more fat tissue, the lower the BMR.
Fever	Fever raises the BMR.[c]
Stresses	Stresses (including many diseases and certain drugs) raise the BMR.
Environmental temperature	Both heat and cold raise the BMR.
Fasting/starvation	Fasting/starvation lowers the BMR.[d]
Malnutrition	Malnutrition lowers the BMR.
Hormones	The thyroid hormone thyroxin, for example, can speed up or slow down the BMR.[e]
Smoking	Nicotine increases energy expenditure.
Caffeine	Caffeine increases energy expenditure.
Sleep	BMR is lowest when sleeping.

[a]The BMR begins to decrease in early adulthood (after growth and development cease) at a rate of about 2 percent/decade. A reduction in voluntary activity as well brings the total decline in energy expenditure to 5 percent/decade.
[b]If two people weigh the same, the taller, thinner person will have the faster metabolic rate, reflecting the greater skin surface, through which heat is lost by radiation, in proportion to the body's volume (see margin drawing).
[c]Fever raises the BMR by 7 percent for each degree Fahrenheit.
[d]Prolonged starvation reduces the total amount of metabolically active lean tissue in the body, although the decline occurs sooner and to a greater extent than body losses alone can explain. More likely, the neural and hormonal changes that accompany fasting are responsible for changes in the BMR.
[e]The thyroid gland releases hormones that travel to the cells and influence cellular metabolism. Thyroid hormone activity can speed up or slow down the rate of metabolism by as much as 50 percent.

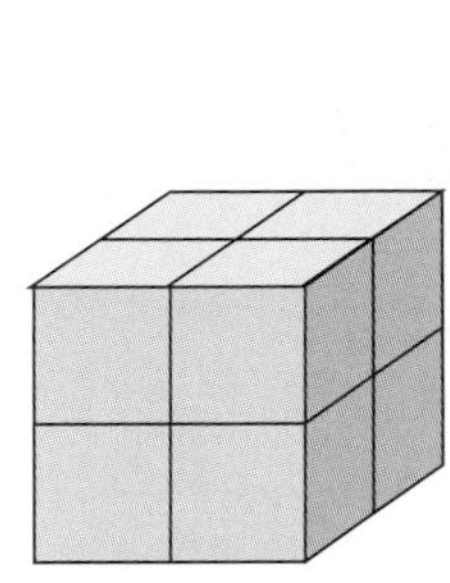
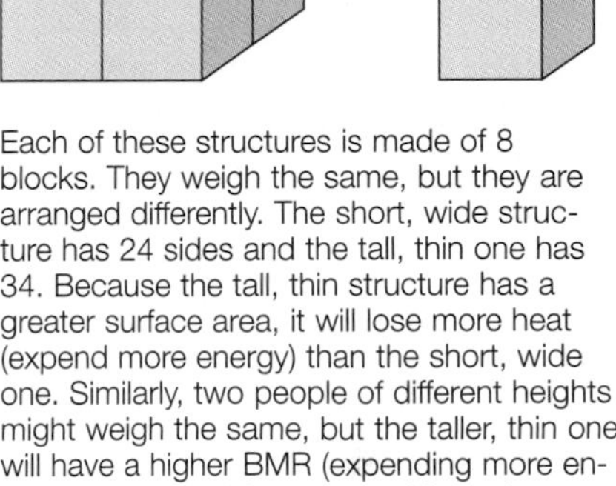

Each of these structures is made of 8 blocks. They weigh the same, but they are arranged differently. The short, wide structure has 24 sides and the tall, thin one has 34. Because the tall, thin structure has a greater surface area, it will lose more heat (expend more energy) than the short, wide one. Similarly, two people of different heights might weigh the same, but the taller, thin one will have a higher BMR (expending more energy) because of the greater skin surface.

lean body mass and may be prevented with ongoing physical activity. The BMR also slows down during fasting and malnutrition.

• **Physical Activity** • The second component of a person's energy output is physical activity: voluntary movement of the skeletal muscles and support systems. Physical activity is the most variable—and the most changeable—component of energy expenditure. Consequently, its influence on both weight gain and weight loss can be significant.

During physical activity, the muscles need extra energy to move, and the heart and lungs need extra energy to deliver nutrients and oxygen and dispose of wastes. The amount of energy needed for any activity, whether playing tennis or studying for an exam, depends on three factors: muscle mass, body weight, and activity. The larger the muscle mass required and the heavier the weight of the body part being moved, the more energy is spent. Table 8-2 on p. 248 gives average energy expenditures for people of different body weights engaged in various activities and shows that a heavy person usually uses more energy per minute to perform a task than a light person does. The activity's duration, frequency, and intensity also influence energy expenditure: the longer, the more frequent, and the more intense the activity, the more kcalories spent.

• **Thermic Effect of Food** • When a person eats, the GI tract muscles speed up their rhythmic contractions, and the cells that manufacture and secrete digestive juices begin their tasks. This acceleration of activity produces heat and is known as the **thermic effect of food (TEF).**

The thermic effect of food is proportional to the food energy taken in and is usually estimated at 10 percent of energy intake. Thus a person who ingests 2000 kcalories in a day probably spends about 200 kcalories on the thermic effect of food. Because the thermic effect of food reflects the body's digestion

thermic effect of food (TEF): an estimation of the energy required to process food (digest, absorb, transport, metabolize, and store ingested nutrients); also called the **specific dynamic effect (SDE)** of food or the **specific dynamic activity (SDA)** of food. The sum of the TEF and any increase in the metabolic rate due to overeating is known as **diet-induced thermogenesis (DIT).**

© 1998 Photo Disc Inc.

It feels like work and it may make you tired, but studying requires only a kcalorie or two per minute.

TABLE 8-2 Energy Spent on Various Activities

The values listed in this table reflect both the energy spent in physical activity *and* the amount used for BMR.

Activity	kCal/lb/min[a]	kCalories per Minute at Different Body Weights				
		110 lb	*125 lb*	*150 lb*	*175 lb*	*200 lb*
Aerobic dance (vigorous)	.062	6.8	7.8	9.3	10.9	12.4
Basketball (vigorous, full court)	.097	10.7	12.1	14.6	17.0	19.4
Bicycling						
13 mph	.045	5.0	5.6	6.8	7.9	9.0
15 mph	.049	5.4	6.1	7.4	8.6	9.8
17 mph	.057	6.3	7.1	8.6	10.0	11.4
19 mph	.076	8.4	9.5	11.4	13.3	15.2
21 mph	.090	9.9	11.3	13.5	15.8	18.0
23 mph	.109	12.0	13.6	16.4	19.0	21.8
25 mph	.139	15.3	17.4	20.9	24.3	27.8
Canoeing, flat water, moderate pace	.045	5.0	5.6	6.8	7.9	9.0
Cross-country skiing						
8 mph	.104	11.4	13.0	15.6	18.2	20.8
Golf (carrying clubs)	.045	5.0	5.6	6.8	7.9	9.0
Handball	.078	8.6	9.8	11.7	13.7	15.6
Horseback riding (trot)	.052	5.7	6.5	7.8	9.1	10.4
Rowing (vigorous)	.097	10.7	12.1	14.6	17.0	19.4
Running						
5 mph	.061	6.7	7.6	9.2	10.7	12.2
6 mph	.074	8.1	9.2	11.1	13.0	14.8
7.5 mph	.094	10.3	11.8	14.1	16.4	18.8
9 mph	.103	11.3	12.9	15.5	18.0	20.6
10 mph	.114	12.5	14.3	17.1	20.0	22.9
11 mph	.131	14.4	16.4	19.7	22.9	26.2
Soccer (vigorous)	.097	10.7	12.1	14.6	17.0	19.4
Studying	.011	1.2	1.4	1.7	1.9	2.2
Swimming						
20 yd/min	.032	3.5	4.0	4.8	5.6	6.4
45 yd/min	.058	6.4	7.3	8.7	10.2	11.6
50 yd/min	.070	7.7	8.8	10.5	12.3	14.0
Table tennis (skilled)	.045	5.0	5.6	6.8	7.9	9.0
Tennis (beginner)	.032	3.5	4.0	4.8	5.6	6.4
Walking (brisk pace)						
3.5 mph	.035	3.9	4.4	5.2	6.1	7.0
4.5 mph	.048	5.3	6.0	7.2	8.4	9.6
Wheelchair basketball	.084	9.2	10.5	12.6	14.7	16.8
Wheeling self in wheelchair	.030	3.3	3.8	4.5	5.3	6.0

[a]To calculate kcalories spent per minute of activity for your own body weight, multiply kcal/lb/min by your exact weight and then multiply that number by the number of minutes spent in the activity. For example, if you weigh 142 pounds, and you want to know how many kcalories you spent doing 30 minutes of vigorous aerobic dance: $0.062 \times 142 = 8.8$ kcalories per minute; 8.8×30 (minutes) $= 264$ total kcalories spent.

and absorption activities, it is influenced by factors such as meal size, frequency, and composition; in general, the thermic effect of food is greater for high-carbohydrate foods than for high-fat foods and for a meal eaten all at once rather than spread out over a couple of hours. For most purposes, however, the thermic effect of food can be ignored when estimating energy expenditure because its contribution to total energy output is smaller than the probable errors involved in estimating overall energy intake and output.

• **Adaptive Thermogenesis** • Some additional energy is spent when a person must adapt to dramatically changed circumstances (**adaptive thermogenesis**). When the body has to adapt to physical conditioning, extreme cold, overfeeding, starvation, trauma, or other types of stress, it has extra work to do, building the tissues and producing the enzymes and hormones necessary to cope with the demand. In some circumstances, this energy makes a considerable difference in the total energy spent. Because this component of energy expenditure is so variable and specific to individuals, it is not included when calculating energy requirements.

TABLE 8-3 Equations for Estimating BMR from Body Weight

Gender and Age (yr)	Equation to Derive BMR in kCal/day
Males	
0–3	(60.9 × wt[a]) − 54
3–10	(22.7 × wt) + 495
10–18	(17.5 × wt) + 651
18–30	(15.3 × wt) + 679
30–60	(11.6 × wt) + 879
>60	(13.5 × wt) + 487
Females	
0–3	(61.0 × wt) − 51
3–10	(22.5 × wt) + 499
10–18	(12.2 × wt) + 746
18–30	(14.7 × wt) + 496
30–60	(8.7 × wt) + 829
>60	(10.5 × wt) + 596

[a]Weight expressed in kilograms.
SOURCE: Reprinted with permission from *Recommended Dietary Allowances*, 10th edition. Copyright 1989 by National Academy of Sciences. Published by the National Academy Press, Washington, D.C.

Estimating Energy Requirements

In calculating energy requirements,◆ the following components of energy expenditure are considered:

- Energy spent on basal metabolism.
- Energy spent on physical activities.
- Energy spent on digesting and metabolizing food.

First, energy spent on basal metabolism for each age-gender group is estimated using equations that consider age, gender, and weight, as shown in Table 8-3. (The "How to" on p. 250 shows a sample calculation.) Next, increments for physical activity are added, assuming the average person will be lightly to moderately active. Finally, increments for the influence of food are added, assuming that each person will meet energy needs by eating a mixed diet of ordinary foods.

To estimate the energy spent on physical activity, individual values such as those presented in Table 8-2 cannot be used. This process is too time-consuming and impractical for estimating the energy needs of a population. Instead, various activities are clustered according to the typical intensity of a day's efforts. Then an "activity factor" for each level of intensity for each gender is determined (see Table 8-4 on p. 250). This activity factor is multiplied by the BMR to yield a total that reflects physical activity *and* basal metabolic activity combined. Again, the "How to" on p. 250 shows a sample calculation; the margin◆ presents quick and easy formulas for estimating energy needs.

Average energy allowance: ◆
• Men: 2300–2900 kcal/day.
• Women: 1900–2200 kcal/day.

Quick and easy estimate for energy needs: ◆
• Men: kg x 24 = kcal/day.
• Women: kg x 23 = kcal/day.

IN SUMMARY

A person takes in energy from food and, on average, spends most of it on basal metabolic activities, some of it on physical activities, and a little on the thermic effect of food. Because energy requirements vary from person to person, such factors as age, gender, and weight must be considered when calculating energy spent on basal metabolism, and the intensity and duration of the activity must be taken into account when calculating expenditures on physical activities.

BODY WEIGHT, BODY COMPOSITION, AND HEALTH

A person 5 feet 10 inches tall who weighs 150 pounds may carry only about 30 of those pounds as fat. The rest is mostly water and lean tissues—muscles, organs such as the heart and liver, and the bones of the skeleton. Direct measures of **body composition** are impossible in living human beings; instead, researchers assess body composition indirectly based on the following assumption:

$$\text{Body weight} = \text{fat} + \text{lean tissue (including water)}.$$

Weight gains and losses tell us nothing about how the body's composition may have changed, yet weight is the measure most people use to judge their

adaptive thermogenesis: adjustments in energy expenditure related to changes in environment such as extreme cold and to physiological events such as overfeeding, trauma, and changes in hormone status.

body composition: the proportions of muscle, bone, fat, and other tissue that make up a person's total body weight.

HOW TO Estimate Daily Energy Output

Basal Metabolism

One way to estimate your energy output for basal metabolism is to use Table 8-3 on p. 249.[a] For example, a 20-year-old male who weighed 160 pounds would select the equation appropriate for his gender and age range:

$$(15.3 \times \text{wt}) + 679.$$

First, he would convert his weight from pounds to kilograms, if necessary:

$$160 \text{ lb} \div 2.2 \text{ lb/kg} = 72.7 \text{ kg}.$$

Then, he would insert his weight into the equation:

$$(15.3 \times 72.7 \text{ kg}) + 679 = 1791 \text{ kcal/day}.$$

The estimated energy expenditure to cover basal metabolism for a 20-year-old male who weighs 160 pounds is 1791 kcalories/day.

A shortcut method uses an easy-to-remember formula for estimating basal energy needs: 1 kcal/kg/hr for men (or 0.9 for women). For example, rounding off 72.7 kilograms to 73:

$$1 \text{ kcal} \times 73 \text{ kg} \times 24 \text{ hr} = 1752 \text{ kcal/day}.$$

The difference between 1791 and 1752 is insignificant and acceptable in estimations such as these.

Basal Metabolism and Voluntary Physical Activity

To account for the energy used in physical activities as well, review the activities listed in Table 8-4 and determine which level of intensity typifies your average daily activity. Then multiply the selected activity factor by your value for basal metabolism. For example, if the man introduced above engages in mostly light activity, his activity factor would be 1.6. Multiply this factor by his basal metabolism kcalories:

$$1.6 \times 1791 = 2866 \text{ kcal/day}.$$

The result, 2866 kcalories/day, expresses his *total* daily energy needs.

Alternatively, total energy expenditure can be estimated in one step based on body weight as shown in the last column of Table 8-4. As an example, for a 160-pound man engaged in mostly light activity:

$$38 \text{ kcal} \times 73 \text{ kg} = 2774 \text{ kcal/day}.$$

Again, the difference between 2866 and 2774 is insignificant and acceptable. Either way, the man's total energy needs are about 2800 kcalories/day. Keep in mind that these estimates of daily energy expenditure are just that—*estimates.*

[a] In the United States, many researchers use another set of equations known as the Harris-Benedict method to estimate energy output for BMR. The values calculated from the two sets of equations do not differ significantly. Harris-Benedict equations:

For men: BMR = 66 + (13.7 × wt in kg) + (5 × ht in cm) − (6.8 × age in yr).
For women: BMR = 655 + (9.6 × wt in kg) + (1.8 × ht in cm) − (4.7 × age in yr).

For equations to convert kg and cm, turn to the Aids to Calculation section on the last two pages of the book.

© Rick Schiff

At 6 feet 3 inches tall and 245 pounds, Mike O'Hearn would be considered overweight by most height-weight standards. Yet he is clearly not overfat.

TABLE 8-4 Estimating Daily Energy Expenditure at Various Levels of Physical Activity

Level of Intensity	Type of Activity	Activity Factor (× BMR)	Energy Expenditure (kcal/kg/day)
Very light	Seated and standing activities, painting trades, driving, laboratory work, typing, sewing, ironing, cooking, playing cards, playing a musical instrument	1.3 (men) 1.3 (women)	31 30
Light	Walking on a level surface at 2.5 to 3 mph, garage work, electrical trades, carpentry, restaurant trades, housecleaning, child care, golf, sailing, table tennis	1.6 (men) 1.5 (women)	38 35
Moderate	Walking 3.5 to 4 mph, weeding and hoeing, carrying a load, cycling, skiing, tennis, dancing	1.7 (men) 1.6 (women)	41 37
Heavy	Walking with a load uphill, tree felling, heavy manual digging, basketball, climbing, football, soccer	2.1 (men) 1.9 (women)	50 44
Exceptional	Training in professional or world-class athletic events	2.4 (men) 2.2 (women)	58 51

Note: The second section of the "How to" above describes how to use this table and explains that the estimate reflects both physical activity *and* basal metabolic activity.
SOURCE: Adapted with permission from *Recommended Dietary Allowances,* 10th edition. Copyright 1989 by the National Academy of Sciences. Published by the National Academy Press, Washington, D.C.

"fatness." For many people, overweight means overfat, but this is not always the case. Athletes with dense bones and well-developed muscles may be overweight by some standards, but have little body fat. Conversely, inactive people may seem to have acceptable weights, when, in fact, they may have too much body fat.

Defining Healthy Body Weight

How much should a person weigh? How can a person know if her weight is appropriate for her height? How can a person know if his weight is jeopardizing his health? Such questions seem so simple, yet even the experts can't agree on the answers. Most often, they try to identify the weights associated with good health and longevity. With this in mind, healthy body weight is defined by three criteria:[5]

- A weight within the suggested range for height.
- A fat distribution pattern that is associated with a low risk of illness and premature death.
- A medical history that reflects an absence of risk factors associated with obesity, such as elevated blood cholesterol, blood glucose, or blood pressure.

People who meet all of these criteria may not gain any health advantage by changing their weights. Anyone who does not meet all of the above criteria may want to consult with a health care professional. The rest of the chapter examines these three criteria in more detail.

• **Body Weight and Its Standards** • Health care professionals often compare people's weights with standard weight-for-height tables, such as Table 8-5. Weight-for-height tables, though commonly used, are not the recommended method of evaluation. Instead, body mass index is used to assess a person's body weight and to monitor changes over time.

• **Body Mass Index** • The **body mass index (BMI)** describes relative weight for height:◆

$$\text{BMI} = \frac{\text{weight (kg)}}{\text{height (m)}^2}.$$

Weight classifications based on BMI are presented in Figure 8-5 on p. 252. Notice that healthy weight falls between a BMI of 18.5 and 24.9, with **underweight** below 18.5 and **overweight** above 25. The average BMI of adults in the United States is 26.5.[6]

The lower end of each weight range in Table 8-5 was calculated at a BMI of 19, and the upper end at a BMI of 25 for adults; the midpoint reflects a BMI of 22. The upper end of the range represents a healthy target either for overweight people to achieve or for others to not exceed. Obesity-related diseases and increased mortality become evident beyond this upper limit.[7] The lower end of the range may be a reasonable target for severely underweight people to achieve. BMI values slightly below the healthy range may be compatible with good health if food intake is adequate, but below a BMI of 17, signs of illness, reduced work capacity, and poor reproductive function become apparent.[8] The inside back cover presents weights and visual images associated with various BMI values. The "How to" on p. 254 describes how to determine an appropriate body weight based on BMI.

Keep in mind that BMI reflects height and weight measures and not body composition. Consequently, a bodybuilder may be classified as over*weight* by BMI standards and not be over*fat*. At the peak of his bodybuilding career, Arnold Schwarzenegger won the Mr. Olympia competition with a BMI of 31; the model on p. 250 also has a BMI greater than 30. Yet neither would be

TABLE 8-5 Healthy Weights for Adults

Height[a]	Weight (lb)[a]	
	Midpoint	*Range*
4'10"	105	91–119
4'11"	109	94–124
5'0"	112	97–128
5'1"	116	101–132
5'2"	120	104–137
5'3"	124	107–141
5'4"	128	111–146
5'5"	132	114–150
5'6"	136	118–155
5'7"	140	121–160
5'8"	144	125–164
5'9"	149	129–169
5'10"	153	132–174
5'11"	157	136–179
6'0"	162	140–184
6'1"	166	144–189
6'2"	171	148–195
6'3"	176	152–200
6'4"	180	156–205
6'5"	185	160–211
6'6"	190	164–216

Note: The higher weights in the ranges generally apply to men, who tend to have more muscle and bone; the lower weights more often apply to women, who have less muscle and bone.
[a]Without shoes or clothes.
SOURCE: *Report of the Dietary Guidelines Advisory Committee on the Dietary Guidelines for Americans* (Washington, D.C.: Government Printing Office, 1995).

To convert pounds to kilograms: ◆
lb ÷ 2.2 lb/kg = kg.
To convert inches to meters:
in ÷ 39.37 in/m = m.

body mass index (BMI): an index of a person's weight in relation to height; determined by dividing the weight (in kilograms) by the square of the height (in meters).

underweight: body weight below some standard of acceptable weight that is usually defined in relation to height (such as BMI).

overweight: body weight above some standard of acceptable weight that is usually defined in relation to height (such as BMI).

FIGURE 8-5 **BMI Values Used to Assess Weight**

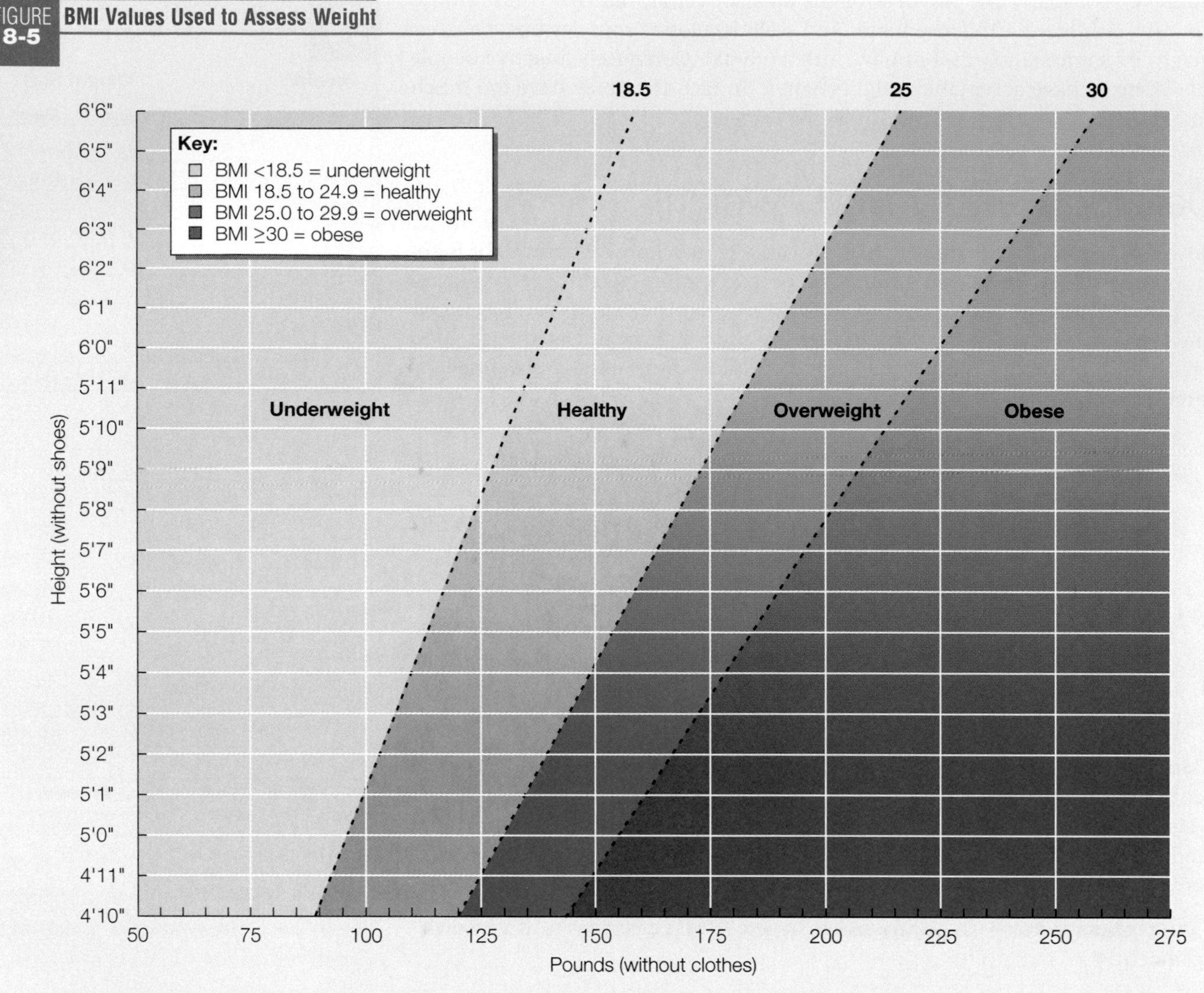

Note: Chapter 15 presents BMI values for children and adolescents age 2 to 20.
SOURCE: U.S. Department of Agriculture and U.S. Department of Health and Human Services, *Nutrition and Your Health: Dietary Guidelines for Americans* (Washington, D.C.: 2000), p. 7.

considered obese. Striking differences in body composition are also apparent among people of various ethnic groups.[9]

IN SUMMARY

Current standards for body weight are based on a person's weight in relation to height, called the body mass index (BMI), and reflect disease risks. Although weight measures are inexpensive, easy to take, and highly accurate, they fail to reveal two valuable pieces of information in assessing disease risk: how much of the weight is fat and where the fat is located.

Body Fat and Its Distribution

The ideal amount of body fat depends partly on the person. A normal-weight man may have from 12 to 20 percent body fat; a woman, because of her greater quantity of essential fat, 20 to 30 percent. Figure 8-6 compares

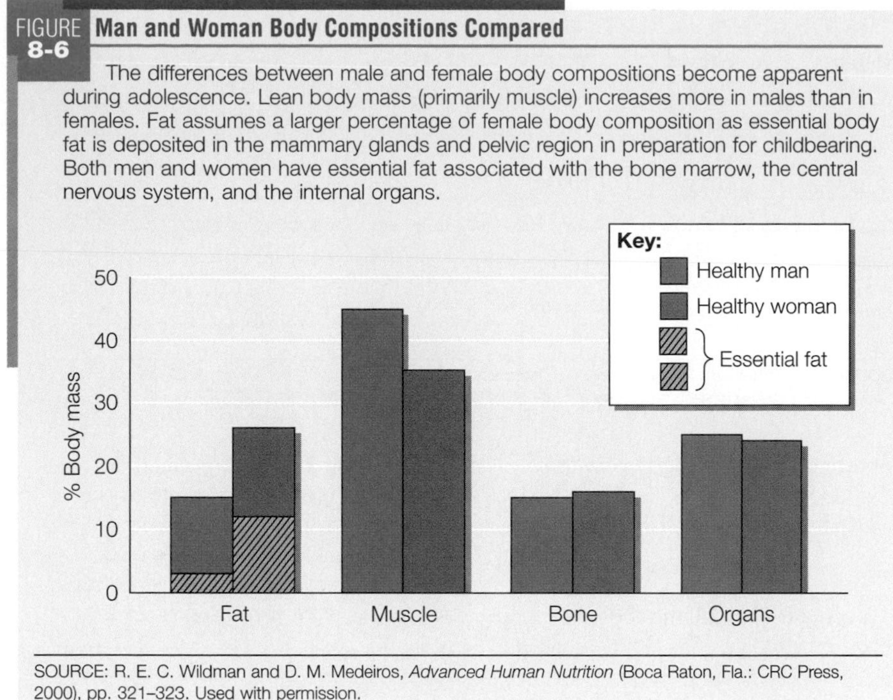

FIGURE 8-6 **Man and Woman Body Compositions Compared**

The differences between male and female body compositions become apparent during adolescence. Lean body mass (primarily muscle) increases more in males than in females. Fat assumes a larger percentage of female body composition as essential body fat is deposited in the mammary glands and pelvic region in preparation for childbearing. Both men and women have essential fat associated with the bone marrow, the central nervous system, and the internal organs.

SOURCE: R. E. C. Wildman and D. M. Medeiros, *Advanced Human Nutrition* (Boca Raton, Fla.: CRC Press, 2000), pp. 321–323. Used with permission.

the body fat and other components of body composition for a healthy man and woman. Figure 8-7 (on p. 254) makes a similar comparison between healthy and obese adults.

• **Some People Need Less** • For many athletes, a lower percentage of body fat may be ideal—just enough fat to provide fuel, insulate and protect the body, assist in nerve impulse transmissions, and support normal hormone activity, but not so much as to burden the body with excess bulk. For some athletes, then, ideal body fat might be 5 to 10 percent for men and 15 to 20 percent for women. (You may want to review the photo on p. 250 to appreciate what 8 percent body fat looks like.)

• **Some People Need More** • For an Alaska fisherman, a higher percentage of body fat is probably beneficial because fat provides an insulating blanket to prevent excessive loss of body heat in cold climates. A woman starting a pregnancy needs sufficient body fat to support conception and fetal growth. Below a certain threshold for body fat, hormone synthesis falters, and individuals may become infertile, develop depression, experience abnormal hunger regulation, or become unable to keep warm. These thresholds differ for each function and for each individual; much remains to be learned about them.

• **The Criterion of Health** • In asking what is ideal, people often mistakenly turn to fashion for the answer. Keep in mind that fashion is fickle; body shapes that our society values change with time and may have little in common with health. Fashion models whose careers depend on body shape often develop eating disorders. Body shapes that a society values also differ from one culture to another.

One important criterion for determining how much a person should weigh and how much body fat a person needs is health. Ideally, a person has enough fat to meet basic needs but not so much as to incur health risks. In general, health problems typically develop when body fat exceeds 22 percent in young men, 25 percent in men over age 40, 32 percent in young women, and 35 percent in women over age 40.

A healthy body contains enough lean tissue to support health and the right amount of fat to meet body needs.

Determine Body Weight Based on BMI

A person whose BMI reflects an unacceptable health risk can choose a desired BMI and then calculate an appropriate body weight. For example, a woman who is 5 feet 5 inches (1.65 meters) tall and weighs 180 pounds (82 kilograms) has a BMI of 30:

$$\text{BMI} = \frac{82 \text{ kg}}{1.65 \text{ m}^2} = 30.$$

A reasonable target for most overweight people is a BMI 2 units below their current one. To determine a desired goal weight based on a BMI of 28, for example, the woman could divide the desired BMI by the factor appropriate for her height from the table above:

Desired BMI ÷ factor = goal weight.

$$28 \div 0.166 = 169 \text{ lb}.$$

To reach a BMI of 28, this woman would need to lose 11 pounds. Such a calculation can help a person to determine realistic weight goals using health risk as a guide. Alternatively, a person could search the table on the inside back cover for the weight that corresponds to his or her height and the desired BMI.

Height	Factor	Height	Factor	Height	Factor
4'7"	0.232	5'3"	0.177	5'11"	0.139
4'8"	0.224	5'4"	0.172	6'0"	0.136
4'9"	0.216	5'5"	0.166	6'1"	0.132
4'10"	0.209	5'6"	0.161	6'2"	0.128
4'11"	0.202	5'7"	0.157	6'3"	0.125
5'0"	0.195	5'8"	0.152	6'4"	0.122
5'1"	0.189	5'9"	0.148	6'5"	0.119
5'2"	0.183	5'10"	0.143	6'6"	0.116

SOURCE: R. P. Abernathy, Body mass index: Determination and use. Copyright the American Dietetic Association. Reprinted by permission from *Journal of the American Dietetic Association* 91 (1991): 843.

• **Fat Distribution** • The distribution of fat on the body may be more critical than the amount of fat alone. **Intra-abdominal fat** that is stored around the organs of the abdomen is referred to as **central obesity** or upper-body fat and, independently of total body fat, is associated with increased risks of heart disease, stroke, diabetes, hypertension, and some types of cancer.

Abdominal fat is common in women past menopause and even more common in men. Even when total body fat is similar, men have more abdominal fat than either premenopausal or postmenopausal women. Regardless of

FIGURE 8-7 **Healthy and Obese Body Compositions Compared**

Body fat may vary from as little as 5 percent in extremely lean adults to as much as 70 percent in excessively obese adults. As the percentage of fat increases, the percentage of other body components decreases.

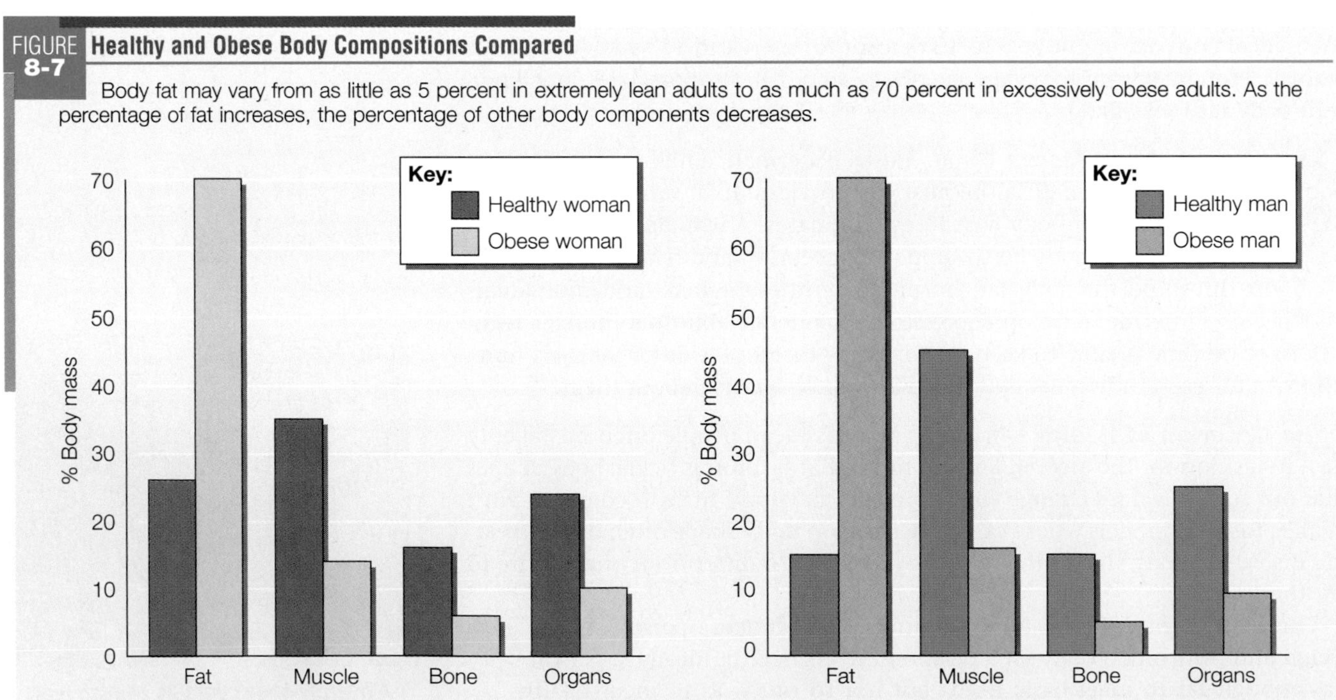

SOURCE: Adapted with permission from R. E. C. Wildman and D. M. Medeiros, *Advanced Human Nutrition* (Boca Raton, Fla.: CRC Press, 2000), pp. 321–323.

menopausal status, the risks of cardiovascular disease and mortality are increased for women with abdominal fat, just as they are for men.[10]

Fat around the hips and thighs, sometimes referred to as lower-body fat,◆ is most common in women during their reproductive years and seems relatively harmless. In fact, people who are overweight, but who do not have excessive fat around the abdomen, are less susceptible to health problems than overweight people with central obesity; theirs is a benign obesity.

◆ Popular articles sometimes call bodies with upper-body fat "apples" and those with lower-body fat, "pears." Researchers sometimes refer to upper-body fat as "android" (manlike) obesity and to lower-body fat as "gynoid" (womanlike) obesity.

• **Waist Circumference** • The most valuable and practical indicator of fat distribution and abdominal fat is a person's **waist circumference.**[11] In general, women with a waist circumference of greater than 35 inches and men with a waist circumference of greater than 40 inches have a high risk of central obesity–related health problems.[12] Appendix E includes instructions for measuring waist circumference.

• **Other Measures of Body Composition** • Health care professionals use several other techniques to estimate body fat and its distribution. These techniques include **fatfold measures, hydrodensitometry,** and **bioelectrical impedance** analysis, among others (see Figure 8-8 on p. 256). Chapter 17 and Appendix E provide additional details and include many of the tables and charts routinely used in assessment procedures.*

IN SUMMARY

The ideal amount of body fat varies from person to person, but researchers have found that body fat in excess of 22 percent for young men and 32 percent for young women (the levels rise slightly with age) poses health risks. Central obesity in which excess fat is distributed around the trunk of the body presents greater health risks than excess fat distributed on the lower body. Researchers use a number of techniques to assess body composition including waist circumference, fatfold measures, hydrodensitometry, and bioelectrical impedance.

Health Risks Associated with Body Weight and Body Fat

Body weight and fat distribution correlate with disease risks.[13] Most people with a BMI between 18.5 and 24.9 have few health risks; risks increase as BMI falls below or rises above this range, indicating that both too little and too much body fat impair health. Factors such as blood pressure or smoking habits raise health risks independently of BMI.

Similarly, epidemiological data show a J-shaped relationship between body weights and mortality (see Figure 8-9 on p. 256).[14] People who are underweight or overweight carry higher risks of early deaths than those whose weights fall within the acceptable range.[15]

• **Health Risks of Underweight** • Some underweight people enjoy an active, healthy life, but others are underweight because of malnutrition, smoking habits, or illnesses.[16] Weight and fat measures alone would not reveal these underlying causes, but a complete assessment that included a diet and medical history, physical examination, and biochemical analysis would.

An underweight person, especially an older adult, may be unable to preserve lean tissue during the fight against a wasting disease such as cancer or a digestive disorder, especially when the disease is accompanied by malnutrition.

*Researchers sometimes estimate body composition using these methods: total body water, radioactive potassium count, dual-energy X-ray absorptiometry, near-infrared spectrophotometry, ultrasound, computed tomography, and magnetic resonance imaging. Each has advantages and disadvantages with respect to cost, technical difficulty, and precision of estimating body fat (see Appendix E for a comparison).

intra-abdominal fat: fat stored within the abdominal cavity in association with the internal abdominal organs, as opposed to the fat stored directly under the skin (subcutaneous fat).

central obesity: excess fat around the trunk of the body; also called **abdominal fat** or **upper-body fat.**

waist circumference: an anthropometric measurement used to assess a person's abdominal fat.

fatfold measures: estimates of total body fatness determined by measuring the thickness of a fold of skin on the back of the arm (over the triceps muscle), below the shoulder blade (subscapular), and in other places as measured with a caliper. (The older, less preferred, term is **skinfold test.**)

hydrodensitometry (HI-dro-DEN-see-TOM-eh-tree): a method of measuring body density in which the person is weighed on land and then weighed again while submerged in water.

bioelectrical impedance (im-PEE-dans): a method of estimating body fat using low-intensity electrical current.

FIGURE 8-8 Methods Used to Assess Body Fat

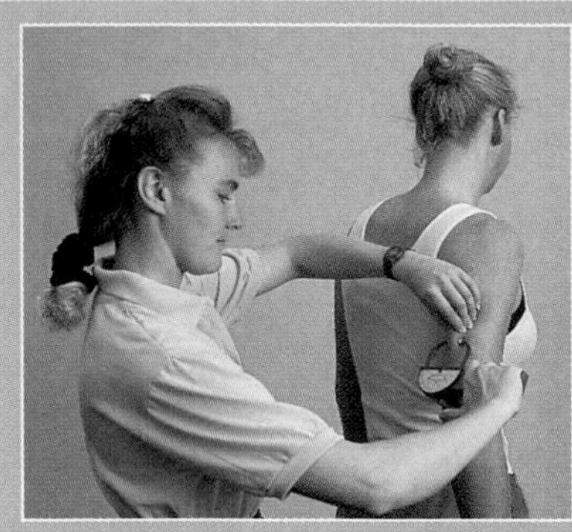

Fatfold measures: The assessor measures body fat by using a caliper to gauge the thickness of a fold of skin on the back of the arm (over the triceps), below the shoulder blade (subscapular), and in other places (including lower-body sites) and then compares these measurements with standards.

Hydrodensitometry: The assessor measures body density by weighing the person first on land and then again while submerged in water. The difference between the person's actual weight and underwater weight provides a measure of the body's volume. A mathematical equation using the two measurements (volume and actual weight) allows the assessor to calculate body density, from which the percentage of body fat can be estimated.

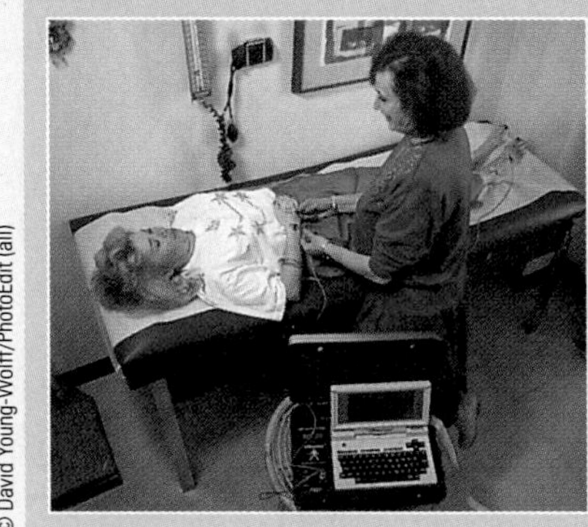

Bioelectrical impedance: The assessor measures body fat by using a low-intensity electrical current. Because electrolyte-containing fluids, which readily conduct an electrical current, are found primarily in lean body tissues, the leaner the person, the less resistance to the current. The measurement of electrical resistance is then used in a mathematical equation to estimate the percentage of body fat.

FIGURE 8-9 Body Mass Index and Mortality

This J-shaped curve describes the relationship between body mass index (BMI) and mortality and shows that both underweight and overweight present risks of a premature death.

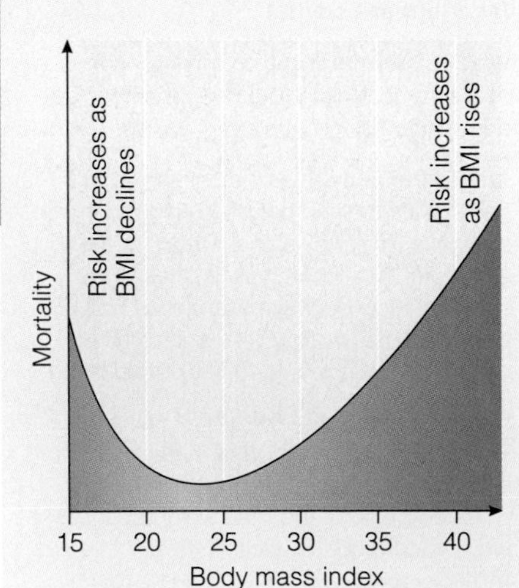

Without adequate nutrient and energy reserves, an underweight person will have a particularly tough battle against such medical stresses. In fact, many people with cancer die, not from the cancer itself, but from malnutrition. Underweight women develop menstrual irregularities and become infertile. Exactly how infertility develops is unclear, but contributing factors include not only body weight but also restricted energy and fat intake and depleted body fat stores. Those who do conceive may give birth to unhealthy infants. An underweight woman can improve her chances of having a healthy infant by gaining weight prior to conception, during pregnancy, or both. Underweight and significant weight loss are also associated with osteoporosis and bone fractures.[17] For all these reasons, underweight people may benefit from enough of a weight gain to provide an energy reserve and protective amounts of all the nutrients that can be stored.

• **Health Risks of Overweight** • As for excessive body fat, the health risks are so many that it has been declared a disease: obesity. Among the health risks associated with obesity are diabetes, hypertension, cardiovascular disease, sleep apnea (abnormal ceasing of breathing during sleep), osteoarthritis, some cancers, gallbladder disease, respiratory problems (including Pickwickian syndrome, a breathing blockage linked with sudden death), and complications in pregnancy and surgery. Each year, these obesity-related illnesses cost our nation billions of dollars.[18]

The cost in terms of lives is also great: an estimated 300,000 people die each year from obesity-related diseases.[19] In fact, obesity is second only to tobacco in causing preventable illnesses and premature deaths. Mortality increases as excess weight increases; people with a BMI greater than 35 are twice as likely to die prematurely as others. The risks associated with a high BMI appear to be greater for whites than for blacks.[20]

Equally important, both central obesity and weight gains of more than 20 pounds between early and middle adulthood correlate with increased mortality.[21] For health's sake, people should limit their weight gains during adulthood to 10 or so pounds. Fluctuations in body weight, as typically occur with "yo-yo" dieting, also increase the risks of chronic diseases and premature death (Chapter 9 provides more details). In contrast, sustained weight loss reduces disease risks and increases life expectancy.[22]

Importantly, BMI and weight gains and losses do not tell all of the story. Cardiorespiratory fitness also plays a major role in health.[23] Normal-weight men who are unfit have more than twice the risk of all-cause mortality than normal-weight men who are fit do. Furthermore, overweight but fit men have lower risks than normal-weight, unfit men. Clearly, a healthy life depends on both body weight and fitness.

• **Cardiovascular Disease** • The relationship between obesity and cardiovascular disease risk is strong, with links to both blood cholesterol and blood pressure.[24] Central obesity may raise the risk of heart attack and stroke as much as the three leading risk factors◆ (high blood cholesterol, hypertension, and smoking) do. In addition to body fat and its distribution, weight gain also increases the risk of cardiovascular disease.[25] Weight loss, on the other hand, can effectively lower both blood cholesterol and blood pressure in obese people. Of course, lean and normal-weight people may also have high blood cholesterol and blood pressure, and these factors are just as dangerous in lean people as in obese people.

◆ **Cardiovascular disease risk factors associated with obesity:**
- High LDL cholesterol.
- Low HDL cholesterol.
- High blood pressure (hypertension).
- Diabetes.

Chapter 26 provides many more details.

• **Diabetes** • Diabetes (type 2) is three times more likely to develop in an obese person than in a nonobese person. Furthermore, the person with type 2 diabetes often has central obesity.[26] Central-body fat cells appear to be larger and more insulin-resistant than lower-body fat cells. The association between **insulin resistance**◆ and obesity is strong.[27] Both are major risk factors for the development of type 2 diabetes.

Diabetes appears to be influenced not only by body weight but by weight gains as well. A weight gain of 11 to 24 pounds since age 18 doubles the risk of developing diabetes, even in women of average weight. In contrast, weight loss is effective in improving glucose tolerance and insulin resistance.

◆ **Highlight 26 describes how the combination of obesity and insulin resistance, along with elevated cholesterol and hypertension—collectively known as the metabolic syndrome or Syndrome X—raises health risks dramatically.**

• **Cancer** • The risk of some cancers increases with both body weight and weight gain, but researchers do not fully understand the relationship.[28] One possible explanation may be that obese people have elevated levels of hormones that could influence cancer development. For example, adipose tissue is the major site of estrogen synthesis in women, obese women have elevated levels of estrogen, and estrogen has been implicated in the development of cancers of the female reproductive system—cancers that account for half of all cancers in women.

insulin resistance: the reduced ability of insulin to regulate glucose metabolism.

IN SUMMARY

The weight appropriate for an individual depends largely on factors specific to that individual, including body fat distribution, family health history, and current health status. At the extremes, both overweight and underweight carry clear risks to health. The next chapter explores weight control and the benefits of achieving and maintaining a healthy weight.

How are you doing?

- Do you balance your food intake with physical activity?
- Is your BMI between 18.5 and 24.9?
- Is your waist circumference less than 35 inches for a woman or 40 inches for a man?

Nutrition on the Net

WEBSITES

Access these websites for further study of topics covered in this chapter.

- Find updates and quick links to these and other nutrition-related sites at our website: **www.wadsworth.com/nutrition**
- Obtain food composition data from the USDA Nutrient Data Laboratory: **www.nal.usda.gov/fnic/foodcomp**
- Calculate your BMI at Shape Up America: **www.shapeup.org**

INTERNET ACTIVITIES

Consumers are bombarded with unrealistic images of body size and shape. Do you know how to determine whether a body size is healthy?

- Go to: **http://www.cyberdiet.com**
- Click on "Self-Assessment" in the upper left corner.
- Click on "What is your Waist-Hip-Ratio?" Measure yourself and calculate your ratio.
- Go back to "Assessment Tools" and click on "Do you know what your Body Mass Index is?" Calculate your BMI.
- Click on "Body Fat Distribution Chart" and read that information.

Calculating your body mass index (BMI) and wait-hip ratio (WHR) gives you valuable personal information about how the issues of body weight and body composition relate to you. How do your numbers stack up?

Making it Click

These problems give you practice in estimating energy needs. Once you have mastered these examples, you will be prepared to examine your own energy intakes and energy expenditures. Be sure to show your calculations for each problem and check p. 261 for answers.

1. Estimate various people's basal metabolic energy needs per day using Table 8-3 on p. 249. For example, calculate the energy needs of a 10-year-old boy who weighs 75 pounds (34 kilograms):

   Using the range for ages 10–18:
   $(17.5 \times 34) + 651 = 1246$ kcal/day.

   Using the range for ages 3–10:
   $(22.7 \times 34) + 495 = 1267$ kcal/day.

   These two answers are similar. The boy needs about 1250 kcal/day for basal metabolism.

a. A 10-year-old girl of the same weight. (Use either range. If you try them both, you'll find about a 100-kcalorie difference, but 10-year-old girls' energy needs vary by even more than this.)
b. An 18-year-old man who weighs 150 pounds (68 kilograms). Again, use either range, and again, if you use both, you'll find about a 100-kcalorie difference.
c. A 35-year-old man who weighs 200 pounds (91 kilograms).
d. A 50-year-old woman who weighs 115 pounds (52 kilograms).

2. Compare the energy a person might spend on various physical activities. Refer to Table 8-2 on p. 248, and compute how much energy a person who weighs 142 pounds would spend doing each of the following. You may want to compare various activities based on your weight.

30 min vigorous aerobic dance:

$$0.062 \text{ kcal/lb/min} \times 142 \text{ lb} = 8.8 \text{ kcal/min.}$$

$$8.8 \text{ kcal/min} \times 30 \text{ min} = 264 \text{ kcal.}$$

a. 2 hours golf, carrying clubs.
b. 20 minutes running at 9 mph.
c. 45 minutes swimming at 20 yd/min.
d. 1 hour walking at 3.5 mph.

3. Consider the effect of age on BMR. An infant who weighs 20 pounds has a BMR of 500 kcalories/day; an adult who weighs 170 pounds has a BMR of about 1500. Based on body weight, who has the faster BMR?
4. Compute daily energy needs for a woman, age 20, who is 5 feet 6 inches tall, weighs 130 pounds, and is lightly active.
   a. From Table 8-3 on p. 249, what is her energy need for basal metabolism (show your calculations)?
   b. From Table 8-4 on p. 250, estimate her daily energy expenditure, using her activity factor and using her weight.
   c. What percentage of her daily energy expenditure is used for basal metabolism?
5. Discover what weight is needed to achieve a desired BMI. Refer to the table on p. 254 and consider a person who is 5 feet 4 inches tall. Suppose this person wants to have a BMI of 21. What should this person weigh? Does this agree with the table on the inside back cover?
6. Calculate safe weight-loss rates for people of different sizes using the recommended rate of 1 percent of body weight per week.
   a. What would the safe rate be for a person who weighs 120 pounds? Show your calculations.
   b. What would the safe rate be for a person who weighs 250 pounds? Show your calculations.
7. Calculate the daily energy intakes appropriate for weight loss using the suggested minimum energy intake of 10 kcalories per pound of body weight per day.
   a. How many kcalories for a person who weighs 130 pounds?
   b. How many kcalories for a person who weighs 250 pounds?
8. Convert body fat into kcalorie values. Suppose a man is 5 feet 3 inches tall and weighs 150 pounds, and 30 percent of that weight is fat.
   a. How many pounds of fat does he have?
   b. Assuming that the energy value of body fat is 3500 kcalories per pound, how many kcalories does this person have stored in his body fat?
   c. Suppose he loses 15 pounds. For purposes of this question, assume that he exercised enough to retain virtually all of his lean tissue. Now he weighs 135 pounds. How many pounds of this is fat? What percentage of his body weight is now fat (show your calculations)?
   d. How many kcalories does a 15-pound loss of fat represent?
   e. If a diet and exercise plan provides a deficit of 500 kcalories/day (that is, 500 kcalories less each day is eaten than is spent), how many days (or weeks or months) will it take to lose that much fat?

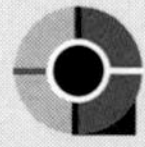

Study Questions

These questions will help you review the chapter. You will find the answers in the discussions on the pages provided.

1. What are the consequences of an unbalanced energy budget? (p. 241)
2. Define hunger, appetite, satiation, and satiety and describe how each influences food intake. (pp. 242–245)
3. Describe each component of energy expenditure. What factors influence each? How can energy expenditure be estimated? (pp. 246–249)
4. Distinguish between body weight and body composition. What assessment techniques are used to measure each? (pp. 249–250, 252–254)

5. What problems are involved in defining "ideal" body weight? (pp. 250–251)
6. What is central obesity, and what is its relationship to disease? (pp. 252–253)
7. What risks are associated with excess body weight and excess body fat? (pp. 255–257)

These questions will help you prepare for an exam. Answers can be found on p. 261.

1. A person who consistently consumes 1700 kcalories a day and spends 2200 kcalories a day for a month would be expected to:
   a. lose ½ to 1 pound.
   b. gain ½ to 1 pound.
   c. lose 4 to 5 pounds.
   d. gain 4 to 5 pounds.
2. A bomb calorimeter measures:
   a. physiological fuel.
   b. energy available from foods.
   c. kcalories a person derives from foods.
   d. heat a person releases in basal metabolism.
3. The psychological desire to eat that accompanies the sight, smell, or thought of food is known as:
   a. hunger.
   b. satiety.
   c. appetite.
   d. palatability.
4. A person watching television after dinner reaches for a snack during a commercial in response to:
   a. external cues.
   b. hunger signals.
   c. stress arousal.
   d. satiety factors.
5. The largest component of energy expenditure is:
   a. basal metabolism.
   b. physical activity.
   c. indirect calorimetry.
   d. thermic effect of food.
6. The major factor influencing BMR is:
   a. gender.
   b. food intake.
   c. body composition.
   d. physical activity.
7. The thermic effect of an 800-kcalorie meal is about:
   a. 8 kcalories.
   b. 80 kcalories.
   c. 160 kcalories.
   d. 200 kcalories.
8. For health's sake, a person with a BMI of 21 might want to:
   a. lose weight.
   b. maintain weight.
   c. gain weight.
9. Which of the following reflects height and weight?
   a. body mass index
   b. fatfold measures
   c. waist-to-hip ratio
   d. bioelectrical impedance
10. Which of the following increases disease risks?
    a. BMI 19–21
    b. BMI 22–25
    c. lower-body fat
    d. central obesity

References

1 A. S. Levine and C. J. Billington, Why do we eat? A neural systems approach, *Annual Review of Nutrition* 17 (1997): 597–619.

2 J. E. Blundell and J. I. Macdiarmid, Fat as a risk factor for overconsumption: Satiation, satiety, and patterns of eating, *Journal of the American Dietetic Association* 97 (1997): S63–S69; Stubbs and coauthors, 1996; Blundell and coauthors, 1996.

3 M. W. Schwartz and coauthors, Model for the regulation of energy balance and adiposity, *American Journal of Clinical Nutrition* 69 (1999): 584–596; A. L. Hirschberg, Hormonal regulation of appetite and food intake, *Annals of Medicine* 30 (1998): 7–20; G. Wolf, Orexins: A newly discovered family of hypothalamic regulators of food intake, *Nutrition Reviews* 56 (1998): 172–189.

4 F. X. Pi-Sunyer, Overnutrition and undernutrition as modifiers of metabolic processes in disease states, *American Journal of Clinical Nutrition* 72 (2000): 533S–537S.

5 National Institutes of Health Obesity Education Initiative, *Clinical Guidelines on the Identification, Evaluation, and Treatment of Overweight and Obesity in Adults* (Washington, D.C.: U.S. Department of Health and Human Services, 1998).

6 R. J. Kuczmarski and coauthors, Varying body mass index cutoff points to describe overweight prevalence among U.S. adults: NHANES III (1988 to 1994), *Obesity Research* 5 (1997): 542–548.

7 J. Stevens and coauthors, Evaluation of WHO and NHANES II standards for overweight using mortality rates, *Journal of the American Dietetic Association* 100 (2000): 825–827; A. Must and coauthors, The disease burden associated with overweight and obesity, *Journal of the American Medical Association* 282 (1999): 1523–1529.

8 L. J. Hoffer, Metabolic consequences of starvation, in M. E. Shils and coeditors, *Modern Nutrition in Health and Disease* (Baltimore: Williams & Wilkins, 1999), pp. 645–665.

9 K. J. Ellis, Body composition of a young, multiethnic, male population, *American Journal of Clinical Nutrition* 66 (1997): 1323–1331.

10 A. C. Perry and coauthors, Relation between anthropometric measures of fat distribution and cardiovascular risk factors in overweight pre- and postmenopausal women, *American Journal of Clinical Nutrition* 66 (1997): 829–836; M. J. Williams and coauthors, Regional fat distribution in women and risk of cardiovascular disease, *American Journal of Clinical Nutrition* 65 (1997): 855–860.

11 T. B. VanItallie, Waist circumference: A useful index in clinical care and health promotion, *Nutrition Reviews* 56 (1998): 300–313.

12 M. E. J. Lean, T. S. Han, and J. C. Seidell, Impairment of health and quality of life in people with large waist circumference, *The Lancet* 351 (1998): 853–856; National Institutes of Health Obesity Education Initiative, 1998.

13 Must and coauthors, 1999.

14 E. E. Calle and coauthors, Body-mass index and mortality in a prospective cohort of U.S. adults, *New England Journal of Medicine* 341 (1999): 1097–1105; K. R. Fontaine and coauthors, Body mass index, smoking, and mortality in older women, *Journal of Women's Health* 7 (1998): 1257–1261; C. T. Sempos and coauthors, The influence of cigarette smoking on the association between body weight and mortality, The Framingham Heart Study revisited, *Annals of Epidemiology* 8 (1998): 289–300; K. D. Lindsted, and P. N. Singh, Body mass and 26-year risk of mortality among women who never smoked: Findings from the Adventist Mortality Study, *American Journal of Epidemiology* 146 (1997): 1–11.

15 R. Bender and coauthors, Assessment of excess mortality in obesity, *American Journal of Epidemiology* 147 (1998): 42–48.

16 Calle and coauthors, 1999; W. C. Willett, Weight loss in the elderly: Cause or effect of poor health? *American Journal of Clinical Nutrition* 66 (1997): 737–738.

17 L. M. Salamone and coauthors, Effect of a lifestyle intervention on bone mineral density in premenopausal women: A randomized trial, *American Journal of Clinical Nutrition* 70 (1999): 97–103; L. W. Turner, M. Q. Wang, and Q. Fu, Risk factors for hip fracture among Southern older women, *Southern Medical Journal* 91 (1998): 533–540; M. E. Mussolino and coauthors, Risk

ANSWERS

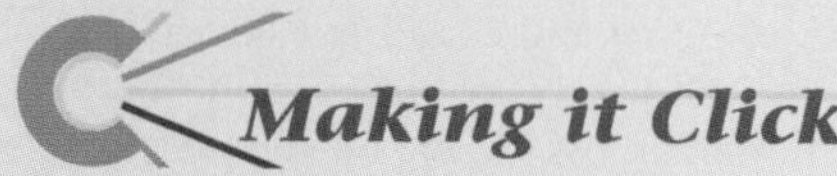

Making it Click

1. a. For ages 10–18: (12.2 × 34) + 746 = 1161 kcal/day.
   For ages 3–10: (22.5 × 34) + 499 = 1264 kcal/day.
   A 10-year-old girl who weighs 75 pounds needs about 1200 kcal/day.
   b. For ages 18–30: (15.3 × 68) + 679 = 1719 kcal/day.
   For ages 10–18: (17.5 × 68) + 651 = 1841 kcal/day.
   An 18-year-old man who weighs 150 pounds needs about 1780 kcal/day.
   c. (11.6 × 91) + 879 = 1935 kcal/day.
   d. (8.7 × 52) + 829 = 1281 kcal/day.

2. a. 0.045 kcal/lb/min × 142 lb = 6.4 kcal/min.
   6.4 kcal/min × 120 min = 768 kcal.
   b. 0.103 kcal/lb/min × 142 lb = 14.6 kcal/min.
   14.6 kcal/min × 20 min = 292 kcal.
   c. 0.032 kcal/lb/min × 142 lb = 4.5 kcal/min.
   4.5 kcal/min × 45 min = 202.5 kcal.
   d. 0.035 kcal/lb/min × 142 lb = 5 kcal/min.
   5 kcal/min × 60 min = 300 kcal.

3. The infant has the faster BMR (500 kcal/day ÷ 20 lb = 25 kcal/lb/day and 1500 kcal/day ÷ 170 lb = 8.8 kcal/lb/day). Because the infant has a BMR of 25 kcal/lb, whereas the adult has a BMR of 8.8 kcal/lb, the infant's BMR is almost 3 times faster than the adult's based on body weight.

4. a. BMR = (14.7 × 59) + 496 = 1363 kcal/day.
   b. With an activity factor of 1.5, her daily energy need is 1.5 × 1365 kcal/day = 2048 kcal/day or, with a weight of 59 kg, her daily energy need is 35 kcal/kg/day × 59 kg = 2065 kcal/day. Either way, she needs about 2050 kcal/day.
   c. 1363 kcal ÷ 2050 kcal = 0.66 or 66%.

5. 21 ÷ 0.172 = 122 lb. Yes.

6. a. 120 lb × 0.01 = 1.2 lb/week.
   b. 250 lb × 0.01 = 2.5 lb/week.

7. a. 130 lb × 10 kcal/lb = 1300 kcal.
   b. 250 lb × 10 kcal/lb = 2500 kcal.

8. a. 150 lb × 0.30 = 45 lb fat.
   b. 3500 kcal/lb × 45 lb = 157,500 kcal of body fat.
   c. 45 lb fat − 15 lb fat = 30 lb fat.
   d. 30 lb fat ÷ 135 lb total body weight = 0.22 or 22% fat.
   e. 15 lb × 3500 kcal/lb = 52,500 kcal.
   52,500 kcal ÷ 500 kcal/day = 105 days (about 3½ months).

Multiple Choice

1. c	2. b	3. c	4. a	5. a
6. c	7. b	8. b	9. a	10. d

factors for hip fracture in white men: The NHANES I Epidemiologic follow-up study, *Journal of Bone Mineral Research* 13 (1998): 918–924; T. V. Nguyen, P. N. Sambrook, and J. A. Eisman, Bone loss, physical activity, and weight change in elderly women: The Dubbo Osteoporosis Epidemiology Study, *Journal of Bone Mineral Research* 13 (1998): 1458–1467.

18 D. Thompson and coauthors, Estimated economic costs of obesity to US business, *American Journal of Health Promotion* 13 (1998): 120–127; A. M. Wolf and G. A. Colditz, Current estimates of the economic cost of obesity in the United States, *Obesity Research* 6 (1998): 97–106.

19 D. B. Allison and coauthors, Annual deaths attributable to obesity in the United States, *Journal of the American Medical Association* 282 (1999): 1530–1538.

20 Calle and coauthors, 1999.

21 C. G. Solomon and J. E. Manson, Obesity and mortality: A review of the epidemiologic data, *American Journal of Clinical Nutrition* 66 (1997): 1044S–1050S.

22 G. Oster and coauthors, Lifetime health and economic benefits of weight loss among obese persons, *American Journal of Public Health* 89 (1999): 1536–1542.

23 C. D. Lee, A. S. Jackson, and S. N. Blair, US weight guidelines: Is it also important to consider cardiorespiratory fitness? *International Journal of Obesity* 22 (1998): S2–S7.

24 R. H. Eckel and R. M. Krauss (for the AHA Nutrition Committee), American Heart Association call to action: Obesity as a major risk factor for coronary heart disease, *Circulation* 97 (1998): 2099–2100.

25 T. B. Harris and coauthors, Carrying the burden of cardiovascular risk in old age: Associations of weight and weight change with prevalent cardiovascular disease, risk factors, and health status in the Cardiovascular Health Study, *American Journal of Clinical Nutrition* 66 (1997): 837–844; K. M. Roxrode and coauthors, A prospective study of body mass index, weight change, and risk of stroke in women, *Journal of the American Medical Association* 277 (1997): 1539–1545.

26 V. J. Carey and coauthors, Body fat distribution and risk of non-insulin-dependent diabetes mellitus in women, *American Journal of Epidemiology* 145 (1997): 614–619.

27 D. H. Bessesen, Obesity as a factor, *Nutrition Reviews* 58 (2000): S12–S15.

28 S. D. Li and S. Mobarhan, Association between body mass index and adenocarcinoma of the esophagus and gastric cardia, *Nutrition Reviews* 58 (2000): 54–56; Z. Huang and coauthors, Dual effects of weight and weight gain on breast cancer risk, *Journal of the American Medical Association* 278 (1997): 1407–1411.

THE LATEST AND GREATEST WEIGHT-LOSS DIET—AGAIN

TO PARAPHRASE William Shakespeare, "a fad diet by any other name would still be a fad diet." And the names are legion: the Atkins New Diet Revolution, the Calories Don't Count Diet, the Protein Power diet, the Carbohydrate Addict's Diet, the Lo-Carbo Diet, the Healthy for Life Diet, the Zone Diet.* Year after year, "new and improved" diets appear on bookstore shelves and circulate among friends. People of all sizes eagerly try the best diet on the market ever, hoping that this one will really work. And sometimes it seems to work for a while, but more often than not, its success is short-lived. And then another fad diet takes the spotlight. Here's how Dr. K. Brownell, an obesity researcher at Yale University, describes this phenomenon: "When I get calls about the latest diet fad, I imagine a trick birthday cake candle that keeps lighting up and we have to keep blowing it out."

Realizing that fad diets do not offer a safe and effective plan for weight loss, health professionals speak out, but they never get the candle blown out permanently. New fad diets can keep making outrageous claims because no one requires their advocates to prove what they say. They do not have to conduct credible research on the benefits or dangers of their diets. They can simply make recommendations and then later, if questioned, search for bits and pieces of research that support the conclusions they have already reached. That's backwards. Diet and health recommendations should *follow* years of sound research that has been reviewed by panels of scientists *before* being offered to the public.

Because anyone can publish anything—in books or on the Internet—peddlers of fad diets can make unsubstantiated statements that fall far short of the truth, but sound impressive to the uninformed. They often offer distorted bits of legitimate research. They may start with one or more actual facts, but then leap from one erroneous conclusion to the next. Anyone who wants to believe them is forced to wonder how the thousands of scientists working on obesity research over the past century could possibly have missed such obvious connections. Table H8-1 presents some of the lies and truths of fad diets.

No matter what their names are, most fad diets espouse essentially the same high-protein, low-carbohydrate diet. After all, diets may come in all flavors, but only in three proportions: high fat, high carbohydrate, or high protein. Few consumers would believe that high-fat diets could lead to weight loss; contrary to such a claim, dietary fat does not promote fat oxidation. Consumers already hear from many free sources that high-carbohydrate diets support good health, so peddling that idea would not be a profitable venture. That leaves high-protein diets, and they surface regularly in various guises as the best way to lose weight. High-protein diets are by design relatively low in carbohydrate. This highlight examines some of the science and the science fiction behind high-protein, low-carbohydrate fad diets.

*The following sources offer comparisons and evaluations of various fad diets for your review: G. L. Blackburn and V. H. He. The changing nature of obesity in the U.S.: How serious is the problem? *Nutrition & the M.D.*, June 1999, pp. 3–7; J. Stein, The low-carb diet craze, *Time*, November 1, 1999, pp. 72–80.

THE DIET'S APPEAL

Perhaps the greatest appeal of a high-protein, low-carbohydrate diet is that it turns current diet recommendations upside down. Foods such as meats and milk products that need to be selected and measured carefully to limit fat can now be eaten with abandon. Grains, legumes, vegetables, and fruits that we are told to eat in abundance can now be ignored. For some people, this is a dream come true: steaks without the potatoes, ribs without the coleslaw, and meatballs without the pasta. Who can resist the promise of weight loss when allowed to eat freely from a list of favorite foods?

To lure dieters in, proponents of high-protein diets often blame the currently recommended high-carbohydrate, low-fat diet for our obesity troubles. They claim that the incidence of obesity is rising because we are eating less fat. Such a claim may impress the naive, but it sends skeptical people running for the facts. True, the incidence of obesity has risen dramatically over the past two decades.[1] True, our intake of fat has dropped from 36 to

The wise consumer distinguishes between loss of fat and loss of weight.

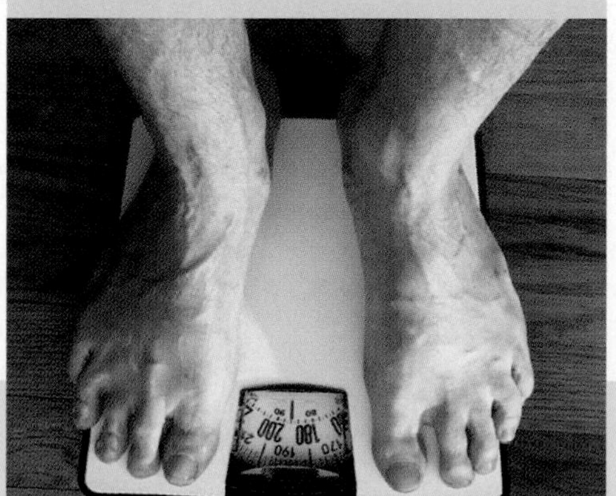

TABLE H8-1 The Lies and Truths of Fad Diets

The Lie:	You can lose weight with "exceptionally easy rules."
The Truth:	Most fad diet plans have complicated rules that require you to calculate protein requirements, count carbohydrate grams, combine certain foods, time meal intervals, purchase special products, plan daily menus, and measure serving sizes.
The Lie:	You can lose weight by eating a specific ratio of carbohydrates, protein, and fat.
The Truth:	Weight loss depends on spending more energy than you take in.
The Lie:	This "revolutionary diet" can "reset your genetic code."
The Truth:	You inherited your genes and cannot alter your genetic code.
The Lie:	High-protein diets are popular, selling more than 20 million books, because they work.
The Truth:	Weight-loss books are popular because people grasp for quick fixes and simple solutions to their weight problems. If book sales were an indication of weight-loss success, we would be a lean nation—but they're not, and neither are we.
The Lie:	People gain weight on low-fat diets.
The Truth:	People can gain weight on low-fat diets if they overindulge in carbohydrates and proteins while cutting fat; low-fat diets are not necessarily low-kcalorie diets. But people can also lose weight on low-fat diets if they cut kcalories as well as fat.
The Lie:	High-protein diets energize the brain.
The Truth:	The brain depends on glucose for its energy; the primary dietary source of glucose is carbohydrate, not protein.
The Lie:	Thousands of people have been successful with this plan.
The Truth:	None of the authors of fad diets have published their research findings in scientific journals. Success stories are anecdotal and failures are not reported.
The Lie:	Carbohydrates raise blood glucose levels, triggering insulin production and fat storage.
The Truth:	Insulin promotes fat storage when energy intake exceeds energy needs. Furthermore, insulin is only one hormone involved in the complex processes of maintaining the body's energy balance and health.
The Lie:	Eat protein and lose weight.
The Truth:	For every complicated problem, there is a simple—and wrong solution.

34 percent of daily energy intake. Such facts might seem to imply that lowering fat intake leads to obesity, but this is an erroneous conclusion. The *percentage* declined only because average energy intakes increased by 200 kcalories a day (from about 2000 kcalories a day to 2200).[2] Actual fat intake *increased* by 4 grams a day (from 82 grams to 86). Furthermore, fewer than half of us engage in regular physical activity.[3] Obesity experts blame our high energy intakes and low energy outputs for the increase in obesity. Weight loss, after all, depends on a negative energy balance. To their credit, some of these diet plans recommend exercise, which helps their success.

Dieters are also lured into fad diets by sophisticated—yet often erroneous—explanations of the metabolic consequences of eating certain foods. Terms such as *eicosanoids* and *de novo lipogenesis* are scattered about, intimidating readers into believing that the authors must be right given their brilliance in understanding the body. One common misconception currently circulating amongst fad diets focuses on insulin. High-protein diet proponents claim that carbohydrates are bad. Some go so far as to equate carbohydrates with toxic poisons or addictive drugs. Starches and sugars are considered evil because they are absorbed easily and raise blood glucose. The pancreas then responds by secreting insulin—and insulin is touted as the real villain responsible for our nation's epidemic of obesity.

What does insulin do? Among its roles, insulin facilitates the transport of glucose into the cells, the storage of fatty acids as fat, and the synthesis of cholesterol. It is an anabolic hormone that builds and stores. True—but not the whole truth and nothing but the truth. Insulin is only one of many factors involved in the body's metabolism of nutrients and regulation of body weight. Furthermore, as Chapter 4's discussion of the glycemic index pointed out, insulin does not always respond to foods as might be expected. Whole-wheat pasta (a high-carbohydrate food), for example, produces less of an insulin response than beef (a high-protein food). Many carbohydrates—fruits, vegetables, legumes, and whole grains—are rich in fibers that slow glucose absorption and moderate insulin response. Most importantly, insulin is critical to maintaining health, as any person with type 1 diabetes can attest. Insulin causes problems only when a person develops insulin resistance—that is, when the body's cells do not respond to the large quantities of insulin that the pancreas continues to pump out in an effort to get a response. Insulin resistance is a major health problem—but it is not caused by carbohydrate, or by protein, or by fat. It accompanies being overweight. When a person loses weight, insulin response improves.

If high-protein diets were as successful as some people claim, then consumers who tried them would lose lots of weight, and our obesity problems would be solved. Obviously, this is not the case. Similarly, if high-protein diets were as worthless as others claim, then consumers would eventually stop pursuing them. Clearly, this is not happening either. These diets have enough going for them that they work for some people at least for a short time, but they fail to produce long-lasting results for most people. The following sections examine some of the apparent achievements and shortcomings of high-protein diets.

THE DIET'S ACHIEVEMENTS

With over half of our nation's adults overweight and many more concerned about their weight, the market for a weight-loss book, product, or program is huge (no pun intended). Americans spend an estimated $33 billion a year on weight-loss books and products. Even a plan that offers only minimal weight-loss success easily attracts a following. High-protein diet plans offer a little success to some people for a short time. Here's why.

Don't Count kCalories

Who wants to count kcalories? Even experienced dieters find counting kcalories burdensome, not to mention timeworn. They want a new, easy way to lose weight, and high-protein diet plans seem to offer this boon. But while these diets often claim to disregard kcalories, their design typically ensures a low energy intake. They advise dieters to stop counting kcalories, but then recommend three meals "not to exceed 500 kcalories each and two snacks of less than 100 kcalories each." Most of the sample menu plans provided by these diets are designed to deliver 800 to 1200 kcalories a day.

Even when it is truly not necessary to count kcalories, the total tends to be low simply because food intake is so limited. Without its refried beans, tortilla wrapping, and chopped vegetables, a burrito is reduced to a pile of ground beef. Weight loss occurs because of the low energy intake—not the proportion of energy nutrients.[4] Success, then, depends on the restricted intake, not on protein's magical powers or carbohydrate's evil forces. This is an important point. Any diet can produce weight loss, at least temporarily, if intake is restrictive. The real value of a diet is determined by its ability to maintain weight loss and support good health over the long term. The goal is not simply weight loss, but health gains—and high-protein, low-carbohydrate diets do not support optimal health over time.

Satisfy Hunger

As Chapter 8 mentioned, of the three energy-yielding nutrients, protein produces the strongest feelings of satiety.[5] People feel full after eating even small quantities of high-protein foods. When a diet leaves a person feeling satisfied instead of deprived, the overall quality of life improves.[6] All meals—whether designed for weight loss or not—should include enough protein to satisfy hunger.

Follow a Plan

People need specific instructions and examples to make dietary changes.[7] Fad diets offer dieters a plan. The user doesn't have to decide what foods to eat, how to prepare them, or how much to eat. Unfortunately, these instructions serve short-term weight-loss needs only. They do not provide for long-term weight maintenance or health goals.

THE DIET'S SHORTCOMINGS

People who have followed high-protein diet plans for several months have lost weight. But can these diets also be harmful?

Too Much Fat

Some fad diets focus so intently on promoting protein and curbing carbohydrate that they fail to account for the fat that accompanies many high-protein foods. A breakfast of bacon and eggs, lunch of ham and cheese, and dinner of barbecued short ribs would provide 100 grams of protein—and 121 grams of fat! Yet this day's meals, even with a snack of peanuts, provide only 1600 kcalories. Without careful selection, protein-rich diets can be extraordinarily high in fat, saturated fat, and cholesterol—all dietary risk factors for heart disease.

People have reported that their blood cholesterol dropped while they were on a high-protein diet, but because no studies have been conducted, such comments cannot be accepted as evidence. Only scientific research can determine if, and how much, cholesterol declines. Furthermore, if there is a decline, is it significant? (High-protein diet proponents commonly report a decline of 5 percent, yet those consuming a low-fat diet typically experience a decline of 40 percent.) And is the decline in cholesterol due to the diet or to the weight loss? If research were to conclude that a high-protein diet is a factor in reducing cholesterol, it would be important to determine how that compares with other types of diets.

Unbalanced Nutrition

Without fruits, vegetables, and whole grains, high-protein diets lack not only carbohydrate, but fiber, vitamins, minerals, and phytochemicals as well—all dietary factors protective against disease. To help shore up some of these inadequacies, fad diets often recommend a daily supplement. Conveniently, many of the companies selling fad diets also peddle these supplements. But as Highlights 10 and 11 explain, foods offer many more health benefits than any supplement can provide. Quite simply, if the diet is inadequate, it needs to be improved, not supplemented.

Too Little Variety

Diets that omit hundreds of foods and several food groups lack variety. Some people lose interest in eating, which further reduces energy intake. Others "cheat" to experience a broader array of flavors. Even if the allowed foods are favorites, eating the same foods week after week can become monotonous.

THE BODY'S PERSPECTIVE

When a person consumes a low-carbohydrate diet, a metabolism similar to that of fasting prevails (see Chapter 7 for a review of fasting). With little dietary carbohydrate coming in, the body uses its glycogen stores to provide glucose for the cells of the brain, central nervous system, and blood. Once the body depletes its glycogen reserves, it turns to its only significant remaining source of glucose—protein. A low-carbohydrate diet may provide abundant protein from food, but the body still uses some protein from body tissues.

Dieters can know this wasting process has begun by monitoring their urine. Whenever glycogen or protein is broken down, water is released and urine production

TABLE H8-2 Adverse Side Effects of Low-Carbohydrate, Ketogenic Diets

- Nausea
- Fatigue (especially if physically active)
- Constipation
- Low blood pressure
- Elevated uric acid (which may exacerbate kidney disease and cause inflammation of the joints in those predisposed to gout)
- Stale, foul taste in the mouth (bad breath)
- In pregnant women, fetal harm and stillbirth

increases. Low-carbohydrate diets also induce ketosis, and ketones can be detected in the urine. Ketones form whenever glucose is lacking and fat breakdown is incomplete.

Many fad diets regard ketosis as the key to losing weight. People in ketosis may experience a loss of appetite and a dramatic weight loss within the first few days. They would be disillusioned if they were aware that much of this weight loss reflects the loss of glycogen and protein together with large quantities of body fluids and important minerals. They need to learn to appreciate the difference between loss of *fat* and loss of *weight.* Fat losses on ketogenic diets are no greater than on other diets providing the same number of kcalories. Once the dieter returns to well-balanced meals that provide adequate energy, carbohydrate, fat, protein, vitamins, and minerals, the body avidly retains these needed nutrients. The weight will return, quite often to a level higher than the starting point. Table H8-2 lists other consequences of a ketogenic diet.

Table H8-3 offers guidelines for identifying fad diets and other weight-loss scams. Diets that fall short, if used for only a little while, may not harm healthy people, but they cannot support optimal health for long. Chapter 9 includes reasonable approaches to weight management and concludes that the ideal diet is one you can live with for the rest of your life. Keep that criterion in mind when you evaluate the next "latest and greatest weight-loss diet" that comes along.

TABLE H8-3 Guidelines for Identifying Fad Diets and Other Weight-Loss Scams

1. They promise dramatic, rapid weight loss. Weight loss should be gradual and not exceed 2 pounds per week.
2. They promote diets that are nutritionally unbalanced or extremely low in kcalories. Diets should provide:
   - A reasonable number of kcalories (not fewer than 1200 kcalories per day).
   - Enough, but not too much, protein (between the RDA and twice the RDA).
   - Enough, but not too much fat (between 20 and 30 percent of daily energy intake from fat).
   - Enough carbohydrate to spare protein and prevent ketosis (at least 100 grams per day) and 20 to 30 grams of fiber from food sources.
   - A balanced assortment of vitamins and minerals from a variety of foods from each of the food groups.
   - At least 1 liter (about 1 quart) of water daily or 1 milliliter per kcalorie daily—whichever is more.
3. They use liquid formulas rather than foods. Foods should accommodate a person's ethnic background, taste preferences, and financial means.
4. They attempt to make clients dependent upon special foods or devices. Programs should teach clients how to make good choices from the conventional food supply.
5. They fail to encourage permanent, realistic lifestyle changes. Programs should provide physical activity plans that involve spending at least 300 kcalories a day and behavior-modification strategies that help to correct poor eating habits.
6. They misrepresent salespeople as "counselors" supposedly qualified to give guidance in nutrition and/or general health. Even if adequately trained, such "counselors" would still be objectionable because of the obvious conflict of interest that exists when providers profit directly from products they recommend and sell.
7. They collect large sums of money at the start or require that clients sign contracts for expensive, long-term programs. Programs should be reasonably priced and run on a pay-as-you-go basis.
8. They fail to inform clients of the risks associated with weight loss in general or the specific program being promoted. They should provide information about dropout rates, the long-term success of their clients, and possible side effects.
9. They promote unproven or spurious weight-loss aids such as human chorionic gonadotropin hormone (HCG), starch blockers, diuretics, sauna belts, body wraps, passive exercise, ear stapling, acupuncture, electric muscle-stimulating (EMS) devices, spirulina, amino acid supplements (e.g., arginine, ornithine), glucomannan, methylcellulose (a "bulking agent"), "unique" ingredients, and so forth.
10. They fail to provide for weight maintenance after the program ends.

SOURCES: Adapted from American College of Sports Medicine, *ACSM's Guidelines for Exercise Testing and Prescription* (Baltimore: Williams & Wilkins, 1995), pp. 218–219; J. T. Dwyer, Treatment of obesity: Conventional programs and fad diets, in *Obesity,* ed. P. Björntorp and B. N. Brodoff (Philadelphia: J. B. Lippincott, 1992), p. 668; *National Council Against Health Fraud Newsletter,* March/April 1987, National Council Against Health Fraud, Inc.

Nutrition on the Net

•WEBSITES•

Access these websites for further study of topics covered in this highlight.

- Find updates and quick links to these and other nutrition-related sites at our website: **www.wadsworth.com/nutrition**
- Review a transcript of presentations and panel discussions of leading obesity experts and fad diet authors by searching for Symposium on the Great Nutrition Debate at USDA's site: **www.usda.gov**

References

1 Update: Prevalence of overweight among children, adolescents, and adults—United States, 1988–1994, *Morbidity and Mortality Weekly Report* 46 (1997): 199–202.

2 N. D. Ernst and coauthors, Consistency between US dietary fat intake and serum total cholesterol concentrations: The National Health and Nutrition Examination Surveys, *American Journal of Clinical Nutrition* 66 (1997): S965–S972.

3 *Physical Activity and Health—A Report of the Surgeon General Executive Summary* (Washington, D.C.: Government Printing Office, 1996).

4 A. Golay and coauthors, Similar weight loss with low- or high-carbohydrate diets, *American Journal of Clinical Nutrition* 63 (1996): 174–178; B. B. Alford, A. C. Blankenship, and R. D. Hagen, The effects of variations in carbohydrate, protein and fat content of the diet upon weight loss, blood values and nutrient intake of adult obese women, *Journal of the American Dietetic Association* 90 (1990): 534–540.

5 J. E. Blundell and coauthors, Control of human appetite: Implications for the intake of dietary fat, *Annual Review of Nutrition* 16 (1996): 285–319.

6 M. Shah, Comparison of a low-fat, ad libitum complex-carbohydrate diet with a low-energy diet in moderately obese women, *American Journal of Clinical Nutrition* 59 (1994): 980–984.

7 R. R. Wing, Food provision in dietary intervention studies, *American Journal of Clinical Nutrition* 66 (1997): 421–422.

CHAPTER 9

Weight Management: Overweight and Underweight

Are you pleased with your body weight? If so, you are a rare individual. Nearly all people in our society think they should weigh more or less (mostly less) than they do. Usually, their primary reason is appearance, but they often understand that physical health is also somehow related to body weight. At the extremes, both overweight and underweight present health risks.

This chapter emphasizes overweight, partly because it has been more intensively studied and partly because it is a widespread health problem in developed countries and a growing concern in developing countries. Information on underweight is presented wherever appropriate. The highlight that follows this chapter delves into the eating disorders anorexia nervosa and bulimia nervosa.

OVERWEIGHT

Despite our preoccupation with body image and weight loss, the prevalence of overweight and obesity in the United States continues to rise dramatically.[1] In the past decade, obesity increased in every state, in both genders, and across all ages, races, and education levels (see Figure 9-1 on p. 270). Over half of the adults◆ in the United States are now considered overweight, as defined by a BMI of 25 or greater (see Figure 9-2 on p. 271).[2] The prevalence of overweight is especially high among women, the poor, blacks, and Hispanics. Obesity is so widespread and its prevalence is rising so rapidly that many refer to it as an **epidemic.**[3] Before examining the suspected causes of obesity and the myriad treatments used to overcome it, it may be helpful to understand the development and metabolism of body fat.

Chapter 15 presents information on overweight and obesity during childhood and adolescence. ◆

◆ Adults gain an average of ½ pound per year between the ages of 25 and 55.◆

HEALTHY PEOPLE 2010

Increase the proportion of adults who are at healthy weight. Reduce the proportion of adults who are obese.

Fat Cell Development

When more energy is consumed than is spent, much of the excess energy is stored in the fat cells of adipose tissue. The amount of fat in a person's body reflects both the *number* and the *size* of the fat cells. The number of fat cells increases most rapidly during the growing years of late childhood and early puberty. Fat cell number increases more rapidly in obese children than in lean children, and obese children entering their teen years may already have as many fat cells as do adults of normal weight.

The fat cells expand in size as they fill with fat droplets (review Figure 5-18 on p. 147). When the cells reach their maximum size, they may also divide. Thus obesity develops◆ when a person's fat cells increase in number, in size, or quite often both. Figure 9-3 (on p. 271) illustrates fat cell development.

With fat loss, the size of fat cells dwindles, but not their number. People can shrink their fat cells, but they cannot make the cells disappear. People with extra fat cells tend to regain lost weight rapidly; with weight gain, their many fat cells readily fill. In contrast, people with an average number of enlarged fat cells may be more successful in maintaining weight losses; when their cells shrink, both cell size and number are normal. Prevention of obesity is most critical, then, during the growing years when fat cells increase in number.

Obesity due to an increase in the *number* of fat cells is **hyperplastic obesity.** Obesity due to an increse in the *size* of fat cells is **hypertrophic obesity.** ◆

Fat Cell Metabolism

The enzyme lipoprotein lipase (LPL) promotes fat storage in both adipose and muscle cells. People with high LPL activity store fat especially efficiently.

epidemic (EP-ee-DEM-ick): the appearance of a disease (usually infectious) or condition that attacks many people at the same time in the same region.
- **epi** = upon
- **demos** = people

FIGURE 9-1 **Prevalence of Obesity (BMI ≥30) Among Adults across the United States**

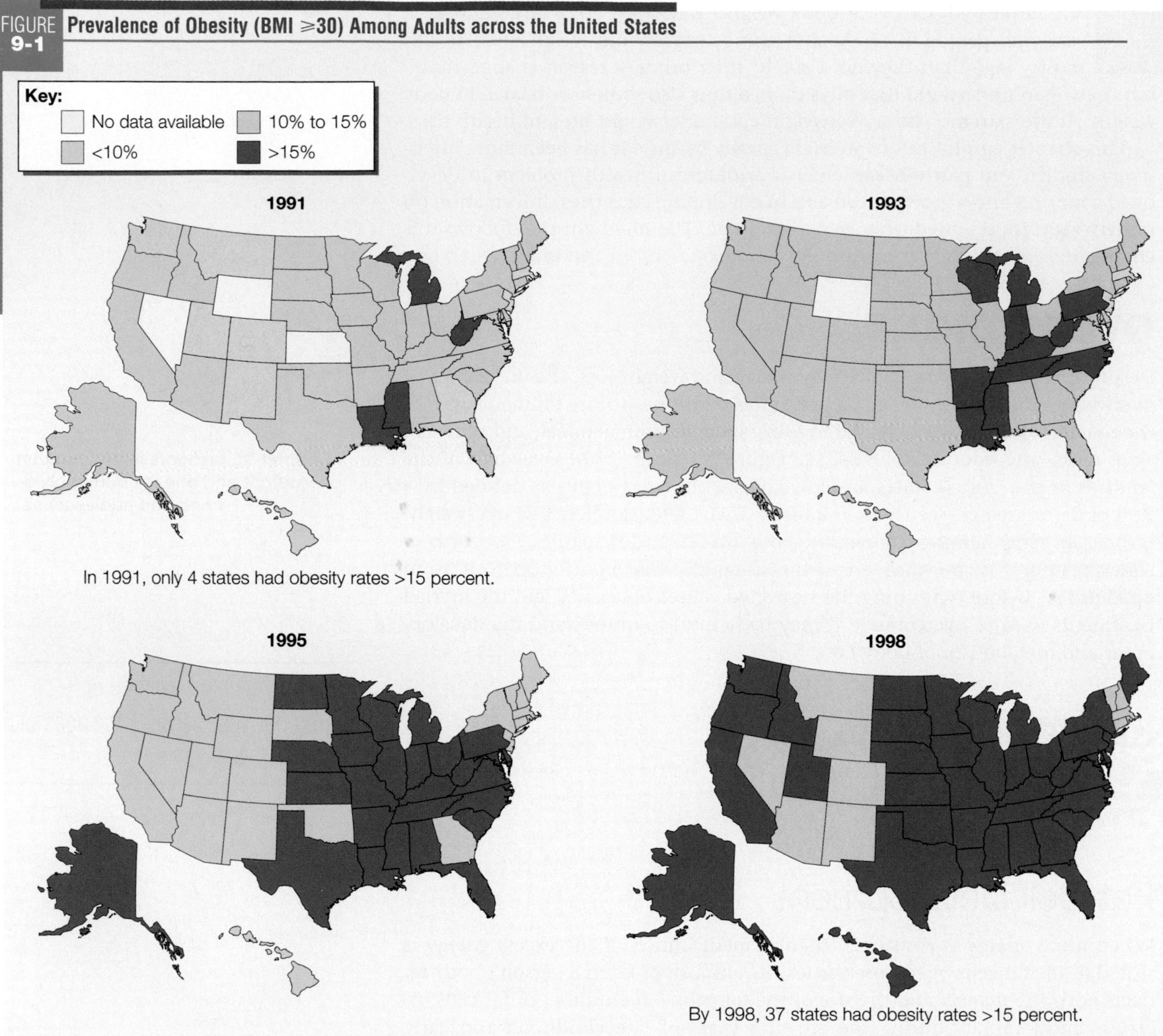

SOURCE: A. H. Mokdad and coauthors, The spread of the obesity epidemic in the United States, 1991–1998, *Journal of the American Medical Association* 282 (1999): 1519–1522.

Because LPL is mounted on fat cell membranes and obese people have many fat cells, they generally have much more LPL activity in their fat cells than lean people do (their muscle cell LPL activity is similar, though).[4] Consequently, even modest excesses in energy intake have a more dramatic impact on obese people than on lean people.

The activity of LPL is partially regulated by gender-specific hormones—estrogen in women and testosterone in men. In women, fat cells in the breasts, hips, and thighs produce abundant LPL, putting fat away in those body sites; in men, fat cells in the abdomen produce abundant LPL. This enzyme activity explains why men tend to develop central obesity whereas women more readily develop lower-body fat.

Differences are also apparent in the activity of the enzymes controlling fat breakdown in various parts of the body. The lower body is less active than the upper body in releasing fat from storage.[5] Consequently, people tend to have a more difficult time losing fat from the hips and thighs than from around the chest and abdomen.

Enzyme activity may also explain why people who lose weight so easily regain it. After weight loss, LPL activity increases, and it does so most dramatically in people who had been fattest prior to weight loss. Apparently, weight loss serves as a signal to the gene that produces the LPL enzyme, saying "Make more enzyme to store fat." People easily regain weight after having lost it because they are battling against enzymes that want to store fat. The activity of these enzymes provides an explanation for the observation that some inner mechanism seems to set a person's weight or body composition at a fixed point; the body will adjust to restore that **set point.**

FIGURE 9-2 **Distribution of Body Weights in U.S. Adults**

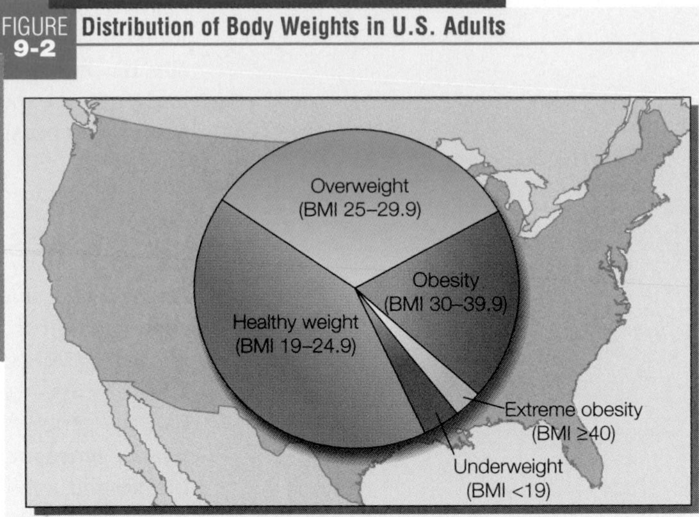

SOURCE: Adapted from R. J. Kuczmarski and coauthors, Varying body mass index cutoff points to describe overweight prevalence among U.S. adults: NHANES III (1988 to 1994), *Obesity Research* 5 (1997): 542–548.

Set-Point Theory

Many internal physiological variables, such as blood glucose, blood pH, and body temperature, remain fairly stable under a variety of conditions. The hypothalamus and other regulatory centers constantly monitor and delicately adjust conditions so as to maintain homeostasis. The stability of such complex systems may depend on set-point regulators that maintain variables within specified limits.

Researchers have confirmed that after weight gains or losses, the body adjusts its metabolism so as to restore the original weight. Energy expenditure increases after weight gain and decreases after weight loss. These changes in energy expenditure differ from those expected based on body composition and help to explain why it is so difficult for an underweight person to maintain weight gains and an overweight person to maintain weight losses.

IN SUMMARY

Fat cells develop by increasing in number and size. Prevention of excess weight gain depends on maintaining a reasonable number of fat cells. With gains or losses, the body attempts to return to its previous status.

CAUSES OF OBESITY

Why do people accumulate excess body fat? The obvious answer is that they take in more food energy than they spend. But that answer falls short of

FIGURE 9-3 **Fat Cell Development**

Fat cells are capable of increasing their size by 20-fold and their number by several thousandfold.

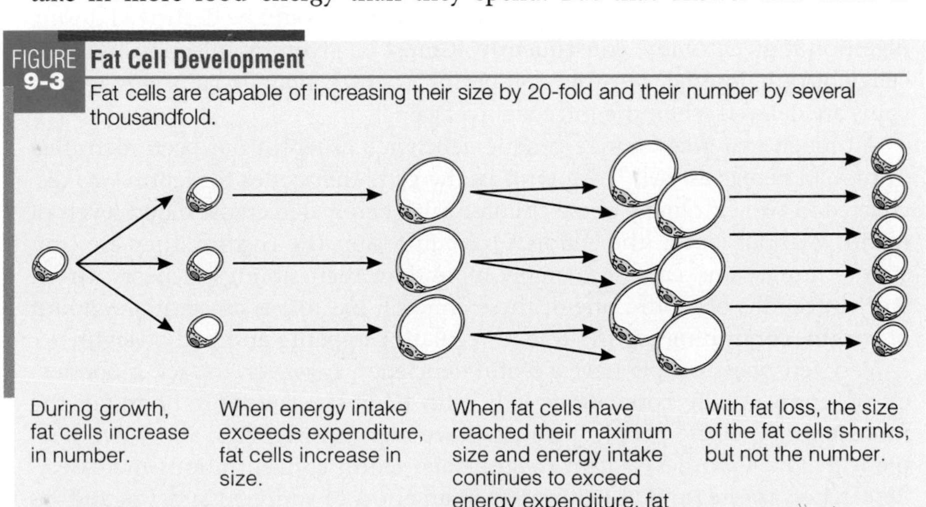

During growth, fat cells increase in number. When energy intake exceeds expenditure, fat cells increase in size. When fat cells have reached their maximum size and energy intake continues to exceed energy expenditure, fat cells increase in number again. With fat loss, the size of the fat cells shrinks, but not the number.

set point: the point at which controls are set (for example, on a thermostat). The set-point theory that relates to body weight proposes that the body tends to maintain a certain weight by means of its own internal controls.

explaining why they do this. Is it genetic? Environmental? Cultural? Behavioral? Socioeconomic? Psychological? Metabolic? All of these? Most likely, obesity has many interrelated causes. Why an imbalance between energy intake and energy expenditure occurs remains a bit of a mystery; the next sections summarize possible explanations.

Genetics

Researchers have found that adopted children tend to be similar in weight to their biological parents, not to their adoptive parents. Studies of twins yield similar findings: identical twins are twice as likely to weigh the same as fraternal twins—even when reared apart. These findings suggest an important role for genetics in determining a person's *susceptibility* to obesity.[6] In other words, genes may not *cause* obesity, but genetic factors may influence the food intake and activity patterns that lead to it and the metabolic pathways that maintain it.[7]

Clearly, something genetic makes a person more or less likely to gain or lose weight when overeating or undereating.[8] Some people gain more weight than others on comparable energy intakes. Given an extra 1000 kcalories a day for 100 days, some pairs of identical twins gain less than 10 pounds while others gain up to 30 pounds. Within each pair, the amounts of weight gained, percentages of body fat, and locations of fat deposits are similar. Similarly, some people lose more weight than others following comparable exercise routines.

Researchers have been examining several genes in search of answers to obesity questions. As Chapter 6's section on protein synthesis described, each cell expresses only the genes for the proteins it needs, and each protein performs a unique function. The following paragraphs describe some recent research involving proteins that might influence obesity development.

• Leptin • Researchers have identified an obesity gene, called *ob,* that is expressed in the fat cells and codes for the protein **leptin.** Leptin acts as a hormone, primarily in the hypothalamus. Preliminary research suggests that leptin promotes a negative energy balance by suppressing appetite and increasing energy expenditure (see Figure 9-4). Changes in energy expenditure primarily reflect changes in basal metabolism, but may also include changes in physical activity patterns.

Mice with a defective *ob* gene do not produce leptin and can weigh up to three times as much as normal mice and have five times as much body fat (see Figure 9-5 on p. 274). When injected with a synthetic form of leptin, the mice rapidly lose body fat. (Because leptin is a protein, it would be destroyed during digestion if given orally; consequently, it must be given by injection.) The fat cells not only lose fat, but they self-destruct, which may explain why weight gains are delayed when the mice are fed again.[9]

Although extremely rare, a genetic deficiency of leptin has been identified in human beings as well.[10] An error in the gene that codes for leptin was discovered in two extremely obese children with barely detectable blood levels of leptin. Without leptin, the children have little appetite control; they are constantly hungry and eat considerably more than their siblings or peers. Given daily injections of leptin, one of these children has lost a substantial amount of weight, confirming leptin's role in regulating appetite and body weight.[11]

Very few obese people have a leptin deficiency, however. In fact, blood levels of leptin usually correlate directly with body fat: the more body fat, the more leptin.[12] Obese people generally have high leptin levels, and when people with low leptin levels gain weight, their leptin concentrations increase.[13] Researchers speculate that leptin rises in an effort to suppress appetite and inhibit fat storage, but its action is ineffective in obesity; obesity appears to be associated with an insensitivity or resistance to leptin.[14] Perhaps leptin or its

leptin: a protein produced by fat cells under direction of the *ob* gene that decreases appetite and increases energy expenditure; sometimes called the ***ob* protein.**
- **leptos** = thin

FIGURE 9-4 Leptin's Action in the Body

Leptin maintains energy homeostasis. When the body gains fat, the increase in leptin shifts energy balance toward the negative: eat less and spend more energy. Such a scenario would ensure that all fat gains were followed by losses, but this is not the case in reality, of course. Most obese people have high levels of leptin, but their energy balance does not automatically shift to the negative, suggesting a resistance to leptin's action in obesity.

When the body loses fat, the decrease in leptin shifts energy balance toward the positive: eat more and spend less energy.

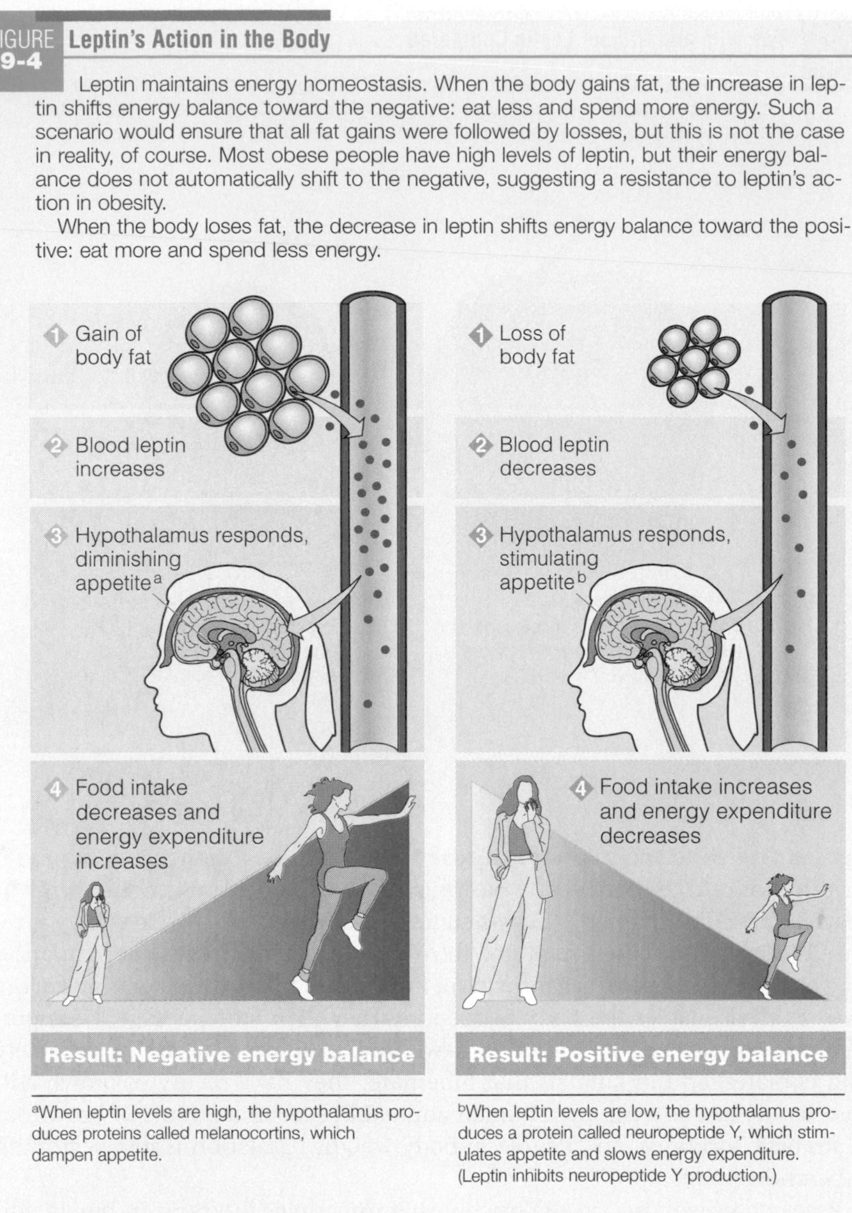

[a]When leptin levels are high, the hypothalamus produces proteins called melanocortins, which dampen appetite.

[b]When leptin levels are low, the hypothalamus produces a protein called neuropeptide Y, which stimulates appetite and slows energy expenditure. (Leptin inhibits neuropeptide Y production.)

receptors are defective or other signals override its action; all these possibilities and more are being considered by researchers.[15] The picture that seems to be developing is that leptin may appear in two scenarios in obesity as insulin does in diabetes. Relatively few people have an insulin deficiency (type 1 diabetes); many others have elevated insulin but are resistant to its action (type 2 diabetes). Just as injections of insulin help to control blood glucose, injections of leptin may help to control obesity. Preliminary research suggests that leptin injections may induce weight loss in obese people.[16] Interestingly, leptin may selectively reduce centrally located fat.[17]

Leptin research has uncovered some of its other regulatory roles around the body. For example, leptin may inform the female reproductive system about body fat reserves; stimulate growth of new blood vessels, especially in the cornea of the eye; enhance the maturation of bone marrow cells; promote formation of red blood cells; and help support a normal immune response.[18]

• **Uncoupling Proteins** • Other genes code for proteins involved in energy metabolism. These proteins may influence obesity by storing or spending energy with different efficiencies or in different types of fat. The body has two

FIGURE 9-5 **Mice with and without Leptin Compared**

Both of these mice have a defective ob gene. Consequently, they do not produce leptin. They both became obese, but the one on the right received daily injections of leptin, which suppressed food intake and increased energy expenditure, resulting in weight loss.

Without leptin, this mouse weighs almost three times as much as a normal mouse.

With leptin treatment, this mouse lost a significant amount of weight, but still weighs almost one and a half times as much as a normal mouse.

types of fat: white and **brown adipose tissue.** White adipose tissue stores fat for other cells to use for energy; brown adipose tissue releases stored energy as heat. Recall from Chapter 7 that when fat is oxidized, some of the energy is released in heat and some is captured in ATP. In brown adipose tissue, oxidation may be uncoupled from ATP formation; it produces heat only. Radiating energy away as heat enables the body to spend, rather than store, energy. Heat production is particularly important in newborns, in adults who live in extremely cold climates, and in animals that hibernate; they have plenty of brown adipose tissue. In contrast, most human adults have small amounts of brown fat in strategic locations, and its role in body weight regulation is just beginning to be understood.

Researchers working on the protein that uncouples reactions in brown adipose tissue discovered a gene that codes for a second uncoupling protein.[19] This protein is active not only in brown fat, but in white fat and many other tissues as well.[20] Its actions seem to influence the basal metabolic rate (BMR) and oppose the development of obesity.[21] Animals with abundant amounts of this uncoupling protein resist weight gain, whereas those with minimal amounts gain weight easily. Similarly, children with a genetic variant of this uncoupling protein are more overweight than others.[22] Whether the body dissipates the energy from an ice cream sundae as heat or stores it in body fat has major consequences, of course, for a person's body weight.

Environment

While genetic studies indicate that body weight is heritable, they do not fully explain obesity. This reality is evident in identical twins who have dramatically different body weights.[23] With obesity rates rising and the gene pool remaining unchanged, environment must play a role as well. The environment includes all of the circumstances that we encounter daily that push us toward fatness or thinness. Keep in mind that genetic and environmental factors are not mutually exclusive; genes can influence eating behaviors, for example.

brown adipose tissue: masses of specialized fat cells packed with pigmented mitochondria that produce heat instead of ATP.

• Overeating • One obvious, although not necessarily satisfactory, explanation for obesity is that overweight people overeat. Whether this is true has been difficult to determine, however. Diet histories from obese people often report energy intakes that are similar to, or even less than, those of others. Diet histories may not be accurate records of actual intakes, though; both normal-weight and obese people commonly underreport their dietary intakes.[24] Most importantly, dietary intakes during the time of study may not reflect the eating habits that lead to obesity. Obese people who had a positive energy balance for years and accumulated excess body fat may not currently have a positive energy balance. This reality highlights an important point: the energy balance equation must consider time. Both present *and* past eating and activity patterns influence current body weight.[25]

© Frank Whitney/The Image Bank

Lack of physical activity fosters obesity.

Studies on rats do not depend on diet histories, of course, and reveal interesting findings. Genetically obese rats eat much more than their nonobese littermates during their early months of growth and rapidly gain weight. When their weights stabilize, however, all rats' intakes become similar, even though the obese rats continue to weigh twice as much as the others. When researchers control the rats' energy intakes, the genetically obese ones gain less than when allowed to eat freely, but they still gain much more weight than the lean rats. Clearly, overeating contributes to obesity, but does not fully explain it.

One obesity educator suggests that Americans live in a **"toxic food environment"** that exposes us to high-kcalorie, high-fat foods that are readily available, relatively inexpensive, heavily advertised, and wonderfully delicious.[26] Food is available everywhere, all the time—thanks largely to fast food. Our highways are lined with fast-food restaurants, and convenience stores and service stations offer fast food as well. Fast food is available in our schools, malls, and airports. It's convenient and it's available morning, noon, and night—and all times in between. Most alarming are the extraordinarily large serving sizes and ready-to-go meals that offer supersize combinations. People buy the large sizes and combinations, perceiving them to be a good value, but then they eat more than they need—a bad deal.

Fast food is a major player in the development of obesity.[27] Fast food is often high in fat, and a high-fat diet promotes obesity,◆ especially in people with a genetic predisposition. Because high-fat foods taste delicious, people may eat more, but fats have little satiating power—so people eat even more.[28] Fat's 9 kcalories per gram quickly add up, amplifying people's energy intakes and enlarging their body fat stores.[29]

◆ Dietary fat promotes obesity because:
- It is palatable (increases food intake).
- It produces little satiety (increases food intake).
- It provides 9 kcalories per gram (increases energy intake).
- It induces efficient metabolism (increases body fat stores).

• Physical Inactivity • The "toxic environment" fosters physical inactivity as well. Life requires little exertion—escalators carry us up stairs, automobiles take us across town, buttons roll down windows, and remote controls change television channels. One hundred years ago, 30 percent of the energy used in farm and factory work came from muscle power; today, only 1 percent does. Modern technology has replaced physical activity at home, at work, and in transportation. Inactivity contributes to weight gain and poor health.[30] In turn, television watching may contribute most to physical inactivity.

Watching television contributes to obesity in several ways. First, television viewing requires little energy beyond the resting metabolic rate. Second, it replaces time spent in more vigorous activities. Third, watching television influences food purchases and correlates with between-meal snacking on the high-kcalorie, high-fat foods most heavily advertised on programs. Nonnutritious foods and beverages appear not only in commercials, but also within the television programs themselves. People, especially children, who see television stars eating and drinking these foods and remaining thin may miss the message that such behavior will bring about weight gain.

Clearly then, people may be obese, not because they eat too much, but because they move too little. Some obese people are so extraordinarily inactive that even when they eat less than lean people, they still have an energy

toxic food environment: a term coined to refer to the easy access to and overabundance of high-fat, high-kcalorie foods in our society. It does *not* refer to the contamination of the food supply with poisonous toxins or infectious microbes.

surplus. Reducing their food intake further would jeopardize health and incur nutrient deficiencies. Physical activity is a necessary component of nutritional health. People must be physically active if they are to eat enough food to deliver all the nutrients they need without unhealthy weight gain. In fact, there seems to be a threshold of physical activity that protects against weight gain.[31] This threshold appears near the moderate-to-heavy level of activity intensity (as defined in Table 8-4 on p. 250).

IN SUMMARY In recent years, the view has been gaining ground that obesity is much more complicated than mere undisciplined gluttony. Most likely, obesity has many causes and different combinations of causes in different people. Some causes, such as overeating and physical inactivity, may be within a person's control, and some, such as genetics, may be beyond it.

CONTROVERSIES IN OBESITY TREATMENT

An estimated 30 to 40 percent of all U.S. women (and 20 to 25 percent of U.S. men) are trying to lose weight at any given time, spending up to $40 billion each year to do so. Some of these people do not even need to lose weight. Others may need to lose weight, but are not successful; few succeed, and even fewer succeed permanently.

© George Sesota/Newsmakers

Camryn Manheim, award-winning TV actress on* The Practice *and author of* Wake Up, I'm Fat, *prefers epicurean joy to ascetic longevity and hopes that one day our society can accept people of all shapes and sizes.

Elusive Goals

Many people assume that every overweight person can achieve slenderness and should pursue that goal. First consider that most overweight people cannot—for whatever reason—become slender: only 5 percent of all people who successfully lose weight maintain their losses for at least a year. Then consider the prejudice involved in that assumption. People come with varying weight tendencies, just as they come with varying potentials for height and degrees of health, yet we do not expect tall people to shrink or healthy people to get sick in an effort to become "normal."

Large segments of our society place such enormous value on thinness that obese people face prejudice and discrimination: they are judged on their appearance more than on their character. Socially, obese people are stereotyped as lazy, stupid, and lacking in self-control. Psychologically, they may suffer embarrassment when others treat them with hostility and contempt, and some have even learned to view their own bodies as grotesque and loathsome. Parents and friends may scold them for lacking the discipline to resolve their weight problems. Health care professionals, including dietitians, are among the chief offenders. All of this hurts self-esteem. Such a critical view of overweight is not prevalent in many other cultures, including segments of our society. Instead, overweight is embraced as a sign of robust health and beauty. Many overweight people today are tired of our obsession with weight control and simply want to be accepted as they are. To free our society of its obsession with body weight and prejudice against obesity, we must first learn to judge others for who they are and not for what they weigh.

Encouraging weight loss through sensible food choices and regular physical activity may be justified when health benefits are clear. For example, a 30-year-old man with a body mass index (BMI) of 40◆ might be able to prevent or control the diabetes that runs in his family by losing 75 pounds. The effort required to do so may be great, but it is far less than the effort and conse-

◆ For reference, a man with a BMI of 40 might be:
- 5 ft 8 in, 265 lb.
- 5 ft 10 in, 280 lb.
- 6 ft, 295 lb.

See the inside back cover for additional examples of heights and weights for various BMI.

quences of living with diabetes. Sometimes health benefits appear with even less weight loss. A 60-year-old man with a BMI of 40 might gain relief from the arthritis in his knees by losing just 25 pounds. In this case, losing more weight might not bring added health benefits.

Often a person's motivations for weight loss have nothing to do with health. A young woman with a BMI of 23◆ might want to lose a few pounds for spring break, but might be healthier *not* losing weight. In the case of a person with anorexia nervosa with a BMI of 18, losing weight might incur devastating medical consequences (see Highlight 9).

For reference, a woman with a BMI of 23 might be: ◆
- 5 ft 3 in, 130 lb.
- 5 ft 5 in, 139 lb.
- 5 ft 7 in, 146 lb.

And a woman with a BMI of 18 might be:
- 5 ft 4 in, 105 lb.
- 5 ft 6 in, 112 lb.
- 5 ft 8 in, 118 lb.

IN SUMMARY

The question whether a person should lose weight depends on many factors: the extent of overweight, age, health, and genetic makeup among them. Not all obesity will cause disease or shorten life expectancy. Just as there are unhealthy, normal-weight people, there are healthy, obese people.

Dangers of Weight Loss

People attach so many dreams of happiness to weight loss that they willingly risk huge sums of money for the slightest chance of success. As a result, weight-loss schemes flourish. Of tens of thousands of claims, treatments, and theories for losing weight, few are effective—and many are downright dangerous. The negative effects must be carefully considered before embarking on any weight-loss program. Physical problems may arise from fad diets and "yo-yo" dieting, and psychological problems may emerge from repeated "failures."

Many states have developed consumer bills of rights to help protect potential weight-loss clients. Such documents explain the risks associated with weight-loss programs and provide honest predictions of success (see Table 9-1).

• Fad Diets • Fad diets often sound good, but typically fall short of delivering on their promises. They espouse exaggerated or false theories of weight loss and advise consumers to follow inadequate diets. Some fad diets are hazardous to health. Adverse reactions can be as minor as headaches, nausea, and dizziness or as serious as death. Highlight 8 offers guidelines for identifying unsound weight-loss schemes and diets.

• Weight Cycling • Many people who try to lose weight become trapped in **weight cycling**, the endless repeating rounds of weight loss and regain from

TABLE 9-1 Weight-Loss Consumer Bill of Rights (An Example)

1. *WARNING:* Rapid weight loss may cause serious health problems. Rapid weight loss is weight loss of more than 1½ to 2 pounds per week or weight loss of more than 1 percent of body weight per week after the second week of participation in a weight-loss program.
2. Consult your personal physician before starting any weight-loss program.
3. Only permanent lifestyle changes, such as making healthful food choices and increasing physical activity, promote long-term weight loss and successful maintenance.
4. Qualifications of this provider are available upon request.
5. *YOU HAVE A RIGHT TO:*
   - Ask questions about the potential health risks of this program and its nutritional content, psychological support, and educational components.
   - Receive an itemized statement of the actual or estimated price of the weight-loss program, including extra products, services, supplements, examinations, and laboratory tests.
   - Know the actual or estimated duration of the program.
   - Know the name, address, and qualifications of the dietitian or nutritionist who has reviewed and approved the weight-loss program.

weight cycling: repeated cycles of weight loss and gain. The weight-cycling pattern is popularly called the **ratchet effect** or **yo-yo effect** of dieting.

FIGURE 9-6 The Weight-Cyclng Effect of Repeated Dieting

Each round of dieting is followed by a rebound of weight to a higher level than before.

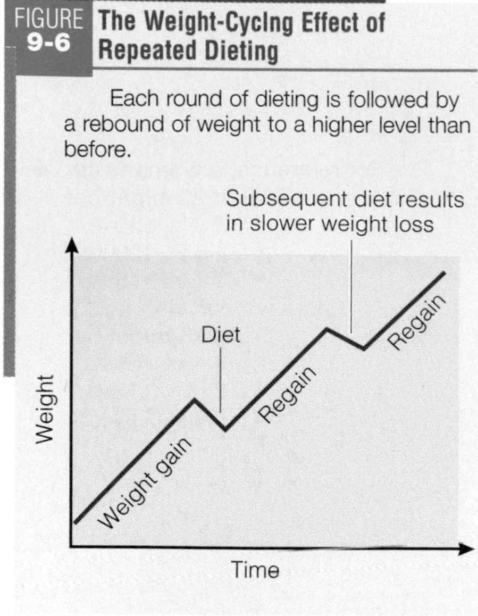

"yo-yo" dieting (see Figure 9-6). When people repeatedly lose weight only to regain it, their bodies become very efficient at making and storing fat. This increased efficiency shows itself in a way familiar to dieters who have lost and gained—and lost and gained again. With each attempt, it becomes harder and takes longer to lose weight and easier and quicker to gain it back. In fact, previous weight-cycling history can predict a person's success (or lack thereof) in maintaining weight loss.

Such fluctuations in body weight appear to increase the risks of chronic diseases and even premature death, independently of obesity itself. Maintaining a stable weight,◆ even if it is overweight, may be less harmful to health than repeated bouts of weight gain and loss. Such concerns should not deter obese people who want to lose weight from trying, but rather should encourage them to commit to lifelong changes in food choices and activity patterns that will support good health.

◆ **Losing 10 lb and keeping them off may be beneficial, but losing 100 lb and regaining them may be quite harmful.**

• **Psychological Problems** • Some of the nation's most popular diet books and weight-loss programs have misled consumers with unsubstantiated claims and deceptive testimonials. Furthermore, they fail to provide an assessment of the short- and long-term results of their treatment plans, even though such evaluations are possible and would permit consumers to make informed decisions. Of course, some weight-loss programs are better than others in terms of cost, approach, and customer satisfaction, but few are particularly successful in helping people keep lost weight off.

Most programs assume that the problem can be solved simply by applying willpower and hard work. If determination were the only factor involved, though, the success rate would be far greater than 5 percent. Overweight people may readily assume the blame for their failures to lose weight and maintain the losses when, in fact, the programs have failed. Ineffective treatment and its associated sense of failure add to a person's psychological burden. Figure 9-7 illustrates how the devastating psychological effects of obesity and dieting perpetuate themselves.

IN SUMMARY

Weight-loss advice does not apply equally to all overweight people. Some people may risk more in the process of losing weight than in remaining overweight. Weight-loss efforts are often misguided. When pursued via unwise weight-loss techniques, they can be physically and psychologically damaging.

FIGURE 9-7 The Psychology of Weight Cycling

I am fat and unhappy.

I want to be happy.

If I lose weight, I will be happy.

I try too hard to reach an unrealistic goal.

I lose a little weight, but then regain it (and sometimes more).

SOURCE: Adapted with permission from J. P. Foreyt and G. K. Goodrick, *Living without Dieting* (Houston: Harrison Publishing, 1992).

AGGRESSIVE TREATMENTS OF OBESITY

The degree of obesity and the risk of disease guide which strategies might be most appropriate in weight reduction. A healthy person with a BMI of 25 may need to improve eating habits and increase physical activity, but someone with hyptertension and a BMI greater than 30 may need more aggressive treatment options.

Drugs

With new understandings of obesity's genetic basis and acceptance of its classification as a chronic disease, drug treatment research has gained much ground in recent years. Experts reason that if obesity is a chronic disease, it should be treated as such—and the treatment of most chronic diseases includes drugs.[32] The challenge, then, is to develop an effective drug that can be used over time without adverse side effects or the potential for abuse. No such drug currently exists.[33]

Several drugs for weight loss have been tried over the years. When used as part of a long-term, comprehensive, weight-loss program, drugs can help obese people to lose approximately 10 percent of their initial weight and maintain that loss for at least a year. Because weight regain commonly occurs with the discontinuation of drug therapy, treatment must be long term. Yet the long-term use of drugs poses risks. We don't yet know whether a person would benefit more from maintaining a 20-pound excess or from taking a drug for a decade to keep the 20 pounds off. Physicians must prescribe drugs appropriately, inform clients of the potential risks, and monitor side effects carefully. Two prescription drugs are currently on the market: sibutramine and orlistat.

• Sibutramine • **Sibutramine** suppresses appetite by inhibiting the uptake of **serotonin**. The drug is most effective when used in combination with a reduced-kcalorie diet and increased physical activity. Side effects include dry mouth, headache, constipation, insomnia, and high blood pressure. The Food and Drug Administration (FDA) advises those with high blood pressure not to use sibutramine and others to monitor their blood pressure.

• Orlistat • **Orlistat** takes a different approach to weight control. It inhibits pancreatic lipase activity, thus blocking dietary fat digestion and absorption by about 30 percent. The drug is taken with meals and is most effective when accompanied by a reduced-kcalorie, low-fat diet. Side effects include gas, frequent bowel movements, and reduced absorption of fat-soluble vitamins.

• Other Drugs • Several other drugs are currently under study, including some that block neuropeptide Y and others that stimulate uncoupling proteins.[34] The use of leptin to treat obesity is also promising.

• Over-the-Counter Drugs • Only one over-the-counter medication to help with weight loss has FDA approval. It contains benzocaine◆ (in a candy or gum form), which anesthetizes the tongue, reducing taste sensations.

In 2000, the FDA recommended banning over-the-counter products containing phenylpropanolamine,◆ an ingredient that had been used in products to suppress appetite. Reported side effects include dry mouth, rapid pulse, nervousness, sleeplessness, hypertension, irregular heartbeats, kidney failure, seizures, and strokes.

• Herbal Products • In their search for weight-loss magic, some consumers turn to "natural" herbal products.◆ St. John's wort, for example, contains substances that enhance serotonin and thus suppress appetite as does the drug sibutramine mentioned earlier. In addition to the many cautions that accompany the use of any herbal remedies, consumers should be aware that St. John's wort is often prepared in combination with the herbal stimulant ephedrine. Ephedrine, in combination with caffeine and/or aspirin, has shown some promise as a treatment for obesity in a limited number of research studies.[35] Without further ado, manufacturers began to market "natural" combinations of ephedrine extracted from the Chinese plant ma huang, caffeine from coffee beans, and acetosalicyclic acid (aspirin) from willow bark.* Some of these ephedrine-containing supplements have been implicated in several cases of heart attacks and seizures and have been linked to about three dozen deaths. Table 9-2 (on p. 280) presents the facts and the fiction behind ephedrine and several other weight-loss supplements.

Herbal laxatives containing senna, aloe, rhubarb root, cascara, castor oil, and buckthorn (or various combinations) are commonly sold as "dieter's tea." Such concoctions commonly cause nausea, vomiting, diarrhea, cramping, and fainting and may have contributed to the deaths of four women who had drastically reduced their food intakes.[36] Consumers mistakenly believe that laxatives will diminish nutrient absorption and save kcalories, but remember that

*Ma huang (ephedrine) is illegal in Canada.

© Anne Dowie

So many promises, so little success.

◆ Benzocaine is marketed under the trade names:
- Diet Ayds (candy).
- Slim Mint (gum).

◆ Phenylpropanolamine was marketed under the trade names:
- Acutrim.
- Dex-A-Diet.
- Dexatrim.
- Permathene.
- Phenoxine.
- Phenyldrine.
- Prolamine.
- Super Ordinex.
- Thin Z.
- Unitrol.

◆ Chapter 18 explores the possible benefits and potential dangers of herbal products and other alternative therapies.

sibutramine (sigh-BYOO-tra-mean): a drug used in the treatment of obesity that slows the reabsorption of serotonin in the brain, thus suppressing appetite and creating a feeling of fullness; marketed under the trade name *Meridia*.

serotonin (SER-oh-tone-in): a neurotransmitter important in sleep regulation, appetite control, and sensory perception among other roles. Serotonin is synthesized in the body from the amino acid tryptophan with the help of vitamin B_6.

orlistat (OR-leh-stat): a drug used in the treatment of obesity that inhibits the absorption of fat in the GI tract, thus limiting kcaloric intake; marketed under the trade name *Xenical.*

TABLE 9-2 Selected Weight-Loss Supplements

Product	Manufacturers' Claims	Research Results	Other Comments
Cellasene (a mixture of herbs, lecithin, and fish oils)	Banishes cellulite	Produced ⅝-inch decrease in thigh measurements (manufacturer's unpublished study)	Retail cost to achieve these results: $275 for two months
Chitosan[a] (pronouned KITE-oh-san; derived from chitin, the substance that forms the hard shells of lobsters, crabs, and other shellfish)	Binds to fat and substances that are soluble in fat	Failed to produce significant weight loss[b]	May block the absorption of fat-soluble vitamins and fat-soluble phytochemicals
Conjugated linoleic acid (CLA)	Increases muscles and decreases fat	Failed to produce weight loss (one unpublished study)	May help add more muscle and less fat during regain of lost weight
Ephedrine[c] (amphetamine-like substance derived from the Chinese ephedra herb ma huang)	Speeds up body's metabolism	Produces weight loss with dangerous side effects	Over 1000 reports of dizziness, headaches, chest pain, psychosis, seizures, and strokes and 35 deaths
Hydroxycitric acid[d] (derived from the tropical fruit *garcinia cambogia*)	Inhibits the enzyme that converts citric acid to fat	Failed to produce significant weight loss[e]	Pharmaceutical company abandoned efforts to develop drug when research found toxicity symptoms in animals
Pyruvate[f]	Contributes to greater rate of fat loss and greater rate of weight loss	Produced weight loss of 3½ pounds more than placebo	Retail cost to achieve these results: $300 a month
Triiodothyroacetic acid[g] (TRIAC, a potent thyroid hormone)	Speeds up body's metabolism	Produces weight loss with dangerous side effects	May cause diarrhea, fatigue, drowsiness, insomnia, nervousness, sweating, heart attacks, and strokes

Note: The FDA has not approved the use of any of these products; a warning not to use any product containing triiodothyroacetic acid (TRIAC) has been issued. Most products are used in conjunction with a 1000- to 1800-kcalorie diet.
[a]Marketed under the trade name: Fat Trapper.
[b]M. H. Pittler and coauthors, Randomized, double-blind trial of chitosan for body weight reduction, *European Journal of Clinical Nutrition* 53 (1999): 379–381.
[c]Marketed under the trade names: Diet Fuel and Metabolife.
[d]Marketed under the trade names: Ultra Burn, Citralean, CitriMax, Citrin.
[e]S. B. Heymsfield and coauthors, *Garcinia cambogia* (hydroxycitric acid) as a potential antiobesity agent: A randomized controlled trial, *Journal of the American Medical Association* 280 (1998): 1596–1600; L. J. C. van Loon and coauthors, Effects of acute (−)-hydroxycitrate supplementation on substrate metabolism at rest and during exercise in humans, *American Journal of Clinical Nutrition* 72 (2000): 1445–1450.
[f]Marketed under the trade names: Exercise in a Bottle, Pyruvate Punch, Pyruvate-c, and Provate.
[g]Marketed under the trade name: Triax Metabolic Accelerator.
SOURCE: Adapted from D. Schardt, Fat burners, *Nutrition Action Healthletter,* July/August 1999, pp. 9–11.

absorption occurs primarily in the upper small intestine and these laxatives act on the lower large intestine.

• **Other Gimmicks** • Other gimmicks don't help with weight loss either. Hot baths do not speed up metabolism so that pounds can be lost in hours. Steam and sauna baths do not melt the fat off the body, although they may dehydrate people so that they lose water weight. Brushes, sponges, wraps, creams, and massages intended to move, burn, or break up **"cellulite"** do nothing of the kind, because there is no such thing as cellulite.

Surgery

Surgery as an approach to weight loss is justified in some specific cases of **clinically severe obesity.** Surgical procedures effectively limit food intake by reducing the size of the stomach. They reduce the size of the outlet as well, so they delay the passage of food from the stomach into the intestine for digestion and absorption.

The long-term safety and effectiveness of gastric surgery depend, in large part, on compliance with dietary instructions. Common immediate postsurgical complications include infections, nausea, vomiting, and dehydration; in the long term, vitamin and mineral deficiencies and psychological problems are common. Lifelong medical supervision is necessary for those who choose

cellulite (SELL-you-light or SELL-you-leet): supposedly, a lumpy form of fat; actually, a fraud. Fatty areas of the body may appear lumpy when the strands of connective tissue that attach the skin to underlying muscles pull tight where the fat is thick. The fat itself is the same as fat anywhere else in the body. If the fat in these areas is lost, the lumpy appearance disappears.

clinically severe obesity: a BMI of 40 or greater or 100 lb or more overweight for an average adult. A less preferred term used to describe the same condition is *morbid obesity.*

the surgical route, but in suitable candidates the benefits of weight loss may prove worth the risks.

Another surgical procedure is used, not to treat obesity, but to remove the evidence. Plastic surgeons can extract some fat deposits by suction lipectomy, or "liposuction." This cosmetic procedure has little effect on body weight, but can alter body shape slightly in specific areas. Liposuction is a popular procedure in part because of its perceived safety, but, in fact, there can be serious complications that occasionally result in death.[37]

IN SUMMARY

Obese people with high risks of medical problems may need aggressive treatment, including drugs or surgery. Others may benefit most from improving eating and exercise habits.

REASONABLE TREATMENTS OF OBESITY

Reasonable treatments of obesity embrace small changes, moderate losses, and reasonable goals. A 200-pound woman who loses 10 to 20 pounds in a year is much more likely to maintain losses and reap health benefits than if she were to drop down to 130 pounds in that same time. In keeping with this philosophy, the *Dietary Guidelines* suggest that for good health, a person should "aim for a healthy weight." The focus is not on weight loss per se, but on health gains. In fact, the *Guidelines* go on to say, "If you are already overweight, first aim to prevent further weight gain, and then lose weight to improve your health."

Health promotion

Modest weight loss, even when a person is still overweight, can improve control of diabetes and reduce the risks of heart disease by lowering blood pressure and blood cholesterol, especially for those with upper-body fat. Improvements in physical capabilities and bodily pain become evident with even a 5-pound weight loss.[38] For these reasons, parameters such as blood pressure, blood cholesterol, or even vitality are more useful than body weight in marking success. People less concerned with disease risks may prefer to set goals for personal fitness, such as being able to play with children or climb stairs without becoming short of breath.

Whether the goal is health or fitness, expectations◆ need to be reasonable. Unreachable targets ensure frustration and failure. If goals are achieved or exceeded, there will be rewards instead of disappointments.

◆ Weight-management tip: Adopt reasonable expectations about health and weight goals and about how long it will take to achieve them.

Findings from a research study highlight the great disparity between lofty expectations and reasonable success.[39] Before beginning a weight-loss program, obese women identified the weights they would describe as "dream," "happy," "acceptable," and "disappointing" (see Figure 9-8 on p. 282). All of these weights were below their starting weight. Their goal weights reflected a 32 percent loss of initial weight, far exceeding the 5 to 10 percent recommended by experts, or even the 15 percent reported by the most successful weight-loss studies. Even the "disappointing" weights exceeded recommended goals. Close to a year later, and after an average loss of 35 pounds, almost half of the women did not achieve even their "disappointing" weights. They did, however, experience more physical, social, and psychological benefits than they had predicted for that weight. Still, in a culture that overvalues thinness, these women were not satisfied with a 16 percent reduction in weight—not because their efforts were unsuccessful, but because their expectations were unrealistic.

Realistic goals for successful weight loss include reasonable time frames—at least six months for a 10 percent loss of initial weight. Keep in mind that pursuing good health is a lifelong journey. Most adults are keenly aware of their body weights and shapes and realize that what they eat and what they do can make a difference to some extent. Those who are most successful at weight

FIGURE 9-8 **Reasonable Weight Goals and Expectations Compared**

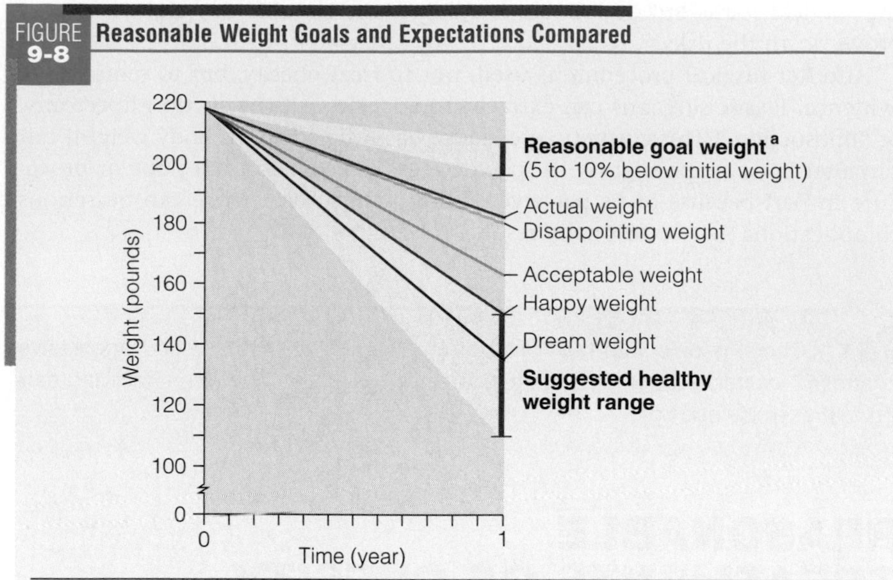

[a]Reasonable goal weights reflect pounds lost over time. Given more time, reasonable goals may eventually fall within the suggested healthy weight range.
SOURCE: Adapted from G. D. Foster and coauthors, What is a reasonable weight loss? Patients' expectations and evaluations of obesity treatment outcomes, *Journal of Consulting and Clinical Psychology* 65 (1997): 79–85.

management seem to have fully incorporated healthful eating and physical activity into their daily lives.[40] Such advice—to reduce kcalorie intake and increase physical activity—would hardly surprise anyone, yet only one in five people trying to control their weight follows these recommendations.[41]

Eating Plans

Contrary to the claims of fad diets, no one food plan is magical, and no specific food must be included or avoided in a weight-management plan. In designing a plan,◆ people need only consider foods that they like or can learn to like, that are available, and that are within their means.

◆ **Weight-management tip: Be involved in planning.**

• Be Realistic about Energy Intake • The main characteristic of a weight-loss diet is that it provides less energy than the person needs to maintain present body weight. Restricting energy intake too severely, however, may produce the devastating physical consequences commonly seen in anorexia nervosa (described in Highlight 9). It may also be counterproductive. Rapid weight loss usually means excessive loss of lean tissue and a rapid weight gain to follow.

Energy intake should provide nutritional adequacy without excess—that is, somewhere between deprivation and complete freedom to eat everything in sight. A reasonable suggestion is that an adult needs to increase activity and reduce food intake◆ enough to create a 500-kcalorie-per-day deficit.[42] Such a deficit will produce a weight loss of about 1 pound per week—a rate that supports the loss of fat efficiently while retaining lean tissue.

◆ **Weight-management tip: "Energy out" should exceed "energy in" by about 500 kcal/day.**

◆ **Weight-management tip: Make nutritional adequacy a high priority.**

• Emphasize Nutritional Adequacy • Nutritional adequacy◆ is difficult to achieve on fewer than 1200 kcalories a day, and most healthy adults need never consume any less than that. A plan that provides an adequate intake supports a healthier and more successful weight loss than a restrictive plan that creates feelings of starvation and deprivation, which can lead to an irresistible urge to binge.

Take a look at Table 2-2 on p. 37 and notice that the 1600-kcalorie food plan represents the minimum servings suggested and supplies about 30 percent of kcalories as fat. Such an intake would allow most people to lose weight and still meet their nutrient needs with careful, nutrient-dense◆ food selections. (Women might need an iron supplement.) Keep in mind, too, that well-balanced diets that emphasize fruits, vegetables, whole grains, lean meats or meat alternates, and low-fat milk products offer many health rewards even when they don't result in weight loss.[43]

◆ Weight-management tip: Emphasize nutrient-dense foods.

• **Eat Small Portions** • Overweight people may need to learn to eat less food◆ at each meal—one piece of chicken for dinner instead of two, a teaspoon of butter on the vegetables instead of a tablespoon, and one cookie for dessert instead of six. The goal is to eat enough food for energy, nutrients, and pleasure, but not more. This amount should leave a person feeling satisfied—not necessarily full.

◆ Weight-management tip: Eat small portions. Share a restaurant meal with a friend or take home half for lunch tomorrow.

Keep in mind that even nonfat and low-fat foods◆ can deliver a lot of kcalories when a person eats large quantities. A low-fat cookie or two can be a sweet treat even on a weight-loss diet, but a whole box is clearly excessive.

◆ Weight-management tip: Limit low-fat treats to the serving size on the label.

• **Focus on Complex Carbohydrates** • Healthy meals and snacks center on complex carbohydrate foods. Fresh fruits, vegetables, legumes, and whole grains◆ offer abundant vitamins, minerals, and fiber but little fat. High-fiber foods also require effort to eat—an added bonus. People who eat these foods in abundance spontaneously eat for longer times and take in fewer kcalories than when eating foods of high energy density. The satiety signal indicating fullness is sent after a 20-minute lag,◆ so a person who slows down and savors each bite eats less before the signal reaches the brain.

◆ Weight-management tip: Make legumes, whole grains, vegetables, and fruits central to your diet plan.

◆ Weight-management tip: Eat slowly.

• **Choose Fats Sensibly** • Recent research on the effects of leptin on food intake and energy expenditure provide possible clues as to why high-fat diets foster weight gain: high-fat meals lower blood leptin levels.[44] When leptin levels drop, people feel hungry.[45] (Review Figure 9-4 on p. 273 to see how decreased leptin levels stimulate appetite and diminish energy expenditure.) Satiety plays a key role in determining food intake, and fat has a weak satiating effect. Consequently, a person eating a high-fat diet tends to overeat, which raises energy intake by adding both more food and more fat kcalories.

Measure fat◆ with extra caution. A slip of the butter knife adds more kcalories than a slip of the sugar spoon. Less fat in the diet means less fat in the body (review p. 152 for strategies to lower fat in the diet). Be careful not to take this advice to the extreme, however; too little fat in the diet or in the body carries health risks as well, as Chapter 5 explained.

◆ Weight-management tip: Limit high-fat foods.

• **Watch for Other Empty kCalories** • Speaking of empty kcalories, a person trying to achieve or maintain a healthy weight needs to pay attention not only to fat, but to sugar◆ and alcohol, too. Using them for pleasure on occasion is compatible with health as long as most daily choices are of nutrient-dense foods. Not only does alcohol add kcalories, but accompanying mixers can also add both kcalories and fat, especially in creamy drinks such as piña coladas (review Table H7-3 on p. 235). Furthermore, drinking alcohol reduces a person's inhibitions, which can sabotage weight-control efforts—at least temporarily.

◆ Weight-management tip: Limit concentrated sweets and alcoholic beverages.

• **Drink Adequate Water** • Learn to satisfy thirst with water.◆ Water fills the stomach between meals and dilutes the metabolic wastes generated from the breakdown of fat. It meets the water need that was formerly met by eating extra food (remember that foods provide water). Water also helps the GI tract adapt to a high-fiber diet.

◆ Weight-management tip: Drink a glass of water before you begin to eat and another while you eat. Drink plenty of water throughout the day (8 glasses or more a day).

Drinking water is a healthy habit.

© Anne Dowie

IN SUMMARY

A person who adopts a lifelong◆ "eating plan for good health" rather than a "diet for weight loss" will be more likely to keep the lost weight off. The margins provide several tips for successful weight management.

◆ Weight-management tip: Learn, practice, and follow a healthful eating plan for the rest of your life.

Physical Activity

◆ Fitness tip: Participate in some form of physical activity regularly.

The best approach to weight management combines diet and physical activity. People who combine diet and exercise are more likely to lose more fat, retain more muscle, and regain less weight than those who only diet. Even when people who include physical activity◆ in their weight-management program do not lose more weight, they seem to follow their diet plans more closely and maintain their losses better than those who do not exercise. Consequently, they benefit both from taking in a little less energy and from expending a little more energy in physical activity. Importantly, they improve their cardiorespiratory fitness, regardless of weight loss.[46]

• **Activity and Energy Expenditure** • Table 8-2 (on p. 248) shows how much energy each of several activities uses. The table also shows that the number of kcalories spent in an activity depends more upon body weight than on how fast the exercise is done. For example, a person who weighs 150 pounds and runs a mile in 6 minutes spends about 103 kcalories. That same person walking a mile in 15 minutes uses almost the same amount—about 92 kcalories. Similarly, a 220-pound person spends about 150 kcalories on the 6-minute mile, and only a little less—about 135 kcalories—on the 15-minute walk. Whether a person chooses to run or walk the distance, a similar amount of energy will be spent; walking will just take longer. To lose fat, exercise as intensely, and expend as much energy, as your time allows. And be careful not to compensate for the energy spent in exercise by eating more food. Otherwise, energy balance won't shift and fat loss will be less significant.

Remember to adjust estimates of energy costs of activities after losing weight (review Table 8-2 on p. 248). After a significant weight loss, formerly obese people spend less energy on the same activity than they spent previously. The new expenditure is even less than would be expected given their lower body weight.

© Myrleen Ferguson Cate/PhotoEdit

The key to good health is to combine sensible eating with regular exercise.

• **Activity and Basal Metabolism** • Activity also contributes to energy expenditure in an indirect way—by speeding up basal metabolism. It does this both immediately and over the long term. On any given day, basal metabolism remains elevated for several hours after intense and prolonged exercise. Over the long term, a person who engages in daily vigorous activity for many weeks gradually develops more lean tissue. Metabolic rate rises accordingly, and this supports continued weight loss or maintenance.

• **Activity and Body Composition** • Physical activity also changes body composition: body fat decreases and lean body mass increases.[47] Furthermore, exercise specifically decreases abdominal fat.[48]

• **Activity and Appetite Control** • Physical activity also helps to control appetite. Many people think that exercising will make them eat more, but this is not entirely true.[49] Active people do have healthy appetites, but immediately after an intense workout, most people do not feel like eating. They may be thirsty and want to shower, but they are not hungry. The reason is that the body has released fuels from storage to support the exercise, so glucose and fatty acids are abundant in the blood. At the same time, the body has suppressed its digestive functions. Hard physical work and eating are not compatible. A person must calm down, put energy fuels back in storage, and relax before eating. Thus exercise may actually help curb appetite, especially the inappropriate appetite that accompanies boredom, anxiety, or depression. Weight-management programs encourage people who feel the urge to eat when not hungry to go out and exercise instead. The activity passes time, relieves anxiety, and prevents inappropriate eating.

• **Activity and Psychological Benefits** • Activity also helps reduce stress. Since stress itself cues inappropriate eating for many people, activity can help here, too. Activity offers still more psychological advantages. The fit person looks and feels healthy and, as a result, gains self-esteem. High self-esteem motivates a person to persist in seeking good health and fitness, which keeps the beneficial◆ cycle going.

Benefits of physical activity in a weight-management program: ◆
- Short-term increase in energy expenditure (from exercise and from a slight rise in BMR).
- Long-term increase (slight) in BMR.
- Improves body composition.
- Appetite control.
- Stress reduction and control of stress eating.
- Physical, and therefore psychological, well-being.
- High self-esteem.

• **Choosing Activities** • Clearly, physical activity is a plus in a weight-management program. What kind of physical activity is best? People should choose activities that they enjoy and are willing to do regularly. Sustained physical activities of moderate intensity (aerobic exercises) are more effective in weight control than short bursts of vigorous exercise.

In addition to exercise, a person can incorporate hundreds of energy-spending activities into daily routines: take the stairs instead of the elevator, walk to the neighbor's apartment instead of making a phone call, and rake the leaves instead of using a blower. Remember that sitting uses more kcalories than lying down, standing uses more kcalories than sitting, and moving uses more kcalories than standing. A 175-pound person who replaces a 30-minute television program with a 2-mile walk a day can spend enough energy to lose (or not gain) 18 pounds in a year. Even chewing gum or fidgeting increases the kcalories spent.[50] The point is: be active. Walk a mile. Run a race. Swim a lap. Dance a jig. Ride a bike. Climb a mountain. Skip a rope. Do whatever you enjoy doing—and do it often.

• **Spot Reducing** • People sometimes ask about "spot reducing." Unfortunately, muscles do not "own" the fat that surrounds them. Fat cells all over the body release fat in response to the demand of physical activity for use by whatever muscles are active. No exercise can remove the fat from any one particular area.

Exercise can help with trouble spots in another way, though. The "trouble spot" for most men is the abdomen, their primary site of fat storage. During aerobic exercise, abdominal fat readily releases its stores, providing fuel to the physically active body. With regular exercise and weight loss, men will deplete these abdominal fat stores before those in the lower body. Women may also deplete abdominal fat with exercise, but their "trouble spots" are more likely to be their hips and thighs.

In addition to aerobic activity, strength training can help to improve the tone of muscles in a trouble area, and stretching to gain flexibility can help with associated posture problems. A combination of aerobic, strength, and flexibility workouts best improves fitness and physical appearance.

IN SUMMARY

Physical activity should be an integral part of a weight-control program. Physical activity can increase energy expenditure, improve body composition, help control appetite, reduce stress and stress eating, and enhance physical and psychological well-being.

Behavior and Attitude

Behavior modification once held a key position in weight-loss programs, but its status has diminished in recent years.[51] Still, behavior and attitude play an important role in supporting efforts to achieve and maintain appropriate body weight and composition. Changing the hundreds of small behaviors of overeating and underexercising that lead to, and perpetuate, obesity requires time and effort. A person must commit to take action.

It also helps to adopt a positive, matter-of-fact attitude. Healthy eating and activity choices are an essential part of healthy living and should simply be incorporated into the day—much like brushing one's teeth or wearing a safety belt.

behavior modification: the changing of behavior by the manipulation of antecedents (cues or environmental factors that trigger behavior), the behavior itself, and consequences (the penalties or rewards attached to behavior).

FIGURE 9-9 Food Record

The entries in a food record should include the times and places of meals and snacks, the types and amounts of foods eaten, and a description of the individual's feelings when eating. The diary should also record physical activities: the kind, the intensity level, the duration, and the person's feelings about them.

Time	Place	Activity or food eaten	People present	Mood
10:30–10:40	School vending machine	6 peanut butter crackers and 12 oz. cola	by myself	Starved
12:15–12:30	Restaurant	Sub sandwich and 12 oz. cola	friends	relaxed & friendly
3:00–3:45	Gym	Weight training	work out partner	tired
4:00–4:10	Snack bar	Small frozen yogurt	by myself	OK

◆ **Behavior change tip: Keep a record of diet and exercise habits; it reveals problem areas, the first step toward improving behaviors.**

◆ **Examples of behavioral strategies to support weight change:**
- Do not grocery shop when hungry.
- Eat slowly (pause during meals, chew thoroughly, put down utensils between bites).
- Exercise when watching television.

◆ **Behavior change tip: Learn alternative ways to deal with emotions and stresses.**

• **Become Aware of Behaviors** • To solve a problem, a person must first identify all the behaviors that created the problem. Keeping a record◆ will help to identify eating and exercise behaviors that may need changing (see Figure 9-9). It will also establish a baseline against which to measure future progress.

• **Change Behaviors** • Strategies◆ focus on learning desired eating and exercise behaviors and eliminating unwanted behaviors. With so many possible behavior changes, a person can choose where to begin. Start simply and don't try to master them all at once. Attempting too many changes at one time invites failure. Pick one trouble area that is manageable and start there. Practice a desired behavior until it becomes routine. Then select another trouble area to work on, and so on. Another bit of advice along the same lines: don't try to tackle major changes during a particularly stressful time of life.

• **Personal Attitude** • For many people, overeating and being overweight have become an integral part of their identity. Those who fully understand their personal relationships with food are best prepared to make healthful changes in eating and exercise behaviors.

Sometimes habitual behaviors that are hazardous to health, such as smoking or drinking alcohol, contribute positively by helping people adapt to stressful situations. Similarly, many people overeat to cope with the stresses of life. To break out of that pattern, they must first identify the particular stressors that trigger the urge to overeat. Then, when faced with these situations, they must learn◆ and practice problem-solving skills that will help them to respond appropriately.

All this is not to imply that psychological therapy holds the magic answer to a weight problem. Still, efforts to improve one's general well-being may result in healthy eating and exercising habits even when weight loss is not the primary goal. When the problems that trigger the urge to overeat are resolved in alternative ways, people may find they eat less. They may begin to respond appropriately to internal cues of hunger rather than inappropriately to external cues of stress. Sound emotional health supports a person's ability to take care of physical health in all ways—including nutrition, weight management, and fitness.

• Support Groups • Group support◆ is important when making life changes. Some people find it helpful to join a group that provides support in efforts to lose weight, such as Take Off Pounds Sensibly (TOPS), Weight Watchers (WW), Overeaters Anonymous (OA), or others. A modest expenditure for health is well worthwhile, but people need to avoid rip-offs, of course. Many dieters find it helpful to form their own self-help groups.

Behavior change tip: Attend support groups regularly or develop supportive relationships with others. ◆

HEALTHY PEOPLE 2010

Increase the proportion of worksites that offer nutrition or weight management classes or counseling.

IN SUMMARY

A surefire remedy for obesity has yet to be found, although many people find a combination of the approaches just described to be most effective. Diet and exercise shift energy balance so that more energy is being spent than is taken in. Physical activity maintains or even builds the lean body so that fat is preferentially lost and metabolic energy needs remain high. Behavior modification retrains habits so that once weight is lost, it will not return; and an improvement in inner self helps a person to manage life without a dependency on food. This treatment package requires time, individualization, and sometimes the assistance of a registered dietitian.

Weight Maintenance

People who are successful often lose much of their weight within half a year and then reach a plateau.[52] This slowdown can be disappointing, but should be recognized as an opportunity for the body to adjust to its new weight. It also gives the person relief from the distraction of weight-loss dieting. An appropriate goal at this point is to continue eating and activity behaviors that will maintain weight. Attempting to lose additional weight at this point would require heroic efforts and would almost certainly meet with failure.

Be aware that obesity is not "cured" by simply attaining a reasonable body weight; eating wisely and staying active must continue to be part of life's daily routines. If, on arriving at goal weight after months of self-discipline, the person resumes old habits, then maintenance will not be successful.

Those who are successful in maintaining◆ their weight loss have established vigorous exercise regimens and careful eating patterns, consuming less energy and a lower percentage of kcalories from fat than the national average.[53] They do not have the same flexibility in their food and activity habits as their friends who have never been overweight; they are more efficient at storing fat.[54] With weight loss, metabolism shifts downward so that formerly overweight people require less energy than might be expected given their current body weight and body composition.[55] Consequently, to keep weight off, they must either eat less or exercise more than people the same size who have never been obese.

Behavior change tip: Adopt permanent lifestyle changes to achieve and maintain a healthy weight. ◆

Physical activity plays a key role in maintaining weight loss. Those who continue exercising vigorously are far more successful than those who remain inactive.[56] On average, weight maintenance requires a person to expend 1500 to 2000 kcalories in physical activity per week.[57] To accomplish this, a person might exercise either moderately (such as brisk walking) for 80 minutes a day or vigorously (such as fast bicycling) for 35 minutes a day, for example.

TABLE 9-3 Suggested Public Health Strategies

Strategies	Examples of Suggested Nutritional Strategies	Examples of Successful Nonnutritional Strategies
Impose safety standards to reduce the potential for harm.	• Regulate the kcalorie or fat density of foods. • Regulate the size of packages of high-fat foods.	• Mandate safety glass in automobiles. • Regulate the lead content of paint.
Control commercial advertising to limit the influence of harmful products.	• Improve nutrition labeling and product packaging. • Restrict the promotion of high-fat foods (especially when directed at children).	• Restrict cigarette advertising (especially when directed at children). • Add health warnings to alcoholic beverages.
Control the conditions under which products are sold to limit exposure to hazardous substances.	• Remove high-fat, low–nutrient density foods from school vending machines. • Restrict the number of vendors licensed to sell high-fat foods.	• Mandate minimum-age laws for the use of tobacco, alcohol, and automobiles. • Restrict the number of vendors licensed to sell alcohol.
Control prices to reduce consumption.	• Tax fat.	• Tax alcohol and tobacco.

SOURCE: Adapted from R. W. Jeffery, Public health approaches to the management of obesity, in K. B. Brownell and C. G. Fairburn, eds., *Eating Disorder and Obesity—A Comprehensive Handbook* (New York: Guilford Press, 1995), pp. 558–563.

Prevention

Given the information presented up to this point in the chapter, the adage "An ounce of prevention is worth a pound of cure" seems particularly apropos. Being obese is unhealthy, and losing weight is challenging and often temporary.[58] Strategies for preventing weight gain◆ are very similar to those for losing weight, with one exception: they begin early and continue throughout life. Over the years, they become an integral part of a person's life. It is much easier for a person to resist doughnuts for breakfast if he rarely eats them. Similarly, a person will have little trouble walking each morning if she has always been active.

◆ **To prevent excessive weight gain:**
- Eat regular meals and limit snacking.
- Drink water instead of high-kcalorie beverages.
- Select low-fat foods regularly and limit dietary fat to 30% of daily kcalorie intake.
- Become physically active and limit television viewing time.

Public Health Programs

Is there anyone in the United States who hasn't heard the message that obesity raises the risks of chronic diseases and that overweight people should aim for a healthy weight by eating sensibly and becoming physically active? Not likely. Yet implementing such advice is difficult in a "toxic environment" of abundant food and physical inactivity. To successfully treat obesity we may have to treat the environment in which we live. Table 9-3 provides examples of public health strategies that might effectively improve our nation's nutrition environment. Some of these strategies may seem radical, but dramatic measures may be needed if we are to curb the obesity epidemic that is sweeping across the nation.

IN SUMMARY Preventing weight gains and maintaining weight losses require vigilant attention to diet and physical activity. Taking care of oneself is a lifelong responsibility.

UNDERWEIGHT

◆ **Reminder: *Underweight* is a body weight so low as to have adverse health effects; it is generally defined as BMI <18.5.**

Underweight◆ is a far less prevalent problem than overweight, affecting no more than 5 percent of U.S. adults (review Figure 9-2 on p. 271). Whether the underweight person needs to gain weight is a question of health and, like weight loss, a highly individual matter. People who are healthy at their present weights may stay there; there are no compelling reasons to try to gain weight. Those who are thin because of malnourishment or illness, however, might benefit from a diet that supports weight gain. Medical advice can help make the distinction.

Thin people may find gaining weight difficult. Those who wish to gain weight for appearance's sake or to improve their athletic performance need to be aware that healthful weight gains can be achieved only by physical conditioning combined with high energy intakes. On a high-kcalorie diet alone, a person will gain weight, but it will be mostly fat. Even if the gain improves appearance, it can be detrimental to health and might impair athletic performance. Therefore, in weight gain, as in weight loss, physical activity and energy intake are essential components of a sound plan.

Problems of Underweight

The causes of underweight may be as diverse as those of overweight—hunger, appetite, and satiety irregularities; psychological traits; metabolic factors; and hereditary tendencies. Habits learned early in childhood, especially food aversions, may perpetuate themselves.

The demand for energy to support physical activity and growth often contributes to underweight. An active, growing boy may need more than 4000 kcalories a day to maintain his weight and may be too busy to take time to eat. Underweight people find it hard to gain weight due, in part, to their expenditure of energy in adaptive thermogenesis. So much energy may be spent adapting to a higher food intake that at first as many as 750 to 800 extra kcalories a day may be needed to gain◆ a pound a week. Like those who want to lose weight, people who want to gain must learn new habits and learn to like new foods. They are also similarly vulnerable to potentially harmful schemes and would be wise to review the consumer bill of rights on p. 277, using "weight gain" instead of "weight loss" where appropriate.

◆ Weight-gain tip: Expect weight gain to take time (1 lb per month would be reasonable).

An extreme underweight condition known as anorexia nervosa sometimes develops in people who employ heroic self-denial to control their weight. They go to such extremes that they become severely undernourished, achieving final body weights of 70 pounds or even less. The distinguishing feature of a person with anorexia nervosa, as opposed to other underweight people, is that the starvation is intentional. Anorexia nervosa is a major eating disorder seen in our society today. Another is bulimia nervosa—compulsive overeating and purging. Eating disorders are the subject of the highlight that follows this chapter.

Weight-Gain Strategies

Weight-gain strategies center on eating foods that provide many kcalories in a small volume and exercising to build muscle. Table 2-2 (on p. 37) provides a diet pattern for 2800 kcalories a day and incorporates many of the principles for planning a healthy diet.

• Energy-Dense Foods • Energy-dense foods◆ (the very ones eliminated from a successful weight-loss diet) hold the key to weight gain. Pick the highest-kcalorie items from each food group—that is, milk shakes instead of nonfat milk, salmon instead of snapper, avocados instead of cucumbers, a cup of grape juice instead of a small apple, and whole-wheat muffins instead of whole-wheat bread. Because fat provides more than twice as many kcalories per teaspoon as sugar does, fat adds kcalories without adding much bulk.

◆ Weight-gain tip: Eat energy-dense foods regularly.

Be aware that health experts routinely recommend a low-fat diet because the biggest health problems in the United States involve overweight and heart disease. Eating high-kcalorie, high-fat foods is not healthy for most people, but may be essential for an underweight individual who needs to gain weight. An underweight person who is physically active and eating a nutritionally adequate diet can afford a few extra kcalories from fat. Well-trained endurance athletes can maintain a favorable lipid profile on a diet providing 42 percent of kcalories from fat.[59] For health's sake, it would be wise to select foods with monounsaturated and polyunsaturated fats instead of those with saturated fats: for example, sautéeing vegetables in olive oil instead of butter.

◆ Weight-gain tip: Eat at least three meals a day.

• **Regular Meals Daily** • People who are underweight need to make meals a priority and take the time to plan, prepare, and eat each meal. They should eat at least three healthy meals◆ every day and learn to eat more food within the first 20 minutes of a meal. Another suggestion is to eat meaty appetizers or the main course first and leave the soup or salad until later.

◆ Weight-gain tip: Eat large portions of foods and expect to feel full.

• **Large Portions** • Underweight people need to learn to eat more food at each meal. Put extra slices of ham and cheese on the sandwich for lunch, drink milk from a larger glass, and eat cereal from a larger bowl.

The person should expect to feel full.◆ Most underweight individuals are accustomed to small quantities of food. When they begin eating significantly more, they feel uncomfortable. This is normal and passes over time.

◆ Weight-gain tip: Eat snacks between meals.

• **Extra Snacks** • Since a substantially higher energy intake is needed each day, in addition to eating more food at each meal, it is necessary to eat more frequently. Between-meal snacking◆ offers a solution. For example, a student might make three sandwiches in the morning and eat them between classes in addition to the day's three regular meals. Snacking on dried fruit, nuts, and seeds will also add kcalories easily.

◆ Weight-gain tip: Drink plenty of juice and milk.

• **Juice and Milk** • Beverages◆ provide an easy way to increase energy intake. Consider that 6 cups of cranberry juice add almost 1000 kcalories to the day's intake. kCalories can be added to milk by mixing in powdered milk or packets of instant breakfast.

For people who are underweight due to illness, concentrated liquid formulas are often recommended because a weak person can swallow them easily. A physician or registered dietitian can recommend high-protein, high-kcalorie formulas to help an underweight person maintain or gain. Used in addition to regular meals, these can help considerably.

◆ Weight-gain tip: Exercise and eat to build muscles.

• **Exercising to Build Muscles** • To gain weight, use strength training◆ primarily and increase energy intake to support that exercise. Eating extra food will then support a gain of both muscle and fat. About 700 to 1000 kcalories a day above normal energy needs is enough to support both the exercise and the building of muscle.

IN SUMMARY

Both the incidence of underweight and the health problems associated with it are less prevalent than overweight and its associated problems. To gain weight, a person must train physically and increase energy intake by selecting energy-dense foods, eating regular meals, taking larger portions, and consuming extra snacks and beverages.

How are you doing?

- Do you try to lose weight even though your BMI falls between 18.5 and 24.9?
- Does your weight fluctuate up and down dramatically over time?
- Do you try to lose weight by following fad diets or taking over-the-counter drugs or herbal supplements?

Nutrition on the Net

WEBSITES

Access these websites for further study of topics covered in this chapter.

- Find updates and quick links to these and other nutrition-related sites at our website: **www.wadsworth.com/nutrition**
- Search for "obesity" and "weight control" at the U.S. Government health information site: **www.healthfinder.gov**
- Review the Clinical Guidelines on the Identification, Evaluation, and Treatment of Overweight and Obesity in Adults: **www.nhlbi.nih.gov/guidelines/obesity/ob_home.htm**
- Learn about the drugs used for weight loss from the Center for Drug Evaluation and Research: **www.fda.gov/cder**
- Learn about weight control and the WIN program from the Weight-control Information Network: **www.niddk.nih.gov/health/nutrit/win.htm**
- Visit weight-loss support groups, such as Take Off Pounds Sensibly, Overeaters Anonymous, and Weight Watchers: **www.tops.org**, **www.overeatersanonymous.org**, and **www.weightwatchers.com**
- See what the obesity professionals think at the North American Association for the Study of Obesity and the American Society for Bariatric Surgery: **www.naaso.org** and **www.asbs.org**
- Consider the nondietary approaches of HUGS International: **www.hugs.com**

INTERNET ACTIVITIES

Weight management is a major health concern in the United States. Many people want to lose weight; some want to gain. Either way, this tool will help you plan a healthy diet while keeping an eye on the kcalorie counter.

- Go to: **http://www.nhlbi.nih.gov/index.htm**
- Click on "Special Web Pages and Interactive Applications."
- Click on "Aim for a Healthy Weight."
- Click on "Menu Planner" and then on "About the Menu Planner."
- After reading the directions, begin planning your menu for a day's meals.

Use this interactive menu planner to plan kcalorie-controlled meals, or to keep track of the kcalories in the foods you normally consume. Did the number of kcalories in the foods you eat surprise you?

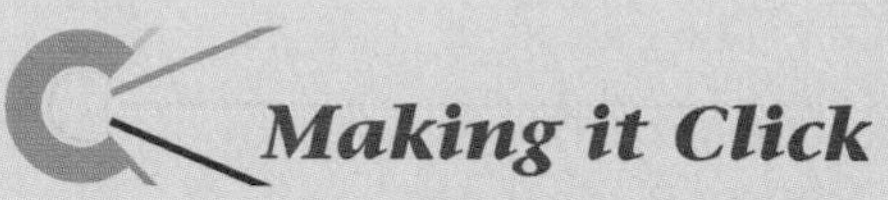

Making it Click

These problems give you practice in doing simple energy-balance calculations (see p. 294 for answers). Once you have mastered these examples, you will be prepared to examine your own food choices. Be sure to show your calculations for each problem.

1. Critique a commercial weight-loss plan. Consumers spend billions of dollars a year on weight-loss programs such as Slim-Fast, Sweet Success, Weight Watchers, Nutri/System, Jenny Craig, Optifast, Medifast, and Formula One. One such plan calls for a milk shake in the morning, at noon, and as an afternoon snack and "a sensible, balanced, low-fat dinner" in the evening. One shake mixed in 8 ounces of vitamin A– and D–fortified nonfat milk offers 190 kcalories; 32 grams of carbohydrate, 13 grams of protein, and 1 gram of fat; at least one-third of the Daily Value for all vitamins and minerals; plus 2 grams of fiber.
   a. Calculate the kcalories and grams of carbohydrate, protein, and fat that three shakes provide.
   b. How do these values compare with the criteria listed in item 2 in Table H8-3 on p. 266?
   c. Plan "a sensible, balanced, low-fat dinner" that will help make this weight-loss plan adequate and balanced. Now, how do the day's totals compare with the criteria in item 2 in Table H8-3 on p. 266?
   d. Critique this plan using the other criteria described in Table H8-3 on p. 266 as a guide.
2. Evaluate a weight-gain attempt. People attempting to gain weight sometimes have a hard time because they choose low-kcalorie, high-bulk foods that make it hard to consume enough energy. Consider the following

lunch: a chef's salad consisting of 2 cups iceberg lettuce, 1 whole tomato, 1 ounce swiss cheese, 1 ounce roasted ham (lean and fat), 1 hard-boiled egg, ½ cucumber, and ¼ cup mayonnaise-type salad dressing. If you weighed these foods, you'd find that they totaled 552 grams. This is a pretty filling meal.

a. How much does this meal weigh in pounds?
b. The meal provides 541 kcalories. What is the energy density of this meal, expressed in kcalories per gram?
c. To gain weight, this person is advised to eat an additional 500 kcalories at this meal. Using foods with this same energy density, how much more chef's salad will this person have to eat?
d. Suppose a person simply can't do this. Try to reduce the bulk of this meal by replacing some of the lettuce with more energy-dense foods. Delete 1 cup lettuce from the salad and add 1 ounce roast beef and 1 ounce cheddar cheese. Show how these changes influence the weight and kcalories of this meal. (Use Appendix H.)

Item No./Food	Weight (g)	Energy (kcal)
Original totals:	552	541
Minus:		
#867 Lettuce, 1 c	−	−
Plus:		
#603 Roast beef, 1 oz	+	+
#37 Cheddar cheese, 1 oz	+	+
Totals:		

e. How many kcalories did the changes add?
f. How much more *weight* of food did these changes add?

This exercise should reveal why people attempting to gain weight are advised to add high-fat items, within reason, to their daily meals.

Study Questions

These questions will help you review the chapter. You will find the answers in the discussions on the pages provided.

1. Describe how body fat develops, and suggest some reasons why it is difficult for an obese person to maintain weight loss. (pp. 269–271)
2. What factors contribute to obesity? (pp. 271–276)
3. List several aggressive ways to treat obesity, and explain why such methods are not recommended for every overweight person. (pp. 276–281)
4. Discuss reasonable dietary strategies for achieving and maintaining a healthy body weight. (pp. 281–283)
5. What are the benefits of increased physical activity in a weight-loss program? (pp. 284–285)
6. Describe the behavioral strategies for changing an individual's dietary habits. What role does personal attitude play? (pp. 285–287)
7. Describe strategies for successful weight gain. (pp. 289–290)

These questions will help you prepare for an exam. Answers can be found on p. 294.

1. With weight loss, fat cells:
   a. decrease in size only.
   b. decrease in number only.
   c. decrease in both number and size.
   d. decrease in number, but increase in size.
2. Obesity is caused by:
   a. overeating.
   b. inactivity.
   c. defective genes.
   d. multiple factors.
3. The protein produced by the fat cells under the direction of the *ob* gene is called:
   a. leptin.
   b. serotonin.
   c. sibutramine.
   d. phentermine.
4. The biggest problem associated with the use of drugs in the treatment of obesity is:
   a. cost.
   b. chronic dosage.
   c. ineffectiveness.
   d. adverse side effects.
5. A realistic goal for weight loss is to reduce body weight:
   a. down to the weight a person was at age 25.
   b. down to the ideal weight in the weight-for-height tables.
   c. by 10 percent over six months.
   d. by 15 percent over three months.
6. A nutritionally sound weight-loss diet might restrict daily energy intake to create a:
   a. 1000-kcalorie-per-month deficit.
   b. 500-kcalorie-per-month deficit.

c. 500-kcalorie-per-day deficit.
d. 1000-kcalorie-per-day deficit.

7. Successful weight loss depends on:
   a. avoiding fats and limiting water.
   b. taking supplements and drinking water.
   c. increasing proteins and restricting carbohydrates.
   d. reducing energy intake and increasing physical activity.
8. Physical activity does not help a person to:
   a. lose weight.
   b. retain muscle.
   c. maintain weight loss.
   d. lose fat in trouble spots.
9. Which strategy would *not* help an overweight person to lose weight?
   a. Exercise.
   b. Eat slowly.
   c. Limit high-fat foods.
   d. Eat energy-dense foods regularly.
10. Which strategy would *not* help an underweight person to gain weight?
   a. Exercise.
   b. Drink plenty of water.
   c. Eat snacks between meals.
   d. Eat large portions of foods.

References

1 A. H. Mokdad and coauthors, The spread of the obesity epidemic in the United States, 1991–1998, *Journal of the American Medical Association* 282 (1999): 1519–1522.

2 R. J. Kuczmarski and coauthors, Varying body mass index cutoff points to describe overweight prevalence among U.S. adults: NHANES III (1988 to 1994), *Obesity Research* 5 (1997): 542–548.

3 J. P. Koplan and W. H. Dietz, Caloric imbalance and public health policy, *Journal of the American Medical Association* 282 (1999): 1579; Mokdad and coauthors, 1999; R. W. Jeffery and S. A. French, Epidemic obesity in the United States: Are fast foods and television viewing contributing? *American Journal of Public Health* 88 (1998): 277–280; A. M. Coulston, Obesity as an epidemic: Facing the challenge, *Journal of the American Dietetic Association* 98 (1998): S6–S8; J. M. Rippe, The obesity epidemic: Challenges and opportunities, *Journal of the American Dietetic Association* 98 (1998): S5.

4 P. A. Kern, Potential role of TNFa and lipoprotein lipase as candidate genes for obesity, *Journal of Nutrition* 127 (1997): 1917S–1922S.

5 M. D. Jensen, Lipolysis: Contribution from regional fat, *Annual Review of Nutrition* 17 (1997): 127–139.

6 J. P. Foreyt and W. S. C. Poston II, Diet, genetics, and obesity, *Food Technology* 51 (1997): 70–73; C. Bouchard, Human variation in body mass: Evidence for a role of the genes, *Nutrition Reviews* 55 (1997): S21–S30.

7 B. L. Heitmann and coauthors, Are genetic determinants of weight gain modified by leisure-time physical activity? A prospective study of Finnish twins, *American Journal of Clinical Nutrition* 66 (1997): 672–678.

8 C. Bouchard and A. Tremblay, Genetic influences on the response of body fat and fat distribution to positive and negative energy balances in human identical twins, *Journal of Nutrition* 127 (1997): 943S–947S.

9 H. Qian and coauthors, Brain administration of leptin causes deletion of adipocytes by apoptosis, *Endocrinology* 139 (1998): 791–794.

10 C. T. Montague and coauthors, Congenital leptin deficiency is associated with severe early-onset obesity in humans, *Nature* 387 (1997): 903–908.

11 I. S. Farooqi and coauthors, Effects of recombinant leptin therapy in a child with congenital leptin deficiency, *New England Journal of Medicine* 341 (1999): 879–884.

12 H. Fors and coauthors, Serum leptin levels correlate with growth hormone secretion and body fat in children, *Journal of Clinical Endocrinology & Metabolism* 84 (1999): 3586–3590; G. Marchini and coauthors, Plasma leptin in infants: Relations to birth weight and weight loss, *Pediatrics* 101 (1998): 429–432; C. S. Fox and coauthors, Is a low leptin concentration, a low resting metabolic rate, or both the expression of the "thrifty genotype"? Results from Mexican Pima Indians, *American Journal of Clinical Nutrition* 68 (1998): 1053–1057; J. M. Friedman, Leptin, leptin receptors, and the control of body weight, *Nutrition Reviews* 56 (1998): S38–S46.

13 E. Ravussin and coauthors, Relatively low plasma leptin concentrations precede weight gain in Pima Indians, *Nature Medicine* 3 (1997): 238–240.

14 J. Albu and coauthors, Obesity solutions: Report of a meeting, *Nutrition Reviews* 55 (1997): 150–156.

15 M. Rosenbaum, R. L. Leibel, and J. Hirsch, Obesity, *New England Journal of Medicine* 337 (1997): 396–407; G. Wolf, Neuropeptides responding to leptin, *Nutrition Reviews* 55 (1997): 85–88.

16 S. B. Heymsfield and coauthors, Recombinant leptin for weight loss in obese and lean adults: A randomized, controlled, dose-escalation trial, *Journal of the American Medical Association* 282 (1999): 1568–1575.

17 Nir Barzilai and coauthors, Leptin selectively decreases visceral adiposity and enhances insulin action, *Journal of Clinical Investigation* 100 (1997): 3105–3110.

18 P. Trayhurn and coauthors, Hormonal and neuroendocrine regulation of energy balance—The role of leptin, *Arch Tierernahr* 51 (1998): 177–185; G. Frunbeck, S. A. Jebb, and A. M. Prentice, Leptin: Physiology and pathophysiology, *Clinical Physiology* 18 (1998): 399–419; M. R. Sierra-Honigmann and coauthors, Biological action of leptin as an angiogenic factor, *Science* 281 (1998): 1683–1686.

19 C. Fleury and coauthors, Uncoupling protein-2: A novel gene linked to obesity and hyperinsulinemia, *Nature Genetics* 15 (1997): 269–272.

20 J. S. Flier and B. B. Lowell, Obesity research springs a proton leak, *Nature Genetics* 15 (1997): 223–224.

21 G. Wolf, A new uncoupling protein: A potential component of the human body weight regulation system, *Nutrition Reviews* 55 (1997): 178–179.

22 J. A. Yanovski and coauthors, Associations between uncoupling protein 2, body composition, and resting energy expenditure in lean and obese African American, white, and Asian children, *American Journal of Clinical Nutrition* 71 (2000): 1405–1412.

23 P. Hakala and coauthors, Environmental factors in the development of obesity in identical twins, *American Journal of Obesity and Related Metabolic Disorders* 23 (1999): 746–753.

24 A. H. C. Goris, M. S. Westerterp-Plantenga, and K. R. Westerterp, Undereating and underrecording of habitual food intake in obese men: Selective underreporting of fat intake, *American Journal of Clinical Nutrition* 71 (2000): 130–134; L. A. Braam and coauthors, Determinants of obesity-related underreporting of energy intake, *American Journal of Epidemiology* 147 (1998): 1081–1086; L. Johansson and coauthors, Under- and overreporting of energy intake related to weight status and lifestyle, *American Journal of Clinical Nutrition* 68 (1998): 266–274; L. M. Carter and S. J. Whiting, Underreporting of energy intake, socioeconomic status, and expression of nutrient intake, *Nutrition Reviews* 56 (1998): 179–182.

25 J. P. Flatt, How not to approach the obesity problem, *Obesity Research* 5 (1997): 632–633.

26 K. Brownell, The pressure to eat—Why we're getting fatter, *Nutrition Action Healthletter,* July/August 1998, pp. 3–6.

27 Jeffery and French, 1998.

28 J. E. Blundell and J. I. Macdiarmid, Fat as a risk factor for overconsumption: Satiation, satiety, and patterns of eating, *Journal of the American Dietetic Association* 97 (1997): S63–S69.

29 C. Proserpi and coauthors, Ad libitum intake of a high-carbohydrate or high-

ANSWERS

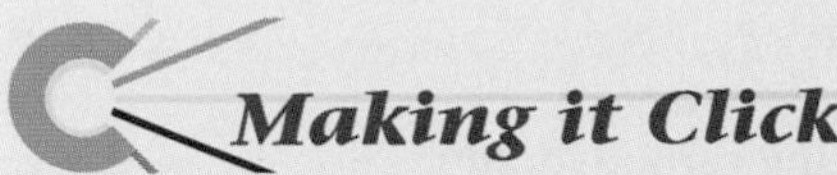

Making it Click

1. a. Three milk shakes provide: 3 × 190 kcal = 570 kcal; 3 × 32 g carbohydrate = 96 g carbohydrate; 3 × 13 g protein = 39 g protein; and 3 × 1 g fat = 3 g fat.
   b. To meet this criteria, the plan needs *at least* an additional 630 kcalories (1200 kcal − 570 kcal = 630 kcal); an additional 5 to 24 grams of protein, depending on the person's RDA based on gender and age (63 g − 39 g = 24 g and 44 g − 39 g = 5 g); an additional 4 grams of carbohydrate (100 g − 96 g = 4 g); and some additional fat.
   c. Of course, there are many possible dinners that you could plan. One might be:
      Salad made with 1 c lettuce, 1 c chopped tomatoes and onions, ¼ c garbanzo beans, and 2 tbs low-fat dressing
      4 oz grilled chicken
      1 medium baked potato
      1 c summer squash and zucchini
      1 c melon cubes
   This meal brings the day's totals to 1215 kcalories, 90 g of protein, 192 g of carbohydrate, and 13 g of fat, which meets the goals for kcalories, protein, and carbohydrate. Because the milk shake has been fortified, all vitamin and mineral needs are covered as well. The only possible dietary shortcoming is that the day's percent kcalories from fat is low (only 10%), but because energy and nutrient recommendations have been met and the goal is weight loss, this may be acceptable.
   d. This weight-loss plan uses a liquid formula rather than foods, making clients dependent on a special device (the formula) rather than teaching them how to make good choices from the conventional food supply. It provides no information about dropout rates, the long-term success of clients, or weight maintenance after the program ends.
2. a. More than a pound (552 g ÷ 454 g/lb = 1.2 lb).
   b. 541 kcal ÷ 552 g = 0.98 kcal/g.
   c. More than another whole pound (0.98 kcal/g × 500 kcal = 490 g; 490 g ÷ 454 g/lb = 1.1 lb).
   d.

Item No./Food	Weight (g)	Energy (kcal)
Original totals:	552	541
Minus:		
#867 Lettuce, 1 c	−55	−7
Plus:		
#603 Roast beef, 1 oz	+28	+68
#37 Cheddar cheese, 1 oz	+28	+113
Totals:	552 g	715 kcal

   e. 715 kcal − 541 kcal = 174 kcal added.
   f. 552 g − 552 g = 0 grams added.

Multiple Choice

1. a	2. d	3. a	4. d	5. c
6. c	7. d	8. d	9. d	10. b

fat diet in young men: Effects on nutrient balances, *American Journal of Clinical Nutrition* 66 (1997): 539–545.

30 M. Wei and coauthors, The association between cardiorespiratory fitness and impaired glucose and type 2 diabetes mellitus in men, *Annals of Internal Medicine* 130 (1999): 89–96; U.S. Department of Health and Human Services, *Physical Activity and Health—A Report of the Surgeon General Executive Summary,* 1996.

31 D. A. Schoeller, Balancing energy expenditure and body weight, *American Journal of Clinical Nutrition* 68 (1998): 956S–961S.

32 R. L. Atkinson, Use of drugs in the treatment of obesity, *Annual Review of Nutrition* 17 (1997): 383–403.

33 C. H. Halsted, Is blockade of pancreatic lipase the answer? *American Journal of Clinical Nutrition* 69 (1999): 1059–1060.

34 L. J. Aronne, Modern medical management of obesity: The role of pharmaceutical intervention, *Journal of the American Dietetic Association* 98 (1998): S23–S26.

35 J. J. Ramsey and coauthors, Energy expenditure, body composition, and glucose metabolism in lean and obese rhesus monkeys treated with ephedrine and caffeine, *American Journal of Clinical Nutrition* 68 (1998): 42–51.

36 P. Kurtzweil, Dieter's brews make tea time a dangerous affair, *FDA Consumer,* July/August 1997, pp. 6–11.

37 R. B. Rao, S. F. Ely, and R. S. Hoffman, Deaths related to liposuction, *New England Journal of Medicine* 340 (1999): 1471–1475.

38 J. T. Fine and coauthors, A prospective study of weight change and health-related quality of life in women, *Journal of the American Medical Association* 282 (1999): 2136–2142.

39 G. D. Foster and coauthors, What is a reasonable weight loss? Patients' expectations and evaluations of obesity treatment outcomes, *Journal of Consulting and Clinical Psychology* 65 (1997): 79–85.

40 Position of The American Dietetic Association: Weight management, *Journal of the American Dietetic Association* 97 (1997): 71–74.

41 M. K. Serdula and coauthors, Prevalence of attempting weight loss and strategies for controlling weight, *Journal of the American Medical Association* 282 (1999): 1353–1358.

42 National Institutes of Health Obesity Education Initiative, *Clinical Guidelines on the Identification, Evaluation, and Treatment of Overweight and Obesity in Adults* (Washington, D.C.: U.S. Department of Health and Human Services, 1998).

43 A. K. Kant and coauthors, A prospective study of diet quality and mortality in women, *Journal of the American Medical Association* 283 (2000) 2109–2115.

44 D. A. Ainslie and coauthoers, Short-term, high-fat diets lower circulating leptin concentrations in rats, *American Journal of Clinical Nutrition* 71 (2000): 438–442; P. J. Havel and coauthors, High-fat meals reduce 24-h circulating leptin concentrations in women, *Diabetes* 48 (1999): 334–341.

45 D. Chapelot and coauthors, An endocrine and metabolic definition of the intermeal interval in humans: Evidence for a role of leptin on the prandial pattern through fatty acid disposal, *American Journal of Clinical Nutrition* 72 (2000): 421–431; N. L. Keim, J. S. Stern, and P. J. Havel, Relation between circulating leptin concentrations and appetite during a prolonged, moderate energy deficit in women, *American Journal of Clinical Nutrition* 68 (1998): 794–801.

46 J. M. Jakicic and coauthors, Effects of intermittent exercise and use of home exercise equipment on adherence, weight loss, and fitness in overweight women—A randomized trial, *Journal of the American Medical Association* 282 (1999): 1554–1560.

47 K. R. Westerterp, Alterations in energy balance with exercise, *American Journal of Clinical Nutrition* 68 (1998): 970S–974S.

48 K. N. Frayn, Regulation of fatty acid delivery in vivo, in *Skeletal Muscle Metabolism in Exercise and Diabetes,* ed. E. A. Richter (New York: Plenum Press, 1998), pp. 171–179.

49 N. A. King, A. Tremblay, and J. E. Blundell, Effects of exercise on appetite control: Implications for energy balance, *Medicine and Science in Sports and Exercise* 29 (1997): 1076–1089.

50 J. Levine, P. Baukol, and I. Pavlidis, The energy expended in chewing gum, *New England Journal of Medicine* 341 (1999): 2100; J. A. Levine, S. J. Schleusner, and M. D. Jensen, Energy expenditure of nonexercise activity, *American Journal of Clinical Nutrition* 72 (2000): 1451–1454.

51 Albu and coauthors, 1997.

52 A. Astrup and coauthors, The role of low-fat diets and fat substitutes in body weight management: What have we learned from clinical studies? *Journal of the American Dietetic Association* 97 (1997): S82–S87.

53 M. T. McGuire and coauthors, Long-term maintenance of weight loss: Do people who lose weight through various weight loss methods use different behaviors to maintain their weight? *International Journal of Obesity and Related Metabolic Disorders* 22 (1998): 572–577; S. M. Shick and coauthors, Persons successful at long-term weight loss and maintenance continue to consume a low-energy, low-fat diet, *Journal of the American Dietetic Association* 98 (1998): 408–413; M. L. Klem and coauthors, A descriptive study of individuals successful at long-term maintenance of substantial weight loss, *American Journal of Clinical Nutrition* 66 (1997): 239–246; Albu and coauthors, 1997.

54 A. Raben and coauthors, Diurnal metabolic profiles after 14d of an ad libitum high-starch, high-sucrose, or high-fat diet in normal-weight, never-obese and postobese women, *American Journal of Clinical Nutrition* 73 (2001): 177–189.

55 A. Astrup and coauthors, Meta-analysis of resting metabolic rate in formerly obese subjects, *American Journal of Clinical Nutrition* 69 (1999): 1117–1122.

56 D. A. Schoeller, K. Shay, and R. F. Kushner, How much physical activity is needed to minimize weight gain in previously obese women? *American Journal of Clinical Nutrition* 66 (1997): 551–556.

57 J. M. Rippe and S. Hess, The role of physical activity in the prevention and management of obesity, *Journal of the American Dietetic Association* 98 (1998) S31–S38.

58 National Institutes of Health Obesity Education Initiative, 1998.

59 J. Leddy and coauthors, Effect of a high or a low fat diet on cardiovascular risk factors in male and female runners, *Medicine and Science in Sports and Exercise* 29 (1997): 17–25.

EATING DISORDERS

FOR SOME people, dieting to lose weight progresses to a dangerous and obsessive point. An estimated 5 million people in the United States, primarily girls and young women, suffer from the **eating disorders** anorexia nervosa and bulimia nervosa (the accompanying glossary defines these and related terms). Many more suffer from binge eating disorders or other unspecified conditions that do not meet the strict criteria for anorexia nervosa or bulimia nervosa, but still imperil a person's well-being.

Why do so many people in our society suffer from eating disorders? Most experts agree that the causes are multifactorial: sociocultural, psychological, and perhaps neurochemical. Excessive pressure to be thin is at least partly to blame. When low body weight becomes an important goal, people begin to view normal healthy body weight as being too fat, and they take unhealthy actions to lose weight.

Young people who attempt extreme weight loss may have learned to identify discomforts such as anger, jealousy, or disappointment as "feeling fat."[1] They may also be depressed or suffer social anxiety. As weight loss and maintenance become more of a focus, psychological problems worsen and the likelihood of developing eating disorders intensifies. Athletes are particularly likely to develop eating disorders.

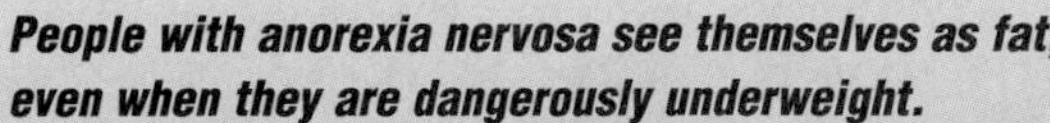

People with anorexia nervosa see themselves as fat, even when they are dangerously underweight.

© Steve Niedorf Photography/The Image Bank

THE FEMALE ATHLETE TRIAD

At age 14, Suzanne was a top contender for a spot on the state gymnastics team. Each day her coach reminded team members that they must weigh no more than their assigned weights in order to qualify for competition. The coach chastised gymnasts who gained weight, and Suzanne was terrified of being singled out. Convinced that the less she weighed the better she would perform, Suzanne weighed herself several times a day to confirm that she had not exceeded her 80-pound limit. Driven to excel in her sport, Suzanne kept her weight down by eating very little and training very hard. Unlike many of her friends, Suzanne never began to menstruate. A few months before her fifteenth birthday, Suzanne's coach dropped her back to the second-level team. Suzanne blamed her poor performance on a slow-healing stress fracture. Mentally stressed and physically exhausted, she quit gymnastics and began overeating between her periods of self-starvation. Suzanne had developed the dangerous combination of problems that characterize the **female athlete triad**—disordered eating, amenorrhea, and osteoporosis (see Figure H9-1).[2]

Disordered Eating

Part of the reason many athletes engage in disordered eating behaviors may be that they and their coaches have embraced unsuitable weight standards. An athlete's body must be heavier for a given height than a nonathlete's body because the athlete's body is dense, containing more healthy bone and muscle and less fat. When athletes rely on scales, they may mistakenly believe they are too fat because weight standards, such as the BMI, are inadequate in providing information about body composition.

Many young athletes severely restrict energy intakes to improve performance, enhance the aesthetic appeal of their performance, or meet the weight guidelines of their specific sports. They fail to realize that the loss of lean tissue that accompanies energy restriction actually impairs their physical performance. The increasing incidence of abnormal eating habits among athletes is causing concern. Male athletes, especially wrestlers and gymnasts, are affected by these disorders as well, but research shows that females are most vulnerable. Risk factors for eating disorders among athletes include the following:

- Young age (adolescence).
- Pressure to excel at a chosen sport.
- Focus on achieving or maintaining an "ideal" body weight or body fat percentage.
- Being a wrestler, jockey, or competitor in a sport where performance is judged on aesthetic appeal such as gymnastics, figure skating, or dance.

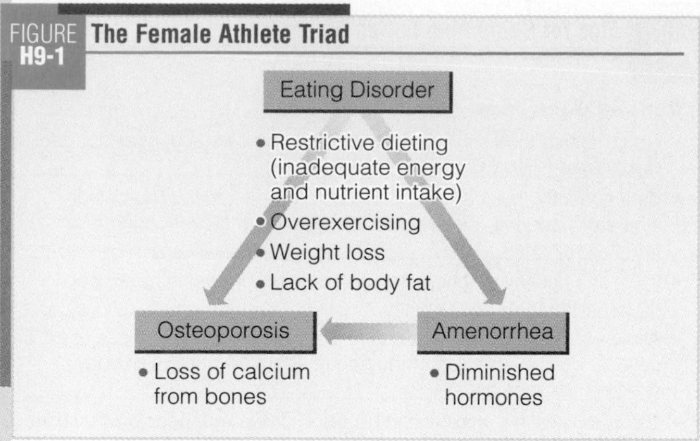

FIGURE H9-1 The Female Athlete Triad

- Dieting at an early age.
- Unsupervised dieting.

Amenorrhea

The prevalence of **amenorrhea** among premenopausal women in the United States is about 2 to 5 percent overall, but among female athletes, it may be as high as 66 percent. Contrary to previous notions, amenorrhea is *not* a normal adaptation to strenuous physical training: it is a symptom of something going wrong. Amenorrhea is characterized by low blood estrogen, infertility, and often bone mineral losses. Some research indicates that depleted body fat contributes to amenorrhea; other studies indicate that the percentage of body fat is not associated with normal menstruation in athletes. However amenorrhea develops, it threatens the integrity of the bones. Bone losses remain significant even after menstruation resumes.

Osteoporosis

For most people, weight-bearing physical activity, dietary calcium, and the hormone estrogen protect against the bone loss of osteoporosis. For young women with disordered eating and amenorrhea, strenuous activity can impair bone health.[3] Vigorous training combined with inadequate food intake and other life stresses may reduce estrogen levels and greatly increase the risks of **stress fractures** today and of osteoporosis in later life.[4] Stress fracture, a serious form of bone injury, commonly occurs among dancers and other athletes with low body weights and disordered eating. Many underweight young athletes have bones like those of postmenopausal women, and they may never again recover their lost bone even after diagnosis and treatment—which makes prevention critical.[5] Athletes should be encouraged to consume at least 1300 milligrams of calcium each day, to eat nutrient-dense foods, and to obtain enough energy to support both weight gain and the energy expended in physical activity. Women with bulimia rarely cease menstruating and so may be spared this loss of bone integrity.[6]

OTHER DANGEROUS PRACTICES OF ATHLETES

Only females face the threats of the female athlete triad, of course, but many male athletes face pressure to achieve a certain body weight and may develop eating disorders. Each week throughout the season, David drastically restricts his food and fluid intake before a wrestling match in an effort to "make weight." For at least three wrestlers in 1997, the consequences were deadly.[7] Wrestlers and their coaches believe that competing in a lower weight class will give them a competitive advantage over smaller opponents. To that end, David practices in rubber suits, sits in steam rooms, and takes diuretics to lose 4 to 6 pounds. He hopes to replenish the lost fluids, glycogen, and lean tissue during the hours between his weigh-in and competition, but the body needs days to correct this metabolic mayhem. Reestablishing fluid and electrolyte balances may take a day or two, replenishing glycogen stores may take two to three days, and replacing lean tissue may take even longer.

GLOSSARY

amenorrhea (ay-MEN-oh-REE-ah): the absence of or cessation of menstruation. **Primary amenorrhea** is menarche delayed beyond 16 years of age. **Secondary amenorrhea** is the absence of three to six consecutive menstrual cycles.

anorexia (an-oh-RECK-see-ah) **nervosa:** an eating disorder characterized by a refusal to maintain a minimally normal body weight and a distortion in perception of body shape and weight.
- **an** = without
- **orex** = mouth
- **nervos** = of nervous origin

binge-eating disorder: an eating disorder whose criteria are similar to those of bulimia nervosa, excluding purging or other compensatory behaviors.

bulimia (byoo-LEEM-ee-ah) **nervosa:** an eating disorder characterized by repeated episodes of binge eating usually followed by self-induced vomiting, misuse of laxatives or diuretics, fasting, or excessive exercise.
- **buli** = ox

cathartic (ka-THAR-tik): a strong laxative.

eating disorders: disturbances in eating behavior that jeopardize a person's physical or psychological health.

emetic (em-ETT-ic): an agent that causes vomiting.

female athlete triad: a potentially fatal combination of three medical problems: disordered eating, amenorrhea, and osteoporosis.

muscle dysmorphia (dis-MORE-fee-ah): a newly coined psychiatric disorder characterized by a preoccupation with building body mass.

stress fractures: bone damage or breaks caused by stress on bone surfaces during exercise.

unspecified eating disorders: eating disorders that do not meet the defined criteria for specific eating disorders.

Ironically, the combination of food deprivation and dehydration impairs physical performance by reducing muscle strength, decreasing anaerobic power, reducing endurance capacity, and lowering oxygen consumption.[8] For optimal performance, wrestlers need to first achieve their competitive weight during the off-season and then eat well-balanced meals and drink plenty of fluids during the competitive season.[9]

Some athletes go to extreme measures to bulk up and *gain* weight. People afflicted with **muscle dysmorphia** eat high-protein diets, take dietary supplements, weight train for hours at a time, and often abuse steroids in an attempt to bulk up. Their bodies are large and muscular, yet they see themselves as puny 90-pound weaklings. They are preoccupied with the idea that their bodies are too small or inadequately muscular. Like others with distorted body images, people with muscle dysmorphia weigh themselves frequently and center their lives on diet and exercise. Paying attention to diet and pumping iron for fitness is admirable, but obsessing over it can cause serious social, occupational, and physical problems.[10]

TABLE H9-1 Tips for Combating Eating Disorders

General Guidelines

- Never restrict food servings to below the numbers suggested for adequacy by the Food Guide Pyramid.
- Eat frequently. Include healthy snacks between meals. The person who eats frequently never gets so hungry as to allow hunger to dictate food choices.
- If not at a healthy weight, establish a reasonable weight goal based on a healthy body composition.
- Allow a reasonable time to achieve the goal. A reasonable loss of excess fat can be achieved at the rate of about 10 percent of body weight in six months.
- Establish a weight-maintenance support group with people who share interests.

Specific Guidelines for Athletes and Dancers

- Remember that eating disorders impair physical performance. Seek confidential help in obtaining treatment if needed.
- Focus on proper nutrition as an important facet of your training, as important as proper technique.

PREVENTING EATING DISORDERS IN ATHLETES

To prevent eating disorders in athletes and dancers, the performers, their coaches, and their parents must learn about inappropriate body weight ideals, improper weight-loss techniques, eating disorder development, proper nutrition, and safe weight-control methods.[11] Young people naturally search for identity and will often follow the advice of a person in authority without question. Therefore, coaches and dance instructors should never encourage unhealthy weight loss to qualify for competition or to conform with distorted artistic ideals. Athletes who truly need to lose weight should try to do so during the off-season and under the supervision of a health care professional.[12] Frequent weighings can push young people who are striving to lose weight into a cycle of starving to confront the scale, then bingeing uncontrollably afterward. The erosion of self-esteem that accompanies these events can interfere with normal psychological development and set the stage for serious problems later on.

Table H9-1 provides some suggestions to help athletes and dancers protect themselves against developing eating disorders. The remaining sections describe eating disorders that anyone, athlete or nonathlete, may experience.

ANOREXIA NERVOSA

Julie is 18 years old and a superachiever in school. She watches her diet with great care, and she exercises daily, maintaining a heroic schedule of self-discipline. She is thin, but she is determined to lose more weight. She is 5 feet 6 inches tall and weighs 85 pounds. She has **anorexia nervosa.**

Characteristics of Anorexia Nervosa

Julie is unaware that she is undernourished, and she sees no need to obtain treatment. She developed amenorrhea several months ago and has become moody and chronically depressed. She insists that she is too fat, although her eyes are sunk in deep hollows in her face. Julie denies that she is ever tired, although she is close to physical exhaustion and no longer sleeps easily. Her family is concerned, and though reluctant to push her, they have finally insisted that she see a psychiatrist. Julie's psychiatrist has diagnosed anorexia nervosa (see Table H9-2 on p. 299) and prescribed group therapy as a start. If she does not begin to gain weight soon, she may need to be hospitalized.

As mentioned in the introduction, most anorexia nervosa victims are females; males account for only about 1 in 20 reported cases. Central to the diagnosis of anorexia nervosa is a distorted body image that overestimates personal body fatness.[13] When Julie looks at herself in the mirror, she sees a "fat" 85-pound body. The more Julie overestimates her body size, the more resistant she is to treatment, and the more unwilling to examine her faulty values and misconceptions. Malnutrition is known to affect brain functioning and judgment in this way, causing lethargy, confusion, and delirium.

Anorexia nervosa cannot be self-diagnosed. Nearly everyone in our society is engaged in the pursuit of thinness, and denial runs high among people with anorexia nervosa. Some women have all the attitudes and behaviors associated with the condition, but without the dramatic weight loss.

• The Role of the Family • Most families normally experience conflicts from time to time, especially when adolescents are trying to establish their identities and

TABLE H9-2 **Criteria for Diagnosis of Anorexia Nervosa**

A person with anorexia nervosa demonstrates the following:

A. Refusal to maintain body weight at or above a minimal normal weight for age and height (e.g., weight loss leading to maintenance of body weight less than 85 percent of that expected; or failure to make expected weight gain during period of growth, leading to body weight less than 85 percent of that expected).

B. Intense fear of gaining weight or becoming fat, even though underweight.

C. Disturbance in the way in which one's body weight or shape is experienced, undue influence of body weight or shape on self-evaluation, or denial of the seriousness of the current low body weight.

D. In females past puberty, amenorrhea, i.e., the absence of at least three consecutive menstrual cycles. (A woman is considered to have amenorrhea if her periods occur only following hormone, e.g., estrogen, administration.)

Two types:

Restricting type: During the episode of anorexia nervosa, the person does not regularly engage in binge eating or purging behavior (i.e., self-induced vomiting or the misuse of laxatives, diuretics, or enemas).

Binge eating/purging type: During the episode of anorexia nervosa, the person regularly engages in binge eating or purging behavior (i.e., self-induced vomiting or the misuse of laxatives, diuretics, or enemas).

SOURCE: Reprinted with permission from American Psychiatric Association, *Diagnostic and Statistical Manual of Mental Disorders,* 4th ed. Text Revision. (Washington, D.C.: American Psychiatric Association, 2000).

independence. Family members who are unable to resolve their conflicts through effective communication sometimes adopt unhealthy coping behaviors. Such unresolved family conflicts may play an important role in the development of anorexia nervosa. Parents may oppose one another's authority and vacillate between defending and condemning the anorexic child's behavior. Parental control is often rigid and overprotective. In the extreme, parents may even be abusive.[14] Julie is a perfectionist, and her parents expect perfection. She identifies so strongly with her parents' ideals and goals that she cannot get in touch with her own identity. She is respectful of authority, but sometimes feels like a robot, and she may act that way, too: polite but controlled, rigid, and unspontaneous.[15] For Julie, rejecting food is a way of gaining control.

Although families of children with anorexia nervosa often have problems, blame is a useless concept. The parents may suffer deeply from being blamed for their child's illness when, in truth, no one knows what causes anorexia nervosa. Rather than judging parents, a more useful tactic is to identify the family's strengths and resources and to prepare them for the job of helping their ill child benefit from treatment.

People with anorexia nervosa may reluctantly eat small amounts of low-kcalorie foods.

© Omni Photo Communications, Inc./Index Stock Imagery

• **Self-Starvation** • How can a person as thin as Julie continue to starve herself? Julie uses tremendous discipline against her hunger to strictly limit her portions of low-kcalorie foods. She will deny her hunger, and having adapted to so little food, she feels full after eating only a half-dozen carrot sticks. She can recite the kcalorie contents of dozens of foods and the kcalorie costs of as many exercises. If she feels that she has gained an ounce of weight, she runs or jumps rope until she is sure she has exercised it off. If she fears that the food she has eaten outweighs the exercise, she takes laxatives to hasten the passage of food from her system. She drinks water incessantly to fill her stomach, risking dangerous mineral imbalances.[16] She is desperately hungry. In fact, she is starving, but she doesn't eat because her need for self-control dominates.

Many people, on learning of this disorder, say they wish they had "a touch" of it to get thin. They mistakenly think that people with anorexia nervosa feel no hunger. They also fail to recognize the pain of the associated psychological and physical trauma.

• **Physical Consequences** • The starvation of anorexia nervosa damages the body just as the starvation of war and poverty does. In fact, after a few months, most people with anorexia nervosa have protein-energy malnutrition (PEM) that is similar to marasmus (described in Chapter 6). Victims are dying to be thin—quite literally. In young people, growth ceases and normal development falters. They lose so much lean tissue that basal metabolic rate slows, an effect that may remain even after treatment and regain of weight. In addition, the heart pumps inefficiently and irregularly, the heart muscle becomes weak and thin, the chambers diminish in size, and the blood pressure falls.[17] Minerals that help to regulate heartbeat become unbalanced. Many deaths occur due to multiple organ system failure.

Starvation brings other physical consequences as well: loss of brain tissue, impaired immune response, anemia, and a loss of digestive functions that worsens malnutrition.[18] Peristalsis becomes sluggish, the stomach empties slowly, and the lining of the intestinal tract atrophies. The deteriorated GI tract fails to provide sufficient digestive enzymes and absorptive surfaces for handling any food the victim may eat. The pancreas slows its production of digestive enzymes. The person may suffer from diarrhea, further worsening malnutrition.

Other effects of starvation include altered blood lipids, high blood vitamin A and vitamin E, low blood proteins, dry thin skin, abnormal nerve functioning, reduced bone density, low body temperature, low blood pressure, and the development of fine body hair (the body's attempt to keep warm). The electrical activity of the brain becomes abnormal, and insomnia is common. Both women and men lose their sex drives.

Women with anorexia nervosa develop amenorrhea (it is one of the diagnostic criteria). Sometimes that symptom precedes the weight loss. Anorexia nervosa delays the onset of menstruation in young girls. Menstrual periods typically resume with recovery, although some women never restart even after they have gained weight. Should an underweight woman with anorexia nervosa become pregnant, she is likely to give birth to an underweight baby—and low-birthweight babies face many health problems (as Chapter 14 explains). Mothers with anorexia nervosa may severely underfeed their children who then fail to grow and suffer the other harms typical of starvation.[19]

Treatment of Anorexia Nervosa

Treatment of anorexia nervosa requires a multidisciplinary approach. Teams of physicians, nurses, psychiatrists, family therapists, and dietitians work together to resolve two sets of issues and behaviors: those relating to food and weight, and those involving relationships with oneself and others. The first dietary objective is to stop weight loss while establishing regular eating patterns. Appropriate diet is crucial to recovery and must be tailored individually to each client's needs.[20] Because body weight is low and fear of weight gain is high, initial food intake may be small. As eating becomes more comfortable, energy intake should increase gradually. Initially, clients may be unwilling to eat for themselves; those who will eat have a good chance of recovering without other interventions. Table H9-3 lists principles of nutrition intervention in anorexia nervosa.

TABLE H9-3 Principles of Nutrition Intervention in Anorexia Nervosa

- Include foods from each of the food groups, with portion sizes increasing as energy intake increases.
- Increase food energy intake gradually (adding 200 kcalories/week).
- Prescribe well-balanced diets, with *some* individual variations according to client preferences (e.g., vegetarian).
- Give multiple vitamin-mineral supplements to restore nutrient losses.
- Enhance elimination with dietary fiber from grain sources.
- Reduce sensations of bloating with small, frequent feedings.
- In behavioral programs, link rewards to food energy intake, not to weight gain.
- Use liquid supplements when the client cannot achieve desired intake with solid food.
- Reduce satiety sensations by offering cold or room-temperature foods and finger foods (e.g., snacks).
- Provide interactive nutrition counseling as an ongoing process.
- Reduce excessive caffeine intake.
- Provide IV nutritional support only in severe states of ill health, malnutrition, and wasting.

SOURCE: Adapted from C. L. Rock and J. Yager, Nutrition and eating disorders: A primer for clinicians, *International Journal of Eating Disorders* 6 (1987): 276, as cited in *Nutrition and the M.D.*, July 1988, with permission. Reprinted with permission of John Wiley & Sons, Inc., copyright 1987.

Because anorexia nervosa is like starvation physically, health care professionals classify clients based on indicators of PEM.* Low-risk clients need nutrition counseling. Intermediate-risk clients may need supplements such as high-kcalorie, high-protein formulas in addition to regular meals. High-risk clients may require hospitalization and may need to be fed by tube at first to prevent death. This step may cause psychological trauma. Drugs are commonly prescribed, but play a limited role in treatment.

Physicians and clients need to realize that weight gain may be difficult, especially during the first week of treatment, perhaps because both the resting metabolic rate and the thermic effect of food in people with anorexia nervosa are so high.[21] Further, as small gains of body fat occur, blood values of the appetite-suppressing hormone leptin begin creeping up, too, which may also contribute to difficulties in gaining weight.[22]

Denial runs high among those with anorexia nervosa. Few seek treatment on their own. Almost half of the women who are treated can maintain their body weight at 85 percent or more of a healthy weight; at that weight, many of them begin menstruating again. The other half have poor to fair treatment outcomes; many relapse into abnormal eating behaviors to some extent. An estimated 1000 women die each year of anorexia nervosa—most commonly from starvation, nutrient imbalances, or suicide.

Before drawing conclusions about someone who is extremely thin or who eats very little, remember that diagnosis requires professional assessment. Several national organizations offer information for people who are seeking help with anorexia nervosa, either for themselves or for others.†

BULIMIA NERVOSA

Kelly is a charming, intelligent, 20-year-old flight attendant of normal weight who thinks constantly about food. She alternatively starves herself and secretly binges; when she has eaten too much, she makes herself vomit. Most readers recognize these symptoms as those of **bulimia nervosa.**

Characteristics of Bulimia Nervosa

Bulimia nervosa is distinct from anorexia nervosa and is more prevalent, although the true incidence is difficult to

*Indicators of protein-energy malnutrition: a low percentage of body fat, low serum albumin, low serum transferrin, and impaired immune reactions.

†Internet sites are listed at the end of this highlight, and phone numbers and addresses are in Appendix F.

establish because bulimia nervosa is not as physically apparent. More men suffer from bulimia nervosa than from anorexia nervosa, but bulimia nervosa is still more common in women than in men. The secretive nature of bulimic behaviors makes recognition of the problem difficult, but once it is recognized, diagnosis is based on the criteria listed in Table H9-4.

Like the typical person with bulimia nervosa, Kelly is single, female, and white. She is well educated and close to her ideal body weight, although her weight fluctuates over a range of 10 pounds or so every few weeks. As a flight attendant, she prefers to weigh less than the weight that her body maintains naturally.

Kelly seldom lets her eating disorder interfere with work or other activities, although a third of all bulimics do. From early childhood she has been a high achiever and emotionally dependent on her parents. As a young teen, Kelly cycled on and off fad diets, but could never maintain an appropriate weight. Kelly feels anxious at social events and cannot easily establish close personal relationships. She is usually depressed, is often impulsive, and has low self-esteem. When crisis hits, Kelly responds by replaying events, worrying excessively, and blaming herself but never asking for help—behaviors that interfere with effective coping.[23]

• The Role of the Family • Parents of bulimic people have often suffered from depression and alcohol abuse. While their expectations for their child's achievements are high, parental involvement is minimal.[24] Any changes in the family structure often meet with resistance, even when the changes would greatly benefit the person with bulimia nervosa. In addition, the family may have secrets that are hidden from outsiders; many people with bulimia nervosa report having been abused sexually or physically by family members or family friends.

• Binge Eating • Like the person with anorexia nervosa, the person with bulimia nervosa spends much time thinking about her body weight and food. Her preoccupation with food manifests itself in secret binge-eating episodes, which usually progress through several emotional stages: anticipation and planning, anxiety, urgency to begin, rapid and uncontrollable consumption of food, relief and relaxation, disappointment, and finally shame or disgust.

A bulimic binge is characterized by a sense of lacking control over eating. During a binge, the person consumes food for its emotional comfort and cannot stop eating or control what or how much is eaten. A typical binge occurs periodically, in secret, usually at night, and lasts an hour or more. Because a binge frequently follows a period of rigid dieting, eating is accelerated by intense hunger. Energy restriction followed by bingeing can set in motion a pattern of weight cycling, which may make weight loss and maintenance more difficult over time.

During a binge, Kelly consumes thousands of kcalories of easy-to-eat, low-fiber, high-fat, and, especially, high-carbohydrate foods. Typically, she chooses cookies, cakes, and ice cream—and she eats the entire bag of cookies, the whole cake, and every last spoonful in a carton of ice cream. After the binge, Kelly pays the price with swollen hands and feet, bloating, fatigue, headache, nausea, and pain.

• Purging • To purge the food from her body, Kelly may use a **cathartic**—a strong laxative that can injure the lower intestinal tract. Or she may induce vomiting, using an **emetic**—a drug intended as first aid for poisoning. These purging behaviors are often accompanied by feelings of shame or guilt. Hence a vicious cycle develops: negative self-perceptions followed by dieting, bingeing, and purging, which in turn lead to negative self-perceptions (see Figure H9-2 on p. 302).

On first glance, purging seems to offer a quick and easy solution to the problems of unwanted kcalories and body weight. Many people perceive such behavior as neutral or even positive, when, in fact, binge eating and purging have serious physical consequences. Signs of subclinical malnutrition are evident in a compromised immune system. Fluid and mineral imbalances caused by vomiting or diarrhea can lead to abnormal heart rhythms and injury to the kidneys, which have to cope with the imbalances. Urinary tract infections can lead to kidney failure. Vomiting causes irritation and infection of the pharynx, esophagus, and salivary glands; erosion of the teeth; and dental caries. The esophagus may rupture or tear, as may the stomach. Sometimes the eyes become red from pressure during vomiting.

TABLE H9-4 Criteria for Diagnosis of Bulimia Nervosa

A person with bulimia nervosa demonstrates the following:

A. Recurrent episodes of binge eating. An episode of binge eating is characterized by both of the following:
   1. Eating, in a discrete period of time (e.g., within any two-hour period), an amount of food that is definitely larger than most people would eat during a similar period of time and under similar circumstances.
   2. A sense of lack of control over eating during the episode (e.g., a feeling that one cannot stop eating or control what or how much one is eating).

B. Recurrent inappropriate compensatory behavior in order to prevent weight gain, such as self-induced vomiting; misuse of laxatives, diuretics, enemas, or other medications; fasting; or excessive exercise.

C. Binge eating and inappropriate compensatory behaviors both occur, on average, at least twice a week for three months.

D. Self-evaluation unduly influenced by body shape and weight.

E. The disturbance does not occur exclusively during episodes of anorexia nervosa.

Two types:

Purging type: The person regularly engages in self-induced vomiting or the misuse of laxatives, diuretics, or enemas.

Nonpurging type: The person uses other inappropriate compensatory behaviors, such as fasting or excessive exercise, but does not regularly engage in self-induced vomiting or the misuse of laxatives, diuretics, or enemas.

SOURCE: Reprinted with permission from American Psychiatric Association, *Diagnostic and Statistical Manual of Mental Disorders*, 4th ed. Text Revision. (Washington, D.C.: American Psychiatric Association, 2000).

Bulimic binges are often followed by self-induced vomiting and feelings of shame or disgust.

© Michael Newman/PhotoEdit

The hands may be calloused or cut by the teeth while inducing vomiting.[25] Overuse of emetics depletes potassium concentrations and can lead to death by heart failure.

Unlike Julie, Kelly is aware that her behavior is abnormal, and she is deeply ashamed of it. She wants to recover, and this makes recovery more likely for her than for Julie, who clings to denial. Feeling inadequate ("I can't even control my eating"), Kelly tends to be passive and to look to others for confirmation of her sense of worth. When she experiences rejection, either in reality or in her imagination, her bulimia nervosa becomes worse. If Kelly's depression deepens, she may seek solace in drug or alcohol abuse or other addictive behaviors. Clinical depression is common in people with bulimia nervosa, and the rates of alcohol, marijuana, and cigarette abuse are high.

Treatment of Bulimia Nervosa

To gain control over food and establish regular eating patterns, Kelly needs to adhere to a structured eating plan. Weight maintenance, rather than cyclic weight gains and losses, is the treatment goal. Many a victim has taken a major step toward recovery by learning to eat enough food to satisfy hunger needs (at least 1600 kcalories a day). Table H9-5 offers diet strategies to correct the eating problems of bulimia nervosa. About half of the women diagnosed with bulimia recover completely after five to ten years, with or without treatment, but treatment probably speeds the recovery process.[26]

A mental health professional should be on the treatment team to help clients with their depression and addictive behaviors. Some physicians prescribe the antidepressant drug fluoxetine in the treatment of bulimia nervosa.* Another drug that may be useful in the management of bulimia nervosa is naloxone, an opiate antagonist that suppresses the consumption of sweet and high-fat foods in binge-eaters.

Anorexia nervosa and bulimia nervosa are distinct eating disorders, yet they sometimes overlap in important ways. Anorexia victims may purge, and victims of both disorders share an overconcern with body weight and the tendency to drastically undereat. Many perceive foods as "forbidden" and "give in" to an eating binge. The two disorders can also appear in the same person, or one can lead to the other. Treatment is challenging and relapses are not

*Fluoxetine is marketed under the trade name Prozac.

FIGURE H9-2 The Vicious Cycle of Restrictive Dieting and Binge Eating

Negative self-perceptions
Restrictive dieting
Binge eating
Purging

TABLE H9-5 Diet Strategies for Combating Bulimia Nervosa

- Avoid finger foods; eat foods that require the use of utensils.
- Enhance satiety by eating warm foods.
- Include vegetables, salad, and/or fruit at meals to prolong eating time.
- Choose whole-grain and high-fiber breads and cereals to maximize bulk.
- Eat a well-balanced diet and meals consisting of a variety of foods.
- Use foods that are naturally divided into portions, such as potatoes (rather than rice or pasta); 4- and 8-ounce containers of yogurt or cottage cheese; precut steak or chicken parts; and frozen entreés.
- Include foods containing ample complex carbohydrates (for satiety) and some fat (to slow gastric emptying).
- Eat meals and snacks sitting down.
- Plan meals and snacks, and record plans in a food diary prior to eating.

SOURCE: Adapted from C. L. Rock and J. Yager, Nutrition and eating disorders: A primer for clinicians, *International Journal of Eating Disorders* 6 (1987): 276, as cited in *Nutrition and the M.D.*, July 1988, with permission. Reprinted with permission of John Wiley & Sons, Inc., copyright 1987.

unusual. Other people have **unspecified eating disorders** that fall short of the criteria for anorexia nervosa or bulimia nervosa, but share some of their features. One such condition is binge eating disorder.

HEALTHY PEOPLE 2010

Reduce the relapse rates for persons with eating disorders including anorexia nervosa and bulimia nervosa.

BINGE-EATING DISORDER

Charlie is a 40-year-old schoolteacher who has been overweight all his life. His friends and family are forever encouraging him to lose weight, and he has come to believe that if he only had more willpower, dieting would work. He periodically gives dieting his best shot—restricting energy intake for a day or two only to succumb to uncontrollable cravings, especially for high-fat foods. Like Charlie, up to half of the obese people who try to lose weight periodically binge; unlike people with bulimia nervosa, however, they typically do not purge. Such an eating disorder does not meet the criteria for either anorexia nervosa or bulimia nervosa—yet such compulsive overeating is a problem and occurs in people of normal weight as well as those who are severely overweight. Table H9-6 lists criteria for unspecified eating disorders, including binge eating. Obesity alone is not an eating disorder.

Clinicians note differences between people with bulimia nervosa and those with **binge-eating disorder.** People with binge-eating disorder consume less during a binge, rarely purge, and exert less restraint during times of dieting. Similarities also exist, including feeling out of control, disgusted, depressed, embarrassed, guilty, or distressed because of their self-perceived gluttony.

There are also differences between obese binge-eaters and obese people who do not binge. Those with the binge-eating disorder report higher rates of self-loathing, disgust about body size, depression, and anxiety. Their eating habits differ as well. Obese binge-eaters tend to consume more kcalories and more dessert and snack-type foods during regular meals and binges than obese people who do not binge.

Binge eating is a behavioral disorder that can be resolved with treatment—even a placebo. Resolving such behavior may not bring weight loss, but it may make participation in weight-control programs easier. It also improves physical health, mental health, and the chances of success in breaking the cycle of rapid weight losses and gains.

TABLE H9-6 Unspecified Eating Disorders, including Binge-Eating Disorder

Criteria for Diagnosis of Unspecified Eating Disorders, in General

Many people have eating disorders but do not meet all the criteria to be classified as having anorexia nervosa or bulimia nervosa. Some examples include those who:

A. Meet all of the criteria for anorexia nervosa, except irregular menses.

B. Meet all of the criteria for anorexia nervosa, except that their current weights fall within the normal ranges.

C. Meet all of the criteria for bulimia nervosa, except that binges occur less frequently than stated in the criteria.

D. Are of normal body weight and who compensate inappropriately for eating small amounts of food (example: self-induced vomiting after eating two cookies).

E. Repeatedly chew food, but spit it out without swallowing.

F. Have recurrent episodes of binge eating but who do not compensate as do those with bulimia nervosa.

Criteria for Diagnosis of Binge-Eating Disorder, Specifically

A person with a binge-eating disorder demonstrates the following:

A. Recurrent episodes of binge eating. An episode of binge eating is characterized by both of the following:
   1. Eating, in a discrete period of time (e.g., within any two-hour period) an amount of food that is definitely larger than most people would eat in a similar period of time under similar circumstances.
   2. A sense of lack of control over eating during the episode (e.g., a feeling that one cannot stop eating or control what or how much one is eating).

B. Binge-eating episodes are associated with at least three of the following:
   1. Eating much more rapidly than normal.
   2. Eating until feeling uncomfortably full.
   3. Eating large amounts of food when not feeling physically hungry.
   4. Eating alone because of being embarrassed by how much one is eating.
   5. Feeling disgusted with oneself, depressed, or very guilty after overeating.

C. The binge eating causes marked distress.

D. The binge eating occurs, on average, at least twice a week for six months.

E. The binge eating is not associated with the regular use of inappropriate compensatory behaviors (e.g., purging, fasting, excessive exercise) and does not occur exclusively during the course of anorexia nervosa or bulimia nervosa.

SOURCE: Reprinted with permission from American Psychiatric Association, *Diagnostic and Statistical Manual of Mental Disorders*, 4th ed. Text Revision. (Washington, D.C.: American Psychiatric Association, 2000).

EATING DISORDERS IN SOCIETY

Proof that society plays a role in eating disorders is found in their demographic distribution—they are known only in developed nations, and they become more prevalent as wealth increases and food becomes plentiful. Some people point to the vomitoriums of ancient times and claim that bulimia nervosa is not new, but the two are actually distinct. Ancient people were eating for pleasure, without guilt, and in the company of others; they vomited so that they could rejoin the feast. Bulimia nervosa is a disorder of isolation and is often accompanied by low self-esteem.

A food-centered society that favors thinness puts people in a bind. Families may encourage hearty eating and socializing around the dinner table. Party hosts take pride in the delicacies they serve, and guests are obliged to indulge. A child raised in such a setting and also encouraged to aspire to a thin ideal may see little alternative but to celebrate by indulging in food and then to vomit, crash diet, or starve to "undo" possible weight gain. Then, starving and guilty, but still reluctant to appear to be a glutton, the child may begin eating uncontrollably to relieve a desperate hunger.

FIGURE H9-3 BMI of Miss America

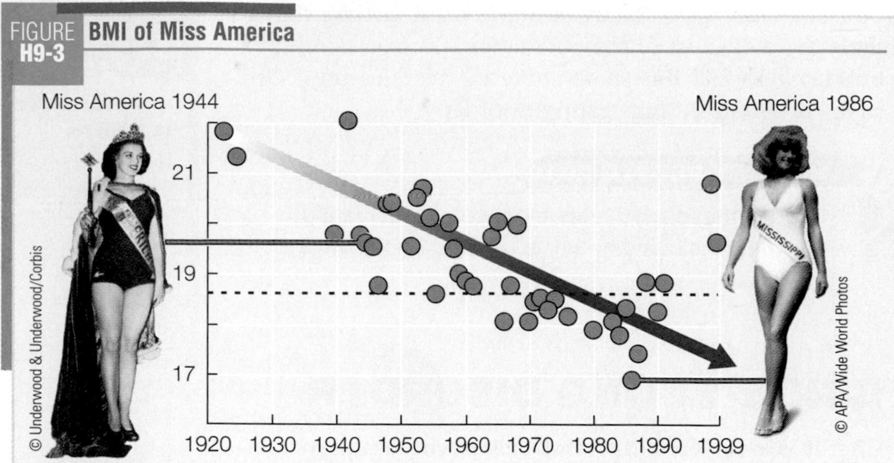

The BMI of Miss America has declined steadily over the years. Since the mid-1960s, most have fallen below 18.5, the cutoff point indicating underweight with its associated health problems.

SOURCE: S. Rubenstein and B. Caballero, Is Miss America an undernourished role model? *Journal of the American Medical Association* 283 (2000): 1569. Used with permission.

No doubt our society sets unrealistic ideals for body weight, especially in women, and devalues those who do not conform to them. Miss America, our nation's icon of beauty, has never been overweight, and she has grown progressively thinner over the years (see Figure H9-3).[27] Magazines, newspapers, and television all convey the message that to be thin is to be beautiful and happy. Anorexia nervosa and bulimia nervosa are not a form of rebellion against these unreasonable expectations, but rather an exaggerated acceptance of them. Even professionals, including physicians and dietitians, are prone to praise people for losing weight and to suggest weight loss to people who do not need it for their health. As a result, even beautiful, normal-weight preteen girls are already worried that they are too fat. Two-thirds of adolescent girls and one-third of adolescent boys are dissatisfied with their body weight and shape. Characteristics of disordered eating such as restrained eating, fasting, binge eating, purging, fear of fatness, and distortion of body image are extraordinarily common among young, white, middle- and upper-class girls. Most are "on diets," and many are poorly nourished.[28] Some eat too little food to support normal growth; thus they miss out on their adolescent growth spurts and may never catch up. Many eat so little that hunger propels them into binge-purge cycles.

Perhaps a person's best defense against these disorders is to learn to appreciate his or her own uniqueness. When people discover and honor the body's real physical needs, they become unwilling to sacrifice health for conformity. To respect and value oneself may be lifesaving.

Nutrition on the Net

•WEBSITES•

Access these websites for further study of topics covered in this highlight.

- Find updates and quick links to these and other nutrition-related sites at our website: **www.wadsworth.com/nutrition**
- Search for "anorexia," "bulimia," and "eating disorders" at the U.S. Government health information site: **www.healthfinder.gov**
- Request fact sheets and resource lists from the Office on Women's Health in the U.S. Department of Health and Human Services' nationwide educational campaign Bodywise: **bodywise@esilsg.org**
- Learn more about anorexia nervosa and related eating disorders from Anorexia Nervosa and Related Eating Disorders: **www.anred.com**
- Find out about help-lines, referral networks, support groups, and prevention programs from the American Anorexia Bulimia Association: **www.aabainc.org**

References

1 Girls in the 90's: Working to undo stereotypes, *Eating Dirorders Review*, September/October 1997, pp. 6–7.

2 Committee on Sports Medicine and Fitness, Medical concerns in the female athlete, *Pediatrics* 106 (2000) 610–613; American College of Sports Medicine, Position stand: The female athlete triad, *Medicine and Science in Sports and Exercise* 29 (1997): i–ix.

3 N. A. Armsey, Stress injury to bone in the female athlete, *Clinical Sports Medicine* 16 (1997): 197–224.

4 M. Hotta and coauthors, The importance of body weight history in the occurrence and recovery of osteoporosis in patients with anorexia nervosa: Evaluation by dual X-ray absorptiometry and bone metabolic markers, *European Journal of Endocrinology* 139 (1998): 276–283.

5 E. R. Brooks, B. W. Ogden, and D. S. Cavalier, Compromised bone density 11.4 years after diagnosis of anorexia nervosa, *Journal of Women's Health* 7 (1998): 567–574.

6 J. Sundgot-Borgen and coauthors, Normal bone mass in bulimic women, *Journal of Clinical Endocrinology and Metabolism* 83 (1998): 3144–3149.

7 Hyperthermia and dehydration-related deaths associated with intentional rapid weight loss in three collegiate wrestlers—North Carolina, Wisconsin, and Michigan, November–December 1997, *Morbidity and Mortality Weekly Report* 47 (1998): 105–108.

8 American College of Sports Medicine, Position stand: Weight loss in wrestlers, *Medicine and Science in Sports and Exercise* 28 (1996): ix–xii.

9 E. Coleman, "Making weight" for wrestling: Six recommendations, *Sports Medicine Digest* 18 (1996): 13, 22–23.

10 H. G. Pope, Jr. and coauthors, Muscle dysmorphia—An underrecognized form of body dysmorphic disorder, *Psychosomatics* 38 (1997): 548–557.

11 R. V. West, The female athlete: The triad of disordered eating, amenorrhoea and osteoporosis, *Sports Medicine* 26 (1998): 63–71.

12 Committee on Sports Medicine and Fitness, Promotion of healthy weight-control practices in young athletes, *Pediatrics* 97 (1996): 752–753.

13 A. Gila and coauthors, Subjective body-image dimensions in normal and anorexic adolescents, *British Journal of Medical Psychology* 71 (1998): 175–184.

14 G. Waller, Perceived control in eating disorders: Relationship with reported sexual abuse, *International Journal of Eating Disorders* 23 (1998): 213–216.

15 T. Pryor and M. W. Weiderman, Personality features and expressed concerns of adolescents with eating disorders, *Adolescence* 33 (1998): 291–300.

16 P. Santonastaso, A. Sala, and A. Favaro, Water intoxication in anorexia nervosa: A case report, *International Journal of Eating Disorders* 24 (1998): 439–442.

17 C. Panagiotopoulos and coauthors, Electrocardiographic findings in adolescents with eating disorders, *Pediatrics* 105 (2000): 1100–1105.

18 G. Addolorato and coauthors, A case of marked cerebellar atrophy in a woman with anorexia nervosa and cerebral atrophy and a review of the literature, *International Journal of Eating Disorders* 24 (1998): 443–447; L. M. Allende and coauthors, Immunodeficiency associated with anorexia nervosa is secondary and improves after refeeding, *Immunology* 94 (1998): 543–551; V. W. Swayze, Brain imaging and eating disorders, *Eating Disorders Review,* May/June 1997, pp. 1–4.

19 G. F. Russell, J. Treasure, and I. Eisler, Mothers with anorexia nervosa who underfeed their children: Their recognition and management, *Psychological Medicine* 28 (1998): 93–108.

20 A. E. Becker and coauthors, Eating disorders, *New England Journal of Medicine* 340 (1999): 1092–1098.

21 M. Moukaddem and coauthors, Increase in diet-induced thermogenesis at the start of refeeding in severely malnourished anorexia nervosa patients, *American Journal of Clinical Nutrition* 66 (1997): 133–140.

22 E. D. Eckert and coauthors, Leptin in anorexia nervosa, *Journal of Clinical Endocrinology and Metabolism* 83 (1998): 791–795.

23 N. A. Troop, A. Holbrey, and J. L. Treasure, Stress, coping, and crisis support in eating disorders, *International Journal of Eating Disorders* 24 (1998): 157–166.

24 C. G. Fairburn and coauthors, Risk factors for bulimia nervosa: A community-based case-control study, *Archives of General Psychiatry* 54 (1997): 509–517.

25 A. Daluiski, B. Rahbar, and R. A. Meals, Russell's sign: Subtle hand changes in patients with bulimia nervosa, *Clinical Orthopaedics and Related Research* 343 (1997): 107–109.

26 P. K. Keel and J. E. Mitchell, Outcome in bulimia nervosa, *American Journal of Psychiatry* 154 (1997): 313–321.

27 S. Rubinstein and B. Caballero, Is Miss America an undernourished role model? *Journal of the American Medical Association* 283 (2000): 1569.

28 Federal Interagency Forum on Child and Family Statistics, *America's Children: Key National Indicators of Well-Being,* 1999, a report from the National Institutes of Health, available from National Maternal and Child Health Clearinghouse, 2070 Chain Bridge Road, Suite 450, Vienna, VA 22182 or on the Internet at http://childstats.gov.

The Water-Soluble Vitamins: B Vitamins and Vitamin C

CHAPTER 10

Earlier chapters focused on the energy-yielding nutrients, which play leading roles in the body. The vitamins and minerals are their supporting cast. This chapter begins with an overview of the vitamins and then examines each of the water-soluble vitamins; the next chapter features the fat-soluble vitamins. Chapters 12 and 13 present the minerals.

THE VITAMINS—AN OVERVIEW

Researchers first recognized that there were substances in foods that were "vital to life" in the early 1900s. Since then, the world of vitamins has opened up dramatically. The vitamins◆ are powerful substances, as their *absence* attests. Vitamin A deficiency can cause blindness; a lack of the B vitamin niacin can cause dementia; and a lack of vitamin D can retard bone growth. The consequences of deficiencies are so dire, and the effects of restoring the needed vitamins so dramatic, that people spend billions of dollars every year in the belief that vitamin pills will cure a host of ailments (see Highlight 10). Vitamins certainly support nutritional health, but they do not cure all ills. Furthermore, vitamin supplements do not offer the many benefits that come from vitamin-rich foods.

◆ Reminder: The *vitamins* are organic, essential nutrients required in tiny amounts to perform specific functions that promote growth, reproduction, or the maintenance of health and life.
- **vita** = life
- **amine** = containing nitrogen (the first vitamins discovered contained nitrogen)

The *presence* of the vitamins also attests to their power. Vitamin C not only prevents the deficiency disease scurvy, but also seems to protect against certain types of cancer. Similarly, vitamin E seems to help protect against some facets of cardiovascular disease. The B vitamin folate helps to prevent birth defects. As you will see, the vitamins' roles in supporting optimal health extend far beyond preventing deficiency diseases. In fact, some of the credit given to low-fat diets in preventing disease actually belongs to the vitamins that diets rich in vegetables, fruits, and whole grains deliver (see Highlight 11 for more on vitamins in disease prevention). Highlight 18 considers the roles of vitamins in supporting a strong immune system.

Health promotion

The vitamins differ from carbohydrates, fats, and proteins in the following ways:

- *Structure.* Vitamins are individual units; they are not linked together (as are molecules of glucose, fatty acids, or amino acids).
- *Function.* Vitamins do not yield usable energy when broken down; they assist the enzymes that release energy from carbohydrates, fats, and proteins.
- *Food contents.* The amounts of vitamins people ingest daily from foods and the amounts they require are measured in *micrograms* (μg) or *milligrams* (mg), rather than grams (g).◆

◆ 1 g = 1000 mg. 1 mg = 1000 μg. For perspective, a dollar bill weighs about 1 g; 28 g equal approximately 1 oz.

The vitamins are similar to the energy-yielding nutrients, though, in that they are vital to life, organic, and available from foods.

• Bioavailability • The availability of vitamins from foods depends on two factors: the quantity provided by a food and the amount absorbed and used by the body (the vitamins' **bioavailability**). Researchers analyze foods to determine their vitamin contents and publish the results in tables of food composition such as Appendix H. Determining the bioavailability of a vitamin is a more complex task because it depends on many factors, including:[1]

- Efficiency of digestion and time of transit through the GI tract.
- Previous nutrient intake and nutrition status.
- Other foods consumed at the same time.◆
- Method of food preparation (raw, cooked, or processed).
- Source of the nutrient (synthetic, fortified, or naturally occurring).

◆ Chapters 10–13 describe factors that inhibit or enhance the absorption of individual vitamins and minerals.

Experts consider these factors when estimating recommended intakes.

• Precursors • Some of the vitamins are available from foods in inactive forms known as **precursors,** or provitamins. Once inside the body, the precursor is

bioavailability: the rate at and the extent to which a nutrient is absorbed and used.

precursors: substances that precede others; with regard to vitamins, compounds that can be converted into active vitamins; also known as **provitamins.**

changed chemically to an active form of the vitamin. Thus, in measuring a person's vitamin intake, it is important to count both the amount of the active vitamin and the potential amount available from its precursors. The summary tables throughout this chapter and the next indicate which vitamins have precursors.

• **Organic Nature** • Being organic, vitamins can be destroyed and left unable to perform their duties. Therefore, they must be handled with care during storage and in cooking. Prolonged heating may destroy much of the thiamin in food. Because riboflavin can be destroyed by the ultraviolet rays of the sun or by fluorescent light, foods stored in transparent glass containers are most likely to lose riboflavin. Oxygen destroys vitamin C, so losses occur when foods are cut or broken and thereby exposed to air. Table 10-1 summarizes ways to minimize nutrient losses in the kitchen.

• **Solubility** • As you may recall, carbohydrates and proteins are hydrophilic and lipids are hydrophobic. The vitamins divide along the same lines—the hydrophilic, water-soluble ones are the B vitamins and vitamin C; the hydrophobic, fat-soluble ones are vitamins A, D, E, and K. As each vitamin was discovered, it was given a name and sometimes a letter and number as well. Many of the water-soluble vitamins have multiple names, which has led to some confusion. The margin◆ lists the standard names; summary tables throughout this chapter provide the common alternative names.

◆ **Water-soluble vitamins:**
- B Vitamins:
  - Thiamin.
  - Riboflavin.
  - Niacin.
  - Biotin.
  - Pantothenic acid.
  - Vitamin B_6.
  - Folate.
  - Vitamin B_{12}.
- Vitamin C.

Fat-soluble vitamins:
- Vitamin A.
- Vitamin D.
- Vitamin E.
- Vitamin K.

Solubility is apparent in the food sources of the different vitamins, and it affects their absorption, transport, storage, and excretion by the body. The water-soluble vitamins are found in the watery compartments of foods; the fat-soluble vitamins usually occur together in the fats and oils of foods. On being absorbed, the water-soluble vitamins move directly into the blood; like fats, the fat-soluble vitamins must first enter the lymph, then the blood. Once in the blood, many of the water-soluble vitamins travel freely; many of the fat-soluble vitamins require protein carriers for transport. Upon reaching the cells, water-soluble vitamins freely circulate in the water-filled compartments of the body; fat-soluble vitamins are held in fatty tissues and the liver until needed. The kidneys, monitoring the blood that flows through them, detect and remove small excesses of water-soluble vitamins (large excesses, however, may overwhelm the system, creating adverse effects); fat-soluble vitamins tend to remain in fat-storage sites in the body rather than being excreted, and so are more likely to reach toxic levels when consumed in excess.

Because the body stores fat-soluble vitamins, they can be eaten in large amounts once in a while and still meet the body's needs over time. Water-soluble vitamins are retained for varying periods in the body; a single day's omission from the diet does not bring on a deficiency, but still, the water-soluble vitamins must be eaten more regularly than the fat-soluble vitamins.

• **Toxicity** • Knowledge about some of the amazing roles of vitamins has prompted many people to begin taking supplements, assuming that more is

TABLE 10-1 Minimizing Nutrient Losses

- To slow the degradation of vitamins, refrigerate (most) fruits and vegetables.
- To minimize the oxidation of vitamins, store fruits and vegetables that have been cut in airtight wrappers and juices that have been opened in closed containers (and refrigerate them).
- To prevent losses during washing, rinse fruits and vegetables before cutting.
- To minimize losses during cooking, use a microwave oven or steam vegetables in a small amount of water. Add vegetables after water has come to a boil. Use the cooking water in mixed dishes such as casseroles and soups. Avoid high temperatures and long cooking times.

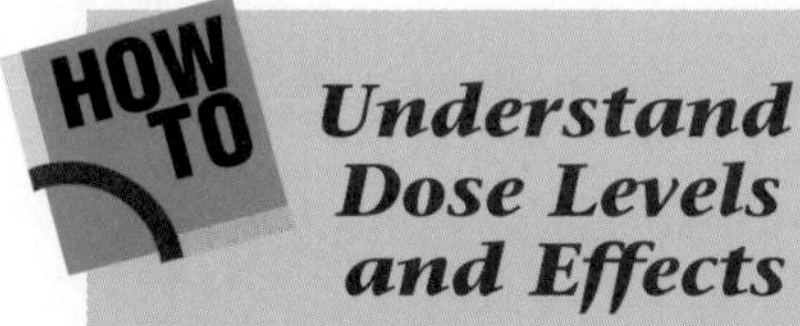

Understand Dose Levels and Effects

A substance may have a beneficial or harmful effect, but a critical thinker would not conclude that the substance itself was beneficial or harmful without first asking what dose was used. The accompanying figure shows three possible relationships between dose levels and effects. The third diagram represents the situation with nutrients—more is better up to a point, but beyond that point, still more is harmful.

Dose Levels and Effects

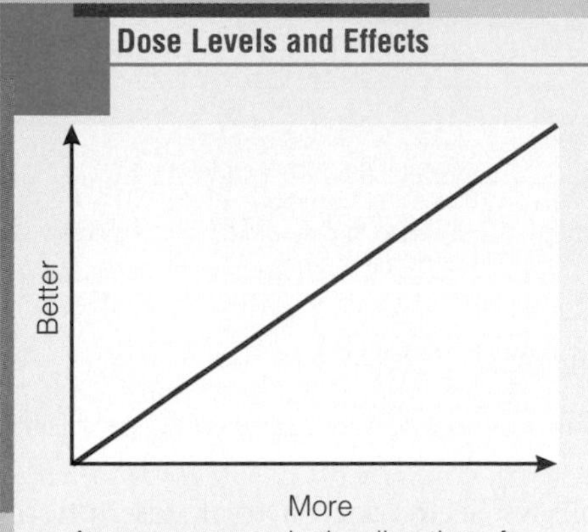

As you progress in the direction of more, the effect gets better and better, with no end in sight (real life is seldom, if ever, like this).

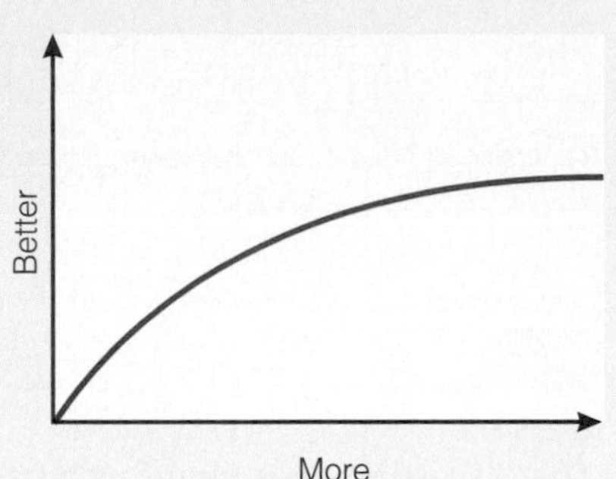

As you progress in the direction of more, the effect reaches a maximum and then a plateau, becoming no better with higher doses.

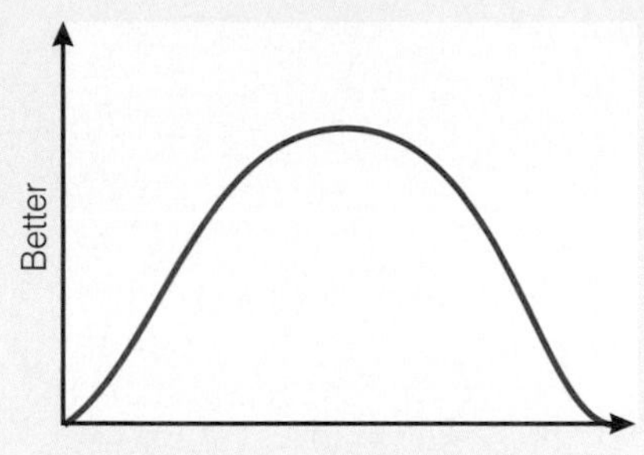

As you progress in the direction of more, the effect reaches an optimum at some intermediate dose and then declines, showing that more is better up to a point and then harmful. That too much is as harmful as too little represents the situation with nutrients.

better. But just as an inadequate intake can cause harm, so can an excessive intake. As mentioned, even some of the water-soluble vitamins have adverse effects when taken in large doses.

That a vitamin can be both essential and harmful may seem surprising, but the same is true of most nutrients. The effects of every substance depend on its dose, and this is one reason consumers should not self-prescribe supplements for their ailments. See the accompanying "How to" for a perspective on doses.

The Committee on Dietary Reference Intakes (DRI) addresses the possibility of adverse effects from high doses of nutrients by establishing Tolerable Upper Intake Levels. An Upper Level defines the highest amount of a nutrient that is likely not to cause harm for most healthy people when consumed daily. The risk of harm increases as intakes rise above the Upper Level. Of the nutrients discussed in this chapter, niacin, vitamin B_6, folate, choline, and vitamin C have Upper Levels set,◆ and these values are presented in their respective summary tables. Data are lacking to establish Upper Levels for the remaining B vitamins, but this does not mean that excessively high intakes would be without risk.

◆ The inside front cover presents Upper Levels for the vitamins and minerals.

IN SUMMARY

The vitamins are essential nutrients that are needed in tiny amounts in the diet both to prevent deficiency diseases and to support optimal health. The water-soluble vitamins are the B vitamins and vitamin C; the fat-soluble vitamins are vitamins A, D, E, and K. The table that follows summarizes the differences between the water-soluble and fat-soluble vitamins.

	Water-Soluble Vitamins: B Vitamins and Vitamin C	Fat-Soluble Vitamins: Vitamins A, D, E, and K
Absorption	Directly into the blood.	First into the lymph, then the blood.
Transport	Travel freely.	Many require protein carriers.
Storage	Circulate freely in water-filled parts of the body.	Stored in the cells associated with fat.
Excretion	Kidneys detect and remove excess in urine.	Less readily excreted; tend to remain in fat-storage sites.
Toxicity	Possible to reach toxic levels when consumed from supplements.	Likely to reach toxic levels when consumed from supplements.
Requirements	Needed in frequent doses (perhaps 1 to 3 days).	Needed in periodic doses (perhaps weeks or even months).

Note: Exceptions occur, but these differences between the water-soluble and fat-soluble vitamins are valid generalizations.

The discussion of B vitamins that follows begins with a brief description of each of them, then offers a look at the ways they work together. Thus a preview of the individuals is followed by a survey of them all together, in concert.

THE B VITAMINS—AS INDIVIDUALS

Despite supplement advertisements that claim otherwise, the vitamins do not provide the body with fuel for energy. The energy-yielding nutrients—carbohydrate, fat, and protein—are used for fuel; the B vitamins help the body to use that fuel. It is true, though, that without B vitamins the body would lack energy. Several of the B vitamins—thiamin, riboflavin, niacin, pantothenic acid, and biotin—form part of the coenzymes◆ that assist certain enzymes in the release of energy from carbohydrate, fat, and protein. Other B vitamins play other indispensable roles in metabolism. Vitamin B_6 assists enzymes that metabolize amino acids; folate and vitamin B_{12} help cells to multiply. Among these cells are the red blood cells and the cells lining the GI tract—cells that deliver energy to all the others.

◆ **Reminder: A *coenzyme* is a small organic molecule that associates closely with certain enzymes; many B vitamins form an integral part of coenzymes.**

The vitamin portion of a coenzyme allows a chemical reaction to occur; the remaining portion of the coenzyme binds to the enzyme. Without its coenzyme, an enzyme cannot function. Thus symptoms of B vitamin deficiencies directly reflect the disturbances of metabolism incurred by a lack of coenzymes. Figure 10-1 illustrates coenzyme action.

The following sections describe individual B vitamins and note many coenzymes and metabolic pathways. Keep in mind that a later section will assemble these pieces of information into a whole picture.

The following sections also present the recommendations, deficiency and toxicity symptoms, and food sources for each vitamin. The recommendations for the B vitamins and vitamin C reflect the 1998 and 2000 DRI, respectively.[2] For thiamin, riboflavin, niacin, vitamin B_6, folate, vitamin B_{12}, and vitamin C, sufficient data were available to establish an RDA; for biotin, pantothenic acid, and choline, an Adequate Intake (AI) was set; only niacin, vitamin B_6, folate, choline, and vitamin C have Tolerable Upper Intake Levels. These values appear in the summary tables and figures that follow, as well as on the inside front covers.

Thiamin

thiamin (THIGH-ah-min): a B vitamin. The coenzyme form is TPP (thiamin pyrophosphate).

beriberi: the thiamin-deficiency disease.
- **beri** = weakness
- **beriberi** = "I can't, I can't"

Thiamin is the vitamin part of the coenzyme TPP, which assists in energy metabolism. The TPP coenzyme participates in the conversion of pyruvate to

FIGURE 10-1 **Coenzyme Action**

Some vitamins form part of the coenzymes that enable enzymes either to synthesize compounds (as illustrated by the lower enzymes in this figure) or to dismantle compounds (as illustrated by the upper enzymes).

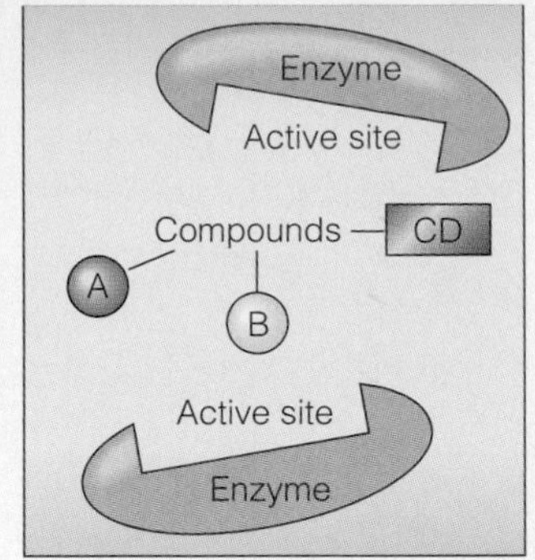

Without coenzymes, compounds A, B, and CD don't respond to their enzymes.

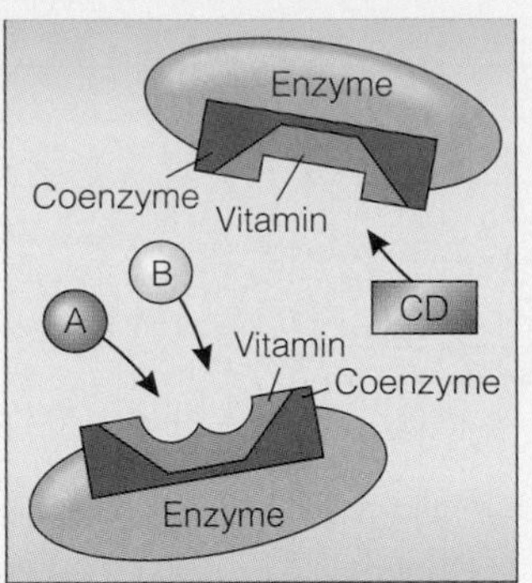

With the coenzymes in place, compounds are attracted to their sites on the enzymes . . .

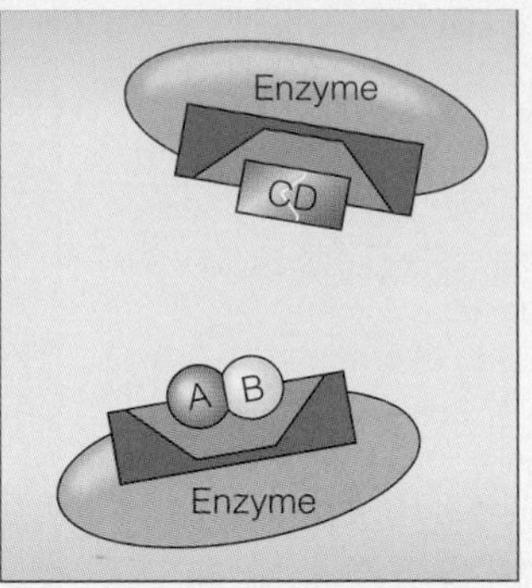

. . . and the reactions proceed instantaneously. The coenzymes often donate or accept electrons, atoms, or groups of atoms.

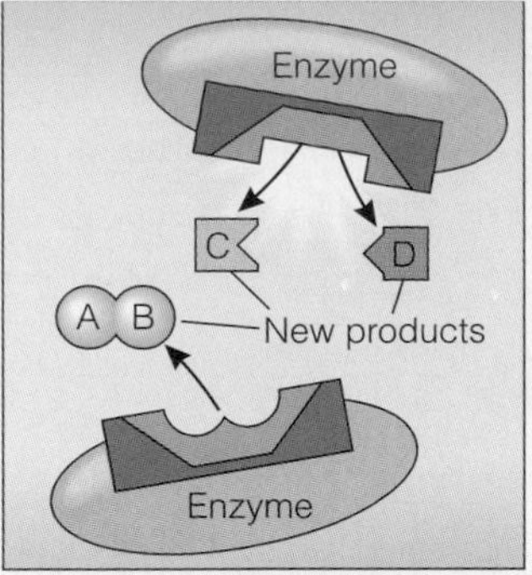

The reactions are completed with either the formation of a new product, AB, or the breaking apart of a compound into two new products, C and D, and the release of energy.

acetyl CoA (described in Chapter 7). The reaction removes one carbon from the 3-carbon pyruvate to make the 2-carbon acetyl CoA and carbon dioxide (CO_2). Later TPP participates in a similar step in the TCA cycle where it helps convert a 5-carbon compound to a 4-carbon compound. Besides playing these pivotal roles in the energy metabolism of all cells, thiamin occupies a special site on the membranes of nerve cells. Consequently, processes in nerves and in their responding tissues, the muscles, depend heavily on thiamin.

◆ Severe thiamin deficiency in alcohol abusers is called the **Wernicke-Korsakoff** (VER-nee-key KORE-sah-kof) **syndrome.** Symptoms include disorientation, loss of short-term memory, jerky eye movements, and staggering gait.

• Thiamin Recommendations • Dietary recommendations are based primarily on thiamin's role in enzyme activity. Generally, if a person eats enough food to meet energy needs and obtains that energy from nutritious foods, thiamin needs will be met. The average thiamin intake in the United States and Canada meets or exceeds recommendations.

• Thiamin Deficiency • People who fail to eat enough food to meet energy needs risk nutrient deficiencies, including thiamin deficiency. Inadequate thiamin intakes have been reported among the nation's malnourished and homeless people.[3] Similarly, people who derive most of their energy from empty-kcalorie items, like alcohol,◆ risk thiamin deficiency. Alcohol contributes energy, but provides few, if any, nutrients and often displaces food. In addition, alcohol impairs thiamin absorption and enhances thiamin excretion in the urine, doubling the risk of deficiency. An estimated four out of five alcoholics are thiamin deficient.

Prolonged thiamin deficiency can result in the disease **beriberi,** which was first observed in East Asia when the custom of polishing rice became widespread. Rice provided 80 percent of the energy intake of the people of that area, and rice germ and bran were their principal source of thiamin. When the germ and bran were removed, beriberi spread like wildfire. The symptoms of beriberi include damage to the nervous system as well as to the heart and other muscles. Figure 10-2 presents one of the symptoms of beriberi.

FIGURE 10-2 **Thiamin-Deficiency Symptom—The Edema of Beriberi**

Beriberi may be characterized as "wet" (referring to edema) or "dry" (with muscle wasting, but no edema). Physical examination confirms that this woman has wet beriberi. Notice how the impression of the physician's thumb remains on her leg.

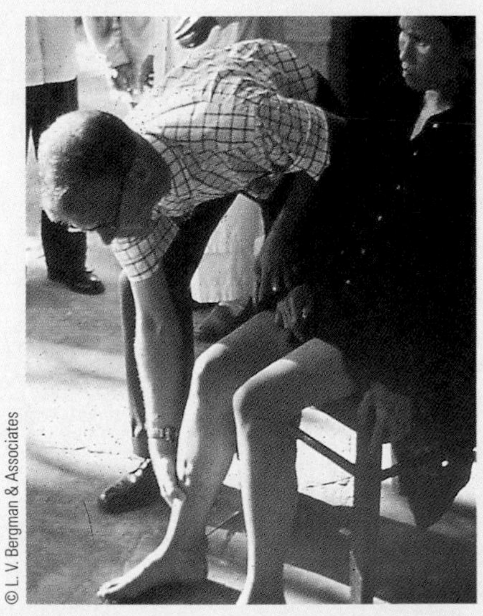

• Thiamin Food Sources • Before examining Figure 10-3 on p. 312, you may want to read the "How to" on p. 313, which describes the many features found in this and similar figures in this chapter and the next three chapters. When you look at Figure 10-3, notice that thiamin occurs in small quantities in many nutritious foods; highly refined foods contain almost no thiamin. The long red bars near the bottom of the graph represent meats that are exceptionally rich in thiamin—those in the pork family.

FIGURE 10-3 Thiamin in Selected Foods

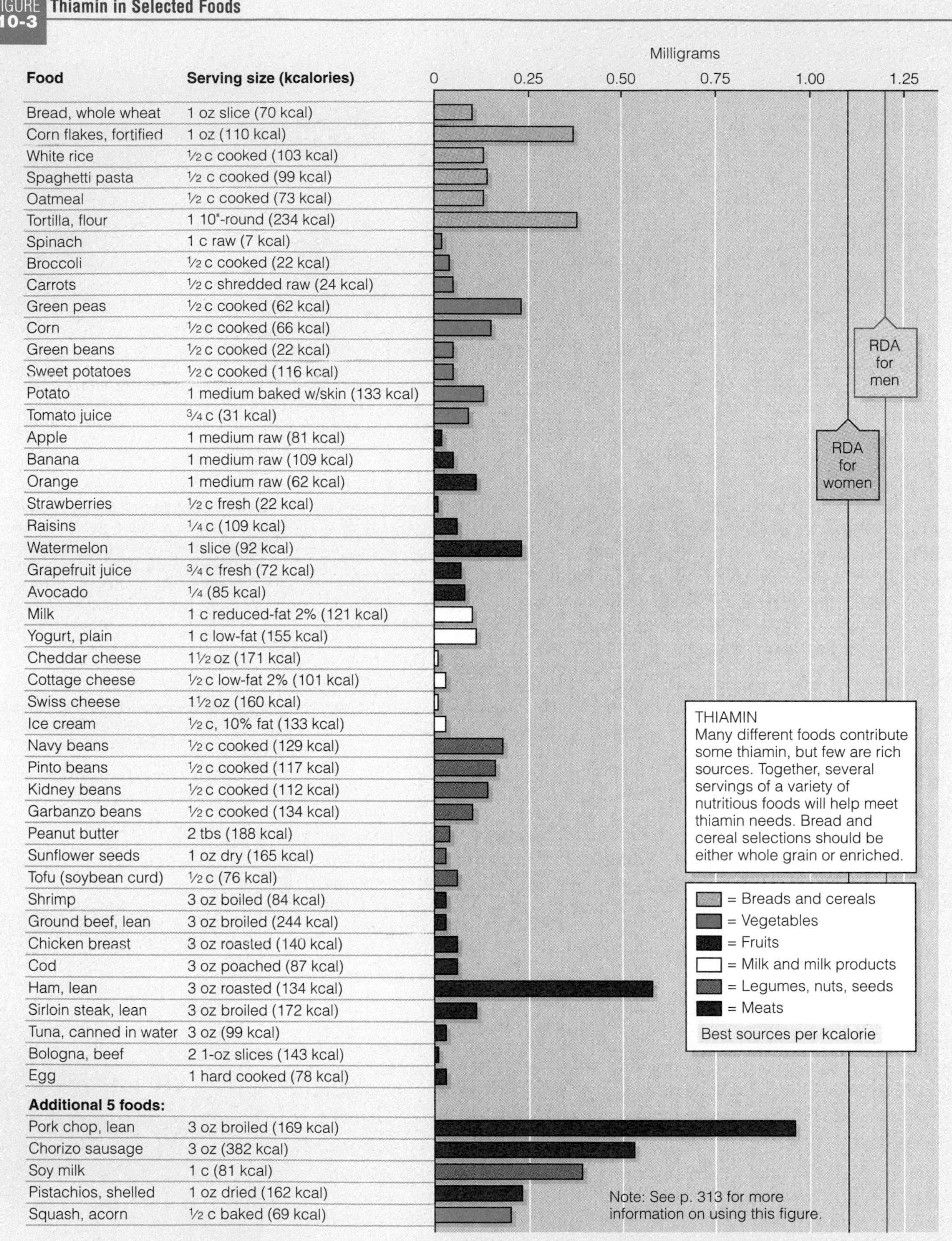

Food	Serving size (kcalories)
Bread, whole wheat	1 oz slice (70 kcal)
Corn flakes, fortified	1 oz (110 kcal)
White rice	½ c cooked (103 kcal)
Spaghetti pasta	½ c cooked (99 kcal)
Oatmeal	½ c cooked (73 kcal)
Tortilla, flour	1 10"-round (234 kcal)
Spinach	1 c raw (7 kcal)
Broccoli	½ c cooked (22 kcal)
Carrots	½ c shredded raw (24 kcal)
Green peas	½ c cooked (62 kcal)
Corn	½ c cooked (66 kcal)
Green beans	½ c cooked (22 kcal)
Sweet potatoes	½ c cooked (116 kcal)
Potato	1 medium baked w/skin (133 kcal)
Tomato juice	¾ c (31 kcal)
Apple	1 medium raw (81 kcal)
Banana	1 medium raw (109 kcal)
Orange	1 medium raw (62 kcal)
Strawberries	½ c fresh (22 kcal)
Raisins	¼ c (109 kcal)
Watermelon	1 slice (92 kcal)
Grapefruit juice	¾ c fresh (72 kcal)
Avocado	¼ (85 kcal)
Milk	1 c reduced-fat 2% (121 kcal)
Yogurt, plain	1 c low-fat (155 kcal)
Cheddar cheese	1½ oz (171 kcal)
Cottage cheese	½ c low-fat 2% (101 kcal)
Swiss cheese	1½ oz (160 kcal)
Ice cream	½ c, 10% fat (133 kcal)
Navy beans	½ c cooked (129 kcal)
Pinto beans	½ c cooked (117 kcal)
Kidney beans	½ c cooked (112 kcal)
Garbanzo beans	½ c cooked (134 kcal)
Peanut butter	2 tbs (188 kcal)
Sunflower seeds	1 oz dry (165 kcal)
Tofu (soybean curd)	½ c (76 kcal)
Shrimp	3 oz boiled (84 kcal)
Ground beef, lean	3 oz broiled (244 kcal)
Chicken breast	3 oz roasted (140 kcal)
Cod	3 oz poached (87 kcal)
Ham, lean	3 oz roasted (134 kcal)
Sirloin steak, lean	3 oz broiled (172 kcal)
Tuna, canned in water	3 oz (99 kcal)
Bologna, beef	2 1-oz slices (143 kcal)
Egg	1 hard cooked (78 kcal)
Additional 5 foods:	
Pork chop, lean	3 oz broiled (169 kcal)
Chorizo sausage	3 oz (382 kcal)
Soy milk	1 c (81 kcal)
Pistachios, shelled	1 oz dried (162 kcal)
Squash, acorn	½ c baked (69 kcal)

Note: See p. 313 for more information on using this figure.

HOW TO Evaluate Foods for Their Nutrient Contributions

Figure 10-3 is the first of a series of figures in this and the next three chapters that present the vitamins and minerals in foods. Each figure presents the same 45 foods, which were selected to ensure a variety of choices representative of each of the food groups as suggested by the Food Guide Pyramid. From its base, for example, a bread, a cereal, a rice, and a pasta were chosen. The suggestion to include a variety of vegetables was also considered: dark green, leafy vegetables (spinach, broccoli); deep yellow vegetables (carrots, sweet potatoes); starchy vegetables (potatoes, corn, green peas); legumes (navy, pinto, kidney, and garbanzo beans); and other vegetables (green beans). The selection of fruits followed the Food Guide Pyramid's suggestions to use whole fruits (apples, bananas); citrus fruits (oranges, grapefruit juice); melons (watermelon); and berries (strawberries). Items were selected from the milk and meat groups in a similar way. In addition to the 45 foods that appear in all of the figures, five different foods were selected for each of the nutrients. These five foods were chosen to add variety and often reflect excellent, and sometimes unusual, sources of the specific nutrient.

Notice that the figures list the food, the serving size, and the food energy (kcalories) on the left and graph the amount of the nutrient per serving on the right along with the RDA (or AI) for adults, so you can see how many servings would be needed to meet recommendations. Serving sizes reflect those used by the Food Guide Pyramid. In some cases, the Pyramid specifies ambiguous serving sizes, recommending "1 medium potato," "1 slice melon," and "¼ avocado." For these foods, serving sizes reflect those presented in Appendix H.

The colored bars show at a glance which food groups best provide a nutrient: yellow for breads and cereals; green for vegetables; purple for fruits; white for milk and milk products; brown for legumes; and red for meat, fish, and poultry. Because the Food Guide Pyramid mentions legumes with both the meat group and the vegetable group and because legumes are especially rich in many vitamins and minerals, they have been given their own color to highlight their nutrient contributions.

Notice how the bar graphs shift in the various figures. Careful study of all of the figures taken together will confirm that variety is the key to nutrient adequacy.

Another way to evaluate foods for their nutrient contributions is to consider their nutrient density (their thiamin *per 100 kcalories,* for example). Quite often, vegetables rank higher on a nutrient-per-kcalorie list than they do on a nutrient-per-serving list (see p. 34 to review how to evaluate foods based on nutrient density). The left column in the figure highlights in yellow the foods that offer the best deal for your energy "dollar" (the kcalorie). Notice how many of them are vegetables.

Realistically, people cannot eat for single nutrients. Fortunately, most foods deliver more than one nutrient, allowing people to combine foods into nourishing meals.

As mentioned earlier, prolonged cooking can destroy thiamin. Also, like other water-soluble vitamins, thiamin leaches into water when foods are boiled or blanched. Cooking methods that require little or no water such as steaming and microwave heating conserve thiamin and other water-soluble vitamins. The accompanying table summarizes thiamin's main functions, food sources, and deficiency symptoms.

IN SUMMARY

Thiamin

Other Names
Vitamin B_1

1998 RDA
Men: 1.2 mg/day
Women: 1.1 mg/day

Chief Functions in the Body
Part of coenzyme TPP (thiamin pyrophosphate) used in energy metabolism

Significant Sources
Whole-grain, fortified, or enriched grain products; moderate amounts in all nutritious food; pork
Easily destroyed by heat

Deficiency Disease
Beriberi (wet, with edema; dry, with muscle wasting)

Deficiency Symptoms*

Cardiovascular Symptoms
Enlarged heart, cardiac failure

Muscular Symptoms
Weakness

Neurological Symptoms
Apathy, poor short-term memory, confusion, irritability

Other
Anorexia and weight loss

Toxicity Symptoms
None reported

*Severe thiamin deficiency is often related to heavy alcohol consumption.

Pork is the richest source of thiamin, but enriched or whole-grain products typically make the greatest contribution to a day's intake because of the quantities eaten. Legumes such as split peas are also valuable sources of thiamin.

All of these foods are rich in riboflavin, but milk and milk products provide much of the riboflavin in the diets of most people.

◆ Turn to p. 34 for a review of how to evaluate foods based on nutrient density (per kcalorie).

Riboflavin

Like thiamin, **riboflavin** serves as a coenzyme in many reactions, most notably in the release of energy from nutrients in all body cells. The coenzyme forms of riboflavin are FMN and FAD; both can accept and then donate two hydrogens (see Figure 10-4). During energy metabolism, FAD picks up two hydrogens (with their electrons) from the TCA cycle and delivers them to the electron transport chain (described in Chapter 7).

• **Riboflavin Recommendations** • Like thiamin's RDA, riboflavin's RDA is based primarily on its role in enzyme activity. Most people in the United States and Canada meet or exceed riboflavin recommendations.[4]

• **Riboflavin Deficiency** • No one disease is associated with riboflavin deficiency. Lack of the vitamin causes inflammation of the membranes of the mouth, skin, eyes, and GI tract.

• **Riboflavin Food Sources** • The greatest contributions of riboflavin come from milk and milk products (see Figure 10-5). Whole-grain or enriched bread and cereal products are also valuable sources because of the quantities typically consumed. When riboflavin sources are ranked by nutrient density (per kcalorie),◆ many dark green, leafy vegetables (such as broccoli, turnip greens, asparagus, and spinach) appear high on the list (notice those foods highlighted in yellow in the left column of Figure 10-5). Vegans and others who don't use milk must rely on ample servings of dark greens and enriched grains for riboflavin. Nutritional yeast is another good source.

Ultraviolet light and irradiation destroy riboflavin. For these reasons, milk is sold in cardboard or opaque plastic containers, and precautions are taken when vitamin D is added to milk by irradiation.* In contrast, riboflavin is stable to heat, so cooking does not destroy it. The summary table on p. 316 lists riboflavin's chief functions, food sources, and deficiency symptoms.

*Vitamin D can be added to milk by feeding cows irradiated yeast or by irradiating the milk itself.

FIGURE 10-4 Riboflavin Coenzyme, Accepting and Donating Hydrogens

This figure shows the chemical structure of the riboflavin portion of the coenzyme only; the remainder of the coenzyme structure is represented by dotted lines (see Appendix C for the complete chemical structures of FAD and FMN). The reactive sites that accept and donate hydrogens are highlighted in white.

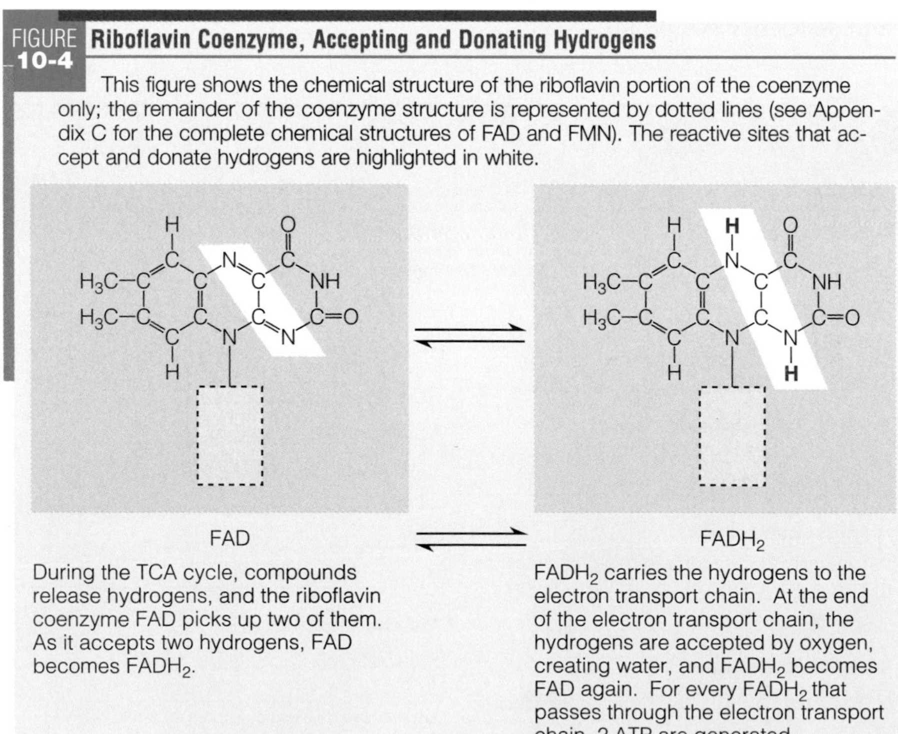

During the TCA cycle, compounds release hydrogens, and the riboflavin coenzyme FAD picks up two of them. As it accepts two hydrogens, FAD becomes $FADH_2$.

$FADH_2$ carries the hydrogens to the electron transport chain. At the end of the electron transport chain, the hydrogens are accepted by oxygen, creating water, and $FADH_2$ becomes FAD again. For every $FADH_2$ that passes through the electron transport chain, 2 ATP are generated.

riboflavin (RYE-boh-flay-vin): a B vitamin. The coenzyme forms are FMN (flavin mononucleotide) and FAD (flavin adenine dinucleotide).

FIGURE 10-5 Riboflavin in Selected Foods

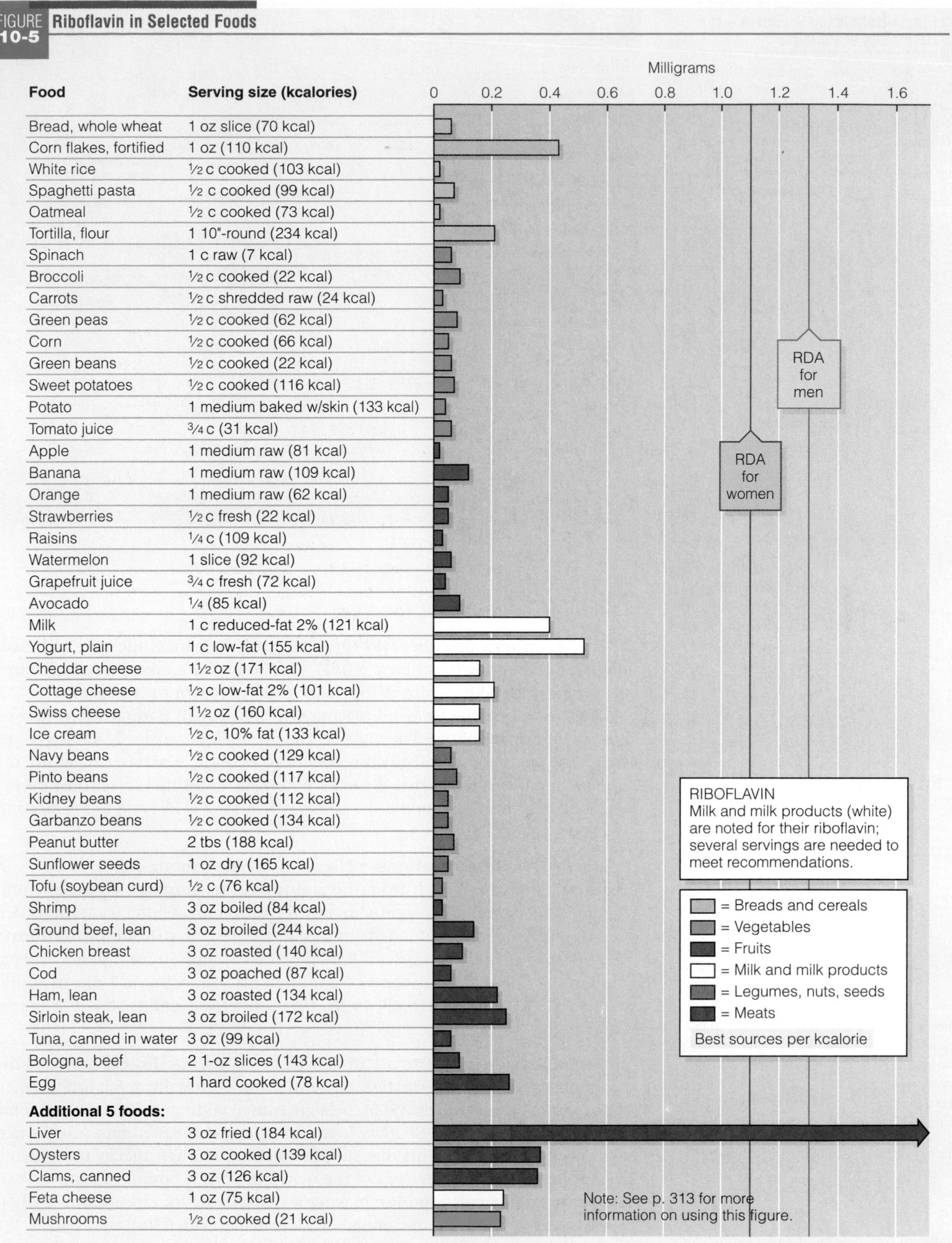

Food	Serving size (kcalories)
Bread, whole wheat	1 oz slice (70 kcal)
Corn flakes, fortified	1 oz (110 kcal)
White rice	½ c cooked (103 kcal)
Spaghetti pasta	½ c cooked (99 kcal)
Oatmeal	½ c cooked (73 kcal)
Tortilla, flour	1 10"-round (234 kcal)
Spinach	1 c raw (7 kcal)
Broccoli	½ c cooked (22 kcal)
Carrots	½ c shredded raw (24 kcal)
Green peas	½ c cooked (62 kcal)
Corn	½ c cooked (66 kcal)
Green beans	½ c cooked (22 kcal)
Sweet potatoes	½ c cooked (116 kcal)
Potato	1 medium baked w/skin (133 kcal)
Tomato juice	¾ c (31 kcal)
Apple	1 medium raw (81 kcal)
Banana	1 medium raw (109 kcal)
Orange	1 medium raw (62 kcal)
Strawberries	½ c fresh (22 kcal)
Raisins	¼ c (109 kcal)
Watermelon	1 slice (92 kcal)
Grapefruit juice	¾ c fresh (72 kcal)
Avocado	¼ (85 kcal)
Milk	1 c reduced-fat 2% (121 kcal)
Yogurt, plain	1 c low-fat (155 kcal)
Cheddar cheese	1½ oz (171 kcal)
Cottage cheese	½ c low-fat 2% (101 kcal)
Swiss cheese	1½ oz (160 kcal)
Ice cream	½ c, 10% fat (133 kcal)
Navy beans	½ c cooked (129 kcal)
Pinto beans	½ c cooked (117 kcal)
Kidney beans	½ c cooked (112 kcal)
Garbanzo beans	½ c cooked (134 kcal)
Peanut butter	2 tbs (188 kcal)
Sunflower seeds	1 oz dry (165 kcal)
Tofu (soybean curd)	½ c (76 kcal)
Shrimp	3 oz boiled (84 kcal)
Ground beef, lean	3 oz broiled (244 kcal)
Chicken breast	3 oz roasted (140 kcal)
Cod	3 oz poached (87 kcal)
Ham, lean	3 oz roasted (134 kcal)
Sirloin steak, lean	3 oz broiled (172 kcal)
Tuna, canned in water	3 oz (99 kcal)
Bologna, beef	2 1-oz slices (143 kcal)
Egg	1 hard cooked (78 kcal)
Additional 5 foods:	
Liver	3 oz fried (184 kcal)
Oysters	3 oz cooked (139 kcal)
Clams, canned	3 oz (126 kcal)
Feta cheese	1 oz (75 kcal)
Mushrooms	½ c cooked (21 kcal)

FIGURE 10-6 Niacin-Deficiency Symptom—The Dermatitis of Pellagra

In the dermatitis of pellagra, the skin darkens and flakes away as if it were sunburned. The protein-deficiency disease kwashiorkor also produces a "flaky paint" dermatitis, but the two are easily distinguished. The dermatitis of pellagra is bilateral and symmetrical and occurs only on those parts of the body exposed to the sun.

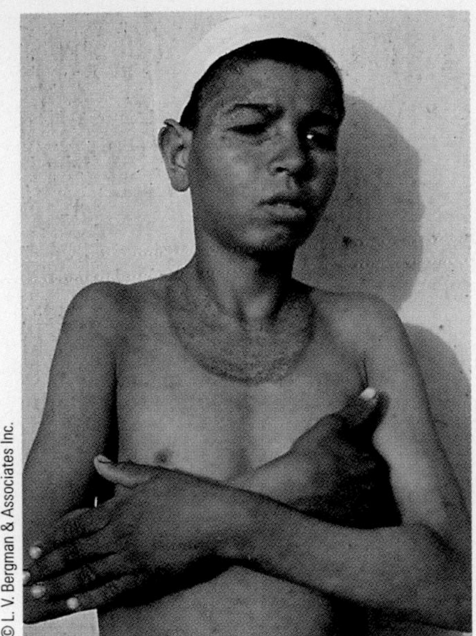

IN SUMMARY

Riboflavin

Other Names

Vitamin B_2

1998 RDA

Men: 1.3 mg/day

Women: 1.1 mg/day

Chief Functions in the Body

Part of coenzymes FMN (flavin mononucleotide) and FAD (flavin adenine dinucleotide) used in energy metabolism

Significant Sources

Milk products (yogurt, cheese); enriched or whole grains; liver

Easily destroyed by ultraviolet light and irradiation

Deficiency Disease

Ariboflavinosis (ay-RYE-boh-FLAY-vin-oh-sis)

Deficiency Symptoms

Eye Symptoms

Inflamed eyelids and sensitivity to light,[a] reddening of cornea

Mouth Symptoms

Sore throat; cracks and redness at corners of mouth;[b] painful, smooth, purplish red tongue[c]

Skin Symptoms

Inflammation characterized by lesions covered with greasy scales

Toxicity Symptoms

None reported

[a]Hypersensitivity to light is *photophobia* (FOE-toe-FOE-bee-ah).

[b]Cracks at the corners of the mouth are termed *cheilosis* (kee-LOH-sis).

[c]Smoothness of the tongue is caused by loss of its surface structures and is termed *glossitis* (gloss-EYE-tis).

Niacin

The name **niacin** describes two chemical structures: nicotinic acid and nicotinamide (also known as niacinamide). The body can easily convert nicotinic acid to nicotinamide, which is the major form of niacin in the blood.

The two coenzyme forms of niacin, NAD and NADP, participate in numerous metabolic reactions. They are central in energy-transfer reactions, especially the metabolism of glucose, fat, and alcohol. NAD is similar to the riboflavin coenzymes in that it carries hydrogens (and their electrons) during metabolic reactions, including the pathway from the TCA cycle to the electron transport chain.

◆ 1 NE = 1 mg niacin or 60 mg tryptophan.

• Niacin Recommendations • Niacin is unique among the B vitamins in that the body can make it from the amino acid tryptophan. To make 1 milligram of niacin requires approximately 60 milligrams of dietary tryptophan. For this reason, recommended intakes are stated in **niacin equivalents (NE).**◆ A food containing 1 milligram of niacin and 60 milligrams of tryptophan provides the equivalent of 2 milligrams of niacin, or 2 niacin equivalents. The RDA for niacin allows for this conversion and is stated in niacin equivalents.

• Niacin Deficiency • The niacin-deficiency disease, **pellagra,** produces the symptoms of diarrhea, dermatitis, dementia, and eventually death (often called "the four Ds"). In the early 1900s, pellagra caused widespread misery and some 87,000 deaths in the U.S. South, where many people subsisted on a low-protein diet centered on corn. This diet supplied neither enough niacin nor enough tryptophan. At least 70 percent of the niacin in corn is bound to complex carbohydrates and small peptides, making it unavailable for absorption. Furthermore, corn is high in the amino acid leucine, which interferes with the tryptophan-to-niacin conversion, thus further contributing to the development of pellagra. Figure 10-6 illustrates the dermatitis of pellagra.

niacin (NIGH-a-sin): a B vitamin. The coenzyme forms are NAD (nicotinamide adenine dinucleotide) and NADP (the phosphate form of NAD). Niacin can be eaten preformed or made in the body from its precursor, tryptophan, one of the amino acids.

niacin equivalents (NE): the amount of niacin present in food, including the niacin that can theoretically be made from its precursor, tryptophan, present in the food.

pellagra (pell-AY-gra): the niacin-deficiency disease.
- **pellis** = skin
- **agra** = rough

Pellagra was first believed to be caused by an infection. Medical researchers spent many years and much effort searching for infectious microbes until they

Estimate Niacin Equivalents

To obtain a rough approximation of niacin equivalents:

- Calculate total protein consumed (grams).
- Assuming that the RDA amount of protein will be used first to make body protein, subtract the RDA to obtain "leftover" protein available to make niacin (grams). (Actually, the RDA provides a generous protein allowance, so "leftover" protein may be even greater than this.)
- About 1 gram of every 100 grams of protein is tryptophan, so divide by 100 to obtain the tryptophan in this leftover protein (grams).
- Multiply by 1000 to express this amount of tryptophan in milligrams.
- Divide by 60 to get niacin equivalents (milligrams).
- Finally, add the amount of preformed niacin obtained in the diet (milligrams).

For example, suppose that a 19-year-old woman who weighs 130 pounds consumes 75 grams of protein in a day. Her protein RDA is 47 grams (calculated from Table 6-4 on p. 189), so she has 28 grams of leftover protein available to make tryptotophan:

$$75 \text{ g protein} - 47 \text{ g RDA} = 28 \text{ g leftover protein.}$$

Next calculate the amount of tryptophan in this leftover protein:

$$28 \text{ g protein} \div 100 = 0.28 \text{ g tryptophan.}$$

$$0.28 \text{ g tryptophan} \times 1000 = 280 \text{ mg tryptophan.}$$

Then convert milligrams of tryptophan to niacin equivalents:

$$280 \text{ mg tryptophan} \div 60 = 4.7 \text{ mg NE.}$$

To determine the total amount of niacin available from the diet, add the amount available from tryptophan (4.7 mg NE) to the amount of preformed niacin obtained from the diet.

realized that the problem was not what was present in the food, but what was *absent* from it. That a disease such as pellagra could be caused by diet—and not by germs—was a groundbreaking discovery. It contradicted commonly held medical opinions that diseases were caused only by infectious agents and advanced the science of nutrition dramatically.*

• Niacin Toxicity • Naturally occurring niacin from foods◆ causes no harm, but large doses from supplements or drugs produce a variety of adverse effects, most notably **"niacin flush."** Niacin flush occurs when niacin in the form of nicotinic acid is taken in doses only three to four times the RDA. It dilates the capillaries and causes a tingling sensation that can be painful. The nicotinamide form does not produce this effect—nor does it lower blood cholesterol.

Large doses of nicotinic acid (3000 milligrams or more a day), however, have been used to lower blood cholesterol. Such therapy must be closely monitored because of its adverse side effects (causes liver damage and aggravates peptic ulcers, among others). People with the following conditions may be particularly susceptible to the toxic effects of niacin: liver disease, diabetes, peptic ulcers, gout, irregular heartbeats, inflammatory bowel disease, migraine headaches, and alcoholism.

• Niacin Food Sources • Tables of food composition typically list preformed niacin only, but as mentioned, niacin can also be made in the body from the amino acid tryptophan. Hence diets that are high in protein are never lacking niacin. The "How to" above shows how to estimate the total amount of niacin available from the diet. Dietary tryptophan could meet about half the daily need for most people, but the average diet easily supplies enough preformed niacin.

The predominance of red bars in Figure 10-7 (see p. 318) explains why meat, poultry, and fish contribute about half the niacin most people need. An additional fourth of most people's niacin comes from enriched and whole grains. Mushrooms, asparagus, and leafy green vegetables are among the richest vegetable sources (per kcalorie) and can provide abundant niacin when eaten in generous amounts.

Niacin is less vulnerable to losses during food preparation and storage than other water-soluble vitamins. Being fairly heat-resistant, niacin can withstand reasonable cooking times, but like other water-soluble vitamins, it will leach

◆ When a normal dose of a nutrient (levels commonly found in foods) provides a normal blood concentration, the nutrient is having a *physiological* effect. When a large dose (levels commonly available only from supplements) overwhelms some body system and acts like a drug, the nutrient is having a *pharmacological* effect.
- **physio** = natural
- **pharma** = drug

Protein-rich foods such as meat, fish, poultry, and peanut butter contribute much of the niacin in people's diets. Enriched breads and cereals and a few vegetables are also rich in niacin.

niacin flush: a temporary burning, tingling, and itching sensation that occurs when a person takes a large dose of nicotinic acid; often accompanied by a headache and reddened face, arms, and chest.

*Dr. Joseph Goldberger, a physician for the U.S. government, headed the investigations that determined that pellagra was a dietary disorder, not an infectious disease. He died several years before Conrad Elevjhem discovered that a deficiency of niacin caused pellagra.

FIGURE 10-7 **Niacin in Selected Foods**

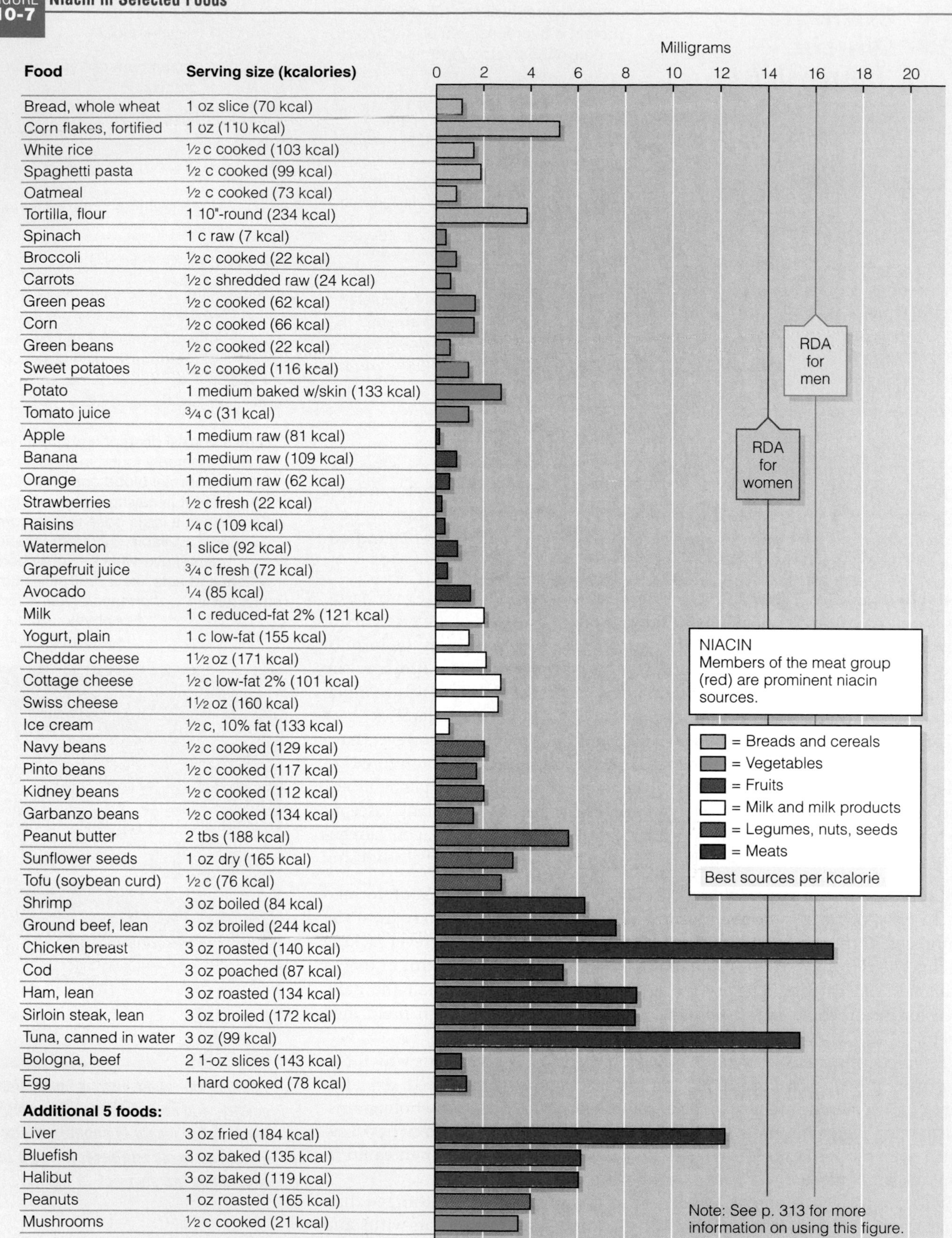

into cooking water. The summary table includes food sources as well as niacin's various names, functions, and deficiency and toxicity symptoms.

IN SUMMARY

Niacin

Other Names

Nicotinic acid, nicotinamide, niacinamide, vitamin B_3; precursor is dietary tryptophan (an amino acid)

1998 RDA

Men: 16 mg NE/day

Women: 14 mg NE/day

Upper Level

Adults: 35 mg/day

Chief Functions in the Body

Part of coenzymes NAD (nicotinamide adenine dinucleotide) and NADP (its phosphate form) used in energy metabolism

Significant Sources

Milk, eggs, meat, poultry, fish, whole-grain and enriched breads and cereals, nuts, and all protein-containing foods

Deficiency Disease

Pellagra

Deficiency Symptoms	Toxicity Symptoms
Eye Symptoms	
	Blurred vision
Digestive Symptoms	
Diarrhea, abdominal pain, vomiting	Nausea, vomiting
Mouth Symptoms	
Inflamed, swollen, smooth, bright red tongue[a]	
Neurological Symptoms	
Depression, apathy, fatigue, loss of memory, headache	
Skin Symptoms	
Bilateral symmetrical rash on areas exposed to sunlight	Painful flush, hives, and rash ("niacin flush"); excessive sweating
Other	
	Liver damage, impaired glucose tolerance

[a]Smoothness of the tongue is caused by loss of its surface structures and is termed *glossitis* (gloss-EYE-tis).

Biotin

Biotin plays an important role in metabolism as a coenzyme that carries activated carbon dioxide. This role is critical in the TCA cycle: biotin delivers a carbon to 3-carbon pyruvate, thus replenishing the 4-carbon compound needed to combine with acetyl CoA to keep the TCA cycle turning.* The biotin coenzyme also serves crucial roles in gluconeogenesis,◆ fatty acid synthesis, and the breakdown of certain fatty acids and amino acids.

◆ Reminder: The synthesis of glucose from noncarbohydrate sources such as amino acids or glycerol is called *gluconeogenesis.*

• Biotin Recommendations • Biotin is needed in very small amounts. Instead of an RDA, an Adequate Intake (AI) has been determined.

• Biotin Deficiency • Biotin deficiencies rarely occur. Researchers can induce a biotin deficiency in animals or human beings by feeding them raw egg whites, which contain a protein◆ that binds biotin and thus prevents its absorption. Biotin-deficiency symptoms include skin rash, hair loss, and neurological impairment. More than two dozen egg whites must be consumed daily for several months to produce these effects, however, and the eggs have to be raw; cooking denatures the binding protein.

◆ The protein **avidin** (AV-eh-din) in egg whites binds biotin.
• **avid** = greedy

• Biotin Food Sources • Biotin is widespread in foods (including egg yolks), so eating a variety of foods protects against deficiencies. Biotin is also synthesized by GI tract bacteria, but how much of it is absorbed is unknown. A review of biotin facts is provided in the summary table that follows.

biotin (BY-oh-tin): a B vitamin that functions as a coenzyme in metabolism.

*This 4-carbon intermediate of the TCA cycle is called *oxaloacetate.*

IN SUMMARY

Biotin

1998 Adequate Intake (AI)	Deficiency Symptoms
Adults: 30 μg/day	**Neurological Symptoms**
Chief Functions in the Body	Depression, lethargy, hallucinations, numb or tingling sensation in the arms and legs
Part of a coenzyme used in energy metabolism, fat synthesis, amino acid metabolism, and glycogen synthesis	**Skin Symptoms**
Significant Sources	Red, scaly rash around the eyes, nose, and mouth; hair loss
Widespread in foods; organ meats, egg yolks, soybeans, fish, whole grains; also produced by GI bacteria	**Toxicity Symptoms**
	None reported

Pantothenic Acid

Pantothenic acid is involved in more than 100 different steps in the synthesis of lipids, neurotransmitters, steroid hormones, and hemoglobin. It serves as part of coenzyme A—the same CoA that forms acetyl CoA, the "crossroads" compound in several metabolic pathways, including the TCA cycle. (Appendix C presents the chemical structures of these two molecules and shows that coenzyme A is made up in part of pantothenic acid.)

• **Pantothenic Acid Recommendations** • An Adequate Intake (AI) for pantothenic acid has been set. It reflects the amount needed to replace daily losses.

• **Pantothenic Acid Deficiency** • Pantothenic acid deficiency is rare. Its symptoms involve a general failure of all the body's systems and include fatigue, GI distress, and neurological disturbances. The "burning feet" syndrome that affected prisoners of war in Asia during World War II was thought to be caused by pantothenic acid deficiency.

• **Pantothenic Acid Food Sources** • Pantothenic acid is widespread in foods, and typical diets seem to provide adequate intakes. Beef, poultry, whole grains, potatoes, tomatoes, and broccoli are particularly good sources. Losses of pantothenic acid during food production can be substantial because it is readily destroyed by the freezing, canning, and refining processes. The summary table presents pantothenic acid facts.

IN SUMMARY

Pantothenic Acid

1998 Adequate Intake (AI)	Deficiency Symptoms
Adults: 5 mg/day	**Digestive Symptoms**
Chief Functions in the Body	Vomiting, nausea, stomach cramps
Part of coenzyme A, used in energy metabolism	**Neurological Symptoms**
Significant Sources	Insomnia, fatigue, depression, irritability, restlessness, apathy
Widespread in foods; organ meats, mushrooms, avocados, broccoli, whole grains	**Other**
Easily destroyed by food processing	Hypoglycemia, increased sensitivity to insulin
	Toxicity Symptoms
	None reported

pantothenic (PAN-toe-THEN-ick) **acid:** a B vitamin. The principal active form is part of coenzyme A, called "CoA" throughout Chapter 7.

- **pantos** = everywhere

Vitamin B_6

Vitamin B_6 occurs in three forms—pyridoxal, pyridoxine, and pyridoxamine. All three can be converted to the coenzyme PLP, which is active in amino acid metabolism. Because PLP can transfer amino groups, the body can make nonessential amino acids when amino groups are available (review Figure 7-15 on p. 218). The ability to add and remove amino groups makes PLP valuable in protein and urea metabolism as well. The conversions of the amino acid tryptophan to niacin or to the neurotransmitter serotonin◆ also depend on PLP as does the synthesis of heme (the nonprotein portion of hemoglobin), nucleic acids (such as DNA and RNA), and lecithin.

◆ Reminder: *Serotonin* is a neurotransmitter important in appetite control, sleep regulation, and sensory perception, among other roles; it is synthesized from the amino acid tryptophan with the help of vitamin B_6.

A surge of research in the last decade has revealed that vitamin B_6 influences cognitive performance, immune function, and steroid hormone activity.[5] Unlike other water-soluble vitamins, vitamin B_6 is stored extensively in muscle tissue.

Among the many drugs that interact with vitamin B_6, alcohol stands out. As Highlight 7 described, when the body breaks down alcohol, it produces acetaldehyde. If allowed to accumulate, acetaldehyde dislodges the PLP coenzyme from its enzymes; once loose, PLP breaks down and is excreted. Thus alcohol contributes to the destruction and loss of vitamin B_6 from the body.

Another drug that acts as a vitamin B_6 **antagonist** is INH, a medication that inhibits the growth of the tuberculosis bacterium.* INH has saved countless lives, but as a vitamin B_6 antagonist, it binds and inactivates the vitamin, inducing a deficiency. Whenever INH is used to treat tuberculosis, vitamin B_6 supplements must be given to protect against deficiency.

Oral contraceptives have raised concerns, but they may be unwarranted. Early studies reported signs of vitamin B_6 deficiency in oral contraceptive users, but that was when the pills contained estrogen at three to five times the quantities used today. Estrogen creates a shortage of vitamin B_6 by stimulating the breakdown of tryptophan, a process that requires the vitamin.

• Vitamin B_6 Recommendations • Because the vitamin B_6 coenzymes play many roles in amino acid metabolism, previous RDA were expressed in terms of protein intakes; the current RDA for vitamin B_6, however, is not. Research does not support claims that large doses of vitamin B_6 enhance muscle strength or physical endurance. In short, vitamin supplements cannot compete with a nutritious diet and physical training.

• Vitamin B_6 Deficiency • Without adequate vitamin B_6, synthesis of key neurotransmitters diminishes, and abnormal compounds produced during tryptophan metabolism accumulate in the brain. Early symptoms of vitamin B_6 deficiency include depression and confusion; advanced symptoms include abnormal brain wave patterns and convulsions.

• Vitamin B_6 Toxicity • The first major report of vitamin B_6 toxicity appeared in 1983. Until that time, everyone (including researchers and dietitians) believed that, like the other water-soluble vitamins, vitamin B_6 could not reach toxic concentrations in the body. The report described neurological damage in people who had been taking more than 2 grams of vitamin B_6 daily (20 times the current Upper Level) for two months or more.

Some women use vitamin B_6 supplements in an attempt to treat premenstrual syndrome (PMS), the cluster of physical, emotional, and psychological symptoms that some women experience seven to ten days prior to menstruation. The cause of PMS remains undefined, although researchers generally agree that the hormonal changes of the menstrual cycle must be responsible. Without a full understanding of PMS causes, medical treatments flounder, and quack treatments abound. Among nutritional approaches, the taking of vitamin B_6 has received much attention, but seems to have done more harm than good.

*INH stands for isonicotinic acid hydrazide.

vitamin B_6: a family of compounds—pyridoxal, pyridoxine, and pyridoxamine. The primary active coenzyme form is PLP (pyridoxal phosphate).

antagonist: a competing factor that counteracts the action of another factor. When a drug displaces a vitamin from its site of action, the drug renders the vitamin ineffective and thus acts as a vitamin antagonist.

© Polara Studios Inc.

Most protein-rich foods such as meat, fish, and poultry provide ample vitamin B_6; some vegetables and fruits are good sources, too.

Some people have taken vitamin B_6 supplements in an attempt to cure carpal tunnel syndrome and sleep disorders. Findings from one study suggest no correlation between blood levels of vitamin B_6 and symptoms of **carpal tunnel syndrome,** but those from another indicate that vitamin B_6 may be effective in treating some people with carpal tunnel syndrome.[6] Self-prescribing is ill-advised, however, because large doses of vitamin B_6 taken for months or years may cause irreversible nerve degeneration.

• **Vitamin B_6 Food Sources** • As you can see from the colored bars in Figure 10-8, meats, fish, and poultry (red), potatoes and a few other vegetables (green), and fruits (purple) offer vitamin B_6. As is true of most of the other vitamins, vegetables would rank considerably higher if foods were ranked by nutrient density (vitamin B_6 per kcalorie). Several servings of vitamin B_6–rich foods are needed to meet recommended intakes.

Foods lose vitamin B_6 when heated. Information is limited, but vitamin B_6 bioavailability from plant-derived foods seems to be lower than from animal-derived foods; fiber does not appear to interfere with absorption. The summary table lists food sources of vitamin B_6 as well as its chief functions in the body and common symptoms of both deficiency and toxicity.

IN SUMMARY

Vitamin B_6

Other Names

Pyridoxine, pyridoxal, pyridoxamine

1998 RDA

Adults (19–50 yr): 1.3 mg/day

Upper Level

Adults: 100 mg/day

Chief Functions in the Body

Part of coenzymes PLP (pyridoxal phosphate) and PMP (pyridoxamine phosphate) used in amino acid and fatty acid metabolism; helps to convert tryptophan to niacin and to serotonin; helps to make red blood cells

Significant Sources

Meats, fish, poultry, potatoes, legumes, noncitrus fruits, fortified cereals, liver, soy products

Easily destroyed by heat

Deficiency Symptoms	Toxicity Symptoms
Blood Symptoms	
Anemia (small-cell type)[a]	
Neurological Symptoms	
Depression, confusion, abnormal brain wave pattern, convulsions	Depression, fatigue, irritability, headaches, nerve damage causing numbness and muscle weakness leading to an inability to walk and convulsions
Skin Symptoms	
Scaly dermatitis	Lesions

[a]Small-cell-type anemia is *microcytic anemia.*

Folate

Folate, also known as folacin or folic acid, has a chemical name that would fit a flying dinosaur: pteroylglutamic acid (PGA for short). Its primary coenzyme form, THF, serves as part of an enzyme complex that transfers one-carbon compounds that arise during metabolism. This action helps convert vitamin B_{12} to one of its coenzyme forms and helps synthesize the DNA required for all rapidly growing cells.

Foods deliver folate mostly in the "bound" form—that is, combined with a string of amino acids (glutamate), known as polyglutamate (see Appendix C for the chemical structure). The intestine prefers to absorb the "free" folate form—folate with only one glutamate attached (the monoglutamate form). Enzymes on the intestinal cell surfaces hydrolyze the polyglutamate to monoglutamate and several glutamates. Then the monoglutamate is attached to a methyl group

carpal tunnel syndrome: a pinched nerve at the wrist, causing pain or numbness in the hand. It is often caused by repetitive motion of the wrist.

folate (FOLE-ate): a B vitamin; also known as folic acid, folacin, or pteroylglutamic (tare-o-EEL-glue-TAM-ick) acid (PGA). The coenzyme forms are DHF (dihydrofolate) and THF (tetrahydrofolate).

FIGURE 10-8 Vitamin B_6 in Selected Foods

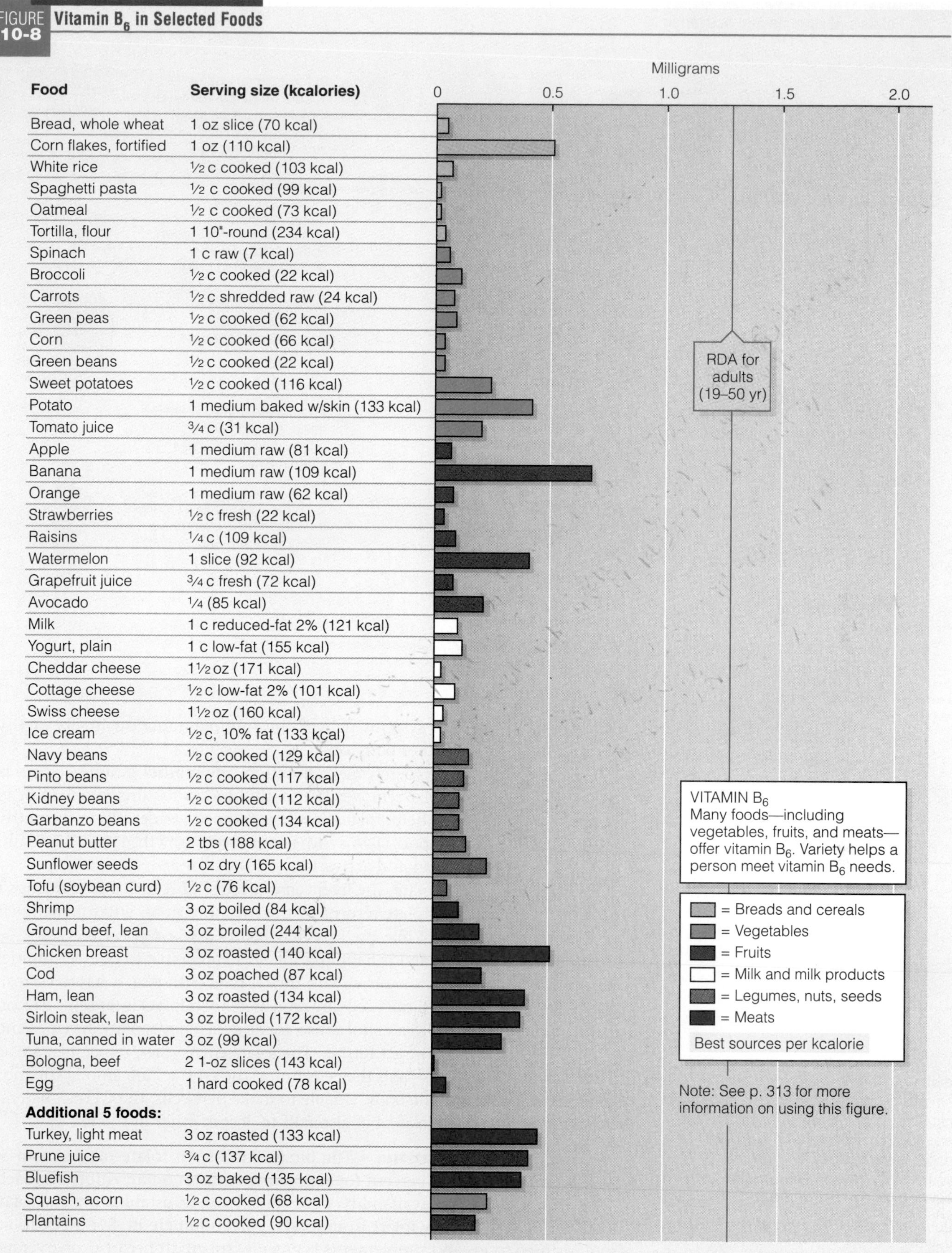

Food	Serving size (kcalories)
Bread, whole wheat	1 oz slice (70 kcal)
Corn flakes, fortified	1 oz (110 kcal)
White rice	½ c cooked (103 kcal)
Spaghetti pasta	½ c cooked (99 kcal)
Oatmeal	½ c cooked (73 kcal)
Tortilla, flour	1 10"-round (234 kcal)
Spinach	1 c raw (7 kcal)
Broccoli	½ c cooked (22 kcal)
Carrots	½ c shredded raw (24 kcal)
Green peas	½ c cooked (62 kcal)
Corn	½ c cooked (66 kcal)
Green beans	½ c cooked (22 kcal)
Sweet potatoes	½ c cooked (116 kcal)
Potato	1 medium baked w/skin (133 kcal)
Tomato juice	¾ c (31 kcal)
Apple	1 medium raw (81 kcal)
Banana	1 medium raw (109 kcal)
Orange	1 medium raw (62 kcal)
Strawberries	½ c fresh (22 kcal)
Raisins	¼ c (109 kcal)
Watermelon	1 slice (92 kcal)
Grapefruit juice	¾ c fresh (72 kcal)
Avocado	¼ (85 kcal)
Milk	1 c reduced-fat 2% (121 kcal)
Yogurt, plain	1 c low-fat (155 kcal)
Cheddar cheese	1½ oz (171 kcal)
Cottage cheese	½ c low-fat 2% (101 kcal)
Swiss cheese	1½ oz (160 kcal)
Ice cream	½ c, 10% fat (133 kcal)
Navy beans	½ c cooked (129 kcal)
Pinto beans	½ c cooked (117 kcal)
Kidney beans	½ c cooked (112 kcal)
Garbanzo beans	½ c cooked (134 kcal)
Peanut butter	2 tbs (188 kcal)
Sunflower seeds	1 oz dry (165 kcal)
Tofu (soybean curd)	½ c (76 kcal)
Shrimp	3 oz boiled (84 kcal)
Ground beef, lean	3 oz broiled (244 kcal)
Chicken breast	3 oz roasted (140 kcal)
Cod	3 oz poached (87 kcal)
Ham, lean	3 oz roasted (134 kcal)
Sirloin steak, lean	3 oz broiled (172 kcal)
Tuna, canned in water	3 oz (99 kcal)
Bologna, beef	2 1-oz slices (143 kcal)
Egg	1 hard cooked (78 kcal)
Additional 5 foods:	
Turkey, light meat	3 oz roasted (133 kcal)
Prune juice	¾ c (137 kcal)
Bluefish	3 oz baked (135 kcal)
Squash, acorn	½ c cooked (68 kcal)
Plantains	½ c cooked (90 kcal)

FIGURE 10-9 Folate's Absorption and Activation

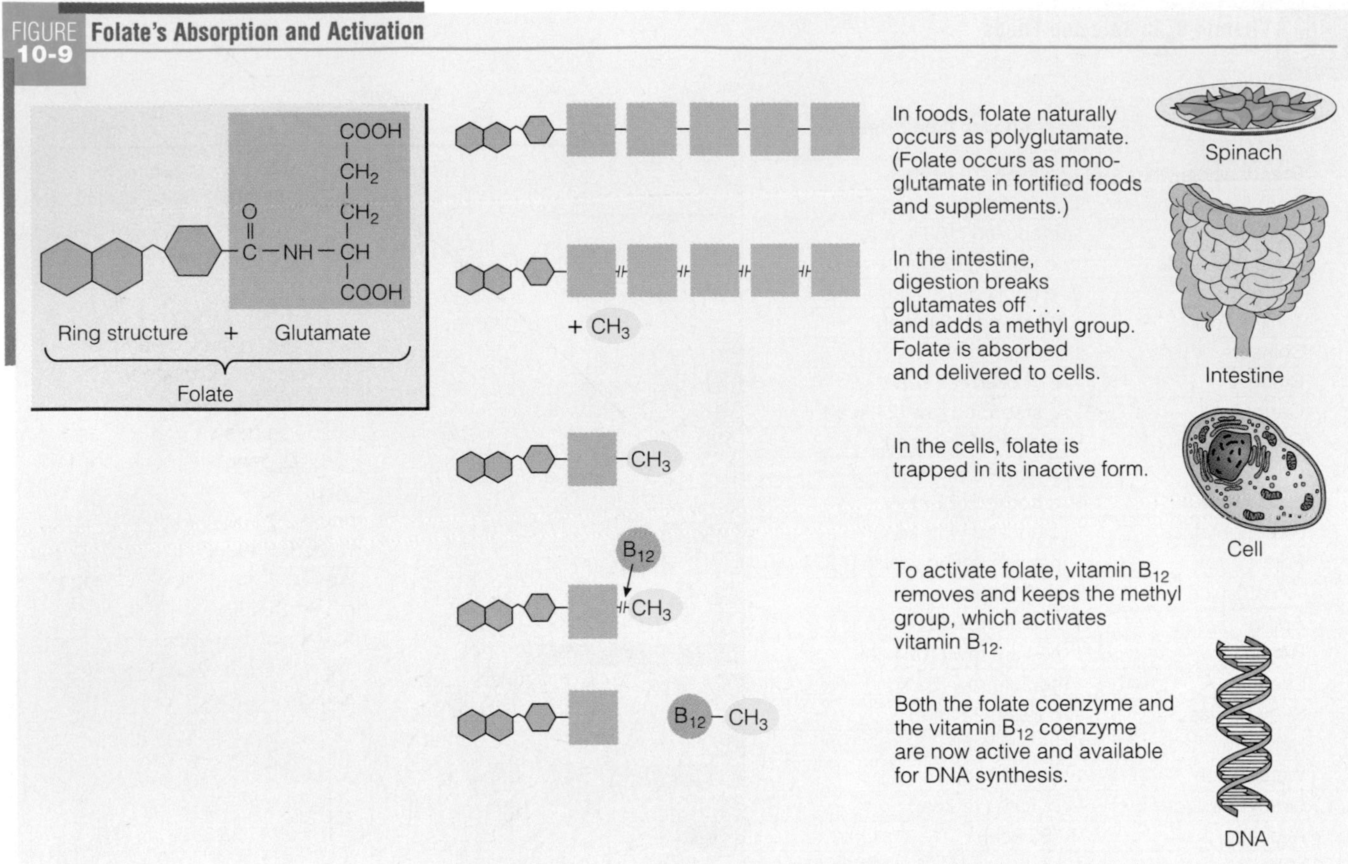

(CH_3). Special transport systems deliver the monoglutamate with its methyl group to the liver and other body cells.

In order for the folate coenzyme to function, the methyl group needs to be removed. The enzyme that removes the methyl group requires the help of vitamin B_{12}. Without that help, folate becomes trapped inside cells in its methyl form, unavailable to support DNA synthesis and cell growth. Figure 10-9 summarizes folate's absorption and activation.

To dispose of excess folate, the liver secretes most of it into bile and ships it to the gallbladder. Thus folate returns to the intestine in an enterohepatic circulation route like that of bile itself (review Figure 5-15 on p. 142).

This complicated system for handling folate is vulnerable to GI tract injuries. Since folate is actively secreted back into the GI tract with bile, it has to be reabsorbed repeatedly. If the GI tract cells are damaged, then folate is rapidly lost from the body. Such is the case in alcohol abuse; folate deficiency rapidly develops and, ironically, damages the GI tract further. The folate coenzymes, remember, are active in cell multiplication—and the cells lining the GI tract are among the most rapidly renewed cells in the body. Unable to make new cells, the GI tract deteriorates and not only loses folate, but also fails to absorb other nutrients.

dietary folate equivalents (DFE): the amount of folate available to the body from naturally occurring sources, fortified foods, and supplements, accounting for differences in the bioavailability from each source. Use the following equation to calculate DFE:

μg food folate + (1.7 × μg synthetic folate).

Using the example in the text:

100 μg food
+ 170 μg supplement (1.7 × 100 μg)
270 μg DFE

• Folate Recommendations • The bioavailability of folate ranges from 50 percent for foods to 100 percent for supplements taken on an empty stomach. These differences in bioavailability were considered in establishing the folate RDA. Naturally occurring folate from foods is given full credit. Synthetic folate from fortified foods and supplements is given extra credit because, on average, it is 1.7 times more available than naturally occurring food folate. Thus a person consuming 100 micrograms of folate from foods and 100 micrograms from a supplement receives 270 **dietary folate equivalents (DFE).** (The accompanying "How to" feature describes how to estimate dietary folate equivalents.) The need for folate rises considerably during pregnancy and whenever cells are multiplying, so the recommendations for pregnant women are considerably higher than for other adults.

Estimate Dietary Folate Equivalents

Folate is expressed in terms of DFE (dietary folate equivalents) because synthetic folate from supplements and fortified foods is absorbed at almost twice (1.7 times) the rate of naturally occurring folate from other foods. Use the following equation to calculate:

DFE = μg food folate + (1.7 × μg synthetic folate).

Consider, for example, a pregnant woman who takes a supplement and eats a bowl of fortified cornflakes, 2 slices of fortified bread, and a cup of fortified pasta. From the supplement and fortified foods, she obtains synthetic folate:

Supplement	100 μg folate
Fortified cornflakes	100 μg folate
Fortified bread	40 μg folate
Fortified pasta	60 μg folate
	300 μg folate

To calculate the DFE, multiply the amount of synthetic folate by 1.7:

300 μg × 1.7 = 510 μg DFE.

Now add the naturally occurring folate from the other foods in her diet—in this example, another 90 μg of folate.

510 μg DFE + 90 μg = 600 μg DFE.

Notice that if we had not converted synthetic folate from supplements and fortified foods to DFE, then this woman's intake would appear to fall short of the 600 μg recommendation for pregnancy (300 μg + 90 μg = 390 μg). But as our example shows, her intake does meet the recommendation. At this time, supplement and fortified food labels list folate in μg only, not μg DFE, making such calculations necessary.

• Folate and Neural Tube Defects • Several research studies have confirmed the importance of folate in reducing the risks of **neural tube defects.**[7] The brain and spinal cord develop◆ from the neural tube, and defects in its orderly formation during the early weeks of pregnancy may result in various central nervous system disorders and death.

Folate supplements taken one month before conception and continued throughout the first trimester of pregnancy can help prevent neural tube defects.[8] For this reason, all women of childbearing age◆ who are capable of becoming pregnant should take 0.4 milligrams (400 micrograms) of folate daily,◆ although only one-third of them do.[9] This amount of folate can be met through a diet that includes at least five servings of fruits and vegetables daily, but many women typically fail to do so and receive only half this amount from foods. Furthermore, because of the enhanced bioavailability of synthetic folate, supplementation or fortification improves folate status significantly.[10] Women who have given birth to infants with neural tube defects previously should take 4 milligrams of folate daily before conception and throughout the first trimester of pregnancy.[11]

Because many pregnancies are unplanned and because neural tube defects occur early in development before most women realize they are pregnant, the Food and Drug Administration (FDA) has mandated that grain products be fortified to deliver folate to the U.S. population.* Labels on fortified products may claim that "adequate intake of folate has been shown to reduce the risk of neural tube defects." Fortification improves folate status and is expected to prevent half of the 4000 neural tube defects that have occurred each year, but folate fortification raises safety concerns as well.[12] Because high intakes of folate complicate the diagnosis of a vitamin B_{12} deficiency, folate consumption should not exceed 1 milligram daily without close medical supervision.[13] Whether it is wise to fortify our food supply with folate, but not vitamin B_{12}, continues to be the subject of much debate.[14]

Recently, researchers have found a genetic flaw that induces a folate deficiency and predisposes women to bear infants with neural tube defects and Down syndrome.[15] Folate's exact role in preventing these birth defects, however, remains unclear. Some women whose infants develop these defects are *not* deficient in folate, and others with severe folate deficiencies do *not* give birth to infants with birth defects. Researchers continue to look for other factors that must also be involved.

◆ **Chapter 14 provides photos of neural tube development.**

◆ **Women of childbearing age (15 to 45 yr) should:**
- Eat folate-rich foods.
- Eat folate-fortified foods.
- Take a multivitamin daily (most provide 400 μg folate).

◆ **Reminder: A milligram (mg) is one-thousandth of a gram. A microgram (μg) is one-thousandth of a milligram (or one-millionth of a gram).**
- 0.4 mg = 400 μg.

neural tube defects: malformations of the brain, spinal cord, or both during embryonic development. The two main types of neural tube defects are spina bifida (literally, "split spine") and anencephaly ("no brain").

*Bread products, flour, corn grits, cornmeal, farina, rice, macaroni, and noodles must be fortified with 140 micrograms of folate per 100 grams of food. For perspective, 100 grams is roughly 3 slices of bread; 1 cup of flour; ½ cup corn grits, cornmeal, farina, or rice; or ¾ cup of macaroni or noodles.

◆ Large-cell anemia is known as **macrocytic** or **megaloblastic** anemia.
- **macro** = large
- **cyte** = cell
- **mega** = large

◆ Chapter 18 discusses nutrient-drug interactions and includes a figure illustrating the similarities between the vitamin folate and the anticancer drug methotrexate.

• Folate and Heart Disease • The FDA's decision to fortify grain products with folate was strengthened by research indicating an important role for folate in defending against heart disease.[16] As Chapter 6 mentioned, research indicates that high levels of the amino acid homocysteine and low levels of folate increase the risk of fatal heart disease.[17] One of folate's key roles in the body is to break down homocysteine. Without folate, homocysteine accumulates, which seems to enhance blood clot formation and arterial wall deterioration. Fortified foods and folate supplements raise blood folate and reduce blood homocysteine levels, but whether these strategies help to prevent heart disease remains to be seen.[18]

• Folate and Cancer • Folate may also play a role in preventing cancer.[19] Notably, folate may be most effective in protecting those most likely to develop cancers: men who smoke (against pancreatic cancer) and women who drink alcohol (against breast cancer).[20]

• Folate Deficiency • Folate deficiency impairs cell division and protein synthesis—processes critical to growing tissues. In a folate deficiency, the replacement of red blood cells and GI tract cells falters. Not surprisingly, then, two of the first symptoms of a folate deficiency are **anemia** and GI tract deterioration.

The anemia of folate deficiency is characterized by large,◆ immature blood cells. Without folate, DNA synthesis slows and the cells lose their ability to divide. The nucleus of the cell is not released as normally occurs during development. As a result, the immature blood cells are enlarged and oval-shaped. They cannot carry oxygen or travel through the capillaries as efficiently as normal red blood cells.

Folate deficiencies may develop from inadequate intake and have been reported in infants fed goat's milk, which is notoriously low in folate. Folate deficiency may also result from impaired absorption or an unusual metabolic need for the vitamin. Metabolic needs increase wherever cell multiplication must speed up: in pregnancies involving twins and triplets; in cancer; in skin-destroying diseases such as chicken pox and measles; and in burns, blood loss, GI tract damage, and the like.

Of all the vitamins, folate appears to be most vulnerable to interactions with drugs, which can lead to a secondary deficiency. Some medications, notably anticancer drugs, have a chemical structure similar to folate's structure◆ and can displace the vitamin from enzymes and interfere with normal metabolism. Cancer cells, like all cells, need the real vitamin to multiply; without it, they die. Unfortunately, these drugs create a deficiency for the other cells in the body that also need folate.

Aspirin and antacids also interfere with the body's handling of folate. Healthy adults who use these drugs to relieve an occasional headache or upset stomach need not be concerned, but people who rely heavily on aspirin or antacids should be aware of the nutrition consequences. Oral contraceptives may also impair folate status, as may smoking.

Leafy green vegetables (such as spinach and broccoli), legumes (such as black beans, kidney beans, and black-eyed peas), liver, and some fruits (notably oranges) are naturally rich in folate.

• Folate Food Sources • Figure 10-10 shows that folate is especially abundant in legumes and vegetables. The vitamin's name suggests the word *foliage,* and indeed, leafy green vegetables are outstanding sources. With fortification, grain products also contribute folate; the bioavailability of added folate is good, too, suggesting that consumers will benefit from the enrichment program.[21] The lack of red and white bars in Figure 10-10 indicates that meats, milk, and milk products are poor folate sources. Heat and oxidation during cooking and storage can destroy as much as half of the folate in foods. The table on p. 328 provides a summary of folate information.

anemia (ah-NEE-me-ah): literally, "too little blood." Anemia is any condition in which too few red blood cells are present, or the red blood cells are immature (and therefore large) or too small or contain too little hemoglobin to carry the normal amount of oxygen to the tissues. It is not a disease itself, but can be a symptom of many different disease conditions, including many nutrient deficiencies, bleeding, excessive red blood cell destruction, and defective red blood cell formation.
- **an** = without
- **emia** = blood

FIGURE 10-10 Folate in Selected Foods

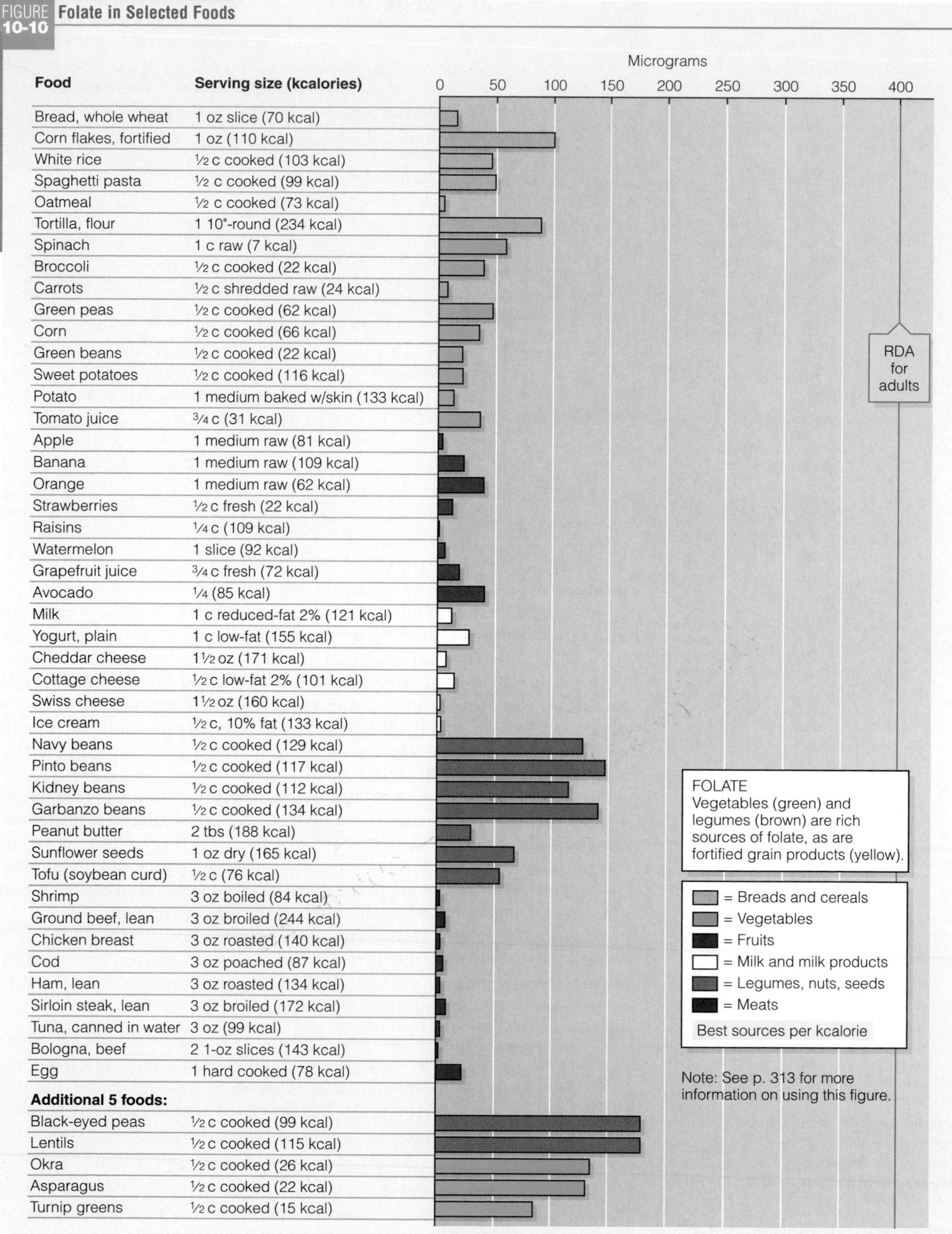

IN SUMMARY

Folate

Other Names	Deficiency Symptoms	Toxicity Symptoms
Folic acid, folacin, pteroylglutamic acid (PGA)	**Blood Symptoms**	
1998 RDA	Anemia (large-cell type)[a]	
Adults: 400 µg/day	**Mouth Symptoms**	
Upper Level	Smooth, red tongue[b]	
Adults: 1000 µg/day	**Neurological Symptoms**	
Chief Functions in the Body	Mental confusion, weakness, fatigue, irritability, headache	
Part of coenzymes THF (tetrahydrofolate) and DHF (dihydrofolate) used in DNA synthesis and therefore important in new cell formation	**Other**	
Significant Sources	Elevated homocysteine	Masks vitamin B_{12}–deficiency symptoms
Fortified grains, leafy green vegetables, legumes, seeds, liver		
Easily destroyed by heat and oxygen		

[a]Large-cell–type anemia is known as either *macrocytic* or *megaloblastic anemia.*
[b]Smoothness of the tongue is caused by loss of its surface structures and is termed *glossitis* (gloss-EYE-tis).

Vitamin B_{12}

Vitamin B_{12} and folate are closely related: each depends on the other for activation. Recall that vitamin B_{12} removes a methyl group to activate the folate coenzyme; when folate gives up its methyl group, the vitamin B_{12} coenzyme becomes activated (review Figure 10-9 on p. 324).

The regeneration of the amino acid methionine and the synthesis of DNA and RNA depend on both folate and vitamin B_{12}.* In addition, without any help from folate, vitamin B_{12} maintains the sheath that surrounds and protects nerve fibers and promotes their normal growth. Bone cell activity and metabolism also depend on vitamin B_{12}.

In the stomach, hydrochloric acid and the digestive enzyme pepsin release vitamin B_{12} from the proteins to which it is attached in foods. Then the vitamin binds with an **"intrinsic factor"** that is made in the stomach. After the intrinsic factor attaches to vitamin B_{12}, the complex passes to the small intestine, where the vitamin is gradually absorbed into the bloodstream. Transport of vitamin B_{12} in the blood depends on specific binding proteins.

Like folate, vitamin B_{12} follows the enterohepatic circulation route. It is continually secreted into bile and delivered to the intestine, where it is reabsorbed. Because most vitamin B_{12} is reabsorbed, healthy people rarely develop a deficiency even when their intake is minimal.

• **Vitamin B_{12} Recommendations** • The RDA for adults is only 2.4 micrograms of vitamin B_{12} a day—just over two-millionths of a gram. The ink in the period at the end of this sentence may weigh about 2.4 micrograms. But tiny though this amount appears to the human eye, it contains billions of molecules of vitamin B_{12}, enough to provide coenzymes for all the enzymes that need its help.

• **Vitamin B_{12} Deficiency** • Most vitamin B_{12} deficiencies reflect inadequate absorption, not poor intake. Inadequate absorption typically occurs for one of two reasons: a lack of hydrochloric acid or a lack of intrinsic factor. Without

*In the body, methionine serves as a methyl (CH_3) donor. In doing so, methionine can be converted to other amino acids. Some of these amino acids can regenerate methionine, but methionine is still considered an essential amino acid that is needed in the diet.

vitamin B_{12}: a B vitamin characterized by the presence of cobalt (see Figure 13-10 in Chapter 13). The active forms of coenzyme B_{12} are methylcobalamin and deoxyadenosylcobalamin.

intrinsic factor: a glycoprotein (a protein with short polysaccharide chains attached) manufactured in the stomach that aids in the absorption of vitamin B_{12}.
- **intrinsic** = on the inside

hydrochloric acid, the vitamin is not released from the dietary proteins and so is not available for binding with the intrinsic factor. Without the intrinsic factor, the vitamin cannot be absorbed.

Many people, especially those over 60, develop **atrophic gastritis,** a common condition in older people that damages the cells of the stomach. Atrophic gastritis may also develop in response to iron deficiency or infection with *Helicobacter pylori,* the bacterium implicated in ulcer formation. Without healthy stomach cells, production of hydrochloric acid and intrinsic factor diminishes. Even with an adequate intake from foods, vitamin B_{12} status suffers. The vitamin B_{12} deficiency caused by atrophic gastritis and a lack of intrinsic factor is known as **pernicious anemia.**[22]

Some people inherit a defective gene for the intrinsic factor. In such cases, or when the stomach has been injured and cannot produce enough of the intrinsic factor, vitamin B_{12} must be injected to bypass the need for intestinal absorption. Alternatively, the vitamin may be delivered by nasal spray; absorption is rapid, high, and well-tolerated.[23]

A prolonged inadequate intake, as can occur with a vegetarian diet,◆ may also create a vitamin B_{12} deficiency.[24] People who stop eating foods containing vitamin B_{12} may take several years to develop deficiency symptoms because the body recycles much of its vitamin B_{12}, reabsorbing it over and over again. Even when the body fails to absorb vitamin B_{12}, deficiency may take up to three years to develop because the body conserves its supply.

◆ Vitamin B_{12} is found primarily in foods derived from animals.

Because vitamin B_{12} is required to convert folate to its active form, one of the most obvious vitamin B_{12}–deficiency symptoms is the anemia of folate deficiency. This anemia is characterized by large, immature red blood cells, which are indicative of slow DNA synthesis and an inability to divide (see Figure 10-11). When folate is trapped in its inactive (methyl folate) form due to vitamin B_{12} deficiency, or is unavailable due to folate deficiency itself, DNA synthesis slows.

First to be affected in a vitamin B_{12} or folate deficiency are the rapidly growing blood cells. Either vitamin B_{12} or folate will clear up the anemia, but if folate is given when vitamin B_{12} is needed, the result is disastrous: devastating neurological symptoms. Remember that vitamin B_{12}, but not folate, maintains the sheath that surrounds and protects nerve fibers and promotes their normal growth. Folate "cures" the *blood* symptoms of a vitamin B_{12} deficiency, but cannot stop the *nerve* symptoms from progressing. By doing so, folate "masks" a

FIGURE 10-11 Normal and Anemic Blood Cells

The anemia of folate deficiency is indistinguishable from that of vitamin B_{12} deficiency. Appendix E describes the biochemical tests used to differentiate the two conditions.

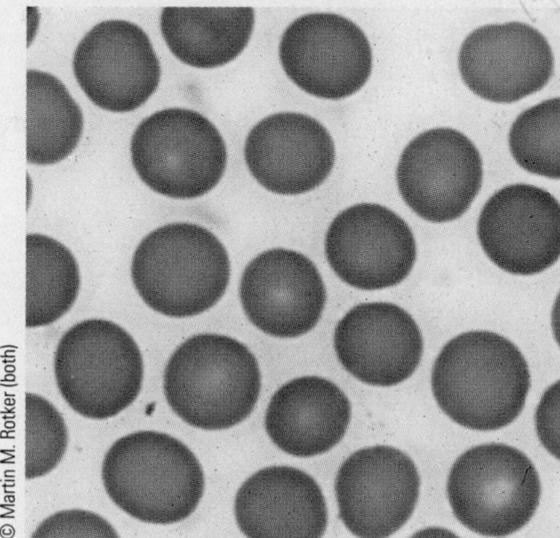

Normal blood cells. The size, shape, and color of the red blood cells show that they are normal.

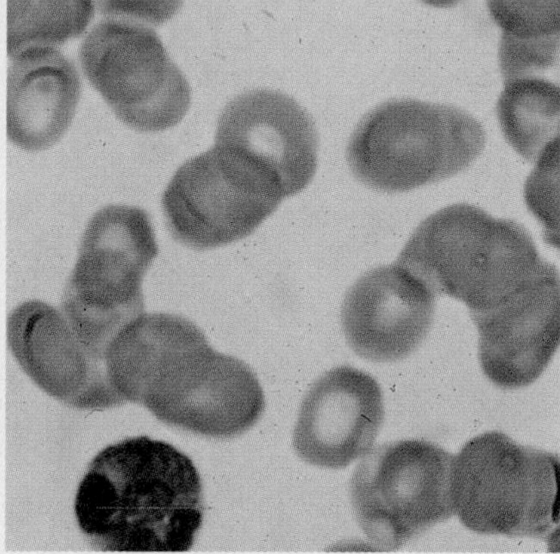

Blood cells in pernicious anemia (megaloblastic). Megaloblastic blood cells are slightly larger than normal red blood cells, and their shapes are irregular.

© Martin M. Rotker (both)

atrophic (a-TRO-fik) **gastritis** (gas-TRY-tis): chronic inflammation of the stomach accompanied by a diminished size and functioning of the mucous membranes and glands.
- **atrophy** = wasting
- **gastro** = stomach
- **itis** = inflammation

pernicious (per-NISH-us) **anemia:** a blood disorder that reflects a vitamin B_{12} deficiency caused by lack of intrinsic factor and characterized by abnormally large and immature red blood cells. Other symptoms include muscle weakness and irreversible neurological damage.
- **pernicious** = destructive

vitamin B_{12} deficiency. The neurological symptoms of a vitamin B_{12} deficiency include a creeping paralysis that begins at the extremities and works inward and up the spine. Early detection and correction are necessary to prevent permanent nerve damage and paralysis. With sufficient folate in the diet, the neurological symptoms of vitamin B_{12} deficiency can develop without evidence of anemia. Such interactions between folate and vitamin B_{12} highlight some of the safety issues surrounding the use of supplements and the fortification of foods.

• Vitamin B_{12} Food Sources • Vitamin B_{12} is unique among the vitamins in being found almost exclusively in foods derived from animals. Anyone who eats reasonable amounts of meat is guaranteed an adequate intake, and vegetarians who use milk products or eggs are also protected from deficiency. Vegans, who restrict all foods derived from animals, need a reliable source, such as vitamin B_{12}–fortified soy milk or vitamin B_{12} supplements. Yeast grown on a vitamin B_{12}–enriched medium and mixed with that medium provides some vitamin B_{12}, but yeast itself does not contain active vitamin B_{12}. Fermented soy products such as miso (a soybean paste) and sea algae such as spirulina also do *not* provide active vitamin B_{12}. Extensive research shows that the amounts listed on the labels of these plant products are inaccurate and misleading because the vitamin B_{12} is in an inactive, unavailable form.

As mentioned earlier, the water-soluble vitamins are particularly vulnerable to losses in cooking. For most of these nutrients, microwave heating minimizes losses as well as, or better than, traditional cooking methods. Such is not the case for vitamin B_{12}, however. Microwave heating inactivates vitamin B_{12}.[25] To preserve this vitamin, use the oven or stovetop instead of a microwave to cook meats and milk products (major sources of vitamin B_{12}). The table below provides a summary of information about vitamin B_{12}.

IN SUMMARY

Vitamin B_{12}

Other Names

Cobalamin (and related forms)

1998 RDA

Adults: 2.4 µg/day

Chief Functions in the Body

Part of coenzymes methylcobalamin and deoxyadenosylcobalamin used in new cell synthesis; helps to maintain nerve cells; reforms folate coenzyme; helps to break down some fatty acids and amino acids

Significant Sources

Animal products (meat, fish, poultry, shellfish, milk, cheese, eggs), fortified cereals

Easily destroyed by microwave cooking

Deficiency Disease

Pernicious anemia[a]

Deficiency Symptoms

Blood Symptoms

Anemia (large-cell type)[b]

Mouth Symptoms

Smooth, sore tongue[c]

Neurological Symptoms

Fatigue, degeneration of peripheral nerves progressing to paralysis

Skin symptoms

Hypersensitivity

Toxicity Symptoms

None reported

[a]The name *pernicious anemia* refers to the vitamin B_{12} deficiency caused by atrophic gastritis and a lack of intrinsic factor, but not to that caused by inadequate dietary intake.

[b]Large-cell–type anemia is known as either *macrocytic* or *megaloblastic anemia*.

[c]Smoothness of the tongue is caused by loss of its surface structures and is termed *glossitis* (gloss-EYE-tis).

Non-B Vitamins

Nutrition scientists debate whether other dietary compounds might also be considered vitamins. In some cases, the compounds may be conditionally essential—that is, needed by the body from foods when synthesis becomes insufficient to support normal growth and metabolism. In other cases, the compounds may be vitamin impostors—not needed under any circumstances.

• **Choline** • The essentiality of **choline** has been blurry for decades, in part because the body can make choline from the amino acid methionine.[26] Furthermore, choline is commonly found in many foods as part of the lecithin molecule (review Figure 5-9 on p. 136). Choline deficiencies are rare. Without any dietary choline, however, synthesis alone appears to be insufficient to meet the body's needs, making choline a conditionally essential nutrient. For this reason, the 1998 DRI report established an Adequate Intake (AI) for choline. The body uses choline to make the neurotransmitter acetylcholine and the phospholipid lecithin. The accompanying table summarizes key choline facts.

IN SUMMARY

Choline

1998 Adequate Intake (AI)	**Deficiency Symptoms**
Men: 550 mg/day	Liver damage
Women: 425 mg/day	
	Toxicity Symptoms
Upper Level	Body odor, sweating, salivation, reduced growth rate, low blood pressure, liver damage
Adults: 3500 mg/day	
Chief Functions in the Body	**Significant Sources**
Needed for the synthesis of the neurotransmitter acetylcholine and the phospholipid lecithin	Milk, liver, eggs, peanuts

• **Inositol and Carnitine** • Like choline, **inositol** and **carnitine** can be made by the body, but unlike choline, no recommendations have been established. Researchers continue to explore the possibility that these substances may be essential. Even if they are essential, though, supplements are unnecessary because these compounds are widespread in foods.

Some vitamin companies include choline, inositol, and carnitine in their formulations to make their vitamin pills look more "complete" than others, but this strategy offers no real advantage. For a rational way to compare vitamin-mineral supplements, read Highlight 10.

• **Vitamin Impostors** • Other substances have been mistaken for essential nutrients for human beings because they are needed for growth by bacteria or other forms of life. Among them are PABA (para-aminobenzoic acid, a component of folate's ring structure), the bioflavonoids (vitamin P or hesperidin), pyrroloquinoline quinone (methoxatin), orotic acid, lipoic acid, and ubiquinone (coenzyme Q_{10}). Other names erroneously associated with vitamins are "vitamin O" (oxygenated salt water), "vitamin B_5" (another name for pantothenic acid), "vitamin B_{15}" (also called "pangamic acid," a hoax), and "vitamin B_{17}" (laetrile, an alleged "cancer cure" and not a vitamin or a cure by any stretch of the imagination—in fact, laetrile is a potentially dangerous substance).

choline (KOH-leen): a nitrogen-containing compound found in foods and made in the body from the amino acid methionine. Choline is used to make the phospholipid lecithin and the neurotransmitter acetylcholine.

inositol (in-OSS-ih-tall): a nonessential nutrient that can be made in the body from glucose. Inositol is used in cell membranes.

carnitine (CAR-neh-teen): a nonessential nutrient made in the body from the amino acid lysine.

IN SUMMARY The B vitamins serve as coenzymes that facilitate the work of every cell. They are active in carbohydrate, fat, and protein metabolism and in the making of DNA and thus new cells. Historically famous B vitamin–deficiency diseases are beriberi (thiamin), pellagra (niacin), and pernicious anemia (vitamin B_{12}). Pellagra can be prevented by adequate protein because the amino acid tryptophan can be converted to niacin in the body. A high intake of folate can mask the blood symptom of a vitamin B_{12} deficiency, but it will not prevent the associated nerve damage. Vitamin B_6 participates in amino acid metabolism and can be harmful in excess. Biotin and pantothenic acid serve important roles in energy metabolism and are common in a variety of foods. Many substances that people claim as B vitamins are not.

THE B VITAMINS—IN CONCERT

This chapter has described some of the impressive ways that vitamins work individually, as if their many actions in the body could easily be disentangled. In fact, oftentimes it is difficult to tell which vitamin is truly responsible for a given effect because the nutrients are interdependent; the presence or absence of one affects another's absorption, metabolism, and excretion. You have already seen this interdependence with folate and vitamin B_{12}.

Riboflavin and vitamin B_6 provide another example. One of the riboflavin coenzymes, FMN, assists the enzyme that converts vitamin B_6 to its coenzyme form PLP. Consequently, a severe riboflavin deficiency can impair vitamin B_6 activity. Thus a deficiency of one nutrient may alter the action of another. Furthermore, a deficiency of one nutrient may create a deficiency of another. For example, both riboflavin and vitamin B_6 (as well as iron) are required for the conversion of tryptophan to niacin. Consequently an inadequate intake of either riboflavin or vitamin B_6 can diminish the body's niacin supply. These interdependent relationships are evident in many of the roles B vitamins play in the body.

B Vitamin Roles

Figure 10-12 is intended to convey an *impression* of the many ways B vitamins busily work in metabolic pathways all over the body. Metabolism is the body's work, and the B vitamin coenzymes are indispensable to every step. In scanning the pathways of metabolism depicted in the figure, note the many abbreviations for the coenzymes that keep the processes going.

Look at the first step in the now-familiar pathway of glucose breakdown. To break down glucose to pyruvate, the cells must have certain enzymes. For the enzymes to work, they must have the niacin coenzyme NAD. To make NAD, the cells must be supplied with niacin (or enough of the amino acid tryptophan to make niacin). They can make the rest of the coenzyme without dietary help.

The next step is the breakdown of pyruvate to acetyl CoA. The enzymes involved in this step require both NAD and the thiamin coenzyme TPP. The cells can manufacture the TPP they need from thiamin, if thiamin is in the diet.

Another coenzyme needed for this step is CoA. Predictably, the cells can make CoA except for an essential part that must be obtained in the diet—pantothenic acid. Another coenzyme requiring biotin serves the enzyme complex involved in converting pyruvate to a compound that can combine with acetyl CoA to start the TCA cycle.

These and other coenzymes participate throughout all the metabolic pathways. When the diet provides riboflavin, the body synthesizes FAD—a needed

FIGURE 10-12 **Metabolic Pathways Involving B Vitamins**

These metabolic pathways were introduced in Chapter 7 and are presented here to highlight the many coenzymes that facilitate the reactions. These coenzymes depend on the following vitamins:

- NAD and NADP: niacin.
- TPP: thiamin.
- CoA: pantothenic acid.
- B_{12}: vitamin B_{12}.
- FMN and FAD: riboflavin.
- THF: folate.
- PLP: vitamin B_6.
- Biotin.

Pathways leading toward acetyl CoA and the TCA cycle are catabolic, and those leading toward amino acids, glycogen, and fat are anabolic. For further details, see Appendix C.

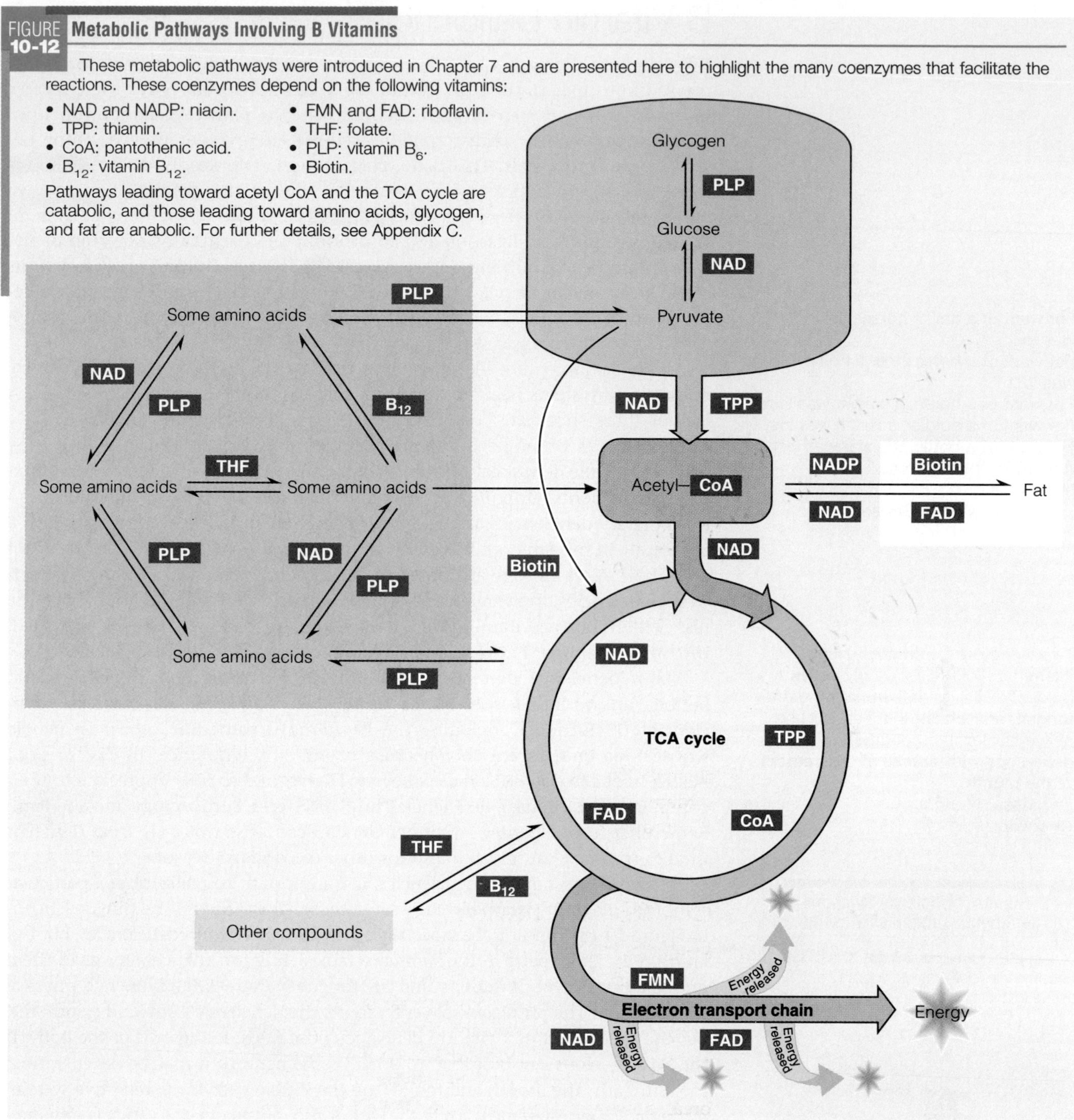

coenzyme in the TCA cycle. Vitamin B_6 is an indispensable part of PLP—a coenzyme required for many amino acid conversions, for a crucial step in the making of the iron-containing portion of hemoglobin for red blood cells, and for many other reactions. Folate becomes THF—the coenzyme required for the synthesis of new genetic material and therefore new cells. The vitamin B_{12} coenzyme, in turn, regenerates THF to its active form; thus vitamin B_{12} is also necessary for the formation of new cells.

Thus each of the B vitamin coenzymes is involved, directly or indirectly, in energy metabolism. Some facilitate the energy-releasing reactions themselves; others help build new cells to deliver the oxygen and nutrients that allow the energy reactions to occur.

B Vitamin Deficiencies

Now suppose the body's cells lack one of these B vitamins—niacin, for example. Without niacin, the cells cannot make NAD. Without NAD, the enzymes involved in every step of the glucose-to-energy pathway cannot function. Then, because all the body's activities require energy, literally everything begins to grind to a halt. This is no exaggeration. The deadly disease pellagra, caused by niacin deficiency, produces the "devastating four *D*s": dermatitis, which reflects a failure of the skin; dementia, a failure of the nervous system; diarrhea, a failure of digestion and absorption; and eventually, as would be the case for any severe nutrient deficiency, death. These symptoms are the obvious ones, but a niacin deficiency affects all other organs, too, because all are dependent on the energy pathways. In short, niacin is like the horseshoe nail◆ for want of which a war was lost.

◆ **For want of a nail, a horseshoe was lost.**
For want of a horseshoe, a horse was lost.
For want of a horse, a soldier was lost.
For want of a soldier, a battle was lost.
For want of a battle, the war was lost,
And all for the want of a horseshoe nail!
—Mother Goose

All the vitamins are like horseshoe nails. With any B vitamin deficiency, many body systems become deranged, and similar symptoms may appear. A lack of "horseshoe nails" can have disastrous and far-reaching effects.

Deficiencies of single B vitamins seldom show up in isolation, however. After all, people do not eat nutrients singly; they eat foods, which contain mixtures of nutrients. Only in two cases described earlier—beriberi and pellagra—have dietary deficiencies associated with single B vitamins been observed on a large scale in human populations. Even in these cases, the deficiencies were not pure. Both diseases were attributed to deficiencies of single vitamins, but both were deficiencies of several vitamins in which one vitamin stood out above the rest. When foods containing the vitamin known to be needed were provided, the other vitamins that were in short supply came as part of the package.

Major deficiency diseases of epidemic proportions such as pellagra and beriberi are no longer seen in the United States and Canada, but lesser deficiencies of nutrients, including the B vitamins, sometimes occur in people whose food choices are poor because of poverty, ignorance, illness, or poor health habits like alcohol abuse. (Review Highlight 7 to fully appreciate how alcohol induces vitamin deficiencies and interferes with energy metabolism.) Remember from Chapter 1 that deficiencies can arise not only from deficient intakes (primary causes), but also for other (secondary) reasons.

In identifying nutrient deficiencies, it is important to realize that a particular symptom may not always have the same cause. The skin and the tongue (shown in Figure 10-13) appear to be especially sensitive to B vitamin deficiencies, but isolating these body parts in the summary tables earlier in this chapter gives them undue emphasis. Both the skin and the tongue◆ are readily visible in a physical examination. The physician sees and reports the deficiency's outward symptoms, but the full impact of a vitamin deficiency occurs inside the cells of the body. If the skin develops a rash or lesions, other tissues beneath it may be degenerating, too. Similarly, the mouth and tongue are the visible part of the digestive system; if they are abnormal, most likely the rest of the GI tract is, too. The accompanying "How to" offers other insights into symptoms and their causes.

◆ **Two symptoms commonly seen in B vitamin deficiencies are glossitis** (gloss-EYE-tis), **an inflammation of the tongue, and cheilosis** (kee-LOH-sis or kye-LOH-sis), **a condition of reddened lips with cracks at the corners of the mouth.**
- **glossa** = tongue
- **cheilos** = lip

FIGURE 10-13 B Vitamin–Deficiency Symptom—The Smooth Tongue of Glossitis

In a B vitamin deficiency, the tongue becomes smooth due to atrophy of the tissue (glossitis).

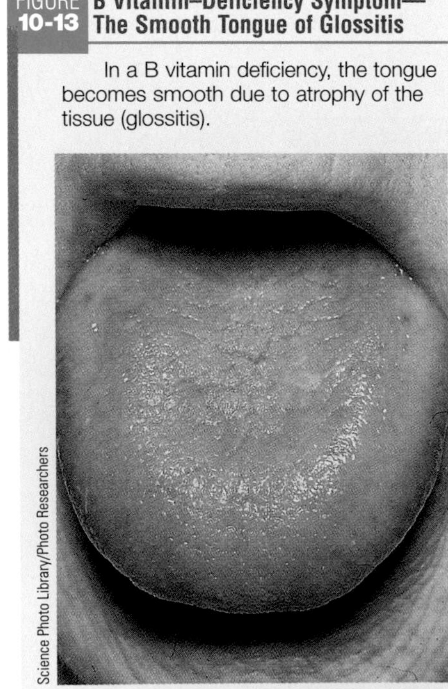

Science Photo Library/Photo Researchers

B Vitamin Toxicities

Toxicities of the B vitamins from foods alone are unknown, but they can occur when people overuse supplements. With supplements, the quantities can quickly overwhelm the cells. Consider that one small capsule can easily deliver 2 milligrams of vitamin B_6, but it would take more than 3000 bananas, 6600 cups of rice, or 3600 chicken breasts to supply an equivalent amount. When the cells become oversaturated with a vitamin, they must work to eliminate the excess. The cells dispatch water-soluble vitamins to the urine for excretion, but sometimes they cannot keep pace with the onslaught. Homeostasis becomes disturbed and symptoms of toxicity develop.

B Vitamin Food Sources

Significantly, the deficiency diseases of beriberi and pellagra were eliminated by supplying foods—not pills. Vitamin pill advertisements make much of the fact that vitamins are indispensable to life, but human beings obtained their nourishment from foods for centuries before vitamin pills existed. If the diet lacks a vitamin, the first solution is to adjust food intake to obtain that vitamin.

Manufacturers of so-called *natural* vitamins boast that their pills are purified from real foods rather than synthesized in a laboratory. Think back on the course of human evolution; it is not *natural* to take any kind of pill. In reality, the finest, most natural vitamin "supplements" available are whole grains, vegetables, fruits, meat, fish, poultry, eggs, legumes, nuts, and milk and milk products.

The food figures presented in this chapter, taken together, sing the praises of a balanced diet. The cereal and bread group delivers thiamin, riboflavin, niacin, and folate. The fruit and vegetable groups excel in folate. The meat group serves thiamin, niacin, vitamin B_6, and vitamin B_{12} well. The milk group stands out for riboflavin and vitamin B_{12}. A diet that offers a variety of foods from each group, prepared with reasonable care, serves up ample B vitamins.

Distinguish Symptoms and Causes

The cause of a symptom is not always apparent. The summary tables in this chapter show that deficiencies of riboflavin, niacin, biotin, and vitamin B_6 can all cause skin rashes. But so can a deficiency of protein, linoleic acid, or vitamin A. Because skin is on the outside and easy to see, it is a useful indicator of things-going-wrong-inside-cells. But by itself, a skin symptom says nothing about its possible cause.

The same is true of anemia. Anemia is often caused by iron deficiency, but it can also be caused by a folate or vitamin B_{12} deficiency; by digestive tract failure to absorb any of these nutrients; or by such nonnutritional causes as infections, parasites, cancer, or loss of blood. No one specific nutrient will always cure a given symptom.

A person who feels chronically tired may be tempted to self-diagnose iron-deficiency anemia and self-prescribe an iron supplement. But this will relieve tiredness only if the cause is indeed iron-deficiency anemia. If the cause is a folate deficiency, taking iron will only prolong the fatigue. A person who is better informed may decide to take a vitamin supplement with iron, covering the possibility of a vitamin deficiency. But the symptom may have a nonnutritional cause. If the cause of the tiredness is actually hidden blood loss due to cancer, the postponement of a diagnosis may be fatal. When fatigue is caused by a lack of sleep, of course, no nutrient or combination of nutrients can replace a good night's rest. A person who is chronically tired should see a physician rather than self-prescribe. If the condition is nutrition related, a registered dietitian should be consulted as well.

IN SUMMARY The B vitamin coenzymes work together in energy metabolism. Some facilitate the energy-releasing reactions themselves; others help build cells to deliver the oxygen and nutrients that permit the energy pathways to run. These vitamins depend on each other to function optimally; a deficiency of any of them creates multiple problems. Fortunately, a variety of foods from each of the food groups will provide an adequate supply of all of the B vitamins.

VITAMIN C

Two hundred and fifty years ago, any man who joined the crew of a seagoing ship knew he had only half a chance of returning alive—not because he might be slain by pirates or die in a storm, but because he might contract the dread disease **scurvy.** As many as two-thirds of a ship's crew might die of scurvy on a long voyage. Only men on short voyages, especially around the Mediterranean Sea, were free of scurvy. No one knew the reason: that on long ocean voyages, the ship's cook used up the fresh fruits and vegetables early and then served cereals and meats until the return to port.

The first nutrition experiment ever performed on human beings was devised in the mid-1700s to find a cure for scurvy. James Lind, a British physician, divided 12 sailors with scurvy into six pairs. Each pair received a different supplemental ration: cider, vinegar, sulfuric acid, seawater, oranges and lemons, or a strong laxative mixed with spices. Those receiving the citrus fruits quickly recovered, but sadly, it was 50 years before the British navy required all vessels to provide every sailor with lime juice daily.

◆ The tradition of providing British sailors with citrus juice daily to prevent scurvy gave them the nickname "limeys."◆

scurvy: the vitamin C–deficiency disease.

The antiscurvy "something" in limes and other foods was dubbed the **antiscorbutic** factor. Nearly 200 years later, the factor was isolated and found to be a six-carbon compound similar to glucose; it was named **ascorbic acid.** Shortly thereafter, it was synthesized, and today hundreds of millions of vitamin C pills are produced in pharmaceutical laboratories each year.

Vitamin C Roles

Vitamin C parts company with the B vitamins in its mode of action. In some settings, vitamin C serves as a **cofactor** helping a specific enzyme perform its job, but in others, it acts as an **antioxidant** participating in more general ways.

• As an Antioxidant • Vitamin C loses electrons easily, which allows it to perform as an antioxidant. In the body, antioxidants defend against free radicals. Free radicals are discussed in Highlight 11, but for now, a simple definition will suffice. A free radical is a molecule with one or more unpaired electrons, which makes it unstable and highly reactive. By donating an electron or two, antioxidants neutralize free radicals and protect other substances from their damage. Figure 10-14 illustrates how vitamin C can give up electrons to stop free-radical damage and then receive them again to become reactivated. This recycling of vitamin C is key to limiting losses and maintaining a reserve of antioxidants in the body.[27]

Vitamin C is like a bodyguard for water-soluble substances; it stands ready to sacrifice its own life to save theirs. In the cells and body fluids, vitamin C protects tissues from **oxidative stress** and thus may play an important role in preventing diseases. In the intestines, vitamin C enhances iron absorption◆ by protecting iron from oxidation.

◆ **Chapter 13 provides more details on the relationship between vitamin C and iron.**

• As a Cofactor in Collagen Formation • Vitamin C helps to form the fibrous structural protein of connective tissues known as collagen. Collagen serves as the matrix on which bones and teeth are formed. When a person is wounded, collagen glues the separated tissues together, forming scars. Cells are held together largely by collagen; this is especially important in the artery walls, which must expand and contract with each beat of the heart, and in the thin capillary walls, which must withstand a pulse of blood every second or so without giving way.

Chapter 6 described how the body makes proteins by stringing together chains of amino acids. During the synthesis of collagen, each time a proline or lysine is added to the growing protein chain, an enzyme hydroxylates it (adds an OH group

antiscorbutic (AN-tee-skor-BUE-tik) **factor:** the original name for vitamin C.
- **anti** = against
- **scorbutic** = causing scurvy

ascorbic acid: one of the two active forms of vitamin C (see Figure 10-14). Many people refer to vitamin C by this name.
- **a** = without
- **scorbic** = having scurvy

cofactor: a small inorganic or organic substance that works with an enzyme to facilitate a chemical reaction.

antioxidant: a substance in foods that significantly decreases the adverse effects of free radicals on normal physiological functions in the human body.

oxidative stress: an imbalance between the production of free radicals and the body's ability to handle them and prevent damage.

FIGURE 10-14 Active Forms of Vitamin C

The two hydrogens highlighted in yellow give vitamin C its acidity and its ability to act as an antioxidant.

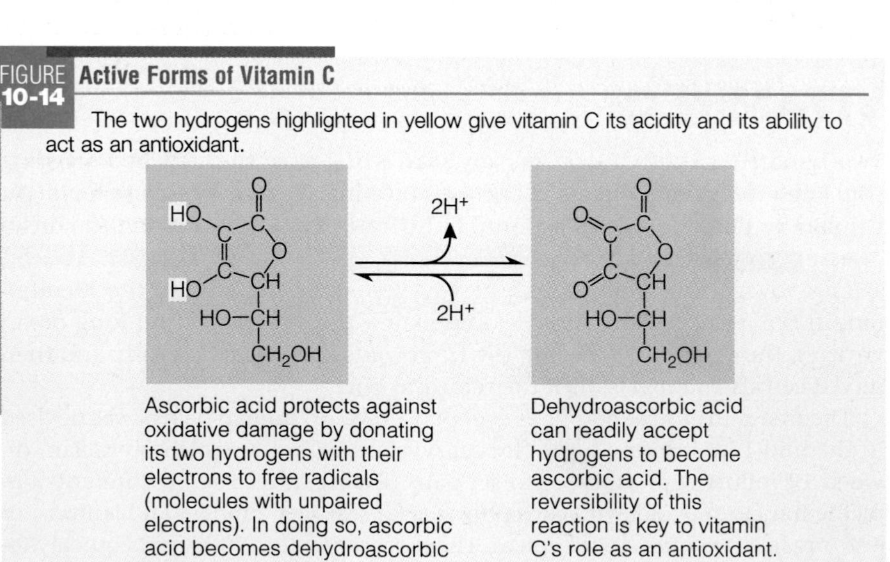

Ascorbic acid protects against oxidative damage by donating its two hydrogens with their electrons to free radicals (molecules with unpaired electrons). In doing so, ascorbic acid becomes dehydroascorbic acid.

Dehydroascorbic acid can readily accept hydrogens to become ascorbic acid. The reversibility of this reaction is key to vitamin C's role as an antioxidant.

to it), making the amino acid hydroxyproline or hydroxylysine, respectively. These two special amino acids facilitate the binding together of collagen fibers to make strong, ropelike structures. The conversion of proline to hydroxyproline requires both vitamin C and iron. Iron works as a cofactor in the reaction, and vitamin C protects iron from oxidation, thereby allowing iron to perform its duty. Without vitamin C and iron, the hydroxylation step does not occur.

• **As a Cofactor in Other Reactions** • Vitamin C also serves as a cofactor in the synthesis of several other compounds. As in collagen formation, vitamin C helps in the hydroxylation of carnitine, a compound that transports long-chain fatty acids into the mitochondria of a cell for energy metabolism. It participates in the conversions of the amino acid tryptophan to the neurotransmitters serotonin and norepinephrine. Vitamin C also assists in the making of hormones, including thyroxin, which regulates the metabolic rate; metabolism speeds up under times of extreme physical stress.

• **In Stress** • The adrenal glands contain more vitamin C than any other organ in the body, and during stress, these glands release the vitamin, together with hormones, into the blood. The vitamin's exact role in the stress reaction remains unclear, but physical stresses raise vitamin C needs. Among the stresses known to increase vitamin C needs are infections; burns; extremely high or low temperatures; intakes of toxic heavy metals such as lead, mercury, and cadmium; the chronic use of certain medications, including aspirin, barbiturates, and oral contraceptives; and cigarette smoking. When immune system cells are called into action, they use a lot of oxygen and produce free radicals. In this case, free radicals are helpful. They act as ammunition in an "oxidative burst" that demolishes the offending viruses and bacteria and destroys the damaged cells. Vitamin C steps in as an antioxidant to control this oxidative activity.

• **As a Cure for the Common Cold** • Newspaper headlines touting vitamin C as a cure for colds have appeared frequently over the years, but research supporting such claims has been conflicting and controversial. A major review of the research on vitamin C in the treatment and prevention of the common cold revealed a significant difference in duration of less than a day per cold in favor of those taking a daily dose of at least 1 gram of vitamin C.[28] The term *significant* means that *statistical* analysis suggests that the findings probably didn't arise by chance, but from the experimental treatment being tested. Is a day enough savings to warrant routine daily supplementation? Supplement users seem to think so.

Interestingly, those who received the placebo *but thought they were receiving vitamin C* had fewer colds than the group who received vitamin C *but thought they were receiving the placebo.* (Never underestimate the healing power of faith!)

Discoveries of the ways vitamin C works in the body provide possible links between the vitamin and the common cold. Anyone who has ever had a cold knows the discomfort of a runny or stuffed-up nose. Nasal congestion develops in response to elevated blood **histamine,** and people commonly take antihistamines for relief. Like an antihistamine, vitamin C comes to the rescue and deactivates histamine.

• **In Disease Prevention** • Whether vitamin C may help in preventing or treating cancer, heart disease, cataract, and other diseases is still being studied, and findings are presented in Highlight 11. Conducting research in the United States and Canada can be difficult, however, because diets typically contribute enough vitamin C to provide optimal health benefits.

Vitamin C Recommendations

How much vitamin C does a person need? As Figure 10-15 illustrates, recommendations are set generously above the minimum requirement to prevent

histamine (HISS-tah-mean or HISS-tah-men): a substance produced by cells of the immune system as part of a local immune reaction to an antigen; participates in causing inflammation.

FIGURE 10-15 **Vitamin C Intake (mg/day)**

Recommendations are generously above the minimum requirement and below the toxicity level.

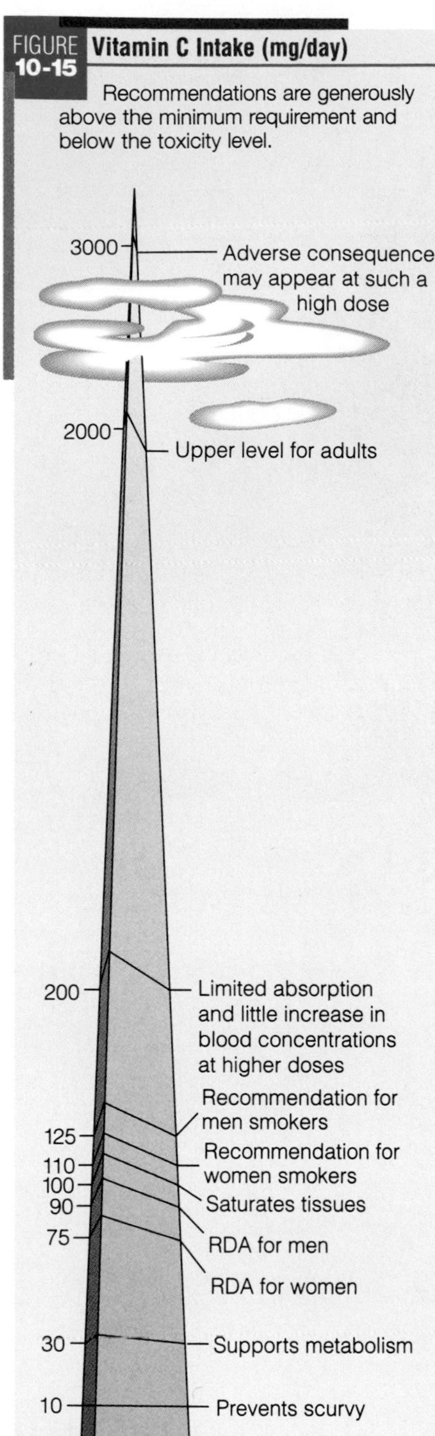

scurvy and well below the toxicity level.[29] Current recommendations are higher than the previous RDA, but not as high as some experts had proposed.[30]

The requirement—the amount needed to prevent the overt symptoms of scurvy—is only 10 milligrams daily. However, 10 milligrams a day does not saturate all the body tissues; higher intakes will increase the body's total vitamin C. At about 100 milligrams per day, 95 percent of the population probably reaches tissue saturation. At about 200 milligrams, absorption reaches a maximum, and there is little, if any, increase in blood concentrations at higher doses. Excess vitamin C is readily excreted.

As mentioned earlier, cigarette smoking increases the need for vitamin C. Cigarette smoke contains oxidants, which greedily deplete this potent antioxidant.[31] Exposure to cigarette smoke, especially when accompanied by low intakes of vitamin C, depletes the body's pool in both active and passive smokers; similarly, people who chew tobacco have low levels of vitamin C as well.[32] Because people who smoke cigarettes regularly suffer significant oxidative stress, their requirement for vitamin C is increased an additional 35 milligrams; nonsmokers regularly exposed to cigarette smoke should be sure to meet their RDA for vitamin C.

After oral surgery, dentists may prescribe supplemental vitamin C to hasten healing. After major operations or extensive burns, when scar tissue is forming, a physician may prescribe 1000 milligrams (1 gram) a day or even more. Self-medication is not recommended.

Vitamin C Deficiency

Two of the most notable signs of a vitamin C deficiency reflect its role in maintaining the integrity of blood vessels. The gums bleed easily around the teeth, and capillaries under the skin break spontaneously, producing pinpoint hemorrhages (see Figure 10-16).

When the vitamin C pool falls to about a fifth of its optimal size (this may take more than a month on a diet lacking vitamin C), scurvy symptoms begin to appear. Inadequate collagen synthesis causes further hemorrhaging. Muscles, including the heart muscle, degenerate. The skin becomes rough, brown, scaly, and dry. Wounds fail to heal because scar tissue will not form. Bone rebuilding falters; the ends of the long bones become softened, malformed, and painful, and fractures develop. The teeth become loose as the cartilage around them weakens. Anemia and infections are common. There are also characteristic psychological signs, including hysteria and depression. Sudden death is likely, caused by massive internal bleeding.

FIGURE 10-16 **Vitamin C–Deficiency Symptoms—Scorbutic Gums and Pinpoint Hemorrhages**

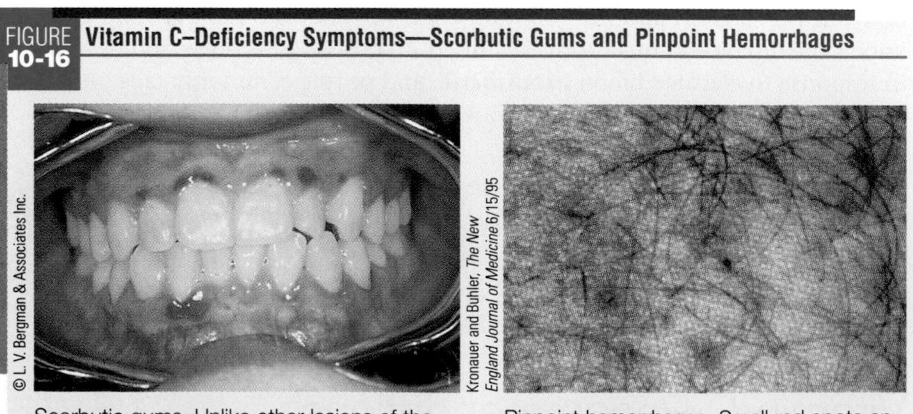

Scorbutic gums. Unlike other lesions of the mouth, scurvy presents a symmetrical appearance without infection.

Pinpoint hemorrhages. Small red spots appear in the skin, indicating spontaneous bleeding internally.

Once diagnosed, scurvy is readily reversed by vitamin C. Moderate doses in the neighborhood of 100 milligrams per day are sufficient, curing the scurvy within about five days. Such an intake is easily achieved by including vitamin C–rich foods in the diet.

Vitamin C Toxicity

The easy availability of vitamin C supplements and the publication of books recommending vitamin C to prevent colds and cancer have led thousands of people to take large doses of vitamin C. Not surprisingly, toxic effects such as nausea, abdominal cramps, and diarrhea are often reported.

Several instances of interference with medical regimens are also known. Large amounts of vitamin C excreted in the urine obscure the results of tests used to detect diabetes, giving a **false positive** result in some instances and a **false negative** in others. People taking anticlotting medications may unwittingly counteract the effect if they also take massive doses of vitamin C.* Those with kidney disease, a tendency toward gout,◆ or a genetic abnormality that alters vitamin C's breakdown to its excretion products are prone to forming kidney stones if they take large doses of vitamin C.† Vitamin C supplements are particularly dangerous for people with iron overload.◆ Vitamin C enhances iron absorption and releases iron from body stores; free iron causes the kind of cellular damage typical of free radicals. These events illustrate how vitamin C can act as a *pro*oxidant when quantities exceed the body's needs.

◆ *Gout* is a metabolic disease in which uric acid crystals precipitate in the joints.

◆ Chapter 13 describes the damaging effects of too much iron.

A person who takes large doses of vitamin C for a long time (say, ten times the RDA daily for several weeks) may adapt by accelerating the metabolism and excretion of the vitamin. If the person then abruptly discontinues supplementation or reduces intake to normal, the clearance system may not be able to slow down fast enough to prevent a deficiency. Evidence supporting this scenario is scanty, but a person discontinuing vitamin C supplementation might be wise to do so gradually.

Few instances warrant consuming more than 200 milligrams of vitamin C a day. For adults who dose themselves with up to 2 grams a day, the risks may not be great; those taking more should be aware of the distinct possibility of adverse effects.

Vitamin C Food Sources

Fruits and vegetables can easily provide a generous amount of vitamin C. A cup of orange juice at breakfast, a salad for lunch, and a stalk of broccoli and a potato for dinner alone provide more than 300 milligrams. Clearly, a person making such food choices needs no vitamin C pills.

Figure 10-17 (on p. 340) shows the amounts of vitamin C in various common foods. The overwhelming abundance of purple and green bars reveals not only that the citrus fruits are justly famous for being rich in vitamin C, but that other fruits and vegetables are in the same league. A single serving of broccoli, bell pepper, or strawberries provides more than 50 milligrams of the vitamin (and an array of other nutrients). Because vitamin C is vulnerable to heat, raw fruits and vegetables usually have a higher nutrient density than their cooked counterparts.

The potato is an important source of vitamin C, not because one potato by itself meets the daily need, but because potatoes are such a common staple that

false positive: a test result indicating that a condition is present (positive) when in fact it is not (therefore false).

false negative: a test result indicating that a condition is not present (negative) when in fact it is present (therefore false).

*Vitamin C interferes with such anticoagulant drugs as warfarin, dicumarol, heparin, and coumadin. It is unclear whether vitamin C inhibits the absorption or the action of these drugs.

†Vitamin C is inactivated and degraded by several routes, and sometimes oxalate, which can form kidney stones, is produced along the way. People may also develop oxalate crystals in their kidneys regardless of vitamin C status.

FIGURE 10-17 **Vitamin C in Selected Foods**

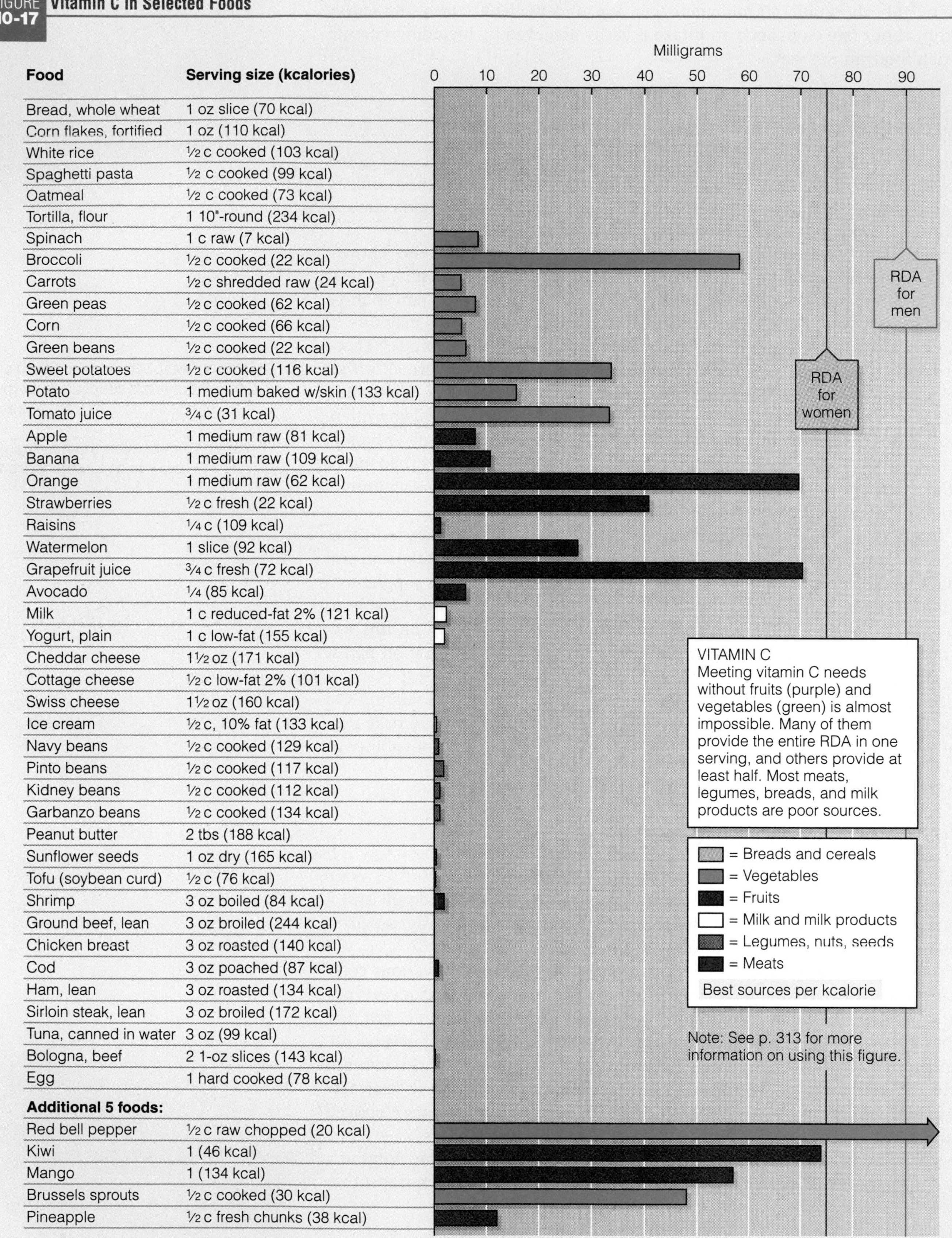

they make significant contributions. In fact, scurvy was unknown in Ireland until the potato blight of the mid-1840s when some two million people died of malnutrition and infection.

The lack of yellow, white, brown, and red bars in Figure 10-17 confirms that grains, milk (except breast milk), legumes, and meats are notoriously poor sources of vitamin C. Organ meats (liver, kidneys, and others) and raw meats contain some vitamin C, but most people don't eat large quantities of these. Raw meats and fish contribute enough vitamin C to be significant in parts of Alaska, Canada, and Japan, but elsewhere fruits and vegetables are necessary to supply sufficient vitamin C.

Because of vitamin C's antioxidant property, food manufacturers sometimes add a variation of vitamin C to some beverages and most cured meats, such as luncheon meats, to prevent oxidation and spoilage. This compound safely preserves these foods, but it does not have vitamin C activity in the body.[33] Simply put, "ham and bacon cannot replace fruits and vegetables."[34] See the accompanying table for a summary of vitamin C.

© PhotoDisc Inc.

When dietitians say "vitamin C," people think "oranges."

© Polara Studios Inc.

But these foods are also rich in vitamin C.

IN SUMMARY

Vitamin C

Other Names

Ascorbic acid

2000 RDA

Men: 90 mg/day

Women: 75 mg/day

Smokers: +35 mg/day

Upper Level

Adults: 2000 mg/day

Chief Functions in the Body

Collagen synthesis (strengthens blood vessel walls, forms scar tissue, provides matrix for bone growth), antioxidant, thyroxin synthesis, amino acid metabolism, strengthens resistance to infection, helps in absorption of iron

Significant Sources

Citrus fruits, cabbage-type vegetables, dark green vegetables (such as bell peppers and broccoli), cantaloupe, strawberries, lettuce, tomatoes, potatoes, papayas, mangoes

Easily destroyed by heat and oxygen

Deficiency Disease

Scurvy

Deficiency Symptoms	Toxicity Symptoms
Blood Symptoms	
Anemia (small-cell type),[a] atherosclerotic plaques, pinpoint hemorrhages	
Bone Symptoms	
Bone fragility, joint pain	
Digestive Symptoms	
	Nausea, abdominal cramps, diarrhea
Immune System Symptoms	
Poor wound healing, frequent infections	
Mouth Symptoms	
Bleeding gums, loosened teeth	
Neurological Symptoms	
Muscle degeneration and pain, hysteria, depression	Headache, fatigue, insomnia
Skin Symptoms	
Rough skin, blotchy bruises	Hot flashes, rashes
Other	
	Interference with medical tests, aggravation of gout symptoms, urinary tract problems, kidney stones[b]

[a]Small-cell–type anemia is *microcytic anemia.*

[b]People with kidney disease, a tendency toward gout, or a genetic abnormality that alters the breakdown of vitamin C are prone to forming kidney stones. Vitamin C is inactivated and degraded by several routes, sometimes producing oxalate, which can form stones in the kidneys.

Vita means life. After this discourse on the vitamins, who could dispute that they deserve their name? Their regulation of metabolic processes makes them vital to the normal growth, development, and maintenance of the body. The accompanying summary table condenses the information provided in this chapter for a quick review. The remarkable roles of the vitamins continue in the next chapter.

IN SUMMARY

The Water-Soluble Vitamins

Vitamin and Chief Functions	Deficiency Symptoms	Toxicity Symptoms	Food Sources
Thiamin Part of coenzyme TPP in energy metabolism	Beriberi (edema or muscle wasting), anorexia and weight loss, neurological disturbances, muscular weakness, heart enlargement and failure	None reported	Enriched, fortified, or whole-grain products; pork
Riboflavin Part of coenzymes FAD and FMN in energy metabolism	Inflammation of the mouth, skin, and eyelids; sensitivity to light	None reported	Milk products; enriched, fortified, or whole-grain products; liver
Niacin Part of coenzymes NAD and NADP in energy metabolism	Pellagra (diarrhea, dermatitis, and dementia)	Niacin flush, liver damage, impaired glucose tolerance	Protein-rich foods
Biotin Part of coenzyme in energy metabolism	Skin rash, hair loss, neurological disturbances	None reported	Widespread in foods; GI bacteria synthesis
Pantothenic acid Part of coenzyme A in energy metabolism	Digestive and neurological disturbances	None reported	Widespread in foods
Vitamin B_6 Part of coenzymes used in amino acid and fatty acid metabolism	Depression, confusion, convulsions, anemia, scaly dermatitis	Nerve degeneration, skin lesions	Protein-rich foods
Folate Activates vitamin B_{12}; helps synthesize DNA for new cell growth	Anemia, glossitis, neurological disturbances, elevated homocysteine	Masks vitamin B_{12} deficiency	Legumes, vegetables, fortified grain products
Vitamin B_{12} Activates folate; helps synthesize DNA for new cell growth; protects nerve cells	Anemia; nerve damage and paralysis	None reported	Foods derived from animals
Vitamin C Synthesis of collagen, carnitine, hormones, neurotransmitters; antioxidant	Scurvy (bleeding gums, pinpoint hemorrhages, abnormal bone growth, and joint pain)	Diarrhea, GI distress	Fruits and vegetables

How are you doing?

- Do you often choose whole or enriched grains, dark green leafy vegetables, citrus fruits, and legumes?
- If you are a woman of childbearing age, do you eat folate-rich foods or take supplements regularly?
- Do you take supplements that provide more than the upper limit of the vitamins?

Nutrition on the Net

WEBSITES

Access these websites for further study of topics covered in this chapter. Be aware that many websites on the Internet are peddling vitamin supplements, not accurate information.

- Find updates and quick links to these and other nutrition-related sites at our website: **www.wadsworth.com/nutrition**
- Search for "vitamins" at the American Dietetic Association: **www.eatright.org**
- Review the Dietary Reference Intakes for the water-soluble vitamins: **www.nap.edu/readingroom**
- Visit the World Health Organization to learn about "vitamin deficiencies" around the world: **www.who.int**
- Search for "vitamins" at the U.S. Government health information site: **www.healthfinder.gov**
- Learn more about neural tube defects from the Spina Bifida Association of America: **www.sbaa.org**
- Read about Dr. Joseph Goldberger and his groundbreaking discovery linking pellagra to diet by searching for his name at: **www.nih.gov** or **www.pbs.org**
- Learn how fruits and vegetables support a healthy diet rich in vitamins from the 5 A Day for Better Health program: **www.5aday.com**

INTERNET ACTIVITIES

Consumers encounter an abundance of information about dietary supplements. Some of this information is reliable, but much of it is designed only to increase sales. How do you know if taking a dietary supplement is a healthy choice?

- ▼ Go to: **www.eatright.org**
- ▼ Click on "Healthy Lifestyle" at the top of the screen.
- ▼ Click on "Nutrition Fact Sheets" on the drop-down menu that appears.
- ▼ Scroll down and click on "Do you need a multivitamin/mineral supplement?"
- ▼ Take the quiz and total your score.

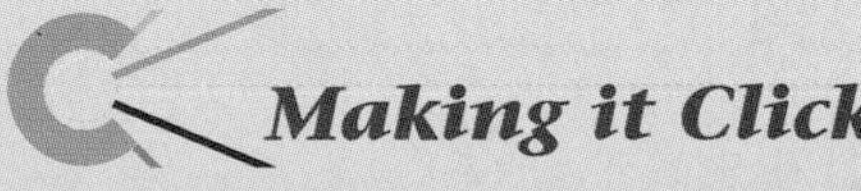

Making it Click

These problems give you practice in doing simple vitamin-related calculations (answers are provided on p. 345). Be sure to show your calculations for each problem.

1. Review the units in which vitamins are measured (a spot check).
   a. For each of these vitamins, note the unit of measure:

Thiamin	Folate
Riboflavin	Vitamin B_{12}
Niacin	Vitamin C
Vitamin B_6	

   b. Recall from the chapter's description of people's self-dosing with vitamin B_6 that people who suffer toxicity symptoms may be taking more than 2 grams a day, whereas the RDA is less than 2 *milli*grams. How much higher than 2 milligrams is 2 grams?

   c. Vitamin B_{12} is measured in micrograms. How many micrograms are in a gram? How many grams are in a teaspoon of a granular powder? How many micrograms does that represent? What is your RDA for vitamin B_{12}?

   This exercise should convince you that the amount of vitamins a person needs is indeed quite small—yet still essential.

2. Be aware of how niacin intakes are affected by dietary protein availability.
   a. Refer to the "How to" on p. 317, and calculate how much niacin a woman receives from a diet that delivers 90 grams protein and 9 milligrams niacin. (Assume her RDA for protein is 46 grams/day.)
   b. Is this woman getting her RDA of niacin (14 milligrams NE)?

   This exercise should demonstrate that protein helps meet niacin needs.

Study Questions

These questions will help you review the chapter. You will find the answers in the discussions on the pages provided.

1. How do the vitamins differ from the energy nutrients? (p. 307)
2. Describe some general differences between fat-soluble and water-soluble vitamins. (p. 308)
3. Which B vitamins are involved in energy metabolism? Protein metabolism? Cell division? (p. 310)
4. For thiamin, riboflavin, niacin, biotin, pantothenic acid, vitamin B_6, folate, vitamin B_{12}, and vitamin C, state:
   - Its chief function in the body.
   - Its characteristic deficiency symptoms.
   - Its significant food sources. (See respective summary tables)
5. What is the relationship of tryptophan to niacin? (pp. 316–317)
6. Describe the relationship between folate and vitamin B_{12}. (pp. 322, 324, 328, 329)
7. What risks are associated with high doses of niacin? Vitamin B_6? Vitamin C? (pp. 317, 321–322, 339)

These questions will help you prepare for an exam. Answers can be found on p. 345.

1. Vitamins:
   a. are inorganic compounds.
   b. yield energy when broken down.
   c. are soluble in either water or fat.
   d. perform best when linked in long chains.
2. The rate at and the extent to which a vitamin is absorbed and used in the body is known as its:
   a. bioavailability.
   b. intrinsic factor.
   c. physiological effect.
   d. pharmacological effect.
3. Many of the B vitamins serve as:
   a. coenzymes.
   b. antagonists.
   c. antioxidants.
   d. serotonin precursors.
4. With respect to thiamin, which of the following is the most nutrient dense?
   a. 1 slice whole-wheat bread (69 kcalories and 0.1 milligram thiamin)
   b. 1 cup yogurt (144 kcalories and 0.1 milligram thiamin)
   c. 1 cup snow peas (69 kcalories and 0.22 milligram thiamin)
   d. 1 chicken breast (141 kcalories and 0.06 milligram thiamin)
5. The body can make niacin from:
   a. tyrosine.
   b. serotonin.
   c. carnitine.
   d. tryptophan.
6. The vitamin that protects against neural tube defects is:
   a. niacin.
   b. folate.
   c. riboflavin.
   d. vitamin B_{12}.
7. A lack of intrinsic factor may lead to:
   a. beriberi.
   b. pellagra.
   c. pernicious anemia.
   d. atrophic gastritis.
8. Which of the following is a B vitamin?
   a. inositol
   b. carnitine
   c. vitamin B_{15}
   d. pantothenic acid
9. Vitamin C serves as a(n):
   a. coenzyme.
   b. antagonist.
   c. antioxidant.
   d. intrinsic factor.
10. The requirement for vitamin C is highest for:
    a. smokers.
    b. athletes.
    c. alcoholics.
    d. the elderly.

References

1 M. J. Jackson, The assessment of bioavailability of micronutrients: Introduction, *European Journal of Clinical Nutrition* 51 (1997): S1–S2.

2 Committee on Dietary Reference Intakes, *Dietary Reference Intakes for Vitamin C, Vitamin E, Selenium, and Carotenoids* (Washington, D.C.: National Academy Press, 2000); Committee on Dietary Reference Intakes, *Dietary Reference Intakes for Thiamin, Riboflavin, Niacin, Vitamin B_6, Folate, Vi-*

tamin B_{12}, Pantothenic Acid, Biotin, and Choline (Washington, D.C.: National Academy Press, 1998).

3 C. K. Austin, C. E. Goodman, and L. L. Van Halderen, Absence of malnutrition in a population of homeless veterans, *Journal of the American Dietetic Association* 96 (1996): 1283–1285.

4 Committee on Dietary Reference Intakes, 1998, p. 113.

5 J. E. Leklem, Vitamin B-6, in *Present Knowledge in Nutrition*, 7th ed., E. E. Ziegler and L. J. Filer (Washington, D.C.: International Life Sciences Institute Press, 1996), pp. 174–183; K. M. Riggs and coauthors, Relations of vitamin B-12, vitamin B-6, folate, and homocysteine to cognitive performance in the Normative Aging Study, *American Journal of Clinical Nutrition* 63 (1996): 306–314.

6 A. Franzblau and coauthors, The relationship of vitamin B_6 status to median nerve function and carpal tunnel syndrome among active industrial workers, *Journal of Occupational and Environmental Medicine* 38 (1996): 485–491; A. L. Bernstein and J. S. Dineson, Effect of pharmacologic doses of vitamin B_6 on carpal tunnel syndrome, electroencephalographic results, and pain, *Journal of the American College of Nutrition* 12 (1993): 73–76.

7 Committee on Genetics, Folic acid for the prevention of neural tube defects, *Pediatrics* 104 (1999): 325–327; L. D. Botto and coauthors, Neural-tube defects, *New England Journal of Medicine* 341 (1999): 1509–1519; C. E. Butterworth, Jr., and A. Bendich, Folic acid and the prevention of birth defects, *Annual Review of Nutrition* 16 (1996): 73–97.

8 R. J. Berry and coauthors, Prevention of neural-tube defects with folic acid in China, *New England Journal of Medicine* 341 (1999): 1485–1490; Committee on Genetics, 1999.

9 Knowledge and use of folic acid by women of childbearing age—United States, 1995 and 1998, *Morbidity and Mortality Weekly Report* 48 (1999): 327–328; C. J. Lewis and coauthors, Estimated folate intakes: Data updated to reflect food fortification, increased bioavailability, and dietary supplement use, *American Journal of Clinical Nutrition* 70 (1999): 198–207; From the Centers for Disease Control and Prevention, Recommendations for use of folic acid to reduce number of spina bifida cases and other neural tube defects, *Journal of the American Medical Association* 269 (1993): 1233–1238.

10 G. J. Cuskelly, H. McNulty, and J. M. Scott, Effect of increasing dietary folate on red-cell folate: Implications for prevention of neural tube defects, *Lancet* 347 (1996): 657–659.

11 *Preventing Neural Tube Birth Defects: A Prevention Model and Resource Guide* (Centers for Disease Control and Prevention, 1998).

12 G. J. Cuskelly, H. McNulty, and J. M. Scott, Fortification with low amounts of folic acid makes a significant difference in folate status in young women: Implications for the prevention of neural tube defects, *American Journal of Clinical Nutrition* 70 (1999): 234–239; P. F. Jacques and coauthors, The effect of folic acid fortification on plasma folate and total homocysteine concentrations, *New England Journal of Medicine* 340 (1999): 1449–1454.

13 Committee on Dietary Reference Intakes, 1998.

14 S. P. Rothenberg, Increasing the dietary intake of folate: Pros and cons, *Seminars in Hematology* 36 (1999): 65–74; V. Herbert and J. Bigaouette, Call for endorsement of a petition to the Food and Drug Administration to always add vitamin B-12 to any folate fortification or supplement, *American Journal of Clinical Nutrition* 65 (1997): 572–573.

15 S. J. James and coauthors, Abnormal folate metabolism and mutation in the methylenetetrahydrofolate reductase gene may be maternal risk factors for Down syndrome, *American Journal of Clinical Nutrition* 70 (1999): 495–501; D. S. Rosenblatt, Folate and homocysteine metabolism and gene polymorphisms in the etiology of Down syndrome, *American Journal of Clinical*

ANSWERS

Making it Click

1. a. Thiamin: mg.
   Riboflavin: mg.
   Niacin: mg NE.
   Vitamin B_6: mg.
   Folate: µg DFE.
   Viatmin B_{12}: µg.
   Vitamin C: mg.

   b. A thousand times higher (2 g × 1000 mg/g = 2000 mg; 2000 mg ÷ 2 mg = 1000).

   c. 1 g = 1000 mg; 1 mg = 1000 µg (1000 × 1000 = 1,000,000); 1 million µg = 1 g.
   1 tsp = 5 g.
   5 × 1,000,000 µg = 5,000,000 µg/tsp.
   See inside front cover for your RDA based on age and gender.

2. a. She eats 90 g protein. Assume she uses 46 g as protein. This leaves 90 g − 46 g = 44 g protein "leftover."
   44 g protein ÷ 100 = 0.44 g tryptophan.
   0.44 g tryptophan × 1000 = 440 mg tryptophan.
   440 mg tryptophan ÷ 60 = 7.3 mg NE.
   7.3 mg NE + 9 mg niacin = 16.3 mg NE.

   b. Yes.

Multiple Choice

1. c	2. a	3. a	4. c	5. d
6. b	7. c	8. d	9. c	10. a

Nutrition 70 (1999): 429–430; A. M. Molloy and coauthors, Thermolabile variant of 5,10methylenetetrahydrofolate reductase associated with low red-cell folates: Implications for folate intake recommendations, *Lancet* 349 (1997): 1591–1593.

16 J. B. Ubbink, P. J. Becker, and W. J. H. Vermaak, Will an increased dietary folate intake reduce the incidence of cardiovascular disease? *Nutrition Reviews* 54 (1996): 213–216.

17 M. L. Bots and coauthors, Homocysteine and short-term risk of myocardial infarction and stroke in the elderly: The Rotterdam Study, *Archives of Internal Medicine* 159 (1999): 38–44; K. Robinson and coauthors, Low circulating folate and vitamin B6 concentrations: Risk factors for stroke, peripheral vascular disease, and coronary artery disease, *Circulation* 97 (1998): 437–443; O. Nygard and coauthors, Plasma homocysteine levels and mortality in patients with coronary artery disease, *New England Journal of Medicine* 337 (1997): 230–236; H. I. Morrison and coauthors, Serum folate and risk of fatal coronary heart disease, *Journal of the American Medical Association* 275 (1996): 1893–1896; J. S. Stamler and A. Slivka, Biological chemistry of thiols in the vasculature and in vascular-related disease, *Nutrition Reviews* 54 (1996): 1–30; J. Selhub and coauthors, Association between plasma homocysteine concentrations and extracranial carotid-artery stenosis, *New England Journal of Medicine* 332 (1995): 286–291; M. J. Stampfer and M. R. Malinow, Can lowering homocysteine levels reduce cardiovascular risk? *New England Journal of Medicine* 332 (1995): 328–329; J. B. Ubbink, Homocysteine—An atherogenic and a thrombogenic factor? *Nutrition Reviews* 53 (1995): 323–332.

18 Jacques and coauthors, 1999; I. A. Brouwer and coauthors, Low-dose folic acid supplementation decreases plasma homocysteine concentrations: A randomized trial, *American Journal of Clinical Nutrition* 69 (1999): 99–104; M. R. Malinow and coauthors, Reduction of plasma homocyst(e)ine levels by breakfast cereal fortified with folic acid in patients with coronary heart disease, *New England Journal of Medicine* 338 (1998): 1009–1015; R. Clarke, Lowering blood homocysteine with folic acid based supplements: Meta-analysis of randomised trials, *British Medical Journal* 316 (1998): 894–898; J. V. Woodside and coauthors, Effect of B-group vitamins and antioxidant vitamins on hyperhomocysteinemia: A double-blind, randomized, factorial-design, controlled trial, *American Journal of Clinical Nutrition* 67 (1998): 858–866.

19 Y. Kim, Folate and cancer prevention: A new medical application of folate beyond hyperhomocysteinemia and neural tube defects, *Nutrition Reviews* 57 (1999): 314–321.

20 Kim, 1999; S. Zhang and coauthors, A prospective study of folate intake and the risk of breast cancer, *Journal of the American Medical Association* 281 (1999): 1632–1637.

21 C. M. Pfeiffer and coauthors, Absorption of folate from fortified cereal-grain products and of supplemental folate consumed with or without food determined by using a dual-label stable-isotope protocol, *American Journal of Clinical Nutrition* 66 (1997): 1388–1397.

22 B. H. Toh, I. R. van Driel, and P. A. Gleeson, Pernicious anemia, *New England Journal of Medicine* 337 (1997): 1441–1448.

23 D. Z. Asselt and coauthors, Nasal absorption of hydrocobalamin in healthy elderly adults, *British Journal of Clinical Pharmacology* 45 (1998): 83–86.

24 B. D. Hokin and T. Butler, Cyanocobalamin (vitamin B-12) status in Seventh-day Adventist ministers in Australia, *American Journal of Clinical Nutrition* 70 (1999): 576S–578S.

25 F. Watanabe and coauthors, Effects of microwave heating on the loss of vitamin B_{12} in foods, *Journal of Agricultural and Food Chemistry* 46 (1998): 206–210.

26 E. P. Shronts, Essential nature of choline with implications for total parenteral nutrition, *Journal of the American Dietetic Association* 97 (1997): 646–649; D. J. Canty and S. H. Zeisel, Lecithin and choline in human health and disease, *Nutrition Reviews* 52 (1994): 327–339.

27 S. Mendiratta, Z. C. Qu, and J. M. May, Erythrocyte ascorbate recycling: Antioxidant effects in blood, *Free Radicals in Biological Medicine* 24 (1998): 789–797.

28 H. Hemilä and Z. S. Herman, Vitamin C and the common cold: A retrospective analysis of Chalmer's review, *Journal of the American College of Nutrition* 14 (1995): 116–123.

29 Committee on Dietary Reference Intakes, 2000.

30 Researchers proposed 120 mg/day. M. Levine and coauthors, Criteria and recommendations for vitamin C intake, *Journal of the American Medical Association* 281 (1999): 1415–1423; A. C. Carr and B. Frei, Toward a new recommended dietary allowance for vitamin C based on antioxidant and health effects in humans, *American Journal of Clinical Nutrition* 69 (1999): 1086–1107.

31 J. Lykkesfeldt and coauthors, Ascorbic acid and dehydroascorbic acid as biomarkers of oxidative stress caused by smoking, *American Journal of Clinical Nutrition* 65 (1997): 959–963.

32 R. S. Strauss, Environmental tobacco smoke and serum vitamin C levels in children, *Pediatrics* 107 (2001): 540–542; D. W. Giraud, H. D. Martin, and J. A. Driskell, Plasma and dietary vitamin C and E levels of tobacco chewers, smokers, and nonusers, *Journal of the American Dietetic Association* 95 (1995): 798–800.

33 H. E. Sauberlich and coauthors, Effects of erythorbic acid on vitamin C metabolism in young women, *American Journal of Clinical Nutrition* 64 (1996): 336–346.

34 M. Levine, Fruits and vegetables: There is no substitute, *American Journal of Clinical Nutrition* 64 (1996): 381–382.

VITAMIN AND MINERAL SUPPLEMENTS

MORE THAN half of the U.S. population takes vitamin and mineral supplements regularly, spending billions of dollars on them each year.[1] Many people take supplements as dietary insurance—in case they are not meeting their nutrient needs from foods alone. Others take supplements as health insurance—to protect against certain diseases.

One out of every five people takes multinutrient pills daily. Others take large doses of single nutrients, most commonly, vitamin C, vitamin E, beta-carotene, iron, and calcium. In many cases, taking supplements is a costly but harmless practice; sometimes, it is both costly and harmful to health.

For the most part, people self-prescribe supplements, taking them on the advice of friends, television, websites, or books that may or may not be reliable. Sometimes, they take supplements on the recommendation of a physician. When such advice follows a valid nutrition assessment, supplementation may be warranted, but even then the preferred course of action is to improve food choices and eating habits. Without an assessment, the advice to take supplements may be inappropriate. A registered dietitian can help with the decision.

When people think of supplements, they often think of vitamins, but minerals are important, too, of course. People whose diets lack vitamins, for whatever reason, probably lack several minerals as well. This highlight asks several questions related to vitamin-mineral **supplements** (the glossary on p. 348 defines supplements and related terms). What are the arguments *for* taking supplements? What are the arguments *against* taking them? Finally, if people do take supplements, how can they choose the appropriate ones? (In addition to vitamins and minerals, supplements may also provide amino acids or herbs, which are discussed in Chapter 6 and Chapter 18, respectively.)

Over 3000 different vitamin and mineral supplements are available in a number of formulations, shapes, and flavors, but none offers the full array of nutrients that a variety of foods can provide.

© L. Share/Tony Stone Images

ARGUMENTS FOR SUPPLEMENTS

Vitamin-mineral supplements may be appropriate in some circumstances. In some cases, they can correct deficiencies; in others, they can reduce the risk of diseases.

Correct Overt Deficiencies

In the United States and Canada, adults rarely suffer nutrient deficiency diseases such as scurvy, pellagra, and beriberi, but they do still occur. To correct an overt deficiency disease, a physician may prescribe therapeutic doses two to ten times the RDA (or AI) of a nutrient. At such high doses, the supplement is acting as a drug.

Improve Nutrition Status

In contrast to the classical deficiencies, which present a multitude of symptoms and are relatively easy to recognize, subclinical deficiencies are subtle and easy to overlook—and they are also more likely to occur. People who do not eat enough food to deliver the needed amounts of nutrients, such as habitual dieters and the elderly, risk developing subclinical deficiencies. Similarly, vegetarians who restrict their use of entire food groups without appropriate substitutions may fail to fully meet their nutrient needs. If there is no way for these people to eat

GLOSSARY

FDA (Food and Drug Administration): a federal agency that is responsible for, among other things, supplement safety, manufacturing, and information, including product labeling, package inserts, and accompanying literature. Another federal agency, the **FTC (Federal Trade Commission)**, is responsible for, among other things, supplement advertising.

high potency: 100% or more of the Daily Value for the nutrient in a single supplement and for at least two-thirds of the nutrients in a multinutrient supplement.

structure-function claims: statements that describe how a product may affect the structure or function of the body (do not require FDA authorization); for example, "calcium builds strong bones."

supplements: pills, capsules, tablets, liquids, or powders that contain vitamins, minerals, herbs, or amino acids; intended to increase dietary intake of these substances.

enough nutritious foods to meet their needs, then vitamin-mineral supplements may be appropriate to help prevent nutrient deficiencies.

Reduce Disease Risks

Highlight 11 reviews the relationships between supplement use and disease prevention. It describes some of the accumulating evidence suggesting that intakes of certain nutrients at levels much higher than can be attained from foods alone may be beneficial in reducing disease risks. It also presents research confirming the associated risks. Clearly, consumers must be cautious in taking supplements to prevent disease.

Many people, especially postmenopausal women and those who are intolerant to lactose or allergic to milk, may not receive enough calcium to forestall the bone degeneration of old age, osteoporosis. For them, nonmilk calcium-rich foods are especially valuable, but calcium supplements may also be appropriate (Highlight 12 provides more details).

Support Increased Nutrient Needs

As Chapters 14–16 explain, nutrient needs increase during certain stages of life, making it difficult to meet some of those needs without supplementation. For example, women who lose a lot of blood and therefore a lot of iron during menstruation each month may need an iron supplement. Women of childbearing age may need folate supplements to reduce the risks of neural tube defects. Similarly, pregnant women and women who are breastfeeding their infants have exceptionally high nutrient needs and so usually need special supplements. Newborns routinely receive a single dose of vitamin K at birth to prevent abnormal bleeding. Infants may need other supplements as well, depending on whether they are breastfed or receiving formula, and on whether their water contains fluoride.

Improve the Body's Defenses

Health care professionals may provide special supplementation to people being treated for addictions to alcohol or other drugs and to people with prolonged illnesses, extensive injuries, or other severe stresses such as surgery. Illnesses that interfere with appetite, eating, or nutrient absorption limit nutrient intakes, yet nutrient needs are often heightened by diseases or medications. In all these cases, supplements are appropriate.

Who Needs Supplements?

In summary, the following list acknowledges that in these specific conditions, these people may need to take supplements:[2]

- People with nutrient deficiencies.
- People with low food energy intakes (fewer than 1200 kcalories per day) need a multivitamin and mineral supplement.
- People who eat all-plant diets (vegans) and those with atrophic gastritis need vitamin B_{12}.
- Women who bleed excessively during menstruation need iron.
- People with lactose intolerance, milk allergies, or who otherwise do not consume enough dairy products to forestall extensive bone loss need calcium.
- People in certain stages of the life cycle who have increased nutrient needs (for example, infants need iron and fluoride, women of childbearing age need folate, pregnant women need iron, and the elderly need vitamin D).
- People with limited milk intake and sun exposure need vitamin D.
- People who have diseases, infections, or injuries or who have undergone surgery that interferes with the intake, absorption, metabolism, or excretion of nutrients.
- People taking medications that interfere with the body's use of specific nutrients.

Except for people in these circumstances, most adults can normally get all the nutrients they need by eating a varied diet of nutrient-dense foods. Even athletes can meet their nutrient needs without the help of supplements.

ARGUMENTS AGAINST SUPPLEMENTS

Foods rarely cause nutrient imbalances or toxicities, but supplements can. The higher the dose, the greater the risk of harm. People's tolerances for high doses of nutrients vary, just as their risks of deficiencies do. Amounts that some can tolerate may be harmful for others, and no one knows who falls where along the spectrum. It is difficult to determine just how much of a nutrient is enough—or too much. The Tolerable Upper Intake Levels of the DRI answer the question how much is too much by defining the highest amount that appears safe for most healthy people. Table H10-1 presents these suggested Upper Levels and Daily Values for selected vitamins and minerals and the quantities typically found in supplements.

Toxicity

The extent and severity of supplement toxicity remain unclear. Only a few alert health care professionals can recognize toxicity, even when it is acute. When it is chronic, with the effects developing subtly and progressing slowly, it often goes unrecognized. In view of the potential hazards, some authorities believe supplements should bear warning labels, advising consumers that large doses may be toxic.

Toxic overdoses of vitamins and minerals in children are more readily recognized and, unfortunately, fairly common. In 1996, poison control centers received more than 50,000 reports of children under the age of six swallowing excessively large doses of supplements.[3] Fruit-flavored, chewable vitamins shaped like cartoon characters entice young children to eat them like candy in amounts that can cause poisoning. High-potency iron supplements (30 milligrams of iron or more per tablet) are especially toxic and are the leading cause of accidental ingestion fatalities among children. Even mild overdoses

TABLE H10-1 Vitamin and Mineral Intakes for Adults

Nutrient	Tolerable Upper Intake Levels[a]	Daily Values	Typical Multivitamin-Mineral Supplement	Average Single-Nutrient Supplement
Vitamins				
Vitamin A	3000 µg (10,000 IU)	5000 IU	5000 IU	8000 to 10,000 IU
Vitamin D	50 µg (2000 IU)	400 IU	400 IU	400 IU
Vitamin E	1000 mg (1500 to 2200 IU)[b]	30 IU	30 IU	100 to 1000 IU
Vitamin K	—[c]	80 µg	40 µg	—[e]
Thiamin	—[c]	1.5 mg	1.5 mg	50 mg
Riboflavin	—[c]	1.7 mg	1.7 mg	25 mg
Niacin (as niacinamide)	35 mg[b]	20 mg	20 mg	100 to 500 mg
Vitamin B_6	100 mg	2 mg	2 mg	100 to 200 mg
Folate	1000 µg[b]	400 µg	400 µg	400 µg
Vitamin B_{12}	—[c]	6 µg	6 µg	100 to 1000 µg
Pantothenic acid	—[c]	10 mg	10 mg	100 to 500 mg
Biotin	—[c]	300 µg	30 µg	300 to 600 µg
Vitamin C	2000 mg	60 mg	10 mg	500 to 2000 mg
Choline	3500 mg	—	10 mg	250 mg
Minerals				
Calcium	2500 mg	1000 mg	160 mg	250 to 600 mg
Phosphorus	4000 mg	1000 mg	110 mg	—[e]
Magnesium	350 mg[d]	400 mg	100 mg	250 mg
Iron	45 mg	18 mg	18 mg	18 to 30 mg
Zinc	40 mg	15 mg	15 mg	10 to 100 mg
Iodine	1100 µg	150 µg	150 µg	—[e]
Selenium	400 µg	70 µg	10 µg	50 to 200 µg
Fluoride	10 mg	—	—	—[e]
Copper	10 mg	2 mg	0.5 mg	—[e]
Manganese	11 mg	2 mg	5 mg	—[e]
Chromium	—[c]	120 µg	25 µg	200 to 400 µg
Molybdenum	2000 µg	75 µg	25 µg	—[e]

[a]Unless otherwise noted, Upper Levels represent total intakes from food, water, and supplements.
[b]Upper Levels represent intakes from supplements, fortified foods, or both.
[c]These nutrients have been evaluated by the DRI committee for Upper Tolerable Intake Levels, but none were established because of insufficient data. No adverse effects have been reported with intakes of these nutrients at levels typical of supplements, but caution is still advised, given the potential for harm that accompanies excessive intakes.
[d]Upper Levels represent intakes from supplements only.
[e]Available as a single supplement by prescription.

cause GI distress, nausea, and black diarrhea that reflects gastric bleeding. Severe overdoses result in bloody diarrhea, shock, liver damage, coma, and death.

Life-Threatening Misinformation

Another problem arises when people who are ill come to believe that high doses of vitamins or minerals can be therapeutic. Not only can high doses be toxic, but the person may take them instead of seeking medical help. Furthermore, there are no guarantees that the supplements will be effective. Marketing materials for supplements often make health statements that are required to be "truthful and not misleading," but often fall far short of both. Chapter 18 revisits this topic and includes a discussion of herbal preparations and other alternative therapies.

Unknown Needs

Another argument against the use of supplements is that no one knows exactly how to formulate the "ideal" supplement. What nutrients should be included? How much of each? On whose needs should the choices be based? Surveys have repeatedly shown little relationship between the supplements people take and the nutrients they actually need.

False Sense of Security

Another argument against supplement use is that it may lull people into a false sense of security. A person might eat irresponsibly, thinking, "My supplement will cover my needs." Or, experiencing a warning symptom of a disease, a person might postpone seeking a diagnosis, thinking, "I probably just need a supplement to make this go away." Such self-diagnosis is potentially dangerous.

Other Invalid Reasons

Other invalid reasons why people might take supplements include:

- The belief that the food supply or soil contains inadequate nutrients.
- The belief that supplements can provide energy.
- The belief that supplements can enhance athletic performance or build lean body tissues without physical work or faster than work alone.
- The belief that supplements will help a person cope with stress.
- The belief that supplements can prevent, treat, or cure conditions ranging from the common cold to cancer.

Ironically, people with health problems are more likely to take supplements than other people, yet today's health problems are more likely to be due to overnutrition and poor lifestyle choices than to nutrient deficiencies. The truth—that most people would benefit from improving their eating and exercise habits—is harder to swallow than a supplement pill.

Bioavailability and Antagonistic Actions

In general, the body absorbs nutrients best from foods in which the nutrients are diluted and dispersed among other substances that may facilitate their absorption. Taken in pure, concentrated form, nutrients are likely to interfere with one another's absorption or with the absorption of nutrients in foods eaten at the same time. Documentation of these effects is particularly extensive for minerals: zinc hinders copper and calcium absorption, iron hinders zinc absorption, calcium hinders magnesium and iron absorption, and magnesium hinders the absorption of calcium and iron. Similarly, binding agents in supplements limit mineral absorption.

Although minerals provide the most familiar and best-documented examples, interference among vitamins is now being seen as supplement use increases. The vitamin A precursor beta-carotene, long thought to be nontoxic, interferes with vitamin E metabolism when taken over the long term as a dietary supplement. Vitamin E, on the other hand, antagonizes vitamin K activity and so should not be used by people being treated for blood-clotting disorders. Consumers who want the benefits of optimal absorption of nutrients should use ordinary foods, selected for nutrient density and variety.

In view of all the negatives associated with supplement taking, nutrition professionals advise that most people need *not* use supplements routinely. Whenever the diet is inadequate, the person should first attempt to improve it so as to obtain the needed nutrients from foods. If that is truly impossible, then the person needs a multivitamin-mineral supplement that supplies between 50 and 150 percent of the Daily Value for each of the nutrients. These amounts reflect the ranges commonly found in foods and therefore are compatible with the body's normal handling of nutrients (its physiologic tolerance). The next section provides some pointers to assist in the selection of an appropriate supplement.

SELECTION OF SUPPLEMENTS

Whenever a physician or registered dietitian recommends a supplement, follow the directions carefully. When selecting a supplement yourself, look for a single, balanced vitamin-mineral supplement.

If you decide to take a vitamin-mineral supplement, ignore the eye-catching art and meaningless claims. Pay attention to the form the supplements are in, the list of ingredients, and the price. Here's where the truth lies, and from it you can make a rational decision based on facts. You have two basic questions to answer.

Form

The first question: What form do you want—chewable, liquid, or pills? If you'd rather drink your supplements than chew them, fine. (If you choose a chewable form, though, be aware that chewable vitamin C can dissolve tooth enamel.) If you choose pills, look for statements about the disintegration time. The U.S. Pharmacopeia (USP) suggests that supplements should completely disintegrate within 30 to 45 minutes.* Obviously, supplements that don't dissolve have little chance of entering the bloodstream, so look for a brand that claims to meet USP disintegration standards.

Contents

The second question: What vitamins and minerals do *you* need? Generally, an appropriate supplement provides vitamins and minerals in amounts smaller than, equal to, or very close to recommended intakes. Avoid supplements that, in a daily dose, provide more than the Tolerable Upper Intake Level for *any* nutrient. Avoid preparations with more than 10 milligrams of iron per dose, except as prescribed by a physician. Iron is hard to get rid of once it's in the body, and an excess of iron can cause problems, just as a deficiency can (see Chapter 13).

Misleading Claims

Ignore "organic" or "natural" claims. Such supplements are no better than others and often cost more. The word *synthetic* may sound like "fake," but to synthesize just means to put together. Whether vitamins are synthesized in a laboratory or synthesized by plants and animals, your body uses them similarly. Only your wallet can tell the difference.

Avoid products that make **"high potency"** claims. More is not better (review the "How to" on p. 309). Remember that foods are also providing these nutrients. Nutrients can build up and cause unexpected problems. For example, a man who takes vitamins and begins to lose his hair may think his hair loss means he needs *more* vitamins, when in fact it may be the early sign of a vitamin A overdose. (Of course, it may be completely unrelated to nutrition as well.)

Be wise to fake vitamins and preparations that contain items not needed in human nutrition, such as carnitine and inositol. Such ingredients reveal a marketing strategy aimed at your pocket, not at your health. The manufacturer wants you to believe that its pills contain the latest "new" nutrient that other brands omit, but in reality, these substances are not known to be needed by human beings.

Realize that the claim that supplements "relieve stress" is another marketing ploy. If you give even passing thought to what people mean by "stress," you'll realize manufacturers could never design a supplement to meet everyone's needs. Is it stressful to take an exam? Well, yes. Is it stressful to survive a major car wreck with third-degree burns and multiple bone fractures? Definitely yes. The body's responses to these stresses are different. The body does use vitamins and minerals in mounting a stress response, but a body fed a well-balanced diet can meet the needs of most minor stresses. As for the major ones, medical intervention is needed. In any case, taking a vitamin supplement won't make life any less stressful.

Other marketing tricks to sidestep are "green" pills that contain dehydrated, crushed parsley, alfalfa, and other fruit and vegetable extracts. The nutrients and phytochemicals advertised can be obtained from a serving of vegetables more easily and for less money. Such pills may also provide enzymes, but these are inactivated in the stomach during protein digestion.

Be aware that some geriatric "tonics" are low in vitamins and minerals, yet so high in alcohol as to threaten inebriation. The liquids designed for infants are more complete.

Recognize the latest nutrition buzzwords. Manufacturers were marketing "antioxidant" supplements before the print had time to dry on the first scientific reports of antioxidant vitamins' action in preventing cancer and cardiovascular disease. Remember, too, that high doses can alter a nutrient's action in the body. An antioxidant in physiological quantities may be beneficial, but in pharmacological quantities, it may act as a prooxidant and produce harmful by-products. Highlight 11 explores antioxidants and supplement use in more detail.

Cost

When shopping for supplements, remember that local or store brands may be just as good as nationally advertised brands. If they are less expensive, it may be because the price does not have to cover the cost of national advertising.

REGULATION OF SUPPLEMENTS

The Dietary Supplement Health and Education Act of 1994 was intended to enable consumers to make informed choices about nutrient supplements. The act subjects supplements to the same general labeling requirements that apply to foods. Specifically:

- Nutrition labeling for dietary supplements is required.
- Labels may make nutrient claims (as "high" or "low") according to specific criteria (for example, "an excellent source of vitamin C").

*The USP establishes standards for quality, strength, and purity of supplements.

Structure-function claims do not need FDA authorization, but they must be accompanied by a disclaimer.

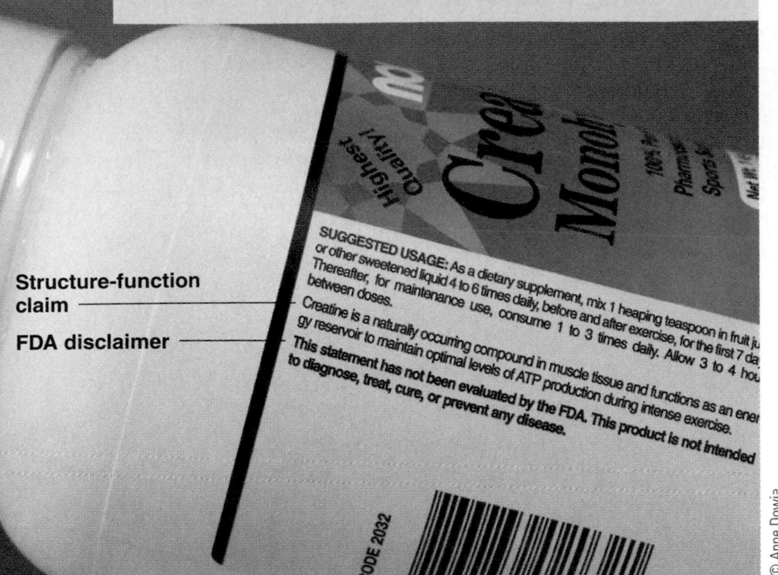

© Anne Dowie

- Labels may claim that the lack of a nutrient can cause a deficiency disease, but if they do, they must also include the prevalence of that deficiency disease in the United States.
- Labels may make health claims that are supported by significant scientific agreement and are not brand specific (for example, "folate protects against neural tube defects"). To date, the following health claims have been approved for supplements: folate and neural tube defects, calcium and osteoporosis, soluble fiber from oat bran and from psyllium husks and cardiovascular disease, and omega-3 fatty acids and cardiovascular disease.
- Labels may claim to diagnose, treat, cure, or relieve common complaints such as menstrual cramps or memory loss, but may *not* make claims about specific diseases (except as noted above).
- Labels may make **structure-function claims** about the role a nutrient plays in the body, how the nutrient performs its function, and how consuming the nutrient is associated with general well-being. These claims must be accompanied by an **FDA** disclaimer statement: "This statement has not been evaluated by the Food and Drug Administration. This product is not intended to diagnose, treat, cure or prevent any disease."

Figure H10-1 provides an example of a supplement label that complies with the mandatory requirements.

In effect, the Dietary Supplement Health and Education Act resulted in a deregulation of the supplement industry.[4] Unlike food additives or drugs, supplements do not need to be proved safe and effective nor do they need the FDA's approval before being marketed. Furthermore, there are no standards for potency or dosage. Should a problem arise, the burden falls to the FDA to prove that the supplement poses an unreasonable risk and should be removed from the market.

FIGURE H10-1 An Example of a Supplement Label

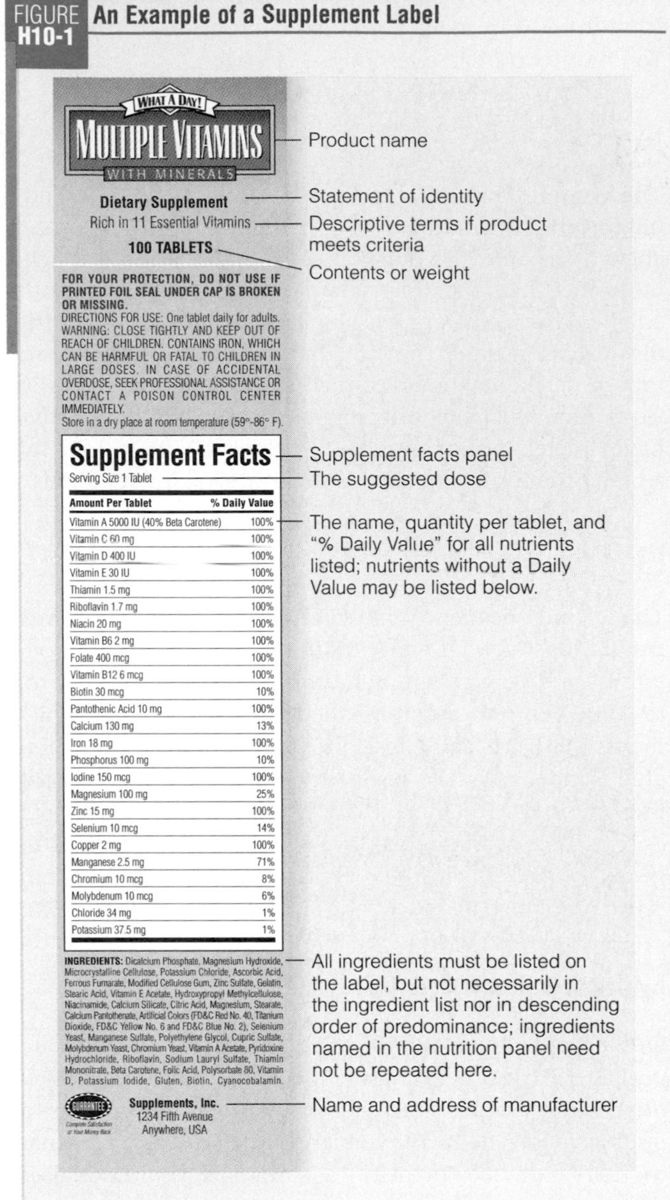
WHAT A DAY!
MULTIPLE VITAMINS
WITH MINERALS

Dietary Supplement
Rich in 11 Essential Vitamins
100 TABLETS

FOR YOUR PROTECTION, DO NOT USE IF PRINTED FOIL SEAL UNDER CAP IS BROKEN OR MISSING.
DIRECTIONS FOR USE: One tablet daily for adults.
WARNING: CLOSE TIGHTLY AND KEEP OUT OF REACH OF CHILDREN. CONTAINS IRON, WHICH CAN BE HARMFUL OR FATAL TO CHILDREN IN LARGE DOSES. IN CASE OF ACCIDENTAL OVERDOSE, SEEK PROFESSIONAL ASSISTANCE OR CONTACT A POISON CONTROL CENTER IMMEDIATELY.
Store in a dry place at room temperature (59°-86° F).

Supplement Facts
Serving Size 1 Tablet

Amount Per Tablet	% Daily Value
Vitamin A 5000 IU (40% Beta Carotene)	100%
Vitamin C 60 mg	100%
Vitamin D 400 IU	100%
Vitamin E 30 IU	100%
Thiamin 1.5 mg	100%
Riboflavin 1.7 mg	100%
Niacin 20 mg	100%
Vitamin B6 2 mg	100%
Folate 400 mcg	100%
Vitamin B12 6 mcg	100%
Biotin 30 mcg	10%
Pantothenic Acid 10 mg	100%
Calcium 130 mg	13%
Iron 18 mg	100%
Phosphorus 100 mg	10%
Iodine 150 mcg	100%
Magnesium 100 mg	25%
Zinc 15 mg	100%
Selenium 10 mcg	14%
Copper 2 mg	100%
Manganese 2.5 mg	71%
Chromium 10 mcg	8%
Molybdenum 10 mcg	6%
Chloride 34 mg	1%
Potassium 37.5 mg	1%

INGREDIENTS: Dicalcium Phosphate, Magnesium Hydroxide, Microcrystalline Cellulose, Potassium Chloride, Ascorbic Acid, Ferrous Fumarate, Modified Cellulose Gum, Zinc Sulfate, Gelatin, Stearic Acid, Vitamin E Acetate, Hydroxypropyl Methylcellulose, Niacinamide, Calcium Silicate, Citric Acid, Magnesium, Stearate, Calcium Pantothenate, Artificial Colors (FD&C Red No. 40, Titanium Dioxide, FD&C Yellow No. 6 and FD&C Blue No. 2), Selenium Yeast, Manganese Sulfate, Polyethylene Glycol, Cupric Sulfate, Molybdenum Yeast, Chromium Yeast, Vitamin A Acetate, Pyridoxine Hydrochloride, Riboflavin, Sodium Lauryl Sulfate, Thiamin Mononitrate, Beta Carotene, Folic Acid, Polysorbate 80, Vitamin D, Potassium Iodide, Gluten, Biotin, Cyanocobalamin.

GUARANTEE
Complete Satisfaction or Your Money Back

Supplements, Inc.
1234 Fifth Avenue
Anywhere, USA

If all the nutrients we need can come from food, why not just eat food? Foods have so much more to offer than supplements do. Nutrients in foods come in an infinite variety of combinations with a multitude of different carriers and absorption enhancers. They come with water, fiber, and an array of beneficial phytochemicals. Foods stimulate the GI tract to keep it healthy. They provide energy, and since you need energy each day, why not have nutritious foods deliver it? They offer pleasure, satiety, and opportunities for socializing while eating. In no way can nutrient supplements hold a candle to foods as a means of meeting human health needs. For further proof, read Highlight 11.

Nutrition on the Net

•WEBSITES•

Access these websites for further study of topics covered in this highlight.

- Find updates and quick links to these and other nutrition-related sites at our website: **www.wadsworth.com/nutrition**
- Review abstracts from more than 250,000 scientific reports on 50 of the most popular dietary supplements: **odp.od.nih.gov/ods/databases/ibids.html**
- Gather information from the Office of Dietary Supplements: **dietary-supplements.info.nih.gov**
- Report adverse reactions associated with dietary supplements to the FDA's MedWatch program: **www.fda.gov/medwatch**
- Search for "supplements" at the American Dietetic Association: **www.eatright.org**
- Learn more about supplements from the FDA Center for Food Safety and Applied Nutrition: **www.cfsan.fda.gov/~dms/supplmnt.html**
- Obtain consumer information on dietary supplements from the U.S. Pharmacopeia: **www.usp.org**
- Review the Federal Trade Commission policies for dietary supplement advertising: **www.ftc.gov/bcp/conline/pubs/buspubs/dietsupp.htm**

References

1 Report of the Commission on Dietary Supplement Labels (Washington, D.C.: Department of Health and Human Services, 1997).

2 Position of The American Dietetic Association: Food fortification and dietary supplements, *Journal of the American Dietetic Association* 101 (2001): 115–125.

3 T. L. Litovitz and coauthors, 1996 Annual Report of the American Association of Poison Control Center Toxic Exposure Surveillance System, *American Journal of Emergency Medicine* 15 (1997): 447–500.

4 M. C. Nesheim, Regulation of dietary supplements, *Nutrition Today* 33 (1998): 62–68.

CHAPTER 11 The Fat-Soluble Vitamins: A, D, E, and K

The fat-soluble vitamins A, D, E, and K◆ differ from the water-soluble vitamins in several significant ways (review the table on p. 310). Being insoluble in water, the fat-soluble vitamins require bile for their absorption. Upon absorption, fat-soluble vitamins travel through the lymphatic system within chylomicrons before entering the bloodstream, where many of them require protein carriers for transport. The fat-soluble vitamins participate in numerous activities all over the body, but excesses are stored primarily in the liver and adipose tissue. The body maintains blood concentrations by retrieving these vitamins from storage as needed; thus people can eat less than their daily need for days, weeks, or even months or years without ill effects. They need only ensure that over time *average* daily intakes approximate recommendations. By the same token, because fat-soluble vitamins are not readily excreted, the risk of toxicity is greater than it is for the water-soluble vitamins.

◆ **The fat-soluble vitamins:**
- Vitamin A.
- Vitamin D.
- Vitamin E.
- Vitamin K.

VITAMIN A AND BETA-CAROTENE

Vitamin A was the first fat-soluble vitamin to be recognized. Almost a century later, vitamin A and its precursor, **beta-carotene,** continue to intrigue researchers with their diverse roles and profound effects on health.

Three different forms of vitamin A are active in the body: retinol, retinal, and retinoic acid. Collectively, these compounds are known as **retinoids.** Foods derived from animals provide compounds (retinyl esters) that are easily converted to retinol in the intestine. Foods derived from plants provide **carotenoids,**◆ some of which have **vitamin A activity.*** The most studied of the carotenoids is beta-carotene, which can be split to form retinol in the intestine and liver. Beta-carotene's absorption and conversion are less efficient than those of the retinoids. Figure 11-1 (on p. 356) illustrates the structural similarities and differences of these vitamin A compounds.

◆ **Carotenoids are among the best-known phytochemicals (see Highlight 11).**

The cells can convert retinol and retinal to the other active forms of vitamin A as needed. The conversion of retinol to retinal is reversible, but the conversion of retinal to retinoic acid is irreversible (see Figure 11-2 on p. 356). This irreversibility is significant because each form of vitamin A performs a function that the others cannot.

A special transport protein, **retinol-binding protein (RBP),** picks up vitamin A from the liver, where it is stored, and carries it in the blood. Cells that will use vitamin A have special protein receptors for it, as if the vitamin were fragile and had to be passed carefully from hand to hand without being dropped. Each form of vitamin A has its own receptor protein (retinol has several) within the cells.[1]

vitamin A: all naturally occurring compounds with the biological activity of retinol (RET-ih-nol), the alcohol form of vitamin A.

beta-carotene (BAY-tah KARE-oh-teen): one of the carotenoids; an orange pigment and vitamin A precursor found in plants. A **precursor** is a compound that can be converted into an active vitamin.

retinoids (RET-ih-noyds): chemically related compounds with biological activity similar to that of retinol.

carotenoids (kah-ROT-eh-noyds): pigments commonly found in plants and animals, some of which have vitamin A activity. The carotenoid with the greatest vitamin A activity is beta-carotene.

vitamin A activity: a term referring to both the active forms of vitamin A and the precursor forms in foods without distinguishing between them.

retinol-binding protein (RBP): the specific protein responsible for transporting retinol.

Roles in the Body

Vitamin A is a versatile vitamin. Its major roles include:

- Promoting vision.
- Participating in protein synthesis and cell differentiation (and thereby maintaining the health of epithelial tissues and skin).
- Supporting reproduction and growth.

As mentioned, each form of vitamin A performs specific tasks. Retinol supports reproduction and is the major transport and storage form of the vitamin. Retinal is active in vision and is also an intermediate in the conversion of retinol to retinoic acid. Retinoic acid acts like a hormone, regulating cell differentiation, growth, and embryonic development. Animals raised on retinoic acid as their sole source of vitamin A can grow normally, but they become blind because retinoic acid cannot be converted to retinal.

*Carotenoids with vitamin A activity include alpha-carotene, beta-carotene, and beta-cryptoxanthin; carotenoids with no vitamin A activity include lycopene, lutein, and zeaxanthin.

FIGURE 11-1 Forms of Vitamin A

In this diagram, corners represent carbon atoms, as in all previous diagrams in this book. A further simplification here is that methyl groups (CH_3) are understood to be at the ends of the lines extending from corners. (See Appendix C for complete structures.)

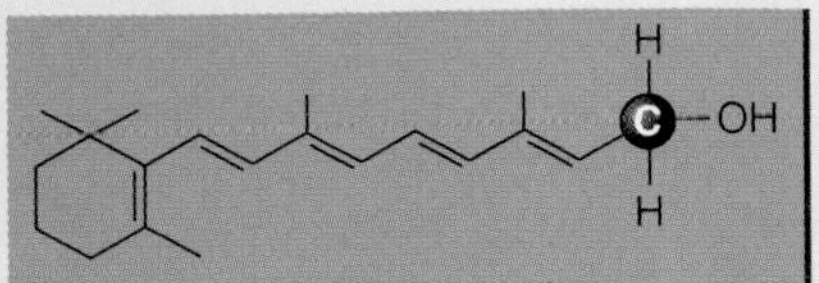

Retinol, the alcohol form

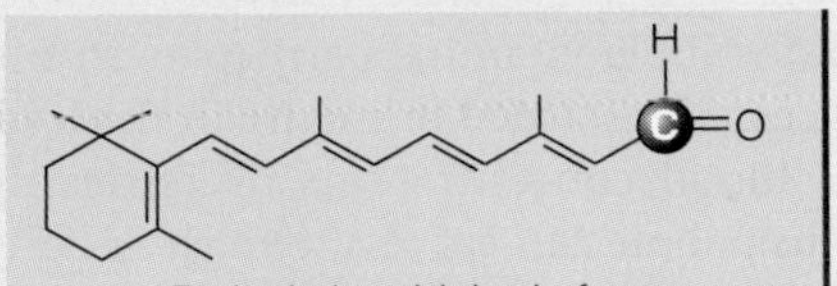

Retinal, the aldehyde form

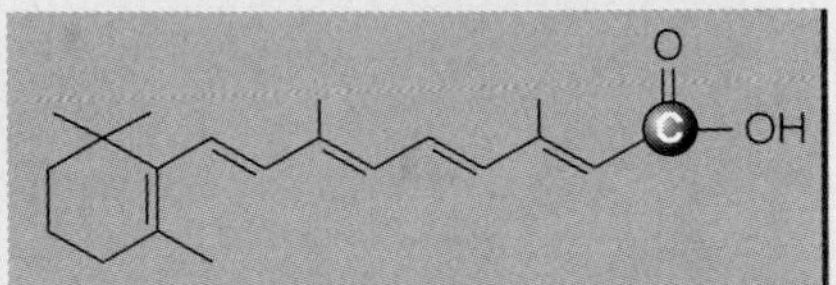

Retinoic acid, the acid form

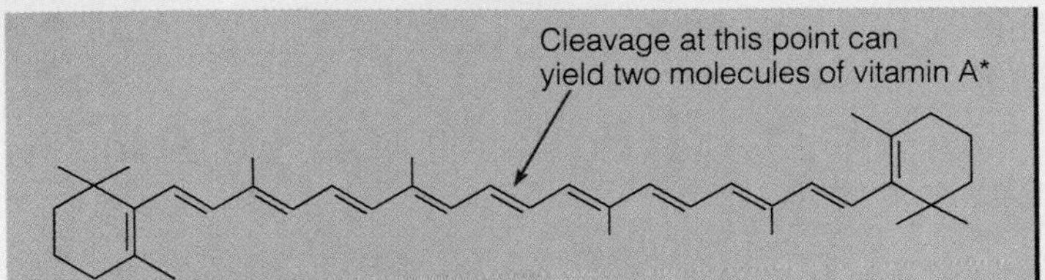

Cleavage at this point can yield two molecules of vitamin A*

Beta-carotene, a precursor

*Sometimes cleavage occurs at other points as well, so that one molecule of beta-carotene may yield only one molecule of vitamin A. Furthermore, not all beta-carotene is converted to vitamin A, and absorption of beta-carotene is not as efficient as vitamin A. For these reasons, 12 µg of beta-carotene are equivalent to 1 µg of vitamin A. Conversion of other carotenoids to vitamin A is even less efficient.

• **Vitamin A in Vision** • Vitamin A plays two indispensable roles in the eye: it helps maintain a crystal-clear outer window, the **cornea,** and it participates in the conversion of light energy into nerve impulses at the **retina** (see Figure 11-3 for details). The cells of the retina contain **pigment** molecules called **rhodopsin;** each rhodopsin molecule is composed of a protein called **opsin** bonded to a molecule of retinal. When light passes through the cornea of the eye and strikes the cells of the retina, rhodopsin responds by changing shape and becoming bleached. As it does, the retinal shifts from a *cis* to a *trans* configuration, just as fatty acids do during hydrogenation (see p. 135). The *trans*-retinal cannot remain bonded to opsin. When retinal is released, opsin changes shape, thereby disturbing the membrane of the cell and generating an electrical impulse that travels along the cell's length. At the other end of the cell, the impulse is transmitted to a nerve cell, which conveys the message to the brain. Much of the retinal is then converted back to its active *cis* form and combined with the opsin protein to regenerate the pigment rhodopsin. Some retinal, however, may be oxidized to retinoic acid, a biochemical dead end for the visual process.

Visual activity leads to repeated small losses of retinal and necessitates its constant replenishment from retinol in the blood, which brings a new supply from the body stores. Ultimately, foods supply all the retinal in the pigments of the eye.

◆ Over 100 million cells reside in the retina, and each contains about 30 million molecules of vitamin A–containing visual pigments.◆

cornea (KOR-nee-uh): the transparent membrane covering the outside of the eye.

retina (RET-in-uh): the layer of light-sensitive nerve cells lining the back of the inside of the eye; consists of rods and cones.

pigment: a molecule capable of absorbing certain wavelengths of light so that it reflects only those that we perceive as a certain color.

rhodopsin (ro-DOP-sin): a light-sensitive pigment of the retina. It contains the retinal form of vitamin A and the protein opsin.

- **rhod** = red (pigment)
- **opsin** = visual protein

opsin (OP-sin): the protein portion of the visual pigment molecule.

FIGURE 11-2 Conversion of Vitamin A Compounds

Notice that the conversion from retinol to retinal is reversible, whereas the pathway from retinal to retinoic acid is not.

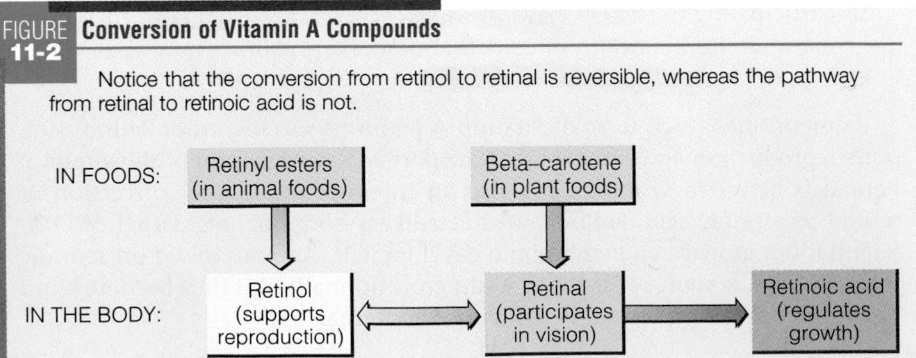

FIGURE 11-3 Vitamin A's Role in Vision

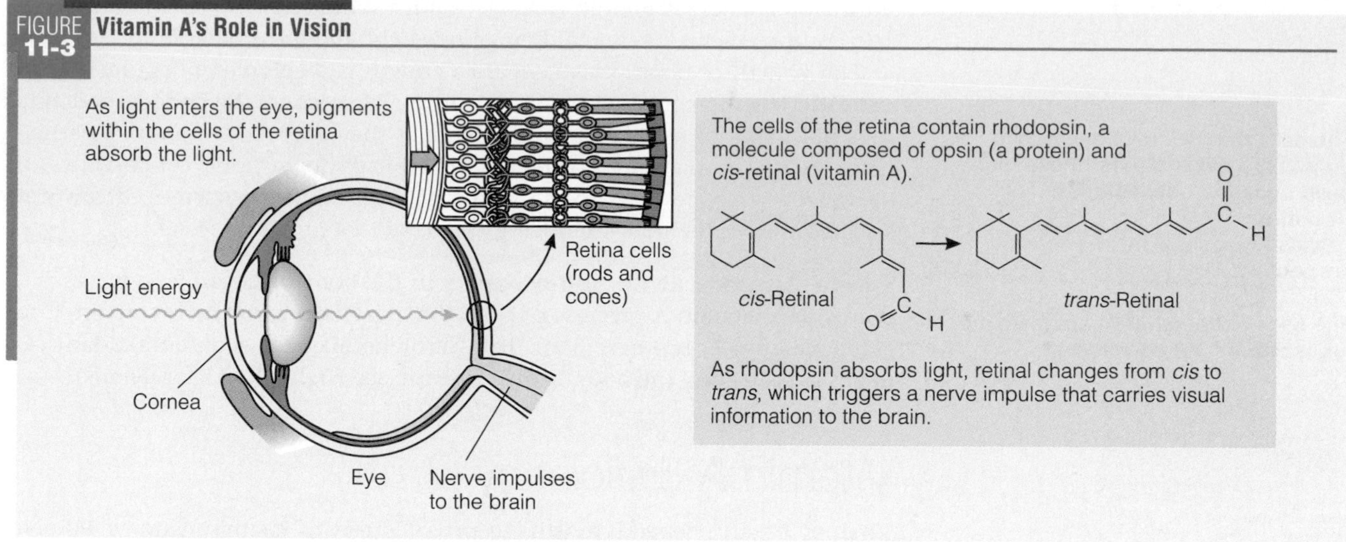

• Vitamin A in Protein Synthesis and Cell Differentiation • Despite its important role in vision, only one-thousandth of the body's vitamin A is in the retina. Much more is in the cells lining the body's surfaces, where the vitamin participates in protein synthesis and cell **differentiation.**

All body surfaces, both inside and out, are covered by layers of cells known as **epithelial cells.** The **epithelial tissue** on the outside of the body is, of course, the skin. The epithelial tissues that line the inside of the body are the **mucous membranes:** the linings of the mouth, stomach, and intestines; the linings of the lungs and the passages leading to them; the linings of the urinary bladder and urethra; the linings of the uterus and vagina; and the linings of the eyelids and sinus passageways. Within the body, the mucous membranes of the GI tract alone line an area larger than a quarter of a football field, and vitamin A helps to maintain their integrity (see Figure 11-4).

Vitamin A promotes differentiation of both epithelial cells and goblet cells, one-celled glands that synthesize and secrete mucus. Mucus coats and protects the epithelial cells from invasive microorganisms and other harmful substances, such as gastric juices.

• Vitamin A in Reproduction and Growth • As mentioned, vitamin A also supports reproduction and growth. In men, retinol participates in sperm development, and in women, vitamin A supports normal fetal development

FIGURE 11-4 Mucous Membrane Integrity

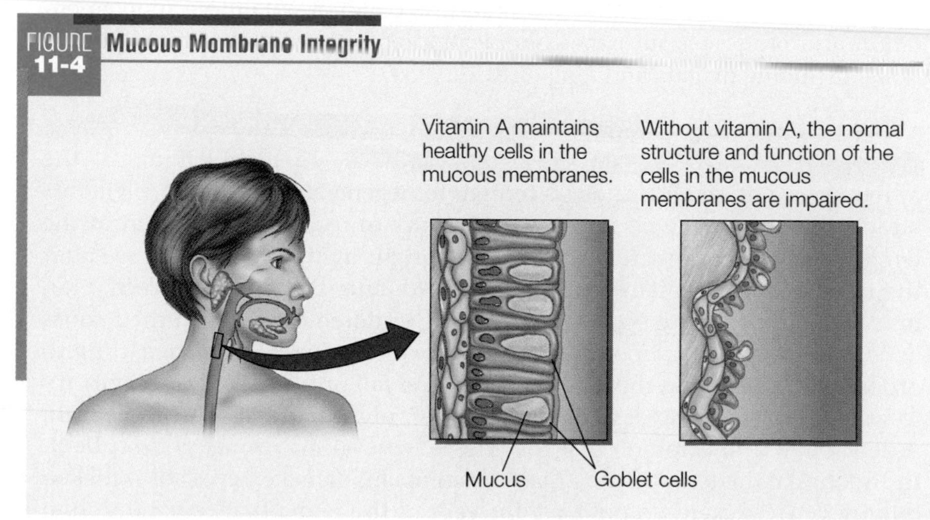

differentiation: the development of specific functions different from those of the original.

epithelial (ep-i-THEE-lee-ul) **cells:** cells on the surface of the skin and mucous membranes.

epithelial tissue: the layer of the body that serves as a selective barrier between the body's interior and the environment (examples are the cornea, the skin, the respiratory lining, and the lining of the digestive tract).

mucous (MYOO-kus) **membranes:** the membranes, composed of mucus-secreting cells, that line the surfaces of body tissues.

during pregnancy.[2] Children lacking vitamin A fail to grow. When given vitamin A supplements, these children gain weight and grow taller.[3]

The growth of bones illustrates that growth is a complex phenomenon of **remodeling.** To convert a small bone into a large bone, the bone-remodeling cells must "undo" some parts of the bone as they go,◆ and vitamin A participates in the dismantling.[4] The cells that break down bone contain sacs of degradative enzymes.◆ With the help of vitamin A, these enzymes eat away at selected sites in the bone, removing the parts that are not needed.

◆ The cells that destroy bone during growth are **osteoclasts;** those that build bone are **osteoblasts.**
- **osteo** = bone
- **clast** = break
- **blast** = build

◆ The sacs of degradative enzymes are **lysosomes** (LYE-so-zomes).

• **Beta-Carotene as an Antioxidant** • In the body, beta-carotene serves primarily as a vitamin A precursor.[5] Not all dietary beta-carotene is converted to active vitamin A, however. Some beta-carotene may act as an antioxidant capable of protecting the body against disease (see Highlight 11 for details).

Vitamin A Deficiency

Vitamin A status depends mostly on the adequacy of vitamin A stores, 90 percent of which are in the liver. Vitamin A status also depends on a person's protein status because retinol-binding proteins serve as the vitamin's transport carriers inside the body.

If a person were to stop eating vitamin A–rich foods, deficiency symptoms would not begin to appear until after stores were depleted—one to two years for a healthy adult but much sooner for a growing child. Then the consequences would be profound and severe. Vitamin A deficiency is one of the developing world's major nutrition problems. More than 100 million children worldwide have some degree of vitamin A deficiency, and so are vulnerable to infectious diseases and blindness.

• **Infectious Diseases** • In developing countries around the world, measles is a devastating infectious disease, killing as many as 2 million children each year. The severity of the illness often correlates with the degree of vitamin A deficiency; deaths are usually due to related infections such as pneumonia and severe diarrhea.[6] Providing large doses of vitamin A reduces the risk of dying from these infections.

The World Health Organization (WHO) and UNICEF (the United Nations International Children's Emergency Fund) have made the control of vitamin A deficiency a major goal in their quest to improve child health and survival throughout the developing world. They recommend routine vitamin A supplementation for all children with measles in areas where vitamin A deficiency is a problem or where the measles death rate is high. In the United States, the American Academy of Pediatrics recommends vitamin A supplementation for certain groups of measles-infected infants and children. Vitamin A supplementation also protects against the complications of other life-threatening infections, including malaria and lung diseases.[7]

• **Night Blindness** • **Night blindness** is one of the first detectable signs of vitamin A deficiency and permits early diagnosis. In night blindness, the retina does not receive enough retinal to regenerate the visual pigments bleached by light. The person loses the ability to recover promptly from the temporary blinding that follows a flash of bright light at night or to see after the lights go out. In many parts of the world, after the sun goes down, vitamin A–deficient people become night-blind: children cannot find their shoes or toys, and women cannot fetch water or wash dishes.[8] They often cling to others or sit still, afraid that they may trip and fall or lose their way if they try to walk alone. In many developing countries, night blindness due to vitamin A deficiency is so common that the people have special words to describe it. In Indonesia, the term is *buta ayam,* which means "chicken eyes" or "chicken blindness." (Chickens do not have the cells of the retina that respond to dim

remodeling: the dismantling and re-formation of a structure, in this case, bone.

night blindness: slow recovery of vision after flashes of bright light at night or an inability to see in dim light; an early symptom of vitamin A deficiency.

FIGURE 11-5 Vitamin A–Deficiency Symptom—Night Blindness

These photographs illustrate the eyes' slow recovery in response to a flash of bright light at night. In animal research studies, the response rate is measured with electrodes.

In dim light, you can make out the details in this room. You are using your rods for vision.

A flash of bright light momentarily blinds you as the pigment in the rods is bleached.

You quickly recover and can see the details again in a few seconds.

With inadequate vitamin A, you do not recover but remain blinded for many seconds.

© David Farr (all)

light and therefore cannot see at night.) Figure 11-5 shows the eyes' slow recovery in response to a flash of bright light in night blindness.

• **Blindness (Xerophthalmia)** • Beyond night blindness is total blindness—failure to see at all. Night blindness is caused by a lack of vitamin A at the back of the eye, the retina; total blindness is caused by a lack at the front of the eye, the cornea. Severe vitamin A deficiency is the major cause of childhood blindness in the world, causing more than half a million preschool children to lose their sight each year.[9] Blindness due to vitamin A deficiency, known as **xerophthalmia,** develops in stages. First, the cornea becomes dry and hard, a condition known as **xerosis.** Corneal xerosis can quickly progress to **keratomalacia,** the softening of the cornea that leads to irreversible blindness.

• **Keratinization** • Elsewhere in the body, vitamin A deficiency affects other surfaces. Without vitamin A, the goblet cells in the GI tract diminish in number and activity, limiting the secretion of mucus. With less mucus, normal digestion and absorption of nutrients falter, and this, in turn, worsens malnutrition by limiting the absorption of whatever nutrients the diet may deliver. Similar changes in the cells of other epithelial tissues weaken defenses, making infections of the respiratory tract, the GI tract, the urinary tract, the vagina, and possibly the inner ear likely. On the body's outer surface, the epithelial cells change shape and begin to secrete the protein **keratin**—the hard, inflexible protein of hair and nails. As Figure 11-6 shows, the skin becomes dry, rough, and scaly as lumps of keratin accumulate **(keratinization).**

FIGURE 11-6 Vitamin A–Deficiency Symptom—The Rough Skin of Keratinization

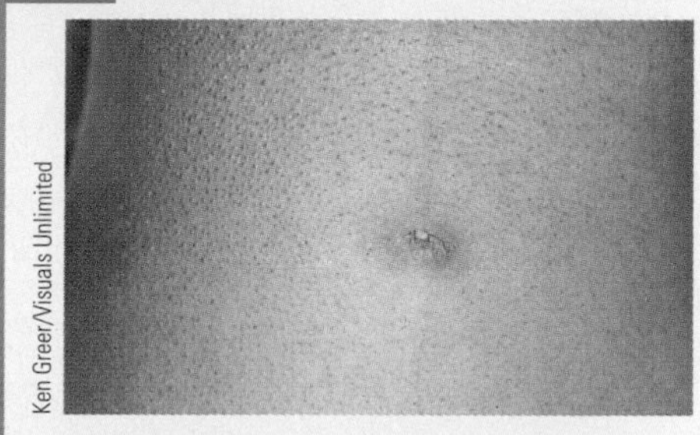

Ken Greer/Visuals Unlimited

In vitamin A deficiency, the epithelial cells secrete the protein keratin in a process known as *keratinization.* (Keratinization doesn't occur in the GI tract, but mucus-producing cells dwindle, and mucus production declines.) The progression of this condition to the extreme is *hyperkeratinization* or *hyperkeratosis.* When keratin accumulates around each hair follicle, the condition is known as *follicular hyperkeratosis.*

xerophthalmia (zer-off-THAL-mee-uh): progressive blindness caused by severe vitamin A deficiency.
- **xero** = dry
- **ophthalm** = eye

xerosis (zee-ROW-sis): abnormal drying of the skin and mucous membranes; a sign of vitamin A deficiency.

keratomalacia (KARE-ah-toe-ma-LAY-shuh): softening of the cornea that leads to irreversible blindness; seen in severe vitamin A deficiency.

keratin (KERR-uh-tin): a water-insoluble protein; the normal protein of hair and nails. Keratin-producing cells may replace mucus-producing cells in vitamin A deficiency.

keratinization: accumulation of keratin in a tissue; a sign of vitamin A deficiency.

Vitamin A Toxicity

Just as a deficiency of vitamin A affects all body systems, so does a toxicity. Symptoms of toxicity begin to develop when all the binding proteins are swamped, and free vitamin A damages the cells. Such effects are unlikely when a person depends on a balanced diet for nutrients, but with concentrated amounts from foods derived from animals or from supplements, toxicity is a real possibility. Children are most vulnerable to toxicity because they need less and are more sensitive to overdoses.

Beta-carotene, which is found in a wide variety of fruits and vegetables, is not converted efficiently enough in the body to cause vitamin A toxicity; instead, it is stored in the fat just under the skin. Overconsumption of beta-carotene from foods may turn the skin yellow, but this is not harmful. In contrast, overconsumption of beta-carotene from supplements may be quite harmful. In excess, this antioxidant may act as a prooxidant, promoting cell division and destroying vitamin A.[10] Furthermore, the adverse effects of beta-carotene supplements are most evident in people who drink alcohol and smoke cigarettes.[11]

• **Birth Defects** • Excessive vitamin A poses a **teratogenic** risk.[12] Among women who took more than 10,000 IU◆ of supplemental vitamin A daily, approximately 1 out of every 57 infants was born with a malformation attributable to vitamin A toxicity.[13] High intakes before the seventh week appear to be the most damaging. For this reason, vitamin A is not given as a supplement in the first trimester of pregnancy unless there is specific evidence of deficiency, which is rare.

◆ For perspective, 10,000 IU ≈ 3000 μg vitamin A, roughly four times the RDA for women.

• **Not for Acne** • Adolescents need to know that massive doses of vitamin A have no beneficial effect on **acne.** The prescription medicine Accutane is made from vitamin A but is chemically different. Taken orally, Accutane is effective against the deep lesions of cystic acne. It is highly toxic, however, especially during growth, and has caused birth defects in infants when women have taken it during their pregnancies. For this reason, women taking Accutane must begin using an effective form of contraception at least one month before taking the drug and continue using contraception at least one month after discontinuing its use.

Another vitamin A relative, Retin-A, fights acne, the wrinkles of aging, and other skin disorders. Applied topically, this ointment smooths and softens skin; it also lightens skin that has become darkly pigmented after inflammation. During treatment, the skin becomes red and tender and peels.

© Polara Studios Limited

The carotenoids in foods bring colors to meals; the retinoids in our eyes allow us to see them.

Vitamin A Recommendations

Because the body can derive vitamin A from various retinoids and carotenoids, its contents in foods and its recommendations are expressed as **retinol activity equivalents (RAE).** A microgram of retinol counts as 1 RAE,◆ as does 12 micrograms of beta-carotene. Some supplements report their vitamin A contents using international units (IU),◆ an old measure of vitamin activity used before direct chemical analysis was possible.

◆ 1 RAE = 1 μg retinol.
= 12 μg beta-carotene.
= 24 μg of other vitamin A precursor carotenoids.

◆ 1 IU = 0.3 μg retinol.
= 3.6 μg beta-carotene.
= 7.2 μg of other vitamin A precursor carotenoids.

Vitamin A in Foods

The richest sources of the retinoids are foods derived from animals—liver, fish liver oils, milk and milk products, butter, and eggs. Since vitamin A is fat soluble, it is lost when milk is skimmed. To compensate, nonfat milk is often fortified so as to supply about 40 percent of the RDA per quart.* Margarine is usually fortified so as to provide the same amount of vitamin A as butter.

Plants contain no retinoids, but many vegetables and some fruits contain vitamin A precursors—the carotenoids, red and yellow pigments of plants. Only a few carotenoids have vitamin A activity; the carotenoid with the greatest vitamin A activity is beta-carotene.

• **The Colors of Vitamin A Foods** • The dark leafy greens (like spinach and broccoli—not celery or cabbage) and the rich yellow or deep orange vegetables and fruits (such as winter squash, cantaloupe, carrots, and sweet potatoes—not corn or bananas) help people meet their vitamin A needs (see Figure 11-7). A diet including several servings of such carotene-rich sources helps to ensure a sufficient intake.

An attractive meal that includes foods of different colors most likely supplies vitamin A as well. Most foods with vitamin A activity are brightly colored—

*Similarly, in Canada all milk that has had fat removed must be fortified with vitamin A.

teratogenic (ter-AT-oh-jen-ik): causing abnormal fetal development and birth defects.
- **terato** = monster
- **genic** = to produce

acne: a chronic inflammation of the skin's follicles and oil-producing glands, which leads to an accumulation of oils inside the ducts that surround hairs; usually associated with the maturation of young adults.

retinol activity equivalents (RAE): a measure of vitamin A activity; the amount of retinol that the body will derive from a food containing preformed retinol or its precursor beta-carotene.

FIGURE 11-7 **Vitamin A in Selected Foods**

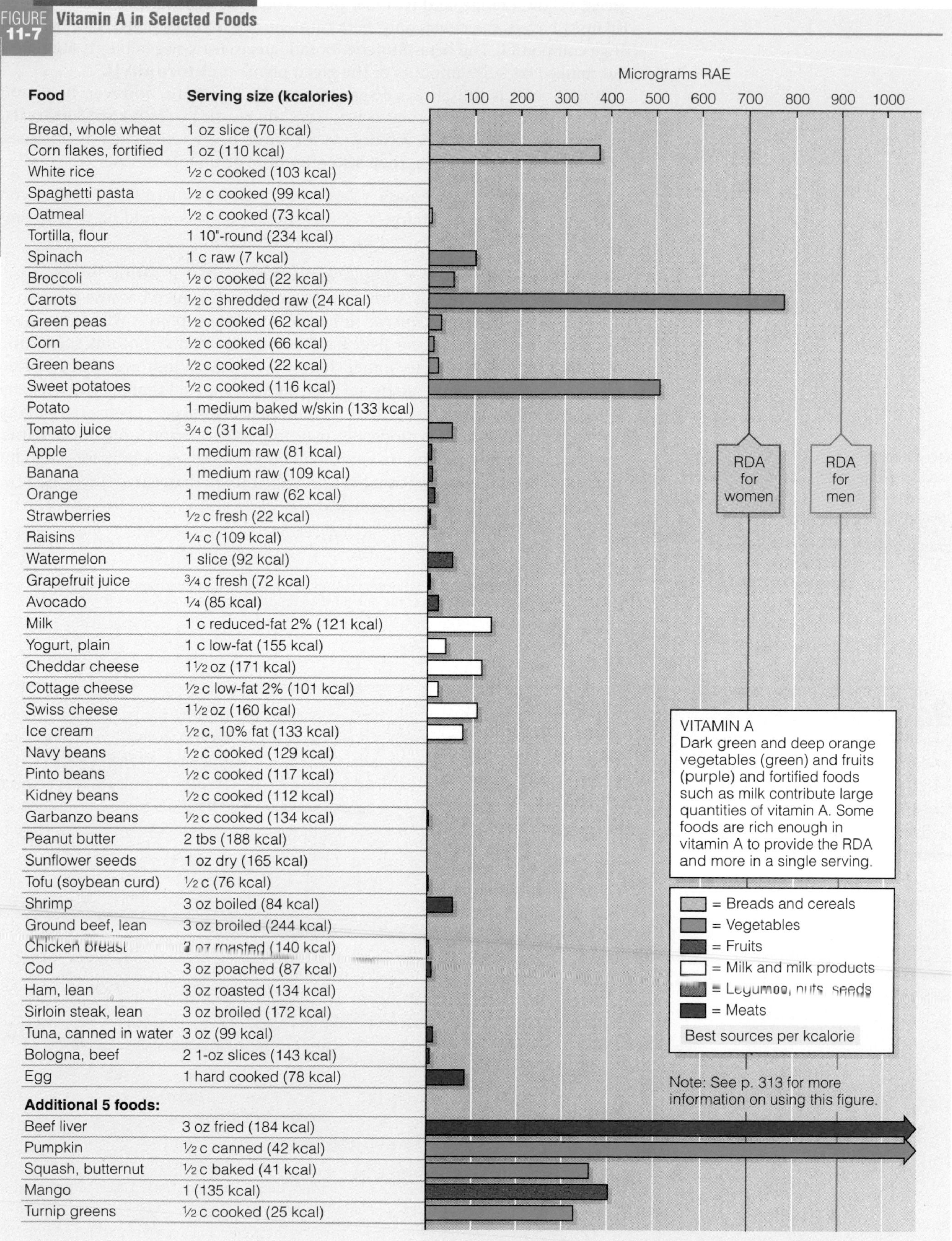

green, yellow, orange, and red. Any plant food with significant vitamin A activity must have some color, since beta-carotene is a rich, deep yellow, almost orange compound. The beta-carotene in dark green, leafy vegetables is abundant but masked by large amounts of the green pigment **chlorophyll.**

Bright color is not always a sign of vitamin A activity, however. Beets and corn, for example, derive their colors from the red and yellow **xanthophylls,** which have no vitamin A activity. As for white plant foods such as potatoes, cauliflower, pasta, and rice, they also offer little or no vitamin A.

• Vitamin A–Poor Fast Foods • Fast foods often lack vitamin A. Anyone who dines frequently on hamburgers, french fries, and colas would be wise to emphasize colorful vegetables and fruits at other meals.

• Vitamin A–Rich Liver • People sometimes wonder if eating liver too frequently can cause vitamin A toxicity. Liver is a rich source because vitamin A is stored there in animals, just as in humans.* Arctic explorers who have eaten large quantities of polar bear liver have become ill with symptoms suggesting vitamin A toxicity. Closer to home, vitamin A toxicity symptoms were reported in young children who regularly ate a chicken liver spread that provided a daily average of up to three times their recommended intake.[14] Liver offers many nutrients, and eating it periodically may improve a person's nutrition status, but caution is warranted not to eat too much too often, especially for pregnant women. With one ounce of beef liver providing more than three times the RDA for vitamin A, intakes can rise quickly.

chlorophyll (KLO-row-fil): the green pigment of plants, which absorbs light and transfers the energy to other molecules, thereby initiating photosynthesis.

xanthophylls (ZAN-tho-fills): pigments found in plants; responsible for the color changes seen in autumn leaves.

*The liver is not the only organ that stores vitamin A. The kidneys, adrenals, and other organs do, too, but the liver stores the most and is the one most commonly eaten.

IN SUMMARY

Vitamin A is found in the body in three forms: retinol, retinal, and retinoic acid. Together, they are essential to vision, healthy epithelial tissues, and growth. Vitamin A deficiency is a major health problem worldwide, leading to infections, blindness, and keratinization. Toxicity can also cause problems and is most often associated with supplement abuse. Animal-derived foods such as liver and milk provide retinoids, whereas brightly colored plant-derived foods such as spinach, carrots, and pumpkins provide beta-carotene and other carotenoids. In addition to serving as a precursor for vitamin A, beta-carotene may act as an antioxidant in the body. The accompanying table summarizes vitamin A's functions in the body, deficiency symptoms, toxicity symptoms, and food sources.

Vitamin A

Other Names	**Deficiency Disease**	**Toxicity Disease**
Retinol, retinal, retinoic acid; precursors are carotenoids such as beta-carotene	Hypovitaminosis A	Hypervitaminosis A[a]
2001 RDA	**Deficiency Symptoms**	**Toxicity Symptoms**
	Blood Symptoms	
Men: 900 µg RAE/day Women: 700 µg RAE/day	Anemia, often masked by dehydration	Loss of hemoglobin and potassium by red blood cells, cessation of menstruation, slowed clotting time, easily induced bleeding
Upper Level	**Deficiency Symptoms**	**Toxicity Symptoms**
	Bones/Teeth Symptoms	
Adults: 3000 µg/day	Cessation of bone growth, painful joints, impaired enamel formation, cracks in teeth, tendency to decay, atrophy of dentin-forming cells	Increased activity of osteoclasts[b] causing decalcification, joint pain, fragility, stunted growth, and thickening of long bones; increase of pressure inside skull, mimicking brain tumor; headaches

[a] A related condition, *hypercarotenemia,* is caused by the accumulation of too much of the vitamin A precursor beta-carotene in the blood, which turns the skin noticeably yellow. Hypercarotenemia is not, strictly speaking, a toxicity symptom.

Vitamin A—(continued)

Chief Functions in the Body

Vision; maintenance of cornea, epithelial cells, mucous membranes, skin; bone and tooth growth; reproduction; immunity

Significant Sources

Retinol: fortified milk, cheese, cream, butter, fortified margarine, eggs, liver

Beta-carotene: spinach and other dark leafy greens; broccoli, deep orange fruits (apricots, cantaloupe) and vegetables (squash, carrots, sweet potatoes, pumpkin)

Deficiency Symptoms	Toxicity Symptoms
Digestive Symptoms	
Changes in lining, diarrhea	Nausea, vomiting, abdominal pain, diarrhea, weight loss
Eye Symptoms[c]	
Night blindness, changes in epithelial tissue (hyperkeratinization), drying (xerosis), triangular gray spots on eye (Bitot's spots), softening of the cornea (keratomalacia), and corneal degeneration and blindness (xerophthalmia)	
Immune System Symptoms	
Suppression of immune reactions; frequent respiratory, digestive, bladder, vaginal, and kidney infections	Overstimulation of immune reactions
Neurological Symptoms	
Brain and spinal cord growth too fast for stunted skull and spine	Loss of appetite, irritability, fatigue, insomnia, restlessness, headaches, blurred vision, nausea, vomiting, muscle weakness
Skin Symptoms	
Plugging of hair follicles with keratin, forming white lumps (hyperkeratosis)	Dryness; itching; peeling; rashes; dry, scaling lips; cracking and bleeding of lips; nosebleeds; loss of hair; brittle nails
Other	
Kidney stones	Amenorrhea,[d] jaundice,[e] enlargement of liver[f] and spleen, massive accumulation of fat and vitamin A in liver, birth defects

[b]*Osteoclasts* are the cells that destroy bone during its growth. Those that build bone are *osteoblasts*.
[c]The eye symptoms of vitamin A deficiency are collectively known as *xerophthalmia*.
[d]Elevated serum carotene concentrations are associated with amenorrhea.
[e]*Jaundice* (JAWN-dice) is a symptom of liver disease, in which bile and related pigments spill into the bloodstream and the skin yellows.
[f]If liver impairment is severe, the "classic" signs seen in skin and hair may be masked.

VITAMIN D

Vitamin D (calciferol)◆ is different from all the other nutrients in that the body can synthesize it, with the help of sunlight, from a precursor that the body makes from cholesterol. Therefore, vitamin D is not an essential nutrient: given enough time in the sun, people need no vitamin D from foods.[15]

◆ Vitamin D comes in many forms, the two most important being a plant version called **vitamin D_2** or **ergocalciferol** (ER-go-kal-SIF-er-ol) and an animal version called **vitamin D_3** or **cholecalciferol** (KO-lee-kal-SIF-er-ol).

Figure 11-8 (on p. 364) diagrams the pathway for making and activating vitamin D in the body. Ultraviolet rays from the sun hit the precursor in the skin and convert it to previtamin D_3. This compound works its way into the body and slowly, over the next 36 hours, is converted to its active form with the help of the body's heat. The biological activity of the active vitamin is 500- to 1000-fold greater than that of its precursor.

Regardless of whether the body manufactures vitamin D_3 or obtains it directly from foods, two reactions must occur before the vitamin becomes fully active. First, the liver adds an OH group, and then the kidneys add another OH group to produce the active vitamin. A review of Figure 11-8 reveals how diseases affecting either the liver or the kidneys can interfere with the activation of vitamin D and produce symptoms of deficiency.

FIGURE 11-8 Vitamin D Synthesis and Activation

The precursor of vitamin D is made in the liver from cholesterol (see Figure 5-10 on p. 137 and Appendix C). The activation of vitamin D is a closely regulated process. The final product, active vitamin D, is also known as1,25-dihydroxycholecalciferol (or calcitriol).

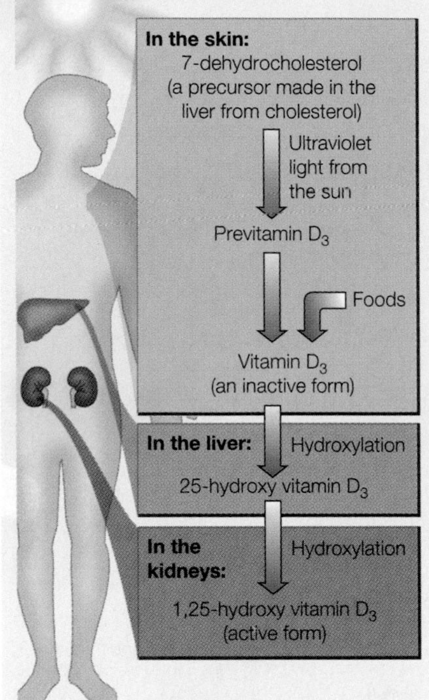

Roles in the Body

Though called a vitamin, vitamin D is actually a hormone—a compound manufactured by one part of the body that causes another part to respond.[16] Like vitamin A, vitamin D has a binding protein that carries it to the target organs—most notably, the intestines, the kidneys, and the bones. All respond to vitamin D by making calcium available for bone growth.

• **Vitamin D in Bone Growth** • Vitamin D is a member of a large and cooperative bone-making and maintenance team composed of nutrients and other compounds, including vitamins A, C, and K; the hormones parathormone and calcitonin; the protein collagen; and the minerals calcium, phosphorus, magnesium, and fluoride. Vitamin D's special role in bone growth is to maintain blood concentrations of calcium and phosphorus. The bones grow denser and stronger as they absorb and deposit these minerals.

Vitamin D raises blood concentrations of these minerals in three ways. It enhances their absorption from the GI tract, their reabsorption by the kidneys, and their mobilization from the bones into the blood. The vitamin may work alone, as it does in the GI tract, or in combination with parathormone, as it does in the bones and kidneys. Vitamin D is the director, but the star of the show is calcium. Details of calcium balance appear in Chapter 12.

• **Vitamin D in Other Roles** • Scientists have discovered many other vitamin D target tissues, including the brain and nervous system, pancreas, skin, muscles and cartilage, reproductive organs, and many cancer cells.[17] These discoveries suggest that vitamin D has numerous functions and may be valuable in treating a number of disorders, including cancer.

Vitamin D Deficiency

In vitamin D deficiency, production of the protein that binds calcium in the intestinal cells slows. Thus, even when calcium in the diet is adequate, it passes through the GI tract unabsorbed, leaving the bones undersupplied. Consequently, a vitamin D deficiency creates a calcium deficiency.

• **Rickets** • Worldwide, the vitamin D–deficiency disease **rickets** still afflicts many children. The bones fail to calcify normally, causing growth retardation and skeletal abnormalities. The bones become so weak that they bend when they have to support the body's weight (see Figure 11-9). A child with rickets who is old enough to walk characteristically develops bowed legs, often the most obvious sign of the disease. Another sign is the protruding belly that results from lax abdominal muscles.

• **Osteomalacia** • The adult form of rickets, **osteomalacia,** occurs most often in women who have low calcium intakes and little exposure to sun and who go through repeated pregnancies and periods of lactation. Given this combination of risk factors, the leg bones may soften to such an extent that a young woman who is tall and straight at 20 may become bent, bowlegged, and stooped before she is 30.

• **Osteoporosis** • Any failure to synthesize adequate vitamin D or obtain enough from foods sets the stage for a loss of calcium from the bones, which can result in fractures. In a group of women with osteoporosis hospitalized for hip fractures, half had an undetected vitamin D deficiency.[18] Highlight 12 describes the many factors that lead to osteoporosis, a condition of reduced bone density.

• **The Elderly** • Vitamin D deficiency is especially likely in older adults for several reasons. For one, the skin, liver, and kidneys lose their capacity to make and activate vitamin D with advancing age. For another, older adults typically

rickets: the vitamin D–deficiency disease in children characterized by inadequate mineralization of bone (manifested in bowed legs or knock-knees, outward-bowed chest, and knobs on ribs). A rare type of rickets, not caused by vitamin D deficiency, is known as *vitamin D–refractory rickets.*

osteomalacia (OS-tee-oh-ma-LAY-shuh): a bone disease characterized by softening of the bones. Symptoms include bending of the spine and bowing of the legs. The disease occurs most often in adult women.
- **osteo** = bone
- **malacia** = softening

drink little or no milk—the main dietary source of vitamin D. And finally, older adults typically spend much of the day indoors, and when they do venture outside, many of them cautiously apply sunscreen to all sun-exposed areas of their skin. All of these factors increase the likelihood of vitamin D deficiency and its consequences: bone losses and fractures.

FIGURE 11-9 **Vitamin D–Deficiency Symptom—The Bowed Legs of Rickets**

The child has the bowed legs commonly seen in rickets.

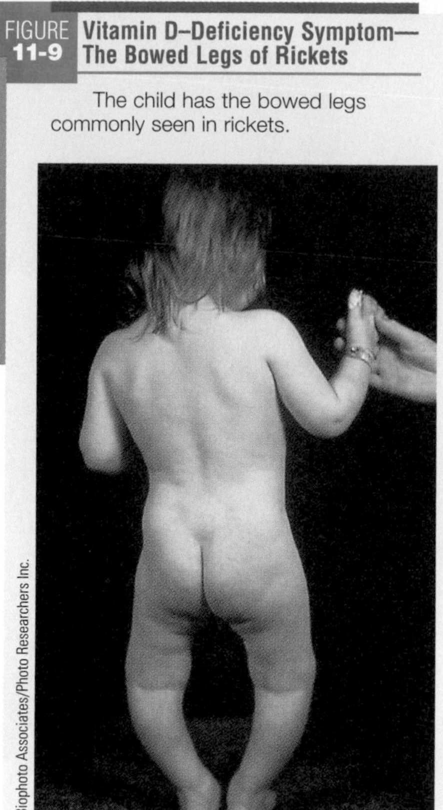

Biophoto Associates/Photo Researchers Inc.

Vitamin D Toxicity

Vitamin D clearly illustrates how nutrients in optimal amounts support health, but both inadequacies and excesses cause trouble. Vitamin D is the most likely of the vitamins to have toxic effects when consumed in excessive amounts. The amounts of vitamin D in foods are well within safe limits, but supplements containing the vitamin in concentrated form should be kept out of the reach of children and used cautiously, if at all, by adults. Vitamin D toxicity has also been reported in people who drank milk that was accidentally fortified with too much vitamin D.[19]

An excess of vitamin D raises the concentration of blood calcium.◆ Excess blood calcium tends to precipitate in the soft tissue, forming stones, especially in the kidneys where calcium is concentrated in the effort to excrete it. Calcification may also harden the blood vessels and is especially dangerous in the major arteries of the heart and lungs, where it can cause death.

◆ High blood calcium is known as **hypercalcemia** and may develop from a variety of disorders, including vitamin D toxicity. It does *not* develop from a high calcium intake.

Vitamin D Recommendations and Sources

Only a few foods contain vitamin D naturally. Fortunately, the body can make all the vitamin D it needs with the help of a little sunshine. In setting dietary recommendations, however, the committee assumed that no vitamin D was available from this source.

• Vitamin D in Foods • Most adults, especially in sunny regions, need not make special efforts to obtain vitamin D from food. People who are not outdoors much or who live in northern or predominantly cloudy or smoggy areas are advised to drink at least 2 cups of vitamin D–fortified milk a day. The fortification of milk with vitamin D is the best guarantee that people will meet their needs and underscores the importance of milk in a well-balanced diet.* For those who use margarine in place of butter, fortified margarine is a significant source. A plant version of vitamin D may yield an active compound on irradiation, but its contribution is minor. Without adequate sunshine, fortification, or supplementation, a vegan diet cannot meet vitamin D needs.

• Vitamin D from the Sun • Most of the world's population relies on natural exposure to sunlight to maintain adequate vitamin D nutrition. The sun imposes no risk of vitamin D toxicity; prolonged exposure to sunlight degrades the vitamin D precursor in the skin, preventing its conversion to the active vitamin. Even lifeguards on southern beaches are safe from vitamin D toxicity from the sun.

Prolonged exposure to sunlight does, however, prematurely wrinkle the skin and present the risk of skin cancer. Sunscreens help reduce these risks, but unfortunately, sunscreens with sun protection factors (SPF) of 8 and above also prevent vitamin D synthesis. A strategy to avoid this dilemma is

© Francisco Cruz/Super Stock

A cool glass of milk refreshes as it replenishes vitamin D and other bone-building nutrients.

*Vitamin D fortification of milk in the United States is 10 micrograms cholecalciferol (400 IU) per quart; in Canada, 9.6 micrograms (385 IU) per liter.

© Corbis/Fotographia

The sunshine vitamin: vitamin D.

to apply sunscreen after enough time has elapsed to provide sufficient vitamin D synthesis. For most people, exposing hands, face, and arms on a clear summer day for 10 to 15 minutes a few times a week should be sufficient to maintain vitamin D nutrition.

The pigments of dark skin provide some protection from the sun's damage, but they also reduce vitamin D synthesis.[20] Dark-skinned people require longer sunlight exposure than light-skinned people: heavily pigmented skin achieves the same amount of vitamin D synthesis in three hours as fair skin in 30 minutes. Latitude, season, and time of day◆ also have dramatic effects on vitamin D synthesis (see Figure 11-10). The ultraviolet (UV) rays of the sun that promote vitamin D synthesis are blocked by heavy clouds, smoke, or smog. Differences in skin pigmentation, latitude, and smog may account for the finding that dark-skinned people in northern, smoggy cities are most likely to develop rickets. For these people, and for those who are unable to go outdoors frequently, dietary vitamin D is essential.

Depending on the radiation used, the UV rays from tanning lamps and tanning booths may also stimulate vitamin D synthesis, but the hazards outweigh any possible benefits.* The Food and Drug Administration (FDA) warns that if the lamps are not properly filtered, people using tanning booths risk burns, damage to the eyes and blood vessels, and skin cancer.

*The best wavelengths for vitamin D synthesis are UV-B rays between 290 and 310 nanometers. Some tanning parlors advertise "UV-A rays only, for a tan without the burn," but in fact, UV-A rays can damage the skin.

◆ **Factors that may limit sun exposure and, therefore, vitamin D synthesis:**
- Geographic location.
- Season of the year.
- Time of day.
- Air pollution.
- Clothing.
- Tall buildings.
- Indoor living.
- Sunscreens.

FIGURE 11-10 Vitamin D Synthesis and Latitude

Above 40° north latitude (and below 40° south latitude in the southern hemisphere), vitamin D synthesis essentially ceases for the four months of winter. Synthesis increases as spring approaches, peaks in summer, and declines again in the fall. People living in regions of extreme northern (or extreme southern) latitudes may miss as much as six months of vitamin D production.

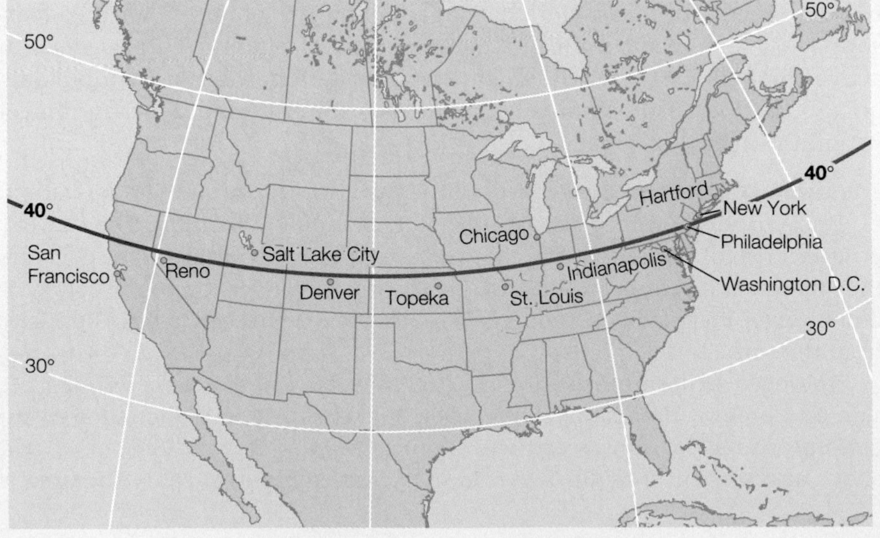

IN SUMMARY Vitamin D can be synthesized in the body with the help of sunlight or obtained from foods derived from animals. It sends signals to three primary target sites: the GI tract to absorb more calcium and phosphorus, the bones to release more, and the kidneys to retain more. These actions maintain blood calcium concentrations and support bone formation. A deficiency causes rickets in childhood and osteomalacia in later life. Fortified milk is an important food source. The accompanying table summarizes vitamin D facts.

◆ **Vitamin D activity was previously expressed in international units (IU), but is now expressed in micrograms of cholecalciferol. To convert, use the following factor: 1 IU = 0.025 µg cholecalciferol. For example:**

- 100 IU = 2.5 µg (100 IU × 0.025 µg).
- 400 IU = 10 µg (400 IU × 0.025 µg).

Vitamin D

Other Names

Calciferol (kal-SIF-er-ol), 1,25-dihyroxy vitamin D (calcitriol); the animal version is vitamin D_3 or cholecalciferol; the plant version is vitamin D_2 or ergocalciferol; precursor is the body's own cholesterol

1997 Adequate Intake (AI)◆

Adults: 5 µg/day (19–50 yr)
10 µg/day (51 –70 yr)
15 µg/day (>70 yr)

Upper Level

Adults: 50 µg/day

Chief Functions in the Body

Mineralization of bones (raises blood calcium and phosphorus by increasing absorption from digestive tract, withdrawing calcium from bones, stimulating retention by kidneys)

Significant Sources

Synthesized in the body with the help of sunlight; fortified milk, margarine, butter, cereals, and chocolate mixes; veal, beef, egg yolks, liver, fatty fish (herring, salmon, sardines) and their oils

Deficiency Diseases	Toxicity Disease
Rickets, osteomalacia	Hypervitaminosis D

Deficiency Symptoms	Toxicity Symptoms
Blood Symptoms	
Decreased calcium and/or phosphorus concentration, increased alkaline phosphatase[a]	Increased calcium and phosphorus concentration
Bones/Teeth Symptoms	
Rickets in Children: Faulty calcification, resulting in misshapen bones (bowing of legs) and retarded growth; enlargement of ends of long bones (knees, wrists); deformities of ribs (bowed, with beads or knobs);[b] delayed closing of fontanel, resulting in rapid enlargement of head (see figure); slow eruption of teeth; malformed, decay-prone teeth *Osteomalacia in Adults*: Softening effect: deformities of limbs, spine, thorax, and pelvis; demineralization; pain in pelvis, lower back, and legs; bone fractures	Enamel hypoplasia, pulp calcification
	Digestive Symptoms
	Nausea, vomiting
Excretory Symptoms	
Increased excretion of calcium in stools, decreased calcium in urine	Increased excretion of calcium in urine, kidney stones, irreversible kidney damage
Neurological/Muscular Symptoms	
Lax muscles resulting in protrusion of abdomen; muscle spasms (rickets); Involuntary twitching, muscle spasms (osteomalacia)	Loss of appetite, headache, muscle weakness, joint pain, fatigue, excessive thirst, irritability, apathy
Other	
Abnormally high secretion of parathormone	Calcification of soft tissues (blood vessels, kidneys, heart, lungs, tissues around joints), frequent urination, death

Fontanel
A fontanel is an open space in the top of a baby's skull before the bones have grown together. In rickets, closing of the fontanel is delayed.

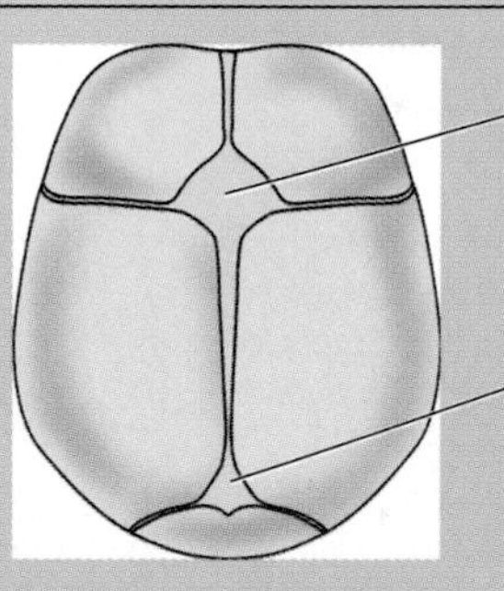

[a]Alkaline phosphatase is an enzyme in the blood that rises during bone resorption.
[b]Bowing of the ribs causes the symptoms known as *pigeon breast.* The beads that form on the ribs resemble rosary beads; thus this symptom is known as *rachitic* (ra-KIT-ik) *rosary* ("the rosary of rickets").

VITAMIN E

Researchers discovered a component of vegetable oils necessary for reproduction in rats and named this antisterility factor **tocopherol,** which means "to bring forth offspring." When chemists isolated four different tocopherol compounds, they designated them by the first four letters of the Greek alphabet: alpha, beta, gamma, and delta. The tocopherols consist of a complex ring structure and a long saturated side chain (Appendix C provides the chemical structures). The positions of methyl groups (CH_3) on the side chain and their chemical rotations distinguish one tocopherol from another. **Alpha-tocopherol** is the only one with vitamin E activity in the human body.[21] The other tocopherols are not readily converted to alpha-tocopherol in the body, nor do they perform the same roles.

Vitamin E as an Antioxidant

Vitamin E is a fat-soluble antioxidant and one of the body's primary defenders against the adverse effects of free radicals. Its main action is to stop the chain reaction of free radicals producing more free radicals (see Figure 11-11). In doing so, vitamin E protects the vulnerable components of the cells and their membranes from destruction. Most notably, vitamin E prevents the oxidation of the polyunsaturated fatty acids (PUFA), but it protects other lipids and related compounds (for example, vitamin A) as well.

Accumulating evidence suggests that vitamin E may reduce the risk of heart disease by protecting low-density lipoproteins (LDL) against oxidation.[22] The oxidation of LDL has been implicated as a key factor in the development of heart disease. Highlight 11 provides many more details on how vitamin E and other antioxidants protect against chronic diseases, such as heart disease and cancer.[23]

While research continues to reveal possible roles for vitamin E, it also has clearly discredited claims that vitamin E improves physical performance, enhances sexual performance, or cures sexual dysfunction in males. Vitamin E does not slow or prevent the processes of aging such as hair turning gray or skin wrinkling. Nor does it slow the progression of Parkinson's disease.

Vitamin E Deficiency

In human beings, a primary deficiency of vitamin E (from a poor intake) is rare; deficiency is usually associated with diseases of fat malabsorption such as cystic fibrosis. Without vitamin E, the red blood cells break open and spill their contents, probably due to oxidation of the PUFA in their membranes. This classic sign of vitamin E deficiency, known as **erythrocyte hemolysis,** is seen in premature infants, born before the transfer of vitamin E from the mother to the infant that takes place in the last weeks of pregnancy. Vitamin E treatment corrects **hemolytic anemia.**

Prolonged vitamin E deficiency also causes neuromuscular dysfunction involving the spinal cord and retina of the eye. Common symptoms include loss of muscle coordination and reflexes and impaired vision and speech. Vitamin E treatment corrects these neurological symptoms of vitamin E deficiency, but it does *not* prevent or cure the hereditary **muscular dystrophy** that afflicts children. Children with this condition do not benefit from vitamin E treatment and usually die at an early age when their respiratory muscles deteriorate.

Two other conditions seem to respond to vitamin E therapy, although results are inconsistent. One is a nonmalignant breast disease **(fibrocystic breast disease),** and the other is an abnormality of blood flow that causes cramping in the legs **(intermittent claudication).**

tocopherol (tuh-KOFF-er-ol): a general term for several chemically related compounds, one of which has vitamin E activity (see Appendix C for chemical structures).

alpha-tocopherol: the active vitamin E compound.

erythrocyte (eh-RITH-ro-cite) **hemolysis** (he-MOLL-uh-sis): the breaking open of red blood cells (erythrocytes); a symptom of vitamin E–deficiency disease in human beings.
- **erythro** = red
- **cyte** = cell
- **hemo** = blood
- **lysis** = breaking

hemolytic (HE-moh-LIT-ick) **anemia:** the condition of having too few red blood cells as a result of erythrocyte hemolysis.

muscular dystrophy (DIS-tro-fee): a hereditary disease in which the muscles gradually weaken. Its most debilitating effects arise in the lungs.

fibrocystic (FYE-bro-SIS-tik) **breast disease:** a harmless condition in which the breasts develop lumps, sometimes associated with caffeine consumption. In some, it responds to abstinence from caffeine; in others, it can be treated with vitamin E.
- **fibro** = fibrous tissue
- **cyst** = closed sac

intermittent claudication (klaw-dih-KAY-shun): severe calf pain caused by inadequate blood supply. It occurs when walking and subsides during rest.
- **intermittent** = at intervals
- **claudicare** = to limp

FIGURE 11-11 **Free Radical Formation and Antioxidant Protection**

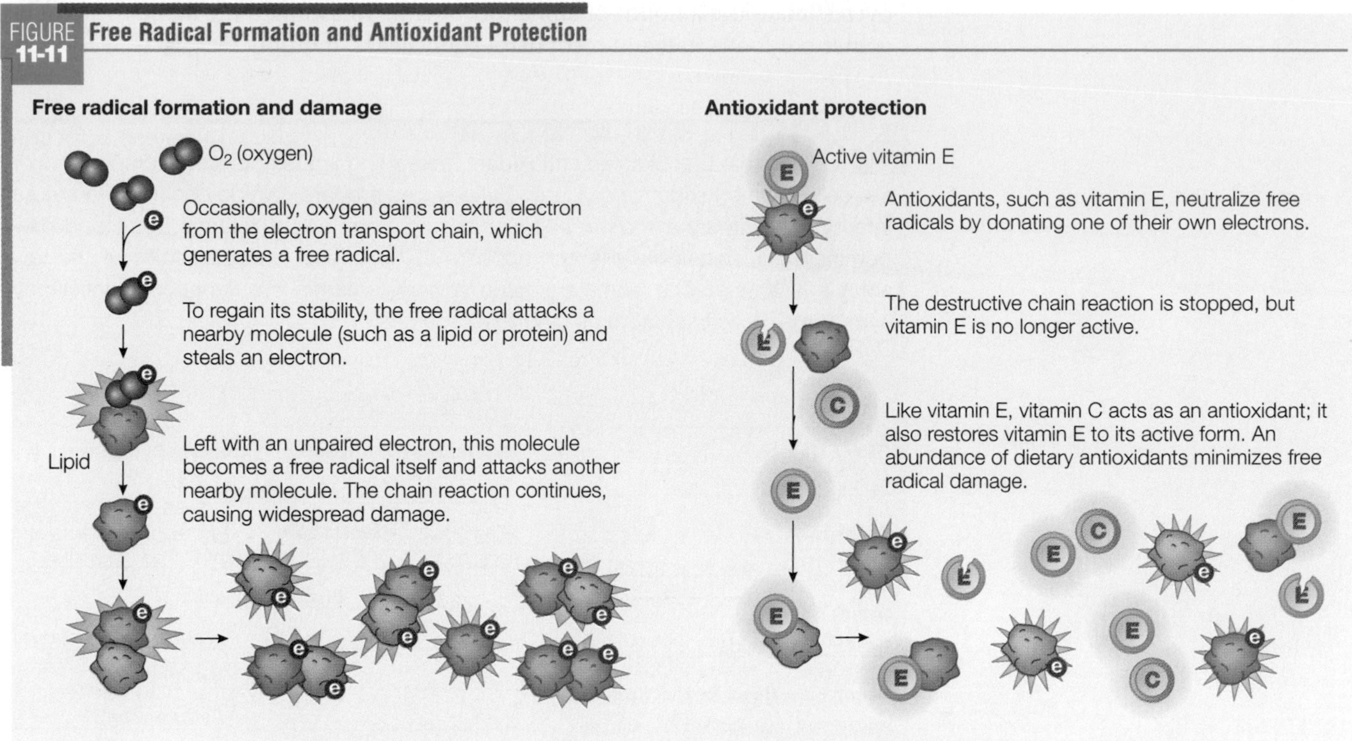

Vitamin E Toxicity

Vitamin E supplement use has risen in recent years as its protective actions against chronic diseases have been recognized. Still, toxicity is rare, and its effects are not as detrimental as with vitamins A and D. The upper level for vitamin E (1000 milligrams) is more than 65 times greater than the recommended intake for adults (15 milligrams). Extremely high doses of vitamin E may interfere with the blood-clotting action of vitamin K and enhance the effects of drugs used to oppose blood clotting, causing hemorrhage.

◆ Measures of vitamin E activity that consider all the various forms of tocopherols are expressed in milligrams of **tocopherol equivalents (mg TE).**

Vitamin E Recommendations

The current RDA for vitamin E differs from previous recommendations in being based on the alpha-tocopherol form only. As mentioned earlier, the other tocopherols cannot be converted to alpha-tocopherol nor can they perform the same metabolic roles in the body. A person who consumes large quantities of PUFA needs more vitamin E. Fortunately, vitamin E and PUFA tend to occur together in the same foods.

Vitamin E in Foods

Vitamin E is widespread in foods. Much of the vitamin E in the diet comes from vegetable oils and products made from them, such as margarine and salad dressings. Wheat germ oil is especially rich in vitamin E.

Vitamin E is readily destroyed by heat processing (such as deep-fat frying) and oxidation, so fresh or lightly processed foods are preferable sources. Most processed and convenience foods do not contribute enough vitamin E to ensure an adequate intake.

Published values of the vitamin E in foods reflect all of the various tocopherols and are expressed in "milligrams of tocopherol equivalents."◆ These measures

© Craig M. Walker

Fat-soluble vitamin E is found predominantly in vegetable oils, seeds, and nuts.

overestimate the amount of alpha-tocopherol. To estimate the alpha-tocopherol content of foods stated in tocopherol equivalents, multiply by 0.8.[24]

IN SUMMARY

Vitamin E acts as an antioxidant, defending lipids and other components of the cells against oxidative damage. Deficiencies are rare, but do occur in premature infants, the primary symptom being erythrocyte hemolysis. Vitamin E is found predominantly in vegetable oils and appears to be one of the least toxic of the fat-soluble vitamins. The summary table reviews vitamin E's functions, deficiency symptoms, toxicity symptoms, and food sources.

Vitamin E

Other Names
Alpha-tocopherol

2000 RDA
Adults: 15 mg/day

Upper Level
Adults: 1000 mg/day

Chief Functions in the Body
Antioxidant (stabilization of cell membranes, regulation of oxidation reactions, protection of polyunsaturated fatty acids [PUFA] and vitamin A)

Significant Sources
Polyunsaturated plant oils (margarine, salad dressings, shortenings), leafy green vegetables, wheat germ, whole-grains, liver, egg yolks, nuts, seeds

Easily destroyed by heat and oxygen

Deficiency Symptoms	Toxicity Symptoms
Blood Symptoms	
Red blood cell breakage,[a] anemia	Augments the effects of anticlotting medication
Digestive Symptoms	
	Nausea, stomach cramps
Eye Symptoms	
	Blurred vision
Neurological/Muscular Symptoms	
Degeneration, weakness, difficulty walking, leg cramps	Fatigue

[a]The breaking of red blood cells is called *erythrocyte hemolysis.*

VITAMIN K

◆ *K* stands for the Danish word *koagulation* ("coagulation" or "clotting").

Like vitamin D, vitamin K can be obtained from a nonfood source. Bacteria in the GI tract synthesize vitamin K that the body can absorb. Vitamin K◆ acts primarily in blood clotting, where its presence can make the difference between life and death. Blood has a remarkable ability to remain a liquid, but can turn solid within seconds when the integrity of that system is disturbed. (If blood did not clot, a single pinprick could drain the entire body of all its blood, just as a tiny hole in a bucket makes the bucket forever useless for holding water.)

Roles in the Body

hemorrhagic (hem-oh-RAJ-ik) **disease:** a disease characterized by excessive bleeding.

hemophilia (HE-moh-FEEL-ee-ah): a hereditary disease that is caused by a genetic defect and has no relation to vitamin K. The blood is unable to clot because it lacks the ability to synthesize certain clotting factors.

More than a dozen different proteins and the mineral calcium are involved in making a blood clot. Vitamin K is essential for the activation of several of these proteins, among them prothrombin, made by the liver as a precursor of the protein thrombin (see Figure 11-12).[25] When any of the blood-clotting factors is lacking, **hemorrhagic disease** results. If an artery or vein is cut or broken, bleeding goes unchecked. (Of course, this is not to say that hemorrhaging is always caused by vitamin K deficiency. Another cause is **hemophilia,** which is not curable with vitamin K.)

Vitamin K also participates in the synthesis of bone proteins. Without vitamin K, the bones produce an abnormal protein that cannot bind to the min-

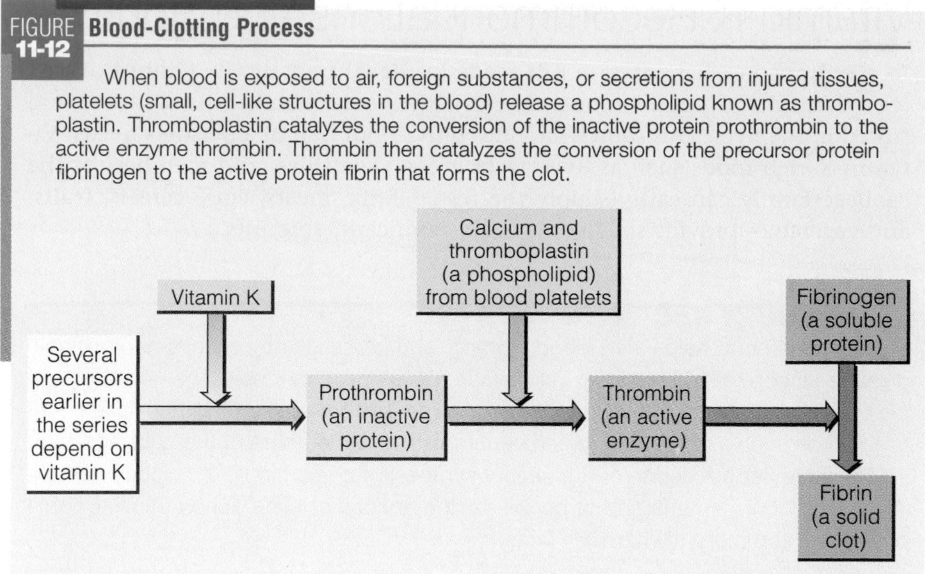

FIGURE 11-12 **Blood-Clotting Process**

When blood is exposed to air, foreign substances, or secretions from injured tissues, platelets (small, cell-like structures in the blood) release a phospholipid known as thromboplastin. Thromboplastin catalyzes the conversion of the inactive protein prothrombin to the active enzyme thrombin. Thrombin then catalyzes the conversion of the precursor protein fibrinogen to the active protein fibrin that forms the clot.

erals that normally form bones. An adequate intake of vitamin K may help protect against hip fractures.[26]

Vitamin K is historically known for its role in blood clotting, and more recently for its participation in bone building, but researchers continue to discover proteins needing vitamin K's assistance. These proteins have been identified in the plaques of atherosclerosis, the kidneys, and the nervous system.[27]

Vitamin K Deficiency

A primary deficiency◆ of vitamin K is rare, but a secondary deficiency may occur in two circumstances. First, whenever fat absorption falters, as occurs when bile production fails, vitamin K absorption diminishes. Second, some drugs disrupt vitamin K's synthesis and action in the body: antibiotics kill the vitamin K–producing bacteria in the intestine, and anticoagulant drugs interfere with vitamin K metabolism and activity. When vitamin K deficiency does occur, it can be fatal.

Newborn infants present a unique case of vitamin K nutrition because they are born with a **sterile** intestinal tract, and the vitamin K–producing bacteria take weeks to establish themselves. At the same time, plasma prothrombin concentrations are low (this reduces the likelihood of fatal blood clotting during the stress of birth). To prevent hemorrhagic disease in the newborn, a single dose of vitamin K◆ (usually as the naturally occurring form, phylloquinone) is given at birth either orally or by intramuscular injection. Concerns that vitamin K given at birth raises the risks of childhood cancer are unproved and unlikely.

Reminder: A *primary deficiency* develops in response to an inadequate dietary intake whereas a *secondary deficiency* occurs for other reasons. ◆

The natural form of vitamin K is phylloquinone (FILL-oh-KWIN-own); the synthetic form is menadione (men-uh-DYE-own). See Appendix C for the chemistry of these structures. ◆

Vitamin K Toxicity

Toxicity is not common but can result when vitamin K supplements are prescribed, especially to infants or pregnant women. High doses of vitamin K can reduce the effectiveness of anticoagulant drugs used to prevent blood clotting. People taking these drugs should eat vitamin K–rich foods in moderation and keep their intakes consistent from day to day. Toxicity symptoms include red blood cell hemolysis, **jaundice**, and brain damage.

sterile: free of microorganisms, such as bacteria.

jaundice (JAWN-dis): yellowing of the skin due to spillover of the bile pigment **bilirubin** (bill-ee-ROO-bin) from the liver into the general circulation; also known as **hyperbilirubinemia** (HIGH-per-BILL-eh-roo-bin-EE-me-ah). When these pigments invade the brain, the condition is **kernicterus** (ker-NICK-ter-us). Jaundice may be caused by obstruction of bile passageways, hemolysis, or dysfunctional liver cells.

© Polara Studios Inc.

Notable food sources of vitamin K include milk, eggs, brussels sprouts, collards, liver, cabbage, spinach, and broccoli.

Vitamin K Recommendations and Sources

As mentioned earlier, vitamin K is made in the GI tract by the billions of bacteria that normally reside there. Once synthesized, vitamin K is absorbed and stored in the liver. This source provides only about half of a person's needs. Vitamin K–rich foods such as liver, leafy green vegetables, and members of the cabbage family can easily supply the rest.[28] Milk, meats, eggs, cereals, fruits, and vegetables provide smaller, but still significant, amounts.

IN SUMMARY

Vitamin K helps with blood clotting, and its deficiency causes hemorrhagic disease (uncontrolled bleeding). Bacteria in the GI tract can make the vitamin; people typically receive about half of their requirements from bacterial synthesis and half from foods such as liver, leafy green vegetables, and members of the cabbage family. Because people depend on bacterial synthesis for vitamin K, deficiency is most likely in newborn infants and in people taking antibiotics. The accompanying table provides a summary of vitamin K facts.

Vitamin K

Other Names	
Phylloquinone, menaquinone, menadione, naphthoquinone	
2001 AI	
Men:	120 μg/day
Women:	90 μg/day

Deficiency Symptoms	Toxicity Symptoms
Blood Symptoms	
Hemorrhaging	Interference with anticlotting medication; vitamin K analogues may cause jaundice, red blood cell hemolysis, and brain damage
Bone Symptoms	
Skeletal weakness	

Chief Functions in the Body

Synthesis of blood-clotting proteins and bone proteins that regulate blood calcium

Significant Sources

Bacterial synthesis in the digestive tract;[a] liver; leafy green vegetables, cabbage-type vegetables; milk

[a]Vitamin K needs cannot be met from bacterial synthesis alone; however, it is a potentially important source in the small intestine, where absorption efficiency ranges from 40 to 70 percent.

THE FAT-SOLUBLE VITAMINS—IN SUMMARY

The four fat-soluble vitamins play many specific roles in the growth and maintenance of the body. Their presence affects the health and function of the eyes, skin, GI tract, lungs, bones, teeth, nervous system, and blood; their deficiencies become apparent in these same areas. Toxicities of the fat-soluble vitamins are possible, especially when people use supplements, because the body stores excesses.

As with the water-soluble vitamins, the function of one fat-soluble vitamin often depends on the presence of another. Recall that vitamin E protects vitamin A from oxidation. In vitamin E deficiency, vitamin A absorption and storage are impaired. Three of the four fat-soluble vitamins—A, D, and K—play important roles in bone growth and remodeling. As mentioned, vitamin K helps synthesize a specific bone protein, and vitamin D regulates that synthesis. Vitamin A, in turn, may control which bone-building genes respond to vitamin D.

Fat-soluble vitamins also interact with minerals: vitamin D and calcium cooperate in bone formation; and zinc is required for the synthesis of vitamin A's transport protein, retinol-binding protein. Zinc also assists the enzyme that regenerates retinal from retinol in the eye.

The roles of the fat-soluble vitamins differ from those of the water-soluble vitamins, and they appear in different foods, yet they are just as essential to life. The need for them underlines the importance of eating a wide variety of nourishing foods daily. The accompanying table condenses the information on fat-soluble vitamins into a short summary.

IN SUMMARY

The Fat-Soluble Vitamins

Vitamin and Chief Functions	Deficiency Symptoms	Toxicity Symptoms	Significant Sources
Vitamin A Vision; maintenance of cornea, epithelial cells, mucous membranes, skin; bone and tooth growth; reproduction; immunity	Infectious diseases, night blindness, blindness (xerophthalmia), keratinization	Bone abnormalities, skin rashes, hair loss, birth defects	Retinol: milk and milk products Beta-carotene: dark green leafy and deep yellow/orange vegetables
Vitamin D Mineralization of bones (raises blood calcium and phosphorus by increasing absorption from digestive tract, withdrawing calcium from bones, stimulating retention by kidneys)	Rickets, osteomalacia	Calcium imbalance (calcification of soft tissues and formation of stones)	Synthesized in the body with the help of sunshine; fortified milk
Vitamin E Antioxidant (stabilization of cell membranes, regulation of oxidation reactions, protection of polyunsaturated fatty acids [PUFA] and vitamin A)	Erythrocyte hemolysis	Not common	Vegetable oils
Vitamin K Synthesis of blood-clotting proteins and bone proteins that regulate blood calcium	Hemorrhage	Not common	Synthesized in the body by GI bacteria; green, leafy vegetables

How are you doing?

- Do you eat dark green, leafy or deep yellow vegetables daily?
- Do you drink vitamin D–fortified milk or go outside in the sunshine regularly?
- Do you use vegetable oils when you cook?

Nutrition on the Net

WEBSITES

Access these websites for further study of topics covered in this chapter. Be aware that many websites on the Internet are peddling vitamin supplements, not accurate information.

- Find updates and quick links to these and other nutrition-related sites at our website: **www.wadsworth.com/nutrition**

- Search for "vitamins" at the American Dietetic Association: **www.eatright.org**
- Review the Dietary Reference Intakes for vitamins A, D, E, and K and the carotenoids by searching for "DRI": **www.nap.edu**
- Visit the World Health Organization to learn about "vitamin deficiencies" around the world: **www.who.int**
- Search for "vitamins" at the U.S. Government health information site: **www.healthfinder.gov**
- Learn how fruits and vegetables support a healthy diet rich in vitamins from the 5 A Day for Better Health program: **www.5aday.com**

INTERNET ACTIVITIES

Interest in alternative medicine and herbal supplements has grown rapidly, as consumers begin to include a variety of alternative treatments in their health routines. Recent surveys show that supplement use is increasing in the college-age population. Have you ever considered taking an herbal product? Are you taking one at the present time?

- Click on: **http://altmedicine.com**
- Click on "Diet & Nutrition," on the left side bar.
- Click on "The Facts about Herbs."
- Click on the active links to learn more about the herbs.

As this information points out, there are many issues to be considered when making the decision to take an herbal product. What did you learn about herbal supplements from this activity?

Making it Click

These exercises will help you learn the best food sources for the vitamins and prepare you to examine your own food choices. See p. 376 for answers.

1. Review the units in which vitamins are measured (a spot check). For each of these vitamins, note the unit of measure:

Vitamin A	Vitamin D
Vitamin E	Vitamin K

2. Analyze the vitamin contents of foods. Review the figures, photos, and food sources sections in Chapters 10 and 11 and list the food group(s) that contributed the most of each vitamin. Which food groups offer the most thiamin? The most riboflavin? The most niacin? The most vitamin B_6? The most folate? The most vitamin B_{12}? The most vitamin C? The most vitamin A? The most vitamin D? The most vitamin E?

   List the groups that provided "the most" and compare them with the Food Guide Pyramid in Chapter 2.

This exercise should convince you that each of the food groups provides some, but not all, of the vitamins needed daily. For a full array, a person needs to eat a variety of foods from each of the food groups regularly.

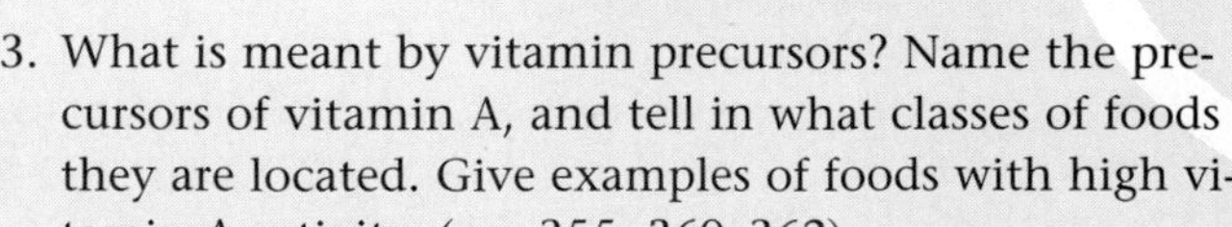

Study Questions

These questions will help you review the chapter. You will find the answers in the discussions on the pages provided.

1. List the fat-soluble vitamins. What characteristics do they have in common? How do they differ from the water-soluble vitamins? (p. 355)
2. Summarize the roles of vitamin A and the symptoms of its deficiency. (pp. 355–359)
3. What is meant by vitamin precursors? Name the precursors of vitamin A, and tell in what classes of foods they are located. Give examples of foods with high vitamin A activity. (pp. 355, 360–362)
4. How is vitamin D unique among the vitamins? What is its chief function? What are the richest sources of this vitamin? (pp. 363–364, 365–366)
5. Describe vitamin E's role as an antioxidant. What are the chief symptoms of vitamin E deficiency? (p. 368)

6. What is vitamin K's primary role in the body? What conditions may lead to vitamin K deficiency? (pp. 370–371)

These questions will help you prepare for an exam. Answers can be found on p. 376.

1. Fat-soluble vitamins:
   a. are easily excreted.
   b. seldom reach toxic levels.
   c. require bile for absorption.
   d. are not stored in the body's tissues.
2. The form of vitamin A active in vision is:
   a. retinal.
   b. retinol.
   c. rhodopsin.
   d. retinoic acid.
3. Vitamin A–deficiency symptoms include:
   a. rickets and osteomalacia.
   b. hemorrhaging and jaundice.
   c. night blindness and keratomalacia.
   d. fibrocystic breast disease and erythrocyte hemolysis.
4. Good sources of vitamin A include:
   a. oatmeal, pinto beans, and ham.
   b. apricots, turnip greens, and liver.
   c. whole-wheat bread, green peas, and tuna.
   d. corn, grapefruit juice, and sunflower seeds.
5. To keep minerals available in the blood, vitamin D targets:
   a. the skin, the muscles, and the bones.
   b. the kidneys, the liver, and the bones.
   c. the intestines, the kidneys, and the bones.
   d. the intestines, the pancreas, and the liver.
6. Vitamin D can be synthesized from a precursor that the body makes from:
   a. bilirubin.
   b. tocopherol.
   c. cholesterol.
   d. beta-carotene.
7. Vitamin E's most notable role is to:
   a. protect lipids against oxidation.
   b. activate blood-clotting proteins.
   c. support protein and DNA synthesis.
   d. enhance calcium deposits in the bones.
8. The classic sign of vitamin E deficiency is:
   a. rickets.
   b. xeropthalmia.
   c. muscular dystrophy.
   d. erythrocyte hemolysis.
9. Without vitamin K:
   a. muscles atrophy.
   b. bones become soft.
   c. skin rashes develop.
   d. blood fails to clot.
10. A significant amount of vitamin K comes from:
    a. vegetable oils.
    b. sunlight exposure.
    c. bacterial synthesis.
    d. fortified grain products.

References

1 E. Li and A. W. Norris, Structure/function of cytoplasmic viatmin A–binding proteins, *Annual Review of Nutrition* 16 (1996): 205–234.

2 D. R. Soprano and K. J. Soprano, Retinoids as teratogens, *Annual Review of Nutrition* 15 (1995): 111–132.

3 H. Hadi and coauthors, Vitamin A supplementation selectively improves the linear growth of Indonesian preschool children: Results from a randomized controlled trial, *American Journal of Clinical Nutrition* 71 (2000): 507–513.

4 S. L. Teitelbaum and coauthors, Cellular and molecular mechanisms of bone resorption, *Mineral and Electrolyte Metabolism* 21 (1995): 193–196.

5 Committee on Dietary Reference Intakes, *Dietary Reference Intakes for Vitamin C, Vitamin E, Selenium, and Carotenoids* (Washington, D.C.: National Academy Press, 2000).

6 C. E. West, Vitamin A and measles, *Nutrition Reviews* 58 (2000): S46–S54.

7 A. H. Shankar and coauthors, Effect of vitamin A supplementation on morbidity due to Plasmodium falciparum in young children in Papua New Guinea randomised trial, *Lancet* 354 (1999): 203–209; J. E. Tyson and coauthors, Vitamin A supplementation for extremely-low-birth-weight infants, *New England Journal of Medicine* 340 (1999): 1962–1968; F. Sempértegui and coauthors, The beneficial effects of weekly low-dose vitamin A supplementation on acute lower respiratory infections and diarrhea in Ecuadorian children, *Pediatrics* 104 (1999): 101 (www.pediatrics.org/cgi/content/full/104/l/el).

8 P. Christian and coauthors, Working after the sun goes down: Exploring how night blindness impairs women's work activities in rural Nepal, *European Journal of Clinical Nutrition* 52 (1998): 519–524.

9 A. Sommer, Xerophthalmia and vitamin A status, *Progress in Retinal and Eye Research* 17 (1998): 9–31.

10 X.-D. Wang and coauthors, Retinoid signaling and activator protein-1 expression in ferrets given β-carotene supplements and exposed to tobacco smoke, *Journal of the National Cancer Institute* 91 (1999): 60–66.

11 M. A. Leo and C. S. Lieber, Alcohol, vitamin A, and β-carotene: Adverse interactions, including hepatotoxicity and carcinogenicity, *American Journal of Clinical Nutrition* 69 (1999): 1071–1085.

12 Soprano and Soprano, 1995.

13 K. J. Rothman and coauthors, Teratogenicity of high vitamin A intake, *New England Journal of Medicine* 333 (1995): 1369–1373.

14 T. O. Carpenter and coauthors, Severe hypervitaminosis A in siblings: Evidence of variable tolerance to retinol intake, *Journal of Pediatrics* 111 (1987): 507–512.

15 R. P. Heaney, Lessons for nutritional science from vitamin D, *American Journal of Clinical Nutrition* 69 (1999): 825–826.

16 A. W. Norman, Sunlight, season, skin pigmentation, vitamin D, and 25-hydroxyvitamin D: Integral components of the vitamin D endocrine system, *American Journal of Clinical Nutrition* 67 (1998): 1108–1110.

17 A. W. Norman and coauthors, Differing shapes of 1α,25-dihydroxyvitamin D_3 function as ligands for the D-binding protein, nuclear receptor and membrane receptor: A status report, *Journal of Steroid Biochemistry and Molecular Biology* 56 (1996): 13–22.

18 M. S. LeBoff and coauthors, Occult vitamin D deficiency in postmenopausal US women with acute hip fracture, *Journal of the American Medical Association* 281 (1999): 1505–1511.

19 J. L. Giunta, Dental changes in hypervitaminosis D, *Oral Surgery, Oral Medicine, Oral Pathology, Oral Radiology, and Endodontics* 85 (1998): 410–413; S. Blank and coauthors, An outbreak of hypervitaminosis D associated with the overfortification of milk from a home-delivery dairy, *American Journal of Health Promotion* 85 (1995): 656–659; C. H. Jacobus and coauthors, Hypervitaminosis D associated with drinking milk, *New England Journal of Medicine* 326 (1992): 1173–1177.

20 S. S. Harris and B. Dawson-Hughes, Seasonal changes in plasma 25-hydroxyvitamin D concentrations of young American black and white women, *American Journal of Clinical Nutrition* 67 (1998): 1232–1236.

21 Committee on Dietary Reference Intakes, 2000.

22 K. G. Losonczy, T. B. Harris, and R. J. Havlik, Vitamin E and vitamin C supplement use and risk of all-cause and coronary heart disease mortality in older persons: The Established Populations for Epidemiologic Studies of the elderly, *American Journal of Clinical Nutrition* 64 (1996): 190–196; J. Regnström and coauthors, Inverse relation between the concentration of low-density lipoprotein vitamin E and severity of coronary artery disease, *American Journal of Clinical Nutrition* 63 (1996): 377–385; L. H. Kushi and coauthors, Dietary antioxidant vitamins and death from coronary heart disease in postmenopausal women, *New England Journal of Medicine* 334 (1996): 1156–1162; J. P. Flaather and coauthors, The antioxidant vitamins and cardiovascular disease: A critical review of epidemiologic and clinical trial data, *Annals of Internal Medicine* 123 (1995): 860–872.

23 M. G. Traber, Vitamin E in humans: Demand and delivery, *Annual Review of Nutrition* 16 (1996): 321–347.

24 Committee on Dietary Reference Intakes, 2000.

25 P. Dowd and coauthors, The mechanism of action of vitamin K, *Annual Review of Nutrition* 15 (1995): 419–440.

26 D. Feskanich and coauthors, Vitamin K intake and hip fractures in women: A prospective study, *American Journal of Clinical Nutrition* 69 (1999): 74–79.

27 K. I. Tsaioun, Vitamin K–dependent proteins in the developing and aging system, *Nutrition Reviews* 57 (1999): 231–240; G. Ferland, The vitamin K–dependent proteins: An update, *Nutrition Reviews* 56 (1998): 223–230.

28 J. W. Suttie, The importance of menaquinones in human nutrition, *Annual Review of Nutrition* 15 (1995): 399–417.

ANSWERS

Making it Click

1. Vitamin A: μg RAE. Vitamin D: μg.
   Vitamin E: mg. Vitamin K: μg.
2. Thiamin: Legumes and grains
   Riboflavin: Milks, grains, and meats
   Niacin: Meats and grains
   Vitamin B_6: Meats
   Folate: Legumes and vegetables
   Vitamin B_{12}: Meats and milks
   Vitamin C: Vegetables and fruits
   Vitamin A: Vegetables, fruits, and milks
   Vitamin D: Milks
   Vitamin E: Legumes and oils

Taken together, "the most" groups form the Pyramid—grains, vegetables, legumes, fruits, milks, meats, and oils.

Multiple Choice

1. c	2. a	3. c	4. b	5. c
6. c	7. a	8. d	9. d	10. c

ANTIOXIDANT NUTRIENTS AND PHYTOCHEMICALS IN DISEASE PREVENTION

COUNT ON supplement manufacturers to exploit the day's hot topics in nutrition. The moment bits of research news surface, new supplements appear—and terms like "antioxidants" and "phytochemicals" become household words. Friendly faces in TV commercials try to persuade us that these supplements hold the magic in the fight against aging and disease. New supplements hit the market and cash registers ring. Vitamin C, for years the leading single nutrient supplement, gains new popularity, and sales of beta-carotene and vitamin E supplements soar as well. A variety of phytochemicals seem to spontaneously appear on market shelves.

In the meantime, scientists and medical experts around the world continue their work to clarify and confirm the roles of antioxidant nutrients and phytochemicals in preventing chronic diseases.[1] This highlight summarizes some of the accumulating evidence. It also revisits the advantages of foods over supplements. But first it is important to introduce the troublemakers—the **free radicals** (the glossary on p. 378 defines free radicals and related terms).

FREE RADICALS AND DISEASE

Chapter 7 described how the body's cells use oxygen in metabolic reactions. In the process, oxygen sometimes reacts with body compounds and produces highly unstable molecules known as free radicals. In addition to normal body processes, environmental factors such as ultraviolet radiation, air pollution, and tobacco smoke generate free radicals.[2]

A free radical is a molecule with one or more unpaired electrons.* An electron without a partner is unstable and highly reactive. To regain its stability, the free radical quickly finds a stable but vulnerable compound from which to steal an electron (see Figure H11-1).

With the loss of an electron, the formerly stable molecule becomes a free radical itself and steals an electron from another nearby molecule. Thus, an electron-snatching chain reaction is under way with free radicals producing more free radicals. Antioxidants neutralize free radicals by donating one of their own electrons, thus ending the chain reaction. When they lose electrons, antioxidants do not become free radicals because they are stable in either form. (Review Figure 10-14 on p. 336 to see how ascorbic acid can give up two hydrogens with their electrons and become dehydroascorbic acid.)

Once formed, free radicals attack. Occasionally, these free-radical attacks are helpful. For example, cells of the immune system use free radicals as ammunition in an "oxidative burst" that demolishes disease-causing viruses and bacteria. Most often, however, free-radical attacks cause widespread damage. They commonly damage the polyunsaturated fatty acids in lipoproteins and in cell membranes, disrupting the transport of substances into and out of cells. Free radicals also damage cell proteins (altering their functions) and DNA (creating mutations).

The body's natural defenses and repair systems try to control the destruction caused by free radicals, but these systems are not 100 percent effective. In fact, they become less effective with age, and the unrepaired damage accumulates. To some extent, dietary antioxidants defend the body against **oxidative stress**, but if antioxidants are unavailable, or if free-radical production becomes excessive, health problems may develop.[3] Oxygen-derived free radicals may cause diseases not only by indiscriminately destroying the valuable components of cells, but also by serving as signals for specific activities within the cells.[4] Scientists have implicated oxidative stress in the aging process and in the development of diseases such as cancer, arthritis, cataracts, and heart disease.

People who eat generous amounts of fruits and vegetables daily are helping their bodies to fight disease.

*Many free radicals exist, but the oxygen-derived ones are most common in the human body. Examples of oxygen-derived free radicals include superoxide radical ($O2\cdot^-$), hydroxyl radical ($OH\cdot$), and nitric oxide ($NO\cdot$). (The dots in the symbols represent the unpaired electrons.) Technically, hydrogen peroxide (H_2O_2) and singlet oxygen are not free radicals because they contain paired electrons, but the unstable conformation of their electrons makes radical-producing reactions likely. Scientists sometimes use the term *reactive oxygen species (ROS)* to describe all of these compounds.

GLOSSARY

free radicals: unstable and highly reactive atoms or molecules that have one or more unpaired electrons in the outer orbital (see Appendix B for a review of basic chemistry concepts).

oxidants (OK-see-dants): compounds (such as oxygen itself) that oxidize other compounds. Compounds that prevent oxidation are called *anti*oxidants, whereas those that promote it are called *pro*oxidants.

- **anti** = against
- **pro** = for

oxidative stress: a condition in which the production of oxidants and free radicals exceeds the body's ability to defend itself.

prooxidants: substances that significantly induce oxidative stress.

Reminder: *Dietary antioxidants* are substances typically found in foods that significantly decrease the adverse effects of free radicals on normal functions in the body. *Nonnutrients* are compounds in foods that do not fit into the six classes of nutrients. *Phytochemicals* are nonnutrient compounds found in plant-derived foods that have biological activity in the body.

FIGURE H11-1 **The Actions of Free Radicals and Antioxidants**

1 FREE-RADICAL FORMATION
During normal energy metabolism, hydrogens and electrons are added to oxygen in a series of reactions known as the electron transport chain (introduced in Chapter 7). This sequence eventually produces water, but some of the intermediate compounds inevitably created during the process are free radicals. Reminder: the dot in the symbols represents the unpaired electrons.

$O_2 \xrightarrow{e^-} O_2^{\bullet -} \xrightarrow{e^-} \xrightarrow{H^+} \xrightarrow{H^+} H_2O_2 \xrightarrow{e^-} \xrightarrow{H^+} OH^\bullet + H_2O$

Oxygen; Superoxide radical; Hydrogen peroxide; Hydroxyl radical; Water

Occasionally, oxygen gains an extra electron from the electron transport chain . . .

. . . which generates the free radical called superoxide radical (a molecule of oxygen with an extra, unpaired electron).

The superoxide radical can gain another electron (again, from the electron transport chain) and react with two hydrogen ions . . .

. . . to form hydrogen peroxide. Hydrogen peroxide can react with an electron and hydrogen . . .

. . . to form another free radical called a hydroxyl radical . . .

. . . and water.

2 FREE-RADICAL CHAIN REACTION AND DAMAGE
Hydroxyl radicals are highly reactive, wanting to match their unpaired electrons. For example, they might take electrons from the lipids in a cell membrane, which causes damage that gives rise to degenerative diseases.

$\text{Lipid} + OH^\bullet \longrightarrow \text{Lipid}^\bullet + H_2O$

Hydroxyl radical; Lipid radical; water

When a hydroxyl radical takes a hydrogen atom from a lipid (such as a polyunsaturated fatty acid) . . .

. . . it generates a lipid radical . . .

. . . and water.

The lipid radical can, in turn, react with oxygen to form another lipid radical, which can, in turn, remove hydrogen atoms from other lipids, producing new radicals, thereby initiating a chain reaction.

3 ANTIOXIDANT PROTECTION
Antioxidants interact with free radicals and break the destructive chain reaction that damages tissues.

Active vitamin E + Lipid$^\bullet$ ⟶ Inactive vitamin E$^\bullet$ + Lipid

Inactive vitamin E$^\bullet$ + Vitamin C (ascorbic acid with its H atoms) = Active vitamin E (with its H atom) + Vitamin C (dehydroascorbic acid without its H atoms)

Vitamin E gives up one of its hydrogens to a lipid radical.*

The result is that vitamin E is no longer active, but it has successfully stopped the radicals from causing more damage and generating more radicals.

Vitamin E can be reactivated by accepting a hydrogen atom from fellow antioxidant vitamin C. Vitamin C's two structures are presented in Figure 10-14 on p. 336.

*The compound is actually a lipid peroxyl radical.

DEFENDING AGAINST FREE RADICALS

The body maintains a couple lines of defense against free-radical damage. A system of enzymes disarms the most harmful **oxidants.*** The action of these enzymes depends on the minerals selenium, copper, manganese, and zinc. If the diet fails to provide adequate supplies of these minerals, this line of defense weakens. The body also uses the antioxidant vitamins: vitamin E and vitamin C. Vitamin E defends the body's lipids (cell membranes and lipoproteins, for example) by efficiently stopping the free-radical chain reaction. Vitamin C protects the body's watery components, such as the fluid of the blood, against free-radical attacks. Vitamin C seems especially adept at neutralizing free radicals from polluted air and cigarette smoke; it may also restore oxidized vitamin E to its active state.

Dietary antioxidants may also include *non*nutrients—some of the phytochemicals. Chapter 1 introduced the phytochemicals as nonnutrient compounds found in plant-derived foods (*phyto* means plant) that have biological activity in the body. Research on phytochemicals is unfolding daily, adding to our knowledge of their roles in human health, but there are still many questions and only tentative answers. Just a few of the tens of thousands of phytochemicals have been researched at all and only a sampling are mentioned in this highlight—enough to illustrate their wide variety of food sources and effects.

Together, nutrients and phytochemicals with antioxidant activity minimize damage by:

- Limiting free-radical formation.
- Destroying free radicals or their precursors.
- Stimulating antioxidant enzyme activity.
- Repairing oxidative damage.
- Stimulating repair enzyme activity.

These actions play key roles in defending the body against cancer and heart disease.

DEFENDING AGAINST CANCER

Cancers arise when cellular DNA is damaged—sometimes by free-radical attacks. Antioxidants may reduce cancer risks by protecting DNA from this damage. Many researchers conducting epidemiological studies have reported low rates of cancer in people whose diets include abundant vegetables and fruits, rich in antioxidants. Preliminary reports suggest an inverse relationship between DNA damage and vegetable intake and a positive relationship with beef and pork intake.[5] Laboratory studies with animals and with cells in tissue culture also seem to support such findings.

*These enzymes include glutathione peroxidase, thioredoxin reductase, superoxide dismutase, and catalase.

The Antioxidant Nutrients

Foods rich in vitamin C seem to protect against certain types of cancers, especially those of the mouth, larynx, and esophagus. Such a correlation may reflect the benefits of a diet rich in fruits and vegetables and low in fat; it does not necessarily support taking vitamin C supplements to treat or prevent cancer.

Evidence that vitamin E helps guard against cancer is less consistent than for vitamin C. Still, people with low blood levels of vitamin E have high rates of some cancers. Several studies report a cancer-preventing benefit of vegetables and fruits rich in beta-carotene and the other carotenoids as well.[6]

The Phytochemicals

A variety of phytochemicals from a variety of foods appear active in defending the body against cancer. A few examples follow.

Soybeans and products made from them correlate with low rates of cancer, especially cancers of the breast and prostate.[7] Soybeans are a rich source of an array of phytochemicals, among them the phytosterols. Phytosterols are plant compounds that weakly mimic or modulate the effects of the steroid hormones estrogen and progesterone in the body. These phytosterols appear to slow the growth of certain cancers.

Tomatoes seem to offer protection against cancers of the esophagus, prostate, and stomach.[8] Among the phytochemicals responsible for this effect is lycopene, one of beta-carotene's many carotenoid relatives. Lycopene is the pigment that gives guava, papaya, pink grapefruits, and watermelon their red color—and it is especially abundant in tomatoes and cooked tomato products.[9] Lycopene is a powerful antioxidant that seems to inhibit the reproduction of cancer cells.[10]

Soybeans and tomatoes are only two of the many fruits and vegetables credited with providing anticancer activity (see Figure H11-2). Researchers speculate that people might cut their risks of cancers in half simply by meeting current recommendations to eat five or more servings of fruits and vegetables a day.[11]

DEFENDING AGAINST HEART DISEASE

High blood cholesterol carried in LDL is a major risk factor for cardiovascular disease, but how do LDL exert their damage? One scenario is that free radicals within the arterial walls oxidize LDL, changing their structure and function. The oxidized LDL then accelerate the formation of artery-clogging plaques.[12] These free radicals also oxidize the polyunsaturated fatty acids of the cell membranes, sparking additional

FIGURE H11-2 **An Array of Phytochemicals in a Variety of Fruits and Vegetables**

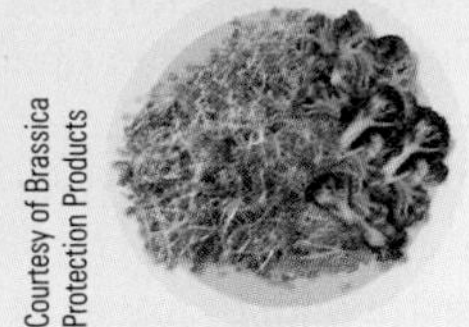
Courtesy of Brassica Protection Products

Broccoli sprouts contain an abundance of the cancer-fighting phytochemical sulforaphane.

Digital Imagery © 2001 PhotoDisc, Inc.

An apple a day—rich in flavonoids—may protect against lung cancer.

Digital Imagery © 2001 PhotoDisc, Inc.

The isoflavones of soybeans seem to starve cancer cells and inhibit tumor growth; they may also lower blood cholesterol and protect cardiac arteries.

© EyeWire, Inc.

Garlic, with its abundant allicin, may lower blood cholesterol and protect against stomach cancer.

Digital Imagery © 2001 PhotoDisc, Inc.

The phytochemical resveratrol found in grapes (and peanuts) protects against cancer by inhibiting cell growth and against heart disease by limiting clot formation.

Digital Imagery © 2001 PhotoDisc, Inc.

The ellagic acid of strawberries may inhibit certain types of cancer.

© EyeWire, Inc.

The limonene of citrus fruits may inhibit cancer growth.

Digital Imagery © 2001 PhotoDisc, Inc.

The flavonoids in black tea may protect against heart disease, whereas those in green tea may defend against cancer.

Digital Imagery © 2001 PhotoDisc, Inc.

Tomatoes, with their abundant lycopene, may defend against cancer by protecting DNA from oxidative damage.

Courtesy of Flax Council of Canada

Flaxseed, the richest source of lignans, may prevent the spread of cancer.

Digital Imagery © 2001 PhotoDisc, Inc.

Blueberries, a rich source of flavonoids, improves memory in animals.

changes in the arterial walls, which impede the flow of blood. Susceptibility to such oxidative damage within the arterial walls is heightened by a high-fat diet or cigarette smoke.[13] In contrast, diets that include plenty of fruits and vegetables, especially when combined with little fat, strengthen antioxidant defenses against LDL oxidation.[14]

The Antioxidant Nutrients

Antioxidants, especially vitamin E, may protect against cardiovascular disease.[15] Epidemiological studies suggest that people who eat foods rich in vitamin E have low rates of death from heart disease. Similarly, large doses of vitamin E supplements are associated with reduced risks of heart disease.

Some studies suggest that vitamin C protects against LDL oxidation, raises HDL, lowers total cholesterol, and improves blood pressure.[16] Vitamin C may also protect the arteries against oxidative damage.[17] Supplementation with both vitamins E and C minimizes the free-radical action within the arterial wall that typically follows a high-fat meal; in fact, blood flow through the arteries is similar to that seen after a low-fat meal.[18]

The Phytochemicals

Diets based primarily on unprocessed foods appear to support heart health better than those founded on highly refined foods—perhaps because of the abundance of nutrients, fiber, or phytochemicals such as the flavonoids. Flavonoids, a large group of phytochemicals known for their health-promoting qualities, are found in whole grains, vegetables, fruits, herbs, spices, teas, and red wines.[19] Researchers studying people in several countries report that deaths from heart disease are low when flavonoid-containing foods are plentiful.[20] Flavonoids are powerful antioxidants that may help to protect LDL against oxidation and reduce blood platelet stickiness, making blood clots less likely.[21]

In addition to flavonoids, fruits and vegetables are rich in carotenoids. Studies suggest that a diet rich in carotenoids is also associated with a lower risk of heart disease.[22] Notable among the carotenoids that may defend against heart disease is lycopene.[23]

FOODS VERSUS SUPPLEMENTS

In the process of scavenging and quenching free radicals, antioxidants themselves become oxidized. To some extent, they can be regenerated, but still, losses occur and free radicals attack continuously. To maintain defenses, a person must replenish dietary antioxidants regularly. But should antioxidants be replenished from foods or from supplements?

Some research suggests a protective effect from as little as a daily glass of orange juice and another of carrot juice (rich sources of vitamin C and beta-carotene, respectively).[24] Other intervention studies, however, have used levels of nutrients that far exceed current recommendations and can be achieved only by taking supplements. In making their recommendations for the antioxidant nutrients, members of the DRI committee considered whether these studies support substantially higher intakes to help protect against chronic diseases. They did raise the recommendations for vitamins C and E, but do not support taking vitamin pills over eating a healthy diet.

While awaiting additional research, should people anticipate the go-ahead and start taking antioxidant supplements now? Most scientists agree that it is too early to make such a recommendation. Though fruits and vegetables containing many antioxidant nutrients and phytochemicals have been associated with a diminished risk of many cancers, supplements have not always proved beneficial.[25] In fact, sometimes the benefits are more apparent when the vitamins come from foods rather than from supplements.[26] Without data to confirm the benefits of supplements, we cannot accept the potential risks. And the risks are real.

Consider the findings from a study to determine whether daily supplements of vitamin E, beta-carotene, or both would reduce the incidence of lung cancer among smokers.[27] After five to eight years of supplementation, there was no reduction in the incidence of lung cancer; in fact, the researchers found a *higher* incidence of lung cancer among smokers receiving the beta-carotene. Another group of researchers reported similar findings: smokers and asbestos workers receiving beta-carotene and vitamin A supplements for four years had a higher incidence of lung cancer and risk of death than those taking a placebo.[28] These findings brought the study to an end much earlier than planned. Given the association between high intakes of *foods* rich in beta-carotene and low rates of lung cancer reported in earlier epidemiological studies, findings of increased risk were surprising, to say the least. Clearly, remedies to life-threatening diseases such as lung cancer are not as simple as taking supplements. Smokers are much wiser to stop smoking than to rely on pills to protect them from lung cancer.

Even if research clearly proves that a particular nutrient is the ultimate protective ingredient in foods, supplements would not be the answer because their contents are limited. Vitamin E supplements, for example, usually contain alpha-tocopherol, but foods provide an assortment of tocopherols among other nutrients, many of which provide valuable protection against free-radical damage.[29] Supplements shortchange users.

Furthermore, much more research is needed to define optimal and dangerous levels of intake. This much we know: antioxidants behave differently under various conditions.[30] At physiological levels typical of a healthy diet, they act as antioxidants, but at pharmacological doses typical of supplements, they may act as **prooxidants,** stimulating the production of free radicals.[31] This is especially likely in the presence of other antioxidants or minerals such as iron. Until the optimum intake of these nutrients can be determined, the risks of supplement use remain unclear. The best way to add antioxidants to the diet is to eat generous servings of fruits and vegetables daily.

FOODS MAKING HEALTH CLAIMS

Results of clinical studies may one day support the use of selected supplements to prevent disease, but until then people will want to select foods rich in all of the vitamins and minerals—particularly the antioxidants. Which foods to select?

The Food and Drug Administration (FDA) examined the available scientific evidence concerning antioxidant vitamins and cancer to determine whether a health claim on food labels was appropriate. The agency concluded that *diets* high in fruits and vegetables, which are good sources of two antioxidant vitamins (vitamin A as beta-carotene and vitamin C), are strongly associated with reduced risks of several types of cancer. Still, the FDA rejected the antioxidant health claim, stating that the reduction in risk could not be attributed directly and solely to the antioxidant effect of the vitamins. Therefore labels may not claim an association between antioxidant vitamins and cancer; the health claim must be stated in terms of "fruits and vegetables and cancer." National campaigns to "Eat 5 a Day" in the United States and "Reach for It" in Canada encourage consumers to select several servings of fruits and vegetables daily.

PHYTOCHEMICALS IN DISEASE PREVENTION

Clearly, the protective effect of fruits and vegetables must depend on more than the antioxidant nutrients alone.[32] Other, nonnutrient compounds must also be involved.

As mentioned earlier, the nonnutrient compounds found in plants are called phytochemicals, and they have been the focus of much recent research. In foods, phytochemicals impart tastes, aromas, colors, and other characteristics. They give hot peppers their burning sensation, garlic its pungent flavor, and tomatoes their dark red color. In the body, phytochemicals can have profound physiological effects, acting as antioxidants, mimicking hormones, and suppressing the development of diseases. Table H11-1 presents the names, possible effects, and food sources of some of the better-known phytochemicals.

Because foods deliver thousands of phytochemicals in addition to dozens of nutrients, researchers must be careful in giving credit for particular health benefits to any one compound. Diets rich in whole grains, legumes, vegetables, and fruits seem to be protective against heart disease and cancer, but identifying *the* specific foods or components of foods that are responsible is difficult.[33] Each food possesses a unique array of phytochemicals—citrus fruits provide limonene; grapes, resveratrol; and flaxseed, lignans. Broccoli may contain as many as 10,000 different phytochemicals—each with the potential to influence some action in the body.[34] Beverages such as wine, spices such as oregano, and oils such as olive oil contain phytochemicals that may explain, in part, why people who live in the Mediterranean region have reduced risks of heart disease.[35] Even identifying all of the phytochemicals and their effects doesn't answer all the questions because the actions of phytochemicals may be complementary or overlapping—which reinforces the principle of variety in

TABLE H11-1 Phytochemicals—Their Food Sources and Actions

Name	Possible Effects	Food Sources
Capsaicin	Modulates blood clotting, possibly reducing the risk of fatal clots in heart and artery disease.	Hot peppers
Carotenoids (include beta-carotene, lycopene, and hundreds of related compounds)[a]	Act as antioxidants, possibly reducing risks of cancer and other diseases.	Deeply pigmented fruits and vegetables (apricots, broccoli, cantaloupe, carrots, pumpkin, spinach, sweet potatoes, tomatoes)
Curcumin	May inhibit enzymes that activate carcinogens.	Tumeric, a yellow-colored spice
Flavonoids (include flavones, flavonols, isoflavones, catechin, and others)[b,c]	Act as antioxidants; scavenge carcinogens; bind to nitrates in the stomach, preventing conversion to nitrosamines; inhibit cell proliferation.	Berries, black tea, celery, citrus fruits, green tea, olives, onions, oregano, purple grapes, purple grape juice, soybeans and soy products, vegetables, whole wheat, wine
Indoles[d]	May trigger production of enzymes that block DNA damage from carcinogens; may inhibit estrogen action.	Broccoli and other cruciferous vegetables (brussels sprouts, cabbage, cauliflower), horseradish, mustard greens
Isothiocyanates (including sulforaphane)	Inhibit enzymes that activate carcinogens; trigger production of enzymes that detoxify carcinogens.	Broccoli and other cruciferous vegetables (brussels sprouts, cabbage, cauliflower), horseradish, mustard greens
Lignans[e]	Block estrogen activity in cells, possibly reducing the risk of cancer of the breast, colon, ovaries, and prostate.	Flaxseed and its oil, whole grains
Monoterpenes (include limonene)	May trigger enzyme production to detoxify carcinogens; inhibit cancer promotion and cell proliferation.	Citrus fruit peels and oils
Organosulfur compounds	May speed production of carcinogen-destroying enzymes; slow production of carcinogen-activating enzymes.	Chives, garlic, leeks, onions
Phenolic acids[c]	May trigger enzyme production to make carcinogens water soluble, facilitating excretion.	Coffee beans, fruits (apples, blueberries, cherries, grapes, oranges, pears, prunes), oats, potatoes, soybeans
Phytic acid	Binds to minerals, preventing free-radical formation, possibly reducing cancer risk.	Whole grains
Phytosterols (genistein and diadzein)	Estrogen inhibition may produce these actions: inhibit cell replication in GI tract; reduce risk of breast, colon, ovarian, prostate, and other estrogen-sensitive cancers; reduce cancer cell survival. Estrogen mimicking may reduce risk of osteoporosis.	Soybeans, soy flour, soy milk, tofu, textured vegetable protein, other legume products
Protease inhibitors	May suppress enzyme production in cancer cells, slowing tumor growth; inhibit hormone binding; inhibit malignant changes in cells.	Broccoli sprouts, potatoes, soybeans and other legumes, soy products
Resveratrol	Offsets artery-damaging effects of high-fat diets.	Red wine, peanuts
Saponins	May interfere with DNA replication, preventing cancer cells from multiplying; stimulate immune response.	Alfalfa sprouts, other sprouts, green vegetables, potatoes, tomatoes
Tannins[c]	May inhibit carcinogen activation and cancer promotion; act as antioxidants.	Black-eyed peas, grapes, lentils, red and white wine, tea

[a]Other carotenoids include alpha-carotene, beta-cryptoxanthin, lutein, and zeaxanthin.
[b]Other flavonoids of interest include ellagic acid and ferulic acid; see also *phytosterols.*
[c]A subset of the larger group *phenolic phytochemicals.*
[d]Indoles include dithiothiones, isothiocyantes, and others.
[e]Lignans act as phytosterols, but their food sources are limited.

Many cancer-fighting products are available now at your local produce counter.

© Jeffry Myers/Stock, Boston/PictureQuest

diet planning.[36] Review Figure H11-2 for an appreciation of the array of phytochemicals offered by a variety of fruits and vegetables.

Everyone eats an assortment of phytochemicals every day. This approach may be more beneficial than taking large doses of any one phytochemical. In large doses, some phytochemicals can be toxic. The regulation of phytochemicals in the food supply depends on how they are used. Consider garlic, for example. A clove of garlic is a food. The FDA classifies dehydrated garlic and garlic extracts as generally recognized as safe (GRAS) substances. A product derived from garlic that makes a special health claim, on the other hand, is classified as a drug.

Of course, manufacturers have already begun to market phytochemicals as dietary supplements. It should be clear by now, though, that we cannot know the identity and action of every chemical in every food. Even if we did, why create a supplement to replicate a food? Why not eat foods and enjoy the pleasure, nourishment, and health benefits they provide? The beneficial constituents in foods are widespread among plants.[37] Don't try to single out one particular food for its magic phytochemical. Instead, eat a wide variety of fruits and vegetables in generous quantities every day—and get *all* the magic compounds these foods have to offer.

References

1. J. E. Buring and C. H. Hennekens, Antioxidant vitamins and cardiovascular disease, *Nutrition Reviews* 55 (1997): S53–S60; B. Halliwell, Antioxidants and human disease: A general introduction, *Nutrition Reviews* 55 (1997): S44–S52; C. L. Rock, R. A. Jacob, and P. E. Bowen, Update on the biological characteristics of the antioxidant micronutrients: Vitamin C, vitamin E, and the carotenoids, *Journal of the American Dietetic Association* 96 (1996): 693–702; Health promotion and disease prevention: The role of antioxidant vitamins, *American Journal of Medicine* 97 (supplement 3A) (1994): 1S–28S.
2. B. Halliwell, Antioxidants in human health and disease, *Annual Review of Nutrition* 16 (1996): 35–50; J. D. Morrow and coauthors, Increase in circulating products of lipid peroxidation (F_2-isoprostanes) in smokers—Smoking as a cause of oxidative damage, *New England Journal of Medicine* 332 (1995): 1198–1203; L. Langseth, *Oxidants, Antioxidants, and Disease Prevention* (Brussels: International Life Sciences Institute, 1995).
3. Halliwell, 1996; R. A. Jacob and B. J. Burri, Oxidative damage and defense, *American Journal of Clinical Nutrition* 63 (1996): 985S–990S.
4. H. J. Palmer and K. E. Paulson, Reactive oxygen species and antioxidants in signal transduction and gene expression, *Nutrition Reviews* 55 (1997): 353–361.
5. Z. Djuric and coauthors, Oxidative DNA damage levels in blood from women at high risk for breast cancer are associated with dietary intakes of meats, vegetables, and fruits, *Journal of the American Dietetic Association* 98 (1998): 524–528.
6. D. S. Michaud and coauthors, Intake of specific carotenoids and risk of lung cancer in 2 prospective US cohorts, *American Journal of Clinical Nutrition,* 72 (2000): 990–997; M. L. Slattery and coauthors, Carotenoids and colon cancer, *American Journal of Clinical Nutrition* 71 (2000): 575–582; D. A. Cooper, A. L. Eldridge, and J. C. Peters, Dietary carotenoids and lung cancer: A review of recent research, *Nutrition Reviews* 57 (1999): 133–134; P. Riso and coauthors, Does tomato consumption effectively increase the resistance of lymphocyte DNA to oxidative damage? *American Journal of Clinical Nutrition* 69 (1999): 712–718; E. Giovannucci, Tomatoes, tomato-based products, lycopene, and cancer: Review of the epidemiologic literature, *Journal of the National Cancer Institute* 91 (1999): 317–331; S. T. Mayne, C. A. Redlich, and M. R. Cullen, Dietary vitamin A and prevalence of bronchial metaplasia in asbestos-exposed workers, *American Journal of Clinical Nutrition* 68 (1998): 630–635; P. A. Kantesky and coauthors, Dietary intake and blood levels of lycopene: Association with cervical dysplasia among non-Hispanic, black women, *Nutrition and Cancer* 31 (1998): 31–40; H. Nishino, Cancer prevention by natural carotenoids, *Journal of Cellular Biochemistry* (supplement) 27 (1997): 86–91; M. S. Tallman and coauthors, All-*trans*-retinoic acid in acute promyelocytic leukemia, *New England Journal of Medicine* 337 (1997): 1021–1028; S. Zhang and coauthors, Measurement of retinoids and carotenoids in breast adipose

tissue and a comparison of concentrations in breast cancer cases and control subjects, *American Journal of Clinical Nutrition* 66 (1997): 626–632.

7 C. A. Lamartiniere, Protection against breast cancer with genistein: A component of soy, *American Journal of Clinical Nutrition* 71 (2000): 1705S–1707S; D. Ingram and coauthors, Case-control study of phytoestrogens and breast cancer, *Lancet* 350 (1997): 990–994; H. Adlercreutz and W. Mazur, Phyto-estrogens and western diets, *Annals of Medicine* 29 (1997): 95–120.

8 S. K. Clinton, Lycopene: Chemistry, biology, and implications for human health and disease, *Nutrition Reviews* 56 (1998): 35–51.

9 G. R. Beecher, Nutrient content of tomatoes and tomato products, *Proceedings of the Society for Experimental and Biological Medicine* 218 (1998): 98–100.

10 H. Gerster, The potential role of lycopene for human health, *Journal of the American College of Nutrition* 16 (1997): 109–126.

11 K. Steinmetz and J. D. Potter, Vegetables, fruit, and cancer prevention: A review, *Journal of the American Dietetic Association* 96 (1996): 1027–1039.

12 P. D. Reaven and J. L. Witztum, Oxidized low density lipoproteins in atherogenesis: Role of dietary modification, *Annual Review of Nutrition* 16 (1996): 51–71; B. Halliwell, Oxidation of low-density lipoproteins: Questions of initiation, propagation, and the effect of antioxidants, *American Journal of Clinical Nutrition* 61 (1995): 670S–677S.

13 C. E. Cross and M. G. Traber, Cigarette smoking and antioxidant vitamins: The smoke screen continues to clear but has a way to go, *American Journal of Clinical Nutrition* 65 (1997): 562–563; Reaven and Witztum, 1996.

14 R. A. Jacob, Evidence that diet modification reduces in vivo oxidant damage, *Nutrient Reviews* 57 (1999): 255–258.

15 M Meydani, Effect of functional food ingredients: Vitamin E modulation of cardiovascular diseases and immune status in the elderly, *American Journal of Clinical Nutrition* 71 (2000): 1665S–1668S; C. Bonithon-Kopp and coauthors, Combined effects of lipid peroxidation and antioxidant status on carotid atherosclerosis in a population aged 59–71 y: The EVA study, *American Journal of Clinical Nutrition* 65 (1997): 121–127; C. C. Tangney, Vitamin E and cardiovascular disease, *Nutrition Today* 32 (1997): 13–22; L. H. Kushi and coauthors, Dietary antioxidant vitamins and death from coronary heart disease in postmenopausal women, *New England Journal of Medicine* 334 (1996): 1156–1162; M. Abbey, The importance of vitamin E in reducing cardiovascular risk, *Nutrition Reviews* 53 (1995): S28–S32; T. Byers, Vitamin E supplements and coronary heart disease, *Nutrition Reviews* 51 (1993): 333–336.

16 D. Harats and coauthors, Citrus fruit supplementation reduces lipoprotein oxidation in young men ingesting a diet high in saturated fat: Presumptive evidence for an interaction between vitamins C and E in vivo, *American Journal of Clinical Nutrition* 67 (1998): 240–245; J. P. Moran and coauthors, Plasma ascorbic acid concentrations relate inversely to blood pressure in human subjects, *American Journal of Clinical Nutrition* 57 (1993): 213–217.

17 G. N. Levine and coauthors, Ascorbic acid reverses endothelial vasomotor dysfunction in patients with coronary artery disease, *Circulation* 93 (1996): 1107–1113.

18 G. D. Plotnick, M. C. Corretti, and R. A. Vogel, Effect of antioxidant vitamins on the transient impairment of endothelium-dependent brachial artery vasoactivity following a single high-fat meal, *Journal of the American Medical Association* 278 (1997): 1682–1686.

19 M. A. Wagstaff and coauthors, Oregano flavonoids as lipid antioxidants, *Journal of the American Dietetic Association* 93 (1993): 1217.

20 M. G. Hertog and coauthors, Flavonoid intake and long-term risk of coronary heart disease and cancer in the seven countries study, *Archives of Internal Medicine* 155 (1995): 381–386.

21 B. Fuhrman and M. Aviram, Flavonoids protect LDL from oxidation and attenuate atherosclerosis, *Current Opinion in Lipidology* 12 (2001): 41–48; S. V. Nigdikar and coauthors, Consumption of red wine polyphenols reduces the susceptibility of low-density lipoproteins to oxidation in vivo, *American Journal of Clinical Nutrition* 68 (1998): 258–265.

22 S. Liu and coauthors, Intake of vegetables rich in carotenoids and risk of coronary heart disease in men: The Physicians' Heart Study, *International Journal of Epidemiology* 30 (2001): 130–135; S. B. Kritchevsky, beta-Carotene, carotenoids and the prevention of coronary heart disease, *Journal of Nutrition* 129 (1999): 5–8.

23 L. Arab and S. Steck, Lycopene and cardiovascular disease, *American Journal of Clinical Nutrition* 71 (2000): 1691S–1695S.

24 M. Abbey, M. Noakes, and P. J. Nestel, Dietary supplementation with orange and carrot juice in cigarette smokers lowers oxidation products in copper-oxidized low-density lipoproteins, *Journal of the American Dietetic Association* 95 (1995): 671–675.

25 J. M. Rapola and coauthors, Effect of vitamin E and beta carotene on the incidence of angina pectoris: A randomized, double-blind, controlled trial, *Journal of the American Medical Association* 275 (1996): 693–698; E. R. Greenberg and coauthors, Mortality associated with low plasma concentration of beta carotene and the effect of oral supplementation, *Journal of the American Medical Association* 275 (1996): 699–703; C. H. Hennekens and coauthors, Lack of effect of long-term supplementation with beta carotene on the incidence of malignant neoplasms and cardiovascular disease, *New England Journal of Medicine* 334 (1996): 1145–1149; E. R. Greenberg and coauthors, A clinical trial of antioxidant vitamins to prevent colorectal adenoma, *New England Journal of Medicine* 331 (1994): 141–147.

26 Kushi and coauthors, 1996.

27 O. P. Heinonen and D. Albanes (and other participants in The Alpha-Tocopherol, Beta Carotene Cancer Prevention Study Group), The effect of vitamin E and beta carotene on the incidence of lung cancer and other cancers in male smokers, *New England Journal of Medicine* 330 (1994): 1029–1035.

28 G. S. Omenn and coauthors, Effects of a combination of beta carotene and vitamin A on lung cancer and cardiovascular disease, *New England Journal of Medicine* 334 (1996): 1150–1155.

29 S. Christen and coauthors, τ-Tocopherol traps mutagenic electrophiles such as NO_x and complements α-tocopherol: Physiological implications, *Proceedings of the National Academy of Sciences* 94 (1997): 3217–3222.

30 E. A Decker, Phenolics: Prooxidants or antioxidants? *Nutrition Reviews* 55 (1997): 396–407.

31 P. Palozza, Prooxidant actions of carotenoids in biologic systems, *Nutrition Reviews* 56 (1998): 257–265; K. M. Brown, P. C. Morrice, and G. G. Duthie, Erythrocyte vitamin E and plasma ascorbate concentrations in relation to erythrocyte peroxidation in smokers and nonsmokers: Dose response to vitamin E supplementation, *American Journal of Clinical Nutrition* 65 (1997): 496–502.

32 H. Priemé and coauthors, No effect of supplementation with vitamin E, ascorbic acid, or coenzyme Q10 on oxidative DNA damage estimated by 8-oxo-7,8-dihydro-2′-deoxyguanosine excretion in smokers, *American Journal of Clinical Nutrition* 65 (1997): 503–507.

33 C. M. Steinmaus, S. Nunez, and A. H. Smith, Diet and bladder cancer: A meta-analysis of six dietary variables, *American Journal of Epidemiology* 151 (2000): 693–702; M. R. Law and J. K. Morris, By how much does fruit and vegetable consumption reduce the risk of ischaemic heart diseaes? *European Journal of Nutrition* 52 (1998): 549–556; F. O. Stephens, Breast cancer: Aetiological factors and associations (a possible protective role of phytoestrogens), *Australian and New Zealand Journal of Surgery* 67 (1997): 755–760.

34 M. Nestle, Broccoli sprouts in cancer prevention, *Nutrition Reviews* 56 (1998): 127–130.

35 A. Trichopoulou, E. Vasilopoulou, and A. Lagiou, Mediterranean diet and coronary heart disease: Are antioxidants critical? *Nutrition Reviews* 57 (1999): 253–255; F. Visioli and C. Galli, The effect of minor constituents of olive oil on cardiovascular disease: New findings, *Nutrition Reviews* 56 (1998): 142–147; Dietary flavonoids and risk of coronary heart disease, *Nutrition Reviews* 52 (1994): 59–61.

36 J. W. Lampe, Health effects of vegetables and fruit: Assessing mechanism of action in human experimental studies, *American Journal of Clinical Nutrition* 70 (1999): 475S–490S.

37 Position of The American Dietetic Association: Phytochemicals and functional foods, *Journal of the American Dietetic Association* 95 (1995): 493–496; E. A. Decker, The role of phenolics, conjugated linoleic acid, carnosine, and pyrroloquinoline quinone as nonessential dietary antioxidants, *Nutrition Reviews* 53 (1995): 49–58.

Water and the Major Minerals

CHAPTER 12

Water is an essential nutrient, as important to life as any of the others. In fact, you can survive only a few days without water, whereas a deficiency of the other nutrients may take weeks, months, or even years to develop.

This chapter begins with a look at water and the body's fluids. The body maintains an appropriate balance and distribution of water with the help of another class of nutrients—the minerals. In addition to introducing the minerals that help regulate body fluids, the chapter describes many of the other important functions minerals perform in the body.

© Barbara Stitzen/PhotoEdit/PictureQuest

Water is the most indispensable nutrient.

WATER AND THE BODY FLUIDS

In the body, water becomes the fluid in which all life processes occur. Every cell contains fluid of the exact composition that is best for that cell **(intracellular fluid)** and is bathed externally in another such fluid **(interstitial fluid).** Figure 12-1 (on p. 388) illustrates a cell and its associated fluids. These fluids continually lose and replace their components, yet the composition in each compartment remains remarkably constant under normal conditions. Whenever disturbances occur, the body quickly responds. Consequently, the entire system of cells and fluids remains in a delicate but controlled state of homeostasis.

The water in the body fluids:

- Carries nutrients and waste products throughout the body.
- Maintains the structure of large molecules such as proteins and glycogen.
- Participates in metabolic reactions.
- Serves as the solvent for minerals, vitamins, amino acids, glucose, and many other small molecules.
- Acts as a lubricant and cushion around joints and inside the eyes, the spinal cord, and, in pregnancy, the amniotic sac surrounding the fetus in the womb.
- Aids in the regulation of body temperature.
- Maintains blood volume.

To support these and other vital functions, the body actively maintains an appropriate **water balance.**

Water Balance and Recommended Intakes

Water constitutes about 60 percent of an adult's body weight and a higher percentage of a child's. Because water makes up about three-fourths of the weight of lean tissue and less than one-fourth of the weight of fat, a person's body composition influences how much of the body's weight is water. The proportion of water is generally smaller in females, obese people, and the elderly because of their smaller proportion of lean tissue. Because imbalances can be devastating, the body attempts to restore homeostasis as promptly as possible, adjusting both water intake and excretion as needed.

• Water Intake • Thirst and satiety influence water intake, apparently in response to changes sensed by the mouth, hypothalamus,◆ and nerves. When the blood becomes concentrated (having lost water, but not the dissolved

◆ Reminder: The *hypothalamus* is a brain center that controls activities such as maintenance of water balance, regulation of body temperature, and control of appetite.

intracellular fluid: fluid within the cells, usually high in potassium and phosphate. Intracellular fluid accounts for approximately two-thirds of the body's water.
- **intra** = within

interstitial (IN-ter-STISH-al) **fluid:** fluid between the cells, usually high in sodium and chloride. Interstitial fluid is a large component of **extracellular fluid** (fluid outside the cells), which also includes plasma and the water of structures such as the skin and bones. Extracellular fluid accounts for approximately one-third of the body's water.
- **inter** = in the midst, between
- **extra** = outside

water balance: the balance between water intake and output (losses).

Water balance = intake − output.

thirst: a conscious desire to drink.

FIGURE 12-1 One Cell and Its Associated Fluids

Fluids are found within the cells (intracellular) or outside the cells (extracellular). Extracellular fluids include plasma (the fluid portion of blood in the intravascular spaces of blood vessels) and interstitial fluids (the tissue fluid that fills the intercellular spaces between the cells).

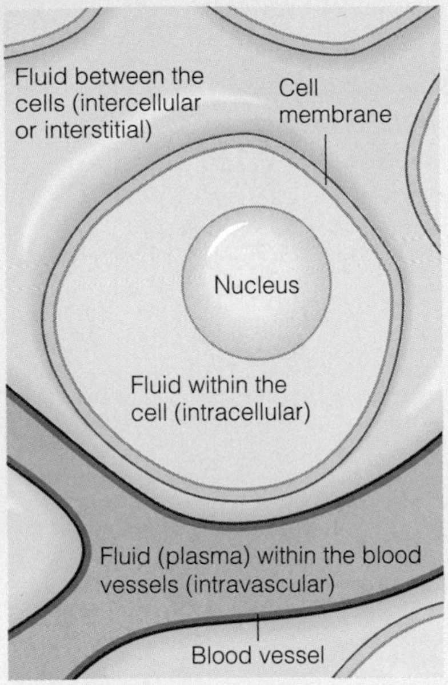

◆ **The amount of water the body has to excrete each day to dispose of its wastes is the obligatory (ah-BLIG-ah-TORE-ee) water excretion—about 500 ml (about 2 c, or a pint).**

◆ A runner can sweat off 50 ounces (about 6 cups) of fluid in an hour.◆

dehydration: the condition in which body water output exceeds water input. Symptoms include thirst, dry skin and mucous membranes, rapid heartbeat, low blood pressure, and weakness.

water intoxication: the rare condition in which body water contents are too high.

TABLE 12-1 Adverse Effects of Dehydration

Body Weight Lost (%)	Symptoms
1–2	Thirst, fatigue, weakness, vague discomfort, loss of appetite
3–4	Impaired physical performance, dry mouth, reduction in urine, flushed skin, impatience, apathy
5–6	Difficulty in concentrating, headache, irritability, sleepiness, impaired temperature regulation, increased respiratory rate
7–10	Dizziness, spastic muscles, loss of balance, delirium, exhaustion, collapse

Note: The onset and severity of symptoms at various percentages of body weight lost depend on the activity, fitness level, degree of acclimation, temperature, and humidity.

substances within it), the mouth becomes dry, and the hypothalamus initiates drinking behavior. Stretch receptors in the stomach send signals to stop drinking as do receptors in the heart that monitor blood volume.

Thirst drives a person to seek water, but it lags behind the body's need. A water deficiency that develops slowly can switch on drinking behavior in time to prevent serious **dehydration,** but a deficiency that develops quickly may not. Also, thirst itself does not remedy a water deficiency; a person must pay attention to the thirst signal and take the time to get a drink. The long-distance runner, the gardener in hot weather, the child busy playing, and the elderly person whose thirst sensation may be blunted can experience serious dehydration if they fail to drink promptly in response to their need for water.

Dehydration may easily develop with either water deprivation or excessive water losses. The symptoms progress rapidly from thirst to weakness, exhaustion, and delirium and end in death if not corrected (see Table 12-1).

Water intoxication, on the other hand, is rare but can occur with excessive water ingestion and kidney disorders that reduce urine production. The symptoms may include confusion, convulsions, and even death in extreme cases.

• **Water Sources** • The obvious dietary sources of water are water itself and other beverages, but nearly all foods also contain water. Most fruits and vegetables contain up to 90 percent water; many meats and cheeses contain at least 50 percent (see Table 12-2 for selected foods and Appendix H for many more). Water is also generated during metabolism. Recall that when the energy-yielding nutrients break down, their carbons and hydrogens combine with oxygen to yield carbon dioxide (CO_2)—and water (H_2O). As Table 12-3 shows, the water derived daily from these three sources averages about 2½ liters (roughly 2½ quarts).

• **Water Losses** • The body must excrete a minimum of about 500 milliliters of water each day◆ as urine—enough to carry away the waste products generated by a day's metabolic activities. Above this amount, excretion adjusts to balance intake. If a person drinks more water, the kidneys excrete more urine, and the urine becomes more dilute. In addition to urine, water is lost from the lungs as vapor and from the skin as sweat; some is also lost in feces.* The amount of fluid lost from each source varies, depending on the environment (such as heat or humidity) and physical conditions (such as exercise or fever). On the average, daily losses total about 2½ liters. Table 12-3 shows how water loss balances intake; maintaining this balance requires healthy kidneys and an adequate intake of fluids.

*Water lost from the lungs and skin accounts for almost one-half of the daily losses even when a person is not visibly perspiring; these losses are commonly referred to as *insensible water losses.*

• Water Recommendations • Because water needs vary depending on diet, activity, environmental temperature, and humidity, a general water requirement is difficult to establish. Recommendations◆ for adults are expressed in proportion to the amount of energy expended under average environmental conditions.[1] A person who expends 2000 kcalories a day needs about 2 to 3 liters of water (about 7 to 11 cups).

Fluid needs are best met by water, but milk and juices can account for part of the day's recommended intake.[2] In addition to their high water content, these beverages deliver valuable nutrients. Alcoholic beverages and those containing caffeine, such as coffee, tea, and sodas, however, are not good substitutes for water. Both alcohol and caffeine act as diuretics, causing the body to lose about half of the liquid consumed from the beverage.

In addition to meeting the body's fluid needs, drinking plenty of water may protect the bladder against cancer by diluting the urine and reducing its holding time. The risk of bladder cancer decreases when fluid intake is high.[3]

TABLE 12-2 Percentage of Water in Selected Foods

100%	Water, diet sodas
90–99%	Nonfat milk, strawberries, watermelon, lettuce, cabbage, celery, spinach, broccoli
80–89%	Fruit juice, yogurt, apples, grapes, oranges, carrots
70–79%	Shrimp, bananas, corn, potatoes, avocados, cottage cheese, ricotta cheese
60–69%	Pasta, legumes, salmon, ice cream, chicken breast
50–59%	Ground beef, hot dogs, feta cheese
40–49%	Pizza
30–39%	Cheddar cheese, bagels, bread
20–29%	Pepperoni sausage, cake, biscuits
10–19%	Butter, margarine, raisins
1–9%	Crackers, cereals, pretzels, taco shells, peanut butter, nuts
0%	Oils

IN SUMMARY

Water makes up about 60 percent of the body's weight. It assists with the transport of nutrients and waste products throughout the body, participates in chemical reactions, acts as a solvent, serves as a shock absorber, and regulates body temperature. To maintain water balance, intake from liquids, foods, and metabolism must equal losses from kidneys, skin, lungs, and feces.

Blood Volume and Blood Pressure

Water maintains the blood volume, which in turn influences blood pressure. Central to the regulation of blood volume and blood pressure are the kidneys. All day, every day, the kidneys reabsorb needed substances and water and excrete wastes with some water in the urine (see Figure 12-2 on p. 390). The kidneys meticulously adjust the volume and the consistency of the urine to accommodate changes in the body, including variations in the day's food and fluid intakes. Instructions on whether to retain or release substances or water come from ADH, renin, angiotensin, and aldosterone.

• ADH and Water Retention • Whenever blood volume or blood pressure falls too low, or whenever the extracellular fluid becomes too concentrated, the hypothalamus signals the pituitary gland to release **antidiuretic hormone (ADH).** ADH is a water-conserving hormone◆ that stimulates the kidneys to reabsorb water. Consequently, the more water you need, the less your kidneys excrete. These events also trigger thirst. Drinking water raises the blood volume and dilutes the concentrated fluids, thus helping to restore homeostasis.

◆ Water recommendation for adults:
- 1.0 to 1.5 ml/kcal expended.
- 4.2 to 6.3 ml/kJ expended.

Water recommendation for infants:
- 1.5 ml/kcal expended.

Note: 1 ml = 0.03 fluid ounce. 125 ml ≈ ½ c.
Easy estimation: ½ c per 100 kcal expended.

◆ Recall from Highlight 7 that alcohol depresses ADH activity, thus promoting fluid losses and dehydration. In addition to its antidiuretic effect, ADH elevates blood pressure, and so it is also called **vasopressin** (VAS-oh-PRES-in).
- **vaso** = vessel
- **press** = pressure

TABLE 12-3 Water Balance

Water Sources	Amount (ml)	Water Losses	Amount (ml)
Liquids	550 to 1500	Kidneys	500 to 1400
Foods	700 to 1000	Skin	450 to 900
Metabolic water	200 to 300	Lungs	350
		Feces	150
	1450 to 2800		1450 to 2800

Note: These values reflect data from several sources and are compatible with those cited in many other references. For further information, see L. Sherwood, *Fundamentals of Physiology: A Human Perspective* (St. Paul, Minn.: West Publishing, 1995), pp. 396–417; Committee on Dietary Allowances, *Recommended Dietary Allowances*, 10th ed. (Washington, D.C.: National Academy Press, 1989), pp. 247–261; J. L. Groff, S. S. Gropper, and S. M. Hunt, *Advanced Nutrition and Human Metabolism* (St. Paul, Minn.: West Publishing, 1995), pp. 423–438.

antidiuretic hormone (ADH): a hormone released by the pituitary gland in response to highly concentrated blood. The kidneys respond by reabsorbing water, thus preventing water loss.
- **anti** = against
- **dia** = through
- **ure** = urine

FIGURE 12-2 A Nephron, One of the Kidney's Many Functioning Units

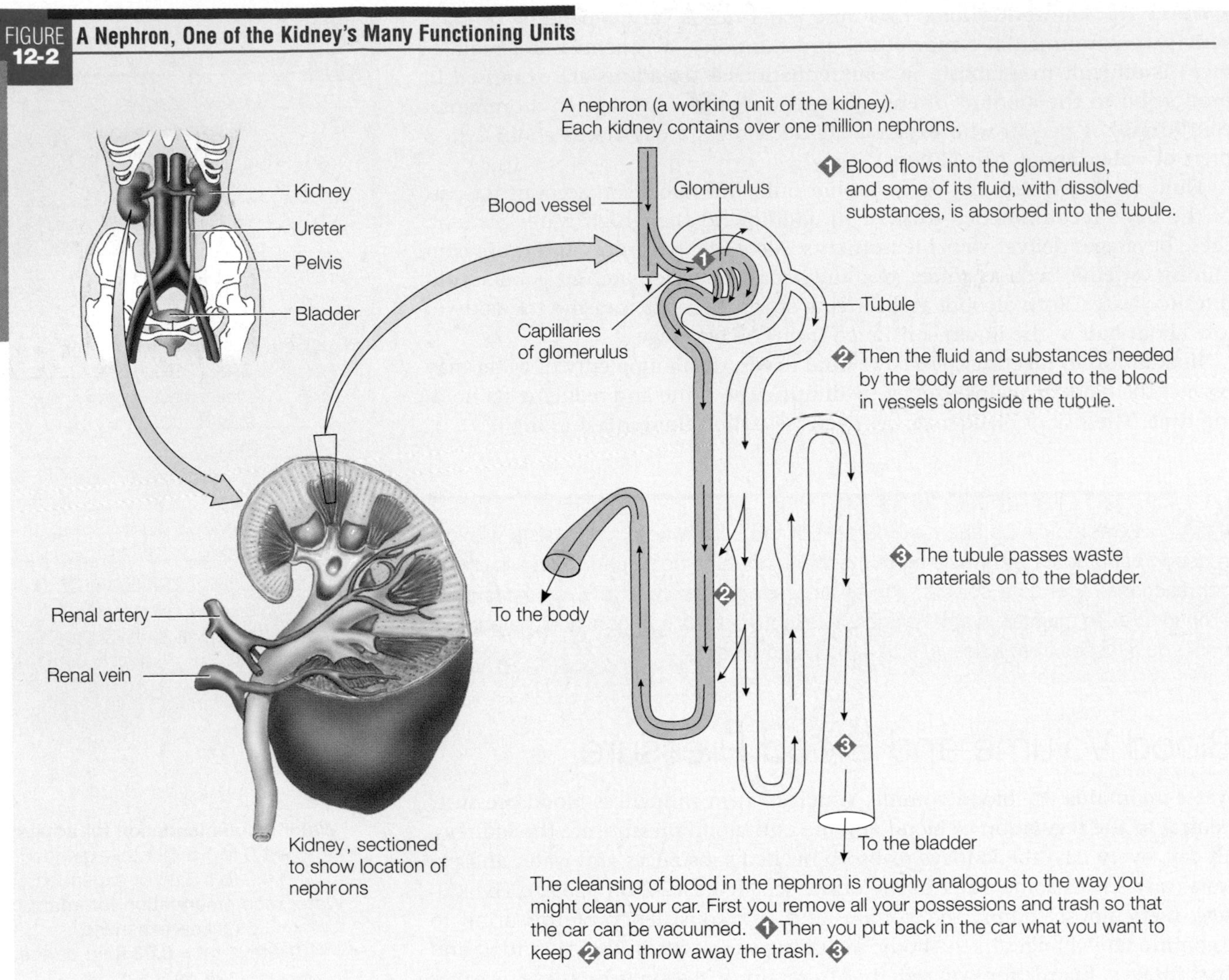

renin (REN-in): an enzyme from the kidneys that activates angiotensin.

angiotensin (AN-gee-oh-TEN-sin): a hormone involved in blood pressure regulation. Its precursor protein is called **angiotensinogen.**

vasoconstrictor (VAS-oh-kon-STRIK-tor): a substance that constricts or narrows the blood vessels.

aldosterone (al-DOS-ter-own): a hormone secreted by the adrenal glands that stimulates the reabsorption of sodium by the kidneys. Aldosterone also regulates chloride and potassium concentrations.

adrenal glands: glands adjacent to, and just above, each kidney.

- **Renin and Sodium Retention** • Cells in the kidneys respond to low blood pressure by releasing an enzyme called **renin.** Through a complex series of events, renin causes the kidneys to reabsorb sodium. Sodium reabsorption, in turn, is always accompanied by water retention, which helps to restore blood volume and blood pressure.

- **Angiotensin and Blood Vessel Constriction** • Renin also activates the blood protein angiotensinogen to **angiotensin.** Angiotensin is a powerful **vasoconstrictor:** it narrows the diameters of blood vessels, thereby raising the blood pressure.

- **Aldosterone and Sodium Retention** • Angiotensin also mediates the release of the hormone **aldosterone** from the **adrenal glands.** Aldosterone signals the kidneys to retain more sodium and therefore water because when sodium moves, fluids follow. Again, the effect is that when more water is needed, less is excreted.

All of these actions help to explain why high-sodium diets aggravate conditions such as hypertension or edema. Too much sodium causes water retention and an accompanying rise in blood pressure or swelling in the interstitial spaces.

IN SUMMARY

In response to low blood volume, low blood pressure, or highly concentrated body fluids, these three actions combine to effectively restore homeostasis (see Figure 12-3):

- *ADH causes water retention.*
- *Renin causes sodium retention.*
- *Angiotensin constricts blood vessels.*
- *Aldosterone causes sodium retention.*

These actions can maintain water balance only if a person drinks enough water.

Fluid and Electrolyte Balance

Maintaining a balance of about two-thirds of the body fluids inside the cells and one-third outside is vital to the life of the cells. If too much water were to enter the cells, it might rupture them; if too much water were to leave, they would collapse. To control the movement of water, the cells direct the movement of the major minerals.

• **Dissociation of Salt in Water** • When a mineral **salt** such as sodium chloride (NaCl) dissolves in water, it separates **(dissociates)** into **ions**—positively and negatively charged particles (Na^+ and Cl^-). The positive ions are **cations;** the negative ones are **anions.**◆ Unlike pure water, which conducts electricity poorly, ions dissolved in water carry electrical current. For this reason, salts that dissociate into ions are called **electrolytes,** and the fluids of the body, which contain them, are **electrolyte solutions.**

◆ **To help you remember the difference, think of the "t" in cations as a "plus" (+) sign.**

FIGURE 12-3 **How the Body Regulates Blood Volume**

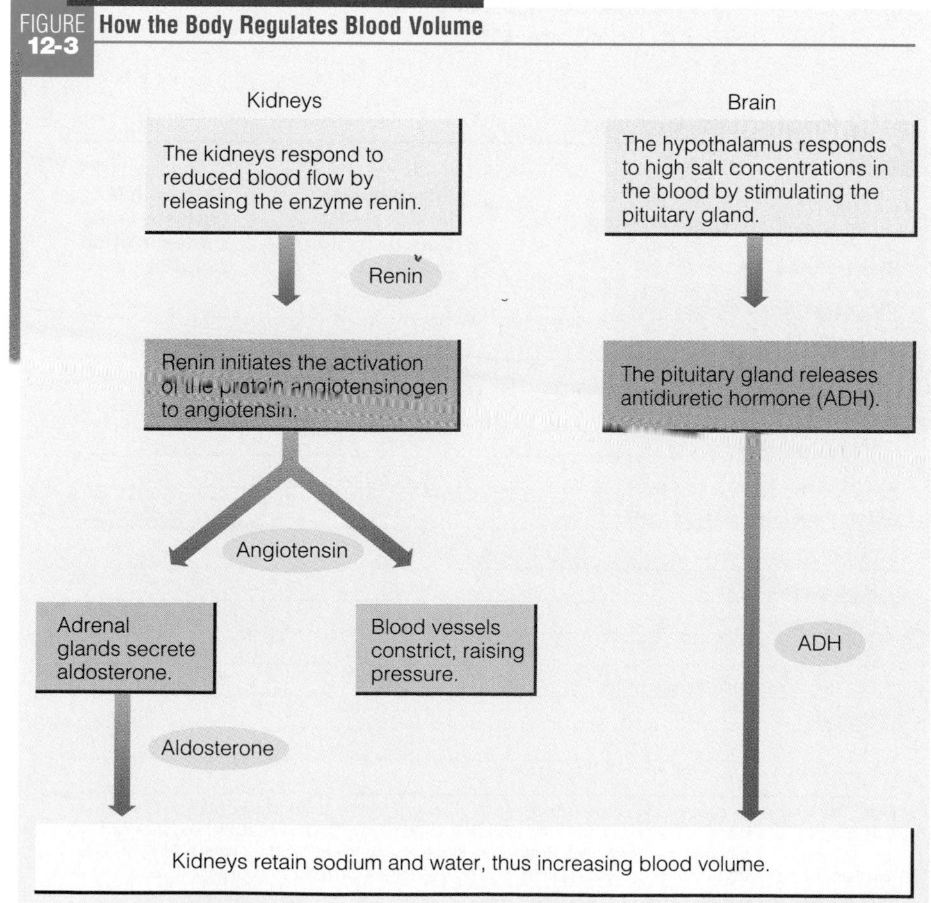

salt: a compound composed of a positive ion other than H^+ and a negative ion other than OH^-. An example is sodium chloride (Na^+Cl^-).
Na = sodium.
Cl = chloride.

dissociates (dis-SO-see-ates): physically separates.

ions (EYE uns): atoms or molecules that have gained or lost electrons and therefore have electrical charges. Examples include the positively charged sodium ion (Na^+) and the negatively charged chloride ion (Cl^-). For a closer look at ions, see Appendix B.

cations (CAT-eye-uns): positively charged ions.

anions (AN-eye-uns): negatively charged ions.

electrolytes: salts that dissolve in water and dissociate into charged particles called ions.

electrolyte solutions: solutions that can conduct electricity.

In all electrolyte solutions, anion and cation concentrations balance (the numbers of negative and positive charges are equal). If a fluid contains 1000 negative charges, it must contain 1000 positive charges, too. If an anion enters the fluid, a cation must accompany it or another anion must leave so that electrical neutrality will be maintained. Whenever Na^+ ions leave a cell, other positive ions enter: potassium (K^+) ions, for example. (In fact, it's a good bet that whenever Na^+ and K^+ ions are moving, they are going in opposite directions.)

Table 12-4 shows that, indeed, the positive and negative charges inside and outside cells are perfectly balanced even though the numbers of each kind of ion differ over a wide range. Inside the cells, the positive charges total 202 and the negative charges balance these perfectly. Outside the cells, the amounts and proportions of the ions differ from those inside, but again the positive and negative charges balance. (Scientists count these charges in **milliequivalents, mEq.**)

• **Electrolytes Attract Water** • Electrolytes attract water. Each water molecule has a net charge of zero,◆ but the oxygen side of the molecule is slightly negatively charged, and the hydrogens are slightly positively charged. Figure 12-4 shows the result in an electrolyte solution: both positive and negative ions attract clusters of water molecules around them. This attraction dissolves salts in water and enables the body to move fluids into appropriate compartments.

◆ A neutral molecule, such as water, that has opposite charges spatially separated within the molecule is **polar;** see Appendix B for more details.

• **Water Follows Electrolytes** • Some electrolytes reside primarily outside the cells (notably, sodium and chloride), while others are predominantly inside the cells (notably, potassium, magnesium, phosphate,◆ and sulfate). Cell membranes are selectively permeable, meaning that they allow the passage of some molecules, but not of others. Whenever electrolytes move across the membrane, water follows.

◆ The word ending *-ate* denotes a salt of the mineral. Thus phosphate is the salt form of the mineral phosphorus, and sulfate is the salt form of sulfur.

The movement of water across a membrane toward the more concentrated **solutes** is called **osmosis.** Osmosis exerts a force, known as osmotic pressure. Figure 12-5 provides familiar examples of osmosis.

TABLE 12-4 Important Body Electrolytes

Electrolytes	Intracellular (inside cells) Concentration (mEq/L)	Extracellular (outside cells) Concentration (mEq/L)
Cations (positively charged ions)		
Sodium (Na^+)	10	142
Potassium (K^+)	150	5
Calcium (Ca^{++})	2	5
Magnesium (Mg^{++})	40	3
	202	155
Anions (negatively charged ions)		
Chloride (Cl^-)	2	103
Bicarbonate (HCO_3^-)	10	27
Phosphate ($HPO_4^=$)	103	2
Sulfate ($SO_4^=$)	20	1
Organic acids (lactate, pyruvate)	10	6
Proteins	57	16
	202	155

Note: The numbers of positive and negative charges in a given fluid are the same. For example, in extracellular fluid, the cations and anions both equal 155 milliequivalents per liter (mEq/L). Of the cations, sodium ions make up 142 mEq/L; and potassium, calcium, and magnesium ions make up the remainder. Of the anions, chloride ions number 103 mEq/L; bicarbonate ions number 27; and the rest are provided by phosphate ions, sulfate ions, organic acids, and protein.

milliequivalents (mEq): the concentration of electrolytes in a volume of solution. Milliequivalents are a useful measure when considering ions because the number of charges reveals characteristics about the solution that are not evident when the concentration is expressed in terms of weight.

solutes (SOLL-yutes): the substances that are dissolved in a solution. The number of molecules in a given volume of fluid is the **solute concentration.**

osmosis: the movement of water across a membrane *toward* the side where the solutes are more concentrated.

FIGURE 12-4 **Water Dissolves Salts and Follows Electrolytes**

The structural arrangement of the two hydrogen atoms and one oxygen atom enables water to dissolve salts. Water's role as a solvent is one of its most valuable characteristics.

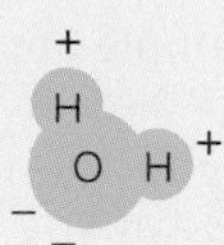

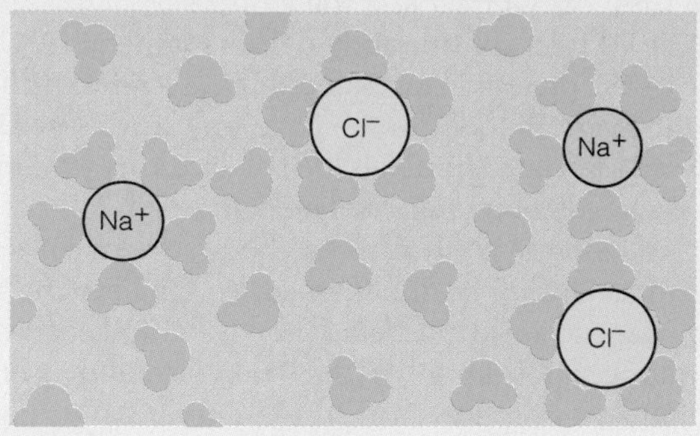

The negatively charged electrons that bond the hydrogens to the oxygen spend most of their time near the oxygen atom. As a result, the oxygen is slightly negative, and the hydrogen is slightly positive (see Appendix B).

In an electrolyte solution, water molecules are attracted to both anions and cations. Notice that the negative oxygen atoms of the water molecules are drawn to the sodium cation (Na^+), while the positive hydrogen atoms of the water molecules are drawn to the chloride ions (Cl^-).

• **Proteins Regulate Flow of Fluids and Ions** • Chapter 6 described how proteins attract water and help to regulate fluid movement. In addition, transport proteins in the cell membranes regulate the passage of positive ions and other substances from one side of the membrane to the other. Negative ions follow positive ions, and water flows toward the more concentrated solution.

A well-understood protein that regulates the flow of fluids and ions in and out of cells is the sodium-potassium pump. The pump actively exchanges

FIGURE 12-5 **Examples of Osmosis**

Water flows in the direction of the higher concentration of solute.

When immersed in water, ... raisins get plump because water moves toward the higher concentration of sugar inside the raisin.

When sprinkled with salt, ... vegetables "sweat" because water moves toward the higher concentration of salt outside the eggplant.

Physically active people must remember to replace their body fluids.

sodium for potassium across the cell membrane, using ATP as an energy source. Figure 6-10 on p. 179 illustrates this action.

• **Regulation of Fluid and Electrolyte Balance** • The amounts of various minerals in the body must remain nearly constant. Regulation occurs chiefly at two sites: the GI tract and the kidneys.

The GI tract continuously delivers minerals to the stomach and small intestine in the digestive juices and bile it secretes. It then reabsorbs these minerals and those from foods in the large intestine as needed. Each day, 8 liters of fluids and associated minerals are recycled this way, providing ample opportunity for the regulation of electrolyte balance.

The kidneys' control of the body's *water* content by way of the hormone ADH has already been described. To regulate the *electrolyte* contents, the kidneys depend on the adrenal glands, which send out messages by way of the hormone aldosterone. If the body's sodium is low, aldosterone stimulates sodium reabsorption from the kidneys. As sodium is reabsorbed, potassium (another positive ion) is excreted in accordance with the rule that total positive charges must remain in balance with total negative charges.

Fluid and Electrolyte Imbalance

Normally, the body defends itself successfully against fluid and electrolyte imbalances. Certain situations, however, may overwhelm the body's ability to compensate. Vomiting, diarrhea, heavy sweating, burns, and wounds may incur such great fluid and electrolyte losses as to precipitate a medical emergency.

• **Sodium and Chloride Most Easily Lost** • Because sodium and chloride are the body's principal extracellular cation and anion, they are first to be lost when fluid is lost by sweating, bleeding, or excretion. It is no coincidence that after sweating excessively or losing fluid in other ways, a person craves salty foods and refreshing drinks.

• **Different Solutes Lost by Different Routes** • If fluid is lost by vomiting or diarrhea,◆ sodium is lost indiscriminately. If the adrenal glands oversecrete aldosterone, as occurs when a tumor develops, the kidneys may excrete too much potassium. And the person with uncontrolled diabetes may lose a solute not normally excreted: glucose and, with it, large amounts of water. All three situations bring on dehydration, but drinking water alone cannot restore electrolyte balance. In each case, medical intervention is required.

◆ Highlight 3 discusses vomiting and diarrhea.

• **Replacing Lost Fluids and Electrolytes** • In many cases, people can replace the fluids and minerals lost in sweat or in a temporary bout of diarrhea by drinking plain cool water and eating regular foods. Some cases, however, demand rapid replacement of fluids and electrolytes—for example, when diarrhea threatens the life of a malnourished child. Caregivers around the world have learned to use simple formulas◆ to treat mild-to-moderate cases of diarrhea. These lifesaving formulas do not require hospitalization and can be prepared from ingredients available locally. Caregivers need only learn to measure ingredients carefully and use sanitary water. Once rehydrated, a person can begin eating foods.

◆ Health care workers use **oral rehydration therapy (ORT)**—a simple solution of sugar, salt, and water, taken by mouth—to treat dehydration caused by diarrhea. A simple ORT recipe:
1 c boiling water.
2 tsp sugar.
A pinch of salt.

Acid-Base Balance

The body uses its ions not only to help maintain fluid and electrolyte balance, but also to regulate the acidity **(pH)** of its fluids. The pH scale of Chapter 3 is repeated here, in Figure 12-6, with the normal and abnormal pH ranges of the blood added. As you can see, the body must maintain the pH within a narrow range to avoid life-threatening consequences. Slight deviations in either direction can damage proteins, causing metabolic mayhem. Enzymes couldn't catalyze reactions and hemoglobin couldn't carry oxygen—to name just two examples.

pH: a measure of the concentration of H^+ ions (see Appendix B). The lower the pH, the higher the H^+ ion concentration and the stronger the acid. A pH above 7 is alkaline, or base (a solution in which OH^- ions predominate).

FIGURE 12-6 The pH Scale

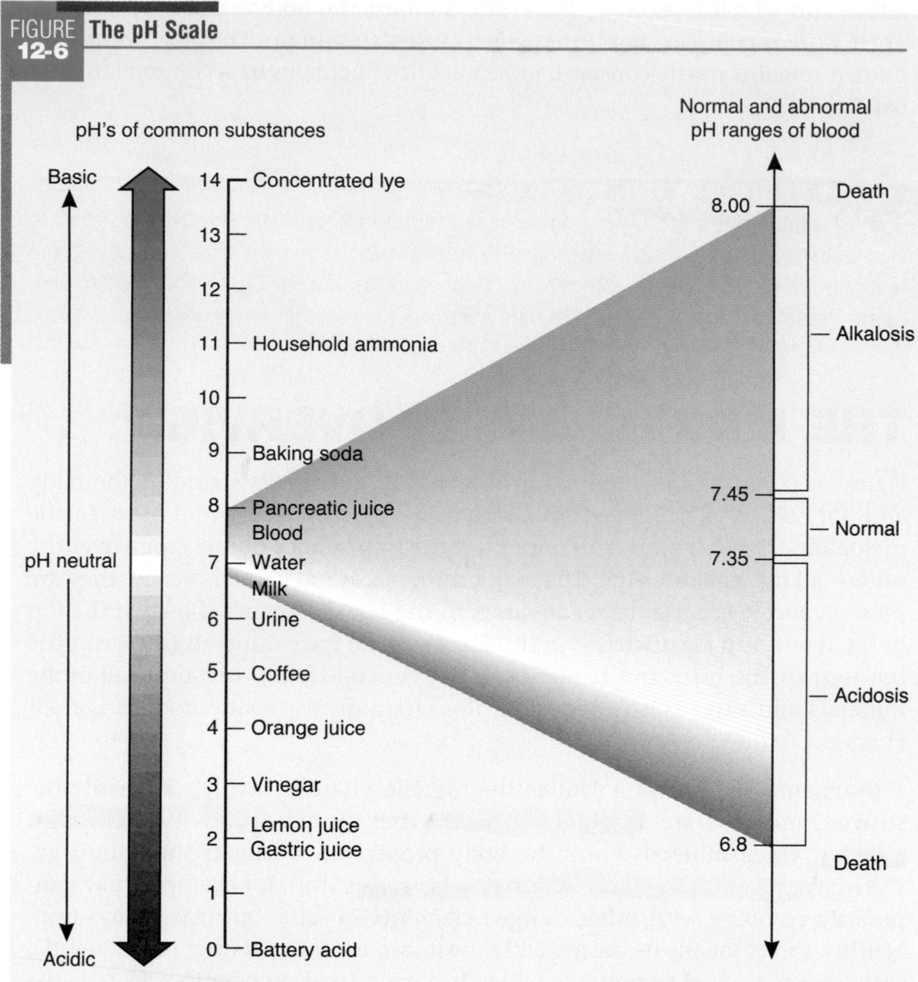

Note: Each step is ten times as concentrated in base (1/10 as much acid, or H^+) as the one below it.

The acidity of the body's fluids is determined by its concentration of hydrogen ions (H^+). A high concentration of hydrogen ions would be very acidic. Normal energy metabolism generates hydrogen ions, as well as many other acids, that must be neutralized. Three systems defend the body against fluctuations in pH—buffers in the blood, respiration in the lungs, and excretion in the kidneys.

• Regulation by the Buffers • **Bicarbonate** (a base) and **carbonic acid** (an acid) in the body fluids, as well as some proteins,◆ protect the body against changes in acidity by acting as buffers—substances that can neutralize acids or bases. These buffer systems serve as a first line of defense against changes in the fluids' acid-base balance.

◆ The buffering action of proteins is described in Chapter 6

• Regulation by the Lungs • Respiration provides another defense. Carbon dioxide, which is formed all the time during cellular metabolism, forms carbonic acid in the blood, pushing the balance toward acid. If too much acid builds up, the respiration rate speeds up; this hyperventilation increases the amount of carbon dioxide exhaled, thereby lowering the carbonic acid concentration and restoring homeostasis. Conversely, if base builds up, the respiration rate slows; carbon dioxide is retained and forms more carbonic acid. Again, homeostasis is restored.

• Regulation by the Kidneys • The kidneys play the primary role in maintaining long-term control of acid-base balance. By selecting which ions to

bicarbonate: a compound with the formula HCO_3 that results from the dissociation of carbonic acid; of particular importance in maintaining the body's acid-base balance. (Bicarbonate is also an alkaline secretion of the pancreas, part of the pancreatic juice.)

carbonic acid: a compound with the formula H_2CO_3 that results from the combination of carbon dioxide (CO_2) and water (H_2O); of particular importance in maintaining the body's acid-base balance.

retain and which to excrete, the kidneys adjust the body's acid-base balance. Their work is complex, but its net effect is easy to sum up. The *body's* total acid burden remains nearly constant; *urine's* acidity fluctuates to accommodate that balance.

IN SUMMARY
Electrolytes (charged minerals) in the fluids help distribute the fluids inside and outside the cells, thus ensuring the appropriate water balance and acid-base balance to support all life processes. Excessive losses of fluids and electrolytes upset these balances; the kidneys play a key role in restoring homeostasis.

THE MINERALS—AN OVERVIEW

Figure 12-7 shows the amounts of the **major minerals** found in the body and, for comparison, some of the trace minerals. The distinction between the major and trace minerals does not reflect the importance of one group over the other—all minerals are vital. The major minerals are so named because they are present, and needed, in larger amounts in the body. They are shown at the top of the figure and are discussed in this chapter. The trace minerals (shown at the bottom) are discussed in Chapter 13. A few generalizations pertain to all of the minerals and distinguish them from the vitamins. Especially notable is their chemical nature.

• **Inorganic Elements** • Unlike the organic vitamins, which are easily destroyed, minerals are inorganic elements that always retain their chemical identity. Once minerals enter the body proper, they remain there until excreted; they cannot be changed into anything else. Iron, for example, may temporarily combine with other charged elements in salts, but it is always iron. Neither can minerals be destroyed by heat, air, acid, or mixing; consequently, little care is needed to preserve minerals during food preparation. In fact, the ash that remains when a food is burned contains all the minerals that were in

FIGURE 12-7 Minerals in a 60-kilogram (132-pound) Human Body

Not only are the major minerals present in the body in larger amounts than the trace minerals, but they are also needed by the body in larger amounts. Recommended intakes for the major minerals are stated in *hundreds of milligrams* or *grams*, whereas those for the trace minerals are listed in *tens of milligrams* or even *micrograms*.

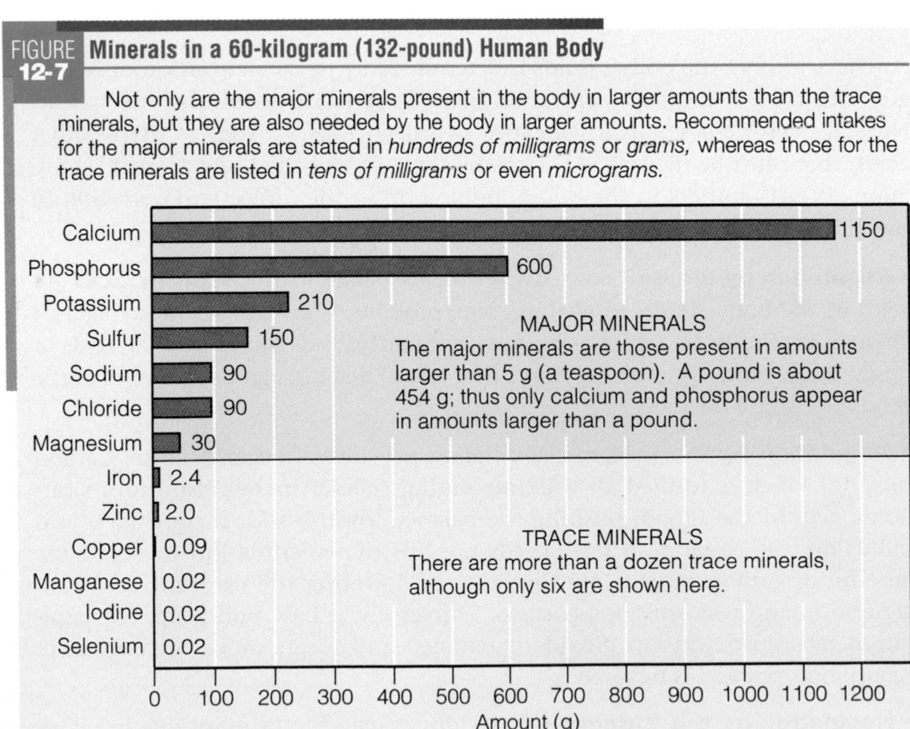

major minerals: essential mineral nutrients found in the human body in amounts larger than 5 g; sometimes called **macrominerals.**

the food originally. Minerals can be lost from food only when they leach into water that is then poured down the drain.

• The Body's Handling of Minerals • The minerals also differ from the vitamins in the amounts the body can absorb and in the extent to which they must be specially handled. Some minerals, such as potassium, are easily absorbed into the blood, transported freely, and readily excreted by the kidneys, much like the water-soluble vitamins. Other minerals, such as calcium, are more like fat-soluble vitamins in that they must have carriers to be absorbed and transported. And, like the fat-soluble vitamins, minerals taken in excess can be toxic.

• Variable Bioavailability • The bioavailability◆ of minerals varies. Some foods contain **binders** that combine chemically with minerals, preventing their absorption and carrying them out of the body with other wastes. Examples of binders include phytates, which are found primarily in legumes and grains, and oxalates, which are present in rhubarb and spinach, among other foods. These foods contain more minerals than the body actually receives for use.

Reminder: *Bioavailability* refers to the rate at and the extent to which a nutrient is absorbed and used. ◆

• Nutrient Interactions • Chapter 10 described how the presence or absence of one vitamin can affect another's absorption, metabolism, and excretion. The same is true of the minerals. The interactions between sodium and calcium, for example, cause both to be excreted when sodium intakes are high. Phosphorus binds with magnesium in the GI tract, so magnesium absorption is limited when phosphorus intakes are high. These are just two examples of interactions involving minerals featured in this chapter. Discussions in both this chapter and the next point out additional problems that arise from such interactions. Notice how often they reflect an excess of one mineral creating an inadequacy of another and how supplements—not foods—are most often to blame.

• Varied Roles • While all the major minerals help to maintain the body's fluid balance described earlier, sodium, chloride, and potassium are most noted for that role. For this reason, these three minerals are discussed first here. Later sections describe the minerals most noted for their roles in bone growth and health—calcium, phosphorus, and magnesium.

IN SUMMARY

The major minerals are found in larger quantities in the body, whereas the trace minerals occur in smaller amounts. Minerals are inorganic elements that retain their chemical identities; they usually receive special handling and regulation in the body; and they may bind with other substances or interact with other minerals, thus limiting their absorption.

SODIUM

People have held salt (sodium chloride) in high regard throughout recorded history. We say "you are the salt of the earth" to someone we admire and "you are not worth your salt" to someone we consider worthless. Even the word *salary* comes from the Latin word for salt.

Cultures vary in their use of salt, but most people find its taste innately appealing.[4] Salt brings its own tangy taste and enhances other flavors, most likely by suppressing the bitter flavors.[5] You can taste this effect for yourself: tonic water with its bitter quinine tastes sweeter with a little salt added.

• Sodium Roles in the Body • Sodium is the principal cation of the extracellular fluid and the primary regulator of its volume. Sodium also helps maintain acid-base balance and is essential to nerve transmission and muscle contraction.*

*One of the ways the kidneys regulate acid-base balance is by excreting hydrogen ions (H^+) in exchange for sodium ions (Na^+).

binders: chemical compounds in foods that combine with nutrients (especially minerals) to form complexes the body cannot absorb. Examples include **phytates** (FYE-tates) and **oxalates** (OK-sa-lates).

sodium: the principal cation in the extracellular fluids of the body; critical to the maintenance of fluid balance, nerve transmissions, and muscle contractions.

Foods usually provide more sodium than the body needs. Sodium is readily absorbed by the intestinal tract and travels freely in the blood until it reaches the kidneys, which filter all the sodium out of the blood; then, with great precision, they return to the bloodstream the exact amount the body needs. Normally, the amount excreted is approximately equal to the amount ingested on a given day. When blood sodium rises, as when a person eats salted foods, thirst signals the person to drink until the appropriate sodium-to-water ratio is restored. Then the kidneys excrete both the excess water and the excess sodium together.

• **Sodium Recommendations** • Diets rarely lack sodium. The *minimum* sodium requirement for adults is set at 500 milligrams in the United States and at 115 milligrams in Canada. Such differences between countries' recommendations typically reflect differences in judgment more than differences in research data. The lower recommendation reflects the minimum average requirement for adults without active sweating, whereas the higher recommendation accommodates a wider variety of physical activities and climates. As for the *maximum,* the Daily Value used on food labels is 2400 milligrams of sodium per day.

HEALTHY PEOPLE 2010

Increase the proportion of persons aged 2 years and older who consume 2400 mg or less of sodium daily.

Health promotion

• **Sodium and Hypertension** • For years, a high *sodium* intake was considered the primary factor responsible for high blood pressure. Then research pointed to *salt* (sodium chloride) as the dietary culprit. Salt has a greater effect on blood pressure than either sodium or chloride alone or in combination with other ions.[6]

Some individuals respond sensitively to excesses in salt intake and experience high blood pressure.[7] People most likely to have a **salt sensitivity** include those with chronic renal disease, diabetes, or hypertension; African Americans; and people over 50 years of age.*[8] Overweight people also appear to be particularly sensitive to the effect of salt on blood pressure. For them, a high salt intake correlates strongly with heart disease and death.[9] Salt restriction may help lower blood pressure in these salt-sensitive individuals, but both a diet abundant in fruits and vegetables and weight loss, especially in combination with salt restriction, may be even more effective.[10]

Whether people with normal blood pressure benefit from a salt-restricted diet remains controversial.[11] Because reducing salt intake causes no harm and diminishes the risk of hypertension and heart disease for some people, health recommendations advise limiting daily *salt* intake to less than 6 grams◆ (the equivalent of 2.4 grams or 2400 milligrams of *sodium*).[12] Chapter 26 offers a complete discussion of hypertension and the dietary recommendations for its prevention and treatment.

◆ Salt (sodium chloride) is about 40% sodium.
1 g salt contributes 400 mg sodium.
5 g salt = 1 tsp.
1 tsp salt contributes 2000 mg sodium.

• **Sodium and Osteoporosis** • A high sodium intake is also associated with calcium excretion, but whether it influences bone loss is less clear.[13] Dietary advice to prevent osteoporosis might suggest eating more calcium-rich foods while eating fewer high-sodium foods.

• **Sodium in Foods** • In general, processed foods have the most sodium, while unprocessed foods such as fresh fruits, vegetables, milk, and meats have the least. In fact, as much as 75 percent of the sodium in people's diets comes from salt added to foods by manufacturers; about 15 percent comes from salt added during cooking and at the table; and only 10 percent comes from the natural content in foods.

salt sensitivity: a characteristic of individuals who respond to a high salt intake with an increase in blood pressure.

*Compared with others, salt-sensitive individuals have elevated concentrations of renin in their blood.

FIGURE 12-8 What Processing Does to the Sodium and Potassium Contents of Foods

People who eat foods high in salt often happen to be eating fewer potassium-containing foods at the same time. Note how the same food loses potassium and gains sodium as it goes through processing, causing its potassium-to-sodium ratio to fall dramatically. Even when potassium isn't lost, the addition of sodium still lowers the potassium-to-sodium ratio. Limiting sodium intake may help in two ways, then—by lowering blood pressure in salt-sensitive individuals and by indirectly raising potassium intakes in all individuals.

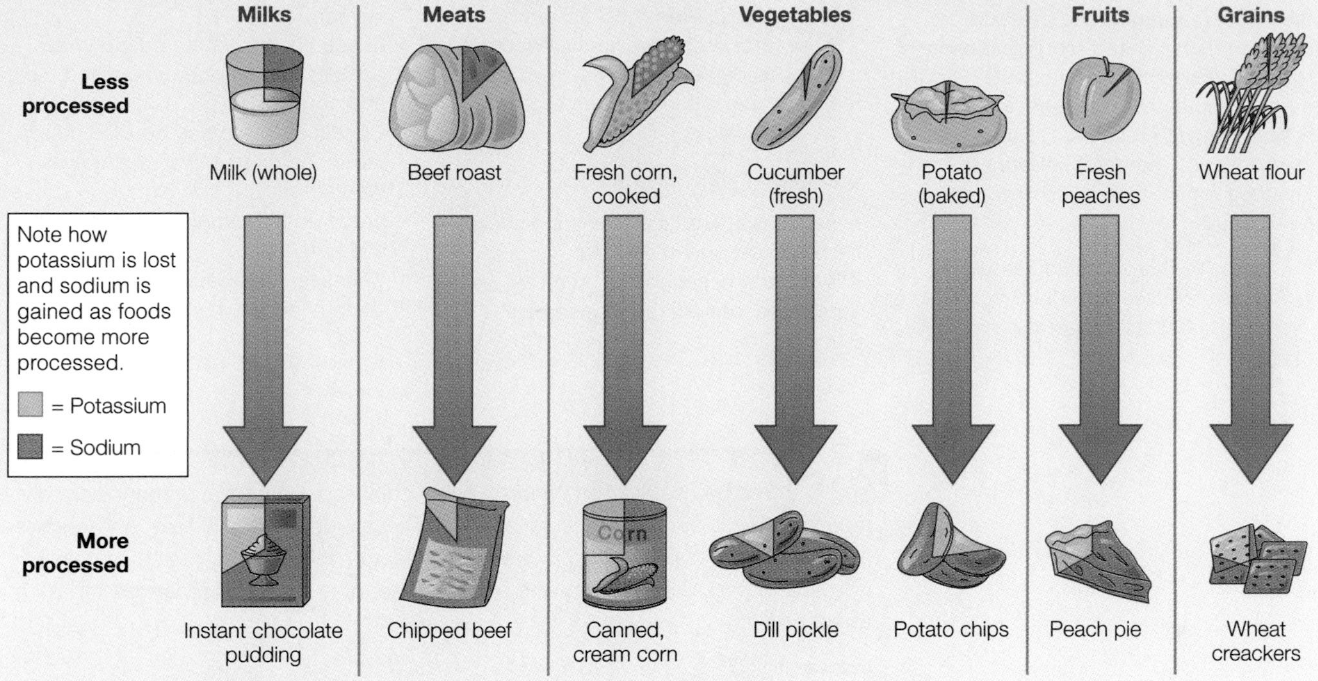

	Milks	Meats	Vegetables			Fruits	Grains
Less processed	Milk (whole)	Beef roast	Fresh corn, cooked	Cucumber (fresh)	Potato (baked)	Fresh peaches	Wheat flour
More processed	Instant chocolate pudding	Chipped beef	Canned, cream corn	Dill pickle	Potato chips	Peach pie	Wheat creackers

Note how potassium is lost and sodium is gained as foods become more processed.

▢ = Potassium

■ = Sodium

Because processed foods may contain sodium without chloride, as in additives such as sodium bicarbonate or sodium saccharin, they do not always taste salty. Most people are surprised to learn that 1 ounce of cornflakes contains more sodium than 1 ounce of salted peanuts—and that ½ cup of instant chocolate pudding contains still more. (A reason the peanuts taste saltier is that the salt is all on the surface, where the tongue's sensors immediately pick it up.)

Figure 12-8 shows that processed foods contain not only more sodium but also less potassium than their less processed counterparts. Low potassium may be as significant as high sodium when it comes to blood pressure regulation, so these foods have two strikes against them. The "How to" on p. 400 offers strategies for cutting salt (and therefore sodium) intake.

Fresh herbs, such as basil, add flavor to a recipe without adding salt.

• Sodium Deficiency • If blood sodium drops, as may occur in vomiting, diarrhea, or heavy sweating, both sodium and water must be replenished. Under normal conditions of sweating due to exercise, salt losses can easily be replaced later in the day with ordinary foods. Salt tablets are not recommended because too much salt, especially if taken with too little water, can induce dehydration.

During intense activities, such as ultra-endurance events, athletes can lose so much sodium and drink so much water that they develop hyponatremia—too little sodium in the blood. Beverages that contain sodium and glucose and salty foods will help restore sodium balance.

• Sodium Toxicity and Excessive Intakes • The immediate symptoms of acute sodium toxicity are edema and hypertension, but such toxicity poses no problem as long as water needs are met. Prolonged excessive sodium intake may contribute to hypertension in some people, as explained earlier.

HOW TO Cut Salt Intake

Most people eat more salt (and therefore sodium) than they need, and some people can lower their blood pressure by avoiding highly salted foods and removing the saltshaker from the table. Foods eaten without salt may seem less tasty at first, but with repetition, people can learn to enjoy the natural flavors of many unsalted foods. Strategies to cut salt intake include:

- Cook with little or no added salt.
- Prepare foods with sodium-free spices such as basil, bay leaves, curry, garlic, ginger, mint, oregano, pepper, rosemary, and thyme; lemon juice; vinegar; or wine.
- Add little or no salt at the table; taste foods before adding salt.
- Read labels with an eye open for salt. (See Table 2–9 on p. 54 for terms used to describe the sodium contents of foods on labels.)
- Select low-salt or salt-free products when available.

Use these foods sparingly:

- Foods prepared in brine, such as pickles, olives, and sauerkraut.
- Salty or smoked meats, such as bologna, corned or chipped beef, bacon, frankfurters, ham, lunch meats, salt pork, sausage, and smoked tongue.
- Salty or smoked fish, such as anchovies, caviar, salted and dried cod, herring, sardines, and smoked salmon.
- Snack items such as potato chips, pretzels, salted popcorn, salted nuts, and crackers.
- Condiments such as bouillon cubes; seasoned salts; MSG: soy, teriyaki, Worcestershire, and barbeque sauces; prepared horseradish, catsup, and mustard.
- Cheeses, especially processed types.
- Canned and instant soups.

IN SUMMARY

Sodium is the main cation outside cells and one of the primary electrolytes responsible for maintaining fluid balance. Dietary deficiency is rare, and excesses may aggravate hypertension in some people. For this reason, health professionals advise a diet moderate in salt and sodium. The table below summarizes information about sodium.

Sodium

1989 Estimated Minimum Requirement	Chief Functions in the Body	Deficiency Symptoms	Toxicity Symptoms	Significant Sources
Adults: 500 mg/day	Maintains normal fluid and electrolyte balance; assists in nerve impulse transmission and muscle contraction	Muscle cramps, mental apathy, loss of appetite	Edema, acute hypertension	Table salt, soy sauce; moderate amounts in meats, milks, breads, and vegetables; large amounts in processed foods

CHLORIDE

The element *chlorine* (Cl_2) is a poisonous gas. When chlorine reacts with sodium or hydrogen, however, it forms the negative chloride ion (Cl^-). *Chloride* is an essential nutrient, required in the diet.

• Chloride Roles in the Body • Chloride is the major anion of the extracellular fluids, where it occurs mostly in association with sodium. Chloride can move freely across membranes and so also associates with potassium inside cells. Like sodium and potassium, chloride maintains fluid and electrolyte balance.

In the stomach, the chloride ion is part of hydrochloric acid, which maintains the strong acidity of the gastric juice. One of the most serious consequences of vomiting is the loss of this acid◆ from the stomach, which upsets the acid-base balance.* Such imbalances are commonly seen in bulimia nervosa, as Highlight 9 describes.

◆ **Reminder: The loss of acid can lead to *alkalosis,* an above-normal alkalinity in the blood and body fluids.**

chloride (KLO-ride): the major anion in the extracellular fluids of the body. Chloride is the ionic form of chlorine, Cl^-; see Appendix B for a description of the chlorine-to-chloride conversion.

*Hydrochloric acid secretion into the stomach involves the addition of bicarbonate ions (base) to the plasma. These bicarbonate ions (HCO_3^-) are neutralized by hydrogen ions (H^+) from the gastric secretions that are reabsorbed into the plasma. When hydrochloric acid is lost during vomiting, these hydrogen ions are no longer available for reabsorption, and so, in effect, the concentrations of bicarbonate ions in the plasma are increased. In this way, excessive vomiting of acidic gastric juices leads to *metabolic alkalosis.*

• **Chloride Recommendations and Intakes** • Chloride is abundant in foods (especially processed foods) as part of sodium chloride and other salts. A recommended intake has not been established for chloride; instead a minimum requirement for adults has been estimated.

• **Chloride Deficiency and Toxicity** • Diets rarely lack chloride. Chloride losses may occur in conditions such as heavy sweating, chronic diarrhea, and vomiting. The only known cause of high blood chloride concentrations is dehydration due to water deficiency. In both cases, consuming ordinary foods and beverages can restore chloride balance.

IN SUMMARY

Chloride is the major anion outside cells, and it associates closely with sodium. In addition to its role in fluid balance, chloride is part of the stomach's hydrochloric acid. The accompanying table summarizes information on chloride.

Chloride

1989 Estimated Minimum Requirement	Chief Functions in the Body	Deficiency Symptoms	Toxicity Symptoms	Significant Sources
Adults: 750 mg/day	Maintains normal fluid and electrolyte balance; part of hydrochloric acid found in the stomach, necessary for proper digestion	Do not occur under normal circumstances	Vomiting	Table salt, soy sauce; moderate amounts in meats, milks,eggs; large amounts in processed foods

POTASSIUM

Like sodium, **potassium** is a positively charged ion. In contrast to sodium, potassium is the body's principal cation *inside* the body cells.

• **Potassium Roles in the Body** • Potassium plays a major role in maintaining fluid and electrolyte balance and cell integrity. During nerve transmission and muscle contraction, potassium and sodium briefly trade places across the cell membrane. The cell then quickly pumps them back into place. Controlling potassium distribution is a high priority for the body because it affects many aspects of homeostasis, including a steady heartbeat.

• **Potassium Recommendations and Intakes** • As for sodium and chloride, a minimum potassium requirement for adults has been estimated. Potassium is abundant in all living cells, both plant and animal. Because cells remain intact unless foods are processed, the richest sources of potassium are *fresh* foods of all kinds—as Figure 12-9 (on p. 402) shows. In contrast, most processed foods such as canned vegetables, ready-to-eat cereals, and luncheon meats contain less potassium—and more sodium (recall Figure 12-8). Toxicity is not a concern when the source is foods.

• **Potassium and Hypertension** • Diets low in potassium seem to play an important role in the development of high blood pressure. Low potassium intakes raise blood pressure, whereas high potassium intakes appear to both prevent and correct hypertension.[14] Potassium-rich fruits and vegetables also appear to reduce the risk of stroke—more so than can be explained by the reduction in blood pressure alone.

• **Potassium Deficiency** • Potassium deficiency is the most common electrolyte imbalance.[15] It is more often caused by excessive losses than by deficient intakes. Conditions such as diabetic acidosis, dehydration, or prolonged vomiting or diarrhea can create a potassium deficiency, as can the regular use

Fresh foods, especially fruits and vegetables, provide potassium in abundance.

potassium: the principal cation within the body's cells; critical to the maintenance of fluid balance, nerve transmissions, and muscle contractions.

FIGURE 12-9 **Potassium in Selected Foods**

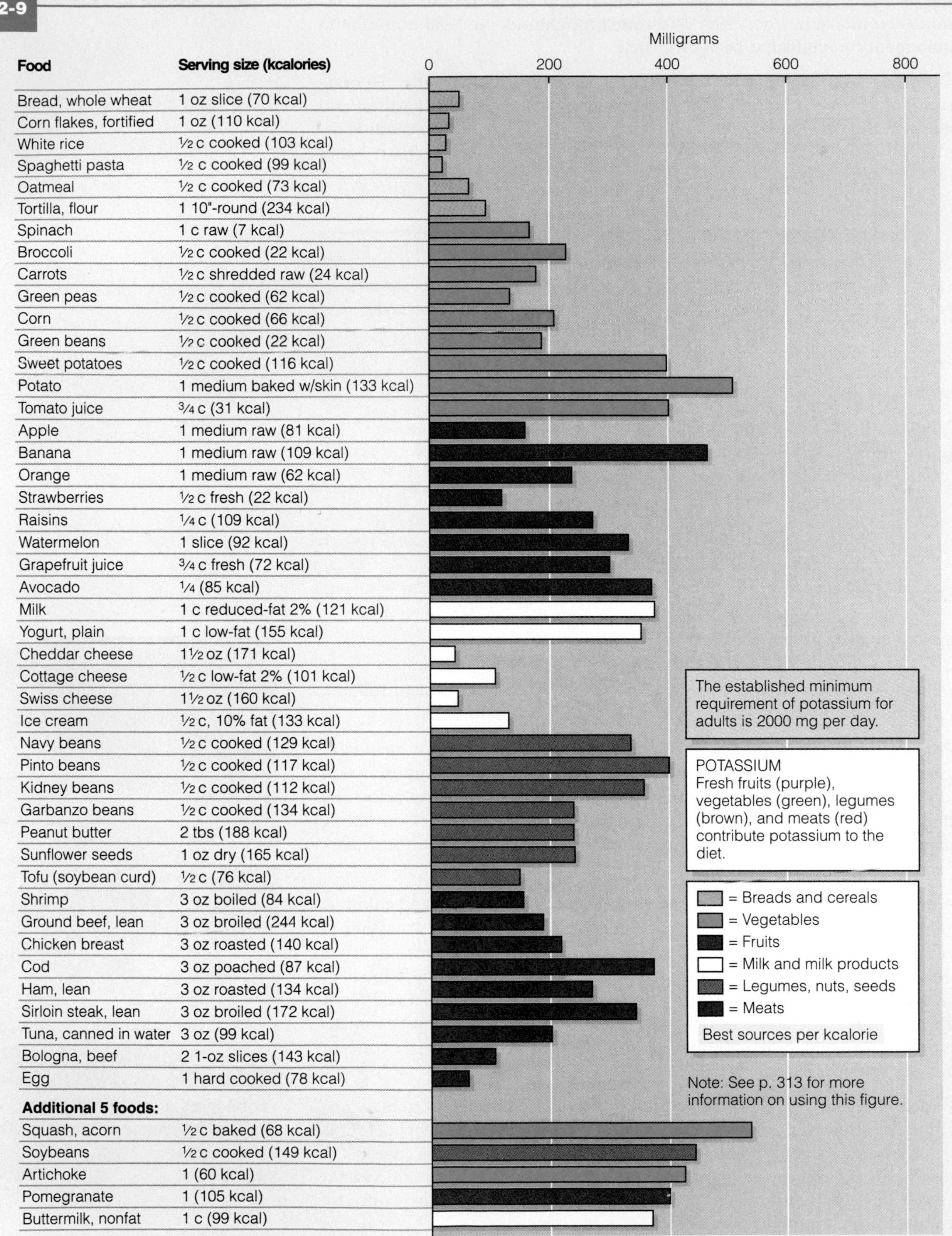

of certain drugs, including diuretics, steroids, and strong laxatives.* For this reason, many physicians prescribe potassium supplements along with these potassium-wasting drugs. One of the earliest symptoms of deficiency is muscle weakness.

• Potassium Toxicity • Potassium toxicity does not result from overeating foods high in potassium. It can result from overconsumption of potassium salts or supplements (including some "energy fitness shakes") and from certain diseases or treatments.[16] Given more potassium than the body needs, the kidneys accelerate their excretion. If the GI tract is bypassed, however, and potassium is injected directly into a vein, it can stop the heart.

IN SUMMARY

Potassium, like sodium and chloride, is an electrolyte that plays an important role in maintaining fluid balance. Potassium is the primary cation inside cells; fresh foods, notably fruits and vegetables, are its best sources. The table below summarizes facts about potassium.

Potassium

1989 Estimated Minimum Requirement	Chief Functions in the Body	Deficiency Symptoms[a]	Toxicity Symptoms	Significant Sources
Adults: 2000 mg/day	Maintains normal fluid and electrolyte balance; facilitates many reactions; supports cell integrity; assists in nerve impulse transmission and muscle contractions	Muscular weakness, paralysis, confusion	Muscular weakness; vomiting; if given into a vein, can stop the heart	All whole foods: meats, milks, fruits, vegetables, grains, legumes

[a]Deficiency accompanies dehydration.

CALCIUM

Calcium is the most abundant mineral in the body. It receives much emphasis in this chapter and in the highlight that follows because an adequate intake helps grow a healthy skeleton in early life and minimize bone loss in later life.

Calcium Roles in the Body

Ninety-nine percent of the body's calcium is in the bones (and teeth), where it plays two roles. First, it is an integral part of bone structure, providing a rigid frame that holds the body upright and serves as attachment points for muscles, making motion possible. Second, it serves as a calcium bank, offering a readily available source of the mineral to the body fluids should a drop in blood calcium occur.

• Calcium in Bones • As bones begin to form, calcium salts form crystals, called **hydroxyapatite,** on a matrix of the protein collagen. During **mineralization,** as the crystals become denser, they give strength and rigidity to the maturing bones. As a result, the long leg bones of children can support their weight by the time they have learned to walk.

Many people have the idea that once a bone is built, it is inert like a rock. Actually, the bones are gaining and losing minerals continuously in an ongoing

*People using diuretics to control hypertension should know that some cause potassium excretion and can induce a deficiency. Those using these drugs must be particularly careful to include rich sources of potassium in their daily diets. (Some diuretics are designed to spare potassium.)

calcium: the most abundant mineral in the body; found primarily in the body's bones and teeth.

hydroxyapatite (high-drox-ee-APP-ah-tite): crystals made of calcium and phosphorus.

mineralization: the process in which calcium, phosphorus, and other minerals crystallize on the collagen matrix of a growing bone, hardening the bone.

process of remodeling. Growing children gain more bone than they lose, and healthy adults maintain a reasonable balance. When withdrawals substantially exceed deposits, problems such as osteoporosis develop (see Highlight 12).

The formation of teeth follows a pattern similar to that of bones. The turnover of minerals in teeth is not as rapid as in bone, however; fluoride hardens and stabilizes the crystals of teeth, opposing the withdrawal of minerals from them.

• Calcium in Body Fluids • The 1 percent of the body's calcium that circulates in the fluids as ionized calcium is vital to life. The calcium ion participates in the regulation of muscle contractions, the clotting of blood, the transmission of nerve impulses, the secretion of hormones, and the activation of some enzyme reactions.

Calcium also activates a protein called **calmodulin.** This protein relays messages from the cell surface to the inside of the cell. Several of these messages help to maintain normal blood pressure.

• Calcium and Disease Prevention • Calcium may be useful in protecting against hypertension, even in salt-sensitive people.[17] An adequate calcium intake can lower blood pressure, superseding the effects of a high sodium intake.[18] For this reason, restricting sodium to treat hypertension is narrow advice, especially considering that low-sodium diets are often low in calcium as well. Instead, recommendations should also focus on raising calcium intakes.[19] Some research also suggests protective relationships between dietary calcium and blood cholesterol, diabetes, and colon cancer. Highlight 12 explores calcium's role in preventing osteoporosis.

HEALTHY PEOPLE 2010

Reduce the prevalence of osteoporosis, among people aged 50 and over.

• Calcium Balance • Calcium homeostasis is one of the body's highest priorities and involves a system of hormones and vitamin D. Whenever blood calcium falls too low or rises too high, three organ systems respond: the intestines, bones, and kidneys. Figure 12-10 illustrates how vitamin D and the hormones **parathormone** and **calcitonin** return blood calcium to normal.

The calcium in bone provides a nearly inexhaustible bank of calcium for the blood. The blood borrows and returns calcium as needed, so that even with a dietary deficiency, *blood* calcium remains normal—even as *bone* calcium diminishes. Blood calcium changes only in response to abnormal regulatory control, not to diet. A person can have an inadequate calcium intake for years and suffer no noticeable symptoms. Only late in life does it become apparent that the integrity of the bones has been compromised.

Blood calcium above normal results in **calcium rigor:** the muscles contract and cannot relax. Similarly, blood calcium below normal causes **calcium tetany**—also characterized by uncontrolled muscle contraction. These conditions do *not* reflect a *dietary* excess or lack of calcium; they are caused by a lack of vitamin D or by abnormal secretion of the regulatory hormones. A chronic *dietary* deficiency of calcium, or a chronic deficiency due to poor absorption over the years, depletes the savings account in the bones. Again: it is the *bones,* not the blood, that are robbed by a calcium deficiency.

• Calcium Absorption • Many factors affect calcium absorption, but on the average, adults absorb about 30 percent of the calcium they ingest. The stomach's acidity helps to keep calcium soluble, and vitamin D helps to make the **calcium-binding protein** needed for absorption. (This explains why calcium-rich milk is the best food for fortification with vitamin D.)

Whenever calcium is needed, the body increases its production of the calcium-binding protein to improve calcium absorption. The result is obvious

calmodulin (cal-MOD-you-lin): an inactive protein that becomes active when bound to calcium. Once activated, it becomes a messenger that tells other proteins what to do. The system serves as an interpreter for hormone- and nerve-mediated messages arriving at cells.

parathormone (PAIR-ah-THOR-moan): a hormone from the parathyroid glands that regulates blood calcium by raising it when levels fall too low; also known as **parathyroid hormone.**

calcitonin (KAL-see-TOE-nin): a hormone from the thyroid gland that regulates blood calcium by lowering it when levels rise too high.

calcium rigor: hardness or stiffness of the muscles caused by high blood calcium concentrations.

calcium tetany (TET-ah-nee): intermittent spasm of the extremities due to nervous and muscular excitability caused by low blood calcium concentrations.

calcium-binding protein: a protein in the intestinal cells, made with the help of vitamin D, that facilitates calcium absorption.

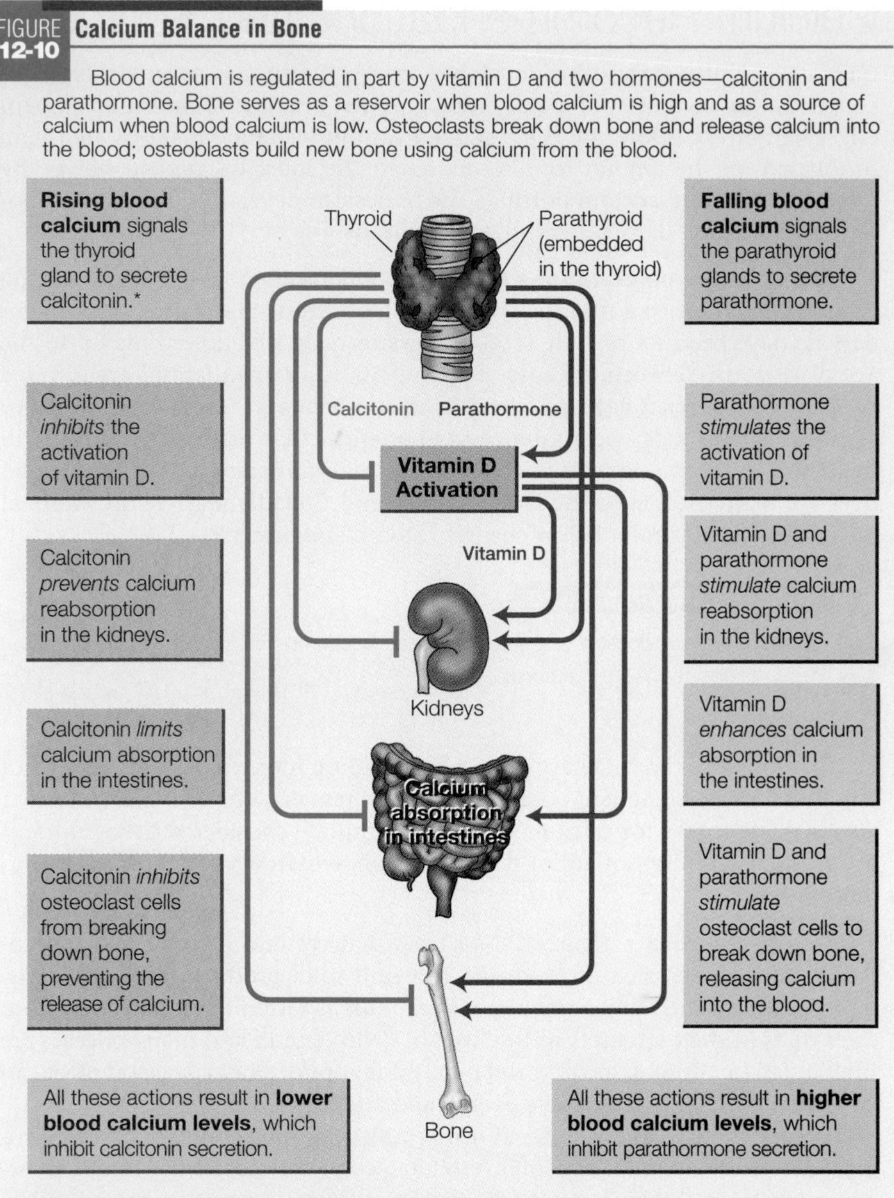

FIGURE 12-10 Calcium Balance in Bone

Blood calcium is regulated in part by vitamin D and two hormones—calcitonin and parathormone. Bone serves as a reservoir when blood calcium is high and as a source of calcium when blood calcium is low. Osteoclasts break down bone and release calcium into the blood; osteoblasts build new bone using calcium from the blood.

*Calcitonin plays a major role in defending infants and young children against the dangers of rising blood calcium that can occur when regular feedings of milk deliver large quanities of calcium to a small body. In contrast, calcitonin plays a relatively minor role in adults because their absorption of calcium is less efficient and their bodies are larger, making elevated blood calcium unlikely.

in the case of a pregnant woman, who absorbs 50 percent of the calcium from the milk she drinks. Similarly, growing children absorb 50 to 60 percent of the calcium they consume. Then, when bone growth slows or stops, absorption falls to the adult level of about 30 percent. In addition, absorption becomes more efficient during times of inadequate intakes.[20]

Many of the conditions that enhance calcium absorption inhibit its absorption when they are absent. For example, sufficient vitamin D supports absorption, while a deficiency impairs it. In addition, fiber, in general, and the binders phytate and oxalate, in particular, interfere with calcium absorption, but their effects are relatively minor at intakes typical of U.S. diets. Vegetables with oxalates and whole grains with phytates are nutritious foods, of course, but they are not useful calcium sources. The margin◆ presents factors that influence calcium balance.

◆ **Factors that *enhance* calcium absorption:**
- Stomach acid.
- Vitamin D.
- Lactose.
- Growth hormones.

Factors that *inhibit* calcium absorption:
- Lack of stomach acid.
- Vitamin D deficiency.
- High phosphorus intake.
- High-fiber diet.
- Phytates (in seeds, nuts, grains).
- Oxalates (in beets, rhubarb, spinach).

Milk and milk products are rightly famous for their calcium contents.

Calcium Recommendations and Sources

Calcium is unlike most other nutrients, in that hormones maintain its *blood* concentration regardless of dietary intake. When intake is high, the *bones* benefit; when intake is low, the *bones* suffer. Calcium recommendations are therefore based on the amount needed to retain the most calcium in bones. By retaining the most calcium possible, the bones can develop to their fullest potential in size and density, within genetic limits.

• **Calcium Recommendations** • Because obtaining enough calcium during growth helps to ensure that the skeleton will be strong and dense, recommendations have been set high at 1300 milligrams daily for adolescents up to the age of 18 years. Between the ages of 19 and 50, recommendations are lowered to 1000 milligrams a day; for later life, recommendations are raised again to 1200 milligrams a day to minimize the bone loss that tends to occur later in life. Some authorities advocate as much as 1500 milligrams a day for women over 50. Many people in the United States and Canada, particularly women, have calcium intakes far below current recommendations.

HEALTHY PEOPLE 2010

Increase the proportion of persons aged 2 years and older who meet dietary recommendations for calcium.

High intakes of both dietary protein and sodium increase calcium losses, but whether these losses impair bone development remains unclear. In establishing an Adequate Intake for calcium, the DRI committee considered these nutrient interactions, but did not adjust dietary recommendations based on this information.

◆ **Suggested minimum daily milk servings:**
- Children: 2 c.
- Teenagers: 3 c.
- Adults: 2 c.
- Pregnant or lactating women: 3 c.
- Pregnant or lactating teens: 4 c.

◆ **People with lactose intolerance may be able to consume small quantities of milk, as Chapter 4 explains.**

• **Calcium Sources** • Figure 12-11 shows that calcium is found most abundantly in a single class of foods—milk◆ and milk products. Unfortunately, many people, especially women, perceive milk as fattening and omit it from their diets in their attempts to lose weight. Whole milk and many cheeses are high in fat, but nonfat, low-fat, and reduced-fat options help a person meet calcium needs within a reasonable energy and fat allowance.

The person who doesn't like to drink milk may prefer to eat cheese or yogurt. Alternatively, milk and milk products can be concealed in foods. Powdered nonfat milk can be added to casseroles, soups, and other mixed dishes during preparation; 5 heaping tablespoons offer the equivalent of a cup of milk. This simple step is an excellent way for older women to obtain not only extra calcium, but more protein, vitamins, and minerals as well.[21]

Bioavailability of Calcium from Selected Foods

Absorption	Foods
≥50% absorbed	Cauliflower, watercress, brussels sprouts, rutabaga, kale, mustard greens, bok choy, broccoli, turnip greens, calcium-fortified foods and beverages
≈30% absorbed	Milk, calcium-fortified soy milk, calcium-set tofu, cheese, yogurt
≈20% absorbed	Almonds, sesame seeds, pinto beans, sweet potatoes
≤5% absorbed	Spinach, rhubarb, Swiss chard

• **Nonmilk Sources** • Some cultures do not use milk in their cuisines; some vegetarians exclude milk as well as meat; and some people are allergic to milk protein or are lactose intolerant.◆ These people need to find nonmilk sources of calcium to help meet their calcium needs. Some brands of tofu, corn tortillas, some nuts (such as almonds), and some seeds (such as sesame seeds) can supply calcium for the person who doesn't use milk products. A slice of most breads contains only about 5 to 10 percent of the calcium found in milk, but can be a major source for people who eat many slices because the calcium is well absorbed. Among the vegetables, mustard and turnip greens, bok choy, kale, parsley, watercress, and broccoli are good sources of available calcium. So are some seaweeds such as the nori popular in Japanese cooking. Some dark green, leafy vegetables—notably spinach and Swiss chard—appear to be calcium-rich but actually provide little, if any, calcium to the body because of the binders they contain. It would take 8 cups of spinach—containing six times as much calcium as 1 cup of milk—to deliver the equivalent in *absorbable* calcium.[22] The margin drawing ranks selected foods according to their calcium bioavailability.

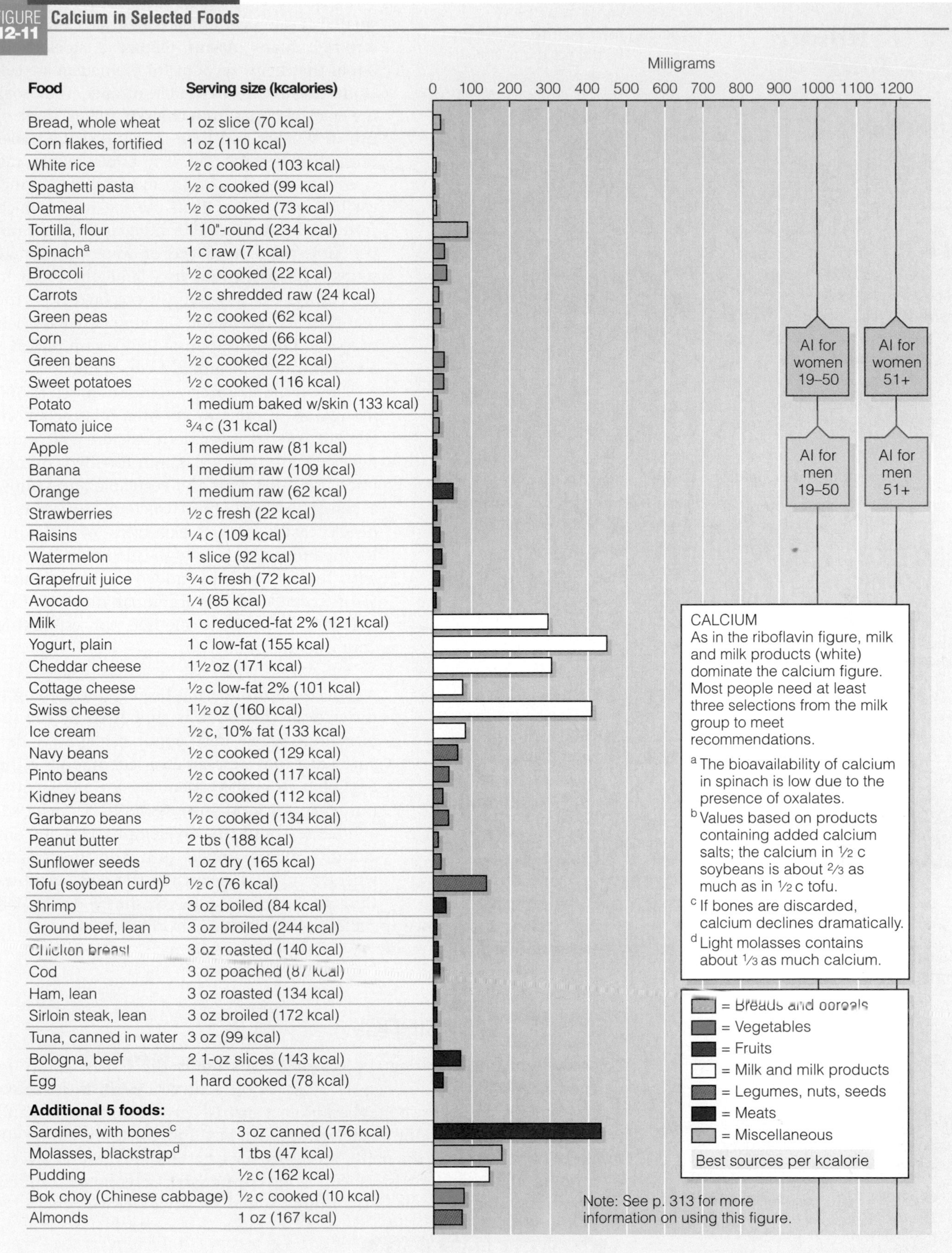

FIGURE 12-11 **Calcium in Selected Foods**

Estimate Your Calcium Intake

The exchange system offers a convenient way to estimate carbohydrate, fat, and protein, but not vitamins or minerals. Most dietitians have developed their own useful shortcuts to help them estimate nutrient intakes and "see" inadequacies in the diet. They can tell at a glance whether a day's meals fall short of calcium recommendations, for example.

To estimate calcium intakes, keep two bits of information in mind:

- A cup of milk provides about 300 milligrams of calcium.
- Adults need between 1000 and 1200 milligrams of calcium per day, which represents 3 to 4 cups of milk—or the equivalent:

$$1000 \text{ mg} \div 300 \text{ mg/c} = 3\tfrac{1}{3} \text{ c.}$$
$$1200 \text{ mg} \div 300 \text{ mg/c} = 4 \text{ c.}$$

If a person drinks 3 to 4 cups of milk a day, it's easy to see that calcium needs are being met. If not, it takes some detective work to identify the other sources and estimate total calcium intake.

To estimate a person's daily calcium intake, use this shortcut, which compares the calcium in calcium-rich foods to the calcium content of milk. The calcium in a cup of milk is assigned 1 point, and the goal is to attain 3 to 4 points per day. Foods are given points as follows:

- 1 c milk, yogurt, or fortified soy milk or 1½ oz cheese = 1 point.
- 4 oz canned fish with bones (sardines or salmon) = 1 point.
- 1 c ice cream, cottage cheese, or calcium-rich vegetable (see the text) = ½ point.

Then, because other foods also contribute small amounts of calcium, together they are given a point.

- Well-balanced diet containing a variety of foods = 1 point.

Now consider a day's meals with calcium in mind. Cereal with 1 cup of milk for breakfast (1 point for milk), a ham and cheese sub sandwich for lunch (1 point for cheese), and a cup of broccoli and lasagna for dinner (½ point for calcium-rich vegetable and 1 point for cheese in lasagna)—plus 1 point for all other foods eaten that day—adds up to 4½ points. This shortcut estimate indicates that calcium recommendations have been met, and a diet analysis of these few foods reveals a calcium intake of over 1000 milligrams. By knowing the best sources of each nutrient, you can learn to scan the day's meals and quickly see if you are meeting your daily goals.

Oysters are also a rich source of calcium, as are small fish eaten with their bones, such as canned sardines. Many Asians prepare a stock from bones that helps account for their adequate calcium intake without the use of milk. They soak the cracked bones from chicken, turkey, pork, or fish in vinegar and then slowly boil the bones until they become soft. The bones release calcium into the acidic broth, and most of the vinegar boils off. Cooks then use the stock, which contains more than 100 milligrams of calcium per tablespoon, in place of water to prepare soups, vegetables, and rice. Similarly, cooks in the Navajo tribe use an ash prepared from the branches and needles of the juniper tree in their recipes.[23] One teaspoon of juniper ash provides about as much calcium as a cup of milk.

Some mineral waters provide as much as 500 milligrams of calcium per liter, offering a convenient way to meet both calcium and water needs.[24] Similarly, calcium-fortified orange juice and other fruit and vegetable juices allow a person to meet both calcium and vitamin needs easily. Other examples of calcium-fortified foods include high-calcium milk (milk with extra calcium added) and calcium-fortified cereals.◆ The accompanying "How to" describes a shortcut method for estimating your calcium intake.

◆ Calcium supplements are discussed in Highlight 12.

A generalization that has been gaining strength throughout this book is supported by the information given here about calcium. A balanced diet that supplies a variety of foods is the best plan to ensure adequacy for all essential nutrients. All food groups should be included, and none should be overemphasized. In our culture, calcium intake is usually inadequate wherever milk is lacking in the diet—whether through ignorance, poverty, simple dislike, fad dieting, lactose intolerance, or allergy. By contrast, iron is usually lacking whenever milk is overemphasized, as Chapter 13 explains.

Calcium Deficiency

A low calcium intake during the growing years limits the bones' ability to achieve an optimal mass and density. Most people achieve a **peak bone mass** by their late 20s, and dense bones best protect against age-related bone loss and fractures. All adults lose bone as they grow older, beginning before they are 40. Should bone losses reach the point of causing fractures under common, everyday stresses, the condition is known as **osteoporosis.** Osteoporosis afflicts more than 25 million people in the United States, mostly older women.

peak bone mass: the highest attainable bone density for an individual, developed during the first three decades of life.

osteoporosis (OS-tee-oh-pore-OH-sis): a condition of older persons in which the bones become porous and fragile due to a loss of minerals; also called **adult bone loss.**
- **osteo** = bone
- **porosis** = porous

HEALTHY PEOPLE 2010

Reduce the proportion of adults with osteoporosis.

Unlike many diseases that make themselves known through symptoms such as pain, shortness of breath, skin lesions, tiredness, and the like, osteoporosis is silent. The body sends no signals saying bones are losing their calcium and, as a result, their integrity. Blood samples offer no clues because blood calcium remains normal regardless of bone content, and measures of bone density are not routinely taken. Highlight 12 suggests strategies to protect against bone loss, of which eating calcium-rich foods is only one.

As important as calcium may be to bone health, osteoporosis is not a calcium-deficiency disease comparable to scurvy. In scurvy, adequate vitamin C reliably reverses the condition; in osteoporosis, high calcium intakes alone during adulthood may prevent further deterioration, but they do little or nothing to reverse bone loss.

IN SUMMARY

Most of the body's calcium is in the bones where it provides a rigid structure and a reservoir of calcium for the blood. Blood calcium participates in muscle contraction, blood clotting, and nerve impulses and is closely regulated by a system of hormones and vitamin D. Calcium is found predominantly in milk and milk products, but some other foods including certain vegetables and tofu also provide calcium. Even when calcium intake is inadequate, blood calcium remains normal, but at the expense of bone loss, which can lead to osteoporosis. Calcium's roles, deficiency symptoms, and food sources are summarized below.

Calcium

1997 Adequate Intake (AI)	Upper Level	Chief Functions in the Body	Deficiency Symptoms	Toxicity Symptoms	Significant Sources
Adults (19–50 yr): 1000 mg/day Adults (51 and older): 1200 mg/day	Adults: 2500 mg/day	Mineralization of bones and teeth; also involved in muscle contraction and relaxation, nerve functioning, blood clotting, blood pressure, and immune defenses	Stunted growth in children; bone loss (osteoporosis) in adults	Constipation; increased risk of urinary stone formation and kidney dysfunction; interference with absorption of other minerals	Milk and milk products, small fish (with bones), tofu (bean curd), greens (broccoli, chard), legumes

PHOSPHORUS

Phosphorus is the second most abundant mineral in the body. About 85 percent of it is found combined with calcium in the hydroxyapatite crystals of bones and teeth.

• Phosphorus Roles in the Body • Phosphorus salts (phosphates) are found not only in bones and teeth, but in all body cells as part of a major buffer system (phosphoric acid and its salts). Phosphorus is also part of DNA and RNA and is therefore necessary for all growth.

Phosphorus assists in energy metabolism. Many enzymes and the B vitamins become active only when a phosphate group is attached. ATP itself, the energy currency of the cells, uses three phosphate groups to do its work.

Lipids containing phosphorus as part of their structures (phospholipids) help to transport other lipids in the blood. Phospholipids are also the major structural components of cell membranes, where they control the transport of nutrients into and out of the cells. Some proteins, such as the casein in milk, contain phosphorus as part of their structures (phosphoproteins).

• Phosphorus Recommendations and Intakes • Diets that provide adequate energy and protein also supply adequate phosphorus. Dietary deficiencies of phosphorus are unknown. As Figure 12-12 (on p. 410) shows, foods rich in proteins are the best sources of phosphorus. In addition to legumes and foods from the milk and meat groups, processed foods (including soft drinks) are usually high in phosphorus (from the additives).

phosphorus: a major mineral found mostly in the body's bones and teeth.

FIGURE 12-12 Phosphorus in Selected Foods

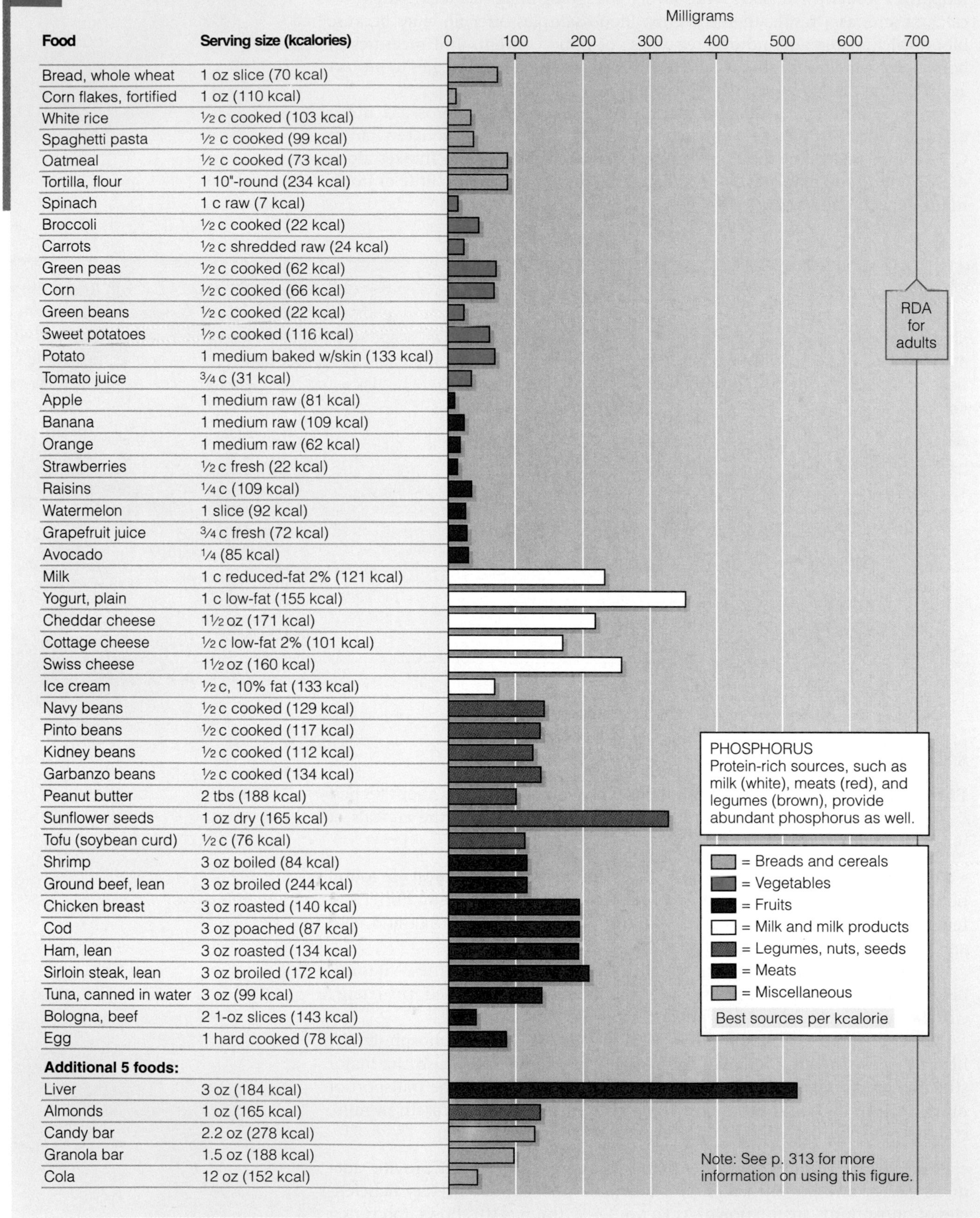
Milligrams
Food
Serving size (kcalories)
0
100
200
300
400
500
600
700
Bread, whole wheat
1 oz slice (70 kcal)
Corn flakes, fortified
1 oz (110 kcal)
White rice
½ c cooked (103 kcal)
Spaghetti pasta
½ c cooked (99 kcal)
Oatmeal
½ c cooked (73 kcal)
Tortilla, flour
1 10"-round (234 kcal)
Spinach
1 c raw (7 kcal)
Broccoli
½ c cooked (22 kcal)
Carrots
½ c shredded raw (24 kcal)
Green peas
½ c cooked (62 kcal)
Corn
½ c cooked (66 kcal)
Green beans
½ c cooked (22 kcal)
Sweet potatoes
½ c cooked (116 kcal)
Potato
1 medium baked w/skin (133 kcal)
Tomato juice
¾ c (31 kcal)
Apple
1 medium raw (81 kcal)
Banana
1 medium raw (109 kcal)
Orange
1 medium raw (62 kcal)
Strawberries
½ c fresh (22 kcal)
Raisins
¼ c (109 kcal)
Watermelon
1 slice (92 kcal)
Grapefruit juice
¾ c fresh (72 kcal)
Avocado
¼ (85 kcal)
Milk
1 c reduced-fat 2% (121 kcal)
Yogurt, plain
1 c low-fat (155 kcal)
Cheddar cheese
1½ oz (171 kcal)
Cottage cheese
½ c low-fat 2% (101 kcal)
Swiss cheese
1½ oz (160 kcal)
Ice cream
½ c, 10% fat (133 kcal)
Navy beans
½ c cooked (129 kcal)
Pinto beans
½ c cooked (117 kcal)
Kidney beans
½ c cooked (112 kcal)
Garbanzo beans
½ c cooked (134 kcal)
Peanut butter
2 tbs (188 kcal)
Sunflower seeds
1 oz dry (165 kcal)
Tofu (soybean curd)
½ c (76 kcal)
Shrimp
3 oz boiled (84 kcal)
Ground beef, lean
3 oz broiled (244 kcal)
Chicken breast
3 oz roasted (140 kcal)
Cod
3 oz poached (87 kcal)
Ham, lean
3 oz roasted (134 kcal)
Sirloin steak, lean
3 oz broiled (172 kcal)
Tuna, canned in water
3 oz (99 kcal)
Bologna, beef
2 1-oz slices (143 kcal)
Egg
1 hard cooked (78 kcal)
Additional 5 foods:
Liver
3 oz (184 kcal)
Almonds
1 oz (165 kcal)
Candy bar
2.2 oz (278 kcal)
Granola bar
1.5 oz (188 kcal)
Cola
12 oz (152 kcal)
RDA for adults
PHOSPHORUS
Protein-rich sources, such as milk (white), meats (red), and legumes (brown), provide abundant phosphorus as well.
= Breads and cereals
= Vegetables
= Fruits
= Milk and milk products
= Legumes, nuts, seeds
= Meats
= Miscellaneous
Best sources per kcalorie
Note: See p. 313 for more information on using this figure.

In the past, researchers emphasized the importance of an ideal calcium-to-phosphorus ratio from the diet to support calcium metabolism, but there is little or no evidence to support this concept.[25] The quantities of calcium and phosphorus in the diet are far more important than their ratio to each other. A high phosphorus intake has been blamed for bone loss when in fact a low calcium intake—not a phosphorus toxicity or an improper ratio—is responsible.[26]

IN SUMMARY

Phosphorus accompanies calcium both in the crystals of bone and in many foods such as milk. Phosphorus is also important in energy metabolism, as part of phospholipids, and as part of the genetic materials DNA and RNA. The summary table below lists functions and other information about phosphorus.

Phosphorus

1997 RDA	Upper Level	Chief Functions in the Body	Deficiency Symptoms	Toxicity Symptoms	Significant Sources
Adults: 700 mg/day	Adults (19–70 yr): 4000 mg/day	Mineralization of bones and teeth; part of every cell; important in genetic material, part of phospholipids, used in energy transfer and in buffer systems that maintain acid-base balance	Weakness, bone pain[a]	Low blood calcium levels	All animal tissues (meat, fish, poultry, eggs, milk)

[a]Dietary deficiency rarely occurs, but some drugs can bind with phosphorus making it unavailable and resulting in bone loss that is characterized by weakness and pain.

MAGNESIUM

Magnesium barely qualifies as a major mineral: only about 1 ounce of magnesium is present in the body of a 130-pound person. Over half of the body's magnesium is in the bones. Most of the rest is in the muscles and soft tissues, with only 1 percent in the extracellular fluid. As with calcium, bone magnesium may serve as a reservoir to ensure normal blood concentrations.

• Magnesium Roles in the Body • Magnesium acts in all the cells of the soft tissues, where it forms part of the protein-making machinery and is necessary for energy metabolism. It participates in hundreds of enzyme systems. A major role is as a catalyst in the reaction that adds the last phosphate to the high-energy compound ATP. As a required component for ATP metabolism, magnesium is essential to the body's use of glucose; the synthesis of protein, fat, and nucleic acids; and the cells' membrane transport systems. Together with calcium, magnesium is involved in muscle contraction and blood clotting: calcium promotes the processes, whereas magnesium inhibits them. This dynamic interaction between the two minerals helps regulate blood pressure and the functioning of the lungs. Magnesium also helps prevent dental caries by holding calcium in tooth enamel. Like many other nutrients, magnesium supports the normal functioning of the immune system.

• Magnesium Intakes • Average dietary magnesium estimates for U.S. adults fall below recommendations. Dietary intake data, however, do not include the contribution made by water. In some areas, the water contains both calcium and magnesium ("hard" water) and contributes significantly to intakes.

The brown bars in Figure 12-13 (on p. 412) indicate that legumes, seeds, and nuts make significant magnesium contributions. Magnesium is part of the chlorophyll molecule, so leafy green vegetables are also good sources.

• Magnesium Deficiency • Even with average magnesium intakes below recommendations, deficiency symptoms rarely appear except with diseases. Magnesium deficiency may develop in cases of alcohol abuse, protein malnutrition, kidney disorders, and prolonged vomiting or diarrhea. People using diuretics may also show symptoms. A severe magnesium deficiency causes a tetany

magnesium: a cation within the body's cells, active in many enzyme systems.

FIGURE 12-13 Magnesium in Selected Foods

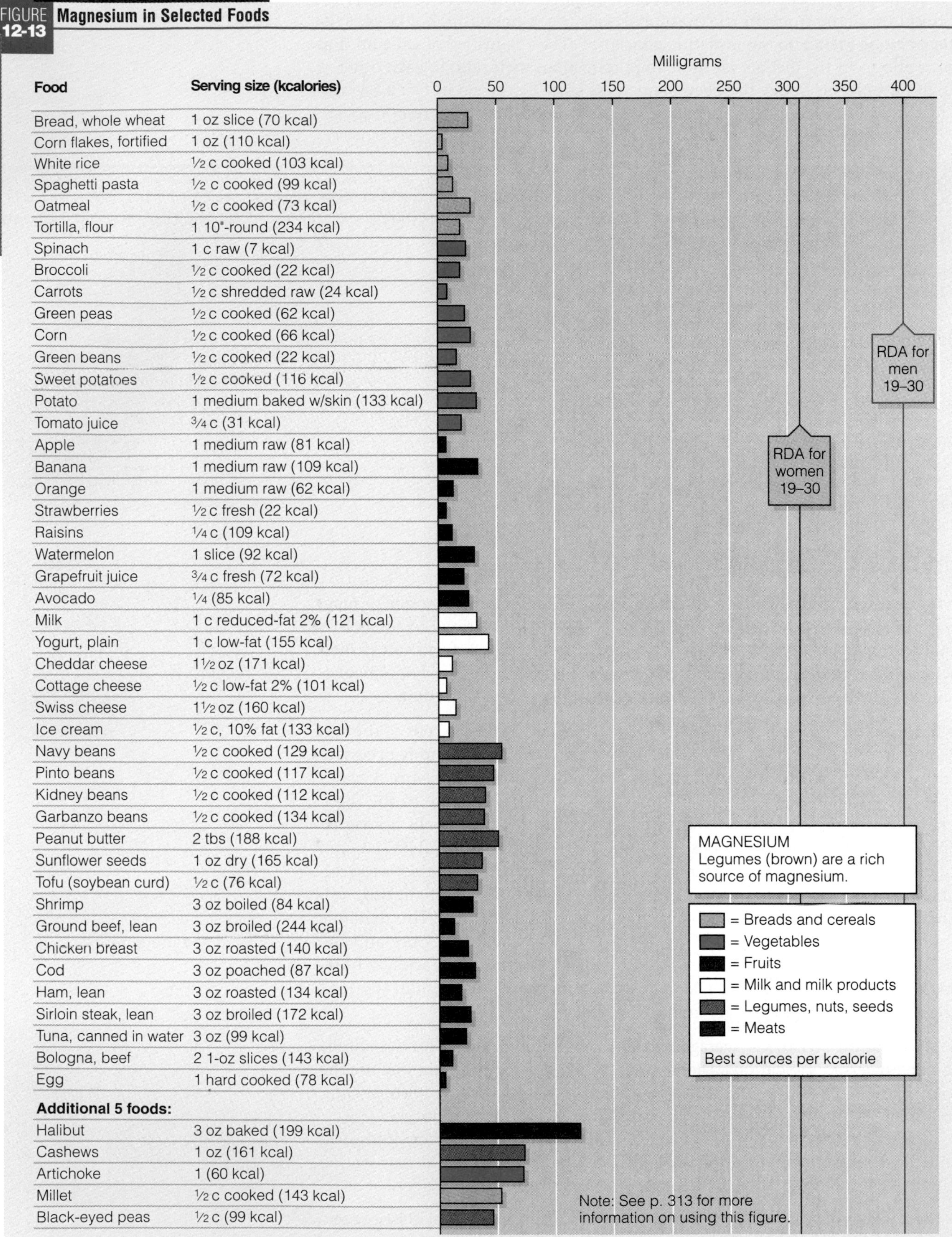

similar to the calcium tetany described earlier. Magnesium deficiencies also impair central nervous system activity and may be responsible for the hallucinations experienced during alcohol withdrawal.

• Magnesium and Hypertension • Magnesium is critical to heart function and seems to protect against hypertension and heart disease. Interestingly, people living in areas of the country with "hard" water, which contains high concentrations of calcium and magnesium, tend to have low rates of heart disease. With magnesium deficiency, the walls of the arteries and capillaries tend to constrict, a possible explanation for the hypertensive effect.

• Magnesium Toxicity • Magnesium toxicity is rare, but it can be fatal.[27] Toxicity occurs only with high intakes from nonfood sources such as supplements or magnesium salts.

IN SUMMARY

Like calcium and phosphorus, magnesium supports bone mineralization. Magnesium is also involved in numerous enzyme systems and in heart function. It is found abundantly in legumes and leafy green vegetables and, in some areas, in water. The table below offers a summary.

Magnesium

1997 RDA	Upper Level	Chief Functions in the Body	Deficiency Symptoms	Toxicity Symptoms	Significant Sources
Men (19–30 yr): 400 mg/day Women (19–30 yr): 310 mg/day	Adults: 350 mg nonfood magnesium/day	Bone mineralization, building of protein, enzyme action, normal muscle contraction, nerve impulse transmission, maintenance of teeth, and functioning of immune system	Weakness; confusion; if extreme, convulsions, bizarre muscle movements (especially of eye and face muscles), hallucinations, and difficulty in swallowing; in children, growth failure[a]	From nonfood sources only; diarrhea, alkalosis, dehydration	Nuts, legumes, whole grains, dark green vegetables, seafood, chocolate, cocoa

[a]A still more severe deficiency causes tetany, an extreme, prolonged contraction of the muscles similar to that caused by low blood calcium.

SULFUR

The body does not use **sulfur** by itself as a nutrient. Sulfur is mentioned here because it is a major mineral that occurs in essential nutrients such as the B vitamin thiamin and the amino acids methionine and cysteine. Sulfur plays a well-known role in determining the contour of protein molecules. The sulfur-containing side chains in cysteine molecules can link to each other, forming disulfide bridges, which stabilize the protein structure (see the drawing of insulin with its disulfide bridges on p. 170). Skin, hair, and nails contain some of the body's more rigid proteins, which have a high sulfur content.

There is no recommended intake for sulfur, and no deficiencies are known. Only when people lack protein to the point of severe deficiency will they lack the sulfur-containing amino acids.

IN SUMMARY

Like the other nutrients, the minerals' actions are coordinated to get the body's work done. The major minerals, especially sodium, chloride, and potassium, influence the body's fluid balance; whenever an anion moves, a cation moves—always maintaining homeostasis. Sodium, chloride, potassium, calcium, and magnesium are key members of the team of nutrients that direct nerve transmission and muscle contraction; they are also the primary nutrients involved in regulating blood pressure. Phosphorus and magnesium participate in many reactions involving glucose, fatty acids, amino acids, and the vitamins. Calcium, phosphorus, and magnesium combine to form the structure of the bones and teeth. Each major mineral also plays other specific roles in the body. (See the summary table on the next page.)

sulfur: a mineral present in the body as part of some proteins.

The Major Minerals

Mineral and Chief Functions	Deficiency Symptoms	Toxicity Symptoms	Significant Sources
Sodium			
Maintains normal fluid and electrolyte balance; assists in nerve impulse transmission and muscle contraction	Muscle cramps, mental apathy, loss of appetite	Edema, acute hypertension	Table salt, soy sauce; moderate amounts in meats, milks, breads, and vegetables; large amounts in processed foods
Chloride			
Maintains normal fluid and electrolyte balance; part of hydrochloric acid found in the stomach, necessary for proper digestion	Do not occur under normal circumstances	Vomiting	Table salt, soy sauce; moderate amounts in meats, milks, eggs; large amounts in processed foods
Potassium			
Maintains normal fluid and electrolyte balance; facilitates many reactions; supports cell integrity; assists in nerve impulse transmission and muscle contractions	Muscular weakness, paralysis, confusion	Muscular weakness; vomiting; if given into a vein; can stop the heart	All whole foods; meats, milks, fruits, vegetables, grains, legumes
Calcium			
Mineralization of bones and teeth; also involved in muscle contraction and relaxation, nerve functioning, blood clotting, blood pressure, and immune defenses	Stunted growth in children; bone loss (osteoporosis) in adults	Constipation; increased risk of urinary stone formation and kidney dysfunction; interference with absorption of other minerals	Milk and milk products, small fish (with bones), tofu (bean curd), greens (broccoli, chard), legumes
Phosphorus			
Mineralization of bones and teeth; part of every cell; important in genetic material, part of phospholipids, used in energy transfer and in buffer systems that maintain acid-base balance	Weakness, bone pain[a]	Low blood calcium levels	All animal tissues (meat, fish, poultry, eggs, milk)
Magnesium			
Bone mineralization, building of protein, enzyme action, normal muscle contraction, nerve impulse transmission, maintenance of teeth, and functioning of immune system	Weakness; confusion; if extreme, convulsions, bizarre muscle movements (especially of eye and face muscles), hallucinations, and difficulty in swallowing; in children, growth failure[b]	From nonfood sources only; diarrhea, alkalosis, dehydration	Nuts, legumes, whole grains, dark green vegetables, seafood, chocolate, cocoa
Sulfur			
As part of proteins, stabilizes their shape by forming disulfide bridges; part of the vitamins biotin and thiamin and the hormone insulin	None known; protein deficiency would occur first	Toxicity would occur only if sulfur-containing amino acids were eaten in excess; this (in animals) depresses growth	All protein-containing foods (meats, fish, poultry, eggs, milk, legumes, nuts)

[a]Dietary deficiency rarely occurs, but some drugs can bind with phosphorus making it unavailable and resulting in bone loss that is characterized by weakness and pain.

[b]A still more severe deficiency causes tetany, an extreme, prolonged contraction of the muscles similar to that caused by low blood calcium.

With all of the tasks these minerals perform, they are of great importance to life. Consuming enough of each of them every day is easy, given a variety of foods from each of the food groups. Whole-grain breads supply magnesium; fruits, vegetables, and legumes also provide magnesium and potassium, too; milks offer calcium and phosphorus; meats also offer phosphorus and sulfur as well; all foods provide sodium and chloride, excesses being more problematic than inadequacies. The message is quite simple and has been repeated throughout this text: for an adequate intake of all the nutrients, including the major minerals, choose different foods from each of the five food groups. And drink plenty of water.

How are you doing?

- Do you drink plenty of water—about 8 glasses—every day?
- Do you select and prepare foods with less salt?
- Do you drink at least 3 glasses of milk—or get the equivalent in calcium—every day?

Nutrition on the Net

WEBSITES

Access these websites for further study of topics covered in this chapter.

- Find updates and quick links to these and other nutrition-related sites at our website: **www.wadsworth.com/nutrition**
- Find out more about the importance of water from the International Food Information Council: **ificinfo.health.org/insight/waterref.htm**
- Search for "minerals" at the American Dietetic Association site: **www.eatright.org**
- Learn about sodium in foods and on food labels from the Food and Drug Administration: **www.fda.gov/fdac/foodlabel/sodium.html**
- Find tips and recipes for including more milk in the diet: **www.whymilk.com**
- Learn about the benefits of calcium from the National Dairy Council: **www.nationaldairycouncil.org**

INTERNET ACTIVITIES

Osteoporosis affects primarily older individuals, so it's sometimes difficult for college students to become interested in learning about its prevention. But if young adults learn how to maximize bone density, they may prevent this disease as they age. Are you aware of the factors that help to maintain a strong skeleton?

- Go to: **www.nof.org**
- Click on "Take our Spring 2001 Better Bone Health Quiz!"
- Click on each active link to learn how to keep your bones strong and healthy.
- Keep track of your "yes" and "no" answers, and total your score.

Learning about the many lifestyle factors that influence the development of osteoporosis is an important first step in maintaining a healthy skeleton in the years ahead. How does your bone health score rate?

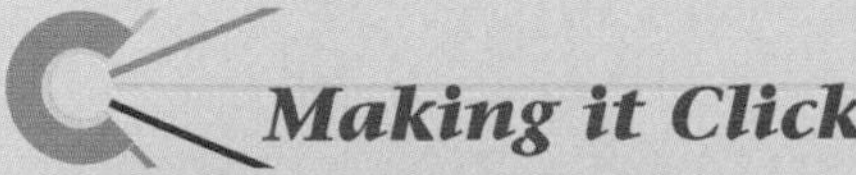

Making it Click

These problems give you an appreciation for the minerals in foods. Be sure to show your calculations (see p. 418 for answers).

1. For each of these minerals, note the unit of measure:
   Calcium Magnesium Phosphorus
   Potassium Sodium
2. Learn to appreciate calcium-dense foods. The list of foods below is ranked in order of their calcium contents per serving.
   a. Which foods offer the most calcium per kcalorie? To calculate calcium density, divide calcium (mg) by energy (kcal). Record your answer in the table (round your answers); the first one is done for you.
   b. The top five items ranked in order of calcium contents per serving are sardines > milk > cheese > salmon > broccoli. What are the top five items in order of calcium contents per kcalorie?

Food	Calcium (mg)	Energy (kcal)	Calcium Density (mg/kcal)
Sardines, 3 oz canned	325	176	1.85
Milk, nonfat, 1 c	301	85	
Cheddar cheese, 1 oz	204	114	
Salmon, 3 oz canned	182	118	
Broccoli, cooked from fresh, chopped, ½ c	36	22	
Sweet potato, baked in skin, 1 ea	32	140	
Cantaloupe melon, ½	29	93	
Whole-wheat bread, 1 slice	21	64	
Apple, 1 medium	15	125	
Sirloin steak, lean, 3 oz	9	171	

This information should convince you that milk, milk products, fish eaten with their bones, and dark green vegetables are the best choices for calcium.

3. a. Consider how the rate of absorption influences the amount of calcium available for the body's use. Use the drawing on p. 406 to determine how much calcium the body actually receives from the foods listed in the table below by multiplying the milligrams of calcium in the food by the percentage absorbed. The first one is done for you.
   b. To appreciate how the absorption rate influences the amount of calcium available to the body, compare broccoli with almonds. Which provides more calcium in foods and to the body?
   c. To appreciate how the calcium content of foods influences the amount of calcium available to the body, compare cauliflower with milk. How much cauliflower would a person have to eat to receive an equivalent amount of calcium as from 1 cup of milk? How does your answer change when you account for differences in their absorption rates?

Food	Calcium in the Food (mg)	Absorption Rate (%)	Calcium in the Body (mg)
Cauliflower, ½ c cooked, fresh	10	≥50	≥5
Broccoli, ½ c cooked, fresh	36		
Milk, 1 c 1% low-fat	300		
Almonds, 1 oz	75		
Spinach, 1 c raw	55		

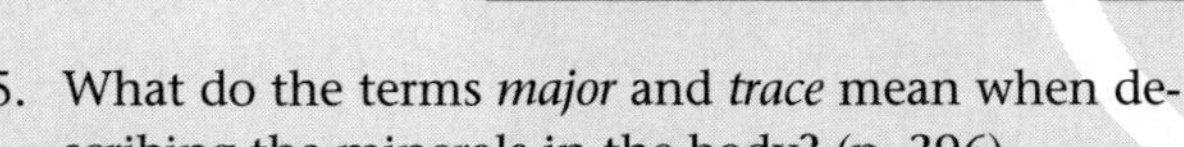

Study Questions

These questions will help you review the chapter. You will find the answers in the discussions on the pages provided.

1. List the roles of water in the body. (p. 387)
2. List the sources of water intake and routes of water excretion. (pp. 387–388)
3. What is ADH? Where does it exert its action? What is aldosterone? How does it work? (pp. 389–390)
4. How does the body use electrolytes to regulate fluid balance? (pp. 391–392, 394)
5. What do the terms *major* and *trace* mean when describing the minerals in the body? (p. 396)
6. Describe some characteristics of minerals that distinguish them from vitamins. (pp. 396–397)
7. What is the major function of sodium in the body? Describe how the kidneys regulate blood sodium. Is a dietary deficiency of sodium likely? Why? (pp. 397–399)
8. List calcium's roles in the body. How does the body keep blood calcium constant regardless of intake? (pp. 403–405)

9. Name significant food sources of calcium. What are the consequences of inadequate intakes? (pp. 406–409)
10. List the roles of phosphorus in the body. Discuss the relationships between calcium and phosphorus. Is a dietary deficiency of phosphorus likely? Why? (pp. 409, 411)
11. State the major functions of chloride, potassium, magnesium, and sulfur in the body. Are deficiencies of these nutrients likely to occur in your own diet? Why? (pp. 400, 401–403, 411, 413)

These questions will help you prepare for an exam. Answers can be found on p. 418.

1. The body generates water during the:
   a. buffering of acids.
   b. dismantling of bone.
   c. metabolism of minerals.
   d. breakdown of energy nutrients.
2. Regulation of fluid and electrolyte balance and acid-base balance depends primarily on the:
   a. kidneys.
   b. intestines.
   c. sweat glands.
   d. specialized tear ducts.
3. The distinction between the major and trace minerals reflects the:
   a. ability of their ions to form salts.
   b. amounts of their contents in the body.
   c. importance of their functions in the body.
   d. capacity to retain their identity after absorption.
4. The principal cation in extracellular fluids is:
   a. sodium.
   b. chloride.
   c. potassium.
   d. phosphorus.
5. The role of chloride in the stomach is to help:
   a. support nerve impulses.
   b. convey hormonal messages.
   c. maintain a strong acidity.
   d. assist in muscular contractions.
6. Which would provide the most potassium?
   a. bologna
   b. potatoes
   c. pickles
   d. whole-wheat bread
7. Calcium homeostasis depends on:
   a. vitamin K, aldosterone, and renin.
   b. vitamin K, parathormone, and renin.
   c. vitamin D, aldosterone, and calcitonin.
   d. vitamin D, calcitonin, and parathormone.
8. Calcium absorption is hindered by:
   a. lactose.
   b. oxalates.
   c. vitamin D.
   d. stomach acid.
9. Phosphorus assists in many activities in the body, but *not:*
   a. energy metabolism.
   b. the clotting of blood.
   c. the transport of lipids.
   d. bone and teeth formation.
10. Most of the body's magnesium can be found in the:
   a. bones.
   b. nerves.
   c. muscles.
   d. extracellular fluids.

References

1 Committee on Dietary Allowances, *Recommended Dietary Allowances,* 10th ed. (Washington, D.C.: National Academy Press, 1989), pp. 247–261.

2 Pamphlet from The American Dietetic Association, Water: The beverage of life, 1994.

3 D. S. Michaud and coauthors, Fluid intake and the risk of bladder cancer in men, *New England Journal of Medicine* 340 (1999); 1390–1397.

4 R. D. Mattes, The taste of salt in humans, *American Journal of Clinical Nutrition* 65 (1997): 692S–697S.

5 P. A. S. Breslin and G. K. Beauchamp, Salt enhances flavour by suppressing bitterness, *Nature* 387 (1997): 563.

6 T. A. Kotchen and J. M. Kotchen, Dietary sodium and blood pressure: Interactions with other nutrients, *American Journal of Clinical Nutrition* 65 (1997): 708S–711S.

7 A. W. Cowley, Jr., Genetic and nongenetic determinants of salt sensitivity and blood pressure, *American Journal of Clinical Nutrition* 65 (1997): 587S–593S.

8 F. C. Luft and M. H. Weinberger, Heterogeneous responses to changes in dietary salt intake: The salt-sensitivity paradigm, *American Journal of Clinical Nutrition* 65 (1997): 612S–617S.

9 J. He and coauthors, Dietary sodium intake and subsequent risk of cardiovascular disease in overweight adults, *Journal of the American Medical Association* 282 (1999): 2027–2034.

10 He and coauthors, 1999; P. K. Whelton and coauthors, Sodium restriction and weight loss in the treatment of hypertension in older persons: A randomized controlled Trial of Nonpharmacologic Interventions in the Elderly

(TONE), *Journal of the American Medical Association* 279 (1998): 839–846; P. K. Whelton and coauthors, Efficacy of nonpharmacologic interventions in adults with high-normal blood pressure: Results from phase 1 of the Trials of Hypertension Prevention, *American Journal of Clinical Nutrition* 65 (1997): 652S–660S.

11 D. A. McCarron, The dietary guideline for sodium: Should we shake it up? Yes! *American Journal of Clinical Nutrition* 71 (2000): 1013–1019; N. M. Kaplan, The dietary guideline for sodium: Should we shake it up? No! *American Journal of Clinical Nutrition* 71 (2000): 1020–1026; J. P. Midgley and coauthors, Effect of reduced dietary sodium on blood pressure: A meta-analysis of randomized controlled trials, *Journal of the American Medical Association* 275 (1996): 1590–1597; Letters: Dietary sodium and blood pressure, *Journal of the American Medical Association* 276 (1996): 1467–1470.

12 J. A. Cutlet, D. Follmann, and P. S. Allender, Randomized trials of sodium reduction: An overview, *American Journal of Clinical Nutrition* 65 (1997): 643S–651S; T. C. Beard and coauthors, Association between blood pressure and dietary factors in the Dietary and National Survey of British Adults, *Archives of Internal Medicine* 157 (1997): 234–238; The Sixth Report of the Joint National Committee on Prevention, Detection, Evaluation, and Treatment of High Blood Pressure (NIH publication, November 1997), p. 20.

13 F. Ginty, A. Flynn, and K. D. Cashman, The effect of dietary sodium intake on biochemical markers of bone metabolism in young women, *British Journal of Nutrition* 79 (1998): 343–350; R. Itoh and Y. Suyama, Sodium excretion in relation to calcium and hydroxyproline excretion in a healthy Japanese population, *American Journal of Clinical Nutrition* 63 (1996): 735–740; A. Devine and coauthors, A longitudinal study of the effect of sodium and calcium intakes on regional bone density in postmenopausal women, *American Journal of Clinical Nutrition* 62 (1995): 740–745.

14 P. K. Whelton and coauthors, Effects of oral potassium on blood pressure: Meta-analysis of randomized con-

ANSWERS

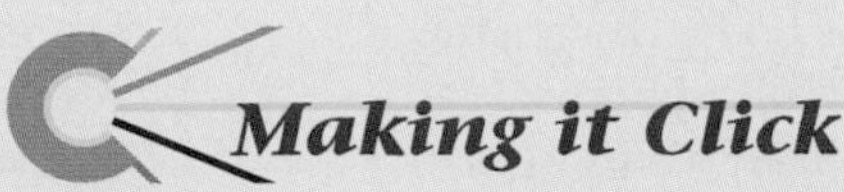

Making it Click

1. Calcium: mg. Magnesium: mg. Phosphorus: mg.
   Potassium: mg. Sodium: mg.
2. a.

Food	Calcium density (mg/kcal)
Sardines, 3 oz canned	325 mg ÷ 176 kcal = 1.85 mg/kcal
Milk, nonfat, 1 c	301 mg ÷ 85 kcal = 3.54 mg/kcal
Cheddar cheese, 1 oz	204 mg ÷ 114 kcal = 1.79 mg/kcal
Salmon, 3 oz canned	182 mg ÷ 118 kcal = 1.54 mg/kcal
Broccoli, cooked from fresh, chopped, ½ c	36 mg ÷ 22 kcal = 1.64 mg/kcal
Sweet potato, baked in skin, 1 ea	32 mg ÷ 140 kcal = 0.23 mg/kcal
Cantaloupe melon, ½	29 mg ÷ 93 kcal = 0.31 mg/kcal
Whole-wheat bread, 1 slice	21 mg ÷ 64 kcal = 0.33 mg/kcal
Apple, 1 medium	15 mg ÷ 125 kcal = 0.12 mg/kcal
Sirloin steak, lean, 3 oz	9 mg ÷ 171 kcal = 0.05 mg/kcal

   b. Milk > sardines > cheese > broccoli > salmon.
3. a.

Food	Calcium in Food (mg) × Absorption rate (%) = Calcium in the Body (mg)
Cauliflower, ½ c cooked, fresh	10 mg × 0.50 = 5 mg (or more)
Broccoli, ½ c cooked, fresh	36 mg × 0.50 = 18 mg (or more)
Milk, 1 c 1% low-fat	300 mg × 0.30 = 90 mg
Almonds, 1 oz	75 mg × 0.20 = 15 mg
Spinach, 1 c raw	55 mg × 0.05 = 3 mg (or less)

   b. The almonds offer more than twice as much calcium per serving, but an equivalent amount after absorption.
   c. To equal the 300 milligrams provided by milk, a person would need to eat 15 cups of cauliflower (300 mg/c milk ÷ 10 mg/½ c cauliflower = 30 ½ c or 15 c). After considering the better absorption rate of cauliflower, a person would need to eat 9 cups of cauliflower (5 mg/½ c or 10 mg/c; 90 mg ÷ 10 mg/c = 9 c) to match the 90 milligrams available to the body from milk after absorption. The better absorption rate reduced the quantity of cauliflower significantly, but that's still a lot of cauliflower.

Multiple Choice

1. d	2. a	3. b	4. a	5. c
6. b	7. d	8. b	9. b	10. a

trolled clinical trial, *Journal of the American Medical Association* 277 (1997): 1624–1632.

15 F. J. Gennari, Hypokalemia, *New England Journal of Medicine* 339 (1998): 451–458.

16 K. Kathleen, Hyperkalemia, *American Journal of Nursing* 100 (2000): 55–56.

17 J. H. Dwyer and coauthors, Dietary calcium, calcium supplementation, and blood pressure in African American adolescents, *American Journal of Clinical Nutrition* 68 (1998): 648–655; H. C. Bucher and coauthors, Effects of dietary calcium supplementation on blood pressure: A meta-analysis of randomized controlled trials, *Journal of the American Medical Association* 275 (1996): 1016–1022; D. A. McCarron and D. Hatton, Dietary calcium and lower blood pressure—We can all benefit, *Journal of the American Medical Association* 275 (1996): 1128–1129; C. G. Osborne and coauthors, Evidence for the relationship of calcium to blood pressure, *Nutrition Reviews* 54 (1996): 365–381.

18 D. A. McCarron, Role of adequate dietary calcium intake in the prevention and management of salt-sensitive hypertension, *American Journal of Clinical Nutrition* 65 (1997): 712S–716S.

19 W. A. Levey and coauthors, Blood pressure responses of white men with hypertension to two low-sodium metabolic diets with different levels of dietary calcium, *Journal of the American Dietetic Association* 95 (1995): 1280–1287.

20 K. O. O'Brien and coauthors, Increased efficiency of calcium absorption during short periods of inadequate calcium intake in girls, *American Journal of Clinical Nutrition* 63 (1996): 579–583.

21 A. Devine, R. L. Prince, and R. Bell, Nutritional effect of calcium supplementation by skim milk powder or calcium tablets on total nutrient intake in postmenopausal women, *American Journal of Clinical Nutrition* 64 (1996): 731–737.

22 C. M. Weaver, W. R. Proulx, and R. Heaney, Choices for achieving adequate dietary calcium with a vegetarian diet, *American Journal of Clinical Nutrition* 70 (1999): 543S–548S.

23 N. K. Christensen and coauthors, Juniper ash as a source of calcium in the Navajo diet, *Journal of the American Dietetic Association* 98 (1998): 333–334.

24 F. Couzy and coauthors, Calcium bioavailability from calcium- and sulfate-rich mineral water, compared with milk, in young adult women, *American Journal of Clinical Nutrition* 62 (1995): 1239–1244.

25 Committee on Dietary Reference Intakes, *Dietary Reference Intakes for Calcium, Phosphorus, Magnesium, Vitamin D, and Fluoride* (Washington, D.C.: National Academy Press, 1997), pp. 5-6–5-7.

26 Committee on Dietary Reference Intakes, 1997, p. 5-31.

27 J. K. McGuire, M. S. Kulkarni, and H. P. Baden, Fatal hypermagnesemia in a child treated with megavitamin/megamineral therapy, *Pediatrics* 105 (2000): 414 [www.pediatrics.org/cgi/content/full/105/2/e18].

OSTEOPOROSIS AND CALCIUM

GLOSSARY

bone density: a measure of bone strength. When minerals fill the bone matrix (making it dense), they give it strength.

cortical bone: the very dense bone tissue that forms the outer shell surrounding trabecular bone and comprises the shaft of a long bone.

trabecular (tra-BECK-you-lar) **bone:** the lacy inner structure of calcium crystals that supports the bone's structure and provides a calcium storage bank.

type I osteoporosis: osteoporosis characterized by rapid bone losses, primarily of trabecular bone.

type II osteoporosis: osteoporosis characterized by gradual losses of both trabecular and cortical bone.

OSTEOPOROSIS SETS in during the later years, but it develops much earlier—and without warning. Few people are aware that their bones are being robbed of their strength. The problem often first becomes apparent when someone's hip suddenly gives way. People say, "She fell and broke her hip," but in fact the hip may have been so fragile that it broke *before* she fell. Even bumping into a table may be enough to shatter a porous bone into fragments so numerous and scattered that they cannot be reassembled. Removing them and replacing them with an artificial joint requires major surgery. An estimated 300,000 people in the United States are hospitalized each year because of hip fractures. About a fifth die of complications within a year. Half of those who survive will never walk or live independently again. Their quality of life slips downward.[1]

This highlight examines osteoporosis, one of the most prevalent diseases of aging, affecting more than 25 million people in the United States—most of them women. It reviews the many factors that contribute to the 1 million breaks in the bones of the hips, wrists, arms, and ankles each year.[2] And it presents strategies to reduce the risks, paying special attention to the role of dietary calcium.

Trabecular bone is the lacy network of calcium-containing crystals that fills the interior. Cortical bone is the dense, ivorylike bone that forms the exterior shell.

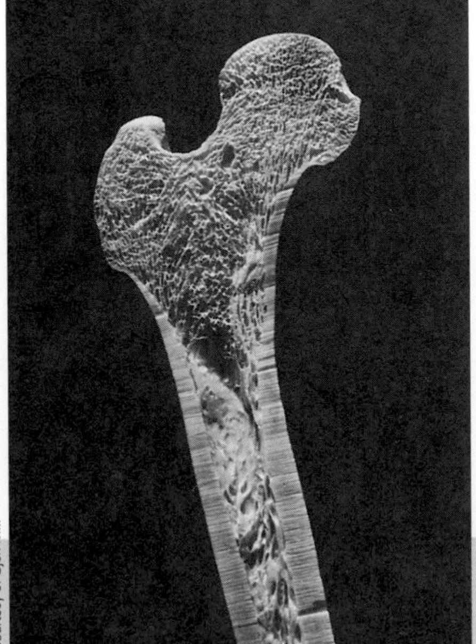

Courtesy of Gjon Mill

BONE DEVELOPMENT AND DISINTEGRATION

Bone has two compartments: the outer, hard shell of cortical bone, and the inner, lacy matrix of trabecular bone. Both can lose minerals, but in different ways and at different rates. The accompanying photograph shows a human leg bone sliced lengthwise, exposing the lacy, calcium-containing crystals of **trabecular bone.** (The glossary defines this and other bone-related terms.) These crystals give up calcium to the blood when the diet runs short, and they take up calcium again when the supply is plentiful. For people who have eaten calcium-rich foods throughout the bone-forming years of their youth, these deposits provide a rich reservoir of calcium.

Surrounding and protecting the trabecular bone is a dense, ivorylike exterior shell—the **cortical bone.** Cortical bone composes the shafts of the long bones, and a thin cortical shell caps the end of the bone, too. Both compartments confer strength on bone: cortical bone provides the sturdy outer wall, while trabecular bone provides support along the lines of stress.

The two types of bone play different roles in calcium balance and osteoporosis. Supplied with blood vessels and metabolically active, trabecular bone is sensitive to hormones that govern day-to-day deposits and withdrawals of calcium. It readily gives up minerals whenever blood calcium needs replenishing. Losses of trabecular bone start becoming significant for men and women in their 30s, although losses can occur whenever calcium withdrawals exceed deposits.

Cortical bone also gives up calcium, but slowly and at a steady pace. Cortical bone losses typically

FIGURE H12-1 **Healthy and Osteoporotic Trabecular Bones**

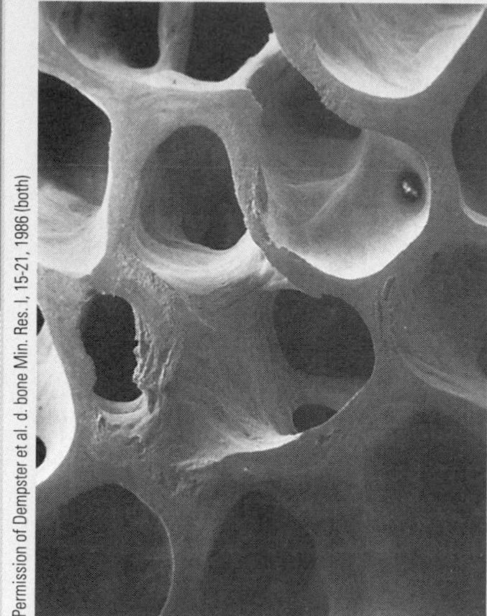

Electron micrograph of healthy trabecular bone.

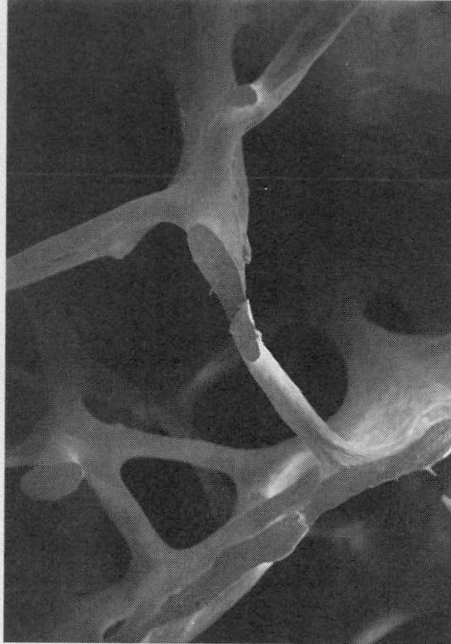

Electron micrograph of trabecular bone affected by osteoporosis.

Permission of Dempster et al. d. bone Min. Res. l, 15-21, 1986 (both)

FIGURE H12-2 **Loss of Height in a Woman Caused by Osteoporosis**

The woman on the left is about 50 years old. On the right, she is 80 years old. Her legs have not grown shorter: only her back has lost length, due to collapse of her spinal bones (vertebrae). Collapsed vertebrae cannot protect the spinal nerves from pressure that causes excruciating pain.

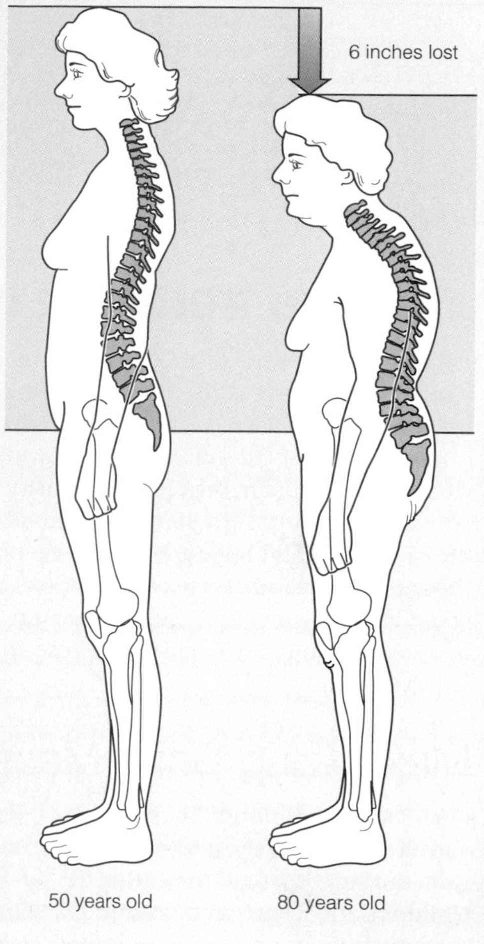

begin at about 40 years of age and continue slowly but surely thereafter.

Losses of trabecular and cortical bone reflect two types of osteoporosis, which cause two types of bone breaks. **Type I osteoporosis** involves losses of trabecular bone (see Figure H12-1). These losses sometimes exceed three times the expected rate, and bone breaks may occur suddenly. Trabecular bone becomes so fragile that even the body's own weight can overburden the spine—vertebrae may suddenly disintegrate and crush down, painfully pinching major nerves. Wrists may break as bone ends weaken, and teeth may loosen or fall out as the trabecular bone of the jaw recedes.[3] Women are most often the victims of this type of osteoporosis, outnumbering men six to one. Taking estrogen after menopause is the most effective preventive measure against this type of osteoporosis.

© Photo Disc Inc.

Young adults wisely drink milk now to defend against bone loss later.

In **type II osteoporosis,** the calcium of both cortical and trabecular bone is drawn out of storage, but slowly over the years. As old age approaches, the vertebrae may compress into wedge shapes, forming what is often called "dowager's hump," the posture many older people assume as they "grow shorter." Figure H12-2 shows the effect of compressed spinal bone on a woman's height and posture. Because both the cortical shell and the trabecular interior weaken, breaks most often occur in the hip, as mentioned in the introductory paragraph. A woman is twice as likely as a man to suffer type II osteoporosis.

Table H12-1 (on p. 422) summarizes the differences between the two types of osteoporosis. Physicians can diagnose osteoporosis and assess the risk of bone fractures by measuring **bone density** using dual-energy X-ray absorptiometry or ultrasound. They also consider risk factors that predict bone fractures, including age, race, low BMI, and physical inactivity.[4]

Whether a person develops osteoporosis seems to depend partly on genetics and partly on other factors, including nutrition. The strongest predictor of bone density is age: 90 percent of the hip fractures in the United States occur in people over the age of 50.[5]

TABLE H12-1 Types of Osteoporosis Compared

	Type I	Type II
Other name	Postmenopausal osteoporosis	Senile osteoporosis
Age of onset	50 to 70 years old	70 years and older
Bone loss	Trabecular bone	Both trabecular and cortical bone
Fracture sites	Wrist and spine	Hip
Gender incidence	6 women to 1 man	2 women to 1 man
Primary causes	Rapid loss of estrogen in women following menopause; loss of testosterone in men with advancing age	Reduced calcium absorption, increased bone mineral loss, increased propensity to fall

AGE AND BONE CALCIUM

Two major stages of life are critical in the development of osteoporosis.[6] First is the bone-acquiring stage of childhood and adolescence. The second stage is the bone-losing decades of late adulthood (following menopause in women). Bones gain strength and density all through the growing years and into young adulthood. As people age, the cells that build bone gradually become less active, but those that dismantle bone continue working. Because the body depends on bone to supply calcium throughout life, bone mass diminishes and bones lose their density and strength.

Maximizing Bone Mass

One factor affecting the balance of bone withdrawal and deposition is calcium nutrition during the first three decades of life. Children and teens who consume adequate calcium and vitamin D deposit more calcium into their bones than those with inadequate intakes. When people reach the bone-losing years of middle age, those who formed dense bones during their youth have the advantage. They simply have more bone starting out and can lose more before suffering ill effects. Figure H12-3 demonstrates this effect.

Minimizing Bone Loss

Factors such as calcium that build strong bones in youth remain important in protecting against losses in the later years. Calcium intakes of older adults are typically low, and calcium absorption declines after about the age of 65 years. The kidneys do not activate vitamin D as well as they did earlier (recall that active vitamin D enhances calcium absorption). Also, sunlight is needed to form vitamin D, and many older people spend little or no time outdoors in the sunshine. For these reasons, and because intakes of vitamin D are typically low anyway, blood vitamin D declines.

Some of the hormones that regulate bone and calcium metabolism also change with age and accelerate bone mineral withdrawal.* Together, these age-related factors probably contribute to bone loss: inefficient bone remodeling, reduced calcium intakes, impaired calcium absorption, poor vitamin D status, and hormonal changes that favor bone mineral withdrawal.

GENDER AND HORMONES

After age, gender is the next strongest predictor of osteoporosis: men have greater bone density than women at maturity, and women have greater losses than men in later life. Consequently, women account for four out of five cases of osteoporosis. Menopause imperils women's bones. Bone dwindles rapidly when the hormone estrogen diminishes and menstruation ceases. Accelerated losses continue for six to eight years following menopause, then taper off so that women again lose bone at the same rate as men their age. Losses of bone minerals continue throughout the remainder of a woman's lifetime, but not at the free-fall pace of the menopause years (review Figure H12-3).

*Among the hormones suggested as influential are parathormone, calcitonin, and estrogen.

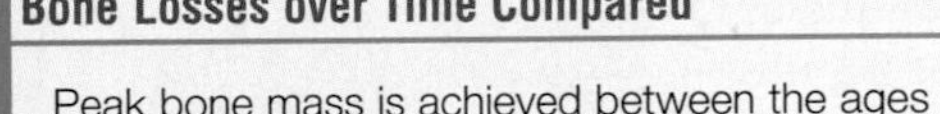

FIGURE H12-3 Bone Losses over Time Compared

Peak bone mass is achieved between the ages of 19 and 30. Women gradually lose bone mass until menopause, when losses accelerate dramatically and then gradually taper off.

Woman A entered adulthood with enough calcium in her bones to last a lifetime.
Woman B had less bone mass starting out and so suffered ill effects from bone loss later on.

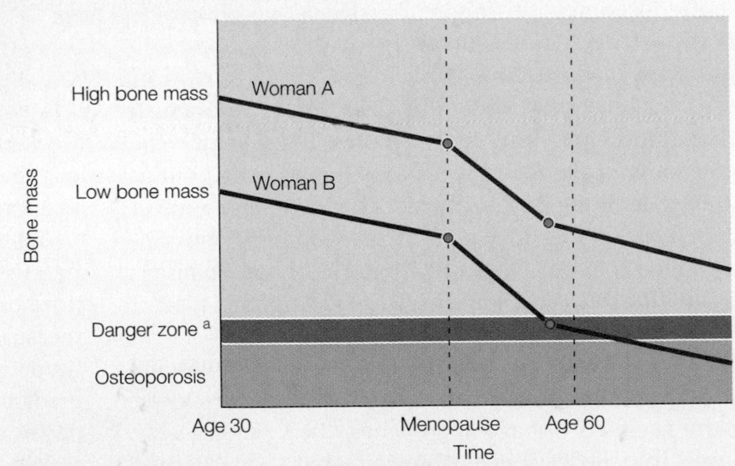

[a]People with a moderate degree of bone mass reduction are said to have *osteopenia* and are at increased risk of fractures.

SOURCE: Data from Committee on Dietary Reference Intakes, *Dietary Reference Intakes for Calcium, Phosphorus, Magnesium, Vitamin D, and Fluoride* (Washington, D.C.: National Academy Press, 1997), pp. 71–145.

When *young* women's ovaries fail to produce enough estrogen, causing menstruation to cease, these women, too, lose bone rapidly. In some, diseased ovaries are to blame and must be removed; in others, the ovaries fail to produce sufficient estrogen because the women suffer from anorexia nervosa and have unreasonably restricted their body weight (see Highlight 9). The amenorrhea and low body weights explain much of the bone loss seen in these young women, even years after diagnosis.[7] Estrogen therapy can help nonmenstruating women prevent further bone loss and reduce the incidence of fractures.[8] In fact, estrogen is the preferred therapy for the prevention and treatment of osteoporosis after menopause.[9]

Interestingly, the phytosterols commonly found in soybeans mimic the actions of estrogen in the body. When natural estrogen is lacking, as after menopause, phytosterols may step in to stimulate estrogen-sensitive tissues. By way of this action, phytosterols may help to prevent the rapid bone losses of the menopause years. Recent research suggests that soybeans, and products made from them, may indeed offer some protection.[10]

If estrogen deficiency is a major cause of osteoporosis in women, what is the cause of bone loss in men? Men produce only a little estrogen, yet they resist osteoporosis better than women. Does the male sex hormone testosterone play a role? Perhaps so, because men suffer more fractures after removal of diseased testes or when their testes lose function with aging. Treatment for men with osteoporosis includes testosterone replacement therapy.[11] Thus both male and female sex hormones appear to play roles in the development and treatment of osteoporosis.

Hormone therapy may slow calcium losses from bones, but a new drug reverses bone loss to some degree.*[12] It works in both men and women by inhibiting the activities of the bone-dismantling cells, thus allowing the bone-building cells to slowly shore up bone tissue with new calcium deposits. The hope now is for a drug that can stimulate the bone-building cells to work faster to restore bone strength.

GENETICS

Genetics plays an influential role in bone density, most likely by influencing both the peak bone mass achieved during growth and the extent of bone loss during the later years. The extent to which a given genetic potential is realized, however, depends on many outside factors.[13] Diet and physical activity, for example, can maximize peak bone density during growth, whereas alcohol and tobacco abuse can accelerate bone losses later in life.

Risks of osteoporosis appear to run along racial lines and reflect genetic differences in bone development between African Americans and Caucasians.[14] African Americans use and conserve calcium more efficiently than Caucasians. Consequently, even though their calcium intakes are typically lower, black people have denser bones than white people do. Greater bone density expresses itself in a lower rate of osteoporosis among blacks. Hip fractures, for example, are about three times more likely at 80 years of age in white women than in black women.

Other ethnic groups have *less* dense bones than do many people of European descent. Asians from China and Japan, Mexican Americans, Hispanic people from Central and South America, and Inuit people from St. Lawrence Island all have lower bone density than do people with a northern European background. One might expect that these groups would suffer more bone fractures, but this is not always the case. Again, genetic differences may explain why. Asians, for example, generally have small, compact hips, which makes them less susceptible to fractures.[15]

Genetics alone does not explain how certain ethnic groups can use no dairy products and have low calcium intakes, yet still maintain calcium balance. Part of the answer is the body's adaptation to low calcium intakes. Calcium absorption is exceptionally efficient in Chinese women whose intakes are notably low.[16] Another part of the answer is that bone loss is not always apparent. Women in regions of China where calcium intakes are high have greater bone densities than women in other areas.

Findings from around the world demonstrate that although a person's genes may lay the groundwork, environmental factors influence the genes' ultimate expression—and calcium nutrition is one of those environmental factors. Others include physical activity, body weight, smoking, and alcohol. Importantly, all of these factors are within a person's control.

PHYSICAL ACTIVITY AND BODY WEIGHT

When people lie idle—for example, when they are confined to bed—their bones lose strength just as their muscles do. Astronauts who live with reduced gravity for weeks or months at a time also lose bone and muscle strength.[17]

Muscle strength and bone strength go together. When muscles work, they pull on the bones, stimulating them to develop more trabeculae and grow denser. Also, when muscles work, the hormones that promote new muscle growth also favor the building of bone. As a result, active bones are denser than sedentary bones.[18] Regular physical activity combined with an adequate calcium intake helps to maximize bone gains in young adulthood and minimize further losses that occur with inactivity.[19]

Heavier body weights and weight gains place a similar stress on the bones and promote their density. In fact, underweight and weight losses are significant and consistent predictors of bone density losses and risk of fractures.[20] As mentioned in Highlight 9, the combination of under-

*The generic name of the drug is aldendronate, marketed as Fosamax.

weight, severely restricted energy intake, extreme daily exercise, and amenorrhea reliably predicts bone loss. Interestingly, recent evidence suggests that the bone density associated with overweight may be due not to body weight alone but to the lack of, or inability to respond to, leptin.[21]

Cells respond, with the help of the necessary regulators, to the demands put upon them. Then they select the nutrients they need from what is offered. To increase bone density, put a demand on the bones, make them work, and then provide the raw materials from which they can grow strong: calcium, other minerals, all the nutrients in the right balance. The combination of regular physical activity and adequate calcium intake best supports bone density.[22]

SMOKING AND ALCOHOL

Add bone damage to the list of ill consequences associated with smoking. Smokers experience more fractures from slight injury than do nonsmokers. Women who smoke a pack of cigarettes a day throughout adulthood lose an extra 5 to 10 percent of their bone density by menopause. Compared with nonsmokers, smokers suffer more bone degeneration and spinal injuries, and surgeries to correct these problems are often less successful.[23] Researchers speculate that both the lower body weights of smokers and the earlier menopause of female smokers may be factors.

People who abuse alcohol often suffer from osteoporosis and experience more bone breaks than others. Several factors appear to be involved: alcohol enhances fluid excretion, which leads to excessive calcium losses in the urine; upsets the hormonal balance required for healthy bones; slows bone formation, leading to lower bone density; stimulates bone breakdown; and increases the risk of falling.[24] Research is less clear as to whether alcohol in moderate amounts is as damaging to the bones.[25]

Table H12-2 summarizes the risk factors and protective factors covered so far and includes some others, among them nutrition factors, discussed next. The more risk factors that apply to a person, the greater the chances of bone loss. Notice that several factors are more influential than diet in the development of osteoporosis.

TABLE H12-2 Risk and Protective Factors for Osteoporosis

Risk Factors	Protective Factors
High Correlation	
Advanced age; postmenopausal	African American
Alcohol abuse	Estrogens, long-term use
Anorexia nervosa	
Caucasian or East Asian	
Chronic steroid use	
Female gender	
Rheumatoid arthritis	
Surgical removal of ovaries or testes	
Underweight (<120 lb) or weight loss	
Moderate Correlation	
Chronic thyroid hormone use	Having given birth
Cigarette smoking	High body weight
Diabetes (type 1)	High-calcium diet (meets recommendations)
Early menopause	Regular physical activity
Excessive antacid use	
Family history of osteoporosis	
Low-calcium diet	
Sedentary lifestyle or immobility	
Vitamin D deficiency	
Probably Important But Not Yet Proved	
Caffeine use	Adequate vitamin K
High-fiber diet	Adequate magnesium
High-protein diet	
High-sodium diet	

CALCIUM, THE KEY TO PREVENTION

Bone strength later in life depends most on how well the bones were developed and maintained during youth. Adequate calcium nutrition during the growing years is essential to achieving optimal peak bone mass.[26] To that end, the Committee on Dietary Reference Intakes recommends 1300 milligrams of calcium per day for everyone 9 through 18 years of age. Unfortunately, few girls meet the recommendations for calcium during these bone-forming years. (Boys generally obtain intakes close to those recommended because they eat more food.) Consequently, most girls start their adult years with less than optimal bone density. As for adults, women rarely meet their recommended intakes of 1000 to 1200 milligrams from food. Some authorities suggest 1500 milligrams of calcium for postmenopausal women who are not receiving estrogen, but warn that intakes exceeding 2500 milligrams a day could cause health problems.

Much research has focused on calcium, but other nutrients support bone health, too. Vitamin D was mentioned earlier. Supplementation with calcium and vitamin D reduces bone loss and the risk of fractures.[27] One leading researcher estimates that as much as half of our osteoporosis problem could be eliminated with adequate calcium and vitamin D intakes.[28] Vitamin K, magnesium, and potassium also help to maintain bone mineral density.[29] Vitamin A is needed in the bone-remodeling process, but too much may be associated with osteoporosis.[30] Additional research points to the benefits not of a specific nutrient, but of a diet rich in fruits and vegetables.[31] Clearly, a well-balanced diet that depends on all the food groups to supply a full array of nutrients is central to bone health.

A PERSPECTIVE ON SUPPLEMENTS

Bone health improves when people increase their intake of calcium-rich foods.[32] But for people who do not consume milk products or other calcium-rich foods in amounts that provide even half the recommended calcium, supplements may be needed. During the menopausal years, calcium supplements of 1 gram may slow, but cannot fully prevent, the inevitable bone loss. Supplements are commonly used as a part of therapy for osteoporosis, along with gentle exercise and, for women, estrogen replacement. (Supplements should not, however, be used as a substitute for estrogen.)

Anyone contemplating the use of calcium supplements should do so only on the advice of a registered dietitian or physician. Taking calcium supplements may present risks, as described in Table H12-3. If these risks are deemed acceptable, the consumer still has several decisions to make when selecting a calcium supplement. Incidentally, multivitamin-mineral pills contain little or no calcium and cannot be used as calcium supplements. The label may list a few milligrams of calcium, but remember that the recommended intake is a gram or more for adults.

Calcium supplements are available in three forms. Simplest are the purified calcium compounds, such as calcium carbonate, citrate, gluconate, lactate, malate, or phosphate, and compounds of calcium with amino acids (called amino acid chelates). Also available are mixtures of calcium with other compounds, such as calcium carbonate with magnesium carbonate, with aluminum salts (as in some antacids), or with vitamin D. Then there are powdered, calcium-rich materials such as bone meal, powdered bone, oyster shell, or dolomite (limestone). See Table H12-4 for a description of various calcium supplements.

The first question to ask is how well the body absorbs and uses the calcium from various supplements. Most healthy people absorb calcium equally well—and as well as from milk—from any of these supplements: amino acids chelated with calcium; calcium phosphate dibasic; or calcium acetate, carbonate, citrate, gluconate, or lactate. People absorb calcium less well from a mixture of calcium and magnesium carbonate, oyster shell calcium fortified with inorganic magnesium, a chelated calcium-magnesium combination, or calcium carbonate fortified with vitamins and iron.

The next question to ask is how much calcium the supplement provides. To be safe, total calcium intake from both foods and supplements should not exceed 2500 milligrams a day. Read the label to find out how much a dose supplies. Select a low-dose supplement and take it several times a day rather than taking a large-dose supplement all at once. Taking supplements in doses of 500 milligrams or less improves absorption.

Then consider that when manufacturers compress large quantities of calcium into small pills, the stomach acid has difficulty penetrating the pill. To test a supplement's ability to dissolve, drop it into a 6-ounce cup of vinegar, and stir occasionally. A high-quality formulation will dissolve within half an hour.

TABLE H12-3 Problems Arising from Calcium Supplementation

People who take calcium supplements risk:

- Impaired iron status. (Calcium inhibits iron absorption.)
- Accelerated calcium loss. (Calcium-containing antacids that also contain aluminum and magnesium hydroxide cause a net calcium loss.)
- Urinary tract stones or kidney damage in susceptible individuals. (People who have a history of kidney stones need to be monitored by a physician and to use calcium citrate supplements, which are most soluble.)
- Exposure to contaminants. (Some preparations of bone meal and dolomites are contaminated with hazardous amounts of arsenic, cadmium, mercury, and lead.)
- Vitamin D toxicity. (Vitamin D is needed to enhance calcium absorption, but continued high intakes of vitamin D, which is present in many calcium supplements, can be toxic. Users must eliminate other concentrated vitamin D sources and take enough, but not too much, vitamin D to normalize calcium absorption.)
- Excess blood calcium. (This complication is seen only with doses of calcium fourfold or more greater than customarily prescribed.)
- Milk alkali syndrome. (This condition is characterized by high blood calcium, metabolic alkalosis, and renal failure. Early symptoms include irritability, headaches, and apathy.)
- Other nutrient interactions. (Calcium inhibits absorption of magnesium, phosphorus, and zinc.)
- Drug interactions. (Calcium and tetracycline form an insoluble complex that impairs both mineral and drug absorption.)
- GI distress. (Constipation, intestinal bloating, and excess gas are especially common in older people.)

TABLE H12-4 Calcium Supplements

- **Amino acid chelates** (KEY-lates) are compounds of minerals (such as calcium) combined with amino acids in a form that favors their absorption. A *chelating agent* is a molecule that surrounds another molecule and can then either promote or prevent its movement from place to place; *chele* means claw.
- **Antacids** are acid-buffering agents used to counter excess acidity in the stomach. Calcium-containing preparations (such as Tums) contain available calcium. Antacids with aluminum or magnesium hydroxides (such as Rolaids) can accelerate calcium losses.
- **Bone meal** or **powdered bones** are crushed or ground bone preparations intended to supply calcium to the diet. Calcium from bone is not well absorbed and is often contaminated with toxic minerals such as arsenic, mercury, lead, and cadmium.
- **Calcium compounds** such as calcium carbonate, citrate, gluconate, lactate, malate, or phosphate are the simplest forms of purified calcium. These supplements vary in the amount of calcium they contain, so read the labels carefully. A 500-milligram tablet of calcium gluconate may provide only 45 milligrams of calcium, for example.
- **Dolomite** is a compound of minerals (calcium magnesium carbonate) found in limestone and marble. Dolomite is powdered and is sold as a calcium-magnesium supplement, but may be contaminated with toxic minerals, is not well absorbed, and interacts adversely with absorption of other essential minerals.
- **Oyster shell** is a product made from the powdered shells of oysters that is sold as a calcium supplement, but is not well absorbed.

Finally, having chosen a supplement, a person must take it regularly, but when should you take it? To circumvent adverse nutrient interactions, take calcium supplements between, not with, meals. To enhance calcium absorption, take supplements with meals. If such contradictory advice drives you crazy, reconsider the benefits of food sources of calcium. Most experts agree that foods are the best source of calcium.

SOME CLOSING THOUGHTS

Unfortunately, many of the strongest risk factors for osteoporosis are beyond people's control: age, gender, and genetics. But there are still several effective strategies for prevention.[33] First, ensure an optimal peak bone mass during childhood and adolescence by eating a balanced diet rich in calcium and exercising regularly. Then maintain that bone mass by continuing those healthy diet and activity habits and abstaining from cigarette smoking and practicing moderation in, or abstinence from, alcohol use. Finally, minimize bone loss by maintaining an adequate nutrition and exercise regimen and, for women, considering estrogen replacement therapy. Other drug therapies may also be effective both in preventing bone loss and in restoring lost bone.[34] The reward is the best possible chance of preserving bone health throughout life.

Nutrition on the Net

•WEBSITES•

Access these websites for further study of topics covered in this highlight.

- Find updates and quick links to these and other nutrition-related sites at our website: **www.wadsworth.com/nutrition**
- Learn how to prevent falls and fractures from the National Institute on Aging: **www.nih.gov/nia**
- Visit the Osteoporosis and Related Bone Diseases' National Resource Center: **www.osteo.org**
- Obtain additional information from the National Osteoporosis Foundation: **www.nof.org**

References

1 T. D. Galsworthy and P. L. Wilson, Osteoporosis: It steals more than bone, *American Journal of Nursing* 96 (1996): 27–32.

2 Incidence and costs to Medicare of fractures among Medicare beneficiaries aged ≥65 years—United States, July 1991–June 1992, *Morbidity and Mortality Weekly Report* 45 (1996): 877–883.

3 M. K. Jeffcoat, Osteoporosis: A possible modifying factor in oral bone loss, *Annals of Periodontology* 3 (1998): 12–21.

4 L. W. Turner, P. A. Faile, and R. Tomlinson, Jr., Osteoporosis diagnosis and fracture, *Orthopaedic Nursing,* September/October 1999, pp. 21–27.

5 J. D. Zuckerman, Hip fracture, *New England Journal of Medicine* 334 (1996): 1519–1525.

6 C. J. Rosen and L. R. Donahue, Insulin-like growth factor and bone: The osteoporosis connection revisited, *Proceedings of the Society for Experimental Biological Medicine* 219 (1998): 1–7.

7 E. R. Brooks, B. W. Ogden, and D. S. Cavalier, Compromised bone density 11.4 years after diagnosis of anorexia nervosa, *Journal of Women's Health* 7 (1998): 567–574.

8 D. L. Schneider, E. L. Barrett-Connor, and D. J. Morton, Timing of postmenopausal estrogen for optimal bone mineral density, *Journal of the American Medical Association* 277 (1997): 543–547; L. Speroff and coauthors, The comparative effect on bone density, endometrium, and lipids of continuous hormones as replacement therapy (CHART Study): A randomized controlled trial, *Journal of the American Medical Association* 276 (1996): 1397–1403; The Writing Group for the PEPI Trial, Effects of hormone therapy on bone mineral density: Results from the Postmenopausal Estrogen/Progestin Interventions (PEPI) Trial, *Journal of the American Medical Association* 276 (1996): 1389–1396.

9 P. A. Rochon and J. H. Gurwitz, Prescribing for seniors: Neither too much nor too little, *Journal of the American Medical Association* 282 (1999): 113–115.

10 B. H. Arjmandi and coauthors, Role of soy protein with normal or reduced isoflavone content in reversing bone loss induced by ovarian hormone deficiency in rats, *American Journal of Clinical Nutrition* 68 (1998): 1358S–1363S; B. H. Arjmandi and coauthors, Bone-sparing effect of soy protein in ovarian-hormone-deficient rats is related to its isoflavone content, *American Journal of Clinical Nutrition* 68 (1998): 1364S–1368S; J. P. Williams and coauthors, Tyrosine kinase inhibitor effects on avian osteoclastic and transport, *American Journal of Clinical Nutrition* 68 (1998): 1369S–1374S.

11 F. H. Anderson, Osteoporosis in men, *International Journal of Clinical Practice* 52 (1998): 176–180; E. Velazquez and G. Bellabarba Arata, Testosterone replacement therapy, *Archives of Andrology* 41 (1998): 79–90.

12 C. J. Strange, Boning up on osteoporosis, *FDA Consumer,* September 1996, pp. 15–20.

13 J. A. Eisman, Genetics, calcium intake and osteoporosis, *Proceedings of the Nutrition Society* 57 (1998): 187–193; C. M. Ulrich and coauthors, Bone mineral density in mother-daughter pairs: Relations to lifetime exercise, lifetime milk consumption, and calcium supplements, *American Journal of Clinical Nutrition* 63 (1996): 72–79.

14 A. M. Parfitt, Genetic effects on bone mass and turnover-relevance to black/white differences, *Journal of the American College of Nutrition* 16 (1997): 325–333.

15 R. P. Heaney, Osteoporosis, in *Nutrition in Women's Health,* ed. D. A. Kummel and P. M. Kris-Etherton (Gaithersburg, Md.: Aspen Publishers, 1996), pp. 418–439.

16 A. W. C. Kung and coauthors, Age-related osteoporosis in Chinese: An evaluation of the response of intestinal calcium absorption and calcitropic hormones to dietary calcium deprivation, *American Journal of Clinical Nutrition* 68 (1998): 1291–1297.

17 S. M. Smith and coauthors, Nutrition in space, *Nutrition Today* 32 (1997): 6–12.

18 I. Vuori, Peak bone mass and physical activity: A short review, *Nutrition Reviews* 54 (1996): S11–S14; L. Alekel and coauthors, Contributions of exercise, body composition, and age to bone mineral density in premenopausal women, *Medicine and Science in Sports and Exercise* 27 (1995): 1477–1485.

19 ACSM Position stand on osteoporosis and exercise, *Medicine and Science in Sports and Exercise* 27 (1995): i–vii.

20 T. A. Ricci and coauthors, Moderate energy restriction increases bone resorption in obese postmenopausal women, *American Journal of Clinical Nutrition* 73 (2001): 347–352; L. M. Salamone and coauthors, Effect of a lifestyle intervention on bone mineral density in premenopausal women: A randomized trial, *American Journal of Clinical Nutrition* 70 (1999): 97–103; L. W. Turner, M. Q. Wang, and Q. Fu, Risk factors for hip fracture among Southern older women, *Southern Medical Journal* 91 (1998): 533–540; M. E. Mussolino and coauthors, Risk factors for hip fracture in white men: The NHANES I Epidemiologic follow-up study, *Journal of Bone Mineral Research* 13 (1998): 918–924; T. V. Nguyen, P. N. Sambrook, and J. A. Eisman, Bone loss, physical activity, and weight change in elderly women: The Dubbo Osteoporosis Epidemiology Study, *Journal of Bone Mineral Research* 13 (1998): 1458–1467.

21 J. C. Fleet, Leptin and bone: Does the brain control bone biology? *Nutrition Reviews* 58 (2000): 209–211.

22 S. Suleiman and coauthors, Effect of calcium intake and physical activity level on bone mass and turnover in healthy, white, postmenopausal women, *American Journal of Clinical Nutrition* 66 (1997): 937–943.

23 M. N. Hadley and S. V. Reddy, Smoking and the human vertebral column: A review of the impact of cigarette use on vertebral bone metabolism and spinal fusion, *Neurosurgery* 41 (1997): 116–124.

24 E. T. Keller, J. Zhang, and W. B. Ershler, Ethanol activates the interleukin-6 promoter in a human bone marrow stromal cell line, *Journal of Gerontology, Series A, Biological Sciences and Medical Sciences* 52 (1997): B311–317; H. W. Sampson, Alcohol, osteoporosis, and bone regulating hormones, *Alcoholism: Clinical and Experimental Research* 21 (1997): 400–403; R. F. Klein, Alcohol-induced bone disease: Impact of ethanol on osteoblast proliferation, *Alcoholism: Clinical and Experimental Research* 21 (1997): 392–399.

25 H. W. Sampson and D. Shipley, Moderate alcohol consumption does not augment bone density in ovariectomized rats, *Alcohol: Clinical and Experimental Research* 21 (1997): 1165–1168; Klein, 1997.

26 D. Teegarden and coauthors, Previous milk consumption is associated with greater bone density in young women, *American Journal of Clinical Nutrition* 69 (1999): 1014–1017.

27 I. R. Reid, The roles of calcium and vitamin D in the prevention of osteoporosis, *Endocrinology and Metabolism Clinics of North America* 27 (1998): 389–398; K. O. O'Brien, Combined calcium and vitamin D supplementation reduces bone loss and fracture incidence in older men and women, *Nutrition Reviews* 56 (1998): 148–150; B. Dawson-Hughes and coauthors, Effect of calcium and vitamin D supplementation on bone density in men and women 65 years of age or older, *New England Journal of Medicine* 337 (1997): 670–676.

28 R. P. Heaney, Bone mass, nutrition, and other lifestyle factors, *Nutrition Reviews* 54 (1996): S3–S10.

29 D. Feskanich and coauthors, Vitamin K intake and hip fractures in women: A prospective study, *American Journal of Clinical Nutrition* 69 (1999): 74–79; K. L. Tucker and coauthors, Potassium, magnesium, and fruit and vegetable intakes are associated with greater bone mineral density in elderly men and women, *American Journal of Clinical Nutrition* 69 (1999): 727–736; R. K. Rude and coauthors, Magnesium deficiency induces bone loss in the rat, *Mineral and Electrolyte Metabolism* 24 (1998): 314–320.

30 J. Melhus and coauthors, Excessive dietary intake of vitamin A is associated with reduced bone mineral density and increased risk for hip fracture, *Annals of Internal Medicine* 129 (1998): 770–778.

31 S. A. New and coauthors, Dietary influences on bone mass and bone metabolism: Further evidence of a positive link between fruit and vegetable consumption and bone health? *American Journal of Clinical Nutrition* 71 (2000): 142–151; J. J. B. Anderson, Plant-based diets and bone health: Nutritional implications, *American Journal of Clinical Nutrition* 70 (1999) 539S–542S; Tucker and coauthors, 1999.

32 R. P. Heaney and coauthors, Dietary changes favorably affect bone remodeling in older adults, *Journal of the American Dietetic Association* 99 (1999): 1228–1233.

33 NIH Consensus Development Panel on Osteoporosis Prevention, Diagnosis, and Therapy, Osteoporosis prevention, diagnosis, and therapy, *Journal of the American Medical Association* 285 (2001): 785–795; C. A. Kulak and J. P. Bilezikian, Osteoporosis: Preventive strategies, *International Journal of Fertility and Womens Medicine* 43 (1998): 56–64.

34 D. Hosking and coauthors, Prevention of bone loss with alendronate in postmenopausal women under 60 years of age, *New England Journal of Medicine* 338 (1998): 485–492; P. D. Delmas and coauthors, Effects of raloxifene on bone mineral density, serum cholesterol concentrations, and uterine endometrium in postmenopausal women, *New England Journal of Medicine* 337 (1997): 1641–1647.

CHAPTER 13 The Trace Minerals

Figure 12-7 in the last chapter (p. 396) showed the tiny quantities of **trace minerals** in the human body. The trace minerals are so named because they are present, and needed, in relatively small amounts in the body. All together, they would produce only a bit of dust, hardly enough to fill a teaspoon. Yet they are no less important than the major minerals or any of the other nutrients. Each of the trace minerals performs a vital role. A deficiency of any of them may be fatal, and an excess of many is equally deadly. Remarkably, people's diets normally supply just enough of these minerals to maintain health.

THE TRACE MINERALS—AN OVERVIEW

The body requires the trace minerals in minuscule quantities. They participate in diverse tasks all over the body, each having special duties that only it can perform.

• Food Sources • The trace mineral contents of foods depend on soil and water composition and on how foods are processed. Furthermore, many factors in the diet and within the body affect the minerals' bioavailability.◆ Still, outstanding food sources for each of the trace minerals, just like those for the other nutrients, would include a wide variety of foods, especially unprocessed, whole foods.

Reminder: *Bioavailability* refers to the rate at and the extent to which a nutrient is absorbed and used. ◆

• Deficiencies • Severe deficiencies of the better-known minerals are easy to recognize. Deficiencies of the others may be harder to diagnose, and for all minerals, mild deficiencies are easy to overlook. In general, the most common result of a deficiency is failure of children to grow and thrive, for the minerals are active in all the body systems—the GI tract, cardiovascular system, blood, muscles, bones, and central nervous system.

• Toxicities • Some of the trace minerals are toxic at intakes not far above the estimated requirements. Thus it is important not to habitually exceed the upper level of recommended intakes. Many vitamin-mineral pills contain trace minerals, making it easy for users to exceed their needs. The Food and Drug Administration (FDA) is not permitted to limit the amounts of trace minerals in supplements; consumers have demanded the freedom to choose their own doses of nutrients.* Individuals who take supplements must therefore be aware of the possible dangers and select supplements that contain no more than 100 percent of the Daily Value. It would be easier and safer to meet nutrient needs by selecting a variety of foods than by combining an assortment of pills without causing toxicity (see Highlight 10).

• Interactions • Interactions among the trace minerals are common and often lead to nutrient imbalances. An excess of one may cause a deficiency of another. (A slight manganese overload, for example, may aggravate an iron deficiency.) A deficiency of one may open the way for another to cause a toxic reaction. (Iron deficiency, for example, makes the body vulnerable to lead poisoning.) A deficiency of one may exacerbate the problems associated with the deficiency of another. (A combined iodine and selenium deficiency, for example, reduces thyroid hormone production more than an iodine deficiency alone.)[1] These examples highlight the need to balance intakes and to steer clear of supplement use. A good food source of one nutrient may be a poor food source of another; and factors that enhance the action of some trace minerals may hinder others. (Meats are a good source of iron, but a poor source of calcium; vitamin C enhances the absorption of iron but hinders that of copper.) Research on the trace minerals is active, suggesting that we have much more to learn about them.

*Canada limits the amounts of trace minerals in supplements.

trace minerals: essential mineral nutrients found in the human body in amounts smaller than 5 g; sometimes called **microminerals.**

IN SUMMARY
To recap, although the body uses only tiny amounts of the trace minerals, they are vital to health. Because so little is required, the trace minerals can be toxic at levels not far above estimated requirements—a consideration for supplement users. Like the other nutrients, the trace minerals are best obtained by eating a variety of whole foods.

IRON

Iron is an essential nutrient, vital to many of the cells' activities, but it poses a problem for millions of people: some people simply don't eat enough iron-containing foods to support their health optimally, while others have so much iron that it threatens their well-being. The principle that both too little or too much of a nutrient is harmful seems particularly apropos for iron.

Iron Roles in the Body

◆ **Iron's two ionic states:**
- Ferrous iron (reduced): Fe^{++}.
- Ferric iron (oxidized): Fe^{+++}.

◆ **Reminder: A *cofactor* is a mineral element that works with an enzyme to facilitate a chemical reaction.**

◆ **For details about ions, oxidation, and reduction, see Appendix B.**

◆ **Reminder: *Hemoglobin* is the oxygen-carrying protein of the red blood cells that transports oxygen from the lungs to tissues throughout the body; hemoglobin accounts for 80% of the body's iron.**

Iron has the knack of switching back and forth between two ionic states.◆ In the reduced state, iron has lost two electrons and therefore has a net positive charge of two; it is known as *ferrous iron*. In the oxidized state, iron has lost a third electron, has a net positive charge of three, and is known as *ferric iron*. Because it can exist in these different ionic states, iron can serve as a cofactor◆ to enzymes involved in oxidation-reduction◆ reactions. Iron forms a part of the electron carriers that participate in the electron transport chain (see Chapter 7).* In the final steps of this energy-yielding metabolic pathway, these carriers transport hydrogens and electrons from energy-yielding nutrients to oxygen, forming water, and in the process make ATP for the cell's energy use.

Most of the body's iron is found in two proteins: hemoglobin ◆ in the red blood cells and **myoglobin** in the muscle cells. In both, iron helps accept, carry, and then release oxygen. Many enzymes that oxidize compounds—reactions so widespread in metabolism that they occur in all cells—depend on iron. Iron is also required by enzymes involved in the making of amino acids, collagen, hormones, and neurotransmitters.

Iron Absorption and Metabolism

The body conserves iron zealously and handles it carefully (see Figure 13-1). Because it is difficult to excrete iron once it is in the body, balance is maintained primarily through absorption: more iron is absorbed when stores are empty and less when they are full.[2]

◆ **A mucous membrane such as the one that lines the GI tract is sometimes called the mucosa (mu-KO-sah). The adjective of mucosa is mucosal (mu-KO-sal).**

• Iron Absorption • Special proteins help the body absorb iron from food. One protein, called *mucosal ferritin,* receives iron from the GI tract and stores it in the mucosal cells◆ of the small intestine. When the body needs iron, mucosal ferritin releases some iron to another protein, called *mucosal transferrin.* Mucosal transferrin transfers the iron to another protein, *blood transferrin,* which transports the iron to the rest of the body. Intestinal cells are replaced about every three days; when the cells are shed and excreted in the feces, they carry some iron out with them. By holding iron temporarily, these cells can either deliver iron when the day's intake falls short or dispose of it when intakes exceed needs.[3]

myoglobin: the oxygen-holding protein of the muscle cells.
- **myo** = muscle

heme (HEEM): the iron-holding part of the hemoglobin and myoglobin proteins. About 40% of the iron in meat, fish, and poultry is bound into heme; the other 60% is **nonheme** iron.

• Heme and Nonheme Iron • Iron absorption depends in part on its source. Iron occurs in two forms in foods: as **heme** iron, which is found only in foods

*The iron-containing electron carriers of the electron transport chain are known as *cytochromes*. See Appendix C for these pathways.

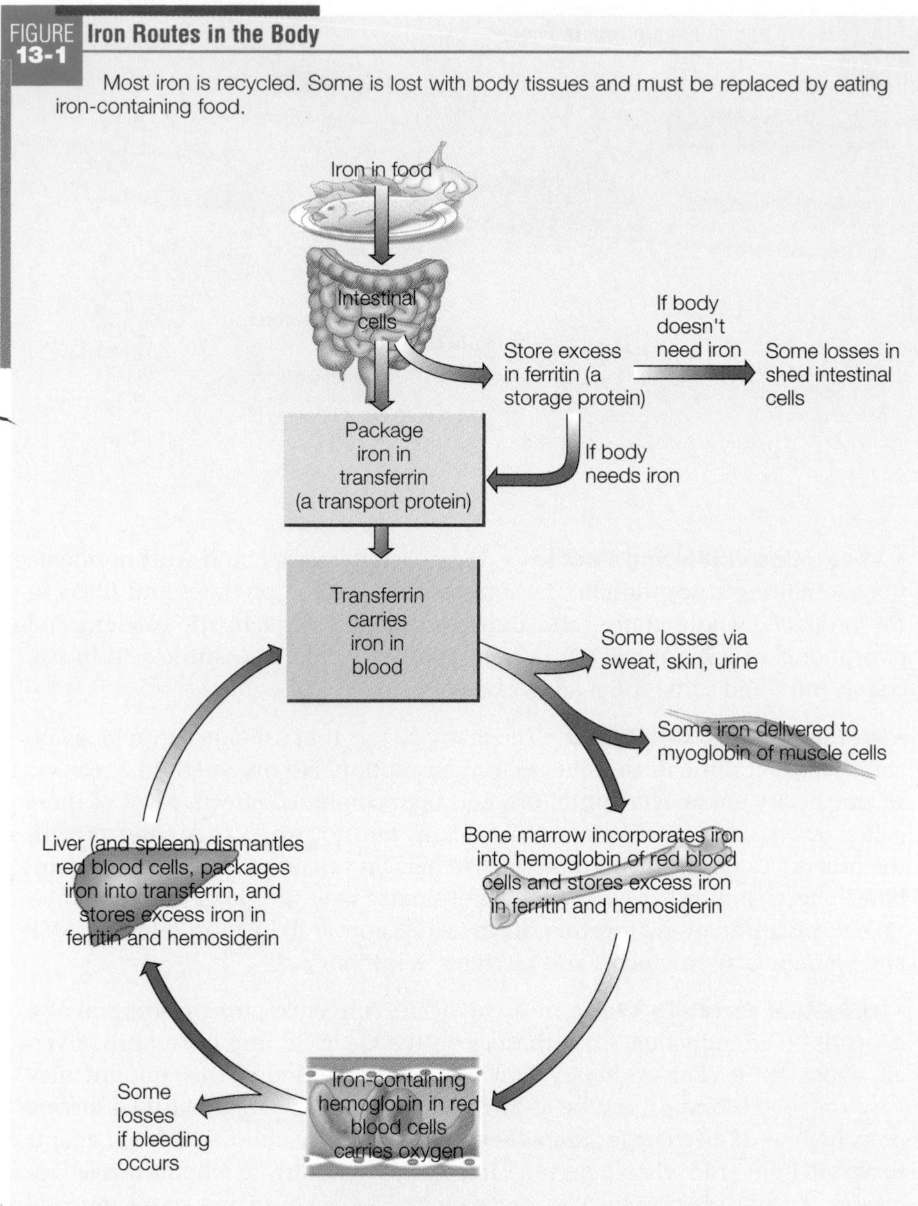

FIGURE 13-1 **Iron Routes in the Body**

Most iron is recycled. Some is lost with body tissues and must be replaced by eating iron-containing food.

derived from the flesh of animals, such as meats, poultry, and fish; and as nonheme iron, which is found in both plant-derived and animal-derived foods (see Figure 13-2 on p. 432). On the average, about 10 percent of the iron a person consumes in a day comes from heme iron. Even though heme iron accounts for only a small proportion of the intake, it is so well absorbed that it contributes significant iron: about 25 percent of heme iron is absorbed. By comparison, only 10 percent of nonheme iron is absorbed, depending on other dietary factors and the body's iron stores. People with severe iron deficiencies absorb both heme and nonheme iron more efficiently and are more sensitive to absorption-enhancing factors than people with adequate iron status.

• Absorption-Enhancing Factors • Meat, fish, and poultry contain not only the well-absorbed heme iron, but also a factor **(MFP factor)** that promotes the absorption of nonheme iron◆ from other foods eaten at the same meal. Vitamin C also enhances nonheme iron absorption from foods eaten in the same meal by capturing the iron and keeping it in the reduced ferrous form, ready for absorption. Some acids and sugars also enhance nonheme iron absorption.

◆ **Factors that *enhance* nonheme iron absorption:**

- MFP factor.
- Vitamin C (ascorbic acid).
- Citric acid and lactic acid from foods and HCl acid from the stomach.
- Sugars (including the sugars in wine).

MFP factor: a factor associated with the digestion of **M**eat, **F**ish, and **P**oultry that enhances iron absorption.

FIGURE 13-2 Heme and Nonheme Iron in Foods

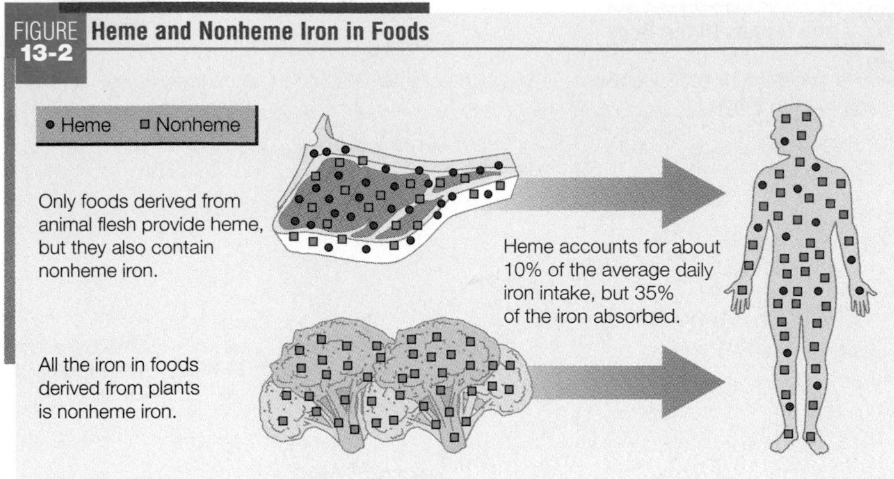

◆ Factors that *inhibit* iron absorption:
- Phytates and fibers (grains and vegetables).
- Oxalates (spinach).
- Calcium and phosphorus (milk).
- EDTA (food additives).
- Tannic acid (and other polyphenols in tea and coffee).

© Photo Disc Inc.

This chili dinner provides several factors that may enhance iron absorption: heme and nonheme iron and MFP from meat, nonheme iron from legumes, and vitamin C from tomatoes.

◆ The iron storage proteins are **ferritin** (FERR-ih-tin) and **hemosiderin** (heem-oh-SID-er-in).

• **Absorption-Inhibiting Factors** • Some dietary factors bind with nonheme iron, inhibiting absorption.◆ These factors include the phytates and fibers in soy products, whole grains, and nuts; oxalates in spinach; the calcium and phosphorus in milk; the EDTA in food additives;* and the tannic acid in tea, coffee, nuts, and some fruits and vegetables.

• **Dietary Factors Combined** • The many factors that influence iron bioavailability make it difficult to estimate iron absorption. No one method considers all the dietary enhancers, inhibitors, and their combined effects. Most of these factors exert a strong influence individually, but not when they interact with the others in a meal. Furthermore, researchers find that the impact of the combined effects diminishes when a diet is evaluated over several days. When multiple meals are analyzed together, three factors appear to be most relevant: MFP and vitamin C as enhancers and phytates as inhibitors.[4]

• **Individual Variation** • In addition to dietary influences, iron absorption also depends on an individual's health, stage in the life cycle, and iron status. Overall, about 18 percent of dietary iron is absorbed, although this amount may vary widely. Absorption can be as low as 2 percent in a person with GI disease or as high as 35 percent in a rapidly growing, healthy child. The body adapts to absorb more iron when a person's iron stores fall short, or when the need increases for any reason (such as pregnancy). The body makes more mucosal transferrin to absorb more iron from the intestines and more blood transferrin to carry more iron around the body. Similarly, when iron stores are sufficient, the body adapts to absorb less iron.[5]

• **Iron Transport and Storage** • Blood transferrin delivers iron to the bone marrow and other tissues. The bone marrow uses large quantities to make new red blood cells, whereas other tissues use less. Surplus iron is stored in the protein ferritin, primarily in the liver, but also in the bone marrow and spleen. When dietary iron has been plentiful, ferritin is constantly and rapidly made and broken down, providing an ever-ready supply of iron. When iron concentrations become abnormally high, the liver converts some ferritin into another storage protein◆ called hemosiderin. Hemosiderin releases iron more slowly than ferritin does. By storing excess iron, the body protects itself: free iron acts as a free radical, attacking cell lipids, DNA, and protein. (See Highlight 11 for more information on free radicals and the damage they can cause.)

• **Iron Recycling** • The average red blood cell lives about four months; then the spleen and liver cells remove it from the blood, take it apart, and prepare

*EDTA is ethylenediamine tetra acetate, a chelating agent that is used in food processing to retard crystal formation and promote color retention.

the degradation products for excretion or recycling. The iron is salvaged: the liver attaches it to blood transferrin, which transports it back to the bone marrow to be reused in making new red blood cells. Thus, although red blood cells live for only about four months, the iron recycles through each new generation of cells. The body loses some iron daily via the GI tract and, if bleeding occurs, in blood; only tiny amounts of iron are lost in feces, urine, sweat, and shed skin.*

Iron Deficiency

Worldwide, **iron deficiency** is the most common nutrient deficiency, affecting more than one billion people. In developing countries, one-third of the children and women of childbearing age suffer from **iron-deficiency anemia.**[6] In the United States, iron deficiency is less prevalent, but still affects 10 percent of toddlers, adolescent girls, and women of childbearing age; preventing and correcting iron deficiency are high priorities.[7]

• Vulnerable Stages of Life • Some stages of life◆ both demand more iron and provide less, making deficiency likely. Women are especially prone to iron deficiency during their reproductive years because of repeated blood losses during menstruation. Pregnancy demands additional iron to support the added blood volume, growth of the fetus, and blood loss during childbirth. Infants and young children receive little iron from their high-milk diets, yet need extra iron to support their rapid growth. The rapid growth of adolescence, especially for males, and the menstrual losses of females also demand extra iron that a typical teen diet may not provide. An adequate iron intake is especially important during these stages of life.

High risk for iron deficiency: ◆
- Women in their reproductive years.
- Pregnant women.
- Infants and young children.
- Teenagers.

HEALTHY PEOPLE 2010

Reduce iron deficiency among young children, females of childbearing age, and pregnant females.

• Blood Losses • Bleeding ◆ from any site incurs iron losses. In some cases, as in an active ulcer, the bleeding may not be obvious, but even small chronic blood losses significantly deplete iron reserves.[8] In developing countries, blood loss is often brought on by parasitic infections of the GI tract. People who donate blood regularly also incur losses and may benefit from iron supplements.[9] As mentioned, menstrual losses can be considerable as they tap women's iron stores regularly.

The iron content of blood is about 0.5 mg/100 mL blood. A person donating a pint of blood (approximately 500 mL) loses about 2.5 mg of iron. ◆

• Assessment of Iron Deficiency • Iron deficiency develops in stages.◆ This section provides a brief overview of how to detect these stages, and Appendix E provides more details. In the first stage of iron deficiency, iron stores diminish. Measures of serum ferritin reflect iron stores and are most valuable in assessing iron status.

Stages of iron deficiency: ◆
- Iron stores diminish.
- Transport iron decreases.
- Hemoglobin production declines.

The second stage of iron deficiency is characterized by a decrease in transport iron: serum iron falls, and the iron-carrying protein transferrin *increases* (an adaptation that enhances iron absorption). Together, these two measures can indicate the severity of the deficiency—the more transferrin and the less iron in the blood, the more advanced the deficiency is.

The third stage of iron deficiency occurs when the lack of iron limits hemoglobin production. Now the hemoglobin precursor, **erythrocyte protoporphyrin,** begins to accumulate as hemoglobin and **hematocrit** values decline.

Hemoglobin and hematocrit tests are easy, quick, and inexpensive, so they are the tests most commonly used in evaluating iron status; their usefulness is

iron deficiency: the state of having depleted iron stores.

iron-deficiency anemia: severe depletion of iron stores that results in low hemoglobin and small, pale, red blood cells.

erythrocyte protoporphyrin (PRO-toe-PORE-fe-rin): a precursor to hemoglobin.

hematocrit (hee-MAT-oh-krit): measurement of the volume of the red blood cells packed by centrifuge in a given volume of blood.

*The adult male loses about 1.0 milligram of iron per day. Women lose additional iron in menses. Menstrual losses vary considerably, but over a month, they average about 0.5 milligram per day.

FIGURE 13-3 **Normal and Anemic Blood Cells**

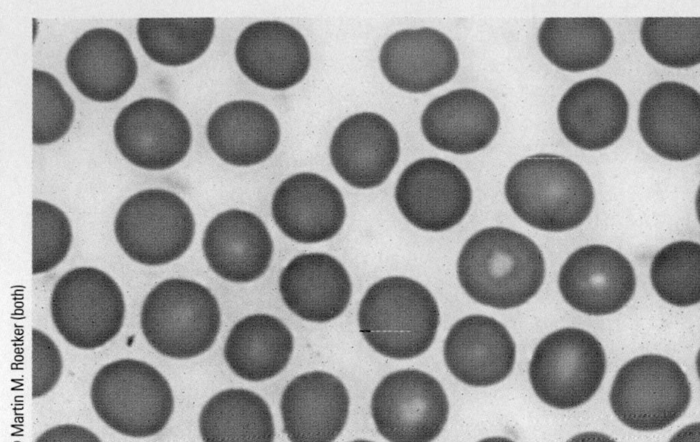

Both size and color are normal in these blood cells.

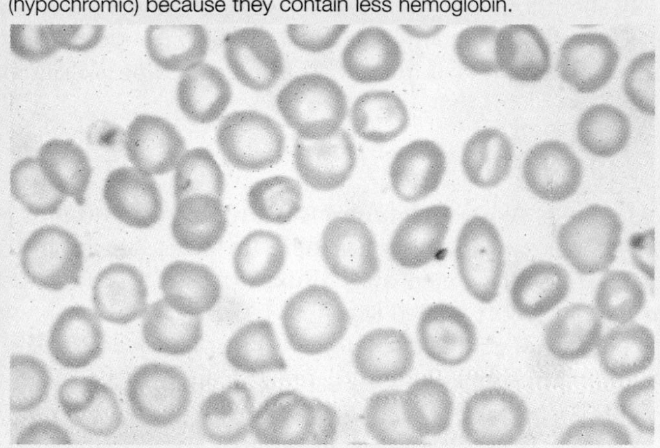

Blood cells in iron-deficiency anemia are small (microcytic) and pale (hypochromic) because they contain less hemoglobin.

© Martin M. Roetker (both)

limited, however, because they are late indicators of iron deficiency. Furthermore, other nutrient deficiencies and medical conditions can influence their values.

• **Iron Deficiency and Anemia** • Iron deficiency and iron-deficiency anemia are not the same: people may be iron deficient without being anemic. The term *iron deficiency* refers to depleted body iron stores without regard to the degree of depletion or to the presence of anemia. The term *iron-deficiency anemia* refers to the severe depletion of iron stores that results in a low hemoglobin concentration. In iron-deficiency anemia, red blood cells are pale and small◆ (see Figure 13-3). They can't carry enough oxygen from the lungs to the tissues, so energy metabolism in the cells falters. The result is fatigue, weakness, headaches, apathy, pallor, and poor resistance to cold temperatures. Since hemoglobin is the bright red pigment of the blood, the skin of a fair person who is anemic may become noticeably pale. In a dark-skinned person, the eye lining, normally pink, will be very pale.

◆ Iron-deficiency anemia is a **microcytic** (my-cro-SIT-ic) **hypochromic** (high-po-KROME-ic) **anemia.**
- **micro** = small
- **cytic** = cells
- **hypo** = too little
- **chrom** = color

• **Iron Deficiency and Behavior** • Long before the red blood cells are affected and anemia is diagnosed, a developing iron deficiency affects behavior. Even at slightly lowered iron levels, the complete oxidation of pyruvate is impaired, reducing physical work capacity and productivity. With reduced energy available to work, plan, think, play, sing, or learn, people simply do these things less. They have no obvious deficiency symptoms; they just appear unmotivated, apathetic, and less physically fit.

Many of the symptoms associated with iron deficiency are easily mistaken for behavioral or motivational problems. A restless child ◆ who fails to pay attention in class might be thought contrary. An apathetic homemaker who has let housework pile up might be thought lazy. No responsible dietitian would ever claim that all behavioral problems are caused by nutrient deficiencies, but poor nutrition is always a possible contributor to problems like these. When investigating a behavioral problem, check the adequacy of the diet and seek a routine physical examination before undertaking more expensive, and possibly harmful, treatment options.

◆ **The effects of iron deficiency on children's behavior are discussed in Chapter 15.**

• **Iron Deficiency and Pica** • A curious behavior seen in some iron-deficient people, especially in women and children of low-income groups, is **pica**—an appetite for ice, clay, paste, and other nonfood substances. These substances contain no iron and cannot remedy a deficiency; in fact, clay actually inhibits iron absorption, which may explain the iron deficiency that accompanies such behavior.

pica (PIE-ka): a craving for nonfood substances. Also known as **geophagia** (gee-oh-FAY-gee-uh) when referring to clay eating and **pagophagia** (pag-oh-FAY-gee-uh) when referring to ice craving.
- **picus** = woodpecker or magpie
- **geo** = earth
- **phagein** = to eat
- **pago** = frost

Iron Toxicity

In general, even a diet that includes fortified foods poses no special risk for iron toxicity.[10] The body normally absorbs less iron when its stores are full, but some individuals are poorly defended against excess iron.[11] Once considered rare, **iron overload** has emerged as an important disorder of iron metabolism and regulation.[12]

• Iron Overload • Iron overload is known as **hemochromatosis** and is usually caused by a genetic disorder that enhances iron absorption.[13] Hereditary hemochromatosis is the most common genetic disorder in the United States, affecting some 1.5 million people.[14] Other causes of iron overload include repeated blood transfusions (which bypass the intestinal defense), massive doses of supplementary iron (which overwhelm the intestinal defense), and other rare metabolic disorders. Long-term overconsumption of iron may cause **hemosiderosis,** a condition characterized by large deposits of the iron storage protein hemosiderin in the liver and other tissues.

Some of the signs and symptoms of iron overload are similar to those of iron deficiency: apathy, lethargy, and fatigue. Therefore, taking iron supplements before assessing iron status is clearly unwise; hemoglobin tests alone would fail to make the distinction.

Iron overload is characterized by tissue damage, especially in iron-storing organs such as the liver. Infections are likely because bacteria thrive on iron-rich blood. Symptoms are most severe in alcohol abusers because alcohol damages the intestine, further impairing its defenses against absorbing excess iron. Untreated hemochromatosis aggravates the risks of diabetes, liver cancer, heart disease, and arthritis.

Iron overload is more common in men than in women and is twice as prevalent among men as iron deficiency. The widespread fortification of foods with iron makes it difficult for people with hemochromatosis to follow a low-iron diet, but greater dangers lie in the indiscriminate use of iron and vitamin C supplements. Vitamin C not only enhances iron absorption, but releases iron from ferritin, allowing free iron to wreak the damage typical of free radicals.[15] This is an example of how vitamin C causes *pro*oxidant damage when taken in high doses (see Highlight 11 for a discussion of free radicals and their effects on disease development).

• Iron and Heart Disease • Some research suggests a link between heart disease and elevated iron stores, especially among older adults.[16] Too much iron may pose no problem for healthy people, but in certain diseases, free radicals attack ferritin, causing it to release iron from storage.[17] Free iron, in turn, acts as an oxidant that can generate more free radicals. Iron's oxidation of LDL may explain its role in the development of heart disease.[18] Researchers report a reduced risk of heart attacks in blood donors, perhaps because of their repeated iron losses.[19]

Health promotion

Prior to these findings, scientists had speculated that premenopausal women were protected against heart disease by their estrogen (women's rate of heart disease is lower and begins to approach that of men only after menopause, when estrogen production ceases). Now scientists are asking whether low iron stores from repeated menstrual losses might be the protective factor. (Women's iron stores tend to catch up with men's after menopause.)

• Iron and Cancer • There also appears to be an association between iron and cancer. Explanations for how iron might be involved in causing cancer focus on its free-radical activity, which can damage DNA (see Highlight 11). One of the benefits of a high-fiber diet may be that its phytates bind iron, making it less available for such reactions.

• Iron Poisoning • Ingestion of iron-containing supplements remains a leading cause of accidental poisoning in small children.[20] Symptoms of intoxication

iron overload: toxicity from excess iron.

hemochromatosis (HE-moh-KRO-ma-toe-sis): a hereditary defect in iron metabolism characterized by deposits of iron-containing pigment in many tissues, with tissue damage.

hemosiderosis (HE-mo-sid-er-OH-sis): a condition characterized by the deposition of hemosiderin in the liver and other tissues.

HOW TO Estimate the Recommended Daily Intake for Iron

To calculate the recommended daily iron intake, the DRI Committee considers a number of factors. For example, for a woman of childbearing age (19 to 50):

- Losses from feces, urine, sweat, and shed skin: 1.0 milligram.
- Losses through menstruation (about 14 milligrams total averaged over 28 days): 0.5 milligram.

These losses reflect an average daily need (total) of 1.5 milligrams of *absorbed* iron.

An estimated average requirement is determined based on the daily need and the assumption that an average of 18 percent of ingested iron is absorbed:

1.5 mg iron (needed) ÷ 0.18 (percent iron absorbed) = 8 mg iron (estimated average requirement).

Then a margin of safety is added to cover the needs of essentially all women of childbearing age and the RDA is set at 18 milligrams.

include nausea, vomiting, diarrhea, a rapid heartbeat, a weak pulse, dizziness, shock, and confusion. As few as 5 iron tablets containing as little as 200 milligrams of iron have caused the deaths of dozens of young children. The exact cause of death is uncertain, but excessive free-radical damage is thought to play a role in heart failure and respiratory distress; autopsy reports reveal iron deposits and cell death in the stomach, small intestine, liver, and blood vessels (which can cause internal bleeding).[21] Keep iron-containing tablets out of the reach of children. If you suspect iron poisoning, call the nearest poison control center or a physician immediately.

Iron Recommendations and Sources

To obtain enough iron, people must first select iron-rich foods and then eat so as to maximize iron absorption. This discussion begins by identifying iron-rich foods, then reviews factors affecting absorption.

• Recommended Iron Intakes • The usual diet in the United States provides about 6 to 7 milligrams of iron for every 1000 kcalories. The recommended daily intake for men is 8 milligrams, and most men eat more than 2000 kcalories a day, so they can meet their iron needs with little effort. Women during their childbearing years, however, need 18 milligrams a day. (The accompanying "How to" explains how to calculate the recommended intake.) Because women have higher iron needs and lower energy needs, they sometimes have trouble obtaining enough iron. On the average, women receive only 12 to 13 milligrams of iron per day, not enough until after menopause. To meet their iron needs from foods, premenopausal women need to select iron-rich foods at every meal.

© Craig M. Moore

When the label on a grain product says "enriched," it means iron and several B vitamins have been added.

• Iron in Foods • Figure 13-4 shows the amounts of iron in selected foods. Meats, fish, and poultry contribute the most iron; other protein-rich foods such as legumes and eggs are also good sources. Foods in the milk group are as poor in iron as they are rich in calcium. Although an indispensable part of the diet, milk products should not be overemphasized. Grain foods vary, with whole-grain and enriched breads and cereals providing the most iron. Finally, dark greens (such as broccoli) and some fruits (especially when dried) contribute some iron.

• Iron-Enriched Foods • Iron is one of the enrichment nutrients for grain products. One serving of enriched bread or cereal provides only a little iron, but because people eat many servings of these foods, the contribution can be significant. Iron added to foods is not absorbed as well as naturally occurring iron, but when eaten with absorption-enhancing foods, enrichment iron can make a difference. In cases of iron overload, enrichment may exacerbate the problem.

• Maximizing Iron Absorption • In general, the bioavailability of iron in meats, fish, and poultry is high; in grains and legumes, intermediate; and in most vegetables, especially those high in oxalates such as spinach, low. As

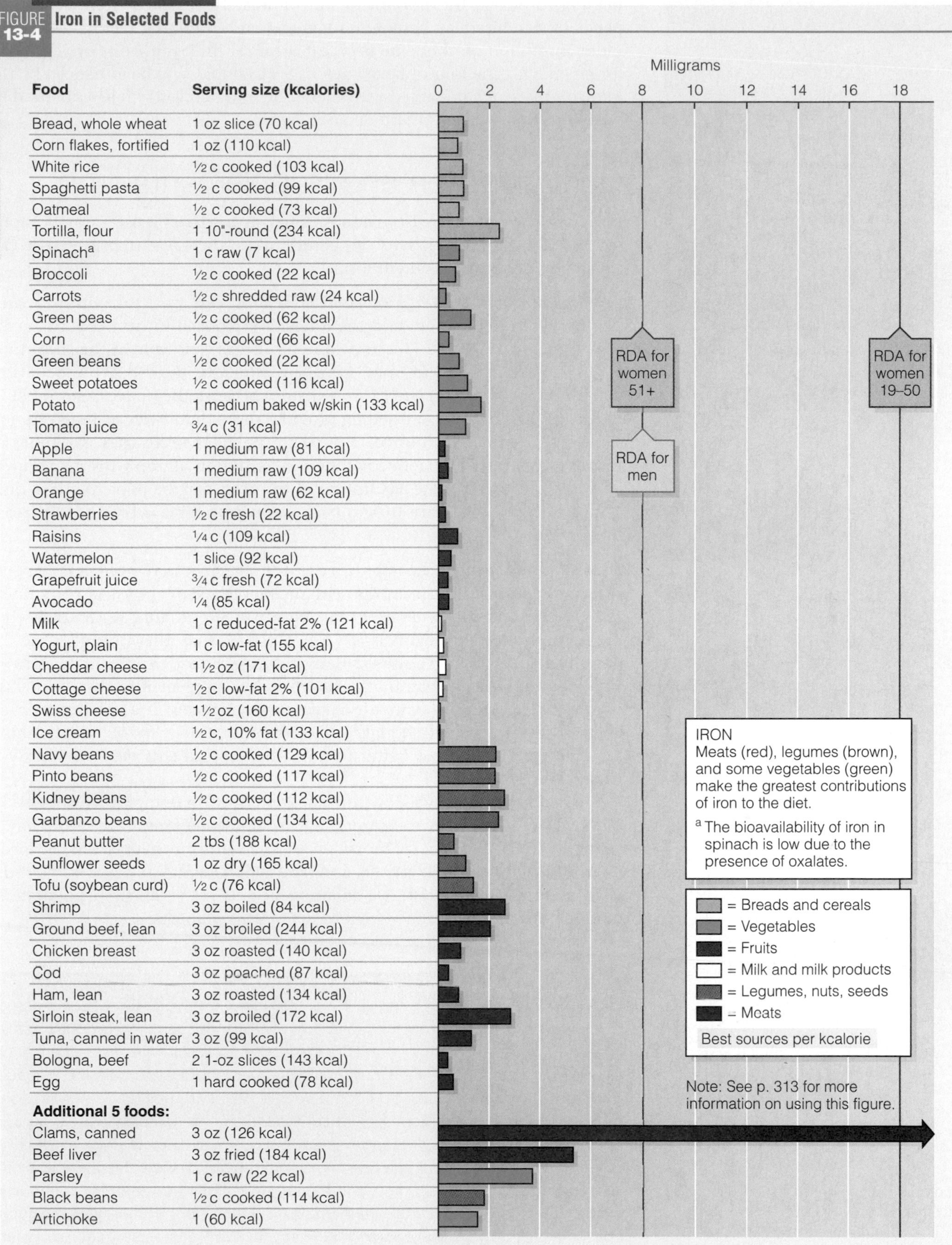

FIGURE 13-4 Iron in Selected Foods

© Polara Studios Inc.

An old-fashioned iron skillet adds iron to foods.

mentioned earlier, the amount of iron ultimately absorbed from a meal depends on the interplay between enhancing and inhibiting factors. For maximum absorption of nonheme iron, eat meat for MFP and fruits or vegetables for vitamin C. The iron of baked beans, for example, will be enhanced by the MFP in a piece of ham served with them; the iron of bread will be enhanced by vitamin C in a slice of tomato on a sandwich.

Contamination and Supplemental Iron

In addition to the iron from foods, **contamination iron** from nonfood sources of inorganic iron salts can contribute to the day's intakes. People can also ingest iron in supplement form.

• **Contamination Iron** • Foods cooked in iron cookware take up iron salts. The more acidic the food, and the longer it is cooked in iron cookware, the higher the iron content. The iron content of eggs can triple in the short time it takes to scramble them in an iron pan. Similarly, dried peaches or raisins contain more iron than the fresh fruits do, because they have been dried in iron pans. For example, a dried fig provides more than twice as much iron as a fresh fig. (Ounce for ounce, a dried fig provides more than six times as much iron as a fresh fig, reflecting the concentration of iron with the removal of water.) Admittedly, the absorption of this iron may be poor (perhaps only 1 to 2 percent), but every little bit helps a person who is trying to increase iron intake.

• **Iron Supplements** • People who are iron deficient may need supplements as well as an iron-rich, absorption-enhancing diet. Many physicians routinely recommend iron supplements to pregnant women, infants, and young children. Iron from supplements is less well absorbed than that from food, so the doses have to be high. The absorption of iron taken as ferrous sulfate or as an iron **chelate** is better than that from other iron supplements. Absorption also improves when supplements are taken between meals or at bedtime on an empty stomach, and with liquids other than milk, tea, or coffee, which inhibit absorption. Whether it is better to take iron supplements daily or weekly is debatable.[22]

There is no benefit to taking iron supplements with orange juice because vitamin C does not enhance absorption from supplements as it does from foods. (Vitamin C enhances iron absorption by converting insoluble ferric iron in foods to the more soluble ferrous iron, and supplemental iron is already in the ferrous form.) Constipation is a common side effect of iron supplementation; drinking plenty of water may help to relieve this problem.

IN SUMMARY

Most of the body's iron is in hemoglobin and myoglobin where it carries oxygen for use in energy metabolism; some iron is also required for enzymes involved in a variety of reactions. Special proteins assist with iron absorption, transport, and storage—all helping to maintain an appropriate balance, because both too little and too much iron can be damaging. Iron deficiency is most common among infants and young children, teenagers, women of childbearing age, and pregnant women; symptoms include fatigue and anemia. Iron overload is most common in men and has been linked with heart disease. Heme iron, which is found only in meat, fish, and poultry, is better absorbed than nonheme iron, which occurs in most foods. Nonheme iron absorption is improved by eating iron-containing foods with foods containing the MFP factor and vitamin C; absorption is limited by phytates and oxalates. The summary table presents a few iron facts.

contamination iron: iron found in foods as the result of contamination by inorganic iron salts from iron cookware, iron-containing soils, and the like.

chelate (KEY-late): a substance that can grasp the positive ions of a metal.
- **chele** = claw

Iron

2001 RDA	Deficiency Symptoms	Toxicity Symptoms
Men: 8 mg/day	**Eye Symptoms**	
Women: 18 mg/day (19–50 yr)	Blue sclera[a]	
8 mg/day (51+)	**Immune System Symptoms**	
Upper Level	Reduced resistance to infection (lowered immunity)	Infections
Adults: 45 mg/day	**Neurological/Muscular Symptoms**	
Chief Functions in the Body	Reduced work productivity, tolerance to work, and voluntary work; reduced physical fitness; weakness; fatigue; impaired cognitive function (children); reduced learning ability; increased distractibility (inability to pay attention); impaired visual discrimination; impaired reactivity and coordination (infants)	Lethargy, joint disease
Part of the protein hemoglobin, which carries oxygen from place to place in the body; part of the protein myoglobin in muscles, which makes oxygen available for muscle contraction; necessary for the utilization of energy as part of the cells' metabolic machinery	**Skin Symptoms**	
Significant Sources	Itching; pale nailbeds, eye membranes, and palm creases; concave nails; impaired wound healing	Pigmentation, loss of hair
Red meats, fish, poultry, shellfish, eggs, legumes, dried fruits	**Other**	
	Reduced resistance to cold, inability to regulate body temperature, pica (clay eating, ice eating)	Death by accidental poisoning in children; organ damage,enlarged liver, amenorrhea, impotence

[a]*Sclera* is a tough fibrous tissue that covers the "white" of the eye; *blue sclera* has an abnormal degree of blueness.

ZINC

Zinc is a versatile trace element required as a cofactor by more than 100 enzymes. Virtually all cells contain zinc, but the highest concentrations are in muscle and bone. Tissues do not readily give up their zinc when blood levels fall, so a person must eat zinc-rich foods frequently.

Zinc Roles in the Body

Zinc supports the work of numerous proteins◆ in the body, including the **metalloenzymes,** which are involved in a variety of metabolic processes.* In addition, zinc stabilizes cell membranes, helping to strengthen their defense against free-radical attacks. Zinc also assists in immune function and in growth and development.[23] Zinc participates in the synthesis, storage, and release of the hormone insulin in the pancreas, although it does not appear to play a direct role in insulin's action. Zinc interacts with platelets in blood clotting, affects thyroid hormone function, and influences behavior and learning performance. It is needed to produce the active form of vitamin A (retinal) in visual pigments and the retinol-binding protein that transports vitamin A. It is essential to normal taste perception, wound healing, the making of sperm, and fetal development. A zinc deficiency impairs all these and other functions, underlining the vast importance of proteins as the body's working machines.

Enzymes that require zinc: ◆
- Help make parts of the genetic materials DNA and RNA.
- Manufacture heme for hemoglobin.
- Participate in essential fatty acid metabolism.
- Release vitamin A from liver stores.
- Metabolize carbohydrates.
- Synthesize proteins.
- Metabolize alcohol in the liver.
- Dispose of damaging free radicals.

Zinc Absorption and Metabolism

The body's handling of zinc resembles that of iron in some ways and differs in others. A key difference is the circular passage of zinc from the intestine to the body and back again.

metalloenzymes (meh-TAL-oh-EN-zimes): enzymes that contain one or more minerals as part of their structures.

*Among the metalloenzymes requiring zinc are carbonic anhydrase, deoxythymidine kinase, DNA and RNA polymerase, and alkaline phosphatase.

• **Zinc Absorption** • The rate of zinc absorption varies from about 15 to 40 percent, depending on a person's zinc status: if more is needed, more is absorbed. Also, dietary factors influence zinc absorption. For example, fiber and phytates bind zinc, thus limiting its bioavailability.

Upon absorption into an intestinal cell, zinc has several options. It may become involved in the metabolic functions of the cell itself. Alternatively, it may be retained within the cell by **metallothionein,** a special binding protein similar to the iron storage protein, mucosal ferritin.

The synthesis of metallothionein in the intestinal cells helps to regulate zinc absorption. When zinc intakes are high, more metallothionein is made; it holds zinc in reserve, thus limiting absorption. (Similarly, metallothionein in the liver binds zinc until other body tissues signal a need for it.) When the body needs zinc, metallothionein releases it into the blood where it can be transported around the body. Some zinc eventually reaches the pancreas.

• **Enteropancreatic Circulation** • Many of the digestive enzymes released from the pancreas into the intestine at mealtimes contain zinc. The intestine thus receives two doses of zinc with each meal—one from foods and the other from the zinc-rich pancreatic secretions. The circulation of zinc in the body from the pancreas to the intestine and back to the pancreas is referred to as the **enteropancreatic circulation** of zinc. As this zinc circulates through the intestine, it may be refused entry by the intestinal cells or retained in them on any of its times around (see Figure 13-5).

• **Zinc Transport by Albumin** • Zinc's main transport vehicle in the blood is the protein albumin, which is a major determinant of zinc absorption. This may account for observations that zinc absorption declines in conditions that lower albumin concentrations—for example, pregnancy and protein-energy malnutrition.

• **Zinc Interactions with Iron and Copper** • Some plasma zinc also binds to transferrin—the same transferrin that carries iron in the blood. In healthy individuals, transferrin is usually less than 50 percent saturated with iron, but in iron overload, it is more saturated. Dietary iron-to-zinc ratios of greater than 2 to 1 leave too few transferrin-binding sites available for zinc and thereby interfere with zinc absorption. The converse is also true: large doses of zinc inhibit iron absorption.

Large doses of zinc create a similar problem with another essential mineral, copper. One possible explanation is that when zinc intakes are high, the intestinal cells synthesize large amounts of the binding protein metallothionein, which binds copper more strongly than zinc. With copper captured in a nonabsorbable form, absorption is limited and a deficiency may result. Another possible explanation is that copper and zinc compete for receptors on the intestinal cells. These nutrient interactions highlight one of the many reasons why people should use supplements conservatively, if at all: supplementation can easily create imbalances.

• **Zinc Losses** • Zinc exits the body primarily in feces. Smaller losses occur in urine, shed skin, hair, sweat, menstrual fluids, and semen.

Zinc Deficiency

Human zinc deficiency was first reported in the 1960s in children and adolescent boys in Egypt, Iran, and Turkey. Children have especially high zinc needs because they are growing rapidly and synthesizing many zinc-containing proteins; the native diets among those populations were not meeting these needs. Middle Eastern diets are typically low in the richest zinc source, meats, and the staple foods are legumes, unleavened breads, and other whole-grain foods—all high in fiber and phytates, which inhibit zinc absorption.*

*Unleavened bread contains no yeast, which normally breaks down phytates during fermentation.

metallothionein (meh-TAL-oh-THIGH-oh-neen): a sulfur-rich protein that avidly binds with metals such as zinc.

- **metallo** = containing a metal
- **thio** = containing sulfur
- **ein** = a protein

enteropancreatic (EN-ter-oh-PAN-kree-AT-ik) **circulation:** the circulatory route from the pancreas to the intestine and back to the pancreas.

FIGURE 13-5 Zinc's Routes in the Body

Notice the enteropancreatic circulation of zinc from the intestines through the blood to the pancreas and back to the intestines.

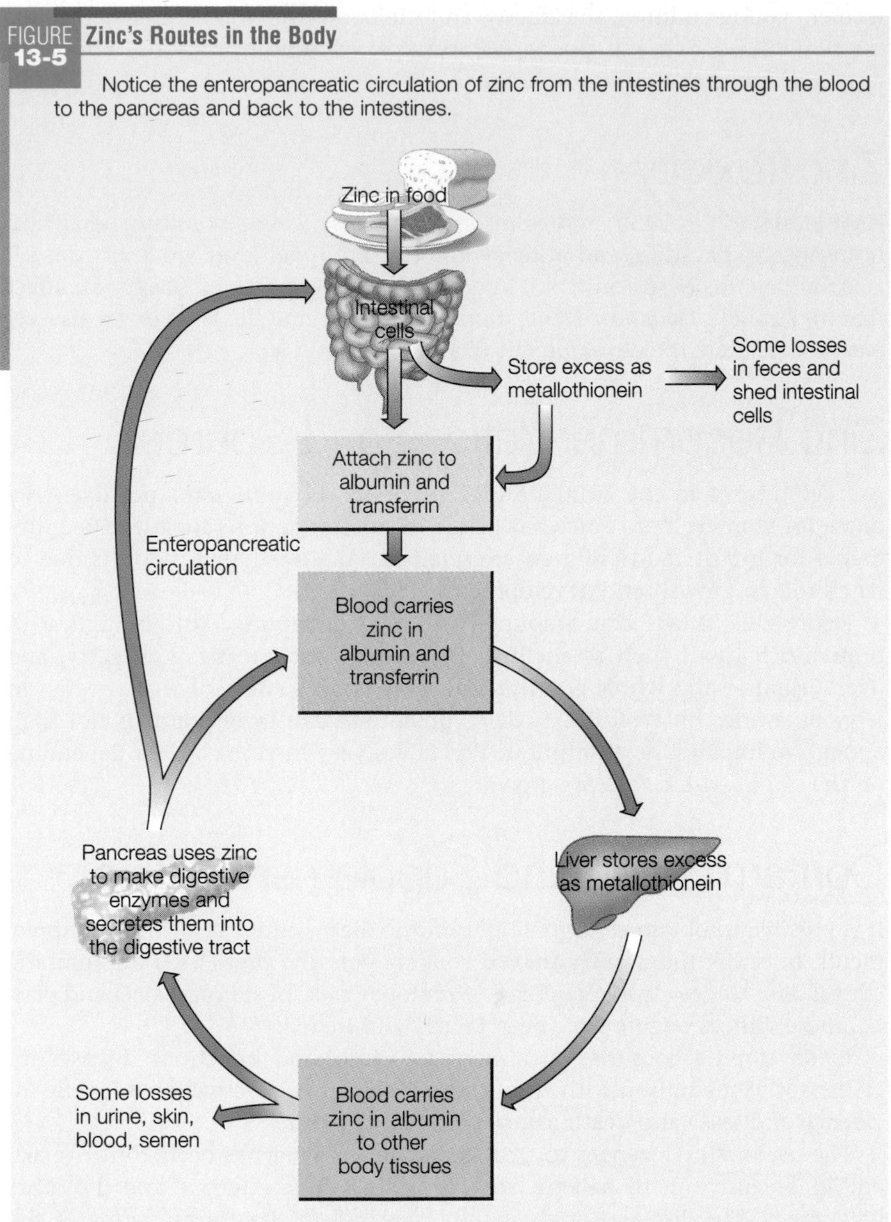

FIGURE 13-6 Zinc-Deficiency Symptom—The Stunted Growth of Dwarfism

The Egyptian man on the right is an adult of average height. The Egyptian boy on the left is 17 years old but is only 4 feet tall, like a 7-year-old in the United States. His genitalia are like those of a 6-year-old. The growth retardation, known as dwarfism, is rightly ascribed to zinc deficiency because it is partially reversible when zinc is restored to the diet.

• **Zinc-Deficiency Symptoms** • Figure 13-6 shows the severe growth retardation and arrested sexual maturation that characterize zinc deficiency. In addition, zinc deficiency hinders digestion and absorption, causing diarrhea, which worsens malnutrition not only for zinc, but for all nutrients. It impairs the immune response, making infections likely—among them, infections of the GI tract, which worsen malnutrition, including zinc malnutrition (a classic downward spiral of events). Chronic zinc deficiency damages the central nervous system and brain and may lead to poor motor development and cognitive performance.[24] Because zinc deficiency directly impairs vitamin A metabolism, vitamin A–deficiency symptoms often appear. Zinc deficiency also disturbs thyroid function and the metabolic rate. It alters taste, causes anorexia, and slows wound healing—in fact, its symptoms are so all-pervasive that generalized malnutrition and sickness are more likely to be the diagnosis than simple zinc deficiency.

• **Vulnerable Stages of Life** • Severe zinc deficiencies are not widespread in developed countries, but they do occur in vulnerable groups—pregnant

© Polara Studios Inc.

Zinc is highest in protein-rich foods such as oysters, beef, poultry, legumes, and nuts.

women, young children, the elderly, and the poor. Even a mild zinc deficiency can result in poor growth, poor appetite, impaired immune response, abnormal taste, and abnormal vision in darkness.

Zinc Toxicity

High doses (50 to 450 milligrams) of zinc may cause vomiting, diarrhea, headaches, exhaustion, and other symptoms. An upper level for adults was set at 40 milligrams based on zinc's interference in copper metabolism—an effect that, in animals, leads to degeneration of the heart muscle. High doses also appear to accelerate the development of atherosclerosis.

Zinc Recommendations and Sources

Average intakes in the United States are about 11 milligrams per day, adequate for women, but somewhat below recommendations for men. Requirements for infants and children are relatively higher than for adults due to zinc's role in growth and development.

Figure 13-7 shows zinc amounts in foods per serving. Zinc is highest in protein-rich foods such as shellfish (especially oysters), meats, poultry, and liver. Legumes and whole-grain products are good sources of zinc if eaten in large quantities; in typical U.S. diets, phytate intake from grains is not high enough to impair zinc absorption. Vegetables vary in zinc content depending on the soil in which they are grown.

Contamination and Supplemental Zinc

It is possible to obtain enough zinc from the diet without resorting to supplements. In earlier times, **galvanized** cooking pots and pipes used in plumbing contributed zinc to people's intakes. With today's use of stainless steel and plastic, these sources of zinc have been largely eliminated.

In developing countries, treatment for childhood infectious diseases includes supplementation with zinc. Zinc supplements effectively reduce the incidence of disease and death associated with diarrhea.[25]

The use of zinc lozenges to treat the common cold has been controversial and inconclusive, with half the studies finding them effective and the other half not.[26] The different study results may reflect the formulations of the lozenges more than the effectiveness of zinc itself. In general, studies using zinc gluconate report positive findings, whereas those using other combinations of zinc or zinc gluconate bound to a flavor-enhancing chelator report negative results. Zinc gluconate shortens the duration of cold symptoms such as coughing, nasal congestion, headache, and sore throat.[27] Common side effects of zinc lozenges include nausea and bad taste reactions.

IN SUMMARY

Zinc-requiring enzymes participate in a multitude of reactions affecting growth, vitamin A activity, and pancreatic digestive enzyme synthesis, among others. Both dietary zinc and zinc-rich pancreatic secretions (via enteropancreatic circulation) are available for absorption. Absorption is monitored by a special binding protein (metallothionein) in the intestine. Protein-rich foods derived from animals are the best sources of bioavailable zinc. Fiber and phytates in cereals bind zinc, limiting absorption. Growth retardation and sexual immaturity are hallmark symptoms of zinc deficiency. These facts and others are included in the accompanying table.

galvanized: a term referring to metals that have been treated with a zinc-containing coating to prevent rust.

Zinc

2001 RDA

Men: 11 mg/day

Women: 8 mg/day

Upper Level

Adults: 40 mg/day

Chief Functions in the Body

Part of many enzymes; associated with the hormone insulin; involved in making genetic material and proteins, immune reactions, transport of vitamin A, taste perception, wound healing, the making of sperm, and the normal development of the fetus

Significant Sources

Protein-containing foods: meats, fish, poultry, whole grains, vegetables

Deficiency Symptoms[a]	Toxicity Symptoms
Blood Symptoms	
High ammonia, low alkaline phosphatase, low insulin	Anemia: reduced hemoglobin production
Bone Symptoms	
Growth retardation, abnormal collagen synthesis	Growth in length, but without normal zinc content
Cellular/Metabolic Symptoms	
Slow DNA synthesis, impaired cell division and protein synthesis	Raised LDL, lowered HDL
Digestive Symptoms	
Weak sense of smell, poor sensitivity to the taste of salt, weight loss, delayed glucose absorption, diarrhea, nausea, impaired appetite	Diarrhea, vomiting, loss of appetite; decreased calcium and copper absorption
Eye Symptoms	
Night blindness; lesions	
Glandular Symptoms	
Delayed onset of puberty, small gonads in males, decreased synthesis and release of testosterone, abnormal glucose tolerance, reduced synthesis of adrenocortical hormones, altered thyroid function	
Immune System Symptoms	
Altered skin test responses, low white blood cell count, few antibody-forming cells, thymus atrophy, susceptibility to infection	Fever, elevated white blood cell count; suppressed response
Kidney Symptoms	
	Renal failure
Liver/Spleen Symptoms	
Enlargement	
Neurological/Muscular Symptoms	
Anorexia (poor appetite), mental lethargy, irritability	Muscular pain and incoordination, heart muscle degeneration, exhaustion, dizziness, drowsiness, headaches
Reproductive Symptoms	
Impaired reproductive function (rats), low sperm counts	
Skin Symptoms	
Generalized hair loss; lesions; rough, dry appearance; slow healing of wounds and burns	

[a]A rare inherited disease of zinc malabsorption, *acrodermatitis* (AK-roh-der-ma-TIE-tis) *enteropathica* (EN-ter-oh-PATH-ick-ah), causes additional and more severe symptoms.

IODINE

Traces of the iodine ion (called iodide)◆ are indispensable to life. In the GI tract, iodine from foods is converted to iodide; this chapter uses *iodine* when referring to the nutrient in foods and *iodide* when referring to it in the body. Iodide occurs in the body in a tiny quantity, but its principal role in human nutrition is well known, and the amount needed is well established.

• Iodide Roles in the Body • Iodide is an integral part of the thyroid hormone◆ that regulates body temperature, metabolic rate, reproduction, growth,

◆ The ion form of *iodine* is called *iodide.*

◆ The thyroid gland releases tetraiodothyronine (T_4), commonly known as thyroxin (thigh-ROCKS-in), to its target tissues. Upon reaching the cells, T_4 is deiodinated to triiodothyronine (T_3), which is the active form of the hormone.

FIGURE 13-7 Zinc in Selected Foods

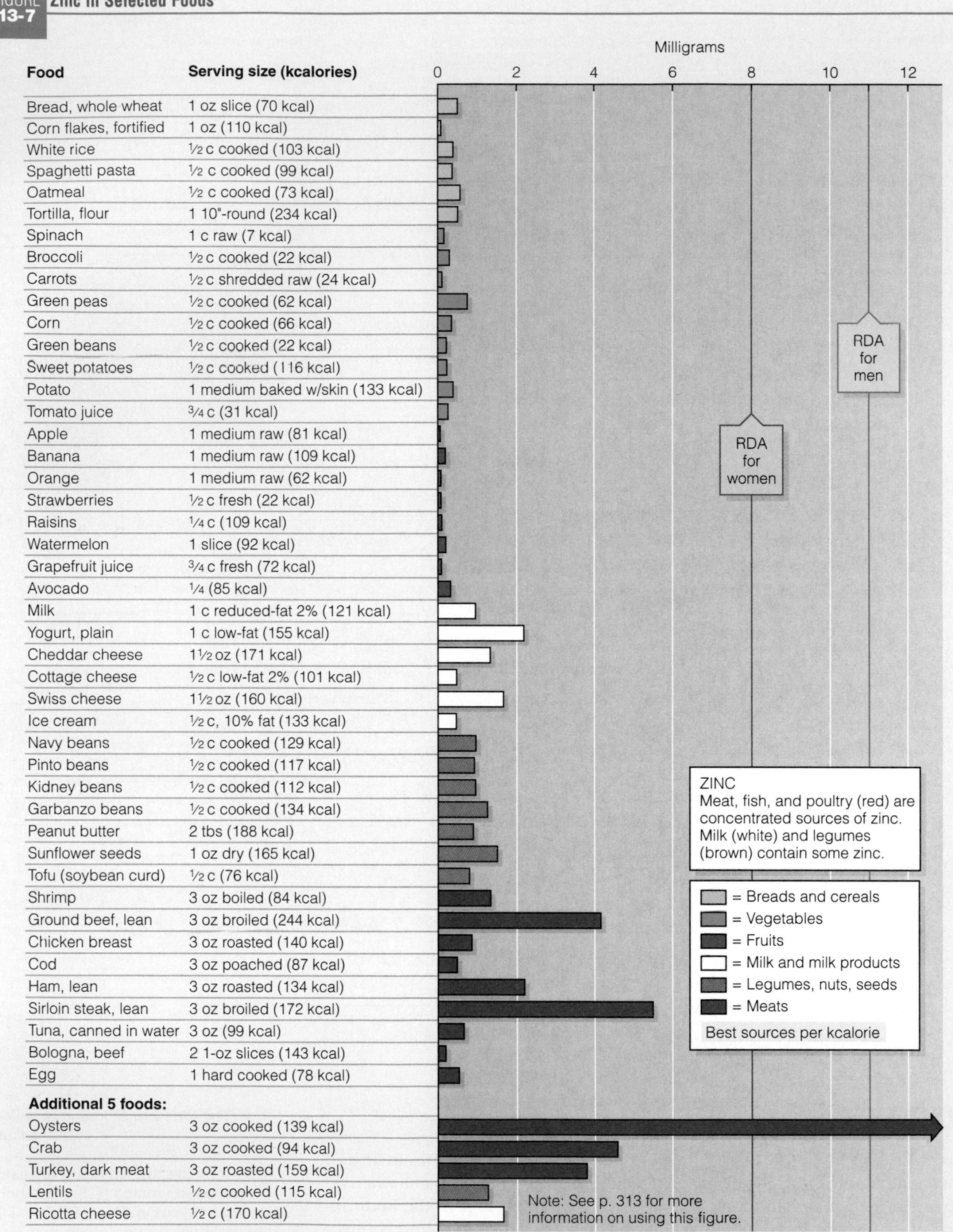

Food	Serving size (kcalories)
Bread, whole wheat	1 oz slice (70 kcal)
Corn flakes, fortified	1 oz (110 kcal)
White rice	½ c cooked (103 kcal)
Spaghetti pasta	½ c cooked (99 kcal)
Oatmeal	½ c cooked (73 kcal)
Tortilla, flour	1 10"-round (234 kcal)
Spinach	1 c raw (7 kcal)
Broccoli	½ c cooked (22 kcal)
Carrots	½ c shredded raw (24 kcal)
Green peas	½ c cooked (62 kcal)
Corn	½ c cooked (66 kcal)
Green beans	½ c cooked (22 kcal)
Sweet potatoes	½ c cooked (116 kcal)
Potato	1 medium baked w/skin (133 kcal)
Tomato juice	¾ c (31 kcal)
Apple	1 medium raw (81 kcal)
Banana	1 medium raw (109 kcal)
Orange	1 medium raw (62 kcal)
Strawberries	½ c fresh (22 kcal)
Raisins	¼ c (109 kcal)
Watermelon	1 slice (92 kcal)
Grapefruit juice	¾ c fresh (72 kcal)
Avocado	¼ (85 kcal)
Milk	1 c reduced-fat 2% (121 kcal)
Yogurt, plain	1 c low-fat (155 kcal)
Cheddar cheese	1½ oz (171 kcal)
Cottage cheese	½ c low-fat 2% (101 kcal)
Swiss cheese	1½ oz (160 kcal)
Ice cream	½ c, 10% fat (133 kcal)
Navy beans	½ c cooked (129 kcal)
Pinto beans	½ c cooked (117 kcal)
Kidney beans	½ c cooked (112 kcal)
Garbanzo beans	½ c cooked (134 kcal)
Peanut butter	2 tbs (188 kcal)
Sunflower seeds	1 oz dry (165 kcal)
Tofu (soybean curd)	½ c (76 kcal)
Shrimp	3 oz boiled (84 kcal)
Ground beef, lean	3 oz broiled (244 kcal)
Chicken breast	3 oz roasted (140 kcal)
Cod	3 oz poached (87 kcal)
Ham, lean	3 oz roasted (134 kcal)
Sirloin steak, lean	3 oz broiled (172 kcal)
Tuna, canned in water	3 oz (99 kcal)
Bologna, beef	2 1-oz slices (143 kcal)
Egg	1 hard cooked (78 kcal)
Additional 5 foods:	
Oysters	3 oz cooked (139 kcal)
Crab	3 oz cooked (94 kcal)
Turkey, dark meat	3 oz roasted (159 kcal)
Lentils	½ c cooked (115 kcal)
Ricotta cheese	½ c (170 kcal)

Note: See p. 313 for more information on using this figure.

blood cell production, nerve and muscle function, and more. By controlling the rate at which the cells use oxygen, this hormone influences the amount of energy released during basal metabolism.

• Iodine Deficiency • The hypothalamus regulates thyroid hormone production by controlling the release of the pituitary's thyroid-stimulating hormone (TSH). With iodine deficiency, thyroid hormone production declines, and the body responds by secreting more TSH in a futile attempt to accelerate iodide uptake by the thyroid gland. If a deficiency persists, the cells of the thyroid gland enlarge, so as to trap as much iodide as possible. Sometimes the gland enlarges until it makes a visible lump in the neck, a simple **goiter** (shown in Figure 13-8).

Goiter afflicts about 200 million people the world over, many of them in South America, Asia, and Africa. In all but 4 percent of these cases, the cause is iodine deficiency. As for the 4 percent (8 million), most have goiter because they overconsume plants of the cabbage family and other foods that contain an antithyroid substance **(goitrogen)** whose effect is not counteracted by dietary iodine. The goitrogens present in plants remind us that even natural components of foods can cause harm when eaten in excess.

An iodine deficiency causes sluggishness and weight gain. During pregnancy, it impairs fetal development, causing the extreme and irreversible mental and physical retardation known as **cretinism.**◆ Cretinism affects approximately 6 million people worldwide and can be averted by the early diagnosis and treatment of maternal iodine deficiency. If treatment comes too late or not at all, the child may live his or her entire life with an IQ as low as 20 (100 is average).[28] Children with even a mild iodine deficiency typically have goiters and perform poorly in school.[29]

• Iodine Toxicity • Excessive intakes of iodine can enlarge the thyroid gland, just as deficiency can. During pregnancy, exposure to excessive iodine from foods, prenatal supplements, or medications is especially damaging to the developing infant. An infant exposed to toxic amounts of iodine during gestation may develop a goiter so severe as to block the airways and cause suffocation. The upper level is 1000 micrograms per day for an adult—several times higher than average intakes.

• Iodine Sources • The ocean is the world's major source of iodine. In coastal areas, seafood, water, and even iodine-containing sea mist are dependable iodine sources. Further inland, the amount of iodine in foods is variable and generally reflects the amount present in the soil in which plants are grown or on which animals graze. Landmasses that were once under the ocean have soils rich in iodine; those in flood-prone areas where water leaches iodine from the soil are poor in iodine. In the United States and Canada, the iodization of salt has eliminated the widespread misery caused by iodine deficiency during the 1930s, but iodized salt is not available in many parts of the world. Some countries add iodine to bread, fish paste, or drinking water instead.

• Iodine Intakes • Average consumption of iodine in the United States exceeds recommendations, but falls below toxic levels as well. Some of the excess iodine in the U.S. diet stems from fast foods, which use iodized salt liberally. Some iodine comes from bakery products and from milk. The baking industry uses iodates (iodine salts) as dough conditioners, and most dairies feed cows iodine-containing medications and use iodine to disinfect milking equipment. Now that these sources have been identified, food industries have reduced their use of these compounds, but the sudden emergence of this problem points to a need for continued surveillance of the food supply. Processed foods in the United States do not use iodized salt.

• Iodine Recommendations • The recommended intake of iodine for adults is a minuscule amount. The need for iodine is easily met by consuming

FIGURE 13-8 Iodine-Deficiency Symptom—The Enlarged Thyroid of Goiter

In iodine deficiency, the thyroid gland enlarges—a condition known as simple goiter.

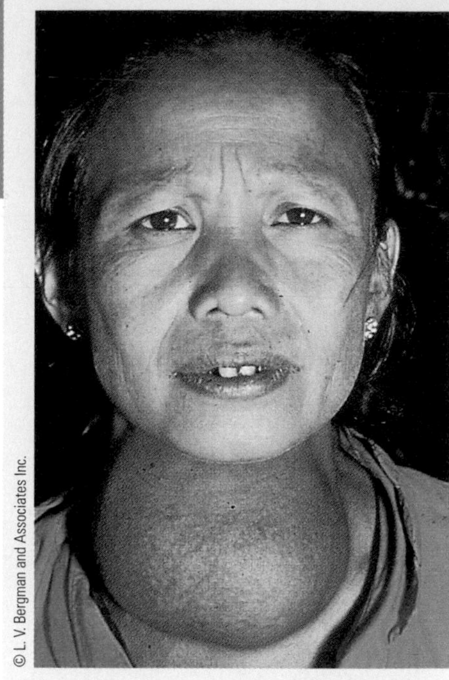

© L. V. Bergman and Associates Inc.

◆ **The underactivity of the thyroid gland is known as *hypothyroidism* and may be caused by iodine deficiency or any number of other causes. Without treatment, an infant with *congenital hypothyroidism* will develop the physical and mental retardation of *cretinism.***

goiter (GOY-ter): an enlargement of the thyroid gland due to an iodine deficiency, malfunction of the gland, or overconsumption of a goitrogen. Goiter caused by iodine deficiency is **simple goiter.**

goitrogen (GOY-troh-jen): a thyroid antagonist found in food; causes **toxic goiter.** Goitrogens are found in such foods as cabbage, kale, brussels sprouts, cauliflower, broccoli, and kohlrabi.

cretinism (CREE-tin-ism): a congenital disease characterized by mental and physical retardation and commonly caused by maternal iodine deficiency during pregnancy.

Only "iodized salt" has had iodine added.

seafood, vegetables grown in iodine-rich soil, and iodized salt.◆ In the United States, labels indicate whether salt is iodized; in Canada, all table salt is iodized.

◆ The FDA requires that iodized salt contain 45 to 76 µg iodine per gram salt. On average, ½ tsp iodized salt provides the RDA for iodine.

IN SUMMARY

Iodide, the ion of the mineral iodine, is an essential component of the thyroid hormone. An iodine deficiency can lead to simple goiter—enlargement of the thyroid gland—and can impair fetal development, causing cretinism. Iodization of salt has largely eliminated iodine deficiency in the United States and Canada. The table provides a summary of iodine.

Iodine

2001 RDA	Deficiency Disease	
Adults: 150 µg/day	Simple goiter, cretinism	
Upper Level	**Deficiency Symptoms**	**Toxicity Symptoms**
1.1 mg/day	Enlargement of the thyroid gland, weight gain, mental and physical retardation of an infant	Enlargement of the thyroid gland, depressed thyroid activity
Chief Functions in the Body		
A component of two thyroid hormones that help to regulate growth, development, and metabolic rate		
Significant Sources		
Iodized salt, seafood, bread, dairy products, plants grown in iodine-rich soil and animals fed those plants		

SELENIUM

The essential mineral **selenium** is one of the body's antioxidant nutrients, working primarily as a part of the enzyme glutathione peroxidase. Glutathione peroxidase and vitamin E work in concert. Glutathione peroxidase prevents free-radical formation, thus blocking the chain reaction before it begins; if free radicals do form and a chain reaction starts, vitamin E stops it. (Highlight 11 describes free-radical formation, chain reactions, and antioxidant action in detail.) An enzyme that converts the thyroid hormone to its active form also contains selenium.

• Selenium Deficiency • Selenium deficiency is associated with a heart disease◆ that is prevalent in regions of China where the soil and foods are lacking selenium. The primary cause of this heart disease is probably a virus, but selenium deficiency appears to predispose people to it, and adequate selenium seems to prevent it.

◆ The heart disease associated with selenium deficiency is named **Keshan** (KESH-an or ka-SHAWN) **disease** for one of the provinces of China where it was studied. Keshan disease is characterized by heart enlargement and insufficiency; fibrous tissue replaces the muscle tissue that normally composes the middle layer of the walls of the heart.

• Selenium and Cancer • In other parts of the world, selenium-poor soil correlates with a high incidence of certain kinds of cancer. Some research suggests that selenium supplements may reduce the incidence of some types of cancers, but given the potential for harm and the lack of additional evidence, recommendations to take selenium supplements would be premature.[30]

• Selenium Intakes • The soil in many regions of the United States and Canada contains selenium. People living in regions with selenium-poor soil may still get enough selenium, partly because they eat vegetables and grains transported from other regions and partly because they eat meats and other animal products, which are reliable sources of selenium.

• Selenium Toxicity • High doses (a milligram or more daily) of selenium are toxic. Selenium toxicity causes vomiting, diarrhea, loss of hair and nails, and lesions of the skin and nervous system.

selenium (se-LEEN-ee-um): a trace element.

IN SUMMARY

Selenium is an antioxidant nutrient that works closely with the glutathione peroxidase enzyme and vitamin E. Selenium is found in association with protein in foods. Deficiencies are associated with a predisposition to a type of heart disease and possibly with some kinds of cancer. See the table below for a summary of selenium.

Selenium

2000 RDA	Chief Functions in the Body	Deficiency Symptoms	Toxicity Symptoms	Significant Sources
Adults: 55 µg/day	Defends against oxidation; regulates thyroid hormone	Predisposition to heart disease characterized by cardiac tissue becoming fibrous (Keshan disease)	Digestive system disorders, loss of hair and nails, skin lesions, nervous system disorders	Seafood, meat, whole grains, vegetables (depending on soil content)
Upper Level				
Adults: 400 µg/day				

COPPER

The body contains about 100 milligrams of copper. It is found in a variety of cells and tissues.

• Copper Roles in the Body • Copper serves as a constituent of several enzymes.[31] The copper-containing enzymes have diverse metabolic roles with one common characteristic: all involve reactions that consume oxygen or oxygen radicals. For example, copper-containing enzymes catalyze the oxidation of ferrous iron to ferric iron.* Copper's role in iron metabolism makes it a key factor in hemoglobin synthesis. Two copper- and zinc-containing enzymes participate in the body's natural defense against free radicals.† Still another copper enzyme helps to manufacture collagen and heal wounds.‡ Copper, like iron, is needed in many of the metabolic reactions related to the release of energy.§

• Copper Deficiency and Toxicity • Copper deficiency is rare, but has been seen in premature infants and malnourished children.[32] Dietary factors such as vitamin C interfere with copper absorption and can lead to deficiency.[33] Copper deficiency in animals raises blood cholesterol and damages blood vessels, raising questions about whether low dietary copper might contribute to cardiovascular disease in humans. Typical U.S. diets provide adequate amounts. Some genetic disorders create a copper toxicity, but excessive intakes from foods are unlikely.[34]

Two rare genetic disorders affect copper status in opposite directions. In Menkes disease, the intestinal cells absorb copper, but cannot release it into circulation, causing a life-threatening deficiency. In Wilson's disease, copper accumulates in the liver and brain, creating a life-threatening toxicity. Wilson's disease can be controlled by reducing copper intake, using chelating agents such as penicillamine, and taking zinc supplements, which interfere with copper absorption.

• Copper Recommendations and Intakes • The richest food sources of copper are legumes, whole grains, nuts, shellfish, organ meats, and seeds. Over half of the copper from foods is absorbed, and the major route of elimination appears to be bile.[35] Water may also provide copper, depending on the type of plumbing pipe and the hardness of the water.[36]

---

*The copper-containing enzyme *ceruloplasmin* participates in the oxidation of ferrous iron to ferric iron.
†Two copper-containing *superoxide dismutase* enzymes defend against free radicals.
‡The copper-containing enzyme *lysyl oxidase* helps synthesize connective tissues.
§The copper-containing enzyme *cytochrome C oxidase* participates in the electron transport chain.

IN SUMMARY

Copper is a component of several enzymes, all of which are involved in some way with oxygen or oxidation. Some act as antioxidants; others are essential to iron metabolism. Legumes, whole grains, and shellfish are good sources of copper. See the table for a summary of copper facts.

Copper

2001 RDA	Chief Functions in the Body	Deficiency Symptoms	Toxicity Symptoms	Significant Sources
Adults: 900 µg/day **Upper Level** Adults: 10 mg/day	Necessary for the absorption and use of iron in the formation of hemoglobin; part of several enzymes	Anemia, bone abnormalities	Vomiting, diarrhea; liver damage	Seafood, nuts, whole grains, seeds, legumes

MANGANESE

The human body contains a tiny 20 milligrams of manganese, mostly in the bones and metabolically active organs such as the liver, kidneys, and pancreas. Manganese acts as a cofactor for many enzymes that facilitate dozens of different metabolic processes. In addition, manganese-containing metalloenzymes assist in urea synthesis, the conversion of pyruvate to a TCA cycle compound, and the prevention of damage by free radicals.

• **Manganese Deficiency** • Manganese requirements are low, and many plant foods contain significant amounts of this trace mineral, so deficiencies are rare.[37] As is true of other trace minerals, however, dietary factors such as phytates inhibit its absorption.[38] In addition, high intakes of iron and calcium limit manganese absorption, so people who use supplements of these minerals regularly may impair their manganese status.

• **Manganese Toxicity** • Toxicity is more likely to occur from an environment contaminated with manganese than from dietary intake. Miners who inhale large quantities of manganese dust on the job over prolonged periods show symptoms of a brain disease, along with abnormalities in appearance and behavior.

IN SUMMARY

Manganese-dependent enzymes are involved in various metabolic processes. Manganese is widespread in plant foods, so deficiencies are rare, although regular use of calcium and iron supplements may limit manganese absorption. A summary of manganese appears in the table below.

Manganese

2001 AI	Chief Functions in the Body	Deficiency Symptoms	Toxicity Symptoms	Significant Sources
Men: 2.3 mg/day Women: 1.8 mg/day **Upper Level** Adults: 11 mg/day	Cofactor for several enzymes	In experimental animals: poor growth, nervous system disorders, reproductive abnormalities	Nervous system disorders	Nuts, whole grains, leafy vegetables

FLUORIDE

Fluoride is present in virtually all soils, water supplies, plants, and animals. Only a trace of fluoride occurs in the human body, but with this amount, the crystalline deposits in bones and teeth are larger and more perfectly formed.

• Fluoride Roles in the Body • During the mineralization of bones and teeth, calcium and phosphorus form crystals called hydroxyapatite. Then fluoride replaces the hydroxyl (OH) portions of the hydroxyapatite crystal, forming **fluorapatite,** which makes the bones stronger and the teeth more resistant to decay.

• Fluoridation and Dental Caries • Dental caries ranks as the nation's most widespread health problem: an estimated 95 percent of the population have decayed, missing, or filled teeth. By interfering with a person's ability to chew and eat a wide variety of foods, these dental problems can quickly lead to a multitude of nutrition problems. Where fluoride is lacking, dental decay is common. Drinking water is usually the best source of fluoride. Fluoridation of drinking water to raise the concentration to 1 part fluoride per 1 million◆ parts water offers the greatest protection against dental caries at virtually no risk of toxicity.[39] By fluoridating the drinking water, a community offers its residents, particularly the children, a safe, economical, practical, and effective way to defend against dental caries.[40] The average cost, per person, of fluoridating drinking water for a lifetime is less than $40, with an estimated savings of as much as $3000 in treatment costs.

• Fluoride Toxicity • Fluoride poisoning has been reported in communities where the public water system failed and allowed fluoride concentrations to reach 150 parts per million. Symptoms of fluoride poisoning include nausea, vomiting, diarrhea, abdominal pain, and numbness or tingling of the face and extremities.

• Fluorosis • Too much fluoride can damage the teeth, causing **fluorosis.** In mild cases, the teeth develop small white specks; in severe cases, the enamel becomes pitted and permanently stained (as shown in Figure 13-9). Fluorosis occurs only during tooth development and cannot be reversed, making its prevention◆ a high priority.

• Fluoride Intakes • Over half of the U.S. population has access to water with an optimal fluoride concentration, which typically delivers about 1 milligram per person per day. Fish and most teas contain appreciable amounts of natural fluoride.

HEALTHY PEOPLE 2010

Increase the proportion of the U.S. population served by community water systems with optimally fluoridated water.

FIGURE 13-9 Fluoride-Toxicity Symptom—The Mottled Teeth of Fluorosis

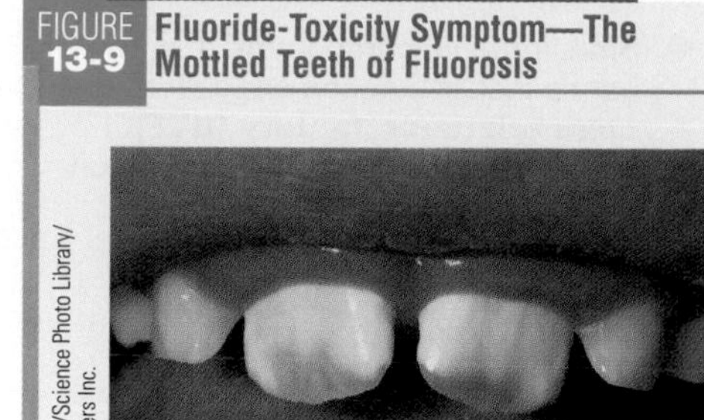

© Dr. P. Marrazi/Science Photo Library/Photo Researchers Inc.

◆ For perspective, 1 part per million (1 ppm) is approximately 1 mg per liter.

◆ To prevent fluorosis:
- Monitor the fluoride content of the local water supply.
- Supervise toddlers when they brush their teeth and use only a little toothpaste (pea-size amount).
- Use fluoride supplements only as prescribed by a physician.

fluorosis (floor-OH-sis): discoloration and pitting of tooth enamel caused by excess fluoride during tooth development.

IN SUMMARY

Fluoride makes bones stronger and teeth more resistant to decay. Fluoridation of public water supplies can significantly reduce the incidence of dental caries, but an excess of fluoride during tooth development can cause fluorosis—discolored and pitted tooth enamel. The table below summarizes fluoride information.

Fluoride

1997 AI	Upper Level	Chief Functions in the Body	Deficiency Symptoms	Toxicity Symptoms	Significant Sources
Men: 3.8 mg/day Women: 3.1 mg/day	Adults: 10 mg/day	Involved in the formation of bones and teeth; helps to make teeth resistant to decay	Susceptibility to tooth decay	Fluorosis (discoloration of teeth), nausea, diarrhea, chest pain, itching, vomiting	Drinking water (if fluoride containing or fluoridated), tea, seafood

CHROMIUM

Chromium is an essential mineral that participates in carbohydrate and lipid metabolism. Like iron, chromium can have different charges. In the case of chromium, the Cr^{+++} ion seems to be the most effective.

fluorapatite (floor-APP-uh-tite): the stabilized form of bone and tooth crystal, in which fluoride has replaced the hydroxyl groups of hydroxyapatite.

◆ Small organic compounds that enhance insulin's action are called **glucose tolerance factors (GTF).** Some glucose tolerance factors contain chromium.

• Chromium Roles in the Body • Chromium helps maintain glucose homeostasis by enhancing the activity of the hormone insulin.[41]◆ Consequently, less insulin is needed to control blood glucose. When chromium is lacking, a diabeteslike condition may develop with elevated blood glucose and impaired glucose tolerance, insulin response, and glucagon response.

• Chromium Recommendations and Intakes • Chromium is present in a variety of foods. The best sources are unrefined foods, particularly liver, brewer's yeast, whole grains, nuts, and cheeses. The more refined foods people eat, the less chromium they ingest.

• Chromium Picolinate Supplements • Supplement advertisements have succeeded in convincing consumers that they can lose fat and build muscle by taking chromium picolinate. Whether chromium—picolinate or plain—supplements reduce body fat or improve muscle strength remains controversial.[42]

IN SUMMARY

Chromium enhances insulin's action. A deficiency can result in a diabetes-like condition. Chromium is widely available in unrefined foods including brewer's yeast, whole grains, and liver. The table below provides a summary of chromium.

Chromium

2001 AI	Chief Functions in the Body	Deficiency Symptoms	Toxicity Symptoms	Significant Sources
Men: 35 µg/day Women: 25 µg/day	Associated with insulin and required for the release of energy from glucose	Diabeteslike condition marked by an inability to use glucose normally	None reported	Meat, unrefined foods, fats, vegetable oils

MOLYBDENUM

Molybdenum acts as a working part of several metalloenzymes. Dietary deficiencies of molybdenum are unknown because the amounts needed are minuscule—as little as 0.1 part per million parts of body tissue. Legumes, breads and other grain products, leafy green vegetables, milk, and liver are molybdenum-rich foods. Average daily intakes fall within the suggested range of intakes.

Molybdenum toxicity is rare, but has been reported in animal studies. Characteristics include kidney damage and reproductive abnormalities. For a summary of molybdenum facts, see the accompanying table.

IN SUMMARY

Molybdenum

2001 RDA	Chief Functions in the Body	Deficiency Symptoms	Toxicity Symptoms	Significant Sources
Adults: 45 µg/day	Cofactor for several enzymes	Unknown	None reported; reproductive effects in animals	Legumes, cereals, organ meats
Upper Level				
Adults: 2 mg/day				

OTHER TRACE MINERALS

Research to determine whether other trace minerals are essential is difficult, both because their quantities in the body are so small and because human deficiencies

molybdenum (mo-LIB-duh-num): a trace element.

are unknown. Guessing their functions in the body can be particularly problematic. Much of the available knowledge comes from research using animals.

Nickel may serve as a cofactor for certain enzymes, but deficiencies are unknown. Silicon is involved in the formation of bones and collagen. Vanadium, too, is necessary for growth and bone development and also for normal reproduction. Cobalt is a key mineral in the large vitamin B_{12} molecule (see Figure 13-10), but it is not an essential nutrient and no recommendation has been established. Boron may play a key role in brain activities. In the future many other trace minerals may turn out to play key nutritional roles. Even arsenic—famous as a poison used by murderers and known to be a carcinogen—may turn out to be essential for human beings in tiny quantities; it has already proved useful in the treatment of some types of leukemia.[43]

FIGURE 13-10 Cobalt with Vitamin B_{12}

The intricate vitamin B_{12} molecule contains one atom of the mineral cobalt. The alternative name for vitamin B_{12}, cobalamin, reflects the presence of cobalt in its structure.

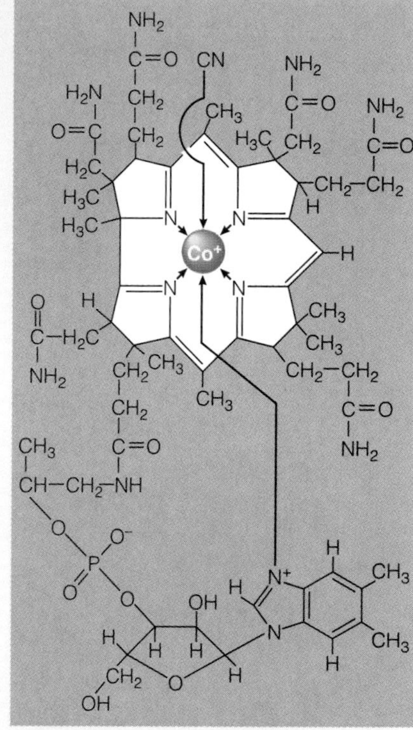

CONTAMINANT MINERALS

Chapter 12 and this chapter have told of the many ways minerals serve the body—maintaining fluid and electrolyte balance, providing structural support to the bones, transporting oxygen, and assisting enzymes. In contrast to those minerals that the body requires, contaminant minerals impair the body's growth, work capacity, and general health. Contaminant minerals include the **heavy metals** lead, mercury, and others that enter the food supply by way of soil, water, and air pollution. This section focuses on lead poisoning because it is the most serious environmental threat to young children, but all contaminant minerals disrupt body processes and impair nutrition status similarly.

Like other minerals, lead is indestructible; the body cannot change its chemistry. Chemically similar to nutrient minerals like iron, calcium, and zinc (cations with two positive charges), lead displaces them from some of the metabolic sites they normally occupy, but is then unable to perform their roles. For example, lead competes with iron in heme, but then cannot carry oxygen; similarly, lead competes with calcium in the brain, but then cannot signal messages from nerve cells. Excess lead in the blood also deranges the structure of red blood cell membranes, making them leaky and fragile. Lead interacts with white blood cells, too, impairing their ability to fight infection, and it binds to antibodies, thereby impairing the body's resistance to disease.

In addition to its effects on the blood, lead damages many body systems, particularly the vulnerable nervous system, kidneys, and bone marrow. It impairs such normal activities as growth by interfering with hormone activity. It interferes with tooth development and causes dental caries in rats, and it has been associated with an increased prevalence of dental caries in people as well.[44] In short, lead's interactions in the body have profound adverse effects.[45] The greater the exposure, the more damaging the effects. Table 13-1 lists symptoms of lead toxicity.

Lead typifies the ways all heavy metals behave in the body: they interfere with nutrients that are trying to do their jobs. The "good guy" nutrients are shoved aside by the "bad guy" contaminants. Then the contaminants cannot perform the roles of the nutrients, and health declines. To safeguard our health, we must defend ourselves against contamination by eating nutrient-rich foods and preserving a clean environment.

TABLE 13-1 Symptoms of Lead Toxicity

Symptoms of Lead Toxicity
In children:
• Learning disabilities (reduced short-term memory; impaired concentration)
• Low IQ
• Behavior problems
• Slow growth
• Iron-deficiency anemia
• Dental caries
• Sleep disturbances (night waking, restlessness, head banging)
• Nervous system disorders; seizures
• Slow reaction time; poor coordination
• Impaired hearing
In adults:
• Hypertension
• Reproductive complications
• Kidney failure

heavy metals: any of a number of mineral ions such as mercury and lead, so called because they are of relatively high atomic weight. Many heavy metals are poisonous.

CLOSING THOUGHTS ON THE NUTRIENTS

This chapter completes the introductory lessons on the nutrients. Each nutrient from the amino acids to zinc has been described rather thoroughly—its chemistry, roles in the body, sources in the diet, symptoms of deficiency and toxicity, and influences on health and disease. Such a detailed examination is

INTERNET ACTIVITIES

Iron-deficiency anemia is the world's most common nutrition deficiency disease, affecting about 15 percent of the world's population. The most vulnerable groups of individuals are women of childbearing age, infants, children, and teenagers. Many people have personal questions about iron and anemia. Perhaps some of these questions are ones that you have also.

- Go to: **www.dietitian.com**
- Scroll down and click on "Iron & Anemia."
- Read the questions submitted by interested consumers and the answers provided by registered dietitians.

Since anemia is such a prevalent problem, there is often confusion about its causes and preventive strategies. What did you learn about iron-deficiency anemia?

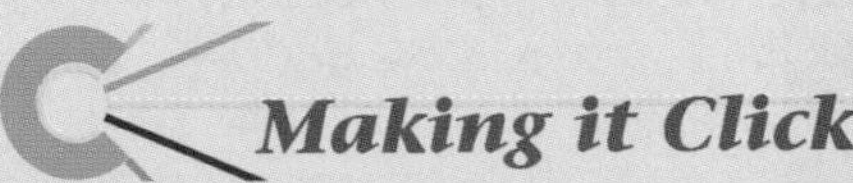

Making it Click

Once you have mastered these examples, you will understand minerals a little better and be prepared to examine your own food choices. Be sure to show your calculations for each problem. (see p. 456 for answers.)

1. For each of these minerals, note the unit of measure:
   Iron
   Zinc
   Iodine
   Selenium
   Copper
   Manganese
   Fluoride
   Chromium
   Molybdenum
2. Appreciate foods for their iron density. Following is a list of foods with the energy amount and the iron content per serving.
   a. Rank these foods by iron per serving.
   b. Calculate the iron density (divide milligrams by kcalories) for these foods and rank them by their iron per kcalorie.
   c. Name three foods that are higher on the second list than they were on the first list.
   d. What do these foods have in common?

Food	Iron (mg)	Energy (kcal)	Iron Density (mg/kcal)
Milk, nonfat, 1 c	0.10	85	
Cheddar cheese, 1 oz	0.19	114	
Broccoli, cooked from fresh, chopped, 1 c	1.31	44	
Sweet potato, baked in skin, 1 ea	0.51	117	
Cantaloupe melon, ½	0.56	93	
Carrots, from fresh, ½ c	0.48	35	
Whole-wheat bread, 1 slice	0.87	64	
Green peas, cooked from frozen, ½ c	1.26	62	
Apple, medium	0.38	125	
Sirloin steak, lean, 4 oz	3.81	228	
Pork chop, lean, broiled, 1 ea	0.66	166	

Study Questions

These questions will help you review the chapter. You will find the answers in the discussions on the pages provided.

1. Distinguish between heme and nonheme iron. Discuss the factors that enhance iron absorption. (pp. 430–431)
2. Distinguish between iron deficiency and iron-deficiency anemia. What are the symptoms of iron-deficiency anemia? (pp. 433–434)
3. What causes iron overload? What are its symptoms? (p. 435)
4. Describe the similarities and differences in the absorption and regulation of iron and zinc. (pp. 430–431, 439–440)

5. Discuss possible reasons for a low intake of zinc. What factors affect the bioavailability of zinc? (pp. 440, 442)
6. Describe the principal functions of iodide, selenium, copper, manganese, fluoride, chromium, and molybdenum in the body. (pp. 443, 446, 447, 448, 449, 450)
7. What public health measure has been used in preventing simple goiter? What measure has been recommended for protection against tooth decay? (pp. 445, 449)
8. Discuss the importance of balanced and varied diets in obtaining the essential minerals and avoiding toxicities. (pp. 429, 451–453)
9. Describe some of the ways trace minerals interact with each other and with other nutrients. (pp. 429, 451–453)

These questions will help you prepare for an exam. Answers can be found on p. 456.

1. Iron absorption is impaired by:
   a. heme.
   b. phytates.
   c. vitamin C.
   d. MFP factor.
2. Which of these people is *least* likely to develop an iron deficiency?
   a. 3-year-old boy
   b. 52-year-old man
   c. 17-year-old girl
   d. 24-year-old woman
3. Which of the following would *not* describe the blood cells of a severe iron deficiency?
   a. anemic
   b. microcytic
   c. pernicious
   d. hypochromic
4. Which provides the most absorbable iron?
   a. 1 apple
   b. 1 c milk
   c. 3 oz steak
   d. ½ c spinach
5. The intestinal protein that helps to regulate zinc absorption is:
   a. albumin.
   b. ferritin.
   c. hemosiderin.
   d. metallothionein.
6. A classic sign of zinc deficiency is:
   a. anemia.
   b. goiter.
   c. mottled teeth.
   d. growth retardation.
7. Cretinism is caused by a deficiency of:
   a. iron.
   b. zinc.
   c. iodine.
   d. selenium.
8. The mineral best known for its role as an antioxidant is:
   a. copper.
   b. selenium.
   c. manganese.
   d. molybdenum.
9. Fluorosis occurs when fluoride:
   a. is excessive.
   b. is inadequate.
   c. binds with phosphorus.
   d. interacts with calcium.
10. Which mineral enhances insulin activity?
    a. zinc
    b. iodine
    c. chromium
    d. manganese

References

1 G. J. Beckett and coauthors, Effects of combined iodine and selenium deficiency on thyroid hormone metabolism in rats, *American Journal of Clinical Nutrition* 57 (1993): 240S–243S.

2 N. C. Andrews, Disorders of iron metabolism, *New England Journal of Medicine* 341 (1999): 1986–1995; E. Beutler, How little we know about the absorption of iron, *American Journal of Clinical Nutrition* 66 (1997): 419–420.

3 J. L. Beard, H. Dawson, and D. J. Piñero, Iron metabolism: A comprehensive review, *Nutrition Reviews* 54 (1996): 295–317.

4 M. B. Reddy, R. F. Hurrell, and J. D. Cook, Estimation of nonheme-iron bioavailability from meal composition, *American Journal of Clinical Nutrition* 71 (2000): 937–943.

5 J. R. Hunt and Z. K. Roughead, Adaptation of iron absorption in men consuming diets with high or low iron bioavailability, *American Journal of Clinical Nutrition* 71 (2000): 94–102.

6 C. E. West, Strategies to control nutritional anemia, *American Journal of Clinical Nutrition* 64 (1996): 789–790.

7 Recommendations to prevent and control iron deficiency in the United States, *Morbidity and Mortality Weekly Report* 47 (1998): supplement; A. C. Looker and coauthors, Prevalence of iron deficiency in the United States, *Journal of the American Medical Association* 277 (1997): 973–976.

8 R. Yip and coauthors, Pervasive occult gastrointestinal bleeding in an Alaska

native population with prevalent iron deficiency: Role of *helicobacter pylori* gastritis, *Journal of the American Medical Association* 277 (1997): 1135–1139.

9 P. J. Garry, K. M. Koehler, and T. L. Simon, Iron stores and iron absorption: Effects of repeated blood donations, *American Journal of Clinical Nutrition* 62 (1995): 611–620.

10 L. Hallberg, L. Hultén, and E. Gramatkovski, Iron absorption from the whole diet in men: How effective is the regulation of iron absorption? *American Journal of Clinical Nutrition* 66 (1997): 347–356.

11 I. E. Roeckel and L. G. Dickson, Understanding iron absorption and metabolism, aided by studies of hemochromatosis, *Annals of Clinical Laboratory Sciences* 28 (1998): 30–33; R. S. Eisenstein, Interaction of the hemochromatosis gene product HFE with transferrin receptor modulates cellular iron metabolism, *Nutrition Reviews* 56 (1998): 356–358.

12 J. C. Fleet, Discovery of the hemochromatosis gene will require rethinking the regulation of iron metabolism, *Nutrition Reviews* 54 (1996): 285–292.

13 M. J. Nowicki and B. R. Bacon, Hereditary hemochromatosis in siblings: Diagnosis by genotyping, *Pediatrics* 105 (2000): 426–429; A. S. Tavill, Clinical implications of the hemochromatosis gene, *New England Journal of Medicine* 341 (1999): 755–757.

14 Iron overload disorders among Hispanics—San Diego, California, 1995, *Morbidity and Mortality Weekly Report* 45 (1996): 991–993; D. H. G. Crawford and coauthors, Factors influencing disease expression in hemochromatosis, *Annual Review of Nutrition* 16 (1996): 139–160.

15 V. Herbert, S. Shaw, and E. Jayatilleke, Vitamin C supplements are harmful to lethal for over 10% of Americans with high iron stores, *FASEB Journal* 8 (1994): A678.

16 K. Klipstein-Grobusch and coauthors, Serum ferritin and risk of myocardial infarction in the elderly: The Rotterdam Study, *American Journal of Clinical Nutrition* 69 (1999): 1231–1236; C. T. Sempos, A. C. Looker, and R. F. Gillum, Iron and heart disease: The epidemiologic data, *Nutrition Reviews* 54 (1996): 73–84.

17 J. M. McCord, Effects of positive iron status at a cellular level, *Nutrition Reviews* 54 (1996): 85–88.

18 K. Klipstein-Grobusch and coauthors, Dietary iron and risk of myocardial infarction in the Rotterdam Study, *American Journal of Epidemiology* 149 (1999): 421–428; H. van Jaarsveld, G. F. Pool, and H. C. Barnard, Dietary iron concentration alters LDL oxidatively: The effect of antioxidants, *Research Communications in Molecular Pathology and Pharmacology* 99 (1998): 69–80; A. Tzonou and coauthors, Dietary iron and coronary heart disease risk: A study from Greece, *American Journal of Epidemiology* 147 (1998): 161–166.

19 T. P. Tuomainen and coauthors, Cohort study of relation between donating blood and risk of myocardial

ANSWERS

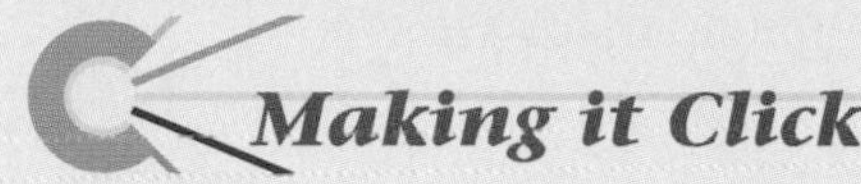

Making it Click

1. Iron: mg.
   Zinc: mg.
   Iodine: µg.
   Selenium: µg.
   Copper: mg.
   Manganese: mg.
   Fluoride: mg.
   Chromium: µg.
   Molybdenum: µg.
2. a. Sirloin steak > broccoli > green peas > bread > pork chop > cantaloupe > sweet potato > carrots > apple > cheese > milk.
   b.

Food	Iron Density (mg/kcal)
Milk, nonfat, 1 c	0.10 mg ÷ 85 kcal = 0.0012 mg/kcal
Cheddar cheese, 1 oz	0.19 mg ÷ 114 kcal = 0.0017 mg/kcal
Broccoli, cooked from fresh, chopped, 1 c	1.31 mg ÷ 44 kcal = 0.0298 mg/kcal
Sweet potato, baked in skin, 1 ea	0.51 mg ÷ 117 kcal = 0.0044 mg/kcal
Cantaloupe melon, ½	0.56 mg ÷ 93 kcal = 0.0060 mg/kcal
Carrots, from fresh, ½ c	0.48 mg ÷ 35 kcal = 0.0137 mg/kcal
Whole-wheat bread, 1 slice	0.87 mg ÷ 64 kcal = 0.0136 mg/kcal
Green peas, cooked from frozen, ½ c	1.26 mg ÷ 62 kcal = 0.0203 mg/kcal
Apple, medium	0.38 mg ÷ 125 kcal = 0.0030 mg/kcal
Sirloin steak, lean, 4 oz	3.81 mg ÷ 228 kcal = 0.0167 mg/kcal
Pork chop, lean broiled, 1 ea	0.66 mg ÷ 166 kcal = 0.0040 mg/kcal

   Broccoli > green peas > sirloin steak > carrots > bread > cantaloupe > sweet potato > pork chop > apple > cheese > milk.
   c. Broccoli, green peas, and carrots are all higher on the per-kcalorie list.
   d. They are all vegetables.

Multiple Choice

1. b	2. b	3. c	4. c	5. d
6. d	7. c	8. b	9. a	10. c

infarction in 2682 men in eastern Finland, *British Medical Journal* 314 (1997): 793–794.

20 M. Shannon, Ingestion of toxic substances by children, *New England Journal of Medicine* 342 (2000): 186–191; C. C. Morris, Pediatric iron poisonings in the United States, *Southern Medicine Journal* 93 (2000): 352–358.

21 W. J. Bartfay and coauthors, Cytotoxic aldehyde generation in heart following acute iron-loading, *Journal of Trace Elements in Medicine and Biology* 14 (2000): 14–20; A. S. Ioannides and J. M. Panisello, Acute respiratory distress syndrome in children with acute iron poisoning: The role of intravenous desferrioxamine, *European Journal of Pediatrics* 159 (2000): 158–159; J. P. Pestaner and coauthors, Ferrous sulfate toxicity: A review of autopsy findings, *Biological Trace Element Research* 69 (1999): 191–198.

22 J. L. Beard, Weekly iron intervention: The case for intermittent iron supplementation, *American Journal of Clinical Nutrition* 68 (1998): 209–212; L. Hallberg, Combating iron deficiency: Daily administration of iron is far superior to weekly administration, *American Journal of Clinical Nutrition* 68 (1998): 213–217.

23 A. H. Shankar and A. S. Prasad, Zinc and immune function: The biological basis of altered resistance to infection, *American Journal of Clinical Nutrition* 68 (1998): 447S–463S.

24 M. M. Black, Zinc deficiency and child development, *American Journal of Clinical Nutrition* 68 (1998): 464S–469S.

25 The Zinc Investigators' Collaborative Group, Therapeutic effects of oral zinc in acute and persistent diarrhea in children in developing countries: Pooled analysis of randomized controlled trials, *American Journal of Clinical Nutrition* 72 (2000): 1516–1522; R. B. Costello and J. Grumstrup-Scott, Zinc: What role might supplements play? *Journal of the American Dietetic Association* 100 (2000): 371–375; R. E. Black, Therapeutic and preventive effects of zinc on serious childhood infectious diseases in developing countries, *American Journal of Clinical Nutrition* 68 (1998): 476S–479S; G. J. Fuchs, Possibilities for zinc in the treatment of acute diarrhea, *American Journal of Clinical Nutrition* 68 (1998): 480S–483S; J. L. Rosado and coauthors, Zinc supplementation reduced morbidity, but neither zinc nor iron supplementation affected growth or body composition of Mexican preschoolers, *American Journal of Clinical Nutrition* 65 (1997): 13–19.

26 A. Gadomski, A cure for the common cold? Zinc again, *Journal of the American Medical Association* 279 (1998): 1999–2000; Zinc lozenges reduce the duration of common cold symptoms, *Nutrition Reviews* 55 (1997): 82–88.

27 S. B. Mossad and coauthors, Zinc gluconate lozenges for treating the common cold: A randomized, double-blind, placebo-controlled study, *Annals of Internal Medicine* 125 (1996): 81–88.

28 N. Bleichrodt and coauthors, The benefits of adequate iodine intake, *Nutrition Reviews* 54 (1996): S72–S78; C. Xue-Yi and coauthors, Timing of vulnerability of the brain to iodine deficiency in endemic cretinism, *New England Journal of Medicine* 331 (1994): 1739–1744.

29 B. D. Tiwari and coauthors, Learning disabilities and poor motivation to achieve due to prolonged iodine deficiency, *American Journal of Clinical Nutrition* 63 (1996): 782–786.

30 Letters from V. Herbert, L. H. Kuller, J. S. Parker, and L. C. Clark, Selenium supplementation and cancer rates, *Journal of the American Medical Association* 277 (1997): 880–881; L. C. Clark and coauthors, Effects of selenium supplementation for cancer prevention in patients with carcinoma of the skin—A randomized controlled trial, *Journal of the American Medical Association* 276 (1996): 1957–1963; G. A. Colditz, Selenium and cancer prevention—Promising results indicate further trials required, *Journal of the American Medical Association* 276 (1996): 1984–1985.

31 R. Uauy, M. Olivares, and M. Gonzalez, Essentiality of copper in humans, *American Journal of Clinical Nutrition* 67 (1998): 952S–959S.

32 A. Cordano, Clinical manifestations of nutritional copper deficiency in infants and children, *American Journal of Clinical Nutrition* 67 (1998): 1012S–1016S.

33 R. A. Wapnir, Copper absorption and bioavailability, *American Journal of Clinical Nutrition* 67 (1998): 1054S–1060S; B. Lönnerdal, Bioavailability of copper, *American Journal of Clinical Nutrition* 63 (1996): 821S–829S.

34 Z. L. Harris and J. D. Gitlin, Genetic and molecular basis for copper toxicity, *American Journal of Clinical Nutrition* 63 (1996): 836S–841S.

35 M. C. Linder and M. Hazegh-Azam, Copper biochemistry and molecular biology, *American Journal of Clinical Nutrition* 63 (1996): 797S–811S.

36 D. J. Fitzgerald, Safety guidelines for copper in water, *American Journal of Clinical Nutrition* 67 (1998): 1098S–1102S.

37 S. Fairweather-Tait and R. F. Hurrell, Bioavailability of minerals and trace elements, *Nutrition Research Reviews* 9 (1996): 295–324.

38 L. Davidson and coauthors, Manganese absorption in humans: The effect of phytic acid and ascorbic acid in soy formula, *American Journal of Clinical Nutrition* 62 (1995): 984–987.

39 Position of the American Dietetic Association: The impact of fluoride on health, *Journal of the American Dietetic Association* 100 (2000): 1208–1213.

40 Achievements in Public Health, 1900–1999: Fluoridation of drinking water to prevent dental caries, *Morbidity and Mortality Weekly Report* 48 (1999): 933–940.

41 W. Mertz, Interaction of chromium with insulin: A progress report, *Nutrition Reviews* 56 (1998): 174–177.

42 R. A. Anderson, Effects of chromium on body composition and weight loss, *Nutrition Reviews* 56 (1998): 266–270.

43 S. L. Soignet, Complete remission after treatment of acute promyelocytic leukemia with arsenic trioxide, *New England Journal of Medicine* 339 (1998): 1341–1348.

44 M. E. Moss, B. P. Lanphear, and P. A. Auinger, Association of dental caries and blood lead levels, *Journal of the American Medical Association* 281 (1999): 2294–2298.

45 P. Mushak and A. Crocetti, Lead and nutrition: Biologic interactions of lead with nutrients, *Nutrition Today* 31 (1996): 12–17.

FUNCTIONAL FOODS—ARE THEY REALLY DIFFERENT?

GLOSSARY

Federal Trade Commission (FTC): a federal agency that is responsible for, among other things, food advertising and industry competition.
www.ftc.gov

functional foods: foods that contain physiologically active compounds that provide health benefits beyond basic nutrition; also called *designer foods* or *nutraceuticals.*

probiotics: microbial food ingredients that are beneficial to health. Nondigestible food ingredients that encourage the growth of favorable bacteria are called **prebiotics.**
- **pro** = for
- **bios** = life
- **pre** = before

yogurt: milk fermented by specific bacterial cultures.

UNTIL RELATIVELY recently, people have always obtained their nourishment from foods provided directly from nature. Only in the last hundred years or so has technology taken nature's foods and reinvented them into a seemingly unlimited number of products. Today, a simple potato is available grated and frozen as hash browns, dehydrated and flaked as potential mashed potatoes, or sliced and deep fried as chips—to name just a few of its many disguises. In the last decade or so, the public's health consciousness has prompted the food industry to generate a whole new line of low-fat, reduced-fat, no-fat, no-salt, no-sugar, and high-fiber foods. In addition, recent knowledge of antioxidants and phytochemicals has sparked a trend to design foods with added ingredients thought to promote optimal health. Foods that provide health benefits beyond those of the traditional nutrients are called **functional foods** (the accompanying glossary defines this and related terms).

This highlight focuses on the most controversial of the functional foods—novel foods to which specific, physiologically active compounds, compounds that act somewhat like drugs, have been added to promote health.[1] Many other chapters and Highlight 11 tout the benefits of nature's functional foods. But first, a look at the past will help put things into perspective.

Nature offers a variety of "functional foods" that provide us with many health benefits.

© Craig M. Moore

TRADITIONAL FOODS

A hundred years ago, the number of options people had when choosing foods was limited. They ate meat, fish, and poultry—not deboned and restructured luncheon meats, fish sticks, and chicken nuggets. They ate fruits and vegetables—not fruit chewies and corn chips. Their milk came with a layer of cream on top, and they did not have carbonated milk, imitation milk, or filled milk.* Bread took hours to bake, was made with whole-wheat flour, and came in whole loaves that were sliced as they were served. People living in the early 1900s would be amazed to see the thousands of foods offered by grocery stores today. Figure H13-1 presents a few selected innovations that have influenced our food options over the past century.

The development of new food products often reflects the industrial and social changes of the time. The widespread use of refrigerators in the 1920s, for example, dramatically changed the way people purchased, used, and stored foods. After World War II, the proportion of women working outside the home increased, and the demand for convenience foods and fast-food restaurants did, too. Baking from scratch became unnecessary when manufacturers began developing mixes for cakes and muffins in the late 1950s. Then further technological advances made frozen foods a common grocery item. With the advent of the microwave oven in the 1980s, "from freezer to table in 5 minutes" became a common boast in advertisements for foods. Convenience took priority in the kitchen, and the food industry responded with prewashed and precut vegetables, boil-in-a-pouch single servings, and ready-to-eat meals. These many changes in the food supply were possible because of new technologies, but they were driven by consumers' demands for foods that were convenient and tasty.

*Carbonated milk has had carbonation added to make it fizz like a soda; imitation milk is not a dairy product and has been made with water, corn syrup solids or sugar to replace the lactose, vegetable oils to replace the milk fat, protein from soybeans, stabilizers, emulsifiers, and flavoring agents; filled milk has had all or part of its milk fat replaced by vegetable fat.

FIGURE H13-1 A Century of Food

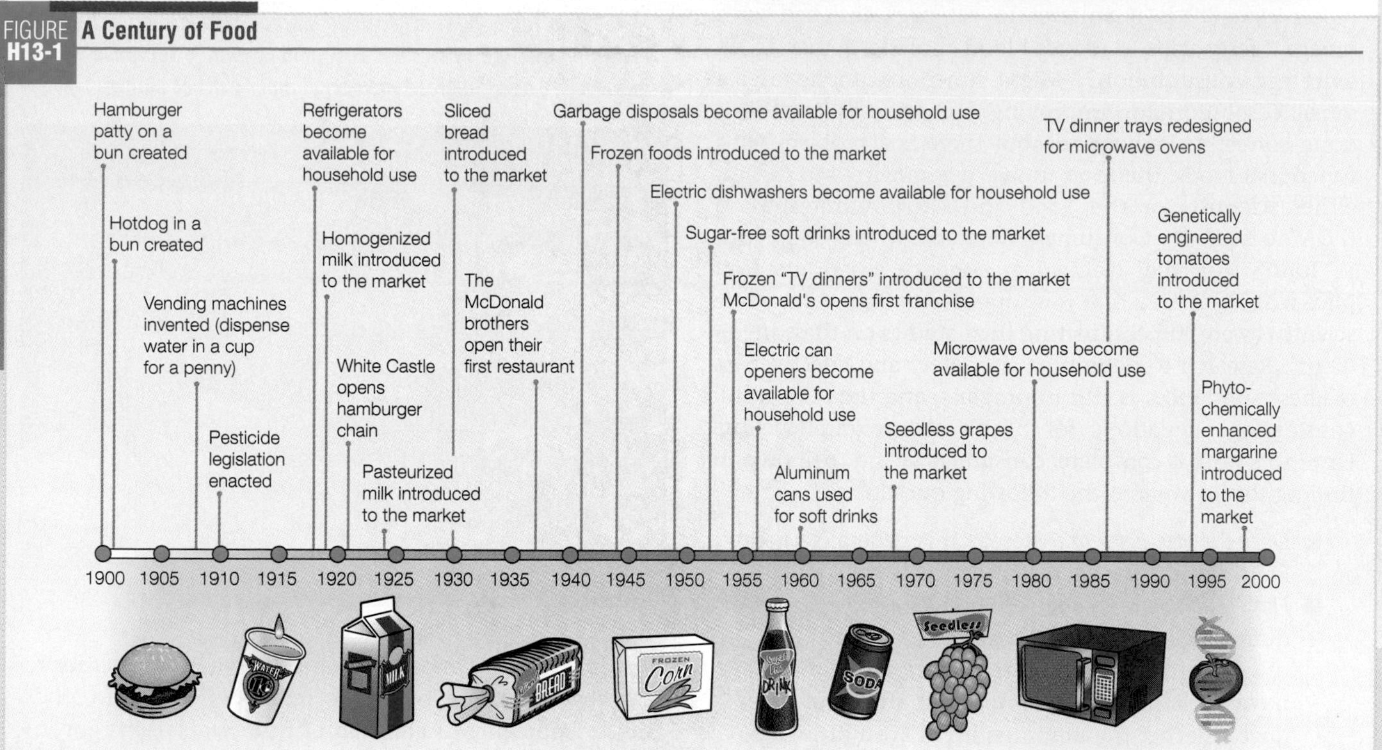

SOURCE: Adapted from Public health nutrition and food safety, 1900–1999, *Nutrition Reviews* 57 (1999): 368–372; The American century in food, *Bon Appétit*, September 1999, entire issue.

FUNCTIONAL FOODS

Some people are looking for more than taste and convenience in their foods—they also want protection against disease. Indeed, the food industry has always catered to consumers' demands for health-promoting foods. Already, food manufacturers fortify margarines and juices with vitamins, reduce the fat in salad dressings and cookies, and increase the fiber in breads and cereals. Now consumers want healing, medicine-like ingredients as well.

Folklore throughout the ages has supported the idea that foods have healing powers. **Yogurt,** for example, has long been regarded as a "health food"—and in fact, it does promote health, although for centuries no one understood why. Today, scientists have explained some of the health benefits of yogurt. Yogurt contains **probiotics**—live, microbial food constituents that influence body functions.[2] These probiotics may alleviate lactose intolerance, enhance immune function, protect against GI cancer, and lower blood cholesterol.[3]

Now that this discovery has been made, probiotics that are normally found only in fermented milk products such as yogurt can be introduced into other foods such as fermented vegetables and meats. This addition gives these new foods added health benefits—hence the term *functional foods.*

As we enter the twenty-first century, the creation of more functional foods is the fastest-growing trend and the greatest influence transforming the American food supply.[4] Consumers may even be led to believe that this is a new category of foods. All foods contain thousands of nonnutrient compounds that are biologically active in the body, however, so virtually all of them have some special value in supporting health.[5] Even simple, whole foods such as fruits and vegetables, in reality, can be considered functional foods. So can many processed foods that have been fortified with nutrients or enhanced with phytochemicals or herbs. Because farm-fresh tomatoes, iodized salt, and kava kava chips all can claim to be "functional foods," consumers may be a little confused. Shoppers who are looking specifically for functional foods in the grocery store will have trouble finding them. They may be able to find "canned vegetables" on aisle 4 and "snack foods" on aisle 6, but there is no aisle devoted to "functional foods." To worsen the consumer's plight, as is always the case with new claims made for foods, some are valid and some are not. Confronted with thousands of food choices, consumers may find it more challenging than ever to make healthy selections.

Unanswered Questions

To achieve a desired health effect, which is the better choice: to eat a food designed to affect some body function or simply to adjust the diet? Does it make more sense to use a margarine enhanced with a phytochemical that lowers blood cholesterol than simply to limit the amount of butter eaten?*[6] Is it smarter to eat eggs enriched with omega-3 fatty acids to reduce blood cholesterol than to restrict egg consumption?[7] Might functional foods offer a sensible solution for improving our nation's health—if done correctly? Perhaps so—but there is a problem with functional foods: the food industry is moving too fast for either scientists or the Food and Drug Administration (FDA) to keep up. Consumers were able to buy soup with St. John's wort that claimed to enhance mood and fruit juice with echinacea that was supposed to fight colds while scientists were still conducting their studies on these ingredients. Research to determine the safety and effectiveness of these substances is still in progress, and the FDA is still considering regulations for health claims and labeling. Until this work is complete, consumers are on their own in finding the answers to the following questions:[8]

- *Does the product work?* Research is generally lacking. Findings to date are often inconclusive.
- *How much does it contain?* Food labels are not required to list the quantities of added herbs and phytochemicals. Even if they were, consumers have no standard for comparison and cannot deduce whether the amounts listed are a little or a lot. Most importantly, until research is complete, food manufacturers do not know what amounts (if any) are most effective.
- *Is it safe?* Functional foods can act like drugs. They contain ingredients that can alter body functions. When consumed in large doses, they may cause allergies, drug interactions, drowsiness, and other side effects. Yet unlike drug labels, food labels do not provide instructions for the dosage, frequency, or duration of treatment.
- *Is it healthy?* A candy bar may be fortified with herbs and phytochemicals, but it is still made mostly of sugar and fat.

Critics suggest that the designation "functional foods" may be nothing more than a marketing tool. After all, even the most experienced researchers cannot yet identify the perfect combination of nutrients, much less phytochemicals, to support optimal health. Yet manufacturers are freely experimenting with various concoctions as if they possessed that knowledge. Is it okay for them to sprinkle phytochemicals on fried snack foods and label them "functional," thus implying health benefits? Do we want our children receiving their nourishment from fortified caramel candies and chocolate cakes?

Regulations

The laws currently in force allow *foods* and *dietary supplements* to appear on market shelves without approval by the FDA. Food *additives,* on the other hand, are required to prove their safety and effectiveness through numerous research studies before being allowed as ingredients in foods. Dietary supplements are exempt from regulations governing food additives. Thus functional foods that use dietary supplements as food additives escape regulation. Consequently, some food manufacturers market functional foods in ways that are not allowed for other foods.

*Margarine products that lower blood cholesterol contain either sterol esters from vegetable oils, soybeans, and corn or stanol esters from wood pulp.

Functional foods currently on the market promise to "enhance mood," "promote relaxation and good karma," "increase alertness," and "improve memory," among other claims.

© Craig M. Moore

The multibillion-dollar-a-year food industry spends much money and effort influencing these regulations. Lobbyists for the food industry have tried to urge the FDA to loosen its criteria for health claims and to allow manufacturers to make exclusive claims based on their own research—without being required to make the research public. Such exclusivity of health claims on products based on private research would be contrary to the open sharing of information that is fundamental to nutrition science. It would also be confusing and misleading for consumers. If a company could claim that its Brand X prevents lung cancer, then consumers would need to be aware that other brands have a similar effect and that many foods provide the same nutrients and phytochemicals as well. Furthermore, consumers would need to know that such a claim is backed by sound scientific research—and that's what FDA approval provides.

• **Health Claims** • Some functional foods qualify to make a health claim (see Chapter 2 for criteria). Health claims may be found on foods that naturally provide compounds that benefit health (oats contain soluble fiber that may reduce the risk of heart disease, for example). Similarly, health claims may be found on foods that have been enhanced with compounds that benefit health (calcium-fortified orange juice may reduce the risk of osteoporosis, for example). Health claims signify that there is substantial scientific agreement that the food protects against disease.

Health claims, however, are not allowed on many functional foods. The benefits of some functional foods, such as garlic (which may lower blood cholesterol), have scientific support, but the manufacturers are not yet allowed to make health claims because they have yet to petition the FDA. The benefits of other foods have enough scientific support to spur market interest, but not enough to warrant a health claim. For example, cooked tomatoes rich in the phytochemical lycopene may reduce the risks of prostate cancer and cardiovascular disease.[9]

Ironically, most of the novel functional foods that have been specifically manufactured to promote health do not meet the FDA's criteria for making health claims. These products typically use "structure-function" claims instead.

• Structure-Function Claims • Unlike health claims, which are tightly regulated, structure-function claims can be made without FDA approval. The food need not be healthy; the ingredients need not be effective. Products can claim to "slow aging," "improve memory," and "build strong bones" without any proof. The only criterion for a structure-function claim is that it must not mention a disease, whereas a health claim may. Unfortunately, most consumers cannot distinguish between these two types of claims. They believe the health claim "may reduce the risk of heart disease" has the same meaning as the structure-function claim "promotes a healthy heart." They mistakenly assume that all claims on all food labels have been scientifically tested and approved by the FDA. Table 2-10 on p. 55 presented approved health claims; Table H13-1 lists examples of structure-function claims; and Figure H13-2 illustrates the differences between various claims.

• Advertising • Manufacturers also skirt around the FDA's regulations in advertising their products. Advertising is under the jurisdiction of the Federal Trade Commission (FTC), which regulates all advertisements for all products in all media. The FTC does not distinguish between various claims or products (whether the product is a food, drug, or dietary supplement). Claims that would not be allowed on product labels are unrestricted in television commercials and magazine advertisements. The FDA has not approved any health claims for the phytochemical lycopene commonly found in tomatoes, for example, yet magazine advertisements for catsup boldly state that lycopene "may help to reduce the risk of prostate and cervical cancer."

FIGURE H13-2 Examples of Various Claims Compared

Health claims and nutrient content claims need FDA approval, but structure-function claims do not.

© Photo Disc

Health claim: May reduce the risk of osteoporosis.
Structure-function claim: Helps promote bone health.
Nutrient content claim: Contains calcium.

© Photo Disc

Health claim: May reduce the risk of cancer.
Structure-function claim: Helps promote immune health.
Nutrient content claim: Contains antioxidants.

© Photo Disc

Health claim: May reduce the risk of heart disease.
Structure-function claim: Helps promote heart health.
Nutrient content claim: Contains soy.

TABLE H13-1 Examples of Structure-Function Claims

- Builds strong bones
- Defends your health
- Promotes relaxation
- Slows aging
- Improves memory
- Guards against colds
- Boosts the immune system
- Lifts your spirits
- Supports heart health

Note: Structure-function claims cannot make statements about diseases. See Table 2-10 on p. 55 for examples of health claims.

Foods as Pharmacy

Not too long ago, most of us could agree on what was a food and what was a drug. Today, functional foods blur the distinctions between foods and drugs. The food industry tried to create a new category called "nutraceuticals" to identify foods that have been engineered to promote health. But this term was never legally or scientifically defined, and it did not catch on with consumers either. "Functional foods" seems to be the preferred term.

Eating margarine sparingly instead of butter generously may gradually lower blood cholesterol slightly over several months and clearly falls into the diet category. Taking the medication Lipitor, on the other hand, dramatically lowers blood cholesterol significantly within weeks and clearly falls into the drug category. But margarine enhanced with a phytochemical that lowers blood cholesterol is in a gray area between the two. The margarine looks and tastes like a food, but it acts like a drug.

Accepting functional foods as drugs creates a whole new set of diet-planning challenges. Not only must foods provide an adequate intake of all the nutrients to support good health, but they must also deliver druglike ingredients to protect against disease. Like drugs used to treat chronic diseases, functional foods may need to be eaten several times a day for several months or years to have a beneficial effect.

Sporadic users may be disappointed in the results. If used three times a day for two weeks, margarine enriched with phytochemicals might reduce cholesterol by 10 percent, much more than regular margarine would, but not nearly as much as the 40 percent reduction seen with cholesterol-lowering drugs. For this reason, functional foods may be more useful for prevention and mild cases of heart disease than for intervention and more severe cases.

Foods and drugs differ dramatically in cost as well. Naturally functional foods such as fruits and vegetables incur no added costs, of course, but foods that have been manufactured with added phytochemicals or herbs can be expensive, costing up to six times as much as their conventional counterparts. Generally, the price of functional foods falls between that of foods and medicines.

FUTURE FOODS

Nature has elegantly designed foods to provide us with a complex array of dozens of nutrients and thousands of additional compounds that may benefit health—most of which we have yet to identify or understand. Over the years, we have taken those foods and first deconstructed them and then reconstructed them in an effort to "improve" them. Although the art of food designing is in a relatively early and experimental stage today, we may someday be able to design foods to meet the *exact* health needs of *each* individual. Indeed, our knowledge of the human genome and of human nutrition may well merge to allow for specific recommendations for individuals.[10] If the present trend continues, then someday physicians may be able to prescribe the perfect foods to enhance your health, and farmers will be able to grow them. Scientists have already developed gene technology to alter the composition of food crops. They can grow rice enriched with vitamin A and tomatoes containing a hepatitis vaccine. It seems quite likely that foods can be created to meet every possible human need. But then, in a sense, that was largely true 100 years ago when we relied on the bounty of nature.

Nutrition on the Net

•WEBSITES•

Access these websites for further study of topics covered in this highlight.

- Find updates and quick links to these and other nutrition-related sites at our website: **www.wadsworth.com/nutrition**
- Search for "functional foods" at the International Food Information Council: **www.ificinfo.health.org**
- Search for "functional foods" at the Center for Science in the Public Interest: **www.cspinet.org**

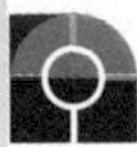

References

1 J. A. Milner, Functional foods: The US perspective, *American Journal of Clinical Nutrition* 71 (2000): 1654S–1659S.

2 Probiotics and prebiotics, *American Journal of Clinical Nutrition* (supplement) 73 (2001): entire issue.

3 L. Kopp-Hoolihan, Prophylactic and therapeutic uses of probiotics: A review, *Journal of the American Dietetic Association* 101 (2001): 299–238; M. B. Roberfroid, Prebiotics and probiotics: Are they functional foods? *American Journal of Clinical Nutrition* 71 (2000): 1682S–1687S.

4 Position of The American Dietetic Association: Functional foods, *Journal of the American Dietetic Association* 99 (1999): 1278–1285.

5 Position of The American Dietetic Association: The role of nutrition in health promotion and disease prevention programs, *Journal of the American Dietetic Association* 98 (1998): 205–208.

6 P. J. H. Jones and F. Ntanios, Comparable efficacy of hydrogenated versus nonhydrogenated plant sterol esters on circulating cholesterol levels in humans, *Nutrition Reviews* 56 (1998): 245–252.

7 D. J. Farrell, Enrichment of hen eggs with n-3 long-chain fatty acids and evaluation of enriched eggs in humans, *American Journal of Clinical Nutrition* 68 (1998): 538–544.

8 B. Brophy and D. Schardt, Functional foods, *Nutrition Action Healthletter,* April 1999, pp. 3–7.

9 L. Arab and S. Steck, Lycopene and cardiovascular disease, *American Journal of Clinical Nutrition* 71 (2000): 1691S–1695S; E. Giovannucci, Tomatoes, tomato-based products, lycopene, and cancer: Review of the epidemiologic literature, *Journal of the National Cancer Institute* 91 (1999): 317–331.

10 I. H. Rosenberg, *What Is a Nutrient? Defining the Food-Drug Continuum* (Washington, D.C.: Center for Food and Nutrition Policy, 1999).

Life Cycle Nutrition: Pregnancy and Lactation

CHAPTER 14

VCG/FPG

All people—pregnant and lactating women, infants, children, adolescents, and adults—need the same nutrients, but the amounts they need vary depending on their stage of life. This chapter focuses on nutrition in preparation for, and support of, pregnancy and lactation. The next two chapters address the needs of infants, children, adolescents, and older adults.

GROWTH AND DEVELOPMENT DURING PREGNANCY

A whole new life begins at **conception.** Organ systems develop rapidly, and nutrition plays many supportive roles. This section describes placental development and fetal growth, paying close attention to times of intense developmental activity.

Placental Development

In the early days of pregnancy, a spongy structure known as the **placenta** develops in the **uterus.** Two associated structures also form (see Figure 14-1). One is the **amniotic sac,** a fluid-filled balloonlike structure that houses the developing fetus. The other is the **umbilical cord,** a ropelike structure containing fetal blood vessels that extends through the fetus's "belly button" (the umbilicus) to the placenta. These three structures play crucial roles during pregnancy and then are expelled from the uterus during childbirth.

The placenta develops as an interweaving of fetal and maternal blood vessels embedded in the uterine wall. The maternal blood transfers oxygen and nutrients to the fetus's blood and picks up fetal waste products. By exchanging oxygen, nutrients, and waste products, the placenta performs the respiratory, absorptive, and excretory functions that the fetus's lungs, digestive system, and kidneys will provide after birth.

The placenta is a versatile, metabolically active organ. Like all body tissues, the placenta uses energy and nutrients to support its work. Like a gland, it produces an array of hormones that maintain pregnancy and prepare the mother's breasts for lactation (making milk). A healthy placenta is essential for the developing fetus to attain its full potential.[1]

Fetal Growth and Development

Fetal development begins with the fertilization of an **ovum** by a **sperm.** Three stages follow: the zygote, the embryo, and the fetus (see Figure 14-2).

• **The Zygote** • The newly fertilized ovum, or **zygote,** begins as a single cell and divides to become many cells during the days after fertilization. Within two weeks, the zygote embeds itself in the uterine wall—a process known as **implantation.** Cell division continues—each set of cells divides into many other cells. As development proceeds, the zygote becomes an embryo.

• **The Embryo** • The **embryo** develops at an amazing rate. At first, the number of cells in the embryo doubles approximately every 24 hours; later the rate slows, and only one doubling occurs during the final ten weeks of pregnancy. The embryo's size changes very little, but at eight weeks, the 1¼-inch embryo has a complete central nervous system, a beating heart, a digestive system, well-defined fingers and toes, and the beginnings of facial features.

• **The Fetus** • The **fetus** continues to grow during the next seven months. Each organ grows to maturity according to its own schedule, with greater intensity at some times than at others. As Figure 14-2 shows, fetal growth is phenomenal: weight increases from less than an ounce to about 7½ pounds (3500 grams).

conception: the union of the male sperm and the female ovum; fertilization.

placenta (plah-SEN-tuh): the organ that develops inside the uterus early in pregnancy, through which the fetus receives nutrients and oxygen and returns carbon dioxide and other waste products to be excreted.

uterus (YOU-ter-us): the muscular organ within which the infant develops before birth.

amniotic (am-nee-OTT-ic) **sac:** the "bag of waters" in the uterus, in which the fetus floats.

umbilical (um-BILL-ih-cul) **cord:** the ropelike structure through which the fetus's veins and arteries reach the placenta; the route of nourishment and oxygen to the fetus and the route of waste disposal from the fetus. The scar in the middle of the abdomen that marks the former attachment of the umbilical cord is the **umbilicus** (um-BILL-ih-cus), commonly known as the "belly button."

ovum (OH-vum): the female reproductive cell, capable of developing into a new organism upon fertilization; commonly referred to as an egg.

sperm: the male reproductive cell, capable of fertilizing an ovum.

zygote (ZY-goat): the product of the union of ovum and sperm; so-called for the first two weeks after fertilization.

implantation: the stage of development in which the zygote embeds itself in the wall of the uterus and begins to develop; occurs during the first two weeks after conception.

embryo (EM-bree-oh): the developing infant from two to eight weeks after conception.

fetus (FEET-us): the developing infant from eight weeks after conception until term.

FIGURE 14-1 **The Placenta and Associated Structures**

To understand how placental villi absorb nutrients without maternal and fetal blood interacting directly, think of how the intestinal villi work. The GI side of the intestinal villi is bathed in a nutrient-rich fluid (chyme). The intestinal villi absorb the nutrient molecules and release them into the body via capillaries. Similarly, the maternal side of the placental villi is bathed in nutrient-rich maternal blood. The placental villi absorb the nutrient molecules and release them to the fetus via fetal capillaries.

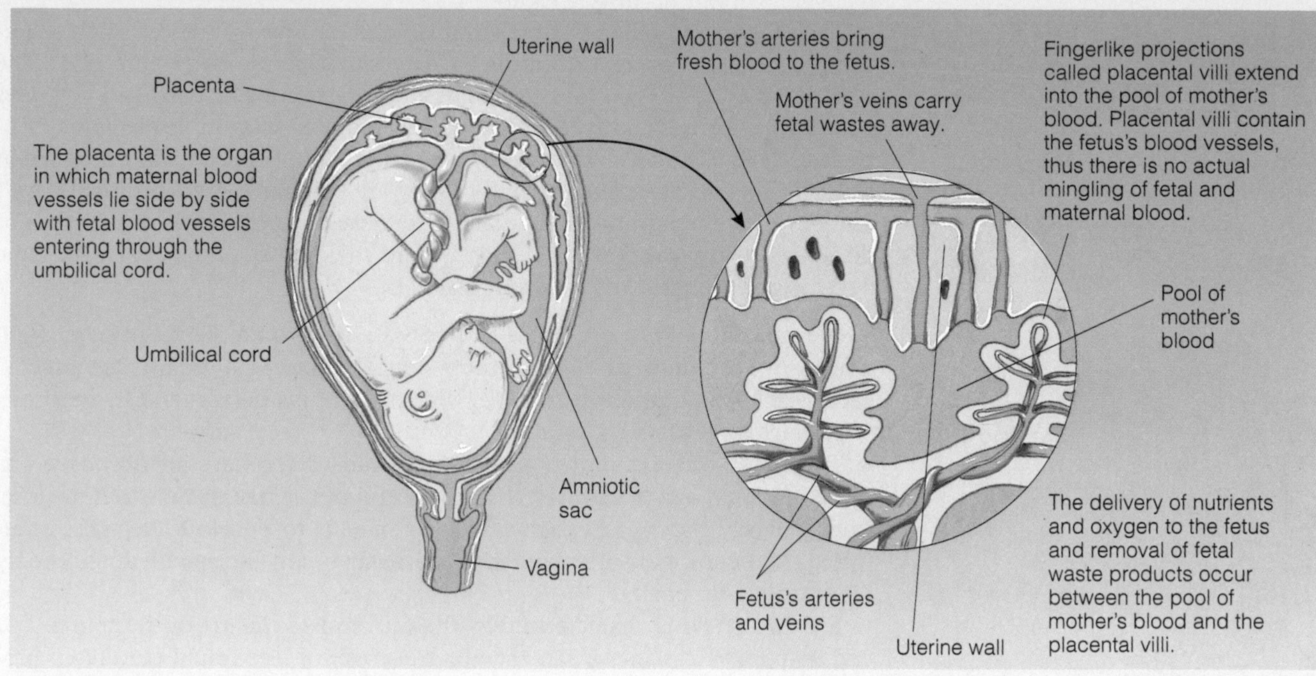

FIGURE 14-2 **Stages of Embryonic and Fetal Development**

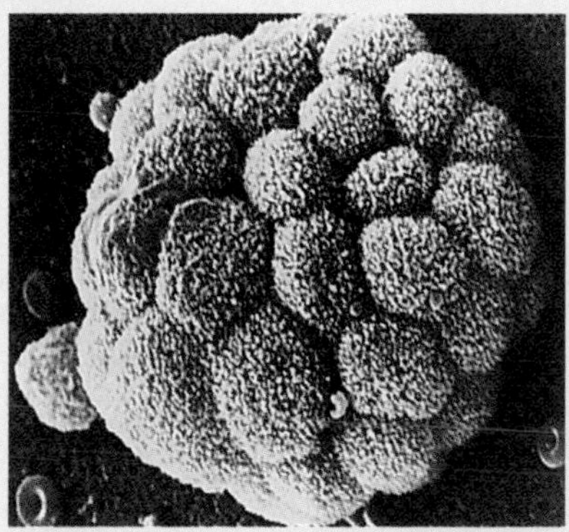

1 A newly fertilized ovum is about the size of the period at the end of this sentence. This zygote at less than one week after fertilization is not much bigger and is ready for implantation.

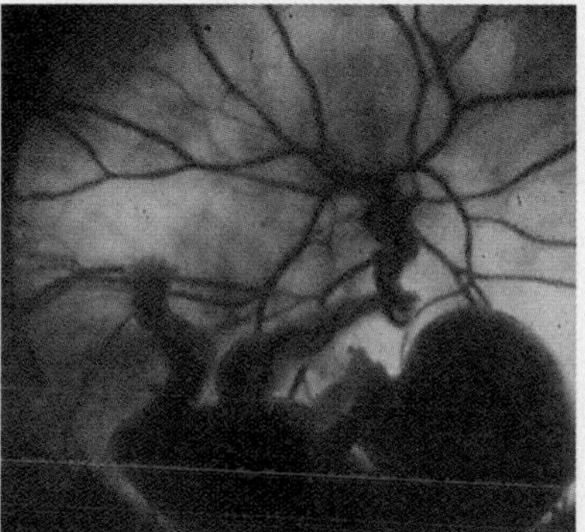

3 A fetus after 11 weeks of development is just over an inch long. Notice the umbilical cord and blood vessels connecting the fetus with the placenta.

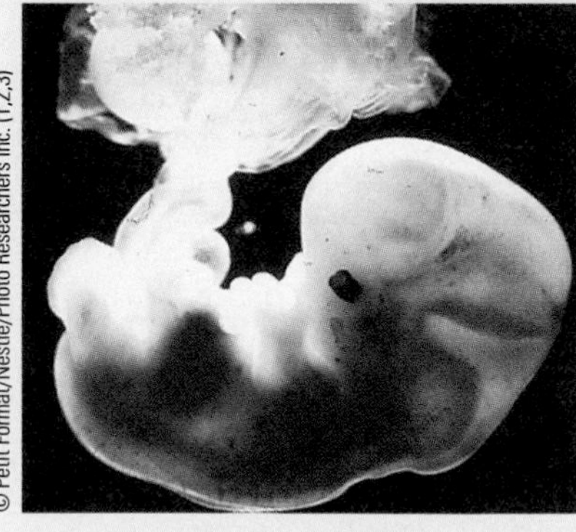

© Petit Format/Nestle/Photo Researchers Inc. (1,2,3)

2 After implantation, the placenta develops and begins to provide nourishment to the developing embryo. An embryo five weeks after fertilization is about ½ inch long.

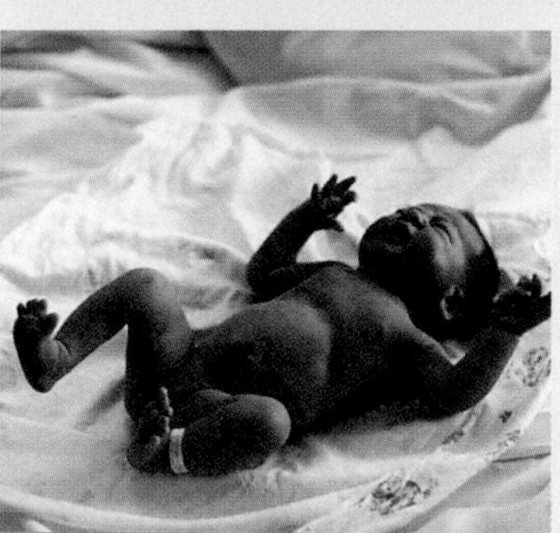

© Anthony M. Vanelli

4 A newborn infant after nine months of development measures close to 20 inches in length. From eight weeks to term, this infant grew 20 times longer and 50 times heavier.

FIGURE 14-3 **The Concept of Critical Periods**

Critical periods occur early in development. An adverse influence felt early can have a much more severe and prolonged impact than one felt later on.

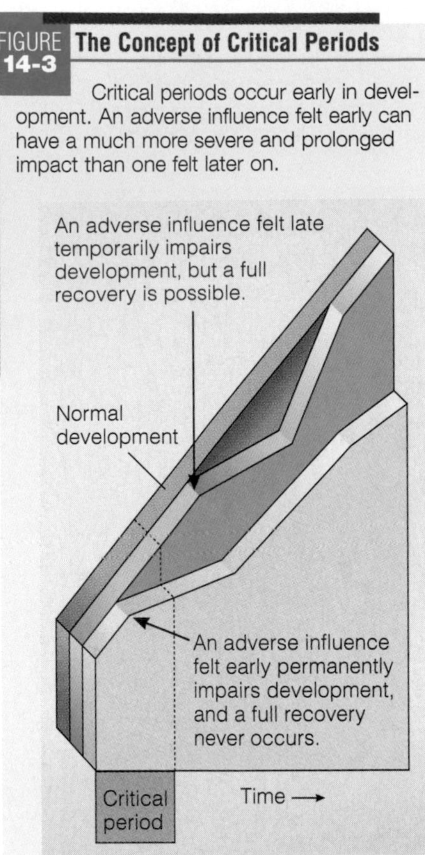

Critical Periods

Times of intense development and rapid cell division are called **critical periods**—critical in the sense that those cellular activities can occur only at those times. If cell division and number are limited during a critical period, full recovery will not occur (see Figure 14-3).

Each organ and tissue is most vulnerable to adverse influences (such as nutrient deficiencies or toxins) during its own critical period (see Figure 14-4). The critical period for neural tube development, for example, is from 17 to 30 days **gestation.** Consequently, neural tube development is most vulnerable to nutrient deficiencies, nutrient excesses, or toxins during this critical time—when most women do not even realize that they are pregnant. Any abnormal development of the neural tube or its failure to close completely can cause a major defect in the central nervous system. Figure 14-5 shows photos of neural tube development in the early weeks of gestation.

• Neural Tube Defects • In the United States, approximately 1 of every 1000 newborns has a **neural tube defect;** some 2500 to 3000 infants are affected each year.* Many other pregnancies with neural tube defects end in abortions or stillbirths.

The two most common types of neural tube defects are anencephaly and spina bifida. In **anencephaly,** the upper end of the neural tube fails to close. Consequently the brain is either missing or fails to develop. Pregnancies affected by anencephaly often end in miscarriage; infants born with anencephaly die shortly after birth.

Spina bifida is characterized by the incomplete closure of the spinal cord and its bony encasement. The membranes covering the spinal cord often protrude as a sac, which may rupture and lead to meningitis, a life-threatening inflammation of the membranes. Spina bifida is accompanied by varying degrees of paralysis, depending on the extent of the spinal cord damage. Mild cases may not even be noticed, but severe cases lead to death. Common problems include clubfoot, dislocated hip, kidney disorders, curvature of the spine, muscle weakness, mental handicaps, and motor and sensory losses.

*Worldwide, some 300,000 to 400,000 infants are born with neural tube defects each year.

critical periods: finite periods during development in which certain events occur that will have irreversible effects on later developmental stages; usually a period of rapid cell division.

gestation (jes-TAY-shun): the period from conception to birth. For human beings, gestation lasts from 38 to 42 weeks. Pregnancy is often divided into thirds, called **trimesters.**

neural tube defect: a serious central nervous system birth defect that often results in lifelong disability or death.

anencephaly (AN-en-SEF-a-lee): an uncommon and always fatal type of neural tube defect, characterized by the absence of a brain.
- **an** = not (without)
- **encephalus** = brain

spina (SPY-nah) **bifida** (BIFF-ih-dah): one of the most common types of neural tube defects, characterized by the incomplete closure of the spinal cord and its bony encasement.
- **spina** = spine
- **bifida** = split

FIGURE 14-4 **Critical Periods of Development**

During embryonic development (between 2 to 8 weeks), many of the tissues are in their critical periods (red area of the bars); events occur that will have irreversible effects on the development of those tissues. In the later stages of development (gray area of the bars), the tissues continue to grow and change, but the events are less critical in that they are relatively minor or reversible.

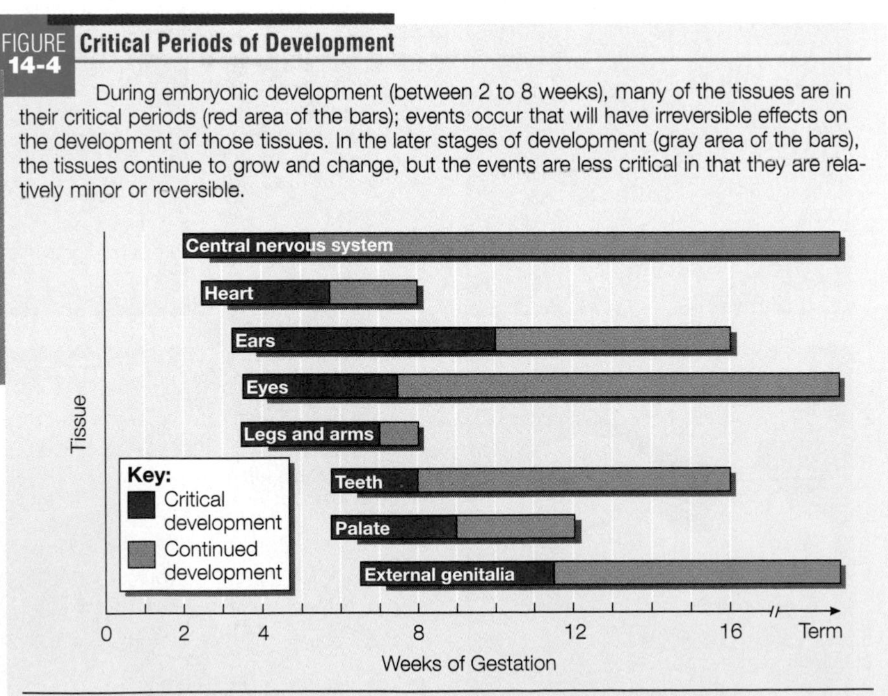

SOURCE: Adapted with permission from *Before We Were Born* by K. L. Moore. W. B. Saunders.

FIGURE 14-5 **Neural Tube Development**

The neural tube is the beginning structure of the brain and spinal cord. Any failure of the neural tube to close or to develop normally results in central nervous system disorders such as spina bifida and anencephaly. Successful development of the neural tube depends, in part, on the vitamin folate.

At four weeks, the neural tube has yet to close (notice the gap at the top).

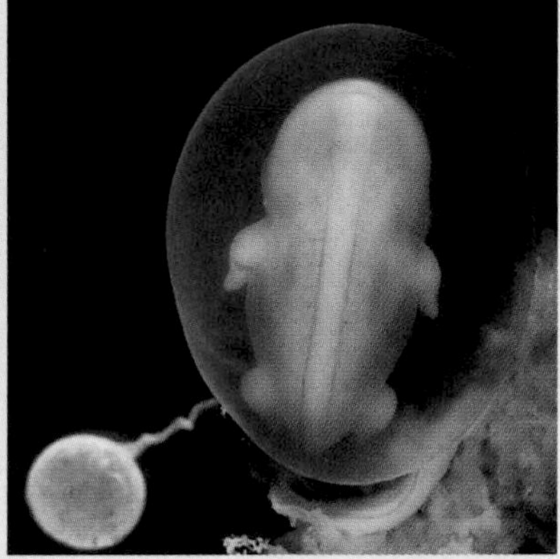

At six weeks, the neural tube (outlined by the delicate red vertebral arteries) has successfully closed.

A pregnancy affected by a neural tube defect can occur in any woman, but these factors increase the risk:

- A previous pregnancy affected by a neural tube defect.
- Maternal diabetes (type 1).
- Maternal use of antiseizure medications.
- Maternal obesity.
- Exposure to high temperatures early in pregnancy (prolonged fever or hot-tub use).
- Race/ethnicity (neural tube defects are more common among whites and Hispanics than others).
- Low socioeconomic status.

Folate supplementation reduces the risk.

• **Folate Supplementation** • Chapter 10 described how folate supplements taken one month before conception and continued throughout the first trimester can reduce the risks of neural tube defects. For this reason, all women of childbearing age◆ who are capable of becoming pregnant are advised to consume 0.4 milligram of folate daily. Supplements offer women a convenient way to ingest sufficient folate regularly and continuously enough to benefit pregnancy.[2] Most over-the-counter multivitamin supplements contain 0.4 milligram of folate; prenatal supplements usually contain at least 0.8 milligram. A woman who has previously had an infant with a neural tube defect may be advised by her physician to take folate supplements in doses ten times larger—4 milligrams daily. Because high doses of folate can mask the pernicious anemia of a vitamin B_{12} deficiency, quantities of 1 milligram or more require a prescription.

Folate RDA: ◆
- For women: 400 μg (0.4 mg)/day.
- During pregnancy: 600 μg (0.6 mg)/day.

To increase folate intake in the United States, the Food and Drug Administration (FDA) has mandated fortification of grain products.* This decision carefully

*Cereal, pasta, flour, rolls, buns, farina, grits, cornmeal, and rice must be fortified with 1.4 milligrams of folate per 100 grams of food.

weighed the benefits of fortification against the risks of overconsumption. On the one hand, an adequate folate intake is expected to reduce the incidence of neural tube defects by 50 percent. On the other hand, if vitamin B_{12} deficiency is masked by folate and left untreated, irreversible nerve damage may occur.

HEALTHY PEOPLE 2010

Reduce the occurrence of spina bifida and other neural tube defects. Increase the proportion of pregnancies begun with an optimum folate level.

Health promotion

• **Chronic Diseases** • Some research suggests that adverse influences at critical times set the stage for chronic diseases in adult life. Poor maternal diet during critical periods may permanently alter body functions such as blood pressure, cholesterol metabolism, and immune functions that influence disease development.[3] For example, maternal anemia during the critical period of placental growth alters the pattern of blood vessel growth, which may affect the cardiovascular health of the infant.[4] Malnutrition during the critical period of pancreatic cell growth provides an example of how diabetes may develop in adulthood. The pancreatic cells responsible for producing insulin (the beta cells) normally increase more than 130-fold between 12 weeks gestation and five months after birth. Nutrition is a primary determinant of beta cell growth, and infants who have suffered prenatal malnutrition have significantly fewer beta cells than well-nourished infants. One hypothesis suggests that diabetes may develop from the interaction of inadequate nutrition early in life with abundant nutrition later in life: the small mass of beta cells developed in lean times during fetal development may be insufficient in times of overnutrition during adulthood when the body needs more insulin.

IN SUMMARY

Maternal nutrition before and during pregnancy affects both the mother's health and the infant's growth. As the infant develops through its three stages—the zygote, embryo, and fetus—its organs and tissues grow, each on its own schedule. Times of intense development are critical periods that depend on nutrients to proceed smoothly. Without folate, for example, the neural tube fails to develop completely during the first month of pregnancy, prompting recommendations that all women of childbearing age take folate daily.

Because critical periods occur throughout pregnancy, a woman should continuously take good care of her health. That care should include achieving and maintaining a healthy body weight prior to pregnancy and gaining sufficient weight during pregnancy to support a healthy infant.

MATERNAL WEIGHT

Birthweight is the most reliable indicator of an infant's health. As a later section of this chapter explains, an underweight infant is more likely to have physical and mental defects, become ill, and die than a normal-weight infant. In general, higher birthweights present fewer risks for infants. Two characteristics of the mother's weight influence an infant's birthweight: her weight *prior* to conception and her weight gain *during* pregnancy.

Weight prior to Conception

A woman's weight prior to conception influences fetal growth. Even with the same weight gain during pregnancy, underweight women tend to have smaller babies than heavier women.

• Underweight • An underweight◆ woman has a high risk of having a low-birthweight infant, especially if she is unable to gain sufficient weight during pregnancy. In addition, the rates of **preterm** births and infant deaths are higher for underweight women.[5] An underweight woman improves her chances of having a healthy infant by gaining sufficient weight prior to conception or by gaining extra pounds during pregnancy. To increase food energy intake, an underweight woman can follow the dietary recommendations for pregnant women (described in Figure 14-10 on p. 474).

• Overweight • Overweight◆ also creates problems related to pregnancy and childbirth. Overweight women have an especially high risk of medical complications such as hypertension, gestational diabetes, and postpartum infections.[6] Compared with other women, overweight women are also more likely to have stillbirths and other complications of labor and delivery.[7]

Compared with other women, overweight women have the lowest rate of low-birthweight infants. In fact, infants of overweight women are more likely to be born **post term** and to weigh more than 9 pounds.◆ Large newborns increase the likelihood of a difficult labor and delivery, birth trauma, and **cesarean section.** Consequently, these infants have a greater risk of poor health and death than infants of normal weight. Overweight women are less likely to have premature infants, but if they do, the infants may be large for their gestational age.

Of greater concern than infant birthweight is the poor development of infants born to obese mothers. Some research suggests that obesity may double the risk for neural tube defects.[8]

Weight-loss dieting during pregnancy is never advisable. Overweight women should try to achieve a healthy body weight before becoming pregnant, avoid excessive weight gain during pregnancy, and postpone weight loss until after childbirth. Weight loss is best achieved by eating moderate amounts of nutrient-dense foods and exercising to lose body fat.

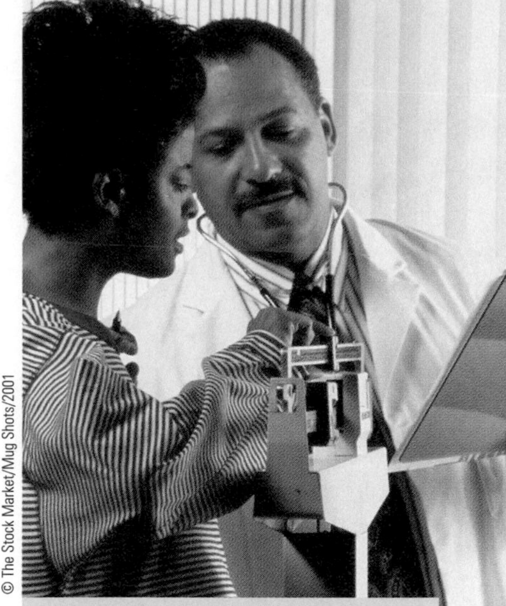

Fetal growth and maternal health depend on a sufficient weight gain during pregnancy.

Underweight is defined as BMI <19.8. ◆

Overweight is defined as BMI 26 to 29. Obese is defined as BMI >29. ◆

The term **macrosomia** (mak-roh-SO-me-ah) describes high-birthweight infants (roughly 9 lb, or 4000 g, or more); macrosomia results from prepregnancy obesity, excessive weight gain during pregnancy, or uncontrolled diabetes. ◆
- **macro** = large
- **soma** = body

Weight Gain during Pregnancy

All pregnant women must gain weight—fetal growth and maternal health depend on it. Maternal weight gain during pregnancy correlates closely with infant birthweight, which is a strong predictor of the health and subsequent development of the infant.

• Recommended Weight Gains • Table 14-1 presents recommended weight gains for various prepregnancy weights. The recommended gain for a woman who begins pregnancy at a healthy weight and is carrying a single fetus is 25 to 35 pounds. An underweight woman needs to gain between 28 and 40 pounds; and an overweight woman, between 15 and 25 pounds. Some women should strive for gains at the upper end of the target range, notably, black women and adolescents who are still growing themselves. Short women (5 feet 2 inches and under) should strive for gains at the lower end of the target range. Women who

TABLE 14-1 Recommended Weight Gains Based on Prepregnancy Weight Status

Prepregnancy Weight Status	Recommended Weight Gain
Underweight (BMI <19.8)	28 to 40 lb (12.5 to 18.0 kg)
Normal weight (BMI 19.8 to 26)	25 to 35 lb (11.5 to 16.0 kg)
Overweight (BMI 26 to 29)	15 to 25 lb (7.0 to 11.5 kg)
Obese (BMI >29)	15 lb minimum (6.8 kg minimum)

SOURCE: Committee on Nutritional Status during Pregnancy and Lactation, Food and Nutrition Board, *Nutrition during Pregnancy* (Washington, D.C.: National Academy Press, 1990), pp. 10, 12.

preterm (infant): an infant born prior to the 38th week of pregnancy; also called a **premature infant.** A **term** infant is born between the 38th and 42nd week of pregnancy.

post term (infant): an infant born after the 42nd week of pregnancy.

cesarean section: a surgically assisted birth involving removal of the fetus by an incision into the uterus, usually by way of the abdominal wall.

are carrying twins should aim for a weight gain of 35 to 45 pounds. If a woman gains more than is recommended early in pregnancy, she should not restrict her energy intake later in order to lose weight. A large weight gain over a short time, however, indicates excessive fluid retention and may be the first sign of the serious medical complication preeclampsia, discussed later.

HEALTHY PEOPLE 2010

Increase the proportion of mothers who achieve a recommended weight gain during their pregnancies.

• Weight-Gain Patterns • For the normal-weight woman, weight gain ideally follows a pattern of 3½ pounds during the first trimester and 1 pound per week thereafter. Health care professionals monitor weight gain using a prenatal weight-gain grid (see Figure 14-6).

• Components of Weight Gain • Women often express concern about the weight gain that accompanies a healthy pregnancy. They may find comfort in a reminder that most of the gain supports the growth and development of the placenta, uterus, blood, and breasts, as well as an optimally healthy 7½-pound

FIGURE 14-6 **Recommended Prenatal Weight Gain Based on Prepregnancy Weight**

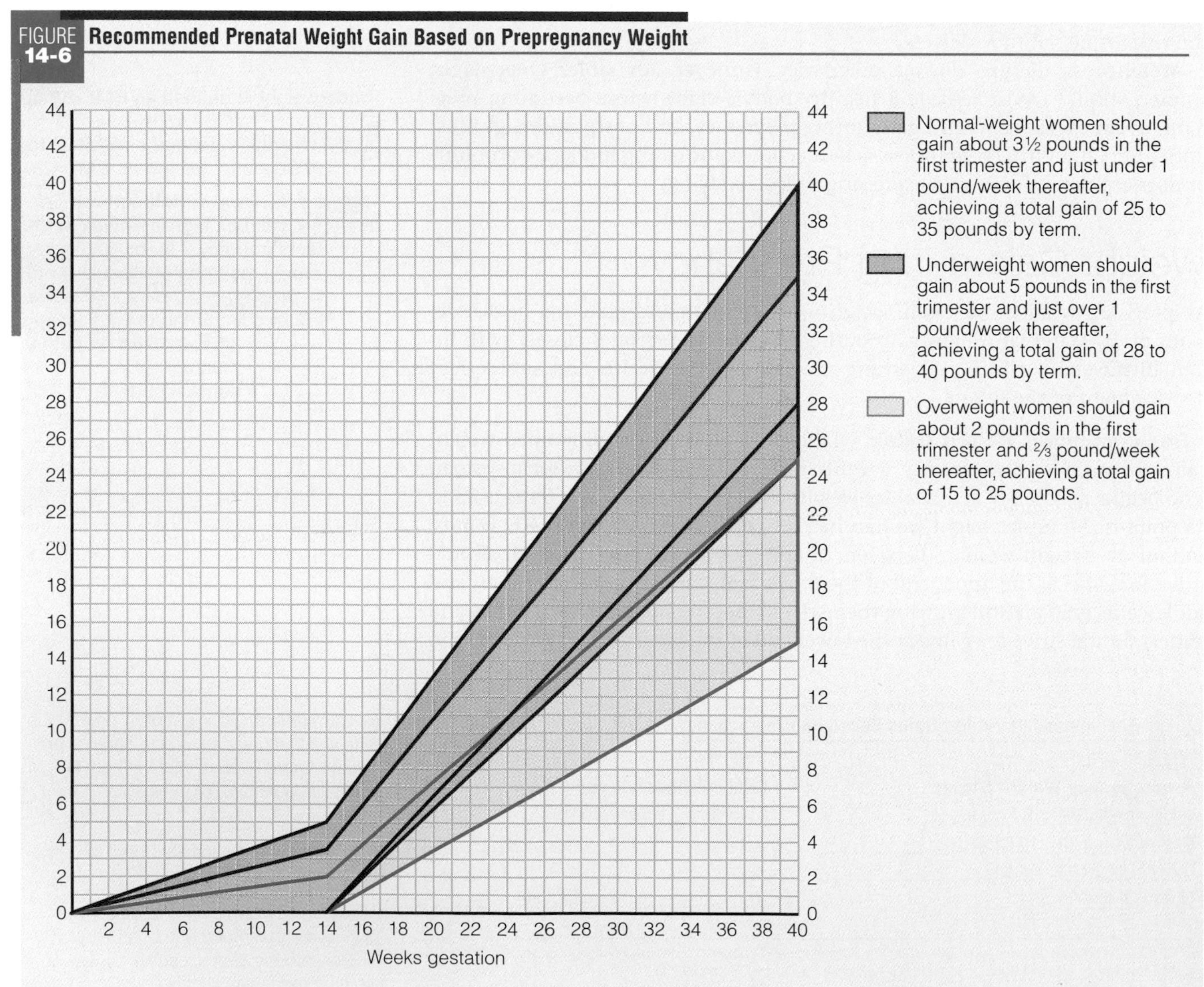

infant. A small amount goes into maternal fat stores, and even that fat is there for a special purpose: to provide energy for labor and lactation. Figure 14-7 shows the components of a typical 30-pound weight gain.

• Weight Loss after Pregnancy • The pregnant woman loses some weight at delivery. In the following weeks, she loses more as her blood volume returns to normal and she sheds accumulated fluids. The typical woman does not, however, return to her prepregnancy weight. In general, the more weight a woman gains beyond the needs of pregnancy, the more she will retain. Even with an average weight gain, though, most women tend to retain a couple of pounds with each pregnancy.

Exercise during Pregnancy

An active, physically fit woman experiencing a normal pregnancy can continue to exercise throughout pregnancy, adjusting the duration and intensity as the pregnancy progresses. Staying active can improve fitness, prevent gestational diabetes, facilitate labor, and reduce stress. Women who exercise during pregnancy report fewer discomforts throughout their pregnancies.[9] Regular exercise develops the strength and endurance a woman needs to carry the extra weight through pregnancy and to labor through an intense delivery. It also maintains the habits that help a woman lose excess weight and get back into shape after the birth.

A pregnant woman should participate in "low-impact" activities and avoid sports in which she might fall or be hit by other people or objects. For example,

FIGURE 14-7 **Components of Weight Gain during Pregnancy**

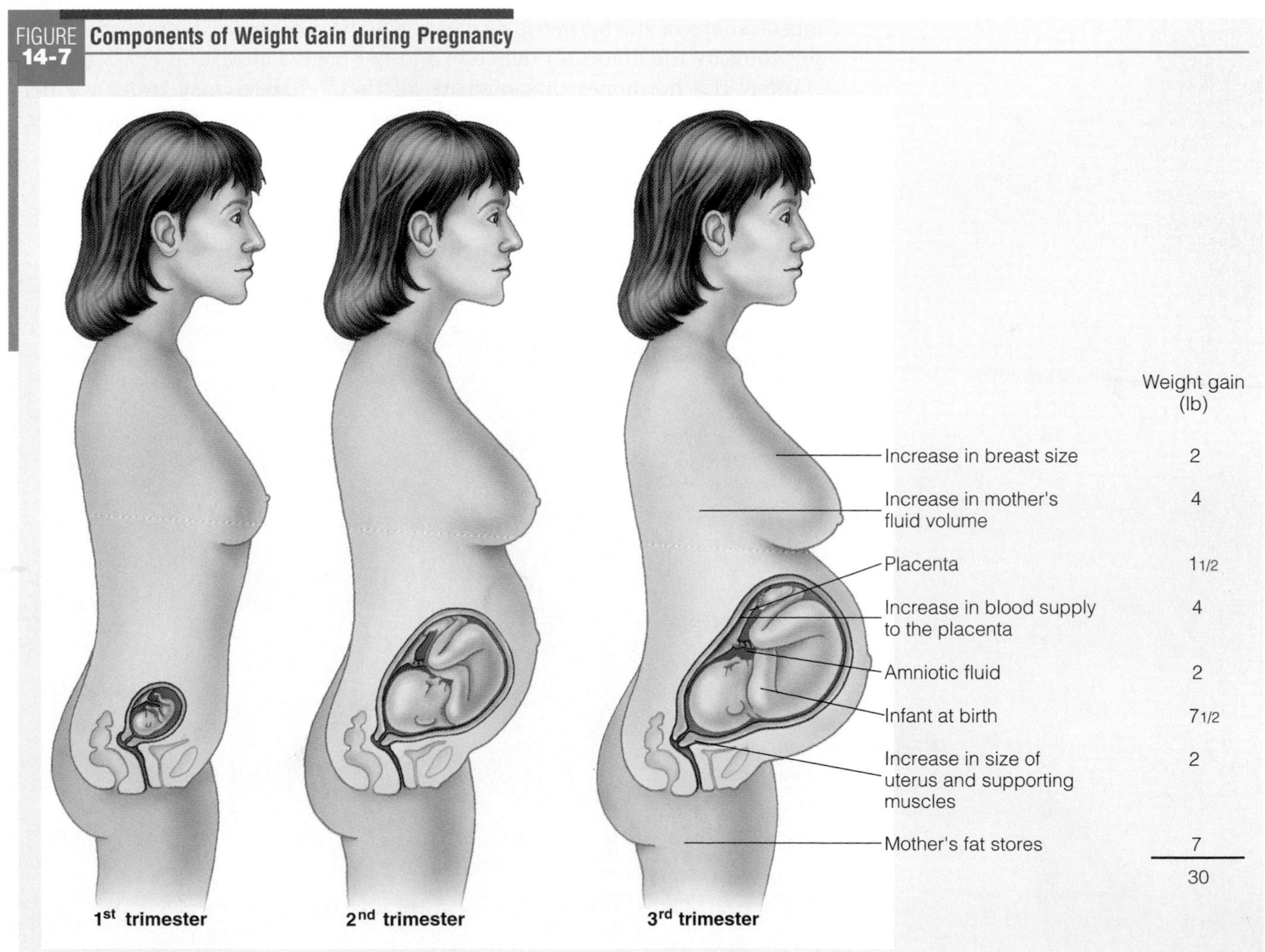

playing singles tennis with one person on each side of the net is safer than a fast-moving game of racquetball in which the two competitors can collide. Swimming and water aerobics are particularly beneficial because they allow the body to remain cool and move freely with the water's support, thus reducing back pain.[10] Figure 14-8 provides some guidelines for exercise during pregnancy.[11] Several of the guidelines are aimed at preventing excessively high internal body temperature and dehydration, both of which can harm fetal development. To this end, pregnant women should also stay out of saunas, steam rooms, and hot whirlpools.

IN SUMMARY

A healthy pregnancy depends on a sufficient weight gain. Women who begin their pregnancies at a healthy weight need to gain about 30 pounds, which covers the growth and development of the placenta, uterus, blood, breasts, and infant. By remaining active throughout pregnancy, a woman can develop the strength she needs to carry the extra weight and maintain habits that will help her lose it after the birth.

NUTRITION DURING PREGNANCY

A woman's body changes dramatically during pregnancy. Her uterus and its supporting muscles increase in size and strength; her blood volume increases by half to carry the additional nutrients and other materials; her joints become more flexible in preparation for childbirth; her feet swell in response to high concentrations of the hormone estrogen, which promotes water retention and helps to ready the uterus for delivery; and her breasts grow in preparation for lactation. The hormones that mediate all these changes may influence her mood. She can best prepare to handle these changes given a nutritious diet,

FIGURE 14-8 Exercise Guidelines during Pregnancy

DO

- Do exercise regularly (at least three times a week).
- Do warm up with 5 to 10 minutes of light activity.
- Do exercise for 20 to 30 minutes at your target heart rate.
- Do cool down with 5 to 10 minutes of slow activity and gentle stretching.
- Do drink water before, after, and during exercise.
- Do eat enough to support the additional needs of pregnancy plus exercise.

© Michael Salas/The Image Bank

Pregnant women can enjoy the benefits of exercise.

DON'T

- Don't exercise vigorously after long periods of inactivity.
- Don't exercise in hot, humid weather.
- Don't exercise when sick with fever.
- Don't exercise while lying on your back after the first trimester of pregnancy or stand motionless for prolonged periods.
- Don't exercise if you experience any pain or discomfort.
- Don't participate in activities that may harm the abdomen or involve jerky, bouncy movements.

regular physical activity, plenty of rest, and caring companions. This section highlights the role of nutrition.

Energy and Nutrient Needs during Pregnancy

From conception to birth, all parts of the infant—bones, muscles, organs, blood cells, skin, and other tissues—are made from nutrients in the foods the mother eats. For most women, nutrient needs during pregnancy and lactation◆ are higher than at any other time (see Figure 14-9).

◆ The table on the inside front cover provides separate listings for women during pregnancy and lactation, reflecting their heightened nutrient needs.

FIGURE 14-9 **Comparison of Nutrient Recommendations for Nonpregnant, Pregnant, and Lactating Women**

For actual values, turn to the table on the inside front cover.

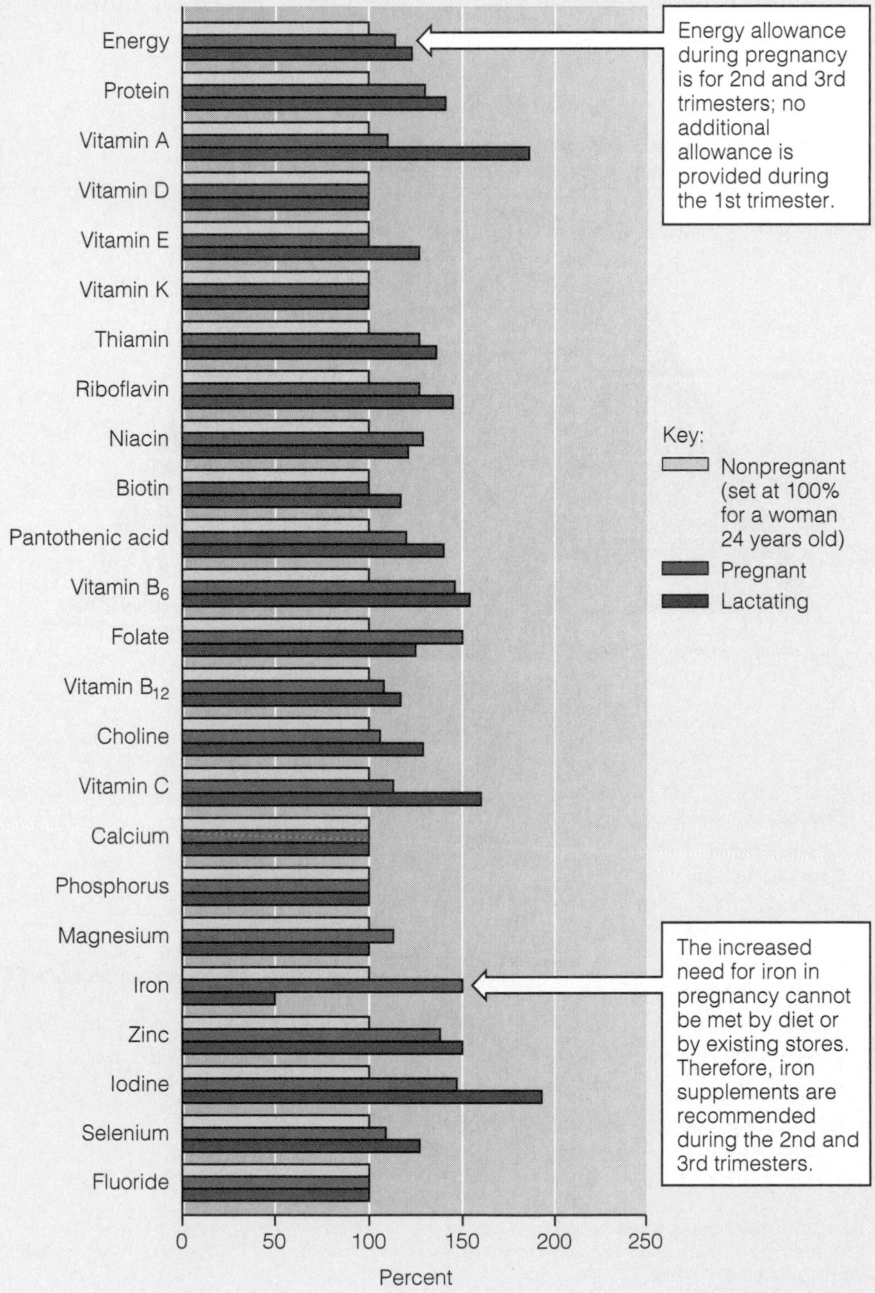

Energy
Protein
Vitamin A
Vitamin D
Vitamin E
Vitamin K
Thiamin
Riboflavin
Niacin
Biotin
Pantothenic acid
Vitamin B_6
Folate
Vitamin B_{12}
Choline
Vitamin C
Calcium
Phosphorus
Magnesium
Iron
Zinc
Iodine
Selenium
Fluoride

0 50 100 150 200 250

Percent

Energy allowance during pregnancy is for 2nd and 3rd trimesters; no additional allowance is provided during the 1st trimester.

The increased need for iron in pregnancy cannot be met by diet or by existing stores. Therefore, iron supplements are recommended during the 2nd and 3rd trimesters.

Key:
Nonpregnant (set at 100% for a woman 24 years old)
Pregnant
Lactating

A pregnant woman's food choices support both her health and her infant's growth and development.

◆ **Energy RDA during pregnancy (2nd and 3rd trimesters):**
+300 kcal/day.
Canadian RNI during pregnancy:
+100 to 300 kcal/day.*

*The lower value indicates recommendations for the first trimester, and the higher value indicates those for the second and third trimesters.

• **Energy** • A pregnant woman needs extra food energy,◆ but only a little extra—300 kcalories above the allowance for nonpregnant women—and only during the second and third trimesters. Pregnant teenagers, underweight women, and physically active women may require more. A woman can easily get 300 kcalories with just one extra serving from each of the five food groups—a slice of bread, a serving of vegetables, an ounce of lean meat, a piece of fruit, and a cup of nonfat milk (see Figure 14-10). Alternatively, a woman who has been neglecting her calcium needs may want to use her additional 300 kcalories for milk and milk products. A variety of strategies are appropriate in meeting the energy demands of pregnancy.[12]

For a 2000-kcalorie daily intake, 300 kcalories represent about 15 percent more food energy than before pregnancy. The increase in nutrient needs is greater than this, so nutrient-dense foods should supply the 300 kcalories: foods such as whole-grain breads and cereals, legumes, dark green vegetables, citrus fruits, nonfat milk and milk products, and lean meats, fish, poultry, and eggs. Ample carbohydrate (ideally, 250 grams or more per day and certainly no less than 100 grams) is necessary to spare the protein needed for growth.

FIGURE 14-10 **Daily Food Choices for Pregnant and Lactating Women**

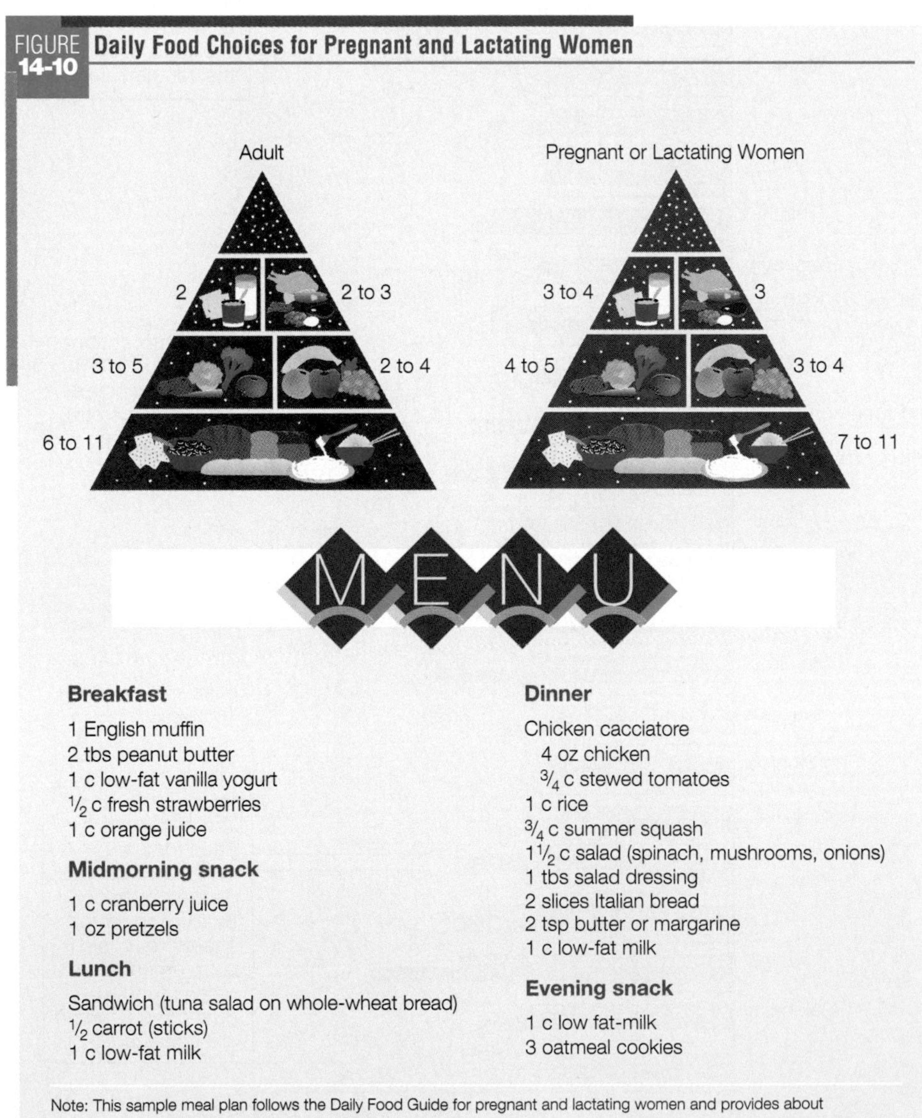

Note: This sample meal plan follows the Daily Food Guide for pregnant and lactating women and provides about 2500 kcalories (55% from carbohydrate, 20% from protein, and 25% from fat). Figure 2-1 in Chapter 2 provides a detailed summary of the Daily Food Guide.

• **Protein** • The protein RDA◆ for pregnancy is 10 grams per day higher than for nonpregnant women. Because people in the United States typically exceed the RDA, most women need not add 10 grams of protein to their diets. In fact, pregnant women in the United States—even those with low incomes who are not participating in food assistance programs—generally receive between 75 and 110 grams of protein a day. Pregnant vegetarian women who meet their energy needs by eating ample servings of protein-containing plant foods such as legumes, whole grains, nuts, and seeds meet their protein needs as well. Use of high-protein supplements during pregnancy can be harmful and is discouraged. Among the problems associated with using high-protein supplements during pregnancy are high rates of low birthweights, preterm births, and deaths.

◆ Protein RDA during pregnancy: +10 g/day.
Canadian RNI during pregnancy: +5 to 24 g/day.*

*For the first trimester, the RNI is an additional 5 g/day; for the second trimester, it is an additional 20 g/day; and for the third trimester, it is 24 g/day.

• **Essential Fatty Acids** • The high nutrient requirements of pregnancy leave little room in the diet for excess fat, but the essential fatty acids are important to the growth of the fetus. Of greatest interest is how the long-chain omega-3 and omega-6 fatty acids influence early development.[13] The brain is largely made of lipid material, and it depends heavily on the long-chain omega-3 and omega-6 fatty acids for its growth, function, and structure. If a pregnant woman regularly eats a diet that includes seafood, she receives a balance of these essential fatty acids◆ and their derivatives. Supplements of fish oil are not recommended, however, both because they may contain concentrated toxins and because their effects on pregnancy remain unknown.

◆ Table 5-2 on p. 145 lists good food sources of the omega fatty acids.

• **Nutrients for Blood Production and Cell Growth** • New cells are laid down at a tremendous pace as the fetus grows and develops. At the same time, the mother's red blood cell mass expands. All nutrients are important in these processes, but for folate, vitamin B_{12}, iron, and zinc, the needs are especially great due to their key roles in the synthesis of DNA and new cells.

The requirement for folate increases dramatically during pregnancy.◆ It is best to obtain sufficient folate from a combination of supplements, fortified foods, and a diet that includes fruits, juices, green vegetables, and whole grains.[14] The "How to" featured in Chapter 10 on p. 325 describes how folate from each of these sources contributes to a day's intake.

◆ Folate RDA during pregnancy: 600 µg/day.

The pregnant woman also has a slightly greater need for the B vitamin that activates the folate enzyme—vitamin B_{12}.◆ Generally, even modest amounts of meat, fish, eggs, or milk products together with body stores easily meet the need for vitamin B_{12}. Vegans who exclude all foods of animal origin, however, need daily supplements of vitamin B_{12} or vitamin B_{12}-fortified foods to prevent deficiency.

◆ Vitamin B_{12} RDA during pregnancy: 2.6 µg/day.

Pregnant women need iron◆ to support their enlarged blood volume and to provide for placental and fetal needs. The developing fetus draws on maternal iron stores to create stores of its own to last through the first four to six months after birth when milk, which is poor in iron, will be its sole food. Also, the blood losses inevitable at birth, especially during a cesarean section, can further drain the mother's supply.*

◆ Iron RDA during pregnancy: 27 mg/day.

During pregnancy, the body makes several adaptations to help meet the exceptionally high need for iron. Menstruation, the major route of iron loss in women, ceases, and iron absorption improves thanks to an increase in blood transferrin, the body's iron-absorbing and iron-carrying protein. Without sufficient intake, though, iron stores would quickly dwindle.

Few women enter pregnancy with adequate iron stores, so a daily iron supplement is recommended during the second and third trimesters for all pregnant women. To enhance absorption, the supplement should be taken between meals or at bedtime on an empty stomach and with liquids other than milk, coffee, or tea, which inhibit iron absorption. (Drinking orange juice does not enhance iron

*On average, almost twice as much blood is lost during a cesarean delivery as during the vaginal delivery of a single fetus.

absorption from supplements as it does from foods; vitamin C enhances iron absorption by converting iron from ferric to ferrous, but supplemental iron is already in the ferrous form.)

HEALTHY PEOPLE 2010

Reduce iron deficiency among pregnant females. Reduce anemia among low-income pregnant females in their third trimester.

◆ Zinc RDA during pregnancy:
13 mg/day (≤18 years).
11 mg/day (19–50 years).

Zinc◆ is required for DNA and RNA synthesis and thus for protein synthesis and cell development. Typical zinc intakes for pregnant women are lower than recommendations, but routine supplementation is not advised. Women taking iron supplements (more than 30 milligrams per day), however, may need zinc supplementation because large doses of iron can interfere with the body's absorption and use of zinc.

• **Nutrients for Bone Development** • Vitamin D and the bone-building minerals calcium, phosphorus, magnesium, and fluoride are in great demand during pregnancy. Insufficient intakes may produce abnormal fetal bones and teeth.

Vitamin D plays a vital role in calcium absorption and utilization. Consequently, severe maternal vitamin D deficiency interferes with normal calcium metabolism, resulting in rickets in the fetus and osteomalacia in the mother. Exposure to sunlight and consumption of vitamin D–fortified milk are usually sufficient to provide the recommended amount of vitamin D during pregnancy. Routine supplementation is not recommended because of the toxicity risk. Vegans who avoid milk, eggs, and fish may receive enough vitamin D from daily exposure to sunlight or from fortified soy milk.

Calcium absorption more than doubles early in pregnancy, helping the mother to meet the calcium needs of pregnancy. During the last trimester, as the fetal bones begin to calcify, over 300 milligrams a day are transferred to the fetus. Recommendations to ensure an adequate calcium intake during pregnancy are aimed at conserving maternal bone while supplying fetal needs.

◆ The Food Guide Pyramid suggests 3 servings of milk for women who are pregnant or breastfeeding.

Calcium intakes for pregnant women◆ typically fall below recommendations. Because bones are still actively depositing minerals until about age 25 or so, adequate calcium is especially important for young women. Pregnant women under age 25 who receive less than 600 milligrams of dietary calcium daily need to increase their intake of milk, cheese, yogurt, and other calcium-rich foods. Alternatively, and less preferably, they may need a daily supplement of 600 milligrams of calcium.

• **Other Nutrients** • The nutrients mentioned here are those most intensely involved in blood production, cell growth, and bone growth. Of course, other nutrients are also needed during pregnancy to support the growth and health of both fetus and mother. Even with adequate nutrition, repeated pregnancies less than a year apart deplete nutrient reserves: fetal growth may be compromised and maternal health may decline. The optimal interval between pregnancies is between 18 and 23 months.[15]

• **Nutrient Supplements** • Women who make wise food choices during pregnancy can meet most of their nutrient needs, except for iron. As mentioned, iron supplements (30 milligrams per day) are recommended during the second and third trimesters of pregnancy. Daily multivitamin-mineral supplements are recommended for women who do not eat adequately and for those in high-risk groups: women carrying multiple fetuses, cigarette smokers, and alcohol and drug abusers. The use of prenatal supplements may help reduce the risks of preterm delivery, low infant birthweights, and birth defects.[16] Table 14-2 lists recommended amounts for supplements.

TABLE 14-2 Nutrient Supplements during Pregnancy[a]

Nutrient	Amount
Folate	400 µg
Vitamin B_6	2 mg
Vitamin C	50 mg
Vitamin D	5 µg
Calcium	600 mg
Copper	2 mg
Iron	30 mg
Zinc	15 mg

[a]For pregnant women at nutritional risk (see Table 14-4 on p. 478).
SOURCE: Reprinted with permission from *Nutrition during Pregnancy* © by the National Academy of Sciences. Published by the National Academy Press, Washington, D.C., 1990.

TABLE 14-3 Strategies to Alleviate Maternal Discomforts

To Alleviate the Nausea of Pregnancy	To Prevent or Alleviate Constipation	To Prevent or Relieve Heartburn
• On waking, arise slowly.	• Eat foods high in fiber (fruits, vegetables, and whole-grain cereals).	• Relax and eat slowly.
• Eat dry toast or crackers.	• Exercise regularly.	• Chew food thoroughly.
• Chew gum or suck hard candies.	• Drink at least eight glasses of liquids a day.	• Eat small, frequent meals.
• Eat small, frequent meals.	• Respond promptly to the urge to defecate.	• Drink liquids between meals.
• Avoid foods with offensive odors.	• Use laxatives only as prescribed by a physician; do not use mineral oil, because it interferes with absorption of fat-soluble vitamins.	• Avoid spicy or greasy foods.
• When nauseated, do not drink citrus juice, water, milk, coffee, or tea.		• Sit up while eating; elevate the head while sleeping.
		• Wait an hour after eating before lying down.
		• Wait two hours after eating before exercising.

Common Nutrition-Related Concerns of Pregnancy

Nausea, constipation, heartburn, and food sensitivities are common nutrition-related concerns during pregnancy. A few simple strategies can help alleviate maternal discomforts (see Table 14-3).

• Nausea • Not all women have uneasy stomachs in the early months of pregnancy, but many do. The nausea of "morning" (actually, anytime) sickness ranges from mild queasiness to debilitating nausea and vomiting. Severe and continued vomiting may require hospitalization if it results in acidosis, dehydration, or excessive weight loss. The hormonal changes of early pregnancy seem to be responsible for a woman's sensitivities to the appearance, texture, or smell of foods. Traditional strategies for quelling nausea are listed in Table 14-3, but some women benefit most from simply eating the foods they want when they feel like eating. They may also find comfort in a cleaner, quieter, and more temperate environment.[17]

• Constipation and Hemorrhoids • As the hormones of pregnancy alter muscle tone and the growing fetus crowds intestinal organs, an expectant mother may experience constipation. She may also develop hemorrhoids (swollen veins of the rectum). These can be painful, and straining during bowel movements may cause bleeding. She can gain relief by following the strategies listed in Table 14-3.

• Heartburn • Heartburn is another common complaint during pregnancy. The hormones of pregnancy relax the digestive muscles, and the growing fetus puts increasing pressure on the mother's stomach. This combination allows stomach acid to back up into the lower esophagus and create a burning sensation near the heart. Tips to help relieve heartburn are included in Table 14-3.

• Food Cravings and Aversions • Some women develop cravings for, or aversions to, particular foods and beverages during pregnancy. These **food cravings** and **food aversions** are fairly common, but do not seem to reflect real physiological needs. In other words, a woman who craves pickles does not necessarily need salt, nor does a woman who craves chocolate need caffeine or fat. Similarly, cravings for ice cream are common in pregnancy, but do not signify a calcium deficiency. Cravings and aversions that arise during pregnancy are most likely due to hormone-induced changes in sensitivity to taste and smell.

• Nonfood Cravings • Some pregnant women develop cravings for nonfood items◆ such as laundry starch, clay, dirt, or ice—a practice known as pica. Pica is especially common among African American women and is often associated with iron-deficiency anemia.[18] Pica is a cultural phenomenon that reflects a

◆ Reminder: The general term for eating nonfood items is *pica.* The specific craving for nonfood items that come from the earth, such as clay or dirt, is known as *geophagia.*

food cravings: strong desires to eat particular foods.

food aversions: strong desires to avoid particular foods.

society's folklore, not a response to the physiological need for a nutrient such as iron; neither clay nor ice provides iron. Pica may lead to anemia by interfering with iron absorption and displacing iron-rich foods from the diet.

IN SUMMARY Energy and nutrient needs are high during pregnancy. A balanced diet that includes an extra serving from each of the five food groups can usually meet these needs, with the exception of iron (supplements are recommended). The nausea, constipation, and heartburn that sometimes accompany pregnancy can usually be alleviated with a few simple strategies. Food cravings do not typically reflect physiological needs.

HIGH-RISK PREGNANCIES

Some pregnancies jeopardize the life and health of the mother and infant. Table 14-4 identifies several characteristics of a **high-risk pregnancy.** A woman with none of these risk factors is said to have a **low-risk pregnancy.** The more factors that apply, the higher the risk. All pregnant women, especially those in high-risk categories, need prenatal care, including dietary◆ advice.[19] The section at the top of the next page describes government efforts to provide food assistance to pregnant women in the United States.

◆ **Nutrition advice in prenatal care:**
- Eat well-balanced meals.
- Take prenatal supplements as prescribed.
- Stop drinking alcohol.
- Gain enough weight to support fetal growth.

HEALTHY PEOPLE 2010

Increase the proportion of pregnant women who receive early and adequate prenatal care.

TABLE 14-4 High-Risk Pregnancy Factors

Factor	Condition That Raises Risk
Maternal weight	
Prior to pregnancy	Prepregnancy BMI either <19.8 or >26.0
During pregnancy	Insufficient or excessive pregnancy weight gain
Maternal nutrition	Nutrient deficiencies or toxicities; eating disorders
Socioeconomic status	Poverty, lack of family support, low level of education, limited food available
Lifestyle habits	Smoking, alcohol or other drug use
Age	Teens, especially 15 years or younger; women 35 years or older
Previous pregnancies	
Number	Many previous pregnancies (3 or more to mothers under age 20; 4 or more to mothers age 20 or older)
Interval	Short intervals between pregnancies (<18 months)
Outcomes	Previous history of problems
Multiple births	Twins or triplets
Birthweight	Low- or high-birthweight infants
Maternal health	
High blood pressure	Development of pregnancy-related hypertension
Diabetes	Development of gestational diabetes
Chronic diseases	Diabetes; heart, respiratory, and kidney disease; certain genetic disorders; special diets and drugs

high-risk pregnancy: a pregnancy characterized by indicators that make it likely the birth will be surrounded by problems such as premature delivery, difficult birth, retarded growth, birth defects, and early infant death.

low-risk pregnancy: a pregnancy characterized by indicators that make a normal outcome likely.

Malnutrition and Pregnancy

Good nutrition clearly supports a pregnancy. In contrast, malnutrition interferes with the ability to conceive, the likelihood of implantation, and the subsequent development of a fetus should conception and implantation occur.

• Malnutrition and Fertility • The nutrition habits and lifestyle choices people make can influence the course of a pregnancy they are not even planning at the time. Severe malnutrition and food deprivation can reduce **fertility:** women may develop amenorrhea,◆ and men may lose their ability to produce viable sperm. Furthermore, both men and women lose sexual interest during times of starvation. Starvation arises predictably during famines, wars, and droughts, but can also occur amidst peace and plenty. Many young women who diet excessively are starving and suffering from malnutrition (see Highlight 9).

Food Assistance Programs for Pregnant Women, Infants, and Children

WIC (the Special Supplemental Nutrition Program for Women, Infants, and Children) provides nutrition education and nutritious foods to low-income pregnant women and their young children. WIC provides eggs, milk, iron-fortified cereal, vitamin C–rich juice, cheese, legumes, peanut butter, and infant formula to infants, children up to age five, and pregnant and breastfeeding women who qualify financially and have a high risk of medical or nutritional problems. The program is both remedial and preventive: services include health care referrals, nutrition education, and food packages or vouchers for specific foods to supply nutrients known to be lacking in the diets of the target population. Most notably, these foods provide protein, calcium, iron, vitamin A, and vitamin C.

Over 7 million people receive WIC benefits each month. Prenatal WIC participation can effectively reduce infant mortality, low birthweight, and maternal and newborn medical costs. For every dollar spent on WIC, an estimated three dollars in medical costs are saved in the first two months after birth.

WIC Program
www.fns.usda.gov/fns
Visit the WIC Program

• Malnutrition and Early Pregnancy • If a malnourished woman does become pregnant, she faces the challenge of supporting both the growth of a baby and her own health with inadequate nutrient stores. Malnutrition prior to and around conception prevents the placenta from developing fully. A poorly developed placenta cannot deliver optimum nourishment to the fetus, and the infant will be born small and possibly with physical and cognitive abnormalities.[20] If this small infant is a female, she may develop poorly and have an elevated risk of developing a chronic condition that could impair her ability to give birth to a healthy infant. Thus a woman's malnutrition can adversely affect not only her children but her *grandchildren.*

◆ The temporary or permanent absence of menstrual periods is called *amenorrhea.* Amenorrhea is normal before puberty, after menopause, during pregnancy, and during lactation; otherwise it is abnormal.

• Malnutrition and Fetal Development • Without adequate nutrition during pregnancy, fetal growth and infant health are compromised. In general, consequences of malnutrition during pregnancy include fetal growth retardation, congenital malformations (birth defects), spontaneous abortion and stillbirth, premature birth, and low infant birthweight. Malnutrition, coupled with low birthweight, is a factor in more than half of all deaths of children under four years of age worldwide.

◆ Some preterm infants are of a weight **appropriate for gestational age (AGA);** others are **small for gestational age (SGA),** often reflecting malnutrition.

The Infant's Birthweight

A high-risk pregnancy is likely to produce an infant with **low birthweight.** Low-birthweight infants, defined as infants who weigh 5½ pounds or less, are classified according to their gestational age. Preterm infants are born before they are fully developed; they are often underweight and have trouble breathing because their lungs are immature. Preterm infants may be small, but if their size and weight are appropriate for their age,◆ they can catch up in growth given adequate nutrition support. In contrast, small-for-gestational-age infants have suffered growth failure in the uterus and do not catch up as well. For the most part, survival improves with increased gestational age and birthweight.

Low-birthweight infants are more likely to experience complications during delivery than normal-weight babies. They also have a statistically greater chance of having physical and mental birth defects, contracting diseases, and

fertility: the capacity of a woman to produce a normal ovum periodically and of a man to produce normal sperm; the ability to reproduce.

low birthweight (LBW): a birthweight of 5½ lb (2500 g) or less; indicates probable poor health in the newborn and poor nutrition status in the mother during pregnancy, before pregnancy, or both. Normal birthweight for a full-term baby is 6½ to 8¾ lb (about 3000 to 4000 g).

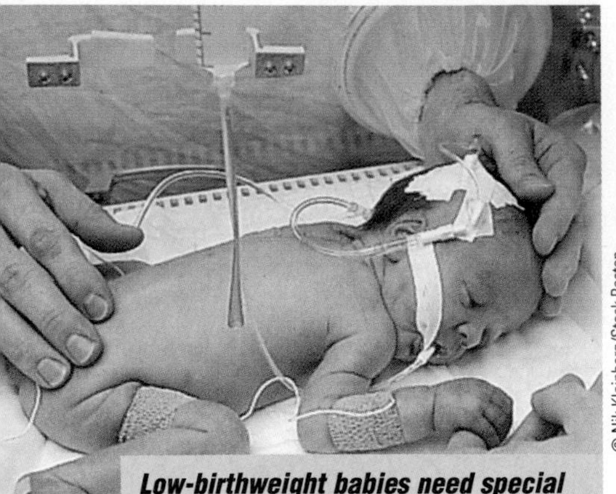
© Nik Kleinberg/Stock Boston

Low-birthweight babies need special care and nourishment.

dying early in life. Of infants who die before their first birthdays, about two-thirds were low-birthweight newborns. Very-low-birthweight infants struggle not only for their immediate physical health and survival, but for their future cognitive development and abilities as well.[21]

A strong relationship is evident between socioeconomic disadvantage and low birthweight. Low socioeconomic status impairs fetal development by causing stress and by limiting access to medical care and to nutritious foods. Low socioeconomic status often accompanies teen pregnancies, smoking, and alcohol and drug abuse—all predictors of low birthweight.

HEALTHY PEOPLE 2010

Reduce low birthweight (LBW) and very low birthweight (VLBW).

Maternal Health

Medical disorders can threaten the life and health of both mother and fetus. If diagnosed and treated early, many diseases can be managed to ensure a healthy outcome—another strong argument for early prenatal care.

• Preexisting Diabetes • Whether diabetes presents risks depends on how well it is controlled before and during pregnancy. Without proper management of maternal diabetes, women face high infertility rates, and those who do conceive may experience episodes of severe hypoglycemia or hyperglycemia, spontaneous abortions, and pregnancy-related hypertension. Infants may be large, suffer physical and mental abnormalities, and experience other complications such as severe hypoglycemia or respiratory distress, both of which can be fatal. Ideally, a woman with diabetes will receive the prenatal care needed to achieve glucose control before conception and continued glucose control throughout pregnancy.

• Gestational Diabetes • Approximately 1 in 25 women who does not have diabetes develops a condition known as **gestational diabetes** during pregnancy. Gestational diabetes usually develops during the second half of pregnancy, with subsequent return to normal after childbirth. Almost one-third of all women with gestational diabetes, however, develop diabetes (type 2) later in life, especially if they are overweight. For this reason, health care professionals advise against excessive weight gain.

The most common consequences of gestational diabetes are complications during labor and delivery and a high infant birthweight. Birth defects associated with gestational diabetes include heart damage, limb deformities, and neural tube defects. To ensure that the problems of gestational diabetes are dealt with promptly, physicians look for the risk factors◆ listed in the margin and screen for glucose intolerance between 24 and 28 weeks gestation.[22] Dietary recommendations encourage small, frequent meals and a bedtime snack (plus additional snacks if blood glucose falls below normal). To maintain normal blood glucose levels and limit excessive weight gain, foods should be high in fiber and other complex carbohydrates and low in fat and sugar. Diet and exercise may control gestational diabetes, but if blood glucose fails to normalize, insulin may be required.

◆ **Risk factors for gestational diabetes:**
- Age 35 or older.
- BMI >25 or excessive weight gain.
- Complications in previous pregnancies, including high-birthweight infants.
- Symptoms of diabetes.
- Family history of diabetes.
- Hispanic, black, Native American, South or East Asian, Pacific Islander, or indigenous Australian.

• Preexisting Hypertension • Hypertension complicates pregnancy and affects its outcome in different ways, depending on when the hypertension first develops and on how severe it becomes.[23] In addition to the threats hypertension always carries (such as heart attack and stroke), high blood pressure increases the risks of a low-birthweight infant or the separation of the placenta from the wall of the uterus before the birth, resulting in stillbirth. Ideally, before a woman with hypertension becomes pregnant, her blood pressure will be under control.

gestational diabetes: abnormal glucose tolerance that is first detected during pregnancy.

• Transient Hypertension of Pregnancy • Some women develop hypertension during the second half of pregnancy.* Most often, the rise in blood pressure is mild and does not affect the pregnancy adversely. Blood pressure usually returns to normal during the first few weeks after childbirth. This **transient hypertension of pregnancy** differs from the life-threatening hypertensive diseases◆ of pregnancy—preeclampsia and eclampsia.

• Preeclampsia and Eclampsia • Hypertension may signal the onset of **preeclampsia,** a condition characterized not only by high blood pressure but also by protein in the urine and fluid retention (edema). The edema◆ of preeclampsia is a whole-body edema, distinct from the localized fluid retention women normally experience late in pregnancy.

Preeclampsia usually occurs with first pregnancies◆ and most often after 20 weeks gestation. Symptoms typically regress within two days of delivery. Black women have a much greater risk of preeclampsia than white women.[24]

Preeclampsia affects almost all of the mother's organs—the circulatory system, liver, kidneys, and brain. Blood flow through the vessels that supply oxygen and nutrients to the placenta diminishes. For this reason, preeclampsia often retards fetal growth. In some cases, the placenta separates from the uterus, resulting in stillbirth.

Preeclampsia can progress rapidly to **eclampsia**—a condition characterized by convulsive seizures and coma. Maternal death during pregnancy and childbirth is extremely rare in developed countries, but when it does occur, eclampsia is a common cause. The rate of death for black women with eclampsia is over four times the rate for white women.[25]

Preeclampsia demands prompt medical attention. Treatment focuses on controlling blood pressure and preventing convulsions. If preeclampsia develops early and is severe, induced labor or cesarean section may be necessary, regardless of gestational age. The infant will be preterm, with all of the associated problems, including poor lung development and special care needs. Several dietary factors have been studied, but none have proved conclusive in preventing preeclampsia.[26]

◆ The hypertensive diseases of pregnancy are sometimes called **toxemia.**

◆ The normal edema of pregnancy responds to gravity; fluid pools in the ankles. The edema of preeclampsia is a generalized edema. The differences between these two types of edema help with the diagnosis of preeclampsia.

◆ Warning signs of preeclampsia:
- Hypertension.
- Protein in the urine.
- Upper abdominal pain.
- Severe and constant headaches.
- Swelling, especially of the face.
- Dizziness.
- Blurred vision.
- Sudden weight gain (1 lb/day).
- Fetal growth retardation.

The Mother's Age

Maternal age also influences the course of a pregnancy. Compared with women of the physically ideal childbearing age of 20 to 25, both younger and older women face more complications of pregnancy.

• Pregnancy in Adolescents • Many adolescents become sexually active before age 19, and over 700,000 adolescent girls face pregnancies each year in the United States; about two-thirds of them give birth.[27] Put another way, about one out of every eight babies is born to a teenager.[28] Nourishing a growing fetus adds to a teenage girl's nutrition burden, especially if her growth is still incomplete.[29] Simply being young increases the risks of pregnancy complications independently of important socioeconomic factors.[30]

Common complications among adolescent mothers include iron-deficiency anemia (which may reflect poor diet and inadequate prenatal care) and prolonged labor (which reflects the mother's physical immaturity). On a positive note, maternal death is lowest for mothers under age 20.

Pregnant teenagers have higher rates of stillbirths, preterm births, and low-birthweight infants than do adult women.[31] Many of these infants suffer physical problems, require intensive care, and die within the first year. The care of

*Blood pressure of 140/90 millimeters mercury during the second half of pregnancy in a woman who has not previously exhibited hypertension indicates high blood pressure. So does a rise in systolic blood pressure of 30 millimeters or in diastolic blood pressure of 15 millimeters on at least two occasions more than six hours apart. By this rule, an apparently "normal" blood pressure of 120/85 would be high for a woman whose normal value was 90/70.

transient hypertension of pregnancy: high blood pressure that develops in the second half of pregnancy and resolves after childbirth, usually without affecting the outcome of the pregnancy.

preeclampsia (PRE-ee-KLAMP-see-ah): a condition characterized by hypertension, fluid retention, and protein in the urine; formerly known as *pregnancy-induced hypertension.**

eclampsia (eh-KLAMP-see-ah): a severe stage of preeclampsia characterized by convulsions.

*The Working Group on High Blood Pressure in Pregnancy, convened by the National High Blood Pressure Education Program of the National Heart, Lung, and Blood Institute, suggested abandoning the term *pregnancy-induced hypertension* because it failed to differentiate between the mild, transient hypertension of pregnancy and the life-threatening hypertension of preeclampsia.

infants born to teenagers costs our society an estimated $1 billion annually. Because teenagers have few financial resources, they cannot pay these costs. Furthermore, their low economic status contributes significantly to the complications surrounding their pregnancies. At a time when prenatal care is most important, it is less accessible. And the pattern of teenage pregnancies continues from generation to generation, with almost 40 percent of the daughters born to teenage mothers becoming teenage mothers themselves.[32] Clearly, teenage pregnancy is a major public health problem.

HEALTHY PEOPLE 2010

Reduce pregnancies among adolescent females.

To support the needs of both mother and fetus, young teenagers (13 to 16 years old) are encouraged to strive for the highest weight gains recommended for pregnancy. For a teen who enters pregnancy at a healthy body weight, a weight gain of approximately 35 pounds is recommended; this amount minimizes the risk of delivering a low-birthweight infant. Gaining less weight may limit fetal growth. Pregnant and lactating teenagers can use the food guide presented in Figure 14-10 (on p. 474), making sure to select at least 4 servings of milk or milk products daily.

Without the appropriate economic, social, and physical support, a young mother will not be able to care for herself during her pregnancy and for her child after the birth. To improve her chances for a successful pregnancy and a healthy infant, she must seek prenatal care. WIC helps pregnant teenagers obtain adequate food for themselves and their infants (WIC was introduced on p. 479).

• **Pregnancy in Older Women** • In the last several decades, many women have delayed childbearing while they pursue education and careers. As a result, the number of first births to women 35 and older has increased dramatically.

Each year, 994 out of every 1000 pregnant women over the age of 35 have healthy pregnancies.[33] The few complications associated with later childbearing reflect chronic conditions such as hypertension and diabetes, which can complicate an otherwise healthy pregnancy. These complications often result in a cesarean section, which is twice as common in women over 35 as among younger women. For all these reasons, maternal death rates are higher in women over 35 than in younger women.

The babies of older mothers face problems of their own. Because 1 out of 50 pregnancies in older women produces an infant with genetic abnormalities, obstetricians routinely screen women older than 35. For a 40-year-old mother, the risk of having a child with **Down syndrome,** for example, is about 1 in 100 compared with 1 in 300 for a 35-year-old and 1 in 10,000 for a 20-year-old. Fetal death is twice as high for women 35 years and older than for younger women.[34] Why this is so remains a bit of a mystery. One possibility is that the uterine blood vessels of older women may not fully adapt to the increased demands of pregnancy.

Practices Incompatible with Pregnancy

Besides malnutrition, a variety of lifestyle factors can have adverse effects on pregnancy; and some may be teratogenic.◆ People who are planning to have children can make the choice to practice healthy behaviors.

◆ **Reminder: The word *teratogenic* describes a factor that causes abnormal fetal development and birth defects.**

Down syndrome: a genetic abnormality that causes mental retardation, short stature, and flattened facial features.

• **Alcohol** • Alcohol consumption during pregnancy can cause irreversible mental and physical retardation of the fetus—fetal alcohol syndrome (FAS). Of the leading causes of mental retardation, FAS is the only one that is totally *preventable.* To that end, the surgeon general has issued a statement that pregnant women should drink absolutely no alcohol. Fetal alcohol syndrome is the topic of Highlight 14, which includes mention of how alcohol consumption by men may also affect fertility and fetal development.

HEALTHY PEOPLE 2010

Increase abstinence from alcohol among pregnant women. Reduce the occurrence of fetal alcohol syndrome (FAS).

© The Stock Market/Jose I. Pelaez/R, 1999

Young adults can prepare for a healthy pregnancy by taking care of themselves today.

• Medicinal Drugs • Drugs other than alcohol can also cause complications during pregnancy, problems in labor, and serious birth defects.[35] For these reasons, pregnant women should not take any medicines without consulting their physicians.

• Herbal Supplements • Similarly, pregnant women should seek a physician's advice before using herbal supplements as well. Women sometimes seek herbal preparations during their pregnancies to induce labor, aid digestion, promote water loss, support restful sleep, and fight depression. Some herbs may be safe, but others are definitely harmful.

• Illicit Drugs • The recommendation to avoid drugs during pregnancy also includes illicit drugs, of course. Unfortunately, use of illicit drugs, such as cocaine and marijuana, is common among some pregnant women.*

Drugs of abuse, such as cocaine, easily cross the placenta and impair fetal growth and development.[36] Furthermore, they are responsible for preterm births, low-birthweight infants, perinatal deaths, and sudden infant deaths.[37] If these newborns survive, central nervous system damage is evident: their cries, sleep, and behaviors early in life are abnormal, and their cognitive development later in life is impaired.[38] They may be hypersensitive or underaroused; those who test positive for drugs suffer the greatest effects of toxicity and withdrawal.[39]

• Smoking and Chewing Tobacco • Smoking cigarettes and chewing tobacco at any time exert harmful effects, and pregnancy dramatically magnifies the hazards of these practices. Smoking restricts the blood supply to the growing fetus and so limits oxygen and nutrient delivery and waste removal. Unfortunately, an estimated 12 percent of pregnant women smoke, with higher rates for unmarried women and those who have not graduated from high school.[40] Smokers tend to eat less nutritious foods during their pregnancies than do nonsmokers, which further impairs fetal development.

Of all preventable causes of low birthweight in the United States, smoking has the greatest impact. A mother who smokes is more likely to have a complicated birth and a low-birthweight infant.[41] Furthermore, smoking causes death in an otherwise healthy fetus or newborn. A positive relationship exists between **sudden infant death syndrome (SIDS)** and both cigarette smoking during pregnancy and postnatal exposure to passive smoke.[42] Smoking during pregnancy may even harm the intellectual and behavioral development of the child later in life.[43] The margin◆ lists other complications of smoking during pregnancy.

◆ **Complications associated with smoking during pregnancy:**
- Fetal growth retardation.
- Low birthweight.
- Complications at birth (prolonged final stage of labor).
- Mislocation of the placenta.
- Premature separation of the placenta.
- Vaginal bleeding.
- Spontaneous abortion.
- Fetal death.
- Sudden infant death syndrome (SIDS).
- Middle ear diseases.
- Cardiac and respiratory diseases.

HEALTHY PEOPLE 2010

Increase smoking cessation during pregnancy. Increase abstinence from cigarettes among pregnant women.

Infants of mothers who chew tobacco also have low birthweights and high rates of fetal death. Any woman who smokes cigarettes or chews tobacco and is considering pregnancy or who is already pregnant should try to quit.

sudden infant death syndrome (SIDS): the unexpected and unexplained death of an apparently well infant; the most common cause of death of infants between the second week and the end of the first year of life; also called *crib death.*

*It is estimated that 17 percent of pregnant women use marijuana and at least 6 percent use cocaine.

• **Environmental Contaminants** • Infants and young children of pregnant women exposed to environmental contaminants such as lead show signs of impaired cognitive development. During pregnancy, lead readily moves across the placenta, inflicting severe damage on the developing fetal nervous system. In addition, infants exposed to even low levels of lead during gestation weigh less at birth and consequently struggle to survive.[44] For these reasons, it is particularly important that pregnant women receive foods and beverages grown and prepared in environments free of contamination.◆

◆ Chapter 13 includes a discussion of contaminant minerals.

• **Vitamin-Mineral Megadoses** • The pregnant woman who is trying to eat well may mistakenly assume that more is better when it comes to vitamin-mineral supplements. This is simply not true; many vitamins and minerals are toxic when taken in excess. Researchers found that among women who took more than 10,000 IU of supplemental vitamin A daily, approximately 1 out of every 57 infants was born with a malformation of the cranial nervous system that was attributable to high vitamin A intake.[45] Intakes before the seventh week appeared to be the most damaging. (Review Figure 14-4 on p. 466 to see how many tissues are in their critical periods prior to the seventh week.) For this reason, vitamin A is not given as a supplement in the first trimester of pregnancy unless there is specific evidence of deficiency, which is rare. A pregnant woman can obtain all the vitamin A and most of the other vitamins and minerals she needs by making wise food choices. She should take supplements only on the advice of a registered dietitian or physician.

• **Caffeine** • Caffeine◆ crosses the placenta, and the developing fetus has a limited ability to metabolize it. Research studies have not proved that caffeine (even in high doses) causes birth defects in human infants (as it does in animals), but some evidence suggests that moderate to heavy use may increase the risk of spontaneous abortion.[46] (Heavy caffeine use was defined as the equivalent of 3 to 6 cups of coffee a day.) All things considered, it might be most sensible to limit caffeine consumption to the equivalent of a cup of coffee or two 12-ounce cola beverages a day.

◆ The caffeine contents of selected beverages, foods, and drugs are listed at the beginning of Appendix H.

• **Weight-Loss Dieting** • Weight-loss dieting, even for short periods, is hazardous during pregnancy. Low-carbohydrate diets or fasts that cause ketosis deprive the fetal brain of needed glucose and may impair cognitive development. Such diets are also likely to lack other nutrients vital to fetal growth. Regardless of prepregnancy weight, pregnant women should never intentionally lose weight.

• **Sugar Substitutes** • Artificial sweeteners have been extensively investigated and found to be acceptable during pregnancy if used within FDA guidelines.[47] (Women with phenylketonuria should not use aspartame, as Highlight 4 explains.) Still, it would be prudent for pregnant women to use sweeteners in moderation and within an otherwise nutritious and well-balanced diet.

IN SUMMARY

High-risk pregnancies, especially for teenagers, threaten the life and health of both mother and infant. Proper nutrition and abstinence from smoking, alcohol, and other drugs improve the outcome. In addition, prenatal care includes monitoring pregnant women for gestational diabetes and preeclampsia.

NUTRITION DURING LACTATION

Before the end of her pregnancy, a woman will need to consider whether to feed her infant breast milk,◆ infant formula, or both. These options are the only recommended foods for an infant during the first four to six months of life.

◆ To learn about breastfeeding, a pregnant woman can read at least one of the many books available. Appendix F provides a list of other nutrition resources, including LaLeche League International.

In many countries around the world, a woman breastfeeds her newborn without considering the alternatives or consciously making a decision. In other parts of the world, a woman feeds her newborn formula simply because she knows so little about breastfeeding. She may have misconceptions or feel uncomfortable about a process she has never seen or experienced. Breastfeeding offers many health benefits to both mother and infant, and every pregnant woman should seriously consider it (see Table 14-5).[48] Still, there are valid reasons for not breastfeeding, and formula-fed infants grow and develop into healthy children.

A woman who decides to breastfeed offers her infant a full array of nutrients and protective factors to support optimal health and development.

HEALTHY PEOPLE 2010

Increase the proportion of mothers who breastfeed their babies.

Lactation: A Physiological Process

Lactation naturally follows pregnancy—and the mother's body continues to nourish the infant. The **mammary glands** secrete milk for this purpose. The mammary glands develop during puberty, but remain fairly inactive until pregnancy. During pregnancy, hormones promote the growth and branching of a duct system in the breasts and the development of the milk-producing cells.

The hormones **prolactin** and **oxytocin** finely coordinate lactation. The infant's demand for milk stimulates the release of these hormones, which signal the mammary glands to supply milk. Prolactin is responsible for milk production. Prolactin concentrations remain high and milk production continues as long as the infant is nursing.

The hormone oxytocin causes the mammary glands to eject milk into the ducts, a response known as the **let-down reflex.** The mother feels this reflex as a contraction of the breast, followed by the flow of milk and the release of pressure. By relaxing and eating well, the nursing mother promotes easy let-down of milk and greatly enhances her chances of successful lactation.

Breastfeeding: A Learned Behavior

Lactation is an automatic physiological process that virtually all mothers are capable of doing. Breastfeeding, on the other hand, is a learned behavior that

TABLE 14-5 Benefits of Breastfeeding

For infants:

- Provides a favorable balance of nutrients with high bioavailability.
- Provides hormones that promote physiological development.
- Improves cognitive development.
- Protects against a variety of infections.
- May protect against some chronic diseases, such as diabetes (type 1) and hypertension, later in life.
- Protects against food allergies.

For mothers:

- Contracts the uterus.
- Delays the return of regular ovulation, thus lengthening birth intervals. (It is not, however, a dependable method of contraception.)
- Conserves iron stores (by prolonging amenorrhea).
- May protect against breast cancer.

SOURCES: Committee on Nutritional Status during Pregnancy and Lactation, *Nutrition during Lactation* (Washington, D.C.: National Academy Press, 1991), pp. 153–212; J. W. Anderson, B. M. Johnstone, and D. T. Remley, Breastfeeding and cognitive development: A meta-analysis. *American Journal of Clinical Nutrition* 70 (1999): 525–535.

lactation: production and secretion of breast milk for the purpose of nourishing an infant.

mammary glands: glands of the female breast that secrete milk.

prolactin (pro-LAK-tin): a hormone secreted from the anterior pituitary gland that acts on the mammary glands to initiate and sustain milk production.
- **pro** = promote
- **lacto** = milk

oxytocin (OK-see-TOH-sin): a hormone that stimulates the mammary glands to eject milk during lactation and the uterus to contract during childbirth.

let-down reflex: the reflex that forces milk to the front of the breast when the infant begins to nurse.

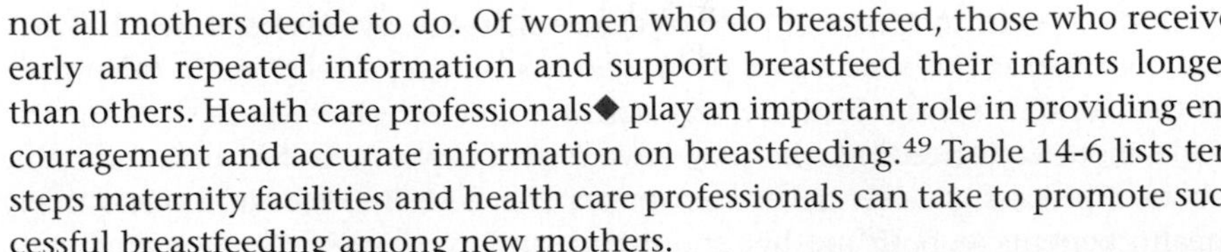

not all mothers decide to do. Of women who do breastfeed, those who receive early and repeated information and support breastfeed their infants longer than others. Health care professionals◆ play an important role in providing encouragement and accurate information on breastfeeding.[49] Table 14-6 lists ten steps maternity facilities and health care professionals can take to promote successful breastfeeding among new mothers.

◆ Some hospitals employ *certified lactation consultants* who specialize in helping new mothers to establish a healthy breastfeeding relationship with their newborn. These consultants are often registered nurses with specialized training in breast and infant anatomy and physiology.

Fathers also play an important role in encouraging breastfeeding.[50] Studies report that most fathers whose partners plan to breastfeed support that decision and respect breastfeeding women. By comparison, fathers whose partners plan to formula feed believe that breastfeeding makes the breasts ugly and interferes with sexual relations. Clearly, educating fathers could change attitudes and promote breastfeeding.

Most healthy women who want to breastfeed can do so with a little preparation; physical obstacles to breastfeeding are rare, although overweight mothers seem to have less success initiating breastfeeding than others.[51] Successful breastfeeding requires adequate nutrition and rest. This, plus the support of all who care, will help to enhance the well-being of mother and infant.

The Mother's Nutrient Needs

Ideally, the mother who chooses to breastfeed her infant will continue to eat nutrient-dense foods throughout lactation. An adequate diet is needed to support the stamina, patience, and self-confidence that nursing an infant demands.

• **Energy Intake and Exercise** • A nursing mother produces about 25 ounces of milk per day, with considerable variation from woman to woman and in the same woman from time to time, depending primarily on the infant's demand for milk. To produce an adequate supply of milk, a woman needs extra energy—almost 650 kcalories a day above her regular need during the first six months of lactation. To meet this energy need,◆ the woman is advised to eat an extra 500 kcalories of food each day and let the fat reserves she accumulated during pregnancy provide the rest. Energy needs for women who are breastfeeding exclusively range from 2500 to 3300 kcalories a day, depending on

◆ Energy RDA during lactation: +500 kcal/day (1800 kcal/day minimum).
Canadian RNI during lactation: +450 kcal/day.

TABLE 14-6 Ten Steps to Successful Breastfeeding

To promote breastfeeding, every maternity facility should:

- Develop a written breastfeeding policy that is routinely communicated to all health care staff.
- Train all health care staff in the skills necessary to implement the breastfeeding policy.
- Inform all pregnant women about the benefits and management of breastfeeding.
- Help mothers initiate breastfeeding within ½ hour of birth.
- Show mothers how to breastfeed and how to maintain lactation, even if they need to be separated from their infants.
- Give newborn infants no food or drink other than breast milk, unless medically indicated.
- Practice rooming-in, allowing mothers and infants to remain together 24 hours a day.
- Encourage breastfeeding on demand.
- Give no artificial nipples or pacifiers to breastfeeding infants.[a]
- Foster the establishment of breastfeeding support groups and refer mothers to them at discharge from the facility.

[a]Compared with nonusers, infants who use pacifiers breastfeed less frequently and stop breastfeeding at a younger age. C. G. Victora and coauthors, Pacifier use and short breastfeeding duration: Cause, consequence, or coincidence? *Pediatrics* 99 (1997): 445–453.
SOURCE: United Nations Children's Fund and World Health Organization, *Barriers and Solutions to the Global Ten Steps to Successful Breast-feeding*, 1994.

Nutritious foods support successful lactation.

physical activity.[52] Lower intakes may support more rapid weight loss, but may not meet vitamin and mineral needs.[53] Most women need at least 1800 kcalories a day to receive all the nutrients required for successful lactation. Severe energy restriction may hinder milk production.

After the birth of the infant, many women are in a hurry to lose the extra body fat they accumulated during pregnancy. Opinions differ as to whether breastfeeding helps with postpartum weight loss. In general, most women lose 1 to 2 pounds a month during the first four to six months of lactation; some may lose more, and others may maintain or even gain weight. Neither the quality nor the quantity of breast milk is adversely affected by moderate weight loss.

Women often exercise to lose weight and improve fitness, and this is compatible with breastfeeding and infant growth.[54] Intense physical activity can raise the lactic acid concentration of breast milk, however, which influences the milk's taste. Infants may prefer milk produced prior to exercise (which has a lower lactic acid content). For this reason, mothers may want to breastfeed their infants before exercise or express their milk before exercise for use afterward.

◆ A nursing mother produces over 35 gallons of milk during the first six months, saving her roughly $450 in formula costs.◆

• **Vitamins and Minerals** • Will a mother's milk lack a nutrient if she fails to get enough in her diet? The answer differs from one nutrient to the next, but in general, nutritional inadequacies reduce the *quantity,* not the *quality,* of breast milk. Women can produce milk with adequate protein, carbohydrate, fat, and most minerals, even when their own supplies are limited. For these nutrients and for the vitamin folate as well, milk quality is maintained at the expense of maternal stores. This is most evident in the case of calcium: dietary calcium has no effect on the calcium concentration of breast milk, but maternal bones lose some of their density during lactation.[55] (Bone density increases again after lactation ceases.) Nutrients in breast milk most likely to decline in response to prolonged inadequate intakes are the vitamins—especially vitamins B_6, B_{12}, A, and D. Review Figure 14-9 (on p. 473) to compare a lactating woman's nutrient needs with those of pregnant and nonpregnant women.

• **Water** • Despite misconceptions, a mother who drinks more fluid does not produce more breast milk. To protect herself from dehydration, however, a lactating woman needs to drink plenty of fluids. A sensible guideline is to drink a glass of milk, juice, or water at each meal and each time the baby nurses.

• **Nutrient Supplements** • Most lactating women can obtain all the nutrients they need from a well-balanced diet without taking vitamin-mineral supplements; some, however, may need iron supplements. Maternal iron stores dwindle during pregnancy when the fetus takes iron to meet its own needs during the first four to six months after birth. In addition, childbirth may have incurred blood losses. A woman may therefore need iron supplements during lactation, not to enhance the iron in her breast milk, but to refill her depleted iron stores. Until menstruation resumes, a lactating woman's iron requirement is about half that of other nonpregnant women her age.

• **Particular Foods** • Foods with strong or spicy flavors (such as garlic) may alter the flavor of breast milk. A sudden change in the taste of the milk may annoy some infants. Infants who are sensitive to particular foods such as cow's milk protein may become uncomfortable when the mother's diet includes these foods. Only a few infants exhibit this sensitivity, so only a few nursing mothers need avoid cow's milk. Generally, nutrients from milk products support both the infant's and the mother's health.

A nursing mother can usually eat whatever nutritious foods she chooses. If she suspects a particular food is causing the infant discomfort, her physician may recommend a dietary challenge: eliminate the food from the diet to see if the infant's reactions subside; then return the food to the diet, and again monitor the infant's reactions. If a food must be eliminated for an extended time, appropriate substitutions must be made to ensure nutrient adequacy.

Practices Incompatible with Lactation

Some substances impair milk production or enter breast milk and interfere with infant development. Some medical conditions prohibit breastfeeding. This section describes these circumstances.

• **Alcohol** • Alcohol easily enters breast milk, and its concentration peaks within an hour of ingestion. Infants drink less breast milk when their mothers have consumed even small amounts of alcohol (equivalent to a can of beer). Three possible reasons, acting separately or together, may explain why. For one, the alcohol may have altered the flavor of the breast milk and thereby the infants' acceptance of it. For another, because infants metabolize alcohol inefficiently, even low doses may be potent enough to suppress their feeding and cause sleepiness. Third, the alcohol may have interfered with lactation by inhibiting the hormone oxytocin.

In the past, alcohol has been recommended to mothers to facilitate lactation despite a lack of scientific evidence that it does so. The research summarized here suggests that alcohol actually hinders breastfeeding. An occasional glass of wine or beer is considered within safe limits, but in general, lactating women should consume little or no alcohol.

• **Medicinal Drugs** • Many drugs are compatible with breastfeeding, but some medicines are contraindicated, either because they suppress lactation or because they are secreted into breast milk and can harm the infant. As a precaution, a nursing mother should consult with her physician prior to taking any drug, including herbal supplements.

• **Illicit Drugs** • Illicit drugs, of course, are harmful to the physical and emotional health of both the mother and the nursing infant. Breast milk can deliver such high doses of illicit drugs as to cause irritability, tremors, hallucinations, and even death in infants.

• **Smoking** • Cigarette smoking reduces milk volume, so smokers may produce too little milk to meet their infants' energy needs. The milk they do produce contains nicotine, which alters its smell and flavor.[56] Consequently, infants of breastfeeding mothers who smoke gain less weight than infants of those who do not smoke. Furthermore, infant exposure to passive smoke negates the protective effect breastfeeding offers against SIDS and increases the risks dramatically.

• **Environmental Contaminants** • Environmental contaminants such as DDT, PCB, and dioxin can find their way into breast milk. Inuit mothers living in Arctic Québec who eat seal and beluga whale blubber have high concentrations of DDT and PCB in their breast milk, but the impact on infant development is unclear. Preliminary studies indicate that the children of these Inuit mothers are developing normally. Researchers speculate that the abundant omega-3 fatty acids of the Inuit diet may protect against damage to the central nervous system. Breast milk tainted with dioxins interferes with tooth development during early infancy, producing soft, mottled teeth that are vulnerable to dental caries.[57]

• **Caffeine** • Caffeine taken during lactation may make a breastfed infant irritable and wakeful. As during pregnancy, caffeine consumption should be moderate—the equivalent of 1 to 2 cups of coffee a day. Larger doses of caffeine may interfere with the bioavailability of iron from breast milk and impair the infant's iron status.

Maternal Health

If a woman has an ordinary cold, she can continue nursing without worry. If susceptible, the infant will catch it from her anyway. (Thanks to the immunological protection of breast milk, the baby may be less susceptible than a formula-fed baby would be.) If a woman has a communicable disease such as tuberculosis or hepatitis that could threaten the infant's health, then mother and baby have to be separated; the mother can pump her breasts several times a day so that the infant can be fed breast milk by bottle.

• HIV Infection and AIDS • Mothers with HIV infections can transmit the virus to their infants through breast milk, especially during the early months of breastfeeding.[58] Where safe alternatives are available, HIV-positive women should *not* breastfeed their infants. In developing countries, where the feeding of inappropriate or contaminated formulas causes 1.5 million infant deaths each year, the decision is less obvious.[59] To prevent the mother-to-child transmission of HIV, WHO and UNICEF also urge mothers in developing countries *not* to breastfeed, but stress the importance of finding suitable feeding alternatives to prevent the malnutrition, disease, and death that commonly occur when women in these countries do not breastfeed.

• Diabetes • Women with diabetes (type 1) may need careful monitoring and counseling to ensure successful lactation. These women need to adjust their energy intakes and insulin doses to meet the heightened needs of lactation. Maintaining good glucose control helps to initiate lactation and support milk production.

• Postpartum Amenorrhea • Women who breastfeed experience prolonged **postpartum amenorrhea.** Absent menstrual periods, however, do not protect a woman from pregnancy. To prevent pregnancy, a couple must use some form of contraception—but not oral contraceptive agents. Standard oral contraceptives contain estrogen, which reduces milk volume and the protein content of breast milk.

• Breast Health • Some women fear that breastfeeding will cause their breasts to sag. The breasts do swell and become heavy and large immediately after the birth, but even when they are producing enough milk to nourish a thriving infant, they eventually shrink back to their prepregnant size. Given proper support, diet, and exercise, breasts often return to their former shape and size after weaning. Breasts change their shape as the body ages, but breastfeeding does not accelerate this process.

Whether the physical and hormonal events of lactation protect women from later breast cancer is an area of active research. Some research suggests no association between breastfeeding and breast cancer, whereas other research suggests a protective effect.[60] The reduction in breast cancer risk is most apparent for premenopausal women who were young when they breastfed and who breastfed for a long time.

IN SUMMARY

The lactating woman needs extra fluid and enough energy and nutrients to produce about 25 ounces of milk a day. Alcohol, other drugs, smoking, and contaminants may reduce milk production or enter breast milk and impair infant development.

This chapter has focused on the nutrition needs of the mother during pregnancy and lactation. The next chapter explores the dietary needs of infants, children, and adolescents.

How are you doing?

- For women of childbearing age, do you consume at least 0.4 milligram of folate daily?

postpartum amenorrhea: the normal temporary absence of menstrual periods immediately following childbirth.

- For women who are pregnant, are you paying attention to your nutrition needs and gaining the amount of weight recommended?
- For women who are about to give birth, have you carefully considered all the advantages of breastfeeding your infant and received the advice and support you need to be successful?

Nutrition on the Net

WEBSITES

Access these websites for further study of topics covered in this chapter.

- Find updates and quick links to these and other nutrition-related sites at our website: **www.wadsworth.com/nutrition**
- Visit the pregnancy and child health center of the Mayo Clinic: **www.mayohealth.org**
- Learn more about having a healthy baby and about birth defects from the March of Dimes: **www.modimes.org**
- Learn more about neural tube defects from the Spina Bifida Association of America: **www.sbaa.org**
- Search for "birth defects," "pregnancy," "adolescent pregnancy," "maternal and infant health," and "breastfeeding" at the U.S. Government health information site: **www.healthfinder.gov**
- Search for "pregnancy" at the American Dietetic Association site: **www.eatright.org**
- Learn more about the WIC Program: **www.fns.usda.gov/fns**
- Visit the American College of Obstetricians and Gynecologists: **www.acog.org**
- Learn more about gestational diabetes from the American Diabetes Association: **www.diabetes.org**
- Learn more about breastfeeding from LaLeche League International: **www.lalecheleague.org**
- Read *A Woman's Guide to Breastfeeding* at the American Academy of Pediatrics site: **www.aap.org**

INTERNET ACTIVITIES

Proper nutrition during pregnancy is vital for the growth and development of a healthy baby. An adequate daily intake of folate during the first three months of pregnancy is necessary to help prevent neural tube defects. Have you ever wondered what an unborn baby looks like at various stages of development?

- Go to: **www.intelihealth.com**
- Click on "Cool Tools" on the left side bar.
- Click on "Baby's Growth" under "Interactive Health Tools."
- Click on the yellow flower to see the baby's development month by month.

All women of childbearing age should take a daily supplement containing folate to ensure an adequate folate intake. If you are a female, have you had your supplement today?

Study Questions

These questions will help you review the chapter. You will find the answers in the discussions on the pages provided.

1. Describe the placenta and its function. (pp. 464–465)
2. Describe the normal events of fetal development. How does malnutrition impair fetal development? (pp. 464–465, 479)
3. Define the term *critical period.* How do adverse influences during critical periods affect later health? (pp. 466–468)
4. Explain why women of childbearing age need folate in their diets. How much is recommended, and how can women ensure that these needs are met? (pp. 467–468, 475)
5. What is the recommended pattern of weight gain during pregnancy for a woman at a healthy

weight? For an underweight woman? For an overweight woman? (pp. 468–469)

6. What does a pregnant woman need to know about exercise? (pp. 471–472)
7. Which nutrients are needed in the greatest amounts during pregnancy? Why are they so important? Describe wise food choices for the pregnant woman. (pp. 473–476)
8. Define low-risk and high-risk pregnancies. What is the significance of infant birthweight in terms of the child's future health? (pp. 478–480)
9. Describe some of the special problems of the pregnant adolescent. Which nutrients are needed in increased amounts? (pp. 481–482)
10. What practices should be avoided during pregnancy? Why? (pp. 482–484)
11. How do nutrient needs during lactation differ from nutrient needs during pregnancy? (pp. 486–487)

These questions will help you prepare for an exam. Answers can be found on p. 492).

1. The spongy structure that delivers nutrients to the fetus and returns waste products to the mother is called the:
   a. embryo.
   b. uterus.
   c. placenta.
   d. amniotic sac.
2. Which of these strategies is *not* a healthy option for an overweight woman?
   a. Limit weight gain during pregnancy.
   b. Postpone weight loss until after pregnancy.
   c. Follow a weight-loss diet during pregnancy.
   d. Try to achieve a healthy weight before becoming pregnant.
3. A reasonable weight gain during pregnancy for a normal-weight woman is about:
   a. 10 pounds.
   b. 20 pounds.
   c. 30 pounds.
   d. 40 pounds.
4. Energy needs during pregnancy increase by about:
   a. 100 kcalories/day.
   b. 300 kcalories/day.
   c. 500 kcalories/day.
   d. 700 kcalories/day.
5. To help prevent neural tube defects, grain products are now fortified with:
   a. iron.
   b. folate.
   c. protein.
   d. vitamin C.
6. Pregnant women should *not* take supplements of:
   a. iron.
   b. folate.
   c. vitamin A.
   d. vitamin C.
7. The combination of high blood pressure, protein in the urine, and edema signals:
   a. jaundice.
   b. preeclampsia.
   c. gestational diabetes.
   d. gestational hypertension.
8. To facilitate lactation, a mother needs:
   a. about 5000 kcalories a day.
   b. adequate nutrition and rest.
   c. vitamin and mineral supplements.
   d. a glass of wine or beer before each feeding.
9. A breastfeeding woman should drink plenty of water to:
   a. produce more milk.
   b. suppress lactation.
   c. prevent dehydration.
   d. dilute nutrient concentration.
10. A woman may need iron supplements during lactation:
    a. to enhance the iron in her breast milk.
    b. to provide iron for the infant's growth.
    c. to replace the iron in her body's stores.
    d. to support the increase in her blood volume.

References

1 W. W. Hay and coauthors, Workshop summary: Fetal growth: Its regulation and disorders, *Pediatrics* 99 (1997): 585–591.

2 J. E. Brown and coauthors, Predictors of red cell folate level in women attempting pregnancy, *Journal of the American Medical Association* 277 (1997): 548–552.

3 J. Newnham, Consequences of fetal growth restriction, *Current Opinion in Obstetrics and Gynecology* 10 (1998): 145–149; K. M. Godfrey, Maternal regulation of fetal development and health in adult life, *European Journal of Obstetrics, Gynecology, and Reproductive Biology* 78 (1998): 141–150; S. C. Langley-Evans, D. S. Gardner, and S. J. Welham, Intrauterine programming of cardiovascular disease by maternal nutritional status, *Nutrition* 14 (1998): 39–47; D. J. Barker and P. M. Clark, Fetal undernutrition and disease in later life, *Reviews of Reproduction* 2 (1997): 105–112; D. J. P. Barker, Growth in utero and coronary heart disease, *Nutrition Reviews* 54 (1996): S1–S7.

4 M. Kadyrov and coauthors, Increased fetoplacental angiogenesis during first trimester in anaemic women, *Lancet* 352 (1998): 1747–1749.

5 R. L. Goldenberg and T. Tamura, Prepregnancy weight and pregnancy outcome, *Journal of the American*

Medical Association 275 (1996): 1127–1128.

6 Goldenberg and Tamura, 1996.

7 S. Cnattingius and coauthors, Prepregnancy weight and the risk of adverse pregnancy outcomes, *New England Journal of Medicine* 338 (1998): 147–152; M. M. Werler and coauthors, Prepregnant weight in relation to risk of neural tube defects, *Journal of the American Medical Association* 275 (1996): 1089–1092.

8 Werler and coauthors, 1996; G. M. Shaw, E. M. Velie, and D. Schaffer, Risk of neural tube defect—Affected pregnancies among obese women, *Journal of the American Medical Association* 275 (1996): 1093–1096.

9 B. Sternfeld and coauthors, Exercise during pregnancy and pregnancy outcome, *Medicine and Science in Sports and Exercise* 27 (1995): 634–640.

10 M. Kihlstrand and coauthors, Water gymnastics reduced the intensity of back/low back pain in pregnant women, *Acta Obstetrica Gynecologica Scandinavica* 78 (1999): 180–185.

11 American College of Obstetricians and Gynecologists, *Planning for Pregnancy, Birth, and Beyond* (Washington, D.C.: The American College of Obstetricians and Gynecologists, 1995), pp. 88–90; American College of Obstetricians and Gynecologists, *ACOG Technical Bulletin 189: Exercise during Pregnancy and the Postpartum Period* (Washington, D.C., 1994).

12 D. L. Dufour, J. C. Reina, and G. B. Spurr, Energy intake and expenditure of free-living, pregnant Colombian women in an urban setting, *American Journal of Clinical Nutrition* 70 (1999): 269–276; L. E. Kopp-Hoolihan and coauthors, Longitudinal assessment of energy balance in well-nourished pregnant women, *American Journal of Clinical Nutrition* 69 (1999): 697–704.

13 M. Makrides and R. A. Gibson, Long-chain polyunsaturated fatty acid requirements during pregnancy and lactation, *American Journal of Clinical Nutrition* 71 (2000): 307S–311S; A. K. Dutta-Roy, Transport mechanisms for long-chain polyunsaturated fatty acids in the human placenta, *American Journal of Clinical Nutrition* 71 (2000): 315S–322S; S. M. Innis, Maternal diet, length of gestation, and long-chain polyunsaturated fatty acid status of infants at birth, *American Journal of Clinical Nutrition* 70 (1999): 181–182; R. Uauy and coauthors, Role of essential fatty acids in the function of the developing nervous system, *Lipids* 31 (1996) S167–S176.

14 Committee on Dietary Reference Intakes, *Dietary Reference Intakes for Thiamin, Riboflavin, Niacin, Vitamin B_6, Folate, Vitamin B_{12}, Pantothenic Acid, Biotin, and Choline* (Washington, D.C.: National Academy Press, 1998), pp. 196–305.

15 B. P. Zhu and coauthors, Effect of the interval between pregnancies on perinatal outcomes, *New England Journal of Medicine* 340 (1999): 589–594.

16 M. M. Werler and coauthors, Multivitamin supplementation and risk of birth defects, *American Journal of Epidemiology* 150 (1999): 675–682; T. O. Scholl and coauthors, Use of multivitamin/mineral prenatal supplements: Influence on the outcome of pregnancy, *American Journal of Epidemiology* 146 (1997): 134–141.

17 M. Erick, Hyperolfaction and hyperemesis gravidarum: What is the relationship? *Nutrition Reviews* 53 (1995): 289–295.

18 A. J. Rainville, Pica practices of pregnant women are associated with lower maternal hemoglobin level at delivery, *Journal of the American Dietetic Association* 98 (1998): 293–296.

19 S. M. Yu and R. T. Jackson, Need for nutrition advice in prenatal care, *Journal of the American Dietetic Association* 95 (1995): 1027–1029.

20 Hay and coauthors, 1997.

21 S. Saigal and coauthors, School difficulties at adolescence in a regional cohort of children who were extremely low birth weight, *Pediatrics* 105 (2000): 325–331.

22 S. L. Kjos and T. A. Buchanan, Gestational diabetes mellitus, *New England Journal of Medicine* 341 (1999): 1749–1756; C. G. Solomon and coauthors, A prospective study of pregravid determinants of gestational diabetes, *Journal of the American Medical Association* 278 (1997): 1078–1083; C. D. Naylor and coauthors, Selective screening for gestational diabetes mellitus, *New England Journal of Medicine* 337 (1997): 1591–1596.

23 B. M. Sibai, Treatment of hypertension in pregnant women, *New England Journal of Medicine* 335 (1996): 257–265.

24 M. Knuist and coauthors, Risk factors for preeclampsia in nulliparous women in distinct ethnic groups: A prospective cohort study, *Obstetrics & Gynecology* 92 (1998): 174–178.

25 *National Vital Statistics Reports* 47 (1999): 14.

26 D. Maine, Role of nutrition in the prevention of toxemia, *American Journal of Clinical Nutrition* 72 (2000): 298S–300S.

27 R. B. Kaufmann and coauthors, The decline in US teen pregnancy rates, 1990–1995, *Pediatrics* 102 (1998): 1141–1147.

28 Kaufmann, 1998; B. Guyer and coauthors, Annual summary of vital statistics—1997, *Pediatrics* 102 (1998): 1333–1349.

29 M. Story and I. Alton, Nutrition issues and adolescent pregnancy, *Nutrition Today* 30 (1995): 142–151.

30 A. M. Fraser, J. E. Brockert, and R. H. Ward, Association of young maternal age with adverse reproductive outcomes, *New England Journal of Medicine* 332 (1995): 1113–1117.

31 J. M. Rees, S. A. Lederman, and J. L. Kiely, Birthweight associated with lowest neonatal mortality: Infants of adolescent and adult mothers, *Pediatrics* 98 (1996): 1161–1166; Fraser, Brockert, and Ward, 1995.

32 J. B. Hardy and coauthors, Adolescent childbearing revisited: The age of

ANSWERS

Multiple Choice

1. c	2. c	3. c	4. b	5. b
6. c	7. b	8. b	9. c	10. c

inner-city mothers at delivery is a determinant of their children's self-sufficiency at age 27 to 33, *Pediatrics* 100 (1997): 802–809.

33 F. G. Cunningham and K. J. Leveno, Childbearing among old women—The message is cautiously optimistic, *New England Journal of Medicine* 333 (1995): 1002–1004.

34 R. C. Fretts and coauthors, Increased maternal age and the risk of fetal death, *New England Journal of Medicine* 333 (1995): 953–957; Cunningham and Leveno, 1995.

35 G. Koren, A. Pastuszak, and S. Ito, Drugs in pregnancy, *New England Journal of Medicine* 338 (1998): 1128–1137.

36 D. A. Frank and coauthors, Level of in utero cocaine exposure and neonatal ultrasound findings, *Pediatrics* 104 (1999): 1101–1105; G. A. Richardson and coauthors, Growth of infants prenatally exposed to cocaine/crack: Comparison of a prenatal care and a no prenatal care sample, *Pediatrics* 104 (1999): 104 [http://www.pediatrics.org/cgi/content/full/104/2/e18]; F. D. Eyler and coauthors, Birth outcome from a prospective, matched study of prenatal crack/cocaine use: I. Interactive and dose effects on health and growth, *Pediatrics* 101 (1998): 229–237; V. Delaney-Black and coauthors, Prenatal cocaine and neonatal outcome: Evaluation of dose-response relationship, *Pediatrics* 98 (1996): 735–740; F. A. Scafidi and coauthors, Cocaine-exposed preterm neonates show behavioral and hormonal differences, *Pediatrics* 97 (1996): 851–855; E. Z. Tronick and coauthors, Late dose-response effects of prenatal cocaine exposure on newborn neurobehavioral performance, *Pediatrics* 98 (1996): 76–83.

37 E. M. Ostrea and coauthors, Mortality within the first 2 years in infants exposed to cocaine, opiate, or cannabinoid during gestation, *Pediatrics* 100 (1997): 79–83; W. T. Weathers and coauthors, Cocaine use in women from a defined population: Prevalence at delivery and effects on growth in infants, *Pediatrics* 91 (1993): 350–354.

38 M. S. Scher, G. A. Richardson, and N. L. Day, Effects of prenatal cocaine/crack and other drug exposure on electroencephalographic sleep studies at birth and one year, *Pediatrics* 105 (2000): 39–48; V. Delaney-Black and coauthors, Prenatal cocaine exposure and child behavior, *Pediatrics* 102 (1998): 945–950; F. D. Eyler and coauthors, Birth outcome from a prospective, matched study of prenatal crack/cocaine use: II. Interactive and dose effects on neurobehavioral assessment, *Pediatrics* 101 (1998): 237–241; L. Fetters and E. Z. Tronick, Neuromotor development of cocaine-exposed and control infants from birth through 15 months: Poor and poorer outcomes, *Pediatrics* 98 (1996): 938–943; C. A. Chiriboga and coauthors, Neurological correlates of fetal cocaine exposure: Transient hypertonia of infancy and early childhood, *Pediatrics* 96 (1995): 1070–1077.

39 Committee on Drugs, American Academy of Pediatrics, Neonatal drug withdrawal, *Pediatrics* 101 (1998): 1079–1088; T. A. King and coauthors, Neurologic manifestations of in utero cocaine exposure in near-term and term infants, *Pediatrics* 96 (1995): 259–264.

40 S. H. Ebrahim and coauthors, Trends in pregnancy-related smoking rates in the United States, 1987–1996, *Journal of the American Medical Association* 283 (2000): 361–366.

41 J. M. Lightwood, C. S. Phibbs, and S. A. Glantz, Short-term health and economic benefits of smoking cessation: Low birth weight, *Pediatrics* 104 (1999): 1312–1320; S. Cnattingius and coauthors, The influence of gestational age and smoking habits on the risk of subsequent preterm deliveries, *New England Journal of Medicine* 341 (1999): 943–948; Medical-care expenditures attributable to cigarette smoking during pregnancy—United States, 1995, *Morbidity and Mortality Weekly Report* 46 (1997): 1048–1050; S. K. Muscati, K. G. Koski, and K. Gray-Donald, Increased energy intake in pregnant smokers does not prevent human fetal growth retardation, *Journal of Nutrition* 126 (1996): 2984–2989.

42 E. Cutz and coauthors, Maternal smoking and pulmonary neuroendocrine cells in sudden infant death syndrome, *Pediatrics* 98 (1996): 668–672; H. S. Klonoff-Cohen and coauthors, The effect of passive smoking and tobacco exposure through breast milk on sudden infant death syndrome, *Journal of the American Medical Association* 237 (1995): 795–798.

43 C. D. Drews and coauthors, The relationship between idiopathic mental retardation and maternal smoking during pregnancy, *Pediatrics* 97 (1996): 547–553; D. L. Olds, C. R. Henderson, Jr., and R. Tatelbaum, Intellectual impairment in children of women who smoke cigarettes during pregnancy, *Pediatrics* 93 (1994): 221–227; D. M. Fergusson, L. J. Horwood, and M. T. Lynskey, Maternal smoking before and after pregnancy: Effects on behavioral outcomes in middle childhood, *Pediatrics* 92 (1993): 815–822.

44 T. González-Cossío and coauthors, Decrease in birth weight in relation to maternal bone-lead burden, *Pediatrics* 100 (1997): 856–862.

45 K. J. Rothman and coauthors, Teratogenicity of high vitamin A intake, *New England Journal of Medicine* 333 (1995): 1369–1373.

46 S. Cnattingius and coauthors, Caffeine intake and the risk of first-trimester spontaneous abortion, *New England Journal of Medicine* 343 (2000): 1839–1845; M. A. Klebanoff and coauthors, Maternal serum paraxanthine, a caffeine metabolite, and the risk of spontaneous abortion, *New England Journal of Medicine* 341 (1999): 1639–1644.

47 Position of The American Dietetic Association: Use of nutritive and nonnutritive sweeteners, *Journal of the American Dietetic Association* 98 (1998): 580–587.

48 Work Group on Breastfeeding, American Academy of Pediatrics, Breastfeeding and the use of human milk, *Pediatrics* 100 (1997): 1035–1039; Position of The American Dietetic Association: Promotion of breast-feeding, *Journal of the American Dietetic Association* 97 (1997): 662–666; M. J. Heinig and K. G. Dewey, Health effects of breast feeding for mothers: A critical review, *Nutrition Research Reviews* 10 (1997): 35–56; M. J. Heinig and K. G. Dewey, Health advantages of breast feeding for infants: A critical review, *Nutrition Research Reviews* 9 (1996): 89–110.

49 Physicians and breastfeeding promotion in the United States: A call to action, *Pediatrics* 107 (2001): 584–588; E. C. Kieffer and coauthors, Health practitioners should consider parity when counseling mothers on decisions about infant feeding methods, *Journal of the American Dietetic Association* 97 (1997): 1313–1316.

50 S. Arora and coauthors, Major factors influencing breastfeeding rates: Mother's perception of father's attitude and milk supply, *Pediatrics* 106 (2000): 1129 [http://www.pediatrics.org/cgi/content/fall/106/5/e67]; M. Sharma and R. Petosa, Impact of expectant fathers in breast-feeding decisions, *Journal of the American Dietetic Association* 97 (1997): 1311–1312; J. P. Sciacca and coauthors, Influences on breastfeeding by lower-income women: An incentive-based, partner-supported educational program, *Journal of the American Dietetic Association* 95 (1995): 323–328.

51 J. A. Hilson, K. M. Rasmussen, and C. L. Kjolhede, Maternal obesity and breastfeeding success in a rural population of white women, *American Journal of Clinical Nutrition* 66 (1997): 1371–1378.

52 K. G. Dewey, Energy and protein requirements during lactation, *Annual Review of Nutrition* 17 (1997): 19–36.

53 L. Doran and S. Evers, Energy and nutrient inadequacies in the diets of low-income women who breast-feed, *Journal of the American Dietetic Association* 97 (1997): 1283–1287.

54 C. A. Lovelady and coauthors, The effect of weight loss in overweight, lactating women on the growth of their infants, *New England Journal of Medicine* 342 (2000): 449–453; M. A. McCrory and coauthors, Randomized trial of the short-term effects of dieting compared with dieting plus aerobic exercise on lactation performance, *American Journal of Clinical Nutrition* 69 (1999): 959–967.

55 M. A. Laskey and coauthors, Bone changes after 3 mo of lactation: Influence of calcium intake, breast-milk output, and vitamin D–receptor genotype, *American Journal of Clinical Nutrition* 67 (1998): 685–692; L. D. Ritchie and coauthors, A longitudinal study of calcium homeostasis during human pregnancy and lactation and after resumption of menses, *American Journal of Clinical Nutrition* 67 (1998): 693–701; H. J. Kalkwarf and coauthors, The effect of calcium supplementation on bone density during lactation and after weaning, *New England Journal of Medicine* 337 (1997): 523–528.

56 J. A. Mennella and G. K. Beauchamp, Smoking and the flavor of breast milk, *New England Journal of Medicine* 339 (1998): 1559–1560.

57 S. Alaluusua and coauthors, Developing teeth as biomarker of dioxin exposure, *Lancet* 353 (1999): 206.

58 R. Nduati and coauthors, Effect of breastfeeding and formula feeding on transmission of HIV-1: A randomized clinical trial, *Journal of the American Medical Association* 283 (2000): 1167–1174; P. G. Miotti and coauthors, HIV transmission through breastfeeding: A study in Malawi, *Journal of the American Medical Association* 282 (1999): 744–749.

59 E. Hormann, Breast-feeding and HIV: What choices does a mother really have? *Nutrition Today* 34 (1999): 189–196; M. G. Fowler, J. Bertolli, and P. Nieburg, When is breastfeeding not best? The dilemma facing HIV-infected women in resource poor settings, *Journal of the American Medical Association* 282 (1999): 781–783.

60 K. B. Michels and coauthors, Prospective assessment of breastfeeding and breast cancer incidence among 89,887 women, *Lancet* 347 (1996): 431–436; United Kingdom National Case-Control Study Group, 1993.

FOURTEEN

FETAL ALCOHOL SYNDROME

As CHAPTER 14 mentioned, drinking alcohol during pregnancy endangers the fetus. Alcohol crosses the placenta freely and deprives the developing fetal brain of both nutrients and oxygen. The result may be **fetal alcohol syndrome (FAS;** see the glossary on p. 496), a cluster of symptoms that includes:

- Prenatal and postnatal growth retardation.
- Impairment of the brain and nerves, with consequent mental retardation, poor coordination, and hyperactivity.
- Abnormalities of the face and skull (see Figure H14-1 on p. 496).
- Increased frequency of major birth defects: cleft palate, heart defects, and defects in ears, eyes, genitals, and urinary system.

Tragically, the damage evident at birth persists: children with FAS never fully recover.

Each year, more than 5000 infants are born with FAS because their mothers drank alcohol during pregnancy.[1] In addition, an estimated 50,000 infants are born with the less serious, yet still significant, damage some clinicians describe as **fetal alcohol effects (FAE).**[2] Some children with FAE have no outward signs; others may be short or have only minor facial abnormalities. Often children with FAE go undiagnosed even when problems develop in the early school years. Learning disabilities, behavioral abnormalities, and motor impairments are common symptoms of FAE.

The surgeon general states that pregnant women should drink absolutely no alcohol. Abstinence from alcohol is the best policy for pregnant women both because alcohol consumption during pregnancy has such severe consequences and because FAS can only be prevented—it cannot be treated. Further, because the most severe damage occurs around the time of conception—*before a woman may even realize that she is pregnant*—even a woman planning to conceive should abstain.

DRINKING DURING PREGNANCY

When a woman drinks during pregnancy, she causes damage in two ways: directly, by intoxication, and indirectly, by malnutrition. Prior to the complete formation of the placenta (approximately 12 weeks), alcohol diffuses directly into the tissues of the developing embryo, causing incredible damage. (Review Figure 14-4 on p. 466 and note that the critical periods for most tissues occur during embryonic development.) Alcohol interferes with the orderly development of tissues during their critical periods, reducing the number of cells and damaging those that are produced.

When alcohol crosses the placenta, fetal blood alcohol rises until it reaches an equilibrium with maternal blood alcohol. The mother may not even appear drunk, but the fetus may be poisoned. The fetus's body is small, its detoxification system is immature, and alcohol remains in fetal blood long after it has disappeared from maternal blood.

A pregnant woman harms her unborn child not only by consuming alcohol but also by not consuming food. This combination enhances the likelihood of malnutrition and a poorly developed infant. It is important to realize, however, that malnutrition is not the cause of FAS. It is true that mothers of FAS children often have unbalanced diets and nutrient deficiencies. It is also true that malnutrition may augment the clinical signs seen in these children, but it is the *alcohol* that is the damaging factor. An adequate diet alone will not prevent FAS if alcohol abuse continues throughout the pregnancy.

The most obvious symptoms of FAS are the abnormal facial features, but the most tragic ones are the mental disabilities.

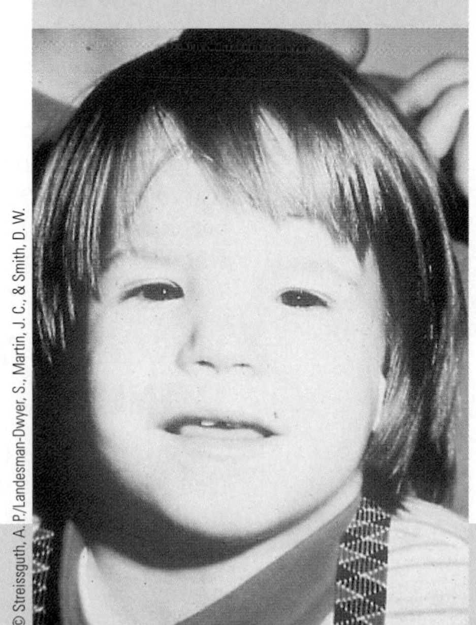

HOW MUCH ALCOHOL IS TOO MUCH?

A pregnant woman need not have an alcohol-abuse problem to give

FIGURE H14-1 **Typical Facial Characteristics of FAS**

The severe facial abnormalities shown here are just outward signs of the severe mental impairments within. The internal organs also suffer irreversible damage that, while hidden, may create major problems for a child's health.

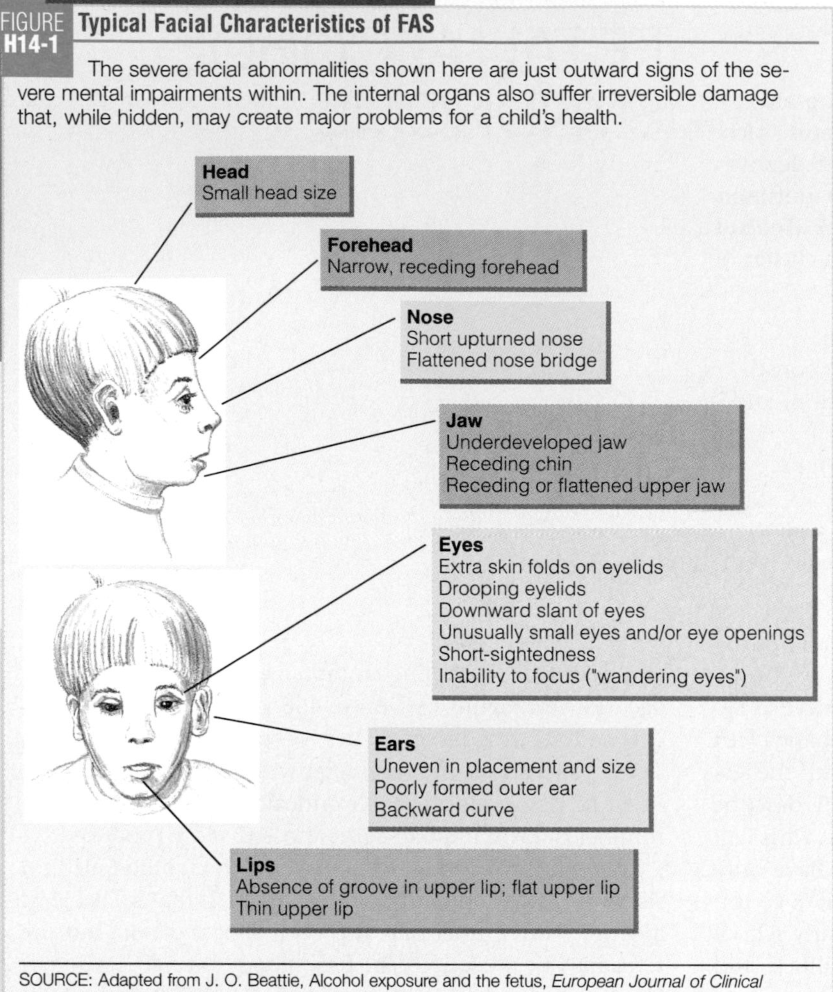

SOURCE: Adapted from J. O. Beattie, Alcohol exposure and the fetus, *European Journal of Clinical Nutrition* 46 (1992): S7–S17.

Characteristic facial features may diminish with time, but children with FAS typically continue to be short and underweight for their age.

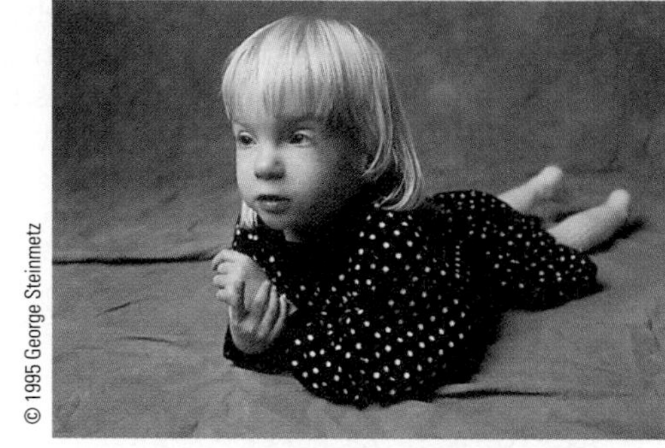

Children born with FAS must live with the long-term consequences of prenatal brain damage.

birth to a baby with FAS. She need only drink in excess of her liver's capacity to detoxify alcohol. Four drinks a day dramatically worsens the risk of having an infant with physical malformations. Even one to two drinks a day threatens to retard growth.

In addition to total alcohol intake, drinking patterns play an important role. Most FAS studies report their findings in terms of average intake per day, but people usually drink more heavily on some days than on others. For example, a woman who drinks an *average* of 1 ounce of alcohol (2 drinks) a day may not drink at all during the week, but then have 10 drinks on Saturday night, exposing the fetus to extremely toxic quantities of alcohol. Whether various drinking patterns incur damage depends on the frequency of consumption, the quantity consumed, and the stage of fetal development at the time of each drinking episode.

An occasional drink may be innocuous, but researchers are unable to say how much alcohol is safe to consume during pregnancy. For this reason, health care professionals urge women to stop drinking alcohol as soon as they realize they are pregnant or, better, as soon as they *plan* to become pregnant. Why take any risk? Only the woman who abstains is sure of protecting her infant from FAS.

GLOSSARY

fetal alcohol effects (FAE): a subclinical version of fetal alcohol syndrome, with hidden defects including learning disabilities, behavioral abnormalities, and motor impairments; also called **alcohol-related birth defects (ARBD).**

fetal alcohol syndrome (FAS): the cluster of symptoms seen in an infant or child whose mother consumed excess alcohol during pregnancy, including retarded growth, impaired development of the central nervous system, and facial malformations.

See Highlight 7 for other alcohol-related terms and information.

WHEN IS THE DAMAGE DONE?

The first month or two of pregnancy is a critical period of fetal development. Because pregnancy confirmation usually requires five to six weeks, a woman may not even realize she is pregnant during that critical time. Therefore, it is advisable for women who are trying to conceive, or who suspect they might be pregnant, to abstain or curtail their alcohol intakes to ensure a healthy start.

The type of abnormality observed in an FAS infant depends on the developmental events occurring at the times of alcohol exposure. During the first trimester, developing organs such as the brain, heart, and kidneys may be malformed. During the second trimester, the risk of spontaneous abortion increases. During the third trimester, body and brain growth may be retarded.

Male alcohol ingestion may also affect fertility and fetal development. Animal studies have found smaller litter sizes, lower birthweights, reduced survival rates, and impaired learning ability in the offspring of males consuming alcohol prior to conception. An association between paternal alcohol intake one month prior to conception and low infant birthweight is also apparent in human beings. (Paternal alcohol intake was defined as an average of two or more drinks daily or at least five drinks on one occasion.) This relationship was independent of either parent's smoking and of the mother's use of alcohol, caffeine, or other drugs.

In view of these findings, it is important to advise women not to drink during pregnancy. Everyone should know of the potential dangers. Heavy drinkers who are sexually active urgently need effective contraception to prevent pregnancy.

All containers of beer, wine, and liquor carry the warning: "Women should not drink alcoholic beverages during pregnancy because of the risk of birth defects." Everyone should hear the message loud and clear: Don't drink alcohol prior to conception or during pregnancy.

Nutrition on the Net

•WEBSITES•

Access these websites for further study of topics covered in this highlight.

- Find updates and quick links to these and other nutrition-related sites at our website: **www.wadsworth.com/nutrition**
- Visit the National Organization on Fetal Alcohol Syndrome: **www.nofas.org**
- Search for "fetal alcohol syndrome" at the U.S. Government health information site: **www.healthfinder.gov**
- Request information on fetal alcohol syndrome from the National Clearinghouse for Alcohol and Drug Information: **www.health.org**
- Request information on drinking during pregnancy from the National Institute on Alcohol Abuse and Alcoholism: **www.niaaa.nih.gov**
- Gather facts on fetal alcohol syndrome from the March of Dimes: **www.modimes.org**

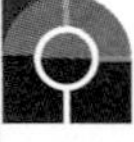

References

1 Update: Trends in fetal alcohol syndrome—United States, 1979–1993, *Morbidity and Mortality Weekly Report* 44 (1995): 249–251.

2 J. M. Aase, K. L. Jones, and S. K. Clarren, Do we need the term "FAE"? *Pediatrics* 95 (1995): 428–430.

CHAPTER 15 Life Cycle Nutrition: Infancy, Childhood, and Adolescence

Infocus, The Image Bank 2001

The first year of life is a time of phenomenal growth and development. After the first year, a child continues to grow and change, but more slowly. Still, the cumulative effects over the next decade are remarkable. Then, as the child enters the teen years, the pace toward adulthood accelerates dramatically. This chapter examines the special nutrient needs of infants, children, and adolescents.

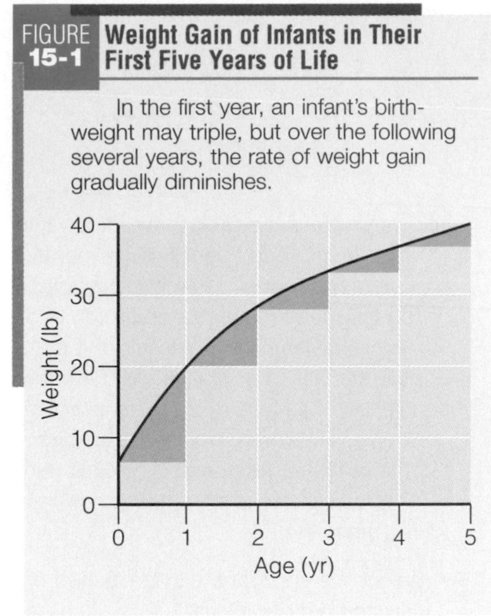

FIGURE 15-1 **Weight Gain of Infants in Their First Five Years of Life**

In the first year, an infant's birthweight may triple, but over the following several years, the rate of weight gain gradually diminishes.

NUTRITION DURING INFANCY

Initially, the infant drinks only breast milk or formula, but later begins to eat some foods, as appropriate. Common sense in the selection of infant foods and a nurturing, relaxed environment go far to promote an infant's health and well-being.

Energy and Nutrient Needs

An infant grows fast during the first year, as Figure 15-1 shows. Growth directly reflects nutrient intake and is an important parameter in assessing the nutrition status of infants and children. Health care professionals measure the heights and weights of infants and children at intervals and compare measures both with standard growth curves for gender and age and with previous measures of each child (see the "How to" on p. 500).[1]

• **Energy Intake and Activity** • A healthy infant's birthweight doubles by about five months of age and triples by one year, typically reaching 20 to 25 pounds. (If an adult starting at 150 pounds were to do this, the person's weight would increase to 450 pounds in a single year.) By the end of the first year, infant growth slows considerably; an infant typically gains less than 10 pounds during the second year.

Not only do infants grow rapidly, but their basal metabolic rate is remarkably high—about twice that of an adult, based on body weight. A newborn baby requires about 650 kcalories per day, whereas most adults require about 2000 kcalories per day. In terms of body weight, the difference is remarkable. Infants require about 100 kcalories per kilogram of body weight per day, whereas most adults need fewer than 40.◆ After six months, metabolic needs decline as the growth rate slows, but some of the energy saved by slower growth is spent in increased activity.

◆

	Infants	Adults
Heart rate (beats/minute)	120 to 140	70 to 80
Respiration rate (breaths/minute)	20 to 40	15 to 20
Energy needs (kcal/body weight)	45/lb (100/kg)	<18/lb (<40/kg)

• **Protein** • No single nutrient is more essential to growth than protein. All of the body's cells and most of its fluids contain protein; it is the basic building material of the body's tissues.

The protein RDA is based on the amount of breast milk that supports adequate growth in healthy, full-term infants. Chapter 6 detailed the problems inadequate protein can cause. Excess dietary protein can cause problems, too, especially in a small infant. Too much protein stresses the kidneys and liver, which have to metabolize and excrete the excess nitrogen. Signs of protein overload include acidosis, dehydration, diarrhea, elevated blood ammonia, elevated blood urea, and fever. Such problems are not common but have been observed in infants fed inappropriate foods, such as nonfat milk or concentrated formula.

After six months, energy saved by slower growth is spent in increased activity.

• **Vitamins and Minerals** • Vitamin and mineral recommendations are based on the average amount of nutrients consumed by thriving infants breastfed by well-nourished mothers. An infant's needs for most of these nutrients, in proportion to body weight, are more than double those of an adult, as Figure 15-2 (on p. 501) illustrates by comparing a five-month-old infant's needs per unit of body weight with those of an adult man. Some of the differences are extraordinary.

• **Water** • One of the most essential nutrients for infants, as for everyone, is water. The younger the infant, the greater the percentage of body weight that is water. During early infancy, breast milk or infant formula normally provides

Plot Measures on a Growth Chart

You can assess the growth of infants and children by plotting their measurements on a percentile graph. Percentile graphs divide the measures of a population into 100 equal divisions so that half of the population falls at or above the 50th percentile, and half falls below. Using percentiles allows for comparisons among people of the same age and gender.

To plot measures on a growth chart, follow these steps:

- Select the appropriate chart based on age and gender. For this example, use the accompanying chart, which gives percentiles for weight for girls from birth to 36 months. (Appendix E provides other growth charts for both boys and girls of various ages.)
- Locate the infant's age along the horizontal axis at the bottom of the chart (in this example, 6 months).
- Locate the infant's weight in pounds or kilograms along the vertical axis of the chart (in this example, 17 pounds or 7.7 kilograms).
- Mark the chart where the age and weight lines intersect (shown here with a red dot), and read off the percentile.

This six-month-old infant is at the 75th percentile. Her pediatrician will weigh her again over the next few months and expect the growth curve to follow the same percentile throughout the first year. In general, dramatic changes or measures much above the 80th percentile or much below the 10th percentile may be cause for concern.

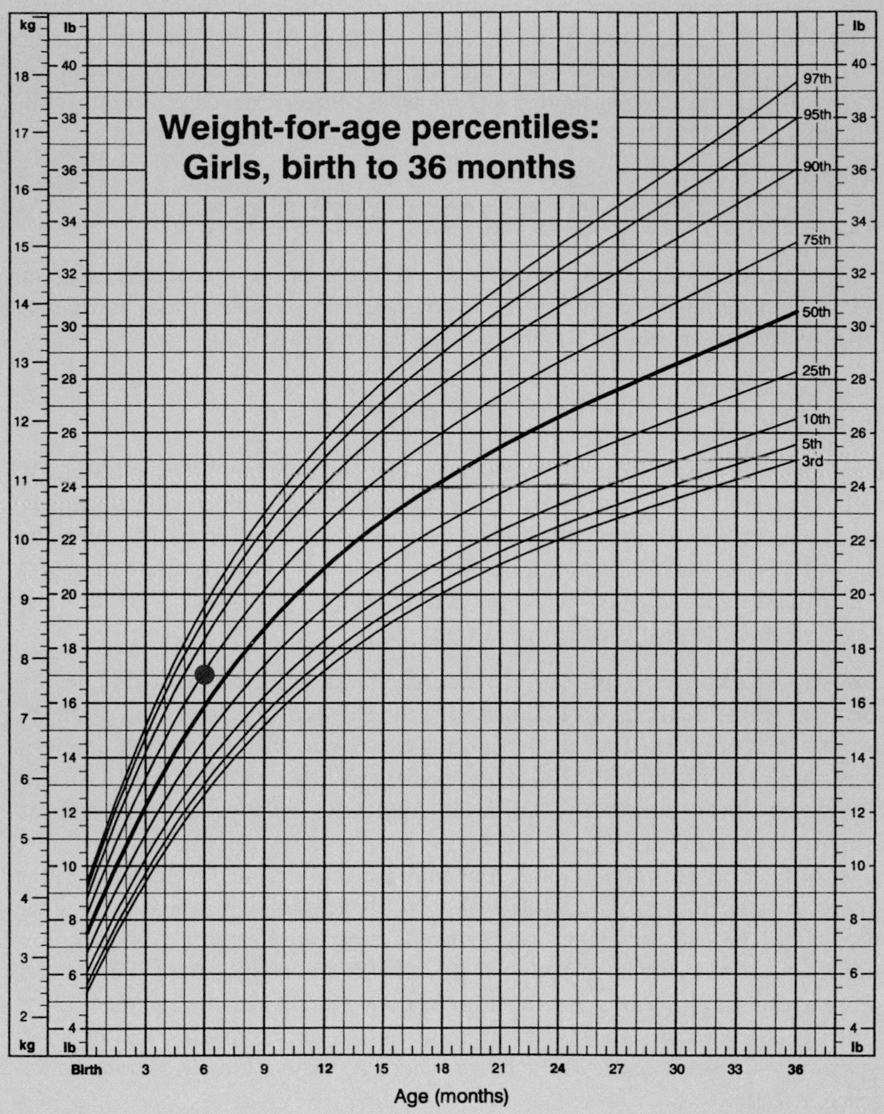

SOURCE: Developed by the National Center for Health Statistics in collaboration with the National Center for Chronic Disease Prevention and Health Promotion (2000).

◆ If an infant's energy needs were superimposed on an adult, a 170-pound adult would require over 7000 kcalories a day.◆

enough water to replace fluid losses in a healthy infant. The water in an infant's body is easily lost because much of it is located *outside* the cells—between the cells and in the vascular space. Conditions that cause fluid loss, such as hot weather, diarrhea, or vomiting, may require supplemental water to prevent life-threatening dehydration.[2] Supplemental water must be free of contamination; to disinfect water, bring it to a rolling boil for five minutes.

Breast Milk

In the United States and Canada, the two dietary practices that have the most effect on an infant's nutrition status are the milk the infant receives and the age at which solid foods are introduced. A later section discusses the introduction of solid food, but as to the milk, both the American Academy of Pediatrics (AAP)

FIGURE 15-2 **Recommended Intakes of an Infant and an Adult Compared on the Basis of Body Weight**

Because infants are small, they need smaller total amounts of the nutrients than adults do, but when comparisons are based on body weight, infants need over twice as much of many nutrients. Infants use large amounts of energy and nutrients, in proportion to their body size, to keep all their metabolic processes going.

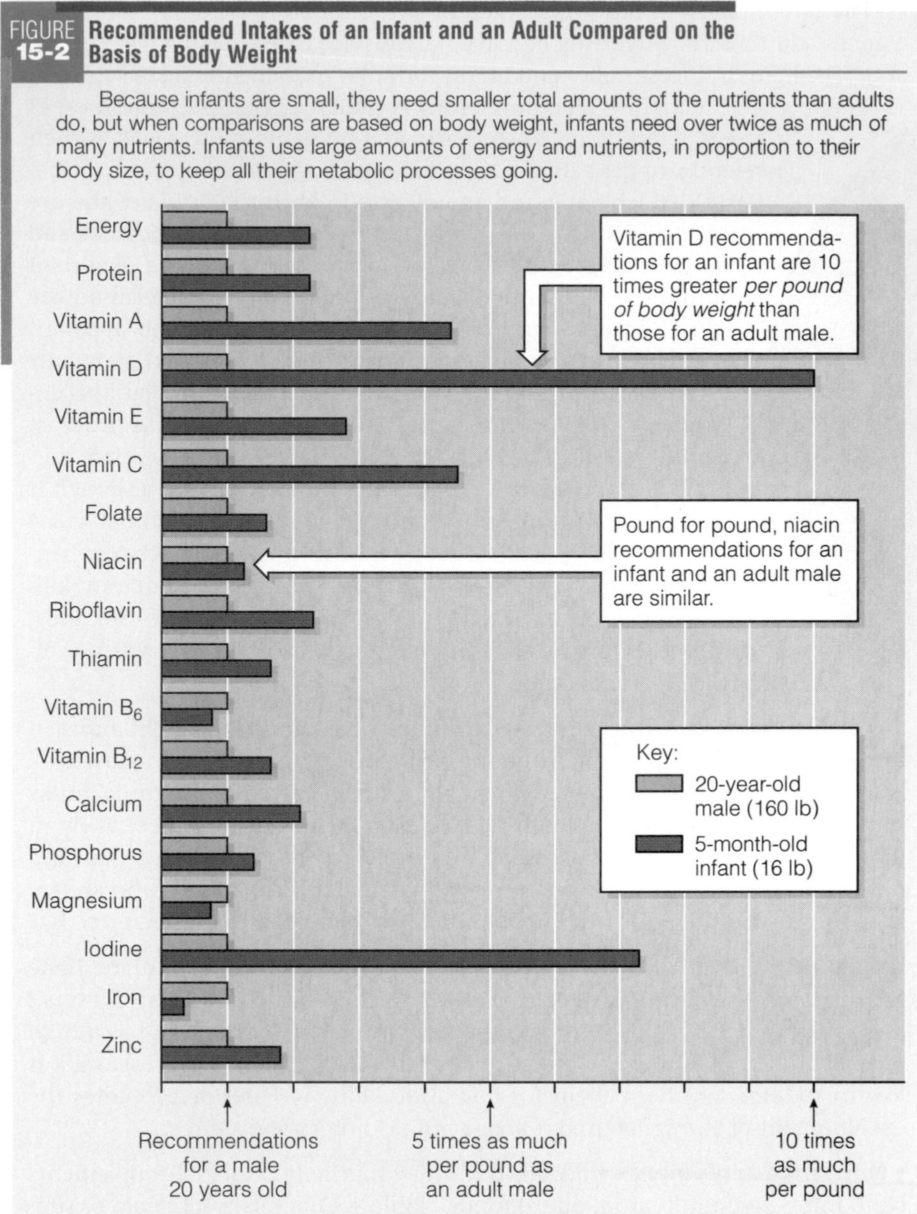

and the Canadian Paediatric Society strongly recommend breastfeeding for full-term infants, except where specific contraindications exist. Breastfeeding offers many benefits to both infant and mother (review Table 14-5 on p. 485).[3]

Breast milk excels as a source of nutrients for the young infant. Its unique nutrient composition and protective factors promote optimal infant health and development throughout the first year of life. Experts add, though, that iron-fortified formula, which imitates the nutrient composition of breast milk, is an acceptable alternative. After all, the primary goal is to provide the infant optimal nourishment in a relaxed and loving environment.

Even two to three months of breastfeeding give the infant immunological protection during the most critical period after birth—protection that persists beyond the breastfeeding period itself. The mother can then shift to formula, if necessary, knowing that she has given her infant those benefits.

• **Energy Nutrients** • The energy-nutrient composition of breast milk differs dramatically from that recommended for adult diets (see Figure 15-3). Yet for infants, breast milk is nature's most nearly perfect food, providing the clear lesson that people at different stages of life have different nutrient needs.

FIGURE 15-3 **Percentages of Energy-Yielding Nutrients in Breast Milk and in Recommended Adult Diets**

The proportions of energy-yielding nutrients in human breast milk differ from those recommended for adults.

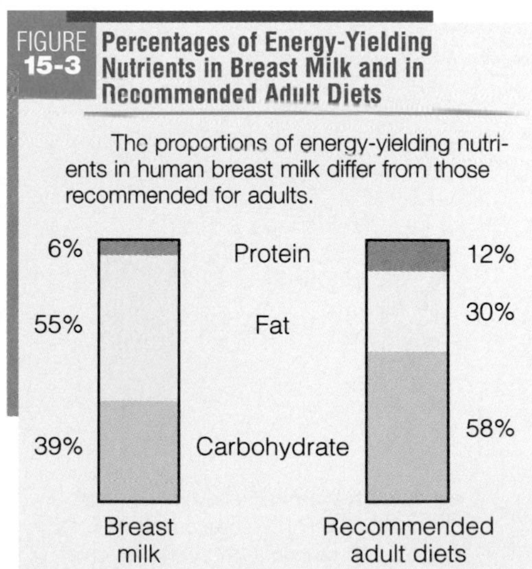

Breast milk offers infants many nutrient and health advantages.

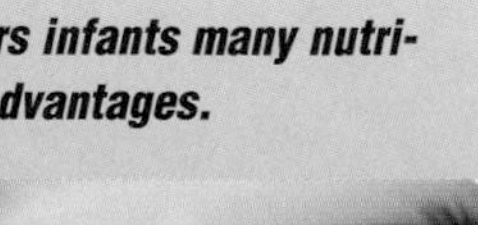

© 2001 PhotoDisc, Inc.

The carbohydrate in breast milk (and infant formula) is the disaccharide lactose. In addition to being easily digested, lactose enhances calcium absorption. The lipids in breast milk—and infant formula—provide the main source of energy in the infant's diet. Breast milk contains a generous proportion of the essential fatty acids linoleic acid and linolenic acid, as well as their longer-chain derivatives arachidonic acid and docosahexaenoic acid (DHA), which are found abundantly in both the retina of the eye and the brain; infant formula contains only linoleic acid and linolenic acid. Because breastfed infants receive more DHA than formula-fed infants, research is under way to determine the physiological significance of this difference.[4] One apparent benefit is that young children who were breastfed as infants have sharper vision than those who were fed formulas; this enhanced visual development is attributed to the DHA in breast milk.[5] Including all the long-chain fatty acids in infant formula may be desirable for optimum development, although it does not seem to influence infant growth.[6] The protein of breast milk is less than in cow's milk, but this quantity is actually beneficial because it places less stress on the infant's immature kidneys to excrete the major end product of protein metabolism, urea. The main protein in breast milk is **alpha-lactalbumin,** which is efficiently digested and absorbed.

• **Vitamins** • With the possible exception of vitamin D, the vitamins in breast milk are ample to support infant growth. The vitamin D in breast milk is low, and vitamin D deficiency impairs bone mineralization. Manufacturers fortify infant formulas with vitamin D. Vitamin D deficiency is most likely in infants who are not exposed to sunlight daily, have darkly pigmented skin, and receive breast milk without vitamin D supplementation.[7] Infants who are exposed to sunlight regularly can make enough vitamin D to meet their needs.

• **Minerals** • The calcium content of breast milk is ideal for infant bone growth, and the calcium is well absorbed. Breast milk contains relatively small amounts of iron, but this iron has a high bioavailability. Zinc also has a high bioavailability, thanks to the presence of a zinc-binding protein. Breast milk is low in sodium, another benefit for immature kidneys. Fluoride promotes the development of strong teeth, but breast milk is not a good source.

• **Nutrient Supplements** • Pediatricians may routinely prescribe supplements containing vitamin D, iron, and fluoride. Table 15-1 offers a schedule of sup-

TABLE 15-1 Supplements for Full-Term Infants

	Vitamin D[a]	Iron[b]	Fluoride[c]
Breastfed infants:			
Birth to six months of age	✔		
Six months to one year	✔	✔	✔
Formula-fed infants:			
Birth to six months of age			
Six months to one year		✔	✔

[a]Vitamin D supplements are recommended for infants whose mothers are vitamin D deficient or those who do not receive adequate exposure to sunlight.
[b]Infants four to six months of age need additional iron, preferably in the form of iron-fortified cereal for both breastfed and formula-fed infants and iron-fortified infant formula for formula-fed infants.
[c]At six months of age, breastfed infants and formula-fed infants who receive ready-to-use formulas (these are prepared with water low in fluoride) or formula mixed with water that contains little or no fluoride (less than 0.3 ppm) need supplements.
SOURCE: Adapted from American Academy of Pediatrics, *Pediatric Nutrition Handbook*, 4th ed., ed. R. E. Kleinman (Elk Grove Village, Ill.: American Academy of Pediatrics, 1998).

alpha-lactalbumin (lact-AL-byoo-min): the chief protein in human breast milk, as opposed to **casein** (CAY-seen), the chief protein in cow's milk.

plements during infancy. In addition, the AAP recommends that a single dose of vitamin K be given to infants at birth to protect them from bleeding to death (see Chapter 11 for a description of vitamin K's role in blood clotting).

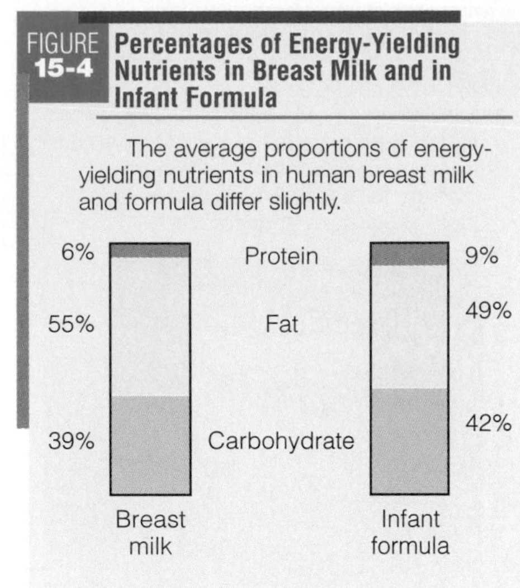

FIGURE 15-4 **Percentages of Energy-Yielding Nutrients in Breast Milk and in Infant Formula**

The average proportions of energy-yielding nutrients in human breast milk and formula differ slightly.

• **Immunological Protection** • In addition to nutritional benefits, breast milk also offers immunological protection. Not only is breast milk sterile, but it actively fights disease and protects infants from illnesses.[8] Such protection is most valuable during the first year, when the infant's immune system is not fully prepared to mount a response against infection.

During the first two or three days after delivery, the breasts produce **colostrum,** a premilk substance containing mostly serum with antibodies and white blood cells. Colostrum (like breast milk) helps protect the newborn from infections against which the mother has developed immunity. The maternal antibodies swallowed with the milk inactivate disease-causing bacteria within the digestive tract before they can start infections. This explains, in part, why breastfed infants have fewer intestinal infections than formula-fed infants.

In addition to antibodies, colostrum and breast milk provide other powerful agents◆ that help to fight against bacterial infection. Among them are **bifidus factors,** which favor the growth of the "friendly" bacterium *Lactobacillus bifidus* in the infant's digestive tract, so that other, harmful bacteria cannot gain a foothold there. An iron-binding protein in breast milk, **lactoferrin,** keeps bacteria from getting the iron they need to grow, helps absorb iron into the infant's bloodstream, and kills some bacteria directly. The protein **lactadherin** in breast milk fights off the virus that causes most infant diarrhea.[9] Also present is a growth factor that stimulates the development and maintenance of the infant's digestive tract and its protective factors. Several breast milk enzymes such as lipase also help protect the infant against infection. Clearly, breast milk is a very special substance.

Protective factors in breast milk: ◆
- Antibodies.
- Bifidus factors.
- Lactoferrin.
- Lactadherin.
- Growth factor.
- Lipase enzyme.

• **Allergy and Disease Protection** • In addition to protection against infection, breast milk may offer protection against the development of allergies and diseases as well. Compared with formula-fed infants, breastfed infants have a lower incidence of allergic reactions, such as asthma, recurrent wheezing, and skin rash. This protection is especially noticeable among infants with a family history of allergies. Similarly, breast milk may offer protection against the development of cardiovascular disease. Compared with formula-fed infants, breastfed infants have a lower blood pressure as adolescents.[10]

Health promotion

Infant Formula

A woman who breastfeeds for a year can **wean** her infant to cow's milk, bypassing the need for infant formula. However, a woman who decides to feed her infant formula from birth, to wean to formula after a short time, or to substitute formula for breastfeeding on occasion must select an appropriate infant formula and learn to prepare it.

• **Infant Formula Composition** • Formula manufacturers attempt to copy the nutrient composition of breast milk as closely as possible. Figure 15-4 illustrates the energy-nutrient balance of both.

The AAP recommends that all formula-fed infants receive iron-fortified infant formulas.[11] The increasing use of iron-fortified formulas during the past few decades is a major reason for the decline in iron-deficiency anemia among U.S. infants.

• **Risks of Formula Feeding** • Infant formulas contain no protective antibodies for infants, but in general, vaccinations, purified water, and clean environments in developed countries help protect infants from infections. Formulas can be prepared safely by following the rules of proper food handling and using water that is free of contamination. Of particular concern is lead-contaminated water, a major source of lead poisoning in infants. Because the

colostrum (ko-LAHS-trum): a milklike secretion from the breast, present during the first day or so after delivery before milk appears; rich in protective factors.

bifidus (BIFF-id-us, by-FEED-us) **factors:** factors in colostrum and breast milk that favor the growth of the "friendly" bacterium *Lactobacillus* (lack-toh-ba-SILL-us) *bifidus* in the infant's intestinal tract, so that other, less desirable intestinal inhabitants will not flourish.

lactoferrin (lack-toh-FERR-in): a protein in breast milk that binds iron and keeps it from supporting the growth of the infant's intestinal bacteria.

lactadherin (lack-tad-HAIR-in): a protein in breast milk that attacks diarrhea-causing viruses.

wean: to gradually replace breast milk with infant formula or other foods appropriate to an infant's diet.

FIGURE 15-5 **Nursing Bottle Tooth Decay**

This child was frequently put to bed sucking on a bottle filled with apple juice, so the teeth were bathed in carbohydrate for long periods of time—a perfect medium for bacterial growth. The upper teeth show signs of decay.

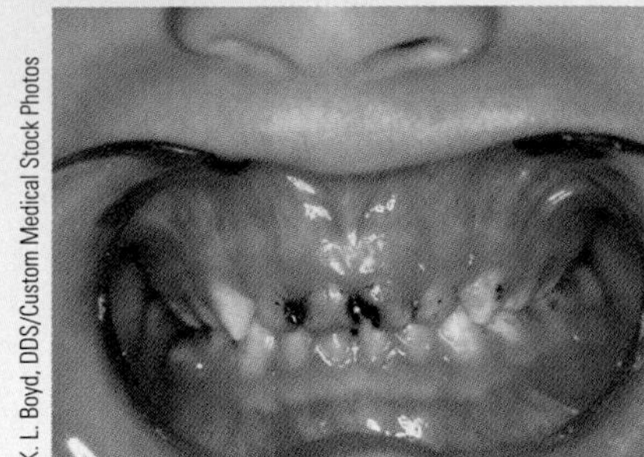

K. L. Boyd, DDS/Custom Medical Stock Photos

first water drawn from the tap each day is highest in lead, a person living in a house with old, lead-soldered plumbing should let the water run a few minutes before drinking or using it to prepare formula.

In developing countries and in poor areas of the United States, formula may be unavailable, prepared with contaminated water, or overdiluted in an attempt to save money. More than 1.2 billion people in developing countries do not have access to safe drinking water. Contaminated formulas often cause infections, leading to diarrhea, dehydration, and malabsorption. Without sterilization and refrigeration, bottles of formula are an ideal breeding ground for bacteria. Whenever such risks are present, breastfeeding can be a life-saving option: breast milk is sterile, and its antibodies enhance an infant's resistance to infections.

• **Infant Formula Standards** • National and international standards have been set for the nutrient contents of infant formulas. In the United States, the standard developed by the AAP reflects "human milk taken from well-nourished mothers during the first or second month of lactation, when the infant's growth rate is high." The Food and Drug Administration (FDA) mandates the safety and nutrition quality of infant formulas.[12] Formulas meeting these standards have similar nutrient compositions; small differences are sometimes confusing, but usually unimportant.

• **Special Formulas** • Standard formulas are inappropriate for some infants. Special formulas have been designed to meet the dietary needs of infants with specific conditions such as prematurity or inherited diseases. Soy formulas use soy for the protein source and cornstarch and sucrose instead of lactose, for example, and so are recommended for infants with milk allergy or lactose intolerance.[13] They are also useful as an alternative to milk-based formulas for vegan families. Despite these limited uses, soy formulas account for one-fourth of the infant formulas sold today.[14] While soy formulas support the normal growth and development of infants, they offer no advantage over milk formulas.

• **Inappropriate Formulas** • Caregivers must use only products designed for infants; soy *beverages,* for example, are nutritionally incomplete and inappropriate for infants. Goat's milk is also inappropriate for infants in part because of its low folate content. An infant receiving goat's milk is likely to develop "goat's milk anemia," an anemia characteristic of folate deficiency.

• **Nursing Bottle Tooth Decay** • An infant cannot be allowed to sleep with a bottle because of the potential damage to developing teeth. Salivary flow, which normally cleanses the mouth, diminishes as the infant falls asleep. Prolonged sucking on a bottle of formula, milk, or juice bathes the upper teeth in a carbohydrate-rich fluid that nourishes decay-producing bacteria. (The tongue covers and protects most of the lower teeth, but they, too, may be affected.) The result is extensive and rapid tooth decay (see Figure 15-5). To prevent **nursing bottle tooth decay,** no infant should be put to bed with a bottle of nourishing fluid.

nursing bottle tooth decay: extensive tooth decay due to prolonged tooth contact with formula, milk, fruit juice, or other carbohydrate-rich liquid offered to an infant in a bottle.

Special Needs of Preterm Infants

An estimated one out of ten pregnancies results in a preterm birth.[15] The terms *preterm* and *premature* imply incomplete fetal development, or immaturity, of many body systems. As might be expected, preterm birth is a leading cause of infant deaths. Preterm infants face physical independence before some of their organs and body tissues are ready. The rate of weight gain in the fetus is greater during the last trimester of gestation than at any other time. Therefore, a preterm infant is most often a low-birthweight infant as well. With a premature birth, the infant is deprived of the nutritional support of the placenta during a time of maximal growth.

The last trimester of gestation is a time of building nutrient stores. Being born with limited nutrient stores intensifies the precarious situation for the infant. Further compromising the nutrition status of preterm infants is their physical and metabolic immaturity. Nutrient absorption, especially of fat and calcium, from an immature GI tract is limited. In short, preterm, low-birthweight infants are candidates for nutrient imbalances. Deficiencies of the fat-soluble vitamins, calcium, iron, and zinc are common.

Infants who are born eight to ten weeks before term miss out on the normal mineralization of bone that takes place during the last trimester of gestation. As a result, they often develop the metabolic bone disease **osteopenia,** the rickets of prematurity. The probability that this condition will occur varies directly with the infant's weight: the smaller the infant, the greater the risk.

Preterm infants may also miss out on the transfer of the long-chain fatty acids arachidonic acid and DHA, so critical to the healthy growth and development of the blood vessels and brain.[16] Supplementing breast milk or enriching infant formulas with these fatty acids may be beneficial for preterm infants.[17]

Preterm breast milk is well suited to meet a preterm infant's needs. During early lactation, preterm milk contains higher concentrations of protein and is lower in volume than term milk. The low milk volume is advantageous because preterm infants consume small quantities of milk per feeding, and the higher protein concentration allows for better growth. In many instances, supplements of nutrients specifically designed for preterm infants are added to the mother's expressed breast milk and fed to the infant from a bottle. When fortified with a preterm supplement, preterm breast milk supports growth at a rate that approximates the growth rate that would have occurred within the uterus.

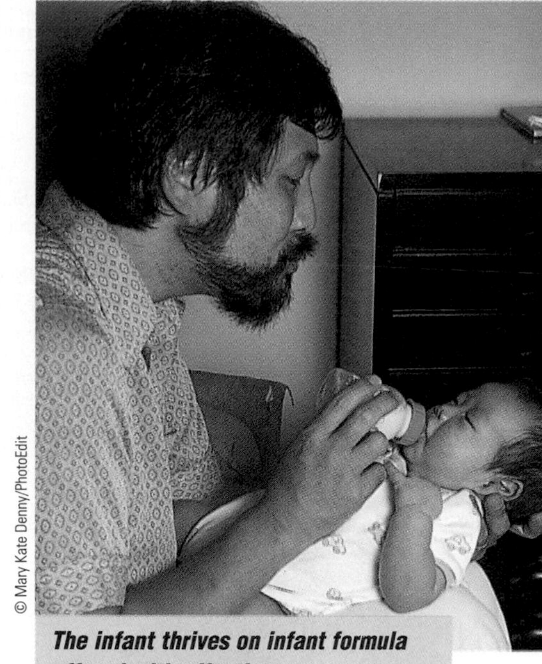

The infant thrives on infant formula offered with affection.

Introducing Cow's Milk

The timing of the introduction of whole cow's milk to the infant's diet has long been a source of controversy. The AAP currently recommends not introducing whole cow's milk before 12 months of age.[18]

In some infants, particularly those younger than six months of age, the consumption of whole cow's milk is associated with intestinal bleeding and iron deficiency. Whole cow's milk is also a poor source of iron. Consequently, cow's milk both causes iron loss and fails to replace iron. Furthermore, the bioavailability of iron from infant cereal and other foods is reduced when cow's milk replaces breast milk or iron-fortified formula during the first year. Compared to breast milk or iron-fortified formula, cow's milk is higher in calcium and lower in vitamin C, characteristics that inhibit iron absorption. In short, cow's milk is a poor choice during the first year of life; infants need breast milk or infant formula.

Research examining the relationships between early exposure to cow's milk (or formula using cow's milk) and the development of type 1 diabetes (insulin-dependent) has been inconclusive and contradictory.[19] Families with a strong history of type 1 diabetes may want to breastfeed and avoid products containing cow's milk protein during the first year. Otherwise, it would be premature to change infant feeding guidelines until, and unless, additional research concludes that cow's milk causes type 1 diabetes.

Introducing Solid Foods

The high nutrient needs of infancy are met first by breast milk or formula only and then by a limited diet to which foods◆ are gradually added. Infants gradually develop the ability to chew, swallow, and digest the wide variety of foods available to adults. The caregiver's selection of appropriate foods at the appropriate stages of development is prerequisite to the infant's optimal growth and health (see Table 15-2 on p. 506).

◆ The German word **beikost** (BYE-cost) describes any nonmilk foods given to an infant.

◆ Infant feeding tip: Introduce solid foods between 4 and 6 mo.

• When to Begin • In addition to breast milk or formula, an infant needs to begin eating solid foods between four and six months.◆ Infants who do not receive solid foods before the end of the first year may suffer delayed growth.

osteopenia (OS-tee-oh-PEE-nee-ah): a metabolic bone disease common in preterm infants; also called **rickets of prematurity.**

TABLE 15-2 Infant Feeding Skills and Recommended Foods

Note: Because each stage of development builds on the previous stage, the foods from an earlier stage continue to be included in all later stages.

Age (mo)	Feeding Skill	Appropriate Foods Added to the Diet
0–4	Turns head toward any object that brushes cheek.	Feed breast milk or infant formula.
	Initially swallows using back of tongue; gradually begins to swallow using front of tongue as well. Strong reflex (extrusion) to push food out during first 2 to 3 months.	
4–6	Extrusion reflex diminishes, and the ability to swallow nonliquid foods develops. Indicates desire for food by opening mouth and leaning forward. Indicates satiety or disinterest by turning away and leaning back. Sits erect with support at 6 months. Begins chewing action. Brings hand to mouth. Grasps objects with palm of hand.	Begin iron-fortified cereal mixed with breast milk, formula, or water. Begin pureed vegetables and fruits.
6–8	Able to feed self finger foods. Develops pincer (finger to thumb) grasp. Begins to drink from cup.	Begin breads and other cereals. Begin textured vegetables and fruits. Begin plain, unsweetened fruit juices from cup.
8–10	Begins to hold own bottle. Reaches for and grabs food and spoon. Sits unsupported.	Begin breads and cereals from table. Begin yogurt. Begin pieces of soft, cooked vegetables and fruit from table. Gradually begin finely cut meats, fish, casseroles, cheese, eggs, and legumes.
10–12	Begins to master spoon, but still spills some.	Include at least 4 servings of breads and cereals from table, in addition to infant cereal.[a] Include at least 2 servings of fruits and 3 servings of vegetables.[a] Include 2 servings of meat, fish, poultry, eggs, or legumes.[a]

[a]Serving sizes for infants and young children are smaller than those for an adult. For example, a serving might be ½ slice of bread instead of 1 slice, or ¼ cup rice instead of ½ cup.
SOURCE: Adapted in part from Committee on Nutrition, American Academy of Pediatrics, *Pediatric Nutrition Handbook,* 4th ed., ed. R. E. Kleinman (Elk Grove Village, Ill.: American Academy of Pediatrics, 1998), pp. 43–53.

The main purpose of introducing solid foods to infants is to provide nutrients that are no longer supplied adequately by breast milk or formula alone. The foods chosen must be foods that the infant is developmentally capable of handling both physically and metabolically. The exact timing depends on the individual infant's needs and developmental readiness (review Table 15-2 above), which vary from infant to infant because of differences in growth rates, activity, and environmental conditions.

◆ **Infant feeding tip: Offer new foods one at a time.**

• Food Allergies • To prevent allergy and to facilitate its prompt identification should it occur, experts recommend introducing single-ingredient foods, one at a time,◆ in small portions, and waiting four to five days before introducing the next new food. For example, rice cereal is usually the first cereal introduced because it is least allergenic. When it is clear that rice cereal is not causing an allergy, another grain, perhaps barley or oats, is introduced. Wheat cereal is offered last because it is the most common offender. If a cereal causes an allergic reaction such as a skin rash, digestive upset, or respiratory discomfort, its use should be discontinued before introducing the next food. A later section in this chapter offers more on food allergies.

• Choice of Infant Foods • Infant foods should be selected to provide variety, balance, and moderation. Commercial baby foods offer a wide variety of palatable, nutritious foods in a safe and convenient form. Homemade infant foods can be as nutritious as commercially prepared ones, as long as the preparer minimizes nutrient losses during preparation. Ingredients for homemade foods should be fresh, whole foods without added salt, sugar, or seasonings. Pureed food can be frozen in ice cube trays, providing convenient-sized blocks of food that can be thawed, warmed, and fed to the infant. To guard against food-borne illnesses, the preparer should be careful to keep hands and equipment clean.

Because recommendations to restrict fat do not apply to children under age two, labels on foods for children under two (such as infant meats and cereals) cannot carry information about fat. Fat information is omitted from infant food labels to prevent parents from restricting fat in infants' diets. Fearing that their infant will become overweight, parents may unintentionally malnourish the infant by limiting fat. In fact, infants and young children, because of their rapid growth, need more fat than older children and adults.

© Polara Studios Inc.

Foods such as iron-fortified cereals and formulas, mashed legumes, and strained meats provide iron.

• Foods to Provide Iron • Rapid growth demands iron. At about four to six months, the infant begins to need more iron than stores plus breast milk or iron-fortified formula can provide. In addition to breast milk or iron-fortified formula, infants can receive iron◆ from iron-fortified cereals and, later, from meat or meat alternates such as legumes. Iron-fortified cereals contribute a significant amount of iron to an infant's diet, but the iron's bioavailability is poor.[20] Caregivers can enhance iron absorption from iron-fortified cereals by serving vitamin C–rich foods and juices with meals.

◆ Infant feeding tip: To provide additional iron, offer iron-fortified cereal, mashed legumes, and infant meats.

• Foods to Provide Vitamin C • The best sources of vitamin C◆ are fruits and vegetables. Some authorities suggest that an infant who is introduced to fruits before vegetables may develop a preference for sweets and find the vegetables less palatable. To prevent this, introduce vegetables first, fruits later.

◆ Infant feeding tip: To provide vitamin C, offer citrus juices, melon pieces, and chopped berries.

Fruit juices should be diluted and served in a cup, not a bottle, once the infant is six months of age or older. Juices provide valuable nutrients, but should be used moderately◆ in the infant diet, so as not to displace other foods.[21] Children fail to grow and thrive when they drink so much juice each day that other, more energy- and nutrient-dense foods are displaced from their diets.[22] Such findings prove that any one food—even a healthful and nutritious one—can create nutrient imbalances, impair growth, or foster obesity when consumed in excess.

◆ Infant (and young child) feeding tip: Limit fruit juice to 8 oz/day.

• Foods to Omit • Concentrated sweets, including baby food "desserts," have no place in an infant's diet.◆ They convey no nutrients to support growth, and the extra food energy can promote obesity. Products containing sugar alcohols such as sorbitol should also be limited, as they may cause diarrhea. Canned vegetables are also inappropriate for infants, as they often contain too much sodium. Honey and corn syrup should never be fed to infants because of the risk of **botulism.*** Infants and even young children cannot safely chew and swallow popcorn, whole grapes, whole beans, hot dog slices, hard candies, and nuts; they can easily choke on these foods, a risk not worth taking.

◆ Infant feeding tip: Limit concentrated sweets, sugar alcohols, and canned vegetables, and avoid foods that may cause choking.

• Foods at One Year • At one year of age, whole cow's milk becomes the primary source of most of the nutrients an infant needs; 2 to 3½ cups a day meets those needs sufficiently. More milk than this displaces iron-rich foods and can lead to **milk anemia.** Children one to two years old should drink whole milk, not reduced-fat, low-fat, or nonfat milk. If powdered milk is used, it should be one of the fat-containing varieties. Children can gradually switch from whole milk to lower-fat milk between the ages of two and five.

Other foods—meats, iron-fortified cereals, enriched or whole-grain breads, fruits, and vegetables—should be supplied in variety and in amounts sufficient to round out total energy needs. Ideally, a one-year-old will sit at the table, eat many of the same foods everyone else eats, and drink liquids from a cup, not a bottle. Figure 15-6 shows a meal plan that meets a one-year-old's requirements.

botulism (BOT-chew-lism): an often fatal food-borne illness caused by the ingestion of foods containing a toxin produced by bacteria that grow without oxygen.

milk anemia: iron-deficiency anemia that develops when an excessive milk intake displaces iron-rich foods from the diet.

*In infants, but not older individuals, ingestion of *Clostridium botulinum* spores can cause illness when the spores germinate in the intestine and produce toxin, which is absorbed. Symptoms include poor feeding, constipation, loss of tension in the arteries and muscles, weakness, and respiratory compromise. Infant botulism has been implicated in 5 percent of cases of sudden infant death syndrome (SIDS).

© Tony Freeman/PhotoEdit

Toddlers need vitamin A– and vitamin D–fortified whole milk.

FIGURE **15-6**
Meal Plan for a One-Year-Old

Breakfast
¼ c whole milk (with cereal)
½ c iron-fortified breakfast cereal
½ c orange juice

Morning snack
½ c vitamin C–fortified fruit juice
1 to 2 oz cheese cubes
Teething crackers

Lunch
1 c whole milk
½ c vegetables[a] (steamed carrots)
½ sandwich: 1 slice bread with 2 tbs tuna salad or egg salad

Afternoon snack
½ c whole milk
1 slice toast
1 to 2 tbs peanut butter

Dinner
1 c whole milk
2 to 3 oz chopped meat or well-cooked mashed legumes
¼ c potato, rice, or pasta
¼ c vegetables[a] (chopped broccoli)
¼ c fruit[b] (sliced strawberries)

[a]Include dark green, leafy and deep yellow vegetables.
[b]Include citrus fruits, melons, and berries.

Mealtimes with Toddlers

Eating habits acquired during infancy and early childhood influence the individual's overall food attitudes throughout life. The nurturing of a young child, however, involves more than nutrition. Those who care for young children are responsible for providing not only nutritious milk, foods, and water, but also a safe, loving, secure environment in which the children may grow and develop. In light of toddlers' developmental and nutrient needs and their often contrary and willful behavior, a few feeding guidelines may be helpful:

- Discourage unacceptable behavior, such as standing at the table or throwing food, by removing the young child from the table to wait until later to eat. Be consistent and firm, not punitive. The child will soon learn to sit and eat.
- Let toddlers explore and enjoy food, even if this means eating with fingers for a while. Use of the spoon will come in time.
- Don't force food on children. Rejecting new foods is normal; acceptance is more likely as children become familiar with new foods through repeated opportunities to taste them.
- Provide nutritious foods, and let children choose which ones and how much they will eat. Gradually, they will acquire a taste for different foods.
- Limit sweets. Infants and young children have little room for empty-kcalorie foods in their daily energy allowance. Do not use sweets as a reward for eating meals.
- Don't turn the dining table into a battleground. Make mealtimes enjoyable. Teach healthy food choices and eating habits in a pleasant environment.

© Stephanie Rausser/FPG International

Ideally, a one-year-old eats many of the same foods as the rest of the family.

IN SUMMARY

The primary food for infants during the first 12 months is either breast milk or iron-fortified formula. In addition to nutrients, breast milk also offers immunological protection. At about four to six months, infants should gradually begin eating solid foods. By one year, they are drinking from a cup and eating many of the same foods as the rest of the family.

NUTRITION DURING CHILDHOOD

Each year from age one to adolescence, a child typically grows taller by 2 to 3 inches and heavier by 5 to 6 pounds. Growth charts provide valuable clues to a child's health. Weight gains out of proportion to height gains may reflect overeating and inactivity, whereas measures significantly below the standard suggest malnutrition.

Increases in height and weight are only two of the many changes growing children experience (see Figure 15-7). At age one, children can stand alone and are beginning to toddle; by two, they can walk and are learning to run; by three, they can jump and are climbing with confidence. Bones and muscles increase in mass and density to make these accomplishments possible. Thereafter, further lengthening of the long bones and increases in musculature proceed unevenly and more slowly until adolescence.

FIGURE 15-7 Body Shape of One-Year-Old and Two-Year-Old Compared

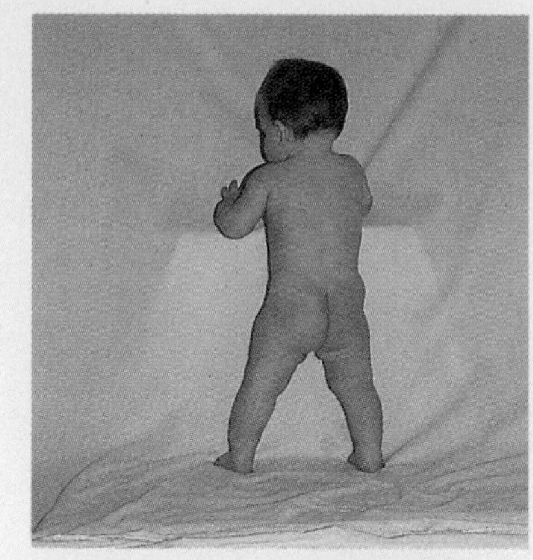

The body shape of a one-year-old (above) changes dramatically by age two (below). The two-year-old has lost much of the baby fat; the muscles (especially in the back, buttocks, and legs) have firmed and strengthened; and the leg bones have lengthened.

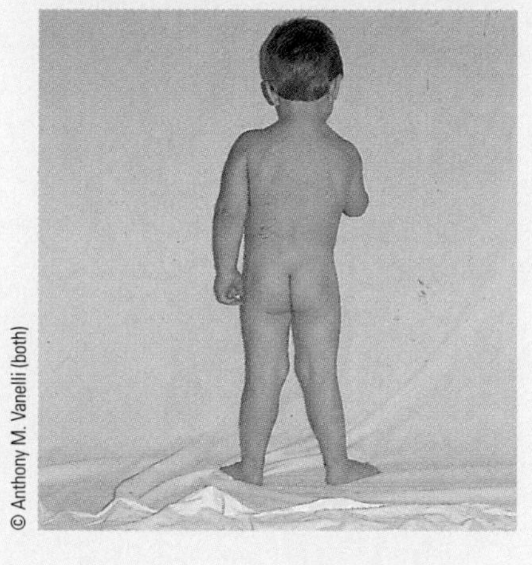

© Anthony M. Vanelli (both)

Energy and Nutrient Needs

Children's appetites begin to diminish around one year, consistent with the slowing growth. Thereafter, children spontaneously vary their food intakes to coincide with their growth patterns; they demand more food during periods of rapid growth than during slow growth. Sometimes they seem insatiable; other times they seem to live on air and water.

Children's energy intakes also vary widely from meal to meal. Even so, their total daily intakes remain remarkably constant. If children eat less at one meal, they typically eat more at the next, and vice versa. Overweight children are an exception: they do not always adjust their energy intakes appropriately and may eat in response to external cues, disregarding hunger and satiety signals.

• Energy Intake and Activity • Individual children's energy needs vary widely, depending on their growth and physical activity. A one-year-old child needs about 1000 kcalories a day; a three-year-old needs only 300 kcalories more. By age ten, a child needs about 2000 kcalories a day. Total energy needs increase slightly with age, but energy needs per kilogram of body weight actually decline gradually.

Inactive children can become obese even when they eat less food than the average. Unfortunately, our nation's children are becoming less and less active, with young girls showing a marked reduction in their physical activity.[23] Schools would serve our children well by offering activities to promote physical fitness.[24] Children who learn to enjoy physical play and exercise are best prepared to maintain active lifestyles as adults.

Some children, notably those adhering to a vegan diet, have difficulty meeting their energy needs. Grains, vegetables, and fruits provide plenty of fiber, adding bulk, but too few kcalories to support growth. A reasonable fiber◆ intake for children is "age plus 5 grams."[25] For example, a 3-year-old might need about 8 grams of fiber a day, and an 18-year-old, 23 grams.

Children's fiber recommendation: ◆
Age + 5 g.

• Protein • Like energy needs, total protein needs increase slightly with age, but when the child's body weight is considered, the protein requirement actually declines gradually (see Table 6-4 on p. 189). The estimation of protein needs considers the requirements for maintaining nitrogen balance, the quality of protein consumed, and the added needs of growth.

• Vitamins and Minerals • The vitamin and mineral needs of children increase with their ages (see inside front cover). A balanced diet of nutritious foods can meet children's needs for these nutrients, with the notable exception of iron. Iron-deficiency anemia is a major problem worldwide, as well as being the most prevalent nutrient deficiency among U.S. and Canadian children. Iron deficiency is most prevalent in young children (under age five).[26]

HEALTHY PEOPLE 2010

Reduce iron deficiency among young children.

To prevent iron deficiency, children's foods must deliver approximately 10 milligrams of iron per day. To achieve this goal, snacks and meals should include iron-rich foods,◆ and milk intake should be reasonable so that it will not displace lean meats, fish, poultry, eggs, legumes, and whole-grain or enriched products.

◆ Chapter 13 describes iron-rich foods and ways to maximize iron absorption.

• Planning Children's Meals • To provide all the needed nutrients, children's meals should include a variety of foods from each food group—in amounts suited to their appetites and needs. Figure 15-8 presents the Food Guide Pyramid for young children. Notice that the number of servings from each group is the same as the lower number for adults. For two- to three-year-olds, serving sizes are smaller, about two-thirds the portion for a child over four years old.

Children whose diets follow the Food Guide Pyramid pattern meet their nutrient needs fully, but few children eat according to these recommendations. In fact, only 5 percent consume the suggested servings from even four of the five food groups.[27] Almost half of them meet none or only one of the food group recommendations. Consequently, intakes of several nutrients, notably calcium, iron, and zinc, fall far below recommendations.[28]

Hunger and Malnutrition in Children

Most children in the United States and Canada are adequately nourished. Malnutrition does appear, however, in certain circumstances. Low-income children, for example, may be hungry and malnourished. An estimated 11 million U.S. children under age 12 are hungry and living in poverty. Highlight 16 examines the causes and consequences of hunger in the United States.

When hunger is chronic, children become malnourished and suffer growth retardation. Worldwide, malnutrition takes a devastating toll on children, contributing to nearly half of the deaths of children under four years old. Vitamin A deficiency afflicts more than 5 million children worldwide, inducing blindness, stunted growth, and infections. Zinc deficiency also retards growth and typically accompanies protein-energy malnutrition and vitamin A deficiency.

HEALTHY PEOPLE 2010

Reduce growth retardation among low-income children under age 5 years.

• Hunger and Behavior • Even when hunger is temporary, as when a child misses one meal, behavior and academic performance are affected. Children who eat nutritious breakfasts improve their school performance and are tardy or absent significantly less often than their peers who do not.[29] Without breakfast, children perform poorly in tasks requiring concentration, their attention spans are shorter, and they even score lower on intelligence tests than their well-fed peers; malnourished children are particularly vulnerable.[30] Unfortunately, an estimated 4 out of 30 students miss breakfast each day. Common sense dictates that it is unreasonable to expect anyone to learn and perform without fuel. For the child who hasn't had breakfast, the morning's lessons may be lost altogether. Even if a child has eaten breakfast, discomfort from hunger may become distracting by late morning.

The problem children face when attempting morning schoolwork on an empty stomach appears to be at least partly due to low blood glucose. The average child up to age ten or so needs to eat about every four hours to maintain

FIGURE 15-8 **Food Guide Pyramid for Young Children**

What counts as one serving?

GRAIN GROUP
1 slice bread
1/2 c cooked rice or pasta
1/2 c cooked cereal
1 oz ready-to-eat cereal

VEGETABLE GROUP
1/2 c chopped raw or cooked vegetables
1 c raw leafy vegetables

FRUIT GROUP
1 piece of fruit or melon wedge
3/4 c juice
1/2 c canned fruit
3/4 c dried fruit

MILK GROUP
1 c milk or yogurt
2 oz cheese

MEAT GROUP
2 to 3 oz cooked lean meat, poultry, or fish
1/2 c cooked dry beans, or 1 egg counts as 1 oz lean meat; 2 tbs of peanut butter count as 1 oz meat

FATS AND SWEETS
Limit kcalories from these.

Four- to six-year-olds can eat these serving sizes. Offer two- to three-year-olds less, except for milk. Two- to six-year-old children need a total of 2 servings from the milk group each day.

SOURCE: USDA Center for Nutrition and Policy Promotion, March 1999, Program AID 1649.

a blood glucose concentration high enough to support the activity of the brain and the rest of the nervous system. A child's brain is as big as an adult's, and the brain is the body's chief glucose consumer, using about three times as much glucose per day as the rest of the body. A child's liver is much smaller than an adult's, however, and the liver is responsible for storing glucose as glycogen and releasing it into the blood as needed. A child's liver can store only about four hours' worth of glycogen—hence the need to eat fairly often. Teachers aware of the late-morning slump in their classrooms wisely request that midmorning snacks be provided; snacks improve classroom performance all the way to lunchtime.

Eating breakfast also helps children to meet their nutrient needs each day.[31] Children who skip breakfast typically do not make up the deficits at later meals—they simply have lower intakes of energy, vitamins, and minerals than those who eat breakfast.

• Iron Deficiency and Behavior • Iron deficiency has well-known and widespread effects on children's behavior.[32] In addition to carrying oxygen in the blood, iron transports oxygen within cells, which use it to help produce energy.

Healthy, well-nourished children are alert in the classroom and energetic at play.

Iron is also used to make neurotransmitters—most notably, those that regulate the ability to pay attention, which is crucial to learning. Consequently, iron deficiency not only causes an energy crisis, but also directly affects mood, attention span, and learning ability.

Iron deficiency is usually diagnosed by a deficit of iron in the *blood,* after the deficiency has progressed all the way to anemia. A child's *brain,* however, is sensitive to low iron concentrations long before the blood effects appear. Research has shown that iron deficiency lowers the "motivation to persist in intellectually challenging tasks," shortens the attention span, and impairs overall intellectual performance. Anemic children perform less well on tests and are more disruptive than their nonanemic classmates; iron supplementation improves learning and memory. When combined with other nutrient deficiencies, iron-deficiency anemia has synergistic effects that are especially detrimental to learning.[33] Furthermore, children who had iron-deficiency anemia *as infants* continue to perform poorly as they grow older, even if their iron status improves.[34] The long-term damaging effects on mental development make prevention of iron deficiency during infancy and early childhood a high priority.

• Other Nutrient Deficiencies and Behavior • A child with any of several nutrient deficiencies may be irritable, aggressive, disagreeable, or sad and withdrawn. Such a child may be labeled "hyperactive," "depressed," or "unlikable," when in fact these traits may arise from simple, even marginal, malnutrition. Parents and medical practitioners often overlook the possibility that malnutrition may account for abnormalities of appearance and behavior. Any departure from normal healthy appearance and behavior is a sign of possible poor nutrition (see Table 15-3). In any such case, inspection of the child's diet by a registered dietitian or other qualified health care professional is clearly in order. Should suspicion of dietary inadequacies be raised, no matter what other causes may be implicated, the people responsible for feeding the child should take steps to correct those inadequacies promptly.

TABLE 15-3 Physical Signs of Malnutrition in Children

	Well-Nourished	Malnourished	Possible Nutrient Deficiencies
Hair	Shiny, firm in the scalp	Dull, brittle, dry, loose; falls out	PEM
Eyes	Bright, clear pink membranes; adjust easily to light	Pale membranes; spots; redness; adjust slowly to darkness	Vitamin A, the B vitamins, zinc, and iron
Teeth and gums	No pain or caries, gums firm, teeth bright	Missing, discolored, decayed teeth; gums bleed easily and are swollen and spongy	Minerals and vitamin C
Face	Clear complexion without dryness or scaliness	Off-color, scaly, flaky, cracked skin	PEM, vitamin A, and iron
Glands	No lumps	Swollen at front of neck, cheeks	PEM and iodine
Tongue	Red, bumpy, rough	Sore, smooth, purplish, swollen	B vitamins
Skin	Smooth, firm, good color	Dry, rough, spotty; "sandpaper" feel or sores; lack of fat under skin	PEM, essential fatty acid, vitamin A, B vitamins, and vitamin C
Nails	Firm, pink	Spoon-shaped, brittle, ridged	Iron
Internal systems	Regular heart rhythm, heart rate, and blood pressure; no impairment of digestive function, reflexes, or mental status	Abnormal heart rate, heart rhythm, or blood pressure; enlarged liver, spleen; abnormal digestion; burning, tingling of hands, feet; loss of balance, coordination; mental confusion, irritability, fatigue	PEM and minerals
Muscles and bones	Muscle tone; posture, long bone development appropriate for age	"Wasted" appearance of muscles; swollen bumps on skull or ends of bones; small bumps on ribs; bowed legs or knock-knees	PEM and vitamin D

The Malnutrition-Lead Connection

© Tony Freeman/PhotoEdit

Old, lead-based paint threatens the health of an exploring child.

Children who are malnourished are vulnerable to lead poisoning. They absorb more lead if their stomachs are empty; if they have low intakes of calcium, zinc, vitamin C, or vitamin D; and, of greatest concern because it is so common, if they have iron deficiencies.[35] Iron deficiency weakens the body's defenses against lead absorption, and lead poisoning can cause iron deficiency. Common to both iron deficiency and lead poisoning are a low socioeconomic background and a lack of immunizations against infectious diseases.[36] Another common factor is pica—a craving for nonfood items. Many children with lead poisoning eat dirt or chew on newspapers, two common sources of lead.

The anemia brought on by lead poisoning may be mistaken for a simple iron deficiency and therefore may be incorrectly treated. Like iron deficiency, mild lead toxicity has nonspecific symptoms, including diarrhea, irritability, and fatigue. The symptoms are not reversible by adding iron to the diet; exposure to lead must stop. With further exposure, the signs become more pronounced: children lose cognitive, verbal, and perceptual abilities and develop learning disabilities and behavioral problems. Still more severe lead toxicity can cause irreversible nerve damage, paralysis, mental retardation, and death.

Lead toxicity is most prevalent among children under six—as many as 1.7 million children (10 to 15 percent of all preschoolers) may have blood lead concentrations high enough to cause mental, behavioral, and other health problems.[37] Lead intoxication in young children comes from their own behaviors and activities—putting their hands in their mouths, playing in dirt, and chewing on nonfood items. Unfortunately, the body readily absorbs lead during times of rapid growth and hoards it possessively thereafter. Lead is not easily excreted and accumulates mainly in the bones, but also in the brain, teeth, and kidneys. Tragically, a child's neuromuscular system is also maturing during these first few years of life. No wonder children with elevated lead levels experience impairment of balance, motor development, and the relaying of nerve messages to and from the brain. Unfortunately, deficits in intellectual development are only partially reversed when lead levels decline.[38]

HEALTHY PEOPLE 2010

Eliminate elevated blood lead levels in children. Increase the proportion of persons living in pre-1950s housing that has been tested for the presence of lead-based paint.

Federal laws mandating reductions in leaded gasolines, lead-based solder, and other products over the past three decades have helped to reduce the amounts of lead in food and in the environment in the United States. As a consequence, the prevalence of lead toxicity in children has declined dramatically for most of the United States, but lead exposure is still a threat in certain communities.[39] The "How to" on p. 514 presents strategies for defending children against lead toxicity.

Hyperactivity and "Hyper" Behavior

All children are naturally active, and many of them become overly active on occasion—for example, in anticipation of a birthday party. Such behavior is markedly different from true **hyperactivity.**

hyperactivity: inattentive and impulsive behavior that is more frequent and severe than is typical of others a similar age; professionally called **attention-deficit/hyperactivity disorder (ADHD).**

• **Hyperactivity** • Hyperactive children have trouble sleeping, cannot sit still for more than a few minutes at a time, act impulsively, and have difficulty paying attention. These behaviors interfere with social development and academic progress. The cause of hyperactivity remains unknown, but it affects about 5

HOW TO Protect against Lead Toxicity

Researchers simultaneously made three major discoveries about lead toxicity: lead poisoning has *subtle* effects, the effects are *permanent*, and they occur at *low levels of exposure*. The amount of lead recognized to cause harm is only 10 micrograms per 100 milliliters of blood. Consequently, consumers should take ultraconservative measures to protect themselves, and especially their infants and young children, from lead poisoning. The American Academy of Pediatrics and the Centers for Disease Control recommend screening in communities with a substantial number of houses built before 1950 and in those with a substantial number of children with elevated lead levels.[a] In addition to screening children most likely to be exposed, pediatricians should alert all parents to the possible dangers of lead exposure and explain prevention strategies.

Defensive strategies include:

- In contaminated environments, keep small children from putting dirty or old painted objects in their mouths, and make sure children wash their hands before eating. Similarly, keep small children from eating any nonfood items. Lead poisoning has been reported in young children who have eaten crayons or pool cue chalk.
- Wet-mop floors and damp-sponge walls regularly. Children's blood lead levels decline when the homes they live in are cleaned regularly.[b]
- Be aware that other countries do not have the same regulations protecting consumers against lead. Children have been poisoned by eating crayons made in China and drinking fruit juice canned in Mexico.
- Do not use lead-contaminated water to make infant formula.
- Once you have opened canned food, immediately move it to a lead-free storage container to prevent lead migration into the food.
- Do not store acidic foods or beverages (such as vinegar or orange juice) in ceramic dishware or alcoholic beverages in pewter or crystal decanters.
- Many manufacturers are now making lead-safe products.[c] Old, handmade, or imported ceramic cups and bowls may contain lead and should not be used to heat coffee or tea or acidic foods such as tomato soup.
- U.S. wineries have stopped using lead in their foil seals, but older bottles may still be around and other countries may still use lead; to be safe, wipe the foil-sealed rim of a wine bottle with a clean wet cloth before removing the cork.
- Feed children nutritious meals regularly.
- Before using your newspaper to wrap food, mulch garden plants, or add to your compost, confirm with the publisher that the paper uses no lead in its ink.

The Environmental Protection Agency (EPA) also publishes a booklet, *Lead and Your Drinking Water,* in which the following cautions appear:

- Have the water in your home tested by a competent laboratory.
- Use only cold water for drinking, cooking, and making formula (cold water absorbs less lead).
- When water has been standing in pipes for more than two hours, flush the cold-water pipes by running water through them for 30 seconds before using it for drinking, cooking, or mixing formulas.
- If lead contamination of your water supply seems probable, obtain additional information and advice from the EPA and your local public health agency.

By taking these steps, parents can protect themselves and their children from this preventable danger.[d]

[a]Committee on Environmental Health, American Academy of Pediatrics, Screening for elevated blood lead levels, *Pediatrics* 101 (1998): 1072–1078.

[b]G. G. Rhoads and coauthors, The effect of dust lead control on blood lead in toddlers: A randomized trial, *Pediatrics* 103 (1999): 551–555.

[c]*A Shopper's Guide to Low-Lead China* is available from the Environmental Defense Fund, 257 Park Avenue South, New York, NY 10010; telephone (800) 284-3322.

[d]The National Lead Information Center provides two hotlines; call (800) LEAD-FYI (532-3394) for general information or (800) 424-LEAD (424-5323) with specific questions.

percent of young school-age children. To resolve the problems surrounding hyperactivity, physicians often recommend specific behavioral strategies, special educational programs, and psychological counseling; in many cases, they prescribe medication.[40]

Parents of hyperactive children sometimes seek help from alternative therapies, including special diets. They mistakenly believe a solution may lie in manipulating the diet—most commonly, by excluding sugar or food additives. Adding carrots or eliminating candy is such a simple solution that many parents eagerly give such diet advice a try. These dietary changes will not solve the problem of true hyperactivity. Studies have consistently found no convincing evidence that sugar causes hyperactivity or worsens behavior; in fact, sugar may actually have a sedative effect.[41] Recommendations to restrict sugar in children's diets to prevent or treat behavioral problems are groundless.

• **Misbehaving** • Even a child who is not truly hyperactive can be difficult to manage at times. Michael may act unruly out of a desire for attention, Jessica may be cranky because of a lack of sleep, Christopher may react violently after watching too much television, and Sheila may be unable to sit still in class due to a lack of exercise. All of these children may benefit from more consistent care—regular hours of sleep, regular mealtimes, and regular outdoor activity.

© Thomas Harm, Tom Peterson/Quest Photographic

Almost 75 percent of adverse reactions are caused by three foods—eggs, peanuts, and milk.

Adverse Reactions to Foods

Adverse reactions to foods can threaten nutritional health to varying extents, depending on the severity and duration of the reactions and the foods they involve. Temporary reactions may lead to permanent avoidance of foods; permanent reactions, if not detected and treated, can cause chronic illness. In rare cases, adverse reactions can result in death.

HEALTHY PEOPLE 2010

Reduce deaths from anaphylaxis caused by food allergies.

• Food Intolerances • Not all adverse reactions to foods are food allergies, although even physicians may describe them as such. Signs of adverse reactions to foods include stomachaches, headaches, rapid pulse rate, nausea, wheezing, hives, bronchial irritation, coughs, and other such discomforts. Among the causes may be reactions to chemicals in foods, such as the flavor enhancer monosodium glutamate (MSG), the natural laxative in prunes, or the mineral sulfur; digestive diseases, such as obstructions or injuries; enzyme deficiencies, such as lactose intolerance; and even psychological aversions. These reactions involve symptoms but no antibody production. Therefore, they are **food intolerances,** not allergies.

Pesticides on produce may also cause adverse reactions. Pesticides may linger on the foods to which they were applied in the field. Health risks from pesticide exposure are probably small for healthy adults, but children may be vulnerable.

• Food Allergies • A true **food allergy** occurs when protein fractions of a food or other large molecules are absorbed into the blood and elicit an immunologic response. (Recall that proteins and other large molecules of food are normally dismantled in the digestive tract to smaller ones that are absorbed without such a reaction.) The body's immune system reacts to these large food molecules as it does to other antigens—by producing antibodies, histamines, or other defensive agents.

Allergies may have one or two components. They always involve antibodies; they may or may not involve symptoms.◆ This means that allergies can be diagnosed only by testing for antibodies. Even symptoms exactly like those of an allergy may not be caused by one.

◆ A person who produces antibodies *without* having any symptoms has an **asymptomatic allergy;** a person who produces antibodies *and* has symptoms has a **symptomatic allergy.**

Allergic reactions to food may be immediate or delayed. In both cases, the antigen interacts immediately with the immune system, but the timing of symptoms varies from minutes to 24 hours after consumption of the antigen. Identifying the food that causes an immediate allergic reaction is fairly easy because the symptoms appear within minutes after eating the food. Identifying the food that causes a delayed reaction is more difficult because the symptoms may not appear until a day later. By this time, many other foods have been eaten, complicating the picture.

Identifying a true food allergy requires a thorough health history, physical examination, and diagnostic tests to eliminate other diseases. Skin pricks with food extracts are one of the most common tests for food allergies, even though the high incidence of false positive results can complicate diagnosis. Physicians also conduct dietary trials that first eliminate the offending food and then reintroduce it in small quantities to substantiate that reactions occur only when that particular food is eaten. Once a food allergy has been diagnosed, therapy requires strict elimination of the offending food.

adverse reactions: unusual responses to food (including intolerances and allergies).

food intolerances: adverse reactions to foods that do not involve the immune system.

food allergy: an adverse reaction to food that involves an immune response; also called **food-hypersensitivity reaction.**

Food allergies are most common during the first few years of life, but then children typically outgrow (become tolerant to) their hypersensitivity. Approximately 5 percent of young children are allergic to certain foods, whereas less than 2 percent of adults have food allergies.[42] Developing tolerance is most likely if the offending food can be identified and eliminated from the diet for at least a year or two.

When parents stop serving a suspected food to their child, they risk the child's suffering growth deficits and nutrient deficiencies.[43] They should be sure to include other foods that offer the same nutrients as the omitted food. Children with allergies, like all children, need all their nutrients.

Hunger, lead poisoning, hyperactivity, and allergic reactions can all adversely affect a child's nutrition status and health. Each of these problems has solutions. They may not be easy solutions, but we have a reasonably good understanding of the problems and ways to correct them. Such is not the case with the most pervasive health problem for children in the United States—obesity.

Childhood Obesity

The incidence of childhood obesity has increased dramatically over the past three decades. Like their parents, children in the United States are becoming fatter. An estimated one out of every five U.S. children is overweight. Figure 15-9 presents the BMI for children and adolescents, indicating cutoff points for overweight and obese.

HEALTHY PEOPLE 2010

Reduce the proportion of children and adolescents who are overweight or obese.

Parental obesity predicts an early increase in a young child's BMI, and it more than doubles the chances that a young child will become an obese adult.[44] As children grow older, their own obesity also becomes an important factor in determining their obesity as adults. Nonobese children with neither parent obese have a less than 10 percent chance of becoming obese in adulthood, whereas obese teens with at least one obese parent have a greater than 80 percent chance of being obese adults.

That children are heavier today than they were 20 or so years ago cannot be explained by genetics, however. Diet and physical inactivity must be responsible. Children's dietary fat intakes vary, of course, but children who prefer high-fat foods tend to be more overweight than their peers.[45] Particularly noteworthy is the finding that children's fat preferences and consumption correlate with their parents' obesity as well. Such findings confirm the significant roles parents play—teaching children about healthy food choices, providing children with low-fat selections, and serving as role models.

Most likely, children have grown more overweight because of their lack of physical activity.[46] An inactive child can become obese even while eating less food than an active child. Today's children are more sedentary and less physically fit than children were even 20 years ago.

◆ **TV fosters obesity because it:**
- Requires no energy beyond basal metabolism.
- Replaces vigorous activities.
- Encourages snacking.
- Promotes a sedentary lifestyle.
- Playing computer games influences children's activity patterns similarly.

Television watching◆ may contribute most to physical inactivity. Children who watch a lot of television (four or more hours a day) are most likely to be obese and least likely to eat fruits and vegetables.[47] They are often snacking on the fattening foods that are advertised. The average child sees an estimated 30,000 TV commercials a year—many peddling foods high in sugar, fat, and salt such as sugar-coated breakfast cereals, candy bars, chips, fast foods, and carbonated beverages.

Watching television accounts for some 24 hours a week of sedentary behavior. Beyond these 24 hours, children spend more time sitting at computers and

FIGURE 15-9 Body Mass Index-for-Age Percentiles: Boys and Girls, Age 2 to 20.

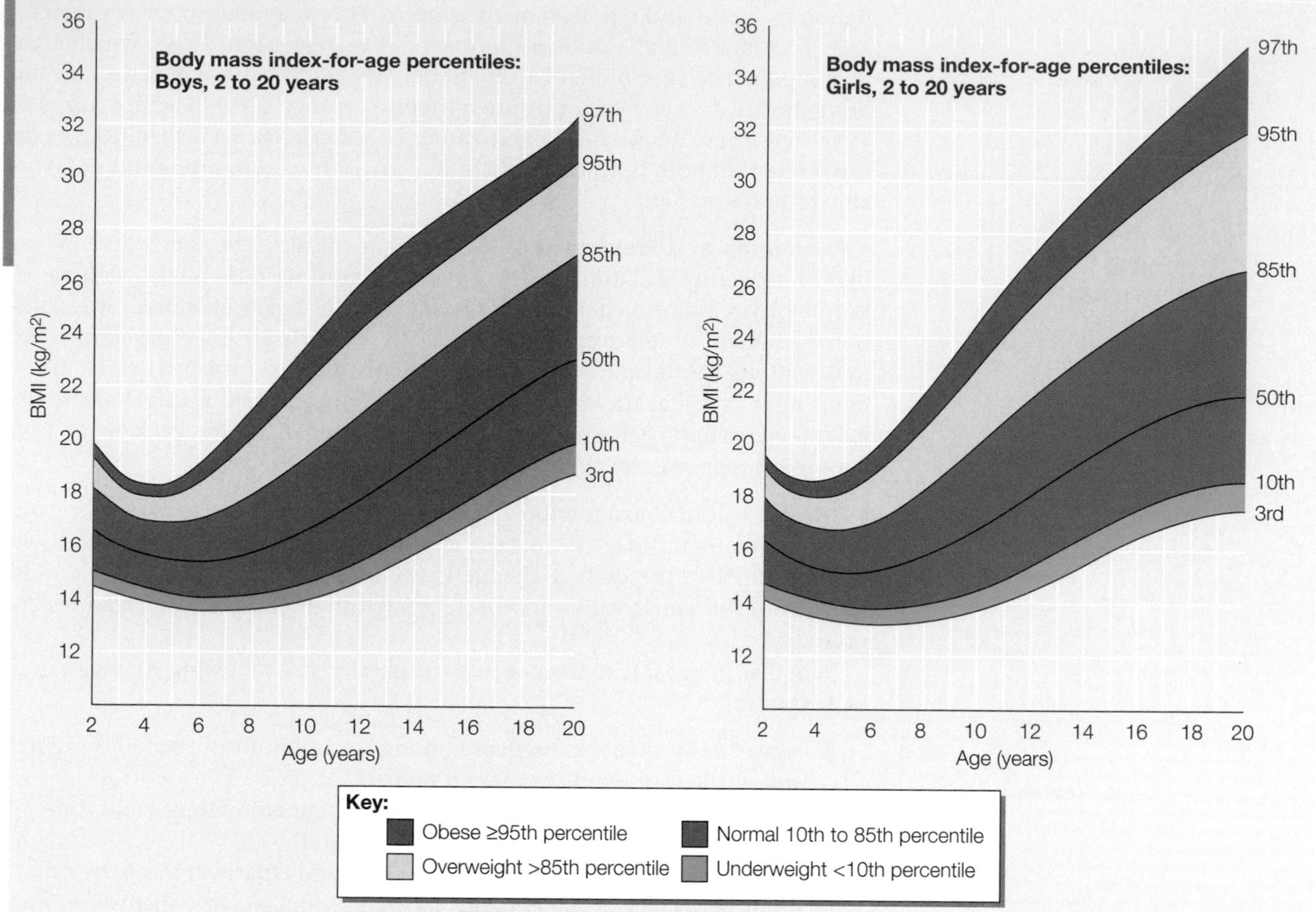

Body mass index-for-age percentiles: Boys, 2 to 20 years

Body mass index-for-age percentiles: Girls, 2 to 20 years

BMI (kg/m²) — Age (years) — 97th, 95th, 85th, 50th, 10th, 3rd

Key:
- Obese ≥95th percentile
- Overweight >85th percentile
- Normal 10th to 85th percentile
- Underweight <10th percentile

playing video games. These activities use no more energy than resting, displace participation in more vigorous activities, and foster snacking on high-fat foods. Simply reducing the amount of time spent watching television (and playing video games) can improve a child's BMI.[48]

The single most important problem for obese children is the potential of becoming obese adults with all the social, economic, and medical ramifications that often accompany obesity. They have additional problems, too, arising from differences in their growth, physical health, and psychological development.

• **Growth** • Obese children develop a characteristic set of physical traits. They typically begin puberty earlier and so grow taller than their peers at first, but then stop growing at a shorter height. They develop greater bone and muscle mass in response to the demand of having to carry more weight—both fat and lean weight. Consequently, they appear "stocky" even when they lose their excess fat. Obese children also have a faster metabolic rate, apparently due to their abundant lean body mass.[49]

© Donna Day/Tony Stone Images

Television watching influences children's eating habits and activity patterns.

• **Physical Health** • Like obese adults, obese children display a blood lipid profile indicative that atherosclerosis is beginning to develop: high levels of total cholesterol, triglycerides, low-density lipoprotein (LDL) cholesterol, and very-low-density lipoprotein (VLDL) cholesterol. Obese children also tend to have high blood pressure; in fact, obesity is the leading cause of pediatric hypertension. Their risks for developing diabetes and respiratory diseases (such as asthma) are also exceptionally high. These relationships between childhood obesity and chronic diseases are discussed fully in Highlight 15.

• Psychological Development • Obesity often causes psychological problems. Because people frequently judge others on appearance more than on character, obese children are often victims of prejudice. Many suffer discrimination by adults and rejection by their peers. They may have poor self-images, a sense of failure, and a passive approach to life. Television shows, which are a major influence in children's lives, frequently portray the fat person as the bumbling misfit. Overweight children themselves may come to accept this negative stereotype. Researchers investigating children's reactions to various body types find that both normal-weight and overweight children respond unfavorably to bulky bodies.

• Prevention and Treatment of Obesity • Medical science has worked wonders in preventing or curing many of even the most serious childhood diseases, but obesity remains a challenge.[50] Once excess fat has been stored, it is stubbornly difficult to remove. In light of all this, parents are encouraged to make major efforts to prevent childhood obesity or to begin treatment early. Treatment must consider the many aspects of the problem and possible solutions. An integrated approach is recommended, involving diet, physical activity, psychological support, and behavioral changes.

• Diet • The initial goal for obese children is to reduce the rate of weight gain; that is, to maintain weight while the child grows taller. Continued growth will then accomplish the desired change in weight for height. Treatment should begin with this conservative approach before other, more drastic measures are taken.

Whether the goal is to treat or prevent obesity, the following strategies may be helpful:

- Serve family meals that reflect kcalorie control both in the foods offered and in the ways foods have been prepared.
- Encourage children to eat slowly, to pause and enjoy their table companions, and to stop eating when they are full.
- Teach them how to select nutrient-dense foods that will meet their nutrient needs within their energy allowances and to serve themselves appropriate portions at meals; the amount of food offered influences the amount of food eaten.[51]
- Never force children to clean their plates.

• Physical Activity • The many benefits of physical activity are well known, but often are not incentive enough to motivate overweight people, especially children. Yet regular vigorous activity can improve a child's weight, body composition, and physical fitness.[52] Ideally, parents will encourage daily physical activity to promote strong skeletal, muscular, and cardiovascular development and instill in their children the desire to be physically active throughout life. Most importantly, parents need to set a good example. Physical activity is a natural and lifelong behavior of healthy living. It can be as simple as riding a bike, playing tag, jumping rope, or doing chores. It need not be an organized sport; it just needs to be some activity on a regular basis.

HEALTHY PEOPLE 2010

Increase the proportion of adolescents who engage in moderate physical activity for at least 30 minutes on 5 or more days per week. Increase the proportion of adolescents who engage in vigorous physical activity that promotes cardiorespiratory fitness 3 or more days per week for 20 or more minutes per occasion. Increase the proportion of the nation's public and private schools that require daily school physical education for all students. Increase the proportion of adolescents who participate in daily physical education being physically active.

• **Psychological Support** • Programs that involve parents in treatment report greater weight loss than programs in which parents are not involved.[53] Because obesity in parents and children tends to be positively correlated, both benefit from a weight-loss program. Parental attitudes about food greatly influence their children's eating behavior, so it is important that the influence be positive. Otherwise, eating problems may become exacerbated. Unaware that they are teaching their children, parents pass on lessons at the dinner table, on television trays, and in drive-through restaurants. Overweight parents may model for their children the behaviors that have led to their weight gains—eating too much, dieting inappropriately, exercising too little. This pattern is especially evident between mothers and daughters.[54]

• **Behavioral Changes** • In contrast to traditional weight-loss programs that focus on *what* to eat, behavioral programs focus on *how* to eat. These techniques involve changing learned habits that lead a child to eat excessively.

Obesity is prevalent in our society. Its far-reaching effects lend urgency to the need to find a remedy. Because treatment of obesity is frequently unsuccessful, it is most important to prevent its onset. Above all, be sensible in teaching children how to maintain appropriate body weight. Parents and the media are most influential in shaping self-concept, weight concerns, and dieting practices.[55] Children can easily get the impression that their worth is tied to their body weight. Some parents fail to realize that society's ideal of slimness can be perilously close to starvation and that a child encouraged to "diet" cannot obtain the energy and nutrients required for normal growth and development. Even healthy children without diagnosable eating disorders have been observed to limit their growth through "dieting."[56] Weight gain in truly overweight children can be managed without compromising growth, but should be overseen by a health care professional.

Mealtimes at Home

Traditionally, parents served as **gatekeepers,** determining what foods and activities were available in their children's lives. Then the children made their own selections. Gatekeepers who wanted to promote nutritious choices and healthful habits provided access to nutrient-dense, delicious foods and opportunities for active play at home.

In today's consumer-oriented society, children have greater influence over family decisions concerning food—the fast-food restaurant the family chooses when eating out, the type of food the family eats at home, and the specific brands the family purchases at the grocery store. Parental guidance in food choices is still necessary, but equally important is teaching children consumer skills to help them make informed choices.[57]

• **Honoring Children's Preferences** • Researchers attempting to explain children's food preferences encounter contradictions. Children say they like colorful foods, yet most often reject green and yellow vegetables while favoring brown peanut butter and white potatoes, apple wedges, and bread. They do like raw vegetables◆ better than cooked ones, so it is wise to offer vegetables that are raw or slightly undercooked, served separately, and easy to eat. Foods should be warm, not hot, because a child's mouth is much more sensitive than an adult's. The flavor should be mild because a child has more taste buds, and smooth foods such as mashed potatoes or split-pea soup should contain no lumps (a child wonders, with some disgust, what the lumps might be). Children prefer foods that are familiar, so offer various foods regularly.

Make mealtimes fun for children. Young children like to eat at little tables and to be served small portions◆ of food. They like sandwiches cut in different geometric shapes and common foods called silly names. They also like to eat

Child feeding tip: Serve vegetables raw or slightly undercooked and crunchy. ◆

Child feeding tip: Provide child-sized portions and utensils. ◆

gatekeepers: with respect to nutrition, key people who control other people's access to foods and thereby exert profound impacts on their nutrition. Examples are the spouse who buys and cooks the food, the parent who feeds the children, and the caregiver in a day-care center.

with other children, and they tend to eat more when in the company of their friends. Children are also more likely to give up their prejudices against foods when they see their peers eating them.

◆ **Child feeding tip: Encourage children to help plan and prepare meals.**

• **Learning through Participation** • Allowing children to help plan and prepare the family's meals◆ provides enjoyable learning experiences and encourages children to eat the foods they have prepared. Vegetables are pretty, especially when fresh, and provide opportunities for children to learn about color, growing vegetables and their seeds, and shapes and textures—all of which are fascinating to young children. Measuring, stirring, washing, and arranging foods are skills that even a young child can practice with enjoyment and pride.

• **Avoiding Power Struggles** • Problems over food often arise during the second or third year, when children begin asserting their independence. Many of these problems stem from the conflict between children's developmental stages and capabilities and parents who, in attempting to do what they think is best for their children, try to control every aspect of eating. Such conflicts can disrupt children's abilities to regulate their own food intakes or to determine their own likes and dislikes. For example, many people share the misconception that children must be persuaded or coerced to try new foods. In fact, the opposite is true. When children are forced to try new foods, even by way of rewards,◆ they are less likely to try those foods again than are children who are left to decide for themselves. Similarly, when children are restricted from eating their favorite foods, they are more likely to want those foods.[58] The parent is responsible for providing healthful foods, but the child is responsible for *how much* and even *whether* to eat.[59]

◆ **Child feeding tip: Don't use food as a reward for good behavior.**

When introducing new foods at the table, offer them one at a time and only in small amounts at first. The more often a food is presented to a young child, the more likely the child will accept that food. Offer the new food at the beginning of the meal, when the child is hungry, and allow the child to make the decision to accept or reject it. Never make an issue of food acceptance. A power struggle almost invariably sets a firm pattern of resistance and permanently closes the child's mind.

◆ **Child feeding tip: To prevent choking, watch children eat and enforce a "sit-down" rule.**

◆ **Young children can easily choke on:**
- Peanut butter (by spoonful).
- Popcorn.
- Whole grapes and cherries.
- Whole beans.
- Hot dog pieces and chunks of meat.
- Hard candies and marshmallows.
- Nuts.
- Raw celery and carrots.

• **Choking Prevention** • Parents must always be alert to the dangers of choking. A choking child is silent, so an adult should be present whenever a child is eating. Make sure the child sits when eating;◆ choking is more likely when a child is running or falling. Round foods◆ such as grapes and hot dog pieces are difficult to control in a mouth with few teeth, and they can easily become lodged in the small opening of a child's trachea.

◆ **Child feeding tip: Play first, then eat.**

• **Play First** • Children may be more relaxed and attentive at mealtime if outdoor play or other fun activities are scheduled before,◆ rather than immediately after, mealtime. Otherwise children "hurry up and eat" so that they can go play.

◆ **Child feeding tip: Provide healthful snacks.**

• **Snacks** • Parents may find that their children snack so much that they aren't hungry at mealtimes. Instead of teaching children *not* to snack, parents might be wise to teach them *how* to snack. Provide snacks◆ that are as nutritious as the foods served at mealtime. Snacks can even be mealtime foods served individually over time, instead of all at once on one plate. When providing snacks to children, think of the five food groups and offer such snacks as pieces of cheese, tangerine slices, carrot sticks, and peanut butter on whole-wheat crackers (see Table 15-4). Snacks that are easy to prepare should be readily available to children, especially if they arrive home from school before their parents.

To ensure that children have healthy appetites and plenty of room for nutritious foods when they are hungry, parents and teachers must limit access to candy, cola, and other concentrated sweets. If such foods are permitted in large quantities, the only possible outcomes are nutrient deficiencies, obesity, or both. The preference for sweets is innate; most children do not naturally select nutritious foods on the basis of taste. When children are allowed to create

Children enjoy eating the foods they help to prepare.

TABLE 15-4 Healthful Snack Ideas—Think Food Groups, Alone and in Combination

Selecting two or more foods from different food groups adds variety and nutrient balance to snacks. The combinations are endless, so be creative.

Grain Products

Grain products are filling snacks, especially when combined with other foods:

- Cereal with fruit and milk
- Crackers and cheese
- Wheat toast with peanut butter
- Popcorn with grated cheese
- Oatmeal raisin cookies with milk

Vegetables

Cut-up fresh, raw vegetables make great snacks alone or in combination with foods from other food groups:

- Celery with peanut butter
- Broccoli, cauliflower, and carrot sticks with a flavored cottage cheese dip

Fruits

Fruits are delicious snacks and can be eaten alone—fresh, dried, or juiced—or combined with other foods:

- Apples and cheese
- Bananas and peanut butter
- Peaches with yogurt
- Raisins mixed with sunflower seeds or nuts

Meats and Meat Alternates

Meat and meat alternates add protein to snacks:

- Refried beans with nachos and cheese
- Tuna on crackers
- Luncheon meat on wheat bread

Milk and Milk Products

Milk can be used as a beverage with any snack, and many other milk products, such as yogurt and cheese, can be eaten alone or with other foods as listed above.

meals freely from a variety of foods, they typically select foods that provide a lot of sugar. When their parents are watching, or even when they think their parents are watching, children improve their selections.

Sweets need not be banned altogether. Children who are exceptionally active can enjoy high-kcalorie foods such as ice cream or pudding from the milk group or pancakes from the bread group. As for sedentary children, they need to become more active, so they can also enjoy some of these foods without unhealthy weight gain.

• **Preventing Dental Caries** • Children frequently snack on sticky, sugary foods◆ that stay on the teeth and provide an ideal environment for the growth of bacteria that cause dental caries. Teach children to brush and floss after meals, to brush or rinse after eating snacks, to avoid sticky foods, and to select crisp or fibrous foods frequently.

◆ Child feeding tip: To protect against dental caries, serve fresh fruits more often than dried fruits or juices.

• **Serving as Role Models** • In an effort to practice these many tips, parents may overlook perhaps the single most important influence on their children's food habits—themselves. Parents who don't eat carrots shouldn't be surprised when their children refuse to eat carrots. Likewise, parents who dislike the smell of brussels sprouts may not be able to persuade children to try them. Children learn much through imitation. It is not surprising that children prefer the foods other family members enjoy and dislike foods that are never offered to them.[60] Parents, older siblings, and other caregivers set an irresistible

◆ Child feeding tip: Set a good example—enjoy nutritious foods.

example◆ by sitting with younger children, eating the same foods, and having pleasant conversations during mealtime.[61]

While serving and enjoying food, caregivers can promote both physical and emotional growth at every stage of a child's life. They can help their children to develop both a positive self-concept and a positive attitude toward food. If the beginnings are right, children will grow without the conflicts and confusions over food that can lead to nutrition and health problems.

Nutrition at School

While parents are doing what they can to establish good eating habits in their children at home, others are preparing and serving foods to their children at day-care centers and schools. In addition, children begin to learn about food and nutrition in the classroom. Meeting the nutrition and education needs of children is critical to supporting their healthy growth and development.[62]

• Meals at School • The U.S. government funds several programs to provide nutritious meals for children at school. Both the School Breakfast Program and the National School Lunch Program provide meals at a reasonable cost to children from families with the financial means to pay. Meals are available free or at reduced cost to children from low-income families. (School lunches in Canada are administered locally and therefore vary from area to area.) Several studies have reported that children who participate in school food programs show improvements in learning. The next page describes food assistance programs for children, and Table 15-5 shows school lunch patterns for children of different ages.

TABLE 15-5 School Lunch Patterns for Different Ages[a]

Food Group	Preschool (Age)		Grade School through High School (Grade)		
	1 to 2	*3 to 4*	*K to 3*	*4 to 6*	*7 to 12*
Meat or meat alternate					
1 serving:					
Lean meat, poultry, or fish	1 oz	1½ oz	1½ oz	2 oz	3 oz
Cheese	1 oz	1½ oz	1½ oz	2 oz	3 oz
Large egg(s)	½	¾	¾	1	1½
Cooked dry beans or peas	¼ c	⅜ c	⅜ c	½ c	¾ c
Peanut butter	2 tbs	3 tbs	3 tbs	4 tbs	6 tbs
Yogurt	½ c	¾ c	¾ c	1 c	1½ c
Peanuts, soynuts, tree nuts, or seeds[b]	½ oz	¾ oz	¾ oz	1 oz	1½ oz
Vegetable and/or fruit					
2 or more servings, both to total	½ c	½ c	½ c	¾ c	¾ c
Bread or bread alternate[c]					
Servings	5 per week	8 per week	8 per week	8 per week	10 per week
Milk					
1 serving of fluid milk	¾ c	¾ c	1 c	1 c	1 c

[a]The quantities listed represent per-lunch minimums for each age and grade except those for the oldest group, which are recommendations. Schools unable to serve the recommended quantities for grades 7 to 12 must provide at least the amount shown for grades 4 to 6.
[b]These meat alternates may be used to meet no more than half of the meat or meat alternate requirement; therefore, they must be used in a meal with another meat or meat alternate.
[c]Schools must serve daily at least ½ serving of bread or bread alternate to the youngest age group and at least 1 serving to older children.
SOURCE: U.S. Department of Agriculture, National School Lunch Program Regulations, revised January 1, 1998.

Food Assistance Programs for Children

The federal School Lunch and School Breakfast Programs assist schools financially so that every student can receive a nutritious lunch, breakfast, or both. These programs enable schools to provide low-income students with meals at no cost while charging other students somewhat less than the full costs of their meals. In addition, schools that participate in the programs can obtain food commodities. Nationally, the U.S. Department of Agriculture (USDA) administers the programs; on the state level, state departments of education operate them. The programs usually cost school districts little.

More than 26 million children receive lunches through the National School Lunch Program—half of them free or at a reduced price. School lunches are designed to provide at least a third of the RDA for energy, protein, vitamin A, vitamin C, iron, and calcium. They must also meet the *Dietary Guidelines* and include specified numbers of servings of milk, protein-rich foods (meat, poultry, fish, cheese, eggs, legumes, or peanut butter), vegetables, fruits, and breads or other grain foods.

The School Breakfast Program is available in slightly more than half of the nation's schools, and about 5 million children participate in it. The school breakfast must provide at least a fourth of the RDA for each of many nutrients and contain at least one serving of milk; one serving of fruit, juice, or vegetable; and either two servings of bread (or bread alternates), two servings of meat (or meat alternates), or one serving of each.

Another federal program, the Child Care Food Program, operates similarly and provides funds to organized child-care programs. All eligible children, centers, and family day-care homes have the right to participate. Meal reimbursements cover most of the meal and administration costs. Sponsors may also receive USDA commodity foods.

HEALTHY PEOPLE 2010

Increase the proportion of children and adolescents aged 6 to 19 years whose intakes of meals and snacks at school contribute to good overall dietary quality.

School lunches offer a variety of food choices and help children meet at least one-third of their needs for selected nutrients. Over a week's menus, these lunches are also required to meet the *Dietary Guidelines*. Schools making special efforts to lower fat in school lunches typically have trouble providing enough energy and nutrients, especially iron, to meet specifications. The American Dietetic Association (ADA) advocates the development of dietary guidelines specifically for children to ensure that school lunches will both provide adequate energy and nutrients and support health.[63] Other health professionals agree that there is a need for separate guidelines that address children's unique needs.[64]

• Competing Influences at School • Serving healthful lunches is only half the battle; students need to eat them, too. Short lunch periods and long waiting lines prevent some students from eating a school lunch and leave others with too little time to complete their meals.[65] Nutrition efforts at schools are also undermined when students can buy meals from fast-food restaurants or a la carte foods such as pizza or snack foods and beverages from snack bars, school stores, and vending machines.[66] These items compete with nutritious school lunches and are often high in fat and sugar.[67]

• Nutrition Education at School • Coincident with the school breakfast and lunch programs is a program of nutrition education and training (NET) in all public schools. This program is minimally funded, but program administrators are ingenious and creative in accomplishing its highest-priority objectives. School health clinics offer another

School lunches provide children with nourishing foods and fun times with friends.

opportunity to provide nutrition education and intervention.[68] Children need to be fed well *and* to learn enough about nutrition to make healthful food choices when the choices become theirs to make.

HEALTHY PEOPLE 2010

Increase the proportion of middle, junior high, and senior high schools that provide school health education to prevent health problems in several topics (including unhealthy dietary patterns and inadequate physical activity).

IN SUMMARY

Children's appetites and nutrient needs reflect their stage of growth. Those who are chronically hungry and malnourished suffer growth retardation; when hunger is temporary and nutrient deficiencies are mild, the problems are usually more subtle—such as poor academic performance. Iron deficiency is widespread and has many physical and behavioral consequences. "Hyper" behavior is not caused by poor nutrition; misbehavior may reflect inconsistent care. Childhood obesity has become a major health problem. Adults at home and at school need to provide children with nutrient-dense foods and teach them how to make healthful diet and activity choices.

NUTRITION DURING ADOLESCENCE

Teenagers make many more choices for themselves than they did as children. They are not fed, they eat; they are not sent out to play, they choose to go. At the same time, social pressures thrust choices at them: whether to drink alcoholic beverages and whether to develop their bodies to meet extreme ideals of slimness or athletic prowess. Their interest in nutrition—both valid information and misinformation—derives from personal, immediate experiences. They are concerned with how diet can improve their lives now—they engage in fad dieting in order to fit into a new bathing suit, avoid greasy foods in an effort to clear acne, or eat a pile of spaghetti to prepare for a big sporting event. In presenting information on the nutrition and health of adolescents, this section includes these many topics of interest to teens.

Growth and Development

With the onset of **adolescence,** the steady growth of childhood speeds up abruptly and dramatically, and the growth patterns of female and male become distinct. Hormones direct the intensity of the adolescent growth spurt, profoundly affecting every organ of the body, including the brain. After two to three years of intense growth and a few more at a slower pace, physically mature adults emerge.

In general, the adolescent growth spurt begins at age 10 or 11 for females and at 12 or 13 for males. It lasts about two and a half years. Before **puberty,** male and female body compositions differ only slightly, but during the adolescent spurt, differences between the genders become apparent in the skeletal system, lean body mass, and fat stores. In females, fat assumes a larger percentage of the total body weight, and in males, the lean body mass—principally muscle and bone—increases much more than in females (review Figure 8-6 on p. 253). On the average, males grow 8 inches taller, and females, 6 inches taller. Males gain approximately 45 pounds, and females, about 35 pounds.

adolescence: the period from the beginning of puberty until maturity.

puberty: the period in life in which a person becomes physically capable of reproduction.

Energy and Nutrient Needs

Energy and nutrient needs are greater during adolescence than at any other time of life, except pregnancy and lactation. In general, nutrient needs rise throughout childhood, peak in adolescence, and then level off or even diminish as the teen becomes an adult.

• Energy Intake and Activity • The energy needs of adolescents vary greatly, depending on their current rate of growth, gender, body composition, and physical activity. Boys' energy needs may be especially high; they typically grow faster than girls and, as mentioned, develop a greater proportion of lean body mass. An active boy of 15 may need 4000 kcalories or more a day just to maintain his weight. Girls start growing earlier than boys and attain shorter heights and lower weights, so their energy needs peak sooner and decline earlier than those of their male peers. An inactive girl of 15 whose growth is nearly at a standstill may need fewer than 2000 kcalories a day if she is to avoid excessive weight gain. Thus adolescent girls need to pay special attention to being physically active and selecting foods of high nutrient density so as to meet their nutrient needs without exceeding their energy needs.

The insidious problem of obesity becomes ever more apparent in adolescence and often continues into adulthood. The problem is most evident in females, especially those of African American descent. Without intervention, overweight adolescents will face numerous physical and socioeconomic consequences for years to come. The consequences of obesity are so dramatic and our society's attitude toward obese people is so negative that even teens of normal weight may perceive a need to lose weight. Those with low self-esteem may diet to lose weight even though their weight is normal or even below normal.[69] When taken to extremes, restrictive diets bring dramatic physical consequences of their own, as Highlight 9 explains.

• Vitamins • The RDA (or AI) for most vitamins increase during the adolescent years (see the table on the inside front cover). Several of the vitamin recommendations for adolescents are similar to those for adults, including the recommendation for vitamin D. During puberty, both the activation of vitamin D and the absorption of calcium are enhanced, thus supporting the intense skeletal growth of the adolescent years without additional vitamin D.

• Iron • The need for iron increases during adolescence for both females and males, but for different reasons. For females, iron needs increase as they start to menstruate, and for males, as their lean body mass develops. Iron intakes often fail to keep pace with increasing needs, especially for females, who typically consume less iron-rich foods such as meat and fewer total kcalories than males. For females, the RDA rises at adolescence and remains high into late adulthood. For males, the RDA returns to preadolescent values in early adulthood.

• Calcium • Adolescence is a crucial time for bone development, and the requirement for calcium reaches its peak during these years.[70] Unfortunately, many adolescents have calcium intakes below current recommendations. Low calcium intakes during times of active growth, especially if paired with physical inactivity, may compromise the development of peak bone mass, which is considered the best protection against adolescent fractures and adulthood osteoporosis.[71] Increasing milk products◆ in the diet to meet calcium recommendations greatly increases bone density.[72] Once again, however, teenage girls are most vulnerable, for their milk—and therefore their calcium—intakes begin to decline at the time when their calcium needs are greatest. Furthermore, women have much greater bone losses than men in later life. In addition to dietary calcium, sports activities during adolescence build strong bones.

◆ Teenagers need to select at least 4 servings from the milk group daily to meet their calcium goal of 1300 mg/day. Chapter 12 presents other calcium-rich food choices.

© Phyllis Picardi/Stock Boston

Nutritious snacks play an important role in an active teen's diet.

Food Choices and Health Habits

Teenagers like the freedom to come and go as they choose. They eat what they want if it is convenient and if they have the time.[73] With a multitude of after-school, social, and job activities, they almost inevitably fall into irregular eating habits. At any given time on any given day, a teenager may be skipping a meal, eating a snack, preparing a meal, or consuming food prepared by a parent or restaurant.

• Snacks • Snacks typically provide at least a fourth of the average teenager's daily food energy intake. Most often, favorite snacks are high in fat and low in calcium, iron, vitamin A, vitamin C, and folate. Most adolescents need to eat a greater variety of foods to obtain these nutrients. Table 15-4 on p. 521 shows how to combine foods from different food groups to create healthy snacks. Vending machines rarely offer nutrient-dense options, and nutrition information alone does not convince people to make healthy choices.

• Beverages • Most frequently, adolescents drink soft drinks instead of fruit juice or milk with lunch, supper, and snacks. About the only time they select fruit juices is at breakfast. When they drink milk, they are more likely to consume it with a meal (especially breakfast) than as a snack. Because of their greater food intakes, boys are more likely than girls to drink enough milk to meet their calcium needs.

Adolescents who drink soft drinks regularly have a higher energy intake and a lower calcium intake than those who do not; they are also likely to be overweight.[74] For adolescents who can afford the kcalories and are meeting their calcium needs, soft drinks are an acceptable part of the diet. Soft drinks may present a problem, however, when caffeine◆ intake becomes excessive. Caffeine seems to be relatively harmless when used in moderate doses (the equivalent of fewer than four 12-ounce cola beverages a day). In greater amounts, it can cause the symptoms associated with anxiety—sweating, tenseness, and inability to concentrate.

◆ For perspective, caffeine-containing soft drinks typically deliver between 30 and 55 mg of caffeine per 12-ounce can. A pharmacologically active dose of caffeine is defined as 200 mg. Appendix H starts with a table listing the caffeine contents of selected foods, beverages, and drugs.

• Eating Away from Home • Adolescents eat about one-third of their meals away from home, and their nutritional welfare is enhanced or hindered by the choices they make. A lunch of a hamburger, a chocolate shake, and french fries supplies substantial quantities of many nutrients at a kcalorie cost of about 800, an energy intake many adolescents can afford. When they eat this sort of lunch, teens can adjust their breakfast and dinner choices to include fruits and vegetables for vitamin A, vitamin C, folate, and fiber and lean meats and legumes for iron and zinc. (See Appendix H for the nutrient contents of fast foods.)

• Peer Influence • Many of the food and health choices adolescents make reflect the opinions and actions of their peers. When others perceive milk as "babyish," a teen will choose soft drinks instead; when others skip lunch and hang out in the parking lot, a teen may join in for the camaraderie, regardless of hunger. Adults need to remember that adolescents have the right to make their own decisions—even if they are contrary to the adults' views. Gatekeepers can set up the environment so that nutritious foods are available and can stand by with reliable nutrition information and advice, but the rest is up to the adolescents. Ultimately, they make the choices. (Highlight 9 examines the influence of social pressures on the development of eating disorders.)

Problems Adolescents Face

Physical maturity and growing independence present adolescents with new choices to make. The consequences of those choices will influence their nutritional health both today and throughout life. Some teenagers begin using drugs, alcohol, and tobacco; others wisely refrain. Information about the use of these substances is presented here because most people are first exposed to them during adolescence, but it actually applies to people of all ages.

• **Marijuana** • Almost half of the high school students in the United States report having at least tried marijuana.[75] Marijuana is unique among drugs in that it seems to enhance the enjoyment of eating, especially of sweets, a phenomenon commonly known as "the munchies." Why or how this effect occurs is not known; it may be a social effect induced by suggestibility, or perhaps the drug stimulates appetite. Whatever the reason, prolonged use of marijuana does not seem to bring about a weight gain.

• **Cocaine** • One in 12 high school seniors reports having used cocaine at least once.[76] Cocaine stimulates the nervous system and elicits the stress response—constricted blood vessels, raised blood pressure, widened pupils of the eyes, and increased body temperature. It also drives away feelings of fatigue. Cocaine occasionally causes immediate death—usually by heart attack, stroke, or seizure in an already damaged body system.

Weight loss is common, and cocaine abusers often develop eating disorders. Notably, the craving for cocaine replaces hunger; rats given unlimited cocaine will choose it over food until they starve to death. Thus, unlike marijuana use, cocaine use has major nutritional consequences.

• **Ecstasy** • The designer drug ecstasy has also lured 1 in 12 high school seniors at least once. Ecstasy signals the nerve cells to dump all their stored serotonin◆ at once and then prevents its reabsorption. The rush of serotonin flooding the gap between the nerve cells (synapse) alters a person's mood, but may also damage nerve cells and impair memory. Because serotonin helps to regulate body temperature, overheating is a common and potentially dangerous side effect. People who use ecstasy regularly lose weight for many of the reasons listed in the margin.◆

◆ **Reminder: *Serotonin* is a neurotransmitter important in the regulation of appetite, sleep, and body temperature.**

◆ **Nutrition problems of drug abusers:**
- They buy drugs with money that could be spent on food.
- They lose interest in food during "highs."
- They use drugs that depress appetite.
- Their lifestyle fails to promote good eating habits.
- If they use intravenous (IV) drugs, they may contract AIDS, hepatitis, or other infectious diseases, which increase their nutrient needs. Hepatitis also causes taste changes and loss of appetite.
- Medicines used to treat drug abuse may alter nutrition status.

• **Drug Abuse, in General** • The nutrition problems associated with other drugs vary in degree, but drug abusers in general face multiple nutrition problems. During withdrawal from drugs, an important part of treatment is to identify and correct these nutrition problems.

• **Alcohol Abuse** • Sooner or later all teenagers face the decision of whether to drink alcohol. The law forbids the sale of alcohol to people under 21, but most adolescents who seek alcohol can obtain it. Four out of five high school students have had at least one alcoholic beverage; about half drink regularly; and one in three students drinks heavily (defined as five or more drinks on at least one occasion in the previous month).[77]

Highlight 7 describes how alcohol affects nutrition status. To sum it up, alcohol provides energy but no nutrients, and it can displace nutritious foods from the diet. Alcohol alters nutrient absorption and metabolism, so imbalances develop. People who cannot keep their alcohol use moderate must abstain to maintain their health. Highlight 7 and Appendix F list resources for people with alcohol-related problems.

• **Smoking** • Almost half of U.S. teens report frequent cigarette smoking. Cigarette smoking is a pervasive health problem causing thousands of people to suffer from cancer and diseases of the cardiovascular, digestive, and respiratory systems. These effects are beyond the scope of nutrition, but smoking cigarettes does influence hunger, body weight, and nutrient status.[78]

HEALTHY PEOPLE 2010

Reduce tobacco use by adolescents. Increase tobacco use cessation attempts by adolescent smokers.

Smoking a cigarette eases feelings of hunger. When smokers receive a hunger signal, they can quiet it with cigarettes instead of food. Such behavior ignores body signals and postpones energy and nutrient intake. In rats, nicotine reduces food intake and increases the rate of energy expenditure, causing weight loss. In humans, cigarette smoking is associated with higher levels of leptin, which may explain the lower body weight.[79] (Recall from Chapter 9 that leptin suppresses appetite and increases energy expenditure.)

Indeed, smokers tend to weigh less than nonsmokers and to gain weight when they stop smoking. People contemplating giving up cigarettes should know that the average person who quits smoking gains about 10 pounds in the first year.[80] Interestingly, and contrary to what might be expected, leptin levels do not increase with the weight gain that accompanies smoking cessation.[81] Smokers wanting to quit should prepare for the possibility of weight gain and adjust their diet and activity habits so as to maintain weight during and after quitting. Smoking cessation programs need to include strategies for weight management.[82]

Nutrient intakes of smokers and nonsmokers differ. Smokers tend to have lower intakes of dietary fiber, vitamin A, beta-carotene, folate, and vitamin C. The association between smoking and low intakes of fruits and vegetables rich in these nutrients may be noteworthy, considering their protective effect against lung cancer (see Highlight 11).

◆ The vitamin C requirement for people who regularly smoke cigarettes is an additional 35 mg/day.

Compared to nonsmokers, smokers require more vitamin C◆ to maintain steady body pools. Oxidants in cigarette smoke accelerate vitamin C metabolism and deplete smokers' body stores of this antioxidant.[83] This depletion is even evident to some degree in nonsmokers who are exposed to passive smoke.

Beta-carotene enhances the immune response and protects against some cancer activity. Specifically, the risk of lung cancer is greatest for smokers who have the lowest intakes. Of course, such evidence should not be misinterpreted. It does not mean that as long as people eat their carrots, they can safely use tobacco. Nor does it mean that beta-carotene *supplements* would be beneficial; smokers taking beta-carotene supplements actually had a higher incidence of lung cancer and risk of death than those taking a placebo (see Highlight 11 for more details). Smokers are ten times more likely to get lung cancer than nonsmokers. Both smokers and nonsmokers, however, can reduce their cancer risks by eating fruits and vegetables rich in carotene (see Highlight 11 for details on antioxidant nutrients and disease prevention).

• Smokeless Tobacco • Nationwide, one in ten high school students reports having used smokeless tobacco products.[84] Like cigarettes, smokeless tobacco use is linked to many health problems, from minor mouth sores to tumors in the nasal cavities, cheeks, gums, and throat. The risk of mouth and throat cancers is even greater than for smoking tobacco. Other drawbacks to tobacco chewing and snuff dipping include bad breath, stained teeth, and blunted senses of smell and taste. Tobacco chewing also damages the gums, tooth surfaces, and jawbones, making it likely that users will lose their teeth in later life.

IN SUMMARY

Nutrient needs rise dramatically as children enter the rapid growth phase of the teen years. The busy lifestyles of adolescents add to the challenge of meeting their nutrient needs—especially for iron and calcium. In addition to making wise food choices, adolescents need to refrain from using substances that will impair their health—including illicit drugs, alcohol, and tobacco.

The nutrition and lifestyle choices people make as children and adolescents have long-term, as well as immediate, effects on their health. Highlight 15 describes how sound choices and good habits can help prevent chronic diseases later in life.

How are you doing?

- If there are children in your life, do they receive enough food for healthy growth, but not so much as to lead to obesity?
- Are they physically active at home and at school?
- Do they get enough calcium and iron?

Nutrition on the Net

WEBSITES

Access these websites for further study of topics covered in this chapter.

- Find updates and quick links to these and other nutrition-related sites at our website: **www.wadsworth.com/nutrition**
- Search for "infants," "baby bottle tooth decay," "premature birth," "hyperactivity," "food allergies," and "adolescent health," at the U.S. Government health information site: **www.healthfinder.gov**
- Learn how to care for infants, children, and adolescents from the American Academy of Pediatrics and the Canadian Paediatric Society: **www.aap.org** and **www.cps.ca**
- Download the current growth charts and learn about their most recent revision: **www.cdc.gov/growthcharts**
- Get information on the Food Guide Pyramid for young children from the USDA: **www.usda.gov/cnpp**
- Get tips for feeding children from the American Dietetic Association, the I Am Your Child Program, and the Kids Food Cyber Club: **www.eatright.org**, **www.iamyourchild.org**, and **www.kidsfood.org**
- Get tips for keeping children healthy from the Nemours Foundation: **www.kidshealth.org**
- Visit the National Center for Education in Maternal & Child Health and the National Institute of Child Health and Development: **www.ncemch.org** and **www.nih.gov/nichd**
- Learn about the Child Nutrition Programs: **www.fns.usda.gov/fns**
- Learn how to reduce lead exposure in your home from the U.S. Department of Housing and Urban Development Office of Lead Hazard Control: **www.hud.gov/lead**
- Visit the Lead Program of the Centers for Disease Control: **www.cdc.gov/nceh/programs/lead**
- Learn more about food allergies from the American Academy of Allergy, Asthma, and Immunology; the Food Allergy Network; and the International Food Information Council: **www.aaaai.org**, **www.foodallergy.org**, and **www.ificinfo.org**
- Learn more about hyperactivity from Children and Adults with Attention Deficit Disorders: **www.chadd.org**
- Visit the Public Health Service's Office of Women's Health site for messages on positive self-images, good nutrition, and fitness for girls between the ages of 9 and 14: **www.health.org/gpower/girlarea/bodywise**
- Visit the Milk Matters section of the National Institute of Child Health and Development (NICHD): **www.nichd.nih.gov**
- Learn more about caffeine from the International Food Information Council: **www.ificinfo.health.org**
- Get weight-loss tips for children and adolescents: **www.shapedown.com**
- Learn about nondietary approaches to weight loss from HUGS International: **www.hugs.com**
- Read the message for parents and teens on the risks of tobacco use from the American Academy of Pediatrics: **www.aap.org**
- Visit the Tobacco Information and Prevention Source (TIPS) of the Centers for Disease Control and Prevention: **www.cdc/gov/tobacco**

INTERNET ACTIVITIES

Children begin to form their attitudes and feelings about food at early ages, and enjoy participating in lessons about food that are fun.

Review this interactive site that is directed at teaching children about good food choices.

- Go to: **www.kidfood.com**
- Click on "Kids only . . . Grown-ups keep out!"
- Click on "Choices, choices, choices."
- Click on "Cyber Food Shopper."
- Follow the instructions to play the game.

Even though this simple game is directed at children, it has lessons for people of all ages. Everyone can benefit from knowing how to get the best nutritional buy for the money spent. According to this game, how did your shopping skills rate?

Study Questions

These questions will help you review the chapter. You will find the answers in the discussions on the pages provided.

1. Describe some of the nutrient and immunological attributes of breast milk. (pp. 500–503)
2. What are the appropriate uses of formula feeding? What criteria would you use in selecting an infant formula? (pp. 503–505)
3. Why are solid foods not recommended for an infant during the first few months of life? When is an infant ready to start eating solid food? (pp. 505–507)
4. Name foods that are inappropriate for infants and explain why they are inappropriate. (p. 507)
5. What nutrition problems are most common in children? What strategies can help prevent these problems? (pp. 509–510)
6. Describe the relationships between nutrition and behavior. How does television influence nutrition? (pp. 510–512, 516–517)
7. Describe a true food allergy. Which foods most often cause allergic reactions? How do food allergies influence nutrition status? (pp. 515–516)
8. Describe the problems associated with childhood obesity and the strategies for prevention and treatment. (pp. 516–519)
9. List strategies for introducing nutritious foods to children. (pp. 519–522)
10. What impact do school meal programs have on the nutrition status of children? (pp. 522–524)
11. Describe the changes in nutrient needs from childhood to adolescence. Why is an adolescent girl more likely to develop an iron deficiency than is a boy? (pp. 524–525)
12. How do adolescents' eating habits influence their nutrient intakes? (p. 526)
13. How does the use of illicit drugs influence nutrition status? (pp. 526–527)
14. How do the nutrient intakes of smokers differ from those of nonsmokers? What impacts can those differences exert on health? (pp. 527–528)

These questions will help you prepare for an exam. Answers can be found on p. 532.

1. A reasonable weight for a healthy five-month-old infant who weighed 8 pounds at birth might be:
   a. 12 pounds.
   b. 16 pounds.
   c. 20 pounds.
   d. 24 pounds.
2. Dehydration can develop quickly in infants because:
   a. much of their body water is extracellular.
   b. they lose a lot of water through urination and tears.
   c. only a small percentage of their body weight is water.
   d. they drink lots of breast milk or formula, but little water.
3. An infant should begin eating solid foods between:
   a. 2 and 4 weeks.
   b. 1 and 3 months.
   c. 4 and 6 months.
   d. 8 and 10 months.
4. Among U.S. and Canadian children, the most prevalent nutrient deficiency is of:
   a. iron.
   b. folate.
   c. protein.
   d. vitamin D.
5. A true food allergy always:
   a. elicits an immune response.
   b. causes an immediate reaction.
   c. creates an aversion to the offending food.
   d. involves symptoms such as headaches or hives.
6. Which of the following strategies is *not* effective?
   a. Play first, eat later.
   b. Provide small portions.
   c. Encourage children to help prepare meals.
   d. Use dessert as a reward for eating vegetables.

7. To help teenagers consume a balanced diet, parents can:
   a. monitor the teens' food intake.
   b. give up—parents can't influence teenagers.
   c. keep the pantry and refrigerator well stocked.
   d. forbid snacking and insist on regular, well-balanced meals.
8. During adolescence, energy and nutrient needs:
   a. reach a peak.
   b. fall dramatically.
   c. rise, but do not peak until adulthood.
   d. fluctuate so much that generalizations can't be made.
9. The nutrients most likely to fall short in the adolescent diet are:
   a. sodium and fat.
   b. folate and zinc.
   c. iron and calcium.
   d. protein and vitamin A.
10. To balance the day's intake, an adolescent who eats a hamburger, fries, and cola at lunch might benefit most from a dinner of:
    a. fried chicken, rice, and banana.
    b. ribeye steak, baked potato, and salad.
    c. pork chop, mashed potatoes, and apple juice.
    d. spaghetti with meat sauce, broccoli, and milk.

References

1 R. J. Kuczmarski and coauthors, CDC growth charts: United States, *Advanced Data* 314 (2000): 1–28.

2 Committee on Nutrition, American Academy of Pediatrics, *Pediatric Nutrition Handbook,* 4th ed., ed. R. E. Kleinman (Elk Grove Village, Ill.: American Academy of Pediatrics, 1998), pp. 49–50.

3 Work Group on Breastfeeding, American Academy of Pediatrics, Breastfeeding and the use of human milk, *Pediatrics* 100 (1997): 1035–1037; Position of The American Dietetic Association: Promotion of breast feeding, *Journal of the American Dietetic Association* 97 (1997): 662–665.

4 R. A. Gibson and M. Makrides, n-3 Polyunsaturated fatty acid requirements of term infants, *American Journal of Clinical Nutrition* 71 (2000): 251S–255S; M. Neuringer, Infant vision and retinal function in studies of dietary long-chain polyunsaturated fatty acids: Methods, results, and implications, *American Journal of Clinical Nutrition* 71 (2000): 256S–267S.

5 C. Williams and coauthors, Stereoacuity at age 3.5 y in children born full-term is associated with prenatal and postnatal dietary factors: A report from a population-based cohort study, *American Journal of Clinical Nutrition* 73 (2001): 316–322.

6 M. Makrides and coauthors, Dietary long-chain polyunsaturated fatty acids do not influence growth of term infants: A randomized clinical trial, *Pediatrics* 104 (1999): 468–475.

7 S. Fitzpatrick and coauthors, Vitamin D–deficient rickets: A multifactorial disease, *Nutrition Reviews* 58 (2000): 218–222.

8 C. G. Victora and coauthors, Potential interventions for the prevention of childhood pneumonia in developing countries: Improving nutrition, *American Journal of Clinical Nutrition* 70 (1999): 309–320; J. Raisler, C. Alexander, and P. O' Campo, Breast-feeding and infant illness: A dose-response relationship? *American Journal of Public Health* 89 (1999): 25–30; A. L. Wright and coauthors, Increasing breastfeeding rates to reduce infant illness at the community level, *Pediatrics* 101 (1998): 837–844; A. H. Cushing and coauthors, Breastfeeding reduces risk of respiratory illness in infants, *American Journal of Epidemiology* 147 (1998): 863–870; M. A. Hylander, D. M. Strobino, and R. Dhanireddy, Human milk feedings and infection among very low birth weight infants, *Pediatrics* 102 (1998): 630 [www.pediatrics.org/cgi/content/full/102/3/e38].

9 Role of human-milk lactadherin in protection against symptomatic rotavirus infection, *Lancet* 351 (1998): 1160–1164.

10 A. Singhal, T. J. Cole, and A. Lucas, Early nutrition in preterm infants and later blood pressure: Two cohorts after randomised trials, *Lancet* 357 (2001): 413–419.

11 Committee on Nutrition, American Academy of Pediatrics, Iron fortification of infant formulas, *Pediatrics* 104 (1999): 119–123.

12 D. J. Raiten, J. M. Talbot, and J. H. Waters, LSRO Report: Assessment of nutrient requirements for infant formulas, *Journal of Nutrition* 128 (1998): 2059S–2293S.

13 L. Businco, G. Bruno, and P. G. Giampietro, Soy protein for the prevention and treatment of children with cow-milk allergy, *American Journal of Clinical Nutrition* 68 (1998): 1447S–1452S.

14 Committee on Nutrition, American Academy of Pediatrics, Soy protein–based formulas: Recommendations for use in infant feeding, *Pediatrics* 101 (1998): 148–153.

15 Preterm singleton births—United States, 1989–1996, *Morbidity and Mortality Weekly Report* 48 (1999): 185–189.

16 M. A. Crawford, Placental delivery of arachidonic and docosahexaenoic acids: Implications for the lipid nutrition of preterm infants, *American Journal of Clinical Nutrition* 71 (2000): 275S–284S.

17 R. Uauy and D. R. Hoffman, Essential fat requirements of preterm infants, *American Journal of Clinical Nutrition* 71 (2000): 245S–250S.

18 Work Group on Breastfeeding, 1997.

19 M. A. Atkinson and T. M. Ellis, Infants' diets and insulin-dependent diabetes: Evaluating the "cows' milk hypothesis" and a role for anti-bovine serum albumin immunity, *Journal of the American College of Nutrition* 16 (1997): 334–340.

20 J. D. Cook and coauthors, The influence of different cereal grains on iron absorption from infant cereal foods, *American Journal of Clinical Nutrition* 65 (1997): 964–969.

21 R. E. Doucette, Is fruit juice a "no-no" in children's diets? *Nutrition Reviews* 58 (2000): 180–183.

22 B. A. Dennison and coauthors, Children's growth parameters vary by type of fruit juice consumed, *Journal of the American College of Nutrition* 18 (1999): 346–352; B. A. Dennison, H. L. Rockwell, and S. L. Baker, Excess fruit juice consumption by preschool-aged children is associated with short stature and obesity, *Pediatrics* 99 (1997): 15–22.

23 M. I. Goran and coauthors, Developmental changes in energy expenditure and physical activity in children: Evidence for a decline in physical activity in girls before puberty, *Pediatrics* 101 (1998): 887–891.

24 Committee on Sports Medicine and Fitness and Committee on School Health, Physical fitness and activity in schools, *Pediatrics* 105 (2000): 1156–1157.

25 Position of The American Dietetic Association: Health implications of dietary fiber, *Journal of the American Dietetic Association* 97 (1997): 1157–1159.

26 A. C. Looker and coauthors, Prevalence of iron deficiency in the United States, *Journal of the American Medical Association* 277 (1997): 973–976.

27 K. A. Muñoz and coauthors, Food intakes of US children and adolescents compared with recommendations, *Pediatrics* 100 (1997): 323–329.

28 S. B. Roberts and M. B. Heyman, Micronutrient shortfalls in young children's diets: Common, and owing to inadequate intakes both at home and at child care centers, *Nutrition Reviews* 58 (2000): 27–29.

29 C. A. Powell and coauthors, Nutrition and education: A randomized trial of the effects of breakfast in rural primary school children, *American Journal of Clinical Nutrition* 68 (1998): 873–879; E. Kennedy and C. Davis, US Department of Agriculture School Breakfast Program, *American Journal of Clinical Nutrition* 67 (1998): 798S–803S.

30 E. Pollitt and R. Mathews, Breakfast and cognition: An integrative summary, *American Journal of Clinical Nutrition* 67 (1998): 804S–813S.

31 P. P. Basiotis, M. Lino, and R. S. Anand, Eating breakfast greatly improves schoolchildren's diet quality, *Family Economics and Nutrition Review* 12 (1999): 81–84; J. T. Dwyer and coauthors, Do third graders eat healthful breakfasts? *Family Economics and Nutrition Review* 11 (1998): 3–18; C. H. Ruxton and T. R. Kirk, Breakfast: A review of associations with measures of dietary intake, physiology and biochemistry, *British Journal of Nutrition* 78 (1997): 199–213.

32 I. de Andraca, M. Castillo, and T. Walter, Psychomotor development and behavior in iron-deficient anemic infants, *Nutrition Reviews* 55 (1997): 125–132.

33 E. Pollitt, Iron deficiency and educational deficiency, *Nutrition Reviews* 55 (1997): 133–141.

34 E. K. Hurtado, A. H. Claussen, and K. G. Scott, Early childhood anemia and mild or moderate mental retardation, *American Journal of Clinical Nutrition* 69 (1999): 115–119.

35 J. A. Simon and E. S. Hudes, Relationship of ascorbic acid to blood lead levels, *Journal of the American Medical Association* 281 (1999): 2289–2293; Y. Cheng and coauthors, Relation of nutrition to bone lead and blood lead levels in middle-aged to elderly men: The Normative Aging Study, *American Journal of Epidemiology* 147 (1998): 1162–1174; R. A. Goyer, Toxic and essential metal interactions, *Annual Review of Nutrition* 17 (1997): 37–50.

36 W. G. Adams and coauthors, Anemia and elevated lead levels in underimmunized inner-city children, *Pediatrics* 101 (1998): 462 [available in full online at http://www.pediatrics.org/cgi/content/full/101/3/e6].

37 Update: Blood lead levels—United States, 1991–1994, *Morbidity and Mortality Weekly Report* 46 (1997): 141–146.

38 S. Tong and coauthors, Declining blood lead levels and changes in cognitive function during childhood: The Port Pirie Cohort Study, *Journal of the American Medical Association* 280 (1998): 1915–1919.

39 Blood lead levels in young children—United States and selected states, 1996–1999, *Morbidity and Mortality Weekly Report* 49 (2000): 1133–1137; S. T. Melman, J. W. Nimeh, and R. D. Anbar, Prevalence of elevated blood lead levels in an inner-city pediatric clinic population, *Environmental Health Perspectives* 106 (1998): 655–657.

40 The MTA Cooperative Group, A 14-month randomized clinical trial of treatment strategies for attention-deficit/hyperactivity disorder: Multimodal treatment study of children with ADHD, *Archives of General Psychiatry* 56 (1999): 1073–1086.

41 J. W. White and M. Wolraich, Effect of sugar and mental performance, *American Journal of Clinical Nutrition* (supplement) 62 (1995): 242–249; M. L. Wolraich, D. B. Wilson, and J. W. White, The effect of sugar on the behavior or cognition in children, *Journal of the American Medical Association* 274 (1995): 1617–1621.

42 H. A. Sampson, Food allergy, *Journal of the American Medical Association* 278 (1997): 1888–1894.

43 E. Isolauri and coauthors, Elimination diet in cow's milk allergy: Risk for impaired growth in young children, *Journal of Pediatrics* 132 (1998): 1004–1009.

44 A. R. Dorosty and coauthors, Factors associated with early adiposity rebound, *Pediatrics* 105 (2000): 1115–1118; R. C. Whitaker and coauthors, Predicting obesity in young adulthood from childhood and parental obesity, *New England Journal of Medicine* 337 (1997): 869–873.

45 S. M. Robertson and coauthors, Factors related to adiposity among children aged 3 to 7 years, *Journal of the American Dietetic Association* 99 (1999): 938–943.

ANSWERS

Multiple Choice

1. b	2. a	3. c	4. a	5. a
6. d	7. c	8. a	9. c	10. d

46 R. P. Troiano and coauthors, Energy and fat intakes of children and adolescents in the United States: Data from the National Health and Nutrition Examination Surveys, *American Journal of Clinical Nutrition* 72 (2000): 1343S–1353S; J. P. DeLany, Role of energy expenditure in the development of pediatric obesity, *American Journal of Clinical Nutrition* 68 (1998): 950S–955S.

47 K. A. Coon and coauthors, Relationships between use of television during meals and children's food consumption patterns, *Pediatrics* 107 (2001): 167 [http://www.pediatrics.org/cgi/content/full107/1/e7]; R. E. Anderson and coauthors, Relationship of physical activity and television watching with body weight and level of fatness among children: Results from the third National Health and Nutrition Examination Survey, *Journal of the American Medical Association* 279 (1998): 938–942.

48 T. N. Robinson, Reducing children's television viewing to prevent obesity: A randomized controlled trial, *Journal of the American Medical Association* 282 (1999): 1561–1567.

49 Y. Schutz and coauthors, Whole-body protein turnover and resting energy expenditure in obese, prepubertal children, *American Journal of Clinical Nutrition* 69 (1999): 857–862.

50 M. I. Goran, Metabolic precursors and effects of obesity in children: A decade of progress, 1990–1999, *American Journal of Clinical Nutrition* 73 (2001): 158–171; S. E. Barlow and W. H. Dietz, Obesity evaluation and treatment: Expert Committee Recommendations, *Pediatrics* 102 (1998): 626 [http://www.pediatrics.org/cgi/content/full/102/3/e29].

51 B. J. Rolls, D. Engell, and L. L. Birch, Serving portion size influences 5-year-old but not 3-year-old children's food intakes, *Journal of the American Dietetic Association* 100 (2000): 232–234.

52 P. Barbeau and coauthors, Correlates of individual differences in body-composition changes resulting from physical training in obese children, *American Journal of Clinical Nutrition* 69 (1999): 705–711; C. B. Ebbeling and N. R. Rodriguez, Effects of exercise combined with diet therapy on protein utilization in obese children, *Medicine and Science in Sports and Exercise* 31 (1999): 378–385.

53 M. Golan and coauthors, Parents as the exclusive agents of change in the treatment of childhood obesity, *American Journal of Clinical Nutrition* 67 (1998): 1130–1135.

54 T. M. Cutting and coauthors, Like mother, like daughter: Familiar patterns of overweight are mediated by mothers' dietary disinhibition, *American Journal of Clinical Nutrition* 69 (1999): 608–613.

55 A. E. Field and coauthors, Peer, parent, and media influences on the development of weight concerns and frequent dieting among preadolescent and adolescent girls and boys, *Pediatrics* 107 (2001): 54–60; K. K. Davison and L. L. Birch, Weight status, parent reaction, and self-concept in five-year-old girls, *Pediatrics* 107 (2001): 46–53.

56 F. Lifshitz and N. Moses, Nutritional dwarfing: Growth, dieting, and fear of obesity, *Journal of the American College of Nutrition* 7 (1998): 367–376.

57 V. Kraak and D. L. Pelletier, The influence of commercialism on the food purchasing behavior of children and teenage youth, *Family Economics and Nutrition Review* 11 (1998): 15–24.

58 J. O. Fisher and L. L. Birch, Restricting access to palatable foods affects children's behavioral response, food selection, and intake, *American Journal of Clinical Nutrition* 69 (1999): 1264–1272.

59 C. Evers, Empower children to develop healthful eating habits, *Journal of the American Dietetic Association* 97 (1997): S116.

60 J. Skinner and coauthors, Toddlers' food preferences: Concordance with family members' preferences, *Journal of Nutrition Education* 30 (1998): 17–22.

61 M. Nahikian Nelms, Influential factors of caregiver behavior at mealtime: A study of child-care programs, *Journal of the American Dietetic Association* 97 (1997): 505–509.

62 Position of The American Dietetic Association: Nutrition standards for child-care programs, *Journal of the American Dietetic Association* 99 (1999): 981–988.

63 Position of The American Dietetic Association: Dietary guidance for healthy children aged 2 to 11 years, *Journal of the American Dietetic Association* 99 (1999): 93–101.

64 M. F. Picciano, L. D. McBean, and V. A. Stallings, How to grow a healthy child: A conference report, *Nutrition Today* 34 (1999): 6–14.

65 E. A. Bergman and coauthors, Time spent by schoolchildren to eat lunch, *Journal of the American Dietetic Association* 100 (2000): 696–698.

66 Position of The American Dietetic Association: Local support for nutrition integrity in schools, *Journal of the American Dietetic Association* 100 (2000): 108–111; K. W. Cullen and coauthors, Effect of a la carte and snack bar foods at school on children's lunchtime intake of fruits and vegetables, *Journal of the American Dietetic Association* 100 (2000): 1482–1486.

67 M. B. Wildey and coauthors, Fat and sugar levels are high in snacks purchased from student stores in middle schools, *Journal of the American Dietetic Association* 100 (2000): 319–322; L. Harnack and coauthors, Availability of a la carte food items in junior and senior high schools: A needs assessment, *Journal of the American Dietetic Association* 100 (2000): 701–703.

68 A. Jasaitis, School-based health clinics: The role for nutrition, *Journal of the American Dietetic Association* 97 (1997): S117.

69 J. Pesa, Psychosocial factors associated with dieting behaviors among female adolescents, *Journal of School Health* 69 (1999): 196–201.

70 A. D. Martin and coauthors, Bone mineral and calcium accretion during puberty, *American Journal of Clinical Nutrition* 66 (1997): 611–615.

71 Committee on Nutrition, American Academy of Pediatrics, Calcium requirements of infants, children, and adolescents, *Pediatrics* 104 (1999): 1152–1157.

72 J. Cadogan and coauthors, Milk intake and bone mineral acquisition in adolescent girls: Randomised, controlled intervention trial, *British Medical Journal* 315 (1997): 1255–1260.

73 D. Neumark-Sztainer and coauthors, Factors influencing food choices of adolescents: Findings from focus-group discussions with adolescents, *Journal of the American Dietetic Association* 99 (1999): 929–934, 937.

74 D. S. Ludwig, K. E. Peterson, and S. L. Gortmaker, Relation between consumption of sugar-sweetened drinks and childhood obesity: A prospective, observational analysis, *Lancet* 357 (2001): 505–508; L. Harnack, J. Stang, and M. Story, Soft drink consumption among US children and adolescents: Nutritional consequences, *Journal of the American Dietetic Association* 99 (1999): 436–441.

75 L. Kann and coauthors, Youth risk behavior surveillance—United States, 1997, *Journal of School Health* 68 (1998): 355–369.

76 Kann and coauthors, 1998.

77 Kann and coauthors, 1998.

78 J. S. Hampl and N. M. Betts, Cigarette use during adolescence: Effects on nutritional status, *Nutrition Reviews* 57 (1999): 215–221.

79 B. J. Nicklas and coauthors, Effects of cigarette smoking and its cessation on body weight and plasma leptin levels, *Metabolism: Clinical and Experimental* 48 (1999): 804–808.

80 P. O'Hara and coauthors, Early and late weight gain following smoking cessation in the Lung Health Study, *American Journal of Epidemiology* 148 (1998): 821–830.

81 Nicklas and coauthors, 1999.

82 L. M. Varner, Impact of combined weight-control and smoking-cessation interventions on body weight: Review of the literature, *Journal of the American Dietetic Association* 99 (1999): 1272–1275.

83 J. Lykkesfeldt and coauthors, Ascorbate is depleted by smoking and repleted by moderate supplementation: A study in male smokers and nonsmokers with matched dietary antioxidant intakes, *American Journal of Clinical Nutrition* 71 (2000): 530–536.

84 Kann and coauthors, 1998.

THE EARLY DEVELOPMENT OF CHRONIC DISEASES

When people think of the health problems of children and adolescents, they typically think of measles and acne, not heart disease. They think of heart disease as the number one killer of adults in the United States and Canada, but it begins in childhood.

Over the past three decades, researchers have been observing how changes in body weight, blood lipids, blood pressure, and individual behaviors correlate with the development of heart disease over time—from infancy to childhood through adolescence and into young adulthood. Some major findings have emerged from this research:

- Changes inside the arteries—changes predictive of heart disease—are evident in childhood.
- Obesity in children affects these changes.
- Behaviors that influence the development of obesity and of heart disease are learned and begin early in life. These behaviors include overeating, eating high-fat foods, physical inactivity, and cigarette smoking.

This highlight focuses on efforts to prevent childhood obesity and heart disease, but the benefits extend to diabetes and other chronic diseases as well. Diabetes (type 2), a chronic disease closely linked with obesity, has been on the rise among children as the prevalence of childhood obesity in the United States has increased.[1] The years of childhood (ages 2 to 18 years) are emphasized here, for the earlier in life health-promoting habits become established, the better they will stick. Later chapters fill in the rest of the story of nutrition's role in reducing chronic disease risk.

Children who are obese and who have high blood lipids and high blood pressure are often from families with a history of heart disease. Invariably, questions arise as to what extent genetics is involved in disease development. Genetics does not appear to play a *determining* role in heart disease; that is, a person is not simply destined at birth to develop heart disease. Instead, genetics appears to play a *permissive* role—the potential is inherited and then will develop, if given a push by poor health choices such as excessive weight gain, poor diet, sedentary lifestyle, and cigarette smoking.

Many experts agree that preventing or treating obesity in childhood will reduce the rate of chronic diseases in adulthood. Without intervention, overweight children become overweight adolescents who become overweight adults, and being overweight exacerbates every chronic disease that adults face.[2]

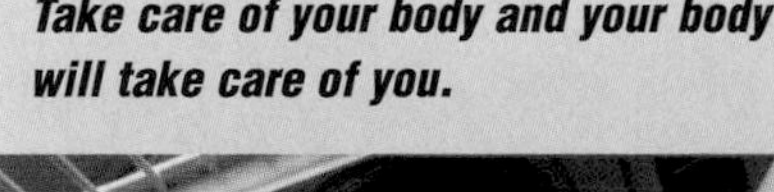

Take care of your body and your body will take care of you.

© The Stock Market/LWA/Sharie Kennedy 2001

EARLY DEVELOPMENT OF HEART DISEASE

Most people consider heart disease to be an adult disease because its incidence rises with advancing age, and symptoms rarely appear before age 30. In actuality, the disease process begins much earlier.

Atherosclerosis

Most **cardiovascular disease** involves **atherosclerosis**—the accumulation of cholesterol and other blood lipids along the walls of the arteries (see the glossary on p. 536). If it progresses, atherosclerosis may eventually block the flow of blood to the heart and cause a heart attack or cut off blood flow to the brain and cause a stroke. Infants are born with healthy, smooth, clear arteries, but within the first decade of life, **fatty streaks** may begin to appear (see Figure H15-1 on p. 536). During adolescence, these fatty streaks may begin to turn to **fibrous plaques.** By early adulthood, the fibrous plaques may begin to calcify and become raised lesions, especially in boys and young men. As the lesions grow more numerous and enlarge, the heart disease rate begins to rise, most dramatically at about age 45 in men and 55 in women. From this point on, arterial damage and blockage progress rapidly, and heart attacks and strokes threaten life. In short, the

GLOSSARY

atherosclerosis (ath-er-oh-scler-OH-sis): a type of artery disease characterized by accumulations of lipid-containing material on the inner walls of the arteries (see Chapter 26).
- **athero** = porridge or soft
- **scleros** = hard
- **osis** = condition

cardiovascular disease (CVD): a general term for all diseases of the heart and blood vessels. Atherosclerosis is the main cause of CVD. When the arteries that carry blood to the heart muscle become blocked, the heart suffers damage known as **coronary heart disease (CHD).**
- **cardio** = heart
- **vascular** = blood vessels

fatty streaks: accumulations of cholesterol and other lipids along the walls of the arteries.

fibrous plaques (PLACKS): mounds of lipid material, mixed with smooth muscle cells and calcium, which develop in the artery walls in atherosclerosis.

consequences of atherosclerosis, which become apparent only in adulthood, have their beginnings in the first decades of life.[3]

Atherosclerosis is not inevitable; people can grow old with relatively clear arteries. Early lesions may either progress or regress, depending on several factors, many of which reflect lifestyle behaviors. Smoking, for example, is strongly associated with the prevalence of fatty streaks and raised lesions, even in young adults.[4]

FIGURE H15-1 **The Formation of Fibrous Plaques in Atherosclerosis**

When fibrous plaques have covered 60 percent of the coronary artery walls, the critical phase of heart disease begins.

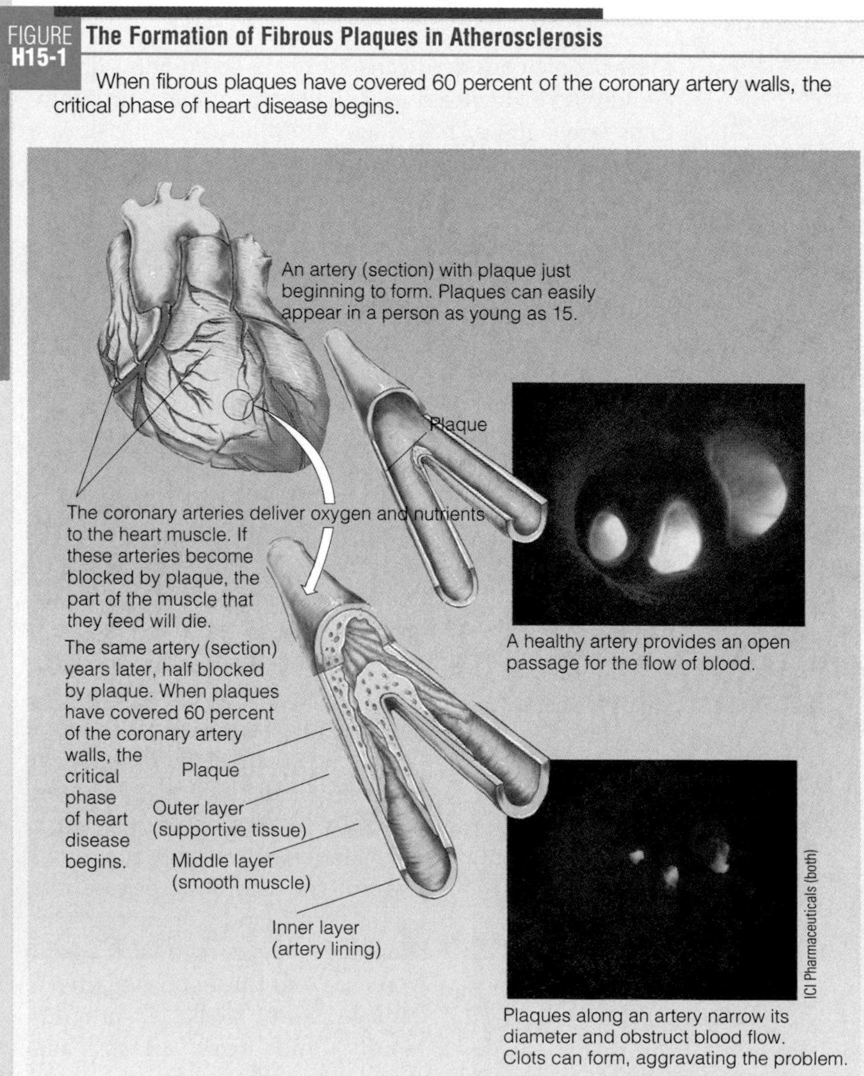

Plaques along an artery narrow its diameter and obstruct blood flow. Clots can form, aggravating the problem.

Blood Cholesterol

Atherosclerotic lesions reflect blood cholesterol: as blood cholesterol increases, lesion coverage increases. Cholesterol values at birth are similar in all populations; differences emerge in early childhood. In countries where the adults have high blood cholesterol and high rates of heart disease, the children also tend to have high blood cholesterol. Conversely, in countries where the adults have low blood cholesterol and low rates of heart disease, the children tend to have low blood cholesterol. These findings suggest that adult heart disease tracks early trends and that early preventive efforts might reduce the incidence of later heart disease.

Such is the case among populations, but individual cholesterol status also becomes established in childhood, as early as one year. The best predictor of a person's blood cholesterol is that person's earlier baseline values: childhood values correlate with values in young adulthood. Quite simply, if you want to know a child's future cholesterol, measure it now. Standard values for cholesterol screening in children and adolescents (ages 2 to 18 years) are listed in Table H15-1.[5]

Blood cholesterol also correlates with childhood obesity, especially central obesity.[6] LDL cholesterol rises with obesity, and HDL declines. These relationships are apparent throughout childhood, and their magnitude increases with age.

Children who are both overweight and have high blood cholesterol are quite likely to have parents who develop heart disease early.[7] For this reason, selective screening is recommended for children and adolescents whose parents or grandparents have heart disease; those whose parents have elevated blood cholesterol; and those whose family history is unavailable, especially if other risk factors are evident.[8] Since blood cholesterol in children is a good predictor of adult values, some experts recommend universal screening to identify all children with high blood cholesterol.[9] They note that many children who have high blood cholesterol would be missed under current screening criteria.

Among those children who may have high blood cholesterol, but may not meet screening criteria are those who are

TABLE H15-1 Cholesterol Values for Children and Adolescents

Disease Risk	Total Cholesterol (mg/dL)	LDL Cholesterol (mg/dL)
Acceptable	<170	<110
Borderline	170–199	110–129
High	≥200	≥130

Note: Adult values appear in Table 26-1 on p. 778.

overweight. The incidence of high blood cholesterol in obese children with no other criteria is similar to that of nonobese children with family histories of heart disease. In addition to overweight, health care professionals should consider whether children smoke or consume a high-fat diet.[10]

Early—but not advanced—atherosclerotic lesions are reversible, making screening and education a high priority. Both those with family histories of heart disease and those with multiple risk factors need intervention. Children with the highest risks of developing heart disease are sedentary and obese, with high blood pressure and high blood cholesterol. In contrast, children with the lowest risks of heart disease are physically active and of normal weight, with low blood pressure and favorable lipid profiles. Routine pediatric care should identify these known risk factors and provide intervention when needed.

Blood Pressure

Pediatricians routinely monitor blood pressure in children and adolescents. High blood pressure may signal an underlying disease or the early onset of hypertension. Hypertension accelerates the development of atheroscerlosis. Standard values for hypertension screening in children and adolescents are given in Table H15-2.[11]

Like atherosclerosis and high blood cholesterol, hypertension may develop in the first decades of life, especially among obese children. Children can control their hypertension by participating in regular aerobic activity and by losing weight or maintaining their weight as they grow taller. No evidence suggests that restricting sodium lowers blood pressure in children and adolescents.[12]

TABLE H15-2 Hypertension Standards for Children and Adolescents

	Systolic over Diastolic Pressure (mm Hg)			
	6 to 9 yr	*10 to 12 yr*	*13 to 15 yr*	*16 to 18 yr*
Mild hypertension	111–121 over 70–77	117–125 over 75–81	124–135 over 77–85	127–141 over 80–91
Moderate hypertension	122–129 over 70–85	126–133 over 82–89	136–143 over 86–91	142–149 over 92–97
Severe hypertension	>129 over >85	>133 over >89	>143 over >91	>149 over >97

PHYSICAL ACTIVITY

Research has also confirmed an association between blood lipids and physical activity in children, similar to that seen in adults. Active children have a better lipid profile than physically inactive children.

Just as blood cholesterol and obesity track over the years, so does a youngster's level of physical activity.[13] A study of almost 1000 teenagers found that over half of those who were initially described as inactive were still inactive six years later. Similarly, almost half of those who were physically active remained so. Compared with inactive teens, those who were physically active weighed less, smoked less, ate a diet lower in saturated fats, and had better blood lipid profiles. Both obesity and blood cholesterol also correlate with the inactive pasttime of watching television.[14] The message is clear: physical activity offers numerous health benefits, and children who are active today are most likely to be active for years to come.

DIETARY RECOMMENDATIONS FOR CHILDREN

Regardless of family history, experts agree that all children should eat a variety of foods and maintain desirable weight. There is less agreement, however, as to whether it is wise to restrict fat in children's diets.[15] Still, health experts recommend that children over age two receive at least 20 percent and no more than 30 percent of total energy from fat, less than 10 percent from saturated fat, and less than 300 milligrams of cholesterol per day.[16] Limiting a child's fat intake to 30 percent appears to improve blood lipids without compromising nutrient adequacy or physical growth.[17]

Not Before Two

Recommendations limiting fat and cholesterol are not intended for infants or children under two years old. Infants and toddlers may need a higher percentage of fat to support their rapid growth.

Moderation, Not Deprivation

Healthy children over age two can begin the transition to eating according to recommendations by eating fewer high-fat foods, replacing some high-fat foods with low-fat choices, and selecting more fruits and vegetables. All high-fat foods need not be eliminated, though. Healthy meals can still include moderate amounts of a child's favorite foods, even if they are high-fat selections such as french fries and ice cream. Without such additions, diets might be too low in fat, not to mention unappetizing and boring.

Balanced meals need to provide lean meat, poultry, fish, and vegetable sources of protein; fruits and vegetables; whole grains; and low-fat milk products. Such meals can provide enough energy and nutrients to support growth and maintain blood cholesterol within a healthy range.

Pediatricians warn parents to avoid extremes; they caution that while intentions may be good, excessive food restriction may create nutrient deficiencies and impair growth. Furthermore, parental control over eating may instigate battles and foster attitudes about foods that can lead to inappropriate eating behaviors.

Diet First, Drugs Later

Experts agree that children with high blood cholesterol should first be treated with diet. If blood cholesterol remains high in children ten years and older after 6 to 12 months of dietary intervention, then drugs may be necessary to lower blood cholesterol.

SMOKING

Even though the focus of this text is nutrition, another risk factor for heart disease that starts in childhood and carries over into adulthood must also be addressed—cigarette smoking. Each day 3000 children light up for the first time—typically in grade school. Seven out of ten high school students have tried smoking, and one in six smokes regularly.[18] Approximately 90 percent of all adult smokers began smoking before the age of 18.

Of those teenagers who continue smoking, half will eventually die of smoking-related causes. Efforts to teach children about the dangers of smoking need to be aggressive. Children are not likely to consider the long-term health consequences of tobacco use. They are more likely to be struck by the immediate health consequences, such as shortness of breath when playing sports, or social consequences, such as having bad breath. Whatever the context, the message to all children and teens should be clear: don't start smoking. If you've already started, quit.

Cigarette smoking is the number one preventable cause of deaths.

In conclusion, *adult* heart disease is a major *pediatric* problem.[19] Without intervention, some 60 million children are destined to suffer its consequences within the next 30 years. Optimal prevention efforts focus on children, especially on those who are overweight.

Just as young children receive vaccinations against infectious diseases, they need screening for, and education about, chronic diseases. Many health education programs have been implemented in schools around the country. These programs are most effective when they include education in the classroom, heart-healthy meals in the lunchroom, fitness activities on the playground, and parental involvement at home.

Nutrition on the Net

•WEBSITES•

Access these websites for further study of topics covered in this highlight.

- Find updates and quick links to these and other nutrition-related sites at our website: **www.wadsworth.com/nutrition**
- Get weight-loss tips for children and adolescents: **www.shapedown.com**
- Learn about nondietary approaches to weight loss from HUGS International: **www.hugs.com**
- Visit the Nemours Foundation: **www.kidshealth.org**

References

1 American Diabetes Association, Type 2 diabetes in children and adolescents, *Pediatrics* 105 (2000): 671–680.

2 D. J. Gunnell and coauthors, Childhood obesity and adult cardiovascular mortality: A 57-y follow-up study based on the Boyd Orr cohort, *American Journal of Clinical Nutrition* 67 (1998): 1111–1118; R. C. Whitaker and coauthors, Predicting obesity in young adulthood from childhood and parental obesity, *New England Journal of Medicine* 337 (1997): 869–873.

3 G. S. Berenson and coauthors, Association between multiple cardiovascular risk factors and atherosclerosis in children and young adults, *New England Journal of Medicine* 338 (1998): 1650–1656; H. C. McGill, Childhood nutrition and adult cardiovascular disease, *Nutrition Reviews* 55 (1997): S2–S11.

4 Berenson and coauthors, 1998.

5 Committee on Nutrition, American Academy of Pediatrics, Cholesterol in childhood, *Pediatrics* 101 (1998): 141–147.

6 N-F Chu and coauthors, Clustering of cardiovascular disease risk factors among obese schoolchildren: The Taipei Children Heart Study, *American Journal of Clinical Nutrition* 67 (1998): 1141–1146; F. J. van Lenthe and coauthors, Association of a central pattern of body fat with blood pressure and lipoproteins from adolescence into adulthood: The Amsterdam Growth and Health Study, *American Journal of Epidemiology* 147 (1998): 686–693.

7 W. Bao and coauthors, Longitudinal changes in cardiovascular risk from childhood to young adulthood in offspring of parents with coronary artery disease: The Bogalusa Heart Study, *Journal of the American Medical Association* 278 (1997): 1749–1754.

8 Committee on Nutrition, 1998.

9 L. Van Horn and P. Greenland, Prevention of coronary artery disease is a pediatric problem, *Journal of the American Medical Association* 278 (1997): 1779–1780; G. S. Berenson, S. R. Srinivasan, and L. S. Webber, Cardiovascular risk prevention in children: A challenge or a poor idea? *Nutrition, Metabolism and Cardiovascular Diseases* 4 (1994); 46–52..

10 Committee on Nutrition, 1998.

11 Committee on Sports Medicine and Fitness, American Academy of Pediatrics, Athletic participation by children and adolescents who have systemic hypertension, *Pediatrics* 99 (1997): 637–638.

12 B. Falkner and S. Michel, Blood pressure response to sodium in children and adolescents, *American Journal of Clinical Nutrition* 65 (1997): 618S–621S.

13 S. J. Marshall and coauthors, Tracking of health-related fitness components in youth age 9 to 12, *Medicine and Science in Sports and Exercise* 30 (1998): 910–916.

14 R. E. Anderson and coauthors, Relationship of physical activity and television watching with body weight and levels of fatness among children: Results from the third National Health and Nutrition Examination Survey, *Journal of the American Medical Association* 279 (1998): 938–942.

15 R. E. Olson, Is it wise to restrict fat in the diets of children? *Journal of the American Dietetic Association* 100 (2000): 28–32; E. Satter, A moderate view on fat restriction, *Journal of the American Dietetic Association* 100 (2000): 32–36; L. A. Lytle, In defense of a low-fat diet for healthy children, *Journal of the American Dietetic Association* 100 (2000): 39–41.

16 Committee on Nutrition, 1998.

17 R. M. Lauer and coauthors, Efficacy and safety of lowering dietary intake of total fat, saturated fat, and cholesterol in children with elevated LDL cholesterol: The Dietary Intervention Study in Children, *American Journal of Clinical Nutrition* 72 (2000): 1332S–1242S; N. F. Butte, Fat intake of children in relation to energy requirements, *American Journal of Clinical Nutrition* 72 (2000): 1246S–1252S.

18 Tobacco use among high school students—United States, 1997, *Morbidity and Mortality Weekly Report* 47 (1998): 229–233.

19 Van Horn and Greenland, 1997.

CHAPTER 16 Life Cycle Nutrition: Adulthood and the Later Years

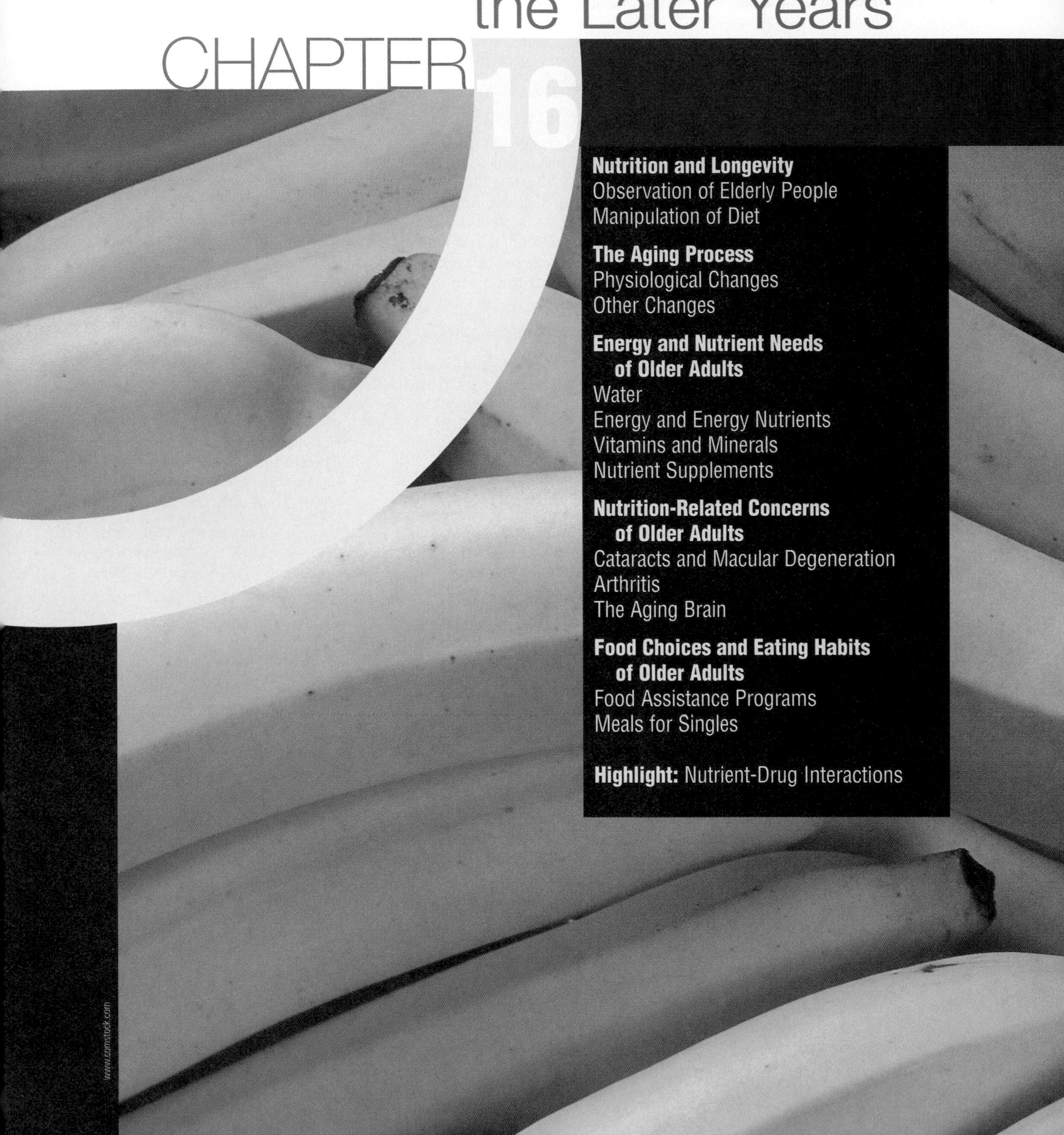

Wise food choices, made throughout adulthood, can support a person's ability to meet physical, emotional, and mental challenges and to enjoy freedom from disease. Two goals motivate adults to pay attention to their diets: promoting health and slowing aging. Much of this text has focused on nutrition to support health; this chapter focuses on aging and the nutrition needs of older adults.

The U.S. population is growing older. The majority is now middle-aged, and the ratio of old people to young is increasing, as Figure 16-1 shows. Our society uses the arbitrary age of 65 years to define the transition point between middle age and old age, but growing "old" happens day by day, with changes occurring gradually over time. Since 1950 the population of those over 65 has more than doubled. Remarkably, the fastest-growing age group today is people over 85 years (see Figure 16-2 on p. 542).

FIGURE 16-1 The Aging of the U.S. Population

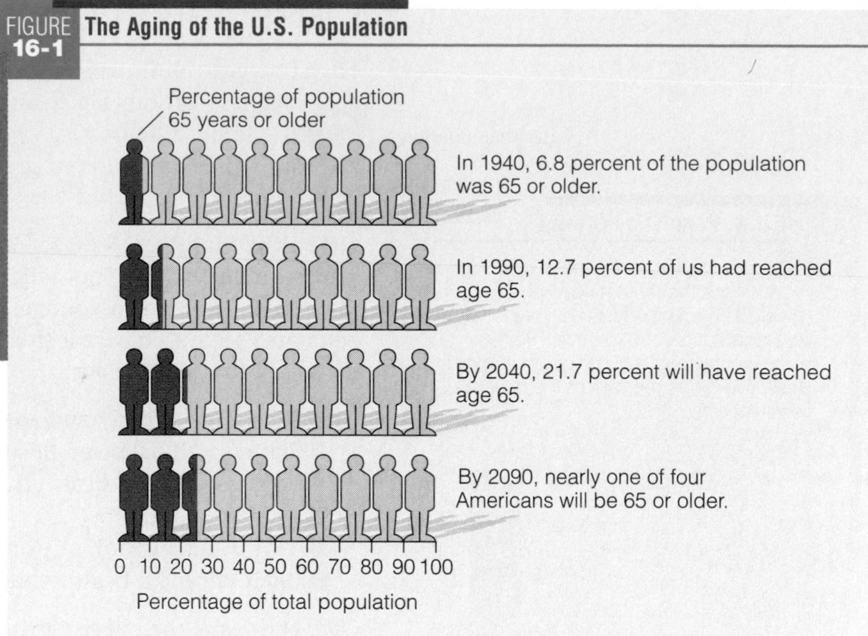

Life expectancy in the United States for white women is 80 years and for black women, 75 years; for white men, it is 75 years and for black men, 68 years—all record highs and much higher than the life expectancy of 47 years in 1900.[1] Women who live to 80 can expect to survive an additional nine years, on average; men, an additional seven. Advances in medical science—antibiotics and other treatments—are largely responsible for almost doubling the life expectancy in the twentieth century. Improved nutrition and an abundant supply of food have also contributed to lengthening life expectancy. The **life span** has not lengthened as dramatically; human **longevity** appears to have an upper limit. The potential human life span is currently 130 years. With recent advances in medical technology and genetic knowledge, however, researchers may one day be able to extend the life span even further by slowing, or perhaps preventing, aging and its accompanying diseases.[2]

◆ In 2000, an estimated 70,000 people in the United States were 100 years or older; by the year 2050, the number of centenarians is expected to grow more than tenfold.◆

NUTRITION AND LONGEVITY

Research in the field of aging is active—and difficult. Researchers are challenged by the diversity of older adults. When older adults experience health problems, it is hard to know whether to attribute these problems to normal, age-related processes or to other factors. The idea that nutrition can influence the aging process is particularly appealing because people can control and change their eating habits. The questions researchers are asking include:

- To what extent is aging inevitable, and can it be slowed through changes in lifestyle and environment?
- What role does nutrition play in the aging process, and what role can it play in slowing aging?

With respect to the first question, it seems that aging is an inevitable, natural process, programmed into the genes at conception. People can, however, slow the process within the natural limits set by genetics. They need to adopt healthy lifestyle habits such as eating nutritious food and engaging in physical activity.

With respect to the second question, good nutrition helps to maintain a healthy body and can therefore ease the aging process in many significant ways. Clearly, nutrition can improve the quality of life in the later years.

life expectancy: the average number of years lived by people in a given society.

life span: the maximum number of years of life attainable by a member of a species.

longevity: long duration of life.

Observation of Elderly People

Health promotion

The strategies adults use to meet the two goals mentioned at the start of this chapter—promoting health and slowing aging—are actually very much the same. What to eat, when to sleep, how physically active to be, and other lifestyle choices greatly influence both physical health and the aging process.

FIGURE 16-2 **U.S. Population Growth**

The "oldest old"—those over 85 years—are the fastest-growing age group in the United States. Since 1960, the population of those over 85 has more than doubled, increasing at more than five times the rate of the general population.

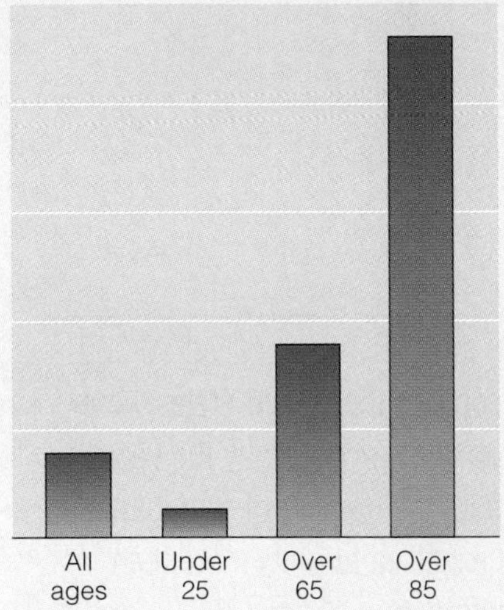

All ages | Under 25 | Over 65 | Over 85

• Healthy Habits • A person's **physiological age** reflects his or her health status and may or may not reflect the person's **chronological age.** Quite simply, some people seem younger, and others older, than their years. Six lifestyle behaviors seem to have the greatest influence on people's health and therefore on their physiological age:

- Sleeping regularly and adequately.
- Eating well-balanced meals, including breakfast, regularly.
- Engaging in physical activity regularly.
- Not smoking.
- Not using alcohol, or using it in moderation.
- Maintaining a healthy body weight.

Over the years, the effects of these lifestyle choices accumulate—that is, people who follow most of these practices live longer and have fewer disabilities as they age.[3] They are in better health, even when older in chronological age, than people who do not adopt these behaviors. Even though people cannot change their birth dates, they may be able to add years to, and enhance the quality of, their lives. Physical activity seems to be most influential in preventing or slowing the many changes that define a stereotypical "old" person.[4]

• Especially Physical Activity • The many and remarkable benefits of regular physical activity are not limited to the young: older adults who are active weigh less and have greater flexibility, more endurance, better balance, and better health than those who are inactive.[5] They reap additional benefits from various activities as well: aerobic activities improve cardiorespiratory endurance, blood pressure, and blood lipid concentrations; moderate endurance activities improve the quality of sleep; and strength training improves posture and mobility.[6] In fact, regular physical activity is the most powerful predictor of a person's mobility in the later years. Physical activity also increases blood flow to the brain, thereby preserving mental ability, alleviating depression, and supporting independence.

Muscle mass and muscle strength tend to decline with aging, making older people vulnerable to falls and immobility. Falls are a major cause of fear, injury, disability, and even death among older adults. Many lose their independence as a result of falls.[7] Regular physical activity tones, firms, and strengthens muscles, helping to improve confidence, reduce the risk of falling, and lessen the risk of injury should a fall occur.

Even without a fall, older adults may become so weak that they can no longer perform life's daily tasks, such as climbing stairs, carrying packages, and opening jars. By improving muscle strength, which allows a person to perform these tasks, strength training helps to maintain independence.[8] Even in frail, elderly people over 85 years of age, strength training not only improves balance, muscle strength, and mobility, but also increases energy expenditure and energy intake, thereby enhancing nutrient intakes.[9] This finding highlights another reason to be physically active: a person spending energy can afford to eat more food and thus receives more nutrients as a result. People who are committed to an ongoing fitness program can benefit from higher energy and nutrient intakes and still maintain their body weights.

Ideally, physical activity should be part of each day's schedule and should be intense enough to prevent muscle atrophy and to speed up the heartbeat and

physiological age: a person's age as estimated from her or his body's health and probable life expectancy.

chronological age: a person's age in years from his or her date of birth.

respiration rate. Although aging reduces both speed and endurance to some degree, older adults can still train and achieve exceptional performances. Healthy older adults who have not been active can ease into a suitable routine. They can start by walking short distances until they are walking at least one mile three times a week, and then gradually increase their pace to achieve a 20- to 25-minute mile.

© R. W. Jones/Corbis

Regular physical activity promotes a healthy, independent lifestyle.

Manipulation of Diet

In their efforts to understand longevity, researchers have not only observed people, but have manipulated influencing factors, such as diet, in animals. This research has given rise to some interesting and suggestive findings.

• Energy Restriction in Animals • Animals live longer and have fewer age-related diseases when their food intakes are restricted.[10] These life-prolonging benefits become evident when the diet provides enough food to prevent malnutrition and an energy intake of about 70 percent of normal.

Exactly how energy restriction prolongs life remains largely unexplained, although genes that play a key role in aging seem to become less active.[11] The consequences of energy restriction include a delay in the onset, or prevention, of diseases such as atherosclerosis; prolonged growth and development; and improved blood glucose, insulin sensitivity, and blood lipids.[12] In addition, energy metabolism slows and body temperature drops—indications of a reduced rate of oxygen consumption.[13] As Highlight 11 explained, the use of oxygen during energy metabolism produces free radicals, which have been implicated in the aging process. Restricting energy intake not only produces fewer free radicals, but increases antioxidant activity and enhances DNA repair.[14] This reduction of oxidative stress may at least partially explain how restricting energy intake may extend the life span.[15]

Interestingly, extending life expectancy appears to depend on restricting energy intake and not on the amount of body fat. Genetically obese rats live longer when given a restricted diet even though their body fat is similar to that of nonobese rats allowed to eat freely.[16]

• Energy Restriction in Human Beings • Research on a variety of animals◆ confirms that restricting energy intake extends the life span, but many more years of research are needed before such findings can be applied to human beings. Applying the results of animal studies to human beings is often unrealistic and potentially dangerous. Extreme starvation to extend life, like any extreme, is rarely, if ever, worth the price. Moderation, on the other hand, may be valuable.

◆ kCalorie-restricted research has been conducted on various species, including mice, rats, rhesus monkeys, cynomolgus monkeys, spiders, and fish.

Many of the physiological responses to energy restriction seen in animals also occur in people whose intakes are *moderately* restricted. When people cut back on their usual energy intake by 10 to 20 percent, body weight, body fat, and blood pressure drop, and HDL cholesterol rises—favorable changes for preventing chronic diseases. Such a moderate restriction of energy intake has no adverse effects on the mental or physical performances of most people.

IN SUMMARY

Life expectancy in the United States increased dramatically in the twentieth century. Factors that enhance longevity include limited or no alcohol use, regular balanced meals, weight control, adequate sleep, abstinence from smoking, and regular physical activity. Nutrition alone, even if ideal, cannot guarantee a long and robust life. At the very least, however, nutrition—especially when combined with regular physical activity—can influence aging and longevity in human beings by supporting good health and preventing disease.

THE AGING PROCESS

As people get older, each person becomes less and less like anyone else. The older people are, the more time has elapsed for such factors as nutrition, genetics, physical activity, and everyday **stress** to influence physical and psychological aging.

Both physical **stressors** (such as alcohol abuse, other drug abuse, smoking, pain, and illness) and psychological stressors (such as exams, divorce, moving, and the death of a loved one) elicit the body's **stress response.** The body responds to such stressors with an elaborate series of physiological steps, as the nervous and hormonal systems bring about defensive readiness in every body part. These effects favor physical action—the classic fight-or-flight response. Stress that is prolonged or severe can drain the body of its reserves and leave it weakened, aged, and vulnerable to illness, especially if physical action is not taken. As people age, they lose their ability to adapt to both external and internal disturbances. When disease strikes, the reduced ability to adapt makes the aging individual more vulnerable to death than a younger person.

Highlight 11 described the oxidative stresses that occur when free radicals exceed the body's ability to defend itself. Increased free-radical activity and decreased antioxidant protection are common, but not inevitable, features of aging.[17] Healthy people over 100 years old who had higher intakes of vegetables showed less evidence of oxidative stress than people 70 to 99 years old.[18] Such findings seem to suggest that the fountain of youth may actually be a cornucopia of fruits and vegetables rich in antioxidants. (Return to Highlight 11 for more details on the antioxidant action of fruits and vegetables in defending against oxidative stress.)

Physiological Changes

As aging progresses, inevitable changes in each of the body's organs contribute to the body's declining function. These physiological changes influence nutrition status, just as growth and development do in the earlier stages of the life cycle.

• **Body Composition** • In general, older people tend to lose bone and muscle and gain body fat.[19] Many of these changes occur because some hormones that regulate appetite and metabolism become less active with age, while others become more active.*[20] Loss of muscle, known as **sarcopenia,** can be significant in the later years and its consequences, quite dramatic.[21] As muscles diminish and weaken, people lose their ability to move and maintain balance, making falls likely. The limitations that accompany the loss of muscle and its strength play a key role in the diminishing health that often accompanies aging. Optimal nutrition and regular physical activity can help maintain muscle mass and strength and minimize the changes in body composition associated with aging.

• **Immune System** • Changes in the immune system also bring declining function with age. In addition, the immune system is compromised by nutrient deficiencies. Thus the combination of old age and malnutrition makes older people vulnerable to infectious diseases.[22] Adding insult to injury, antibiotics often are not effective against infections in people with compromised immune systems. Consequently, infectious diseases are a major cause of death in older adults.

stress: any threat to a person's well-being; a demand placed on the body to adapt.

stressors: environmental elements, physical or psychological, that cause stress.

stress response: the body's response to stress, mediated by both nerves and hormones.

sarcopenia (SAR-koh-PEE-nee-ah): loss of skeletal muscle mass, strength, and quality.
- **sarco** = flesh
- **penia** = loss or lack

*Examples of hormones that change with age include growth hormone and androgens, which decline with advancing age, thus contributing to the decrease in lean body mass, and prolactin, which increases with age, helping to maintain body fat. Insulin sensitivity also diminishes as people grow older, most likely because of increases in body fat and decreases in physical activity.

• GI Tract • In the GI tract, the intestinal wall loses strength and elasticity with age, and GI hormone secretions change.[23] All of these actions slow motility. Constipation is much more common in the elderly than in the young.

Atrophic gastritis, a condition that affects almost one-third of those over 60, is characterized◆ by an inflamed stomach, bacterial overgrowth, and a lack of hydrochloric acid and intrinsic factor—all of which can impair the digestion and absorption of nutrients, most notably, vitamin B_{12}, but also biotin, folate, calcium, iron, and zinc.

Consequences of atrophic gastritis: ◆
- Inflamed stomach.
- Increased bacterial growth.
- Reduced hydrochloric acid.
- Reduced intrinsic factor.
- Increased risk of nutrient deficiencies, notably of vitamin B_{12}.

• Tooth Loss • Regular dental care over a lifetime protects against tooth loss and gum disease, which are common in old age. These conditions make chewing difficult or painful. Dentures, even when they fit properly, are less effective than natural teeth, and inefficient chewing can cause choking. People with tooth loss, gum disease, and ill-fitting dentures tend to limit their food selections to soft foods. If foods such as corn on the cob, apples, and hard rolls are replaced by creamed corn, applesauce, and rice, then nutrition status may not be greatly affected, but when food groups are eliminated and variety is limited, poor nutrition follows. People without teeth typically have low intakes of fiber and vitamins and high intakes of saturated fat and energy. To determine whether a visit to the dentist is needed, an older adult can check the conditions listed in the margin.◆

Conditions requiring dental care: ◆
Dry mouth.
Eating difficulty.
No dental care within 2 years.
Tooth or mouth pain.
Altered food selections.
Lesions, sores, or lumps in mouth.

• Sensory Losses and Other Physical Problems • A multitude of sensory losses and other physical problems can also interfere with an older person's ability to obtain adequate nourishment. Failing eyesight, for example, can make driving to the grocery store impossible and shopping for food a frustrating experience. It may become so difficult to read food labels and count money that the person doesn't buy needed foods. Carrying bags of groceries may be an unmanageable task. Similarly, a person with limited mobility may find cooking and cleaning up too hard to do. Not too surprisingly, the prevalence of undernutrition is high among those who are homebound.[24]

Sensory losses can also interfere with a person's ability or willingness to eat. Taste and smell sensitivities tend to diminish with age and may make eating less enjoyable. Consequently, food intake may diminish and nutrient deficiencies follow.[25] Loss of vision and hearing may contribute to social isolation, and eating alone may lead to poor intake.

Other Changes

In addition to the physiological changes that accompany aging, adults are changing in many other ways that influence their nutrition status. Psychological, economic, and social factors play big roles in a person's ability and willingness to eat.

• Psychological Changes • Although not an inevitable component of aging, depression is common, affecting an estimated 6 million older adults.[26] Depressed people, even those without disabilities, lose their ability to perform simple physical tasks.[27] They frequently lose their appetite and the motivation to cook or even to eat. An overwhelming sense of grief and sadness at the death of a spouse, friend, or family member may leave a person, especially an elderly person, feeling powerless to overcome depression. When a person is suffering the heartache and loneliness of bereavement, cooking meals may not seem worthwhile. The support and companionship of family and friends, especially at mealtimes, can help overcome depression and enhance appetite.

Shared meals can brighten the day and enhance the appetite.

• Economic Changes • Overall, older adults today have higher incomes than their cohorts of previous generations. Still, poverty is a major problem for about 20 percent of the people over age 65. Factors such as living arrangements and income make significant differences in the food choices, eating habits, and

Growing old can be enjoyable for people who take care of their health and live each day fully.

© Kwame Zikomo/Super Stock

nutrition status of older adults, especially those over age 80. People of low socioeconomic means are likely to have inadequate food and nutrient intakes. Only about one-third of the needy elderly receive assistance from federal programs.[28]

• **Social Changes** • Malnutrition among older adults is most likely to occur among those living alone, especially men; those with the least education; those living in federally funded housing (an indicator of low income); and those who have recently experienced a change in lifestyle. One study of home-delivered meals confirmed that men living alone eat less than men living with others; interestingly, women living alone eat more than women living with others. Adults who live alone do not necessarily make poor food choices, but they often consume too little food: loneliness is directly related to nutritional inadequacies, especially of energy intake.

IN SUMMARY

Many changes that accompany aging can impair nutrition status. Among physiological changes, hormone activity alters body composition, immune system changes raise the risk of infections, atrophic gastritis interferes with digestion and absorption, and tooth loss limits food choices. Psychological changes such as depression, economic changes such as loss of income, and social changes such as loneliness contribute to poor food intake.

ENERGY AND NUTRIENT NEEDS OF OLDER ADULTS

Knowledge about the nutrient needs and nutrition status of older adults has grown considerably in recent years. The Dietary Reference Intakes (DRI) cluster people over 50 into two age categories—one group of 51 to 70 years and one of 71 and older.[29] Increasingly, research is showing that the nutrition needs of people 50 to 60 years old may be very different from those of people over 80.

Setting standards for older people is difficult because individual differences become more pronounced as people grow older. People start out with different genetic predispositions and ways of handling nutrients, and the effects of these differences become magnified with years of unique dietary habits. For example, one person may tend to omit fruits and vegetables from his diet, and by the time he is old, he may have a set of nutrition problems associated with a lack of fiber and antioxidants. Another person may have omitted milk and milk products all her life—her nutrition problems may be related to a lack of calcium. Also, as people age, they suffer different chronic diseases and take various medicines—both of which will affect nutrient needs. For all of these reasons, researchers have difficulty even defining "healthy aging," a prerequisite to developing recommendations to meet the "needs of practically all healthy persons." The following discussion gives special attention to a few nutrients of concern.

Water

◆ Older adult eating tip: Drink plenty of water.

Dehydration is a risk for older adults,◆ who may find it difficult and bothersome to get a drink or to get to a bathroom. Older adults who have lost bladder control may be afraid to drink too much water. Despite real fluid needs, many older people do not seem to feel thirsty or notice mouth dryness. Many nursing home employees say it is hard to persuade their elderly clients to drink enough water and fruit juices.

Total body water decreases as people age, so even mild stresses such as fever or hot weather can precipitate rapid dehydration in older adults. Dehydrated older adults seem to be more susceptible to urinary tract infections, pneumonia, pressure ulcers, and confusion and disorientation.[30] To prevent dehydration, older adults need to drink at least 6 glasses of water a day.[31]◆ Milk and juices may replace some of this water, but beverages containing alcohol or caffeine should be limited because of their diuretic effect.

Water recommendation for adults: ◆
- 1 oz/kg actual body weight, with a minimum of 50 oz (6¼ c) daily.

Energy and Energy Nutrients

On average, adult energy needs decline an estimated 5 percent per decade. One reason is that people usually reduce their physical activity as they age, although they need not do so. Another reason is that lean body mass diminishes, slowing the basal metabolic rate.

The lower energy expenditure of older adults requires that they eat less food to maintain their weights. Accordingly, the energy RDA for adults decreases slightly after age 50. Energy intakes typically decline in parallel with needs. Still, many older adults are overweight, indicating that their food intakes do not decline enough to compensate for their sedentary lifestyles.[32] Overweight among older adults increases the risks for numerous diseases and for disabilities as well.[33]

On limited energy allowances, people must select mostly nutrient-dense foods.◆ There is little leeway for sugars, fats, oils, or alcohol. The Daily Food Guide (on pp. 38–39) offers a dietary framework for adults of all ages. Most older adults would do well to select the lower number of recommended servings from each food group. Those who need additional food energy can choose extra servings.

Older adult eating tip: Select nutrient-dense foods, low in fats, sugars, and alcohol. ◆

Not all older adults are overweight, of course. In fact, the prevalence of overweight decreases with increasing age after age 55.[34] Many older adults experience unintentional weight loss, in large part because of an inadequate food intake.[35] Such weight loss is often associated with illness and premature death, making every meal a life-saving event.[36]

• Protein • Because energy needs decrease, protein must be obtained from low-kcalorie sources of high-quality protein, such as lean meats, poultry, fish, and eggs; nonfat and low-fat milk products; and legumes. Protein is especially important for the elderly to support a healthy immune system and to prevent muscle wasting. Some research suggests that current protein recommendations may be inadequate and that a daily intake of 1 gram of high-quality protein per kilogram of body weight would be more appropriate for older adults.[37] Additional research is needed to clarify the protein needs of older people, especially those over 70 years old.

• Carbohydrate and Fiber • As always, abundant carbohydrate is needed to protect protein from being used as an energy source. Sources of complex carbohydrates such as legumes, vegetables, whole grains, and fruits are also rich in fiber and essential vitamins and minerals. Average fiber intakes among older adults are lower than current recommendations (10 to 13 grams per 1000 kcalories).[38] Eating high-fiber foods and drinking water can alleviate constipation—a condition common among older adults, and especially among nursing home residents. Physical inactivity and medications also contribute to the high incidence of constipation.

Many nursing home residents are malnourished and underweight. For these people, a diet that emphasizes fiber-rich foods such as whole grains, fruits, and vegetables may be too low in protein and energy. Protein- and energy-dense snacks such as hard-boiled eggs, tuna fish and crackers, peanut butter on graham crackers, and hearty soups are valuable additions to the diets of underweight or malnourished older adults. Liquid nutritional formulas can also be used to supplement intake.

Myrleen Ferguson Cate/PhotoEdit

Taking time to nourish your body well is a gift you give yourself.

• **Fat** • As is true for people of all ages, fat needs to be limited in the diets of most older adults. Cutting fat may help prevent or delay the development of cancer, atherosclerosis, and other degenerative diseases. This recommendation should not be taken too far; for some older adults, limiting fat too severely may lead to nutrient deficiencies and weight loss—two problems that carry greater health risks in the elderly than overweight.

Vitamins and Minerals

Most people can achieve adequate vitamin and mineral intakes simply by including foods from all food groups in their diets, but older adults often omit fruits and vegetables. Similarly, few older adults consume the recommended amounts of milk or milk products.

• **Vitamin B_{12}** • An estimated 15 percent of the elderly population is deficient in vitamin B_{12}.[39] As Chapter 10 explained, people with atrophic gastritis are particularly vulnerable to vitamin B_{12} deficiency: the bacterial overgrowth that accompanies this condition uses up the vitamin and without hydrochloric acid and intrinsic factor, digestion and absorption of vitamin B_{12} are inefficient. Given the devastating neurological effects of a vitamin B_{12} deficiency, an adequate intake is imperative.[40]

• **Vitamin D** • Vitamin D deficiency is a problem among older adults.[41] Only vitamin D–fortified milk provides significant vitamin D, and many older adults drink little or no milk. Further compromising the vitamin D status of many older people, especially those in nursing homes, is their limited exposure to sunlight. Finally, aging reduces the skin's capacity to make vitamin D and the kidneys' ability to convert it to its active form. Not only are older adults not getting enough vitamin D, but they may actually need more. An intake of 10 micrograms daily is recommended to prevent bone loss and to maintain vitamin D status, especially in those who engage in minimal outdoor activity.[42]

• **Calcium** • Chapter and Highlight 12 emphasized the importance of abundant dietary calcium throughout life, and especially for women after menopause, to protect against osteoporosis. The recommended intake for older adults is 1200 milligrams of calcium daily, but the calcium intakes of older people in the United States are well below recommendations.[43] Some older adults avoid milk and milk products because they dislike these foods or associate them with stomach discomfort. One simple solution is to add powdered nonfat milk to recipes; Chapter 12 offers many other strategies for including nonmilk sources of calcium for those who do not drink milk.

• **Iron** • Iron-deficiency anemia is less common in older adults than in younger people, but still occurs in some, especially in people with low food energy intakes. Aside from diet, other factors make iron deficiency likely in older people: chronic blood loss from diseases and medicines, and poor iron absorption due to reduced stomach acid secretion and antacid use. For older people with infectious diseases, the consequences of iron-deficiency anemia can be life-threatening.[44] Anyone concerned with older people's nutrition should keep these possibilities in mind.

Nutrient Supplements

People judge for themselves how to manage their nutrition, and some turn to supplements. Advertisers target older people with appeals to take supplements and eat "health" foods, claiming that these products prevent disease and promote longevity. About half of all women over 65 take some type of dietary supplement, while about one-fifth of older men do. Certain diseases or health problems may necessitate the taking of supplements, but often supplements have not been prescribed by health care professionals and are unnecessary.

When recommended by a physician or registered dietitian, vitamin D and calcium supplements for osteoporosis or iron for iron-deficiency anemia may be beneficial. In most cases, though, the money spent on supplements would be better spent on nutritious foods. Older adults with food energy intakes less than about 1500 kcalories per day, however, should probably take a once-daily type of multivitamin-mineral supplement.

People with small energy allowances would do well to become more active so they can afford to eat more food. Food is the best source of nutrients for everybody. Supplements are just that—supplements to foods, not substitutes for them. For anyone who is motivated to obtain the best possible health, it is never too late to learn to eat well, drink water, exercise regularly, and adopt other lifestyle habits such as quitting smoking and moderating alcohol use.

IN SUMMARY

The accompanying table summarizes the nutrient concerns of aging. Although some nutrients need special attention in the diet, supplements are not routinely recommended. The ever-growing number of older people creates an urgent need to learn more about how their nutrient requirements differ from those of others and how such knowledge can enhance their health.

Nutrient	Effect of Aging	Comments
Water	Lack of thirst and decreased total body water make dehydration likely.	Mild dehydration is a common cause of confusion. Difficulty obtaining water or getting to the bathroom may compound the problem.
Energy	Need decreases as muscle mass decreases (sarcopenia).	Physical activity moderates the decline.
Fiber	Likelihood of constipation increases with low intakes and changes in the GI tract.	Inadequate water intakes and lack of physical activity, along with some medications, compound the problem.
Protein	Needs may stay the same or increase slightly.	Low-fat, high-fiber legumes and grains meet both protein and other nutrient needs.
Vitamin B_{12}	Atrophic gastritis is common.	Deficiency causes neurological damage.
Vitamin D	Increased likelihood of inadequate intake; skin synthesis declines.	Daily sunlight exposure in moderation may be of benefit.
Calcium	Intakes may be low; osteoporosis is common.	Stomach discomfort commonly limits milk intake; calcium substitutes are needed.
Iron	In women, status improves after menopause; deficiencies are linked to chronic blood losses and low stomach acid output.	Adequate stomach acid is required for absorption; antacid or other medicine use may aggravate iron deficiency; vitamin C and meat increase absorption.

NUTRITION-RELATED CONCERNS OF OLDER ADULTS

Nutrition may play a greater role than has been realized in preventing many changes once thought to be inevitable consequences of growing older. The following discussions of cataracts and macular degeneration, arthritis, and the aging brain show that nutrition may provide at least some protection against some of the conditions associated with aging.

Cataracts and Macular Degeneration

Cataracts are age-related thickenings in the lenses of the eyes that impair vision. If not surgically removed, they ultimately lead to blindness. Cataracts occur even in well-nourished individuals as a result of ultraviolet light exposure, oxidative stress, injury, viral infections, toxic substances, and genetic disorders. Many cataracts, however, are vaguely called senile cataracts—meaning "caused by aging." In the United States, more than half of all adults 65 and older have a cataract.

Oxidative stress appears to play a significant role in the development of cataracts, and the antioxidant nutrients may help minimize the damage. Studies have reported an inverse relationship between cataracts and dietary intakes of vitamin C, vitamin E, and carotenoids; taking supplements of vitamins C and E seems to reduce the likelihood of developing age-related cataracts.[45]

One other diet-related factor may play a role in the development of cataracts: overweight.[46] Overweight appears to be associated with cataracts, but its role has not been identified. Risk factors that typically accompany overweight, such as inactivity, diabetes, or hypertension, do not explain the association.

Another common cause of visual loss among older people is **macular degeneration,** a deterioration of the macular region of the retina. Like cataracts, risk factors for macular degeneration include oxidative stress from sunlight. Dietary fat may also be a risk factor for macular degeneration, but the omega fatty acids of fish oils may be protective.[47] Similarly, foods rich in antioxidant nutrients and wine, with its protective phytochemicals, seem to reduce the risk of developing macular degeneration.[48]

Arthritis

Over 40 million people in the United States have some form of **arthritis.**[49] As the population ages, it is expected that the prevalence will continue to increase.

• Osteoarthritis • The most common type of arthritis that disables older people is **osteoarthritis,** a painful swelling of the joints. During movement, the ends of bones are normally protected from wear by cartilage and by small sacs of fluid that act as a lubricant. With age, bones sometimes disintegrate, and the joints become malformed and painful to move.

One known connection between osteoarthritis◆ and nutrition is overweight. Weight loss may relieve some of the pain for overweight persons with osteoarthritis, partly because the joints affected are often weight-bearing joints that are stressed and irritated by having to carry excess poundage. Interestingly, though, weight loss often relieves the worst of the pain of arthritis in the hands as well, even though they are not weight-bearing joints. Jogging and other weight-bearing exercises do not worsen arthritis. In fact, both aerobic activity and strength training offer modest improvements in physical performance and pain relief.[50]

◆ Risk factors for osteoarthritis:
- Age.
- High BMI at age 40.
- Smoking.
- Lack of hormone therapy (in women).

• Rheumatoid Arthritis • Another type of arthritis known as **rheumatoid arthritis** has a possible link to diet through the immune system. In rheumatoid arthritis, the immune system mistakenly attacks the bone coverings as if they were made of foreign tissue. The integrity of the immune system depends

cataracts (KAT-ah-rakts): thickenings of the eye lenses that impair vision and can lead to blindness.

macular (MACK-you-lar) **degeneration:** deterioration of the macular area of the eye that can lead to loss of central vision and eventual blindness. The **macula** is a small, oval, yellowish region in the center of the retina that provides the sharp, straight-ahead vision so critical to reading and driving.

arthritis: inflammation of a joint, usually accompanied by pain, swelling, and structural changes.

osteoarthritis: a painful, chronic disease of the joints that occurs when the cushioning cartilage in a joint breaks down; joint structure is usually altered, with loss of function; also called **degenerative arthritis.**

rheumatoid (ROO-ma-toyd) **arthritis:** a disease of the immune system involving painful inflammation of the joints and related structures.

on adequate nutrition, and a poor diet may worsen this type of arthritis. It is also possible that in some individuals, certain foods may stimulate the immune system to attack. For example, milk and milk products seem to aggravate rheumatoid arthritis in some people.

Nutrition is also linked to rheumatoid arthritis through EPA and DHA, the omega-3 fatty acids commonly found in fish oil.[51] The same diet recommended for heart health—one low in saturated fat from meats and milk products and high in oils from fish—helps prevent or reduce the inflammation in the joints that makes arthritis so painful. Most likely, these omega-3 fatty acids interfere with the action of prostaglandins, chemicals involved in inflammation.

Another possible link between nutrition and rheumatoid arthritis involves the oxidative damage to the membranes within joints that causes inflammation and swelling. Vitamin E helps to prevent oxidation, but it does not improve active cases of rheumatoid arthritis. This is not surprising, though, because the vitamin's role as an antioxidant is preventive, not restorative.

• **Treatment** • Traditional medical intervention for arthritis includes medication and surgery. Alternative therapies to treat arthritis abound, but none have proved themselves safe and effective in scientific studies. Two currently popular supplements—glucosamine and chondroitin—may relieve pain and improve mobility as well as over-the-counter pain relievers, but stronger research studies are needed to confirm reports.[52] Drugs and supplements used to relieve arthritis can impose nutrition risks; many affect appetite and alter the body's use of nutrients, as Chapter 18 explains.

Both foods and mental challenges nourish the brain.

The Aging Brain

The brain, like all of the body's organs, responds to both genetic and environmental factors that can enhance or diminish its amazing capacities. One of the challenges researchers face when studying the human brain is to distinguish among normal age-related physiological changes, changes caused by diseases, and changes that result from cumulative, extrinsic factors such as diet.

The brain normally changes in some characteristic ways as it ages. For one thing, its blood supply decreases. For another, the number of **neurons,** the brain cells that specialize in transmitting information, diminishes as people age. When the number of nerve cells in one part of the cerebral cortex diminishes, hearing and speech are affected. Losses of neurons in other parts of the cortex can impair memory and cognitive function. When the number of neurons in the hindbrain diminishes, balance and posture are affected. Losses of neurons in other parts of the brain affect still other functions.

Clinicians now recognize that some of the cognitive loss and forgetfulness generally attributed to aging may be due in part to extrinsic, and therefore controllable, factors—including nutrient deficiencies. In some instances, the degree of cognitive loss is extensive. Such **senile dementia** may be attributable to a specific disorder such as a brain tumor or Alzheimer's disease.

• **Alzheimer's Disease** • Much attention has focused on the *abnormal* deterioration of the brain called **Alzheimer's disease,** which affects 10 percent of U.S. adults by age 65 and 30 percent of those over 85.[53] Diagnosis of Alzheimer's disease depends on its characteristic symptoms: the victim gradually loses memory and reasoning, the ability to communicate, physical capabilities, and eventually life itself. Nerve cells in the brain die, and communication between the cells breaks down.

Researchers are closing in on the exact cause of Alzheimer's disease.* Clearly, genetic factors are involved.[54] Free radicals may also be involved.[55] Nerve cells in the brains of people with Alzheimer's disease show evidence of free-radical

*A report on the genetic and other aspects of Alzheimer's is available from Alzheimer's Disease Education and Referral Center, P.O. Box 8250, Silver Springs, MD 20907-8250.

neurons: nerve cells; the structural and functional units of the nervous system. Neurons initiate and conduct nerve transmissions.

senile dementia: the loss of brain function beyond the normal loss of physical adeptness and memory that occurs with aging.

Alzheimer's disease: a degenerative disease of the brain involving memory loss and major structural changes in neuron networks; also known as *senile dementia of the Alzheimer's type (SDAT), primary degenerative dementia of senile onset,* or *chronic brain syndrome.*

attack—damage to DNA, cell membranes, and proteins. They also show evidence of the minerals that trigger free-radical attacks—iron, copper, zinc, and aluminum.[56]

In Alzheimer's disease, the brain is littered with clumps of a protein fragment called beta-amyloid. Free radicals and beta-amyloid have a sinister relationship: free radicals accelerate the clumping of beta-amyloid, and beta-amyloid produces more free radicals. Scientists believe beta-amyloid clogs the brain and damages or kills certain nerve cells, causing memory loss. Dietary research is examining whether the antioxidant nutrients can limit free-radical damage and delay or prevent Alzheimer's disease.[57] Drug research is focusing on developing an immunization for beta-amyloid or an enzyme to block its production.[58]

Late in the course of the disease there is a decline in the activity of the enzyme that assists in the production of the neurotransmitter acetylcholine from choline and acetyl CoA.[59] Acetylcholine is essential to memory, but supplements of choline (or of lecithin, which contains choline) have no effect on memory or on the progression of the disease. Drugs that inhibit the breakdown of acetylcholine, on the other hand, have proved beneficial.[60]

Research suggests that cardiovascular disease risk factors such as high blood pressure, diabetes, and elevated levels of homocysteine may be related to the development of Alzheimer's disease. [61] Diets designed to support a healthy heart may benefit a healthy brain as well.

Treatment for Alzheimer's disease involves providing care to clients and support to their families. Drugs are used to improve or at least to slow the loss of short-term memory and cognition, but they do not treat the disease.[62] Other drugs may be used to control depression, anxiety, and behavior problems.

Maintaining appropriate body weight may be the most important nutrition concern for the person with Alzheimer's disease.[63] Depression and forgetfulness can lead to changes in eating behaviors and poor food intake. Furthermore, changes in the body's weight-regulation system may contribute to weight loss. Perhaps the best that a caregiver can do nutritionally for a person with Alzheimer's disease is to supervise food planning and mealtimes.[64] Providing well-liked and well-balanced meals and snacks in a cheerful atmosphere encourages food consumption. To minimize confusion, offer a few ready-to-eat foods, in bite-size pieces, with seasonings and sauces. To avoid mealtime disruptions, control distractions such as music, television, children, and the telephone.

• Nutrient Deficiencies and Brain Function • Nutrients influence the development and activities of the brain.[65] The ability of neurons to synthesize specific neurotransmitters, for example, depends in part on the availability of precursor nutrients that are obtained from the diet. The neurotransmitter serotonin derives from the amino acid tryptophan. To function properly, the enzymes involved in neurotransmitter synthesis require vitamins and minerals. Severe dietary deficiencies of the B vitamins impair mental ability, including memory.[66] Similarly, long-term, moderate nutrient deficiencies may contribute to the loss of memory and cognition that some older adults experience.[67] If long-term, moderate nutrient deficiencies influence the loss of cognitive function that accompanies aging, then the loss may be preventable or at least diminished or delayed through diet. Table 16-1 summarizes some of the better-known connections between brain function and nutrients.

IN SUMMARY

Senile dementia and other losses of brain function afflict millions of older adults, while others face loss of vision due to cataracts or macular degeneration, or cope with the pain of arthritis. As the number of people over age 65 continues to grow, the need for solutions to these problems becomes urgent. Some problems may be inevitable, but others are preventable and good nutrition may play a key preventive role.

TABLE 16-1 Summary of Nutrient-Brain Relationships

Brain Function	Depends on an Adequate Intake of:
Short-term memory	Vitamin B_{12}, vitamin C, vitamin E
Performance in problem-solving tests	Riboflavin, folate, vitamin B_{12}, vitamin C
Mental health	Thiamin, niacin, zinc, folate
Cognition	Folate, vitamin B_6, vitamin B_{12}, iron, vitamin E
Vision	Essential fatty acids, vitamin A
Neurotransmitter synthesis	Tyrosine, tryptophan, choline

We can now state with certainty that a person's nutrition status affects the health and functioning of the whole body. Eating a nutritious, balanced diet throughout life seems a small effort in light of the rewards of continued health and enjoyment in later life.

Social interactions at a congregate meal site can be as nourishing as the foods served.

FOOD CHOICES AND EATING HABITS OF OLDER ADULTS

Older people are an incredibly diverse group, and for the most part they are independent, socially sophisticated, mentally lucid, fully participating members of society who report themselves to be happy and healthy. In fact, the quality of life among the elderly has improved and their chronic disabilities have declined dramatically in recent years.[68] By practicing stress-management skills, maintaining physical fitness, participating in activities of interest, and cultivating spiritual health, as well as obtaining adequate nourishment, people can support a high quality of life into old age (see Table 16-2 on p. 554 for some strategies).

Older people spend more money per person on foods to eat at home than other age groups and less money on foods away from home. Manufacturers would be wise to cater to the preference of older adults by providing good-tasting, nutritious foods in easy-to-open, single-serving packages with labels that are easy to read. Such services enable older adults to maintain their independence◆ and to feel a sense of control and involvement in their own lives. Another way older adults can take care of themselves is by remaining or becoming physically active. As mentioned earlier, physical activity helps preserve one's ability to perform daily tasks and so promotes independence.

Familiarity,◆ taste, and health beliefs are most influential on older people's food choices. Eating foods that are familiar, especially those that recall family meals and pleasant times, can be comforting. People 65 and over are less likely to diet to lose weight than younger people are, but are more likely to diet in pursuit of medical goals such as controlling blood glucose and cholesterol.

◆ Older adult eating tip: Try to maintain independence.

◆ Older adult eating tip: Select familiar foods, especially ethnic favorites.

Food Assistance Programs

The Nutrition Screening Initiative is part of a national effort to identify and treat nutrition problems in older persons; it uses a screening checklist. To *determine*◆ the risk of malnutrition in older clients, health care professionals can keep in mind the characteristics listed in the margin.

Nutrition services are an integral part of health care, and different subgroups of the aging population need different programs designed to meet their specific needs.[69] People living alone can benefit from **congregate meals;** people confined to their homes need meals delivered.

◆ Risk factors for malnutrition in older adults:

Disease.
Eating poorly.
Tooth loss or oral pain.
Economic hardship.
Reduced social contact.
Multiple medications.
Involuntary weight loss or gain.
Needs assistance with self-care.
Elderly person older than 80 years.

congregate meals: nutrition programs that provide food for the elderly in a conveniently located setting such as a community center.

TABLE 16-2 Strategies for Growing Old Healthfully

- Choose nutrient-dense foods.
- Be physically active. Walk, run, dance, swim, bike, or row for aerobic activity. Lift weights, do calisthenics, or pursue some other activity to tone, firm, and strengthen muscles. Modify activities to suit changing abilities and tastes.
- Maintain appropriate body weight.
- Reduce stress (cultivate self-esteem, maintain a positive attitude, manage time wisely, know your limits, practice assertiveness, release tension, and take action).
- For women, see a physician about estrogen replacement.
- For people who smoke, quit.
- Expect to enjoy sex, and learn new ways of enhancing it.
- Use alcohol only moderately, if at all; use drugs only as prescribed.
- Take care to prevent accidents.
- Expect good vision and hearing throughout life; obtain glasses and hearing aids if necessary.
- Take care of your teeth; obtain dentures if necessary.
- Be alert to confusion as a disease symptom, and seek diagnosis.
- Take medications as prescribed; see a physician before self-prescribing medicines or herbal remedies and a registered dietitian before self-prescribing supplements.
- Control depression through activities and friendships; seek professional help if necessary.
- Drink 6 to 8 glasses of water every day.
- Practice mental skills. Keep on solving math problems and crossword puzzles, playing cards or other games, reading, writing, imagining, and creating.
- Make financial plans early to ensure security.
- Accept change. Work at recovering from losses; make new friends.
- Cultivate spiritual health. Cherish personal values. Make life meaningful.
- Go outside for sunshine and fresh air as often as possible.
- Be socially active—play bridge, join an exercise group, take a class, teach a class, eat with friends, volunteer time to help others.
- Stay interested in life—pursue a hobby, spend time with grandchildren, take a trip, read, grow a garden, or go to the movies.
- Enjoy life.

HEALTHY PEOPLE 2010

Increase the receipt of home foodservices by people aged 65 and older who have difficulty in preparing their own meals or are otherwise in need of home-delivered meals.

The U.S. government funds programs to provide nutritious meals to older adults at congregate meal sites. These congregate meals are a valuable source of nutrients for more than 3 million older adults.[70] Like school lunches, though, congregate meals typically do not meet current dietary recommendations to limit sodium, fat, and cholesterol. The section on p. 555 describes food assistance programs for older adults and Highlight 16 discusses community nutrition programs.

Meals for Singles

Many older adults live alone, and singles of all ages face difficulties in purchasing, storing, and preparing food. Large packages of meat and vegetables are often intended for families of four or more, and even a head of lettuce can spoil before one person can use it all. Many singles live in small dwellings and have little storage space for foods. A limited income presents additional obstacles. This section offers suggestions that can help to solve some of these problems.

• **Spend Wisely** • People who have the means to shop and cook for themselves can cut their food bills just by being wise shoppers. Large supermarkets

Food Assistance Programs for Older Adults

The federal Elderly Nutrition Program is intended to improve older people's nutrition status and enable them to avoid medical problems, continue living in communities of their own choice, and stay out of institutions. Its specific goals are to provide low-cost, nutritious meals; opportunities for social interaction; homemaker education and shopping assistance; counseling and referral to social services; and transportation.

The Elderly Nutrition Program provides for congregate meal programs. Administrators try to select sites for congregate meals so as to feed as many eligible people as possible. Volunteers may also deliver meals to those who are homebound either permanently or temporarily; these efforts are known as Meals on Wheels. The home-delivery program ensures nutrition, but its recipients miss out on the social benefit of the congregate meal sites; every effort is made to persuade older people to come to the shared meals, if they can. All persons aged 60 years and older and their spouses are eligible to receive meals from these programs, regardless of their income. Priority is given to those who are economically and socially needy. An estimated 25 percent of our nation's elderly poor benefit from these meals.[a]

These programs provide at least one meal a day that meets a third of the RDA for this age group; they must operate five or more days a week. Many programs voluntarily offer additional services: provisions for special diets, food pantries, ethnic meals, and delivery of meals to the homeless.

Older adults can learn about the available programs in their communities by looking in the yellow pages of the telephone book under "Social Services" or "Senior Citizens' Organizations." In addition, the local senior center and hospital can usually direct people to programs providing nutrition and other health-related services.

[a]Federal program nourishes poor elderly, *Journal of the American Medical Association* 278 (1997): 1301.

are usually less expensive than convenience stores. A grocery list helps reduce impulse buying, and specials and coupons can save money when the items featured are those that the shopper needs and uses.

Buying the right amount so as not to waste any food is a challenge for people eating alone. They can buy fresh milk in the size best suited for personal needs. Pint-size and even cup-size boxes◆ of milk are available and can be stored unopened on a shelf for up to three months without refrigeration.

◆ Boxes of milk kept at room temperature on the shelves of grocery stores have been exposed to temperatures above those of pasteurization just long enough to sterilize the milk—a process called **ultrahigh temperature (UHT).**

Many foods that offer a variety of nutrients for practically pennies have a long shelf life; staples such as rice, pastas, nonfat dry powdered milk, and dried beans and peas can be purchased in bulk and stored on a shelf for months at room temperature. Other foods that are usually a good buy include whole pieces of cheese rather than sliced or shredded cheese, fresh produce in season, variety meats such as chicken livers, and cereals that require cooking instead of ready-to-serve cereals.

A person who has ample freezing space can buy large packages of meat, such as pork chops, ground beef, or chicken, when they are on sale. Then the meat can be immediately divided into individual servings and wrapped in aluminum foil, not freezer paper: the foil can become the liner for the pan in which to bake or broil the meat, thus saving work. All the individual servings can be put in a bag marked appropriately with the contents and the date.

Frozen vegetables are more economical in large bags than in small boxes. The amount needed can be taken out, and the bag closed tightly with a twist tie or rubber band. If the package is returned quickly to the freezer each time, the vegetables will stay fresh for a long time.

Finally, breads and cereals usually must be purchased in larger quantities. Again the amount needed for a few days can be taken out and the rest stored in the freezer.

Grocers will break open a package of wrapped meat and rewrap the portion needed. Similarly, eggs can be purchased by the half-dozen. Eggs do keep for long periods, though, if stored properly in the refrigerator.

Fresh fruits and vegetables can be purchased individually. A person can buy fresh fruit at various stages of ripeness: a ripe one to eat right away, a semiripe one to eat soon after, and a green one to ripen on the windowsill. If vegetables are packaged in large quantities, the grocer can break open the package so that a

Buy only what you will use.

Invite guests to share a meal.

© Elena Rooraid/PhotoEdit

smaller amount can be purchased. Small cans of fruits and vegetables, even though they are more expensive per unit, are a reasonable alternative, considering that it is expensive to buy a regular-size can and let the unused portion spoil.

• **Be Creative** • Creative chefs think of various ways to use foods when only large amounts are available. For example, a head of cauliflower can be divided into thirds. Then one-third is cooked and eaten hot. Another third is put into a vinegar and oil marinade for use in a salad. And the last third can be used in a casserole or stew.

Chefs also experiment with stir-fried foods. A large frying pan often works as well or better than a wok on most stovetops. A variety of vegetables and meats can be enjoyed this way; inexpensive vegetables such as cabbage, celery, and onion are delicious when crisp cooked in a little oil with herbs or lemon added. Interesting frozen vegetable mixtures are available in larger grocery stores. Cooked, leftover vegetables can be dropped in at the last minute. A bonus of a stir-fried meal is that there is only one pan to wash. Similarly, a microwave oven allows a chef to use fewer pots and pans. Meals and leftovers can also be frozen or refrigerated in microwavable containers to reheat as needed.

Many frozen dinners that are commercially available are now low in fat and nutritious. Adding a fresh salad, a whole-wheat roll, and a glass of milk can make a nutritionally balanced meal.

Also, single people shouldn't hesitate to invite someone to share meals with them whenever there is a lot of food. It's likely that the person will return the invitation, and both parties will get to enjoy companionship and a meal prepared by others.

IN SUMMARY Older people can benefit from both the nutrients provided and the social interaction available at congregate meals. Other government programs deliver meals to those who are homebound. With creativity and careful shopping, those living alone can prepare nutritious, inexpensive meals. Physical activity, mental challenges, stress management, and social activities can also help people grow old comfortably.

How are you doing?

- If there are older adults in your life, do they have the financial means, physical ability, and social support they need to eat adequately?
- Have they experienced an unintentional loss of weight recently?
- Are they active physically, socially, and mentally?

Nutrition on the Net

WEBSITES

Access these websites for further study of topics covered in this chapter.

- Find updates and quick links to these and other nutrition-related sites at our website: **www.wadsworth.com/nutrition**
- Search for "aging," "arthritis," and "Alzheimer's" on the U.S. Government health information site: **www.healthfinder.gov**
- Visit the National Aging Information Center of the Administration on Aging: **www.aoa.gov**
- Visit the American Geriatrics Society: **www.americangeriatrics.org**
- Visit the National Institute on Aging: **www.nih.gov.nia**

- Visit the American Association of Retired Persons: **www.aarp.org**
- Get nutrition tips for growing older in good health from the American Dietetic Association: **www.eatright.org**
- Learn more about cataracts and macular degeneration from the National Eye Institute and the American Society of Cataract and Refractive Surgery: **www.nei.nih.gov** and **www.ascrs.org**
- Learn more about arthritis from the Arthritis Foundation and the National Institute of Arthritis and Musculoskeletal and Skin Diseases Information Clearinghouse of the National Institutes of Health: **www.arthritis.org** and **www.nih.gov/niams**
- Learn more about Alzheimer's disease from the NIA Alzheimer's Disease Education and Referral Center and the Alzheimer's Association: **www.alzheimers.org/adear** and **www.alz.org**
- Learn more about malnutrition in older adults and the Nutrition Screening Initiative from the American Academy of Family Physicians: **www.aafp.org**
- Find out about federal government programs designed to help senior citizens maintain good health: **www.seniors.gov**

INTERNET ACTIVITIES

The number of elderly individuals in our society is growing larger each year. As people age, they encounter many factors that may impair their nutritional status. Do you know how to recognize the warning signs of nutritional risk?

- Go to: **www.agingwell.state.ny.us**
- Click on the "Eating Well" icon.
- Click on "The Warning Signs" and read that information.
- Click on "Determine your Nutritional Health" at the bottom of the screen.
- Take the quiz on behalf of an elderly friend or relative and click "Submit."

This survey is used to *determine* which elderly individuals face nutritional problems. According to this survey, how does your elderly friend or relative score?

Study Questions

These questions will help you review the chapter. You will find the answers in the discussions on the pages provided.

1. What roles does nutrition play in aging, and what roles can it play in retarding aging? (pp. 541–543)
2. What are some of the physiological changes that occur in the body's systems with aging? To what extent can aging be prevented? (pp. 544–546)
3. Why does the risk of dehydration increase as people age? (pp. 546–547)
4. Why do energy needs usually decline with advancing age? (p. 547)
5. Which vitamins and minerals need special consideration for the elderly? Explain why. Identify some factors that complicate the task of setting nutrient standards for older adults. (pp. 548–549)
6. Discuss the relationships between nutrition and cataracts and between nutrition and arthritis. (pp. 550–551)
7. What characteristics contribute to malnutrition in older people? (pp. 553–556)

These questions will help you prepare for an exam. Answers can be found on p. 559.

1. Life expectancy in the United States is:
   a. 48 to 60 years.
   b. 58 to 70 years.
   c. 68 to 80 years.
   d. 78 to 90 years.
2. The human life span is about:
   a. 85 years.
   b. 100 years.
   c. 115 years.
   d. 130 years.
3. A 72-year-old person whose physical health is similar to that of people 10 years younger has a(n):
   a. chronological age of 62.
   b. physiological age of 72.
   c. physiological age of 62.
   d. absolute age of minus 10.
4. Rats live longest when given diets that:
   a. eliminate all fat.
   b. provide lots of protein.
   c. allow them to eat freely.
   d. restrict their energy intakes.

5. Which characteristic is *not* commonly associated with atrophic gastritis?
   a. inflamed stomach
   b. vitamin B_{12} toxicity
   c. bacterial overgrowth
   d. lack of intrinsic factor
6. On average, adult energy needs:
   a. decline 5 percent per year.
   b. decline 5 percent per decade.
   c. remain stable throughout life.
   d. rise gradually throughout life.
7. Which nutrients seem to protect against cataract development?
   a. minerals
   b. lecithins
   c. antioxidants
   d. amino acids
8. The best dietary advice for a person with osteoarthritis might be to:
   a. avoid milk products.
   b. take fish oil supplements.
   c. take vitamin E supplements.
   d. lose weight, if overweight.
9. Congregate meal programs are preferable to Meals on Wheels because they provide:
   a. nutritious meals.
   b. referral services.
   c. social interactions.
   d. financial assistance.
10. The Elderly Nutrition Program is available to:
   a. all people 65 years and older.
   b. all people 60 years and older.
   c. homebound people only, 60 years and older.
   d. low-income people only, 60 years and older.

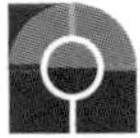

References

1 B. Guyer and coauthors, Annual summary of vital statistics—1998, *Pediatrics* 104 (1999): 1229–1246.

2 L. Guarente, G. Ruvkun, and R. Amasino, Aging, life span, and senescence, *Proceedings of the National Academy of Sciences* 95 (1998): 11034–11036; D. A. Banks and M. Fossel, Telomeres, cancer, and aging: Altering the human life span, *Journal of the American Medical Association* 278 (1997): 1345–1348.

3 A. J. Vita and coauthors, Aging, health risks, and cumulative disability, *New England Journal of Medicine* 338 (1998): 1035–1041.

4 American College of Sports Medicine, *ACSM Fitness Book* (Hong Kong: Paramount Printing Ltd., 1998), pp. 6–8.

5 American College of Sports Medicine, Position stand: Exercise and physical activity for older adults, *Medicine and Science in Sports and Exercise* 30 (1998): 992–1008.

6 M. Motoyama and coauthors, Blood pressure lowering effect of low intensity aerobic training in elderly hypertensive patients, *Medicine and Science in Sports and Exercise* 30 (1998): 818–823.

7 C. Dutta, Significance of sarcopenia in the elderly, *Journal of Nutrition*127 (1997): 992S–997S.

8 W. J. Evans and D. Cyr-Campbell, Nutrition, exercise, and healthy aging, *Journal of the American Dietetic Association* 97 (1997): 632–638.

9 A. C. King and coauthors, Moderate intense exercise and self-rated quality of sleep in older adults: A randomized controlled trial, *Journal of the American Medical Association* 277 (1997): 32–37.

10 R. Weindruch and R. S. Sohal, Caloric intake and aging, *New England Journal of Medicine* 337 (1997): 986–994.

11 C. K. Lee and coauthors, Gene expression profile of aging and its retardation by calorie restriction, *Science* 285 (1999): 1390–1393.

12 A. C. Gazdag and coauthors, Effect of long-term calorie restriction on GLUT4, phosphatidylinositol-3 kinase p85 subunit, and insulin receptor substrate-1 protein levels in rhesus monkey skeletal muscle, *Journals of Gerontology: Series A, Biological Sciences and Medical Sciences* 55 (2000): B44–B46; W. T. Cefalu and coauthors, Influence of caloric restriction on the development of atherosclerosis in nonhuman primates: Progress to date, *Toxicological Sciences* 52 (1999): 49–55; I. J. Edwards and coauthors, Caloric restriction in rhesus monkeys reduces low density lipoprotein interaction with arterial proteoglycans, *Journals of Gerontology: Series A, Biological Sciences and Medical Sciences* 53 (1998): B443–B448; R. B. Verdery and coauthors, Calorie restriction increases HDL2 levels in rhesus monkeys (Macaca mulatta), *American Journal of Physiology* 273 (1997): E714–E719.

13 J. P. DeLany and coauthors, Long-term calorie restriction reduces energy expenditure in aging monkeys, *Journals of Gerontology: Series A, Biological Sciences and Medical Sciences* 54 (1999): B5–B11; J. J. Ramsey and coauthors, Energy expenditure of adult male rhesus monkeys during the first 30 mo of dietary restriction, *American Journal of Physiology* 272 (1997): E901–E907.

14 D. Kritchevsky, Caloric restriction and experimental mammary carcinogenesis, *Breast Cancer Research and Treatment* 46 (1997): 161–167.

15 J. Wanagat, D. B. Allison, and R. Weindruch, Calorie intake and aging: Mechanisms in rodents and a study in nonhuman primates, *Toxicological Sciences* 52 (1999): 35–40.

16 P. R. Johnson and coauthors, Longevity in obese and lean male and female rats of the Zucker strain: Prevention of hyperphagia, *American Journal of Clinical Nutrition* 66 (1997): 890–903.

17 L. E. Rikans and K. R. Hronbrook, Lipid peroxidation, antioxidant protection and aging, *Biochimica et Biophysica Acta* 1362 (1997): 116–127.

18 G. Paolisso and coauthors, Oxidative stress and advancing age: Results in

healthy centenarians, *Journal of the American Geriatrics Society* 46 (1998): 833–838.

19 J. J. Kehayias and coauthors, Total body potassium and body fat: Relevance to aging, *American Journal of Clinical Nutrition* 66 (1997) 904–910.

20 J. E. Morley, Anorexia of aging: Physiologic and pathologic, *American Journal of Clinical Nutrition* 66 (1997): 760–763.

21 I. H. Rosenberg, Sarcopenia: Origins and clinical relevance, *Journal of Nutrition* 127 (1997): 990S–991S.

22 G. Ravaglia and coauthors, Effect of micronutrient status on natural killer cell immune function in healthy free-living subjects aged ≥90 y, *American Journal of Clinical Nutrition* 71 (2000): 590–598; K. Buzina-Suboticanec and coauthors, Ageing, nutritional status and immune response, *International Journal of Vitamin and Nutrition Research* 68 (1998): 133–141.

23 C. G. MacIntosh and coauthors, Effects of age on concentrations of plasma cholecystokinin, glucagon-like peptide 1, and peptide YY and their relation to appetite and pyloric motility, *American Journal of Clinical Nutrition* 69 (1999): 999–1006.

24 C. S. Ritchie and coauthors, Nutritional status of urban homebound older adults, *American Journal of Clinical Nutrition* 66 (1997): 815–818.

25 S. S. Schiffman, Taste and smell losses in normal aging and disease, *Journal of the American Medical Association* 278 (1997): 1357–1362.

26 J. Unützer and coauthors, Depressive symptoms and the cost of health services in HMO patients aged 65 years and older, *Journal of the American Medical Association* 277 (1997): 1618–1623; Depressing statistics, *Journal of the American Medical Association* 277 (1997): 1584.

27 B. W. J. H. Penninx, and coauthors, Depressive symptoms and physical decline in community-dwelling older persons, *Journal of the American Medical Association* 279 (1998): 1720–1726.

28 N. S. Wellman and coauthors, Elder insecurities: Poverty, hunger, and malnutrition, *Journal of the American Dietetic Association* 97 (1997): S120–S122.

29 Committee on Dietary Reference Intakes, *Dietary Reference Intakes for Calcium, Phosphorus, Magnesium, Vitamin D, and Fluoride* (Washington, D.C.: National Academy Press, 1997); R. M. Russell, New views on the RDAs for older adults, *Journal of the American Dietetic Association* 97 (1997): 515–518.

30 J. C. Chidester and A. A. Spangler, Fluid intake in the institutionalized elderly, *Journal of the American Dietetic Association* 97 (1997): 23–28.

31 D. H. Holben and coauthors, Fluid intake compared with established standards and symptoms of dehydration among elderly residents of a long-term-care facility, *Journal of the American Dietetic Association* 99 (1999): 1447–1450.

32 G. L. Jensen and J. Rogers, Obesity in older persons, *Journal of the American Dietetic Association* 98 (1998): 1308–1311.

33 M. Visser and coauthors, High body fatness, but not low fat-free mass, predicts disability in older men and women: The Cardiovascular Health Study, *American Journal of Clinical Nutrition* 68 (1998): 584–590.

34 Surveillance for selected public health indicators affecting older adults—United States, *Morbidity and Mortality Weekly Report* 48 (1999): 94–95.

35 S. B. Roberts, Energy regulation and aging: Recent findings and their implications, *Nutrition Reviews* 58 (2000): 91–97; M. J. Toth and E. T. Poehlman, Energetic adaptation to chronic disease in the elderly, *Nutrition Reviews* 58 (2000): 61–66.

36 F. Landi and coauthors, Body mass index and mortality among older people living in the community, *Journal of the American Geriatrics Society* 47 (1999): 1072–1076.

37 W. J. Carter, Macronutrient requirements for elderly persons, in *Geriatric Nutrition,* ed. R. Chernoff (Gaithersburg, Md.: Aspen Publishers, 1999), pp. 13–26; D. J. Millward and coauthors, Aging, protein requirements, and protein turnover, *American Journal of Clinical Nutrition* 66 (1997): 774–786.

38 Position of The American Dietetic Association: Health implications of dietary fiber, *Journal of the American Dietetic Association* 97 (1997): 1157–1159.

39 S. P. Stabler, J. Lindenbaum, and R. H. Allen, Vitamin B-12 deficiency in the elderly: Current dilemmas, *American Journal of Clinical Nutrition* 66 (1997): 741–749.

40 C. Ho, G. P. A. Kauwell, and L. B. Bailey, Practitioners' guide to meeting the vitamin B-12 Recommended Dietary Allowance for people aged 51 years and older, *Journal of the American Dietetic Association* 99 (1999): 725–727; H. W. Baik and R. M. Russell, Vitamin B_{12} deficiency in the elderly, *Annual Review of Nutrition* 19 (1999): 357–377.

41 P. F. Jacques and coauthors, Plasma 25-hydroxyvitamin D and its determinants in an elderly population sample, *American Journal of Clinical Nutrition* 66 (1997): 929–936.

42 Committee on Dietary Reference Intakes, 1997.

43 Committee on Dietary Reference Intakes, 1997; C. Marwick, NHANES III health data relevant for aging nation, *Journal of the American Medical Association* 277 (1997): 100–102.

44 G. J. Izaks, R. G. J. Westendorp, and D. L. Knook, The definition of anemia in older persons, *Journal of the American Medical Association* 281 (1999): 1714–1717.

45 L. Brown and coauthors, A prospective study of carotenoid intake and risk of cataract extraction in US men, *American Journal of Clinical Nutrition* 70 (1999): 517–524; L. Chason-Taber and coauthors, A prospective study of carotenoid and vitamin A intake and

ANSWERS

Multiple Choice

1. c	2. d	3. c	4. d	5. b
6. b	7. c	8. d	9. c	10. b

risk of cataract extraction in US women, *American Journal of Clinical Nutrition* 70 (1999): 509–516; M. C. Leske and coauthors, Antioxidant vitamins and nuclear opacities: The longitudinal study of cataract, *Ophthalmology* 105 (1998): 831–836; P. F. Jacques and coauthors, Long-term vitamin C supplement use and prevalence of early age-related lens opacities, *American Journal of Clinical Nutrition* 66 (1997): 911–916.

46 D. A. Schaumberg and coauthors, Relations of body fat distribution and height with cataract in men, *American Journal of Clinical Nutrition* 72 (2000): 1495–1502.

47 E. Cho and coauthors, Propspective study of dietary fat and the risk of age-related macular degeneration, *American Journal of Clinical Nutrition* 73 (2000): 209–218.

48 B. R. Hammond and coauthors, Dietary modification of human macular pigment density, *Investigative Opthalmology and Visual Science* 38 (1997): 1795–1801; T. O. Obisesan and coauthors, Moderate wine consumption is associated with decreased odds of developing age-related macular degeneration in NHANES-1, *Journal of the American Geriatrics Society* 46 (1998): 1–7.

49 R. C. Lawrence and coauthors, Estimates of the prevalence of arthritis and selected musculoskeletal disorders in the United States, *Arthritis & Rheumatism* 41 (1998): 778–799.

50 W. H. Ettinger and coauthors, A randomized trial comparing aerobic exercise and resistance exercise with a health education program in older adults with knee osteoarthritis: The Fitness Arthritis and Seniors Trial (FAST), *Journal of the American Medical Association* 277 (1997): 25–31.

51 J. M. Kremer, n-3 Fatty acid supplements in rheumatoid arthritis, *American Journal of Clinical Nutrition* 71 (2000): 349S–351S.

52 T. E. McAlindon and coauthors, Glucosamine and chondroitin for treatment of osteoarthritis: A systematic quality assessment and meta-analysis, *Journal of the American Medical Association* 283 (2000): 1469–1475.

53 C. M. Cullum and R. N. Rosenberg, Memory loss—When is it Alzheimer disease? *Journal of the American Medical Association* 279 (1998): 1689–1690.

54 E. Rogaeva and coauthors, Evidence for an Alzheimer disease susceptibility locus on chromosome 12 and further locus heterogeneity, *Journal of the American Medical Association* 280 (1998): 614–618; W. S. Wu and coauthors, Genetic studies on chromosome 12 in late-onset Alzheimer disease, *Journal of the American Medical Association* 280 (1998): 619–622; National Institute of Aging, *Progress Report on Alzheimer's Disease,* 1994, NIH publication no. 94-3885 (Washington, D.C.: Government Printing Office, 1994).

55 Y. Christen, Oxidative stress and Alzheimer disease, *American Journal of Clinical Nutrition* 71 (2000): 621S–629S.

56 M. A. Lovell and coauthors, Copper, iron and zinc in Alzheimer's disease senile plaques, *Journal of Neurological Sciences* 158 (1998): 47–52; C. R. Cornett, W. R. Markesbery, and W. D. Ehmann, Imbalances of trace elements related to oxidative damage in Alzheimer's disease brain, *Neurotoxicology* 19 (1998): 339–345.

57 M. Grundman, Vitamin E and Alzheimer disease: The basis for additional clinical trials, *American Journal of Clinical Nutrition* 71 (2000): 630S–636S.

58 D. Schenk and coauthors, Immunization with amyloid-beta attenuates Alzheimer-disease-like pathology in the PDAPP mouse, *Nature* 400 (1999): 173–177; R. Vassar and coauthors, Beta-secretase cleavage of Alzheimer's amyloid precursor protein by the transmembrane aspartic protease BACE, *Science* 286 (1999): 735; I. Hussain and coauthors, Identification of novel aspartic protease (Asp 2) as beta-secretase, *Molecular and Cellular Neuroscience* 14 (1999): 419–427.

59 K. L. Davis and coauthors, Cholinergic markers in elderly patients with early signs of Alzheimer disease, *Journal of the American Medical Association* 281 (1999): 1401–1406.

60 E. R. Peskind, Pharmacologic approaches to cognitive deficits in Alzheimer's disease, *Journal of Clinical Psychiatry* 59 (1998): 22–27.

61 D. Snowdon and coauthors, Serum folate and the severity of atrophy of the neocortex in Alzheimer disease: Findings from the Nun Study, *American Journal of Clinical Nutrition* 71 (2000): 993–998; D. G. Weir and A. M. Molloy, Microvascular disease and dementia in the elderly: Are they related to hyperhomocysteinemia? *American Journal of Clinical Nutrition* 71 (2000): 859–860; J. W. Miller, Homocysteine and Alzheimer's disease, *Nutrition Reviews* 57 (1999): 126–129.

62 R. Mayeux and M. Sand, Treatment of Alzheimer's disease, *New England Journal of Medicine* 341 (1999): 1670–1679; G. W. Small, Treatment of Alzheimer's disease: Current approaches and promising developments, *American Journal of Medicine* 104 (1998): 32S–38S.

63 S. Gillette-Guyonnet and coauthors, Weight loss in Alzheimer disease, *American Journal of Clinical Nutrition* 71 (2000): 637S–642S; E. T. Poehlman and R. V. Dvorak, Energy expenditure, energy intake, and weight loss in Alzheimer disease, *American Journal of Clinical Nutrition* 71 (2000): 650S–655S; S. Rivière and coauthors, Nutrition and Alzheimer's disease, *Nutrition Reviews* 57 (1999): 363–367.

64 B. Finley, Nutritional needs of the person with Alzheimer's disease: Practical approaches to quality care, *Journal of the American Dietetic Association* 97 (1997): S177–S180.

65 J. D. Fernstrom, Can nutrient supplements modify brain function? *American Journal of Clinical Nutrition* 71 (2000): 1669S–1673S.

66 J. Selhub and coauthors, B vitamins, homocysteine, and neurocognitive function in the elderly, *American Journal of Clinical Nutrition* 71 (2000): 614S–620S.

67 A. La Rue and coauthors, Nutritional status and cognitive functioning in a normally aging sample: A 6-y reassessment, *American Journal of Clinical Nutrition* 65 (1997): 20–29.

68 Y. Liao and coauthors, Quality of the last year of life of older adults: 1986 vs 1993, *Journal of the American Medical Association* 283 (2000): 512–518; K. G. Manton, L. Corder, and E. Stallard, Chronic disability trends in elderly United States population: 1982–1994, *Proceedings of the National Academy of Sciences* 94 (1997): 2593–2598.

69 Position of the American Dietetic Association: Nutrition, aging, and the continuum of care, *Journal of the American Dietetic Association* 100 (2000): 580–595.

70 Federal program nourishes poor elderly, *Journal of the American Medical Association* 278 (1997): 1301.

HUNGER AND COMMUNITY NUTRITION

ONE PERSON in every five worldwide experiences persistent hunger—not the healthy appetite triggered in anticipation of a hearty meal, but the painful sensation caused by a lack of food. Tens of thousands die of starvation each day—one every two seconds.

In the United States, an estimated 36 million people including one out of every five children, live in poverty and cannot afford to buy enough food to maintain good health. Given the enormous wealth and economic growth in this country, do these numbers surprise you? The limited or uncertain availability of nutritionally adequate and safe foods is known as food insecurity and is a major problem in our nation today (see the glossary on p. 562).[1] The "How to" on p. 562 describes how national surveys identify **food insecurity** in the United States, and Figure H16-1 presents the most recent findings. Questions like these provide crude, but necessary, data to estimate the degree of hunger in this country.

HEALTHY PEOPLE 2010

Increase the prevalence of food security among U.S. households.

DEFINING HUNGER IN THE UNITED STATES

At its most extreme, people experience hunger because they have absolutely no food. More often, they have too little food and try to stretch their limited resources by eating small meals or skipping meals—often for days at a time. Sometimes hungry people obtain enough food to satisfy their hunger, but only through socially unacceptable ways—begging from strangers, stealing from markets, or digging through garbage cans, for example.

Hunger has many causes, but in developed countries, the primary cause is **food poverty.** People are hungry not because there is no food nearby to purchase, but because they lack money. An estimated one out of eight people in the United States lives in poverty. Even those above the poverty line may not have food security. Physical and mental illnesses and disabilities, sudden job losses, and high living expenses threaten their financial stability. Further contributing to food poverty are other problems such as abuse of alcohol and other drugs; lack of awareness of available food assistance programs; and the reluctance of people, particularly the elderly, to accept what they perceive as "welfare" or "charity." Lack of resources remains the major cause of food poverty, and solving this problem would do a lot to relieve hunger.

In the United States, food poverty and hunger reach across various segments of society, touching Hispanics and African Americans, those living in the inner cities, and those living in households with children more than others. People living in poverty are simply unable to buy sufficient amounts of nourishing foods, even if they are wise shoppers. For many of the children in these families, school lunch is their only meal of the day. Otherwise they go hungry, waiting for an adult to find money for food.

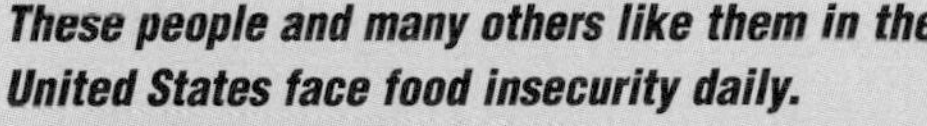

These people and many others like them in the United States face food insecurity daily.

COMMUNITY NUTRITION PROGRAMS TO RELIEVE HUNGER IN THE UNITED STATES

The American Dietetic Association (ADA) calls for aggressive action to bring an end to domestic hunger and to achieve food and nutrition security.[2] Many federal and local programs aim to prevent or relieve malnutrition and hunger in the United States.

• Federal Food Assistance Programs • An extensive network of federal assistance programs provides life-giving food daily to millions of U.S. citizens.[3] One out of every six

GLOSSARY

food insecurity: limited or uncertain access to foods of sufficient quality or quantity to sustain a healthy and active life.

food poverty: hunger occurring when enough food exists in an area but some of the people cannot obtain it because they lack money, are being deprived for political reasons, live in a country at war, or suffer from other problems such as lack of transportation.

food recovery: collecting wholesome food for distribution to low-income people who are hungry. Four common methods of food recovery are:

- Field gleaning: collecting crops from fields that either have already been harvested or are not profitable to harvest.
- Perishable food rescue or salvage: collecting perishable produce from wholesalers and markets.
- Prepared food rescue: collecting prepared foods from commercial kitchens.
- Nonperishable food collection: collecting processed foods from wholesalers and markets.

food bank: a central source for the donation and distribution of food to local charities feeding the hungry.

FIGURE H16-1 **Prevalence of Food Security and Hunger in the U.S. Households, 1999**

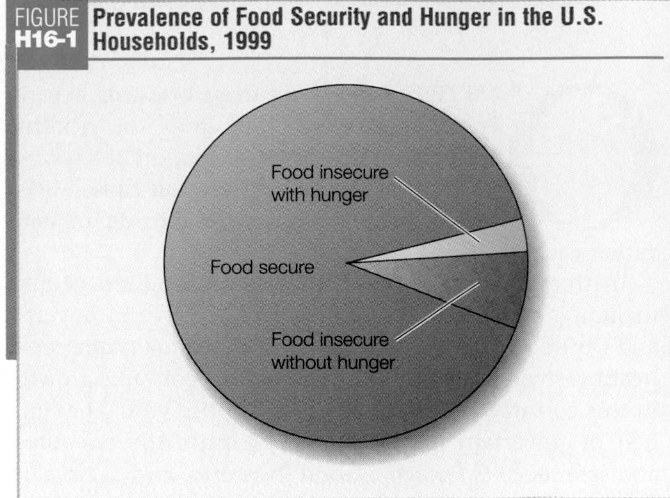

Americans receives food assistance of some kind, at a total cost of almost $40 billion per year.[4] Even so, the programs are not fully successful in preventing hunger, even among those who receive their benefits.[5] Programs described in the life cycle chapters preceding this highlight include the WIC program for low-income pregnant women, breastfeeding mothers, and their young children (Chapter 14); the school lunch, breakfast, and child care food programs for children (Chapter 15); and the food assistance programs for older adults such as congregate meals and Meals on Wheels (Chapter 16).

HOW TO *Identify Food Insecurity in a U.S. Household*

Questions like these are asked on surveys to determine the extent of food insecurity in a household. The more questions answered "Yes," the more intense the hunger the household is experiencing.

- Do you often go hungry?
- Do you often have too little food to eat because you have no money, transportation, or kitchen appliances (stove, refrigerator)?
- Do you ever rely on nutritionally inferior foods to feed yourself or your children because you lack any of these resources?
- Do you ever eat less than you feel you should because you lack any of these resources?
- Do you ever skip meals or cut the size of meals because you lack any of these resources?
- Do you ever rely on neighbors, friends, relatives, or schools to feed any of your children because there is not enough food in the house?
- Do your children ever say they are hungry because there is not enough food in the house?
- Do you or any of your children ever go to bed hungry because there is not enough food in the house?

The centerpiece of food programs for low-income people in the United States is the Food Stamp Program, administered by the U.S. Department of Agriculture (USDA). The USDA issues food stamp coupons or debit cards through state agencies to households—people who buy and prepare food together. The amount a household receives depends on its size and income. Recipients may use the coupons or cards like cash to purchase food and food-bearing plants and seeds, but not to buy tobacco, cleaning items, alcohol, or other nonfood items. The "How to" on p. 563 features shopping tips for those on a limited budget.

The Food Stamp Program is the largest of the federal food assistance programs, both in amount of money spent and in number of people participating. More than 17 million people receive food stamps at a cost of over $21 billion per year; over half of the recipients are children.[6]

The Food Stamp Program improves nutrient intakes significantly, but hunger continues to plague the United States.[7] Of the estimated 2 million homeless people in the United States who are eligible for food assistance, only 15 percent of single adults and 50 percent of families receive food stamps. Health care professionals working with clients who may be having financial problems can encourage such clients to talk with a social worker who can assess their eligibility for food assistance programs.

• National Food Recovery Programs • Efforts to resolve the problem of hunger in the United States do not depend solely on federal assis-

Plan Healthy, Thrifty Meals

Chapter 2 introduced the Food Guide Pyramid and principles for planning a healthy diet. Meeting that goal on a limited budget adds to the challenge. To save money and spend wisely, plan and shop for healthy meals with the following tips in mind:

Planning

- Make a grocery list before going to the store to avoid expensive "impulse" items. Do not shop when hungry.
- Use leftovers.
- Center meals on rice, noodles, and other grains.
- Use small quantities of meat, poultry, fish, or eggs.
- Use legumes instead of meat, poultry, fish, or eggs several times a week.
- Use cooked cereals such as oatmeal instead of ready-to-eat breakfast cereals.
- Cook large quantities when time and money allow.
- Check for sales and clip coupons for products you need; plan meals to take advantage of sale items.

Shopping

- Buy day-old bread and other products from the bakery outlet.
- Select whole foods instead of convenience foods (potatoes instead of instant mashed potatoes, for example).
- Try store brands.
- Buy fresh produce that is in season; buy canned or frozen items at other times.
- Buy only the amount of fresh foods that you will eat before it spoils. Buy large bags of frozen items or dry goods; when cooking, take out the amount needed and store the remainder.
- Buy nonfat dry milk; mix and refrigerate quantities needed for a day or two. Buy fresh milk by the gallon or half-gallon.
- Buy less expensive cuts of meat. Chuck and bottom round roast are usually inexpensive; cover during cooking and cook long enough to make meat tender. Buy whole chickens instead of pieces.
- Compare the unit price (cost per ounce, for example) of similar foods so that you can select the least expensive brand or size.
- Buy nonfood items such as toilet paper and laundry detergent at discount stores instead of grocery stores.

For daily menus and recipes for healthy, thrifty meals, visit the USDA Center for Nutrition Policy and Promotion: www.usda.gov/cnpp

tance programs. National **food recovery** programs have made a dramatic difference; the largest program, Second Harvest, coordinates the efforts of almost 200 food banks in providing more than 1 billion pounds of food to 45,000 local agencies that feed 26 million people a year. Table H16-1 lists addresses, phone numbers, and websites for Second Harvest and other national and international hunger relief organizations.

Each year, an estimated one-fifth of our food supply is wasted in fields, commercial kitchens, grocery stores, and restaurants—that's enough food to feed 49 million people. Food recovery programs collect and distribute good food that would otherwise go to waste.[8] Each year, about 27 percent of America's food supply ends up as waste in landfills. Volunteers might pick corn left in an already harvested field, a grocer might deliver ripe bananas to a local **food bank**, and a caterer might take leftover chicken salad to a community shelter, for example. All of these efforts help to feed the hungry in the United States.

• Local Efforts • Food recovery programs depend on volunteers. Concerned citizens work through local agencies and churches to feed the hungry. Community-based food pantries provide groceries, and soup kitchens serve

TABLE H16-1 Hunger-Relief Organizations

Action without Borders
350 Fifth Ave., Suite 6614
New York, NY 10118
(212) 843-3973
www.idealist.org

Bread for the World Institute
1100 Wayne Ave., Suite 1000
Silver Spring, MD 20910
(301) 608-2400
www.bread.org

Congressional Hunger Center
229½ Pennsylvania Ave.
Washington, DC 20003
(202) 547-7022
www.hungercenter.org

Foodchain
912 Baltimore, #300
Kansas City, MO 64105
(800) 845-3008
www.foodchain.org

OXFAM America
26 West St.
Boston, MA 02111-1206
(617) 482-1211
www.oxfam.org

Pan American Health Organization
525 23 St. NW
Washington, DC 20037
(202) 974-3000
www.paho.org

Second Harvest
116 S. Michigan Ave., #4
Chicago, IL 60603
(800) 771-2303
www.secondharvest.org

Society of St. Andrew
3383 Sweet Hollow Rd.
Big Island, VA 24526
(800) 333-4597
www.endhunger.org

United Nations Food and Agriculture Organization (FAO)
1001 22nd St. NW, Suite 300
Washington, DC 20437
(202) 653-2400
www.fao.org

United Nations International Children's Emergency Fund (UNICEF)
3 United Nations Plaza
New York, NY 10017-4414
(212) 326-7035
www.unicef.org

World Food Program
Via Cesare Giulio Viola, 68
Parco dé Medici
Rome, Italy 00148
www.wfp.org

World Health Organization (WHO)
525 23rd St. NW
Washington, DC 20037
(202) 861-3200
www.who.org

World Hunger Program
Brown University
Box 1831
Providence, RI 02912
(401) 863-2700
www.brown.edu/Departments/World_Hunger_Program/hungerweb/WHP/overview.html

World Hunger Year
505 Eighth Ave., 21st Floor
New York, NY 10018-6582
(800) GleanIt
www.worldhungeryear.org

Community-based efforts to feed citizens include food pantries that provide groceries.

© AP/Wide World Photos

Feeding the hungry in the United States.

© Thomas Muscianico/Contact Press/PictureQuest

prepared meals. Meals often deliver adequate nourishment, but most homeless people receive fewer than one and a half meals a day, so many are still inadequately nourished. Health care professionals can serve as valuable members of community groups seeking to provide food assistance. Table H16-2 shows how communities can address their hunger problems.

In closing, although the problems of food insecurity and hunger among those living in poverty are apparent, the solutions may be less clear. Government assistance programs help to relieve some poverty and hunger, but equally important are food recovery programs and other local efforts. Individuals can help in the fight against hunger and poverty by joining and working for hunger-relief organizations (review Table H16-1). The actions of many individuals add up to a powerful force. In the wisdom of Margaret Mead, "Never doubt that a small group of thoughtful, committed people can change the world. Indeed, it is the only thing that ever has."

TABLE H16-2 Fourteen Ways Communities Can Address Their Hunger Problems

1. Establish a community-based emergency food delivery network.
2. Assess community hunger problems and evaluate community services. Create strategies for responding to unmet needs.
3. Establish a group of individuals, including low-income participants, to develop and implement policies and programs to combat hunger and the threat of hunger; monitor responsiveness of existing services; and address underlying causes of hunger.
4. Participate in federally assisted nutrition programs that are easily accessible to targeted populations.
5. Integrate public and private resources, including local businesses, to relieve hunger.
6. Establish an education program that addresses the food needs of the community and the need for increased local citizen participation in activities to alleviate hunger.
7. Provide information and referral services for accessing both public and private programs and services.
8. Support programs to provide transportation and assistance in food shopping, where needed.
9. Identify high-risk populations and target services to meet their needs.
10. Provide adequate transportation and distribution of food from all resources.
11. Coordinate food services with parks and recreation programs and other community-based outlets to which residents of the area have easy access.
12. Improve public transportation to human service agencies and food resources.
13. Establish nutrition education programs for low-income citizens to enhance their food purchasing and preparation skills and make them aware of the connections between diet and health.
14. Establish a program for collecting and distributing nutritious foods—either agricultural commodities in farmers' fields or prepared foods that would have been wasted.

References

1 Position of The American Dietetic Association: Domestic food and nutrition security, *Journal of the American Dietetic Association* 98 (1998): 337–342.

2 Position of The American Dietetic Association: World hunger, *Journal of the American Dietetic Association* 95 (1995): 1160–1162.

3 C. S. Kramer-LeBlanc and K. McMurry, Discussion paper on domestic food security, *Family Economics and Nutrition Review* 11 (1998): 49–78.

4 P. P. Basiotis, C. S. Kramer-LeBlanc, and E. T. Kennedy, Maintaining nutrition security and diet quality: The role of the Food Stamp Program and WIC, *Family Economics and Nutrition Review* 11 (1998): 4–16.

5 USDA Center for Nutrition Policy and Promotion, Could there be hunger in America? *Nutrition Insight,* September 1998.

6 USDA Food and Nutrition Service website, www.fns.usda.gov/fns, visited March 19, 2001.

7 D. Rose, J.-P. Habicht, and B. Devaney, Household participation in the food stamp and WIC programs increases the nutrient intakes of preschool children, *Journal of Nutrition* 128 (1998): 548–555.

8 U.S. Department of Agriculture, *A Citizen's Guide to Food Recovery,* December 1996.

CHAPTER 17 Illness and Nutrition Status

The earlier chapters of this book have shown how nutrition supports a physically fit body and an alert mind throughout life. Chapter 1 described the basics of nutrition assessments and how they help to create a picture of a person's nutrition status. Turning now to clinical nutrition, the remaining chapters describe how changes in health affect nutrition status and nutrient needs and how **medical nutrition therapy** addresses those changes.

Even a person in good nutrition status can develop nutrient imbalances as a consequence of illness. Health care professionals who understand the impact that illnesses have on nutrition status and make a conscientious effort to think "nutrition" are best prepared to help clients recover from disease and maintain an optimal quality of life. They are alert to signs of poor nutrition status and quickly implement corrective measures whenever such signs become apparent. This chapter discusses the effects of illness on nutrition status, defines the nutrition care process, and describes the signs that point to nutrition problems.

NUTRITION IN HEALTH CARE

Poor nutrition can lead to health problems and make existing problems worse. Excessive energy intakes, for example, contribute to the development of some chronic diseases including heart disease, hypertension, and diabetes. Once these disorders develop, excessive energy intakes aggravate the problems. Other illnesses, however, lead to inadequate energy and protein intakes and multiple nutrient deficiencies. Without adequate energy, protein, vitamins, and minerals, the body has difficulty maintaining immune defenses, mending broken bones, healing wounds, utilizing medications, and supporting organ function. The compromised person may then fall victim to additional complications, and a downward spiral is set in motion (see Figure 17-1). The following sections describe the ways that health problems can directly and indirectly affect nutrition status. Highlight 17 discusses the interrelationships between mental health and nutrition status.

FIGURE 17-1 Illness, Malnutrition and Immunity

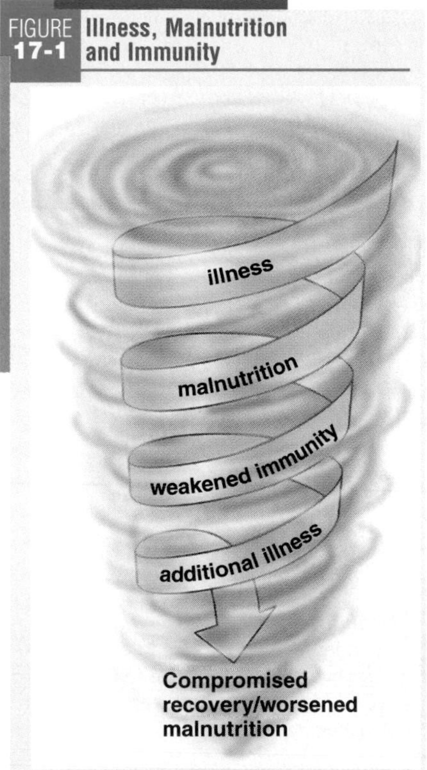

Regardless of where a person enters the spiral, the effects of illness, malnutrition, and impaired immunity can interact to compromise recovery and worsen malnutrition.

Health Problems

As the remaining chapters of this book show, health problems, symptoms, and treatments can lead to malnutrition by impairing a person's ability to eat, interfering with digestion and absorption, and altering metabolism and excretion. Regardless of which function is initially affected, once a problem intensifies, all functions are affected. When problems with food intake progress unchecked, for example, the body's ability to digest and absorb nutrients declines, and the body alters its metabolism in an effort to maintain homeostasis. Figure 17-2 on p. 568 provides examples of the effects of illness on nutrition status.

Medications and other treatments can also affect nutrition status. The potential effects of medications are extensive and are described in the next chapter. Preparation for diagnostic tests and surgeries and the treatment of some disorders may require that a person not eat. For the person who needs many such tests and procedures or many days of bowel rest, food intake may be very poor at a time when nutrient needs may be especially high due to illness. Consequences of some surgeries, especially those of the GI tract, often include limited food intake or altered digestion and absorption of nutrients. The location of a wound (burns to the hands, for example) or surgical incision may make eating difficult or painful. When treatment of a medical condition includes a highly restrictive diet, the client's motivation to eat may be impaired by limited food choices.

• Immobility and Pressure Sores • Bed rest (a treatment for some conditions) and immobility (a consequence of some medical conditions) further

medical nutrition therapy: the provision of a client's nutrient, dietary, and nutrition education needs based on a complete nutrition assessment.

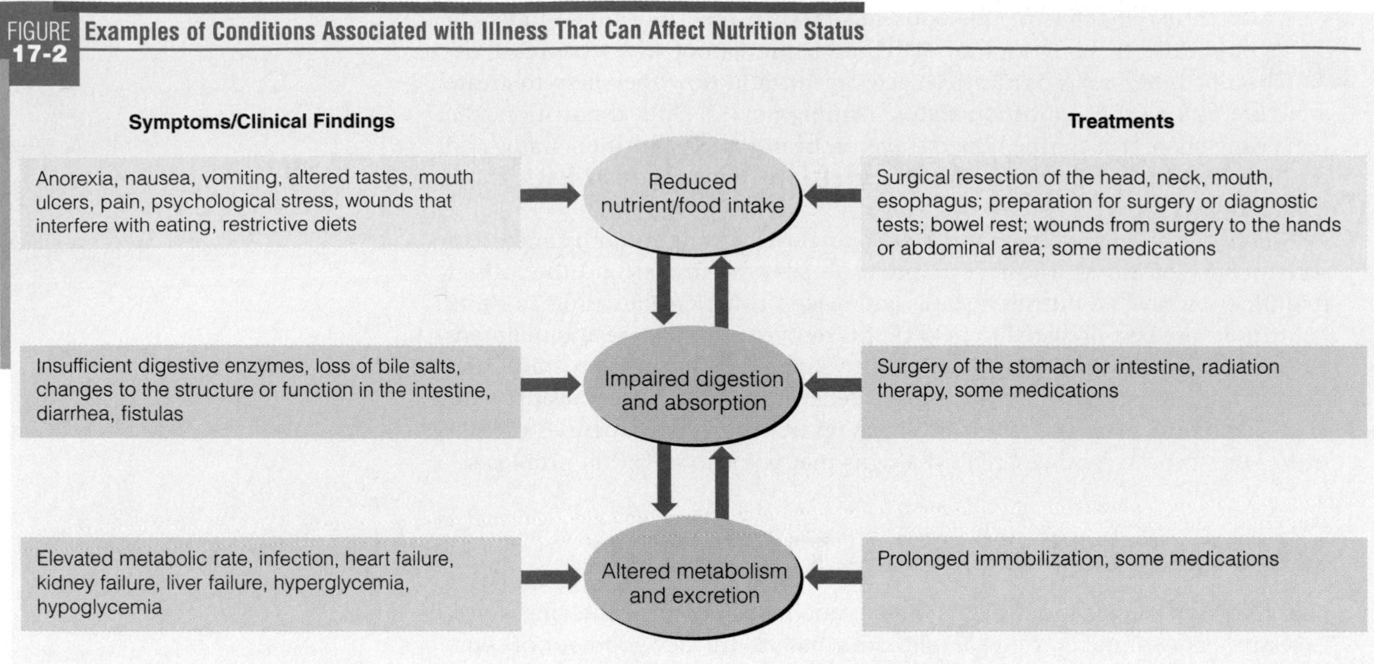

FIGURE 17-2 Examples of Conditions Associated with Illness That Can Affect Nutrition Status

compromise nutrition status. Without physical activity, the muscles and bones begin to lose nitrogen and calcium, respectively, which can further contribute to malnutrition.

Immobility, poor food intake, and protein-energy malnutrition (PEM) are all associated with the development of **pressure sores.** Pressure sores can form wherever there is constant pressure on the skin. The elderly, people who are unable to respond to pain or change body positions, and people who are malnourished are most likely to develop pressure sores. Pressure sores can be extremely painful and can easily become infected, raising the body's need for nutrients.

• **Indirect Effects** • In addition to the direct effects of medical conditions on nutrition status, the costs of health care can seriously affect economic status and lifestyle choices, limiting a person's ability to obtain adequate amounts of foods and the space and equipment necessary to store and prepare them. Health problems, especially chronic diseases or terminal illnesses, can have significant effects on mental health, which can also affect nutrition status (see Highlight 17). With chronic disorders that demand a substantial commitment of time and financial resources for management, nutrition problems are quite likely.

pressure sores: the breakdown of skin and underlying tissues due to constant pressure and lack of oxygen to the affected area; also called **decubitus** (dee-CUE-bih-tus) **ulcers** or bedsores.

nutrition care process: an organized approach to nutrition intervention that consists of assessing, planning, implementing, and evaluating. The nutrition care process parallels the nursing care process, but focuses on nutrition concerns.

individualized: based on consideration of a person as a unique being rather than one of many.

holistic (hoe-LIS-tik): based on consideration of a person as a whole, including physical, emotional, intellectual, social, and spiritual needs.

The Nutrition Care Process

Considering the many ways illness can affect nutrition status, how do health care professionals identify potential or actual nutrition-related problems and take measures to prevent or remedy them? Clients come in all shapes and sizes, and in addition to their physical needs, they have unique emotional, intellectual, social, and spiritual needs that both contribute to their health and nutrition status and determine their abilities and motivation to follow recommended therapies. Dietitians use a systematic and logical approach to nutrition care—the **nutrition care process**—to focus in on each person's unique needs and address them in an **individualized** and **holistic** manner. Other health care professionals who understand the basics of nutrition care are best prepared to actively contribute to the process.

The client is central to the nutrition care process. When clients cannot be active participants in the process, as occurs with infants and young children, people who are very ill or unconscious, people with mental disabilities, or people who are uncooperative, health care professionals enlist the involvement of family members or other support people.

• Steps of the Nutrition Care Process • To use the nutrition care process, dietitians follow five steps:

1. Assess nutrition status.
2. Analyze assessment data.
3. Develop a plan of action **(nutrition care plan).**
4. Implement the nutrition care plan.
5. Evaluate the effectiveness of the nutrition care plan and make necessary changes.

Although the process is easiest to visualize as five distinct steps (see Figure 17-3), in reality the process is continuous and the steps overlap.

FIGURE 17-3 **The Nutrition Care Process**

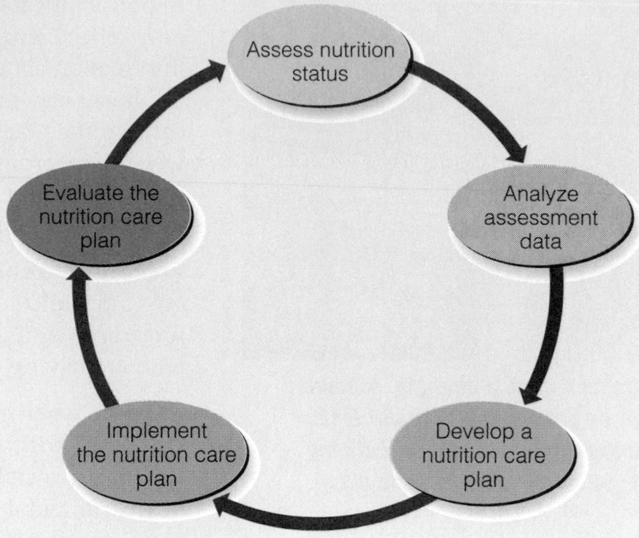

• Applications • The nutrition care process can be applied to clients in any setting—healthy clients visiting wellness clinics, pregnant women receiving prenatal care, elderly clients in assisted living, and people being treated for illness in private physicians' offices or in the hospital, to name a few. Most often, however, clients referred for nutrition care have health problems.

Roles of Health Care Professionals

For busy health care professionals◆ with many responsibilities, it may be easy to put clients' nutrition needs on the back burner. After all, the effects of nutrition therapy are not always as immediate or apparent as the effects of other treatments. Yet poor nutrition can significantly impair clients' health and affect their responses to other treatments, delay recovery, and complicate the tasks of all health care professionals.

◆ Highlight 23 describes the members and functions of nutrition support teams.

• Physicians • Physicians hold the ultimate responsibility for ensuring that all the client's medical needs, including nutrition, are met. The physician prescribes the client's diet and writes the **diet order** in the medical record. The physician may also write orders that relate to nutrition care, including orders for comprehensive nutrition assessments, diet counseling, and evaluation of the client's food intake. Physicians, in turn, rely on registered dietitians, nurses, and other health care professionals to alert them to nutrition problems, suggest strategies for handling nutrition problems, and provide requested nutrition services.

• Clinical Dietitians • Clinical dietitians have the primary responsibility for ensuring that clients receive optimal nutrition care. Clinical dietitians conduct complete nutrition assessments, analyze nutrient needs, and develop, implement, and evaluate nutrition care plans. Not all facilities employ dietitians, however, and even in those that do, the dietitian cannot see every client. Most often, the dietitian will routinely monitor clients with written orders for special diets or other nutrition services. Sometimes nutrition problems are overlooked when a client seeks health care, or nutrition status gradually deteriorates over the course of care; in these circumstances, the physician's written orders may not reflect the need for a dietitian's services. Consequently, nutrition problems may not be addressed in a timely manner—or at all. Thus dietitians depend on other health care professionals to alert them to clients who are having nutrition-related problems. Dietitians also rely on other health care professionals to provide information about clients' lifestyles and beliefs that affect their abilities to follow nutrition care plans.

nutrition care plan: a plan that translates nutrition assessment data into a strategy for meeting a client's nutrient and nutrition education needs.

diet order: a statement of the client's diet prescription that the physician writes in the medical record.

© Michelle DelGuercio/Photo Researchers, Inc.

Optimal muscle strength, supported by an optimal intake of nutrients, makes the job easier for both the physical therapist and the client.

• **Nurses** • Nurses, with their frequent and intimate contact with clients, are in a unique position to identify clients who need nutrition services and to provide information that can ensure successful nutrition care. Nurses and dietitians share responsibility for identifying diet-drug interactions, which can alter a medication's effectiveness or affect a client's nutrition status. Chapter 18 provides more information about diet-medication interactions.

Nurses also provide direct nutrition services, such as encouraging clients to eat, finding practical solutions to food-related problems, measuring height and weight, recording information about a client's food intake, and answering questions about special diets. In facilities that do not employ dietitians, the nurse often assumes the primary responsibility for nutrition care.

• **Dietetic Technicians** • Dietetic technicians directly assist dietitians. Dietetic technicians may take diet histories, collect information for nutrition screenings and assessments, calculate special diets, work directly with clients who are having problems with foods, and help with nutrition education.

• **Other Health Care Professionals** • Other health care professionals may also assist with nutrition care. Pharmacists, physical therapists, occupational therapists, social workers, nursing assistants, and home health care aides can be instrumental in alerting dietitians or nurses to nutrition problems and sharing relevant information about clients' health status or personal histories or concerns.

IN SUMMARY

Illnesses and their treatments have both direct and indirect effects on nutrition status. Either or both can interfere with the intake, digestion, absorption, metabolism, and excretion of nutrients or any combination of these. The nutrition care process is a logical system for uncovering and addressing nutrition-related problems. The first step of the nutrition care process, assessment, is described next.

NUTRITION ASSESSMENT

Nutrition assessment provides the information needed to evaluate a client's current nutrition status and nutrient needs and determine the most realistic and effective way to meet those needs. From the information, the assessor can develop a plan of action to prevent or correct any imbalances. Follow-up assessments help determine if the care plan is working.

• **Complete Nutrition Assessments** • Registered dietitians most often conduct complete nutrition assessments, sometimes with the assistance of dietetic technicians. In some cases, nurses and physicians skilled in clinical nutrition may also perform complete assessments, which draw on many sources of information including:

- Health, drug, personal, and diet histories.
- Anthropometric measurements.
- Laboratory tests.
- Physical examinations.

A meaningful assessment depends on both accurate information and the careful interpretation of each finding in relation to the others. Time constraints make it difficult to conduct complete assessments for all clients, and many facilities use the screening techniques described in the next section to identify clients who are most likely to benefit from complete assessments.

nutrition screening: a tool for quickly identifying clients at risk for malnutrition so that they can receive complete nutrition assessments.

• **Nutrition Screening** • Although health care professionals can help identify clients at risk for malnutrition in any health care setting, formal procedures for **nutrition screening** are most likely to be used routinely in hospitals and extended care facilities. Nurses, nursing assistants, and dietetic technicians most often conduct nutrition screenings. The Joint Commission on Accreditation of Healthcare Organizations (JCAHO) recommends that nutrition screenings be completed on each client shortly after admission, ideally within 24 hours, and at regular intervals thereafter. A person may be admitted to a health care facility in good nutrition status, but nutrient stores may decline as a consequence of the client's medical condition and treatments (**iatrogenic malnutrition**).

The exact criteria that determine nutrition risk vary from facility to facility and are based, in part, on the client population. The screening criteria in a facility that provides rehabilitative services, for example, differ from those in a hospital that serves a geriatric population. In general, people most likely to develop malnutrition include:

- Those with compromised nutrition status prior to developing a health problem.
- Those with one or more health problems associated with PEM (described later).
- Those with no appetites or poor appetites for more than a few days.
- Those with severe, persistent diarrhea or vomiting.
- Those who are significantly underweight or who have lost significant amounts of weight (described later in this chapter).

The remainder of this chapter describes the most common components of nutrition assessments. Many other techniques and tests are available for complete nutrition assessments; their use varies from facility to facility.

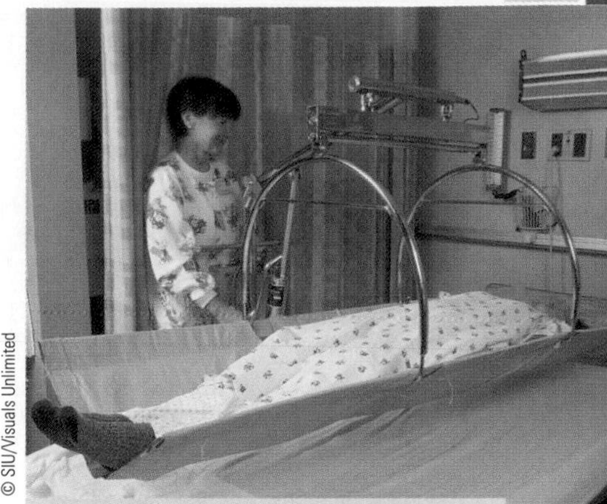

Measuring a client's weight is one component of nutrition screening.

IN SUMMARY

Dietitians conduct complete nutrition assessments to obtain the information needed to evaluate a client's nutrition status and develop a realistic and effective nutrition care plan. Complete assessments require a great deal of time and are often reserved for clients at risk for malnutrition. Nutrition screenings conducted when clients are admitted to health care facilities and at regular intervals thereafter, help identify clients who need complete nutrition assessments.

HISTORICAL INFORMATION

Table 17-1 on p. 572 summarizes the types of information information that provide clues about a client's nutrition status, the client's nutrient needs, and the many lifestyle and personal factors that shape the nutrition care plan. As Figure 17-4 on p. 572 shows, health, medications, nutrition status, and personal factors are interrelated; all must be considered to extract meaningful information. A thorough history points out potential problems requiring further investigation.

• Health History • A review of a client's **health history** identifies past, current, and potential health problems. Table 17-2 on p. 573 lists health problems associated with a high risk of PEM. Clients found to have any of these conditions during nutrition screening often receive complete nutrition assessments. Besides identifying risk for PEM, a review of the health history uncovers other nutrition-related information including:

- The need for diet modifications, including weight reduction.
- Potential health problems that modified diets might help to prevent or delay.
- Symptoms and clinical findings that can affect food intake or alter nutrient needs.
- Disorders and treatments that demand a great deal of time, motivation, or financial resources.
- Physical disabilities that interfere with a person's ability to purchase, prepare, and/or eat adequate amounts of food.

iatrogenic (EYE-at-row-JEN-ick) **malnutrition:** malnutrition that develops as a consequence of a treatment. Malnutrition that occurs because nutrient needs are not met during the course of hospitalization is an example.

health history: an account of the client's current and past health status and risk factors for disease. Traditionally, the health history has been called the *medical history.* The term *health history* now seems more appropriate, however, because the contents describe the client's health status, and the goal of medical care is health promotion and disease prevention.

FIGURE 17-4 **Interrelationships among Health, Medications, Nutrition Status, and Personal Factors**

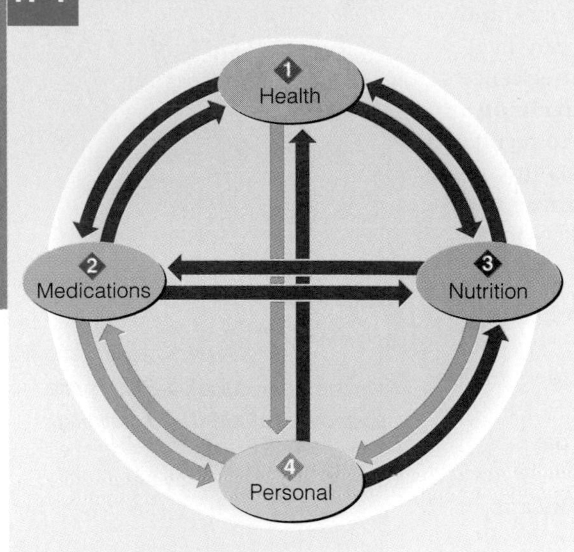

1. Health status can alter nutrient requirements, alter nutrition status, and directly affect the need for medications. Health care costs and the effects of health status on quality of life can alter lifestyle and seriously affect economic status.

2. Medications can alter nutrient requirements and affect nutrition status. Medications can have positive effects, negative effects, or both positive and negative effects on health. The costs of medications and their side effects can affect economic status, lifestyle, and quality of life.

3. Nutrition status can affect a client's health status, which, in turn, can affect the need for medications. Nutrition status and food intake can also alter the way medications are digested, absorbed, metabolized, and excreted. Although many personal factors are independent of nutrition status, poor nutrition and health status can affect a person's quality of life and have an impact on lifestyle. The costs of some nutrition therapies as well as a person's quality of life can interfere with a person's ability to maintain an optimal economic status.

4. Personal factors can have an impact on nutrient needs, food availability, and food choices. Personal factors can affect health status by altering nutrition status. Socioeconomic factors can also increase the predisposition to some diseases, such as hypertension and diabetes, and can affect people's ability to pay for health care, as well as their attitudes and beliefs about seeking health care and/or following health care recommendations.

A review of health problems helps direct the assessment process. For example, when a person reports problems sleeping at night, the astute assessor checks the client's diet for sources of caffeine and the times of the day that they are consumed.

TABLE 17-1 **Historical Information and Nutrition Assessments**

Type of History and Significant Information	What It Identifies
Health history	
• Current health problem(s) • Past health problem(s) • Family health history • Previous surgeries • Potential health problem(s)	Health factors that affect nutrient or nutrition education needs or place the client at risk for poor nutrition status.
Medication history	
• Prescription medications • Over-the-counter medications • Herbal supplements • Dietary supplements • Illegal drugs	Medications, alternative therapies, and illegal drug use that can affect nutrient needs or alter nutrition status.
Personal history	
• Age • Gender • Cultural/ethnic identity • Occupation • Role in family • Educational level • Motivational level • Economic status	Factors that affect nutrient needs, influence food choices, or limit diet therapy options.
Diet history	
• Food intake • Eating habits and patterns • Lifestyle patterns	Nutrient intake and imbalances, reasons for potential nutrition problems, and dietary factors important to shaping a nutrition care plan.

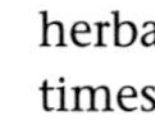
THINK NUTRITION whenever clients have disorders that raise their risk for PEM or have nutrition-related implications.

• Medication History • The nutrition assessment includes a record of all medications the client takes (prescription and over-the-counter), the dose, and the number of times each medication is taken. This drug history includes any type of nutrient or herbal supplements and the types, amounts, and number of times these preparations are taken as well. The next chapter describes the potential effect of medications and dietary supplements on nutrition status as well the effects of diet on drug absorption and metabolism.

THINK NUTRITION whenever clients are on medications or herbal supplements that can alter nutrient needs and nutrition status as well as medications whose absorption and metabolism are affected by diet.

• Personal History • To develop a realistic and attainable nutrition care plan, the dietitian considers intellectual, spiritual, social, psychological, and financial factors to help identify influences on food choices, resources for dealing with health and nutrition problems, and constraints that may interfere with prescribed interventions. Some of these personal factors include age and gender, number of people in the household, education level, occupation, income, ethnic background, cultural orientation, and religious affiliation, as well as perception of health status, level of psychological stress, and coping skills.

THINK NUTRITION whenever clients have personal histories that can influence their ability to follow health care advice and share this information with the dietitian.

Personal factors help direct the assessment process. Age, for example, helps pinpoint nutrient needs and also cues the assessor to look for nutrition problems common to a specific age. For another example, if the assessor discovers that a person depends on a significant other to make health care decisions, then the assessor includes that person in the remainder of the nutrition care process. If the assessor suspects that a client's limited financial resources may be placing a nutritious diet out of reach, then the assessor remains alert to the possibility of malnutrition throughout the remainder of the assessment.

• Diet History • Dietitians or dietetic technicians conduct complete diet histories, which include information about food intake, eating patterns, and lifestyle habits. The **diet history** uncovers current or potential nutrient imbalances and patterns and habits that support health or contribute to health problems. Table 17-3 on p. 574 shows examples of questions included in diet histories. An important component of the diet history is an account of how much and what a person is eating, as described in the next section.

TABLE 17-2 Examples of Health Conditions Associated With a High Risk of Protein-Energy Malnutrition (PEM)[a]

Acquired immune deficiency syndrome (AIDS)
Alcoholism
Anorexia nervosa
Bone marrow transplants
Burns, second or third degree covering more than 20% of body surface area
Cancer, some types
Chronic obstructive pulmonary disease, later stages
Chronic pancreatitis
Congestive heart failure, later stages
Crohn's disease
Cystic fibrosis
Depression, especially in the elderly
Diarrhea, prolonged, severe
Dysphagia
Feeding disabilities
Infections
Kidney failure, later stages
Liver failure, later stages
Malabsorption
Nonhealing wounds
Pressure sores
Septicemia
Surgical resection of the mouth, esophagus, and small intestine
Trauma (tissue damage)
Vomiting, prolonged, severe

[a]The conditions listed here frequently lead to PEM. Many other conditions can lead to malnutrition, especially if nutrition-related problems are not addressed and corrected in a timely manner.

IN SUMMARY

Health, medication, personal, and diet histories provide a wealth of information about a person's nutrient needs and socioeconomic and psychosocial factors that influence the person's ability to follow nutrition advice. All of these factors assist the health care professional in understanding the person's needs so that the nutrition care plan can be realistic and attainable.

FOOD INTAKE DATA

Finding out how much food a person eats and how an illness has affected the amount of food eaten is important for both complete nutrition assessments and nutrition screenings. For people who need to make changes in diet, finding out what a person currently eats helps the dietitian evaluate the general quality of the diet and provides information necessary to devise a nutrition care plan.

Identifying Potential Food Intake Problems

Although guidelines vary, the risks for malnutrition increase when clients eat less than half their usual intakes for five or more days. Assessment of weight, rate of weight loss (described later in this chapter), and symptoms and clinical findings help clarify the severity of food intake problems and assist in identifying appropriate interventions.

THINK NUTRITION whenever clients have very poor appetites for more than a few days.

• Information from Clients • Nutrition screenings generally rely on clients' perceptions and accounts of their food intakes.◆ For example, the nurse conducting a nutrition screening might ask the client, "Have you noticed any changes in the amount of food you are eating?" If the response indicates that the client has not been eating very well, the nurse then asks, "How long have you had the problem?" The nurse might also inquire about how much less the client has been eating. If the nurse determines that the problem is significant, the client is referred for a complete nutrition assessment.

To obtain accurate information about food intakes, health care professionals must conduct interviews and ask questions in a nonjudgmental manner. ◆

diet history: a comprehensive record of eating-related behaviors and the foods a person eats.

TABLE 17-3 Examples of Questions to Uncover Dietary Patterns or Habits

- Do you have any favorite foods?
- Do you dislike any foods?
- Are there any foods that you do not eat for any reason?
- Are you allergic or intolerant to any foods?
- Are you on a special diet?
- How many times a day do you typically eat meals and snacks?
- Where do you typically eat?
- How many times a week do you typically eat out?
- How is your appetite?
- Who does the shopping in your household?

Note: In addition to questions such as these, a food intake tool would be used to evaluate actual food intake.

• Observation of Food Intake in Health Care Facilities • For clients already admitted to facilities that serve foods, health care professionals can use direct observation to identify problems with nutrient intake by taking these steps:

- Regularly check the client's tray to be sure that foods are being eaten. Remember that appetites can change during the course of an illness, and catching problems early can help prevent serious problems later.
- If the client is to receive no food or is unable to eat, find out how long it has been since the client has eaten. Ask if the client is expected to be able to eat soon.
- If the client is unable to eat, check to see if nutrients are being delivered by tube (Chapter 22) or by vein (Chapter 23).

Communicate problems to the dietitian or physician, record the problems in the medical record, and follow up to make sure problems are being addressed.

For some clients who are having problems with appetite, the physician may order a **kcalorie count**—a procedure that evaluates what and how much clients are eating. To conduct a kcalorie count, nurses or nursing assistants (or, in some cases, the client) write down all the foods and beverages, as well as the amounts of each, the client receives. After the client finishes the food, beverage, or meal, the amount that is left is also recorded. Dietitians or dietetic technicians retrieve the record, deduce what has been eaten, and use the information to estimate energy and nutrient intake.

Other Tools for Gathering Food Intake Data

When health care professionals need information about what and how much a client eats at home, they can use simple tools to get a picture of the client's diet. The tools described here are the 24-hour recall, the usual intake method, food frequency questionnaires, and food records. The best tool to use depends on the purpose of the record.

Dietitians and dietetic technicians use food models, photos, pictures, and measuring utensils to help clients estimate portion sizes.

• The 24-Hour Recall and Usual Intake Method • The **24-hour recall** asks the client to recount everything eaten or drunk during the previous day. The **usual intake method** is similar, but asks the client to recount everything eaten or drunk in a typical day. For both methods, the assessor asks about the times when meals or snacks are eaten, the amounts of foods eaten, and the ways foods are prepared.

To obtain food intake data using a 24-hour recall or usual intake method, the assessor might begin by asking, "What is the first thing you ate or drank yesterday?" (Or "What is the first thing you usually eat or drink during the day?") After the client responds, the next question might be, "What time was that?" For each food the client recounts, the health care professional asks how it was prepared or eaten. Similar questions follow until the intake for the previous day or typical day is complete. When using the 24-hour recall, health care professionals ask clients if their intake that day was fairly typical. If not, the assessor finds out how it varied from the usual intake. Another approach might be to gather information from three or more 24-hour periods, being sure to include at least one weekend day.

Clients may forget to mention the beverages they drink and the fats, sweeteners, condiments, and other foods they may use, unless specifically prompted to do so. Clients may state they ate 2 slices of toast, for example, and may not think to mention that they spread 4 pats of butter and 2 tablespoons of jelly on the toast. For another example, they may mention they drank a cup of coffee, but not mention that the "coffee" they drink is half milk or cream.

Dietitians and dietetic technicians skilled at gathering food intake data can use the information to identify usual eating habits or to see how well a client

kcalorie count: a determination of a client's food intake from a direct observation of how much the client eats.

24-hour recall: a record of foods eaten by a person in the previous 24 hours.

usual intake method: a record of the foods eaten by a person in a typical day.

is able to follow a prescribed diet. For example, one person may always eat an afternoon snack, another may eat very large portions of food, and still another may avoid milk or vegetables. Or the record may reveal that a person on a special diet does not understand it and is eating a food that is generally restricted or avoiding a whole group of foods that are allowed.

Dietitians or nurses can also use information from food intake records to help clients plan when and how to take their medications relative to food intake. More information about the timing of medication doses with food intake is provided in Chapter 18.

• Food Frequency Questionnaire • A **food frequency questionnaire** is typically used to compare a client's usual food intake with the Daily Food Guide and thus helps to identify energy intakes, type and amount of fat in the diet, and specific nutrient imbalances. Clients may be asked how many servings of breads, cereals, or grain products; vegetables; fruits; meat, poultry, fish, and alternatives; milk, cheese, yogurt, and milk products; fats, oils, and sweets they eat in a typical day, week, or month. Clients are often also asked specific questions about the type of fat in their diets. For example, "How often do you eat red meat in a typical week?" or "What type of milk do you drink?" Food frequency questionnaires can also be used in addition to a 24-hour recall or usual intake method to verify the accuracy of a client's recollection of food intake.

• Food Records • Food records maintained over several days provide valuable information about a client's food intake as well as the person's response to and compliance with medical nutrition therapy or tolerance for foods. The client writes down all foods and beverages consumed, times foods are eaten, amounts consumed, and methods of preparation.◆ Depending on the purpose of the food record, the person may be asked to record other information as well. If the record is to be used to help a person change eating behaviors and lose weight, it might also include information about the person's mood, the occasion (party, holiday, family meal), behaviors associated with eating food (watching TV, driving in the car, sitting at the table with the family), and physical activity. When the purpose of the food record is to establish blood glucose control (see Chapter 25), records include details of medication administration, physical activity, illness, and the results of blood glucose monitoring. When the purpose of the record is to establish food tolerances (such as the amount of lactose a person can handle), food records also include symptoms associated with eating (for example, cramps, diarrhea, nausea, or hives).

◆ Figure 9-9 on p. 286 provides an example of a food record.

Food records help pinpoint problem food patterns so that solutions can be implemented. Unfortunately, they require a great deal of time to complete, and clients must be motivated to keep them accurately. Another drawback is that clients may either consciously or unconsciously change their eating behaviors while keeping the records. If clients understand their diets but are not following them, the clients may record what they believe they should be eating, rather than what they are actually eating.

• Applying Food Intake Information • In addition to helping evaluate a client's appetite, information from 24-hour recalls, usual intakes, food frequencies, and food records helps identify dietary patterns that suggest nutrition-related problems or patterns that must be considered when nutrition care plans are devised. To estimate the energy and energy nutrients in a client's diet, dietitians often use the exchange system described in Chapter 2. To check for nutrition-related problems, dietitians compare the client's intake to the Daily Food Guide and other dietary guidelines. Are all food groups included in appropriate amounts? If not, which nutrient may be excessive or deficient (see Figure 2-2 on pp. 38–39)? Are food choices varied, or is the diet monotonous? Does the diet provide adequate fiber or excessive amounts of refined sugars? Are the total amount of fat and types of fat appropriate? Is caffeine, alcohol, or salt use excessive?

food frequency questionnaire: a tool for gathering food intake data that asks clients about the types and amounts of foods they routinely eat.

food records: logs of all the food eaten over a period of time that may also include records of behaviors and symptoms, physical activity and medications; also called **eating** and/or **food diaries.**

Measure Length and Height

Tips for measuring length and height include:

- Always measure—never ask! Self-reported heights are often greater than measured heights. If height is not measured, document that the height is self-reported.

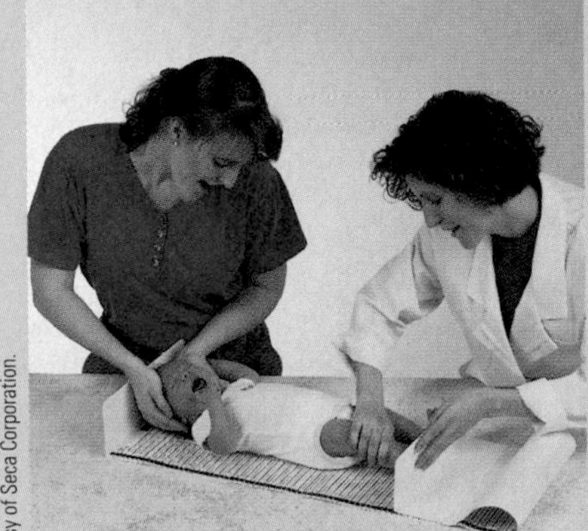

Courtesy of Seca Corporation.

Lying with legs straight for a length measurement can be a trying experience for an infant.

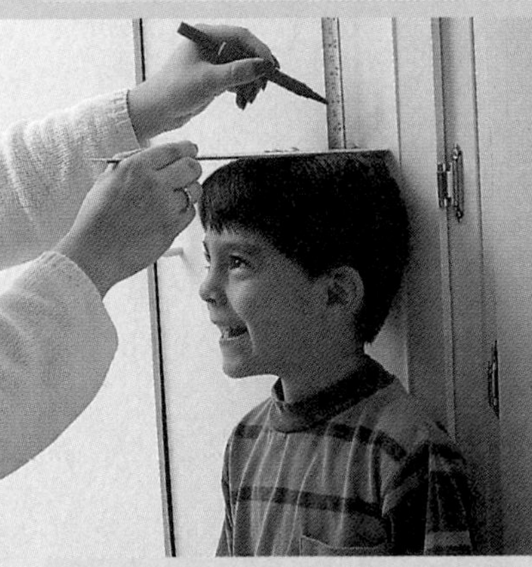

© Tony Freeman/PhotoEdit

Health care professionals obtain accurate height measurements using a measuring tape or a board fixed to a wall.

- Measure length for infants and young children using a measuring board with a fixed headboard and movable footboard. It often takes two people to measure length. One person gently holds the infant's head against the headboard; the other straightens the infant's legs and moves the footboard to the bottom of the infant's feet.

  Using a measuring tape is a less exacting way to measure length in infants and is the method used to measure length for others who cannot stand erect. To use this method, straighten out the infant's or person's body, make a mark at the top of the head, make another mark at the bottom of the heel, and then measure the distance between the two marks.
- Measure height against a wall to which a nonstretchable measuring tape or a board has been fixed. Ask the person to stand erect without shoes and with heels together. The person's line of sight should be horizontal, with the heels, buttocks, shoulders, and head touching the wall. Place a ruler or other flat, stiff object on the top of the head at a right angle to the wall and carefully note the height measurement. Although less accurate, the measuring rod of a scale can also be used to measure height. To use this method, follow the same general procedure, asking the person to face away from the scale. Take extra care to ensure that the client is standing erect and that the line of sight is horizontal.
- Immediately record length and measurements to the nearest ¼ inch or 0.5 centimeter.

In assessing food intake data, the dietitian keeps the client's health problems in mind. Consider a client who reports problems with constipation, for example. The problem alerts the dietitian to look for sources of fiber and the volume of fluid in the client's diet.

IN SUMMARY

Finding out about a client's food intake is an important component of both nutrition screenings and nutrition assessments. Although the client's description of food intake is often a good starting point for identifying nutrition-related problems, several tools can be used to get a clearer picture of what and how much a person is eating. These tools include the 24-hour recall, the usual intake method, food questionnaires, and food records.

ANTHROPOMETRIC MEASUREMENTS

Anthropometric measurements, introduced in Chapter 1, measure physical characteristics of the body. Height and weight are the most common anthropometric measurements, and both are important components of nutrition screenings and complete nutrition assessments. As Table 17-4 on p. 578 shows, other anthropometric measurements include head circumference (described in this chapter), fatfold tests (see Chapter 8 and Appendix E), abdominal girth (see Chapter 28), and waist circumference (see Chapter 8 and Appendix E).

Other measurements do not measure the physical characteristics of the body itself, but instead use measurements to evaluate the body's composition. Examples include hydrodensitometry (underwater weighing) and bioelectrical impedence. Figure 8-8 on p. 256 illustrates methods used to assess body fat and Appendix E provides more information about these techniques. For another example, bioelectrical impedance analysis (see Chapter 8) measures the body's resistance to an electrical current to assess body composition.

Length or height, weight, and head circumference are routinely measured in children, as are height and weight in adults. Dietitians use anthropometric measurements to evaluate current nutrition status and to look for patterns—obviously high or low measurements and upward or downward trends that occur in the individual over time. They also use height and weight measurements to estimate energy and energy-nutrient needs and then use the measurements again to see whether their estimates require further refinements.

◆ Measure *length* for children under age three when using growth charts from birth to 36 months to plot growth data; measure *height* for children two or older when using growth charts for children over two years.

• Length, Height, and Weight • Length measurements for infants and children up to age two or three and height measurements for older children◆ help evaluate growth, which depends on adequate nutrition. Poor growth in children is an important indicator of malnutrition. For adults, height measurements alone do not reflect current nutrition status but help to estimate desirable weight and energy needs. Length may also be measured in adults and children who are unable to stand for physical or medical reasons. Recumbent length measurements can be significantly greater than height measurements, however, and when length measurements are used to estimate desirable weight or energy needs, the estimates must be carefully interpreted. The "How to" features on pp. 576 and 577 describe the proper techniques for measuring length, height, and weight.

• Head Circumference • Health professionals may also measure head circumference to confirm that an infant's growth is proceeding normally. To measure head circumference, the assessor places a nonstretchable tape so that it encircles the largest part of the infant's head: just above the eyebrows, just above the point where the ears attach, and around the occipital prominence at the back of the head. The measurement is recorded to the nearest ⅛ inch or 0.5 centimeter.

• Analysis of Measures in Infants and Children • Health professionals evaluate physical development by periodically measuring length or height, weight, and head circumference (as appropriate), plotting the results on growth charts (see the "How to" on p. 500), and analyzing the results. For children two and over, height and weight can also be used to calculate the body mass index (BMI)◆ and can be compared to percentile standards (see p. 517). Growth charts consist of percentiles, which are used to compare weight to age, length or height to age, weight to length or height, and BMI to age. Although individual growth patterns vary, ideally a child's length or height, weight, and BMI should fall roughly in the same percentile and remain at about the same percentile throughout childhood.

HOW TO Measure Weight

To improve the accuracy of weight measurements, keep in mind these tips:

- Always measure—never ask! Self-reported weights are often inaccurate and are frequently lower than measured weights.
- Use the right equipment. Beam balance or electronic scales that have been calibrated and checked for accuracy at regular intervals provide the most reliable weight measurements. Infant scales are equipped with platforms that allow infants to sit or lie down while being weighed. Once children can stand erect, regular scales provide accurate weight measurements. Special scales and hospital beds with built-in scales assist in weighing clients who are unable to stand on regular scales.
- Follow standard procedures. For children under two or three (depending on the growth chart used), weigh the child without clothing or diapers. For all others, try to use the same scale and take weight measurements at about the same time of the day (preferably before breakfast), in about the same amount of clothing (without shoes), and after the person has voided.
- Record the measurement immediately either to the nearest ¼ pound or 0.1 kilogram.

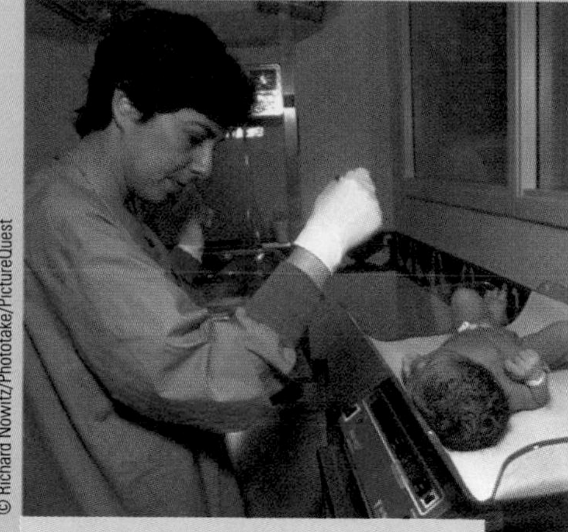

© Richard Nowitz/Phototake/PictureQuest

Infants are weighed on scales equipped with platforms that allow them to sit or lie down during the measurement.

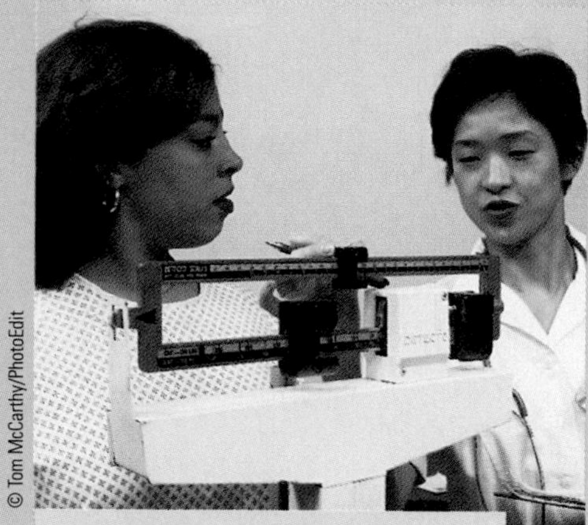

© Tom McCarthy/PhotoEdit

Beam balance or electronic scales provide accurate weight measurements for older children and adults.

In addition to nutrition, genetic factors influence height;◆ therefore, when analyzing the measurements, the health care professional considers the heights of family members and disorders that affect growth. State of hydration influences weight measurements. Children who are retaining fluids may have weights that are deceptively high; children who are dehydrated may have weights that are deceptively low.

Length or height and weight below the 25th percentile or a sudden drop in a previously steady growth pattern suggests growth retardation, an important sign of malnutrition. A weight percentile significantly lower than the height percentile—for example, weight in the 25th percentile and height in the 95th percentile—or a BMI-for-age below the 5th percentile indicates that the child is underweight. Conversely, a weight percentile significantly higher than the height percentile—for example, weight in the 95th percentile and height in

◆ Reminder: The *body mass index (BMI)* is an index of a person's weight in relation to height, determined as follows:

$$BMI = \frac{\text{weight (kg)}}{\text{height (m)}^2}$$

◆ Size and growth differences have been identified between various ethnic and racial groups. These differences are small and inconsistent, and the recently revised growth charts (see Appendix E) provide an acceptable tool for assessing growth in all infants and children in the United States.

TABLE 17-4 Anthropometric and Other Body Measurements Useful in Assessing Nutrition Status

Measurement	What It Reflects
Abdominal girth measurement	Abdominal fluid retention and abdominal organ size
Hand grip strength	Ability of muscles to perform work (protein status)
Height-weight	Overnutrition and undernutrition; growth in children
%IBW, %UBW,[a] recent weight change	Overnutrition and undernutrition
Head circumference	Brain growth and development in infants and children under age two
Fatfold	Subcutaneous and total body fat
Midarm muscle circumference	Skeletal muscle mass (protein status)
Size of reaction to skin test	Immune function (protein status)
Waist circumference	Fat distribution

[a]%IBW = percent ideal body weight; %UBW = percent usual body weight.

the 25th percentile—or a BMI in the 95th percentile or above indicates that the child is overweight. A BMI at or greater than the 85th percentile indicates risk for overweight.

THINK NUTRITION whenever children are underweight, overweight, or have risk factors for either.

In children whose growth has been retarded, nutrition rehabilitation will ideally induce height and weight to increase to higher percentiles. In overweight children, the goal is for weight to remain stable as height increases, until weight becomes appropriate for height.

Head circumference reflects brain growth, which occurs rapidly before birth and during early infancy and is nearly complete by age seven. Malnutrition before birth and during early childhood can impair brain development. In addition to nutrition factors, genetic variations and certain disorders can also influence head circumference.

◆ Figure 8-5 on page 253 shows BMI values for adults.

◆ Reminder: Weight-for-height tables suggest a weight range, rather than pinpoint one ideal weight—a helpful reminder that several weights at any given height can suggest good health.

◆ Although the term *ideal body weight* is a misnomer, it is the term most likely to be used in health care settings, and so it is used here.

• **Analysis of Measures for Adults** • For people who are healthy or whose weights have not been affected by illness, the BMI◆ can be used to assess health risks in relation to height and weight. For people with illnesses that affect their weights, practitioners typically use weight-for-height tables◆ (see p. 252) or a quick estimate of desirable body weight (see Table 17-5) and compare a person's actual weight with desirable weight, a figure commonly known as the percent ideal body weight (%IBW).◆ Actual weight can also be compared with usual body weight to generate the percent usual body weight (%UBW), which considers what is normal for a particular individual. Both nutrition screenings and complete nutrition assessments generally include an evaluation of the %IBW and %UBW. The %UBW is particularly useful for evaluating the degree of nutrition risk associated with illness. In cases where an overweight individual becomes acutely ill and is rapidly losing weight, the health care professional relying on the %IBW may inadvertently overlook significant weight loss. Conversely, for clients who have been underweight throughout life, %IBW may overstate the degree of weight loss. The "How to" and Table 17-6 on p. 580 show how to calculate and evaluate the %IBW and the %UBW.

In assessing weight loss, consider not only the amount of loss, but also the rate. A person who loses less than 5 percent UBW over a six-month period is at

TABLE 17-5 Quick Estimate of Desirable Body Weight	
Men	**Women**
For 5 ft, consider 106 lb a reasonable weight.	For 5 ft, consider 100 lb a reasonable weight.
For each inch over 5 ft, add 6 lb.	For each inch over 5 ft, add 5 lb.
Subtract 6 lb for each inch under 5 ft.	Subtract 5 lb for each inch under 5 ft.
Add 10% for a large-framed individual; subtract 10% for a small-framed individual.	Add 10% for a large-framed individual; subtract 10% for a small-framed individual.
Example: A man 5 ft 8 in tall (medium frame) would start at 106 lb, add 48, and arrive at a reasonable weight of 154 lb.	*Example:* A woman 5 ft 6 in tall (medium frame) would start at 100 lb, add 30, and arrive at a reasonable weight of 130 lb.

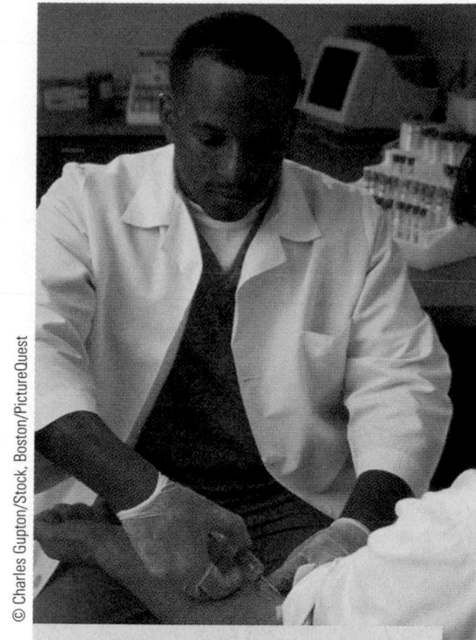

© Charles Gupton/Stock, Boston/PictureQuest

Lab tests are indirect body measurements—they help determine what is happening inside the body.

minimal risk for poor nutrition status. A loss of 5 to 10 percent UBW over six months is significant, and a loss of greater than 10 percent usual body weight is highly significant.

THINK NUTRITION whenever clients are underweight or overweight or have lost or gained significant amounts of weight over short periods of time.

Keep in mind that a person's state of hydration can significantly influence body weight measurements. Disorders or medications that cause fluid retention can mask significant weight loss. Conversely, a person who is dehydrated may have a body weight that is deceptively low.

LABORATORY TESTS OF NUTRITION STATUS

Laboratory tests measure levels of metabolites to evaluate the body's state of health or its response to various treatments. Laboratory tests can provide information about protein-energy balance, vitamin-mineral status, fluid balance, body composition, organ function, and metabolic status. They can also help determine if nutrition therapy is appropriate or if a person is complying with a special diet. This section describes some tests useful in assessing nutrition status and response to diet therapy. Other tests, such as serum◆ glucose, cholesterol, and tests that define fluid and electrolyte balance, acid-base balance, and organ function, help pinpoint disorders or problems with nutrition implications. Table 17-7 lists some common laboratory tests with nutrition implications and shows what these tests reflect. Tests relevant to specific disorders will be discussed in the appropriate chapters.

◆ The *serum* is the watery portion of the blood that remains after removal of the cells and clot-forming material; *plasma* is the fluid that remains when unclotted blood is centrifuged. Usually, serum and plasma concentrations are similar, but plasma samples are more likely to clog mechanical blood analyzers, so serum samples are preferred.

To interpret laboratory tests that fall outside the normal range requires consideration of all the factors that influence the test results. For laboratory tests that suggest malnutrition, the person's state of health and hydration can influence test results independently of nutrition factors. With dehydration, lab results may be deceptively high; with fluid retention, lab results may be deceptively low.

The body's way of handling nutrients also affects lab tests. The low blood concentration of a nutrient, for example, may reflect a primary deficiency◆ of that nutrient, but it may also be secondary to the deficiency of one or several other nutrients involved in the transport or metabolism of that nutrient. Nutrient concentrations in the blood and urine sometimes reflect recent intakes rather than long-term intakes. Thus blood concentrations of a nutrient may be normal, even when tissue levels are deficient.

◆ Chapter 1 described primary, secondary, and subclinical nutrient deficiencies and Figure 1-5 on p. 17 shows how they can be detected by various assessment techniques.

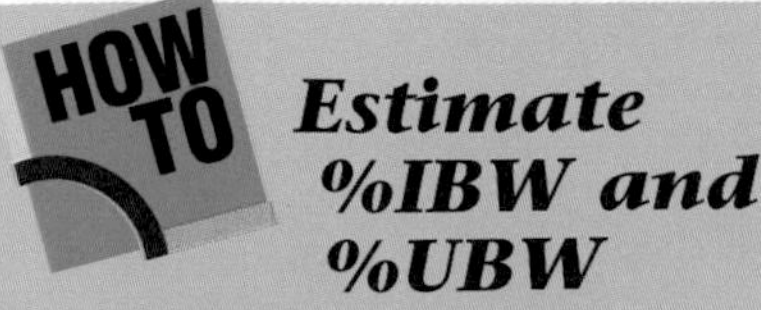

Estimate %IBW and %UBW

To estimate %IBW, compare the client's current weight with the desirable body weight from standard weight-for-height tables:

$$\%\text{IBW} = \frac{\text{actual body weight}}{\text{desirable (ideal) weight}} \times 100.$$

For example, suppose you wish to calculate %IBW for a man who is 5 feet 8 inches tall and weighs 123 pounds. Use Table 8-5 on p. 252 to estimate desirable (ideal) weight. The desirable weight range for someone of this height is 125 to 164 pounds. Because the client is a man, begin with the upper half of the weight range (144.5 to 164 pounds). Use the midpoint (154 pounds) as the desirable weight estimate. In this example, the desirable weight is 154 pounds.

$$\%\text{IBW} = \frac{123 \text{ lb}}{154 \text{ lb}} \times 100 = 80\%.$$

The man in this example is at 80 percent of his ideal body weight. Table 17-6 indicates that at 80 percent IBW he is mildly underweight.

This man has lost 15 pounds in the last month. To calculate his %UBW, use this formula:

$$\%\text{UBW} = \frac{\text{actual weight}}{\text{usual weight}} \times 100.$$

Calculate the usual weight (138 pounds) by adding the weight loss (15 pounds) to the current body weight (123 pounds).

$$\%\text{UBW} = \frac{123 \text{ lb}}{138 \text{ lb}} \times 100 = 89\%.$$

The man is at 89 percent of his usual body weight. A look at Table 17-6 reveals that a person at 89 percent UBW is mildly underweight. Based on %UBW, the degree of underweight is less severe than the %IBW implied, because this man has consistently weighed less than standard weight. Nevertheless, his recent rate of weight change is very significant: he has lost almost 4 pounds per week.

It is beyond the scope of this text to describe all lab tests used to assess nutrition status, define organ function, and develop nutrition care plans. Instead, the emphasis is on lab tests used to evaluate protein status—serum proteins and the total lymphocyte count. Appendix E provides information about nitrogen balance studies and tables of lab tests useful in detecting various vitamin and mineral deficiencies, including those associated with nutrition-related anemias.

• Factors Affecting Serum Protein Levels • Serum protein levels reflect protein intake (the availability of amino acids) and the body's metabolism, degradation, and distribution of the specific protein. The larger the body's pool of a serum protein, and the slower its rate of degradation, the longer it takes for the protein to be affected by changes in diet. During acute stresses, such as an extensive burn injury, some serum proteins shift from the blood to the extracellular fluid. Thus lab tests of these serum proteins will reveal low levels, even though the body's pool may not be low. Furthermore, serum proteins are synthesized in the liver, so low levels can also reflect altered liver function. As Table 17-8 on p. 582 shows, a variety of medical conditions and treatments can alter levels of serum proteins and the total lymphocyte count independently of nutrition factors.

TABLE 17-6 Weight Parameters as an Indicator of Nutrition Status

%IBW	%UBW	Nutrition Status
>120	—	Obese
110–120	—	Overweight
90–109	—	Adequate
80–89	85–95	Mildly underweight
70–79	75–84	Moderately underweight
<70	<75	Severely underweight

• Albumin • Albumin is the most abundant serum protein, and its serum levels are inexpensive to measure. Serum albumin is affected by many medical conditions and is slow to reflect changes in nutrition status because of its large body pools and slow rate of degradation. In people with chronic PEM, serum albumin levels remain normal for long periods of time despite a depletion of body proteins; the levels fall only after prolonged malnutrition. Likewise, albumin concentrations increase slowly with appropriate nutrition support, so albumin is not a sensitive indicator of response to nutrition therapy.

• Serum Transferrin • Transferrin transports iron, and its concentrations reflect both protein-energy and iron status. As with albumin, markedly reduced transferrin levels indicate severe PEM. Transferrin breaks down in the body more rapidly than albumin, but is still relatively slow to respond to nutrition therapy. Thus it may not be a sensitive indicator of response to nutrition therapy. Furthermore, interpreting transferrin levels as an indicator of protein-energy status is difficult when iron deficiency is present. Transferrin rises as iron deficiency grows worse and falls as iron status improves.

THINK NUTRITION whenever clients have low levels of albumin◆ and transferrin, but remember to consider the many other factors that can influence the lab test results.

◆ Albumin transports many nutrients through the blood to the tissues. When albumin levels are low, the tissues may be unable to receive the nutrients they need, creating secondary deficiencies.

◆ *Transthyretin* is also known as prealbumin or thyroxin-binding protein.

• Transthyretin and Retinol-Binding Protein • Transthyretin◆ and retinol-binding protein levels decrease rapidly during PEM and respond quickly to changes in protein intake. Thus lab tests of transthyretin and retinol-binding protein are more sensitive tests of protein status, but they are also more expensive than serum albumin and less commonly measured in clients. Measurement of transthyretin and retinol-binding protein is often reserved for

TABLE 17-7 Examples of Routine Laboratory Tests with Nutrition Implications

Test	Uses
Hematology	
Hemoglobin (Hg)	To detect anemia and determine state of hydration.
Hematocrit (Hct)	To detect anemia and determine state of hydration.
White blood cells (WBC)	To detect infection and determine total lymphocyte count.
Mean corpuscular volume (MCV)	To detect anemia and determine its causes.
Mean corpuscular hemoglobin (MCH)	To detect anemia and determine its causes.
Mean corpuscular hemoglobin concentration (MCHC)	To detect anemia and determine its causes.
Blood Chemistry	
Proteins	
Total protein[a]	To detect PEM and various nutrient imbalances.
Albumin	To detect PEM and determine state of hydration.
Transferrin	To detect PEM and iron status, and monitor response to feeding.
Transthyretin (prealbumin)	To detect PEM and monitor response to feeding.
Electrolytes	
Sodium	To check state of hydration.
Potassium	To monitor acid-base balance and renal function and detect imbalances.
Chloride	To monitor acid-base balance and detect GI losses of chloride (from vomiting or nasogastric suctioning).
Carbon dioxide	To monitor acid-base balance.
Other	
Glucose	To detect diabetes mellitus, pancreatic tumors, and hypoglycemia and monitor glucose intolerance.
Glycated hemoglobin	To monitor blood glucose control over the past 2–3 months.
Blood urea nitrogen	To monitor renal function and determine state of hydration.
Calcium	To detect hormonal imbalances, certain malignancies, and calcium imbalances.
Phosphorus	To detect imbalances and PEM and monitor renal function and response to feeding.
Magnesium	To monitor renal function and response to feeding and detect PEM.
Cholesterol	To assess risk of heart disease and detect possibility of obstructive jaundice.
Uric acid	To detect gout and determine state of hydration.
Serum creatinine	To monitor renal function and determine state of hydration.
Serum enzymes	
Creatinine phosphokinase (CPK)	To monitor heart function and muscle damage.
Lactic dehydrogenase (LDH)	To monitor heart and renal function.
Alanine transaminase (ALT, formerly SGPT)	To monitor heart and liver function.
Aspartate transaminase (AST, formerly SGOT)	To monitor heart and liver function.
Alkaline phosphatase	To monitor liver function.
Serum amylase	To monitor pancreatic function.
Serum lipase	To monitor pancreatic function.

Note: This table presents a partial listing of the major uses of certain commonly performed lab tests that have implications for nutrition.
[a]More than half of the total protein is albumin.

clients with disorders that markedly raise metabolic rates because such disorders can rapidly lead to malnutrition.

• **Total Lymphocyte Count** • PEM compromises the immune system, reducing the number of white blood cells (lymphocytes), which are important in resisting and fighting infections. Two laboratory tests, the number of white blood cells and the percentage of lymphocytes, are used to derive the total lymphocyte count.◆ The total lymphocyte count is inexpensive and easy to obtain, but its value in nutrition assessment is limited because so many variables affect its level. Table 17-9 on p. 582 provides standards for evaluating serum proteins and the total lymphocyte count.

◆ **Total lymphocyte count (mm^3) = white blood cells (mm^3) × percent lymphocytes.**

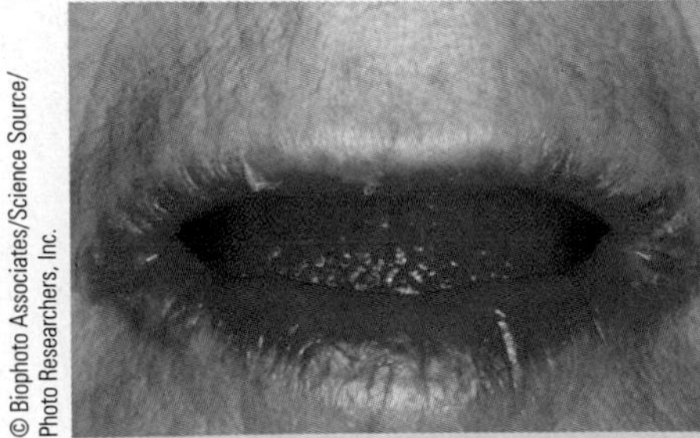

Physical signs of B vitamin deficiencies include dry, cracked lips and sores in the corners of the lips.

TABLE 17-8 Factors That Affect Serum Protein Levels and the Total Lymphocyte Count

Lab Test	Factors That Influence Values
Albumin	Chronic PEM, metabolic stress, liver disease, kidney disease (nephrotic syndrome), and eclampsia may lower serum levels.
Transferrin	Chronic PEM, metabolic stress, liver disease, kidney disease (nephrotic syndrome), and the use of some antibiotics lower serum levels; pregnancy, iron deficiency, and use of oral contraceptives raise serum levels.
Transthyretin	PEM, metabolic stress, hemodialysis, and hypothyroidism lower serum levels; kidney disease and the use of corticosteroids may raise serum levels.
Retinol-binding protein	PEM, metabolic stress, liver disease, hyperthyroidism, vitamin A deficiency, and cystic fibrosis may lower serum levels; kidney disease may raise serum levels.
Total lymphocyte count	PEM, metabolic stress, and the use of chemotherapy, immunosuppressants, and corticosteroids lower levels; infections raise levels.

IN SUMMARY Anthropometric measurements and laboratory tests provide valuable information regarding nutrition status. Single measurements allow for a comparison with norms. Repeated measurements allow health care professionals to evaluate how nutrition status is affected by illness or by medical nutrition therapy. Body measurements, together with historical information and physical findings, described next, help define nutrition status and nutrient needs.

PHYSICAL SIGNS OF MALNUTRITION

◆ For the purposes of this discussion, physical signs of malnutrition include clinical signs such as elevated blood pressure and changes in taste perception.

One clinician astutely summarized the role of the physical examination◆ in nutrition assessment this way: "To me, physical examination proves the saying 'A picture is worth a thousand words.'"[1] Indeed, health care professionals can simply look at people to see if they are overweight, underweight, lethargic, confused, or unable to feed themselves, to give a few examples.

With closer examination, a skilled health care professional can learn to detect physical signs of nutrient deficiencies and excesses. Signs of malnutrition

TABLE 17-9 Relationship among Degree of Overnutrition, Serum Proteins, and the Total Lymphocyte Count

Indicator	Degree of Depletion			
	Normal	Mild	Moderate	Severe
Albumin (g/100 ml)	3.5–5.4	2.8–3.4	2.1–2.7	<2.1
Transferrin (mg/100 ml)	200–400	150–200	100–149	<100
Transthyretin (mg/100 ml)	23–43	10–15	5–9	<5
Retinol-binding protein[a] (mg/100 ml)	3–7	—	—	—
Total lymphocyte count (mm³)	2500	<1500	<1200	<800

Note: To convert albumin (g/100 ml) to standard international units (g/L), multiply by 100. To convert transferrin (mg/100 ml) to standard international units (g/L), multiply by 0.01.
[a]Levels less than normal suggest compromised protein status. The actual degree of depletion (mild, moderate, and severe) has not been defined.

TABLE 17-10 Physical Signs of Nutrient Imbalances

Body System	Acceptable	Signs of Malnutrition	Other Possible Causes
Hair	Shiny, firm in scalp	Dull, brittle, dry, loose; falls out (PEM) corkscrew hair (copper)	Excessive hair bleaching; hair loss from aging, chemotherapy, or radiation therapy
Eyes	Bright, clear pink membranes; adjust easily to light	Pale membranes (iron); spots, dryness, nightblindness (vitamin A); redness at corners of eyes (B vitamins)	Anemia, unrelated to nutrition; eye disorders; allergies
Lips	Smooth	Dry, cracked, or with sores in the corner of the lips (B vitamins)	Sunburn, windburn, excessive salivation from ill-fitting dentures or other disorders
Mouth and gums	Red tongue without swelling, normal sense of taste; teeth without caries; gums without bleeding, swelling, or pain	Smooth or magenta tongue (B vitamins), decreased taste sensations (zinc); swollen, bleeding gums (vitamin C)	Medications, periodontal disease (poor oral hygiene)
Skin	Smooth, firm, good color	Poor wound healing, (PEM, vitamin C, zinc); dry, rough, lack of fat under skin (essential fatty acids, PEM, B vitamins); bruising, bleeding under skin (vitamins C and K)	Poor skin care, diabetes mellitus, aging, medications
Nails	Smooth, firm, pink	Ridged (PEM), spoon shaped, pale (iron)	
Other		Dementia, peripheral neuropathy (B vitamins); swollen glands at front of neck (PEM, iodine); bowed legs (vitamin D)	Disorders of aging (dementia), diabetes mellitus (peripheral neuropathy)

appear most rapidly in parts of the body where cell replacement occurs at a rapid rate, such as the hair, skin, and digestive tract (including the mouth and tongue). The summary tables in Chapters 10 through 13 include physical signs of specific vitamin and mineral imbalances. Table 17-10 lists physical signs of PEM and of vitamin and mineral malnutrition in general.

• Fluid Balance • Among the most useful physical signs of nutrition status are those that reflect dehydration and fluid retention (see Table 17-11). Along with laboratory tests, physical signs of hydration help guide medical and nutrition therapy and are important in the interpretation of weight measurements. Various medical conditions and treatments (including medications) can upset fluid balances, and signs of fluid imbalance can vary as well.

THINK NUTRITION whenever clients have physical or clinical signs of nutrient or fluid imbalances.

• Limitations of Physical Findings • Identifying and interpreting physical signs of malnutrition require knowledge, skill, and clinical judgment. Many physical signs are nonspecific; they can reflect any of several nutrient deficiencies as well as conditions not related to nutrition (see Table 17-10). For example, cracked lips may be caused by any of several B vitamin deficiencies or sunburn, windburn, or dehydration. Food intake information and laboratory tests can provide further support for a suspected nutrient deficiency.

IN SUMMARY

The information gathered from nutrition assessments conducted at regular intervals allows health care professionals to construct nutrition care plans and see how well they are working. Complete assessments include historical information, body measurements, and physical examinations. Health care professionals who understand the basics of nutrition assessment are in the best position to identify clients with nutrition problems and contribute to the nutrition care of clients. The case study that follows provides practice in identifying and evaluating factors that suggest malnutrition or shape future nutrition therapy.

TABLE 17-11 Physical Signs of Dehydration and Fluid Retention

Dehydration

- Sunken eyes
- Hollow cheekbones
- Dry mucous membranes
- Loss of skin turgor (elasticity)[a]
- Weak cry[b]
- Depression of the anterior fontanel[b]
- Deep, gasping respirations
- Weak, rapid pulse
- Thirst
- Reduced urinary output
- Weight loss

Fluid Retention

- Edema
- Ascites (abdominal fluid retention)
- Elevated blood pressure
- Increased urinary output
- Weight gain

[a]May not be a useful parameter in the elderly.
[b]Findings specific to infants.

CASE STUDY

Nutrition Screening and Assessment

Mrs. Genosa is an 85-year-old retired schoolteacher who has been a widow for 15 years. She has been admitted to the hospital with pneumonia and has congestive heart failure and diabetes. She routinely takes several medications, and, in addition to these, the physician has ordered antibiotics to treat the pneumonia. During an initial nutrition screening, Mrs. Genosa states that she has been eating very poorly over the past two weeks. She says that she usually weighs about 125 pounds, and that was her weight at her last physician's visit one month ago. Although she feels she has been losing weight, she doesn't know how much weight she has lost or when she started losing weight. Mrs. Genosa currently weighs 115 pounds and is 5 feet, 2 inches tall. A physical exam reveals edema, and laboratory tests confirm that she is retaining fluid. As a result of the nutrition screening, Mrs. Genosa has been referred to the dietitian for a complete nutrition assessment.

1. From the brief description provided, what factors in Mrs. Genosa's health, medication, personal, and diet histories might alert the nurse to risk for malnutrition?
2. Identify a desirable body weight for Mrs. Genosa and calculate her %IBW and %UBW. What do the results reveal? What effect does fluid retention have on Mrs. Genosa's weight?
3. How might fluid retention alter Mrs. Genosa's serum protein levels?
4. What tools might the dietitian conducting a complete nutrition assessment use to estimate what and how much Mrs. Genosa has been eating?
5. Describe some other types of assessment information the dietitian would need to develop a nutrition care plan.

Nutrition on the Net

WEBSITES

Assess these websites for further study of topics covered in this chapter.

- Find updates and quick links to these and other nutrition-related sites at our website: **www.wadsworth.com/nutrition**
- To view a food frequency questionnaire designed for women, visit the Women's Health Center: **www.womens-health.com/health_center/nutrition/nta_1.html**
- To view a food frequency questionnaire and assessment tool for children, visit the National Network for Child Care: **www.nncc.org/Nutrition/health.quiz.html**
- To use a nutrient analysis program, visit NAT Tools for good health: **www.nat.umc.edu**
- To use a health assessment tool that includes nutrition, visit Ask the Dietitian: **www.dietitian.com/ibw/ibw.html**
- Find more information about body composition analysis, including underwater weighing and bioelectrical impedance, at the "Virtual" Nutrition Center: **www.sci.lib.ucl.edu/HSG/Nutrition.html HYDRO**

INTERNET ACTIVITIES

Food frequency questionnaires are often designed to uncover eating habits that raise the risk of chronic diseases. This site provides an example of a food frequency questionnaire and how it can be used to assess eating habits.

- Go to **www.womens-health.com/health_center/nutrition/nta_1.html**
- Complete the food frequency questionnaire.
- Click on the "Submit Form" button.

Although this site is designed for women, the questionnaire applies to everyone. How did you score? What did the food frequency questionnaire reveal about your dietary habits and how can you improve them?

Study Questions

These questions will help you review the chapter. You will find the answers in the discussions on the pages provided.

1. In what ways can illnesses affect nutrition status? (pp. 567–569)
2. What roles do physicians, dietitians, nurses, dietetic technicians, and other health care professionals play in the nutrition care of clients? (pp. 569–570)
3. What is the difference between complete nutrition assessments and nutrition screenings? (pp. 570–571)
4. Provide examples of how information from health, medication, personal, and diet histories can alert health care professionals to potential nutrition problems or the need for nutrition education. (p. 571)
5. Describe ways of gathering food intake data and briefly discuss how each method works. (pp. 573–575)
6. Which anthropometric measurements are commonly used in both complete nutrition assessments and nutrition screenings? How do these measurements help define nutrition status? (p. 576)
7. What information about nutrition status do laboratory tests reveal? What factors influence lab test results? (p. 579)
8. How can health care professionals use physical findings to help pinpoint nutrition problems? (pp. 582–583)

These questions will help you prepare for an exam. Answers can be found on p. 586.

1. Mr. Salpingo has experienced a loss of appetite, difficulty swallowing, and mouth pain as a consequence of illness. Mr. Salpingo is at risk for malnutrition due to:
   a. altered metabolism.
   b. reduced food intake.
   c. altered excretion of nutrients.
   d. altered digestion and absorption.
2. Health care professionals can assist dietitians in the nutrition care of clients by:
   a. calculating clients' nutrient needs.
   b. providing medical nutrition therapy.
   c. conducting complete nutrition assessments.
   d. identifying clients at risk for malnutrition.
3. The nutrition care process is an organized approach for:
   a. identifying the nutrition contents of foods.
   b. ordering special diets.
   c. conducting a nutrition screening.
   d. meeting the nutrient and nutrition education needs of clients.
4. To conduct complete nutrition assessments, dietitians rely on several sources of information, which include all of the following *except:*
   a. nutrition care plans.
   b. body measurements.
   c. health, medication, personal, and diet histories.
   d. physical findings.
5. All of the following factors place a client at risk for poor nutrition status, *except:*
   a. a health problem frequently associated with PEM.
   b. use of several prescription medications that may affect nutrient needs.
   c. a personal history that reveals that the client lives with a spouse in a middle-income neighborhood.
   d. a significant reduction in food intake over five or more days.
6. After a nutrition screening revealed that a client had been eating very poorly during the past several weeks and had lost a considerable amount of weight, the client was referred to the dietitian for a complete nutrition assessment. Which food intake tool would the dietitian most likely use to get a clearer picture of how much the client has been eating?
   a. kcalorie count
   b. 24-hour recall or usual intake method
   c. food frequency questionnaire
   d. food record
7. Both height and weight measurements:
   a. are affected by fluid status.
   b. cannot be taken on bedridden clients.
   c. are routine measurements in health care facilities.
   d. require equipment that is not readily available in most health care facilities.
8. The %IBW of a person who weighs 185 pounds and has a desirable body weight of 150 pounds is:
   a. 123 percent.
   b. 150 percent.
   c. 23 percent.
   d. 50 percent.
9. A client has just begun to eat after days without significant amounts of food. Which of the following laboratory tests would be expected to respond most quickly to changes in energy and protein intakes?
   a. albumin
   b. transferrin
   c. total lymphocyte count
   d. retinol-binding protein
10. Physical signs of PEM might include all of the following *except:*
   a. low serum albumin.
   b. dull, brittle hair.
   c. poor grip strength.
   d. wasted appearance.

Clinical Applications

1. Considering the many ways that health problems can affect nutrient intake, explain why it is important to follow a systematic approach to nutrition care. Provide some specific examples of how the physician, nurse, dietetic technician, pharmacist, and social worker might assist the dietitian in the assessment process.

2. Describe the possible nutrition implications of these findings from a client's history and physical examination: age 73, lives alone, recently lost spouse, uses a walker, no teeth, pale skin, lack of energy, history of hypertension and diabetes, several medications prescribed.

3. Nurses and nurse's aides frequently shoulder much of the responsibility for collecting food intake data for kcalorie counts because they often deliver food trays and snacks and later retrieve them. Why is it important for a nurse or aide to verify and record both what the client receives (both the foods and the amounts) and the foods that remain uneaten? When might clients be enlisted to aid in the collection of food intake data, and when might such a course be unwise?

References

1 K. Hammond, Nutrition focused physical assessment, *Support Line,* August 1996, pp. 1–4.

ANSWERS

Multiple Choice

1. b	2. d	3. d	4. a	5. c
6. b	7. c	8. a	9. d	10. a

NUTRITION AND MENTAL HEALTH

MENTAL AND nutritional health go together. Mentally healthy people have the capacity to feed themselves well. Well-nourished people experience none of the nutrient deficiencies that might contribute to poor mental health. By the same token, when either type of health is impaired, both may be affected. People with mental and emotional problems often have poor diets, and people who are malnourished are often in poor mental health. The health care professional who recognizes the relationships between nutrition and mental health is in a better position to offer effective care.

IMPACT OF EMOTIONS ON EATING

To understand the connections between nutrition and emotional health, consider what ordinary anxiety does to your own eating habits. Do you lose your appetite? Overeat? Eat "junk" foods instead of balanced meals? Your reaction depends on your personality type. Temporary emotional stresses may have little effect on nutrition, but if such stresses become prolonged or chronic, the resulting changes in eating habits can lead to underweight, overweight, or nutrient imbalances.

Mental health problems that are difficult to overcome, such as depression, are more likely to lead to nutrition problems. Quite often people who are depressed lose interest in caring for themselves and in participating in usual activities such as eating, socializing, or pursuing hobbies. When people fail to care for themselves and cut themselves off from pleasurable activities and friendships, depression deepens and becomes a self-aggravating condition. People with depression may feel worthless, hopeless, drained of energy, and unable to sleep and concentrate. Thus people who are depressed are likely to have little interest in preparing and eating food.

DEPRESSION AND MALNUTRITION

Not surprisingly, many conditions increase the risk of both depression and malnutrition. The pain, loss of physical independence, and economic hardships imposed by serious illness as well as some medications used to treat illness can lead to both depression and malnutrition. The emotional adjustments necessary to accept and handle a terminal illness, such as an HIV infection, can also lead to both depression and malnutrition. Furthermore, when people become ill and lose significant amounts of weight, they may develop depression as their physical appearance deteriorates, and they become unable to perform routine tasks.

Depression and loneliness can profoundly affect nutrition status.

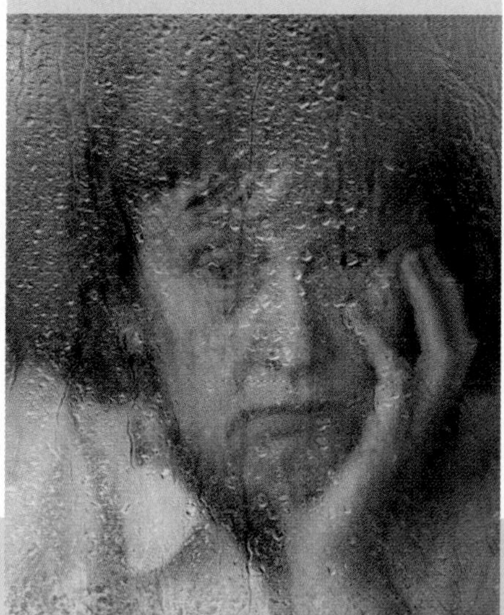

Depression in the Elderly

Although depression and its consequences affect people of all ages, depression is pervasive among the elderly and is a common cause of weight loss, especially for those in nursing homes.[1] Depression may be overlooked in the elderly because health care professionals may mistakenly regard depression as a normal consequence of aging.[2] Thus depression may deepen until it leads to serious problems.

Depression in the elderly often results from loneliness associated with social isolation and the loss of loved ones, mobility, or sense of purpose. Many authorities believe that among the elderly, loneliness is particularly relevant to depression and malnutrition. For human beings, eating is as much a social and psychological event as a biological one. Without companionship, appetite diminishes. Some 6 million adults over age 65 live alone. Their most pressing need seems to be for companionship; food takes second place. Social interaction is important to mental health, and elderly people of all classes in our society, both the financially secure and the poverty-stricken, tend to become isolated. Jack

Weinberg, professor of psychiatry at the University of Illinois, wrote perceptively of this problem:

> In our efforts to provide the aged with a proper diet, we often fail to perceive it is not what the older person eats but with whom that will be the deciding factor in proper care for him. The oft-repeated complaint of the older patient that he has little incentive to prepare food for only himself is not merely a statement of fact but also a rebuke to the questioner for failing to perceive his isolation and aloneness and to realize that food . . . for one's self lacks the condiment of another's presence which can transform the simplest fare to the ceremonial act with all its shared meaning.[3]

A sad spiral can set in when a lonely person begins to neglect to eat well. Malnutrition worsens the apathy felt due to loneliness. Then the person has even less energy with which to secure nourishment. Watch for this spiral in all people, and especially in elderly people who live alone and in those who have recently lost a spouse or other loved one and are grieving.

Preventing and Addressing Problems

Caring health care professionals remain alert for signs of depression in people of all ages. When depression is recognized early, the person can receive appropriate treatment before health and nutrition status markedly deteriorate. Nurses, counselors, and social workers can help elderly clients work through depression and find solutions to their loneliness. Nurses and dietitians can help clients understand how depression affects food intake and health and how eating a well-balanced diet can prevent health and nutrition problems. Meal plans that emphasize foods that are easy to prepare and eat can help some clients meet their nutrient needs when they lack the motivation to eat. Encouraging clients to eat with family or friends or at congregate meal sites can help combat loneliness.

NUTRIENT DEFICIENCIES AND MENTAL HEALTH

Among all age groups, nutrients and mental health can interact. Mental retardation in a child may be a consequence of protein-energy malnutrition during pregnancy. Children malnourished early in life often exhibit behavioral and social deficits as well as physical retardation. Conversely, children who are neglected early in life show a greater tendency to suffer from severe malnutrition than children who receive love and attention. Wherever you see a malnourished child, offer the child emotional as well as physical support. And wherever you see emotional illness, look to the child's nutrition, too.

Individual nutrient deficiencies can also affect brain function. Table 16-1 on p. 553 lists some brain functions affected by nutrient deficiencies. People with B vitamin deficiencies often exhibit symptoms ranging from confusion, apathy, fatigue, memory deficits, and irritability to delirium and psychoses. Severe niacin deficiencies, for example, can lead to **dementia.**[4] (The glossary defines dementia and other terms related to mental illness.) Deficiencies of folate and vitamin B_{12} are also associated with memory loss, depression and dementia.[5] Depression is a common manifestation of folate deficiency, and conversely, people diagnosed with depressive disorders frequently have low serum or red blood cell folate levels.[6] In the latter case, it is not clear whether folate deficiency leads to depression or depression leads to low folate levels by altering food intake.

NUTRIENT DEFICIENCIES AND SENILITY

Sometimes the confusion caused by nutrient deficiencies is incorrectly diagnosed as **senility.** An elderly person may even be wrongly confined to a nursing home. The story is told of a woman who exhibited the classic signs of senility—mental confusion, inability to make decisions, and forgetting to perform important tasks, such as turning off a stove burner. The woman's family decided to move her into a nursing home. While she was waiting for a place there, her family took her into their own home. After several weeks of eating good meals and enjoying social stimulation, the woman became her old self again and was able to return to her home. This story has been repeated with many variations and serves to remind us to think about loneliness and nutrition before concluding that a person is senile and needs institutional care. What harm could there be in first trying good, balanced meals served with tender, loving care?

GLOSSARY

Alzheimer's (ALTZ-high-merz) **disease:** see p. 551.

delusions (dee-LOO-shuns): inappropriate beliefs not consistent with the individual's own knowledge and experience.

dementia (dee-MEN-she-ah): irreversible loss of mental function.

mood disorders: mental illness characterized by episodes of severe depression or excessive excitement (mania) or both.

paranoia (PAR-ah-NOY-ah): mental illness characterized by delusions of persecution.

schizophrenia (SKITZ-oh-FREN-ee-ah): mental illness characterized by an altered concept of reality and, in some cases, delusions and hallucinations.

senility (see-NIL-ih-tee): mental or physical weakness associated with old age.

NUTRITION AND MENTAL ILLNESS

Nutrition in psychiatric care is a specialty all its own because there are so many connections. Mental illnesses characterized by depression, illogical thinking, dementia, **paranoia, delusions,** and inappropriate eating habits can alter food intake and thus interfere with nutrition status. Some of these disorders include **schizophrenia, Alzheimer's disease, mood disorders,** substance abuse, and eating disorders.

People with mental disorders characterized by depression risk poor nutrition status for reasons already described. People with mental illnesses characterized by illogical thinking or dementia may have little interest in food or may be unable to make appropriate food choices. Those who are paranoid may believe that foods are being used to poison them. People suffering from delusions may attribute magical powers to certain foods and insist on eating only those foods. Medications used in the treatment of mental illnesses can also interact with nutrients and alter nutrition status.

Among the most common mental disorders with considerable effects on nutrition status are alcoholism and eating disorders. The relationships between alcoholism and nutrition are treated in Highlight 7 and eating disorders are discussed in Highlight 9.

Nutrition affects the brain and the mind, and the brain and the mind affect the way people eat. All are interrelated, and the wise health care professional will keep these interrelationships in focus.

References

1 G. J. Kennedy, The geriatric syndrome of late-life depression, *Psychiatric Services* 46 (1995): 43–48.

2 C. Ryan and M. E. Shea, Recognizing depression in older adults: The role of the dietitian, *Journal of the American Dietetic Association* 96 (1996): 1042–1044.

3 J. Weinberg, Psychological implications of the nutritional needs of the elderly, *Journal of the American Dietetic Association* 60 (1972): 293–296.

4 J. E. Morley, Nutritional modulation of behavior and immunocompetence, *Nutrition Reviews* (supplement 2) 52 (1994): 6–8.

5 T. Bottiglieri, Folate, vitamin B_{12}, and neuropsychiatric disorders, *Nutrition Reviews* 54 (1996): 382–390.

6 J. E. Alpert and M. Fava, Nutrition and depression: The role of folate, *Nutrition Reviews* 55 (1997): 145–149.

Nutrition, Medications, and Complementary Therapies

CHAPTER 18

People today increasingly rely on **medications** and dietary supplements◆ to prevent and treat health problems. Conventional medical therapy includes a myriad of prescription and over-the-counter medications that reach the market each year. As the population expands and people live longer, the number of people taking medications is growing rapidly. At the same time, people are using more and more complementary therapies, especially dietary supplements. In 1997, for example, Americans spent an estimated $12 billion for dietary supplements (including herbs).[1] Whether a person uses a vitamin C supplement to prevent a cold, an herbal remedy to treat indigestion, an over-the-counter medication to relieve pain, or a prescription medication to lower blood pressure, the potential for adverse side effects exists. The chemicals that compose medications and dietary supplements can affect metabolism and body processes just as nutrients from foods can. Some of the most serious side effects occur when medications interact with each other (drug-drug interactions). Adverse side effects that can affect nutrition status include those that interfere with the ingestion, digestion, absorption, metabolism, or excretion of nutrients. This chapter explains the differences between medications and dietary supplements and explores the dynamics and consequences of diet-drug interactions.

Reminder: *Dietary supplements* are chemicals (drugs) taken by mouth that contain ingredients such as vitamins, minerals, herbs, amino acids, enzymes, organ tissues, metabolites, extracts, or concentrates. Examples include vitamin-mineral supplements, creatine, melatonin, tryptophan, shark cartilage, and coenzyme Q. ◆

OVERVIEW OF CONVENTIONAL AND COMPLEMENTARY THERAPIES

What distinguishes conventional from complementary therapies? Conventional therapies are widely taught in required courses in medical schools and practiced routinely in health care facilities. Conversely, many practicing health care professionals have no formal training in complementary therapies, although more recent graduates are more likely to have some training.

Whereas research often supports the value and safety of conventional treatments, evidence to support complementary therapies often comes from folklore, tradition and testimonial accounts. When clinical trials of complementary therapies have been conducted, the studies are often poorly-controlled, include only small numbers of people, or have not been repeated. In fairness, however, not all conventional therapies are based on controlled clinical trials. Even some of the conventional diet therapies presented in this book are based largely on logic and clinical experience rather than research. An example is the clear-liquid diet traditionally provided following surgery. Such a diet is believed to be easier to tolerate after surgery than solid foods. Some clinicians question this assumption and advocate the use of easy-to-digest solid foods instead. Research to back either position is minimal. Conversely, just because a complementary therapy is based on hearsay or folklore doesn't mean it isn't effective. A complementary therapy may become a conventional therapy if enough research and experience become available to support its use.

One thing is clear—people are using complementary therapies regardless of whether conventional health care professionals approve. In recognition of this reality, more medical schools offer elective courses in complementary therapies or discuss these therapies in required courses. More health care professionals are learning about complementary therapies and helping clients to use them along with conventional therapy. Some major hospitals now have complementary medicine centers and some health plans cover the costs of complementary therapies.

The term *alternative therapy* is frequently used interchangeably with the term *complementary therapy*. Table 18-1 presents examples of alternative therapies. We prefer the term *complementary therapy* because most people use alternative therapies in addition to, rather than in place of, conventional therapies.

medications: chemicals (drugs) that alter one or more body functions that are marketed only with approval of the Food and Drug Administration and only after research shows that they are safe and effective.

TABLE 18-1 Fields of Alternative Medicine and Selected Examples

Mind-body interventions	Manual healing methods
Biofeedback	Biofield therapeutics
Faith healing	Chiropractic
Hypnotherapy	Massage therapy
Imagery	Pharmacological and biological treatments
Meditation	Cartilage therapy
Bioelectromagnetic applications in medicine	Chelation therapy
Electroacupuncture	Ozone therapy
Microwave resonance therapy	Herbal medicine
Alternative systems of medical practice	Diet and nutrition in the prevention and treatment of chronic disease
Acupuncture	Macrobiotic diets
Ayurveda	Orthomolecular medicine
Homeopathic medicine	
Naturopathic medicine	

In this chapter, the discussion of complementary therapies includes only dietary supplements. Like prescription and over-the-counter medications, these substances are composed of chemicals taken to treat and prevent health problems and have similar effects on body processes. As you read through the remainder of the chapter, keep in mind that people may use other chemicals, including alcohol (see Highlight 7) and illicit drugs, that can also interact with diet and nutrients, medical drugs, and dietary supplements.

THINK NUTRITION whenever clients use prescription medications, over-the-counter medications, dietary supplements, alcohol, or illicit drugs.

Medications

Over-the-counter and prescription medications fall within conventional medical therapies. Medications make a claim to affect a health problem, and that claim must be backed up by research. Medications reach the market only after they have been extensively studied and approved by the Food and Drug Administration (FDA). Thus the benefits and risks of medications are often clear, and information about proper dosing is available. Medication labels, even for medicated throat lozenges, must carry general precautions and dosing information. The amount of active ingredients in each dose is also carefully controlled. Even with these safeguards in place, however, serious side effects may become evident after a product is marketed, and a previously approved medication may be withdrawn from the market.

Many new medications become available each year. Often these medications are stronger and act for longer periods of time than older medications.

Health care professionals face a challenge in helping clients prevent and recognize adverse side effects and interactions associated with medications and dietary supplements.

• Prescription Medications • Prescription medications are often more potent and used for longer periods of time than over-the-counter medications, and they also carry a greater risk of serious side effects. Thus they require a prescription, which ensures that a physician evaluates the client's health status and determines that the benefits from using a prescription medication outweigh the risks of potential or actual side effects. Physicians, pharmacists, nurses, and dietitians share responsibility for educating clients about proper use of prescription medications and how to prevent, recognize, and handle side effects.

• Over-the-Counter Medications • Many medications, once available only by prescription, become available over-the-counter. Over-the-counter medications are most often intended for temporary use to relieve common but relatively minor health problems. Examples include aspirin to treat headaches or pain, decongestants to relieve stuffy noses, and antacids to combat indigestion. Although they are less potent and are used for shorter periods of time than prescription medications, over-the-counter medications may present a hidden danger by masking symptoms that require a physician's care. Because people often self-medicate with over-the-counter medications, they may use the medications for the wrong reasons, in the wrong amounts, or for longer-than-recommended periods of time. People who self-medicate may also be unaware of potential interactions with prescription medications and dietary supplements.

People who use herbs often use them along with conventional therapies to treat or prevent minor health problems.

Dietary Supplements

Before medications became widely available, people often relied on herbs and plants to cure aches and ills with varying degrees of success. Upon scientific scrutiny, some of these folk remedies reveal ingredients that explain their use for a medical problem. The herb valerian, for example, which has long been used as a tranquilizer, contains oils that have a sedative effect. The compounds that some plants make are so beneficial, in fact, that they are isolated and used in many modern medicines. The quantities of active ingredients in the plants, however, are often much lower than in the medications, where the ingredients are isolated. Other dietary supplements have failed to show benefits or simply have not been thoroughly studied. Many of these dietary supplements are safe, even though they may not be effective.

Highlight 10 provided information about dietary supplements, including examples of risks associated with their use. When nutrient supplements, such as antioxidant supplements, are taken at levels that greatly exceed recommended intakes, the supplement's action is pharmacological rather than physiological; that is, the supplement acts more like a drug than a nutrient or other food component.

• Marketing of Dietary Supplements • Unlike medications, dietary supplements can be marketed without studies to document their effectiveness and safety and without prior approval of the FDA.◆ Although labels of dietary supplements can make no direct claim to affect a health problem, manufacturers avoid this restriction with creative wording. The label of an herbal product cannot claim that an herb alleviates insomnia, for example, but it can claim that the herb promotes restful sleep. Such a claim may be made without research to support the claim provided that the label carries the disclaimer: "Has not been evaluated by the Food and Drug Administration."

◆ The regulations that govern the labeling and marketing of dietary supplements are contained in the *Dietary Supplement Health and Education Act* (DSHEA).

Even when the active ingredients in a dietary supplement have been shown to be safe and effective, the dose suggested on the label may not provide the amount of active ingredients found to be effective. Preliminary clinical testing of the herb saw palmetto, for example, has shown some benefits over the placebo for treating some types of prostate-related problems. In a recent test, however, only 8 of 13 brands of saw palmetto provided effective amounts of the active ingredient when used as suggested on the label; the cost ranged from 44 cents to $1.44 per day.[2] Of the remaining 5 brands, the label-recommended doses supplied as little as 1 percent of the effective dose. Thus people may spend a great deal of money on products that may or may not be safe, effective, or reliably labeled.

In contrast to medications, which cannot be marketed unless they are proved to be safe, the FDA has the burden of proving that a dietary supplement is not safe. And such proof often must be substantial. Thus some products remain on the market despite warnings from the FDA. Such is the case with ephedra (commonly known as ma huang), a product promoted for weight loss, which remains on the market even though its use has been linked to 38 deaths.

• Other Problems Associated with Dietary Supplements • Health care professionals and clients are faced with a countless array of herbs and dietary supplements, which come in a multitude of brand names, strengths, and formulations of the same product.[3] Although some clients consult traditional health care professionals or licensed, alternative practitioners, others simply self-medicate or ask the advice of store clerks. In so doing, clients may delay or discontinue effective conventional treatments for remedies that may have little merit. The consequences can sometimes be serious and irreversible.

Furthermore, clients who use herbs and dietary supplements often fail to tell health care professionals about them. Some clients may simply forget to mention the products, especially if health care professionals don't ask. Others mistakenly believe that herbs and dietary supplements are natural and safe and pose no health risks. Still others may consciously withhold information about herb or dietary supplement use because they feel health care professionals will disapprove or have little knowledge about these products, which is true in many cases.

Clients may also be unaware that dietary supplements can interact with medications—sometimes with potentially serious consequences. Because research on dietary supplements is often lacking, assessing potential interactions is difficult for health care professionals and clients alike.[4] Even less is known about the effects of herbs and other dietary supplements on nutrition status. In the United States, about 44 percent of people take prescription drugs, and of these, more than 18 percent say they also use an herbal product or a megadose of a nutrient supplement.[5] Of the18 percent who use herbal products or dietary supplements, 46 percent do so with no medical supervision.

IN SUMMARY

Clearly, people use a wide variety of chemicals to prevent and treat health problems, and all of these pose risks, some known, some unknown. The sheer number of medications, herbs, and dietary supplements makes it difficult for health care professionals and consumers to evaluate risks, and undoubtedly, new problems will surface in the future. The next section looks at factors that raise the risk of serious side effects and offers suggestions for minimizing risks.

WEIGHING RISKS AND BENEFITS

Physicians decide what medication to prescribe based on the client's health problems, any drugs currently being taken, and the potential benefits of the medication weighed against the risks. When clients self-medicate with over-the-counter medications or dietary supplements, they may not have all the information needed to make educated decisions.

The remainder of this chapter specifically addresses interactions between medications and other remedies that affect nutrient needs or nutrition status or the medication's effectiveness. Keep in mind, however, that many of the risks for diet-drug interactions also apply to medication-medication interactions. For simplicity, we use the term *diet-drug interactions* to include interactions between prescription medications, over-the-counter medications, herbal products, and dietary supplements and foods, individual nutrients, and non-nutrient food components.

• Learning about Benefits and Risks • With hundreds of prescription medications, over-the-counter medications, herbs, and dietary supplements available, how do health care professionals learn about potential diet-drug interactions? This chapter provides a basic introduction and some examples, but is by no means complete. Learning about diet-drug interactions is difficult at first, but becomes easier with clinical experience. At first, looking up and learning about medications and dietary supplements encountered in clinical practice is an inescapable task. Drug guides that are available for purchase or use in health care facilities or libraries provide information about prescription and over-the-counter

TABLE 18-2 **Popular Herbs, Their Common Uses, and Risks**

Common Name	Claims and Uses[a]	Risks[b]
Echinacea (root)	Boosts immunity; promotes wound healing; shortens duration of colds and flus	Generally considered safe; may potentiate the effects of warfarin and interfere with immunosuppressants
Evening primrose (oil)	Relieves symptoms of premenstrual syndrome	Few studies to support safety and effectiveness
Garlic (bulb)	Lowers blood lipids and prevents atherosclerosis; lowers blood pressure	Generally considered safe; may cause mouth odors, gas, and heartburn; may potentiate the effects of warfarin, antiplatelet agents, and antidiabetic agents
Ginkgo (leaf)	Enhances memory; improves peripheral circulation	Generally considered safe; may cause headaches and GI distress; may potentiate the effects of warfarin and antiplatelet agents
Ginseng (root)	Combats fatigue; restores stamina and impaired concentration; improves sexual performance	May cause hypertension, insomnia, headaches, euphoria, diarrhea, and edema; may interact with monoamine oxidase (MAO) inhibitors, aspirin, caffeine, warfarin, heparin, anti-inflammatory agents, and antidiabetic agents
Kava (root)	Promotes sleep; relaxes muscles; reduces stress and anxiety	Generally considered safe; may interfere with the actions of central nervous system (CNS) stimulants and potentiate the effects of CNS depressants (including alcohol)
Milk thistle (fruit)	Protects liver tissue	Generally considered safe; may cause diarrhea
Saw palmetto (fruit)	Relieves symptoms of enlarged prostate; enhances sexual performance; enlarges mammary glands	Generally considered safe; may cause GI distress; may interact with medications used to treat prostate problems and steriod hormones (testosterone, oral contraceptives)
St. John's wort (leaf and flower)	Relieves depression and anxiety	Considered to be safer than prescription antidepressants; may cause GI distress, dizziness, confusion, and sedation; may interact with antidepressants, cyclosporine, digoxin, oral contraceptives, antiviral agents, theophylline, warfarin, and calcium channel blockers (anti-hypertensive agents)
Valerian (root)	Promotes sleep; relieves anxiety	Generally considered safe; may cause headaches, insomnia, and restlessness; long-term use not recommended; may potentiate the effects of CNS depressants and limit the effects of CNS stimulants

[a]Reminder: Although some dietary supplements have undergone clinical studies to support health-related claims, research is limited.
[b]Allergies are always a possible side effect.
SOURCES: J. E. Robbers and V. E. Tyler, *Herbs of Choice: The Therapeutic Use of Phytochemicals* (New York: The Haworth Herbal Press, 1999); P. Shah and K. L. Grant, An overview of common herbal supplements, *Support Line,* October 2000, pp. 3–7; S. M. Riddle, Drug interactions: Examining the impact of botanicals and dietary supplements, *Support Line,* October 2000, pp. 9–13, 16–18.

medications. It is vital that the guide be current—many new medications reach the market each year, and new information about side effects and precautions also becomes available. Many drug guides include information about interactions between medications and herbs and dietary supplements. Several Internet sources for information about drugs and complementary therapies are listed at the end of this chapter. Table 18-2 lists popular herbs and their common uses and risks. Remember, however, that information about the safety and effectiveness of herbs and dietary supplements is often scarce.

• **Evaluating Risks** • Health care professionals should be aware that some clients are at higher risk for adverse nutrition-related side effects from medications, herbs, and dietary supplements. These include women who are pregnant or nursing and clients who:

- Take medications, herbs, or other dietary supplements for long periods of time.
- Regularly take several medications, herbs, or other dietary supplements.
- Regularly use alcohol or illegal drugs, either of which can affect nutrition status or interact with other drugs.
- Fail to use medications or other remedies as directed.
- Are in poor health, especially if they have altered organ function.
- Are in poor nutrition status.
- Have medical conditions that markedly raise nutrient needs.

Several of these conditions may coexist, especially in the elderly. Consider that elderly people are more likely than others to have:

- Chronic diseases that require the use of multiple medications over long periods of time.
- Difficulty taking medications as prescribed, either due to changes in cognitive function or due to physical or financial problems that affect their ability to acquire and pay for medications.
- Malnutrition due to chronic diseases, altered organ function, or limited food intake as a consequence of poor mental and emotional health, physical disabilities, or financial problems.

Not surprisingly, studies of institutionalized elderly people show that multiple medication use may significantly affect their nutrition status.[6]

• Limiting Risks • The most important steps health care professionals can take to limit the risks of drug-related side effects is to know what and how much prescription medications, over-the-counter medications, herbs, other dietary supplements, alcohol, and illegal drugs clients are taking. Always ask! Clients may be taking medications prescribed by different physicians; taking health remedies recommended by alternative medicine practitioners, store clerks, trainers, or friends; or simply trying a remedy they read about. Health care professionals who use a nonjudgmental, culturally sensitive, respectful, and open-minded approach are most likely to learn the facts about what drugs clients take and why they are taking them. This information provides the basis for clients and health care professionals to work out acceptable medical care plans that respect clients' rights to make informed health care decisions with the fewest risks of side effects. For clients with limited knowledge of the products they use, it is wise to ask them to bring in their medications and dietary supplements.

Once health care professionals know what products clients use, the next step is to evaluate what and how much the client is taking and limit the total number of products or doses the client uses whenever possible. Ideally, the client's primary physician evaluates prescription medications and determines if all are safe and necessary for the client to take. Health care professionals including physicians, nurses, pharmacists, and dietitians share responsibility for educating clients about the safety, effectiveness, side effects, precautions, and interactions of all the products they use.

For clients who regularly use over-the-counter medications and dietary supplements, find out the reasons why. Perhaps a problem needs to be investigated, or the client may be able to make lifestyle changes that eliminate the problem without the use of a drug. A client who regularly uses an over-the counter medication or herb to relieve indigestion, for example, may benefit from nutrition counseling that helps to identify foods and eating habits that contribute to indigestion (see Chapter 20). In some cases, a client who takes a vitamin or mineral supplement may be able to include more food sources of the nutrient and eliminate the need for the supplement.

For clients contemplating the use of a dietary supplement, discuss what is known about the therapy's safety and effectiveness. Recommend that the client try only one new therapy at a time, so that the client can determine if the therapy is working. One clinician recommends that the client keep a symptom record, logging the severity and frequency of symptoms as well as factors that make the symptoms worse or better.[7]

Health care professionals can also encourage clients to purchase all medications at the same pharmacy and to tell the pharmacist about over-the-counter medications, herbs, and dietary supplements they are taking. With a complete account of clients' medication use, the pharmacist will be able to alert physicians and clients to potential problems. Many pharmacies have computer programs that track potential interactions between medications, herbs, and supplements.

IN SUMMARY

Health care professionals limit the likelihood of serious side effects from medications and dietary supplements by evaluating their clients' risks for serious side effects, knowing what clients are taking, and eliminating the use of unnecessary products. Health care professionals share responsibility for explaining what is known about each product's safety and effectiveness, precautions, side effects, and interactions.

DIET-DRUG INTERRELATIONSHIPS

Current information about diet-drug interactions involving dietary supplements is very limited, and this discussion, therefore, focuses mainly on diet-medication interactions. As more and more people use dietary supplements, however, the likelihood of additional interactions will grow. Health care professionals who understand the ways nutrients and drugs can interact will be best prepared to understand and recognize such interactions.

Clinicians frequently overlook or fail to recognize diet-drug interactions, yet these interactions can raise health care costs and result in serious, and even fatal, complications.[8] With hundreds of diet-drug interactions known to exist and many more to be identified in the future, health care professionals must learn how to prevent, recognize, and handle them. The "How to" on p. 598 offers practical suggestions.

Foods and food components, including nutrients and nonnutrients, can alter the absorption, metabolism, and excretion of medications. Likewise, medications can alter food intake and the absorption, metabolism, and excretion of nutrients. Table 18-3 on p. 599 lists the general classes of medications notable for their interactions with foods and food components. The following sections provide examples of different types of interactions, and Table 18-4 on p. 600 summarizes this information and provides specific examples.

Medications and Food Intake

Medications can reduce food intake by directly suppressing the appetite or by causing complications that make eating difficult. Conversely, some medications heighten the appetite and lead to weight gain. Still other medications provide relief from complications that interfere with the appetite, so the person will be able to eat.

THINK NUTRITION whenever clients take medications or dietary supplements that can reduce food intake.

• **Altering the Appetite** • Most medications prescribed for obesity work by intentionally suppressing the appetite. Amphetamines, sibutramine (Meridia), and the herb ephedra are examples. Sometimes, however, appetite suppression is unintentional. Such is the case when amphetamines are prescribed to treat attention deficit hyperactivity disorder. In this case, amphetamines are used to improve concentration and behavior, and appetite suppression and weight loss are undesirable side effects.

Medications such as megestrol acetate, dronabinol, growth hormone, and testosterone can improve the appetite and help people gain weight. Unintentional weight gains can result from the use of some antianxiety agents, antidepressants, and antipsychotics.

• **Causing or Alleviating Complications** • Medications can lead to symptoms that make it difficult to eat. Sedatives, for example, can make a person too tired to eat. Many medications including some antibiotics, many medications used in

HOW TO Manage Diet-Drug Interactions

Begin with a list of medications prescribed for a particular client. Ask the client about the types and amounts of over-the-counter medications and dietary supplements used on a regular basis. Using a drug guide, look up each medication and make a note of:

- The appropriate method of administering the medication (twice daily or at bedtime, for example).
- How the medication should be given with respect to foods or specific nutrients (give on an empty stomach, give with food, do not give with milk, give iron supplements at least two hours apart from the medication dose, for example).
- How the medication should be given with respect to other medications.
- Side effects that can affect food intake (nausea, vomiting, or sedation, for example), nutrient needs (hypokalemia or hyperglycemia, for example), or dietary recommendations (constipation or flatulence, for example).

Standards set by the Joint Committee on Accreditation of Healthcare Organizations (JCAHO) include the provision that all clients be educated about potential diet-drug interactions. The physician, nurse, pharmacist, or dietitian should review all precautions related to medications and dietary supplements with the client, explain signs of potential nutrient deficiencies or nutrition-related problems, and advise the client of actions to take if problems arise.

Use a similar process for any dietary supplements the client is taking. Using a reliable reference, study what is known about the dietary supplements, how they should be taken, and their potential interactions and side effects.

Clients, particularly those who must take multiple medications, may need help figuring out when to take each medication to avoid medication-medication or diet-drug interactions. Chapter 17 described ways to uncover information about a client's usual eating habits. Use this information to coordinate medications that must be administered with regard to food intake or specific dietary components. For additional information about a medication or potential interaction, remember to ask the pharmacist.

For medications that can lead to nutrient deficiencies or alter nutrient needs, remain alert for signs of nutrient imbalances, especially when:

- Imbalances are commonly noted with the use of the medication.
- The adverse effect persists over time.
- The client is in a high-risk group (see p. 595–596).
- The client will need to take the medication for a long period of time.

Remember to alert the dietitian if you suspect a problem or think that the client can benefit from nutrition counseling.

the treatment of cancer, garlic, ginkgo, saw palmetto, St. John's wort, and iron supplements can lead to indigestion and nausea and thus limit food intake. Table 18-5 on p. 601 lists many other symptoms and complications associated with drugs that can lead to a poor appetite. Complications that limit food intake are significant only when they persist over time. Almost all medications, for example, can cause nausea for some people. Often nausea subsides after the first few doses of the medication. If nausea persists, however, weight loss and malnutrition can follow, especially if the medication must be taken over a long period of time.

Some medications treat symptoms and complications that can reduce food intake. Antinauseants and antiemetics, for example, help reduce nausea and vomiting. The herbs peppermint and chamomile are promoted as aids to alleviate indigestion.

Diet-Drug Interactions and Absorption

Some foods and food components can affect the ways medications are absorbed and consequently how much of a medication becomes available to the body. Medications can also affect the ways some nutrients are absorbed.

THINK NUTRITION whenever clients take medications that can interfere with nutrient absorption or that require dietary considerations for proper absorption.

• Diet Effects on Medication Absorption • Foods frequently affect medication absorption. Foods reduce the absorption of one antihypertensive drug, captopril, and improve the absorption of another, hydralazine. Thus captopril should be taken on an empty stomach, while hydralazine should be taken with food. In some cases, foods delay, but do not reduce, a medication's absorption. Aspirin works faster when taken on an empty stomach than when given with food, but because aspirin can irritate the GI tract, taking it with food can reduce nausea.

Individual nutrients and nonnutrients in foods can also affect drug absorption. Among the more common substances in foods that can bind with medications and reduce their absorption are minerals, fiber, phytates,◆ and oxalates. Note that antacids contain a variety of minerals, which may include aluminum, calcium, magnesium, and sodium. Thus, when any of these minerals interferes with the absorption of a medication, antacids or other medications (including mineral supplements) that contain the offending mineral must be restricted.

◆ Reminder: *Phytates* are nonnutrients found in the husks of grains, legumes, and seeds. *Oxalates* are also nonnutrients; they are found in significant amounts in rhubarb, spinach, beets, nuts, chocolate, tea, wheat bran, and strawberries.

• Medication Effects on Nutrient Absorption • Laxatives provide an example of how medications can interfere with nutrient absorption. Some laxa-

TABLE 18-3 **Classes of Medications That Can Affect Nutrition Status**

Classification	Possible Side Effects That Can Affect Nutrition Status
Analgesics, narcotic	Sedation, nausea and vomiting, reduced motility of GI tract
Antacids	Constipation, diarrhea
Antibiotics	Nausea, vomiting, diarrhea
Anticonvulsants	Nausea, vomiting, GI distress
Antidepressants	Weight changes, dry mouth, nausea and vomiting, diarrhea, constipation
Antidiabetic agents	GI distress, diarrhea
Antidiarrheals	Nausea, constipation
Antifungal agents	Depressed appetite, nausea and vomiting, GI distress, diarrhea
Antihypertensives	Nausea, drowsiness, dry mouth, constipation, dizziness
Antilipemics	Nausea, GI distress, constipation
Antineoplastics	Depressed appetite, nausea and vomiting, dry mouth, taste alterations, mouth ulcers, mouth inflammation, fatigue, diarrhea, fever
Antiulcer agents	Reduced absorption of iron and vitamin B_{12}
Antiviral agents	Depressed appetite, nausea and vomiting, GI distress
Central nervous system stimulants	Depressed appetite, dry mouth, taste alterations
Corticosteroids	Nausea and vomiting, insulin resistance, altered calcium and vitamin D metabolism, negative nitrogen balance, sodium and fluid retention
Diuretics	Altered excretion of sodium, potassium, mangesium, phosphorus, calcium, and zinc
Hormonal agents	Appetite and weight changes; various other side effects depending on agent
Immunosuppressants	Nausea and vomiting, diarrhea, constipation, impaired renal function
Laxatives	GI gas, laxative dependency, nutrient malabsorption

tives move foods rapidly through the intestine, reducing the time available for nutrient absorption. Other laxatives reduce nutrient absorption for different reasons. For example, fat-soluble vitamins (notably, vitamin D) dissolve in and are excreted along with mineral oil, an indigestible oil that is sometimes used as a laxative. Calcium, too, is excreted. An added danger with all laxatives is that a person who uses them daily for a long time may find that the intestines can no longer function without them. The more often laxatives are used, the more likely that nutrient deficiencies will develop.

Other medications improve nutrient absorption. Enzyme replacements, for example, help clients who lack digestive enzymes absorb protein, fat, and carbohydrate. Lactase enzyme replacements help clients with lactose intolerance to absorb lactose.

• Other Absorption-Related Interactions • Nutrients and medications can also interact and reduce the absorption of both. A classic example is the interaction between the antibiotic tetracycline and the minerals calcium and iron. When either of these minerals and tetracycline are taken at the same time, the mineral binds to the tetracycline, and both are excreted. To circumvent this problem, clients are instructed to take tetracycline on an empty stomach at least one hour before or two hours after meals or after using milk, milk products, calcium-containing antacids, and mineral supplements.

Diet-Drug Interactions and Metabolism

Medications taken orally and absorbed through the GI tract or those delivered directly into the circulation can interact with the components from foods that enter the bloodstream or from intravenous nutrients. The alterations in

TABLE 18-4 Mechanisms and Examples of Diet-Drug Interactions

Drugs Can Alter Food Intake by:	
• Altering the appetite (amphetamines suppress the appetite).	
• Interfering with taste or smell (amphetamines change taste perceptions).	
• Inducing nausea or vomiting (digitalis can do both).	
• Changing the oral environment (phenobarbital can cause dry mouth).	
• Causing sores or inflammation of the mouth (methotrexate can cause painful mouth ulcers).	
Drugs Can Alter Nutrient Absorption by:	**Foods Can Alter Drug Absorption by:**
• Changing the acidity of the digestive tract (antisecretory agents can interfere with iron absorption).	• Changing the acidity of the digestive tract (candy can change the acidity, thereby causing slow-acting asthma medication to dissolve too quickly).
• Altering motility of the digestive tract (laxatives speed motility, causing the malabsorption of many nutrients).	• Stimulating secretion of digestive juices (griseofulvin is absorbed better when taken with foods that stimulate the release of digestive enzymes).
• Damaging mucosal cells (chemotherapy can damage mucosal cells).	• Altering rate of absorption (aspirin is absorbed more slowly when taken with food).
• Binding to nutrients (some antacids bind phosphorus).	• Binding to drugs (calcium binds to tetracycline, limiting drug absorption).
	• Competing for absorption sites in the intestines (dietary amino acids interefere with levodopa absorption this way).
Drugs and Nutrients Can Interact and Alter Metabolism by:	
• Acting as structural analogs (as warfarin and vitamin K do).	
• Competing with each other for metabolic enzyme systems (as phenobarbital and folate do).	
• Altering enzyme activity and contributing pharmacologically active substances (as monoamine oxidase inhibitors and tyramine do).	
Drugs Can Alter Nutrient Excretion by:	**Nutrients Can Alter Medication Excretion by:**
• Altering reabsorption in the kidneys (some diuretics increase the excretion of sodium and potassium).	• Changing the acidity of the urine (vitamin C can alter urinary pH and limit the excretion of aspirin).
• Displacing nutrients from their plasma protein carriers (aspirin displaces folate).	

metabolism can affect the ways medications work, affect the availability of nutrients, or in other ways exert negative effects on body processes.

THINK NUTRITION whenever clients take medications that affect, or are affected by, the absorption of nutrients.

• Interactions Affecting Medications • Vitamin K and the anticlotting medication warfarin (Coumadin) provide an example of how a nutrient can affect the way a medication works. The chemical structure of warfarin resembles vitamin K, and it is this property that makes it an effective anticlotting agent. Warfarin interferes with the synthesis of clotting factors that require vitamin K. The prescribed warfarin dose depends, in part, on how much vitamin K is in the diet. If a person's vitamin K intake changes, as may happen in summer when lettuce and greens are in season, then the physician has to alter the medication dose.◆ A note of caution: many herbs and dietary supplements can interact with warfarin and either reduce or potentiate its effectiveness. In either case, clients should be advised to avoid these herbs and dietary supplements or to use them consistently so that the proper warfarin dose can be determined. Herbs that can interact with warfarin include danshen, don quai, echinacea, feverfew, garlic, ginger, ginkgo, ginseng, and St. John's wort.[9] Other dietary supplements that can interact with warfarin include coenzyme Q, omega-3 fatty acids, and vitamin E (doses greater than 400 IU per day).

◆ To avoid potential problems, people taking warfarin should try to consume a consistent intake of vitamin K every day, and that amount should meet current dietary recommendations.

In some cases, nutrients must be available to maximize a medication's effectiveness. The medication aldendronate sodium (Fosamax), used to increase bone mass and prevent osteoporosis, for example, depends on an adequate supply of vitamin D and calcium, either from the diet or from supplements, for maximum effectiveness.

Among the notable foods and food components that can affect medication metabolism are grapefruit juice (but not other citrus juices), caffeine, and natural licorice. An area currently generating a great deal of interest and research is the interaction between grapefruit juice and a variety of medications. So far, grapefruit juice has been shown to raise blood levels of some medications, including some calcium channel blockers used to treat hypertension, some cholesterol-lowering medications (statins), the antianxiety agent buspirone, and the immunosuppressant cyclosporine to name a few.[10]

Caffeine, which acts as a central nervous system stimulant, can potentiate the actions of other central nervous stimulants, such as amphetamines or ephedra, and limit the effectiveness of some central nervous system depressants, such as barbituates. Natural licorice and the herb licorice root can complicate drug therapy that includes diuretics and antihypertensive agents because they promote sodium retention and potassium excretion. Most licorice sold as candy or breath fresheners in the United States is not natural licorice, however, but a flavored substitute that does not interact with medications.

• **Interactions Affecting Nutrients** • Corticosteroids, which act as anti-inflammatory agents and immunosuppressants, provide an example of how a medication can affect nutrient metabolism. Corticosteroids alter hormones that affect the way the body uses calcium and vitamin D, and in so doing, they raise the risk of osteoporosis.

Methotrexate, a medication used to treat certain cancers, provides another example. Methotrexate resembles folate in structure (see Figure 18–1). Because of this similarity, methotrexate prevents the conversion of folate to its active form, and signs of folate deficiency develop. Methotrexate may also be used to treat rheumatoid arthritis, psoriasis, and inflammatory bowel diseases, but in these cases, lower doses of the medication are prescribed, and signs of folate deficiencies are less common.

Aspirin can also alter folate metabolism but in a different way. Aspirin competes with folate for its protein carrier, thus hindering the body's use of the vitamin. When aspirin is used over long periods of time, health care professionals should ensure that either the diet or supplements are supplying sufficient folate to meet the added needs.

TABLE 18-5 Examples of Drug-Induced Side Effects That Can Limit Food Intake

Altered tastes
Anorexia
Belching
Bloating
Blurred vision
Chest pain
Confusion
Congestion, nasal
Constipation
Coughing
Cramps, abdominal
Diarrhea
Dizziness
Dry mouth
Epigastric pain
Fatigue
GI distress
Indigestion
Inflammation of mouth tissue
Intestinal gas
Mouth ulcers or lesions
Nausea
Pain
Sedation
Shortness of breath
Throat irritation

FIGURE 18-1 Folate and Methotrexate

By competing for the enzyme that activates folate, methotrexate prevents cancer cells from obtaining the folate they need to multiply. In the process, normal cells are also deprived of the folate they need.

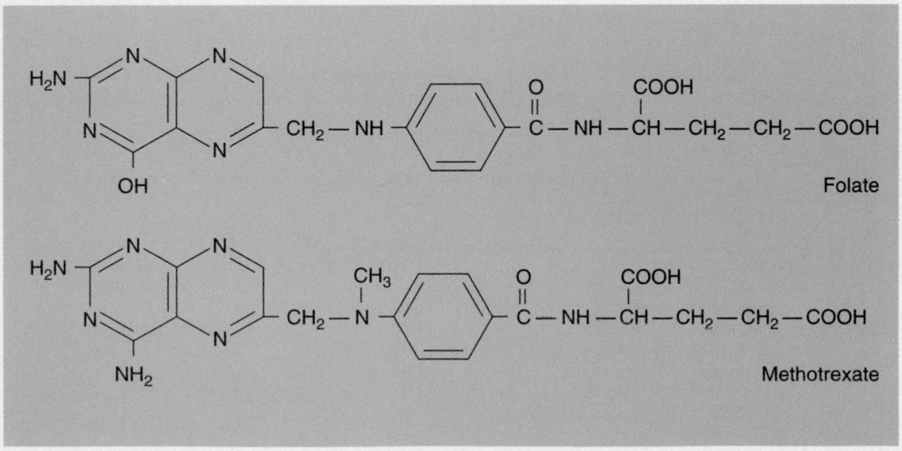

• Other Interactions • Tyramine, a substance found in some foods, and monoamine oxidase (MAO) inhibitors, medications prescribed to treat certain forms of severe depression, provide an example of how foods and medications can alter metabolism and lead to a potentially fatal outcome. MAO inhibitors block the action of the enzyme in the brain that normally inactivates tyramine. When people who take MAO inhibitors consume large amounts of tyramine, tyramine remains active and stimulates the release of the neurotransmitter norepinephrine. Severe headaches and hypertension can result, and if blood pressure rises high enough, it can be fatal. For this reason, people taking MAO inhibitors are advised to restrict their intakes of foods rich in tyramine (see Table 18-6).

Diet-Drug Interactions and Excretion

Nutrients and medications can also interact and affect the way one or the other is excreted. When diet-drug interactions cause nutrients to be excreted in greater-than-normal amounts, deficiencies can develop. When diet-drug interactions cause medications to be excreted in greater-than-normal amounts, the medication may not remain available to the body for as long as intended.

THINK NUTRITION whenever clients take medications that affect nutrient excretion or those whose excretion can be affected by diet.

• Nutrient Effects on Medication Excretion • Nutrients can alter urinary acidity, which, in turn, can affect the reabsorption of a medication from the kidneys back into the blood. An acidic urine limits the excretion of acidic drugs like aspirin, and megadoses of vitamin C contribute to urinary acidity. Consequently, when a person takes aspirin along with megadoses of vitamin C, the aspirin remains available to the body for longer periods of time. Dietary sodium intake greatly affects the reabsorption of the medication lithium, used to treat certain psychiatric disorders. People taking lithium are advised to maintain a consistent intake of sodium from day to day in order to maintain a stable blood level of lithium.

• Medication Effects on Nutrient Excretion • Medications can also alter the urinary excretion of nutrients. For example, some diuretics accelerate the ex-

TABLE 18-6 Foods Restricted in a Tyramine-Controlled Diet

Beverages	Red wines including chianti, sherry[a]
Cheeses	Aged cheeses, American, camembert, cheddar, gouda, gruyère, mozzarella, parmesan, provolone, romano, roquefort, stilton[b]
Meats	Liver; dried, salted, smoked, or pickled fish; sausage; pepperoni; salami; dried meats
Vegetables	Fava beans; Italian broad beans; sauerkraut; snow peas; fermented pickles and olives
Other	Brewer's yeast;[c] all aged and fermented products; soy sauce in large amounts; cheese-filled breads, crackers, and desserts; salad dressings containing cheese

Note: The tyramine contents of foods vary from product to product depending on the methods used to prepare, process, and store the food. In some cases, as little as 1 ounce of cheese can cause a severe hypertensive reaction in people taking monoamine oxidase inhibitors. In general, the following foods contain small enough amounts of tyramine that they can be consumed in small quantities: ripe avocado, banana, yogurt, sour cream, acidophilus milk, buttermilk, raspberries, and peanuts.

[a]Most wine and domestic beer can be consumed in small quantities.

[b]Unfermented cheeses, such as ricotta, cottage cheese, and cream cheese, are allowed.

[c]Products made with baker's yeast are allowed.

cretion of calcium, potassium, magnesium, and zinc. Use of the antifungal agent amphotericin B leads to loss of potassium and magnesium in the urine.

Other Ingredients in Medications

Besides the active ingredients, medications may contain other substances such as sugar, sorbitol, lactose, sodium, and caffeine. For most people who use medications on occasion and in small amounts, such ingredients pose no problems. When medications are taken regularly or in large doses, however, people with specific problems may need to be aware of these additional ingredients and their effects.

Many liquid preparations contain sugar or sorbitol to make them taste better. People who must regulate their intakes of carbohydrates, such as people with diabetes, need to consider the amount of sugar these medications contribute to their diets. Large doses of liquids containing sorbitol may result in diarrhea. This can be a problem for adults who must use a pediatric liquid formulation to take a medication. Because they must use more of the liquid than a child would use, they may develop diarrhea with repeated use. The lactose added as a filler to some medications may cause problems for people who cannot digest lactose or those who cannot metabolize galactose.◆

◆ Reminder: Lactose is composed of two simple sugars, glucose and galactose.

Antibiotics and antacids often contain sodium. People who take Alka-Seltzer may not realize that a single two-tablet dose may exceed their safe sodium intakes for a whole day. Medications given by vein provide water and frequently provide sodium, potassium, and other electrolytes, or dextrose (the name for glucose in intravenous solutions). The contributions these nutrients make must be considered when a client's diet must be restricted in any of these nutrients. Interactions between medications, tube feedings, and intravenous feedings can also occur; these will be described in Chapters 22 and 23.

IN SUMMARY

Considering the many ways that medications and nutrients can interact and the number of medications available to clients, it is no wonder that serious side effects are increasingly recognized. Health care professionals are challenged to understand the mechanisms of diet-drug interactions, identify them when they occur, and prevent them whenever possible.

Nutrition on the Net

WEBSITES

Access these websites for further study of topics covered in this chapter.

- Find updates and quick links to these and other nutrition-related sites at our website: **www.wadsworth.com/nutrition**
- To search for specific drug information, visit RxList: **www.rxlist.com**
- Medscape DrugInfo: **promini.medscape.com/drugdb/search.asp**
- To investigate dietary supplements, visit the U.S. Food and Drug Administration, the National Center for Complementary and Alternative Medicine, and the National Institutes of Health's Office of Dietary Supplements and the Natural Pharmacist: **www.fda.gov** and **nccam.nih.gov** and **odp.od.nih.gov/ods** and **www.TNP.com**

INTERNET ACTIVITIES

People use thousands of different medications to treat hundreds of different illnesses. Diet-drug interactions can range from mild to severe. Learn how medications interact with foods or other drugs.

▼ Go to: **promini.medscape.com/drugsdb/search.asp**

- Click on the "drug" button.
- Type in the name of the drug you are interested in.
- Click on "Drug Interactions."
- Click on "Drug-Food."

This site provides a wealth of information about specific medications—including information about interactions, adverse effects, uses and dosages, precautions, and much more. If and when you take medications, how does each one interact with the foods you eat, or other medications you take?

Study Questions

These questions will help you review the chapter. You will find the answers in the discussions on the pages provided.

1. In what ways are medications and dietary supplements alike? In what ways do they differ? (pp. 592–594)
2. Identify conditions that place clients at high risk for diet-drug interactions. (pp. 595–596)
3. Why is it important for health care professionals to have a full account of all drugs clients take? What steps can health care professionals take to help clients limit their risks of diet-drug interactions? (p. 596)
4. Discuss some ways medications can affect food intake. (pp. 597–598)
5. What effects can nutrients have on medication absorption? What effects can medications have on nutrient absorption? (pp. 598–599)
6. How can nutrients and medications interact and affect metabolism? How can diet-drug interactions affect the excretion of nutrients and drugs? Provide some examples. (pp. 600–602)

These questions will help you prepare for an exam. Answers can be found on p. 606.

1. The health care professional recognizes that a client who exercises daily, adheres to a low-fat diet, and uses a garlic extract to prevent heart disease is following:
   a. conventional medical therapies.
   b. complementary therapies.
   c. alternative therapies.
   d. medical nutrition therapy.
2. Over-the-counter and prescription medications:
   a. must be used under the supervision of a physician.
   b. cannot make health claims.
   c. do not require approval of the FDA prior to reaching the market.
   d. require research to support their safety and effectiveness before they reach the market.
3. Which of the following have the greatest risk of serious adverse effects if used inappropriately?
   a. prescription medications
   b. over-the-counter medications
   c. herbs
   d. dietary supplements
4. Over-the-counter medications:
   a. are not required to list precautions on their labels.
   b. are often used for longer periods of time than prescription medications.
   c. may be used for the wrong reasons and in the wrong amounts.
   d. are unlikely to interact with prescription medications.
5. An important difference between medications and herbs and dietary supplements that reach the marketplace is that:
   a. medications are not required to contain a specific amount of active ingredients per dose.
   b. medications cannot make health claims.
   c. herbs and dietary supplements are not required to have proof of safety and effectiveness.
   d. herbs and dietary supplements must provide standard amounts of active ingredients.
6. Adverse diet-drug interactions are more likely to occur if:
   a. the person taking a medication is malnourished.
   b. one medication or herb is taken exclusively.
   c. all organ systems are fully functional.
   d. medications or dietary supplements are taken for a few days.
7. The most important step health care professionals can take to limit the risk of medication-related side effects is to:
   a. encourage clients to take as many medications, herbs, and dietary supplements as possible.
   b. encourage clients to communicate openly and honestly about what medications, herbs, and dietary supplements they take and how much of each they are taking.

c. encourage clients to use more over-the-counter medications, herbs, and dietary supplements and fewer prescription medications.
d. encourage clients who want to try using herbs and dietary supplements to begin by using several products at the same time.

8. Examples of medication-related symptoms that will most likely significantly limit food intake include:
   a. skin rash and ringing in the ears.
   b. insomnia and sensitivity to sunlight.
   c. nasal congestion and hair loss.
   d. persistent nausea and vomiting.
9. For a client taking the antibiotic tetracycline, which foods and medications should not be taken at the same time as the tetracycline?
   a. milk, calcium-containing antacids, mineral supplements
   b. milk, calcium-containing antacids, folate supplements
   c. grapefruit juice, caffeine, mineral supplements
   d. grapefruit juice, milk, and calcium-containing antacids
10. A client is taking the anticoagulant warfarin. The health care professional should periodically evaluate the client's intake of:
   a. folate.
   b. vitamin K, herbs, and dietary supplements.
   c. vitamin D, caffeine, oxalates, and phytates.
   d. calcium, vitamin D, caffeine, and licorice.

Clinical Applications

1. An elderly person uses mineral oil (an over-the-counter medication) to treat chronic constipation. Drawing on the discussion of laxatives on pp. 598–599, discuss why elderly people are at particular risk for developing both diet-drug interactions and serious consequences from taking mineral oil. Consider the reasons and implications for the following statements to guide your thinking: elderly people are more likely to use laxatives to treat chronic constipation, elderly people are more likely to have chronic diseases and take many medications, elderly people are more likely to be malnourished, and elderly people are more likely to develop osteoporosis.

2. A client confides in you that he takes more than ten herbs and dietary supplements and that he has not told the physician that he takes them. His prescription medications include an antihypertensive agent (to reduce blood pressure) and warfarin. What approach might you use to uncover exactly what he is taking and his reasons for taking them? If you discover that some of his herbs and supplements might pose a risk for interactions with his prescription medications, what steps should you take?

3. A client states that she has a hard time sleeping even though she is using three different herbs and a dietary supplement to help her sleep. She says she drinks 8 to 10 cups of coffee each day, the last of which she drinks after dinner. The client's job often requires that she work into the evening, and she seldom engages in physical activity. After she finishes working, she "unwinds" for a few hours by finding interesting chat rooms and surfing the Net. What dietary and lifestyle factors might contribute to the client's inability to fall asleep? Can you suggest lifestyle changes she might try? What might you suggest that she do to determine if any of the herbs or the dietary supplement she uses is helping her sleep?

References

1 As cited in Herbal Rx: The promises and pitfalls, *Consumer Reports Online,* www.consumerreports.org, site visited on November 28, 2000.

2 Herbal Rx for prostate problems, *Consumer Reports Online,* www. consumerreports.org,site visited on November 28, 2000.

3 R. Chang, Quality and standards in dietary supplements, *Support Line,* August 1998, pp. 1–4.

4 S. M. Riddle, Drug interactions: Examining the impact of botanicals and dietary supplements, *Support Line,* October 2000, pp. 8–18.

5 D. M. Eisenberg, Trends in alternative medicine use in the United States: 1990–1997, *Journal of the American Medical Association* 280 (1998): 1569–1575.

6 C. W. Lewis, E. A. Frogillo, and D. A. Roe, Drug-nutrient interactions in long-term care facilities, *Journal of the American Dietetic Association* 95 (1995): 309–315; R. N. Varma, Risk for drug-induced malnutrition is unchecked in elderly patients in nursing homes, *Journal of the American Dietetic Association* 94 (1994): 192–194.

7 D. M. Eisenberg, Advising patients who seek alternative medical therapies, *Annals of Internal Medicine* 127 (1997): 61–69.

8 L. Chan, Redefining drug-nutrient interactions, *Nutrition in Clinical Practice* 15 (2000): 249–252.

9 Riddle, 2000.

10 T. Kantola, K. T. Kivisto, and P. J. Neuvonen, Grapefruit juice greatly increases serum concentrations of lovastatin and lovastatin acid, *Clinical Pharmacology and Therapeutics* 63 (1998): 397–402; J. J. Lilja and coauthors, Grapefruit juice-simvastatin interaction: Effect on serum concentrations of simvastatin, simvastatin acid, and HMG-CoA reductase inhibitors, *Clinical Pharmacology and Therapeutics* 64 (1998): 477–483; J. J. Lilja and coauthors, Grapefruit juice substantially increases plasma concentration of buspirone, *Clinical Pharmacology and Therapeutics* 64 (1998): 655–660; E. B. Feldman, How grapefruit juice potentiates drug bioavailability, *Nutrition Reviews* 55 (1997): 398–400.

ANSWERS

Multiple Choice

1. b	2. d	3. a	4. c	5. c
6. a	7. b	8. d	9. a	10. b

NUTRITION AND IMMUNITY

WHILE PEOPLE are often quite aware of how medications and dietary supplements can be used to prevent and treat diseases, they may know far less about their body's natural defense system. **The immune system** protects the body and enables it to fight off infectious agents and rid itself of abnormal and worn-out cells. Like drugs, the immune system works, in part, by inducing chemical changes that alter body processes. This highlight describes the immune system and examines its relationships to nutrition. The accompanying glossary defines terms used in this discussion.

COMPONENTS OF THE IMMUNE SYSTEM

The immune defenses operate so adeptly that few healthy people realize how many tasks they perform each day. The immune system's highly elaborate and intricate network of organs and cells works to block invading organisms from entering the body or destroy those that do gain entry. Any substance that triggers an immune response is called an **antigen.** Antigens include infectious agents (including bacteria, viruses, fungi, and parasites), worn-out and malignant cells, and tissues or cells from another person. Harmless proteins, such as milk protein, can trigger immune responses if they enter the circulation as a protein rather than amino acids. In such a case, the antigen is called an **allergen,** and the immune response is called an allergic response. In some medical disorders, immune responses are mounted against the body's own cells. Such **autoimmune disorders** include rheumatoid arthritis and lupus and some cases of diabetes mellitus.

Organs of the Immune System

The immune system resides in no single organ, but depends on physical and chemical interactions between many organs and cells. Strategically scattered throughout the body, **lymph tissues** house **lymphocytes**—the white blood cells central to immune defenses. Lymph tissues include the bone marrow, thymus, lymph nodes, spleen, tonsils, adenoids, appendix, and Peyer's patches (clumps of lymph tissue in the intestine). Lymph tissues can also be found interspersed in the mucous membranes. Lymphocytes circulate throughout the body by way of the blood and lymphatic system, where they vigilantly search for foreign substances.

GLOSSARY OF IMMUNE SYSTEM TERMS

allergen: any substance that triggers an inappropriate immune response to a substance not normally harmful to the body.

antibodies: proteins produced by B-cells in response to invasion of the body by specific antigens.

antigen: any substance that triggers an immune system response, including bacteria, viruses, fungi, parasites, worn-out cells, and malignant cells.

autoimmune disorders: disorders that result from immune system defenses attacking the body's own cells.

B-cells: lymphocytes that produce antibodies.

cell-mediated immunity: immunity conferred by T-cells traveling to the invasion site to fight specific antigens.

complement: a group of blood proteins that assist the activities of antibodies.

cytokines: special proteins that direct immune and inflammatory responses.

humoral immunity: immunity conferred by B-cells, which produce antibodies that travel through the blood to the invasion site.

immune system: the body's system of defense against harmful substances.

lymphocytes (limb-FOE-sites): cells made in lymph tissues that travel throughout the lymphatic and circulatory systems.

lymph tissues: tissues that contain lymphocytes.

natural killer cells: lymphocytes that confer nonspecific immunity. Natural killer cells destroy viruses and tumor cells.

nonspecific immunity: immunity directed at foreign substances in general, rather than specific antigens.

phagocytes: large white blood cells that confer nonspecific immunity. Phagocytes engulf and destroy foreign substances. Phagocytes that travel in the blood are called *monocytes;* when monocytes embed themselves in tissues, they grow larger and are called *macrophages.* Other types of phagocytes include *neutrophils, polymorphonuclear leukocytes,* and *basophils.*

specific immunity: immunity directed at specific organisms. The B-cells and T-cells confer this type of immunity.

T-cells: lymphocytes that react to specific antigens by traveling directly to the invasion site. Some T-cells (cytotoxic T-cells) kill invaders; others (helper/inducer T-cells) activate immune responses; still others (suppressor T-cells) turn off immune responses.

• The Skin and Mucous Membranes • The world around us teems with infectious agents that can gain entry through the skin and other gateways to the body, including the eyes, nose, mouth, lungs, GI tract, and genitourinary tract. The skin provides a physical barrier to infectious agents because it is thick and coated with protective waxes that hinder attempts by infectious agents to penetrate it. Glands found within the skin also secrete chemicals that destroy some microbes.

To protect other gateways into the body, mucous membranes produce mucus—a secretion containing chemicals and enzymes that destroys invading organisms and forms a slippery coat that prevents them from sticking to the cells of the mucous membranes. The body traps microbes in mucus, and both are eventually excreted.

• The GI Tract • Microbes that reach the stomach face destruction from highly acidic gastric juices and enzymes. Those that do survive enter the intestine, where further defenses are in place. Recall from Chapter 5 that the suface of the intestine is lined with fingerlike projections called villi. Healthy villi are crowded close together, forming a physical barrier that hinders passage between them. Interspersed among the villi are mucus-secreting cells and lymph tissue that house immune cells to fend off invaders (see Figure H18-1). Consequently, substances can pass from the intestine to the inside of the body only by crossing the cells' membranes, and the cells are remarkably efficient at keeping antigens out.

Many species of harmless bacteria normally reside in the large intestine, and this bacterial flora helps prevent the growth of harmful bacteria by competing with them for nutrients and space. These harmless bacteria also produce short-chain fatty acids that prevent harmful bacteria from sticking to the intestinal surface.

• Cells of the Immune System • If an invading organism manages to penetrate all these barriers, as may happen as a consequence of a wound or malnutrition, the cells of the immune system play a vital role in defending the body. Four types of white blood cells, the **phagocytes** and three types of lymphocytes, shoulder much of the responsibility for immune responses. Figure 18-2 describes immune cells, their actions, and the results of their actions.

• Nonspecific Immunity: Phagocytes and Natural Killer Cells • The actions of phagocytes and natural killer cells are not directed at specific cells, but rather at any foreign invader. Thus the immunity they confer is called **nonspecific immunity.**

Phagocytes, large white blood cells, act as scavengers in the blood and in tissues. Phagocytes rid the body of worn-out cells and debris. When a phagocyte recognizes an antigen, it engulfs and digests it in a process called phagocytosis.

FIGURE H18-1 Immune Cells of the Intestinal Villi

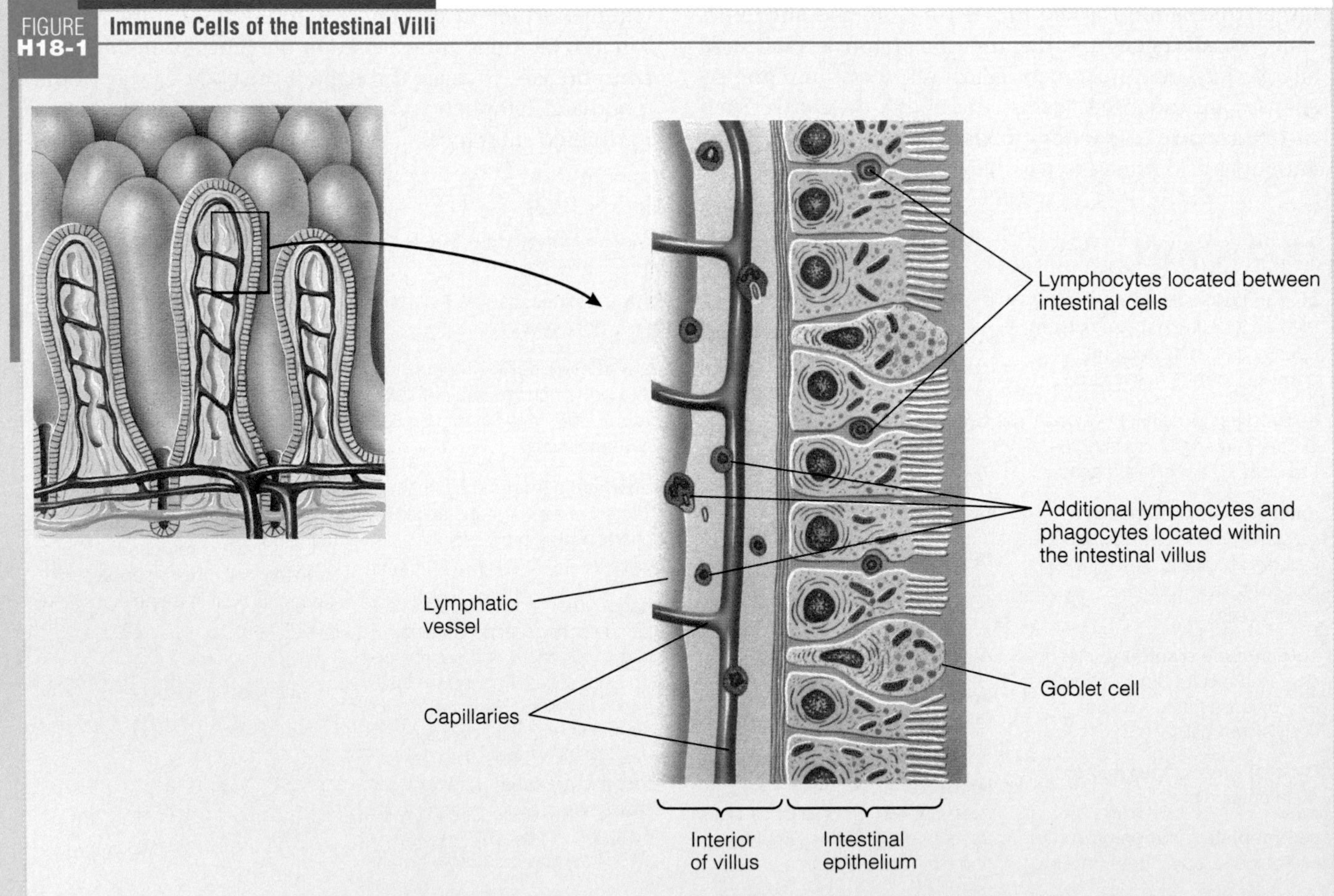

FIGURE H18-2 **Immune System Cells**

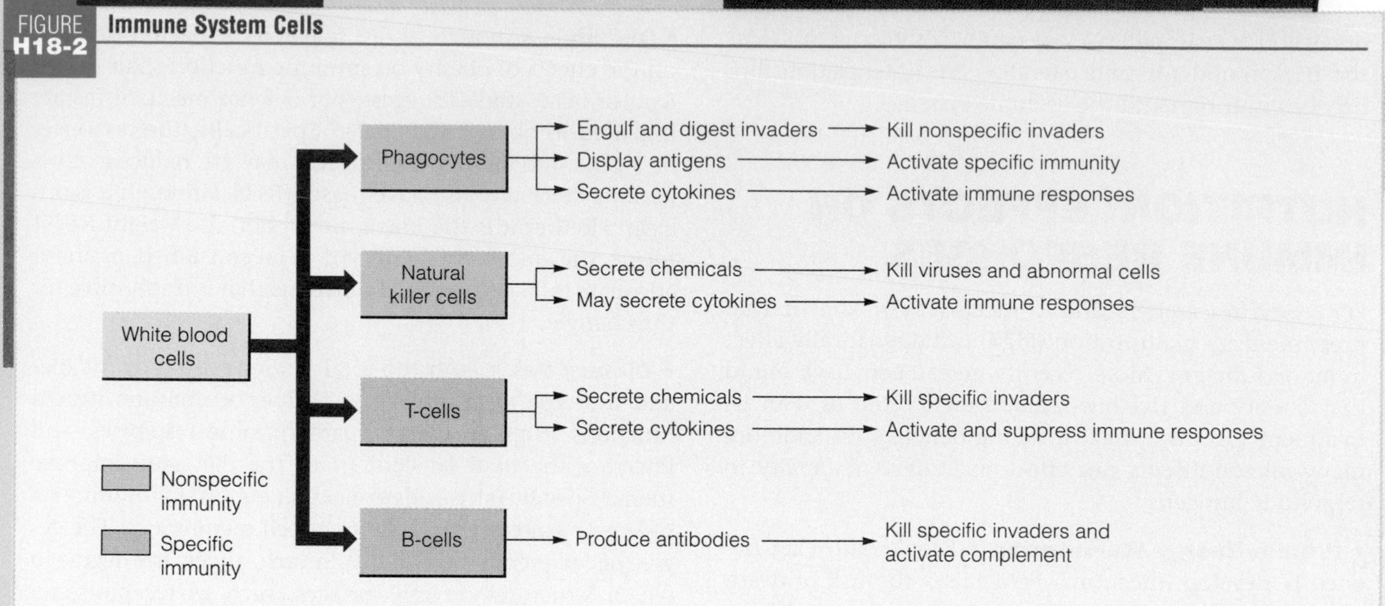

Some phagocytes play a crucial role in initiating immune responses. As these phagocytes engulf a foreign particle, they display a portion of the antigen on their own cell surface. This display triggers lymphocytes into action.

Still other phagocytes contain granules filled with potent chemicals that destroy microbes and trigger inflammatory responses. Inflammatory responses help limit the spread of infectious agents (Chapter 24 provides more information).

Phagocytes also secrete enyzmes, **complement** proteins (described later), and special proteins called **cytokines.** Cytokines bind to receptors on target cells and direct the actions of many other cells and substances. They stimulate cell growth, activate cells, and destroy target cells, to name a few. Cytokines also result in many clinical findings associated with infections including fever and appetite suppression.

Natural killer cells are lymphocytes capable of recognizing and destroying abnormal cells and many infectious agents, especially viruses. Natural killer cells bind to their target and secrete chemicals that puncture the target cell's membrane. They may also help modulate immune responses by secreting cytokines.

• Specific Immunity: T-Cells and B-Cells • Unlike phagocytes and natural killer cells, two other types of lymphocytes, the **T-cells** and **B-cells,** recognize specific antigens. Thus the immunity they confer is called **specific immunity.** T-cells and B-cells can attack only one type of antigen. After making enough cells to destroy a particular antigen, some lymphocytes retain the necessary information to serve as memory cells so that the immune system can rapidly respond if the same infection should recur.

T-cells participate in **cell-mediated immunity,** so named because the cells themselves travel to the invasion site to wage a battle against foreign organisms. Like some phagocytes, regulatory T-cells secrete cytokines that orchestrate immune responses. Some of the regulatory T-cells activate immune responses, while others suppress responses when an infection is under control. Still other T-cells work like natural killer cells and release powerful chemicals; the difference is that T-cells destroy specific foreign particles. T-cells are responsible for the rejection of newly transplanted tissue, which is why physicians use drugs to inactivate them when an organ transplant has been performed.

The B-cells confer **humoral immunity,** so named because the cells' secretions, not the cells themselves, mount the defensive effort. B-cells respond to antigens by rapidly making cells that produce **antibodies.** Each cell derived from a given B-cell synthesizes millions of identical antibodies and pours them into the bloodstream. Some antibodies directly inactivate microbes. Other stick to the surfaces of antigens and make them easy prey for attack by phagocytes. Still others block viruses from entering cells. Often an antigen-antibody combination triggers the synthesis of complement, described next.

• Complement • Complement is a group of about 25 enzymes, so named because they "complement" the activity of antibodies. Complement disarms microbes either by rendering them more susceptible to phagocytosis or by puncturing the target cell's membrane. Complement also helps rid the body of antigen-antibody complexes and stimulates inflammatory responses (see Chapter 24).

In summary, the immune system prevents foreign materials from entering the body through physical and chemical barriers provided by the skin, mucous membranes, and the GI tract. If disease-causing agents survive these barriers, phagocytes, natural killer cells, T-cells, B-cells, and complement spring into action, depending on the agent. Although this discussion oversimplifies a very complex subject, it provides the background for understanding the interrelationships between nutrition and the immune system.

NUTRITION: EFFECTS ON IMMUNE RESPONSES

For years, researchers and clinicians have known that protein-energy malnutrition (PEM) can dramatically affect immune function. More recently, researchers have found that obesity and the low-kcalorie diets often used in its treatment can also affect immune defenses. In addition, many micronutrients can affect an individual's ability to respond to antigens.

• **Protein-Energy Malnutrition** • People with PEM frequently develop infections. PEM raises the risk of death from infection for children in developing countries.[1] Even moderate PEM impairs all components of immune defenses, especially the T-cells and cell-mediated immunity. These changes are summarized in Table H18-1. They are highly significant because some T-cells play important roles in directing further immune responses.

PEM can weaken the physical barriers to infectious agents and alter a barrier's function. The skin, for example, loses collagen and connective tissue and becomes thinner as a consequence of malnutrition. PEM may also reduce the quantity of antibodies in mucus, which may also raise the risk of infection. The microvilli of the intestines flatten and shrink as a consequence of malnutrition, and these physical changes may allow pathogens to cross the intestinal barrier—a process called translocation. These changes may be responsible for the repeated lung and GI tract infections that frequently develop in children with PEM.

• **Obesity** • Although far less research has been conducted on the effects of obesity on immune function than the effects of PEM, studies suggest that defense mechanisms are impaired in obese individuals.[2] Specifically, the responses of T-cells and B-cells to antigens may be reduced. Low-kcalorie diets may also have these effects, although it is unclear whether it is the low-kcalorie diet, the weight loss it incurs, the amount of fat or type of fat consumed, or an inadequate intake of essential nutrients that is responsible for the changes.

• **Dietary Fat** • Both the total amount of fat in the diet and the type of fat appear to influence immune system functions. High-fat diets impair immune responses, and lowering the total fat content of the diet may improve them.[3] Specific fatty acids exert their effects on immune responses by altering the fluidity of cell membranes. The degree of cell membrane fluidity, in turn, affects the degree to which structures on cell surfaces, such as receptors, respond to stimulation, and receptors play an important role in immune responses. Preliminary research suggests that diets rich in omega-3 fatty acids (such as fish oils) have beneficial effects on immune system responses not conferred by diets rich in omega-6 fatty acids (such as corn and safflower oils).

• **Micronutrients** • Among the micronutrients, vitamin A appears to have a particularly strong relationship with immunity. This strong relationship may help explain why the distribution of vitamin A capsules in developing countries has had a major impact on reducing child mortality.[4] All body surfaces, both inside and out, maintain their integrity with the help of vitamin A. These body surfaces include the skin and mucous membranes. All lymphocytes require vitamin A to develop and function properly. Vitamin A deficiency can alter the response of some antibodies to antigens and may also exert effects on the network of cytokines secreted during immune responses.

Other nutrient deficiencies that may impair immune functions include those of vitamin E, vitamin C, vitamin B_6, zinc, and selenium. Table 18-2 summarizes the effects of micronutrient deficiencies on immune functions.

TABLE H18-1 Effects of Malnutrition on the Immune System

Immune System Component	Effects of Malnutrition
Lymph tissues	Thymus gland atrophied; lymph nodes and spleen smaller; T-cell areas depleted of lymphocytes
Skin	Thinner, with less connective tissue and collagen
Mucus	Reduced in quantity and amount of antibodies it contains
Intestine	Microvilli flattened; reduced number of T-cells in lymph tissue, reduced quantity of mucus with fewer antibodies
Phagocytosis	Kill time delayed
Cell-mediated immunity	Circulating T-cells and responsiveness of T-cells reduced
Humoral immunity	Responsiveness of B-Cells reduced

TABLE H18-2 Effects of Selected Micronutrient Deficiencies on Immune Function

Deficiency	Impairments
Vitamin A	T-cell and antibody production; response of lymphocytes, resistance to infections
Vitamin B_6	Antibody production and response of lymphocytes
Vitamin E	Phagocytosis, antibody production, response of lymphocytes, increased activity of viruses
Zinc	T-cell production, response of lymphocytes, resistance to infection

IMMUNE FUNCTION AND NUTRITION STATUS

Adequate nutrition status supports a healthy immune system, and once immune defenses are called into play, they can affect nutrition status. The most severe effects occur as a consequence of severe medical stress—multiple broken bones, deep penetrating wounds, or extensive and serious infections. The details of these effects are reserved for Chapter 24, but a few general points are relevant here.

• Energy Reserves and Protein • To mount an attack against infectious agents, the immune system needs energy and amino acids to synthesize the cells and chemicals it needs to function. A serious threat can rapidly and dramatically drain energy reserves and result in the loss of vital protein.

• Limited Food Intake • At the same time that nutrient and energy needs are high, cytokines produced during immune and inflammatory response depress the appetite, produce a feeling of discomfort, and cause pain at the site of injury. All of these effects make it difficult for the person to eat.

• Consequences • Severe stresses can lead to malnutrition in a previously well-nourished person or worsen preexisting malnutrition. A self-aggravating cycle then develops. Infection aggravates malnutrition, and malnutrition reduces the body's immune defenses. Careful administration of nutrients during stress aims to provide enough nutrients to support immune responses without providing an excess, which can further stress the body.

The healthy immune system is both fascinating and remarkable in its ability to distinguish the body's own cells from microbes and malignant cells. Occasionally, the immune system is defective or fails, and serious consequences follow. Some of the many medical conditions and treatments with direct connections to the immune system that you will encounter in the remaining chapters of this book include inflammatory bowel diseases, severe stress, atherosclerosis, cancer, HIV infections, and organ transplants. This highlight will prepare you for a deeper understanding of these disorders and treatments and their connections to nutrition.

References

1 D. L. Pelletier and coauthors, The effects of malnutrition on child mortality in developing countries, *Bulletin of the World Health Organization* 73 (1995): 443–448.

2 L. Langseth, Dietary factors which alter immune responses, in *Nutrition and Immunity in Man* (ILSI Europe Concise Monographs, ILSI Press, 1999).

3 P. C. Calder, Dietary fatty acids and the immune system, *Nutrition Reviews* (supplement) 56 (1998): 70–73.

CHAPTER 19 Nutrition Intervention

Chapters 17 and 18 described how illnesses and treatments can affect nutrient needs and how dietitians use assessments to evaluate clients' health and nutrition status. This chapter describes the process used to correct the nutrition problems identified through nutrition assessments. Assuring that a person's nutrient needs are met is a key part of this process, so this chapter also describes modified diets and looks at how the client's needs are communicated among health care team members.

NUTRITION CARE PLANS

After completing a nutrition assessment the dietitian identifies actual and potential nutrition problems. Does the client's history indicate the need for changes in diet? Is body weight appropriate? Do laboratory tests or physical signs suggest malnutrition? Are nutrient needs altered due to growth, illness, or medications? Will long-term dietary adjustments be necessary? Answers to questions like these enable the dietitian to generate a nutrition problem list—the basis of the nutrition care plan.

• Goals and Interventions • The dietitian reviews the problem list and considers solutions to the problems. The plan sets goals and identifies interventions that aim to resolve the client's immediate and long-term nutrition problems.◆ To achieve the desired outcome, care plans must take into account the client's food habits, lifestyle, and other personal factors. The care plan should also be consistent with the plans of other members of the health care team to ensure high-quality and cost-effective implementation. Nursing care plans, for example, should include nursing diagnoses that address nutrition-related problems and nursing interventions relevant to the diagnoses. Table 19-1 lists examples of nursing diagnoses with nutrition implications. In some cases, plans for addressing nutrition problems are incorporated into a total medical care plan developed by the health care team. Such plans, called **clinical pathways, critical pathways,** or **care maps,** are charts or tables that outline a coordinated plan of care for a specific medical diagnosis, treatment, or procedure. Figure 19-1 on p. 614 provides an example of such a plan.

• Addressing Nutrition-Related Problems • Goals for nutrition care plans are stated in terms of measurable outcomes, such as target ranges for body weight or blood glucose levels. **Outcome measures** provide a means for measuring the success of therapy. Consider, for example, a problem of diarrhea. If a medication is causing the diarrhea, an appropriate nutrition strategy might be to prevent dehydration and electrolyte imbalances by encouraging the client to consume ample fluids and electrolytes while the physician determines if another medication might be appropriate. Measurable goals might include a target range for serum electrolytes, urinary output, and blood pressure, as well as reduced frequency of bowel elimination and the production of stools of normal volume and consistency. If the diarrhea results from lactose intolerance, the goals might be the same, but the problem-solving strategy includes a temporary elimination of milk and milk products.

• Addressing Nutrition Education Needs • Nutrition education is a key part of the nutrition care plan. Goals for nutrition education can also be defined in terms of measurable goals. For the client experiencing diarrhea as a result of lactose intolerance, for example, the dietitian plans counseling sessions to teach the client how to plan a nutritionally adequate diet that limits lactose from milk and milk products. Examples of measurable goals for nutrition education might include the client's ability to verbally identify foods containing milk or milk products and plan a day's menus appropriate for the restrictions. Keep in mind that the nutrition education plan must be flexible to accommodate the client's

TABLE 19-1 Examples of Nursing Diagnoses with Nutrition Implications[a]

- Altered nutrition: Less than body requirements
- Altered nutrition: More than body requirements
- Altered nutrition: Risk for more than body requirements
- Altered growth and development
- Altered oral mucous membrane
- Body image disturbance
- Chronic confusion
- Chronic pain
- Diarrhea
- Feeding self-care deficit
- Hopelessness
- Impaired memory
- Impaired physical mobility
- Impaired skin integrity
- Impaired social interaction
- Impaired swallowing
- Ineffective breastfeeding
- Ineffective individual coping
- Ineffective infant feeding pattern
- Powerlessness
- Risk for loneliness
- Self-esteem disturbance
- Social isolation
- Unilateral neglect

[a]North American Nursing Diagnosis Association–approved nursing diagnoses.

◆ The goals of medical nutrition therapy are:
- To meet the client's nutrient needs.
- To meet the client's needs for nutrition education.

clinical pathways, critical pathways, or **care maps:** charts or tables that outline a plan of care for a specific diagnosis, treatment, or procedure, with a goal of providing the best possible outcome at the lowest cost. The plan, developed by the health care team after a careful study of each facility's unique client population, is regularly reassessed and improved.

outcome measures: indicators that describe an observable change that are used to evaluate the effects of interventions.

FIGURE 19-1 **Critical Pathway for Ventilator Dependency—The First Seven Days**

	Day 1	Day 2	Day 3	Day 4	Day 5	Day 6	Day 7
Discharge Planning	**Social Service** Obtain phone numbers, next of kin. Identify spokesperson from patient/family. Review DNR discussion.	Review power of attorney vs. legal guardianship. Evaluate insurance coverage for: • home health • nursing home • rehab hospital Evaluate the need for assistance: • Medicare • Medicaid • Social Security		Clarify patient/family wishes regarding DNR status. Review goals of therapy with patient/family: • short-term • long-term			Identify potential discharge plans, document. Discuss discharge plans with family.
Ventilator Dependency	Check weaning parameters BID. Identify conditions to correct before weaning: • fever • malnutrition • anemia	Identify wean method and time. Evaluate need and plan for trach.	Wean as directed by plan.			Evaluate need and plan for trach.	
Fluid Balance	Identify dry weight. Measure I & O q shift. Assess need for diuretic. Draw Chem 7. Assess LV function from previous cardiac cath data.	Weigh patient every other day. Draw Chem 7.	Draw Chem 7. Assess need for ECHO.	Weigh patient. Draw Chem 7. Review Chem 7 frequency.		Weigh patient.	Review Chem 7 frequency.
Nutrition	**Registered dietitian** Evaluate: • calorie/nutrient • PO diet requirement • tolerance • enteral formula Insert Flexiflo. If DB or if on TPN, check 24-hour insulin need. Draw Chem 23, prealbumin, 24-hour urine urea nitrogen (UUN). Check finger stick glucose q 6h x 3 days if on TPN or diabetic. Begin SCU enteral feed protocol, unless alternate plan identified.	Progress tube feeding to goal rate per protocol or identified plan. Begin calorie counts. Consider PO diet. • assess gag/cough/swallow • begin full fluids Review 24-hour insulin needs; adjust if needed.	If patient has trach, assess cough and gag reflexes. Evaluate plan for PO food. Assess need for Broviac/central line for TPN. Make sure patient is receiving 100% caloric requirement. Calorie count.	Calorie count.			Identify deficits and address as necessary.
Elimination	Assess need for Foley/urine C & S. Document last BM; if no BM x 3 days, rectal check. If patient has diarrhea, start protocol. If patient is constipated, add stool softeners; if no response in 24 hours, Fleet's enema x 2, notify MD.		Assess BM.		Assess BM.		Assess BM.
Sleep Pattern	Identify sleep history. Review last 48 hours' sleep pattern. Correlate sleep pattern & current meds that may affect sleep.	Document sleep periods on special care unit flow sheet. Establish nighttime ritual. Review appropriate hypnotic with PharmD.	Follow nightime ritual. Effective hypnotic agent established. Does patient consistently sleep three or more hours?				
Activity	Assess mobility. Consult with PT and OT.	Determine acitvity plan OOB 2x day. Review ADL plan to avoid weaning times. Check pulse ox with activity.					

Note: DNR, do not resuscitate; LV, left ventricular; TPN, total parenteral nutrition; BM, bowel movement; PT, physical therapy; OT, occupational therapy; ADL, activities of daily living; SCU, special care unit; BID, twice a day; PO, orally; DB, diabetic; q, every; I & O, input and output; OOB, out of bed; ABG, arterial blood

SOURCE: Reprinted with permission from Lippincott Williams & Wilkins. K. S. Thompson and coauthors, Building a critical path for ventilator dependency, *American Journal of Nursing* 91 (1991): 28–31.

personal goals, understanding of the information presented, and motivation to practice the suggestions offered.

Nurses, thanks to their frequent daily contact with clients, can offer important support in the educational aspect of client care. Clients often think of questions long after the dietitian has left, and they most often ask the nurse. The nurse who is confident of the answers should provide them. If not sure of the answers, the nurse should express honest uncertainty and reassure the client that the information will be provided. The nurse should then inform the dietitian that the client needs a follow-up visit.

CASE STUDY

Nutrition Care Plans

Sam is a nine-year-old Native American who was admitted to the hospital after he passed out while playing with friends. Tests confirm a diagnosis of diabetes mellitus. Sam will be in the hospital for several days until his blood glucose levels are under control. During this time, he and his family will begin learning about diabetes mellitus, the diet Sam will have to follow, how to use insulin, how to monitor blood glucose levels, and how to coordinate diet, insulin, and physical activity.

The details of diabetes mellitus are reserved for Chapter 25, but for now consider what steps will be necessary to develop a nutrition care plan.

1. What is the first step for the dietitian to take before a care plan can be developed?
2. From the limited information available, what personal factors will be important in devising a realistic care plan? Consider how these factors might affect the diet prescribed as well as nutrition education.
3. The extensive amount of information that Sam and his family will need to handle his diabetes will require the combined expertise of many health care professionals. Describe ways these health care professionals can communicate with each other to provide the most effective care.
4. Sam will need follow-up to learn more about diabetes and make the adjustments that will allow him to cope with diabetes. Why is it important that plans for follow-up be addressed before Sam leaves the hospital?

• **Ongoing Evaluation** • As the planned strategies are implemented, the dietitian regularly evaluates their effectiveness in achieving goals. If, for example, a client on a weight-reduction diet fails to lose weight, a change may be needed. Is the client eating too much? Is the client too inactive? Perhaps the client should keep a food and activity record to help identify problems with the weight-loss plan.

If a client's situation changes, so may nutrition status and nutrient needs. For example, when a pregnant woman delivers her baby, she will need instructions on how to feed her infant. She will also need information on how to revise her diet to support lactation (if she is breastfeeding) and return to a healthy weight. The accompanying case study offers an opportunity to apply some of the planning principles introduced here.

A care plan may be ideal, but still fall short of meeting goals if a client is unable or unwilling to comply with it. If the client is unable to comply, reassessing communication techniques may help. Perhaps the level of instruction needs to be simplified or cultural differences addressed. If the client is unwilling, despite the best efforts of health care professionals, little can be done except to try later when the client may be more receptive.

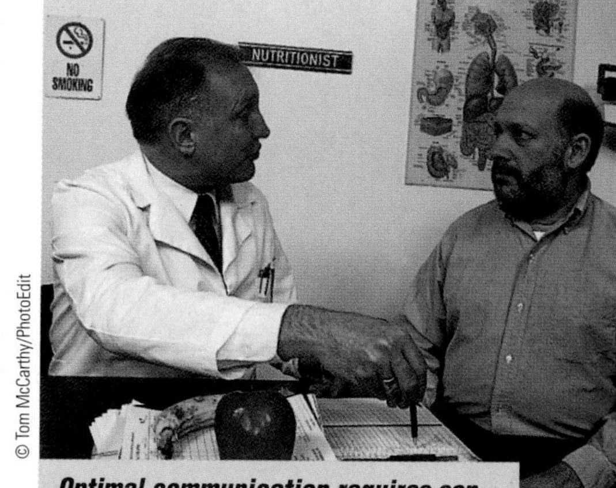

Optimal communication requires sensitivity to cultural orientation, education, and motivation.

IN SUMMARY Once nutrient and nutrition education needs have been identified, the dietitian develops a nutrition care plan to meet those needs. The plan includes measurable goals and interventions to meet those goals. Once implemented, the plan is regularly evaluated to see how well it is working. Often a nutrition care plan includes diet modifications, described in the next section, to help clients meet their needs.

STANDARD AND MODIFIED DIETS

An essential part of any nutrition care plan is to provide the appropriate amounts of energy, protein, carbohydrate, fat, vitamins, major minerals, trace elements, and water in whatever form best meets the client's needs. For **standard** or **regular diets,** any food can serve as a source of energy or nutrients. **Modified diets** vary from standard diets in the types or amounts of nutrients, foods, or food components they provide.

standard or **regular diets:** diets that include all foods and meet the nutrient needs of healthy people.

modified diet: a diet that is adjusted to meet medical needs. Such diets may be adjusted in consistency, level of energy and nutrients, amount of fluid, or number of meals, or by the inclusion or elimination of certain foods.

Feeding Routes

Nutrients can be delivered using the GI tract **(enteral nutrition)** or by bypassing the GI tract and supplying nutrients intravenously **(parenteral nutrition).** Enteral diets include both oral diets and tube feedings. Most often, people meet their nutrient needs by eating regular foods. If their nutrient needs are high or their appetites are poor, liquid formulas taken orally can help meet some or all of their nutrient needs. Sometimes, however, a person's medical condition may make it difficult to meet nutrient needs orally. Two options remain: **tube feedings** or **intravenous** feedings.

• Tube Feedings • With a tube feeding, nutritionally complete formulas are delivered through a tube placed in the stomach or intestine (Chapter 22 provides the details). Tube feedings are preferred to intravenous feedings, but they can be used only when clients are able to digest and absorb enough nutrients to meet their needs. A person in a coma, for example, though unable to eat, may be able to digest foods and absorb nutrients. In such a case, a tube feeding would be an appropriate option.

• Intravenous Feedings • In some cases, however, a person's medical condition prohibits the use of the GI tract to deliver nutrients. If the person is malnourished and the GI tract cannot be used for long periods of time, then intravenous feedings can supply nutrients (see Chapter 23).

Diet Modifications

Modifying the standard diet is much like tailoring a suit. A tailored suit is the same suit after alterations—only it fits better. In the case of a modified diet, the tailoring may involve changing the consistency; adjusting the amounts of energy, individual nutrients, or fluid; altering the number of meals; or including or eliminating certain foods. For example, a person who cannot chew easily needs soft foods; a person who is retaining fluid may need to limit sodium; a person with diabetes mellitus needs a diet that provides consistent amounts of carbohydrate at regular intervals.

• Treating Symptoms and Clinical Findings • It is helpful to think about modified diets in terms of the symptoms or conditions they prevent or relieve rather than in terms of disorders. Two people with the same disorder may need two different diets. Consider two people with an HIV (human immunodeficiency virus) infection, for example: one may need a diet that will help control nausea; the other may need a fat-modified diet to control elevated blood lipids. Conversely, people with two different disorders may develop similar problems, and both may benefit from the same diet. Many neurological disorders, for example, can lead to problems with swallowing (dysphagia). Whatever the cause, the diet aims to ease the task of swallowing.

• Diet Orders • As Chapter 17 described, the physician prescribes the client's diet and writes the diet order in the medical record. The physician often relies on the dietitian or nurse to suggest a diet prescription or make recommendations when changes in the diet order appear warranted or when clarity is lacking.

To avoid confusion, physicians should provide clear and precise diet orders. For example, a "low-sodium diet" order should specify the amount of sodium; otherwise, "low-sodium" could be interpreted to mean any amount from 500 to 4000 milligrams. For uncomplicated diets, such as low-sodium diets, the foodservice department often sends a preselected diet **(house diet)** to the client until orders are clarified, and the level of restriction is recorded in the medical record. For more complicated diets, such as renal diets, meals will not be sent until the order is clarified.

enteral (EN-ter-all) **nutrition:** the provision of nutrients using the GI tract. Enteral nutrition includes both oral diets and tube feedings.
- **enteron** = intestine

parenteral (par-EN-ter-all) **nutrition:** the provision of nutrients bypassing the intestine.
- **par** = beside

tube feedings: liquid formulas delivered through a tube placed in the stomach or intestine.

intravenous: through a vein.

house diet: a menu preselected by the dietary department.

Occasionally, diet orders may be inappropriate. For example, the physician may describe an obese individual as "well-nourished" and order a regular diet. Unless the nurse identifies the problem, the client will not be referred to a dietitian and will receive an inappropriate diet and no nutrition advice. As another example, a physician may order that a client receive no food or fluids◆ after midnight for a lab test to be conducted in the morning. The doctor assumes that the order will be discontinued after the test, but it may not be. The client may miss several meals before someone notices the error. These examples illustrate situations where communication between health care professionals can make a difference in client care. Whenever you notice inappropriate diet orders, contact the dietitian or alert the physician.

An order to give a client nothing orally (including food, beverages, and medications) reads *NPO*, the abbreviation for *non per os*, which means "nothing by mouth." *PO* stands for *per os*, which means "by mouth" or "orally." ◆

• Diets in Facilities That Serve Foods • In an in-patient health care facility (such as a hospital, nursing home, or in-patient mental health care facility), providing an appropriate diet appears deceptively simple: appropriate foods are delivered to clients. Behind the scenes, however, much work goes into providing appropriate foods.

In large facilities, the staff of dietitians may compile a **diet manual,** subject to approval by the hospital administrator, several physicians, and representatives of the nursing service. Small facilities may adopt the diet manual of another hospital or an organization such as the state dietetic association. The diet manual describes the foods allowed and not allowed on each diet, outlines the rationale and indications for use of each diet, provides information on the nutritional adequacy of the diets, and offers sample menus. The dietary department uses the manual to design menus for each diet.

The exact foods excluded from or included on a specific modified diet, and even the name given to the diet, may differ among health care facilities, generally in minor ways. These variations reflect different schools of thought regarding diet. Health care professionals consult diet manuals as a standard of practice whenever they have questions about the foods provided for a modified diet.

Aside from planning appropriate diets, the foodservice department assures that the appropriate foods are carefully prepared and delivered. Highlight 19 describes how foodservice systems operate. Equally important, once delivered, the food must be eaten by the client. As Chapter 17 described, many medical conditions and treatments can depress the appetite. Dietitians, nurses, and their assistants play a central role in helping clients to eat. The "How to" on p. 618 offers suggestions for improving clients' food intakes. The chapters that follow provide many examples of modified diets, and the next section introduces diets that are routinely served in health care facilities.

© Craig M. Moore

Health care professionals consult diet manuals to clarify what foods are included on or excluded from different diets.

Routine Progressive Diets

In health care facilities that serve meals, a common diet order reads, "progress diet from clear liquids to a regular diet as tolerated." Diets that progress from liquids to solids are typical for clients after uncomplicated surgeries, for clients experiencing nausea or diarrhea, or for clients who have not been eating foods orally and are beginning to eat again. At each step of a **progressive diet,** other diet modifications may also apply. For example, sodium may be restricted. Then the diet order might read, "progress diet from low-sodium clear liquids to a 4-gram sodium diet as tolerated." Progressive diets frequently begin with clear-liquid diets.

• Clear-Liquid Diets • Clear liquids are often the first foods offered to clients on progressive diets. A **clear-liquid diet** provides fluids and electrolytes and consists of foods that are liquid at room temperature and leave little residue in the intestine. Clear liquids provide minimal stimulation to the GI tract, and they are easily and almost completely absorbed; thus they help determine if the digestive system is working well enough to handle more complex foods. Table 19-2 lists the

diet manual: a book that describes the foods allowed and restricted on a diet, outlines the rationale and indications for use of each diet, and provides sample menus.

progressive diet: a diet that changes as the client's tolerances permit.

clear-liquid diet: a diet that consists of foods that are liquid at room temperature and leave little residue in the intestine.

Help Clients Improve Their Food Intakes

1. Empathize. If the person is frightened, angry, or confused, show that you care and are there to help. Imagine feeling too sick to move or too tired to sit up. Show that you understand how difficult eating may be.
2. Motivate. Be sure the client understands how important nutrition is to recovery.
3. Help clients select foods they like and mark menus appropriately. Call the dietitian if the client needs extra help. When appropriate and permissible, let a friend or family member bring in favorite foods from outside the hospital. This may be especially helpful for clients with strong ethnic, religious, or personal food preferences.
4. Solve eating problems. Encourage clients who feel full after a short time to eat the most nutritious foods first and save liquids until after meals. For clients who are weak or tired, suggest foods that require little effort to eat. Eating a roast beef sandwich, for example, requires less effort than cutting and eating a steak; drinking soup is easier than eating it with a spoon. For clients who either fill up quickly when eating or are weak or tired, smaller meals combined with snacks, such as a sandwich at bedtime and milk shakes or instant breakfast drinks between meals, can improve intake considerably.
5. Help clients prepare for meals. Encourage clients to wash their hands and faces and to brush their teeth or rinse their mouths before eating. Help them get comfortable, either in bed or in a chair. Adjust the extension table to a comfortable distance and height, and make sure it is clean. A clean, odor-free room also helps. Take these steps before the tray arrives, so the meal can be served promptly and at the right temperature.
6. Check for accuracy and appearance. When the food cart arrives, check the client's tray. Confirm that the client is receiving the right diet, that the foods on the tray are the ones the client marked on the menu, and that the foods look appealing. Order a new tray if foods are not appropriate.
7. Help with eating. Help clients who need assistance in opening containers or cutting foods and those who are unable to feed themselves.
8. Take a positive attitude toward the hospital's food. Never say something like "I couldn't eat this stuff either." Instead, say, "The foodservice department really tries to make foods appetizing. I'm sure we can find a solution."

foods allowed on clear-liquid diets, and the sample menu shows an example of a day's meals. Once the person tolerates clear liquids, the diet may progress to full liquids.

• **Full-Liquid Diets** • A **full-liquid diet** includes both clear and opaque liquid foods and some semiliquid foods (see Table 19-2 and the sample menu). A

Routine progressive diets help test a client's ability to digest food.

TABLE 19-2 Foods Included on Liquid Diets

Clear-Liquid Diets	Full-Liquid Diets
Bouillon	All clear liquids
Broth, clear	Butter
Carbonated beverages	Commercially prepared liquid formulas (all)
Coffee, regular and decaffeinated	Cooked cereals, strained
Commercially prepared clear-liquid formulas	Cream
Fruit drinks	Custard
Fruit ices	Ice cream, plain
Fruit juices, strained	Instant breakfast drinks
Gelatin	Margarine
Hard candy	Milk, all types
Honey	Pudding
Lemonade	Sherbet
Popsicles	Soups, strained vegetable, meat, or cream
Salt	Sour cream
Salt substitutes	Vegetable juices, strained
Sugar	Vegetable purees, diluted in cream soups
Sugar substitutes	Yogurt
Tea, regular and decaffeinated	

full-liquid diet: a diet that consists of both clear and opaque liquid foods and near-liquid foods.

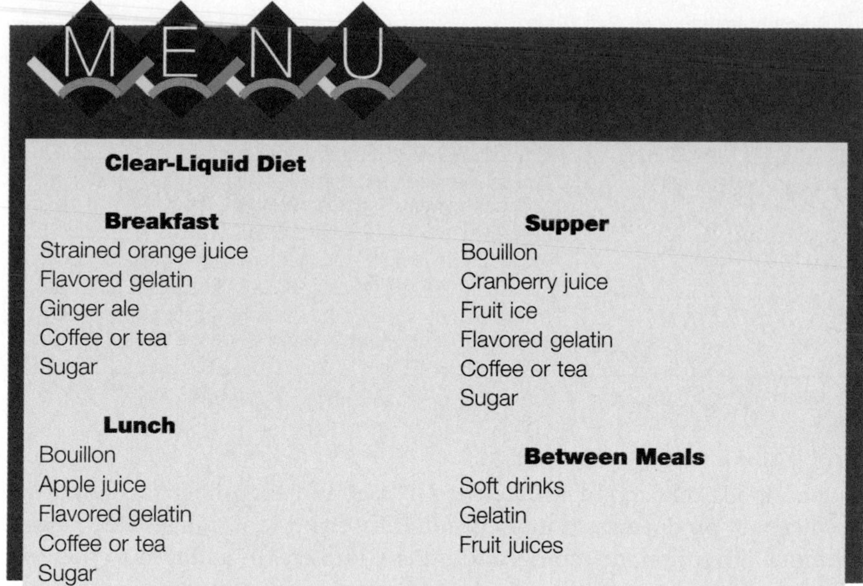

Clear-Liquid Diet

Breakfast
Strained orange juice
Flavored gelatin
Ginger ale
Coffee or tea
Sugar

Lunch
Bouillon
Apple juice
Flavored gelatin
Coffee or tea
Sugar

Supper
Bouillon
Cranberry juice
Fruit ice
Flavored gelatin
Coffee or tea
Sugar

Between Meals
Soft drinks
Gelatin
Fruit juices

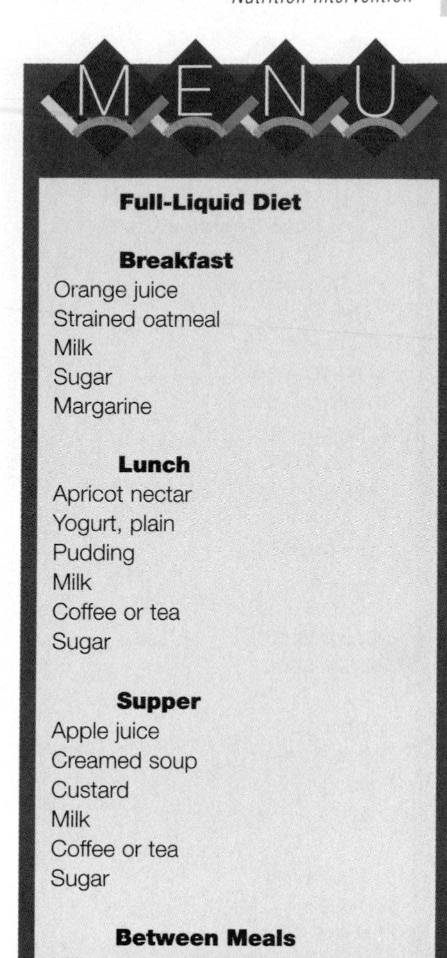

Full-Liquid Diet

Breakfast
Orange juice
Strained oatmeal
Milk
Sugar
Margarine

Lunch
Apricot nectar
Yogurt, plain
Pudding
Milk
Coffee or tea
Sugar

Supper
Apple juice
Creamed soup
Custard
Milk
Coffee or tea
Sugar

Between Meals
Milk shakes (made with plain ice cream, ice cream, eggnog, pudding, custard, or gelatin

full-liquid diet may serve as the second diet as a person progresses from clear liquids to regular foods, or as a diet for people who are unable to chew or swallow regular foods for medical reasons or because they are too ill to eat.

Note that many foods provided on full-liquid diets contain lactose. The same conditions for which liquid diets are prescribed often result in temporary lactose intolerance. When potential problems with lactose intolerance are suspected, lactose-containing liquids cannot be provided. In such cases, lactose-free formulas can meet nutrient needs, or the diet may progress directly from clear liquids to small, frequent servings of soft foods.

• Cautious Use of Liquid Diets • Liquids are offered in small amounts at first to make sure the person can tolerate them. Many clients willingly accept liquid diets because they are too ill to eat solid foods. When left on the diet for any length of time, however, people may understandably find these foods unappetizing and boring. More importantly, both standard clear-liquid and full-liquid diets are lacking many nutrients and, therefore, should be used only temporarily. When clients are unable to tolerate solid foods for more than a few days, liquid formulas (see Chapter 22) provide an option. Formulas are of a known nutrient composition, and they can be selected to meet nutrient needs and reduce the likelihood of GI problems, including lactose intolerance.

THINK NUTRITION whenever clients are placed on unsupplemented liquid diets for more than a few days.

• Solid Foods and Diet Progression • Once a person is tolerating liquids, the next step is solid foods. The exact diet used as the first solid diet depends on the client's medical condition. Some clients begin receiving regular foods, but others better tolerate the shift from liquid diets to solid diets if they first receive soft, low-fiber, or low-residue diets.

• Soft Diets • Soft diets provide soft but solid foods (such as tender meats and soft fresh fruit like bananas) that are lightly seasoned and moderate in fiber. Generally, soft foods are offered in frequent, small meals. As Table 19-3 shows, soft diets include foods that are easy to chew, digest, and absorb. Soft diets limit high-fiber foods, nuts, foods with seeds, and coconut. By limiting highly

Soft/Low-Fiber/ Low-Residue Diet

Breakfast
Orange juice
Soft-cooked egg
Puffed rice cereal
White bread toast
Coffee or tea
Milk for cereal
Creamer

Lunch
Baked fish
White rice
Green beans
Small banana
Roll
Margarine
Coffee or tea
Sugar
Creamer

Supper
Roast beef
Mashed potatoes
Cooked carrots
Canned peaches
Roll
Margarine
Coffee or tea
Sugar
Creamer

Snack
Applesauce
Vanilla wafers

◆ **Soft, low-fiber, and low-residue diets are similar, and in facilities that serve meals, menus for these diets are often the same. All three diets restrict high-fiber foods. Soft diets limit highly seasoned foods, low-residue diets limit milk and milk products, and both low-fiber and low-residue diets limit serving sizes of allowed fruits and vegetables. A soft, low-fiber or soft, low-residue diet would also limit highly seasoned foods.**

residue: the total amount of material in the colon; includes dietary fiber and undigested food, intestinal secretions, bacterial cell bodies, and cells shed from the intestinal mucosa.

TABLE 19-3 Foods Included on Soft Diets

Meat and Meat Alternates	Tender, moist meats, fish, or poultry; mild cheeses, creamy peanut butter, and eggs
Milk and Milk Products	Milk, milk products, yogurt without seeds or nuts
Fruits and Vegetables	Cooked, canned, or soft fresh fruits such as melons; fruit juice; soft-cooked vegetables except those likely to produce gas (see Table 21-1 in Chapter 21)
Grains	Refined white or light rye bread, rolls, or crackers; cooked or ready-to-eat cereals without nuts or seeds
Other	Mild condiments; salt; sugar; mildly seasoned broths or soups; all nonalcoholic beverages; desserts without nuts, seeds, or coconut

seasoned foods that might irritate the GI tract and high-fiber foods that are more likely to produce gas, soft diets minimize the risk of indigestion, nausea, abdominal distention, cramping, and other GI upsets. In addition to their use in a progressive diet, soft diets also benefit people with temporary indigestion or nausea. Clients tolerating soft diets can then progress to regular diets that provide the full spectrum of nutrients and include all foods.

• Low-Fiber and Low-Residue Diets • Low-fiber and low-residue diets◆ serve to reduce the total fecal volume **(residue)** and minimize the risk of an intestinal obstruction when GI tract motility is slow or when the intestinal tract is narrowed by inflammation or scarring. Thus a physician may prescribe a low-fiber or low-residue diet either before or after surgery of the intestinal tract. Low-fiber diets restrict high-fiber foods and tough meats. Low-residue diets limit these foods and also limit milk and milk products. Table 19-4 lists foods low in fiber. The sample menu shows a day's meals for a soft/low-fiber/low-residue diet.

• Advancing the Diet • The nurse is often responsible for monitoring the client's tolerance and readiness to advance through the stages of a progressive diet. Indigestion, nausea, vomiting, diarrhea, cramping, or other GI upsets indicate intolerance. At each progressive step, a client may be intolerant to particular foods (orange juice or milk, for example), rather than to the diet itself. In such a case, the offending food is withheld temporarily.

IN SUMMARY

Providing appropriate nutrients in an appropriate form is an important component of medical nutrition therapy. When standard diets fail to meet these needs, diet modifications are necessary. The physician prescribes each client's diet. The exact foods included on and excluded from modified diets, as well as serving sizes, are specified in each facility's diet manual. Routine progressive diets test clients' tolerance for foods and begin with liquid foods and progress to solid foods. The medical record, described next, provides a vehicle for health care professionals to document a client's medical and nutrition needs and responses to interventions.

THE MEDICAL RECORD

Maintaining strong professional communication networks benefits both health care professionals and their clients. Conversely, miscommunication between professionals can result in inappropriate therapy or ineffective care with serious consequences for clients' health.

TABLE 19-4 Low-Fiber Foods

Meat	Tender meat, poultry, seafood, and eggs
Breads and Cereals	Refined breads, cereals, rice, and pasta
Fruits	Cooked or canned peeled apples, apricots, peaches, pineapple, plums; bananas, cherries, cranberry sauce, mandarin oranges, pomegranates, tangerines; fruit juice
Vegetables	Cooked bean sprouts, cabbage, onions, summer squash; celery, endive, lettuce, tomato paste, tomato puree, vegetable juice, water chestnuts, watercress
Other	Avocado, nuts

Medical records are legal documents that record a client's health history; the assessment, diagnosis, and prognosis of medical problems; the measures being taken to treat those problems; and the results of tests and therapy. Writing in the medical records allows health care professionals to document the actions taken to comply with physicians' orders, the client's responses to those actions, and recommendations. This information helps determine if medical orders are being followed and directs future care.

• Types of Medical Records • Medical records can be organized in many ways. In today's cost-conscious health care environment, time constraints compel health care professionals to minimize the time spent in documenting medical care.[1] Traditional problem-oriented medical records (POMR), which focus on a client's medical problems and the strategies being used to address those problems, are gradually being replaced by outcome- or goal-oriented medical records, which focus on observable medical goals. A problem-oriented medical record for a person with diabetes mellitus, for example, would list diabetes mellitus as a problem and then define the strategies used to control the diabetes, such as a 1600-kcalorie diet, oral hypoglycemic agents, and the like. A goal-oriented medical record for the same client would list the client's goal weight and target blood glucose measurements, and the health care team would document how these values have responded to therapy.

• Recording Nutrition Information • Learn how to effectively use the record in the facility where you work. Regardless of the approach used, be sure the client's medical record includes important nutrition-related information. Examples of important information include:

- Documentation of nutrition screening.
- Assessment data relevant to nutrition status.
- Recommended medical nutrition therapy, including goals.
- Acceptance and tolerance of current diet.
- Problems with food intake.
- Documentation of diet counseling, including an assessment of the client's understanding of the diet.
- Any planned follow-up or referral to another person or agency.

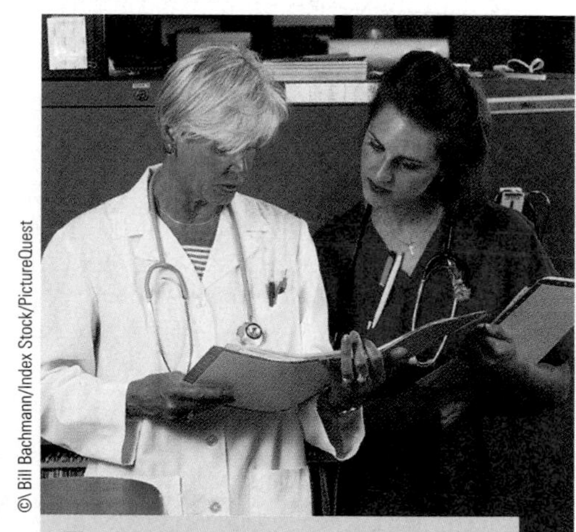

The medical record serves as a tool for communicating information about a client's health and responses to treatments.

IN SUMMARY

Effective nutrition care addresses the unique needs of the individual, framing nutrition needs in the context of the client's personal and medical needs. This chapter has described techniques for building effective nutrition care plans. The remaining chapters describe how nutrition care can support recovery when health problems arise.

Nutrition on the Net

WEBSITES

Access these websites for further study of topics covered in this chapter.

- Learn more about medical nutrition therapy by visiting Ask the Dietitian: **www.dietitian.com/medtherapy.html**
- To find more information about nursing diagnoses, visit the North American Nursing Diagnosis Association: **www.nanda.org**

INTERNET ACTIVITIES

Health care professionals encounter clients with a variety of disorders that may have nutrition implications. These clients need specific medical nutrition therapies.

- ▼ Go to **www.dietitian.com/medtherapy.html**
- ▼ Read the questions and answers about medical nutrition therapy from nurses, dietitians, and clients themselves.
- ▼ Click on "Ask the Dietitian" in the upper left corner and scroll down the list of other nutrition-related topics.
- ▼ Click on a topic of interest to you and read the questions and answers.

This site offers questions and answers on many different nutrition topics. What did you learn about nutrition intervention?

Study Questions

These questions will help you review the chapter. You will find the answers in the discussions on the pages provided.

1. How does the dietitian formulate a nutrition care plan? What two services should the care plan deliver to the client? (pp. 613–614)
2. Why is it important to define goals for nutrition care in terms of measurable outcomes? Discuss the value of ensuring that nutrition care plans are consistent with the care plans of other health care professionals. (p. 613)
3. Who prescribes the client's diet? How do health care professionals know what foods are included on and excluded from specific diets provided in a health care facility? (pp. 616–617)
4. What are progressive diets? Describe the steps of a routine progressive diet that tests a client's tolerance for food. (pp. 617–620)
5. Describe some ways that health care professionals can help clients improve their food intakes in facilities that serve foods. (p. 618)
6. What ways can medical records be used to improve the quality of care? What kinds of information about nutrition care should be recorded in the medical record? (pp. 620–621)

These questions will help you prepare for an exam. Answers can be found on p. 623.

1. At a minimum, the nutrition problem list of a client with a very poor appetite and weight of less than 80 percent IBW would include:
   a. difficulty chewing and swallowing foods.
   b. risk for malnutrition.
   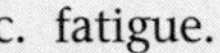
   c. fatigue.
   d. fluid and electrolyte imbalance.
2. Based solely on the information presented in question 1 above, which of the following interventions is appropriate for the client?
   a. a high-kcalorie clear-liquid diet
   b. a fluid and electrolyte-restriced diet
   c. a soft diet
   d. referral for a complete nutrition assessment
3. Measurable goals for a client who must lose weight might include:
   a. the number of pounds the client is expected to lose each week.
   b. the type of diet the client is expected to follow.
   c. the family members' attitudes toward the client's weight.
   d. the client's food intake record.
4. The most important factor(s) that affect how nutrition information is presented to a client is/are:
   a. the client's nutrient needs and nutrition status.
   b. the client's ability level and motivation.
   c. the client's health history.
   d. the medical record.
5. A dietitian instructs a client on a weight-reduction diet and exercise program and sets a goal for weight loss of 1 pound per week. The next step for the dietitian is to:
   a. assume the plan is working.
   b. formulate a new nutrition care plan.
   c. find a new strategy for meeting weight-loss goals.
   d. plan a follow-up meeting to weigh the client and see how well the plan is working.
6. All of the following statements describe diet orders *except:*
   a. they are prescribed by the physician.

b. they should be written clearly and precisely.
c. they should never be questioned.
d. the physician may write an order that reads, "progress diet from clear liquids to a regular diet as tolerated."

7. A nurse notices a food on a client's tray and is not sure if the food is allowed on the client's diet. An appropriate action for the nurse to take would be to:
a. check the care plan.
b. check the diet order.
c. check the diet manual.
d. check the medical record.

8. Which of the following statements describes both standard clear- and full-liquid diets?
a. They can be used for long periods of time.
b. They are deficient in nutrients and are temporary diets.
c. They are contraindicated for people who are unable to chew or swallow regular foods.
d. They are contraindicated for people experiencing gastrointestinal disturbances.

9. A client's diet order reads, "progress diet from clear liquids to a regular diet as tolerated" following surgery. What signs and symptoms suggest the client is not ready to advance to the next step of the diet progression?
a. nausea, vomiting, diarrhea, and cramping
b. headache or fever
c. dehydration or overhydration
d. pain and confusion

10. All of the following are true about medical records *except:*
a. they are legal documents.
b. they provide information about a client's medical problems and the measures being taken to address those problems.
c. they all conform to the same format.
d. they demand the time of health care professionals.

Clinical Applications

1. A client who recently suffered a heart attack is on a special diet. The client tells the nurse that the dietitian has talked to him about his diet and that he is totally confused. He confides that diet is the last thing on his mind right now. What actions should the nurse and/or dietitian take?

2. An initial nutrition screening of an elderly client admitted to the hospital for surgery revealed that the client was not at risk for poor nutrition status. The client was not on a modified diet and was not referred for nutrition assessment, and so has not seen a dietitian. Following surgery, ongoing assessment has revealed that the client has developed several complications, and recovery has been slower than expected. The nurse working with the client notices that the client has eaten only minimal amounts of food for several days. Describe several steps that can be taken to uncover and address problems the client might be having with food.

3. Clients respond differently to treatments, and usually there is more than one treatment option. Consider the following scenario. The physician orders a routine progressive diet for a client who has not eaten an oral diet for two weeks. The client has problems tolerating a clear-liquid diet at first, but by the end of two days is ready to advance to a full-liquid diet. (Note: this is a longer than usual amount of time for a client to adjust to a clear-liquid diet.) The physician suspects that the client may experience temporary lactose intolerance and have even more problems tolerating a full-liquid diet. Describe the potential benefits and drawbacks of each of these options:
   - Advance the client to an unsupplemented, lactose-free full-liquid diet for a day, and monitor the client's response.
   - Provide a lactose-free full-liquid diet supplemented with a formula, and monitor the client's response.
   - Skip the full-liquid step in the diet progression, and offer lactose-free, soft foods in very limited amounts.

References

1 C. J. Klein, J. B. Bosworth, and C. E. Wiles, Physicians prefer goal-oriented note format more than three to one over other outcome-focused documentation, *Journal of the American Dietetic Association* 97 (1997): 1306–1310.

ANSWERS

Multiple Choice:

1. b	2. d	3. a	4. b	5. d
6. c	7. c	8. b	9. a	10. c

HIGHLIGHT

FOOD AND FOODSERVICE IN HEALTH CARE FACILITIES

HOSPITALS AND long-term care facilities prepare and provide foods for clients with a variety of medical conditions. The health care facility's foodservice department faces a challenge in planning, producing, and delivering appetizing, nutritious meals designed to accommodate dozens of special diets and food preferences. Although this discussion focuses on foodservice in hospitals, much of the information applies to foodservice in any health care facility, including nursing homes, assisted living centers, rehabilitation centers, and residential mental health care facilities. An important difference between hospitals and long-term health care facilities deserves attention, however. When clients in hospitals eat poorly, they can make up for nutrient deficits by eating well when they return home. Residents of a long-term care facility do not have this option. For this reason, foodservice departments in long-term care facilities must make even greater efforts to ensure that their clients receive nutritious foods and eat them.

THE CLIENT'S PERSPECTIVE

As Chapter 17 described, people in hospitals may lose their appetites either as a direct result of their medical condition or through emotional distress. Even when appetites are healthy, comments about hospital foods—both positive and negative—are common. Many obvious and not-so-obvious factors shape a client's perceptions of hospital foods.

Sometimes complaints about hospital foods are valid objections to the ways foods are prepared and served. A client may be expecting to enjoy a favorite food for dinner, only to be disappointed with the way the food is prepared. Unfortunately, hot foods may not be hot and cold foods may not be cold by the time they arrive in the client's room, or foods may have been left in the room while the client was gone for tests or therapy. In addition, the client receives meals at specified times regardless of hunger and often must eat in bed without companionship, which can be more of a chore than a pleasurable experience. Many disorders, medications, and treatments can dramatically alter taste perceptions and lead to complaints about food.[1] Meals may also be unwelcome if the person is in pain or has been sedated.

In the hospital, eating offers clients familiarity in an otherwise strange environment. Most people generally look forward to eating, and for many people in the hospital, a healthy appetite signals a return to health. Eating may become even more enjoyable than usual. It is also one of the few hospital experiences where clients have a choice. Consider that clients usually cannot choose when they will receive tests, how many times blood will be drawn, what nurse will care for them, or what time they will have surgery. But they usually can select their meals, and they can use those meals to exercise control or express their feelings: they can eat or refuse to eat!

Complaints about hospital foods may have little to do with the food itself, but instead serve as a way for clients to vent fear, frustration, anger, and physical pain. Clients need opportunities to express their feelings, and often you may find that a problem can be resolved simply by listening and providing emotional support.[2]

HANDLING FOOD-RELATED PROBLEMS

The majority of people in the hospital will eat adequate amounts of foods, even though they complain about them. If they fail to eat enough food to meet their nutrient needs, the deficit will be easy to correct once they are at home eating familiar foods. Chapter 19 provides suggestions for helping people in the hospital to eat. For people in pain, administering pain medications so that they will be effective during meals can be helpful. For people with altered taste perceptions, the dietitian can work closely with them to uncover the

Foodservice departments strive to prepare appetizing and nutritious foods designed to accommodate dozens of special diets and hundreds of food preferences.

tastes and food preferences that they can tolerate and enjoy.

Some problems can be handled directly by the person caring for the client. For example, the nurse or assistant providing trays to clients should distribute the trays as soon as possible after the food carts arrive. That person can also make sure that foods and utensils are arranged attractively before they are served. In some cases, the foodservice department must be contacted to solve food-related problems. For example, the foodservice department should be contacted if a client receives the wrong diet or consistently receives foods that differ from those that were requested. Foodservice departments often conduct periodic surveys to uncover problems clients may have with foods or food service.

HOW FOODSERVICE SYSTEMS WORK

The responsibility of budgeting, planning, preparing, and serving appropriate meals rests with either a chief administrative dietitian or a foodservice manager. In some facilities, foodservice companies from outside the hospital perform these duties.

Clinical dietitians work directly with clients to assess their nutrition status, plan appropriate diets, and provide nutrition education. Clinical dietitians may also assist in menu planning, especially for special diets. In some facilities, dietetic technicians assist dietitians in both administrative and clinical responsibilities. Other dietary employees include clerks, porters, and other assistants. Keep in mind that many dietary employees do not have formal education in nutrition, and their ability to interpret diet orders and provide accurate information is limited.

Menu Procedures

Most hospitals provide menus from which clients can select their meals. A client who must follow a modified diet receives menus that include only foods specified in the hospital's diet manual for that particular diet (see Figure H19-1 on p. 626). By allowing a choice, this system helps to ensure that clients receive foods they prefer and will eat. An added advantage for people on special diets is that they become familiar with their diets by marking appropriate menus.

Although procedures vary between hospitals, generally dietary employees deliver menus to each client's room early in the day and pick them up again later in the day. Each menu shows the client's name and room number, as well as the name of the meal (breakfast, lunch, or supper), the type of diet, and the day the menu will be served. Often clients make selections for the next few days to give the foodservice department time to collect the menus and estimate the amount and type of food to prepare. Menus are usually color coded by diet. Color coding helps ensure that foodservice employees put the right types of foods on food trays and helps the person delivering the tray to quickly determine if the right diet has been delivered.

Clients typically select one or more items from each food category on the menu. Clients may not receive foods they enjoy if they fail to mark the menus or inadvertently make the wrong food selections. If menus are not marked or if a menu is lost, the client receives meals selected by the foodservice department. Consider these potential problems:

- Clients may have difficulty seeing, reading, understanding, or physically marking menus.
- Clients may not understand that their selections will be for the next (or another) day.
- Clients may be out of their rooms (for tests, procedures, or physical activity) or asleep when the menus arrive; when the clients return or wake up, they may not see the menus or may have missed the menu pickup time.
- Clients may be too ill or too disinterested in food to make menu selections.

Occasional problems with menu selections can usually be corrected simply by explaining the menu system to clients or taking time to help them mark menus. If clients continue to complain about food selections, contact the dietetic technician or dietitian.

Once food selections have been made and menus collected, menus are often checked by a member of the foodservice staff (usually, a dietetic technician or dietitian) to ensure that selections are appropriate. Completed menus can provide valuable clues about a person's usual eating habits or understanding of a modified diet. In checking menus, for example, the technician may notice that one person on a regular diet is selecting very little or that another is selecting too much. In another case, the technician may notice that a person on a kcalorie-restricted diet is not selecting the appropriate number of servings of allowed foods. Such problems suggest the need for further intervention.

Some hospitals do not offer selective menus. Instead, they serve a standard house diet, adjusting the menu for individual food preferences. For example, clients can request simple changes, such as the substitution of one vegetable for another. Similarly if the regular menu offers fried fish, a person on a low-fat or low-kcalorie diet would receive baked fish.

Food Preparation

The logistics of preparing foods tailored to each modified diet can be overwhelming. For this reason, foodservice departments use systems designed to limit costs and minimize errors. Foods prepared for regular and soft/bland/low-fiber/low-residue diets are prepared with some fat and salt, because these dietary components are not restricted on such diets. Note that the other diet menus shown in

Figure H19-1 provide a number of low-fat (LF) or low-sodium (LS) or low-sodium, low-fat (LSLF) foods.

If the foodservice department were to prepare a food (green beans, for example) for each different diet, it would have to prepare green beans made with some fat and salt, green beans made without fat, green beans made without salt, and green beans made without fat or salt. Instead, only two types of green beans are prepared—one with some fat and salt, and the other without fat or salt. Clients can then add allowed ingredients to food. For example, the person on a low-salt diet could add margarine to the green beans; the person on a low-fat diet could add salt.

Keep in mind that green beans are only one of many menu items in a day, and you can see why preparing individual foods for each diet is not feasible. However, you can also see how such a system can lead to complaints about food. A person on a low-fat diet, for example, may not realize that the food has been prepared without any salt. In such a case, the client should be advised to add other seasonings to enliven the flavor of the food.

Food Delivery

Sometimes foods are prepared in a main kitchen, assembled on trays, and delivered to the nursing unit; then foods intended to be eaten hot are heated in areas close to the clients' rooms. In other cases, foods are delivered directly from the main kitchen, using serving pieces that help keep hot foods hot and cold foods cold. In either case, foodservice personnel deliver food carts directly to the nursing unit; then nursing or foodservice personnel take a tray to

FIGURE H19-1 Sample Lunch Menus

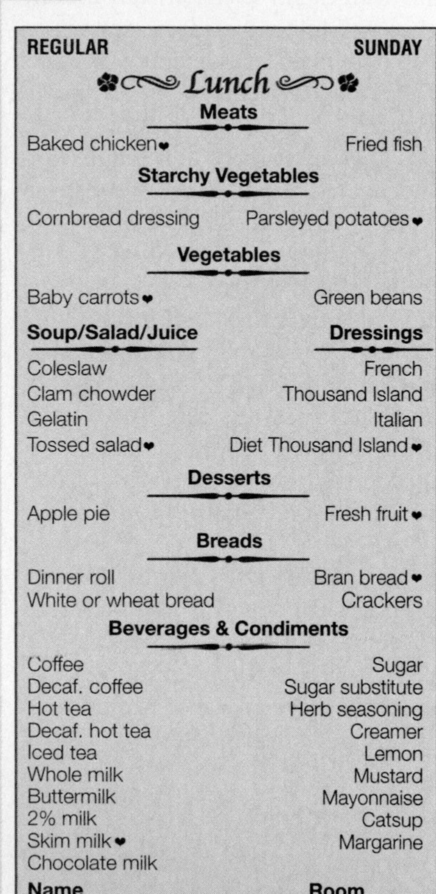

REGULAR — **SUNDAY**

Lunch

Meats

Baked chicken♥ — Fried fish

Starchy Vegetables

Cornbread dressing — Parsleyed potatoes♥

Vegetables

Baby carrots♥ — Green beans

Soup/Salad/Juice — **Dressings**

Coleslaw — French
Clam chowder — Thousand Island
Gelatin — Italian
Tossed salad♥ — Diet Thousand Island♥

Desserts

Apple pie — Fresh fruit♥

Breads

Dinner roll — Bran bread♥
White or wheat bread — Crackers

Beverages & Condiments

Coffee — Sugar
Decaf. coffee — Sugar substitute
Hot tea — Herb seasoning
Decaf. hot tea — Creamer
Iced tea — Lemon
Whole milk — Mustard
Buttermilk — Mayonnaise
2% milk — Catsup
Skim milk♥ — Margarine
Chocolate milk

Name ________ **Room** ____

People on regular diets select the foods of their choice. The regular menu may also be used for high-kcalorie, high-protein diets. The menu items marked with a heart guide people in selecting foods that are lower in fat, cholesterol, sodium, and caffeine or higher in fiber than other menu choices.

SOFT/BLAND/LOW RESIDUE — **SUNDAY**

Lunch

Meats

Baked chicken — Baked fish (cod)

Starchy Vegetables

Rice — Boiled potatoes

Vegetables

Baby carrots — Green beans

Soup/Salad/Juice — **Dressings**

Gelatin — Mayonnaise
Tomato soup

Desserts

Apple pie — Pears

Breads

Dinner roll — Crackers
White bread

Beverages & Condiments

Decaf. coffee — Sugar
Decaf. hot tea — Sugar substitute
Decaf. iced tea — Creamer
Whole milk — Lemon
2% milk — Margarine
Buttermilk — Catsup
Skim milk
Lemonade

No pepper

Name ________ **Room** ____

Foods for soft/bland/low-residue diets are similar to those for regular diets. Foods from the regular menu that are not appropriate have been eliminated from the menu, and substitutes have been made. For clients on bland diets, decaffeinated coffee and tea would be crossed off the menu. For low-residue diets, milk would be restricted to two servings a day.

KCALORIE RESTRICTED, DIABETIC — **SUNDAY**
_1200_ **CALORIES**

Lunch

LF = Low Fat LSLF = Low Sodium, Low Fat

Meat Exchange (Select _1_)

LSLF Baked chicken (2 oz) — LSLF Baked fish (2 oz)

Starch Exchange (Select _1_)

Clam chowder (1 c) — LF dinner roll (1)
LSLF Rice (1/3 c) — White bread (1 slice)
LSLF Boiled potatoes (1/2 c) — Wheat bread (1 slice)
Bran bread (1 slice)
Crackers (6)

Vegetable Exchange (Select _2_)

LSLF Baby carrots (1/2 c) — LSLF green beans (1/2 c)

Fruit Exchange (Select _1_)

Diet pears (1/2 c) — Fresh fruit

Milk Exchange (Select _1_)

Whole milk (1 c) omit 2 fats — Buttermilk (1 c)
2% milk (1 c) omit 1 fat — Skim milk (1 c)

Fat Exchange (Select _1_)

Margarine (1 tsp) — Creamer (1 = 1/2 fat)
Diet mayonnaise (1/2 oz)

Calorie-free foods

Coffee — LSLF Coleslaw (1/2 c)
Decaf. coffee — Tossed salad (1/2 c)
Hot tea — Diet gelatin (1/2 c)
Decaf. hot tea — Diet French
Iced tea — Diet Thousand Island
Sugar substitute — Diet Italian
Lemon — Mustard
Herb seasoning — Diet catsup

Name ________ **Room** ____

For kcalorie-restricted and diabetic diets, the number of exchanges allowed is written on the menu beforehand. (This example uses a 1200-kcalorie diet.) Note that the meat exchange is written in 2-ounce portions so that 1 serving = 2 exchanges.

each client. Efficient delivery of foods to the nursing unit and then to the client helps ensure that clients receive foods at the appropriate temperature.

Once the client is finished eating, the tray is returned to the food cart. Foodservice personnel pick up the carts and return them to the foodservice department.

WORKING WITH THE SYSTEM

Learning about the foodservice system in the health care facility where you work can help save both you and your clients needless aggravation. Ask to spend a few hours or a day working with different foodservice employees to see firsthand how the department operates and what problems they encounter. If that is not possible, learn the facility's procedures for ordering diets, making diet changes, reporting problems with a client's tray, or making special requests. Remember that requests are not simply made by one individual to another. Often many people are involved in processing a single request, and the number of requests made during any one meal can be considerable. Translating requests (for example, preparing another tray) takes time, and delays are unavoidable. The best strategy is prevention—make sure that clients mark menus and that you call in requests as early as possible to allow the foodservice department the time to process the request.

One of the most important things to know about your facility's foodservice system is the time when meals are actually assembled, so that you can call in requests well before this time. Once tray assembly begins, dietary

FIGURE H19-1 **Sample Lunch Menus**

LOW-FAT/LOW CHOLESTEROL/CARDIAC SUNDAY

Lunch

LF = Low Fat LSLF = Low Sodium, Low Fat

Meats

LSLF Baked chicken LSLF Baked fish (cod)

Starchy Vegetables

LSLF Rice LSLF Boiled potatoes

Vegetables

LSLF Baby carrots LSLF Green beans

Soup/Salad/Juice

LSLF Coleslaw
Gelatin
Tomato soup
Tossed salad

Dressings

Diet French
Diet Thousand Island
Diet Italian

Desserts

Pears Fresh fruit

Breads

LF Dinner roll
White bread
Wheat bread
Bran bread
LS Crackers

Beverages & Condiments

Coffee
Decaf. coffee
Hot tea
Decaf. hot tea
Iced tea
Buttermilk
Skim milk
Creamer
Sugar
Sugar substitute
Herb seasoning
Lemon
Margarine
Mustard
Diet mayonnaise
Catsup

Name ______ Room ______

People on low-fat, low-cholesterol diets who also need kcalorie restriction receive a kcalorie-restricted menu to control portion sizes and number of servings. Both menus provide low-fat, low-cholesterol foods. Foods not appropriate for a low-fat, low-cholesterol diet, such as whole milk, would be crossed off the menu beforehand.

LOW SODIUM SUNDAY

Lunch

LF = Low Fat LSLF = Low Sodium, Low Fat

Meats

LSLF Baked chicken LSLF Baked fish (cod)

Starchy Vegetables

LSLF Rice LSLF Boiled potatoes

Vegetables

LSLF Baby carrots LSLF Green beans

Soup/Salad/Juice

LSLF Coleslaw
LS Chicken broth
Apple juice
Tossed salad

Dressings

Diet French
Diet Thousand Island
Diet Italian

Desserts

Pears Fresh fruit

Breads

Dinner roll
White bread
Wheat bread
Bran bread
LS Crackers

Beverages & Condiments

Coffee
Decaf. coffee
Hot tea
Decaf. hot tea
Iced tea
Whole milk
2% milk
Skim milk
Sugar
Sugar substitute
Creamer
Lemon
Herb seasoning
Margarine
Diet mustard
Diet mayonnaise
Diet catsup

No salt

Name ______ Room ______

Low-sodium menus are similar to those provided for low-fat, low-cholesterol diets, but they eliminate high-sodium foods, such as tomato soup. The person on a low-sodium, low-fat, low-cholesterol diet selects foods from a low-fat menu with high-sodium foods crossed off beforehand. If the person is also on a low-kcalorie diet, foods would by selected from a low-kcalorie menu with high-sodium foods crossed off the menu beforehand.

RENAL SUNDAY

Lunch

LF = Low Fat LSLF = Low Sodium, Low Fat

Meats

LSLF Baked chicken LSLF Baked fish

Starchy Vegetables

LSLF Rice LSLF Dialyzed potatoes

Vegetables

LSLF Baby carrots LSLF Green beans

Soup/Salad/Juice

Lemonade
LSLF Coleslaw
Tossed salad (no tomato)

Dressings

Diet French
Diet Thousand Island
Diet Italian

Desserts

Pears Apple pie

Breads

Dinner roll
White bread
Wheat bread
Bran bread
LS Crackers

Beverages & Condiments

Coffee
Decaf. coffee
Hot tea
Decaf. hot tea
Iced tea
Sugar
Sugar substitute
Creamer
Lemon
Margarine
Diet mustard
Mayonnaise

No salt

Name ______ Room ______

Renal diets must be highly individualized, and the person checking the menu has to carefully consider the client's selections and make appropriate changes when necessary.

employees are extremely busy, and requests will be difficult to process.

With so many people and steps involved in foodservice, and so many clients with individual dietary needs, food preferences, and emotional responses, it is easy to see many opportunities for problems to arise. Once you understand the many factors that affect clients' appetites and perceptions of hospital foods as well as how the foodservice department operates, you can begin to tackle problems efficiently and avoid needless frustration for your clients and yourself.

References

1 M. A. Hess, Taste: The neglected nutritional factor, *Journal of the American Dietetic Association* (supplement 2) 97 (1997): 205–207.

2 M. Bélanger and L. Dubé, The emotional experience of hospitalization: Its moderation and its role in patient satisfaction with foodservice, *Journal of the American Dietetic Association* 96 (1996): 354–360.

Nutrition and Disorders of the Upper GI Tract

CHAPTER 20

The remarkable GI tract serves as a passageway through which the body assimilates nutrients and other food components from the external world. The diet of a healthy newborn gradually adjusts to include a wide variety of foods as its GI tract matures and the infant becomes progressively more efficient at ingesting, digesting, and absorbing nutrients. With aging, some GI tract functions decline, and again, dietary adjustments to accommodate these normal physiologic changes help meet nutrient needs. At any stage of development, medical conditions can affect the functions of the GI tract, and dietary adjustments can ease symptoms and help prevent malnutrition. This chapter describes how medical nutrition therapy serves in the treatment of some upper GI tract symptoms and disorders. The next chapter presents disorders of the lower GI tract and their relationships to diet and nutrition status.

DISORDERS OF THE MOUTH AND ESOPHAGUS

Figure 20-1 illustrates the upper GI tract and reviews the functions of its various components. In the mouth, the teeth and jaw muscles work together to break down food to a consistency that can be easily swallowed and digested. The upper esophagus assists in the swallowing process, and the remaining esophagus transports food from the mouth to the stomach. The muscles controlling the esophagus push foods down and prevent them from moving back toward the throat.

Difficulties Chewing

People may have difficulties chewing◆ from many reasons ranging from sedation and pain to neurological disorders, facial injuries, mouth ulcers, missing teeth,◆ or ill-fitting dentures. People with persistent problems with chewing may eat too little, lose too much weight, and suffer the consequences of a deteriorating nutrition status.

◆ Reminder: The process of chewing is also called **mastication.**

◆ A person without teeth is **edentulous** (ee-DENT-you-lus).
- **e** = without
- **dens** = teeth

THINK NUTRITION whenever clients have persistent problems with chewing food for whatever reason.

• **Diet Options** • People's tolerances of food consistencies vary greatly. Following surgery to repair a broken jaw, for example, a person may be able to consume liquids only. With time, however, the person gradually progresses to foods that are modified in texture to make them easy to chew and swallow—a **mechanical soft diet.** Unlike the soft diet described in Chapter 19, mechanical soft diets include all foods and seasonings. Nurses and dietitians work together with the client, family, and caregivers to identify which foods the client can safely handle; such foods may be naturally soft foods such as fish or bananas, tender cooked whole foods, ground or chopped foods, or pureed foods. The goal is to provide a wide variety of foods that are as similar as possible to those of a regular diet. Such a strategy enhances the appetite and minimizes the likelihood of nutrient deficiencies. Therefore, soft natural foods, tender-cooked foods, and chopped foods are provided whenever possible; only when the person cannot chew or swallow adequate amount of these foods are pureed foods provided.

The person recovering from a broken jaw, described above, may progress from liquids to pureed foods, then to ground or chopped foods, and then to regular foods. In some cases, however, a person may need to follow a mechanical

mechanical soft diet: a diet that excludes only those foods that a person cannot chew; also called a **dental soft diet** or an **edentulous diet.**

FIGURE 20-1 The Upper GI Tract

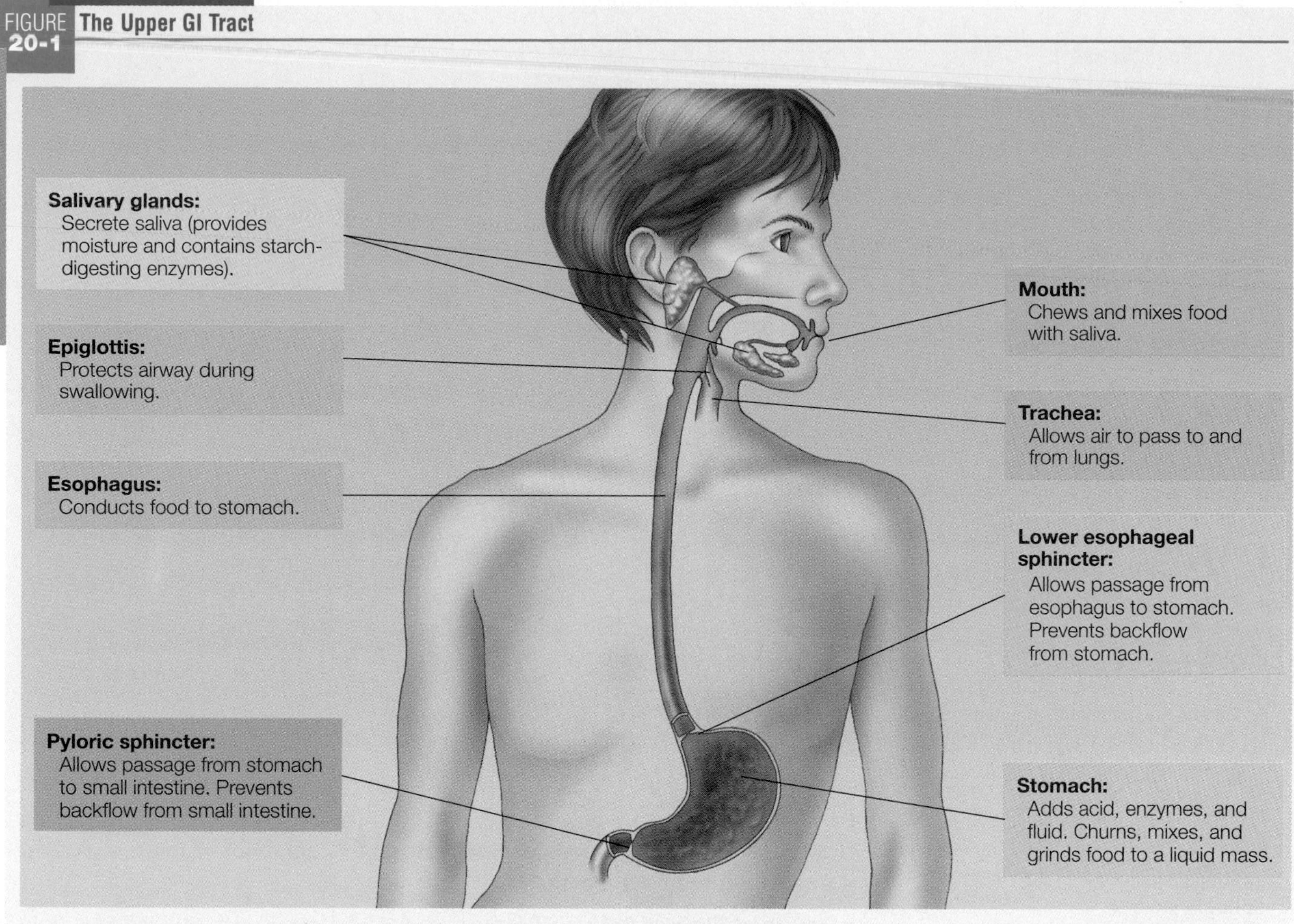

soft diet permanently. Moist, soft-textured foods, such as foods prepared with sauces and gravies, often work well. Drinking liquids along with meals makes it easier to chew and swallow foods.

• Cautious Use of Pureed Foods • Pureed foods can meet all nutrient needs and may taste good, but the monotony of eating foods of the same consistency for every meal, day after day, can create a psychological block to eating. Commercial products that thicken pureed foods and allow them to be shaped attractively can make a remarkable difference in enhancing the appetite. The "How to" on p. 632 offers other suggestions for clients and caregivers to improve the acceptance of pureed foods.

Health care professionals employed in long-term care facilities such as nursing homes and rehabilitation centers face a formidable challenge in ensuring that clients receiving pureed foods over long periods of time continue to eat enough to meet their nutrient needs. In addition to the problems associated with pureed diets, clients in long-term care facilities lack the comforts of home and may be overwhelmed by their medical conditions, isolation, and loss of control. Addressing the client's emotional needs is critical to stimulating the appetite.

• Mouth and Throat Pain • The mechanical soft diet for people with **mouth ulcers** or inflammation of the lips or throat (following a **tonsillectomy,** for example) provides moist, soft-textured foods and eliminates spicy, salty, or acidic foods (such as citrus fruits and juices and tomato products) that may be painful to eat. In addition, nuts or seeds in foods (such as sesame or poppy seeds in breads) and sticky foods (like peanut butter) can cause discomfort.

mouth ulcers: lesions or sores in the lining of the mouth. Certain medications, radiation therapy, and some disorders, such as oral herpes infections, can cause mouth ulcers.

tonsillectomy (tawn-sill-ECK-tah-me): surgical removal of the tonsils.

HOW TO Improve Acceptance of Pureed Foods

Take a moment to think about a meal of pureed foods. A typical dinner of baked chicken, mashed potatoes, and green beans is blenderized to white mush, more white mush, and a green blob. The foods may taste great, but on seeing the plate, the person may be reluctant to try the first bite. To stimulate the appetite, use creative techniques such as these for preparing and serving food:

- Encourage clients and their caregivers to prepare a variety of favorite foods and blenderize them to a tolerable consistency. The smell of favorite foods and the thought of consuming them stimulate the appetite.
- Consider color when planning meals. The meal of baked chicken, mashed potatoes, and green beans, described above, can be made more appealing by substituting mashed sweet potatoes for the white potatoes. Arranging foods attractively on a plate with appropriate garnishes also adds color and eye appeal.
- Serve foods at the right temperature and puree them so that they are smooth and thick—not watery and thin. Commercially available thickeners add shape, texture, and even nutrients to food.
- Experiment with seasonings and spices to enliven food flavors, excluding only those that the person cannot tolerate for personal or medical reasons. Seasoning food to accommodate personal tastes adds flavor, allows for variety, and improves the appetite.
- Supplement the diet with nutritious liquids such as milk, instant breakfast drinks, or liquid formulas (see Chapter 22).

Courtesy Diamond Crystal Specialty Foods

Can you tell that the foods in this photo are pureed foods shaped with commercial thickeners?

Efforts to improve the acceptance of pureed foods can go a long way toward helping people to eat and maintain or improve their weights. When efforts to improve a client's intake of pureed foods are unsuccessful, feeding the person by tube becomes an option (see Chapter 22).

Heat may intensify pain, too. Many clients prefer cold foods or foods served no warmer than room temperature.

• **Reduced Flow of Saliva** • People with mouth dryness due to a reduced flow of saliva often prefer moist, soft-textured foods. Clients can moisten foods with sauces and gravies. Salty foods and snacks dry the mouth and should be avoided. Encourage clients to practice good oral hygiene; when salivary flow is reduced, the mouth is poorly defended against dental caries. Clients can increase salivary secretions by sucking on sugarless candy, chewing sugarless gum, or using drugs that stimulate the flow of saliva.

Difficulties Swallowing

Problems with swallowing, known as **dysphagia,** can arise from many causes, including aging, neurological disorders, developmental disabilities, and strokes.◆ Dysphagia may occur when the muscles of the mouth or esophagus fail to function properly and interfere with a person's ability to push foods to the back of the throat to initiate swallowing or to propel foods through the esophagus to the stomach. In **achalasia,** the lower esophageal sphincter◆ fails to relax in response to the presence of foods in the esophagus, and foods and liquids accumulate in the esophagus until the pressure of the food forces the lower esophageal sphincter to open, or the food is regurgitated.

◆ Conditions associated with dysphagia include:
- Acquired immune deficiency syndrome (AIDS).
- Aging.
- Alzheimer's disease.
- Brain tumors.
- Cancers of the head and neck.
- Developmental feeding disorders.
- Guillain-Barré syndrome.
- Head injuries.
- Lou Gehrig's disease (amyotrophic lateral sclerosis).
- Multiple sclerosis.
- Myasthenia gravis.
- Parkinson's disease.
- Polio.
- Reflux esophagitis.
- Strokes.

◆ The **lower esophageal sphincter (LES)** is also called the cardiac sphincter or **gastroesophageal sphincter.**

dysphagia (dis-FAY-gee-ah): difficulty in swallowing.

achalasia (ack-ah-LAY-zee-ah): failure of the lower esophageal sphincter to relax and allow foods to pass from the esophagus to the stomach; formally called **achalasia of the cardia** or **cardiospasm.**
- **a** = not
- **chalasis** = relaxation

THINK NUTRITION when a client has dysphagia. Consider the client to be at risk for protein-energy malnutrition as well as for aspiration.

• **Subtle and Dangerous** • Dysphagia often goes undiagnosed, especially when associated with aging, because symptoms may be sporadic and mild at first. All people "catch food in their throat" at one time or another, so a person who experiences this feeling more frequently may consider the condition normal and ignore it. Eventually, the person may eat less, experience marked weight loss, and develop nutrient deficiencies.

When undetected, dysphagia can be dangerous for other reasons as well. Unlike healthy people, people with dysphagia may fail to cough in response to the presence of food in the trachea. Food particles may then pass unnoticed into the lungs **(aspiration),** allowing bacteria to multiply. Such "silent" aspiration, which has been observed in people following strokes, carries with it the risk of serious pneumonia and death.

People with achalasia generally experience only mild dysphagia at first. Eventually, the esophagus may enlarge and regurgitation may occur frequently. Nutrients lost through regurgitation as well as a fear of eating◆ to avoid the discomfort of vomiting can markedly interfere with nutrition status.

◆ The fear of eating is called **sitophobia (SIGH-toe-FOE-bee-ah).**
- **sitos** = food
- **phobia** = fear

• Signs of Dysphagia • Health care professionals should be alert to subtle symptoms of dysphagia including an unexplained decline in food intake or repeated bouts of pneumonia. Other symptoms include pain upon swallowing, weight loss, a fear of eating certain foods or any food at all, a feeling that food is sticking in the throat, a tendency to hold food in the mouth rather than swallowing it, coughing or choking during meals, frequent throat clearing, drooling, or a change in voice quality. Depending on the cause of dysphagia, the voice may be hoarse, nasal, or have a "wet" sound. Diagnosis is based on extensive testing that may include cranial nerve assessment, X rays, fluoroscopy, and measurements of esophageal sphincter pressure and esophageal peristalsis.

• Dietary Interventions for Dysphagia • The mechanical soft diet for dysphagia leaves little room for error or experimentation. Speech therapists, dietitians, physicians, and nurses work together to assess the client's swallowing abilities and design an individualized diet. Many clients with dysphagia tolerate mildly spiced and moderately sweet foods that are served at room temperature. Sticky foods and small pieces of food or foods that break into small pieces when eaten (such as rice or chopped meats) can be difficult to handle. Foods with more than one texture, such as yogurt with fruit or dry cereal with milk, are also difficult for many clients to handle. Clients often experience the most difficulty swallowing true liquids. Thickened liquids or smooth solids, such as milk shakes or puddings, are good choices; these foods flow slowly enough to allow time to coordinate swallowing movements. Commercial thickeners, puddings, yogurt, or baby cereal can be used to thicken liquids.

Appropriate body positioning during meals minimizes the risk of choking. Advise the client to sit upright at a 90-degree angle with hips flexed, feet flat on the floor, and the head tilted slightly forward.

With time, swallowing function may improve. The health care team continuously monitors the person and expands the diet to include additional foods as tolerated. Ideally, the diet will progress to a regular diet, although this is not always possible.

• Tube Feedings • Feedings by tube (described in Chapter 22) may be necessary for people who are unable to eat adequate amounts of foods orally, particularly for those who are severely malnourished or those whose swallowing function continues to deteriorate. Tube feedings delivered into the stomach, however, may be contraindicated due to the high risk of aspiration in people with dysphagia. Intestinal tube feedings often provide a safer alternative.

IN SUMMARY

Many disorders that affect the mouth and esophagus can interfere with chewing and swallowing and may require a modification in food consistency. Dysphagia, a serious swallowing disorder that frequently goes undiagnosed, can lead to repeated bouts of pneumonia and even death. Medical nutrition therapy for chewing and swallowing disorders includes a highly individualized mechanical soft diet based on the client's individual tolerances.

aspiration: the drawing of food, gastric secretions, or liquid into the lungs.

DISORDERS OF THE STOMACH

Once swallowed, food travels down the esophagus into the stomach. The stomach retains the bolus for a while, adding acids, fluids, and enzymes to the mixture before slowly releasing its contents into the intestine. Disorders of the stomach range from occasional bouts of indigestion◆ to serious conditions that require surgical resections.

Indigestion and Reflux Esophagitis

Indigestion, or **dyspepsia,** describes abdominal pain associated with eating. Highlight 3 (pp. 85–91) describes the causes and consequences of occasional bouts of indigestion including **heartburn,** belching, and hiccups. Many people suffer from indigestion caused by the backflow of the stomach's acid fluids into the esophagus and mouth (acid indigestion). Acid indigestion occurs when the lower esophageal sphincter fails to close tightly enough to keep the stomach's contents from backing up into the esophagus **(esophageal reflux).**

• Causes of Esophageal Reflux • Conditions and substances that weaken the lower esophageal sphincter itself or raise the pressure in the stomach increase the likelihood of reflux (see Table 20-1). Aging, certain medications, and the use of tubes that pass from the nose through the stomach for more than a few days increase the risk of reflux. Esophageal reflux frequently occurs in people with asthma, peptic ulcers (described later in this chapter), irritable bowel syndrome (described in Chapter 21), developmental disabilities (see Highlight 20), and sliding hiatal hernias. A sliding **hiatal hernia** occurs when the diaphragm weakens, allowing the lower esophageal sphincter and a portion of the stomach to slip through the esophageal **hiatus** and protrude above the diaphragm (see Figure 20-2). Once a sliding hiatal hernia forms, the lower esophageal sphincter is no longer reinforced by the surrounding diaphragm, and it becomes easy for gastric juices to reflux into the esophagus.

• Consequences of Esophageal Reflux • In some cases reflux causes no symptoms or problems. When reflux becomes a chronic problem, however, the highly acidic gastric fluids can irritate the esophagus and cause a painful inflammation called **reflux esophagitis.** Heartburn and a feeling of fluid accumulation in the throat and mouth accompanied by a sour or bitter taste are common symptoms. Clients with reflux esophagitis may also report symptoms not associated with the GI tract, including chest pain, chronic coughing, hoarseness, and sore throat. Severe and chronic inflammation may lead to **esophageal ulcers,** and the consequent bleeding, inflammation, and scarring can narrow the inner diameter of the esophagus **(esophageal stricture).** Chronic blood loss can lead to anemia. Dysphagia may develop. Chronic reflux is also associated with **Barrett's esophagus,** a condition linked to an increased risk of cancer of the esophagus.[1] Chronic pulmonary disease can develop if the gastric contents are frequently aspirated into the lungs through the throat.

• Prevention and Treatment • Treatment for active reflux esophagitis aims to alleviate reflux and reduce inflamma-

TABLE 20-1 Conditions and Substances Associated with Esophageal Reflux

Conditions That Raise the Likelihood of Reflux
Ascites (accumulation of fluid in the abdomen)
Delayed gastric emptying
Eating large meals
Lying flat after eating
Obesity
Pregnancy
Wearing clothes that fit tightly across the waist or abdomen
Substances That Weaken Lower Esophageal Sphincter Pressure
Alcohol
Anticholinergic agents
Caffeine
Calcium channel blockers
Chocolate
Cigarette smoking
Diazepam
Garlic
High-fat foods
Meperidine
Onions
Peppermint oils
Spearmint oils
Theophylline

◆ Indigestion and other common digestive problems are discussed in Highlight 3.

dyspepsia: vague abdominal pain; a symptom, not a disease.
- **dys** = bad
- **pepteln** = to digest

heartburn: a burning sensation felt behind the sternum caused by the presence of gastric juices in the esophagus; also called **pyrosis** (pie-ROE-sis).

esophageal reflux: the backflow of the gastric contents of the esophagus.

hiatal hernia: a protrusion of a portion of the stomach through the esophageal hiatus of the diaphragm. There are several types of hiatal hernias, but the type most commonly associated with reflux is a **sliding hiatal hernia.**

hiatus (high-AY-tus): the opening in the diaphragm through which the esophagus passes.
- **hiatus** = to yawn

reflux esophagitis (eh-sof-ah-JYE-tis): inflammation of the esophagus caused by esophageal reflux; also called **gastroesophageal reflux disease (GERD).**
- **re** = back
- **fluxus** = flow

esophageal ulcers: lesions or sores in the lining of the esophagus.

esophageal stricture: narrowing of the inner diameter of the esophagus from inflammation and scarring.

Barrett's esophagus: changes in the cells of the esophagus associated with chronic reflux that raise the risk of cancer of the esophagus.

FIGURE 20-2 **The Upper GI Tract, Acid Reflux, and Hiatal Hernia**

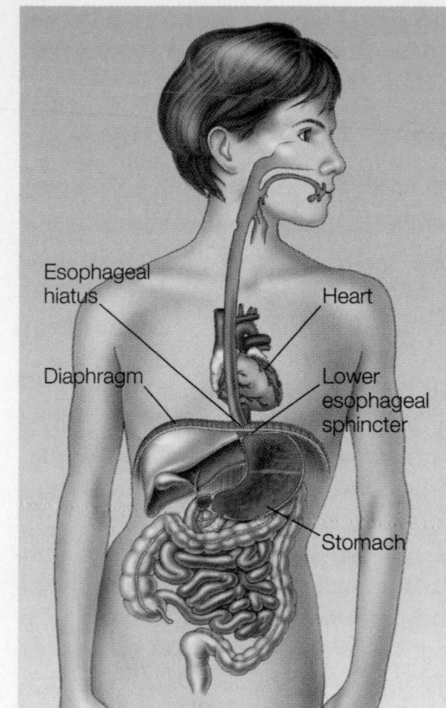

The stomach normally lies below the diaphragm, and the esophagus passes through the esophageal hiatus. The lower esophageal sphincter prevents reflux of stomach contents.

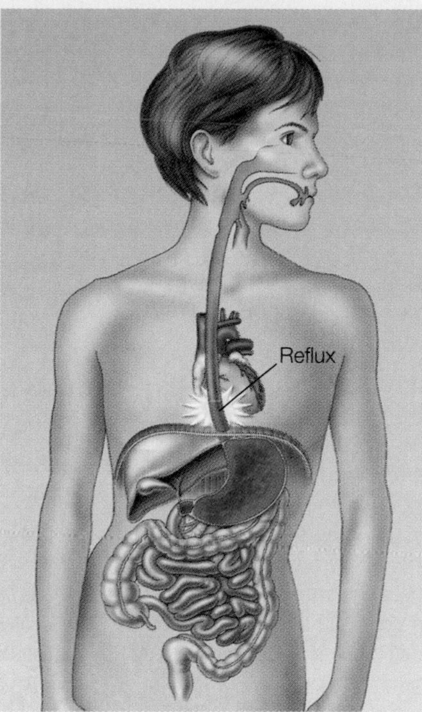

Whenever the pressure in the stomach exceeds the pressure in the esophagus, as can occur with overeating and overdrinking, the chance of reflux increases. The resulting "heartburn" is so-named because it is felt in the area of the heart.

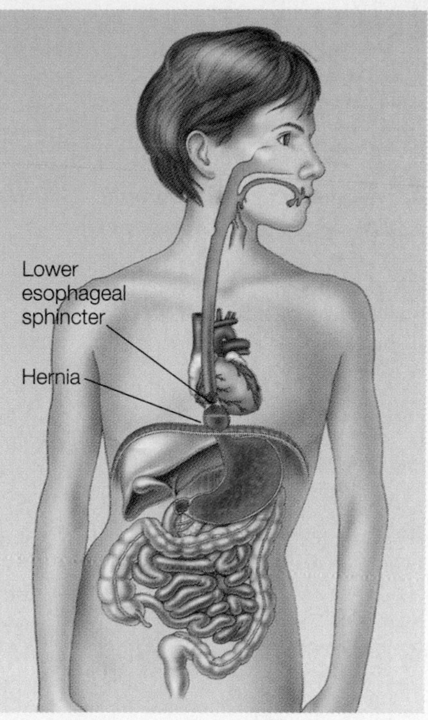

Acid reflux often occurs as a consequence of a hiatal hernia. A sliding hiatal hernia occurs when part of the stomach, with the lower esophageal sphincter, slips through the diaphragm.

tion of the esophagus and its associated pain by reducing gastric acidity, reducing pressure in the stomach, and eliminating foods, substances, or activities that weaken the lower esophageal sphincter. The "How to" on p. 636 offers tips for preventing reflux and treating its symptoms.

THINK NUTRITION whenever clients have a diagnosis of reflux esophagitis. Medical nutrition therapy can help ease symptoms and reduce the risk that pain will keep the client from eating adequate amounts of food.

• **Drug Therapy** • Turn on the television set and quite likely you will see several advertisements touting antacids and a new generation of over-the-counter medications that are highly effective in both preventing and relieving acid indigestion. The ads imply that rather than eat sensibly and listen to the body's signals about what kinds and amounts of foods to eat and the right way to eat them, people can eat anything, anyway they like, and then simply take medication to avoid the discomforts of acid indigestion. All medications carry risks, however, and everyone should be encouraged to take steps to prevent acid indigestion without medication. People who must rely on medications to treat acid indigestion or reflux esophagitis should do so with the advice of their physicians. Physicians frequently prescribe antacids to neutralize gastric acidity and antisecretory agents◆ to suppress or inhibit gastric acid secretion (see

Antisecretory agents are also called ◆ **anti-GERD (GastroEsophageal Reflux Disease) agents.**

Prevent and Treat Reflux Esophagitis

To prevent and treat reflux esophagitis and its associated discomfort, recommend that clients:

- Eat small meals and drink liquids one hour before or one hour after meals to avoid distending the stomach. A distended stomach exerts pressure on the lower esophageal sphincter.
- Relax during mealtimes, eat foods slowly, and chew them thoroughly to avoid swallowing air and distending the stomach.
- Limit foods that weaken lower esophageal sphincter pressure or increase gastric acid secretion, including fat, alcohol, caffeine, decaffeinated coffee and tea, chocolate, spearmint, and peppermint.
- Lose weight, if necessary. Overweight tends to increase abdominal pressure.
- Refrain from lying down, bending over, and wearing tight-fitting clothing or belts, particularly after eating, to avoid increasing pressure in the stomach.
- Elevate the head of the bed by 4 to 6 inches. Keeping the chest higher than the stomach helps prevent reflux.
- Refrain from smoking cigarettes. Smoking relaxes the lower esophageal sphincter.
- During periods of active reflux esophagitis, avoid foods and beverages that irritate the esophagus, such as citrus fruits and juices, tomatoes and tomato-based products, pepper, spices, and very hot or very cold foods, according to individual tolerances.

Individual tolerances of types and amounts of foods and spices vary markedly especially during periods of active esophagitis. Health care professionals can help clients pinpoint food intolerances by advising them to keep a record of the types and amounts of foods and beverages consumed, the time of consumption, GI symptoms, and time of occurrence. Assessment of the record by a dietitian or health care professional will help determine the types and amounts of food the client can handle without discomfort. If a client cannot tolerate citrus fruits and juices, ensure that the diet provides adequate vitamin C from other foods or supplements.

the Diet-Drug Interactions box on p. 643). Medications that strengthen lower esophageal sphincter pressure (cholinergics) may be used when other measures prove unsuccessful in controlling symptoms. Surgery may be necessary in some cases. The accompanying case study provides questions that review the nutrition needs of a client with reflux esophagitis.

Nausea and Vomiting

Nausea describes the feeling that one is about to vomit. Nausea is a symptom of many medical conditions,◆ medications, and treatments and can be triggered by emotional stress, certain odors, motions, and disturbing sights.

As Highlight 3 describes, simple vomiting is certainly unpleasant and wearying for the nauseated person, but is not cause for alarm. Prolonged vomiting, however, can be serious and requires professional medical care. With vomiting, foods and medications fail to reach the intestine and are unavailable to the body. Large amounts of fluids and electrolytes are also expelled, raising the risk of dehydration and nutrient imbalances. Infants and children can quickly become seriously dehydrated as a consequence of vomiting, because their bodies contain a greater percentage of water than an adult body. Both nausea and vomiting lead to reduced food intake, and when they persist over time, both can lead to malnutrition.

• Treatment of Nausea and Vomiting • Treatment of the medical condition that causes nausea and vomiting often alleviates the problem. Nausea and vomiting associated with medications are sometimes temporary and often resolve after a few doses. If not, the physician may change the medication, if possible. The suggestions in the "How to" on p. 638 can help relieve nausea and vomiting. Physicians may also prescribe medications to relieve nausea (antinauseants) and vomiting (antiemetics). The nutrition-related interactions of these medications are provided in the Diet-Drug Interactions box on p. 643).

◆ Some conditions that can lead to nausea and vomiting include pregnancy, reflux esophagitis, delayed gastric emptying, gastritis, peptic ulcers, gallbladder disorders, pancreatic disorders, liver disorders, kidney disorders, cancer, HIV infections, and food poisoning.

• Dietary Interventions for Vomiting • If efforts to control nausea fail, foods are withheld for a few hours. If tolerated, clear liquids can be provided to replace fluids and electrolytes. If oral liquids cannot be tolerated, intravenous (IV) fluids may be given until vomiting resolves. Providing all nutrients by vein (see Chapter 23) may be indicated in cases of prolonged or severe vomiting, especially for clients who are malnourished or those who have high nutrient needs.

Gastritis

Gastritis is an inflammation of the lining of the stomach that can develop suddenly (acute gastritis) or over time (chronic gastritis). Unresolved gastritis can lead to ulcers (described in the next section), hemorrhage, shock, obstruction, and perforation and raises the risk of gastric cancer.

gastritis: inflammation of the stomach lining.

CASE STUDY

Accountant with Reflux Esophagitis

Mrs. Scarlatti, a 49-year-old accountant, recently underwent a complete physical examination. She told her physician that she had been feeling fairly well until she began experiencing heartburn, which has become more frequent and more painful. The heartburn usually occurs after she has eaten a large meal, particularly when she lies down after eating. After inspecting the interior of her esophagus, stomach, and duodenum with a long tube equipped with an optical device called a gastroscope, the physician diagnosed a sliding hiatal hernia and reflux esophagitis.

Mrs. Scarlatti's past health history shows no indication of significant health problems. During her last physical, the physician did advise her to stop smoking cigarettes and to lose 20 pounds, which she has yet to do. The nutrition assessment reveals that Mrs. Scarlatti is experiencing a great deal of stress because it is the middle of the tax season. She usually does not have time for breakfast, eats a lunch of fast foods hurriedly while continuing to work, and eats a large dinner around 8:00 P.M. She generally drinks one or two alcoholic beverages before going to sleep. Her current height and weight are 5 feet 6 inches and 170 pounds.

Explain to Mrs. Scarlatti what a sliding hiatal hernia is, and describe how the hernia leads to gastric reflux and heartburn.

From the brief history provided, list the factors and behaviors that increase Mrs. Scarlatti's chances of experiencing reflux and the pain of heartburn. What recommendations can you make to help her change these behaviors? What medications might the physician prescribe and why?

• **Acute Gastritis** • Acute gastritis most often follows the repeated use of aspirin or other medications that irritate the gastric mucosa. ***Helicobacter pylori*** infections, and other bacterial infections, alcohol abuse, food irritants, food allergies, food poisoning, radiation therapy, and metabolic stress can also cause gastritis. Symptoms of acute gastritis may include anorexia, nausea and vomiting, stomach pain, and fever.

Treatment for acute gastritis depends on the cause, but in all cases any food or substance that irritates the gastric mucosa is withheld. For the person with acute gastritis who cannot eat because of nausea or vomiting, foods may be withheld for a day or two. Then, the diet progresses from liquids to a highly individualized diet that aims to eliminate foods that stimulate gastric acid secretion, irritate the gastric mucosa, or cause discomfort for the individual. Table 20-2 describes foods and food components that may cause gastric irritation.

TABLE 20-2 Foods and Food Components That May Cause Gastric Irritation

- Alcohol
- Caffeine and caffeine-containing beverages (including cola beverages, cocoa, coffee, and tea)
- Decaffeinated coffee and tea
- Pepper and spicy foods except as tolerated

Note: An individual may find other foods to be irritating, and, if so, those foods should be eliminated.

• **Chronic Gastritis** • Chronic gastritis may be associated with aging, *Helicobacter pylori* infections, and conditions that cause the chronic reflux of basic fluids from the duodenum into the stomach (gastric reflux). With time, gastric acid secretions diminish, and the production of intrinsic factor falters. Vitamin B_{12} malabsorption and pernicious anemia◆ follow. People with chronic gastritis may have mild symptoms similar to those of acute gastritis or no symptoms at all. For this reason, a diagnosis of pernicious anemia may actually lead to the diagnosis of chronic gastritis. In the later stages of unresolved chronic gastritis, the gastric cells gradually shrink and lose their function (atrophic gastritis).◆

Chronic gastritis requires diagnosis and treatment before damage progresses too far. Interventions need to begin early enough to prevent complications such as dehydration, malnutrition, or damage to the esophagus. Foods and substances that irritate the gastric mucosa must be avoided (see Table 20-2). For clients with pernicious anemia, vitamin B_{12} is given by injection or using a prescription nasal spray to bypass the need for absorption. Antacids, antisecretory agents, antiulcer agents, and antibiotics may also be prescribed.

Reminder: Pernicious (per-NISH-us) anemia is a reduced number of red blood cells caused by a lack of intrinsic factor and the consequent malabsorption of vitamin B_{12}. ◆

Reminder: Atrophic gastritis is the severe form of gastritis in which chronic inflammation diminishes the size and function of the stomach's mucosal cells and glands. ◆

Ulcers

Ulcers can develop both inside and outside the body, but the term **ulcer** used alone generally refers to a peptic ulcer—an erosion of the top layer of cells from the lining of the esophagus, stomach, or small intestine. This erosion leaves the

ulcer: an open sore or lesion. A **peptic ulcer** is an erosion of the cells of the mucosa of the lower esophagus, the stomach, or small intestine. Ulcers may also develop in the mouth, upper esophagus, and large intestine, or on the skin.

Helicobacter pylori: a bacterium that may lead to gastritis and peptic ulcers and may raise the risk of cancer of the stomach.

HOW TO Minimize Nausea and Vomiting

The following suggestions can help clients alleviate nausea:

- Relax before eating, chew foods thoroughly, and avoid overeating. Breathe deeply and slowly if you begin to experience nausea. Eat small meals and save liquids for between meals to prevent distention of the stomach.
- Drink cold or carbonated liquids when you first experience nausea.
- Identify individual food intolerances and aromas that precipitate nausea; avoid these foods and aromas, if possible. High-fat foods and highly spiced foods are often poorly tolerated. Cold or room-temperature foods may be better tolerated than hot foods, especially if food aromas trigger nausea. Ask others to prepare foods, if possible.
- Eat carbohydrate-rich, low-fat foods before getting out of bed if you feel nauseated in the morning. Crackers or bread can be kept at bedside. Dry, salty foods like crackers, pretzels, and popcorn can sometimes help relieve nausea.
- If nausea is a problem at specific times of the day, avoid eating or drinking immediately before, during, or after those times.
- Try drinking ginger or peppermint teas, which provide relief for some people.
- Relax after eating, but don't lie down. Get fresh air after meals and wear nonconstrictive clothing.

Physicians may also prescribe antinauseants or antiemetics to help control nausea and vomiting. Counsel clients to take these medications at least 30 minutes before eating to ensure effectiveness during mealtimes.

underlying layers of cells exposed to gastric juices. When the erosion reaches the capillaries, the ulcer bleeds, and when gastric juices reach the nerves, they cause pain.

• Causes of Ulcers • Highlight 3 introduced the three major causes of ulcers: *Helicobacter pylori* infections, the use of certain anti-inflammatory drugs,◆ and disorders that cause excessive gastric acid secretion. One such disorder, the **Zollinger-Ellison syndrome,** results from a tumor of the pancreas that produces gastrin, which, in turn, stimulates the production of gastric acid. Severe peptic ulcer disease follows.

A study of more than 45,000 men suggests that higher intakes of vitamin A, fruits and vegetables, and soluble dietary fibers are associated with a lower incidence of duodenal ulcers.[2] Further research is necessary to determine if other closely related nutrition (or other) factors may actually account for the protective effects the researchers observed.

• Treatment for Ulcers • As mentioned in Highlight 3, treatment aims at relieving pain, healing the ulcer, and minimizing the likelihood of recurrence. Antibiotics are used to treat bacterial infections. Antiulcer and antisecretory agents may be used to suppress or inhibit gastric acid secretion and protect the esophagus, stomach, or intestinal lining from acid erosion. As for diet, advise the client to relax during mealtimes, eat slowly, chew foods throughly, avoid overeating, and follow the suggestions in Table 20-2 on p. 637.

For clients with Zollinger-Ellison syndrome, surgical removal of the tumor, when possible, reduces gastric acid production. Surgery to sever the nerves that stimulate gastric acid production or to remove all or part of the stomach, described next, may be indicated.

◆ The medications associated with ulcers are *nonsteroidal anti-inflammatory agents (NSAIDS)* such as ibuprofen and naproxen.

Gastric Surgery

Several surgical procedures affect the functions of the stomach. During a **gastrectomy,** the surgeon removes either a portion or all of the stomach. Figure 20-3 on p. 639 illustrates three common gastrectomy procedures. Another type of gastric surgery, **pyloroplasty,** enlarges the pyloric sphincter (the sphincter that joins the stomach and small intestine) so that the intestinal fluids (with a higher pH) reflux into the stomach and neutralize gastric acidity. During a **vagotomy,** the surgeon severs the nerves that stimulate gastric acid production. A vagotomy may accompany either a gastrectomy or a pyloroplasty in some cases. In **gastric partitioning,** a treatment for severe obesity, the stomach remains intact, but all or a portion of the stomach is bypassed.

• Dumping Syndrome • When the portion of the stomach containing the pyloric sphincter has been removed, bypassed, or disrupted, one problem that may occur is **dumping syndrome** (see Figure 20-4 on p. 640). A typical scenario goes something like this: Mr. Clark had a fairly extensive gastric resection about a week ago and has just begun to eat solid foods. He swallows the food and about 15 minutes later begins to feel weak and dizzy. He looks pale, his

Zollinger-Ellison syndrome: marked hypersecretion of gastric acid and consequent peptic ulcers caused by a tumor of the pancreas.

gastrectomy (gas-TREK-tah-mee): surgery to remove all (total gastrectomy) or part (subtotal or partial gastrectomy) of the stomach.

pyloroplasty (pie-LOOR-oh-PLAS-tee): surgery that enlarges the pyloric sphincter.

vagotomy (vay-GOT-oh-mee): surgery that severs the nerves that stimulate gastric acid secretion.

gastric partitioning: surgery for severe obesity that limits the functional size of the stomach.

dumping syndrome: the symptoms that result from the rapid entry of undigested food into the jejunum: sweating, weakness, and diarrhea shortly after eating and hypoglycemia later.

FIGURE 20-3 **Typical Gastric Surgery Resections**

In a gastric resection, part or all of the stomach is surgically removed. the dashed lines show the removed section.

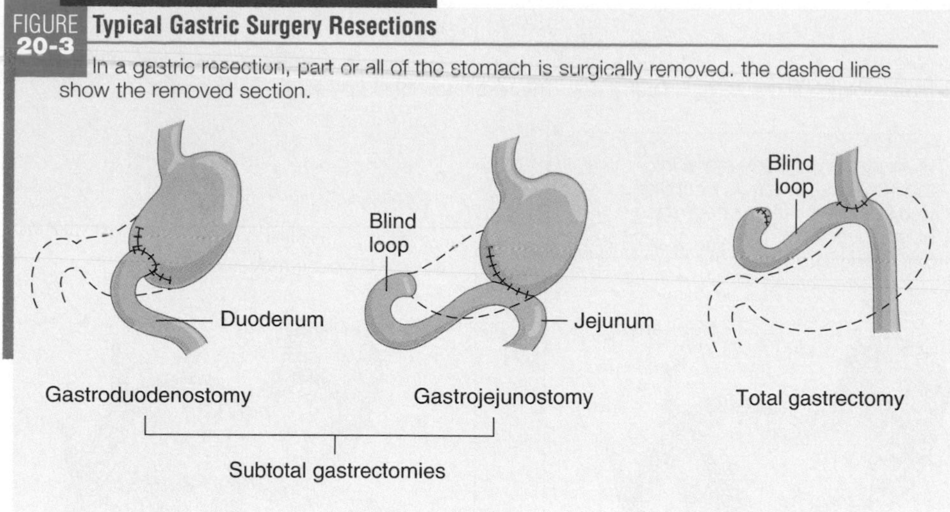

heart beats rapidly, and he breaks out in a sweat. Shortly thereafter, he develops diarrhea. What causes this sequence of events?

Without a functional pyloric sphincter, Mr. Clark's stomach has lost control over the rate at which food enters the intestine. Now, food gets "dumped" rapidly into the jejunum. (The duodenum is short, and even if it had not been bypassed during surgery, large amounts of food would pass quickly through it into the jejunum.) As the mass of food is digested, the intestinal contents rapidly become concentrated (hypertonic). Fluids from the body move into the intestinal lumen to dilute the concentration. Consequently, the volume of circulating blood falls rapidly, causing weakness, dizziness, and a rapid heartbeat. The large volume of hypertonic fluid and unabsorbed material in the jejunum causes pain and **hyperperistalsis,** and diarrhea results.

Two to three hours later, Mr. Clark experiences many of the same symptoms again: dizziness, fainting, nausea, and sweating. This time the cause is different. Mr. Clark's intestines efficiently absorbed so much glucose from the meal that his blood glucose rose quickly. His pancreas responded by overproducing insulin, which made his blood glucose fall quickly. Now, low blood glucose (hypoglycemia)◆ is causing the symptoms.

◆ The type of hypoglycemia that occurs following gastric surgery is called **reactive** or **postgastrectomy hypoglycemia.**

Not all people experience the diarrhea of dumping syndrome following gastric surgery or vagotomies. Even fewer develop hypoglycemia. Most people who initially experience the dumping syndrome gradually adapt to a fairly regular diet. Nevertheless, a postgastrectomy diet serves as a preventive measure following the immediate postsurgical period and benefits clients with prolonged or severe problems.

• Postsurgical Care • Immediately following surgery, the client receives no foods or fluids by mouth. Health care professionals monitor fluid and electrolyte balances carefully, and the physician corrects any imbalances promptly. After several days, liquids and solids are gradually introduced in very small amounts offered frequently.

• The Postgastrectomy Diet • The postgastrectomy diet limits the total amount of carbohydrate as well as the amount of simple sugars. To provide energy, slow the passage of foods through the stomach, and minimize diarrhea, the diet emphasizes foods high in protein and fat. When digested, proteins and fats produce fewer particles than carbohydrates and do not attract fluid as rapidly as carbohydrates do, thereby minimizing diarrhea. Table 20-3 on p. 641 lists foods included on, and excluded from, the postgastrectomy diet. Diet advice to offer with the postgastrectomy diet is provided in the "How to" on p. 642.

hyperperistalsis: rapid movement through the intestine.

FIGURE 20-4 **The Dumping Syndrome**

When partially digested food rapidly enters the jejunum, it is quickly digested and creates a concentrated mass. Fluid from the intestinal capillaries enters the jejunum, diminishing blood volume and stimulating peristalsis. The result: low blood pressure and diarrhea.

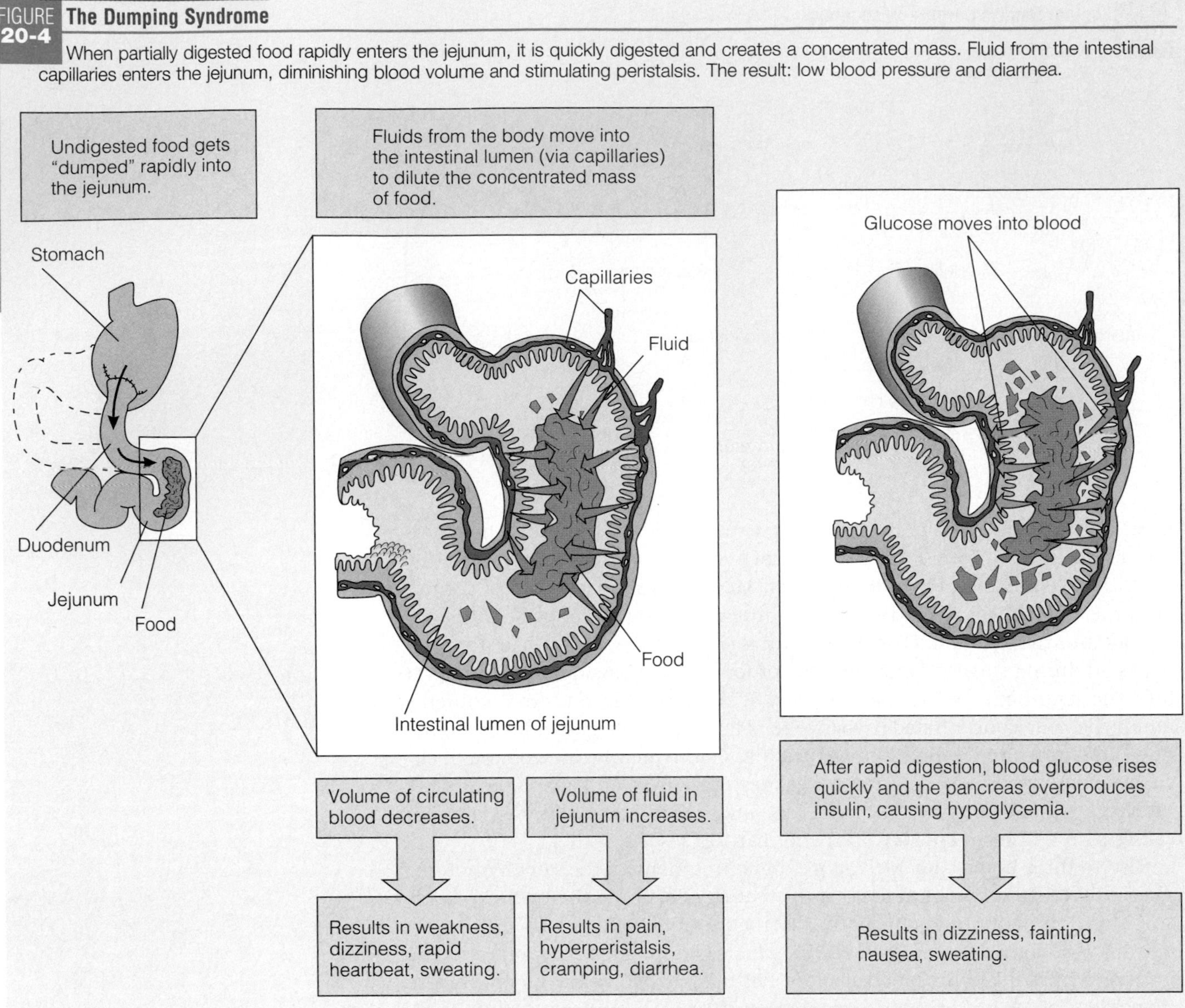

Dietitians carefully tailor the postgastrectomy diet to meet the person's tolerances. Initially, the dietitian visits the client after each meal to check the person's tolerance for different foods. With time, the symptoms of dumping syndrome resolve or improve. Gradually, most people begin to tolerate limited amounts of concentrated sweets, larger quantities of food, and some liquids with meals. Sometimes adding pectin and guar gum (types of dietary fibers) to the diet can help control dumping syndrome. If dietary and medical management of dumping syndrome fail to resolve the problem, additional surgery may be necessary.

• **Weight Loss and Nutrient Deficiencies** • While weight loss is the goal for people undergoing gastric partitioning, people undergoing other gastric surgeries have often lost significant amounts of weight before surgery, and they risk serious malnutrition if they do not receive adequate energy and nutrients. Unfortunately, many people continue to experience weight loss and develop nutrient deficiencies following surgery. Reflux esophagitis commonly occurs following gastric surgery, and its associated **epigastric** pain can contribute to poor food intake. Other potential problems that can contribute to poor food intake include early satiety, postsurgical pain, fear that eating will bring on the symptoms of the dumping syndrome, and sometimes dysphagia.

epigastric: the region of the body just above the stomach.
- **epi** = above
- **gastric** = stomach

TABLE 20-3 **Postgastrectomy Diet**[a]

Meat and Meat Alternatives

Any type allowed.

Milk and Milk Products

Withheld initially and then gradually introduced as tolerated.

Grains and Starchy Vegetables

Allowed (up to 5 servings per day): Plain breads, crackers, rolls, unsweetened cereal, rice, pasta, corn, lima beans, parsnips, peas, white potatoes, sweet potatoes, pumpkin, yams, winter squash.

Excluded: Sweetened cereal; cereal containing dates, raisins, or brown sugar.

Nonstarchy Vegetables

Allowed (unlimited): Cabbage, Chinese cabbage, celery, cucumbers, lettuce, parsley, radishes, watercress.

Allowed (up to two ½ c servings per day as individual tolerances permit): Asparagus, bean sprouts, beets, broccoli, brussels sprouts, carrots, cauliflower, eggplant, green pepper, greens, mushrooms, okra, onions, rhubarb, sauerkraut, string beans, summer squash, tomatoes, turnips, zucchini.

Excluded: Vegetables prepared with sugar or creamed.

Fruits

Allowed (up to 3 servings per day): Unsweetened fruits and fruit juices.

Excluded: Sweetened fruits and fruit juices, dates, raisins.

Fats

Any type allowed.

Beverages

Allowed: Coffee, tea, artificially sweetened drinks.

Excluded: Alcohol; sweetened milk, beverages, and fruit drinks; cocoa.

Other

Excluded: Cakes, cookies, ice cream, sherbet, honey, jam, jelly, syrup, and sugar.

Note: [a]Clients with dumping syndrome who are unable to tolerate a sufficient variety or volume of foods over long periods of time often require nutrient supplements.

In addition to the malabsorption associated with the dumping syndrome, fat malabsorption (see Chapter 21) can occur whenever clients have undergone surgery that bypasses the duodenum. When the food bolus bypasses the duodenum, the trigger for the release of the hormones that mediate the secretion of enzymes and bile to aid fat digestion and absorption is also bypassed. Fat malabsorption results. Fat malabsorption can also occur as a consequence of reduced gastric acid secretion, which can lead to bacterial overgrowth in the stomach or upper small intestine (described in detail in Chapter 21). Bacterial overgrowth contributes to the malabsorption not only of fat but also of fat-soluble vitamins (especially vitamin D), as well as folate, vitamin B_{12}, and calcium.

• Anemia • Iron-deficiency anemia frequently occurs following gastric surgery, although it may take several years to develop. When iron's exposure to gastric acid is limited, less iron is converted to its absorbable form. In addition, iron absorption normally occurs more efficiently in the duodenum than in the rest of the small intestine, and following gastric surgery, the food bolus either bypasses the duodenum altogether or moves rapidly through it. Inadequate intake of iron from foods and blood loss also contribute to the problem. An iron supplement helps correct iron deficiencies.

Inadequate intake and malabsorption can also lead to anemia caused by folate and, less often, vitamin B_{12} deficiencies. Although one might expect vitamin B_{12} deficiencies to be common after gastric surgery because intrinsic factor production could be impaired, surgeons often avert this problem by leaving intact a

HOW TO Adjust Meals to Prevent the Dumping Syndrome

The person on a postgastrectomy diet benefits from the following suggestions:

- Avoid concentrated sweets (sugar, cookies, cakes, pies, or soft drinks) because the body digests and absorbs these carbohydrates rapidly and breaks them down into many particles that draw fluids from the body into the intestine.
- Eat frequent, small meals to fit the reduced storage capacity of the stomach.
- Drink liquids in small amounts about 45 minutes before or after meals, not with them. This precaution prevents overloading the stomach's reduced storage capacity and slows the transit of food from the stomach to the intestine.
- Lie down immediately after eating to help slow the transit of food to the intestine. Clients who experience esophageal reflux, however, should not lie down after eating.
- Be aware that lactose intolerance (described in Chapter 21) may develop and add to the problem of diarrhea and abdominal pain. Enzyme-treated milk and milk products should also be avoided because the enzymes break down lactose to glucose and galactose—simple sugars that further concentrate the intestinal contents and promote dumping. Discontinue all types of milk and milk products until recovery is under way. Then try them gradually and in small amounts.

small area of the stomach where intrinsic factor production occurs.◆ To correct deficiencies, clients receive supplements.

◆ While most gastric surgeries do not lead to vitamin B_{12} deficiencies, such deficiencies are common when the entire stomach (total gastrectomy) is removed.

• **Bone Disease** • People who experience fat malabsorption following gastric surgery also malabsorb vitamin D and calcium. After many years, a significant number of people who have undergone gastric surgery develop a bone disease similar to osteomalacia.◆ Although the bone disease usually does not respond well to treatment, vitamin D and calcium supplements are often provided. The case study reviews the needs of a client following gastric surgery.

◆ Reminder: *Osteomalacia* is a bone disease characterized by softening of the bones.

• **Gastric Partitioning** • Unlike with other gastric surgeries, weight loss is a goal following gastric partitioning. Most clients lose weight for about 12 to 18 weeks after surgery and lose 50 to 60 percent of their excess body weight.[3]

The safety and effectiveness of gastric partitioning depend, in large part, on compliance with medical nutrition therapy, which aims to prevent nutrient deficiencies and promote eating and lifestyle habits that will help the client lose weight and maintain a desirable weight loss. Poor dietary habits may prevent weight loss, rupture staples, or obstruct the small opening left for food to pass into the lower stomach. Other potential complications include infections, nausea, vomiting, dehydration, dumping syndrome, esophageal reflux, and, as a result of all this and more, depression.

• **Diet following Gastric Partitioning** • After surgery, clients typically follow a mechanical soft diet and are cautioned to chew foods thoroughly to reduce the risk of dislodging surgical staples and to prevent foods from obstructing the opening into the lower stomach. After about three months, clients can generally begin to eat solid foods. Regardless of food consistency, clients tolerate only small amounts of food (about a cup of food at a time); overeating or overdrinking can cause nausea, reflux, and vomiting.

• **Food Selections** • Clients must understand that the foods and beverages they select influence the extent of their weight loss. If they regularly drink high-kcalorie liquids or eat high-kcalorie foods, even if only in small quantities,

CASE STUDY

Commercial Artist Requiring Gastric Surgery

Mr. Miyamotto, a 58-year-old commercial artist, was admitted to the hospital for gastric surgery after numerous attempts to medically manage his severe peptic ulcer disease had failed. A gastrojejunostomy and vagotomy were performed, and Mr. Miyamotto is recovering as expected. The health care team members anticipate nutrition-related problems and are taking measures to prevent them.

1. Review Figure 20-3 to understand the surgical procedure that Mr. Miyamotto underwent. Consider the possibilities that he might experience early satiety, nausea, vomiting, reflux esophagitis, weight loss, dumping syndrome, malabsorption, anemia, and bone disease. Describe how these conditions might occur.
2. What type of diet will the physician prescribe for Mr. Miyamotto after he begins eating orally? Describe the diet and how it progresses. What advice can you give Mr. Miyamotto to prevent dumping syndrome?
3. Describe the nutrition-related concerns associated with fat malabsorption and anemia. How can these concerns be handled?

ANTACIDS

Aluminum-containing antacids contribute aluminum to the diet and may cause constipation or lead to phosphorus deficiency. Long-term or inappropriate use can lead to aluminum toxicity.

Calcium-containing antacids contribute calcium to the diet and may cause constipation. Concurrent use with vitamin D supplements or foods containing large amounts of vitamin D can lead to elevated blood calcium levels (hypercalcemia).

Magnesium-containing antacids contribute magnesium to the diet and may cause diarrhea; long-term use may lead to magnesium toxicity.

Sodium bicarbonate, an antacid that contributes sodium to the diet, can alter serum electrolyte levels and raise the pH of the blood. Sodium bicarbonate should not be taken with milk—hypercalcemia can result.

ANTIBIOTICS (FOR *HELICOBACTER PYLORI* INFECTIONS)

When *amoxicillin* is given without regard to food, nausea and diarrhea are common side effects. Encourage clients to take with food to reduce nausea.

Metronidazole may cause taste alterations (metallic taste), and no alcohol should be used during treatment and for 24 hours afterward. Alcohol can react with metronidazole and result in a disulfram-like reaction with symptoms that include nausea, vomiting, headache, cramps, and flushing of the skin.

Tetracycline and its classic interactions with nutrients were mentioned in Chapter 18. Calcium, magnesium, zinc, aluminum, antacids that contain any of these nutrients, and vitamin-mineral supplements should not be given from one hour before to two hours after a tetracycline dose. Iron supplements should not be given from two hours before to three hours after a tetracycline dose. Tetracycline can also cause nausea and diarrhea.

ANTIEMETICS, ANTINAUSEANTS

Antiemetics frequently lead to drowsiness. Two antiemetics, *dronabinol* (a derivative of marijuana) and *prochlorperazine,* stimulate the appetite and can result in weight gain. Prochlorperazine also increases the urinary excretion of riboflavin. Prochlorperazine, *prochloramide* and *triethylperazine maleate* can lead to mouth dryness. Prochlorperazine, *prochloramide, triethylperazine maleate, granisetron,* and *odansteron* may cause constipation. *Metoclopramide* rarely causes nutrition-related side effects.

ANTISECRETORY AGENTS

Antisecretory agents may interfere with iron absorption. When iron supplements are necessary, they should be given two hours before or after taking these medications. People taking some antisecretory agents should avoid the use of the herb pennyroyal; antisecretory agents may increase the formation of toxic metabolites from pennyroyal.

they will not lose weight. Their selections also affect the nutritional quality of the diet. Nutrient deficiencies, particularly of vitamin B_{12}, iron, and vitamin D, are common. Careful planning, diligent compliance with the prescribed diet, and vitamin-mineral supplements help ensure that nutrient needs are met within a strictly limited energy intake.

IN SUMMARY

Indigestion, reflux esophagitis, nausea, vomiting, gastritis, and ulcers are frequent GI problems that can lead to a limited food intake, weight loss, and malnutrition. When treatment for a medical condition requires gastric surgery, the impact on nutrition status can be severe. Medical nutrition therapy for conditions that affect the upper GI tract can relieve uncomfortable symptoms and minimize the risk of malnutrition. The accompanying Diet-Drug Interactions box describes the nutrition implications of medications used in the treatment of upper GI tract disorders, and the Nutrition Assessment Checklist summarizes information that health care professionals can use to assess and monitor the nutrition status of clients with upper GI tract disorders.

As the next chapter shows, medical nutrition therapy also plays a major role in the treatment of lower GI tract disorders. Nutrition problems are especially likely in those with fat malabsorption—a problem that can arise from a variety of conditions.

Nutrition Assessment Checklist for People with Upper GI Tract Disorders

MEDICAL HISTORY

Does the client's health history reveal medical conditions or treatments that:

- ❑ Interfere with chewing and swallowing?
- ❑ Result in indigestion, nausea, or vomiting?

Does the client have a medical diagnosis of:

- ❑ Hiatal hernia or reflux esophagitis?
- ❑ Gastritis?
- ❑ Pernicious anemia?
- ❑ Peptic ulcers?

For a client who has undergone gastric surgery, check for the following complications:

- ❑ Dumping syndrome
- ❑ Malabsorption
- ❑ Anemia
- ❑ Bone disease

MEDICATIONS

Check the medication and dosing schedule for:

- ❑ Medications that may cause indigestion, nausea, and vomiting. Note that many medications can cause nausea, especially the first few doses. Remember to give medications with food when possible, to help alleviate nausea.
- ❑ Clients receiving tetracycline for *Helicobacter pylori* infections. Do not give tetracycline one hour before or two hours after giving milk and milk products or antacids. Do not give tetracycline two hours before or three hours after giving iron supplements.
- ❑ Clients receiving antisecretory agents and iron supplements. Give the antisecretory agent two hours before or after the iron supplement.

FOOD/NUTRIENT INTAKE

Confer with clients and the dietitian to help pinpoint food intolerances for clients with:

- ❑ Indigestion/nausea
- ❑ Active reflux esophagitis, gastritis, or peptic ulcers
- ❑ Dumping syndrome

For clients on long-term mechanical soft diets, regularly note:

- ❑ Appetite
- ❑ Variety of food being offered and eaten
- ❑ Consistencies the client can handle

ANTHROPOMETRICS

Measure baseline height and weight. Address weight loss early to prevent malnutrition for clients with:

- ❑ Any condition requiring mechanical soft diets for long periods of time
- ❑ Dysphagia
- ❑ Indigestion and nausea of long duration
- ❑ Dumping syndrome/malabsorption

LABORATORY TESTS

Check laboratory tests for signs of dehydration for clients with:

- ❑ Persistent vomiting
- ❑ Dumping syndrome

Check laboratory tests for nutrition-related anemias (see Appendix E) for people with:

- ❑ Gastritis
- ❑ Gastric surgeries
- ❑ Conditions that require long term use of antisecretory agents

PHYSICAL SIGNS

Look for physical signs of:

- ❑ Dehydration (especially for people with persistent vomiting or dumping syndrome)
- ❑ Iron deficiency (especially for people who have had gastric surgery or those on tetracycline or antisecretory agents)
- ❑ Vitamin B_{12} deficiency (especially for people with chronic gastritis or those who have undergone gastric partitioning)

Nutrition on the Net

WEBSITES

Access these websites for further study of topics covered in this chapter.

- Search for specific upper GI tract disorders at Gut Feelings and the American College of Gastroenterology: **www.gutfeelings.com** and **www.acg.gi.org**
- Find more information about dysphagia at the Dysphagia Resource Center: **www.dysphagia.com**
- Review information about reflux esophagitis at the GERD Information Center: **www.gerd.com**
- Review information about *Helicobacter pylori* infections and their consequences at the Helicobacter Foundation: **www.helico.com**
- Find more information about peptic ulcers at the U.S. Department of Health and Human Services: **www.hoptechno.com/book35.htm**

INTERNET ACTIVITY

People seldom think about chewing and swallowing, much less about what happens when the food they eat reaches the stomach. When something goes wrong and these processes are disrupted, health and nutrition can be compromised. You may be interested to know about conditions that can impair the health of the upper GI tract.

- Go to **www.gutfeelings.com**
- Click on "Diseases."
- Click on a disease of interest and learn about the symptoms, diagnostic tests, prevention, and treatment.

What did you learn about the disease and its symptoms, prevention, and treatment?

Study Questions

These questions will help you review the chapter. You will find the answers in the discussions on the pages provided.

1. Provide examples of conditions that can interfere with chewing. Describe ways that diets can be adjusted to meet the needs of people who are having problems chewing foods. (pp. 630–632)
2. Discuss ways to provide appetizing foods for people on pureed diets. (p. 632)
3. What is dysphagia, and how is the diet managed to ease its symptoms? What are the potential consequences of dysphagia? (pp. 632–633)
4. What is reflux esophagitis, and what are some of its complications? What advice can benefit clients to prevent and treat symptoms of reflux esophagitis? (pp. 634–636)
5. Under what conditions can nausea and vomiting present a risk to nutrition status? (p. 636)
6. What are the differences between acute and chronic gastritis? Describe the role of diet therapy for both types of gastritis. (p. 637)
7. What are the possible nutrition consequences of gastric surgery? Describe the relationship of gastric surgery to those consequences. What dietary interventions help prevent those consequences? (pp. 638–642)

These questions will help you prepare for an exam. Answers can be found on p. 646.

1. The health care professional working with a client on a mechanical soft diet recognizes that:
   a. only pureed foods should be given to minimize the risk of aspiration.
   b. the client can have any food that can be comfortably and safely chewed and swallowed.
   c. highly seasoned foods are always restricted.
   d. such diets cannot be planned to meet total nutrient needs and supplements are always necessary.
2. For a person with dysphagia:
   a. the diet requires a great deal of experimentation to uncover which foods the person can tolerate.
   b. meals should be eaten in bed with the head tilted back.
   c. the diet is based on a health care team's assessment of the person's swallowing abilities and close monitoring of the person's ability to handle different foods.
   d. coughing during meals indicates that the person is able to clear the throat and is not at risk for aspiration.
3. Reflux esophagitis is:
   a. an inflammation of the esophagus caused by the backflow of acidic gastric juices from the stomach.
   b. a protuberance of a portion of the stomach above the cardiac sphincter.
   c. an erosion of the lining of the stomach caused by excess acid in gastric juices.
   d. an obstruction of the lower esophagus that results in dysphagia.
4. The health care professional working with a client with reflux esophagitis recognizes that the client understands her diet when she says:
   a. "I need to eat three meals a day and drink liquids with my meals."
   b. "I need to eat food slowly, relax during meals, and lie down after meals."
   c. "I need to drink more citrus juices and tomato-based products at the first sign of heartburn."
   d. "I need to limit my intake of fat, alcohol, caffeine, and decaffeinated coffee, and tea."
5. For the client with persistent vomiting, the major nutrition-related concern(s) is/are:
   a. dehydration and malnutrition.
   b. reflux esophagitis.
   c. emotional distress.
   d. peptic ulcers.
6. Dietary suggestions that help clients with gastritis or active ulcers discourage foods that:
   a. are high in fiber.
   b. irritate the gastric mucosa.
   c. are easy to chew and swallow.
   d. contain simple sugars.

7. Nutrition concerns most commonly associated with chronic gastritis include:
   a. malnutrition and pernicious anemia.
   b. iron-deficiency anemia and protein malabsorption.
   c. dumping syndrome and pernicious anemia.
   d. dehydration and electrolyte imbalances.
8. Antisecretory agents and the suggestions in Table 20-2 are recommended for a client with peptic ulcers. The client complies with the recommendations but continues to complain of indigestion and gastric pain. The most appropriate advice would be to:
   a. try a mechanical soft diet.
   b. eat large meals.
   c. keep a record of food intake, gastric symptoms, and when the symptoms occur.
   d. drink milk with each meal and snack.
9. Which of the following snacks would be an appropriate choice for a client on a postgastrectomy diet?
   a. milk shake
   b. saltine crackers with peanut butter
   c. cookies and milk
   d. eggnog
10. The health care professional assessing a client who underwent a gastrectomy three years ago should be alert to signs of:
   a. problems with chewing and swallowing.
   b. vitamin C deficiency.
   c. zinc deficiency.
   d. iron-deficiency anemia.

Clinical Applications

1. People on mechanical soft diets differ in the kinds of foods they can handle, in the lengths of time that diet modifications remain necessary, and in the help they need from health care professionals. Think about the difference between working with a person who has had no teeth for years and a person who recently had mouth surgery and is just beginning to eat again. Describe some nutrition-related concerns you might have for the person who has been following a mechanical soft diet for years. How would these concerns differ for a person who needs the mechanical soft diet only temporarily? Contrast the amounts of time a nurse or dietitian might need to spend with the two types of clients.
2. Many of the diets described in this chapter are highly individualized. A particular food may cause discomfort for one person and have no effect on another. Describe practical ways to keep track of food intolerances.
3. Review the chapter to find disorders that often occur as a consequence of aging. Referring to Chapter 16, describe the effects of aging on the upper GI tract and relate these changes to the disorders you find.

References

1 J. Lagergren and coauthors, Symptomatic gastroesophageal reflux as a risk factor for esophageal adenocarcinoma, *New England Journal of Medicine* 340 (1999): 825–831.

2 W. H. Aldoori and coauthors, Prospective study of diet and the risk of duodenal ulcer in men, *American Journal of Epidemiology* 145 (1997): 42–50.

3 F. P. Kennedy, Medical management of patients following bariatric surgery, presented at the Eighth Annual Advances and Controversies in Clinical Nutrition, sponsored by the Mayo Clinic Foundation, Dallas, Texas, April 5–7, 1998.

ANSWERS

Multiple Choice

1. b 2. c 3. a 4. d 5. a
6. b 7. a 8. c 9. b 10. d

HELPING PEOPLE WITH FEEDING DISABILITIES

CHAPTER 20 described problems encountered when people have difficulties with chewing and swallowing. This highlight discusses a broader problem faced by thousands of people who must cope with disabilities that interfere with the process of eating, including some that interfere with chewing and swallowing. These obstacles can arise at any time during a person's life and from any number of causes. An infant may be born with a physical impairment such as cleft palate; an adolescent may suffer nerve damage from injuries sustained in a car accident; a middle-aged adult may lose motor control following a stroke; an older adult may struggle with the pain of arthritis or the mental deterioration of dementia. Table H20-1 on page 648 lists some of the conditions that may lead to feeding problems.

EFFECTS OF DISABILITIES ON NUTRITION STATUS

To get food from the table to the stomach requires an amazing number of individual coordinated motions. Consider an infant learning to feed himself. Every single step—sitting, grasping and bringing utensils or food to the mouth, biting and chewing, and swallowing—requires coordinated movements. Any injury or disability that interferes with these movements can lead to feeding problems.

Disabilities may have nutrition-related consequences beyond their effects on the physical process of eating. For example, a person who has problems with sight, or who cannot drive or walk or carry groceries, or who cannot plan meals and think through what to buy, has a disability that affects eating. Disabilities of any type can cause people to have trouble maintaining adequate nutrition status.[1] Their number one problem is inadequate food intake, which leads to malnutrition, underweight, and, in children, poor growth.[2] Children and adolescents with severe disabilities are especially vulnerable to these problems.

Effects on Energy Needs

Disabilities can also affect energy needs. For example, a disability may make it difficult for a person to engage in enough physical activity to support a healthy appetite or, conversely, to burn energy. Then the person's intake may either fail to meet needs or result in excessive weight gain. As another example, a person who has lost a limb to amputation has altered energy needs. Energy needs are reduced in proportion to the weight and metabolism represented by the missing limb, but may be increased if the person needs extra energy to perform tasks such as pushing a wheelchair. As still another example, people with involuntary motor activity may have exceedingly high energy needs.

Secondary Effects of Disabilities on Nutrition Status

Depending on the disability, many symptoms can interfere with eating. Some of these include nausea, frequent coughing, choking, reflux esophagitis, and language and hearing disabilities. Frequent coughing and choking can lead to respiratory infections, which raise nutrient needs. People with speech and hearing problems may have a difficult time communicating with caregivers about thirst and hunger. People with disabilities that interfere with the ability to walk and move are at risk for bone demineralization and pressure sores (see Chapter 17).

Many conditions that lead to feeding problems also require the use of multiple medications, which can also have a significant impact on nutrition status (see Chapter 18). Furthermore, people with feeding problems often encounter emotional and social problems. For example, children may fail to receive the social training that mealtimes provide, and any person may miss the social stimulation that goes with eating in the company of others. The process of eating

Adaptive feeding equipment can help clients with feeding disabilities gain independence.

TABLE H20-1 Conditions That May Lead to Feeding Problems

The following conditions may lead to feeding problems by interfering with a person's ability to suck, bite, chew, swallow, or coordinate hand-to-mouth movements.

• Accidents	• Language, visual, or hearing impairment
• Amputations	• Microcephalia
• Arthritis	• Multiple sclerosis
• Birth defects	• Muscle weakness
• Cerebral palsy	• Muscular dystrophy
• Cleft palate	• Neuromotor dysfunction
• Down syndrome	• Parkinson's disease
• Head injuries	• Polio
• Huntington's chorea	• Spinal cord injuries
• Hydrocephalia	• Stroke

may take such a long time and be so difficult that the person fails to eat adequate amounts of food.

INDEPENDENT EATING FOR PEOPLE WITH DISABILITIES

The evaluation and treatment of a feeding problem ideally involve the joint efforts of several health care professionals, possibly including a dietitian, a psychologist, an occupational therapist, a physical therapist, a speech-language pathologist, a dentist, and one or several nurses. Together, health care professionals evaluate each client's ability to perform eating-related tasks such as chewing, sipping, swallowing, grasping utensils, using utensils to pick up foods, and bringing foods from the plate to the mouth.

Roles of Team Members

The physician diagnoses the client's medical problems, prescribes treatments (including medications), and coordinates the health care team. The nurse identifies a nursing plan to address the client's needs and educates the client and caregivers in accordance with that plan. The dietitian assesses the client's nutrition status, plans a diet, monitors medications for potential diet-drug interactions, and provides nutrition counseling.[3] Table H20-2 lists important nutrition assessment parameters. A speech-language pathologist most often evaluates the client's ability to chew and swallow and trains the client to use lips, tongue, and throat for eating and speaking. The occupational therapist may work with the client to evaluate the need for special feeding devices and then show the client how to use them. A dentist evaluates dental health and provides instructions for maintaining oral hygiene. The health care team and client face many challenges in solving feeding problems, but the personal satisfaction of seeing someone gain control of eating can be very rewarding.

TABLE H20-2 Nutrition Assessment for People with Feeding Disabilities

Diet History
Feeding environment
Feeding position
Symptoms associated with eating (coughing, gagging)
Amount of food lost from feeding utensil or mouth during feeding
Person responsible for feeding
Client/caregiver attitudes toward feeding
Food Intake Data
Length of feeding
Types of foods consumed
Food aversions, intolerances, or allergies
Total fluid intake
Anthropometrics
Height
Weight
For children, plot data on growth charts
Laboratory Data
Hemoglobin
Hematocrit
Albumin
Prealbumin
Physical Findings
Physical signs of nutrient imbalances and poor dental health

Teaching Clients to Eat

Direct observation of the client during mealtimes allows the health care team to evaluate current eating behaviors, demonstrate feeding techniques, monitor the client's and caregiver's understanding of the techniques, and evaluate how well the care plan is working. To illustrate, consider an example of a child with a feeding problem caused by a hypersensitivity to oral stimulation. The health care professional may start by teaching the caregiver to gently and playfully stroke the child's face with a hand, wash cloth, or soft toy. Once the child tolerates touch on less-sensitive areas of the face, the health care professional may encourage the caregiver to slowly begin to rub the child's lips, gums, palate, and tongue. With time, the child may be better able to tolerate the presence of food in the mouth.

Adaptive Feeding Equipment

Figure H20-1 shows a few of many special feeding devices and describes their uses. These devices can make a remark-

FIGURE H20-1 Examples of Adaptive Feeding Devices

Utensils

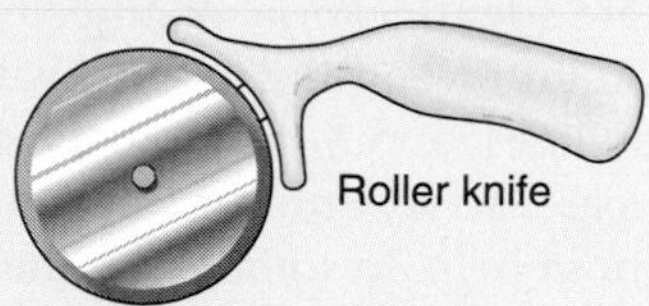

People with only one arm or hand may have difficulty cutting foods and may appreciate using a *rocker knife* or a *roller knife*.

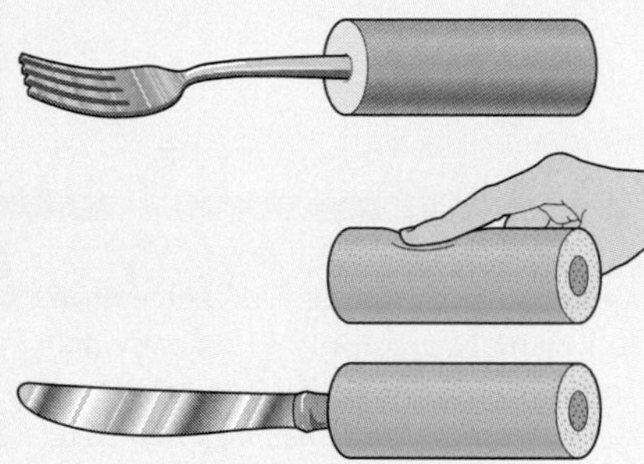

People with a limited range of motion can feed themselves better when they use *flatware with built-up handles.*

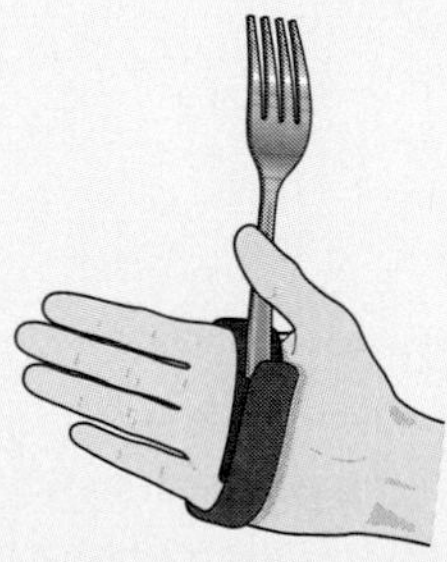

People with extreme muscle weakness may be able to eat with a *utensil holder.*

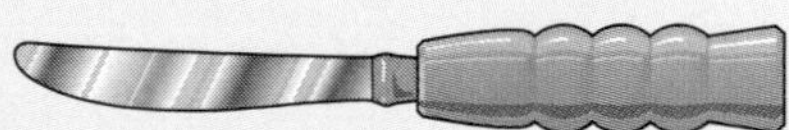

For people with tremors, spasticity, and uneven jerky movements, *weighted utensils* can aid the feeding process.

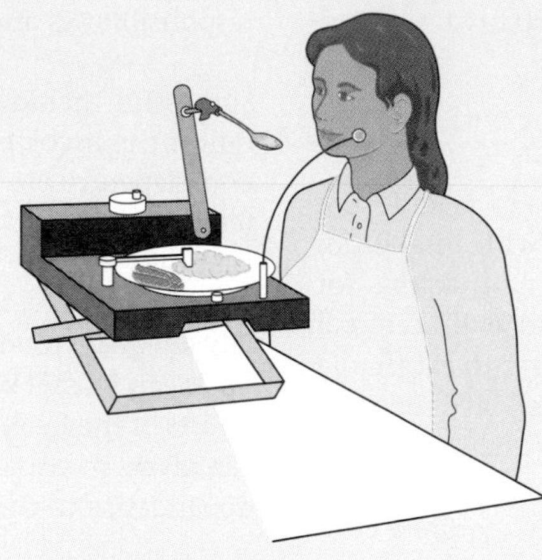

Battery-powered feeding machines enable people with severe limitations to eat with less assistance from others.

Plates

People who have limited dexterity and difficulty maneuvering food find *scoop dishes* or *food guards* useful.

People with uncontrolled or excessive movements might move dishes around while eating and may benefit from using *unbreakable dishes with suction cups.*

Cups

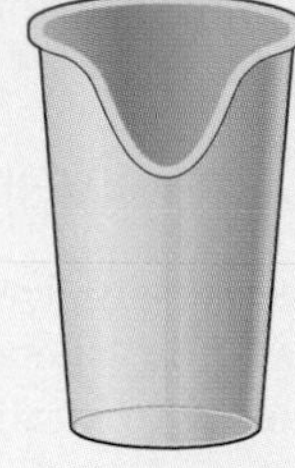

People with limited neck motion can use a *cutout plastic cup.*

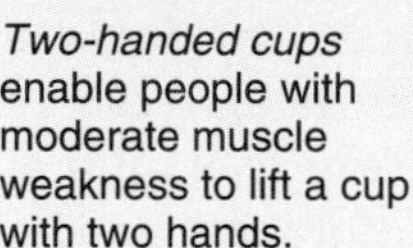

Two-handed cups enable people with moderate muscle weakness to lift a cup with two hands.

People with uncontrolled or excessive movements might prefer to drink liquids from a *covered cup* or glass with a *slotted opening* or *spout.*

A soft, flexible long plastic straw may also ease the task of drinking.

able difference in a person's ability to eat independently. For a person who cannot grasp an ordinary fork, for example, a special fork may be the key to future health.

Sometimes, despite the best efforts of all involved, the client still can't eat enough food by mouth. In these cases, tube feedings can help improve nutrition status.[4]

A Note about Caregivers

The time and patience needed to learn and handle the many tasks required to care for a person with disabilities often result in a great deal of frustration and distress for both the client and the caregiver. A child with cerebral palsy, for example, may take ten times longer to eat than a child without cerebral palsy. Some mothers have reported spending up to seven hours a day feeding these children. Besides feeding, the caregiver must often help the disabled person with many other tasks, and all may require a considerable amount of time. Caregivers may feel overwhelmed with responsibility and have little time to care for themselves and other family members. Psychologists can offer counseling to clients or caregivers to help them adjust; all members of the health care team can offer emotional support and practical suggestions to ease caregivers' responsibilities and frustrations.

Successful therapy for people with feeding disabilities requires the involvement of many health care professionals and depends on the accurate identification of impaired feeding skills and appropriate interventions. Ideally, with training, people with disabilities attain total independence—they are able to prepare, serve, and eat nutritionally adequate food daily without help. In some cases, these goals can be met with the help of caregivers. The combined efforts of the health care team can support both clients and caregivers in enhancing quality of life and in achieving independence to the greatest degree possible.

References

1 Position of The American Dietetic Association: Nutrition services for children with specific needs, *Journal of the American Dietetic Association* 95 (1995): 800–812.

2 R. D. Stevenson, Feeding and nutrition in children with developmental disabilities, *Pediatric Annals* 24 (1995): 225–260.

3 H. H. Cloud, Expanding roles for dietitians working with persons with developmental disabilities, *Journal of the American Dietetic Association* 97 (1997): 129–130.

4 B. Björnestam and coauthors, Long-term home jejunostomy feeding of young children, *Nutrition in Clinical Practice* 14 (1999): 247–248; R. Tawfik and coauthors, Caregivers' perceptions following gastrostomy in severely disabled children with feeding problems, *Developmental Medicine and Child Neurology* 39 (1997): 746–751.

Nutrition and Lower GI Tract Disorders

CHAPTER 21

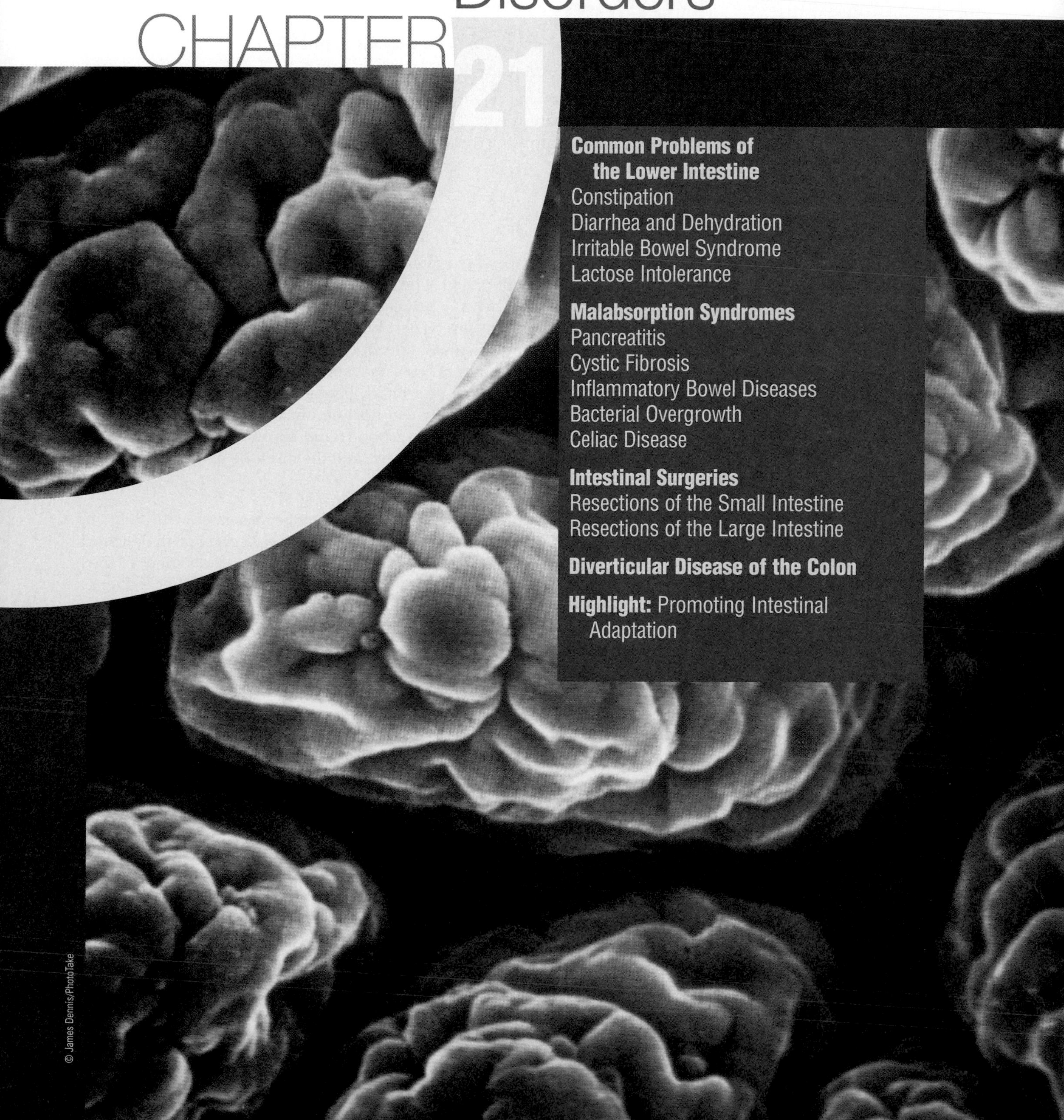

As Chapter 3 noted, the intestine is "the" organ of digestion and absorption. The intestine also provides a physical barrier against invading organisms and contains many immune cells to help protect against disease (see Highlight 18). Figure 21-1 shows the lower GI tract and related organs and reviews some of their functions. Many health problems and related treatments can affect the rate at which a food bolus passes through the intestine. Others affect the absorption of nutrients.

COMMON PROBLEMS OF THE LOWER INTESTINE

Several health problems—constipation, diarrhea, irritable bowel syndrome, and lactose intolerance—are common GI tract disorders characterized by changes in the rate at which a food bolus passes through the intestine. In each case, medical nutrition therapy may help alleviate the discomforts clients experience.

Constipation

◆ Reminder: Constipation is the condition of having infrequent or difficult bowel movements.

Constipation◆ describes a symptom, not a disease. Each person's GI tract has its own cycle of waste elimination, which depends on its owner's physical makeup, the type of food eaten, when it was eaten, and when the person makes time to defecate. For some people, several days typically pass between movements. Only when people pass stools that are difficult or painful to expel or when they experience a reduced frequency of bowel movements from their typical pattern are they constipated. Abdominal discomfort, headaches, backaches, and the passing of gas sometimes accompany constipation. In severe cases, hardened stool in the intestine may form an obstruction **(fecal impaction).** Left untreated, pressure may build above the impaction, and the intestine may rupture, which can lead to **peritonitis.**

© Peter Kane/The Stock Shop

High-fiber foods promote normal bowel movements.

• Causes of Constipation • Constipation may be associated with fluid and electrolyte imbalances, hormonal imbalances, many diseases of the GI tract, chronic laxative abuse, stress, lack of physical activity, and a variety of medications, including narcotic analgesics, anticholinergics (drugs that reduce GI tract motility), aluminum- and magnesium-containing antacids, and some antihypertensives. In addition, dietary supplements, including chondrotin sulfate, glucosamine, calcium, and magnesium can lead to constipation.[1] Expectant mothers frequently experience constipation as the growing fetus crowds intestinal organs and hormonal changes alter intestinal muscle tone. Constipation is a problem for many elderly people, particularly those in institutions. With aging, gastrointestinal motility slows, and elderly people are less likely than others to engage in regular physical activity, drink adequate amounts of fluids, and eat foods high in fiber. Elderly people may also use several medications that contribute to constipation. Some people, especially elderly people, report problems with constipation and take laxatives to treat it, although objective measurements of stool consistency and frequency fail to confirm constipation.[2]

• Treatment of Constipation • When constipation occurs as a result of an underlying medical condition, treatment of that disorder may alleviate the problem. Medical nutrition therapy includes a diet containing at least 25 grams of fiber, drinking plenty of fluids, eating prunes, drinking prune juice, and engaging in regular physical activity. Laxatives, often **bulk-forming agents,** are prescribed only when other measures fail to relieve constipation (see the Diet-Drug Interactions box on p. 673).

fecal impaction: a compacted mass of fecal material in the colon or rectum.

peritonitis (PARE-ih-toe-NYE-tis): infection and inflammation of the membrane lining the abdominal cavity caused by leakage of infectious organisms through a perforation (hole) in an abdominal organ.

bulk-forming agents: laxatives composed of fibers that work like dietary fibers. They attract water in the intestine to form a bulky stool, which then stimulates peristalsis. Metamucil and Fiberall are examples.

FIGURE 21-1 The Lower GI Tract and Related Organs

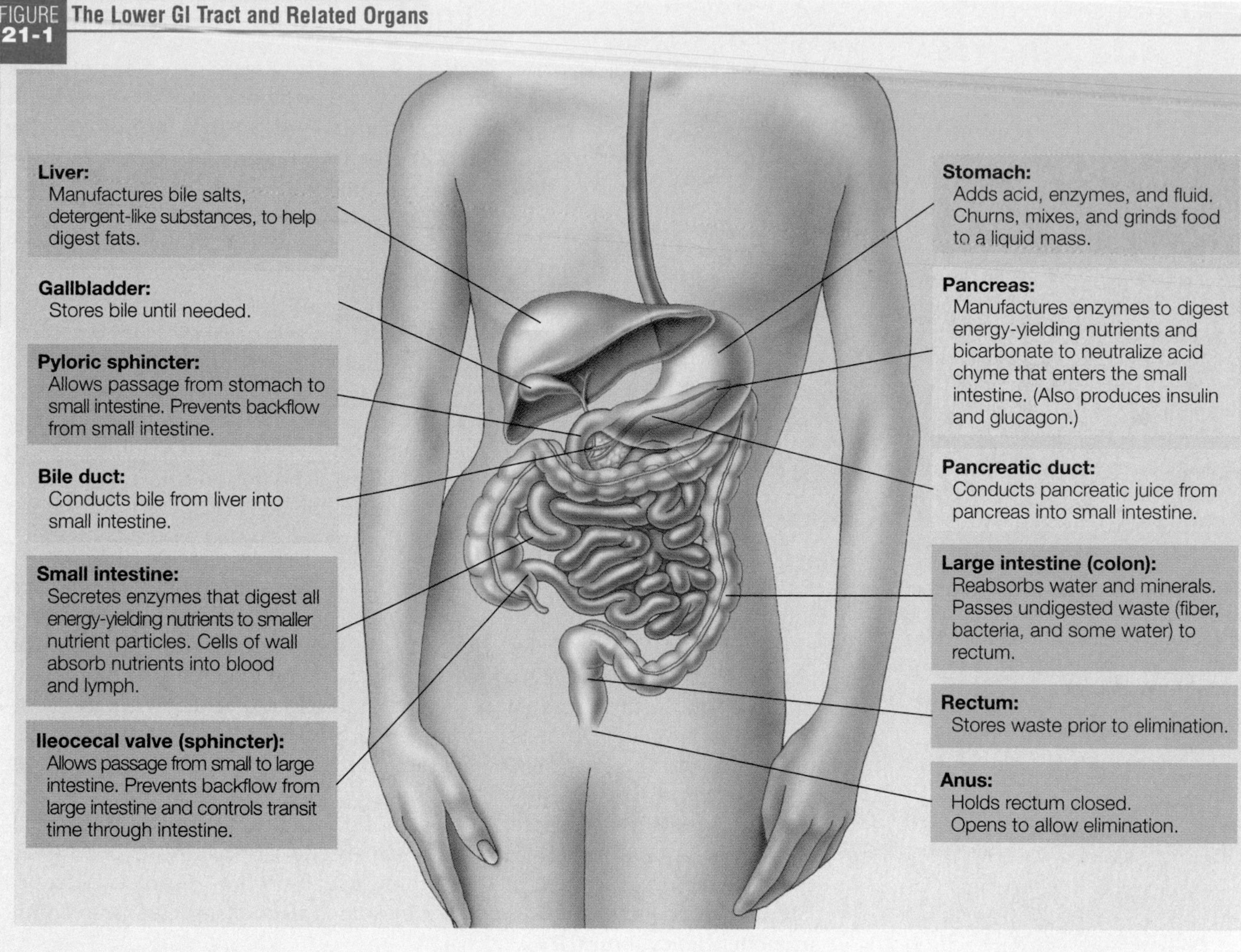

• Fiber in Foods • Foods containing fiber include whole-grain breads and cereals, fruits, vegetables, and legumes (see Table 4-5 on p. 117). Fiber-containing foods help maintain GI tract function by adding volume and weight to the stool, normalizing the transit of undigested materials through the intestine, and minimizing pressure within the colon. The "How to" box on p. 654 offers suggestions for helping clients adjust their fiber intakes with minimal discomfort from gas.

• Intestinal Gas • Excessive gas production can be an uncomfortable side effect of high-fiber diets. Undigested and unabsorbed dietary fibers may pass into the colon where bacteria may metabolize them and produce gas in the process. People with excessive gas may feel bloated and experience abdominal pain and **flatus.** Thus many fiber-containing foods, including legumes, some grains, fruits, and vegetables, are also gas-forming foods (see Table 21-1). Note that in addition to fibers, any undigested and unabsorbed food can cause intestinal gas. Thus high-fat foods, fructose, and sugar alcohols (sorbitol, mannitol, xylitol, and maltitol) taken in large amounts may be incompletely absorbed and cause gas for some people. Unabsorbed lactose can cause gas for people with lactose intolerance (described later in this chapter). People troubled by gas need to determine which foods bother them and then avoid those foods or eat them in moderation. Medications that help reduce intestinal gas production, called antiflatulents, are also available.

TABLE 21-1 Foods or Substances that May Produce Gas

Apples	High-fat meats
Asparagus	Honey
Beer	Legumes
Bran	Mannitol
Broccoli	Milk and milk products
Brussels sprouts	Nuts
Cabbage	Onions
Carbonated beverages	Peppers, green
Cauliflower	Prunes
Corn	Radishes
Cream sauces	Raisins
Cucumber	Sorbitol
Fried foods	Soybeans
Fructose	Wheat
Gravy	

flatus (FLAY-tuss): gas in the intestinal tract or the expelling of gas from the intestinal tract, especially through the anus.

Add Fiber to the Diet

During the first few weeks on a high-fiber diet, a person may feel bloated, pass gas frequently, or experience heartburn. Provide these suggestions to help:

- *Go slow.* Add high-fiber foods gradually and in small portions at first. Increase portion sizes and add foods as tolerance improves.
- *Add fluids.* Fiber attracts water as it moves through the intestine. Aim to drink 64 ounces of water each day. If you have trouble remembering to drink fluids, set up a schedule of when you will drink fluids each day and follow the schedule until drinking becomes automatic.
- *Experiment.* Try small servings of various fiber-containing foods at first and adopt those that are most pleasing.
- *Mix high-fiber foods with other foods.* Sprinkle bran flakes, wheat germ, or raisins on salads or applesauce. Add bran or mashed legumes to meat loaf. Add legumes and other high-fiber vegetables to soups and salads.

Diarrhea and Dehydration

Diarrhea◆ is not a disease, but a symptom of many medical conditions and a complication of some medical treatments, including many medications. Diarrhea is characterized by the passage of frequent, watery bowel movements. It can be acute, lasting less than two weeks, or chronic, lasting longer. Mild diarrhea that remits in 24 to 48 hours is seldom a cause for concern unless the person is already dehydrated. A person with severe, persistent diarrhea may rapidly become dehydrated, lose weight, and develop multiple nutrient deficiencies. A child or infant can lose proportionately more fluid and weight than an adult and can develop severe dehydration and malnutrition in a short time. Table 17-11 in Chapter 17 listed physical findings associated with dehydration.

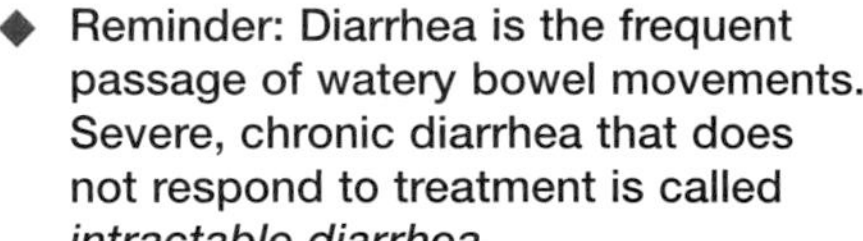

◆ Reminder: Diarrhea is the frequent passage of watery bowel movements. Severe, chronic diarrhea that does not respond to treatment is called *intractable diarrhea.*

• Causes of Diarrhea • Acute diarrhea that occurs abruptly in a healthy person frequently results from viral, bacterial, or protozoal infections or as a side effect of medications or dietary supplements. It can also occur in a person who begins to eat foods or begins a tube feeding after a period of fasting, starvation, or intravenous nutrition. Infants frequently develop diarrhea when given formulas their immature GI tracts cannot tolerate or when they are ill. When used in large quantities, food ingredients such as sorbitol and olestra may cause diarrhea in some people.

Chronic diarrhea can occur as a result of disorders that alter GI tract motility, such as irritable bowel syndrome. It can also develop as a result of disorders that alter absorption, such as the chronic malabsorption disorders described later in this chapter, and those caused by some infections, such as human immunodeficiency virus (HIV).

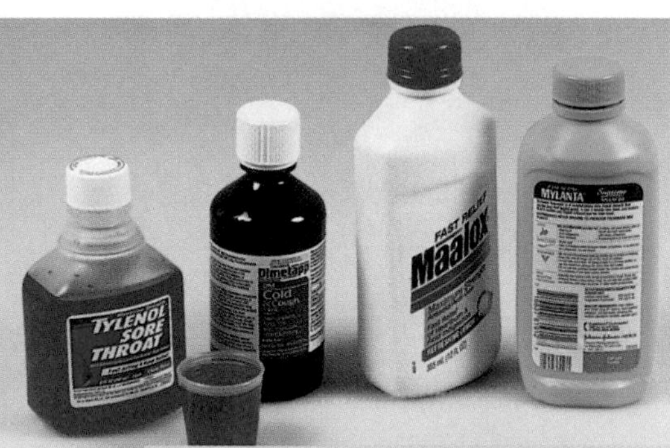

© Craig M. Moore

People who use high doses of sugar-free liquid medications, such as adults who have trouble swallowing pills, may develop diarrhea from the sorbitol these medications contain.

• Treatment of Diarrhea • The treatment of diarrhea includes treatment of the primary medical condition. If a food is responsible, that food must be omitted from the diet. If a medication is responsible, a different medication or form of medication (injectable versus oral, for example) may alleviate the problem. Infections are treated with appropriate anti-infective agents. Antidiarrheals slow GI motility and are often recommended along with other therapies to treat diarrhea (see the Diet-Drug Interactions box on p. 673).

• Medical Nutrition Therapy • Until the diarrhea resolves, treatment includes replacement of lost fluids and electrolytes to prevent dehydration. Fluids that provide both glucose and sodium help maximize the body's absorption of fluids and electrolytes.[3] Good choices for mild cases of diarrhea include diluted fruit juices, sports drinks, or caffeine-free carbonated beverages. For mild-to-moderate cases of diarrhea, oral rehydration formulas—simple solutions of water, salts, and sugar—provide needed fluids and electrolytes. Salty broths, soups, bouillon, flavored ades, and jello can be used to supplement replacement fluids. For severe cases, especially if the client is dehydrated or experiencing persistent vomiting, intravenous fluids and electrolytes replace losses.

If eating aggravates diarrhea, clients may be advised to drink only clear liquids (including oral rehydration formulas). In severe cases, it may become necessary to stop placing demands on the GI tract by withholding all foods and beverages until the diarrhea remits, usually in a day or two. For people who can tolerate solid foods, the liberal use of foods with soluble fibers can often help control diarrhea (see Table 21-2). Yogurt, which contains helpful bacteria,◆ can help to control diarrhea in some cases. Foods with insoluble fibers should be

◆ Yogurt contains *lactobacillus,* a bacterium that helps to establish a healthy bacterial flora and secretes lactase to help digest lactose.

avoided temporarily. The diet also temporarily excludes lactose, caffeine, highly seasoned foods, foods high in fat, foods that cause gas, and any food that aggravates the diarrhea. Frequent, small meals are easiest to tolerate at first. Permanent dietary changes may be necessary for diarrhea caused by food sensitivities or allergies.

TABLE 21-2 Dietary Suggestions for Treating Diarrhea

- Drink plenty of fluids including diluted fruit juices, sports drinks, salty soups and broths, or oral rehydration formulas.
- Add salt to foods.
- Use foods containing soluble fibers including: bananas, peeled apples, applesauce, peeled pears, oranges, peeled white or sweet potatoes, oatmeal, barley, millet, couscous, noodles, rice, white bread, and saltine crackers.
- Avoid foods with insoluble fibers including: whole-wheat breads and cereals, bran, raw vegetables, corn, peas, nuts, seeds, and skins of fruits and vegetables.
- Avoid caffeine and lactose, highly seasoned foods, high-fat foods, foods that cause gas, and any food that aggravates the diarrhea.
- Eat frequent, small meals.

Irritable Bowel Syndrome

Irritable bowel syndrome is a common motility disorder characterized by abdominal pain associated with diarrhea, constipation, or alternating episodes of both.[4] The person may also experience bloating, flatulence, the passage of mucus with stools, indigestion, and nausea. Symptoms frequently occur shortly after a person eats and resolve temporarily following a bowel movement. For many, the symptoms are mild, but in some cases, the need to be close to a bathroom can interfere with work or prevent participation in social activities.

In the United States, irritable bowel syndrome affects about 5 million people annually and occurs more frequently in women.[5] Symptoms often begin in childhood or young adulthood and decrease with age. Even though symptoms persist for several months, people with irritable bowel syndrome and no other health problems generally remain in good health and seldom experience malnutrition.

• Causes of Irritable Bowel Syndrome • The causes of irritable bowel syndrome remain elusive, but stress and anxiety are believed to be contributing factors. Symptoms often worsen during periods of psychological stress or, for women, during menstruation. Other contributing factors may include abnormal GI tract motility and hypersensitivity to intestinal distention. Some studies suggest that bacterial overgrowth in the small intestine (described later in this chapter) may contribute to diarrhea and abdominal pain in some people with irritable bowel syndrome.[6] Foods and food components, including gas-forming foods (review Table 21-1), high-fat foods, fat substitutes, lactose, fructose, sorbitol, caffeine, and alcohol, may aggravate the symptoms of irritable bowel syndrome, but are not believed to cause it. Medications and herbal remedies can also aggravate diarrhea or lead to constipation.

• Treatment of Irritable Bowel Syndrome • Medical therapy for irritable bowel syndrome often includes stress management along with medical nutrition therapy. Drug therapy, often reserved for cases that do not respond to other treatments may include antidepressants, anticholinergics, antidiarrheals, and laxatives (see the Diet-Drug Interactions box on p. 673).

• Medical Nutrition Therapy • To help minimize abdominal distention, improve digestion, and reduce the chance of swallowing air with foods (which can contribute to abdominal distention and gas in the intestine), clients with irritable bowel syndrome are advised to avoid eating too much or too fast and to chew foods thoroughly. Clients respond uniquely to specific dietary interventions for irritable bowel syndrome, and no one diet is suitable for all clients. Dietitians often rely on food records that include food intake, fluid intake, stool consistencies, other GI symptoms, and stress levels to help uncover individual food intolerances.

Many clients benefit from a fat-restricted diet with a liberal fiber and fluid intake. The dietitian often adjusts types of fiber (soluble and insoluble) depending on whether the client primarily experiences constipation or diarrhea. Because many fiber-containing foods cause gas for some people, however, eating too much fiber can also aggravate symptoms. Some clients tolerate the fiber from bulk-forming agents better than dietary fibers for this reason. Highly seasoned foods, lactose, fructose, sorbitol, caffeine, and alcohol are restricted to the extent that they cause discomfort for the individual. The case study on p. 656 provides questions to help you apply information about irritable syndrome to a clinical situation.

irritable bowel syndrome: an intestinal disorder of unknown cause characterized by abdominal discomfort, cramping, diarrhea or constipation, or alternating diarrhea and constipation.

CASE STUDY

Marketing Professional with Irritable Bowel Syndrome

Sudah Patel is a 22-year-old recent college graduate who began her first professional job in a marketing firm one month ago. As a college student, she occasionally experienced abdominal pain and cramping after eating. She also experienced frequent bouts of diarrhea and noticed that she felt better for a while after a bowel movement. Once Sudah began her new job, she noticed that her symptoms were occurring more frequently. At first she attributed her symptoms to the stress related to her new job, but when the symptoms continued for several months, she decided to see her physician. Sudah is 5 feet 3 inches tall and weighs 118 pounds. After taking a careful history and conducting several tests to rule out other bowel disorders, the physician diagnosed irritable bowel syndrome. The physician prescribed bulk-forming agents and advised Sudah to keep a food intake and symptoms record for one week. Sudah was then referred to a dietitian to review the records and recommend appropriate dietary suggestions. In reviewing Sudah's food intake records, the dietitian notices that Sudah eats many traditional Indian foods, which are highly seasoned and often include legumes. Sudah tells the dietitian that she began to drink coffee and caffeinated cola drinks in college so that she could study for longer periods of time. She has continued to drink coffee since that time.

1. How would you explain irritable bowel syndrome to Sudah: What role does stress play in irritable bowel syndrome?
2. Can Sudah's diet be responsible for causing irritable bowel syndrome? Can any of these foods or food components aggravate her symptoms? How will the dietitian use the food intake and symptoms record to devise medical nutrition therapy for Sudah?
3. What type of diet benefits people with irritable bowel syndrome? What problem(s) can this diet cause?

Lactose Intolerance

Unlike diarrhea and constipation, which can occur as a consequence of many health-related problems, the cause of lactose intolerance is clear. It results from a deficiency of lactase, the enzyme that splits lactose to glucose and galactose.

• Causes of Lactase Deficiency • In rare cases, a person is born with a lactase deficiency; more often, lactase activity gradually diminishes with age. Lactose intolerance is prevalent among people in certain ethnic groups including those of Mediterranean origin, African Americans, Asians, Jews, Mexicans, and Native Americans. Permanent or temporary lactase deficiencies can develop as a consequence of any disorder or condition that damages the delicate intestinal microvilli including malnutrition, radiation therapy, and many of the conditions described later in this chapter.

Vigorous advertising campaigns to promote products for lactose intolerance have led many people to believe they have lactose intolerance, when, in fact, they do not.[7] Some believe they are intolerant to even the smallest amounts of lactose. These people may unnecessarily eliminate milk and milk products from the diet and inadvertently develop calcium deficiencies in the process.

© Bard Martin/The Image Bank

Most people with lactose intolerance can drink milk, if they drink it along with other foods and limit the amount they have at any one time.

• Symptoms of Lactose Intolerance • Undigested lactose causes the intestinal contents to become hypertonic, and the result is cramps, distention, and diarrhea. Bacteria in the large intestine metabolize the undigested sugars to irritating acids and gases, which further contribute to cramping and diarrhea and cause flatulence. The severity of the symptoms varies among individuals.[8]

• Treatment of Lactose Intolerance • Medical nutrition therapy provides relief from the symptoms of lactose intolerance. Lactose-restricted diets are highly individualized diets that most often limit, but do not exclude, milk and milk products. Dietitians advise clients to test their tolerance for lactose by gradually increasing consumption of lactose-containing foods to the point that precipitates symptoms of lactose intolerance. Many lactose-intolerant individuals can tolerate up to 1 or 2 cups of milk a day, provided that the milk is taken with food and intake is divided throughout the day.[9] Often people can eat cheeses, particularly aged cheeses.[10] Some tolerate chocolate milk better than plain milk. Most tolerate yogurt well because yogurt contains bacteria that secrete lactase and thus help digest lactose.

People can also add a lactase enzyme preparation to milk before they drink it or take enzyme tablets whenever they eat lactose-containing foods. Lactose-free milk and milk products that have been treated with lactase are also available. Products to aid lactose digestion are often unnecessary, however, because most people tolerate a fair amount of lactose.

People with temporary lactose intolerance are advised to temporarily restrict all milk and milk products. Then lactose-containing foods are gradually reintroduced in small amounts.

• Preventing Calcium and Vitamin D Deficiencies • People who restrict milk and milk products because they have lactose intolerance, or believe they have it, risk calcium and vitamin D deficiencies. To help prevent such deficiencies, encourage people to include milk and milk products in their diets to the extent that they can tolerate these foods, with or without the use of enzyme preparations. People who fail to receive adequate amounts of calcium from milk and milk products should be encouraged to eat other food sources of calcium, including calcium-fortified juices, cereals, and snacks; broccoli; mustard greens; kale; and sardines. For people who cannot get enough calcium from conventional foods, calcium supplements are indicated, particularly for children, adolescents, and pregnant, lactating, or postmenopausal women. Vitamin D is not a problem if the person gets regular exposure to sunlight; otherwise, a supplement may be necessary.

IN SUMMARY

Constipation, diarrhea, irritable bowel syndrome, and lactose intolerance are common GI problems that may cause a great deal of discomfort for people who experience them. Medical nutrition therapy can alleviate these problems in some cases and lessen undesirable side effects in others. A diet adequate in fiber and fluid and physical activity can often alleviate constipation. Diets high in soluble fibers can also help treat diarrhea, but the best dietary advice depends on the cause of the diarrhea. In all cases, attention to fluids and electrolytes helps prevent dehydration. Because people with irritable bowel syndrome can experience either diarrhea or constipation as a primary symptom, diet treatments are tailored to each person's unique needs. People with lactose intolerance frequently tolerate moderate amounts of lactose provided that they eat lactose-containing foods with other foods and spread out their intake of lactose-containing foods throughout the day.

MALABSORPTION SYNDROMES

Many disorders lead to malabsorption, and any nutrient can be involved. Chapter 20 described vitamin B_{12} malabsorption resulting from chronic gastritis and general malabsorption resulting from the dumping syndrome. The next few sections show how disorders of the pancreas and intestine can interfere with the absorption of nutrients, particularly fat.◆ Malabsorption syndromes often affect fat absorption more profoundly than carbohydrate or protein absorption because fat digestion and absorption is more complex. Table 21-3 on p. 658 lists examples of laboratory tests useful in assessing the absorption of nutrients.

◆ **Disorders of the liver (see Chapter 28) that disrupt the liver's production or secretion of bile, necessary for the absorption of fat, can also lead to fat malabsorption.**

• Fat Malabsorption • The body excretes unabsorbed fat in the stools, causing the type of diarrhea called **steatorrhea.** The loss of fat in the stools means that valuable food energy, essential fatty acids, fat-soluble vitamins, and some minerals are lost as well (see Figure 21-2 on p. 658). The absorption of protein, carbohydrate, other vitamins and minerals, and water may also be impaired, depending on the cause of the malabsorption. Disorders that cause malabsorption and their treatments further tax nutrition status by leading to complications that reduce nutrient intakes, incur further nutrient losses, and/or raise nutrient needs (see Table 21-4 on p. 659).

steatorrhea (STEE-ah-toe-REE-ah): fatty diarrhea characterized by loose, foamy, foul-smelling stools.

TABLE 21-3 Examples of Tests Used to Evaluate Nutrient Absorption

- Direct stool examinations: Stool checked for weight and fat (greater than normal weight or excess fat suggests malabsorption).
- Chemical analysis of fecal fat: Fecal fat of greater than 7 grams/day when the diet includes 100 grams of fat/day indicates fat malabsorption.
- Chemical analysis of fecal nitrogen: Normal fecal nitrogen is less than 2 grams/day.
- D-xylose test: Test of carbohydrate absorption.
- Serum calcium: Low levels seen in calcium or vitamin D malabsorption.
- Serum carotene: Low levels accompany steatorrhea.
- Schilling test: Identifies vitamin B_{12} malabsorption.

• Essential Fatty Acid Deficiencies • Among the fats lost in steatorrhea are the essential fatty acids. Studies show that many people with severe fat malabsorption develop essential fatty acid deficiencies.[11]

• Vitamin and Mineral Malabsorption • Fat-soluble vitamins are excreted in the stools along with unabsorbed fat. Some minerals, including calcium and magnesium, form **soaps** with unabsorbed fatty acids and are not available for absorption. Vitamin D losses further aggravate calcium malabsorption. Bone loss is a possible consequence.

soaps: chemical compounds formed between a basic mineral (such as calcium) and unabsorbed fatty acids. Soaps give steatorrhea its foamy appearance.

• Oxalate Stones • The binding of calcium to fatty acids can cause another problem. Oxalate, which is present in some foods, normally binds with some of the calcium in the gut and is excreted with it. But when fatty acids bind with calcium, the oxalate remains unbound. The intestine absorbs the unbound oxalate, but the body cannot metabolize it, and so excretes it in the urine. High urinary oxalate favors the formation of kidney stones (see Chapter 27).

FIGURE 21-2 The Consequences of Fat Malabsorption

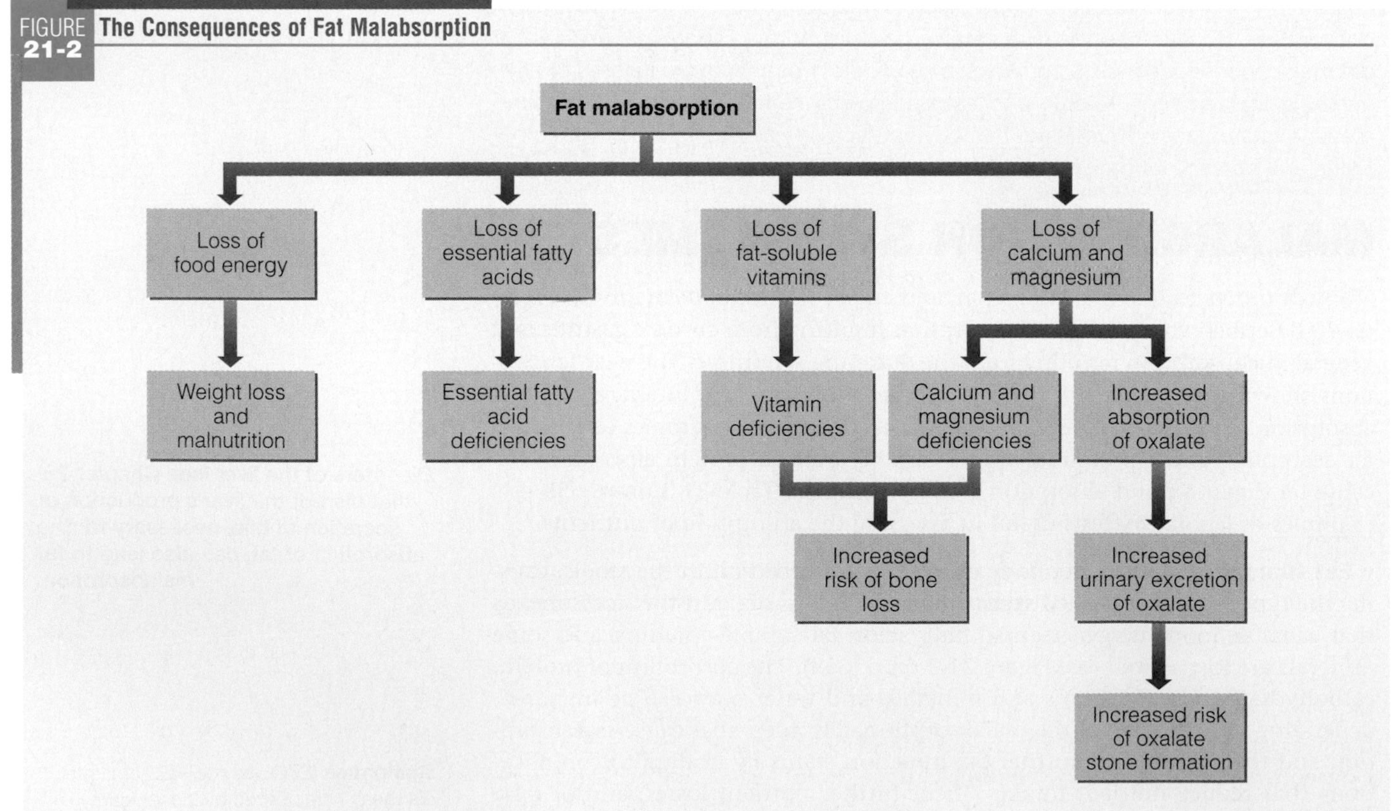

TABLE 21-4 Possible Causes of Malnutrition in Malabsorption Syndromes

Reduced Nutrient Intake	Excessive Nutrient Losses	Raised Nutrient Needs
Abdominal pain	Blood loss	High basal energy expenditure
Anorexia	Diarrhea	Infection
Bowel rest	Fistulas	Inflammation
Emotional stress	General malabsorption	Medications
Food intolerance	Medications	Surgery
Indigestion	Steatorrhea	
Medications	Vomiting	
Nausea		
Obstructions		

• **Treatments for Malabsorption** • To treat steatorrhea successfully, the underlying disorder must be diagnosed and treated. Drug therapy, and sometimes surgery, may be necessary. High-kcalorie, high-protein diets meet energy and protein needs. Dietary fat may be modified in type or amount; otherwise enzyme replacements (described later) are given to aid absorption. People following a fat-restricted diet are often advised to avoid foods high in oxalate to reduce the risk of kidney stones.

• **Fat-Restricted Diets** • Individual tolerances determine the specific amount of fat prescribed, but usually it ranges from 20 to 30 percent of total energy. Typically, the person begins with a diet containing about 35 to 45 grams of fat. Fat intake gradually increases as tolerance permits. Table 21-5 on p. 660 provides instructions for a fat-restricted diet. Whenever fat is restricted, the following diet-planning principles apply:

- Fat is not restricted any more than is necessary to prevent steatorrhea because the person needs food energy.
- Part of the fat may be given as **medium-chain triglycerides (MCT)** because these fats are easier to digest and absorb than long-chain triglycerides (LCT). MCT supply almost as many kcalories as regular fats, but they do not provide essential fatty acids, so some LCT are required.
- Foods are provided in frequent, small meals because fat is best tolerated in small amounts at a time.
- Fat-soluble vitamins are provided in a water-miscible form when fat malabsorption is severe. Water-miscible, fat-soluble vitamins mix readily with water and can be absorbed without fat.

The "How to" on p. 661 offers suggestions for helping people accept fat-restricted diets and includes tips for using MCT.

• **Enzyme Replacement Therapy** • **Pancreatic enzyme replacements** serve to treat malabsorption when the pancreas's ability to produce and secrete digestive enyzmes is altered as a consequence of severe and chronic damage. Because a severely damaged pancreas also fails to secrete enough digestive enzymes to provide an optimal pH for enzymes to function, people with pancreatic insufficiency often take antisecretory agents to limit gastric acid production. At best, enzyme replacements taken with meals lessen the malabsorption of fat and protein, but may not fully correct it.

• **Oxalate-Restricted Diet** • Oxalate in the urine derives primarily from the body's synthesis of it, and a smaller amount comes from food. One way the body synthesizes oxalate is through a pathway that begins with vitamin C. Consequently, megadoses of vitamin C over prolonged periods of time raise urinary oxalate concentrations. Thus people at risk for oxalate stone formation

medium-chain triglycerides (MCT): triglycerides containing fatty acids with 8 to 12 carbon atoms; they require minimal lipase and no bile for absorption.

pancreatic enzyme replacements: extracts of pork or beef pancreatic enzymes that are taken as supplements to aid digestion.

TABLE 21-5 Fat-Restricted Diet (35 grams)

Use:

1. Fat-free milk, fat-free cheeses, fat-free yogurt, sherbet, and fruit ices.
2. Low-fat egg substitutes and up to three regular eggs per week.
3. Up to 6 oz of lean meats and poultry without skin daily.
4. Up to 3 servings of fat daily. One serving is any one of the following:
   - 1 tsp butter, margarine, shortening, oil, or mayonnaise
   - 1 tbs reduced-fat butter, margarine, or mayonnaise
   - 1 strip crisp bacon
   - 1 tbs salad dressing or 2 tbs reduced-fat salad dressing
   - ⅛ avocado
   - 2 tbs cream (half and half)
   - 10 small nuts
   - 8 large black olives or 10 large green olives
   - If fat is used to cook or season food, it must be taken from this allowance.
5. All vegetables prepared without fat.
6. All fruits prepared without fat.
7. Plain white or whole-grain bread; fat-free cereals, pasta, rice, noodles, and macaroni.
8. Clear soups.
9. Angel food cake and fruit whips made with gelatin, sugar, and egg-white meringues.
10. Jelly, jam, honey, gumdrops, jelly beans, and marshmallows.

Do not use:

1. Whole milk, chocolate milk, whole-milk cheeses, and ice cream.
2. Pastries, cakes, pies, sweet rolls, breads, or vegetables made with fat.
3. More than one egg a day, fried or fatty meats (sausage, luncheon meats, spareribs, frankfurters), duck, goose, or tuna packed in oil (unless well drained).
4. More than 3 servings of fat.
5. Desserts, candy, or anything made with chocolate, nuts, or foods not allowed.
6. Creamed soups made with whole milk.

Suggestions:

1. To make the diet still lower in fat, reduce the fat and meat (and egg) servings.
2. To raise the fat content, give additional fat or meat servings.
3. To improve acceptance of the diet, check the fat content of a well-liked food and allow that food if possible. Use the exchange system fat list for alternate suggestions for fat servings (see Appendixes G and I).

Note: A "How to" on p. 152 provides additional tips for lowering fat in the diet.

are advised to avoid vitamin supplements as well as foods high in oxalate (see Chapter 27).

Pancreatitis

Pancreatic secretions contain many enzymes necessary for the digestion of protein, fat, and carbohydrate, together with bicarbonate-rich juices that provide the optimal pH necessary to activate these enzymes. Normally, the pancreas stores digestive enzymes in an inactive form to protect itself from digestion. In pancreatitis, however, the pancreas becomes inflamed, digestive enzymes are activated within the pancreas, and the enzymes damage the pancreas itself.◆ The blood picks up some of these enzymes; thus serum amylase and lipase rise and serve as indicators of pancreatitis.

◆ **Inflammation sets in motion processes that heal and repair damaged tissues, but when inflammation is extensive or persistent (chronic), additional tissue damage results. Any disorder that results in extensive inflammation can lead to malnutrition and multiple organ failure. Chapter 24 and Highlights 18 and 24 provide more information.**

• **Acute Pancreatitis** • Pancreatitis most often develops as a consequence of gallstones (Highlight 28) or alcoholism (Highlight 7); sometimes, though, the reasons are unclear because a variety of medical conditions and some drugs can also precipitate pancreatitis. Sudden, severe abdominal pain, nausea, vomiting, and diarrhea are common symptoms. While mild cases of pancreatitis may subside in a few days, in other cases, impaired pancreatic function may persist for weeks or months. In severe cases, life-threatening complications including shock and multiple organ failure (Chapter and Highlight 24), pancreatic hemorrhages, **fistulas, abscesses,** and peritonitis may develop. Malnutrition, chronic pancreatitis, and diabetes mellitus (Chapter 25) can also occur.

• Medical Nutrition Therapy for Acute Pancreatitis • Initially, food is withheld because foods stimulate pancreatic secretions. Fluids and electrolytes are provided intravenously. In some cases, a tube is inserted into the stomach to suction gastric secretions and help relieve pain and distention (gastric decompression). Oral intake begins when abdominal pain subsides and serum amylase returns to normal or near-normal levels. The diet progresses from liquids to a fat-restricted diet and, eventually, a regular diet as tolerated. Alcohol is restricted.

Severe acute pancreatitis can lead to secondary infections, shock, and multiple organ failure. Evidence suggests that these infections may arise from bacteria in the intestine, and that delivering tube feedings of easy-to-absorb formulas directly into the jejunum early after severe pancreatitis develops may limit the risks of infection and multiple organ failure.[12] In some cases, intravenous nutrition may be necessary, especially when fistulas develop or when the GI tract becomes immobile **(adynamic ileus)** for long periods of time.

• Chronic Pancreatitis • When severe pancreatitis or repeated episodes of pancreatitis permanently damage the pancreas, absorption, especially of fat, becomes permanently impaired. Chronic pancreatitis is most commonly associated with alcoholism. Abdominal pain is often severe and unrelenting, vomiting is frequent, and severe weight loss and malnutrition are common.

• Medical Nutrition Therapy for Chronic Pancreatitis • Medical nutrition therapy aims to maintain optimal nutrition status, reduce malabsorption, and avoid subsequent attacks of acute pancreatitis. High-kcalorie, high-protein diets containing normal amounts of fat are frequently prescribed. Enzyme replacements taken with meals help the person digest and absorb protein and fat while minimizing steatorrhea. Clients are strongly cautioned to avoid alcohol. Clients who develop glucose intolerance, diabetes, or hypoglycemia require additional dietary adjustments (see Chapter 25). Furthermore, clients who develop chronic pancreatitis as a consequence of alcohol abuse may have multiple nutrient deficiencies and alterations in metabolism, and these require additional dietary adjustments. The case study on p. 662 describes a client with chronic pancreatitis.

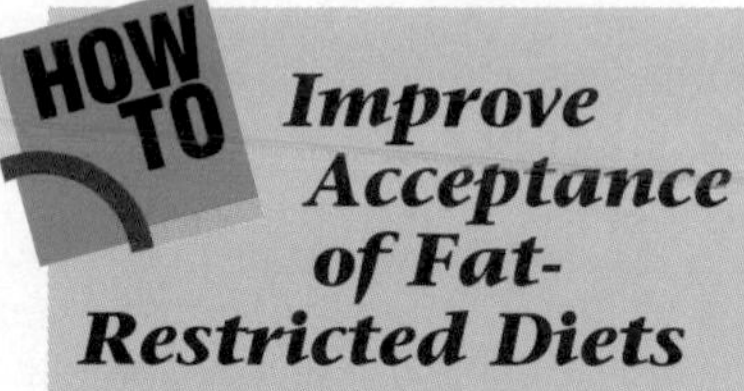

Improve Acceptance of Fat-Restricted Diets

Fat-restricted diets can be difficult to follow. Fats give flavors, aromas, and textures to foods—characteristics that people may miss. Unlike some diets that can be introduced gradually, the fat-restricted diet for malabsorption must be implemented immediately without giving the person time to adapt to the changes. These suggestions may help:

- Provide clients with tips for making foods palatable while lowering fat intake, such as those found in the box on p. 152.
- Remind clients that new fat-free and low-fat products appear on market shelves daily, and most people find these products very acceptable. Caution clients to avoid products containing fat substitutes (such as olestra), however. While healthy people can use fat substitutes in appropriate amounts, people with digestive problems and fat-soluble vitamin malabsorption may find that fat substitutes aggravate their conditions.

People who use MCT need additional advice:

- Advise clients to add MCT to the diet gradually. Nausea, vomiting, diarrhea, abdominal pain, and distention can result from using too much MCT all at once.
- Recommend that clients improve the palatability of MCT oil by substituting it for regular oil in salad dressing and for cooking and baking and by adding it to beverages, desserts, and other dishes.
- Warn clients that MCT products are expensive, and explain that these products can be purchased at pharmacies and are sometimes covered by medical insurance.

Cystic Fibrosis

People with **cystic fibrosis,** the most common fatal genetic disorder of the white population, produce secretions of thick, sticky mucus that may seriously impair the function of many organs, notably the lungs and pancreas.[13] Until

fistulas (FIS-chew-lahs): abnormal openings formed between two organs or between an internal organ and the skin.

abscesses: accumulations of pus, caused by local infection, that contain live microorganisms and immune system cells. Treatment involves draining the abscess.

adynamic ileus (ILL-ee-us): obstruction of the intestine caused by paralysis of the intestinal muscles.

cystic fibrosis: a hereditary disorder characterized by the production of thick mucus that affects many organs, including the pancreas, lungs, liver, heart, gallbladder, and small intestine.

CASE STUDY

Retired Executive with Chronic Pancreatitis

Mr. McDonald is a 62-year-old alcohol abuser with chronic pancreatitis and malnutrition. Mr. McDonald was forced to resign his position as president of an import/export company in his early fifties, when his problems with alcohol and declining health seriously impaired his abilities to run the company. His wife divorced him shortly thereafter, and he currently lives alone. At 5 feet 11 inches tall, Mr. McDonald weighs 145 pounds. He continues to experience frequent, severe abdominal pain and vomiting. Although Mr. McDonald has been advised to follow a high-kcalorie, high-protein diet with no fat restrictions, he has a great deal of difficulty eating enough food. He uses enzyme replacements with meals to aid digestion and has not used alcohol for eight months. Mr. McDonald takes a water-miscible form of fat-soluble vitamins and uses a supplement to meet his other vitamin and mineral needs.

1. What is the reason for malabsorption in chronic pancreatitis? How does chronic pancreatitis differ from acute pancreatitis?
2. Would Mr. McDonald benefit from following a fat-restricted diet? Why or why not?
3. What role do enzyme replacements play in the management of chronic pancreatitis?
4. Why would Mr. McDonald benefit from fat-soluble vitamins in a water-miscible form?
5. What benefits might Mr. McDonald realize from eating frequent, small meals rather than three regular meals? Do you think Mr. McDonald might benefit from using an enteral formula supplement? Why or why not?
6. Why is it vitally important that Mr. McDonald refrain from drinking any alcohol?

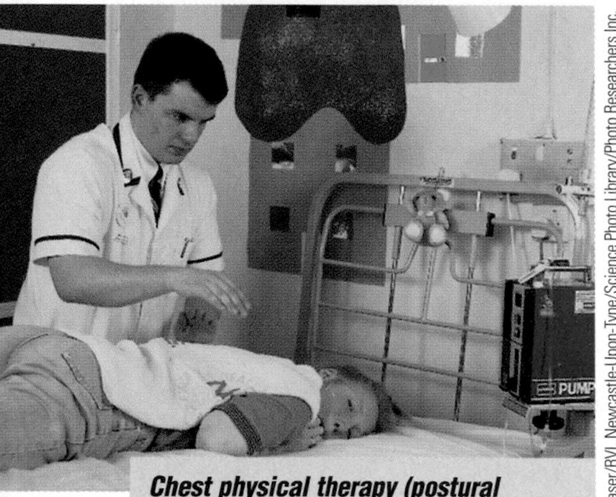

Chest physical therapy (postural drainage) promotes maximal lung function for people with cystic fibrosis.

recently, few infants born with cystic fibrosis survived to adulthood. Now, with early detection and advances in medical therapy and nutrition care, the outlook is much brighter, with some surviving into their forties and even fifties.

• Consequences of Cystic Fibrosis • Cystic fibrosis has three major consequences with nutrition implications: chronic lung disease, malabsorption, and the loss of electrolytes in the sweat. People with cystic fibrosis produce secretions of thick, sticky mucus that clog the lungs' airways and the ducts of the pancreas and liver. Labored breathing and stagnation of the thick mucus in the bronchial tubes provide an ideal environment for bacteria to multiply. Lung infections occur frequently, and they further damage lung tissue and raise nutrient needs in people with cystic fibrosis.[14]

Damage to the pancreas and bile duct interferes with the secretion of digestive enzymes, pancreatic juices, pancreatic hormones, and bile. With aging, damage to the pancreas worsens. Malabsorption of many nutrients and diabetes mellitus are frequent complications. Cystic fibrosis is believed to cause some degree of pancreatic insufficiency in all cases, with about 85 to 90 percent of cases serious enough to require enzyme replacement therapy.

• Medical Nutrition Therapy for Cystic Fibrosis • Although the person with cystic fibrosis may have a hearty appetite, meeting the high energy and nutrient needs imposed by labored breathing, repeated infections, and the loss of nutrients through malabsorption represents a considerable challenge. With such high energy needs, fat restrictions are inappropriate. An unrestricted high-kcalorie, high-protein diet carefully tailored to individual needs and tolerances supports nutritional health, while enzyme replacements help control steatorrhea and relieve abdominal pain. Multivitamin and fat-soluble vitamin supplements are routinely recommended. The liberal use of table salt is encouraged to make up for losses of electrolytes in the sweat. Encouraging fluid intake is also important. Fluids help liquify thick secretions and prevent dehydration.

Breast milk, standard infant formulas, and hydrolyzed infant formulas can all meet the nutrient needs of infants with cystic fibrosis provided that enzyme replacements are given as well. The breastfed infant is also given additional table salt mixed with water to replace electrolytes lost through sweat. Infant formulas contain adequate salt to prevent depletion.

CASE STUDY

Child with Cystic Fibrosis

Meredith is a 7-year-old girl diagnosed with cystic fibrosis. Symptoms of steatorrhea and failure to gain weight during infancy prompted the test that led to the diagnosis. She is 46 inches tall and weighs 45 pounds. Her height-for-age and weight-for-age fall just below the 25th percentile (see Appendix E). When she is at home, Meredith eats table foods during the day and receives additional nutrients through tube feedings delivered at night.

1. Explain cystic fibrosis to Meredith and her parents. Why are growth failure and repeated respiratory infections common in people with cystic fibrosis?
2. What do Meredith's height and weight percentiles tell you about her nutrition status? Why is it important to reassess Meredith's nutrition status at regular intervals?
3. What type of diet should Meredith follow? How can enzyme replacements be used effectively?
4. How can tube feedings delivered at night help Meredith meet her nutrient needs? Do you think this measure is necessary?
5. How will Meredith's nutrient needs change if she develops a respiratory infection?

For all people with cystic fibrosis, every effort is made to maintain an appropriate weight for height. If weight falls between 85 and 90 percent of desirable weight for height, the diet should also include high-kcalorie snacks and formula supplements to meet energy needs.[15] Clients whose weights fall below 75 percent of desirable weight for height may benefit from home nutrition programs that include oral diets during waking hours and tube feedings at night. Use the accompanying case study to see how medical nutrition therapy applies to a child with cystic fibrosis.

Inflammatory Bowel Diseases

Two of the most prevalent **inflammatory bowel diseases (IBD)** are **Crohn's disease** and **ulcerative colitis.** IBD share some clinical features but are distinct disorders. Research suggests that altered intestinal immune responses◆ coupled with genetic and environmental factors contribute to the development of IBD.[16]

In Crohn's disease, cracklike ulcers and sometimes **granulomas** accompany intestinal inflammation. Crohn's disease most often affects the ileum and colon, but can occur throughout the GI tract. The person often experiences intermittent fatigue, abdominal pain, diarrhea, and weight loss.

Ulcerative colitis develops only in the large intestine. During active episodes, ulcerative colitis causes an almost continuous diarhhea with malabsorption and great losses of fluids and electrolytes. Bloody diarrhea, cramping, abdominal pain, anorexia, and weight loss are clinical manifestations of the disorder. Anemia may develop as a consequence of bleeding and malabsorption.

◆ Evidence suggests that IBD result from exaggerated intestinal immune responses to the normal intestinal flora. IBD occurs more frequently in people who have relatives (especially siblings) with IBD and American Jews of European descent. Adolescents and adults between the ages of 15 and 35 and those older than 50 years (to a lesser degree) most often develop IBD.

• Complications Associated with Inflammatory Bowel Diseases • As the inflammatory bowel disease progresses, fibrous tissue (scar tissue) forms in the intestine, reducing its absorptive ability, narrowing the intestinal lumen, and sometimes creating an obstruction. Localized infections can develop. The intestine may also rupture, which can lead to peritonitis.

The area and extent of the intestine affected by IBD determine the type of and degree to which malabsorption occurs. Malnutrition is more likely to be a problem in people with Crohn's disease of the small intestine than in people with Crohn's disease of the colon or ulcerative colitis. Many people with Crohn's disease malabsorb carbohydrates, especially lactose, and this malabsorption can lead to gas and bloating, which makes abdominal pain worse. Fat and vitamin B_{12} malabsorption occurs more often in people who have had surgery that removes the last part of the small intestine (terminal ileum), because bile salts and vitamin B_{12} are absorbed in this portion of the intestine.

inflammatory bowel diseases (IBD): diseases characterized by inflammation of the bowel.

Crohn's disease: inflammation and ulceration along the length of the GI tract, often with granulomas.

ulcerative colitis (ko-LYE-tis): inflammation and ulceration of the colon.

granulomas (gran-you-LOH-mahs): tumors or growths that are covered with a fibrous coat and contain foreign organisms surrounded by immune system cells.

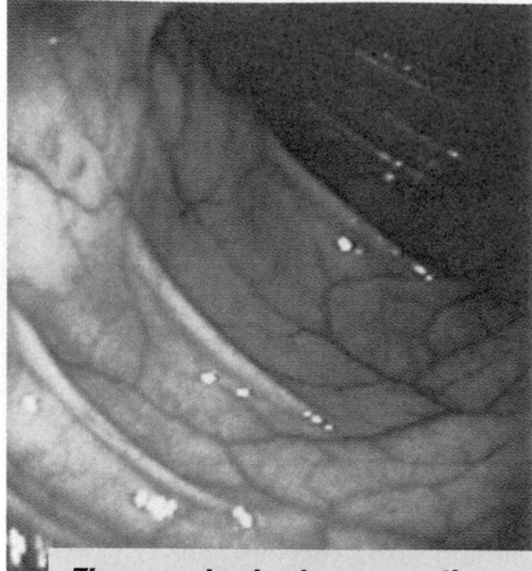

The normal colon has a smooth, shiny surface with a visible pattern of fine blood vessels.

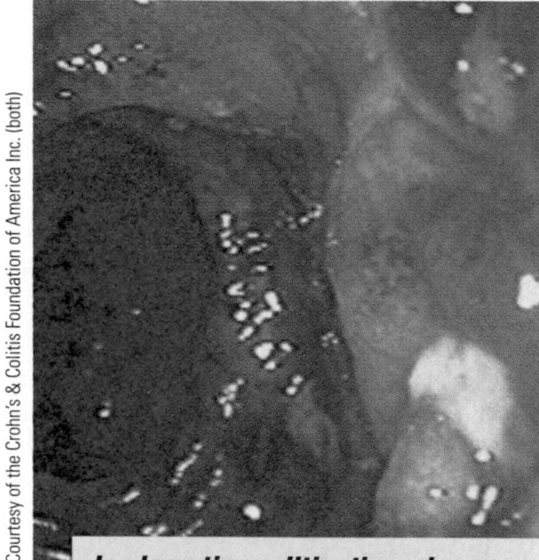

In ulcerative colitis, the colon appears inflamed and reddened, and ulcers are visible.

Courtesy of the Crohn's & Colitis Foundation of America Inc. (both)

◆ Reminder: Probiotics are live microorganisms that survive passage through the GI tract and become part of the normal bacterial flora. They are believed to help modulate immune responses.

Iron deficiencies are more common in IBD affecting the colon because small amounts of blood are lost regularly.

Fistulas may develop if an inflamed loop of intestine sticks to another loop of intestine, to another organ, or to the skin and gradually erodes. If a fistula forms between the stomach or the upper portion of the small intestine and the colon, ingested food is shunted directly into the colon, and malabsorption worsens. Bacteria from the colon can then invade the stomach or upper small intestine, contributing to further malabsorption (see "Bacterial Overgrowth" later in this chapter), increasing the risk of serious infections, and causing severe inflammation, nausea, and vomiting. If a fistula forms between the small intestine and the skin, significant malabsorption, dehydration, and sepsis can follow.[17]

• Treatments for Inflammatory Bowel Diseases • Treatments for inflammatory bowel diseases aim to reduce inflammation, control symptoms, and prevent malnutrition. People with IBD often rely on medications to reduce inflammation and control symptoms. They may remain symptom-free for long periods of time, but the symptoms tend to recur. Medications used in the treatment of IBD may include analgesics, antidiarrheals, anti-infective agents, and immune system suppressors and modulators; their potential nutrition-related side effects are shown in the Diet-Drug Interactions box on p. 673. Probiotics◆ have also been used in studies to reduce intestinal inflammation. Surgery may be necessary to remove a diseased or obstructed portion of the intestine or to repair a fistula.

• Medical Nutrition Therapy for Inflammatory Bowel Diseases • For people with Crohn's disease, restoring and maintaining nutrition status can be a challenging task. Growth failure in children, protein-energy malnutrition (PEM), and vitamin and mineral deficiencies, including deficiencies of antioxidant nutrients, may occur.[18] Poor nutrition status and nutrient deficiencies threaten immune function and may reduce the effectiveness of medications and impair recovery.

Some people with Crohn's disease meet their nutrient needs through unsupplemented oral diets. Those who cannot maintain a desirable weight need high-kcalorie, high-protein diets to help prevent or treat malnutrition. Liquid formulas (Chapter 22) may be recommended, especially for children with inflammatory bowel diseases. Children sometimes eat table foods during the day and receive tube feedings at night.

During active periods of disease, eating often intensifies uncomfortable symptoms. No specific dietary restrictions apply to all clients with Crohn's disease, although some people may find that certain foods make their symptoms worse. The most common offenders include lactose-containing foods and high-fiber foods, and these foods are restricted to the degree necessary to control symptoms. Sometimes eating smaller amounts of food more often is helpful. Low-fiber diets may be recommended for people with partial intestinal obstructions or those at high risk for obstructions (see Table 19-4 in Chapter 19). Vitamin and mineral supplements are generally recommended. Fish oils have been reported to help prevent recurrences of active Crohn's disease and, to a lesser extent, ulcerative colitis, although further research is necessary to confirm a potential benefit.[19]

For people with active ulcerative colitis, no dietary interventions seem to lessen symptoms. People with severe abdominal pain and diarrhea need complete bowel rest.

For people with active Crohn's disease who have intestinal obstructions or fistulas or for whom eating significantly worsens symptoms, foods and fluids may temporarily be withheld while fluids and electrolytes are replaced intravenously. For people who are already severely malnourished or for those who have gone for several days without eating, feeding by tube (Chapter 22) should be considered. In some cases, feeding tubes can be placed so as to bypass a fistula or partial obstruction. Intravenous nutrition (Chapter 23) can provide nutrients when oral or tube feedings significantly aggravate pain and diarrhea, when the bowel is obstructed, when complete bowel rest might help a fistula to close, or when oral or tube feedings cannot meet nutrient requirements.

Bacterial Overgrowth

Although the colon normally houses a considerable bacterial population, the small intestine is protected from bacterial overgrowth by gastric acid, which kills bacteria, and peristalsis, which flushes microorganisms through the small intestine before they can multiply. Conditions that disrupt these protective mechanisms can result in bacterial overgrowth in the stomach and small intestine. Bacteria in the small intestine partly dismantle bile salts, which are essential for fat digestion and absorption. Malabsorption of fat and fat-soluble vitamins can occur as a consequence. The bacteria also disrupt the absorption of vitamin B_{12} and can lead to vitamin B_{12} deficiencies.

• **Causes of Bacterial Overgrowth** • In some types of gastric surgery, a portion of the small intestine is bypassed, causing stasis in that portion of the intestine and allowing bacteria to flourish (see Figure 20-3 on p. 639). The bypassed portion is called a blind loop, and the symptoms associated with bacterial overgrowth are called the **blind loop syndrome.**

Other conditions that can lead to bacterial overgrowth by significantly reducing gastric acid secretions include chronic gastritis (see p. 637), medications (antisecretory agents), chronic pancreatitis, and HIV infections (see Chapter 29). Small bowel obstructions and nerve dysfunction associated with diabetes (Chapter 25) can lead to bacterial overgrowth by altering peristalsis.

• **Consequences of Bacterial Overgrowth** • Typical symptoms of bacterial overgrowth include chronic diarrhea, intestinal gas, malnutrition, and weakness. In some cases, signs of fat-soluble vitamin deficiencies may develop including dry, scaly skin (vitamin A deficiency), bruising (vitamin K deficiency), and numbness, tingling, muscle spasms, and bone pain (vitamin D and calcium deficiencies). Pernicious anemia can develop as a consequence of vitamin B_{12} deficiency.

• **Treatments for Bacterial Overgrowth** • The primary treatment for bacterial overgrowth is antibiotics. In some cases, antibiotics correct the problem, but symptoms recur weeks to months later. During active periods of bacterial overgrowth, clients benefit from fat-restricted diets. Oral vitamin and mineral supplements serve to correct nutrient deficiencies with the exception of vitamin B_{12}. Because the absorption of vitamin B_{12} is impaired, vitamin B_{12} must be given by injection or using prescription vitamin B_{12} nasal sprays.

Celiac Disease

The incidence of **celiac disease,** a hereditary intestinal disorder, appears to be increasing; currently, it affects about 1 in every 250 people.[20] In celiac disease, the intestinal cells become sensitive to certain fractions of proteins found in some grains, including wheat, rye, and barley. **Gluten** is a protein found in wheat, and **gliadin** is the fraction of gluten that causes sensitivity in celiac disease. Other grains have corresponding protein fractions that cause sensitivity in celiac disease. These protein fractions act as toxic substances, triggering immune system responses (inflammation) that damage the intestinal villi and lead to malabsorption.

• **Consequences of Celiac Disease** • People with celiac disease may malabsorb many nutrients, notably, fat, protein, carbohydrate, vitamin K, folate, vitamin B_{12}, iron, and calcium. Lactose intolerance is common. Clinical

blind loop syndrome: the problems of fat and vitamin B_{12} malabsorption that result from the overgrowth of bacteria in a bypassed segment of the intestine.

celiac (SEE-lee-ack) **disease:** a sensitivity to a part of the protein gluten that causes flattening of the intestinal villi and malabsorption; also called *gluten-sensitive enteropathy* or *celiac sprue.*

gluten (GLUE-ten): a protein found in wheat.

gliadin (GLY-ah-din): the fraction of gluten that causes the toxic effects in celiac disease. Corresponding protein fractions in barley, rye, and possibly oats also have these effects.

manifestations of the disease vary and may range from simple iron and folate deficiencies to serious malnutrition, fatigue, and severe anemia.[21] A vitamin K deficiency may precipitate clotting abnormalities, and the person may bleed easily. Reduced bone mineral density and bone diseases are common.[22] People with unexplainable bone disease may actually have celiac disease that has gone undiagnosed because GI symptoms are either mild or absent.[23]

• **Medical Nutrition Therapy for Celiac Disease** • Lifelong adherence to a gluten-free diet serves as the primary treatment for celiac disease. Once the person follows a gluten-free diet for a few weeks, the intestinal changes reverse almost completely. With early diagnosis and treatment, children and adolescents with celiac disease who adhere to a gluten-free diet can achieve normal bone mass and avoid bone diseases.[24] In adults, bone demineralization is not always reversible.[25] Lactose intolerance may be permanent.

This international symbol identifies gluten-free foods.

• **Gluten-Free Diets** • The gluten-free diet strictly eliminates all foods that contain wheat, rye, and barley. Even the smallest amounts of these grains can cause problems for people with celiac disease. Therefore, people with celiac disease must be vigilant in preparing and storing foods. They must be careful to avoid contaminating the foods they eat with bread crumbs from toasters, cutting boards, grills, or those that may be in margarine or jelly or in fats used to fry foods. As Table 21-6 shows, many commonly used foods contain wheat, rye, and barley in various forms, many of which are not obvious. To further complicate matters, not all authorities agree on what foods are safe to include on gluten-free diets.[26] In the United States, all celiac organizations oppose the inclusion of oats, wheat starch, malt, and malt flavorings on the gluten-free diet, whereas so[illegible]nizations in Europe consider these products safe.

TABLE 21-6 Glute[illegible]Diet[a]

Meat and [illegible]ternates	Any allowed [illegible]ose that are breaded, prepared with bread crumbs or ingredients that are not allowed, [illegible]med.
Milk and Milk Products	Any allowed if client is not intolerant to lactose except milk mixed with Ovaltine, commercial chocolate milk with a cereal additive, milk beverages flavored with malt, pudding thickened with wheat flour, ice cream or sherbet containing gluten stabilizers, processed cheese foods and spreads, Roquefort and ricotta cheese.
Fruits and Vegetables	Any allowed except those that are breaded, prepared with bread crumbs, or creamed.
Starches and Grains	*Allowed:* Bread, cereal, or dessert products made from buckwheat, cornmeal, soybean flour, rice flour, or potato flour; tapioca; cornmeal, popcorn, and hominy; rice, cream of rice, puffed rice, and rice flakes; potato chips.
	Not allowed:[b] Bread, cereal, or dessert products made from wheat, rye, barley, and oats; wheat starch; low-gluten flours; commercially prepared mixes for biscuits, cornbread, muffins, pancakes, cakes, cookies, or waffles; bran; pasta, macaroni, and noodles; malt; pretzels; wheat germ; doughnuts; ice cream cones; matzo.
Other	*Not allowed:* Beer; ale; certain whiskeys (Canadian rye); alcohol-based extracts; cereal beverages (Postum); root beer; commercial salad dressings that contain gluten stabilizers; soy sauce; soups containing any ingredients not allowed (such as barley or noodles); products made with hydrolyzed vegetable protein.

[a]This is a partial list. Celiac organizations can provide additional information.
[b]Note that many special products made with allowed ingredients are available. Gluten-free pastas and macaroni, for example, are acceptable substitutes for regular pastas and macaroni.

CASE STUDY

Physical Therapist with Celiac Disease

Anna Kovacs was diagnosed with celiac disease at age 31. Anna sought medical help after she began to experience fatigue and weight loss. Laboratory tests revealed iron-deficiency anemia, and Anna was placed on iron supplements. The iron supplements improved her anemia, but she continued to lose weight. After a series of tests, the physician diagnosed celiac disease, and Anna was referred to the dietitian for consultation regarding a gluten-free diet.

1. Describe celiac disease and its potential nutrition problems.
2. How would you explain the gluten-free diet to Anna? What benefits will she realize by strictly adhering to the diet?
3. What precautions should Anna consider with respect to the gluten-free diet and her food choices?

If people with celiac disease do not take care to vary their food choices and select appropriate foods, malnutrition can be a problem.[27] Food manufacturers make many gluten-free products that serve as acceptable substitutes for foods not permitted on gluten-free diets. These products help people with celiac disease expand their food choices and enjoy foods that would otherwise be forbidden. However, clients must carefully select the products they use; some products do not contain the same levels of B vitamins, iron, and fiber as the foods they are intended to replace.[28] The accompanying case study presents a client with celiac disease.

IN SUMMARY

Clearly, the disorders that lead to fat malabsorption can seriously impair nutrition status. With fat malabsorption, needed energy, essential fatty acids, fat-soluble vitamins, and some minerals are lost as well. Malnutrition and bone diseases are frequent complications. Oxalate stones can also form. To help people with fat malabsorption get enough energy from foods, some of the fat can be supplied as MCT, which are easier to digest and absorb than LCT. Enzyme replacements may be given along with antisecretory agents to help people digest and absorb fat so that greater quantities of fat can be consumed. Fat malabsorption in people with pancreatitis (especially chronic pancreatitis) and cystic fibrosis results from the inability of the pancreas to make and secrete enough digestive enzymes to allow for adequate fat absorption. Some people with Crohn's disease (especially when it affects the small intestine) and people with short-bowel syndrome (described in the next section) experience fat malabsorption because their intestinal surface area for absorbing fat is reduced. In bacterial overgrowth, fat malabsorption results from the dismantling of bile salts by bacteria. People with celiac disease are sensitive to gliadin, a portion of a protein found in wheat and corresponding protein portions found in rye and barley and possibly other grains as well. Strict adherence to a gluten-free diet reverses much of the intestinal damage and provides the primary treatment for celiac disease.

INTESTINAL SURGERIES

The treatment of inflammatory bowel diseases, cancer of the intestine, intestinal obstructions, fistulas, diverticulitis, or impaired blood supply to the intestine may include surgery to remove a portion of the intestine. Surgical resections of the small intestine have a greater impact on nutrient absorption than surgical resections of the large intestine, because most nutrient absorption is complete by the time the intestinal contents reach the colon.

Resections of the Small Intestine

When portions of the small intestine must be removed, the length, location, and health of the remaining intestine determine if the person will be able to meet nutrient needs using the GI tract (enterally). Figure 21-3 reviews nutrient absorption in the GI tract and describes how absorption is affected by surgical resection. Generally, up to 50 percent of the intestine can be resected without serious nutrition consequences. When the absorptive surface of the small intestine is significantly reduced, **short-bowel syndrome**—characterized by severe diarrhea and malabsorption following surgery—results. Gradually, diarrhea and malabsorption lessen as adaptation occurs in the remaining bowel.

• Adaptation • After an intestinal resection, a remarkable adaptive response occurs in the healthy portions that remain: they get longer, thicker, and wider, and they absorb nutrients more efficiently.[29] Adaptation begins soon after surgery, but may take as long as a year following surgery. The presence of nutrients in the remaining gut stimulates this adaptation. Specific dietary and other factors, such as the amino acid glutamine, short-chain fatty acids, fiber, and growth hormone, may also aid in this adaptation. Highlight 21 describes the effects of these factors in more detail. Intestinal disease in the remaining portions of the intestine (such as active Crohn's disease or inflammation caused by radiation therapy) may impede adaptation. Following extensive small bowel resections, adaptation is most likely to allow clients to meet their nutrient needs enterally if the ileum, the ileocecal valve, and the colon remain intact.

The ileum helps prevent malabsorption because it is the site where bile salts and vitamin B_{12} are absorbed. When the ileum has been resected, the absorption of vitamin B_{12} and bile salts may be impaired. In addition to vitamin B_{12} deficiencies, the body's pool of bile salts diminishes, and this loss can lead to fat malabsorption and its consequences.

With surgical removal of the ileocecal valve, the contents of the small intestine empty rapidly into the colon, fluid and electrolyte losses are compounded, and bacteria from the colon can invade the small intestine, leading

short-bowel or **short-gut syndrome**: severe malabsorption that may occur when the absorptive surface of the small bowel is reduced, resulting in diarrhea, weight loss, bone disease, hypocalcemia, hypomagnesemia, and anemia.

FIGURE 21-3 **Nutrient Absorption and Consequences of Intestinal Surgeries**

About 90 to 95 percent of nutrient absorption takes place in the first half of the small intestine. After a resection, nutrient absorption may be reduced.

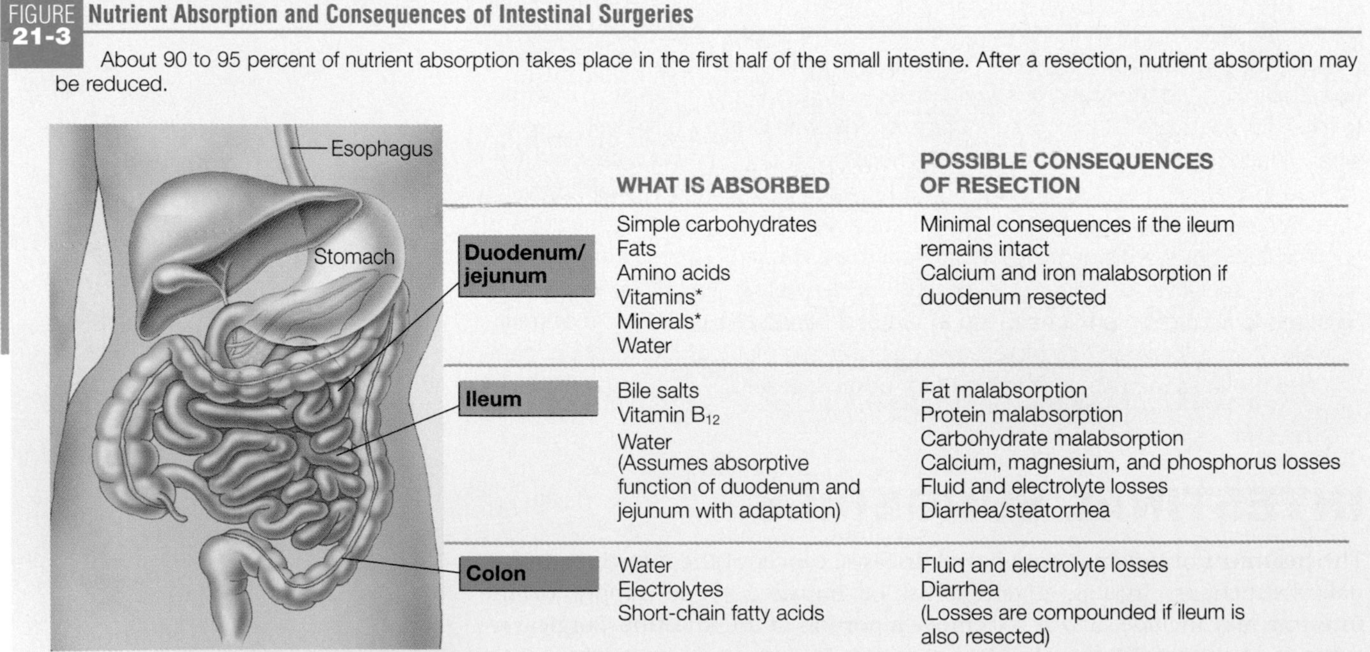

	WHAT IS ABSORBED	POSSIBLE CONSEQUENCES OF RESECTION
Duodenum/ jejunum	Simple carbohydrates Fats Amino acids Vitamins* Minerals* Water	Minimal consequences if the ileum remains intact Calcium and iron malabsorption if duodenum resected
Ileum	Bile salts Vitamin B_{12} Water (Assumes absorptive function of duodenum and jejunum with adaptation)	Fat malabsorption Protein malabsorption Carbohydrate malabsorption Calcium, magnesium, and phosphorus losses Fluid and electrolyte losses Diarrhea/steatorrhea
Colon	Water Electrolytes Short-chain fatty acids	Fluid and electrolyte losses Diarrhea (Losses are compounded if ileum is also resected)

*The absorption of vitamins and minerals begins in the duodenum and continues throughout the length of the small intestine.

to bacterial overgrowth. Bacterial overgrowth further damages bile salts and worsens fat and vitamin B_{12} malabsorption.

For the person with an extensive resection of the small intestine, the presence of an intact colon helps reduce losses of fluids and electrolytes. Furthermore, the colon houses bacteria that can metabolize unabsorbed carbohydrate to short-chain fatty acids. Short-chain fatty acids, in turn, can provide energy the body can use and help promote adaptation in the colon.

As a guideline, enteral nutrition is possible with a functional (healthy) bowel length of greater than 40 inches (100 centimeters) with the colon removed, or greater than 24 inches (60 centimeters) with the colon intact.[30] When people cannot meet nutrient needs enterally, two options remain: permanent intravenous nutrition or intestinal transplantation. Due to the major risks associated with intestinal transplantation, this procedure is reserved for clients who require lifelong intravenous nutrition support, but who develop complications that make intravenous support extremely difficult or impossible.

• Medical Nutrition Therapy for Short-Bowel Syndrome • Immediately following surgery, fluid and electrolyte balances are markedly upset, and intravenous fluids serve to correct these imbalances. For resections of less than 50 percent of the intestine, oral nutrition often begins after a few days. With extensive intestinal resections, severe diarrhea, malabsorption, and lack of GI motility prevent use of the GI tract to provide nutrients. Once fluid and electrolyte balances are stabilized, intravenous nutrition is cautiously initiated. Once adaptation is under way, generally in about one to three weeks, diarrhea and malabsorption begin to lessen, and enteral nutrition (usually by tube feeding) can be started in limited amounts. If tolerance for enteral feeding improves, intravenous feedings are tapered and gradually discontinued. Some people may be unable to discontinue intravenous nutrition and will require lifelong intravenous feedings to supply all or some of their nutrient needs.

Once oral diets are possible, clients whose colons remain intact benefit from diets high in complex carbohydrates, restricted in fat, and low in oxalates.[31] Bacteria in the colon break down complex carbohydrates to short-chain fatty acids, which provide a source of energy (up to 500 kcalories per day) for the body. For people without intact colons following surgery, carbohydrates contribute to further intestinal losses, and diets with a high percentage of fat and a low percentage of carbohydrate are recommended. Dietitians work closely with clients to help them identify individual food intolerances and to offer suggestions to help control the rate at which foods pass through the intestinal tract.[32] Vitamin and mineral supplements are routinely recommended. For those with vitamin B_{12} malabsorption, vitamin B_{12} injections or prescription nasal sprays are necessary to correct deficiencies and maintain vitamin B_{12} status. Use the case study on p. 670 to apply the concepts described in this section to a client with short-bowel syndrome.

Resections of the Large Intestine

When the small bowel remains intact, resections of the colon do not result in major nutrition consequences. However, because water is absorbed along the length of the colon, additional fluid will be lost in the stool, and stool consistency will vary depending on both the length and the portion of the resected bowel. In general, the greater the length of colon left intact, the more fluid that will be absorbed and the more formed the stool.

Treatment for some medical conditions and surgeries affecting the large intestine necessitates that the feces be diverted from all or portions of the colon. A **colostomy** is a surgical procedure that creates an opening **(stoma)** from the colon through the abdominal wall to the surface of the skin to allow for defecation when feces cannot pass through the rectum and anus. A pouch placed over the stoma collects the feces. Colostomies may be permanent or temporary.

colostomy (co-LOSS-toe-me): surgery that creates an opening from any portion of the colon through the abdominal wall and out through the skin.

stoma (STOH-mah): a surgically formed opening.

stoma = window

CASE STUDY

Artist with Short-Bowel Syndrome

Adam Goldstein is a 25-year-old artist with an 8-year history of Crohn's disease. Adam is 5 feet, 9 inches tall. Three years ago, Adam underwent a small bowel resection and remained free of active disease for two years. During that time, Adam's symptoms subsided, he gained weight, and with the exception of milk products that aggravate his lactose intolerance, he was able to eat many foods. Adam has been able to maintain a modest income from selling his paintings and working part-time in an art gallery. Several months ago, Adam experienced a severe flare-up of his Crohn's disease. He lost 10 pounds and currently weighs 158 pounds. He experienced fatigue and severe abdominal pain that persisted despite aggressive medical management that included intravenous nutrition. Adam underwent another resection five days ago, which left him with 36 percent of healthy small intestine. His colon is intact. He is experiencing extensive diarrhea.

1. What is Crohn's disease, and why is surgery sometimes necessary in its treatment?
2. Calculate Adam's IBW. What nutrition concerns do his current weight and recent weight loss suggest? What other nutrition problems might Adam be experiencing as a consequence of Crohn's disease? Describe some complications and other factors that can affect nutrient needs in people with Crohn's disease.
3. What factors determine how well a person will be able to meet nutrient needs enterally following an extensive intestinal resection?
4. How are nutrient needs met following extensive intestinal resections? Why? What medical conditions indicate that enteral feeding is possible? How are enteral feedings usually delivered at first?
5. What diet will be advised for Adam once he is able to eat table foods? How would this advice differ if Adam's medical condition had necessitated removal of his colon?

In an **ileostomy,** the entire colon is bypassed, and the stoma is created from the ileum. In an alternative to an ileostomy, the ileal pouch/anal anastomosis, the surgeon removes the diseased colon and rectal tissue and connects the ileum to the anus. A temporary ileostomy allows time for the tissue to heal; thereafter, defecation occurs through the anus, rather than a stoma. Figure 21-4 shows examples of a colostomy and ileostomy. In this section, the word *ostomy* is used to refer to both colostomies and ileostomies.

FIGURE 21-4 Colostomy and Ileostomy

Colostomy

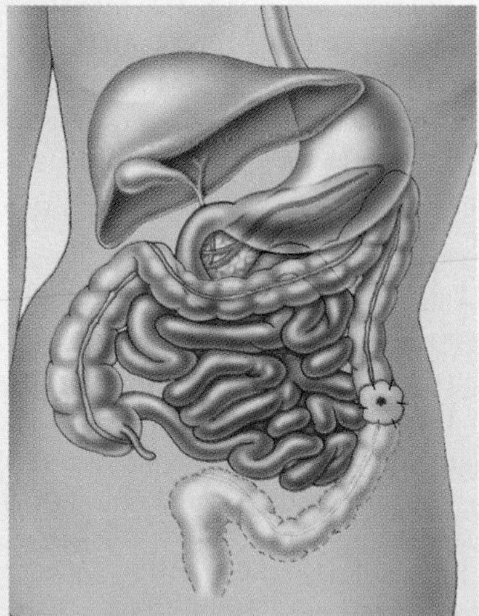

In a colostomy, the rectum and anus are removed, and the stoma is formed from the remaining colon.

Ileostomy

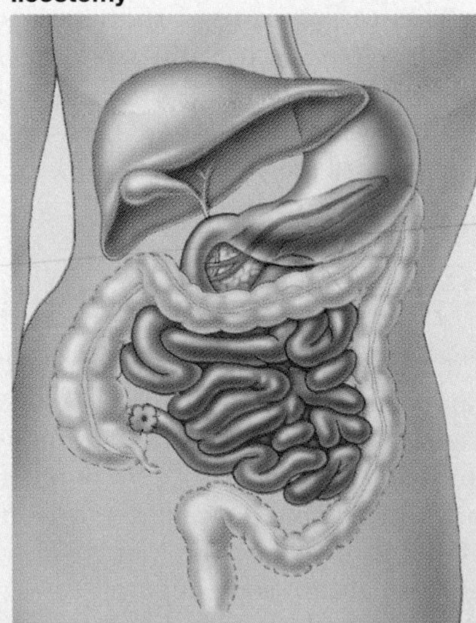

In an ileostomy, the entire colon, rectum, and anus are removed, and the stoma is formed from the ileum.

ileostomy (ILL-ee-OSS-toe-me): surgery that creates a stoma from the ileum through the abdominal wall and out through the skin.

• Medical Nutrition Therapy • Once fluid and electrolyte balances have been restored and solid foods are permitted following surgery, people who have undergone surgery of the large intestine often receive low-fiber, soft diets to prevent obstructing the colon or ostomy site, to help promote healing, and to prevent GI upsets. Encourage people to judiciously include other foods as soon as possible, however. Foods are added one at a time and in small amounts so that their effects can be assessed. If the added food presents problems, the person can try it again in a few weeks or months.

• Preventing Obstructions • People with ostomies may find that some foods are more likely than others to be incompletely digested and obstruct the stoma. People with ileostomies are more likely to experience this problem than people with colostomies. Some foods that might be incompletely digested include stringy foods such as celery, spinach, coconut, and bean sprouts; foods with tough skins such as dried fruits, raw apples, and corn; foods with seeds; mushrooms; and nuts. Practitioners report that some of these foods can be used if the client cuts the food into small pieces and chews them thoroughly. A large portion of an undigested mushroom, for example, may act as a plug and obstruct a stoma, but if it is cut into small pieces and chewed thoroughly, it may be tolerated.

• Encouraging Fluids • People who have resections of the large intestine, especially those whose entire colons have been removed, need extra fluids because they are absorbing less fluid from the large intestine. They may tend to restrict their fluid intakes, however, for fear of aggravating diarrhea.

• Controlling Diarrhea • People who experience diarrhea as a consequence of their surgery—a frequent problem for those with ileostomies—may benefit from eating foods that thicken the stool. These foods include applesauce, bananas, cheese, creamy peanut butter, and starchy foods such as white breads, rice, and potatoes without skins. Foods that may aggravate diarrhea include apple, grape, and prune juices; highly seasoned foods; foods that cause gas; and alcohol and caffeine. The foods listed here are suggestions only; what works for the individual is determined by trial and error.

• Reducing Gas and Odors • People with ostomies are often concerned about gas and odors. Gas-forming foods in general were listed in Table 21-1, but certain foods in particular seem to cause gas and odors for people with ostomies: asparagus, beans, beer, broccoli, brussels sprouts, cabbage, carbonated beverages, cauliflower, eggs, fish, garlic, and onions. Foods thought to reduce gas and odors include buttermilk, cranberry juice, parsley, and yogurt.

IN SUMMARY

Surgery of the intestines can have a variety of effects on nutrient absorption depending on the length, portion, and health of the intestine that remains. Severe problems with malabsorption are most likely when significant portions of the small intestine have been resected, especially if the ileum, ileocecal valve, and colon do not remain intact. With intestinal adaptation, some people can meet their nutrient needs enterally. Others will need to meet some or all of their nutrient needs using intravenous nutrition. For those with extensive small bowel resections and intact colons, diets high in complex carbohydrates and restricted in fat can help minimize malabsorption. For those without intact colons, diets lower in carbohydrate and higher in fat are recommended. People who have all or portions of their large intestines removed can maintain nutrition status, provided their small intestines remain healthy and intact. For these clients, encouraging fluids is a primary concern. Other dietary advice depends on problems the client may experience, which can include foods obstucting a stoma, diarrhea, and gas and odors.

FIGURE 21-5 Diverticula

This figure is repeated from Highlight 3, which briefly introduced diverticulosis.

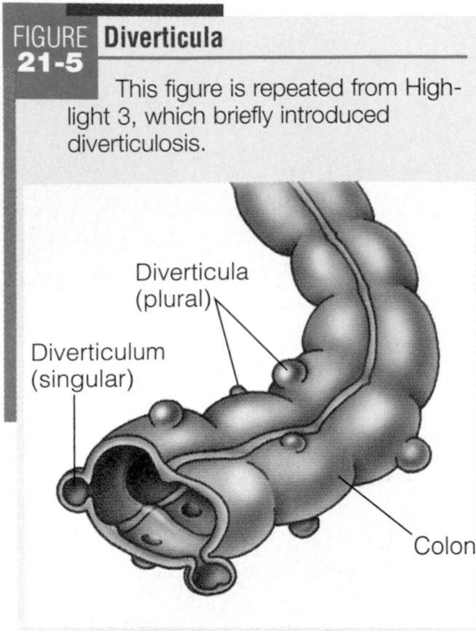

DIVERTICULAR DISEASE OF THE COLON

Sometimes pouches of the intestinal wall (called diverticula)◆ bulge out through the muscles surrounding the large intestine, often at points where blood vessels enter the muscles (see Figure 21-5). Most authorities believe that low-fiber diets and constipation lead to the development of diverticula (diverticulosis). Without adequate fiber, stools become hard and difficult to pass. The person must then strain the intestinal muscles to defecate, and this raises pressure in the colon. As pressure builds, parts of the intestinal membrane balloon outward through the muscle layer. People with diverticulosis are frequently symptom-free and unaware of the disorder. Others may develop cramps, bloating, and constipation. Diverticulosis often develops with age. About half of all people between the ages of 60 and 80 have diverticulosis; virtually all people over age 80 have it.

◆ **Reminder: Diverticula (dye-ver-TIC-you-lah) are sacs or pouches that develop in weakened areas of the intestinal wall (like bulges in an inner tube where the tire wall has been punctured). The condition of having diverticula is called diverticulosis. When diverticula become infected the condition is called diverticulitis.**

• **Diverticulitis** • In about 10 to 25 percent of people with diverticulosis, a localized area of inflammation and infection develops around a diverticulum, a condition called diverticulitis. Although the cause is unknown, it is believed that bacteria or fecal matter gets trapped in a diverticulum and leads to infection. People with diverticulitis may suffer from abdominal pain and distention, alternating episodes of diarrhea and constipation, indigestion, flatus, and fever. Occasionally, a diverticulum ruptures, causing a localized or sometimes life-threatening infection (peritonitis). Bleeding from a diverticulum may also occur. If the diverticula become inflamed repeatedly, the intestinal wall can form scar tissue and thicken (fibrosis), narrowing the intestinal lumen and creating an obstruction. An inflamed bowel segment can also stick to other pelvic organs, forming a fistula.

• **Treatments for Diverticular Diseases** • People with diverticulosis require no medications, but are advised to eat fiber-containing foods to reduce pressure in the colon and stimulate peristalsis. A study of over 45,000 people suggests that high-fiber, low-fat diets are associated with a lower incidence of symptomatic diverticular disease.[33] Traditionally, diet advice for diverticulosis has included a recommendation to avoid foods with seeds such as okra and strawberries because seeds were believed to get trapped in the diverticula and cause irritation. However, evidence to support this theory is lacking.[34]

During periods of active diverticulitis, antibiotics serve to treat the infection, and analgesics provide pain relief and reduce fever. Clients may be advised to follow liquid diets until symptoms subside. If bouts of diverticulitis are severe and occur repeatedly, or if complications occur, surgery may be necessary to remove the affected portion of the colon.

IN SUMMARY

Diverticular disease, a frequent consequence of aging, is believed to develop as a consequence of rising pressure in the colon. High-fiber diets can help prevent diverticular disease and diverticulitis.

This chapter has shown how disorders of the intestine can have effects that range from discomfort to serious nutrition problems. Compared to disorders of the small intestine, disorders of the large intestine are less likely to lead to severe nutrition problems because most nutrients are absorbed by the time the food bolus reaches the colon. The accompanying Diet-Drug Interactions box shows examples of nutrition-related concerns related to medications used to treat disorders of the lower GI tract. The Nutrition Assessment Checklist on p. 674 highlights nutrition-related findings important to consider for clients with disorders of the lower GI tract.

ANALGESICS

All *analgesics* can cause nausea, vomiting, and GI upsets. When oral intake is permitted, giving the medication along with food can minimize these side effects. *Narcotic analgesics* can lead to constipation; these medications can also lead to lethargy, which can contribute to reduced food intake.

ANTIANXIETY AGENTS

Antianxiety agents can cause drowsiness, which can interfere with food intake. *Alprazolam* and *choldiazepoxide* can stimulate the appetite and lead to weight gain. *Diazepam, lorazepam,* and *oxazepam* can cause constipation, diarrhea, or dry mouth. *Meprobamate* can cause nausea and vomiting and drowsiness. Caution clients on antianxiety agents to avoid alcohol.

ANTIDEPRESSANTS

Antidepressants frequently cause mouth dryness, nausea, vomiting, and constipation. *Nephazone* should be administered without food to ensure proper drug absorption. The herb *belladonna* can enhance the effects of antidepressants. People taking *St. John's wort* and antidepressants should do so only under medical supervision to minimize potential interactions.

ANTIDIARRHEALS

Antidiarrheals seldom result in significant nutrition-related side effects, with the exception of *opium* and *paregoric,* which can cause nausea and vomiting, constipation, and lethargy.

ANTIEMETICS

Antiemetics frequently lead to drowsiness. *Prochloramide* and *triethyperaine maleate* can also lead to mouth dryness and constipation.

ANTI-INFECTIVE AGENTS

Numerous *anti-infective agents* with a variety of nutrition-related interactions may be used in the treatment of diarrhea, pulmonary infections (cystic fibrosis), bacterial overgrowth of the intestine, and infections that arise in clients recovering from surgery of the intestine. Consult a drug guide for specific examples.

IMMUNOSUPPRESSANTS (FOR INFLAMMATORY BOWEL DISEASES)

Prednisone can stimulate the appetite and lead to fluid retention—both can lead to weight gain. Long-term use also leads to negative nitrogen and calcium balances and osteoporosis. For clients who take prednisone, high-protein, high-calcium, high-potassium diets are recommended. If lactose intolerance is a problem, clients may need to use digestive aids or calcium supplements. Clients who are not losing fluids through diarrhea and malabsorption may need to limit sodium. The dietary supplement *melatonin* can antagonize the effects of prednisone. Two other immunosuppressants, *azathioprine* and *6-mercaptopurine,* may lead to nausea and vomiting.

LAXATIVES

Common side effects of *laxatives* include nausea and cramps. Mineral oils can reduce the absorption of fat-soluble vitamins. Clients on laxatives are encouraged to eat high-fiber foods and drink generous amounts of fluids.

PANCREATIC ENZYME REPLACEMENTS

Enzyme replacements sometimes cause nausea and may interfere with iron absorption. Enzyme replacements must be taken with meals and snacks. Enteric-coated forms should not be crushed or chewed. Capsules containing enteric-coated microspheres (tiny spheres of medication) may be sprinkled on soft foods (such as applesauce), provided they can be swallowed without chewing and are followed by a glass of water or juice.

UNCATEGORIZED DRUGS (FOR INFLAMMATORY BOWEL DISEASES)

Sulfasalazine may cause nausea, vomiting, and diarrhea, and it may also lead to folate deficiency. The drug should be taken with food at regular intervals during the day. *Mesalamine* may lead to anorexia and weight loss, but nutrition related side effects are less common with mesalamine than sulfasalazine. Clients unable to eat enough folate from foods may need supplements. Encourage the client to drink fluids. *Infliximab* may cause nausea and abdominal pain and is given by injection.

Nutrition Assessment Checklist for People with Lower GI Tract Disorders

MEDICAL HISTORY

Does the client have a medical diagnosis of:

- ☐ Irritable bowel syndrome?
- ☐ Lactose intolerance?
- ☐ Pancreatitis?
- ☐ Cystic fibrosis?
- ☐ Crohn's disease?
- ☐ Ulcerative colitis?
- ☐ Short-bowel syndrome?
- ☐ Celiac disease?
- ☐ Diverticular disease?

Has the client had surgery that includes:

- ☐ Intestinal resection?
- ☐ Colostomy?
- ☐ Ileostomy?

Does the client have the following symptoms or complications:

- ☐ Constipation?
- ☐ Diarrhea/dehydration?
- ☐ Lactose intolerance?
- ☐ Malabsorption?
- ☐ Essential fatty acid deficiency?
- ☐ Bone disease?
- ☐ Oxalate stones?
- ☐ Infection?
- ☐ Fistulas?
- ☐ Obstructions?
- ☐ Bacterial overgrowth of the intestine?

MEDICATIONS

Check for medications or herbal remedies that may:

- ☐ Cause constipation or diarrhea
- ☐ Interfere with food intake including those that cause nausea, vomiting, cramps, dry mouth, or drowsiness
- ☐ Alter nutrient needs, including prednisone, sulfasalazine, and antidepressants

FOOD/NUTRIENT INTAKE

Note the following conditions and contact the dietitian if you suspect a problem with:

- ☐ Poor appetite or limited food intake
- ☐ Food intolerances (including amount of fat for those with malabsorption, and amount of lactose for those with lactose intolerance)
- ☐ Inadequate fiber/fluid intake for those with constipation
- ☐ Inadequate calcium intake for those with lactose intolerance
- ☐ Fluid intake

ANTHROPOMETRICS

Measure baseline height and weight. Address weight loss early to prevent malnutrition for clients with:

- ☐ Severe or persistent diarrhea
- ☐ Malabsorption

LABORATORY TESTS

Check laboratory tests for signs of dehydration for clients with:

- ☐ Severe or persistent diarrhea
- ☐ Malabsorption
- ☐ Intestinal resections

Check laboratory tests for signs of nutrition-related anemias and nutrient deficiencies (see Appendix F) for clients with:

- ☐ Severe or persistent diarrhea
- ☐ Malabsorption

PHYSICAL SIGNS

Look for physical signs of:

- ☐ Dehydration (especially for people with severe or persistent diarrhea or malabsorption)
- ☐ PEM
- ☐ Essential fatty acid deficiencies (especially for people with fat malabsorption)
- ☐ Folate and vitamin B_{12} deficiencies (especially for people with Crohn's disease or bacterial overgrowth)
- ☐ Mineral deficiencies (especially for people with severe and persistent diarrhea, lactose intolerance, or fat malabsorption)

Nutrition on the Net

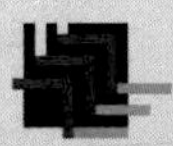

WEBSITES

Access these websites for further study of topics covered in this chapter.

- Find updates and quick links to these and other nutrition-related sites at our website: **www.wadsworth.com/nutrition**
- Find more information on many diseases of the lower GI tract at:

  National Institute of Diabetes & Digestive & Kidney Diseases **www.niddk.nih.gov/health/digest/digest.html**

  American College of Gastroenterology **www.acg.gi.org**

  Medscape Gastroenterology **www.medscape.com/medscape/gastro/journal/public/MG-journal.html**

  InteliHealth site: **www.intelihealth.com**
- Find more information about celiac disease by visiting the

  Celiac Diseases Foundation at **www.celiac.org**

  Canadian Celiac Association at **www.celiac.ca**

  Celiac Sprue Association at **www.csaceliacs.org**

  Gluten Intolerance Group of North America at **www.gluten.net**

- Search for more information on Crohn's disease and ulcerative colitis at the Crohn's and Colitis Foundation site: **www.ccfa.org**

INTERNET ACTIVITIES

Diseases and disorders of the digestive tract are second only to colds in causing people to miss days of work. People diagnosed with certain digestive diseases may be embarrassed to ask questions about their disease. Here is one way to become more knowledgeable about digestive diseases.

- Go to: **www.niddk.nih.gov/health/digest/digest.htm**
- Click on the specific disease you want information about.
- Learn about the disease and its causes and treatments.

This site provides information about many different digestive diseases—from constipation to lactose intolerance to pancreatitis. Do you or does anyone close to you have symptoms that might be related to a disease of the digestive tract? How can you find out whether symptoms are related to a specific disease?

Study Questions

These questions will help you review the chapter material. You will find the answers in the discussions on the pages provided.

1. What measures can help to prevent and treat constipation? (pp. 652–653)
2. Describe dietary measures that may be useful in the treatment of diarrhea. When is diarrhea a cause for alarm? (pp. 654–655)
3. Describe the irritable bowel syndrome and discuss the ways diet can be used in its treatment. (p. 655)
4. What is lactose intolerance? What nutrients may be deficient in the diet of people with lactose intolerance? (pp. 656–657)
5. Explain why disorders of the stomach, pancreas, liver, and intestine can all lead to fat malabsorption. What are the primary nutrition problems that accompany fat malabsorption? (pp. 657–660)
6. In what ways are the dietary treatments for chronic pancreatitis and cystic fibrosis similar? (pp. 661–662)
7. What types of inflammatory bowel diseases are most likely to lead to severe malnutrition? Why? (p. 663)
8. How can bacteria in the stomach or small intestine lead to fat malabsorption? What other nutrients can be affected by bacterial overgrowth? (p. 665)
9. Discuss the cause of celiac disease and describe the diet used in its treatment. (pp. 665–667)
10. What is the short-bowel syndrome, and what factors determine the likelihood of its development following intestinal resections? (pp. 668–669)
11. Describe the primary nutrition-related concerns of people who have undergone colostomies and ileostomies. (p. 671)
12. What theory explains the development of diverticular disease, and what diet is most useful in its prevention and treatment? (p. 672)

These questions will help you prepare for an exam. Answers can be found on p. 677.

1. The health care professional advising an elderly client with constipation encourages the client to eat a:
   a. low-fat diet rich in potassium.
   b. low-fiber diet rich in calcium.
   c. diet rich in fiber and drink plenty of fluids.
   d. gluten-free diet rich in protein.
2. For people with permanent lactose intolerance, lactose-restricted diets:
   a. include lactose-containing foods according to individual tolerances.
   b. strictly limit lactose from all sources.
   c. require the use of lactase-containing digestive aids.
   d. require careful review of the client's intake of folate and vitamin B_{12}.
3. Nutrition problems associated with fat malabsorption syndromes typically include all of the following *except:*
   a. weight loss and PEM.
   b. essential amino acid deficiencies.
   c. bone diseases.
   d. oxalate kidney stones.

4. The client with chronic pancreatitis benefits from the following diet-related advice:
   a. follow a very-low-fat diet.
   b. moderately restrict alcohol.
   c. eat three meals a day and avoid snacks.
   d. use enzyme replacements to improve fat absorption.
5. Dietary recommendations for people with cystic fibrosis include:
   a. limited use of table salt.
   b. a high-kcalorie, high-protein diet.
   c. a fat-restricted diet.
   d. fluid restrictions.
6. The dietitian working with a client with an inflammatory bowel disease recognizes that all of the following can affect nutrient needs *except:*
   a. the portion of the intestinal tract affected by disease.
   b. medications.
   c. dumping syndrome.
   d. fistulas.
7. The common nutrition problems associated with bacterial overgrowth in the stomach and small intestine include:
   a. sensitivity to gluten and gliaden.
   b. fat malabsorption and vitamin B_{12} deficiencies.
   c. increased absorption of bile salts and constipation.
   d. permanent loss of digestive enzymes.
8. The health care professional recognizes that a client understands the basics of a gluten-free diet when the client states:
   a. "I must avoid all products containing wheat, corn, and rice."
   b. "I must avoid all products containing barley, soybeans, and corn."
   c. "I must avoid all products containing wheat, barley, and rye."
   d. "I must limit my use of products containing wheat, barley, rye, and oats."
9. Diets for people who have undergone extensive intestinal resections but whose colons remain intact should be:
   a. gluten-free.
   b. low in simple sugars and high in protein and fat.
   c. low in carbohydrate to prevent lactose intolerance.
   d. high in complex carbohydrate and restricted in fat.
10. Long-term management of diverticular disease includes a:
   a. high-fiber diet.
   b. low-fiber diet that omits foods with seeds.
   c. high-fiber, lactose-free diet.
   d. low-fiber, lactose-free diet.

Clinical Applications

1. Using Table 21-5 on p. 660 as a guide, plan a day's menus for a diet containing 35 grams of fat. Take care to make the menus both palatable and nutritious. How can these menus be improved using the suggestions in the box on p. 661?

2. As stated in this chapter, treatment of celiac disease is deceptively simple—eliminate gluten. Take a trip to the grocery store and randomly select ten of your favorite snack and convenience foods. Check the labels of these products and see if they are allowed on gluten-restricted diets. (As you complete this part of the assignment, keep in mind that the labels may not list all offending ingredients.) Find acceptable substitutes for the products that are not allowed, either by substituting other foods or by checking for gluten-free products in the grocery store. (If you have access to a computer, you may want to look up sites that sell gluten-free products to get an idea of what's available.) No doubt, this will be a challenging assignment.

References

1 J. E. Suneson, Irritable bowel syndrome: A practical approach to medical nutrition therapy, *Support Line*, February 1999, pp. 11–15.

2 D. Harari and coauthors, Bowel habit relation to age and gender: Findings from the National Health Interview Survey and clinical implications, *Archives of Internal Medicine* 156 (1996): 315–320; C. S. Probert and coauthors, Evidence for the ambiguity of the term constipation: The role of irritable bowel syndrome, *Gut* 35 (1994): 1455–1458.

3 W. McCray and B. Krevsky, Diarrhea in adults: When is intervention necessary? *Hospital Medicine* 35 (1999): 39–46.

4 L. R. Schiller, Irritable bowel syndrome: One physician's perspective, *Support Line,* October 1998, pp. 3–8.

5 Suneson, 1999.

6 M. Pimentel and coauthors, Eradication of small intestinal bacterial overgrowth reduces symptoms of irritable

bowel syndrome, *American Journal of Gastroenterology* 95 (2000): 3503–3506.

7 F. L. Suarez, D. A. Savaiano, and M. D. Levitt, A comparison of symptoms after the consumption of milk or lactose-hydrolyzed milk by people with self-reported severe lactose intolerance, *New England Journal of Medicine* 333 (1995): 1–4; F. L. Suarez and coauthors, Tolerance to the daily ingestion of two cups of milk by individuals claiming lactose intolerance, *American Journal of Clinical Nutrition* 65 (1997): 1502–1506.

8 M. Lee and S. D. Kransinski, Human adult-onset lactase decline: An update, *Nutrition Reviews* 56 (1998): 1–8.

9 Suarez and coauthors, 1997.

10 S. R. Hertzler and coauthors, How much lactose is low lactose? *Journal of the American Dietetic Association* 96 (1996): 243–246.

11 P. B. Jeppesen and coauthors, Essential fatty acid deficiency in patients with severe fat malabsorption, *American Journal of Clinical Nutrition* 65 (1997): 837–843.

12 P. Lehocky and M. G. Sarr, Early enteral feeding in severe acute pancreatitis: Can it prevent secondary pancreatic (super) infection? *Digestive Surgery* 17 (2000): 571–577; R. Andersson and X. D. Wang, Gut barrier dysfunction in experimental acute pancreatitis, *Annals of the Academy of Medicine, Singapore* 28 (1999): 141–146.

13 C. E. Beck and coauthors, Improvement in the nutritional and pulmonary profiles of cystic fibrosis patients undergoing bilateral sequential lung and heart-lung transplantation, *Nutrition in Clinical Practice* 12 (1997): 45–53.

14 L. G. Tolstoi and C. L. Smith, Human Genome Project and cystic fibrosis—A symbiotic relationship, *Journal of the American Dietetic Association* 99 (1999): 1421–1427.

15 A.S.P.E.N. Board of Directors, Practice guidelines: Cystic fibrosis, *Journal of Parenteral and Enteral Nutrition* (supplement) 17 (1993): 44.

16 S. B. Hanauer, Update on inflammatory bowel disease therapy, *Digestive Disease Week 2000*, May 23, 2000, www.medscape.com/medscape/cno/2000/DDW/Story.cfm?story_id=1312, site visited January 3, 2001.

17 S. Fukuchi and coauthors, Nutrition support of patients with enterocutaneous fistulas, *Nutrition in Clinical Practice* 13 (1998): 59–65.

18 B. J. Geerling and coauthors, Comprehensive nutritional status in recently diagnosed patients with inflammatory bowel disease compared with population controls, *European Journal of Clinical Nutrition* 54 (2000): 514–521; E. Levy, Altered lipid profile, lipoprotein composition, and oxidant and antioxidant status in pediatric Crohn's disease, *American Journal of Clinical Nutrition* 71 (2000): 807–815.

19 A. Belluzzi and coauthors, Effect of an enteric-coated fish-oil preparation on relapses in Crohn's disease, *New England Journal of Medicine* 334 (1996): 1557–1560; Y. Kim, Can fish oil supplements maintain Crohn's disease in remission? *Nutrition Reviews* 54 (1996): 248–257.

20 R. P. Anderson and coauthors, In vivo antigen challenge in celiac disease identifies a single transglutaminase-modified peptide as the dominant A-gliadin T-cell epitope, *Nature Medicine* 6 (2000): 337–342; T. Not and coauthors, Celiac disease risk in the USA: High prevalence of antiendomysium antibodies in healthy blood donors, *Scandinavian Journal of Gastroenterology* 33 (1998): 494–498.

21 J. R. Saltzman and B. D. Clifford, Identification of the triggers of celiac sprue, *Nutrition Reviews* 52 (1994): 317–319.

22 S. Mora and coauthors, Reversal of low bone density with a gluten-free diet in children and adolescents with celiac disease, *American Journal of Clinical Nutrition* 67 (1998): 477–481; G. R. Corazza and coauthors, Propeptide of type I procollagen is predicative of posttreatment bone mass gain in adult celiac disease, *Gastroenterology* 113 (1997): 67–71.

23 J. L. Shaker and coauthors, Hypocalcemia and skeletal disease as presenting features of celiac diseases, *Archives of Internal Medicine* 157 (1997): 1013–1016.

24 Mora and coauthors, 1998.

25 Corazza and coauthors, 1997.

26 T. Thompson, Questionable foods and the gluten-free diet: Survey of current recommendations, *Journal of the American Dietetic Association* 100 (2000): 463–465.

27 M. T. Bardella and coauthors, Body composition and dietary intakes in adult celiac disease patients consuming a strict gluten-free diet, *American Journal of Clinical Nutrition* 72 (2001): 937–939.

28 T. Thompson, Folate, iron, and dietary fiber contents of the gluten-free diet, *Journal of the American Dietetic Association* 100 (2000): 1389–1396; T. Thompson, Thiamin, riboflavin, and niacin contents of the gluten-free diet: Is there a cause for concern? *Journal of the American Dietetic Association* 99 (1999): 858–862.

29 M. D. Johnson, Management of short bowel syndrome—A review, *Support Line*, December 2000, pp. 11–13, 16–23.

30 Johnson, 2000.

31 T. A. Byrne and coauthors, A new treatment option for patients with short-bowel syndrome: Bowel rehabilitation with growth hormone, glutamine, and a modified diet, *Support Line*, February 1996, pp. 1–7.

32 T. C. Lykins and J. Stockwell, Comprehensive modified diet simplifies nutrition management of adults with short-bowel syndrome, *Journal of the American Dietetic Association* 98 (1998): 309–315.

33 W. H. Aldoori and coauthors, A prospective study of diet and risk of symptomatic diverticular disease in men, *American Journal of Clinical Nutrition* 60 (1994): 757–764.

34 National Institute of Diabetes & Digestive & Kidney Diseases, Diverticulosis and diverticulitis, November 1998, www.niddk.nih.gov/health/digest/pubs/divert/divert.htm, site visited on January 5, 2001.

ANSWERS

Multiple Choice

1. c	2. a	3. b	4. d	5. b
6. c	7. b	8. c	9. d	10. a

PROMOTING INTESTINAL ADAPTATION

AS CHAPTER 21 described, people who must undergo extensive resections of the small intestine can have grievous problems meeting their nutrient needs. Imagine for a moment how difficult survival might be without intravenous feeding to provide nutrients to support recovery. Even a person who might eventually be able to meet nutrient needs enterally might not survive the process of adaptation. Without parenteral nutrition, the medical team could only provide simple intravenous solutions of fluids, electrolytes, and minimal energy and hope that the person would be able to eat before malnutrition was irreversible. Indeed, the ability to survive a massive intestinal resection was spurred largely by the development and improvement of techniques for providing all needed nutrients by vein and the adaptation of these techniques for use at home.

As experience with long-term parenteral nutrition grew, however, it became clear that this lifesaving technique carried with it a risk of serious complications (Chapter 23 provides more details). In addition, the costs of home parenteral nutrition programs are extremely high and can exceed $400 per day or over $145,000 per year.[1] Consequently, clinicians began to search for ways to foster intestinal adaptation to the fullest extent possible so that clients might be able to maintain their nutrition status with an enteral diet or at least reduce their dependence on parenteral nutrition. This highlight focuses on research and techniques aimed at promoting optimal adaptation in the remaining small bowel, or bowel rehabilitation. Controversies remain, of course, partly because much of the research has been conducted on animals and partly because studies in human beings require further elucidation and confirmation.

The need for intravenous feedings after extensive intestinal resections may be eliminated or reduced by taking full advantage of the remaining intestinal cells' ability to adapt.

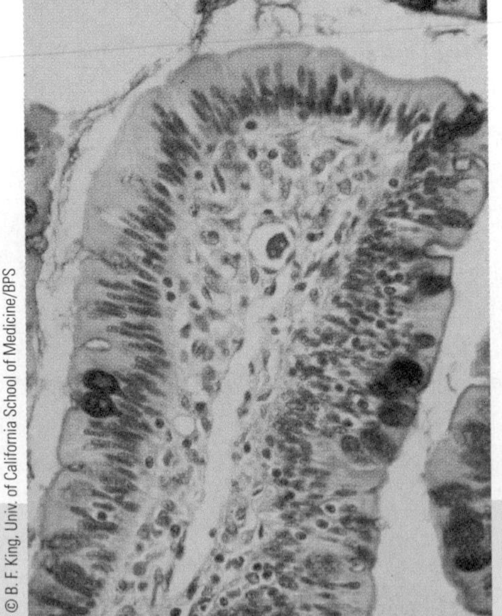

© B. F. King, Univ. of California School of Medicine/BPS

DIETARY AND GROWTH FACTORS AND INTESTINAL ADAPTATION

For some time, clinicians have accepted that intestinal adaptation can occur only when the GI tract is stimulated by enteral nutrients.[2] Providing enteral nutrients as early as possible after an intestinal resection improves the chances that the person will be able to tolerate enteral nutrition and will not be dependent on parenteral nutrition. In addition, researchers and clinicians have begun to identify specific factors, including dietary components, that might optimally stimulate adaptation.

Glutamine

The amino acid glutamine is common in food and is the most abundant amino acid in the blood. Glutamine provides fuel for rapidly dividing cells and is the major fuel for the intestinal cells and immune system cells. After the intestinal cells have metabolized glutamine, the liver uses the end products, alanine and ammonia, to make glucose and urea, respectively. Glutamine is also important for the replication of all body cells, which use it to synthesize purines, pyrimidines, and nucleotides, as well as other amino acids.

In healthy individuals, glutamine is a nonessential amino acid. If food sources fail to meet the body's needs, the body can synthesize more glutamine from the branched-chain amino acids of skeletal muscle. Following severe stresses, including intestinal resection, however, the body may need more glutamine than it receives from typical nutrient sources or than it can synthesize. Under such conditions, glutamine becomes a conditionally essential amino acid.

Various animal studies show that adding glutamine to intravenous feeding solutions helps to promote adaptation following intestinal resections.[3] Human studies suggest that oral glutamine raises growth hormone levels, which may further promote adaptation, as described in a later section.

Short-Chain Fatty Acids

As Chapter 21 noted, bacteria in the colon metabolize soluble fibers and undigested dietary fibers to short-chain fatty acids. Short-chain fatty acids possess several characteristics that enhance their importance following intestinal resections. First, they stimulate intestinal cell growth, enhance intestinal blood flow, bolster secretion of pancreatic enzymes, and promote sodium and water absorption in the colon. In animals, short-chain fatty acids stimulate intestinal cell growth following resections, even when they are provided intravenously.

Short-chain fatty acids can also provide the body with usable energy. Although they may not be a significant source of kcalories when malabsorption is not present, with malabsorption there is more substrate for bacteria to metabolize, and bacteria become more efficient at metabolizing it.

In healthy people with full-length intestinal tracts, short-chain fatty acids normally provide about 5 to 10 percent of the total daily energy needs (about 100 to 200 kcalories per day). Following extensive small bowel resections in which the colon remains intact, however, adaptation occurs in the colon so that it assumes a greater role in providing energy. When the bacteria in the colon are confronted with more substrates in the form of dietary fiber, unabsorbed carbohydrate, and, to a lesser extent, unabsorbed protein, they step up production of short-chain fatty acids and can provide up to 1000 kcalories per day of usable energy.[4] For people with malabsorption who have intact colons, the remarkable ability of colonic bacteria to salvage energy following extensive small bowel resections provides a source of much-needed energy.

Thus, for people whose colons remain intact following intestinal resections, providing a diet high in complex carbohydrates supports adaptation and improves energy conservation. Because the colonic bacteria cannot convert fat to short-chain fatty acids, unabsorbed fat reaching the colon provides no additional energy. When bile salts are malabsorbed (most likely when the ileum is resected), the unabsorbed bile salts and undigested fatty acids contribute further to diarrhea and malabsorption.

People without intact colons appear to absorb fats better than carbohydrates. For these people, high-carbohydrate diets may increase diarrhea.

Growth Factors

Researchers are currently exploring other factors that might foster adaptation. Growth hormone and insulin-like growth factor-1 (IGF-1) are two such factors. As its name implies, growth hormone stimulates the growth of tissues. When growth hormone reaches its target tissues and attaches to receptors, the tissues produce IGF-1. The cells of the intestinal tract are among the tissues that contain receptors for growth hormone. Serum IGF-1 levels fall with malnutrition and rise with the reintroduction of nutrients.[5] Mineral deficiencies, especially zinc and magnesium deficiencies, may also have a negative effect on growth hormone and IGF-1 levels. When either growth hormone or IGF-1 is provided, nutrient uptake and utilization by intestinal cells, skeletal muscle, and other organs increase. Animal studies suggest that administering either growth hormone or IGF-1 raises serum IGF-1 levels, but if growth hormone alone is administered, the primary effects are on skeletal muscle mass. IGF-1 administration alone appears to directly stimulate intestinal cell growth.[6]

CURRENT RESEARCH

Following extensive small bowel resections, clinicians strive to promote maximum absorptive capacity in the remaining bowel. The most desirable outcome is for the person to be able to meet all nutrient needs orally. When that is not possible, the goal is to minimize the need for parenteral nutrition, thus reducing the complications associated with it.

Preliminary clinical studies designed to assess the effectiveness of a combination therapy in promoting bowel rehabilitation yielded impressive results. The therapy consisted of growth hormone administration, glutamine supplementation, and a diet high in complex carbohydrates, low in fat, and supplemented with fiber. Eight clients with severe short-bowel syndrome comprised the first study group; all were believed to be dependent on parenteral nutrition for life and past the period of intestinal adaptation. They were able to consume oral foods as tolerated, but were unable to maintain nutrition status or adequate fluid balance without parenteral nutrition. After only three weeks of treatment, the clients' energy intake and abilities to absorb protein, carbohydrates, water, and sodium showed significant improvements.

The positive study results prompted a subsequent study of 47 clients to determine if the combination therapy could eliminate or reduce parenteral nutrition requirements.[7] Again, all clients were considered to be dependent on parenteral nutrition and past the period of intestinal adaptation. At the time of the study, 39 clients were consuming oral diets and receiving parenteral nutrition. Eight clients were receiving oral diets only—this was because parenteral nutrition had to be discontinued for medical reasons. All clients received the combination therapy for a minimum of 26 days, after which time growth hormone was discontinued. Clients were discharged from the research facility and instructed to continue oral glutamine supplements and the modified diet. After about a year, 40 percent of the clients were able to maintain nutrition status with an oral diet alone, and another 40 percent were able to reduce their parenteral nutrition requirements.

Two additional studies of eight clients with short-bowel syndrome failed to confirm the positive results.[8] In these studies, all clients ate a high-carbohydrate, low-fat diet. Four of the eight clients received growth hormone and glutamine, while the other four received placebos. After three weeks, the groups reversed—those who had received the study treatment were switched to placebo, and those who had received the placebo were switched to the study treatment for the same length of time. In the first study, intestinal transit time slowed in the treatment group, and those in this group with intact colons (two clients) showed modest improvements in protein absorption; however, the absorption of nutrients, amount of stool produced, and changes in the structure of intestinal cells were not statistically different in the two groups. The second study used the same subjects and design, but looked at body composition as well. In this study, the treatment group experienced temporary increases in body weight and lean body mass, but body weight returned to prestudy levels when the treatment was discontinued. Nutrient and fluid absorption was not affected by the treatment.

A variety of factors may have affected the outcomes of these studies. In the first two studies, for example, clients continued to take medications to help reduce diarrhea. In the last two studies, clients were taken off these medications during the study period. Other possible factors, to name a few, might include the clients' nutrition status prior to the study (nutrition status can affect a person's response to growth hormone) and how many clients in each study had intact colons. Perhaps some clients benefit from the treatment strategies more than others, and the response may depend, in part, on the medical condition that necessitated an intestinal resection.

Aside from these possible discrepancies, the studies raise additional questions for future research to address. Would even better results occur if combination therapy is provided soon after surgery, before adaptation has occurred? Might an individual component of the combination therapy prove just as effective as all three in minimizing or reducing the need for parenteral nutrition? Might IGF-1 administration have benefits not demonstrated with growth hormone administration? Might other nutrients also have positive effects?

Medical research is engaged in a continuously evolving quest to lengthen life and improve its quality. Often research in one area overlaps with research in another. For example, severe stresses cause atrophy of the intestinal cells and significantly reduce their absorptive capacity. Factors that stimulate intestinal cell growth and adaptation for short-bowel syndrome might also help correct the intestinal atrophy that accompanies stress and the administration of parenteral nutrition. Indeed, much current research examines the roles of glutamine and growth factors in severe medical stresses (Chapter 24 provides more information).

References

1 P. Reddy and M. Malone, Cost and outcome analysis of home parenteral and enteral nutrition, *Journal of Parenteral and Enteral Nutrition* 22 (1998): 302–310.

2 T. R. Ziegler and coauthors, Interactions between nutrients and peptide growth factors in intestinal growth, repair, and function, *Journal of Parenteral and Enteral Nutrition* (supplement) 23 (1999): 174–183.

3 J. Schroder and coauthors, Glutamine dipeptide-supplemented parenteral nutrition reverses gut atrophy, disaccharidase enzyme activity, and absorption in rats, *Journal of Parenteral and Enteral Nutrition* 19 (1995): 502–506; H. Tamada and coauthors, Alanyl glutamine-enriched total parenteral nutrition restores intestinal adaptation after either proximal or distal massive resection in rats, *Journal of Parenteral and Enteral Nutrition* 17 (1993): 236–242; M. C. Gouttebel and coauthors, Influence of N-acetylglutamine or glutamine infusion on plasma amino acid concentrations during the early phase of small-bowel adaptation in the dog, *Journal of Parenteral and Enteral Nutrition* 16 (1992): 117–121.

4 I. Norgaard, B. S. Hansen, and P. B. Mortensen, Importance of colonic support for energy absorption as small-bowel failure proceeds, *American Journal of Clinical Nutrition* 64 (1996): 222–231.

5 Ziegler and coauthors, 1999.

6 D. M. Ney, Effects of insulin-like growth factor-I and growth hormone in models of parenteral nutrition, *Journal of Parenteral and Enteral Nutrition* (supplement) 23 (1999): 184–189.

7 T. A. Byrne and coauthors, A new treatment for patients with short bowel syndrome: Growth hormone, glutamine, and a modified diet, *Annals of Surgery* 222 (1995): 243–255.

8 J. S. Scolapio, Effect of growth hormone, glutamine, and diet on body composition in short bowel syndrome: A randomized, controlled study, *Journal of Parenteral and Enteral Nutrition* 23 (1999): 309–313; J. S. Scolapio and coauthors, Effect of growth hormone, glutamine, and diet on adaptation in short-bowel syndrome: A randomized, controlled study, *Gastroenterology* 113 (1997): 1074–1081.

CHAPTER 22 Enteral Formulas

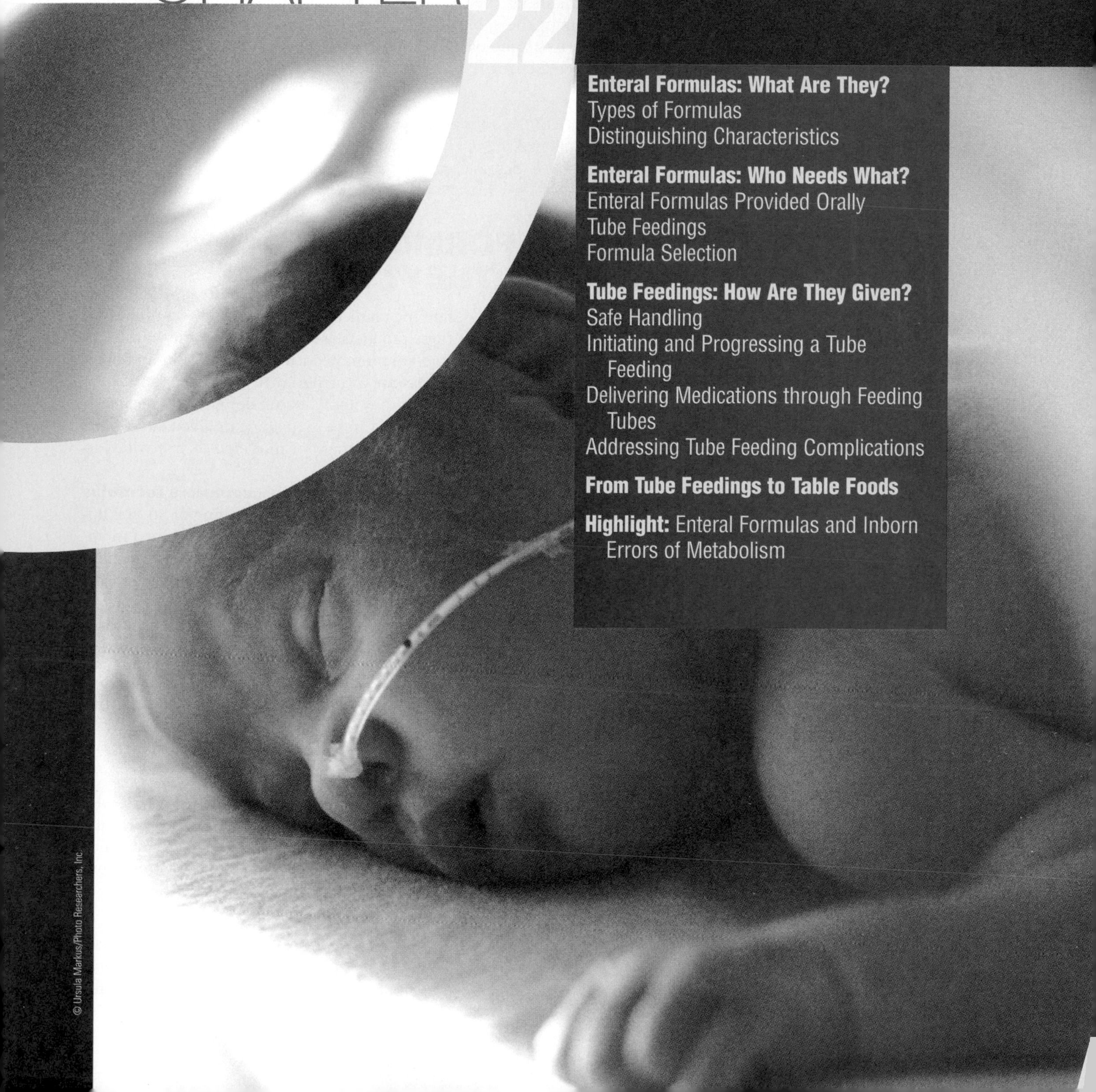

© Ursula Markus/Photo Researchers, Inc

To meet nutrient needs with conventional foods, a person must be able to chew and swallow and digest and absorb nutrients in the amounts necessary to satisfy metabolic demands. As Chapters 20 and 21 have shown, however, illnesses may interfere with eating, digestion, and absorption to such a degree that conventional foods fail to deliver necessary nutrients. If a poor appetite is the primary nutrition problem, **enteral formulas** given orally can help clients meet nutrient needs, if the clients can drink them in sufficient amounts.

For clients who cannot eat enough food or drink enough formula to meet nutrient needs, however, it may be necessary to deliver nutrients by tube or by vein. Formulas provided either orally or by tube are enteral feedings, the subject of this chapter. Enteral feedings are possible whenever a client can digest and absorb nutrients via the GI tract. Otherwise, nutrients are given by vein as parenteral feedings, the subject of the next chapter. (Figure 23-1 in the next chapter summarizes some of the factors involved in deciding the most appropriate way to feed a client.)

THINK NUTRITION whenever clients cannot meet their nutrient needs using conventional foods. Such clients may benefit from enteral formulas provided orally, whenever possible, or by tube when necessary.

ENTERAL FORMULAS: WHAT ARE THEY?

The number of enteral formulas on the market is staggering (some examples are listed in Appendix J). Some of these products are available over-the-counter in pharmacies and grocery stores and are promoted through television and magazine advertisements. Most formulas are available in ready-to-use form or in powdered form that must be mixed with water. They are designed to meet a variety of medical and nutrition needs and can be used alone or given along with other foods. Thus a formula is simply a standard or modified diet provided in liquid form.

Whenever formula is the primary source of nutrients, **complete formulas** are necessary. Such is the case when a client is on a tube feeding or an oral liquid diet for more than a few days. Complete formulas, when consumed in appropriate amounts, supply all the nutrients a client needs. Complete formulas can also be (and often are) used in smaller quantities to supplement table foods.

Types of Formulas

Formulas are classified in many ways, but for purposes of this book, it is reasonable to think of two major kinds categorized by the type of protein they supply. Standard formulas contain complete proteins, whereas hydrolyzed formulas contain small fragments of proteins, which may include free amino acids, dipeptides, and tripeptides.

• **Standard Formulas** • **Standard formulas** are appropriate for people who are able to digest and absorb nutrients without difficulty. Standard formulas contain whole proteins or one or a combination of **protein isolates** (purified proteins). A few formulas, called *blenderized formulas,* derive their protein primarily from pureed meat (a mixture of whole proteins).

• **Hydrolyzed Formulas** • To simplify the body's digestive work, a complete protein can be hydrolyzed—that is, partially broken down to yield small peptides. Alternatively, a formula can be made from free amino acids. In this text, we call both types *hydrolyzed* for simplicity. **Hydrolyzed formulas** are often

enteral formulas: liquid diets designed to be delivered through the GI tract, either orally or by tube.

complete formulas: liquid diets designed to supply all needed nutrients when consumed in sufficient volume.

standard formulas: liquid diets that contain complete molecules of protein; also called **intact** or **polymeric formulas.**

protein isolates: proteins that have been separated from a food. Examples include casein from milk and albumin from egg.

hydrolyzed formulas: liquid diets that contain broken-down molecules of protein, such as amino acids and short peptide chains; also called **monomeric formulas.**

very low in fat or provide some fat from medium-chain triglycerides (MCT) to ease digestion and absorption. People who cannot digest nutrients well may benefit from hydrolyzed formulas.

• Modular Formulas • Unlike complete formulas, a few formulas, called **modules,** provide essentially a single nutrient (protein, carbohydrate, or fat). In addition to commercial modular formulas, intravenous nutrients◆ and even table foods (vegetable oil or corn syrup, for example) can serve as modules; modules can then be added to enteral formulas to alter nutrient composition (for example, to add kcalories or protein). Modules can also be combined with other modules and liquid vitamin and mineral preparations to construct individualized formulas for clients with unique nutrient needs. Designing, preparing, and delivering such a formula is a challenge that requires in-depth nutrition knowledge and skills.

Distinguishing Characteristics

Formulas differ not only in the form of protein they contain but also in the amount of energy they provide and the source of energy nutrients. Although formulas differ, most fit into general categories and within each category can often be used interchangeably. For example, standard formulas provide similar amounts of energy and nutrients and require the same degree of digestive function. They may derive their nutrients from different sources, but the sources are all of similar quality. Thus the physician or dietitian has many choices in selecting a formula for an individual client. Table 22-1 lists examples of protein, carbohydrate, and fat sources in formulas.

• Nutrient Density • Standard formulas provide about 1.0 kcalorie per milliliter.◆ Nutrient-dense formulas provide 1.2 to 2.0 kcalories per milliliter and meet energy and nutrient needs in a smaller volume. Thus nutrient-dense formulas often benefit clients with high nutrient needs or those with fluid restrictions. Formulas also vary in the percentage of energy from protein, fat, and carbohydrate.

• Residue and Fiber • The positive health effects of dietary fibers suggest that fiber-enriched formulas would be the best choice for most people. Why, then, do many standard formulas have a low-to-moderate residue content?◆ The answer is that formulas are most often used for relatively short periods of time, and low-to-moderate residue formulas are least likely to cause gas and abdominal distention and thus are often well tolerated by many who need them: people with GI tract disorders (inflammatory bowel diseases or partial obstructions, for example), those who have undergone surgeries of the GI tract, or those beginning enteral nutrition after periods of GI tract disuse.

Many fiber-enriched formulas are available for people who need them, however. People who depend on tube feedings for long periods of time, those with constipation, and some people with short-bowel syndrome (see Chapter 21) are more likely than others to benefit from fiber-enriched formulas. Soluble fibers may be effective in controlling diarrhea in some people receiving enteral formulas.[1]

• Osmolality • Osmolality is a measure of the concentration of molecular and ionic particles in a solution. A formula that approximates the osmolality of blood serum (about 300 milliosmoles per kilogram) is referred to as an **isotonic formula.** A **hypertonic formula** has a higher osmolality than serum.

Most people tolerate both isotonic and hypertonic formulas without difficulty. When hypertonic formulas are delivered directly into the intestine, however, the hyperosmolar load can result in diarrhea, a situation analogous to the dumping syndrome (see Chapter 20). To prevent this problem, hypertonic formulas delivered into the intestine are initially delivered at a slow, even rate and gradually increased as tolerated.

TABLE 22-1 Examples of Protein, Carbohydrate, and Fat Sources in Enteral Formulas

Protein sources
Casein
Egg white
Free amino acids
Hydrolyzed casein, whey, or soy protein
Soy
Whey
Carbohydrate sources
Corn starch
Corn syrup
Fructose
Guar gum
Maltodextrin
Sucrose
Fat sources
Canola oil
Fish oil
MCT oil
Safflower oil
Soybean oil
Sunflower oil

◆ **Caution:** Although intravenous nutrients can be used enterally, the reverse is not true. Enteral formulas cannot be delivered by vein without serious consequences.

◆ For practical purposes, 1 ml (milliliter) is equivalent to 1 cc (cubic centimeter).

◆ Since hydrolyzed formulas are almost completely absorbed, they leave little residue in the intestine.

modules: formulas or foods that provide primarily a single nutrient and are designed to be added to other formulas or foods to alter nutrient composition. Modules can also be combined to create a highly individualized formula.

osmolality (OZ-moh-LAL-eh-tee): a measure of the concentration of particles in a solution, expressed as the number of milliosmoles (mOsm) per kilogram.

isotonic formula: a formula with an osmolality similar to that of blood serum (300 mOsm/kg).
- **iso** = the same
- **ton** = tension

hypertonic formula: a formula with an osmolality greater than that of blood serum.
- **hyper** = greater, more

• **Costs** • Costs of individual products vary greatly. As a general rule, however, hydrolyzed formulas and products formulated for specific disorders (renal or respiratory failure, for example) are more costly than standard formulas.

IN SUMMARY

Enteral formulas are liquid mixtures of nutrients compounded to meet a variety of medical and nutrition needs. Standard formulas meet the nutrient needs of people who can digest and absorb nutrients without difficulty, whereas hydrolyzed formulas meet the nutrient needs of people who may have some difficulty digesting and absorbing complete proteins and fats. Formulas differ in the amount of energy they deliver, sources of energy nutrients, and percentages of energy nutrients. Some contain fiber, while others are designed to provide minimal residue. A formula's osmolality can also vary. In some cases, hypertonic formulas may cause diarrhea, which can often be prevented by delivering the formula at a slow, even rate. Like table foods, enteral formulas can meet a variety of dietary needs. With that in mind, it is important to identify which people benefit from such formulas.

ENTERAL FORMULAS: WHO NEEDS WHAT?

All people with functional GI tracts who cannot get the nutrients they need from table foods can potentially benefit from enteral formulas. They could also get nutrients from parenteral nutrition, but enteral nutrition is preferable whenever it is possible. Compared with parenteral nutrition, enteral nutrition helps maintain normal gut function better, causes fewer complications, and is less costly. Enteral feedings help stimulate intestinal adaptation following intestinal resections and long periods of GI tract disuse.

Similarly, oral feedings are preferred to tube feedings whenever a person can drink the formula and drink enough of it. In so doing, the client avoids the stress of the procedure to insert a feeding tube, and nurses save valuable time. Orally provided formulas are also less costly than feeding by tube and less likely to result in complications.

Enteral Formulas Provided Orally

For people who can tolerate only liquids for long periods and those who need hydrolyzed formulas, formulas provided orally can meet all nutrient needs in ways that foods cannot. If people can drink enough of the formula, they can avoid being fed by tube.

More often, formulas provided orally supplement a conventional diet. Some people can eat table foods, but not in the quantities they need to meet nutrient needs. Enteral formulas provide a reliable source of nutrients and work particularly well for adding energy and protein to the diet. Psychologically, liquids seem less filling than foods, and they are easier for debilitated, weak, or tired clients to handle. Highlight 22 describes how enteral formulas provided orally meet the special needs of people with inborn errors of metabolism.

Formula supplements often help meet nutrient needs when clients cannot eat enough conventional foods.

When a client uses an oral formula, taste becomes an important consideration. Allowing clients to sample different products and flavors and select the ones they like best helps promote acceptance.[2] The "How to" offers suggestions for helping clients accept oral formulas.

Tube Feedings

Tube feedings are simply complete formulas delivered through a tube into the stomach or intestine. An individual who has a functional GI tract but is unable to ingest enough nutrients (or the appropriate type) by mouth to meet nutrient needs may need a tube feeding. Candidates for tube feedings include:

- People with physical problems that seriously interfere with chewing and swallowing.
- People with no appetite for an extended time, especially if they are malnourished.
- People with a partial obstruction, some types of fistulas, or altered motility in the upper GI tract.
- People in a coma.
- People with high nutrient requirements.
- People who have undergone extensive intestinal resections and are just beginning enteral feedings.
- People who are unable to ingest a hydrolyzed formula orally.

Help Clients Accept Oral Formulas

People on enteral formulas are often quite ill and frequently have poor appetites. Even when a person enjoys a formula, palatability can become a problem after a while. Hydrolyzed formulas are often less palatable than standard formulas, and clients may find them difficult to accept. Caring professionals can help by using these suggestions:

- Let the client try both different flavors and different formulas appropriate for the client's needs; use those the client likes best.
- Serve formulas attractively and remind clients to drink them. Formulas offered in a glass are more appealing than those served from a can with an unfamiliar name. Some people find the smell of formulas unappealing. Covering the top of the glass with plastic wrap or a lid, leaving just enough room for a straw, can help.
- Provide easy access. Keep the formula close to the client's bed where little effort is required to reach it, and within sight to remind the client to drink it. Clients who are very ill may lack the motivation even to reach for formula, let alone drink it. In such cases, offer the formula ready-to-drink and in small amounts frequently through the day.
- Keep formula in an ice bath so that it will be cool and refreshing when the client drinks it.
- If the client stops drinking the formula after a while, recommend different flavors or another formula to help relieve boredom.

• Feeding Tube Placement • Feeding tubes are inserted into different locations along the GI tract depending on the client's medical problems and the estimated length of time that the feeding will be required. Figure 22-1 shows feeding tube placement sites, and the glossary on p. 686 describes these sites.

When clients are expected to be on tube feedings for less than about four weeks, feeding tubes are frequently inserted through the nose and passed into the stomach or intestine (**transnasal** placement). The client often remains fully alert during the procedure and helps pass the tube by swallowing. For an infant, the feeding tube may be inserted through the mouth and to the stomach (**orogastric** placement) before each feeding and removed immediately after the feeding to allow the infant to breathe easily and reduce the risk of aspiration.

When a client will be on a tube feeding for a longer period, or when a feeding tube cannot be passed through the nose, esophagus, or stomach due to an

FIGURE 22-1 Feeding Tube Placement Sites

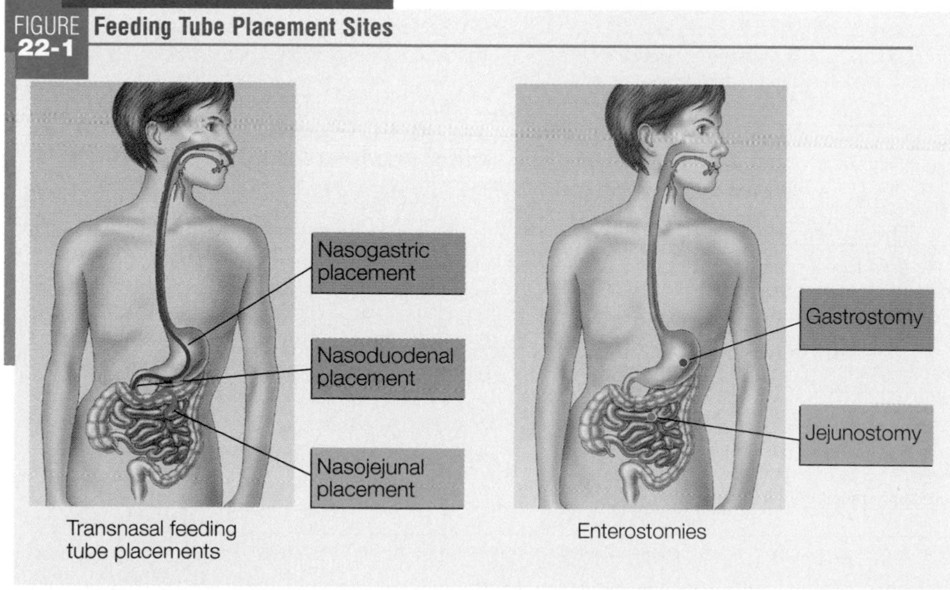

GLOSSARY OF TUBE FEEDING PLACEMENT SITES

These terms are listed in order from the nose to lower organs of the digestive system.

transnasal: through the nose. A **transnasal feeding tube** is one that is inserted through the nose.
- **naso** = nose

nasogastric (NG): from the nose to the stomach.

nasoenteric: from the nose to the stomach or intestine. *Nasoenteric feedings* include nasogastric, nasoduodenal, and nasojejunal feedings. Most clinicians use nasoenteric to refer to nasoduodenal and nasojejunal feedings only.

nasoduodenal (ND): from the nose to the duodenum.

nasojejunal (NJ): from the nose to the jejunum.

orogastric: from the mouth to the stomach. This method is often used to feed infants because they breathe through their noses, and a nasogastric tube can hinder the infant's breathing.

enterostomy (EN-ter-OSS-toe-mee): an opening into the stomach or jejunum through which a feeding tube can be passed.

gastrostomy (gas-TROSS-toe-mee): an opening in the stomach made surgically or under local anesthesia through which a feeding tube can be passed. The technique for creating a gastrostomy under local anesthesia is called **percutaneous endoscopic gastrostomy,** or **PEG** for short. When the feeding tube is guided from such an opening into the jejunum, the procedure is called **percutaneous endoscopic jejunostomy (PEJ),** a misnomer because the enterostomy is in the stomach rather than the jejunum.

jejunostomy (JEE-ju-NOSS-toe-mee): an opening in the jejunum made surgically or under local anesthesia through which a feeding tube can be passed. The technique for creating a jejunostomy under local anesthesia is called a **direct endoscopic jejunostomy (DEJ).** Note: Some clinicians also refer to this procedure as a PEJ, which is a more accurate use of the term than the more common use described above.

obstruction or for other medical reasons, an opening can be made into the stomach or jejunum. A tube **enterostomy** can be made either surgically or nonsurgically using local anesthesia. Table 22-2 compares some of the features of various tube feeding sites. The "How to" suggests ways to reduce anxiety for clients beginning a tube feeding.

• Gastric Feedings • When formulas are delivered into the stomach, either through a **nasogastric** tube or a **gastrostomy,** the digestive process begins in the stomach, just as it does with an oral diet. The stomach empties its contents at a controlled rate and delivers small volumes of nutrients into the intestine. Thus gastric feedings are often preferred whenever they are possible. Gastric feedings are not possible for people with gastric obstructions or condi-

TABLE 22-2 Comparison of Feeding Tube Sites[a]

Insertion Method and Feeding Site	Advantages	Disadvantages
Transnasal	Does not require surgery or incisions for placement.	Easy to remove by disoriented clients; long-term use may irritate the nasal passages, throat, and esophagus.
Nasogastric	Easiest to insert and confirm placement; feedings can often be given intermittently and without an infusion pump.	Highest risk of aspiration in compromised clients.
Nasoduodenal and nasojejunal	Lower risk of aspiration in compromised clients; allow for enteral nutrition earlier than gastric feedings following severe stress; may allow for enteral feeding when partial obstructions, fistulas, or other medical conditions prevent gastric feeding.	More difficult to insert and confirm placement; feedings require an infusion pump for administration; may take longer to reach nutrition goals.
Tube enterostomies	Allow lower esophageal sphincter to remain closed, reducing the risk of aspiration; more comfortable than transnasal insertion for long-term use; site is not visible under clothing.	May require general anesthesia for insertion; require incisions; greater risk of complications from the insertion procedure; greater risk of infection; may cause skin irritation around the insertion site.
Gastrostomy	Feedings can often be given intermittently and without a pump; easier to insert than a jejunostomy.	Moderate risk of aspiration in high-risk clients.
Jejunostomy	Lowest risk of aspiration; allows for enteral nutrition earlier following severe stress; may allow for enteral feeding when partial obstructions, fistulas, or medical conditions prevent gastric feeding.	Most difficult to insert; feedings require an infusion pump for administration; may take longer to reach nutrition goals.

[a]Relative to each tube feeding site. The actual advantages and disadvantages of different insertion procedures depend on the person's medical condition.

Help Clients Cope with Tube Feedings

The thought of being "force-fed" is frightening to many people. One person may envision a large feeding tube and fear that the procedure will be extremely painful. Another may have heard about tube feedings only from the popular press and associate them with irreversible comas. All clients benefit when they understand the insertion procedure, the expected duration of the tube feeding, and the strategic role that nutrition plays in recovery from disease. These pointers can help health care professionals prepare clients for transnasal tube feedings:

- Allow clients to see and touch the feeding tube. Seeing firsthand that the tube is soft and narrow (only about half the diameter of a pencil) often alleviates anxiety. Show clients how the feeding apparatus is attached to the feeding tube, and explain how the feeding will work. Use dolls or stuffed toys to demonstrate tube insertion and feeding procedures to young children.
- Explain that the client remains fully alert during the procedure and helps pass the tube by swallowing. A numbing solution sprayed on the back of the throat minimizes discomfort and prevents gagging during the procedure.
- Tell the client that once the tube has been inserted, most people become accustomed to its presence within a few hours. In most cases, the client can easily swallow foods and liquids with the tube in place. If permitted, favorite foods or beverages can still be enjoyed.
- Assure the client that the tube feeding will be temporary, if such assurance is appropriate.

Although a tube feeding may be frightening for some, for others, it is a relief. People who understand that they should eat, but cannot do so, may be relieved to receive sound nutrition without any effort. As they feel better and begin to eat again, the volume of the tube feeding can often be reduced and then discontinued when oral intake is adequate.

Some people feel a loss of control over their lives; others feel self-conscious about how the feeding tube looks or awkward about moving around with the equipment. A few simple measures can help:

- Involve older children, teens, and adults in the decision-making and care process whenever possible. Clients can help arrange daily feeding schedules, and some can also perform many of the feeding procedures themselves.
- Show clients how to manipulate the feeding equipment so that they can get out of bed and move around.
- Recommend that clients walk around and socialize with others, if permissible.
- Encourage clients to maintain contact with friends and keep busy with hobbies and activities they enjoy. This measure is especially important for children, teens, and those on long-term feedings.
- For infants and children, keep the developmental age of the child in mind and work with parents to ensure that appropriate feeding skills are mastered. For infants, providing a pacifier during feedings helps maintain the associations between sucking, swallowing, eating, and fullness. When possible, some of the tube feeding formula may be provided by bottle or by spoon to further develop skills.

The more complex the procedure, the easier it becomes for health care professionals to focus on the procedure and forget about the client's emotions. No matter how many technicalities you have to keep in mind, remember to stay focused on the person receiving your care.

tions that significantly interfere with the stomach's ability to empty. Following a severe stress, such as intestinal surgery, for example, GI motility may be temporarily disrupted. Activity resumes more quickly in the small intestine than in the stomach, however. Thus, after severe stress, the delivery of formulas into the small intestine◆ can be initiated earlier than gastric feedings.

◆ The final location of the feeding tube determines the type of feeding. If a feeding tube is passed through a gastrostomy into the duodenum or jejunum, the feeding is intestinal rather than gastric.

Gastric feedings may also be a problem for people at risk for aspiration. In these clients, formula may reflux from the stomach into the esophagus, and the client may aspirate the formula into the lungs. **Aspiration pneumonia,** a lung infection that can be fatal, may develop. To minimize the possibility of aspiration, clinicians may prefer a **nasoenteric** feeding for clients at risk. Alternatively, for a client with a very high risk of aspiration, some clinicians prefer a gastrostomy or **jejunostomy,** which allows the lower esophageal sphincter to remain tightly closed.

The major disadvantage of intestinal feedings is loss of the controlled emptying action of the stomach. Assuring passage of the feeding tube into the appropriate location is also more difficult, and placement of the tube must be confirmed before the feeding begins. Formulas must be carefully administered to avoid diarrhea and dehydration.

• Feeding Tubes • Feeding tubes are soft and flexible and come in a variety of diameters◆ and lengths. Many have special characteristics that make them desirable for specific purposes. For example, tubes with double lumens allow for gastric decompression and intestinal feedings at the same time.

◆ The outer diameter of a feeding tube is measured using the French scale, where each unit is about one-third of a millimeter. Thus the outer diameter of a 10 French feeding tube is a little over 3 mm. The inner diameter varies depending on the thickness of the material used to construct the tube.

aspiration pneumonia: an infection of the lungs caused by inhaling fluids regurgitated from the stomach.

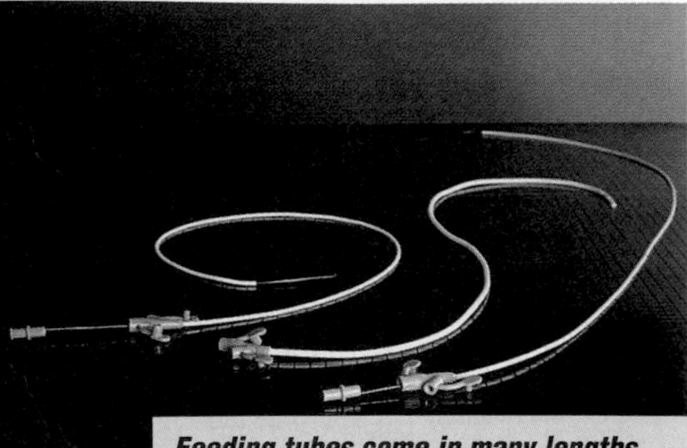

Feeding tubes come in many lengths and diameters. The thin wires protruding from the end of the feeding tubes are stylets, which stiffen the tube to ease insertion and are discarded thereafter. The Y connector (shown here in orange) provides a port for administering water or medications without disrupting the feeding.

Which feeding tube is appropriate depends on the client's age and size, medical condition, how the tube will be placed (transnasally or through an enterostomy), how far the final placement will be from the insertion site (from the nose to the stomach or intestine, for example), and its inner diameter. Once the appropriate length is selected, the smallest tube through which the formula will flow without clogging the tube is selected. Unclogging a tube is a difficult procedure that interrupts the feeding schedule and may require insertion of a new tube, causing stress and anxiety for the client and raising the cost of the feeding.

Formula Selection

To select an appropriate formula requires a logical approach. Figure 22-2 shows some considerations involved. In a nutshell, the formula that meets the client's medical and nutrient needs with the lowest risk of complications and at the lowest cost is the best choice. If no formula can be found that meets the client's needs, then modules can be used to create an appropriate formula.

Selecting a formula that meets the client's nutrient needs is paramount. Nutrient requirements are estimated based on a careful assessment of the client's age, nutrition status, medical condition, ability to digest and absorb nutrients, and metabolic rate. Standard formulas are appropriate for the vast majority of clients. However, the person with a functional, but impaired, GI tract may benefit from hydrolyzed formulas. Besides the client's ability to digest and absorb nutrients, other nutrition-related factors that affect the selection of formula include:

- The client's energy, protein, fluid, and nutrient needs. High nutrient needs must be met in the volume of formula the client can tolerate. If nutrients (including water) must be restricted, the selected formula should deliver the prescribed amount of nutrients in the volume prescribed.
- The need for residue or fiber modifications. The choice of formulas is narrowed when a person needs a low-residue or a high-fiber diet.
- Individual tolerances (food allergies and sensitivities). Most formulas are lactose-free because temporary and permanent lactose intolerances are common problems for people who need enteral formulas. Many formulas are also gluten-free and can accommodate the needs of people with celiac disease.
- The size of the inner lumen of the feeding tube. When a small-diameter tube has been placed, the formula selected must flow readily through the tube to prevent it from becoming obstructed.

In addition, health care facilities cannot stock all formulas, so formula selection is limited by availability. In the final analysis, the dietitian or physician can make only an educated guess in selecting the best formula for an individual. The health care team monitors each person's nutrition status and responses to the formula to help ensure that individual needs are being met.

IN SUMMARY

People who can digest and absorb foods but cannot get the nutrients they need from conventional foods can benefit from enteral formulas. Whenever possible, people are encouraged to drink the formula, but sometimes the formula must be delivered through feeding tubes inserted through the nose to the stomach or intestine, or through enterostomies made directly into the stomach or intestine. Transnasal insertions are often used for short-term feedings, provided that the client's medical condition permits the passage of the tube through the nose, esophagus, or stomach. Otherwise, enterostomies are appropriate. Selection of an appropriate feeding tube depends on the client's age and size, medical condition, how the tube will be placed, and how far the final placement of the tube will be from the insertion site. The health care team considers the client's medical and nutrient needs and selects a formula that meets those needs with the least risk of complications.

FIGURE 22-2 Selecting a Formula

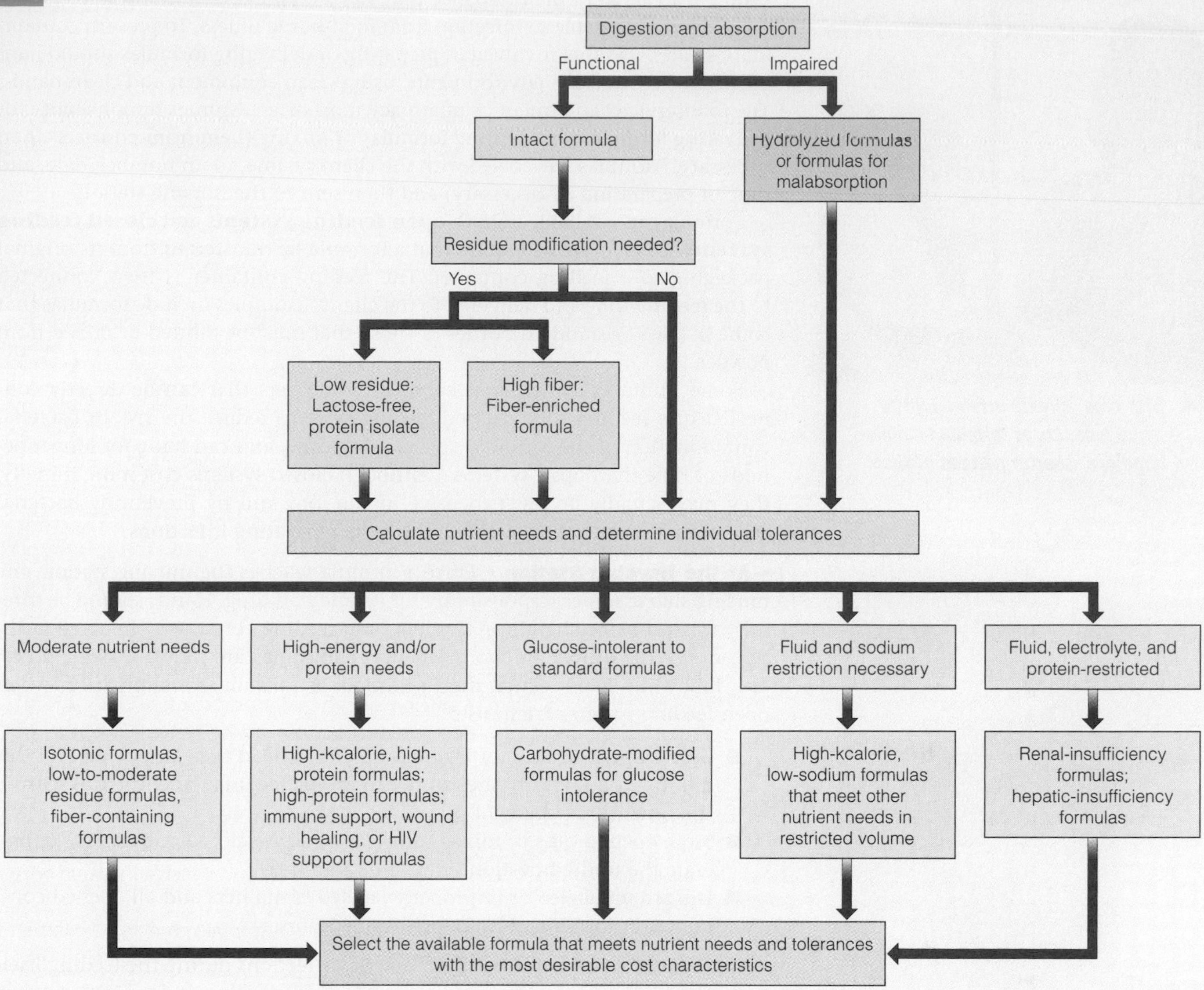

Digestion and absorption

Functional → Intact formula

Impaired → Hydrolyzed formulas or formulas for malabsorption

Intact formula → Residue modification needed?

Yes → Low residue: Lactose-free, protein isolate formula

Yes → High fiber: Fiber-enriched formula

No → Calculate nutrient needs and determine individual tolerances

Low residue / High fiber / Hydrolyzed formulas or formulas for malabsorption → Calculate nutrient needs and determine individual tolerances

Calculate nutrient needs and determine individual tolerances →

- Moderate nutrient needs → Isotonic formulas, low-to-moderate residue formulas, fiber-containing formulas
- High-energy and/or protein needs → High-kcalorie, high-protein formulas; high-protein formulas; immune support, wound healing, or HIV support formulas
- Glucose-intolerant to standard formulas → Carbohydrate-modified formulas for glucose intolerance
- Fluid and sodium restriction necessary → High-kcalorie, low-sodium formulas that meet other nutrient needs in restricted volume
- Fluid, electrolyte, and protein-restricted → Renal-insufficiency formulas; hepatic-insufficiency formulas

All → Select the available formula that meets nutrient needs and tolerances with the most desirable cost characteristics

TUBE FEEDINGS: HOW ARE THEY GIVEN?

Once a feeding route has been selected and the feeding tube inserted, attention turns to delivering the formula. Just as the actual foods provided on special diets vary from one health care facility to the next, the exact procedures for safely handling tube feedings and providing them to clients vary as well. The procedures presented in the following sections are suggested guidelines.

Many people beginning a tube feeding are seriously ill or malnourished. Using the safest methods of preparing and administering the formula helps minimize the risk of complications, which can delay or prevent the attainment of medical and nutrition goals. Of all health care professionals, nurses hold the greatest responsibility for ensuring that formulas are safely administered. Although the following sections specifically address the preparation and administration of tube feedings in health care institutions, the principles apply to people who receive tube feedings at home as well.

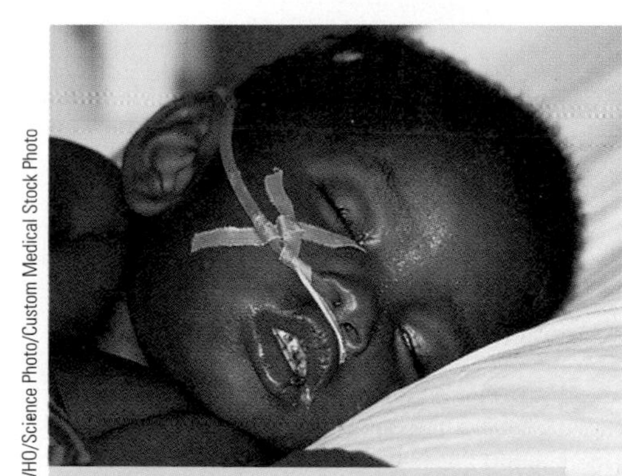

The safe delivery of a tube feeding helps clients meet their medical and nutrition goals with minimal discomfort.

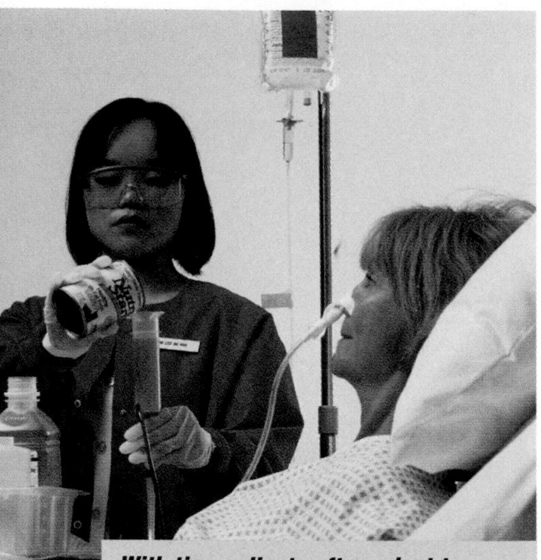

With time, clients often adapt to larger amounts of formula administered over shorter periods of time.

Safe Handling

People who are ill or malnourished may have suppressed immune systems that make them vulnerable to infection from food-borne illness. To prevent contamination, all personnel involved in preparing or delivering formulas should handle them only in clean environments, using clean equipment and clean hands. The foodservice department or pharmacy most often assumes responsibility for delivering formulas and diluting formulas or mixing them from powders when necessary. Formulas are labeled with the client's name, room number, date, and time of preparation (if necessary) and then sent to the nursing station.

Formulas are available in both **open feeding systems** and **closed feeding systems.** Open systems require that a formula be transferred from its original packaging to a feeding container. The feeding container is then connected to the feeding tube and delivered to the client. Examples include formulas that come in cans or standard bottles or those that must be diluted or mixed from powder.

Some formulas come prepackaged in containers that can be directly connected to a feeding tube. Closed feeding systems reduce the risk of bacterial contamination of the formula, save nursing time, and can hang for longer periods of time than open systems.[3] Although closed systems cost more initially, they may actually be less expensive in the long run by preventing bacterial contamination and thus avoiding the costs of treating infections.

• At the Nursing Station • Once a formula reaches the nursing station, the nursing staff assumes responsibility for its safe handling. Hands should be carefully washed before handling formulas and feeding containers. In some facilities, nonsterile gloves are worn whenever formulas are handled (see photo). The following steps reduce the likelihood of formula contamination when open feeding systems are in use:

- Before opening a can of formula, carefully clean the can opener and the lid. If you do not use the entire can at one feeding, label the can with the time it was opened.
- Store opened cans or mixed formulas in clean, closed containers. Refrigerate the unused portion of formula promptly.
- Discard unlabeled or improperly labeled containers and all opened containers of formula not used within 24 hours.

• At Bedside • Prevent the risk of bacterial infections during the feeding itself by following these procedures:

- Hang no more than an 8- to 12-hour supply of formula administered through an open system or no more than a 36- to 48-hour supply of formula administered through a closed system.
- For open systems, discard any formula that remains after 8 to 12 hours, rinse out the feeding bag and tubing, and add fresh formula to the feeding bag. Use a new feeding container and tubing (except the feeding tube itself) every 24 hours.

Initiating and Progressing a Tube Feeding

Two of the most serious complications associated with tube feedings are the inadvertent placement of a transnasal tube into the respiratory tract and the aspiration of formula from the stomach into the lungs. In both cases, potentially fatal complications can follow. To minimize the risk of such complications, most clinicians use X rays to verify the position of the feeding tube before a feeding is initiated, and nurses confirm the tube's position several times a day. Lying flat raises the risk of aspiration; therefore, the client's upper body is elevated to at least a 30- to 45-degree angle during the feeding and for 30 minutes

open feeding systems: enteral formula delivery systems that require formula to be transferred from its original packaging to a feeding container before it can be administered through a feeding tube.

closed feeding systems: enteral formula delivery systems in which the formula comes prepackaged in a container that is ready to be attached to a feeding tube for administration.

after the feeding whenever possible.[4] To assist in the early identification of aspiration, sterile blue food coloring is sometimes added to formulas so that the formula can easily be distinguished from pulmonary secretions.

• Formula Delivery Techniques • A day's volume of formula can be given in relatively large amounts at intervals or in relatively small amounts continuously throughout the day. A client may start on a continuous feeding and gradually be converted to an intermittent feeding. Each method has specific uses, advantages, and disadvantages.

• Intermittent Feedings • Intermittent feedings are best tolerated when they are delivered into the stomach (not intestine) and no more than 250 to 400 milliliters is given over 30 minutes or more using the gravity drip method or an infusion pump. The larger the volume of formula needed to meet nutrient needs, the more frequently feedings are delivered. (People who have very high nutrient needs benefit from either high-nutrient-density formulas or continuous feedings.) Rapid delivery (in 10 minutes or less) of a large volume of formula (300 to 400 milliliters)—called a **bolus feeding**—often leads to complaints of abdominal discomfort, nausea, fullness, and cramping, especially when a feeding is first initiated. This makes sense. After all, people do not gobble down a meal in just a few minutes, especially if they have not been eating for a while.

Intermittent feedings may be difficult for a client to tolerate, and because there is a greater volume of formula in the stomach at one time, the risk of aspiration is higher than with continuous feedings. Intermittent feedings work well for clients able to tolerate them, however. Often clients gradually adapt to larger volumes of formula given over shorter periods of time. Such feedings mimic the usual pattern of eating and allow the client freedom of movement between meals. They also require less time, making them less costly and easier for people to use at home.

• Continuous Feedings • Continuous feedings are delivered slowly and in constant amounts over a period of 8 to 24 hours. Such feedings benefit people who have received no food through the GI tract for a long time, those who are hypermetabolic, and those who are receiving intestinal feedings. Although continuous feedings are often well tolerated and are associated with a relatively low risk of aspiration, they require infusion pumps◆ to help ensure accurate and constant flow rates; the pump makes the feedings more costly to provide and limits the client's freedom of movement. The "How to" on p. 692 explains several ways to plan tube feeding schedules.

• Formula Volume and Strength • Recommended formula administration schedules vary between institutions; the ones provided here are examples. Almost all people can receive undiluted formula (either isotonic or hypertonic) at the start of a feeding. Formulas, especially hypertonic formulas, are given slowly at first, and the volume is gradually increased. In rare cases, formulas may need to be diluted initially, the strength increased gradually, and then the rate advanced.

Intermittent feedings may start at about 100 to 150 milliliters at each feeding and then be increased by 50 to 100 milliliters each day until the goal volume is reached.[5] Continuous feedings may start at about 25 to 50 milliliters per hour. If the person tolerates the formula, the rate of feeding can be increased by about 25 milliliters per hour every 4 to 12 hours,◆ depending on the location of the feeding tube (gastric or intestinal), the person's medical condition, and the nutrient density of the formula. For both intermittent and continuous feedings, if the new rate is not tolerated, back up and proceed more slowly, giving the person additional time to adapt. If the person on an intermittent feeding cannot tolerate the feeding, a continuous feeding may be a better choice than intermittent feeding.

◆ **Caution: The bright lights, beeps, and controls of an infusion pump may make it irresistible to young children. Keep the infusion pump at a safe distance from children to prevent them from toppling the pump or IV pole and possibly injuring themselves or damaging the equipment.**

◆ **As an example of a volume progression for a continuous tube feeding, start the feeding at 50 ml/hr at full strength and then progress as follows:**
- After 6 hours: 75ml/hr.
- After 12 hours: 100 ml/hr.
- After 18 hours: 125ml/hr.

intermittent feedings: delivery of about 250 to 400 ml of formula over 30 minutes or more.

bolus feeding: delivery of about 300 to 400 ml of formula over 10 minutes or less.

continuous feedings: slow delivery of formula in constant amounts over an 8- to 24-hour period.

HOW TO Plan a Tube Feeding Administration Schedule

After selecting a formula that meets the client's medical and nutrient needs, the planner, usually a dietitian, determines the volume of formula per day that meets those needs. Consider a client who needs 2000 milliliters of formula per day. If the client is to receive the formula intermittently six times a day, he needs about 330 milliliters of formula at each feeding (2000 ml ÷ 6 feedings = 333 ml/feeding). Alternatively, if he is to receive the same volume of formula eight times a day, then he needs 250 milliliters (or about one can of ready-to-feed formula) at each feeding (2000 ml ÷ 8 feedings = 250 ml/feeding). He will probably tolerate this volume of formula best if it is given to him over 30 minutes or more at each feeding. If the client is to receive the formula continuously over 24 hours, he needs about 85 milliliters of formula each hour (2000 ml ÷ 24 hr = 83 ml/hr).

• Supplemental Water • Formulas contain a considerable amount of water, and they often meet a substantial portion of clients' water needs.◆ Standard formulas contain about 850 milliliters of water per liter of formula. Nutrient-dense formulas provide less water. In addition to the formula itself, water provided through the feeding tube helps meet additional fluid needs and helps prevent clogged feeding tubes. Most adults require about 2000 milliliters (approximately 2 quarts) of water daily. For people with kidney, liver, and heart diseases, less water may be necessary. Clients with fever, excessive sweating, severe vomiting, diarrhea, fistula drainage, high-output ostomies, blood loss, or open wounds require additional water.

Attention to indicators of body water balance can help determine how much additional water an individual needs. In alert adults, thirst is a good indicator of water needs; a person complaining of thirst generally needs water. In the elderly, however, thirst may be slow to develop in response to dehydration. Health care professionals monitor clients' weight changes, record their intake and output, and measure urine specific gravity to evaluate hydration status. Table 17-11 on p. 583 listed physical and laboratory indices of dehydration.

◆ **The following formula is one of several used to estimate fluid requirements for adults and children:**
- Allow 100 ml/kg for each of the first 10 kg of body weight.
- Allow 50 ml/kg for each of the next 10 kg of body weight.
- Allow 20 ml/kg for each kilogram of body weight above 20 kg.

Water is often given by syringe or gravity bag at room (for gastric feedings) or body (for intestinal feedings) temperature.[6] Automatic flush pumps that deliver a selected volume of water each hour are also available.[7] For clients on continuous feedings, the feeding tube is generally flushed with water every four hours. For clients on intermittent feedings, the feeding tube is generally flushed with water after each feeding.

• Gastric Residuals • Nurses measure the **gastric residual** to ensure that the stomach is emptying properly and to prevent nausea, vomiting, and possible aspiration of formula into the lungs. The gastric residual is the volume of formula that remains in the stomach from a previous feeding. It is measured by gently withdrawing the gastric contents through the feeding tube using a syringe. The gastric residual is measured before each feeding for intermittent feedings and every four hours for continuous feedings. The gastric residual that is considered excessive varies among facilities, but ranges from about 100 to 200 milliliters. If the residual is excessive, the feeding is held for an hour or two, and then the residual is rechecked. If excessive residuals persist, the physician may withhold the feeding, reduce the rate of administration, or begin drug therapy to stimulate gastric emptying.

gastric residual: the volume of formula that remains in the stomach from a previous feeding.

Delivering Medications through Feeding Tubes

Clients receiving tube feedings are also likely to be receiving numerous medications. Often these medicines are delivered through feeding tubes, and in some cases, complications can occur.

Keep in mind that enteral formulas can interact with medications in the same ways that foods can. For example, the vitamin K in a formula can interact with warfarin (see Chapter 18) just as the vitamin K in foods can. The health care team must consider the effects of drug therapy on nutrient requirements, the effects of the formula on medication absorption, the effects of medications on physical properties of formulas, and the development and prevention of complications.

• Medication Forms • A medication may come in any of several forms including tablets, liquid, injectable, and intravenous. People on tube feedings have functional GI tracts, and oral medications are less costly and easier to deliver than injectable or intravenous forms. Thus clinicians often prefer to use oral medications for clients on tube feedings. The following guidelines may be helpful in preventing medication-medication interactions, medication-formula interactions, or clogged feeding tubes when medications are delivered through feeding tubes:

- Give medications by mouth instead of by tube whenever possible.
- Use the liquid form of medications, whenever available. Dilute hypertonic liquid medications◆ with water before administering them through the feeding tube.
- Use injectable or intravenous forms of medications if they are not available in liquid form. Ideally, tablets should not be crushed and administered through feeding tubes. If using tablets is unavoidable, crush the tablets to a fine powder and mix with water before administering them.
- Never crush tablets intended to release their contents slowly (time-released tablets) or enteric-coated medications. In these cases, another medication form must be given.
- Unless the compatibility of multiple medications is known, do not mix medications together or mix medications with the formula hanging in the feeding container. Instead, give each medication individually using the separate port whenever available.
- Flush the feeding tube with warm (body temperature) water before and after administering each medication.
- Avoid medications known to be incompatible with formulas. Bulk-forming agents, for example, can clog the feeding tube. Other medications that may be incompatible with some formulas are listed in Table 22-3.

◆ Like hypertonic formulas, hypertonic liquid medications can lead to diarrhea. Examples of hypertonic medications include potassium chloride elixir (a potassium supplement), liquid multivitamin preparations, Tagamet liquid, and theophylline elixir.

• GI Side Effects • Medications can trigger GI side effects, including nausea, vomiting, and diarrhea, that are also frequent complications of tube feedings. Medications are a frequent culprit when diarrhea is associated with tube feedings. In one study, the most common cause of diarrhea (responsible for about half the cases) in clients receiving tube feeding was sorbitol-containing medications.[8]

◆ Some clients have a specific type of jejunostomy, called a *needle catheter jejunostomy,* through which phenytoin cannot be delivered. In such cases, phenytoin is given intravenously.

• Additional Considerations • The location of the feeding tube (whether gastric or intestinal) is also relevant when administering medications. Medications designed to dissolve in the mouth should not be given through a feeding tube. Medications that depend on the stomach's acidic environment for absorption may be poorly absorbed if delivered directly into the duodenum or jejunum. Similarly, a medication that is optimally absorbed in the duodenum may be poorly absorbed in the jejunum. In such cases, oral, intravenous, or injectable forms of the medication should be provided.

In some cases, formulas can alter medication absorption. One example is phenytoin, a medication used to control seizures. Absorption of phenytoin may be markedly reduced for a person who is on continuous tube feedings. Although opinions of the best way to handle this problem vary, some clinicians suggest that for most clients on either intermittent or continuous feedings, the feeding should be stopped for two hours before and two hours after giving phenytoin.[9]◆ For clients requiring continuous feedings, the rate of delivery is increased during the times the feeding is given to ensure that nutrient needs are met.

TABLE 22-3 Selected Medications that Are Incompatible with Formulas

Aluminum and magnesium hydroxide	MCT oil
Chlorpromazine concentrate	Mellaril concentrate
Ciballth-S syrup	Mellaril oral solution
Cimetidine	Paregoric elixir
Dimetane elixir	Potassium chloride
Dimetapp elixir	Reglan syrup
Feosol elixir	Riopan
Fleet's phosphosoda	Robitussin expectorant
Gevrabon liquid	Sudafed syrup
Klorvess syrup	Thorazine concentrate
Mandelamine Forte suspension	Zinc sulfate capsules

Note: These substances may be compatible with some formulas and not others.
SOURCES: Z. M. Pronsky, *Food-Medication Interactions*, 11th ed. (Pottstown, Pa.: Food-Medication Interactions, 2000); F. C. Thompson, M. R. Naysmith, and A. Lindsay, Managing drug therapy in patients receiving enteral and parenteral nutrition, *Hospital Pharmacist* 7 (2000): 155–164.

Addressing Tube Feeding Complications

Table 22-4 summarizes complications associated with tube feedings and shows that many problems can be prevented or corrected by selecting the formula and feeding route wisely, preparing the formula correctly, and delivering it appropriately. Attention to the person's primary medical condition and medications is important as well. Table 22-5 provides a monitoring schedule that helps ensure early detection of problems that may be encountered when clients are fed by tube.

• Types of Complications • Failure to estimate nutrient needs correctly or to ensure that the selected formula meets these needs limits the client's ability to achieve or maintain adequate nutrition status. Mechanical problems, such as a

TABLE 22-4 Causes and Prevention or Correction of Tube Feeding Complications

Complications	Possible Causes	Preventive/Corrective Measures
Aspiration of formula	Compromised lower esophageal sphincter, delayed gastric emptying	Use nasoenteric, gastrostomy, or jejunostomy feedings in high-risk clients; check tube placement; elevate head of bed during and for 45 minutes after feeding; check gastric residuals.
Clogged feeding tube	Formula too thick for tube	Select appropriate tube size; flush tubing with water before and after giving formula; use infusion pump to deliver thick formulas; remedies reported to help unclog feeding tubes include cola, cranberry juice, meat tenderizer, and pancreatic enzymes.
	Medications delivered through feeding tube	Use oral, liquid, or injectable medications whenever possible; dilute thick or sticky liquid medications with water before administering; crush tablets to a fine powder and mix with water; flush tubing with water before and after medications are given; give medications individually; do not add medications to the feeding container.
Constipation	Low-fiber formula	Provide additional fluids; use high-fiber formula.
	Lack of exercise	Encourage walking and other activities, if appropriate.
Dehydration and electrolyte imbalance	Excessive diarrhea	See items under *Diarrhea*.
	Inadequate fluid intake	Provide additional fluid.
	Carbohydrate intolerance	Use continuous drip administration of formula; monitor blood glucose; select a formula with a lower amount or different type of carbohydrate.
	Excessive protein intake	Monitor blood electrolyte levels; reduce protein intake.
Diarrhea, cramps, abdominal distention	Bacterial contamination	Use fresh formula every 24 hours; store opened or mixed formula in a refrigerator; rinse feeding bag and tubing before adding fresh formula; change feeding apparatus every 24 hours; prepare formula with clean hands using clean equipment in a clean environment.
	Lactose intolerance	Use lactose-free formula in lactose-intolerant and high-risk clients.
	Hypertonic formula	Use small volume of formula and increase volume gradually.
	Rapid formula administration	Slow administration rate or use continuous drip feedings.
	Malnutrition/low serum albumin	Use small volume of dilute formula and increase volume and concentration gradually.
Hyperglycemia	Diabetes, hypermetabolism, drug therapy	Check blood glucose; slow administration rate; provide adequate fluids; select a formula with a lower amount or different type of carbohydrate.
Nausea and vomiting	Obstruction	Discontinue tube feeding.
	Delayed gastric emptying	Check gastric residual; slow administration rate, use continuous drip feedings, or discontinue tube feeding.
	Intolerance to concentration or volume of formula	Use small volume of formula and increase volume and concentration gradually; use continuous drip feedings.
	Psychological reaction to tube feeding	Address client's concerns.
Skin irritation at enterostomy site	Leakage of GI secretions and friction caused by the tube	Keep site clean; inspect area for redness, tenderness, and drainage; use protective skin cream.

Note: Many of the complications presented here can be caused by the client's primary disorder or drug therapy rather than the tube feeding itself. In such a case, the corrective measure would include treatment of the disorder or a change in drug therapy. Additionally, other corrective measures that require a physician's order are not shown here.

TABLE 22-5 Suggested Guidelines for Monitoring Clients on Tube Feedings

Before starting a new feeding:	Complete a nutrition assessment. Check tube placement.
Before each intermittent feeding:	Check client position. Check gastric residual. Check tube placement.
After each intermittent feeding:	Flush feeding tube with water.
Every half hour:	Check gravity drip rate, when applicable.
Every hour:	Check pump drip rate, when applicable.
Every 4 hours:	Check vital signs, including blood pressure, temperature, pulse, and respiration.
Every 6 hours:	Check blood glucose; monitoring blood glucose can be discontinued after 48 hours if test results are consistently negative in a nondiabetic client.
Every 4 to 6 hours of continuous feeding:	Check client position. Check gastric residual. Flush feeding tube with water.
Every 8 hours:	Check intake and output. Check specific gravity of urine. Check tube placement. Chart client's total intake of, acceptance of, and tolerance to tube feeding.
Every day:	Weigh client. Change feeding container and attached tubing. Clean feeding equipment.
Every 7 to 10 days:	Reassess nutrition status.
As needed:	Observe client for any undesirable responses to tube feeding; for example, delayed gastric emptying, nausea, vomiting, or diarrhea. Check nitrogen balance. Check laboratory data. Chart significant details.

clogged feeding tube, a malfunctioning feeding pump, or a tube that becomes dislodged from its appropriate location, can interrupt the feeding schedule. Other complications related to the formula or its administration can result in nausea, vomiting, diarrhea, cramps, constipation, delayed gastric emptying, abdominal distention, and aspiration.

Metabolic complications such as dehydration, fluid and electrolyte imbalances, and elevated blood glucose can also occur. When complications arise, they can place further stress on the client and make it more difficult to achieve medical and nutrition goals.

• **What to Chart** • Chapter 19 emphasized the importance of the medical record as a legal document and communication tool. Before reimbursing for the cost of tube feedings, many types of insurance (including Medicare) and managed care organizations require that the physician appropriately document the client's need for a tube feeding as well as the justification for using an infusion pump or a special formula, when appropriate. Medicare does not cover the cost of tube feeding unless the physician believes the condition necessitating the tube feeding will last for at least three months. Furthermore, Medicare does not cover the cost of tube feeding for clients with functioning GI tracts who are unable to eat due to a lack of appetite.

The health care team should routinely document the following information for clients on tube feedings:

- The nutrition goals for the client.
- The condition necessitating the tube feeding.
- The placement—both where (gastric or intestinal) and how (transnasal or enterostomy) the feeding tube is placed—and the type and size of the feeding tube.
- The formula selected to meet nutrition goals and its nutrient composition.
- The recommended administration schedule (concentration and rate) and method of delivery (intermittent or continuous, gravity drip or infusion pump).
- The client's education regarding the nutrition goals and the tube feeding procedure.
- The client's physical and emotional responses to the tube insertion and tube feeding procedure.
- The client's tolerances to the formula and administration schedule, complications (if any), and corrective actions recommended.
- All substances delivered through the feeding tube including the formula, dye, additional water, medications, and any substances used to unclog the feeding tube (review Table 22-4).
- Confirmation that measures for monitoring the client have been taken.
- The reasons why a tube feeding was interrupted or could not be delivered, if necessary.

The medical record serves as both a legal record and an invaluable source of information for investigating problems so that corrective measures can be taken promptly.

IN SUMMARY

To maximize the benefits of tube feedings, formulas must be prepared and delivered using techniques that minimize the risk of complications. Formulas must be prepared safely to minimize the risk of bacterial contamination and its consequent effects on health. Formulas are given slowly at first and gradually increased until the client is meeting nutrition goals. The delivery of a formula may be intermittent or continuous, depending on the client's medical condition. Supplemental water is given through the tube to prevent dehydration and clogged feeding tubes. Confirming tube placement, elevating the head of the bed during a feeding, adding dye to the enteral formula, and measuring gastric residuals can reduce the risk of aspiration. When feeding tubes are used to deliver medications, special care must be taken to ensure that the client does not develop complications as a consequence. Throughout a tube feeding, health care professionals should chart the client's progress in the medical record to ensure that the client obtains the most benefits from the procedure.

FROM TUBE FEEDINGS TO TABLE FOODS

Once the medical condition that required a tube feeding resolves, the client can gradually shift to an oral diet as the volume of formula is tapered off. The client should be eating about two-thirds of the estimated nutrient needs by mouth before the tube feeding is discontinued.[10] In many cases, the person can begin to drink the same formula that is being delivered by tube. As clients begin to take more food or formula orally, they receive less of the formula by tube. Some people cannot make the transition to oral intake for medical reasons and go home on tube feedings. Chapter 23 discusses how home tube feedings work. The accompanying case study helps you to consider the many factors involved in tube feedings, and the Nutrition Assessment Checklist reviews key points for assessing nutrition status in people receiving tube feedings.

CASE STUDY

Graphics Designer Requiring Enteral Nutrition

Mrs. Innis is a 24-year-old graphics designer who suffered multiple fractures when she fell from a cliff while hiking. She has been in the hospital for seven days and has no appetite. Mrs. Innis has lost 8 pounds over the course of her hospitalization. Due to the nature of her injuries, Mrs. Innis is in traction and is immobile, although the head of her bed can be elevated to 45 degrees. From the history, the dietitian determined that Mrs. Innis's nutrition status was adequate prior to hospitalization. The health care team agrees that a nasoduodenal tube feeding should be instituted before nutrition status deteriorates further. The standard formula selected for the feeding is lactose-free, and Mrs. Innis's nutrient requirements can be met with 2200 milliliters of the formula per day.

1. What steps can be taken to prepare Mrs. Innis for tube feeding? Why might nasoduodenal placement of the feeding tube be preferred to nasogastric placement?
2. The physician's orders specify that the feeding should be given continuously over 18 hours. Develop a tube feeding schedule.
3. What parameters should be monitored to ensure that Mrs. Innis's fluid needs are being met? How can additional fluids be given? Describe precautions that should be taken if Mrs. Innis is to receive medications through the feeding tube.
4. After three days of feeding, Mrs. Innis develops diarrhea. Look at Table 22-4 on p. 694 to determine the possible causes. What measures can be taken to correct the various causes of diarrhea?
5. What information should be charted in Mrs. Innis's medical record? When Mrs. Innis is ready to eat table foods again, what steps will the health care team take?

Nutrition Assessment Checklist for People Receiving Tube Feedings

MEDICAL HISTORY

Check the medical record for medical conditions that:

- ☐ Alter nutrient needs (including malabsorption) and narrow the selection of formula
- ☐ Suggest how long the feeding will be required and how the tube will be inserted (transnasal versus enterostomy)
- ☐ Narrow the choice of placement sites (gastric versus intestinal)

Regularly review the medical record for complications that may suggest the need to alter the tube feeding formula or delivery technique, including:

- ☐ Aspiration of formula
- ☐ Constipation
- ☐ Dehydration/electrolyte imbalance
- ☐ Diarrhea
- ☐ Hyperglycemia
- ☐ Nausea and vomiting
- ☐ Skin irritation

MEDICATIONS

Check medications for those that can cause side effects similar to those associated with tube feedings including:

- ☐ Nausea/vomiting
- ☐ Diarrhea
- ☐ Constipation
- ☐ GI discomfort/cramps

For medications being delivered through the feeding tube, check:

- ☐ Form of medication (avoid crushing tablets; use other drug forms, if possible)
- ☐ Consistency of liquid medications (dilute thick or sticky liquids)
- ☐ Compatibility with formulas (see Table 22-3 on p. 693)
- ☐ Type of medication (do not crush time-released tablets)

FOOD/NUTRIENT INTAKE

To assess the adequacy of a tube feeding, check to see if:

- ☐ Nutrition goals are consistent with nutrient needs
- ☐ Formula is being administered as prescribed
- ☐ Table foods are meeting nutrient needs before stopping the feeding
- ☐ Supplemental water is being given to meet needs

ANTHROPOMETRICS

Measure baseline height and weight. If weight is not consistent with meeting nutrition goals:

- ☐ Reestimate energy needs and determine if energy needs have been correctly calculated
- ☐ Check to see if formula is being delivered as prescribed
- ☐ Check for physical and laboratory signs of dehydration or overhydration

LABORATORY TESTS

Check serum and urine tests for signs of:

- ☐ Dehydration
- ☐ Electrolyte imbalances
- ☐ Glucose intolerance
- ☐ Adequacy of protein intake (prealbumin and retinol-binding protein), when available
- ☐ Improvement or deterioration of medical condition

PHYSICAL SIGNS

Look for physical signs of:

- ☐ Dehydration
- ☐ Overhydration
- ☐ Delayed gastric emptying (gastric residuals)
- ☐ Malnutrition

IN SUMMARY Tube feeding is a practical solution to feeding a person who is unable to consume adequate nutrients by mouth. However, a person without a functional GI tract cannot benefit from a tube feeding. In such a case, intravenous nutrition (the subject of the next chapter) can be a lifesaving treatment option.

Nutrition on the Net

WEBSITES

Access these websites for further study of topics covered in this chapter.

- Find updates and quick links to these and other nutrition-related sites at our website: **www.wadsworth.com/nutrition**
- To find out more about organizations that promote the appropriate use of enteral and parenteral nutrition, visit the:

  American Society for Parenteral and Enteral Nutrition site: **www.clinnutri.org**

  Canadian Parenteral-Enteral Nutrition Association site: **www.magi.com/~cpena**

  European Society of Parenteral and Enteral Nutrition site: **www.espen99.org**
- To view illustrations of how PEG tubes are inserted, visit Healthgate: **www.healthgate.com/sym/surg68.shtml**
- To find information about caring for feeding tubes and other information for caregivers, visit the Children's Hospital Medical Center: **www.cincinnatichildrens.org/family/pep/homecare**

INTERNET ACTIVITIES

When people cannot eat, nutrients need to be delivered by way of formulas—either orally or sometimes, by tube. You may have questions about specific formulas, how and when feeding tubes are used, and what the latest research on enteral nutrition is.

- ▼ Go to: **www.ross.com**
- ▼ Scroll down and click on topics of interest, including those listed under Adult Nutritionals and Enteral Feeding Devices.

This site helps to fill in knowledge gaps in the area of enteral nutrition. What did you learn about enteral feedings?

Study Questions

These questions will help you review the chapter. You will find the answers in the discussions on the pages provided.

1. Describe standard formulas, hydrolyzed formulas, complete formulas, and modular formulas, explaining the characteristics of each and how they differ. (pp. 682–683)
2. Who can benefit from enteral formulas? When can a person drink a formula, and when must the formula be delivered by tube? (pp. 684–685)
3. Suggest ways for improving acceptance of enteral formulas for people who can drink them. (p. 685)
4. In what ways and in what locations can feeding tubes be placed? What factors must be considered in selecting an appropriate formula for a tube feeding? (pp. 685–688)
5. How are tube feedings usually initiated? Discuss the ways that tube feedings can be delivered to clients. (pp. 690–691)
6. Describe the problems that can occur when medications are delivered through feeding tubes. What guidelines can be used to prevent these problems? (pp. 692–693)
7. What complications are associated with tube feedings? What steps help to identify and prevent complications? (pp. 694–695)

These questions will help you prepare for an exam. Answers can be found on p. 700.

1. Which of the following statements is correct?
   a. Standard formulas contain whole proteins or protein isolates.
   b. Standard formulas contain free amino acids or small peptide chains.
   c. Hydrolyzed formulas are made from pureed meats.
   d. Hydrolyzed formulas may contain protein isolates or whole proteins.

2. When a client cannot meet nutrient needs from table foods due to a poor appetite:
   a. parenteral nutrition is preferred to tube feedings.
   b. parenteral nutrition is preferred to enteral formulas provided orally.
   c. enteral formulas are inappropriate.
   d. enteral formulas provided orally are preferred to formulas provided by tube.
3. For a client expected to be able to eat table foods in about a month, but with a high risk of aspiration, an appropriate placement of a feeding tube would most likely be:
   a. nasogastric.
   b. nasoenteric.
   c. gastrostomy.
   d. jejunostomy.
4. In selecting an appropriate enteral formula for a client, the primary consideration is:
   a. the formula's osmolality.
   b. the client's nutrient needs.
   c. the availability of infusion pumps.
   d. the formula's cost and availability.
5. What step can health care professionals take to prevent bacterial contamination of tube feeding formulas?
   a. Deliver the formula continuously.
   b. Do not change the feeding bag and attached tubing.
   c. Discard all opened containers of formula not used in 24 hours.
   d. Add formula to the feeding container before it empties completely.
6. When compared to intermittent feedings, continuous feedings:
   a. require a pump for infusion.
   b. allow greater freedom of movement.
   c. are more like normal patterns of eating.
   d. are associated with more GI side effects.
7. A client needs 1800 milliliters of formula a day. If the client is to receive the formula intermittently every 4 hours, he will need _____ milliliters of formula at each feeding.
   a. 225
   b. 300
   c. 400
   d. 425
8. The term that describes the volume of formula that remains in the stomach from a previous feeding is:
   a. residue.
   b. osmolar load.
   c. gastric residual.
   d. intermittent feeding.
9. The nurse using the feeding tube to deliver medications recognizes that:
   a. medications generally do not result in GI complaints.
   b. medications can be added directly to the feeding container.
   c. thick or sticky liquid medications and crushed tablets can clog feeding tubes.
   d. enteral formulas do not interact with medications in the same way that foods do.
10. Tube feedings can gradually be discontinued when:
    a. discharge planning begins.
    b. the client experiences hunger.
    c. the medical condition resolves.
    d. the client is able to eat foods or drink formula in sufficient amounts.

Clinical Applications

1. Complex procedures, such as those necessary to insert feeding tubes and deliver tube feedings, require attention to many technical details, making it easy to focus on the procedure and forget about the client. Imagine that you need a transnasal tube feeding. How might you react to the news that you need the feeding and to the insertion procedure? What things would you miss most about eating table foods? Think about ways a nurse might help you deal with these feelings.

2. Review Chapters 20 and 21 and note symptoms or disorders that may require the use of tube feedings. For each, consider when and why a tube feeding might be appropriate and which conditions, if any, might require hydrolyzed formulas.

References

1 E. A. Emergy and coauthors, Banana flakes control diarrhea in enterally fed patients, *Nutrition in Clinical Practice* 12 (1997): 72–75.

2 A. Skipper, C. Bohac, and M. B. Gregoire, Knowing brand name affects patient preferences for enteral supplements, *Journal of the American Dietetic Association* 99 (1999): 91–92.

3 V. Vanek, Closed *versus* open enteral delivery systems: A quality improvement study, *Nutrition in Clinical Practice* 15 (2000): 234–243; M. E. Rupp and coauthors, Evaluation of bacterial contamination of a sterile, non-air-dependent enteral feeding system in immunocompromised patients, *Nutrition in Clinical Practice* 14 (1999): 135–137.

4 E. H. Elpern, Pulmonary aspiration in hospitalized adults, *Nutrition in Clinical Practice* 12 (1997): 286–289.

5 M. H. DeLegge and B. M. Rhodes, Continous versus intermittent feedings: Slow and steady or fast and furious? *Support Line,* October 1998, pp. 11–15.

6 J. Lipp, L. M. Lord, and L. H. Scholer, Fluid management in enteral nutrition, *Nutrition in Clinical Practice* 14 (1999): 232–237.

7 M. D. Coleman and K. A. Tappenden, Enteral feeding equipment: From the 15th century to the millennium, *Support Line,* October 1999, pp. 22–27.

8 M. S. Williams and coauthors, Diarrhea management in enterally fed patients, *Nutrition in Clinical Practice* 13 (1998): 225–229.

9 J. Hatton and B. Magnuson, How to minimize interaction between phenytoin and enteral nutrition: Two approaches, *Nutrition in Clinical Practice* 11 (1996): 28–31.

10 T. C. Lykins, Nutrition support clinical pathways, *Nutrition in Clinical Practice* 11 (1996) 16–20.

ANSWERS

Multiple Choice

1. a	2. d	3. b	4. b	5. c
6. a	7. b	8. c	9. c	10. d

ENTERAL FORMULAS AND INBORN ERRORS OF METABOLISM

CHAPTER 22 described the use of enteral formulas for people who have nutrient needs that cannot be met using table foods. Such is the case for people with some inborn errors of metabolism, and in these cases, enteral formulas play a vital role in management. This highlight describes inborn errors of metabolism and discusses the role of diet in two of these disorders: phenylketonuria and galactosemia. The accompanying glossary defines terms related to inborn errors of metabolism.

INBORN ERRORS OF METABOLISM

An **inborn error of metabolism** is a genetic error **(mutation)** that alters the production of a protein. In many cases, the protein is an enzyme. When the body fails to make an enzyme, makes an enzyme in insufficient amounts, or makes an enzyme with an abnormal structure, body functions that depend on that enzyme cannot proceed as they normally do. For example, if an enzyme is missing or malfunctioning in the metabolic pathway that converts compound A to compound B, then compound A accumulates and compound B becomes deficient. Both the excess of compound A and the lack of compound B can lead to a variety of problems. Furthermore, these imbalances alter other metabolic pathways, creating another array of problems.

Not all inborn errors of metabolism are serious. In some instances, the accumulated compound is not toxic, and the deficient compound is not essential, so individuals experience no problem. They most likely do not even know about the error. In other cases, however, inborn errors have severe consequences, including mental retardation. Without prompt diagnosis and treatment, they can be lethal.

Treatments for Inborn Errors of Metabolism

Medical nutrition therapy comprises the primary treatment for many inborn errors. With an understanding of the biochemical pathway involved, a clinician can often manipulate the diet to compensate for excesses and inadequacies. Management involves restricting dietary precursors that occur prior to the error in the metabolic pathway, replacing needed products that fail to be produced, or both. The goal of therapy is to:

- Prevent the accumulation of toxic metabolites.
- Replace essential nutrients that are deficient as a result of the defective metabolic pathway.
- Provide a diet that supports normal growth, development, and maintenance.

Meeting these three objectives is a major challenge that was previously unattainable. New knowledge about the body's many biochemical pathways, coupled with current technology for synthesizing formulas with specific nutrient compositions, has greatly enhanced the treatment of inborn errors.

GLOSSARY OF TERMS RELATED TO INBORN ERRORS OF METABOLISM

carriers: individuals who possess one **dominant** and one **recessive gene** for a genetic trait, such as an inborn error of metabolism. When a trait is recessive, a carrier may show no signs of the trait but can pass it on.

dominant gene: a gene that has an observable effect on an organism. If an altered gene is dominant, it has an observable effect even when it is paired with a normal gene; see also *recessive gene.*

galactosemia (ga-LAK-toe-SEE-me-ah): an inborn error of metabolism in which enzymes that normally metabolize galactose to compounds the body can handle are missing and an alternative metabolite accumulates in the tissues, causing damage.

genes: the basic units of hereditary information, made of DNA, that are passed from parent to offspring in the chromosomes. A pair of genes codes for each genetic trait.

inborn error of metabolism: an inherited flaw evident as a metabolic disorder or disease present from birth.

mutation: an alteration in a gene such that an altered protein is produced.
muta = change

PKU, phenylketonuria (FEN-el-KEY-toe-NEW-ree-ah): an inborn error of metabolism in which phenylalanine, an essential amino acid, cannot be converted to tyrosine. Alternative metabolites of phenylalanine (phenylketones) accumulate in the tissues, causing damage, and overflow into the urine.

recessive gene: a gene that has no observable effect as long as it is paired with a normal gene that can produce a normal product. In this case, the normal gene is said to be *dominant.*

Other Interventions

Therapy for inborn errors goes beyond nutrition. Many disorders result in medical problems that need specific treatments. Psychological and genetic counseling for people who are affected and their families is also important. Psychological counseling can help families cope with adjustments their disorders entail. Genetic counseling can help family members determine if they carry **genes** for the disorder. **Carriers** (those who possess one **dominant** and one **recessive gene** for the disorder) need to learn about their risks for passing on the disorder should they have children. A classic example that illustrates the principles of treatment is the most common inborn error of metabolism: **phenylketonuria (PKU).**

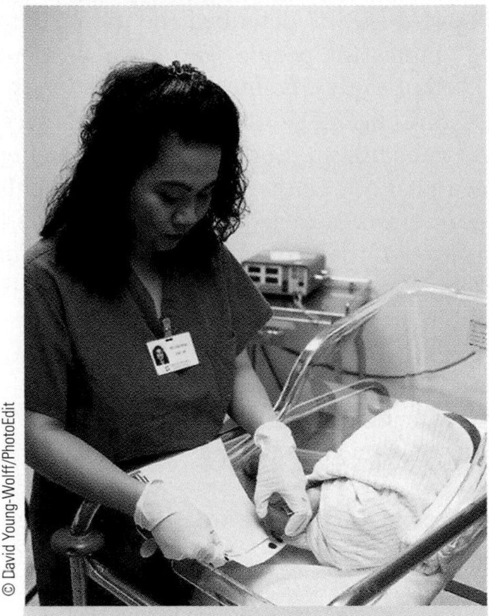

© David Young-Wolff/PhotoEdit

A simple blood test screens newborns for PKU—the most common inborn error of metabolism.

PHENYLKETONURIA

PKU is one of many inborn errors that affect amino acid metabolism. Other disorders can affect carbohydrate, lipid, vitamin, and mineral metabolism. PKU affects approximately 1 out of every 10,000 newborns in the United States each year. The ability to detect and treat PKU has saved and significantly improved the lives of many people; its example offers hope to those suffering from other inborn errors.

The Error in PKU

Classic PKU results from a deficiency of an enzyme that converts the essential amino acid phenylalanine to tyrosine (see Figure H22-1). Without the enzyme, abnormally high concentrations of phenylalanine and other related compounds accumulate and damage the developing nervous system. Simultaneously, the body cannot make tyrosine or other compounds (such as the neurotransmitter epinephrine) that normally derive from tyrosine. Under these conditions, tyrosine becomes an essential amino acid; that is, the body cannot make it, and so the diet must supply it.

Detecting PKU

PKU is a hidden disease that cannot be seen at birth, yet diagnosis and treatment beginning in the first few days of life can prevent its devastating effects. For these reasons, and because PKU is the most common inborn error of metabolism, all newborns in the United States receive a screening test for PKU. The test must be conducted after the infant has consumed several meals containing protein (usually after 24 hours and before seven days).

FIGURE H22-1 **Biochemical Alterations in PKU**

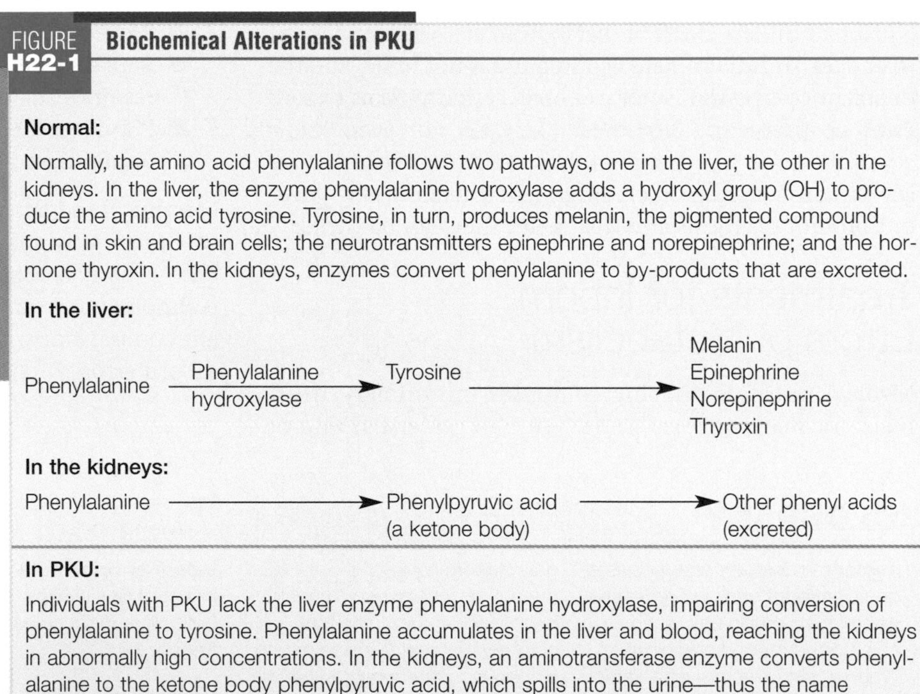

Before screening became routine, an infant with PKU would suffer the dire consequences of uncorrected high phenylalanine concentrations. At first, the only signs are a skin rash and light skin pigmentation. Between three and six months, signs of developmental delay begin to appear. By one year, irreversible brain damage is clearly evident.

MEDICAL NUTRITION THERAPY FOR PKU

Essentially, the diet restricts phenylalanine and supplements tyrosine to maintain blood concentrations within a safe range. The diet's effectiveness is remarkable; in almost every case, it can prevent the devastating array of symptoms described. As most dietitians can attest, though, the diet is more easily described than designed.

Because phenylalanine is an essential amino acid, the diet cannot exclude it completely. Children with PKU do not require less phenylalanine than other children, but they cannot handle excesses without detrimental effects. If phenylalanine intake is too low, children suffer bone, skin, and blood disorders; growth and mental retardation; and death. Failure to eat the recommended amount of phenylalanine, along with inadequate energy intakes, can lead to negative nitrogen balance and a consequent *rise* in blood phenylalanine levels.[1] Therefore, the diet must strike a balance, providing enough phenylalanine to support normal growth and health and prevent negative nitrogen balance, but not enough to cause harm. To ensure that blood phenylalanine and tyrosine concentrations remain within a safe range, children with PKU receive blood tests periodically and alter their diets when necessary.

Sample Phenylalanine-Restricted Menu for a Child with PKU

Breakfast
2 tbs raisins
5 tbs cream of rice
2 tsp sugar
Low-phenylalanine formula

Snack
4 oz orange juice

Lunch
½ small banana
2 tbs tomato soup (without milk)
3 tbs rice
1½ tsp margarine
Low-phenylalanine formula

Snack
Low-phenylalanine formula
5 round butter crackers

Supper
2 tbs instant potatoes (without milk)
3 tbs green beans
4 tbs vegetable and beef broth
1½ tsp margarine
¾ c sliced peaches
Low-phenylalanine formula

Snack
2 tbs raisins
Low-phenylalanine formula

Diet for PKU

Phenylalanine is widespread in foods that contain protein, and therefore, the diet for PKU includes very little natural protein. The diet excludes high-protein foods such as meat, fish, poultry, cheese, eggs, milk, nuts, and legumes. Also excluded are commercial breads and pastries made from regular flour, which has a high phenylalanine content. Basically, the diet allows foods that contain some phenylalanine, such as fruits, vegetables, and cereals, and those that contain none, such as fats, sugars, jellies, and some candies. Clearly, it is impossible to create such a diet that is adequate in energy and all other nutrients using only whole, natural foods.

Enteral Formulas for PKU

Formulas supply energy, protein, vitamins, and minerals for people with PKU. Some formulas are phenylalanine-free; others supply small amounts of phenylalanine. The less phenylalanine in the formula, the more phenylalanine the person can eat from foods. Children who depend primarily on a formula for their nourishment risk multiple trace mineral deficiencies. Health care professionals monitor trace mineral status and supplement as needed. The accompanying menu provides a sample phenylalanine-restricted diet for a child with PKU.

Continuing Diet Restrictions

During the early years of central nervous system development, the diet for PKU is critical to preventing irreversible mental retardation. Until the late 1970s, researchers assumed that the child with PKU could abandon the special diet after the first few years of life when the central nervous system had completed its development. Unfortunately, however, elevated phenylalanine concentrations in the older child do cause problems such as short attention span, poor short-term memory, and poor eye-to-hand coordination, although the damage is less severe than at an earlier age. A child with PKU who discontinues the controlled diet may experience problems in school performance, mood, and behavior. Clinicians generally encourage children to continue the low-phenylalanine diet indefinitely. As children reach their teen years, they begin to assume more responsibility for their food choices, and it becomes very important that they understand their disorder and its treatment.[2] This is especially important for girls, because they may eventually become pregnant.

Diet, Pregnancy, and PKU

High blood phenylalanine in a pregnant woman with PKU presents a hostile environment to fetal development. The fetus's blood concentrations rise even higher than hers, and she may experience a spontaneous abortion; or her infant may suffer mental retardation, congenital heart disease, and low birthweight. For these reasons, women with elevated phenylalanine concentrations need counseling prior to pregnancy, so that they understand the problems their condition may create for their children.

Dietary control of maternal PKU may protect the fetus, at least in part, if implemented early enough. Dietary control does not ensure a successful outcome of pregnancy, but the infants of women who follow a low-phenylalanine diet from at least one to two months prior to conception and continue it throughout pregnancy are more likely to have higher birthweights, larger head circumferences, fewer malformations, and higher scores on intelligence tests than the infants of women who begin diet therapy during their pregnancies or not at all. Again, women rely on formulas to help them meet the nutrient demands of pregnancy.

OTHER INBORN ERRORS OF METABOLISM

Many other inborn errors of metabolism exist, and the effect of diet in treatment varies. As an example, **galactosemia** is an inborn error of carbohydrate metabolism in which any one of three enzymes that convert galactose to glucose is missing or defective. When infants with galactosemia are given infant formula or breast milk (which contains a galactose unit in each molecule of lactose), they vomit and have diarrhea. Alterations in metabolism cause growth failure, liver enlargement, and other neurological abnormalities that can lead to coma and death.

Diet for Galactosemia

The diet for galactosemia is simpler than the diet for PKU for two reasons. First, unlike phenylalanine, galactose is not an essential nutrient. Diets for galactosemia need only to restrict galactose, not to provide a perfectly calculated dose. Second, galactose occurs primarily in lactose (the sugar in milk), so treatment depends chiefly on the careful elimination of all milk and milk products. Some fruits and vegetables contain free galactose; the amounts in fresh blueberries and honeydew melon warrant their exclusion from the diet as well.[3] This is not to say that the diet is easy to follow; many commercially prepared products contain milk. Still, milk is less widespread in the diet than the amino acid phenylalanine, which appears in all proteins.

For infants with galactosemia, lactose-free formulas meet nutrient needs. Once a child is eating table food, special formulas are generally unnecessary. However, care must be taken to ensure that the diet supplies adequate calcium.

Effect of Diet for Galactosemia

Early introduction of lactose-free formulas and galactose-restricted diets prevent or minimize some symptoms, but the long-term effects of diet are not as dramatic as for PKU.[4] Most people with galactosemia experience developmental delays, speech abnormalities, mental disabilities, cataracts, and, in women, ovarian disfunction.

As scientific understanding of human genetics and biochemistry increases, more and more inborn errors affecting enzyme function are being recognized. Understanding the roles of enzymes in metabolism sometimes makes it possible to develop enteral formulas that compensate for these defects, which otherwise would destroy the quality of life. In such cases, diet can make a dramatic difference in people's lives.

References

1 G. P. Duran and coauthors, Necessity of complete intake of phenyalanine-free mixture for metabolic control of phenylketonuria, *Journal of the American Dietetic Association* 99 (1999): 1559–1563.

2 R. Singh and coauthors, Impact of a camp experience on phenylalanine levels, knowledge, attitudes, and health beliefs relevant to nutrition management of phenylketonuria in adolescent girls, *Journal of the American Dietetic Association* 100 (2000): 797–803.

3 S. S. Gropper and coauthors, Free galactose content of fresh fruits and strained fruit and vegetable baby foods: More foods to consider for the galactose-restricted diet, *Journal of the American Dietetic Association* 100 (2000): 573–575.

4 K. Widhalm, B. D. O. M. da Cruz, and M. Koch, Diet does not ensure normal development in galactosemia, *Journal of the American College of Nutrition* 16 (1997): 204–208.

CHAPTER 23 Parenteral Nutrition

Many IV solutions look alike, and the only way to discover their contents is to read the label.

The science of medical nutrition as we know it today was shaped tremendously by the demonstration in 1968 that all nutrient needs could be met by vein.[1] Practitioners now had the means to feed people who otherwise might have died from malnutrition. With time, clinicians learned more and more about the solutions and delivery techniques used to provide parenteral nutrition. They also discovered that although parenteral nutrition is a lifesaving procedure, it is also very costly and is associated with serious complications including liver dysfunction, progressive kidney problems, bone disorders, and many nutrient deficiencies. These findings prompted a renewed appreciation for the GI tract, and clinicians today subscribe to the adage "If the GI tract works, use it."◆ Only when people cannot meet their nutrient needs using the enteral route should they receive total parenteral (or intravenous) nutrition support. Figure 23-1 summarizes the decision-making process in selecting the most appropriate feeding method.

◆ When parenteral nutrition is necessary to meet nutrient needs, some nutrients should be provided enterally (whenever possible) to help maintain the structure and function of the intestinal cells.

THINK NUTRITION whenever clients need nutrients but are unable to use their GI tracts. Parenteral nutrition provides a life-saving alternative.

INTRAVENOUS SOLUTIONS: WHAT ARE THEY?

As is true of all medical nutrition therapy, the decision to use intravenous (IV) nutrition, the method of delivery, and the type and amount of nutrients to provide are based on a thorough assessment of the client's medical condition and nutrient needs. Infusion of IV nutrients immediately changes blood levels of fluids, electrolytes, and other nutrients and, therefore, requires vigilant attention to the individual's responses and adjustments to the formula, if appropriate.

FIGURE 23-1 **Selecting a Feeding Route**

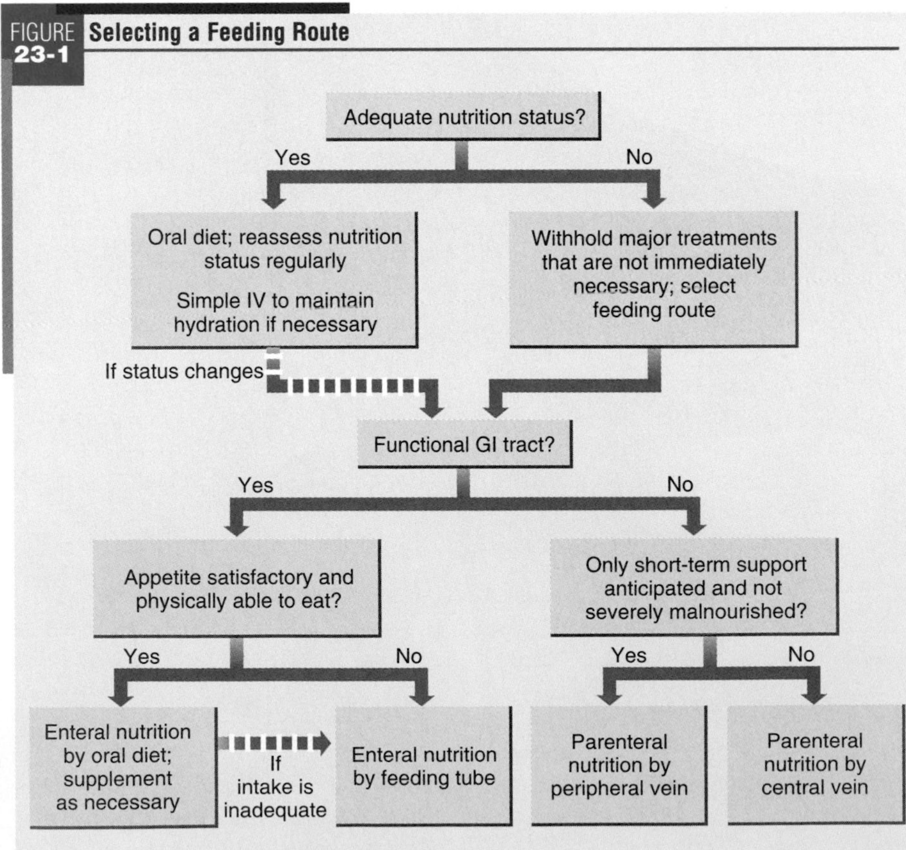

Intravenous Nutrients

A variety of nutrients can be administered by vein in different combinations.◆ These IV solutions may contain any or all of the essential nutrients: water, amino acids, carbohydrate, fat, vitamins, and minerals. The "How to" explains how to read IV solution abbreviations.

Read IV Solution Abbreviations

The labels of IV solutions use several abbreviations, including:

- **D:** dextrose.
- **W:** water.
- **NS:** normal saline (normal saline has a sodium chloride concentration similar to that of blood, which is 0.9 percent).

Interpret these abbreviations as follows:

- **D_5W or D5:** Read as 5 percent dextrose in water (the subscript following the *D* tells you the percentage of dextrose in the solution) or 5 percent dextrose.
- **$D_{10}W$ or D10:** Read as 10 percent dextrose in water or 10 percent dextrose.
- **D_5½ NS:** Read as 5 percent dextrose in a ½ normal saline solution. (Since a normal saline solution contains 0.9 percent sodium chloride, a ½ normal saline solution contains 0.45 percent sodium chloride.)

• Amino Acids • Normally, by the time protein from food reaches the bloodstream, it has been broken down into amino acids. Thus IV solutions supply amino acids rather than whole proteins. Standard IV solutions contain both essential and nonessential amino acids to meet the body's protein needs. Special solutions that contain only essential amino acids or have large amounts of certain amino acids and small amounts of others meet protein needs for specific medical conditions. Products designed for liver failure, for example, may contain more branched-chain amino acids and fewer aromatic amino acids (see Chapter 28).

Sometimes a nonessential amino acid may be omitted from standard parenteral solutions because it does not mix well or is unstable in the solution. For example, glutamine, which may be a conditionally essential amino acid◆ for some clients, is not stable in IV solutions. Providing glutamine as a dipeptide solves the instability problem, and studies suggest that short-chain peptides can be digested to free amino acids by enzymes bound to cell membranes.

◆ Reminder: Formulas intended for enteral use cannot be delivered intravenously.

◆ Reminder: A conditionally essential amino acid is one that is normally nonessential but must be supplied by the diet when the body's need for it exceeds the body's ability to produce it. Glutamine may be a conditionally essential amino acid following intestinal resections or during recovery from severe stress.

• Carbohydrate • Standard IV solutions provide carbohydrate as **dextrose monohydrate.** This form of dextrose contains some water—a form more suitable for IV solutions. Intravenous dextrose provides 3.4 kcalories per gram, whereas glucose provides 4.0.

• Lipid • Lipid emulsions deliver fat in IV solutions. Lipid emulsions are provided either daily or periodically (once or twice a week). If provided daily, IV fat serves as a concentrated source of energy; if offered less often, it serves primarily as a source of essential fatty acids. A 500-milliliter bottle of a 10 percent fat emulsion provides 550 kcalories (1.1 kcalories per milliliter). The same volume of a 20 percent fat emulsion provides 1000 kcalories (2.0 kcalories per milliliter).

Lipid emulsions are contraindicated for some newborns with markedly elevated **bilirubin** levels, some people with elevated blood lipids, people with severe liver disease,◆ and those with severe egg allergies. Cautious use of IV fats is recommended for people with atherosclerosis, moderate liver disease, blood coagulation disorders, pancreatitis, and some lung problems. After long-term administration, brown pigments may accumulate in certain liver cells, but these pigments disappear after parenteral therapy is discontinued; their effects on liver function are unknown. Prolonged IV lipid use may also enlarge the liver and spleen and reduce the number of blood platelets and white blood cells.

◆ Hyperlipidemia and atherosclerosis are discussed in Chapter 26, and liver disorders are the subject of Chapter 28.

• Micronutrients • Vitamins, electrolytes (minerals), and trace elements may all be used in IV solutions. The U.S. Food and Drug Administration (FDA) recently issued new specifications for IV multivitamin formulations.[2] In the past, vitamin K was not added to adult IV multivitamin formulations, but once the new specifications take effect, vitamin K will be included. Pediatric multivitamin formulations, on the other hand, contain vitamin K and will continue to do so.

Failure to include vitamins in parenteral nutrition solutions can result in serious nutrient deficiencies and even death. During 1988, for example, a national shortage of parenteral multivitamin formulations led to the omission

dextrose monohydrate: a form of glucose that contains a molecule of water and is stable in IV solutions. IV dextrose solutions provide 3.4 kcal/g.

bilirubin: a pigment in the bile whose blood concentration may rise as a result of medical conditions that affect the liver.

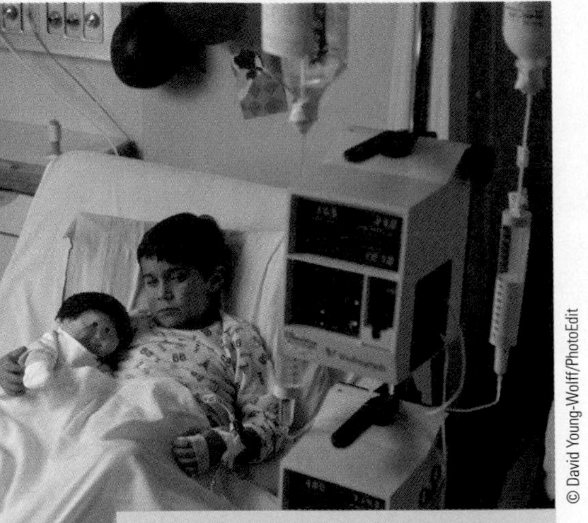

© David Young-Wolff/PhotoEdit

Intravenous lipid emulsions, made from plant-derived oils and egg phospholipids, provide energy and essential fatty acids and can be easily identified by their milky white color.

of vitamins in parenteral solutions and resulted in severe thiamin deficiencies and three deaths.[3] Conversely, long-term parenteral nutrition can sometimes lead to elevated manganese (a trace mineral) levels; these levels return to normal when manganese is removed from the IV solution.[4]

Some electrolytes (particularly, calcium and phosphorus) can precipitate with other IV solution components, potentially resulting in life-threatening complications. As an indication of the seriousness of this problem, the FDA warned health care professionals that a precipitate of calcium phosphate might have been responsible for two deaths and at least two cases of respiratory distress in one institution. Pharmacists skilled in formulating IV solutions follow specific mixing procedures to minimize the risk of precipitation and protect the stability of IV solutions.

• **Other Additives** • To avoid the need for a separate infusion site, intravenous medications are sometimes added directly to IV solutions or infused along with IV solutions through a separate port (Y-connector). These practices are not recommended, however, because interactions between medications and IV solutions can and do occur. Only medications proved to be physically compatible with, and biochemically stable in IV solutions are safe to administer together with the solution. Intravenous catheters with separate lumens allow medications to be delivered separately from the IV solution and limit the risks of adverse interactions.

Other problems may occur when medications are added directly to the IV solution. If the total volume of the solution is not infused, the client does not receive the full dose of the medication. Conversely, if the addition of a medication to an IV solution is not noted clearly in the medical record, the physician may inadvertently reorder the medication, and the client may suffer potentially severe consequences.

Types of Intravenous Solutions

Intravenous solutions can be compounded in many ways. The client's medical condition and nutrient needs determine what goes into an IV solution. The concentrations of energy nutrients and electrolytes are often expressed as percentages (see the accompanying "How to"), although the recommended practice is for physicians' orders for IV solutions and labels for their containers to state the formula's composition in both grams per day and grams per liter.[5]

• **Simple Intravenous Solutions** • Simple IV solutions typically contain 5 percent dextrose and normal saline. Other electrolytes or salts may be added as needed. Often 3 liters of the solution are provided daily and deliver about 150 grams of dextrose or about 500 kcalories per day. Although simple IV solutions provide some energy, they fall far short of meeting energy and nutrient needs. The volume of simple IV solutions required to meet energy needs exceeds the volume of fluid the body can safely handle.

• **Complete Nutrient Solutions** • Complete nutrient solutions provide amino acids, dextrose, fatty acids, vitamins, minerals, and trace elements. Many facilities use standard forms to assist physicians in writing orders for complete nutrient solutions. Solutions delivered through the **peripheral veins**◆ typically contain from 5 to 10 percent dextrose (like simple IVs) and 3 to 5 percent amino acids and provide about 1500 to 2000 kcalories per day; lipid emulsions contribute more than half of the total kcalories.

◆ The dextrose concentrations of solutions infused through peripheral veins should not exceed 10%, and the osmolarity of the solution should not exceed 600 milliosmoles per liter (mOsm/L).

peripheral veins: the small-diameter veins that carry blood from the arms and legs.

central veins: the large-diameter veins located close to the heart.

Complete nutrient solutions delivered through the **central veins** meet energy needs primarily from concentrated dextrose solutions. These solutions ideally provide about 70 to 85 percent of the nonprotein energy from dextrose and 15 to 30 percent from fat.[6] If lipids are not used as an energy source, however, essential fatty acid requirements may be met by giving IV lipids periodically (two to three times per week).

HOW TO Calculate the Energy Nutrient Content of IV Solutions Expressed as Percentages

The percentage of a substance in an IV solution tells you how many grams of that substance are present in 100 milliliters[a]. For example, a 5 percent dextrose solution contains 5 grams of dextrose per 100 milliliters. A 0.9 percent normal saline solution contains 0.9 gram sodium per 100 milliliters.

Suppose a person is receiving 1800 milliliters of a complete nutrient IV solution containing 16 percent dextrose, 5 percent amino acids, and 2.5 percent lipid. First, determine the number of grams of each energy nutrient:

- For dextrose:

$$\frac{16\text{ g dextrose}}{100\text{ ml}} = \frac{X\text{ g dextrose.}}{1800\text{ ml}}$$

$$\frac{16\text{ g} \times 1800\text{ ml}}{100\text{ ml}} = 288\text{ g dextrose.}$$

- For amino acids:

$$\frac{5\text{ g amino acids}}{100\text{ ml}} = \frac{X\text{ g amino acids.}}{1800\text{ ml}}$$

$$\frac{5\text{ g} \times 1800\text{ ml}}{100\text{ ml}} = 90\text{ g amino acids.}$$

- For lipid:

$$\frac{2.5\text{ g lipid}}{100\text{ ml}} = \frac{X\text{ g lipid.}}{1800\text{ ml}}$$

$$\frac{2.5\text{ g lipid} \times 1800\text{ ml}}{100\text{ ml}} = 45\text{ g lipid.}$$

Next, calculate the total kcalories in 1800 milliliters of the solution by multiplying the kcalories per gram and then adding the totals:

$$288\text{ g dextrose} \times 3.4\text{ kcal/g} = 979\text{ kcal.}$$
$$90\text{ g amino acids} \times 4.0\text{ kcal/g} = 360\text{ kcal.}$$
$$45\text{ g lipid} \times 9.0\text{ kcal/g}^{b} = 405\text{ kcal.}$$
$$\text{Total} = 1744\text{ kcal.}$$

Note that this solution provides 71 percent of the nonprotein kcalories from carbohydrate and 29 percent from fat.

[a]The preferred method for ordering energy nutrients in IV solutions and labeling their containers is to state the amount of each energy nutrient in grams per day and grams per liter. National Advisory Group on Standards of Practice Guidelines for Parenteral Nutrition, *Journal of Parenteral and Enteral Nutrition* 22 (1998): 49–66.
[b]The number of kcalories per gram of IV lipid is actually more than 9.0 kcalories per gram, but 9.0 kcalories per gram is an acceptable estimate.

IN SUMMARY Intravenous solutions contain all or a combination of nutrients and can include water, amino acids, dextrose, lipid, and micronutrients. The solutions may be given as simple IV solutions that contain primarily water, dextrose, and minerals or complete nutrient solutions that meet total nutrient needs.

INTRAVENOUS NUTRITION: WHO NEEDS WHAT?

The different types of intravenous solutions deliver nutrients for different purposes. The composition of the solution and the delivery method depend on the person's immediate medical and nutrient needs, nutrition status, and anticipated length of time on IV nutrition support. When enteral feeding is not possible and complete nutrient solutions are required, they should be initiated before nutrition status is severely compromised. It is much easier to maintain nutrition status than to try to replenish lost nutrient stores.

• Simple Intravenous Infusions • People with medical conditions that disrupt fluid and electrolyte or acid-base balances may need simple IV solutions to help them maintain or restore their body's normal homeostasis. Simple IV solutions are delivered through a peripheral **IV catheter** inserted into a small-diameter peripheral vein, such as those in the forearm and back of the hand. For people who are expected to be unable to eat for more than a few days, especially those who are malnourished or have high nutrient requirements, simple IV solutions are insufficient.

• Peripheral Parenteral Nutrition (PPN) • For some people, nutrient needs can be met using the peripheral veins. **Peripheral parenteral nutrition (PPN)** relies on IV lipid emulsions to provide a concentrated source of kcalories in a form that is isotonic and less irritating to the blood than concentrated dextrose solutions, which irritate the peripheral veins and cause them to collapse.

The relatively large volume of fluid necessary to deliver a limited number of kcalories and the need for accessible and strong peripheral veins limit the use

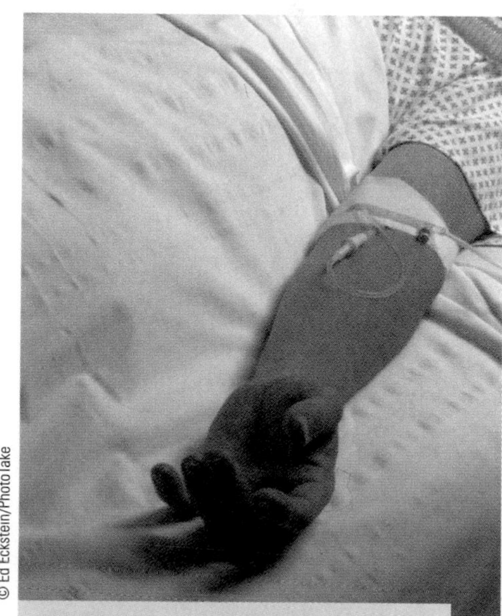

The peripheral veins can provide access to the blood for the delivery of simple IV solutions, PPN, and central TPN.

IV catheter: a thin tube inserted into a peripheral or central vein. Additional tubing connects the IV solution to the catheter.

peripheral parenteral nutrition (PPN): the provision of an IV solution that meets nutrient needs delivered into the peripheral veins.

of PPN. Peripheral parenteral nutrition best suits people with normal renal function who need only short-term nutrition support (about 7 to 14 days), people who need additional nutrients temporarily to supplement an oral diet or tube feeding, and those in whom inserting an IV catheter into a central vein is medically unsound. People who need fluid restrictions, people who have high energy requirements, and people for whom lipid emulsions are contraindicated are not candidates for PPN. In addition, the need for strong peripheral veins limits the use of PPN for people with weak veins that collapse easily.

• Central Total Parenteral Nutrition (Central TPN) • Another method used to meet all nutrient needs by vein is **central total parenteral nutrition,** or **central TPN** for short. In central TPN, the tip of a central venous catheter is either placed directly into a large-diameter central vein or threaded into a central vein through a peripheral vein (see Figure 23-2). The advantage of TPN is that it allows the infusion of concentrated IV solutions, which deliver larger amounts of energy and nutrients in smaller volumes than are possible with PPN solutions. The central veins lie close to the heart, where a large volume of blood rapidly dilutes the TPN solution. Consider that even under resting conditions the heart pumps more than a gallon of blood out of its chambers and into the arteries each minute. By the time the TPN solution reaches the peripheral veins, it is no longer concentrated enough to irritate the blood vessels.

Central TPN is indicated whenever parenteral nutrition will be required for long periods of time, when nutrient requirements are high, or when people are severely malnourished (see Table 23-1). A **peripherally inserted central catheter** is associated with fewer insertion- and infection-related complications

FIGURE 23-2 Central TPN

❶ Traditionally, central TPN catheters enter the circulation at the right subclavian vein and are threaded into the superior vena cava with the tip of the catheter lying close to the heart. Sometimes catheters are threaded into the superior vena cava from the left subclavian vein, the internal jugular vein, or the external jugular vein.

❷ Peripherally inserted central catheters usually enter the circulation at the basillic or cephalic vein and are guided up toward the heart so that the catheter tip rests in the superior vena cava.

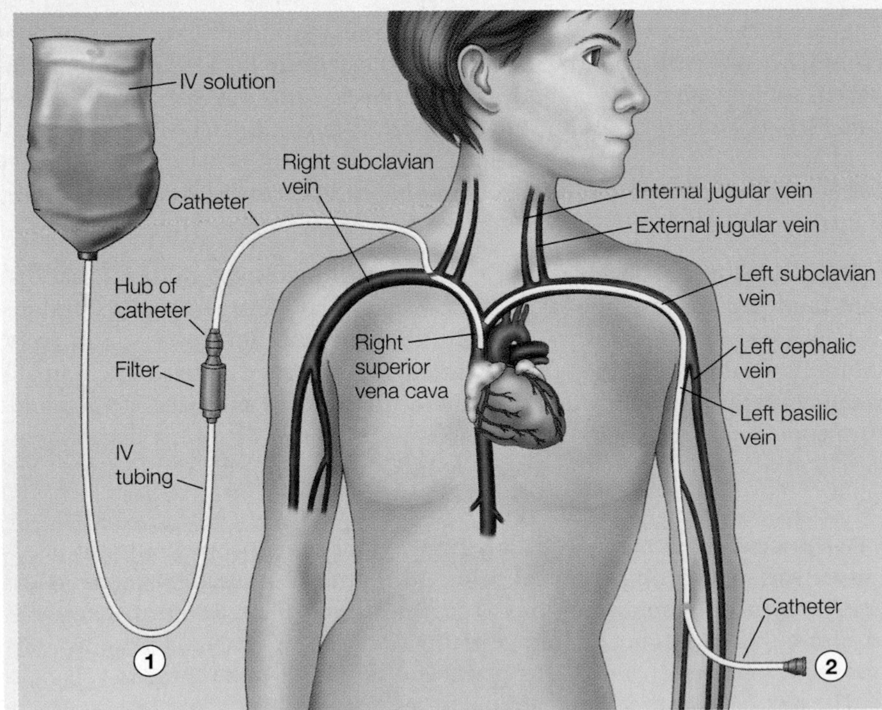

central total parenteral nutrition (central TPN): the provision of an IV solution that meets nutrient needs delivered into a central vein.

peripherally inserted central catheter (PICC): a catheter inserted into a peripheral vein and advanced into a central vein.

TABLE 23-1 Possible Indications for Central TPN

Acquired immune deficiency syndrome (AIDS)
Bone marrow transplants
Extensive small bowel resections
GI tract obstructions
High-output enterocutaneous fistulas
Hypermetabolic disorders, or major surgery, when it is anticipated that the GI tract will be unusable for more than 2 weeks
Intractable diarrhea
Intractable vomiting
Low birthweight with necrotizing enterocolitis (severe GI inflammatory disease) or bronchopulmonary dysplasia (chronic lung disease)
Radiation enteritis (inflammation of intestine caused by radiation)
Severe acute pancreatitis
Severe malnutrition if surgical or intensive medical intervention is necessary
Severe nausea and vomiting associated with pregnancy (hyperemesis gravidarum) when they last more than 14 days
When it is anticipated that adequate enteral nutrition cannot be established within 14 days of hospitalization

Note: Parenteral nutrition is indicated only when enteral nutrition is contraindicated. If short-term parenteral nutrition support is anticipated (less than 14 days), PPN might be preferred.
SOURCE: Adapted from A.S.P.E.N. Board of Directors, Guidelines for the use of parenteral and enteral nutrition in adult and pediatric patients, *Journal of Parenteral and Enteral Nutrition* (supplement) 17 (1993): 1–49.

than a catheter inserted directly into a central vein and is also less costly. Although access to a strong peripheral vein is still required, once inserted, a peripherally inserted central catheter generally remains usable longer than a standard peripheral IV catheter. Thus, when people need TPN to meet nutrient needs for weeks or months or when they need nutrition support but have few strong veins remaining, a peripherally inserted central catheter is an option.[7]

IN SUMMARY

Simple IV solutions serve to maintain fluid, electrolyte, and acid-base balances, but they provide little energy and few nutrients. To meet all nutrient needs by vein, complete nutrient solutions can be delivered using both peripheral and central veins. PPN works best for people with strong peripheral veins who will not need parenteral nutrition for more than a week or two. They must be able to tolerate lipid emulsions and large amounts of fluids. Central TPN meets nutrient needs when clients need long-term parenteral nutrition, when they lack the strong veins necessary for PPN, when they are malnourished, when their nutrient needs are high, or when they have medical conditions that require fluid restrictions.

INTRAVENOUS SOLUTIONS: HOW ARE THEY GIVEN?

Intravenous feedings are like tube feedings in that careful attention to formula selection, preparation, and delivery helps support nutrition status and minimize the risks of complications. To prevent bacterial contamination and ensure the stability of IV solutions, they are carefully compounded in the pharmacy and shielded from light and refrigerated. Prior to infusion, the IV solution is removed from the refrigerator and allowed to reach room temperature. As Table 23-2 on p. 712 shows, many of the risks associated with IV nutrition are more serious than those associated with enteral nutrition. Some of the complications

TABLE 23-2 Complications Associated with Intravenous Nutrition

Catheter- or Care-Related Complications
Air leaking into catheter, obstructing blood flow (air embolism)
Air or gas in the chest (pneumothorax)[a]
Blood clot (thrombosis)
Blood in the chest (hemothorax)[a]
Catheter inadvertently placed in subclavian artery (arterial puncture)[a]
Catheter occlusion
Catheter tip broken off, obstructing blood flow (catheter embolism)[a]
Fluid in the chest (hydrothorax)[a]
Hole or tear in heart made by catheter tip (myocardial perforation)[a]
Improperly positioned catheter tip
Infection
Inflammation of vein (phlebitis)
Infusion pump malfunctions
Sepsis
Metabolic or Nutrition-Related Complications
Acid-base imbalances
Bone demineralization
Coma from excessive glucose load (hyperosmolar, hyperglycemic, nonketotic coma)[a]
Dehydration
Electrolyte imbalances
Elevated blood glucose (hyperglycemia)
Elevated liver enzymes
Essential fatty acid deficiency[a]
Fatty liver
Fluid overload
High blood ammonia levels (hyperammonemia)
Low blood glucose (hypoglycemia)
Trace element deficiencies
Vitamin, mineral and trace element deficiencies

[a]Central TPN.

directly involve the catheter or the delivery of IV solutions; others occur as a consequence of altered metabolism.

Insertion and Care of Intravenous Catheters

Insertion of a peripheral catheter for PPN is the same as for simple IV solutions. Skilled nurses can place peripherally inserted catheters for simple IVs, PPN, or TPN, but a catheter for direct central vein access requires surgical insertion by a qualified physician. The client may be awake for the procedure, but is given a local anesthetic. Unnecessary apprehension can be avoided by explaining the procedure to the client beforehand.

The difficulty of maintaining the integrity of peripheral veins frequently interferes with the ability to sustain PPN. Veins may become inflamed and sometimes infected. Often the catheter must be removed and reinserted at a new site; consequently, long-term feedings are difficult and rarely indicated. Peripherally inserted central catheters are less irritating to the veins and can often remain in place longer than PPN catheters.

Infections can develop at the catheter site in both PPN and central TPN. Compared with peripherally inserted catheters (for either PPN or TPN), though, central TPN presents a greater risk of introducing disease-causing microorganisms into the bloodstream◆ because the catheter is inserted so near the heart. Nurses inspect the catheter site regularly and use aseptic techniques when changing the dressing covering a catheter insertion site.

◆ Reminder: The presence of disease-causing organisms in the blood is *sepsis*—a life-threatening complication that may occur as a consequence of parenteral nutrition.

Administration of Complete Nutrient Solutions

Just as a tube feeding is started slowly to give the GI tract time to adapt to the formula, a complete nutrient solution is started slowly to allow time for the body to adapt to the glucose concentration and osmolality of the solution. An infusion pump◆ ensures an accurate and steady delivery rate. Rapid changes in the infusion rate, especially for central TPN, can cause severe hyperglycemia or hypoglycemia, which can lead to coma, convulsions, and even death. When the administration of solution gets behind or ahead of schedule, the drip rate should be adjusted to the correct hourly infusion rate, but no attempt should be made to speed up or slow down the rate to meet the originally ordered volume. All changes to the infusion rate must be made gradually and cautiously, and infusion pumps must be checked regularly for malfunctions. Problems are more likely to occur in people with organ dysfunction or in infants with immature organ systems.

◆ Reminder: Keeping IV poles and infusion pumps at a safe distance from young children helps to prevent the children from toppling the pole or pump, potentially injuring themselves and damaging the equipment.

Electrolytes and blood glucose are monitored vigilantly initially and periodically thereafter. If tests indicate electrolyte imbalances or unacceptably high blood glucose,◆ the causes are investigated and treated. Table 23-3 provides guidelines for monitoring clients on PPN and TPN.

◆ To prevent hyperglycemia, the final volume of a TPN solution designed for a client should provide no more than 7 g of carbohydrate per kilogram of body weight per day.

• **Intravenous Lipid Infusion** • Traditionally, IV fat emulsions are infused separately from the base solution containing amino acids, dextrose, and micronutrients. People sometimes experience adverse reactions to IV lipid emulsions, particularly when the IV lipids are given in large amounts or administered too rapidly. Immediate reactions may include fever, warmth, chills, backache, chest pain, allergic reactions, palpitations, rapid breathing, wheezing, cyanosis, nausea, and an unpleasant taste in the mouth. To guard against adverse reactions, the client receives only small amounts of lipid emulsion over the first 15 to 30 minutes. After that time, the rate can be increased.

TABLE 23-3 Guidelines for Monitoring Clients Receiving Parenteral Nutrition

Before starting TPN:	Complete nutrition assessment.
	Record weight.
	Confirm placement of catheter tip by X ray.
	Check blood glucose, electrolytes, chemical profile, and complete blood count.
Every 4 to 6 hours:	Check blood glucose.[a]
	Monitor vital signs.
	Check pump infusion rate.
Daily:	Monitor weight changes.[b]
	Record intake and output.
	Inspect catheter site for signs of inflammation and infection.
Weekly:	Reassess nutrition status.
	Monitor serum proteins, ammonia, enzymes, triglycerides, cholesterol, and bleeding indices.
	Check the complete blood count.
As needed:	Monitor serum transferrin, electrolytes, calcium, magnesium, phosphorus, blood urea nitrogen, and creatinine.

[a]If blood glucose levels are stable, they can be checked once daily after the first week.
[b]After the first week, weight can usually be measured less often (about twice a week).

When IV lipid emulsions are used as an energy source, they are often added directly to the base solution and infused along with it. The use of **total nutrient admixtures** for clients in the hospital as well as at home has grown dramatically. Total nutrient admixtures must be compounded carefully, refrigerated prior to use, and mixed gently before they are infused.

• Cyclic Infusion • A person on **cyclic parenteral nutrition** receives a TPN solution at a constant rate for 8 to 12 hours a day with breaks in the infusion during the rest of the day. The infusion can be given during the night to allow freedom for routine daytime activities. Consequently, cyclic parenteral nutrition is suited to long-term TPN, especially for clients who receive TPN at home. When a person receives a TPN solution continuously, insulin levels stay high. As a result, the person cannot mobilize fat stores for energy or for essential fatty acids; eventually, fat may be deposited in the liver. Cyclic TPN reverses these problems. Additionally, fewer kcalories seem to be effective in maintaining nitrogen balance, probably because the person uses body fat for energy. Some people, however, cannot tolerate the delivery of a day's volume of solution over a short period of time.

Discontinuing Intravenous Nutrition

Although some clients will need parenteral nutrition for the rest of their lives, most often the medical problem causing the need for IV nutrition resolves, and the client can gradually shift to an enteral diet. The transition requires careful planning, especially when the client has been receiving no nutrients enterally for a long time. During long periods of disuse, the intestinal villi shrink and lose some of their function. Reintroducing nutrients to the GI tract at the appropriate rate and volume will stimulate the progressive restoration of the villi's normal structure and function and prevent malabsorption and other GI discomforts.

total nutrient admixtures: intravenous solutions that contain all nutrients, including lipid emulsions; also called **three-in-one (3-in-1) admixtures** or **all-in-one admixtures.**

cyclic parenteral nutrition: the continuous administration of a parenteral solution for 8 to 12 hours with time periods when no nutrients are infused.

To prevent hypoglycemia when a person is being taken off central TPN, the infusion rate of the solution must be tapered off gradually. PPN can be discontinued without tapering.

• **Transitional Feedings** • The transition from IV feeding to an enteral diet can be accomplished in different ways and often involves a combination of feeding methods. One way is to start an oral diet while the person is still on IV nutrition. The diet is often progressive, beginning with liquids provided in small amounts. If the person cannot eat enough food to meet at least 50 percent of daily nutrient needs within a few days, and intake does not seem to be improving, a tube feeding may be considered.

Whether a person is given a tube feeding or provided an oral diet, the volume of the IV solution is reduced as the volume of enteral feeding is increased. The person who cannot tolerate enteral feedings can still rely on TPN to meet nutrient needs. Parenteral nutrition can be discontinued when at least 70 to 75 percent of estimated energy needs are being met by oral intake, tube feeding, or a combination of the two.[8] Chapter 22 described the transition from tube feedings to table foods.

• **Psychological Effects** • Returning to oral intake after having been fed either intravenously or by tube can have a variety of psychological effects. Some people may be extremely eager to eat again, and food can be an important morale booster. Others may be apprehensive about eating, particularly if they have had extensive GI problems. Appetite may be slow to return for some. In such circumstances, all members of the health care team can support the successful reintroduction of food. Recognize the person's concerns, and provide reassurances that help will be available throughout the process. The many decisions surrounding the provision of TPN require careful consideration. The case study presents an example for review.

IN SUMMARY

Complete nutrient solutions are given slowly at first to allow the body time to adapt to the glucose concentration and osmolality of the solution. Changes are made carefully and slowly. Parenteral nutrition can be given either continuously for 24 hours or over 8 to 12 hours with breaks in the infusion during the rest of the day. As the need for parenteral nutrition resolves, clients are gradually shifted to an enteral diet (either an oral diet or tube feeding), while the volume of parenteral nutrition is gradually reduced.

SPECIALIZED NUTRITION SUPPORT AT HOME

Occasionally, a client must continue to receive nutrition support (tube feedings or parenteral nutrition) after the primary medical condition has stabilized. In such a case, home nutrition support might be an option.

Since the first report of a person sent home successfully on TPN in 1969,[9] the use of home nutrition support has expanded rapidly. According to recent estimates, about 40,000 people receive parenteral nutrition at home, and more than 150,000 receive enteral nutrition at home.[10] As the number of people benefiting from home nutrition programs continues to grow, health care professionals who work with these programs are gaining valuable experience and improving the quality of home nutrition support. Medical supply companies provide the equipment, formulas, and service necessary to support home nutrition care.

CASE STUDY

Artist Requiring Parenteral Nutrition

Adam Goldstein, a 25-year-old artist with Crohn's disease underwent an extensive small bowel resection five days ago (see p. 670 in Chapter 21). Adam had received central TPN prior to surgery and continues to receive it. After ten days, a tube feeding was begun in very small amounts.

1. Review possible reasons for the use of central TPN in the management of active Crohn's disease. Why would Adam require central TPN following surgery? How would you explain the need for central TPN to Adam?
2. Describe the components of a typical TPN solution. Calculate the energy content of 1 liter of a solution that provides 140 grams dextrose, 45 grams amino acids, and 20 grams lipid. If Adam's energy requirements are 2100 kcalories per day, how many liters of solution will he need each day?
3. Why are enteral feedings important for Adam, even though he relies on parenteral nutrition to meet his nutrient needs? Assuming that Adam will eventually be able to tolerate the tube feeding, how will the health care team help Adam make the transition from parenteral feedings to tube feeding? How will the health care team know when it is safe to take Adam off TPN? Consider some of the physiological and psychological problems Adam might face when he begins eating an oral diet.
4. If Adam is unable to meet nutrient needs orally, he may need to continue a tube feeding or central TPN at home. As you read through the following sections, think about the factors health care professionals might look for in determining if Adam will be a candidate for a home nutrition program. Consider some of the benefits of a home program for Adam. What are some of the problems he might encounter on a home program?

The Basics of Home Programs

As with tube feedings and TPN in the hospital, the main objective of home nutrition support is to maintain or achieve adequate nutrition status. Nutrition support at home, however, has an added dimension: it permits a person to continue nutrition care in familiar surroundings. If you have ever been in the hospital, or taken a long trip for that matter, you probably remember the comfort you experienced when you returned to your own bed, knew where things were, and could get things when you needed them.

• Candidates for Home Nutrition Support • The nutrition support team most frequently determines whether a client is a candidate for a home nutrition program. In addition to medical considerations, the candidate for home nutrition support and those who care for that person must have rational, stable personalities so that they can successfully handle the problems that arise. They must be capable of learning the necessary techniques and of dealing with complications. They must be willing to comply with recommendations. They must have a home that is clean and has electricity, refrigeration, formula preparation and storage areas, a telephone for emergencies, and a safe water supply. They must also have adequate financial resources and access to the equipment, supplies, and professional support that are integral components of a successful home program.

• Roles of Health Care Professionals • The health care team not only assesses the client's medical and nutrition needs, but also evaluates the individual's home environment so that they can plan realistic feeding and care schedules and find practical solutions to the unique problems the client will face at home. Health care teams who work with home nutrition programs must also understand the regulations that govern payment for home nutrition programs by both government and private insurance. They must use this knowledge to document the need for services to ensure maximum reimbursement for the client.

Once a home nutrition program is initiated, the physician directs the program and monitors the client's care. Nurses train the client in the appropriate techniques; dietitians monitor nutrition status; pharmacists coordinate the delivery of formulas, medications, and supplies with the home care company; and social workers provide insurance assistance. All team members are expected to answer questions and provide emotional support. Home visits by a nurse, and sometimes the dietitian, help ensure that the client is able to comply with the demands of the home care program. A qualified nurse, dietitian, or physician must be available to answer questions and handle problems as they arise.

• **Cost Considerations** • One recent analysis found that the average cost of maintaining a home parenteral nutrition program was $70,000 per year with a range of $15,000 to $169,000 per year.[11] Costs for enteral nutrition programs are lower (average $18,000 per year with a range of $5,000 to $50,000 per year), but are still substantial.◆ Insurance and Medicare often pay for a portion of the costs, but people on home programs are also more likely than others to have high medical costs for managing their diseases as well. It is easy to see why the economic impact of home nutrition programs is often the primary concern for people on these programs.

◆ To compare the expense of receiving nutrients from parenteral or enteral solutions with the cost of table foods, consider that the average monthly "grocery bill" for parenteral and enteral nutrition is $5800 and $1500, respectively, per person.

• **Adjustments for Clients** • Although home nutrition programs can extend and improve the quality of life, clients may struggle with the many lifestyle adjustments the programs entail. In addition to the economic impact of the therapy, clients may find the demands of the home schedule to be frustrating, inconvenient, and monotonous. Among physical complaints, clients cite frequent urination (due to high volumes of fluids), disturbed sleep (often as a consequence of waking up to go to the bathroom), and feeling weak as common problems.[12] Clients say that the lifestyle areas most affected by home nutrition programs are sleep, travel, exercise, and social life.[13] Despite the big impact a home program can have on lifestyle, clients report that their lives have improved with home nutrition therapy.[14] Many people who require nutrition support at home can also eat some foods by mouth, and some resume activities, such as going to work, driving, and playing sports.◆

◆ Return to part four of the Case Study on p. 715 and provide responses.

Portable pumps and convenient carrying cases allow people who require nutrition support at home to move about freely.

How Home Enteral and Parenteral Nutrition Programs Work

People on home enteral nutrition programs commonly have cancer or neurological disorders that interfere with swallowing. People on home TPN often have acquired immune deficiency syndrome (AIDS), cancer, Crohn's disease, short-bowel syndrome, or other intestinal disorders. Sometimes home TPN is relatively short term, lasting for weeks or months as opposed to lifelong.

• **Home Enteral Nutrition** • Both transnasal tubes and tube enterostomies provide access to the GI tract for home tube feedings (see Chapter 22). When possible, intermittent feeding schedules are arranged so that clients are free to move around between meals. Some people are able to meet some of their nutrient needs by eating during the day and using tube feedings at night. Clients on continuous feedings who must use pumps can obtain small, lightweight pumps that are easily concealed in carrying cases and allow freedom of movement.

• **Home TPN** • Different types of home TPN programs are currently in use. Ideally, clients are given as much responsibility for their own care as they can handle. For example, a client who is capable of changing the catheter dressings is trained to do so. Typically, caregivers also learn the techniques so that they can assist as needed.

Special catheters designed for long-term use are often inserted for home TPN. Many times these catheters are tunneled under the skin so that the catheter exit site lies in the abdominal area where the client can access and care for it easily. The day's volume of TPN solution is frequently delivered within 8 to 12 hours using an infusion pump. The client infuses the solution while sleeping or at some other convenient time and thus is free to move about unencumbered for much of the day. For those who need continuous feedings, a lightweight carrying case that holds a small pump and IV bags enables the client to move around freely.

IN SUMMARY

Unquestionably, tube feedings and parenteral nutrition provide a lifesaving feeding alternative for people who cannot eat traditional diets either in the hospital or at home. The Nutrition Assessment Checklist reviews areas of concern for people on parenteral nutrition support.

Nutrition Assessment Checklist for People Receiving Parenteral Nutrition

MEDICAL HISTORY

Check the medical record for medical conditions that:

- ☐ Prevent the use of enteral nutrition
- ☐ Suggest how long parenteral nutrition will be required
- ☐ Indicate the appropriate feeding route (peripheral versus central)

Review the medical record for complications that may suggest the need to alter the IV solution or the administration technique:

- ☐ Hyperglycemia/hypoglycemia
- ☐ Dehydration/overhydration
- ☐ Electrolyte imbalances
- ☐ Acid-base imbalances
- ☐ Essential fatty acid deficiencies
- ☐ Vitamin, mineral, and trace element deficiencies
- ☐ Fatty liver

MEDICATIONS

Check medications and TPN solution for:

- ☐ Physical compatibility with the TPN solution
- ☐ Effects of TPN solution on medication absorption and availability

When a medication is added directly to the TPN solution:

- ☐ Note times that the TPN solution is not delivered
- ☐ Note that the full amount of medication is not delivered when the feeding is stopped

FOOD/NUTRIENT INTAKE

If the client is not meeting nutrition goals, check to see if the:

- ☐ TPN solution is being delivered as prescribed
- ☐ Infusion pump is working correctly
- ☐ Client's nutrient needs were correctly determined

When the client is ready to begin an enteral feeding program:

- ☐ Record all amounts of foods or formulas eaten or provided enterally
- ☐ Reduce volume of TPN solution as enteral intake increases

ANTHROPOMETRICS

Measure baseline height and weight and daily weights. For clients who are not meeting goals for weight:

- ☐ Reestimate energy needs and verify that the TPN solution is delivering an amount sufficient to meet those needs
- ☐ Check to see if formula is being delivered as prescribed
- ☐ For rapid weight changes, check for physical and laboratory signs of dehydration or overhydration

LABORATORY TESTS

Check serum and urine tests for signs of:

- ☐ Dehydration/overhydration
- ☐ Hyperglycemia/hypoglycemia
- ☐ Electrolyte imbalances
- ☐ Acid-base imbalances
- ☐ Vitamin and trace element deficiencies
- ☐ Essential fatty acid deficiencies
- ☐ Adequacy of protein intake (prealbumin and retinol-binding protein), when available
- ☐ Improvement or deterioration of medical condition, which can alter nutrient needs

PHYSICAL SIGNS

Check:

- ☐ Catheter insertion site for signs of infection or inflammation
- ☐ Blood pressure, temperature, pulse, and respiration for signs of fluid, electrolyte, acid, and base imbalances

Look for physical signs of:

- ☐ Dehydration/overhydration
- ☐ PEM
- ☐ Essential fatty acid deficiencies
- ☐ Vitamin, mineral, and trace element deficiencies

Nutrition on the Net

WEBSITES

Access these websites for further study of the topics covered in this chapter.

- Find updates and quick links to these and other nutrition-related sites at our website: **www.wadsworth.com/nutrition**

Find more information about organizations that support the appropriate use of parenteral and enteral nutrition by visiting:

American Society for Parenteral and Enteral Nutrition: **www.nutritioncare.org**

Canadian Parenteral-Enteral Nutrition Association: **www.magi.com/~cpena**

European Society of Parenteral and Enteral Nutrition: **www.espen99.org**

Explore resources available to clients on home nutrition programs by visiting the Oley Foundation site: **www.wizvax.net/oleyfdn**

INTERNET ACTIVITIES

Sometimes nutrients must be delivered directly into the bloodstream, bypassing the GI tract. For example, people with electrolyte imbalances may need simple intravenous solutions to help restore fluid-electrolyte balance. In people with more serious conditions, however, total nutrient needs may have to be met by way of intravenous nutrition. You may want to learn more about parenteral nutrition and the people who need it.

- Go to **www.wizvax.net/oleyfdn**
- Click on "The Lifeline Letter."
- Scroll down through the article listed, click on one that interests you, and read it.

After reading one or more of the articles, consider how you think you would cope with long-term parenteral nutrition. How would it affect your lifestyle?

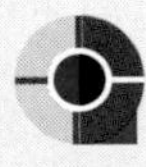

Study Questions

These questions will help you review the chapter. You will find the answers in the discussions on the pages provided.

1. How are protein, carbohydrate, and fat provided in IV solutions? What other nutrients or other additives may be present? (pp. 707–708)
2. Describe the characteristics of simple IV solutions, complete nutrient solutions delivered through peripheral veins, and complete nutrient solutions delivered through central veins. (p. 708)
3. Who can benefit from PPN? In what cases would PPN be inappropriate? (pp. 709–710)
4. When is central TPN preferred to PPN? In what ways is central TPN different from PPN? (pp. 710–711)
5. What precautions must be taken when PPN, central TPN, and IV fat emulsions are initiated? (p. 712)
6. How do clients make the transition from parenteral to enteral nutrition? (pp. 713–714)
7. What are the advantages of home enteral and parenteral nutrition programs? What are some factors that help determine if a person is a candidate for a home nutrition program? What problems do people on home programs face? (pp. 715–716)

The questions will help you prepare for an exam. Answers can be found on p. 720.

1. Which of the following cannot be delivered intravenously?
   a. dextrose
   b. amino acids
   c. lipid emulsions
   d. hydrolyzed enteral formulas
2. A simple IV solution is most appropriate for people who:
   a. are malnourished.
   b. have high nutrient needs.
   c. can't eat for long periods of time.
   d. are well nourished and are expected to eat in a few days.
3. For a client receiving central TPN who also receives IV lipid emulsions two or three times a week, the lipid emulsions serve primarily as a source of:
   a. vitamin C.
   b. essential fatty acids.
   c. fat-soluble vitamins.
   d. concentrated energy.

4. Compared to solutions delivered by peripheral vein, solutions delivered by central vein provide:
   a. more fat, less dextrose.
   b. more dextrose, less fat.
   c. a lower osmolality.
   d. more vitamins and minerals.
5. Among the following, the people least likely to benefit from PPN are those:
   a. who need long-term IV nutrition support.
   b. on oral or tube feedings who need additional nutrients temporarily.
   c. with normal renal function who need short-term IV support.
   d. in whom inserting an IV catheter into a central vein would be difficult.
6. The person who is a good candidate for central TPN rather than PPN:
   a. is well-nourished.
   b. does not have high nutrient requirements.
   c. needs long-term parenteral nutrition support.
   d. has strong peripheral veins and moderate nutrient needs.
7. The solutions infused through peripherally inserted central catheters are the same as those infused for:
   a. PPN.
   b. simple IVs.
   c. central TPN.
   d. enteral feedings.
8. Which type of IV insertion procedure is associated with the highest risk of sepsis?
   a. central TPN catheter
   b. peripheral catheter for PPN
   c. peripheral catheter for simple IVs
   d. peripherally inserted central catheter
9. A gradual transition from an IV feeding to an oral diet is primarily designed to:
   a. improve the appetite.
   b. prevent hypoglycemia.
   c. prevent apprehension about eating.
   d. ensure that nutrient needs will continue to be met.
10. The health care team evaluating a client's ability to manage a home enteral or parenteral nutrition program would be least concerned with:
    a. the client's financial resources.
    b. the client's psychological status.
    c. the exact composition of the enteral formula or IV solution.
    d. the client and caregivers' abilities to learn the necessary techniques and follow medical instructions.

Clinical Applications

1. One liter of a TPN solution contains 500 milliliters of 50 percent dextrose and 500 milliliters of 5 percent amino acids. Determine the daily kcalorie and protein intakes of a person who receives 2 liters of such a TPN solution. Calculate the average daily energy intake if the person also receives 500 milliliters of a 20 percent fat emulsion three times a week.

2. Consider what it must be like to be on a home TPN program with no foods allowed by mouth. What would be the advantages of being at home instead of in the hospital? What would be the disadvantages? Think of how you would manage feedings. How would you feel about the time, costs, and commitment required to maintain this therapy? If you were not permitted to take any foods by mouth, how would you feel about not being able to eat after a long time? How would you handle holidays and special occasions that often center around food?

References

1 D. W. Wilmore and S. J. Dudrick, Growth and development of an infant receiving all nutrients exclusively by way of the vein, *Journal of the American Medical Association* 203 (1968): 860–864.

2 U.S. Food and Drug Administration, Parenteral multivitamin products; Drugs for human use; Drug efficacy study implementation; Amendment, *Federal Register* 65 (77; April 20, 2000): 21200–21201.

3 National Advisory Group on Standards of Practice Guidelines for Parenteral Nutrition, *Journal of Parenteral and Enteral Nutrition* 22 (1998): 49–66.

4 As cited in C. A. Wardle and coauthors, Hypermanganesemia in long-term intravenous nutrition and chronic liver disease, *Journal of Parenteral and Enteral Nutrition* 23 (1999): 350–355.

5 National Advisory Group on Standards of Practice Guidelines for Parenteral Nutrition, 1998.

6 National Advisory Group on Standards of Practice Guidelines for Parenteral Nutrition, 1998.

7 J. Z. Rogers, K. McKee, and E. McDermott, Peripherally inserted central venous catheters, *Support Line,* October 1995, pp. 6–9; S. C. Loughran and

M. Borzatta, Peripherally inserted central catheters: A report of 2506 catheter days, *Journal of Parenteral and Enteral Nutrition* 19 (1995): 133–136.

8 T. C. Clark, Nutrition support clinical pathways, *Nutrition in Clinical Practice* 11 (1996): 16–20; The 1995 A.S.P.E.N. standards for nutrition support: Hospitalized patients, *Nutrition in Clinical Practice* 10 (1995): 206–207.

9 M. E. Shils and coauthors, Long-term parenteral nutrition through an external arteriovenous shunt, *New England Journal of Medicine* 283 (1970): 314–343.

10 P. Reddy and M. Malone, Cost and outcome analysis of home parenteral and enteral nutrition, *Journal of Parenteral and Enteral Nutrition* 22 (1998): 302–310.

11 Reddy and Malone, 1998.

12 M. F. Winkler, Nutrition support from hospital to home, Presented at the 82nd Annual Meeting of the American Dietetic Association, October 21, 1999.

13 Reddy and Malone, 1998.

14 M. Malone, Effect of home nutrition support on patient's lifestyle (abstract), *Journal of Parenteral and Enteral Nutrition* (supplement) 19 (1995): 23.

ANSWERS

Multiple Choice

1. d	2. d	3. b	4. b	5. a
6. c	7. c	8. a	9. d	10. c

NUTRITION AND COST-CONSCIOUS HEALTH CARE

DECADES OF medical research have resulted in an explosion of knowledge and technologies to diagnose and treat diseases. Among the benefits spurred by this explosion are a greater understanding of the GI tract and the ability to safely feed people by tube and by vein. This astounding progress, however, has come at a tremendous financial cost. The United States spends more money on health care than any other nation, yet the high cost of health care has put medical care out of the reach of many citizens. Without attention to cost containment, the health status of the nation is threatened; sophisticated medical services do little good if people cannot use them. The medical community, which once embodied the idealistic approach of sparing no cost when it came to health care, has embraced the reality that cost is an element of quality.[1] To discuss all the ramifications of cost containment for health care delivery is beyond the scope of this highlight. Instead, this discussion identifies major trends affecting medical services, including nutrition services, and points out some ways nutrition can help reduce health care costs.

MEASURES TO REDUCE HEALTH CARE COSTS

Among other things, efforts to control health care costs aim to eliminate duplication of services, reduce the number of hours spent in client care, maximize the use of health care professionals, and limit access to unnecessary services. The task is to cut costs without compromising the quality of care.

A major change in health care delivery that has evolved from efforts to control costs is the shift from the traditional **indemnity insurance** (fee-for-service) system to **managed care** (see the glossary on p. 722). A traditional fee-for-service health insurance plan allows clients to use the physicians and health care facilities of their choice and then reimburses the clients for a percentage of the services covered by the plan. Managed care organizations, which include **health maintenance organizations (HMOs)** and **preferred provider organizations (PPOs),** strive to provide high-quality, low-cost care by controlling the access to and cost of services. Each client selects a primary care physician, who then determines what services the client needs. Managed care is moving toward a capitated system. **Capitation** is a system where health care providers agree to supply all the services a client needs for a set monthly fee. Because the providers assume the financial responsibility if they fail to operate within a predetermined budget, they have a strong incentive to control costs. Managed care organizations, which were rare just 15 years ago, are increasingly replacing traditional insurance plans. While managed care organizations have been effective at controlling costs, many consumers and health care professionals worry that quality of care has taken a backseat.

QUALITY OF CARE

What is quality health care? Obviously, obtaining the best treatments to maintain optimal health and treat illness is an important part of quality health care. To deliver such care requires the services of competent, skilled, and caring health care professionals who attend to clients' physical, emotional, intellectual, social, and spiritual needs. In today's cost-conscious health care environment, an added requirement is "at the least cost."

Cost-cutting measures have redefined the responsibilities of health care professionals as well as the settings in which they work.

Measuring Quality of Care in Terms of Costs

In cost-conscious health care, quality of care is most often defined in terms of outcomes.[2] Examples of desirable outcomes include the person's ability to function independently, reduced number or length of hospital stays,

GLOSSARY OF HEALTH CARE INSURANCE TERMS

capitation: prepayment of a set fee per client in exchange for medical services.

health maintenance organizations (HMOs): managed care organizations that limit the subscriber's choice of health care professionals to those affiliated with the organization and control access to services by directing care through a primary care physician.

indemnity insurance: traditional fee-for-service insurance.

managed care: a health care delivery system that aims to provide cost-effective health care by coordinating services and limiting access to services.

preferred provider organizations (PPOs): managed care organizations that encourage subscribers to select health care providers from a group that has contracted with the organization to provide services at lower costs.

prevention of diseases and complications, and extended survival time. Desirable outcomes include those that improve a person's quality of life, those that save money, or both. People who are able to function independently, for example, enjoy the freedom from dependence on others and also save the expense of having others care for them.

Health care services are cost justified when they have a reasonable chance of improving a client's outcome. Services that produce the best outcomes at the lowest costs are the most cost-effective. The more expensive the procedure or test, the more critical justifying its cost becomes. Thus it may not be cost-effective to perform a nutrition assessment on every client admitted to a hospital, whereas nutrition screening is cost-effective. For another example, providing enteral formulas orally to clients who can drink them is more cost-effective than providing the formula by tube.

Nutrition and Quality of Care

A review of the desirable outcomes listed above reveals many ways that nutrition can affect quality of care and reduce health care costs. By providing clients with the nutrients they need to maintain both physical and mental health, nutrition can significantly improve the client's ability to function independently, prevent or forestall diseases, limit complications, and maintain quality of life. All of these factors can help prevent the need for medical treatments, hospitalization, or lengthy hospital stays. Both timely identification of nutrition problems and appropriate intervention are essential to improving outcomes and cutting costs. Malnutrition, a persistent and common problem in hospitalized clients, is associated with significantly longer hospital stays, higher costs, and the need for home health care.[3] Based on one analysis of data from several sources, researchers estimate that early attention to malnutrition in hospitalized clients can reduce length of stay and may result in cost savings of approximately $8300 per hospital bed per year.[4] A recent study found that clients whose nutrition status deteriorated during the course of hospitalization experienced higher rates of complications and incurred higher hospital costs than clients who maintained their nutrition status.[5]

Outside the hospital, attention to nutrition can be cost-effective as well. The authors of a review article estimate a potential savings of billions of health care dollars from an increased public awareness of the roles of nutrition in preventing several common disorders (type 2 diabetes, hypertension, coronary heart disease, stroke, cancer, osteoporosis, and depression) and consequent changes in eating habits.[6] Medical nutrition therapy can also have a significant impact on the costs of treating common chronic diseases such as diabetes and cardiovascular diseases.[7]

Medical nutrition therapy is a relatively inexpensive service, and, as the examples above indicate, it can also be cost-effective. (Note: In most managed care settings, medical nutrition therapy is treated as specialty care that requires referral by the primary care physician.) In representing nutrition professionals, the American Dietetic Association takes the position that managed care organizations and integrated health delivery systems should provide medical nutrition therapy as an essential component of health care and that it should be provided by qualified nutrition professionals.[8] The task is to document the cost savings of nutrition intervention and ensure that nutrition is addressed in standards that guide the care of clients.

Standards of Care and Nutrition

Health care professionals and the organizations that represent them are working together to define outcome-oriented standards of practice and measures of care. The Joint Commission on Accreditation of Healthcare Organizations (JCAHO), an agency that oversees the accreditation of health care facilities, requires compliance with nutrition care standards to identify, address, and monitor each client's nutrition needs.[9] The standards promote coordination and communication among disciplines in an effort to improve quality of care in a cost-efficient manner.[10]

CHANGING ROLES OF HEALTH CARE PROFESSIONALS

One of the positive trends spurred by cost-cutting measures is the replacement of traditional health care delivery, which is discipline- or department-specific, with a multidisciplinary or team approach. By working together, health care professionals can coordinate their services, which eliminates unnecessary services and solves problems in a timely and efficient manner. Traditionally, for example, nurses performed nursing functions and left the responsibility for the nutrition care of clients to the dietitian. When a client was admitted to the hospital, the nurse would perform a nursing assessment, and the dietitian, a nutrition assessment. Yet much of the information from these assessments overlaps. As Chapter 17 described, a nurse can use the nursing assessment to determine the client's nutrition risk and make appropriate referrals to the dietitian. The dietitian, in turn, can rely on much of the information the nurse collects during the nursing assessment to complete a nutrition assessment.

Nutrition Support Teams

Nutrition support teams often include a physician, nurse, dietitian, and pharmacist. Team members have a special interest in and in-depth knowledge of nutrition assessment, tube feedings, and parenteral nutrition. As Chapters 17, 22, and 23 show, the appropriate use of these techniques requires skill and considerable amounts of time.

• Responsibilities of the Nutrition Support Team • The team members develop standards and protocols for nutrition screening, nutrition assessment, tube feedings, and parenteral nutrition; identify and monitor the care of people who need nutrients from enteral formulas or parenteral nutrition; and serve as consultants to the rest of the institution's staff. Team members also identify and solve problems regarding individual clients, problems with procedures or equipment, or issues appropriate to their specialty area.

The team analyzes new products and reviews current research to determine which products and services might serve their clients. As an example, the nutrition support team might evaluate open and closed enteral feeding systems to see if the higher initial costs of a closed feeding system might be offset by a lower incidence of bacterial contamination of formulas and reduced hospital costs for treating infections. Figure H23-1 shows the responsibilities of individual team members as well as those shared by the team. Nutrition support teams play an important role in ensuring quality nutrition care while reducing health care costs.

• Coordinating Services to Save Costs • To understand the ways that communication among health care professionals and coordination of services can save costs, consider this example. During nutrition support team rounds, the dietitian mentions a recurring problem with the collection of urine for nitrogen balance studies, which has invalidated several test results, wasted time (and money), and eliminated a source of information that could improve client care. The nurse recalls that many of the nurses are having difficulty with the procedure and suggests an in-service session on the proper techniques for nitrogen balance studies. The nurse agrees to schedule and conduct the session with the help of the dietitian. In an alternative scenario, the pharmacist (reacting to the dietitian's comments) relays information he read about a cost-effective and efficient procedure for conducting nitrogen balance studies. The team agrees that each team member will review the article and decide at the next meeting whether to test the new procedure. Clients benefit from many eyes noting problems before they become serious and many brains searching for better ideas and solutions.

In a similar manner, the team relies on the expertise of its members to develop protocols for cost-effective, high-quality care. One example might be the development of a clinical pathway (see Chapter 17). For the nutrition

FIGURE H23-1 The Nutrition Support Team

The physician
- Diagnoses medical problems
- Performs medical procedures
- Coordinates and prescribes therapy
- Directs and supervises team
- Approves guidelines and protocols
- Consults with other physicians

The nurse
- Assesses nursing needs
- Performs direct client care
- Explains medical procedures and treatment plans
- Instructs clients regarding medical care
- Acts as a liaison between team and nursing staff
- Coordinates discharge plans

All team members
- Review current research
- Analyze new products
- Develop guidelines
- Provide in-service training
- Monitor clients
- Correct problems
- Educate clients
- Evaluate the outcome of the care provided and cost savings
- Promote the appropriate use of nutrition support
- Improve communications among team members and between the team and other health care professionals

The dietitian
- Assesses nutrition status
- Determines clients' nutrient needs
- Recommends appropriate diet therapy
- Reevaluates clients regularly
- Instructs clients about their diets
- Acts as a liaison between the team and the dietary department

The pharmacist
- Recommends appropriate drug therapy
- Identifies medication-medication and diet-medication interactions
- Identifies medication-related complications
- Educates clients about their medications
- Acts as a liaison between the team and the pharmacy

support team, examples might include clinical pathways for enteral and parenteral nutrition. Such a plan defines a time frame for each intervention with a goal of providing the best outcome at the lowest cost. Once the plan is in place, each team member assists in studying unexpected outcomes or variances from the expected course of treatment and fine-tuning the plan as necessary to achieve the best outcomes.

Specific examples of cost savings generated by nutrition support teams include:

- Preventing overuse of parenteral nutrition, which is costly. In one hospital that implemented procedures to reduce the inappropriate use of parenteral nutrition, quality nutrition care could be achieved with an estimated cost savings of over $225,000 for one year.[11] In another study, reducing the inappropriate use of parenteral nutrition generated potential cost savings estimated at $500,000 per year.[12]
- Selecting cost-effective enteral and parenteral formulas and supplies. In one hospital seeking to reduce formula waste and the incidence of clogged feeding tubes and encourage appropriate formula administration techniques, clinicians studied the costs and benefits of a new type of pump (automatic flush pump) that automatically delivers water through feeding tubes at intervals. The potential cost savings generated by limiting the number of clogged feeding tubes was estimated at $43,350 per month.[13] In addition, the cost savings due to the reduced time nurses spent manually flushing feeding tubes was estimated at $40,150 per year.

These are but a few of a growing number of examples that suggest nutrition support teams improve the quality of care and generate cost savings at the same time.

Skilled Health Care Assistants

Another change in the way health care professionals operate as a result of cost containment is an increased use of skilled assistants in health care. A dietetic technician assisting a clinical dietitian, for example, may collect the data for a nutrition assessment, and the dietitian may analyze the data. Because the dietitian's time is reserved for tasks that require a dietitian's unique skills, the dietitian can manage more clients, and money is saved. Assistants for physicians, nurses, and pharmacists perform similar roles.

The more health care professionals involved in nutrition care—or any medical service for that matter—the more important effective communications become. In the case of nutrition care, all persons involved must understand their roles and perform their functions effectively to ensure quality of care.

Greater Use of Home Care

Another major change affecting health care professionals has been a great reduction in the length of hospital stays and the growth of home health care. Early discharges lower hospital costs. As a result, hospital staffs have been reduced, and there is less time to identify nutrition problems. Clients may be discharged before they have regained health, and they may require additional medical and nutrition care in outpatient settings and at home. Estimates suggest that more than 8 million Americans currently receive home care to meet both long- and short-term medical needs.[14] Nurses most often provide nutrition services that are part of a home care treatment plan, although the number of dietitians in home health care is expected to grow.[15] Health care professionals are challenged to ensure that home care agencies understand the cost-effectiveness of nutrition services in helping clients maintain health and quality of life.

The full implications of cost containment and its impact on the nation's health remain to be seen. In the words of one clinician, "Our American society is in the midst of the most far-reaching and profoundly disturbing uncontrolled study in the history of health care. We are experiencing major changes in the way we practice and pay for health care with very little evidence that these changes will achieve the desired outcomes."[16] In the midst of these changes, health care professionals have often been pushed to provide better outcomes with fewer personnel and less time. Little wonder, then, that the changes are often met with resistance. Yet significant improvements in quality of care continue to be realized.[17] Health care professionals who react positively to the inevitable changes have the opportunity to make a significant difference in ensuring the success of the health care system for the future.

References

1. A. Bothe, Consensus: We should not lower quality to cut costs, *Nutrition in Clinical Practice* (supplement) 10 (1995): 1–7.
2. D. A. August, Outcomes research, nutrition care, and the provider-payer relationship, *Nutrition in Clinical Practice* (supplement) 13 (1998): S8–S11.
3. C. S. Chima and coauthors, Relationship of nutritional status to length of stay, hospital costs, and discharge status of patients hospitalized in the medicine service, *Journal of the American Dietetic Association* 97 (1997): 975–978.
4. H. N. Tucker and S. G. Miguel, Cost containment through nutrition intervention, *Nutrition Reviews* 54 (1996): 975–978.

5 C. Braunschweig, S. Gomez, and P. M. Sheean, Impact of declines in nutritional status on outcomes in adult patients hospitalized for more than 7 days, *Journal of the American Dietetic Association* 2000 (100): 1316–1322.

6 As cited in D. J. Rodriquez, Managed care: Key concepts and nutritional implications, *Support Line,* October 1999, pp. 8–15.

7 G. Sikand and coauthors, Dietitian intervention improves lipid values and saves medication costs in men with combined hyperlipidemia and a history of niacin noncompliance, *Journal of the American Dietetic Association* 2000 (100): 218–224; J. Sheils, R. Rubin, and D. C. Stapleton, The estimated costs and savings of medical nutrition therapy: The Medicare population, *Journal of the American Dietetic Association* 99 (1999): 428–435.

8 Position of The American Dietetic Association: Nutrition services in managed care, *Journal of the American Dietetic Association* 96 (1996): 391–395.

9 D. Dougherty and coauthors, Nutrition care given new importance in JCAHO standards, *Nutrition in Clinical Practice* 10 (1995): 26–31.

10 Bothe, 1995.

11 N. M. Pace and coauthors, Performance model anchors successful nutrition support protocol, *Nutrition in Clinical Practice* 12 (1997): 274–279.

12 E. B. Trujillo and coauthors, Metabolic and monetary costs of avoidable parenteral nutrition use, *Journal of Parenteral and Enteral Nutrition* 23 (1999): 109–113.

13 M. P. Petnicki, Cost savings and improved patient care with use of a flush enteral feeding pump, *Nutrition in Clinical Practice* (supplement) 13 (1998): 29–41.

14 National Association of Home Care web page (www.nahc.org), visited on November 3, 1999.

15 P. S. Anthony, The business of home care, *Support Line,* October 1999, pp. 16–21; M. S. Schiller, M. B. Arensberg, and B. Kantor, Administratiors' perceptions of nutrition services in home health care agencies, *Journal of the American Dietetic Association* 98 (1998): 56–61.

16 J. R. Wesley, Managing the future of nutrition support, *Journal of Parenteral and Enteral Nutrition* 20 (1996): 391–395.

17 P. J. Schneider, A. Bothe, and M. Bisognago, Improving the nutrition support process: Assuring that more patients receive optimal nutrition support, *Nutrition in Clinical Practice* 14 (1999): 221–226.

CHAPTER 24 Nutrition in Severe Stress

Chapters 20 and 21 described how diseases of the GI tract compromise nutrition status by interfering with the intake, digestion, or absorption of nutrients. This chapter focuses on medical conditions that can alter nutrition status by creating **stress**—a disruption of the body's internal balance. (Highlight 17 describes emotional stress and its connections to nutrition.)

The body routinely experiences some degree of stress and makes adjustments to restore balance. These stresses fall within the body's normal and healthy functioning and are known as **physiological stresses. Pathological stresses,** on the other hand, push the body beyond these limits. The amount of stress the body tolerates without significant disruption varies from individual to individual depending on many factors, including nutrition status.

Conditions that cause acute, severe stress—the subject of this chapter—lead to major disruptions of the body's normal internal balance and rapidly and markedly raise its metabolic rate. These conditions include uncontrolled infections or extensive tissue damage, such as deep, penetrating wounds or multiple broken bones.◆ When confronted with such a severe stress, the body employs complicated mechanisms to reestablish its balance. Once in play, these mechanisms have both helpful and harmful effects.

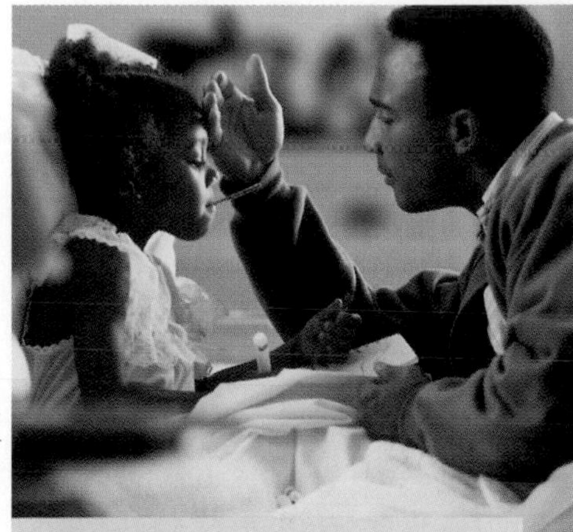

© Bruce Ayres/Stone

Although all illnesses threaten the body's internal balances, the healthy body adapts to minor stresses quickly and efficiently.

THE BODY'S RESPONSES TO STRESS

To regain homeostasis during severe stress, the body speeds up its metabolic rate (hypermetabolism) to mobilize glucose and amino acids. From these nutrients the body can synthesize the special factors it needs to limit and repair damage. The exact factors and the amount the body makes depend, in part, on what condition precipitates the stress and the severity of the stress. Mending a broken bone requires different factors than does healing a wound or fighting an infection. Healing a blister produces less stress than mending a ruptured organ.

Researchers are working to elucidate the body's complex and interrelated responses to stress, which ultimately affects many body systems. Much of that research focuses on **cytokines**—proteins that direct changes in the cardiovascular and nervous systems and stimulate the production of the many immune system cells◆ and chemicals necessary to recover from stress (see Highlight 18).[1]

• Immune System Responses • The body's natural system of defense against pathogens—the immune system—enables the body to fight off infectious agents and repair tissue damage. The immune system defends the body so alertly and silently that most healthy people are unaware that thousands of microbes mount attacks against them every day. Occasionally, though, an infection succeeds in making a person ill, and the immune system must then mount a vigorous counterattack. Serious infections greatly tax the body, and if the counterattack fails, death follows.

Tissues damaged by trauma, heat, chemicals, loss of blood flow, or infectious agents can disrupt organ function and render the body vulnerable to invading organisms. The body's response to tissue damage—the **inflammatory response**—serves to isolate and destroy foreign particles and repair tissue damage. At the site of injury, the walls of the blood vessels swell, which shrinks the diameter of the blood vessels in the affected area, slows blood flow, and allows immune system factors and albumin to flow into the injured area. Figure 24-1 on p. 728 illustrates the changes in blood vessels that occur during inflammatory responses. Some eicosanoids—derivatives of omega-6 and omega-3 fatty acids direct changes that alter the permeability of blood vessels and cause immune system

◆ Examples of stressors that can lead to severe stress are infection; surgery; burns; fractures; deep, penetrating wounds (gunshot wounds, fistulas, or surgical incisions); and extensive bleeding. These stressors may lead to shock, tissue injury, or tissue death (necrosis).

◆ The cells of the immune system—white blood cells—include a variety of phagocytes and lymphocytes (see Highlight 18).

stress: the state in which a body's internal balance (homeostasis) is upset by a threat to a person's physical well-being **(stressor).** The terms **acute stress** and **severe stress** are used in this chapter to refer to pathological stresses that rapidly and markedly raise the body's metabolic rate and significantly upset its normal internal balance.

physiological stresses: disruptions to the body's internal balance caused by processes necessary to sustain life.

pathological stresses: disruptions to the body's internal balance that lie beyond its normal and healthy functioning.

cytokines (SIGH-toe-kynes): proteins that help regulate immune system responses. Cytokines trigger hypermetabolism and cause anorexia, fever, and discomfort.

inflammatory response: the changes orchestrated by the immune system when tissues are injured by such forces as blows, wounds, foreign bodies (chemicals, microorganisms), loss of blood flow, heat, cold, electricity, or radiation.

FIGURE 24-1 **Inflammatory Responses**

1 Normal cells lining the blood vessels (red) lie close together, and immune system cells (blue) and albumin (green) do not cross into tissue.
2 With inflammation, the cells lining the blood vessels swell (pink), gaps form between the cells, and immune system cells and albumin (and fluid along with it) enter tissues in the area and cause edema.
3 Along with changes in blood vessels, inflammatory responses also result in the recruitment and adhesion of immune system cells. These responses cause blood flow to slow through the inflamed area.
4 Immune system cells produce toxins (including free radicals and enzymes) that kill foreign invaders and destroy healthy tissue in the process.

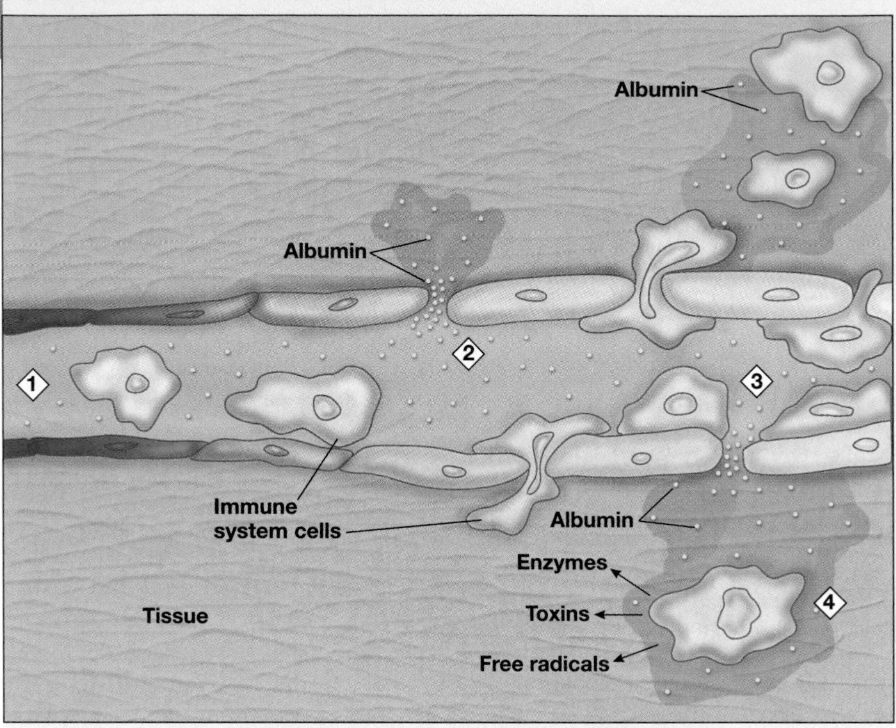

◆ Reminder: Eicosanoids, which include prostaglandins and thromboxanes, are powerful, biologic compounds made from 20-carbon fatty acids (see pp. 145–146). Some eicosanoids are **proinflammatory**—they direct changes that heighten inflammatory responses. Other eicosanoids are **anti-inflammatory**—they direct changes that slow inflammatory responses.

◆ An understanding of inflammatory responses helps explain why disorders characterized by chronic inflammation, such as chronic pancreatitis and Crohn's disease (see Chapter 21), lead to tissue damage and ongoing problems. The body's response to acute, severe stress, however, is far more exaggerated and pronounced than its response to chronic inflammation.

shock: a sudden drop in blood volume that disrupts the supply of oxygen to the tissues and organs and the return of blood to the heart. Shock is a critical event that requires immediate correction.

cells to adhere to the area.◆ The immune system factors then destroy foreign particles and remove dead cells. Together these responses protect the body from the spread of infection. In the final stage of the inflammatory response, scar tissue forms to replace damaged tissues.

Unfortunately, the benefits conferred by inflammatory responses come with a cost to healthy tissue. Acute, severe stresses result in exaggerated or prolonged inflammatory responses that significantly alter blood flow and can lead to **shock.** Blood flow can be further affected by the formation of tiny blood clots, which form under the direction of certain eicosanoids to help immune system cells "stick" to the damaged area. The altered blood flow and the release of toxins to destroy certain foreign particles also damage healthy tissues.◆ Among the damaging toxins produced are free radicals, and extensive tissue injuries can lead to oxidative stress (see p. 380). Furthermore, the replacement of healthy tissue with scar tissue (fibrosis) can interfere with organ function and significantly impair health. Figure 24-2 shows how inflammatory responses can lead to tissue damage during severe stress.

If the immune factors fail to disarm the invaders but succeed in preventing an infection from spreading, an abscess or granuloma may form. If local defenses are overwhelmed and the invaders gain entry into the bloodstream, sepsis develops. Sepsis triggers exaggerated inflammatory responses and may lead to multiple organ failure—the most common cause of life-threatening compli-

FIGURE 24-2 Inflammation and Tissue Damage

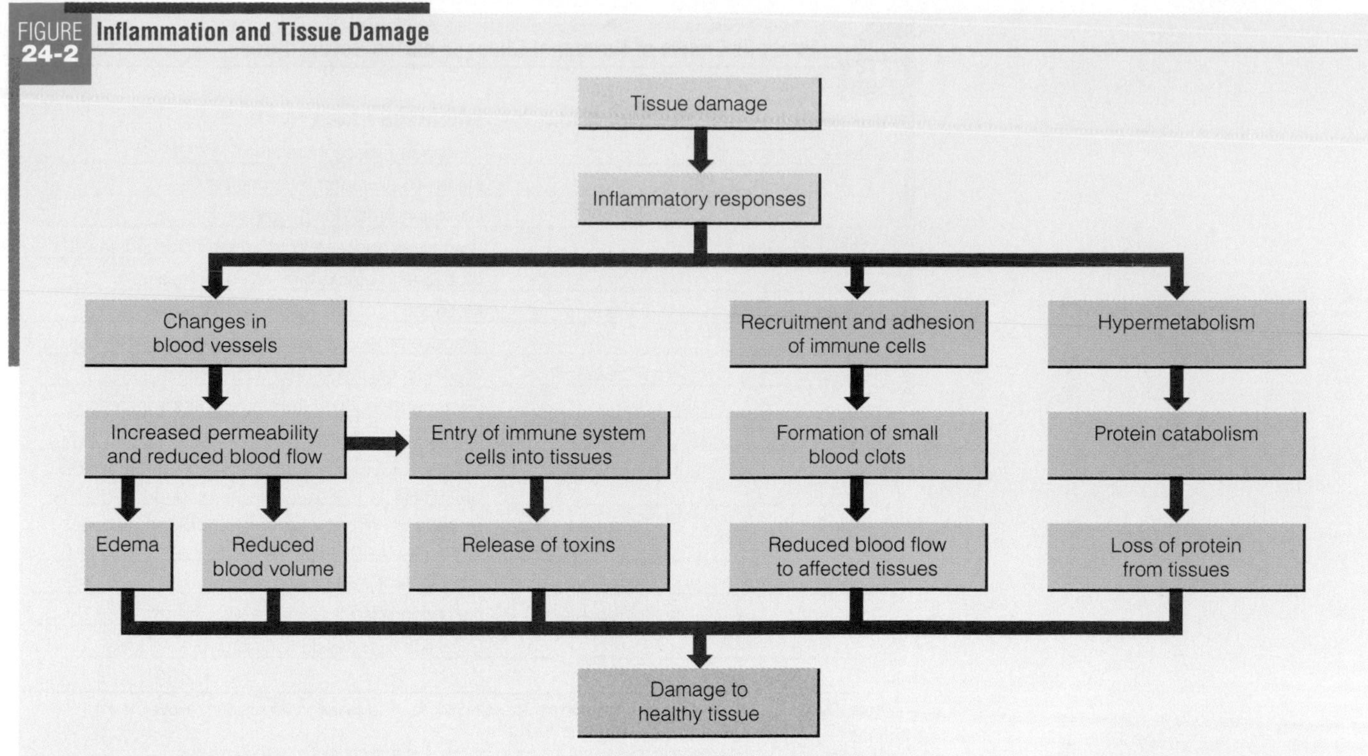

cations and death in critically ill people. (Highlight 24 provides more information about multiple organ failure.)

• **Hormonal Responses** • The hormonal changes characteristic of the immediate stress response shift the balance between insulin, which promotes the storage of carbohydrate and lipids and the synthesis of protein, and the **counterregulatory hormones,** which promote the breakdown of glycogen, the mobilization of fatty acids from lipids, and the synthesis of glucose from protein (see Table 24-1 on p. 730). Although insulin levels rise, the rise in the level of counterregulatory hormones is greater still. These changes result in negative nitrogen balance; that is, the body breaks down and uses more protein than it receives and synthesizes. The protein loss comes primarily from skeletal muscle, connective tissue, and the gut; protein synthesis in the liver and the tissues that produce immune system cells actually increases. Thus skeletal muscle, connective tissue, and gut tissue are sacrificed to supply the glucose and amino acids necessary to respond to stress.[2]

As a result of the hormonal changes, the metabolic rate and blood glucose levels rise. High insulin levels suppress the mobilization of essential fatty acids from body stores, and clinical signs of essential fatty acid deficiencies may develop. Figure 24-3 on p. 731 illustrates the differences between the body's responses to simple fasting and to severe stress to show why the severely stressed body rapidly consumes energy and loses vital nutrients.

Other hormonal changes promote the retention of water and sodium and the excretion of potassium. Hypermetabolism generally peaks at about 3 to 4 days and subsides in about 7 to 10 days.◆ With recovery, hormone levels gradually return to normal.

• **Clinical Signs and Symptoms** • The inflammatory and immune responses to stressors result in redness, swelling, heat, and pain at the injury site. Body temperature, heart rate, respiratory rate, and the white blood cell count increase. The person experiences anorexia and malaise—a feeling of discomfort and uneasiness. The symptoms associated with the inflammatory response are collectively called the **systemic inflammatory response syndrome (SIRS).** In addition to an elevated white blood cell count, laboratory tests reveal elevated blood glucose (hyperglycemia), blood urea nitrogen (from protein

◆ The **acute phase** of the stress response occurs immediately following an injury or infection and is characterized by hypermetabolism. During the **adaptive phase,** the body begins to recover from stress. If adaptation succeeds, **recovery** follows. If adaptation fails, **exhaustion** leads to death.

counterregulatory hormones: hormones such as glucagon, cortisol, and catecholamines that oppose insulin's actions and promote catabolism.

systemic inflammatory response syndrome (SIRS): the complex of symptoms (see the text) that occur as a result of immune and inflammatory factors in response to tissue damage. In severe cases, SIRS may progress to multiple organ failure.

TABLE 24-1 Metabolic Effects of Hormonal Changes during Severe Stress

Hormone	Alteration	Metabolic Effect
Catecholamines	Increase	Glucagon release increases.
		Insulin-to-glucagon ratio decreases.[a]
		Glycogen breakdown increases.
		Glucose production from amino acids increases.
		Mobilization of free fatty acids increases.
Cortisol	Increases	Mobilization of free fatty acids increases.
		Glucose production from amino acids increases.
Glucagon	Increases	Insulin-to-glucagon ratio decreases.[a]
		Glucose production from amino acids increases.
		Glycogen breakdown increases.
		Storage of glucose, amino acids, and fatty acids decreases.
Insulin	Increases	Breakdown of body fat inhibited; blood glucose levels rise despite the increase.
Antidiuretic hormone	Increases	Retention of water increases.
Aldosterone	Increases	Retention of sodium increases.
		Excretion of potassium increases.

Note: These changes are part of the immediate stress response. As adaptation occurs and recovery is in progress, hormone levels gradually return to normal.
[a]The net effect of a decrease in the ratio of insulin to glucagon is that catabolism predominates.

catabolism), and triglycerides; negative nitrogen balance; an increased retention of fluid and sodium; and an increased excretion of potassium. Blood concentrations of albumin, iron, and zinc fall.

• GI Tract Responses • During stress, blood flow to the GI tract diminishes, and GI tract motility slows. Altered blood flow to the stomach may hinder the stomach's ability to protect itself from gastric acid. Consequently, ulcers may form in the stomach or small intestine. The cells of the intestinal tract gradually shrink and may lose some of their absorptive and immune functions as gut proteins are sacrificed to support the body's defenses. Highlight 24 reviews the immune functions of the GI tract and discusses the ways that changes in the GI tract during severe stress raise the likelihood of infection and multiple organ failure.

• Effects on Nutrition Status • For a well-nourished person with a short-term stress, the duration of negative nitrogen balance (loss of protein) is acceptable because sufficient protein remains to support defense systems and maintain vital functions. Acute stresses that markedly elevate the metabolic rate over longer periods of time, however, rapidly deplete energy reserves and break down protein tissues. Consequently, they may lead to **acute malnutrition** in a previously healthy person or worsen preexisting malnutrition.◆ When the chronically malnourished person suffers an acute stress, or when acute malnutrition extends over long periods of time, the loss of lean body mass and the consequent depletion of vital proteins compromise the function of the immune system, heart, lungs, kidneys, and GI tract. Once organ systems begin to fail, recovery is severely compromised. Table 24-2 lists the consequences of stress that can lead to malnutrition.

◆ **Reminder: Some effects of illness on nutrition status arise not from the stress itself, but rather from its consequences or treatment. Some of these effects can arise from emotional stress; immobility, which can lead to pressure sores, negative nitrogen and calcium balance, and the formation of calcium stones; the location of injuries, which can make it difficult or painful to eat; diagnostic tests and medical procedures, which can interfere with nutrient intake (see Chapter 17); and medications.**

acute malnutrition: protein-energy malnutrition (PEM) that develops rapidly due to a sudden and dramatic demand for nutrients. The person suffering from acute malnutrition may be of normal weight or may be overweight, but serum protein levels are low. **Chronic malnutrition,** on the other hand, develops as a consequence of insufficient intake of energy and protein over long periods of time and is characterized by underweight, depleted fat stores, and normal serum protein levels.

• Effects on Medications • Severe stress, and the malnutrition that may accompany it, can alter the body's use of medications and worsen health and nutrition status. Many medications are transported in the blood bound to blood proteins such as albumin, and low blood albumin is a symptom of both stress and PEM. Without sufficient carriers, medications may be slow to reach their sites of action. Once medications do arrive at their target cells, the lack of carriers may delay the medications' transport to the liver and kidneys, where they are often detoxified and excreted. Thus medications may take a long time to

FIGURE 24-3 Metabolic Responses to Fasting and Stress

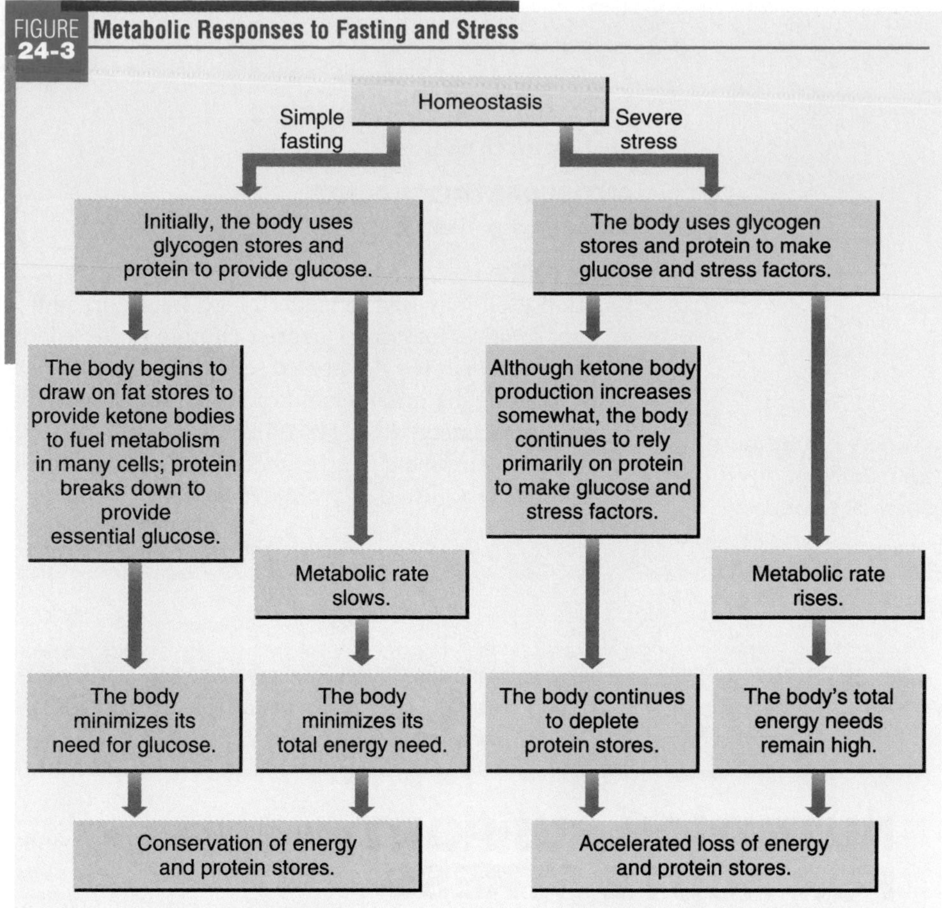

work and then may remain active for longer than intended. In addition, people with acute stresses often receive multiple medications, which complicates decisions about the correct dosage and raises the risk of diet-drug interactions. Furthermore, when medications are given in an oral form, the effects of both stress and malnutrition on GI tract function can alter the absorption of both medications and nutrients. The box on p. 732 describes diet-drug interactions of medications commonly used in the treatment of acute stresses.

IN SUMMARY

Severe stress sparks a series of hormonal and metabolic responses to reestablish balance. These changes demand energy and drain the body of nutrients, which can lead a well-nourished person rapidly into PEM or lead a malnourished body perilously close to death, if not to death itself.

TABLE 24-2 Possible Causes of Malnutrition Associated with Acute Stress

Reduced Nutrient Intake	Excessive Nutrient Losses	Raised Nutrient Needs
Anorexia	Blood loss	Fever
Emotional stress	Immobility	Hypermetabolism
Immobility	Malabsorption	Infection
Location of injury	Medications	Medications
Malaise	Urinary losses	Pressure sores
Medications	Vomiting	
Nausea/vomiting		

ANALGESICS
See box on p. 673.

ANTIANXIETY AGENTS
See box on p. 673.

ANTIDIARRHEALS
See box on p. 673.

ANTIEMETICS/ANTINAUSEANTS
See box on p. 644.

ANTI-INFECTIVE AGENTS
Numerous *anti-infective agents* with a variety of nutrition-related interactions may be used to treat infections related to acute stresses. Consult a pharmacist or drug guide for specific interactions.

ANTI-INFLAMMATORY AGENTS
See box on p. 673.

ANTISECRETORY AGENTS
See box on p. 644.

SEDATIVES
Sedation itself may lead to reduced food intake. In addition, *sedatives* may cause nausea. Of note is the sedative *propofol,* which is a lipid-based sedative that contains 1.1 kcalorie per milliliter and can provide a significant amount of energy from fat. When propofol is provided to acutely stressed clients, enteral and parenteral formulas must be adjusted to avoid overfeeding.

THINK NUTRITION whenever clients are faced with severe stress, particularly if they are in marginal or poor nutrition status. The metabolic demands of severe stress can rapidly lead to serious nutrition problems.

NUTRITION SUPPORT FOR ACUTE STRESS

To support recovery from acute stress, the immediate concerns of the health care team are to restore blood flow and oxygen transport and to prevent or treat infection. Possible measures include providing intravenous (IV) solutions to correct fluid and electrolyte imbalances, giving transfusions, removing dead tissues **(debridement),** draining abscesses, and administering antibiotics. Supplying adequate energy and nutrients following stress helps to minimize nutrient losses, preserve organ function, promote wound healing, restore blood flow and oxygen transport, and maintain immune defenses—all of which support recovery. Once hypermetabolism subsides, nutrition support aims to restore nutrient deficits and promote positive energy and nitrogen balance.

Nutrient Needs

Feeding the acutely stressed individual is a formidable challenge. Nourishment must be introduced cautiously, and adjustments must often be made throughout the course of the stress. Both underfeeding and introducing nutrients too rapidly to people who are severely stressed can result in serious complications that compromise recovery. Underfeeding hinders the functions of vital organs and impedes the production of the special factors necessary to respond to stress. Reintroducing nutrients too rapidly can overtax organ systems and lead to many complications, including malabsorption, cardiac insufficiency, respiratory distress, congestive heart failure, convulsions, coma, and even death. Collectively, the complications that occur as a consequence of reintroducing nutrients too rapidly are called the **refeeding syndrome.**

Because the course of critical illness may involve failure of one or more organs, nutrient needs must often be adjusted to an even greater degree than described here. Remember to keep the considerations of critical illness in mind when you read about acute respiratory failure in Chapter 26, acute renal failure in Chapter 27, and liver failure in Chapter 28.

debridement (dee-BREED-ment): the removal of dead tissues resulting from burns and other wounds to speed healing and prevent infection.

refeeding syndrome: the physiologic and metabolic complications associated with reintroducing nutrients too rapidly in people with depleted nutrient stores due to chronic malnutrition or in those who have been underfed for several days. These complications can include malabsorption, cardiac insufficiency, respiratory distress, congestive heart failure, convulsions, coma, and possibly death.

• **Fluids and Electrolytes** • To restore circulation and prevent dehydration and electrolyte imbalances, the medical team must act quickly to stabilize the body's blood volume and electrolyte balance. This step is critical for, without adequate blood volume, oxygen, nutrients, and medications cannot be delivered to the cells, organ systems cannot function properly, and waste products cannot be eliminated from the body. Conversely, providing too much fluid stresses the heart, which must pump a greater blood volume, and the kidneys, which must work harder to excrete the extra fluid.

The physician determines the person's fluid needs based on clinical measures such as blood pressure, heart rate, respiratory rate, urinary output, level of consciousness, and body temperature.◆ Serum electrolytes are closely monitored, and adjustments are made as necessary.

◆ An estimate of fluid requirements can be made using the formula on p. 692.

With catabolism, blood levels of electrolytes normally concentrated in the intracellular fluids (potassium, phosphorus, and magnesium) rise. Once catabolism subsides, these electrolytes, along with glucose and amino acids, move back into the cells to begin rebuilding tissues. Without careful attention to replacement, blood levels of these electrolytes can plummet, resulting in life-threatening complications.

• **Energy** • Supplying too little energy further compromises nutrition status, raises the risk of infections, and interferes with wound healing and lung function.[3] Supplying too much energy elevates the metabolic rate, which, in turn, increases oxygen use and carbon dioxide production. Then the heart and lungs, already working hard as a consequence of stress, must work even harder to keep the body's gases in balance. If these vital organs are weakened by preexisting malnutrition or if their function is already compromised by stress, the heart and/or lungs may fail. Providing too much energy can also lead to fatty liver (see Chapter 28), which enlarges the liver and can interfere with its function.

Energy needs during severe stress depend on both the type and the degree of stress, as well as the individual's metabolic rate and nutrition status. Indirect calorimetry◆ is widely believed to provide the best estimate of energy needs during acute stress, especially when the stress is particularly severe.[4] Use of indirect calorimetry is limited in many facilities, however, because it requires special equipment (metabolic cart) and skilled clinicians to perform the measurement and interpret the results.[5] Instead, clinicians often rely on estimates, such as the Harris-Benedict equation to estimate basal energy needs (see Table 24-3 on p. 734).◆ Additional kcalories are added to the basal energy estimate to meet the demands of stress. For most stresses, energy needs range from 100 to 120 percent of the basal energy expenditure, a range that is not significantly greater than a typical diet. Head and severe burn injuries, however, raise energy needs to a greater extent than other stresses.

◆ Reminder: *Indirect calorimetry* is an indirect estimate of resting energy needs made by measuring the amount of carbon dioxide expired (carbon dioxide production) and the amount of oxygen inspired (oxygen consumption).

◆ Many clinicians estimate energy needs in critically ill people based on body weight and provide 25 to 35 kcal/kg body weight for people who are not obese, and 21 kcal/kg body weight for people who are obese.

The effects of obesity on the heart, lungs, and immune function, as well as the insulin resistance typical of obesity, add to the problems of the obese person who suffers a severe stress. Although it is critical to provide nutrition support to those who are obese, clinical research to determine the best approach for feeding obese people with severe stress is lacking. Some researchers report that obese people may benefit by providing energy at about half of their estimated needs, provided that protein needs are met.[6] Table 24-3 includes one method for adjusting energy needs for obesity.

• **Protein** • The greater the stress, the more body protein is broken down (up to the limit of the body's capabilities), the more nitrogen is excreted in the urine, and the greater the need for protein. To meet protein needs, stressed people with normal kidney and liver function generally need about 1.5 and 2.0 grams of protein per kilogram of body weight per day. People with severe burns may need more—their exact needs depend on the extent of the burn.

Even when adequate amounts of protein are supplied, negative nitrogen balance cannot be fully corrected. Providing too much protein places an additional stress on the liver, which must increase its production of urea to handle

◆ Basal metabolic rate (BMR, described on p. 246) and BEE express the same thing: basal energy need. The equation for BMR is traditionally used in physiology and fitness laboratories; that for BEE, in hospitals. The two equations yield slightly different results, each suitable for the purposes intended. All results are approximations; all require judgment in their application.

TABLE 24-3 The Harris-Benedict Equation for Estimating Basal Energy Expenditure (BEE)◆

Women:

BEE = 655 + (9.6 × wt[a] in kg) + (1.8 × ht in cm) − (4.7 × age in yr).

Men:

BEE = 66.5 + (13.8 × wt[a] in kg) + (5 × ht in cm) − (6.8 × age in yr).

Add to BEE for stress:

0–20% for most stresses

30–50% for head injuries

Energy needs for burn injuries depend on the percentage of the total body surface area (TBSA) that is burned.[b] Add to BEE for burns:

20–40% (<20% TBSA)

60% (20–25% TBSA)

70% (25–30% TBSA)

80% (30–35% TBSA)

90% (35–40% TBSA)

100% (40 45% TBSA)

110% (>45% TBSA)

For people with a %IBW greater than 125, instead of actual weight, use an adjusted body weight in the BEE equation. To determine adjusted body weight, use this equation:[c]

(actual body weight − IBW) × 25%[d] + IBW = adjusted body weight

Note: wt = weight; ht = height; yr = years; kg = kilograms; cm = centimeters.

[a]Use actual body weight, not ideal body weight.

[b]Many formulas are available for estimating energy needs for burn injuries. A variety of formulas are described and referenced in D. J. Rodriguez, Nutrition in major burn patients: State of the art, *Support Line,* August 1995, pp. 1–8.

[c]From J. M. Karkeck, Adjustment for obesity, *American Dietetic Association Renal Practice Group Newsletter,* Winter, 1984.

[d]Approximately 25 percent of body fat tissue is metabolically active.

SOURCES: M. M. McMahon, Nutrition support of hospitalized patients, presented at the Eighth Annual Advances and Controversies in Clinical Nutrition, Mayo Clinic Foundation, April 5–7, 1998; E. Weeks and M. Elia, Observations on the patterns of 24-hour energy expenditure changes in body composition and gastric emptying in head-injured patients receiving nasogastric tube feeding, *Journal of Parenteral and Enteral Nutrition* 20 (1996): 31–37.

the excess nitrogen, and the kidneys, which then must excrete it. Only after the hypermetabolic stage of severe stress subsides can negative nitrogen balance be fully corrected. The "How to" shows an example of how to estimate energy and protein needs following severe stress.

◆ Highlight 24 provides more information about glutamine, arginine, and necleotides.

• Amino Acids • An amino acid receiving wide attention in relation to stress is glutamine.◆ Glutamine is a nonessential amino acid in healthy people, but the body's demands for this amino acid during stress may exceed its ability to synthesize glutamine in adequate amounts. Consequently, glutamine may become a conditionally essential amino acid. Glutamine provides fuel for the intestinal cells and plays important roles in maintaining intestinal immune function and promoting wound healing.[7] Glutamine supplementation may reduce the incidence of infection in acutely stressed people and people undergoing bone marrow transplants.[8]

Another nonessential amino acid, arginine, and nucleotides (nitrogen-containing components of RNA and DNA) may also be important during severe stress. Arginine plays an important role in wound healing and may help minimize negative nitrogen balance, improve blood flow, and reduce clot formation. Nucleotides have been shown to improve immune responses in animals. However, further research is necessary to determine if supplementing arginine and nucleotides provides benefits for people experiencing severe stress.

• Carbohydrates and Lipids • Nonprotein energy sources spare protein, and the amount of energy provided as carbohydrate or lipid has ramifications for stressed people. Carbohydrates provide a readily usable source of energy. During stress, however, the body can metabolize only a fixed amount (about 500 grams per day) of glucose. Excess glucose serves no benefit, but contributes to hyperglycemia, which can adversely affect fluid and electrolyte balances and may also increase the risk of infections in stressed people.[9]

Lipids provide energy and essential fatty acids, but given in excess, they can also tax metabolic functions and hamper immune responses. Clinicians often supply nonprotein kcalories through a mixture of about 70 percent carbohydrates and 30 percent lipids.[10] For clients with burns, restricting fat to 15 to 20 percent of the nonprotein kcalories may help prevent respiratory infections and speed healing.[11]

HOW TO Estimate Energy and Protein Needs Following Severe Stress

Bernadette is a 39-year-old female, who is 5 feet 3 inches tall and weighs 130 pounds. She recently underwent extensive surgery. Her energy needs can be estimated using the Harris-Benedict equation as follows:

Weight in kilograms =
130 lb ÷ 2.2 kg/lb = 59 kg.

Height in centimeters =
63 in × 2.54 cm/in = 160 cm.

BEE = 655 + (9.6 × wt in kg)
+ (1.8 × ht in cm) − (4.7 × age in yr).
655 + (9.6 × 59 kg) + (1.8 × 160 cm)
− (4.7 × 39) = 655 + 566 + 288 − 183
= 1326 kcal.

Next add 20% × BEE for surgery:[a]

1326 kcal × 20% = 262 kcal.
1326 + 265 = 1591 kcal.

Bernadette needs about 1591 kcalories to meet her BEE and additional energy needs due to surgery; clinicians monitor weight changes to determine if actual needs are higher or lower. Her energy needs will change as stress resolves.

Protein needs for Bernadette can be estimated at 1.5 to 2.0 grams of protein per kilogram of body weight per day. Use her weight of 59 kilograms to make the calculation:

59 kg × 1.5 g/kg = 89 g protein.
59 kg × 20 g/kg = 118 g protein.

Bernadette needs an estimated 89 to 118 grams of protein daily. Some clinicians subtract the kcalories provided by protein from the total energy needs and then provide the remaining calories as carbohydrate and fat.[b] Others meet the energy needs with carbohydrate and fat only and do not consider the kcalories contributed by protein. In either case, clinicians can monitor serum proteins (see Chapter 17) or use nitrogen balance studies to determine whether the estimate is meeting actual protein needs.

[a]Alternatively, multiply the BEE by 120%, (× 1.20); which is the mathematical equivalent.
[b]J. M. Miles and J. A. Klein, Should protein be included in calorie calculations for a TPN prescription? *Nutrition in Clinical Practice* 11 (1996): 204–206.

• Fatty Acids • Much research today focuses on the best source of fatty acids for clients experiencing severe stress. Standard IV lipid emulsions and many enteral formulas are rich sources of omega-6 fatty acids. When given in excess of essential fatty acid requirements, however, omega-6 fatty acids lead to the production of eicosanoids that accelerate inflammatory responses.[12] Alternative lipid sources such as fish oils (a rich source of omega-3 fatty acids) may be a better choice. Omega 3-fatty acids lead to the production of factors that slow inflammatory responses.[13] Still other research suggests that olive oil–based lipids (a rich source of monounsaturated fatty acids) may also be beneficial in modulating inflammatory responses.[14]

The type of fat may also have an effect on the altered lipid metabolism that accompanies severe stress. Lipid emulsions containing both long-chain and medium-chain triglycerides (LCT and MCT) may help normalize lipid levels more quickly than lipid emulsions containing only LCT.[15] For people experiencing severe stress, MCT provide a readily available source of energy.

• Micronutrients • Vitamin and mineral needs during stress are highly variable, and specific requirements are unknown. The needs for many B vitamins increase when energy and protein intakes increase, and these vitamins are important cofactors in many metabolic reactions. Other micronutrients that play specific roles in repairing tissues include vitamin A, vitamin C, iron, zinc, copper, and selenium. With oxidative stress, antioxidant nutrients become depleted and supplementing these nutrients may prove to be beneficial. To date, little research has been done to determine if supplementing antioxidants can help reverse oxidative stress.[16]

During stress, blood iron levels may fall as iron moves from the circulation into tissues for storage. Unless a true iron deficiency is identified, iron is not given to compensate, however, because the shift robs invading organisms of the iron they need to grow and thus helps prevent the spread of infection.

• Growth Hormone and Insulin-like Growth Factor • Researchers have begun to study nondietary factors that might improve nitrogen balance and lessen the impact of severe stress on the host. Two such factors, growth hormone and insulin-like growth factor-1 (IGF-1), stimulate the growth of intestinal cells and may play an important role in protecting the intestine during stress.

Delivery of Nutrients following Stress

Selecting the appropriate amounts and types of nutrients to help people recover from stress is only part of medical nutrition therapy. Just as important is supplying nutrients in a form that best serves the body's ability to recover.

• **Oral Diets** • Well-nourished clients with mild stresses who are expected to eat within 7 to 10 days following stress receive simple IV solutions to maintain fluid and electrolyte balances and provide minimal kcalories. Once GI tract motility returns, they begin an oral diet that often progresses from clear liquids to regular foods as tolerated (see Chapter 19). Many people have great difficulty eating enough food to meet nutrient needs following an acute stress, however. Use the suggestions in the box on p. 618 in Chapter 19 to help improve the client's intake. Enteral formulas provided orally often help supplement energy and nutrient intake.

• **Tube Feedings and Parenteral Nutrition** • People with preexisting malnutrition, those who are not expected to be able to eat within 7 to 10 days, or those who have especially high nutrient needs benefit from tube feedings or parenteral nutrition. To prevent abdominal distention, nausea, vomiting, and the possible aspiration of foods or formula into the lungs, oral or gastric feedings have to wait until gastric motility is restored. Because peristalsis returns more quickly to the small intestine than to the stomach, feeding formula directly into the small intestine through a tube is not only possible, but provides advantages over parenteral nutrition.[17] Early feeding (initiated within 48 hours following stress) stimulates intestinal blood flow, function, and adaptation and may minimize hypermetabolism and help maintain the GI tract's immune and absorptive functions (see Highlight 24). Most significantly, however, early enteral feeding improves recovery following stress by reducing septic complications.[18] Early intestinal feedings are not possible in all cases, however, particularly when blood flow to the intestine is disrupted either because of the person's medical condition or as a consequence of medications.[19] Additionally, some clients may need both enteral feedings and parenteral nutrition until they are able to meet all nutrient needs enterally. Once oral feeding is possible, tube feedings or parenteral nutrition is gradually discontinued (see Chapters 22 and 23).

• **Stress Formulas** • Clinicians eager to improve a client's outcome often rely on enteral formulas designed to meet nutrient needs during stress. Many such formulas are of high nutrient density. Some contain extra vitamins A and C, zinc, and other nutrients designed to promote wound healing. Formulas designed to improve immune function often contain added glutamine, arginine, nucleotides, and omega-3 fatty acids. Although such formulas appear to be beneficial for specific situations, further research is necessary to determine if their impact on recovery justifies their use.

IN SUMMARY Recovery depends, in part, on the body's receiving the energy and nutrients required to mount a defense, repair damaged tissues, and replenish nutrient reserves. Providing nutrients too rapidly, however, can stress the body and worsen stress. The accompanying case study of a client with burns tests your knowledge of nutrition and severe stress, and the Nutrition Assessment Checklist summarizes information necessary to monitor the nutrition care of stressed clients.

CASE STUDY

Journalist with a Third-Degree Burn

Mr. Sampson, a 48-year-old journalist, has been admitted to the emergency room. He suffered a severe burn covering over 40 percent of his body when he was trapped in a burning building. His wife told the nurse that Mr. Sampson's height is 6 feet and that he weighs about 175 pounds. The physician ordered lab work, including serum proteins; the results are not back yet.

1. Identify Mr. Sampson's immediate postinjury needs. What measures might be taken to meet those needs? What additional concerns might you have if Mr. Sampson was malnourished before experiencing the burn?
2. Considering Mr. Sampson's condition, what problems might the health care team encounter in getting information from him about his preburn nutrition status?
3. Calculate Mr. Sampson's energy and protein needs to support burn healing (use 2 to 3 grams of protein per kilogram of body weight). What other nutrients might be of concern (consider proportions of energy nutrients and also nutrients that might bolster immune responses or speed wound healing)?
4. What are the possible benefits of early enteral nutrition to Mr. Sampson. How might these benefits be particularly important following a severe burn injury?

Nutrition Assessment Checklist For People Experiencing Severe Stress

MEDICAL HISTORY

Check the medical record regularly to determine:

- ☐ Type of stress
- ☐ Severity of stress
- ☐ Stage of stress
- ☐ Route of feeding (oral/tube feeding/parenteral)
- ☐ If any organ system is compromised

Review the medical record for complications that may be related to underfeeding or overfeeding:

- ☐ Dehydration/fluid overload
- ☐ Hyperglycemia
- ☐ Electrolyte imbalances
- ☐ Acid-base imbalances
- ☐ Fatty liver

MEDICATIONS

Record all medications and note:

- ☐ Signs that medication dosage may be inappropriate
- ☐ Signs of nutrient deficiencies
- ☐ kCalories provided from propofol, if prescribed

For people on medications who are able to eat an oral diet:

- ☐ Note any medications that may result in reduced food intake or nutrient imbalances
- ☐ Provide pain medications at times when they will be effective during meals, when appropriate
- ☐ Provide antinauseants or antiemetics at times when they will be effective during meals, when appropriate

FOOD/NUTRIENT INTAKE

If the client is not meeting nutrition goals:

- ☐ Calculate nutrient intake from table foods, enteral formulas, and/or parenteral solutions
- ☐ Determine the contribution of nutrients from the sedative propofol, if prescribed
- ☐ Investigate causes, including incorrect determination of nutrient needs or need to alter prescription based on client's unique responses

For clients eating an oral diet:

- ☐ Note the problems client is having eating
- ☐ Consider interventions to correct problems with eating and take action

ANTHROPOMETRICS

Measure baseline height and weight and daily weights. Note that body weight changes erratically in acutely ill clients due to the large amounts of fluids required for resuscitation. Once clients are in the adaptive stage of stress, if their weights are not meeting goals:

- ☐ Reestimate energy needs
- ☐ Check to see if client is receiving the diet (or formula) that has been prescribed
- ☐ Consider the need to change the energy prescription to meet weight goals
- ☐ Consider kcalories contributed from propofol, if prescribed

LABORATORY TESTS

Laboratory tests that often change as a consequence of stress itself and require careful interpretation as indicators of nutrition status during acute stress include:

- ☐ Albumin
- ☐ Transferrin
- ☐ Prealbumin
- ☐ Total lymphocyte count (white blood cell counts are often elevated)

Check laboratory tests for signs of:

- ☐ Dehydration/fluid overload
- ☐ Hypertriglyceridemia
- ☐ Electrolyte imbalances
- ☐ Acid-base imbalances
- ☐ Vitamin and trace element deficiencies
- ☐ Essential fatty acid deficiencies
- ☐ Response to protein intake (nitrogen balance studies, prealbumin, and retinol-binding protein), when available
- ☐ Organ dysfunction or return to normal organ function

PHYSICAL SIGNS

Regularly assess vital signs including:

- ☐ Blood pressure
- ☐ Pulse
- ☐ Temperature
- ☐ Respiration

Look for physical signs of:

- ☐ PEM (muscle mass and strength; body fat)
- ☐ Essential fatty acid deficiencies
- ☐ Dehydration/fluid overload
- ☐ Nutrient deficiencies and excesses

Nutrition on the Net

WEBSITES

Access these websites for further study of topics covered in this chapter.

- Find updates and quick links to these and other nutrition-related sites at our website: **www.wadsworth.com/nutrition**
- To read complete articles or uncover more information relevant to critical care, visit these sites: American Association of Critical Care Nurses: **www.aacn.org**
- Canadian Association of Critical Care Nurses: **www.execulink.com/~caccn**

These sites provide many resources regarding burns for clients and families:

- American Burn Association: **www.ameriburn.org**
- Alberta Burn Rehabilitation Society: **www.burnrehab.com**

INTERNET ACTIVITIES

When a person experiences a sudden, severe stress such as a deep penetrating wound or a serious burn, feeding the person is but one of the challenges health care professionals face. You can read questions and answers about caring for people with burns, and even contribute questions or comments.

- Go to **www.ameriburn.org**
- Click on "Advocacy and Communications."
- Click on "ABA On-Line Forum."
- Scroll down the left side of the screen and click on a topic of interest.
- Read the comments and questions and then click on "Next" (at the top of the screen) to read replies from different contributors. To contribute your own comments, click on "Reply."

What would be the most challenging aspect of feeding people with severe burns?

Study Questions

These questions will help you review the chapter. You will find the answers in the discussions on the pages provided.

1. What is severe stress? List some medical conditions that can lead to severe stress. (p. 727)
2. How do immune system factors and hormones mediate the stress response? (pp. 727–729)
3. Describe the effect of nutrition status on stress. How can stress rapidly lead to malnutrition? How do stress and PEM affect vital organ systems, GI tract motility, and GI tract absorptive and immune functions? (p. 730)
4. Describe how stress affects nutrient needs. Discuss the precautions that must be taken when feeding the person who suffers a severe stress. (p. 732)
5. Why might enteral feedings be preferable to parenteral nutrition after a severe stress? Why can enteral formulas be delivered by tube into the intestine but not be taken by mouth in the early period following a severe stress? (p. 736)

These questions will help you prepare for an exam. Answers can be found on p. 740.

1. The severe stresses described in this chapter are characterized by:
   a. chronic malnutrition.
   b. tissue damage and hypermetabolism.
   c. reduced protein synthesis in the liver.
   d. hormonal changes that protect skeletal muscle.
2. Cytokines are proteins that:
   a. repair damaged tissues.
   b. destroy microorganisms.
   c. direct stress responses.
   d. oppose insulin's action and result in the catabolism of protein in skeletal muscle, connective tissue, and the gut.
3. If the immune system fails to control an infection following a severe stress, the most serious complication that can occur from the list below is:
   a. sepsis.
   b. catabolism of protein.
   c. essential fatty acid deficiency.
   d. low serum albumin and transferrin.
4. Which of the following metabolic changes accompany acute stress?
   a. Proteins are broken down in the liver.
   b. Protein synthesis in the liver decreases.

c. The body conserves protein as it does in simple fasting.
d. Proteins are broken down in skeletal muscle, connective tissue, and the gut.

5. Which of the following statements with respect to nutrition and severe stress is true?
a. A person with preexisting malnutrition who suffers an acute stress does not require immediate attention to energy and nutrient needs.
b. A previously well-nourished person can develop acute malnutrition if the stress is extreme or prolonged.
c. A person with either acute or preexisting malnutrition has the energy reserves and protein needed to respond successfully to stress.
d. A person with malnutrition who suffers an acute stress generally does not need tube feedings or TPN unless he or she will be unable to eat for 7 to 10 days.

6. All of the following parameters help to assess a person's fluid needs during stress *except:*
a. urinary output.
b. nitrogen balance.
c. blood pressure.
d. body temperature.

7. Which of the following statements correctly describes an appropriate diet for severe stress?
a. The diet is always high in kcalories.
b. The diet has little effect on blood glucose levels.
c. The amounts of carbohydrates, lipids, and proteins that supply energy make little difference.
d. Although the diet must supply adequate energy, energy needs are generally not high, except for some people with extensive burns or head injuries.

8. Depending on the type of stress, the amount of protein a stressed person who weighs 150 pounds needs can range from about ________ grams of protein per day.
a. 100 to 130
b. 100 to 200
c. 130 to 200
d. 200 to 300

9. The use of oral diets in the immediate poststress period is not possible because severe stress:
a. slows gastric motility.
b. decreases the appetite.
c. alters plasma amino acid levels.
d. results in increased blood flow to the GI tract.

10. The major reason early enteral nutrition must be introduced by tube following a severe stress is that:
a. appetite is depressed.
b. hydrolyzed diets are necessary.
c. gastric feeding is not medically possible.
d. oral diets cannot meet energy and nutrient needs.

Clinical Applications

1. Returning to Bernadette from the "How to" on p. 735 and assuming that she can tolerate an intact enteral formula, find at least three formulas in Appendix J that the health care team might select if she needed a tube feeding. Determine the volume of each formula that would be needed to meet Bernadette's energy and protein needs. Would this volume also meet the recommendations for vitamins and minerals?

2. Bennie is a well-nourished seven-year-old who develops the flu and has a fever of 101°F for two days. Describe how this stress could temporarily affect his nutrition status. How would your concerns differ if Bennie were a seven-year-old hospitalized for injuries suffered in a car accident, who develops the flu and a fever and is unable to eat for several days?

3. Susan Griff is a 28-year-old woman admitted to the hospital following a car accident in which she broke several bones, ruptured a portion of her small intestine, and suffered a severe burn. She has been in the hospital for several weeks and is now eating table foods. Aside from the nutrient demands imposed by the stresses she withstood, describe how the following factors can impair her nutrition status:

- Susan's injuries are painful.
- Susan's medications cause extreme drowsiness.
- Susan is depressed.
- Susan is often out of her room for X rays and other diagnostic tests when her menus and food trays arrive.
- Susan's food intake is often restricted for diagnostic tests she will be receiving.

How might these problems be resolved to improve Susan's ability to eat?

4. Reviewing the information about the metabolic changes directed by hormones in response to acute stress (p. 729), suggest a reason why a previously well-nourished person who develops acute malnutrition as a consequence of stress might have normal or excessive body fat stores.

References

1 H. R. Chang and B. Bistrian, The role of cytokines in the catabolic consequences of infection and injury, *Journal of Parenteral and Enteral Nutrition* 22 (1998): 156–166.

2 L. L. Moldawer, Cytokines and the cachexia response to acute inflammation, *Support Line,* April 1996, pp.1–6.

3 S. A. McClave and coauthors, Are patients fed appropriately according to their caloric requirements? *Journal of Parenteral and Enteral Nutrition* 22 (1998): 375–381.

4 S. A. McClave and D. A. Spain, Indirect calorimetry should be used, *Nutrition in Clinical Practice* 13 (1998): 143–145.

5 C. S. Ireton-Jones and J. D. Jones, Should predictive equations or indirect calorimetry be used to design nutrition support regimens? *Nutrition in Clinical Practice* 13 (1998): 141–145.

6 P. S. Choban, J. C. Burge, and L. Flancbaum, Nutrition support of obese hospitalized patients, *Nutrition in Clinical Practice* 12 (1997): 149–154; J. C. Burge and coauthors, Efficacy of hypocaloric total parenteral nutrition in hospitalized obese patients: A prospective, double-blind, randomized trial, *Journal of Parenteral and Enteral Nutrition* 18 (1994): 203–207.

7 H. Saito, S. Furukawa, and T. Matsuda, Glutamine as an immunoenhancing nutrient, *Journal of Parenteral and Enteral Nutrition* (supplement) 23 (1999): S59–S61.

8 A. P. J. Houdijk, R. J. Nijveldt, and P. A. M. van Leeuwen, Glutamine-enriched enteral feeding in trauma patients: Reduced infectious morbidity is not related to changes in endocrine and metabolic responses, *Journal of Parenteral and Enteral Nutrition* (supplement) 23 (1999): S52–S58; A. P. J. Houdijk and coauthors, Randomised trial of glutamine-enriched enteral nutrition on infectious morbidity in patients with multiple trauma, *Lancet* 352 (1998): 772–776; P. R. Schloerb and M. Amare, TPN with glutamine in bone marrow transplantation and other clinical applications (a randomized double blind study), *Journal of Parenteral and Enteral Nutrition* 17 (1993): 407–413; T. R. Ziegler and coauthors, Clinical and metabolic efficacy of glutamine-supplemented parenteral nutrition after one marrow transplantation, *Annals of Internal Medicine* 116 (1992): 821–828.

9 M. M. McMahon, Nutrition support of hospitalized patients, presented at the Eighth Annual Advances and Controversies in Clinical Nutrition, Mayo Clinic Foundation, April 5–7, 1998; J. J. Pomposelli and coauthors, Early postoperative glucose control predicts nosocomial infection rate in diabetic patients, *Journal of Parenteral and Enteral Nutrition* 22 (1998): 77–81.

10 McMahon, 1998.

11 D. R. Garrel and coauthors, Improved clinical status and length of care with low-fat nutrition support in burn patients, *Journal of Parenteral and Enteral Nutrition* 19 (1995): 505–507.

12 Z. A. Gonzalez, Practical aspects of nutritional modulation of the immune response, *Nutrition in Clinical Practice* 15 (2000): 45–47; C. Lo and coauthors, Fish oil modulates macrophage P44/P42 mitogen-activated protein kinase activity by lipopolysaccharide, *Journal of Parenteral and Enteral Nutrition* 24 (2000): 159–163.

13 Lo and coauthors, 2000.

14 D. Granato and coauthors, Effects of parenteral lipid emulsions with different fatty acid composition on immune cell functions *in vitro, Journal of Parenteral and Enteral Nutrition* 24 (2000): 113–118.

15 S. Hailer, K. W. Jauch, and G. Wolfram, Influence of different fat emulsions with 10 or 20% MCT/LCT or LCT on lipoproteins in plasma of patients after abdominal surgery, *Annals of Nutrition and Metabolism* 42 (1998): 170–180.

16 K. M. Oldham and P. E. Bowen, Oxidative stress in critical care: Is antioxidant supplementation beneficial? *Journal of the American Dietetic Association* 98 (1998): 1001–1008.

17 W. W. Souba, Nutritional support, *New England Journal of Medicine* 336 (1997): 41–48; B. Beier, E. A. Bergman, and M. J. Morrissey, Factors related to the use of early postoperative enteral feedings in thoracic and abdominal surgery patients in the United States, *Journal of the American Dietetic Association* 97 (1997): 293–295; S. Trice, G. Melnik, and C. P. Page, Complications or costs of early postoperative parenteral versus enteral nutrition in trauma patients, *Nutrition in Clinical Practice* 12 (1997): 114–119.

18 K. A. Kudsk, Immunologic support: Enteral vs parenteral feeding, in *Enteral Nutrition Support,* Report of the First Ross Conference on Enteral Devices, Ross Laboratories, 1996, pp. 70–74.

19 G. Minard, Early enteral feeding—Is it safe for every patient? *Nutrition in Clinical Practice* 13 (1998): 79–80.

ANSWERS

Multiple Choice

1. b	2. c	3. a	4. d	5. b
6. b	7. d	8. a	9. a	10. c

NUTRITION AND MULTIPLE ORGAN FAILURE

Multiple organ failure is a critical, and often fatal, complication related to the inflammatory responses that occur as a consequence of uncontrolled infection or severe tissue injury. It generally develops days or weeks after the initial stress and may develop in organs not related to the site of injury. Often, the lungs fail first, followed by the liver and kidneys. This highlight describes the causes of multiple organ failure and discusses the potential role of the GI tract and nutrition in protecting against its development.

THE GI TRACT AND MULTIPLE ORGAN SYSTEM FAILURE

As Chapter 24 described, the inflammatory responses to tissue injury serve to protect the body from invading organisms and repair tissue damage, but the responses can also lead to further tissue damage. Most often, the body's inflammatory responses are controlled and self-limiting; the responses gradually diminish and eventually end once the danger is over. If the insult to the body is severe, however, or if further injuries occur or infection develops, inflammatory responses may be prolonged or exaggerated, and the harmful effects of the responses overwhelm the beneficial effects. Tissues especially vulnerable to the damaging effects of inflammatory responses include the lungs, liver, kidneys, and GI tract.[1] Further damage to these organs results from the breakdown of protein that accompanies hypermetabolism.

The exact reasons why inflammatory responses run amok and lead to multiple organ failure remains unclear. However, a growing body of evidence suggests that changes that occur in the GI tract as a consequence of severe stress disrupt the GI tract's immune defenses and allow bacteria to enter the body. According to this theory, the bacteria then trigger further inflammatory responses, thereby setting in motion a vicious cycle of further infection and tissue damage. A short review of GI tract immune functions provides the basis for understanding the connections between the GI tract and multiple organ failure.

GI Tract Immune Functions

As Highlight 18 described, the GI tract plays a pivotal role in preventing microbes from entering the body. The defense mechanisms in the intestines include the physical barrier to microbial invasion provided by the closely packed intestinal villi and immune system cells that destroy invaders and stimulate immune responses. The intestinal tract houses 70 to 80 percent of the immune tissue in the entire body.[2] In addition, the normal bacterial flora inhibit the growth of harmful bacteria by competing with them for nutrients and space. The normal bacterial flora do not stimulate inflammatory responses. Recent evidence suggests that the normal bacteria in the gut may also directly affect intestinal functions, including nutrient absorption and immune functions.[3] Researchers found that introducing one strain of normally harmless bacteria into mice with germ-free GI tracts enhanced the production of genes important in nutrient absorption and processing as well as genes that limit damage resulting from activation of inflammatory processes. Furthermore, the introduction of different strains of bacteria produced variations in the genes that were produced, suggesting that the composition of the bacterial flora plays an important role in providing optimal protection to the GI tract.

Inflammatory Response Effect on GI Immune Function

Conditions that compromise the intestinal tract's physical barrier, alter its immune cell responses, or alter the colon's normal bacterial flora raise the likelihood that infectious agents can cross the intestinal barrier and enter the body—a process called translocation. Table H24-1 lists conditions that increase the likelihood of translocation. In small amounts, the translocation of infectious agents may stimulate the immune system,

In multiple organ failure, the harmful effects of inflammatory responses overwhelm the beneficial effects.

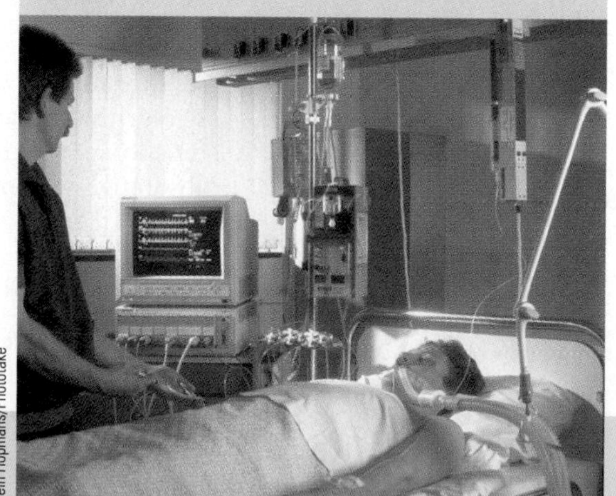

but extensive translocation may cause serious infection and even death.

Severe Stress and the GI Tract

As already described, inflammatory responses render the GI tract susceptible to tissue damage. During severe stress, changes in the circulatory system divert blood flow away from the GI tract. The disruption of the delivery of oxygen and nutrients to the intestine, coupled with the loss of protein due to hypermetabolism, impairs the structure and function of intestinal cells. The provision of fluid and electrolytes following stress (reperfusion) restores blood flow to the intestine. While reperfusion provides needed oxygen and nutrients, it also spurs inflammatory responses in the intestine, which further damages intestinal cells and stimulates the production of cytokines.[4] The cytokines, in turn, exaggerate and prolong inflammatory responses and have effects throughout the body. Figure H24-1 illustrates the effects and consequences of severe stress on the GI tract and the potential for the development of multiple organ failure.

In addition to these effects, intestinal edema (from increased permeability of blood vessels) and reduced peristalsis provide conditions that favor the overgrowth of bacteria from the colon to the small intestine. The net effect of all these consequences is an increased risk of bacterial translocation. Animal studies suggest that bacterial translocation from the intestine can lead to peritonitis, pneumonia, and widespread sepsis.[5] A study of people undergoing surgery found that bacterial overgrowth of the small intestine increased the incidence of both bacterial translocation and sepsis.[6] Recall from Chapter 21 that bacterial translocation may also be a cause of shock and multiple organ failure following severe acute pancreatitis.

Although much of the evidence is compelling, clinical evidence to support a role for bacterial translocation as the cause of, or even one of the major factors in, multiple organ failure is lacking.[7] It may be that bacterial translocation leads to multiple organ failure in some cases but not in others. Furthermore, it may be the type of intestinal bacteria present, rather than the inflammatory responses, that lead to bacterial translocation.[8]

FIGURE H24-1 Effects and Potential Consequences of Severe Stress on the GI Tract

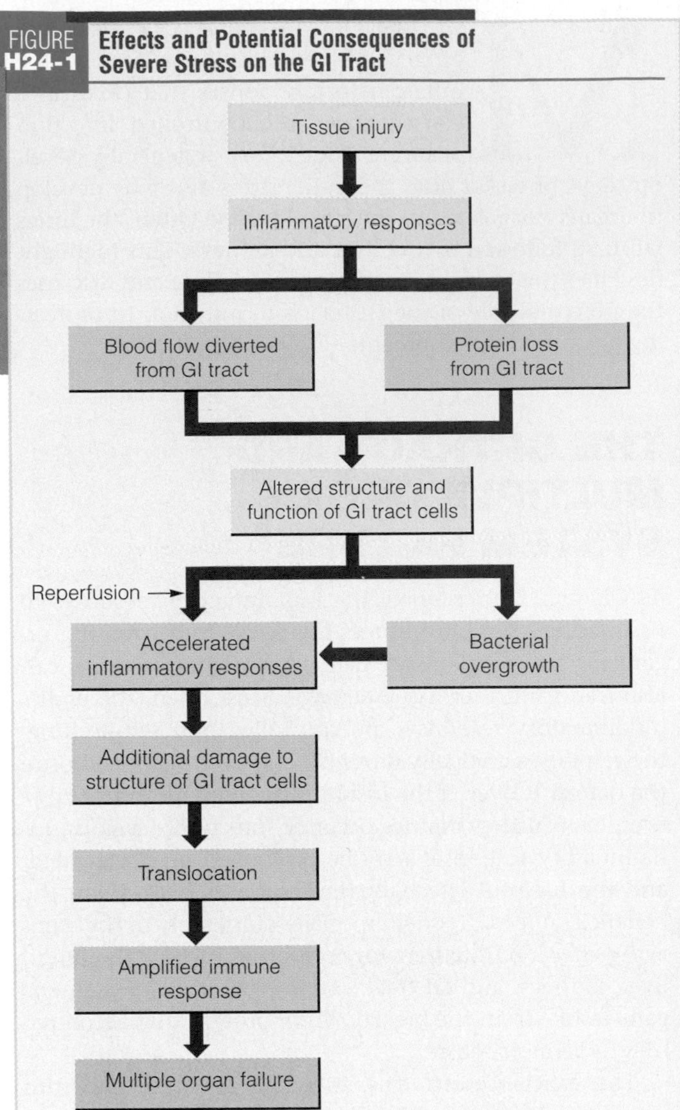

TABLE H24-1 Conditions That Increase the Likelihood of Translocation

Conditions
Altered structure and function of GI tract barrier
Prolonged fasting or lack of enteral nutrients
Injury to the GI tract
Altered blood flow to GI tract
Inflammatory responses
Malnutrition
Changes in bacterial flora
Lack of enteral nutrients
Decreased GI tract motility
Use of broad-spectrum antibiotics
Compromised function of immune factors
Malnutrition
Hypermetabolism

SOURCE: Adapted from M. T. DeMeo, The role of enteral nutrition in maintaining the structural and functional integrity of the GI tract, in *Enteral Nutrition Support*, Report of the First Ross Conference on Enteral Devices, Ross Laboratories, 1996, pp. 4–8.

NUTRITION AND GI TRACT INTEGRITY

In clinical practice, measures to support GI integrity during severe stress are widely accepted and used by practitioners. These measures are designed to maintain the structure and function of the GI tract cells in order to preserve both their absorptive and immune functions.

Enteral versus Parenteral Nutrition

As Chapter 24 described, without the stimulation of nutrients delivered enterally, the cells of the GI tract shrink and lose some of their functions. Providing at least some nutrients enterally provides benefits when compared to parenteral nutrition, including lower rates of infection and improved blood flow to the gut.[9] With improved blood flow, the risk of bacterial overgrowth—and, therefore, translocation—may also be reduced.

Timing and Components of Nutrition Support

Just as important as the method of feeding is its timing. Although negative nitrogen balance occurs regardless of nutrient availability, delivery of nutrients within a few days after the stress occurs helps prevent acute protein malnutrition, preserves lean body mass, and may preserve immune function.

As mentioned in Chapter 24, during stress the amino acid glutamine, which provides fuel to both the intestinal cells and immune system cells, may become conditionally essential. Glutamine added to parenteral nutrition solutions has been shown to help preserve intestinal structure. More recently, animal studies show that adding glutamine to parenteral solutions helps preserve some immune functions both within and outside the intestine.[10] Glutamine-enriched enteral diets resulted in fewer incidences of pneumonia, bacteremia, and sepsis than a similar diet without glutamine in people with multiple trauma.[11] Enteral formulas supplemented with arginine, nucleotides, and omega-3 fatty acids have also been shown to have protective effects on immune function following severe stress.[12]

Evidence suggesting that a breach in the intestinal barrier can lead to multiple organ failure is mounting, but the theory remains unproven. While waiting for more information, clinicians often rely on nutrition measures to help maintain the integrity of the GI tract. These measures include enteral feeding in general, and especially early enteral feeding. Many also use immune-enhancing formulas, which may include glutamine, arginine, nucleotides, and omega-3 fatty acids, although the benefits of such formulas remain unclear as well. The prospects of supporting GI immune system functions and preventing further decline in severely stressed people are inviting, and the potential benefits may prove lifesaving.

References

1 K. H. Cheever, Early enteral feeding of patients with multiple trauma, *Critical Care Nurse* 19 (1999): 40–51.

2 J. S. Thompson, The intestinal response to critical illness, *American Journal of Gastroenterology* 90 (1996): 190–200.

3 L. V. Hooper and coauthors, Molecular analysis of commensal host-microbial relationships in the intestine, *Science* 291 (2001): 881–884.

4 F. A. Moore, The role of the gastrointestinal tract in postinjury multiple organ failure, *American Journal of Surgery* 178 (1999): 449–453.

5 T. J. Babineau and G. L. Blackburn, Time to consider early gut feeding, *Critical Care Medicine* 22 (1994): 191–193.

6 J. MacFie and coauthors, Gut origin of sepsis: A prospective study investigating associations between bacterial translocation, gastric microflora, and septic morbidity, *Gut* 45 (1999): 223–228.

7 S. Kanwar and coauthors, Lack of correlation between failure of gut barrier function and septic complications after major upper gastrointestinal surgery, *Annals of Surgery* 231 (2000): 88–95; G. A. Nieuwenhuijzen and R. J. Goris, The gut: The "motor" of multiple organ dysfunction syndrome? *Current Options in Nutritional and Metabolic Care* 2 (1999): 399–404.

8 M. Ljungdahl and coauthors, Bacterial translocation in experimental shock is dependent on strains in intestinal flora, *Scandinavian Journal of Gastroenterology* 35 (2000): 389–397.

9 K. Takagi and coauthors, Modulating effects of the feeding route on stress response and endotoxin translocation in severely stressed patients receiving thoracic esophagectomy, *Nutrition* 16 (2000): 355–360; T. R. Ziegler, C. Gatzen, and D. W. Wilmore, Strategies for attenuating protein-catabolic responses in the critically ill, *Annual Review of Medicine* 45 (1994): 459–480; T. J. Babineau and G. L. Blackburn, Time to consider early gut feeding, *Critical Care Medicine* 22 (1994): 191–193; D. K. Heyland, D. J. Cook, and D. H. Guyatt, Does the formulation of enteral feeding influence infectious morbidity and mortality rates in the critically ill patient? A critical review of the evidence, *Critical Care Medicine* 22 (1994): 1192–1202.

10 K. A. Kudsk and coauthors, Glutamine-enriched parenteral nutrition maintains intestinal interleukin-4 and muscosal immunoglobulin A levels, *Journal of Parenteral and Enteral Nutrition* 24 (2000): 270–275.

11 A. P. J. Houdijk and coauthors, Randomised trial of glutamine-enriched enteral nutrition on infectious morbidity in patients with multiple trauma, *Lancet* 352 (1998): 772–776.

12 L. Gianotti and coauthors, A prospective, randomized clinical trial on perioperative feeding with an arginine-, omega-3 fatty acid-, and RNA-enriched enteral diet: Effect on host response and nutritional status, *Journal of Parenteral and Enteral Nutrition* 23 (1999): 314–320; M. Braga and coauthors, Immune and nutritional effects of early enteral nutrition after major abdominal operations, *European Journal of Surgery* 162 (1996): 105–112; M. Kemen and coauthors, Early postoperative enteral nutrition with a diet enriched with arginine-, omega-3 fatty acids-, and ribonucleic acid-supplemented diet *versus* control in cancer patients: An immunologic evaluation of impact, *Critical Care Medicine* 23 (1995): 652–659.

CHAPTER 25
Nutrition and Diabetes Mellitus

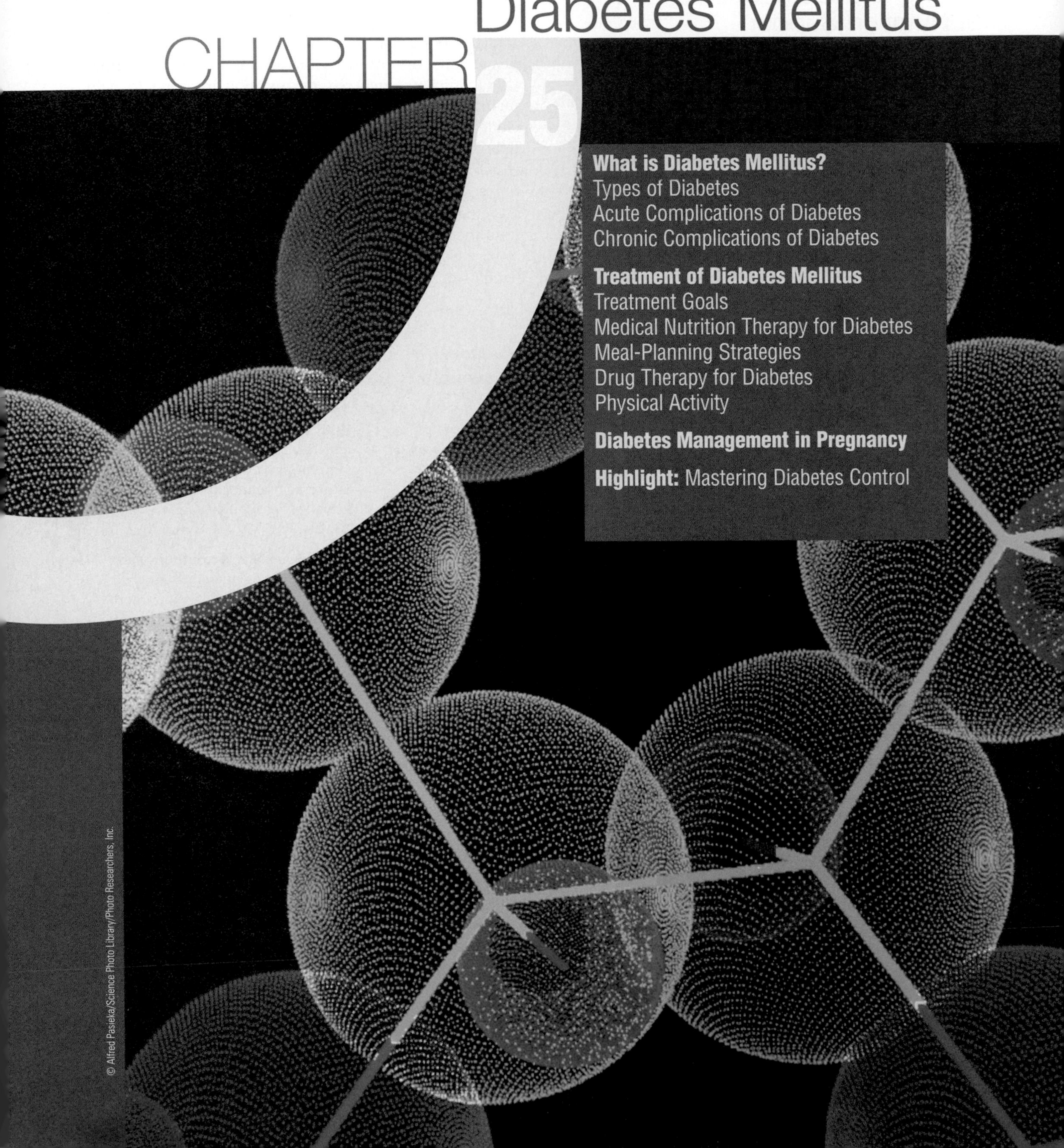

As Chapter 24 showed, medical conditions that lead to severe stress upset the body's metabolic activities to such an extent that their consequences can be extensive and even fatal. Part of the body's response to severe stress includes alterations in the body's normal balance between insulin◆ and the counterregulatory hormones; the result is an increase in blood glucose and altered energy metabolism. Alterations in the balance between insulin and the counterregulatory hormones also occur in **diabetes mellitus,** but for different reasons. In this case, the body's production or use of insulin falters.

◆ Reminder: Insulin is the hormone that, among other things, enables many cells to take up glucose from the blood and store energy fuels. The counterregulatory hormones, which include glucagon, epinephrine, norepinephrine, cortisol, and growth hormone, oppose insulin's actions.

WHAT IS DIABETES MELLITUS?

Diabetes mellitus describes a group of metabolic disorders characterized by elevated blood glucose and altered energy metabolism and caused by defective insulin secretion, defective insulin action, or a combination of the two. There are two major types of diabetes, and their distinguishing features are summarized in Table 25-1. A later section of this chapter describes the special case of gestational diabetes. Other types of diabetes can occur as a consequence of genetic disorders, diseases of the exocrine pancreas, hormonal imbalances, drugs or chemicals, certain infections, and immune system disorders. Chapter 20, for example, noted that the destruction of pancreatic cells in both pancreatitis and cystic fibrosis can lead to diabetes. Some people who develop hyperglycemia as a consequence of severe stress may also develop diabetes permanently as a consequence.

The diagnosis of diabetes is based on blood glucose levels, which can be measured under different conditions and include:

- A random blood glucose sample (one taken without regard to food intake) that exceeds 200 milligrams per deciliter in a person with symptoms of hyperglycemia (described in a later section).
- A blood glucose level of 126 milligrams per deciliter or greater in a person who has been fasting for at least 8 hours.
- A blood glucose level of 200 milligrams per deciliter or greater at any time during a glucose tolerance test—a test that measures a person's fasting blood glucose and then retests it several times after a specific amount of glucose is given orally.

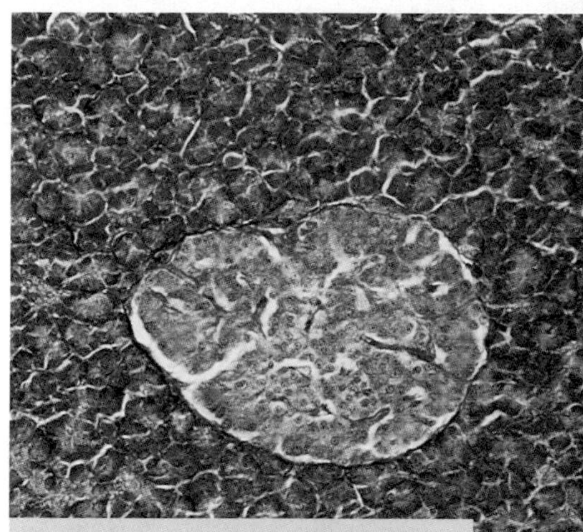

Specialized endocrine cells in the pancreas, the islets of Langerhans, produce hormones. Among these cells are the alpha cells, which produce glucagon, and the beta cells, which produce insulin.

TABLE 25-1 Features of Type 1 and Type 2 Diabetes

	Type 1	Type 2
Age of onset	<20 (mean age, 12)	10–19, >40
Associated conditions	Viral infection, heredity	Obesity, heredity, aging
Insulin required?	Yes	Sometimes
Cell response to insulin	Normal	Resistant
Clinical findings (generally)	Hyperglycemia with ketoacidosis	Hyperglycemia without ketoacidosis
Prevalence in diabetic population	5 to 10%	90 to 95%
Other names	IDDM[a]	NIDDM[a]
	Juvenile-onset diabetes	Adult-onset diabetes
	Ketosis-prone diabetes	Ketosis-resistant diabetes
	Brittle diabetes	Lipoplethoric diabetes
		Stable diabetes

[a]The names IDDM (insulin-dependent diabetes mellitus) and NIDDM (noninsulin-dependent diabetes mellitus) frequently appear in references and generally describe type 1 and type 2 diabetes, respectively.

diabetes (DYE-uh-BEET-eez) **mellitus** (MELL-ih-tus or mell-EYE-tus): a group of metabolic disorders of glucose regulation and utilization.
- **diabetes** = passing through (the body)
- **mellitus** = honey-sweet (sugar)

Any one of these tests can suggest diabetes, but a positive result must be confirmed by performing a second test (using any of the three criteria) on a subsequent day. Physicians determine the type of diabetes by assessing a variety of parameters including medical history, body weight, clinical findings, and laboratory tests.

Types of Diabetes

The incidence of diabetes is reaching epidemic proportions. Over 16 million people in the United States have diabetes, and that number is expected to rise to 22 million by 2025. Of the 16 million people who currently have diabetes, about one-third do not know they have it.[1] Undiagnosed diabetes is especially dangerous because its damaging effects (described later) may begin to develop years before the symptoms of diabetes appear.[2]

• Type 1 Diabetes • In **type 1 diabetes,** the less common of the two major types of diabetes (about 5 to 10 percent of diagnosed cases), the pancreas gradually loses its ability to synthesize insulin. In most cases, the individual inherits a defect in which immune cells mistakenly attack and destroy the insulin-producing pancreatic cells.◆ Usually, by the time type 1 diabetes is diagnosed, the destruction of beta cells has reached a point where insulin must be supplied. Most often, type 1 diabetes begins in childhood or adolescence, although it can occur at any age. People with siblings or parents with type 1 diabetes have a high risk of developing the disease themselves. Studies are currently under way to determine if the destruction of beta cells in type 1 diabetes can be delayed or prevented.[3]

◆ Immune system disorders in which the body destroys its own cells or tissues are called **autoimmune disorders.** Laboratory tests that detect antibodies to insulin can help clinicians predict which individuals will develop type 1 diabetes and also help them distinguish type 1 from type 2 diabetes.

• Type 2 Diabetes • The predominant type of diabetes mellitus (90 to 95 percent of cases), and the type most likely to go undiagnosed, is **type 2 diabetes.** In type 2 diabetes, the pancreas produces insulin, and the cells respond to it, but with less sensitivity **(insulin resistance).** As blood glucose rises, the pancreas makes more insulin, and blood insulin rises to abnormally high levels (hyperinsulinemia). During this period of **impaired glucose tolerance,** the body is able to maintain blood glucose within a fairly normal range but with a cost. The chronic demand for insulin gradually exhausts the beta cells of the pancreas, and finally insulin production falters as the disease progresses.

Impaired glucose tolerance and type 2 diabetes are associated with excess body fat, especially abdominal fat; physical inactivity; and aging. Obesity◆ aggravates insulin resistance: as body fat increases, body tissues become less and less able to respond to insulin. About 20 percent of people over 65 have diabetes mellitus.[4] The incidence in people over 80 may be as high as 40 percent.[5] People most likely to develop type 2 diabetes include:

- Those who are obese.
- Those who have immediate family members with diabetes.
- Members of high-risk ethnic populations (African Americans, Asian Americans, Pacific Islanders, Hispanic Americans, and Native Americans).
- Those who are over age 45.
- Women who have given birth to babies weighing over 9 pounds or who have been diagnosed with diabetes while pregnant.

◆ The Expert Committee on the Diagnosis and Classification of Diabetes Mellitus defines obesity as greater than 120% desirable weight or a body mass index (BMI) of 27 or higher.

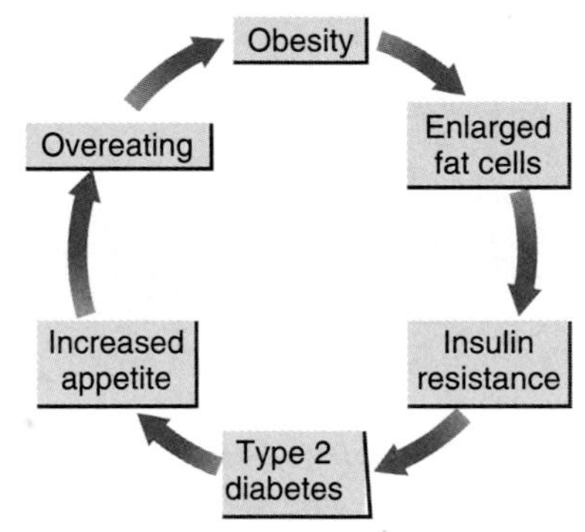

Although type 2 diabetes frequently occurs in obese, middle-aged or older adults, children and adolescents represent a sizable and growing number of cases. Children and adolescents with type 2 diabetes are generally obese (body mass index of 27 or greater) and between 10 and 19 years old, and they have a strong family history of diabetes.

Most experts believe that the rising rates of diabetes and its destructive consequences can be eased or even halted by adoption of healthy behaviors that include weight management and regular physical activity. A nationwide multicenter study is currently under way to determine whether these interventions or others can prevent type 2 diabetes in people with impaired glucose tolerance.[6]

type 1 diabetes: the less common type of diabetes in which the person produces no insulin at all.

type 2 diabetes: the more common type of diabetes that develops gradually and is associated with obesity and insulin resistance.

insulin resistance: the condition in which the cells fail to respond to insulin as they do in healthy people.

impaired glucose tolerance: inability to maintain normal blood glucose levels without excessive insulin production. Some people with impaired glucose tolerance have fasting glucose levels somewhat higher than normal but not high enough to diagnose diabetes. Others have normal blood glucose levels most of the time, but when given a large amount of glucose, their blood glucose rises too high.

Acute Complications of Diabetes

Elevated blood glucose leads to immediate complications. Figure 25-1 presents an overview of these acute metabolic changes and clinical manifestations. The acute metabolic consequences of type 1 diabetes are more immediate and severe than those of type 2 diabetes because people with type 1 diabetes produce little or no insulin, so no glucose enters the cells. The glossary on p. 748 defines diabetes-related symptoms and complications.

• **Hyperglycemia and Glycosuria** • Normally when a person absorbs carbohydrate from a meal or snack, blood glucose rises. The rising blood glucose triggers the release of insulin, which allows glucose to enter the cells and effectively lowers blood glucose. With insufficient or ineffective insulin, blood glucose remains high, and the person develops symptoms of **hyperglycemia.**◆ Despite the high blood glucose, the cells are deprived of energy. The body responds as it would to starvation—levels of counterregulatory hormones rise, insulin resistance develops or worsens, and the body mobilizes its glycogen stores and uses protein to make more glucose. Blood glucose rises higher still. Some people experience hunger and eat excessively **(polyphagia),** which aggravates hyperglycemia.

Symptoms of hyperglycemia: ◆
- Intense thirst and, sometimes, hunger.
- Increased urination.
- Blurred vision.
- Fatigue.
- Acetone breath.
- Labored breathing.

Under normal conditions, glucose is not excreted in the urine, but eventually the concentration of glucose in the blood exceeds the kidneys' ability to

FIGURE 25-1 Acute Metabolic Consequences and Clinical Manifestations of Untreated or Uncontrolled Diabetes

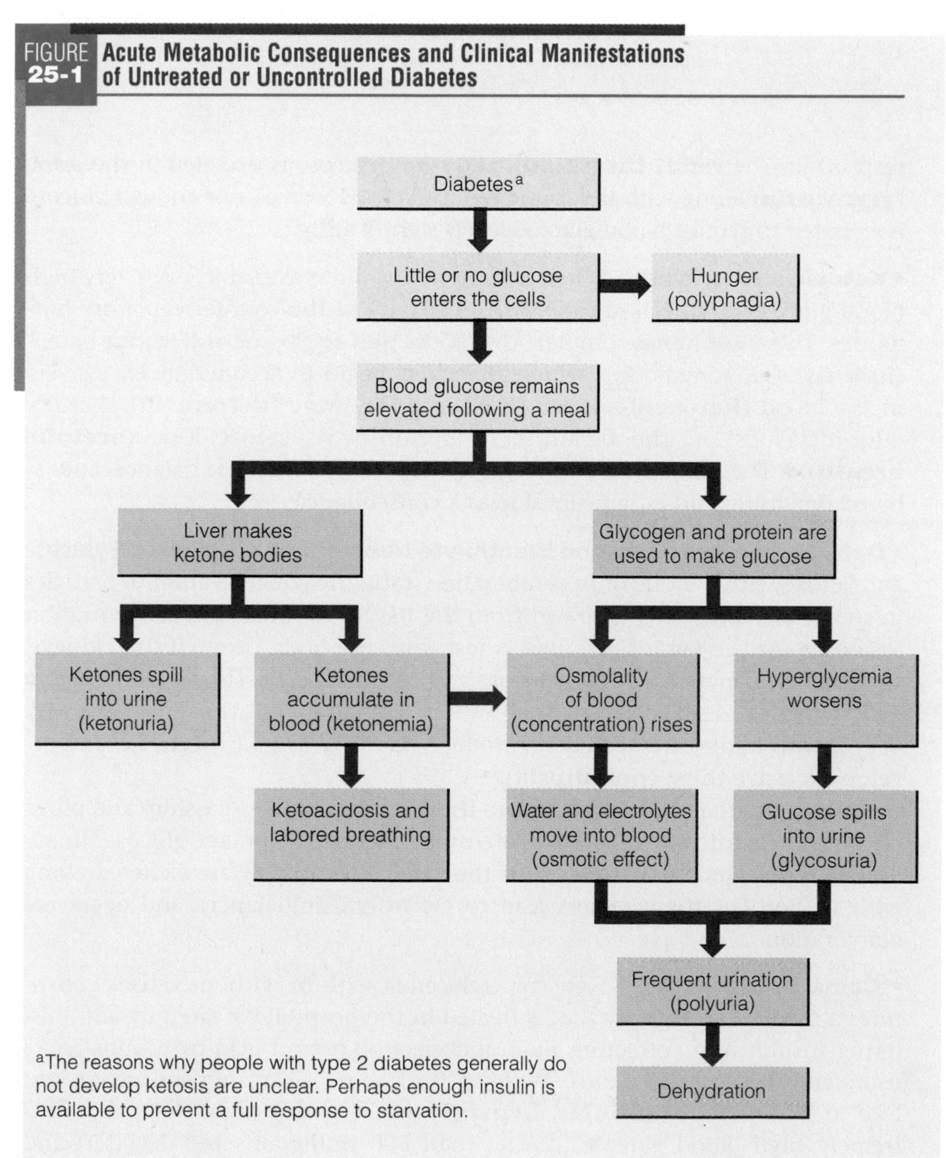

[a]The reasons why people with type 2 diabetes generally do not develop ketosis are unclear. Perhaps enough insulin is available to prevent a full response to starvation.

GLOSSARY OF DIABETES-RELATED SYMPTOMS AND COMPLICATIONS

acetone breath: a distinctive fruity odor that can be detected on the breath of a person who is experiencing ketosis.

dawn phenomenon: early morning hyperglycemia that develops in response to elevated levels of counterregulatory hormones that act to raise blood glucose after an overnight fast.

diabetic coma: unconsciousness precipitated by hyperglycemia, dehydration, ketosis, and acidosis in people with diabetes.

gangrene: death of tissue due to a deficient blood supply and/or infection.

gastroparesis: delayed gastric emptying.

glycosuria (GLY-ko-SUE-ree-ah) or **glucosuria** (GLUE-ko-SUE-ree-ah): glucose in the urine, which generally occurs when blood glucose exceeds 180 mg/dL.

hyperglycemia: elevated blood glucose. Normal fasting blood glucose is less than 110 mg/dL. Fasting blood glucose between 110 and 125 mg/dL suggests impaired glucose tolerance; values of 126 mg/dL or higher suggest diabetes.

hyperosmolar, hyperglycemic coma: coma that occurs in uncontrolled type 2 diabetes precipitated by the presence of hypertonic blood and dehydration.

hypoglycemia: low blood glucose.

ketoacidosis: lowering of the blood's normal pH due to the accumulation of acidic ketones.

ketonuria: ketones in the urine.

ketosis: accelerated production of ketones.

macroangiopathies: disorders of the large blood vessels.

microangiopathies: disorders of the capillaries.

nephropathy: a disorder of the kidneys.

neuropathy: a disorder of the nerves.

nocturnal hypoglycemia: hypoglycemia that occurs while a person is sleeping.

polydipsia (POLL-ee-DIP-see-ah): excessive thirst.

polyphagia (POLL-ee-FAY-gee-ah): excessive eating.

polyuria (POLL-ee-YOU-ree-ah): excessive urine production.

rebound hyperglycemia: hyperglycemia resulting from excessive secretion of counterregulatory hormones in response to excessive insulin and consequent low blood glucose levels; also called the **Somogyi** (so-MOHG-yee) **effect.**

renal threshold: the point at which a blood constituent that is normally reabsorbed by the kidneys reaches a level so high the kidneys cannot reabsorb it. The renal threshold for glucose is generally reached when blood glucose rises above 180 mg/dL.

retinopathy: a disorder of the retina.

reabsorb it (the **renal threshold**), and some glucose is excreted in the urine **(glycosuria)** along with fluids and electrolytes. However, not enough glucose is excreted to reduce blood glucose levels significantly.

• Ketosis and Acidosis• When little or no insulin is available, even very high blood glucose levels fail to suppress the activity of the counterregulatory hormones. These hormones stimulate the production of glucose and ketone bodies **(ketosis).◆** Ketone bodies, which are acidic, begin to accumulate excessively in the blood **(ketoacidosis)** and spill into the urine **(ketonuria).** A fruity odor develops on the breath of a person with ketoacidosis **(acetone breath).◆** The lungs must work hard to help restore acid-base balance, and labored breathing and fatigue develop as a consequence.

◆ **Reminder: Ketone bodies are produced by the incomplete breakdown of fat when glucose is not available to the cells (see p. 225).**

◆ **The odor associated with *acetone breath* is typically described as "fruity" for lack of a better description, but it is actually a sharper, more distinctive smell similar to that of nail polish remover, which contains acetone.**

• Dehydration and Fluid and Electrolyte Imbalances • High blood glucose and ketosis, either alone or in combination, raise the concentration of particles in the blood, and water is drawn from the tissues to dilute the concentration (osmosis). At the same time, fluid is lost from the body through the kidneys, which must excrete the excess glucose and ketone bodies. Thus hyperglycemia and ketosis can lead to severe dehydration. This series of events explains why people with undiagnosed diabetes produce excessive urine **(polyuria)** and develop excessive thirst **(polydipsia).**

Electrolytes that normally reside in the cells, especially potassium and phosphorus, move into the blood as proteins are sacrificed to make glucose. In addition, when glucose is excreted in the urine, electrolytes are excreted along with it. Both of these factors lead to electrolyte imbalances and aggravate dehydration.

• Coma in Diabetes • Severe hyperglycemia with or without ketosis constitutes a medical emergency that is treated in the hospital by carefully administering insulin and correcting fluid and electrolyte and acid-base imbalances using intravenous (IV) fluids. Without treatment, severe hyperglycemia can lead to coma. **Hyperosmolar, hyperglycemic coma** can develop from extremely high blood glucose (greater than 600 milligrams per deciliter) and severe dehydration without ketoacidosis or with only minimal ketoacidosis. Coma that occurs when ketoacidosis is also present is called **diabetic coma.** In this type of coma, blood glucose does not rise as high as in hyperosmolar, hypoglycemic coma, but acidosis becomes a serious problem.

• Hyperglycemia in Untreated Diabetes • Hyperglycemia, glycosuria, and associated symptoms may be the only clinical findings present at the time diabetes is diagnosed, especially for people with type 2 diabetes. Ketoacidosis and its clinical findings are more likely to be present at the time of diagnosis in type 1 diabetes than in type 2 diabetes, because insulin production is markedly reduced by the time type 1 diabetes is identified.

• Causes of Hyperglycemia in Treated Diabetes • Once diabetes has been diagnosed and treatment is under way, hyperglycemia can develop when a person eats too much carbohydrate or during times when counterregulatory hormone levels rise.◆ Counterregulatory hormones raise blood glucose, but with no insulin or ineffective insulin, glucose cannot enter the cells. Thus hyperglycemia can be present when the person with diabetes arises after an overnight fast **(dawn phenomenon);** when blood glucose levels are high and the person engages in strenuous physical activity; when a person injects too much insulin **(rebound hyperglycemia);** or when the person develops an illness or infection. When these problems occur, insulin doses must be adjusted.

Reminder: When counterregulatory hormones rise, they cause insulin resistance and signal the liver to mobilize glycogen and protein to provide glucose. ◆

Even a minor illness such as a cold or flu can cause blood glucose to rise dramatically. Infections and injuries that evoke inflammatory responses are the most common causes of severe hyperglycemia in people with diabetes.

• Hypoglycemia • Hypoglycemia, or low blood glucose, arises from the inappropriate management of diabetes, rather than from the disease itself. It can result from taking too much insulin◆ or glucose-lowering medications, strenuous physical activity, skipped or delayed meals, inadequate food intake, vomiting, or severe diarrhea. Some people develop hypoglycemia while they are sleeping **(nocturnal hypoglycemia).** Among the symptoms of hypoglycemia,◆ mental confusion and shakiness may make it difficult for the person to recognize the problem and take corrective actions.

Hypoglycemia in a person who uses insulin is also called an **insulin reaction** or **insulin shock.** ◆

Symptoms of hypoglycemia: ◆

- Hunger.
- Headache.
- Sweating.
- Shakiness.
- Nervousness.
- Confusion.
- Disorientation.
- Slurred speech.

Left untreated, hypoglycemia can lead to coma and death. Repeated episodes of hypoglycemia may permanently impair cognitive function.[7] Avoiding hypoglycemia is especially important for children, particularly children under seven years old, because hypoglycemia can interfere with normal brain development.[8] Children and people who have had diabetes for a long time may not notice the warning signs, however, and thus risk severe hypoglycemia. A later section describes the dietary treatment of hypoglycemia.

Notice that many of the symptoms of hypoglycemia are those of alcohol intoxication. If the true problem goes unrecognized, the person may die. To prevent such a tragic mistake, advise every person with diabetes to wear identification in the form of a bracelet or necklace.

Chronic Complications of Diabetes

Exposure of the tissues to high glucose concentrations over time results in the chronic complications of diabetes. When blood glucose concentrations remain high, glucose attaches to proteins in the blood and cells lining the blood vessels. These glycated proteins damage the structures of the blood vessels and nerves.◆ The cells lining the blood vessels thicken, circulation becomes poor, and nerve function falters. Infections are likely to occur due to poor circulation coupled with glucose-rich blood and urine. Infections often go undetected due to impaired nerve function; **gangrene** may follow. People with diabetes must pay special attention to hygiene and keep alert for early signs of infection.

Reminder: Insulin resistance may have damaging effects on blood vessels and nerves even before a diagnosis of diabetes is made. ◆

• Cardiovascular Diseases • Disorders of the large blood vessels **(macroangiopathies),** including coronary heart disease and hypertension (see Chapter 26), commonly develop in people with diabetes. Cardiovascular disease tends to develop early, progress rapidly, and be more advanced at the time of diagnosis in people with type 2 diabetes.[9] More than 80 percent of people with diabetes die as a consequence of cardiovascular diseases, especially heart attacks. If nerve function is also impaired, the person may suffer a heart attack◆ and not even realize it.

◆ A heart attack that goes unnoticed is called a **silent heart attack.**

As Chapter 26 will show, people with diabetes often have many risk factors for coronary heart disease including altered blood lipid levels, hypertension, and obesity.◆ Furthermore, as described above, hyperglycemia damages tissues. Tissue damage elicits inflammatory responses, which alter blood flow and bring about further tissue damage by leading to oxidative stress and an environment that favors the formation of blood clots (procoagulant state).

◆ The combination of insulin resistance, secretion of more and more insulin to maintain blood glucose levels, obesity, hypertension, elevated LDL and triglycerides, and lowered HDL (see Chapter 26) frequently leads to both type 2 diabetes and cardiovascular disease and is called the **metabolic syndrome** (see Highlight 26).

• **Small Blood Vessel Disorders** • Disorders of the smallest blood vessels **(microangiopathies)** also occur as a consequence of diabetes. Hyperglycemia damages the capillaries, which may lead to loss of kidney function and retinal degeneration and loss of vision. About 85 percent of people with diabetes have **nephropathy, retinopathy,** or both. Consequently, diabetes is a leading cause of both kidney failure and blindness.

• **Neuropathy** • Nerve tissues may also deteriorate, resulting in **neuropathy.** At first, the person may experience a painful prickling sensation, often in the arms and legs, which progresses until the person loses sensations in the hands and feet. Injuries to these areas may go unnoticed, and infections can progress rapidly. If tissues die as a consequence, amputation of the affected limb (usually the toes, feet, or legs) may be necessary. Neuropathy can also delay gastric emptying **(gastroparesis).** When the stomach empties slowly after a meal, the person may experience a premature feeling of fullness, nausea, vomiting, weight loss, and poor blood glucose control due to irregular nutrient absorption.[10]

IN SUMMARY

Diabetes develops when little or no insulin is produced (type 1 diabetes) or when insulin is ineffective (type 2 diabetes). In either case, undiagnosed or uncontrolled diabetes can lead to acute complications, including hyperglycemia and hypoglycemia. If left untreated, both of these complications can lead to coma and death. The chronic complications of diabetes include disorders of the large and small blood vessels and nerve damage.

TREATMENT OF DIABETES MELLITUS

A diagnosis of diabetes can be devastating. An adult with recently diagnosed diabetes may fear possible complications and be overwhelmed by the lifestyle changes necessary to control the disorder. The parents of a child with diabetes may feel overwhelmed, angry, anxious, and even guilty. Preteens and teenagers often have intense difficulty accepting a diagnosis of diabetes. As young adults strive for independence from their families, they become increasingly self-conscious and reject those things that set them apart from peer groups—including the tasks necessary to control their diabetes. Once children reach school age, they should be active participants in their treatment plans. Adolescents need to know they can manage their disease themselves and achieve the independence they strive for.

metabolic syndrome: the combination of insulin resistance, hyperinsulinemia, obesity, hypertension, elevated LDL and triglycerides, and reduced HDL that is frequently associated with type 2 diabetes and cardiovascular disease; also called *syndrome X* and *insulin-resistance syndrome.*

Treatment Goals

The goals of both medical and nutrition therapy for diabetes are to maintain blood glucose within a fairly normal range, achieve optimal blood lipid levels (see Chapter 26), control blood pressure (Chapter 26), support health and well-being, and prevent and treat complications. The most important of these goals is blood glucose control. Table 25-2 on p. 752 compares two treatment plans for controlling blood glucose—traditional therapy and intensive therapy.

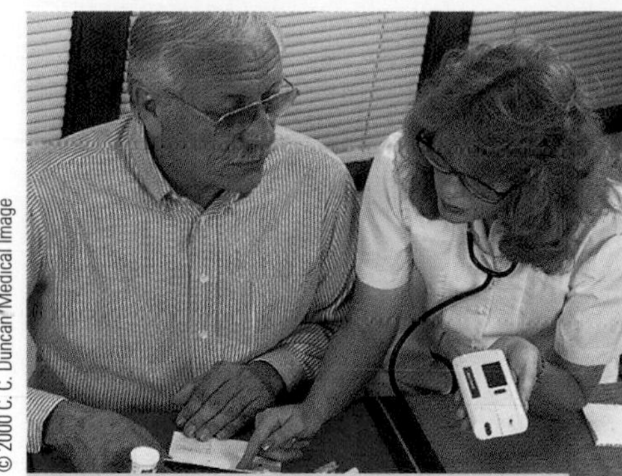

© 2000 C. C. Duncan/Medical Image

Blood glucose monitoring helps people with diabetes learn how their blood glucose responds to carbohydrate, physical activity, and illness and helps them maintain it in a safe range.

• Benefits of Intensive Therapy • A major multicenter clinical trial (the Diabetes Control and Complications Trial, or DCCT) clearly showed that compared to traditional therapy, intensive therapy reduces the risks of onset and progression of nephropathy, retinopathy, and neuropathy by about 50 percent. The study included and found comparable benefits in both adults and children age 13 and older. A follow-up study found that the benefits of intensive therapy persist for at least four years (the time period of the study), even though blood glucose levels tend to rise with time.[11] The United Kingdom Prospective Diabetes Study showed similar benefits for people with type 2 diabetes.[12]

• Evaluating Diabetes Control • The health care team routinely evaluates the client's blood glucose levels, blood lipid levels, blood pressure, weight, and reflexes to see how well the treatment plan is working and to check for signs of complications. Urine tests can help detect the early signs of kidney disease, eye exams help detect the early signs of retinopathy, and foot exams help detect early signs of infection.

Clients learn to monitor their blood glucose levels at home, often using computerized blood glucose meters. Clients who practice intensive therapy check their blood glucose about four times a day, and those who practice traditional therapy check it once or twice a day. At medical appointments, the health care team reviews the client's test results and records to monitor the client's progress and look for patterns that suggest the need for adjustments in the treatment plan. Highlight 25 describes how the health care team works with each client to develop an individualized treatment plan.

Health care professionals often advise clients to monitor ketones in the urine when they find that blood glucose levels are consistently high or during illness. As described earlier, high blood glucose may predispose the person to ketosis and coma.

Physicians periodically evaluate blood glucose control through laboratory measurement of the percentage of **glycated hemoglobin.**◆ As blood glucose rises, glucose attaches to amino acids on hemoglobin molecules and remains there until the red blood cells that carry the hemoglobin die (about 120 days). The percentage of glycated hemoglobin reflects diabetes control over the past two to three months, rather than just prior to the test.

◆ In the Diabetes Control and Complications Trial, groups receiving intensive treatment and those following traditional therapy achieved average glycated hemoglobin (GHb) of 7.2% and 9.1%, respectively (normal <6%). In the follow-up study, average GHb in groups receiving intensive treatment and those following traditional therapy averaged 7.9% and 8.2%, respectively.

• Treatment Plans • An accurate and thorough assessment and monitoring of a client's medical condition, developmental stage, and psychosocial and economic needs form the basis for an acceptable diet, physical activity, and medication plan. Successful management plans provide the fewest disruptions to the client's lifestyle and family (or caregiver) relationships.

Medical Nutrition Therapy for Diabetes

Medical nutrition therapy for diabetes is most appropriately designed and implemented by a skilled dietitian. A complete nutrition assessment provides the foundation. The dietitian uses the assessment to learn about the client's usual eating habits, analyzes the diet's composition and adequacy, and devises a plan that considers the client's age and daily routines that affect food intake. After the client tries the plan for a while, adjustments often need to be made. Sometimes nutrient needs change (as a child grows, for example), or the client may have problems with blood glucose control or simply find it hard to follow certain parts of the plan.

glycated (GLIGH-kate-id) **hemoglobin:** hemoglobin with glucose molecules attached to its amino acids; also called *glycosylated hemoglobin.* The type of glycated hemoglobin most commonly measured is hemoglobin A_{1c}.

Effective treatment plans for diabetes must consider individual needs. A plan for a child, for example, must be flexible enough to accommodate a typical child's activity levels (and appetite), which often vary widely from day to day.

TABLE 25-2 Comparision of Traditional and Intensive Therapy for Diabetes

Diet	Traditional	Intensive
Diet	Consistent intake of energy and carbohydrates at set times each day.	Consistent intake of energy and carbohydrates at set times as often as possible.
Physical Activity	Regular schedule of physical activity at set times.	Regular schedule of physical activity at set times and as often as possible.
Insulin	One or two injections at same time each day.	Multiple daily injections or insulin pump. Adjust insulin doses to accommodate deviations from meal plan or physical activity schedule.
Blood Glucose Monitoring	Once or twice a day.	At least three or four times a day.
Advantages	Easier to learn, requires less time to follow, fewer injections and finger pricks, less risk of hypoglycemia and weight gain.	Better blood glucose control, fewer microvascular complications, greater flexibility.
Disadvantages/ Contraindications	Poorer blood glucose control (blood glucose stays higher), greater risk of complications, inflexible—deviations from diet/physical activity plan not recommended.	Greater risk of hypoglycemia and excessive weight gain; not recommended for children under age 2 or older people with significant cardiovascular disease.

THINK NUTRITION whenever clients have diabetes of any type. Medical nutrition therapy is an integral part of the treatment plan.

The diet for diabetes parallels a healthy diet for all people in both amounts and types of nutrients. Attention to all energy nutrients is important: controlling carbohydrate prevents hyper- and hypoglycemia; controlling protein may help preserve kidney function; controlling fat helps prevent cardiovascular complications.

• Energy • Medical nutrition therapy for diabetes first focuses on providing food energy in the amount necessary to achieve or maintain a healthy and realistic body weight and to support growth in children and pregnant women. The diet planner calculates the person's average daily energy intake and uses it to determine an appropriate energy level based on body weight, monitors weight periodically, and adjusts the diet as necessary.

People with newly diagnosed or poorly controlled type 1 diabetes are likely to be thin despite eating apparently adequate amounts of food. However, with treatment, especially intensive therapy, excessive weight can be a problem. Teenagers with type 1 diabetes who practice intensive therapy are often heavier than teenagers without diabetes. Children and adults with type 2 diabetes are likely to be overweight and often benefit from weight loss. Even moderate weight loss (10 to 20 pounds) can help reverse insulin resistance, improve the blood lipid profile, and reduce blood pressure. A diet plan that moderately restricts energy intake (250 to 500 kcalories less than the current average intake) provides for a realistic and gradual weight loss.[13] Elderly people with diabetes, even those with type 2 diabetes, are often underweight.

◆ An early sign of impending kidney disease is the excretion of small amounts of albumin in the urine or **microalbuminemia.**

• Protein • Protein provides about 10 to 20 percent of the total kcalories in the diet for diabetes. Limited and inconclusive data suggest that providing adequate, but not excessive, protein may help delay the onset or progression of kidney disease (see Chapter 27). At the first sign of kidney disease,◆ people with diabetes may need to restrict protein to no more than 0.8 gram per kilogram of body weight (the same as the RDA).[14] If kidney disease progresses, protein may need to be restricted further.

• **Carbohydrate** • To control blood glucose, the person with diabetes needs to have glucose available throughout the day, but not so much at one time that blood glucose levels rise too high. Of all the energy nutrients, carbohydrates have the greatest effect on blood glucose. Once eaten, carbohydrates raise blood glucose in about an hour. Some protein and fat from a meal may also raise blood glucose, but to a lesser degree and more slowly than carbohydrates. Most diet plans provide from 45 to 60 percent of the total kcalories from carbohydrate.

A person with a regular physical activity program who takes a prescribed dose of medication at a set time and then eats about the same amount of carbohydrate at about the same time each day is likely to have a safe amount of glucose available to the body when it is needed. Thus, for people with diabetes, consistent timing and composition of meals and snacks from day to day improve glucose control. Without medication adjustments, too much carbohydrate at a meal or snack can cause hyperglycemia. Skipping meals or eating too little can lead to hypoglycemia. An evening snack is especially important because it helps sustain the person's blood glucose through the night and helps prevent nocturnal hypoglycemia.

• **Carbohydrate Sources** • Encourage clients with diabetes to select whole-grain breads and cereals, legumes, fruits, and vegetables. In addition to carbohydrates, these foods provide fiber,◆ vitamins, and minerals and offer many health benefits (see Chapter 4).

◆ Reminder: Authorities recommend an intake of 20 to 35 g of dietary fiber a day.

Traditionally, concentrated sweets were strictly excluded from the diet for diabetes, but now they are restricted only to the same extent as they are for all people. The total amount of carbohydrate is of greater concern in diabetes than the type of carbohydrate. The person with diabetes can consume concentrated sweets and alternative nutritive sweeteners◆ as a limited part of a healthy diet, as long as they are counted as part of the carbohydrate allowance. Artificial sweeteners (such as saccharin and aspartame), and products made from them, contain no carbohydrate and minimal kcalories and can be used in place of sugar.

◆ Reminder: Nutritive sweeteners are carbohydrates and include sucrose, fructose, sorbitol, mannitol, and xylitol.

• **Glycemic Index** • As Chapter 4 described, carbohydrates from different sources have varying effects on blood glucose after a meal. Foods with a high glycemic index raise blood glucose faster and to a greater extent than foods with a low glycemic index (see Table 4-2 on p. 109). Controversy abounds regarding the value of the glycemic index of foods and the diet for diabetes.[15] Some feel enough evidence exists to support the use of glycemic indexing in the diet for diabetes. Others feel that, at present, too little information is available to warrant the imposition of additional restrictions. Although various health organizations around the world and many clinicians endorse the use of the glycemic index, the American Diabetes Association currently does not.

• **Fat** • People with diabetes who have acceptable blood lipid concentrations benefit from a fat intake consistent with the *Dietary Guidelines for Americans* (30 percent or less of kcalories from fat, less than 10 percent of kcalories from saturated fat, and less than 300 milligrams of cholesterol). People with diabetes who have elevated LDL cholesterol◆ may need to restrict saturated fat intake to 7 percent or less of total kcalories and cholesterol to less than 200 milligrams daily. To lower fat and cholesterol intakes, clients can use low-fat and nonfat milk and milk products and lean meats, among other strategies (see Chapter 26 for more information).

◆ Reminder: Cholesterol is transported through the blood packaged with lipoproteins. LDL are low-density lipoproteins and VLDL are very low-density lipoproteins. Chapter 26 provides more information on the relationships of lipoproteins to cardiovascular disease.

For people with elevated triglycerides◆ and VLDL, another diet approach may be useful. These clients may be able to control their blood glucose and improve their blood lipid levels with a moderately higher fat intake and a lower

◆ People with very high triglyceride levels (≥1000 mg/dL) need to restrict all types of dietary fat to less than 10% of kcalories.

carbohydrate intake, provided that the additional fat kcalories come from monounsaturated fat.[16]

People with diabetes who use reduced-fat products◆ must use them cautiously.[17] People may overeat when they believe they are saving kcalories by eating reduced-fat products, and this is not always the case. Additionally, carbohydrate often replaces the fat in these products. The energy and carbohydrate that the reduced-fat products contribute to the person's food intake must be considered as part of the energy and carbohydrate allowance in the diet plan.

◆ The fat substitute, olestra, does not contribute kcalories or carbohydrate.

• Sodium • People with diabetes frequently develop hypertension, and limiting sodium can help reduce blood pressure. An intake of sodium of 2400 milligrams of sodium (6 grams salt) or less per day is appropriate for all people, including those with diabetes. People with diabetes and kidney disease may need to restrict sodium even more.[18]

• Alcohol • The person whose blood glucose is well controlled can usually include some alcoholic beverages with the consent of the physician. Alcohol can cause hypoglycemia in any person, and people with diabetes who take medications to lower blood glucose are particularly likely to develop this complication. People who take insulin are advised to drink only moderate amounts (no more than one drink per day for women or two drinks per day for men),◆ with meals, and in addition to the usual meal plan. To protect against hypoglycemia, no foods should be omitted. (Remember, too, that the person with hypoglycemia may appear to be intoxicated and alcohol use can add confusion to a potentially dangerous situation.)

◆ One alcoholic drink is defined as 1½ oz of distilled liquor, 12 oz of beer, or 5 oz of wine. Note: Light beer contains the same amount of alcohol as regular beer, with half the carbohydrate. For people who need to lose weight, 1 drink = 2 fat exchanges.

People with a history of alcohol abuse, pancreatitis, abnormal blood lipids, or neuropathy and women who are pregnant are strongly cautioned to avoid alcohol completely. Alcohol use is also discouraged for people who are overweight; if it is used, alcohol should be considered as part of the fat allowance. Drinks that contain simple sugars (mixers, sweet wines, and liqueurs) are best avoided. If they are used, the person must count their carbohydrate contents as part of the daily carbohydrate allowance. The combination of alcohol and some **oral antidiabetic agents** may cause nausea, vomiting, headache, cramps, flushing of the skin, and a rapid heartbeat **(disulfiram-like reaction).**

• Micronutrients • Unless the person with diabetes has a vitamin, mineral, or trace element deficiency, micronutrient◆ needs for people with diabetes do not differ from those of the healthy population. People who take some types of diuretics (see Chapter 26), however, may need to take potassium supplements.

◆ Some people take chromium to lower blood glucose levels. However, this strategy is ineffective unless the person has a chromium deficiency.

• Missed Meals and Illness • When people with diabetes are ill, their blood glucose levels often rise dramatically. During illness, clients may be advised to increase their doses of medication, reduce carbohydrate intakes somewhat, or a combination of both. However, they need some carbohydrate to forestall hypoglycemia; so do people who must miss a meal for any reason. If appetite is poor, people can use juice, flavored gelatin, soft drinks, or frozen juice bars to meet their carbohydrate needs.

Hospitals employ different procedures for people with diabetes who miss a meal. One procedure provides at least half the prescribed carbohydrate and kcalories within three hours of the missed meal. If this cannot be done, the physician may change the insulin schedule, give IV dextrose, and/or change the diet prescription to include more simple carbohydrates.

• Treating Hypoglycemia • If a client develops hypoglycemia for any reason, a judicious approach prevents overtreatment and subsequent hyperglycemia. As soon as the symptoms are observed, the person needs 10 to 15 grams of carbohydrate. People who take oral agents that interfere with the digestion of sucrose and complex carbohydrates need to take glucose to treat hypoglycemia; for all others, any carbohydrate source that is readily available and easy to eat is a good choice.◆ Advise clients to avoid foods that also contain fat, which slows the absorption of carbohydrate. Blood glucose is then checked within 15

◆ Easy-to-eat sources of carbohydrate (10 to 15 g per serving):
- 2 to 3 tsp honey.
- 4 to 5 hard candies (such as Lifesavers).
- 5 to 6 large jelly beans.
- 4 to 6 oz sweetened soft drink.
- 4 oz orange or other fruit juice.
- 1 tbs icing from a can or tube.
- Glucose gel or tablets (check label for amount).

oral antidiabetic agents: medications taken by mouth to lower blood glucose levels in people with type 2 diabetes.

disufiram-like reaction: nausea, vomiting, headache, cramps, flushing of the skin, and a rapid heartbeat that can occur when some medications are taken along with alcohol. The medication disulfiram produces these effects when combined with alcohol to discourage alcohol abusers from using alcohol.

to 20 minutes to see if it has risen to an acceptable level. If not, an additional 10 to 15 grams of carbohydrate are given, and blood glucose is rechecked. The procedure continues until blood glucose returns to an acceptable range. Advise clients to carry some convenient source of carbohydrates with them at all times, so they can act immediately when hypoglycemic symptoms occur.

If hypoglycemia becomes severe, the person may be disoriented, unable to recognize a hypoglycemic reaction, and unable to swallow safely. In such cases, the person needs to receive IV glucose or the hormone glucagon or both to counteract the insulin reaction. Without treatment, the person may lapse into shock and die.

• Enteral and Parenteral Formulas • The indications for enteral and parenteral nutrition for people with diabetes are the same as for other people (see Chapters 22 and 23). When people with diabetes require enteral or parenteral formulas, however, they may also be experiencing insulin resistance, so adjustments may need to be made. Often health care professionals provide or adjust insulin doses to cover the higher blood glucose levels. If insulin fails to control hyperglycemia, people on parenteral nutrition may need to receive less energy from dextrose and more from IV lipid emulsions. People who cannot tolerate the carbohydrate in standard enteral formulas may benefit from specially designed formulas that contain less total carbohydrate (see Appendix J).

Meal-Planning Strategies

No single approach to medical nutrition therapy meets everyone's needs, and dietitians use several approaches to help clients follow a consistent diet plan and maintain their target blood glucose levels.[19] Some diet strategies use food guides or simple menus to teach clients to plan diets. Traditionally, however, diet planners use the exchange system described briefly here and more completely in Chapter 2 (see pp. 41–43).

• Exchange Lists • Recall that the exchange system◆ sorts foods into three main groups by their proportions of carbohydrate, fat, and protein and further divides each group into exchange lists. By strictly defining portion sizes, all of the foods on a given exchange list provide approximately the same amounts of energy (kcalories) and energy nutrients (carbohydrate, fat, and protein). Thus any food on a list can be exchanged, or traded, for any other food on that same list without affecting a plan's energy balance. Figure 2-4 on pp. 44–45 shows examples of foods and portion sizes for each group of foods, and Table 2-3 on p. 42 shows the energy and energy nutrients found in each exchange list. Clients also learn how to use food labels to determine the number of exchanges for foods not shown on the exchange lists. Review the "How to" to learn the basics of using exchange lists to plan a diet for diabetes.

◆ Appendix G includes the complete U.S. exchange system, and Appendix I includes the Canadian choice system.

• Carbohydrate Counting • Another widely used strategy for planning diets for diabetes is called carbohydrate counting.[20] Carbohydrate counting allows clients more flexibility in adjusting their diets while still maintaining blood glucose control. To learn carbohydrate counting, clients may first learn the exchange list system to help them establish healthy eating habits, understand portion sizes, control their intake of energy and energy nutrients, and learn to eat consistent amounts of carbohydrates at regular times. Clients must also learn to perform the mathematical operations necessary to calculate their carbohydrate intakes.

Once clients understand how to maintain a healthy diet and control their blood glucose levels, they can focus on the carbohydrate they eat to simplify meal planning and manage times when their intakes vary. Clients who monitor their blood glucose and keep records of their glucose levels, the time and amount of carbohydrate they eat at each meal, and the time and amount of insulin or oral antidiabetic agents they use gain the most benefits from carbohydrate counting. The "How to" on p. 758 shows how carbohydrate counting works.

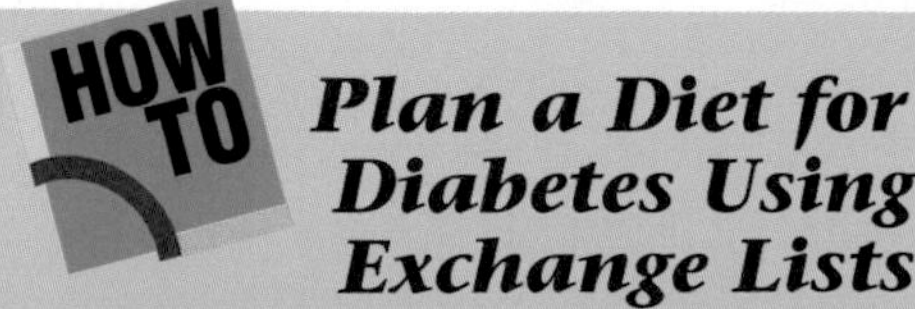

Plan a Diet for Diabetes Using Exchange Lists

Using exchange lists to plan diets takes time at first, but with practice, the planning process becomes routine. This box describes a simplified diet plan.

The first step is to assess each individual to determine a reasonable body weight and the energy intake necessary to achieve or maintain that body weight. For this example, we will use a man who is 6 feet tall and is comfortable with the weight of 178 pounds that he has maintained throughout his adult life. From an assessment of food intake, the dietitian estimates that the man has maintained his weight on about 2900 kcalories per day with 25 percent of kcalories from protein, 45 percent from carbohydrate, and 30 percent from fat.

1. The next step is to determine the recommended grams of protein, carbohydrate, and fat based on 2900 kcalories:

   - Protein (10 to 20% of kcalories):

     $.10 \times 2900$ kcal $= 290$ kcal.
     290 kcal $\div$ 4 kcal/g $= 73$ g.

     $.20 \times 2900$ kcal $= 580$ kcal.
     580 kcal $\div$ 4 kcal/g $= 145$ g.

     Thus the man needs between 73 and 145 grams of protein.

   - Carbohydrate (45 to 60% of kcalories):

     $.45 \times 2900$ kcal $= 1305$ kcal.
     1305 kcal $\div$ 4 kcal/g $= 326$ g.

     $.60 \times 2900$ kcal $= 1740$ kcal.
     1740 kcal $\div$ 4 kcal/g $= 435$ g.

     Thus the man needs between 326 and 435 grams of carbohydrate.

   - Fat (30% or less of kcalories):

     $.30 \times 2900$ kcal $= 870$ kcal.
     870 kcal $\div$ 9 kcal/g $= 97$ g.

     Thus the man needs about 97 grams of fat or less.

2. The dietitian recognizes that the client will need to make dietary changes to conform to a healthy eating plan. To minimize the changes the client must make, the dietitian plans the diet to include 20 percent protein (580 kcalories). Thus 80 percent or 2320 kcalories remain for carbohydrate and fat. After reviewing information about the man's blood lipids, which are within acceptable limits, the dietitian plans the diet to keep fat at the current level of 30 percent (870 kcalories). This means that 50 percent of the kcalories (1450 kcalories) remain for carbohydrate.

   2900 total kcal − 580 protein kcal
   − 870 fat kcal
   = 1450 carbohydrate kcal.

3. To translate the kcalories from fat and carbohydrate to grams:

   870 fat kcal $\div$ 9 kcal/g $= 96$ g.*
   1450 carbohydrate kcal $\div$
   4 kcal/g $= 363$ g.*

   Thus the diet will provide 2900 kcalories with a distribution of 20 percent protein (145 grams or 580 kcalories), 50 percent carbohydrate (363 grams or 1450 kcalories), and 30 percent fat (96 grams or 870 kcalories).

4. Now it is time to translate the diet prescription into a meal plan. Table 2-3 on p. 42 shows the grams of carbohydrate, protein, and fat and the energy value in each serving on an exchange list. Using this table and the client's food intake record as a guide, the dietitian first plans servings of foods that contain carbohydrate, then protein, and finally fat, trying to match foods as closely as possible to the client's usual food intake. This process takes practice and requires some adjusting based on trial and error. Most often, the final result does not fit the meal plan exactly, but comes close. Table 25-3 shows how the dietitian might plan a day's exchanges for the man in this example. Note that the plan falls within the guidelines of the Daily Food Guide on pp. 38–39. The plan uses fat-free milk and lean meat exchanges for calculations. Lower-fat foods are encouraged. If the client occasionally

*To limit fat, round down the actual 96.6 grams to 96. Round up the 362.5 grams carbohydrate to 363.

TABLE 25-3 A Day's Exchanges for a Sample 2900-kCalorie Diet

Exchange Group/List	Number of Exchanges	Carbohydrate (g)	Protein (g)	Fat[a] (g)
Carbohydrate group[b]				
Starch	13	195	39	0
Fruit	7	105	—	—
Milk	3	36	24	0
Vegetable	6	30	12	—
Meat and meat substitutes group				
Lean	10	—	70	30
Fat group	13	—	—	65
Total grams		366	145	95
Total kcalories		1464	580	855
% kcalories		50.5	20	29.5

[a]To ease calculation, exchanges from the carbohydrate groups are assumed to have 0 grams fat. If the client uses a fat-containing exchange, the fat can be deducted from the daily fat allowance.
[b]Foods from the "other carbohydrates" list can be substituted for a starch, fruit, or milk list exchange. Any fat in the selected food is then deducted from the daily fat allowance.

Drug Therapy for Diabetes

People with type 1 diabetes need insulin to control blood glucose and utilize energy. People with type 2 diabetes can sometimes control their blood glucose without medications by using a combination of diet and physical activity.

HOW TO

chooses to use another type of milk or meat, the number of fat servings must be adjusted accordingly. For example, if the man eats 4 ounces of a high-fat meat (32 grams of fat) instead of lean meat (12 grams of fat), he must then use four fewer fat exchanges during the day (20 grams of fat). The plan shown in Table 25-3 does not include the "other carbohydrates" list; starches and other foods (described later) can be substituted for foods on this list.

5. Distribute foods into meals that fit the client's usual eating patterns. Table 25-4 shows how the day's exchanges might be divided for the man in this example. With this information in hand, the dietitian and client can begin to fill in the plan with real foods to create a sample menu such as the one shown in Table 25-5. The client is reminded to eat about the same amount of carbohydrate at about the same time each day.

6. Teach clients how to tailor the diet to meet their own preferences. For example, foods from the starch, fruit, milk, and other carbohydrate lists contain similar amounts of energy and carbohydrate and can be substituted for one another from time to time. Regular substitution is discouraged, however, because each list makes unique contributions to other nutrient needs. A client who regularly substitutes fruit for milk, for example, may not be getting enough calcium. The client who regularly substitutes milk for a starch or fruit may not be getting enough fiber.

7. Three servings of free foods can be included as long as they are spread throughout the day. Free foods contain up to 20 kcalories and 5 grams of carbohydrate per serving. A serving of food that contains a carbohydrate-based fat substitute that provides 5 grams or less carbohydrate counts as a free food.

TABLE 25-4 Dividing a Day's Exchanges between Meals and Snacks

Exchange Group/List	Number of Exchanges	Breakfast	Lunch	Midafternoon Snack	Supper	Bedtime Snack
Carbohydrate group						
Starch	13	3	3	2	3	2
Fruit	7	2	1	1	1	2
Milk	3	1	1			1
Vegetable	6		3		3	
Meat and meat substitutes group						
Lean	10		3	1	4	2
Fat group	13	3	3	2	3	2

TABLE 25-5 Translating a Day's Exchanges into a Day's Meals

Breakfast

½ c bran cereal (1 starch)
1 c fat-free milk (1 milk)
1 bagel, 2 oz (2 starch)
2 tsp margarine (2 fat)
1 banana (2 fruit)
Coffee

Lunch

1 cup pasta (2 starch) served with:
½ c spaghetti sauce (1 starch, 1 fat)
2 oz meatballs, medium fat (2 meat, 2 fat)
2 tbs grated parmesan cheese (1 meat, ½ fat)
1½ c broccoli (3 vegetables)
17 small grapes (1 fruit)
1 c fat-free milk (1 milk)

Midafternoon Snack

1 fat-free granola bar (2 starch)
2 tbs peanut butter (1 meat, 1½ fat) spread over slices of 1 small apple (1 fruit)

Supper

4 oz grilled salmon (4 meat)
⅔ c rice (2 starch)
1 whole-wheat roll (1 starch)
1 c cooked carrots (2 vegetables)
1 tsp margarine (1 fat)
1 c mixed green salad (1 vegetable)
Salad dressing made with 2 tsp olive oil (2 fat) and wine vinegar
1 c diced cantaloupe (1 fruit)
Iced tea

Bedtime Snack

½ c lowfat cottage cheese (2 meat)
1 c fruit cocktail (2 fruit)
1 small frosted cupcake (2 starch, 1 fat)
1 c fat-free milk (1 milk)

Note: Compared with Table 25-4, this menu provides one fewer fat exchange at breakfast and one-half fat exchange more at lunch.

When these measures fail, one or a combination of oral antidiabetic agents may be prescribed. Oral antidiabetic agents have a maximum dose; if blood glucose cannot be controlled at the maximum dose, the physician may prescribe insulin or insulin analogs, alone or in combination with oral antidiabetic agents. Medications do not replace diet and physical activity in the management of

Help Clients Count Carbohydrates

1. Begin by calculating how much carbohydrate the person usually eats at each meal and snack. For example, assume that the client is a woman who maintains her weight and controls her blood glucose levels on an intake of 1700 kcalories. Use exchanges to determine her usual carbohydrate intake. In this example, the client and dietitian determined that her usual eating pattern includes:

- Breakfast: 2 starch, 1 fruit, 1 milk.
- Lunch: 2 starch, 1 fruit, 2 vegetables.
- Midafternoon snack: 1 fruit, 1 milk.
- Supper: 2 starch, 1 fruit, 2 vegetables.
- Bedtime snack: 1 starch, 1 fruit.

2. Next, use Table 2-3 on p. 42 to determine the grams of carbohydrate in each meal. Breakfast, for example, contains:

2 starch = 2 × 15 g carbohydrate = 30 g
1 fruit = 1 × 15 g carbohydrate = 15 g
1 milk = 1 × 12 g carbohydrate = 12 g
Total = 57 g carbohydrate.

Using the same method to calculate the carbohydrate from other meals and snacks, the carbohydrate contents are as follows:

- Lunch: 55 g.
- Midafternoon snack: 27 g.
- Supper: 55 g.
- Bedtime snack: 30 g.

3. Teach the client to include these same amounts of carbohydrate consistently at each meal and snack. Clients can use exchanges, food labels, and food composition tables to determine the carbohydrate contents of the foods they eat.
4. Encourage clients to carefully weigh and measure all foods initially. Once clients are familiar with portion sizes, they can weigh and measure foods less often, but should still check portion sizes occasionally.
5. Remind clients that they still need to be aware of the total amount of food they eat as well as the type and amount of fat they are eating.

Clients who use insulin can learn how much insulin they need to cover their carbohydrate intakes (see p. 760). Then, if they change their carbohydrate intakes on occasion, they can adjust their insulin dose accordingly.

diabetes; advise clients to continue these therapies. The Diet-Drug Interactions box on p. 759 includes the agents described here. Medications used to treat cardiovascular and renal complications that may accompany diabetes are described in Chapters 26 and 27, respectively.

• Oral Antidiabetic Agents • The number of oral agents available to treat people with type 2 diabetes has grown markedly in recent years. Some oral agents work by stimulating the release of insulin from the beta cells; these include sulfonylureas, replaglinide, and nateglinide. Sulfonylureas (mainly chlorpropamide) may react with alcohol (see p. 754) and can cause a disulfiram-like reaction. Nateglinide stimulates insulin release for only short periods following meals to help minimize post-meal hyperglycemia without overtaxing the pancreas. Metformin and the thioglitazones (rosiglitazone, and piolitazone) work primarily by lessening peripheral insulin resistance. The alpha-glucosidase inhibitors, acarbose and meglitol, block the digestion of starches and slow the digestion of disaccharides, limiting the rise in blood glucose after a meal. People who use alpha-glucosidase inhibitors must use glucose to treat episodes of hypoglycemia.

• Insulin and Insulin Analogs • For people who need insulin, commercial insulin comes in different forms that act with different timings so that it can be delivered in a manner that mimics the body's normal insulin actions as closely as possible.◆ As Figure 25-2 shows, insulin can be either rapid acting (regular), intermediate acting (NPH and lente), or long acting (ultralente). Insulin analog (lispro) is a rapid-acting

◆ Although the availability of insulin has been lifesaving, insulin therapy cannot achieve the same degree of blood glucose control as a body that produces its own insulin.

FIGURE 25-2 Actions of Insulin Types

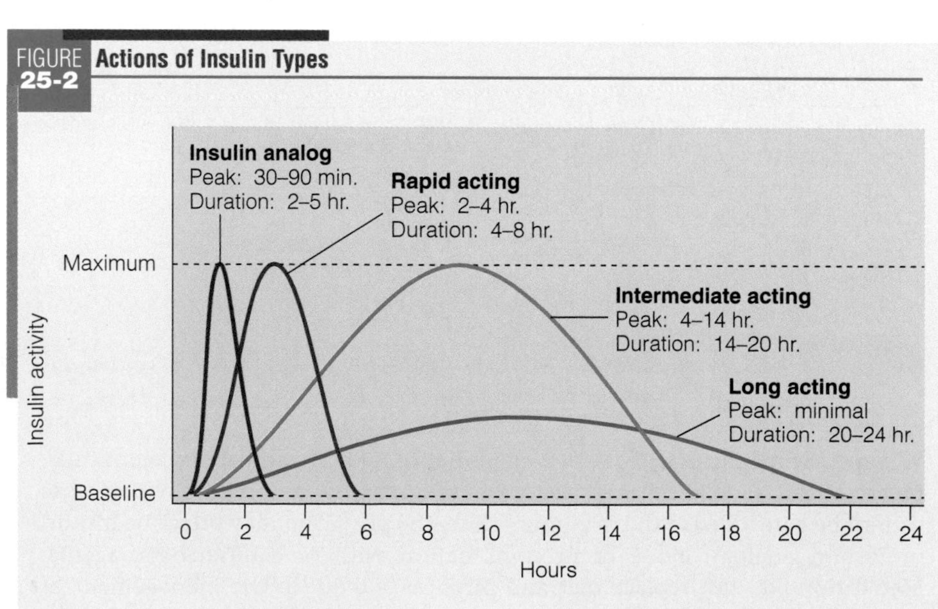

Diet-Drug Interactions

ALPHA-GLUCOSIDASE INHIBITORS (ACARBOSE AND MEGLITOL)

Alpha-glucosidase inhibitors are taken at the start of each meal. Clients using these agents must use glucose to treat hypoglycemia. Nutrition-related side effects include abdominal pain, gas, and diarrhea.

INSULIN

The timing of *insulin* administration varies, depending on the action of the insulin prescribed. Insulin may cause hypoglycemia.

LISPRO (INSULIN ANALOG)

Lispro is taken 5 to 10 minutes before meals. Lispro lowers the risk of hypoglycemia compared to insulin.

METFORMIN

Metformin is taken once or twice a day before meals (breakfast or breakfast and dinner). GI side effects are uncommon, but metformin may cause diarrhea or leave a metallic taste in the mouth.

NATEGLINIDE

Nateglinide is taken immediately before eating. Nateglinide can cause hypoglycemia and sometimes diarrhea.

REPAGLINIDE

Repaglinide is taken before meals. Repaglinide can cause hypoglycemia and diarrhea.

SULFONYLUREAS (CHLORPROPAMIDE, GLIPZIDE, GLYBURIDE, GLIMEPIRIDE)

Sulfonylureas are generally taken one or two times a day, before meals. Disulfiram-like reactions can occur when large amounts of alcohol are taken (especially with *chlorpropamide*). These medications can also cause hypoglycemia.

THIOGLITAZONES (ROSIGLITAZONE, PIOLITAZONE)

Thioglitazones are taken once or twice a day before meals. Nutrition-related side effects are uncommon. One medication of this type, *troglitazone,* was recently taken off the market due to concerns that its use might be a cause of liver failure.

Note: Ask clients who take insulin or antidiabetic agents about their use of the following dietary supplements, which may affect blood glucose: chromium, ephedra, garlic, ginger, ginseng, and niacin.

insulin whose amino acid composition has been modified so that it works faster and has a shorter duration of action.[21] As a result, lispro reduces after-meal hyperglycemia to a greater extent than regular insulin and is also associated with a lower risk of hypoglycemia between meals and during the night. Some insulins come mixed together so that only one injection is required to provide both types of insulin.

Some people experience a temporary remission from diabetes after their initial treatment with insulin—a time referred to as the "honeymoon phase." Recall that some beta cell function may remain when type 1 diabetes is diagnosed. With insulin treatment and relief from hyperglycemia, the beta cells of the pancreas regain their function again—but only temporarily.

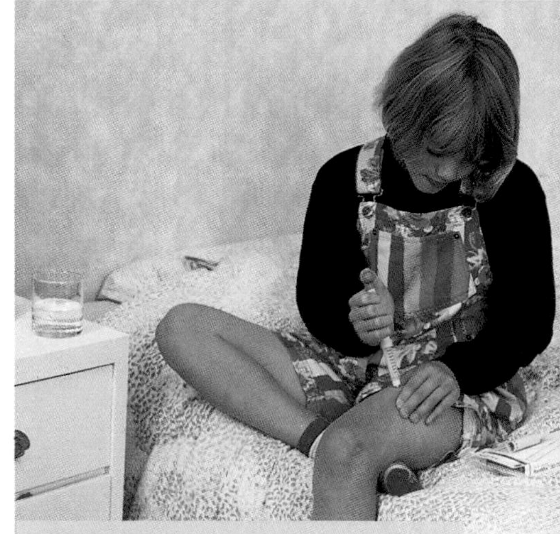

Injections are one option for delivering insulin to people with diabetes.

• Insulin Therapy • Normally, the body secretes a constant, baseline amount of insulin at all times and more after blood glucose rises following a meal. The person with type 1 diabetes often receives NPH (intermediate-acting) insulin to meet baseline needs and regular (rapid-acting) insulin and/or insulin analogs to process energy nutrients after meals. Clients who practice intensive therapy use **multiple daily injections** (a mixture of two or more types of insulin three to four times daily) to meet their insulin needs. People who practice traditional therapy use insulin less often (one to two times daily). Single injections are seldom effective.

People with type 2 diabetes may be treated with insulin alone, or they may use it in combination with oral antidiabetic agents. Often a single injection of NPH (intermediate-acting) insulin is given at bedtime. Insulin analogs may also be used in the treatment of type 2 diabetes.

multiple daily injections: delivery of a mixture of insulins by injection three or more times daily.

• Insulin Delivery • Currently, people who need insulin must inject it or use external pumps to deliver the insulin they need. For those who choose pumps to deliver insulin, external pumps, about the size of a beeper, hold enough insulin to meet needs for two or three days. From the pump, insulin enters the

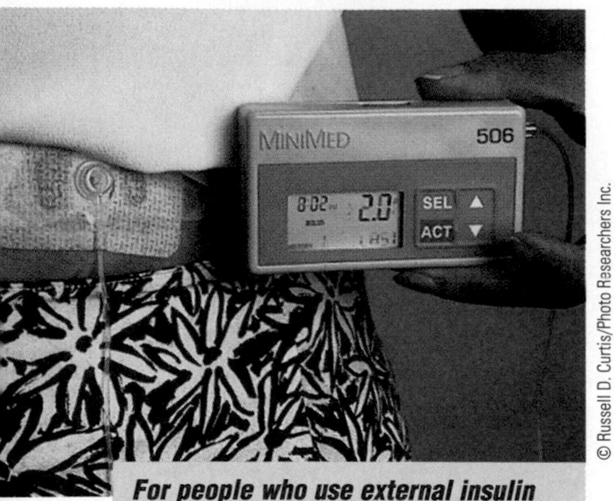

© Russell D. Curtis/Photo Researchers Inc.

For people who use external insulin pumps, insulin passes from the pump through tubing that enters the body through the abdomen as shown.

body through tubing and a needle inserted into the abdominal area. Personal preferences, motivational level, and financial considerations guide clients in deciding which delivery system works best for them. In the future; people may also have the option of inhaling insulin or using it as a mouth spray. Implanted pumps with built-in glucose sensors that automatically dispense insulin in response to blood glucose levels may also be an option for insulin delivery in the future.[22]

• Intentional Misuse of Insulin • As described earlier, teenagers with diabetes are often heavier than teenagers without diabetes, and they are also very self-conscious about their weights. While the incidence of overt eating disorders is no greater among teenagers with diabetes than among their peers, overweight teens with diabetes sometimes use excessive measures to control their weights.[23] Adolescents, particularly girls, may intentionally use less than recommended amounts or omit insulin doses as a means of weight control (recall that hyperglycemia leads to weight loss). Such practices can lead to both acute and chronic complications.

• Carbohydrate-to-Insulin Ratios• Clients who practice intensive therapy can work with health care professionals to determine the unique way their body uses insulin in response to carbohydrate. To uncover their **carbohydrate-to-insulin ratio,** clients eat a consistent amount of carbohydrate, monitor their blood glucose, and keep records of their food intake, insulin use, physical activity, and illness. From these records, health care professionals deduce the number of units of insulin a person needs per gram of carbohydrate. Then clients can use their carbohydrate-to-insulin ratio to adjust their insulin doses when they eat more or less carbohydrate than their meal plan provides.

• Pancreas Transplants • For people with type 1 diabetes who encounter serious problems managing their diseases with insulin, pancreas transplants have been successful in providing functional, insulin-producing beta cells. Most often, a pancreas transplant is combined with a kidney transplant because the person often has significant kidney problems as well. A combination pancreas-kidney transplant can eliminate the need for insulin and for dialysis (see Chapter 27) and greatly enhance the person's quality of life.

Very promising clinical trials have shown that transplanted beta cells infused into the portal vein can attach to the blood vessels and function as normal beta cells do. Early studies show that beta cell transplantation can eliminate the need for insulin injections or infusions in people with type 1 diabetes, suggesting that a cure for type 1 diabetes may be possible.

Physical Activity

Together with diet and insulin, physical activity plays an important role in the management of diabetes. The client with diabetes should be carefully evaluated to determine a safe type and amount of physical activity. For people with type 2 diabetes, a regular program of physical activity improves blood glucose control, contributes to weight loss, improves blood lipid levels, and lowers blood pressure. Although physical activity has not been shown to aid blood glucose control in people with type 1 diabetes, its value lies in its benefits to the cardiovascular system. People who use insulin, however, need to take special precautions when exercising.

carbohydrate-to-insulin ratio: the number of units of insulin needed to cover 1 g of carbohydrate. The lower the ratio, the more insulin is needed to cover carbohydrate intake.

• Physical Activity and Insulin • Generally, insulin should be taken more than an hour before physical activity. Vigorous physical activity and warm temperatures speed blood flow, increase the rate of insulin absorption, and set the stage for a hypoglycemic reaction, which may occur after several hours. Reducing the insulin dose before and after the activity by up to 30, or even 50, percent can help to prevent this sequence of events. People prone to nocturnal hypoglycemia who engage in strenuous physical activity are often advised to undertake the activity early in the day or to reduce the insulin dose when the activity is undertaken late in the day.

People who need insulin should check their blood glucose before and after engaging in physical activity to help them maintain their blood glucose levels. Physical activity should not be undertaken if blood glucose is too low (less than 100 milligrams per deciliter) or too high (greater than 300 milligrams per deciliter). If blood glucose is too low, hypoglycemia can quickly develop. If blood glucose is too high, exercising can cause blood glucose levels to rise even higher. Activity raises the body's need for energy, and the body responds by calling into action counterregulatory hormones, which raise blood glucose by breaking down glycogen stores and synthesizing glucose from protein.

• Physical Activity and Food Intake • All persons, especially those with diabetes, need to make sure they are adequately hydrated before and during physical activity by drinking liquids throughout the day. The person who takes insulin or hypoglycemic agents may need to eat before, during, and after vigorous physical activity. Especially important is carbohydrate, which is readily available from fruits, fruit juices, yogurt, crackers, and other starches. The amount of carbohydrate the client needs depends on the type of activity, its duration, the client's individual responses, and the results of blood glucose tests.

• Making Adjustments • Once a client implements a diet, medication, and physical activity plan, adjustments may be necessary from time-to-time. Table 25-6 on p. 762 summarizes strategies for correcting problems with hyperglycemia and hypoglycemia.

Physical activity plays an important role in the management of diabetes.

IN SUMMARY

The goals of medical and nutrition management of diabetes are to control blood glucose levels, limit and treat complications, and support health and well-being. To reach these goals, clients must learn the complex task of coordinating diet, medications, and physical activity. The diet for diabetes parallels a healthy diet for all people, but emphasizes a consistent intake of carbohydrate spaced evenly throughout the day. The case study describes a child with type 1 diabetes and poses questions to help you sort through the many factors that affect treatment plans.

CASE STUDY

Child with Type 1 Diabetes

Roscoe is a 12-year-old boy diagnosed with type 1 diabetes two years ago. Like most children, Roscoe's appetite varies from day to day. Roscoe frequently joins his friends for a neighborhood basketball or soccer game, although sometimes he prefers to watch TV or surf the Net. Roscoe practices intensive therapy and has the support of both his parents and an excellent diabetes management team. With their help, Roscoe has been able to assume the bulk of the responsibility for his diabetes care, and he has managed to control his blood glucose remarkably well.

In the last few months, however, Roscoe has begun to complain bitterly about the impositions diabetes has placed on his life and his interactions with friends. Sometime he refuses to perform blood glucose tests, and he has skipped insulin injections a few times. Recently, Roscoe was admitted to the emergency room, complaining of fever, nausea, vomiting, and intense thirst. The physician noted that Roscoe was confused and breathing with difficulty. A urine test was positive for ketones, and his blood glucose was 400 milligrams per 100 milliliters. The diagnosis was diabetic ketoacidosis.

1. Describe the metabolic events associated with diabetes that lead to ketoacidosis. Were Roscoe's symptoms and laboratory tests consistent with the diagnosis?
2. Think about Roscoe's age and discuss the influence of his age on his ability to cope with and manage his diabetes. Consider how his age affects his food intake and physical activity. Why might he feel that diabetes is disrupting his life?
3. Review Table 25-2 and consider what advantages or disadvantages intensive therapy might have for roscoe. How might carbohydrate counting be a useful diet-planning strategy for a teenager?

TABLE 25-6 Strategies for Managing Hyperglycemia and Hypoglycemia

Problem	Possible Solutions[a]
	Hyperglycemia
Before breakfast	• Adjust dose of intermediate-acting insulin at bedtime.[b]
Before lunch	• Adjust morning dose of rapid-acting insulin.[b]
	• Reduce amount of carbohydrate at breakfast.
	• Reduce or omit midmorning snack.
	• Change time of breakfast or midmorning snack.
	• Add physical activity after breakfast.
Before dinner	• Adjust afternoon dose of rapid-acting insulin.[b]
	• Reduce carbohydrate at lunch.
	• Reduce or omit midafternoon snack.
	• Change time of lunch or midafternoon snack.
	• Add physical activity between lunch and dinner.
At bedtime	• Adjust insulin dose before dinner.[b]
	• Reduce amount of carbohydrate at dinner.
	• Reduce or omit evening snack.
	• Add physical activity after dinner.
	Hypoglycemia
Before breakfast	• Adjust dose of intermediate- or long-acting insulin at bedtime.[b]
	• Add carbohydrate at evening snack.
	• Avoid strenuous activity late in the day.
Before lunch	• Adjust morning dose of rapid-acting insulin.[b]
	• Add carbohydrate at breakfast.
	• Add a morning snack.
	• Change time of breakfast, lunch, or morning snack.
	• Adjust physical activity schedule.
Before dinner	• Adjust afternoon dose of rapid-acting insulin.[b]
	• Add carbohydrate at lunch.
	• Add an afternoon snack.
	• Change time of lunch, dinner, or afternoon snack.
	• Adjust physical activity schedule.
At bedtime	• Adjust insulin dose before dinner.[b]
	• Add carbohydrate at dinner.
	• Add an evening snack.
	• Change time of dinner or evening snack.

[a]Skilled health care professionals gather additional data to find the best solution for problems with blood glucose control. Is the problem an isolated occurrence or a pattern? Has food intake changed? If yes, why? Has the activity level changed? Has illness been a problem? Whenever altering the diet or physical activity plan is difficult for the client, insulin is adjusted to correct problems, if possible.

[b]Insulin doses can be adjusted in amount or timing or both.

DIABETES MANAGEMENT IN PREGNANCY

All women who become pregnant face new challenges that include altering their diets to meet the demands of the growing fetus and addressing complications that occur as a consequence of pregnancy. Pregnancy in all women elevates blood insulin and alters insulin resistance.◆ Blood insulin begins to rise soon after conception, and the cells respond by storing energy nutrients to provide for the developing fetus. Later in pregnancy, insulin remains high, but the

◆ The hormones that oppose the action of insulin during late pregnancy are *placental lactogen, cortisol, prolactin,* and *progesterone*.

cells become insulin resistant. Levels of hormones that act antagonistically to insulin rise. This hormonal shift signals the body to stop storing energy fuels and to begin allowing the fetus to rapidly take up energy nutrients. Because pregnancy stresses the glucose regulatory system in these ways, women with diabetes should expect control to become more difficult during pregnancy.

Digital Imagery © 2001 PhotoDisc, Inc.

Careful control of blood glucose during pregnancy offers the best chance of a safe delivery and healthy infant for women with diabetes.

• **Health Risks Associated with Diabetes during Pregnancy** • Women with either type 1 or type 2 diabetes who are contemplating pregnancy should know that uncontrolled diabetes in early pregnancy raises the risk of spontaneous abortions.[24] Women with gestational diabetes also have a higher risk than other women of experiencing pregnancy-related hypertension (see Chapter 14).

A fetus exposed to high blood glucose and ketones may develop birth defects. High blood glucose levels also "overfeed" the growing fetus, resulting in large infants◆ that are difficult to deliver. As a consequence, more women with gestational diabetes require a cesarean delivery. The extra glucose also means that the fetus must make extra insulin to handle the load, which may lead to severe hypoglycemia in the infant after birth.

Women with diabetes who are contemplating pregnancy need to receive preconception care, which aims to achieve the best possible blood glucose control before pregnancy, and continued care to maintain control during pregnancy. Women in the Diabetes Control and Complications Trial who tightly controlled their blood glucose levels and became pregnant experienced rates of spontaneous abortion and birth defects similar to those of women without diabetes.[25]

◆ Reminder: The term that describes high-birthweight infants is *macrosomia.*

◆ Chaper 14 provides additional informaton about gestational diabetes.

• **Gestational Diabetes** • Women who have never had diabetes or never knew that they had it may be diagnosed with diabetes during pregnancy (gestational diabetes).◆ Gestational diabetes is the most common medical complication of pregnancy.[26] The American Diabetes Association recommends that women be screened for diabetes between 24 and 28 weeks of gestation, with the exception of women who meet all of these criteria:

- Less than 25 years of age.
- Normal body weight.
- No first-degree relatives with diabetes.
- Not of a high-risk ethnic origin including Hispanic American, Native American, Asian American, Pacific Islander, and African American.[27]

• **Blood Glucose Monitoring** • Obstetricians recommend blood glucose monitoring for all pregnant women with any type of diabetes. Establishing blood glucose control is important to the health of both mother and infant.

• **Medical Nutrition Therapy** • For pregnant women with diabetes, individualized diets tailored to meet the added nutrient demands of pregnancy and carefully coordinated with insulin therapy (when necessary) are central to therapy.◆ The diet plan aims to provide adequate but not excessive kcalories to support appropriate weight gain (see the weight-gain recommendations in Table 14-1). Table 25-7 on p. 764 summarizes guidelines for energy and energy nutrients for pregnant women with diabetes. Careful and ongoing nutrition assessments help dietitians determine if the diet should be adjusted. Carbohydrate is moderately restricted and spaced throughout the day to help keep blood glucose levels from rising too high after meals.◆ Limiting carbohydrate (less than 20 percent of the day's total carbohydrate) at breakfast helps maintain morning blood glucose levels in an acceptable range until counterregulatory hormone levels diminish. Meals and snacks help assure an ongoing supply of glucose without inducing hyperglycemia. Concentrated sweets are discouraged because they quickly raise blood glucose. A bedtime snack is recommended to prevent

◆ Oral antidiabetic agents are not currently recommended to treat diabetes during pregnancy, although some studies suggest they may be safe to use.

◆ Carbohydrates distributed among small meal snacks serves to maintain blood glucose levels of pregnant women who do not require insulin. For pregnant women who use insulin, carbohydrate distribution is adjusted to accommodate each woman's life style and insulin administration schedule.

TABLE 25-7 Energy and Energy Nutrient Recommendations for Diabetes during Pregnancy

Energy:
- For women who have a desirable prepregnancy weight (BMI 20–26):
  30 kcal/kg actual weight
- For women who are obese prior to pregnancy (BMI >30):
  24 kcal/kg actual weight

Carbohydrate:
- 35–45% of total kcalories
- <20% of total carbohydrates at breakfast

Protein: 20–25% of total kcalories

Fat: 35–40% of total kcalories

SOURCE: B. E. Metzger and D. R. Coustan, Summary and recommendations of the Fourth International Workshop-Conference on Gestational Diabetes Mellitus, *Diabetes Care* (supplement 2) 21 (1998): 161–167.

nocturnal hypoglycemia and ketosis in the mother and to provide fuel to the developing fetus.

• **Preventive Measures after Gestational Diabetes** • For most women with gestational diabetes, glucose tolerance returns to normal after pregnancy, but women with gestational diabetes and their offspring risk developing permanent diabetes (usually type 2 diabetes) later in life, especially if they are overweight.[28] The offspring of women with gestational diabetes are also likely to develop obesity and impaired glucose tolerance. For these reasons, health care professionals closely monitor women who have experienced gestational diabetes and their offspring and recommend that they seek medical attention if they develop symptoms suggestive of diabetes. Education focuses on strategies to achieve and maintain a healthy weight and to develop a regular program of physical activity. Use the case study to review the connections between gestational and type 2 diabetes.

IN SUMMARY Careful management of blood glucose levels before and during pregnancy helps women with diabetes limit the risks of complications to mother and infant. The diet plan for diabetes during pregnancy carefully controls energy intakes, limits total carbohydrate, and carefully spaces out carbohydrate intake throughout the day. Women with gestational diabetes and their offspring may develop diabetes at a later time, and health care professionals check for diabetes at regular intervals following the pregnancy. The Nutrition Assessment Checklist highlights areas of concern for people with diabetes.

CASE STUDY

School Counselor with Type 2 Diabetes

Mrs. Lopez is a 41-year-old Hispanic American woman recently diagnosed with type 2 diabetes. Mrs. Lopez developed gestational diabetes while she was pregnant with her second child. Her blood glucose returned to normal following pregnancy, and she was advised to get regular checkups, maintain a desirable weight, and engage in regular physical activity. Although she visits her physician at least once a year, she has been unable to maintain a healthy weight. At 5 feet 3 inches tall, Mrs. Lopez currently weighs 155 pounds. She is determined to lose weight and begin an activity plan because she fears that she might need to use insulin injections. She is also concerned about her husband and children because they are overweight as well. The physician refers Mrs. Lopez to a dietitian to help her plan a diet.

1. What factors in Mrs. Lopez's history increase her risk for diabetes? Are her husband and children also at risk?
2. Describe the general characteristics of the diet that will be appropriate for Mrs. Lopez to follow. In what ways can weight loss and physical activity benefit Mrs. Lopez? How will the dietitian determine what diet-planning strategy will work best for Mrs. Lopez?
3. If Mrs. Lopez is unable to control her blood glucose with diet and physical activity, what type of medication would the physician most likely prescribe? Can you explain to Mrs. Lopez why she would probably not require insulin at this time?
4. Why might the dietitian suggest nutrition counseling for the entire family?

Nutrition Assessment Checklist for People with Diabetes

MEDICAL HISTORY

Check the medical record to determine:

- ☐ Type of diabetes
- ☐ Duration of diabetes
- ☐ Acute and chronic complications
- ☐ Other medical conditions, including pregnancy, that alter nutrient needs

MEDICATIONS

For clients with preexisting diabetes who use antidiabetic agents, insulin, or both, note:

- ☐ Type(s) of antidiabetic agent or insulin
- ☐ Administration schedule

Check for other medications and note possible diet-drug interactions, including:

- ☐ Antilipemics (to lower blood lipids, see Chapter 26)
- ☐ Antihypertensive agents (to reduce blood pressure, see Chapter 26)
- ☐ Diuretics (to reduce blood pressure, see Chapter 26)
- ☐ Dietary supplements

FOOD/NUTRIENT INTAKE

To devise an acceptable meal plan and coordinate medications, obtain:

- ☐ An accurate and thorough record of food intake and usual eating habits
- ☐ An account of usual physical activities

At medical checkups, reassess the client's ability to:

- ☐ Maintain an appropriate energy intake
- ☐ Maintain a consistent intake of carbohydrate
- ☐ Adjust the diet for missed meals due to illness
- ☐ Use appropriate amounts and types of foods to treat hypoglycemia

ANTHROPOMETRICS

Take accurate baseline height and weight measurements as a basis for:

- ☐ An appropriate energy intake
- ☐ Initial insulin therapy

Reassess height and weight for children and weight for adults periodically to ensure that the meal plan provides an appropriate energy intake.

LABORATORY TESTS

Check the following tests to monitor the success of diabetes therapy:

- ☐ Results of home blood glucose monitoring
- ☐ Glycated hemoglobin
- ☐ Blood lipids
- ☐ Ketones in urine
- ☐ Microalbuminemia, when available

PHYSICAL SIGNS

Look for signs of:

- ☐ Dehydration, especially in the elderly
- ☐ Nutrient deficiencies and excesses

Nutrition on the Net

WEBSITES

Access these websites for further study of topics covered in this chapter.

- Find updates and quick links to these and other nutrition-related sites at our website: **www.wadsworth.com/nutrition**
- Visit the American Diabetes Association: **www.diabetes.org** and the Joslin Diabetes Center: **www.joslin.org** to find information on a wide range of topics related to diabetes.
- Find out what it takes to become a diabetes educator by visiting the American Association of Diabetes Educators site: **www.aadenet.org**
- Learn more about alternative therapies for diabetes by visiting the Natural Pharmacist encyclopedia: **www.alternativediabetes.com**

INTERNET ACTIVITIES

More than 10 million people in the United States have diabetes. Many acute and chronic conditions are associated with diabetes and successful management can reduce these complications. Complementary and alternative treatments may benefit some people with diabetes. Perhaps you would like to know more about such treatments.

- ▼ Go to: **www.alternativediabetes.com**
- ▼ Click on "The Natural Pharmacist encyclopedia."
- ▼ Click on "Conditions A–Z."
- ▼ Click on the letter "D" and then click on "Diabetes."
- ▼ Read about alternative treatments for diabetes including safety issues, dosages, and scientific evidence about each treatment.

What factors should be considered in deciding whether an alternative therapy is appropriate?

Study Questions

These questions will help you review the chapter. You will find the answers in the discussions on the pages provided.

1. Name the two major types of diabetes. Which type is more common? Describe the differences between the two types of diabetes. (p. 746)
2. What metabolic changes and clinical findings can occur as a consequence of hyperglycemia? What are the differences between severe hyperglycemia with and without ketosis? (pp. 747–749)
3. What causes hypoglycemia in diabetes, and what are its consequences? (p. 749)
4. Discuss the chronic complications that can arise as a result of diabetes. (pp. 749–750)
5. What are the goals of medical and nutrition therapy for people with diabetes? Which goal is the most important for preventing acute and chronic complications? (pp. 750–751)
6. Describe the diet for diabetes. Why is a consistent intake of carbohydrate spaced evenly throughout the day important? (pp. 752–755)
7. What are oral antidiabetic agents? Describe the ways antidiabetic agents work. (p. 758)
8. In what ways does the body normally secrete insulin, and how is commerically available insulin given to simulate these actions? (p. 759)
9. Describe how physical activity affects insulin and carbohydrate needs in people with diabetes. (pp. 760–761)
10. What are the risks of poorly controlled diabetes during pregnancy? How are treatment plans for pregnant women with diabetes managed to control blood glucose? (pp. 763–764)

These questions will help you review for an exam. Answers can be found on p. 768.

1. Which of the following is characteristic of type 1 diabetes?
   a. The pancreas makes little or no insulin.
   b. It frequently goes undiagnosed.
   c. It is the predominant form of diabetes.
   d. Insulin secretion is ineffective in preventing hyperglycemia.
2. Which of the following describes type 2 diabetes?
   a. Immune factors play a role.
   b. The pancreas makes little or no insulin.
   c. Hyperglycemia with ketoacidosis is a common complication.
   d. Chronic complications may have begun to develop before it is diagnosed.
3. Sudden hyperglycemia in a person who has consistently maintained good blood glucose control can be precipitated by:
   a. infections or illnesses.
   b. chronic alcohol ingestion.
   c. undertreatment of hypoglycemia.
   d. conditions that lower levels of counterregulatory hormones.
4. The chronic complications associated with diabetes result from:
   a. alterations in kidney function.
   b. weight gain and hypertension.
   c. damage to blood vessels and nerves.
   d. infections that deplete nutrient reserves.
5. The health care professional working with a client with diabetes emphasizes that the diet should provide:
   a. a very low intake of fat.
   b. more protein than regular diets.
   c. a restricted intake of simple sugars and concentrated sweets.
   d. a consistent carbohydrate intake from day to day and at each meal and snack.
6. Which of the following is true regarding the use of alcohol in a diet for diabetes?
   a. A serving of alcohol is considered part of the carbohydrate allowance.
   b. Alcohol can cause hypoglycemia in all people, including those with diabetes.
   c. People with well-controlled blood glucose levels should refrain from alcohol use.
   d. In combination with alcohol, some types of insulin can cause a disulfirim-like reaction in people with type 2 diabetes.
7. The meal-planning strategy that is most effective for the person with diabetes is:
   a. carbohydrate counting.
   b. the exchange list system.
   c. food guides and sample menus.
   d. the one that best helps the client control blood glucose levels.
8. Which of the following best describes insulin therapy in a person receiving both NPH insulin and an insulin analog?
   a. Since NPH insulin is of long duration, there is no need for the insulin analog.
   b. NPH insulin covers basal insulin needs while the insulin analog covers the carbohydrate from meals.
   c. NPH insulin covers the carbohydrate from meals while the insulin analog covers basal insulin needs.
   d. Since glucose is not available from food between meals, the insulin analog alone would cover the person's insulin needs.

9. Which of the following describes the treatment plans for a person with type 1 diabetes who practices intensive therapy?
   a. The person cannot use a pump to deliver insulin.
   b. The person uses one type of insulin one or two times daily.
   c. The person must monitor blood glucose several times a day.
   d. After learning to control blood glucose levels, the person can quit monitoring blood glucose levels.

10. Women with pregnancies complicated by diabetes:
   a. often need less carbohydrate at breakfast.
   b. generally benefit from larger meals and a snack at bedtime.
   c. need more carbohydrate than women with diabetes who are not pregnant.
   d. need more kcalories to support the pregnancy than women without diabetes.

Clinical Applications

1. Using the box on pp. 756–757, plan a diet using the exchange lists for a sedentary woman with type 1 diabetes who is 5 feet 9 inches tall and weighs 160 pounds. Assume that the distribution of kcalories will be 55 percent from carbohydrate, 20 percent from protein, and 25 percent from fat. Develop a sample menu.

2. An important part of learning is being able to apply knowledge and guidelines to real-life situations. Using Table 25-6 on p. 762 as a guide, think about the possible remedies for either hyperglycemia or hypoglycemia. Describe at least one situation when it might be preferable to alter the insulin dose and one situation when it might be preferable to alter the carbohydrate intake.

3. Take a trip to a pharmacy and price these items: blood glucose meter, test strips appropriate for the glucose meter you select, lancets, insulin, and syringes. Determine the approximate cost of insulin for a person who uses 14 units of regular insulin and 26 units of NPH insulin in three injections daily. Then estimate the cost of testing blood glucose four times daily. Consider how an external pump might affect the total cost of managing diabetes. How does the need for a well-balanced diet influence the cost of diabetes care?

References

1 A. H. Mokdad and coauthors, The continuing increase of diabetes in the U.S., *Diabetes Care* 24 (2001): 1278–1283.

2 The Expert Committee on the Diagnosis and Classification of Diabetes Mellitus, Report of the Expert Committee on the diagnosis and classification of diabetes mellitus, *Diabetes Care* (supplement 1) 24 (2001): 5–19.

3 E. A. Simone, D. R. Wegmann, and G. S. Eisenbarth, Immunologic "vaccination" for the prevention of autoimmune diabetes (type 1a), *Diabetes Care* (supplement 2) 22 (1999): B7– B15.

4 A. D. Mooradian and coauthors, Diabetes care for older adults, *Diabetes Spectrum* 12 (1999): 70–77.

5 S. Saffel-Shrier, Carbohydrate counting for older patients, *Diabetes Spectrum* 13 (2000): 158–162.

6 A. I. Adler, The Diabetes Prevention Program, *Diabetes Care* 22 (1999): 543–544; W. Y. Fujimoto, A national multicenter study to learn whether type II diabetes can be prevented: The Diabetes Prevention Program, *Clinical Diabetes* 15 (1997): 13–15.

7 I. J. Deary, Hypoglycemia-induced cognitive decrements in adults with type 1: A case to answer? *Diabetes Spectrum* 10 (1997): 13–15.

8 American Diabetes Association, Implications of the Diabetes Control and Complications Trial, *Diabetes Care* (supplement 1) 24 (2001): 25–27.

9 B. Janand-Delenne and coauthors, Silent myocardial ischemia in patients with diabetes, *Diabetes Care* 22 (1999): 1396–1400.

10 V. Valentine, J. A. Barone, and J. V. C. Hill, Gastropathy in patients with diabetes: Current concepts and treatment recommendations, *Diabetes Spectrum* 11 (1998): 248–253.

11 The Diabetes Control and Complications Trial/Epidemiology of Diabetes Interventions and Complications Research Group, Retinopathy and nephropathy in patients with type 1 diabetes four years after a trial of intensive therapy, *New England Journal of Medicine* 342 (2000): 381–389.

12 American Diabetes Association, Implications of the United Kingdom Prospective Diabetes Study, *Diabetes Care* (supplement 1) 24 (2001): 28–32.

13 American Diabetes Association, Nutrition recommendations and principles for people with diabetes mellitus, *Diabetes Care* (supplement 1) 24 (2001): 44–47.

14 American Diabetes Association, Nutrition recommendations and principles for people with diabetes mellitus, 2001.

15 S. Carden, The glycemic index in diabetes meal planning, 60th Scientific

Sessions of the American Diabetes Association, June 13, 2000, www.medscape.com/medscape/cno/2000/ADA/Story.cfm?story_id=1393, site visited on February 14, 2001.

16 A. Garg, High-monunsaturated diets for patients with diabetes mellitus: A meta-analysis, *American Journal of Clinical Nutrition* (supplement) 67 (1998): 577S–582S; J. P. Parnett and A. Garg, Medical nutrition therapy for patients with diabetes mellitus: Role of dietary fats, *On the Cutting Edge: Diabetes Care and Education,* Fall 1997, pp. 5–6; A. M. Coulston, Monounsaturated fats for people with diabetes, *On the Cutting Edge: Diabetes Care and Education,* Fall 1997, pp. 14–16; A. Garg and coauthors, Effects of varying carbohydrate content of diet in patients with non-insulin-dependent diabetes mellitus, *Journal of the American Medical Association* 271 (1994): 1421–1428.

17 American Diabetes Association, Role of fat replacers in diabetes medical nutrition therapy, *Diabetes Care* (supplement 1) 24 (2001): 104–105.

18 American Diabetes Association, Nutrition recommendations and principles for people with diabetes mellitus, 2001.

19 M. D. Maryniuk, The new shape of medical nutrition therapy, *Diabetes Spectrum* 13 (2000): 122–124.

20 Carbohydrate counting: Back to basics, *Diabetes Spectrum* 13 (2000): 149; S. J. Gillespie, K. Kulkarni, and A. E. Daly, Using carbohydrate counting in diabetes clinical practice, *Journal of the American Dietetic Association* 98 (1998): 897–905.

21 American Diabetes Association, Lispro: A new fast-acting insulin option, *Diabetes Spectrum* 9 (1996): 253.

22 ABCNews.com, Inhaled insulin study results promising, www.diabetes.org, site visited February 9, 2001; Z. T. Bloomgarden, New technologies for diabetes: Continuous glucose monitoring and new strategies for insulin administration, www.medscape.com, site visited April 9, 2001.

23 K. S. Bryden and coauthors, Eating habits, body weight, and insulin misuse, *Diabetes Care* 22 (1999): 1956–1960.

24 American Diabetes Association, Position statement: Preconception care of women with diabetes, *Diabetes Care* (supplement 1) 24 (2001): 61–68.

25 The Diabetes Control and Complications Trial Research Group: Pregnancy outcomes in the Diabetes Control and Complications Trial, *American Journal of Obstetrics and Gynecology* 174 (1996): 1343–1353.

26 D. B. Carr and S. Gabbe, Gestational diabetes: Detection, management, and implications, *Clinical Diabetes* 16 (1998): 4–11.

27 The Expert Committee on the Diagnosis and Classification of Diabetes Mellitus, 2001; American Diabetes Association website, www.diabetes.org, visited November 16, 1999.

28 American Diabetes Association, Position statement: Gestational diabetes mellitus, *Diabetes Care* (supplement) 24 (2001): 77–79.

ANSWERS

Multiple Choice

1. a	2. d	3. a	4. c	5. d
6. b	7. d	8. b	9. c	10. a

MASTERING DIABETES CONTROL

A HEALTHY PERSON goes about daily activities with little thought to how the body will react to everyday routines or disruptions to those routines. If you usually eat breakfast at 8:00 A.M., you may sleep in and choose not to eat breakfast on weekends without a second thought. If your friend asks you to play tennis and the match interferes with dinner, you simply eat later. If you get hungry during the day, you eat a snack. If you're not hungry at your usual dinner hour, you wait and eat later. For people with diabetes, even such simple variations in a daily schedule require thought and adjustments. They need to learn facts, master techniques, and develop new attitudes and behaviors that will promote a healthy life. With assistance from the health care team, clients with the motivation can learn to manage their diabetes, rather than allow their disease to control their lives.

CLIENT-CENTERED CARE

Health care professionals who simply "prescribe" remedies and then expect their clients to comply with those remedies fail to consider the impact that lifestyle changes have on a person's quality of life. Clients can easily be overwhelmed, and their motivation and compliance may be poor. This highlight describes a different education approach—client empowerment. To use this approach, health care professionals provide clients with the information and encouragement they need to make decisions about their treatment plans. When clients are allowed to make the final decisions about their health care, the treatment plan has a better chance for success.

Clients with diabetes decide what overall approach to care—traditional or intensive—best suits their lifestyles and goals. They also make many decisions about managing their diabetes on a day-to-day basis.

The Health Care Team

Together with the client, the diabetes management team includes physicians, nurses, dietitians, and counselors with training in diabetes management.[1] The diabetes management team provides clients with the information about options for care and the skills and support they will need to follow treatment plans. The Diabetes Control and Complications Trial (DCCT) identified a team approach as an important strategy in the management of diabetes. Clients who practiced intensive therapy in the DCCT spent more than 80 percent of their time during follow-up visits with nonphysician team members; about 60 percent of their time was spent with nurses.[2]

For a person with diabetes, even simple changes in routine require planning and adjustments.

© Ann Dowie

Weighing Treatment Options

Health care professionals cannot knowingly encourage clients to practice behaviors that are medically harmful. Nor can they force clients to follow advice. Some aspects of diabetes management are critical for survival. Other aspects may be beneficial but less critical. Still other aspects may be ideal but less pressing when considered in the total context of a treatment plan.

A person with type 1 diabetes, for example, must take insulin or face death, but in many cases, the person's goals, motivation, finances, and ability level help determine if traditional therapy or intensive therapy is more appropriate. Health care professionals explain the ramifications of treatment choices and may continue to encourage clients to consider different options throughout the course of care, but ultimately they respect the individual's right to make health care decisions without judgment.

One client with diabetes may choose traditional therapy even though it means less than the best blood glucose control, while another may be eager to learn intensive therapy. (Table 25-2 on p. 752 compares traditional and intensive therapy.) Traditional therapy requires fewer insulin injections, fewer finger pricks for blood glucose testing, and fewer calculations. It is also less flexible and results in poorer control of blood glucose. Intensive therapy is more difficult to master and requires more time, but it also offers greater flexibility in food choices, schedule changes, and activity levels and results in better control of blood glucose.

DIABETES EDUCATION

All newly diagnosed clients and their families need diabetes management training and counseling, and the learning process takes time. Even highly motivated clients who listen attentively need several counseling sessions to learn the basics of diabetes management. Diabetes education is an ongoing process that is a routine part of diabetes management. To manage their diabetes, clients must learn:

- *Diet planning.* Clients need to know how to control their energy intakes so that they eat a consistent amount of carbohydrate at regular times throughout the day. Clients help the dietitian determine which meal-planning strategy works best for them (see Chapter 25).
- *Medication administrations.* Clients need to learn how to use their medications and how to take them in relation to foods. Clients who use insulin need to know how to store, draw, and administer insulin or how to use their external pumps.
- *Blood glucose monitoring.* Clients need to learn how to use blood glucose meters, draw blood samples, and record and interpret results.

People who are contemplating intensive therapy must have a thorough understanding of their diabetes and have the skills to take that knowledge a step further. To practice intensive therapy, clients also learn about:

- *Carbohydrate-to-insulin ratios.* Clients need to know how much insulin their bodies need to cover the carbohydrate in meals and snacks so that they can make adjustments to their plans.
- *Self-management.* Clients need to learn how to keep an accurate record of blood glucose test results, food intake, activity, illness, and medication administration so that they can keep their blood glucose tightly controlled.

Individualizing Treatment Plans

Regardless of whether a client chooses traditional or intensive therapy, health care professionals facilitate the learning process by working out a highly individualized plan that carefully considers the client's current lifestyle, motivation, goals, and ability to grasp new concepts and make lifestyle changes. The plan must be flexible to accommodate changing needs; a person who is highly motivated at one point, for example, may become totally discouraged at another time and need to restructure goals temporarily. At each step, the health care team balances medical needs with the individual's goals and motivation.

As Chapter 25 described, a complete assessment serves as the first step in formulating a treatment plan. The more detailed the assessment, the more closely the plan can be tailored to the individual's existing lifestyle and the better the chances for success.[3] Using assessment data, the health care team sets long-term medical goals. To enable the team and the client to measure the success of therapy, the goals are stated in terms of measurable outcomes such as target ranges for blood glucose, hemoglobin A_{1c}, blood lipids, and body weight. The health care team then works with the client to set short-term goals and determine the strategies (or interventions) that will help the client meet medical goals. For example, consider the process for a 55-year-old woman just diagnosed with type 2 diabetes. The health care team conducts a complete assessment and sets medical goals. The client and health care team agree to a trial of diet and physical activity to reduce blood glucose without medications. The nurse explains type 2 diabetes and the consequences of high blood glucose to the woman, and the client is referred to the dietitian for nutrition counseling.

Setting Short-Term Goals

From the assessment, the dietitian finds that in addition to elevated blood glucose, the client's blood lipids are elevated, and she is 30 pounds overweight and does not have a physical activity program. A review of the client's eating habits reveals that she often skips breakfast, eats a large dinner, eats many fried and other high-fat foods, and seldom eats vegetables. The dietitian recognizes that several dietary changes are warranted, but talks with the client to find out what she thinks about her health situation and the steps she can take to improve her diet. The client's primary concern is her weight, but she feels overwhelmed and doesn't know where to begin. With the dietitian's guidance, the woman sets these goals: to eat a consistent amount of carbohydrate three times a day, with careful attention to portion sizes; to limit high-fat foods; and to begin a physical activity program.

Determining Appropriate Interventions

Once goals have been set, the next set is to plan interventions. What specific activities can help the client meet her goals? Returning to our example, the dietitian and client might discuss a diet plan that includes appropriate amounts of foods for breakfast, lunch, and dinner; review several menu options; practice weighing and measuring foods; and develop a physical activity plan. The client feels that she could walk after dinner for 20 minutes five times a week. The dietitian would like the client to keep records of her food intake, activity, and blood glucose, but the client feels she cannot handle that right now.

Evaluating the Interventions

Once the client tries the plan, the next step is to see how well the plan is working. Which strategies were successful and which were not? Suppose the client in our example returns for her next appointment and relates that she has been successful at eating breakfast and limiting high-fat foods, has been inconsistent about reducing her portion sizes, and has managed to walk after dinner only twice a week. From the client's medical record, the dietitian sees that her blood glucose is somewhat lower and that she has lost one pound. The dietitian reinforces the value of the positive changes the client has made and praises her efforts. At this point, the dietitian must reassess the client's motivation and decide what strategies to suggest next. The client may want to continue the plan and renew her commitment to control portion sizes and walk more frequently. Alternatively, she may be so pleased with how she is feeling that she is ready to do more. She may agree not only to continue her original plan, but also to keep food and activity records, for example.

To help people adjust to the demands imposed by diabetes, health care professionals guide clients through management plans with measurable goals set by the clients. Clients' responses and level of motivation dictate future actions.

DIABETES AND EMOTIONAL HEALTH

In addition to their diabetes-related tasks, clients have many other responsibilities related to work, family, and community. It should come as little surprise that even when clients know what to do and why they should do it, they may be unable to carry out the plan at times. Many people with diabetes report feeling overwhelmed and frustrated by the multitude of self-care demands.[4] In the words of one diabetes educator: "No one but another person who has diabetes can fully appreciate the demands of diabetes. It is 24 hours a day, 365 days a year (except on Leap Year when you get an extra day of diabetes). It involves all manner of imposition and deprivation. And even when you do everything right, there are no guarantees."[5]

Clients' Perceptions

People with diabetes often feel that health care professionals, family, and friends "blame" them for having diabetes-related problems and complications.[6] They may feel guilt and remorse because their efforts at diabetes control were not good enough to prevent complications. Health care professionals are wise to remain nonjudgmental when working with clients with diabetes and to recognize that diabetes control must be balanced with quality of life. Clients also need to be reassured that they may experience complications even when they are doing everything possible. Insisting on perfection can only lead to failure.

Caregivers' Perceptions

Parents and other family members also face the challenge of living with diabetes. The intensity of the situation can either reinforce or disrupt family unity. Parents may resent the demands of caring for a child with a chronic illness and then may experience guilt for having those feelings. They may feel anxious and be reluctant to allow their child to follow the diabetes care plan without their constant assistance. Especially in the case of an older child, they may press their care and control on a child who needs to develop autonomy and self-care. Parents may also become emotionally upset when they see their child feeling anxious, depressed, or withdrawn. Parents need time to work through these feelings. They might want to attend meetings for parents of children with diabetes. Such meetings offer opportunities to share feelings, ideas, and frustrations with others in similar situations. Sometimes just knowing that you're not alone helps.

Support Networks

After an initial introduction to the world of diabetes, many hospitals and medical centers offer clients educational programs designed to expand their knowledge and promote independence. Some programs may encourage children to bring friends, which makes the experience more comfortable and fun. Some programs are designed specifically for parents, grandparents, and other caregivers.

Children can combine education and summer vacation at camps designed especially for children with diabetes. These camps offer the chance to learn more about diabetes while "living" the lifestyle with companions under supervision. Children trade snack ideas, try new recipes, and help prepare meals. Older children assist younger ones, and all benefit.

The results of the Diabetes Control and Complications Trial clearly show that tightly managing diabetes can dramatically reduce the complications associated with it. Helping clients with diabetes make the necessary adjustments is a continuous process that balances medical and individual needs.

References

1 M. Bayless and C. Martin, The team approach to intensive diabetes management, *Diabetes Spectrum* 11 (1998): 33–37.

2 M. Bayless and Martin, 1998.

3 S. M. Strowig, Improved methods of therapy in diabetes, *Diabetes Spectrum* 11 (1998): 16–17.

4 W. H. Polonsky, Listening to our patients' concerns: Understanding and addressing diabetes-specific emotional distress, *Diabetes Spectrum* 9 (1996): 8–11; R. R. Rubin, Life's work they have not chosen, *Diabetes Spectrum* 8 (1995): 308.

5 Rubin, 1995.

6 K. F. McFarland, The power of words, *Diabetes Spectrum* 8 (1995): 308.

Nutrition and Disorders of the Heart, Blood Vessels, and Lungs

CHAPTER 26

cardiovascular diseases (CVD): diseases of the heart and blood vessels.
- **cardio** = heart
- **vascular** = blood vessels

coronary heart disease (CHD): heart damage that results from an inadequate supply of blood to the heart.
- **corona** = crown or circle

atherosclerosis (ATH-er-oh-skler-OH-sis): a condition characterized by the buildup of plaque along the inner walls of the arteries, which narrows the lumen of the artery and restricts blood flow to the tissue it supplies.

hypertension: elevated blood pressure.

plaques (PLACKS): mounds of lipid material (mostly cholesterol) with some macrophages (a type of white blood cell) covered with fibrous connective tissue and embedded in artery walls. With time, the plaques may harden as the fibrous coat thickens and calcium is deposited in the plaque.
- **placken** = patch or plate

The heart receives oxygen-rich blood from the lungs and delivers this oxygen (as well as nutrients) to the body's tissues by way of the blood vessels (see Figure 26-1). A disrupted supply of oxygen and nutrients alters the body's ability to carry out its metabolic functions. Such a disruption can occur as a consequence of either cardiovascular or lung disorders.

CARDIOVASCULAR DISEASES

Cardiovascular diseases (CVD) are the leading cause of death in the United States and around the world. Cardiovascular diseases claim over 10,000 more lives in the United States than the next six leading causes of death combined.[1] Men have a greater risk of heart disease and are at risk at an earlier age than women, but this gap closes with age. **Coronary heart disease (CHD),** the most common form of cardiovascular disease, usually involves **atherosclerosis** and **hypertension.** Thus treatment strategies to reduce coronary heart disease aim to prevent or treat both atherosclerosis and hypertension.

Atherosclerosis

Atherosclerosis or "hardening of the arteries" occurs when fibrous **plaques,** which are composed primarily of cholesterol, build up in the arteries, especially at branch points (see Figure H15-1 on p. 536). The first sign of atherosclerosis

FIGURE 26-1 The Heart, Blood Vessels, and Lungs

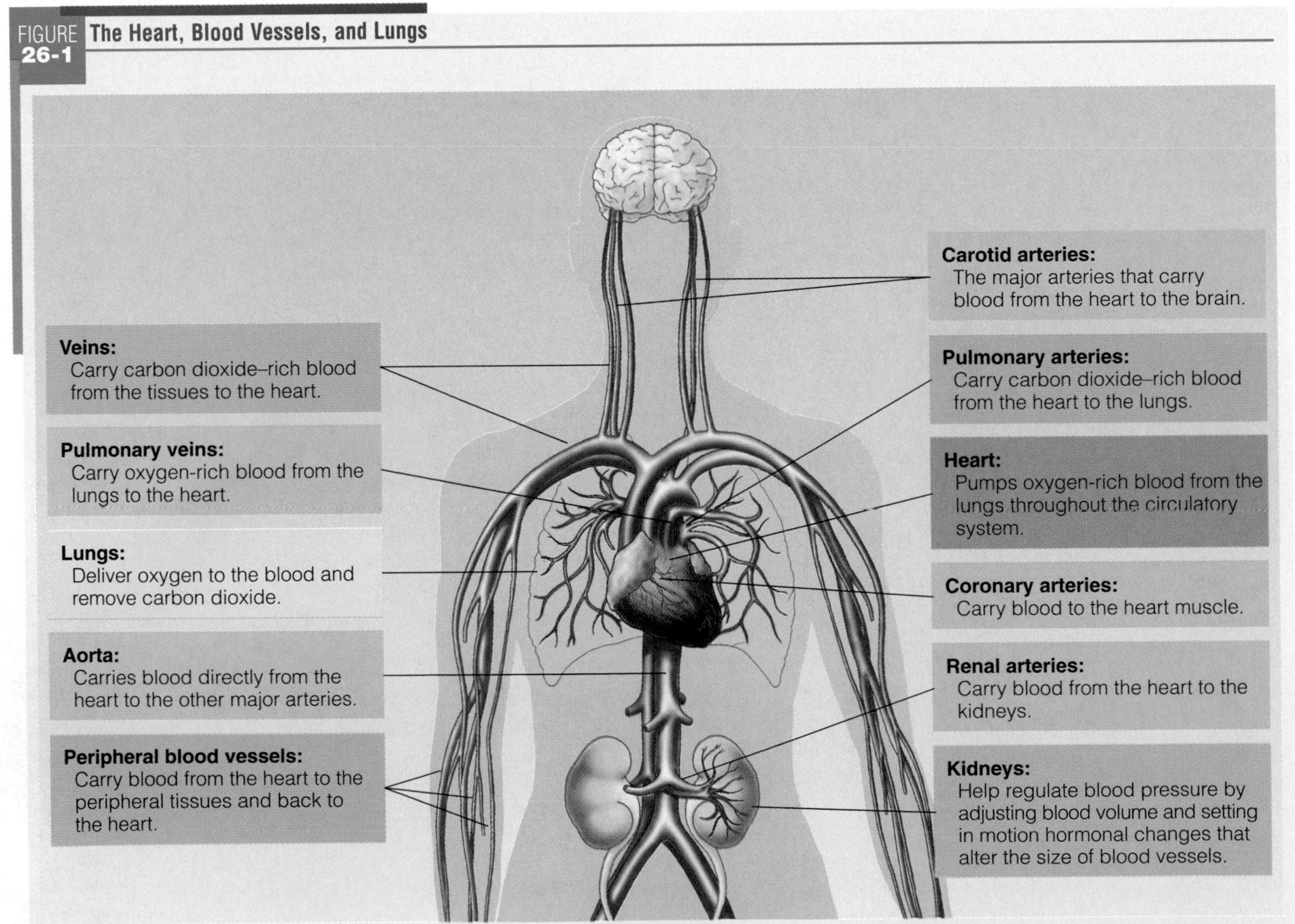

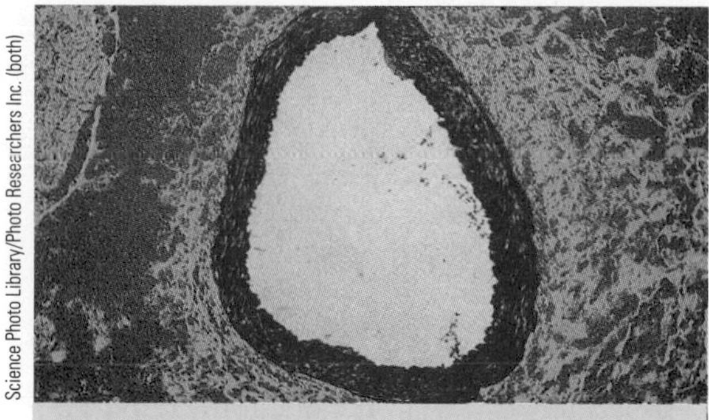

Blood flows unencumbered through a normal artery.

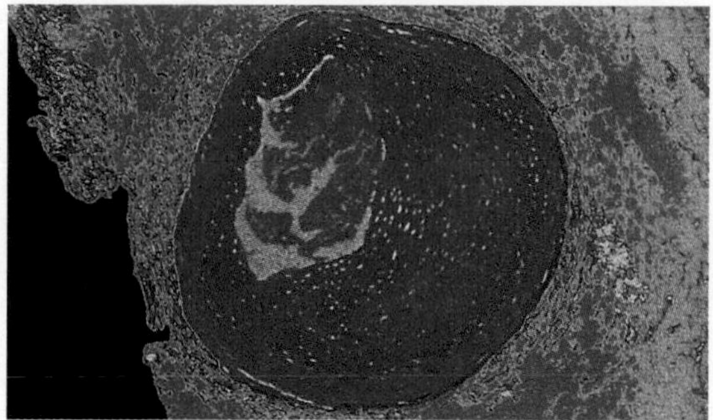

The progressive buildup of plaque can seriously impede blood flow through an artery.

is soft fatty streaks visible along the walls of the arteries. Often these fatty streaks gradually enlarge and harden as they fill with lipids and minerals (especially calcium) and become encased in fibrous connective tissue (scar tissue). Plaque stiffens the arteries and narrows their diameter. Although the plaques of atherosclerosis may obstruct blood flow through any blood vessel, the coronary arteries are frequently affected.

• Causes of Atherosclerosis • What causes plaque to develop is unknown, but many scientists believe that tissue damage and inflammatory responses play a role. Tissues may be damaged by high cholesterol levels, hypertension, toxins from tobacco products, or some viral and bacterial infections. Inflammatory responses increase the permeability of the blood vessels and allow immune system cells (macrophages) and LDL cholesterol◆ to deposit in the blood vessel walls. Free radicals produced during inflammatory responses oxidize LDL cholesterol, and this allows the macrophages to engulf it. Macrophages swell with large quantities of oxidized LDL cholesterol and eventually become the cells that comprise plaque. Inflammatory responses also direct changes that allow minerals to harden plaque and form the fibrous connective tissue that encapsulates plaque.

• Blood Lipids and Atherosclerosis • The blood cholesterol most clearly linked to atherosclerosis is LDL cholesterol. If excess LDL cholesterol◆ remains after the body's cells take up the amount they need, then the excess becomes available for oxidation. Some people are genetically programmed to make an excess of a variant form of LDL cholesterol called **lipoprotein(a).** People with elevated levels of lipoprotein(a) often develop atherosclerosis and coronary heart disease at an early age. Currently available medications and diet are ineffective in lowering lipoprotein(a).

HDL also carry cholesterol in the blood, but it is believed they carry cholesterol away from the arteries and back to the liver. Although the reasons are unclear, high levels of HDL may protect against the development of plaque. Conversely, low HDL may favor the development of plaque.

To a lesser extent, elevated triglycerides◆ are also linked to atherosclerosis. Elevated triglycerides often occur together with elevated LDL, low HDL, and other conditions that favor plaque development (overweight, diabetes), however, so it is difficult to determine if elevated triglycerides alone are problematic.[2]

• Consequences of Atherosclerosis • When progression of atherosclerosis in the coronary arteries restricts blood flow and damages the heart muscle, coronary heart disease, or CHD, develops. The person with CHD often experiences pain and pressure in and around the area of the heart **(angina).** If blood flow to the heart is cut off, that area of heart muscle dies, and a **myocardial infarction** (heart attack) results.

◆ Reminder: Cholesterol is carried in several lipoproteins, including chylomicrons, very-low-density lipoprotein (VLDL), low-density lipoprotein (LDL), and high-density lipoprotein (HDL). Cholesterol carried on LDL is the type that oxidizes and is found in plaques.

◆ The technical term for abnormal blood lipids is **dyslipidemia;** elevated LDL and low HDL are examples. Elevated blood lipids may also be called **hyperlipidemia.** The laboratory measurement of each type of lipoprotein is called a **blood lipid profile.**

◆ Reminder: Triglycerides are carried on VLDL.

lipoprotein(a): a variant form of LDL cholesterol associated with a high risk of CHD.

angina (an-JYE-nah or AN-ji-nah): a painful feeling of tightness or pressure in and around the heart, often radiating to the back, neck, and arms; caused by a lack of oxygen to an area of heart muscle.

myocardial (my-oh-CAR-dee-al) **infarction** (in-FARK-shun) or **MI:** sudden tissue death caused by blockages of vessels that feed the heart muscle; also called **heart attack, cardiac arrest,** or **acute heart failure.**

- **myo** = muscle
- **cardial** = heart
- **infarct** = tissue death

A sudden spasm or surge in blood pressure in an artery can tear away part of the fibrous coat covering a plaque. When this happens, the body responds to the damage as it would to other tissue injuries. **Platelets,** tiny disc-shaped bodies cover the damaged area, and together with other factors, they form a clot **(thrombosis).** A blood clot **(thrombus)** on a plaque may grow large enough to close off a blood vessel. A portion of a clot may also break free from the plaque **(embolus)** and travel through the circulatory system until it lodges in a small artery and suddenly shuts off the blood flow to that area **(embolism).** The gradual or sudden loss of blood flow to the portion of tissue supplied by the clotted artery robs the tissue of oxygen and nutrients, and the tissue may eventually die.

Sometimes atherosclerosis obstructs blood flow to the brain; the result may include a **transient ischemic attack** or **stroke** (described later in this chapter). Blood flow to the kidneys can also be affected. Renal failure is described in Chapter 27.

Hypertension

Chronic elevated blood pressure, or hypertension, is estimated to affect some 50 million people in the United States.[3] Although one out of four adults in the United States may have high blood pressure, almost a third of these people do not know they have it. Thus hypertension is sometimes called the "silent killer"; it may go undetected until the person experiences a potentially fatal complication such as a heart attack or stroke. Sometimes hypertension develops as a consequence of another disorder, but most often the cause is unknown. The higher the blood pressure above normal,◆ the greater the risk of heart disease. People cannot feel the physical effects of high blood pressure, but it can impair life's quality and end life prematurely.

The body's ability to maintain blood pressure is vital to life. The heart's pumping action must create enough force to push the blood through the major arteries into the smaller arteries and finally into tiny capillaries,◆ whose thin porous walls permit fluid exchange between the blood and tissues (see Figure 26-2). The nervous system helps maintain blood pressure by adjusting the size of the blood vessels and by influencing the heart's pumping action. The kidneys also help regulate blood pressure by setting in motion mechanisms that change the blood volume. The narrower the blood vessels or the greater the volume of blood in the circulatory system, the harder the heart must pump (and the more pressure the heart must create) to feed the tissues.

◆ **Optimal resting blood pressure for adults averages about 120 over 80 millimeters of mercury (mmHg). At a blood pressure of 140 over 90 mmHg or higher, the risks of heart attacks and strokes increase in direct proportion to rising blood pressure (see Table 26-1 later in the chapter).**

◆ **The resistance to the flow of blood caused by the reduced diameter of the smallest arteries and capillaries at the periphery of the body is called peripheral resistance.**

• Consequences of Hypertension • Hypertension stiffens the arteries and restricts blood flow through them. Thus hypertension can lead to many of the same complications as atherosclerosis—heart attacks, transient ischemic attacks, strokes, and renal failure. Constant high pressure damages the arterial walls and may cause them to balloon out **(aneurysm).** Aneurysms that go undetected can burst and lead to massive bleeding and death, particularly when a large vessel such as the aorta is affected. In the small arteries of the brain, an aneurysm may lead to stroke, and in the eye, it may lead to blindness. Likewise the kidneys may be damaged (kidney disease) when the heart is unable to adequately pump blood through them. Strain on the heart's pump, the left ventricle, can enlarge and weaken it until finally it fails (heart failure is described later).

• Interrelationships between Hypertension and Atherosclerosis • Hypertension injures the arterial walls, and plaques and clots are especially likely to form at damage points. Thus hypertension sets the stage for atherosclerosis or makes it worse. Once plaques or clots develop, they may reduce the diameter of an artery and raise blood pressure even higher. Hypertension and atherosclerosis are mutually aggravating conditions, and both contribute to acute and chronic heart failure and strokes.

platelets: tiny, disc-shaped bodies in the blood that are important in clot formation.

thrombosis (throm-BOH-sis): the formation or presence of a blood clot in the vascular system. A *coronary thrombosis* occurs in a coronary artery, and a *cerebral thrombosis* occurs in an artery that feeds the brain.
- **cerebro** = brain

thrombus (THROM-bus): a blood clot that blocks a blood vessel or cavity of the heart.

embolus (EM-boh-lus): a traveling blood clot.

embolism (EM-boh-lizm): the obstruction of a blood vessel by an embolus, causing sudden tissue death.

transient ischemic attack (TIA): a temporary reduction in blood flow to the brain, which causes temporary symptoms that vary depending on the part of the brain that is affected. Common symptoms include light-headedness, visual disturbances, paralysis, staggering, numbness, or dysphagia.

stroke: an event in which the blood flow to a part of the brain is cut off; also called a **cerebral vascular accident (CVA).**

aneurysm (AN-you-riz-um): a ballooning out of a portion of a blood vessel (usually an artery) due to weakness of the vessel's wall.

FIGURE 26-2 **Blood Pressure and Fluid Exchange**

At the same time the heart pushes blood into an artery, the small-diameter arteries and capillaries at its other end resist the blood's flow (peripheral resistance). Both actions contribute to the pressure inside the artery. Another determining factor is the volume of fluid in the circulatory system, which depends in turn on the number of dissolved particles in that fluid.

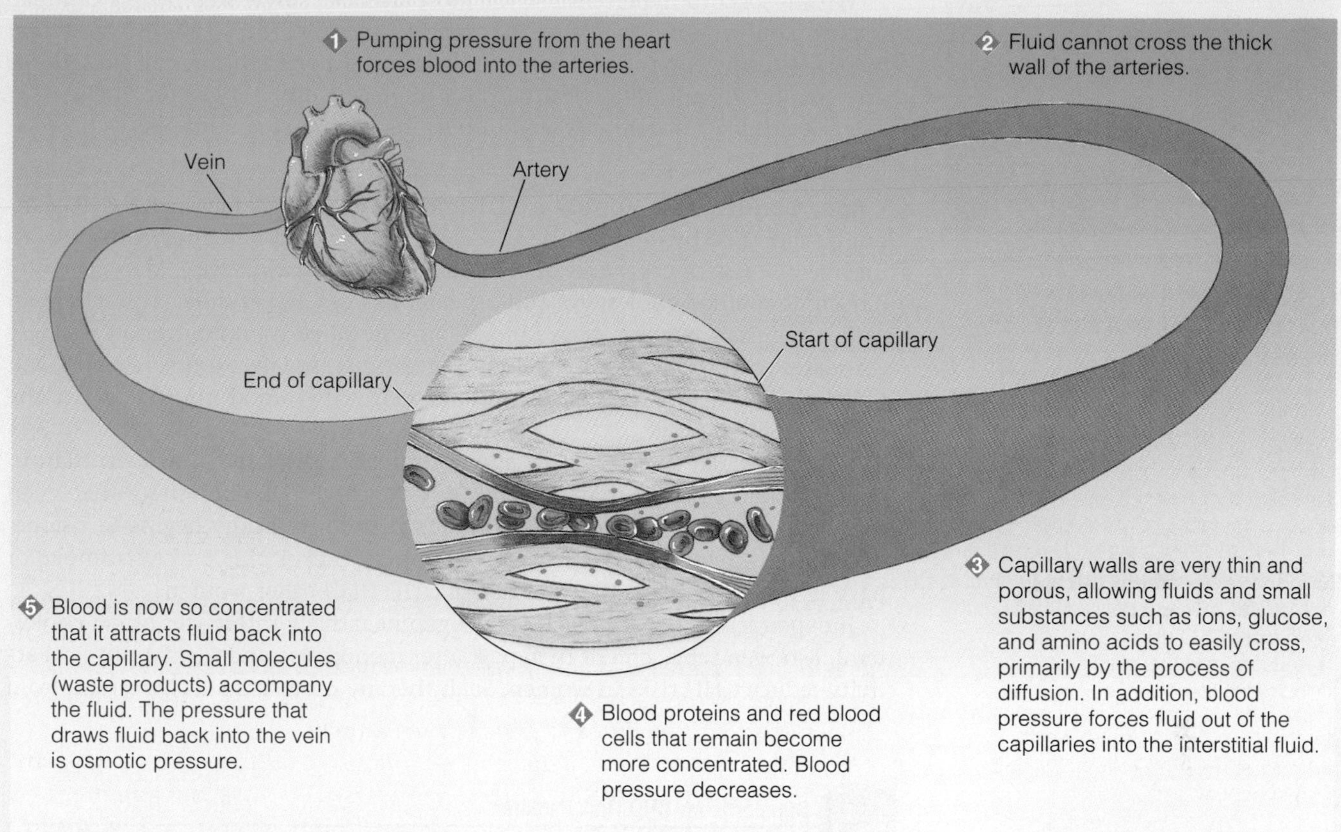

IN SUMMARY Most CHD involves atherosclerosis, hypertension, or a combination of the two. In atherosclerosis, cholesterol-filled plaques develop on the inner walls of the arteries. The plaques gradually enlarge and harden and may block the flow of blood to organs. A plaque may also rupture, and the clot that forms to repair the damage may further restrict blood flow or may break away and cut off blood flow through a smaller artery. Hypertension obstructs blood flow and strains the heart. Atherosclerosis and hypertension often occur together, and both are self-aggravating conditions that can lead to heart failure, transient ischemic attacks, strokes, and renal disease.

RISK FACTORS FOR CORONARY HEART DISEASE

Efforts to reduce the incidence of coronary heart disease (CHD) focus on normalizing blood lipids and hypertension. The margin◆ lists the major risk factors for CHD identified by the American Heart Association (AHA).[4] Many risk factors are interrelated—lack of physical activity, overweight, hypertension, elevated blood cholesterol, and type 2 diabetes frequently occur together, for example. Other factors may contribute to the risk of CHD, including emotional stress and drinking too much alcohol.

• Prevalence of Risk Factors • By middle age, most adults have at least one risk factor for CHD, and many have more than one.[5] Both the United States and Canada recommend screening to identify individuals at risk so as to offer

◆ **Major risk factors for CHD that cannot be modified:**
- Increasing age.
- Male gender.
- Heredity.

Modifiable risk factors for CHD:
- Tobacco smoke.
- **Elevated blood cholesterol (LDL).**
- **Hypertension.**
- Physical inactivity.
- **Excess body weight and body fat, especially abdominal fat.**
- **Diabetes mellitus.**

Note: Risk factors in bold type have relationships with diet.

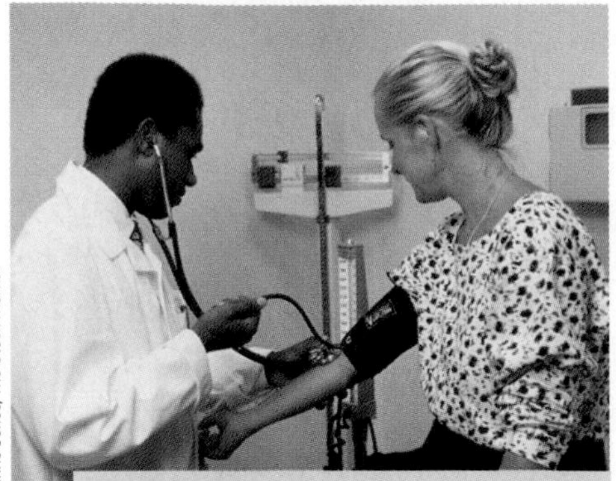
© Chris Jones/The Stock Market

Screening people for hypertension is a first step toward early detection of hypertension and prevention of complications.

preventive advice and treatment. Such public health programs are proving successful: since 1960, both blood cholesterol levels and deaths from cardiovascular disease among U.S. adults have shown a continuous and substantial downward trend.[6]

With respect to hypertension, the most effective single step people can take is to find out whether they have it. A major national effort to identify and treat hypertension is currently under way. Even mild hypertension can be serious; early treatment promotes health and a higher-quality, longer life. Table 26-1 lists the criteria for defining blood lipids, blood pressure, and obesity in relation to CHD risk.

• Age, Gender, and Heredity • Three of the major risk factors for CHD cannot be modified by diet or otherwise: age, gender, and heredity. With increasing age, the risk of CHD and its complications increases as well. Men generally have higher blood cholesterol and a greater risk of CHD and heart attacks at an earlier age than women. Men's blood cholesterol early in adulthood strongly correlates with their risk of developing heart disease later in life.[7] Almost half of all deaths from CHD occur among men with blood cholesterol in the borderline-high range.

Cardiovascular disease occurs about 10 to 12 years later in women than in men. Women younger than 45 tend to have lower LDL cholesterol than men of the same age, but a woman's blood cholesterol typically begins to rise between ages 45 and 55. Women who use birth control pills and also smoke or have hypertension are at greater risk of CHD than other women.

◆ Estrogen replacement therapy:
- Alleviates menopausal symptoms.
- Reduces CHD risks.
- Reduces osteoporosis risks.
- Increases some cancer risks.

Independently of age, a lack of estrogen negatively influences blood cholesterol. Estrogen replacement therapy◆ after menopause and regular physical activity reduce CHD risks in women. Such therapy protects LDL cholesterol from

TABLE 26-1 Standards for CHD Risk Factors[a]

	Desirable	Borderline	High
Lipids (mg/dL)[b]			
Total cholesterol	<200	200–239	≥240
LDL cholesterol[c]	<130	130–159	≥160
Triglycerides	<200	200–399	≥400
Body Mass Index (BMI)	18.5–24.9	25–29.9	≥30

	Systolic	Diastolic
Blood Pressure (mm/Hg)		
Optimal	<120	<80
Normal	120–129	80–84
High-normal	130–139	85–89
Mild	140–159	90–99
Moderate	160–179	100–109
Severe	≥180	≥110

[a] See Highlight 15 for lipid and blood pressure standards for children.
[b] HDL cholesterol is considered to be protective against CHD when it is ≥60 mg/dL and low when it is <40 mg/dL. An LDL-to-HDL ratio of >5 in men and >4.5 in women indicates risk for CHD.
[c] For people with two or more CHD risk factors, LDL cholesterol of ≥130 mg/dL is high. For people with preexisting CHD or diabetes, LDL cholesterol of ≥100 mg/dL is high.
SOURCES: Blood lipid standards adapted from The Expert Panel, Executive Summary of the third report of the National Cholesterol Education Program (NCEP) Expert Panel on Detection, Evaluation and Treatment of High Blood Cholesterol in Adults (Adult Treatment Panel III), available from the National Heart, Lung and Blood Institute, www.nhlbi.nih.gov/guidelines/cholesterol/index.htm hypertension standards adapted from the Sixth Report of the Joint National Committee on Prevention, Detection, Evaluation, and Treatment of High Blood Pressure, National High Blood Pressure Education Program, National Heart, Lung, and Blood Institute, National Institutes of Health, November 1997, p. 11.

oxidation, as does vitamin E supplementation.[8] Estrogen replacement therapy increases the risk of some cancers, however, so a woman must carefully consider her health history when balancing the apparent benefits against the potential risks of such treatment.[9]

Risks of CHD are higher for children of parents with heart disease and for people belonging to certain ethnic groups. These groups include African Americans, Mexican Americans, Native Americans, Native Hawaiians, and some Asian Americans.

© Ronnie Kaufman/The Stock Market

Regular aerobic exercise can help to defend against heart disease by strengthening the cardiovascular system, promoting weight loss, reducing blood pressure, and improving blood lipid and blood glucose levels.

• Tobacco Smoke • Smoking stresses the cardiovascular system by depriving the heart of oxygen and raising the blood pressure. It also damages platelets, making blood clot formation more likely. Toxins in tobacco smoke damage the blood vessels, setting the stage for plaque formation. People who smoke cigars and pipes have an increased risk of CHD, but not as high as the risk for people who smoke cigarettes. Finally, even nonsmokers have a higher risk of CHD if they are regularly exposed to other people's smoke (secondhand or environmental tobacco smoke).

• Physical Inactivity • Physical inactivity raises the risk of CHD because regular exercise confers many benefits to the cardiovascular system and helps control other risk factors for CHD. Aerobic, endurance-type activities,◆ undertaken faithfully for 30 minutes or more as a daily or every-other-day routine, provide the most benefits. Such a physical activity program can strengthen the heart and blood vessels; expand the volume of oxygen the heart can deliver to the tissues at each beat and so reduce the heart's workload; and bring about a redistribution of body water that eases the transit of blood through the peripheral arteries.[10] In addition regular aerobic physical activity lowers LDL, raises HDL, lowers blood pressure, speeds weight loss and loss of body fat, improves blood glucose control in people with type 2 diabetes, and reduces emotional stress. These changes are so beneficial that some experts believe that physical activity should be *the* primary focus of cardiovascular disease prevention efforts.[11]

◆ Examples of aerobic, endurance-type physical activities that can improve cardiovascular fitness include brisk walking, jogging, running, cycling, and rowing.

Some researchers wonder if physical activity itself raises blood HDL or if the weight loss that often accompanies exercise is the real protective factor. For women, weight loss through diet alone appears to *lower* HDL, but when diet is combined with moderate aerobic activity, HDL do not decline. In fact, HDL increase substantially in women who exercise regularly.[12] In men, diet raises HDL, and the combination of activity and diet results in a significantly greater rise in HDL than diet alone.

If heart and artery disease has already set in, a monitored program of physical activity may actually help to reverse it. Activity may stimulate development of new arteries to feed the heart muscle, which may account for the excellent recovery seen in some heart attack victims who exercise regularly.

• Diet-Related Risk Factors • It befits a nutrition book to focus on dietary strategies to prevent heart disease, but it should be noted that only four of the nine risk factors shown in the margin on p. 777 can be modified by diet. Individually, each of these diet-related risk factors—abnormal blood lipids, hypertension, diabetes (and insulin resistance), and obesity—increases the likelihood of developing CHD, but when they occur together (the metabolic syndrome), they synergistically raise the risk.[13] Highlight 26 presents more information about the metabolic syndrome and its relationship to cardiovascular disease.

As appealing as the solution may sound, diet may not reduce the risk of heart disease and stroke as successfully as other interventions do. Regular physical activity, quitting smoking, and taking medications to lower blood lipids or blood pressure, for example, may be more effective than dietary changes alone, but dietary changes confer additional benefits when they are combined with other strategies.

IN SUMMARY
Some risk factors for CHD, such as age, gender, and heredity, cannot be modified, but many other risk factors including smoking and exposure to tobacco smoke, elevated LDL cholesterol, hypertension, physical inactivity, obesity and overweight, and diabetes mellitus can be modified or controlled. Diet can play a role in lowering blood lipids, blood pressure, and weight and also in controlling blood glucose levels. Diet strategies for reducing the risk of CHD are most effective when combined with other strategies.

PREVENTION AND TREATMENT OF CORONARY HEART DISEASE

Treatment plans for preventing and treating CHD, including medical nutrition therapy, aim to normalize blood lipids and blood pressure, alter modifiable risk factors, and prevent complications. Treatment focuses on lifestyle changes first; clients are encouraged to increase physical activity, implement dietary changes, and reduce exposure to tobacco smoke either by quitting smoking or by avoiding secondhand smoke. If these changes fail to normalize lipid levels or reduce blood pressure, then lipid-lowering and/or antihypertensive medications are added to the treatment plan. For people with diabetes, controlling blood glucose levels is an important strategy. Many physicians advise clients to take low doses of aspirin daily. For people with advanced cardiovascular disease, therapy may also include surgery to restore blood flow to the affected organ.

THINK NUTRITION whenever you work with healthy clients as well as those with risk factors for CHD. Dietary strategies can often help to modify the risks of heart disease and strokes, especially when they are combined with other preventive strategies.

Diet Strategies

Diet strategies to both prevent and treat CHD focus on four main goals: a healthy eating pattern, a healthy body weight, a desirable blood cholesterol and lipoprotein profile, and a desirable blood pressure.[14] The specific strategies for achieving a desirable blood cholesterol and lipoprotein profile vary somewhat, depending on whether the objective is to prevent CHD in a healthy person with desirable LDL cholesterol levels or to improve blood lipid levels in high-risk groups—people with elevated LDL cholesterol, preexisting CHD, insulin resistance, or diabetes mellitus. Table 26-2 summarizes diet strategies for preventing CHD and reducing elevated LDL cholesterol; the "How to" on p. 782 offers practical suggestions. Dietitians often use the exchange system described in Chapter 25 or similar lists to help clients plan the diet.◆

◆ The box on pp. 154–155 describes how to use the exchange system to estimate fat intake.

© Michael Newman/PhotoEdit/PictureQuest

A varied, nutritious diet confers many benefits to support the health of the heart and blood vessels.

• **A Healthy Eating Pattern** • In the past, dietary recommendations for preventing and treating CHD centered specifically on the total energy and the amount of fat in the diet. While guidelines for energy and fat remain a pivotal component of the current recommendations, the new recommendations have a broader focus on eating patterns that foster both general and cardiovascular health. Adherence to the guidelines may also reduce the risks of other chronic health problems, including type 2 diabetes, osteoporosis, and certain types of cancer.[15]

Besides limiting fats (described later), the heart-healthy diet encourages consumption of carbohydrate-rich foods that provide nutrients and other components that benefit health and may protect against CHD. The fiber found in oats, barley, and pectin-rich fruits and vegetables, for example, helps to reduce blood lipids.[16] Carbohydrate-rich foods also provide minerals, which may help control blood pressure (described later); antioxidant nutrients, which may pro-

TABLE 26-2 **Diet Strategies for Meeting American Heart Association Dietary Goals**

Goals:

A healthy eating pattern, a healthy body weight, a desirable blood cholesterol and lipoprotein profile, and a desirable blood pressure.

Diet Strategies:

- Balance energy intake with energy needs to maintain weight, or limit energy to lose weight.
- Achieve a level of physical activity that balances with energy intake (to maintain weight) or exceeds energy intake (to lose weight).
- Limit foods high in saturated fatty acids (<10 percent of total energy intake) *trans*-fatty acids, and cholesterol (less than 300 milligrams).[a,b]
- Limit total fat to <30 percent of total energy.
- Replace saturated fats with carbohydrate from grains, legumes, fruits, and vegetables or with unsaturated fats (both long-chain omega-3 polyunsaturated fats and monounsaturated fats) from fish, vegetable oils, and nuts.
- Eat 5 or more servings of a variety of fruits and vegetables each day.
- Include 6 or more servings of a variety of starchy vegetables and grain products, including legumes and whole grains, each day.
- Use 6 ounces or less of lean meat, skinless poultry, or fish each day. Limit the use of whole eggs to 4 or less per week (including those used in cooking).[c]
- Include fish in at least two meals per week.
- Include at least 2 servings of nonfat or low-fat milk products each day for children 1 to 8 years old; 3 servings a day for adults aged 19 to 50 years; 3 to 4 servings for women who are pregnant or breastfeeding; and 4 servings for children and adolescents 9 to 18 years old and adults 51 years and older.
- Limit the intake of salt (sodium chloride) to <6 grams per day.
- Limit alcohol consumption (no more than 1 drink a day for women or 2 drinks a day for men).

[a] For high-risk individuals—those with elevated LDL cholesterol, preexisting cardiovascular disease, or diabetes—the saturated fat and cholesterol should be lower: <7 percent of total energy intake and <200 milligrams, respectively.

[b] See p. 152 in Chapter 5 for ways to lower saturated fat and cholesterol intake and the margin on p. 151 for a list of foods high in *trans*-fatty acids.

[c] High-risk individuals are advised to eat fewer than 5 ounces of lean meat, skinless poultry, and fish per day and fewer than 2 eggs per week.

SOURCES: Adapted from R. M. Krauss and coauthors, AHA dietary guidelines revision 2000: A statement for healthcare professionals from the nutrition committee of the American Heart Association; American Heart Association, An eating plan for a healthy America: The new 2000 food guidelines, available at www.americanheart.org, site visited February 21, 2001; National Heart, Lung, and Blood Institute, Tipsheet: Step II diet daily food guide, available at www.nhlbi.nih.gov/chd/Tipsheets/daily.htm, site visited February 21, 2001.

tect against LDL cholesterol oxidation (see Highlight 11); and B vitamins, which may help lower blood homocysteine levels. Studies have shown a positive association between elevated blood homocysteine and the risk of CHD.[17] Although an adequate supply of folate, vitamin B_6, and vitamin B_{12} lowers homocysteine levels, whether such a diet-induced reduction also reduces CHD risk remains to be answered.[18] Table 26-3 on p. 783 summarizes many of these factors, and earlier chapters and highlights have provided more details.

• **A Healthy Body Weight** • The dietary guidelines recommend energy intakes to achieve and maintain a desirable weight. Obesity, especially central obesity,◆ is associated with elevated blood lipids, hypertension, insulin resistance, and diabetes. For people who are overweight, weight loss reduces LDL cholesterol, triglycerides, blood pressure, and insulin resistance. Even a modest weight loss can help reduce the risk of CHD and stroke.

◆ Chapter 8 provides more information about central fat and its effects on the body.

• **A Desirable Blood Cholesterol and Lipoprotein Profile** • Too much total fat, saturated fatty acids, *trans*-fatty acids, and, to a lesser extent, dietary cholesterol raise blood LDL cholesterol levels. To achieve a healthy blood cholesterol and lipoprotein profile, the dietary guidelines recommend that all people limit total fat to less than 30 percent of daily kcalories and avoid *trans*-fatty acids.◆ For people without CHD or elevated LDL cholesterol, saturated fat should comprise less than 10 percent of the total kcalories, and dietary

◆ Notice that the low-fat diet described here controls both the amount of fat and the amount of saturated, monounsaturated, and polyunsaturated fat. By comparison, the low-fat diet described in Chapter 21's discussion of fat malabsorption controls the amount of fat and sometimes includes fat from medium-chain triglycerides (MCT).

Help Clients Implement Heart-Healthy Diets

For many people, following a heart-healthy diet translates into major changes in the foods they eat. They may find it easier to adapt to the diet if changes are made gradually. The following suggestions can help:

- Gradually increase servings of fruits, vegetables, and grains to reduce gas and bloating that may accompany a higher-than-normal fiber intake. Advise clients to drink plenty of water as well.
- For a client who usually eats only 1 or two servings of fruits or vegetables a day, start by including at least 1 serving of a fruit or vegetable at each meal (3 servings). Then add an additional serving at two meals or have one fruit and one vegetable as a snack (total 5 servings). Use a similar approach to add starchy vegetables and grains and include at least 6 servings daily. Choose extra servings of fruits, vegetables, starchy vegetables and grains if additional kcalories are needed to round out the meal plan.
- Cut back on servings of meat, poultry, and fish by eating one-third to one-half the amount usually eaten at each meal. The goal is to include no more than 5 to 6 ounces of meat, poultry, and fish a day.
- For a client who doesn't drink milk or has only one serving, start by adding a glass of nonfat or very-low fat (1%) milk at one meal. then add a glass at another meal. A third glass of milk can be used as a snack or taken with a third meal.
- Reduce salt intake by gradually cutting down on the salt added to food and used in cooking. The goal is to reduce salt intake to about 1 teaspoon a day.

To help clients implement their diet, recommend they use:

- Whole-grain breads and cereals instead of refined breads and cereals.
- Whole fruits and vegetables, rather than juice, as often as possible.
- Fresh, frozen, dried, or canned fruit packed in its own juice.
- Fresh, frozen, or canned vegetables that contain no added salt.
- Nonfat, ½%, or 1% fat milk and low-fat and fat-free yogurt.
- Casseroles, pasta meals, and stir-fry dishes to limit meat and increase servings of vegetables and grains.
- Two or more vegetarian meals per week.
- Fats that contain 2 grams or less of saturated fat per serving including liquid and tub margarines and canola, corn, olive, safflower, sesame, soybean, and sunflower oils.
- Liquid or tub margarines with no *trans*-fatty acids.
- Low-salt or salt-free products.
- Products containing fat substitutes in moderate amounts.
- Sodium-free spices such as basil, bay leaves, curry, garlic, ginger, lemon, mint, oregano, pepper, rosemary, and thyme to season foods.

(The box on page 152 offered many tips for reducing fat and the type of fat in the diet.)

Offer clients these suggestions for snacks:

- Frozen low-fat or fat-free yogurt.
- Fruit.
- Angel food cake.
- Low-fat ice cream (no more than 3 grams fat per ½ cup), fruit ices, sherbets, or sorbets.
- Raw vegetables.
- Unsalted pretzels and nuts.
- Plain popcorn without butter, margarine, or salt.

Suggest that clients limit or avoid these foods:

- Foods of low nutrient density including high-fat foods, foods high in sugar, and alcohol.
- Foods prepared with hydrogenated fat, which contains *trans*-fatty acids, including donuts, crackers, commercially prepared baked goods, and fried foods in restaurants.

Recommend that clients also avoid these high-salt foods:

- Pickles, olives, and sauerkraut.
- Cured or smoked meats, such as beef jerky, bologna, corned or chipped beef, frankfurters, ham, luncheon meats, salt pork, and sausage. Many of these foods are also high in fat.
- Salty or smoked fish, such as anchovies, caviar, salted or dried cod, herring, sardines, and smoked salmon.
- Salted snack foods like potato chips, pretzels, popcorn, nuts, and crackers. Many of these foods are also high in fat.
- Bouillon cubes, seasoned salts, and soy, steak, Worcestershire, and barbecue sauces.
- Cheeses, especially processed cheeses. Many cheeses are also high in fat.
- Salted canned and instant soups.
- Prepared horseradish, catsup, and mustard.

Clients with lactose intolerance unable to tolerate the recommended servings of dairy foods can use the digestive aids described in Chapter 21 (see p. 657).

cholesterol should be less than 300 milligrams a day.* In most cases, foods high in cholesterol are also high in saturated fat. Eggs and some vegetable oils (coconut, palm, and palm kernel) are exceptions. Eggs are low in saturated fats and high in cholesterol; coconut, palm, and palm kernel oils are high in saturated fats and low in cholesterol.

For people with elevated LDL cholesterol,◆ prexisting cardiovascular disease (including those who have experienced a heart attack or stroke), or diabetes, the recommended diet further reduces saturated fat to 7 percent of daily kcalo-

◆ **Target levels for LDL cholesterol vary depending on each person's health status. As Table 26-1 shows, for the general population, LDL cholesterol is elevated at ≥160 mg/dL. However, for people who have two or more risk factors for CHD, LDL cholesterol is considered to be elevated at ≥130 mg/dL. Finally, for people who already have CHD, LDL cholesterol is considered to be elevated at ≥100 mg/dL.**

* These recommendations for fat and cholesterol correspond to the Step I diet that had been recommended in the past.

TABLE 26-3 Dietary Factors Protecting against CHD

In addition to reducing total fat and saturated fat, these dietary factors may also protect against CHD.

Dietary Factor	Protection against CHD
Soluble fiber (apples and other fruits, oats, soy, barley, legumes)	• Lowers blood cholesterol, especially in those with high cholesterol • Lowers risk of heart attack • Improves LDL-to-HDL ratio
Omega-3 fatty acids (fish and some plant oils)	• Limit clot formation • Prevent irregular heartbeats • Lower risk of heart attack
Alcohol (in moderation)	• Raises HDL • Prevents clot formation
Folate, vitamin B_6, vitamin B_{12}	• Reduce homocysteine
Vitamin E (vegetable oils and margarines, some nuts, wheat germ)	• Slows progression of plaque formation • Lowers risk of heart attack in people with CHD • Limits LDL oxidation
Soy (protein and isoflavones)	• Lowers blood cholesterol • Raises HDL cholesterol • Improves LDL-to-HDL ratio

ries and cholesterol to less than 200 milligrams per day.* Compared to the diet for healthy people, the diet for high-risk groups provides less lean meat, skinless poultry, fish, and eggs (see Table 26-2) and strictly limits high-fat meats, especially organ meats (liver, brain, and kidney). Dietitians can help plan the diet to ensure that saturated fats and cholesterol are reduced without sacrificing the nutritional quality of the diet.

Polyunsaturated fatty acids, monounsaturated fatty acids, and, to a lesser extent, soluble fibers lower LDL cholesterol. Either unsaturated fat◆ or carbohydrate can replace saturated fats in a heart-healthy diet. As Chapter 25 described, however, high-carbohydrate diets can elevate blood triglycerides◆ and reduce HDL cholesterol, especially for some people with insulin resistance or type 2 diabetes. In these cases, replacing saturated fats with unsaturated fats instead of carbohydrate may help improve the lipoprotein profile.

Reminder: Fat from all sources should not exceed 30% of the total kcalories. ◆

People with elevated triglycerides should avoid simple sugars, which often cause triglyceride levels to rise. ◆

Fish oils, rich in omega-3 polyunsaturated fatty acids, improve blood lipids (primarily by lowering triglycerides), prevent blood clots, and reduce the risk of sudden death associated with CHD.[19] For these reasons, the dietary guidelines recommend at least two servings of fish per week. Plant sources of omega-3 fatty acids may also confer benefits.[20] Plant sources of omega-3 fatty acids include flaxseed and flaxseed oil, canola oil, soybean oil, and nuts. For people with preexisting CHD, long-term use of omega-3 fatty acids may reduce overall mortality as well as sudden death from CHD.[21] To achieve the intake of omega-3 fatty acids found to be effective requires that people either eat one fish meal each day or take fish oil supplements daily.[22]

• Achieve and Maintain a Normal Blood Pressure • Diet strategies to reduce hypertension have traditionally included recommendations to control weight, reduce salt intake, increase potassium intake, and limit alcohol. While health authorities continue to recommend these modifications, they also recommend a diet that contains adequate amounts of calcium and magnesium and that limits saturated fats and cholesterol. Results of the Dietary Approaches to Stop Hypertension (DASH) trial show that a diet rich in fruits, vegetables, and low-fat dairy products and with reduced total fat and saturated fat can lower blood pressure to a significant degree.[23] When the DASH diet is combined with a limited intake of sodium, the effects on blood pressure are greater still.[24] Thus the heart-healthy dietary guidelines embrace these strategies in an overall diet to prevent and treat CHD.

* These recommendations for fat and cholesterol correspond to the Step II diet that had been recommended in the past.

For many years, controversy surrounded recommendations to limit salt to prevent or reduce hypertension. The second study of the DASH diet, which looked at the effects of sodium restriction on blood pressure control, however, provides strong evidence that sodium restriction plays an important role. Lowering sodium intakes reduced blood pressure regardless of gender, race, presence or absence of preexisting hypertension, or whether people followed the DASH diet or a typical American diet. However, the most benefits to blood pressure occurred in the groups who followed the DASH diet and lowered their sodium intakes. Furthermore, the lower the sodium intake, the greater the drop in blood pressure.

• **Alcohol** • Moderate consumption of alcohol may reduce the risk of CHD by raising HDL cholesterol and preventing clot formation.[25] Beneficial effects of alcohol are most apparent in people over age 50, those with other risk factors, and those with high LDL.[26] For others, the benefits may not be apparent. At least one recent study reports that abstainers and moderate drinkers shared similar risks of dying from heart disease. [27]

These findings pose a dilemma for health care professionals who are well aware of the potentially damaging effects of alcohol on many body systems (see Highlight 7). Too much alcohol can raise blood pressure. Alcohol can also raise triglycerides, and people with elevated triglycerides are advised to restrict alcohol. Most authorities do not advise clients who do not drink alcohol to begin to do so to reduce their risk of CHD. If clients do drink alcohol, clinicians stress that moderation is the key.

Alcohol from any source—red or white wine, beer, or distilled liquor—appears to be equally effective, and alcohol itself may be a protective factor.[28] In addition to alcohol, wine contains phytochemicals that may also protect against cardiovascular disease.[29] These substances may act as antioxidants, reducing LDL oxidation, and may alter prostaglandin metabolism, reducing blood clot formation. These protective effects may explain the so-called French paradox: the wine-drinking people of France enjoy a lower incidence of CHD even though they have many of the same risk factors as people in the United States.

• **Plant Sterols** • Margarines made from plant sterols are being marketed as aids to lower cholesterol. Plant sterols, which are found in oil derived from some complex carbohydrates (including soy), resemble cholesterol in structure; they reduce the absorption of dietary cholesterol and lower blood cholesterol levels.[30] Even vegetarians, however, may find it difficult to consume enough plant sterols from food to lower blood cholesterol. Thus the margarines serve to supplement plant sterols in the diet. The long-term safety of these products is unclear, however, and the AHA currently recommends that margarines made from plant sterols be used only by people with elevated LDL cholesterol or preexisting heart disease.

• **Soy Protein and Isoflavones** • When soy proteins are substituted for animal proteins, blood levels of LDL cholesterol and triglycerides fall, and HDL levels do not.[31] The cholesterol-lowering effect may be due to soy isoflavones—phytochemicals that have an estrogen-like effect.[32] For people who have elevated cholesterol levels and are following low-fat, low–saturated fat diets, adding soy protein daily can significantly lower LDL cholesterol.[33] The Food and Drug Administration (FDA) allows foods that contain 6.25 grams of soy protein per serving to carry a health claim for reduced risk of heart disease.

Drug Therapy

When used together with diet therapy and a physical activity program, drug therapy can effectively lower blood lipids. Lipid-lowering medications carry health risks, however, and are costly. Therefore, physicians as a rule do not prescribe medications to treat hyperlipidemia until after a six-month trial of intensive diet therapy and physical activity has proved unsuccessful in lowering blood lipid concentrations.

• **Medications** • For people with very high LDL cholesterol (greater than 220 milligrams per deciliter), a shorter diet trial may be considered. Besides lipid-lowering medications, medications used in the treatment of cardiovascular diseases may include antihypertensives◆ and diuretics to reduce blood pressure, aspirin and anticoagulants to prevent clot formation, and nitroglycerin to alleviate angina. (Clients with diabetes may also be taking oral antidiabetic agents or insulin.) All of these medications are associated with significant risks and nutrition-related side effects (see the Diet-Drug Interactions box on p. 791), a problem compounded by the fact that drug therapy often includes multiple medications and continues for many years or even life. Elderly clients especially risk diet-medication interactions.

Some antihypertensive agents can lead to potassium deficiencies; others can lead to elevated blood potassium. ◆

• **Dietary Supplements** • People may use dietary supplements including antioxidant nutrients and folate (described earlier in this chapter) and garlic and black and green teas to protect against cardiovascular disease. Garlic is a relatively safe remedy that, when used along with dietary changes, may counter the tendency of blood to clot and modestly lower serum cholesterol and blood pressure. Black and green teas are sources of a type of antioxidant that may play a role in CHD prevention. The Diet-Drug Interactions box describes herbs used for other reasons that may interact with medications prescribed for cardiovascular diseases.

IN SUMMARY

Lifestyle changes that involve diet, physical activity, and avoidance of tobacco smoke often form the first strategy for reducing the risk of CHD. A heart-healthy diet focuses on an overall healthy eating pattern, an appropriate body weight, a desirable cholesterol and blood lipid profile, and a desirable blood pressure. Among the many medications that may be used in the treatment of CHD, lipid-lowering agents (antilipemics) and antihypertensives are commonly prescribed.

HEART FAILURE AND STROKES

When atherosclerosis and hypertension run their course, the consequences can be fatal. Most often, these consequences include heart failure or strokes.

Heart Attacks

The heart receives nutrients and oxygen, not from inside its chambers, but from arteries on its surface (see Figure 26-1 on p. 774). As mentioned earlier, a heart attack, or myocardial infarction (MI), occurs when CHD robs the heart muscle of its blood supply. Treatment aims to relieve pain, stabilize the heart rhythm, and reduce the heart's workload.

• **Immediate Care** • Like an accident victim, a heart attack victim is in shock at first, and diet therapy cannot begin until shock resolves. After several hours of observation, the person can usually begin to eat again, as outlined in the "How to" on p. 786. Medical nutrition therapy aims to reduce the work of the heart and therefore restricts total energy, the amount of food or drink at each feeding, sodium, and caffeine. Because nausea is a common problem following an MI, liquids are offered first. Low-sodium foods prevent fluid retention, and soft foods can help prevent nausea and abdominal distention, which can push the diaphragm up toward the heart and stress the heart muscle. Temperature extremes can stimulate nerves that slow the heart rate, so foods are offered at moderate temperatures. Caffeine stimulates the metabolic rate and is usually restricted.

HOW TO Manage Diets after a Myocardial Infarction (MI)

- Offer nothing by mouth until shock resolves.
- After several hours, give a 1000- to 1200-kcalorie diet that progresses from low-sodium liquids to low-sodium soft foods of moderate temperature in frequent, small feedings.
- After five to ten days, adjust the diet to meet individual needs, generally to three meals a day.
- Although somewhat controversial, many practitioners restrict caffeine completely during the first few days after an MI and generally recommend a moderate restriction (no more than 3 cups of a caffeine-containing beverage per day) thereafter.
- Gradually progress diet to a heart-healthy diet with less than 7 percent saturated fat and less than 200 milligrams cholesterol with three meals a day.

• Long-Term Diet Therapy • After the person is out of immediate danger (in about five to ten days), the diet is tailored to meet individual needs and to deal with conditions such as hyperlipidemia, hypertension, obesity, and diabetes. Because the person has CHD, the heart-healthy diet described earlier is appropriate. Such a diet can be planned to provide three meals a day, but people who still have chest pain after a heart attack may continue to benefit from eating frequent, small meals. Advise clients to eat slowly and to avoid strenuous physical activity before and after meals.

• Encouraging Lifestyle Changes • Often a person who has experienced an MI is eager to apply strategies to reduce the risks of further CHD, and health care professionals can use the opportunity to offer sound and useful counsel. Continue to offer support and encouragement for as long as possible. Clients may readily return to their old habits when their symptoms disappear if they haven't fully incorporated new healthful behaviors into their lives.

Chronic Heart Failure

While heart attacks represent acute heart failure, heart failure can also occur gradually. Coronary heart disease and hypertension are common causes. A person may develop chronic heart failure following a heart attack or a severe stress. The conditions that lead to **chronic,** or **congestive, heart failure (CHF)** cause the heart muscle to work unusually hard. As a result, the heart muscle enlarges **(cardiomegaly)** and gradually weakens as it strains to supply adequate blood to the tissues.

• Consequences of CHF • As heart failure progresses, reduced blood flow impairs the function of all organs. Reduced blood flow to the kidneys triggers the retention of fluid, further stressing the heart and compounding the stagnation of fluid in the body. Peripheral, pulmonary, and hepatic edema may develop as the person becomes increasingly "congested" with excess fluids. Pulmonary edema increases the likelihood of respiratory infections, which can further stress the heart and lungs.

• CHF and Nutrition Status • As CHF progresses, energy needs increase because organ systems, particularly the heart and lungs, must work extra hard to maintain their functions. At the same time, the disrupted blood flow limits the supply of nutrients and oxygen to the organs and tissues. Repeated respiratory infections can further tax nutrition status. People in the later stages of CHF are often unable to eat enough to meet energy and protein demands; oral intake may be limited due to anorexia, altered taste sensitivity, intolerance to food odors, physical exhaustion, the low-sodium diet used for treatment (described later), and medications. Weight loss in people with CHF may go unnoticed until it has progressed considerably, because edema masks their underweight condition. Thus protein-energy malnutrition (PEM) can occur as a consequence of CHF **(cardiac cachexia),** particularly when the disorder progresses to the later stages. Malnutrition then further contributes to the weakness of the heart muscle and raises the likelihood of respiratory infections.

THINK NUTRITION whenever clients are in the late stages of CHF. Fluid retention may mask the client's underweight condition, and PEM can progress unnoticed.

chronic or **congestive heart failure (CHF):** a syndrome in which the heart gradually weakens and can no longer adequately pump blood through the circulatory system.

cardiomegaly (CAR-dee-oh-MEG-ah-lee): enlargement of the heart.
- **mega** = large

cardiac cachexia (ka-KEKS-ee-ah): chronic PEM that develops as a consequence of heart failure. Research suggests that cytokines play a role in the development of PEM in the late stages of CHF.
- **kakos** = bad
- **hexia** = condition

• Treatment of CHF • Drug therapy for CHF includes diuretics to reduce the fluid volume and cardiac glycosides to increase the strength of heart muscle contractions. People taking thiazide or loop diuretics and cardiac glycosides together are at high risk for potassium deficiency and may be prescribed a potassium supplement. Stool softeners may be prescribed, particularly for elderly clients who frequently experience constipation, because straining to empty the bowels can stress the heart. Initially, bed rest helps reduce the heart's workload. Once recovery is under way, the person must rest frequently and avoid overexertion.

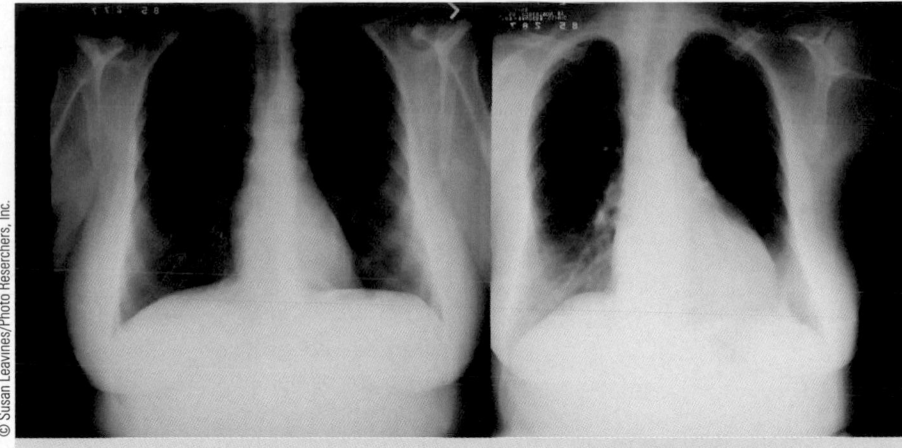

An overburdened heart enlarges in an effort to supply blood to the body's tissues.

• Medical Nutrition Therapy • The person newly diagnosed with CHF who is overweight benefits from a safe weight-loss program. Medical nutrition therapy for the wasting associated with later stages of CHF aims to preserve or restore nutrition status and reduce the work of the heart. Providing enough energy to maintain body weight◆ is vital, but providing too much energy increases the body's metabolic rate, stressing the heart. Likewise, giving too much fluid and sodium expands the body's fluid volume, which also taxes the heart. When clients are unable to eat adequate amounts of table foods, formulas of high nutrient density, which provide energy and protein with less fluid, may help clients meet their nutrient needs.

The heart-healthy diet described earlier is also appropriate. As CHF progresses, the prescribed diet limits sodium (2 to 3 grams or less per day).◆ Dietary fiber is carefully adjusted: the goal is to provide some fiber to prevent constipation, but to avoid amounts and types of fibers that produce gas and abdominal distention.

◆ Energy needs for people with severe heart failure may be as high as 20 to 30 percent above basal energy needs (see Table 24-3 on p. 734).

◆ In most cases, diuretics serve to eliminate excess fluids from the body. In severe cases of heart failure, however, people may need to restrict their intake of fluid to 1,000 to 2,000 milliliters per day. Chapter 27 provides more information about fluid- and sodium-restricted diets.

Strokes

Temporary interference with blood flow to the brain may result in a transient ischemic attack (TIA), a condition that causes changes in mental status that may last for a few minutes or a few hours. People who experience a TIA may or may not develop a total blockage of blood flow to a portion of the brain that results in a stroke, or cerebral vascular accident (CVA). Most strokes occur as a consequence of atherosclerosis, hypertension, or a combination of the two.

• Complications Affecting Food Intake • For some stroke victims, recovery is unremarkable; others may require months of rehabilitative therapy. Victims may suffer from temporary or permanent problems that interfere with the ability to communicate and sometimes develop dysphagia (see Chapter 20).◆ The inability to communicate effectively makes it difficult for them to tell health care professionals about foods they can or would like to eat or about problems they may be having with swallowing foods.

◆ Reminder: *Dysphagia* refers to an inability to coordinate swallowing movements appropriately.

Following a stroke, some people develop physical problems that interfere with their ability to prepare foods or to engage in the physical process of eating. Such problems can include an inability to grasp utensils or coordinate movements that bring foods or liquids from the table to the mouth. Highlight 20 described some ways to handle such problems.

• Medical Nutrition Therapy • Clients with significant problems with chewing and swallowing may require tube feedings until they have recovered from the stroke and a speech pathologist determines which foods they can safely chew and swallow. Long-term diet therapy for stroke victims depends on the underlying medical condition, but often includes the heart-healthy diet to help lower blood lipids and blood pressure. For people with limited activity, food energy may need to be limited. A client who must relearn to walk or use other muscle groups can best do so if not overweight. Underweight can also hinder physical rehabilitation—another reason to ensure that the person does not become malnourished.

IN SUMMARY

Some of the major consequences of atherosclerosis and hypertension include heart failure and strokes. The heart can fail suddenly (myocardial infarction) or gradually (congestive heart failure). For both types of heart failure, heart-healthy diets are appropriate once clients are out of immediate danger. In the later stages of congestive heart failure, clients may need to restrict sodium, and health care professionals must remain alert for signs of PEM. The dietary needs of people who experience a stroke vary, depending on the extent and area of the brain affected. The accompanying case study presents a client with CHD. Carefully consider the questions posed to review the information presented so far in this chapter.

DISORDERS OF THE LUNGS

The heart and lungs work together to ensure the delivery of oxygen to the cells and the elimination of carbon dioxide generated by metabolic processes. Lung disorders can be acute or chronic; both can affect nutrition status.

Acute Respiratory Failure

Recall from Chapter 24 that the lungs are often the first organ to fail in people following a severe stress or in those with sepsis. With hypermetabolism, the body avidly consumes oxygen and produces carbon dioxide, and the lungs must work hard to keep the body's gases in balance. At the same time, the inflammatory responses to stress increase the permeability of the lungs' **alveoli,** fluids accumulate in the lungs' interstitial spaces, and pulmonary edema may

alveoli (al-VEE-oh-lie): air sacs in the lungs. One sac is an *alveolus.*

CASE STUDY

History Professor with Cardiovascular Disease

Mr. Jablonski, a 48-year-old history professor, has a blood lipid profile that includes elevated LDL cholesterol. He is 5 feet 7 inches tall and weighs 200 pounds. Mr. Jablonski has a family history of CHD. His diet history shows excessive intakes of food energy, cholesterol, total fat, saturated fat, and salt. He smokes a pack of cigarettes a day, and his lifestyle leaves him little time for physical activity. Mr. Jablonski also has hypertension, for which antihypertensive agents have been prescribed. He frequently forgets to take his pills, though, and his blood pressure is often quite high.

1. Name the risk factors for CHD and hypertension in Mr. Jablonski's history. Which of them can he control? Which can be helped by diet? What complications might you expect if his condition goes untreated?
2. What type of diet, if any, would you recommend to treat Mr. Jablonski's high LDL cholesterol and hypertension? Explain the rationale for each diet change. How will his current diet change? Prepare a day's menus for Mr. Jablonski. What suggestions might you offer to help him make the necessary diet changes?
3. What laboratory and clinical tests would you expect to see monitored regularly? Why?
4. Name at least three ways in which Mr. Jablonski could benefit from losing weight. How does physical activity fit into a weight-loss plan? Describe other ways Mr. Jablonski might benefit from a physical activity program.
5. Discuss nutrition considerations if Mr. Jablonski should suffer a heart attack or stroke. Describe the relationships of these disorders to elevated blood lipids and hypertension.

develop. If the problem becomes severe enough, the lungs lose their elasticity and ability to exchange gases, and **respiratory failure** results. Inflammatory responses can also lead to acute respiratory failure in people with severe allergic reactions.

Acute respiratory failure can also occur when an embolus lodges in the lungs or when toxic substances such as gastric contents (aspiration pneumonia), smoke, or some inhaled toxic drug directly damage the lung tissue. Premature infants may develop acute respiratory failure because their lungs have not fully developed.

• Consequences of Acute Respiratory Failure • To compensate for altered gas exchange, the person breathes faster, the heart rate increases, breathing becomes labored, and the person becomes restless.◆ Lack of oxygen in the blood **(hypoxemia)** causes the person's skin to become pale and develop a bluish tint **(cyanosis)**. If respiratory failure progresses to a severe stage, mental status diminishes, and the person may lapse into a coma. The weakened lungs may fail to push oxygenated blood to the heart, which may also fail. The combination of poor respiration, poor circulation, and poor local nutrition raises the likelihood of respiratory infections. Clients who recover from respiratory failure, however, often regain normal lung function.

The cluster of symptoms associated with acute respiratory failure is called the acute respiratory distress syndrome (ARDS). ◆

• Treatment of Acute Respiratory Failure • Treatment focuses on correcting the underlying disorder and preventing the progression of acute respiratory failure. The person often requires a **mechanical ventilator** in order to breathe. Intravenous solutions are used to correct fluid and electrolyte and acid-base imbalances. Diuretics help mobilize fluids.

• Medical Nutrition Therapy • Medical nutrition therapy aims to provide enough energy and protein to support lung function and prevent infections without overtaxing the compromised respiratory system. Fluid and sodium restrictions may be necessary to ease pulmonary edema. Overfeeding generates excess carbon dioxide, so total kcalories must be carefully controlled. Glucose metabolism generates more carbon dioxide than does fat metabolism, but this effect is of lesser importance when total energy needs are not exceeded.[34] Actual energy needs are highly variable. Ideally, clinicians use indirect calorimetry to determine the ratio of carbon dioxide produced to oxygen consumed and then use that information to estimate total energy needs and the appropriate mix of carbohydrate and fat to meet those needs. Alternatively, clinicians use formula estimates, such as the Harris-Benedict equation, to predict energy needs (see Table 24-3 on p. 734) and prevent overfeeding.

THINK NUTRITION whenever clients are in acute respiratory failure. Providing too little or too much energy and nutrients can further tax the respiratory system.

People recovering from respiratory failure who are being weaned from mechanical ventilators also have high energy and protein needs. The person's lungs were weak before mechanical ventilation began, and they become weaker with disuse. As mechanical ventilation decreases, the lungs must do more work—a stressful and energy-consuming process.

• Tube Feedings and Parenteral Nutrition • Often people with acute respiratory failure are unable to meet their nutrient needs with an oral diet. For people in respiratory failure, tube feedings are preferred to parenteral nutrition, and intestinal feedings are preferred to gastric feedings because they reduce the risk of aspiration. Special nutrient-dense pulmonary formulas that provide less carbohydrate and more fat are available, but other nutrient-dense formulas can effectively meet nutrient needs.[35]

respiratory failure: failure of the lungs to exchange gases. In *acute* respiratory failure, the lungs fail over a short period of time.

hypoxemia (high-pox-EE-me-ah): lack of oxygen in the blood.

cyanosis (sigh-ah-NOH-sis): bluish discoloration of the skin caused by a lack of oxygen.

mechanical ventilator: a machine that "breathes" for the person who can't. In a normal respiration, the lungs expand, which draws air into the lungs. With mechanical ventilation, air is forced into the lungs at regular intervals using pressure.

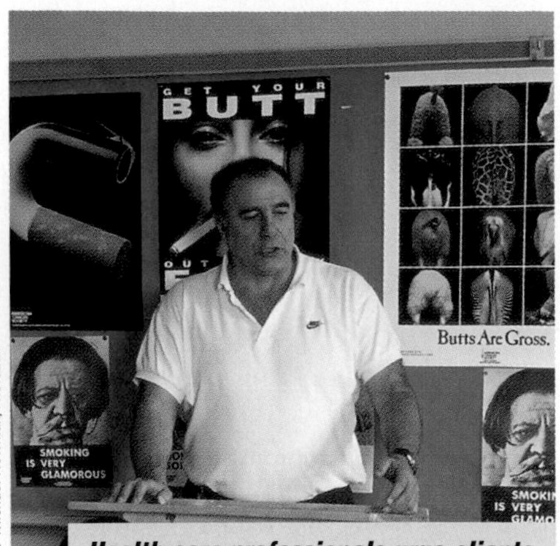

Health care professionals urge clients not to smoke to reduce the incidence of COPD or preserve lung function in those who already have the disorder.

Chronic Obstructive Pulmonary Disease

Chronic obstructive pulmonary disease (COPD) describes several conditions characterized by persistent obstruction of air flow through the lungs. Unlike acute respiratory failure, which may occur suddenly, lung failure resulting from COPD is gradual. The two major types of COPD are **emphysema** and chronic **bronchitis.** COPD ranks as the fourth leading cause of death in the United States.[36] Smoking is the primary risk factor for COPD, and most people with COPD are smokers. Other risk factors include exposure to environmental pollution (including exposure of nonsmokers to cigarette smoke) and, possibly, repeated respiratory tract infections.

Experts believe that COPD may be largely preventable by encouraging people not to smoke and controlling environmental pollution. Although COPD has no cure, people with COPD who smoke can help preserve lung function if they quit smoking before the disorder has progressed too far.

• Consequences of COPD • Regardless of the type of COPD, the lungs gradually lose their functional surface area and strength, making it difficult for them to deliver oxygen to the blood and to remove carbon dioxide from it. As lung function becomes increasingly compromised, pulmonary infections, respiratory failure, and heart failure can follow.

• COPD and Nutrition Status • People with COPD frequently experience weight loss, PEM, and infections and account for many cases of malnutrition in hospitals. The extent of malnutrition is associated with the severity of the pulmonary disease. Body weight is a predictor of survival in people with COPD, and low body weight is associated with reduced muscle mass, respiratory function, and immune competence.[37] Weight loss, which may be rapid and dramatic, may occur for many reasons, including the following:

- Anorexia and poor food intake. These may be caused by depression and anxiety; chronic mouth breathing, which can alter tastes for foods; difficulty breathing while preparing and eating food; and gastric discomfort from swallowing air while eating.[38]
- High energy expenditures associated with labored breathing.
- Medications, including anti-inflammatory agents, diuretics, and antibiotics, which alter nutrient requirements and compromise nutrition status.
- Use of oxygen masks, because people cannot eat while using them.
- Repeated infections, which raise nutrient needs and deplete nutrient stores. Poor nutrition status, in turn, opens the way for further infection—a vicious cycle.

THINK NUTRITION whenever clients have COPD. PEM becomes common as the disorder progresses.

• Treatment of COPD • People with COPD are encouraged to quit smoking, receive vaccinations to prevent influenza (flu) and pneumonia, and take antibiotics at the first sign of bacterial infection. Medications for COPD frequently include bronchodilators to limit respiratory muscle spasms and corticosteroids to reduce inflammation of lung tissue (see the Diet-Drug Interactions box). People with severe COPD may need long-term oxygen therapy to help relieve symptoms.

• Medical Nutrition Therapy • Like people with CHF, people newly diagnosed with COPD or those with mild cases who are also overweight benefit from safe weight-loss programs. Excessive body weight increases the work of the lungs in maintaining respiration. As the disorder progresses, unintentional weight loss is more likely to occur. At this stage, maintaining or repleting nutrient stores helps

chronic obstructive pulmonary disease (COPD): one of several disorders, including emphysema and bronchitis, that interfere with respiration.

emphysema (EM-fih-SEE-mah): a type of COPD in which the lungs lose their elasticity and the victim has difficulty breathing; often occurs along with bronchitis.

bronchitis (bron-KYE-tis): inflammation of the lungs' air passages.

Diet-Drug Interactions

ANTICOAGULANTS (INCLUDING ASPIRIN)

Aspirin can be given with food to reduce nausea and GI distress. Long-term use of aspirin may lead to folate and vitamin C deficiencies. *Ticlopidine* is best absorbed along with foods. Ticlopidine may cause nausea, GI distress, and diarrhea. For people taking *warfarin,* maintaining a consistent vitamin K intake from day to day helps assure the medication's effectiveness. Warfarin may also cause diarrhea. The herbs *danshen, dong quai, ginkgo, cayenne, feverfew, echinacea, ginger, St. John's wort, garlic* and coenzyme Q, omega-3 fatty acids, and high doses of vitamin E can affect clotting times and must be used cautiously by people who are also taking anticoagulants.

ANTIHYPERTENSIVES

All clients on *antihypertensives* should avoid natural licorice and limit alcohol. People on *ACE inhibitors* should avoid salt substitutes containing potassium and use potassium supplements (if prescribed) cautiously because these agents can raise blood potassium levels. One ACE inhibitor, *fosinopril,* should not be given along with calcium or magnesium supplements or calcium- or magnesium-containing antacids because calcium and magnesium reduce the drug's absorption. *Alpha-adrenergic blockers* can lead to weight gain and fatigue. *Beta-blockers,* which also act as antianginals, should be given with foods to reduce nausea and GI distress. Grapefruit juice should not be given along with *calcium channel blockers* because it can potentiate the effects of the drugs. *Clonidine* may dry the mouth and lead to constipation and drowsiness.

ANTILIPEMICS

Cholestyramine can cause nausea, GI distress, and constipation and can lead to fat-soluble vitamin deficiencies because it reduces their absorption. *Colestipol* is less likely than cholestyramine to cause nausea and lead to fat-soluble vitamin deficiencies, but it can lead to constipation. *Gemfibrozil* can lead to nausea and GI distress. The *statins (lovastatin, pravastatin, simvastin)* should not be given with grapefruit juice. People taking statins should limit alcohol use.

BRONCHODILATORS

Bronchodilators can dry the mouth and throat and lead to nausea and vomiting.

CARDIAC GLYCOSIDES

Digitoxin, digoxin, and *digitalis* can lead to anorexia and nausea. Low blood potassium and elevated blood calcium levels can lead to drug toxicity, and clients are advised to eat a high-potassium diet. If they use calcium and vitamin D supplements, they must be used cautiously. Magnesium-containing antacids or supplements given along with *cardiac glycosides* can reduce drug absorption. The herb *hawthorn* can potentiate the effects of cardiac glycosides.

CORTICOSTEROIDS, *SEE ANTI-INFLAMMATORY AGENTS* ON P. 673.

DIURETICS

Diuretics can cause potassium imbalances and lead to muscle weakness, unexplained numbness and tingling sensations, irregular heartbeats, and cardiac arrest. *Thiazide* and *loop diuretics* increase the urinary excretion of potassium, and clients on these diuretics are encouraged to include rich sources of potassium daily and may be prescribed potassium supplements. When taken internally, *aloe vera* can potentiate the loss of potassium in people taking potassium-wasting diuretics. Potassium-sparing diuretics *(amiloride, spironolactone,* and *triamterene),* on the other hand, lead to potassium retention. Clients on potassium-sparing diuretics should avoid excessive potassium intakes, potassium supplements, and salt substitutes that contain potassium. Sometimes a combination of both types of diuretics is used to avoid potassium imbalances.

to maintain lung function and prevent lung infections. Malnourished clients replete nutrient stores best when refed gradually, with a goal of providing a high-kcalorie, high-protein diet without stressing the lungs. Overfeeding people with COPD and PEM, however, can be as harmful as underfeeding them. Overfeeding and excessive carbohydrate intakes produce high levels of carbon dioxide, which the already stressed lungs must work hard to expel.

Meeting energy and protein needs from easy-to-eat foods provided in frequent, small meals works best.◆ When appetite is poor, clients may benefit from enteral formulas provided either orally or by tube. Pulmonary formulas that provide more kcalories from fat and fewer from carbohydrate than standard formulas are available, but there is little evidence to suggest that pulmonary formulas are superior to standard enteral formulas for managing COPD.[39]

Advise clients to relax during meals, eat slowly, and chew foods thoroughly to prevent swallowing air while eating. ◆

IN SUMMARY

Chronic diseases of the heart and lungs gradually lead to loss of organ function and can eventually progress to acute events that are fatal. Health care professionals seek to identify people at risk for heart and lung disorders and encourage clients to modify behaviors known to increase that risk. The Nutrition Assessment Checklist helps to identify nutrition-related factors that may help to prevent or treat heart and lung disorders.

Nutrition Assessment Checklist for People with Disorders of the Heart, Blood Vessels, and Lungs

MEDICAL HISTORY

Check the medical record for a diagnosis of:

- ☐ Coronary heart disease
- ☐ Hypertension
- ☐ Congestive heart failure
- ☐ Acute respiratory failure
- ☐ COPD

Review the medical record for complications related to CVD or COPD:

- ☐ Heart attacks
- ☐ Stroke
- ☐ Weight loss related to congestive heart failure or COPD

Note risk factors for CHD related to diet, including:

- ☐ Elevated LDL cholesterol
- ☐ Obesity or overweight
- ☐ Diabetes

MEDICATIONS

For people who are using drug therapy for cardiovascular and respiratory diseases, note:

- ☐ Side effects that may alter food intake
- ☐ Potential for folate and vitamin C deficiencies for people taking aspirin
- ☐ Consistency of vitamin K intake from day to day for people taking warfarin
- ☐ Consistency of use of denshen, dong quai, ginkgo, cayenne, feverfew, echinacea, ginger, St. John's wort, garlic, coenzyme Q, omega-3 fatty acids, and vitamin E for people taking anticoagulants
- ☐ Intake of potassium (including potassium from salt substitutes) for people taking ACE inhibitors, cardiac glycosides, and diuretics
- ☐ Intake of fat-soluble vitamins for people taking cholestyramine
- ☐ Use of alcohol and natural licorice for people taking antihypertensive agents
- ☐ Use of grapefruit juice for people taking calcium channel blockers and statins
- ☐ Use of calcium- and magnesium-containing antacids and supplements for people taking fosinopril and cardiac glycosides
- ☐ Use of hawthorn for people taking cardiac glycosides

FOOD/NUTRIENT INTAKE

For all clients, especially those with CHD or hypertension, or those with risk factors for CHD or hypertension, assess the diet for:

- ☐ Total energy
- ☐ Total fat and sources of saturated fat, monounsaturated fat, polyunsaturated fat, and cholesterol
- ☐ *Trans*-fatty acids
- ☐ Fiber
- ☐ Salt, potassium, calcium, and magnesium
- ☐ Folate and other B vitamins
- ☐ Alcohol

Considerations for clients with complications of the heart or lungs include:

- ☐ Energy and nutrient intake for clients in the later stages of congestive heart failure or COPD or those with acute respiratory failure
- ☐ Physical disabilities that interfere with clients' abilities to prepare and eat foods and/or dysphagia following a stroke

ANTHROPOMETRICS

Measure baseline height and weight and reassess weight at each medical checkup.

Note whether clients are meeting weight goals, including:

- ☐ Weight loss for clients who are overweight, especially those who have or are at risk for CHD, hypertension, congestive heart failure, or COPD
- ☐ Weight maintenance for people in the later stages of congestive heart failure or COPD

Remember that weight may be deceptively high in people who are retaining fluids, especially those in the later stages of:

- ☐ Congestive heart failure
- ☐ COPD

LABORATORY TESTS

Monitor the following laboratory tests for people with cardiovascular diseases or those at risk for CHD and hypertension:

- ☐ Blood lipids
- ☐ Blood glucose for people with diabetes
- ☐ Blood potassium for people taking diuretics and/or cardiac glycosides
- ☐ Indicators of fluid retention for people with heart or respiratory failure

PHYSICAL SIGNS

Blood pressure measurement is routine in physical exams, but is especially important for people:

- ☐ With cardiovascular diseases or following a heart attack or stroke
- ☐ At risk for CHD or hypertension

Look for physical signs of:

- ☐ Folate and vitamin C deficiencies in people taking aspirin
- ☐ Potassium imbalances (muscle weakness, numbness and tingling sensations, irregular heartbeats) in people taking diuretics or cardiac glycosides
- ☐ Fat-soluble vitamin deficiencies in people taking cholestyramine
- ☐ Fluid overload in people in the later stages of congestive heart failure, acute respiratory failure, and COPD

Nutrition on the Net

WEBSITES

Access these websites for further study of topics covered in this chapter.

- Find updates and quick links to these and other nutrition-related sites at our website: **www.wadsworth.com/nutrition**
- To search for more information about CHD and its consequences, including the complete American Heart Association dietary guidelines, planning guides to implement the dietary guidelines, the DASH diet, tools for assessing your personal risk of heart disease, links to related websites, and much more, visit the American Heart Association: **www.americanheart.org**
- To review other CHD risk assessment tools, more information about CHD and its consequences, and a variety of suggestions for implementing heart-healthy diets, visit the National Heart, Lung, and Blood Institute: **www.nhlbi.nih.gov/nhilbi/nhlbi.htm** and the Heart and Stroke Foundation of Canada at: **www.hsf.ca**
- To find more information about alternative therapies for heart disease, visit The Natural Pharmacist: **www.TNP.com**
- To learn more about lung diseases, visit the American Lung Association: **www.lungusa.org** and the Canadian Lung Association at: **www.lung.ca**

INTERNET ACTIVITIES

Cardiovascular diseases are the leading cause of death in the United States and around the world. Smoking is the primary cause of heart and lung disease. If you are a smoker, or if you know someone who is, help in quitting may be just what you or your friend needs.

- Go to **www.lung.ca**
- Scroll down headings on the left side of the page. Click on "Smoking and Tobacco."
- Read the paragraph and click on "Smoking Cessation Guide."
- Read the information and then click on a topic of interest such as "I want to quit," or "I'm a friend of a smoker."
- Follow the directions for your topic.

What did you learn about quitting smoking, or helping someone else to quit? What other suggestions do you have?

Study Questions

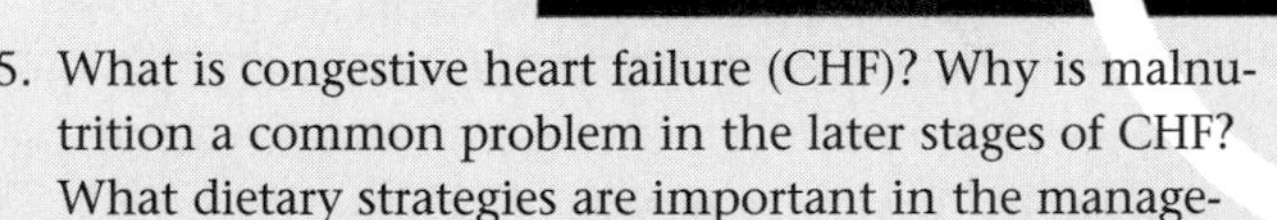

These questions will help you review the chapter. You will find the answers in the discussions on the pages provided.

1. What is atherosclerosis? How can atherosclerosis lead to hypertension, thrombosis, heart attacks, and strokes? (pp. 774–776)
2. What risk factors for CHD can be helped by diet? What diet strategies are recommended to reduce the risk of CHD? (pp. 779–780)
3. How do the diet strategies to prevent the risk of CHD in healthy people differ from those recommended for people with elevated LDL cholesterol, preexisting CHD, and diabetes? (pp. 781–782)
4. What is a myocardial infarction (MI), and what is its cause? Describe medical nutrition therapy immediately after an MI as well as after the immediate danger is over. (pp. 785–786)
5. What is congestive heart failure (CHF)? Why is malnutrition a common problem in the later stages of CHF? What dietary strategies are important in the management of CHF? (pp. 786–787)
6. Describe some ways that a person's nutrition needs might be affected by a stroke. How can attention to nutrient needs affect a person's responses to physical therapy? (pp. 787–788)
7. What is respiratory failure? How can malnutrition affect the course of respiratory failure? (pp. 788–789)
8. What is COPD? What circumstances can alter nutrition status in people with COPD? (p. 790)

These questions will help you review for an exam. Answers can be found on p. 796.

1. The form of blood cholesterol associated with an increased risk of atherosclerosis is:
   a. triglycerides.

b. chylomicrons.
c. LDL cholesterol.
d. HDL cholesterol.

2. Which of the following statements about hypertension is true?
 a. Low blood pressure reduces life expectancy.
 b. The narrower the diameter of the arteries, the greater the blood pressure.
 c. People can tell when their blood pressure rises.
 d. Hypertension aggravates atherosclerosis, but atherosclerosis does not aggravate hypertension.
3. Modifiable risk factors for CHD include:
 a. age and heredity.
 b. heredity, smoking, and salt intake.
 c. obesity, diabetes, and LDL cholesterol.
 d. physical activity, hypertension, gender, and heredity.
4. Which of the following diet modifications would be recommended by a health care professional suggesting diet strategies to help a client lower the risk of CHD?
 a. high complex carbohydrate, low total fat, low saturated fat, and low cholesterol
 b. high complex carbohydrate, low total fat, low saturated fat, and high cholesterol
 c. low carbohydrate, low fiber, low total fat, low saturated fat, and low cholesterol
 d. low carbohydrate, low fiber, low total fat, low monounsaturated fat, and low cholesterol
5. Compared to the heart-healthy diet recommended for healthy people, the diet recommended for people with elevated LDL cholesterol, preexisting CHD, and diabetes is:
 a. lower in kcalories.
 b. higher in complex carbohydrates.
 c. lower in total fat and cholesterol.
 d. lower in saturated fat and cholesterol.
6. Which of the following statements about alcohol and cardiovascular diseases might a health care professional relay to a client who wishes to start using alcohol to lower the risks of CHD?
 a. Alcohol lowers blood pressure.
 b. Alcohol intake should not exceed more than 4 drinks a day.
 c. A reduced risk of CHD is observed at all levels of alcohol intake.
 d. Risks associated with alcohol use may outweigh its benefits in protecting against CHD.
7. For people receiving drug therapy for either hyperlipidemia or hypertension:
 a. physical activity must be restricted.
 b. the risk of diet-drug interactions is high.
 c. diet restrictions and physical activity programs are no longer effective.
 d. the risk of potassium imbalances is great for people taking anticoagulants.
8. Diet strategies to help a person recover from a myocardial infarction include:
 a. low-fiber foods to prevent constipation.
 b. foods of moderate temperatures to avoid slowing the heart rate.
 c. three larger meals to allow the heart more time to rest between meals.
 d. high-kcalorie, high-protein, low-sodium liquids at first to speed repair of the heart and prevent fluid retention.
9. Medical nutrition therapy for a person in acute respiratory failure includes:
 a. careful attention to providing enough, but not too much, energy.
 b. high volumes of fluids and electrolytes to clear the lungs of mucus.
 c. foods of moderate temperature to allow the lungs to ventilate fully.
 d. high-carbohydrate, high-fiber foods to prevent constipation and limit carbon dioxide production.
10. Factors that frequently contribute to wasting in people with either congestive heart failure or COPD include:
 a. anorexia, medications, and respiratory infections.
 b. dehydration, fat malabsorption, and lactose intolerance.
 c. essential fatty acid deficiencies and potassium imbalances.
 d. vitamin K deficiencies, dysphagia, and low-fiber, low-potassium diets.

Clinical Applications

1. Consider the risk factors for CHD. Describe possible interrelationships among the factors. For example, a female over age 55 is also at higher risk of diabetes; a person with diabetes is more likely to have hypertension.
2. Plan a week's menus that incorporate the diet strategies listed in Table 26-2. Begin by calculating how much energy you need. If you are at a desirable weight, you need about 15 kcalories per pound of body weight to maintain your weight if you are active and 13 kcalories, if you are less active. If you are overweight, you can use 10 kcalories per pound of body weight to estimate your energy needs. Use the exchange lists to determine the number of servings of fats, starchy vegetables and grains, lean meat, and alcohol that you can include in your diet.

References

1 American Heart Association, 2001 Heart and Stroke Statistical Update, www.americanheart.org/statistics/cvd.html, site visited on February 19, 2001.

2 S. M. Grundy and coauthors, Primary prevention of coronary heart disease: Guidance from Framingham, *Circulation* 97 (1998): 1876–1887.

3 American Heart Association, 2001 Heart and Stroke Statistical Update, 2001.

4 American Heart Association, Risk factors and coronary heart disease: AHA Scientific position, 2000, www.americanheart.org, site visited on February 20, 2001.

5 G. S. Berenson and coauthors, Association between multiple cardiovascular risk factors and atherosclerosis in children and young adults, *New England Journal of Medicine* 338 (1998): 1650–1656.

6 W. D. Rosamond and coauthors, Trends in the incidence of myocardial infarction and in mortality due to coronary heart disease 1987 to 1994, *New England Journal of Medicine* 339 (1998): 861–867.

7 J. Stamler and coauthors, Relationship of baseline serum cholesterol levels in 3 large cohorts of younger men to long-term coronary, cardiovascular, and all-cause mortality and to longevity, *Journal of the American Medical Association* 284 (2000): 311–318; D. Steinberg and A. M. Gotto, Preventing coronary artery disease by lowering cholesterol levels, *Journal of the American Medical Association* 282 (1999): 2043–2050.

8 R. C. Wander and coauthors, Effects of interaction of RRR-alpha-tocopheryl acetate and fish oil on low-density-lipoprotein oxidation in postmenopausal women with and without hormone-replacement therapy, *American Journal of Clinical Nutrition* 63 (1996): 184–193.

9 H. Jernstrom and E. Barrett-Connor, Obesity, weight change, fasting insulin, proinsulin, C-peptide, and insulin-like growth factor-1 levels in women with and without breast cancer: The Rancho Bernardo Study, *Journal of Women's Health and Gender-Based Medicine* 8 (1999): 1265–1272; N. E. Davidson, Hormone-replacement therapy—Breast versus heart versus bone, *New England Journal of Medicine* 332 (1995): 1638–1639.

10 G. F. Fletcher and coauthors, Statement on exercise: Benefits and recommendations for physical activity programs for all Americans, *Circulation* 94 (1996): 857–862.

11 F. W. Booth and coauthors, Waging war on modern chronic diseases: Primary prevention through exercise biology, *Journal of Applied Physiology* 88 (2000): 774–787; F. B. Hu and coauthors, Physical activity and risk of stroke in women, *Journal of the American Medical Association* 283 (2000): 2961–2967; F. W. Farrell and coauthors, Influences of cardiorespiratory fitness levels and other predictors on cardiovascular disease mortality in men, *Medicine and Science in Sports and Exercise* 30 (1998): 899–905.

12 P. T. Williams, High-density lipoprotein cholesterol and other risk factors for coronary heart disease in female runners, *New England Journal of Medicine* 334 (1996): 1298–1303.

13 M. E. Daly and coauthors, Dietary carbohydrates and insulin sensitivity: A review of the evidence and clinical implications, *American Journal of Clinical Nutrition* 66 (1997): 1072–1085.

14 R. M. Krauss and coauthors, AHA dietary guidelines, revision 2000: A statement for healthcare professionals from the nutrition committee of the American Heart Association, *Circulation* 102 (2000): 2284–2299.

15 Krauss and coauthors, 2000.

16 L. Van Horn, Fiber, lipids, and coronary heart disease, *Circulation* 95 (1997): 2701–2704.

17 E. Arnesen and coauthors, Serum total homocysteine and coronary heart disease, *International Journal of Epidemiology* 24 (1995): 704–709; M. J. Stampfer and coauthors, A prospective study of plasma homocyst(e)ine and risk of myocardial infarction in US physicians, *Journal of the American Medical Association* 268 (1992): 877–880.

18 M. J. Stampfer and E. B. Rimm, Folate and cardiovascular disease, *Journal of the American Medical Association* 275 (1995):1929–1930.

19 C. M. Albert and coauthors, Fish consumption and risk of sudden cardiac death, *Journal of the American Medical Association* 279 (1998): 23–28; W. S. Harris, n-3 Fatty acids and serum lipoproteins: Human studies, *American Journal of Clinical Nutrition* (supplement) 65 (1997): 1645–1654; T. A. Mori and coauthors, Interactions between dietary fat, fish, and fish oils and their effects on platelet function in men at risk of cardiovascular disease, *Atherosclerosis, Thrombosis and Vascular Biology* 17 (1997): 279–286.

20 C. Von Shacky and coauthors, The effect of dietary omega-3 fatty acids on coronary atherosclerosis: A randomized, double-blind, placebo-controlled trial, *Annals of Internal Medicine* 130 (1999): 554–562.

21 GISSI-Prevenzione Investigators, Dietary supplementation with n-3 polyunsaturated fatty acids and vitamin E after myocardial infarction, *Lancet* 354 (1999): 447–455.

22 Krauss and coauthors, 2000.

23 L. J. Appel and coauthors, A clinical trial of the effects of dietary patterns on blood pressure, *New England Journal of Medicine* 336 (1997): 1117–11124.

24 F. M. Sacks and coauthors, Effects on blood pressure of reduced sodium and the Dietary Approaches to Stop Hypertension (DASH) diet. DASH-Sodium Collaborative Research Group, *New England Journal of Medicine* 344 (2001): 3–10.

25 P. R. Ridker and coauthors, Association of moderate alcohol consumption and plasma concentrations of endogenous tissue-type plasminogen activator, *Journal of the American Medical Association* 272 (1994): 929–933; J. M. Gaziano and coauthors, Moderate alcohol intake, increased levels of high-density lipoprotein and its subfractions, and decreased risk of myocardial infarction, *New England Journal of Medicine* 329 (1993): 1829–1834.

26 H. O. Hein, P. Suadiacani, and F. Gyntelberg, Alcohol consumption, serum low density lipoprotein choles-

terol concentration, and risk of ischaemic heart disease: six year follow up in the Copenhagen male study, *British Medical Journal* 312 (1996): 731–736; C. S. Fuchs and coauthors, Alcohol consumption and mortality among women, *New England Journal of Medicine* 332 (1995): 1245–1250.

27 C. L. Hart and coauthors, Alcohol consumption and mortality from all causes, coronary heart disease, and stroke: Results from a prospective cohort study of Scottish men with 21 years of follow up, *British Medical Journal* 318 (1999): 1725–1729.

28 E. B. Rimm and coauthors, Review of moderate alcohol consumption and reduced risk of coronary heart disease: is the effect due to beer, wine, or spirits? *British Medical Journal* 312 (1996): 731–736.

29 S. V. Nigdikar and coauthors, Consumption of red wine polyphenols reduces the susceptibility of low-density lipoproteins to oxidation in vivo, *American Journal of Clinical Nutrition* 68 (1998): 258–265.

30 H. Gylling, Reduction of serum cholesterol in postmenopausal women with previous myocardial infarction and cholesterol malabsorption induced by dietary sitostanol ester margarine, *Circulation* 95 (1997): 4226–4231; B. V. Howard and D. Kritchevsky, Phytochemicals and cardiovascular disease: A statement for healthcare professionals from the American Heart Association, *Circulation* 95 (1997): 2591–2593.

31 J. W. Anderson, B. M. Johnstone, and M. E. Cook-Newell, Meta-analysis of the effects of soy protein intake on serum lipids, *New England Journal of Medicine* 333 (1995): 276–282.

32 J. R. Crouse and coauthors, A randomized trial comparing the effect of casein with that of soy protein containing varying amounts of isoflavones on plasma concentrations of lipids and lipoproteins, *Archives of Internal Medicine* 159 (1999): 2070–2072.

33 S. R. Teixeira and coauthors, Effects of feeding 4 levels of soy protein for 3 and 6 weeks on blood lipids and apolipoproteins in moderately hypercholesterolemic men, *American Journal of Clinical Nutrition* 71 (2000): 1077–1084.

34 A. M. Malone, Is a pulmonary enteral formula warranted for patients with pulmonary dysfunction? *Nutrition in Clinical Practice* 12 (1997): 168–171.

35 Malone, 1997.

36 D. L. Hoyert, K. D. Kochanek, and S. L. Murphy, Deaths: Final data for 1997, *National Vital Statistics Report,* June 30, 1999.

37 K. Gray-Donald and coauthors, Nutritional status and mortality in chronic obstructive pulmonary disease, *American Journal of Respiratory and Critical Care Medicine* 153 (1996): 961–966.

38 K. M. Chapman and L. Winter, COPD: Using nutrition to prevent respiratory function decline, *Geriatrics* 51 (1996): 37–42.

39 Malone, 1997.

ANSWERS

Multiple Choice

1. c	2. b	3. c	4. a	5. d
6. d	7. b	8. b	9. a	10. a

THE METABOLIC SYNDROME

As CHAPTERS 25 and 26 described, several risk factors for cardiovascular diseases (CVD) and their complications frequently occur together. These risk factors—insulin resistance, high triglycerides, low HDL cholesterol, and hypertension—collectively comprise the metabolic syndrome, which is sometimes called syndrome X or the insulin resistance syndrome. Often obesity, especially central obesity, triggers insulin resistance, which then leads to a cascade of metabolic events that impairs the health of the cardiovascular system. The metabolic syndrome frequently accompanies diabetes, especially type 2 diabetes. This may explain why CVD account for over 75 percent of hospital admissions and over 80 percent of deaths in people with diabetes. Conversely, because the risk factors that comprise the metabolic syndrome are modifiable, it might be possible to prevent their development and reduce the risks associated with them. While researchers continue to unravel the mechanisms of the metabolic syndrome, many pieces of the puzzle have emerged. This highlight summarizes some of the most important findings. The accompanying glossary defines terms related to the metabolic syndrome.

INSULIN RESISTANCE

In a person with insulin resistance, insulin fails to lower blood glucose as it would in a healthy, insulin-sensitive person. The beta cells of the pancreas respond by secreting more insulin. Thus people with insulin resistance have elevated levels of both glucose and insulin (hyperinsulemia). While chronic elevated blood glucose levels correspond to the acute and **microvascular** complications associated with diabetes, insulin resistance may play a bigger role in the cardiovascular complications. Figure H26-1 on p. 798 shows how insulin resistance leads to problems typical of the metabolic syndrome.

Who Develops Insulin Resistance?

People with type 2 diabetes develop insulin resistance, but the opposite is not true. Not all people with insulin resistance have type 2 diabetes or develop it later. For some people, the extra insulin secretion compensates for insulin resistance, and they are able to maintain normal blood glucose levels. People who may be insulin resistant and not have diabetes include some who have hypertension or have survived a heart attack and women who have **polycystic ovary syndrome.**

Obesity and Insulin Resistance

Obesity, especially central obesity, leads to insulin resistance and hyperinsulinemia and is a risk factor for both type 2 diabetes and cardiovascular disease. Many people with the metabolic syndrome are obese, but others are not. For people with central obesity, high insulin levels represent a response both to insulin resistance and to a reduced clearance rate of insulin by the liver. With central obesity, less insulin reaches the liver, and more is diverted through the peripheral circulation, so insulin levels remain high.

ALTERED LIPIDS

The hyperinsulinemia that accompanies insulin resistance ultimately leads to altered blood lipid patterns. Triglyceride levels rise, and with this rise, the size of the LDL particles is

GLOSSARY

fibrinogen (fie-BRIN-oh-jen): a protein produced by the liver that is essential to blood clotting.

insulin resistance of adipose tissue: failure of the enzyme that releases free fatty acids from fat cells to adequately reduce its activity in the presence of insulin.

insulin resistance of skeletal muscle: failure of the muscle cells to take up glucose from the blood in response to insulin as they normally would.

microvascular: pertaining to the capillaries. Retinopathy, nephropathy, and neuropathy are microvascular complications.

nitric oxide (NO): a substance produced by the vascular endothelium that causes blood vessels to dilate and inhibits clot formation.

plasminogen (plaz-MIN-oh-jen) **activator inhibitor-1 (PAL-1):** a substance important in blood clotting.

polycystic (POL-ee-SIS-tik) **ovary syndrome:** a disorder of women characterized by ovaries enlarged with fluid-filled sacs and elevated levels of male hormones (androgens).

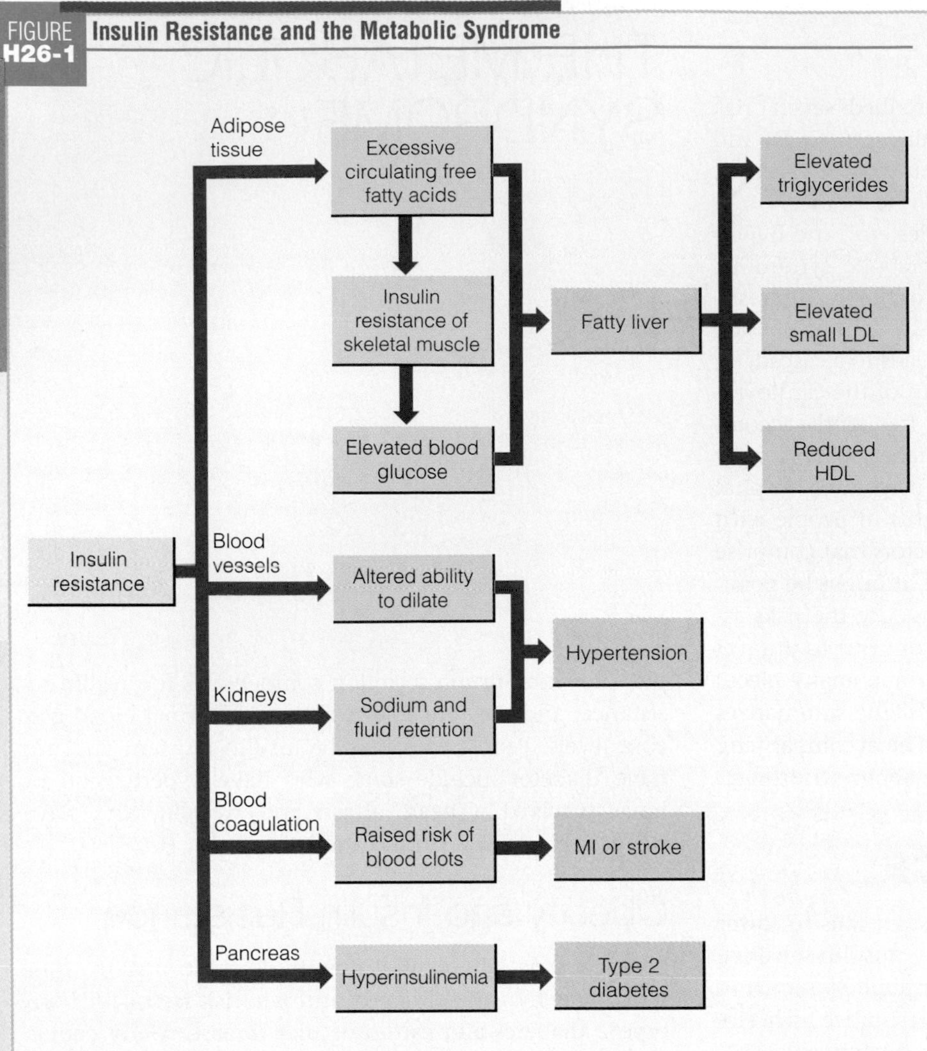

FIGURE H26-1 **Insulin Resistance and the Metabolic Syndrome**

reduced. Among the LDL, small LDL have a particularly strong correlation with CHD risk.[1] Compared to regular-size LDL, small LDL may cross through the arterial wall and be oxidized more readily. In addition to elevated triglycerides and small LDL, HDL levels fall.

Hyperinsulinemia and Fatty Acids

After a meal, insulin levels rise and suppress the activity of an enzyme that releases free fatty acids from adipose cells. In a fasting state, insulin levels fall, the enzyme is activated, and free fatty acids are released. Skeletal muscle cells use these free fatty acids for energy during the fast. Most of the rest reach the liver, where they may be used for energy or packaged in lipoproteins and returned to the circulation.

The enzyme that releases fatty acids is highly sensitive to insulin in some cases and less sensitive in others. The enzyme is often less sensitive in people who are obese, especially those with central obesity, and in people with a genetic predisposition. People with enzymes that are not sensitive to insulin tend to have high circulating fatty acid levels despite high levels of insulin. This condition is described as **insulin resistance of adipose tissue.**[2] In these people, the high fatty acid levels can overload skeletal muscle and the liver. An overload of lipid in skeletal muscle impairs the uptake of glucose and results in **insulin resistance of skeletal muscle.** The potential effects of lipid overload on the liver are described next.

Fatty Liver

The excess accumulation of lipid in the liver can lead to fatty liver (see Chapter 28). Although the reasons are unclear, fatty liver results in various changes in normal fat metabolism. Some of these changes include:

- Increased production of VLDL (triglyceride-rich lipoproteins). Serum triglycerides and VLDL rise, and small LDL rise as well.
- Impaired action of the enzyme (lipoprotein lipase) that allows fat cells to take up fatty acids from lipoprotein. Triglycerides and triglyceride-rich lipoproteins remain high.
- Overactivity of the enzyme that degrades HDL. HDL levels fall.

Thus, once insulin resistance and hyperinsulinemia are set in motion, blood glucose rises higher (less is taken up by skeletal muscle), and lipid abnormalities that favor the development of CVD ensue.

OTHER CONSEQUENCES

Insulin resistance and hyperinsulinemia result in still other changes that raise the risk of CVD. Some of these changes affect the volume of blood and the ability of the blood vessels to dilate and constrict. Still other changes foster the formation of clots.

Fluids and Electrolytes

Insulin influences sodium reabsorption in the kidneys. People who are insulin resistant reabsorb more sodium and fluid along with it, and the increased blood volume contributes to hypertension.

Blood Vessel Changes

Like many other cells, healthy blood vessel cells also respond to insulin. For healthy people, the rise in insulin following a meal relaxes blood vessels, and the vessels dilate. Once a blood vessel is dilated, blood flow through the vessel increases. Insulin resistance changes the ability of the blood vessels to relax so that blood flow can increase. Whereas blood flow to muscle cells increases rapidly and significantly in insulin-sensitive individuals, in obese people blood flow takes longer to increase to the same level; in people with type 2 diabetes, blood flow takes even longer to increase, and it never rises to the same level as it does in insulin-sensitive and obese people.[3] These changes may also contribute to the hypertension that frequently accompanies insulin resistance and diabetes.[4]

Blood Coagulation

Insulin resistance alters the body's coagulation system in a way that favors the formation of clots. Among the blood factors that favor the formation of clots are **fibrinogen** and **plasminogen activator inhibitor-1,** and both of these are elevated in people with insulin resistance.[5] Improved blood glucose control may lower plasma fibrinogen, suggesting that prompt treatment of the insulin resistance may help reduce the progression of the syndrome and the development of cardiovascular complications.[6]

Nitric Oxide

An interesting hypothesis that may explain some of the effect of insulin resistance on the ability of the blood vessels to dilate and on the changes in blood coagulation suggests that insulin resistance interferes with the production of **nitric oxide (NO).**[7] NO is continuously produced from arginine (a nonessential amino acid) by the cells lining the blood vessels. NO relaxes the smooth muscle cells within the blood vessels and also inhibits factors that favor clot formation. In people with insulin resistance, NO production appears to falter.[8]

A number of contributing factors raise the risk for CVD in people who develop insulin resistance and the clinical findings associated with the metabolic syndrome. Abnormal blood lipids, changes in the blood vessels, hypertension, and an increased risk of clot formation are some of these clinical findings. Equally as important as determining how insulin resistance exerts its negative effects on the cardiovascular system is learning if the cascade of events can be prevented with lifestyle changes or other interventions.

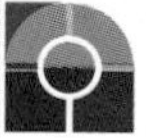

References

1 R. M. Krauss, Triglycerides and atherogenic lipoproteins: Rationale for lipid management, *American Journal of Medicine* (supplement) 105 (1998): 58–62.

2 S. M. Grundy, Pathogenesis of atherogenic dyslipidemia, *Drug Benefit Trends* 15 (2000): 22–27.

3 A. D. Baron, Insulin and the vasculature—Old actors, new roles, *Journal of Investigative Medicine* 44 (1996): 406–412.

4 D. B. Corry and M. L. Tuck, Pathogenesis of hypertension in diabetes, *Journal of Cardiovascular Pharmacology* (supplement 4) 28 (1996): 6–15.

5 Grundy, 2000; G. Imperatore and coauthors, Plasma fibrinogen: A new factor of the metabolic syndrome, *Diabetes Care* 21 (1998): 649–654.

6 G. Bruno and coauthors, Association of fibrinogen with glycemia control and albumin excretion rate in patients with non-insulin-dependent diabetes mellitus, *Annals of Internal Medicine* 125 (1996): 653–657.

7 Type 2 diabetes mellitus: New perspectives of an old problem, *Clinical Courier,* November 1999, pp. 1–7.

8 Baron, 1996.

Nutrition and Renal Diseases

CHAPTER 27

The kidneys produce urine, which travels through the ureters to the bladder where it is stored temporarily and then excreted from the body (see Figure 27-1). The kidneys certainly illustrate the old adage that good things come in small packages. Each bean-shaped kidney is only about the size of a fist, yet the kidneys carry out many critical functions. The kidneys shoulder much of the responsibility for maintaining the body's chemical balance. Figure 12-2 on p. 390 shows a **nephron**—one of the kidneys' functional units. Within the nephron, the **glomerulus** serves as a gate through which fluids and other blood components enter the nephron and form **filtrate.** As the filtrate passes through the **tubule,** some of its components are returned to the body, and others are excreted in the urine. Healthy kidneys, by continuously filtering the blood, maintain the body's fluid and electrolyte and acid-base balances and eliminate metabolic waste products.

In addition the kidneys:

- Help regulate blood pressure by secreting the enzyme **renin.** Renin triggers the release of the hormone **aldosterone,** which, in turn, signals the kidneys to retain sodium and water and raises blood pressure.
- Produce the hormone **erythropoietin,** which stimulates red blood cell production.
- Convert vitamin D to its most active form **(1,25-dihydroxy vitamin D)** and play an important role in maintaining healthy bone tissue.

The kidneys' functions are so vital to life that if the kidneys fail, life cannot continue for more than a few days without medical intervention. The glossary on p. 802 defines terms related to the kidneys and their functions.

KIDNEY STONES

Of the disorders that affect the kidneys and urinary tract, kidney or **renal** stones◆ are the most common. About 500,000 people in the United States develop kidney stones each year. Most of these people are men over 20 years old. Kidney stones are often quite painful, although sometimes they produce no symptoms. Stones often recur, but recurrences may be preventable.

◆ Kidney stones are also called **renal calculi.** Stones that have passed from the kidney to the ureter are technically urethral stones, but most people use the term *kidney stone* to describe stones anywhere in the urinary tract.
• **calculi** = pebble

renal: pertaining to the kidneys.

FIGURE 27-1 The Kidneys and Urinary Tract

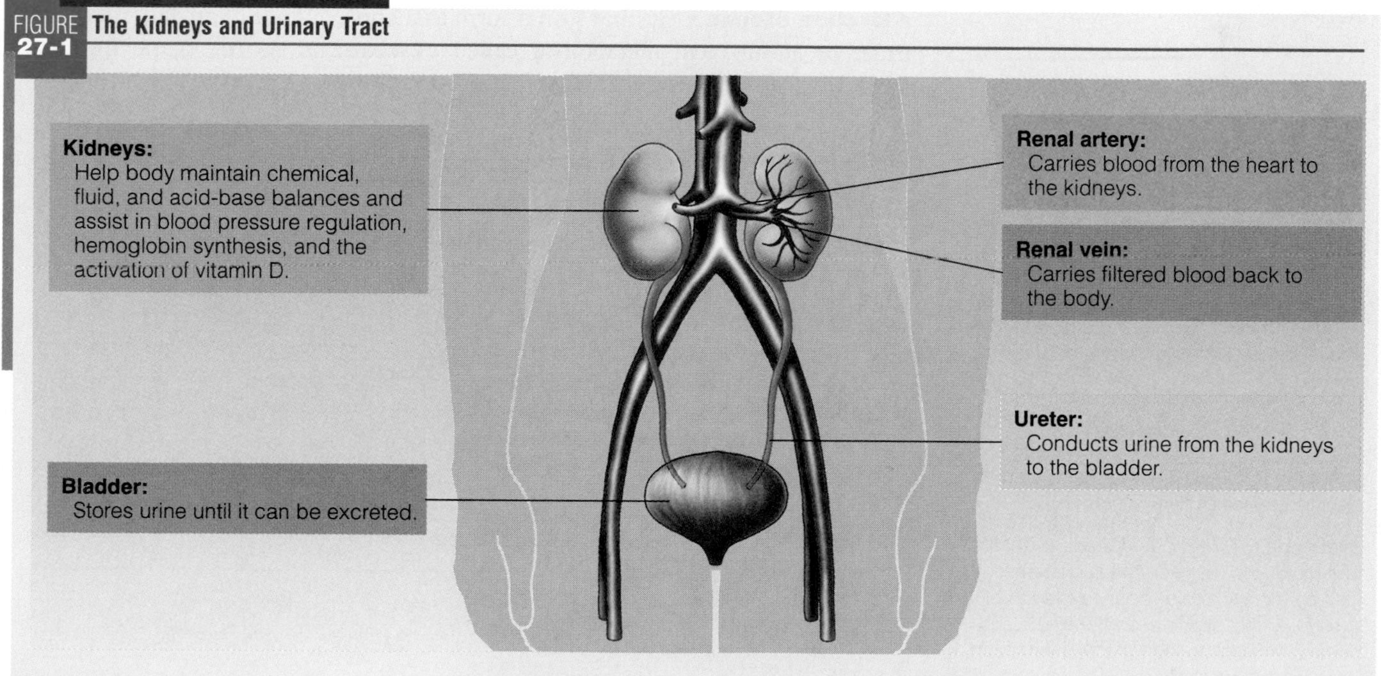

GLOSSARY OF KIDNEY-RELATED TERMS

aldosterone (al-DOS-ter-own or AL-dough-STEER-own): a hormone secreted from the adrenal glands that signals the kidneys to retain sodium and fluid.

erythropoietin (eh-RITH-row-POY-eh-tin): a hormone secreted by the kidneys in response to oxygen depletion or anemia that stimulates the bone marrow to produce red blood cells.
- **erthro** = red (blood cell)
- **poi** = to make

filtrate: in the kidneys, the fluid that passes from the blood through the capillary walls of the glomeruli, eventually forming urine.

glomerulus (glow-MARE-you-lus): a cup-shaped membrane enclosing a tuft of capillaries within a nephron. (The plural is *glomeruli.*)

nephron (NEF-ron): the working unit of the kidneys. Each nephron consists of a glomerulus and a tubule.

1,25-dihydroxy vitamin D: the active form of vitamin D (see Figure 11-8 on p. 364).

renin (REN-in): an enzyme secreted by the kidneys in response to a reduced blood flow that triggers the release of the hormone aldosterone.

tubule: a tubelike structure that surrounds the glomerulus and descends through the nephron. A pressure gradient between the glomerular capillaries and the tubule returns needed materials to the blood and moves wastes into the tubule to be excreted in the urine.

The New England Journal of Medicine October, 1992, p. 1142 with permission

The most common type of kidney stone is composed of calcium oxalate crystals, shown here.

◆ **Chapter 17 described how immobilization can lead to kidney stones, and Chapter 21 described how fat malabsorption can lead to kidney stones.**

Causes of Kidney Stones

Kidney stones develop when stone constituents become concentrated in the urine and form crystals that grow. The stone constituents vary, but more than three-fourths of them contain calcium◆ as calcium oxalate, calcium phosphate, or a combination of calcium, oxalate, and phosphate. Stones that contain calcium oxalate or a combination of calcium oxalate and phosphate are by far the most common. Less commonly, stones are composed of uric acid, the amino acid cystine, or magnesium ammonium phosphate. Table 27-1 lists some conditions associated with stone formation.

• **Calcium Stones** • Some people with calcium stones excrete normal amounts of calcium in the urine; others excrete excess calcium **(hypercalciuria).** The reason why people with normal amounts of calcium in their urine form calcium stones is unclear, although they may lack factors that normally inhibit stone formation in most people. People with hypercalciuria are either more efficient at absorbing calcium from the intestine or more wasteful in their excretion of calcium than most people.

• **Uric Acid Stones** • Uric acid stones are frequently associated with **gout,** a metabolic disorder characterized by elevated levels of uric acid in the blood and urine. Uric acid stones form when the urine becomes persistently acid, contains excessive uric acid, or both.

• **Cystine Stones** • Cystine stones form as a consequence of an inherited disorder of amino acid metabolism called **cystinuria.** As the name implies, cystinuria causes the abnormal excretion of cystine in the urine.

TABLE 27-1 Conditions Associated with Kidney Stones

Condition
Cystinuria
Fat malabsorption
Glucocorticoid excess
Gout
Hyperparathyroidism
Hyperthyroidism
Immobilization (prolonged bed rest or paralysis)
Malignancies (some types)
Osteoporosis
Paget's disease
Recurrent urinary tract infections
Renal tubular acidosis
Vitamin D toxicity

hypercalciuria (HIGH-per-kal-see-YOU-ree-ah): excessive urinary excretion of calcium.

gout: an inherited metabolic disorder that results in excessive uric acid in the blood and urine and the deposition of uric acid in and around the joints, which causes acute arthritis and joint inflammation.

cystinuria (SIS-tin-NEW-ree-ah): an inherited metabolic disorder that is characterized by the excessive urinary excretion of cystine, lysine, arginine, and ornithine and commonly leads to kidney stone formation.

• Magnesium Stones • Stones composed of magnesium ammonium phosphate, or **struvite,** are associated with recurring urinary tract infections. Urinary tract infections and struvite stones are more common in women than men.

Consequences of Kidney Stones

In most cases, kidney stones pose no serious medical problems, especially when they are few and small. Small stones (less than one-fifth of an inch in diameter) may readily pass through the ureters (see Figure 27-1) and out of the body via the urine with minimal treatment.

• Renal Colic • Large stones cannot pass easily through the ureters. When a large stone enters a ureter, it produces a sharp, stabbing pain, called **renal colic.** Typically, the pain starts suddenly in the back and intensifies as the stone follows the ureter's course down the abdomen toward the groin. The intense pain often occurs when the person moves about and may be accompanied by nausea and vomiting. When the stone reaches the bladder, the pain subsides abruptly.

• Urinary Tract Complications • Symptoms associated with kidney stones may include frequent urination, urgency of urination, **dysuria,** and **hematuria.** Stones that cannot pass through the ureter may cause a urinary tract obstruction or infection and serious bleeding.

www.comstock.com

Drinking plenty of water regularly throughout the day is the most important measure a person can take to prevent kidney stones.

Prevention and Treatment of Kidney Stones

Treatment of the underlying medical condition is necessary to help prevent recurrences of kidney stones. Specific treatment measures vary according to the composition of the stone, but advice to prevent stones always includes this recommendation: increase fluid intake to dilute the urine. People who have had kidney stones need to drink enough fluid (mostly water) to maintain a urine volume of at least 2 liters per day.◆ Generally, this requires a total intake of about 3 to 4 liters of fluid throughout the day. People who are physically active or live in warm climates may need additional fluids. People with fevers, diarrhea, or vomiting also need additional fluids until these conditions resolve. Once a stone has formed, drinking plenty of fluids (more than 3 liters a day) can sometimes help a small kidney stone to pass through the ureters.

◆ Water should comprise at least 50 percent of the fluid intake. The choice of other beverages may also be important. Coffee (both caffeinated and decaffeinated), tea, and wine may lower the likelihood of stone formation; grapefruit juice may raise the likelihood.[1]

THINK NUTRITION whenever a client has had a kidney stone. Drinking plenty of fluids and altering the diet may help prevent recurrences.

For people who have never had a kidney stone, a high intake of calcium from foods may actually lower the risk of developing kidney stones.[2] Using calcium supplements, especially if they are taken without meals or with foods low in oxalate,◆ may raise the risk of kidney stones. Whether it is the calcium from foods or some other constituent in the calcium-containing food that provides protection remains to be determined.

◆ Reminder: When calcium and oxalate are eaten together, they bind together and both are excreted, so eating calcium-containing foods and oxalate-containing foods together may benefit the person who forms calcium oxalate stones.

Preventing urinary tract infections is an important strategy for preventing struvite stones. Limited studies suggest that cranberry juice may help prevent urinary tract infections in women by preventing bacteria from adhering to the inner lining of the urinary tract.[3] People with urinary tract infections may also need to take anti-infective agents.

• Drug Therapy • For people who excrete too much calcium, thiazide diuretics help to reduce the amount of calcium excreted in the urine. People with uric acid stones are often treated with allopurinol, a medication that reduces urinary uric acid concentrations. Other medications may include agents that reduce urinary acidity. The Diet-Drug Interactions box on pp. 817–818 includes

struvite (STREW-vite): crystals of magnesium ammonium phosphate formed by the action of bacterial enzymes.

renal colic: the severe pain that accompanies the movement of a kidney stone through the ureter to the bladder.

dysuria (dis-YOU-ree-ah): painful or difficult urination.

hematuria (HE-mah-TOO-ree-ah): blood in the urine.

more information about medications used to treat kidney stones. Large stones that block the flow of urine or cause an infection require removal either surgically or, more commonly, by using shock waves to break the stone into pieces small enough to pass through the urinary tract (lithotripsy).

• Medical Nutrition Therapy for Calcium and Oxalate Stones • Medical nutrition therapy for people with hypercalciuria includes the recommendation to avoid excessive calcium intakes, but not to let calcium intakes fall below recommended intakes.◆ People with hypercalciuria who follow a low-calcium diet generally excrete more calcium than they ingest, indicating that they are losing calcium from their bones. Calcium from food sources should be encouraged. A person who cannot eat sufficient dietary calcium should be advised to use appropriate amounts of calcium supplements cautiously and to take them with meals. (Calcium restriction is not appropriate in the treatment of calcium stones unrelated to hypercalciuria, including stones related to fat malabsorption.)

◆ **For people with hypercalciuria who absorb too much calcium, calcium is restricted as follows:**
- 800 mg/day for most adults.
- 1,200 mg/day for pregnant and lactating women.
- 1,200 to 1,500 mg/day for postmenopausal women.

For people with hypercalciuria and normal calcium absorption, calcium is restricted to 1,000 mg/day. (See Figure 12-11 on p. 407 for the calcium content of selected foods.)

People with calcium oxalate stones are advised to limit their intakes of foods high in oxalate (see Table 27-2). **Hyperoxaluria** increases the likelihood of calcium oxalate stone formation more than hypercalciuria does. Megadoses of vitamin C over long times raise urinary oxalate concentrations, so people at risk for oxalate stones are advised to avoid vitamin C in excess of recommended amounts.

In addition to avoiding excessive calcium and restricting oxalate, limiting salt and including foods high in potassium may also be important in preventing calcium oxalate stones. Excess salt increases urinary calcium excretion in all people, but causes a proportionately greater amount of calcium to be excreted in people with hypercalciuria.[4] For people taking thiazide diuretics, moderate salt restriction and inclusion of high-potassium foods help limit the urinary excretion of both calcium and potassium. Some clinicians recommend a moderate protein intake (0.8 to 1.0 gram protein per kilogram of body weight per day) primarily from vegetable sources to limit urinary acidity.

• Uric Acid Stones • Diets restricted in **purine** are commonly prescribed to prevent uric acid stones. A purine-restricted diet limits red meats, particularly

TABLE 27-2 Foods High in Oxalate

Vegetables	Fruits	Other
Beans, green and wax	Blackberries	Chocolate and chocolate beverages*
Beets*	Blueberries	Cocoa
Celery	Currants, red	Coffee
Chard, swiss	Gooseberries	Draft beer
Collard greens	Grapes, Concord	Fruit cake
Dandelion greens	Lemon peel	Grits
Eggplant	Lime peel	Nuts, nut butters*
Endive	Orange peel	Peanut butter*
Escarole	Raspberries	Pepper
Leeks	Rhubarb*	Soybean crackers
Legumes	Strawberries*	Tea*
Okra		Tofu
Parsley		Wheat bran*
Potatoes, sweet		Wheat germ
Spinach*		
Squash, summer		

Note: The oxalate content of many foods has not been analyzed and even fewer studies have been conducted to determine which foods raise urinary oxalate. The foods marked with an asterisk have been documented to raise urinary oxalate and should be avoided by people who form calcium stones.

hyperoxaluria (HIGH-per-OX-all-YOU-ree-ah): excessive urinary excretion of oxalate.

purine (PU-reen): an end product of nucleotide metabolism that eventually breaks down to form uric acid.

organ meats, anchovies, sardines, and meat extracts. The benefits of such a diet are unproven, but avoiding excessive protein may be useful, and foods high in purine are also high in protein. Alcohol intake is also limited.

• **Cystine Stones** • The body synthesizes cystine, a nonessential amino acid, from methionine, an essential amino acid. Therefore, people with cystinuria benefit from a diet that provides enough methionine to meet the body's needs without providing too much. Medications to reduce urinary acidity may also be beneficial.

IN SUMMARY Kidney stones form when stone constituents become very concentrated in the urine. The majority of kidney stones contain calcium, usually as calcium oxalate. Drinking plenty of water throughout the day is the most important dietary measure for preventing kidney stones of any type. People who form calcium oxalate stones may need to moderately limit calcium and protein, avoid foods high in oxalate and sodium, and include high-potassium foods. Treatment of uric acid stones includes purine-restricted or moderate-protein diets, while the treatment of cystine stones includes a methionine-restricted diet.

THE NEPHROTIC SYNDROME

The **nephrotic syndrome** is not a disease, but rather a distinct cluster of symptoms caused by damage to the glomerular capillaries. The damage may occur as a consequence of diabetes mellitus, hypertension, infections (either of the kidneys or elsewhere in the body), immunological and hereditary disorders, chemicals (medications, illicit drugs, or contaminants), and some cancers to name a few. The damage alters the permeability of the glomerular capillaries and allows plasma proteins to escape in the urine **(proteinuria).** Along with proteinuria, low serum albumin, edema, and elevated blood lipids are typical clinical findings. The nephrotic syndrome is often an early sign of renal failure, especially in people with diabetes (see Chapter 25). In some cases, treatment of the underlying condition corrects the disorder.

Consequences of the Nephrotic Syndrome

The consequences of the nephrotic syndrome include protein-energy malnutrition (PEM), anemia, infection, blood coagulation disorders, and accelerated atherosclerosis. Many of the consequences of this disorder are the same as those of malnutrition, which is not surprising because both alter protein status. If the nephrotic syndrome progresses to renal failure, the person develops uremia and other manifestations, as a later section describes.

THINK NUTRITION whenever clients have the nephrotic syndrome. Attention to diet can help prevent PEM, reduce edema, and limit kidney damage.

• **Loss of Blood Proteins** • As plasma proteins escape through the urine, blood proteins plummet. Albumin, the major plasma protein, is also the major protein lost in the urine, and its blood level is markedly reduced. Because albumin acts as a carrier for a variety of nutrients, hormones, and mediations, blood levels of these may drop as well. Other proteins that are lost in the urine include immunoglobulins, transferrin, and vitamin D–binding protein. Losses of immunoglobulins render the person prone to infections. Loss of transferrin, the iron-carrying protein, may lead to anemia. When vitamin D–binding protein is lost in the urine, vitamin D deficiency may develop, which impairs

nephrotic (neh-FRAUT-ic) **syndrome:** the cluster of clinical findings that occur when glomerular function falters, including proteinuria, low serum albumin, edema, and elevated blood lipids.

proteinuria (pro-teen-YOUR-ee-ah): the loss of protein in the urine. The loss of albumin in the urine is *albuminuria.*

calcium absorption. Some calcium may also be lost in the urine along with the albumin that carries it. Consequently, rickets may develop, especially in children. If protein loss continues without replacement, lean body tissues break down, and PEM and general malnutrition follow. Figure 27-2 shows the consequences of falling levels of various proteins.

• **Edema** • For many years, clinicians believed that edema occurred in the nephrotic syndrome due to the loss of albumin, which helps maintain the pressure that draws fluid from tissues to the bloodstream. However, while plasma proteins are lost as a consequence of the nephrotic syndrome, sodium is reabsorbed by the nephrons in greater amounts than normal, and this change is believed to cause the edema.[5]

• **Altered Blood Lipids** • Elevated cholesterol, triglycerides, LDL, and VLDL are characteristic of the nephrotic syndrome. Blood coagulation disorders are also common, and people with the nephrotic syndrome frequently develop blood clots. All of these factors combine to raise the risk of cardiovascular disease and stroke. Blood clots can also form in the renal veins and further injure the kidneys.

Treatment of the Nephrotic Syndrome

Medical treatment of the nephrotic syndrome first requires treatment of the underlying disorder and then drug and medical nutrition therapy to resolve symptoms. Medications may include anti-infective agents, anticoagulants, antihypertensives, anti-inflammatory agents, antilipemics, and diuretics (see the Diet-Drug Interactions box on pp. 817–818). Diet is central to preventing PEM and alleviating edema.

• **Energy** • A diet adequate in energy (35 kcalories per kilogram body weight per day) sustains weight and spares protein. Weight loss or infections signal the need for additional kcalories. People who are obese may benefit from lowering their energy intakes to help control blood lipids and blood glucose (when necessary).

FIGURE 27-2 **Consequences of Urinary Protein Losses in the Nephrotic Syndrome**

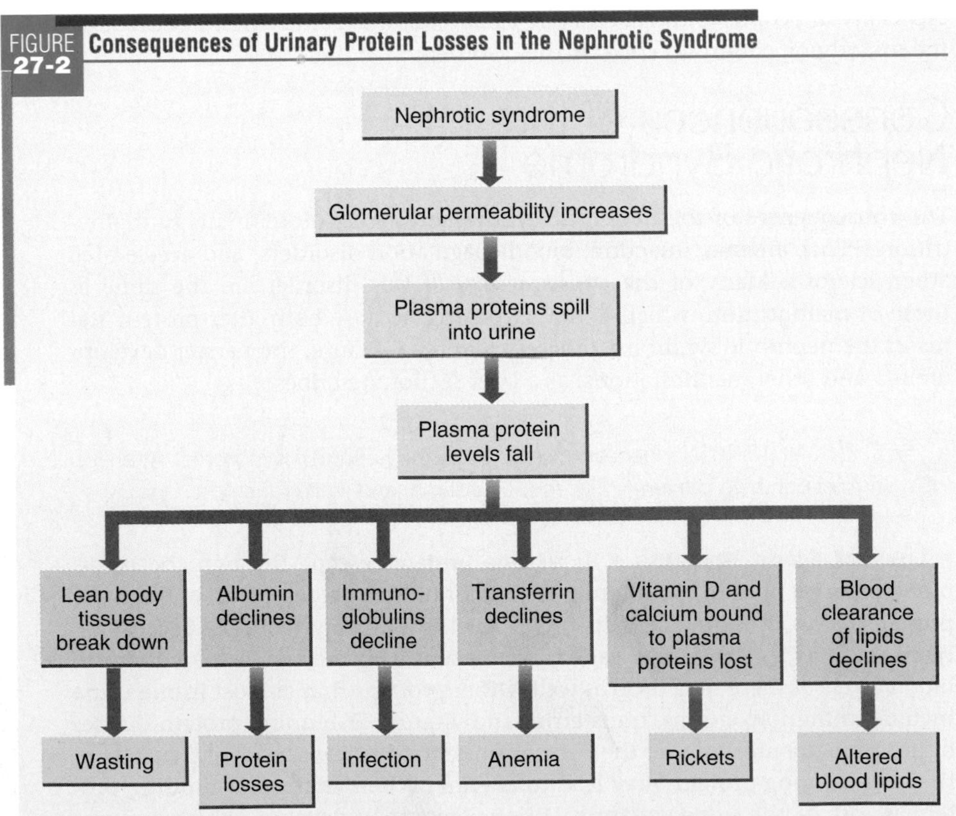

• **Protein** • In the past, high-protein diets (about 120 grams per day) were often prescribed for people with the nephrotic syndrome, stemming from the belief that high dietary protein would compensate for protein losses. High-protein diets accelerate protein synthesis, but they also incur greater urinary protein losses.[6] Furthermore, extra dietary protein may accelerate deterioration of renal function. For these reasons, protein is provided at about the RDA (0.8 to 1.0 gram per kilogram of body weight per day).

• **Fat** • A diet low in saturated fat, and cholesterol,◆ such as the plans described in Chapter 26, can help control the elevated blood lipids associated with the nephrotic syndrome. Often, however, people with the nephrotic syndrome are unable to control blood lipids adequately using diet alone, and physicians may prescribe antilipemic agents as well.

◆ Reminder: If blood lipids are elevated, an appropriate diet would provide less than 30% of the total energy from fat, with less than 7% saturated fat and less than 200 mg cholesterol per day.

• **Sodium** • A person with the nephrotic syndrome avidly retains sodium, and dietary sodium is generally restricted. Although the level of sodium restriction varies depending on the individual's response to diuretics, the diet often provides about 2 grams of sodium (see Table 27-3 on p. 808). Clients placed on thiazide or loop diuretics should be encouraged to select foods rich in potassium (see p. 405).

IN SUMMARY

The nephrotic syndrome describes the symptoms of proteinuria, edema, altered blood lipids, and altered blood clotting associated with damage to the glomerular capillaries. Losses of protein in the urine can lead to infections, anemia, rickets, PEM, and general malnutrition. Medications and a diet adequate in energy, moderate in protein, and low in sodium are the primary treatments.

ACUTE RENAL FAILURE

The kidneys may fail suddenly or deteriorate gradually. In **acute renal failure,** the nephrons suddenly lose function and are unable to maintain homeostasis. The degree of renal dysfunction varies from mild to severe. With prompt treatment, acute renal failure is often reversible; less often, the damage is permanent.

Acute renal failure frequently develops when blood flow to the kidneys suddenly declines, often as a result of a severe stress such as heart failure, shock, or severe blood loss after surgery or trauma.◆ In these cases, acute renal failure is imposed on catabolic illness, and mortality rates are high. When acute renal failure develops for other reasons, such as urinary tract obstructions, recovery rates are much better. Infections, toxins, and some drugs can directly damage the kidneys' cells and lead to acute renal failure.

◆ Reduced blood flow to the kidneys is a **prerenal** cause of acute renal failure. A urinary tract obstruction, which can be a consequence of a kidney stone, is a **postrenal** cause of acute renal failure. Direct damage to the kidneys' cells is an **intrarenal** cause of acute renal failure.

- **pre** = before
- **post** = after
- **intra** = within

Consequences of Acute Renal Failure

When acute renal failure accompanies severe stress, the complications of renal failure occur in addition to those incurred by the stress itself. Hyperglycemia, insulin resistance, and elevated triglycerides are common clinical findings. As renal function falters, the composition of the blood and urine changes. The rate at which the kidneys form filtrate, the **glomerular filtration rate (GFR),** declines, upsetting fluid and electrolyte and acid-base balances. Acute renal failure is characterized by a sudden and precipitous drop in the GFR and urine output. As the nephrons fail, the body's principal nitrogen-containing metabolic waste products—blood urea nitrogen (BUN),◆ creatinine, and uric acid—accumulate in the blood **(uremia).** The person who is also catabolic is in negative nitrogen balance and has large amounts of nitrogen to excrete. It is easy to see how uremia and its clinical manifestations—the **uremic syndrome**—can quickly develop under such conditions.

◆ Normal blood urea nitrogen (BUN) is 10 to 20 mg/dL. A BUN of 50 to 150 mg/dL indicates serious impairment of renal function. BUN may rise as high as 150 to 250 mg/dL in end-stage renal disease.

acute renal failure: the sudden loss of the kidneys' ability to function.

glomerular filtration rate (GFR): the rate at which the kidneys form filtrate, usually measured by determining the amount of creatinine excreted in 24 hours. The normal GFR is about 130 ml/min for males and 120 ml/min for females.

uremia (you-REE-me-ah): abnormal accumulation of nitrogen-containing substances, especially urea, in the blood; also called *azotemia* (AZE-oh-TEE-me-ah).

uremic syndrome: the cluster of clinical findings associated with the buildup of nitrogen-containing waste products in the blood, which may include fatigue, diminished mental alertness, agitation, muscle twitches, cramps, anorexia, nausea, vomiting, inflammation of the membranes of the mouth, unpleasant taste in the mouth, itchy skin, skin hemorrhages, gastritis, GI bleeding, and diarrhea.

TABLE 27-3 Two-Gram Sodium-Restricted Diet[a]

General Guidelines

About 75 percent of sodium in the typical diet comes from processed foods, and about 10 percent comes from unprocessed natural foods. About 15 percent of sodium in a typical diet comes from table salt. With this in mind:

- Choose fresh foods and foods frozen or canned without added salt.
- Avoid adding salt to foods while cooking.
- Avoid adding salt to foods at the table.
- When eating out, ask that meals be prepared without salt.

Sodium in Foods

All foods contain sodium, but some contain more than others. Use the information about the average sodium contents of foods to tailor the diet to the individual's preferences.

Food Group	Serving Size	Sodium (mg) per Serving
Fresh meats, poultry, freshwater fish; low-sodium canned meats and fish; low-sodium peanut butter and cheese; unsalted cottage cheese, soybeans, and textured vegetable protein	1 oz	20
Regular nonfat, low-fat, and whole milk and yogurt	8 oz	120
Eggs	1	60
Fresh artichokes, beets, carrots, and celery; beet, collard, dandelion, mustard, and turnip greens	½ c	50
Regular canned vegetables	½ c	300
Regular white and whole-grain bread	1 slice	150
Butter and margarine	1 tsp	50
Salt	½ tsp	1000

Other Foods

These foods can be used freely or with some limits with respect to sodium, although some of these foods may need to be restricted for weight or blood lipid control:

- All fruits and fruit juices.
- Low-sodium canned or frozen vegetables without added salt except those listed above; low-sodium vegetable juices.
- Low-sodium bread and bread products; puffed rice and wheat and shredded wheat cereals; rice; pasta.
- Soups, casseroles, and recipes made with allowed foods and ingredients.
- Unsalted butter, margarine, nuts, and gravy; low-sodium mayonnaise and salad dressing; shortening.
- Low-sodium catsup, mustard, tabasco sauce, and other condiments; low-sodium baking powder.

These foods and dishes prepared with them are high in sodium and should be avoided:

- Cured, canned, salted, or smoked meats, poultry, and fish such as bacon, luncheon meats, corned beef, kosher meats, and canned tuna and salmon; imitation fish products, salted textured vegetable protein, peanut butter, and nuts.
- Buttermilk, regular cheeses.
- Maraschino cherries; crystallized or glazed fruits; dried fruits with sodium sulfite added.
- Pickles, pickled vegetables, sauerkraut, and regular vegetable juices.
- Instant and quick-cooking hot cereals; commercial bread products made from self-rising flour or cornmeal; salted snack foods.
- Salt pork and bacon; commercial salad dressing; olives; regular gravy; catsup, baking powder, soy sauce, bouillon.

A Sample Two-Gram Sodium-Restricted Diet

Using the information above, many diet plans that meet individual needs are possible. Using the guidelines for a heart-healthy diet, a typical plan for a day might look like this:

Food Group	Sodium (mg)
Meat, 6 oz (6 × 25 mg)	120
Milk, 3 c (3 × 120 mg)	360
Fruit, 3 servings	negligible
Vegetables, ½ c vegetables with some sodium (1 × 50)	50
Vegetables, other vegetables and legumes, 2 servings	negligible
Whole-grain bread, 4 slices (4 × 125 mg)	600
Salted margarine, 6 servings (6 × 50 mg)	300
Total	1430

Clients can use the remainder of the sodium allowance for whatever foods they choose. The sodium content of other foods can be determined by reading food labels or using food composition tables. One client may choose to use some (¼ teaspoon) table salt or a favorite food that contains sodium.

[a]The 2400-milligram sodium diet described in Chapters 25 and 26 is more liberal than the diet shown here. The 2400-milligram sodium diet requires that a person choose unprocessed foods as much as possible and avoid table salt and highly salted foods.

• **Electrolyte Levels Rise** • When blood electrolyte levels rise and the kidneys are unable to restore balance, serious consequences can follow. In a person who is catabolic, electrolyte levels rise rapidly as the body's cells use protein for energy and release potassium, phosphorus, and magnesium in the process. Elevated potassium **(hyperkalemia)** is of particular concern because potassium imbalances can alter the heart rate and lead to acute heart failure.

• **Blood Volume Changes** • In the early stages of acute renal failure, blood pressure may rise dangerously high as fluids rapidly accumulate.◆ During this stage, clients, especially elderly clients, may develop fluid overload. Later in the course of acute renal failure, the kidneys cannot conserve water, and the person begins to excrete large amounts of fluid and electrolytes. If recovery occurs, kidney function gradually normalizes.

◆ The early phase of acute renal failure, when urine volume is reduced, is the **oliguric phase.** The phase characterized by large fluid and electrolyte losses in the urine is the **diuretic phase.** The gradual return of renal function marks the **recovery phase.**

Treatment of Acute Renal Failure

The primary goal in acute renal failure is to treat the underlying cause in order to prevent permanent or further damage to the kidneys. For example, if severe blood loss is the problem, a blood transfusion may be needed to restore blood volume. A combination of medical nutrition therapy, drug therapy, and **dialysis** may be undertaken to restore fluid and electrolyte balances and minimize blood concentrations of toxic waste products. (Highlight 27 describes dialysis and **continuous renal replacement therapy (CRRT),** a type of dialysis used specifically in the treatment of acute renal failure, and discusses why nutrient needs are affected by various dialysis procedures.)

THINK NUTRITION whenever clients develop acute renal failure. Attention to nutrition maximizes the chance of recovery and prevents complications.

• **Energy** • If the person in acute renal failure receives too little energy, body proteins break down, raising blood urea and potassium even higher and taxing the kidneys further. Thus wasting and malnutrition complicate recovery. The person's energy needs are highly variable and depend on the rate of catabolism, so indirect calorimetry provides the best estimate of energy needs.[7] When indirect calorimetry is not available, clinicians estimate energy needs and monitor the client's responses. One approach is to provide 30 to 45 kcalories per kilogram of body weight per day.[8]

• **Protein** • Although protein contributes nitrogen and can tax the kidneys, protein is necessary to prevent complications such as impaired wound healing, infections, muscle wasting, and negative nitrogen balance—complications that could prove fatal. Accordingly, clinicians often provide a higher protein intake, even if it means the person will need dialysis. Actual protein needs of people with acute renal failure depend on the degree of renal function, the metabolic rate, and nutrition status. Dietitians may use measurements of urea nitrogen◆ and body weight to determine a client's nitrogen balance and specific protein needs. In general, people who are not on dialysis receive about 0.6 to 1.0 gram of protein per kilogram of body weight per day. If dialysis is begun, a more liberal protein intake (1.0 to 1.5 grams per kilogram body weight) can be provided because nitrogenous waste products are removed during the procedure.[9] Protein needs are higher (1.5 to 2.0 grams per kilogram body weight) for people on continuous renal replacement therapy.

◆ To evaluate dialysis and guide diet therapy, renal teams often use a mathematical model that takes into account the kidneys' ability to clear urea and the person's protein catabolic rate. The technique is called **urea kinetic modeling.**

• **Fluids** • Fluid balances are carefully restored in clients who are either overhydrated or dehydrated. Thereafter, health care professionals determine fluid needs by measuring urine output and then adding about 500 milliliters for water lost through the skin, lungs, and perspiration. The person who is vomiting, has diarrhea, has a high fever, has a high-output fistula, or otherwise loses fluids has greater fluid needs. In the oliguric stage, the person needs small

hyperkalemia (HIGH-per-kay-LEE-me-ah): excessive potassium in the blood.

dialysis (dye-AL-ih-sis): removal of waste from the blood through a semipermeable membrane using the principles of simple diffusion and osmosis. The two main types are **hemodialysis** and **peritoneal dialysis** (see Highlight 27).

continuous renal replacement therapy (CRRT): a slow and continuous type of dialysis used in the treatment of acute renal failure.

amounts of fluids. In the diuretic stage, urine volume may increase significantly, and large amounts of fluids may have to be provided.

• **Electrolytes** • Sodium may be restricted to about 2 to 3 grams daily in the oliguric phase, but this may change as the person enters the diuretic phase. Generally, potassium and phosphorus are restricted as well; however, lab values must be monitored carefully, especially in those who are catabolic. Once the catabolic period of stress subsides, potassium and phosphorus may shift back into the intracellular fluid, and blood levels can plummet.

• **Enteral and Parenteral Nutrition** • Because clients with acute renal failure are often severely stressed, they frequently receive their nutrients from tube feedings or TPN. Special enteral and parenteral formulas for renal failure meet energy and protein needs in small volumes. Compared with standard enteral formulas, renal formulas have less protein, fewer electrolytes, and more kcalories per milliliter (see Appendix J). TPN formulas are compounded with mixtures of both nonessential and essential amino acids at lower concentrations and dextrose at higher concentrations than in standard TPN solutions. Electrolytes are added in appropriate amounts.

The carbohydrate-dense enteral and parenteral formulas used for acute renal failure may exacerbate the hyperglycemia and insulin resistance that frequently accompany both severe stress and renal failure. Fat can be used to add kcalories, reduce the need for carbohydrate, and limit the glucose load. Insulin must often be provided to lower blood glucose.

• **Drug Therapy** • In the oliguric phase of acute renal failure, diuretics may be used to mobilize fluids. Medications called potassium exchange resins must sometimes be used to treat hyperkalemia. These medications, provided by mouth or through an enema, cause sodium to be exchanged for potassium in the colon, and the potassium is then excreted in the stool.

As mentioned, insulin may be provided to help lower blood glucose. Insulin also lowers blood potassium and phosphorus in two ways. First, as insulin moves glucose into the cells, potassium and phosphorus follow. Second, as an anabolic hormone, insulin minimizes tissue breakdown, and consequently, the cells retain potassium and phosphorus. The Diet-Drug Interactions box on pp. 817–818 provides more information about nutrition and medications used in the treatments of renal diseases. The accompanying case study helps direct your thoughts toward the needs of a client with acute renal failure.

CASE STUDY

Store Manager with Acute Renal Failure

Mrs. Calley is a 35-year-old store manager admitted to the hospital's intensive care unit. She was first seen in the emergency room after she sustained multiple and severe injuries in an auto accident. She had lost so much blood she almost died before reaching the hospital. Her injuries include a fractured leg, broken ribs, a collapsed lung, and internal bleeding. Following emergency surgery to stop the internal bleeding and repair injuries, she developed acute renal failure. Mrs. Calley is 5 feet 3 inches tall and weighs 125 pounds.

Mrs. Calley has a urine volume of less than 50 ml/day and a BUN of 75 mg/dL. A test of GFR could not be performed due to the low volume of urine she was excreting.

1. Describe the most probable reason why Mrs. Calley developed acute renal failure. What other problems can cause acute renal failure? Describe the phases of acute renal failure.
2. What are Mrs. Calley's dietary needs in the early phase? What waste products and electrolytes are of greatest concern? Why? What factors do you have to keep in mind in determining Mrs. Calley's energy, protein, and fluid needs during acute renal failure? How will these needs change if dialysis is begun?
3. How will Mrs. Calley's nutrient needs change as she progresses to the second stage of acute renal failure? Why? As you read through the discussion of chronic renal failure, consider how Mrs. Calley's needs would change if her renal failure became chronic.

IN SUMMARY In acute renal failure, which most often occurs as a consequence of severe stress, the glomerular filtration rate falls suddenly. Nitrogen-containing metabolic waste products, fluids, sodium, potassium, phosphorus, and magnesium often rise in the blood as urine output falls. Nutrient needs depend on the degree of catabolism and on whether dialysis is a treatment.

CHRONIC RENAL FAILURE

Unlike acute renal failure, in which the GFR drops suddenly and sharply, in chronic renal failure the GFR gradually deteriorates, and renal function is irreversibly altered. Recall that many people who develop the nephrotic syndrome eventually develop chronic renal failure. Thus diabetic nephropathy, hypertension, certain infections,◆ and immunological and hereditary diseases can lead to chronic renal failure. Other causes include renal artery obstructions, atherosclerosis of the renal arteries, and congestive heart failure. In a few cases, renal function is permanently damaged following acute renal failure.

◆ Upper respiratory tract infections, especially those caused by streptococci, and viral infections, including hepatitis B virus (HBV), hepatitis C virus (HCV), and immunodeficiency virus (HIV), can lead to chronic renal failure.

Consequences of Chronic Renal Failure

In the early stages of chronic renal failure, the body compensates for the loss of some nephron function by enlarging the remaining functional nephrons. The hypertrophied nephrons work so efficiently that the GFR may fall to 75 percent of its normal rate before the symptoms of renal failure appear.◆

◆ The capacity of the kidneys to function despite loss of some nephrons is referred to as **renal reserve.** Normal values for GFR in males and females are about 130 and 120 ml/min, respectively. In either gender, a GFR of 56 to 100 ml/min constitutes mild renal failure, 25 to 55 ml/min constitutes moderate renal failure, and 24 ml/min or less constitutes severe renal failure.

The body eventually exhausts the overworked nephrons, and renal function deteriorates. (This effort is similar to that of the pancreatic beta cells in type 2 diabetes, which at first produce more and more insulin in response to high blood glucose and later become exhausted and unable to produce adequate insulin.) In **end-stage renal disease (ESRD),** the GFR drops below 20 percent of normal. **Renal insufficiency** describes the period in which kidney function has deteriorated but not to the point of ESRD. The consequences described next develop gradually and are pronounced in ESRD.

• Uremic Syndrome • As renal function deteriorates, nitrogen-containing waste products◆ accumulate in the blood, and the uremic syndrome (see p. 807) develops. The skin becomes dry and scaly, and the person may itch uncomfortably. Skin hemorrhages may be visible. In the later stages, urea (which can be excreted through the sweat) may crystallize on the skin, a symptom known as **uremic frost.** Nausea, vomiting, diarrhea, gastritis, and GI bleeding frequently accompany the uremic syndrome.

◆ Reminder: The nitrogen-containing waste products that accumulate in renal failure include blood urea nitrogen (BUN), creatinine, and uric acid.

• Blood Chemistry Alterations • In addition to retaining nitrogen-containing waste products, the body retains excess fluids, electrolytes, and acids◆ produced during metabolism and normally excreted in the urine. As a consequence, the person develops edema, electrolyte imbalances, and acidosis, which stress the cardiovascular and pulmonary systems. Elevated blood potassium can trigger irregular heartbeats (arrhythmias) and heart failure. Elevated phosphorus upsets the body's balance of phosphorus and calcium, often leading to bone disease.◆

◆ Reminder: *Acidosis* is the condition of having too much acid in the blood.

◆ Bone disorders resulting from calcium and phosphorus imbalances in renal disease are called **renal osteodystrophies** (OS-tee-oh-DIS-tro-fees).

Hormonal balances are also upset. Altered hormone levels together with the retention of waste products lead to hypertension, hyperglycemia, elevated blood lipids (especially triglycerides), low serum albumin, anemia, and altered bone metabolism.

• Other Complications • Accelerated atherosclerosis and cardiovascular diseases frequently accompany renal failure, and they can aggravate hypertension and raise the risk of congestive heart failure, heart attacks, and pulmonary edema. Anemia may result from depressed erythropoietin synthesis by the

end-stage renal disease (ESRD): the severe stage of chronic renal failure in which dialysis or a kidney transplant is necessary to sustain life. In ESRD the GRF falls to less than about 25 ml/min, and the BUN may rise as high as 150–250 mg/dL.

renal insufficiency: the stage of renal failure in which renal function is reduced but not to a life-threatening degree.

uremic frost: the appearance of urea crystals on the skin.

damaged kidneys, restrictive diets, nausea and vomiting, GI blood losses, and blood losses through dialysis and from frequent blood testing. Inability to activate vitamin D and poor nutrient intakes of calcium contribute to the bone disease that often develops.

• **Growth Failure and Wasting** • Both children and adults with chronic renal disease frequently develop wasting and PEM. Nutrition status becomes more difficult to maintain as renal disease progresses. Table 27-4 summarizes the causes of wasting associated with renal failure. Children with renal disease need nutrition intervention before the end of puberty if they are to make up growth deficits. Adults with renal disease can maintain or restore their nutrition status, avoid complications, and improve quality of life by attending to diet as well.

THINK NUTRITION **whenever clients develop chronic renal failure. Early nutrition intervention can help delay the progression of renal failure and control its complications.**

Treatment of Chronic Renal Failure

Treatment of chronic renal failure aims to delay the progression of the disorder for as long as possible and improve quality of life. Some of the measures include medications to control blood pressure (antihypertensives), blood lipids (antilipemics), blood glucose (in people with diabetes), and phosphorus (phosphate binders) and dietary protein and phosphorus restrictions. Other medications may be used to control symptoms or prevent complications; these medications include antiemetics, antisecretory agents, cardiac glycosides, diuretics, and potassium exchange resins. Once renal failure reaches its end stage, dialysis or a kidney transplant is necessary to sustain life.

Table 27-5 summarizes the changes in nutrient needs as renal insufficiency progresses to ESRD. Note that the two dialysis procedures (hemodialysis and peritoneal dialysis) affect nutrient needs differently. The physician monitors the client's renal function and medical status to determine the appropriate diet prescription.

The complexity of the renal diet, as well as its critical role in the treatment of renal disease, underscores the need for a specialist, a renal dietitian, to educate clients and provide diet plans. Other health care professionals do not need to know the specifics of medical nutrition therapy, but they must understand the general concepts in order to communicate effectively with clients.

• **Energy** • All people with renal failure need adequate energy to maintain a desirable weight and prevent protein catabolism. As renal failure progresses, growth failure and wasting become more likely, and consuming adequate en-

TABLE 27-4 Possible Causes of Wasting in Renal Failure

Reduced Nutrient Intake	Excessive Nutrient Losses	Raised Nutrient Needs
Anorexia	Dialysis	Hormonal alterations
Fatigue	Diarrhea	Infection
Medications	GI bleeding	Inflammation
Nausea	Medications	Medications
Pain	Numerous blood tests	
Restrictive diet	Poor absorption	
Taste alterations	Vomiting	

TABLE 27-5 Nutrient Needs in Chronic Renal Failure

Nutrients[a]	Renal Insufficiency (Predialysis)	Hemodialysis	Peritoneal Dialysis
Energy (kcal/kg)[b]	30–40	30–35	25–35
Protein (g/kg)	0.6–0.8	1.2–1.4	1.2–1.5
Fluid (ml)	Typically not restricted	500–750 plus daily urine output, or 1000 if anuric	≥2000
Sodium (g)	2–4	2–3	3–4
Potassium (mg/kg)	Typically not restricted	40	Typically not restricted
Phosphorus (mg)	<1200[c]	800–1200[c]	1200[c]
Supplements			
Calcium (mg)	1000–1500	1000–1500	1000–1500
Folate (mg)	1	1	1
Vitamin B_6 (mg)	5	10	10
Vitamin D	As appropriate	As appropriate	As appropriate

Note: The actual amounts of these nutrients in the diet must be highly individualized based on each person's responses. For example, energy needs in renal insufficiency may be lower for people who are overweight or higher for people who are underweight.
[a]Besides the specific nutrients listed, all others should meet recommended amounts.
[b]For children, an intake of 100 kcalories per kilogram of body weight is desirable, but 80 kcalories per kilogram of body weight is good. At a minimum, energy intake should not fall below the RDA.
[c]The extent of phosphorus restriction depends on serum phosphorus. The goal is to maintain serum phosphorus between 4.5 and 6.0 milligrams per deciliter. Often, phosphate binders are useful for this purpose.
SOURCES: Adapted from Meeting the challenge of the renal diet: A preview of the "National Renal Diet" educational series, *Journal of the American Dietetic Association* 93 (1993): 637–639; J. A. Beto, Which diet for which renal failure: Making sense of the options, *Journal of the American Dietetic Association* 95 (1995): 898–903.

ergy becomes more difficult. Nausea, vomiting, anorexia, taste alterations, and fatigue that may occur as a consequence of the uremic syndrome can all reduce food intake. As additional dietary restrictions are imposed, people are challenged to find enough foods they can eat and enjoy.

Notice from Table 27-5 that energy needs are slightly lower once dialysis, especially peritoneal dialysis, begins. This is because the person obtains some glucose from fluids **(dialysate)** used to remove waste products (Highlight 27 provides the details). Energy needs are based on body weight, and people on dialysis retain fluid between treatments. For people on dialysis, body weight immediately following a dialysis treatment **(dry weight)** provides the best estimate of true body weight.

• Protein • Controlling protein (nitrogen) intake is a primary goal in the treatment of chronic renal failure. Providing the right amount of protein to the person in renal failure, however, is like walking a tightrope. Too little protein, and the person develops malnutrition. Too much protein, and blood urea (the toxic waste product of protein metabolism) rises. For people with renal insufficiency, restricting protein may help protect the remaining nephrons, although human studies have not clearly demonstrated a beneficial effect.[10] As renal insufficiency progresses,◆ dietary protein restrictions tighten.[11] When protein intakes fall below the RDA recommendation, consuming adequate energy to spare protein and eating high-quality protein become vitally important. Clinicians generally recommend a protein-restricted diet that derives at least half of its protein from sources such as eggs, milk, meat, poultry, and fish. The remaining protein is derived from plant sources. A diet that includes protein from both animal and plant sources may be best because it combines the high quality of animal proteins with the low–saturated fat, low-cholesterol nature of plant proteins.

As renal failure progresses, some clinicians prescribe very-low-protein diets (0.3 grams protein per kilogram of body weight per day) supplemented with essential amino acids or essential amino acid precursors (keto acids). Once the

◆ For people with early renal insufficiency (GFR less than 55 but greater than 25 ml/min), protein is restricted to 0.8 g/kg/day (the RDA). For people with more advanced renal insufficiency (GFR less than 25 ml/min), protein is restricted to 0.6 g/kg/day until dialysis begins (see Table 27-5).

dialysate (dye-AL-ih-SATE): a solution used during dialysis to draw wastes and fluids from the blood.

dry weight: weight after excess fluids are removed from the body.

© Craig M. Moore

When a diet must be restricted in protein, it is important that the protein that is consumed be of high quality, including protein from eggs, milk, meat, poultry, and fish.

◆ Foods and other substances that are considered part of the fluid allowance on fluid-restricted diets include:
- Cream.
- Frozen yogurt.
- Fruit ice.
- Gelatin.
- Ice.
- Ice cream.
- Ice milk.
- Liquid medications.
- Popsicles.
- Sherbet.
- Soup.

◆ Caution: People with renal failure should avoid salt substitutes that contain potassium and products made from such salt substitutes, including many low-sodium soups and low-sodium baking powder. Most fruits and vegetables provide potassium as well. Lists that separate fruits and vegetables according to their potassium contents enable clients to know which fruits and vegetables to avoid or limit. A few examples include:
- Apricots
- Bananas.
- Chard.
- Kiwi.
- Melons.
- Oranges.
- Potatoes, especially white potatoes.
- Pumpkin.
- Spinach.
- Squash.

◆ Phosphorus appears in many high-protein foods including milk and milk products, eggs, cheese, peanut butter, sardines, and legumes. Bran cereal is also high in phosphorus.

◆ The active form of supplemental vitamin D, or **calcitriol,** can be given orally or intravenously.

client begins dialysis, protein restrictions can be relaxed, because dialysis removes nitrogenous waste products and incurs protein losses.

• **Lipid and Carbohydrate** • The ideal renal diet restricts total fat, saturated fat, and cholesterol to help control elevated blood lipid levels. A diet rich in complex carbohydrates helps minimize elevated blood glucose and triglycerides. People on peritoneal dialysis (see Highlight 27) need to further restrict their intake of total carbohydrate, and especially simple carbohydrates, because they absorb a considerable amount of glucose as a consequence of the dialysis procedure.

• **Sodium and Fluids** • As renal failure progresses, the person excretes less urine and cannot handle normal amounts of sodium and fluids. At this point, limiting sodium and fluids helps to prevent hypertension, edema, and heart failure. Individual needs for sodium and fluids are determined by carefully monitoring each person's weight, blood pressure, urine output, and blood electrolyte levels. A rapid rise in body weight and blood pressure suggests that the person is retaining sodium and fluid; conversely, a rapid decline in body weight and blood pressure (a desirable outcome of dialysis) indicates fluid loss.

Fluids are not restricted in renal insufficiency until urine output decreases. After that time, fluids are restricted.◆ For the person who is neither dehydrated nor overhydrated, daily fluid needs amount to the daily urine output plus 500 to 750 milliliters to provide for obligatory water losses. Once a person is on dialysis, sodium and fluid intakes are controlled to allow a weight gain of about 2 pounds (of fluid) between dialysis treatments, although larger weight gains are common.[12]

• **Potassium** • Many people with renal insufficiency and those on peritoneal dialysis can handle typical, but not excessive, intakes of potassium (about 2.5 to 3.5 grams per day)◆ until urine output falls below one liter per day. Some people with diabetic nephropathy may experience hyperkalemia earlier in the course of renal failure. People on hemodialysis may develop hyperkalemia between dialysis treatments. In such cases, potassium may be moderately restricted to about 1.5 to 3.0 grams per day. Remember, however, that individual needs vary. People taking thiazide and loop diuretics may need to adjust their potassium intakes accordingly.

• **Phosphorus, Calcium, and Vitamin D** • Controlling blood phosphorus and calcium in renal failure may help slow its progression and help prevent bone diseases. As renal function declines, blood phosphorus levels rise and levels of active vitamin D fall. The low amount of active vitamin D limits calcium absorption, and the combination of high blood phosphorus and limited availability of calcium leads to the loss of calcium from the bones and consequent bone disorders.

Dietary phosphorus restrictions◆ are instituted early in the course of renal failure to help control rising blood phosphorus. Fortunately, protein-restricted diets are restricted in phosphorus as well. As renal failure progresses and especially when protein is liberalized, physicians often prescribe medications that bind phosphorus in the GI tract and make it unavailable for absorption. Calcium and aluminum salts, and most recently a binder that contains neither calcium nor aluminum, can be used to bind phosphorus. The use of aluminum-containing phosphate binders is discouraged because aluminum toxicity is a problem for many people on dialysis.

In addition to restricting phosphorus, most people with renal disease need calcium supplements. Supplemental vitamin D in its active form◆ can help maintain blood calcium and prevent bone disease. Some people, however, develop hypercalcemia. The renal team monitors serum calcium closely to prevent both low and high blood calcium and prescribes calcium and vitamin D supplements accordingly. Calcium-containing phosphate binders provide some calcium, but the absorption of calcium from these products varies widely.

For people who tend to develop hypercalcemia, the phosphate binders that do not contain calcium may be useful.

• Other Vitamins • People with renal failure frequently develop folate and vitamin B_6 deficiencies because of restrictive diets, loss of vitamins during dialysis, drug therapy, and altered metabolism. Medical nutrition therapy often includes folate and vitamin B_6 in generous amounts, along with the recommended amounts of the remaining water-soluble vitamins.[13] Adequate amounts of folate and vitamins B_6 and B_{12} may also help to lower serum homocysteine levels (see Chapter 26), which may help to protect against cardiovascular disease in people with renal failure. Furthermore, these vitamins are important in preventing anemia.

Because many people with renal failure have high blood oxalate levels, intakes of vitamin C from both the diet and supplements are often limited to less than 100 milligrams per day.[14] When oxalate levels are elevated, oxalate crystals can become deposited in soft tissues and result in complications including kidney stones and heart attacks. Supplementation of fat-soluble vitamins other than vitamin D is usually not necessary.

© Craig M. Moore

In limited amounts, the fruits and vegetables pictured here provide a level of potassium acceptable for renal diets.

• Trace Minerals • The administration of human erythropoietin along with the B vitamins and iron needed to synthesize hemoglobin is effective in treating iron-deficiency anemia, once a common and persistent problem in people with chronic renal disease. Clients should be cautioned to avoid iron supplements that also contain vitamin C.

For people with renal failure, aluminum and magnesium can reach toxic levels. Therefore, clients should avoid aluminum- and magnesium-containing antacids as well as supplements, laxatives, and enemas containing magnesium.

People on dialysis frequently complain of anorexia, altered taste perceptions (dysgeusia), and a decreased sexual drive, symptoms typical of zinc deficiency. Clients who have these symptoms may need zinc supplements if their serum zinc levels are inadequate. Recall that children have particularly high needs for zinc, and zinc deficiencies can contribute to growth retardation.

• Enteral and Parenteral Nutrition • Enteral formulas and parenteral nutrition can provide nutrients to people with chronic renal failure who are unable to eat adequate amounts of foods. Formulas designed for use in renal failure are described on p. 810.

• Diet Planning • The ideal renal diet presents a challenge: provide adequate energy, but restrict protein, fat, and, sometimes, simple carbohydrate. Complex carbohydrates, which might appear to be the ideal energy source, are often also rich sources of potassium. Because potassium must be restricted, complex carbohydrates must be used cautiously. Diet planners must accept that under such circumstances, no diet is truly ideal. They must recognize that the need to adjust protein and electrolytes outweighs the need to restrict fat and simple carbohydrate.

To help meet energy needs, clients include as many complex carbohydrates as their diet plans allow. They supplement their meals with formulas high in kcalories but restricted in protein and electrolytes and use foods such as sugars (hard candy and jelly) and fats (margarine and oil) freely. The person with elevated blood lipids is advised to restrict fat and modify the type of fat to whatever extent is possible. The person with diabetes or hyperglycemia is advised to eat a consistent carbohydrate intake at regular intervals and to adjust insulin to cover carbohydrate intake.

To help individuals on renal diets find foods they will accept and enjoy, food lists similar to the exchange system are available. Whereas the exchange system for diabetes groups foods by their energy, carbohydrate, protein, and fat contents, renal food lists group foods by their energy, protein, sodium, potassium, and phosphorus contents. The "How to" on p. 816 provides suggestions to ease the task of complying with a renal diet.

Help Clients Comply with a Renal Diet

The following suggestions can assist clients in complying with the renal diet:

1. To keep track of fluid intake:
   - Fill a container with an amount of water equal to your total fluid allowance. Each time you use a liquid food or beverage, discard an equivalent amount of water from the container. The amount remaining in the container will show you how much fluid you have left for the day.
   - Be sure to save enough fluid to take medications.
2. To help control thirst:
   - Chew gum or suck hard candy.
   - Freeze fluids so they take longer to consume.
   - Add lemon juice to water to make it more refreshing.
   - Gargle with refrigerated mouthwash.
3. To prevent the diet from becoming monotonous:
   - Experiment with new combinations of allowed foods.
   - Use favorite foods whenever possible.
   - Substitute nondairy products for regular dairy products. Nondairy products are lower in protein, phosphorus, and potassium than regular dairy foods, and they can substitute for milk and add energy to the diet.
   - Add zest to foods by seasoning with garlic, onion, chili, curry powder, oregano, pepper, or lemon juice.
   - Consult a dietitian when you want to eat restricted foods. Many restricted foods can be used occasionally and in small amounts if the diet is carefully adjusted.

• Diet Compliance • The challenges dietitians face in designing renal diets pale in comparison with those encountered by clients and caregivers who must follow a complicated medical plan of which diet is only a part. To help clients understand their medical plans and support their efforts, the members of the health care team must effectively communicate with their clients and, just as important, listen to them. Within a tangled web of grossly altered metabolic processes, dialysis lines, and toxic waste products is a person, often a frightened or discouraged one. All members of the health care team need to understand the plan if they are to offer the most effective support.

A nutrition education approach that emphasizes self-management appears to be a useful strategy for helping clients comply with a renal diet.[15] This approach guides clients in identifying goals, selecting strategies to meet goals, and evaluating progress (see Highlight 25).

Kidney Transplants and Diet

A preferable alternative to dialysis in end-stage renal disease is a kidney transplant. Kidney transplants can successfully restore kidney function and promote normal growth. For this reason, transplants are particularly desirable in children. Given a choice, many would prefer transplants, but suitable kidney donors cannot always be found. Of the more than 35,000 people awaiting transplants, only about 11,000 will receive them.[16]

• Immunosuppressive Drug Therapy • After receiving a new kidney, the person must take very large doses of immunosuppressive medications to prevent tissue rejection (see the Diet-Drug Interactions box). Infections and increased susceptibility to malignant tumors are common adverse effects of these medications. Muscular weakness, GI bleeding, protein catabolism, carbohydrate intolerance, sodium retention, fluid retention, hypertension, weight gain, and a characteristic puffy-faced appearance commonly accompany immunosuppressant therapy. Diuretics are frequently prescribed to promote the excretion of fluid and sodium and to prevent hypertension.

• Diet Interventions • Once recovery is under way, the degree of renal function guides diet therapy. Typical post-transplant diet modifications appear in Table 27-6. Protein is provided in amounts adequate to prevent the protein catabolism that immunosuppressants may incur, but not so much as to tax renal function. Because blood lipids are frequently elevated, clients are advised to follow a fat-modified diet. Sodium restrictions help prevent fluid retention and hypertension. Depending on the type of diuretic prescribed and the person's responses, potassium intake may have to be adjusted.

A person with a kidney transplant may reject the new kidney either temporarily or permanently. During these times, dialysis must be reinstituted, and the person must return to the prescribed diet for renal failure. Clients may find this regression difficult to accept and should be prepared for the possibility before it occurs. The accompanying case study helps you apply information about chronic renal disease and kidney transplants.

TABLE 27-6 Dietary Guidelines following a Kidney Transplant

Nutrient	Guideline
Energy	Adequate to achieve or maintain desirable body weight.
Protein	1 gram per kilogram of body weight. (Adjust based on renal function tests.)
Fat	≤30 percent of total kcalories; ≤10 percent saturated fat; ≤300 milligrams cholesterol.
Sodium	3 to 4 grams per day.
Potassium	Adjust according to individual needs.

SOURCE: Adapted from J. A. Beto, Which diet for which renal failure: Making sense of the options, *Journal of the American Dietetic Association* 95 (1995): 898–903.

CASE STUDY

Child with Chronic Renal Failure

Jason is a nine-year-old child who developed chronic renal failure after suffering from a streptococcal infection (strep throat). Jason's renal function has declined steadily over the last three years, and he is currently receiving hemodialysis three times a week. A search is on for a suitable kidney donor. Jason is 4 feet 3 inches tall and weighs 55 pounds. He follows a fluid-, sodium-, potassium-, and phosphorus-restricted diet. His typical energy intake is about 1000 kcalories per day.

1. Describe chronic renal failure. What happens to the GFR and BUN as renal function declines? What determines when renal failure reaches its end stage?
2. Look at Jason's height and weight. Use a growth chart to determine how Jason's height and weight compare to those of other children of the same age. (see Appendix E.) Discuss reasons why growth may be compromised in a child with renal failure.
3. Describe the reasons for each of Jason's dietary restrictions. Calculate Jason's energy needs and compare them to his typical energy intake. Consider the effect of Jason's energy intake on his growth. What other nutrients are important to consider in Jason's diet?
4. Consider the ways Jason's diet will change if he undergoes a successful kidney transplant.
5. Discuss some diet strategies you can suggest to Jason and his parents to help him comply with his diet. Consider the impact of renal disease on Jason, his family, and his interactions with friends.

IN SUMMARY

Chronic renal failure develops gradually from a variety of disorders that permanently damage the kidneys. Diet plans may control energy, protein, fluid, sodium, potassium, phosphorus, and calcium. When end-stage renal disease develops, dialysis or a kidney transplant is necessary to sustain life. The Diet-Drug Interactions box describes nutrition-related concerns of medications used in the treatment of disorders of the kidneys and urinary tract. Use the Nutrition Assessment Checklist to review assessment findings relevant to people with disorders of the urinary tract and kidneys.

Diet-Drug Interactions

ANTICOAGULANTS

For *anticoagulants,* see p. 794.

ANTIEMETICS

For *antiemetics,* see p. 644.

ANTIGOUT

Allopurinol and *colchicine* should both be given with meals. Nutrition-related side effects of allopurinol are uncommon, but colchicine can cause nausea, vomiting, GI distress, and diarrhea. Colchicine reduces the absorption of vitamin B_{12}, and the person may need to take vitamin B_{12} supplements.

ANTIHYPERTENSIVES

For *antihypertensives,* see p. 791.

ANTI-INFECTIVE AGENTS

Numerous *anti-infective agents* may be used in the treatment of kidney stones, nephrotic syndrome, and acute or chronic renal failure. Consult a current drug guide for specific interactions.

ANTILIPEMICS

For *antilipemics,* see p. 791.

ANTISECRETORY AGENTS

For *antisecretory agents,* see p. 644

CARDIAC GLYCOSIDES

For *cardiac glycosides,* see p. 791.

DIURETICS

For *diuretics,* see p. 791.

EXCHANGE RESINS

Cellulose sodium phosphate (prescribed for people with hypercalciuria who absorb too much calcium) can cause altered tastes, GI distress, and diarrhea and can lead to low blood levels of magnesium. Clients should take the medication with meals and should not take a magnesium supplement for at least one hour before or after taking the resin. *Sodium polystyrene* (prescribed to reduce blood potassium levels) can cause anorexia, nausea, vomiting, and constipation and low blood levels of potassium (intentional) and calcium. The powder can be mixed with sorbitol-containing syrup to combat constipation. Calcium-containing antacids or calcium supplements should not be taken within several hours of taking sodium polystyrene.

IMMUNOSUPPRESSANTS

Immunosuppressants have multiple effects on many organ systems, may be toxic to the kidneys and liver, and can lead to many problems including nausea, vomiting, and a high risk of infections. *Cyclosporine* can elevate blood potassium, and it should not be given with grapefruit or grapefruit juice (unless prescribed). The person taking cyclosporine should not use potassium supplements or salt substitutes containing potassium. *Azathioprine* can also lead to pancreatitis.

Lymphocyte immune globulin and muromonab-cd3 can lead to pulmonary edema.

See also *anti-inflammatory agents* on p. 673.

PHOSPHATE BINDERS

For *calcium acetate, calcium carbonate, calcium citrate, aluminum carbonate,* and *aluminum hydroxide,* see calcium- and aluminum-containing antacids on p. 644. *Sevelamer hydrochloride* does not contain calcium or aluminum. It is given with meals and can cause nausea, flatulence, GI distress, diarrhea, and, less frequently, constipation.

Nutrition Assessment Checklist for People with Renal and Urinary Tract Disorders

MEDICAL HISTORY

Check the medical record to determine:

- ☐ Type of kidney stone
- ☐ Degree of renal function
- ☐ Cause of nephrotic syndrome or renal failure
- ☐ Type of dialysis, if appropriate
- ☐ If client has received a kidney transplant

Review medical record for complications that may alter nutrient needs:

- ☐ PEM
- ☐ Severe stress
- ☐ Infection
- ☐ Anemia
- ☐ Diabetes mellitus
- ☐ Hyperlipidemia
- ☐ Hypertension
- ☐ Heart failure (acute or chronic)

MEDICATIONS

For clients with kidney stones, note diet-medication interactions for medications taken by the client, including:

- ☐ Anti-infective agents
- ☐ Diuretics (see Chapter 26)
- ☐ Exchange resins (cellulose sodium phosphate)
- ☐ Antigout agents

Note that clients with nephrotic syndrome, renal insufficiency, or renal failure and those who have had a kidney transplant risk medication-related malnutrition for many reasons, including:

- ☐ Long-term use of medications
- ☐ Multiple medication use with many of the medications having significant effects on nutrition status
- ☐ Altered renal function, which compounds nutrition risks
- ☐ Preexisting malnutrition due to the disorder itself and complications of the disorder
- ☐ Reduced food intake, altered digestion and absorption, altered metabolism, and the altered excretion of nutrients due to the medications as well as the disorders themselves

For all clients with kidney diseases, note:

- ☐ Use of over-the-counter medications and supplements that may contain electrolytes that must be controlled
- ☐ Use of herbs and other remedies, which can have a significant impact on clients who suffer from malnutrition and renal insufficiency or renal failure and who use multiple medications

FOOD/NUTRIENT INTAKE

For people with kidney stones or a past history of kidney stones:

- ☐ Stress the importance of drinking plenty of fluids regularly throughout the day.
- ☐ Assess intake of calcium, oxalate, salt, and protein as appropriate for type of stone.

For people who wish to try cranberry juice cocktail to prevent urinary tract infections, explain that:

- ☐ Studies are limited and they were conducted on older women; but, at worst, cranberry juice cocktail is not harmful.
- ☐ The amount of cranberry juice cocktail used by study participants was 10 ounces per day.
- ☐ A 10-ounce serving of regular cranberry juice cocktail has about 180 kcalories, while the same serving of low-kcalorie cranberry juice has about 60 kcalories.

For clients with the nephrotic syndrome, renal insufficiency, or renal failure, or those who have undergone kidney transplants, regularly assess intakes of:

- ☐ Energy
- ☐ Protein
- ☐ Sodium
- ☐ Potassium

In addition, for clients with renal insufficiency or renal failure, assess intakes of:

- ☐ Fluid
- ☐ Phosphorus
- ☐ Calcium
- ☐ Vitamins
- ☐ Minerals

ANTHROPOMETRICS

Take accurate baseline height and weight measurements. Keep in mind that:

- ☐ Fluid retention in people with the nephrotic syndrome or renal failure can mask malnutrition.
- ☐ For people on dialysis, the weight measured immediately following a dialysis treatment, called the "dry weight," most accurately reflects the person's true weight.
- ☐ Rapid weight gain between dialysis treatments often reflects fluid retention. For clients who regularly have problems with fluid retention, review fluid intake to ensure that the client understands and is complying with diet recommendations.
- ☐ Weight loss is expected and intentional following a dialysis treatment.

LABORATORY TESTS

Note that serum protein levels are often low in people with the nephrotic syndrome or renal failure. Review the following laboratory test results to assess degree of renal function and response to treatments:

- ☐ Glomerular filtration rate (GFR)
- ☐ Creatinine
- ☐ Blood urea nitrogen (BUN)
- ☐ Electrolytes

Check laboratory test results for complications associated with renal disease including:

- ☐ Anemia
- ☐ Hyperglycemia
- ☐ Hyperlipidemia
- ☐ Hyperparathyroidism (bone diseases)

PHYSICAL SIGNS

For clients with the nephrotic syndrome and renal insufficiency or renal failure, look for physical signs of:

- ☐ Dehydration and fluid retention
- ☐ Iron deficiency
- ☐ Uremia
- ☐ Bone diseases
- ☐ Hyperkalemia
- ☐ Zinc deficiencies

Nutrition on the Net

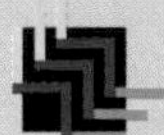

WEBSITES

Access these websites for further study of topics covered in this chapter.

- Find updates and quick links to these and other nutrition-related sites at our website: **www.wadsworth.com/nutrition**
- To find out more about kidney stones, visit these sites:

  Oxalosis and Hyperoxaluria Foundation: **www.ohf.org**

  American Foundation for Urologic Disease: **www.afud.org**.
- To search for specific topics related to kidney diseases, dialysis, and kidney transplants, including materials available for clients with kidney diseases, visit these sites:

  National Institute of Diabetes and Digestive and Kidney Diseases: **www.niddk.nih.gov**

  National Kidney Foundation: **www.kidney.org**

  Kidney Foundation of Canada: **www.kidney.ca**

  Renalnet Kidney Information Clearinghouse: **www.renalnet.org/renalnet/renalnet.cfm**
- To learn more about support for people with kidney diseases, visit The American Association of Kidney Patients: **www.aakp.org**
- To uncover more information about dialysis, visit the Kidney Dialysis Foundation: **www.kdf.sg**

INTERNET ACTIVITIES

The kidneys are vital to the health and proper functioning of the body. Diseases that damage the kidneys or interfere with their function can have devastating effects on health. People with kidney disease may need special diets, medical treatment, dialysis, and even kidney transplants. You may have questions about different diseases of the kidney.

- ▼ Go to **www.niddk.nih.gov**
- ▼ Under the heading "Health Information," click on "Kidney."
- ▼ Scroll down the list of "Publications online" and click on one that interests you.
- ▼ Read about causes, symptoms, and treatments (including nutrition).

What did you learn about kidney disease and its treatment? What role does nutrition play in treatment?

Study Questions

These questions will help you review the chapter. You will find the answers in the discussions on the pages provided.

1. What are kidney stones, and how do they develop? What is the composition of the most common type of kidney stone? What dietary adjustments can help prevent the recurrence of this type of stone? (pp. 802–804)
2. What is the nephrotic syndrome? What are its consequences? Describe the role of protein in the diet for the nephrotic syndrome. Why is a low-salt, low-fat, low-cholesterol diet recommended for its treatment? (pp. 805–807)
3. Describe acute renal failure and list some of its causes. What symptoms are associated with the buildup of toxic metabolic waste products in the blood? What electrolytes are of concern? How do nutrient needs change as the person progresses through the oliguric and diuretic phases of acute renal failure? (pp. 807–810)
4. How does chronic renal failure differ from acute renal failure? What happens as renal failure gradually progresses? Describe end-stage renal disease. (pp. 811–812)
5. What are the objectives of the treatment of chronic renal failure? Why are the protein and phosphorus intakes of the person with chronic renal failure of particular concern? What other measures can be taken to help lower phosphorus levels? (p. 812)
6. What fluid and electrolyte modifications are made to the diet of the person with chronic renal failure? What modifications are made concerning calcium, vitamin D, other vitamins, minerals, and trace elements? Why? (pp. 814–815)
7. Discuss the nutrient needs of the person with a kidney transplant. Can these nutrient needs change? Why or why not? (pp. 816–817)

These questions will help you prepare for an exam. Answers can be found on p. 821.

1. Which of the following is not a function of the kidneys?
   a. activation of vitamin K
   b. maintenance of acid-base balance
   c. elimination of metabolic waste products
   d. maintenance of fluid and electrolyte balance
2. Treatment for all kidney stones includes:
   a. a protein-restricted diet.
   b. a methionine-restricted diet.
   c. a calcium intake that meets but does not exceed the DRI.
   d. a fluid intake to maintain a urine volume of at least 2 liters a day.
3. People with calcium oxalate stones may benefit from diets that restrict:
   a. oxalate and sodium.
   b. calcium and potassium.
   c. protein and methionine.
   d. calcium and phosphorus.
4. A person with the nephrotic syndrome is prone to infections due to losses of:
   a. albumin.
   b. transferrin.
   c. lean body mass.
   d. immunoglobulins.
5. Diet recommendations for the nephrotic syndrome include:
   a. no diet restrictions.
   b. sodium restrictions.
   c. protein intakes that are less than the RDA.
   d. protein intakes that are 1.5 to 2.0 times the RDA.
6. The electrolytes that may rise rapidly in people with acute renal failure who are catabolic include:
   a. sodium, phosphorus, and calcium.
   b. sodium, potassium, and phosphorus.
   c. potassium and phosphorus.
   d. potassium, phosphorus, calcium, and magnesium.
7. The health care team estimates fluid requirements for clients with acute renal failure by adding _____ milliliters to the amount of urine output.
   a. 100
   b. 300
   c. 500
   d. 750
8. Complications commonly associated with chronic renal failure may include:
   a. growth failure and renal colic.
   b. nausea, vomiting, and reflux esophagitis.
   c. anemia, edema, and potassium deficiencies.
   d. anemia, bone disease, cardiovascular disease, and malnutrition.
9. Which of the following nutrients may be unintentionally restricted when a person follows a renal diet?
   a. fluid
   b. calcium
   c. potassium
   d. phosphorus
10. The health care professional recognizes that compared to the person with renal failure who is not on dialysis, the diet of a person on dialysis is:
    a. lower in protein.
    b. higher in protein.
    c. lower in potassium and phosphorus.
    d. higher in potassium and phosphorus.

Clinical Applications

1. Consider that a person with chronic renal failure may need multiple medications to control disease progression and treat symptoms and complications. For people with diabetes and hyperlipidemia who develop renal failure, medications might include insulin, antihypertensives, diuretics, antilipemics, antiemetics (for nausea), antisecretory agents, and phosphate binders. Review the nutrition-related side effects of these medications. Describe the ways that medications can make it harder for people to maintain nutrition status.

2. Think about the case of a person with type 2 diabetes, elevated triglycerides and LDL cholesterol, and diabetic nephropathy. First consider the recommended diet for type 2 diabetes and heart health. In what ways might careful adherence to such a diet prevent or delay the development of diabetic nephropathy? How would the diet change if the person developed the nephrotic syndrome and eventually kidney failure? Using the box on p. 816 as a guide, give practical suggestions for helping a person adjust to a renal diet.

References

1. G. C. Curhan and coauthors, Beverage use and risk of kidney stones in women, *Annals of Internal Medicine* 128 (1998): 534–540.

2 G. C. Curhan and coauthors, Comparison of dietary calcium with supplemental calcium and other nutrients as factors affecting the risk for kidney stones in women, *Annals of Internal Medicine* 126 (1997): 497–504.

3 S. Ahuja, B. Kaack, and J. Roberts, Loss of fimbrial adhesion with the addition of Vaccinum macorcarpon to the growth medium of P-fimbriated Escheria coli, *Journal of Urology* 159 (1998): 559–562; J. Avorn and coauthors, Reduction of bacteriuria and pyuria after ingestion of cranberry juice, *Journal of the American Medical Association* 271 (1994): 751–754.

4 W. J. Burtis and coauthors, Dietary hypercalciuria in patients with calcium oxalate stones, *American Journal of Clinical Nutrition* 60 (1994): 424–429.

5 S. R. Orth and E. Ritz, The nephrotic syndrome, *New England Journal of Medicine* 338 (1998): 1202–1211.

6 R. Rodrigo and M. Pino, Proteinuria and albumin homeostasis in the nephrotic syndrome: Effect of dietary protein intake, *Nutrition Reviews* 54 (1996): 337–347.

7 D. J. Rodriguez and W. M. Sandoval, Nutrition support in acute renal failure patients: Current perspectives, *Support Line,* December 1997, pp. 3–7.

8 Rodriguez and Sandoval, 1997.

9 J. D. Kopple, The nutrition management of the patient with acute renal failure, *Journal of Parenteral and Enteral Nutrition* 20 (1996): 3–12.

10 The Modification of Diet in Renal Disease Study Group, The effects of dietary protein restriction and blood pressure control on the progression of chronic renal disease, *New England Journal of Medicine* 330 (1994): 877–884.

11 S. Kobrin and S. Aradhye, Preventing progression and complications of renal disease, *Hospital Medicine* 33 (1997): 11–12, 17–18, 20, 29–31, 35–36, 39–40.

12 J. A. Beto, Which diet for which renal failure: Making sense of the options, *Journal of the American Dietetic Association* 95 (1995): 898–903.

13 R. Makoff and H. Gonick, Renal failure and concomitant derangement of micronutrient metabolism, *Nutrition in Clinical Practice* 14 (1999): 238–246.

14 Beto, 1995.

15 B. P. Gillis and coauthors, Nutrition intervention program of the Modification of Diet in Renal Disease Study: A self-management approach, *Journal of the American Dietetic Association* 95 (1995): 1288–1294.

16 National Kidney Foundation, About kidney disease, www.kidney.org, website visited November 30, 1999.

ANSWERS

Multiple Choice

1. a	2. d	3. a	4. d	5. b
6. c	7. c	8. d	9. b	10. b

DIALYSIS AND NUTRITION

Although there is no perfect substitute for one's own kidneys, dialysis offers a life-sustaining treatment option for people with end-stage renal disease. Dialysis can serve as a permanent treatment for kidney failure or as a temporary measure to sustain life until a suitable kidney donor can be found. Dialysis also benefits the person with acute renal failure who needs immediate help in restoring normal blood balances. All health care professionals need to understand that renal diseases alter nutrient needs and that dialysis affects those needs. Those who routinely work with clients with renal diseases, however, need to understand more about dialysis procedures. This highlight describes the different types of dialysis procedures and explains why different procedures affect nutrient needs in different ways.

THE BASICS OF DIALYSIS

Dialysis removes excess fluids and wastes from the blood, in part, by employing the principles of simple **diffusion** and **osmosis** across a **semipermeable membrane** (see the accompanying glossary). For **hemodialysis** and **peritoneal dialysis,** a solution similar in composition to normal blood plasma, called the dialysate, is placed on one side of the semipermeable membrane; the person's blood flows by on the other side.

The Semipermeable Membrane

A semipermeable membrane acts like a filter. Different types of membranes have pores of different sizes, and the size of the pores determines which molecules will be able to pass through the membrane and which will not. Small molecules like urea and electrolytes cross the membrane freely; large molecules like proteins are less likely to cross the membrane.

The Dialysate

Wastes are removed from the blood by altering the composition of the dialysate. When the dialysate contains a lower concentration of a substance than the blood, the substance, provided it can cross the membrane, will diffuse out of the blood. To maximally remove waste products like urea from the blood, the dialysate contains no urea. In some cases, however, the dialysate must be adjusted so that only excesses will be removed. Thus potassium can be removed from the blood, for example, by providing a dialysate with a lower concentration of potassium than the person's blood. However, some potassium must remain in the dialysate, or blood potassium will fall too low.

Dialysis can filter excess substances only from the blood.[1] Thus electrolytes like potassium and phosphorus that reside mainly in the intracellular fluid are more difficult to control with dialysis.

The pressure created by proteins that cannot cross the membrane tends to hold excess fluid in the blood. Thus, to remove excess fluid (and sodium along with it), pressure gradients must be created between the blood and the dialysate, in such a way that water and small molecules are "pushed" through the pores of the membrane **(ultrafiltration).** Figure H27-1 illustrates the different forces that work to remove wastes and fluids during dialysis.

The dialysate can also be used to add needed components back into the blood. For a person with acidosis, for example, bases such as bicarbonate can be added to the dialysate. The base moves by diffusion into the person's blood to ease acidosis.

TYPES OF DIALYSIS

Three main types of procedures are available to remove fluids and wastes from the body. These include hemodialysis, peritoneal dialysis, and continuous renal replacement therapy (CRRT). Each of these has advantages and disadvantages (see Table H27-1 on p. 824).

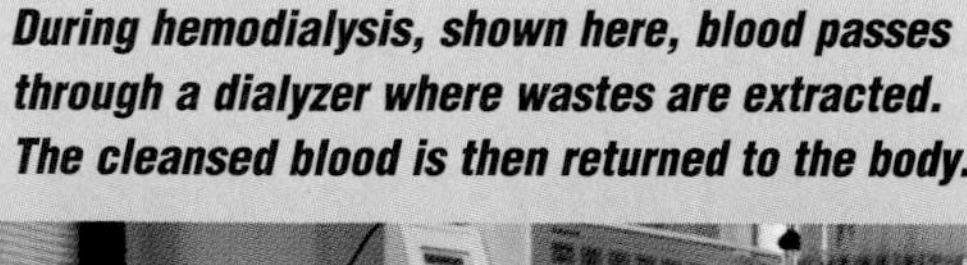

During hemodialysis, shown here, blood passes through a dialyzer where wastes are extracted. The cleansed blood is then returned to the body.

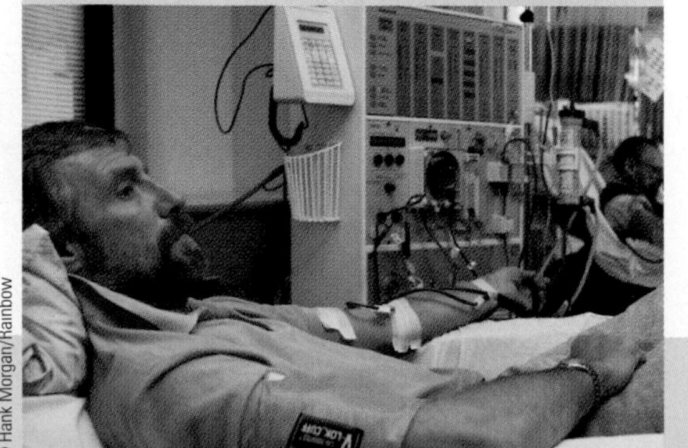

Hemodialysis

In hemodialysis, the blood enters a machine called a **dialyzer** or artificial kidney. The dialyzer houses a series of tubes made

GLOSSARY

continuous ambulatory peritoneal dialysis (CAPD): the most common type of peritoneal dialysis that people use at home.

dialyzer (dye-ah-LYES-er): the machine used for hemodialysis; also called an **artificial kidney.**

diffusion: movement of solutes from an area of high concentration to one of low concentration.

hemodialysis: removal of fluids and wastes from the blood by passing it through a dialyzer.

hemofiltration: removal of fluids and wastes from the blood by using ultrafiltration and fluid replacement.

osmosis: movement of water from an area of low solute concentration to one of high solute concentration.

peritoneal dialysis: removal of wastes and fluids from the body by using the peritoneal membrane as a semipermeable membrane.

semipermeable membrane: a membrane with pores that allow some particles to pass through the membrane but not others.

ultrafiltration: removal of fluids and small- to medium-size molecules from the blood by using pressure to transfer the blood across a semipermeable membrane.

from synthetic semipermeable membranes. Blood is pumped out of the body and into the dialyzer where it flows between the tubes that carry the dialysate. The dialysis machine applies a negative pressure to the dialysate side of the membrane so that the excess water and sodium can be removed from the blood. The dialysate extracts wastes from the blood, and then the blood is returned to the body.

Hemodialysis can take place in a hospital or clinic or in the home. Often the person receives three treatments a week, and each treatment takes from two to four hours.

Peritoneal Dialysis

In peritoneal dialysis, the peritoneal membrane (the membrane that covers the abdominal organs) serves as the semipermeable membrane. The dialysate is infused directly through a tube into the peritoneal space—the space within the person's abdomen that overlays the intestine. There are several different peritoneal dialysis techniques, and the technique used determines how often treatments are given and how long fluid remains in the abdomen. With the most common technique, **continuous ambulatory peritoneal dialysis (CAPD),** the dialysate is placed in the abdomen where it remains for four to six hours. The dialysate is then drained from the abdomen through a tube using gravity and replaced with fresh dialysate. It takes about 30 to 40 minutes to drain the dialysate and replace it, and generally the solution is changed four times a day.

Because negative pressure cannot be created in the peritoneal cavity as it can in a dialyzer, glucose (and sometimes amino acids) is added to the dialysate. Increasing the concentration of glucose creates osmotic pressure so that fluids and sodium can be removed through osmosis.

Unlike hemodialysis, peritoneal dialysis does not require that blood exit the body. In addition, the removal of fluids and wastes is more gradual.

Continuous Renal Replacement Therapy (CRRT)

For people in acute renal failure, another procedure, called continuous renal replacement therapy (CRRT), allows removal of fluids or wastes.[2] CRRT is usually reserved for people with acute renal insufficiency or renal failure who cannot tolerate hemodialysis or peritoneal dialysis for medical reasons.

Several different methods can be used to perform CRRT, depending on the individual's needs. If the purpose is to remove fluids only, no dialysate is used. Instead blood is circulated through a filter where pressure gradients force all blood components small enough to pass through the pores of a semipermeable membrane out of the blood (ultrafiltration). The final composition of the filtrate is similar to that of the blood but without plasma proteins and blood cells, which are too large to pass

FIGURE H27-1 **Diffusion, Osmosis, and Ultrafiltration**

Diffusion

Small molecules (electrolytes and waste products) move from an area of high concentration to an area of low concentration by diffusion.

Osmosis

Water moves from an area of low concentration to an area of high concentration by osmosis. In other words, water moves from an area that has fewer particles dissolved in it to an area with more particles dissolved in it.

Ultrafiltration

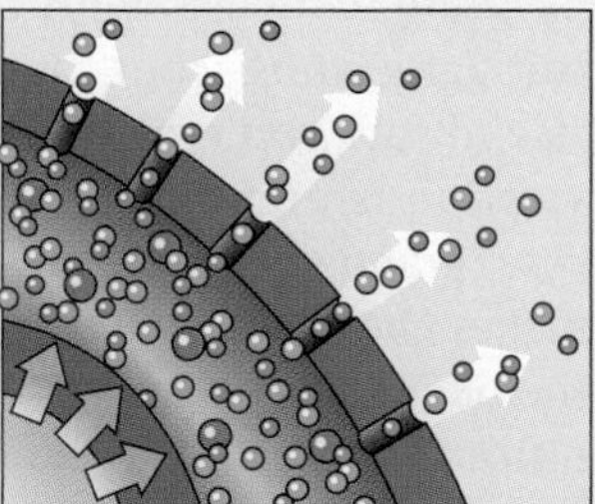

Pressure squeezes water and small molecules through the pores of a semipermeable membrane during ultrafiltration.

TABLE H27-1 **Advantages and Disadvantages of Various Dialysis Procedures**

	Advantages	Disadvantages
Hemodialysis	Acts quickly; shorter treatment time; minimal absorption of glucose; can be done at home or in clinic.	Requires access to blood; risk of hypotension, hypokalemia, arrhythmias, blood clots, and anemia; requires a dialyzer.
Peritoneal dialysis	Gradual removal of fluids and wastes; low risk of hypotension, hypokalemia, and blood clots, fewer dietary restrictions; requires little equipment; easy to do at home.	Requires access to peritoneal cavity; takes a great deal of time; risk of peritonitis; greater loss of protein.
Continuous renal replacement therapy (CRRT)	Gradual removal of fluids and wastes; useful only for acute renal failure; done only in hospital.	Requires access to blood; less risk of hypotension.

through the membrane. Thus excess electrolytes and waste products are not removed, but the total volume is reduced. In some cases, part of the fluid removed during ultrafiltration is replaced with IV fluids that contain only those components needed by the client **(hemofiltration).** If blood containing electrolytes and wastes is removed and some of the fluid is replaced with a very dilute dextrose solution, for example, eventually the concentration of electrolytes and wastes in the blood is lowered, as is the amount of fluids. The filter can also be adjusted to perform dialysis as well as filtration, making it possible to provide slow, continuous dialysis. The methods can also be combined so that ultrafiltration and dialysis can take place together.

EFFECTS OF DIALYSIS ON NUTRIENT NEEDS

Both the dialysis itself and the type of dialysis affect nutrient needs. For some nutrients, the differences can be substantial.

Energy

For people on dialysis, dextrose (glucose) is added to the dialysate. Glucose helps to draw fluid into the dialysate through osmosis, but the person undergoing dialysis may also absorb some of it. During hemodialysis, glucose absorption is minimal. In peritoneal dialysis, however, glucose gradually crosses the peritoneal membrane. This makes the dialysate less efficient at drawing fluid out of the blood and also means that the person may absorb significant amounts of glucose. The glucose may provide as many as 600 to 800 kcalories per day. For people on CRRT, the absorption of glucose is also considerable and can range from about 35 to 45 percent to as high as 60 to 70 percent, depending on the type of CRRT and many other factors.[3] Because people undergoing CRRT are often catabolic, they frequently experience insulin resistance, and the carbohydrate provided must be carefully adjusted to avoid serious problems with hyperglycemia.

Protein

Compared to people with renal failure who are not on dialysis, people on dialysis receive diets higher in protein. In part, this is because dialysis can remove the nitrogen-containing waste products of protein metabolism. In addition, some amino acids and smaller proteins may be lost in the dialysate. In hemodialysis, protein losses average about 5 to 8 grams per treatment. Protein losses from peritoneal dialysis are slightly higher (about 10 to 20 grams per day) because more blood proteins pass into the dialysate through the peritoneal membrane than through the synthetic membrane used for hemodialysis. Clients with acute renal failure often receive TPN solutions during the time that they are undergoing CRRT. People undergoing CRRT retain about 90 percent of infused amino acids.[4]

Potassium

Another difference between hemodialysis and peritoneal dialysis with respect to nutrition concerns potassium. In hemodialysis, the dialysate contains potassium; because a relatively large volume of blood is pumped through the dialyzer, potassium levels can drop rapidly, and hypokalemia can follow and interfere with heart function. In peritoneal dialysis and CRRT, the exchange of potassium between the blood and the dialysate is much slower, and blood potassium does not change rapidly. In peritoneal dialysis and CRRT, potassium may be added in varying amounts or left out of the dialysate altogether, depending on the person's blood potassium levels.

Since peritoneal dialysis can remove more potassium than hemodialysis, the person on peritoneal dialysis generally does not have to restrict dietary sources of potassium. The person on hemodialysis, however, does need to moderately restrict potassium between dialysis treatments.

Dialysis and CRRT help remove wastes and fluids normally removed by functional kidneys. Although these procedures cannot restore the hormonal functions of the kidneys, they provide a lifesaving means for alleviating the symptoms of uremia, hypertension, and edema and limiting the risk of heart failure.

References

1 D. I. Charney, Medical treatment in renal disease: Basic concepts in dialysis, *Support Line,* February 1998, pp. 3–7.

2 D. J. Rodriguez and W. M. Sandoval, Nutrition support in acute renal failure patients: Current perspectives, *Support Line,* December 1997, pp. 3–7.

3 As cited in D. C. Kaufman and coauthors, Adjustment of nutrition support with continuous hemodiafiltration in a critically ill patient, *Nutrition in Clinical Practice* 14 (1999): 120–123.

4 Kaufman and coauthors, 1999; Rodriguez and Sandoval, 1997.

CHAPTER 28 Nutrition and Liver Disorders

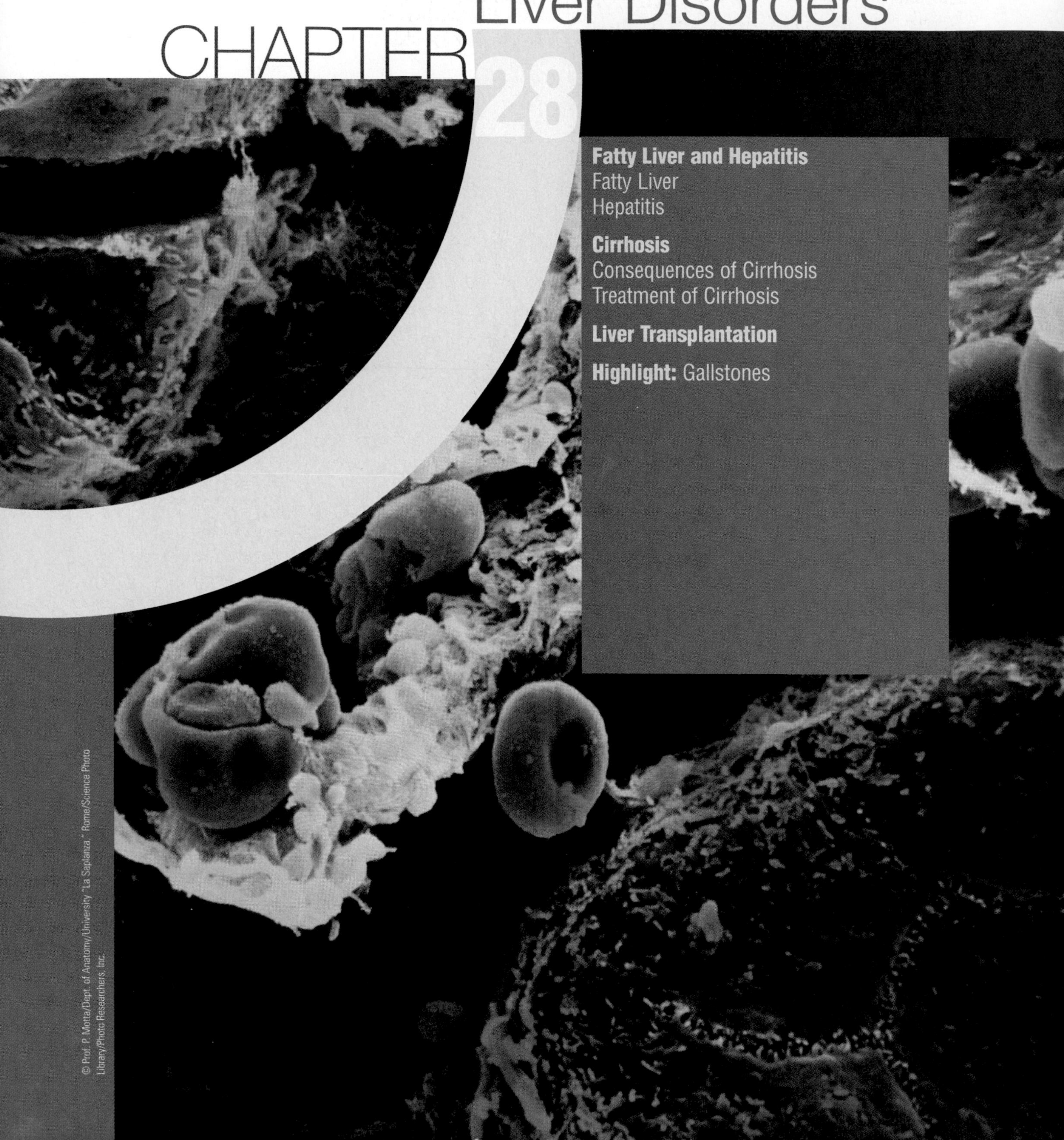

The liver is the metabolic crossroads of the body, and its health is crucial to every body function. The liver receives nutrients and then metabolizes, packages, stores, or ships them out for use by other organs.◆ It manufactures bile, which the body uses to prepare fat for digestion and absorption. It synthesizes albumin, cholesterol, lipoproteins, carrier proteins, and some clotting and stress response factors. The liver also detoxifies drugs, packages excess nitrogen in urea so that it can be excreted by the kidneys, and participates in iron recycling, the manufacture of red blood cells, and the activation of vitamin D. No wonder then that disorders that affect the liver can profoundly affect both nutrition and general health status. The liver conditions described in this chapter are not diseases, but rather forms of liver damage that can occur as a consequence of many disorders.

Review Table 7-1 on p. 208 to refresh your memory regarding the major metabolic functions of the liver. ◆

As Figure 28-1 shows, after the liver processes nutrient-rich blood from the intestine, the blood is returned to the heart through the hepatic vein and distributed throughout the body. The liver excretes excess cholesterol in bile through the **biliary tract** and into the intestine, where it is eliminated in the feces. Highlight 28 provides additional information about the gallbladder and the connections between gallstones and diet.

biliary tract: the gallbladder and bile ducts.

fatty liver: an accumulation of triglycerides in the liver resulting from many disorders, including exposure to excessive alcohol, excessive weight gain, and diabetes mellitus; also called **hepatic steatosis, steatohepatitis,** and **fatty infiltration of the liver.**

hepatitis (hep-ah-TIE-tis): inflammation of the liver.
- **hepatic** = liver

FATTY LIVER AND HEPATITIS

Fatty liver and **hepatitis** are two of the more common signs of liver dysfunction. Both inflame and enlarge the liver. Dietary factors may play a role in the development of both disorders, although both may also arise from causes unrelated to diet.

FIGURE 28-1 The Liver, Biliary Tract, and Associated Blood Vessels

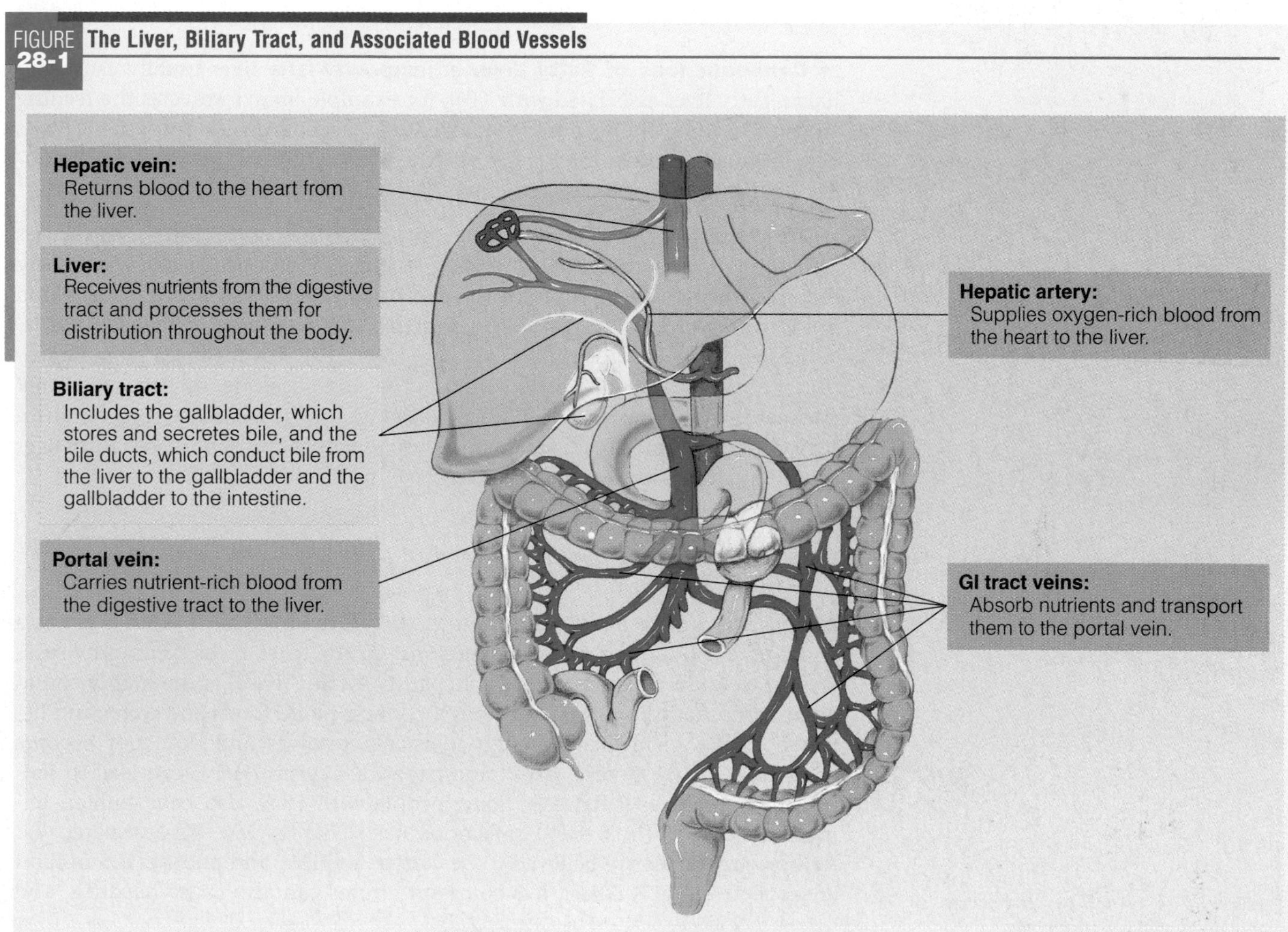

Fatty Liver

Fatty liver is a clinical finding common to many conditions associated with liver dysfunction. Fatty liver may develop from damage to liver cells or from altered nutrient availability to the liver. Most commonly, fatty liver develops from the liver's exposure to toxic substances such as alcohol (Highlight 7) or as a consequence of excessive weight gain, insulin resistance, or diabetes mellitus.◆ Fatty liver can also develop as a consequence of inadequate intake of protein (as in protein-energy malnutrition, or PEM), an infection or malignant disease, drug therapy (such as therapy with corticosteroids or tetracycline), long-term TPN, or small bowel bypass surgery. Fatty liver does not arise from eating too much fat alone; excess total kcalories (from either fat or carbohydrate) and high blood glucose levels (as found in insulin resistance and poorly controlled diabetes) appear to have a greater association with fatty liver.

◆ Highlight 26 describes the development of fatty liver as a consequence of obesity and insulin resistance.

The exact reasons why fats (triglycerides) accumulate in the liver are unknown, but the liver may either synthesize too much fat, use too little for energy, take up too much from the blood, release too little back to the blood, or make a combination of these errors. In severe cases, liver weight may increase from a typical weight of about 3 pounds to as much as 11 pounds, with triglycerides increasing from 5 percent to as much as 40 percent of liver weight. Laboratory findings associated with fatty liver may include elevated serum transaminases (ALT and AST),◆ alkaline phosphatase, **bilirubin,** glucose, triglycerides, and lipids.

◆ The two transaminase enzymes that may be elevated in liver disease are **alanine transaminase (ALT)** and **aspartate transaminase (AST).**

THINK NUTRITION whenever clients have fatty liver. Nutrition can play a role in both the development and the treatment of fatty liver.

• **Consequences of Fatty Liver** • Temporary fatty liver usually causes no harm. Fatty liver associated with TPN, for example, may resolve as the feeding continues, when the feeding is changed to a cyclic infusion, or when TPN is discontinued. In other cases, however, the liver's accumulation of fat can damage liver cells and result in permanent liver damage.

• **Treatment of Fatty Liver** • The appropriate therapy for fatty liver focuses on eliminating the cause and reversing its effects. Fatty liver related to obesity requires weight reduction and treatment of elevated blood glucose and blood lipids, when appropriate. Fatty liver related to diabetes mellitus and insulin resistance requires control of blood glucose levels. Fatty liver caused by malnutrition requires a gradual introduction of a high-kcalorie, high-protein diet adequate in all other nutrients. Fatty liver due to alcohol abuse requires abstinence from alcohol and an adequate diet to replenish nutrient stores. Fatty liver caused by drug therapy requires alternative medications or other therapies.

Hepatitis

In hepatitis, inflammation and enlargement of the liver most often occur as a consequence of one of six viral infections. Of the three most common viruses known to cause hepatitis (A, B, C), hepatitis A virus (HAV) is the highly contagious form that can be spread through contaminated foods and water and between family members, although it usually resolves and does not become chronic. Hepatitis B virus (HBV) and hepatitis C virus (HCV) can lead to serious, permanent liver damage. Some people with HCV also have human immunodeficiency virus (HIV) infections (see Chapter 29). An estimated 1.2 million Americans are believed to be carriers of HBV, and another 3.5 million are carriers of HCV. Other, less common viruses can also cause hepatitis, and

bilirubin (bill-ih-REW-bin): an orange- or yellow-colored pigment in bile.

hepatitis can occur as a consequence of damage to liver cells by chronic and excessive alcohol ingestion, certain medications and illicit drugs, or other chemical toxins. The dietary supplements chaparral, bee pollen, germander, jin bu huan, ma huang, skullcap, mistletoe, senna, and valerian root have also been reported to cause hepatitis.[1]

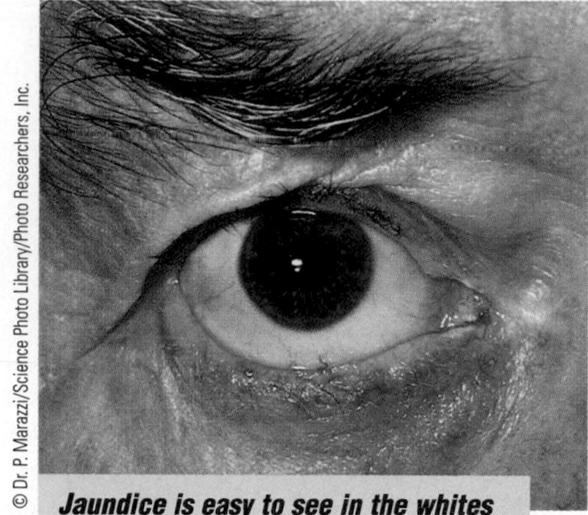

© Dr. P. Marazzi/Science Photo Library/Photo Researchers, Inc.

Jaundice is easy to see in the whites of the eyes.

• Symptoms of Hepatitis • For the person with hepatitis, the symptoms may be mild, and the disease may go undiagnosed. In some cases, the person may develop flu-like symptoms including fatigue, anorexia, nausea, vomiting, diarrhea or constipation, fever, and pain in the area of the liver. If hepatitis progresses to a severe stage, bilirubin accumulates in the inflamed liver and spills into the blood, causing **jaundice** and producing a dark urine. Serum transaminase levels (AST and ALT) rise, and the liver becomes enlarged and tender.

• Consequences of Hepatitis • The origin and type of hepatitis, the extent of liver damage, and the person's response to treatment all determine how seriously the disease will affect health. In many cases, liver cells gradually regenerate and liver function recovers. Recovery from HAV generally occurs over several weeks. Ten percent of people with HBV and 85 percent of those with HCV develop chronic hepatitis, which increases the risk of cirrhosis (described next), liver cancer, and liver failure. Viral hepatitis can also lead to both acute and chronic renal failure.[2] Far less frequently, severe hepatitis progresses rapidly and leads to acute liver failure **(fulminant liver failure)** and death. This is more likely to occur when hepatitis results from medications or chemical agents that are toxic to liver cells.

• Treatment of Hepatitis • Vaccines to prevent HAV and HBV, but not HCV, infections are available. When hepatitis does occur, treatment aims to reduce further insult to the liver cells; thus the person must abstain from alcohol and avoid taking any unnecessary medications and dietary supplements. An HAV infection generally resolves without medications. HBV and HBC infections generally require treatment with antiviral agents (lamivudine and ribavirin, respectively) and interferon (an immune system modulator). Other cases of severe hepatitis may be treated with anti-inflammatory agents (see the Diet-Drug Interactions box on p. 835).

• Medical Nutrition Therapy • Liver cells need nutrients to help them recover from hepatitis. The person with hepatitis who is in good nutrition status receives a regular, well-balanced diet. The malnourished person with hepatitis receives a high-kcalorie, high-protein diet to replenish nutrient stores. For the person with mild anorexia, frequent small meals, enteral formula supplements, or both may be helpful. For the person with vomiting, intravenous replacement of fluids is important. Parenteral nutrition is an alternative if the person is malnourished and vomiting persists.

THINK NUTRITION whenever clients have hepatitis. Attention to nutrition can help restore liver function.

IN SUMMARY

Fatty liver develops when the liver is unable to process fats as it normally does. The treatment of fatty liver depends on the cause, and many of the causes are nutrition related. Hepatitis can develop as a consequence of several viral infections or from alcohol abuse or the ingestion of chemicals, including some medications and illicit drugs, that are toxic to liver cells. The goal of nutrition support for hepatitis is to provide the nutrients liver cells need to regenerate.

jaundice (JON-dis): a characteristic yellowing of the skin, whites of the eyes, mucous membranes, and body fluids resulting from the accumulation of bilirubin in the blood.

fulminant (FULL-mih-nant) **liver failure:** liver failure that rapidly progresses to a life-threatening stage.

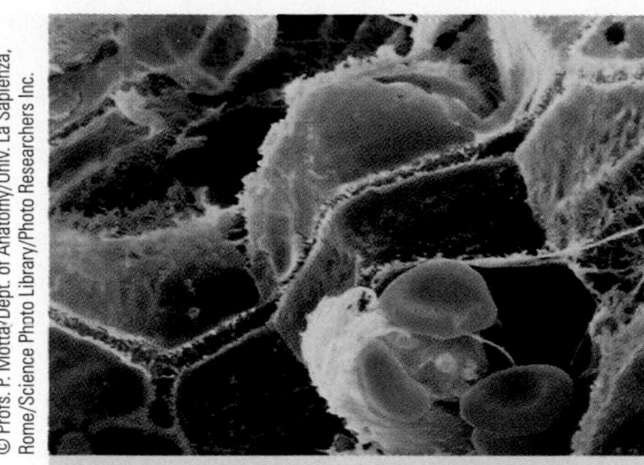
© Profs. P. Motta/Dept. of Anatomy/Univ. La Sapienza, Rome/Science Photo Library/Photo Researchers Inc.

This micrograph of a normal liver shows several hepatocytes (dark reddish-brown), bile ducts (yellowish green), red blood cells (bright red), and a macrophage (yellow).

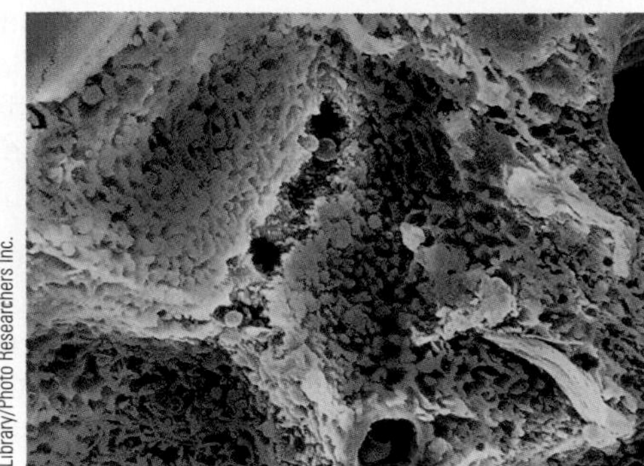
© Profs. P. Motta, M. Nishi, J. Fujita/Science Photo Library/Photo Researchers Inc.

In a liver damaged by cirrhosis, the hepatocytes lose their structure and flexibility, obstructing the flow of blood and bile through the liver and ultimately preventing the liver from carrying out its functions.

CIRRHOSIS

In **cirrhosis,** a serious consequence of some chronic liver diseases, liver cells are damaged to the point that they cannot regenerate, and liver function falters and, finally, fails. Scar tissue (fibrosis) forms within the liver, altering the structure of the liver cells and interrupting the blood flow through it. Most often the damage occurs over time, but sometimes the liver can fail suddenly.

Chronic alcohol abuse is the most common cause of cirrhosis **(Laennec's cirrhosis)** in the United States, although not all people with cirrhosis are alcohol abusers, and not all alcohol abusers develop cirrhosis. Other causes of cirrhosis include:

- Infections, including those that lead to chronic hepatitis **(postnecrotic cirrhosis).**
- Progression from fatty liver in people who do not consume excessive alcohol **(nonalcoholic fatty cirrhosis).**
- Diseases of the bile ducts, including **biliary atresia** and **primary biliary cirrhosis,** or obstructions of the bile ducts due to a stone or tumor **(obstructive cirrhosis).**
- Later stages of congestive heart failure **(cardiac cirrhosis).**
- Severe reactions to medications and prolonged exposure to toxic chemicals **(toxic cirrhosis).**◆
- Inherited metabolic disorders, including those that cause excessive accumulation of iron **(hemochromatosis)** or copper **(Wilson's disease)** in the liver or alter the liver's ability to use glucose **(glycogen storage disease).**

◆ **Ingestion of chaparral tea, mentioned earlier, has led to liver failure requiring transplantation. Another herb, pennyroyal (used to treat coughs and upset stomachs), may also be toxic to liver cells.**

Sometimes cirrhosis develops for no identifiable cause **(idiopathic cirrhosis).** The glossary on p. 831 defines terms related to cirrhosis.

Consequences of Cirrhosis

Unlike healthy liver tissue, which is soft and flexible, scar tissue is unyielding—a difference with major implications. Many people have no symptoms at first, but as the disorder progresses, symptoms and clinical findings may include nausea, vomiting, weight loss, liver enlargement, edema, jaundice, mental disturbances, and itching from the accumulation of bile pigments under the skin. The liver progressively loses function and, in the end stages, fails. Table 28-1 shows standards for laboratory tests often used to diagnose and monitor the extent of liver disease. Figure 28-2 on p. 832 shows many of the consequences of cirrhosis described in the sections that follow.

cirrhosis (sih-ROW-sis): an advanced form of liver disease in which scar tissue replaces liver cells that have permanently lost their function.

GLOSSARY OF TERMS RELATED TO CIRRHOSIS

biliary (BILL-ee-air-ee) **atresia** (ah-TREE-zee-ah): a disorder of infants characterized by absent or injured bile ducts.

cardiac cirrhosis: severe liver damage associated with the later stages of congestive heart failure (see Chapter 26).

glycogen storage disease: one of several inherited disorders in which a person lacks one of the enzymes that allow glycogen stores to be utilized efficiently. The accumulation of glycogen in the liver leads to severe liver damage.

hemochromatosis (HE-moe-CROW-mah-toe-sis): an inherited disorder in which a person absorbs too much iron from the intestine and stores too much iron in the liver. The accumulation of iron in the liver leads to severe liver damage.

idiopathic (id-ee-oh-PATH-ic) **cirrhosis:** severe liver damage for which the cause cannot be identified.

Laennec's (LAY-eh-necks) **cirrhosis:** severe liver damage related to alcohol abuse.

obstructive cirrhosis: severe liver damage caused by obstruction of the bile ducts due to a stone or tumor.

nonalcoholic fatty cirrhosis: severe liver damage that develops from fatty liver that is unrelated to alcohol abuse.

postnecrotic cirrhosis: severe liver damage that develops as a complication of chronic hepatitis.

primary biliary cirrhosis: severe liver damage from the gradual destruction of the bile ducts; can lead to chronic pancreatitis and fat malabsorption.

toxic cirrhosis: severe liver damage that results from toxic levels of chemicals.

Wilson's disease: an inherited disorder that can lead to severe liver damage. People with Wilson's disease absorb too much copper from the intestine and have too little of the protein that transports copper from the liver to the sites where it is needed.

• Portal Hypertension • The portal vein◆ and the hepatic artery carry 11½ quarts of blood every minute to the miles of intermeshed capillaries and arterioles within the liver. This huge volume of blood cannot pulse easily through the scarred tissue of a cirrhotic liver. Consequently, blood flow to and through the liver decreases, blood backs up, and pressure in the portal vein rises sharply, causing **portal hypertension.**

◆ **Reminder: The *portal vein* carries nutrients from the GI tract to the liver. Normally, almost all the blood that flows from the intestine passes through the liver. The *hepatic vein* returns blood from the liver to the heart. The *hepatic artery* delivers oxygen-rich blood from the heart back to the liver.**

• Collaterals and Esophageal Varices • With normal blood flow through the portal vein obstructed, pressure forces some of the blood to take a detour from the portal vein through the smaller vessels around the liver. These **collaterals,** or **shunts,** can develop throughout the GI tract, but often occur in the area of the esophagus. As pressure builds, the collaterals become enlarged and twisted, forming **varices.** Esophageal varices bulge into the lumen of the esophagus, and they can rupture and lead to massive bleeding that can be fatal. In addition, blood from ruptured varices anywhere in the GI tract can travel to the intestine and serve as a source of ammonia (described later).

• Ascites and Edema • The rising pressure in the portal vein forces plasma out of the liver's blood vessels into the abdominal cavity, causing the abdomen to swell. This accumulation of fluid, or edema, in the abdominal cavity is called **ascites.** At the same time, the diseased liver is unable to synthesize adequate amounts of albumin, which normally creates a pressure that draws fluids from

TABLE 28-1 Standards for Laboratory Tests Used to Diagnose and Monitor Liver Diseases

Test	Normal Values	Values in Liver Disease
Albumin	3.5–5.0 g/dL	Decreased
Alkaline phosphatase	Varies[a]	Normal or elevated
ALT (formerly SGPT)[b]	Varies[a]	Elevated
Ammonia	<50 μg/dL	Elevated
AST (formerly SGOT)[b]	Varies[a]	Elevated
Bilirubin (direct)	0.1–0.3 mg/dL	Elevated
Prothrombin time	10–13 seconds	Prolonged

Note: To convert albumin (g/dL) to standard international (SI) units, multiply by 10; to convert ammonia (μg/dL) to SI units, multiply by 0.5872; to convert bilirubin (μg/dL) to SI units (μmol/L), multiply by 17.10.
[a]Reference ranges vary depending on the test used. Consult laboratory report for normal ranges.
[b]ALT = alanine transaminase; SGPT = serum glutamic pyruvic transaminase; AST = aspartate transaminase; SGOT = serum glutamic oxaloacetic transaminase.

portal hypertension: elevated blood pressure in the portal vein caused by obstructed blood flow through the liver.

collaterals: small branches of a blood vessel that develop when blood flow through the liver is obstructed; also called **shunts.**

varices (VAIR-ih-seez): blood vessels that have become twisted and distended.

ascites (ah-SIGH-teez): a type of edema characterized by the accumulation of fluid in the abdominal cavity.

FIGURE 28-2 The Consequences of Cirrhosis

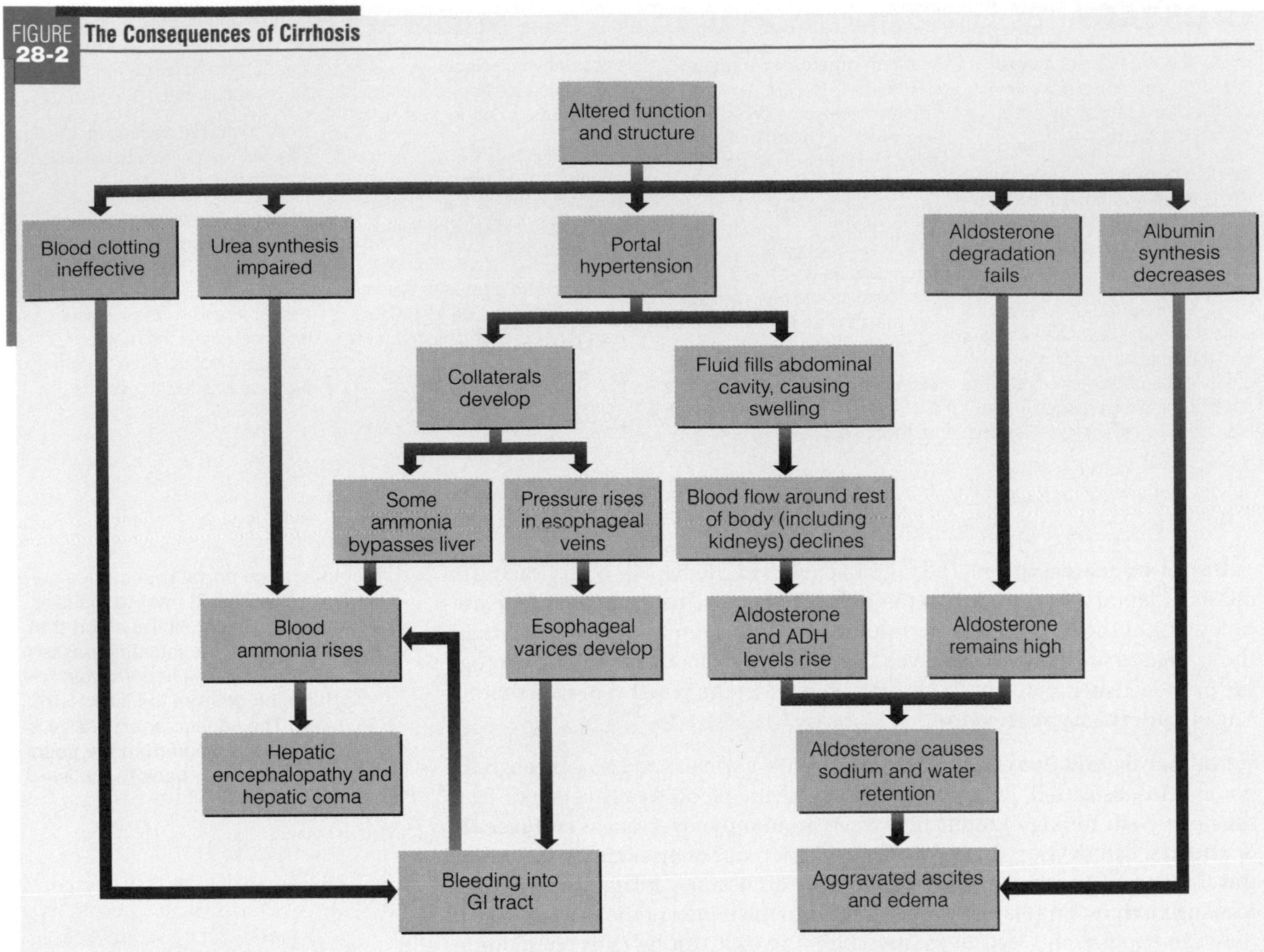

tissues back into the blood. As fluid accumulates in the abdomen, less blood reaches the kidneys, a sign the body interprets as fluid depletion. In response, the body makes more aldosterone◆ and antidiuretic hormone (ADH), hormones that expand the body's fluid volume by triggering retention of sodium and water in the kidneys. As a result of sodium and water retention, ascites worsens, and edema may spread to all body compartments (peripheral edema). To make matters worse, the diseased liver cannot dispose of aldosterone as it usually does, so aldosterone levels remain high. Thus ascites is a self-aggravating condition. Ascites frequently causes early satiety and nausea and also raises the basal metabolic rate.

◆ **Reminder:** ***Aldosterone*** **is the hormone secreted by the adrenal gland that acts on the kidneys to increase sodium and fluid retention.** ***Antidiuretic hormone ADH*** **is the hormone secreted by the hypothalamus that increases the reabsorption of water by the kidneys.**

• **Elevated Blood Ammonia Levels** • Ammonia is a normal but toxic product of protein digestion and metabolism (see Figure 28-3). The healthy liver detoxifies ammonia by removing it from circulation and converting it to urea. As liver disease progresses, ammonia-laden blood bypasses the liver by way of the collaterals, the liver may be unable to detoxify the ammonia it does retrieve, and blood ammonia rises. **Hyperammonemia** disrupts central nervous system function.

hyperammonemia (HIGH-per-AM-moe-KNEE-me-ah): elevated blood ammonia.

hepatic encephalopathy (en-SEF-ah-LOP-ah-thee): mental changes associated with liver disease that may include irritability, short-term memory loss, and an inability to concentrate.
- **encephalo** = brain
- **pathy** = disease

hepatic coma: a state of unconsciousness that results from severe liver disease.

• **Hepatic Encephalopathy and Hepatic Coma** • Liver diseases can alter mental function **(hepatic encephalopathy)** and lead to **hepatic coma**—a dangerous complication that develops in some people when liver function deteriorates significantly. The causes of encephalopathy and hepatic coma remain elusive, although elevated blood ammonia appears to play a role. As Figure 28-3 shows, most of the ammonia in the body comes from the GI tract. Thus events

FIGURE 28-3 **Ammonia Production in the Body**

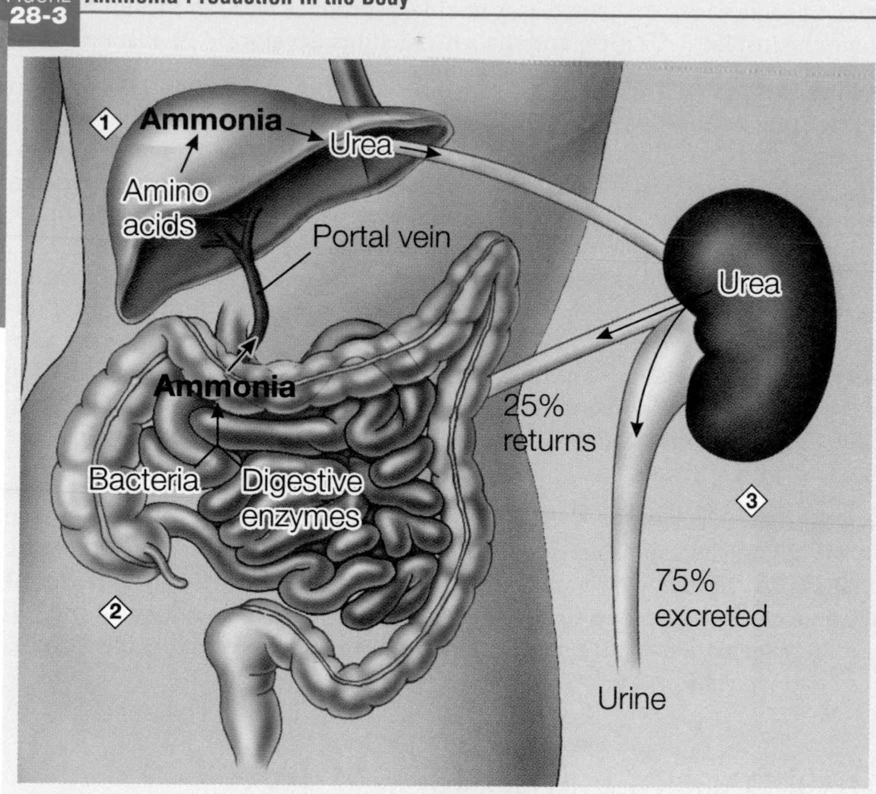

1 The liver makes urea mainly from ammonia formed in the GI tract and also makes some during amino acid metabolism.

2 Most ammonia in the body derives from the GI tract. Intestinal bacteria make ammonia from undigested protein and urea. Digestive enzymes make ammonia as they dismantle protein from the diet, blood in the GI tract, and shed intestinal cells. Some foods directly contribute to ammonia.

3 The kidneys excrete about 75 percent of ammonia as urea, but about 25 percent of urea is returned to the intestine, where it can form ammonia again.

that contribute protein to the GI tract (such as GI bleeding) or lengthen the time that protein remains in the GI tract (such as constipation)◆ may raise blood ammonia and the risk of coma. Infections and any acute stress result in protein catabolism, thereby producing more ammonia and raising the risk of coma.

◆ **Sedatives can lead to constipation, and sedative use can also precipitate hepatic coma in some cases.**

Elevated blood ammonia is associated with encephalopathy and hepatic coma, but the degree of ammonia elevation does not correlate with the severity of symptoms. A possible explanation for this poor correlation is that blood ammonia does not always parallel brain ammonia concentration. When brain ammonia is high, the body produces greater quantities of two substances (glutamine and ketoglutarate), and the degree of their elevation tends to correlate with the severity of symptoms.

Blood amino acid patterns also change in the later stages of liver disease: aromatic amino acid levels rise,◆ and branched-chain amino acid levels fall. These alterations may alter chemicals in the brain or add ammonia to the blood and contribute to encephalopathy and hepatic coma in some people.

◆ **Reminder: Phenylalanine and tyrosine are aromatic amino acids; leucine, lysine, and valine are the branched-chain amino acids.**

The person with hepatic encephalopathy may suffer from irritability, short-term memory loss, and an inability to concentrate. The person with impending hepatic coma exhibits changes in judgment, personality, or mood. The person may be unable to draw even a simple shape, such as a star. A sweet, musty, or pungent odor **(fetor hepaticus)** may develop on the breath. **Flapping tremor** may also develop in the precoma state. Just before passing into a coma, the person becomes very difficult to arouse.

fetor hepaticus (FEE-tor he-PAT-eh-cuss): a pungent odor of the breath that may develop in people with impending hepatic coma.

flapping tremor: uncontrolled movement of the muscle group that causes the outstretched arm and hand to flap like a wing; occurs in disorders that cause encephalopathy; also called **asterixis** (AS-ter-ICK-sis).

• Clotting Abnormalities • The person with cirrhosis bruises and bleeds easily because the liver is unable to make an adequate supply of clotting factors. Problems with clotting raise the risk of a fatal hemorrhage if varices develop and bleed. Blood loss in the GI tract also contributes to rising blood ammonia.

• Insulin Resistance • Insulin resistance can lead to fatty liver, which may eventually progress to cirrhosis, and people with cirrhosis often develop insulin resistance, hyperinsulinemia, and hyperglycemia, although the reasons are unclear. In some cases, the pancreas is unable to produce enough insulin to overcome the insulin resistance, and diabetes mellitus develops.

• Malnutrition and Wasting • As liver deterioration progresses, malnutrition and wasting become evident. Table 28-2 summarizes possible causes of wasting in people with liver disorders, which can include anorexia and reduced food intake, altered metabolism, and malabsorption. Fat malabsorption occurs when a liver disorder interferes with bile salt production, when the flow of bile from the liver is obstructed, or when the person develops pancreatic insufficiency.◆

◆ Reminder: Alcoholism is a common cause of chronic pancreatitis.

In advanced liver disorders, the liver's inability to activate nutrients can result in deficiencies despite an adequate intake. In that case, providing the nutrient in its active form may prevent deficiencies. One example involves methionine, an essential amino acid. To perform its functions, methionine needs to be activated to S-adenosylmethionine (SAMe) in the liver. Researchers have shown that providing methionine as SAMe can reduce mortality in some children with cirrhosis.[3]

• Other Consequences • When liver function deteriorates significantly, the alterations in fluid and electrolyte balances, as well as the buildup of waste products in the blood, can precipitate kidney failure and the **hepatorenal syndrome.** Bone disorders that are generally unresponsive to vitamin D and calcium supplementation can also become a problem, especially for people with cirrhosis and malabsorption.

Treatment of Cirrhosis

Treatment of cirrhosis aims to preserve remaining organ function, reversing the damage that has occurred to whatever extent is possible, and to prevent and control complications. Treatment depends, in part, on addressing the underlying medical condition leading to cirrhosis.◆ When cirrhosis has progressed to a severe stage, liver transplantation becomes a treatment option in some cases.

◆ For people with hemochromatosis, treatment may include surgically opening a vein to withdraw blood (phlebotomy), which reduces the body's stores of iron by stimulating red blood cell production.

• Drug Therapy • One goal of drug therapy in cirrhosis is to control blood ammonia levels. About two-thirds of the body's ammonia comes from the intestine (see Figure 28-3) where bacteria make ammonia from undigested dietary protein, protein from shed intestinal cells, and protein from GI bleeding. Digestive enzymes also produce some ammonia as they dismantle proteins. Thus drug therapy often includes antibiotics to limit the growth of intestinal bacteria and laxatives to speed intestinal transit time and provide less time for ammonia absorption.

Other medications may include antihypertensives to control portal hypertension and diuretics to reduce fluid retention and prevent ascites and edema.[4]

TABLE 28-2 Possible Causes of Wasting in Liver Disease

Reduced Nutrient Intake	Excessive Nutrient Losses	Raised Nutrient Needs
Abdominal pain	Blood loss through GI bleeding	Ascites (raises metabolic rate)
Anorexia	Diarrhea	Infections
Early satiety	Malabsorption	Inflammation
Esophageal varices	Medications	Malnutrition
Medications	Steatorrhea	Medications
Nausea	Vomiting	
Restrictive diet		
Vomiting		

hepatorenal syndrome: the combined symptoms of liver and renal failure that occur as a consequence of severe liver damage.

Diet-Drug Interactions

ANTIDIABETIC AGENTS

For *antidiabetic agents,* see p. 760.

ANTIHYPERTENSIVES

For *antihypertensives,* see p. 791.

ANTI-INFECTIVES

The antibiotic frequently used to control blood ammonia levels in liver disease is *neomycin.* Because neomycin kills bacteria in the intestine, the person may need a vitamin K supplement. The person may be prone to other intestinal infections that ordinarily are suppressed by normal bacterial flora. Neomycin can also be toxic to the kidneys. The antiviral agents *ribavirin* and *lamivudine* may be used in the treatment of liver diseases. These will be described in the box in Chapter 29.

ANTI-INFLAMMATORY AGENTS

For *anti-inflammatory agents,* see p. 673.

ANTILIPEMICS

For *antilipemics,* see p. 791.

APPETITE STIMULATES

Megestrol acetate and *dronabinol* (a derivative of marijuana) stimulate the appetite and result in weight gain. GI side effects are uncommon, although some people develop nausea, vomiting, and diarrhea. Dronabinol produces euphoria in some people and can reach toxic levels in people with liver dysfunction.

DIURETICS

Spironolactone is a potassium-sparing diuretic frequently used to treat ascites. People taking spironolactone should avoid foods high in potassium, potassium supplements, and salt substitutes that contain potassium. See also p. 791.

IMMUNOSUPPRESSANTS

Immunosuppressants are used for people who undergo liver transplants. For *cyclosporine* and a general discussion of immunosuppressants, see pp. 816 and 818. Another immunosuppressant sometimes used to prevent tissue rejection following a liver transplant is *tacrolimus.* Tacrolimus can be given with food to reduce nausea, vomiting, and GI upsets. Other nutrition-related side effects include anorexia, diarrhea, constipation, anemia, hyperglycemia, potassium imbalances, low blood magnesium levels, and ascites.

INTERFERON

Interferon can lead to nausea, vomiting, weight loss, fever, fatigue, and depression.

LAXATIVES

The laxative commonly used in the treatment of liver disease is *lactulose.* Lactulose can cause belching, cramps, and diarrhea. Encourage clients to replace fluid and electrolytes lost through diarrhea.

PENICILLAMINE

Penicillamine is used in the treatment of Wilson's disease. Foods decrease the absorption of penicillamine, and it should be taken on an empty stomach. Nutrition-related side effects may include altered taste perceptions, anorexia, nausea, vomiting, diarrhea, vitamin B_6 deficiency, iron deficiency, and loss of protein in the urine. The person should avoid foods and supplements that contain copper. Iron supplements should be taken at least two hours before or after penicillamine is given.

Oral antidiabetic agents may be used to control hyperglycemia. Cholestyramine, an antilipemic, may be used to alleviate the excessive itching that may accompany liver disease. Megestrol acetate and dronabinol may help improve the appetite and stimulate weight gain.[5] Interferon and ribavirin may be used to improve immune responses in people with cirrhosis caused by viral hepatitis. People with Wilson's disease may be treated with penicillamine, which binds with copper and allows it to be excreted in the urine. The Diet-Drug Interactions box describes the nutrition-related concerns associated with these medications.

Promising treatments for liver disease include the dietary supplement milk thistle (silymarin), which may prove to have a protective effect on liver cells.[6] Milk thistle increases protein synthesis in liver cells and acts as an antioxidant and free-radical scavenger. An extract from soybeans, polyenylphosphatidylcholine (PPC), may also protect the livers of alcoholics with liver disease, even if they continue to consume alcohol.[7]

THINK NUTRITION whenever clients have cirrhosis. As liver failure progresses, malnutrition and wasting are frequent complications, and diet can help alleviate many symptoms and prevent complications.

• **Medical Nutrition Therapy** • Medical nutrition therapy for people with cirrhosis carefully considers each client's needs for energy, protein, sodium, and fluid (see Table 28-3). Needs vary considerably and depend on the complications present in each case. In all cases, clients with cirrhosis must abstain completely from alcohol and avoid using unprescribed medications, herbal supplements, and drugs to protect the liver from further injury.

• **Energy** • The compromised liver needs energy to protect its functioning cells and to prevent protein catabolism. People with ascites, malabsorption, or infections have higher energy needs than people without these complications. For people with ascites, energy needs are based on the person's desirable or estimated dry weight (weight without ascites) to avoid overestimating energy needs.[8] Energy needs range from 120 to 175 percent of the basal energy expenditure (see Table 24-3 on p. 734).

TABLE 28-3 Diet Guidelines for Liver Failure

Energy

- Without ascites, malnutrition, or infection: 120% of basal energy expenditure (BEE); see Table 24-3 on p. 734.
- With ascites, malnutrition, or infection: 150–175% of BEE.

Protein

- Not sensitive to protein: 1.0 to 1.5 g protein/kg body weight/day.
- Protein sensitive: Start with 0.5 to 0.7 g protein/kg body weight/day; increase gradually to level of tolerance with a goal of at least 1.0 g protein/kg body weight/day and up to 1.5 g protein/kg body weight/day if the person is malnourished.
- Very sensitive to protein (unable to tolerate at least 0.8 g protein/kg body weight/day): Consider vegetable-based diets or enteral supplements enriched with branched-chain amino acids and low in aromatic amino acids.

Carbohydrate

- Not restricted.
- For people with insulin resistance or diabetes, provide up to 50 to 60% of kcalories from carbohydrates (mainly complex carbohydrates) with a consistent carbohydrate intake from day to day and at each meal and snack.

Fat

- Not restricted unless fat malabsorption is present.
- With fat malabsorption: Restrict fat only as necessary to control steatorrhea (see Chapter 21); use medium-chain triglycerides (MCT) to increase kcalories as necessary.

Sodium

- Restrict only as necessary to control ascites, but not less than 2 g sodium/day in most cases.

Fluid

- Restrict to 1.0 to 1.5 liters per day for people with ascites who also have low blood sodium levels. In severe cases, limit fluids to 500–750 milliliters plus urinary output/day.

Vitamins and minerals

- Ensure adequate intake from diet or supplements based on individual needs.

SOURCE: Adapted from J. Hasse and coauthors, Nutrition therapy for end-stage liver disease: A practical approach, *Support Line,* August 1997, pp. 8–15.

• Protein • For people with cirrhosis who do not show signs of impending coma, protein restrictions are unnecessary, and a high-protein diet (1.0 to 1.5 grams of protein per kilogram of body weight per day) is appropriate.[9] For people who develop signs of impending coma, the cause is investigated, and the underlying condition is treated. In a small number of people (7 to 9 percent), dietary protein can precipitate hepatic encephalopathy.[10] For people who are protein sensitive, the diet plan is low in protein at first, and protein is gradually increased. The final amount of protein that is optimal depends on the person's nutrition status and individual tolerance to protein. For a person who is malnourished, up to 1.5 grams of protein per kilogram of body weight is the goal. The final level of protein the person tolerates may be less than this amount, however. Sometimes medications can be adjusted to control symptoms, allowing a higher protein intake.

The foods shown here provide 60 grams of protein—a day's protein allowance for some people with liver disease.

In some cases, the person who is protein sensitive can tolerate only very low protein intakes (0.5 to 0.7 gram of protein per kilogram of body weight per day). Although research is limited and controversial, many clinicians recommend special enteral (or parenteral, when needed) formulas that are low in aromatic amino acids and high in branched-chain amino acids for these clients (see Appendix J).[11] Some studies suggest that vegetable proteins may be better tolerated than meat proteins, perhaps because vegetables contain fewer ammonia-forming constituents and aromatic amino acids and more branched-chain amino acids than meats do. In addition, the fiber from plant foods speeds intestinal transit time, reducing the amount of time for ammonia absorption from the intestine.

If a person lapses into a coma, all protein is withheld temporarily until the cause is determined and corrective measures are taken. For protein-sensitive people, protein is then gradually reintroduced in small amounts and increased gradually as tolerated.

• Carbohydrate and Fat • In a diet for liver disease, carbohydrate generally provides 50 to 60 percent of energy needs. A high-carbohydrate (50 grams) snack at bedtime may help prevent excessive breakdown of fat and protein during an overnight fast.[12] People with hyperglycemia should follow the guidelines for diets in diabetes; that is, eat mostly complex carbohydrates and eat them at consistent times throughout the day.

Because fat helps make foods appetizing and delivers energy efficiently, fat plays an important role in the diet of a person with cirrhosis. Fat is restricted only if the person develops fat malabsorption. In these cases, medium-chain triglycerides (MCT) can provide additional kcalories. People with liver disease related to the biliary tract or those in the later stages of cirrhosis are most likely to malabsorb fat.

• Sodium and Fluid • For people with ascites, the diet restricts sodium. The lower the sodium intake, the more quickly ascites resolves. Very-low-sodium diets are unpalatable to many people, however, so most clinicians recommend a diet that allows from 2 to 4 grams of sodium and depend on diuretics to mobilize excess fluids.

In people with ascites, the amount of sodium in the body is excessive, even though blood levels are sometimes low. In these cases, the low blood sodium levels **(hyponatremia)** signify fluid overload. Thus, when people with ascites have low blood sodium, adding sodium to the diet is inappropriate; instead, fluids may be restricted to about 1.0 to 1.5 liters per day. To assess changes in fluid balance, health care professionals monitor weight and measure abdominal girth.◆ Rapid weight gain and an increasing abdominal girth indicate fluid retention; sudden weight loss and decreasing abdominal girth indicate successful fluid excretion.

◆ Abdominal girth is measured by placing a tape measure around the back and over the person's umbilicus.

• Vitamins • The liver's central role in the metabolism and storage of vitamins and minerals, combined with coexisting conditions (such as malabsorption, alcoholism, and malnutrition), explains why nutrient deficiencies commonly occur in people with liver diseases. Virtually all people with advanced liver

hyponatremia (HIGH-poe-nay-TREE-mee-ah): low levels of sodium in the blood.

HOW TO Help the Person with Cirrhosis Eat Enough Food

People with cirrhosis often have difficulty eating enough food to prevent malnutrition and its consequences. Ascites and gastrointestinal symptoms such as nausea and vomiting can interfere with food intake. The person with encephalopathy may be confused about what to eat or have little interest in eating. People with sodium restrictions may have difficulty adjusting to a low-salt diet. To facilitate diet compliance:

- Individualize the diet plan based on each person's symptoms and responses. The diet should restrict protein, fat, sodium, or fluid only when such restrictions are warranted.
- Use the suggestions for improving food intake in the "How to" box in Chapter 17.
- Advise clients to eat frequent small meals and to eat a high-carbohydrate bedtime snack. Eating frequent small meals can help with nausea and can also improve glucose tolerance. For people who are protein sensitive, eating small amounts of protein frequently throughout the day improves tolerance.
- Recommend the suggestions in the box on p. 816 for people who must restrict fluids and proteins.
- Point out that there is probably no food that cannot be incorporated into the diet on special occasions and in limited amounts. Advise clients to talk with the dietitian about how to change their meal plans so that favorite foods can be incorporated from time to time.

The low-sodium diet used in the treatment of cirrhosis may be more restrictive than the low-salt diet recommended for hypertension. Table 27-3 on p. 808 describes a 2-gram sodium-restricted diet. Some suggestions that can help clients adhere to their sodium restrictions include:

- Suggest that clients replace the salt they use for cooking and seasoning with herbs or spices like basil, bay leaves, curry, garlic, ginger, lemon, mint, oregano, rosemary, and thyme.
- Suggest that clients experiment with low-sodium products to find the ones they like. (People on potassium-sparing diuretics should be cautioned to avoid salt substitutes that replace sodium with potassium.)
- Advise clients to check food labels to learn the sodium content of the foods they eat. They may be able to find similar products with lower sodium contents.

Continue to offer support and encouragement to the client with cirrhosis. Severe weight loss is less likely to occur if nutrition intervention is provided before problems progress too far.

disease require supplements of some vitamins, as well as minerals. Physicians determine which nutrients to supplement by monitoring serum levels and checking for clinical signs of deficiencies.

The B vitamins serve as cofactors for the liver's many metabolic reactions and repair work; deficiencies of thiamin, vitamin B_6, riboflavin, and folate are common. Fat-soluble vitamins may need to be supplemented if fat malabsorption develops. If the diseased liver fails to synthesize adequate amounts of retinol-binding protein, body tissues may not receive the vitamin A they need. Vitamin D deficiencies may develop if the liver is unable to perform its roles in the activation of vitamin D. Although the bone diseases associated with liver disease are generally not responsive to vitamin D supplements, supplements are provided to ensure adequate vitamin D intakes.◆ Vitamin K deficiencies can prolong the time it takes for blood to clot,◆ a dangerous complication that increases the risk of massive bleeding from the GI tract. (Recall that liver damage itself can interfere with blood coagulation.)

• Other Minerals • Calcium deficiencies can develop from three causes in people with cirrhosis: steatorrhea, low serum albumin (albumin, which carries calcium in the blood, is manufactured in the liver), and impaired vitamin D metabolism. Thus it is important that the diet or supplements provide adequate calcium to prevent deficiencies. Magnesium deficiencies are also common. Zinc stores may also be depleted as a consequence of liver damage, and limited research suggests that zinc deficiencies may be implicated in the development of hepatic encephalopathy.[13] Zinc may also help prevent the muscle cramps frequently experienced by people with liver disorders.[14]

• Enteral and Parenteral Nutrition • People with cirrhosis face many difficulties in eating enough food to maintain nutrition status. The accompanying box offers tips for maintaining an adequate intake. If the person with cirrhosis cannot take enough food or formula by mouth, tube feedings or TPN is indicated. As mentioned earlier, enteral and parenteral formulas designed for liver disease provide fewer aromatic and more branched-chain amino acids than standard formulas (see Appendix J). These formulas are generally reserved for people highly-sensitive to protein. Both enteral and parenteral nutrition have been used successfully in people with cirrhosis.

People with bleeding esophageal varices will be unable to consume food by mouth and are often given simple intravenous solutions to maintain fluid and electrolyte balances. When they begin to eat, they are frequently given liquid, and then soft, foods. Parenteral nutrition should be considered if the person is malnourished or if the health care team anticipates that oral intake will not resume for an extended time. The case study on p. 839 asks you to use clinical knowledge and judgment in answering questions about cirrhosis.

◆ Vitamin D is supplied as 25-hydroxyvitamin D (ergocalciferol).

◆ One laboratory test that evaluates the time it takes for blood to clot is called the **prothrombin time.** Both vitamin K deficiency and liver disease can prolong the prothrombin time.

CASE STUDY

Carpenter with Cirrhosis

Mr. Sloan, a 48-year-old carpenter, has been hospitalized many times and has been diagnosed with alcoholic cirrhosis. He recognizes his problem with alcohol abuse and has entered alcohol rehabilitation programs several times over the last few years. Nevertheless, he is still drinking. Mr. Sloan was recently admitted to the hospital with advanced liver diseae and signs of impending hepatic coma. At 5 feet 7 inches tall, Mr. Sloan, who once weighed 150 pounds, now weighs 120 pounds. He is jaundiced and looks thin, although his abdomen is distended with ascites. Laboratory findings include elevated AST, ALT, alkaline phosphatase, and blood ammonia. Compare these findings with Table 28-1 to determine if they are consistent with liver disease.

1. Can you explain to Mr. Sloan what cirrhosis is and what its consequences are? From the limited information available, what can you determine about Mr. Sloan's nutrition status? What medical problem makes it difficult to interpret Mr. Sloan's actual weight? How can his weight measurements help determine if his condition is improving? Calculate Mr. Sloan's energy needs. What factors influence protein needs for a person with impending hepatic coma? What signs suggest that a person is in a precoma state?
2. Why is Mr. Sloan's abdomen distended? Explain the development of ascites in liver disease and how diet is adjusted.
3. Would you expect Mr. Sloan's blood ammonia levels to be high? Why or why not?
4. Describe portal hypertension, jaundice, and esophageal varices. How would Mr. Sloan's diet be changed if he were found to have bleeding esophageal varices?

IN SUMMARY

In cirrhosis, a complication of a variety of disorders, liver cells undergo permanent changes, and function falters. Consequences of liver failure may include portal hypertension, varices in the esophagus and GI tract, ascites and edema, elevated blood ammonia levels, hepatic encephalopathy, hepatic coma, clotting abnormalities, insulin resistance, malnutrition, renal failure, and bone disease. Therapy often includes medications and a carefully tailored diet that considers the client's energy, protein, sodium, fluid, vitamin, and mineral needs. Once liver failure reaches its final stages, liver transplantation is necessary to sustain life.

LIVER TRANSPLANTATION

If liver failure progresses to a severe and irreversible state, liver transplantation becomes an option in some cases. A liver transplant candidate must often wait for a liver donor before surgery is possible. Wise health care professionals use this time to identify and correct nutrient imbalances whenever possible.

THINK NUTRITION whenever clients are awaiting liver transplants. The waiting period provides an opportunity to improve nutrition status and minimize the risk of complications.

• **Nutrition before Transplantation** • In severe liver failure, malnutrition has often progressed for some time. Clinicians report malnutrition in over 70 percent of liver transplant recipients and note that malnutrition increases the risk of complications following a liver transplant.[15] The person equipped with adequate nutrient stores faces the transplant better prepared to fight infections, heal wounds, and mount a stress response.

Difficulties arise in assessing nutrition status in liver transplant candidates because the metabolic effects of liver dysfunction and those of malnutrition are difficult to distinguish. Edema may mask weight loss and alter other anthropometric measurements. Fluid retention can also interfere with laboratory tests.

• **Nutrition following Transplantation** • Following liver transplantation, liver function determines nutrient needs. All people are hypermetabolic after surgery, and energy needs must be met. Immunosuppressant drugs given to prevent tissue rejection can contribute to nutrient imbalances by causing nausea, vomiting, diarrhea, and mouth sores. The nutrition support team may use indirect calorimetry to estimate energy needs and carefully monitors clinical and laboratory data so that specific nutrient recommendations can be made.

Although TPN has been the traditional source of nutrients following transplantation, researchers report equal success with intestinal tube feedings.[16] Early enteral nutrition may reduce the incidence of infection, a particularly important consideration for people with suppressed immune systems.

IN SUMMARY

Halting or delaying the progression of liver dysfunction depends in large part on attention to nutrition and nutrition assessment parameters. Once liver failure progresses, a liver transplant becomes a life-sustaining option. Attention to nutrition needs prior to and after a liver transplant provides the best assurance that a person will be able to heal wounds and fight infections. The accompanying Nutrition Assessment Checklist reviews important points to keep in mind when assessing the nutrition status of people with liver disorders.

Nutrition Assessment Checklist for People with Liver Disorders

MEDICAL HISTORY

Check the medical record to determine:

- ☐ Type of liver disorder
- ☐ Cause of liver disorder
- ☐ If the client has received a liver transplant

Review the medical record for complications that may alter medical nutrition therapy including:

- ☐ Malnutrition
- ☐ Esophageal varices
- ☐ Ascites
- ☐ Hepatic encephalopathy
- ☐ Hepatic coma
- ☐ Insulin resistance/diabetes mellitus
- ☐ Heart failure
- ☐ Pancreatitis
- ☐ Malabsorption
- ☐ Renal failure

MEDICATIONS

For clients with liver dysfunction, note that risk of diet-drug interactions is very high because many drugs are metabolized in the liver. The risk of interactions is intensified for clients with:

- ☐ Ascites (medications may take a long time to reach the liver)
- ☐ Renal failure (medications are often metabolized further in the kidneys and excreted in the urine)
- ☐ Malnutrition
- ☐ Multiple medication prescriptions and long-term medication use

FOOD/NUTRIENT INTAKE

For clients with fatty liver, pay special attention to the client's intake of:

- ☐ Energy, if the client is overweight or malnourished, has diabetes, or is receiving TPN
- ☐ Carbohydrate, if the client has diabetes or is receiving TPN
- ☐ Alcohol

For clients with hepatitis and cirrhosis, make a note of:

- ☐ Appetite
- ☐ Adequacy of energy and nutrient intake
- ☐ Alcohol use

For clients with protein-sensitive encephalopathy:

- ☐ Ensure that total energy intake is adequate.
- ☐ Find the level of protein restriction that is appropriate for the individual.
- ☐ Base energy needs on desirable or estimated dry weight to avoid overfeeding.

ANTHROPOMETRICS

Take baseline height and weight measurements and monitor weight regularly. For clients with ascites and edema:

- ☐ Use weight to monitor the degree of fluid retention.
- ☐ Remember that the client may be malnourished even though weight may be deceptively high.

LABORATORY TESTS

Note that albumin and serum proteins are often reduced in people with liver disease and are difficult to interpret as an indicator of nutrition status. Review the following laboratory test results to assess liver function (see Table 28-1):

- ☐ Albumin
- ☐ Alkaline phosphatase
- ☐ ALT
- ☐ Ammonia
- ☐ AST
- ☐ Bilirubin
- ☐ Prothrombin time

Check laboratory test results for complications associated with liver failure including:

- ☐ Anemia
- ☐ Fluid retention
- ☐ Hyperglycemia
- ☐ Renal function tests

PHYSICAL SIGNS

Look for physical signs of:

- ☐ Fluid retention (ascites and edema)
- ☐ PEM (muscle wasting and unintentional weight loss)
- ☐ B vitamin deficiencies
- ☐ Fat-soluble vitamin deficiencies
- ☐ Calcium deficiency
- ☐ Magnesium deficiency
- ☐ Potassium imbalances
- ☐ Zinc deficiency

Nutrition on the Net

WEBSITE
Access these websites for further study of topics covered in this chapter.

- Find updates and quick links to these and other nutrition-related sites at our website: **www.wadsworth.com/nutrition**
- To learn more about preventing liver disorders, as well as obtain information about fatty liver, hepatitis, cirrhosis, diseases related to these conditions, and links to other websites, visit the American Liver Foundation: **www.liverfoundation.org** and the Canadian Liver Foundation: **www.liver.ca.**
- To find out more about resources and support for children with liver diseases and liver transplants, visit the Children's Liver Alliance: **www.livertx.org.**
- To review information about hepatitis, visit the Hepatitis Education Project: **www.scn.org/health/hepatitis/index.htm**, the Hepatitis Foundation International: **www.hepfi.org**, and the Hepatitis B Foundation: **www2.hepb.org.**
- To uncover more information about liver transplants, search the Center Span Transplant News Network: **www.centerspan.org.**

INTERNET ACTIVITIES
Damage to the liver can have profound effects on nutrition and health. Liver damage can occur as a consequence of many different disorders, including diabetes, alcohol toxicity, and viral infections. In the best interest of your health and the health of those you care about, you may want to know more about liver damage, its consequences, and most importantly, prevention.

▼ Go to: **www.liver.ca**

▼ Look at the left-hand column on the screen and click on "Liver Disease."

▼ Click on a type of liver disease or related topic and read questions and answers about the disease or topic.

What lifestyle factors can affect the risk of liver damage? What changes in your own diet or lifestyle might you make to reduce your risk of liver damage?

Study Questions

These questions will help you review the chapter. You will find the answers in the discussions on the pages provided.

1. What is fatty liver? Describe the ways fatty liver can develop. What is the treatment for fatty liver? What role does nutrition play in the treatment? (p. 828)
2. What is hepatitis, and what are its causes? What nutrition concerns arise in the person with hepatitis? (pp. 828–829)
3. Discuss cirrhosis. How is it associated with portal hypertension, esophageal varices, ascites, elevated blood ammonia, changes in mental status, clotting abnormalities, insulin resistance, malnutrition, renal failure, and bone disease? (pp. 830–834)
4. Describe the dietary treatment of the person with cirrhosis and hepatic coma. Review the special dietary concerns of the person with ascites and esophageal varices. (pp. 836–838)
5. How does nutrition status influence recovery from a liver transplant? (pp. 839–840)

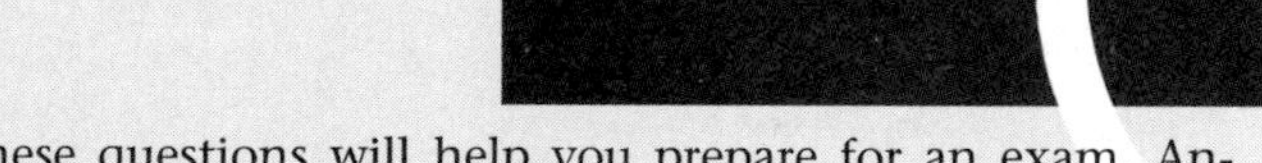

These questions will help you prepare for an exam. Answers can be found on p. 843.

1. Which of the following diet strategies would be most appropriate for helping to reverse fatty liver associated with diabetes mellitus?
   a. low-protein diet
   b. fat-restricted diet
   c. fluid- and sodium-restricted diet
   d. energy to achieve or maintain a desirable weight with a consistent intake of carbohydrate
2. Which of the following statements about hepatitis is true?
   a. Chronic hepatitis can progress to cirrhosis.
   b. Regardless of the type of hepatitis, symptoms are severe.
   c. People with hepatitis always need high-kcalorie, high-protein diets.
   d. HCV infections are often mild and can be spread through contaminated foods and water.

3. The consequences of cirrhosis are primarily due to:
   a. chronic malnutrition.
   b. chronic alcohol abuse.
   c. liver cell damage and altered hepatic blood flow.
   d. elevated blood ammonia, amino acid, and sodium levels.
4. Esophageal varices are a dangerous complication of liver disease primarily because they:
   a. interfere with food intake.
   b. can lead to massive bleeding.
   c. divert blood flow from the GI tract.
   d. cause portal hypertension and collateral development.
5. The condition(s) in liver disease most likely to lead to ascites is (are):
   a. portal hypertension.
   b. rising blood ammonia levels.
   c. elevated serum albumin levels.
   d. insulin resistance and diabetes mellitus.
6. For the person with cirrhosis, short-term memory loss and an inability to concentrate are signs of:
   a. coma.
   b. encephalopathy.
   c. hyperammonemia.
   d. hepatorenal syndrome.
7. Medical nutrition therapy for people with cirrhosis includes diets that are:
   a. restricted in fat.
   b. high in kcalories and protein.
   c. based on symptoms and responses to treatment.
   d. restricted in protein, carbohydrate, sodium, and fluid.
8. Which of the following statements regarding diet and cirrhosis is false?
   a. People with ascites need to restrict sodium.
   b. People with steatorrhea may need to restrict fat.
   c. People with ascites and low blood sodium levels need to restrict sodium and fluid.
   d. People who are protein sensitive always need to restrict protein to 0.5 to 0.7 gram of protein per kilogram of body weight per day.
9. With respect to vitamins and minerals, people with cirrhosis:
   a. may develop calcium deficiencies.
   b. seldom require nutrient supplements.
   c. frequently develop vitamin C deficiencies.
   d. may develop clotting abnormalities associated with vitamin A deficiency.
10. For the person undergoing a liver transplant:
    a. immunosuppressant drugs seldom alter nutrient needs.
    b. enteral nutrition is contraindicated following the transplant.
    c. attention to nutrition before a transplant improves chances of recovery.
    d. provided adequate nutrition is given following a transplant, nutrition status before surgery has little impact on recovery.

1. Think about the problems a person might encounter when following an 1800-kcalorie diet that is restricted to 40 grams of protein. On such a diet, the total protein allowance could be used up by just one scrambled egg, 3 ounces of meat, a cup of milk, and two slices of bread. Using Appendix H, Figure 2-4 on pp. 44–45, and Table 2-3 on p. 42 for reference, calculate the kcalories these foods provide. Then consider what other foods to add to the diet to meet energy needs without contributing additional protein.

   Now compare the foods listed with the Daily Food Guide on pp. 38–39. Which food groups have you offered in the recommended quantities? Which food groups are in short supply on such a diet? Which nutrients might be low? How might fats and sugars be useful in such a diet?

2. Chapter 24 described acute stresses and discussed how the combination of hypermetabolism and infection can lead to multiple organ failure. The respiratory system, kidneys, and liver may all fail as a consequence of multiple organ failure. On a sheet of paper make four columns, one for stress, one for respiratory failure, one for kidney failure, and one for liver failure. Under each column, list the energy, protein, fat, fluid, and other nutrient considerations for each disorder individually. Which nutrient modifications are common to all disorders? Do some of the modifications necessary for one disorder conflict with those for another? Consider why providing adequate nutrients, but not providing too much, is important for supporting the function of each organ.

3. Think about the relationship of ammonia to urea (see pp. 218–219) and explain why you would expect blood ammonia levels to rise in liver disease and blood urea nitrogen levels to rise in renal disease.

References

1 J. A. Shad, C. G. Chinn, and O. S. Brann, Acute hepatitis after ingestion of herbs, *Southern Medical Journal* 92 (1999): 1095–1097.

2 N. K. Krane and P. Gaglio, Viral hepatitis as a cause of renal disease, *Southern Medical Journal* 92 (1999): 354–360.

3 C. S. Lieber, Alcoholic liver disease: New insights in pathogenesis lead to new treatments, *Journal of Hepatology* (supplement 1) 32 (2000): 113–128.

4 J. S. Crippin, Medical management of end-stage liver disease: A bridge to transplantation, *Support Line,* August 1997, pp. 3–7.

5 G. Gurk-Turner, Management of the metabolic complications of liver disease: An overview of commonly used pharmacologic agents, *Support Line,* August 1997, pp. 17–19.

6 J. E. Robbers and V. E. Tyler, *Tyler's Herbs of Choice* (Binghamton, N.Y.: Hawthorn Herbal Press, 1999), pp. 76–79.

7 C. S. Lieber, Alcohol: Its metabolism and interaction with nutrients, *Annual Reviews of Nutrition* 20 (2000): 395–430; Lieber, Alcoholic liver disease: New insights in pathogenesis lead to new treatments, 2000.

8 J. Hasse and coauthors, Nutrition therapy for end-stage liver disease: A practical approach, *Support Line,* August 1997, pp. 8–15.

9 G. L. Braden, Clinical Gastroenterology: Highlights of the Annual Postgraduate Course, Part II, 1999 American College of Gastroenterolgy Annual Scientific Meeting, available from Medscape at http://gastroenterology.medscape.com/Medscape/CNO/1999/ACG/ACG-06.html, site visited March 17, 2001; C. Corish, Nutrition and liver disease, *Nutrition Reviews* 55 (1997): 17–20.

10 As cited in Hasse and coauthors, 1997.

11 A. Fabri and coauthors, Overview of randomized clinical trials of oral branched-chain amino acid treatment in chronic hepatic encephalopathy, *Journal of Parenteral and Enteral Nutrition* 20 (1996): 159–164.

12 W. Chang and coauthors, Effects of extra-carbohydrate supplementation in the late evening on energy expenditure and substrate oxidation in patients with liver cirrhosis, *Journal of Parenteral and Enteral Nutrition* 21 (1997): 96–99.

13 Hasse and coauthors, 1997.

14 M. Kugelmas, Preliminary observation: Oral zinc sulfate replacement is effective in treating muscle cramps in cirrhotic patients, *Journal of the American College of Nutrition* 19 (2000): 13–15.

15 J. Pikul and coauthors, Degree of preoperative malnutrition is predictive of postoperative morbidity and mortality in liver transplant recipients, *Transplantation* 57 (1994): 469–472; J. Hasse, Nutrition and transplantation, *Nutrition in Clinical Practice* 8 (1993): 3–4.

16 J. M. Hasse, Early enteral nutrition support in patients undergoing liver transplantation, *Journal of Parenteral and Enteral Nutrition* 19 (1995): 437–443; C. Wicks and coauthors, Comparison of enteral feeding and total parenteral nutrition after liver transplantation, *Lancet* 344 (1994) 837–840.

ANSWERS

Multiple Choice

1. d	2. a	3. c	4. b	5. a
6. b	7. c	8. d	9. a	10. c.

GALLSTONES

AFTER THE liver produces bile, it travels to the gallbladder via the bile ducts, where it remains in storage until it is needed for the digestion of fat. As Chapter 28 discussed, disorders that affect the liver's production of bile or its secretion into the biliary tract or intestine can damage the liver and result in serious liver disease. A far more common and often less serious disorder related to bile and the biliary tract is **gallstones.** Gallstones affect an estimated 20 million Americans, or about 10 percent of the population. This highlight provides an overview of gallstones and the role of diet in their prevention. The glossary defines terms related to disorders of the biliary tract.

WHAT ARE GALLSTONES?

Gallstones are solid masses of material that form in the gallbladder from bile. Bile provides the vehicle for excreting excess cholesterol from the body. In addition to cholesterol, bile contains water, bile salts, and bile pigments (bilirubin). While stored in the gallbladder, bile becomes more concentrated as water is slowly extracted from it. Sometimes the hardened material eventually clumps together and forms stones. The formation or presence of stones in the gallbladder is called **cholelithiasis.** The vast majority of gallstones (about 80 percent of all cases) contain primarily cholesterol; the rest (20 percent) contain mainly calcium salts and bilirubin.

Most gallstones are made primarily of cholesterol; they can be as small as a grain of sand or as large as a ping-pong ball.

© C. James Webb/PhotoTake

• Cholesterol Gallstones • Most clinicians believe that cholesterol gallstones form when bile contains too much cholesterol, when substances present in bile allow cholesterol to crystalize, and when the gallbladder does not contract forcibly enough to regularly empty its contents. Sluggish motility in the large intestine may also play a role.[1] The cholesterol in bile precipitates as small crystals, which gradually fuse together and form stones. Stones can be tiny specks or as large as ping-pong balls. Most stones are less than about an inch in diameter.

• Pigment Stones • Pigment stones form when excess bile pigments crystalize in bile and grow. Pigment stones are more likely to develop in people with cirrhosis, biliary tract infections, or some hereditary blood disorders, including sickle-cell anemia. The remainder of this discussion focuses on cholesterol stones because they are more common than pigment stones and more is known about their causes.

Consequences of Gallstones

Most often, gallstones cause no symptoms **(silent stones)** and require no treatment. Once symptoms develop, however, the symptoms persist and occur more frequently. The longer a stone remains lodged in the gallbladder, the greater the likelihood that symptoms will develop.

• Common Symptoms • Steady, intense pain in the upper abdomen that increases in intensity and lasts from 30 minutes to several hours develops when gallstones become symptomatic. Sometimes the pain spreads to the chest and shoulders and mimics the symptoms of a heart attack. Indigestion, nausea, vomiting, gas, and bloating may also occur. For some, symptoms begin following a fatty meal; for others, the pain occurs at night and awakens them from sleep. When a stone enters the common bile duct, the pain that results is called **biliary colic.**

Immediate medical attention is required for anyone who has pain that does not resolve over time; develops fever, chills, severe nausea and vomiting, or jaundice; or excretes clay-colored stools. In such cases, the gallstones may be causing serious problems.

• Complications • A serious complication can occur if a gallstone lodges in the bile duct **(choledocholithiasis)** as it is expelled from the gallbladder during a contraction. A stone can obstruct the flow of bile or lead to inflammation of the gallbladder **(cholecystitis)** or the bile duct itself **(cholangitis).** Acute pancreatitis can follow if the stone obstructs the pancreatic duct.

GLOSSARY

biliary colic: pain associated with gallstones that have entered the common bile duct.

cholangitis (KOH-lan-JYE-tis): inflammation of the bile ducts.

cholecystectomy (KOH-lee-sis-TEK-toe-mee): surgical removal of the gallbladder.

cholecystitis (KOH-lee-sis TYE-tis): inflammation of the gallbladder.

choledocholithiasis (koh-LED-oh-koh-lih-THIGH-ah-sis): the presence of gallstones in the common bile duct.

cholelithiasis (KOH-lee-lih-THIGH-ah-sis): the formation or presence of stones in the gallbladder or common bile duct.
- **chole** = bile
- **liti** = stone
- **iasis** = condition

gallstones: crystals of cholesterol or bile pigments that precipitate together and form a hard mass.

silent stones: gallstones that cause no symptoms.

Gallstones are the most common cause of acute pancreatitis. If the bile ducts are blocked over the course of years, the liver cells are progressively destroyed, and liver failure can follow. Elderly people and people with diabetes may develop serious complications without first experiencing the symptoms of gallstones.[2] Figure H28-1 shows the location of the gallbladder and the bile ducts.

Treatments for Gallstones

As described earlier, no treatment is necessary for gallstones that do not produce symptoms. Medications and surgery can be used to treat gallstones once they become a problem.

• Medications • In some cases, a medication called ursodiol, a natural bile salt, can dissolve a gallstone. Ursodiol reduces the amount of cholesterol the liver adds to the bile. Ursodiol works best for small stones. Although symptoms may resolve shortly after a person begins taking this medication, it takes about six months for the stone to dissolve. Other medications used in the treatment of gallstones include antiemetics to reduce nausea and vomiting, analgesics to relieve pain, and sometimes antibiotics.

• Surgery • The most common treatment for gallstones is removal of the gallbladder **(cholecystectomy).** Cholecystectomy is the most common surgery performed in the United States with over 500,000 people undergoing the procedure each year.[3]

Once the gallbladder is removed, bile flows directly from the liver into the intestine. Although most people experience no symptoms once they recover from surgery, others continue to experience nausea, bloating, gas, and abdominal pain. Certain foods can cause pain for some people, and if so, these foods should be avoided. Fatty or spicy foods are most frequently cited as offending foods.

RISK FACTORS FOR GALLSTONES

Are some people more likely to develop gallstones than others? Native Americans and Mexican Americans have high rates of gallstones. The three most important risk factors, however, are body weight, gender, and age.

• Body Weight • Weight has more than one connection with the risk of gallstones. Obesity, especially in women, raises the risk. People who are obese have a three to seven times greater risk of developing gallstones than people of desirable weight. The ways obesity contributes to stone formation are unclear, but some evidence suggests that obese people produce too much cholesterol in their livers and then excrete more in bile.

Weight loss can also raise the risk of gallstones, particularly if the person fasts or loses a lot of weight rapidly. Gallstones are common following intestinal surgeries for obesity and also for people who follow diets of less than 800 kcalories per day. It may be that very-low-kcalorie diets

FIGURE H28-1 The Gallbladder and Bile Ducts

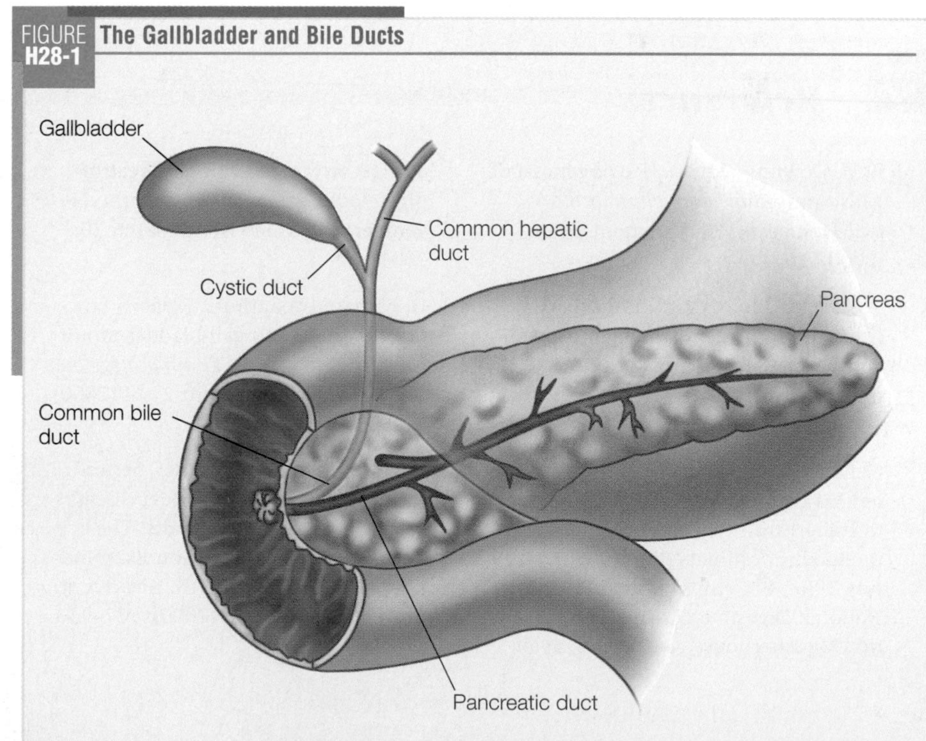

fail to provide sufficient fat to promote efficient emptying of the gallbladder.[4] It is unclear whether rapid weight loss leads to gallstones or increases the likelihood that silent stones become problematic.

Research suggests that dieting may shift the balance of cholesterol and bile salts excreted in bile, such that gallstone formation is more likely. When people go for long periods without eating (as many dieters do), the gallbladder may not contract enough to dispel its contents, favoring the formation of gallstones.

The health risks of obesity (see Chapter 9) often outweigh the risks of gallstones during weight loss. If gallstones do develop, surgery to remove the gallbladder is an option. Nevertheless, it makes sense to avoid unnecessary risks. Although research is still needed to back up this recommendation, it seems prudent for those undertaking a weight-loss program to follow a plan that allows for a gradual weight loss, contains some fat, and provides meals throughout the day with no long periods of fasting.

• **Gender** • Women between the ages of 20 and 60 have twice the risk of developing gallstones as men. Women who are pregnant, use estrogen replacement therapy, or take birth control pills also have a higher risk of gallstones. Extra estrogen increases the amount of cholesterol in bile and may also disrupt the gallbladder's contractions. In women, but not in men, serum levels of ascorbic acid relate inversely to the prevalence of gallbladder disease.[5] (Ascorbic acid plays a role in the catabolism of cholesterol.)

• **Age** • The risk of gallstones increases with age for both men and women. By age 60, 10 percent of men and 20 percent of women have gallstones.[6]

• **Other Risk Factors** • People with diabetes, those who use cholesterol-lowering medications, and those on total parenteral nutrition also have higher risks of gallstones than others. People with short bowel syndrome commonly experience gallstones; often cholecystectomies are performed as a preventive measure when large portions of the intestine must be removed.

Other diet factors have also been implicated in the development of gallstones, but whether altering these factors can prevent gallstones is unknown. Among these factors, soluble dietary fibers may protect against gallstones by reducing the saturation of cholesterol in bile.[7] Factors that may promote gallstones include simple sugars, alcohol, and, possibly, diets rich in animal fats and low in vegetable fats.[8] More research is needed to determine if specific dietary recommendations, other than avoiding obesity, can reduce the risk of gallbladder disease.

Gallstones are the most common disorder affecting the biliary tract. Many people with gallstones have no symptoms, but when symptoms develop, they can be quite painful. Gallstones can become serious if a gallstone lodges in the bile duct and obstructs the flow of bile from an adjoining organ. Inflammation of the bile duct, gallbladder, pancreas, and liver are possible. Obesity and rapid weight loss raise the risk of gallstones. Other nutrition-related conditions and treatments associated with a greater risk of gallstones include diabetes, the use of cholesterol-lowering medications, total parenteral nutrition, and short bowel syndrome. The potential for preventing gallstones, especially symptomatic gallstones, remains a focus of research.

References

1 R. H. Dowling, Review: Pathogenesis of gallstones, *Alimentary Pharmacology and Therapeutics* (Supplement 2) 14 (2000): 39–47.

2 S. Santen, Cholecystitis and biliary colic from emergency medicine, available at www.emedicine.com/EMERG/topic98.htm, updated January 23, 2001, site visited March 20, 2001.

3 Gallstones: A national health problem, available at www.liverfoundation.org/html/livheal.dir/lhimdox.dir/iml3dox.fol/onlmats.dir/_lnh013.htm, July 1996, site visited March 20, 2001; National Digestive Diseases Information Clearinghouse, Gallstones, available at www.niddk.nih.gov/health/digest/pubs/gallstns/gallstns.htm, November 1998, site visited March 20, 2001.

4 D. Festi and coauthors, Review: Low calorie intake and gall-bladder motor function, *Alimentary Pharmacology and Therapeutics* (supplement 2) 14 (2000): 51–53.

5 J. A. Simon and E. S. Hudes, Serum ascorbic acid and gallbladder disease prevalence among US adults: The Third Health and Nutrition Examination Survey (NHANES III), *Archives of Internal Medicine* 160 (2000): 931–936.

6 Gallstones: A national health problem, 1996.

7 W. H. Schwesinger and coauthors, Soluble dietary fiber protects against cholesterol gallstone formation, *American Journal of Surgery* 177 (1999): 307–310.

8 M. Tseng, J. E. Everhart, and R. S. Sandler, Dietary intake and gallbladder disease: A review, *Public Health Nutrition* 2 (1999): 161–172; G. Misciagna and coauthors, Diet, physical activity, and gallstones—A population-based, case-control study in southern Italy, *American Journal of Clinical Nutrition* 69 (1999): 120–126.

CHAPTER 29
Nutrition, Cancer, and HIV Infection

Although cancers and **HIV (human immunodeficiency virus)** infections are distinct disorders, from a nutrition standpoint, they share many similarities. Both disorders can have effects on many organ systems that affect nutrient needs, and both can lead to severe wasting in some cases. Both require medical nutrition therapy that is highly individualized based on the symptoms manifested and the organ systems involved.

© Kent Meireis/The Image Works

People with cancer take comfort from the support of others and from the knowledge that medical science is waging an unrelenting battle in their defense.

CANCER

The thought of cancer often strikes fear in people, and, indeed, cancer ranks just below cardiovascular disease as a cause of death. As with cardiovascular diseases, however, the prognosis for most people with cancer today is far brighter than in the past. Identification of risk factors, new techniques for early detection, and innovative therapies offer hope and encouragement.

Cancer is not a single disorder; instead, there are many different cancers. They have different characteristics, occur in different locations in the body, take different courses, and require different treatments. Whereas an isolated, nonspreading type of skin cancer may be removed in a physician's office with no observable effect on nutrition status, advanced cancers (especially those of the GI tract, pancreas, and liver) can seriously impair nutrition status.

How Cancers Develop

The genes in a healthy body work together to regulate cell division and ensure that each new cell is a replica of the parent cell. In this way, the healthy body grows, replacing dead cells and repairing damaged ones. Cancers develop from mutations in the genes that normally regulate cell division. The mutations silence the genes that ordinarily monitor replicating DNA for chemical errors. The affected cells seemingly have no brakes to halt cell division. As the abnormal mass of cells, called a **tumor,** grows, blood vessels form to supply the tumor with the nutrients it needs to support its growth. Eventually, the tumor invades more and more healthy tissue and may **metastasize** to other parts of the body. In leukemia (cancer of the blood-forming cells of the bone marrow), the cancer cells do not form a tumor, but rather circulate with the blood through other tissues where they can accumulate. Clinicians describe cancers by their type,◆ size, and extent, specifically noting if the tumor has spread to surrounding lymph nodes or to distant sites in the body. Figure 29-1 illustrates how a tumor develops and shows the difference between a **benign** and a **malignant** tumor.

◆ Cancers are classified by the tissues or cells from which they develop:
- **Adenomas** (ADD-eh-NO-mahz) arise from glandular tissues.
- **Carcinomas** (KAR-see-NO-mahz) arise from epithelial tissues.
- **Gliomas** (gly-OH-mahz) arise from the glial cells of the central nervous system.
- **Leukemias** (loo-KEY-mee-ahz) arise from the blood-forming cells of the bone marrow.
- **Lymphomas** (lim-FOE-mahz) arise from lymph tissue.
- **Melanomas** (MEL-ah-NO-mahz) arise from pigmented skin cells.
- **Sarcomas** (sar-KO-mahz) arise from muscle, bone, or connective tissue.

• Genetic Factors • All cancers have a genetic component in that a mutation causes abnormal cell growth, but some cancers have a genetically inherited component. A person with a family history of colon cancer, for example, has a greater risk of colon cancer than a person without such a genetic predisposi-

cancers: diseases that result from the unchecked growth of cells.

HIV (human immunodeficiency virus): a virus that progressively hampers the function of the immune system and leaves its host defenseless against other infections and cancer and eventually causes AIDS. The most common HIV is HIV-1—the virus described in this chapter.

tumor: a new growth of tissue forming an abnormal mass with no function; also called a **neoplasm** (NEE-oh-plazm).

metastasize (meh-TAS-tah-size): to spread by the movement of cancer cells from one part of the body to another.

benign (bee-NINE): describes tumors that stop growing without intervention or can be removed surgically and most often pose no threat to health.
- **benign** = mild

malignant (ma-LIG-nant): describes tumors that multiply out of control, threaten health, and require treatment.
- **malignus** = of bad kind

FIGURE 29-1 Tumor Formation

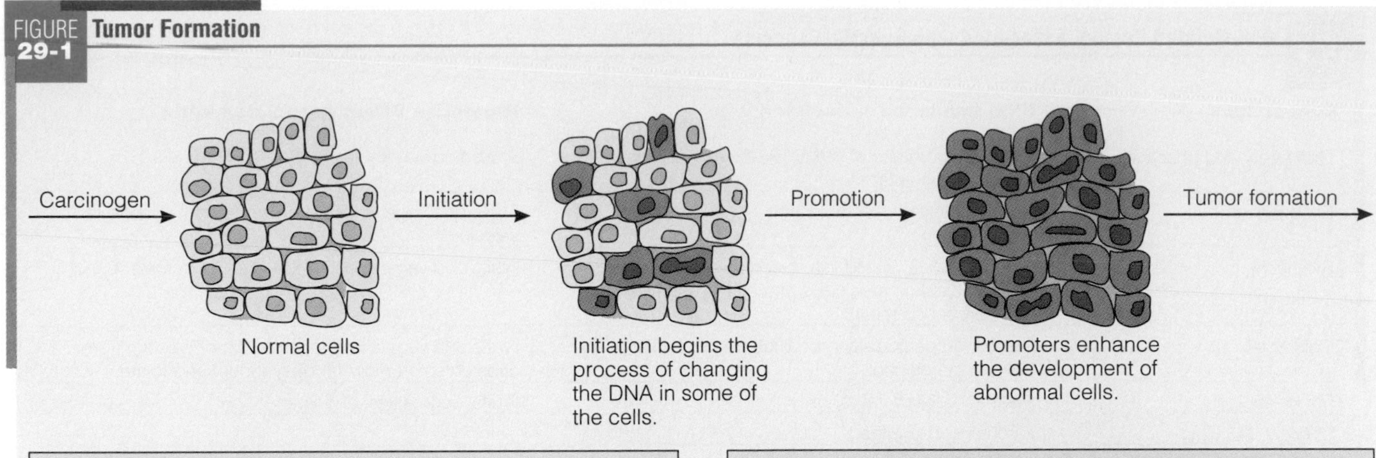

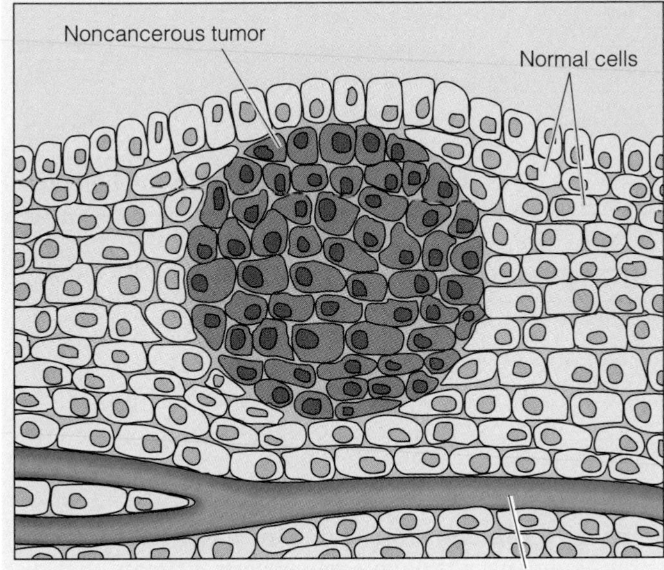

A noncancerous (benign) tumor usually grows within a self-contained capsule. It does not invade nearby tissue, nor does it spread.

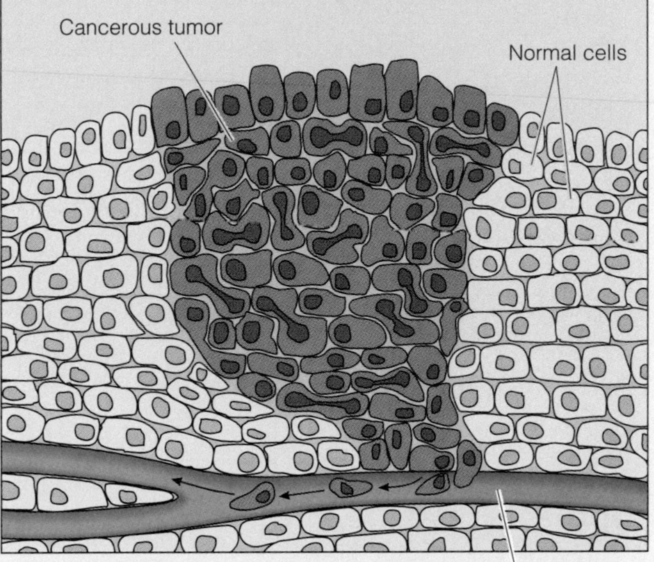

A cancerous (malignant) tumor usually grows out of control and may spread to other parts of the body through the blood or lymph systems.

tion. This does not mean, however, that the person *will* develop cancer, only that the risk is greater.

• **Immune Factors** • A healthy immune system recognizes foreign cells and destroys them. Researchers theorize that an ineffective immune system may not recognize tumor cells as foreign, thus allowing tumor growth. Aging affects immune function, and the incidence of cancer increases with age. Medications that suppress the immune system (immunosuppressants) and viral infections (including HIV infection) and other disorders that severely tax the immune system increase the risk of cancer.

• **Environmental Factors** • Among environmental factors, exposure to radiation and sunlight, water and air pollution, and smoking are known to cause cancer. Lack of physical activity may also play a role in the development of some types of cancer.[1] Men and women whose lifestyles include regular, vigorous physical activity have the lowest risk of colon cancer.[2] As Table 29-1 shows, dietary constituents are also associated with an increased risk of certain cancers. Some dietary factors may initiate cancer development **(initiators),** others may promote cancer development once it has started **(promoters),** and still others may protect against the development of cancer **(antipromoters).**

initiators: factors that cause mutations that give rise to cancer, such as radiation and carcinogens.

promoters: factors that favor the development of cancers once they have begun.

antipromoters: factors that oppose the development of cancers.

TABLE 29-1 Diet-Related Factors Associated with Specific Cancers[a]

Cancer Type	High Incidence Associated with:	Protective Effect Associated with:
Oral cavity and pharynx	Alcohol, low intake of vitamin A; excessive vitamin A from supplements	Fruits and vegetables
Esophageal	Alcohol, pickled vegetables, obesity	Fruits and vegetables, vitamins A and C, riboflavin, selenium
Stomach	Smoked foods, salted fish, cured meats, pickled vegetables, alcohol, pernicious anemia, *possibly* grilled and barbecued meats	Fruits and vegetables, whole grains, vitamins A and C
Colorectal	Dietary fat (particularly animal fat), meat, obesity, central fat, and alcohol	Fruits and vegetables; *possibly* fiber; calcium from supplements or low-fat dairy foods, and folate
Pancreatic	Meat and dietary fat, obesity	Fruits, vegetables, and fiber
Liver	Alcohol, iron overload	
Lung		Fruits and vegetables
Breast	Alcohol, obesity, central fat, *possibly* dietary fat	Fruits and vegetables
Endometrial	Obesity, dietary fat (particularly animal fat)	*Possibly* fruits and vegetables
Cervical		Fruits and vegetables
Ovarian		Fruits and vegetables
Prostate	Dietary fat (particularly animal fat), *possibly* vitamin A supplements and low-calcium, high-fructose diets	Fruits and vegetables, *possibly* selenium and vitamin E supplements
Bladder	The herb Arisocholia fanchic, found in some herbal weight-loss products	Fruits and vegetables, adequate intake of fluids

[a]Factors unrelated to diet are not shown. Often these factors are related more strongly to cancer risk than diet factors. The risk of stomach cancer, for example, is more closely associated with *Heliobacter pylori* infections than with diet factors.
SOURCE: American Cancer Society, Cancer Resource Center, www3.cancer.org/cancerinfo, visited March 27, 2001.

• Dietary Factors—Cancer Initiators • We do not know to what extent diet contributes to cancer development, although some experts estimate that diet may be linked to a third or more of all cases. Consequently, many people think that certain foods are **carcinogenic,** especially those that contain additives or pesticides. However, our food supply is one of the safest in the world. Additives that have been approved for use in foods are not carcinogenic. Some pesticides are carcinogenic at high doses, but not at the concentrations allowed on fruits and vegetables. The benefits of eating fruits and vegetables are far greater than any potential risk.[3]

The incidence of cancers, especially stomach cancers, is high in parts of the world where people eat a lot of heavily pickled or salt-cured foods that produce carcinogenic nitrosamines. Most commercial manufacturers in the United States use different preservative methods, and all are carefully controlled to minimize carcinogenic contamination.

Alcohol has also been associated with a high incidence of some cancers, especially of the mouth, esophagus, and liver. Breast cancer has also been associated with alcohol use in women.[4] Beverages such as beer and scotch may contain damaging nitrosamines as well as alcohol. Other beverages, such as wine and brandy, may contain the carcinogen urethane, which is produced during fermentation. The amounts of these compounds in alcoholic beverages currently on the market are not considered harmful—assuming consumption in moderate amounts.

• Dietary Factors—Cancer Promoters • Unlike carcinogens, which initiate cancers, some dietary components may accelerate cancers that have already begun to develop. Studies suggest that certain dietary fats eaten in excess may promote cancer, in part by contributing to obesity. More specifically, linoleic

carcinogenic (CAR-sin-oh-JEN-ick): producing cancer. A substance that produces or enhances the development of cancer is a **carcinogen.**

acid, the omega-6 fatty acid of vegetable oils, has been implicated in enhancing cancer development in some animals; in contrast, omega-3 fatty acids from fish oils appear to prevent or delay cancer development.[5]

• Dietary Factors—Antipromoters • It seems apparent that foods may also contain antipromoters. Almost without exception, epidemiological studies find a link between eating plenty of fruits and vegetables and a low incidence of cancers. The fiber in fruits and vegetables helps to protect against some cancers by speeding up the transit time of all materials through the colon so that the colon walls are not exposed to cancer-causing substances for long. Many studies strongly support the premise that fiber-rich diets protect against some forms of cancer, including colon cancer.[6] A few studies, however, dispute such findings, suggesting that high-fiber diets do not protect against colon cancer.[7]

In addition to fiber, fruits and vegetables contain both nutrients and phytochemicals that protect against cancer. By acting as scavengers of oxygen-derived free radicals, the antioxidant nutrients beta-carotene, vitamin C, and vitamin E may help to prevent tissue damage that can give rise to cancer. Phytochemicals common to many vegetables, especially cruciferous vegetables, can activate enzymes that destroy carcinogens (see Highlight 11). The recommendations for reducing cancer risks are quite similar to the recommendations for heart health and are summarized in Table 29-2.

Cruciferous vegetables, such as cauliflower, broccoli, and brussels sprouts, contain nutrients and phytochemicals that may inhibit cancer development.

TABLE 29-2 Recommendations for Reducing Cancer Risk[a]

Choose a diet rich in a variety of plant-based foods.

- Eat seven or more servings a day of a variety of whole grains, legumes, and starchy vegetables.
- Eat five or more servings a day of other vegetables and fruits all year round.
- Limit consumption of processed foods and refined sugar.

Maintain a healthy weight and be physically active.

- Avoid being underweight or overweight and limit weight gain during adulthood to less than 11 pounds (5 kilograms).
- If occupational activity is low or moderate, take an hour's brisk walk or participate in a similar exercise daily.
- Exercise vigorously for at least one hour each week.

Drink alcohol in moderation, if at all.

- Avoid alcohol consumption.
- If alcohol is consumed, limit it to less than two drinks a day for men and one for women.

Select foods low in fat and salt.

- Limit consumption of fatty foods, particularly those of animal origin. If red meat is eaten, limit intake to less than 3 ounces daily.
- Choose modest amounts of vegetable oils.
- Limit consumption of salted foods and use of cooking and table salt.
- Use herbs and spices to season foods.

Prepare and store foods safely.

- Use refrigeration and other appropriate methods to preserve perishable foods as purchased and at home.
- Do not eat charred food.
- Consume meat and fish grilled in direct flame and cured and smoked meats only occasionally.

Most importantly, do not smoke or use tobacco in any form.[b]

[a]American Institute for Cancer Research, *Food, Nutrition and the Prevention of Cancer: A Global Perspective,* 1997.
[b]One additional recommendation is in order: vary food choices. This last suggestion is based on an important concept that applies specifically to the prevention of cancer initiation—dilution. Switching from food to food dilutes the negative qualities of a food.

Consequences of Cancer

Once cancer develops, its consequences depend on its location, severity, and treatment. Cancer of the lungs, for example, will have different consequences than cancer of the kidneys. The chances of effective treatment are best in the early stages of cancer, yet during this time many cancers produce no symptoms and the person may be unaware of any threat to health. Efforts to detect cancers early aim to identify cancers before they have progressed too far.

• **Wasting Associated with Cancer** • Loss of appetite, weight loss, depletion of lean body mass and serum proteins, and debilitation typify the **cancer cachexia syndrome,** which occurs in as many as 80 percent of people with cancer before they die.[8] Weight loss is often evident at the time cancer is diagnosed, and severe malnutrition, often found in the later stages of cancer, may be the ultimate cause of death in many cases. Studies have shown that once lean body mass is significantly depleted, regardless of the cause, death will follow.[9] Without adequate energy and nutrients, the body is poorly equipped to maintain immune defenses, support organ function, absorb nutrients, mend damaged tissues, and utilize medications.

Many factors appear to play a role in the wasting associated with cancer, although the exact causes are unknown. Cytokines,◆ produced as inflammatory responses mount a battle to halt tissue damage caused by tumor cells, appear to play an important role.[10] In one study, poor appetite and weight loss occurred more often in people with cancer who had heightened inflammatory responses.[11] Whatever the cause, the combination of poor appetite, accelerated and altered metabolism, and the diversion of nutrients to support tumor growth simultaneously reduces the supply of energy and nutrients and raises the demand for them.

◆ Reminder: Chapter 24 describes immune system responses to tissue damage, which include the secretion of cytokines. Some of the cytokines identified as mediators of cancer cachexia include tumor necrosis factor (cachetin), interleukin-1, interleukin-6, interferon-γ, and differentiation factor.

People who develop cachexia swiftly fall into a downward spiral. Once weight loss and wasting have been set in motion, the debilitation and general poor health that follow make it even more difficult for the person to eat. The body is unable to respond to the reduced nutrient supply as it does during fasting, and it rapidly depletes its nutrient stores. Figure 29-2 summarizes some of the factors that contribute to the cancer cachexia syndrome.

FIGURE 29-2 **The Cancer Cachexia Syndrome: Contributing Factors**

Anorexia and wasting contribute to each other. Cancer and its treatments make both problems worse.

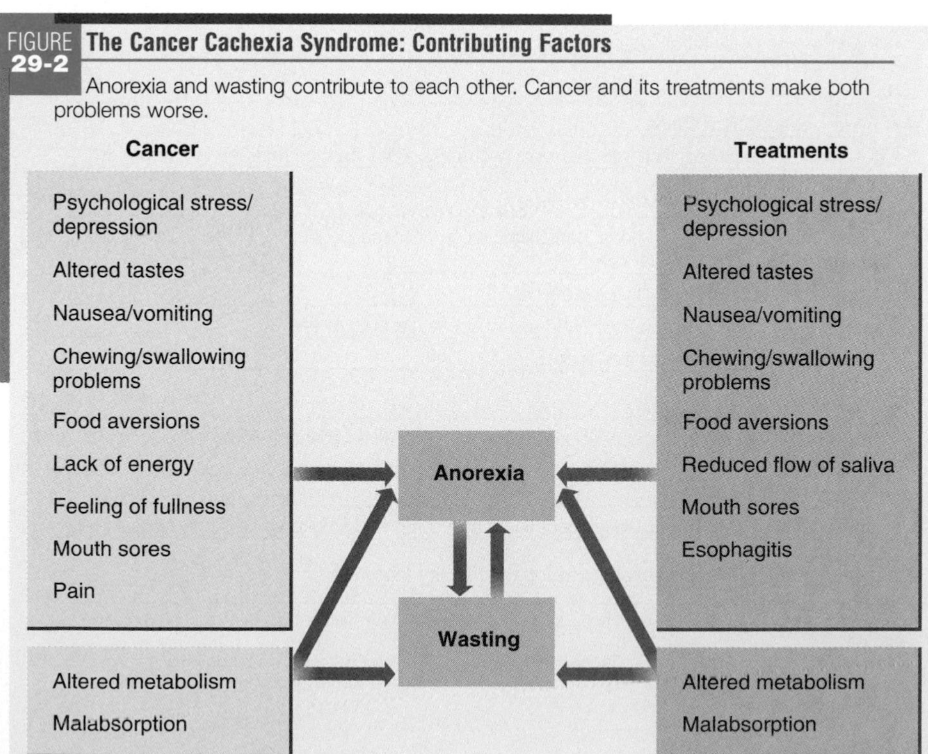

cancer cachexia (ka-KEKS-ee-ah) **syndrome:** loss of lean body mass, depletion of serum proteins, and debilitation that frequently accompany cancer.

• **Anorexia and Reduced Food Intake** • Anorexia is a major contributor to wasting associated with cancer. Some factors that can contribute to anorexia or otherwise reduce food intake in the person with cancer include:

- *Chronic nausea and early satiety.* People with cancer frequently experience nausea and a premature feeling of fullness after eating small amounts of food.
- *Fatigue.* People with cancer often tire easily and lack energy to prepare and eat meals. Once wasting is evident, these tasks become even more difficult for the person to handle.
- *Pain.* People in pain may have little interest in eating, particularly if eating makes the pain worse.
- *Psychological stress.* The very diagnosis of cancer can cause so much distress that eating becomes unimportant. Stress can be compounded by the person's fear and anxiety about the medical, personal, and financial concerns created by the diagnosis. Once wasting is evident, people may become depressed by their inability to perform routine tasks and by their physical appearance, and depression can also lead to reduced food intake.
- *Obstructions.* A tumor may partially or completely obstruct any portion of the GI tract and interfere with chewing and swallowing; cause delayed gastric emptying, early satiety, nausea, or vomiting; or make oral diets impossible.
- *Cancer therapy.* Therapy for cancer including medications, **chemotherapy, radiation therapy,** surgery, and **bone marrow transplants** can dramatically affect food intake by causing nausea, vomiting, altered taste perceptions, diminished taste sensitivity (mouth blindness), inflammation of the mouth (stomatitis) and esophagus (esophagitis), mouth ulcers, mouth dryness, food aversions (strong dislikes for certain foods), and depression. (A later section provides more information about treatments for cancer.)

• **Metabolic Alterations** • Altered metabolism causes people with cancer to use nutrients in inefficient ways that demand more energy and waste vital protein tissues.◆ This explains why some people with cancer fail to regain lean body mass even when they are receiving adequate energy and nutrients. Some people with cancer are hypermetabolic and have high nutrient needs; others may become hypermetabolic as a consequence of surgery or infection. People who are hypermetabolic are also insulin resistant, a condition that interferes with the body's ability to obtain the energy it needs. In addition, cancer and its treatments tax the immune system, raising the likelihood of infections, which, in turn, create additional demands for energy and nutrients. Chemotherapy◆ can interfere with normal metabolic pathways and create nutrient deficiencies as well.

• **Nutrient Losses** • Excessive nutrient losses◆ can develop as a consequence of cancer itself or the treatment for it. Causes of nutrient losses include vomiting, inadequate digestion, diarrhea, and malabsorption, and often several of these conditions occur at the same time. Some tumors, some antineoplastic agents, sometimes radiation therapy, and bone marrow transplants can lead to vomiting and diarrhea, electrolyte imbalances, and dehydration. Cancers of the pancreas, liver, or small intestine often lead to malabsorption. Radiation therapy to the small intestine can cause **radiation enteritis,** which can lead to malabsorption, chronic blood loss, fluid and electrolyte imbalances, and, sometimes, intestinal obstructions and fistulas. Intestinal function may return after radiation therapy ends, but for some the changes are permanent. Severe diarrhea and malabsorption, with fluid losses often exceeding 10 liters a day, can occur in people who undergo bone marrow transplants and then reject the transplanted tissue.

◆ **Altered metabolism may raise nutrient needs in cancer due to:**
- Inefficient use of energy and nutrients.
- Hypermetabolism.
- Insulin resistance.
- Secondary infections.
- Nutrient deficiencies created by chemotherapy.

◆ **Chapter 18 described how one medication used for chemotherapy, methotrexate, resembles folate and works by depriving growing cells (tumor cells and healthy cells alike) of folate.**

◆ **Cancer-induced causes of nutrient losses can include:**
- Inadequate digestion.
- Malabsorption.
- Vomiting.
- Diarrhea.

chemotherapy: the use of drugs to arrest or destroy cancer cells. Drugs used for chemotherapy are called **chemotherapeutic** or **antineoplastic agents.**

radiation therapy: the use of radiation to arrest or destroy cancer cells.

bone marrow transplants: the replacement of diseased bone marrow in a recipient with healthy bone marrow from a donor; sometimes used as a treatment for breast cancer, leukemia, lymphomas, and certain blood disorders.

radiation enteritis: inflammation and scarring of the intestinal cells caused by exposure to radiation.

Treatments for Cancer

The primary treatments for cancer—radiation therapy, chemotherapy, surgery, or any combination of the three—aim to annihilate cancer cells, relieve pain, alleviate symptoms, and prevent tumor growth. Table 29-3 summarizes the nutrition-related side effects of radiation therapy and chemotherapy, and Table 29-4 shows the potential effects of various surgeries for cancers on nutrition status. The use of bone marrow transplants to treat certain cancers has grown markedly.[12] To prepare a person for a bone marrow transplant, high doses of chemotherapy and sometimes whole-body radiation are used to eradicate cancer cells. Thus the nutrition-related side effects of these treatments apply to bone marrow transplants as well. In addition, immunosuppressants, given to help prevent **tissue rejection** following a bone marrow transplant, have multiple effects on nutrition status. The Diet-Drug Interactions box on p. 864 includes interactions for medications used in the treatment of cancer.

• Medications to Combat Anorexia and Wasting • To help people with advanced cancers combat anorexia, medications that stimulate the appetite have gained wide use. One of the most promising medications, megestrol acetate,◆ stimulates the appetite and promotes weight gain. Dronabinol (a medication containing the principal psychoactive ingredient in marijuana) works as both an appetite stimulant and an antiemetic, although side effects including euphoria and confusion limit its use in some cases.

◆ **Megestrol acetate is believed to exert its effects on appetite by altering the synthesis and release of certain cytokines. Megestrol acetate is also used as an antineoplastic agent in the treatment of breast cancer.**

Even when diets supply seemingly adequate amounts of energy and nutrients, wasting may continue. People may be able to regain weight, but the weight gain is often in the form of fat rather than lean body mass. Thus diet alone may be ineffective in treating cachexia, and drug therapy may be necessary.[13] Under investigation are medications, notably anabolic steroids, growth hormone, insulin-like growth factor, and thalidomide, which may help to restore lean body mass.[14]

tissue rejection: destruction of healthy donor cells by the recipient's immune system, which recognizes the donor cells as foreign; also called **graft-versus-host disease (GVHD).**

• Other Medications • Depending on the organ systems affected by cancer and by the side effects of treatments, many medications may be used in treatment. Medications commonly used to treat symptoms of cancer include antiemetics, antidiarrheals, analgesics, and sedatives.

TABLE 29-3 Possible Causes of Wasting Associated with Radiation and Chemotherapy

	Reduced Nutrient Intake	Accelerated Nutrient Losses	Altered Metabolism
Radiation	Anorexia Damage to teeth and jaws Dysphagia Esophagitis Mouth ulcers Nausea Reduced salivary secretions Taste alterations Thick salivary secretions Vomiting	Blood loss from intestine and bladder Diarrhea Fistulas Intestinal obstructions Malabsorption Radiation enteritis Vomiting	Fluid and electrolyte imbalances as a consequence of vomiting, diarrhea or malabsorption. Secondary effects of malnutrition, infection, or tissue damage (inflammation)
Chemotherapy	Abdominal pain Anorexia Mouth ulcers Nausea Taste alterations Vomiting	Diarrhea Intestinal ulcers Malabsorption Vomiting	Fluid and electrolyte imbalances Hyperglycemia Interference with vitamins or other metabolites Negative nitrogen and calcium balance Secondary effects of malnutrition, infection, or tissue damage (inflammation)

TABLE 29-4 Possible Effects of Surgeries for Cancers on Nutrition Status

Head and Neck Resection	
Difficulty in chewing/swallowing	Inability to chew/swallow
Esophageal Resection	
Diarrhea	Reduced gastric motility
Fistula formation	Steatorrhea (fat malabsorption)
Reduced gastric acid secretion	Stenosis (constriction)
Gastric Resection	
Dumping syndrome	Lack of gastric acid
General malabsorption	Vitamin B_{12} malabsorption
Hypoglycemia	
Intestinal Resection	
Blind loop syndrome	Hyperoxaluria
Diarrhea	Malabsorption
Fluid and electrolyte imbalances	Steatorrhea
Pancreatic Resection	
Diabetes mellitus	Malabsorption

• **Alternative Therapies** • Many people with cancer turn to complementary therapies (see Chapter 18) to assist them in their fight against cancer.◆ Others who feel they are making little progress or who think conventional medicine offers little hope of recovery may rely solely on alternative therapies for treatment. Up to 83 percent of people with cancer use at least one complementary therapy; 63 percent use dietary supplements (vitamins and herbs).[15] The majority of people use complementary therapies because they believe such therapies will improve the quality of their lives, boost their immune systems, prolong their lives, or relieve uncomfortable symptoms; 38 percent, however, believe that complementary therapies will cure their cancer. Yet the majority of people (60 percent) do not discuss complementary therapies with their physician.

◆ "Nutrition on the Net" on p. 868 provides a website with more information about complementary therapies for cancer.

Clinical studies are vitally needed to find out what interactions, if any, dietary supplements may have with antineoplastic agents or whether the supplements may have other negative effects on the body. Taking high doses of folate, for example, might interfere with the effects of methotrexate. Even when it seems logical that a supplement might be helpful and carry few risks, the reverse might be true. Consider, for example, the case of beta-carotene and lung cancer. Because a high intake of fruits and vegetables protects against lung cancer, and these same foods are good sources of beta-carotene, logic predicts that beta-carotene might have a protective effect against lung cancer or at least not be harmful. Yet such logic did not hold under the scrutiny of clinical research. Not only did beta-carotene not protect against lung cancer, it increased the risk of developing the disease.[16] Other researchers have voiced concerns over the potential effects of antioxidant supplements on chemotherapy and radiation therapy.[17] These therapies are believed to exert their effects on cancer cells, in part, by generating free radicals. Since antioxidant supplement use is widespread among people with cancer, an answer derived from careful research appears warranted.

The herbs and other dietary supplements often used by people with cancer may have unknown effects on cancer treatments.

Medical Nutrition Therapy

Although nutrition cannot change the ultimate outcome of cancer, attention to nutrition can help people maintain their strength and nutrition status and bolster the immune system while they undergo stressful treatments and cope with their diseases. Compared to malnourished people with cancer, well-nourished people with cancer enjoy a better quality of life—that is, they feel better, function better, are stronger and more active, and eat more.[18] Malnourished people with cancer may also have a poorer response to cancer treatments and a shorter survival time than people without malnutrition.[19] At a minimum, meeting nutrient needs eliminates the additional stresses imposed by malnutrition.

• **Early Nutrition Intervention** • For people with cancer, early nutrition intervention is a high priority. The initial nutrition assessment evaluates the individual's current nutrition status and establishes baseline parameters from which to monitor changes. Early intervention helps to prepare the person for the stresses ahead and helps detect and correct deficiencies before the task becomes monumental.

THINK NUTRITION whenever clients receive a diagnosis of cancer. Early attention to nutrition can help prevent serious malnutrition.

◆ Energy and protein needs may vary considerably and are difficult to predict from equations. Dietitians regularly monitor weight changes and assess food intake information to make energy and protein recommendations.

• **Oral Diets** • For people with cancer, medical nutrition therapy must account for the organ systems involved and the type and severity of the cancer (see Table 29-5.)◆ For people who are having problems maintaining weight, the ideal diet may need to supply up to about 150 percent of the basal energy expenditure (see Table 24-3 on p. 734) and 1.5 or more grams of protein per kilogram of body weight per day. In addition, a thorough nutrition assessment uncovers specific symptoms that each person is experiencing, and the interventions address these symptoms. The "How to" on pp. 858–859 outlines some strategies for alleviating symptoms and improving oral intake. The suggestions are numerous and detailed, reflecting both the complexity of the problems and the importance of offering specific suggestions to deal with individual problems.

Clients who are unable to eat adequate amounts of table foods often benefit from nutrient-dense formula supplements. Limited research suggests that

TABLE 29-5 Dietary Considerations for Specific Cancers

Cancer Sites	Dietary Considerations
Brain/Nervous System	Physical feeding disabilities (see Highlight 20); chewing and swallowing problems (see Chapter 20).
Head/neck	Chewing and swallowing problems.
Mouth/esophagus	Chewing and swallowing problems; vomiting, if obstructed, tube feeding below the obstruction may be necessary.
Stomach	Nausea, vomiting, early satiety; if obstructed, tube feeding below the obstruction or TPN may be necessary; if resection is performed, a postgastrectomy diet (see Chapter 20) may be needed; bacterial overgrowth (Chapter 21) may occur.
Intestine	If obstructed, tube feeding or TPN may be necessary; resections or inflammation may cause multiple nutrition problems (see Chapter 21); fat- and lactose-restricted diet may be useful.
Liver	Protein-, sodium-, and fluid-restricted diet may be necessary (see Chapter 28).
Pancreas	Fat-restricted diet and enzyme replacements may be necessary (see Chapter 21); diabetic diet may be necessary if insulin production is affected (see Chapter 25).
Kidneys	Protein-, electrolyte-, and fluid-controlled diet may be necessary (see Chapter 27).

Note: The considerations listed here are specific to the type of cancer; they do not include other nutrition-related effects of treatment.

omega-3 fatty acids from fish oils may help limit weight loss in people with cancer.[20]

Not all clients with cancer lose weight. Women diagnosed with breast cancer often gain, rather than lose weight, and this weight gain can be distressing. Weight gain may continue for some time after diagnosis. By encouraging physical activity, regularly assessing weight and energy intakes, and recommending strategies to correct problems early, health care professionals can help clients avoid unnecessary weight gain.[21]

• Tube Feedings and TPN • In general, a tube feeding or TPN is not routinely recommended for an adequately nourished or mildly malnourished person with cancer who is unable to eat. Most studies have failed to show that the use of tube feedings or parenteral nutrition reduces complications, lowers mortality rates, or shortens hospital stays for people with cancer. However, nutrition support can maintain functional and nutrition status when anorexia persists or when a person is severely malnourished and is about to undergo aggressive cancer therapy.[22] Each case is decided individually, and the use of special nutrition support is more likely when the person's chances of recovery or of significant response to treatment are good, or when the type of cancer is associated with a high risk of death from malnutrition.[23] People requiring head and neck resections, for example, may need long-term tube feedings and may need to continue tube feedings at home. Considering the many negative effects that cancers can have on GI tract absorptive and immune functions, tube feedings are strongly preferred to parenteral nutrition, whenever possible. People with severe radiation enteritis, however, may require home TPN.

• Nutrition Support and Bone Marrow Transplants • The person undergoing a bone marrow transplant routinely receives TPN before and after the transplant, because the GI tract is severely compromised by the preparatory procedure. Some researchers have found that adding glutamine to the TPN solution results in fewer infections and shorter hospital stays for people undergoing bone marrow transplants.[24] However, other researchers failed to confirm a benefit of glutamine-supplemented TPN on infection rates and hospital stays, but found evidence to suggest that *oral* glutamine may reduce the need for TPN and improve long-term survival.[25]

When GI tract function returns, the person begins to receive foods orally along with TPN, whenever possible. TPN is gradually tapered as oral intake improves. After a bone marrow transplant, nutrition complications can be severe and debilitating, especially for people who reject the transplant and develop GI complications.◆ In some cases, oral intake fails to meet nutrient needs, and TPN is required permanently.

◆ Examples of GI complications that can occur after a bone marrow transplant include anorexia, bleeding, infections, altered taste sensations, mouth dryness, inflammation of the mucous membranes of the mouth and esophagus, gastroesophageal reflux, esophagitis, early satiety, nausea, vomiting, diarrhea, and malabsorption.

Early oral feedings often start with lactose-free, low-residue, low-fat liquids to maximize absorption and minimize nausea, vomiting, and fat malabsorption. Gradually, solid foods are introduced. For about three months after the transplant, the diet may exclude most fresh fruits and vegetables, undercooked meats, poultry and eggs, and ground meats to minimize the risk of food-borne infections. Clients are advised to follow safe food handling practices (see Table 29-6 on p. 860) to minimize the risks of food-borne illnesses. Fiber, lactose, and fat are gradually added to the diet as individual tolerances allow. Because the transplant recipient also receives immunosuppressants, which often incur negative nitrogen and calcium balances, the final goal is to provide a high-kcalorie, high-protein, high-calcium diet. In addition, physicians often prescribe calcium and vitamin D supplements. Individuals with persistent diarrhea are encouraged to eat high-potassium foods.

• Ethical Issues • Every malnourished person with cancer or an HIV infection who cannot eat an adequate diet orally is a potential candidate for a tube feeding or TPN. This chapter describes uses of tube feedings or TPN as they are applied to

Help Clients Handle Food-Related Problems

For people with cancer or HIV infections, many problems can interfere with eating. It is important to find out what problems clients are having with food. The following list offers possible solutions to improve food intake based on answers to the question "Are you having any problems with eating or food?" For all clients, explain how eating can help them feel better. Not all of the suggestions will work for each client; encourage clients to experiment and find the ones that work best.

I just don't have an appetite.

- Eat small meals and snacks at regular times each day.
- Eat the most food at the time of day when you feel the best.
- Use nutrient-dense foods for meals and snacks. (Suggestions are provided later.)
- Eat nutrient-dense foods first.
- Indulge in favorite foods throughout the day.
- Avoid drinking large amounts of liquids with meals.
- Eat in a pleasant and relaxed environment.
- Listen to your favorite music or enjoy a program on TV while you eat.
- Eat with family and friends.
- Serve foods attractively.
- Take a walk before you eat.

I am too tired to fix meals and eat.

- Let friends and family members prepare food for you.
- Use foods that are easy to prepare and eat like sandwiches, frozen dinners, meals from take-out restaurants, instant breakfast drinks, liquid formulas, and supplements in candy bar and pudding form.
- Eat nutrient-dense foods first.

Foods just don't taste right.

- Brush your teeth or use a mouthwash before you eat.
- Add sauces and seasonings to meats.
- Eat meats cold or at room temperature.
- Use eggs, fish, poultry, and dairy products instead of meats.
- Try new foods and experiment with herbs and spices.
- Use plastic, rather than metal, eating utensils.
- Ask your doctor about zinc supplements. If you have a deficiency, your tastes may change.

I am nauseated a lot of the time, and sometimes I throw up.

- If you experience vomiting, use clear liquids like broths, carbonated beverages, juices, jello, or popsicles to replace fluids and electrolytes.
- If you become nauseated from chemotherapy treatments, avoid eating for at least two hours before treatments.
- See also the suggestions in the box on p. 638.

I can't stand some of the foods I really used to like.

- Save your favorite foods for times when you are not feeling nauseated or sick to your stomach.
- Maintain a food-free "window" of an hour or so before and after you have treatments or take medications that cause nausea or vomiting.

I am having problems chewing and swallowing food.

- Experiment with food consistencies to find the ones you can handle best. Thin liquids, true solids, and sticky foods (like peanut butter) are often difficult to swallow.
- Add sauces and gravies to dry foods.
- Drink fluids with meals to ease chewing and swallowing.
- Try using a straw to drink liquids.
- Tilt your head forward and backward to see if you can swallow easier with the head positioned differently.

I have sores in my mouth and they hurt when I eat.

- Use cold or frozen foods; they are often soothing.
- Try soft, soothing foods like ice cream, milk shakes, bananas, applesauce, mashed potatoes, cottage cheese, and macaroni and cheese.
- Avoid foods that irritate mouth sores like citrus fruits and juices, tomatoes and tomato-based products, spicy foods, foods that are very salty, foods with seeds (like poppy seeds and sesame seeds) that can be trapped in

people with cancer and HIV infections who have a chance of recovery (from cancer) or a reasonable life expectancy. When incurable cancer or HIV infection has reached its final stages, however, the person, caregivers, and the health care team need to make some important decisions about the use of a tube feeding or parenteral nutrition support. Highlight 29 provides more information about medical ethics and describes some factors that go into the decision-making process. The case study on p. 861 helps you apply information about nutrition and cancer.

IN SUMMARY Cancers develop when genes that normally regulate cell division fail to function properly. Dietary constituents can act as cancer initiators, promoters, and antipromoters. Once cancers develop, the effects on nutrition status depend on the type of cancer, its severity, and the types of treatment. Wasting is a frequent complication of many types of cancer because cancer and its treatment often result in inadequate nutrient intake, wasteful metabolism of nutrients, and excessive nutrient losses. Attention to nutrition can help prevent wasting and preserve quality of life.

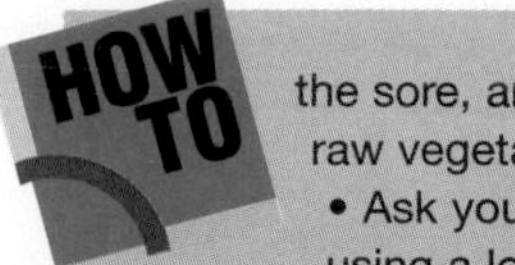

the sore, and coarse foods like raw vegetables and toast.
- Ask your doctor about using a local anesthetic solution like lidocaine before eating to reduce pain.
- Use a straw for drinking liquids, to bypass the sores.

My mouth is really dry.
- Rinse your mouth with warm salt water or mouthwash frequently, and drink liquids between meals.
- Ask your doctor or pharmacist about medications that can help with dry mouth.
- Use sour candy or gum to stimulate the flow of saliva.
- Make sure you brush your teeth and floss regularly to prevent cavities and oral infections.

I am having trouble with diarrhea.
- Drink plenty of fluids. Salty broths and soups, diluted fruit juices, and sports drinks are good choices. For severe diarrhea, try commercially prepared oral rehydration formulas.
- Eat foods with soluble fiber, and temporarily limit foods with insoluble fibers (see Table 21-2 on p. 655).
- Temporarily avoid foods that cause gas like dried beans and peas, broccoli, cucumbers, and onions.
- Take lactase enzyme replacements when you use dairy products because you may also experience lactose intolerance while you are having diarrhea. You may be able to tolerate low-fat yogurt.
- Avoid high-fat foods and foods made with lots of sugar.
- Avoid caffeine.
- Eat smaller meals more often.
- Check with your doctor about using digestive enzyme replacements if you have diarrhea for a long time.

I am having trouble with constipation.
- Drink plenty of fluids. Try warm fluids, especially in the morning.
- Eat whole-grain breads and cereals, nuts, fresh fruits, prunes, prune juice, and raw vegetables. Avoid refined carbohydrates like white bread, white rice, and pasta.
- Exercise regularly.

I need to gain weight, but my blood lipids are elevated.
- To gain weight, you will need to eat more fat, but the type of fat you use is important. Use unhydrogenated margarine in tubs (some are available that contain no *trans*-fatty acids), but don't use the low-kcalorie types.
- Use more monounsaturated fats like olive, canola, and peanut oils for baking and frying and in salad dressings and dips. Snack on avocados, nuts, and peanut butter. Make guacamole dip and spread it on vegetables and low-fat crackers. Spread peanut butter on celery, cucumbers, fruit, and low-fat crackers.
- To increase protein in your diet without adding too much of the fat you need to avoid, eat larger-than-usual servings of chicken and fish, especially salmon. Add chicken, fish, or low-fat and fat-free cheeses to sauces, soups, casseroles, and vegetables. Use instant breakfast drinks made with low-fat milk, or use commercially available supplements that contain no more than 30 percent fat. Add low-fat milk or milk powder to meat loaves, casseroles, soups, puddings, and cereals. Use low-fat yogurt.
- Use plenty of dried fruits. Add nuts and dried fruits to desserts, cereals, and salads.

I need help figuring out how to eat more energy and protein. (Note: All of the suggestions directly above apply here as well, but for some people the need for energy and protein outweighs the need to restrict the type of fat.)
- Use whole milk and regular yogurt instead of the low-fat or fat-free varieties.
- Use plenty of butter, margarine, mayonnaise, cream cheese, oil, and salad dressings on breads, sandwiches, potatoes, vegetables, salads, pasta, and rice.
- Use yogurt, sour cream, or a sour cream dip with vegetables.
- Add whipping cream to desserts and hot chocolate, or use it to lighten coffee.
- Use cream instead of milk with cereal.

HIV INFECTION

For many years, the devastating effects of infection by the human immunodeficiency virus (HIV), the virus that causes **acquired immune deficiency syndrome (AIDS),** seemed unstoppable. Although the disease still has no cure, remarkable progress has been made in extending the life expectancy of people with HIV infections. In the United States, deaths from AIDS declined from a high of over 50,000 in 1995 to just over 16,000 in 1999.[26]

Without a cure, however, the best course is prevention. HIV is transmitted from one person to another by direct contact with contaminated body fluids, most often through sexual intercourse, through contaminated needles or direct blood contact, or from mother to infant during pregnancy, birth, or breastfeeding.

Once a person has been infected, it takes about 6 to 12 weeks before laboratory tests can confirm a diagnosis. Because people remain symptom-free in the early stages of infection, however, they may not be tested for several years following infection. Thus early detection to prevent the spread of HIV infection and to ensure early treatment for the person infected are important health goals.

acquired immune deficiency syndrome (AIDS): the severe complications associated with the end stages of HIV infection. A person with an HIV infection is diagnosed with AIDS when AIDS-defining illnesses (certain infections or severe wasting) develop.

TABLE 29-6 Suggestions for Preventing Food-Borne Illness

Keep hot foods hot.

- When cooking meats or poultry, use a thermometer and cook meat to appropriate temperature.
- Cook stuffing separately or stuff poultry just prior to cooking.
- Avoid cooking large cuts of meat in a microwave.
- Marinate meats, poultry, or fish in the refrigerator. Don't use the marinade that was in contact with raw meat for basting or sauces.
- Never eat raw eggs. Cook eggs until the whites are firmly set and the yolks begin to thicken.
- When serving hot foods, keep them at a temperature of at least 140°F.
- Heat leftovers thoroughly.

Keep cold foods cold.

- Refrigerate perishable food items as soon as you get home from the grocery store.
- Put packages of raw meat, poultry, and fish on a plate before refrigerating to prevent juices from dripping on other foods.
- Buy only frozen foods that are solidly frozen.
- Keep cold foods at 40°F or less.
- Keep frozen foods at 0°F or less.
- Refrigerate leftovers promptly. Use shallow containers to cool foods promptly.
- Thaw meats in the refrigerator, not at room temperature. If you must hasten thawing, use cool running water or a microwave oven.
- Freeze meat, poultry, or fish immediately if not planning to use within a few days.

Keep a clean and safe kitchen.

- Wash fruits and vegetables with a scrub brush in warm water; store washed and unwashed produce separately.
- Use hot, soapy water to wash hands, utensils, dishes, nonporous cutting boards, and countertops between tasks when working with different foods. Use a bleach solution on cutting boards (one capful per gallon of water).
- Cover cuts with clean bandages before preparing food.
- Avoid cross-contamination by washing all surfaces that have been in contact with raw meats, poultry, or eggs before reusing; serving cooked meats on a clean plate; and separating raw foods from those that have been cooked.
- Mix foods with utensils, not hands; keep hands and utensils away from mouth, nose, and hair.
- Do not allow a person with a skin infection or infectious disease to prepare food.
- Wash or replace sponges and towels regularly.
- Clean up food spills and crumb-filled crevices.

Keep food safety in mind when shopping.

- When shopping, don't place packages of meat, poultry, and fish on top of other foods to avoid cross-contamination.
- Do not buy or use items with broken seals or mangled packaging.
- Check safety seals, buttons, and expiration dates.
- Buy only pasteurized milk and milk products.
- Check eggs and buy only those with intact shells.
- Buy only seafood that has been properly refrigerated or iced. Cooked seafood for purchase should be displayed separately from raw seafood to avoid cross-contamination.

Keep these pointers in mind.

- Always wash hands with soap and hot water before preparing food.
- Do not taste food that is suspect. "If in doubt, throw it out."
- Throw out foods with danger-signaling odors. Be aware, though, that most food-poisoning bacteria are odorless, colorless, and tasteless.
- Discard foods that are discolored, moldy, or decayed or that have been contaminated by insects or rodents.

CASE STUDY

Public Relations Consultant with Cancer

Mrs. Barnett is a 54-year-old public relations consultant recently diagnosed with cancer of the colon. Mrs. Barnett was feeling well and was unaware of any problem. A fecal occult (hidden) blood test performed as part of a routine physical exam alerted her physician to a potential problem. Mrs. Barnett then underwent a colonoscopy—a procedure that uses a tube with an optical device that allows the physician to look inside the colon and remove any suspicious areas for examination. Mrs. Barnett is scheduled to have surgery to remove the cancer and determine if it has spread to the surrounding lymph nodes and other organs. Before surgery, the surgeon explains to Mrs. Barnett that most likely a portion of her intestine will be removed, and there is a possibility that she will require a temporary colostomy. The severity and extent of the cancer will determine whether radiation therapy and/or chemotherapy will be required. The dietitian completing a nutrition assessment discovers that Mrs. Barnett is 5 feet 7 inches tall, and weighs 165 pounds. Her typical diet is high in saturated fat and includes meat, poultry, or fish at both lunch (about four ounces) and dinner (six to eight ounces). She eats two to three servings of fruits and vegetables and between three and six servings of grains and starchy vegetables each day. She rarely drinks milk, but often eats cheese.

1. Review Table 29-1 on p. 850 and describe the diet-related factors in Mrs. Barnett's history that might have contributed to the development of colon cancer.
2. What nutrition-related problems can occur as a consequence of surgical resection of the colon? Is malabsorption a common problem following surgery of the colon?
3. Describe some of the effects of radiation and chemotherapy on nutrition status.
4. If Mrs. Barnett is unresponsive to treatments and her cancer progresses, she may develop the cancer cachexia syndrome. What is this syndrome, and what are its causes? What are the benefits of preventing or correcting the wasting associated with cancer?
5. Provide suggestions to help Mrs. Barnett handle these problems should they develop: poor appetite, fatigue, taste alterations, nausea and vomiting, food aversions, chewing and swallowing difficulties, mouth ulcers, mouth dryness, diarrhea, and weight loss.
6. Under what circumstances might a tube feeding or parenteral nutrition be appropriate for Mrs. Barnett if she is unable to eat an oral diet?

How HIV Develops

HIV attacks the immune system and leaves its victims defenseless against **opportunistic illnesses**—illnesses that would cause few, if any, symptoms in a person with a healthy immune system. The infection progresses in stages. The virus gradually destroys cells with a specific protein called CD4+ on their surfaces. Among the cells most affected are the **CD4+ T-cells,** lymphocytes that are essential components of the immune system.

• **HIV Stages** • At first, the number of CD4+ T-cells declines gradually. As the infection continues, though, depletion of CD4+ T-cells progressively impairs immune function, and in the final stages, fatal complications develop.

• **Monitoring HIV Progression** • With improved treatments for HIV infection, the progression of the disorder has been slowed dramatically. On average, it takes about ten years for an HIV infection to progress to AIDS (the final stage of infection). Clinicians monitor the course of HIV infections by measuring concentrations of CD4+ T-cells and circulating virus (viral load) and by monitoring clinical symptoms.

◆ The cluster of symptoms and disorders that sometimes occurs before AIDS develops is called **AIDS-related complex (ARC).**

Consequences of HIV Infection

In the initial stages of HIV-infection, the individual remains symptom-free. As the infection progresses, early symptoms may include fatigue, skin rashes, fevers, diarrhea, joint pain, night sweats, weight loss, oral lesions and infections, and other opportunistic illnesses that are not life-threatening.◆ Infection with hepatitis C, which often leads to chronic hepatitis, is common in people with

opportunistic illnesses: illnesses that normally would not occur or that would cause only minor problems in the healthy population, but can cause great harm when the immune system is compromised.

CD4+ T-cells: a type of white blood cell (lymphocyte) that has a specific protein receptor on its surface and is a necessary component of the immune system. Highlight 18 describes T-cells.

© Lauren Goodsmith/The Image Works

The countless lives touched by AIDS serve as a potent reminder of the need to continue the search for a cure.

HIV infections, especially those who use injectable drugs or those who acquired the infection from contaminated blood products. In the final stages, CD4+ T-cell counts become markedly reduced, and the person develops frequent and eventually fatal complications **(AIDS-defining illnesses),** such as severe weight loss; recurrent bacterial pneumonia; serious infections of the central nervous system, GI tract, and skin; cancers; and severe diarrhea. Improved treatments for HIV infections have significantly reduced the incidence of many opportunistic illnesses and severe malnutrition.[27]

• **Lipodystrophy** • Prior to improved treatments for HIV infections, severe wasting and debilitation were common manifestations of the disease. Although involuntary weight loss and wasting remain a problem for many people with AIDS, people on aggressive therapies (described later) who are not experiencing complications may gain weight.[28] For some people, the weight gain is composed primarily of fat, and the fat has an altered distribution **(lipodystrophy).**[29] People with lipodystrophy associated with HIV infections tend to accumulate fat around the abdominal area (central fat) and lose fat from the face, arms, and legs. Thus they may appear to be quite thin except for a "pot belly." Fat accumulates not only under the skin in the abdominal area, but also interspersed between the organs located there. Breast enlargement, thick necks, fat at the top of the back **(buffalo hump),** and multiple benign growths composed of fat **(lipomas)** have also been observed. Central body fat, whether a consequence of overweight or HIV infection, is associated with hyperlipidemia and insulin resistance and with increased risks of cardiovascular disease and diabetes. The reasons why lipodystrophy develops are unclear, and different factors may lead to lipodystrophy in different cases. Many believe lipodystrophy is related to antiviral medications, but some people develop lipodystrophy even when they are not taking these medications.

• **Weight Loss and Wasting** • Although weight loss is uncommon in the early stages of HIV infection, in the later stage, the person with AIDS may experience severe wasting.◆ People with HIV infections often lose weight rapidly during periods when they are experiencing complications that alter their appetites or interfere with the absorption or metabolism of nutrients; then they regain weight during periods when they are feeling better.[30] Others experience a gradual but steady weight loss, which may be related to inadequate energy intake and gastrointestinal complications. Much as in cancer, the causes of wasting associated with HIV infection are multifactorial: anorexia and inadequate food intake, altered metabolism, excessive nutrient losses, and nutrient-medication interactions.

◆ **Wasting, an AIDS-defining illness, is the involuntary loss of greater than 10 percent of body weight accompanied by chronic diarrhea, lethargy, and/or fever lasting longer than 30 days.**

Vitamin and mineral deficiencies are also common in people with HIV infections. Many deficiencies have been documented, even in people who are consuming recommended amounts of micronutrients.

• **Anorexia and Reduced Nutrient Intake** • Anorexia and reduced food intake play a major role in the development of wasting associated with both HIV infections and cancer. For people with HIV infections, anorexia and inadequate nutrient intakes may occur as a consequence of:

- *Psychological stress and pain.* As in cancer, fear, depression, and anxiety over the HIV diagnosis, the prognosis, and the medical, personal, and financial problems that lie ahead, as well as the pain associated with complications of the disorder, can destroy the appetite.
- *Oral infections.* Infections and fever cause anorexia. In addition, oral infections associated with HIV infection cause further problems. **Thrush,** a common oral infection associated with HIV infection, can alter taste sensitivity, reduce the flow of saliva, and cause pain on swallowing. Oral infections caused by **herpes virus** can cause painful mouth ulcers that interfere with chewing and swallowing.

AIDS-defining illnesses: very low CD4+ T-cell counts, wasting, and other complications that mark the final stages of an HIV infection.

lipodystrophy (LIP-oh-DISS-tro-fee): the redistribution of fat that can occur as a consequence of HIV infections as well as other disorders. The accumulation of abdominal fat associated with HIV infections is sometimes called *protease paunch.*

buffalo hump: the accumulation of fat at the top of the back.

lipomas (lih-POE-mahs): benign tumors composed of fat.

thrush: a fungal infection of the mouth and esophagus, caused by *Candida albicans,* that coats the tongue with a milky film and leads to mouth ulcers, altered taste sensations, and pain on chewing and swallowing. The technical term for this infection is *candidiasis.*

herpes virus: a virus that can lead to mouth lesions and may also affect the lower GI tract, causing diarrhea.

- *Respiratory infections.* Pneumonia and tuberculosis, frequent complications associated with HIV infection, cause fever and pain that contribute to anorexia. The person who must use an oxygen mask to improve breathing may find it difficult to eat.
- *GI tract complications and altered organ function.* In addition to oral infections, people with HIV infection may experience belching, reflux esophagitis, gastritis, and heartburn that may cause nausea and interfere with eating. Intestinal complications, including infections, diarrhea, and constipation, and hepatitis can contribute to reduced food intake and food aversions.
- *Fatigue, lethargy, and dementia.* Fatigue is a common complication of HIV infections, even in the early stages. Fatigue may be a consequence of anemia, which is also a frequent complication of HIV infections, or it may develop for other reasons. In the later stages of HIV infections, lethargy and dementia frequently occur and may interfere with food intake. The individual may not care or even remember to eat.
- *Cancer.* As previously described, cancer often leads to anorexia. **Kaposi's sarcoma,** a cancer associated with HIV infection, can cause lesions and obstructions in the esophagus that make eating painful.
- *Medical treatments.* Medications used to treat HIV infection, associated infections, and cancer often cause anorexia, taste alterations, nausea, and vomiting and reduce food intake. Food aversions may also arise.

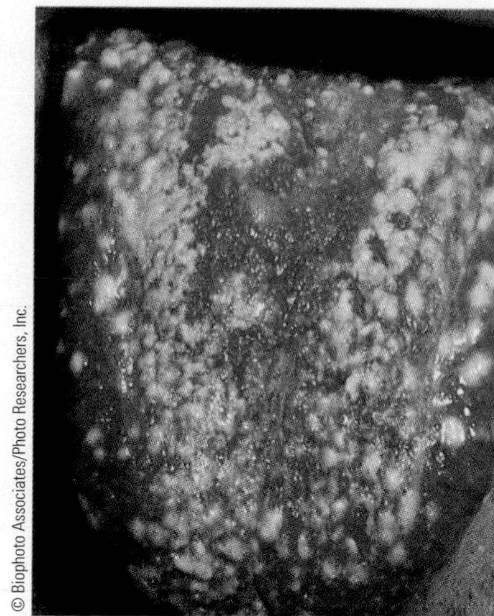

The oral infection thrush is easily identified by the characteristic milky white patches on the tongue.

• Metabolic Alterations and Nutrient Losses • The metabolic alterations that occur as a result of HIV infections and their consequences, including repeated infections and cancer, contribute to wasting in people with HIV infections. In addition, nutrient losses can occur as a consequence of diarrhea and malabsorption,◆ which often develop late in the course of HIV infections. About 60 percent of people with HIV infections in the United States experience significant diarrhea and malabsorption. Both the structure and the function of the intestinal cells are altered as a consequence of GI tract infections. Food, water, and enteral formulas may serve as a source of infectious agents, and people with advanced HIV infections are highly susceptible to food-borne illness. It is unclear whether HIV infection itself leads to malabsorption.

◆ The diarrhea and malabsorption associated with HIV infection are called **HIV enteropathy** (EN-ter-OP-a-thee).

The prolonged use of antibiotics to treat infections and antisecretory agents and antacids to relieve nausea can lead to bacterial overgrowth in the upper small intestine, further contributing to malabsorption. Treatments common among people with HIV infections, especially anti-infective agents, chemotherapy, and radiation therapy, can accelerate nutrient losses due to vomiting, diarrhea, and malabsorption. Megadoses of vitamin C and other home remedies that some people with HIV infections use may also cause diarrhea.

In most cases, diarrhea is recurrent, and typical losses average less than 1 liter of diarrheal fluids daily. Diarrhea caused by parasites, however, may be severe and unresponsive to medications—the person may lose 10 or more liters of diarrheal fluids daily. Once malnutrition is under way, it, too, contributes to malabsorption.

Treatments for HIV Infection

Treatments for HIV infection focus on slowing the course of the infection, controlling symptoms, and alleviating pain. Drug therapy often includes a combination of medications that disrupt HIV at different stages of replication.◆ The combined drug therapy called highly active antiretroviral therapy (HAART) has made a remarkable difference in the treatment of HIV infection. In 1996, for the first time since 1990, AIDS was no longer among the top ten causes of death in the United States, falling from 8th to 14th place.[31] Antiviral agents can have many interactions with diet as the box on p. 864 describes.

◆ **The medications used to treat HIV infections are antiviral agents. Antiviral agents include nonnucleoside reverse transcriptase inhibitors, nucleoside reverse transcriptase inhibitors, and protease inhibitors.**

Kaposi's (cap-OH-seez) **sarcoma:** a type of cancer rare in the general population but common in people with HIV infections.

Diet-Drug Interactions

ANABOLIC AGENTS

Testosterone, testosterone derivatives *(oxandrolone, nandrolone,* and *oxymetholone),* and *growth hormone* may be used to promote weight gain, specifically a gain of lean body mass. These medications can be taken without regard to food, and nutrition-related side effects are uncommon.

ANTIDIABETIC AGENTS

For *antidiabetic agents,* see p. 760.

ANTIDIARRHEALS

For *antidiarrheals,* see p. 673.

ANTI-INFECTIVES

Many medications may be used to treat the infections associated with both cancer and HIV infections (consult a drug guide for specific examples). The antiviral agents specifically used to treat HIV infections are described below.

ANTILIPEMICS

For *antilipemics,* see p. 791.

ANTINAUSEANTS

For *antinauseants,* see p. 644.

ANTINEOPLASTICS

About 80 antineoplastic agents are in use today. Many of these agents can cause nausea and vomiting. Most often nausea and vomiting begin a few hours after a treatment and resolve shortly thereafter. In some cases, the problems may last for a few days. The suggestions in the "How to" on pp. 858–859 can help alleviate these problems. Antinauseants may also be prescribed. Mouth sores, taste alterations, fatigue, anemia, diarrhea, and constipation are also associated with many agents. *Megestrol acetate* is used as both an antineoplastic agent and an appetite stimulant. Megestrol acetate and some antineoplastic agents that are hormones (including testosterone) lead to weight gain. Fluid and sodium retention and edema can also occur, although these side effects are not common.

ANTISECRETORY AGENTS

For *antisecretory agents,* see p. 644.

ANTIVIRALS

The nonnucleoside reverse transcriptase inhibitors include *delavirdine* and *nevirapine,* which must be taken one hour apart from antacids and without regard to food. Neither is associated with significant nutrition-related side effects. The nucleoside reverse transcriptase inhibitors include *lamivudine (3TC), zidovudine (AZT), stavudine, zalcitabine (ddC),* and *didanosine (ddI).* Lamivudine, zidovudine, and stavudine can be taken without regard to food; zidovudine can cause nausea and vomiting. Clients who use zalcitabine and didanosine should avoid aluminum- and magnesium-containing antacids. Didanosine can lead to nausea and vomiting and should be taken one hour before or two hours after meals. The protease inhibitors include *indinavir, saquinavir, ritonavir,* and *nelfinavir.* Indinavir should be taken one hour before or two hours after meals; however, people who experience nausea and abdominal pain can eat a light meal (less than 300 kcalories, 6 grams of protein, and 3 grams of fat). Indinavir can lead to kidney stones, so clients taking this medication are encouraged to drink liquids (6 liters a day). Saquinavir should be taken within two hours of a meal. Ritonavir and nelfinavir are better absorbed along with food and should be taken with meals. Ritonavir can cause nausea, vomiting, diarrhea, taste alterations, and abdominal pain. The most common side effect of nelfinavir is mild-to-moderate diarrhea, although nausea and abdominal pain may also occur.

APPETITE STIMULANTS

For *appetite stimulants,* see p. 835.

IMMUNOSUPPRESANTS

For *immunosuppressants,* see p. 835.

LAXATIVES

For *laxatives,* see p. 835.

◆ Growth hormone and testosterone are called **anabolic agents.**

• **Medications to Combat Anorexia and Wasting** • The medications megestrol acetate and dronabinol (described on p. 854) may be prescribed to stimulate the appetite and help people with HIV infections to gain weight. More recently, human growth hormone has been approved to treat wasting in HIV infection. Growth hormone helps restore lean body mass. Testosterone and its derivatives may also be beneficial and are currently being studied.[32]◆ Testosterone levels often decrease as an HIV infection progresses, and people with low testosterone may lose lean body mass even when their weights remain stable. When testosterone levels are low, administering testosterone can increase lean body mass and improve strength and quality of life. The combination of megestrol acetate and testosterone can improve the appetite and increase weight and lean body mass.

• Other Medications • Still other medications serve to control infections (anti-infective agents) and complications. Antinauseants, antisecretory agents, oral antidiabetic agents, antilipemics, and antidiarrheals are examples. Like people with cancer, people with HIV infections are likely to use complementary therapies,◆ including dietary supplements.

© Gary Conner/PhotoEdit

Nutrition provides an edge in maintaining quality of life and encouraging independence.

◆ "Nutrition on the Net" on p. 868 provides a website with more information about complementary therapies used by people with HIV infections.

◆ Reminder: Monounsaturated fat is found in olive, canola, and peanut oils; avocados; nuts; and peanut butter. Omega-3 fatty acids are found primarily in fish oils.

Medical Nutrition Therapy

Nutrition assessment and counseling should begin as soon as a diagnosis of HIV infection has been made. The initial assessment provides baseline data from which to monitor progress throughout the course of the disease. For people with HIV infections on aggressive combination drug therapies, assessment should include an evaluation of body composition. Clinics that treat clients with HIV infections may use bioelectrical impedance analysis to estimate how much muscle and lean tissue, fat, and water a person's body contains (see Appendix E). While the technique does have drawbacks—it can't tell where fat is, for example—it provides a simple and convenient analysis. Measurements repeated at intervals allow clinicians to make adjustments to diet and drug therapies.

• Oral Diets • For people on aggressive combination drug therapies who have hyperlipidemia, diet advice mimics that for all clients with elevated lipids: achieve or maintain a desirable weight, reduce fat to less than 30 percent of total kcalories, reduce saturated fat, limit *trans*-fatty acids, and replace saturated fats with monounsaturated fats and omega-3 fatty acids.◆ People with elevated triglycerides benefit from limiting simple sugars. People with glucose intolerance or diabetes benefit from eating a consistent intake of carbohydrate throughout the day and using mostly complex carbohydrates. A combination of aerobic activity and resistance training may also play an important role in preventing central obesity.[33] As Chapter 26 described, physical activity can also help lower insulin resistance.

For clients who are losing weight, a high-kcalorie (up to 150 percent of the basal energy expenditure), high-protein (at least 1.5 grams or more per kilogram of body weight) oral diet can help to halt weight loss and restore weight. Enteral formulas and supplements in the form of easy-to-eat bars and pudding can help boost nutrient intake. Liquid formulas may be especially useful for the person who is too tired to eat and prepare meals. The "How to" on pp. 858–859 offers other suggestions for improving the appetite and adding kcalories and protein to the diet.

All people with HIV infections experience anorexia from time to time. Almost all experience bouts of diarrhea and constipation as well. It is important to find out what problems the person is having and make appropriate suggestions.

• Vitamins and Minerals • Vitamin and mineral needs for people with HIV infections are highly variable, and little information is available concerning specific needs. Anemia, fatigue, and neuropathy, common clinical findings in people with HIV infections, are associated with vitamin B and iron deficiencies, and such deficiencies are also common in people with HIV infections. People with fat malabsorption may also have deficiencies of fat-soluble vitamins. Deficiencies of minerals, especially zinc and selenium, are also common. A vitamin and mineral supplement that contains at least 100 percent of recommended intakes is frequently prescribed to ensure an adequate intake. Some clinicians recommend two multivitamin and mineral tablets daily. Some people with HIV infections may have elevated blood levels of iron, and excess iron may be harmful to the liver and interfere with immune function. For women with HIV infections who are not menstruating or pregnant, the physician should determine if iron supplementation is appropriate. Some clinicians recommend a variety of nutrient supplements, including antioxidant nutrients, and supplemental nonessential amino acids, including glutamine, cysteine, and carnitine, although research to support the use of these supplements is lacking.

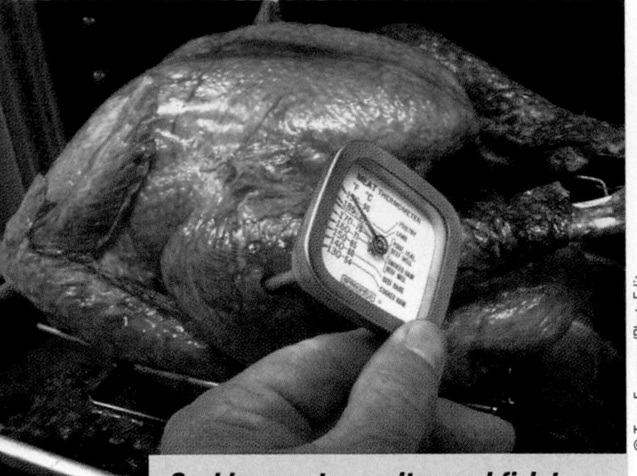

© Tony Freeman/PhotoEdit

Cooking meats, poultry, and fish to the correct internal temperature is an important safeguard against food-borne illness.

• Food Safety • The depressed immune system in people with HIV infections raises the likelihood of infections from microorganisms in food and water. To prevent infections from food-borne microorganisms, clients are provided with instructions for the safe handling and preparation of foods (see Table 29-6 on p. 860). Water can also be a source of food-borne illnesses, especially **cryptosporidiosis.** Water quality varies throughout the United States, and local health departments should be consulted to determine if the local tap water is safe for people with HIV infections to drink. If not, or to take additional safety measures, water used for cooking and making ice cubes should be boiled for one minute. Some types of filtered and bottled waters are also safe, but not all.◆

◆ For specific instructions on safe filtered and bottled waters, visit www.the body.com/cdc/crypto1297.html.

• Tube Feedings and TPN • People with HIV infections need aggressive nutrition support whenever they are unable to consume oral diets that are sufficient to prevent nutrition complications and unintentional weight loss. Tube feedings are preferred whenever the GI tract is functional. Tube feedings can be given at night to supplement oral diets during the day. Preventing bacterial contamination of the formula is particularly important. TPN is generally reserved for people with HIV infections who are unable to tolerate enteral nutrition, but need to maintain their nutrition status while undergoing a therapy that is expected to improve their condition. For people with severe HIV-related malabsorption, orally administered hydrolyzed formulas containing medium-chain triglycerides and supplemented with glutamine may be as effective as TPN in reversing weight loss and wasting.[34]

People with GI tract obstructions, severe vomiting, or GI infections affecting the entire small intestine may benefit from TPN. As with enteral formulas, the concern for infection related to TPN is magnified in people with HIV infections. The accompanying case study discusses nutrition concerns of a person with HIV infection. The Nutrition Assessment Checklist applies to people with cancer or HIV infections.

cryptosporidiosis (KRIP-toe-spo-rid-ee-OH-sis): a food-borne illness caused by the parasite *Cryptosporidium parvum.* Most people develop few or minor problems from this infection, but people with HIV infections, and especially those with AIDS, can develop long-lasting and serious problems.

IN SUMMARY

In some cases, people with HIV infections develop lipodystrophy, hyperlipidemia, and insulin resistance. For others, especially those with AIDS, weight loss and wasting are the primary concerns. Reduced food intake, metabolic alterations, and increased nutrient losses all play a role in wasting associated with HIV infections. Health care professionals serve people with HIV infections best by identifying nutrition problems early and offering solutions before nutrition status seriously deteriorates.

CASE STUDY

Travel Agent with HIV Infection

Three years ago, Mr. Sands, a 34-year-old travel agent, sought medical help when he began feeling run-down and developed a painful white coating over his mouth and tongue. The presence of thrush and anemia alerted Mr. Sands's physician to the possibility of an HIV infection. When Mr. Sands tested positive for an HIV infection, he and his family and friends were devastated by the news, but those closest to him have remained supportive. Mr. Sands has gained weight and developed lipodystrophy and hyperlipidemia during the three years since he began combination drug therapy. Mr. Sands is 6 feet tall and currently weighs 190 pounds. He occasionally develops diarrhea and sometimes anorexia.

1. What is Mr. Sands's percent ideal body weight (%IBW)? What is lipodystrophy, and what is its typical pattern in people with HIV infections? Describe an appropriate diet for Mr. Sands.
2. What suggestions can you give Mr. Sands for the times he has trouble with diarrhea or anorexia? What factors can lead to diarrhea or contribute to anorexia in people with HIV infections?
3. Describe how HIV infection can lead to wasting as the disease progresses to the later stages.
4. How will Mr. Sands's diet change if wasting becomes a problem for him?

This chapter brings to a close your introduction to normal and clinical nutrition. Congratulations! You have received an abundance of information since you first began your study. The normal nutrition chapters of this text provided you with current recommendations to promote optimal health. You learned how the body transforms foods into nutrients and how those nutrients support the body's well-being. The clinical chapters addressed you as a future health care professional, concerned with the well-being of others during times of illness. We hope this text has served you well and that you will remember, when selecting food for yourself or when making recommendations for others, to honor the body.

Nutrition Assessment Checklist for People with Cancer or HIV Infections

MEDICAL HISTORY

Check the medical record to determine:

- ☐ Type and extent of cancer
- ☐ Stage of HIV infection

Review the medical record for complications that may alter medical nutrition therapy including:

- ☐ Malnutrition/wasting
- ☐ Altered organ function
- ☐ Anorexia
- ☐ Nausea/vomiting
- ☐ Mouth ulcers/thrush
- ☐ Taste alterations
- ☐ Dry mouth
- ☐ Diarrhea/malabsorption
- ☐ Constipation

MEDICATIONS

For clients with cancer or HIV infections:

- ☐ Make a note of all medications the client is taking and remain alert for possible diet-medication interactions.
- ☐ Give antinauseants or pain medications at times when they will be most effective during meals.
- ☐ Ask about use of alternative therapies, including herbal preparations and megadoses of vitamins.

For clients with cancer who require chemotherapy:

- ☐ Recommend strategies to prevent food aversions (see the box on pp. 858–859).
- ☐ Offer suggestions for handling complications associated with medications.

For clients on antivirals for HIV infections:

- ☐ Note that some antivirals can be taken without regard to foods, some are better absorbed when taken with foods, and still others must be taken on an empty stomach.
- ☐ Help clients work out an acceptable medication schedule that considers the client's lifestyle and the timing of the medication dose with food intake.
- ☐ Offer suggestions for handling complications associated with medications.

FOOD/NUTRIENT INTAKE

For clients with poor food intakes and weight loss:

- ☐ Determine the reason(s) for reduced food intake.
- ☐ Offer suggestions for improving intake based on specific problems the client is having.
- ☐ Provide interventions before weight loss progresses too far.

For clients with HIV infections who experience weight gain and hyperlipidemia and/or glucose intolerance:

- ☐ Assess diet for total energy; total fat and types and amounts of specific fats; and total carbohydrate, fiber, and simple sugars.
- ☐ Recommend a low-fat, low-saturated fat, energy-controlled diet for hyperlipidemia.
- ☐ Recommend a low-fat, low-saturated fat, energy-controlled diet for glucose intolerance or diabetes with a consistent carbohydrate intake from day to day.

ANTHROPOMETRICS

Take baseline height and weight measurements, monitor weight regularly, and make dietary adjustments promptly, if necessary. Baseline and periodic body composition measurements are important for people with HIV infections who take combination antiviral medications.

LABORATORY TESTS

Note that albumin and serum proteins may be reduced for people with cancer or HIV infections, especially those who are experiencing wasting. Check laboratory tests for indications of:

- ☐ Changes in organ function
- ☐ Anemia
- ☐ Dehydration

For people with HIV infections, the progression of the HIV infection is evaluated by checking:

- ☐ CD4+ T-cell counts
- ☐ Viral loads

PHYSICAL SIGNS

Look for physical signs of:

- ☐ Wasting and PEM
- ☐ Fluid status (especially for those with fever, vomiting, or diarrhea)
- ☐ Mouth ulcers

Nutrition on the Net

Access these websites for further study of topics covered in this chapter.

- Find updates and quick links to these and other nutrition-related sites at our website: **www.wadsworth.com/nutrition**
- To learn more about cancer, including risk factors, prevention, screening, detection, treatments (including nutrition), complementary therapies, support networks, and links to online cancer resources, visit these sites:

  American Cancer Society: **www.cancer.org**

  National Cancer Institute: **www.nci.nih/gov**

  CancerSource: **www.cancersource.com**

  American Institute for Cancer Research: **www.aicr.org**

  American Association for Cancer Research: **www.aacr.org**
- To find virtually anything you would like to know about HIV infections and AIDS, visit the Body: A Multimedia AIDS and HIV Information Resource: **www.thebody.com**

  Be sure to check out the forums on a variety of topics (including wasting, lipodystrophy, and controlling symptoms) to gain valuable insights into what it is like to live with an HIV infection.
- To view statistics about the incidence of HIV infections and AIDS as well as basic facts on the disease and its prevention and detection, visit the Centers for Disease Control: **www.cdc.gov/nehstp/od/nchstp.html.**
- To uncover links to online resources for HIV infection and AIDS, visit the AIDS Education Global Information System: **www.aegis.com.**
- To research complementary therapies for both cancer and AIDS, visit The Natural Pharmacist: **www.TNP.com.**
- To review information about safe food handling, visit Healthfinder: **www.healthfinder.gov/searchoptions/topicsaz.htm** and search for food safety.

INTERNET ACTIVITIES

The foods people eat can positively or negatively affect their risk of developing some types of cancer. When people do develop cancer, especially cancer of the GI tract, pancreas, and liver, nutrition status can be severely impaired. You may be interested in learning more about diet and cancer, as well as about detection, symptoms, treatment, and other information about specific cancers.

- Go to: **www.cancer.org**
- Look at the left side of the screen, under the heading "Cancer Resource Center Overview."
- Use the "Select cancer type" scroll to choose the type of cancer you are interested in. Click on the cancer you want to learn about and then click "Go."

Describe the type of cancer you selected including its incidence, causes, symptoms, progression, and treatments. What role does diet play in the prevention or treatment of this type of cancer?

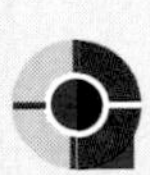

Study Questions

These questions will help you review the chapter. You will find the answers in the discussions on the pages provided.

1. What is cancer? What events are believed to lead to its development? What roles may dietary factors play in cancer prevention or cancer development? (pp. 848–851)
2. What is the cancer cachexia syndrome? What factors contribute to its development? (p. 852)
3. Describe how cancer and its treatments contribute to anorexia, altered metabolism, and increased nutrient losses. (p. 853)
4. Discuss strategies for combating anorexia, bitter or metallic tastes, nausea and vomiting, problems with mouth ulcers, reduced flow of saliva, or other problems with chewing or swallowing. (pp. 858–859)
5. What is an HIV infection, and what happens as it progresses? (p. 861)
6. Describe the pattern of lipodystrophy that develops in some people with HIV infections. What other complications generally occur in these cases? (p. 862)
7. Describe factors that can lead to reduced nutrient intake, altered metabolism, excessive nutrient losses, and wasting in people with HIV infections. (pp. 862–863)
8. Why are people with HIV infections highly susceptible to food-borne illness? Describe some measures that can be taken to prevent food-borne illness. (p. 866)

These questions will help you prepare for an exam. Answers can be found on p. 871.

1. Energy and protein needs for people with wasting due to cancer or HIV infections are often about _____ of the basal energy expenditure (BEE) and _____ grams or more of protein per kilogram of body weight per day.
   a. 100 percent; 1.5
   b. 120 percent; 1.0
   c. 150 percent; 1.5
   d. 200 percent; 1.0
2. Which of the following statements describes wasting associated with cancer and HIV infections?
   a. Anorexia is a major factor in the development of wasting.
   b. Wasting always accompanies cancer and HIV infection.
   c. Altered metabolism plays no role in the development of wasting.
   d. Unlike wasting associated with acute stresses, cytokines do not contribute to wasting in cancer and HIV infections.
3. Practical advice for a person with cancer or HIV infection who has trouble preparing and eating foods due to fatigue might include:
   a. prepare dinner for friends.
   b. prepare fresh vegetables every day.
   c. make homemade ice cream for snacks.
   d. keep premixed breakfast drinks or formula supplements in the refrigerator for snacks.
4. Oral diets after bone marrow transplants may:
   a. limit calcium.
   b. test tolerance for lactose.
   c. restrict high-protein foods.
   d. restrict raw fruits and vegetables.
5. Mouth sores in people with HIV infections are most frequently due to:
   a. dehydration.
   b. oral infections.
   c. malabsorption.
   d. food-borne illnesses.
6. A tube feeding or TPN is most likely to benefit people with cancer or HIV infection if:
   a. they do not wish to prolong their lives.
   b. no further treatments are available to them.
   c. they have been told they have no other alternatives.
   d. malnutrition may have an undesirable effect on their ability to receive additional treatments.
7. Which of the following statements is true with respect to the diarrhea that often occurs as a consequence of HIV infection?
   a. Megadoses of vitamin C can help resolve it.
   b. HIV and secondary infections can play roles in its development.
   c. Fibers from wheat bran and raw vegetables are useful in controlling it.
   d. Diarrhea can always be corrected with appropriate nutrition therapy and fluid replacement.
8. Which people are most prone to infections arising from foods, enteral formulas, and TPN?
   a. people who have undergone radiation therapy
   b. people with oral infections or cancer of the GI tract
   c. people who have undergone chemotherapy or surgery
   d. people with HIV infections or those who have undergone bone marrow transplants
9. The changes in body fat seen in some people with HIV infections include:
   a. increased central and peripheral fat.
   b. decreased central and peripheral fat.
   c. increased central and decreased peripheral fat.
   d. decreased central and increased peripheral fat.
10. Which of the following statements is true?
   a. People with HIV infections may gain weight.
   b. People with cancer never have to worry about gaining weight.
   c. People with cancer never have to worry about losing weight.
   d. People with HIV infections never have to worry about losing weight.

Clinical Applications

1. Many disorders can lead to wasting. For some of these disorders, such as fat malabsorption (Chapter 21), diet is a cornerstone of treatment. For others, such as congestive heart failure (CHF), chronic obstructive pulmonary disease (COPD), cancer, and HIV infection, nutrition plays a supportive role. What determines whether nutrition plays a major or a supportive role in the treatment of a disorder? Review the effects of PEM on pp. 184–187 and p. 567. Carefully consider how severe malnutrition can further debilitate people with wasting due to cancer or HIV infection.

2. The suggestions for handling food-related problems in the "How to" on pp. 858–859 appear simple enough, but many of the suggestions may be

difficult to implement in some cases. What suggestions to control nausea and vomiting contradict suggestions to add kcalories and protein? What other contradictions can you find? How might a health care provider deal with such contradictions?

3. Consider problems associated with nutrition in a 36-year-old woman with a malignant brain tumor affecting her ability to move the right side of her body (including the tongue) and to speak coherently. She has an expected length of survival of six months and is taking a pain medication that makes her nauseated and sleepy. What would be a realistic goal of nutrition support? If she is right-handed, how can her impairment interfere with eating? What suggestions might you have for overcoming this problem? How might nutrition be affected by her problems with communication? Describe ways that the medications she is taking can affect her nutrition status. Would tube feedings or TPN be appropriate for this woman? Why or why not?

References

1 S. A. Oliveria and P. J. Christos, The epidemiology of physical activity and cancer, *Annals of the New York Academy of Sciences* 833 (1997): 79–90.

2 M. L. Slattery and coauthors, Lifestyle and colon cancer: An assessment of factors associated with risk, *American Journal of Epidemiology* 150 (1999): 869–877.

3 Food and Nutrition Science Alliance (FANSA) Statement on diet and cancer prevention in the United States, November 1999.

4 S. A. Smith-Warner and coauthors, Alcohol and breast cancer in women: A pooled analysis of cohort studies, *Journal of the American Medical Association* 279 (1998): 535–540.

5 H. Senzake and coauthors, Dietary effects of fatty acids on growth and metastasis of KPL-1 human breast cancer cells in vivo and in vitro, *Anticancer Research* 18 (1998): 1621–1627; M. W. Pariza, Animal studies: Summary, gaps, and future research, *American Journal of Clinical Nutrition* 66 (1997): 1539–1540.

6 B. S. Reddy, Role of dietary fiber in colon cancer: An overview, *American Journal of Medicine* 106 (1999): S16–S19; D. Kritchevsky, Protective role of wheat bran fiber: Preclinical data, *American Journal of Medicine* 106 (1999): S28–S31; M. C. Jansen and coauthors, Dietary fiber and plant foods in relation to colorectal cancer mortality: The Seven Countries Study, *International Journal of Cancer* 81 (1999): 174–179; M. J. Hill, Cereals, cereal fibre, and colorectal cancer risk: A review of the epidemiological literature, *European Journal of Cancer Prevention* 6 (1997): 219–225; L. Le Marchand and coauthors, Dietary fiber and colorectal cancer risk, *Epidemiology* 8 (1997): 658–665.

7 A. Schatzkin and coauthors, Lack of effect of a low-fat, high-fiber diet on the recurrence of colorectal adenomas, *New England Journal of Medicine* 342 (2000): 1149–1155; C. S. Fuchs and coauthors, Dietary fiber and the risk of colorectal cancer and adenoma in women, *New England Journal of Medicine* 340 (1999): 169–176.

8 As cited in A. M. Herrington, J. D. Herrington, and C. A. Church, Pharmacologic options for the treatment of cachexia, *Nutrition in Clinical Practice* 12 (1997): 101–113.

9 D. P. Kotler and coauthors, Magnitude of body cell mass depletion and timing of death from wasting in AIDS, *American Journal of Clinical Nutrition* 50 (1989): 444–447.

10 M. Puccio and L. Nathanson, The cancer cachexia syndrome, *Seminars in Oncology* 24 (1997): 277–278.

11 P. O'Gorman, D. C. McMillan, C. S. McArdle, Impact of weight loss, appetite, and the inflammatory response on quality of life in gastrointestinal cancer patients, *Nutrition and Cancer* 32 (1998): 76–80.

12 T. Duell and coauthors, Health and functional status of long-term survivors of bone marrow transplantation, *Annals of Internal Medicine* 126 (1997): 182–184.

13 K. Mulligan and A. S. Bloch, Energy expenditure and protein metabolism in human immunodeficiency virus infection and cancer cachexia, *Seminars in Oncology* (supplement 6) 1998: 82–91.

14 T. M. Nash, Use of anabolic agents in patients with cancer cachexia, *Support Line,* June 1999, pp. 14–18; Herrington, Herrington, and Church, 1997.

15 M. A. Richardson and coauthors, Complementary/alternative medicine use in a comprehensive cancer center and implications for oncology, *Journal of Clinical Oncology* 18 (2000): 2505–2514.

16 D. Albanes and coauthors, Alpha-tocopherol and beta-carotene supplements and lung cancer incidence in the Alpha-Tocopherol, Beta-Carotene Cancer Prevention Study: Effects of base-line characteristics and study compliance, *Journal of the National Cancer Institute* 88 (1996): 1560–1570.

17 D. Labriola and R. Livingstone, Possible interactions between dietary antioxidants and chemotherapy, *Oncology* 13 (1999): 1003–1008.

18 O'Gorman, McMillan, and McArdle, 1998.

19 S. Mercadante, Parenteral versus enteral nutrition in cancer patients: Indications and practice, *Supportive Care in Cancer* 6 (1998): 85–93.

20 P. Bougnoux, n-3 polyunsaturated fatty acids and cancer, *Current Opinion in Clinical Nutrition and Metabolic Care* 2 (1999): 121–126; M. D. Barber and coauthors, Fish oil–enriched nutritional supplement attenuates progression of the acute-phase response in weight-losing patients with advanced cancer, *Journal of Nutrition* 129 (1999): 1120–1125; J. M. Daly and coauthors, Enteral nutrition with supplemental arginine, RNA, and omega-3 fatty acids in patients after operation: Immuno-

logic, metabolic, and clinical outcome, *Surgery* 112 (1992): 56–67.

21 C. L. Rock, Factors associated with weight gain in women after diagnosis of breast cancer, *Journal of the American Dietetic Association* 99 (1999): 1212–1218, 1221.

22 Mercadante, 1998.

23 M. Marian, Cancer cachexia: Prevalence, mechanisms, and interventions, *Support Line,* April 1998, pp. 3–12.

24 P. R. Schloerb and M. Amare, Total parenteral nutrition with glutamine in bone marrow transplantation and other clinical applications (randomized, double-blind study), *Journal of Parenteral and Enteral Nutrition* 17 (1993): 407–413; T. R. Ziegler and coauthors, Clinical and metabolic efficacy of glutamine-supplemented parenteral nutrition after bone marrow transplantation, *Annals of Internal Medicine* 116 (1992): 821–828.

25 P. R. Schloerb and B. S. Skikne, Oral and parenteral glutamine in bone marrow transplantation: A randomized, double-blind study, *Journal of Parenteral and Enteral Nutrition* 23 (1999): 117–122.

26 Centers for Disease Control and Prevention, HIV/AIDS Surveillance Report, Vol. 12, No. 1, Midyear 2000.

27 A. Mocroft and coauthors, Changes in AIDS-defining illnesses in a London clinic, 1987–1998, *Journal of Acquired Immune Deficiency Syndrome* 21 (1999): 401–407; Centers for Disease Control and Prevention, Surveillance for AIDS-defining opportunistic illness, 1992–1997, *Morbidity and Mortality Weekly Report,* April 19, 1999.

28 E. Gomez, Wasting and body changes: What do we know so far? *PWA Newsline,* September 1998, pp. 34–36.

29 A. Carr and coauthors, Diagnosis, prediction, and natural course of HIV-1 protease-inhibitor-associated lipodystrophy, hyperlipidemia, and diabetes mellitus: A cohort study, *Lancet* 353 (1999): 2093–2099; M. Mann and coauthors, Unusual distributions of body fat in AIDS patients: A review of adverse events reported to the Food and Drug Administration, *AIDS Patient Care Standards* 13 (1999): 287–295.

30 L. M. Kruse, Nutritional assessment and management for patients with HIV disease, *The AIDS Reader* 8 (1998): 121–130. (Available online from Medscape at www.medscape.com/SCP/TAR/1998/v08.n03/a3013.krus/pnt-a3013.krus.html.)

31 J. Henkel, Attacking AIDS with a "cocktail" therapy, www.thebody.com/fda/cocktail.html, site visited December 9, 1999.

32 Centers for Disease Control and Prevention, The use of testosterone in AIDS wasting syndrome, *AIDS Clinical Care* 11 (1999): 25; J. C. Loss, The use of anabolic agents in HIV disease, *Support Line,* June 1999, pp. 23–28.

33 R. Roubenoff and coauthors, A pilot study of exercise training to reduce trunk fat in adults with HIV-associated fat redistribution, *AIDS* 13 (1999): 1373–1375.

34 D. P. Kotler, L. Fogelman, and A. R. Tierney, Comparison of total parenteral nutrition and an oral, semielemental diet on body composition, physical function, and nutrition-related costs of patients with malabsorption due to acquired immunodeficiency syndrome, *Journal of Parenteral and Enteral Nutrition* 22 (1998): 120–126.

ANSWERS

Multiple Choice

1. c	2. a	3. d	4. d	5. b
6. d	7. b	8. d	9. c	10. a

ETHICAL ISSUES IN NUTRITION CARE

As CHAPTER 29 described, every person with cancer or an HIV infection who cannot eat an oral diet is a potential candidate for a tube feeding or total parenteral nutrition (TPN). Tube feedings and TPN can meet nutrient needs and support recovery at times when table foods cannot. Like other medical technologies, however, the availability of special nutrition support forces health care professionals and society to face **ethical** issues. Such treatments can prolong life by merely delaying death; the remaining life may be of low quality. (Terms relating to ethical issues appear in the accompanying glossary.)

THE SCOPE OF THE PROBLEM

Understanding that the provision of nutrition support may do little to promote recovery raises the question "Is it ever morally and legally appropriate to withhold or withdraw nutrition support?" In attempting to answer a question such as this one, ethics experts weigh such factors as:

- The client's right to make decisions concerning his or her own well-being **(autonomy),** even if withholding treatment will result in death.
- The potential benefits **(beneficence)** of the treatment versus the potential risks **(maleficence)** the treatment poses to the client's health or quality of life.
- The client's right to be fully informed of a treatment's benefits and risks in a fair and honest manner.
- The ability of caregivers or family to promote the client's well-being without selfish intent when the client cannot speak for herself or himself.
- The client's right to expect that health care professionals and caregivers will honor his or her stated wishes about health care.

When sifting through ethical dilemmas, answers rarely come easily. It may be difficult to determine whether clients truly comprehend the complexity and potential finality of their decisions, and it may be equally as difficult to determine whether a treatment will ultimately pose more risks than provide benefits. In reality, the answers often lie tangled in personal values, charged emotions, and legal conflict.

Examples of Ethical Issues

Health care professionals readily recommend whatever form of nutrition is necessary to support clients who have any reasonable chance of recovering from a disease. Clearly, health care professionals cannot rightfully withhold nutrition support because of poor judgment or negligence. The decision of whether to feed a client becomes less clear, however, when clients have a **terminal illness,** when they are in a **persistent vegetative state,** or when they (or their caregivers) simply refuse specialized nutrition support because they feel the quality of their lives is poor.[1] Do we (as a society) allow them such choices? Are health care professionals morally and legally obligated to comply with, or to deny, such requests? Furthermore, who determines when clients are **competent** to speak for themselves? If a client is **comatose** or incompetent to make such a decision, who, if anyone, should be allowed to make life-and-death decisions for the client? These are but a few of the questions that have evolved along with advances in medical technology and nutrition support; a discipline known as "medical ethics" has developed out of the need to discuss and solve problems such as these.

When is it morally and legally appropriate to withhold or withdraw special nutrition support?

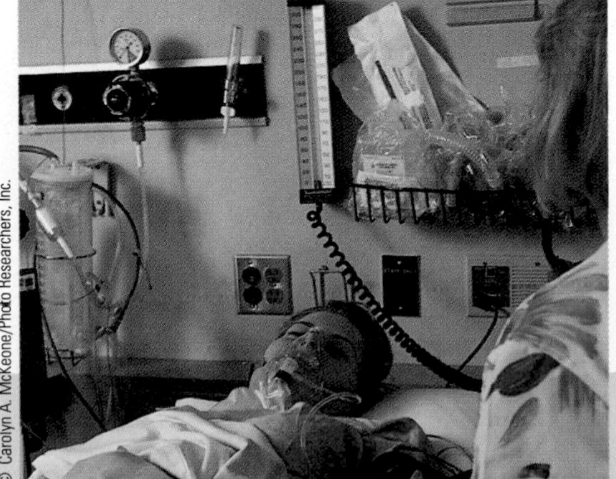

Decision Making

Most often the client (when competent) and the client's family decide what to do in cases where there are ethical dilemmas. They often make their decisions based on extensive consultation with the physician and the health care

GLOSSARY

advance directives: the means by which competent adults record their preferences for future medical interventions. A living will and durable power of attorney are types of advance directives.

artificial feedings: parenteral and enteral nutrition; feeding by a route other than the normal ingestion of food.

autonomy: independence.
- **auto** = self
- **nomos** = law

beneficence (be-NEF-eh-sens): doing good.

comatose: in a state of deep unconsciousness from which the person cannot be aroused.

competent: having sufficient mental ability to understand a treatment, weigh its risks and benefits, and comprehend the consequences of refusing or accepting the treatment.

durable power of attorney: a legal document in which one competent adult authorizes another competent adult to make decisions for her or him in the event of incapacitation. The phrase "durable power" means that the agent's authority survives the client's incompetence; "attorney" refers to an attorney-in-fact (not an attorney-at-law).

ethical: in accordance with moral principles or professional standards. Socrates described *ethics* as "how we ought to live."

legal: established by law.

living will: a document signed by a competent adult that specifically states whether the person wishes aggressive treatment in the event of terminal illness or irreversible coma from which the person is not expected to recover.

maleficence (mah-LEF-eh-sens): doing harm.

persistent vegetative state: exhibiting motor reflexes but without the ability to regain cognitive behavior, communicate, or interact purposefully with the environment.

terminal illness: a progressive, irreversible disease that will lead to death in the near future.

team. Health care professionals must provide decision makers with the information they need to fully understand the disease, its treatments, the potential benefits of treatments, and the risks so that they can make a truly informed choice. When the client or client's family makes a decision that the health care team or the facility housing the client believes may make the team or facility liable for malpractice, the conflict of interests may give rise to court cases. Then the court is charged with defining the problems and solving them.

Court Decisions

One of the most widely publicized cases involving nutrition support issues was that of Nancy Cruzan. Cruzan was a young woman who suffered permanent and irreversible brain damage after a car crash in 1983. For eight years, she was in a persistent vegetative state—awake but unaware. Her physicians and parents held no hope for her recovery, yet given food and water, she might have lived for another 30 years. Her parents requested permission to discontinue tube feeding, but the Missouri Supreme Court rejected their request. The court held that Cruzan never definitively stated her "right to die" wishes and that her parents had no **legal** right to make such a request for her. The court stated that preserving life, no matter what its quality, takes precedence over all other considerations.

Cruzan's parents appealed the decision, and in 1989, the U.S. Supreme Court agreed to hear their arguments. The Cruzans tried to convince the Supreme Court that their once independent and vivacious daughter would not want to live in a vegetative state. No one questioned that Cruzan's parents knew their daughter's wishes better than anyone and had the highest and most loving motives. The question for the Court was whether families (or anyone) can make life-and-death decisions on behalf of incompetent persons. The Supreme Court recognized that competent adults have the right to stop life-sustaining treatment, but upheld the Missouri Supreme Court's decision to require "clear and convincing" evidence that the incompetent person would refuse treatment.[2] It took still another round of court battles before additional evidence convinced the Missouri Supreme Court of Nancy Cruzan's wishes. Finally, her feeding tube was removed, and she died from dehydration two weeks later.

The high court's decision had widespread implications for people with terminal illnesses and for those in a persistent vegetative state, as well as the health professionals who take care of them. Additional ethical concerns are likely to arise as the population of elderly clients grows and new medical advances capable of sustaining life are developed. End-of-life decisions touch all of us, because they influence the extent to which our society views life-sustaining treatment as optional not only for our clients, but for ourselves and our families.

Prevailing Opinions

The emerging ethical, medical, and legal consensus seems to support the view that individual rights outweigh those of the state. Competent individuals have a legal right to refuse medical treatment—including nourishment and hydration—even when medical experts consider that treatment necessary to sustain life. Most people believe that it is acceptable to withhold nourishment and hydration when a competent person desires to forgo it, when an incompetent person has given **advance directives** (described next) about it, or when a person is in a permanently unconscious state.[3] In other words, even when treatment is lifesaving and its refusal may bring an earlier death, clients' rights remain paramount.

ADVANCE DIRECTIVES

Health care professionals should routinely encourage their competent clients to use advance directives to express ahead of time their preferences regarding medical treatments, including **artificial feedings,** should terminal illness, coma, or incompetency develop.[4] One form of an advance directive is a **living will.** Another is the **durable power of attorney.** All states have statutes governing the use of advance directives; health care professionals should be aware of the regulations of the states where they work.[5] Some states' laws specifically address the conditions under which nutrition support can be withheld or withdrawn.

Living Wills

A living will allows a competent adult to express clear directions regarding medical treatment in the event that the person is unable to make the necessary decisions at that time. The living will may specify that no extraordinary treatments should be administered, or alternatively, it may declare that every effort should be made to maintain life. People who prefer that nourishment and hydration be continued or discontinued need to write this specification into their living wills to ensure that their wishes are known.

Durable Power of Attorney

A durable power of attorney allows a competent adult to designate another competent adult (usually a relative or close friend) as an agent to make health care decisions in the event of incapacitation. In essence, it says, "I give this person the right to make health care decisions on my behalf should I become unable to make them."

When people give another person a durable power of attorney, they should also have a living will. They should ensure that the person they appoint clearly understands their health care preferences, knows the contents of their living will (when available), and can be trusted to make decisions that reflect those preferences.

Compliance with Advance Directives

People often fail to complete advance directives, and when they do, compliance with them is not guaranteed. Advance directives are not given to the physician 62 percent of the time, and even when they are, they are frequently misinterpreted.[6] The Study to Understand Prognoses and Preferences for Outcomes and Risks of Treatments (SUPPORT) found that among people with terminal diseases who were hospitalized, only 10 percent had advance directives.[7] Physicians were aware of these directives only 25 percent of the time. Other researchers found that among elderly clients admitted to intensive care, only 5 percent had advance directives.[8] Forty percent of the clients in the study who died had received resuscitation despite advance directives that instructed health care professionals to forgo this procedure.

Promoting Compliance with Advance Directives

People who wish to ensure that they receive medical care consistent with their individual wishes need to complete advance directives. Furthermore, they need to discuss these directives and hypothetical scenarios with their physicians, family members, and the person who holds the durable power of attorney before medical conditions arise that will necessitate others making decisions for them. They must ensure that their physician has a copy of any advanced directive and that the copy is part of their medical record. The person who plans ahead for future care in the case of a terminal illness or irreversible state of unconsciousness relieves others of the guilt and some of the anxiety of having to make decisions. Imagine the anguish a family member may go through in directing the health care team to stop nutrition support, knowing that doing so will hasten death. That decision is a little easier if the family member knows it is what the individual would want, or if a legal document takes the decision out of the family's hands altogether.

Responsibilities of Health Care Professionals

Clients should expect that physicians, other health care professionals, and health care facilities will comply with their preferences and not ignore them or merely tolerate them grudgingly. Health care professionals should reassure clients that their decisions to refuse life-sustaining treatments, including nutrition support, will not mean that other care will be withheld. The client who makes such a choice still deserves meticulous physical care, pain management, and compassionate emotional support.[9] Any health care professional who is unwilling to abide by the client's stated preferences should arrange for continuing care by another equally qualified professional and then withdraw from that client's care.

Ethical questions have no easy answers, yet decisions must be made. Each case requires careful, individualized consideration. Most hospitals and extended care facilities have established ethics committees to deal with issues such as those presented here. Health care team members should ensure that their disciplines are represented on such committees and should become familiar with their profession's ethics policies and guidelines.

Nutrition on the Net

WEBSITES

Access these websites for further study of topics covered in this highlight.

- For more information about medical ethics, visit these websites:

  American Society of Law, Medicine and Ethics: **www.aslme.org**

  Choice in Dying: **www.choices.org**

  American Medical Association: **www.ama-assn.org**

- Links to living wills can be found at this site: **www.midspring.com/~scottr/will.html**

References

1 R. DeChicco, A. Trew, and D. L. Seidner, What to do when the patient refuses a feeding tube, *Nutrition in Clinical Practice* 12 (1997): 228–230.

2 *Cruzan v. Director, Missouri Department of Health,* 497 U.S. 261, 110 S.Ct. 2841, 111 L.Ed.2d 224 (1990).

3 R. Burck, Feeding, withdrawing, and withholding: Ethical perspectives, *Nutrition in Clinical Practice* 11 (1996): 243–253.

4 *Advance Directives: The Role of Health Care Professionals* (Columbus, Ohio: Ross Products Division, Abbott Laboratories, 1996).

5 B. Dorner and coauthors, The "to feed or not to feed" dilemma, *Journal of the American Dietetic Association* (supplement 2) 97 (1997): 172–176.

6 D. R. Lustbader, Medical-legal issues in the ICU: When the court speaks, 30th International Educational and Scientific Symposium of the Society of Critical Care Medicine, February 10, 2001, available from Medscape at www.medscape.com/medscape/cno/2001/SCCM/Story.cfm?story_id=2077, visited March 28, 2001.

7 The SUPPORT Principal Investigators, A controlled trial to improve care for seriously ill hospitalized patients. The Study to Understand Prognoses and Preferences for Outcomes and Risks of Treatment (SUPPORT), *Journal of the American Medical Association* 274 (1995): 1591–1598.

8 M. D. Goodman, M. Tarnoff, and G. J. Slotman, Effect of advance directives on the management of elderly critically ill patients, *Critical Care Medicine* 26 (1998): 701–704.

9 V. M. Herrmann, Ethics in nutrition support, presented at the Eighth Annual Advances and Controversies in Clinical Nutrition, Dallas, Texas, April 5–7, 1998.

Appendixes

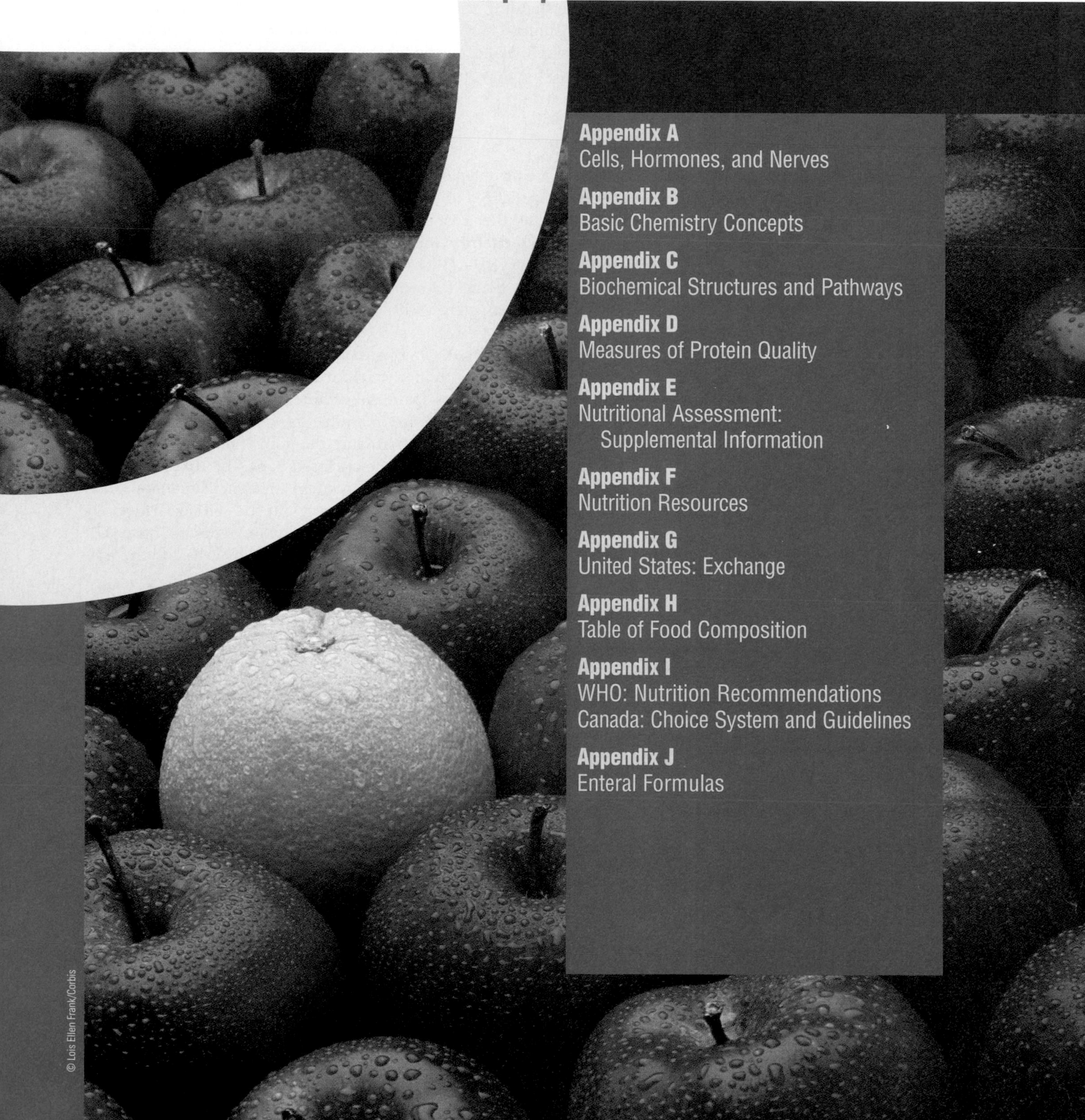

Contents

CELLS, HORMONES, AND NERVES

This appendix is offered as an optional chapter for readers who want to enhance their understanding of how the body coordinates its activities. The text presents a brief summary of the structure and function of the body's basic working unit (the cell) and of the body's two major regulatory systems (the hormonal system and the nervous system).

The Cell

The body's organs are made up of millions of cells and of materials produced by them. Each **cell** is specialized to perform its organ's functions, but all cells have common structures (see the accompanying glossary and Figure A-1). Every cell is contained within a **cell membrane.** The cell membrane assists in moving materials into and out of the cell, and some of its special proteins act as "pumps" (described in Chapter 6). Some features of cell membranes, such as microvilli (Chapter 3), permit cells to interact with other cells and with their environments in highly specific ways.

Inside the membrane lies the **cytoplasm,** or cell "fluid." The cytoplasm contains much more than just fluid, though. It is a highly organized system of fibers, tubes, membranes, particles, and subcellular **organelles** as complex as a city. These parts intercommunicate, manufacture and exchange materials, package and prepare materials for export, and maintain and repair themselves.

Within each cell is another membrane-enclosed body, the **nucleus.** Inside the nucleus are the **chromosomes,** which contain the genetic material, DNA. The DNA encodes all the instructions for carrying out the cell's activities. The role of DNA in coding for cell proteins is summarized in Figure 6-7 on p. 174. Chapter 6 also describes the variety of proteins produced by cells and the ways they perform the body's work.

Among the organelles within a cell are ribosomes, mitochondria, and lysosomes. Figure 6-7 briefly refers to the **ribosomes;** they assemble amino acids into proteins, following directions conveyed to them by RNA copies from the DNA in the chromosomes.

The **mitochondria** are made of intricately folded membranes that bear thousands of highly organized sets of enzymes on their inner and outer surfaces. Although mentioned only briefly in this book's chapters, their presence is implied whenever the enzymes of the TCA cycle and electron transport chain are mentioned because the mitochondria house all these enzymes.* Mitochondria are therefore crucial to aerobic metabolism, described in Chapter 7, and muscles conditioned to work aerobically are packed with them.

The **lysosomes** are membranes that enclose degradative enzymes. When a cell needs to self-destruct or to digest materials in its surroundings, its lysosomes free their enzymes. Lysosomes are active when tissue repair or remodeling is taking place—for example, in cleaning up infections, healing wounds, shaping embryonic organs, and remodeling bones.

*For the reactions of glycolysis, the TCA cycle, and the electron transport chain, see Chapter 7 and Appendix C. The reactions of glycolysis take place in the cytoplasm; the conversion of pyruvate to acetyl CoA takes place in the mitochondria, as do the TCA cycle and electron transport chain reactions. The mitochondria then release carbon dioxide, water, and ATP as their end products.

GLOSSARY OF CELL STRUCTURES

cell: the basic unit of life, of which all living things are composed. Every cell is surrounded by a membrane and contains cytoplasm, within which are organelles and a nucleus; the cell nucleus contains chromosomes.

cell membrane: the membrane that surrounds the cell and encloses its contents; made primarily of lipid and protein.

chromosomes: a set of structures within the nucleus of every cell that contains the cell's genetic material, DNA, associated with other materials (primarily proteins).

cytoplasm (SIGH-toh-plazm): the cell contents, except for the nucleus.
- **cyto** = cell
- **plasm** = a form

endoplasmic reticulum (en-doh-PLAZ-mic reh-TIC-you-lum): a complex network of intracellular membranes. The **rough endoplasmic reticulum** is dotted with ribosomes, where protein synthesis takes place. The **smooth endoplasmic reticulum** bears no ribosomes.
- **endo** = inside
- **plasm** = the cytoplasm

Golgi (GOAL-gee) **apparatus:** a set of membranes within the cell where secretory materials are packaged for export.

lysosomes (LYE-so-zomes): cellular organelles; membrane-enclosed sacs of degradative enzymes.
- **lysis** = dissolution

mitochondria (my-toh-KON-dree-uh); singular **mitochondrion:** the cellular organelles responsible for producing ATP aerobically; made of membranes (lipid and protein) with enzymes mounted on them.
- **mitos** = thread (referring to their slender shape)
- **chondros** = cartilage (referring to their external appearance)

nucleus: a major membrane-enclosed body within every cell, which contains the cell's genetic material, DNA, embedded in chromosomes.
- **nucleus** = a kernel

organelles: subcellular structures such as ribosomes, mitochondria, and lysosomes.
- **organelle** = little organ

FIGURE A-1 **The Structure of a Typical Cell**

The cell shown might be one in a gland (such as the pancreas) that produces secretory products (enzymes) for export (to the intestine). The rough endoplasmic reticulum with its ribosomes produces the enzymes; the smooth reticulum conducts them to the Golgi region; the Golgi membranes merge with the cell membrane, where the enzymes can be released into the extracellular fluid.

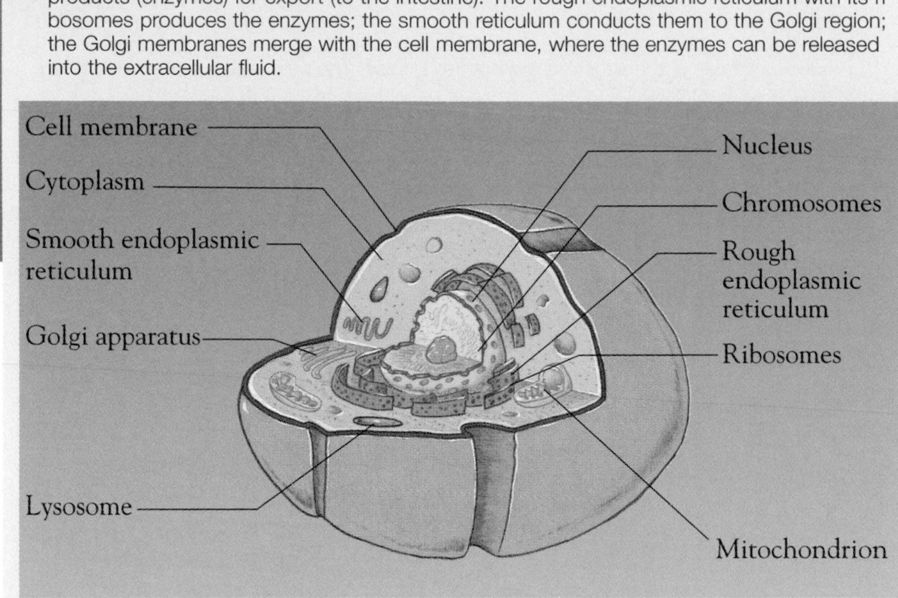

Besides these and other cellular organelles, the cell's cytoplasm contains a highly organized system of membranes, the **endoplasmic reticulum.** The ribosomes may either float free in the cytoplasm or be mounted on these membranes. A membranous surface dotted with ribosomes looks speckled under the microscope and is called "rough" endoplasmic reticulum; such a surface without ribosomes is called "smooth." Some intracellular membranes are organized into tubules that collect cellular materials, merge with the cell membrane, and discharge their contents to the outside of the cell; these membrane systems are named the **Golgi apparatus,** after the scientist who first described them. The rough and smooth endoplasmic reticula and the Golgi apparatus are continuous with one another, so secretions produced deep in the interior of the cell can be efficiently transported to the outside and released. These and other cell structures enable cells to perform the multitudes of functions for which they are specialized.

The actions of cells are coordinated by both hormones and nerves, as the next sections show. Among the types of cellular organelles are receptors for the hormones delivering instructions that originate elsewhere in the body. Some hormones penetrate the cell and its nucleus and attach to receptors on chromosomes, where they activate certain genes to initiate, stop, speed up, or slow down synthesis of certain proteins as needed. Other hormones attach to receptors on the cell surface and transmit their messages from there. The hormones◆ are described in the next section; the nerves, in the one following.

The Hormones

A hormonal message originates in a gland and travels as a chemical compound—a **hormone**—in the bloodstream. The hormone flows everywhere in the body, but only its target organs respond to it, because only they possess the receptors to receive it.

The hormones, the glands they originate in, and their target organs and effects are described in this section. Many of the hormones you might be interested in are included, but only a few are discussed in detail. Figure A-2 identifies the glands that produce the hormones, and the accompanying glossary defines the hormones discussed in this section.

ribosomes (RYE-boh-zomes): protein-making organelles in cells; composed of RNA and protein.
- **ribo** = containing the sugar ribose (in RNA)
- **some** = body

GLOSSARY OF HORMONES

adrenocorticotropin (ad-REE-noh-KORE-tee-koh-TROP-in) **ACTH:** a hormone, so named because it stimulates *(trope)* the adrenal cortex. The adrenal gland, like the pituitary, has two parts, in this case an outer portion *(cortex)* and an inner core *(medulla).* The realease of ACTH is mediated by **corticotropin-releasing hormone (CRH).**

aldosterone: a hormone from the adrenal gland involved in blood pressure regulation
- **aldo** = aldehyde

angiotensin: a hormone involved in blood pressure regulation that is activated by **renin** (REN-in), an enzyme from the kidneys.
- **angio** = blood vessels
- **tensin** = pressure
- **ren** = kidneys

antidiuretic hormone (ADH): the hormone that prevents water loss in urine (also called **vasopressin**).
- **anti** = against
- **di** = through
- **ure** = urine
- **vaso** = blood vessels
- **pressin** = pressure

calcitonin (KAL-see-TOH-nin): a hormone secreted by the thyroid gland that regulates (tones) calcium metabolism.

erythropoietin (eh-REE-throh-POY-eh-tin): a hormone that stimulates red blood cell production.
- **erythro** = red (blood cell)
- **poiesis** = creating (like poetry)

estrogens: hormones responsible for the menstrual cycle and other female characteristics.
- **oestrus** = the egg-making cycle
- **gen** = gives rise to

The study of hormones and their effects is ◆ **endocrinology.**

FIGURE A-2 **The Endocrine System**

These organs and glands release hormones that regulate body processes. An *endocrine gland* secretes its product directly into *(endo)* the blood; for example, the pancreas cells that produce insulin. An *exocrine gland* secretes its product(s) out *(exo)* of the gland through a duct into a cavity; the sweat glands of the skin and the enzyme-producing glands of the pancreas are both examples. The pancreas is therefore both an endocrine and an exocrine gland.

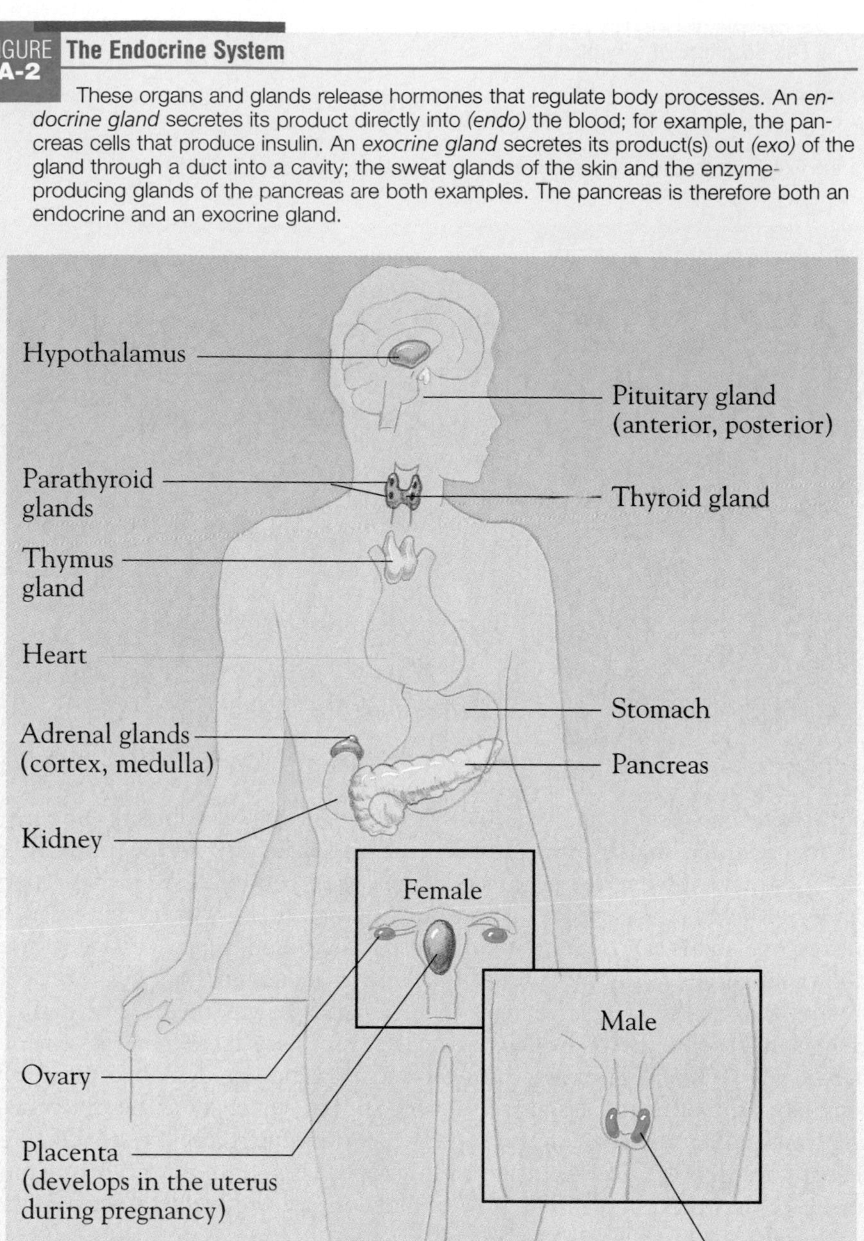

follicle-stimulating hormone (FSH): a hormone that stimulates maturation of the ovarian follicles in females and the production of sperm in males. (The ovarian follicles are part of the female reproductive system where the eggs are produced.) The release of FSH is mediated by **follicle-stimulating hormone releasing hormone (FSH–RH).**

glucocorticoids: hormones from the adrenal cortex that affect the body's management of glucose.
- **gluco** = glucose
- **corticoid** = from the cortex

growth hormone (GH): a hormone secreted by the pituitary tht regulates the cell division and protein synthesis needed for normal growth. The release of GH is mediated by **GH-releasing hormone (GHRH).**

◆ The **pituitary gland** in the brain has two parts—the **anterior** (front) and the **posterior** (hind) parts.

The hormonal system is a complex system in which many of the parts interact with one another. For example, several hormones are produced in the anterior pituitary gland in the brain. All of these hormones are regulated by other hormones produced in another part of the brain, the hypothalamus. Furthermore, each of the pituitary gland hormones has effects on the production of compounds elsewhere in the body. Some of these compounds are also hormones that will affect still other body parts. A hormone may travel far from its point of origin and ultimately have profound, even unexpected, effects.

Hormones of the Pituitary Gland and Hypothalamus

The anterior pituitary gland◆ produces the following hormones, each of which acts on one or more target organs and elicits a characteristic response:

- **Adrenocorticotropin (ACTH)** acts on the adrenal cortex, promoting the making and release of its hormones.

- **Thyroid-stimulating hormone (TSH)** acts on the thyroid gland, promoting the making and release of thyroid hormones.
- **Growth hormone (GH)** works on all tissues, promoting growth, fat breakdown, and the formation of antibodies.
- **Follicle-stimulating hormone (FSH)** works on the ovaries in the female, promoting their maturation, and on the testicles in the male, promoting sperm formation.
- **Luteinizing hormone (LH)** also acts on the ovaries, stimulating their maturation, the making and release of progesterone and estrogens, and ovulation; and on the testicles, promoting the making and release of testosterone.
- **Prolactin,** secreted in the female during pregnancy and lactation, acts on the mammary glands to stimulate their growth and the making of milk.

Each of these hormones has one or more signals that turn it on and another (or others) that turns it off.◆ Among the controlling signals are several hormones from the hypothalamus:

- **Corticotropin-releasing hormone (CRH),** which promotes release of ACTH, is turned on by stress and turned off by ACTH when enough has been released.
- **TSH-releasing hormone (TRH),** which promotes release of TSH, is turned on by large meals or low body temperature.
- **GH-releasing hormone (GHRH),** which stimulates the release of growth hormone, is turned on by insulin.
- **GH-inhibiting hormone (GHIH** or **somatostatin),** which inhibits the release of GH and interferes with the release of TSH, is turned on by hypoglycemia and/or physical activity and is rapidly destroyed by body tissues so that it does not accumulate.
- **FSH/LH–releasing hormone (FSH/LH–RH)** is turned on in the female by nerve messages or low estrogen and in the male by low testosterone.
- **Prolactin-inhibiting hormone (PIH)** is turned on by high prolactin levels and off by estrogen, testosterone, and suckling (by way of nerve messages).

Let's examine some of these controls. PIH, for example, responds to high prolactin levels (remember, prolactin promotes the making of milk). High prolactin levels ensure that milk is made and—by calling forth PIH—ensure that prolactin levels don't get too high. But when the infant is suckling—and creating a demand for milk—PIH is not allowed to work (suckling turns off PIH). The consequence: prolactin remains high, and milk production continues. Demand from the infant thus directly adjusts the supply of milk. This example shows how the need is met through the cooperation of the nerves and hormones.

As another example, consider CRH. Stress, perceived in the brain and relayed to the hypothalamus, switches on CRH. On arriving at the pituitary, CRH switches on ACTH. Then ACTH acts on its target organ, the adrenal cortex, which responds by producing and releasing stress hormones, and the stress response is under way. Events cascading from there involve every body cell and many other hormones.

The numerous steps required to set the stress response in motion make it possible for the body to fine-tune the response; control can be exerted at each step. These two examples illustrate what the body can do in response to two different stimuli—producing milk in response to an infant's need and gearing up for action in an emergency.

The posterior pituitary gland produces two hormones, each of which acts on one or more target cells and elicits a characteristic response:

- **Antidiuretic hormone (ADH),** or **vasopressin,** acts on the arteries, promoting their contraction, and on the kidneys, preventing water excretion. ADH is turned on whenever the blood volume is low, the blood pressure is low, or the salt concentration of the blood is high (see Chapter 12). It is turned off by the return of these conditions to normal.

Hormones that are turned off by their own effects are said to be regulated by **negative feedback.** For example, when a pituitary gland hormone has caused the release of a substance from a target organ, that substance itself switches off the original hormone signal (that is, it feeds back negatively). ◆

A

hormone: a chemical messenger. Hormones are secreted by a variety of endocrine glands in response to altered conditions in the body. Each hormone travels to one or more specific target tissues or organs, where it elicits a specific response to maintain homeostasis.

luteinizing (LOO-tee-in-EYE-zing) **hormone (LH):** a hormone that stimulates ovulation and the development of the corpus luteum (the small tissue that develops from a ruptured ovarian follicle and secretes hormones); so called because the follicle turns yellow as it matures. In men, LH stimulates testosterone secretion. The release of LH is mediated by **luteinizing hormone–releasing hormone (LH–RH).**
- **lutein** = a yellow pigment

oxytocin (OK-see-TOH-sin): a hormone that stimulates the mammary glands to eject milk during lactation and the uterus to contract during childbirth.
- **oxy** = quick
- **tocin** = childbirth

progesterone: the hormone of gestation (pregnancy).
- **pro** = promoting
- **gest** = gestation (pregnancy)
- **sterone** = a steroid hormone

prolactin (proh-LAK-tin): a hormone so named because it promotes *(pro)* the production of milk *(lacto).* The release of prolactin is mediated by **prolactin-inhibiting hormone (PIH).**

relaxin: the hormone of late pregnancy.

somatostatin (GHIH): a hormone that inhibits the release of growth hormone; the opposite of **somatotropin (GH).**
- **somato** = body
- **stat** = keep the same
- **tropin** = make more

testosterone: a steroid hormone from the testicles, or testes. The steroids, as explained in Chapter 5, are chemically related to, and some are derived from, the lipid cholesterol.
- **sterone** = a steroid hormone

thyroid-stimulating hormone (TSH): a hormone secreted by the pituitary that stimulates the thyroid gland to secrete its hormones—thyroxine and triiodothyronine. The release of TSH is mediated by **TSH-releasing hormone (TRH).**

- **Oxytocin** acts on the uterus, inducing contractions, and on the mammary glands, causing milk ejection. Oxytocin is produced in response to reduced progesterone levels, suckling, or the stretching of the cervix.

Hormones That Regulate Energy Metabolism

Hormones produced by a number of different glands have effects on energy metabolism:

- Insulin from the pancreas beta cells is turned on by many stimuli, including raised blood glucose. It acts on cells to increase glucose and amino acid uptake into them and to promote the secretion of GHRH.
- Glucagon from the pancreas alpha cells responds to low blood glucose and acts on the liver to promote the breakdown of glycogen to glucose, the conversion of amino acids to glucose, and the release of glucose.
- Thyroxin from the thyroid gland responds to TSH and acts on many cells to increase their metabolic rate, growth, and heat production.
- Norepinephrine and epinephrine◆ from the adrenal medulla respond to stimulation by sympathetic nerves and produce reactions in many cells that facilitate the body's readiness for fight or flight: increased heart activity, blood vessel constriction, breakdown of glycogen and glucose, raised blood glucose levels, and fat breakdown. Norepinephrine and epinephrine also influence the secretion of the many hormones from the hypothalamus that exert control on the body's other systems.

◆ Norepinephrine and epinephrine were formerly called **noradrenalin** and **adrenalin,** respectively.

- Growth hormone (GH) from the anterior pituitary (already mentioned).
- **Glucocorticoids** from the adrenal cortex become active during times of stress and carbohydrate metabolism.

Every body part is affected by these hormones. Each different hormone has unique effects; and hormones that oppose each other are produced in carefully regulated amounts, so each can respond to the exact degree that is appropriate to the condition.

Hormones That Adjust Other Body Balances

Hormones are involved in moving calcium into and out of the body's storage deposits in the bones:

- **Calcitonin** from the thyroid gland acts on the bones, which respond by storing calcium from the bloodstream whenever blood calcium rises above the normal range. It also acts on the kidneys to increase excretion of both calcium and phosphorus in the urine. Calcitonin plays a major role in infants and young children, but is less active in adults.
- Parathormone (parathyroid hormone or PTH) from the parathyroid gland responds to the opposite condition—lowered blood calcium—and acts on three targets: the bones, which release stored calcium into the blood; the kidneys, which slow the excretion of calcium; and the intestine, which increases calcium absorption.
- Vitamin D from the skin and activated in the kidneys acts with parathormone and is essential for the absorption of calcium in the intestine.

Figure 12-10 on p. 405 diagrams the ways vitamin D and the hormones calcitonin and parathormone regulate calcium homeostasis.

Another hormone has effects on blood-making activity:

- **Erythropoietin** from the kidneys is responsive to oxygen depletion of the blood and to anemia. It acts on the bone marrow to stimulate the making of red blood cells.

Another hormone is special for pregnancy:

- **Relaxin** from the ovaries is secreted in response to the raised progesterone and estrogen levels of late pregnancy. This hormone acts on the cervix and

A

pelvic ligaments to allow them to stretch so that they can accommodate the birth process without strain.

Other agents help regulate blood pressure:

- **Rennin** (an enzyme), from the kidneys, in cooperation with **angiotensin** in the blood responds to a reduced blood supply experienced by the kidneys and acts in several ways to increase blood pressure. Renin and angiotensin also stimulate the adrenal cortex to secrete the hormone aldosterone.
- **Aldosterone,** a hormone from the adrenal cortex, targets the kidneys, which respond by reabsorbing sodium. The effect is to retain more water in the bloodstream—thus, again, raising the blood pressure. Figure 12-3 in Chapter 12 provides more details.

The Gastrointestinal Hormones

Several hormones are produced in the stomach and intestines in response to the presence of food or the components of food:

- Gastrin from the stomach and duodenum stimulates the production and release of gastric acid and other digestive juices and the movement of the GI contents through the system.
- Cholecystokinin from the duodenum signals the gallbladder and pancreas to release their contents into the intestine to aid in digestion.
- Secretin from the duodenum calls forth acid-neutralizing bicarbonate from the pancreas into the intestine and slows the action of the stomach and its secretion of acid and digestive juices.
- Gastric-inhibitory peptide from the duodenum and jejunum inhibits the secretion of gastric acid and slows the process of digestion.

These hormones are defined and presented in more detail in Chapter 3.

The Sex Hormones

There are three major sex hormones:

- **Testosterone** from the testicles is released in response to LH (described earlier). It acts on all the tissues that are involved in male sexuality and promotes their development and maintenance.
- **Estrogens** from the ovary are released in response to both FSH and LH and act similarly in females.
- **Progesterone** from the ovary's corpus luteum and from the placenta acts on the uterus and mammary glands, preparing them for pregnancy and lactation.

The Prostaglandins

The prostaglandins◆ are a group of substances produced by many different cells. They perform a multitude of diverse functions in those cells, including the regulation of blood vessel contractions, nerve impulses, and hormone responses. The prostaglandins are all derived from the 20-carbon polyunsaturated fatty acids and account in part for the necessity for these fatty acids in the diet.

◆ Reminder: *Prostaglandins* are biologically active compounds, derived from the 20-carbon polyunsaturated fatty acids, that regulate blood pressure, blood clotting, and other body functions.

This brief description of the hormones and their functions should suffice to provide an awareness of the enormous impact these compounds have on body processes. The other overall regulating agency is the nervous system.

The Nervous System

The nervous system has a central control system—a sort of computer—that can evaluate information about conditions within and outside the body, and a vast

GLOSSARY OF THE NERVOUS SYSTEM

autonomic nervous system: the division of the nervous system that controls the body's automatic responses. Its two branches are the **sympathetic** branch, which helps the body respond to stressors from the outside environment, and the **parasympathetic** branch, which regulates normal body activities between stressful times.

- **autonomos** = self-governing

central nervous system: the central part of the nervous system; the brain and spinal cord.

peripheral (puh-RIFF-er-ul) **nervous system:** the peripheral (outermost) part of the nervous system; the vast complex of wiring that extends from the central nervous system to the body's outermost areas. It contains both somatic and autonomic components.

somatic (so-MAT-ick) **nervous system:** the division of the nervous system that controls the voluntary muscles, as distinguished from the autonomic nervous system, which controls involuntary functions.

- **soma** = body

system of wiring that receives information and sends instructions. The control unit is the brain and spinal cord, called the **central nervous system;** and the vast complex of wiring between the center and the parts is the **peripheral nervous system.** The smooth functioning that results from the system's adjustments to changing conditions is homeostasis.

The nervous system has two general functions: it controls voluntary muscles in response to sensory stimuli from them, and it controls involuntary, internal muscles and glands in response to nerve-borne and chemical signals about their status. In fact, the nervous system is best understood as two systems that use the same or similar pathways to receive and transmit their messages. The **somatic nervous system** controls the voluntary muscles; the **autonomic nervous system** controls the internal organs.

When scientists were first studying the autonomic nervous system, they noticed that when something hurt one organ of the body, some of the other organs reacted as if in sympathy for the afflicted one. They therefore named the nerve network they were studying the sympathetic nervous system. The term is still used today to refer to that branch of the autonomic nervous system that responds to pain and stress. The other branch is called the parasympathetic nervous system. (Think of the sympathetic branch as the responder when homeostasis needs restoring and the parasympathetic branch as the commander of function during normal times.) Both systems transmit their messages through the brain and spinal cord. Nerves of the two branches travel side by side along the same pathways to transmit their messages, but they oppose each other's actions (see Figure A-3).

An example will show how the sympathetic and parasympathetic nervous systems work to maintain homeostasis. When you go outside in cold weather, your skin's temperature receptors send "cold" messages to the spinal cord and brain. Your conscious mind may intervene at this point to tell you to zip your jacket, but let's say you have no jacket. Your sympathetic nervous system reacts to the external stressor, the cold. It signals your skin-surface capillaries to shut down so that your blood will circulate deeper in your tissues, where it will conserve heat. Your sympathetic nervous system also signals involuntary contractions of the small muscles just under the skin surface. The product of these muscle contractions is heat, and the visible result is goose bumps. If these measures do not raise your body temperature enough, then the sympathetic nerves signal your large muscle groups to shiver; the contractions of these large muscles produce still more heat. All of this activity adds up to a set of adjustments that maintain your homeostasis (with respect to temperature) under conditions of external extremes (cold) that would throw it off balance. The cold was a stressor; the body's response was resistance.

Now let's say you come in and sit by a fire and drink hot cocoa. You are warm and no longer need all that sympathetic activity. At this point, your parasympathetic nerves take over; they signal your skin-surface capillaries to dilate again, your goose bumps to subside, and your muscles to relax. Your body is back to normal. This is recovery.

Putting It Together

The hormonal and nervous systems coordinate body functions by transmitting and receiving messages. The point-to-point messages of the nervous system travel through a central switchboard (the spinal cord and brain), whereas the messages of the hormonal system are broadcast over the airways (the bloodstream), and any organ with the appropriate receptors can pick them up. Nerve impulses travel faster than hormonal messages do—although both are remarkably swift. Whereas your brain's command to wiggle your toes reaches the toes within a fraction of a second and stops as quickly, a gland's message to alter a

FIGURE A-3 **The Organization of the Nervous System**

The brain and spinal cord evaluate information about conditions within and outside the body, and the peripheral nerves receive information and send instructions.

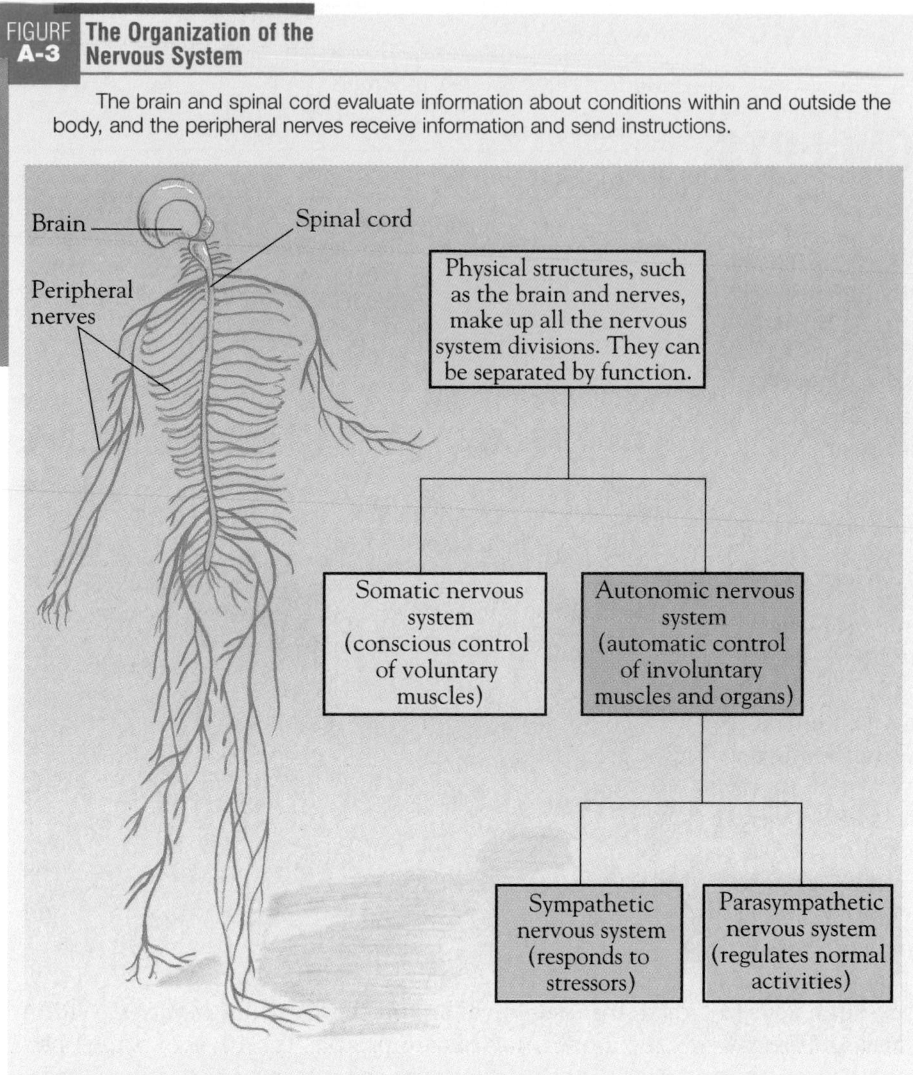

body condition may take several seconds or minutes to get started and may fade away equally slowly.

Together, the two systems possess every characteristic a superb communication network needs: varied speeds of transmission, along with private communication lines or public broadcasting systems, depending on the needs of the moment. The hormonal system, together with the nervous system, integrates the whole body's functioning so that all parts act smoothly together.

B BASIC CHEMISTRY CONCEPTS

This appendix is intended to provide the background in basic chemistry you need to understand the nutrition concepts presented in this book. Chemistry is the branch of natural science that is concerned with the description and classification of **matter,** the changes that matter undergoes, and the **energy** associated with these changes. The accompanying glossary defines matter, energy, and other related terms.

CONTENTS

Matter: The Properties of Atoms

Every substance has characteristics or properties that distinguish it from all other substances and thus give it a unique identity. These properties are both physical and chemical. The physical properties include such characteristics as color, taste, texture, and odor, as well as the temperatures at which a substance changes its state (from a solid to a liquid or from a liquid to a gas) and the weight of a unit volume (its density). The chemical properties of a substance have to do with how it reacts with other substances or responds to a change in its environment so that new substances with different sets of properties are produced.

A physical change does not change a substance's chemical composition. The three states—ice, water, and steam—all consist of two hydrogen atoms and one oxygen atom bound together. However, a chemical change occurs if an electric current passes through water. The water disappears, and two different substances are formed: hydrogen gas, which is flammable, and oxygen gas, which supports life. Chemical changes are also referred to as chemical reactions.

GLOSSARY

atoms: the smallest components of an element that have all of the properties of the element.

compound: a substance composed of two or more different atoms—for example, water (H_2O).

element: a substance composed of atoms that are alike—for example, iron (Fe).

energy: the capacity to do work.

matter: anything that takes up space and has mass.

molecule: two or more atoms of the same or different elements joined by chemical bonds. Examples are molecules of the element oxygen, composed of two oxygen atoms (O_2), and molecules of the compound water, composed of two hydrogen atoms and one oxygen atom (H_2O).

Substances: Elements and Compounds

The smallest part of a substance that can exist separately without losing its physical and chemical properties is a **molecule.** If a molecule is composed of **atoms** that are alike, the substance is an **element** (for example, O_2). If a molecule is composed of two or more different kinds of atoms, the substance is a **compound** (for example, H_2O).

Just over 100 elements are known, and these are listed in Table B-1. A familiar example is hydrogen, whose molecules are composed only of hydrogen atoms linked together in pairs (H_2). On the other hand, over a million compounds are known. An example is the sugar glucose. Each of its molecules is composed of 6 carbon, 6 oxygen, and 12 hydrogen atoms linked together in a specific arrangement (as described in Chapter 4).

The Nature of Atoms

Atoms themselves are made of smaller particles. Within the atomic nucleus are protons (positively charged particles), and surrounding the nucleus are electrons (negatively charged particles). The number of protons (+) in the nucleus of an atom determines the number of electrons (−) around it. The positive charge on a proton is equal to the negative charge on an electron, so the charges cancel each other out and leave the atom neutral to its surroundings.

The nucleus may also include neutrons, subatomic particles that have no charge. Protons and neutrons are of equal mass, and together they give an atom its weight. Electrons bond atoms together to make molecules, and they are involved in chemical reactions.

Each type of atom has a characteristic number of protons in its nucleus. The hydrogen atom is the simplest of all. It possesses a single proton, with a single electron associated with it:

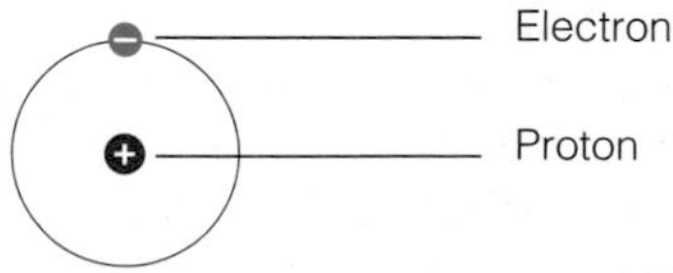

Hydrogen atom (H), atomic number 1.

B

TABLE B-1 Chemical Symbols for the Elements

Number of Protons (Atomic Number)	Element	Number of Electrons in Outer Shell	Number of Protons (Atomic Number)	Element	Number of Electrons in Outer Shell
1	Hydrogen (H)	1	52	Tellurium (Te)	6
2	Helium (He)	2	53	Iodine (I)	7
3	Lithium (Li)	1	54	Xenon (Xe)	8
4	Beryllium (Be)	2	55	Cesium (Cs)	1
5	Boron (B)	3	56	Barium (Ba)	2
6	Carbon (C)	4	57	Lanthanum (La)	2
7	Nitrogen (N)	5	58	Cerium (Ce)	2
8	Oxygen (O)	6	59	Praseodymium (Pr)	2
9	Fluorine (F)	7	60	Neodymium (Nd)	2
10	Neon (Ne)	8	61	Promethium (Pm)	2
11	Sodium (Na)	1	62	Samarium (Sm)	2
12	Magnesium (Mg)	2	63	Europium (Eu)	2
13	Aluminum (Al)	3	64	Gadolinium (Gd)	2
14	Silicon (Si)	4	65	Terbium (Tb)	2
15	Phosphorus (P)	5	66	Dysprosium (Dy)	2
16	Sulfur (S)	6	67	Holmium (Ho)	2
17	Chlorine (Cl)	7	68	Erbium (Er)	2
18	Argon (Ar)	8	69	Thulium (Tm)	2
19	Potassium (K)	1	70	Ytterbium (Yb)	2
20	Calcium (Ca)	2	71	Lutetium (Lu)	2
21	Scandium (Sc)	2	72	Hafnium (Hf)	2
22	Titanium (Ti)	2	73	Tantalum (Ta)	2
23	Vanadium (V)	2	74	Tungsten (W)	2
24	Chromium (Cr)	1	75	Rhenium (Re)	2
25	Manganese (Mn)	2	76	Osmium (Os)	2
26	Iron (Fe)	2	77	Iridium (Ir)	2
27	Cobalt (Co)	2	78	Platinum (Pt)	1
28	Nickel (Ni)	2	79	Gold (Au)	1
29	Copper (Cu)	1	80	Mercury (Hg)	2
30	Zinc (Zn)	2	81	Thallium (Tl)	3
31	Gallium (Ga)	3	82	Lead (Pb)	4
32	Germanium (Ge)	4	83	Bismuth (Bi)	5
33	Arsenic (As)	5	84	Polonium (Po)	6
34	Selenium (Se)	6	85	Astatine (At)	7
35	Bromine (Br)	7	86	Radon (Rn)	8
36	Krypton (Kr)	8	87	Francium (Fr)	1
37	Rubidium (Rb)	1	88	Radium (Ra)	2
38	Strontium (Sr)	2	89	Actinium (Ac)	2
39	Yttrium (Y)	2	90	Thorium (Th)	2
40	Zirconium (Zr)	2	91	Protactinium (Pa)	2
41	Niobium (Nb)	1	92	Uranium (U)	2
42	Molybdenum (Mo)	1	93	Neptunium (Np)	2
43	Technetium (Tc)	1	94	Plutonium (Pu)	2
44	Ruthenium (Ru)	1	95	Americium (Am)	2
45	Rhodium (Rh)	1	96	Curium (Cm)	2
46	Palladium (Pd)	—	97	Berkelium (Bk)	2
47	Silver (Ag)	1	98	Californium (Cf)	2
48	Cadmium (Cd)	2	99	Einsteinium (Es)	2
49	Indium (In)	3	100	Fermium (Fm)	2
50	Tin (Sn)	4	101	Mendelevium (Md)	2
51	Antimony (Sb)	5	102	Nobelium (No)	2

Key:
Elements found in energy-yielding nutrients, vitamins, and water.
Major minerals.
Trace minerals.

Just as hydrogen always has one proton, helium always has two, lithium three, and so on. The atomic number of each element is the number of protons in the nucleus of that atom, and this never changes in a chemical reaction; it gives the atom its identity. The atomic numbers for the known elements are listed in Table B-1.

Besides hydrogen, the atoms most common in living things are carbon (C), nitrogen (N), and oxygen (O), whose atomic numbers are 6, 7, and 8, respectively. Their structures are more complicated than that of hydrogen, but each of them possesses the same number of electrons as there are protons in the nucleus. These electrons are found in orbits, or shells (shown on p. B-2).

Carbon atom (C), atomic number 6. | Nitrogen atom (N), atomic number 7. | Oxygen atom (O), atomic number 8.

In these and all diagrams of atoms that follow, only the protons and electrons are shown. The neutrons, which contribute only to atomic weight, not to charge, are omitted.

The most important structural feature of an atom for determining its chemical behavior is the number of electrons in its outermost shell. The first, or innermost, shell is full when it is occupied by two electrons; so an atom with two or more electrons has a filled first shell. When the first shell is full, electrons begin to fill the second shell.

The second shell is completely full when it has eight electrons. A substance that has a full outer shell tends not to enter into chemical reactions. Atomic number 10, neon, is a chemically inert substance because its outer shell is complete. Fluorine, atomic number 9, has a great tendency to draw an electron from other substances to complete its outer shell, and thus it is highly reactive. Carbon has a half-full outer shell, which helps explain its great versatility; it can combine with other elements in a variety of ways to form a large number of compounds.

Atoms seek to reach a state of maximum stability or of lowest energy in the same way that a ball will roll down a hill until it reaches the lowest place. An atom achieves a state of maximum stability:

- By gaining or losing electrons to either fill or empty its outer shell.
- By sharing its electrons through bonding together with other atoms and thereby completing its outer shell.

The number of electrons determines how the atom will chemically react with other atoms. The atomic number, not the weight, is what gives an atom its chemical nature.

Chemical Bonding

Atoms often complete their outer shells by sharing electrons with other atoms. In order to complete its outer shell, a carbon atom requires four electrons. A hydrogen atom requires one. Thus, when a carbon atom shares electrons with four hydrogen atoms, each completes its outer shell (as shown in the next column). Electron sharing binds the atoms together and satisfies the conditions of maximum stability for the molecule. The outer shell of each atom is complete, since hydrogen effectively has the required two electrons in its first (outer) shell, and carbon has eight electrons in its second (outer) shell; and the molecule is electrically neutral, with a total of ten protons and ten electrons.

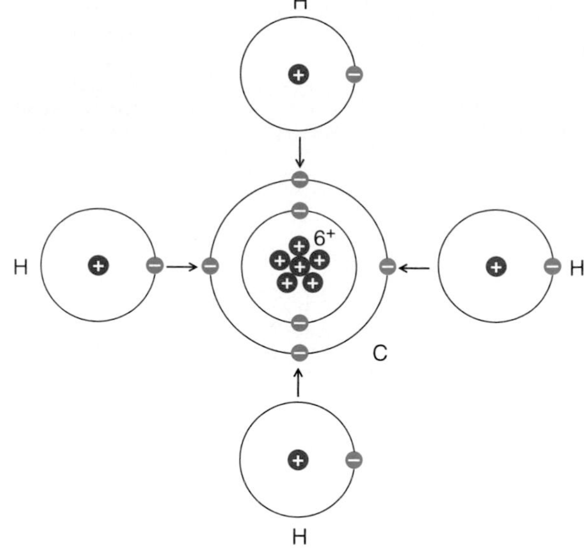

When a carbon atom shares electrons with four hydrogen atoms, a methane molecule is made.

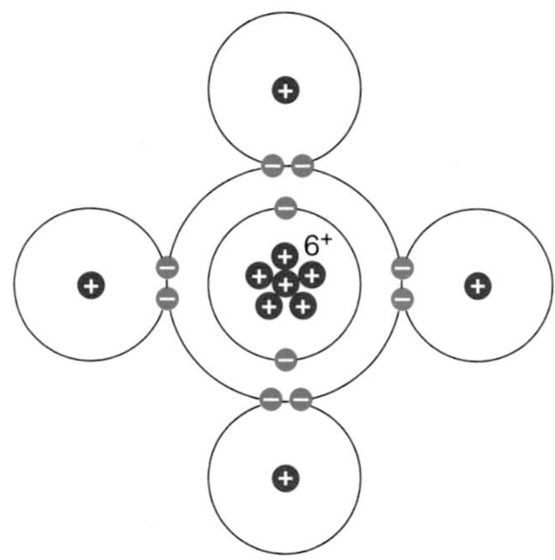

The chemical formula for methane is CH_4. Note that by sharing electrons, every atom achieves a filled outer shell.

Bonds that involve the sharing of electrons, like the bond between carbon and hydrogen, are the most stable kind of association that atoms can form with one another. They are sometimes called covalent bonds, and the resulting combinations of atoms are called molecules. A single pair of shared electrons forms a single bond. A simplified way to represent a single bond is with a single line. Thus the structure of methane (CH_4) could be represented like this (ignoring the inner-shell electrons, which do not participate in bonding):

```
   H
   |
H—C—H
   |
   H
```

Methane (CH_4).

Similarly, one nitrogen atom and three hydrogen atoms can share electrons to form one molecule of ammonia (NH_3):

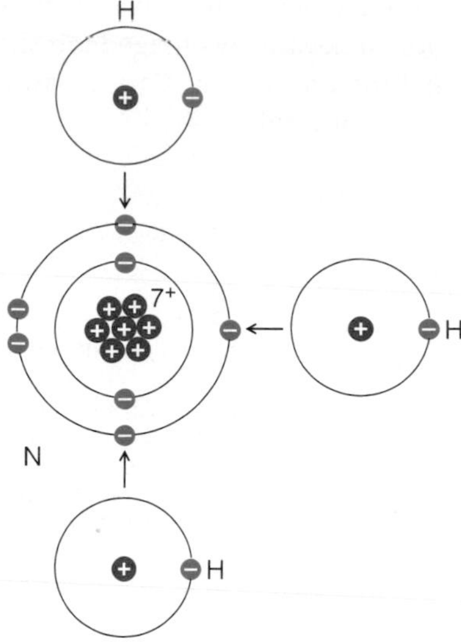

When a nitrogen atom shares electrons with three hydrogen atoms, an ammonia molecule is made.

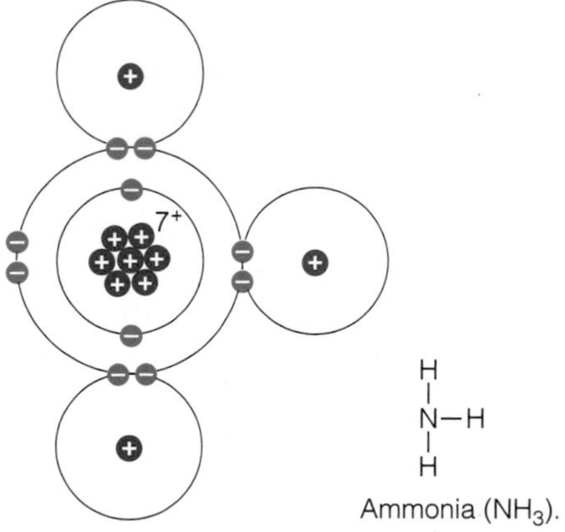

Ammonia (NH_3).

The chemical formula for ammonia is NH_3. Count the electrons in each atom's outer shell to confirm that it is filled.

One oxygen atom may be bonded to two hydrogen atoms to form one molecule of water (H_2O):

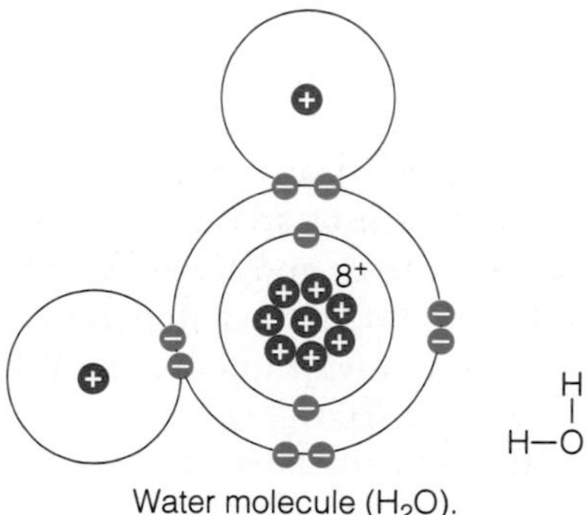

Water molecule (H_2O).

When two oxygen atoms form a molecule of oxygen, they must share two pairs of electrons. This double bond may be represented as two single lines:

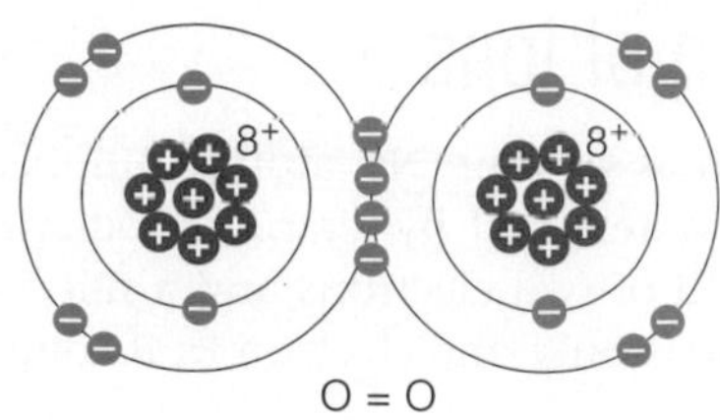

Oxygen molecule (O_2).

Small atoms form the tightest, most stable bonds. H, O, N, and C are the smallest atoms capable of forming one, two, three, and four electron-pair bonds (respectively). This is the basis for the statement in Chapter 4 that in drawings of compounds containing these atoms, hydrogen must always have one, oxygen two, nitrogen three, and carbon four bonds radiating to other atoms:

H— —O— —N— —C—

The stability of the associations between these small atoms and the versatility with which they can combine make them very common in living things. Interestingly, all cells, whether they come from animals, plants, or bacteria, contain the same elements in very nearly the same proportions. The atomic elements commonly found in living things are shown in Table B-2.

TABLE B-2 Elemental Composition of the Human Body

Element	Composition Chemical Symbol	By Weight (%)
Oxygen	O	65
Carbon	C	18
Hydrogen	H	10
Nitrogen	N	3
Calcium	Ca	1.5
Phosphorus	P	1.0
Potassium	K	0.4
Sulfur	S	0.3
Sodium	Na	0.2
Chloride	Cl	0.1
Magnesium	Mg	0.1
Total		99.6[a]

[a]The remaining 0.40 percent by weight is contributed by the trace elements: chromium (Cr), copper (Cu), zinc (Zn), selenium (Se), molybdenum (Mo), fluorine (F), iodine (I), manganese (Mn), and iron (Fe). Cells may also contain variable traces of some of the following: boron (B), cobalt (Co), lithium (Li), strontium (Sr), aluminum (Al), silicon (Si), lead (Pb), vanadium (V), arsenic (As), bromine (Br), and others.

Formation of Ions

An atom such as sodium (Na, atomic number 11) cannot easily fill its outer shell by sharing. Sodium possesses a filled first shell of two electrons and a filled second shell of eight; there is only one electron in its outermost shell:

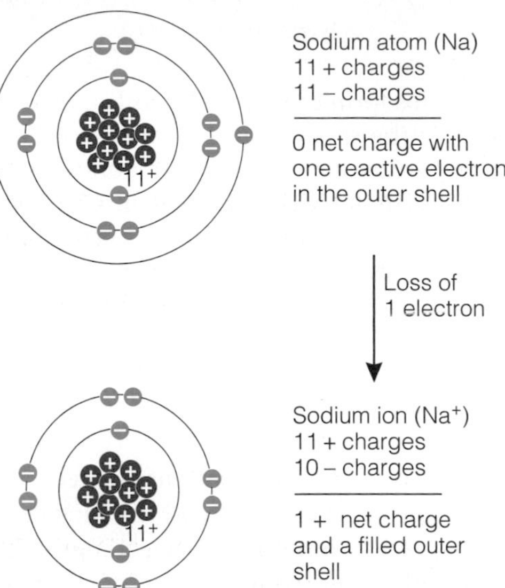

If sodium loses this electron, it satisfies one condition for stability: a filled outer shell (now its second shell counts as the outer shell). However, it is not electrically neutral. It has 11 protons (positive) and only 10 electrons (negative). It therefore has a net positive charge. An atom or molecule that has lost or gained one or more electrons and so is electrically charged is called an ion.

An atom such as chlorine (Cl, atomic number 17), with seven electrons in its outermost shell, can share electrons to fill its outer shell, or it can gain one electron to complete its outer shell and thus give it a negative charge:

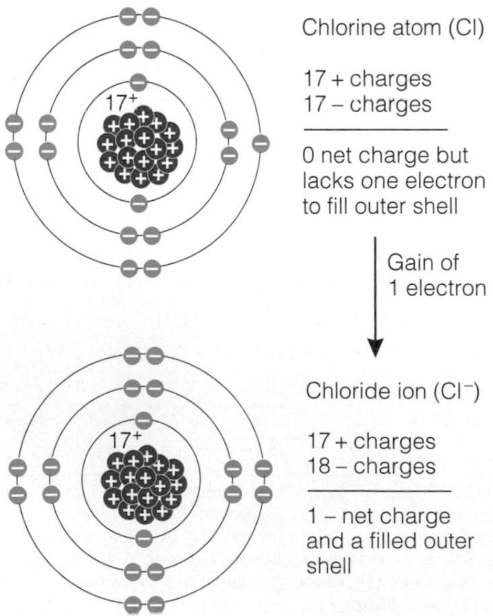

A positively charged ion such as sodium ion (Na^+) is called a cation; a negatively charged ion such as a chloride ion (Cl^-) is called an anion. Cations and anions attract one another to form salts:

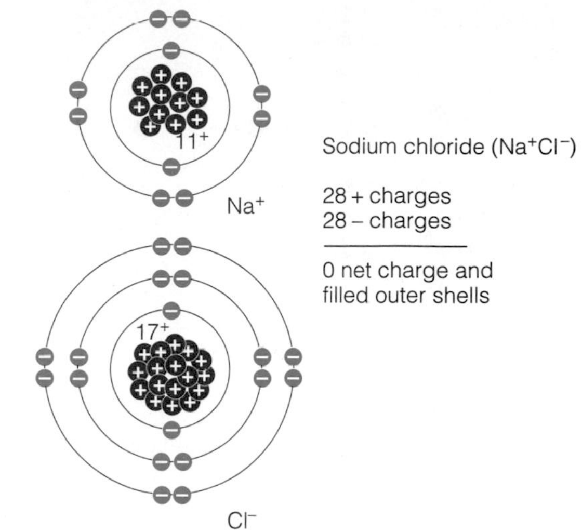

With all its electrons, sodium is a shiny, highly reactive metal; chlorine is the poisonous greenish yellow gas that was used in World War I. But after sodium and chlorine have transferred electrons, they form the stable white salt familiar to you as table salt, or sodium chloride (Na^+Cl^-). The dramatic difference illustrates how profoundly the electron arrangement can influence the nature of a substance. The wide distribution of salt in nature attests to the stability of the union between the ions. Each meets the other's needs (a good marriage).

When dry, salt exists as crystals; its ions are stacked very regularly into a lattice, with positive and negative ions alternating in a three-dimensional checkerboard structure. In water, however, the salt quickly dissolves, and its ions separate from one another, forming an electrolyte solution in which they move about freely. Covalently bonded molecules rarely dissociate like this in a water solution. The most common exception is when they behave like acids and release H^+ ions, as discussed in the next section.

An ion can also be a group of atoms bound together in such a way that the group has a net charge and enters into reactions as a single unit. Many such groups are active in the fluids of the body. The bicarbonate ion is composed of five atoms—one H, one C, and three Os—and has a net charge of −1 (HCO_3^-). Another important ion of this type is a phosphate ion with one H, one P, and four O, and a net charge of −2 (HPO_4^{-2}).

Whereas many elements have only one configuration in the outer shell and thus only one way to bond with other elements, some elements have the possibility of varied configurations. Iron is such an element. Under some conditions iron loses two electrons, and under other circumstances it loses three. If iron loses two electrons, it then has a net charge of +2, and we call it ferrous iron

(Fe^{++}). If it donates three electrons to another atom, it becomes the +3 ion, or ferric iron (Fe^{+++}).

Ferous iron (Fe^{++})	Ferric iron (Fe^{+++})
(had 2 outer-shell electrons but has lost them)	(had 3 outer-shell electrons but has lost them)
26 + charges	26 + charges
24 − charges	23 − charges
2 + net charge	3 + net charge

Remember that a positive charge on an ion means that negative charges—electrons—have been lost and not that positive charges have been added to the nucleus.

Water, Acids, and Bases

• ***Water*** • The water molecule is electrically neutral, having equal numbers of protons and electrons. However, when a hydrogen atom shares its electron with oxygen, that electron will spend most of its time closer to the positively charged oxygen nucleus. This leaves the positive proton (nucleus of the hydrogen atom) exposed on the outer part of the water molecule. We know, too, that the two hydrogens both bond toward the same side of the oxygen. These two facts explain why water molecules are polar: they have regions of more positive and more negative charge.

Polar molecules like water are drawn to one another by the attractive forces between the positive polar areas of one and the negative poles of another. These attractive forces, sometimes known as polar bonds or hydrogen bonds, occur among many molecules and also within the different parts of single large molecules. Although very weak in comparison with covalent bonds, polar bonds may occur in such abundance that they become exceedingly important in determining the structure of such large molecules as proteins and DNA.

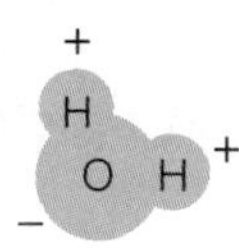

This diagram of the polar water molecule shows displacement of electrons toward the O nucleus; thus the negative region is near the O and the positive regions are near the Hs.

Water molecules have a slight tendency to ionize, separating into positive (H^+) and negative (OH^-) ions. In pure water, a small but constant number of these ions is present, and the number of positive ions exactly equals the number of negative ions.

• ***Acid*** • An acid is a substance that releases H^+ ions (protons) in a water solution. Hydrochloric acid (HC^-) is such a substance because it dissociates in a water solution into H^+ and Cl^- ions. Acetic acid is also an acid because it dissociates in water to acetate ions and free H^+:

$$H-\underset{H}{\overset{H}{\underset{|}{\overset{|}{C}}}}-\overset{O}{\overset{\|}{C}}-O-H \longrightarrow H-\underset{H}{\overset{H}{\underset{|}{\overset{|}{C}}}}-\overset{O}{\overset{\|}{C}}-O^- + H^+$$

Acetic acid dissociates into an acetate ion and a hydrogen ion.

The more H^+ ions released, the stronger the acid.

• ***pH*** • Chemists define degrees of acidity by means of the pH scale, which runs from 0 to 14. The pH expresses the concentration of H^+ ions: a pH of 1 is extremely acidic, 7 is neutral, and 13 is very basic. There is a tenfold difference in the concentration of H^+ ions between points on this scale. A solution with pH 3, for example, has *ten times* as many H^+ ions as a solution with pH 4. At pH 7, the concentrations of free H^+ and OH^- are exactly the same—1/10,000,000 moles per liter (10^{-7} moles per liter).* At pH 4, the concentration of free H^+ ions is 1/10,000 (10^{-4}) moles per liter. This is a higher concentration of H^+ ions, and the solution is therefore acidic. Figure 3-6 on p. 72 presents the pH scale.

• ***Bases*** • A base is a substance that can soak up, or combine with, H^+ ions, thus reducing the acidity of a solution. The compound ammonia is such a substance. The ammonia molecule has two electrons that are not shared with any other atom; a hydrogen ion (H^+) is just a naked proton with no shell of electrons at all. The proton readily combines with the ammonia molecule to form an ammonium ion; thus a free proton is withdrawn from the solution and no longer contributes to its acidity. Many compounds containing nitrogen are important bases in living systems. Acids and bases neutralize each other to produce substances that are neither acid nor base.

$$:\underset{H}{\overset{H}{\underset{|}{\overset{|}{N}}}}-H + H^+ \longrightarrow H-\underset{H}{\overset{H}{\underset{|}{\overset{|}{N^+}}}}-H$$

Ammonia captures a hydrogen ion from water. The two dots here represent the two electrons not shared with another atom. These dots are ordinarily not shown in chemical structure drawings. Compare this drawing with the earlier diagram of an ammonia molecule (p. B-3).

Chemical Reactions

A chemical reaction, or chemical change, results in the breakdown of substances and the formation of new ones. Almost all such reactions involve a change in the bonding of atoms. Old bonds are broken, and new ones are formed. The nuclei of atoms are never involved in chemical reactions—only their outer-shell electrons take part. At the end of a chemical reaction, the number of atoms of

*A mole is a certain number (about 6×10^{23}) of molecules. The pH of a solution is defined as the negative logarithm of the hydrogen ion concentration of the solution. Thus, if the concentration is 10^{-2} (moles per liter), the pH is 2; if 10^{-8}, the pH is 8; and so on.

each type is always the same as at the beginning. For example, two hydrogen molecules ($2H_2$) can react with one oxygen molecule (O_2) to form two water molecules ($2H_2O$). In this reaction two substances (hydrogen and oxygen) disappear, and a new one (water) is formed, but at the end of the reaction there are still four H atoms and two O atoms, just as there were at the beginning. Because the atoms are now linked in a different way, their characteristics or properties have changed.

In many instances chemical reactions involve not the relinking of molecules but the exchanging of electrons or protons among them. In such reactions the molecule that gains one or more electrons (or loses one or more hydrogen ions) is said to be reduced; the molecule that loses electrons (or gains protons) is oxidized. A hydrogen ion is equivalent to a proton. Oxidation and reduction take place simultaneously because an electron or proton that is lost by one molecule is accepted by another. The addition of an atom of oxygen is also oxidation because oxygen (with six electrons in the outer shell) accepts two electrons in becoming bonded. Oxidation, then, is loss of electrons, gain of protons, or addition of oxygen (with six electrons); reduction is the opposite—gain of electrons, loss of protons, or loss of oxygen. The addition of hydrogen atoms to oxygen to form water can thus be described as the reduction of oxygen *or* the oxidation of hydrogen.

If a reaction results in a net increase in the energy of a compound, it is called an endergonic, or "uphill," reaction (energy, *erg,* is added into, *endo,* the compound). An example is the chief result of photosynthesis, the making of sugar in a plant from carbon dioxide and water using the energy of sunlight. Conversely, the oxidation of sugar to carbon dioxide and water is an exergonic, or "downhill," reaction because the end products have less energy than the starting products. Oftentimes, but not always, reduction reactions are endergonic, resulting in an increase in the energy of the products. Oxidation reactions often, but not always, are exergonic.

Chemical reactions tend to occur spontaneously if the end products are in a lower energy state and therefore are more stable than the reacting compounds. These reactions often give off energy in the form of heat as they occur. The generation of heat by wood burning in a fireplace and the maintenance of human body warmth both depend on energy-yielding chemical reactions. These downhill reactions occur easily, although they may require some activation energy to get them started, just as a ball requires a push to start rolling downhill.

Uphill reactions, in which the products contain more energy than the reacting compounds started with, do not occur until an energy source is provided. An example of such an energy source is the sunlight used in photosynthesis, where carbon dioxide and water (low-energy compounds) are combined to form the sugar glucose (a higher-energy compound). Another example is the use of

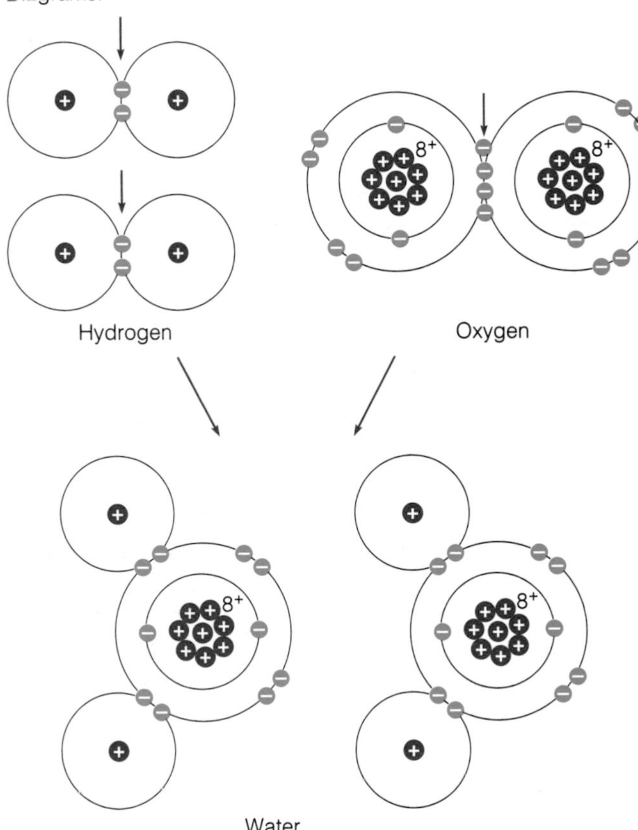

Hydrogen and oxygen react to form water.

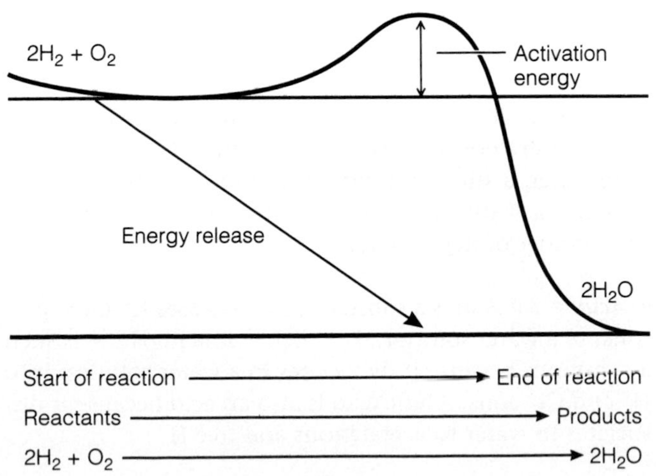

the energy in glucose to combine two low-energy compounds in the body into the high-energy compound ATP (see Chapter 7). The energy in ATP may be used to power many other energy-requiring, uphill reactions. Clearly, any of many different molecules can be used as a temporary storage place for energy.

Neither downhill nor uphill reactions occur until something sets them off (activation) or until a path is provided for them to follow. The body uses enzymes as a means of providing paths and controlling chemical reactions (see Chapter 6). By controlling the availability and the action of its enzymes, the cells can "decide" which chemical reactions to prevent and which to promote.

Formation of Free Radicals

Normally, when a chemical reaction takes place, bonds break and re-form with some redistribution of atoms and rearrangement of bonds to form new, stable compounds. Normally, bonds don't split in such a way as to leave a molecule with an odd, unpaired electron. When they do, free radicals are formed. Free radicals are highly unstable and quickly react with other compounds, forming more free radicals in a chain reaction. A cascade may ensue in which many highly reactive radicals are generated, resulting finally in the disruption of a living structure such as a cell membrane.

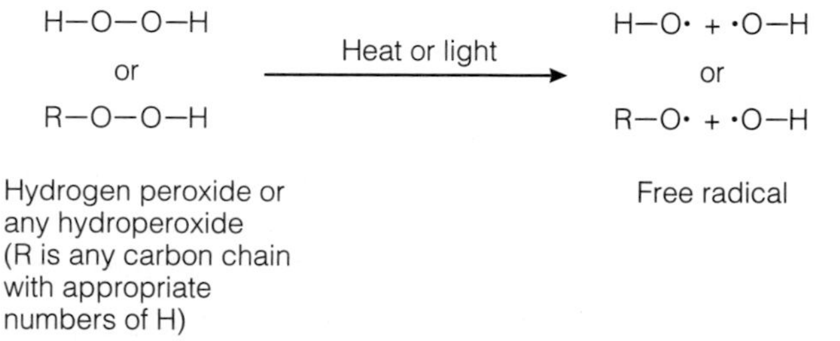

H—O—O—H or R—O—O—H —(Heat or light)→ H—O· + ·O—H or R—O· + ·O—H

Hydrogen peroxide or any hydroperoxide (R is any carbon chain with appropriate numbers of H) → Free radical

Free radicals are formed. The dots represent single electrons that are available for sharing (the atom needs another electron to fill its outer shell).

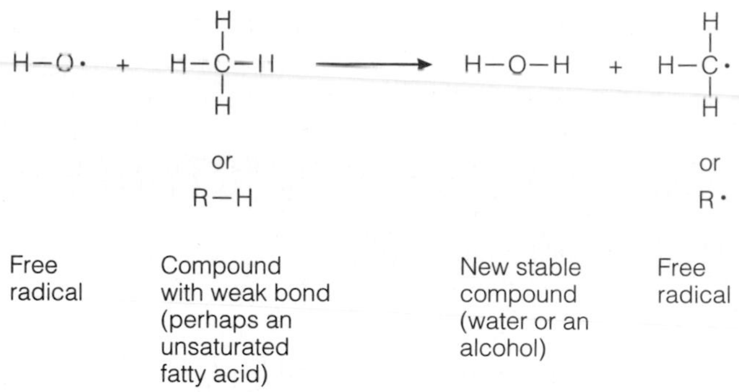

H—O· + CH_4 (H—C—H with H above and below) or R—H → H—O—H + ·CH_3 (H—C· with H above and below) or R·

Free radical + Compound with weak bond (perhaps an unsaturated fatty acid) → New stable compound (water or an alcohol) + Free radical

Destruction of biological compounds by free radicals. The free radical attacks a weak bond in a biological compound, disrupting it and forming a new stable molecule and another free radical. This can attack another biological compound, and so on.

Oxidation of some compounds can be induced by air at room temperature in the presence of light. Such reactions are thought to take place through the formation of compounds called peroxides:

Peroxides:

H—O—O—H	Hydrogen peroxide
R—O—O—H	Hydroperoxides (R is any carbon chain with appropriate numbers of H)
R—O—O—R	Peroxide

Some peroxides readily disintegrate into free radicals, initiating chain reactions like those just described.

Free radicals are of special interest in nutrition because the antioxidant properties of vitamins A, C, and E as well as the mineral selenium are thought to protect against the destructive effects of these free radicals (see Highlight 11). For example, vitamin E on the surface of the lungs reacts with, and is destroyed by, free radicals, thus preventing the radicals from reaching underlying cells and oxidizing the lipids in their membranes.

C BIOCHEMICAL STRUCTURES AND PATHWAYS

The diagrams of nutrients presented here are meant to enhance your understanding of the most important organic molecules in the human diet. Following the diagrams of nutrients are sections on the major metabolic pathways mentioned in Chapter 7—glycolysis, fatty acid oxidation, amino acid degradation, the TCA cycle, and the electron transport chain—and a description of how alcohol interferes with these pathways. Discussions of the urea cycle and the formation of ketone bodies complete the appendix.

CONTENTS

Carbohydrates

Monosaccharides

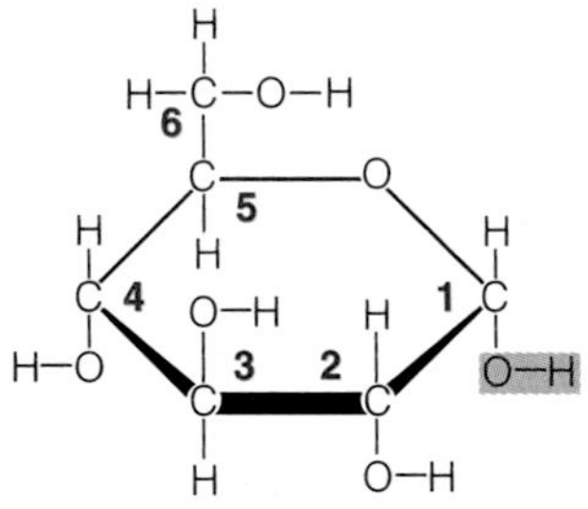

Glucose (alpha form). The ring would be at right angles to the plane of the paper. The bonds directed upward are above the plane; those directed downward are below the plane. This molecule is considered an alpha form because the OH on carbon 1 points downward.

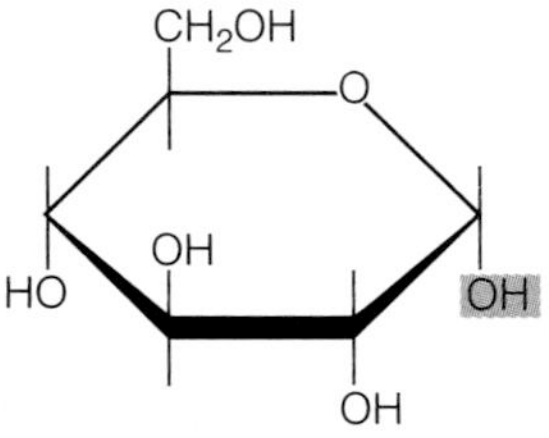

Glucose (alpha form) shorthand notation. This notation, in which the carbons in the ring and single hydrogens have been eliminated, will be used throughout this appendix.

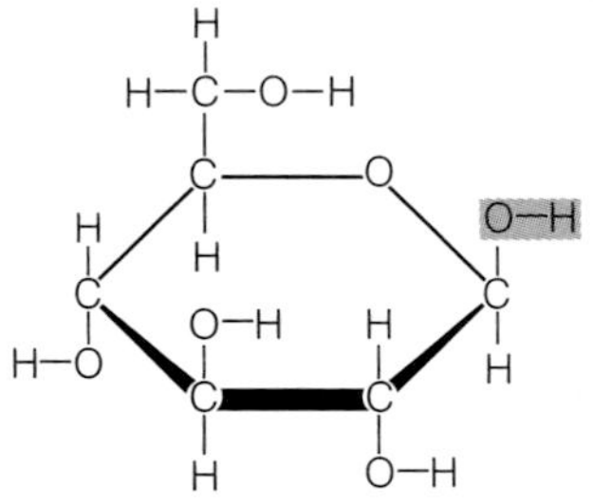

Glucose (beta form). The OH on carbon 1 points upward.
Fructose, galactose: see Chapter 4.

Disaccharides

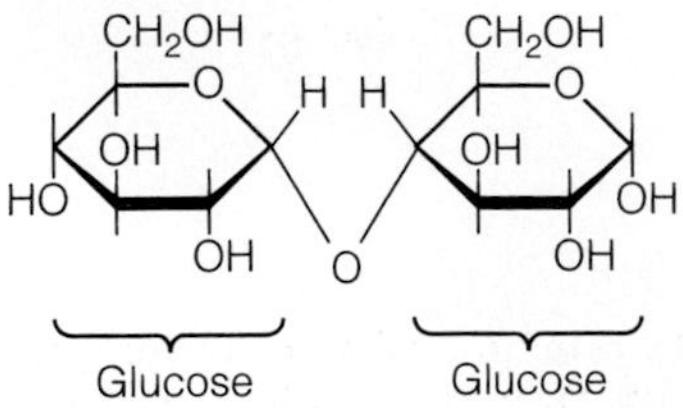

Glucose Glucose
Maltose.

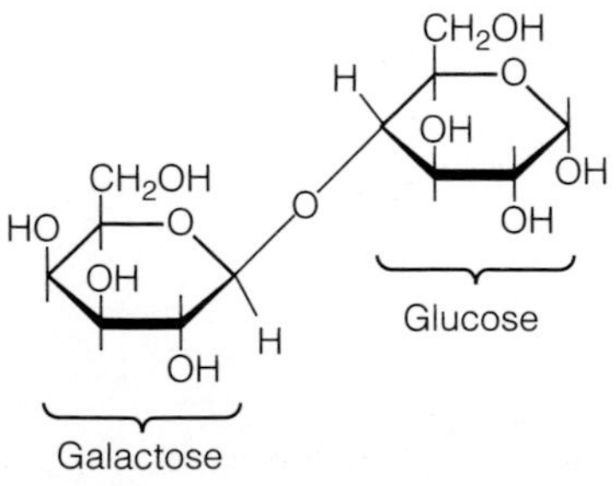

Galactose Glucose
Lactose (alpha form).

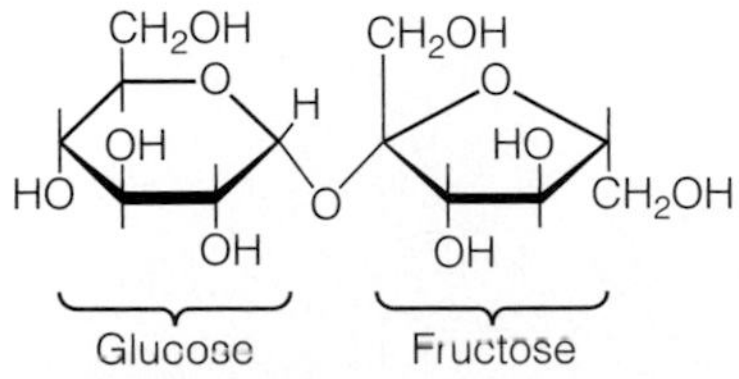

Glucose Fructose
Sucrose.

Polysaccharides

As described in Chapter 4, starch, glycogen, and cellulose are all long chains of glucose molecules covalently linked together.

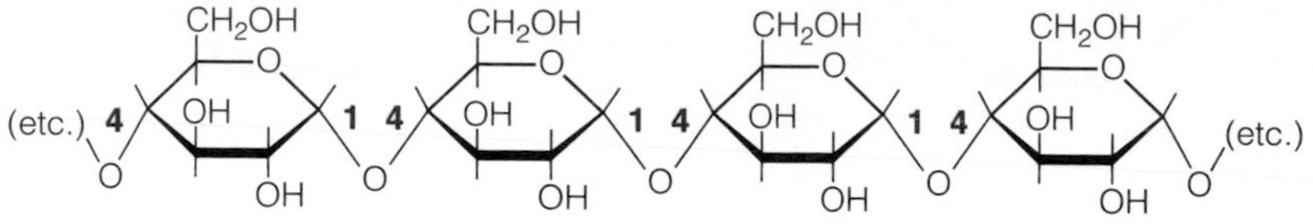

Starch. Two kinds of covalent linkages occur between glucose molecules in starch, giving rise to two kinds of chains. Amylose is composed of straight chains, with carbon 1 of one glucose linked to carbon 4 of the next (α-1,4 linkage). Amylopectin is made up of straight chains like amylose but has occasional branches arising where the carbon 6 of a glucose is also linked to the carbon 1 of another glucose (α-1,6 linkage).

Glycogen. The structure of glycogen is like amylopectin but with many more branches.

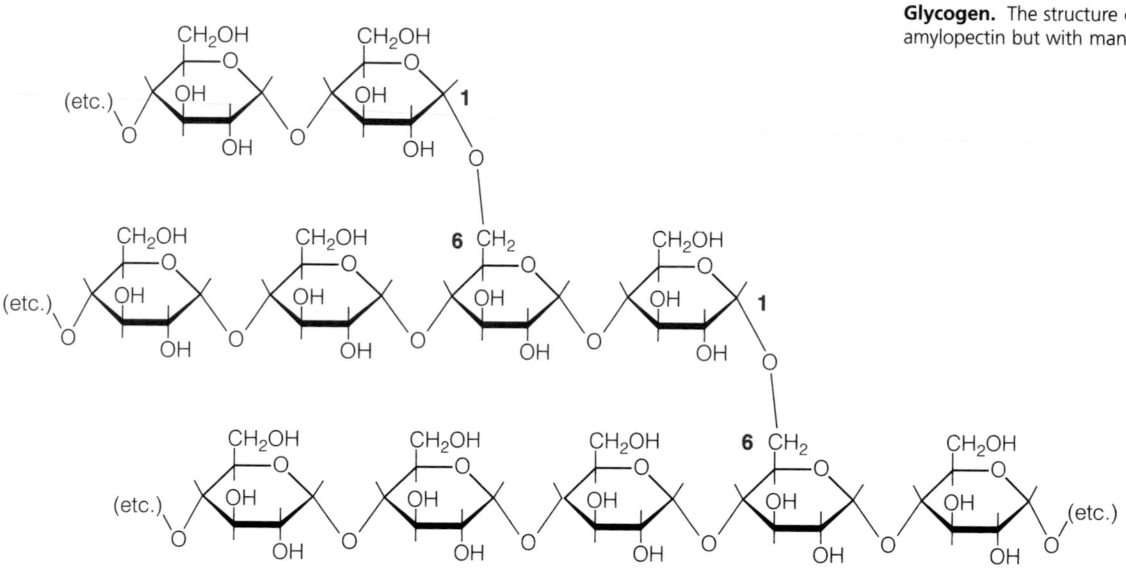

Amylose (unbranched starch)

Amylopectin (branched starch)

Cellulose. Like starch and glycogen, cellulose is also made of chains of glucose units, but there is an important difference: in cellulose, the OH on carbon 1 is in the beta position (see p. C-1). When carbon 1 of one glucose is linked to carbon 4 of the next, it forms a b-1, 4 linkage, which cannot be broken by digestive enzymes in the human GI tract.

Fibers, such as hemicelluloses, consist of long chains of various monosaccharides.

Monosaccharides common in the backbone chain of hemicelluloses:

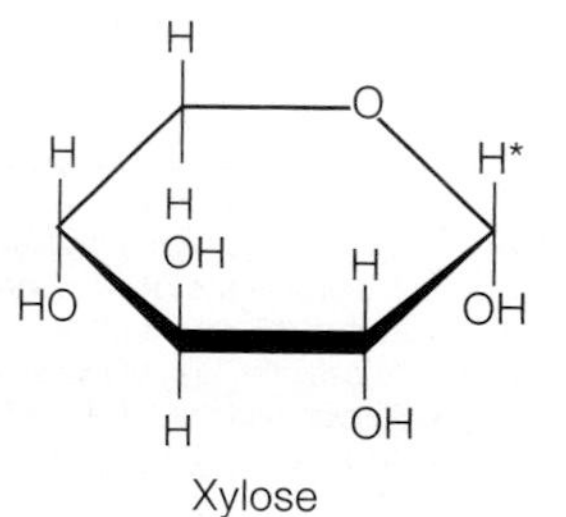

Xylose

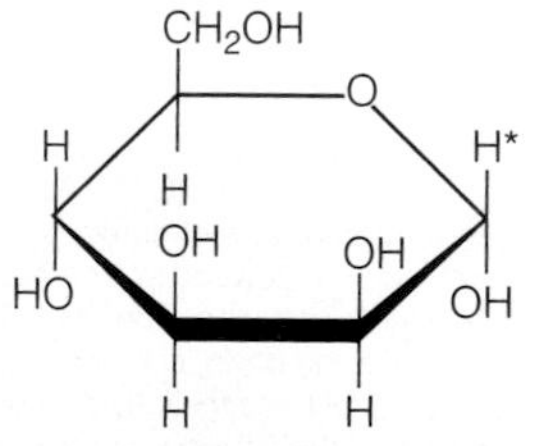

Mannose

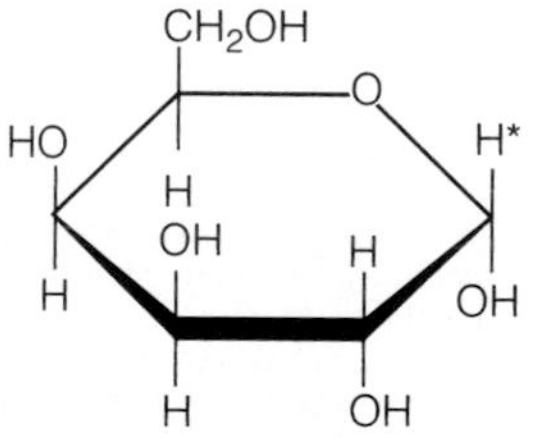

Galactose

*These structures are shown in the alpha form with the H on the carbon pointing upward and the OH pointing downward, but they may also appear in the beta form with the H pointing downward and the OH upward.

Monosaccharides common in the side chains of hemicelluloses:

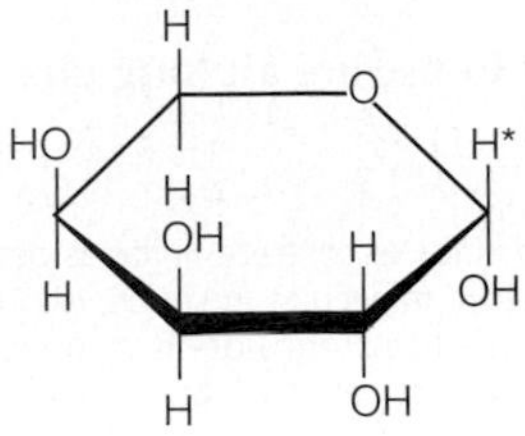

Arabinose

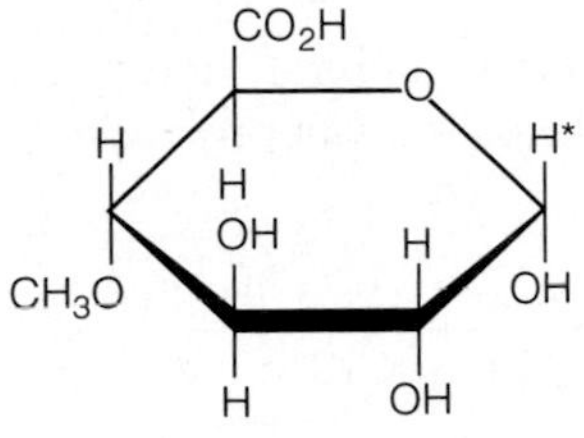

Glucuronic acid

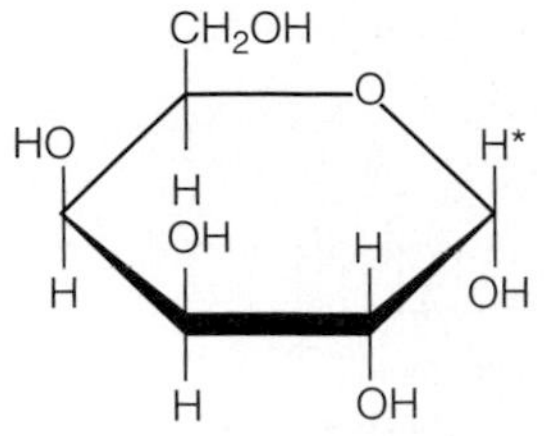

Galactose

Hemicelluloses. The most common hemicelluloses are composed of a backbone chain of xylose, mannose, and galactose, with branching side chains of arabinose, glucuronic acid, and galactose.

Lipids

TABLE C-1 Saturated Fatty Acids Found in Natural Fats

Saturated Fatty Acids	Chemical Formulas	Number of Carbons	Major Food Sources
Butyric	C_3H_7COOH	4	Butterfat
Caproic	$C_5H_{11}COOH$	6	Butterfat
Caprylic	$C_7H_{15}COOH$	8	Coconut oil
Capric	$C_9H_{19}COOH$	10	Palm oil
Lauric	$C_{11}H_{23}COOH$	12	Coconut oil, palm oil
Myristic[a]	$C_{13}H_{27}COOH$	14	Coconut oil, palm oil
Palmitic[a]	$C_{15}H_{31}COOH$	16	Palm oil
Stearic[a]	$C_{17}H_{35}COOH$	18	Most animal fats
Arachidic	$C_{19}H_{39}COOH$	20	Peanut oil
Behenic	$C_{21}H_{43}COOH$	22	Seeds
Lignoceric	$C_{23}H_{47}COOH$	24	Peanut oil

[a]Most common saturated fatty acids.

TABLE C-2 Unsaturated Fatty Acids Found in Natural Fats

Unsaturated Fatty Acids	Chemical Formulas	Number of Carbons	Number of Double Bonds	Standard Notation[a]	Omega Notation[b]	Food Sources
Palmitoleic	$C_{15}H_{29}COOH$	16	1	16:1;9	16:1v7	Seafood, beef
Oleic	$C_{17}H_{33}COOH$	18	1	18:1;9	18:1v9	Olive oil, canola oil
Linoleic	$C_{17}H_{31}COOH$	18	2	18:2;9,12	18:2v6	Sunflower oil, safflower oil
Linolenic	$C_{17}H_{29}COOH$	18	3	18:3;9,12,15	18:3v3	Soybean oil, canola oil
Arachidonic	$C_{19}H_{31}COOH$	20	4	20:4;5,8,11,14	20:4v6	Eggs, most animal fats
Eicosapentaenoic	$C_{19}H_{29}COOH$	20	5	20:5;5,8,11,14,17	20:5v3	Seafood
Docosahexaenoic	$C_{21}H_{31}COOH$	22	6	22:6;4,7,10,13,16,19	22:6v3	Seafood

Note: A fatty acid has two ends; designated the methyl (CH_3) end and the carboxyl, or acid (COOH), end.

[a]Standard chemistry notation begins counting carbons at the acid end. The number of carbons the fatty acid contains comes first, followed by a colon and another number that indicates the number of double bonds; next comes a semicolon followed by a number or numbers indicating the positions of the double bonds. Thus the notation for linoleic acid, an 18-carbon fatty acid with two double bonds between carbons 9 and 10 and between carbons 12 and 13, is 18:2;9,12.

[b]Because fatty acid chains are lengthened by adding carbons at the acid end of the chain, chemists use the omega system of notation to ease the task of identifying them. The omega system begins counting carbons at the methyl end. The number of carbons the fatty acid contains comes first, followed by a colon and the number of double bonds; next come the omega symbol (v) and a number indicating the position of the double bond nearest the methyl end. Thus linoleic acid with its first double bond at the sixth carbon from the methyl end would be noted 18:2v6 in the omega system.

Protein: Amino Acids

The common amino acids may be classified into the seven groups listed on the next page. Amino acids marked with an asterisk (*) are essential because human beings cannot synthesize them.

1. Amino acids with aliphatic side chains, which consist of hydrogen and carbon atoms (hydrocarbons):

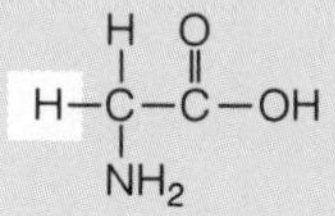

$H-C(H)(NH_2)-C(=O)-OH$ **Glycine (Gly)**

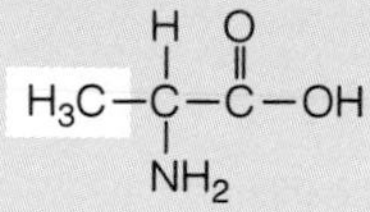

$H_3C-C(H)(NH_2)-C(=O)-OH$ **Alanine (Ala)**

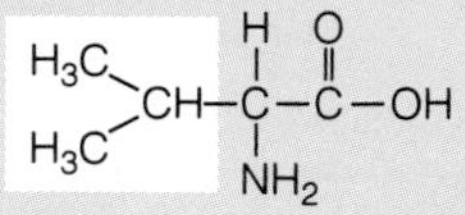

$(H_3C)_2CH-C(H)(NH_2)-C(=O)-OH$ **Valine* (Val)**

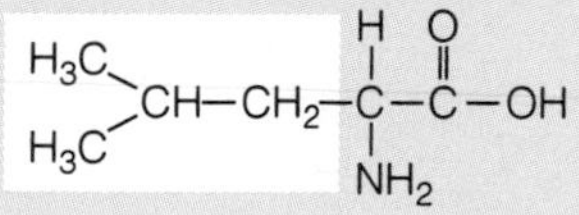

$(H_3C)_2CH-CH_2-C(H)(NH_2)-C(=O)-OH$ **Leucine* (Leu)**

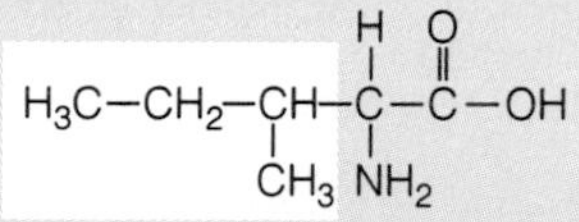

$H_3C-CH_2-CH(CH_3)-C(H)(NH_2)-C(=O)-OH$ **Isoleucine* (Ile)**

2. Amino acids with hydroxyl (OH) side chains:

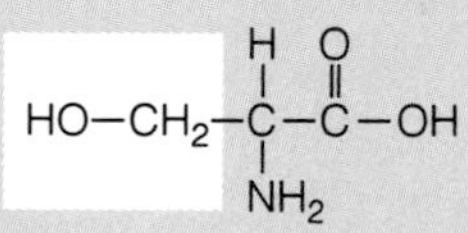

$HO-CH_2-C(H)(NH_2)-C(=O)-OH$ **Serine (Ser)**

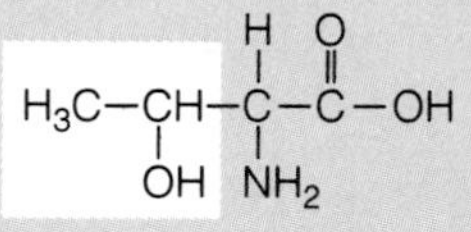

$H_3C-CH(OH)-C(H)(NH_2)-C(=O)-OH$ **Threonine* (Thr)**

3. Amino acids with side chains containing acidic groups or their amides, which contain the group NH_2:

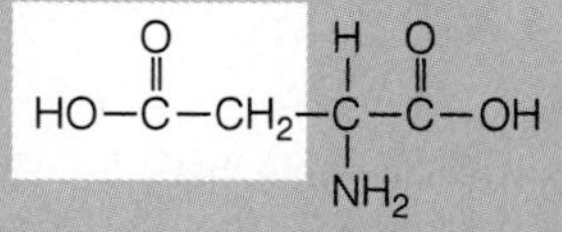

$HO-C(=O)-CH_2-C(H)(NH_2)-C(=O)-OH$ **Aspartic acid (Asp)**

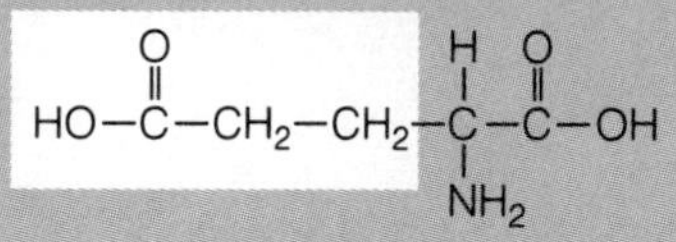

$HO-C(=O)-CH_2-CH_2-C(H)(NH_2)-C(=O)-OH$ **Glutamic acid (Glu)**

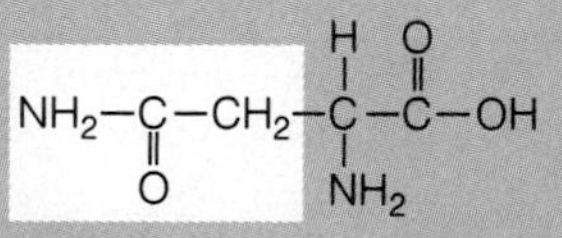

$NH_2-C(=O)-CH_2-C(H)(NH_2)-C(=O)-OH$ **Asparagine (Asn)**

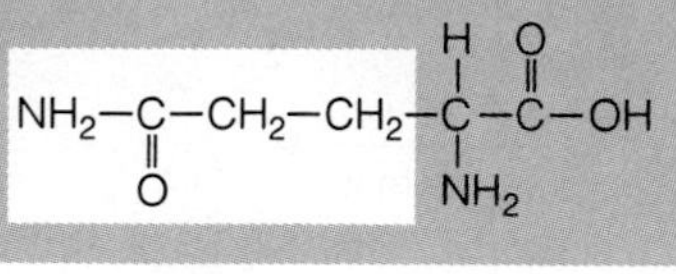

$NH_2-C(=O)-CH_2-CH_2-C(H)(NH_2)-C(=O)-OH$ **Glutamine (Gln)**

4. Amino acids with basic side chains:

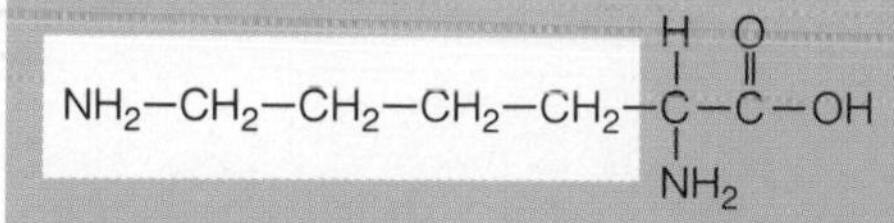

$NH_2-CH_2-CH_2-CH_2-CH_2-C(H)(NH_2)-C(=O)-OH$ **Lysine* (Lys)**

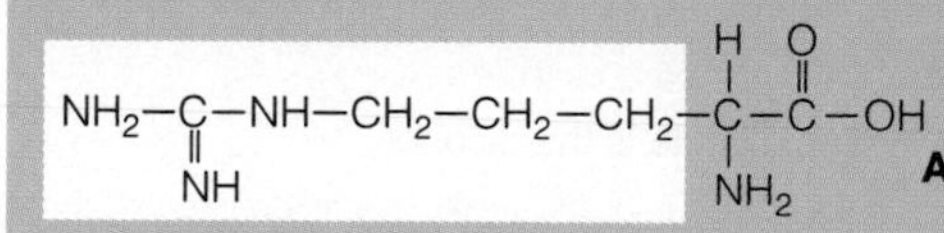

$NH_2-C(=NH)-NH-CH_2-CH_2-CH_2-C(H)(NH_2)-C(=O)-OH$ **Arginine (Arg)**

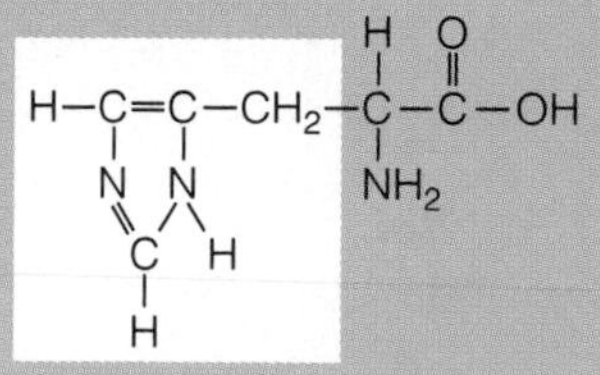

$H-C=C-CH_2-C(H)(NH_2)-C(=O)-OH$ (imidazole ring: N, N, C–H, H) **Histidine* (His)**

5. Amino acids with aromatic side chains, which are characterized by the presence of at least one ring structure:

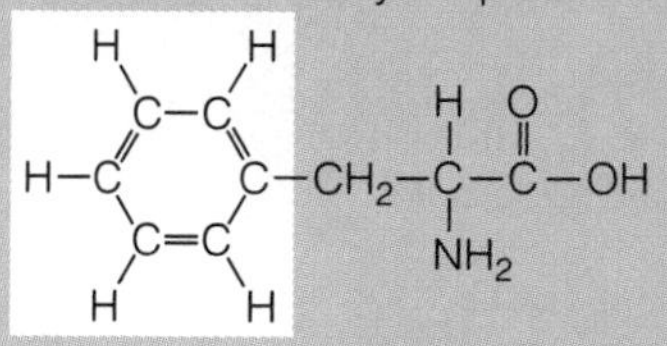

$C_6H_5-CH_2-C(H)(NH_2)-C(=O)-OH$ **Phenylalanine* (Phe)**

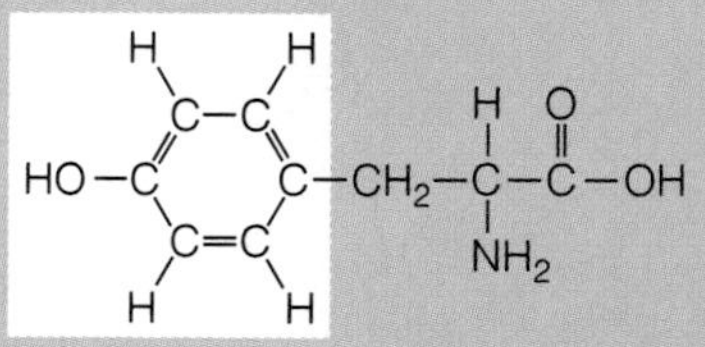

$HO-C_6H_4-CH_2-C(H)(NH_2)-C(=O)-OH$ **Tyrosine (Tyr)**

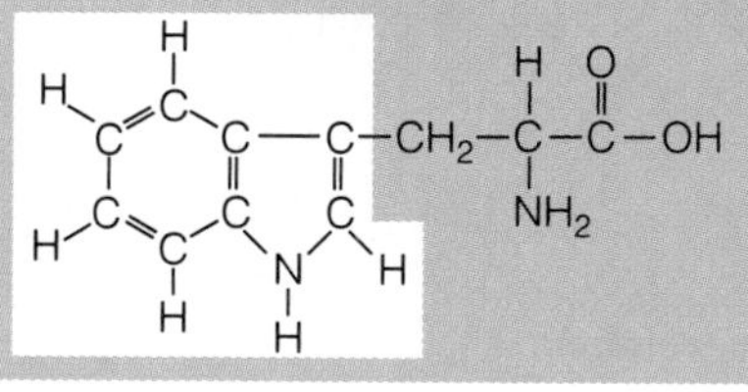

(indole ring)$-CH_2-C(H)(NH_2)-C(=O)-OH$ **Tryptophan* (Trp)**

6. Amino acids with side chains containing sulfur atoms:

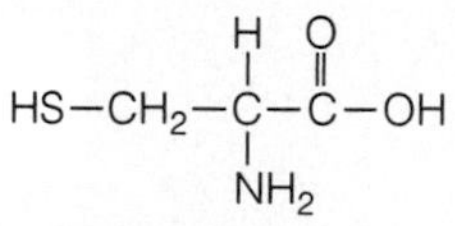

$HS-CH_2-C(H)(NH_2)-C(=O)-OH$

Cysteine (Cys)

$CH_3-S-CH_2-CH_2-C(H)(NH_2)-C(=O)-OH$

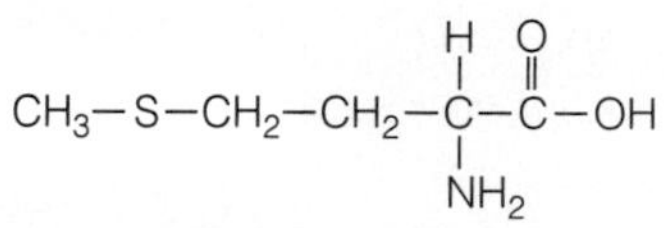

Methionine* (Met)

7. Imino acid:

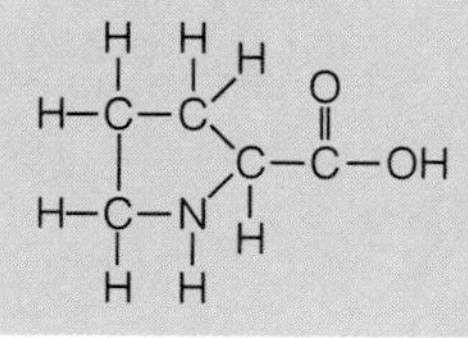

Proline (Pro)

(pyrrolidine ring: H–C–C, H–C–N–H)$-C(H)-C(=O)-OH$

Proline has the same chemical structure as the other amino acids, but its amino group has given up a hydrogen to form a ring.

Vitamins and Coenzymes

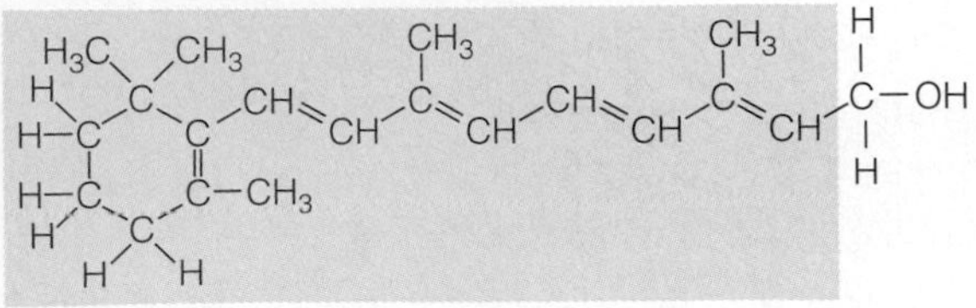

Vitamin A: retinol. This molecule is the alcohol form of vitamin A.

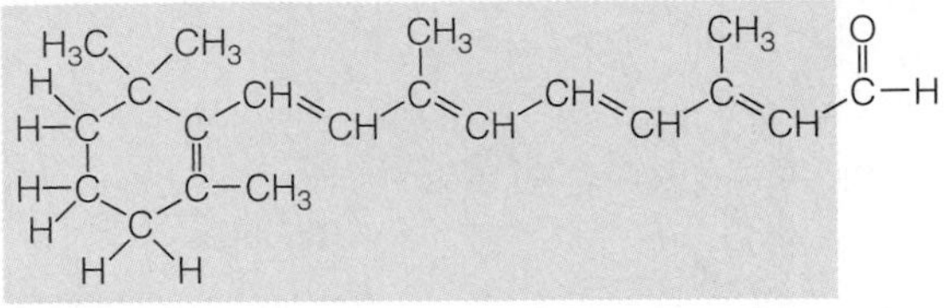

Vitamin A: retinal. This molecule is the aldehyde form of vitamin A.

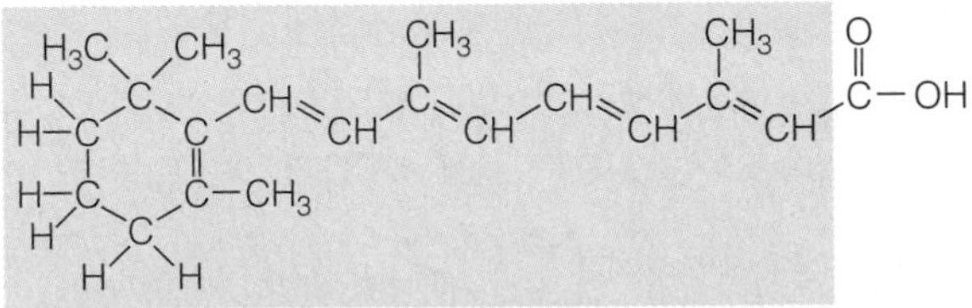

Vitamin A: retinoic acid. This molecule is the acid form of vitamin A.

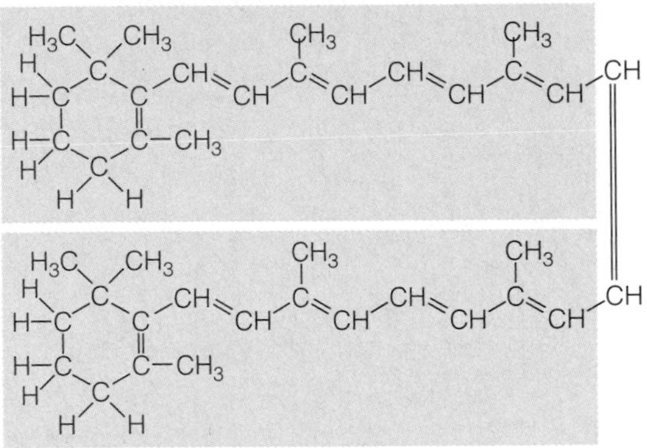

Vitamin A precursor: beta-carotene.

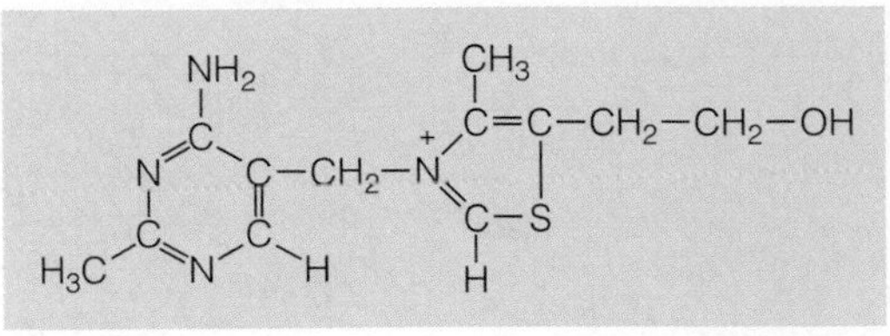

Thiamin. This molecule is part of the coenzyme thiamin pyrophosphate (TPP).

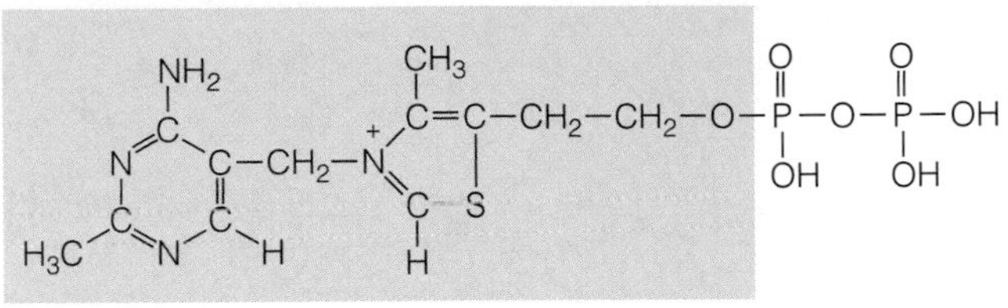

Thiamin pyrophosphate (TPP). TPP is a coenzyme that includes the thiamin molecule as part of its structure.

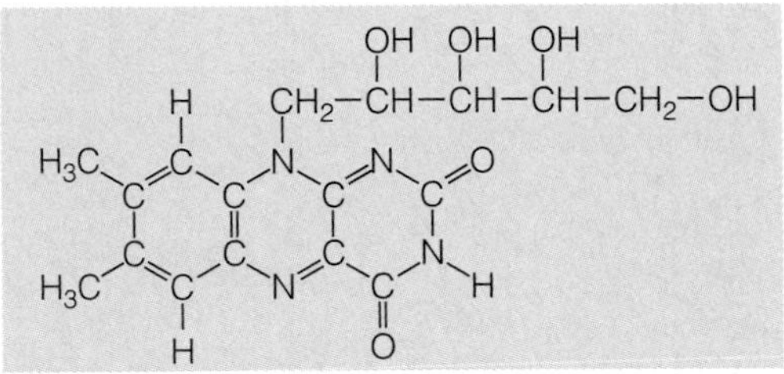

Riboflavin. This molecule is a part of two coenzymes—flavin mononucleotide (FMN) and flavin adenine dinucleotide (FAD).

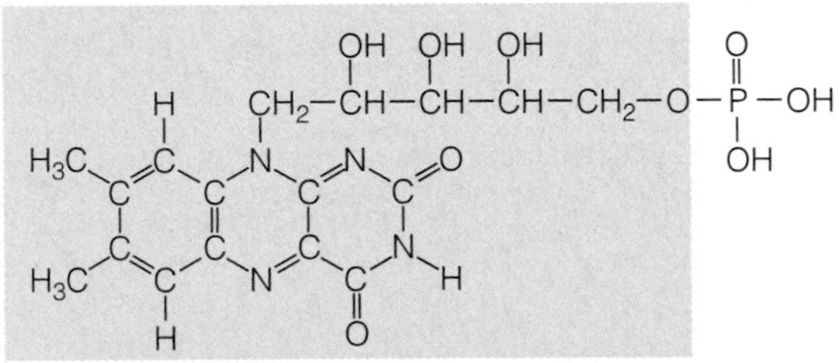

Flavin mononucleotide (FMN). FMN is a coenzyme that includes the riboflavin molecule as part of its structure.

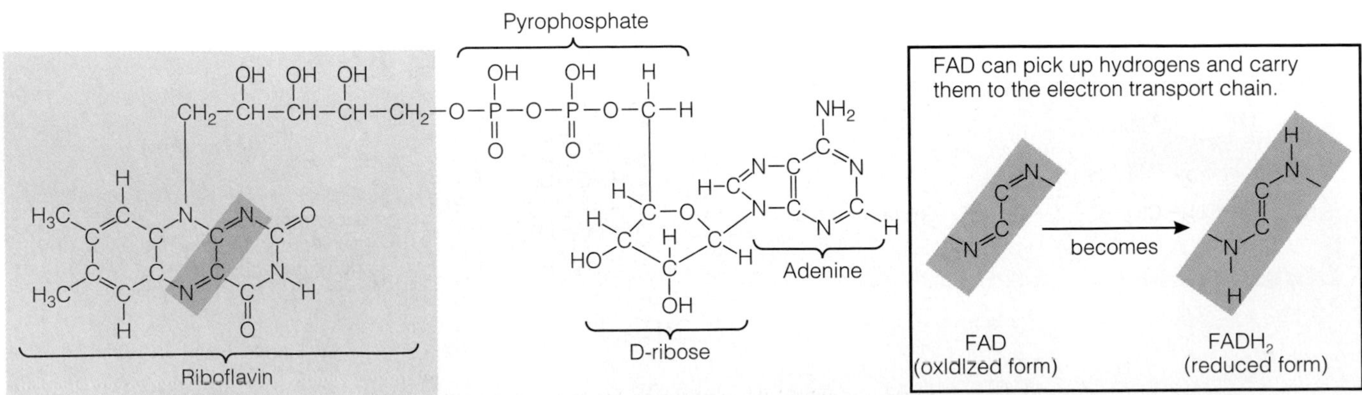

Flavin adenine dinucleotide (FAD). FAD is a coenzyme that includes the riboflavin molecule as part of its structure.

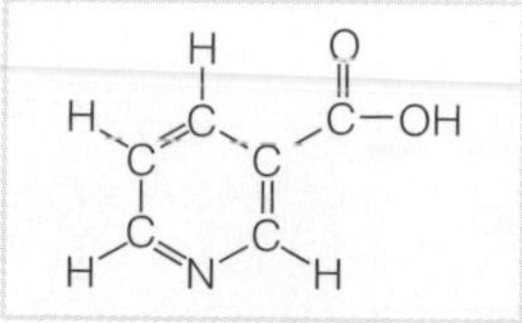

C

Nicotinic acid

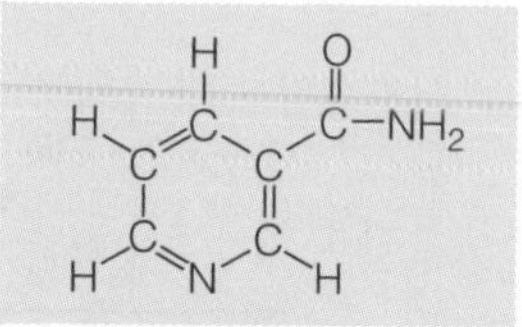

Nicotinamide

Niacin (nicotinic acid and nicotinamide). These molecules are a part of two coenzymes—nicotinamide adenine dinucleotide (NAD^+) and nicotinamide adenine dinucleotide phosphate ($NADP^+$).

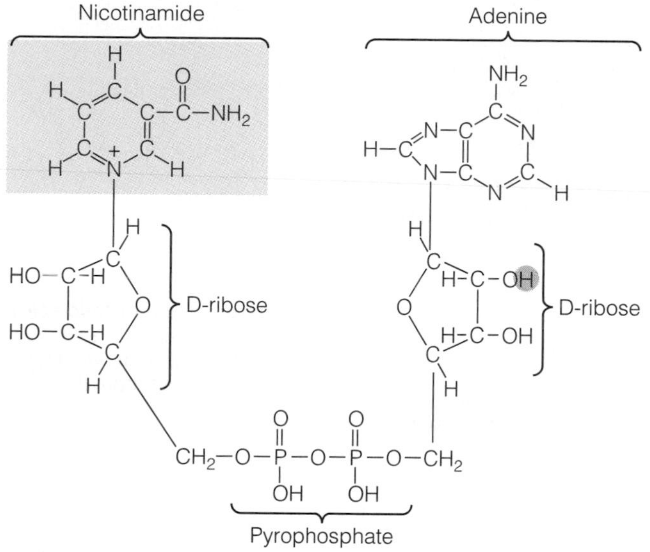

Nicotinamide

Adenine

D-ribose

D-ribose

Pyrophosphate

Nicotinamide adenine dinucleotide (NAD^+) and nicotinamide adenine dinucleotide phosphate ($NADP^+$). NADP has the same structure as NAD but with a phosphate group attached to the O instead of the H.

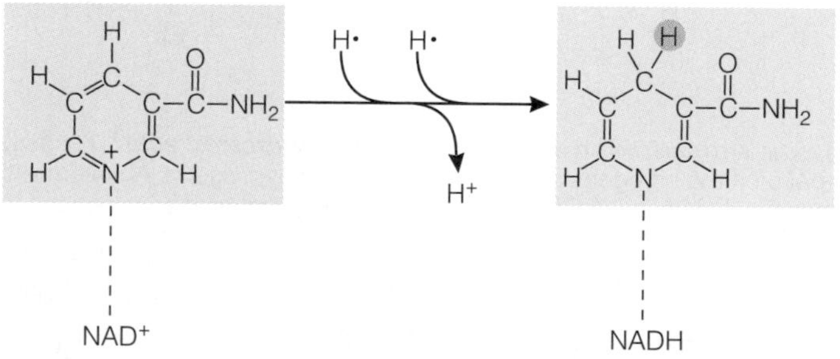

NAD^+

NADH

Reduced NAD^+ (NADH). When NAD^+ is reduced by the addition of H^+ and two electrons, it becomes the coenzyme NADH. (The dots on the H entering this reaction represent electrons—see Appendix B.)

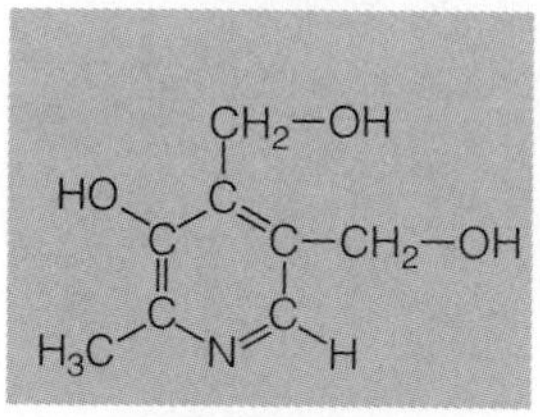

Pyridoxine

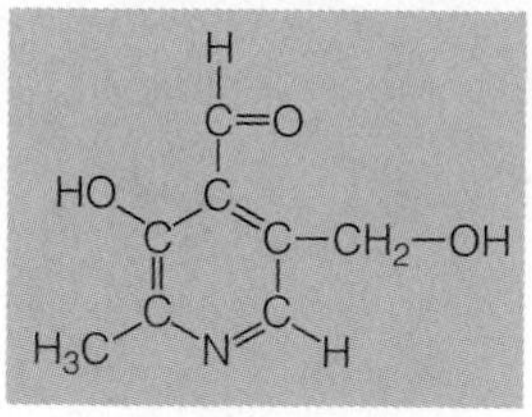

Pyridoxal

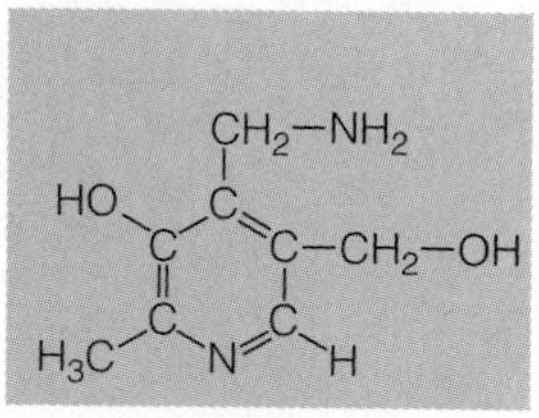

Pyridoxamine

Vitamin B_6 (a general name for three compounds—pyridoxine, pyridoxal, and pyridoxamine). These molecules are a part of two coenzymes—pyridoxal phosphate and pyridoxamine phosphate.

Pyridoxal phosphate

Pyridoxamine phosphate

Pyridoxal phosphate (PLP) and pyridoxamine phosphate. These coenzymes include vitamin B_6 as part of their structures.

Folate (folacin or folic acid). This molecule consists of a double ring combined with a single ring and at least one glutamate (a nonessential amino acid marked in the box). Folate's biologically active form is tetrahydrofolic acid.

Vitamin B_{12} (cyanocobalamin). The arrows in this diagram indicate that the spare electron pairs on the nitrogens attract them to the cobalt.

Tetrahydrofolic acid. This active coenzyme form of folate has four added hydrogens. An intermediate form, dihydrofolate, has two added hydrogens.

Pantothenic acid. This molecule is part of coenzyme A (CoA).

Coenzyme A (CoA). Coenzyme A is a coenzyme that includes pantothenic acid as part of its structure.

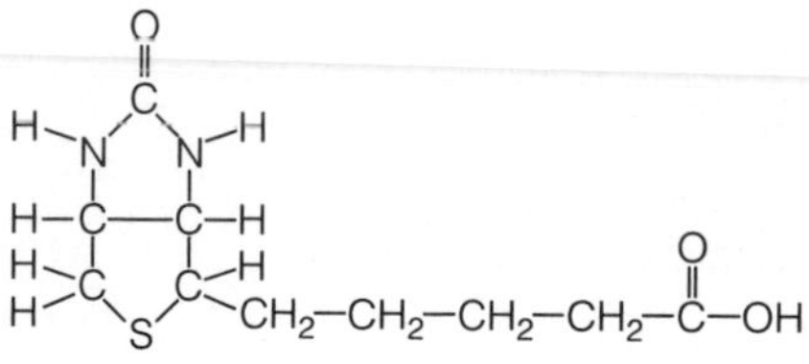

Biotin.

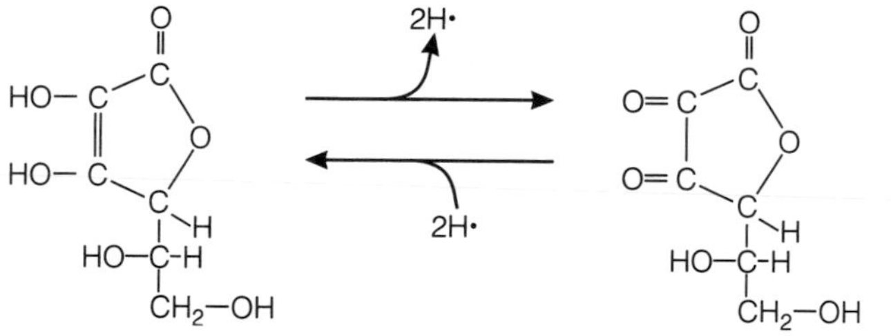

Ascorbic acid
(reduced form)

Dehydroascorbic acid
(oxidized form)

Vitamin C. The dots on the H indicate that two hydrogen atoms, complete with their electrons, are lost when ascorbic acid is oxidized and gained when it is reduced again.

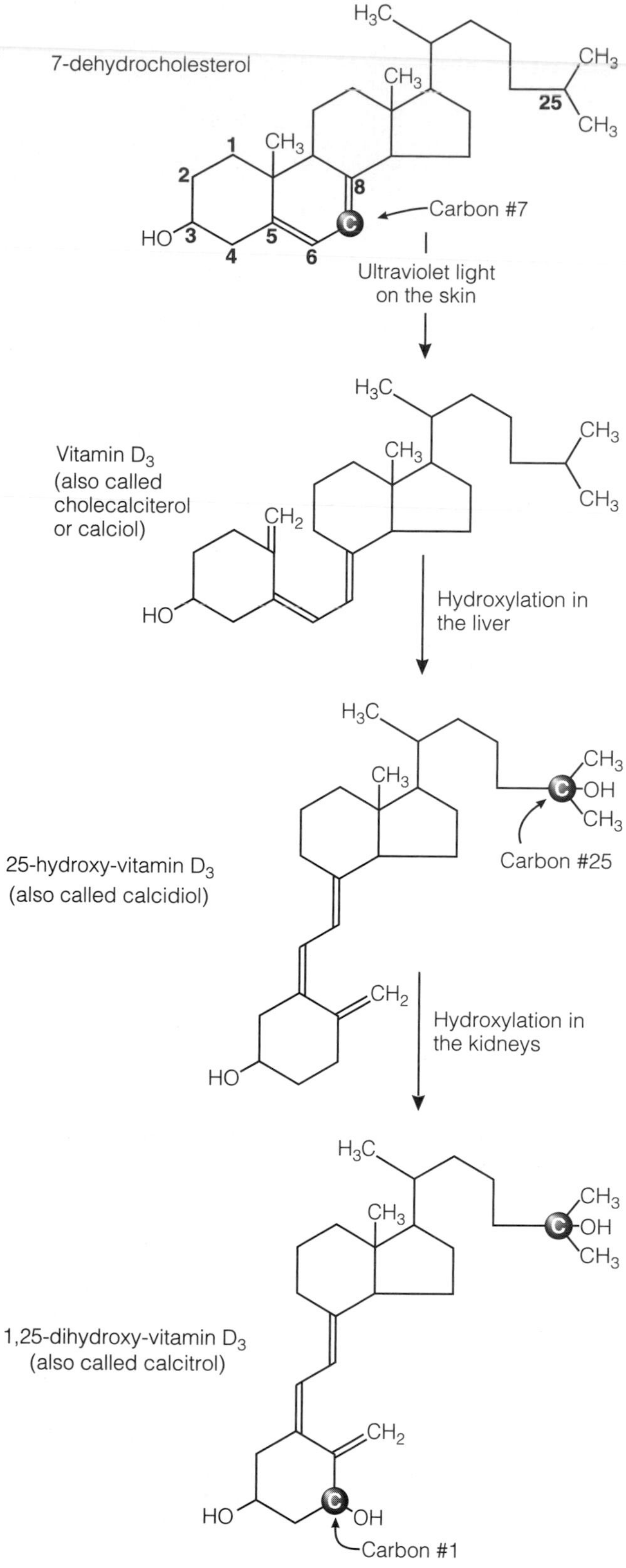

7-dehydrocholesterol

Carbon #7

Ultraviolet light on the skin

Vitamin D_3 (also called cholecalciterol or calciol)

Hydroxylation in the liver

25-hydroxy-vitamin D_3 (also called calcidiol)

Carbon #25

Hydroxylation in the kidneys

1,25-dihydroxy-vitamin D_3 (also called calcitrol)

Carbon #1

Vitamin D. The synthesis of active vitamin D begins with 7-dehydrocholesterol. (The carbon atoms at which changes occur are numbered.)

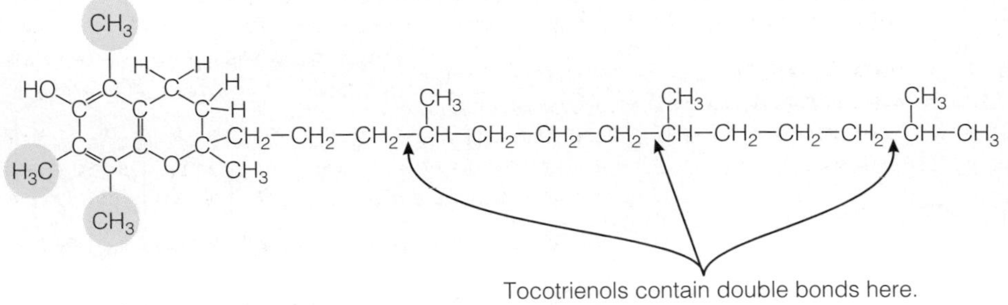

C

Tocotrienols contain double bonds here.

Vitamin E (alpha-tocopherol). The number and position of the methyl groups (CH_3) bonded to the ring structure differentiate among the tocopherols.

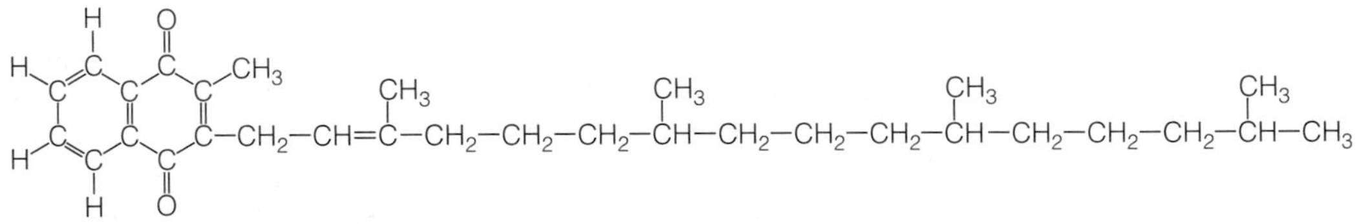

Vitamin K. Naturally occurring compounds with vitamin K activity include phylloquinones (from plants) and menaquinones (from bacteria).

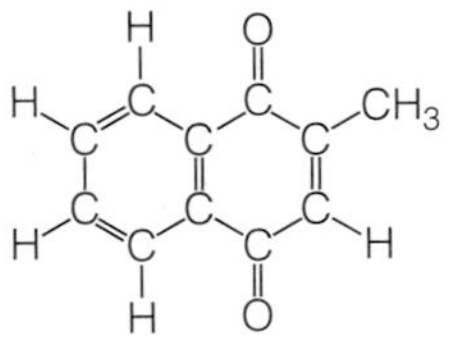

Menadione. This synthetic compound has the same activity as natural vitamin K.

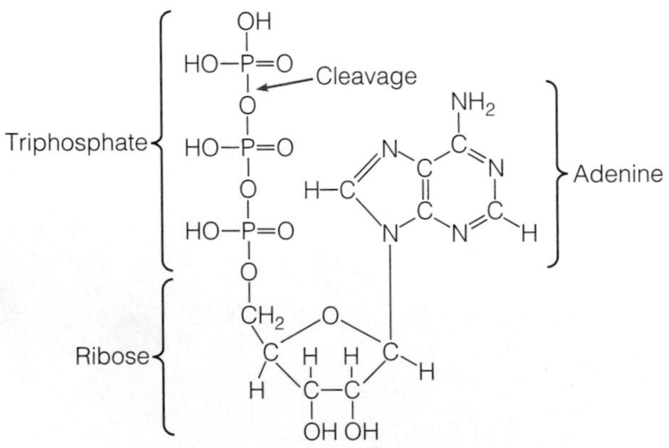

Triphosphate

Cleavage

Adenine

Ribose

Adenosine triphosphate (ATP), the energy carrier. The cleavage point marks the bond that is broken when ATP splits to become ADP + P.

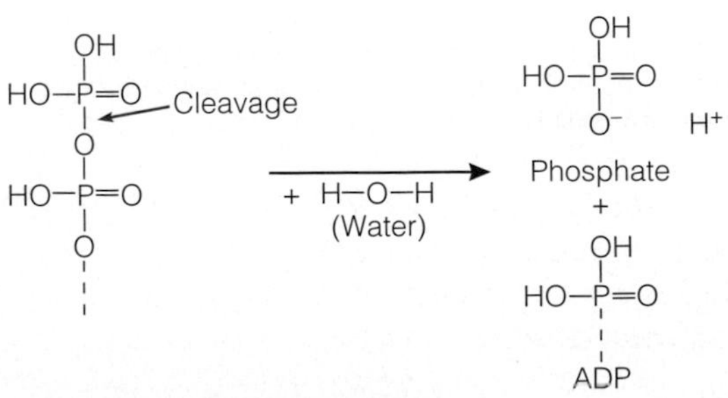

Cleavage

+ H—O—H (Water)

Phosphate

+

ADP

Adenosine diphosphate (ADP).

Glycolysis

Figure C-1 depicts the events of glycolysis. The following text describes key steps as numbered on the figure.

FIGURE C-1 **Glycolysis**

Notice that galactose and fructose enter at different places but all continue on the same pathway.

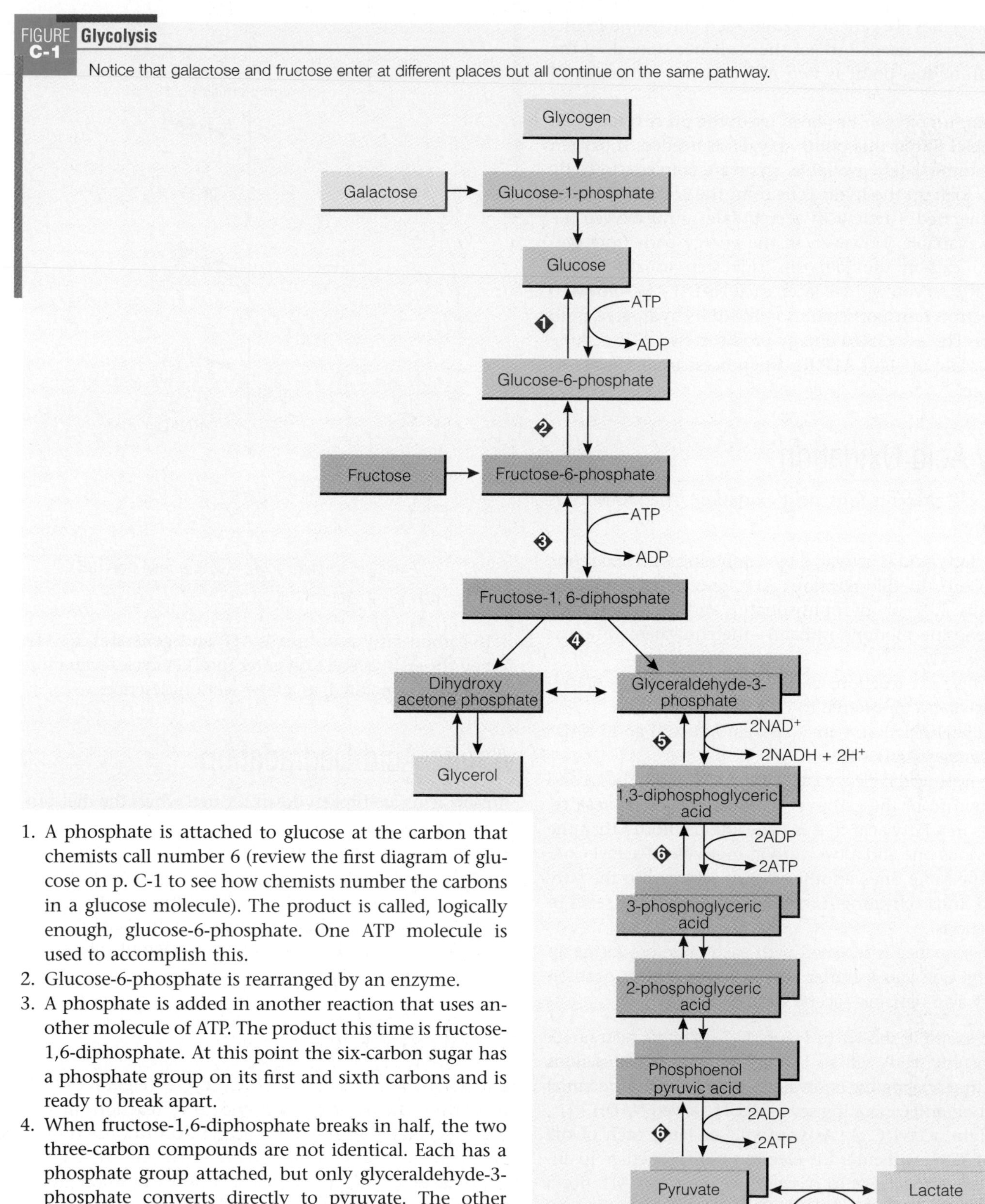

1. A phosphate is attached to glucose at the carbon that chemists call number 6 (review the first diagram of glucose on p. C-1 to see how chemists number the carbons in a glucose molecule). The product is called, logically enough, glucose-6-phosphate. One ATP molecule is used to accomplish this.
2. Glucose-6-phosphate is rearranged by an enzyme.
3. A phosphate is added in another reaction that uses another molecule of ATP. The product this time is fructose-1,6-diphosphate. At this point the six-carbon sugar has a phosphate group on its first and sixth carbons and is ready to break apart.
4. When fructose-1,6-diphosphate breaks in half, the two three-carbon compounds are not identical. Each has a phosphate group attached, but only glyceraldehyde-3-phosphate converts directly to pyruvate. The other compound, however, converts easily to glyceraldehyde-3-phosphate.

5. In the next step, enough energy is released to convert NAD^+ to NADH + H^+.
6. In two of the following steps ATP is regenerated.

Remember that in effect two molecules of glyceraldehyde-3-phosphate are produced from glucose; therefore, four ATP molecules are generated from each glucose molecule. Two ATP were needed to get the sequence started, so the net gain at this point is two ATP and two molecules of NADH + H^+.

So far, no oxygen has been used; the process has been anaerobic. But at this point, oxygen is needed. If oxygen is not immediately available, pyruvate converts to lactic acid to soak up the hydrogens from the NADH + H^+ that was generated. Lactic acid accumulates until oxygen becomes available. However, in the energy path from glucose to carbon dioxide, this side step usually is not necessary. As you will see later, each NADH + H^+ moves to the electron transport chain to unload its hydrogens onto oxygen. The associated energy produces two ATP, making a total yield of eight ATP for the process from glucose to pyruvate.

Fatty Acid Oxidation

Figure C-2 presents fatty acid oxidation. The sequence is as follows.

1. The fatty acid is activated by combining with coenzyme A (CoA). In this reaction, ATP loses two phosphorus atoms (PP, or pyrophosphate) and becomes AMP (adenosine monophosphate)—the equivalent of a loss of two ATP.
2. In the next reaction, two H with their energy are removed and transferred to FAD, forming $FADH_2$.
3. In a later reaction, two H are removed and go to NAD^+ (forming NADH + H^+).
4. The fatty acid is cleaved at the "beta" carbon, the second carbon from the carboxyl (COOH) end. This break results in a fatty acid that is two carbons shorter than the previous one and a two-carbon molecule of acetyl CoA. At the same time, another CoA is attached to the fatty acid, thus activating it for its turn through the series of reactions.
5. The sequence is repeated with each cycle producing an acetyl CoA and a shorter fatty acid until only a 2-carbon fatty acid remains—acetyl CoA.

In the example shown in Figure C-2, palmitic acid (a 16-carbon fatty acid) will go through this series of reactions seven times, using the equivalent of two ATP for the initial activation and generating seven $FADH_2$, seven NADH + H^+, and eight acetyl CoA. As you will see later, each of the seven $FADH_2$ will enter the electron transport chain to unload its hydrogens onto oxygen, yielding two ATP (for a total of 14). Similarly, each NADH + H^+ will enter the electron transport chain to unload its hydrogens onto oxygen, yielding three ATP (for a total of 21). Thus the oxidation of a 16-carbon fatty acid uses 2 ATP and generates 35 ATP. When the eight acetyl CoA enter the TCA cycle, even more ATP will be generated, as a later section describes.

FIGURE C-2 Fatty Acid Oxidation

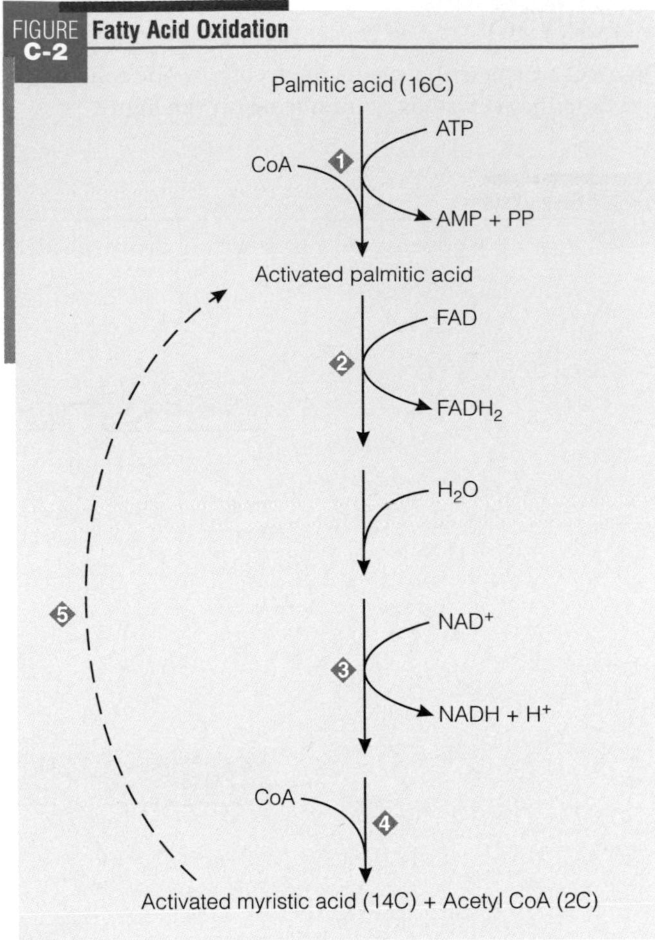

Amino Acid Degradation

Amino acids are broken down for fuel when the diet provides protein foods in excess or carbohydrate foods or energy in inadequate amounts. The first step in amino acid degradation is the removal of the nitrogen-containing amino group through either deamination or transamination reactions. Then the remaining carbon skeletons may enter the metabolic pathways at different places, as shown in Figure C-3.

The TCA Cycle

The tricarboxylic acid, or TCA, cycle (Figure C-4 on p. C-12) is the name given to the set of reactions involving oxygen and leading from acetyl CoA to carbon dioxide (and water). To link glycolysis to the TCA cycle, pyruvate enters the mitochondrion, loses a carbon group, and bonds with a molecule of CoA to become acetyl CoA. The TCA cycle uses the products of carbohydrate, fat, and protein metabolism. Any substance that can be converted

FIGURE C-3 **Amino Acid Degradation**

After losing their amino groups, carbon skeletons can be converted to one of seven molecules that can enter the TCA cycle, presented in Figure C-4.

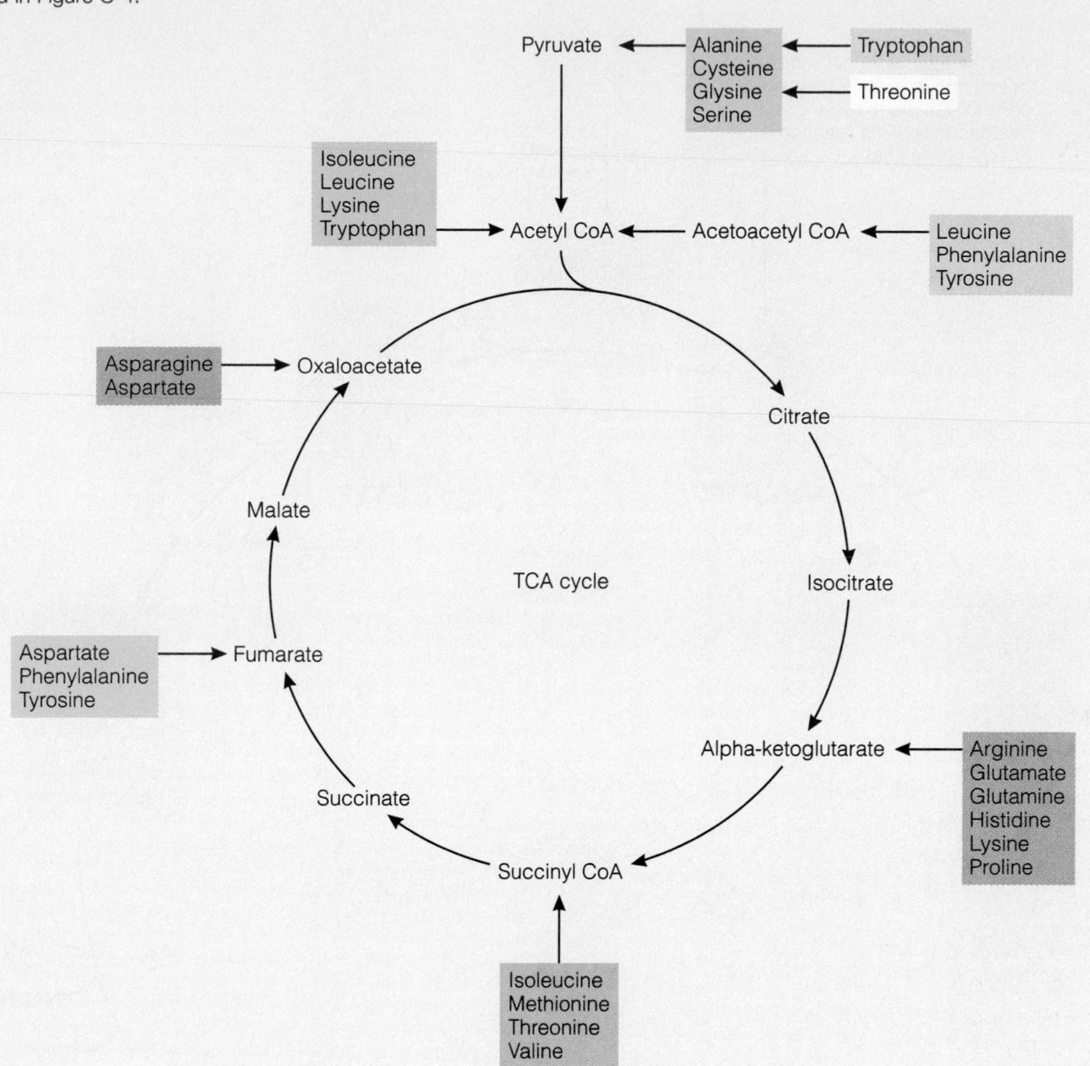

FIGURE C-4 The TCA Cycle

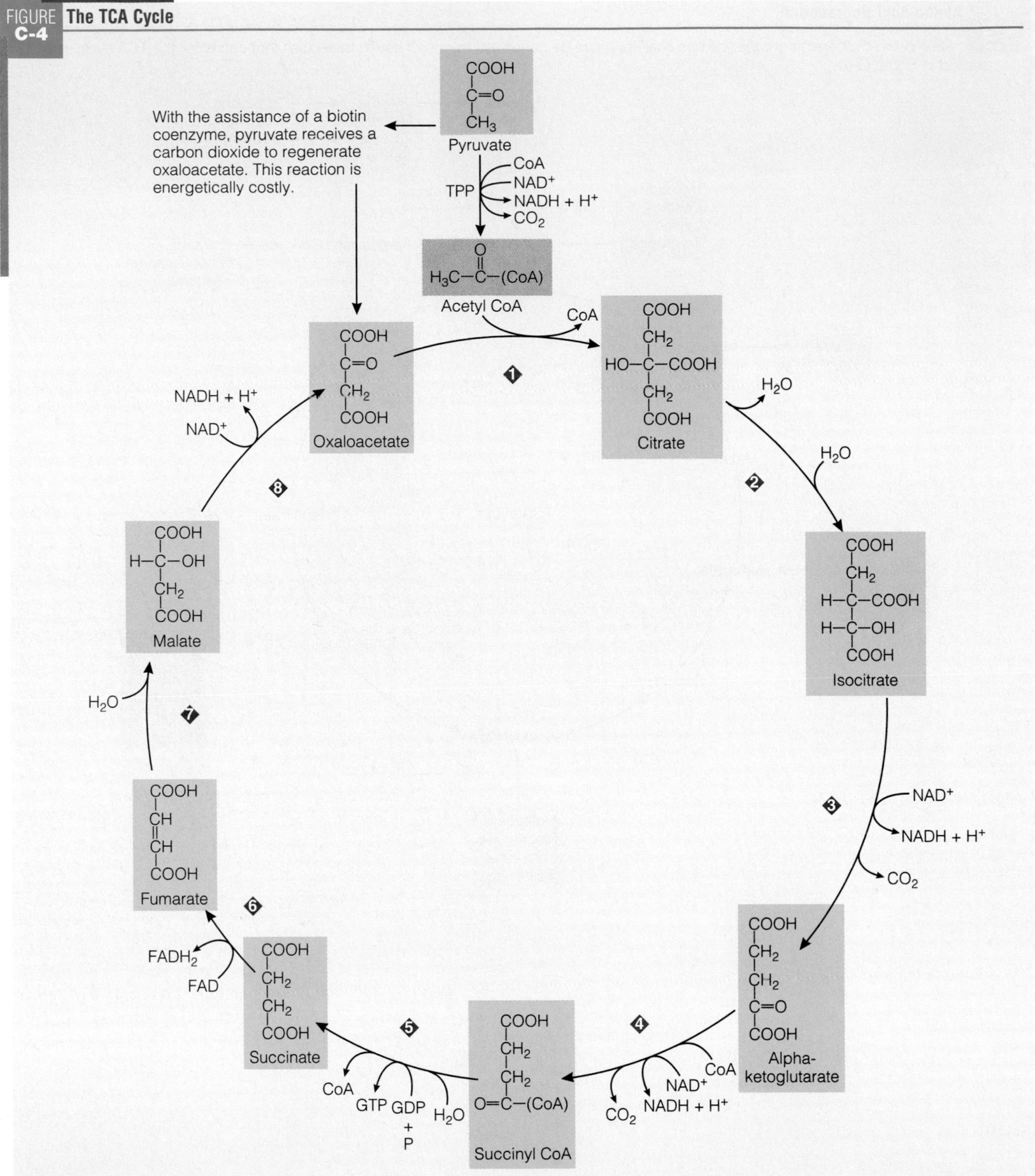

to acetyl CoA directly, or indirectly through pyruvate, may enter the cycle.

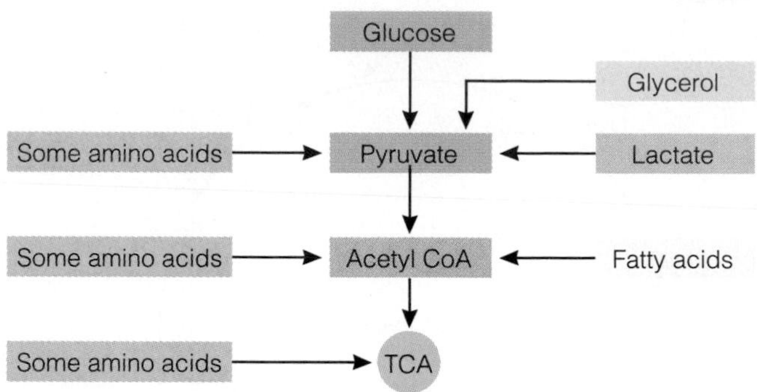

The step from pyruvate to acetyl CoA is exceedingly complex. We have included only those substances that will help you understand the transfer of energy from the nutrients. In the presence of oxygen, pyruvate loses a carbon to carbon dioxide and is attached to a molecule of CoA. In the process, NAD^+ picks up two hydrogens with their associated energy, becoming NADH + H^+.

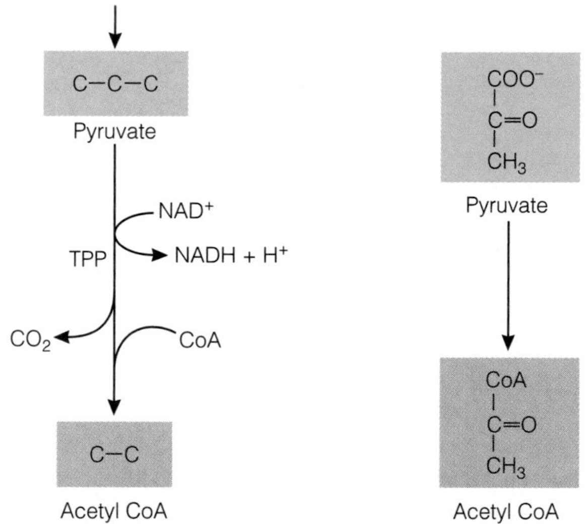

The step from pyruvate to acetyl CoA. (TPP and NAD are coenzymes containing the B vitamins thiamin and niacin, respectively.)

As the acetyl CoA breaks down to carbon dioxide and water, its energy is captured in ATP. Let's follow the steps by which this occurs (see the corresponding numbers in Figure C-4).

1. The two-carbon acetyl CoA combines with a four-carbon compound, oxaloacetate. The CoA comes off, and the product is a six-carbon compound, citrate.
2. The atoms of citrate are rearranged to form isocitrate.
3. Now NAD^+ reacts with isocitrate. Two H and two electrons are removed from the isocitrate. One H becomes attached to the NAD^+ with the two electrons; the other H is released as H^+. Thus NAD^+ becomes NADH + H^+. (Remember this NADH + H^+. It is carrying the H and the energy released from the last reaction. But let's follow the carbons first.) A carbon is combined with two oxygens, forming carbon dioxide (which diffuses away into the blood and is exhaled). What is left is the five-carbon compound alpha-ketoglutarate.
4. Now two compounds interact with alpha-ketoglutarate—a molecule of CoA and a molecule of NAD^+. In this complex reaction, a carbon and two oxygens are removed (forming carbon dioxide); two hydrogens are removed and go to $NAD1^+$ (forming NADH + H^+); and the remaining four-carbon compound is attached to the CoA, forming succinyl CoA. (Remember this NADH + H^+ also. You will see later what happens to it.)
5. Now two molecules react with succinyl CoA—a molecule called GDP and one of phosphate (P). The CoA comes off, the GDP and P combine to form the high-energy compound GTP (similar to ATP), and succinate remains. (Remember this GTP.)
6. In the next reaction, two H with their energy are removed from succinate and are transferred to a molecule called FAD (an electron-hydrogen receiver like NAD^+) to form $FADH_2$. The product that remains is fumarate. (Remember this $FADH_2$.)
7. Next a molecule of water is added to fumarate, forming malate.
8. A molecule of NAD^+ reacts with the malate; two H with their associated energy are removed from the malate and form NADH + H^+. The product that remains is the four-carbon compound oxaloacetate. (Remember this NADH + H^+.)

We are back where we started. The oxaloacetate formed in this process can combine with another molecule of acetyl CoA (step 1), and the cycle can begin again. The whole scheme is shown in Figure C-4.

So far, we have seen two carbons brought in with acetyl CoA and two carbons ending up in carbon dioxide. But where are the energy and the ATP we promised?

Each time a pair of hydrogen atoms is removed from one of the compounds in the cycle, it includes a pair of electrons. Then the energy from this chemical bond is captured in the compound to which the H become attached. A review of the eight steps of the cycle shows that energy is transferred in this way into other compounds in steps 3, 4, 6, and 8. In step 5, energy is stored when GDP and P are bound together to form GTP. Thus the compounds NADH + H^+ (three molecules), $FADH_2$, and GTP store energy originally found in acetyl CoA. To see how this energy ends up in ATP, we must follow the electrons further. Let us take those attached to NAD^+ as an example.

The Electron Transport Chain

The six reactions described here are those of the electron transport chain, which is shown in Figure C-5 on p. C-14. Since oxygen is required for these reactions, and ADP and P are combined to form ATP in several of them (ADP is phosphorylated), these reactions are also called oxidative phosphorylation.

An important concept to remember at this point is that an electron is not a fixed amount of energy. The electrons that bond the H to NAD^+ in NADH have a relatively large amount of energy. In the series of reactions that follow, they lose this energy in small amounts, until at the end they are attached (with H) to oxygen (O) to make water (H_2O). In some of the steps, the energy they lose is captured into ATP in coupled reactions.

1. In the first step of the electron transport chain, NADH reacts with a molecule called a flavoprotein, losing its electrons (and their H). The products are NAD^+ and reduced flavoprotein. A little energy is lost as heat in this reaction.
2. The flavoprotein passes on the electrons to a molecule called coenzyme Q. Again they lose some energy as heat, but ADP and P bond together and form ATP, storing much of the energy. This is a coupled reaction: ADP + P → ATP.
3. Coenzyme Q passes the electrons to cytochrome b. Again the electrons lose energy.
4. Cytochrome *b* passes the electrons to cytochrome c in a coupled reaction in which ATP is formed: ADP + P → ATP.
5. Cytochrome *c* passes the electrons to cytochrome *a*.
6. Cytochrome *a* passes them (with their H) to an atom of oxygen (O), forming water (H_2O). This is a coupled reaction in which ATP is formed: ADP + P → ATP.

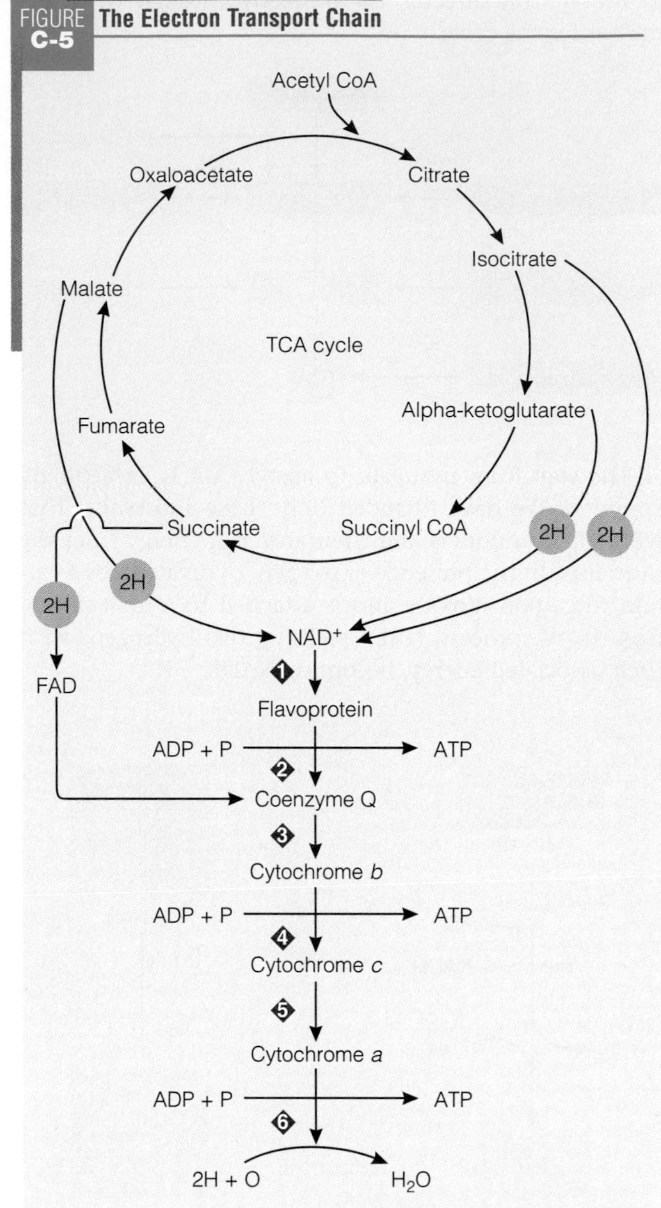

FIGURE C-5 The Electron Transport Chain

As Figure C-5 shows, each time NADH is oxidized (loses its electrons) by this means, the energy it loses is parceled out into three ATP molecules. When the electrons are passed on to water at the end, they are much lower in energy than they were originally. This completes the story of the electrons from NADH.

As for $FADH_2$, its electrons enter the electron transport chain at coenzyme Q. From coenzyme Q to water, ATP is generated in only two steps. Therefore, $FADH_2$ coming out of the TCA cycle yields just two ATP molecules.

One energy-receiving compound of the TCA cycle (GTP) does not enter the electron transport chain but gives its energy directly to ADP in a simple phosphorylation reaction. This reaction yields one ATP.

It is now possible to draw up a balance sheet of glucose metabolism (see Table C-3). Glycolysis has yielded 4 NADH + H^+ and 4 ATP molecules and has spent 2 ATP. The 2 acetyl CoA going through the TCA cycle have yielded 6 NADH + H^+, 2 $FADH_2$, and 2 GTP molecules. After the NADH + H^+ and $FADH_2$ have gone through the electron transport chain, there are 34 ATP. Added to these are the 4 ATP from glycolysis and the 2 ATP from GTP, making the total 40 ATP generated from one molecule of glucose. After the expense of 2 ATP is subtracted, there is a net gain of 38 ATP.*

A similar balance sheet from the complete breakdown of one 16-carbon fatty acid would show a net gain of 129 ATP. As mentioned earlier, 35 ATP were generated from the seven $FADH_2$ and seven NADH + H^+ produced during fatty acid oxidation. The eight acetyl CoA produced will each generate 12 ATP as they go through the TCA cycle and the electron transport chain, for a total of 96 more ATP. After subtracting the 2 ATP needed to activate the fatty acid initially, the net yield from one 16-carbon fatty acid: 35 + 96 − 2 = 129 ATP.

These calculations help explain why fat yields more energy (measured as kcalories) per gram than carbohydrate or protein. The more hydrogen atoms a fuel contains, the more ATP will be generated during oxidation. The 16-carbon fatty acid molecule, with its 32 hydrogen atoms, generates 129 ATP, whereas glucose, with its 12 hydrogen atoms, yields only 38 ATP.

*The total may sometimes be 36 or 37, rather than 38, ATP. The NADH + H^+ generated in the cytoplasm during glycolysis pass their electrons on to shuttle molecules, which move them into the mitochondria. One shuttle, malate, contributes its electrons to the electron transport chain before the first site of ATP synthesis, yielding 3 ATP. Another, glycerol phosphate, adds its electrons into the chain beyond that first site, yielding 2 ATP. Thus sometimes 3, and sometimes only 2, ATP result from the NADH + H^+ that arise from glycolysis. The amount depends on the cell.

TABLE C-3 Balance Sheet for Glucose Metabolism

	Expenditures	Income
Glycolysis:		
1 glucose	2 ATP	4 ATP
1 fructose-1,6-diphosphate		2 NADH + H^+
2 pyruvate		2 NADH + H^+
TCA cycle:		
2 isocitrate		2 NADH + H^+
2 alpha-ketoglutarate		2 NADH + H^+
2 succinyl CoA		2 GTP
2 succinate		2 $FADH_2$
2 malate		2 NADH + H^+
Total ATP collected:		
From glycolysis	2 ATP	4 ATP
From 2 NADH + H^+		4–6 ATP[a]
From 8 NADH + H^+		24 ATP
From 2 GTP		2 ATP
From 2 $FADH_2$		4 ATP
Totals:	2 ATP	38–40 ATP
Balance on hand from 1 molecule of glucose:		36–38 ATP

[a]Each NADH + H^+ from glycolysis can yield 2 or 3 ATP. See the accompanying text.

The TCA cycle and the electron transport chain are the body's major means of capturing the energy from nutrients in ATP molecules. Other means, such as anaerobic glycolysis, contribute, but the aerobic processes are the most efficient. Biologists and chemists understand much more about these processes than has been presented here.

Alcohol's Interference with Energy Metabolism

Highlight 7 provides an overview of how alcohol interferes with energy metabolism. With an understanding of the TCA cycle, a few more details may be appreciated. During alcohol metabolism, the enzyme alcohol dehydrogenase oxidizes alcohol to acetaldehyde while it simultaneously reduces a molecule of NAD^+ to NADH + H^+. The related enzyme acetaldehyde dehydrogenase reduces another NAD^+ to NADH + H^+ while it oxidizes acetaldehyde to acetyl CoA, the compound that enters the TCA cycle to generate energy. Thus, whenever alcohol is being metabolized in the body, NAD^+ diminishes, and NADH + H^+ accumulates. Chemists say that the body's "redox state" is altered, because NAD^+ can oxidize, and NADH + H^+ can reduce, many other body compounds. During alcohol metabolism, NAD^+ becomes unavailable for the multitude of reactions for which it is required.

As the previous sections just explained, for glucose to be completely metabolized, the TCA cycle must be operating, and NAD^+ must be present. If these conditions are not met (and when alcohol is present, they may not be), the pathway will be blocked, and traffic will back up—or an alternate route will be taken. Think about this as you follow the pathway shown in Figure C-6.

In each step of alcohol metabolism in which NAD+ is converted to NADH + H^+, hydrogen ions accumulate, resulting in a dangerous shift of the acid-base balance toward acid (Chapter 12 explains acid-base balance). The accumulation of NADH + H^+ depresses TCA cycle activity, so pyruvate and acetyl CoA build up. This condition favors the conversion of pyruvate to lactic acid, which serves as a temporary storage place for hydrogens from NADH + H^+. The conversion of pyruvate to lactic acid restores some NAD^+, but a lactic acid buildup has serious consequences of its own. It adds to the body's acid burden and interferes with the excretion of uric acid, causing goutlike symptoms. Molecules of acetyl CoA become building blocks for fatty acids or ketone bodies. The making of ketone bodies

FIGURE C-6 Ethanol Enters the Metabolic Path

This is a simplified version of the glucose-to-energy pathway showing the entry of ethanol. The coenzyme NAD (which is the active form of the B vitamin niacin) is the only one shown here; however, many others are involved.

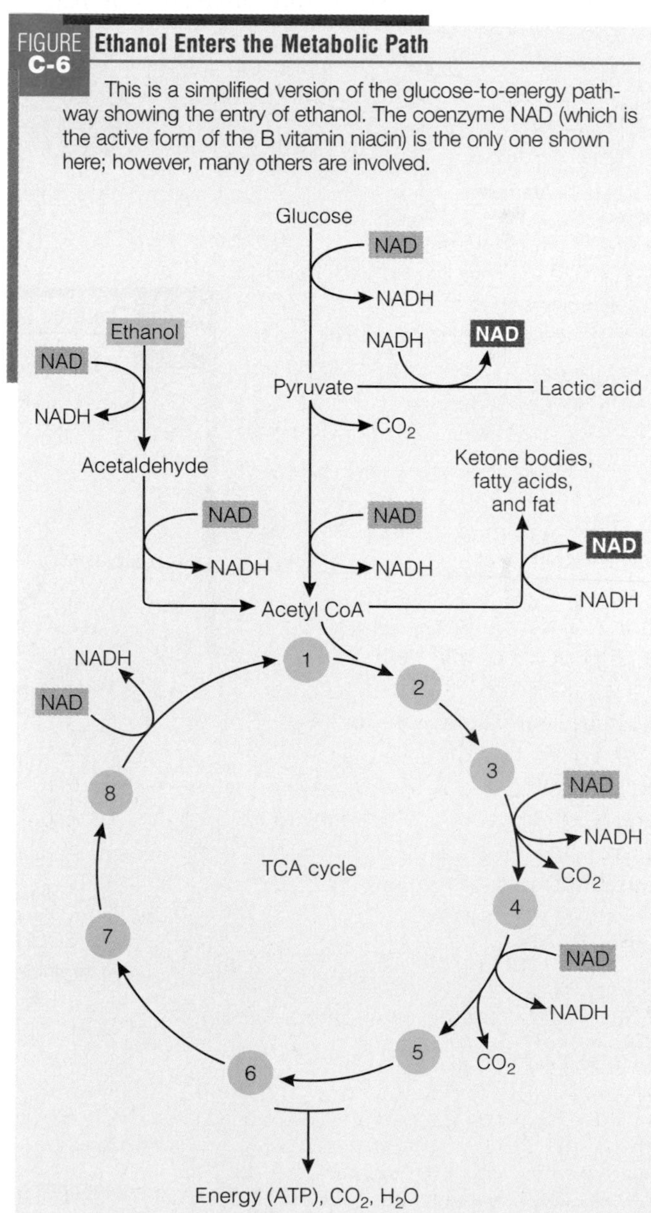

consumes acetyl CoA and generates NAD^+; but some ketone bodies are acids, so they push the acid-base balance further toward acid.

Thus alcohol cascades through the metabolic pathways, wreaking havoc along the way. These consequences have physical effects, which Highlight 7 describes.

The Urea Cycle

Chapter 7 sums up the process by which waste nitrogen is eliminated from the body by stating that ammonia molecules combine with carbon dioxide to produce urea. This is true, but it is not the whole story. Urea is produced in a multistep process within the cells of the liver.

Ammonia, freed from an amino acid or other compound during metabolism anywhere in the body, arrives at the liver by way of the bloodstream and is taken into a liver cell. There, it is first combined with carbon dioxide and a phosphate group from ATP to form carbamyl phosphate:

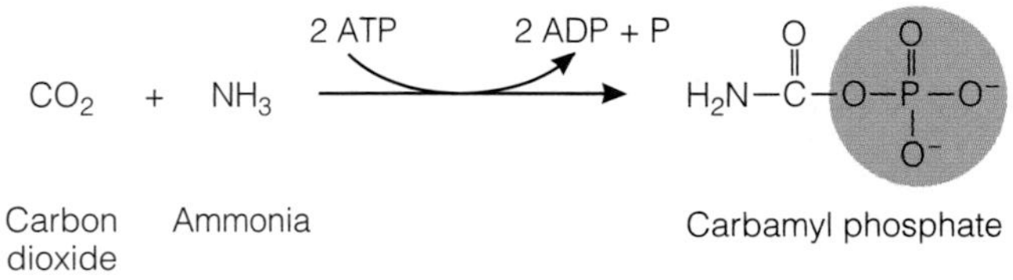

$$\underset{\text{Carbon dioxide}}{CO_2} + \underset{\text{Ammonia}}{NH_3} \xrightarrow{2\,ATP \;\to\; 2\,ADP + P} \underset{\text{Carbamyl phosphate}}{H_2N-\overset{\overset{O}{\|}}{C}-O-\overset{\overset{O}{\|}}{\underset{\underset{O^-}{|}}{P}}-O^-}$$

Figure C-7 shows the cycle of four reactions that follow.

1. Carbamyl phosphate combines with the amino acid ornithine, losing its phosphate group. The compound formed is citrulline.
2. Citrulline combines with the amino acid aspartic acid, to form argininosuccinate. The reaction requires energy from ATP. (ATP was shown earlier losing one phosphorus atom in a phosphate group, P, to become ADP. In this reaction, it loses two phosphorus atoms joined together, PP, and becomes adenosine monophosphate, AMP.)
3. Argininosuccinate is split, forming another acid, fumarate, and the amino acid arginine.
4. Arginine loses its terminal carbon with two attached amino groups and picks up an oxygen from water. The end product is urea, which the kidneys excrete in the urine. The compound that remains is ornithine, identical to the ornithine with which this series of reactions began, and ready to react with another molecule of carbamyl phosphate and turn the cycle again.

Formation of Ketone Bodies

Normally, fatty acid oxidation proceeds all the way to carbon dioxide and water. However, in ketosis (discussed in

FIGURE C-7 The Urea Cycle

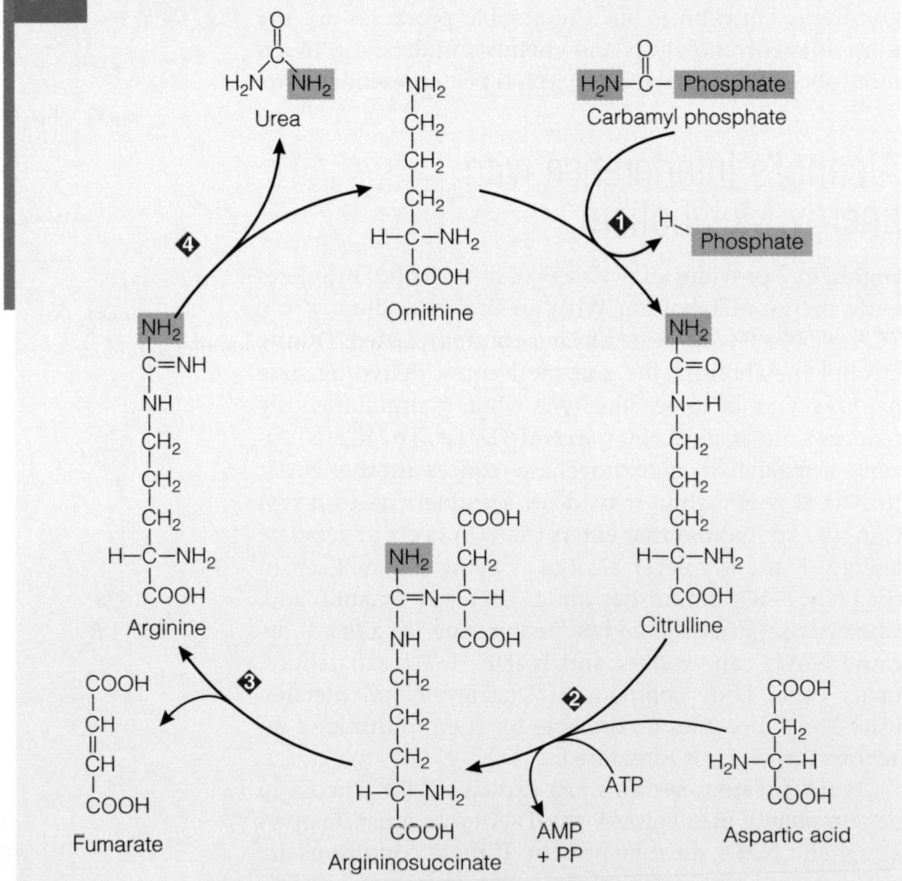

Urea
Ornithine
Carbamyl phosphate
Phosphate
Citrulline
Aspartic acid
ATP
AMP + PP
Argininosuccinate
Fumarate
Arginine

Chapter 7), an intermediate is formed from the condensation of two molecules of acetyl CoA: acetoacetyl CoA. Figure C-8 shows the formation of ketone bodies from that intermediate.

1. Acetoacetyl CoA condenses with acetyl CoA to form a six-carbon intermediate, beta-hydroxy-beta methylglutaryl CoA.
2. This intermediate is cleaved to acetyl CoA and acetoacetic acid.
3. Acetoactate can be metabolized either to beta-hydroxybutyric acid (step 3a) or to acetone (3b).

Acetoacetic acid, beta-hydroxybutyric acid, and acetone are the so-called ketone bodies of ketosis. Two are real ketones (they have a C=O group between two carbons); the other is an alcohol that has been produced during ketone formation—hence the term *ketone bodies,* rather than ketones, to describe the three of them. There are many other ketones in nature; these three are characteristic of ketosis in the body.

FIGURE C-8 The Formation of Ketone

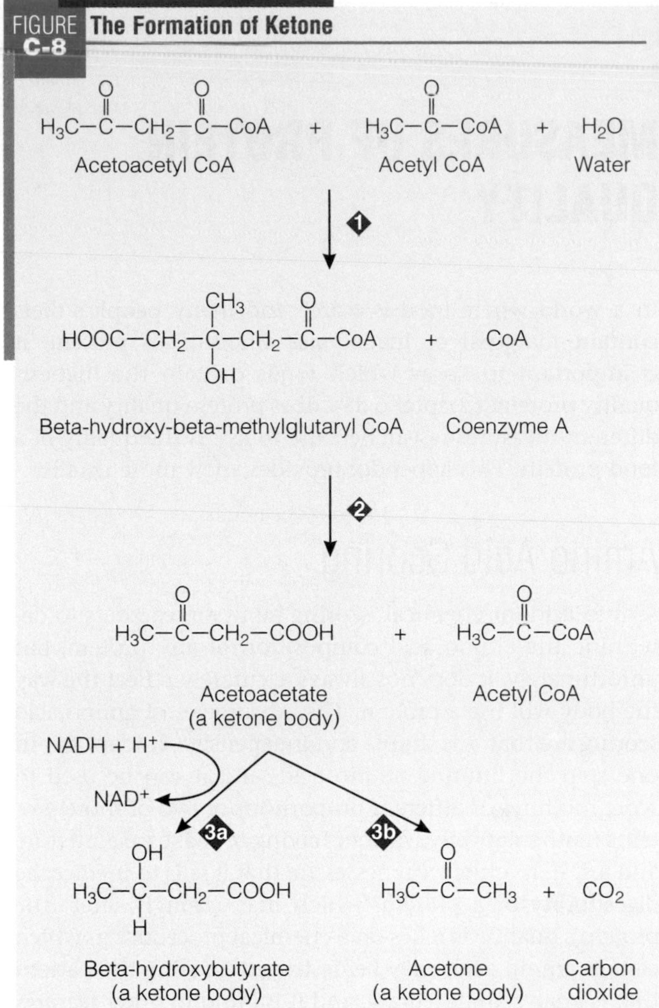

D MEASURES OF PROTEIN QUALITY

CONTENTS

In a world where food is scarce and many people's diets contain marginal or inadequate amounts of protein, it is important to know which foods contain the highest-quality protein. Chapter 6 describes protein quality and the different measures researchers use to assess the quality of a food protein. This appendix provides a few more details.

Amino Acid Scoring

Amino acid, or chemical, scoring allows researchers to determine the amino acid composition of any protein, but unfortunately, it does not always accurately reflect the way the body will use a protein. The advantages of amino acid scoring are that it is simple and inexpensive, it identifies in one step the limiting amino acid, and it can be used to score mixtures of different proportions of two or more proteins mathematically without having to make up a mixture and test it. Its chief weaknesses are that it fails to predict the digestibility of a protein, which may strongly affect the protein's quality; it relies on a chemical procedure in which certain amino acids may be destroyed, making the pattern that is analyzed inaccurate; and it is blind to other features of the protein (such as the presence of substances that may inhibit the digestion or utilization of the protein) that would only be revealed by a test in living animals.

Table D-1 shows the reference pattern for the nine essential amino acids. To interpret the table, read, "For every 3210 units of essential amino acids, 145 must be histidine, 340 must be isoleucine, 540 must be leucine," and so on. To compare a test protein with the reference protein, the experimenter first obtains a chemical analysis of the test protein's amino acids. Then, taking 3210 units of the amino acids, the experimenter compares the amount of each amino acid to the amount found in 3210 units of essential amino acids in egg protein. For example, suppose the test protein contained (per 3210 units) 360 units of isoleucine; 500 units of leucine; 350 of lysine; and for each of the other amino acids, more units than egg protein contains. The two amino acids that are low are leucine (500 as compared with 540 in egg) and lysine (350 versus 440 in egg). The ratio, amino acid in the test protein divided by amino acid in egg, is 500/540 (or about 0.93) for leucine and 350/440 (or about 0.80) for lysine. Lysine is the limiting amino acid (lowest ratio compared with egg), so the test protein receives a chemical score of 80.

PDCAAS

PDCAAS (protein-digestibility–corrected amino acid score) takes the amino acid scoring method a step further by correcting for the digestibility of the protein. To calculate the PDCAAS, researchers first determine the amino acid profile of the test protein (in this example, pinto beans). The sec-

TABLE D-1 A Reference Pattern for Amino Acid Scoring of Proteins

Essential Amino Acids	Reference Protein—Whole Egg (mg amino acid/g nitrogen)
Histidine	145
Isoleucine	340
Leucine	540
Lysine	440
Methionine + cystine[a]	355
Phenylalanine + tyrosine[b]	580
Threonine	294
Tryptophan	106
Valine	410
Total	3210

[a]Methionine is essential and is also used to make cystine. Thus the methionine requirement is lower if cystine is supplied.
[b]Phenylalanine is essential and is also used to make tyrosine if not enough of the latter is available. Thus the phenylalanine requirement is lower if tyrosine is also supplied.

D

TABLE D-2 An Example of PDCAAS

Essential Amino Acids	Amino Acid Profile of Pinto Beans (mg/g protein)	Amino Acid Requirements for 2–5 yr (mg/g protein)	Ratio
Histidine	30.0	19	1.58
Isoleucine	42.5	28	1.52
Leucine	80.4	66	1.22
Lysine	69.0	58	1.19
Methionine + cystine	21.1	25	0.84
Phenylalanine + tyrosine	90.5	63	1.44
Threonine	43.7	34	1.28
Tryptophan	8.8	11	0.80
Valine	50.1	35	1.43

ond column of Table D-2 presents the essential amino acid profile for pinto beans. The third column presents the amino acid requirements of preschool-aged children for comparison. (The rationale behind using the requirements of this age group is that if a protein will effectively support a young child's growth and development, then it will meet or exceed the requirements of older children and adults.) To determine how well the food protein meets human needs, researchers calculate the ratio by dividing the second column by the third column (for example, 30 ÷ 19 = 1.578 or 1.58).

The amino acid with the lowest ratio is the first limiting amino acid—in this case, tryptophan. Its ratio is the amino acid score for the protein—in this case, 0.80. Remember, though, the amino acid score does not account for digestibility. Protein digestibility, as determined by rat balance studies, yields a value of 79 percent for pinto beans. Together, the amino acid score and the digestibility value determine the PDCAAS:

$$\text{PDCAAS} = \text{protein digestibility} \times \text{lowest amino acid ratio.}$$
$$\text{PDCAAS for pinto beans} = 0.79 \times 0.80 = 0.63.$$

Thus the PDCAAS for pinto beans is 0.63. Table D-3 lists the PDCAAS values of selected foods.

The PDCAAS is used to determine the % Daily Value on food labels. To calculate the % Daily Value for protein for canned pinto beans, multiply the number of grams of protein in a standard serving (in this case, 7 grams per ½ cup) by the PDCAAS:

$$7\text{ g} \times 0.63 = 4.41.$$

This value is then divided by the recommended standard for protein (for children over age four and adults, 50 grams):

$$4.41 \div 50 = 0.088 \text{ (or 8.8\%).}$$

The food label for this can of pinto beans would declare that one serving provides 7 grams protein, and if the label included a % Daily Value for protein, the value would be 9 percent.

Biological Value

To determine the actual value of a protein as it is used by the body, it is necessary to measure both urinary and fecal losses of nitrogen when that protein is actually fed to human beings under test conditions. Even then, small additional losses from sweat, shed skin, hair, and fingernails will be missed. This kind of experiment determines the biological value (BV) of proteins.

In a test of biological value, two nitrogen balance studies are done. In the first, no protein is fed, and nitrogen (N) excretions in the urine and feces are measured. It is assumed that under these conditions, N lost in the urine is the amount the body always necessarily loses by filtration into the urine each day, regardless of what protein is fed (endogenous N). The N lost in the feces (called metabolic N) is the amount the body invariably loses into the intestine each day, whether or not food protein is fed. (To help you remember the terms: endogenous N is "urinary N on a zero-protein diet"; metabolic N is "fecal N on a zero-protein diet.")

TABLE D-3 PDCAAS Values of Selected Foods

Food	PDCAAS
Casein (milk protein)	1.00
Egg white	1.00
Soybean (isolate)	.99
Beef	.92
Pea flour	.69
Kidney beans (canned)	.68
Chick peas (canned)	.66
Pinto beans (canned)	.63
Rolled oats	.57
Lentils (canned)	.52
Peanut meal	.52
Whole wheat	.40

Note: 1.0 is the maximum PDCAAS a food protein can receive.

D

TABLE D-4 Biological Values (BV) of Selected Foods

Food	BV
Egg	100
Milk	93
Beef	75
Fish	75
Corn	72

Note: 100 is the maximum BV a food protein can receive.

In the second study, an amount of protein slightly below the requirement is fed. Intake and losses are measured; then the BV is derived using this formula:

$$BV = \frac{\text{N retained}}{\text{N absorbed}} \times 100.$$

The denominator of this equation expresses the amount of nitrogen *absorbed:* food N minus fecal N (excluding the metabolic N the body would lose in the feces anyway, even without food). The numerator expresses the amount of N *retained* from the N absorbed: absorbed N (as in the denominator) minus the N excreted in the urine (excluding the endogenous N the body would lose in the urine anyway, even without food).

For egg protein, the BV is 100 (all the absorbed protein is retained). Supplied in adequate quantity, a protein with a BV of 70 or greater can support human growth as long as energy intake is adequate. Table D-4 presents the BV for selected foods.

This method has the advantages of being based on experiments with human beings (it can be done with animals, too, of course) and of measuring actual nitrogen retention. But it is also cumbersome, expensive, and often impractical, and it is based on several assumptions that may not be valid. For example, the physiology, normal environment, or typical food intake of the subjects used for testing may not be similar to those for whom the test protein may ultimately be used. For another example, the retention of protein in the body does not necessarily mean that it is being well utilized. Considerable exchange of protein among tissues (protein turnover) occurs, but is hidden from view when only N intake and output are measured. The test of biological value wouldn't detect if one tissue were shorted.

Net Protein Utilization

Like measurements of BV, determinations of net protein utilization (NPU) involve two balance studies: one on zero nitrogen intake, and the other on submaximal intake. The formula for NPU is:

$$NPU = \frac{\text{N retained}}{\text{N intake}} \times 100.$$

The numerator is the same as it is for BV, but the denominator represents food N intake only—not N absorbed.

This method offers advantages similar to those of BV determinations and is used more frequently, with animals as the test subjects. A drawback is that if a low NPU is obtained, the test results offer no help in distinguishing between two possible causes: a poor amino acid composition of the test protein or poor digestibility. There is also a limit to the extent to which animal test results can be assumed to be applicable to human beings.

Protein Efficiency Ratio

The protein efficiency ratio (PER) is a widely used procedure for evaluating protein quality. Young rats are fed a measured amount of protein and weighed periodically as they grow. The PER is expressed as:

$$PER = \frac{\text{weight gain (g)}}{\text{protein intake (g)}}.$$

This method has the virtues of economy and simplicity, but it also has many drawbacks. The experiments are time-consuming; the amino acid needs of rats are not the same as those of human beings; and the amino acid needs for growth are not the same as for the maintenance of adult animals (growing animals need more lysine, for example). Table D-5 presents PER values for selected foods.

TABLE D-5 Protein Efficiency Ratio (PER) Values of Selected Proteins

Protein	PER
Casein (milk)	2.8
Soy	2.4
Glutein (wheat)	0.4

NUTRITION ASSESSMENT: SUPPLEMENTAL INFORMATION

CONTENTS

E

Chapter 17 described data from nutrition assessments that help evaluate clients' nutrition status and nutrient needs. This appendix provides additional information that may be useful for complete assessments.

Diet-Drug Interactions

Chapter 18 described diet-drug interactions and later chapters provided a series of diet-drug interaction boxes. Table E-1 shows where you can find examples of diet-drug interactions for different classes of medications.

TABLE E-1 Locating Examples of Diet-Drug Interactions

Medication	Page Number(s)
Anabolic agents	864
Analgesics	673
Antacids	643
Antianxiety agents	673
Antibiotics	643
Anticoagulants	791
Antidepressants	673
Antidiabetic agents	760
Antidiarrheals	673
Antiemetics, antinauseants	643
Anti-GERD, see *antisecretory agents*	
Antigout agents	817
Antihypertensives	791
Anti-infectives	835
Antilipemics	791
Antineoplastic agents	864
Antisecretory agents	643
Antiviral agents	864
Appetite stimulants	835
Bronchodilators	791
Cardiac glycosides	791
Diuretics	791, 835
Immunosuppressants	673, 818, 835
Laxatives	673, 835
Phosphate binders	818
Potassium binders	818
Sedatives	732
Other	
Infliximab	673
Interferon	835
Pancreatic enzyme replacements	673
Penicillamine	835
Sulfasalazine	673

Growth Charts and Body Measurements

Growth charts, shown in Figures E-1 (A and B) through E-6 (A and B), allow health care professionals to evaluate the growth and development of children from birth to 18 years of age. The growth charts for plotting body mass index-for-age are shown in Figure 14-9 on p. 473. Percentile charts divide the measures of a population into 100 equal divisions. Thus, half of the population falls above the 50th percentile, and half falls below. The use of percentile measures allows for comparisons among children of the same age and gender. For example, a six-month-old female infant whose weight is at the 75 percentile weighs more than 75 percent of the female infants her age.

• **Fatfold Measures** • Fatfold measures provide a good estimate of total body fat and a fair assessment of the fat's location. Approximately half the fat in the

FIGURE E-1B Weight-for-Age Percentiles: Girls, Birth to 36 Months

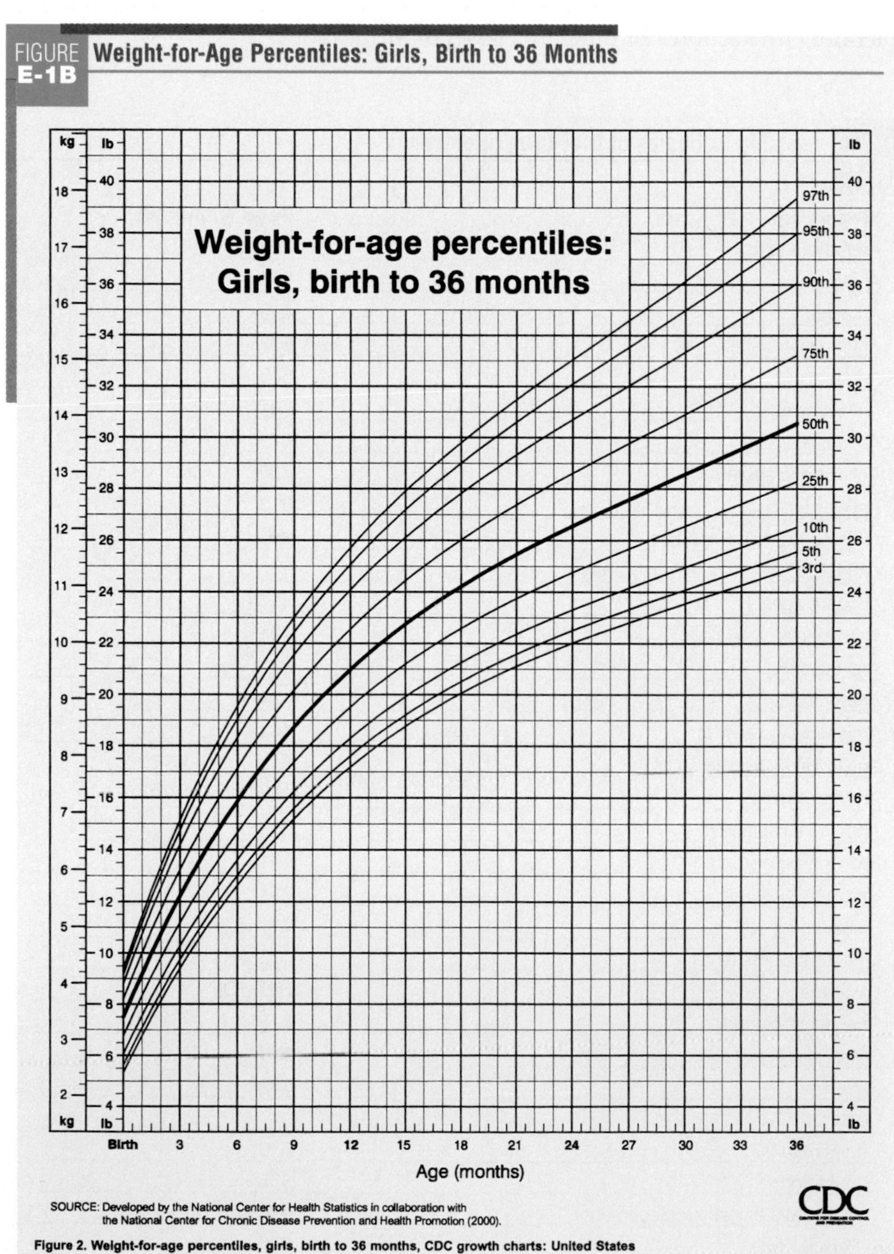

SOURCE: Developed by the National Center for Health Statistics in collaboration with the National Center for Chronic Disease Prevention and Health Promotion (2000).

Figure 2. Weight-for-age percentiles, girls, birth to 36 months, CDC growth charts: United States

body lies directly beneath the skin, and the thickness of this subcutaneous fat reflects total body fat. In some parts of the body, such as the back and the back of the arm over the triceps muscle, this fat is loosely attached;◆ a person can pull it up between the thumb and forefinger to obtain a measure of fatfold thickness. To measure fatfold, a skilled assessor follows a standard procedure using reliable calipers (illustrated in Figure E-7 on p. E-9) and then compares the measurement with standards. Triceps fatfold measures greater than 15 millimeters in men or 25 millimeters in women suggest excessive body fat.

Common sites for fatfold measures: ◆
- Triceps
- Subscapular (below shoulder blade)
- Suprailae (above hip bone)
- Abdomen
- Upper thigh

Fatfold measurements correlate directly with the risk of heart disease. They assess central obesity and its associated risks better than do weight measures alone. If a person gains body fat, the fatfold increases proportionately; if the person loses fat, it decreases. Measurements taken from central-body sites (around the abdomen) better reflect changes in fatness than those taken from upper sites

(continues on p. E-9)

FIGURE E-1A **Weight-for-Age Percentiles: Boys, Birth to 36 Months**

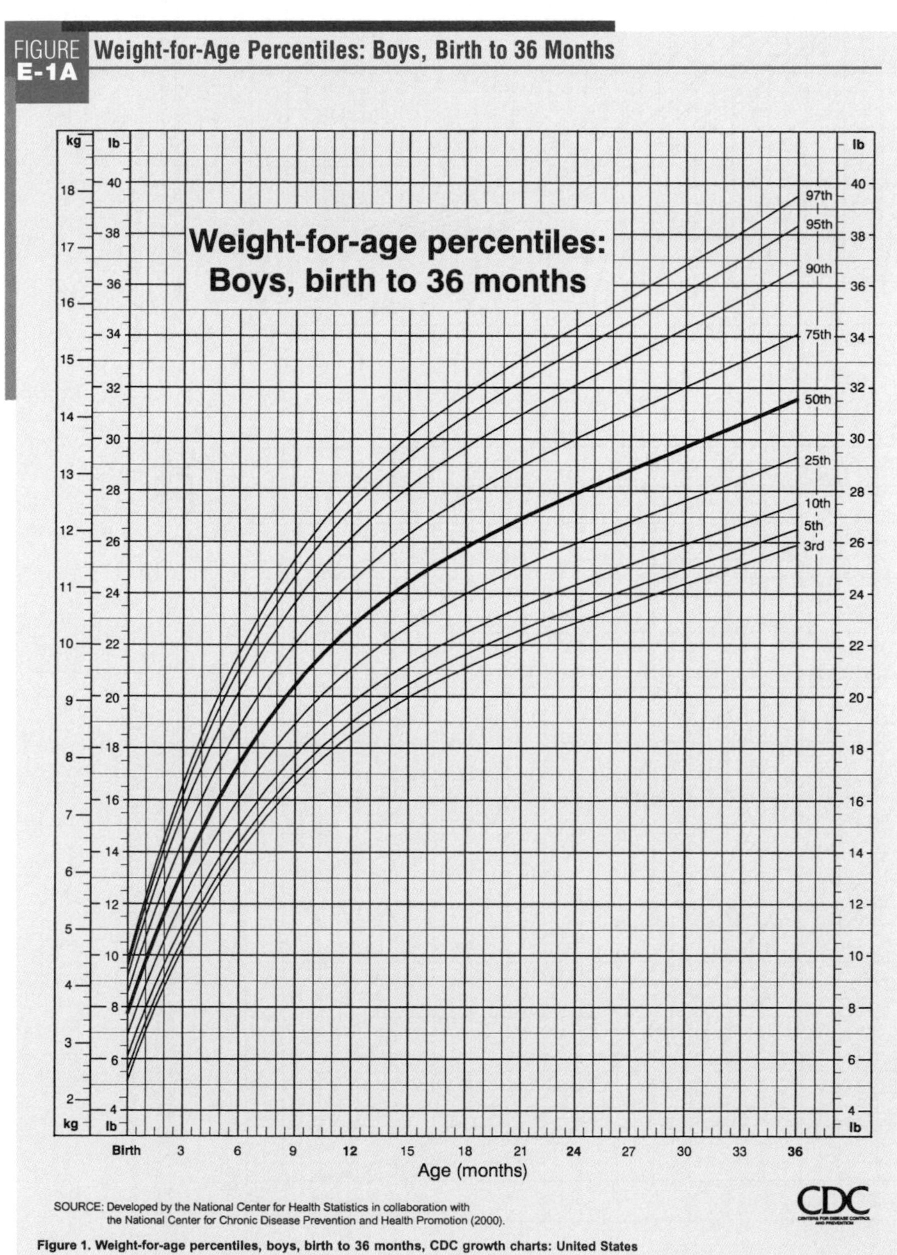

FIGURE E-2A Length-for-Age Percentiles: Boys, Birth to 36 Months

Length-for-age percentiles: Boys, birth to 36 months

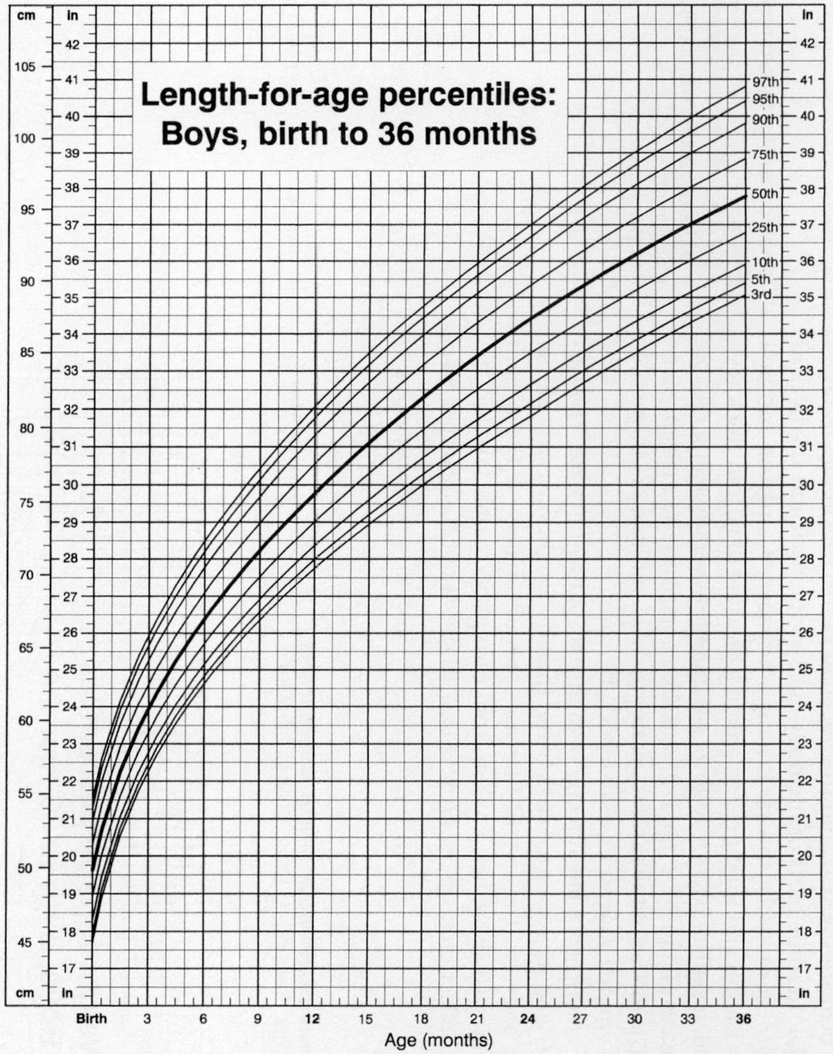

SOURCE: Developed by the National Center for Health Statistics in collaboration with the National Center for Chronic Disease Prevention and Health Promotion (2000).

Figure 3. Length-for-age percentiles, boys, birth to 36 months, CDC growth charts: United States

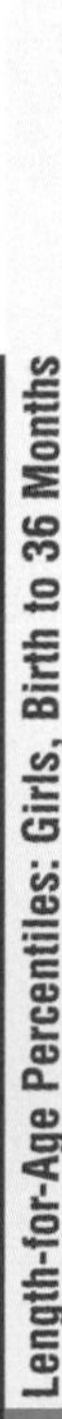

FIGURE E-2B Length-for-Age Percentiles: Girls, Birth to 36 Months

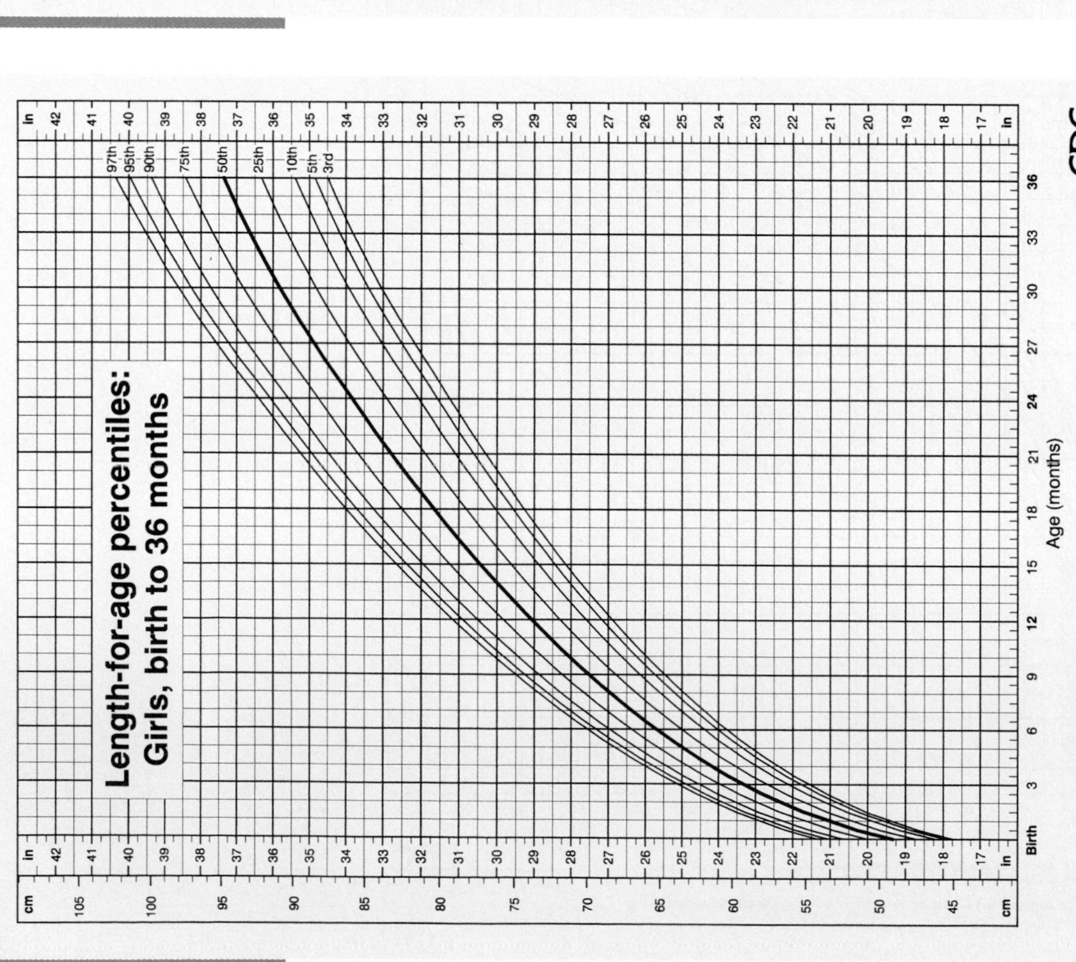

SOURCE: Developed by the National Center for Health Statistics in collaboration with the National Center for Chronic Disease Prevention and Health Promotion (2000).

Figure 4. Length-for-age percentiles, girls, birth to 36 months, CDC growth charts: United States

FIGURE E-3B Weight-for-Length Percentiles: Girls, Birth to 36 Months

Weight-for-length percentiles: Girls, birth to 36 months

97th, 95th, 90th, 75th, 50th, 25th, 10th, 5th, 3rd

Length

Revised and corrected June 8, 2000.
SOURCE: Developed by the National Center for Health Statistics in collaboration with the National Center for Chronic Disease Prevention and Health Promotion (2000).

CDC

Figure 6. Weight-for-length percentiles, girls, birth to 36 months, CDC growth charts: United States

FIGURE E-3A Weight-for-Length Percentiles: Boys, Birth to 36 Months

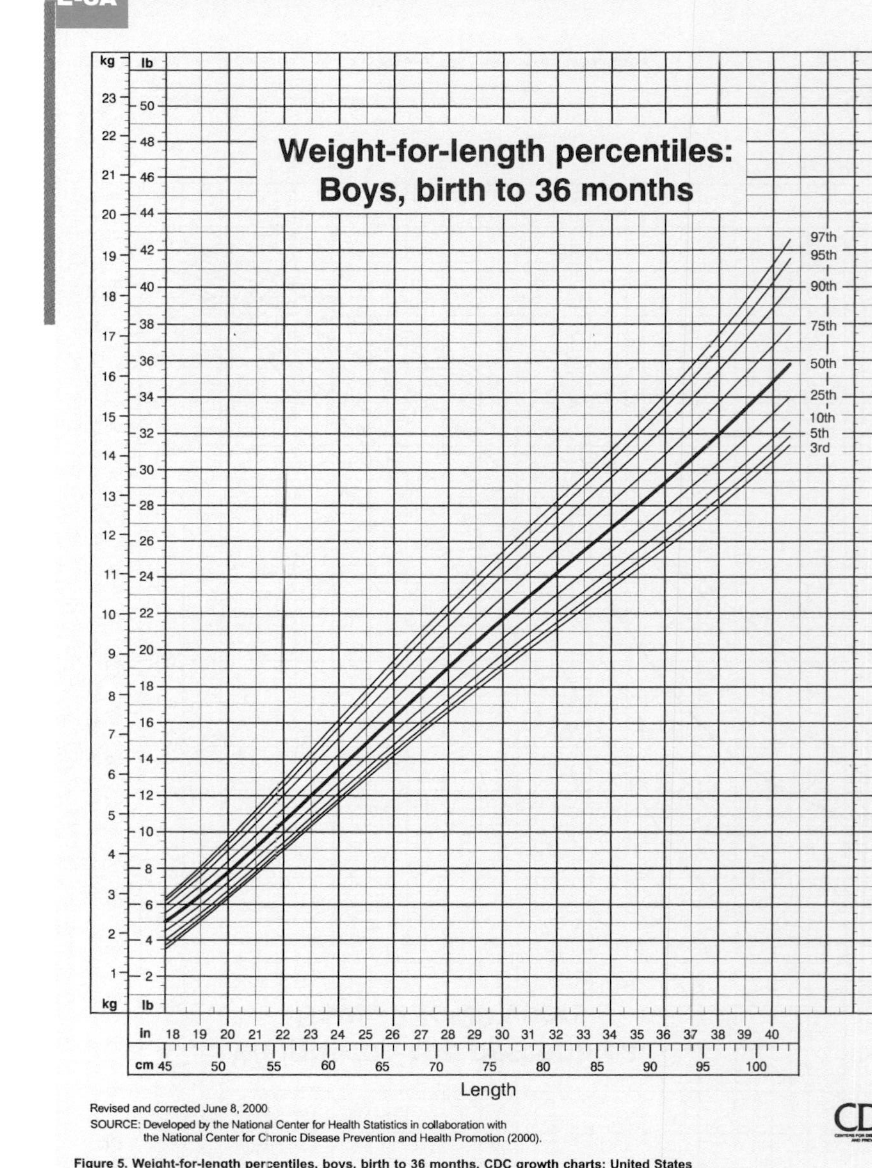

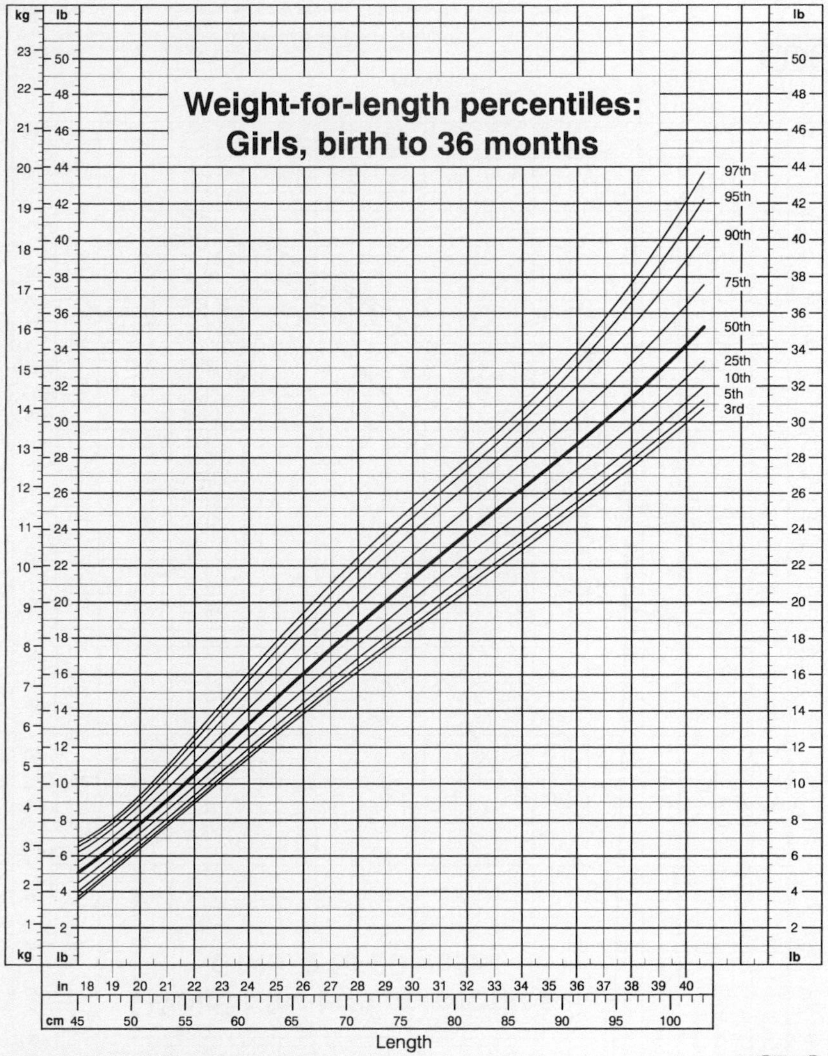

Revised and corrected June 8, 2000
SOURCE: Developed by the National Center for Health Statistics in collaboration with the National Center for Chronic Disease Prevention and Health Promotion (2000).

Figure 5. Weight-for-length percentiles, boys, birth to 36 months, CDC growth charts: United States

FIGURE E-4B Weight-for-Age Percentiles: Girls, 2 to 20 Years

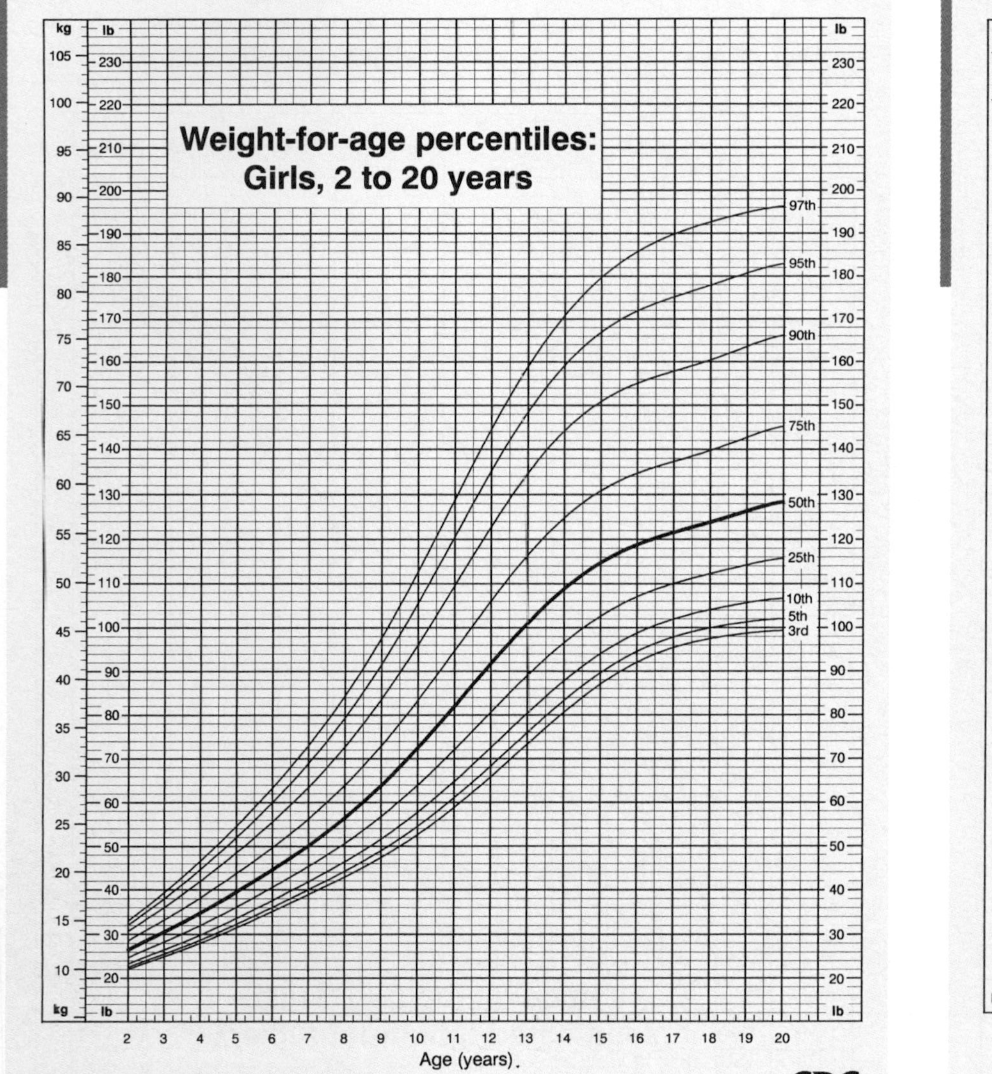

SOURCE: Developed by the National Center for Health Statistics in collaboration with the National Center for Chronic Disease Prevention and Health Promotion (2000).

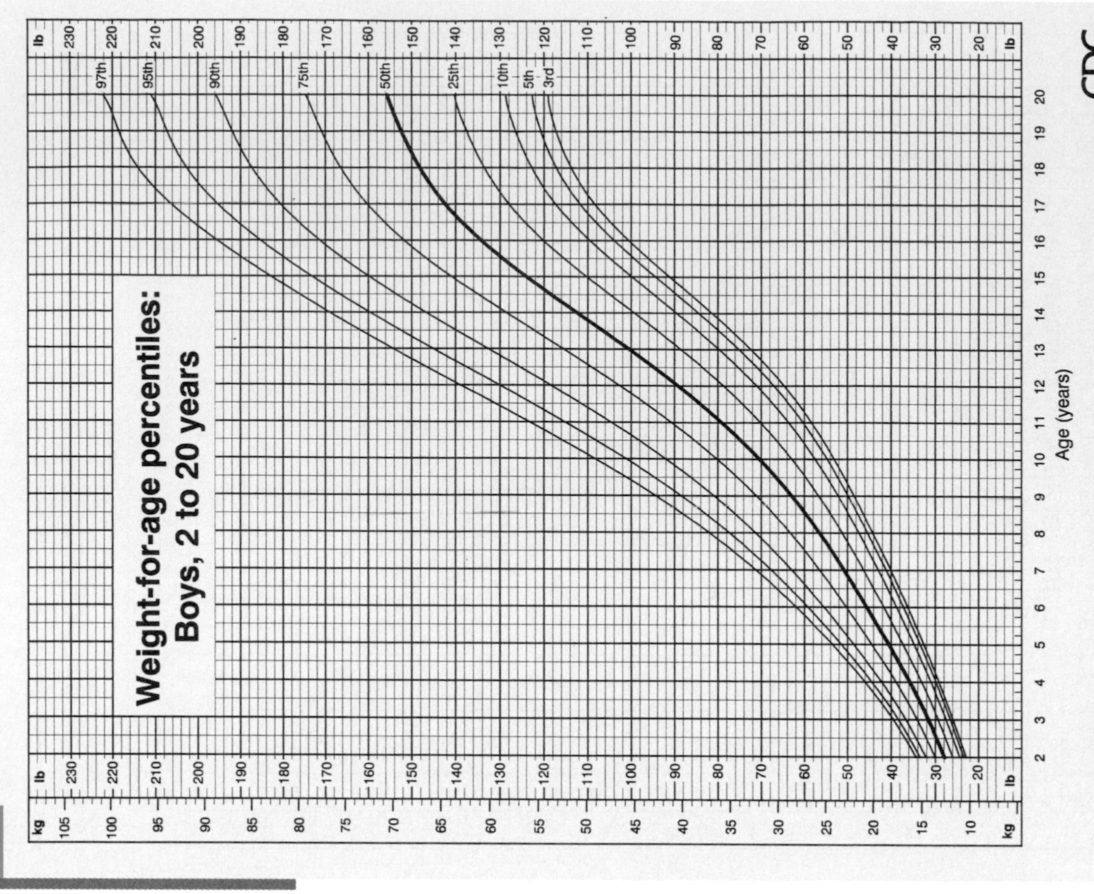

Figure 10. Weight-for-age percentiles, girls, 2 to 20 years, CDC growth charts: United States

FIGURE E-4A Weight-for-Age Percentiles: Boys, 2 to 20 Years

Weight-for-age percentiles:
Boys, 2 to 20 years

SOURCE: Developed by the National Center for Health Statistics in collaboration with the National Center for Chronic Disease Prevention and Health Promotion (2000).

Figure 9. Weight-for-age percentiles, boys, 2 to 20 years, CDC growth charts: United States

FIGURE E-5A Stature-for-Age Percentiles: Boys, 2 to 20 Years

Stature-for-age percentiles: Boys, 2 to 20 years

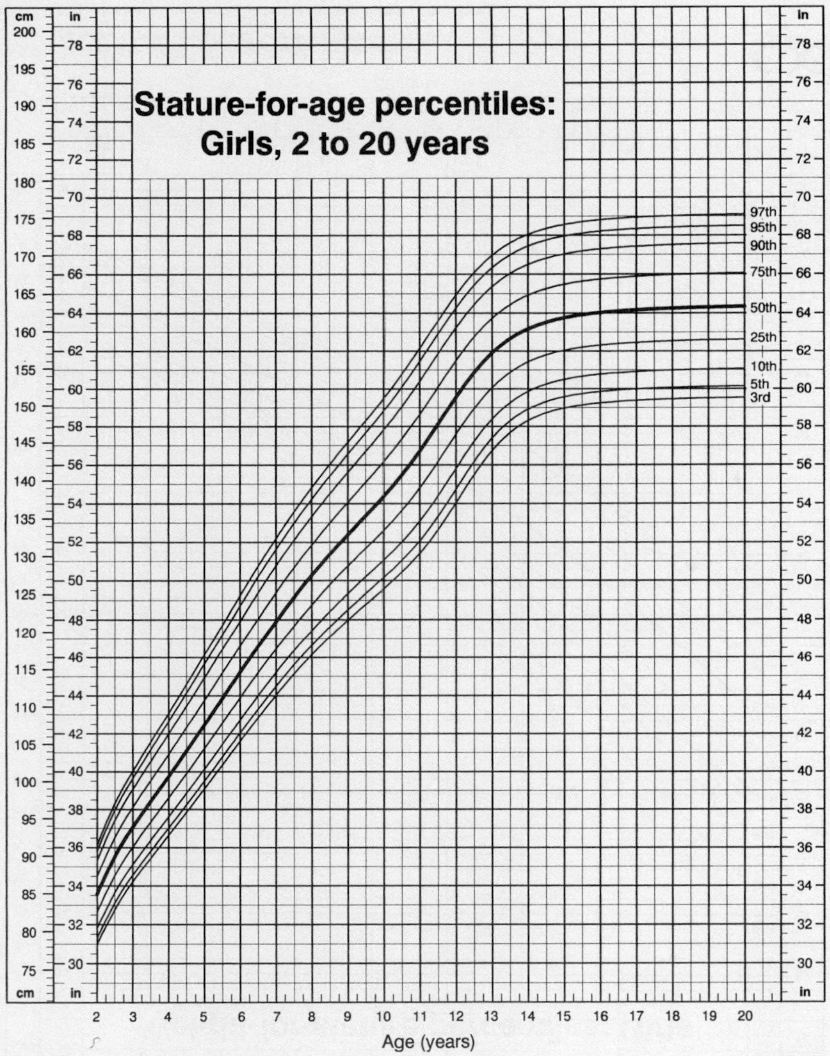

97th, 95th, 90th, 75th, 50th, 25th, 10th, 5th, 3rd

Age (years)

SOURCE: Developed by the National Center for Health Statistics in collaboration with the National Center for Chronic Disease Prevention and Health Promotion (2000).

CDC

Figure 11. Stature-for-age percentiles, boys, 2 to 20 years, CDC growth charts: United States

FIGURE E-5B Stature-for-Age Percentiles: Girls, 2 to 20 Years

Stature-for-age percentiles: Girls, 2 to 20 years

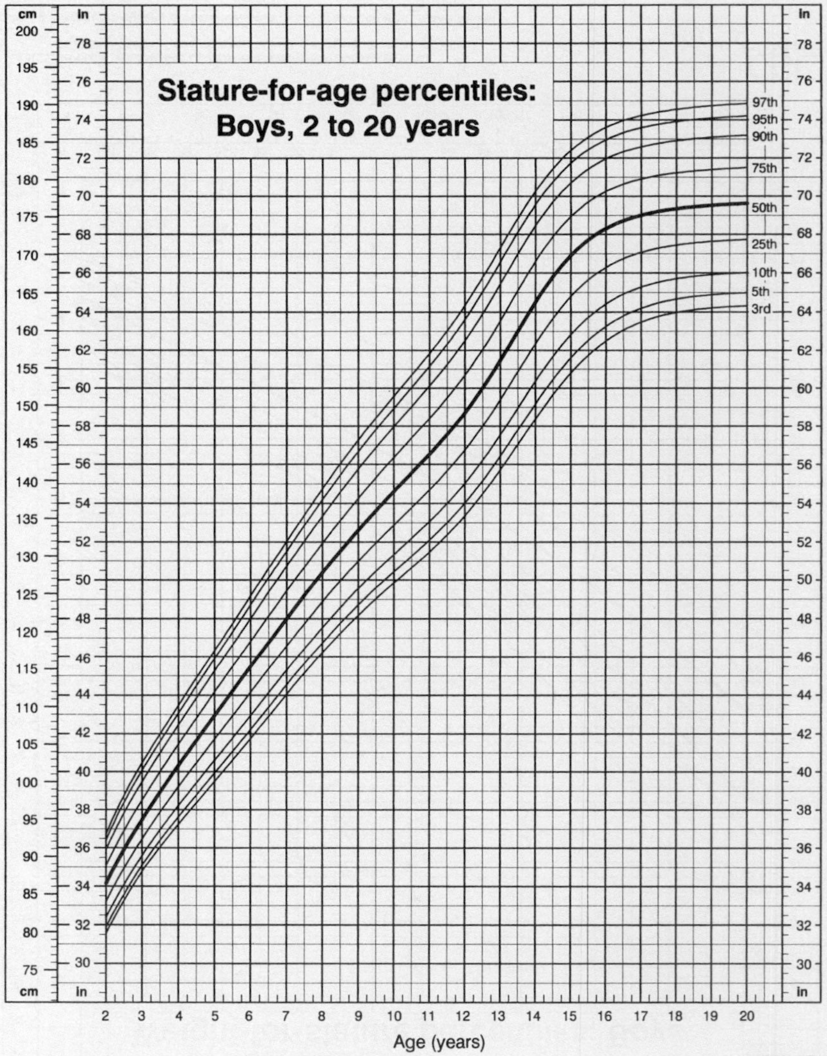

97th, 95th, 90th, 75th, 50th, 25th, 10th, 5th, 3rd

Age (years)

SOURCE: Developed by the National Center for Health Statistics in collaboration with the National Center for Chronic Disease Prevention and Health Promotion (2000).

CDC

Figure 12. Stature-for-age percentiles, girls, 2 to 20 years, CDC growth charts: United States

FIGURE E-6A Weight-for-Stature Percentiles: Boys, 2 to 20 Years

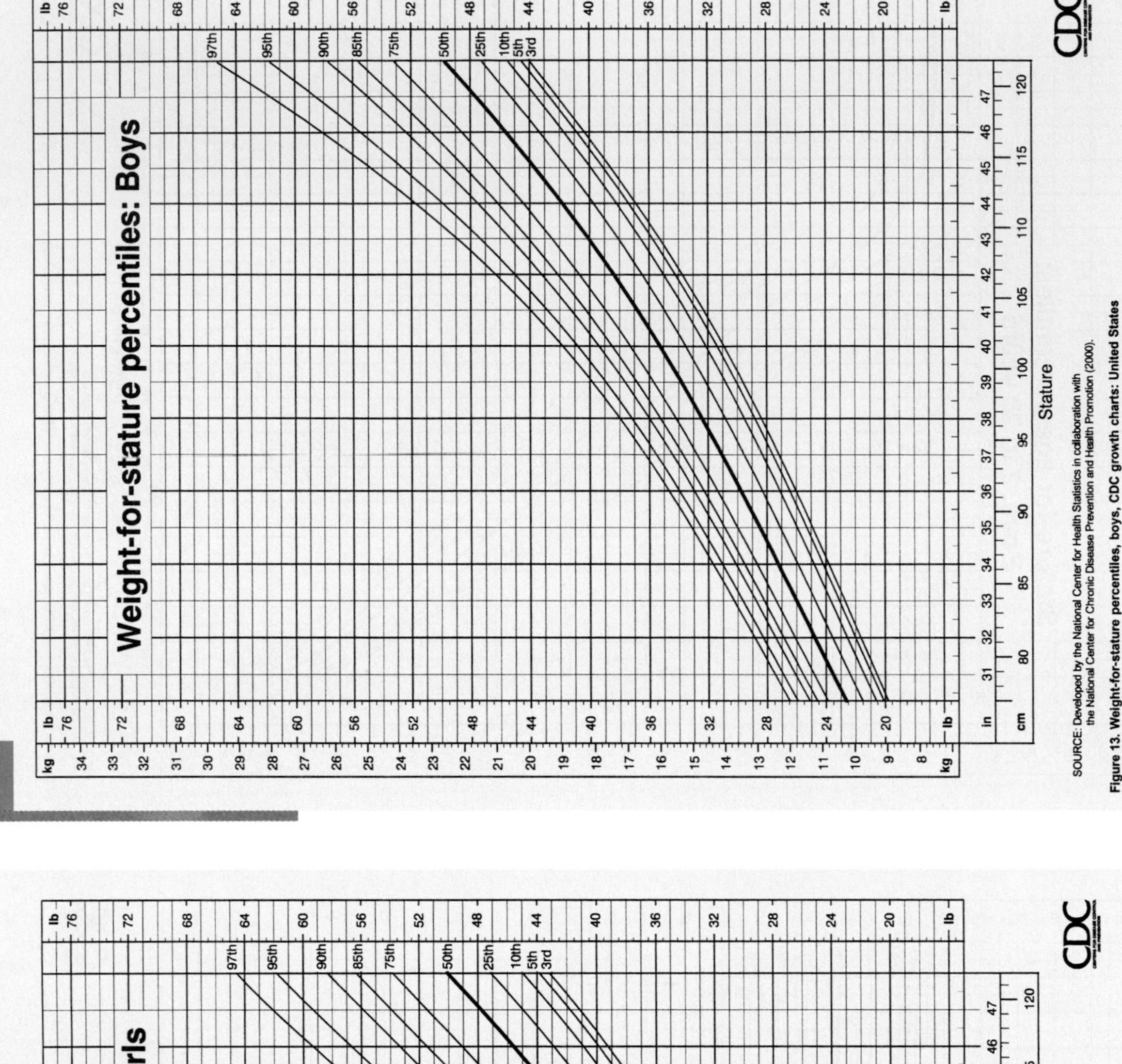

SOURCE: Developed by the National Center for Health Statistics in collaboration with the National Center for Chronic Disease Prevention and Health Promotion (2000).

Figure 13. Weight-for-stature percentiles, boys, CDC growth charts: United States

FIGURE E-6B Weight-for-Stature Percentiles: Girls, 2 to 20 Years

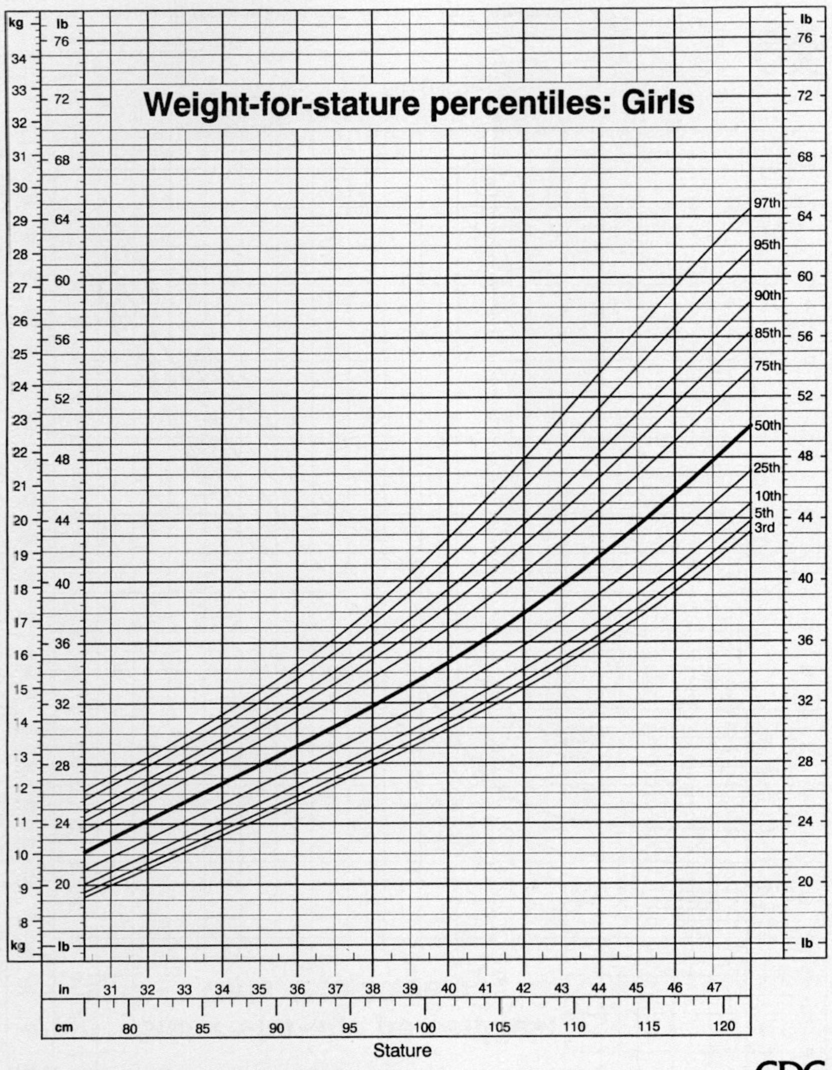

Weight-for-stature percentiles: Girls

SOURCE: Developed by the National Center for Health Statistics in collaboration with the National Center for Chronic Disease Prevention and Health Promotion (2000).

Figure 14. Weight-for-stature percentiles, girls, CDC growth charts: United States

FIGURE E-7 **How to Measure the Triceps Fatfold**

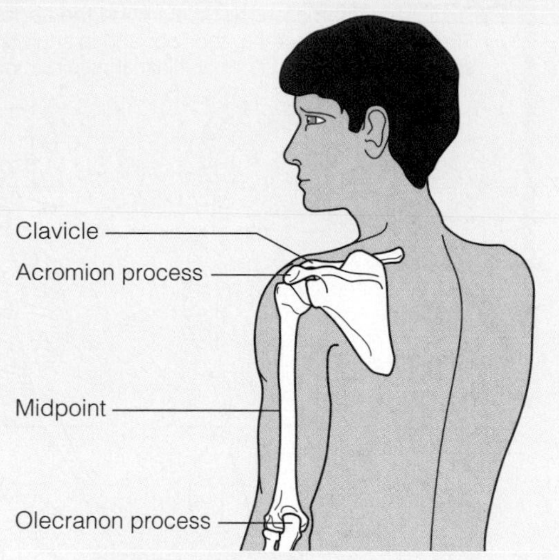

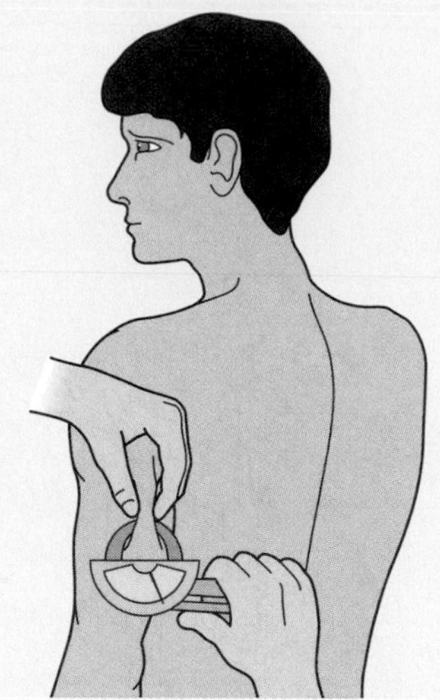

A. Find the midpoint of the arm:
1. Ask the subject to bend his or her arm at the elbow and lay the hand across the stomach. (If he or she is right-handed, measure the left arm, and vice versa.)
2. Feel the shoulder to locate the acromion process. It helps to slide your fingers along the clavicle to find the acromion process. The olecranon process is the tip of the elbow.
3. Place a measuring tape from the acromion process to the tip of the elbow. Divide this measurement by 2, and mark the midpoint of the arm with a pen.

B. Measure the fatfold:
1. Ask the subject to let his or her arm hang loosely to the side.
2. Grasp a fold of skin and subcutaneous fat between the thumb and forefinger slightly above the midpoint mark. Gently pull the skin away from the underlying muscle. (This step takes a lot of practice. If you want to be sure you don't have muscle as well as fat, ask the subject to contract and relax the muscle. You should be able to feel if you are pinching muscle.)
3. Place the calipers over the fatfold at the midpoint mark, and read the measurement to the nearest 1.0 millimeter in two to three seconds. (If using plastic calipers, align pressure lines, and read the measurement to the nearest 1.0 millimeter in two to three seconds.)
4. Repeat steps 2 and 3 twice more. Add the three readings, and then divide by 3 to find the average.

(arm and back). A major limitation of the fatfold test is that fat may be thicker under the skin in one area than in another. A pinch at the side of the waistline may not yield the same measurement as a pinch on the back of the arm. This limitation can be overcome by taking fatfold measurements at several (often three) different places on the body (including upper-, central-, and lower-body sites) and comparing each measurement with standards for that site. Multiple measures are not always practical in clinical settings, however, and most often, the triceps fatfold measurement alone is used because it is easily accessible.◆

◆ **Figure 8-8 on p. 256 includes photos of triceps fatfold measurement, hydro-densitometry, and bioelectrical impedance.**

• **Waist Circumference** • Chapter 8 described how fat distribution correlates with health risks and mentioned that the waist circumference is a valuable indicator of fat distribution. To measure waist circumference, the assessor places a nonstretchable tape around the person's body, crossing just above the upper hip bones and making sure that the tape remains on a level horizontal plane on all sides (see Figure E-8 on p. E-10). The tape is tightened slightly, but without compressing the skin.◆

◆ **Reminder: A waist circumference of > 35 inches in women and > 40 inches in men is associated with a high risk of central obesity-related health problems.**

• **Hydrodensitometry** • To estimate body density using hydrodensitometry, the person is weighed twice—first on land and then again when submerged under water. Underwater weighing usually generates a good estimate of body fat and is useful in research, although the technique has drawbacks: it requires bulky, expensive, and nonportable equipment. Furthermore, submerging some people (especially those who are very young, very old, ill, or fearful) under water is not always practical.

• **Bioelectrical Impedance** • Bioelectrical impedance provides a means to estimate body composition using portable equipment that incurs minimal

FIGURE E-8 How to Measure Waist Circumference

Place the measuring tape around the abdomen just above the bony crest of the hip. The tape runs parallel to the floor and is snug (but does not compress the skin). The measurement is taken at normal minimal respiration.

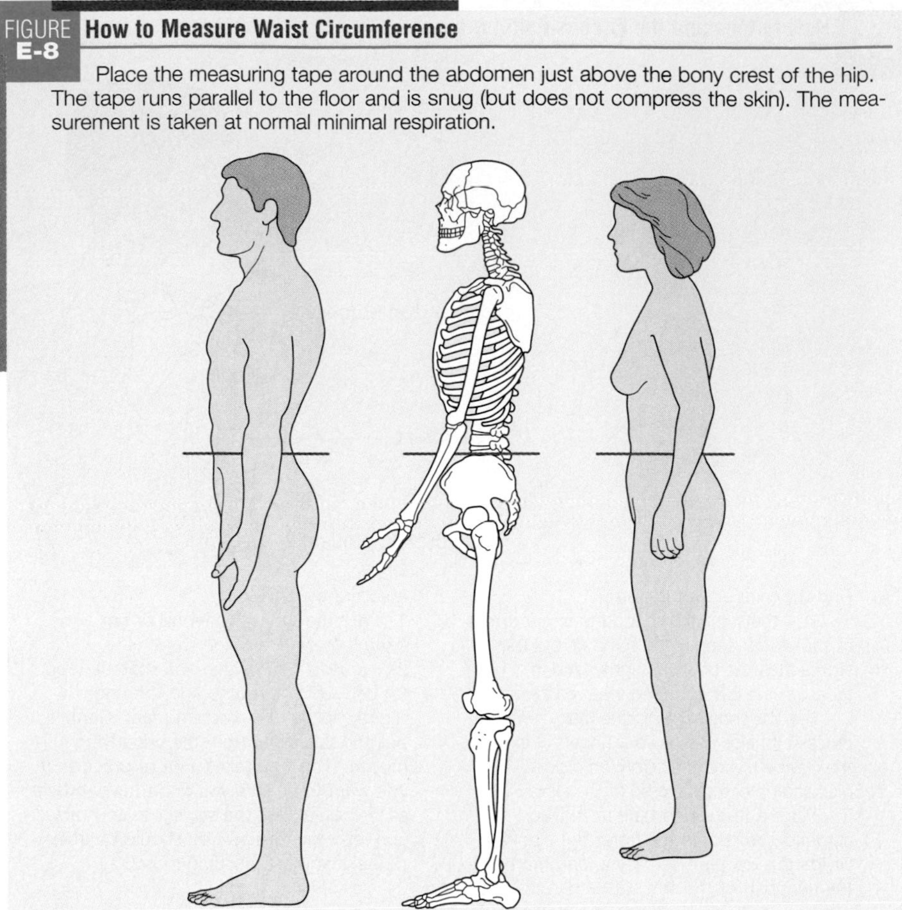

SOURCE: National Institutes of Health Obesity Education Initiative, *Clinical Guidelines on the Identification, Evaluation, and Treatment of Overweight and Obesity in Adults* (Washington, D.C.: U.S. Department of Health and Human Services, 1998), p. 59.

inconvenience to the client. To evaluate body composition using the bioelectrical impedance technique, an electrical current of very low intensity is briefly sent through the body by way of electrodes typically placed on the wrist and ankles. The measurement of electrical resistance is then used in a mathematical equation to estimate the total body water, lean body mass, and body fat. Since electrolyte-containing fluids, which readily conduct electrical current, are found primarily in lean tissues, the leaner the person, the less resistance to the current. To provide reliable results, bioelectrical impedance requires standardized procedures, calibrated instruments, and mathematical equations that are population- and diagnosis-specific.

Clinicians use many other methods to estimate body fat and its distribution. Each has its advantages and disadvantages as Table E-2 summarizes.

Laboratory Tests of Nutrition Status

As Chapter 17 pointed out, blood and urine tests provide valuable information about nutrition status. Table E-3 on p. E-12 shows laboratory tests that help assess vitamin and mineral status.

TABLE E-2 **Methods of Estimating Body Fat and Its Distribution**

Method	Cost	Ease of Use	Accuracy	Measures Fat Distribution
Height and weight	Low	Easy	High	No
Fatfolds	Low	Easy	Low	Yes
Circumferences	Low	Easy	Moderate	Yes
Ultrasound	Moderate	Moderate	Moderate	Yes
Hydrodensitometry	Low	Moderate	High	No
Heavy water tritiated	Moderate	Moderate	High	No
Deuterium oxide, or heavy oxygen	High	Moderate	High	No
Potassium isotope (^{40}K)	Very high	Difficult	High	No
Total body electrical conductivity (TOBEC)	High	Moderate	High	No
Bioelectric impedance (BIA)	Moderate	Easy	High	No
Dual energy X-ray absorptiometry (DEXA)	High	Easy	High	No
Computed tomography (CT)	Very high	Difficult	High	Yes
Magnetic resonance imaging (MRI)	Very high	Difficult	High	Yes

SOURCE: Adapted with permisssion from G. A. Bray, a handout presented at the North American Association for the Study of Obesity and Emory University School of Medicine Conference on Obesity Update: Pathophysiology, Clinical Consequences, and Therapeutic Options, Atlanta, Georgia, August 31–September 2, 1992.

• **Nutrition-Related Anemia** • Anemia, a symptom of a wide variety of nutrition- and nonnutrition-related disorders, is characterized by a reduced number of red blood cells. Iron, folate, and vitamin B_{12} deficiencies caused by inadequate intake, poor absorption, or abnormal metabolism of these nutrients are the most common nutritional anemias. Some nonnutrition-related causes of anemia include massive blood loss, infections, hereditary blood disorders, such as sickle-cell anemia, and chronic liver or kidney disease. Table E-4 on p. E-13 lists laboratory tests that help to define anemia and distinguish among its major nutrition-related causes. Table E-5 on p. E-13 provides standards for tests used to detect iron status (hemoglobin, hematocrit, serum ferritin, total iron binding capacity, serum iron transferrin saturation, erythrocyte protoporphyrin). Table E-6 on p. E-13 shows standards for assessing folate and vitamin B_{12} status.

• **Nitrogen Balance Studies** • Nitrogen balance studies provide information about protein status and helps assessors determine a person's protein requirements. Nitrogen balance studies require that food intake be recorded and urine be collected during the same time period, often for 24 hours. Much effort is wasted if the proper techniques are not conscientiously followed or communications between health care professionals and clients are not clear. Nurses or assistants caring for clients must collect and save all urine samples and record food intake data or instruct clients on how to perform these tasks. Each urine sample is added to a collection container and refrigerated until the collection is complete. If even one urine sample is spilled or discarded, the test is invalid. Furthermore, any protein provided intravenously or by tube must be included as part of the protein intake. Nitrogen intake is calculated from the protein intake, and urine urea nitrogen (UUN) is measured from the urine sample. Nitrogen balance equals the nitrogen intake minus the nitrogen output.

In most cases, protein contains about 16 percent nitrogen. Thus, assessors multiply the protein intake by 16 percent (0.16) or divide by 6.25—the mathematical equivalent—to calculate nitrogen intake. To determine nitrogen output, assessors must consider that in addition to nitrogen lost as UUN, the person also loses nitrogen through the skin, feces, and other sources of urinary nitrogen. To account

TABLE E-3 Biochemical Tests Useful for Assessing Vitamin and Mineral Status

Nutrient	Assessment Tests
Vitamins	
Vitamin A	Serum retinol, retinol-binding protein
Thiamin[a]	Erythrocyte (red blood cell) transketolase activity, erythrocyte thiamin pyrophosphate
Riboflavin[a]	Erythrocyte glutathione reductase activity
Vitamin B_6[a]	Urinary xanthurenic acid excretion after tryptophan load test, erythrocyte transaminase activity, plasma pyridoxal 5′-phosphate (PLP)
Niacin	Plasma or urinary metabolites NMN (N-methyl nicotinamide) or 2-pyridone, or preferably both expressed as a ratio
Folate[b]	Serum folate, erythrocyte folate (reflects liver stores)
Vitamin B_{12}[b]	Serum vitamin B_{12}, serum and urinary methylmalonic acid, Schilling test
Biotin	Urinary biotin, urinary 3-hydroxyisovaleric acid
Vitamin C	Plasma vitamin C[c], leukocyte vitamin C
Vitamin D	Serum vitamin D
Vitamin E	Serum α-tocopherol, erythrocyte hemolysis
Vitamin K	Serum vitamin K, plasma prothrombin; blood-clotting time (prothrombin time) is not an adequate indicator
Minerals	
Phosphorus	Serum phosphate
Sodium	Serum sodium
Chloride	Serum chloride
Potassium	Serum potassium
Magnesium	Serum magnesium, urinary magnesium
Iron	Hemoglobin, hematocrit, serum ferritin, total iron-binding capacity (TIBC), erythrocyte protoporphyrin, serum iron, transferrin saturation
Iodine	Serum thyroxine or thyroid-stimulating hormone (TSH), urinary iodine
Zinc	Plasma zinc, hair zinc
Copper	Erythrocyte superoxide dismutase, serum copper, serum ceruloplasmin
Selenium	Erythrocyte selenium, glutathione peroxidase activity

[a]Urinary measurements for these vitamins are common, but may be of limited use. Urinary measurements reflect recent dietary intakes and may not provide reliable information concerning the severity of a deficiency.
[b]Folate assessments should always be conducted in conjunction with vitamin B_{12} assessments (and vice versa) to help distinguish the cause of common deficiency symptoms.
[c]Vitamin C shifts between the plasma and the white blood cells known as leukocytes; thus a plasma determination may not accurately reflect the body's pool. A measurement of leukocyte vitamin C can provide information about the body's stores of vitamin C. A combination of both tests may be more reliable than either one alone.
SOURCE: Adapted from H. E. Sauberlich, *Laboratory Tests for the Assessment of Nutritional Status* (Boca Raton, Fla.: CRC Press, 1999).

for these losses, the assessor adds 4 grams of nitrogen to the UUN. Thus, nitrogen balance is calculated as follows:

$$\text{Nitrogen balance} = \frac{\text{24-hour protein intake (g)}}{6.25} - [\text{24-hour UUN (g)} + 4 \text{ g}].$$

For example, if a woman's 24-hour protein intake is 90 grams and her 24-hour UUN is 17 grams, nitrogen balance can be calculated as follows:

$$\text{Nitrogen balance} = \frac{90 \text{ g protein}}{6.25} - (17 + 4)$$

$$\text{Nitrogen balance} = 14.4 \text{ g} - 21 \text{ g} = -6.6 \text{ g}.$$

The woman in this example is in negative nitrogen balance; that is, she is losing about 6.5 grams of nitrogen more than she is consuming.

E

TABLE E-4 Laboratory Tests Useful in Evaluating Nutrition-Related Anemias

Test or Test Result	What It Reflects
General Tests for Anemia	
Hemoglobin (Hg)	Total amount of hemoglobin in the red blood cells (RBC)
Hematocrit (Hct)	Percentage of RBC in the total blood volume
Red blood cell (RBC) count	Number of RBC
Mean corpuscular volume (MCV)	RBC size; helps to determine if anemia is microcytic (iron deficiency) or macrocytic (folate or vitamin B_{12} deficiency)
Mean corpuscular hemoglobin concentration (MCHC)	Hemoglobin concentration within the average RBC; helps to determine if anemia is hypochromic (iron deficiency) or normochromic (folate or vitamin B_{12} deficiency)
Bone marrow aspiration	The manufacture of blood cells in different developmental states
Iron Deficiency	
↓ Serum ferritin	Early deficiency state (iron stores diminish)
↓ Transferrin saturation	Progressing deficiency state (transport iron decreases)
↑ Erythrocyte protoporphyrin	Later deficiency state (hemoglobin production falters)
Folate Deficiency	
↓ Serum folate	Progressing deficiency state
↓ RBC folate	Later deficiency state
Vitamin B_{12} Deficiency	
↓ Serum vitamin B_{12}	Progressing deficiency state
Schilling test	Absorption of vitamin B_{12}

TABLE E-5 Criteria for Assessing Iron Status

Test	Age (yr)	Gender	Deficiency Value
Hemoglobin (g/dL)	0.5–10	M–F	<11
	11–15	M	<12
		F	<11.5
	>15	M	<13
		F	<12
	Pregnancy		<11
Hematocrit (%)	0.5–4	M–F	<32
	5–10	M–F	<33
	11–15	M	<35
		F	<34
	>15	M	<40
		F	<36
Serum ferritin (μg/L)	0.5–15	M–F	<10
	>15	M–F	<12
Total iron-binding capacity (μg/dL)	>15	M–F	>400
Serum iron (μg/dL)	>15	M–F	<60
Transferrin saturation (%)	0.5–4	M–F	<12
	5–10	M–F	<14
	>10	M–F	<16
Erythrocyte protoporphyrin (μg/dL RBC)	0.5–4	M–F	>80
	>4	M–F	>70

TABLE E-6 Criteria for Assessing Folate and Vitamin B_{12}

	Deficient	Borderline	Acceptable
Serum folate (ng/ml)[a]	<3.0	3.0–5.9	>6.0
Erythrocyte folate (ng/ml)[a]	<140	140–159	>160
Serum vitamin B_{12} (pg/ml)	<150	150–200	≥201
Serum methylmalonic acid (nmol/L)	<376	—	—

[a]To convert folate values (ng/ml) to international standard units (nmol/L), multiply by 2.266.

Note: A nanogram (ng) is one-billionth of a gram; a picogram (pg) is one-trillionth of a gram.

NUTRITION RESOURCES

CONTENTS

F

People interested in nutrition often want to know where they can find reliable nutrition information. Wherever you live, there are several sources you can turn to:

- The Department of Health may have a nutrition expert.
- The local extension agent is often an expert.
- The food editor of your local paper may be well informed.
- The dietitian at the local hospital had to fulfill a set of qualifications before he or she became an RD (see Highlight 1).
- There may be knowledgeable professors of nutrition or biochemistry at a nearby college or university.

In addition, you may be interested in building a nutrition library of your own. Books you can buy, journals you can subscribe to, and addresses you can contact for general information are given below.

Books

For students seeking to establish a personal library of nutrition references, the authors of this text recommend the following books:

- *Present Knowledge in Nutrition,* 7th ed. (Washington, D.C.: International Life Sciences Institute—Nutrition Foundation, 1996).

This 646-page paperback has a chapter on each of 64 topics, including energy, obesity, each of the nutrients, several diseases, malnutrition, growth and its assessment, immunity, alcohol, fiber, exercise, drugs, and toxins. Watch for an update; new editions come out every few years.

- M. E. Shils and coeditors, *Modern Nutrition in Health and Disease,* 9th ed. (Baltimore: Williams & Wilkins, 1999).

This reference book contains encyclopedic articles on the nutrients, foods, diet, metabolism, malnutrition, age-related needs, and nutrition in disease.

- Committee on Dietary Reference Intakes, *Dietary Reference Intakes for Calcium, Phosphorus, Magnesium, Vitamin D, and Fluoride* (Washington, D.C.: National Academy Press, 1997).
- Committee on Dietary Reference Intakes, *Dietary Reference Intakes for Thiamin, Riboflavin, Niacin, Vitamin B_6, Folate, Vitamin B_{12}, Pantothenic Acid, Biotin, and Choline* (Washington, D.C.: National Academy Press, 1998).
- Committee on Dietary Reference Intakes, *Dietary Reference Intakes for Vitamin C, Vitamin E, Selenium, and Carotenoids* (Washington, D.C.: National Academy Press, 2000).
- Committee on Dietary Reference Intakes, *Dietary Reference Intakes for Vitamin A, Vitamin K, Arsenic, Boron, Chromium, Copper, Iodine, Iron, Manganese, Molybdenum, Nickel, Silicon, Vanadium, and Zinc* (Washington, D.C.: National Academy Press, 2001).

These reports review the function of each nutrient, dietary sources, and deficiency and toxicity symptoms as well as provide recommendations for intakes. Watch for additional reports on the Dietary Reference Intakes for the remaining nutrients and other food components.

- Committee on Diet and Health, *Diet and Health Implications for Reducing Chronic Disease Risk* (Washington, D.C.: National Academy Press, 1989).

This 749-page book presents the integral relationship between diet and chronic disease prevention. Its nutrient chapters provide evidence on how diet influences disease development, and its disease chapters review the dietary patterns implicated in each chronic disease.

- S. S. Gropper, *The Biochemistry of Human Nutrition: A Desk Reference,* 2nd ed. (Belmont, Calif.: Wadsworth/Thomson Learning, 2000).

This 263-page paperback presents the biochemical concepts necessary for an understanding of nutrition. It is a handy reference book for those who need a refresher in the basics of biochemistry or for those who are learning biochemistry for the first time.

We also recommend two of our own books that explore current topics in nutrition, health, and the life span:

- S. R. Rolfes, L. K. DeBruyne, and E. N. Whitney, *Life Span Nutrition: Conception through Life,* 2nd ed. (Belmont, Calif.: West/Wadsworth, 1998).
- E. N. Whitney, C. B. Cataldo, L. K. DeBruyne, and S. R. Rolfes, *Nutrition for Health and Health Care,* 2nd ed. (Belmont, Calif.: Wadsworth/Thomson Learning, 2001).

Journals

Nutrition Today is an excellent magazine for the interested layperson. It makes a point of raising controversial issues and providing a forum for conflicting opinions. Six issues per year are published. Order from Williams & Wilkins, 12107 Insurance Way, Hagerstown, MD 21740.

The *Journal of the American Dietetic Association,* the official publication of the ADA, contains articles of interest to dietitians and nutritionists, news of legislative action on food and nutrition, and a very useful section of abstracts of articles from many other journals of nutrition and related areas. There are 12 issues per year, available from the American Dietetic Association (see "Addresses," later).

Nutrition Reviews, a publication of the International Life Sciences Institute, does much of the work for the library researcher, compiling recent evidence on current topics and presenting extensive bibliographies. Twelve issues per year are available from Nutrition Reviews, P.O. Box 1897, Lawrence, KS 66044-8897.

Nutrition and the M.D. is a monthly newsletter that provides up-to-date, easy-to-read, practical information on nutrition for health care providers. It is available from Lippincott-Williams & Wilkens, 16522 Hunters Green Parkway, Hagerstown, MD 21740.

Other journals that deserve mention here are *Food Technology, Journal of Nutrition, American Journal of Clinical Nutrition, Nutrition Research,* and *Journal of Nutrition Education. FDA Consumer,* a government publication with many articles of interest to the consumer, is available from the Food and Drug Administration (see "Addresses," below). Many other journals of value are referred to throughout this book.

Addresses

Many of the organizations listed below will provide publication lists free on request. Government and international agencies and professional nutrition organizations are listed first, followed by organizations in the following areas: aging, alcohol and drug abuse, consumer organizations, fitness, food safety, health and disease, infancy and childhood, pregnancy and lactation, trade and industry organizations, weight control and eating disorders, and world hunger.

U.S. Government

- Federal Trade Commission (FTC)
  Public Reference Branch
  600 Pennsylvania Avenue NW
  Washington, DC 20580
  (202) 326-2222
  www.ftc.gov
- Food and Drug Administration (FDA)
  Office of Consumer Affairs, HFE 1
  Room 16-85
  5600 Fishers Lane
  Rockville, MD 20857
  (301) 443-1726
  www.fda.gov
- FDA Consumer Information Line
  (888) INFO-FDA
  (888) 463-6332
- FDA Center for Food Safety & Applied Nutrition, HFS 150
  200 C Street SW
  Washington, DC 20204
  (202) 205-4561; fax (202) 205-4564
  www.cfsan.fda.gov
- Food and Nutrition Information Center
  National Agricultural Library, Room 304
  10301 Baltimore Avenue
  Beltsville, MD 20705-2351
  fax (301) 504-6409
  www.nal.usda.gov/fnic
- Superintendent of Documents
  U.S. Government Printing Office
  Washington, DC 20402
  (202) 512-1530
  www.access.gpo.gov/su_docs
- U.S. Department of Agriculture (USDA)
  14th Street and Independence Avenue SW
  Washington, DC 20250
  (202) 720-2791
  www.fns.usda.gov/fncs
- USDA Center for Nutrition Policy and Promotion (Dietary Guidelines and Food Guide Pyramid)
  1120 20th Street NW, Suite 200
  North Lobby
  Washington, DC 20036
  (800) 687-2258 or (202) 418-2312
  www.cnpp.usda.gov
- USDA Food Safety and Inspection Service
  Food Safety Education Office
  1400 Independence Avenue SW,
  Room 2932-S
  Washington, DC 20250
  (202) 690-0351
  www.usda.gov/fsis
- U.S. Department of Education (DOE)
  Accreditation and State Liaison
  Accrediting Agency Evaluation
  1990 K Street NW, #7105
  Washington, DC 20006-8509
  (202) 219-7011
  www.ed.gov/offices/OPE/accreditation
- U.S. Department of Health and Human Services
  200 Independence Avenue SW
  Washington, DC 20201
  (202) 619-0257
  www.os.dhhs.gov
- U.S. Environmental Protection Agency (EPA)
  1200 Pennsylvania Avenue NW
  Washington, DC 20460
  (202) 260-2090
  www.epa.gov
- Assistant Secretary of Health
  Office of Public Health and Science
  Department of Health and Human Sciences
  200 Independence Avenue SW,
  Room 725-H
  Washington, DC 20201
  (202) 690-7694

Canadian Government

Federal

- Bureau of Nutritional Sciences
  Food Directorate
  Health Protection Branch
  3-West
  Sir Frederick Banting Research Centre, AL0904A
  Tunney's Pasture
  Ottawa, Ontario K1A 0K9
  (613) 957-2991
  fax (613) 941-5366
  www.hc-sc.gc.ca
- Canadian Food Inspection Agency
  Agriculture and Agri-Food Canada
  59 Camelot Drive
  Nepean, Ontario K1A 0Y9
  (613) 225-CFIA or (613) 225-2342
  fax (613) 228-6653
  www.agr.ca

- Office of Nutrition Policy and Promotion
  Health Products and Food Branch
  AL 90761
  7th Floor—Jeanne Mance Bldg.
  Tunney's Pasture
  Ottawa, Ontario K1A 1B4
  www.hc-sc.gc.ca/hppb/nutrition

Provincial and Territorial

- Population Health Strategies Branch
  Alberta Health
  23rd Floor, TELUS Plaza, North Tower
  10025 Jasper Avenue
  Edmonton, AB T5J 2N3
- Prevention and Health Promotion Strategies
  Ministry of Health
  1520 Blanshard Street, 2nd Floor
  Victoria BC V8W 3C8
- Executive Director
  Health Programs
  Room 2094-300
  Carlton Street
  Winnipeg, MB R3B 3M9
- Project Manager
  Public Health Management Services
  Health and Community Services
  P.O. Box 5100
  520 King Street
  Fredericton, NB E3B 5G8
- Director, Health Promotion
  Department of Health
  Government of Newfoundland and Labrador
  P.O. Box 8700
  Confederation Building, West Block
  St. John's, NF A1B 4J6
- Consultant, Nutrition
  Health & Wellness Promotion
  Population Health
  Department of Health and Social Services
  Government of the Northwest Territories
  Centre Square Tower, 6th Floor
  P.O. Box 1320
  Yellowknife, NT X1A 2L9
- Public Health Nutritionist
  Capital District Health Authority
  201 Brownlow Avenue, Unit 4
  Dartmouth, NS B3B 1W2
- Senior Consultant, Nutrition
  Public Health Branch
  Ministry of Health, 8th Floor
  5700 Yonge St.
  North York, ON M2M 4K5
- Coordinator, Health Information Resource Centre
  Department of Health and Social Services
  1 Rochford Street, Box 2000
  Charlottetown, PEI C1A 7N8
- Responsables de la santé cardio-vasculaire et de la nutrition
  Ministère de la Santé et des Services sociaux, Service de la Prévention en Santé
  3e étage
  1075, chemin Sainte-Foy
  Quèbec (Quèbec) G1S 2M1
- Health Promotion Unit
  Population Health Branch
  Saskatchewan Health
  3475 Albert Street
  Regina, SK S4S 6X6
- Director, Nutrition Services
  Yukon Hospital Corporation
  #5 Hospital Road
  Whitehorse, YT Y1A 3H7

International Agencies

- Food and Agriculture Organization of the United Nations (FAO)
  Liaison Office for North America
  2175 K Street, Suite 300
  Washington, DC 20437
  (202) 653-2400; fax (202) 653-5760
  www.fao.org
- International Food Information Council Foundation
  1100 Connecticut Avenue NW, Suite 430
  Washington, DC 20036
  (202) 296-6540
  ificinfo.health.org
- UNICEF
  3 United Nations Plaza
  New York, NY 10017
  (212) 326-7000
  www.unicef.org
- World Health Organization (WHO) Regional Office
  525 23rd Street NW
  Washington, DC 20037
  (202) 974-3000
  www.who.org

Professional Nutrition Organizations

- American Society of Nutritional Sciences
  9650 Rockville Pike
  Bethesda, MD 20814
  (301) 530-7050; fax (301) 571-1892
  www.nutrition.org
- American Dietetic Association (ADA)
  216 West Jackson Boulevard, Suite 800
  Chicago, IL 60606-6995
  (800) 877-1600; (312) 899-0040
  www.eatright.org
- ADA, Consumer Nutrition Hotline
  (800) 366-1655
- American Society for Clinical Nutrition
  9650 Rockville Pike
  Bethesda, MD 20814-3998
  (301) 530-7110; fax (301) 571-1863
  www.faseb.org/ascn
- Dietitians of Canada
  480 University Avenue, Suite 604
  Toronto, Ontario, Canada M5G 1V2
  (416) 596-0857; fax (416) 596-0603
  www.dietitians.ca
- International Life Sciences Institute
  1126 Sixteenth Street NW
  Washington, DC 20036
  (202) 659-0074
  www.ilsi.org
- National Academy of Sciences/ National Research Council (NAS/NRC)
  2101 Constitution Avenue, NW
  Washington, DC 20418
  (202) 334-2000
  www.nas.edu
- National Institute of Nutrition
  265 Carling Avenue, Suite 302
  Ottawa, Ontario K1S 2E1
  (613) 235-3355; fax (613) 235-7032
  www.nin.ca

- Society for Nutrition Education
  1001 Connecticut Avenue NW,
  Suite 528
  Washington, DC 20036
  (202) 452-8534; fax (202) 452-8536
  www.sne.org

F

Aging

- Administration on Aging
  330 Independence Avenue SW
  Washington, DC 20201
  (202) 619-0724
  www.aoa.dhhs.gov
- American Association of Retired Persons (AARP)
  601 E Street NW
  Washington, DC 20049
  (800) 424-3410
  www.aarp.org
- National Aging Information Center
  330 Independence Avenue SW
  Washington, DC 20201
  (202) 619-7501
  www.aoa.dhhs.gov/naic
- National Institute on Aging
  Public Information Office
  31 Center Drive, MSC 2292
  Bethesda, MD 20892
  (800) 222-2225
  www.nih.gov/nia

Alcohol and Drug Abuse

- Al-Anon Family Groups, Inc.
  1600 Corporate Landing Parkway
  Virginia Beach, VA 23454-5617
  (888) 4AL-ANON or (888) 425-2666
  (757) 563-1600; fax (757) 563-1655
  www.al-anon.alateen.org
- Alcohol & Drug Abuse Information Line
  Adcare Hospital
  (800) 252-6465
- Alcoholics Anonymous (AA)
  Grand Central Station
  P.O. Box 459
  New York, NY 10163
  (212) 870-3400
  www.alcoholics-anonymous.org
- Narcotics Anonymous (NA)
  P.O. Box 9999
  Van Nuys, CA 91409
  (818) 773-9999; fax (818) 700-0700
  www.wsoinc.com
- National Clearinghouse for Alcohol and Drug Information (NCADI)
  P.O. Box 2345
  Rockville, MD 20847-2345
  (800) 729-6686
  www.health.org
- National Council on Alcoholism and Drug Dependence (NCADD)
  12 West 21st Street
  New York, NY 10010
  (800) NCA-CALL or (800) 622-2255
  (212) 206-6770; fax (212) 645-1690
  www.ncadd.org
- U.S. Center for Substance Abuse Prevention
  1010 Wayne Avenue, Suite 850
  Silver Spring, MD 20910
  (301) 459-1591 ext. 244
  fax (301) 495-2919
  www.covesoft.com/csap.html

Consumer Organizations

- Center for Science in the Public Interest (CSPI)
  1875 Connecticut Avenue NW,
  Suite 300
  Washington, DC 20009-5728
  (202) 332-9110; fax (202) 265-4954
  www.cspinet.org
- Choice in Dying, Inc.
  1035 30th Street NW
  Washington, DC 20007
  (800) 989-WILL or (800) 989-9455
  (202) 338-9790; fax (202) 338-0242
  www.choices.org
- Consumer Information Center
  Pueblo, CO 81009
  (888) 8 PUEBLO or (888) 878-3256
  www.pueblo.gsa.gov
- Consumers Union of US Inc.
  101 Truman Avenue
  Yonkers, NY 10703-1057
  (914) 378-2000
  www.consumersunion.org
- National Council Against Health Fraud, Inc. (NCAHF)
  P.O. Box 141
  Fort Lee, NJ 07024
  (212) 723-2955
  www.ncahf.org

Fitness

- American College of Sports Medicine
  401 West Michigan Street
  Indianapolis, IN 46206-1440
  (317) 637-9200
  www.acsm.org
- American Council on Exercise (ACE)
  5820 Oberlin Drive, Suite 102
  San Diego, CA 92121
  (800) 825-3636
  www.acefitness.org
- Shape Up America!
  6707 Democracy Boulevard,
  Suite 306
  Bethesda, MD 20817
  (301) 493-5368
  www.shapeup.org

Food Safety

- Alliance for Food & Fiber
  Food Safety Hotline
  (800) 266-0200
- FDA Center for Food Safety and Applied Nutrition Outreach and Information
  200 C Street SW
  Washington, DC 20204
  www.cfsan.fda.gov
- National Lead Information Center
  (800) LEAD-FYI or (800) 532-3394
  (800) 424-LEAD or (800) 424-5323
- National Pesticide Telecommunications Network (NPTN)
  Oregon State University
  333 Weniger Hall
  Corvallis, OR 97331-6502
  (800) 858-7378
  www.ace.orst.edu/info/nptn
- USDA Meat and Poultry Hotline
  (800) 535-4555
- U.S. EPA Safe Drinking Water Hotline
  (800) 426-4791

Health and Disease

- Alzheimer's Disease Education and Referral Center
  P. O. Box 8250
  Silver Spring, MD 20907-8250
  (800) 438-4380
  www.alzheimers.org
- Alzheimer's Disease Information and Referral Service
  919 North Michigan Avenue,
  Suite 1100

Chicago, IL 60611
(800) 272-3900
www.alz.org

- American Academy of Allergy, Asthma, and Immunology
611 East Wells Street
Milwaukee, WI 53202
(414) 272-6071; fax (414) 272-6070
www.aaaai.org
- American Cancer Society
National Home Office
1599 Clifton Road NE
Atlanta, GA 30329-4251
(800) ACS-2345 or (800) 227-2345
www.cancer.org
- American Council on Science and Health
1995 Broadway, 2nd Floor
New York, NY 10023-5860
(212) 362-7044; fax (212) 362-4919
www.acsh.org
- American Dental Association
211 East Chicago Avenue
Chicago, IL 60611
(312) 440-2500; fax (312) 440-2800
www.ada.org
- American Diabetes Association
1701 North Beauregard Street
Alexandria, VA 22311
(800) 232-3472 or (703) 549-1500
www.diabetes.org
- American Heart Association
Box BHG, National Center
7272 Greenville Avenue
Dallas, TX 75231
(800) 242-8721
www.americanheart.org
- American Institute for Cancer Research
1759 R Street NW
Washington, DC 20009
(800) 843-8114 or (202) 328-7744
fax (202) 328-7226
www.aicr.org
- American Medical Association
515 North State Street
Chicago, IL 60610
(312) 464-5000
www.ama-assn.org
- American Public Health Association (APHA)
800 I Street NW
Washington, DC 20001-3710
(282) 777-2742
www.apha.org
- American Red Cross
National Headquarters
8111 Gatehouse Road
Falls Church, VA 22042
(703) 206-8143
www.redcross.org
- Arthritis Foundation
(800) 283-7800
www.arthritis.org
- Canadian Diabetes Association
15 Toronto Street, Suite 800
Toronto, ON M5C 2E3
(800) BANTING or (800) 226-8464
(416) 363-3373
www.diabetes.ca
- Canadian Public Health Association
400-1565 Carling Avenue
Ottawa, Ontario K1Z 8R1
(613) 725-3769; fax (613) 725-9826
www.cpha.ca
- Centers for Disease Control and Prevention (CDC)
1600 Clifton Road NE
Atlanta, GA 30333
(404) 639-3311
www.cdc.gov
- The Food Allergy Network
10400 Eaton Place, Suite 107
Fairfax, VA 22030-2208
(800) 929-4040 or (703) 691-2713
www.foodallergy.org
- Internet Health Resources
www.ihr.com
- Mayo Clinic Health Oasis
www.mayohealth.org
- National AIDS Hotline (CDC)
(800) 342-AIDS (English)
(800) 344-SIDA (Spanish)
(800) 2437-TTY (Deaf)
(900) 820-2437
- National Cancer Institute
Office of Cancer Communications
Building 31, Room 10A31
31 Center Drive MSC 2580
Bethesda, MD 20892
(800) 4-CANCER or
(800) 422-6237
www.nci.nih.gov
- National Diabetes Information Clearinghouse
31 Center Drive MSC 2560
Bethesda, MD 20892-2560
(301) 654-3327
www.niddk.nih.gov
- National Digestive Disease Information Clearinghouse (NDDIC)
31 Center Drive MSC 2560
Bethesda, MD 20892-2560
(301) 654-3810
www.niddk.nih.gov
- National Health Information Center (NHIC)
P.O. Box 1133
Washington, DC 20013
(800) 336-4797
nhic-nt.health.org
- National Heart, Lung, and Blood Institute Information Center
P.O. Box 30105
Bethesda, MD 20824-0105
(301) 592-8573
www.nhlbi.nih.gov
- National Institute of Allergy and Infectious Diseases
Office of Communications
Building 31, Room 7A50
31 Center Drive, MSC 2520
Bethesda, MD 20892-2520
(301) 496-5717
www.niaid.nih.gov
- National Institute of Dental Research (NIDR)
National Institute of Health
Bethesda, MD 20892-2190
www.nidr.nih.gov
- National Institutes of Health (NIH)
9000 Rockville Pike
Bethesda, MD 20892
(301) 496-2433
www.nih.gov
- National Osteoporosis Foundation
1232 22 Street NW
Washington, DC 20037
(202) 223-2226
www.nof.org
- Office of Disease Prevention and Health Promotion
odphp.osophs.dhhs.gov
- Osteoporosis and Related Bone Diseases
(800) 624-BONE or (800) 624-2663
www.osteo.org

Infancy and Childhood

- American Academy of Pediatrics
141 Northwest Point Boulevard
Elk Grove Village, IL 60007-1098

F

F

(847) 434-4000; fax (847) 434-8000
www.aap.org

- Birth Defect Research for Children
930 Woodcock Road, Suite 225
Orlando, FL 32803
(407) 895-0802; fax (407) 895-0824
www.birthdefects.org
- Canadian Paediatric Society
100-2204 Walkley Road
Ottawa, ON K1G 4G8
(613) 526-9397; fax (613) 526-3332
www.cps.ca
- National Center for Education in Maternal & Child Health
2000 15th Street North, Suite 701
Arlington, VA 22201-2617
(703) 524-7802
www.ncemch.org

Pregnancy and Lactation

- American College of Obstetricians and Gynecologists Resource Center
409 12th Street SW
Washington, DC 20090
(202) 638-5577
www.acog.org
- La Leche International, Inc.
1400 N. Meacham Road
Schaumburg, IL 60173
(847) 519-7730
www.lalecheleague.org
- March of Dimes Birth Defects Foundation
1275 Mamaroneck Avenue
White Plains, NY 10605
(888) MoDimes or (888) 663-4637
www.modimes.org

Trade and Industry Organizations

- Beech-Nut Nutrition Corporation
100 South 4th Street
St. Louis, MO 63102
(800) 523-6633
www.beechnut.com
- Borden Foods Nutrition Department
180 East Broad Street
Columbus, OH 43215
(800) 426-7336
- Campbell Soup Company
Consumer Response Center
Campbell Place, Box 26B
Camden, NJ 08103-1701
(800) 257-8443
www.campbellssoup.com
- General Mills, Inc.
Number One General Mills Boulevard
Minneapolis, MN 55440
(800) 328-6787
www.generalmills.com
- Kellogg Company
P.O. Box 3599
Battle Creek, MI 49016-3599
(800) 962-1413
www.kelloggs.com
- Kraft Foods
Consumer Response and Information Center
One Kraft Court
Glenview, IL 60025
(800) 323-0768
www.kraftfoods.com
- Mead Johnson Nutritionals
2400 West Lloyd Expressway
Evansville, IN 47721
(800) 247-7893
www.meadjohnson.com
- Nabisco Consumer Affairs
100 DeForest Avenue
East Hanover, NJ 07936
(800) NABISCO or (800) 932-7800
www.nabisco.com
- National Dairy Council
10255 West Higgins Road, Suite 900
Rosemond, IL 60018-5616
(847) 803-2000
www.dairyinfo.com
- NutraSweet/KELCO
P.O. Box 2986
Chicago, IL 60654-0986
www.equal.com
- Pillsbury Company
2866 Pillsbury Center
Minneapolis, MN 55402
(800) 767-4466
www.pillsbury.com
- Procter and Gamble Company
One Procter and Gamble Plaza
Cincinnati, OH 45202
(513) 983-1100
www.pg.com/info
- Ross Laboratories, Abbot Laboratory
625 Cleveland Avenue
Columbus, OH 43215
(800) 227-5767
www.abbot.com
- Sunkist Growers
Consumer Affairs
Fresh Fruit Division
14130 Riverside Drive
Sherman Oaks, CA 91423
www.sunkist.com
- United Fresh Fruit and Vegetable Association
727 North Washington Street
Alexandria, VA 22314
(703) 836-3410
- USA Rice Federation
4301 North Fairfax Drive, Suite 305
Arlington, VA 22203
Phone: (703) 351-8161
www.usarice.com

Weight Control and Eating Disorders

- American Anorexia & Bulimia Association, Inc.
165 West 46th Street #1108
New York, NY 10036
(212) 575-6200
www.aabaine.org
- American Obesity Association
1250 24 Street NW, Suite 300
Washington, DC 20037
(800) 98-OBESE or (800) 986-2373
www.obesity.org
- Anorexia Nervosa and Related Eating Disorders (ANRED)
P.O. Box 5102
Eugene, OR 97405
(541) 344-1144
www.anred.com
- National Association of Anorexia Nervosa and Associated Disorders, Inc. (ANAD)
P.O. Box 7
Highland Park, IL 60035
(847) 831-3438; fax (847) 433-4632
www.anad.org
- National Eating Disorder Information Centre
Toronto General Hospital
200 Elizabeth Street, CW 1-211
Toronto, Ontario M5G 2C4
(416) 340-4156; fax (416) 340-4736
www.nedic.ca
- National Institute of Diabetes and Digestive Diseases Weight-Control

Information Network (WIN)
31 Center Drive MSC 2560
Bethesda, MD 20892
(800) WIN-8098
www.niddk.nih.gov/health/nutrit/win.htm

- Overeaters Anonymous (OA)
World Service Office
6075 Zenith Court NE
Rio Rancho, NM 87124
(505) 891-2664; fax (505) 891-4320
www.overeatersanonymous.org
- TOPS (Take Off Pounds Sensibly)
4575 South Fifth Street
P.O. Box 07360
Milwaukee, WI 53207-0360
(800) 932-8677 or (414) 482-4620
www.tops.org
- Weight Watchers International, Inc.
Consumer Affairs Department/IN
175 Crossways Park West
Woodbury, NY 11797
(800) 651-6000; fax (516) 390-1632
www.weightwatchers.com

World Hunger

- Bread for the World
50 F Street NW, Suite 500
Washington, DC 20001
(800) 82-BREAD or (800) 822-7323
(202) 639-9400; fax (202) 639-9401
www.bread.org
- Center on Hunger, Poverty and Nutrition Policy
Tufts University School of Nutrition
11 Curtis Avenue
Medford, MA 02155
(617) 627-3956
- Freedom from Hunger
P.O. Box 2000
1644 DaVinci Court
Davis, CA 95616
(530) 758-6241
www.freefromhunger.org
- Oxfam America
26 West Street
Boston, MA 02111-1206
(800) 77-OXFAM or (800) 776-9326
www.oxfamamerica.org
- Worldwatch Institute
1776 Massachusetts Avenue NW
Washington, DC 20036
(202) 452-1999
www.worldwatch.org

UNITED STATES: EXCHANGE LISTS

◆ Appendix I presents Canada's Choice System for Meal Planning.

Chapter 2 introduced the exchange system.◆ This appendix provides additional details. The exchange system groups together foods that have about the same amount of carbohydrate, protein, fat, and kcalories. Then any food on a list can be "exchanged" for any other on that same list (see Tables G-1 through G-9).

TABLE G-1 U.S. Exchange System: Starch List

1 starch exchange = 15 g carbohydrate, 3 g protein, 0–1 g fat, and 80 kcal
Note: In general, a starch serving is ½ c cereal, grain, pasta, or starchy vegetable; 1 oz of bread; ¾ to 1 oz snack food.

Serving Size	Food
Bread	
½ (1 oz)	Bagels
2 slices (1½ oz)	Bread, reduced-kcalorie
1 slice (1 oz)	Bread, white (including French and Italian), whole-wheat, pumpernickel, rye
2 (⅔ oz)	Bread sticks, crisp, 40 x ½″
½	English muffins
½ (1 oz)	Hot dog or hamburger buns
½	Pita, 6″ across
1 (1 oz)	Plain rolls, small
1 slice (1 oz)	Raisin bread, unfrosted
1	Tortillas, corn, 6″ across
1	Tortillas, flour, 7–8″ across
1	Waffles, 4½″ square, reduced-fat
Cereals and Grains	
½ c	Bran cereals
½ c	Bulgur, cooked
½ c	Cereals, cooked
¾ c	Cereals, unsweetened, ready-to-eat
3 tbs	Cornmeal (dry)
⅓ c	Couscous
3 tbs	Flour (dry)
¼ c	Granola, low-fat
¼ c	Grape nuts
½ c	Grits, cooked
½ c	Kasha
¼ c	Millet
¼ c	Muesli
½ c	Oats
½ c	Pasta, cooked
1½ c	Puffed cereals
½ c	Rice milk
⅓ c	Rice, white or brown, cooked
½ c	Shredded wheat
½ c	Sugar-frosted cereal
3 tbs	Wheat germ
Starchy Vegetables	
⅓ c	Baked beans
½ c	Corn
1 (5 oz)	Corn on cob, medium
1 c	Mixed vegetables with corn, peas, or pasta
½ c	Peas, green
½ c	Plantains
1 small (3 oz)	Potatoes, baked or boiled
½ c	Potatoes, mashed
1 c	Squash, winter (acorn, butternut)
½ c	Yams, sweet potatoes, plain
Crackers and Snacks	
8	Animal crackers
3	Graham crackers, 2½″ square
¾ oz	Matzoh
4 slices	Melba toast
24	Oyster crackers
3 c	Popcorn (popped, no fat added or low-fat microwave)
¾ oz	Pretzels
2	Rice cakes, 4″ across
6	Saltine-type crackers
15–2″ (¾ oz)	Snack chips, fat-free (tortilla, potato)
2–5 (¾ oz)	Whole-wheat crackers, no fat added
Dried Beans, Peas, and Lentils	
½ c	Beans and peas, cooked (garbanzo, lentils, pinto, kidney, white, split, black-eyed)
⅔ c	Lima beans
3 tbs	Miso 🧂
Starchy Foods Prepared with Fat	
Count as 1 starch + 1 fat exchange.	
1	Biscuit, 2½″ across
½ c	Chow mein noodles
1 (2 oz)	Cornbread, 2″ cube
6	Crackers, round butter type
1 c	Croutons
16–25 (3 oz)	French-fried potatoes
¼ c	Granola
1 (1½ oz)	Muffin, small
2	Pancake, 4″ across
3 c	Popcorn, microwave
3	Sandwich crackers, cheese or peanut butter filling
⅓ c	Stuffing, bread (prepared)
2	Taco shell, 6″ across
1	Waffle, 4½″ square
4–6 (1 oz)	Whole-wheat crackers, fat added

🧂 = 400 mg or more of sodium per serving.

G

TABLE G-2 U.S. Exchange System: Fruit List

1 fruit exchange = 15 g carbohydrate and 60 kcal

Note: In general, a fruit serving is 1 small to medium fresh fruit; ½ c canned or fresh fruit or fruit juice; ¼ c dried fruit.

Serving Size	Food
1 (4 oz)	Apples, unpeeled, small
½ c	Applesauce, unsweetened
4 rings	Apples, dried
4 whole (5½ oz)	Apricots, fresh
8 halves	Apricots, dried
½ c	Apricots, canned
1 (4 oz)	Bananas, small
¾ c	Blackberries
¾ c	Blueberries
⅓ melon (11 oz) or 1 c cubes	Cantaloupe, small
12 (3 oz)	Cherries, sweet, fresh
½ c	Cherries, sweet, canned
3	Dates
1½ large or 2 medium (3½ oz)	Figs, fresh
1½	Figs, dried
½ c	Fruit cocktail
½ (11 oz)	Grapefruit, large
¾ c	Grapefruit sections, canned
17 (3 oz)	Grapes, small
1 slice (10 oz) or 1 c cubes	Honeydew melon
1 (3½ oz)	Kiwi
¾ c	Mandarin oranges, canned
½ (5½ oz) or ½ c	Mangoes, small
1 (5 oz)	Nectarines, small
1 (6½ oz)	Oranges, small
½ (8 oz) or 1 c cubes	Papayas
1 (6 oz)	Peaches, medium, fresh
½ c	Peaches, canned
½ (4 oz)	Pears, large, fresh
½ c	Pears, canned
¾ c	Pineapple, fresh
½ c	Pineapple, canned
2 (5 oz)	Plums, small
½ c	Plums, canned
3	Prunes, dried
2 tbs	Raisins
1 c	Raspberries
1¼ c whole berries	Strawberries
2 (8 oz)	Tangerines, small
1 slice (13½ oz) or 1¼ c cubes	Watermelon
Fruit Juice	
½ c	Apple juice/cider
⅓ c	Cranberry juice cocktail
1 c	Cranberry juice cocktail, reduced-kcalorie
⅓ c	Fruit juice blends, 100% juice
⅓ c	Grape juice
½ c	Grapefruit juice
½ c	Orange juice
½ c	Pineapple juice
⅓ c	Prune juice

TABLE G-3 U.S. Exchange System: Milk List

Serving Size	Food
Nonfat and Low-Fat Milk	
1 nonfat/low-fat milk exchange = 12 g carbohydrate, 8 g protein, 0–3 g fat, 90 kcal	
1 c	Nonfat milk
1 c	½% milk
1 c	1% milk
1 c	Nonfat or low-fat buttermilk
½ c	Evaporated nonfat milk
⅓ c dry	Dry nonfat milk
¾ c	Plain nonfat yogurt
1 c	Nonfat or low-fat fruit-flavored yogurt sweetened with aspartame or with a nonnutritive sweetener
Reduced-Fat Milk	
1 reduced-fat milk exchange = 12 g carbohydrate, 8 g protein, 5 g fat, 120 kcal	
1 c	2% milk
¾ c	Plain low-fat yogurt
1 c	Sweet acidophilus milk
Whole Milk	
1 whole milk exchange = 12 g carbohydrate, 8 g protein, 8 g fat, 150 kcal	
1 c	Whole milk
½c	Evaporated whole milk
1 c	Goat's milk
1 c	Kefir

G

TABLE G-4 **U.S. Exchange System: Other Carbohydrates List**

1 other carbohydrate exchange = 15 g carbohydrate, or 1 starch, or 1 fruit, or 1 milk exchange

Food	Serving Size	Exchanges per Serving
Angel food cake, unfrosted	1⁄12 cake	2 carbohydrates
Brownies, small, unfrosted	2″ square	1 carbohydrate, 1 fat
Cake, unfrosted	2″ square	1 carbohydrate, 1 fat
Cake, frosted	2″ square	2 carbohydrates, 1 fat
Cookie, fat-free	2 small	1 carbohydrate
Cookies or sandwich cookies	2 small	1 carbohydrate, 1 fat
Cupcakes, frosted	1 small	2 carbohydrates, 1 fat
Cranberry sauce, jellied	¼ c	2 carbohydrates
Doughnuts, plain cake	1 medium, (1½ oz)	1½ carbohydrates, 2 fats
Doughnuts, glazed	3¾″ across (2 oz)	2 carbohydrates, 2 fats
Fruit juice bars, frozen, 100% juice	1 bar (3 oz)	1 carbohydrate
Fruit snacks, chewy (pureed fruit concentrate)	1 roll (¾ oz)	1 carbohydrate
Fruit spreads, 100% fruit	1 tbs	1 carbohydrate
Gelatin, regular	½ c	1 carbohydrate
Gingersnaps	3	1 carbohydrate
Granola bars	1 bar	1 carbohydrate, 1 fat
Granola bars, fat-free	1 bar	2 carbohydrates
Hummus	⅓ c	1 carbohydrate, 1 fat
Ice cream	½ c	1 carbohydrate, 2 fats
Ice cream, light	½ c	1 carbohydrate, 1 fat
Ice cream, fat-free, no sugar added	½ c	1 carbohydrate
Jam or jelly, regular	1 tbs	1 carbohydrate
Milk, chocolate, whole	1 c	2 carbohydrates, 1 fat
Pie, fruit, 2 crusts	⅙ pie	3 carbohydrates, 2 fats
Pie, pumpkin or custard	⅛ pie	1 carbohydrate, 2 fats
Potato chips	12–18 (1 oz)	1 carbohydrate, 2 fats
Pudding, regular (made with low-fat milk)	½ c	2 carbohydrates
Pudding, sugar-free (made with low-fat milk)	½ c	1 carbohydrate
Salad dressing, fat-free 🧂	¼ c	1 carbohydrate
Sherbet, sorbet	½ c	2 carbohydrates
Spaghetti or pasta sauce, canned 🧂	½ c	1 carbohydrate, 1 fat
Sweet roll or danish	1 (2½ oz)	2½ carbohydrates, 2 fats
Syrup, light	2 tbs	1 carbohydrate
Syrup, regular	1 tbs	1 carbohydrate
Syrup, regular	¼ c	4 carbohydrates
Tortilla chips	6–12 (1 oz)	1 carbohydrate, 2 fats
Yogurt, frozen, low-fat, fat-free	⅓ c	1 carbohydrate, 0–1 fat
Yogurt, frozen, fat-free, no sugar added	½ c	1 carbohydrate
Yogurt, low-fat with fruit	1 c	3 carbohydrates, 0–1 fat
Vanilla wafers	5	1 carbohydrate, 1 fat

🧂 = 400 mg or more sodium per exchange.

TABLE G-5 U.S. Exchange System: Vegetable List

1 vegetable exchange = 5 g carbohydrate, 2 g protein, 25 kcal
Note: In general, a vegetable serving is ½ c cooked vegetables or vegetable juice; 1 c raw vegetables. Starchy vegetables such as corn, peas, and potatoes are on the starch list.

Artichokes	Mushrooms
Artichoke hearts	Okra
Asparagus	Onions
Beans (green, wax, Italian)	Pea pods
Bean sprouts	Peppers (all varieties)
Beets	Radishes
Broccoli	Salad greens (endive, escarole, lettuce, romaine, spinach)
Brussels sprouts	Sauerkraut 🧂
Cabbage	Spinach
Carrots	Summer squash (crookneck)
Cauliflower	Tomatoes
Celery	Tomatoes, canned
Cucumbers	Tomato sauce 🧂
Eggplant	Tomato/vegetable juice 🧂
Green onions or scallions	Turnips
Greens (collard, kale, mustard, turnip)	Water chestnuts
Kohlrabi	Watercress
Leeks	Zucchini
Mixed vegetables (without corn, peas, or pasta)	

🧂 = 400 mg or more sodium per exchange.

G

G

TABLE G-6 U.S. Exchange System: Meat and Meat Substitutes List

Note: In general, a meat serving is 1 oz meat, poultry, or cheese; ½ c dried beans (weigh meat and poultry and measure beans after cooking).

Serving Size	Food
Very Lean Meat and Substitutes	
1 very lean meat exchange = 7 g protein, 0–1 g fat, 35 kcal	
1 oz	Poultry: Chicken or turkey (white meat, no skin), Cornish hen (no skin)
1 oz	Fish: Fresh or frozen cod, flounder, haddock, halibut, trout; tuna, fresh or canned in water
1 oz	Shellfish: Clams, crab, lobster, scallops, shrimp, imitation shellfish
1 oz	Game: Duck or pheasant (no skin), venison, buffalo, ostrich
	Cheese with $\leq$1g fat/oz:
¼ c	Nonfat or low-fat cottage cheese
1 oz	Fat-free cheese
	Other:
1 oz	Processed sandwich meats with $\leq$1 g fat/oz (such as deli thin, shaved meats, chipped beef, turkey ham)
2	Egg whites
¼ c	Egg substitutes, plain
1 oz	Hot dogs with $\leq$1 g fat/oz
1 oz	Kidney (high in cholesterol)
1 oz	Sausage with $\leq$1 g fat/oz
Count as one very lean meat and one starch exchange:	
½ c	Dried beans, peas, lentils (cooked)
Lean Meat and Substitutes	
1 lean meat exchange = 7 g protein, 3 g fat, 55 kcal	
1 oz	Beef: USDA Select or Choice grades of lean beef trimmed of fat (round, sirloin, and flank steak); tenderloin; roast (rib, chuck, rump); steak (T-bone, porterhouse, cubed), ground round
1 oz	Pork: Lean pork (fresh ham); canned, cured, or boiled ham; Canadian bacon; tenderloin, center loin chop
1 oz	Lamb: Roast, chop, leg
1 oz	Veal: Lean chop, roast
1 oz	Poultry: Chicken, turkey (dark meat, no skin), chicken white meat (with skin), domestic duck or goose (well-drained of fat, no skin)
	Fish:
1 oz	Herring (uncreamed or smoked)
6 medium	Oysters
1 oz	Salmon (fresh or canned), catfish
2 medium	Sardines (canned)
1 oz	Tuna (canned in oil, drained)
1 oz	Game: Goose (no skin), rabbit
	Cheese:
¼ c	4.5%-fat cottage cheese
2 tbs	Grated Parmesan
1 oz	Cheeses with $\leq$3 g fat/oz
	Other:
1½ oz	Hot dogs with $\leq$3 g fat/oz
1 oz	Processed sandwich meat with $\leq$3 g fat/oz (turkey pastrami or kielbasa)
1 oz	Liver, heart (high in cholesterol)
Medium-Fat Meat and Substitutes	
1 medium-fat meat exchange = 7 g protein, 5 g fat, and 75 kcal	
1 oz	Beef: Most beef products (ground beef, meatloaf, corned beef, short ribs, Prime grades of meat trimmed of fat, such as prime rib)
1 oz	Pork: Top loin, chop, Boston butt, cutlet
1 oz	Lamb: Rib roast, ground
1 oz	Veal: Cutlet (ground or cubed, unbreaded)
1 oz	Poultry: Chicken dark meat (with skin), ground turkey or ground chicken, fried chicken (with skin)
1 oz	Fish: Any fried fish product
	Cheese with $\leq$5 g fat/oz:
1 oz	Feta
1 oz	Mozzarella
¼ c (2 oz)	Ricotta
	Other:
1	Egg (high in cholesterol, limit to 3/week)
1 oz	Sausage with $\leq$5 g fat/oz
1 c	Soy milk
¼ c	Tempeh
4 oz or ½ c	Tofu
High-Fat Meat and Substitutes	
1 high-fat meat exchange = 7 g protein, 8 g fat, 100 kcal	
1 oz	Pork: Spareribs, ground pork, pork sausage
1 oz	Cheese: All regular cheeses (American, cheddar, Monterey Jack, swiss)
	Other:
1 oz	Processed sandwich meats with $\leq$8 g fat/oz (bologna, pimento loaf, salami)
1 oz	Sausage (bratwurst, Italian, knockwurst, Polish, smoked)
1 (10/lb)	Hot dog (turkey or chicken)
3 slices (20 slices/lb)	Bacon
Count as one high-fat meat plus one fat exchange:	
1 (10/lb)	Hot dog (beef, pork, or combination)
2 tbs	Peanut butter (contains unsaturated fat)

= 400 mg or more of sodium per serving.

G

TABLE G-7 U.S. Exchange System: Fat List

1 fat exchange = 5 g fat, 45 kcal

Note: In general, a fat serving is 1 tsp regular butter, margarine, or vegetable oil; 1 tbs regular salad dressing. Many fat-free and reduced-fat foods are on the Free Foods List.

Serving Size	Food
Monounsaturated Fats	
⅛ medium (1 oz)	Avocados
1 tsp	Oil (canola, olive, peanut)
8 large	Olives, ripe (black)
10 large	Olives, green, stuffed
6 nuts	Almonds, cashews
6 nuts	Mixed nuts (50% peanuts)
10 nuts	Peanuts
4 halves	Pecans
2 tsp	Peanut butter, smooth or crunchy
1 tbs	Sesame seeds
2 tsp	Tahini paste
Polyunsaturated Fats	
1 tsp	Margarine, stick, tub, or squeeze
1 tbs	Margarine, lower-fat (30% to 50% vegetable oil)
1 tsp	Mayonnaise, regular
1 tbs	Mayonnaise, reduced-fat
4 halves	Nuts, walnuts, English
1 tsp	Oil (corn, safflower, soybean)
1 tbs	Salad dressing, regular
2 tbs	Salad dressing, reduced-fat
2 tsp	Mayonnaise type salad dressing, regular
1 tbs	Mayonnaise type salad dressing, reduced-fat
1 tbs	Seeds: pumpkin, sunflower
Saturated Fats*	
1 slice (20 slices/lb)	Bacon, cooked
1 tsp	Bacon, grease
1 tsp	Butter, stick
2 tsp	Butter, whipped
1 tbs	Butter, reduced-fat
2 tbs (½ oz)	Chitterlings, boiled
2 tbs	Coconut, sweetened, shredded
2 tbs	Cream, half and half
1 tbs (½ oz)	Cream cheese, regular
2 tbs (1 oz)	Cream cheese, reduced-fat
	Fatback or salt pork†
1 tsp	Shortening or lard
2 tbs	Sour cream, regular
3 tbs	Sour cream, reduced-fat

= 400 mg or more sodium per exchange

*Saturated fats can raise blood cholesterol levels.

† Use a piece 1″ × 1″ × ¼″ if you plan to eat the fatback cooked with vegetables. Use a piece 2″ × 1″ × ½″ when eating only the vegetables with the fatback removed.

G

TABLE G-8 U.S. Exchange System: Free Foods List

Note: A serving of free food contains less than 20 kcalories; those with serving sizes should be limited to 3 servings a day whereas those without serving sizes can be eaten freely.

Serving Size	Food
Fat-Free or Reduced-Fat Foods	
1 tbs	Cream cheese, fat-free
1 tbs	Creamers, nondairy, liquid
2 tsp	Creamers, nondairy, powdered
1 tbs	Mayonnaise, fat-free
1 tsp	Mayonnaise, reduced-fat
4 tbs	Margarine, fat-free
1 tsp	Margarine, reduced-fat
1 tbs	Mayonnaise type salad dressing, nonfat
1 tsp	Mayonnaise type salad dressing, reduced-fat
	Nonstick cooking spray
1 tbs	Salad dressing, fat-free
2 tbs	Salad dressing, fat-free, Italian
¼ c	Salsa
1 tbs	Sour cream, fat-free, reduced-fat
2 tbs	Whipped topping, regular or light
Sugar-Free or Low-Sugar Foods	
1 piece	Candy, hard, sugar-free
	Gelatin dessert, sugar-free
	Gelatin, unflavored
	Gum, sugar-free
2 tsp	Jam or jelly, low-sugar or light
	Sugar substitutes
2 tbs	Syrup, sugar-free
Drinks	
	Bouillon, broth, consommé (salt shaker symbol)
	Bouillon or broth, low-sodium
	Carbonated or mineral water
1 tbs	Cocoa powder, unsweetened
	Coffee
	Club soda
	Diet soft drinks, sugar-free
	Drink mixes, sugar-free
	Tea
	Tonic water, sugar-free
Condiments	
1 tbs	Catsup
	Horseradish
	Lemon juice
	Lime juice
	Mustard
1½ large	Pickles, dill (salt shaker symbol)
	Soy sauce, regular or light (salt shaker symbol)
1 tbs	Taco sauce
	Vinegar
Seasonings	
Flavoring extracts	
Garlic	
Herbs, fresh or dried	
Pimento	
Spices	
Hot pepper sauces	
Wine, used in cooking	
Worcestershire sauce	

(salt shaker symbol) = 400 mg or more of sodium per serving.

G

TABLE G-9 U.S. Exchange System: Combination Foods List

Food	Serving Size	Exchanges per Serving
Entrees		
Tuna noodle casserole, lasagna, spaghetti with meatballs, chili with beans, macaroni and cheese ✎	1 c (8 oz)	2 carbohydrates, 2 medium-fat meats
Chow mein (without noodles or rice)	2 c (16 oz)	1 carbohydrate, 2 lean meats
Pizza, cheese, thin crust ✎	¼ of 10" (5 oz)	2 carbohydrates, 2 medium-fat meats, 1 fat
Pizza, meat topping, thin crust ✎	¼ of 10" (5 oz)	2 carbohydrates, 2 medium-fat meats, 2 fats
Pot pie ✎	1 (7 oz)	2 carbohydrates, 1 medium-fat meat, 4 fats
Frozen Entrees		
Salisbury steak with gravy, mashed potato	1 (11 oz)	2 carbohydrates, 3 medium-fat meats, 3–4 fats
Turkey with gravy, mashed potato, dressing ✎	1 (11 oz)	2 carbohydrates, 2 medium-fat meats, 2 fats
Entree with less than 300 kcalories ✎	1 (8 oz)	2 carbohydrates, 3 lean meats
Soups		
Bean ✎	1 c	1 carbohydrate, 1 very lean meat
Cream (made with water) ✎	1 c (8 oz)	1 carbohydrate, 1 fat
Split pea (made with water) ✎	½ c (4 oz)	1 carbohydrate
Tomato (make with water) ✎	1 c (8 oz)	1 carbohydrate
Vegetable beef, chicken noodle, or other broth-type ✎	1 c (8 oz)	1 carbohydrate
Fast Foods		
Burritos with beef ✎	2	4 carbohydrates, 2 medium-fat meats, 2 fats
Chicken nuggets ✎	6	1 carbohydrate, 2 medium-fat meats, 1 fat
Chicken breast and wing, breaded and fried ✎	1	1 carbohydrate, 4 medium-fat meats, 2 fats
Fish sandwich/tartar sauce ✎	1	3 carbohydrates, 1 medium-fat meat, 3 fats
French fries, thin	20–25	2 carbohydrates, 2 fats
Hamburger, regular	1	2 carbohydrates, 2 medium-fat meats
Hamburger, large ✎	1	2 carbohydrates, 3 medium-fat meats, 1 fat
Hot dog with bun ✎	1	1 carbohydrate, 1 high-fat meat, 1 fat
Individual pan pizza ✎	1	5 carbohydrates, 3 medium-fat meats, 3 fats
Soft serve cone	1 medium	2 carbohydrates, 1 fat
Submarine sandwich ✎	1 (60)	3 carbohydrates, 1 vegetable, 2 medium-fat meats, 1 fat
Taco, hard shell ✎	1 (6 oz)	2 carbohydrates, 2 medium-fat meats, 2 fats
Taco, soft shell ✎	1 (3 oz)	1 carbohydrate, 1 medium-fat meat, 1 fat

✎ = 400 mg or more sodium per exchange.

H TABLE OF FOOD COMPOSITION

This edition of the table of food composition includes a wide variety of foods from all food groups. It is updated yearly to reflect nutrient changes for current foods, remove outdated foods, and add foods that are new to the marketplace.*

The nutrient database for this appendix is compiled from a variety of sources, including the USDA Standard Release database (Release 13), literature sources, and manufacturers' data. The USDA database provides data for a wider variety of foods and nutrients than other sources. Because laboratory analysis for each nutrient can be quite costly, manufacturers tend to provide data only for those nutrients mandated on food labels. Consequently, data for their foods are often incomplete; any missing information is designated in this table as a blank space. Keep in mind that a blank space means only that the information is unknown and should not be interpreted as a zero.

Whenever using nutrient data, remember that many factors influence the nutrient contents of foods, including the mineral content of the soil, the diet of the animal or the fertilizer of the plant, the season of harvest, the method of processing, the length and method of storage, the method of cooking, the method of analysis, and the moisture content of the sample analyzed. With so many factors involved, users must view nutrient data as a close approximation of the actual amount.

For updates, corrections, and a list of 3000 additional foods and codes found in the diet analysis software that accompanies this text, visit www.wadsworth.com/nutrition and click on *Diet Analysis*.

• ***Fats*** • Total fats, as well as the breakdown of total fats to saturated, monounsaturated, and polyunsaturated fats, are listed in the table. The fatty acids seldom add up to the total due to rounding and to other fatty acid components that are not included in these basic categories, such as *trans*-fatty acids and glycerol. *Trans*-fatty acids can comprise a large share of the total fat in margarine and shortening (hydrogenated oils) and in any foods that include them as ingredients.

• ***Vitamin A*** • Databases currently report vitamin A in retinol equivalents (RE)—a measure of vitamin A activity based on an older and less accurate estimation of how much vitamin A the body derives from carotenoids. Recent evidence, however, has determined that the vitamin A activity of carotenoids is about half the amount previously thought. Consequently, data using RE overestimate vitamin A activity. The 2001 RDA for vitamin A establishes a new measure of vitamin A activity—retinol activity equivalents (RAE)—to reflect the change in equivalencies. Chapter 11 provides conversion factors and additional details.

• ***Vitamin E*** • Databases currently report vitamin E in α-tocopherol equivalents (α-TE)—a measure of vitamin E activity based on an older and less accurate estimation of how much vitamin E the body derives from all eight naturally occurring forms of the vitamin (the various tocopherols and tocotrienols). Recent evidence, however, has determined that the body derives vitamin E activity only from the α-tocopherol form. Consequently, data using measures of α-TE overestimate vitamin E activity. The 2000 RDA for vitamin E is based only on the α-tocopherol form. Chapter 11 provides conversion factors and additional details.

• ***Bioavailability*** • Keep in mind that the availability of nutrients from foods depends not only on the quantity provided by a food, but also on the amount absorbed and used by the body—the bioavailability. The bioavailability of folate from fortified foods, for example, is greater than from naturally occurring sources. (Note that this appendix has been updated with data to reflect the folate fortification of grain products.) Similarly, the body can make niacin from the amino acid tryptophan, but niacin values in this table (and most databases) report preformed niacin only. Chapter 10 provides conversion factors and additional details.

• ***Using the Table*** • The items in this table have been organized into several categories, which are listed at the head of each right-hand page. Page numbers have been provided, and each group has been color-coded to make it easier to find individual items.

In an effort to conserve space, the following abbreviations have been used in the food descriptions and nutrient breakdowns:

- diam = diameter
- ea = each
- enr = enriched
- f/ = from
- frzn = frozen
- g = grams

*This food composition table has been prepared for Wadsworth Publishing Company and is copyrighted by ESHA Research in Salem, Oregon—the developer and publisher of the Food Processor and Genesis nutritional software programs. The nutritional data are supported by over 1300 references. Because the list of sources is so extensive, it is not provided here, but is available from the publisher.

2 T Peanut butter

14gmCHO x 4cal/gm=56cal
8gmPro x 4cal/gm= 32cal
11gmfat x 9cal/gm= 99cal
Total= 187cal

Carbohydrates provide 4kcal/gm
Protein provides 4kcal/gm
Fat provides 9kcal/gm

Alcohol provides 7kcal/gm

Calculate the calories!
Wendy's Big Bacon Classic
44 grams carbohydrate
36 grams fat
37 grams protein
Add a Regular Beer (12oz)
12 grams alcohol
13 grams carbohydrate

How many calories ?

How many total kcals? ________________

Diet & Health
Family histories of these diseases are most probably the largest risk factor for the development of the following diseases. They are influenced by the interaction between genetic and nutritional factors.
Heart disease/ stroke
Hypertension
Obesity
Diabetes
Certain cancers
Osteoporosis
Cirrhosis of liver
KNOW YOUR HISTORY
1. Name and student # - cover sheet
2. Read pages 19 –20 in text; submit completed genogram & 1 page analysis of your relative risk. Include in your analysis the answers to the 3 questions on page 20…"How are you doing?"
3. DUE- February 2, 2004

Breakfast

250 C Nutritional Supplement

2 Cookies

Lunch

2 chicken drumsticks

Portion of Rice

2 glasses of juice

Dinner

Roast Beef Sandwich

Soup

2 cookies

2 glasses of choc. milk

2 more glasses of juice

Chapter 1

What influences your food choices?

Is it true: you are what you eat?

Food -

Nutrients-

Nutrition-

Diet therapy-

6 Classes of nutrients:

Macronutrients (These are also energy-yielding)

- Carbohydrates
- Protein
- Fat

Micronutrients

- Vitamins
- Minerals
- Water

Kilocalories- unit of measurement, amount needed to raise the temperature of 1 Kg of water 1° Celsius

- liq = liquid
- pce = piece
- pkg = package
- w/ = with
- w/o = without
- t = trace
- 0 = zero (no nutrient value)
- blank space = information not available

• ***Caffeine Sources*** • Caffeine occurs in several plants, including the familiar coffee bean, the tea leaf, and the cocoa bean from which chocolate is made. Most human societies use caffeine regularly, most often in beverages, for its stimulant effect and flavor. Caffeine contents of beverages vary depending on the plants they are made from, the climates and soils where the plants are grown, the grind or cut size, the method and duration of brewing, and the amounts served. The accompanying table shows that in general, a cup of coffee contains the most caffeine; a cup of tea, less than half as much; and cocoa or chocolate, less still. As for cola beverages, they are made from kola nuts, which contain caffeine, but most of their caffeine is added, using the purified compound obtained from decaffeinated coffee beans.

The FDA lists caffeine as a multipurpose GRAS substance that may be added to foods and beverages. Drug manufacturers use caffeine in many kinds of drugs: stimulants, pain relievers, cold remedies, diuretics, and weight-loss aids.

H

TABLE **Caffeine Content of Beverages, Foods, and Over-the-Counter Drugs**

Beverages and Foods	Average (mg)	Range (mg)
Coffee (5-oz cup)		
Brewed, drip method	130	110–150
Brewed, percolator	94	64–124
Instant	74	40–108
Decaffeinated, brewed or instant	3	1–5
Tea (5-oz cup)		
Brewed, major U.S. brand	40	20–90
Brewed, imported brands	60	25–110
Instant	30	25–50
Iced (12-oz can)	70	67–76
Soft drinks (12-oz can)		
Dr. Pepper	40	
Colas and cherry cola		
Regular		30–46
Diet		2–58
Caffeine-free		0–trace
Jolt	72	
Mountain Dew, Mello Yello	52	
Fresca, Hires Root Beer, 7-Up, Sprite, Squirt, Sunkist Orange	0	
Cocoa beverage (5-oz cup)	4	2–20
Chocolate milk beverage (8 oz)	5	2–7
Milk chocolate candy (1 oz)	6	1–15
Dark chocolate, semisweet (1 oz)	20	5–35
Baker's chocolate (1 oz)	26	
Chocolate flavored syrup (1 oz)	4	

Drugs[a]	Average (mg)
Cold remedies (standard dose)	
Dristan	0
Coryban-D, Triaminicin	30
Diuretics (standard dose)	
Aqua-ban, Permathene H_2Off	200
Pre-Mens Forte	100
Pain relievers (standard dose)	
Excedrin	130
Midol, Anacin	65
Aspirin, plain (any brand)	0
Stimulants	
Caffedrin, NoDoz, Vivarin	200
Weight-control aids (daily dose)	
Prolamine	280
Dexatrim, Dietac	200

Note: A pharmacologically active dose of caffeine is defined as 200 milligrams.
[a]Because products change, contact the manufacturer for an update on products you use regularly.

Table H-1

Food Composition (Computer code number is for West Diet Analysis program) (For purposes of calculations, use "0" for t, <1, <.1, <.01, etc.)

Computer Code Number	Food Description	Measure	Wt (g)	H_2O (%)	Ener (kcal)	Prot (g)	Carb (g)	Dietary Fiber (g)	Fat (g)	Fat Breakdown (g) Sat	Mono	Poly
	BEVERAGES											
	Alcoholic:											
	Beer:											
1	Regular (12 fl oz)	1½ c	356	92	146	1	13	1	0	0	0	0
2	Light (12 fl oz)	1½ c	354	95	99	1	5	0	0	0	0	0
1506	Nonalcoholic (12 fl oz)	1½ c	360	98	32	1	5	0	0	0	0	0
	Gin, rum, vodka, whiskey:											
3	80 proof	1½ fl oz	42	67	97	0	0	0	0	0	0	0
4	86 proof	1½ fl oz	42	64	105	0	<1	0	0	0	0	0
5	90 proof	1½ fl oz	42	62	110	0	0	0	0	0	0	0
	Liqueur:											
1359	Coffee liqueur, 53 proof	1½ fl oz	52	31	175	<1	24	0	<1	.1	t	.1
1360	Coffee & cream liqueur, 34 proof	1½ fl oz	47	46	154	1	10	0	7	4.5	2.1	.3
1361	Crème de menthe, 72 proof	1½ fl oz	50	28	186	0	21	0	<1	t	t	.1
	Wine, 4 fl oz:											
6	Dessert, sweet	½ c	118	72	181	<1	14	0	0	0	0	0
7	Red	½ c	118	88	85	<1	2	0	0	0	0	0
8	Rosé	½ c	118	89	84	<1	2	0	0	0	0	0
9	White medium	½ c	118	90	80	<1	1	0	0	0	0	0
1592	Nonalcoholic	1 c	232	98	14	1	3	0	0	0	0	0
1593	Nonalcoholic light	1 c	232	98	14	1	3	0	0	0	0	0
1409	Wine cooler, bottle (12 fl oz)	1½ c	340	90	169	<1	20	<1	<1	t	t	t
1595	Wine cooler, cup	1 c	227	90	113	<1	13	<1	<1	t	t	t
	Carbonated:											
10	Club soda (12 fl oz)	1½ c	355	100	0	0	0	0	0	0	0	0
11	Cola beverage (12 fl oz)	1½ c	372	89	153	0	39	0	0	0	0	0
12	Diet cola w/aspartame (12 fl oz)	1½ c	355	100	4	<1	<1	0	0	0	0	0
13	Diet soda pop w/saccharin (12 fl oz)	1½ c	355	100	0	0	<1	0	0	0	0	0
14	Ginger ale (12 fl oz)	1½ c	366	91	124	0	32	0	0	0	0	0
15	Grape soda (12 fl oz)	1½ c	372	89	160	0	42	0	0	0	0	0
16	Lemon-lime (12 fl oz)	1½ c	368	89	147	0	38	0	0	0	0	0
17	Orange (12 fl oz)	1½ c	372	88	179	0	46	0	0	0	0	0
18	Pepper-type soda (12 fl oz)	1½ c	368	89	151	0	38	0	<1	.1	0	0
19	Root beer (12 fl oz)	1½ c	370	89	152	0	39	0	0	0	0	0
20	Coffee, brewed	1 c	237	99	5	<1	1	0	<1	t	0	t
21	Coffee, prepared from instant	1 c	238	99	5	<1	1	0	<1	t	0	t
	Fruit drinks, noncarbonated:											
22	Fruit punch drink, canned	1 c	248	88	117	0	29	<1	<1	t	t	t
1358	Gatorade	1 c	241	93	60	0	15	0	0	0	0	0
23	Grape drink, canned	1 c	250	87	125	<1	32	<1	0	0	0	0
1304	Koolade sweetened with sugar	1 c	262	90	97	0	25	0	<1	t	t	t
1356	Koolade sweetened with nutrasweet	1 c	240	95	43	0	11	0	0	0	0	0
26	Lemonade,frzn concentrate (6-oz can)	¾ c	219	52	396	1	103	1	<1	.1	t	.1
27	Lemonade, from concentrate	1 c	248	89	99	<1	26	<1	<1	t	t	t
28	Limeade, frzn concentrate (6-oz can)	¾ c	218	50	408	<1	108	1	<1	t	t	.1
29	Limeade, from concentrate	1 c	247	89	101	0	27	<1	<1	t	t	t
24	Pineapple grapefruit, canned	1 c	250	88	118	<1	29	<1	<1	t	t	.1
25	Pineapple orange, canned	1 c	250	87	125	3	29	<1	0	0	0	0
	Fruit and vegetable juices: see Fruit and Vegetable sections											
	Ultra Slim Fast, ready to drink, can:											
30411	Chocolate Royale	1 ea	350	84	220	10	38	5	3	1	1	.5
30415	French Vanilla	1 ea	350	84	220	10	38	5	3	1	1.5	.5
30413	Strawberries n' cream	1 ea	350	83	220	10	42	5	3	1	1.5	.5
1357	Water, bottled: Perrier (6½ fl oz)	1 ea	192	100	0	0	0	0	0	0	0	0
1594	Water, bottled: Tonic water	1½ c	366	91	124	0	32	0	0	0	0	0
	Tea:											
30	Brewed, regular	1 c	237	100	2	0	1	0	<1	t	0	t
1662	Brewed, herbal	1 c	237	100	2	0	<1	0	<1	t	t	t
32	From instant, sweetened	1 c	259	91	88	<1	22	0	<1	t	t	t
31	From instant, unsweetened	1 c	237	100	2	<1	<1	0	0	0	0	0

Chol (mg)	Calc (mg)	Iron (mg)	Magn (mg)	Pota (mg)	Sodi (mg)	Zinc (mg)	VT-A (RE)	Thia (mg)	VT-E (α-TE)	Ribo (mg)	Niac (mg)	V-B6 (mg)	Fola (µg)	VT-C (mg)
0	18	.11	21	89	18	.07	0	.02	0	.09	1.61	.18	21	0
0	18	.14	18	64	11	.11	0	.03	0	.11	1.39	.12	14	0
0	25	.04	32	90	18	.04	0	.02	0	.09	1.63	.18	22	0
0	0	.02	0	1	<1	.02	0	<.01	0	<.01	<.01	0	0	0
0	0	.02	0	1	<1	.02	0	<.01	0	<.01	<.01	0	0	0
0	0	.02	0	1	<1	.02	0	<.01	0	<.01	<.01	0	0	0
0	1	.03	2	16	4	.02	0	<.01	0	.01	.07	0	0	0
7	8	.06	1	15	43	.07	20	0	.12	.03	.04	.01	0	0
0	0	.03	0	0	2	.02	0	0	0	0	<.01	0	0	0
0	9	.28	11	109	11	.08	0	.02	0	.02	.25	0	<1	0
0	9	.51	15	132	6	.11	0	.01	0	.03	.1	.04	2	0
0	9	.45	12	117	6	.07	0	<.01	0	.02	.09	.03	1	0
0	11	.38	12	94	6	.08	0	<.01	0	.01	.08	.02	<1	0
0	21	.93	23	204	16	.19	0	0	0	.02	.23	.05	2	0
0	21	.93	23	204	16	.19	0	0	0	.02	.23	.05	2	0
0	19	.92	18	152	29	.2	<1	.02	.02	.02	.15	.04	4	6
0	13	.61	12	102	19	.13	<1	.01	.02	.02	.1	.03	3	4
0	18	.04	4	7	75	.35	0	0	0	0	0	0	0	0
0	11	.11	4	4	15	.04	0	0	0	0	0	0	0	0
0	14	.11	4	0	21	.28	0	.02	0	.08	0	0	0	0
0	14	.14	4	7	57	.18	0	0	0	0	0	0	0	0
0	11	.66	4	4	26	.18	0	0	0	0	0	0	0	0
0	11	.3	4	4	56	.26	0	0	0	0	0	0	0	0
0	7	.26	4	4	40	.18	0	0	0	0	.05	0	0	0
0	19	.22	4	7	45	.37	0	0	0	0	0	0	0	0
0	11	.15	0	4	37	.15	0	0	0	0	0	0	0	0
0	18	.18	4	4	48	.26	0	0	0	0	0	0	0	0
0	5	.12	12	128	5	.05	0	0	0	0	.53	0	<1	0
0	7	.12	10	86	7	.07	0	0	0	<.01	.67	0	0	0
0	20	.52	5	62	55	.3	3	0	0	.06	.05	0	3	73
0	0	.12	2	26	96	.05	0	.01	0	0	0	0	0	0
0	7	.25	10	87	2	.07	<1	.02	0	.02	.25	.05	2	40
0	42	.13	3	3	37	.08	0	0	0	<.01	<.01	0	<1	31
0	17	.65	5	50	50	.26	2	.02	0	.05	.05	0	5	77
0	15	1.58	11	147	9	.17	22	.06	0	.21	.16	.05	22	39
0	7	.4	5	37	7	.1	5	.01	0	.05	.04	.01	5	10
0	11	.22	9	129	0	.09	0	.02	0	.02	.22	0	9	26
0	7	.07	2	32	5	.05	0	<.01	0	<.01	.05	0	2	7
0	17	.77	15	153	35	.15	9	.07	0	.04	.67	.1	26	115
0	12	.67	15	115	7	.15	13	.07	0	.05	.52	.12	27	56
5	400	2.7	140	530	220	2.24	525	.52	7	.59	7	.7	120	21
5	400	2.8	140	450	460	2.1	525	.52	7	.59	7	.7	120	21
5	400	2.7	140	450	460	2.24	525	.52	7	.59	7	.7	120	21
0	27	0	0	0	2	0	0	0	0	0	0	0	0	0
0	4	.04	0	0	15	.37	0	0	0	0	0	0	0	0
0	0	.05	7	88	7	.05	0	0	0	.03	0	0	12	0
0	5	.19	2	21	2	.09	0	.02	0	.01	0	0	1	0
0	5	.05	5	49	8	.08	0	0	0	.05	.09	<.01	10	0
0	5	.05	5	47	7	.07	0	0	0	<.01	.09	<.01	1	0

Table H-1

Food Composition (Computer code number is for West Diet Analysis program) (For purposes of calculations, use "0" for t, <1, <.1, <.01, etc.)

Computer Code Number	Food Description	Measure	Wt (g)	H_2O (%)	Ener (kcal)	Prot (g)	Carb (g)	Dietary Fiber (g)	Fat (g)	Fat Breakdown (g) Sat	Mono	Poly
	DAIRY											
	Butter: see Fats and Oils, #158,159,160											
	Cheese, natural:											
33	Blue	1 oz	28	42	99	6	1	0	8	5.2	2.2	.2
34	Brick	1 oz	28	41	104	6	1	0	8	5.3	2.4	.2
35	Brie	1 oz	28	48	93	6	<1	0	8	4.9	2.2	.2
36	Camembert	1 oz	28	52	84	6	<1	0	7	4.3	2	.2
37	Cheddar:	1 oz	28	37	113	7	<1	0	9	5.9	2.6	.3
38	1" cube	1 ea	17	37	68	4	<1	0	6	3.6	1.6	.2
39	Shredded	1 c	113	37	455	28	1	0	37	24	10.6	1.1
1406	Low fat, low sodium	1 oz	28	65	48	7	1	0	2	1.2	.6	.1
	Cottage:											
984	Low sodium, low fat	1 c	225	83	162	28	6	0	2	1.4	.7	.1
40	Creamed, large curd	1 c	225	79	232	28	6	0	10	6.4	2.9	.3
41	Creamed, small curd	1 c	210	79	216	26	6	0	9	6	2.7	.3
42	With fruit	1 c	226	72	280	22	30	0	8	4.9	2.2	.2
43	Low fat 2%	1 c	226	79	203	31	8	0	4	2.8	1.2	.1
44	Low fat 1%	1 c	226	82	164	28	6	0	2	1.5	.7	.1
46	Cream	1 tbs	15	54	52	1	<1	0	5	3.3	1.5	.2
983	low fat	1 tbs	15	64	35	2	1	0	3	1.7	.9	.1
47	Edam	1 oz	28	42	100	7	<1	0	8	4.9	2.3	.2
48	Feta	1 oz	28	55	74	4	1	0	6	4.2	1.3	.2
49	Gouda	1 oz	28	41	100	7	1	0	8	4.9	2.2	.2
50	Gruyère	1 oz	28	33	116	8	<1	0	9	5.3	2.8	.5
51	Gorgonzola	1 oz	28	43	97	6	1	0	8	5		
1676	Limburger	1 oz	28	48	92	6	<1	0	8	4.7	2.4	.1
53	Monterey Jack	1 oz	28	41	104	7	<1	0	8	5.3	2.4	.3
54	Mozzarella, whole milk	1 oz	28	54	79	5	1	0	6	3.7	1.8	.2
55	Mozzarella, part-skim milk, low moisture	1 oz	28	49	78	8	1	0	5	3	1.4	.1
56	Muenster	1 oz	28	42	103	7	<1	0	8	5.3	2.4	.2
2422	Neufchatel	1 oz	28	62	73	3	1	0	7	4.1	1.9	.2
1399	Nonfat cheese (Kraft Singles)	1 oz	28	61	44	6	4	0	0	0	0	0
59	Parmesan, grated:	1 oz	28	18	128	12	1	0	8	5.5	2.4	.2
57	Cup, not pressed down	1 c	100	18	456	42	4	0	30	19.7	8.7	.7
58	Tablespoon	1 tbs	6	18	27	2	<1	0	2	1.2	.5	t
60	Provolone	1 oz	28	41	98	7	1	0	7	4.9	2.1	.2
61	Ricotta, whole milk	1 c	246	72	428	28	7	0	32	20.4	8.9	.9
62	Ricotta, part-skim milk	1 c	246	74	339	28	13	0	19	12.1	5.7	.6
63	Romano	1 oz	28	31	108	9	1	0	8	4.8	2.2	.2
64	Swiss	1 oz	28	37	105	8	1	0	8	5	2	.3
976	low fat	1 oz	28	60	50	8	1	0	1	.9	.4	t
	Pasteurized processed cheese products:											
65	American	1 oz	28	39	105	6	<1	0	9	5.5	2.5	.3
66	Swiss	1 oz	28	42	93	7	1	0	7	4.5	2	.2
67	American cheese food, jar	½ c	57	43	187	11	4	0	14	9	4.1	.4
68	American cheese spread	1 tbs	15	48	43	2	1	0	3	2	.9	.1
982	Velveeta cheese spread, low fat, low sodium, slice	1 pce	34	62	61	9	1	0	2	1.5	.7	.1
	Cream, sweet:											
69	Half & half (cream & milk)	1 c	242	81	315	7	10	0	28	17.3	8	1
70	Tablespoon	1 tbs	15	81	19	<1	1	0	2	1.1	.5	.1
71	Light, coffee or table:	1 c	240	74	468	6	9	0	46	28.8	13.4	1.7
72	Tablespoon	1 tbs	15	74	29	<1	1	0	3	1.8	.8	.1
73	Light whipping cream, liquid:	1 c	239	63	698	5	7	0	74	46.1	21.7	2.1
74	Tablespoon	1 tbs	15	63	44	<1	<1	0	5	2.9	1.4	.1
75	Heavy whipping cream, liquid:	1 c	238	58	821	5	7	0	88	54.7	25.5	3.3
76	Tablespoon	1 tbs	15	58	52	<1	<1	0	6	3.4	1.6	.2
77	Whipped cream, pressurized:	1 c	60	61	154	2	7	0	13	8.3	3.8	.5
78	Tablespoon	1 tbs	4	61	10	<1	<1	0	1	.6	.3	t
79	Cream, sour, cultured:	1 c	230	71	492	7	10	0	48	29.9	13.9	1.8
80	Tablespoon	1 tbs	14	71	30	<1	1	0	3	1.8	.8	.1

Chol (mg)	Calc (mg)	Iron (mg)	Magn (mg)	Pota (mg)	Sodi (mg)	Zinc (mg)	VT-A (RE)	Thia (mg)	VT-E (α-TE)	Ribo (mg)	Niac (mg)	V-B6 (mg)	Fola (μg)	VT-C (mg)
21	148	.09	6	72	391	.74	64	.01	.18	.11	.29	.05	10	0
26	189	.12	7	38	157	.73	85	<.01	.14	.1	.03	.02	6	0
28	51	.14	6	43	176	.67	51	.02	.18	.15	.11	.07	18	0
20	109	.09	6	52	236	.67	71	.01	.18	.14	.18	.06	17	0
29	202	.19	8	28	174	.87	85	.01	.1	.1	.02	.02	5	0
18	123	.12	5	17	106	.53	51	<.01	.06	.06	.01	.01	3	0
119	815	.77	31	111	702	3.51	342	.03	.41	.42	.09	.08	21	0
6	197	.2	8	31	6	.86	17	.01	.05	.01	.02	.02	5	0
9	137	.31	11	194	29	.85	25	.04	.25	.36	.29	.16	27	0
33	135	.31	12	190	911	.83	108	.05	.27	.37	.28	.15	27	0
31	126	.29	11	177	851	.78	101	.04	.26	.34	.26	.14	26	0
25	108	.25	9	151	915	.65	81	.04	.21	.29	.23	.12	22	0
19	155	.36	14	217	918	.95	45	.05	.13	.42	.32	.17	30	0
10	138	.32	12	193	918	.86	25	.05	.25	.37	.29	.15	28	0
16	12	.18	1	18	44	.08	66	<.01	.14	.03	.01	.01	2	0
8	17	.25	1	25	44	.11	33	<.01	.07	.04	.02	.01	3	0
25	205	.12	8	53	270	1.05	71	.01	.21	.11	.02	.02	5	0
25	138	.18	5	17	312	.81	36	.04	.01	.24	.28	.12	9	0
32	196	.07	8	34	229	1.09	49	.01	.1	.09	.02	.02	6	0
31	283	.05	10	23	94	1.09	84	.02	.1	.08	.03	.02	3	0
30	170	.18			280		43							0
25	139	.04	6	36	224	.59	88	.02	.18	.14	.04	.02	16	0
25	209	.2	8	23	150	.84	71	<.01	.09	.11	.03	.02	5	0
22	145	.05	5	19	104	.62	67	<.01	.1	.07	.02	.02	2	0
15	205	.07	7	26	148	.88	53	.01	.13	.1	.03	.02	3	0
27	201	.11	8	37	176	.79	88	<.01	.13	.09	.03	.02	3	0
21	21	.08	2	32	112	.15	74	<.01	.26	.05	.03	.01	3	0
4	221	0		81	427		126		0	.1				0
22	385	.27	14	30	521	.89	48	.01	.22	.11	.09	.03	2	0
79	1375	.95	51	107	1861	3.19	173	.04	.8	.39	.31	.1	8	0
5	82	.06	3	6	112	.19	10	<.01	.05	.02	.02	.01	<1	0
19	212	.15	8	39	245	.9	74	<.01	.1	.09	.04	.02	3	0
124	509	.93	28	258	207	2.85	330	.03	.86	.48	.26	.11	30	0
76	669	1.08	36	308	308	3.3	278	.05	.53	.45	.19	.05	32	0
29	298	.22	11	24	336	.72	39	.01	.2	.1	.02	.02	2	0
26	269	.05	10	31	73	1.09	71	.01	.14	.1	.03	.02	2	0
10	269	.05	10	31	73	1.09	18	.01	.05	.1	.02	.02	2	0
26	172	.11	6	45	400	.84	81	.01	.13	.1	.02	.02	2	0
24	216	.17	8	60	384	1.01	64	<.01	.19	.08	.01	.01	2	0
36	327	.48	17	159	678	1.7	125	.02	.4	.25	.08	.08	4	0
8	84	.05	4	36	202	.39	28	.01	.11	.06	.02	.02	1	0
12	233	.15	8	61	2	1.13	22	.01	.17	.13	.03	.03	3	0
89	254	.17	25	315	98	1.23	259	.08	.27	.36	.19	.09	6	2
6	16	.01	2	19	6	.08	16	<.01	.02	.02	.01	.01	<1	<1
159	231	.1	21	293	95	.65	437	.08	.36	.35	.14	.08	6	2
10	14	.01	1	18	6	.04	27	<.01	.02	.02	.01	<.01	<1	<1
265	166	.07	17	231	82	.6	705	.06	1.43	.3	.1	.07	9	1
17	10	<.01	1	14	5	.04	44	<.01	.09	.02	.01	<.01	1	<1
326	154	.07	17	179	89	.55	1001	.05	1.5	.26	.09	.06	9	1
21	10	<.01	1	11	6	.03	63	<.01	.09	.02	.01	<.01	1	<1
46	61	.03	6	88	78	.22	124	.02	.36	.04	.04	.02	2	0
3	4	<.01		6	5	.01	8	<.01	.02	<.01	<.01	<.01	<1	0
102	267	.14	26	331	123	.62	449	.08	1.3	.34	.15	.04	25	2
6	16	.01	2	20	7	.04	27	<.01	.08	.02	.01	<.01	2	<1

Table H-1

Food Composition (Computer code number is for West Diet Analysis program) (For purposes of calculations, use "0" for t, <1, <.1, <.01, etc.)

Computer Code Number	Food Description	Measure	Wt (g)	H_2O (%)	Ener (kcal)	Prot (g)	Carb (g)	Dietary Fiber (g)	Fat (g)	Fat Breakdown (g) Sat	Mono	Poly
	DAIRY—Continued											
	Cream products—imitation and part dairy:											
81	Coffee whitener, frozen or liquid	1 tbs	15	77	20	<1	2	0	1	1.4	t	0
82	Coffee whitener, powdered	1 tsp	2	2	11	<1	1	0	1	.6	t	0
83	Dessert topping, frozen, nondairy:	1 c	75	50	239	1	17	0	19	16.4	1.2	.4
84	Tablespoon	1 tbs	5	50	16	<1	1	0	1	1.1	.1	t
85	Dessert topping, mix with whole milk:	1 c	80	67	151	3	13	0	10	8.6	.7	.2
86	Tablespoon	1 tbs	5	67	9	<1	1	0	1	.5	t	t
88	Dessert topping, pressurized	1 c	70	60	185	1	11	0	16	13.3	1.3	.2
87	Tablespoon	1 tbs	4	60	11	<1	1	0	1	.8	.1	t
91	Sour cream, imitation:	1 c	230	71	478	6	15	0	45	40.9	1.3	.1
92	Tablespoon	1 tbs	14	71	29	<1	1	0	3	2.5	.1	t
89	Sour dressing, part dairy:	1 c	235	75	418	8	11	0	39	31.3	4.6	1.1
90	Tablespoon	1 tbs	15	75	27	1	1	0	2	2	.3	.1
	Milk, fluid:											
93	Whole milk	1 c	244	88	150	8	11	0	8	5.1	2.7	.3
94	2% reduced-fat milk	1 c	244	89	121	8	12	0	5	2.9	1.3	.2
95	2% milk solids added	1 c	245	89	125	9	12	0	5	2.9	1.4	.2
96	1% lowfat milk	1 c	244	90	102	8	12	0	3	1.6	.7	.1
97	1% milk solids added	1 c	245	90	104	9	12	0	2	1.5	.7	.1
98	Nonfat milk, vitamin A added	1 c	245	91	85	8	12	0	<1	.3	.1	t
99	Nonfat milk solids added	1 c	245	90	90	9	12	0	1	.4	.2	t
100	Buttermilk, skim	1 c	245	90	99	8	12	0	2	1.3	.6	.1
	Milk, canned:											
101	Sweetened condensed	1 c	306	27	982	24	166	0	27	16.8	7.4	1
103	Evaporated, nonfat	1 c	256	79	199	19	29	0	1	.3	.2	t
	Milk, dried:											
104	Buttermilk, sweet	1 c	120	3	464	41	59	0	7	4.3	2	.3
105	Instant, nonfat, vit A added (makes 1 qt)	1 ea	91	4	326	32	47	0	1	.4	.2	t
106	Instant nonfat, vit A added	1 c	68	4	243	24	35	0	<1	.3	.1	t
107	Goat milk	1 c	244	87	168	9	11	0	10	6.5	2.7	.4
108	Kefir	1 c	233	88	149	8	11	0	8			
	Milk beverages and powdered mixes:											
	Chocolate:											
109	Whole	1 c	250	82	209	8	26	2	8	5.3	2.5	.3
110	2% fat	1 c	250	84	179	8	26	1	5	3.1	1.5	.2
111	1% fat	1 c	250	84	158	9	26	1	2	1.5	.7	.1
	Chocolate-flavored beverages:											
112	Powder containing nonfat dry milk:	1 oz	28	1	101	3	22	<1	1	.7	.4	t
113	Prepared with water	1 c	275	86	138	4	30	3	2	.9	.5	t
114	Powder without nonfat dry milk:	1 oz	28	1	98	1	25	2	1	.5	.3	t
115	Prepared with whole milk	1 c	266	81	226	9	31	1	9	5.5	2.6	.3
116	Eggnog, commercial	1 c	254	74	343	10	34	0	19	11.3	5.7	.9
974	Eggnog, 2% reduced-fat	1 c	254	85	189	12	17	0	8	3.7	2.7	.7
1027	Instant Breakfast, envelope, powder only:	1 ea	37	7	131	7	24	<1	1	.3	.1	t
1028	Prepared with whole milk	1 c	281	77	280	15	36	<1	9	5.4	2.5	.3
1029	Prepared with 2% milk	1 c	281	78	252	15	36	<1	5	3.3	1.5	.2
1283	Prepared with 1% milk	1 c	281	79	233	15	36	<1	3	1.9	.9	.1
1284	Prepared with nonfat milk	1 c	282	80	216	16	36	<1	1	.7	.3	t
117	Malted milk, chocolate, powder:	3 tsp	21	1	79	1	18	<1	1	.5	.2	.1
118	Prepared with whole milk	1 c	265	81	228	9	30	<1	9	5.5	2.6	.4
1661	Ovaltine with whole milk	1 c	265	81	225	9	29	<1	9	5.5	2.5	.4
119	Malted mix powder, natural:	3 tsp	21	2	87	2	16	<1	2	.9	.4	.3
120	Prepared with whole milk	1 c	265	81	236	10	27	<1	10	6	2.8	.6
121	Milk shakes, chocolate	1 c	166	71	211	6	34	1	6	3.8	1.8	.2
122	Milk shakes, vanilla	1 c	166	75	184	6	30	1	5	3.1	1.4	.2
	Milk desserts:											
134	Custard, baked	1 c	282	79	296	14	30	0	13	6.6	4.3	1
1548	Low-fat frozen dessert bars	1 ea	81	72	88	2	19	0	1	.2	.1	.4
	Ice cream, vanilla (about 10% fat):											
123	Hardened: ½ gallon	1 ea	1064	61	2138	37	251	0	117	72.4	33.8	4.4
124	Cup	1 c	132	61	265	5	31	0	14	9	4.2	.5
126	Soft serve	1 c	172	60	370	7	38	0	22	12.9	6	.8

Chol (mg)	Calc (mg)	Iron (mg)	Magn (mg)	Pota (mg)	Sodi (mg)	Zinc (mg)	VT-A (RE)	Thia (mg)	VT-E (α-TE)	Ribo (mg)	Niac (mg)	V-B6 (mg)	Fola (μg)	VT-C (mg)
0	1	<.01		29	12	<.01	1	0	.24	0	0	0	0	0
0		.02		16	4	.01	<1	0	<.01	<.01	0	0	0	0
0	5	.09	1	14	19	.02	64	0	.14	0	0	0	0	0
0		.01		1	1	<.01	4	0	.01	0	0	0	0	0
8	72	.03	8	121	53	.22	39	.02	.11	.09	.05	.02	3	1
<1	5	<.01		8	3	.01	2	<.01	.01	.01	<.01	<.01	<1	<1
0	4	.01	1	13	43	.01	33	0	.12	0	0	0	0	0
0		<.01		1	2	0	2	0	.01	0	0	0	0	0
0	6	.9	15	370	235	2.71	0	0	.34	0	0	0	0	0
0		.05	1	22	14	.16	0	0	.02	0	0	0	0	0
13	266	.07	23	381	113	.87	5	.09	.29	.38	.17	.04	28	2
1	17	<.01	1	24	7	.06	<1	.01	.02	.02	.01	<.01	2	<1
33	290	.12	33	371	120	.93	76	.09	.24	.39	.2	.1	12	2
18	298	.12	33	376	122	.95	139	.09	.17	.4	.21	.1	12	2
18	314	.12	35	397	128	.98	140	.1	.17	.42	.22	.11	13	2
10	300	.12	34	381	123	.95	144	.09	.1	.41	.21	.1	12	2
10	314	.12	35	397	128	.98	145	.1	.1	.42	.22	.11	13	2
4	301	.1	28	407	126	.98	149	.09	.1	.34	.22	.1	13	2
5	316	.12	35	419	130	1	149	.1	.1	.43	.22	.11	13	2
9	284	.12	27	370	257	1.03	20	.08	.15	.38	.14	.08	12	2
104	869	.58	79	1135	389	2.88	248	.27	.65	1.27	.64	.16	34	8
9	742	.74	69	850	294	2.3	300	.11	.01	.79	.44	.14	22	3
83	1420	.36	132	1910	620	4.82	65	.47	.48	1.9	1.05	.41	57	7
17	1119	.28	106	1551	500	4.01	646	.38	.02	1.58	.81	.31	45	5
12	836	.21	80	1159	373	3	483	.28	.01	1.18	.61	.23	34	4
28	327	.12	34	498	122	.73	137	.12	.22	.34	.68	.11	1	3
		.3	33	373	107									
30	280	.6	32	418	149	1.03	72	.09	.23	.4	.31	.1	12	2
17	285	.6	33	423	151	1.03	143	.09	.13	.41	.31	.1	12	2
7	288	.6	33	425	152	1.03	148	.09	.06	.41	.32	.1	12	2
1	91	.33	23	199	141	.41	1	.03	.04	.16	.16	.03	0	<1
3	129	.47	33	270	198	.6	1	.04	.06	.21	.22	.04	0	1
0	10	.88	27	165	59	.43	1	.01	.11	.04	.14	<.01	2	<1
32	301	.8	53	497	165	1.28	77	.1	.21	.43	.32	.1	12	2
149	330	.51	47	419	138	1.17	203	.09	.58	.48	.27	.13	2	4
194	269	.71	32	367	155	1.26	197	.11	1.01	.55	.21	.15	30	2
4	105	4.74	84	350	142	3.16	554	.31	5.31	.07	5.27	.42	105	28
38	396	4.86	117	719	262	4.09	630	.41	5.51	.47	5.46	.52	118	31
23	401	4.86	118	726	264	4.12	693	.41	5.41	.48	5.46	.53	118	31
14	406	4.86	118	731	266	4.12	698	.41	5.36	.48	5.47	.53	118	31
9	407	4.83	112	755	268	4.14	703	.4	5.3	.42	5.47	.52	118	31
1	13	.48	15	130	53	.17	4	.04	.08	.04	.42	.03	4	<1
34	305	.61	48	498	172	1.09	79	.13	.26	.44	.62	.13	16	3
34	384	3.76	53	620	244	1.17	901	.73	.32	1.26	10.9	1.02	32	34
4	63	.15	19	159	104	.21	18	.11	.08	.19	1.1	.09	10	1
37	355	.26	53	530	223	1.14	95	.2	.32	.59	1.31	.19	22	3
22	188	.51	28	332	161	.68	38	.1	.11	.41	.27	.08	6	1
18	203	.15	20	289	136	.6	53	.07	.1	.3	.31	.09	5	1
245	316	.85	39	431	217	1.49	169	.09	.68	.64	.24	.14	28	1
1	82	.07	10	111	47	.26	38	.03	.07	.11	.06	.03	3	1
468	1361	.96	149	2117	851	7.34	1244	.44	0	2.55	1.23	.51	53	6
58	169	.12	18	263	106	.91	154	.05	0	.32	.15	.06	7	1
157	225	.36	21	304	105	.89	265	.08	.64	.31	.16	.08	15	1

Table H-1

Food Composition (Computer code number is for West Diet Analysis program) (For purposes of calculations, use "0" for t, <1, <.1, <.01, etc.)

Computer Code Number	Food Description	Measure	Wt (g)	H_2O (%)	Ener (kcal)	Prot (g)	Carb (g)	Dietary Fiber (g)	Fat (g)	Fat Breakdown (g) Sat	Mono	Poly
	DAIRY—Continued											
	Ice cream, rich vanilla (16% fat):											
127	Hardened: ½ gallon	1 ea	1188	60	2554	49	264	0	154	88.9	41.5	5.5
128	Cup	1 c	148	57	357	5	33	0	24	14.8	6.9	.9
1724	Ben & Jerry's	½ c	108		230	4	21	0	17	10		
	Ice milk, vanilla (about 4% fat):											
129	Hardened: ½ gallon	1 ea	1048	68	1456	40	238	0	45	27.7	12.9	1.7
130	Cup	1 c	132	68	183	5	30	0	6	3.5	1.6	.2
131	Soft serve (about 3% fat)	1 c	176	70	222	9	38	0	5	2.9	1.3	.2
	Pudding, canned (5 oz can = .55 cup):											
135	Chocolate	1 ea	142	69	189	4	32	1	6	1	2.4	2
136	Tapioca	1 ea	142	74	169	3	27	<1	5	.9	2.3	1.9
137	Vanilla	1 ea	142	71	185	3	31	<1	5	.8	2.2	1.9
	Puddings, dry mix with whole milk:											
138	Chocolate, instant	1 c	294	74	326	9	55	3	9	5.4	2.7	.5
139	Chocolate, regular, cooked	1 c	284	74	315	9	51	3	10	5.9	2.8	.4
140	Rice, cooked	1 c	288	72	351	9	60	<1	8	5.1	2.4	.3
141	Tapioca, cooked	1 c	282	74	321	8	55	0	8	5.1	2.3	.3
142	Vanilla, instant	1 c	284	73	324	8	56	0	8	4.9	2.4	.4
143	Vanilla, regular, cooked	1 c	280	75	311	8	52	0	8	5.1	2.4	.4
	Sherbet (2% fat):											
132	½ gallon	1 ea	1542	66	2127	17	469	8	31	17.9	8.3	1.2
133	Cup	1 c	198	66	273	2	60	1	4	2.3	1.1	.2
144	Soy milk	1 c	245	93	81	7	4	3	5	.7	1	2.6
2301	Soy milk, fortified, fat free	1 c	240	88	110	6	22	1	0	0	0	0
	Yogurt, frozen, low-fat											
1584	Cup	1 c	144	65	229	6	35	0	8	4.9	2.3	.3
1512	Scoop	1 ea	79	74	78	4	15	0	<1	.1	t	t
	Yogurt, lowfat:											
1172	Fruit added with low-calorie sweetener	1 c	241	86	122	12	19	1	<1	.2	.1	t
145	Fruit added	1 c	245	74	250	11	47	<1	3	1.7	.7	.1
146	Plain	1 c	245	85	155	13	17	0	4	2.4	1	.1
147	Vanilla or coffee flavor	1 c	245	79	209	13	34	0	3	2	.8	.1
148	Yogurt, made with nonfat milk	1 c	245	85	137	15	19	0	<1	.3	.1	t
149	Yogurt, made with whole milk	1 c	245	88	150	9	11	0	8	5.1	2.2	.2
	EGGS											
	Raw, large:											
150	Whole, without shell	1 ea	50	75	74	6	1	0	5	1.5	1.9	.7
151	White	1 ea	33	88	16	3	<1	0	0	0	0	0
152	Yolk	1 ea	17	49	61	3	<1	0	5	1.6	2	.7
	Cooked:											
153	Fried in margarine	1 ea	46	69	91	6	1	0	7	1.9	2.8	1.3
154	Hard-cooked, shell removed	1 ea	50	75	77	6	1	0	5	1.6	2	.7
155	Hard-cooked, chopped	1 c	136	75	211	17	2	0	14	4.4	5.5	1.9
156	Poached, no added salt	1 ea	50	75	74	6	1	0	5	1.5	1.9	.7
157	Scrambled with milk & margarine	1 ea	61	73	101	7	1	0	7	2.2	2.9	1.3
1681	Egg substitute, liquid:	½ c	126	83	106	16	1	0	4	.8	1.1	2
1254	Egg Beaters, Fleischmann's	½ c	122		60	12	2	0	0	0	0	0
1262	Egg substitute, liquid, prepared	½ c	105	80	100	14	1	0	4	.8	1.1	1.9
	FATS and OILS											
158	Butter: Stick	½ c	114	16	817	1	<1	0	92	57.7	27.4	3.4
159	Tablespoon:	1 tbs	14	16	100	<1	<1	0	11	7.1	3.4	.4
8025	Unsalted	1 tbs	14	18	100	<1	<1	0	11	7.1	3.4	.4
160	Pat (about 1 tsp)	1 ea	5	16	36	<1	<1	0	4	2.5	1.2	.2
1682	Whipped	1 tsp	3	16	21	<1	<1	0	2	1.5	.7	.1
	Fats, cooking:											
1363	Bacon fat	1 tbs	14		125	0	0	0	14	6.3	5.9	1.1
1362	Beef fat/tallow	1 c	205	0	1849	0	0	0	205	103	87.3	8.2
1364	Chicken fat	1 c	205		1845	0	0	0	205	61.1	91.6	42.8
161	Vegetable shortening:	1 c	205	0	1812	0	0	0	205	52.1	91.2	53.5
162	Tablespoon	1 tbs	13	0	115	0	0	0	13	3.3	5.8	3.4

Chol (mg)	Calc (mg)	Iron (mg)	Magn (mg)	Pota (mg)	Sodi (mg)	Zinc (mg)	VT-A (RE)	Thia (mg)	VT-E (α-TE)	Ribo (mg)	Niac (mg)	V-B6 (mg)	Fola (μg)	VT-C (mg)
1081	1556	2.49	143	2102	725	6.18	1829	.58	4.4	2.16	1.13	.57	107	9
90	173	.07	16	235	83	.59	272	.06	0	.24	.12	.06	7	1
95	150	.36			55		214		0					0
147	1456	1.05	157	2211	891	4.61	493	.61	0	2.78	.94	.68	63	8
18	183	.13	20	279	112	.58	62	.08	0	.35	.12	.09	8	1
21	276	.11	25	389	123	.93	51	.09	0	.35	.21	.08	11	2
4	128	.72	30	256	183	.6	16	.04	.18	.22	.49	.04	4	3
1	119	.33	11	148	168	.38	0	.03	.13	.14	.44	.14	4	1
10	125	.18	11	160	192	.35	9	.03	.18	.2	.36	.02	0	0
32	300	.85	53	488	835	1.23	62	.1	.18	.41	.28	.11	12	3
34	315	1.02	43	463	293	1.28	74	.09	.17	.49	.29	.1	11	2
32	297	1.09	37	372	314	1.09	58	.22	.17	.4	1.28	.1	11	2
34	293	.17	34	372	341	.96	76	.08	.23	.4	.21	.11	11	2
31	287	.2	34	364	812	.94	71	.09	.18	.39	.21	.1	11	2
34	300	.14	36	381	448	.98	76	.08	.17	.4	.21	.09	11	2
77	833	2.16	123	1480	709	7.4	216	.39	.88	1.05	1.48	.52	77	66
10	107	.28	16	190	91	.95	28	.05	.11	.13	.19	.07	10	9
0	10	1.42	47	345	29	.56	7	.39	.02	.17	.36	.1	4	0
0	400	1.44		20	60		0	.07		.1	3			0
3	206	.43	20	304	125	.6	82	.05	.07	.32	.41	.11	9	1
1	137	.07	13	175	53	.67	1	.03	<.01	.16	.08	.04	8	1
3	369	.61	41	550	139	1.83	6	.1	.17	.45	.5	.11	32	26
10	372	.17	36	478	143	1.81	27	.09	.07	.44	.23	.1	23	2
15	448	.2	43	573	172	2.18	39	.11	.1	.52	.28	.12	27	2
12	419	.17	40	537	161	2.03	32	.1	.08	.49	.26	.11	26	2
4	488	.22	47	625	187	2.38	5	.12	.01	.57	.3	.13	30	2
31	296	.12	28	380	114	1.45	73	.07	.22	.35	.18	.08	18	1
213	24	.72	5	60	63	.55	95	.03	.52	.25	.04	.07	23	0
0	2	.01	4	47	54	<.01	0	<.01	0	.15	.03	<.01	1	0
218	23	.6	2	16	7	.53	99	.03	.54	.11	<.01	.07	25	0
211	25	.72	5	61	162	.55	114	.03	.75	.24	.03	.07	17	0
212	25	.59	5	63	62	.52	84	.03	.52	.26	.03	.06	22	0
577	68	1.62	14	171	169	1.43	228	.09	1.43	.7	.09	.16	60	0
212	24	.72	5	60	140	.55	95	.02	.52	.21	.03	.06	17	0
215	43	.73	7	84	171	.61	119	.03	.8	.27	.05	.07	18	<1
1	67	2.65	11	416	223	1.64	272	.14	.61	.38	.14	<.01	19	0
0	80	2.16		170	200		80		.59					
1	63	2.51	10	394	211	1.55	258	.11	.58	.34	.12	<.01	13	0
250	27	.18	2	30	942	.06	860	.01	1.8	.04	.05	<.01	3	0
31	3	.02		4	116	.01	106	<.01	.22	<.01	.01	0	<1	0
31	3	.02		4	2	.01	106	<.01	.22	<.01	.01	0	<1	0
11	1	.01		1	41	<.01	38	0	.08	<.01	<.01	0	<1	0
7	1	<.01		1	25	<.01	23	0	.05	<.01	<.01	0	<1	0
14		0			76	<.01	0	0	.31	0	0	0	0	0
223	0	0	0		<1	0	0	0	3.08	0	0	0	0	0
174	0	0	0	0	0	0	0	0	5.54	0	0	0	0	0
0	0	0	0	0	0	0	0	0	17	0	0	0	0	0
0	0	0	0	0	0	0	0	0	1.08	0	0	0	0	0

Table H-1

Food Composition (Computer code number is for West Diet Analysis program) (For purposes of calculations, use "0" for t, <1, <.1, <.01, etc.)

Computer Code Number	Food Description	Measure	Wt (g)	H_2O (%)	Ener (kcal)	Prot (g)	Carb (g)	Dietary Fiber (g)	Fat (g)	Fat Breakdown (g) Sat	Mono	Poly
	FATS and OILS—Continued											
163	Lard:	1 c	205	0	1849	0	0	0	205	81.1	87	28.3
164	Tablespoon	1 tbs	13	0	117	0	0	0	13	5.1	5.5	1.8
	Margarine:											
165	Imitation (about 40% fat), soft:	1 c	232	58	800	1	1	0	90	17.9	36.4	32
166	Tablespoon	1 tbs	14	58	48	<1	<1	0	5	1.1	2.2	1.9
167	Regular, hard (about 80% fat):	½ c	114	16	820	1	1	0	92	18	40.8	29
168	Tablespoon	1 tbs	14	16	101	<1	<1	0	11	2.2	5	3.6
169	Pat	1 ea	5	16	36	<1	<1	0	4	.8	1.8	1.3
170	Regular, soft (about 80% fat):	1 c	227	16	1625	2	1	0	183	31.3	64.7	78.5
171	Tablespoon	1 tbs	14	16	100	<1	<1	0	11	1.9	4	4.8
2056	Saffola, unsalted	1 tbs	14	20	100	0	0	0	11	2	3	4.5
2057	Saffola, reduced fat	1 tbs	14	37	60	0	0	0	8	1.3	2.7	4.4
172	Spread (about 60% fat), hard:	1 c	227	37	1225	1	0	0	138	32	59	41.1
173	Tablespoon	1 tbs	14	37	76	<1	0	0	9	2	3.6	2.5
174	Pat	1 ea	5	37	27	<1	0	0	3	.7	1.2	1
175	Spread (about 60% fat), soft:	1 c	227	37	1225	1	0	0	138	29.3	71.5	31.3
176	Tablespoon	1 tbs	14	37	76	<1	0	0	9	1.8	4.4	1.9
2160	Touch of Butter (47% fat)	1 tbs	14		60	0	0	0	7	1.5	3.1	1.5
	Oils:											
1585	Canola:	1 c	218	0	1927	0	0	0	218	15.5	128	64.5
1586	Tablespoon	1 tbs	14	0	124	0	0	0	14	1	8.2	4.1
177	Corn:	1 c	218	0	1927	0	0	0	218	29.4	54.1	131
178	Tablespoon	1 tbs	14	0	124	0	0	0	14	1.9	3.5	8.4
179	Olive:	1 c	216	0	1909	0	0	0	216	29.4	159	21.3
180	Tablespoon	1 tbs	14	0	124	0	0	0	14	1.9	10.3	1.4
1683	Olive, extra virgin	1 tbs	14		126	0	0	0	14	2	10.8	1.3
181	Peanut:	1 c	216	0	1909	0	0	0	216	40	99.8	71.3
182	Tablespoon	1 tbs	14	0	124	0	0	0	14	2.6	6.5	4.6
183	Safflower:	1 c	218	0	1927	0	0	0	218	19.8	26.4	162
184	Tablespoon	1 tbs	14	0	124	0	0	0	14	1.3	1.7	10.4
185	Soybean:	1 c	218	0	1927	0	0	0	218	32	50.8	126
186	Tablespoon	1 tbs	14	0	124	0	0	0	14	2.1	3.3	8.1
187	Soybean/cottonseed:	1 c	218	0	1927	0	0	0	218	39.5	64.3	105
188	Tablespoon	1 tbs	14	0	124	0	0	0	14	2.5	4.1	6.7
189	Sunflower:	1 c	218	0	1927	0	0	0	218	25.3	42.5	143
190	Tablespoon	1 tbs	14	0	124	0	0	0	14	1.6	2.7	9.2
	Salad dressings/sandwich spreads:											
191	Blue cheese, regular	1 tbs	15	32	76	1	1	0	8	1.5	1.8	4.2
1040	Low calorie	1 tbs	15	79	15	1	<1	<1	1	.2	.5	.4
1684	Caesar's	1 tbs	12	36	56	1	<1	<1	5	1	3.7	.5
192	French, regular	1 tbs	16	38	69	<1	3	0	7	1.5	1.3	3.5
193	Low calorie	1 tbs	16	71	21	<1	3	0	1	.1	.2	.5
194	Italian, regular	1 tbs	15	40	70	<1	2	0	7	1	1.7	4.2
195	Low calorie	1 tbs	15	84	16	<1	1	<1	1	.2	.3	.9
	Kraft, Deliciously Right:											
2150	1000 Island	1 tbs	16		35	0	4	0	2	.5		
2153	Bacon & tomato	1 tbs	16		31	1	2	0	3	.5		
2154	Cucumber ranch	1 tbs	16		31	0	1	0	3	.5		
2151	French	1 tbs	16		25	0	3	0	1	.2		
2152	Ranch	1 tbs	16		52	0	3	0	5	.8		
199	Mayo type, regular	1 tbs	15	40	58	<1	4	0	5	.7	1.3	2.7
1030	Low calorie	1 tbs	14	54	36	<1	3	0	3	.4	.7	1.4
	Mayonnaise:											
197	Imitation, low calorie	1 tbs	15	63	35	<1	2	0	3	.5	.7	1.6
196	Regular (soybean)	1 tbs	14	17	100	<1	<1	0	11	1.7	3.2	5.8
1488	Regular, low calorie, low sodium	1 tbs	14	63	32	<1	2	0	3	.5	.6	1.4
1493	Regular, low calorie	1 tbs	15	63	35	<1	2	0	3	.5	.7	1.6
198	Ranch, regular	1 tbs	15	39	80	0	<1	0	8	1.2		
2251	Low calorie	1 tbs	14	69	30	<1	1	0	2	.5		
1685	Russian	1 tbs	15	34	74	<1	2	0	8	1.1	1.8	4.4
1502	Salad dressing, low calorie, oil free	1 tbs	15	88	4	<1	1	<1	<1	0	0	0

Chol (mg)	Calc (mg)	Iron (mg)	Magn (mg)	Pota (mg)	Sodi (mg)	Zinc (mg)	VT-A (RE)	Thia (mg)	VT-E (α-TE)	Ribo (mg)	Niac (mg)	V-B6 (mg)	Fola (μg)	VT-C (mg)
195		0			<1	.23	0	0	2.46	0	0	0	0	0
12		0			<1	.01	0	0	.16	0	0	0	0	0
0	41	0	4	59	2227	0	1853	.01	5.41	.05	.03	.01	2	<1
0	2	0		4	134	0	112	<.01	.33	<.01	<.01	<.01	<1	<1
0	34	.07	3	48	1075	0	911	.01	14.6	.04	.03	.01	1	<1
0	4	.01		6	132	0	112	<.01	1.79	<.01	<.01	<.01	<1	<1
0	1	<.01		2	47	0	40	<.01	.64	<.01	<.01	0	<1	<1
0	60	0	5	86	2447	0	1813	.02	27.2	.07	.04	.02	2	<1
0	4	0		5	151	0	112	<.01	1.68	<.01	<.01	<.01	<1	<1
	0	0			0		51							0
	0	0			115		51							0
0	47	0	4	68	2256	0	1813	.02	11.4	.06	.04	.01	2	<1
0	3	0		4	139	0	112	<.01	.7	<.01	<.01	<.01	<1	<1
0	1	0		1	50	0	40	0	.25	<.01	<.01	0	<1	<1
0	47	0	4	68	2256	0	1813	.02	20.5	.06	.04	.01	2	<1
0	3	0		4	139	0	112	<.01	1.26	<.01	<.01	<.01	<1	<1
0	0	0		0	110		100		1.27					0
0	0	0	0	0	0	0	0	0	45.8	0	0	0	0	0
0	0	0	0	0	0	0	0	0	2.94	0	0	0	0	0
0	0	0	0	0	0	0	0	0	46	0	0	0	0	0
0	0	0	0	0	0	0	0	0	2.95	0	0	0	0	0
0		.82		0	<1	.13	0	0	26.8	0	0	0	0	0
0		.05		0	<1	.01	0	0	1.74	0	0	0	0	0
									1.74					
0		.06			<1	.02	0	0	27.9	0	0	0	0	0
0		<.01			<1	<.01	0	0	1.81	0	0	0	0	0
0	0	0	0	0	0	0	0	0	94	0	0	0	0	0
0	0	0	0	0	0	0	0	0	6.03	0	0	0	0	0
0		.04		0	0	0	0	0	39.7	0	0	0	0	0
0		<.01		0	0	0	0	0	2.55	0	0	0	0	0
0	0	0	0	0	0	0	0	0	61.5	0	0	0	0	0
0	0	0	0	0	0	0	0	0	3.95	0	0	0	0	0
0	0	0	0	0	0	0	0	0	110	0	0	0	0	0
0	0	0	0	0	0	0	0	0	7.08	0	0	0	0	0
3	12	.03	0	6	164	.04	10	<.01	1.4	.01	.01	.01	1	<1
<1	13	.07	1	1	180	.04	<1	<.01	.13	.01	.01	<.01	<1	<1
12	23	.2	3	21	207	.13	7	<.01	.72	.02	.5	.01	2	1
0	2	.06	0	13	219	.01	21	<.01	1.35	<.01	0	<.01	1	0
0	2	.06	0	13	126	.03	21	0	.19	0	0	0	0	0
0	1	.03		2	118	.02	4	<.01	1.56	<.01	0	<.01	1	0
1		.03	0	2	118	.02	0	0	.22	0	0	0	0	0
2	0	0		27	160		0		.19					0
1	0	0		21	155		0		.75					0
0	0	0		10	232		0		.73					0
0	0	0		7	130		50		.42					0
0	0	0		5	165		0		1.31					0
4	2	.03		1	107	.03	13	<.01	.6	<.01	<.01	<.01	1	0
4	2	.03		1	99	.02	9	<.01	.6	<.01	0	<.01	1	0
4		0		1	75	.02	0	0	.96	0	0	0	0	0
8	3	.07		5	79	.02	12	0	1.65	0	<.01	.08	1	0
3	0	0	0	1	15	.01	1	0	.53	<.01	0	0	<1	0
4		0		1	75	.02	0	0	.96	0	0	0	0	0
5	0	0			105		0							0
5	5	0			120		0		.7					0
3	3	.09		24	130	.06	31	.01	1.53	.01	.09	<.01	2	1
0	1	.04	2	7	256	<.01	<1	0	0	0	<.01	<.01	<1	<1

Table H-1

Food Composition (Computer code number is for West Diet Analysis program) (For purposes of calculations, use "0" for t, <1, <.1, <.01, etc.)

Computer Code Number	Food Description	Measure	Wt (g)	H_2O (%)	Ener (kcal)	Prot (g)	Carb (g)	Dietary Fiber (g)	Fat (g)	Fat Breakdown (g) Sat	Mono	Poly
	FATS and OILS—Continued											
	Salad dressing, no cholesterol											
1605	Miracle Whip	1 tbs	15	57	48	0	2	0	4	1.1	1.1	2.1
203	Salad dressing, from recipe, cooked	1 tbs	16	69	25	1	2	0	2	.5	.6	.3
200	Tartar sauce, regular	1 tbs	14	34	74	<1	1	<1	8	1.5	2.6	4.1
1503	Low calorie	1 tbs	14	63	31	<1	2	<1	2	.4	.6	1.3
201	Thousand island, regular	1 tbs	16	46	60	<1	2	0	6	1	1.3	3.2
202	Low calorie	1 tbs	15	69	24	<1	2	<1	2	.2	.4	.9
204	Vinegar and oil	1 tbs	16	47	72	0	<1	0	8	1.5	2.4	3.9
	Wishbone:											
2180	Creamy Italian, lite	1 tbs	15		26	<1	2		2	.4	.9	.7
2166	Italian, lite	1 tbs	16	90	6	0	1		<1	0	.2	.1
8427	Ranch, lite	1 tbs	15	56	50	0	2	0	4	.7		
	FRUITS and FRUIT JUICES											
	Apples:											
	Fresh, raw, with peel:											
205	2¾" diam (about 3 per lb w/cores)	1 ea	138	84	81	<1	21	4	<1	.1	t	.1
206	3¼" diam (about 2 per lb w/cores)	1 ea	212	84	125	<1	32	6	1	.1	t	.2
207	Raw, peeled slices	1 c	110	84	63	<1	16	2	<1	.1	t	.1
208	Dried, sulfured	10 ea	64	32	156	1	42	6	<1	t	t	.1
209	Apple juice, bottled or canned	1 c	248	88	117	<1	29	<1	<1	t	t	.1
210	Applesauce, sweetened	1 c	255	80	194	<1	51	3	<1	.1	t	.1
211	Applesauce, unsweetened	1 c	244	88	105	<1	28	3	<1	t	t	t
	Apricots:											
212	Raw, w/o pits (about 12 per lb w/pits)	3 ea	105	86	50	1	12	3	<1	t	.2	.1
	Canned (fruit and liquid):											
213	Heavy syrup	1 c	240	78	199	1	52	4	<1	t	.1	t
214	Halves	3 ea	120	78	100	1	26	2	<1	t	t	t
215	Juice pack	1 c	244	87	117	2	30	4	<1	t	t	t
216	Halves	3 ea	108	87	52	1	13	2	<1	t	t	t
217	Dried, halves	10 ea	35	31	83	1	22	3	<1	t	.1	t
218	Dried, cooked, unsweetened, w/liquid	1 c	250	76	213	3	55	8	<1	t	.2	.1
219	Apricot nectar, canned	1 c	251	85	141	1	36	2	<1	t	.1	t
	Avocados, raw, edible part only:											
220	California	1 ea	173	73	306	4	12	8	30	4.5	19.5	3.5
221	Florida	1 ea	304	80	340	5	27	16	27	5.3	14.8	4.5
222	Mashed, fresh, average	1 c	230	74	370	5	17	11	35	5.6	22.1	4.5
	Bananas, raw, without peel:											
223	Whole, 8¾" long (175g w/peel)	1 ea	118	74	109	1	28	3	1	.2	t	.1
224	Slices	1 c	150	74	138	2	35	4	1	.3	.1	.1
1285	Bananas, dehydrated slices	½ c	50	3	173	2	44	4	1	.3	.1	.2
225	Blackberries, raw	1 c	144	86	75	1	18	8	1	t	.1	.3
	Blueberries:											
226	Fresh	1 c	145	85	81	1	20	4	1	t	.1	.2
227	Frozen, sweetened	10 oz	284	77	230	1	62	6	<1	t	.1	.2
228	Frozen, thawed	1 c	230	77	186	1	51	5	<1	t	t	.1
	Cherries:											
229	Sour, red pitted, canned water pack	1 c	244	90	88	2	22	3	<1	.1	.1	.1
230	Sweet, red pitted, raw	10 ea	68	81	49	1	11	2	1	.1	.2	.2
231	Cranberry juice cocktail,vitamin C added	1 c	253	85	144	0	36	<1	<1	t	t	.1
1411	Cranberry juice, low calorie	1 c	237	95	45	0	11	<1	0	0	0	0
232	Cranberry-apple juice,vitamin C added	1 c	245	83	164	<1	42	<1	0	0	0	0
233	Cranberry sauce, canned, strained	1 c	277	61	418	1	108	3	<1	t	.1	.2
234	Dates, whole, without pits	10 ea	83	22	228	2	61	6	<1	.2	.1	t
235	Dates, chopped	1 c	178	22	490	4	131	13	1	.3	.3	.1
236	Figs, dried	10 ea	190	28	485	6	124	18	2	.4	.5	1.1
	Fruit cocktail, canned, fruit and liq:											
237	Heavy syrup pack	1 c	248	80	181	1	47	2	<1	t	t	.1
238	Juice pack	1 c	237	87	109	1	28	2	<1	t	t	t
	Grapefruit:											
	Raw 3¾" diam (half w/rind = 241g)											
239	Pink/red, half fruit, edible part	1 ea	123	91	37	1	9	2	<1	t	t	t
240	White, half fruit, edible part	1 ea	118	90	39	1	10	1	<1	t	t	t
241	Canned sections with light syrup	1 c	254	84	152	1	39	1	<1	t	t	.1

Chol (mg)	Calc (mg)	Iron (mg)	Magn (mg)	Pota (mg)	Sodi (mg)	Zinc (mg)	VT-A (RE)	Thia (mg)	VT-E (α-TE)	Ribo (mg)	Niac (mg)	V-B6 (mg)	Fola (μg)	VT-C (mg)
0	0	<.01	0	0	102	0	2	0	.64	0	0	0	0	0
9	13	.08	1	19	117	.06	20	.01	.3	.02	.04	<.01	1	<1
7	3	.13		11	99	.02	9	<.01	2.24	<.01	0	<.01	1	<1
3	2	.09		5	83	.02	2	0	.83	<.01	.01	<.01	<1	<1
4	2	.1		18	112	.02	15	<.01	.18	<.01	<.01	<.01	1	0
2	2	.09		17	150	.02	14	<.01	.18	<.01	0	<.01	1	0
0	0	0	0	1	<1	0	0	0	1.41	0	0	0	0	0
<1	0	0			148		0	0	.56	0	0			0
0	1	0			255		2	0	.24	0	0			<1
2	0	0			120		0							0
0	10	.25	7	159	0	.05	7	.02	.44	.02	.11	.07	4	8
0	15	.38	11	244	0	.08	11	.04	.68	.03	.16	.1	6	12
0	4	.08	3	124	0	.04	4	.02	.09	.01	.1	.05	<1	4
0	9	.9	10	288	56	.13	4	0	.35	.1	.59	.08	0	2
0	17	.92	7	295	7	.07	<1	.05	.02	.04	.25	.07	<1	2
0	10	.89	8	156	8	.1	3	.03	.03	.07	.48	.07	2	4
0	7	.29	7	183	5	.07	7	.03	.02	.06	.46	.06	1	3
0	15	.57	8	311	1	.27	274	.03	.93	.04	.63	.06	9	10
0	22	.72	17	336	10	.26	295	.05	2.14	.05	.9	.13	4	7
0	11	.36	8	168	5	.13	148	.02	1.07	.03	.45	.06	2	4
0	29	.73	24	403	10	.27	412	.04	2.17	.05	.84	.13	4	12
0	13	.32	11	178	4	.12	183	.02	.96	.02	.37	.06	2	5
0	16	1.65	16	482	3	.26	253	<.01	.52	.05	1.05	.05	4	1
0	40	4.18	42	1222	7	.65	590	.01	1.25	.07	2.36	.28	0	4
0	18	.95	13	286	8	.23	331	.02	.2	.03	.65	.05	3	2
0	19	2.04	71	1096	21	.73	106	.19	2.32	.21	3.32	.48	113	14
0	33	1.61	103	1483	15	1.28	185	.33	2.37	.37	5.84	.85	162	24
0	25	2.35	90	1377	23	.97	140	.25	3.08	.28	4.42	.64	142	18
0	7	.37	34	467	1	.19	9	.05	.32	.12	.64	.68	22	11
0	9	.46	43	594	1	.24	12	.07	.4	.15	.81	.87	29	14
0	11	.57	54	746	1	.3	15	.09	0	.12	1.4	.22	7	3
0	46	.82	29	282	0	.39	23	.04	1.02	.06	.58	.08	49	30
0	9	.25	7	129	9	.16	14	.07	1.45	.07	.52	.05	9	19
0	17	1.11	6	170	3	.17	11	.06	2.02	.15	.72	.17	19	3
0	14	.9	5	138	2	.14	9	.05	1.63	.12	.58	.14	15	2
0	27	3.34	15	239	17	.17	183	.04	.32	.1	.43	.11	19	5
0	10	.26	7	152	0	.04	14	.03	.09	.04	.27	.02	3	5
0	8	.38	5	45	5	.18	1	.02	0	.02	.09	.05	<1	90
0	21	.09	5	52	7	.05	1	.02	0	.02	.08	.04	<1	76
0	17	.15	5	66	5	.1	1	.01	0	.05	.15	.05	<1	78
0	11	.61	8	72	80	.14	6	.04	.28	.06	.28	.04	3	6
0	27	.95	29	541	2	.24	4	.07	.08	.08	1.83	.16	10	0
0	57	2.05	62	1160	5	.52	9	.16	.18	.18	3.92	.34	22	0
0	274	4.24	112	1352	21	.97	25	.13	9.5	.17	1.32	.43	14	2
0	15	.72	12	218	15	.2	50	.04	.72	.05	.93	.12	6	5
0	19	.5	17	225	9	.21	73	.03	.47	.04	.95	.12	6	6
0	13	.15	10	159	0	.09	32	.04	.31	.02	.23	.05	15	47
0	14	.07	11	175	0	.08	1	.04	.29	.02	.32	.05	12	39
0	36	1.02	25	328	5	.2	0	.1	.63	.05	.62	.05	22	54

Table H-1

Food Composition (Computer code number is for West Diet Analysis program) (For purposes of calculations, use "0" for t, <1, <.1, <.01, etc.)

Computer Code Number	Food Description	Measure	Wt (g)	H_2O (%)	Ener (kcal)	Prot (g)	Carb (g)	Dietary Fiber (g)	Fat (g)	Fat Breakdown (g) Sat	Mono	Poly
	FRUITS and FRUIT JUICES											
	Grapefruit juice:											
242	Fresh, white, raw	1 c	247	90	96	1	23	<1	<1	t	t	.1
243	Canned, unsweetened	1 c	247	90	94	1	22	<1	<1	t	t	.1
244	Sweetened	1 c	250	87	115	1	28	<1	<1	t	t	.1
	Frozen concentrate, unsweetened:											
245	Undiluted, 6-fl-oz can	¾ c	207	62	302	4	72	1	1	.1	.1	.2
246	Diluted with 3 cans water	1 c	247	89	101	1	24	<1	<1	t	t	.1
	Grapes, raw European (adherent skin):											
247	Thompson seedless	10 ea	50	81	35	<1	9	<1	<1	.1	t	.1
248	Tokay/Emperor, seeded types	10 ea	50	81	35	<1	9	<1	<1	.1	t	.1
	Grape juice:											
249	Bottled or canned	1 c	253	84	154	1	38	<1	<1	.1	t	.1
	Frozen concentrate, sweetened:											
250	Undiluted, 6-fl-oz can, vit C added	¾ c	216	54	387	1	96	1	1	.2	t	.2
251	Diluted with 3 cans water, vit C added	1 c	250	87	128	<1	32	<1	<1	.1	t	.1
1410	Low calorie	1 c	253	84	154	1	38	<1	<1	.1	t	.1
252	Kiwi fruit, raw, peeled (88 g with peel)	1 ea	76	83	46	1	11	3	<1	t	t	.2
253	Lemons, raw, without peel and seeds (about 4 per lb whole)	1 ea	58	89	17	1	5	2	<1	t	t	.1
	Lemon juice:											
254	Fresh:	1 c	244	91	61	1	21	1	0	0	0	0
255	Tablespoon	1 tbs	15	91	4	<1	1	<1	0	0	0	0
256	Canned or bottled, unsweetened:	1 c	244	92	51	1	16	1	1	.1	t	.2
257	Tablespoon	1 tbs	15	92	3	<1	1	<1	<1	t	t	t
258	Frozen, single strength, unsweetened:	1 c	244	92	54	1	16	1	1	.1	t	.2
2298	Tablespoon	1 tbs	15	92	3	<1	1	<1	<1	t	t	t
	Lime juice:											
260	Fresh:	1 c	246	90	66	1	22	1	<1	t	t	.1
261	Tablespoon	1 tbs	15	90	4	<1	1	<1	<1	t	t	t
262	Canned or bottled, unsweetened	1 c	246	92	52	1	16	1	1	.1	.1	.2
263	Mangos, raw, edible part (300 g w/skin & seeds)	1 ea	207	82	135	1	35	4	1	.1	.2	.1
	Melons, raw, without rind and contents:											
264	Cantaloupe, 5" diam (2⅓ lb whole with refuse), orange flesh	½ ea	276	90	97	2	23	2	1	.2	t	.3
265	Honeydew, 6½" diam (5¼ lb whole with refuse), slice = ⅒ melon	1 pce	160	90	56	1	15	1	<1	t	t	.1
266	Nectarines, raw, w/o pits, 2¼" diam	1 ea	136	86	67	1	16	2	1	.1	.2	.3
	Oranges, raw:											
267	Whole w/o peel and seeds, 2⅝" diam (180 g with peel and seeds)	1 ea	131	87	62	1	15	3	<1	t	t	t
268	Sections, without membranes	1 c	180	87	85	2	21	4	<1	t	t	t
	Orange juice:											
269	Fresh, all varieties	1 c	248	88	112	2	26	<1	<1	.1	.1	.1
270	Canned, unsweetened	1 c	249	89	105	1	24	<1	<1	t	.1	.1
271	Chilled	1 c	249	88	110	2	25	<1	1	.1	.1	.2
	Frozen concentrate:											
272	Undiluted (6-oz can)	¾ c	213	58	339	5	81	2	<1	.1	.1	.1
273	Diluted w/3 parts water by volume	1 c	249	88	112	2	27	<1	<1	t	t	t
1345	Orange juice, from dry crystals	1 c	248	88	114	0	29	0	0	0	0	0
274	Orange and grapefruit juice, canned	1 c	247	89	106	1	25	<1	<1	t	t	t
	Papayas, raw:											
275	½" slices	1 c	140	89	55	1	14	3	<1	.1	.1	t
276	Whole, 3½" diam by 5⅛" w/o seeds and skin (1 lb w/refuse)	1 ea	304	89	119	2	30	5	<1	.1	.1	.1
1031	Papaya nectar, canned	1 c	250	85	143	<1	36	1	<1	.1	.1	.1
	Peaches:											
277	Raw, whole, 2½" diam, peeled, pitted (about 4 per lb whole)	1 ea	98	88	42	1	11	2	<1	t	t	t
278	Raw, sliced	1 c	170	88	73	1	19	3	<1	t	.1	.1
	Canned, fruit and liquid:											
279	Heavy syrup pack:	1 c	262	79	194	1	52	3	<1	t	.1	.1
280	Half	1 ea	98	79	72	<1	19	1	<1	t	t	t

Chol (mg)	Calc (mg)	Iron (mg)	Magn (mg)	Pota (mg)	Sodi (mg)	Zinc (mg)	VT-A (RE)	Thia (mg)	VT-E (α-TE)	Ribo (mg)	Niac (mg)	V-B6 (mg)	Fola (μg)	VT-C (mg)
0	22	.49	30	400	2	.12	2	.1	.12	.05	.49	.11	25	94
0	17	.49	25	378	2	.22	2	.1	.12	.05	.57	.05	26	72
0	20	.9	25	405	5	.15	0	.1	.12	.06	.8	.05	26	67
0	56	1.01	79	1001	6	.37	6	.3	.37	.16	1.6	.32	26	248
0	20	.35	27	336	2	.12	2	.1	.12	.05	.54	.11	9	83
0	5	.13	3	92	1	.02	3	.05	.35	.03	.15	.05	2	5
0	5	.13	3	92	1	.02	3	.05	.35	.03	.15	.05	2	5
0	23	.61	25	334	8	.13	3	.07	0	.09	.66	.16	7	<1
0	28	.78	32	160	15	.28	6	.11	.38	.2	.93	.32	9	179
0	10	.25	10	52	5	.1	2	.04	.12	.06	.31	.1	3	60
0	23	.61	25	334	8	.13	3	.07	0	.09	.66	.16	7	<1
0	20	.31	23	252	4	.13	14	.01	.85	.04	.38	.07	29	74
0	15	.35	5	80	1	.03	2	.02	.14	.01	.06	.05	6	31
0	17	.07	15	303	2	.12	5	.07	.22	.02	.24	.12	31	112
0	1	<.01	1	19	<1	.01	<1	<.01	.01	<.01	.01	.01	2	7
0	27	.32	19	249	51	.15	5	.1	.22	.02	.48	.1	25	60
0	2	.02	1	15	3	.01	<1	.01	.01	<.01	.03	.01	2	4
0	19	.29	19	217	2	.12	2	.14	.22	.03	.33	.15	23	77
0	1	.02	1	13	<1	.01	<1	.01	.01	<.01	.02	.01	1	5
0	22	.07	15	268	2	.15	2	.05	.22	.02	.25	.11	20	72
0	1	<.01	1	16	<1	.01	<1	<.01	.01	<.01	.01	.01	1	4
0	29	.57	17	185	39	.15	5	.08	.17	.01	.4	.07	19	16
0	21	.27	19	323	4	.08	805	.12	2.32	.12	1.21	.28	29	57
0	30	.58	30	853	25	.44	889	.1	.41	.06	1.58	.32	47	116
0	10	.11	11	434	16	.11	6	.12	.24	.03	.96	.09	10	40
0	7	.2	11	288	0	.12	101	.02	1.21	.06	1.35	.03	5	7
0	52	.13	13	237	0	.09	27	.11	.31	.05	.37	.08	40	70
0	72	.18	18	326	0	.13	38	.16	.43	.07	.51	.11	54	96
0	27	.5	27	496	2	.12	50	.22	.22	.07	.99	.1	75	124
0	20	1.1	27	436	5	.17	45	.15	.22	.07	.78	.22	45	86
0	25	.42	27	473	2	.1	20	.28	.47	.05	.7	.13	45	82
0	68	.75	72	1435	6	.38	60	.6	.68	.14	1.53	.33	330	294
0	22	.25	25	473	2	.12	20	.2	.47	.04	.5	.11	109	97
0	62	.2	2	50	12	.1	551	<.01	0	.04	0	0	143	121
0	20	1.14	25	390	7	.17	30	.14	.17	.07	.83	.06	35	72
0	34	.14	14	360	4	.1	39	.04	1.57	.04	.47	.03	53	86
0	73	.3	30	781	9	.21	85	.08	3.4	.1	1.03	.06	116	188
0	25	.85	7	77	12	.37	27	.01	.05	.01	.37	.02	5	7
0	5	.11	7	193	0	.14	53	.02	.69	.04	.97	.02	3	6
0	8	.19	12	335	0	.24	92	.03	1.19	.07	1.68	.03	6	11
0	8	.71	13	241	16	.24	86	.03	2.33	.06	1.61	.05	8	7
0	3	.26	5	90	6	.09	32	.01	.87	.02	.6	.02	3	3

Table H-1

Food Composition (Computer code number is for West Diet Analysis program) (For purposes of calculations, use "0" for t, <1, <.1, <.01, etc.)

Computer Code Number	Food Description	Measure	Wt (g)	H_2O (%)	Ener (kcal)	Prot (g)	Carb (g)	Dietary Fiber (g)	Fat (g)	Fat Breakdown (g) Sat	Mono	Poly
	FRUITS and FRUIT JUICES—Continued											
281	Juice pack:	1 c	248	87	109	2	29	3	<1	t	t	t
282	Half	1 ea	98	87	43	1	11	1	<1	t	t	t
283	Dried, uncooked	10 ea	130	32	311	5	80	11	1	.1	.4	.5
284	Dried, cooked, fruit and liquid	1 c	258	78	199	3	51	7	1	.1	.2	.3
	Frozen, slice, sweetened:											
285	10-oz package,vitamin C added	1 ea	284	75	267	2	68	5	<1	t	.1	.2
286	Cup, thawed measure, vitamin C added	1 c	250	75	235	2	60	4	<1	t	.1	.2
1032	Peach nectar, canned	1 c	249	86	134	1	35	1	<1	t	t	t
	Pears:											
	Fresh, with skin, cored:											
287	Bartlett, 2½" diam (about 2½ per lb)	1 ea	166	84	98	1	25	4	1	t	.1	.2
288	Bosc, 2⅕" diam (about 3 per lb)	1 ea	139	84	82	1	21	3	1	t	.1	.1
289	D'Anjou, 3" diam (about 2 per lb)	1 ea	209	84	123	1	32	5	1	t	.2	.2
	Canned, fruit and liquid:											
290	Heavy syrup pack:	1 c	266	80	197	1	51	4	<1	t	.1	.1
291	Half	1 ea	76	80	56	<1	15	1	<1	t	t	t
292	Juice pack:	1 c	248	86	124	1	32	4	<1	t	t	t
293	Half	1 ea	76	86	38	<1	10	1	<1	t	t	t
294	Dried halves	10 ea	175	27	459	3	122	13	1	.1	.2	.3
1033	Pear nectar, canned	1 c	250	84	150	<1	39	1	<1	t	t	t
	Pineapple:											
295	Fresh chunks, diced	1 c	155	86	76	1	19	2	1	t	.1	.2
	Canned, fruit and liquid:											
	Heavy syrup pack:											
296	Crushed, chunks, tidbits	½ c	127	79	99	<1	26	1	<1	t	t	.1
297	Slices	1 ea	49	79	38	<1	10	<1	<1	t	t	t
298	Juice pack, crushed, chunks, tidbits	1 c	250	83	150	1	39	2	<1	t	t	.1
299	Juice pack, slices	1 ea	47	83	28	<1	7	<1	<1	t	t	t
300	Pineapple juice, canned, unsweetened	1 c	250	85	140	1	34	<1	<1	t	t	.1
	Plantains, yellow flesh, without peel:											
301	Raw slices (whole=179 g w/o peel)	1 c	148	65	181	2	47	3	1	.2	t	.1
302	Cooked, boiled, sliced	1 c	154	67	179	1	48	4	<1	.1	t	.1
	Plums:											
303	Fresh, medium, 2⅛" diam	1 ea	66	85	36	1	9	1	<1	t	.3	.1
304	Fresh, small, 1½" diam	1 ea	28	85	15	<1	4	<1	<1	t	.1	t
	Canned, purple, with liquid:											
305	Heavy syrup pack:	1 c	258	76	230	1	60	3	<1	t	.2	.1
306	Plums	3 ea	138	76	123	<1	32	1	<1	t	.1	t
307	Juice pack:	1 c	252	84	146	1	38	3	<1	t	t	t
308	Plums	3 ea	138	84	80	1	21	1	<1	t	t	t
1698	Pomegranate, fresh	1 ea	154	81	105	1	26	1	<1	.1	.1	.1
	Prunes, dried, pitted:											
309	Uncooked (10 = 97 g w/pits, 84 g w/o pits)	10 ea	84	32	201	2	53	6	<1	t	.3	.1
310	Cooked, unsweetened, fruit & liq (250 g w/pits)	1 c	248	70	265	3	70	16	1	t	.4	.1
311	Prune juice, bottled or canned	1 c	256	81	182	2	45	3	<1	t	.1	t
	Raisins, seedless:											
312	Cup, not pressed down	1 c	145	15	435	5	115	6	1	.2	t	.2
313	One packet, ½ oz	½ oz	14	15	42	<1	11	1	<1	t	t	t
	Raspberries:											
314	Fresh	1 c	123	87	60	1	14	8	1	t	.1	.4
315	Frozen, sweetened	10 oz	284	73	293	2	74	12	<1	t	t	.3
316	Cup, thawed measure	1 c	250	73	258	2	65	11	<1	t	t	.2
317	Rhubarb, cooked, added sugar	1 c	240	68	278	1	75	5	<1	t	t	.1
	Strawberries:											
318	Fresh, whole, capped	1 c	144	92	43	1	10	3	1	t	.1	.3
	Frozen, sliced, sweetened:											
319	10-oz container	10 oz	284	73	273	2	74	5	<1	t	.1	.2
320	Cup, thawed measure	1 c	255	73	245	1	66	5	<1	t	t	.2
	Tangerines, without peel and seeds:											
321	Fresh (2⅜" whole) 116 g w/refuse	1 ea	84	88	37	1	9	2	<1	t	t	t
322	Canned, light syrup, fruit and liquid	1 c	252	83	154	1	41	2	<1	t	t	t

Chol (mg)	Calc (mg)	Iron (mg)	Magn (mg)	Pota (mg)	Sodi (mg)	Zinc (mg)	VT-A (RE)	Thia (mg)	VT-E (α-TE)	Ribo (mg)	Niac (mg)	V-B6 (mg)	Fola (µg)	VT-C (mg)
0	15	.67	17	317	10	.27	94	.02	3.72	.04	1.44	.05	8	9
0	6	.26	7	125	4	.11	37	.01	1.47	.02	.57	.02	3	4
0	36	5.28	55	1294	9	.74	281	<.01	0	.28	5.69	.09	<1	6
0	23	3.38	33	826	5	.46	52	.01	0	.05	3.92	.1	<1	10
0	9	1.05	14	369	17	.14	79	.04	2.53	.1	1.85	.05	9	268
0	7	.92	12	325	15	.12	70	.03	2.23	.09	1.63	.04	8	236
0	12	.47	10	100	17	.2	65	.01	.2	.03	.72	.02	3	13
0	18	.41	10	208	0	.2	3	.03	.83	.07	.17	.03	12	7
0	15	.35	8	174	0	.17	3	.03	.69	.06	.14	.02	10	6
0	23	.52	12	261	0	.25	4	.04	1.05	.08	.21	.04	15	8
0	13	.58	11	173	13	.21	0	.03	1.33	.06	.64	.04	3	3
0	4	.17	3	49	4	.06	0	.01	.38	.02	.18	.01	1	1
0	22	.72	17	238	10	.22	2	.03	1.24	.03	.5	.03	3	4
0	7	.22	5	73	3	.07	1	.01	.38	.01	.15	.01	1	1
0	59	3.68	58	933	10	.68	1	.01	0	.25	2.4	.13	0	12
0	12	.65	7	32	10	.17	<1	<.01	.25	.03	.32	.03	3	3
0	11	.57	22	175	2	.12	3	.14	.15	.06	.65	.13	16	24
0	18	.48	20	132	1	.15	1	.11	.13	.03	.36	.09	6	9
0	7	.19	8	51	<1	.06	<1	.04	.05	.01	.14	.04	2	4
0	35	.7	35	305	2	.25	10	.24	.25	.05	.71	.18	12	24
0	7	.13	7	57	<1	.05	2	.04	.05	.01	.13	.03	2	4
0	42	.65	32	335	2	.27	1	.14	.05	.05	.64	.24	58	27
0	4	.89	55	739	6	.21	167	.08	.4	.08	1.02	.44	33	27
0	3	.89	49	716	8	.2	140	.07	.22	.08	1.16	.37	40	17
0	3	.07	5	114	0	.07	21	.03	.4	.06	.33	.05	1	6
0	1	.03	2	48	0	.03	9	.01	.17	.03	.14	.02	1	3
0	23	2.17	13	235	49	.18	67	.04	1.81	.1	.75	.07	6	1
0	12	1.16	7	126	26	.1	36	.02	.97	.05	.4	.04	3	1
0	25	.86	20	388	3	.28	255	.06	1.76	.15	1.19	.07	7	7
0	14	.47	11	213	1	.15	139	.03	.97	.08	.65	.04	4	4
0	5	.46	5	399	5	.18	0	.05	.85	.05	.46	.16	9	9
0	43	2.08	38	626	3	.44	167	.07	1.22	.14	1.65	.22	3	3
0	57	2.75	50	828	5	.59	77	.06	<.01	.25	1.79	.54	<1	7
0	31	3.02	36	707	10	.54	1	.04	.03	.18	2.01	.56	1	10
0	71	3.02	48	1088	17	.39	1	.23	1.02	.13	1.19	.36	5	5
0	7	.29	5	105	2	.04	<1	.02	.1	.01	.11	.03	<1	<1
0	27	.7	22	187	0	.57	16	.04	.55	.11	1.11	.07	32	31
0	43	1.85	37	324	3	.51	17	.05	1.28	.13	.65	.1	74	47
0	37	1.63	32	285	2	.45	15	.05	1.13	.11	.57	.08	65	41
0	348	.5	29	230	2	.19	17	.04	.48	.05	.48	.05	13	8
0	20	.55	14	239	1	.19	4	.03	.2	.09	.33	.08	25	82
0	31	1.68	20	278	9	.17	6	.04	.4	.14	1.14	.08	42	118
0	28	1.5	18	250	8	.15	5	.04	.36	.13	1.02	.08	38	106
0	12	.08	10	132	1	.2	77	.09	.2	.02	.13	.06	17	26
0	18	.93	20	197	15	.6	212	.13	.86	.11	1.12	.11	12	50

Table H-1

Food Composition (Computer code number is for West Diet Analysis program) (For purposes of calculations, use "0" for t, <1, <.1, <.01, etc.)

Computer Code Number	Food Description	Measure	Wt (g)	H_2O (%)	Ener (kcal)	Prot (g)	Carb (g)	Dietary Fiber (g)	Fat (g)	Fat Breakdown (g) Sat	Mono	Poly
	FRUITS and FRUIT JUICES—Continued											
323	Tangerine juice, canned, sweetened	1 c	249	87	125	1	30	<1	<1	t	t	.1
	Watermelon, raw, without rind and seeds:											
324	Piece, 1/16 wedge	1 pce	286	91	91	2	20	1	1	.1	.3	.4
325	Diced	1 c	152	91	49	1	11	1	1	.1	.2	.2
	BAKED GOODS: BREADS, CAKES, COOKIES, CRACKERS, PIES											
326	Bagels, plain, enriched, 3½" diam.	1 ea	71	33	195	7	38	2	1	.2	.1	.5
1663	Bagel, oat bran	1 ea	71	33	181	8	38	3	1	.1	.2	.3
	Biscuits:											
327	From home recipe	1 ea	60	29	212	4	27	1	10	2.6	4.2	2.5
328	From mix	1 ea	57	29	191	4	28	1	7	1.6	2.4	2.5
329	From refrigerated dough	1 ea	74	27	276	4	34	1	13	8.7	3.4	.5
330	Bread crumbs, dry, grated (see # 364, 365 for soft crumbs)	1 c	108	6	427	13	78	3	6	1.4	2.3	1.7
2087	Bread sticks, brown & serve	1 ea	57	34	150	7	28	1	1	.5	.5	.5
	Breads:											
331	Boston brown, canned, 3¼" slice	1 pce	45	47	88	2	19	2	1	.1	.1	.3
332	Cracked wheat (¼ cracked-wheat & ¾ enr wheat flour): 1-lb loaf	1 ea	454	36	1180	39	225	25	18	4.2	8.6	3.1
333	Slice (18 per loaf)	1 pce	25	36	65	2	12	1	1	.2	.5	.2
334	Slice, toasted	1 pce	23	30	65	2	12	1	1	.2	.5	.2
335	French/Vienna, enriched: 1-lb loaf	1 ea	454	34	1243	40	236	14	14	2.9	5.5	3.1
337	Slice, 4¾ x 4 x ½"	1 pce	25	34	68	2	13	1	1	.2	.3	.2
336	French, slice, 5 x 2½"	1 pce	25	34	68	2	13	1	1	.2	.3	.2
	French toast: see Mixed Dishes, and Fast Foods, #691											
2083	Honey wheatberry	1 pce	38	38	100	3	18	2	1	0	.5	
338	Italian, enriched: 1-lb loaf	1 ea	454	36	1230	40	227	12	16	3.9	3.7	6.3
339	Slice, 4½ x 3¼ x ¾"	1 pce	30	36	81	3	15	1	1	.3	.2	.4
340	Mixed grain, enriched: 1-lb loaf	1 ea	454	38	1135	45	211	29	17	3.7	6.9	4.2
341	Slice (18 per loaf)	1 pce	26	38	65	3	12	2	1	.2	.4	.2
342	Slice, toasted	1 pce	24	32	65	3	12	2	1	.2	.4	.2
343	Oatmeal, enriched: 1-lb loaf	1 ea	454	37	1221	38	220	18	20	3.2	7.2	7.7
344	Slice (18 per loaf)	1 pce	27	37	73	2	13	1	1	.2	.4	.5
345	Slice, toasted	1 pce	25	31	73	2	13	1	1	.2	.4	.5
346	Pita pocket bread, enr, 6½" round	1 ea	60	32	165	5	33	1	1	.1	.1	.3
	Pumpernickel (⅔ rye & ⅓ enr wheat flr):											
347	1-lb loaf	1 ea	454	38	1135	40	216	29	14	2	4.2	5.6
348	Slice, 5 x 4 x ⅜"	1 pce	26	38	65	2	12	2	1	.1	.2	.3
349	Slice, toasted	1 pce	29	32	80	3	15	2	1	.1	.3	.4
350	Raisin, enriched: 1-lb loaf	1 ea	454	34	1243	36	237	19	20	4.9	10.5	3.1
351	Slice (18 per loaf)	1 pce	26	34	71	2	14	1	1	.3	.6	.2
352	Slice, toasted	1 pce	24	28	71	2	14	1	1	.3	.6	.2
353	Rye, light (⅓ rye & ⅔ enr wheat flr): 1-lb loaf	1 ea	454	37	1175	39	219	26	15	2.9	6	3.6
354	Slice, 4¾ x 3¾ x 7/16"	1 pce	32	37	83	3	15	2	1	.2	.4	.3
355	Slice, toasted	1 pce	24	31	68	2	13	2	1	.2	.3	.2
356	Wheat (enr wheat & whole-wheat flour): 1-lb loaf	1 ea	454	37	1160	43	213	25	19	3.9	7.3	4.5
357	Slice (18 per loaf)	1 pce	25	37	65	2	12	1	1	.2	.4	.2
358	Slice, toasted	1 pce	23	32	65	2	12	1	1	.2	.4	.2
359	White, enriched: 1-lb loaf	1 ea	454	35	1293	36	225	9	26	5.4	5.9	12.6
360	Slice	1 pce	42	35	120	3	21	1	2	.5	.5	1.2
361	Slice, toasted	1 pce	38	29	119	3	21	1	2	.5	.5	1.2
366	Whole-wheat: 1-lb loaf	1 ea	454	38	1116	44	209	31	19	4.2	7.6	4.5
367	Slice (16 per loaf)	1 pce	28	38	69	3	13	2	1	.3	.5	.3
368	Slice, toasted	1 pce	25	30	69	3	13	2	1	.3	.5	.3
	Bread stuffing, prepared from mix:											
369	Dry type	1 c	200	65	356	6	43	6	17	3.5	7.6	5.2
370	Moist type, with egg and margarine	1 c	232	65	390	9	51	5	17	3.4	7.4	4.9

Chol (mg)	Calc (mg)	Iron (mg)	Magn (mg)	Pota (mg)	Sodi (mg)	Zinc (mg)	VT-A (RE)	Thia (mg)	VT-E (α-TE)	Ribo (mg)	Niac (mg)	V-B6 (mg)	Fola (μg)	VT-C (mg)
0	45	.5	20	443	2	.07	105	.15	.22	.05	.25	.08	11	55
0	23	.49	31	332	6	.2	106	.23	.43	.06	.57	.41	6	27
0	12	.26	17	176	3	.11	56	.12	.23	.03	.3	.22	3	15
0	52	2.53	21	72	379	.62	0	.38	.02	.22	3.24	.04	62	0
0	9	2.19	40	145	360	1.48	<1	.23	.17	.24	2.1	.14	57	<1
2	141	1.74	11	73	348	.32	14	.21	1.45	.19	1.77	.02	37	<1
2	105	1.17	14	107	544	.35	15	.2	.23	.2	1.72	.04	3	<1
5	89	1.64	9	87	584	.29	24	.27	.44	.18	1.63	.03	6	0
0	245	6.61	50	239	931	1.32	<1	.83	.95	.47	7.4	.11	118	0
0	60	2.7			290		0	.22		.1	1.6			0
<1	31	.94	28	143	284	.22	5	.01	.13	.05	.5	.04	5	0
														0
0	195	12.8	236	804	2442	5.63	0	1.63	2.56	1.09	16.7	1.38	277	
0	11	.7	13	44	135	.31	0	.09	.14	.06	.92	.08	15	0
0	11	.7	13	44	135	.31	0	.07	.14	.05	.83	.07	7	0
0	341	11.5	123	513	2764	3.95	0	2.36	1.07	1.49	21.6	.19	431	0
0	19	.63	7	28	152	.22	0	.13	.06	.08	1.19	.01	24	0
0	19	.63	7	28	152	.22	0	.13	.06	.08	1.19	.01	24	0
0	20	.72			200		0	.12	.24	.07	.8			0
0	354	13.3	123	499	2651	3.9	0	2.15	1.26	1.33	19.9	.22	431	0
0	23	.88	8	33	175	.26	0	.14	.08	.09	1.31	.01	28	0
0	413	15.8	241	926	2210	5.77	0	1.85	2.79	1.55	19.8	1.51	363	1
0	24	.9	14	53	127	.33	0	.11	.16	.09	1.14	.09	21	<1
0	24	.9	14	53	127	.33	0	.08	.16	.08	1.02	.08	16	<1
0	300	12.3	168	645	2719	4.63	9	1.81	1.56	1.09	14.3	.31	281	2
0	18	.73	10	38	162	.27	1	.11	.09	.06	.85	.02	17	<1
0	18	.73	10	38	163	.28	<1	.09	.09	.06	.77	.02	13	<1
0	52	1.57	16	72	322	.5	0	.36	.02	.2	2.78	.02	57	0
0	309	13	245	944	3046	6.72	0	1.48	2.3	1.38	14	.57	363	0
0	18	.75	14	54	174	.38	0	.08	.13	.08	.8	.03	21	0
0	21	.91	17	66	214	.47	0	.08	.17	.09	.89	.04	20	0
0	300	13.2	118	1030	1770	3.27	1	1.54	3.44	1.81	15.8	.31	395	2
0	17	.75	7	59	101	.19	<1	.09	.2	.1	.9	.02	23	<1
0	17	.76	7	59	102	.19	<1	.07	.2	.09	.81	.02	18	<1
0	331	12.8	182	754	2996	5.18	2	1.97	2.51	1.52	17.3	.34	390	1
0	23	.91	13	53	211	.36	<1	.14	.18	.11	1.22	.02	27	<1
0	19	.74	10	44	174	.3	<1	.09	.15	.08	.9	.02	17	<1
0	572	15.8	209	627	2447	4.77	0	2.09	3	1.45	20.5	.49	204	0
0	26	.83	11	50	133	.26	0	.1	.14	.07	1.03	.02	19	0
0	26	.83	11	50	132	.26	0	.08	.14	.06	.93	.02	15	0
14	259	13.5	86	663	1629	2.91	100	1.84	4.95	1.74	16.3	.23	413	1
1	24	1.25	8	61	151	.27	9	.17	.46	.16	1.51	.02	38	<1
1	24	1.24	8	61	150	.27	8	.13	.46	.14	1.35	.02	12	<1
0	327	15	390	1144	2392	8.81	0	1.59	4.72	.93	17.4	.81	227	0
0	20	.92	24	71	148	.54	0	.1	.29	.06	1.08	.05	14	0
0	20	.93	24	71	148	.54	0	.08	.23	.05	.97	.04	9	0
0	64	2.18	24	148	1086	.56	162	.27	2.8	.21	2.96	.08	202	0
0	148	3.8	35	304	1069	.74	160	.39	2.78	.33	3.69	.12	39	4

H

Table H-1

Food Composition (Computer code number is for West Diet Analysis program) (For purposes of calculations, use "0" for t, <1, <.1, <.01, etc.)

Computer Code Number	Food Description	Measure	Wt (g)	H_2O (%)	Ener (kcal)	Prot (g)	Carb (g)	Dietary Fiber (g)	Fat (g)	Fat Breakdown (g) Sat	Mono	Poly
	BAKED GOODS: BREADS, CAKES, COOKIES, CRACKERS, PIES—Continued											
	Cakes, prepared from mixes using enriched flour and veg shortening, w/frostings made from margarine:											
	Angel food:											
371	Whole cake, 9 ¾" diam tube	1 ea	340	33	877	20	197	5	3	.4	.2	1.2
372	Piece, 1/12 of cake	1 pce	28	33	72	2	16	<1	<1	t	t	.1
373	Boston cream pie, ⅛ of cake	1 pce	123	45	310	3	53	2	10	3.1	5.4	1.2
	Coffee cake:											
374	Whole cake, 7¾ x 5⅝ x 1¼"	1 ea	336	30	1068	18	177	4	32	6.3	13	10.7
375	Piece, ⅙ of cake	1 pce	56	30	178	3	30	1	5	1	2.2	1.8
	Devil's food, chocolate frosting:											
376	Whole cake, 2 layer, 8 or 9" diam	1 ea	1021	23	3747	42	557	29	167	47.9	91.9	19.5
377	Piece, 1/16 of cake	1 pce	64	23	235	3	35	2	10	3	5.8	1.2
378	Cupcake, 2½" diam	1 ea	42	23	154	2	23	1	7	2	3.8	.8
	Gingerbread:											
379	Whole cake, 8" square	1 ea	603	33	1863	24	306	7	61	15.8	34	8.1
380	Piece, 1/9 of cake	1 pce	67	33	207	3	34	1	7	1.8	3.8	.9
	Yellow, chocolate frosting, 2 layer:											
381	Whole cake, 8 or 9" in diam	1 ea	1024	22	3880	39	567	18	178	49	99	21.4
382	Piece, 1/16 of cake	1 pce	64	22	243	2	35	1	11	3.1	6.2	1.3
	Cakes from home recipes w/enr flour:											
	Carrot cake, made with veg oil, cream cheese frosting:											
383	Whole, 9 x 13" cake	1 ea	1776	21	7743	82	838	21	469	86.8	116	242
384	Piece, 1/16 of cake, 2¼ x 3¼" slice	1 pce	111	21	484	5	52	1	29	5.4	7.2	15.1
	Fruitcake, dark:											
385	Whole cake, 7½"diam tube, 2¼"high	1 ea	1376	25	4458	40	848	51	125	15.4	57.4	44.6
386	Piece, 1/32 of cake, ⅔" arc	1 pce	43	25	139	1	26	2	4	.5	1.8	1.4
	Sheet, plain, made w/veg shortening, no frosting:											
387	Whole cake, 9" square	1 ea	774	24	2817	35	433	3	108	29.9	51.5	25.5
388	Piece, 1/9 of cake	1 pce	86	24	313	4	48	<1	12	3.3	5.7	2.8
	Sheet, plain, made w/margarine, uncooked white frosting:											
389	Whole cake, 9" square	1 ea	576	22	2148	20	339	2	83	13.8	35.5	29.5
390	Piece, 1/9 of cake	1 pce	64	22	239	2	38	<1	9	1.5	3.9	3.3
	Cakes, commerical:											
	Cheesecake:											
401	Whole cake, 9" diam	1 ea	960	46	3081	53	245	4	216	111	74.4	13.2
402	Piece, 1/12 of cake	1 pce	80	46	257	4	20	<1	18	9.2	6.2	1.1
	Pound cake:											
393	Loaf, 8½ x 3½ x 3"	1 ea	340	25	1319	19	166	2	68	38.1	19	3.7
394	Slice, 1/17 of loaf, 2" slice	1 pce	28	25	109	2	14	<1	6	3.1	1.6	.3
	Snack cakes:											
395	Chocolate w/creme filling,Ding Dong	1 ea	50	20	188	2	30	<1	7	1.6	2.7	2.1
396	Sponge cake w/creme filling,Twinkie	1 ea	43	20	157	1	27	<1	5	1.2	1.9	1.5
1677	Sponge cake, 1/12 of 12" cake	1 pce	38	30	110	2	23	<1	1	.3	.4	.2
	White, white frosting, 2 layer:											
397	Whole cake, 8 or 9" diam	1 ea	1136	20	4260	37	716	11	153	68.2	60.1	15.4
398	Piece, 1/16 of cake	1 pce	71	20	266	2	45	1	10	4.3	3.8	1
	Yellow, chocolate frosting, 2 layer:											
399	Whole cake, 8 or 9" in diam	1 ea	1024	22	3880	39	567	18	178	49	99	21.4
400	Piece, 1/16 of cake	1 pce	64	22	243	2	35	1	11	3.1	6.2	1.3
1332	Bagel chips	5 pce	70	3	298	6	52	6	7	1.2	2	3.4
2225	Bagel chips, onion garlic, toasted	1 oz	28		193	5	31	3	8	1.7	5.2	0
1035	Cheese puffs/Cheetos	1 c	20	1	111	2	11	<1	7	1.3	4.1	1
	Cookies made with enriched flour:											
	Brownies with nuts:											
403	Commercial w/frosting, 1½ x 1¾ x ⅞"	1 ea	61	14	247	3	39	1	10	2.6	5.1	1.6
1902	Fat free fudge, Entenmann's	1 pce	40	24	110	2	27	1	0	0	0	0

Chol (mg)	Calc (mg)	Iron (mg)	Magn (mg)	Pota (mg)	Sodi (mg)	Zinc (mg)	VT-A (RE)	Thia (mg)	VT-E (α-TE)	Ribo (mg)	Niac (mg)	V-B6 (mg)	Fola (μg)	VT-C (mg)
0	476	1.77	41	316	2546	.24	0	.35	.34	1.67	3	.1	119	0
0	39	.15	3	26	210	.02	0	.03	.03	.14	.25	.01	10	0
45	28	.47	7	48	177	.2	28	.5	1.3	.33	.23	.03	18	<1
165	457	4.8	60	376	1414	1.51	134	.56	5.58	.59	5.11	.17	228	1
27	76	.8	10	63	236	.25	22	.09	.93	.1	.85	.03	38	<1
470	439	22.5	347	2042	3410	7.04	286	.28	17.3	1.36	5.89	.32	174	1
29	27	1.41	22	128	214	.44	18	.02	1.08	.08	.37	.02	11	<1
19	18	.92	14	84	140	.29	12	.01	.71	.06	.24	.01	7	<1
211	416	20	96	1453	2761	2.47	96	1.14	8.26	1.12	9.41	.23	60	1
23	46	2.22	11	161	307	.27	11	.13	.92	.12	1.05	.02	7	<1
563	379	21.3	307	1822	3450	6.35	276	1.23	27.6	1.61	12.8	.3	225	1
35	24	1.33	19	114	216	.4	17	.08	1.73	.1	.8	.02	14	<1
959	444	22.2	320	1989	4368	8.7	6819	2.42	74.9	2.77	17.9	1.35	213	19
60	28	1.39	20	124	273	.54	426	.15	4.68	.17	1.12	.08	13	1
69	454	28.5	220	2105	3715	3.72	261	.69	42.9	1.36	10.9	.63	261	5
2	14	.89	7	66	116	.12	8	.02	1.34	.04	.34	.02	8	<1
503	495	11.7	108	611	2322	2.74	372	1.24	11	1.39	10.1	.26	54	2
56	55	1.3	12	68	258	.3	41	.14	1.22	.15	1.12	.03	6	<1
323	357	6.16	35	305	1981	1.44	109	.58	10.9	.4	2.88	.2	156	1
36	40	.68	4	34	220	.16	12	.06	1.22	.04	.32	.02	17	<1
528	490	6.05	106	864	1987	4.9	1545	.27	10.1	1.85	1.87	.5	173	6
44	41	.5	9	72	166	.41	129	.02	.84	.15	.16	.04	14	<1
751	119	4.69	37	405	1353	1.56	530	.47	2.24	.78	4.45	.12	139	<1
62	10	.39	3	33	111	.13	44	.04	.18	.06	.37	.01	11	<1
8	36	1.68	20	61	213	.28	2	.11	1.01	.15	1.22	.01	14	<1
7	19	.55	3	39	157	.13	2	.07	.83	.06	.52	.01	12	<1
39	27	1.03	4	38	93	.19	17	.09	.17	.1	.73	.02	15	0
91	545	9.09	60	659	2658	1.76	368	1.14	20.4	1.48	10.2	.16	64	1
6	34	.57	4	41	166	.11	23	.07	1.28	.09	.64	.01	4	<1
563	379	21.3	307	1822	3450	6.35	276	1.23	27.6	1.61	12.8	.3	225	1
35	24	1.33	19	114	216	.4	17	.08	1.73	.1	.8	.02	14	<1
0	9	1.38	41	167	419	.9	0	.13	.46	.12	1.57	.19	58	0
0	0	2.52			490		0	.39	<.01	.24	3.5			0
1	12	.47	4	33	210	.08	7	.05	1.02	.07	.65	.03	24	<1
10	18	1.37	19	91	190	.44	12	.16	1.3	.13	1.05	.02	13	<1
0	0	1.08		90	140		0		.01					0

Table H-1

Food Composition (Computer code number is for West Diet Analysis program) (For purposes of calculations, use "0" for t, <1, <.1, <.01, etc.)

Computer Code Number	Food Description	Measure	Wt (g)	H_2O (%)	Ener (kcal)	Prot (g)	Carb (g)	Dietary Fiber (g)	Fat (g)	Fat Breakdown (g) Sat	Mono	Poly
	BAKED GOODS: BREADS, CAKES, COOKIES, CRACKERS, PIES—Continued											
	Chocolate chip cookies:											
405	Commercial, 2¼" diam	4 ea	60	12	275	2	35	2	15	4.5	7.8	1.6
406	Home recipe, 2¼" diam	4 ea	64	6	312	4	37	2	18	5.2	6.7	5.4
407	From refrigerated dough, 2¼" diam	4 ea	64	13	284	3	39	1	13	4.5	6.5	1.3
408	Fig bars	4 ea	64	16	223	2	45	3	5	.9	2.6	.8
2052	Fruit bar, no fat	1 ea	28		90	2	21	0	0	0	0	0
2162	Fudge, fat free, Snackwell	1 ea	16	14	53	1	12	<1	<1	.1	.1	t
409	Oatmeal raisin, 2⅝" diam	4 ea	60	6	261	4	41	2	10	1.9	4.1	3
410	Peanut butter, home recipe, 2⅝"diam	4 ea	80	6	380	7	47	2	19	3.5	8.7	5.8
411	Sandwich-type, all	4 ea	40	2	189	2	28	1	8	1.7	4.7	1.1
412	Shortbread, commercial, small	4 ea	32	4	161	2	21	1	8	2	4.3	1
413	Shortbread, home recipe, large	2 ea	22	3	120	1	12	<1	7	4.5	2.1	.3
414	Sugar from refrigerated dough, 2" diam	4 ea	48	5	232	2	31	<1	11	2.8	6.2	1.4
1874	Vanilla sandwich, Snackwell's	2 ea	26	4	109	1	21	1	2	.5	.8	.2
415	Vanilla wafers	10 ea	40	5	176	2	29	1	6	1.4	2.4	1.5
416	Corn chips	1 c	26	1	140	2	15	1	9	1.2	2.5	4.3
	Crackers (enriched):											
417	Cheese	10 ea	10	3	50	1	6	<1	3	.9	.9	.5
418	Cheese with peanut butter	4 ea	28	4	135	4	16	1	6	1.4	3.4	1.2
	Fat free, enriched:											
2161	Cracked pepper, Snackwell	1 ea	15	2	60	2	12	<1	<1	.1	t	.1
2159	Wheat, Snackwell	7 ea	15	1	60	2	12	1	<1	.1	.1	.1
2075	Whole wheat, herb seasoned	5 ea	14	5	50	2	11	2	0	0	0	0
2077	Whole wheat, onion	5 ea	14	4	50	2	11	2	0	0	0	0
419	Graham, enriched	2 ea	14	4	59	1	11	<1	1	.4	.7	.2
420	Melba toast, plain, enriched	1 pce	5	5	19	1	4	<1	<1	t	t	.1
1514	Rice cakes, unsalted, enriched	2 ea	18	6	70	1	15	1	<1	.1	.2	.2
421	Rye wafer, whole grain	2 ea	22	5	73	2	18	5	<1	t	t	.1
422	Saltine-enriched	4 ea	12	4	52	1	9	<1	1	.3	.8	.2
1971	Saltine, unsalted tops, enriched	2 ea	6		25	1	4	0	<1	0	0	0
423	Snack-type, round like Ritz, enriched	3 ea	9	3	45	1	5	<1	2	.4	1	.7
424	Wheat, thin, enriched	4 ea	8	3	38	1	5	<1	2	.7	.8	.2
425	Whole-wheat wafers	2 ea	8	3	35	1	5	1	1	.2	.8	.2
426	Croissants, 4½ x 4 x 1¾"	1 ea	57	23	231	5	26	1	12	6.7	3.2	.7
1699	Croutons, seasoned	½ c	20	4	93	2	13	1	4	1	1.9	.5
	Danish pastry:											
427	Packaged ring, plain, 12 oz	1 ea	340	21	1349	19	181	1	65	13.5	40.9	6.4
428	Round piece, plain, 4¼" diam, 1" high	1 ea	88	21	349	5	47	<1	17	3.5	10.6	1.6
429	Ounce, plain	1 oz	28	21	111	2	15	<1	5	1.1	3.4	.5
430	Round piece with fruit	1 ea	94	29	335	5	45		16	3.3	10.1	1.6
	Desserts, 3 x 3" piece:											
1348	Apple crisp	1 pce	78	61	127	1	25	1	3	.6	1.2	.9
1353	Apple cobbler	1 pce	104	57	199	2	35	2	6	1.2	2.8	2
1349	Cherry crisp	1 pce	138	77	146	2	24	1	5	.9	2.5	1.8
1352	Cherry cobbler	1 pce	129	66	198	2	34	1	6	1.2	2.8	1.9
1350	Peach crisp	1 pce	139	75	155	2	27	2	5	.9	2.5	1.7
1351	Peach cobbler	1 pce	130	64	204	2	36	2	6	1.2	2.8	1.9
	Doughnuts:											
431	Cake type, plain, 3¼" diam	1 ea	47	21	198	2	23	1	11	1.8	4.5	3.8
432	Yeast-leavened, glazed, 3¾" diam	1 ea	60	25	242	4	27	1	14	3.5	7.8	1.7
	English muffins:											
433	Plain, enriched	1 ea	57	42	134	4	26	2	1	.2	.2	.5
434	Toasted	1 ea	52	37	133	4	26	2	1	.1	.2	.5
1504	Whole wheat	1 ea	66	46	134	6	27	4	1	.2	.3	.6
1414	Granola bar, soft	1 ea	28	6	124	2	19	1	5	2	1.1	1.5
1415	Granola bar, hard	1 ea	25	4	118	3	16	1	5	.6	1.1	3
1985	Granola bar, fat free, all flavors	1 ea	42	10	140	2	35	3	0	0	0	0
	Muffins, 2½" diam, 1½" high:											
	From home recipe:											
435	Blueberry	1 ea	57	39	165	4	23	1	6	1.4	1.6	3.1
436	Bran, wheat	1 ea	57	35	164	4	24	4	7	1.5	1.8	3.6
437	Cornmeal	1 ea	57	32	183	4	25	2	7	1.6	1.8	3.5

Chol (mg)	Calc (mg)	Iron (mg)	Magn (mg)	Pota (mg)	Sodi (mg)	Zinc (mg)	VT-A (RE)	Thia (mg)	VT-E (α-TE)	Ribo (mg)	Niac (mg)	V-B6 (mg)	Fola (μg)	VT-C (mg)
0	9	1.45	21	56	196	.28	1	.07	1.74	.12	.97	.1	23	0
20	25	1.57	35	143	231	.59	105	.12	1.86	.11	.87	.05	21	<1
15	16	1.44	15	115	134	.32	11	.12	1.31	.12	1.27	.03	36	0
0	41	1.86	17	132	224	.25	3	.1	.45	.14	1.2	.05	17	<1
0	0	.36			95		0		.01					0
0	3	.29	5	26	71	.08	<1	.02	<.01	.02	.26	<.01		0
20	60	1.59	25	143	323	.52	98	.15	1.5	.1	.76	.04	18	<1
25	31	1.78	31	185	414	.66	125	.18	3.04	.17	2.81	.07	44	<1
0	10	1.55	18	70	242	.32	<1	.03	1.21	.07	.83	.01	17	0
6	11	.88	5	32	146	.17	4	.11	.98	.1	1.07	.01	19	0
20	4	.58	3	15	102	.09	67	.08	.18	.06	.64	<.01	2	0
15	43	.88	4	78	225	.13	5	.09	1.54	.06	1.16	.01	25	0
<1	17	.61	5	28	95	.16	<1	.05		.07	.69	.01		0
23	19	.95	6	39	125	.14	7	.11	.54	.13	1.24	.03	20	0
0	33	.34	20	37	164	.33	2	.01	.35	.04	.31	.06	5	0
1	15	.48	4	14	99	.11	3	.06	.1	.04	.47	.05	8	0
1	22	.82	16	69	278	.3	10	.11	1.24	.1	1.83	.42	25	0
<1	26	.73	4	19	148	.14	<1	.05		.06	.78	.01		<1
<1	28	.58	7	43	169	.21	<1	.04		.07	.73	.02		0
0	0				80		100							2
0	0	0			80		100							2
0	3	.52	4	19	85	.11	0	.03	.27	.04	.58	.01	8	0
0	5	.18	3	10	41	.1	0	.02	.01	.01	.21	<.01	6	0
0	2	.27	24	52	5	.54	1	.01	.02	.03	1.41	.03	4	0
0	9	1.31	27	109	175	.62	<1	.09	.44	.06	.35	.06	3	<1
0	14	.65	3	15	156	.09	0	.07	.2	.05	.63	<.01	15	0
0		.36		5	50				.1					
0	11	.32	2	12	76	.06	0	.04	.4	.03	.36	<.01	7	0
2	3	.28	5	16	70	.13	<1	.04	.02	.03	.34	.01	1	0
0	4	.25	8	24	53	.17	0	.02	.31	.01	.36	.01	3	0
43	21	1.16	9	67	424	.43	78	.22	.24	.14	1.25	.03	35	<1
1	19	.56	8	36	248	.19	1	.1	.32	.08	.93	.02	18	0
105	143	6.94	54	371	1261	1.87	20	.99	3.06	.75	8.5	.2	211	10
27	37	1.8	14	96	326	.48	5	.25	.79	.19	2.2	.05	55	3
9	12	.57	4	30	104	.15	2	.08	.25	.06	.7	.02	17	1
19	22	1.4	14	110	333	.48	24	.29	.85	.21	1.8	.06	31	2
0	22	.58	5	76	142	.12	24	.07		.06	.6	.03	4	2
1	21	.79	6	106	288	.16	76	.1	1.11	.09	.74	.04	3	<1
0	26	2.14	11	154	74	.15	150	.06	.93	.08	.6	.06	11	3
1	28	1.81	9	133	294	.2	135	.1	1.01	.11	.85	.05	9	2
0	20	.89	12	189	70	.19	108	.06	1.13	.05	1.06	.03	6	5
1	24	.91	10	159	291	.23	105	.09	1.16	.09	1.19	.03	6	3
17	21	.92	9	60	257	.26	8	.1	1.63	.11	.87	.03	22	<1
4	26	1.22	13	65	205	.46	6	.22	1.75	.13	1.71	.03	26	0
0	99	1.43	12	75	264	.4	0	.25	.07	.16	2.21	.02	46	<1
0	98	1.41	11	74	262	.39	0	.2	.07	.14	1.98	.02	38	<1
0	175	1.62	47	139	420	1.06	0	.2	.46	.09	2.25	.11	28	0
<1	29	.72	21	91	78	.42	0	.08	.34	.05	.14	.03	7	0
0	15	.74	24	84	73	.51	4	.07	.33	.03	.39	.02	6	<1
0	0	3.6			5		100							0
22	107	1.29	9	69	251	.31	16	.15	1.03	.16	1.26	.02	7	1
20	106	2.39	44	181	335	1.57	136	.19	1.31	.25	2.29	.18	30	4
26	147	1.49	13	82	333	.35	23	.17	1.08	.18	1.36	.05	10	<1

H

Table H-1

Food Composition (Computer code number is for West Diet Analysis program) (For purposes of calculations, use "0" for t, <1, <.1, <.01, etc.)

H

Computer Code Number	Food Description	Measure	Wt (g)	H_2O (%)	Ener (kcal)	Prot (g)	Carb (g)	Dietary Fiber (g)	Fat (g)	Fat Breakdown (g) Sat	Mono	Poly
	BAKED GOODS: BREADS, CAKES, COOKIES, CRACKERS, PIES—Continued											
	From commercial mix:											
438	Blueberry	1 ea	50	36	150	3	24	1	4	.7	1.8	1.5
439	Bran, wheat	1 ea	50	35	138	3	23	2	5	1.2	2.3	.7
440	Cornmeal	1 ea	50	30	161	4	25	1	5	1.4	2.6	.6
1864	Nabisco Newtons, fat free, all flavors	1 ea	23		69	1	16		0	0	0	0
	Pancakes, 4" diam:											
441	Buckwheat, from mix w/ egg and milk	1 ea	30	54	62	2	8	1	2	.6	.6	.8
442	Plain, from home recipe	1 ea	38	53	86	2	11	1	4	.8	.9	1.7
443	Plain, from mix; egg, milk, oil added	1 ea	38	53	74	2	14	<1	1	.2	.3	.3
1468	Pan dulce, sweet roll w/topping	1 ea	79	21	291	5	48	1	9	2	3.9	2.7
	Piecrust,with enriched flour, vegetable shortening, baked:											
444	Home recipe, 9" shell	1 ea	180	10	949	12	85	3	62	15.5	27.4	16.4
	From mix:											
445	Piecrust for 2-crust pie	1 ea	320	10	1686	21	152	5	111	27.6	48.6	29.2
446	1 pie shell	1 ea	160	11	802	11	81	3	49	12.3	27.7	6.2
	Pies, 9" diam; pie crust made with vegetable shortening, enriched flour:											
447	Apple: Whole pie	1 ea	1000	52	2370	19	340	16	110	21.1	59.4	20.9
448	Piece, ⅙ of pie	1 pce	167	52	396	3	57	3	18	3.5	9.9	3.5
449	Banana cream: Whole pie	1 ea	1152	48	3098	51	379	8	157	43.3	65.9	38
450	Piece, ⅙ of pie	1 pce	192	48	516	8	63	1	26	7.2	11	6.3
451	Blueberry: Whole pie	1 ea	1176	51	2881	32	394	16	140	34.3	60.2	36.2
452	Piece, ⅙ of pie	1 pce	196	51	480	5	66	3	23	5.7	10	6
453	Cherry: Whole pie	1 ea	1140	46	3078	32	439	17	139	34.1	60.5	37.1
454	Piece, ⅙ of pie	1 pce	240	46	648	7	92	4	29	7.2	12.7	7.8
455	Chocolate cream: Whole pie	1 ea	1194	63	2150	49	281	12	97	35.5	38.2	18.6
456	Piece, ⅙ of pie	1 pce	199	63	358	8	47	2	16	5.9	6.4	3.1
457	Custard: Whole pie	1 ea	630	61	1323	35	131	10	73	17.5	36.3	12.1
458	Piece, ⅙ of pie	1 pce	105	61	221	6	22	2	12	2.9	6	2
459	Lemon meringue: Whole pie	1 ea	678	42	1817	10	320	8	59	10.6	24.6	19.6
460	Piece, ⅙ of pie	1 pce	113	42	303	2	53	1	10	1.8	4.1	3.3
461	Peach: Whole pie	1 ea	1111	45	2994	26	443	16	130	31.1	55.7	37.4
462	Piece, ⅙ of pie	1 pce	139	45	375	3	55	2	16	3.9	7	4.7
463	Pecan: Whole pie	1 ea	678	19	2712	27	388	24	125	25.5	73.2	20.1
464	Piece, ⅙ of pie	1 pce	113	19	452	5	65	4	21	4.2	12.2	3.4
465	Pumpkin: Whole pie	1 ea	654	58	1373	25	179	18	62	13.2	32.8	10.5
466	Piece, ⅙ of pie	1 pce	109	58	229	4	30	3	10	2.2	5.5	1.7
467	Pies, fried, commercial: Apple	1 ea	85	40	266	2	33	1	14	6.5	5.8	1.2
468	Pies, fried, commercial: Cherry	1 ea	128	38	404	4	54	3	21	3.1	9.5	6.9
	Pretzels, made with enriched flour:											
469	Thin sticks, 2¼" long	1 oz	28	3	107	3	22	1	1	.2	.4	.3
470	Dutch twists	10 pce	60	3	229	5	47	2	2	.4	.8	.7
471	Thin twists, 3¼ x 2¼ x ¼"	10 pce	60	3	229	5	47	2	2	.4	.8	.7
	Rolls & buns, enriched, commercial:											
472	Cloverleaf rolls, 2½" diam, 2" high	1 ea	28	32	84	2	14	1	2	.5	1	.3
473	Hot dog buns	1 ea	43	34	123	4	22	1	2	.5	1.1	.4
474	Hamburger buns	1 ea	43	34	123	4	22	1	2	.5	1.1	.4
475	Hard roll, white, 3¾" diam, 2" high	1 ea	57	31	167	6	30	1	2	.3	.6	1
476	Submarine rolls/hoagies, 11¼ x 3 x 2½"	1 ea	135	31	392	12	75	4	4	.9	1.3	1.4
	Rolls & buns, enriched, home recipe:											
477	Dinner rolls 2½" diam, 2" high	1 ea	35	29	112	3	19	1	3	.7	1.1	.7
	Sports/fitness bar:											
2043	Forza energy bar	1 ea	70	18	231	10	45	4	1			
2042	Power bar	1 ea	65		230	10	45	3	2			
2041	Tiger sports bar	1 ea	65	17	229	11	40	4	2			
478	Toaster pastries, fortified (Poptarts)	1 ea	52	12	204	2	37	1	5	.8	2.1	2
2132	Toaster strudel pastry—cream cheese	1 ea	54	32	188	3	24	<1	9	2.7		
2134	Toaster strudel pastry—french toast	1 ea	54	32	188	3	24	<1	9	2.9		

Chol (mg)	Calc (mg)	Iron (mg)	Magn (mg)	Pota (mg)	Sodi (mg)	Zinc (mg)	VT-A (RE)	Thia (mg)	VT-E (α-TE)	Ribo (mg)	Niac (mg)	V-B6 (mg)	Fola (μg)	VT-C (mg)
23	12	.56	5	39	219	.19	11	.07	.7	.16	1.12	.04	5	<1
34	16	1.27	28	73	234	.57	15	.1	.75	.12	1.44	.09	8	0
31	37	.97	10	65	398	.32	22	.12	.75	.14	1.05	.05	5	<1
					77									
20	77	.56	17	70	160	.35	20	.05	.62	.08	.4	.04	5	<1
22	83	.68	6	50	167	.21	20	.08	.36	.11	.6	.02	14	<1
5	48	.59	8	66	239	.15	3	.08	.32	.08	.65	.03	3	<1
26	13	1.82	10	57	140	.35	87	.23	1.35	.21	1.98	.04	22	<1
0	18	5.2	25	121	976	.79	0	.7	9.94	.5	5.96	.04	121	0
0	32	9.25	45	214	1734	1.41	0	1.25	17.7	.89	10.6	.08	214	0
0	96	3.44	24	99	1166	.62	0	.48	8.83	.3	3.79	.09	19	0
0	110	4.5	70	650	2660	1.6	300	.28	16.5	.27	2.63	.38	220	32
0	18	.75	12	109	444	.27	50	.05	2.76	.04	.44	.06	37	5
588	864	12	184	1900	2764	5.53	806	1.6	16.9	2.38	12.1	1.53	311	18
98	144	2	31	317	461	.92	134	.27	2.82	.4	2.02	.25	52	3
0	82	14.5	94	588	2175	2.35	47	1.8	24.7	1.55	14	.4	270	8
0	14	2.41	16	98	363	.39	8	.3	4.12	.26	2.33	.07	45	1
0	114	21.1	103	878	2177	2.28	547	1.69	21.7	1.43	14.6	.39	308	11
0	24	4.44	22	185	458	.48	115	.35	4.56	.3	3.07	.08	65	2
109	1028	8.84	170	1705	2085	4.93	235	1.03	11.4	2.43	7.34	.37	59	6
18	171	1.47	28	284	348	.82	39	.17	1.9	.41	1.22	.06	10	1
208	504	3.65	69	668	1512	3.28	315	.25	7.5	1.31	1.84	.3	126	2
35	84	.61	12	111	252	.55	52	.04	1.25	.22	.31	.05	21	<1
305	380	4.14	102	603	990	3.32	353	.42	9.7	1.42	4.4	.2	88	22
51	63	.69	17	101	165	.55	59	.07	1.62	.24	.73	.03	15	4
0	59	12.1	79	1047	2025	1.88	386	1.41	26.2	1.19	15.3	.2	58	589
0	7	1.52	10	131	253	.23	48	.18	3.28	.15	1.91	.02	7	74
217	115	7.05	122	502	2874	3.86	319	.62	17.2	.83	1.69	.14	183	7
36	19	1.18	20	84	479	.64	53	.1	2.86	.14	.28	.02	30	1
131	392	5.17	98	1007	1844	2.94	3139	.36	10.5	1	1.22	.37	131	10
22	65	.86	16	168	307	.49	523	.06	1.75	.17	.2	.06	22	2
13	13	.88	8	51	325	.17	33	.1	.37	.08	.98	.03	4	1
0	28	1.56	13	83	479	.29	22	.18	.55	.14	1.83	.04	23	2
0	10	1.21	10	41	480	.24	0	.13	.06	.17	1.47	.03	48	0
0	22	2.59	21	88	1029	.51	0	.28	.13	.37	3.15	.07	103	0
0	22	2.59	21	88	1029	.51	0	.28	.13	.37	3.15	.07	103	0
<1	33	.88	6	37	146	.22	0	.14	.22	.09	1.13	.01	27	<1
0	60	1.36	9	61	241	.27	0	.21	.2	.13	1.69	.02	41	0
0	60	1.36	9	61	241	.27	0	.21	.2	.13	1.69	.02	41	0
0	54	1.87	15	62	310	.54	0	.27	.1	.19	2.42	.03	54	0
0	122	3.78	27	122	783	.85	0	.54	.1	.33	4.47	.05	40	0
13	21	1.04	7	53	145	.24	28	.14	.35	.14	1.21	.02	15	<1
0	300	6.3	160	220	65	5.25		1.5	20	1.7	20	2	400	60
0	300	5.4	140	150	110	5.25		1.5		1.7	20	2	400	60
	349	4.49	140	279	100		50	1.5	19.9	1.69	19.9	1.99	399	60
0	13	1.81	9	58	218	.34	149	.15	.97	.19	2.05	.2	34	<1
12	12	.97			217		17		1					0
12	12	.97			217		17		1					0

Table H-1

Food Composition (Computer code number is for West Diet Analysis program) (For purposes of calculations, use "0" for t, <1, <.1, <.01, etc.)

Computer Code Number	Food Description	Measure	Wt (g)	H_2O (%)	Ener (kcal)	Prot (g)	Carb (g)	Dietary Fiber (g)	Fat (g)	Fat Breakdown (g) Sat	Mono	Poly
	BAKED GOODS: BREADS, CAKES, COOKIES, CRACKERS, PIES—Continued											
	Tortilla chips:											
1271	Plain	10 pce	18	2	90	1	11	1	5	.9	2.8	.7
1036	Nacho flavor	1 c	26	2	129	2	16	1	7	1.3	3.9	.9
1037	Taco flavor	1 pce	18	2	86	1	11	1	4	.8	2.6	.6
	Tortillas:											
479	Corn, enriched, 6″ diam	1 ea	26	44	58	2	12	1	1	.1	.2	.3
480	Flour, 8″ diam	1 ea	49	27	159	4	27	2	3	.6	1.4	1.4
1301	Flour, 10″ diam	1 ea	72	27	234	6	40	2	5	.8	2.1	2
481	Taco shells	1 ea	14	4	63	1	9	1	3	.4	1.5	.6
	Waffles, 7″ diam:											
482	From home recipe	1 ea	75	42	218	6	25	1	11	2.2	2.6	5.1
483	From mix, egg/milk added	1 ea	75	42	218	5	26	1	10	1.7	2.7	5.2
1510	Whole grain, prepared from frozen	1 ea	39	43	107	4	13	1	5	1.6	1.9	.9
	GRAIN PRODUCTS: CEREAL, FLOUR, GRAIN, PASTA and NOODLES, POPCORN											
484	Barley, pearled, dry, uncooked	1 c	200	10	704	20	155	31	2	.5	.3	1.1
485	Barley, pearled, cooked	1 c	157	69	193	4	44	6	1	.1	.1	.3
2009	Breakfast bars, fat free, all flavors	1 ea	38	25	110	2	26	3	0	0	0	0
	Breakfast bar, Snackwell:											
2165	Apple-cinnamon	1 ea	37	16	119	1	29	1	<1	.1	t	.1
2164	Blueberry	1 ea	37	16	121	1	29	1	<1	t	t	.1
2163	Strawberry	1 ea	37	16	120	1	29	1	<1	t	t	.1
	Breakfast cereals, hot, cooked w/o salt added:											
	Corn grits (hominy) enriched:											
486	Regular/quick prep w/o salt, yellow:	1 c	242	85	145	3	31	<1	<1	.1	.1	.2
487	Instant, prepared from packet, white	1 ea	137	82	89	2	21	1	<1	t	t	.1
	Cream of wheat:											
488	Regular, quick, instant	1 c	239	87	129	4	27	1	<1	.1	.1	.3
489	Mix and eat, plain, packet	1 ea	142	82	102	3	21	<1	<1	t	t	.2
1664	Farina cereal, cooked w/o salt	1 c	233	88	117	3	25	3	<1	t	t	.1
490	Malt-O-Meal, cooked w/o salt	1 c	240	88	122	4	26	1	<1	.1	.1	t
494	Maypo	1 c	216	83	153	5	29	5	2	.4	.7	.8
	Oatmeal or rolled oats:											
491	Regular, quick, instant,nonfortified cooked w/o salt	1 c	234	85	145	6	25	4	2	.4	.7	.9
	Instant, fortified:											
492	Plain, from packet	½ c	118	85	70	4	12	2	1	.2	.4	.4
493	Flavored, from packet	½ c	109	76	106	3	21	2	1	.2	.5	.5
	Breakfast cereals, ready to eat:											
495	All-Bran	1 c	62	3	160	8	46	20	2	.4	.4	1.3
1306	Alpha Bits	1 c	28	1	110	2	24	1	1	.1	.2	.2
1307	Apple Jacks	1 c	33	3	120	2	30	1	<1	.1	.1	.2
1308	Bran Buds	1 c	90	3	240	8	72	36	2	.4	.4	1.4
1305	Bran Chex	1 c	49	2	156	5	39	8	1	.2	.3	.7
1309	Honey BucWheat Crisp	1 c	38	5	147	4	31	3	1	.2	.3	.6
1310	C.W. Post, plain	1 c	97	2	421	9	73	7	13	1.7	6	4.7
1311	C.W. Post, with raisins	1 c	103	4	446	9	74	14	15	11	1.7	1.4
496	Cap'n Crunch	1 c	37	2	147	2	32	1	2	.5	.4	.3
1312	Cap'n Crunchberries	1 c	35	2	140	2	30	1	2	.5	.3	.3
1313	Cap'n Crunch, peanut butter	1 c	35	2	146	3	28	1	3	.7	1.1	.7
497	Cheerios	1 c	23	3	84	2	17	2	1	.3	.5	.2
1314	Cocoa Krispies	1 c	41	2	159	3	36	1	1	.7	.2	.2
1316	Cocoa Pebbles	1 c	32	2	131	1	27	1	2	1.1	.4	.1
1315	Corn Bran	1 c	36	3	120	2	30	6	1	.3	.3	.4
1317	Corn Chex	1 c	28	2	110	2	25	<1	<1	t	t	t
498	Corn Flakes, Kellogg's	1 c	28	3	100	2	24	1	<1	.1	t	.1
499	Corn Flakes, Post Toasties	1 c	24	3	93	2	21	1	<1	t	t	t
1340	Corn Pops	1 c	31	3	120	1	28	<1	<1	.1	.1	t
1318	Cracklin' Oat Bran	1 c	65	4	252	5	48	8	8	3.4	3.8	.9
1038	Crispy Wheat 'N Raisins	1 c	43	7	150	3	35	3	1	.1	.1	.2

Chol (mg)	Calc (mg)	Iron (mg)	Magn (mg)	Pota (mg)	Sodi (mg)	Zinc (mg)	VT-A (RE)	Thia (mg)	VT-E (α-TE)	Ribo (mg)	Niac (mg)	V-B6 (mg)	Fola (μg)	VT-C (mg)
0	28	.27	16	35	95	.27	4	.01	.24	.03	.23	.05	2	0
1	38	.37	21	56	184	.31	11	.03	.35	.05	.37	.07	4	<1
1	28	.36	16	39	142	.23	16	.04	.24	.04	.36	.05	4	<1
0	45	.36	17	40	42	.24	6	.03	.04	.02	.39	.06	30	0
0	61	1.62	13	64	234	.35	0	.26	.62	.14	1.75	.02	60	0
0	90	2.38	19	94	344	.51	0	.38	.91	.21	2.57	.04	89	0
0	35	.36	15	34	25	.19	6	.04	.59	.02	.24	.04	4	0
52	191	1.73	14	119	383	.51	49	.2	1.73	.26	1.55	.04	34	<1
38	93	1.22	15	134	458	.35	19	.15	1.5	.19	1.23	.07	9	<1
39	84	.69	15	91	150	.45	25	.08	.53	.13	.75	.04	7	<1
0	58	5	158	560	18	4.26	4	.38	.26	.23	9.2	.52	46	0
0	17	2.09	34	146	5	1.29	2	.13	.08	.1	3.23	.18	25	0
0	20	.72			25		20							1
<1	17	5	6	68	103	3.88	260	.39		.44	5.2	.52		<1
<1	14	4.83	5	43	107	3.85	260	.39		.44	5.2	.52		<1
<1	14	4.82	6	47	102	3.83	260	.39		.44	5.2	.52		2
0	0	1.55	10	53	0	.17	14	.24	.12	.14	1.96	.06	75	0
0	8	8.19	11	38	289	.21	0	.15	.03	.08	1.38	.05	47	0
0	50	10.3	12	45	139	.33	0	.24	.03	0	1.43	.03	108	0
0	20	8.09	7	38	241	.24	125	.43	.02	.28	4.97	.57	101	0
0	5	1.17	5	30	0	.16	0	.19	.03	.12	1.28	.02	54	0
0	5	9.6	5	31	2	.17	0	.48	.03	.24	5.76	.02	5	0
0	112	7.56	45	190	233	1.34	633	.65	1.51	.65	8.42	.86	9	26
0	19	1.59	56	131	2	1.15	5	.26	.23	.05	.3	.05	9	0
0	109	4.2	28	66	190	.58	302	.35	.14	.19	3.65	.49	100	0
0	112	4.45	34	91	169	.66	306	.35	.14	.25	3.92	.51	100	<1
0	200	9	280	620	560	7.5	450	.75	1.14	.85	10	1	186	30
0	8	2.66	16	54	178	1.48	371	.36	.02	.42	4.93	.5	99	0
0	0	4.5	8	35	150	3.75	225	.38	.05	.43	5	.5	116	15
0	60	13.5	240	809	599	11.3	676	1.17	1.42	1.26	15	1.53	270	45
0	29	14	69	216	345	6.47	11	.64	.56	.26	8.62	.88	173	26
0	54	10.9	43	142	361	.68	913	.9	8.99	1.03	12.1	1.88	11	36
<1	47	15.4	67	198	167	1.64	1284	1.26	.68	1.46	17.1	1.75	342	0
<1	50	16.4	74	261	161	1.64	1363	1.34	.72	1.55	18.1	1.85	364	0
0	7	6.18	13	47	286	5.14	5	.51	.18	.58	6.85	.68	137	0
0	9	6.06	13	49	256	5.39	6	.5	.25	.57	6.72	.67	135	<1
0	3	5.85	24	80	264	4.87	5	.49	.19	.55	6.48	.65	130	0
0	42	6.21	25	68	218	2.88	288	.29	.16	.33	3.84	.38	77	11
0	0	2.38	11	79	278	1.97	298	.49	.19	.57	6.6	.66	123	20
0	5	2.02	13	53	180	1.7	424	.42	.04	.48	5.63	.58	113	0
0	27	10.1	19	75	338	5	5	.1	.19	.56	6.66	.67	134	0
0	3	8.01	4	23	306	.1	14	.36	.07	.07	4.93	.5	99	15
0	0	8.68	3	25	300	.17	225	.36	.03	.43	5	.5	99	15
0	1	.63	4	28	252	.07	318	.31	.06	.36	4.22	.43	85	0
0	0	1.86	2	25	120	1.55	225	.4	.03	.43	5.18	.5	109	15
0	26	2.41	79	305	226	1.95	299	.5	.43	.56	6.63	.66	181	20
0	54	3.52	33	180	223	.85	293	.29	.45	.33	3.91	.39	78	0

Table H-1

Food Composition (Computer code number is for West Diet Analysis program) (For purposes of calculations, use "0" for t, <1, <.1, <.01, etc.)

Computer Code Number	Food Description	Measure	Wt (g)	H_2O (%)	Ener (kcal)	Prot (g)	Carb (g)	Dietary Fiber (g)	Fat (g)	Fat Breakdown (g) Sat	Mono	Poly
	GRAIN PRODUCTS: CEREAL, FLOUR, GRAIN, PASTA and NOODLES, POPCORN—Continued											
1319	Fortified Oat Flakes	1 c	48	3	180	8	36	1	1	.2	.3	.4
500	40% Bran Flakes, Kellogg's	1 c	39	4	121	4	32	7	1	.2	.2	.5
501	40% Bran Flakes, Post	1 c	47	3	152	5	37	9	1	.1	.1	.4
502	Froot Loops	1 c	32	2	120	2	28	1	1	.4	.2	.3
518	Frosted Flakes	1 c	41	3	159	2	37	1	<1	.1	t	.1
1320	Frosted Mini-Wheats	1 c	51	5	170	5	41	5	1	.2	.1	.6
1321	Frosted Rice Krispies	1 c	35	2	135	2	32	<1	<1	.1	.1	.1
1324	Fruit & Fibre w/dates	1 c	57	9	193	5	43	8	3	.4	1.3	1
1322	Fruity Pebbles	1 c	32	3	130	1	28	<1	2	1.4	.1	.1
503	Golden Grahams	1 c	39	3	150	2	33	1	1	.2	.4	.2
504	Granola, homemade	½ c	61	5	285	9	32	6	15	2.9	4.8	6.5
505	Granola, low fat	½ c	47	3	181	5	38	3	3	0		
1670	Granola, low fat, commercial	½ c	45	5	165	4	35	2	2	.6	.7	.9
505	Grape Nuts	½ c	55	3	196	7	45	5	<1	t	t	.1
1326	Grape Nuts Flakes	1 c	39	3	144	4	32	4	1	.6	.1	.2
1665	Heartland Natural with raisins	1 c	110	5	468	11	76	6	16	4	4.2	6.2
1327	Honey & Nut Corn Flakes	1 c	37	2	148	3	31	1	2	.3	.7	.6
506	Honey Nut Cheerios	1 c	33	2	126	3	27	2	1	.3	.5	.2
1328	HoneyBran	1 c	35	2	119	3	29	4	1	.3	.1	.3
1329	HoneyComb	1 c	22	1	86	1	20	1	<1	.2	.1	.1
1330	King Vitaman	1 c	21	2	81	2	18	1	1	.2	.3	.2
1039	Kix	1 c	19	2	72	1	16	1	<1	.1	.1	t
1331	Life	1 c	44	4	167	4	35	3	2	.3	.6	.8
507	Lucky Charms	1 c	32	2	124	2	27	1	1	.2	.4	.2
1323	Mueslix Five Grain	1 c	82	5	279	7	63	7	3	.5	1	1.2
508	Nature Valley Granola	1 c	113	4	510	12	74	7	20	2.6	13.3	3.8
1666	Nutri Grain Almond Raisin	1 c	40	6	147	3	31	3	2	.1	1	1.2
1336	100% Bran	1 c	66	3	178	8	48	19	3	.6	.6	1.9
509	100% Natural cereal, plain	1 c	104	2	462	11	71	8	17	7.5	7.5	2.3
1337	100% Natural with apples & cinnamon	1 c	104	2	477	11	70	7	20	15.5	1.8	1.3
1338	100% Natural with raisins & dates	1 c	110	3	496	12	72	7	20	13.6	3.7	1.7
510	Product 19	1 c	33	4	110	2	28	1	<1	t	.2	.2
1339	Quisp	1 c	30	3	121	2	25	1	2	.5	.4	.2
511	Raisin Bran, Kellogg's	1 c	61	9	200	6	47	8	1	.1	.1	.4
512	Raisin Bran, Post	1 c	56	9	172	5	42	8	1	.2	.1	.5
1667	Raisin Squares	1 c	71	9	241	6	55	7	2	.2	.2	.6
1041	Rice Chex	1 c	33	3	130	2	29	1	<1	t	t	t
513	Rice Krispies, Kellogg's	1 c	28	2	111	2	25	<1	<1	t	t	t
514	Rice, puffed	1 c	14	4	54	1	12	<1	<1	t	t	t
515	Shredded Wheat	1 c	43	5	154	5	35	4	1	.1	.1	.4
516	Special K	1 c	31	3	110	6	22	1	<1	t	t	.2
517	Super Golden Crisp	1 c	33	1	123	2	30	<1	<1	.1	.1	.1
519	Honey Smacks	1 c	36	3	133	3	32	1	1	.4	.1	.3
1341	Tasteeos	1 c	24	2	94	3	19	3	1	.2	.2	.2
1342	Team	1 c	42	4	164	3	36	1	1	.1	.2	.3
520	Total, wheat, with added calcium	1 c	40	3	140	4	32	4	1	.2	.2	.1
521	Trix	1 c	28	2	114	1	24	1	2	.4	.9	.3
1344	Wheat Chex	1 c	46	2	169	5	38	4	1	.2	.1	.5
1043	Wheat cereal, puffed, fortified	1 c	12	4	44	2	9	1	<1	t	t	.1
522	Wheaties	1 c	29	3	106	3	23	2	1	.2	.2	.1
523	Buckwheat flour, dark	1 c	120	11	402	15	85	12	4	.8	1.1	1.1
525	Buckwheat, whole grain, dry	1 c	170	10	583	23	122	17	6	1.3	1.8	1.8
526	Bulgar, dry, uncooked	1 c	140	9	479	17	106	26	2	.3	.2	.8
527	Bulgar, cooked	1 c	182	78	151	6	34	8	<1	.1	.1	.2
	Cornmeal:											
528	Whole-ground, unbolted, dry	1 c	122	10	442	10	94	9	4	.6	1.2	2
530	Degermed, enriched, dry	1 c	138	12	505	12	107	10	2	.3	.6	1
38041	Degermed, enriched, baked	1 c	138	12	505	12	107	10	2	.3	.6	1
	Macaroni, cooked:											
532	Enriched	1 c	140	66	197	7	40	2	1	.1	.1	.4
533	Whole wheat	1 c	140	67	174	7	37	4	1	.1	.1	.3

Chol (mg)	Calc (mg)	Iron (mg)	Magn (mg)	Pota (mg)	Sodi (mg)	Zinc (mg)	VT-A (RE)	Thia (mg)	VT-E (α-TE)	Ribo (mg)	Niac (mg)	V-B6 (mg)	Fola (μg)	VT-C (mg)
0	68	13.7	58	228	220	2.54	636	.62	.34	.72	8.45	.86	169	0
0	0	10.9	81	229	309	5.03	505	.51	7.22	.58	6.71	.66	138	20
0	21	13.4	102	251	431	2.49	622	.61	.54	.7	8.27	.85	166	0
0	0	4.51	8	35	150	3.75	225	.37	.12	.43	5	.5	96	15
0	0	6.15	4	26	264	.2	298	.5	.05	.56	6.61	.66	123	20
0	0	15	60	170	0	1.5	0	.37	.46	.42	4.64	.5	102	0
0	0	2.42	8	27	256	.42	303	.49	.03	.56	6.72	.66	140	20
0	30	10.1	81	335	270	3.02	725	.75	1.32	.85	10.1	1	201	0
0	4	2.02	9	24	178	1.7	424	.42	.03	.48	5.63	.58	113	0
0	19	5.85	12	69	357	4.88	293	.49	.29	.55	6.51	.65	130	19
0	49	2.56	109	328	15	2.48	2	.45	7.87	.17	1.25	.19	52	1
0		2.71	36	143	90	5.64	226	.56	7.57	.64	7.52	.75	151	
0	15	1.35	30	127	101	2.84	169	.27	4.03	.31	3.74	.36	90	0
0	5	15.7	37	184	382	1.21	728	.71	.14	.82	9.68	.99	194	0
0	16	11.2	43	136	220	.78	516	.51	.1	.58	6.86	.7	138	0
0	66	4.02	141	415	226	2.83	7	.32	.77	.14	1.54	.2	44	1
0	0	3.03	3	40	249	.2	152	.26	.09	.3	3.37	.33	74	10
0	22	4.95	32	94	285	4.13	248	.41	.34	.47	5.51	.55	110	16
0	16	5.57	46	151	202	.9	463	.45	.81	.52	6.16	.63	23	19
0	4	2.09	7	25	124	1.17	291	.29	.09	.33	3.87	.4	78	0
0	3	5.92	18	58	176	2.65	212	.26	1.42	.3	3.53	.35	71	8
0	28	5.13	6	26	167	2.38	238	.24	.05	.27	3.17	.32	63	9
0	134	12.3	43	109	240	5.5	2	.55	.22	.62	7.35	.73	147	0
0	35	4.8	21	58	217	4	240	.4	.14	.45	5.34	.53	107	16
0	38	8.94	82	369	107	7.46	747	.75	8.94	.84	9.84	.99	197	1
0	85	3.53	107	375	183	2.27	0	.35	7.97	.12	1.25	.16	17	0
0	122	1.14	13	147	139	3.06	0	.32	4.38	.35	4.08	.41	80	0
0	46	8.12	312	652	457	5.74	0	1.58	1.53	1.78	20.9	2.11	47	63
1	100	3.11	109	457	28	2.5	1	.36	1.19	.17	1.84	.19	26	<1
0	157	2.89	72	514	52	2	6	.33	.73	.57	1.87	.11	17	1
0	160	3.12	124	538	47	2.11	7	.31	.77	.65	2.09	.16	45	0
0	0	19.8	18	55	308	16.5	248	1.65	24.4	1.88	22	2.21	429	66
0	6	5.1	15	40	216	4.26	4	.42	.15	.48	5.67	.56	113	0
0	40	4.5	80	350	390	3.75	225	.37	.56	.43	5	.5	122	0
0	26	8.9	95	345	365	2.97	741	.73	1.3	.84	9.86	1.01	198	0
0	0	21.7	54	335	4	1.99	0	.5	.38	.57	6.67	.64	142	0
0	5	9.44	8	38	276	.45	2	.43	.04	.01	5.81	.59	116	17
0	5	.7	12	27	206	.46	371	.52	.03	.59	6.92	.69	138	15
0	1	.41	4	16	1	.15	0	.06	.01	.01	.87	0	1	0
0	16	1.81	57	155	4	1.42	0	.11	.23	.12	2.26	.11	21	0
0	0	8.4	16	55	250	3.75	225	.53	.08	.59	7.01	.71	93	15
0	7	2.08	20	48	51	1.75	437	.43	.12	.49	5.81	.59	116	0
0	0	2.4	21	53	67	.4	300	.5	.18	.58	6.66	.68	133	20
0	11	6.86	26	71	183	.69	318	.31	.17	.36	4.22	.43	85	13
0	6	12	12	71	260	.58	556	.55	.1	.63	7.39	.76	7	22
0	344	24	43	129	265	20	500	2	31.3	2.27	26.8	2.67	533	80
0	30	4.2	3	16	184	3.5	210	.35	.56	.4	4.68	.47	93	14
0	18	13.2	58	173	308	1.23	0	.6	.17	.17	8.1	.83	162	24
0	3	.56	16	44	1	.37	<1	.05	.08	.03	1.43	.02	4	0
0	53	7.83	31	101	215	.68	218	.36	.36	.41	4.84	.48	97	14
0	49	4.87	301	692	13	3.74	0	.5	1.24	.23	7.38	.7	65	0
0	31	3.74	393	782	2	4.08	0	.17	1.75	.72	11.9	.36	51	0
0	49	3.44	230	574	24	2.7	0	.32	.22	.16	7.15	.48	38	0
0	18	1.75	58	124	9	1.04	0	.1	.05	.05	1.82	.15	33	0
0	7	4.21	155	350	43	2.22	57	.47	.82	.24	4.43	.37	31	0
0	7	5.7	55	224	4	.99	57	.99	.45	.56	6.94	.35	258	0
0	7	5.7	55	224	4	.99	57	.79	.5	.5	6.25	.32	181	0
0	10	1.96	25	43	1	.74	0	.29	.04	.14	2.34	.05	98	0
0	21	1.48	42	62	4	1.13	0	.15	.14	.06	.99	.11	7	0

Table H-1

Food Composition (Computer code number is for West Diet Analysis program) (For purposes of calculations, use "0" for t, <1, <.1, <.01, etc.)

Computer Code Number	Food Description	Measure	Wt (g)	H_2O (%)	Ener (kcal)	Prot (g)	Carb (g)	Dietary Fiber (g)	Fat (g)	Fat Breakdown (g) Sat	Mono	Poly
	GRAIN PRODUCTS: CEREAL, FLOUR, GRAIN, PASTA and NOODLES, POPCORN—Continued											
534	Vegetable, enriched	1 c	134	68	172	6	36	2	<1	t	t	.1
535	Millet, cooked	1 c	240	71	286	8	57	3	2	.4	.4	1.2
	Noodles (see also Pasta and Spaghetti):											
1507	Cellophane noodles, cooked	1 c	190	79	160	<1	39	<1	<1	t	t	t
1995	Cellophane noodles, dry	1 c	140	13	491	<1	121	1	<1	t	t	t
537	Chow mein, dry	1 c	45	1	237	4	26	2	14	2	3.5	7.8
536	Egg noodles, cooked, enriched	1 c	160	69	213	8	40	2	2	.5	.7	.7
538	Spinach noodles, dry	3½ oz	100	8	372	13	75	11	2	.2	.2	.6
1343	Oat bran, dry	¼ c	24	7	59	4	16	4	2	.3	.6	.7
	Pasta, cooked:											
1418	Fresh	2 oz	57	69	75	3	14	1	1	.1	.1	.2
1417	Linguini/Rotini	1 c	140	66	197	7	40	4	1	.1	.1	.4
	Popcorn:											
539	Air popped, plain	1 c	8	4	31	1	6	1	<1	t	.1	.2
1042	Microwaved, low fat, low sodium	1 c	6	3	25	1	4	1	1	.1	.2	.3
540	Popped in vegetable oil/salted	1 c	11	3	55	1	6	1	3	.5	.9	1.5
541	Sugar-syrup coated	1 c	35	3	151	1	28	2	4	1.3	1	1.6
	Rice:											
542	Brown rice, cooked	1 c	195	73	216	5	45	4	2	.4	.6	.6
2215	Mexican rice, cooked	1 c	226		820	16	180	6	30	4	1	1
2216	Spanish rice, cooked	1 c	246	85	130	3	28	2	1			
	White, enriched, all types:											
543	Regular/long grain, dry	1 c	185	12	675	13	148	2	1	.3	.4	.3
544	Regular/long grain, cooked	1 c	158	68	205	4	45	1	<1	.1	.1	.1
545	Instant, prepared without salt	1 c	165	76	162	3	35	1	<1	.1	.1	.1
	Parboiled/converted rice:											
546	Raw, dry	1 c	185	10	686	13	151	3	1	.3	.3	.3
547	Cooked	1 c	175	72	200	4	43	1	<1	.1	.1	.1
1486	Sticky rice (glutinous), cooked	1 c	174	77	169	4	37	2	<1	.1	.1	.1
548	Wild rice, cooked	1 c	164	74	166	7	35	3	1	.1	.1	.4
1700	Rice and pasta (Rice-a-Roni), cooked	1 c	202	72	246	5	43	1	6	1.1	2.3	1.9
549	Rye flour, medium	1 c	102	10	361	10	79	15	2	.2	.2	.8
1044	Soy flour, low-fat	1 c	88	3	325	45	30	9	6	.9	1.3	3.3
	Spaghetti pasta:											
550	Without salt, enriched	1 c	140	66	197	7	40	4	1	.1	.1	.4
551	With salt, enriched	1 c	140	66	197	7	40	2	1	.1	.1	.4
552	Whole-wheat spaghetti, cooked	1 c	140	67	174	7	37	6	1	.1	.1	.3
1302	Tapioca-pearl, dry	1 c	152	11	544	<1	135	1	<1	t	t	t
553	Wheat bran, crude	1 c	58	10	125	9	37	25	2	.4	.4	1.3
554	Wheat germ, raw	1 c	115	11	414	27	60	15	11	1.9	1.6	6.9
555	Wheat germ, toasted	1 c	113	6	432	33	56	15	12	2.1	1.7	7.5
1669	Wheat germ, with brown sugar & honey	1 c	113	3	420	30	66	11	9	1.5	1.2	5.5
556	Rolled wheat, cooked	1 c	240	84	149	5	33	4	1	.1	.1	.5
557	Whole-grain wheat, cooked	1 c	150	86	84	4	20	3	<1	.1	.1	.2
	Wheat flour (unbleached):											
	All-purpose white flour, enriched:											
558	Sifted	1 c	115	12	419	12	88	3	1	.2	.1	.5
559	Unsifted	1 c	125	12	455	13	95	3	1	.2	.1	.5
560	Cake or pastry, enriched, sifted	1 c	96	12	348	8	75	2	1	.1	.1	.4
561	Self-rising, enriched, unsifted	1 c	125	11	443	12	93	3	1	.2	.1	.5
562	Whole wheat, from hard wheats	1 c	120	10	407	16	87	15	2	.4	.3	.9
	MEATS: FISH and SHELLFISH											
1045	Bass, baked or broiled	4 oz	113	69	165	27	0	0	5	1.4	1.5	2.3
1046	Bluefish, baked or broiled	4 oz	113	63	180	29	0	0	6	1.4	2	2.7
1686	Catfish, breaded/flour fried	4 oz	113	49	325	21	14	1	20	5	9	4.7
	Clams:											
563	Raw meat only	1 ea	145	82	107	19	4	0	1	.3	.4	.7
564	Canned, drained	1 c	160	64	237	41	8	0	3	.7	.9	1.5
1290	Steamed, meat only	10 ea	95	64	141	24	5	0	2	.4	.5	.9

Chol (mg)	Calc (mg)	Iron (mg)	Magn (mg)	Pota (mg)	Sodi (mg)	Zinc (mg)	VT-A (RE)	Thia (mg)	VT-E (α-TE)	Ribo (mg)	Niac (mg)	V-B6 (mg)	Fola (μg)	VT-C (mg)
0	15	.66	25	41	8	.59	7	.15	.05	.08	1.43	.03	87	0
0	7	1.51	106	149	5	2.18	0	.25	.43	.2	3.19	.26	46	0
0	14	1	3	5	9	.23	0	.07	.06	0	.09	.02	1	0
0	35	3.04	4	14	14	.57	0	.21	.18	0	.28	.07	3	0
0	9	2.13	23	54	198	.63	4	.26	.07	.19	2.68	.05	40	0
53	19	2.54	30	45	11	.99	10	.3	.08	.13	2.38	.06	102	0
0	58	2.13	174	376	36	2.76	46	.37	.04	.2	4.55	.32	48	0
0	14	1.3	56	136	1	.75	0	.28	.41	.05	.22	.04	12	0
19	3	.65	10	14	3	.32	3	.12	.09	.09	.56	.02	36	0
0	10	1.96	25	43	1	.74	0	.29	.08	.14	2.34	.05	98	0
0	1	.21	10	24	<1	.27	2	.02	.01	.02	.15	.02	2	0
0	1	.14	9	14	29	.23	1	.02	.06	.01	.12	.01	1	0
0	1	.31	12	25	97	.29	2	.01	.03	.01	.17	.02	2	<1
2	15	.61	12	38	72	.2	3	.02	.42	.02	.77	.01	1	0
0	19	.82	84	84	10	1.23	0	.19	.53	.05	2.98	.28	8	0
0	300	9			2700		120							96
0		.72			1340									
0	52	7.97	46	213	9	2.02	0	1.07	.24	.09	7.75	.3	427	0
0	16	1.9	19	55	2	.77	0	.26	.08	.02	2.34	.15	92	0
0	13	1.04	8	7	5	.4	0	.12	.08	.08	1.45	.02	68	0
0	111	6.59	57	222	9	1.78	0	1.1	.24	.13	6.72	.65	427	0
0	33	1.98	21	65	5	.54	0	.44	.09	.03	2.45	.03	87	0
0	3	.24	9	17	9	.71	0	.03	.07	.02	.5	.04	2	0
0	5	.98	52	166	5	2.2	0	.08	.38	.14	2.12	.22	43	0
2	16	1.9	24	85	1147	.57	0	.25	.27	.16	3.6	.2	89	<1
0	24	2.16	76	347	3	2.03	0	.29	1.36	.12	1.76	.27	19	0
0	165	5.27	202	2261	16	1.04	4	.33	.17	.25	1.9	.46	361	0
0	10	1.96	25	43	1	.74	0	.29	.08	.14	2.34	.05	98	0
0	10	1.96	25	43	140	.74	0	.29	.38	.14	2.34	.05	98	0
0	21	1.48	42	62	4	1.13	0	.15	.07	.06	.99	.11	7	0
0	30	2.4	2	17	2	.18	0	.01	0	0	0	.01	6	0
0	42	6.15	354	686	1	4.22	0	.3	1.35	.33	7.89	.75	46	0
0	45	7.2	275	1025	14	14.1	0	2.16	20.7	.57	7.83	1.5	323	0
0	51	10.3	362	1070	5	18.9	0	1.89	20.5	.93	6.32	1.11	398	7
0	56	9.1	307	1089	12	15.7	11	1.51	24.9	.78	5.34	.56	376	0
0	17	1.49	53	170	0	1.15	0	.17	.48	.12	2.14	.17	26	0
0	9	.88	35	99	1	.73	0	.12	.3	.03	1.5	.08	12	0
0	17	5.34	25	123	2	.8	0	.9	.07	.57	6.79	.05	177	0
0	19	5.8	27	134	2	.87	0	.98	.07	.62	7.38	.05	193	0
0	13	7.03	15	101	2	.59	0	.86	.06	.41	6.52	.03	148	0
0	423	5.84	24	155	1587	.77	0	.84	.07	.52	7.29	.06	193	0
0	41	4.66	166	486	6	3.52	0	.54	1.48	.26	7.64	.41	53	0
98	116	2.16	43	515	102	.94	40	.1	.84	.1	1.72	.16	19	2
86	10	.7	47	539	87	1.18	156	.08	.71	.11	8.19	.52	2	0
92	41	1.44	34	376	598	1.05	33	.4	2.48	.18	3.37	.21	19	1
49	67	20.3	13	455	81	1.99	131	.12	1.45	.31	2.57	.09	23	19
107	147	44.8	29	1004	179	4.37	274	.24	3.04	.68	5.36	.18	46	35
64	87	26.6	17	597	106	2.59	162	.14	1.86	.4	3.18	.1	27	21

Table H-1

Food Composition (Computer code number is for West Diet Analysis program) (For purposes of calculations, use "0" for t, <1, <.1, <.01, etc.)

Computer Code Number	Food Description	Measure	Wt (g)	H_2O (%)	Ener (kcal)	Prot (g)	Carb (g)	Dietary Fiber (g)	Fat (g)	Fat Breakdown (g) Sat	Mono	Poly
	MEATS: FISH and SHELLFISH—Continued											
	Cod:											
565	Baked	4 oz	113	76	119	26	0	0	1	.2	.1	.5
566	Batter fried	4 oz	113	67	196	20	8	<1	9	2.2	3.6	2.6
567	Poached, no added fat	4 oz	113	77	116	25	0	0	1	.2	.1	.3
	Crab, meat only:											
1048	Blue crab, cooked	1 c	118	77	120	24	0	0	2	.3	.3	.8
1049	Dungeness crab, cooked	1 c	118	73	130	26	1	0	1	.2	.3	.5
568	Blue crab, canned	1 c	135	76	134	28	0	0	2	.4	.3	.6
1587	Crab, imitation, from surimi	4 oz	113	74	115	14	11	0	1	.3	.2	.8
569	Fish sticks, breaded pollock	2 ea	56	46	152	9	13	<1	7	1.8	2.8	1.8
572	Flounder/sole, baked	4 oz	113	73	132	27	0	0	2	.5	.4	.9
1599	Grouper, baked or broiled	4 oz	113	73	133	28	0	0	1	.4	.4	.6
573	Haddock, breaded, fried	4 oz	113	55	264	22	14	1	13	3.2	5.4	3.3
1050	Haddock, smoked	4 oz	113	71	131	28	0	0	1	.3	.3	.5
	Halibut:											
17291	Baked	4 oz	113	72	158	30	0	0	3	.7	1	1.4
1051	Smoked	4 oz	113	64	203	34			4	.6	1.2	1.5
1054	Raw	4 oz	113	78	124	23	0	0	3	.7	.8	1.1
575	Herring, pickled	4 oz	113	55	296	16	11	0	20	4.4	11	4.8
1052	Lobster meat, cooked w/moist heat	1 c	145	76	142	30	2	0	1	.2	.2	.5
1687	Ocean perch, baked/broiled	4 oz	113	73	137	27	0	0	2	.4	1	.7
576	Ocean perch, breaded/fried	4 oz	113	59	249	22	9	<1	13	3.2	5.7	3.4
1056	Octopus, raw	4 oz	113	80	93	17	2	0	1	.3	.2	.3
	Oysters:											
577	Raw, Eastern	1 c	248	85	169	17	10	0	6	2	.9	2.8
578	Raw, Pacific	1 c	248	82	201	23	12	0	6	1.3	.9	2.2
	Cooked:											
579	Eastern, breaded, fried, medium	5 ea	73	65	144	6	8	<1	9	2.7	1.7	4.6
580	Western, simmered	5 ea	125	64	204	24	12	0	6	1.9	.9	2.7
581	Pollock, baked, broiled, or poached	4 oz	113	74	128	27	0	0	1	.3	.2	.6
	Salmon:											
582	Canned pink, solids and liquid	4 oz	113	69	157	22	0	0	7	1.7	2.1	2.3
583	Broiled or baked	4 oz	113	62	244	31	0	0	12	2.2	6	2.7
584	Smoked	4 oz	113	72	132	21	0	0	5	1.2	2.3	1.1
585	Atlantic sardines, canned, drained, 2 = 24 g	4 oz	113	60	235	28	0	0	13	1.9	4.4	6.4
586	Scallops, breaded, cooked from frozen	6 ea	93	58	200	17	9	<1	10	2	2.5	5.3
1588	Scallops, imitation, from surimi	4 oz	113	74	112	14	12	0	1	.1	.1	.3
1688	Scallops, steamed/boiled	½ c	60	76	64	10	1	0	2	.3	.7	.6
	Shrimp:											
587	Cooked, boiled, 2 large = 11g	16 ea	88	77	87	18	0	0	1	.2	.2	.5
588	Canned, drained	½ c	64	73	77	15	1	0	1	.3	.3	.7
589	Fried, 2 large = 15 g, breaded	12 ea	90	53	218	19	10	<1	11	2	3	5.9
1057	Raw, large, about 7g each	14 ea	98	76	104	20	1	0	2	.3	.3	1
1589	Shrimp, imitation, from surimi	4 oz	113	75	114	14	10	0	2	.3	.2	1
1053	Snapper, baked or broiled	4 oz	113	70	145	30	0	0	2	.4	.4	.7
1060	Squid, fried in flour	4 oz	113	64	198	20	9	0	8	2.1	3.1	2.4
1590	Surimi	4 oz	113	76	112	17	8	0	1	.2	.2	.6
1058	Swordfish, raw	4 oz	113	76	137	22	0	0	5	1.3	1.7	1
1059	Swordfish, baked or broiled	4 oz	113	69	175	29	0	0	6	1.7	2.2	1.3
590	Trout, baked or broiled	4 oz	113	70	170	26	0	0	7	1.8	2	2.2
	Tuna, light, canned, drained solids:											
591	Oil pack	1 c	145	60	287	42	0	0	12	2.2	4.3	4.2
592	Water pack	1 c	154	74	179	39	0	0	1	.4	.2	.5
1061	Bluefin tuna, fresh	4 oz	113	68	163	26	0	0	6	1.4	1.8	1.6
	MEATS: BEEF, LAMB, PORK and others											
	BEEF, cooked, trimmed to ½" outer fat:											
	Braised, simmered, pot roasted:											
	Relatively fat, choice chuck blade:											
593	Lean and fat, piece 2½ x 2½ x ¾"	4 oz	113	47	393	30	0	0	29	13	14.8	1.2
594	Lean only	4 oz	113	55	297	35	0	0	16	7.3	8.2	.7

Chol (mg)	Calc (mg)	Iron (mg)	Magn (mg)	Pota (mg)	Sodi (mg)	Zinc (mg)	VT-A (RE)	Thia (mg)	VT-E (α-TE)	Ribo (mg)	Niac (mg)	V-B6 (mg)	Fola (µg)	VT-C (mg)
62	16	.55	47	276	88	.65	16	.1	.39	.09	2.84	.32	9	1
64	43	.92	36	443	124	.62	17	.13	.92	.13	2.58	.23	10	1
61	23	.54	41	496	69	.63	14	.09	.32	.08	2.48	.28	8	1
118	123	1.07	39	382	329	4.98	2	.12	1.18	.06	3.89	.21	60	4
90	70	.51	68	481	446	6.45	37	.07	1.33	.24	4.27	.2	50	4
120	136	1.13	53	505	450	5.43	2	.11	1.35	.11	1.85	.2	57	4
23	15	.44	49	102	950	.37	23	.04	.12	.03	.2	.03	2	0
63	11	.41	14	146	326	.37	17	.07	.77	.1	1.19	.03	10	0
77	20	.38	65	389	119	.71	12	.09	2.6	.13	2.46	.27	10	0
53	24	1.29	42	537	60	.58	56	.09	.71	.01	.43	.4	11	0
96	63	1.92	46	345	523	.59	33	.08	1.56	.14	4.49	.28	19	<1
87	55	1.58	61	469	862	.56	25	.05	.56	.05	5.73	.45	17	0
46	68	1.21	121	651	78	.6	61	.08	1.23	.1	8.05	.45	16	0
59	87	1.56	154	833	2260	.78	86	.11	1.11	.14	10.8	.64	22	0
36	53	.95	94	509	61	.47	53	.07	.96	.08	6.61	.39	14	0
15	87	1.38	9	78	983	.6	292	.04	1.81	.16	3.73	.19	3	0
104	88	.57	51	510	551	4.23	38	.01	2.1	.1	1.55	.11	16	0
61	155	1.33	44	396	108	.69	16	.15	1.84	.15	2.76	.3	12	1
71	136	1.57	38	323	431	.67	23	.14	2.41	.18	2.68	.24	15	1
54	60	5.99	34	396	260	1.9	51	.03	1.36	.04	2.37	.41	18	6
131	112	16.5	117	387	523	225	74	.25	1.98	.24	3.42	.15	25	9
124	20	12.7	55	417	263	41.2	201	.17	2.11	.58	4.98	.12	25	20
59	45	5.07	42	178	304	63.6	66	.11	1.66	.15	1.2	.05	23	3
125	20	11.5	55	378	265	41.5	183	.16	2.21	.55	4.53	.11	19	16
108	7	.32	82	437	131	.68	26	.08	.32	.09	1.86	.08	4	0
62	241	.95	38	368	626	1.04	19	.03	1.53	.21	7.39	.34	17	0
98	8	.62	35	424	75	.58	71	.24	1.42	.19	7.54	.25	6	0
26	12	.96	20	198	886	.35	29	.03	1.53	.11	5.33	.31	2	0
160	432	3.3	44	449	571	1.48	76	.09	.34	.26	5.93	.19	13	0
57	39	.76	55	310	432	.99	20	.04	1.77	.1	1.4	.13	34	2
25	9	.35	49	116	898	.37	23	.01	.12	.02	.35	.03	2	0
19	15	.15	32	168	246	.55	27	.01	.81	.04	.6	.08	7	1
172	34	2.72	30	160	197	1.37	58	.03	.66	.03	2.28	.11	3	2
111	38	1.75	26	134	108	.81	11	.02	.59	.02	1.77	.07	1	1
159	60	1.13	36	203	310	1.24	50	.12	1.35	.12	2.76	.09	7	1
149	51	2.36	36	181	145	1.09	53	.03	.8	.03	2.5	.1	3	2
41	21	.68	49	101	797	.37	23	.03	.12	.04	.19	.03	2	0
53	45	.27	42	590	64	.5	40	.06	.71	<.01	.39	.52	7	2
294	44	1.14	43	315	346	1.97	12	.06	2.09	.52	2.94	.07	16	5
34	10	.29	49	127	162	.37	23	.02	.28	.02	.25	.03	2	0
44	5	.91	30	325	102	1.3	41	.04	.56	.11	10.9	.37	2	1
56	7	1.18	38	417	130	1.66	46	.05	.71	.13	13.3	.43	3	1
78	97	.43	35	506	63	.58	17	.17	.57	.11	6.52	.39	21	2
26	19	2.02	45	300	513	1.31	33	.05	1.74	.17	18	.16	8	0
46	17	2.36	42	365	521	1.19	26	.05	.82	.11	20.5	.54	6	0
43	9	1.15	56	285	44	.68	740	.27	1.13	.28	9.77	.51	2	0
112	11	3.45	21	275	67	7.57	0	.08	.26	.27	3.54	.32	10	0
120	15	4.16	26	297	80	11.6	0	.09	.16	.32	3.02	.33	7	0

H

Table H-1

Food Composition (Computer code number is for West Diet Analysis program) (For purposes of calculations, use "0" for t, <1, <.1, <.01, etc.)

Computer Code Number	Food Description	Measure	Wt (g)	H_2O (%)	Ener (kcal)	Prot (g)	Carb (g)	Dietary Fiber (g)	Fat (g)	Fat Breakdown (g) Sat	Mono	Poly
	MEATS: BEEF, LAMB, PORK and others—Continued											
	Relatively lean, like choice round:											
595	Lean and fat, pce 4⅛ x 2½ x ¾"	4 oz	113	52	311	32	0	0	19	8.5	9.7	.8
596	Lean only	4 oz	113	57	249	36	0	0	11	4.8	5.4	.5
	Ground beef, broiled, patty 3 x ⅝":											
597	Extra lean, about 16% fat	4 oz	113	54	299	32	0	0	18	8	9	.8
598	Lean, 21% fat	4 oz	113	53	316	32	0	0	20	8.9	10.1	.8
	Roasts, oven cooked, no added liquid:											
	Relatively fat, prime rib:											
601	Lean and fat, piece 4⅛ x 2¼ x ½"	4 oz	113	46	425	25	0	0	35	15.8	17.9	1.5
602	Lean only	4 oz	113	58	271	31	0	0	16	7	7.9	.7
	Relatively lean, choice round:											
603	Lean and fat, piece 2½ x 2½ x ¾"	4 oz	113	59	272	30	0	0	16	7.1	8.1	.7
604	Lean only	4 oz	113	65	198	33	0	0	6	2.9	3.3	.3
1701	Steak, rib, broiled, lean	4 oz	113	58	250	32	0	0	13	5.7	6.4	.5
	Steak, broiled, relatively lean,											
606	choice sirloin, lean only	4 oz	113	62	228	34	0	0	9	4	4.6	.4
	Steak, broiled, relatively fat,											
	choice T-bone:											
1063	Lean and fat	4 oz	113	52	349	26	0	0	26	11.8	13.3	1.1
1064	Lean only	4 oz	113	61	232	30	0	0	11	5.1	5.8	.5
	Variety meats:											
1086	Brains, panfried	4 oz	113	71	221	14	0	0	18	6.7	7	3.9
599	Heart, simmered	4 oz	113	64	198	32	<1	0	6	2.9	1.5	1.5
600	Liver, fried	4 oz	113	56	245	30	9	0	9	3.1	1.8	1.9
1062	Tongue, cooked	4 oz	113	56	320	25	<1	0	23	10.1	10.7	.9
607	Beef, canned, corned	4 oz	113	58	283	31	0	0	17	7.5	8.5	.7
608	Beef, dried, cured	1 oz	28	56	46	8	<1	0	1	.5	.5	.1
	LAMB, domestic, cooked:											
	Chop, arm, braised (5.6 oz raw w/bone):											
609	Lean and fat	1 ea	70	44	242	21	0	0	17	7.8	7.3	1.4
610	Lean only	1 ea	55	49	153	19	0	0	8	3.6	3.4	.6
	Chop, loin, broiled (4.2 oz raw w/bone):											
611	Lean and fat	1 ea	64	52	202	16	0	0	15	6.8	6.4	1.2
612	Lean only	1 ea	46	61	99	14	0	0	4	2.1	1.9	.4
1067	Cutlet, avg of lean cuts, cooked	4 oz	113	54	330	28	0	0	23	10.9	10.2	1.9
	Leg, roasted, 3 oz = 4⅛ x 2¼ x ½":											
613	Lean and fat	4 oz	113	57	292	29	0	0	19	8.7	8.1	1.6
614	Lean only	4 oz	113	64	216	32	0	0	9	4.1	3.8	.7
615	Rib, roasted, lean and fat	4 oz	113	48	406	24	0	0	34	15.7	14.7	2.8
616	Rib, roasted, lean only	4 oz	113	60	262	30	0	0	15	7	6.6	1.2
1065	Shoulder, roasted, lean and fat	4 oz	113	56	312	25	0	0	23	10.5	9.8	1.9
1066	Shoulder, roasted, lean only	4 oz	113	63	231	28	0	0	12	5.7	5.3	1.1
	Variety meats:											
1069	Brains, panfried	4 oz	113	76	164	14	0	0	11	4.4	3.7	1.9
1068	Heart, braised	4 oz	113	64	209	28	2	0	9	3.9	2.7	1.1
1070	Sweetbreads, cooked	4 oz	113	60	264	26	0	0	17	8.1	6.5	1.4
1071	Tongue, cooked	4 oz	113	58	311	24	0	0	23	8.8	11.6	1.4
	PORK, cured, cooked (see also Sausages and Lunch Meats)											
617	Bacon, medium slices	3 pce	19	13	109	6	<1	0	9	3.3	4.5	1.1
1087	Breakfast strips, cooked	2 pce	23	27	106	7	<1	0	8	2.9	3.8	1.3
618	Canadian-style bacon	2 pce	47	62	87	11	1	0	4	1.3	1.9	.4
	Ham, roasted:											
619	Lean and fat, 2 pces 4⅛ x 2¼ x ¼"	4 oz	113	64	201	25	0	0	10	3.4	5	1.7
620	Lean only	4 oz	113	68	164	24	2	0	6	2.1	3	.6
621	Ham, canned, roasted, 8% fat	4 oz	113	69	154	24	1	0	6	1.8	2.8	.5
	PORK, fresh, cooked:											
	Chops, loin (cut 3 per lb with bone):											
1291	Braised, lean and fat	1 ea	89	58	213	24	0	0	12	4.5	5.4	1
1292	Lean only	1 ea	80	61	163	23	0	0	7	2.7	3.3	.6
622	Broiled, lean and fat	1 ea	82	58	197	23	0	0	11	3.9	4.8	.8
623	Broiled, lean only	1 ea	74	61	149	22	0	0	6	2.2	2.7	.4

Chol (mg)	Calc (mg)	Iron (mg)	Magn (mg)	Pota (mg)	Sodi (mg)	Zinc (mg)	VT-A (RE)	Thia (mg)	VT-E (α-TE)	Ribo (mg)	Niac (mg)	V-B6 (mg)	Fola (µg)	VT-C (mg)
108	7	3.53	25	319	56	5.55	0	.08	.21	.27	4.21	.37	11	0
108	6	3.91	28	348	58	6.19	0	.08	.2	.29	4.61	.41	12	0
112	10	3.13	28	417	93	7.27	0	.08	.2	.36	6.61	.36	12	0
114	14	2.77	27	394	101	7.01	0	.07	.23	.27	6.75	.34	12	0
96	12	2.61	21	334	71	5.92	0	.08	.27	.19	3.8	.26	8	0
91	11	2.95	28	425	84	7.84	0	.09	.14	.24	4.64	.34	9	0
81	7	2.07	27	406	67	4.87	0	.09	.23	.18	3.92	.4	7	0
78	6	2.2	30	446	70	5.36	0	.1	.12	.19	4.24	.43	8	0
90	15	2.9	30	445	78	7.9	0	.11	.16	.25	5.42	.45	9	0
101	12	3.8	36	455	75	7.37	0	.15	.16	.33	4.84	.51	11	0
76	9	3.06	26	363	72	5.03	0	.1	.24	.24	4.46	.37	8	0
67	7	3.58	32	427	80	6	0	.12	.16	.28	5.23	.44	9	0
2254	10	2.51	17	400	179	1.53	0	.15	2.37	.29	4.27	.44	7	4
218	7	8.49	28	263	71	3.54	0	.16	.81	1.74	4.6	.24	2	2
545	12	7.1	26	411	120	6.16	12123	.24	.72	4.68	16.3	1.62	249	26
121	8	3.83	19	203	68	5.42	0	.03	.4	.4	2.43	.18	6	1
97	14	2.35	16	154	1136	4.03	0	.02	.17	.17	2.75	.15	10	0
12	2	1.26	9	124	972	1.47	0	.02	.04	.06	1.53	.1	3	0
84	17	1.67	18	214	50	4.26	0	.05	.1	.17	4.66	.08	13	0
67	14	1.49	16	186	42	4.02	0	.04	.1	.15	3.48	.07	12	0
64	13	1.16	15	209	49	2.23	0	.06	.08	.16	4.54	.08	11	0
44	9	.92	13	173	39	1.9	0	.05	.07	.13	3.15	.07	11	0
110	12	2.26	25	340	77	4.67	0	.12	.15	.32	7.48	.16	19	0
105	12	2.24	27	354	75	4.97	0	.11	.17	.3	7.45	.17	23	0
101	9	2.4	29	382	77	5.58	0	.12	.2	.33	7.16	.19	26	0
110	25	1.81	23	306	82	3.94	0	.1	.11	.24	7.63	.12	17	0
99	24	2	26	356	91	5.05	0	.1	.17	.26	6.96	.17	25	0
104	23	2.23	26	284	75	5.91	0	.1	.16	.27	6.95	.15	24	0
98	21	2.41	28	299	77	6.83	0	.1	.2	.29	6.51	.17	28	0
2308	14	1.9	16	232	151	1.54	0	.12	1.73	.27	2.79	.12	6	14
281	16	6.24	27	212	71	4.16	0	.19	.79	1.34	4.93	.34	2	8
452	14	2.4	21	329	59	3.03	0	.02	.78	.24	2.89	.06	15	23
214	11	2.97	18	179	76	3.38	0	.09	.36	.47	4.17	.19	3	8
16	2	.31	5	92	303	.62	0	.13	.1	.05	1.39	.05	1	0
24	3	.45	6	107	483	.85	0	.17	.08	.08	1.75	.08	1	0
27	5	.38	10	183	727	.8	0	.39	.15	.09	3.25	.21	2	0
67	9	1.51	25	462	1695	2.79	0	.82	.45	.37	6.95	.35	3	0
60	9	1.67	16	324	1359	3.25	0	.85	.29	.23	4.54	.45	3	0
34	7	1.04	24	393	1282	2.52	0	1.18	.29	.28	5.53	.51	6	0
71	19	.95	17	333	43	2.12	2	.56	.3	.23	3.93	.33	3	1
63	14	.9	16	310	40	1.98	2	.53	.3	.21	3.67	.31	3	<1
67	27	.66	20	294	48	1.85	2	.88	.27	.24	4.3	.35	5	<1
61	23	.63	20	278	44	1.76	2	.85	.31	.23	4.1	.35	4	<1

Table H-1

Food Composition (Computer code number is for West Diet Analysis program) (For purposes of calculations, use "0" for t, <1, <.1, <.01, etc.)

Computer Code Number	Food Description	Measure	Wt (g)	H_2O (%)	Ener (kcal)	Prot (g)	Carb (g)	Dietary Fiber (g)	Fat (g)	Fat Breakdown (g) Sat	Mono	Poly
	MEATS: BEEF, LAMB, PORK and others—Continued											
624	Panfried, lean and fat	1 ea	78	53	216	23	0	0	13	4.7	5.5	1.5
625	Panfried, lean only	1 ea	63	59	152	16	0	0	10	3.2	3.9	1.2
626	Leg, roasted, lean and fat	4 oz	113	55	308	30	0	0	20	7.3	8.9	1.9
627	Leg, roasted, lean only	4 oz	113	61	233	35	0	0	9	3.2	4.3	.9
628	Rib, roasted, lean and fat	4 oz	113	56	288	31	0	0	17	6.7	7.9	1.4
629	Rib, roasted, lean only	4 oz	113	59	252	32	0	0	13	4.9	5.9	1
630	Shoulder, braised, lean and fat	4 oz	113	48	372	32	0	0	26	9.6	11.8	2.6
631	Shoulder, braised, lean only	4 oz	113	54	280	36	0	0	14	4.7	6.5	1.3
1088	Spareribs, cooked, yield from 1 lb raw with bone	4 oz	113	40	449	33	0	0	34	12.5	15.3	3.1
1095	Rabbit, roasted (1 cup meat = 140 g)	4 oz	113	61	223	33	0	0	9	4	2	3
	VEAL, cooked:											
632	Cutlet, braised or broiled, 4⅛ x 2¼ x ½"	4 oz	113	52	321	34	0	0	19	7.6	7.6	1.3
633	Rib roasted, lean, 2 pieces 4⅛ x 2¼ x ¼"	4 oz	113	60	258	27	0	0	16	6.1	6.1	1.1
634	Liver, panfried	4 oz	113	67	186	24	3	0	8	3.3	1.9	2.2
1096	Venison (deer meat), roasted	4 oz	113	65	179	34	0	0	4	1.4	1	.7
	MEATS: POULTRY and POULTRY PRODUCTS											
	CHICKEN, cooked:											
	Fried, batter dipped:											
635	Breast	1 ea	280	52	728	69	25	1	37	9.9	15.3	8.6
636	Drumstick	1 ea	72	53	193	16	6	<1	11	3	4.7	2.7
637	Thigh	1 ea	86	51	238	19	8	<1	14	3.8	5.9	3.3
638	Wing	1 ea	49	46	159	10	5	<1	11	2.9	4.5	2.5
	Fried, flour coated:											
639	Breast	1 ea	196	57	435	62	3	<1	17	4.9	7.1	3.8
1212	Breast, without skin	1 ea	86	60	161	29	<1	<1	4	1.1	1.5	.9
640	Drumstick	1 ea	49	57	120	13	1	<1	7	1.8	2.7	1.6
641	Thigh	1 ea	62	54	162	17	2	<1	9	2.5	3.7	2.1
1099	Thigh, without skin	1 ea	52	59	113	15	1	<1	5	1.4	2	1.3
642	Wing	1 ea	32	49	103	8	1	<1	7	1.9	2.9	1.6
	Roasted:											
643	All types of meat	1 c	140	64	266	40	0	0	10	2.9	3.8	2.4
644	Dark meat	1 c	140	63	287	38	0	0	14	3.7	5.2	3.2
645	Light meat	1 c	140	65	242	43	0	0	6	1.8	2.2	1.4
646	Breast, without skin	1 ea	172	65	284	53	0	0	6	1.8	2.2	1.3
647	Drumstick, without skin	1 ea	44	67	76	12	0	0	2	.7	.8	.6
1703	Leg, without skin	1 ea	95	65	181	26	0	0	8	2.2	2.9	1.9
648	Thigh	1 ea	62	59	153	16	0	0	10	2.7	3.9	2.1
1100	Thigh, without skin	1 ea	52	63	109	13	0	0	6	1.6	2.2	1.3
649	Stewed, all types	1 c	140	67	248	38	0	0	9	2.6	3.5	2.2
656	Canned, boneless chicken	4 oz	113	69	186	25	0	0	9	2.5	3.6	2
1102	Gizzards, simmered	1 c	145	67	222	39	2	0	5	1.5	1.3	1.5
1101	Hearts, simmered	1 c	145	65	268	38	<1	0	11	3.3	2.9	3.3
2300	Liver, simmered: Ounce	3 oz	85	68	133	21	1	0	5	1.6	1.1	.8
1098	Liver, simmered: Piece = 20 g	6 ea	120	68	188	29	1	0	7	2.2	1.6	1.1
	DUCK, roasted:											
1293	Meat with skin, about 2.7 cups	½ ea	382	52	1287	73	0	0	108	36.9	49.3	13.9
651	Meat only, about 1.5 cups	½ ea	221	64	444	52	0	0	25	9.2	8.2	3.2
	GOOSE, domesticated, roasted:											
1294	Meat only, about 4.2 cups	½ ea	591	57	1406	173	0	0	75	23.6	40.2	10.9
1295	Meat with skin, about 5.5 cups	½ ea	774	52	2360	195	0	0	170	53.2	80.5	24.8
	TURKEY:											
	Roasted, meat only:											
652	Dark meat	4 oz	113	63	211	33	0	0	8	2.7	1.8	2.5
653	Light meat	4 oz	113	66	177	34	0	0	4	1.2	.6	1
654	All types, chopped or diced	1 c	140	65	238	42	0	0	7	2.3	1.5	2
1103	Ground, cooked	4 oz	113	59	266	31	0	0	15	4.1	5.5	3.6
1106	Gizzard, cooked	2 ea	134	65	218	39	1	0	5	1.5	1	1.5
1107	Heart, cooked	4 ea	64	64	113	17	1	0	4	1.1	.8	1.1
1108	Liver, cooked	1 ea	75	66	127	18	3	0	4	1.4	1.1	.8

Chol (mg)	Calc (mg)	Iron (mg)	Magn (mg)	Pota (mg)	Sodi (mg)	Zinc (mg)	VT-A (RE)	Thia (mg)	VT-E (α-TE)	Ribo (mg)	Niac (mg)	V-B6 (mg)	Fola (µg)	VT-C (mg)
72	21	.71	23	332	62	1.8	2	.89	.32	.24	4.37	.37	5	1
52	14	.67	16	230	49	2.44	1	.46	.3	.23	2.8	.26	3	<1
106	16	1.14	25	398	68	3.34	3	.72	.34	.35	5.16	.45	11	<1
108	8	1.29	33	442	73	3.4	3	.91	.46	.4	5.56	.38	3	<1
82	32	1.06	24	476	52	2.33	2	.82	.41	.34	6.92	.37	3	<1
80	29	1.11	25	494	53	2.41	2	.86	.55	.36	7.25	.38	3	<1
123	20	1.82	21	417	99	4.72	3	.61	.5	.35	5.89	.4	5	<1
129	9	2.2	25	458	115	5.62	3	.68	.58	.41	6.71	.46	6	<1
137	53	2.09	27	362	105	5.2	3	.46	.52	.43	6.19	.4	5	0
93	21	2.57	24	433	53	2.57	0	.1	.96	.24	9.53	.53	12	0
133	32	1.23	27	316	90	4.1	0	.04	.45	.34	10.2	.29	16	0
124	12	1.1	25	333	104	4.62	0	.06	.4	.3	7.89	.28	15	0
634	8	2.96	21	232	60	10.8	9095	.15	.42	2.19	9.58	.55	858	35
127	8	5.05	27	379	61	3.11	0	.2	.28	.68	7.58	.42	5	0
238	56	3.5	67	563	770	2.66	56	.32	2.97	.41	29.4	1.2	42	0
62	12	.97	14	134	194	1.68	19	.08	.88	.15	3.67	.19	13	0
80	15	1.25	18	165	248	1.75	25	.1	1.05	.19	4.92	.22	16	0
39	10	.63	8	68	157	.68	17	.05	.52	.07	2.58	.15	9	0
174	31	2.33	59	508	149	2.16	29	.16	1.12	.26	26.9	1.14	12	0
78	14	.98	27	237	68	.93	6	.07	.36	.11	12.7	.55	3	0
44	6	.66	11	112	44	1.42	12	.04	.41	.11	2.96	.17	5	0
60	9	.92	15	147	55	1.56	18	.06	.52	.15	4.31	.2	7	0
53	7	.76	13	135	49	1.45	11	.05	.3	.13	3.7	.2	5	0
26	5	.4	6	57	25	.56	12	.02	.18	.04	2.14	.13	2	0
125	21	1.69	35	340	120	2.94	22	.1	.58	.25	12.8	.66	8	0
130	21	1.86	32	336	130	3.92	31	.1	.81	.32	9.17	.5	11	0
119	21	1.48	38	346	108	1.72	13	.09	.37	.16	17.4	.84	6	0
146	26	1.79	50	440	127	1.72	10	.12	.66	.2	23.6	1.03	7	0
41	5	.57	11	108	42	1.4	8	.03	.25	.1	2.68	.17	4	0
89	11	1.24	23	230	86	2.72	18	.07	.55	.22	6	.35	8	0
58	7	.83	14	138	52	1.46	30	.04	.35	.13	3.95	.19	4	0
49	6	.68	12	124	46	1.34	10	.04	.3	.12	3.4	.18	4	0
116	20	1.64	29	252	98	2.79	21	.07	.42	.23	8.57	.36	8	0
70	16	1.79	14	156	568	1.59	38	.02	.24	.15	7.15	.4	5	2
281	14	6.02	29	260	97	6.35	81	.04	2.29	.35	5.77	.17	77	2
351	28	13.1	29	191	70	10.6	13	.1	2.32	1.07	4.06	.46	116	3
536	12	7.2	18	119	43	3.69	4176	.13	1.45	1.49	3.78	.49	655	13
757	17	10.2	25	168	61	5.21	5895	.18	2.04	2.1	5.34	.7	924	19
321	42	10.3	61	779	225	7.11	241	.66	2.5	1.03	18.5	.69	23	0
197	26	5.97	44	557	144	5.75	51	.57	1.55	1.04	11.3	.55	22	0
567	83	17	148	2293	449	18.7	71	.54	9.16	2.3	24.1	2.78	71	0
704	101	21.9	170	2546	542	20.3	163	.6	13.5	2.5	32.3	2.86	15	0
96	36	2.63	27	328	89	5.04	0	.07	.94	.28	4.12	.41	10	0
78	21	1.53	32	345	72	2.31	0	.07	.12	.15	7.73	.61	7	0
106	35	2.49	36	417	98	4.34	0	.09	.59	.25	7.62	.64	10	0
115	28	2.18	27	305	121	3.23	0	.06	.45	.19	5.45	.44	8	0
311	20	7.29	25	283	72	5.57	74	.04	.27	.44	4.11	.16	70	2
145	8	4.41	14	117	35	3.37	5	.04	.13	.56	2.08	.2	51	1
470	8	5.85	11	146	48	2.32	2805	.04	2.41	1.07	4.46	.39	500	1

Table H-1

Food Composition (Computer code number is for West Diet Analysis program) (For purposes of calculations, use "0" for t, <1, <.1, <.01, etc.)

Computer Code Number	Food Description	Measure	Wt (g)	H_2O (%)	Ener (kcal)	Prot (g)	Carb (g)	Dietary Fiber (g)	Fat (g)	Fat Breakdown (g) Sat	Mono	Poly
	MEATS: POULTRY and POULTRY PRODUCTS—Continued											
	POULTRY FOOD PRODUCTS (see also items in Sausages & Lunchmeats section):											
1567	Chicken patty, breaded, cooked	1 ea	75	49	213	12	11	<1	13	4.1	6.4	1.6
659	Turkey and gravy, frozen package	3 oz	85	85	57	5	4	<1	2	.8	.8	.4
	Turkey breast, Louis Rich:											
1104	Barbecued	2 oz	56	72	58	12	2	0	<1	.2	.2	.1
1943	Hickory smoked	1 pce	80		80	16	2	0	1	0		
1947	Honey roasted	1 pce	80		80	16	3	0	1	.5		
1945	Oven roasted	1 pce	80		70	16		0	1	0		
661	Turkey patty, breaded, fried	2 oz	57	50	161	8	9	<1	10	2.7	4.3	2.7
662	Turkey, frozen, roasted, seasoned	4 oz	113	68	175	24	3	0	7	2.1	1.4	1.9
1704	Turkey roll, light meat	1 pce	28	72	41	5	<1	0	2	.6	.7	.5
	MEATS: SAUSAGES and LUNCHMEATS (see also Poultry Food Products)											
1072	Beerwurst/beer salami, beef	1 oz	28	53	92	3	<1	0	8	3.6	3.9	.3
1074	Beerwurst/beer salami, pork	1 oz	28	61	67	4	1	0	5	1.8	2.5	.7
1075	Berliner sausage	1 oz	28	61	64	4	1	0	5	1.7	2.2	.4
	Bologna:											
1297	Beef	1 pce	23	55	72	3	<1	0	7	2.8	3.2	.3
2115	Beef, light, Oscar Mayer	1 pce	28	65	56	3	2	0	4	1.6	2	.1
663	Beef & pork	1 pce	28	54	88	3	1	0	8	3	3.7	.7
2155	Healthy Favorites	1 pce	23		22	3	1	0	<1	0		
1298	Pork	1 pce	23	61	57	4	<1	0	5	1.6	2.2	.5
2114	Regular, light, Oscar Mayer	1 pce	28	65	56	3	2	0	4	1.6	2	.4
664	Turkey	1 pce	28	65	56	4	<1	0	4	1.4	1.3	1.2
1970	Turkey, Louis Rich	1 pce	28	67	57	3	<1	0	5	1.5	1.8	1.3
665	Braunschweiger sausage	2 pce	57	48	205	8	2	0	18	6.2	8.5	2.1
1073	Bratwurst, link	1 ea	70	51	226	10	2	0	19	6.9	9.3	2
666	Brown & serve sausage links, cooked	2 ea	26	45	102	4	1	0	10	3.4	4.4	1
1089	Cheesefurter/cheese smokie	2 ea	86	52	281	12	1	0	25	9	11.8	2.6
2157	Chicken breast, Healthy Favorites	4 pce	52		40	9	1	0	0	0	0	0
1556	Chorizo, pork & beef	1 ea	60	32	273	15	1	0	23	8.6	11	2.1
1090	Corned beef loaf, jellied	1 pce	28	69	43	6	0	0	2	.7	.7	.1
	Frankfurters:											
1077	Beef, large link, 8/package	1 ea	57	55	180	7	1	0	16	6.9	7.9	.8
1078	Beef and pork, large link, 8/package	1 ea	57	54	182	6	1	0	17	6.2	8	1.6
667	Beef and pork, small link, 10/pkg	1 ea	45	54	144	5	1	0	13	4.9	6.3	1.2
668	Turkey frankfurter, 10/package	1 ea	45	63	102	6	1	0	8	2.7	2.5	2.2
1968	Turkey/chicken frank 8/pkg	1 ea	43		80	6	1	0	6	2		
	Ham:											
669	Ham lunchmeat, canned, 3 x 2 x ½"	1 pce	21	52	70	3	<1	0	6	2.3	3	.7
670	Chopped ham, packaged	2 pce	42	64	96	7	0	0	7	2.4	3.4	.9
2156	Honey ham, Healthy Favorites	4 pce	52	73	55	9	2	0	1	.4	.8	.1
2113	Oscar Mayer lower sodium ham	1 pce	21	73	23	3	1	0	1	.3	.4	.1
673	Turkey ham lunchmeat	2 pce	57	71	73	11	<1	0	3	1	.7	.9
1091	Kielbasa sausage	1 pce	26	54	81	3	1	0	7	2.6	3.4	.8
1092	Knockwurst sausage, link	1 ea	68	55	209	8	1	0	19	6.9	8.7	2
1093	Mortadella lunchmeat	2 pce	30	52	93	5	1	0	8	2.8	3.4	.9
1097	Olive loaf lunchmeat	2 pce	57	58	134	7	5	<1	9	3.3	4.5	1.1
1952	Turkey breast, fat free	1 pce	28	77	22	4	1	0	<1	.1	.1	t
1080	Turkey pastrami	2 pce	57	71	80	10	1	0	4	1	1.2	.9
1969	Turkey salami	1 pce	28	72	41	4	<1	0	3	.9	1	.8
1081	Pepperoni sausage	2 pce	11	27	55	2	<1	0	5	1.8	2.3	.5
1094	Pickle & pimento loaf	2 pce	57	57	149	7	3	<1	12	4.5	5.5	1.5
1082	Polish sausage	1 oz	28	53	91	4	<1	0	8	2.9	3.8	.9
674	Pork sausage, cooked, link, small	2 ea	26	45	96	5	<1	0	8	2.8	4.1	.8
1079	Pork sausage, cooked, patty	4 oz	113	45	417	22	1	0	35	12.1	17.7	3.3
675	Salami, pork and beef	2 pce	57	60	143	8	1	0	11	4.6	5.2	1.1
677	Salami, pork and beef, dry	3 pce	30	35	125	7	1	0	10	3.7	5.1	1
676	Salami, turkey	2 pce	57	66	112	9	<1	0	8	2.3	2.6	2
	Sandwich spreads:											
1300	Ham salad spread	2 tbs	30	63	65	3	3	0	5	1.5	2.2	.8
678	Pork and beef	2 tbs	30	60	70	2	4	<1	5	1.8	2.3	.8
1296	Chicken/turkey	2 tbs	26	66	52	3	2	0	4	.9	.8	1.6

Chol (mg)	Calc (mg)	Iron (mg)	Magn (mg)	Pota (mg)	Sodi (mg)	Zinc (mg)	VT-A (RE)	Thia (mg)	VT-E (α-TE)	Ribo (mg)	Niac (mg)	V-B6 (mg)	Fola (µg)	VT-C (mg)
45	12	.94	15	185	399	.78	22	.07	1.46	.1	5.04	.23	8	<1
15	12	.79	7	52	471	.59	11	.02	.3	.11	1.53	.08	3	0
25	14	.62	16	175	599	.59	0	.02		.06	5.35	.22	2	0
35	0	.72			1060		0							0
35	0	.72			940		0							0
35	0				910		0							0
35	8	1.25	9	157	456	.82	6	.06	1.36	.11	1.31	.11	16	0
60	6	1.84	25	337	768	2.87	0	.05	.43	.18	7.09	.3	6	0
12	11	.36	4	70	137	.44	0	.02	.04	.06	1.96	.09	1	0
17	3	.42	3	49	288	.68	0	.02	.05	.03	.95	.05	1	0
16	2	.21	4	71	347	.48	0	.15	.06	.05	.91	.1	1	0
13	3	.32	4	79	363	.69	0	.11	.06	.06	.87	.06	1	0
13	3	.38	3	36	226	.5	0	.01	.04	.02	.55	.03	1	0
13	4	.34	4	44	314	.53	0							0
15	3	.42	3	50	285	.54	0	.05	.06	.04	.72	.05	1	0
7		.18			255									
14	3	.18	3	65	272	.47	0	.12	.06	.04	.9	.06	1	0
15	14	.39	5	46	312	.45	0							0
28	23	.43	4	56	246	.49	0	.01	.15	.05	.99	.06	2	0
22	34	.45	5	51	242	.57	0	.01		.05	1.08	.05		0
89	5	5.34	6	113	652	1.6	2405	.14	.2	.87	4.77	.19	25	0
44	34	.72	11	197	778	1.47	0	.17	.19	.16	2.31	.09	3	0
16	2	.62	4	70	248	.3	0	.21	.06	.09	.96	.06	1	0
58	50	.93	11	177	931	1.94	33	.21	.27	.14	2.49	.11	3	0
25		.72			620									
53	5	.95	11	239	741	2.05	0	.38	.13	.18	3.08	.32	1	0
13	3	.57	3	28	267	1.15	0	0	.05	.03	.49	.03	2	0
35	11	.81	2	95	585	1.24	0	.03	.11	.06	1.38	.07	2	0
28	6	.66	6	95	638	1.05	0	.11	.14	.07	1.5	.07	2	0
22	5	.52	4	75	504	.83	0	.09	.11	.05	1.18	.06	2	0
48	48	.83	6	81	642	1.4	0	.02	.28	.08	1.86	.1	4	0
40	60	1.08			480		0							0
13	1	.15	2	45	271	.31	0	.08	.05	.04	.66	.04	1	<1
21	3	.35	7	134	576	.81	0	.26	.11	.09	1.63	.15	<1	0
24	6	.7	18	144	635	1.02	0							0
9	1	.3	5	197	174	.42	0							0
32	6	1.57	9	185	568	1.68	0	.03	.36	.14	2.01	.14	3	0
17	11	.38	4	70	280	.52	0	.06	.06	.06	.75	.05	1	0
39	7	.62	7	135	687	1.13	0	.23	.39	.09	1.86	.12	1	0
17	5	.42	3	49	374	.63	0	.04	.07	.05	.8	.04	1	0
22	62	.31	11	169	846	.79	34	.17	.14	.15	1.05	.13	1	0
9	3	.34	8	59	387	.24	0							0
31	5	.95	8	148	596	1.23	0	.03	.12	.14	2.01	.15	3	0
21	11	.35	6	61	281	.65	0							0
9	1	.15	2	38	224	.27	0	.03	.02	.03	.55	.03	<1	0
21	54	.58	10	194	792	.8	4	.17	.14	.14	1.17	.11	3	0
20	3	.4	4	66	245	.54	0	.14	.06	.04	.96	.05	1	<1
22	8	.33	4	94	336	.65	0	.19	.07	.07	1.18	.09	1	<1
94	36	1.42	19	408	1462	2.84	0	.84	.29	.29	5.11	.37	2	2
37	7	1.52	9	113	607	1.22	0	.14	.12	.21	2.02	.12	1	0
24	2	.45	5	113	558	.97	0	.18	.08	.09	1.46	.15	1	0
47	11	.92	9	139	572	1.03	0	.04	.32	.1	2.01	.14	2	0
11	2	.18	3	45	274	.33	0	.13	.52	.04	.63	.04	<1	0
11	4	.24	2	33	304	.31	3	.05	.52	.04	.52	.04	1	0
8	3	.16	3	48	98	.27	11	.01	.57	.02	.43	.03	1	<1

Table H-1

Food Composition (Computer code number is for West Diet Analysis program) (For purposes of calculations, use "0" for t, <1, <.1, <.01, etc.)

Computer Code Number	Food Description	Measure	Wt (g)	H_2O (%)	Ener (kcal)	Prot (g)	Carb (g)	Dietary Fiber (g)	Fat (g)	Fat Breakdown (g) Sat	Mono	Poly
	MEATS: SAUSAGES and LUNCHMEATS—Continued											
1084	Smoked link sausage, beef and pork	1 ea	68	52	228	9	1	0	21	7.2	9.7	2.2
1083	Smoked link sausage, pork	1 ea	68	39	265	15	1	0	22	7.7	9.9	2.6
1085	Summer sausage	2 pce	46	51	154	7	<1	0	14	5.5	6	.6
1076	Turkey breakfast sausage	1 pce	28	60	64	6	0	0	5	1.6	1.8	1.2
679	Vienna sausage, canned	2 ea	32	60	89	3	1	0	8	3	4	.5
	MIXED DISHES and FAST FOODS											
	MIXED DISHES:											
1445	Almond Chicken	1 c	242	77	275	20	18	4	14	2	5.3	5.8
1981	Baked beans, fat free, honey	½ c	120	73	110	7	24	7	0	0	0	0
1454	Bean cake	1 ea	32	23	130	2	16	1	7	1	2.9	2.6
680	Beef stew w/ vegetables, homemade	1 c	245	82	218	16	15	2	10	4.9	4.5	.5
1109	Beef stew w/ vegetables, canned	1 c	245	82	194	14	17	2	8	2.4	3.1	.3
1116	Beef, macaroni, tomato sauce casserole	1 c	226	76	255	16	26	2	10	3.8	4.1	.5
2295	Beef fajita	1 ea	223	63	409	17	46	4	17	5.1	7.6	3.9
1265	Beef flauta	1 ea	113	49	360	16	13	2	27	4.9	11.6	9.1
681	Beef pot pie, homemade	1 pce	210	55	517	21	39	3	30	8.4	14.7	7.3
1898	Broccoli, batter fried	1 c	85	74	123	3	9	2	9	1.3	2.2	4.9
1462	Buffalo wings/spicy chicken wings	2 pce	32	53	98	8	<1	<1	7	1.8	2.8	1.6
1675	Carrot raisin salad	½ c	88	58	204	1	21	2	14	2	3.9	7.3
2248	Cheeseburger deluxe	1 ea	219	52	563	28	38		33	15	12.6	2
682	Chicken à la king, homemade	1 c	245	68	468	27	12	1	34	12.7	14.3	6.2
683	Chicken & noodles, homemade	1 c	240	71	367	22	26	2	18	5.9	7.1	3.5
684	Chicken chow mein, canned	1 c	250	89	95	6	18	2	1	0	.1	.8
685	Chicken chow mein, homemade	1 c	250	78	255	31	10	1	10	2.4	4.3	3.1
1266	Chicken fajita	1 ea	223	61	405	22	50	4	13	2.4	6	3.5
1264	Chicken flauta	1 ea	113	52	343	14	13	2	27	4.3	11.1	9.6
686	Chicken pot pie, homemade (⅓)	1 pce	232	57	545	23	42	3	31	10.9	14.5	5.8
1672	Chili con carne	½ c	127	77	128	12	11	2	4	1.7	1.7	.3
1112	Chicken salad with celery	½ c	78	53	268	11	1	<1	25	3.1	4.5	15.8
1382	Chicken teriyaki, breast	1 ea	128	67	176	26	7	<1	4	.9	1	.9
687	Chili with beans, canned	1 c	256	75	287	15	30	11	14	6	6	.9
1479	Chinese pastry	1 oz	28	46	67	1	13	<1	1	.2	.4	.8
688	Chop suey with beef & pork	1 c	220	63	425	22	31	3	24	5	8.6	9.3
690	Coleslaw	1 c	132	74	195	2	17	2	15	2.1	3.2	8.5
689	Corn pudding	1 c	250	76	273	11	32	4	13	6.4	4.3	1.8
1110	Corned beef hash, canned	1 c	220	67	398	19	23	1	25	11.9	10.9	.9
1255	Deviled egg (½ egg + filling)	1 ea	31	69	62	4	<1	0	5	1.2	1.7	1.5
	Egg foo yung patty:											
1467	Meatless	1 ea	86	78	113	6	3	1	8	1.9	3.3	2.1
1458	With beef	1 ea	86	74	129	9	3	<1	9	2.2	3.2	2.4
1465	With chicken	1 ea	86	74	130	9	4	<1	9	2.1	3.1	2.5
1602	Egg roll, meatless	1 ea	64	70	102	3	10	1	6	1.2	2.5	1.6
1550	Egg roll, with meat	1 ea	64	66	114	5	9	1	6	1.5	2.7	1.5
1113	Egg salad	1 c	183	57	586	17	3	0	56	10.6	17.4	24.2
691	French toast w/wheat bread, homemade	1 pce	65	54	151	5	16	<1	7	2	3	1.7
1355	Green pepper, stuffed	1 ea	172	75	229	11	20	2	11	5	4.9	.5
1487	Hot & sour soup (Chinese)	1 c	244	88	133	12	5	<1	6	2	2.5	1
2242	Hamburger deluxe	1 ea	110	49	279	13	27		13	4.1	5.3	2.6
1997	Hummous/hummus	¼ c	62	65	106	3	12	3	5	.8	2.2	2
	Lasagna:											
1346	With meat, homemade	1 pce	245	67	382	22	39	3	15	7.7	5	.9
1111	Without meat, homemade	1 pce	218	69	298	15	39	3	9	5.4	2.4	.6
1117	Frozen entree	1 ea	340	75	390	24	42	4	14	6.7	5.5	.8
1606	Lo mein, meatless	1 c	200	82	134	6	27	3	1	.1	.1	.3
1607	Lo mein, with meat	1 c	200	70	285	17	31	2	10	1.9	2.9	4.5
692	Macaroni & cheese, canned	1 c	240	80	228	9	26	1	10	4.2	3.1	1.4
693	Macaroni & cheese, homemade	1 c	200	58	430	17	40	1	22	8.9	8.8	3.6
1115	Macaroni salad, no cheese	1 c	177	60	461	5	28	2	37	4	6	25.5
1120	Meat loaf, beef	1 pce	87	63	182	16	4	<1	11	4.4	4.7	.5
1119	Meat loaf, beef and pork (⅓)	1 pce	87	60	205	15	4	<1	14	5.2	6.3	.9
1303	Moussaka (lamb & eggplant)	1 c	250	82	237	16	13	4	13	4.6	5.4	1.9

Chol (mg)	Calc (mg)	Iron (mg)	Magn (mg)	Pota (mg)	Sodi (mg)	Zinc (mg)	VT-A (RE)	Thia (mg)	VT-E (α-TE)	Ribo (mg)	Niac (mg)	V-B6 (mg)	Fola (μg)	VT-C (mg)
48	7	.99	8	129	643	1.43	0	.18	.15	.12	2.2	.12	1	0
46	20	.79	13	228	1020	1.92	0	.48	.17	.17	3.08	.24	3	1
34	6	1.17	6	125	571	1.18	0	.07	.1	.15	1.98	.12	1	0
23	5	.51	6	75	188	.96	0	.03	.14	.08	1.4	.08	1	0
17	3	.28	2	32	305	.51	0	.03	.07	.03	.51	.04	1	0
35	81	2	59	551	615	1.54	75	.08	2.64	.19	8.59	.42	31	10
0	40	2.7			135		450							12
0	3	.67	6	57	55	.16	0	.07	1.14	.05	.55	.02	9	0
64	29	2.94	40	613	292	5.29	568	.15	.49	.17	4.66	.28	37	17
34	29	2.21	39	426	1006	4.24	262	.07	.34	.12	2.45	.2	31	7
39	26	2.7	40	522	862	3.14	97	.22	.57	.22	4.31	.29	20	14
26	76	3.69	38	427	850	2.38	52	.46	2.08	.3	4.73	.32	25	29
45	50	2.15	29	292	187	4.18	15	.07	4	.15	2.13	.25	10	14
44	29	3.78	6	334	596	3.17	519	.29	3.78	.29	4.83	.24	29	6
16	67	.94	20	242	62	.38	102	.08	2.1	.13	.75	.11	43	53
26	5	.4	6	59	61	.56	17	.01	.23	.04	2.06	.13	1	<1
10	26	.75	14	317	118	.19	1452	.08	5.03	.05	.64	.22	9	5
88	206	4.66	44	445	1108	4.6	129	.39	1.18	.46	7.38	.28	81	8
186	127	2.45	20	404	760	1.8	272	.1	.98	.42	5.39	.23	11	12
96	26	2.16	26	149	600	1.53	10	.05		.17	4.32	.19	10	0
7	45	1.25	14	418	725	1.3	28	.05	.05	.1	1	.09	12	12
77	57	2.5	28	473	718	2.12	50	.07	.75	.22	4.25	.41	19	10
41	83	3.7	51	532	439	1.77	55	.48	2.04	.37	6.64	.35	41	22
37	52	.97	27	243	189	1.18	21	.05	4.06	.1	3.21	.22	8	14
72	70	3.02	25	343	594	2	735	.32	3.25	.32	4.87	.46	29	5
67	34	2.6	23	347	505	1.79	84	.06	.81	.57	1.24	.16	23	1
48	16	.62	11	138	201	.79	31	.03	6.27	.07	3.28	.34	8	1
80	27	1.75	36	309	1866	1.94	16	.08	.35	.2	8.69	.46	13	3
43	120	8.78	115	934	1336	5.12	87	.12	1.88	.27	.92	.34	58	4
0	8	.51	6	28	3	.18	<1	.05	.25	<.01	.41	.02	1	0
46	39	4.16	54	515	818	3.52	134	.36	1.82	.37	5.63	.44	44	20
7	45	.96	12	236	356	.26	66	.05	5.28	.04	.11	.14	51	11
250	100	1.4	37	403	138	1.25	90	1.03	.52	.32	2.47	.29	63	7
73	29	4.4	36	440	1188	3.3	0	.02	.48	.2	4.62	.43	20	0
121	15	.35	3	36	94	.3	49	.02	.86	.14	.02	.05	13	0
184	31	1.04	12	118	310	.7	86	.04	1.57	.25	.44	.09	30	5
180	26	1.11	11	145	184	1.16	92	.05	1.79	.24	.74	.15	22	3
182	27	.86	12	144	187	.81	95	.05	1.87	.25	.96	.13	22	3
30	12	.76	9	98	306	.25	15	.08	.81	.1	.81	.05	12	3
37	13	.78	10	124	304	.46	14	.16	.78	.13	1.31	.1	9	2
574	74	1.8	13	180	665	1.42	260	.08	8.87	.66	.09	.46	62	0
76	64	1.09	11	86	311	.44	81	.13	.31	.21	1.06	.05	15	<1
34	16	1.77	20	233	201	2.28	44	.15	.75	.1	2.74	.3	17	55
23	29	1.83	27	351	1562	1.17	2	.19	.12	.22	4.58	.15	12	1
26	63	2.63	22	227	504	2.06	9	.23	.82	.2	3.69	.12	52	2
0	31	.97	18	108	151	.68	1	.06	.62	.03	.25	.25	37	5
56	258	3.22	50	461	745	3.25	158	.23	1.15	.33	3.97	.21	19	15
31	252	2.5	44	375	714	1.77	156	.22	1.07	.27	2.49	.17	17	15
55	263	3.44	64	752	823	3.7	248	.29	3.45	.39	5.07	.32	28	41
0	47	2.06	33	389	623	.92	130	.23	.35	.24	2.83	.19	49	12
30	25	2.11	39	246	276	1.63	6	.37	1.51	.24	4.25	.28	41	8
24	199	.96	31	139	730	1.2	73	.12	.14	.24	.96	.02	8	<1
42	362	1.8	37	240	1086	1.2	234	.2	.12	.4	1.8	.05	10	1
27	31	1.56	20	170	352	.53	44	.18	10.3	.1	1.43	.33	20	4
84	29	1.61	14	187	145	3.23	23	.05	.31	.22	2.61	.15	11	1
84	33	1.42	14	213	381	2.68	23	.19	.32	.22	2.68	.17	10	1
97	68	1.79	40	557	432	2.56	105	.15	.81	.31	4.14	.23	45	6

Table H-1

Food Composition (Computer code number is for West Diet Analysis program) (For purposes of calculations, use "0" for t, <1, <.1, <.01, etc.)

Computer Code Number	Food Description	Measure	Wt (g)	H_2O (%)	Ener (kcal)	Prot (g)	Carb (g)	Dietary Fiber (g)	Fat (g)	Fat Breakdown (g) Sat	Mono	Poly
	MIXED DISHES and FAST FOODS—Continued											
1899	Mushrooms, batter fried	5 ea	70	66	148	2	8	1	12	2.1	3	6.4
715	Potato salad with mayonnaise and eggs	½ c	125	76	179	3	14	2	10	1.8	3.1	4.7
1674	Pizza, combination, 1/12 of 12" round	1 pce	79	48	184	13	21		5	1.5	2.5	.9
1673	Pizza, pepperoni, 1/12 of 12" round	1 pce	71	46	181	10	20		7	2.2	3.1	1.2
694	Quiche Lorraine ⅛ of 8" quiche	1 pce	176	54	508	20	20	1	39	17.6	13.8	4.9
1449	Ramen noodles, cooked	1 c	227	82	156	6	29	3	2	.4	.5	.5
1671	Ravioli, meat	½ c	125	68	194	10	18	1	9	2.9	3.6	1
1597	Fried rice (meatless)	1 c	166	68	264	5	34	1	12	1.7	3	6.3
2142	Roast beef hash	½ c	117	66	230	9	11	1	16	7	5.8	3.2
	Spaghetti (enriched) in tomato sauce											
	With cheese:											
695	Canned	1 c	250	80	190	5	38	2	1	0	.4	.5
696	Home recipe	1 c	250	77	260	9	37	2	9	2	5.4	1.2
	With meatballs:											
697	Canned	1 c	250	78	258	12	28	6	10	2.1	3.9	3.9
698	Home recipe	1 c	248	70	332	19	39	8	12	3.3	6.3	2.2
716	Spinach soufflé	1 c	136	74	219	11	3	3	19	9.5	5.7	2.2
1553	Sweet & sour pork	1 c	226	77	231	15	25	1	8	2.2	2.9	2.4
1263	Sweet & sour chicken breast	1 ea	131	79	117	8	15	1	3	.6	.8	1.5
1515	Three bean salad	1 c	150	82	139	4	13	3	8	1.2	1.9	4.9
717	Tuna salad	1 c	205	63	383	33	19	0	19	3.2	5.9	8.4
1121	Tuna noodle casserole, homemade	1 c	202	75	237	17	25	1	7	1.9	1.5	3.2
1270	Waldorf salad	1 c	137	58	408	4	13	2	40	4.1	7.3	27
	FAST FOODS and SANDWICHES (see end of this appendix for additional Fast Foods)											
699	Burrito, beef & bean	1 ea	116	52	255	11	33	3	9	4.2	3.5	.6
700	Burrito, bean	1 ea	109	52	225	7	36	4	7	3.5	2.4	.6
2106	Burrito, chicken con queso	1 ea	306	76	280	12	53	5	6	1.5		
701	Cheeseburger with bun, regular	1 ea	154	55	359	18	28		20	9.2	7.2	1.5
702	Cheeseburger with bun, 4-oz patty	1 ea	166	51	417	21	35		21	8.7	7.8	2.7
703	Chicken patty sandwich	1 ea	182	47	515	24	39	1	29	8.5	10.4	8.4
704	Corndog	1 ea	175	47	460	17	56		19	5.2	9.1	3.5
1922	Corndog, chicken	1 ea	113	52	271	13	26		13			
705	Enchilada	1 ea	163	63	319	10	28		19	10.6	6.3	.8
706	English muffin with egg, cheese, bacon	1 ea	146	49	383	20	31	1	20	9	6.8	2.1
	Fish sandwich:											
707	Regular, with cheese	1 ea	183	45	523	21	48	<1	28	8.1	8.9	9.4
708	Large, no cheese	1 ea	158	47	431	17	41	<1	23	5.2	7.7	8.2
709	Hamburger with bun, regular	1 ea	107	45	275	14	33	1	10	3.5	3.7	1.8
710	Hamburger with bun, 4-oz patty	1 ea	215	50	576	32	39		32	12	14.1	2.8
711	Hotdog/frankfurter with bun	1 ea	98	54	242	10	18		14	5.1	6.8	1.7
	Lunchables:											
2129	Bologna & American cheese	1 ea	128		450	18	19	0	34	15		
2130	Ham & cheese	1 ea	128		320	22	19	0	17	8		
2117	Honey ham & Amer. w/choc pudding	1 ea	176		390	18	34	<1	20	9		
2118	Honey turkey & cheddar w/Jello	1 ea	163		320	17	27	<1	16	9		
2131	Pepperoni & American cheese	1 ea	128		480	20	19	0	36	17		
2125	Salami & American cheese	1 ea	128		430	18	18	0	32	15		
2127	Turkey & cheddar cheese	1 ea	128		360	20	20	1	22	11		
712	Pizza, cheese, ⅛ of 15" round	1 pce	63	48	140	8	20	1	3	1.5	1	.5
	SANDWICHES:											
	Avocado, chesse, tomato & lettuce:											
1276	On white bread, firm	1 ea	210	58	478	15	41	5	29	8.8	11.3	7.2
1278	On part whole wheat	1 ea	201	59	444	14	34	7	29	8.6	11.4	7.3
1277	On whole wheat	1 ea	214	58	468	16	40	8	30	8.7	11.6	7.5
	Bacon, lettuce & tomato sandwich:											
1137	On white bread, soft	1 ea	124	53	308	10	28	2	18	4.5	6.1	6.1
1139	On part whole wheat	1 ea	124	54	303	10	26	3	17	4.3	6.2	6.1
1138	On whole wheat	1 ea	137	53	328	12	32	4	18	4.4	6.4	6.3

Chol (mg)	Calc (mg)	Iron (mg)	Magn (mg)	Pota (mg)	Sodi (mg)	Zinc (mg)	VT-A (RE)	Thia (mg)	VT-E (α-TE)	Ribo (mg)	Niac (mg)	V-B6 (mg)	Fola (µg)	VT-C (mg)
14	54	.76	8	180	121	.42	10	.07	.92	.22	1.65	.05	8	1
85	24	.81	19	318	661	.39	41	.1	2.33	.07	1.11	.18	8	12
20	101	1.53	18	179	382	1.11	101	.21		.17	1.96	.09	32	2
14	65	.94	9	153	267	.52	55	.13		.23	3.05	.06	37	2
205	201	1.9	27	271	549	1.66	243	.23	1.91	.44	4.71	.19	17	3
38	20	1.89	24	51	1349	.76	204	.22	.09	.1	1.75	.06	9	<1
84	32	2.03	20	259	619	1.67	94	.15	1.52	.22	2.95	.14	14	11
42	30	1.84	24	134	286	.89	21	.21	2.46	.11	2.25	.15	22	4
40	10	.9	22	362	695	2.99	0	.09		.12	2.33	.3	12	0
7	40	2.75	21	303	955	1.12	120	.35	2.13	.27	4.5	.13	6	10
7	80	2.25	26	408	955	1.3	140	.25	2.75	.17	2.25	.2	8	12
22	52	3.25	20	245	1220	2.39	100	.15	1.5	.17	2.25	.12	5	5
74	124	3.72	40	665	1009	2.45	159	.25	1.64	.3	3.97	.2	10	22
184	230	1.35	38	201	763	1.29	675	.09	1.22	.3	.48	.12	80	3
38	28	1.36	34	390	1219	1.46	28	.55	.62	.21	3.6	.41	10	20
23	16	.79	21	187	732	.66	20	.06	.39	.08	3.06	.18	6	12
0	35	1.42	25	224	514	.54	23	.07	1.96	.09	.4	.04	53	4
27	35	2.05	39	365	824	1.15	55	.06	1.95	.14	13.7	.17	16	5
41	34	2.3	30	182	772	1.2	13	.18	1.18	.15	7.78	.2	10	1
21	43	.88	39	270	236	.63	39	.1	8.67	.05	.36	.36	27	6
24	53	2.46	42	329	670	1.93	32	.27	.7	.42	2.71	.19	58	1
2	57	2.27	44	328	495	.76	16	.32	.87	.3	2.04	.15	44	1
10	40	.72			600		40							15
52	182	2.65	26	229	976	2.62	71	.32	1.34	.23	6.38	.15	65	2
60	171	3.42	30	335	1050	3.49	65	.35		.28	8.05	.18	61	2
60	60	4.68	35	353	957	1.87	31	.33	.55	.24	6.81	.2	100	9
79	102	6.18	17	263	973	1.31	37	.28	.7	.7	4.17	.09	103	0
64					668									
44	324	1.32	50	240	784	2.51	186	.08	1.47	.42	1.91	.39	65	1
234	207	3.29	34	213	784	1.81	158	.48	.6	.53	3.93	.16	47	1
68	185	3.5	37	353	939	1.17	97	.46	1.83	.42	4.23	.11	91	3
55	84	2.61	33	340	615	.99	30	.33	.87	.22	3.4	.11	85	3
43	51	2.46	22	215	564	2.05	13	.26	.43	.32	4.7	.13	52	3
103	92	5.55	45	527	742	5.81	4	.34	1.61	.41	6.73	.37	84	1
44	23	2.31	13	143	670	1.98	0	.23	.27	.27	3.65	.05	48	<1
85	300	2.7			1620		60							0
60	300	1.8			1770		80							
55	250	2.7			1540		40							
50	20	6			1360		80							
95	250	2.7			1840		60							
80	250	2.7			1740		60							
70	300	1.8			1650		60							
9	117	.58	16	110	336	.81	74	.18		.16	2.48	.04	35	1
34	294	3.06	54	581	550	1.71	140	.37	4.55	.39	3.77	.32	80	11
31	291	3.1	67	617	525	1.91	140	.35	4.55	.39	3.94	.35	80	11
31	281	3.53	102	679	593	2.68	140	.36	4.17	.37	4.29	.42	92	11
21	52	1.99	20	233	590	.96	31	.35	2.34	.19	3.2	.14	34	12
20	63	2.27	33	283	604	1.19	31	.36	2.7	.21	3.62	.17	37	12
20	55	2.68	64	342	670	1.9	31	.37	2.36	.2	3.97	.24	48	12

Table H-1

Food Composition (Computer code number is for West Diet Analysis program) (For purposes of calculations, use "0" for t, <1, <.1, <.01, etc.)

Computer Code Number	Food Description	Measure	Wt (g)	H_2O (%)	Ener (kcal)	Prot (g)	Carb (g)	Dietary Fiber (g)	Fat (g)	Fat Breakdown (g) Sat	Mono	Poly
	MIXED DISHES and FAST FOODS—Continued											
	Cheese, grilled:											
1140	On white bread, soft	1 ea	119	37	400	18	30	1	24	13.2	7.5	2
1142	On part whole wheat	1 ea	119	37	396	18	28	3	24	13	7.6	2.1
1141	On whole wheat	1 ea	132	38	420	20	33	4	24	13.1	7.8	2.2
1596	Chicken fillet	1 ea	182	47	515	24	39	1	29	8.5	10.4	8.4
	Chicken salad:											
1143	On white bread, soft	1 ea	110	40	369	11	31	1	23	3.7	6.1	12.1
1145	On part whole wheat	1 ea	110	41	364	11	29	4	23	3.5	6.1	12.1
1144	On whole wheat	1 ea	123	41	387	13	34	5	23	3.6	6.3	12.2
1146	Corned beef & swiss on rye	1 ea	156	49	420	28	22	<1	26	9.4	7.4	6.3
	Egg salad:											
1147	On white bread, soft	1 ea	117	43	380	10	31	1	25	4.4	6.8	12
1149	On part whole wheat	1 ea	116	43	374	10	29	3	25	4.1	6.8	12
1148	On whole wheat	1 ea	130	43	400	12	35	5	25	4.3	7.1	12.2
	Ham:											
1279	On rye bread	1 ea	150	60	283	22	21	<1	13	2.4	3.8	6
1151	On white bread, soft	1 ea	157	55	334	22	30	1	14	3	4.4	5.9
1153	On part whole wheat	1 ea	156	55	328	22	28	3	14	2.7	4.4	6
1152	On whole wheat	1 ea	169	54	352	24	34	4	15	2.9	4.7	6.1
	Ham & cheese:											
1280	On white bread, soft	1 ea	157	49	403	23	30	1	22	8.4	5.9	6.1
1282	On part whole wheat	1 ea	156	50	397	23	28	3	21	8.1	5.9	6.2
1281	On whole wheat	1 ea	170	49	424	24	34	4	22	8.3	6.2	6.4
1150	Ham & swiss on rye	1 ea	150	54	339	22	22	<1	19	6.5	5.1	6
	Ham salad:											
1154	On white bread, soft	1 ea	131	47	362	11	37	1	20	4.8	6.7	7.4
1156	On part whole wheat	1 ea	131	48	357	11	35	3	20	4.5	6.8	7.5
1155	On whole wheat	1 ea	144	47	380	12	40	4	20	4.7	7	7.6
1157	Patty melt: Ground beef & cheese on rye	1 ea	182	46	561	37	22	3	37	13.2	11.7	8.4
	Peanut butter & jelly:											
1158	On white bread, soft	1 ea	101	26	351	12	47	3	15	3.1	6.7	3.9
1160	On part whole wheat	1 ea	101	27	346	12	45	5	15	2.9	6.7	4
1159	On whole wheat	1 ea	114	28	370	13	50	6	15	3	7	4.1
1161	Reuben, grilled: Corned beef, swiss cheese, sauerkraut on rye	1 ea	239	64	462	28	25	2	29	9.9	9.5	7.1
	Roast beef:											
713	On a bun	1 ea	139	49	346	21	33		14	3.6	6.8	1.7
1162	On white bread, soft	1 ea	157	46	404	29	34	1	17	3.4	4.2	8.2
1164	On part whole wheat	1 ea	156	47	398	29	32	3	17	3.2	4.3	8.3
1163	On whole wheat	1 ea	169	46	422	31	38	4	17	3.3	4.5	8.4
	Tuna salad:											
1165	On white bread, soft	1 ea	122	46	327	14	35	2	15	2.5	3.8	7.9
1167	On part whole wheat	1 ea	122	47	322	14	33	4	15	2.2	3.8	8
1166	On whole wheat	1 ea	135	46	346	16	39	5	15	2.3	4.1	8.1
	Turkey:											
1168	On white bread, soft	1 ea	156	54	346	24	29	1	15	2.4	3.2	8.3
1170	On part whole wheat	1 ea	155	54	338	24	27	3	14	2.1	3.2	8.3
1169	On whole wheat	1 ea	169	53	365	26	33	4	15	2.3	3.5	8.5
	Turkey ham:											
1272	On rye bread	1 ea	150	60	280	21	20	<1	14	2.5	2.8	6.9
1273	On white bread, soft	1 ea	156	55	331	21	29	1	14	3	3.4	6.8
1275	On part whole wheat	1 ea	156	56	326	21	28	3	14	3	4.2	5.6
1274	On whole wheat	1 ea	169	55	350	23	33	4	15	2.9	3.7	7
714	Taco	1 ea	171	58	369	21	27		20	11.4	6.6	1
	Tostada:											
1114	With refried beans	1 ea	144	66	223	10	26	7	10	5.4	3	.7
1118	With beans & beef	1 ea	225	70	333	16	30	4	17	11.5	3.5	.6
1354	With beans & chicken	1 ea	156	68	248	19	18	3	11	5.3	3.9	1.6
	NUTS, SEEDS, and PRODUCTS											
	Almonds:											
1365	Dry roasted, salted	1 c	138	3	810	22	33	19	71	6.7	46.2	14.9

Chol (mg)	Calc (mg)	Iron (mg)	Magn (mg)	Pota (mg)	Sodi (mg)	Zinc (mg)	VT-A (RE)	Thia (mg)	VT-E (α-TE)	Ribo (mg)	Niac (mg)	V-B6 (mg)	Fola (µg)	VT-C (mg)
55	399	1.81	25	154	1143	2.05	212	.24	1.13	.34	1.91	.06	24	<1
54	412	2.12	39	209	1160	2.31	212	.25	1.53	.36	2.39	.1	28	<1
54	402	2.54	73	271	1226	3.08	212	.26	1.14	.35	2.74	.17	40	<1
60	60	4.68	35	353	957	1.87	31	.33	.55	.24	6.81	.2	100	9
32	60	2.04	18	139	460	.8	24	.25	6.16	.18	3.68	.25	26	1
31	73	2.35	33	195	475	1.06	24	.26	6.57	.2	4.16	.29	30	1
30	63	2.79	68	259	543	1.85	24	.27	6.14	.19	4.5	.36	42	1
82	268	3.12	28	225	1392	3.65	81	.19	2.59	.33	2.72	.17	19	1
157	71	2.18	16	113	526	.74	76	.26	4.52	.31	1.98	.2	37	0
155	83	2.47	31	169	539	1	75	.27	4.91	.33	2.45	.23	41	0
155	73	2.92	66	234	611	1.8	76	.28	4.53	.32	2.82	.3	53	0
47	48	2.3	26	364	1566	2.11	8	.99	2.36	.31	5.45	.47	15	23
47	60	2.39	29	368	1619	2.04	8	1.02	2.34	.33	5.99	.47	24	22
45	71	2.68	43	421	1630	2.29	8	1.03	2.73	.35	6.45	.5	27	22
45	62	3.1	77	483	1696	3.05	8	1.04	2.34	.33	6.78	.57	39	22
61	232	2.28	30	315	1620	2.34	90	.76	2.53	.36	4.64	.36	25	15
59	244	2.58	45	368	1630	2.59	90	.77	2.93	.39	5.09	.39	28	15
59	236	3.02	79	432	1707	3.37	90	.78	2.56	.37	5.47	.46	40	15
57	258	2.25	29	344	1602	2.59	79	.72	2.52	.36	4.06	.35	16	15
30	56	2.07	19	159	921	1.06	8	.51	3.29	.22	3.25	.17	22	4
29	69	2.38	33	216	936	1.32	8	.52	3.69	.24	3.74	.21	26	4
29	59	2.81	69	279	1001	2.11	8	.53	3.29	.23	4.09	.28	38	4
113	222	4.19	36	391	701	7.11	123	.25	3.5	.46	6.14	.35	25	<1
2	60	2.25	56	245	293	1.07	<1	.27	.12	.17	5.33	.13	40	<1
0	72	2.55	70	299	308	1.32	<1	.28	.51	.19	5.8	.17	44	<1
0	63	2.97	104	361	375	2.09	<1	.29	.14	.17	6.14	.24	56	<1
80	288	4.24	38	361	1949	3.73	130	.21	4.46	.34	2.79	.27	38	13
51	54	4.23	31	316	792	3.39	21	.37	.19	.31	5.87	.26	57	2
45	60	3.98	28	432	1595	3.78	12	.29	3.3	.3	6.39	.39	30	12
43	72	4.27	42	485	1607	4.02	12	.31	3.69	.32	6.84	.42	34	12
43	62	4.7	77	547	1672	4.79	12	.31	3.31	.31	7.18	.49	45	12
14	60	2.24	22	161	567	.67	22	.25	2.72	.18	5.53	.11	25	1
13	73	2.55	37	217	582	.93	22	.26	3.12	.2	6	.16	29	1
12	63	2.98	72	280	649	1.72	22	.27	2.71	.19	6.34	.23	41	1
45	56	2	29	302	1585	1.33	12	.26	3.46	.23	8.97	.4	24	0
43	68	2.28	43	354	1589	1.57	11	.27	3.82	.25	9.38	.44	28	0
43	59	2.73	77	418	1665	2.35	12	.28	3.47	.23	9.77	.51	39	0
55	51	4.06	25	342	1185	3	8	.22	2.8	.33	4.3	.29	17	<1
55	62	4.09	28	346	1248	2.9	8	.27	2.75	.35	4.87	.28	25	0
53	74	4.37	42	400	1262	3.15	8	.28	3.57	.37	5.34	.32	29	0
53	65	4.81	76	462	1329	3.91	8	.29	2.76	.35	5.68	.39	41	0
56	221	2.41	70	474	802	3.93	147	.15	1.88	.44	3.21	.24	68	2
30	210	1.89	59	403	543	1.9	85	.1	1.15	.33	1.32	.16	43	1
74	189	2.45	67	491	871	3.17	173	.09	1.8	.49	2.86	.25	85	4
53	168	1.79	47	365	433	2.28	86	.11	1.87	.2	4.52	.32	53	3
0	389	5.24	420	1062	1076	6.76	0	.18	7.66	.83	3.89	.1	88	1

Table H-1

Food Composition (Computer code number is for West Diet Analysis program) (For purposes of calculations, use "0" for t, <1, <.1, <.01, etc.)

Computer Code Number	Food Description	Measure	Wt (g)	H_2O (%)	Ener (kcal)	Prot (g)	Carb (g)	Dietary Fiber (g)	Fat (g)	Fat Breakdown (g) Sat	Mono	Poly
	NUTS, SEEDS, and PRODUCTS—Continued											
	Almonds:											
718	Slivered, packed, unsalted	1 c	108	4	636	22	22	12	56	5.3	36.6	11.9
719	Whole, dried, unsalted	1 c	142	4	836	28	29	15	74	7	48.1	15.6
720	Ounce	1 oz	28	4	165	6	6	3	15	1.4	9.5	3.1
721	Almond butter:	1 tbs	16	1	101	2	3	1	9	.9	6.1	2
4572	Salted	1 tbs	16	1	101	2	3	1	9	.9	6.1	2
722	Brazil nuts, dry (about 7)	1 c	140	3	918	20	18	8	93	22.7	32.2	33.7
	Cashew nuts, dry roasted:											
723	Salted:	1 c	137	2	786	21	45	4	64	12.8	37.4	10.7
724	Ounce	1 oz	28	2	161	4	9	1	13	2.6	7.6	2.2
4621	Unsalted:	1 c	137	2	786	21	45	4	64	12.8	37.4	10.7
4621	Ounce	1 oz	28	2	161	4	9	1	13	2.6	7.6	2.2
725	Oil roasted:	1 c	130	4	749	23	37	5	63	12.6	36.9	10.6
726	Ounce	1 oz	28	4	161	5	8	1	13	2.7	7.9	2.3
4622	Unsalted:	1 c	130	4	749	21	37	5	63	12.6	36.9	10.6
4622	Ounce	1 oz	28	4	161	5	8	1	13	2.7	7.9	2.3
727	Cashew butter, unsalted	1 tbs	16	3	94	3	4	<1	8	1.6	4.7	1.3
4662	Cashew butter, salted	1 tbs	16	3	94	3	4	<1	8	1.6	4.7	1.3
728	Chestnuts, European, roasted (1 cup = approx 17 kernels)	1 c	143	40	350	5	76	7	3	.6	1.1	1.2
	Coconut, raw:											
729	Piece 2 x 2 x ½"	1 pce	45	47	159	2	7	4	15	13.5	.6	.2
730	Shredded/grated, unpacked	½ c	40	47	142	1	6	4	13	12	.6	.1
	Coconut, dried, shredded/grated:											
731	Unsweetened	1 c	78	3	515	6	19	13	50	45.1	2.1	.6
732	Sweetened	1 c	93	13	466	3	44	4	33	29.6	1.4	.4
733	Filberts/hazelnuts, chopped:	1 c	135	5	853	18	21	8	84	6.2	66.3	8.1
734	Ounce	1 oz	28	5	177	4	4	2	17	1.3	13.7	1.7
735	Macadamias, oil roasted, salted:	1 c	134	2	962	10	17	12	103	15.4	80.9	1.8
736	Ounce	1 oz	28	2	201	2	4	3	21	3.2	16.9	.4
1368	Macadamias, oil roasted, unsalted	1 c	134	2	962	10	17	12	103	15.4	80.9	1.8
	Mixed nuts:											
737	Dry roasted, salted	1 c	137	2	814	24	35	12	71	9.4	43	14.8
738	Oil roasted, salted	1 c	142	2	876	24	30	13	80	12.4	45	18.9
1369	Oil roasted, unsalted	1 c	142	2	876	27	30	14	80	12.4	45	18.9
	Peanuts:											
739	Oil roasted, salted	1 c	144	2	837	38	27	13	71	9.8	35.3	22.5
740	Ounce	1 oz	28	2	163	7	5	3	14	1.9	6.9	4.4
1370	Oil roasted, unsalted	1 c	144	2	837	38	27	10	71	9.8	35.3	22.5
741	Dried, salted	1 c	146	2	854	35	31	12	73	10.1	36.1	22.9
742	Ounce	1 oz	28	2	164	7	6	2	14	1.9	6.9	4.4
743	Peanut butter:	½ c	128	1	759	33	25	8	65	14.3	31.1	17.7
1371	Tablespoon	2 tbs	32	1	190	8	6	2	16	3.6	7.8	4.4
744	Pecan halves, dried, unsalted:	1 c	108	5	720	9	20	8	73	5.8	45.6	18
745	Ounce	1 oz	28	5	187	2	5	2	19	1.5	11.8	4.7
1372	Pecan halves, dry roasted, salted	¼ c	28	1	185	2	6	3	18	1.4	11.3	4.5
746	Pine nuts/piñons, dried	1 oz	28	6	176	3	5	3	17	2.6	6.4	7.2
747	Pistachios, dried, shelled	1 oz	28	4	162	6	7	3	14	1.8	9.2	2
1373	Pistachios,dry roasted,salted,shelled	1 c	128	2	776	19	35	14	68	8.8	45.7	10.2
748	Pumpkin kernels, dried, unsalted	1 oz	28	7	151	7	5	1	13	2.4	4	5.8
1374	Pumpkin kernels, roasted, salted	1 c	227	7	1184	75	30	9	96	18.1	29.7	43.6
749	Sesame seeds, hulled, dried	¼ c	38	5	223	10	4	3	21	2.9	7.9	9.1
	Sunflower seed kernels:											
750	Dry	¼ c	36	5	205	8	7	4	18	1.9	3.4	11.8
751	Oil roasted	¼ c	34	3	209	7	5	2	20	2	3.7	12.9
752	Tahini (sesame butter)	1 tbs	15	3	91	3	3	1	8	1.2	3.2	3.7
1334	Trail mix w/chocolate chips	1 c	146	7	707	21	66	8	47	9.3	19.8	16.5
753	Black walnuts, chopped:	1 c	125	4	759	31	15	6	71	4.8	15.9	46.9
754	Ounce	1 oz	28	4	170	7	3	1	16	1.1	3.6	10.5
755	English walnuts, chopped:	1 c	120	4	770	17	22	6	74	7.2	17	46.9
756	Ounce	1 oz	28	4	180	4	5	1	17	1.7	4	10.9

Chol (mg)	Calc (mg)	Iron (mg)	Magn (mg)	Pota (mg)	Sodi (mg)	Zinc (mg)	VT-A (RE)	Thia (mg)	VT-E (α-TE)	Ribo (mg)	Niac (mg)	V-B6 (mg)	Fola (µg)	VT-C (mg)
0	287	3.95	320	791	12	3.15	0	.23	25.9	.84	3.63	.12	63	1
0	378	5.2	420	1039	16	4.15	0	.3	34.1	1.11	4.77	.16	83	1
0	74	1.02	83	205	3	.82	0	.06	6.72	.22	.94	.03	16	<1
0	43	.59	48	121	2	.49	0	.02	3.25	.1	.46	.01	10	<1
0	43	.59	48	121	72	.49	0	.02	3.25	.1	.46	.01	10	<1
0	246	4.76	315	840	3	6.43	0	1.4	10.6	.17	2.27	.35	6	1
0	62	8.22	356	774	877	7.67	0	.27	.78	.27	1.92	.35	95	0
0	13	1.68	73	158	179	1.57	0	.06	.16	.06	.39	.07	19	0
0	62	8.22	356	774	22	7.67	0	.27	.78	.27	1.92	.35	95	0
0	13	1.68	73	158	4	1.57	0	.06	.16	.06	.39	.07	19	0
0	53	5.33	332	689	814	6.18	0	.55	2.03	.23	2.34	.32	88	0
0	11	1.15	71	148	175	1.33	0	.12	.44	.05	.5	.07	19	0
0	53	5.33	332	689	22	6.18	0	.55	2.03	.23	2.34	.32	88	0
0	11	1.15	71	148	5	1.33	0	.12	.44	.05	.5	.07	19	0
0	7	.8	41	87	2	.83	0	.05	.25	.03	.26	.04	11	0
0	7	.8	41	87	98	.83	0	.05	.25	.03	.26	.04	11	0
0	41	1.3	47	847	3	.81	3	.35	1.72	.25	1.92	.71	100	37
0	6	1.09	14	160	9	.49	0	.03	.33	.01	.24	.02	12	1
0	6	.97	13	142	8	.44	0	.03	.29	.01	.22	.02	11	1
0	20	2.59	70	424	29	1.57	0	.05	1.05	.08	.47	.23	7	1
0	14	1.79	46	313	244	1.69	0	.03	1.26	.02	.44	.25	8	1
0	254	4.41	385	601	4	3.24	9	.67	32.3	.15	1.54	.83	97	1
0	53	.92	80	125	1	.67	2	.14	6.69	.03	.32	.17	20	<1
0	60	2.41	157	441	348	1.47	1	.28	.55	.15	2.71	.26	21	0
0	13	.5	33	92	73	.31	<1	.06	.11	.03	.57	.05	4	0
0	60	2.41	157	441	9	1.47	1	.28	.55	.15	2.71	.26	21	0
0	96	5.07	308	818	917	5.21	1	.27	8.22	.27	6.44	.41	69	1
0	153	4.56	334	825	926	7.21	3	.71	8.52	.31	7.19	.34	118	1
0	153	4.56	334	825	16	7.21	3	.71	8.52	.31	7.19	.34	118	1
0	127	2.64	266	982	624	9.55	0	.36	10.7	.16	20.6	.37	181	0
0	25	.51	52	191	121	1.86	0	.07	2.07	.03	4	.07	35	0
0	127	2.64	266	982	9	9.55	0	.36	10.7	.16	20.6	.37	181	0
0	79	3.3	257	961	1186	4.83	0	.64	10.8	.14	19.7	.37	212	0
0	15	.63	49	184	228	.93	0	.12	2.07	.03	3.78	.07	41	0
0	49	2.36	204	856	598	3.74	0	.11	12.8	.13	17.2	.58	95	0
0	12	.59	51	214	149	.93	0	.03	3.2	.03	4.29	.14	24	0
0	39	2.3	138	423	1	5.91	14	.92	3.35	.14	.96	.2	42	2
0	10	.6	36	110	<1	1.53	4	.24	.87	.04	.25	.05	11	1
0	10	.61	37	104	218	1.59	4	.09	.84	.03	.26	.05	11	1
0	2	.86	65	176	20	1.2	1	.35	.98	.06	1.22	.03	16	1
0	38	1.9	44	306	2	.37	6	.23	1.46	.05	.3	.07	16	2
0	90	4.06	166	1241	998	1.74	31	.54	8.26	.31	1.8	.33	76	9
0	12	4.2	150	226	5	2.09	11	.06	.28	.09	.49	.06	16	1
0	98	33.8	1212	1829	1305	16.9	86	.48	2.27	.72	3.95	.2	130	4
0	50	2.96	132	155	15	3.91	3	.27	.86	.03	1.78	.05	36	0
0	42	2.44	127	248	1	1.82	2	.82	18.1	.09	1.62	.28	82	<1
0	19	2.28	43	164	1	1.77	2	.11	17.1	.09	1.4	.27	80	<1
0	21	.95	53	69	<1	1.58	1	.24	.34	.02	.85	.02	15	0
6	159	4.95	235	946	177	4.58	7	.6	15.6	.33	6.44	.38	95	2
0	72	3.84	253	655	1	4.28	37	.27	3.28	.14	.86	.69	82	4
0	16	.86	57	147	<1	.96	8	.06	.73	.03	.19	.15	18	1
0	113	2.93	203	602	12	3.28	14	.46	3.14	.18	1.25	.67	79	4
0	26	.68	47	141	3	.76	3	.11	.73	.04	.29	.16	18	1

Table H-1

Food Composition (Computer code number is for West Diet Analysis program) (For purposes of calculations, use "0" for t, <1, <.1, <.01, etc.)

Computer Code Number	Food Description	Measure	Wt (g)	H_2O (%)	Ener (kcal)	Prot (g)	Carb (g)	Dietary Fiber (g)	Fat (g)	Fat Breakdown (g) Sat	Mono	Poly
	SWEETENERS and SWEETS (see also Dairy [milk desserts] and Baked Goods)											
757	Apple butter	2 tbs	36	52	66	<1	17	<1	<1	t	t	t
1124	Butterscotch topping	2 tbs	41	32	103	1	27	<1	<1	t	t	0
1125	Caramel topping	2 tbs	41	32	103	1	27	<1	<1	t	t	0
	Cake frosting, creamy vanilla:											
1127	Canned	2 tbs	39	13	163	<1	27	<1	7	1.9	3.4	.9
1123	From mix	2 tbs	39	12	165	<1	28	<1	6	1.3	2.6	2.2
	Cake frosting, lite:											
2061	Milk chocolate	1 tbs	16	18	58	<1	11	<1	1	.4		
2062	Vanilla	1 tbs	16	15	60	0	12	<1	1	.4		
	Candy:											
1128	Almond Joy candy bar	1 oz	28	10	131	1	16	1	7	4.8	1.8	.4
2069	Butterscotch morsels	¼ c	43	1	246	0	31	0	12	12.5	0	0
758	Caramel, plain or chocolate	1 pce	10	8	38	<1	8	<1	1	.7	.1	t
1961	Chewing gum, sugarless	1 pce	3		6	0	2		0	0	0	0
	Chocolate (see also #784, 785, 971):											
	Milk chocolate:											
759	Plain	1 oz	28	1	144	2	17	1	9	5.2	2.8	.3
760	With almonds	1 oz	28	1	147	3	15	2	10	4.8	3.8	.6
761	With peanuts	1 oz	28	1	155	5	11	2	11	3.4	5.1	2.5
762	With rice cereal	1 oz	28	2	139	2	18	1	7	4.4	2.4	.2
763	Semisweet chocolate chips	1 c	168	1	805	7	106	10	50	29.9	16.8	1.6
764	Sweet dark chocolate (candy bar)	1 ea	41	1	226	2	25	2	13	8.3	4.6	.4
765	Fondant candy, uncoated (mints, candy corn, other)	1 pce	16	7	57	0	15	0	0	0	0	0
1697	Fruit Roll-Up (small)	1 ea	14	11	49	<1	12	<1	<1	.1	.2	.1
766	Fudge, chocolate	1 pce	17	10	65	<1	13	<1	1	.9	.4	.1
767	Gumdrops	1 c	182	1	703	0	180	0	0	0	0	0
768	Hard candy, all flavors	1 pce	6	1	22	0	6	0	<1	0	0	0
769	Jellybeans	10 pce	11	6	40	0	10	0	<1	t	t	t
1134	M&M's plain chocolate candy	10 pce	7	2	34	<1	5	<1	1	.9	.5	t
1135	M&M's peanut chocolate candy	10 pce	20	2	103	2	12	1	5	2.1	2.2	.8
1130	Mars almond bar	1 ea	50	4	234	4	31	1	11	2.7	5.5	2.8
1129	Milky Way candy bar	1 ea	60	6	254	3	43	1	10	4.7	3.6	.4
1708	Milk chocolate-coated peanuts	1 c	149	2	773	19	74	7	50	21.8	19.4	6.4
1709	Peanut brittle, recipe	1 c	147	2	666	11	102	3	28	7.4	12.5	6.9
1132	Reese's peanut butter cup	2 ea	50	3	271	5	27	2	16	5.5	6.5	2.8
1133	Skor English toffee candy bar	1 ea	39	3	217	2	22	1	13	8.5	4.3	.5
1131	Snickers candy bar (2.2oz)	1 ea	62	5	297	5	37	2	15	5.6	6.5	3
1482	Fruit juice bar (2.5 fl oz)	1 ea	77	78	63	1	16	0	<1	t	0	t
771	Gelatin dessert/Jello, prepared	½ c	135	85	80	2	19	0	0	0	0	0
1702	SugarFree	½ c	117	98	8	1	1	0	0	0	0	0
772	Honey:	1 c	339	17	1030	1	279	1	0	0	0	0
773	Tablespoon	1 tbs	21	17	64	<1	17	<1	0	0	0	0
774	Jams or preserves:	1 tbs	20	29	54	<1	14	<1	<1	0	t	t
775	Packet	1 ea	14	34	34	<1	9	<1	<1	t	t	0
776	Jellies:	1 tbs	19	28	51	<1	13	<1	<1	t	t	t
777	Packet	1 ea	14	28	38	<1	10	<1	<1	t	t	t
1136	Marmalade	1 tbs	20	33	49	<1	13	<1	0	0	0	0
770	Marshmallows	1 ea	7	16	22	<1	6	<1	<1	t	t	t
1126	Marshmallow creme topping	2 tbs	38	18	118	1	30	<1	<1	t	t	t
778	Popsicle/ice pops	1 ea	128	80	92	0	24	0	0	0	0	0
	Sugars:											
779	Brown sugar	1 c	220	2	827	0	214	0	0	0	0	0
780	White sugar, granulated:	1 c	200		774	0	200	0	0	0	0	0
781	Tablespoon	1 tbs	12		46	0	12	0	0	0	0	0
782	Packet	1 ea	6		23	0	6	0	0	0	0	0
783	White sugar, powdered, sifted	1 c	100		389	0	99	0	<1	t	t	t
	Sweeteners:											
1711	Equal, packet	1 ea	1	12	4	<1	1	0	<1	0	t	t
1712	Sweet 'N Low, packet	1 ea	1		0	0	1	0	0	0	0	0

Chol (mg)	Calc (mg)	Iron (mg)	Magn (mg)	Pota (mg)	Sodi (mg)	Zinc (mg)	VT-A (RE)	Thia (mg)	VT-E (α-TE)	Ribo (mg)	Niac (mg)	V-B6 (mg)	Fola (μg)	VT-C (mg)
0	2	.05	1	33	0	.02	0	<.01	.01	<.01	.03	.01	<1	1
<1	22	.08	3	34	143	.08	11	<.01	0	.04	.02	.01	1	<1
<1	22	.08	3	34	143	.08	11	<.01	0	.04	.02	.01	1	<1
0	1	.04		14	35	0	88	0	.79	<.01	<.01	0	0	0
0	4	.09	1	9	87	.04	42	.01	.79	.01	.13	<.01	0	0
0	1	.24			40		<1							0
0		.02			29		0							0
1	17	.39	18	69	41	.22	1	.01	.63	.04	.13	.02		<1
0	0	0		80	46		0	.03		.04	.03			0
1	14	.01	2	21	24	.04	1	<.01	.05	.02	.02	<.01	<1	<1
				0	0									
6	53	.39	17	108	23	.39	15	.02	.35	.08	.09	.01	2	<1
5	63	.46	25	124	21	.37	4	.02	.53	.12	.21	.01	3	<1
3	32	.52	34	150	11	.68	6	.08	1.3	.05	2.12	.04	23	0
5	48	.21	14	96	41	.31	3	.02	.35	.08	.13	.02	3	<1
0	54	5.26	193	613	18	2.72	3	.09	2	.15	.72	.06	5	0
<1	11	.98	47	139	3	.61	1	.01	.41	.1	.27	.02	1	0
0		.01		3	6	.01	<1	0	0	<.01	0	0	0	0
0	4	.14	3	41	9	.03	2	.01	.04	<.01	.01	.04	1	1
2	7	.08	4	17	10	.07	8	<.01	.02	.01	.02	<.01	<1	<1
0	5	.73	2	9	80	0	0	0	0	<.01	<.01	0	0	0
0		.02			2	<.01	0	0	0	0	0	0	0	0
0		.12		4	3	.01	0	0	0	0	0	0	0	0
1	7	.08	3	19	4	.07	4	<.01	.06	.01	.02	<.01	<1	<1
2	20	.23	12	69	10	.27	5	.03	.43	.04	.41	.02	7	<1
4	84	.55	36	163	85	.55	22	.02	.3	.16	.47	.03	9	<1
8	78	.46	20	145	144	.43	34	.02	.39	.13	.21	.03	6	1
13	155	1.95	134	748	61	2.8	0	.17	3.8	.26	6.33	.31	12	0
19	44	2.03	73	306	664	1.43	69	.28	2.41	.08	5.15	.15	103	0
2	39	.6	42	176	159	.7	9	.02	.66	.1	1.99	.04	27	<1
20	51	.02	13	93	108	.3	27	.01	.53	.13	.03	.01		<1
8	58	.47	42	209	165	.88	24	.13	.95	.1	2.26	.07	25	<1
0	4	.15	3	41	3	.04	2	.01	0	.01	.12	.02	5	7
0	3	.04	1	1	57	.04	0	0	0	<.01	<.01	<.01	0	0
0	2	.01	1	0	56	.03	0	0	0	<.01	<.01	<.01	0	0
0	20	1.42	7	176	14	.75	0	0	0	.13	.41	.08	7	2
0	1	.09		11	1	.05	0	0	0	.01	.02	<.01	<1	<1
0	4	.2	1	18	2	.01	<1	<.01	.02	.01	.04	<.01	2	<1
0	3	.07	1	11	6	.01	<1	0	0	<.01	<.01	<.01	5	1
0	2	.04	1	12	7	.01	<1	0	0	<.01	.01	<.01	<1	<1
0	1	.03	1	9	5	.01	<1	0	0	<.01	<.01	<.01	<1	<1
0	8	.03		7	11	.01	1	<.01	0	<.01	.01	<.01	7	1
0		.02			3	<.01	<1	0	0	0	<.01	0	<1	0
0	1	.08	1	2	17	.01	<1	0	0	0	.03	<.01	<1	0
0	0	0	1	5	15	.03	0	0	0	0	0	0	0	0
0	187	4.2	64	761	86	.4	0	.02	0	.01	.18	.06	2	0
0	2	.12	0	4	2	.06	0	0	0	.04	0	0	0	0
0		.01	0		<1	<.01	0	0	0	<.01	0	0	0	0
0		<.01	0		<1	<.01	0	0	0	<.01	0	0	0	0
0	1	.06	0	2	1	.03	0	0	0	0	0	0	0	0
0		<.01			<1	0	0	0	0	0	0	0	0	0
0	0	0			0		0							0

Table H-1

Food Composition (Computer code number is for West Diet Analysis program) (For purposes of calculations, use "0" for t, <1, <.1, <.01, etc.)

Computer Code Number	Food Description	Measure	Wt (g)	H_2O (%)	Ener (kcal)	Prot (g)	Carb (g)	Dietary Fiber (g)	Fat (g)	Fat Breakdown (g) Sat	Mono	Poly
	SWEETENERS and SWEETS—Continued											
	Syrups, chocolate:											
785	Hot fudge type	2 tbs	43	22	149	2	27	1	4	2.4	1.6	1.4
784	Thin type	2 tbs	38	29	93	1	25	1	<1	.3	.2	t
786	Molasses, blackstrap	2 tbs	41	29	96	0	25	0	0	0	0	0
1710	Light cane syrup	2 tbs	41	24	103	0	27	0	0	0	0	0
787	Pancake table syrup (corn and maple)	2 tbs	40	24	115	0	30	0	0	0	0	0
	VEGETABLES and LEGUMES											
788	Alfalfa sprouts	1 c	33	91	10	1	1	1	<1	t	.1	t
1815	Amaranth leaves, raw, chopped	1 c	28	92	7	1	1	<1	<1	t	t	t
1816	Amaranth leaves, raw, each	1 ea	14	92	4	<1	1	<1	<1	t	t	t
1817	Amaranth leaves, cooked	1 c	132	91	28	3	5	2	<1	.1	.1	.1
1987	Arugula, raw, chopped	½ c	10	92	2	<1	<1	<1	<1	t	t	t
789	Artichokes, cooked globe (300 g with refuse)	1 ea	120	84	60	4	13	6	<1	t	t	.1
1177	Artichoke hearts, cooked from frozen	1 c	168	86	76	5	15	8	1	.2	t	.4
1176	Artichoke hearts, marinated	1 c	130	81	128	3	10	6	10	1.5	2.3	5.9
2021	Artichoke hearts, in water	½ c	100	91	37	2	6	0	0	0	0	0
	Asparagus, green, cooked:											
	From fresh:											
790	Cuts and tips	½ c	90	92	22	2	4	1	<1	.1	t	.1
791	Spears, ½" diam at base	4 ea	60	92	14	2	3	1	<1	t	t	.1
	From frozen:											
792	Cuts and tips	½ c	90	91	25	3	4	1	<1	.1	t	.2
793	Spears, ½" diam at base	4 ea	60	91	17	2	3	1	<1	.1	t	.1
794	Canned, spears, ½" diam at base	4 ea	72	94	14	2	2	1	<1	.1	t	.2
795	Bamboo shoots, canned, drained slices	1 c	131	94	25	2	4	2	1	.1	t	.2
1795	Bamboo shoots, raw slices	1 c	151	91	41	4	8	3	<1	.1	t	.2
1798	Bamboo shoots, cooked slices	1 c	120	96	14	2	2	1	<1	.1	t	.1
	Beans (see also alphabetical listing this section):											
1990	Adzuki beans, cooked	½ c	115	66	147	9	28	1	<1	t	t	t
796	Black beans, cooked	½ c	86	66	114	8	20	7	<1	.1	t	.2
	Canned beans (white/navy):											
803	With pork and tomato sauce	½ c	127	73	124	7	25	6	1	.5	.6	.2
804	With sweet sauce	½ c	130	71	144	7	27	7	2	.7	.8	.2
805	With frankfurters	½ c	130	69	185	9	20	9	9	3.1	3.7	1.1
	Lima beans:											
797	Thick seeded (Fordhooks), cooked from frozen	½ c	85	73	85	5	16	5	<1	.1	t	.1
798	Thin seeded (Baby), cooked from frozen	½ c	90	72	94	6	18	5	<1	.1	t	.1
799	Cooked from dry, drained	½ c	94	70	108	7	20	7	<1	.1	t	.2
1998	Red Mexican, cooked f/dry	½ c	112	70	126	8	23	9	<1	.1	.1	.2
	Snap bean/green string beans cuts and french style:											
800	Cooked from fresh	½ c	63	89	22	1	5	2	<1	t	t	.1
801	Cooked from frozen	½ c	68	91	19	1	4	2	<1	t	t	.1
802	Canned, drained	½ c	68	93	14	1	3	1	<1	t	t	t
1713	Snap bean, yellow, cooked f/fresh	½ c	63	89	22	1	5	2	<1	t	t	.1
	Bean sprouts (mung):											
806	Raw	½ c	52	90	16	2	3	1	<1	t	t	t
807	Cooked, stir fried	½ c	62	84	31	3	7	1	<1	t	t	t
808	Cooked, boiled, drained	½ c	62	93	13	1	3	<1	<1	t	t	t
1788	Canned, drained	½ c	63	96	8	1	1	<1	<1	t	t	t
	Beets, cooked from fresh:											
809	Sliced or diced	½ c	85	87	37	1	8	2	<1	t	t	.1
810	Whole beets, 2" diam	2 ea	100	87	44	2	10	2	<1	t	t	.1
	Beets, canned:											
811	Sliced or diced	½ c	79	91	24	1	6	1	<1	t	t	t
812	Pickled slices	½ c	114	82	74	1	19	2	<1	t	t	t
813	Beet greens, cooked, drained	½ c	72	89	19	2	4	2	<1	t	t	.1

Chol (mg)	Calc (mg)	Iron (mg)	Magn (mg)	Pota (mg)	Sodi (mg)	Zinc (mg)	VT-A (RE)	Thia (mg)	VT-E (α-TE)	Ribo (mg)	Niac (mg)	V-B6 (mg)	Fola (μg)	VT-C (mg)
5	43	.52	21	92	56	.34	9	.01	0	.09	.09	.01	2	<1
0	5	5.17	25	183	58	.28	494	<.01	.01	.31	12.8	.01	2	<1
0	353	7.18	88	1021	23	.41	0	.01	0	.02	.44	.29	<1	0
0	68	1.76	100	376	6	.12	0	.03	0	.02	.08	.27	0	0
0		.04	1	1	33	.02	0	<.01	0	<.01	.01	0	0	0
0	11	.32	9	26	2	.3	5	.02	—	.04	.16	.01	12	3
0	60	.65	15	171	6	.25	82	.01	.22	.04	.18	.05	24	12
0	30	.32	8	85	3	.13	41	<.01	.11	.02	.09	.03	12	6
0	276	2.98	73	846	28	1.16	366	.03	.66	.18	.74	.23	75	54
0	16	.15	5	37	3	.05	24	<.01	.04	.01	.03	.01	10	1
0	54	1.55	72	425	114	.59	22	.08	.23	.08	1.2	.13	61	12
0	35	.94	52	444	89	.6	27	.1	.32	.26	1.54	.15	200	8
0	30	1.24	37	335	688	.41	21	.05	1.43	.13	1.06	.11	114	40
0	0	1.35		0	250		12							4
0	18	.66	9	144	10	.38	49	.11	.34	.11	.97	.11	131	10
0	12	.44	6	96	7	.25	32	.07	.23	.08	.65	.07	88	6
0	21	.58	12	196	4	.5	74	.06	1.13	.09	.94	.02	122	22
0	14	.38	8	131	2	.34	49	.04	.75	.06	.62	.01	81	15
0	11	1.32	7	124	207	.29	38	.04	.31	.07	.69	.08	69	13
0	10	.42	5	105	9	.85	1	.03	.5	.03	.18	.18	4	1
0	20	.75	5	805	6	1.66	3	.23	1.51	.11	.91	.36	11	6
0	14	.29	4	640	5	.56	0	.02	.8	.06	.36	.12	3	1
0	32	2.3	60	612	9	2.04	1	.13	.11	.07	.82	.11	139	0
0	23	1.81	60	305	1	.96	1	.21	.07	.05	.43	.06	128	0
9	71	4.17	44	381	559	7.44	15	.07	.69	.06	.63	.09	29	4
9	79	2.16	44	346	437	1.95	14	.06	.7	.08	.46	.11	49	4
8	62	2.25	36	306	559	2.43	19	.07	.61	.07	1.17	.06	39	3
0	19	1.16	29	347	45	.37	16	.06	.25	.05	.91	.1	18	11
0	25	1.76	50	370	26	.49	15	.06	.58	.05	.69	.1	14	5
0	16	2.25	40	478	2	.89	0	.15	.17	.05	.4	.15	78	0
0	42	1.86	48	369	240	.87	<1	.13	.08	.07	.37	.11	94	2
0	29	.81	16	188	2	.23	42	.05	.09	.06	.39	.03	21	6
0	33	.6	16	86	6	.33	27	.02	.09	.06	.26	.04	16	3
0	18	.61	9	74	178	.2	24	.01	.09	.04	.14	.02	22	3
0	29	.81	16	188	2	.23	5	.05	.18	.06	.39	.03	21	6
0	7	.47	11	77	3	.21	1	.04	.02	.06	.39	.05	32	7
0	8	1.18	20	136	6	.56	2	.09	.01	.11	.74	.08	43	10
0	7	.4	9	63	6	.29	1	.03	.01	.06	.51	.03	18	7
0	9	.27	6	17	88	.18	1	.02	.01	.04	.14	.02	6	<1
0	14	.67	20	259	65	.3	3	.02	.25	.03	.28	.06	68	3
0	16	.79	23	305	77	.35	4	.03	.3	.04	.33	.07	80	4
0	12	1.44	13	117	153	.17	1	.01	.24	.03	.12	.04	24	3
0	12	.47	17	169	301	.3	1	.01	.15	.05	.29	.06	30	3
0	82	1.37	49	654	174	.36	367	.08	.22	.21	.36	.09	10	18

Table H-1

Food Composition (Computer code number is for West Diet Analysis program) (For purposes of calculations, use "0" for t, <1, <.1, <.01, etc.)

Computer Code Number	Food Description	Measure	Wt (g)	H_2O (%)	Ener (kcal)	Prot (g)	Carb (g)	Dietary Fiber (g)	Fat (g)	Fat Breakdown (g) Sat	Mono	Poly
	VEGETABLES and LEGUMES—Continued											
	Broccoli, raw:											
817	Chopped	½ c	44	91	12	1	2	1	<1	t	t	.1
818	Spears	1 ea	31	91	9	1	2	1	<1	t	t	.1
	Broccoli, cooked from fresh:											
819	Spears	1 ea	180	91	50	5	9	5	1	.1	t	.3
820	Chopped	½ c	78	91	22	2	4	2	<1	t	t	.1
	Broccoli, cooked from frozen:											
821	Spear, small piece	½ c	92	91	26	3	5	3	<1	t	t	.1
822	Chopped	½ c	92	91	26	3	5	3	<1	t	t	.1
1603	Broccoflower, steamed	½ c	78	90	25	2	5	2	<1	t	t	.1
823	Brussels sprouts, cooked from fresh	½ c	78	87	30	2	7	2	<1	.1	t	.2
824	Brussels sprouts, cooked from frozen	½ c	78	87	33	3	6	3	<1	.1	t	.2
	Cabbage, common varieties:											
825	Raw, shredded or chopped	1 c	70	92	17	1	4	2	<1	t	t	.1
826	Cooked, drained	1 c	150	94	33	2	7	3	1	.1	t	.3
	Cabbage, Chinese:											
1178	Bok choy, raw, shredded	1 c	70	95	9	1	2	1	<1	t	t	.1
827	Bok choy, cooked, drained	1 c	170	96	20	3	3	3	<1	t	t	.1
1937	Kim chee style	1 c	150	92	31	2	6	2	<1	t	t	.2
828	Pe Tsai, raw, chopped	1 c	76	94	12	1	2	2	<1	t	t	.1
1796	Pe Tsai, cooked	1 c	119	95	17	2	3	3	<1	t	t	.1
	Cabbage, red, coarsely chopped:											
829	Raw	1 c	89	92	24	1	5	2	<1	t	t	.1
830	Cooked, drained	1 c	150	94	31	2	7	3	<1	t	t	.1
831	Cabbage, savoy, coarsely chopped, raw	1 c	70	91	19	1	4	2	<1	t	t	t
1785	Cabbage, savoy, cooked	1 c	145	92	35	3	8	4	<1	t	t	.1
1896	Capers	1 ea	5	86		<1	<1	<1	<1			
	Carrots, raw:											
832	Whole, 7½ x 1⅛"	1 ea	72	88	31	1	7	2	<1	t	t	.1
833	Grated	½ c	55	88	24	1	6	2	<1	t	t	t
	Carrots, cooked, sliced, drained:											
834	From raw	½ c	78	87	35	1	8	3	<1	t	t	.1
835	From frozen	½ c	73	90	26	1	6	3	<1	t	t	t
836	Carrots, canned, sliced, drained	½ c	73	93	17	<1	4	1	<1	t	t	.1
837	Carrot juice, canned	1 c	236	89	94	2	22	2	<1	.1	t	.2
	Cauliflower, flowerets:											
838	Raw	½ c	50	92	12	1	3	1	<1	t	t	t
839	Cooked from fresh, drained	½ c	62	93	14	1	3	2	<1	t	t	.1
840	Cooked, from frozen, drained	½ c	90	94	17	1	3	2	<1	t	t	.1
	Celery, pascal type, raw:											
841	Large outer stalk, 8 x 1½"(root end)	1 ea	40	95	6	<1	1	1	<1	t	t	t
842	Diced	1 c	120	95	19	1	4	2	<1	t	t	.1
1789	Celeriac/celery root, cooked	1 c	155	92	39	1	9	2	<1	.1	.1	.2
1179	Chard, swiss, raw, chopped	1 c	36	93	7	1	1	1	<1	t	t	t
1180	Chard, swiss, cooked	1 c	175	93	35	3	7	4	<1	t	t	t
1855	Chayote fruit, raw	1 ea	203	94	39	2	9	3	<1	.1	t	.1
1856	Chayote fruit, cooked	1 c	160	93	38	1	8	4	1	.2	.1	.3
	Chickpeas (see Garbanzo Beans #854)											
	Collards, cooked, drained:											
843	From raw	½ c	95	92	25	2	5	3	<1	t	t	.1
844	From frozen	½ c	85	88	31	3	6	3	<1	.1	t	.2
	Corn, yellow, cooked, drained:											
845	From raw, on cob, 5" long	1 ea	77	73	72	2	17	2	1	.1	.2	.3
846	From frozen, on cob, 3½" long	1 ea	63	73	59	2	14	2	<1	.1	.1	.2
847	Kernels, cooked from frozen	½ c	82	77	66	2	16	2	<1	.1	.1	.2
	Corn, canned:											
848	Cream style	½ c	128	79	92	2	23	2	1	.1	.2	.3
849	Whole kernel, vacuum pack	½ c	105	77	83	3	20	2	1	.1	.2	.2
	Cowpeas (see Black-eyed peas #814-816)											
850	Cucumber slices with peel	7 pce	28	96	4	<1	1	<1	<1	t	t	t
1948	Cucumber, kim chee style	1 c	150	91	31	2	7	2	<1	t	0	.1

Chol (mg)	Calc (mg)	Iron (mg)	Magn (mg)	Pota (mg)	Sodi (mg)	Zinc (mg)	VT-A (RE)	Thia (mg)	VT-E (α-TE)	Ribo (mg)	Niac (mg)	V-B6 (mg)	Fola (μg)	VT-C (mg)
0	21	.39	11	143	12	.18	68	.03	.73	.05	.28	.07	31	41
0	15	.27	8	101	8	.12	48	.02	.51	.04	.2	.05	22	29
0	83	1.51	43	526	47	.68	250	.1	3.04	.2	1.03	.26	90	134
0	36	.65	19	228	20	.3	108	.04	1.32	.09	.45	.11	39	58
0	47	.56	18	166	22	.28	174	.05	.95	.07	.42	.12	28	37
0	47	.56	18	166	22	.28	174	.05	1.52	.07	.42	.12	52	37
0	25	.55	16	251	18	.39	5	.06	.23	.07	.59	.14	38	49
0	28	.94	16	247	16	.26	56	.08	.66	.06	.47	.14	47	48
0	19	.58	19	254	18	.28	46	.08	.45	.09	.42	.22	79	36
0	33	.41	10	172	13	.13	9	.03	.07	.03	.21	.07	30	22
0	46	.25	12	146	12	.13	19	.09	.16	.08	.42	.17	30	30
0	73	.56	13	176	45	.13	210	.03	.08	.05	.35	.14	46	31
0	158	1.77	19	631	58	.29	437	.05	.2	.11	.73	.28	69	44
0	145	1.28	27	375	995	.35	426	.07	.24	.1	.75	.34	88	80
0	58	.24	10	181	7	.17	91	.03	.09	.04	.3	.18	60	20
0	38	.36	12	268	11	.21	115	.05	.14	.05	.59	.21	63	19
0	45	.44	13	183	10	.19	4	.04	.09	.03	.27	.19	18	51
0	55	.52	16	210	12	.22	4	.05	.18	.03	.3	.21	19	52
0	24	.28	20	161	20	.19	70	.05	.07	.02	.21	.13	56	22
0	43	.55	35	267	35	.33	129	.07	.15	.03	.03	.22	67	25
0	2	.05			105		1							0
0	19	.36	11	233	25	.14	2025	.07	.33	.04	.67	.11	10	7
0	15	.27	8	178	19	.11	1547	.05	.25	.03	.51	.08	8	5
0	24	.48	10	177	51	.23	1914	.03	.33	.04	.39	.19	11	2
0	20	.34	7	115	43	.17	1292	.02	.31	.03	.32	.09	8	2
0	18	.47	6	131	177	.19	1005	.01	.31	.02	.4	.08	7	2
0	57	1.09	33	689	68	.42	2584	.22	.02	.13	.91	.51	9	20
0	11	.22	7	152	15	.14	1	.03	.02	.03	.26	.11	28	23
0	10	.2	6	88	9	.11	1	.03	.02	.03	.25	.11	27	27
0	15	.37	8	125	16	.12	2	.03	.04	.05	.28	.08	37	28
0	16	.16	4	115	35	.05	5	.02	.14	.02	.13	.03	11	3
0	48	.48	13	344	104	.16	16	.05	.43	.05	.39	.1	34	8
0	40	.67	19	268	95	.31	0	.04	.31	.06	.66	.16	5	6
0	18	.65	29	136	77	.13	119	.01	.68	.03	.14	.04	5	11
0	102	3.96	151	961	313	.58	550	.06	3.31	.15	.63	.15	15	31
0	34	.69	24	254	4	1.5	12	.05	.24	.06	.95	.15	189	16
0	21	.35	19	277	2	.5	8	.04	.19	.06	.67	.19	29	13
0	113	.44	16	247	9	.4	297	.04	.84	.1	.55	.12	88	17
0	179	.95	25	213	42	.23	508	.04	.42	.1	.54	.1	65	22
0	2	.47	22	193	3	.48	16	.13	.07	.05	1.17	.17	23	4
0	2	.38	18	158	3	.4	13	.11	.06	.04	.96	.14	19	3
0	3	.29	16	121	4	.33	18	.07	.07	.06	1.07	.11	25	3
0	4	.49	22	172	365	.68	13	.03	.11	.07	1.23	.08	57	6
0	5	.44	24	195	286	.48	25	.04	.09	.08	1.23	.06	52	9
0	4	.07	3	40	1	.06	6	.01	.02	.01	.06	.01	4	1
0	13	7.23	12	176	1531	.76	49	.04	.24	.04	.69	.16	34	5

Table H-1

Food Composition (Computer code number is for West Diet Analysis program) (For purposes of calculations, use "0" for t, <1, <.1, <.01, etc.)

Computer Code Number	Food Description	Measure	Wt (g)	H_2O (%)	Ener (kcal)	Prot (g)	Carb (g)	Dietary Fiber (g)	Fat (g)	Fat Breakdown (g) Sat	Mono	Poly
	VEGETABLES and LEGUMES—Continued											
	Dandelion Greens:											
851	Raw	1 c	55	86	25	1	5	2	<1	.1	t	.2
852	Chopped, cooked, drained	1 c	105	90	35	2	7	3	1	.2	t	.3
853	Eggplant, cooked	1 c	99	92	28	1	7	2	<1	t	t	.1
1714	Endive, fresh, chopped	1 c	50	94	8	1	2	2	<1	t	t	t
856	Escarole/curly endive, chopped	1 c	50	94	8	1	2	2	<1	t	t	t
854	Garbanzo beans (chickpeas), cooked	1 c	164	60	269	14	45	12	4	.4	1	1.9
1939	Grape leaf, raw:	1 ea	3	73	3	<1	1	<1	<1	t	t	t
7914	Cup	1 c	14	73	13	1	2	2	<1	t	t	.1
855	Great northern beans, cooked	1 c	177	69	209	15	37	12	1	.2	t	.3
857	Jerusalem artichoke, raw slices	1 c	150	78	114	3	26	2	<1	0	t	t
1794	Jicama	1 c	120	90	46	1	11	6	<1	t	t	t
	Kale, cooked, drained:											
858	From raw	1 c	130	91	36	2	7	3	1	.1	t	.3
859	From frozen	1 c	130	90	39	4	7	3	1	.1	t	.3
860	Kidney beans, canned	1 c	256	77	218	13	40	16	1	.1	.1	.5
1181	Kohlrabi, raw slices	1 c	135	91	36	2	8	5	<1	t	t	.1
861	Kohlrabi, cooked	1 c	165	90	48	3	11	2	<1	t	t	.1
1183	Leeks, raw, chopped	1 c	89	83	54	1	13	2	<1	t	t	.1
1182	Leeks, cooked, chopped	1 c	104	91	32	1	8	1	<1	t	t	.1
862	Lentils, cooked from dry	1 c	198	70	230	18	40	16	1	.1	.1	.3
1288	Lentils, sprouted, stir-fried	1 c	124	69	125	11	26	5	1	.1	.1	.2
1289	Lentils, sprouted, raw	1 c	77	67	82	7	17	3	<1	t	.1	.2
	Lettuce:											
	Butterhead/Boston types:											
863	Head, 5" diameter	¼ ea	41	96	5	1	1	<1	<1	t	t	t
864	Leaves, inner or outer	4 ea	30	96	4	<1	1	<1	<1	t	t	t
	Iceberg/crisphead:											
867	Chopped or shredded	1 c	55	96	7	1	1	1	<1	t	t	.1
865	Head, 6" diameter	1 ea	539	96	65	5	11	8	1	.1	t	.5
866	Wedge, ¼ head	1 ea	135	96	16	1	3	2	<1	t	t	.1
868	Looseleaf, chopped	½ c	28	94	5	<1	1	1	<1	t	t	t
869	Romaine, chopped	½ c	28	95	4	<1	1	<1	<1	t	t	t
870	Romaine, inner leaf	3 pce	30	95	4	<1	1	1	<1	t	t	t
1930	Luffa, cooked (Chinese okra)	1 c	178	89	57	3	13	6	<1	.1	t	.1
	Mushrooms:											
871	Raw, sliced	½ c	35	92	9	1	2	<1	<1	t	t	.1
872	Cooked from fresh, pieces	½ c	78	91	21	2	4	2	<1	t	t	.1
1962	Stir fried, shitake slices	½ c	73	83	40	1	10	2	<1	t	t	t
873	Canned, drained	½ c	78	91	19	1	4	2	<1	t	t	.1
1951	Mushroom caps, pickled	8 ea	47	92	11	1	2	1	<1	t	t	.1
	Mustard greens:											
874	Cooked from raw	½ c	70	94	10	2	1	1	<1	t	.1	t
875	Cooked from frozen	½ c	75	94	14	2	2	2	<1	t	.1	t
876	Navy beans, cooked from dry	1 c	182	63	258	16	48	12	1	.3	.1	.4
	Okra, cooked:											
877	From fresh pods	8 ea	85	90	27	2	6	2	<1	t	t	t
878	From frozen slices	1 c	184	91	51	4	11	5	1	.1	.1	.1
1236	Batter fried from fresh	1 c	92	69	175	3	11	2	13	2.1	3.4	7.1
1930	Chinese, (Luffa), cooked	1 c	178	89	57	3	13	6	<1	.1	t	.1
	Onions:											
879	Raw, chopped	½ c	80	90	30	1	7	1	<1	t	t	t
880	Raw, sliced	½ c	58	90	22	1	5	1	<1	t	t	t
881	Cooked, drained, chopped	½ c	105	88	46	1	11	1	<1	t	t	.1
882	Dehydrated flakes	¼ c	14	4	49	1	12	1	<1	t	t	t
1934	Onions, pearl, cooked	½ c	93	87	41	1	9	1	<1	t	t	.1
883	Spring/green onions, bulb and top, chopped	½ c	50	90	16	1	4	1	<1	t	t	t
884	Onion rings, breaded, heated f/frozen	2 ea	20	28	81	1	8	<1	5	1.7	2.2	1
1917	Palm hearts, cooked slices	1 c	146	69	150	4	39	2	<1	.1	.1	t
885	Parsley, raw, chopped	½ c	30	88	11	1	2	1	<1	t	.1	t
888	Parsnips, sliced, cooked	½ c	78	78	63	1	15	3	<1	t	.1	t

Chol (mg)	Calc (mg)	Iron (mg)	Magn (mg)	Pota (mg)	Sodi (mg)	Zinc (mg)	VT-A (RE)	Thia (mg)	VT-E (α-TE)	Ribo (mg)	Niac (mg)	V-B6 (mg)	Fola (μg)	VT-C (mg)
0	103	1.71	20	218	42	.23	770	.1	1.38	.14	.44	.14	15	19
0	147	1.89	25	244	46	.29	1228	.14	2.63	.18	.54	.17	13	19
0	6	.35	13	246	3	.15	6	.07	.03	.02	.59	.08	14	1
0	26	.41	7	157	11	.39	103	.04	.22	.04	.2	.01	71	3
0	26	.41	7	157	11	.39	103	.04	.22	.04	.2	.01	71	3
0	80	4.74	79	477	11	2.51	5	.19	.57	.1	.86	.23	282	2
0	11	.08	3	8	<1	.02	81	<.01	.06	.01	.07	.01	2	<1
0	51	.37	13	38	1	.09	378	.01	.28	.05	.33	.06	12	2
0	120	3.77	88	692	4	1.56	<1	.28	.53	.1	1.21	.21	181	2
0	21	5.1	25	644	6	.18	3	.3	.28	.09	1.95	.12	20	6
0	14	.72	14	180	5	.19	2	.02	5.48	.03	.24	.05	14	24
0	94	1.17	23	296	30	.31	962	.07	1.11	.09	.65	.18	17	53
0	179	1.22	23	417	19	.23	826	.06	.23	.15	.87	.11	19	33
0	61	3.23	72	658	873	1.41	0	.27	.13	.22	1.17	.06	130	3
0	32	.54	26	473	27	.04	5	.07	.65	.03	.54	.2	22	84
0	41	.66	31	561	35	.51	7	.07	2.76	.03	.64	.25	20	89
0	52	1.87	25	160	18	.11	9	.05	.82	.03	.36	.21	57	11
0	31	1.14	15	90	10	.06	5	.03	.63	.02	.21	.12	25	4
0	38	6.59	71	731	4	2.51	2	.33	.22	.14	2.1	.35	358	3
0	17	3.84	43	352	12	1.98	5	.27	.11	.11	1.49	.2	83	16
0	19	2.47	28	248	8	1.16	4	.18	.07	.1	.87	.15	77	13
0	13	.12	5	105	2	.07	40	.02	.18	.02	.12	.02	30	3
0	10	.09	4	77	1	.05	29	.02	.13	.02	.09	.01	22	2
0	10	.27	5	87	5	.12	18	.02	.15	.02	.1	.02	31	2
0	102	2.7	48	852	48	1.19	178	.25	1.51	.16	1.01	.22	302	21
0	26	.67	12	213	12	.3	45	.06	.38	.04	.25	.05	76	5
0	19	.39	3	74	3	.08	53	.01	.12	.02	.11	.01	14	5
0	10	.31	2	81	2	.07	73	.03	.12	.03	.14	.01	38	7
0	11	.33	2	87	2	.07	78	.03	.13	.03	.15	.01	41	7
0	112	.8	101	570	420	.97	103	.23	1.22	.1	1.54	.33	81	29
0	2	.43	3	130	1	.26	0	.04	.04	.16	1.44	.03	7	1
0	5	1.36	9	278	2	.68	0	.06	.09	.23	3.48	.07	14	3
0	2	.32	10	85	3	.97	0	.03	.09	.12	1.1	.12	15	<1
0	9	.62	12	101	332	.56	0	.07	.09	.02	1.24	.05	10	0
0	2	.5	5	139	95	.28	0	.03	.05	.16	1.42	.03	6	1
0	52	.49	10	141	11	.08	212	.03	1.41	.04	.3	.07	51	18
0	76	.84	10	104	19	.15	335	.03	1.31	.04	.19	.08	52	10
0	127	4.51	107	670	2	1.93	<1	.37	.73	.11	.97	.3	255	2
0	54	.38	48	274	4	.47	49	.11	.59	.05	.74	.16	39	14
0	177	1.23	94	431	6	1.14	94	.18	1.27	.23	1.44	.09	269	22
15	104	.77	37	214	137	.5	43	.13	3.08	.1	.75	.13	37	10
0	112	.8	101	570	420	.97	103	.23	1.22	.1	1.54	.33	81	29
0	16	.18	8	126	2	.15	0	.03	.1	.02	.12	.09	15	5
0	12	.13	6	91	2	.11	0	.02	.07	.01	.09	.07	11	4
0	23	.25	12	174	3	.22	0	.04	.14	.02	.17	.13	16	5
0	36	.22	13	227	3	.26	0	.07	.19	.01	.14	.22	23	10
0	21	.22	10	154	218	.19	0	.04	.12	.02	.15	.12	14	5
0	36	.74	10	138	8	.19	19	.03	.06	.04	.26	.03	32	9
0	6	.34	4	26	75	.08	5	.06	.14	.03	.72	.01	13	<1
0	26	2.47	15	2636	20	5.45	10	.07	.73	.25	1.25	1.06	30	10
0	41	1.86	15	166	17	.32	156	.03	.54	.03	.39	.03	46	40
0	29	.45	23	286	8	.2	0	.06	.78	.04	.56	.07	45	10

Table H-1

Food Composition (Computer code number is for West Diet Analysis program) (For purposes of calculations, use "0" for t, <1, <.1, <.01, etc.)

H

Computer Code Number	Food Description	Measure	Wt (g)	H_2O (%)	Ener (kcal)	Prot (g)	Carb (g)	Dietary Fiber (g)	Fat (g)	Fat Breakdown (g) Sat	Mono	Poly
	VEGETABLES and LEGUMES—Continued											
	Peas:											
	Black-eyed, cooked:											
814	From dry, drained	½ c	86	70	100	7	18	6	<1	.1	t	.2
815	From fresh, drained	½ c	82	75	79	3	17	4	<1	.1	t	.1
816	From frozen, drained	½ c	85	66	112	7	20	5	1	.1	.1	.2
889	Edible pod peas, cooked	½ c	80	89	34	3	6	2	<1	t	t	.1
890	Green, canned, drained:	½ c	85	82	59	4	11	3	<1	.1	t	.1
5267	Unsalted	½ c	124	86	66	4	12	4	<1	.1	t	.2
891	Green, cooked from frozen	½ c	80	79	62	4	11	4	<1	t	t	.1
1786	Snow peas, raw	½ c	49	89	21	1	4	1	<1	t	t	t
1787	Snow peas, raw	10 ea	34	89	14	1	3	1	<1	t	t	t
892	Split, green, cooked from dry	½ c	98	69	116	8	21	8	<1	.1	.1	.2
1187	Peas & carrots, cooked from frozen	½ c	80	86	38	2	8	2	<1	.1	t	.2
1186	Peas & carrots, canned w/liquid	½ c	128	88	49	3	11	3	<1	.1	t	.2
	Peppers, hot:											
893	Hot green chili, canned	½ c	68	92	14	1	3	1	<1	t	t	t
894	Hot green chili, raw	1 ea	45	88	18	1	4	1	<1	t	t	t
1715	Hot red chili, raw, diced	1 tbs	9	88	4	<1	1	<1	<1	t	t	t
1988	Jalapeno, raw	1 ea	45	90	11	<1	2		<1			
895	Jalapeno, chopped, canned	½ c	68	89	18	1	3	2	1	.1	t	.3
1918	Jalapeno wheels, in brine (Ortega)	2 tbs	29		10	0	2		0	0	0	0
	Peppers, sweet, green:											
896	Whole pod (90 g with refuse), raw	1 ea	74	92	20	1	5	1	<1	t	t	.1
897	Cooked, chopped (1 pod cooked = 73g)	½ c	68	92	19	1	5	1	<1	t	t	.1
	Peppers, sweet, red:											
1286	Raw, chopped	½ c	75	92	20	1	5	1	<1	t	t	.1
1807	Raw, each	1 ea	74	92	20	1	5	1	<1	t	t	.1
1287	Cooked, chopped	½ c	68	92	19	1	5	1	<1	t	t	.1
	Peppers, sweet, yellow:											
1872	Raw, large	1 ea	186	92	50	2	12	2	<1	.1	t	.2
1873	Strips	10 pce	52	92	14	1	3	<1	<1	t	t	.1
898	Pinto beans, cooked from dry	½ c	85	64	116	7	22	7	<1	.1	.1	.2
1191	Poi, two finger	½ c	120	72	134	<1	33	<1	<1	t	t	.1
	Potatoes:											
	Baked in oven, 4¾"x2⅓" diam											
899	With skin	1 ea	202	71	220	5	51	5	<1	.1	t	.1
900	Flesh only	1 ea	156	75	145	3	34	2	<1	t	t	.1
901	Skin only	1 ea	58	47	115	2	27	5	<1	t	t	t
	Baked in microwave, 4¾"x 2⅓"dm:											
902	With skin	1 ea	202	72	212	5	49	5	<1	.1	t	.1
903	Flesh only	1 ea	156	74	156	3	36	2	<1	t	t	.1
904	Skin only	1 ea	58	63	77	3	17	3	<1	t	t	t
	Boiled, about 2½" diam:											
905	Peeled after boiling	1 ea	136	77	118	3	27	2	<1	t	t	.1
906	Peeled before boiling	1 ea	135	77	116	2	27	2	<1	t	t	.1
	French fried, strips 2–3½" long:											
907	Oven heated	10 ea	50	35	167	2	20	2	9	3	5.7	.7
908	Fried in vegetable oil	10 ea	50	38	158	2	20	2	8	1.9	4.7	.7
1188	Fried in veg and animal oil	10 ea	50	38	158	2	20	2	8	1.9	4.7	.7
909	Hashed browns from frozen	1 c	156	56	340	5	44	3	18	7	8	2.1
	Mashed:											
910	Home recipe with whole milk	½ c	105	78	81	2	18	2	1	.4	.2	.1
911	Home recipe with milk and marg	½ c	105	76	111	2	17	2	4	1.1	1.9	1.3
912	Prepared from flakes; water, milk, margarine, salt added	½ c	110	76	124	2	16	3	6	1.6	2.5	1.7
	Potato products, prepared:											
	Au gratin:											
913	From dry mix	½ c	123	79	114	3	16	1	5	3.5	1.5	.2
914	From home recipe, using butter	½ c	122	74	161	7	14	2	9	4.8	3.2	1.3
	Scalloped:											
915	From dry mix	½ c	122	79	113	3	16	1	5	3.2	1.5	.2
916	From home recipe, using butter	½ c	123	81	106	4	13	2	5	1.7	1.7	.9

Chol (mg)	Calc (mg)	Iron (mg)	Magn (mg)	Pota (mg)	Sodi (mg)	Zinc (mg)	VT-A (RE)	Thia (mg)	VT-E (α-TE)	Ribo (mg)	Niac (mg)	V-B6 (mg)	Fola (μg)	VT-C (mg)
0	21	2.16	46	239	3	1.11	2	.17	.24	.05	.43	.09	179	<1
0	105	.92	43	343	3	.84	65	.08	.18	.12	1.15	.05	104	2
0	20	1.8	42	319	4	1.21	7	.22	.33	.05	.62	.08	120	2
0	34	1.58	21	192	3	.3	10	.1	.31	.06	.43	.11	23	38
0	17	.81	14	147	214	.6	65	.1	.32	.07	.62	.05	38	8
0	22	1.26	21	124	11	.87	47	.14	.47	.09	1.04	.08	35	12
0	19	1.26	23	134	70	.75	54	.23	.14	.08	1.18	.09	47	8
0	21	1.02	12	98	2	.13	7	.07	.19	.04	.29	.08	20	29
0	15	.71	8	68	1	.09	5	.05	.13	.03	.2	.05	14	20
0	14	1.26	35	355	2	.98	1	.19	.38	.05	.87	.05	64	<1
0	18	.75	13	126	54	.36	621	.18	.26	.05	.92	.07	21	6
0	29	.96	18	128	333	.74	739	.09	.24	.07	.74	.11	23	8
0	5	.34	10	127	798	.12	41	.01	.47	.03	.54	.1	7	46
0	8	.54	11	153	3	.13	35	.04	.31	.04	.43	.12	10	109
0	2	.11	2	31	1	.03	97	.01	.06	.01	.09	.02	2	22
				2	2		30		.37					53
0	16	1.28	10	131	1136	.23	116	.03	.47	.03	.27	.13	10	7
0				55	390		10		.2					21
0	7	.34	7	131	1	.09	47	.05	.51	.02	.38	.18	16	66
0	6	.31	7	113	1	.08	40	.04	.47	.02	.32	.16	11	51
0	7	.34	7	133	1	.09	428	.05	.52	.02	.38	.19	16	143
0	7	.34	7	131	1	.09	422	.05	.51	.02	.38	.18	16	141
0	6	.31	7	113	1	.08	256	.04	.47	.02	.32	.16	11	116
0	20	.86	22	394	4	.32	45	.05	1.28	.05	1.66	.31	48	342
0	6	.24	6	110	1	.09	12	.01	.36	.01	.46	.09	13	96
0	41	2.22	47	398	2	.92	<1	.16	.8	.08	.34	.13	146	2
0	19	1.06	29	220	14	.26	2	.16	.22	.05	1.32	.33	26	5
0	20	2.75	54	844	16	.65	0	.22	.1	.07	3.33	.7	22	26
0	8	.55	39	610	8	.45	0	.16	.06	.03	2.18	.47	14	20
0	20	4.08	25	332	12	.28	0	.07	.02	.06	1.78	.36	12	8
0	22	2.5	54	903	16	.73	0	.24	.1	.06	3.45	.69	24	30
0	8	.64	39	641	11	.51	0	.2	.06	.04	2.54	.5	19	24
0	27	3.45	21	377	9	.3	0	.04	.02	.04	1.29	.28	10	9
0	7	.42	30	515	5	.41	0	.14	.07	.03	1.96	.41	14	18
0	11	.42	27	443	7	.36	0	.13	.07	.03	1.77	.36	12	10
0	6	.83	11	270	307	.2	0	.04	.25	.02	1.34	.11	11	3
0	9	.38	17	366	108	.19	0	.09	.25	.01	1.63	.12	14	5
6	9	.38	17	366	108	.19	0	.09	.25	.01	1.63	.12	14	5
0	23	2.36	26	680	53	.5	0	.17	.3	.03	3.78	.2	10	10
2	27	.28	19	314	318	.3	6	.09	.05	.04	1.18	.24	9	7
2	27	.27	19	303	310	.28	21	.09	.31	.04	1.13	.23	8	6
4	54	.24	20	256	365	.2	23	.12	.77	.05	.74	.01	8	11
18	102	.39	18	269	540	.29	38	.02	1.48	.1	1.15	.05	8	4
18	145	.78	24	483	528	.84	46	.08	.64	.14	1.21	.21	13	12
13	44	.46	17	248	416	.3	26	.02	.18	.07	1.26	.05	12	4
7	70	.7	23	465	412	.49	23	.08	.4	.11	1.29	.22	13	13

H

Table H-1

Food Composition (Computer code number is for West Diet Analysis program) (For purposes of calculations, use "0" for t, <1, <.1, <.01, etc.)

Computer Code Number	Food Description	Measure	Wt (g)	H_2O (%)	Ener (kcal)	Prot (g)	Carb (g)	Dietary Fiber (g)	Fat (g)	Fat Breakdown (g) Sat	Mono	Poly
	VEGETABLES and LEGUMES—Continued											
	Potato Salad (see Mixed Dishes #715)											
1192	Potato puffs, cooked from frozen	½ c	64	53	142	2	19	2	7	3.3	2.8	.5
918	Pumpkin, cooked from fresh, mashed	½ c	123	94	25	1	6	1	<1	t	t	t
919	Pumpkin, canned	½ c	123	90	42	1	10	4	<1	.2	t	t
1891	Radicchio, raw, shredded	½ c	20	93	5	<1	1	<1	<1	t	t	t
1894	Radicchio, raw, leaf	10 ea	80	93	18	1	4	1	<1	t	t	.1
920	Red radishes	10 ea	45	95	8	<1	2	1	<1	t	t	t
1793	Daikon radishes (Chinese) raw	½ c	44	95	8	<1	2	1	<1	t	t	t
921	Refried beans, canned	½ c	126	76	118	7	19	7	2	.6	.7	.2
1375	Rutabaga, cooked cubes	½ c	85	89	33	1	7	2	<1	t	t	.1
922	Sauerkraut, canned with liquid	½ c	118	92	22	1	5	3	<1	t	t	.1
923	Seaweed, kelp, raw	½ c	40	82	17	1	4	1	<1	.1	t	t
924	Seaweed, spirulina, dried	½ c	8	5	23	5	2	<1	1	.2	.1	.2
1866	Shallots, raw, chopped	1 tbs	10	80	7	<1	2	<1	<1	t	t	t
1557	Snow peas, stir-fried	½ c	83	89	35	2	6	2	<1	t	t	.1
925	Soybeans, cooked from dry	½ c	86	63	149	15	9	5	8	1.1	1.7	4.4
1996	Soybeans, dry roasted	½ c	86	1	387	34	28	7	19	2.7	4.1	10.6
	Soybean products:											
926	Miso	½ c	138	41	284	17	39	7	8	1.2	1.9	4.7
	Soy milk (see #144 and #2301 under Dairy)											
	Tofu (soybean curd):											
7540	Extra firm, silken	½ c	126	88	69	9	3	<1	2	.4	.4	1.3
7542	Firm, silken	½ c	126	87	77	9	3	<1	3	.5	.7	1.9
927	Regular	½ c	124	87	76	8	2	<1	5	.7	1	2.6
7541	Soft, silken	½ c	124	89	68	6	4	<1	3	.4	.6	1.9
	Spinach:											
928	Raw, chopped	½ c	28	92	6	1	1	1	<1	t	t	t
929	Cooked, from fresh, drained	½ c	90	91	21	3	3	2	<1	t	t	.1
930	Cooked from frozen (leaf)	½ c	95	90	27	3	5	3	<1	t	t	.1
931	Canned, drained solids:	½ c	107	92	25	3	4	3	1	.1	t	.2
5149	Unsalted	½ c	107	92	25	3	4	3	1	.1	t	.2
	Spinach soufflé (see Mixed Dishes)											
	Squash, summer varieties,cooked w/skin:											
932	Varieties averaged	½ c	90	94	18	1	4	1	<1	.1	t	.1
933	Crookneck	½ c	90	94	18	1	4	1	<1	.1	t	.1
934	Zucchini	½ c	90	95	14	1	4	1	<1	t	t	t
	Squash, winter varieties, cooked:											
	Average of all varieties, baked:											
935	Mashed	1 c	245	89	96	2	21	7	2	.3	.1	.6
936	Cubes	1 c	205	89	80	2	18	6	1	.3	.1	.5
937	Acorn, baked, mashed	½ c	123	83	69	1	18	5	<1	t	t	.1
1218	Acorn, boiled, mashed	½ c	122	90	41	1	11	3	<1	t	t	t
	Butternut squash:											
938	Baked cubes	½ c	103	88	41	1	11	3	<1	t	t	t
1219	Baked, mashed	½ c	103	88	41	1	11	3	<1	t	t	t
1193	Cooked from frozen	½ c	120	88	47	1	12	3	<1	t	t	t
1194	Hubbard, baked, mashed	½ c	120	85	60	3	13	3	1	.2	.1	.3
1195	Hubbard, boiled, mashed	½ c	118	91	35	2	8	3	<1	.1	t	.2
1196	Spaghetti, baked or boiled	½ c	77	92	22	<1	5	1	<1	t	t	.1
1189	Succotash, cooked from frozen	½ c	85	74	79	4	17	3	1	.1	.1	.4
	Sweet potatoes:											
939	Baked in skin, peeled, 5 x 2″ diam	1 ea	114	73	117	2	28	3	<1	t	t	.1
940	Boiled without skin, 5 x 2″ diam	1 ea	151	73	159	2	37	3	<1	.1	t	.2
941	Candied, 2½ x 2″	1 pce	105	67	144	1	29	3	3	1.4	.7	.2
	Canned:											
942	Solid pack	½ c	128	74	129	3	30	2	<1	.1	t	.1
943	Vacuum pack, mashed	½ c	127	76	116	2	27	2	<1	.1	t	.1
944	Vacuum pack, 3¾ x 1″	2 pce	80	76	73	1	17	1	<1	t	t	.1
1940	Taro shoots, cooked slices	1 c	140	95	20	1	4	1	<1	t	t	t
1941	Taro, tahitian, cooked slices	1 c	137	86	60	6	9	1	1	.2	.1	.4
	Tomatillos:											
1877	Raw, each	1 ea	34	92	11	<1	2	1	<1	t	.1	.1
1875	Raw, chopped	1 c	132	92	42	1	8	3	1	.2	.2	.6

Chol (mg)	Calc (mg)	Iron (mg)	Magn (mg)	Pota (mg)	Sodi (mg)	Zinc (mg)	VT-A (RE)	Thia (mg)	VT-E (α-TE)	Ribo (mg)	Niac (mg)	V-B6 (mg)	Fola (μg)	VT-C (mg)
0	19	1	12	243	477	.19	1	.12	.03	.05	1.38	.15	11	4
0	18	.7	11	283	1	.28	1330	.04	1.3	.1	.51	.05	10	6
0	32	1.71	28	253	6	.21	2713	.03	1.3	.07	.45	.07	15	5
0	4	.11	3	60	4	.12	1	<.01	.45	.01	.05	.01	12	2
0	15	.45	10	242	18	.5	2	.01	1.81	.02	.2	.05	48	6
0	9	.13	4	104	11	.13	<1	<.01	0	.02	.13	.03	12	10
0	12	.18	7	100	9	.07	0	.01	0	.01	.09	.02	12	10
10	44	2.09	42	336	377	1.47	0	.03	.39	.02	.4	.18	14	8
0	41	.45	20	277	17	.3	48	.07	.13	.03	.61	.09	13	16
0	35	1.73	15	201	780	.22	2	.02	.12	.03	.17	.15	28	17
0	67	1.14	48	36	93	.49	5	.02	.35	.06	.19	<.01	72	1
0	10	2.28	16	109	84	.16	5	.19	.4	.29	1.02	.03	8	1
0	4	.12	2	33	1	.04	125	.01	.01	<.01	.02	.03	3	1
0	36	1.73	20	166	3	.22	11	.11	.32	.06	.47	.13	28	42
0	88	4.42	74	443	1	.99	1	.13	1.68	.24	.34	.2	46	1
0	120	3.4	196	1173	2	4.1	2	.37	3.96	.65	.91	.19	176	4
0	91	3.78	58	226	5032	4.58	12	.13	.01	.34	1.19	.3	45	0
0	39	1.5	34	195	80	.76	0	.1	.18	.04	.3	.01		0
0	41	1.3	34	244	45	.77	0	.13	.24	.05	.31	.01		0
0	138	1.38	33	149	10	.79	1	.06	.01	.05	.66	.06	55	<1
0	38	1.02	36	223	6	.64	0	.12	.25	.05	.37	.01		0
0	28	.76	22	156	22	.15	188	.02	.53	.05	.2	.05	54	8
0	122	3.21	78	419	63	.68	737	.09	.86	.21	.44	.22	131	9
0	139	1.44	66	283	82	.66	739	.06	.91	.16	.4	.14	103	12
0	136	2.46	81	370	29	.49	939	.02	1.39	.15	.41	.11	105	15
0	136	2.46	81	370	29	.49	939	.02	1.39	.15	.41	.11	105	15
0	24	.32	22	173	1	.35	26	.04	.11	.04	.46	.06	18	5
0	24	.32	22	173	1	.35	26	.04	.11	.04	.46	.08	18	5
0	12	.31	20	228	3	.16	22	.04	.11	.04	.38	.07	15	4
0	34	.81	20	1070	2	.64	872	.21	.29	.06	1.72	.18	69	23
0	29	.68	16	896	2	.53	730	.17	.25	.05	1.44	.15	57	20
0	54	1.14	53	538	5	.21	53	.2	.15	.02	1.08	.24	23	13
0	32	.68	32	321	4	.13	32	.12	.15	.01	.65	.14	14	8
0	42	.62	30	293	4	.13	721	.07	.17	.02	1	.13	20	16
0	42	.62	30	293	4	.13	721	.07	.17	.02	1	.13	20	16
0	23	.7	11	160	2	.14	401	.06	.16	.05	.56	.08	20	4
0	20	.56	26	430	10	.18	725	.09	.14	.06	.67	.21	19	11
0	12	.33	15	253	6	.12	473	.05	.14	.03	.39	.12	11	8
0	16	.26	8	90	14	.15	8	.03	.09	.02	.62	.08	6	3
0	13	.76	20	225	38	.38	20	.06	.31	.06	1.11	.08	28	5
0	32	.51	23	397	11	.33	2487	.08	.32	.14	.69	.27	26	28
0	32	.85	15	278	20	.41	2574	.08	.42	.21	.97	.37	17	26
8	27	1.19	12	198	73	.16	440	.02	3.99	.04	.41	.04	12	7
0	38	1.7	31	269	96	.27	1936	.03	.35	.11	1.22	.3	14	7
0	28	1.13	28	396	67	.23	1013	.05	.32	.07	.94	.24	21	33
0	18	.71	18	250	42	.14	638	.03	.2	.05	.59	.15	13	21
0	20	.57	11	482	3	.76	7	.05	1.4	.07	1.13	.16	4	26
0	204	2.14	70	854	74	.14	241	.06	3.7	.27	.66	.16	10	52
0	2	.21	7	91	<1	.07	4	.01	.13	.01	.63	.02	2	4
0	9	.82	26	354	1	.29	14	.06	.5	.05	2.44	.07	9	15

Table H-1

Food Composition (Computer code number is for West Diet Analysis program) (For purposes of calculations, use "0" for t, <1, <.1, <.01, etc.)

Computer Code Number	Food Description	Measure	Wt (g)	H_2O (%)	Ener (kcal)	Prot (g)	Carb (g)	Dietary Fiber (g)	Fat (g)	Fat Breakdown (g) Sat	Mono	Poly
	VEGETABLES and LEGUMES—Continued											
	Tomatoes:											
945	Raw, whole, 2 ⅗" diam	1 ea	123	94	26	1	6	1	<1	.1	.1	.2
946	Raw, chopped	1 c	180	94	38	2	8	2	1	.1	.1	.2
947	Cooked from raw	1 c	240	92	65	3	14	2	1	.1	.2	.4
948	Canned, solids and liquid:	1 c	240	94	46	2	10	2	<1	t	t	.1
5741	Unsalted	1 c	240	94	46	2	10	2	<1	t	t	.1
1879	Tomatoes, sundried:	1 c	54	15	139	8	30	7	2	.2	.3	.6
1881	Pieces	10 pce	20	15	52	3	11	2	1	.1	.1	.2
1885	Oil pack, drained	10 pce	30	54	64	2	7	2	4	.6	2.6	.6
2020	Tomato, raw	1 ea	123	94	26	1	6	1	<1	.1	.1	.2
949	Tomato juice, canned:	1 c	243	94	41	2	10	1	<1	t	t	.1
5397	Unsalted	1 c	243	94	41	2	10	2	<1	t	t	.1
	Tomato products, canned:											
950	Paste, no added salt	1 c	262	74	215	10	51	11	1	.2	.2	.6
951	Puree, no added salt	1 c	250	87	100	4	24	5	<1	.1	.1	.2
952	Sauce, with salt	1 c	245	89	73	3	18	3	<1	.1	.1	.2
953	Turnips, cubes, cooked from fresh	1 c	156	94	33	1	8	3	<1	t	t	.1
	Turnip greens, cooked:											
954	From fresh, leaves and stems	1 c	144	93	29	2	6	5	<1	.1	t	.1
955	From frozen, chopped	1 c	164	90	49	6	8	6	1	.2	t	.3
956	Vegetable juice cocktail, canned	1 c	242	93	46	2	11	2	<1	t	t	.1
	Vegetables, mixed:											
957	Canned, drained	½ c	81	87	38	2	7	2	<1	t	t	.1
958	Frozen, cooked, drained	½ c	91	83	54	3	12	4	<1	t	t	.1
1818	Water chestnuts, Chinese, raw	½ c	62	73	60	1	15	2	<1	t	t	t
	Water chestnuts, canned:											
959	Slices	½ c	70	86	35	1	9	2	<1	t	t	t
960	Whole	4 ea	28	86	14	<1	3	1	<1	t	t	t
1190	Watercress, fresh, chopped	½ c	17	95	2	<1	<1	<1	<1	t	t	t
	VEGETARIAN FOODS:											
7509	Bacon strips, meatless	3 ea	15	49	46	2	1	<1	4	.7	1.1	2.3
1511	Baked beans, canned	½ c	127	73	118	6	26	6	1	.1	t	.2
7526	Bakon crumbles	¼ c	7	16	28	2	1	<1	2			
7548	Chicken, breaded, fried, meatless	1 pce	57	70	97	6	3	3	7	1	2.9	2.5
7547	Chicken slices, meatless	2 ea	60	59	132	10	4	3	8	1.3	2	4.4
7557	Chili w/meat substitute	½ c	107	64	141	19	15	4	2	.3	.6	.9
7549	Fish stick, meatless	2 ea	57	45	165	13	5	3	10	1.6	2.5	5.4
7550	Frankfurter, meatless	1 ea	51	58	102	10	4	2	5	.8	1.2	2.7
7504	GardenBurger, patty	1 ea	45	53	87	5	13	3	1	.4	.3	.7
7505	GardenSausage, patty	1 ea	35	15	117	4	22	5	1	.7	.3	t
7551	Luncheon slice, meatless	1 sl	67	46	188	17	6	3	11	1.7	2.6	5.6
7560	Meatloaf, meatless	1 ea	71	58	142	15	6	3	6	1	1.5	3.3
1171	Nuteena	1 ea	55	58	162	6	6	2	13	5.1	5.8	1.7
7556	Pot pie, meatless	1 ea	227	59	524	15	41	5	34	9.5	12.6	9.8
7554	Soyburger, patty	1 ea	71	58	142	15	6	3	6	1	1.5	3.3
7562	Soyburger w/cheese, patty	1 ea	135	50	316	21	29	4	13	4.2	3.9	3.7
7564	Tempeh	1 c	166	55	330	31	28	9	13	1.9	2.9	7.2
7670	Vegan burger, patty	1 ea	78	71	75	11	6	4	<1	.1	.3	.2
	Vegetarian foods, Green Giant:											
7677	Breakfast links	3 ea	68	65	114	12	5	4	5	.7	1.2	3.1
7676	Breakfast patties	2 ea	57	65	95	10	5	3	4	.6	1	2.6
	Burger, harvest, patty:											
7673	Italian	1 ea	90	65	139	17	8	5	4	1.4	.3	.4
7674	Original	1 ea	90	65	137	18	8	5	4	1.3	.1	.4
7675	Southwestern	1 ea	90	65	135	16	9	5	4	1.4	.2	.4
	Vegetarian foods, Loma Linda											
7727	Chik nuggets, frozen	5 pce	85	47	245	12	13	5	16	2.5	4	8.8
7753	Chik-fried, frozen	1 pce	57	51	178	11	1	1	15	1.9	3.7	8.7
7744	Franks, big, canned	1 ea	51	59	110	10	2	2	7	1.1	1.7	3.8
7747	Linketts, canned	1 ea	35	60	72	7	1		4	.7	1.2	2.5
1173	Redi-burger, patty	1 ea	85	59	172	16	5		10	1.5	2.4	5.8
7755	Swiss steak w/gravy, canned	1 pce	92	71	120	9	8	4	6	.8	1.5	3.3

Chol (mg)	Calc (mg)	Iron (mg)	Magn (mg)	Pota (mg)	Sodi (mg)	Zinc (mg)	VT-A (RE)	Thia (mg)	VT-E (α-TE)	Ribo (mg)	Niac (mg)	V-B6 (mg)	Fola (µg)	VT-C (mg)
0	6	.55	13	273	11	.11	76	.07	.47	.06	.77	.1	18	23
0	9	.81	20	400	16	.16	112	.11	.68	.09	1.13	.14	27	34
0	14	1.34	34	670	26	.26	178	.17	.91	.14	1.8	.23	31	55
0	72	1.32	29	530	355	.38	144	.11	.77	.07	1.76	.22	19	34
0	72	1.32	29	545	24	.38	144	.11	.91	.07	1.76	.22	19	34
0	59	4.91	105	1850	1131	1.07	47	.28	<.01	.26	4.89	.18	37	21
0	22	1.82	39	685	419	.4	17	.11	<.01	.1	1.81	.07	14	8
0	14	.8	24	470	80	.23	39	.06	.16	.11	1.09	.1	7	31
0	6	.55	13	273	11	.11	76	.07	.47	.06	.77	.1	18	23
0	22	1.41	27	535	877	.34	136	.11	2.21	.07	1.64	.27	48	44
0	22	1.41	27	535	24	.34	136	.11	2.21	.07	1.64	.27	48	44
0	92	5.08	134	2454	231	2.1	639	.41	11.3	.5	8.44	1	59	111
0	42	3.1	60	1065	85	.55	320	.18	6.3	.13	4.3	.38	27	26
0	34	1.89	47	909	1482	.61	240	.16	3.43	.14	2.82	.38	23	32
0	34	.34	12	211	78	.31	0	.04	.05	.04	.47	.1	14	18
0	197	1.15	32	292	42	.2	792	.06	2.48	.1	.59	.26	170	39
0	249	3.18	43	367	25	.67	1308	.09	4.79	.12	.77	.11	65	36
0	27	1.02	27	467	653	.48	283	.1	.77	.07	1.76	.34	51	67
0	22	.85	13	236	121	.33	944	.04	.49	.04	.47	.06	19	4
0	23	.75	20	154	32	.45	389	.06	.33	.11	.77	.07	17	3
0	7	.04	14	362	9	.31	0	.09	.74	.12	.62	.2	10	2
0	3	.61	3	83	6	.27	0	.01	.35	.02	.25	.11	4	1
0	1	.24	1	33	2	.11	0	<.01	.14	.01	.1	.04	2	<1
0	20	.03	4	56	7	.02	80	.01	.17	.02	.03	.02	2	7
0	3	.36	3	25	220	.06	1	.66	1.04	.07	1.13	.07	6	0
0	63	.37	41	376	504	1.78	22	.19	.67	.08	.54	.17	30	4
	8	.44	11	120	172	.25	0	.06		.02	.12	.02	7	0
0	13	.97	7	171	228	.37	0	.4	1.11	.27	2.68	.28	32	0
0	21	.78	10	198	474	.42	0	.66	1.61	.24	3.18	.42	46	0
0	53	4.24	36	362	527	1.26	78	.12	1.25	.07	1.21	.15	82	16
0	54	1.14	13	342	279	.8	0	.63	2.25	.51	6.84	.85	58	0
0	17	.92	9	76	219	.61	0	.56	.98	.61	8.16	.5	40	0
0	36	1.35		129	112		18	.05		.09		.06		<1
0	181	.33		307	78		3	.08		.2		.13		<1
0	27	1.54	15	188	576	1.07	0	.64	2.01	.37	7.37	.74	67	0
0	21	1.49	13	128	391	1.28	0	.64	1.23	.43	7.1	.85	55	0
0	9	.27	33	166	119	.46	0	.1		.35	1.04	.45	49	0
20	66	2.9	31	331	538	1.05	729	.65	4	.4	4.47	.31	40	10
0	21	1.49	13	128	391	1.28	0	.64	1.23	.43	7.1	.85	55	0
13	146	2.71	26	211	931	1.97	45	.77	1.43	.55	8.14	.86	69	1
0	154	3.75	116	609	10	3	115	.22	.03	.18	7.69	.5	86	0
0	79	2.66	15	398	351	.69	0	.23	.01	.51	3.78	.18	225	0
0	65	1.84			340	4.56	0	.18		.09	.27	.18		0
0	54	2			285	3.82	0	.15		.07	2.28	.15		0
0	74	2.61			374	6.93	3	.28		.14	4.05	.28		0
0	76	2.7			378	7.2	0	.29		.14	4.32	.29		0
0	71	2.52			371	6.66	3	.27		.13	4.05	.27		0
2	40	1.4		153	709	.43	0	.67		.3	2.89	.45		0
4	2	.63		76	503	.2	0	.98		.46	2.1	.35		0
2	8	.77		51	243	.89	0	.26		.46	1.98	.14		0
1	4	.39		29	160	.46	0	.13		.22	.64	.29		0
1	12	1.06	16	121	455	1.11	0	.14		.3	1.9	.51	21	0
2	24	.31		225	433	.41	0	1.25		.65	5.41	1		0

Table H-1

Food Composition (Computer code number is for West Diet Analysis program) (For purposes of calculations, use "0" for t, <1, <.1, <.01, etc.)

Computer Code Number	Food Description	Measure	Wt (g)	H_2O (%)	Ener (kcal)	Prot (g)	Carb (g)	Dietary Fiber (g)	Fat (g)	Fat Breakdown (g) Sat	Mono	Poly
	VEGETARIAN FOODS:—Continued											
1174	Vege-Burger, patty	1 ea	55	71	66	10	2	2	2	.4	.6	.5
	Vegetarian foods, Morningstar Farms:											
7672	Better-n-burgers, svg	1 ea	78	71	75	11	6	4	<1	.1	.3	.2
7766	Better-n-eggs	¼ c	57	88	23	5	<1	0	<1	.1	.1	.1
57436	Breakfast links	2 pce	45	60	63	8	2	2	2	.5	.7	1.3
7752	Breakfast strips	2 pce	16	43	56	2	2	<1	4	.7	1.1	2.6
7725	Burger crumbles, svg	1 ea	55	60	116	11	3	3	6	1.6	2.3	2.5
7726	Burger, spicy black bean	1 ea	78	60	113	11	15	5	1	.2	.3	.4
7665	Chik pattie	1 ea	71	51	177	7	15	2	10	1.3	2.6	5.9
7724	Frank, deli	1 ea	45	52	109	10	3	3	7	1	2.1	3.5
7722	Garden vege pattie	1 ea	67	60	104	11	9	4	4	.5	1.1	2.2
7746	Grillers	1 ea	64	55	139	14	5	3	7	1.7	2.2	3
7664	Prime pattie	1 ea	64	64	94	16	4	3	2	.2	.4	.6
	Vegetarian foods, Worthington:											
7634	Beef style, meatless, frzn	3 pce	55	58	113	9	4	3	7	1.2	2.7	2.6
7732	Burger, meatless, patty	¼ c	55	71	60	9	2	1	2	.3	.5	1.1
1846	Chik slices, canned	2 pce	60	78	62	6	1	1	4	.6	.9	2.3
1833	Chili, canned	½ c	106	73	136	9	10	4	7	1.1	1.7	4.1
1835	Choplets, slices, canned	2 pce	92	72	93	17	3	2	2	.9	.3	.3
7608	Corned beef style, meatless, frzn	4 pce	57	55	138	10	5	2	9	1.9	4.1	3.1
1831	Country stew, canned	1 c	240	81	208	13	20	5	9	1.6	2.3	4.8
7632	Egg roll, meatless, frzn	1 ea	85	53	181	6	20	2	8	1.7	4.5	2.3
1838	Numete, slices, canned	1 pce	55	58	132	6	5	3	10	2.4	4.4	2.7
1839	Prime steaks, slices, canned	1 pce	92	71	136	9	4	4	9	1.4	2.9	4.9
1840	Protose, slices, canned	1 pce	55	53	131	13	5	3	7	1	3	2.4
7606	Roast, dinner, meatless, frzn	1 ea	85	63	180	12	5	3	12	2.2	5	5.2
1842	Saucette links, canned	1 pce	38	62	86	6	1		6	1.1	1.6	3.8
1844	Savory slices, canned	1 pce	28	66	48	3	2	1	3	1.2	1.3	.6
7735	Steaklets, frzn	1 pce	71	58	145	12	6	2	8	1.4	2.7	3.9
1847	Turkee slices, canned	1 pce	33	64	68	5	1	1	5	.8	1.9	2.1
	MISCELLANEOUS											
	Baking powders for home use:											
	Sodium aluminum sulfate:											
962	With monocalcium phosphate monohydrate	1 tsp	5	2	6	<1	2	0	0	0	0	0
963	With monocalcium phosphate monohydrate, calcium sulfate	1 tsp	5	5	3	0	1	<1	0	0	0	0
964	Straight phosphate	1 tsp	5	4	3	<1	1	<1	0	0	0	0
965	Low sodium	1 tsp	5	6	5	<1	2	<1	<1	t	0	t
1204	Baking soda	1 tsp	5		0	0	0	0	0	0	0	0
966	Basil, dried	1 tbs	5	6	13	1	3	2	<1	t	t	.1
2068	Cajun seasoning	1 tsp	3	5	6	<1	1	<1	<1			
961	Carob flour	1 c	103	4	185	5	92	41	1	.1	.2	.2
967	Catsup:	¼ c	61	67	63	1	17	1	<1	t	t	.1
968	Tablespoon	1 tbs	15	67	16	<1	4	<1	<1	t	t	t
1200	Cayenne/red pepper	1 tbs	5	8	16	1	3	1	1	.2	.1	.4
969	Celery seed	1 tsp	2	6	8	<1	1	<1	<1	t	.3	.1
1203	Chili powder:	1 tbs	8	8	25	1	4	3	1	.2	.3	.6
970	Teaspoon	1 tsp	3	8	9	<1	2	1	<1	.1	.1	.2
	Chocolate:											
971	Baking, unsweetened, square	1 oz	28	1	146	3	8	4	15	9.1	5.2	.5
	For other chocolate items, see Sweeteners & Sweets											
972	Cilantro/coriander, fresh	1 tbs	1	93		<1	<1	<1	<1	0	t	0
2287	Cinnamon	1 tsp	2	10	5	<1	2	1	<1	t	t	t
1197	Cornstarch	1 tbs	8	8	30	<1	7	<1	<1	t	t	t
2239	Curry powder	1 tsp	2	10	6	<1	1	1	<1	t	.1	.1
1202	Dill weed, dried	1 tbs	3	7	8	1	2	<1	<1	t	.1	t
975	Garlic cloves	1 ea	3	59	4	<1	1	<1	<1	t	0	t
2238	Garlic powder	1 tsp	3	6	10	<1	2	<1	<1	t	t	t
977	Gelatin, dry, unsweetened: Envelope	1 ea	7	13	23	6	0	0	<1	t	t	t
978	Ginger root, slices, raw	2 pce	5	82	3	<1	1	<1	<1	t	t	t

Chol (mg)	Calc (mg)	Iron (mg)	Magn (mg)	Pota (mg)	Sodi (mg)	Zinc (mg)	VT-A (RE)	Thia (mg)	VT-E (α-TE)	Ribo (mg)	Niac (mg)	V-B6 (mg)	Fola (μg)	VT-C (mg)
0	8	.5	12	30	114	.58	0	.2		.25	.78	.31	15	0
0	79	2.66	15	398	351	.69	0	.23	.01	.51	3.78	.18	225	0
2	7	.63		68	90	.51	64	.01		.26	0	.11		0
1	15	2.14	16	59	338	.36	0	6.95		.22	5.19	.33	12	0
<1	7	.27		15	220	.05	0	.75		.04	.6	.07		0
0	40	3.2	1	89	238	.82	0	4.96	.34	.18	1.49	.27		0
1	56	1.84	44	269	499	.93	14	8.03	.36	.14	0	.21		0
1	11	1.02		163	536	.31	0	2.15		.16	1.51	.14		0
2	16	.26	4	50	524	.38	0	.14	1.26	.02	0	.01		0
1	34	.72	29	180	382	.59	20	6.47	.98	.1	0	0	29	0
2	43	1.16		127	256	.49	0	11.7		.24	2.99	.37		0
1	46	2.14		142	247	.74	0	.51		.25	.92	.41		2
0	4	2.63		44	624	.22	0	.89		.34	6.46	.56		0
0	4	1.73		25	269	.38	0	.13		.1	1.96	.24		0
1	9	.73		111	257	.26	0	.06		.05	.37	.08		0
0	20	1.49		195	523	.57	0	.02		.03	1.04	.31		0
0	6	.37		40	500	.65	0	.05		.05	0	.05		0
1	6	1.17		58	524	.26	0	10.6		.07	1.36	.3		0
2	51	5.09		270	826	1.03	216	1.85		.29	4.22	.86		0
1	15	.57		96	384	.31	0	1.22		.19	0	.03		0
0	10	1.12		155	272	.56	0	.08		.06	.54	.2		0
2	12	.38		82	445	.38	0	.12		.13	1.98	.38		0
<1	1	1.84		50	283	.7	0	.18		.13	1.34	.24		0
2	36	2.87		38	566	.64	0	2.13		.25	6.02	.6		0
1	9	1.15		25	205	.26	0	.59		.08	.09	.13		0
<1		.47		14	179	.08	0	.08		.06	.48	.1		0
2	49	.99		95	484	.5	0	1.51		.12	3.1	.26		0
1	3	.47		16	203	.11	0	1.13		.05	.39	.09		0
0	97	0		7	547	0	0	0	0	0	0	0	0	0
0	294	.55	1	1	530	<.01	0	0	0	0	0	0	0	0
0	368	.56	2		395	<.01	0	0	0	0	0	0	0	0
0	217	.41	1	505	4	.04	0	0	<.01	0	0	0	0	0
0	0	0	0	0	1368	0	0	0	0	0	0	0	0	0
0	106	2.1	21	172	2	.29	47	.01	.08	.02	.35	.06	14	3
				29	474									
0	358	3.03	56	852	36	.95	1	.05	.65	.47	1.96	.38	30	<1
0	12	.43	13	293	723	.14	62	.05	.9	.04	.84	.11	9	9
0	3	.1	3	72	178	.03	15	.01	.22	.01	.21	.03	2	2
0	7	.39	8	101	1	.12	208	.02	.24	.05	.43	.1	5	4
0	35	.9	9	28	3	.14	<1	.01	.02	.01	.06	.01	<1	<1
0	22	1.14	14	153	81	.22	279	.03	.08	.06	.63	.15	8	5
0	8	.43	5	57	30	.08	105	.01	.03	.02	.24	.06	3	2
0	21	1.77	87	233	4	1.12	3	.02	.34	.05	.31	.03	2	0
0	1	.02		5	<1	<.01	3	<.01	.02	<.01	.01	<.01	<1	<1
0	25	.76	1	10	1	.04	1	<.01	0	<.01	.03	<.01	1	1
0		.04			1	<.01	0	0	0	0	0	0	0	0
0	10	.59	5	31	1	.08	2	<.01	.01	.01	.07	.01	3	<1
0	53	1.46	13	99	6	.1	18	.01		.01	.08	.04		1
0	5	.05	1	12	1	.03	0	.01	0	<.01	.02	.04	<1	1
0	2	.08	2	33	1	.08	0	.01	0	<.01	.02	.08	<1	1
0	4	.08	2	1	14	.01	0	<.01	0	.02	.01	0	2	0
0	1	.02	2	21	1	.02	0	<.01	.01	<.01	.03	.01	1	<1

Table H-1

Food Composition (Computer code number is for West Diet Analysis program) (For purposes of calculations, use "0" for t, <1, <.1, <.01, etc.)

Computer Code Number	Food Description	Measure	Wt (g)	H_2O (%)	Ener (kcal)	Prot (g)	Carb (g)	Dietary Fiber (g)	Fat (g)	Fat Breakdown (g) Sat	Mono	Poly
	MISCELLANEOUS—Continued											
1198	Horseradish, prepared	1 tbs	15	85	7	<1	2	<1	<1	t	t	.1
1997	Hummous/hummus	1 c	246	65	421	12	50	12	21	3.1	8.8	7.8
1909	Mustard, country dijon	1 tsp	5		5	<1	<1	0	0	0	0	0
2019	Mustard, gai choy Chinese	1 tbs	16	94	3	<1	1		<1			
979	Mustard, prepared (1 packet = 1 tsp)	1 tsp	5	80	4	<1	<1	<1	<1	t	.2	t
	Miso (see #926 under Vegetables and Legumes, Soybean products)											
980	Olives, green	5 ea	20	78	23	<1	<1	<1	3	.3	1.9	.2
981	Olives, ripe, pitted	5 ea	22	80	25	<1	1	1	2	.3	1.7	.2
26008	Onion powder	1 tsp	2	5	7	<1	2	<1	<1	t	t	t
2237	Oregano, ground	1 tsp	2	7	6	<1	1	1	<1	.1	t	.1
2236	Paprika	1 tsp	2	10	6	<1	1	<1	<1	t	t	.2
887	Parsley, freeze dried	¼ c	1	2	3	<1	<1	<1	<1	t	t	t
	Parsley, fresh (see #885 and #886)											
985	Pepper, black	1 tsp	2	10	5	<1	1	1	<1	t	t	t
	Pickles:											
986	Dill, medium, 3¾ x 1¼" diam	1 ea	65	92	12	<1	3	1	<1	t	t	t
987	Fresh pack, slices, 1½" diam x ¼"	2 pce	15	79	11	<1	3	<1	<1	0	0	t
988	Sweet, medium	1 ea	35	65	41	<1	11	<1	<1	t	t	t
989	Pickle relish, sweet	1 tbs	15	63	21	<1	5	<1	<1	t	t	t
	Popcorn (see Grain Products #539-541)											
917	Potato chips:	10 pce	20	2	107	1	11	1	7	2.2	2	2.4
44076	Unsalted	1 oz	28	2	150	2	15	1	10	3.1	2.8	3.4
1201	Sage, ground	1 tsp	1	8	3	<1	1	<1	<1	.1	t	t
1347	Salsa, from recipe	1 tbs	15	93	3	<1	1	<1	<1	t	t	t
2218	Salsa, pico de gallo, medium	1 tbs	15	92	2	0	1	<1	0	0	0	0
990	Salt	1 tsp	6		0	0	0	0	0	0	0	0
	Salt Substitutes:											
1205	Morton, salt substitute	1 tsp	6			0	<1		0	0	0	0
1207	Morton, light salt	1 tsp	6			0	<1		0	0	0	0
2067	Seasoned salt, no MSG	1 tsp	5	5	4	<1	1	<1	<1			
991	Vinegar, cider	½ c	120	94	17	0	7	0	0	0	0	0
2172	Balsamic	1 tbs	15	64	21	0	4	0	0	0	0	0
2176	Malt	1 tbs	15	90	5	0	<1	0	0	0	0	0
2182	Tarragon	1 tbs	15	95	3	0	<1	0	0	0	0	0
2181	White wine	1 tbs	15	89	5	0	<1	0	0	0	0	0
	Yeast:											
992	Baker's, dry, active, package	1 ea	7	8	21	3	3	1	<1	t	.2	t
993	Brewer's, dry	1 tbs	8	5	23	3	3	3	<1	t	t	0
	SOUPS, SAUCES, and GRAVIES											
	SOUPS, canned, condensed:											
	Unprepared, condensed:											
1210	Cream of celery	1 c	251	85	181	3	18	2	11	2.8	2.6	5
1215	Cream of chicken	1 c	251	82	233	7	18	<1	15	4.2	6.5	3
1216	Cream of mushroom	1 c	251	81	259	4	19	1	19	5.1	3.6	8.9
1220	Onion	1 c	246	86	113	8	16	2	3	.5	1.5	1.3
	Prepared w/equal volume of whole milk:											
994	Clam chowder, New England	1 c	248	85	164	9	17	1	7	2.9	2.3	1.1
1209	Cream of celery	1 c	248	86	164	6	14	1	10	3.9	2.5	2.6
995	Cream of chicken	1 c	248	85	191	8	15	<1	11	4.6	4.5	1.6
996	Cream of mushroom	1 c	248	85	203	6	15	<1	14	5.1	3	4.6
1214	Cream of potato	1 c	248	87	149	6	17	<1	6	3.8	1.7	.6
1213	Oyster stew	1 c	245	89	135	6	10	0	8	5	2.1	.3
997	Tomato	1 c	248	85	161	6	22	3	6	2.9	1.6	1.1
	Prepared with equal volume of water:											
998	Bean with bacon	1 c	253	84	172	8	23	9	6	1.5	2.2	1.8
999	Beef broth/bouillon/consommé	1 c	240	98	17	3	<1	0	1	.3	.2	t
1000	Beef noodle	1 c	244	92	83	5	9	1	3	1.1	1.2	.5
1001	Chicken noodle	1 c	241	92	75	4	9	1	2	.7	1.1	.6
1002	Chicken rice	1 c	241	94	60	4	7	1	2	.5	.9	.4
1208	Chili beef	1 c	250	85	170	7	21	9	7	3.3	2.8	.3
1003	Clam chowder, Manhattan	1 c	244	92	78	2	12	1	2	.4	.4	1.3

Chol (mg)	Calc (mg)	Iron (mg)	Magn (mg)	Pota (mg)	Sodi (mg)	Zinc (mg)	VT-A (RE)	Thia (mg)	VT-E (α-TE)	Ribo (mg)	Niac (mg)	V-B6 (mg)	Fola (μg)	VT-C (mg)
0	8	.06	4	37	47	.12	<1	<.01	<.01	<.01	.06	.01	9	4
0	123	3.86	71	428	600	2.71	5	.23	2.46	.13	1.01	.98	146	19
0				10	120									
0	4	.1	2	6	63	.03	0	0	.09	0	0	<.01	0	0
0	12	.32	4	11	480	.01	6	0	.6	0	0	<.01	<1	0
0	19	.73	1	2	192	.05	9	<.01	.66	0	.01	<.01	0	<1
0	7	.05	2	19	1	.05	0	.01	<.01	<.01	.01	.03	3	<1
0	31	.88	5	33	<1	.09	14	.01	.03	.01	.12	.02	5	1
0	4	.47	4	47	1	.08	121	.01	.01	.03	.31	.04	2	1
0	2	.54	4	63	4	.06	63	.01	.06	.02	.1	.01	15	1
0	9	.58	4	25	1	.03	<1	<.01	.02	<.01	.02	.01	<1	<1
0	6	.34	7	75	833	.09	21	.01	.1	.02	.04	.01	1	1
0	5	.27	1	30	101	0	2	0	.02	<.01	0	<.01	0	1
0	1	.21	1	11	329	.03	5	<.01	.06	.01	.06	<.01	<1	<1
0	3	.12	1	30	107	.01	1	0	.02	<.01	0	0	0	1
0	5	.33	13	255	119	.22	0	.03	.98	.04	.77	.13	9	6
0	7	.46	19	357	2	.3	0	.05	1.37	.05	1.07	.18	13	9
0	16	.28	4	11	<1	.05	6	.01	.02	<.01	.06	.01	3	<1
0	1	.06	1	24	58	.02	22	.01	.04	<.01	.06	.01	2	5
0					130									
0	1	.02			2325	.01	0	0	0	0	0	0	0	0
	33			3018	<1									
	2		4	1560	1170									
				15	1542									
0	7	.72	26	120	1	0	0	0	0	0	0	0	0	0
	2	.07		10	3		<1	.07		.07	.07			<1
	2	.07		13	4		<1	.07		.07	.07			1
		.07		2	1		<1	.07		.07	.07			<1
	1	.07		12	1		<1	.07		.07	.07			<1
0	4	1.16	7	140	3	.45	<1	.16	.01	.38	2.79	.11	164	<1
0	17	1.38	18	151	10	.63	0	1.25		.34	3.03	.4	313	0
28	80	1.26	13	246	1900	.3	60	.06	.38	.1	.66	.02	5	<1
20	68	1.2	5	176	1972	1.26	113	.06	.33	.12	1.64	.03	3	<1
3	65	1.05	10	168	1736	1.19	0	.06	2.61	.17	1.62	.02	8	2
0	54	1.35	5	138	2115	1.23	0	.07	.57	.05	1.21	.1	30	2
22	186	1.49	22	300	992	.8	40	.07	.15	.24	1.03	.13	10	3
32	186	.69	22	310	1009	.2	67	.07	.97	.25	.44	.06	8	1
27	181	.67	17	273	1046	.67	94	.07	.24	.26	.92	.07	8	1
20	179	.59	20	270	918	.64	37	.08	1.34	.28	.91	.06	10	2
22	166	.55	17	322	1061	.67	67	.08	.1	.24	.64	.09	9	1
32	167	1.05	20	235	1041	10.3	44	.07	.49	.23	.34	.06	10	4
17	159	1.81	22	449	744	.29	109	.13	2.6	.25	1.52	.16	21	68
3	81	2.05	45	402	951	1.03	89	.09	.08	.03	.57	.04	32	2
0	14	.41	5	130	782	0	0	<.01	0	.05	1.87	.02	5	0
5	15	1.1	5	100	952	1.54	63	.07	<.01	.06	1.07	.04	19	<1
7	17	.77	5	55	1106	.39	72	.05	.07	.06	1.39	.03	22	<1
7	17	.75	0	101	815	.26	65	.02	.05	.02	1.13	.02	1	<1
12	42	2.13	30	525	1035	1.4	150	.06	.17	.07	1.07	.16	17	4
2	27	1.63	12	188	578	.98	98	.03	.73	.04	.82	.1	10	4

Table H-1

Food Composition (Computer code number is for West Diet Analysis program) (For purposes of calculations, use "0" for t, <1, <.1, <.01, etc.)

Computer Code Number	Food Description	Measure	Wt (g)	H_2O (%)	Ener (kcal)	Prot (g)	Carb (g)	Dietary Fiber (g)	Fat (g)	Fat Breakdown (g) Sat	Mono	Poly
	SOUPS, SAUCES, and GRAVIES—Continued											
1004	Cream of chicken	1 c	244	91	117	3	9	<1	7	2.1	3.3	1.5
1005	Cream of mushroom	1 c	244	90	129	2	9	<1	9	2.4	1.7	4.2
1006	Minestrone	1 c	241	91	82	4	11	1	3	.6	.7	1.1
1211	Onion	1 c	241	93	58	4	8	1	2	.3	.7	.7
1007	Split pea & ham	1 c	253	82	190	10	28	2	4	1.8	1.8	.6
1008	Tomato	1 c	244	90	85	2	17	<1	2	.4	.4	1
1009	Vegetable beef	1 c	244	92	78	6	10	<1	2	.9	.8	.1
1010	Vegetarian vegetable	1 c	241	92	72	2	12	<1	2	.3	.8	.7
	Ready to serve:											
1707	Chunky chicken soup	1 c	251	84	178	13	17	2	7	2	3	1.4
	SOUPS, dehydrated:											
	Prepared with water:											
1299	Beef broth/bouillon	1 c	244	97	19	1	2	0	1	.3	.3	t
1376	Chicken broth	1 c	244	97	22	1	1	0	1	.3	.4	.4
1013	Chicken noodle	1 c	252	94	53	3	7	1	1	.3	.5	.4
1122	Cream of chicken	1 c	261	91	107	2	13	<1	5	3.4	1.2	.4
1014	Onion	1 c	246	96	27	1	5	1	1	.1	.3	.1
1217	Split pea	1 c	255	87	125	7	21	3	1	.4	.7	.3
1015	Tomato vegetable	1 c	253	93	56	2	10	<1	1	.4	.3	.1
	Unprepared, dry products:											
1011	Beef bouillon, packet	1 ea	6	3	14	1	1	0	1	.3	.2	t
1012	Onion soup, packet	1 ea	39	4	115	5	21	4	2	.5	1.4	.3
	SAUCES											
	From dry mixes, prepared with milk:											
1016	Cheese sauce	1 c	279	77	307	17	23	1	17	9.3	5.3	1.6
1017	Hollandaise	1 c	259	84	240	5	14	<1	20	11.6	5.9	.9
1018	White sauce	1 c	264	81	240	10	21	<1	13	6.4	4.7	1.7
	From home recipe:											
1206	Lowfat cheese sauce	¼ c	61	73	85	6	4	<1	5	2.1	1.9	.9
1019	White sauce, medium	¼ c	72	77	102	2	6	<1	8	2.3	3.2	2
	Ready to serve:											
2202	Alfredo sauce, reduced fat	¼ c	69		170	5	16	0	10	6		
1020	Barbeque sauce	1 tbs	16	81	12	<1	2	<1	<1	t	.1	.1
1706	Chili sauce, tomato base	1 tbs	17	68	18	<1	4	<1	<1	t	t	t
2126	Creole sauce	¼ c	62	89	25	1	4	1	1	.1	.2	.3
2124	Hoisin sauce	1 tbs	17	47	35	<1	7	0	1	0		
2199	Pesto sauce	2 tbs	16		83	2	1	0	8	1.8	5.4	.7
1021	Soy sauce	1 tbs	16	71	8	1	1	<1	<1	t	t	t
2123	Szechuan sauce	1 tbs	16	71	21	<1	3	<1	1	.1	.3	.4
1380	Teriyaki sauce	1 tbs	18	68	15	1	3	<1	0	0	0	0
	Spaghetti sauce, canned:											
1377	Plain	1 c	249	75	271	5	40	8	12	1.7	6.1	3.3
1378	With meat	1 c	250	74	300	8	37	8	14	2.7	7	3.2
1379	With mushrooms	½ c	123	84	108	2	13	1	3	.4	1.5	.8
	GRAVIES											
	Canned:											
1022	Beef	1 c	233	87	123	9	11	1	5	2.7	2.2	.2
1023	Chicken	1 c	238	85	188	5	13	1	14	3.4	6.1	3.6
1024	Mushroom	1 c	238	89	119	3	13	1	6	1	2.8	2.4
1025	From dry mix, brown	1 c	258	92	75	2	13	<1	2	.8	.7	.1
1026	From dry mix, chicken	1 c	260	91	83	3	14	<1	2	.5	.9	.4
	FAST FOOD RESTAURANTS											
	ARBY'S											
1402	Bac'n cheddar deluxe	1 ea	231	59	512	21	39	<1	31	8.7	12.7	10.1
	Roast beef sandwiches:											
1403	Regular	1 ea	155	47	383	22	35	1	18	7	8	3.5
1404	Junior	1 ea	89	48	233	11	23	<1	11	4.1	5.2	2.5
1405	Super	1 ea	254	58	552	24	54	1	28	7.6	12.2	8.4
1407	Beef 'n cheddar	1 ea	194	50	508	25	43		26	7.7	12	6.8
1408	Chicken breast sandwich	1 ea	204	52	445	22	52	1	22	3	9.7	10.1
1412	Ham'n cheese sandwich	1 ea	169	54	355	25	34	<1	14	5.1	5.8	3.8
1726	Italian sub sandwich	1 ea	297	58	671	34	47		39	12.8	15.7	8.5

Chol (mg)	Calc (mg)	Iron (mg)	Magn (mg)	Pota (mg)	Sodi (mg)	Zinc (mg)	VT-A (RE)	Thia (mg)	VT-E (α-TE)	Ribo (mg)	Niac (mg)	V-B6 (mg)	Fola (µg)	VT-C (mg)
10	34	.61	2	88	986	.63	56	.03	.2	.06	.82	.02	2	<1
2	46	.51	5	100	881	.59	0	.05	1.24	.09	.72	.01	5	1
2	34	.92	7	313	911	.73	234	.05	.07	.04	.94	.1	36	1
0	26	.67	2	67	1053	.61	0	.03	.29	.02	.6	.05	15	1
8	23	2.28	48	400	1006	1.32	45	.15	.15	.08	1.47	.07	3	2
0	12	1.76	7	264	695	.24	68	.09	2.49	.05	1.42	.11	15	66
5	17	1.12	5	173	791	1.54	190	.04	.32	.05	1.03	.08	10	2
0	22	1.08	7	210	822	.46	301	.05	.79	.05	.92	.05	11	1
30	25	1.73	8	176	889	1	131	.08	.18	.17	4.42	.05	5	1
0	10	.02	7	37	1361	.07	<1	<.01	.02	.02	.36	0	0	0
0	15	.07	5	24	1483	.01	12	.01	.02	.03	.19	0	2	0
3	33	.5	8	30	1282	.2	5	.07	.02	.06	.88	.01	18	<1
3	76	.26	5	214	1184	1.57	123	.1	.15	.2	2.61	.05	5	1
0	12	.15	5	64	849	.06	<1	.03	.1	.06	.48	0	1	<1
3	20	.94	43	224	1147	.56	5	.21	.13	.14	1.26	.05	39	0
0	8	.63	20	104	1146	.17	20	.06	.81	.05	.79	.05	10	6
1	4	.06	3	27	1018	0	<1	<.01	.01	.01	.27	.01	2	0
2	55	.58	25	260	3493	.23	1	.11	.42	.24	1.99	.04	6	1
53	569	.28	47	552	1565	.97	117	.15	.33	.56	.32	.14	13	2
52	124	.9	8	124	1564	.7	220	.04	.26	.18	.06	.5	22	<1
34	425	.26	264	444	797	.55	92	.08	1.58	.45	.53	.07	16	3
11	166	.25	10	100	389	.73	58	.03	.55	.14	.16	.03	4	<1
8	75	.21	9	100	82	.26	89	.05	.98	.12	.28	.03	4	1
30	150	0	8	80	600		80	0		.1	0			0
0	3	.14	3	28	130	.03	14	<.01	.18	<.01	.14	.01	1	1
0	3	.14	2	63	227	.05	24	.01	.05	.01	.27	.02	1	3
0	35	.31	9	187	339	.1	24	.03	.61	.02	.53	.07	9	0
0	0	0			250		0							0
4	64	.09	6	15	129	.29	39	.01		.03	0	.02	<1	0
0	3	.32	5	29	914	.06	0	.01	0	.02	.54	.03	2	0
0	2	.12	2	13	218	.02	10	<.01	.07	<.01	.1	.01	1	<1
0	4	.31	11	40	690	.02	0	<.01	0	.01	.23	.02	4	0
0	70	1.62	60	956	1235	.52	306	.14	4.98	.15	3.76	.88	54	28
15	68	1.94	60	952	1179	1.37	577	.13	5.91	.17	4.51	.87	52	26
0	15	1	15	332	494	.34	241	.08	1.35	.08	.93	.16	12	9
7	14	1.63	5	189	1304	2.33	0	.07	.15	.08	1.54	.02	5	0
5	48	1.12	5	259	1373	1.9	264	.04	.37	.1	1.05	.02	5	0
0	17	1.57	5	252	1356	1.67	0	.08	.19	.15	1.6	.05	29	0
3	67	.23	10	57	1075	.31	0	.04	.05	.08	.81	0	0	0
3	39	.26	10	62	1133	.32	0	.05	.05	.15	.78	.03	3	3
38	110	4.32		491	1094	3	40	.34		.46	9.6			11
43	60	4.86	16	422	936	3.75	0	.28		.48	11	.2	14	1
22	40	2.7	8	201	519	1.5		.18		.25	6.6	.1	7	
43	90	6.48	25	533	1174	3.75	30	.39		.58	12.4	.3	21	9
52	150	6.12		321	1166	3		.42		.63	9.8			1
45	60	2.88	30	330	1019	.15		.22		.54	9	.38	18	5
55	170	2.7	31	382	1400	.9	40	.82		.37	7.8	.31	26	24
69	410	4.32		565	2062		100	.91		.49	8.2			11

Table H-1

Food Composition (Computer code number is for West Diet Analysis program) (For purposes of calculations, use "0" for t, <1, <.1, <.01, etc.)

Computer Code Number	Food Description	Measure	Wt (g)	H_2O (%)	Ener (kcal)	Prot (g)	Carb (g)	Dietary Fiber (g)	Fat (g)	Fat Breakdown (g) Sat	Mono	Poly
	FAST FOOD RESTAURANTS											
	ARBY'S—Continued											
1413	Turkey sandwich, deluxe	1 ea	195	69	260	20	33	<1	6	1.6	2.3	2.4
1680	Turkey sub sandwich	1 ea	277	62	486	33	46		19	5.3	6	7
	Milkshakes:											
1419	Chocolate	1 ea	340	74	451	10	76	<1	12	2.8	7	1.7
1420	Jamocha	1 ea	326	75	368	9	59	0	10	2.5	6.4	1.6
1421	Vanilla	1 ea	312	77	330	10	46	0	11	3.9	5.3	2.3
1728	Salad, roast chicken	1 ea	400	88	204	24	12		7	3.3	.9	.9
1729	Sports drink, Upper Ten	1 ea	358	88	169	0	42		0	0	0	0
	Source: Arby's											
	BURGER KING											
1423	Croissant sandwich, egg, sausage & cheese	1 ea	176	46	600	22	25	1	46	16		
	Whopper sandwiches:											
1425	Whopper	1 ea	270	58	640	27	45	3	39	11		
1426	Whopper with cheese	1 ea	294	57	730	33	46	3	46	16		
	Sandwiches:											
1629	BK broiler chicken sandwich	1 ea	248	59	550	30	41	2	29	6		
1432	Cheeseburger	1 ea	138	48	380	23	28	1	19	9		
1434	Chicken sandwich	1 ea	229	45	710	26	54	2	43	9		
1427	Double beef	1 ea	351	57	870	46	45	3	56	19		
1428	Double beef & cheese	1 ea	375	56	960	52	46	3	63	24		
1433	Double cheeseburger with bacon	1 ea	218	48	640	44	28	1	39	18		
1431	Hamburger	1 ea	126	48	330	20	28	1	15	6		
1437	Ocean catch fish fillet	1 ea	255	51	700	26	56	3	41	6		
1435	Chicken tenders	1 ea	88	50	230	16	14	2	12	3		
1439	French fries (salted)	1 svg	116	40	370	5	43	3	20	5		
1630	French toast sticks	1 svg	141	33	500	4	60	1	27	7		
1440	Onion rings	1 svg	124	51	310	4	41	6	14	2	8	4
1441	Milk shakes, chocolate	1 ea	284	75	320	9	54	3	7	4		
1442	Milk shakes, vanilla	1 ea	284	75	300	9	53	1	6	4		
1443	Fried apple pie	1 ea	113	47	300	3	39	2	15	3		
	Source: Burger King Corporation											
	CHICK-FIL-A											
	Sandwiches:											
69153	Chargrilled chicken	1 ea	150	54	280	27	36	1	3	1		
69152	Chicken	1 ea	167	61	290	24	29	1	9	2		
69155	Chicken salad	1 ea	167	55	320	25	42	1	5	2		
69154	Chicken salad club	1 ea	232	62	390	33	38	2	12	5		
	Salads:											
52139	Carrot and raisin	1 ea	76	53	150	5	28	2	2	0		
52136	Chicken plate	1 ea	468	85	290	21	40	6	5	0		
52134	Chicken garden, charbroiled	1 ea	397	89	170	26	10	5	3	1		
52135	Chick-n-strips	1 ea	451	86	290	32	21	5	9	2		
52138	Cole slaw	1 ea	79	70	130	6	11	1	6	1		
52137	Tossed salad	1 ea	130	85	70	5	13	1	0	0	0	0
15263	Chicken nuggets, svg	1 ea	110	51	290	28	12	60	12	3		
15262	Chicken-n-strips, svg	1 ea	119	59	230	29	10	0	8	2		
50885	Hearty breast of chicken soup, svg	1 ea	215	86	110	16	10	1	1	0		
7973	Waffle potato fries, svg	1 ea	85	28	290	1	49	0	10	4		
46489	Cheesecake, svg	1 ea	88	52	270	13	7	0	21	9		
49134	Fudge nut brownie, svg	1 ea	88	8	416	12	49	0	19	3.6		
20601	Icedream, svg	1 ea	127	74	140	11	16	0	4	1		
48214	Lemon pie, svg	1 ea	99	56	280	1	19	0	22	6		
	Source: Chick-Fil-A											

Chol (mg)	Calc (mg)	Iron (mg)	Magn (mg)	Pota (mg)	Sodi (mg)	Zinc (mg)	VT-A (RE)	Thia (mg)	VT-E (α-TE)	Ribo (mg)	Niac (mg)	V-B6 (mg)	Fola (μg)	VT-C (mg)
33	130	3.42	30	353	1262	1.5	40	.08		.41	15.4	.52	20	12
51	400	4.68		500	2033		20	13.2		.54	18.8			
36	250	.72	48	410	341	1.5	60	.12		.68	.8	.14	14	5
35	250	2.7	36	525	262	1.5	60	.12		.68	.8	.14	14	2
32	300	2.7	36	686	281	1.5	60	.12		.68	4	.14	37	2
43	170	1.98		877	508		485	.33		.54	5.6			51
0				0	40									
260	150	3.6			1140		80							0
90	80	4.5			870		100	.33		.41	7	.35		9
115	250	4.5			1350		150	.34		.48	7	.33		9
80	60	5.4			480		60							6
65	100	2.7			770		60							0
60	100	3.6			1400		0							0
170	80	7.2			940		100	.34		.56	10			9
195	250	7.2			1420		150	.35		.63	10			9
145	200	4.5			1240		80	.31		.42	6			0
55	40	1.8			530		20	.28		.31	4.89			0
90	60	2.7			980		20							1
35	0	.72			530		0							0
0	0	1.08			240		0							4
0	60	2.7			490		0							0
0	100	1.44			810		0							0
20	200	1.8			230		60	.13		.55	.13			0
20	300	0			230		60	.11		.57	.13			4
0	0	1.44			230		0							6
40					640									
50					870									
10					810									
70					980									
6					650									
35					570									
25					650									
20					430									
15					430									
0					0									
14					770									
20					380									
45					760									
5					960									
10					510									
36					773									
40					240									
5					550									

Table H-1

Food Composition (Computer code number is for West Diet Analysis program) (For purposes of calculations, use "0" for t, <1, <.1, <.01, etc.)

Computer Code Number	Food Description	Measure	Wt (g)	H_2O (%)	Ener (kcal)	Prot (g)	Carb (g)	Dietary Fiber (g)	Fat (g)	Fat Breakdown (g) Sat	Mono	Poly
	FAST FOOD RESTAURANTS—Continued											
	DAIRY QUEEN											
	Ice cream cones:											
1446	Small vanilla	1 ea	142	63	230	6	38	0	7	4.5		
1447	Regular vanilla	1 ea	213	64	350	8	57	0	10	7		
1448	Large vanilla	1 ea	253	65	410	10	65	0	12	8		
1450	Chocolate dipped	1 ea	234	59	510	9	63	1	25	13		
1453	Chocolate sundae	1 ea	241	62	410	8	73	0	10	6		
1455	Banana split	1 ea	369	67	510	8	96	3	12	8		
1456	Peanut buster parfait	1 ea	305	51	730	16	99	2	31	17		
1457	Hot fudge brownie delight	1 ea	305	52	710	11	102	1	29	14	12	2
1459	Buster bar	1 ea	149	45	450	10	41	2	28	12		
1645	Breeze, strawberry, regular	1 ea	383	70	460	13	99	1	1	1	0	0
1460	Dilly bar	1 ea	85	55	210	3	21	0	13	7	3	3
1461	DQ ice cream sandwich	1 ea	61	46	150	3	24	1	5	2		
1463	Milk shakes, regular	1 ea	397	71	520	12	88	<1	14	8	2	2
1464	Milk shakes, large	1 ea	461	71	600	13	101	<1	16	10	2	2
1466	Milk shakes, malted	1 ea	418	68	610	13	106	<1	14	8	2	2
1470	Misty slush, small	1 ea	454	88	220	0	56	0	0	0	0	0
2250	Starkiss	1 ea	85	75	80	0	21	0	0	0	0	0
	Yogurt:											
1641	Yogurt cone, regular	1 ea	213	66	280	9	59	0	1	.5		
1643	Yogurt sundae, strawberry	1 ea	255	69	300	9	66	1	<1	.5	0	0
	Sandwiches:											
1481	Cheeseburger, double	1 ea	219	55	540	35	30	2	31	16		
1480	Cheeseburger, single	1 ea	152	55	340	20	29	2	17	8		
1474	Chicken	1 ea	191	56	430	24	37	2	20	4		
1647	Chicken fillet, grilled	1 ea	184	64	310	24	30	3	10	2.5		
1475	Fish fillet sandwich	1 ea	170	57	370	16	39	2	16	3.5		
1476	Fish fillet with cheese	1 ea	184	56	420	19	40	2	21	6	7	8
1477	Hamburger, single	1 ea	128	56	269	16	27	2	11	4.6	5.6	.9
1478	Hamburger, double	1 ea	212	62	440	30	29	2	22	10		
	Hotdog:											
1483	Regular	1 ea	99	57	240	9	19	1	14	5		
1484	With cheese	1 ea	113	55	290	12	20	1	18	8	8	2
1485	With chili	1 ea	128	61	280	12	21	2	16	6		
1489	French fries, small	1 ea	71	39	210	3	29	3	10	2	5	3
1490	French fries, large	1 ea	128	40	390	5	52	6	18	4	8	6
1491	Onion rings	1 ea	85	46	240	4	29	2	12	2.5		
	Source: International Dairy Queen											
	HARDEES											
1734	Frisco burger hamburger	1 ea	242		760	36	43		50	18		
1736	Frisco grilled chicken salad	1 ea	278		120	18	2		4	1		
1737	Peach shake	1 ea	345		390	10	77		4	3		
	Source: Hardees											
	JACK IN THE BOX											
	Breakfast items:											
1492	Breakfast Jack sandwich	1 ea	121	49	300	18	30	0	12	5	5	2.5
1494	Sausage crescent	1 ea	156	39	580	22	28	0	43	16		
1495	Supreme crescent	1 ea	153	40	530	23	34	0	33	10	18.9	7.8
1496	Pancake platter	1 ea	231	45	610	15	87	0	22	9	7.6	3.5
1497	Scrambled egg platter	1 ea	213	52	560	18	50	0	32	9	16.6	4.4
	Sandwiches:											
1654	Bacon cheeseburger	1 ea	242	49	710	35	41	0	45	15	15.7	8.7
1499	Cheeseburger	1 ea	110	41	330	16	32	0	15	6	5.9	2.3
1739	Chicken caesar pita sandwich	1 ea	237	59	520	27	44	4	26	6		
1655	Chicken sandwich	1 ea	160	52	400	20	38	0	18	4		
1656	Chicken sandwich, sourdough ranch	1 ea	225		490	29	45	1	21	6		
1505	Chicken supreme	1 ea	245	55	620	25	48	0	36	11	14.8	11.4
1583	Double cheeseburger	1 ea	152	44	450	24	35	0	24	12	11.6	3.1

Chol (mg)	Calc (mg)	Iron (mg)	Magn (mg)	Pota (mg)	Sodi (mg)	Zinc (mg)	VT-A (RE)	Thia (mg)	VT-E (α-TE)	Ribo (mg)	Niac (mg)	V-B6 (mg)	Fola (μg)	VT-C (mg)
20	200	1.08		250	115		122	.05		.28				
30	300	1.8		390	170		150	.09		.38	.16	.13		2
40	350	1.8		451	200		200	.11		.4	.2			2
30	300	1.8		435	200		150	.09		.38	.16	.13		2
30	250	1.44		394	210		150	.08		.35	.4	.19		0
30	250	1.8		860	180		200	.15		.25	.4	.2		15
35	300	1.8		660	400		150	.15		.51	3	.22		1
35	300	5.4		510	340		80	.15		.68	.3	.18		1
15	150	1.08		400	280		80	.09		.17	3	.08		0
10	450	2.7		530	270		0	.13		.73				9
10	100	.36		170	75		60	.03		.14		.06		0
5	60	.72		105	115		40	.03		.25	.4	.05		0
45	400	1.44		570	230		80	.12		.59	.8	.19		<1
50	450	1.44		660	260		200	.15		.68	.8			<1
45	400	1.44		570	230		80	.12		.59	.8	.19		<1
0	0	0			20		0							0
0	0	0			10		0							0
5	300	1.8		285	170		0	.09		.38				2
5	300	1.8		352	180		0	.09		.49				6
115	250	4.5		426	1130		150	.29		.49	6.78			4
55	150	3.6		263	850		60	.29		.33	3.89			4
55	40	1.8		350	760		0	.37		.34	11			0
50	200	2.7		330	1040		0	.3		1.02	12			0
45	40	1.8		280	630		0	.3		.22	3			0
60	100	1.8		290	850		80	.3		.25	5			0
42	56	2.5		234	584		37	.27		.23	3.6			3
90	60	4.5		444	680		60	.32		.45	7.49			6
25	60	1.8		170	730		20	.22		.14	2			4
40	150	1.8		180	950		60	.22		.17	2			4
35	60	1.8		262	870		80	.23		.14	3			4
0	0	.72		430	115		0	.09		.03	2			5
0	0	1.44		780	200		0	.15		.07	3			9
0	0	1.08		90	135		0	.09		.05	.4			0
70					1280									
60					520									
25					290									
185	200	2.7		220	890		80	.47		.41	3			9
185	150	2.7		260	1010		100	.6		.51	4.6			0
210	150	3.6		270	930		150	.65		.54	4.2			12
100	100	1.8		310	890		80	.03		.85	7			6
380	150	4.5		450	1060		150			.66	5			9
110	250	5.4		540	1240		80	.24		.48	8.8	.39		9
35	200	2.7		200	510		60	.23		.23	3			1
55	250	2.7		490	1050		80							2
45	150	1.8		180	1290		40							0
65	150	1.8		340	1060									0
75	200	2.7		190	1520		100	.39		.32	11			2
75	250	3.6		320	900		100	.15		.34	6			0

Table H-1

Food Composition (Computer code number is for West Diet Analysis program) (For purposes of calculations, use "0" for t, <1, <.1, <.01, etc.)

Computer Code Number	Food Description	Measure	Wt (g)	H_2O (%)	Ener (kcal)	Prot (g)	Carb (g)	Dietary Fiber (g)	Fat (g)	Fat Breakdown (g) Sat	Mono	Poly
	FAST FOOD RESTAURANTS—Continued											
	JACK IN THE BOX—Continued											
1651	Grilled sourdough burger	1 ea	223	48	670	32	39	0	43	16	17.8	7.9
1498	Hamburger	1 ea	97	42	280	13	31	0	11	4	4.9	2
1500	Jumbo Jack burger	1 ea	229	55	560	26	41	0	32	10	13	8
1501	Jumbo Jack burger with cheese	1 ea	242	55	610	29	41	0	36	12	15	9
1740	Monterey roast beef sandwich	1 ea	238	57	540	30	40	3	30	9		
1508	Tacos, regular	1 ea	78	57	190	7	15	2	11	4		
1509	Tacos, super	1 ea	126	59	280	12	22	3	17	6		
	Teriyaki bowl:											
1679	Beef	1 ea	440	62	640	28	124	7	3	1		
1668	Chicken	1 ea	440	62	580	28	115	6	1			
1516	French fries	1 ea	109	38	350	4	45	4	17	4		
1517	Hash browns	1 ea	57	53	160	1	14	1	11	2.5	6.8	.3
1518	Onion rings	1 ea	103	34	380	5	38	0	23	6	15.2	.9
	Milkshakes:											
1519	Chocolate	1 ea	322	72	390	9	74	0	6	3.5	2.1	
1520	Strawberry	1 ea	298	74	330	9	60	0	7	4	2	
1521	Vanilla	1 ea	304	73	350	9	62	0	7	4	1.8	
1522	Apple turnover	1 ea	110	34	350	3	48	0	19	4	10.6	1.5
	Source: Jack in the Box Restaurant, Inc											
	KENTUCKY FRIED CHICKEN											
	Rotisserie gold:											
1472	Dark qtr, no skin	1 ea	117	66	217	27	0	0	12	3.5		
1473	Dark qtr, w/skin	1 ea	146	62	333	30	1		24	6.6		
1513	White qtr with wing, w/skin	1 ea	176	65	335	40	1		19	5.4		
1525	White qtr with wing, no skin	1 ea	117	63	199	37	0	0	6	1.7		
	Original Recipe:											
1253	Center breast	1 ea	103	52	260	25	9	<1	14	3.8	7.8	2
1251	Side breast	1 ea	83	47	245	18	9	<1	15	4.2	8.8	2.2
1250	Drumstick	1 ea	57	54	152	14	3	<1	8	2.2	4.1	1.3
1252	Thigh	1 ea	95	49	287	18	8	<1	21	5.3	9.4	3.1
1249	Wing	1 ea	53	45	172	12	5	<1	12	3	6	1.8
	Hot & spicy:											
1451	Center breast	1 ea	125	48	360	28	13		22	5		
1452	Side breast	1 ea	120	43	400	22	16		28	6		
1430	Thigh	1 ea	119	47	370	24	10		27	6		
1471	Wing	1 ea	61	38	220	14	5		16	4		
	Extra crispy recipe:											
1261	Center breast	1 ea	118	48	330	26	14	<1	20	4.8	10.8	2.1
1259	Side breast	1 ea	116	40	400	21	19	<1	27	5.5	12.9	2.3
1258	Drumstick	1 ea	65	49	190	14	6	<1	12	3.4	7.7	1.7
1260	Thigh	1 ea	109	43	380	23	7	<1	30	7.7	16	4.2
1257	Wing	1 ea	59	34	240	13	8	<1	17	4.4	10.7	2.5
1390	Baked beans	½ c	167	70	200	8	36	6	3	1.5	.7	.4
1526	Breadstick	1 ea	33	30	110	3	17	0	3	0		
1388	Buttermilk biscuit	1 ea	65	28	234	5	28	<1	13	3.4	6.2	2.3
1391	Chicken little sandwich	1 ea	47	35	169	6	14	<1	10	2	4.7	3.4
1269	Coleslaw	1 svg	90	75	114	1	13	<1	6	1	1.7	3.4
1527	Cornbread	1 ea	56	26	228	3	25	1	13	2		
1268	Corn-on-the-cob	1 ea	151	70	222	4	27	8	12	2	1	1.5
1429	Chicken, hot wings	1 svg	135	38	471	27	18		33			
1386	Kentucky fries	1 svg	77	42	228	3	26	3	12	3.2		
1381	Kentucky nuggets	6 ea	95	48	284	16	15	<1	18	4		
1534	Macaroni & cheese	1 svg	114	71	162	7	15	0	8	3		
1387	Mashed potatoes & gravy	1 svg	120	80	103	1	16	<1	5	.4	.5	.2
1530	Pasta salad	1 svg	108	78	135	2	14	1	8	1		
1389	Potato salad	½ c	188	74	271	5	27	3	16	3	4.2	7.3
1383	Potato wedges	1 svg	92	59	192	3	25	3	9	3		
1535	Red beans & rice	1 svg	112	76	114	4	18	3	3	1		
1529	Vegetable medley salad	1 ea	114	77	126	1	21	3	4	1		
	Source: Kentucky Fried Chicken Corp											

Chol (mg)	Calc (mg)	Iron (mg)	Magn (mg)	Pota (mg)	Sodi (mg)	Zinc (mg)	VT-A (RE)	Thia (mg)	VT-E (α-TE)	Ribo (mg)	Niac (mg)	V-B6 (mg)	Fola (μg)	VT-C (mg)
110	200	4.5		510	1140		150	.65		.48	8	.33		6
25	100	2.7		190	430		20	.15		.26	2			1
65	100	4.5		450	700		40	.36		.29	1.8			6
80	200	5.4		460	780		60	.36		.44	1.6			6
75	300	3.6		500	1270		80							5
20	100	1.08	35	240	410	1.2	0	.07		.17	1	.13		0
30	150	1.8	45	370	720	1.8	0	.12		.08	1.4	.18		2
25	150	4.5		430	930		1000							6
30	100	1.8		380	1220		1100							9
0	0	1.08		690	190		0	.18		.03	3.8			24
0	0	.36		190	310		0	.05			1			6
0	20	1.8		130	450		0	.29		.17	2.6			2
25	300	.72		680	210		0	.15		.6	.4			0
30	300	0		550	180		0	.15		.43	.4			0
30	300	0		570	180		0	.15		.34	.4			0
0	0	1.8		80	460		0	.2		.12	1.8			9
128	10	.18			772		15							1
163	10	.18			980		15							1
157	10	.18			1104		15							1
97	10	.18			667		15							1
92	30	.72			609		15	.09		.17	11.5			
78	68	1.2			604		15	.06		.13	6.9			
75	21	1.1			269		15	.05		.12	3.2			
112	40	1.08			591		31	.08		.3	5.5			
59	30	.54			383		15	.03		.08	3.7			
80	20	.72			750		15							6
80	40	1.08			850		15							6
100	20	1.08			670		15							6
65	20	.72			440		30							
75	33	.8			740		15	.11		.13	13.1			
75	20	.72			710		15	.09		.1	8.5			
65	20	.36			310		30	.06		.12	3.7			
90	49	1.2			520		30	.1		.21	6.5			
65	20	.36			320		30			.04	.06	3.3		
5	61	2.17	44	348	812	1.96	38	.09		.06	.76	.11	49	3
0	30	.18			15		0							0
3	43	1.92			565		28	.26		.2	2.77			
18	23	1.7			331		5	.16		.12	2.2			
5	30	.36			177		32	.03		.03	.2			27
42	60	.72			194		10							
0	0	.36			76		20	.14		.11	1.8			2
150	40	3.24			1230		15							6
4	11	.98			535		0							0
66	2	.1			865		15	.02		.02	1	.05		<1
16	120	.72			531		190							0
<1	20	.4			388		15			.04	1.2			
1	20	1.08			663		110							7
16	16	3.25	23	385	636	.44	120	.1		.03	.9	.29	11	
3					428		0							
4	10	.72			315									
0	20	.36			240		375							5

Table H-1

Food Composition (Computer code number is for West Diet Analysis program) (For purposes of calculations, use "0" for t, <1, <.1, <.01, etc.)

Computer Code Number	Food Description	Measure	Wt (g)	H_2O (%)	Ener (kcal)	Prot (g)	Carb (g)	Dietary Fiber (g)	Fat (g)	Fat Breakdown (g)		
										Sat	Mono	Poly
	FAST FOOD RESTAURANTS—Continued											
	LONG JOHN SILVER'S											
1528	Chicken plank dinner, 3 piece	1 ea	399	56	890	32	101		44	9.5	24.8	9.4
1531	Clam chowder	1 ea	198	86	140	11	10	1	6	1.8	2.5	1.7
1532	Clam dinner	1 ea	361	46	990	24	114		52	10.9	31.4	9.9
	Fish, batter fried:											
1523	Fish & fryes (fries), 3 piece	1 ea	384	54	980	31	92		50	11.3	28.4	9.7
1524	Fish & fryes, 2 piece	1 ea	261	54	610	27	52		37	7.9	23.5	5.3
2240	Fish and lemon crumb dinner, 3 piece	1 ea	493	71	610	39	86		13	2.2	3.9	5.3
2241	Fish and lemon crumb dinner, 2 piece	1 ea	334	77	330	24	46		5	.9	1.6	1.2
1533	Fish & chicken dinner	1 ea	431	55	950	36	102		49	10.6	28.8	9.5
1537	Shrimp dinner, batter fried	1 ea	331	54	840	18	88		47	9.7	27.2	9.1
	Salads:											
1541	Cole slaw	1 ea	98	70	140	1	20	1	6	1	1.5	3.5
1539	Ocean chef salad	1 ea	234	89	110	12	13	2	1	.4	.4	.2
1540	Seafood salad	1 ea	278	79	380	15	12	2	31	5.1	8.2	17.5
1542	Fryes (fries) serving	1 ea	85	43	250	3	28	1	15	2.5	7.4	5.1
1543	Hush puppies	1 ea	24	40	70	2	10	<1	2	.4	1.3	.2
	Source: Long John Silver's, Lexington KY											
	McDONALD'S											
	Sandwiches:											
1221	Big mac	1 ea	216	53	510	25	46	3	26	9.3	7.5	4.1
1226	Cheeseburger	1 ea	122	46	318	15	36	2	13	5.6	3.8	1.1
1224	Filet-o-fish	1 ea	145	49	364	14	41	1	16	3.7	3.8	5.6
1225	Hamburger	1 ea	108	49	266	12	35	2	9	3.2	2.8	.9
1444	McChicken	1 ea	189	52	491	17	42	2	29	5.4	8.5	10.2
1591	McLean deluxe	1 ea	214	64	345	23	37	2	12	4.4	3.6	1.2
1438	McLean deluxe with cheese	1 ea	228	63	398	26	38	2	16	6.8	4.6	1.3
1222	Quarter-pounder	1 ea	171	52	415	23	36	2	20	7.8	6.7	1.3
1223	Quarter-pounder with cheese	1 ea	199	50	520	28	37	2	29	12.6	8.7	1.6
1227	French fries, small serving	1 ea	68	40	207	3	26	2	10	1.7	3.1	2.5
1228	Chicken McNuggets	4 pce	73	51	198	12	10	0	12	2.5	3.7	2.4
	Sauces (packet):											
1229	Hot mustard	1 ea	30	60	63	1	7	1	4	.5	1.1	2
1230	Barbecue	1 ea	32	58	53	<1	12	<1	<1	.1	.1	.2
1231	Sweet & sour	1 ea	32	57	55	<1	12	<1	<1	.1	.1	.3
	Low-fat (frozen yogurt) milk shakes:											
1232	Chocolate	1 ea	295		348	13	62	1	6	3.5	.1	.7
1233	Strawberry	1 ea	294		343	12	63	<1	5	3.4	.1	.6
1234	Vanilla	1 ea	293		308	12	54	<1	5	3.3	.1	.6
	Low-fat (frozen yogurt) sundaes:											
1237	Hot caramel	1 ea	182	56	307	7	62	1	3	2	.3	1
1235	Hot fudge	1 ea	179	60	293	8	53	2	5	4.7	.1	.4
1267	Strawberry	1 ea	178	65	239	6	51	1	1	.7	.1	.2
1238	Vanilla	1 ea	90	68	118	4	24	<1	1	.5	.2	t
1241	Cookies, McDonaldland	1 ea	56	3	258	4	41	1	9	1.7	6.4	.8
1242	Cookies, chocolaty chip	1 ea	56	3	282	3	36	1	14	3.9	4.4	1
1240	Muffin, apple bran, fat-free	1 ea	75	39	182	4	40	1	1	.2	.1	.3
1239	Pie, apple	1 ea	84	35	289	3	37	1	14	3.7	4.5	.3
	Breakfast items:											
1243	English muffin with spread	1 ea	63	33	189	5	30	2	6	2.4	1.5	1.3
1244	Egg McMuffin	1 ea	137	57	289	17	27	1	13	.7	4.5	1.6
1245	Hotcakes with marg & syrup	1 ea	222	44	557	8	100	2	14	2.4	4.6	5.8
1246	Scrambled eggs	1 ea	102	73	170	13	1	0	12	3.6	5.3	1.7
1247	Pork sausage	1 ea	43	45	173	6	<1	0	16	5.5	6.4	2.1
1248	Hashbrown potatoes	1 ea	53	55	130	1	13	1	8	1.3	2.3	1.9
1392	Sausage McMuffin	1 ea	112	42	361	13	26	1	23	8.3	8.2	2.8
1393	Sausage McMuffin with egg	1 ea	163	52	443	19	27	1	29	10	10.7	3.6
1394	Biscuit with biscuit spread	1 ea	76	32	260	4	32	1	13	3.8	3.7	.8

Chol (mg)	Calc (mg)	Iron (mg)	Magn (mg)	Pota (mg)	Sodi (mg)	Zinc (mg)	VT-A (RE)	Thia (mg)	VT-E (α-TE)	Ribo (mg)	Niac (mg)	V-B6 (mg)	Fola (μg)	VT-C (mg)
55	200	4.5		1170	2000	3	40	.52		.51	16			9
20	200	1.8		380	590	.6	150	.09		.25	2			
75	200	4.5		910	1830	3	40	.75		.42	12			12
70	200	4.5		1120	1530	3	40	.45		.42	8			15
60	40	1.8		900	1480	1.2		.37		.34	8			9
125	200	5.4		990	1420	2.25	700	.75		.59	24			6
75	80	1.8		440	640	.9	1000	.3		.25	14			18
75	200	4.5		1280	2090	3	40	.6		.59	14			9
100	200	3.6		840	1630	3	40	.45		.42	9			9
15	60	.72		190	260	.6	40	.06		.07	2			
40	100	3.6		95	730	.3	500	.12		.14	3			21
55	150	4.5		130	980	.9	200	.15		.25	3			21
0	200	.72		370	500	.3	0	.09			1.6			6
	40	.72		65	25	.3		.06		.03	.8			
76	202	4.32	46	456	932	4.81	66	.49	1.01	.44	6.08	.25	49	3
42	134	2.73	27	281	766	2.62	64	.33	.46	.31	3.81	.15	24	2
37	123	1.85	32	266	708	.7	21	.32	1.52	.23	2.58	.07	30	0
28	126	2.73	24	260	531	2.25	22	.33	.23	.26	3.81	.14	21	2
52	128	2.5	32	319	797	1.06	29	.91	6.16	.24	7.74	.38	37	1
59	131	4.29	40	537	811	4.9	74	.42	.63	.34	7.16	.29	44	8
73	139	4.29	43	559	1046	5.26	115	.42	.85	.39	7.16	.3	47	8
70	127	4.33	33	405	692	4.66	33	.39	.36	.32	6.78	.24	27	3
97	143	4.5			1160		115	.39	.81	.43	6.78	.26	33	3
0	9	.53	26	469	135	.32	0	.05	.83	0	1.94	.24	26	8
42	9	.65	17	210	353	.69	0	.08	.96	.11	5.15	.21		0
3	7	.78		29	85		4	.01		.01	.15			0
0	4	0		51	277		0	.01		.01	.17			4
0	2	.16		7	158		74	0		.01	.08			0
24	372	1.04		543	241		46	.12		.51	.4	.1		3
24	366	.29		542	170		46	.12		.51	.4	.11		3
24	360	.29		533	193		45	.12		.51	.31			3
7	246	.15		344	197		18	.09		.34	.27			2
5	258	.59		441	190		7	.09		.34	.29			2
5	221	.25		325	115		6	.06		.34	.25			2
3	132	.23		175	84		4	<.01		.01	.23			1
0	10	1.73	11	62	267	.38	0	.24	.99	.16	2.01	.03		0
3	28	1.78	24	142	229	.4	0	.14	.92	.16	1.48			0
0	34	1.29	13	77	215	.33	0	.14	0	.14	1.32	.03	5	1
0	17	1.23	7	69	221	.23		.19	1.5	.12	1.55	.04	9	27
13	103	1.59	13	69	386	.42	33	.25	.13	.31	2.61	.04	57	1
234	151	2.44	24	199	730	1.56	100	.49	.85	.45	3.33	.15	33	2
11	108	1.98	27	285	746	.53	119	.24	1.2	.26	1.86	.09	<1	<1
424	50	1.19	10	126	143	1.06	168	.07	.92	.51	.06	.12	44	0
33	7	.5	7	102	292	.78	0	.18	.26	.06	1.7	.09		0
0	7	.3	11	213	332	.15	0	.08	.58	.02	.9	.08	8	3
46	132	2.07	22	191	751	1.51	48	.56	.66	.27	3.76	.14	16	0
257	156	2.8	26	251	821	2.07	117	.59	1.11	.49	3.79	.19	33	0
0	68	1.85	9	105	836	.3	2	.29	.81	.23	2.23	.03	5	0

Table H-1

Food Composition (Computer code number is for West Diet Analysis program) (For purposes of calculations, use "0" for t, <1, <.1, <.01, etc.)

Computer Code Number	Food Description	Measure	Wt (g)	H_2O (%)	Ener (kcal)	Prot (g)	Carb (g)	Dietary Fiber (g)	Fat (g)	Fat Breakdown (g) Sat	Mono	Poly
	FAST FOOD RESTAURANTS—Continued											
	McDONALD'S—Continued											
1395	Biscuit with sausage	1 ea	119	37	433	10	32	1	29	8.6	10.1	2.8
1396	Biscuit with sausage & egg	1 ea	170	48	518	16	33	1	35	10.5	12.7	3.7
1397	Biscuit with bacon, egg, cheese	1 ea	152	46	450	17	33	1	27	8.7	8.9	2.3
	Salads:											
1398	Chef salad	1 ea	313	86	206	19	9	3	11	4.2	3	1.2
1400	Garden salad	1 ea	234	92	84	6	7	3	4	1.1	1.4	.7
1401	Chunky chicken salad	1 ea	296	87	164	23	8	3	5	1.3	1.6	1
	Source: McDonald's Corporation											
	PIZZA HUT											
	Pan pizza:											
1657	Cheese	2 pce	216	51	522	24	56	4	22	10	6.8	3.4
1658	Pepperoni	2 pce	208	49	531	22	56	4	24	8	9.9	3.7
1659	Supreme	2 pce	273	56	622	30	56	6	30	12	12	4.2
1660	Super supreme	2 pce	286	56	645	30	56	6	34	12		
	Thin 'n crispy pizza:											
1649	Cheese	2 pce	174	52	411	22	42	4	16	8	4.4	2.3
1623	Pepperoni	2 pce	168	48	431	22	42	2	20	8		
1622	Supreme	2 pce	232	57	514	28	42	4	26	10		
1620	Super supreme	2 pce	247	57	541	28	44	4	28	12		
	Hand tossed pizza:											
1619	Cheese	2 pce	216	53	470	26	58	4	14	7.9		
1618	Pepperoni	2 pce	208	51	477	24	58	4	16	8		
1648	Supreme	2 pce	273	56	568	32	60	6	24	10		
1617	Super supreme	2 pce	286	57	591	32	60	6	26	10		
	Personal pan pizza:											
1610	Pepperoni	1 ea	255	50	637	27	69	5	28	10	11.8	4.5
1609	Supreme	1 ea	327	57	721	33	70	6	34	12	14.7	5.6
	Source: Pizza Hut											
	SUBWAY											
	Deli style sandwich:											
69104	Bologna	1 ea	171	64	292	10	38	2	12	4		
69102	Ham	1 ea	171	69	234	11	37	2	4	1		
69103	Roast beef	1 ea	180	69	245	13	38	2	4	1		
69105	Seafood and crab:	1 ea	178	66	298	12	37	2	11	2		
69106	With light mayo	1 ea	178	68	256	12	37	2	7	2		
69108	Tuna:	1 ea	178		354	11	37	2	18	3		
69107	With light mayo	1 ea	178	67	279	11	38	2	9	2		
69101	Turkey	1 ea	180	69	235	12	38	2	4	1		
	Sandwiches, 6 inch:											
	B.L.T.:											
69135	On white bread	1 ea	191	67	311	14	38	3	10	3		
69136	On wheat bread	1 ea	198	65	327	14	44	3	10	3		
	Chicken taco sub:											
69131	On white bread	1 ea	286	70	421	24	43	3	16	5		
69132	On wheat bread	1 ea	293	69	436	25	49	4	16	5		
	Club :											
69117	On white bread	1 ea	246	73	297	21	40	3	5	1		
69118	On wheat bread	1 ea	253	71	312	21	46	3	5	1		
	Cold cut trio:											
69113	On white bread	1 ea	246	71	362	19	39	3	13	4		
69114	On wheat bread	1 ea	253	68	378	20	46	3	13	4		
	Ham:											
69115	On white bread	1 ea	232	73	287	18	39	3	5	1		
69115	On wheat bread	1 ea	239	71	302	19	45	3	5	1		

Chol (mg)	Calc (mg)	Iron (mg)	Magn (mg)	Pota (mg)	Sodi (mg)	Zinc (mg)	VT-A (RE)	Thia (mg)	VT-E (α-TE)	Ribo (mg)	Niac (mg)	V-B6 (mg)	Fola (μg)	VT-C (mg)
33	75	2.35	15	207	1128	1.08	2	.48	1.07	.29	3.93	.12	5	0
245	100	2.95	20	271	1199	1.61	59	.51	1.53	.55	3.96	.18	27	0
238	103	2.6	20	245	1315	1.64	99	.39	1.49	.57	3.32	.13	30	0
179	157	1.81	40	605	727	2.16	1179	.33	1.45	.37	4.32	.36	100	22
139	52	1.34	24	407	61	.73	1114	.12	.95	.24	.65	.16	96	22
76	54	1.62	44	673	318	1.52	1973	.51	1.28	.21	8.46	.52	83	30
50	288	3	63	337	1002	4.32	211	.6		.64	5.48	.18		7
48	206	3.21	55	399	1140	4.14	190	.62		.48	5.31	.16	0	8
60	234	4.6	81	620	1529	6	195	.86		.84	6.4	.33		11
68	236	4.39	80	592	1649	5.99	201	.83		.73	7.13			12
50	291	2.06	56	307	1070	4.23	217	.46		.46	5.65	.18		6
50	208	2.2	51	330	1255	4.02	199	.48		.49	5.97			7
62	238	3.6	79	631	1591	5.4	197	.7		.57	6.27			12
70	238	3.41	73	563	1762	5.47	208	.72		.53	6.57			9
50	284	3	71	388	1242	4.6	198	.48		.48	5.3			10
48	202	3.21	84	610	1380	6.01	187	.72		.56	7.59			13
60	232	4.6	87	589	1769	5.48	192	.82		.66	8.45			14
68	232	4.39	89	607	1889	5.65	198	.84		.68	8.71			14
55	250	4	60	406	1338	3.8	233	.56		.66	8.16	.2		10
66	276	5.19	74	603	1757	4.69	240	.73		.82	9.91	.4		14
20	39	3			744		113							14
14	24	3			773		113							14
13	23	3			638		113							14
17	24	3			544		113							14
16	24	3			556		118							14
18	26	3			557		116							14
16	26	3			583		126							14
12	26	3			944		113							14
16	27	3			945		120							15
16	33	3			957		120							15
52	118	4			1264		209							18
52	124	4			1275		209							18
26	29	4			1341		120							15
26	35	4			1352		120							15
64	49	4			1401		130							16
64	55	4			1412		130							16
28	28	3			1308		120							15
28	35	3			1319		120							15

Table H-1

Food Composition (Computer code number is for West Diet Analysis program) (For purposes of calculations, use "0" for t, <1, <.1, <.01, etc.)

Computer Code Number	Food Description	Measure	Wt (g)	H_2O (%)	Ener (kcal)	Prot (g)	Carb (g)	Dietary Fiber (g)	Fat (g)	Fat Breakdown (g) Sat	Mono	Poly
	FAST FOOD RESTAURANTS—Continued											
	SUBWAY—Continued											
	Italian B.M.T.											
69139	On white bread	1 ea	246	66	445	21	39	3	21	8		
69140	On wheat bread	1 ea	253	64	460	21	45	3	22	7		
	Meatball:											
69129	On white bread	1 ea	260	70	404	18	44	3	16	6		
69130	On wheat bread	1 ea	267	67	419	19	51	3	16	6		
	Melt with turkey, ham, bacon, cheese:											
69127	On white bread	1 ea	251	70	366	22	40	3	12	5		
69128	On wheat bread	1 ea	258	68	382	23	46	3	12	5		
	Pizza sub:											
69133	On white bread	1 ea	250	66	448	19	41	3	22	9		
69134	On wheat bread	1 ea	257	65	464	19	48	3	22	9		
	Roast beef:											
69121	On white bread	1 ea	232	72	288	19	39	3	5	1		
69122	On wheat bread	1 ea	239	70	303	20	45	3	5	1		
	Roasted chicken breast:											
69125	On white bread	1 ea	246	70	332	26	41	3	6	1		
69126	On wheat bread	1 ea	253	68	348	27	47	3	6	1		
	Seafood and crab:											
69145	On white bread:	1 ea	246	69	415	19	38	3	19	3		
69147	With light mayo	1 ea	246	72	332	19	39	3	10	2		
69146	On wheat bread:	1 ea	253	67	430	20	44	3	19	3		
69148	With light mayo	1 ea	253	70	347	20	45	3	10	2		
	Spicy italian:											
69123	On white bread	1 ea	232	64	467	20	38	3	24	9		
69124	On wheat bread	1 ea	239	62	482	21	44	3	25	9		
	Steak and cheese:											
69119	On white bread	1 ea	257	68	383	29	41	3	10	6		
69120	On wheat bread	1 ea	264	67	398	30	47	3	10	6		
	Tuna:											
69141	On white bread:	1 ea	246	62	527	18	38	3	32	5		
69143	With light mayo	1 ea	246	70	376	18	39	3	15	2		
69142	On wheat bread:	1 ea	253	62	542	19	44	3	32	5		
69144	With light mayo	1 ea	253	68	391	19	46	3	15	2		
	Turkey:											
69111	On white bread	1 ea	232	73	273	17	40	3	4	1		
69112	On wheat bread	1 ea	239	71	289	18	46	3	4	1		
	Turkey breast and ham:											
69137	On white bread	1 ea	232	73	280	18	39	3	5	1		
69138	On wheat bread	1 ea	239	71	295	18	46	3	5	1		
	Veggie delite:											
69109	On white bread	1 ea	175	71	222	9	38	3	3	0		
69110	On wheat bread	1 ea	182	69	237	9	44	3	3	0		
	Salads:											
52128	B.L.T.	1 ea	276	91	140	7	10	2	8	3		
52124	B.M.T., classic Italian	1 ea	331	86	274	14	11	1	20	7		
52127	Chicken taco	1 ea	370	87	250	18	15	2	14	5		
52115	Club	1 ea	331	91	126	14	12	1	3	1		
52120	Cold cut trio	1 ea	330	89	191	13	11	1	11	3		
52123	Ham	1 ea	316	91	116	12	11	1	3	1		
52129	Meatball	1 ea	345	88	233	12	16	2	14	5		
52131	Melt	1 ea	336	88	195	16	12	1	10	4		
52121	Pizza	1 ea	335	86	277	12	13	2	20	8		
52126	Roast beef	1 ea	316	92	117	12	11	1	3	1		
52119	Roasted chicken breast	1 ea	331	89	162	20	13	1	4	1		
52117	Seafood and crab:	1 5	331	88	244	13	10	2	17	3		
52116	With light mayo	1 5	331	90	161	13	11	2	8	1		
52130	Steak and cheese	1 ea	342	87	212	22	13	1	8	5		
52122	Tuna:	1 ea	331	84	356	12	10	1	30	5		
52118	With light mayo	1 ea	331	89	205	12	11	1	13	2		
52114	Turkey breast	1 ea	316	92	102	11	12	1	2	1		
52125	With ham	1 ea	316	92	109	11	11	1	3	1		

Chol (mg)	Calc (mg)	Iron (mg)	Magn (mg)	Pota (mg)	Sodi (mg)	Zinc (mg)	VT-A (RE)	Thia (mg)	VT-E (α-TE)	Ribo (mg)	Niac (mg)	V-B6 (mg)	Fola (μg)	VT-C (mg)
56	44	4			1652		151							15
56	50	4			1664		151							15
33	32	4			1035		142							16
33	39	4			1046		142							16
42	93	4			1735		155							15
42	100	3			1746		156							15
50	103	4			1609		238							16
50	110	3			1621		238							16
20	25	4			928		120							15
20	32	3			939		120							15
48	35	3			967		123							15
48	42	3			978		123							15
34	28	3			849		121							15
32	28	3			873		131							15
34	34	3			860		121							15
32	34	3			884		131							15
57	40	4			1592		169							15
57	47	4			1604		169							15
70	88	5			1106		175							18
70	95	5			1117		176							18
36	32	3			875		125							15
32	32	3			928		146							15
36	38	3			886		126							15
32	38	3			940		146							15
19	30	4			1391		120							15
19	37	3			1403		120							15
24	29	3			1350		120							15
24	36	3			1361		120							15
0	25	3			582		120							15
0	32	3			593		120							15
16	24	1			672		273							32
56	41	2			1379		303							32
52	115	3			990		361							35
26	26	2			1067		273							32
64	46	2			1127		282							33
28	25	2			1034		273							32
33	30	2			761		295							33
42	90	2			1461		308							32
50	100	2			1336		390							33
20	23	2			654		273							32
48	32	2			693		276							32
34	25	2			575		273							32
32	25	2			599		284							32
70	86	3			832		328							35
36	29	2			601		278							32
32	29	2			654		298							32
19	28	2			1117		273							32
24	27	2			1076		273							32

Table H-1

Food Composition (Computer code number is for West Diet Analysis program) (For purposes of calculations, use "0" for t, <1, <.1, <.01, etc.)

Computer Code Number	Food Description	Measure	Wt (g)	H_2O (%)	Ener (kcal)	Prot (g)	Carb (g)	Dietary Fiber (g)	Fat (g)	Fat Breakdown (g) Sat	Mono	Poly
	FAST FOOD RESTAURANTS—Continued											
	SUBWAY—Continued											
52113	Veggie delite	1 ea	260	94	51	2	10	1	1	0		
	Cookies:											
47662	Brazil nut and chocolate chip	1 ea	48	12	229	3	27	1	12	3.5		
47655	Chocolate chip:	1 ea	48	14	209	2	29	1	10	3.5		
47658	With M&M's	1 ea	48	14	209	2	29	1	10	3		
47659	Chocolate chunk	1 ea	48	14	209	2	29	1	10	3.5		
47656	Oatmeal raisin	1 ea	48	15	199	3	29	1	8	2		
47657	Peanut butter	1 ea	48	13	219	3	26	1	12	2.5		
47660	Sugar	1 ea	48	11	229	2	28	0	12	3		
47661	White chip macademia	1 ea	48	12	229	2	28	1	12	2.5		
	Source: Subway International											
	TACO BELL											
	Breakfast burrito:											
1601	Bacon breakfast burrito	1 ea	99	48	291	11	23		17	4		
1627	Country breakfast burrito	1 ea	113	55	220	8	26	2	14	5		
1626	Fiesta breakfast burrito	1 ea	92	44	280	9	25	2	16	6		
1625	Grande breakfast burrito	1 ea	177	56	420	13	43	3	22	7		
1604	Sausage breakfast burrito	1 ea	106	49	303	11	23		19	6		
	Burritos:											
1544	Bean with red sauce	1 ea	198	58	380	13	55	13	12	4		
1545	Beef with red sauce	1 ea	198	57	432	22	42	4	19	8	6.7	.7
1546	Beef & bean with red sauce	1 ea	198	57	412	17	50	5	16	6	6.1	2.1
1569	Big beef supreme	1 ea	298	64	520	24	54	11	23	10		
1552	Chicken burrito	1 ea	171	58	345	17	41		13	5		
1547	Supreme with red sauce	1 ea	248	64	428	16	50	10	18	7.8		
1571	7 layer burrito	1 ea	234	61	438	13	55	11	19	5.8		
1538	Chilito	1 ea	156	49	391	17	41		18	9		
1549	Chilito, steak	1 ea	257	62	496	26	47		23	10		
	Tacos:											
1551	Taco	1 ea	78	58	180	9	12	3	10	4		
1554	Soft taco	1 ea	99	63	242	10	13	3	11	4.4		
1536	Soft taco supreme	1 ea	128	64	234	11	21	3	13	6.3		
1568	Soft taco, chicken	1 ea	128	63	212	15	22	2	7	2.6		
1572	Soft taco, steak	1 ea	100	63	180	12	16	2	8	1.9		
1555	Tostada with red sauce	1 ea	156	67	264	9	27	11	13	4.4		
1558	Mexican pizza	1 ea	223	53	578	21	43	8	35	10.1		
1559	Taco salad with salsa	1 ea	585	71	923	33	70	17	56	16.3		
1560	Nachos, regular	1 ea	106	40	343	5	36	3	19	4.3		
1561	Nachos, bellgrande	1 ea	287	51	708	19	77	16	36	10.1		
1562	Pintos & cheese with red sauce	1 ea	128	68	203	10	19	11	10	4.3		
1563	Taco sauce, packet	1 ea	9	94	2	<1	<1	<1	<1	0	0	0
1564	Salsa	1 ea	10	28	27	1	6		<1	0	0	0
1565	Cinnamon twists	1 ea	35	6	175	1	24	0	7	0		
1628	Caramel roll	1 ea	85	19	353	6	46		16	4		
	Source: Taco Bell Corporation											
	WENDY'S											
	Hamburgers:											
1566	Single on white bun, no toppings	1 ea	133	44	360	24	31	2	16	6		
1570	Cheeseburger, bacon	1 ea	166	55	380	20	34	2	19	7	10.3	1.4
1730	Chicken sandwich, grilled	1 ea	189	62	310	27	35	2	8	1.5		
	Baked potatoes:											
1573	Plain	1 ea	284	71	310	7	71	7	0	0	0	0
1574	With bacon & cheese	1 ea	380	69	530	17	78	7	18	4	10.7	3.3
1575	With broccoli & cheese	1 ea	411	74	470	9	80	9	14	2.5	8	2.5
1576	With cheese	1 ea	383	68	570	14	78	7	23	8	9.2	4.8

Chol (mg)	Calc (mg)	Iron (mg)	Magn (mg)	Pota (mg)	Sodi (mg)	Zinc (mg)	VT-A (RE)	Thia (mg)	VT-E (α-TE)	Ribo (mg)	Niac (mg)	V-B6 (mg)	Fola (μg)	VT-C (mg)
0	23	1			308		136							32
10	32	1.99			115		0							0
10	16	1.99			139		0							0
15	16	1			139		0							0
10	16	1			139		0							0
15	32	1			159		0							0
0	16	1			179		0							0
20	0	.72			179		0							0
10	16	1			139		0							0
181	80	1.8			652		310							
195	80	1.08			690		250							0
25	80	.72			580		150							0
205	100	1.8			1050		500							0
183	80	1.8			661		320							
10	150	2.7		495	1100		450	.04		2.02	1.98	.31		0
57	160	3.96		380	1303		530	.4		2.14	3.44	.32		1
32	170	3.78	50	442	1221	2.67	450	.49		.41	3.09	.59	38	1
55	150	2.7			1520		600							5
57	140	2.52			854		440							1
34	146	8.75	48	410	1196		486	.39		2.04	2.81	.34		5
21	165	2.98			1058		248							5
47	300	3.06			980		950							
78	200	2.7			1313		970							2
25	80	1.08		159	330		100	.05		.14	1.2	.12		0
27	88	1.19		211	363		110	.42		.24	2.95	1.08		0
31	90	1.62			532		135							3
37	85	1.52			571		63							1
19	62	1.13			797		31							0
13	132	1.59		401	573		441	.05		.17	.63	.26		1
46	253	3.65	80	408	1054	5.37	405	.32		.33	2.96	1.12	60	5
65	326	6.84		1048	1931	1.67	1736	.51		.76	4.8	.56	10	26
5	107	.77		160	610	1.68	64	.17		.16	.68	.19	10	0
32	184	3.31		674	1205		138	.1		.34	2.17			3
16	160	1.92	110	384	693	2.17	267	.05		.15	.43	.21	68	0
0	0	.07		9	75		30	0			.02			<1
0	50	.6		376	709		168	.02		.14	0			10
0	0	.45		27	238		50	.1		.04	.71	.04		0
15	60	1.44			312		330							4
65	110	4.14		296	580		0	.43		.38	6.71			0
60	170	3.42	38	375	850	5.9	80	.3		.31	6.43	.26	28	6
65	100	2.7			790		40							6
0	30	3.78	75	1187	25	.74	0	.31	.14	.12	4.3	.8	31	36
20	180	4.32	87	1498	1390	2.75	100	.24		.19	5.04	.94	36	36
5	210	4.5	93	1745	470	.97	350	.34		.29	4.5	.97	74	72
30	380	4.14	85	1510	640	.67	200	.25		.28	3.6	.88	36	36

Table H–1

Food Composition (Computer code number is for West Diet Analysis program) (For purposes of calculations, use "0" for t, <1, <.1, <.01, etc.)

Computer Code Number	Food Description	Measure	Wt (g)	H_2O (%)	Ener (kcal)	Prot (g)	Carb (g)	Dietary Fiber (g)	Fat (g)	Fat Breakdown (g) Sat	Mono	Poly
	FAST FOOD RESTAURANTS—Continued											
	WENDY'S—Continued											
1577	With chili & cheese	1 ea	439	69	630	20	83	9	24	9		
1578	With sour cream & chives	1 ea	314	71	380	8	74	8	6	4		
1579	Chili	1 ea	227	81	210	15	21	5	7	2.5		
1582	Chocolate chip cookies	1 ea	57	6	270	3	36	1	13	6		
1580	French fries	1 ea	130	41	390	5	50	5	19	3	11.9	2.4
1581	Frosty dairy dessert	1 ea	227	68	330	8	56	0	8	5		
	Source: Wendy's International											
	CONVENIENCE FOODS and MEALS											
	BUDGET GOURMET											
1695	Chicken cacciatore	1 ea	312	80	300	20	27		13			
1692	Linguini & shrimp	1 ea	284	77	330	15	33		15			
1691	Scallops & shrimp	1 ea	326	79	320	16	43		9			
2245	Seafood newburg	1 ea	284	74	350	17	43		12			
1693	Sirloin tips with country gravy	1 ea	284	80	310	16	21		18			
1694	Sweet & sour chicken with rice	1 ea	284	72	350	18	53		7			
1689	Teriyaki chicken	1 ea	340	77	360	20	44		12			
1690	Veal parmigiana	1 ea	340	75	440	26	39		20			
1696	Yankee pot roast	1 ea	312	77	380	27	22		21			
	Source: The All American Gourmet Co.											
	HAAGEN DAZS											
1755	Ice cream bar, vanilla almond	1 ea	107		371	6	26		27	14.1	10	3
	Sorbet:											
1758	Lemon	½ c	113		140	0	35		0	0	0	0
1760	Orange	½ c	113		140	0	36		0	0	0	0
1759	Raspberry	½ c	113		110	0	27		0	0	0	0
	Yogurt, frozen:											
1753	Chocolate	½ c	98		171	8	26		4	2	2	0
1754	Strawberry	½ c	98		171	6	27		4	2	2	0
	Yogurt extra, frozen:											
1752	Brownie nut	½ c	101		220	8	29		9	4	4	1
1751	Raspberry rendezvous	½ c	101		132	4	26		2	1	1	0
	Source: Pillsbury											
	HEALTHY CHOICE											
	Entrees:											
2112	Fish, lemon pepper	1 ea	303	78	290	14	47	7	5	1		
1624	Lasagna	1 ea	383	76	390	26	60	9	5	2		
2111	Meatloaf, traditional	1 ea	340	79	320	16	46	7	8	4		
2104	Zucchini lasagna	1 ea	396	80	329	20	58	11	1	1		
2110	Dinner, pasta shells marinara	1 ea	340	74	360	25	59	5	3	1.5		
	Low-fat ice cream:											
973	Brownie	½ c	71	60	120	3	22	2	2	1	.3	.7
259	Butter pecan	½ c	71	60	120	3	22	1	2	1	.3	.7
650	Chocolate chip	½ c	71	62	120	3	21	<1	2	1	1	0
1608	Cookie & cream	½ c	71	62	120	3	21	<1	2	1.5	.5	0
650	Chocolate chip	½ c	71	62	120	3	21	<1	2	1	1	0
45	Rocky road	½ c	71	53	140	3	28	2	1	1	.5	0
1621	Vanilla	½ c	71	66	100	3	18	1	2	.5	1.5	0
391	Vanilla fudge	½ c	71	62	120	3	21	1	2	1.5		
	Source: ConAgra Frozen Foods, Omaha, NE											

Chol (mg)	Calc (mg)	Iron (mg)	Magn (mg)	Pota (mg)	Sodi (mg)	Zinc (mg)	VT-A (RE)	Thia (mg)	VT-E (α-TE)	Ribo (mg)	Niac (mg)	V-B6 (mg)	Fola (μg)	VT-C (mg)
40	330	5.04	122	1745	770	4.15	200	.33		.29	4.5	.99	55	36
15	80	4.32	71	1438	40	.91	300	.23		.14	3.04	.8	32	48
30	80	2.9		501	800		80	.11		.15	2.66			4
30	10	1.8	13	89	120	.41	0	.05		.06	.36	.03	5	0
0	20	1.08	55	845	120	.62	0	.18		.04	3.6	.33	40	6
35	310	1.08	46	544	200	.97	150	.11		.47	.32	.13	17	0
60	150	1.8			810		40	.23		.51	5			21
75	10	3.6			1250		1000	.3		.17	3			2
70	150	.72			690		150			.26	3			12
70	100	.72			660		40	.23		.26	2			
40	60	.36			570		150	.15		.17	4	.28		2
40	60	.72			640		80	.12		.34	3			2
55	80	1.4			610		300	.15		.34	6			12
165	30	4.5			1160		1000	.45		.6	6			6
70	150	1.8			690		600	.15		.43	7			6
90	161	.38		221	85		161			.18				
0				30	20									7
0				80	20									20
0				60	15									7
40	147	.71		241	45		20			.17				
50	147			141	45		20	.03		.17				5
55	152	.73		250	60		20			.14				
20	81			97	25		0			.1				5
25	20	1.08			360		100							30
15	150	3.6		500	550		100	.3		.26	2			6
35	40	1.8			460		150							54
10	199	2.69			309		249							0
25	400	1.8			390		100							4
2	80	0		268	55		40							0
2	100	0		211	60		40							0
2	100	0		240	50		40							0
2	100			254	90		60	.03		.15				2
2	100	0		240	50		40							0
2	100	0		168	60		40	.03		.15				0
5	100			254	50		60	.05		.22				2
2	100	0		296	50		40							0

Table H-1

Food Composition (Computer code number is for West Diet Analysis program) (For purposes of calculations, use "0" for t, <1, <.1, <.01, etc.)

Computer Code Number	Food Description	Measure	Wt (g)	H_2O (%)	Ener (kcal)	Prot (g)	Carb (g)	Dietary Fiber (g)	Fat (g)	Fat Breakdown (g) Sat	Mono	Poly
	CONVENIENCE FOODS and MEALS—Continued											
	HEALTH VALLEY											
	Soups, fat-free:											
2001	Beef broth, no salt added	1 c	240	98	18	5	0	0	0	0	0	0
2073	Beef broth, w/salt	1 c	240	98	30	5	2	0	0	0	0	0
2016	Black bean & vegetable	1 c	240	85	110	11	24	12	0	0	0	0
2017	Chicken broth	1 c	240	97	30	6	0	0	0	0	0	0
2018	14 garden vegetable	1 c	240	90	80	6	17	4	0	0	0	0
2015	Lentil & carrot	1 c	240	85	90	10	25	14	0	0	0	0
2014	Split pea & carrot	1 c	240	89	110	8	17	4	0	0	0	0
2013	Tomato vegetable	1 c	240	90	80	6	17	5	0	0	0	0
	Source: Health Valley											
	LA CHOY											
2100	Egg rolls, mini, chicken	1 svg	106	53	220	8	35	3	6	1.5		
2099	Egg rolls, mini, shrimp	1 svg	106	56	210	7	35	3	4	1		
	Source: Beatrice/Hunt Wesson											
	LEAN CUISINE											
	Dinners:											
1639	Baked cheese ravioli	1 ea	241	77	250	12	32	4	8	3	2	1
1632	Chicken chow mein	1 ea	255	81	210	13	28	2	5	1	2	1
1633	Lasagna	1 ea	291	79	270	19	34	5	6	2.5	1.5	.5
1634	Macaroni & cheese	1 ea	255	78	270	13	39	2	7	3.5	1.5	.5
1631	Spaghetti w/meatballs	1 ea	269	74	290	17	40	4	7	2	3	1.5
	Pizza:											
1636	French bread sausage pizza	1 ea	170	53	420	19	41	4	20	5	13.9	1.1
	Source: Stouffer's Foods Corp, Solon, OH											
	TASTE ADVENTURE SOUPS											
1905	Black bean	1 c	242		139	6	28	6	1			
1904	Curry lentil	1 c	241		138	6	30	5	1			
1906	Lentil chili	1 c	242		181	11	33	6	1			
1903	Split pea	1 c	244		140	5	27	5	1			
	Source: Taste Adventure Soups											
	WEIGHT WATCHERS											
	Cheese, fat-free slices:											
1978	Cheddar, sharp	2 pce	21	65	30	5	2	0	0	0	0	0
1980	Swiss	2 pce	21	65	30	5	2	0	0	0	0	0
1977	White	2 pce	21	65	30	5	2	0	0	0	0	0
1979	Yellow	2 pce	21	65	30	5	2	0	0	0	0	0
	Dinners:											
2029	Chicken chow mein	1 ea	255	81	200	12	34	3	2	.5		
1646	Oven fried fish	1 ea	218	78	230	15	25	2	8	2.5	5	2
1972	Margarine, reduced fat	1 tbs	14	49	59	0	0	0	7	1.5		
	Pizza:											
1653	Cheese	1 ea	163	48	390	23	49	6	12	4	3	1
1650	Deluxe combination pizza	1 ea	186	56	380	23	47	6	11	3.5	5	2
1652	Pepperoni pizza	1 ea	158	48	390	23	46	4	12	4	5	2
	Desserts:											
1644	Chocolate brownie	1 ea	182	75	190	6	35	4	4	1	2	1
2024	Chocolate eclair	1 ea	60	45	151	3	24	2	5	1.5		
2247	Chocolate mousse	1 ea	78	44	190	6	33	3	4	1.5		
1642	Strawberry cheesecake	1 ea	111	62	180	7	28	2	5	2	1	2

Chol (mg)	Calc (mg)	Iron (mg)	Magn (mg)	Pota (mg)	Sodi (mg)	Zinc (mg)	VT-A (RE)	Thia (mg)	VT-E (α-TE)	Ribo (mg)	Niac (mg)	V-B6 (mg)	Fola (μg)	VT-C (mg)
0				196	74						.98			
0	0	0		196	160		0				.98			5
0	40	3.6		676	280		2000	.34		.11	1.35	.22	135	9
0	20	1.8		147	170		0			.03	2.45			1
0	40	1.8		406	250		2000	.26		.08	2.25	.18	27	15
0	60	5.4		439	220		2000	.1		.16	5.63	.45	27	2
0	40	5.4		439	230		2000	.1		.16	5.63	.45		9
0	40	5.4		609	240		2000	.1		.08	2.25	.13	<1	9
5	20	1.44			460		20							0
5	20	1.44			510		20							0
55	200	1.08	42	400	500	1.5	150	.06		.25	1.2	.2	48	6
35	20	.36	30	300	510	1.1	20	.15		.17	5			6
25	150	1.8	44	620	560	2.9	100	.15		.25	3	.32		12
20	250	.72		170	550		20	.12		.25	1.2			0
30	100	2.7	47	480	520	2.5	80	.15		.25	3	.2		4
35	250	2.7	39	340	900	2.2	80	.45		.51	5	.07		6
				650	565									
				467	584									
				650	448									
				484	591									
0	99	0		64	306		56							0
0	99	0		74	276		56							0
0	99	0		64	306		56							0
0	99	0		64	306		56							0
25	40	.72		360	430		300							36
25	20	1.44		370	450		40	.09		.14	1.6			0
0	0	0		5	128		49							0
35	700	1.8		290	590		80	.3		.51	3	.06		6
40	500	3.6		370	550		150	.3		.51	3	.2		5
45	450	1.8		320	650		80	.23		.51	3			5
5	80	1.08		230	160		0	.06		.03	.2	.03		0
0	40	0		65	151		0							0
5	60	1.8		320	150		0							0
15	80	.36		115	230		40	.06		.07	1.6			2

Table H-1

Food Composition (Computer code number is for West Diet Analysis program) (For purposes of calculations, use "0" for t, <1, <.1, <.01, etc.)

Computer Code Number	Food Description	Measure	Wt (g)	H_2O (%)	Ener (kcal)	Prot (g)	Carb (g)	Dietary Fiber (g)	Fat (g)	Fat Breakdown (g) Sat	Mono	Poly
	CONVENIENCE FOODS and MEALS—Continued											
	WEIGHT WATCHERS—CONTINUED											
2027	Triple chocolate cheesecake	1 ea	89	52	199	7	32	1	5	2.5		
	Source: Weight Watchers											
	SWEET SUCCESS:											
	Drinks, prepared:											
1776	Chocolate chip	1 c	265	81	180	15	30	6	3	1.6		
1777	Chocolate fudge	1 c	265	81	180	15	30	6	2			
1774	Chocolate mocha	1 c	265	81	180	15	30	6	1	1		
1778	Milk chocolate	1 c	265	81	180	15	30	6	2	1		
1775	Vanilla	1 c	265	81	180	15	33	6	1	.6		
	Drinks, ready to drink:											
2147	Chocolate mint	1 c	297	82	187	11	36	6	3	0		
2148	Strawberry	1 c	265	82	167	10	32	5	3	0		
	Shakes:											
1771	Chocolate almond	1 c	250	82	158	9	30	5	2	0	2.1	.2
1773	Chocolate fudge	1 c	250	82	158	9	30	5	2	0	2.1	.2
1768	Chocolate mocha	1 c	250	82	158	9	30	5	2	0	.6	1.8
1769	Chocolate raspberry truffle	1 c	250	82	158	9	30	5	2	0	2.2	.2
1770	Vanilla creme	1 c	250	82	158	9	30	5	2	0	2.1	.3
	Snack bars:											
1767	Chocolate brownie	1 ea	33	9	120	2	23	3	4	2	.5	.6
1766	Chocolate chip	1 ea	33	9	120	2	23	3	4	2	.4	.5
1921	Oatmeal raisin	1 ea	33	9	120	2	23	3	4	2		
1765	Peanut butter	1 ea	33	9	120	2	23	3	4	2	.6	.6
	Source: Foodway National Inc, Boise, ID											
	BABY FOODS											
1720	Apple juice	½ c	125	88	59	0	15	<1	<1	t	t	t
1721	Applesauce, strained	1 tbs	16	89	7	<1	2	<1	<1	t	t	t
1716	Carrots, strained	1 tbs	14	92	4	<1	1	<1	<1	t	t	t
1718	Cereal, mixed, milk added	1 tbs	15	75	17	1	2	<1	1	.3		
1719	Cereal, rice, milk added	1 tbs	15	75	17	<1	3	<1	1	.3		
1723	Chicken and noodles, strained	1 tbs	16	88	8	<1	1	<1	<1	.1	.1	t
1722	Peas, strained	1 tbs	15	87	6	1	1	<1	<1	t	t	t
1717	Teething biscuits	1 ea	11	6	43	1	8	<1	<1	.2	.2	.1

Chol (mg)	Calc (mg)	Iron (mg)	Magn (mg)	Pota (mg)	Sodi (mg)	Zinc (mg)	VT-A (RE)	Thia (mg)	VT-E (α-TE)	Ribo (mg)	Niac (mg)	V-B6 (mg)	Fola (μg)	VT-C (mg)
10	80	1.08		169	199		0							0
6	500	6.3	140	600	288	5.25	350	.52	7.05	.59	7	.7	140	21
6	500	6.3	140	750	336	5.25	350	.52	7.05	.59	7	.7	140	21
6	500	6.3	140	800	336	5.25	350	.52	7.05	.59	7	.7	140	21
6	500	6.3	140	750	336	5.25	350	.52	7.05	.59	7	.7	140	21
6	500	6.3	140	830	312	5.25	250	.52	7.05	.59	7	.7	140	21
6	470	5.94	131	526	226	5.05	329	.5	6.56	.56	6.53	.65	131	20
5	419	5.3	117	310	175	4.51	294	.45	5.86	.5	5.83	.58	117	17
5	396	5	110	443	190	4.25	277	.42	5.53	.47	5.5	.55	110	16
5	396	5	110	443	175	4.25	277	.42	5.53	.47	5.5	.55	110	16
5	396	5	110	403	175	4.25	277	.42	5.53	.47	5.5	.55	110	16
5	383	5	110	443	175	4.25	277	.42	5.53	.47	5.5	.55	110	16
5	396	5	110	293	175	4.25	277	.42	5.53	.47	5.5	.55	110	16
3	150	2.71	60	140	45	.59	150	.22	3.01	.25	3	.3	60	9
3	150	2.71	60	110	40	.59	150	.22	3.01	.25	3	.3	60	9
3	150	2.71	60		30	.59	150	.22	3.01	.25	3	.3	60	9
3	150	2.71	60	125	35	.59	150	.22	3.01	.25	3	.3	60	9
0	5	.71	4	114	4	.04	2	.01	.75	.02	.1	.04	<1	72
0	1	.03		11	<1	<.01	<1	<.01	.1	<.01	.01	<.01	<1	6
0	3	.05	1	27	5	.02	160	<.01	.07	.01	.06	.01	2	1
2	33	1.56	4	30	7	.11	4	.06		.09	.87	.01	2	<1
2	36	1.83	7	28	7	.1	4	.07		.07	.78	.02	1	<1
3	4	.07	1	6	3	.05	18	<.01	.04	.01	.07	<.01	2	<1
0	3	.14	2	17	1	.05	8	.01	.08	.01	.15	.01	4	1
0	29	.39	4	35	40	.1	1	.03	.05	.06	.48	.01	5	1

H

Contents

WHO: NUTRITION RECOMMENDATIONS CANADA: CHOICE SYSTEM AND GUIDELINES

This appendix first presents nutrition recommendations from the World Health Organization (WHO) and then provides details for Canadians on the exchange system (called the Choice System in Canada) and food guides. Appendix F includes addresses of Canadian and international agencies and professional organizations that may provide additional information.◆

◆ For information on Canadian guidelines and food labels, visit: www.hc-sc.gc.ca

Nutrition Recommendations from WHO

The World Health Organization (WHO) has assessed the relationships between diet and the development of chronic diseases. Its recommendations are expressed in average daily ranges that represent the lower and upper limits:

- Total energy: sufficient to support normal growth, physical activity, and body weight (body mass index = 20 to 22).
- Total fat: 15 to 30 percent of total energy.
  - Saturated fatty acids: 0 to 10 percent total energy.
  - Polyunsaturated fatty acids: 3 to 7 percent total energy.
  - Dietary cholesterol: 0 to 300 milligrams per day.
- Total carbohydrate: 55 to 75 percent total energy.
  - Complex carbohydrates: 50 to 75 percent total energy.
  - Dietary fiber: 27 to 40 grams per day.
  - Refined sugars: 0 to 10 percent total energy.
- Protein: 10 to 15 percent total energy.
- Salt: upper limit of 6 grams/day (no lower limit set).

Canada's Choice System for Meal Planning

The *Good Health Eating Guide* is the Canadian choice system of meal planning.[1] It contains several features similar to those of the U.S. exchange system including the following:

- Foods are divided into lists according to carbohydrate, protein, and fat content.
- Foods are interchangeable within a group.
- Most foods are eaten in measured amounts.
- An energy value is given for each food group.

Tables I-1 through I-8 present the Canadian choice system.

[1] The tables for the Canadian choice system are adapted from the *Good Health Eating Guide Resource,* copyright 1994, with permission of the Canadian Diabetes Association.

I

TABLE I-1 Canadian Choice System: Starch Foods

1 starch choice = 15 g carbohydrate (starch), 2 g protein, 290 kJ (68 kcal)

Food	Measure	Mass (Weight)
Breads		
Bagels	½	30 g
Bread crumbs	50 mL (¼ c)	30 g
Bread cubes	250 mL (1 c)	30 g
Bread sticks	2	20 g
Brewis, cooked	50 mL (¼ c)	45 g
Chapati	1	20 g
Cookies, plain	2	20 g
English muffins, crumpets	½	30 g
Flour	40 mL (2½ tbs)	20 g
Hamburger buns	½	30 g
Hot dog buns	½	30 g
Kaiser rolls	½	30 g
Matzo, 15 cm	1	20 g
Melba toast, rectangular	4	15 g
Melba toast, rounds	7	15 g
Pita, 20 cm (8") diameter	¼	30 g
Pita, 15 cm (6") diameter	½	30 g
Plain rolls	1 small	30 g
Pretzels	7	20 g
Raisin bread	1 slice	30 g
Rice cakes	2	30 g
Roti	1	20 g
Rusks	2	20 g
Rye, coarse or pumpernickel	½ slice	30 g
Soda crackers	6	20 g
Tortillas, corn (taco shell)	1	30 g
Tortilla, flour	1	30 g
White (French and Italian)	1 slice	25 g
Whole-wheat, cracked-wheat, rye, white enriched	1 slice	30 g
Cereals		
Bran flakes, 100% bran	125 mL (½ c)	30 g
Cooked cereals, cooked	125 mL (½ c)	125 g
Dry	30 mL (2 tbs)	20 g
Cornmeal, cooked	125 mL (½ c)	125 g
Dry	30 mL (2 tbs)	20 g
Ready-to-eat unsweetened cereals	125 mL (½ c)	20 g
Shredded wheat biscuits, rectangular or round	1	20 g
Shredded wheat, bite size	125 mL (½ c)	20 g
Wheat germ	75 mL (⅓ c)	30 g
Cornflakes	175 mL (⅔ c)	20 g
Rice Krispies	175 mL (⅔ c)	20 g
Cheerios	200 mL (¾ c)	20 g
Muffets	1	20 g
Puffed rice	300 mL (1¼ c)	15 g
Puffed wheat	425 mL (1⅔ c)	20 g
Grains		
Barley, cooked	125 mL (½ c)	120 g
Dry	30 mL (2 tbs)	20 g
Bulgur, kasha, cooked, moist	125 mL (½ c)	70 g

(continued on the next page)

TABLE I-1 **Canadian Choice System: Starch Foods—(continued)**

1 starch choice = 15 g carbohydrate (starch), 2 g protein, 290 kJ (68 kcal)

Food	Measure	Mass (Weight)
Cooked, crumbly	75 mL (⅓ c)	40 g
Dry	30 mL (2 tbs)	20 g
Rice, cooked, brown & white (short & long grain)	125 mL (½ c)	70 g
Rice, cooked, wild	75 mL (⅓ c)	70 g
Tapioca, pearl and granulated, quick cooking, dry	30 mL (2 tbs)	15 g
Couscous, cooked moist	125 mL (½ c)	70 g
Dry	30 mL (tbs)	20 g
Quinoa, cooked moist	125 mL (½ c)	70 g
Dry	30 mL (2 tbs)	20 g
Pastas		
Macaroni, cooked	125 mL (½ c)	70 g
Noodles, cooked	125 mL (½ c)	80 g
Spaghetti, cooked	125 mL (½ c)	70 g
Starchy Vegetables		
Beans and peas, dried, cooked	125 mL (½ c)	80 g
Breadfruit	1 slice	75 g
Corn, canned, whole kernel	125 mL (½ c)	85 g
Corn on the cob	½ medium cob	140 g
Cornstarch	30 mL (2 tbs)	15 g
Plantains	⅓ small	50 g
Popcorn, air-popped, unbuttered	750 mL (3 c)	20 g
Potatoes, whole (with or without skin)	½ medium	95 g
Yams, sweet potatoes (with or without skin)	½	75 g

Food	Choices per Serving	Measure	Mass (Weight)
NOTE: Food items found in this category provide more than 1 starch choice:			
Bran flakes	1 starch + ½ sugar	150 mL (⅔ c)	24 g
Croissant, small	1 starch + 1½ fats	1 small	35 g
Large	1 starch + 1½ fats	½ large	30 g
Corn, canned creamed	1 starch + ½ fruits and vegetables	12 mL (½ c)	113 g
Potato chips	1 starch + 2 fats	15 chips	30 g
Tortilla chips (nachos)	1 starch + 1½ fats	13 chips	20 g
Corn chips	1 starch + 2 fats	30 chips	30 g
Cheese twists	1 starch + 1½ fats	30 chips	30 g
Cheese puffs	1 starch + 2 fats	27 chips	30 g
Tea biscuit	1 starch + 2 fats	1	30 g
Pancakes, homemade using 50 mL (¼ c) batter (6" diameter)	1½ starches + 1 fat	1 medium	50 g
Potatoes, french fried (homemade or frozen)	1 starch + 1 fat	10 regular size	35 g
Soup, canned* (prepared with equal volume of water)	1 starch	250 mL (1 c)	260 g
Waffles, packaged	1 starch + 1 fat	1	35 g

*Soup can vary according to brand and type. Check the label for Food Choice Values and Symbols or the core nutrient listing.

TABLE I-2 **Canadian Choice System: Fruits and Vegetables**

1 fruits and vegetables choice = 10 g carbohydrate, 1 g protein, 190 kJ (44 kcal)

Food	Measure	Mass (Weight)
Fruits (fresh, frozen, without sugar, canned in water)		
Apples, raw (with or without skin)	½ medium	75 g
Sauce unsweetened	125 mL (½ c)	120 g
Sweetened	see *Combined Food Choices*	
Apple butter	20 mL (4 tsp)	20 g
Apricots, raw	2 medium	115 g
Canned, in water	4 halves, plus 30 mL (2 tbs) liquid	110 g
Bake-apples (cloudberries), raw	125 mL (½ c)	120 g
Bananas, with peel	½ small	75 g
Peeled	½ small	50 g
Berries (blackberries, blueberries, boysenberries, huckleberries, loganberries, raspberries)		
Raw	125 mL (½ c)	70 g
Canned, in water	125 mL (½ c), plus 30 mL (2 tbs) liquid	100 g
Cantaloupe, wedge with rind	¼	240 g
Cubed or diced	250 mL (1 c)	160 g
Cherries, raw, with pits	10	75 g
Raw, without pits	10	70 g
Canned, in water, with pits	75 mL (⅓ c), plus 30 mL (2 tbs) liquid	90 g
Canned, in water, without pits	75 mL (⅓ c), plus 30 mL (2 tbs) liquid	85 g
Crabapples, raw	1 small	55 g
Cranberries, raw	250 mL (1 c)	100 g
Figs, raw	1 medium	50 g
Canned, in water	3 medium, plus 30 mL (2 tbs) liquid	100 g
Foxberries, raw	250 mL (1 c)	100 g
Fruit cocktail, canned, in water	125 mL (½ c), plus 30 mL (2 tbs) liquid	120 g
Fruit, mixed, cut-up	125 mL (½ c)	120 g
Gooseberries, raw	250 mL (1 c)	150 g
Canned, in water	250 mL (1 c), plus 30 mL (2 tbs) liquid	230 g
Grapefruit, raw, with rind	½ small	185 g
Raw, sectioned	125 mL (½ c)	100 g
Canned, in water	125 mL (½ c), plus 30 mL (2 tbs) liquid	120 g
Grapes, raw, slip skin	125 mL (½ c)	75 g
Raw, seedless	125 mL (½ c)	75 g
Canned, in water	75 mL (⅓ c), plus 30 mL (2 tbs) liquid	115 g
Guavas, raw	½	50 g
Honeydew melon, raw, with rind	½	225 g
Cubed or diced	250 mL (1 c)	170 g
Kiwis, raw, with skin	2	155 g
Kumquats, raw	3	60 g
Loquats, raw	8	130 g
Lychee fruit, raw	8	120 g
Mandarin oranges, raw, with rind	1	135 g
Raw, sectioned	125 mL (½ c)	100 g
Canned, in water	125 mL (½ c), plus 30 mL (2 tbs) liquid	100 g
Mangoes, raw, without skin and seed	⅓	65 g
Diced	75 mL (⅓ c)	65 g
Nectarines	½ medium	75 g
Oranges, raw, with rind	1 small	130 g
Raw, sectioned	125 mL (½ c)	95 g

(continued on the next page)

TABLE I-2 Canadian Choice System: Fruits and Vegetables—(continued)

1 fruits and vegetables choice = 10 g carbohydrate, 1 g protein, 190 kJ (44 kcal)

Papayas, raw, with skin and seeds	¼ medium	150 g
Raw, without skin and seeds	¼ medium	100 g
Cubed or diced	125 mL (½ c)	100 g
Peaches, raw, with seed and skin	1 large	100 g
Raw, sliced or diced	125 mL (½ c)	100 g
Canned in water, halves or slices	125 mL (½ c), plus 30 mL (2 tbs) liquid	120 g
Pears, raw, with skin and core	½	90 g
Raw, without skin and core	½	85 g
Canned, in water, halves	1 half plus 30 mL (2 tbs) liquid	90 g
Persimmons, raw, native	1	30 g
Raw, Japanese	¼	50 g
Pineapple, raw	1 slice	75 g
Raw, diced	125 mL (½ c)	75 g
Canned, in juice, diced	75 mL (⅓ c), plus 15 mL (1 tbs) liquid	55 g
Canned, in juice, sliced	1 slice, plus 15 mL (1 tbs) liquid	55 g
Canned, in water, diced	125 mL (½ c), plus 30 mL (2 tbs) liquid	100 g
Canned, in water, sliced	2 slices, plus 15 mL (1 tbs) liquid	100 g
Plums, raw	2 small	60 g
Damson	6	65 g
Japanese	1	70 g
Canned, in apple juice	2, plus 30 mL (2 tbs) liquid	70 g
Canned, in water	3, plus 30 mL (2 tbs) liquid	100 g
Pomegranates, raw	½	140 g
Strawberries, raw	250 mL (1 c)	150 g
Frozen/canned, in water	250 mL (1 c), plus 30 mL (2 tbs) liquid	240 g
Rhubarb	250 mL (1 c)	150 g
Tangelos, raw	1	205 g
Tangerines, raw	1 medium	115 g
Raw, sectioned	125 mL (½ c)	100 g
Watermelon, raw, with rind	1 wedge	310 g
Cubed or diced	250 mL (1 c)	160 g
Dried Fruit		
Apples	5 pieces	15 g
Apricots	4 halves	15 g
Banana flakes	30 mL (2 tbs)	15 g
Currants	30 mL (2 tbs)	15 g
Dates, without pits	2	15 g
Peaches	½	15 g
Pears	½	15 g
Prunes, raw, with pits	2	15 g
Raw, without pits	2	10 g
Stewed, no liquid	2	20 g
Stewed, with liquid	2, plus 15 mL (1 tbs) liquid	35 g
Raisins	30 mL (2 tbs)	15 g
Juices (no sugar added or unsweetened)		
Apricot, grape, guava, mango, prune	50 mL (¼ c)	55 g
Apple, carrot, papaya, pear, pineapple, pomegranate	75 mL (⅓ c)	80 g
Cranberry (see Sugars Section)		
Clamato (see Sugars Section)		

(continued on the next page)

TABLE I-2 Canadian Choice System: Fruits and Vegetables—(continued)

1 fruits and vegetables choice = 10 g carbohydrate, 1 g protein, 190 kJ (44 kcal)

Grapefruit, loganberry, orange, raspberry, tangelo, tangerine	125 mL (½ c)	130 g
Tomato, tomato-based mixed vegetables	250 mL (1 c)	255 g
Vegetables (fresh, frozen, or canned)		
Artichokes, French, globe	2 small	50 g
Beets, diced or sliced	125 mL (½ c)	85 g
Carrots, diced, cooked or uncooked	125 mL (½ c)	75 g
Chestnuts, fresh	5	20 g
Parsnips, mashed	125 mL (½ c)	80 g
Peas, fresh or frozen	125 mL (½ c)	80 g
Canned	75 mL (⅓ c)	55 g
Pumpkin, mashed	125 mL (½ c)	45 g
Rutabagas, mashed	125 mL (½ c)	85 g
Sauerkraut	250 mL (1 c)	235 g
Snow peas	250 mL (1 c)	135 g
Squash, yellow or winter, mashed	125 mL (½ c)	115 g
Succotash	75 mL (⅓ c)	55 g
Tomatoes, canned	250 mL (1 c)	240 g
Tomato paste	50 mL (¼ c)	55 g
Tomato sauce*	75 mL (⅓ c)	100 g
Turnips, mashed	125 mL (½ c)	115 g
Vegetables, mixed	125 mL (½ c)	90 g
Water chestnuts	8 medium	50 g

*Tomato sauce varies according to brand name. Check the label or discuss with your dietitian.

TABLE I-3 Canadian Choice System: Milk

Type of Milk	Carbohydrate (g)	Protein (g)	Fat (g)	Energy
Nonfat (0%)	6	4	0	170 kJ (40 kcal)
1%	6	4	1	206 kJ (49 kcal)
2%	6	4	2	244 kJ (58 kcal)
Whole (4%)	6	4	4	319 kJ (76 kcal)

Food	Measure	Mass (Weight)
Buttermilk (higher in salt)	125 mL (½ c)	125 g
Evaporated milk	50 mL (¼ c)	50 g
Milk	125 mL (½ c)	125 g
Powdered milk, regular	30 mL (2 tbs)	15 g
Instant	50 mL (¼ c)	15 g
Plain yogurt	125 mL (½ c)	125 g

Food	Choices per Serving	Measure	Mass (Weight)
NOTE: Food items found in this category provide more than 1 milk choice:			
Milkshake	1 milk + 3 sugars + ½ protein	250 mL (1 c)	300 g
Chocolate milk, 2%	2 milks 2% + 1 sugar	250 mL (1 c)	300 g
Frozen yogurt	1 milk + 1 sugar	125 mL (½ c)	125 g

I

TABLE I-4 Canadian Choice System: Sugars

1 sugar choice = 10 g carbohydrate (sugar), 167 kJ (40 kcal)

Food	Measure	Mass (Weight)
Beverages		
Condensed milk	15 mL (1 tbs)	
Flavoured fruit crystals*	75 mL (⅓ c)	
Iced tea mixes*	75 mL (⅓ c)	
Regular soft drinks	125 mL (½ c)	
Sweet drink mixes*	75 mL (⅓ c)	
Tonic water	125 mL (½ c)	
*These beverages have been made with water.		
Miscellaneous		
Bubble gum (large square)	1 piece	5 g
Cranberry cocktail	75 mL (⅓ c)	80 g
Cranberry cocktail, light	350 mL (1⅓ c)	260 g
Cranberry sauce	30 mL (2 tbs)	
Hard candy mints	2	5 g
Honey, molasses, corn & cane syrup	10 mL (2 tsp)	15 g
Jelly bean	4	10 g
Licorice	1 short stick	10 g
Marshmallows	2 large	15 g
Popsicle	1 stick (½ popsicle)	
Powdered gelatin mix		
(Jello®) (reconstituted)	50 mL (¼ c)	
Regular jam, jelly, marmalade	15 mL (1 tbs)	
Sugar, white, brown, icing, maple	10 mL (2 tsp)	10 g
Sweet pickles	2 small	100 g
Sweet relish	30 mL (2 tbs)	

Food	Choices per Serving	Measures	Mass (Weight)
The following food items provide more than 1 sugar choice:			
Brownie	1 sugar + 1 fat	1	20 g
Clamato juice	1½ sugars	175 mL (⅔ c)	
Fruit salad, light syrup	1 sugar + 1 fruits & vegetables	125 mL (½ c)	130 g
Aero® bar	2½ sugars + 2½ fats	1 bar	43 g
Smarties®	4½ sugars + 2 fats	1 box	60 g
Sherbet	3 sugars + ½ fat	125 mL (½ c)	95 g

TABLE I-5 Canadian Choice System: Protein Foods

1 protein choice = 7 g protein, 3 g fat, 230 kJ (55 kcal)

Food	Measure	Mass (Weight)
Cheese		
Low-fat cheese, about 7% milk fat	1 slice	30 g
Cottage cheese, 2% milkfat or less	50 mL (¼ c)	55 g
Ricotta, about 7% milkfat	50 mL (¼ c)	60 g
Fish		
Anchovies (see *Extras,* Table I-7)		
Canned, drained (e.g., mackerel, salmon, tuna packed in water)	(⅓ of 6.5 oz can)	30 g
Cod tongues, cheeks	75 mL (⅓ c)	50 g
Fillet or steak (e.g., Boston blue, cod, flounder, haddock, halibut, mackerel, orange roughy, perch, pickerel, pike, salmon, shad, snapper, sole, swordfish, trout, tuna, whitefish)	1 piece	30 g
Herring	⅓ fish	30 g
Sardines, smelts	2 medium or 3 small	30 g
Squid, octopus	50 mL (¼ c)	40 g
Shellfish		
Clams, mussels, oysters, scallops, snails	3 medium	30 g
Crab, lobster, flaked	50 mL (¼ c)	30 g
Shrimp, fresh	5 large	30 g
Frozen	10 medium	30 g
Canned	18 small	30 g
Dry pack	50 mL (¼ c)	30 g
Meat and Poultry (e.g., beef, chicken, goat, ham, lamb, pork, turkey, veal, wild game)		
Back, peameal bacon	3 slices, thin	30 g
Chop	½ chop, with bone	40 g
Minced or ground, lean or extra-lean	30 mL (2 tbs)	30 g
Sliced, lean	1 slice	30 g
Steak, lean	1 piece	30 g
Organ Meats		
Hearts, liver	1 slice	30 g
Kidneys, sweetbreads, chopped	50 mL (¼ c)	30 g
Tongue	1 slice	30 g
Tripe	5 pieces	60 g
Soyabean		
Bean curd or tofu	½ block	70 g
Eggs		
In shell, raw or cooked	1 medium	50 g
Without shell, cooked or poached in water	1 medium	45 g
Scrambled	50 mL (¼ c)	55 g

Food	Choices per Serving	Measures	Mass (Weight)
Note: The following choices provide more than 1 protein exchange:			
Cheese			
Cheeses	1 protein + 1 fat	1 piece	25 g
Cheese, coarsely grated (e.g., cheddar)	1 protein + 1 fat	50 mL (¼ c)	25 g
Cheese, dry, finely grated (e.g., parmesan)	1 protein + 1 fat	45 mL	15 g
Cheese, ricotta, high fat	1 protein + 1 fat	50 mL (¼ c)	55 g
Fish			
Eel	1 protein + 1 fat	1 slice	50 g

(continued on the next page)

I

TABLE I-5 Canadian Choice System: Protein Foods—(continued)

1 protein choice = 7 g protein, 3 g fat, 230 kJ (55 kcal)

Meat			
Bologna	1 protein + 1 fat	1 slice	20 g
Canned lunch meats	1 protein + 1 fat	1 slice	20 g
Corned beef, canned	1 protein + 1 fat	1 slice	25 g
Corned beef, fresh	1 protein + 1 fat	1 slice	25 g
Ground beef, medium-fat	1 protein + 1 fat	30 mL (2 tbs)	25 g
Meat spreads, canned	1 protein + 1 fat	45 mL	35 g
Mutton chop	1 protein + 1 fat	½ chop, with bone	35 g
Paté (see *Fats and Oils* group, Table I-6)			
Sausages, garlic, Polish or knockwurst	1 protein + 1 fat	1 slice	50 g
Sausages, pork, links	1 protein + 1 fat	1 link	25 g
Spareribs or shortribs, with bone	1 protein + 1 fat	1 large	65 g
Stewing beef	1 protein + 1 fat	1 cube	25 g
Summer sausage or salami	1 protein + 1 fat	1 slice	40 g
Weiners, hot dog	1 protein + 1 fat	½ medium	25 g
Miscellaneous			
Blood pudding	1 protein + 1 fat	1 slice	25 g
Peanut butter	1 protein + 1 fat	15 mL (1 tbs)	15 g

TABLE I-6 Canadian Choice System: Fats and Oils

1 fat choice = 5 g fat, 190 kJ (45 kcal)

Food	Measure	Mass (Weight)
Avocado*	⅛	30 g
Bacon, side, crisp*	1 slice	5 g
Butter*	5 mL (1 tsp)	5 g
Cheese spread	15 mL (1 tbs)	15 g
Coconut, fresh*	45 mL (3 tbs)	15 g
Coconut, dried*	15 mL (1 tbs)	10 g
Cream, Half and half		
(cereal), 10%*	30 mL (2 tbs)	30 g
Light (coffee), 20%*	15 mL (1 tbs)	15 g
Whipping, 32 to 37%*	15 mL (1 tbs)	15 g
Cream cheese*	15 mL (1 tbs)	15 g
Gravy*	30 mL (2 tbs)	30 g
Lard*	5 mL (1 tsp)	5 g
Margarine	5 mL (1 tsp)	5 g
Nuts, shelled:		
Almonds	8	5 g
Brazil nuts	2	10 g
Cashews	5	10 g
Filberts, hazelnuts	5	10 g
Macadamia	3	5 g
Peanuts	10	10g
Pecans	5 halves	5 g
Pignolias, pine nuts	25 mL (5 tsp)	10 g
Pistachios, shelled	20	10 g
Pistachios, in shell	20	20 g
Pumpkin and squash seeds	20 mL (4 tsp)	10 g
Nuts (continued):		
Sesame seeds	15 mL (1 tbs)	10 g
Sunflower seeds		
Shelled	15 mL (1 tbs)	10 g
In shell	45 mL (3 tbs)	15 g
Walnuts	4 halves	10 g
Oil, cooking and salad	5 mL (1 tsp)	5 g
Olives, green	10	45 g
Ripe black	7	57 g
Pâté, liverwurst, meat spreads	15 mL (1 tbs)	15 g
Salad dressing: blue, French, Italian,	10 mL (2 tsp)	10 g
mayonnaise, Thousand Island	5 mL (1 tsp)	5 g
Salad dressing, low-calorie	30 mL (2 tbs)	30 g
Salt pork, raw or cooked*	5 mL (1 tsp)	5 g
Sesame oil	5 mL (1 tsp)	5 g
Sour cream		
12% milkfat	30 mL (2 tbs)	30 g
7% milkfat	60 mL (4 tbs)	60 g
Shortening*	5 mL (1 tsp)	

*These items contain higher amounts of saturated fat.

TABLE I-7 **Canadian Choice System: Extras**

Extras have no more than 2.5 g carbohydrate, 60 kJ (14 kcal)

Vegetables 125 mL (½ c)
Artichokes
Asparagus
Bamboo shoots
Bean sprouts, mung or soya
Beans, string, green, or yellow
Bitter melon (balsam pear)
Bok choy
Broccoli
Brussels sprouts
Cabbage
Cauliflower
Celery
Chard
Cucumbers
Eggplant
Endive
Fiddleheads
Greens: beet, collard, dandelion, mustard, turnip, etc.
Kale
Kohlrabi
Leeks
Lettuce
Mushrooms
Okra
Onions, green or mature
Parsley
Peppers, green, yellow or red
Radishes
Rapini
Rhubarb
Sauerkraut
Shallots
Spinach
Sprouts: alfalfa, radish, etc.
Tomato wedges
Watercress
Zucchini

Free Foods (may be used without measuring)	
Artificial sweetener, such as cyclamate or aspartame	Lime juice or lime wedges Marjoram, cinnamon, etc.
Baking powder, baking soda	Mineral water
Bouillon from cube, powder, or liquid	Mustard Parsley
Bouillon or clear broth	Pimentos
Chowchow, unsweetened	Salt, pepper, thyme
Coffee, clear	Soda water, club soda
Consommé	Soya sauce
Dulse	Sugar-free Crystal Drink
Flavorings and extracts	Sugar-free Jelly Powder
Garlic	Sugar-free soft drinks
Gelatin, unsweetened	Tea, clear
Ginger root	Vinegar
Herbal teas, unsweetened	Water
Horseradish, uncreamed	Worcestershire sauce
Lemon juice or lemon wedges	

Condiments

Food	**Measure**
Anchovies	2 fillets
Barbecue sauce	15 mL (1 tbs)
Bran, natural	30 mL (2 tbs)
Brewer's yeast	5 mL (1 tsp)
Carob powder	5 mL (1 tsp)
Catsup	5 mL (1 tsp)
Chili sauce	5 mL (1 tsp)
Cocoa powder	5 mL (1 tsp)
Cranberry sauce, unsweetened	15 mL (1 tbs)
Dietetic fruit spreads	5 mL (1 tsp)
Maraschino cherries	1
Nondairy coffee whitener	5 mL (1 tsp)
Nuts, chopped pieces	5 mL (1 tsp)
Pickles	
unsweetened dill	2
sour mixed	11
Sugar substitutes, granular	5 mL (1 tsp)
Whipped toppings	15 mL (1 tbs)

I

Canada's Food and Activity Guides

Canada's Food Guide to Healthy Eating, shown in Figure I-1 (pp. I-12–I-13), gives detailed information for selecting foods to meet the nutritional needs of all Canadians four years of age and older. Like the U.S. Daily Food Guide, Canada's Food Guide also takes a total diet approach, rather than emphasizing a single food, meal, or day's meals and snacks. Figure I-2 (pp. I-14–I-15), presents Canada's Physical Activity Guide.

TABLE I-8 Canadian Choice System: Combined Food Choices

Food	Choices per Serving	Measure	Mass (Weight)
Angel food cake	½ starch + 2½ sugars	1/12 cake	50 g
Apple crisp	½ starch + 1½ fruits & vegetables + 1 sugar + 1–2 fats	125 mL (½ c)	
Applesauce, sweetened	1 fruits & vegetables + 1 sugar	125 mL (½ c)	
Beans and pork in tomato sauce	1 starch + ½ fruits & vegetables + ½ sugar + 1 protein	125 mL (½ c)	135 g
Beef burrito	2 starches + 3 proteins + 3 fats		110 g
Brownie	1 sugar + 1 fat	1	20 g
Cabbage rolls*	1 starch + 2 proteins	3	310 g
Caesar salad	2–4 fats	20 mL dressing (4 tsp)	
Cheesecake	½ starch + 2 sugars + ½ protein + 5 fats	1 piece	80 g
Chicken fingers	1 starch + 2 proteins + 2 fats	6 small	100 g
Chicken and snow pea Oriental	2 starches + ½ fruits & vegetables + 3 proteins + 1 fat	500 mL (2 c)	
Chili	1½ starches + ½ fruits & vegetables + 3½ protein	300 mL (1¼ c)	325 g
Chips			
Potato chips	1 starch + 2 fats	15 chips	30 g
Corn chips	1 starch + 2 fats	30 chips	30 g
Tortilla chips	1 starch + 1½ fats	13 chips	
Cheese twist	1 starch + 1½ fats	30 chips	30 g
Chocolate bar			
Aero®	2½ sugars + 2½ fats	bar	43 g
Smarties®	4½ sugars + 2 fats	package	60 g
Chocolate cake (without icing)	1 starch + 2 sugars + 3 fats	1/10 of a 8" pan	
Chocolate devil's food cake (without icing)	2 starches + 2 sugars + 3 fats	1/12 of a 9" pan	
Chocolate milk	2 milks 2% + 1 sugar	250 mL (1 c)	300 g
Clubhouse (triple-decker) sandwich	3 starches + 3 proteins + 4 fats		
Cookies			
Chocolate chip	½ starch + ½ sugar + 1½ fats	2	22 g
Oatmeal	1 starch + 1 sugar + 1 fat	2	40 g
Donut (chocolate glazed)	1 starch + 1½ sugars + 2 fats	1	65 g
Egg roll	1 starch + ½ protein + 1 fat		75 g
Four bean salad	1 starch + ½ protein + 1 fat	125 mL (½ c)	
French toast	1 starch + ½ protein + 2 fats	1 slice	65 g
Fruit in heavy syrup	1 fruits & vegetables + 1½ sugars	125 mL (½ c)	
Granola bar	½ starch + 1 sugar + 1–2 fats		30 g
Granola cereal	1 starch + 1 sugar + 2 fats	125 mL (½ c)	45 g
Hamburger	2 starches + 3 proteins + 2 fats	junior size	

* If eaten with sauce, add ½ fruits & vegetables exchange.

(continued on the next page)

TABLE 1-8 **Canadian Choice System: Combined Food Choices—(continued)**

Food	Choices per Serving	Measure	Mass (Weight)
Ice cream and cone, plain flavour			
Ice cream	½ milk + 2–3 sugars + 1–2 fats		100 g
Cone	½ sugar		4 g
Lasagna			
Regular cheese	1 starch + 1 fruits & vegetables + 3 proteins + 2 fats	3" × 4" piece	
Low-fat cheese	1 starch + 1 fruits & vegetables + 3 proteins	3" × 4" piece	
Legumes			
Dried beans (kidney, navy, pinto, fava, chick peas)	2 starches + 1 protein	250 mL (1 c)	180 g
Dried peas	2 starches + 1 protein	250 mL (1 c)	210 g
Lentils	2 starches + 1 protein	250 mL (1 c)	210 g
Macaroni and cheese	2 starches + 2 proteins + 2 fats	250 mL (1 c)	210 g
Minestrone soup	1½ starches + ½ fruits & vegetables + ½ fat	250 mL (1 c)	
Muffin	1 starch + ½ sugar + 1 fat	1 small	45 g
Nuts (dry or roasted without any oil added).			
Almonds, dried sliced	½ protein + 2 fats	50 mL (¼ c)	22 g
Brazil nuts, dried unblanched	½ protein + 2½ fats	5 large	23 g
Cashew nuts, dry roasted	½ starch + ½ protein + 2 fats	50 mL (¼ c)	28 g
Filbert hazelnut, dry	½ protein + 3½ fats	50 mL (¼ c)	30 g
Macadamia nuts, dried	½ protein + 4 fats	50 mL (¼ c)	28 g
Peanuts, raw	1 protein + 2 fats	50 mL (¼ c)	30 g
Pecans, dry roasted	½ fruits & vegetables + 3 fats	50 mL (¼ c)	22 g
Pine nuts, pignolia dried	1 protein + 3 fats	50 mL (¼ c)	34 g
Pistachio nuts, dried	½ fruits & vegetables + ½ protein + 2½ fats	50 mL (¼ c)	27 g
Pumpkin seeds, roasted	2 proteins + 2½ fats	50 mL (¼ c)	47 g
Sesame seeds, whole dried	½ fruits & vegetables + ½ protein + 2½ fats	50 mL (¼ c)	30 g
Sunflower kernel, dried	½ protein + 1½ fats	50 mL (¼ c)	17 g
Walnuts, dried chopped	½ protein + 3 fats	50 mL (¼ c)	26 g
Perogies	2 starches + 1 protein + 1 fat	3	
Pie, fruit	1 starch + 1 fruits & vegetables + 2 sugars + 3 fats	1 piece	120 g
Pizza, cheese	1 starch + 1 protein + 1 fat	1 slice (⅛ of a 12")	50 g
Pork stir fry	½ to 1 fruits & vegetables + 3 proteins	200 mL (¾ c)	
Potato salad	1 starch + 1 fat	125 mL (½ c)	130 g
Potatoes, scalloped	2 starches + 1 milk + 1–2 fats	200 mL (¾ c)	210 g
Pudding, bread or rice	1 starch + 1 sugar + 1 fat	125 mL (½ c)	
Pudding, vanilla	1 milk + 2 sugars	125 mL (½ c)	
Raisin bran cereal	1 starch + ½ fruits & vegetables + ½ sugar	175 mL (⅔ c)	40 g
Rice krispie squares	½ starch + 1½ sugars + ½ fat	1 square	30 g
Shepherd's pie	2 starches + 1 fruits & vegetables + 3 proteins	325 mL (1⅓ c)	
Sherbet, orange	3 sugars + ½ fat	125 mL (½ c)	
Spaghetti and meat sauce	2 starches + 1 fruits & vegetables + 2 proteins + 3 fats	250 mL (1 c)	
Stew	2 starches + 2 fruits & vegetables + 3 proteins + ½ fat	200 mL (¾ c)	
Sundae	4 sugars + 3 fats	125 mL (½ c)	
Tuna casserole	1 starch + 2 proteins + ½ fat	125 mL (½ c)	
Yogurt, fruit bottom	1 fruits & vegetables + 1 milk + 1 sugar	125 mL (½ c)	125 g
Yogurt, frozen	1 milk + 1 sugar	125 mL (½ c)	125 g

FIGURE I-1 Canada's Food Guide to Healthy Eating

FIGURE I-1 **Canada's Food Guide to Healthy Eating—continued**

Different People Need Different Amounts of Food

The amount of food you need every day from the 4 food groups and other foods depends on your age, body size, activity level, whether you are male or female and if you are pregnant or breast-feeding. That's why the Food Guide gives a lower and higher number of servings for each food group. For example, young children can choose the lower number of servings, while male teenagers can go to the higher number. Most other people can choose servings somewhere in between.

Grain Products
5–12
SERVINGS PER DAY

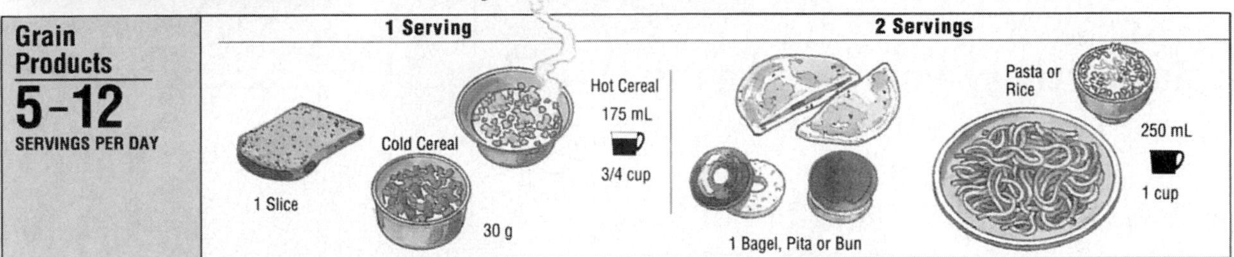

Vegetables & Fruit
5–10
SERVINGS PER DAY

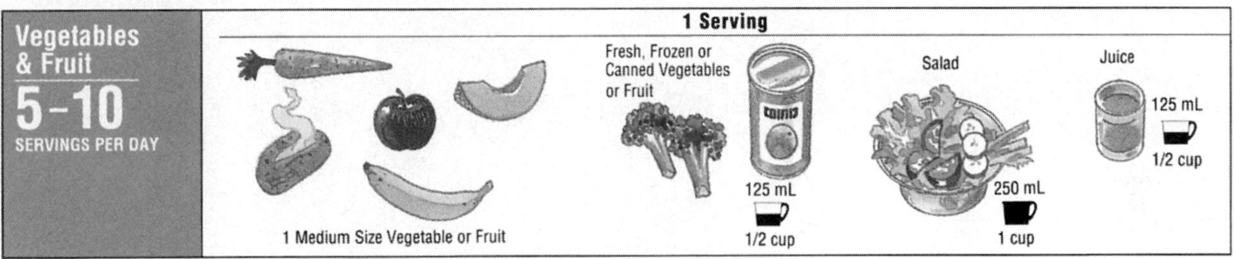

Milk Products
SERVINGS PER DAY
Children 4–9 years: 2–3
Youth 10–16 years: 3–4
Adults: 2–4
Pregnant & Breast-feeding Women: 3–4

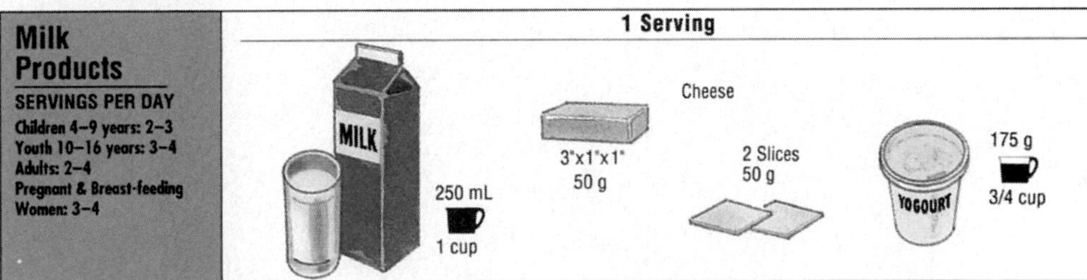

Meat & Alternatives
2–3
SERVINGS PER DAY

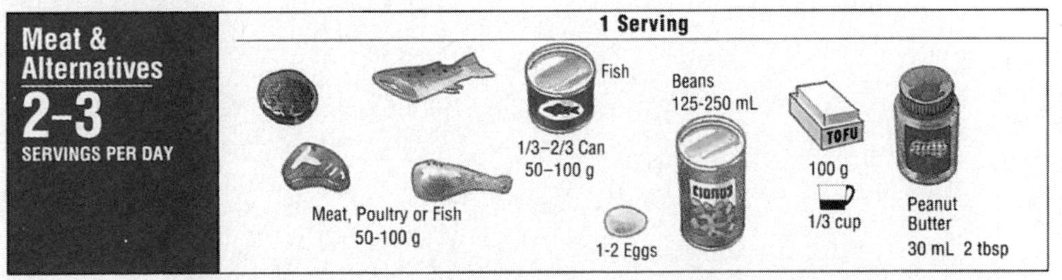

Other Foods

Taste and enjoyment can also come from other foods and beverages that are not part of the 4 food groups. Some of these foods are higher in fat or Calories, so use these foods in moderation.

Enjoy eating well, being active and feeling good about yourself. That's VITALITY

FIGURE I-2 Canada's Physical Activity Guide

Physical activity improves health.

Every little bit counts, but more is even better – everyone can do it!

Get active your way – build physical activity into your daily life...

- at home
- at school
- at work
- at play
- on the way

...that's active living!

Increase Endurance Activities

Increase Flexibility Activities

Increase Strength Activities

Reduce Sitting for long periods

Health Canada Santé Canada

Canadian Society for Exercise Physiology

I

FIGURE I-2 **Canada's Physical Activity Guide—continued**

Choose a variety of activities from these three groups:

Endurance

4-7 days a week
Continuous activities for your heart, lungs and circulatory system.

Flexibility

4-7 days a week
Gentle reaching, bending and stretching activities to keep your muscles relaxed and joints mobile.

Strength

2-4 days a week
Activities against resistance to strengthen muscles and bones and improve posture.

Starting slowly is very safe for most people. Not sure? Consult your health professional.

For a copy of the *Guide Handbook* and more information: **1-888-334-9769**, or **www.paguide.com**

Eating well is also important. Follow *Canada's Food Guide to Healthy Eating* to make wise food choices.

Get Active Your Way, Every Day – For Life!

Scientists say accumulate 60 minutes of physical activity every day to stay healthy or improve your health. As you progress to moderate activities you can cut down to 30 minutes, 4 days a week. Add-up your activities in periods of at least 10 minutes each. Start slowly... and build up.

Time needed depends on effort

Very Light Effort	Light Effort *60 minutes*	Moderate Effort *30-60 minutes*	Vigorous Effort *20-30 minutes*	Maximum Effort
• Strolling • Dusting	• Light walking • Volleyball • Easy gardening • Stretching	• Brisk walking • Biking • Raking leaves • Swimming • Dancing • Water aerobics	• Aerobics • Jogging • Hockey • Basketball • Fast swimming • Fast dancing	• Sprinting • Racing

Range needed to stay healthy

You Can Do It – Getting started is easier than you think

Physical activity doesn't have to be very hard. Build physical activities into your daily routine.

- Walk whenever you can – get off the bus early, use the stairs instead of the elevator.
- Reduce inactivity for long periods, like watching TV.
- Get up from the couch and stretch and bend for a few minutes every hour.
- Play actively with your kids.
- Choose to walk, wheel or cycle for short trips.
- Start with a 10 minute walk – gradually increase the time.
- Find out about walking and cycling paths nearby and use them.
- Observe a physical activity class to see if you want to try it.
- Try one class to start, you don't have to make a long-term commitment.
- Do the activities you are doing now, more often.

Benefits of regular activity:

- better health
- improved fitness
- better posture and balance
- better self-esteem
- weight control
- stronger muscles and bones
- feeling more energetic
- relaxation and reduced stress
- continued independent living in later life

Health risks of inactivity:

- premature death
- heart disease
- obesity
- high blood pressure
- adult-onset diabetes
- osteoporosis
- stroke
- depression
- colon cancer

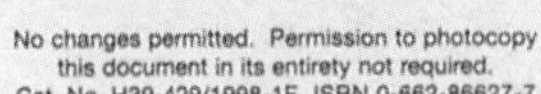

Cat. No. H39-429/1998-1E ISBN 0-662-86627-7

J Enteral Formulas

The staggering number of enteral formulas available allows health care professionals to meet a variety of their clients' medical needs, but also complicates the process of selecting an appropriate formula. The first step in narrowing the choice of formulas is to determine the client's ability to digest and absorb nutrients. Table J-1 on pp. J-1 through J-4 lists examples of intact formulas for clients with the ability to digest and absorb nutrients, and Table J-2 on pp. J-5 through J-6 provides examples of hydrolyzed formulas for clients with limited ability to digest and absorb nutrients. Products promoted to the general public and intended primarily as oral supplements, including Carnation Instant Breakfast® (Nestlé), Boost® (Mead Johnson), and Ensure (Ross) are not included as examples. Each formula is listed only once, although the formula may have more than one use. A high-protein formula, for example, may also be a fiber-containing formula. Tables J-3 through J-5 on p. J-7 list modular formulas. Although this appendix provides many examples, the list of formulas is not complete. The information reflects the manufacturer's literature and does not suggest endorsement by the authors. Manufacturers frequently add new formulas and stop production or rename formulas. Be aware that formula composition changes periodically. Consult manufacturer's literature and web sites for updates. The following products are listed in this appendix:

- B. Braun[a]
  - Hepatic-Aid® II
  - Immun-Aid®
- Mead Johnson Nutritionals[b]
  - Casec®
  - Choice DM®
  - Comply®
  - Criticare HN®
  - Deliver® 2.0
  - Isocal®
  - Isocal® HN
  - Kindercal® TF
  - Kindercal® TF with Fiber
  - Lipisorb®
  - Magnacal® Renal
  - MCT Oil®
  - Microlipid®
  - Moducal®
  - Protain XL®
  - Respalor®
  - TraumaCal®
  - Ultracal®
- Nestlé Clinical Nutrition[c]
  - Crucial®
  - Glytrol®
  - Nutren ®1.0
  - Nutren® 1.0 with Fiber
  - Nutren® 1.5
  - Nutren® 2.0
  - Nutren® Junior
  - Nutren® Junior with Fiber
  - NutriHep®
  - NutriRenal®
  - NutriVent®
  - Peptamen®
  - Peptamen® 1.5
  - Peptamen® Junior
  - Peptamen® VHP
  - ProBalance®
  - Reabilan®
  - Reabilan® HN
  - Renalcal® Diet
  - Replete®
  - Replete® with Fiber
- Novartis Nutrition Corporation[d]
  - Compleat® Modified
  - Compleat® Pediatric
  - Diabetisource®
  - Fibersource®
  - Fibersource® HN
  - Impact®
  - Impact® Glutamine
  - Impact® 1.5
  - Impact® with Fiber
  - Isosource® Standard
  - Isosource® HN
  - Isosource® VHN
  - Isosource® 1.5
  - Novasource® Pulmonary
  - Novasource® Renal
  - Novasource® 2.0
  - Resource® Diabetic
  - Sandosource® Peptide
  - Tolerex®
  - Vivonex® Pediatric
  - Vivonex® Plus
  - Vivonex® T.E.N.
- Ross Laboratories[e]
  - Advera®
  - Alitraq®
  - Glucerna®
  - Jevity®
  - Jevity® Plus
  - Nepro®
  - Optimental®
  - Osmolite®
  - Osmolite® HN
  - Osmolite® HN Plus
  - Oxepa®
  - PediaSure®
  - PediaSure® with Fiber
  - Perative®
  - Polycose®
  - ProMod®
  - Promote®
  - Promote® with Fiber
  - Pulmocare®
  - Suplena®
  - TwoCal® HN
  - Vital® HN

[a] *Partners in Caring* (1998), B. Braun Medical Inc., Irvine, CA 92614.
[b] Mead Johnson Nutritionals, www.meadjohnson.com, visited April 6, 2001.
[c] Nestlé Clinical Nutrition, www.nestleclinicalnutrition.com, visited April 6, 2001.
[d] *Pocket Guide* (2001), Novartis Nutrition, Minneapolis, MN 55440.
[e] Ross Products, www.ross.com, visited April 6, 2001.

TABLE J-1 Intact Protein Formulas

Product[a]	Volume to Meet 100% RDI (ml)	Energy (kcal/ml)	Protein or Amino Acids (g/L)	Carbohydrate (g/L)	Fat (g/L)	Osmolality[b] (mOsm/kg)	Notes
Lactose-Free, Standard Formulas							
Compleat® Modified	1500	1.07	43	140	37	300	4.3 g fiber/L
Isocal®	1890	1.06	34	135	44	270	20% fat from MCT
Isosource Standard®	1165	1.20	43	170	39	490	50% fat from MCT
Nutren® 1.0	1500	1.00	40	127	38	315	25% fat from MCT
Osmolite®	2000	1.06	37	151	35	300	20% fat from MCT
Lactose-Free, Fiber-Containing Formulas							
Fibersource® Standard	1165	1.20	43	170	39	490	10 g fiber/L
Impact® with Fiber	1500	1.00	56	140	28	375	10 g fiber/L, enriched with arginine, nucleotides, and omega-3 fatty acids
Jevity®	1321	1.06	44	155	35	300	14 g fiber/L
Nutren® 1.0 with Fiber	1500	1.00	40	127	38	320	14 g fiber/L
ProBalance®	1000	1.20	54	156	41	350	10 g fiber/L
Promote® with Fiber	1000	1.00	63	138	28	370	14 g fiber/L
Replete® with Fiber	1000	1.00	63	113	34	310	14 g fiber/L
Ultracal®	1120	1.06	45	142	39	360	14.4 g fiber/L

[a]Formulas come in ready-to-use (liquid) form unless specified under "Notes."
[b]Osmolality may vary, depending on the flavorings added to a product.
MCT= Medium-chain triglycerides.

TABLE J-1 Intact Protein Formulas (continued)

Product[a]	Volume to Meet 100% RDI (ml)	Energy (kcal/ml)	Protein or Amino Acids (g/L)	Carbohydrate (g/L)	Fat (g/L)	Osmolality[b] (mOsm/kg)	Notes
Lactose-Free, High-kCalorie Formulas							
Comply®	830	1.50	60	180	61	460	
Deliver® 2	1000	2.00	75	200	102	640	
Isosource® 1.5	933	1.50	68	170	65	650	8 g fiber/L
Novasource® 2.0	948	2.00	90	220	88	790	
Nutren® 1.5	1000	1.50	60	169	68	430	50% fat from MCT
Nutren® 2.0	750	2.00	80	196	106	745	75% fat from MCT
Osmolite® HN Plus	1000	1.20	56	158	39	360	19% fat from MCT
TwoCal HN®	947	2.00	84	219	91	730	
Lactose-Free, High-Protein Formulas							
Fibersource® HN	1165	1.20	53	160	39	490	7 g fiber/L
Isocal® HN	1180	1.06	45	124	45	270	20% fat from MCT
Isocal® HN Plus	1000	1.20	54	156	40	390	Part of fat from MCT
Isosource® HN	1165	1.20	53	160	39	401	
Isosource® VHN	1250	1.00	62	130	29	300	10g fiber/L
Jevity® Plus	1000	1.20	56	173	39	450	12 g fiber/L
Lipisorb®	1180	1.35	57	161	57	630	85% fat from MCT
Osmolite® HN	1321	1.06	44	144	35	300	Low-residue, 20% fat from MCT
Promote®	1000	1.00	63	130	26	340	Fat primarily from polyunsaturated and monounsaturated sources, 20% fat from MCT
Ultracal® HN Plus	1000	1.20	54	156	40	370	30% fat from MCT, 11 g fiber/L
Special-Use Formulas: Pediatric (1 to 10 years)							
Compleat® Pediatric	Varies[c]	1.00	38	130	39	380	Blenderized formula, 4 g fiber/L
Kindercal® TF	Varies[c]	1.06	30	135	44	345	20% fat from MCT
Kindercal® TF with Fiber	Varies[c]	1.06	30	135	44	345	6 g fiber/L
Nutren Junior®	Varies[c]	1.00	30	128	42	350	25% fat from MCT
Nutren Junior® with Fiber	Varies[c]	1.00	30	128	42	350	25% fat from MCT, 6 g fiber/L
PediaSure®	Varies[c]	1.00	30	110	50	335	
PediaSure® Enteral Formula with Fiber	Varies[c]	1.00	30	114	45	345	7 g fiber/L

[c]Depends on age of child.

TABLE J-1 Intact Protein Formulas (continued)

Product[a]	Volume to Meet 100% RDI (ml)	Energy (kcal/ml)	Protein or Amino Acids (g/L)	Carbohydrate (g/L)	Fat (g/L)	Osmolality[b] (mOsm/kg)	Notes
Special-Use Formulas: Glucose Intolerance							
Choice dm® TF	946	1.06	45	119	51	300	14 g fiber/L
Diabetisource®	1500	1.00	50	90	49	360	4 g fiber/L
Glucerna®	1422	1.00	42	96	54	355	14 g fiber/L
Glytrol®	1400	1.00	45	100	48	380	15 g fiber/L; 20% fat from MCT
Resource® Diabetic	1180	1.06	63	100	47	320	13 g fiber/L
Special-Use Formulas: Immune System Support							
Immun-Aid®	2000	1.00	80	120	22	460	Powder form; enriched with arginine, glutamine, branched-chain amino acids, nucleic acids, omega-3 fatty acids, vitamins A, C, E, and B_6 and trace elements
Impact®	1500	1.00	56	130	28	375	Enriched with arginine, nucleic acids, and omega-3 fatty acids
Impact® 1.5	1250	1.50	84	140	69	550	Same as above
Impact® Glutamine	1000	1.30	78	150	43	630	Same as above and enriched with glutamine; 10 g fiber/L
Special-Use Formulas: Renal Insufficiency[d]							
Magnacal® Renal	—	2.00	75	200	101	570	High in monounsaturated fat, 20% fat from MCT; intended for use once hemodialysis has been instituted
Nepro®	—	2.00	70	222	96	665	High-calcium, low-phosphorus; intended for use once dialysis has been instituted
Novasource® Renal	—	2.00	74	200	100	700	Low in electrolytes; intended for use once dialysis has been instituted
NutriRenal®	—	2.00	70	205	104	N/A	50% fat from MCT; enriched with vitamins C and B_6, folate, zinc, and selenium; intended for use once dialysis has been instituted
Renalcal® Diet	—	2.00	35	290	82	NA	Contains histidine, 70% fat from MCT; intended for use before dialysis is instituted
Suplena®	—	2.00	30	255	96	600	Low in electrolytes; intended for use before dialysis is instituted

[d] Renal formulas are intentionally low in essential minerals and thus do not meet the RDI in commonly administered volumes.

TABLE J-1 Intact Protein Formulas (continued)

Product[a]	Volume to Meet 100% RDI (ml)	Energy (kcal/ml)	Protein or Amino Acids (g/L)	Carbohydrate (g/L)	Fat (g/L)	Osmolality[b] (mOsm/kg)	Notes
Special-Use Formulas: Respiratory Insufficiency							
Novasource® Pulmonary	933	1.50	75	150	68	650	8 g fiber/L
NutriVent®	1000	1.50	68	100	94	330	55% kcal from fat, 40% fat from MCT
Oxepa	947	1.50	63	106	94	493	55% kcal from fat, enriched with antioxidant nutrients
Pulmocare®	1136	1.50	63	106	93	475	55% kcal from fat, 20% fat from MCT, enriched with antioxidant nutrients
Respalor®	1000	1.50	75	146	68	400	41% kcal from fat; 30% fat from MCT; enriched with vitamins C and E, B-complex vitamins, zinc and trace elements
Special-Use Formulas: Wound Healing							
Protain XL®	—	1.00	57	145	30	340	9 g fiber/L, 20% fat from MCT, enriched with vitamins A and C and zinc
Replete®	1000	1.00	63	113	34	300	Enriched with vitamins A and C and zinc; 25% fat from MCT
TraumaCal®	2000	1.50	82	142	68	560	Enriched with vitamins C, B-complex, and E and copper and zinc

TABLE J-2 Hydrolyzed Protein Formulas

Product	Volume to Meet 100% RDI (ml)	Energy (kcal/ml)	Protein or Amino Acids (g/L)	Carbohydrate (g/L)	Fat (g/L)	Osmolality (mOsm/kg)	Notes
Special-Use Hydrolyzed Formulas: Hepatic Insufficiency							
Hepatic Aid® II	N/A	1.20	44	169	36	560	Powder form; free amino acids, high in branched-chain amino acids, low in aromatic amino acids, no added vitamins or electrolytes
NutriHep®	1000	1.50	40	290	21	790	Free amino acids, 50% branched-chain amino acids, 50% aromatic amino acids, contains vitamins and minimal electrolytes
Special-Use Hydrolyzed Formulas: HIV Infection or AIDS							
Advera®	1184	1.28	60	216	23	680	78% hydrolyzed and 22% intact protein, low-fat, fiber added, enriched with vitamins E, C, B_6, B_{12}, and folate
Special-Use Hydrolyzed Formulas: Immune System Support							
Alitraq®	1500	1.00	53	165	16	575	Powder form; 47% free amino acids, 42% small peptides, enriched with glutamine and arginine
Crucial®	1000	1.50	94	135	68	490	Enriched with arginine, glutamine, antioxidant nutrients, and zinc
Perative®	1155	1.30	67	177	37	385	Enriched with arginine and b-carotene
Vivonex® Plus	1800	1.00	45	190	7	650	Powder form; 100% free amino acids, enriched with glutamine, arginine, and branched-chain amino acids

TABLE J-2 Hydrolyzed Protein Formulas (continued)

Product	Volume to Meet 100% RDI (ml)	Energy (kcal/ml)	Protein or Amino Acids (g/L)	Carbohydrate (g/L)	Fat (g/L)	Osmolality[a] (mOsm/kg)	Notes
Special-Use Hydrolyzed Formulas: Malabsorption							
Criticare HN®	1890	1.06	38	220	5	650	50% free amino acids, 50% small peptides
Optimental®	1422	1.5	51	139	28	540	Contains MCT and arginine; enriched with vitamins C and E and beta-carotene
Peptamen®	1500	1.00	40	127	39	270	70% fat from MCT; contains glutamine
Peptamen® 1.5	1000	1.5	60	191	59	450	70% fat from MCT; contains glutamine
Peptamen VHP®	1500	1.00	63	105	39	300	70% fat from MCT; contains glutamine
Reabilan®	2000	1.00	32	132	41	350	50% fat from MCT
Reabilan® HN	1500	1.33	58	158	54	490	50% fat from MCT
SandoSource® Peptide	1750	1.00	50	160	17	490	60% free amino acids and small peptides
Tolerex® acids	1800	1.00	21	230	1.5	550	Powder form; 100% free amino
Vital® HN	1500	1.00	42	185	11	500	Powder form; 87% hydrolyzed proteins, 13% essential amino acids
Vivonex® T.E.N.	2000	1.00	38	210	3	630	Powder form; 100% free amino acids, enriched with branched-chain amino acids
Special-Use Hydrolyzed Formulas: Pediatric (1 to 10 years)							
Peptamin Junior®	varies[b]	1.0	30	138	39	260	60% fat from MCT; contains glutamine
Vivonex® Pediatric	varies[b]	0.8	24	130	24	360	Powder form; 100% free amino acids

[a]Osmolality may vary depending on the flavorings added to a product.
[b]Depends on age of child.

TABLE J-3 Protein Modules

Product	Form	Major Protein Source	Energy (kcal/g)	Protein (g/100 g)
Casec®	powder	Calcium caseinate	3.8	90
ProMod®	powder	Whey protein	4.2	75

TABLE J-4 Carbohydrate Modules

Product	Form	Major Carbohydrate Source	Energy (kcal/ml or g)
Moducal®	powder	Hydrolyzed cornstarch	3.8 kcal/g
Polycose Liquid®	liquid	Hydrolyzed cornstarch	2.0 kcal/ml
Polycose Powder®	powder	Hydrolyzed cornstarch	3.8 kcal/g

TABLE J-5 Fat Modules

Product	Form	Major Fat Source	Energy (kcal/ml)	Fat (g/100 ml)
MCT Oil®	liquid	Medium-chain triglycerides	7.7	86
Microlipid®	liquid	Safflower oil	4.5	51

GLOSSARY

MANY MEDICAL TERMS have their origins in Latin or Greek. By learning a few common derivations, you can glean the meaning of words you have never heard of before. For example, once you know that "hyper" means above normal, "glyc" means glucose, and "emia" means blood, you can easily determine that "hyperglycemia" means high blood glucose. The following derivations will help you to learn many terms presented in this glossary.

General

a- or *an-* = not or without
ana- = up
ant- or *anti-* = against
ante- or *pre-* or *pro-* = before
cata- = down
co- = with or together
di- = two, twice
dys- or *mal-* = bad, difficult, painful
endo- = inner or within
epi- = upon
extra- = outside of, beyond, or in addition
exo- = outside of or without
gen- or *-gen* = gives rise to, producing
homeo- = like, similar, constant unchanging state
hyper- = over, above, excessive
hypo- = below, under, beneath
in- = not
inter- = between, in the midst
intra- = within
-itis = infection or inflammation
-lysis = break
macro- = large or long
micro- = small
mono- = one, single
neo- = new, recent
oligo- = few or small
-osis or *-asis* = condition
peri- = around, about
poly- = many or much
-stat or *-stasis-* = stationary
tri- = three

Body

arterio- = artery
cardiac or *cardio-* = heart
-cyte = cell
enteron = intestine
gastro- = stomach
hema- or *-emia* = blood
hepatic = liver
myo- = muscle
osteo- = bone
pulmo- = lung
renal = kidney
ure or *-uria* = urine
vaso- = vessel
vena = vein

Chemistry

-al = aldehyde
-ase = enzyme
-ate = salt
glyc- or *gluc-* = sweet (glucose)
hydro- or *hydrate* = water
lipo- = lipid
-ol = alcohol
-ose = carbohydrate
saccha- = sugar

1,25-dihydroxy vitamin D: the active form of vitamin D (see Figure 11-8 on p. 364).

24-hour recall: a record of foods eaten by a person in the previous 24 hours.

—A—

abscesses: accumulations of pus, caused by local infection, that contain live microorganisms and immune system cells. Treatment involves draining the abscess.

absorption: the passage of nutrients from the GI tract into either the blood or the lymph.

Acceptable Daily Intake (ADI): the estimated amount of a sweetener that individuals can safely consume each day over the course of a lifetime without adverse effect. It includes a 100-fold safety factor.

accredited: approved; in the case of medical centers or universities, certified by an agency recognized by the U.S. Department of Education.

acesulfame (AY-sul-fame) **potassium:** an artificial sweetener composed of an organic salt that has been approved in both the United States and Canada; also known as acesulfame-K because K is the chemical symbol for potassium.

acetaldehyde (ass-et-AL-duh-hide): an intermediate in alcohol metabolism.

acetone breath: a distinctive fruity odor that can be detected on the breath of a person who is experiencing ketosis.

acetyl CoA (ASS-eh-teel, or ah-SEET-il, coh-AY): a 2-carbon compound (**acetate**, or **acetic acid**, shown in Figure 5-1 on p. 130) to which a molecule of CoA is attached.

achalasia (ack-ah-LAY-zee-ah): failure of the lower esophageal sphincter to relax and allow foods to pass from the esophagus to the stomach; formally called achalasia of the cardia or **cardiospasm.**

acid controllers: medications used to prevent or relieve indigestion by suppressing production of acid in the stomach; also called H2 blockers. Common brands include Pepcid AC, Tagamet HB, Zantac 75, and Axid AR.

acid-base balance: the equilibrium in the body between acid and base concentrations; see Chapter 12.

acidophilus (ASS-ih-dof-ih-lus) **milk:** a cultured milk created by adding *Lactobacillus acidophilus,* a bacterium that breaks down lactose to glucose and galactose, producing a sweet, lactose-free product.

acidosis (assi-DOE-sis): above-normal acidity in the blood and body fluids.

acids: compounds that release hydrogen ions in a solution.

acne: a chronic inflammation of the skin's follicles and oil-producing glands, which leads to an accumulation of oils inside the ducts that surround hairs; usually associated with the maturation of young adults.

acquired immune deficiency syndrome (AIDS): the severe complications associated with the end stages of HIV infection. A person with an HIV infection is diagnosed with AIDS when AIDS-defining illnesses (certain infections or severe wasting) develop.

acute malnutrition: protein-energy malnutrition (PEM) that develops rapidly due to a sudden and dramatic demand for nutrients. The person suffering from acute malnutrition may be of normal weight or may be overweight, but serum protein levels are low.

acute PEM: protein-energy malnutrition caused by recent severe food restriction; characterized in children by thinness for height (wasting).

acute renal failure: the sudden loss of the kidneys' ability to function.

adaptive thermogenesis: adjustments in energy expenditure related to changes in environment such as extreme cold and to physiological events such as overfeeding, trauma, and changes in hormone status.

adequacy (dietary): providing all the essential nutrients, fiber, and energy in amounts sufficient to maintain health.

Adequate Intake (AI): the average amount of a nutrient that appears sufficient to maintain a specified criterion; a value used as a guide for nutrient intake when an RDA cannot be determined.

adipose (ADD-ih-poce) **tissue:** the body's fat tissue; consists of masses of fat-storing cells.

adolescence: the period from the beginning of puberty until maturity.

adrenal glands: glands adjacent to, and just above, each kidney.

advance directives: the means by which competent adults record their preferences for future medical interventions. A living will and durable power of attorney are types of advance directives.

adverse reactions: unusual responses to food (including intolerances and allergies).

adynamic ileus (ILL-ee-us): obstruction of the intestine caused by paralysis of the intestinal muscles.

aerobic (air-ROE-bic): requiring oxygen.

AIDS-defining illnesses: very low CD4+ T-cell counts, wasting, and other complications that mark the final stages of an HIV infection.

alcohol: a class of organic compounds containing hydroxyl (OH) groups.

alcohol abuse: a pattern of drinking that includes failure to fulfill work, school, or home responsibilities; drinking in situations that are physically dangerous (as in driving while intoxicated); recurring alcohol-related legal problems (as in aggravated assault charges); or continued drinking despite ongoing social problems that are caused by or worsened by alcohol.

alcohol dehydrogenase (dee-high-DROJ-eh-nayz): an enzyme active in the stomach and the liver that converts ethanol to acetaldehyde.

alcoholism: a pattern of drinking that includes a strong craving for alcohol, a loss of control and an inability to stop drinking once begun, withdrawal symptoms (nausea, sweating, shakiness, and anxiety) after heavy drinking, and the need for increasing amounts of alcohol in order to feel "high."

aldosterone (al-DOS-ter-own or AL-dough-STEER-own): a hormone secreted by the adrenal glands that signals the kidneys to reabsorb sodium and fluid. Aldosterone also regulates chloride and potassium concentrations.

alitame (AL-ih-tame): an artificial sweetener composed of two amino acids (alanine and aspartic acid); FDA approval pending.

alkalosis (alka-LOE-sis): above-normal alkalinity (base) in the blood and body fluids.

allergen: any substance that triggers an inappropriate immune response to a substance not normally harmful to the body.

alpha-lactalbumin (lact-AL-byoo-min): the chief protein in human breast milk, as opposed to **casein** (CAY-seen), the chief protein in cow's milk.

alpha-tocopherol: the active vitamin E compound.

Alzheimer's (ALTZ-high-merz) **disease:** a degenerative disease of the brain involving memory loss and major structural changes in neuron networks; also known as *senile dementia of the Alzheimer's type (SDAT), primary degenerative dementia of senile onset,* or *chronic brain syndrome.*

amenorrhea (ay-MEN-oh-REE-ah): the absence of or cessation of menstruation. **Primary amenorrhea** is menarche delayed beyond 16 years of age. **Secondary amenorrhea** is the absence of three to six consecutive menstrual cycles.

American Dietetic Association (ADA): the professional organization of dietitians in the United States. The Canadian equivalent is Dietitians of Canada, which operates similarly.

amino (a-MEEN-oh) **acids:** building blocks of proteins. Each contains an amino group, an acid group, a hydrogen atom, and a distinctive side group, all attached to a central carbon atom.

amino acid pool: the supply of amino acids derived from either food proteins or body proteins that collect in the cells and circulating blood and stand ready to be incorporated in proteins and other compounds or used for energy.

amino acid scoring: a method of evaluating protein quality by comparing a protein's amino acid pattern with that of a reference protein; sometimes called **chemical scoring.**

ammonia: a compound with the chemical formula NH_3; produced during the deamination of amino acids.

amniotic (am-nee-OTT-ic) **sac:** the "bag of waters" in the uterus, in which the fetus floats.

amylase (AM-ih-lace): an enzyme that hydrolyzes amylose (a form of starch). Amylase is a carbohydrase, an enzyme that breaks down carbohydrates.

anabolism (an-ABB-o-lism): reactions in which small molecules are put together to build larger ones. Anabolic reactions require energy.

anaerobic (AN-air-ROE-bic): not requiring oxygen.

anencephaly (AN-en-SEF-a-lee): an uncommon and always fatal type of neural tube defect, characterized by the absence of a brain.

aneurysm (AN-you-riz-um): a ballooning out of a portion of a blood vessel (usually an artery) due to weakness of the vessel's wall.

angina (an-JYE-nah or AN-ji-nah): a painful feeling of tightness or pressure in and around the heart, often radiating to the back, neck, and arms; caused by a lack of oxygen to an area of heart muscle.

angiotensin (AN-gee-oh-TEN-sin): a hormone involved in blood pressure regulation. Its precursor protein is called **angiotensinogen.**

anions (AN-eye-uns): negatively charged ions.

anorexia (an-oh-RECK-see-ah) **nervosa:** an eating disorder characterized by a refusal to maintain a minimally normal body weight and a distortion in perception of body shape and weight.

antacids: medications used to relieve indigestion by neutralizing acid in the stomach. Common brands include Alka-Seltzer, Maalox, Rolaids, and Tums.

antagonist: a competing factor that counteracts the action of another factor. When a drug displaces a vitamin from its site of action, the drug renders the vitamin ineffective and thus acts as a vitamin antagonist.

anthropometric (AN-throw-poe-MET-rick): relating to measurement of the physical characteristics of the body, such as height and weight.

antibodies: large proteins produced by the B-cells of the immune system in response to the invasion of the body by foreign molecules (usually proteins called antigens). Antibodies combine with and inactivate the foreign invaders, thus protecting the body.

antidiuretic hormone (ADH): a hormone released by the pituitary gland in response to highly concentrated blood. The kidneys respond by reabsorbing water, thus preventing water loss.

antigen: any substance that elicits the formation of antibodies or an inflammation reaction from the immune system. A bacterium, a virus, a toxin, and a protein in food that causes allergy are all examples of antigens.

antioxidant: a substance in foods that protects other substances from oxidation by being oxidized themselves and thus decreases the adverse effects of free radicals on normal physiological functions in the human body.

antipromoters: factors that oppose the development of cancers.

antiscorbutic (AN-tee-skor-BUE-tik) **factor:** the original name for vitamin C.

anus (AY-nus): the terminal outlet of the GI tract.

appendix: a narrow blind sac extending from the beginning of the colon that stores lymph cells.

appetite: the integrated response to the sight, smell, thought, or taste of food that initiates or delays eating.

arachidonic (a-RACK-ih-DON-ic) **acid:** an omega-6 polyunsaturated fatty acid with 20 carbons and four double bonds (20:4); synthesized from linoleic acid.

arteries: vessels that carry blood away from the heart.

arthritis: inflammation of a joint, usually accompanied by pain, swelling, and structural changes.

artificial feedings: parenteral and enteral nutrition; feeding by a route other than the normal ingestion of food.

artificial sweeteners: sugar substitutes that provide negligible, if any, energy; sometimes called **nonnutritive sweeteners.**

ascites (ah-SIGH-teez): a type of edema characterized by the accumulation of fluid in the abdominal cavity.

ascorbic acid: one of the two active forms of vitamin C (see Figure 10-14). Many people refer to vitamin C by this name.

-ase (ACE): a word ending denoting an enzyme. The word beginning often identifies the compounds the enzyme works on.

aspartame (ah-SPAR-tame or ASS-par-tame): an artificial sweetener composed of two amino acids (phenylalanine and aspartic acid); approved in both the United States and Canada.

aspiration pneumonia: an infection of the lungs caused by inhaling fluids regurgitated from the stomach.

aspiration: the drawing of food, gastric secretions, or liquid into the lungs.

atherosclerosis (ATH-er-oh-skler-OH-sis): a type of artery disease characterized by formulations of plaques along the inner walls of the arteries, which narrows the lumen of the artery and restricts blood flow to the tissue it supplies.

ATP or **adenosine** (ah-DEN-oh-seen) **triphosphate** (try-FOS-fate): a common high-energy compound composed of a purine (adenine), a sugar (ribose), and three phosphate groups.

atrophic (a-TRO-fik) **gastritis** (gas-TRY-tis): chronic inflammation of the stomach accompanied by a diminished size and functioning of the mucous membranes and glands.

autoimmune disorders: disorders that result from immune system defenses attacking the body's own cells.

autonomy: independence.

—B—

balance (dietary): providing foods of a number of types in proportion to each other, such that foods rich in some nutrients do not crowd out of the diet foods that are rich in other nutrients.

Barrett's esophagus: changes in the cells of the esophagus associated with chronic reflux that raise the risk of cancer of the esophagus.

basal metabolic rate (BMR): the rate of energy use for metabolism under specified conditions: after a 12-hour fast and restful sleep, without any physical activity or emotional excitement, and in a comfortable setting. It is usually expressed as kcalories per kilogram body weight per hour. (Table 8-3 on p. 249 provides equations for estimating BMR.)

basal metabolism: the energy needed to maintain life when a body is at complete digestive, physical, and emotional rest.

bases: compounds that accept hydrogen ions in a solution.

B-cells: lymphocytes that produce antibodies.

beer: an alcoholic beverage brewed by fermenting malt and hops.

behavior modification: the changing of behavior by the manipulation of antecedents (cues or environmental factors that trigger behavior), the behavior itself, and consequences (the penalties or rewards attached to behavior).

belching: the expulsion of gas from the stomach through the mouth.

beneficence (be-NEF-eh-sens): doing good.

benign (bee-NINE): describes tumors that stop growing without intervention or can be removed surgically and most often pose no threat to health.

beriberi: the thiamin-deficiency disease.

beta-carotene (BAY-tah KARE-oh-teen): one of the carotenoids; an orange pigment and vitamin A precursor found in plants.

bicarbonate: a compound with the formula HCO_3 that results from the dissociation of carbonic acid; of particular importance in maintaining the body's acid-base balance. (Bicarbonate is also an alkaline secretion of the pancreas, part of the pancreatic juice.)

bifidus (BIFF-id-us, by-FEED-us) **factors:** factors in colostrum and breast milk that favor the growth of the "friendly" bacterium *Lactobacillus* (lack-toh-ba-SILL-us) *bifidus* in the infant's intestinal tract, so that other, less desirable intestinal inhabitants will not flourish.

bile: an emulsifier that prepares fats and oils for digestion; an exocrine secretion made by the liver, stored in the gallbladder, and released into the small intestine when needed.

biliary (BILL-ee-air-ee) **atresia** (ah-TREE-zee-ah): a disorder of infants characterized by absent or injured bile ducts.

biliary colic: pain associated with gallstones that have entered the common bile duct.

biliary tract: the gallbladder and bile ducts.

bilirubin (bill-ih-REW-bin): an orange- or yellow-colored pigment in the bile whose blood concentration may rise as a result of medical conditions that affect the liver.

binders: chemical compounds in foods that combine with nutrients (especially minerals) to form complexes the body cannot absorb. Examples include **phytates** (FYE-tates) and **oxalates** (OK-sa-lates).

binge-eating disorder: an eating disorder whose criteria are similar to those of bulimia nervosa, excluding purging or other compensatory behaviors.

bioavailability: the rate at and the extent to which a nutrient is absorbed and used.

bioelectrical impedance (im-PEE-dans): a method of estimating body fat using low-intensity electrical current.

biological value (BV): the amount of protein nitrogen that is retained from a given amount of protein nitrogen absorbed; a measure of protein quality.

biotin (BY-oh-tin): a B vitamin that functions as a coenzyme in metabolism.

blind experiment: an experiment in which the subjects do not know whether they are members of the experimental group or the control group.

blind loop syndrome: the problems of fat and vitamin B_{12} malabsorption that result from the overgrowth of bacteria in a bypassed segment of the intestine.

blood lipid profile: results of blood tests that reveal a person's total cholesterol, triglycerides, and various lipoproteins.

body composition: the proportions of muscle, bone, fat, and other tissue that make up a person's total body weight.

body mass index (BMI): an index of a person's weight in relation to height; determined by dividing the weight (in kilograms) by the square of the height (in meters).

bolus (BOH-lus): a portion; with respect to food, the amount swallowed at one time.

bolus feeding: delivery of about 300 to 400 ml of formula over 10 minutes or less.

bomb calorimeter (KAL-oh-RIM-eh-ter): an instrument that measures the heat energy released when foods are burned, thus providing an estimate of the potential energy of foods.

bone density: a measure of bone strength. When minerals fill the bone matrix (making it dense), they give it strength.

bone marrow transplants: the replacement of diseased bone marrow in a recipient with healthy bone marrow from a donor; sometimes used as a treatment for breast cancer, leukemia, lymphomas, and certain blood disorders.

botulism (BOT-chew-lism): an often fatal food-borne illness caused by the ingestion of foods containing a toxin produced by bacteria that grow without oxygen (see Chapter 19 for details).

bronchitis (bron-KYE-tis): inflammation of the lungs' air passages.

brown adipose tissue: masses of specialized fat cells packed with pigmented mitochondria that produce heat instead of ATP.

buffalo hump: the accumulation of fat at the top of the back.

bulimia (byoo-LEEM-ee-ah) **nervosa:** an eating disorder characterized by repeated episodes of binge eating usually followed by self-induced vomiting, misuse of laxatives or diuretics, fasting, or excessive exercise.

bulk-forming agents: laxatives composed of fibers that work like dietary fibers. They attract water in the intestine to form a bulky stool, which then stimulates peristalsis. Metamucil and Fiberall are examples.

—C—

calcitonin (KAL-see-TOE-nin): a hormone from the thyroid gland that regulates blood calcium by lowering it when levels rise too high.

calcium rigor: hardness or stiffness of the muscles caused by high blood calcium concentrations.

calcium tetany (TET-ah-nee): intermittent spasm of the extremities due to nervous and muscular excitability caused by low blood calcium concentrations.

calcium: the most abundant mineral in the body; found primarily in the body's bones and teeth.

calcium-binding protein: a protein in the intestinal cells, made with the help of vitamin D, that facilitates calcium absorption.

calmodulin (cal-MOD-you-lin): an inactive protein that becomes active when bound to calcium. Once activated, it becomes a messenger that tells other proteins what to do. The system serves as an interpreter for hormone- and nerve-mediated messages arriving at cells.

calories: units by which energy is measured. Food energy is measured in **kilocalories** (1000 calories equal 1 kilocalorie), abbreviated **kcalories** or **kcal.** A capitalized version is also sometimes used: **Calories.** One kcalorie is the amount of heat necessary to raise the temperature of 1 kilogram (kg) of water 1°C.

cancer cachexia (ka-KEKS-ee-ah) **syndrome:** loss of lean body mass, depletion of serum proteins, and debilitation that frequently accompany cancer.

cancers: diseases that result from the unchecked growth of cells.

capillaries (CAP-ill-aries): small vessels that branch from an artery. Capillaries connect arteries to veins. Exchange of oxygen, nutrients, and waste materials takes place across capillary walls.

capitation: prepayment of a set fee per client in exchange for medical services.

carbohydrates: compounds composed of carbon, oxygen, and hydrogen arranged as monosaccharides or multiples of monosaccharides. Most, but not all, carbohydrates have a ratio of one carbon molecule to one water molecule: $(CH_2O)_n$.

carbohydrate-to-insulin ratio: the number of units of insulin needed to cover 1 g of carbohydrate. The lower the ratio, the more insulin is needed to cover carbohydrate intake.

carbonic acid: a compound with the formula H_2CO_3 that results from the combination of carbon dioxide (CO_2) and water (H_2O); of particular importance in maintaining the body's acid-base balance.

carcinogenic (CAR-sin-oh-JEN-ick): producing cancer. A substance that produces or enhances the development of cancer is a **carcinogen.**

cardiac cachexia (ka-KEKS-ee-ah): chronic PEM that develops as a consequence of heart failure.

cardiac cirrhosis (sih-ROW-sis): severe liver damage associated with the later stages of congestive heart failure.

cardiomegaly (CAR-dee-oh-MEG-ah-lee): enlargement of the heart.

cardiovascular disease (CVD): a general term for all diseases of the heart and blood vessels. Atherosclerosis is the main cause of CVD. When the arteries that carry blood to the heart muscle become blocked, the heart suffers damage known as **coronary heart disease (CHD).**

carnitine (CAR-neh-teen): a nonessential nutrient made in the body from the amino acid lysine.

carotenoids (kah-ROT-eh-noyds): pigments commonly found in plants and animals, some of which have vitamin A activity. The carotenoid with the greatest vitamin A activity is beta-carotene.

carpal tunnel syndrome: a pinched nerve at the wrist, causing pain or numbness in the hand. It is often caused by repetitive motion of the wrist.

carriers: individuals who possess one **dominant** and one **recessive gene** for a genetic trait, such as an inborn error of metabolism. When a trait is recessive, a carrier may show no signs of the trait but can pass it on.

catabolism (ca-TAB-o-lism): reactions in which large molecules are broken down to smaller ones. Catabolic reactions usually release energy.

catalyst (CAT-uh-list): a compound that facilitates chemical reactions without itself being changed in the process.

cataracts (KAT-ah-rakts): thickenings of the eye lenses that impair vision and can lead to blindness.

cathartic (ka-THAR-tik): a strong laxative.

cations (CAT-eye-uns): positively charged ions.

CD4+ T-cells: a type of white blood cell (lymphocyte) that has a specific protein receptor on its surface and is a necessary component of the immune system.

celiac (SEE-lee-ack) **disease:** a sensitivity to a part of the protein gluten that causes flattening of the intestinal villi and malabsorption; also called *gluten-sensitive enteropathy* or *celiac sprue.*

cell-mediated immunity: immunity conferred by T-cells traveling to the invasion site to fight specific antigens.

cellulite (SELL-you-light or SELL-you-leet): supposedly, a lumpy form of fat; actually, a fraud. Fatty areas of the body may appear lumpy when the strands of connective tissue that attach the skin to underlying muscles pull tight where the fat is thick. The fat itself is the same as fat anywhere else in the body. If the fat in these areas is lost, the lumpy appearance disappears.

central obesity: excess fat around the trunk of the body; also called **abdominal fat** or **upper-body fat.**

central total parenteral nutrition (central TPN): the provision of an IV solution that meets nutrient needs delivered into a central vein.

central veins: the large-diameter veins located close to the heart.

cesarean section: a surgically assisted birth involving removal of the fetus by an incision into the uterus, usually by way of the abdominal wall.

chelate (KEY-late): a substance that can grasp the positive ions of a metal.

chemotherapy: the use of drugs to arrest or destroy cancer cells. Drugs used for chemotherapy are called **chemotherapeutic** or **antineoplastic agents.**

chloride (KLO-ride): the major anion in the extracellular fluids of the body. Chloride is the ionic form of chlorine, Cl_; see Appendix B for a description of the chlorine-to-chloride conversion.

chlorophyll (KLO-row-fil): the green pigment of plants, which absorbs light and transfers the energy to other molecules, thereby initiating photosynthesis.

cholangitis (KOH-lan-JYE-tis): inflammation of the bile ducts.

cholecystectomy (KOH-lee-sis-TEK-toe-mee): surgical removal of the gallbladder.

cholecystitis (KOH-lee-sis-TYE-tis): inflammation of the gallbladder.

cholecystokinin (coal-ee-sis-toe-KINE-in), or **CCK:** a hormone produced by cells of the intestinal wall. Target organ: the gallbladder. Response: release of bile and slowing of GI motility.

choledocholithiasis (koh-LED-oh-koh-lih-THIGH-ah-sis): the presence of gallstones in the common bile duct.

cholelithiasis (KOH-lee-lih-THIGH-ah-sis): the formation or presence of stones in the gallbladder or common bile duct.

cholesterol (koh-LESS-ter-ol): one of the sterols containing four carbon rings and a carbon side chain.

choline (KOH-leen): a nitrogen-containing compound found in foods and made in the body from the amino acid methionine. Choline is found in foods as part of the phospholipid lecithin and in the body as part of the neurotransmitter acetylcholine.

chronic diseases: long-duration degenerative diseases characterized by deterioration of the body organs. Examples include heart disease, cancer, and diabetes.

chronic malnutrition: protein-energy malnutrition (PEM) that develops as a consequence of insufficient intake of energy and protein over long periods of time and is characterized by underweight, depleted fat stores, and normal serum protein levels.

chronic obstructive pulmonary disease (COPD): one of several disorders, including emphysema and bronchitis, that interfere with respiration.

chronic or **congestive heart failure (CHF):** a syndrome in which the heart gradually weakens and can no longer adequately pump blood through the circulatory system.

chronic PEM: protein-energy malnutrition caused by long-term food deprivation; characterized in children by short height for age (stunting).

chronological age: a person's age in years from his or her date of birth.

chylomicrons (kye-lo-MY-cronz): the class of lipoproteins that transport lipids from the intestinal cells to the rest of the body.

chyme (KIME): the semiliquid mass of partly digested food expelled by the stomach into the duodenum.

cirrhosis (sih-ROW-sis): an advanced form of liver disease in which scar tissue replaces liver cells that have permanently lost their function.

clear-liquid diet: a diet that consists of foods that are liquid at room temperature and leave little residue in the intestine.

clinical pathways, critical pathways, or **care maps:** charts or tables that outline a plan of care for a specific diagnosis, treatment, or procedure, with a goal of providing the best possible outcome at the lowest cost. The plan, developed by the health care team after a careful study of each facility's unique client population, is regularly reassessed and improved.

clinically severe obesity: a BMI of 40 or greater or 100 lb or more overweight for an average adult. A less preferred term used to describe the same condition is *morbid obesity.*

closed feeding systems: enteral formula delivery systems in which the formula comes prepackaged in a container that is ready to be attached to a feeding tube for administration.

CoA (coh-AY): coenzyme A; the coenzyme derived from the B vitamin pantothenic acid and central to energy metabolism.

coenzymes: small organic molecules that work with enzymes to facilitate the enzymes' activity. Many coenzymes have B vitamins as part of their structures (Figure 10-2 in Chapter 10 illustrates coenzyme action).

cofactor: a small inorganic or organic substance that works with an enzyme to facilitate a chemical reaction.

colitis (ko-LYE-tis): inflammation of the colon.

collagen (KOL-ah-jen): the protein from which connective tissues such as scars, tendons, ligaments, and the foundations of bones and teeth are made.

collaterals: small branches of a blood vessel that develop when blood flow through the liver is obstructed; also called **shunts.**

colonic irrigation: the popular, but potentially harmful practice of "washing" the large intestine with a powerful enema machine.

colostomy (co-LOSS-toe-me): surgery that creates an opening from any portion of the colon through the abdominal wall and out through the skin.

colostrum (ko-LAHS-trum): a milklike secretion from the breast, present during the first day or so after delivery before milk appears; rich in protective factors.

comatose: in a state of deep unconsciousness from which the person cannot be aroused.

competent: having sufficient mental ability to understand a treatment, weigh its risks and benefits, and comprehend the consequences of refusing or accepting the treatment.

complement: a group of blood proteins that assist the activities of antibodies.

complementary proteins: two or more proteins whose amino acid assortments complement each other in such a way that the essential amino acids missing from one are supplied by the other.

complete formulas: liquid diets designed to supply all needed nutrients when consumed in sufficient volume.

complete protein: a dietary protein containing all the essential amino acids in relatively the same amounts that human beings require. It may also contain nonessential amino acids.

complex carbohydrates (starches and **fibers):** polysaccharides composed of straight or branched chains of monosaccharides.

conception: the union of the male sperm and the female ovum; fertilization.

condensation: a chemical reaction in which two reactants combine to yield a larger product.

conditionally essential amino acid: an amino acid that is normally nonessential, but must be supplied by the diet in special circumstances when the need for it exceeds the body's ability to produce it.

congregate meals: nutrition programs that provide food for the elderly in a conveniently located setting such as a community center.

constipation: the condition of having infrequent or difficult bowel movements.

contamination iron: iron found in foods as the result of contamination by inorganic iron salts from iron cookware, iron-containing soils, and the like.

continuous ambulatory peritoneal dialysis (CAPD): the most common type of peritoneal dialysis that people use at home.

continuous feedings: slow delivery of formula in constant amounts over an 8- to 24-hour period.

continuous renal replacement therapy (CRRT): a slow and continuous type of dialysis used in the treatment of acute renal failure.

control group: a group of individuals similar in all possible respects to the experimental group except for the treatment. Ideally, the control group receives a placebo while the experimental group receives a real treatment.

Cori cycle: the path from muscle glycogen to glucose to pyruvate to lactic acid (which travels to the liver) to glucose (which can travel back to the muscle) to glycogen; named after the scientist who elucidated this pathway.

cornea (KOR-nee-uh): the transparent membrane covering the outside of the eye.

coronary heart disease (CHD): heart damage that results from an inadequate supply of blood to the heart.

correlation (core-ee-lay-shun): the simultaneous increase, decrease, or change in two variables. If A increases as B increases, or if A decreases as B decreases, the correlation is **positive.** (This does not mean that A causes B or vice versa.) If A increases as B decreases, or if A decreases as B increases, the correlation is **negative.** (This does not mean that A prevents B or vice versa.) Some third factor may account for both A and B.

correspondence schools: schools that offer courses and degrees by mail. Some correspondence schools are accredited; others are *diploma mills.*

cortical bone: the very dense bone tissue that forms the outer shell surrounding trabecular bone and comprises the shaft of a long bone.

counterregulatory hormones: hormones such as glucagon, cortisol, and catecholamines that oppose insulin's actions and promote catabolism.

coupled reactions: pairs of chemical reactions in which energy released from the breakdown of one compound is used to create a bond in the formation of another compound.

covert (KOH-vert): hidden, as if under covers.

cretinism (CREE-tin-ism): a congenital disease characterized by mental and physical retardation and commonly caused by maternal iodine deficiency during pregnancy.

critical periods: finite periods during development in which certain events occur that will have irreversible effects on later developmental stages; usually a period of rapid cell division.

Crohn's disease: inflammation and ulceration along the length of the GI tract, often with granulomas.

cryptosporidiosis (KRIP-toe-spo-rid-ee-OH-sis): a food-borne illness caused by the parasite *Cryptosporidium parvum.* Most people develop few or minor problems from this infection, but people with HIV infections, and especially those with AIDS, can develop long-lasting and serious problems.

crypts (KRIPTS): tubular glands that lie between the intestinal villi and secrete intestinal juices into the small intestine.

cyanosis (sigh-ah-NOH-sis): bluish discoloration of the skin caused by a lack of oxygen.

cyberspace: a term coined by William Gibson referring to the nonphysical place where all Internet activity occurs.

cyclamate (SIGH-kla-mate): an artificial sweetener that is being considered for approval in the United States and is available in Canada as a tabletop sweetener, but not as an additive.

cyclic parenteral nutrition: the continuous administration of a parenteral solution for 8 to 12 hours with time periods when no nutrients are infused.

cystic fibrosis: a hereditary disorder characterized by the production of thick mucus that affects many organs, including the pancreas, lungs, liver, heart, gallbladder, and small intestine.

cystinuria (SIS-tin-NEW-ree-ah): an inherited metabolic disorder that is characterized by the excessive urinary excretion of cystine, lysine, arginine, and ornithine and commonly leads to kidney stone formation.

cytokines (SIGH-toe-kynes): proteins that help regulate immune system responses. Cytokines trigger hypermetabolism and cause anorexia, fever, and discomfort.

—D—

dawn phenomenon: early morning hyperglycemia that develops in response to elevated levels of counterregulatory hormones that act to raise blood glucose after an overnight fast.

deamination (dee-AM-eh-NAY-shun): removal of the amino (NH2) group from a compound such as an amino acid.

debridement (dee-BREED-ment): the removal of dead tissues resulting from burns and other wounds to speed healing and prevent infection.

defecate (DEF-uh-cate): to move the bowels and eliminate waste.

deficient: the amount of a nutrient below which almost all healthy people can be expected, over time, to experience deficiency symptoms.

dehydration: the condition in which body water output exceeds water input. Symptoms include thirst, dry skin and mucous membranes, rapid heartbeat, low blood pressure, and weakness.

delusions (dee-LOO-shuns): inappropriate beliefs not consistent with the individual's own knowledge and experience.

dementia (dee-MEN-she-ah): irreversible loss of mental function.

denaturation (dee-NAY-chur-AY-shun): the change in a protein's shape and consequent loss of its function brought about by heat, agitation, acid, base, alcohol, heavy metals, or other agents.

dental caries: decay of teeth.

dental plaque: a gummy mass of bacteria that grows on teeth and can lead to dental caries and gum disease.

dextrose monohydrate: a form of glucose that contains a molecule of water and is stable in IV solutions. IV dextrose solutions provide 3.4 kcal/g.

diabetes (DYE-uh-BEET-eez) **mellitus** (MELL-ih-tus or mell-EYE-tus): a group of metabolic disorders that result from inadequate or ineffective insulin causing abnormal glucose regulation and utilization.

diabetic coma: unconsciousness precipitated by hyperglycemia, dehydration, ketosis, and acidosis in people with diabetes.

dialysate (dye-AL-ih-SATE): a solution used during dialysis to draw wastes and fluids from the blood.

dialysis (dye-AL-ih-sis): removal of waste from the blood through a semipermeable membrane using the principles of simple diffusion and osmosis. The two main types are **hemodialysis** and **peritoneal dialysis.**

dialyzer (dye-ah-LYES-er): the machine used for hemodialysis; also called an *artificial kidney.*

diarrhea: the frequent passage of watery bowel movements.

diet history: a comprehensive record of eating-related behaviors and the foods a person eats.

diet manual: a book that describes the foods allowed and restricted on a diet, outlines the rationale and indications for use of each diet, and provides sample menus.

diet order: a statement of the client's diet prescription that the physician writes in the medical record.

diet: the foods and beverages a person eats and drinks.

dietary folate equivalents (DFE): the amount of folate available to the body from naturally occurring sources, fortified foods, and supplements, accounting for differences in the bioavailability from each source.

Dietary Reference Intakes (DRI): a set of values for the dietary nutrient intakes of healthy people in the United States and Canada. These values are used for planning and assessing diets.

dietetic technician: a person who has completed a minimum of an associate's degree from an accredited university or college and an approved dietetic technician program that includes a supervised practice experience.

dietetic technician, registered (DTR): a dietetic technician who has passed a national examination and maintains registration through continuing professional education.

dietitian: a person trained in nutrition, food science, and diet planning.

differentiation: the development of specific functions different from those of the original.

diffusion: movement of solutes from an area of high concentration to one of low concentration.

digestion: the process by which food is broken down into absorbable units.

digestive enzymes: proteins found in digestive juices that act on food substances, causing them to break down into simpler compounds.

digestive system: all the organs and glands associated with the ingestion and digestion of food.

dipeptide (dye-PEP-tide): two amino acids bonded together.

disaccharides (dye-SACK-uh-rides): pairs of monosaccharides linked together. See Appendix C for the chemical structures of the disaccharides.

dissociates (dis-SO-see-ates): physically separates.

distilled liquor: an alcoholic beverage made by fermenting and distilling grains; sometimes called *distilled spirits* or *hard liquor.*

disufiram-like reaction: nausea, vomiting, headache, cramps, flushing of the skin, and a rapid heartbeat that can occur when some medications are taken along with alcohol. The medication disulfiram produces these effects when combined with alcohol to discourage alcohol abusers from using alcohol.

diverticula (dye-ver-TIC-you-la): sacs or pouches that develop in the weakened areas of the intestinal wall (like bulges in an inner tube where the tire wall is weak).

diverticulitis (DYE-ver-tic-you-LYE-tis): infected or inflamed diverticula.

diverticulosis (DYE-ver-tic-you-LOH-sis): the condition of having diverticula. About one in every six people in Western countries develops diverticulosis in middle or later life.

docosahexaenoic (DOE-cossa-HEXA-ee-NO-ick) **acid (DHA):** an omega-3 polyunsaturated fatty acid with 22 carbons and six double bonds (22:6); synthesized from linolenic acid.

dominant gene: a gene that has an observable effect on an organism. If an altered gene is dominant, it has an observable effect even when it is paired with a normal gene; see also *recessive gene.*

double-blind experiment: an experiment in which neither the subjects nor the researchers know which subjects are members of the experimental group and which are serving as control subjects, until after the experiment is over.

Down syndrome: a genetic abnormality that causes mental retardation, short stature, and flattened facial features.

drink: a dose of any alcoholic beverage that delivers ½ oz of pure ethanol.

drug: a substance that can modify one or more of the body's functions.

dry weight: weight after excess fluids are removed from the body.

DTR: see *dietetic technician, registered.*

dumping syndrome: the symptoms that result from the rapid entry of undigested food into the jejunum: sweating, weakness, and diarrhea shortly after eating and hypoglycemia later.

duodenum (doo-oh-DEEN-um, doo-ODD-num): the top portion of the small intestine (about "12 fingers' breadth" long in ancient terminology).

durable power of attorney: a legal document in which one competent adult authorizes another competent adult to make decisions for her or him in the event of incapacitation. The phrase "durable power" means that the agent's authority survives the client's incompetence; "attorney" refers to an attorney-in-fact (not an attorney-at-law).

dysentery (DISS-en-terry): an infection of the digestive tract that causes diarrhea.

dyspepsia: vague abdominal pain; a symptom, not a disease.

dysphagia (dis-FAY-gee-ah): difficulty in swallowing.

dysuria (dis-YOU-ree-ah): painful or difficult urination.

—E—

eating disorders: disturbances in eating behavior that jeopardize a person's physical or psychological health.

eclampsia (eh-KLAMP-see-ah): a severe stage of preeclampsia characterized by convulsions.

edema (eh-DEEM-uh): the swelling of body tissue caused by excessive amounts of fluid in the interstitial spaces; seen in protein deficiency (among other conditions).

eicosanoids (eye-COSS-uh-noyds): derivatives of 20-carbon fatty acids; biologically active compounds that regulate blood pressure, blood clotting, and other body functions. They include *prostaglandins* (PROS-tah-GLAND-ins), *thromboxanes* (throm-BOX-ains), and *leukotrienes* (LOO-ko-TRY-eens).

eicosapentaenoic (EYE-cossa-PENTA-ee-NO-ick) **acid (EPA):** an omega-3 polyunsaturated fatty acid with 20 carbons and five double bonds (20:5); synthesized from linolenic acid.

electrolyte solutions: solutions that can conduct electricity.

electrolytes: salts that dissolve in water and dissociate into charged particles called ions.

electron transport chain (ETC): the final pathway in energy metabolism where the electrons from hydrogen are passed to oxygen and the energy released is trapped in the bonds of ATP.

embolism (EM-boh-lizm): the obstruction of a blood vessel by an embolus, causing sudden tissue death.

embolus (EM-boh-lus): a traveling blood clot.

embryo (EM-bree-oh): the developing infant from two to eight weeks after conception.

emetic (em-ETT-ic): an agent that causes vomiting.

emphysema (EM-fih-SEE-mah): a disorder in which the lungs lose their elasticity and the victim has difficulty breathing; often occurs along with bronchitis.

empty-kcalorie foods: a popular term used to denote foods that contribute energy but lack protein, vitamins, and minerals.

emulsifier (ee-MUL-sih-fire): a substance with both water-soluble and fat-soluble portions that promotes the mixing of oils and fats in a watery solution.

end-stage renal disease (ESRD): the severe stage of chronic renal failure in which dialysis or a kidney transplant is necessary to sustain life.

enemas: solutions inserted into the rectum and colon to stimulate a bowel movement and empty the lower large intestine.

energy: the capacity to do work. The energy in food is chemical energy. The body can convert this chemical energy to mechanical, electrical, or heat energy.

energy-yielding nutrients: the nutrients that break down to yield energy the body can use.

enriched: the addition to a food of nutrients that were lost during processing so that the food will meet a specified standard.

enteral (EN-ter-all) **nutrition:** the provision of nutrients using the GI tract. Enteral nutrition includes both oral diets and tube feedings.

enteral formulas: liquid diets designed to be delivered through the GI tract, either orally or by tube.

enteropancreatic (EN-ter-oh-PAN-kree-AT-ik) **circulation:** the circulatory route from the pancreas to the intestine and back to the pancreas.

enterostomy (EN-ter-OSS-toe-mee): an opening into the stomach or jejunum through which a feeding tube can be passed.

enzymes: proteins that facilitate chemical reactions without being changed in the process; protein catalysts.

epidemic (EP-ee-DEM-ick): the appearance of a disease (usually infectious) or condition that attacks many people at the same time in the same region.

epiglottis (epp-ee-GLOTT-iss): cartilage in the throat that guards the entrance to the trachea and prevents fluid or food from entering it when a person swallows.

epinephrine (EP-ih-NEFF-rin): a hormone of the adrenal gland that modulates the stress response; formerly called **adrenaline.**

epithelial (ep-i-THEE-lee-ul) **cells:** cells on the surface of the skin and mucous membranes.

epithelial tissue: the layer of the body that serves as a selective barrier between the body's interior and the environment (examples are the cornea, the skin, the respiratory lining, and the lining of the digestive tract).

erythrocyte (eh-RITH-ro-cite) **hemolysis** (he-MOLL-uh-sis): the breaking open of red blood cells (erythrocytes); a symptom of vitamin E–deficiency disease in human beings.

erythrocyte protoporphyrin (PRO-toe-PORE-fe-rin): a precursor to hemoglobin.

erythropoietin (eh-RITH-row-POY-eh-tin): a hormone secreted by the kidneys in response to oxygen depletion or anemia that stimulates the bone marrow to produce red blood cells.

esophageal reflux: the backflow of the gastric contents of the esophagus.

esophageal sphincters (ee-SOF-ah-GEE-al SFINK-ters): sphincter muscles at the upper and lower ends of the esophagus. The *lower esophageal sphincter* is also called the *cardiac sphincter.*

esophageal stricture: narrowing of the inner diameter of the esophagus from inflammation and scarring.

esophageal ulcers: lesions or sores in the lining of the esophagus.

esophagus (ee-SOFF-ah-gus): the food pipe; the conduit from the mouth to the stomach.

essential amino acids: amino acids that the body cannot synthesize in amounts sufficient to meet physiological needs (see Table 6-1). Some researchers refer to essential amino acids as **indispensable** and to nonessential amino acids as **dispensable.**

essential fatty acids: fatty acids needed by the body, but not made by it in amounts sufficient to meet physiological needs.

essential nutrients: nutrients a person must obtain from food because the body cannot make them for itself in sufficient quantity to meet physiological needs; also called **indispensable nutrients.** About 40 nutrients are known to be essential for human beings.

Estimated Average Requirement: the amount of a nutrient that will maintain a specific biochemical or physiological function in half the people of a given age and gender group.

ethanol: a particular type of alcohol found in beer, wine, and distilled spirits; also called *ethyl alcohol* (see Figure H7-1). Ethanol is the most widely used—and abused—drug in our society. It is also the only legal, nonprescription drug that produces euphoria.

ethical: in accordance with moral principles or professional standards. Socrates described *ethics* as "how we ought to live."

exchange lists: diet-planning tools that organize foods by their proportions of carbohydrate, fat, and protein. Foods on any single list can be used interchangeably.

experimental group: a group of individuals similar in all possible respects to the control group except for the treatment. The experimental group receives the real treatment.

—F—

false negative: a test result indicating that a condition is not present (negative) when in fact it is present (therefore false).

false positive: a test result indicating that a condition is present (positive) when in fact it is not (therefore false).

fat replacers: ingredients that replace some or all of the functions of fat and may or may not provide energy. In this text, the term *fat replacer* is used interchangeably with **fat substitute,** which technically applies only to an ingredient that replaces all of the functions of fat and provides no energy.

fatfold measures: estimates of total body fatness determined by measuring the thickness of a fold of skin on the back of the arm (over the triceps muscle), below the shoulder blade (subscapular), and in other places as measured with a caliper. (The older, less preferred, term is **skinfold test.**)

fats: lipids in foods or the body; composed mostly of triglycerides.

fatty acid oxidation: the metabolic breakdown of fatty acids to acetyl CoA; also called **beta oxidation.**

fatty acid: an organic compound composed of a carbon chain with hydrogens attached and an acid group (COOH) at one end.

fatty liver: an early stage of liver deterioration seen in several diseases, including kwashiorkor and alcoholic liver disease. Fatty liver is characterized by an accumulation of fat in the liver cells also called **hepatic steatosis, steatohepatitis,** and **fatty infiltration of the liver.**

fatty streaks: accumulations of cholesterol and other lipids along the walls of the arteries.

FDA (Food and Drug Administration): a federal agency that is responsible for, among other things, supplement safety, manufacturing, and information, including product labeling, package inserts, and accompanying literature.

fecal impaction: a compacted mass of fecal material in the colon or rectum.

Federal Trade Commission (FTC): a federal agency that is responsible for, among other things, food advertising and industry competition.

female athlete triad: a potentially fatal combination of three medical problems: disordered eating, amenorrhea, and osteoporosis.

ferment: to digest in the absence of oxygen.

fertility: the capacity of a woman to produce a normal ovum periodically and of a man to produce normal sperm; the ability to reproduce.

fetal alcohol effects (FAE): a subclinical version of fetal alcohol syndrome, with hidden defects including learning disabilities, behavioral abnormalities, and motor impairments; also called **alcohol-related birth defects (ARBD).**

fetal alcohol syndrome (FAS): the cluster of symptoms seen in an infant or child whose mother consumed excess alcohol during pregnancy, including retarded growth, impaired development of the central nervous system, and facial malformations.

fetor hepaticus (FEE-tor he-PAT-eh-cuss): a pungent odor of the breath that may develop in people with impending hepatic coma.

fetus (FEET-us): the developing infant from eight weeks after conception until term.

fibers: in plant foods, the *nonstarch polysaccharides* that are not digested by human digestive enzymes, although some are digested by GI tract bacteria. Fibers include cellulose, hemicelluloses, pectins, gums, and mucilages and the nonpolysaccharides lignins, cutins, and tannins.

fibrinogen (fie-BRIN-oh-jen): a protein produced by the liver that is essential to blood clotting.

fibrocystic (FYE-bro-SIS-tik) **breast disease:** a harmless condition in which the breasts develop lumps, sometimes associated with caffeine consumption. In some, it responds to abstinence from caffeine; in others, it can be treated with vitamin E.

fibrosis (fye-BROH-sis): an intermediate stage of liver deterioration seen in several diseases, including viral hepatitis and alcoholic liver disease. In fibrosis, the liver cells lose their function and assume the characteristics of connective tissue cells (fibers).

fibrous plaques (PLACKS): mounds of lipid material, mixed with smooth muscle cells and calcium, which develop in the artery walls in atherosclerosis.

filtrate: in the kidneys, the fluid that passes from the blood through the capillary walls of the glomeruli, eventually forming urine.

fistulas (FIS-chew-lahs): abnormal openings formed between two organs or between an internal organ and the skin.

flapping tremor: uncontrolled movement of the muscle group that causes the outstretched arm and hand to flap like a wing; occurs in disorders that cause encephalopathy; also called **asterixis** (AS-ter-ICK-sis).

flatus (FLAY-tuss): gas in the intestinal tract or the expelling of gas from the intestinal tract, especially through the anus.

fluid balance: maintenance of the proper types and amounts of fluid in each compartment of the body fluids.

fluorosis (floor-OH-sis): discoloration and pitting of tooth enamel caused by excess fluoride during tooth development.

folate (FOLE-ate): a B vitamin; also known as folic acid, folacin, or pteroylglutamic (tare-o-EEL-glue-TAM-ick) acid (PGA). The coenzyme forms are DHF (dihydrofolate) and THF (tetrahydrofolate).

food allergy: an adverse reaction to food that involves an immune response; also called **food-hypersensitivity reaction.**

food aversions: strong desires to avoid particular foods.

food bank: a central source for the donation and distribution of food to local charities feeding the hungry.

food consumption survey: a survey that measures the amounts and kinds of foods people consume (using diet histories), estimates the nutrient intakes, and compares them with a standard.

food cravings: strong desires to eat particular foods.

food frequency questionnaire: a tool for gathering food intake data that asks clients about the types and amounts of foods they routinely eat.

food group plans: diet-planning tools that sort foods of similar origin and nutrient content into groups and then specify that people should eat certain numbers of servings from each group.

food insecurity: limited or uncertain access to foods of sufficient quality or quantity to sustain a healthy and active life.

food intolerances: adverse reactions to foods that do not involve the immune system.

food poverty: hunger occurring when enough food exists in an area but some of the people cannot obtain it because they lack money, are being deprived for political reasons, live in a country at war, or suffer from other problems such as lack of transportation.

food records: logs of all the food eaten over a period of time that may also include records of behaviors, symptoms, physical activity, and medications; also called **eating** and/or **food diaries.**

food recovery: collecting wholesome food for distribution to low-income people who are hungry.

food substitutes: foods that are designed to replace other foods.

foods: products derived from plants or animals that can be taken into the body to yield energy and nutrients for the maintenance of life and the growth and repair of tissues.

fortified: the addition to a food of nutrients that were either not originally present or present in insignificant amounts. Fortification can be used to correct or prevent a widespread nutrient deficiency or to balance the total nutrient profile of a food.

fraud or **quackery:** the promotion, for financial gain, of devices, treatments, services, plans, or products (including diets and supplements) that alter or claim to alter a human condition without proof of safety or effectiveness. (The word *quackery* comes from the term *quacksalver,* meaning a person who quacks loudly about a miracle product—a lotion or a salve.)

free radicals: unstable and highly reactive atoms or molecules that have one or more unpaired electrons in the outer orbital (see Appendix B for a review of basic chemistry concepts).

fructose (FRUK-tose or FROOK-tose): a monosaccharide. Sometimes known as fruit sugar or **levulose,** fructose is found abundantly in fruits, honey, and saps.

fuel: compounds that cells can use for energy. The major fuels include glucose, fatty acids, and amino acids; other fuels include ketone bodies, lactic acid, glycerol, and alcohol.

full-liquid diet: a diet that consists of both clear and opaque liquid foods and near-liquid foods.

fulminant (FULL-mih-nant) **liver failure:** liver failure that rapidly progresses to a life-threatening stage.

functional foods: foods that contain physiologically active compounds that provide health benefits beyond their nutrient contributions; also called *designer foods* or *nutraceuticals.*

—G—

galactose (ga-LAK-tose): a monosaccharide; part of the disaccharide lactose.

galactosemia (ga-LAK-toe-SEE-me-ah): an inborn error of metabolism in which enzymes that normally metabolize galactose to compounds the body can handle are missing and an alternative metabolite accumulates in the tissues, causing damage.

gallbladder: the organ that stores and concentrates bile. When it receives the signal that fat is present in the duodenum, the gallbladder contracts and squirts bile through the bile duct into the duodenum.

gallstones: crystals of cholesterol or bile pigments that precipitate together and form a hard mass.

galvanized: a term referring to metals that have been treated with a zinc-containing coating to prevent rust.

gangrene: death of tissue due to a deficient blood supply and/or infection.

gastrectomy (gas-TREK-tah-mee): surgery to remove all (total gastrectomy) or part (subtotal or partial gastrectomy) of the stomach.

gastric glands: exocrine glands in the stomach wall that secrete gastric juice into the stomach.

gastric juice: the digestive secretion of the gastric glands of the stomach.

gastric partitioning: surgery for severe obesity that limits the functional size of the stomach.

gastric residual: the volume of formula that remains in the stomach from a previous feeding.

gastric-inhibitory peptide: a hormone produced by the intestine. Target organ: the stomach. Response: slowing of the secretion of gastric juices and of GI motility.

gastrin: a hormone secreted by cells in the stomach wall. Target organ: the glands of the stomach. Response: secretion of gastric acid.

gastritis: inflammation of the stomach lining.

gastroesophageal reflux: the backflow of stomach acid into the esophagus, causing damage to the cells of the esophagus and the sensation of heartburn.

gastrointestinal (GI) tract: the digestive tract. The principal organs are the stomach and intestines.

gastroparesis: delayed gastric emptying.

gastrostomy (gas-TROSS-toe-mee): an opening in the stomach made surgically or under local anesthesia through which a feeding tube can be passed. The technique for creating a gastrostomy under local anesthesia is called **percutaneous endoscopic gastrostomy,** or **PEG** for short. When the feeding tube is guided from such an opening into the jejunum, the procedure is called **percutaneous endoscopic jejunostomy (PEJ),** a misnomer because the enterostomy is in the stomach rather than the jejunum.

gatekeepers: with respect to nutrition, key people who control other people's access to foods and thereby exert profound impacts on their nutrition. Examples are the spouse who buys and cooks the food, the parent who feeds the children, and the caregiver in a day-care center.

genes: the basic units of hereditary information, made of DNA, that are passed from parent to offspring in the chromosomes. A pair of genes codes for each genetic trait.

gestation (jes-TAY-shun): the period from conception to birth. For human beings, gestation lasts from 38 to 42 weeks. Pregnancy is often divided into thirds, called **trimesters.**

gestational diabetes: abnormal glucose tolerance that is first detected during pregnancy.

gland: a cell or group of cells that secretes materials for special uses in the body. Glands may be **exocrine** (EKS-oh-crin) **glands,** secreting their materials "out" (into the digestive tract or onto the surface of the skin), or **endocrine** (EN-doe-crin) **glands**, secreting their materials "in" (into the blood).

gliadin (GLY-ah-din): the fraction of gluten that causes the toxic effects in celiac disease. Corresponding protein fractions in barley, rye, and possibly oats also have these effects.

glomerular filtration rate (GFR): the rate at which the kidneys form filtrate, usually measured by determining the amount of creatinine excreted in 24 hours. The normal GFR is about 130 ml/min for males and 120 ml/min for females.

glomerulus (glow-MARE-you-lus): a cup-shaped membrane enclosing a tuft of capillaries within a nephron. (The plural is *glomeruli.*)

glucagon (GLOO-ka-gon): a hormone that is secreted by special cells in the pancreas in response to low blood glucose concentration and elicits release of glucose from storage.

gluconeogenesis (gloo-co-nee-oh-GEN-ih-sis): the making of glucose from a noncarbohydrate source.

glucose (GLOO-kose): a monosaccharide; sometimes known as blood sugar or **dextrose.**

gluten (GLUE-ten): a protein found in wheat.

glycated (GLIGH-kate-id) **hemoglobin:** hemoglobin with glucose molecules attached to its amino acids; also called *glycosylated hemoglobin.* The type of glycated hemoglobin most commonly measured is hemoglobin A_{1c}.

glycemic (gligh-SEEM-ic) **effect:** a measure of the extent to which a food, as compared with pure glucose, raises the blood glucose concentration and elicits an insulin response.

glycemic index: a method used to classify foods according to their potential for raising blood glucose.

glycerol (GLISS-er-ol): an alcohol composed of a three-carbon chain, which can serve as the backbone for a triglyceride.

glycogen (GLY-co-gen): an animal polysaccharide composed of glucose; manufactured and stored in the liver and muscles as a storage form of glucose. Glycogen is not a significant food source of carbohydrate and is not counted as one of the complex carbohydrates in foods.

glycogen storage disease: one of several inherited disorders in which a person lacks one of the enzymes that allow glycogen stores to be utilized efficiently. The accumulation of glycogen in the liver leads to severe liver damage.

glycolysis (gligh-COLL-ih-sis): the metabolic breakdown of glucose to pyruvate. Glycolysis does not require oxygen (anaerobic).

glycosuria (GLY-ko-SUE-ree-ah) or **glucosuria** (GLUE-ko-SUE-ree-ah): glucose in the urine, which generally occurs when blood glucose exceeds 180 mg/dL.

goblet cells: cells of the GI tract (and lungs) that secrete mucus.

goiter (GOY-ter): an enlargement of the thyroid gland due to an iodine deficiency, malfunction of the gland, or overconsumption of a goitrogen. Goiter caused by iodine deficiency is **simple goiter.**

goitrogen (GOY-troh-jen): a thyroid antagonist found in food; causes **toxic goiter.** Goitrogens are found in such foods as cabbage, kale, brussels sprouts, cauliflower, broccoli, and kohlrabi.

gout: an inherited metabolic disorder that results in excessive uric acid in the blood and urine and the deposition of uric acid in and around the joints, which causes acute arthritis and joint inflammation.

granulomas (gran-you-LOH-mahs): tumors or growths that are covered with a fibrous coat and contain foreign organisms surrounded by immune system cells.

—H—

HDL (high-density lipoprotein): the type of lipoprotein that transports cholesterol back to the liver from the cells; composed primarily of protein.

health claims: statements that characterize the relationship between a nutrient or other substance in a food and a disease or health-related condition.

health history: an account of the client's current and past health status and risk factors for disease. Traditionally, the health history has been called the *medical history.* The term *health history* now seems more appropriate, however, because the contents describe the client's health status, and the goal of medical care is health promotion and disease prevention.

health maintenance organizations (HMOs): managed care organizations that limit the subscriber's choice of health care professionals to those affiliated with the organization and control access to services by directing care through a primary care physician.

Healthy People: a national public health initiative under the jurisdiction of the U.S. Department of Health and Human Services (DHHS) that identifies the most significant preventable threats to health and focuses efforts toward eliminating them.

heartburn: a burning sensation felt behind the sternum caused by the backflow of gastric juices into the esophagus; also called **pyrosis** (pie-ROE-sis).

heavy metals: any of a number of mineral ions such as mercury and lead, so called because they are of relatively high atomic weight. Many heavy metals are poisonous.

Heimlich (HIME-lick) **maneuver (abdominal thrust maneuver):** a technique for dislodging an object from the trachea of a choking person (see Figure H3-2); named for the physician who developed it.

Helicobacter pylori: a bacterium that may lead to gastritis and peptic ulcers and may raise the risk of cancer of the stomach.

hematocrit (hee-MAT-oh-krit): measurement of the volume of the red blood cells packed by centrifuge in a given volume of blood.

hematuria (HE-mah-TOO-ree-ah): blood in the urine.

heme (HEEM): the iron-holding part of the hemoglobin and myoglobin proteins. About 40% of the iron in meat, fish, and poultry is bound into heme; the other 60% is **nonheme** iron.

hemochromatosis (HE-moe-CROW-mah-toe-sis): an inherited disorder in which a person absorbs too much iron from the intestine and stores too much iron in the liver. The accumulation of iron in the liver leads to severe liver damage.

hemodialysis: removal of fluids and wastes from the blood by passing it through a dialyzer.

hemofiltration: removal of fluids and wastes from the blood by using ultrafiltration and fluid replacement.

hemoglobin (HE-moh-GLOW-bin): the globular protein of the red blood cells that carries oxygen from the lungs to the cells throughout the body.

hemolytic (HE-moh-LIT-ick) **anemia:** the condition of having too few red blood cells as a result of erythrocyte hemolysis.

hemophilia (HE-moh-FEEL-ee-ah): a hereditary disease that is caused by a genetic defect and has no relation to vitamin K. The blood is unable to clot because it lacks the ability to synthesize certain clotting factors.

hemorrhagic (hem-oh-RAJ-ik) **disease:** a disease characterized by excessive bleeding.

hemorrhoids (HEM-oh-royds): painful swelling of the veins surrounding the rectum.

hemosiderosis (HE-mo-sid-er-OH-sis): a condition characterized by the deposition of hemosiderin in the liver and other tissues.

hepatic coma: a state of unconsciousness that results from severe liver disease.

hepatic encephalopathy (en-SEF-ah-LOP-ah-thee): mental changes associated with liver disease that may include irritability, short-term memory loss, and an inability to concentrate.

hepatitis (hep-ah-TIE-tis): inflammation of the liver.

hepatorenal syndrome: the combined symptoms of liver and renal failure that occur as a consequence of severe liver damage.

herpes virus: a virus that can lead to mouth lesions and may also affect the lower GI tract, causing diarrhea.

hiatal hernia: a protrusion of a portion of the stomach through the esophageal hiatus of the diaphragm. There are several types of hiatal hernias, but the type most commonly associated with reflux is a **sliding hiatal hernia.**

hiatus (high-AY-tus): the opening in the diaphragm through which the esophagus passes.

hiccups (HICK-ups): repeated cough like sounds and jerks that are produced when an involuntary spasm of the diaphragm muscle sucks air down the windpipe; also spelled *hiccoughs*.

high potency: 100% or more of the Daily Value for the nutrient in a single supplement and for at least two-thirds of the nutrients in a multinutrient supplement.

high-quality protein: an easily digestible, complete protein.

high-risk pregnancy: a pregnancy characterized by indicators that make it likely the birth will be surrounded by problems such as premature delivery, difficult birth, retarded growth, birth defects, and early infant death.

histamine (HISS-tah-mean or HISS-tah-men): a substance produced by cells of the immune system as part of a local immune reaction to an antigen; participates in causing inflammation.

HIV (human immunodeficiency virus): a virus that progressively hampers the function of the immune system and leaves its host defenseless against other infections and cancer and eventually causes AIDS. The most common HIV is HIV-1.

holistic (hoe-LIS-tik): based on consideration of a person as a whole, including physical, emotional, intellectual, social, and spiritual needs.

homeostasis (HOME-ee-oh-STAY-sis): the maintenance of constant internal conditions (such as blood chemistry, temperature, and blood pressure) by the body's control systems. A homeostatic system is constantly reacting to external forces so as to maintain limits set by the body's needs.

hormones: chemical messengers. Hormones are secreted by a variety of glands in response to altered conditions in the body. Each hormone travels to one or more specific target tissues or organs, where it elicits a specific response to maintain homeostasis.

hormone-sensitive lipase: an enzyme inside adipose cells that responds to the body's need for fuel by hydrolyzing triglycerides so that their parts (glycerol and fatty acids) escape into the general circulation and thus become available to other cells as fuel. The signals to which this enzyme responds include epinephrine and glucagon, which oppose insulin.

house diet: foods preselected by the dietary department of a health care facility.

humoral immunity: immunity conferred by B-cells, which produce antibodies that travel through the blood to the invasion site.

hunger: the physiological drive for food that initiates food-seeking behavior.

hydrochloric acid: an acid composed of hydrogen and chloride atoms (HCl). The gastric glands normally produce this acid.

hydrodensitometry (HI-dro-DEN-see-TOM-eh-tree): a method of measuring body density in which the person is weighed on land and then weighed again while submerged in water.

hydrogenation (high-dro-gen-AY-shun): a chemical process by which hydrogens are added to monounsaturated or polyunsaturated fats to reduce the number of double bonds, making the fats more saturated (solid) and more resistant to oxidation (protecting against rancidity). Hydrogenation produces *trans*-fatty acids.

hydrolysis (high-DROL-ih-sis): a chemical reaction in which a major reactant is split into two products, with the addition of a hydrogen atom (H) to one and a hydroxyl group (OH) to the other (from water, H_2O). (The noun is **hydrolysis;** the verb is **hydrolyze.**)

hydrolyzed formulas: liquid diets that contain broken-down molecules of protein, such as amino acids and short peptide chains; also called **monomeric formulas.**

hydrophilic (high-dro-FIL-ick): a term referring to water-loving, or water-soluble, substances.

hydrophobic (high-dro-FOE-bick): a term referring to water-fearing, or non-water-soluble, substances; also known as **lipophilic** (fat loving).

hydroxyapatite (high-drox-ee-APP-ah-tite): crystals made of calcium and phosphorus.

hyperactivity: inattentive and impulsive behavior that is more frequent and severe than is typical of others a similar age; professionally called **attention-deficit/hyperactivity disorder (ADHD).**

hyperammonemia (HIGH-per-AM-moe-KNEE-me-ah): elevated blood ammonia.

hypercalciuria (HIGH-per-kal-see-YOU-ree-ah): excessive urinary excretion of calcium.

hyperglycemia: elevated blood glucose. Normal fasting blood glucose is less than 110 mg/dL. Fasting blood glucose between 110 and 125 mg/dL suggests impaired glucose tolerance; values of 126 mg/dL or higher suggest diabetes.

hyperkalemia (HIGH-per-kay-LEE-me-ah): excessive potassium in the blood.

hyperosmolar, hyperglycemic coma: coma that occurs in uncontrolled type 2 diabetes precipitated by the presence of hypertonic blood and dehydration.

hyperoxaluria (HIGH-per-OX-all-YOU-ree-ah): excessive urinary excretion of oxalate.

hyperperistalsis: rapid movement through the intestine.

hypertension: elevated blood pressure.

hypertonic formula: a formula with an osmolality greater than that of blood serum.

hypoglycemia (HIGH-po-gligh-SEE-me-ah): an abnormally low blood glucose concentration.

hyponatremia (HIGH-poe-nay-TREE-mee-ah): low levels of sodium in the blood.

hypothalamus (high-po-THAL-ah-mus): a brain center that controls activities such as maintenance of water balance, regulation of body temperature, and control of appetite.

hypoxemia (high-pox-EE-me-ah): lack of oxygen in the blood.

iatrogenic (EYE-at-row-JEN-ick) **malnutrition:** malnutrition that develops as a consequence of a treatment. Malnutrition that occurs because nutrient needs are not met during the course of hospitalization is an example.

idiopathic (id-ee-oh-PATH-ic) **cirrhosis:** severe liver damage for which the cause cannot be identified.

ileocecal (ill-ee-oh-SEEK-ul) **valve:** the sphincter separating the small and large intestines.

ileostomy (ILL-ee-OSS-toe-me): surgery that creates a stoma from the ileum through the abdominal wall and out through the skin.

ileum (ILL-ee-um): the last segment of the small intestine.

imitation foods: foods that substitute for and resemble another food, but are nutritionally inferior to it with respect to vitamin, mineral, or protein content. If the substitute is not inferior to the food it resembles and if its name provides an accurate description of the product, it need not be labeled "imitation."

immune system: the body's system of defense against harmful substances.

immunity: the body's ability to recognize and eliminate foreign invaders.

impaired glucose tolerance: inability to maintain normal blood glucose levels without excessive insulin production. Some people with impaired glucose tolerance have fasting glucose levels somewhat higher than normal but not high enough to diagnose diabetes. Others have normal blood glucose levels most of the time, but when given a large amount of glucose, their blood glucose rises too high.

implantation: the stage of development in which the zygote embeds itself in the wall of the uterus and begins to develop; occurs during the first two weeks after conception.

inborn error of metabolism: an inherited flaw evident as a metabolic disorder or disease present from birth.

indemnity insurance: traditional fee-for-service insurance.

indigestion: incomplete or uncomfortable digestion, usually accompanied by pain, nausea, vomiting, heartburn, intestinal gas, or belching.

individualized: based on consideration of a person as a unique being rather than one of many.

inflammatory bowel diseases (IBD): diseases characterized by inflammation of the bowel.

inflammatory response: the changes orchestrated by the immune system when tissues are injured by such forces as blows, wounds, foreign bodies (chemicals, microorganisms), loss of blood flow, heat, cold, electricity, or radiation.

initiators: factors that cause mutations that give rise to cancer, such as radiation and carcinogens.

inorganic: not containing carbon or pertaining to living things.

inositol (in-OSS-ih-tall): a nonessential nutrient that can be made in the body from glucose. Inositol is used in cell membranes.

insoluble fibers: indigestible food components that do not dissolve in water. Examples include the tough, fibrous structures found in the strings of celery and the skins of corn kernels.

insulin (IN-suh-lin): a hormone secreted by special cells in the pancreas in response to (among other things) increased blood glucose concentration. The primary role of insulin is to control the transport of glucose from the bloodstream into the muscle and fat cells.

insulin resistance of adipose tissue: failure of the enzyme that releases free fatty acids from fat cells to adequately reduce its activity in the presence of insulin.

insulin resistance of skeletal muscle: failure of the muscle cells to take up glucose from the blood in response to insulin as they normally would.

insulin resistance: the condition in which the cells fail to respond to insulin thus preventing the normal regulation of glucose metabolism.

intermittent claudication (klaw-dih-KAY-shun): severe calf pain caused by inadequate blood supply. It occurs when walking and subsides during rest.

intermittent feedings: delivery of about 250 to 400 ml of formula over 30 minutes or more.

Internet (the Net): a worldwide network of millions of computers linked together to share information.

interstitial (IN-ter-STISH-al) **fluid:** fluid between the cells, usually high in sodium and chloride. Interstitial fluid is a large component of **extracellular fluid** (fluid outside the cells), which also includes plasma and the water of structures such as the skin and bones. Extracellular fluid accounts for approximately one-third of the body's water.

intra-abdominal fat: fat stored within the abdominal cavity in association with the internal abdominal organs, as opposed to the fat stored directly under the skin (subcutaneous fat).

intracellular fluid: fluid within the cells, usually high in potassium and phosphate. Intracellular fluid accounts for approximately two-thirds of the body's water.

intravenous (IV): through a vein.

intrinsic factor: a glycoprotein (a protein with short polysaccharide chains attached) manufactured in the stomach that aids in the absorption of vitamin B_{12}.

ions (EYE-uns): atoms or molecules that have gained or lost electrons and therefore have electrical charges. Examples include the positively charged sodium ion (Na^+) and the negatively charged chloride ion (C_{12}). For a closer look at ions, see Appendix B.

iron deficiency: the state of having depleted iron stores.

iron overload: toxicity from excess iron.

iron-deficiency anemia: severe depletion of iron stores that results in low hemoglobin and small, pale, red blood cells.

irritable bowel syndrome: an intestinal disorder of unknown cause characterized by abdominal discomfort, cramping, diarrhea or constipation, or alternating diarrhea and constipation.

isotonic formula: a formula with an osmolality similar to that of blood serum (300 mOsm/kg).

IV catheter: a thin tube inserted into a peripheral or central vein. Additional tubing connects the IV solution to the catheter.

—J—

jaundice (JAWN-dis): a characteristic yellowing of the skin, whites of the eyes, mucous membranes and body fluids resulting from the spillover of the bile pigment **bilirubin** (bill-ee-ROO-bin) from the liver into the general circulation; also known as **hyperbilirubinemia** (HIGH-per-BILL-eh-roo-bin-EE-me-ah). When these pigments invade the brain, the condition is **kernicterus** (ker-NICK-ter-us). Jaundice may be caused by obstruction of bile passageways, hemolysis, or dysfunctional liver cells.

jejunostomy (JEE-ju-NOSS-toe-mee): an opening in the jejunum made surgically or under local anesthesia through which a feeding tube can be passed. The technique for creating a jejunostomy under local anesthesia is called a **direct endoscopic jejunostomy (DEJ).** Note: Some clinicians also refer to this procedure as a PEJ, which is a more accurate use of the term than the more common use described above.

jejunum (je-JOON-um): the first two-fifths of the small intestine beyond the duodenum.

—K—

Kaposi's (cap-OH-seez) **sarcoma:** a type of cancer rare in the general population but common in people with HIV infections.

kCalories: see *calories*.

kcalorie (energy) control: management of food energy intake.

kcalorie count: a determination of a client's food intake from a direct observation of how much the client eats.

keratin (KERR-uh-tin): a water-insoluble protein; the normal protein of hair and nails. Keratin-producing cells may replace mucus-producing cells in vitamin A deficiency.

keratinization: accumulation of keratin in a tissue; a sign of vitamin A deficiency.

keratomalacia (KARE-ah-toe-ma-LAY-shuh): softening of the cornea that leads to irreversible blindness; seen in severe vitamin A deficiency.

keto (KEY-toe) **acid:** an organic acid that contains a carbonyl group (C=O).

ketoacidosis: lowering of the blood's normal pH due to the accumulation of acidic ketones.

ketone (KEE-tone) **bodies:** the product of the incomplete breakdown of fat when glucose is not available in the cells.

ketonuria: ketones in the urine.

ketosis (kee-TOE-sis): accelerated production of ketones that results in an undesirably high concentration in the blood and urine.

kwashiorkor (kwash-ee-OR-core, kwash-ee-or-CORE): a form of PEM that results either from inadequate protein intake or, more commonly, from infections.

—L—

lactadherin (lack-tad-HAIR-in): a protein in breast milk that attacks diarrhea-causing viruses.

lactase deficiency: a lack of the enzyme required to digest the disaccharide lactose into its component monosaccharides (glucose and galactose).

lactase: an enzyme that hydrolyzes lactose.

lactation: production and secretion of breast milk for the purpose of nourishing an infant.

lactic acid: lactate, a 3-carbon compound produced from pyruvate during anaerobic metabolism.

lactoferrin (lack-toh-FERR-in): a protein in breast milk that binds iron and keeps it from supporting the growth of the infant's intestinal bacteria.

lacto-ovo-vegetarians: people who include milk, milk products, and eggs, but exclude meat, poultry, fish, and seafood from their diets.

lactose (LAK-tose): a disaccharide composed of glucose and galactose; commonly known as milk sugar.

lactose intolerance: a condition that results from inability to digest the milk sugar lactose; characterized by bloating, gas, abdominal discomfort, and diarrhea. Lactose intolerance differs from milk allergy, which is caused by an immune reaction to the protein in milk.

lactovegetarians: people who include milk and milk products, but exclude meat, poultry, fish, seafood, and eggs from their diets.

Laennec's (LAY-eh-necks) **cirrhosis:** severe liver damage related to alcohol abuse.

large intestine or **colon** (COAL-un): the lower portion of intestine that completes the digestive process. Its segments are the ascending colon, the transverse colon, the descending colon, and the sigmoid colon.

larynx: the voice box (see Figure H3-1).

LDL (low-density lipoprotein): the type of lipoprotein derived from very-low-density lipoproteins (VLDL) as cells remove triglycerides from them; composed primarily of cholesterol.

lean body mass: the weight of the body minus the fat content.

lecithin (LESS-uh-thin): one of the phospholipids; a compound of glycerol to which are attached two fatty acids, a phosphate group, and a choline molecule. Both nature and the food industry use lecithin as an emulsifier to combine two ingredients that do not ordinarily mix, such as water and oil.

legal: established by law.

legumes (lay-gyooms, leg-yooms): plants of the bean and pea family, rich in high-quality protein compared with other plant-derived foods.

leptin: a protein produced by fat cells under direction of the *ob* gene that decreases appetite and increases energy expenditure; sometimes called the ***ob* protein.**

let-down reflex: the reflex that forces milk to the front of the breast when the infant begins to nurse.

license to practice: permission under state or federal law, granted on meeting specified criteria, to use a certain title (such as dietitian) and offer certain services. **Licensed dietitians** may use the initials **LD** after their names.

life expectancy: the average number of years lived by people in a given society.

life span: the maximum number of years of life attainable by a member of a species.

limiting amino acid: the essential amino acid found in the shortest supply relative to the amounts needed for protein synthesis in the body.

linoleic (lin-oh-LAY-ick) **acid:** an essential fatty acid with 18 carbons and two double bonds (18:2).

linolenic (lin-oh-LEN-ick) **acid:** an essential fatty acid with 18 carbons and three double bonds (18:3).

lipids: a family of compounds that includes triglycerides (fats and oils), phospholipids, and sterols.

lipodystrophy (LIP-oh-DISS-tro-fee): the redistribution of fat that can occur as a consequence of HIV infections as well as other disorders. The accumulation of abdominal fat associated with HIV infections is sometimes called *protease paunch.*

lipomas (lih-POE-mahs): benign tumors composed of fat.

lipoprotein lipase (LPL): an enzyme mounted on the surface of fat cells (and other cells) that hydrolyzes triglycerides passing by in the bloodstream and directs their parts into the cells, where they can be metabolized or reassembled for storage.

lipoprotein(a): a variant form of LDL cholesterol associated with a high risk of heart disease.

lipoproteins (LIP-oh-PRO-teenz): clusters of lipids associated with proteins that serve as transport vehicles for lipids in the lymph and blood.

liver: the organ that manufactures bile. The liver's many other functions are described in Chapter 7.

living will: a document signed by a competent adult that specifically states whether the person wishes aggressive treatment in the event of terminal illness or irreversible coma from which the person is not expected to recover.

longevity: long duration of life.

low birthweight (LBW): a birthweight of 5½ lb (2500 g) or less; indicates probable poor health in the newborn and poor nutrition status in the mother during pregnancy, before pregnancy, or both. Normal birthweight for a full-term baby is 6½ to 8¾ lb (about 3000 to 4000 g).

low-risk pregnancy: a pregnancy characterized by indicators that make a normal outcome likely.

lumen (LOO-men): the space within a vessel, such as the intestine.

lymph (LIMF): a clear yellowish fluid that is almost identical to blood except that it contains no red blood cells or platelets. Lymph from the GI tract transports fat and fat-soluble vitamins to the bloodstream via lymphatic vessels.

lymph tissues: tissues that contain lymphocytes.

lymphatic (lim-FAT-ic) **system:** a loosely organized system of vessels and ducts that convey fluids toward the heart. The GI part of the lymphatic system carries the products of digestion into the bloodstream.

lymphocytes (limb-FOE-sites): cells made in lymph tissues that travel throughout the lymphatic and circulatory systems.

—M—

macroangiopathies: disorders of the large blood vessels.

macrobiotic diets: extremely restrictive diets limited to a few grains and vegetables; based on metaphysical beliefs and not on nutrition.

macular (MACK-you-lar) **degeneration:** deterioration of the macular area of the eye that can lead to loss of central vision and eventual blindness. The **macula** is a small, oval, yellowish region in the center of the retina that provides the sharp, straight-ahead vision so critical to reading and driving.

magnesium: a cation within the body's cells, active in many enzyme systems.

major minerals: essential mineral nutrients found in the human body in amounts larger than 5 g; sometimes called **macrominerals.**

maleficence (mah-LEF-eh-sens): doing harm.

malignant (ma-LIG-nant): describes tumors that multiply out of control, threaten health, and require treatment.

malnutrition: any condition caused by excess or deficient food energy or nutrient intake or by an imbalance of nutrients.

maltase: an enzyme that hydrolyzes maltose.

maltose (MAWL-tose): a disaccharide composed of two glucose units; sometimes known as malt sugar.

mammary glands: glands of the female breast that secrete milk.

managed care: a health care delivery system that aims to provide cost-effective health care by coordinating services and limiting access to services.

marasmus (ma-RAZ-mus): a form of PEM that results from a severe deprivation, or impaired absorption, of energy, protein, vitamins, and minerals.

matrix (MAY-tricks): the basic substance that gives form to a developing structure; in the body, the formative cells from which teeth and bones grow.

meat replacements: products formulated to look and taste like meat, fish, or poultry; usually made of textured vegetable protein.

mechanical soft diet: a diet that excludes only those foods that a person cannot chew; also called a **dental soft diet** or an **edentulous diet.**

mechanical ventilator: a machine that "breathes" for the person who can't. In a normal respiration, the lungs expand, which draws air into the lungs. With mechanical ventilation, air is forced into the lungs at regular intervals using pressure.

medical nutrition therapy: the provision of a client's nutrient, dietary, and nutrition education needs based on a complete nutrition assessment.

medications: chemicals (drugs) that alter one or more body functions that are marketed only with approval of the Food and Drug Administration and only after research shows that they are safe and effective.

medium-chain triglycerides (MCT): triglycerides containing fatty acids with 8 to 12 carbon atoms; they require minimal lipase and no bile for absorption.

MEOS or **microsomal** (my-krow-SO-mal) **ethanol-oxidizing system:** a system of enzymes in the liver that oxidize not only alcohol, but also several classes of drugs.

metabolic syndrome: the combination of insulin resistance, hyperinsulinemia, obesity, hypertension, elevated LDL and triglycerides, and reduced HDL that is frequently associated with type 2 diabetes and cardiovascular disease; also called *syndrome X* and *insulin-resistance syndrome.*

metabolism: the sum total of all the chemical reactions that go on in living cells. **Energy metabolism** includes all the reactions by which the body obtains and spends the energy from food.

metalloenzymes (meh-tal-oh-EN-zimes): enzymes that contain one or more minerals as part of their structures.

metallothionein (meh-TAL-oh-THIGH-oh-neen): a sulfur-rich protein that avidly binds with metals such as zinc.

metastasize (meh-TAS-tah-size): to spread by the movement of cancer cells from one part of the body to another.

MFP factor: a factor associated with the digestion of Meat, Fish, and Poultry that enhances iron absorption.

micelles (MY-cells): tiny spherical complexes of emulsified fat that arise during digestion. Each carries dozens of molecules of bile and fatty acids and/or monoglycerides.

microangiopathies: disorders of the capillaries.

microvascular: pertaining to the capillaries. Retinopathy, nephropathy, and neuropathy are microvascular complications.

microvilli (MY-cro-VILL-ee, MY-cro-VILL-eye): tiny, hairlike projections on each cell of every villus that can trap nutrient particles and transport them into the cells; singular **microvillus.**

milk anemia: iron-deficiency anemia that develops when an excessive milk intake displaces iron-rich foods from the diet.

milliequivalents (mEq): the concentration of electrolytes in a volume of solution. Milliequivalents are a useful measure when considering ions because the number of charges reveals characteristics about the solution that are not evident when the concentration is expressed in terms of weight.

mineralization: the process in which calcium, phosphorus, and other minerals crystallize on the collagen matrix of a growing bone, hardening the bone.

minerals: inorganic elements. Some minerals are essential nutrients required in small amounts.

misinformation: false or misleading information.

moderation (dietary): providing enough but not too much of a substance.

moderation: in relation to alcohol consumption, not more than two drinks a day for the average-sized man and not more than one drink a day for the average-sized woman.

modified diet: a diet that is adjusted to meet medical needs. Such diets may be adjusted in consistency, level of energy and nutrients, amount of fluid, or number of meals, or by the inclusion or elimination of certain foods.

modules: formulas or foods that provide primarily a single nutrient and are designed to be added to other formulas or foods to alter nutrient composition. Modules can also be combined to create a highly individualized formula.

molybdenum (mo-LIB-duh-num): a trace element.

monoglycerides: molecules of glycerol with one fatty acid attached. A molecule of glycerol with two fatty acids attached is a **diglyceride.**

monosaccharides (mon-oh-SACK-uh-rides): carbohydrates of the general formula $C_nH_{2n}O_n$ that consist of a single ring. See Appendix C for the chemical structures of the monosaccharides.

monounsaturated fatty acid: a fatty acid that lacks two hydrogen atoms and has one double bond between carbons—for example, oleic acid. A **monounsaturated fat** is composed of triglycerides in which most of the fatty acids are monounsaturated.

mood disorders: mental illness characterized by episodes of severe depression or excessive excitement (mania) or both.

mouth: the oral cavity containing the tongue and teeth.

mouth ulcers: lesions or sores in the lining of the mouth. Certain medications, radiation therapy, and some disorders, such as oral herpes infections, can cause mouth ulcers.

mucous (MYOO-kus) **membranes:** the membranes, composed of mucus-secreting cells, that line the surfaces of body tissues.

mucus (MYOO-kus): a slippery substance secreted by goblet cells of the GI lining (and other body linings) that protects the cells from exposure to digestive juices (and other destructive agents). The lining of the GI tract with its coat of mucus is a **mucous membrane.** (The noun is **mucus;** the adjective is **mucous.**)

multiple daily injections: delivery of a mixture of insulins by injection three or more times daily.

muscle dysmorphia (dis-MORE-fee-ah): a newly coined psychiatric disorder characterized by a preoccupation with building body mass.

muscular dystrophy (DIS-tro-fee): a hereditary disease in which the muscles gradually weaken. Its most debilitating effects arise in the lungs.

mutation: an alteration in a gene such that an altered protein is produced.

mutual supplementation: the strategy of combining two protein foods in a meal so that each food provides the essential amino acid(s) lacking in the other. Mutual supplementation is the dietary strategy that brings complementary proteins together in a meal.

myocardial (my-oh-CAR-dee-al) **infarction** (in-FARK-shun) or **MI:** sudden tissue death caused by blockages of vessels that feed the heart muscle; also called **heart attack, cardiac arrest,** or **acute heart failure.**

myoglobin: the oxygen-holding protein of the muscle cells.

—N—

NAD (nicotinamide adenine dinucleotide): the main coenzyme form of the vitamin niacin. Its reduced form is NADH.

narcotic (nar-KOT-ic): drug that dulls the senses, induces sleep, and becomes addictive with prolonged use.

nasoduodenal (ND): from the nose to the duodenum.

nasoenteric: from the nose to the stomach or intestine. *Nasoenteric feedings* include nasogastric, nasoduodenal, and nasojejunal feedings. Most clinicians use nasoenteric to refer to nasoduodenal and nasojejunal feedings only.

nasogastric (NG): from the nose to the stomach.

nasojejunal (NJ): from the nose to the jejunum.

natural killer cells: lymphocytes that confer nonspecific immunity. Natural killer cells destroy viruses and tumor cells.

nephron (NEF-ron): the working unit of the kidneys. Each nephron consists of a glomerulus and a tubule.

nephropathy: a disorder of the kidneys.

nephrotic (neh-FRAUT-ic) **syndrome:** the cluster of clinical findings that occur when glomerular function falters, including proteinuria, low serum albumin, edema, and elevated blood lipids.

net protein utilization (NPU): the amount of protein nitrogen that is retained from a given amount of protein nitrogen eaten; a measure of protein quality.

neural tube defects: malformations of the brain, spinal cord, or both during embryonic development that often results in lifelong disability or death. The two main types of neural tube defects are spina bifida (literally, "split spine") and anencephaly ("no brain").

neurons: nerve cells; the structural and functional units of the nervous system. Neurons initiate and conduct nerve transmissions.

neuropathy: a disorder of the nerves.

neuropeptide Y: a chemical produced in the brain that stimulates appetite, diminishes energy expenditure, and increases fat storage.

neurotransmitters: chemicals that are released at the end of a nerve cell when a nerve impulse arrives there. They diffuse across the gap to the next cell and alter the membrane of that second cell to either inhibit or excite it.

niacin (NIGH-a-sin): a B vitamin. The coenzyme forms are NAD (nicotinamide adenine dinucleotide) and NADP (the phosphate form of NAD). Niacin can be eaten preformed or made in the body from its precursor, tryptophan, one of the amino acids.

niacin equivalents (NE): the amount of niacin present in food, including the niacin that can theoretically be made from its precursor, tryptophan, present in the food.

night blindness: slow recovery of vision after flashes of bright light at night or an inability to see in dim light; an early symptom of vitamin A deficiency.

nitric oxide (NO): a substance produced by the vascular endothelium that causes blood vessels to dilate and inhibits clot formation.

nitrogen balance: the amount of nitrogen consumed (N in) as compared with the amount of nitrogen excreted (N out) in a given period of time.

nocturnal hypoglycemia: hypoglycemia that occurs while a person is sleeping.

nonalcoholic fatty cirrhosis: severe liver damage that develops from fatty liver that is unrelated to alcohol abuse.

nonessential amino acids: amino acids that the body can synthesize (see Table 6-1).

nonnutrients: compounds in foods that do not fit within the six classes of nutrients.

nonspecific immunity: immunity directed at foreign substances in general, rather than specific antigens.

nursing bottle tooth decay: extensive tooth decay due to prolonged tooth contact with formula, milk, fruit juice, or other carbohydrate-rich liquid offered to an infant in a bottle.

nutrient claims: statements that characterize the quantity of a nutrient in a food.

nutrient density: a measure of the nutrients a food provides relative to the energy it provides. The more nutrients and the fewer kcalories, the higher the nutrient density.

nutrients: chemical substances obtained from food and used in the body to provide energy, structural materials, and regulating agents to support growth, maintenance, and repair of the body's tissues. Nutrients may also reduce the risks of some diseases.

nutrition: the science of foods and the nutrients and other substances they contain, and of their actions within the body (including ingestion, digestion, absorption, transport, metabolism, and excretion). A broader definition includes the social, economic, cultural, and psychological implications of food and eating.

nutrition assessment: a comprehensive evaluation of a person's nutrition status, completed by a registered dietitian, using health, socioeconomic, drug, and diet histories; anthropometric measurements; physical examinations; and laboratory tests.

nutrition care plan: a plan that translates nutrition assessment data into a strategy for meeting a client's nutrient and nutrition education needs.

nutrition care process: an organized approach to nutrition intervention that consists of assessing, planning, implementing, and evaluating. The nutrition care process parallels the nursing care process, but focuses on nutrition concerns.

nutrition screening: a tool for quickly identifying clients at risk for malnutrition so that they can receive complete nutrition assessments.

nutrition status survey: a survey that evaluates people's nutrition status using diet histories, anthropometric measures, physical examinations, and laboratory tests.

nutritionist: a person who specializes in the study of nutrition. Some nutritionists are registered dietitians, whereas others are self-described experts whose training is questionable. In states with responsible legislation, the term applies only to people who have MS or PhD degrees from properly accredited institutions.

nutritive sweeteners: sweeteners that yield energy, including both sugars and sugar replacers.

—O—

obstructive cirrhosis: severe liver damage caused by obstruction of the bile ducts due to a stone or tumor.

oils: liquid fats (at room temperature).

olestra: a synthetic fat made from sucrose and fatty acids that provides 0 kcalories per gram; also known as **sucrose polyester.**

omega: the last letter of the Greek alphabet (σ), used by chemists to refer to the position of the first double bond from the methyl end in a fatty acid.

omega-3 fatty acid: a polyunsaturated fatty acid in which the first double bond is three carbons away from the methyl (CH3) end of the carbon chain.

omega-6 fatty acid: a polyunsaturated fatty acid in which the first double bond is six carbons from the methyl (CH3) end of the carbon chain.

omnivores: people who have no formal restriction on the eating of any foods.

open feeding systems: enteral formula delivery systems that require formula to be transferred from its original packaging to a feeding container before it can be administered through a feeding tube.

opportunistic illnesses: illnesses that normally would not occur or that would cause only minor problems in the healthy population, but can cause great harm when the immune system is compromised.

opsin (OP-sin): the protein portion of the visual pigment molecule.

oral antidiabetic agents: medications taken by mouth to lower blood glucose levels in people with type 2 diabetes.

organic: a substance or molecule containing carbon-carbon bonds or carbon-hydrogen bonds. Some farmers call their produce "organic" if it was grown without manufactured fertilizers and pesticides, but by the definition given here, all foods are organic.

orlistat (OR-leh-stat): a drug used in the treatment of obesity that inhibits the absorption of fat in the GI tract, thus limiting kcaloric intake; marketed under the trade name *Xenical.*

orogastric: from the mouth to the stomach. This method is often used to feed infants because they breathe through their noses, and a nasogastric tube can hinder the infant's breathing.

osmolality (OZ-moh-LAL-eh-tee): a measure of the concentration of particles in a solution, expressed as the number of milliosmoles (mOsm) per kilogram.

osmosis: the movement of water across a membrane *toward* the side where the solutes are more concentrated.

osteoarthritis: a painful, chronic disease of the joints that occurs when the cushioning cartilage in a joint breaks down; joint structure is usually altered, with loss of function; also called **degenerative arthritis.**

osteomalacia (OS-tee-oh-ma-LAY-shuh): a bone disease characterized by softening of the bones. Symptoms include bending of the spine and bowing of the legs. The disease occurs most often in adult women.

osteopenia (OS-tee-oh-PEE-nee-ah): a metabolic bone disease common in preterm infants; also called **rickets of prematurity.**

osteoporosis (OS-tee-oh-pore-OH-sis): a condition of older persons in which the bones become porous and fragile due to a loss of minerals; also called **adult bone loss.**

outcome measures: indicators that describe an observable change that are used to evaluate the effects of interventions.

overnutrition: excess energy or nutrients.

overt (oh-VERT): out in the open and easy to observe.

overweight: body weight above some standard of acceptable weight that is usually defined in relation to height (such as BMI).

ovum (OH-vum): the female reproductive cell, capable of developing into a new organism upon fertilization; commonly referred to as an egg.

oxaloacetate (OKS-ah-low-AS-eh-tate): a carbohydrate intermediate of the TCA cycle.

oxidants (OK-see-dants): compounds (such as oxygen itself) that oxidize other compounds. Compounds that prevent oxidation are called *anti*oxidants, whereas those that promote it are called *pro*oxidants.

oxidation (OKS-ee-day-shun): the process of a substance combining with oxygen.

oxidative stress: a condition in which the production of oxidants and free radicals exceeds the body's ability to defend itself and prevent damage.

oxytocin (OK-see-TOH-sin): a hormone that stimulates the mammary glands to eject milk during lactation and the uterus to contract during childbirth.

—P—

pancreas: a gland that secretes digestive enzymes and juices into the duodenum.

pancreatic (pank-ree-AT-ic) **juice:** the exocrine secretion of the pancreas, containing enzymes for the digestion of carbohydrate, fat, and protein as well as bicarbonate, a neutralizing agent. The juice flows from the pancreas into the small intestine through the pancreatic duct. (The pancreas also has an endocrine function, the secretion of insulin and other hormones.)

pancreatic enzyme replacements: extracts of pork or beef pancreatic enzymes that are taken as supplements to aid digestion.

pantothenic (PAN-toe-THEN-ick) **acid:** a B vitamin. The principal active form is part of coenzyme A, called "CoA" throughout Chapter 7.

paranoia (PAR-ah-NOY-ah): mental illness characterized by delusions of persecution.

parathormone (PAIR-ah-THOR-moan): a hormone from the parathyroid glands that regulates blood calcium by raising it when levels fall too low; also known as **parathyroid hormone.**

parenteral (par-EN-ter-all) **nutrition:** the provision of nutrients bypassing the intestine.

pathological stresses: disruptions to the body's internal balance that lie beyond its normal and healthy functioning.

peak bone mass: the highest attainable bone density for an individual, developed during the first three decades of life.

peer review: a process in which a panel of scientists rigorously evaluates a research study to assure that the scientific method was followed.

pellagra (pell-AY-gra): the niacin-deficiency disease.

pepsin: a gastric enzyme that hydrolyzes protein. Pepsin is secreted in an inactive form, **pepsinogen,** which is activated by hydrochloric acid in the stomach.

peptic ulcer: an erosion in the mucous membrane of either the stomach (a gastric ulcer) or the duodenum (a duodenal ulcer).

peptidase: a digestive enzyme that hydrolyzes peptide bonds. *Tripeptidases* cleave tripeptides; *dipeptidases* cleave dipeptides. *Endopeptidases* cleave peptide bonds within the chain to create smaller fragments, whereas *exopeptidases* cleave bonds at the ends to release free amino acids.

peptide bond: a bond that connects the acid end of one amino acid with the amino end of another, forming a link in a protein chain.

peripheral parenteral nutrition (PPN): the provision of an IV solution that meets nutrient needs delivered into the peripheral veins.

peripheral veins: the small-diameter veins that carry blood from the arms and legs.

peripherally inserted central catheter (PICC): a catheter inserted into a peripheral vein and advanced into a central vein.

peristalsis (peri-STALL-sis): wavelike muscular contractions of the GI tract that push its contents along.

peritoneal dialysis: removal of wastes and fluids from the body by using the peritoneal membrane as a semipermeable membrane.

peritonitis (PARE-ih-toe-NYE-tis): infection and inflammation of the membrane lining the abdominal cavity caused by leakage of infectious organisms through a perforation (hole) in an abdominal organ.

pernicious (per-NISH-us) **anemia:** a blood disorder that reflects a vitamin B_{12} deficiency caused by lack of intrinsic factor and characterized by abnormally large and immature red blood cells. Other symptoms include muscle weakness and irreversible neurological damage.

persistent vegetative state: exhibiting motor reflexes but without the ability to regain cognitive behavior, communicate, or interact purposefully with the environment.

pH: a measure of the concentration of H^+ ions (see Appendix B). The lower the pH, the higher the H^+ ion concentration and the stronger the acid. A pH above 7 is alkaline, or base (a solution in which OH_- ions predominate).

phagocytes: large white blood cells that confer nonspecific immunity. Phagocytes engulf and destroy foreign substances. Phagocytes that travel in the blood are called *monocytes;* when monocytes embed themselves in tissues, they grow larger and are called *macrophages.* Other types of phagocytes include *neutrophils, polymorphonuclear leukocytes,* and *basophils.*

pharynx (FAIR-inks): the passageway leading from the nose and mouth to the larynx and esophagus, respectively.

phospholipid (FOS-foe-LIP-id): a compound similar to a triglyceride but having a phosphate group (a phosphorus-containing salt) and choline (or another nitrogen-containing compound) in place of one of the fatty acids.

phosphorus: a major mineral found mostly in the body's bones and teeth.

photosynthesis: the process by which green plants make carbohydrates from carbon dioxide and water using the green pigment chlorophyll to trap the sun's energy.

physiological age: a person's age as estimated from her or his body's health and probable life expectancy.

physiological stresses: disruptions to the body's internal balance caused by processes necessary to sustain life.

phytic (FYE-tick) **acid:** a nonnutrient component of plant seeds; also called **phytate** (FYE-tate). Phytic acid occurs in the husks of grains, legumes, and seeds and is capable of binding minerals such as zinc, iron, calcium, magnesium, and copper in insoluble complexes in the intestine, which the body excretes unused.

phytochemicals (FIE-toe-KEM-ih-cals): nonnutrient compounds found in plant-derived foods that have biological activity in the body.

pica (PIE-ka): a craving for nonfood substances. Also known as **geophagia** (gee-oh-FAY-gee-uh) when referring to clay eating and **pagophagia** (pag-oh-FAY-gee-uh) when referring to ice craving.

pigment: a molecule capable of absorbing certain wavelengths of light so that it reflects only those that we perceive as a certain color.

PKU, phenylketonuria (FEN-el-KEY-toe-NEW-ree-ah): an inborn error of metabolism in which phenylalanine, an essential amino acid, cannot be converted to tyrosine. Alternative metabolites of phenylalanine (phenylketones) accumulate in the tissues, causing damage, and overflow into the urine.

placebo (pla-see-bo): an inert, harmless medication given to provide comfort and hope; a sham treatment used in controlled research studies.

placebo effect: the healing effect that faith in medicine, even medicine without pharmaceutical effects, often has.

placenta (plah-SEN-tuh): the organ that develops inside the uterus early in pregnancy, through which the fetus receives nutrients and oxygen and returns carbon dioxide and other waste products to be excreted.

plaques (PLACKS): mounds of lipid material (mostly cholesterol) with some macrophages (a type of white blood cell) covered with fibrous connective tissue and embedded in artery walls. With time, the plaques may harden as the fibrous coat thickens and calcium is deposited in the plaque.

plasminogen (plaz-MIN-oh-jen) **activator inhibitor-1 (PAL-1):** a substance important in blood clotting.

platelets: tiny, disc-shaped bodies in the blood that are important in clot formation.

point of unsaturation: the double bond of a fatty acid, where hydrogen atoms can easily be added to the structure.

polycystic (POL-ee-SIS-tik) **ovary syndrome:** a disorder of women characterized by ovaries enlarged with fluid-filled sacs and elevated levels of male hormones (androgens).

polydipsia (POLL-ee-DIP-see-ah): excessive thirst.

polypeptide: many (ten or more) amino acids bonded together. An intermediate string of four to nine amino acids is an **oligopeptide** (OL-ee-go-PEP-tide).

polyphagia (POLL-ee-FAY-gee-ah): excessive eating.

polysaccharides: compounds composed of many monosaccharides linked together. An intermediate string of three to ten monosaccharides is an **oligosaccharide.**

polyunsaturated fatty acid (PUFA): a fatty acid that lacks four or more hydrogen atoms and has two or more double bonds between carbons—for example, linoleic acid (two double bonds) and linolenic acid (three double bonds). A **polyunsaturated fat** is composed of triglycerides in which most of the fatty acids are polyunsaturated.

polyuria (POLL-ee-YOU-ree-ah): excessive urine production.

portal hypertension: elevated blood pressure in the portal vein caused by obstructed blood flow through the liver.

post term (infant): an infant born after the 42nd week of pregnancy.

postnecrotic cirrhosis: severe liver damage that develops as a complication of chronic hepatitis.

potassium: the principal cation within the body's cells; critical to the maintenance of fluid balance, nerve transmissions, and muscle contractions.

prebiotics: nondigestible food ingredients that encourage the growth of favorable bacteria.

precursors: substances that precede others; with regard to vitamins, compounds that can be converted into active vitamins; also known as **provitamins.**

preeclampsia (PRE-ee-KLAMP-see-ah): a condition characterized by hypertension, fluid retention, and protein in the urine; formerly known as *pregnancy-induced hypertension.*

preferred provider organizations (PPOs): managed care organizations that encourage subscribers to select health care providers from a group that has contracted with the organization to provide services at lower costs.

pressure sores: the breakdown of skin and underlying tissues due to constant pressure and lack of oxygen to the affected area; also called **decubitus** (dee-CUE-bih-tus) **ulcers** or bedsores.

preterm (infant): an infant born prior to the 38th week of pregnancy; also called a **premature infant.** A **term** infant is born between the 38th and 42nd week of pregnancy.

primary biliary cirrhosis: severe liver damage from the gradual destruction of the bile ducts; can lead to chronic pancreatitis and fat malabsorption.

primary deficiency: a nutrient deficiency caused by inadequate dietary intake of a nutrient.

probiotics: microbial food ingredients that are beneficial to health.

progressive diet: a diet that changes as the client's tolerances permit.

prolactin (pro-LAK-tin): a hormone secreted from the anterior pituitary gland that acts on the mammary glands to initiate and sustain milk production.

promoters: factors that favor the development of cancers once they have begun.

proof: a way of stating the percentage of alcohol in distilled liquor. Liquor that is 100 proof is 50% alcohol; 90 proof is 45%, and so forth.

prooxidants: substances that significantly induce oxidative stress.

proteases (PRO-tee-aces): enzymes that hydrolyze protein.

protein digestibility: a measure of the amount of amino acids absorbed from a given protein intake.

protein digestibility–corrected amino acid score (PDCAAS): a measure of protein quality assessed by comparing the amino acid score of a food protein with the amino acid requirements of preschool-age children and then correcting for the true digestibility of the protein; recommended by the FAO/WHO and used to establish protein quality of foods for Daily Value percentages on food labels.

protein efficiency ratio (PER): a measure of protein quality assessed by determining how well a given protein supports weight gain in growing rats; used to establish the protein quality for infant formulas and baby foods.

protein isolates: proteins that have been separated from a food. Examples include casein from milk and albumin from egg.

protein turnover: the degradation and synthesis of protein.

protein-energy malnutrition (PEM), also called **protein-kcalorie malnutrition (PCM):** a deficiency of protein, energy, or both, including kwashiorkor, marasmus, and instances in which they overlap.

proteins: compounds composed of carbon, hydrogen, oxygen, and nitrogen atoms, arranged into amino acids linked in a chain. Some amino acids also contain sulfur atoms.

protein-sparing action: the action of carbohydrate (and fat) in providing energy that allows protein to be used for other purposes.

proteinuria (pro-teen-YOUR-ee-ah): the loss of protein in the urine. The loss of albumin in the urine is *albuminuria.*

puberty: the period in life in which a person becomes physically capable of reproduction.

public health dietitians: dietitians who specialize in providing nutrition services through organized community efforts.

purine (PU-reen): an end product of nucleotide metabolism that eventually breaks down to form uric acid.

pyloric (pie-LORE-ic) **sphincter:** the circular muscle that separates the stomach from the small intestine and regulates the flow of partially digested food into the small intestine; also called *pylorus* or *pyloric valve.*

pyloroplasty (pie-LOOR-oh-PLAS-tee): surgery that enlarges the pyloric sphincter.

pyruvate (PIE-roo-vate): pyruvic acid, a 3-carbon compound that plays a key role in energy metabolism.

—R—

radiation enteritis: inflammation and scarring of the intestinal cells caused by exposure to radiation.

radiation therapy: the use of radiation to arrest or destroy cancer cells.

randomization (ran-dom-ih-zay-shun): a process of choosing the members of the experimental and control groups without bias.

RD: see *registered dietitian.*

rebound hyperglycemia: hyperglycemia resulting from excessive secretion of counterregulatory hormones in response to excessive insulin and consequent low blood glucose levels; also called the **Somogyi** (so-MOHG-yee) **effect.**

recessive gene: a gene that has no observable effect as long as it is paired with a normal gene that can produce a normal product. In this case, the normal gene is said to be *dominant.*

Recommended Dietary Allowance (RDA): the average daily amount of a nutrient considered adequate to meet the known nutrient needs of practically all healthy people; a goal for dietary intake by individuals.

rectum: the muscular terminal part of the intestine, extending from the sigmoid colon to the anus.

refeeding syndrome: the physiologic and metabolic complications associated with reintroducing nutrients too rapidly in people with depleted nutrient stores due to chronic malnutrition or in those who have been underfed for several days. These complications can include malabsorption, cardiac insufficiency, respiratory distress, congestive heart failure, convulsions, coma, and possibly death.

reference protein: a standard against which to measure the quality of other proteins.

refined: the process by which the coarse parts of a food are removed. When wheat is refined into flour, the bran, germ, and husk are removed, leaving only the endosperm.

reflux esophagitis (eh-sof-ah-JYE-tis): inflammation of the esophagus caused by esophageal reflux; also called **gastroesophageal reflux disease (GERD).**

reflux: a backward flow.

registered dietitian (RD): a person who has completed a minimum of a bachelor's degree from an accredited university or college, has completed approved course work and a supervised practice program, has passed a national examination, and maintains registration through continuing professional education.

registration: listing; with respect to health professionals, listing with a professional organization that requires specific course work, experience, and passing of an examination.

remodeling: the dismantling and re-formation of a structure.

renal colic: the severe pain that accompanies the movement of a kidney stone through the ureter to the bladder.

renal insufficiency: the stage of renal failure in which renal function is reduced but not to a life-threatening degree.

renal threshold: the point at which a blood constituent that is normally reabsorbed by the kidneys reaches a level so high the kidneys cannot reabsorb it. The renal threshold for glucose is generally reached when blood glucose rises above 180 mg/dL.

renal: pertaining to the kidneys.

renin (REN-in): an enzyme secreted by the kidneys in response to a reduced blood flow that activates angiotensin, which triggers the release of the hormone aldosterone.

replication (REP-lee-kay-shun): repeating an experiment and getting the same results. The skeptical scientist, on hearing of a new, exciting finding, will ask, "Has it been replicated yet?" If it hasn't, the scientist will withhold judgment regarding the finding's validity.

requirement: the lowest continuing intake of a nutrient that will maintain a specified criterion of adequacy.

resistant starch: starch that escapes digestion and absorption in the small intestine of healthy people.

respiratory failure: failure of the lungs to exchange gases. In *acute* respiratory failure, the lungs fail over a short period of time.

resting metabolic rate (RMR): similar to the BMR, a measure of a person at rest in a comfortable setting, but with less stringent criteria for the number of hours fasting. Consequently, the RMR is slightly higher than the BMR.

retina (RET-in-uh): the layer of light-sensitive nerve cells lining the back of the inside of the eye; consists of rods and cones.

retinoids (RET-ih-noyds): chemically related compounds with biological activity similar to that of retinol.

retinol activity equivalents (RAE): a measure of vitamin A activity; the amount of retinol that the body will derive from a food containing preformed retinol or its precursor beta-carotene.

retinol-binding protein (RBP): the specific protein responsible for transporting retinol.

retinopathy: a disorder of the retina.

rheumatoid (ROO-ma-toyd) **arthritis:** a disease of the immune system involving painful inflammation of the joints and related structures.

rhodopsin (ro-DOP-sin): a light-sensitive pigment of the retina. It contains the retinal form of vitamin A and the protein opsin.

riboflavin (RYE-boh-flay-vin): a B vitamin. The coenzyme forms are FMN (flavin mononucleotide) and FAD (flavin adenine dinucleotide).

rickets: the vitamin D–deficiency disease in children characterized by inadequate mineralization of bone (manifested in bowed legs or knock-knees, outward-bowed chest, and knobs on ribs). A rare type of rickets, not caused by vitamin D deficiency, is known as *vitamin D–refractory rickets.*

risk factor: a condition or behavior associated with an elevated frequency of a disease but not proved to be causal. Risk factors for disease include overweight, cigarette smoking, alcohol abuse, high blood pressure, high blood cholesterol, high-fat diet, and physical inactivity.

—S—

saccharin (SAK-ah-ren): an artificial sweetener that has been approved in the United States. In Canada, approval for use in foods and beverages is pending; currently available only in pharmacies and only as a tabletop sweetener, not as an additive.

saliva: the secretion of the salivary glands. Its principal enzyme begins carbohydrate digestion.

salivary glands: exocrine glands that secrete saliva into the mouth.

salt sensitivity: a characteristic of individuals who respond to a high salt intake with an increase in blood pressure.

salt: a compound composed of a positive ion other than H^+ and a negative ion other than OH_2. An example is sodium chloride ($Na + C_{12}$). Na = sodium. Cl = chloride.

sarcopenia (SAR-koh-PEE-nee-ah): loss of skeletal muscle mass, strength, and quality.

satiating: having the power to suppress hunger and inhibit eating.

satiation (say-she-AY-shun): the feeling of satisfaction and fullness that occurs during a meal and halts eating. Satiation determines how much food is consumed during a meal.

satiety (sah-TIE-eh-tee): the feeling of satisfaction that occurs after a meal and inhibits eating until the next meal. Satiety determines how much time passes between meals.

saturated fatty acid: a fatty acid carrying the maximum possible number of hydrogen atoms—for example, stearic acid. A **saturated fat** is composed of triglycerides in which most of the fatty acids are saturated.

schizophrenia (SKITZ-oh-FREN-ee-ah): mental illness characterized by an altered concept of reality and, in some cases, delusions and hallucinations.

secondary deficiency: a nutrient deficiency caused by something other than an inadequate intake such as a disease condition that reduces absorption, accelerates use, hastens excretion, or destroys the nutrient.

secretin (see-CREET-in): a hormone produced by cells in the duodenum wall. Target organ: the pancreas. Response: secretion of bicarbonate-rich pancreatic juice.

segmentation (SEG-men-TAY-shun): a periodic squeezing or partitioning of the intestine at intervals along its length by its circular muscles.

selenium (se-LEEN-ee-um): a trace element.

semipermeable membrane: a membrane with pores that allow some particles to pass through the membrane but not others.

senile dementia: the loss of brain function beyond the normal loss of physical adeptness and memory that occurs with aging.

senility (see-NIL-ih-tee): mental or physical weakness associated with old age.

serotonin (SER-oh-tone-in): a neurotransmitter important in sleep regulation, appetite control, and sensory perception among other roles. Serotonin is synthesized in the body from the amino acid tryptophan with the help of vitamin B_6.

shock: a sudden drop in blood volume that disrupts the supply of oxygen to the tissues and organs and the return of blood to the heart. Shock is a critical event that requires immediate correction.

short-bowel or **short-gut syndrome**: severe malabsorption that may occur when the absorptive surface of the small bowel is reduced, resulting in diarrhea, weight loss, bone disease, hypocalcemia, hypomagnesemia, and anemia.

sibutramine (sigh-BYOO-tra-mean): a drug used in the treatment of obesity that slows the reabsorption of serotonin in the brain, thus suppressing appetite and creating a feeling of fullness; marketed under the trade name *Meridia.*

sickle-cell anemia: a hereditary form of anemia characterized by abnormal sickle- or crescent-shaped red blood cells. Sickled cells interfere with oxygen transport and blood flow. Symptoms are precipitated by dehydration and insufficient oxygen (as may occur at high altitudes) and include hemolytic anemia (red blood cells burst), fever, and severe pain in the joints and abdomen.

silent stones: gallstones that cause no symptoms.

simple carbohydrates (sugars): monosaccharides and disaccharides.

small intestine: a 10-foot length of small-diameter intestine that is the major site of digestion of food and absorption of nutrients. Its segments are the duodenum, jejunum, and ileum.

soaps: chemical compounds formed between a basic mineral (such as calcium) and unabsorbed fatty acids. Soaps give steatorrhea its foamy appearance.

sodium: the principal cation in the extracellular fluids of the body; critical to the maintenance of fluid balance, nerve transmissions, and muscle contractions.

soluble fibers: indigestible food components that dissolve in water to form a gel. An example is pectin from fruit, which is used to thicken jellies.

solutes (SOLL-yutes): the substances that are dissolved in a solution. The number of molecules in a given volume of fluid is the **solute concentration.**

specific immunity: immunity directed at specific organisms. The B-cells and T-cells confer this type of immunity.

sperm: the male reproductive cell, capable of fertilizing an ovum.

sphincter (SFINK-ter): a circular muscle surrounding, and able to close, a body opening. Sphincters are found at specific points along the GI tract and regulate the flow of food particles.

spina (SPY-nah) **bifida** (BIFF-ih-dah): one of the most common types of neural tube defects, characterized by the incomplete closure of the spinal cord and its bony encasement.

standard formulas: liquid diets that contain complete molecules of protein; also called **intact** or **polymeric formulas.**

standard or **regular diets:** diets that include all foods and meet the nutrient needs of healthy people.

starches: plant polysaccharides composed of glucose.

steatorrhea (STEE-ah-toe-REE-ah): fatty diarrhea characterized by loose, foamy, foul-smelling stools.

sterile: free of microorganisms, such as bacteria.

sterols (STARE-ols or STEER-ols): compounds composed of C, H, and O atoms arranged in rings, like those of cholesterol, with any of a variety of side chains attached.

stevia (STEE-vee-ah): a South American shrub whose leaves are used as a sweetener; sold in the United States as a dietary supplement that provides sweetness without kcalories.

stoma (STOH-mah): a surgically formed opening.

stomach: a muscular, elastic, saclike portion of the digestive tract that grinds and churns swallowed food, mixing it with acid and enzymes to form chyme.

stools: waste matter discharged from the colon; also called **feces** (FEE-seez).

stress fractures: bone damage or breaks caused by stress on bone surfaces during exercise.

stress response: the body's response to stress, mediated by both nerves and hormones.

stress: the state in which a body's internal balance (homeostasis) is upset by a threat to a person's physical well-being **(stressor).** The terms **acute stress** and **severe stress** are used in this chapter to refer to physiological stresses that rapidly and markedly raise the body's metabolic rate and significantly upset its normal internal balance.

stressors: environmental elements, physical or psychological, that cause stress.

stroke: an event in which the blood flow to a part of the brain is cut off; also called a **cerebral vascular accident (CVA).**

structure-function claims: statements that describe how a product may affect the structure or function of the body (do not require FDA authorization); for example, "calcium builds strong bones."

struvite (STREW-vite): crystals of magnesium ammonium phosphate formed by the action of bacterial enzymes.

subclinical deficiency: a deficiency in the early stages, before the outward signs have appeared.

subjects: the people or animals participating in a research project.

sucralose (SUE-kra-lose): an artificial sweetener approved for use in the United States.

sucrase: an enzyme that hydrolyzes sucrose.

sucrose (SUE-krose): a disaccharide composed of glucose and fructose; commonly known as table sugar, beet sugar, or cane sugar. Sucrose also occurs in many fruits and some vegetables and grains.

sudden infant death syndrome (SIDS): the unexpected and unexplained death of an apparently well infant; the most common cause of death of infants between the second week and the end of the first year of life; also called *crib death.*

sugar replacers: sugarlike compounds that can be derived from fruits or commercially produced from dextrose; also called **sugar alcohols** or **polyols.** Sugar alcohols are absorbed more slowly than other sugars and metabolized differently in the human body; they are not readily utilized by ordinary mouth bacteria. Examples are **maltitol, mannitol, sorbitol, xylitol, isomalt,** and **lactitol.**

sulfur: a mineral present in the body as part of some proteins.

supplements: pills, capsules, tablets, liquids, or powders that contain vitamins, minerals, herbs, or amino acids; intended to increase dietary intake of these substances.

systemic inflammatory response syndrome (SIRS): the complex of symptoms that occur as a result of immune and inflammatory factors in response to tissue damage. In severe cases, SIRS may progress to multiple organ failure.

—T—

TCA cycle: tricarboxylic acid cycle; a series of metabolic reactions that break down molecules of acetyl CoA to carbon dioxide and hydrogen atoms also called the **Krebs cycle** after the biochemist who elucidated its reactions.

T-cells: lymphocytes that react to specific antigens by traveling directly to the invasion site. Some T-cells (cytotoxic T-cells) kill invaders; others (helper/inducer T-cells) activate immune responses; still others (suppressor T-cells) turn off immune responses.

tempeh (TEM-pay): a fermented soybean food, rich in protein and fiber.

teratogenic (ter-AT-oh-jen-ik): causing abnormal fetal development and birth defects.

terminal illness: a progressive, irreversible disease that will lead to death in the near future.

textured vegetable protein: processed soybean protein used in vegetarian products such as soy burgers.

thermic effect of food (TEF): an estimation of the energy required to process food (digest, absorb, transport, metabolize, and store ingested nutrients); also called the **specific dynamic effect (SDE)** of food or the **specific dynamic activity (SDA)** of food. The sum of the TEF and any increase in the metabolic rate due to overeating is known as **diet-induced thermo-genesis (DIT).**

thermogenesis: the generation of heat; used in physiology and nutrition studies as an index of how much energy the body is spending.

thiamin (THIGH-ah-min): a B vitamin. The coenzyme form is TPP (thiamin pyrophosphate).

thirst: a conscious desire to drink.

thrombosis (throm-BOH-sis): the formation or presence of a blood clot in the vascular system. A *coronary thrombosis* occurs in a coronary artery, and a *cerebral thrombosis* occurs in an artery that feeds the brain.

thrombus (THROM-bus): a blood clot that blocks a blood vessel or cavity of the heart.

thrush: a fungal infection of the mouth and esophagus, caused by *Candida albicans,* that coats the tongue with a milky film and leads to mouth ulcers, altered taste sensations, and pain on chewing and swallowing. The technical term for this infection is *candidiasis.*

tissue rejection: destruction of healthy donor cells by the recipient's immune system, which recognizes the donor cells as foreign; also called *graft-versus-host disease (GVHD).*

tocopherol (tuh-KOFF-er-ol): a general term for several chemically related compounds, one of which has vitamin E activity (see Appendix C for chemical structures).

tofu (TOE-foo): a curd made from soybeans, rich in protein and often fortified with calcium; used in many Asian and vegetarian dishes in place of meat.

Tolerable Upper Intake Level: the maximum amount of a nutrient that appears safe for most healthy people and beyond which there is an increased risk of adverse health effects.

tonsillectomy (tawn-sill-ECK-tah-me): surgical removal of the tonsils.

total nutrient admixtures: intravenous solutions that contain all nutrients, including lipid emulsions; also called **three-in-one (3-in-1) admixtures** or **all-in-one admixtures.**

toxic cirrhosis: severe liver damage that results from toxic levels of chemicals.

toxic food environment: a term coined to refer to the easy access to and overabundance of high-fat, high-kcalorie foods in our society. It does *not* refer to the contamination of the food supply with poisonous toxins or infectious microbes.

trabecular (tra-BECK-you-lar) **bone:** the lacy inner structure of calcium crystals that supports the bone's structure and provides a calcium storage bank.

trace minerals: essential mineral nutrients found in the human body in amounts smaller than 5 g; sometimes called **microminerals.**

trachea (TRAKE-ee-uh): the windpipe; the passageway from the mouth and nose to the lungs.

transamination (TRANS-am-ih-NAY-shun): the transfer of an amino group from one amino acid to a keto acid, producing a new nonessential amino acid and a new keto acid.

***trans*-fatty acids:** fatty acids with an unusual configuration around the double bond.

transient hypertension of pregnancy: high blood pressure that develops in the second half of pregnancy and resolves after childbirth, usually without affecting the outcome of the pregnancy.

transient ischemic attack (TIA): a temporary reduction in blood flow to the brain, which causes temporary symptoms that vary depending on the part of the brain that is affected. Common symptoms include light-headedness, visual disturbances, paralysis, staggering, numbness, or dysphagia.

transnasal: through the nose. A **transnasal feeding tube** is one that is inserted through the nose.

triglycerides (try-GLISS-er-rides): the chief form of fat in the diet and the major storage form of fat in the body; composed of a molecule of glycerol with three fatty acids attached; also called **triacylglycerols** (try-ay-seel-GLISS-er-ols).

tripeptide: three amino acids bonded together.

tube feedings: liquid formulas delivered through a tube placed in the stomach or intestine.

tubule: a tubelike structure that surrounds the glomerulus and descends through the nephron. A pressure gradient between the glomerular capillaries and the tubule returns needed materials to the blood and moves wastes into the tubule to be excreted in the urine.

tumor: a new growth of tissue forming an abnormal mass with no function; also called a **neoplasm** (NEE-oh-plazm).

type 1 diabetes: the less common type of diabetes in which the person produces no insulin at all; also known as **insulin-dependent diabetes mellitus (IDDM)** or **juvenile-onset diabetes** (because it frequently develops in childhood), although some cases arise in adulthood.

type 2 diabetes: the more common type of diabetes in which the fat cells resist insulin; also called **noninsulin-dependent diabetes mellitus (NIDDM)** or **adult-onset diabetes.** Type 2 usually progresses more slowly than type 1.

type I osteoporosis: osteoporosis characterized by rapid bone losses, primarily of trabecular bone.

type II osteoporosis: osteoporosis characterized by gradual losses of both trabecular and cortical bone.

—U—

ulcer: an open sore or lesion. A **peptic ulcer** is an erosion of the cells of the mucosa of the lower esophagus, the stomach, or small intestine. Ulcers may also develop in the mouth, upper esophagus, and large intestine, or on the skin.

ulcerative colitis (ko-LYE-tis): inflammation and ulceration of the colon.

ultrafiltration: removal of fluids and small- to medium-size molecules from the blood by using pressure to transfer the blood across a semipermeable membrane.

umbilical (um-BILL-ih-cul) **cord:** the ropelike structure through which the fetus's veins and arteries reach the placenta; the route of nourishment and oxygen to the fetus and the route of waste disposal from the fetus. The scar in the middle of the abdomen that marks the former attachment of the umbilical cord is the **umbilicus** (um-BILL-ih-cus), commonly known as the "belly button."

undernutrition: deficient energy or nutrients.

underweight: body weight below some standard of acceptable weight that is usually defined in relation to height (such as BMI).

unsaturated fatty acid: a fatty acid that lacks hydrogen atoms and has at least one double bond between carbons (includes monounsaturated and polyunsaturated fatty acids). An **unsaturated fat** is composed of triglycerides in which most of the fatty acids are unsaturated.

unspecified eating disorders: eating disorders that do not meet the defined criteria for specific eating disorders.

urea (you-REE-uh): the principal nitrogen-excretion product of metabolism. Two ammonia fragments are combined with carbon dioxide to form urea.

uremia (you-REE-me-ah): abnormal accumulation of nitrogen containing substances, especially urea, in the blood; also called *azotemia* (AZE-oh-TEE-me-ah).

uremic frost: the appearance of urea crystals on the skin.

uremic syndrome: the cluster of clinical findings associated with the buildup of nitrogen-containing waste products in the blood, which may include fatigue, diminished mental alertness, agitation, muscle twitches, cramps, anorexia, nausea, vomiting, inflammation of the membranes of the mouth, unpleasant taste in the mouth, itchy skin, skin hemorrhages, gastritis, GI bleeding, and diarrhea.

usual intake method: a record of the foods eaten by a person in a typical day.

uterus (YOU-ter-us): the muscular organ within which the infant develops before birth.

—V—

vagotomy (vay-GOT-oh-mee): surgery that severs the nerves that stimulate gastric acid secretion.

validity (va-lid-ih-tee): having the quality of being founded on fact or evidence.

variables: factors that change. A variable may depend on another variable (for example, a child's height depends on his age), or it may be independent (for example, a child's height does not depend on the color of her eyes). Sometimes both variables correlate with a third variable (a child's height and eye color both depend on genetics).

varices (VAIR-ih-seez): blood vessels that have become twisted and distended.

variety (dietary): eating a wide selection of foods within and among the major food groups (the opposite of monotony).

vasoconstrictor (VAS-oh-kon-STRIK-tor): a substance that constricts or narrows the blood vessels.

vegans (VEE-guns, VAY-guns, or VEJ-ans): people who exclude all animal-derived foods (including meat, poultry, fish, eggs, and dairy products) from their diets; also called **pure vegetarians, strict vegetarians,** or **total vegetarians.**

vegetarians: a general term used to describe people who exclude meat, poultry, fish, or other animal-derived foods from their diets.

veins (VANES): vessels that carry blood back to the heart.

villi (VILL-ee, VILL-eye): fingerlike projections from the folds of the small intestine; singular **villus.**

vitamin A: all naturally occurring compounds with the biological activity of retinol (RET-ih-nol), the alcohol form of vitamin A.

vitamin A activity: a term referring to both the active forms of vitamin A and the precursor forms in foods without distinguishing between them.

vitamin B_{12}: a B vitamin characterized by the presence of cobalt (see Figure 13-10 in Chapter 13). The active forms of coenzyme B_{12} are methylcobalamin and deoxyadenosylcobalamin.

vitamin B_6: a family of compounds—pyridoxal, pyridoxine, and pyridoxamine.

The primary active coenzyme form is PLP (pyridoxal phosphate).

vitamins: organic, essential nutrients required in small amounts by the body for health.

VLDL (very-low-density lipoprotein): the type of lipoprotein made primarily by liver cells to transport lipids to various tissues in the body; composed primarily of triglycerides.

voluntary activities: conscious and deliberate muscular work—walking, lifting, climbing, and other physical activities. In contrast, involuntary activities occur independently, without conscious will or knowledge—heart beating, lungs breathing, and other activities critical to maintaining life.

vomiting: expulsion of the contents of the stomach up through the esophagus to the mouth.

waist circumference: an anthropometric measurement used to assess a person's abdominal fat.

water balance: the balance between water intake and output (losses).

water intoxication: the rare condition in which body water contents are too high.

wean: to gradually replace breast milk with infant formula or other foods appropriate to an infant's diet.

websites: Internet resources composed of text and graphic files, each with a unique URL (Uniform Resource Locator) that names the site (for example, www.usda.gov).

weight cycling: repeated cycles of weight loss and gain. The weight-cycling pattern is popularly called the **ratchet effect** or **yo-yo effect** of dieting.

whole grain: a grain milled in its entirety (all but the husk), not refined.

Wilson's disease: an inherited disorder that can lead to severe liver damage. People with Wilson's disease absorb too much copper from the intestine and have too little of the protein that transports copper from the liver to the sites where it is needed.

wine: an alcoholic beverage made by fermenting grape juice.

World Wide Web (WWW, the **Web):** a graphical subset of the Internet.

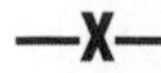

xanthophylls (ZAN-tho-fills): pigments found in plants; responsible for the color changes seen in autumn leaves.

xerophthalmia (zer-off-THAL-mee-uh): progressive blindness caused by severe vitamin A deficiency.

xerosis (zee-ROW-sis): abnormal drying of the skin and mucous membranes; a sign of vitamin A deficiency.

yogurt: milk fermented by specific bacterial cultures.

Zollinger-Ellison syndrome: marked hypersecretion of gastric acid and consequent peptic ulcers caused by a tumor of the pancreas.

zygote (ZY-goat): the product of the union of ovum and sperm; so-called for the first two weeks after fertilization.

CREDITS

1 © www.comstock.com; 2 © SuperStock; 3 © The Stock Market/Chuck Savage 2001; 4 © Bob Daemmrich/Stock Boston; 6 © Felicia Martinez/PhotoEdit (top-both); 6 © Thomas Harm, Tom Peterson/Quest Photographic Inc. (bottom-left); 6 © Tony Freeman/PhotoEdit (bottom-right); 8 © 2001 Stone/Christopher Bissell; 9 © R. Benali/Liaison Gamma (bottom-left); 9 © 2001 PhotoDisc Inc. (bottom-right); 9 © L. V. Bergman and Associates Inc. (top-right); 12 © Tony Freeman/PhotoEdit; 17 © Tom & Dee Ann McCarthy/The Stock Market; 24 © Bill Aron/PhotoEdit/PictureQuest; 27 © Marilyn Herbert Photography; 32 www.comstock.com; 35 © Polara Studios Inc.; 36 © Thomas Harm/Tom Peterson/Quest Photographic Inc.; 38 © Polara Studios Inc. (all); 39 © Polara Studios Inc. (all); 40 © Becky Luigart-Stayner/Corbis (top); 40 © PhotoDisc Inc. (center and bottom); 42 © Polara Studios Inc.; 43 © Polara Studios Inc.; 44 © Polara Studios Inc. (all); 45 © Polara Studios Inc. (all); 47 © Thomas Harm/Tom Peterson/Quest Photographic Inc.; 48 © Polara Studios Inc.; 49 © 1998 PhotoDisc Inc. (left); 49 © 1998 PhotoDisc Inc. (left-center); 49 © Felicia Martinez/PhotoEdit (right-center); 49 © Michael Newman/PhotoEdit (right); 50 © Bob Daemmrich Photography; 52 Courtesy of the FDA; 61 © Craig M. Moore; 63 Photos reprinted with permission of the Pillsbury Company, 2001; 64 © www.comstock.com; 65 © 1998 PhotoDisc Inc.; 75 Photo © Don W. Fawcett; 81 © Michael S. Yamashita/Corbis; 85 © The Stock Market/Ronnie Kaufman 2001; 89 © Polara Studios Inc.; 92 © Richard Hamilton Smith/Corbis; 95 © Wartenberg/Picture Press/Corbis; 98 © Polara Studios Inc.; 100 © Wartenberg/Picture Press/Corbis; 105 © 2001 PhotoDisc, Inc.; 112 © Thomas Harm/Tom Peterson/Quest Photographic Inc. (both); 114 © Felicia Martinez/PhotoEdit; 118 © Polara Studios Inc. (both); 123 Sunette Division, Hoechst Celanese Corp.; 127 © Craig M. Moore; 129 © Michael Pole/Corbis; 133 © Polara Studios Inc.; 144 © Peter Correz/Tony Stone Images; 148 © Bob Daemmrich Photography; 149 © Tony Freeman/PhotoEdit; 150 © Polara Studios Inc.; 151 © Polara Studios Inc., (left and center); 151 © Thomsas Harn and Tom Peterson/Quest Photographic Inc. (right); 154 © Thomas Harm and Tom Peterson/Quest Photographic Inc. (top); 154 © Polara Studios Inc. (bottom); 155 © Thomas Harm and Tom Peterson/Quest Photographic Inc. (top); 155 © Polara Studios Inc. (bottom); 157 © 1998 PhotoDisc, Inc.; 163 © Gerard Fritz/SuperStock; 165 © Polara Studios Inc.; 167 © Richard Fukuhara/Corbis; 176 © IT International/eStock Photography/PictureQuest; 182 © Polara Studios Inc.; 183 © Polara Studios Inc.; 184 AP/Wide World Photos; 185 AP/Wide World Photos; 186 © Paul A. Souders/Corbis; 187 © World Health Organization, Geneva; 190 © Polara Studios Inc. (both); 196 © Polara Studios Inc.; 201 © Courtesy National Cattlemen's Beef Association; 203 © EyeWire, Inc.; 204 © PhotoDisc Inc.; 223 © SW Production/Index Stock Imagery; 230 © 1998 PhotoDisc Inc.; 240 © FoodPix; 245 © Polara Studios Inc. (both); 248 © 1998 PhotoDisc Inc.; 250 © Rick Schaff; 253 © Lori Adamski Peek/Tony Stone Images; 256 © David Young-Wolff/PhotoEdit (all); 263 © 1998 PhotoDisc Inc.; 268 © 2001 PhotoDisc, Inc.; 274 © Courtesy Amgen, Inc.; 275 © Frank Whitney/The Image Bank; 276 © George Sesota/Newsmakers; 279 © Anne Dowie; 283 © Anne Dowie; 284 © Myrleen Ferguson Cate/PhotoEdit; 296 © Steve Niedorf Photography/The Image Bank; 299 © Omni Photo Communications, Inc./Index Stock Imagery; 302 © Michael Newman/PhotoEdit; 304 © Underwood & Underwood/Corbis (left); 304 AP/Wide World Photos (right); 306 © 2001 PhotoDisc, Inc.; 311 © L. V. Bergman & Associates; 313 © Polara Studios Inc.; 314 © Polara Studios Inc.; 316 © L. V. Bergman & Associates Inc.; 317 © Polara Studios Inc.; 322 © Polara Studios Inc.; 326 © Polara Studios Inc.; 329 © Martin M. Rotker (both); 334 © Science Photo Library/Photo Researchers; 338 © L. V. Bergman & Associates Inc. (left); 338 From Kronauer and Buhler, "Skin findings in a patient with scurvy." *The New England Journal of Medicine,* 6/15/95 (right); 341 © PhotoDisc Inc. (top); 341 © Polara Studios Inc. (bottom); 347 © J. Share/Tony Stone Images; 352 © Anne Dowie; 354 © Michael Boys/Corbis; 359 © David Farr/Image Smythe (all); 359 Ken Greer/Visuals Unlimited; 360 © Polara Studios Inc.; 365 © Biophoto Associates/Photo Researchers Inc. (top); 365 © Francisco Cruz/SuperStock (bottom); 366 © Corbis/Fotographia; 369 © Craig M. Moore; 372 © Polara Studios Inc.; 377 © The Stock Market/Michael Keller 2001; 380 (top, left to right) Courtesy of Brassica Protection Products; 380 Digital Imagery © 2001 PhotoDisc, Inc.; 380 Digital Imagery © 2001 PhotoDisc, Inc.; 380 © EyeWire, Inc.; 380 (center, left to right) Digital Imagery © 2001 PhotoDisc, Inc.; 380 Digital Imagery © 2001 PhotoDisc, Inc.; 380 © EyeWire, Inc.; 380 Digital Imagery © 2001 PhotoDisc, Inc.; 380 (bottom, left to right) Digital Imagery © 2001 PhotoDisc, Inc.; 380 Courtesy of Flax Council of Canada; 380 Digital Imagery © 2001 PhotoDisc, Inc.; 383 © Jeffry Myers/Stock, Boston/PictureQuest; 386 © FoodPix; 387 © Barbara Stitzen/PhotoEdit/PictureQuest; 393 © Craig M. Moore (both); 394 © Bob Daemmrich/Stock Boston; 399 © 2001 PhotoDisc, Inc.; 401 © Polara Studios Inc.; 406 © Thomas Harm, Tom Peterson/Quest Photographic Inc.; 420 Courtesy of Gjon Mili; 421 Permission by David Dempster from J Bone Miner Res, 1986; 421 © PhotoDisc Inc.; 428 H. Armstrong Roberts, Inc.; 432 © PhotoDisc Inc.; 434 © Martin M. Roetker (both); 436 © Craig M. Moore; 438 © Polara Studios Inc.; 441 © H. Sanstead, University of Texas at Galveston; 442 © Polara Studios Inc.; 445 © L. V. Bergman and Associates Inc.; 446 © Craig M. Moore; 449 © Dr. P. Marrazi/Science Photo Library/Photo Researchers Inc.; 458 © Craig M. Moore; 460 © Craig M. Moore; 461 © 2001 PhotoDisc, Inc. (all); 463 © VCG/FPG; 465 © Petit Format/Nestle/Photo Researchers Inc. (1,2,3); 465 © Anthony M. Vanelli; 467 © Lennart Nilsson/Albert Bonniers Förlag AB, from *A Child is Born,* Dell Publishing Co. (both); 469 © The Stock Market/Mug Shots 2001; 472 © Michael Salas/The Image Bank; 473 © The Stock Market/Rick Gomez 2001; 480 © Nik Kleinberg/Stock Boston; 483 © The Stock Market/Jose I. Pelaez/R, 1999; 485 © The Stock Market/Ariel Skelley 2001; 486 © Photo Disc Inc.; 495 © Streissguth, A. P.; Landesman-Dwyer, S.; Martin, J. C.; & Smith, D. W. (1980), Teratogenic effects of alcohol in humans and laboratory animals. Science, 209 (18): 353–361; 496 © 1995 George Steinmetz (both); 498 © Infocus, The Image Bank 2001; 499 © 1998 PhotoDisc Inc.; 502 © 2001 PhotoDisc, Inc.; 504

K. L. Boyd, DDS/Custom Medical Stock Photos; 505 © Mary Kate Denny/PhotoEdit; 507 © Polara Studios Inc.; 508 © Tony Freeman/PhotoEdit; 508 © Stephanie Rausser/FPG International; 509 © Anthony M. Vanelli (both); 511 © 2001 PhotoDisc, Inc.; 513 © Tony Freeman/PhotoEdit; 515 © Thomas Harm, Tom Peterson/Quest Photographic; 517 © Donna Day/Tony Stone Images; 523 © Bob Daemmrich/ Stock Boston/PictureQuest; 526 © Phyllis Picardi/Stock Boston; 535 © The Stock Market/LWA/Sharie Kennedy 2001; 536 ICU Pharmaceuticals (both); 538 © M. Greenlar/The Image Works; 540 © www.comstock.com; 543 © R. W. Jones/Corbis; 545 © 2001 Stone/Bob Thomas; 546 © Kwame Zikomo/Super Stock; 548 Myrleen Ferguson Cate/PhotoEdit; 551 © Deborah Davis/PhotoEdit; 553 © Richard Pasley/ Stock Boston/PictureQuest; 555 © Lonnie Duka/Index Stock Imagery; 556 © Elena Rooraid/PhotoEdit; 561 © The Stock Market/David Woods 2001; 564 AP/Wide World Photos (left); 564 © Thomas Muscianico/Contact Press/Picture-Quest (right); 566 © Anthony Johnson/The Image Bank; 570 © Michelle DelGuercio/Photo Researchers, Inc.; 571 © SIU/Visuals Unlimited; 574 © Nathan Benn/Stock, Boston/ PictureQuest; 576 Courtesy of Seca Corporation (top); 576 © Tony Freeman/PhotoEdit (bottom); 577 © Richard Nowitz/ Phototake/PictureQuest; 577 © Tom McCarthy/PhotoEdit; 579 © Charles Gupton/Stock, Boston/PictureQuest; 582 © Biophoto Associates/Science Source/Photo Researchers, Inc.; 587 © 2000 Photo Disc Inc.; 590 © www.comstock.com; 592 © Josh Sher/Science Photo Library/Photo Researchers, Inc.; 593 © 2001 Stone/Christel Rosenfeld; 612 © CORBIS; 615 © Tom McCarthy/PhotoEdit; 617 © Craig M. Moore; 618 © 2001 Stone/Cat Gwynn; 621 © Bill Bachmann/Index Stock/PictureQuest; 624 © Leslie O'Shaughnessy/Medical Images, Inc.; 629 © 2001 PhotoDisc, Inc.; 632 Courtesy Diamond Crystal Specialty Foods; 647 © Charles Gupton/Stock Boston; 651 © James Dennis/ Phototake; 652 © Peter Kane/ The Stock Shop;; 656 © Bard Martin/The Image Bank; 659 © Craig M. Moore; 662 © Simon Fraser/RVI, Newcastle-Upon-Tyne/Science Photo Library/Photo Researchers Inc.; 664 Courtesy of the Crohn's & Colitis Foundation of America Inc. (both); 678 © B. F. King, Univ. of California School of Medicine/BPS; 681 © Ursula Markus/ Photo Researchers, Inc.; 684 © 2000 C. C. Duncan/Medical Images Inc.; 688 © Flexiflo® Over-The-Counter Nasojeunal Feeding Tube, Courtesy of Ross Products Div., Abbott Laboratories, Columbus, OH.; 689 © Andy Crump, TDR, WHO/Science Photo/ Custom Medical Stock Photo; 690 © Ed Eckstein/ Phototake; 702 © David Young-Wolff/PhotoEdit; 705 © Bob Wickley/ SuperStock; 706 © John Greim/Medichrome; 708 © David Young-Wolff/PhotoEdit; 709 © Ed Eckstein/PhotoTake; 716 Courtesy of Kendall Healthcare; 721 © Dan Reynolds/Medical Images, Inc.; 726 © BSI Agency/ Index Stock Imagery/ PictureQuest; 727 © Bruce Ayres/Stone; 741 © Hein Hopmans/Phototake; 744 © Alfred Pasieka/ Science Photo Library/ Photo Researchers, Inc.; 745 © Astrid & Hanns-Frieder Michler/ Science Photo Library/ Photo Researchers, Inc.; 751 © 2000 C. C. Duncan/Medical Image Inc.; 752 © Tony Freeman/PhotoEdit (both); 759 © Mark Clarke/Science Photo Library/ Photo Researchers, Inc.; 760 © Russell D. Curtis/Photo Researchers Inc.; 761 © 2000 PhotoDisc Inc.; 763 Digital Imagery © 2001 PhotoDisc, Inc.; 769 © Ann Dowie; 773 © PictorIntl/ Pictor International/ PictureQuest; 775 Science Photo Library/Photo Researchers Inc. (both); 778 © Chris Jones/The Stock Market; 779 © Ronnie Kaufman/The Stock Market; 780 © Michael Newman/-PhotoEdit/PictureQuest; 787 © Susan Leavines/Photo Reserchers, Inc.; 790 © Michael Newman/PhotoEdit; 800 © Alfred Pasieka/ Science Photo Library/ Photo Researchers, Inc.; 802 Printed with permission from *The New England Journal of Medicine,* October, 1992, p. 1142; 803 www.comstock.com; 812 © Craig M. Moore; **815 © Craig M. Moore; 822 © Hank Morgan/Rainbow; 826 © Prof. P. Motta/ Dept. of Anatomy/ Univ. La Sapienza, Rome/ Science Photo Library/ Photo Researchers, Inc.; 829 © Dr. P. Marazzi/Science Photo Library/Photo Researchers, Inc.; 830 © Profs. P. Motta/Dept. of Anatomy/Univ. La Sapienza, Rome/Science Photo Library/Photo Researchers Inc.; 830 © Profs. P. Motta, M. Nishi, J. Fujita/Science Photo Library/Photo Researchers Inc.; 837 © Polara Studios Inc.; 844 © C. James Webb/Photo-Take; 847 © Will and Deni McIntyre/ Photo Researchers, Inc.; 848 © Kent Meireis/The Image Works; 851 © Polara Studios Inc.; 855 © Erika Craddock/Science Photo Library/ Photo Researchers, Inc.; 862 © Lauren Goodsmith/The Image Works; 863 © Biophoto Associates/Photo Researchers, Inc.; 865 © Gary Conner/PhotoEdit; 866 © Tony Freeman/ PhotoEdit; 872 © Carolyn A. McKeone/Photo Researchers, Inc.

INDEX

Page references in bold indicate definitions of terms.
Page references followed by the letter "f" indicate figures.
Page references followed by the letter "t" indicate tables.
Page references followed by the letter "n" indicate footnotes.
Page references with combined letter and number (A-1) refer to the appendixes.

O

AIDS TO CALCULATION

Many mathematical problems have been worked out in the "How to" and "Making it Click" sections of the text. These pages provide additional help and examples.

Conversion Factors

A conversion factor is a fraction in which the numerator (top) and the denominator (bottom) express the same quantity in different units. For example, 2.2 pounds (lb) and 1 kilogram (kg) are equivalent; they express the same weight. The conversion factors used to change pounds to kilograms and vice versa are:

$$\frac{2.2 \text{ lb}}{1 \text{ kg}} \quad \text{and} \quad \frac{1 \text{ kg}}{2.2 \text{ lb}}.$$

Because a conversion factor equals 1, measurements can be multiplied by the factor to change the *unit* of measure without changing the *value* of the measurement. To change one unit of measurement to another, use the factor with the unit you are seeking in the numerator (top) of the fraction.

Example 1 Convert the weight of 130 pounds to kilograms.

- Choose the conversion factor in which the kilograms are on top and multiply by 130 pounds:

$$\frac{1 \text{ kg}}{2.2 \text{ lb}} \times 130 \text{ lb} = \frac{130 \text{ kg}}{2.2} = 59 \text{ kg}.$$

Example 2 Consider a 4-ounce (oz) hamburger that contains 7 grams (g) of saturated fat. How many grams of saturated fat are contained in a 3-ounce hamburger?

- Because you are seeking grams of saturated fat, the conversion factor is:

$$\frac{7 \text{ g saturated fat}}{4 \text{ oz hamburger}}.$$

- Multiply 3 ounces of hamburger by the conversion factor:

$$3 \text{ oz hamburger} \times \frac{7 \text{ g saturated fat}}{4 \text{ oz hamburger}} = \frac{3 \times 7}{4} = \frac{21}{4} =$$

5 g saturated fat (rounded off).

Percentages

A percentage is a comparison between a number of items (perhaps the number of kcalories in your daily energy intake) and a standard number (perhaps the number of kcalories used for Daily Values on food labels). To find a percentage, first divide by the standard number and then multiply by 100 to state the answer as a percentage (*percent* means "per 100").

Example 3 Suppose your energy intake for the day is 1500 kcalories (kcal): What percentage of the Daily Value (DV) for energy does your intake represent? (Use the Daily Value of 2000 kcalories as the standard.)

- Divide your kcalorie intake by the Daily Value:

1500 kcal (your intake) ÷ 2000 kcal (DV) = 0.75.

- Multiply your answer by 100 to state it as a percentage:

0.75 × 100 = 75% of the Daily Value.

Example 4 Sometimes the percentage is more than 100. Suppose your daily intake of vitamin C is 120 milligrams (mg) and your RDA (male) is 90 milligrams. What percentage of the RDA for vitamin C is your intake?

120 mg (your intake) ÷ 90 mg (RDA) = 1.33.

1.33 ×100 = 133% of the RDA.

Example 5 Sometimes the comparison is between a part of a whole (for example, your kcalories from protein) and the total amount (your total kcalories). In this case, the total is the number you divide by. If you consume 60 grams (g) protein, 80 grams fat, and 310 grams carbohydrate, what percentages of your total kcalories for the day come from protein, fat, and carbohydrate?

- Multiply the number of grams by the number of kcalories from 1 gram of each energy nutrient (conversion factors):

$$60 \text{ g protein} \times \frac{4 \text{ kcal}}{1 \text{ g protein}} = 240 \text{ kcal}.$$

$$80 \text{ g fat} \times \frac{9 \text{ kcal}}{1 \text{ g fat}} = 720 \text{ kcal}.$$